AF443173

I.D.L.I.
INTERNATIONAL DRIVE LINE INTERCHANGE
GUIDE™

WORLD'S MOST COMPLETE HISTORY OF NUMBER ALTERNATIVES FOR DRIVE LINE COMPONENTS

A COMPUTERIZED INTERCHANGE OF UNIVERSAL JOINT CROSSES AND BEARINGS, YOKES, SHAFTS, FLANGES, TUBING, ETC. FOR GROUND AND OTHER EQUIPMENT

FIRST EDITION
1986-87

Published by

INTERCHANGE
INCORPORATED™

P. O. BOX 16012, St. Louis Park, MN 55416, U.S.A.
(612) 929-6669   CABLE CODE: INTERCHANG   TELEX: 29-1081

P.O. Box 16012 • St. Louis Park, Minnesota 55416-0012 U.S.A.
Telephone: (612) 929-6669 • Telex: 29-1081

A MESSAGE TO OUR CUSTOMERS

This Interchange Guide is only one of four (4) unique cross-reference guides now available to you. The Interchange Guides have been published for more than 20 years, and are currently sold in over 100 countries around the world.

The 1986-87 Editions include:

> 9th Edition International Bearing Interchange (I.B.I.) Guide
> 6th Edition International Seal Interchange (I.S.I.) Guide
> 4th Edition International Drive Belt Interchange (I.D.B.I.) Guide
> 1st Edition International Drive Line Interchange (I.D.L.I.) Guide

New numbers and brands are continually added to our previous bank of information, so that these guides serve as encyclopedias of information for our customers throughout the world.

To continue providing this history of cross-reference information, we urge you to share with us any catalogs or other material that should be considered for inclusion in our future editions. Old or new part numbers, manufacturers or Original Equipment Users (O.E.M.) cross-references, or suggestions for improving the listings already shown in our Guides will all be considered.

These are the only reference Guides that include both U.S.A. and foreign interchanges and both producers and users. Many government and industry standard numbering systems are also included.

We would also appreciate your sending us a listing of your branches or outlying affiliates for our mailing list. A quantity of our circulars can be sent on request, and if we have a sales representative or distributor in your area, we will ask him to contact you directly.

Thank you for your interest in our product line. We look forward to your continued support of this exclusive service.

Sincerely,

S.H. Friedman, President
INTERCHANGE, INC.

Worldwide Interchange Guides to
Bearings, Seals, Drive Belts,
and Drive Line Components

AUTHOR AND PUBLISHER
SY H. FRIEDMAN

With over 50 years of business experience, Sy H. Friedman is one of the world's foremost authorities on finding interchangeable part numbers for bearings, seals, drive belts and drive line components.

Mr. Friedman began compiling lists of interchanges in 1935, when he first worked as a clerk for a bearing wholesaler. He felt strongly that his knowledge on how to find fast, accurate replacement parts would save time, save money, and keep customers satisfied.

Mr. Friedman continued to collect interchange data while working in his own bearing distributorship, and while still in that business he began his research in 1955 toward publishing the Interchange Guides.

The International Bearing Interchange (I.B.I.) Guide was first published in 1966.

The success of that first book, and popular demand, found Mr. Friedman researching and publishing additional worldwide Interchange Guides for oil seals and "O" rings (The I.S.I. Guide), drive belts (The I.D.B.I. Guide) and, just released in 1986-87, the Interchange Guide for drive line components (The I.D.L.I. Guide).

Mr. Friedman has designed a computerized system for finding suitable substitutions in the Interchange Guides, making them as easy-to-use as a telephone directory.

The International Interchange Guides are sold in more than 100 countries throughout the world, saving time and money for anyone who uses, buys, sells, or specifies, bearings, oil seals, drive belts or drive line components for cars, trucks, tractors, construction machinery, agricultural or factory and other equipment.

ALPHABETICAL LIST OF MANUFACTURERS AND BRANDS SHOWN IN THE IDLI GUIDE

REFER TO THIS INDEX FOR NAMES YOU MAY NOT FIND IN THE FOLLOWING PAGES.
WE HAVE ALSO ENDEAVORED TO SHOW
ALTERNATE BRANDS AND SUBSTITUTE TRADE-MARKS WHEREVER POSSIBLE.

BUCYR-ERIE . . . . Bucyrus-Erie Co. (Also see HY-DYNAM)
BUFFALO-SP . . . . Buffalo-Springfield Roller Cor. (Also see
KOEHRING, PARSONS & SCHIELD)
BUICK . . . . See GMC
BUSH HOG . . . . Bush Hog Mfg. Co.
BUTLER. . . . . Butler Mfg. Co., Inc.
BWE . . . . BWE, Inc.
BYPY . . . . Bondioli & Pavesi, Inc. (Also see BOND. & PAV.)
CADILLAC . . . . See GMC
CARDWELL . . . . Cardwell Mfg. Co.
CASE, J.I. . . . . Case, J.I.
CATERPILLR . . . . Caterpillar Tractor Co.
CEDAR RAP . . . . Cedar Rapids Mfg. Co. (Also see Iowa Mfg.)
CENTRAL . . . . Central States Power Co.
CHAIN BELT . . . . Chain Belt Mfg. Co.
CHAMPION . . . . Champion Products, Inc.
CHAMPLAIN . . . . Champlain Mfg. Co., Inc.
CHECKER . . . . Checker Cab Mfg. Co., Checker Motor Corp.
CHELSEA . . . . . Chelsea Power Take-Off (Also see SPICER & DANA)
CHEVROLET . . . . See GMC
CHRYSLER . . . . Chrysler Corp.
CHRYSLER-BRITISH . . . . See CHRYSLER
CHRYSLER-FRANCE . . . . See CHRYSLER
CITROEN . . . . S. A. Andre Citroen (France)
CLARK EQU. . . . . Clark Equipment Mfg. Co.
(Also see AUST. WEST. & MICHIGAN)
CLAYSON . . . . See NEW HOLL.
CLEVE. MOT. . . . . Cleveland Motive Products, Inc.
CLEVE. STL. . . . . Cleveland Steel Products Corp.
CMP . . . . CMP Corporation
COCKSHUTT . . . . Cockshutt Farm Equipment MFG. Co.
COLES . . . . . Coles Cranes Ltd., British Crane & Excavatcor Co., Ltd.
COMMER . . . . See CHRYSLER
CONS. MACH. . . . . Construction Machinery Corp.
COOK, J.B. . . . . J.B. Cook Machinery Co., Inc.
CRANE CARR . . . . Crane Carrier Co.
CROSLEY . . . . Crosley Motors Corp.
CROY . . . . Croy Automotive Co., Croy Mfg. Co.
(Also see ALLOY & WESCO)
CURTIS . . . . Curtis Machine Co., Inc.
CURTISS-WR . . . . Curtiss-Wright Corp.
D&D MACH. . . . . D & D Machine (Also see MASTERS)
DAF . . . . DAF Ltd. (Holland)
DAFFIN . . . . Daffin Corp.
DAIHATSU. . . . . Daihatsu Kogyo Co. Ltd. (Japan)
DAIM. RUBE. . . . . Daimler Rubens GmbH
DAIMLER . . . . Daimler-Benz Motor Co. (Also see MERCEDES)
DANA . . . . Dana Corp., Spicer-Dana Corp. (Also see SPICER)
DANLEY . . . . Danley Machine Co.
DANUSER . . . . Danuser Machine Co. (Also see HESSTON)
DARLINGTON . . . . Darlington Dumpers Ltd.
DART . . . . See CHRYSLER
DATSUN . . . . Nissan Motor Co., Ltd. (Also see NISSAN)
DAVEY COM. . . . . Davey Compressor Co.
DEERE, JOHN . . . . John Deere & Co.
DENNIS . . . . Dennis Motors, Ltd.
DE SOTO. . . . . See CHRYSLER
DET. HARV. . . . . Detroit Harvester Co. (Also see DURA-DET.)

DETROIT . . . . Detroit Universal, Inc.
DIAMOND R . . . . Diamond Reo Co. (Also see DIMOND T & WHITE)
DIAMOND T . . . Diamond Reo Co. (Also see Diamond R and WHITE)
DIGIO-LIFT . . . . Di Gio Lift Truck Spa
DIVCO . . . . Divco-Wayne Corp.
DODGE . . . . See CHRYSLER
DOMINION . . . . Dominion Road Machinery
DORMAN . . . . Dorman Products Co., Inc.
DROTT . . . . Drott Mfg. Co., Inc.
DUPONT . . . . E.I. DuPont de Nemours
DURA-DET. . . . . . . Dura-Detroit Harvester Co. (Also see DET. HARV.)
EATON . . . . Eaton Yale & Towne, Inc.
EIMCO. . . . . Eimco Corp.
ELEC. MTR. . . . . Electra-Motor Corp.
ELECT. MOT. . . . . Electro-Motive Div. of GMC (See GMC)
ELG. SWEEP . . . . Elgin Sweeper Co.
ENGLER . . . . Engler Mfg. Corp.
ERF . . . . E.R.F Mfg. Co. (See also FODEN)
ETNYRE . . . . E.D. Etnyre & Co.
EUCLID . . . . Euclid Corp. (See GMC)
EXCEL . . . . Excel Corp.
FALK . . . . Falk & Co.
FARGO . . . . See CHRYSLER
FARMHAND . . . . Farmhand, Inc.
FED. BEAR. . . . . Schatz-Federal Bearing Co., Inc.
FED. MOGUL . . . . Federal Mogul Corp.
FENNER . . . . Listed By Dimensions Only
FERGUSON . . . . Ferguson Co. (See FORD & MASS-FERG.)
FIAT . . . . Fiat S.p.A. & Fiat Ltd.
FIAT-ALLIS . . . . Fiat-Allis, Inc. (See also ALLIS-CHLM)
FIRESTONE . . . . Firestone Tire & Rubber Co.
FLEETRITE . . . . Fleetrite, Inc.
FLXIBLE . . . . Flxible Co., Inc.
FMC . . . . FMC Corp.
FODEN . . . . Foden Co., Inc.
FOR. PARTS . . . . Foreign Parts Distributors, Inc.
FORD . . . . Ford Motor Co.
FOX. . . . . Fox Trailer Manufacturing Co.
FOX RIVER . . . . Fox River Tractor Co.
FWD . . . . Four Wheel Drive Corp.
G & G MFG. . . . . G & G Manufacturing Co.
GALION . . . . Galion Iron Works & Mfg. Co.
GAR WOOD . . . . . . Gar Wood Industries, Inc. (Also see BUCKEYE)
GARD. DENV. Gardner-Denver Corp.
GEHL BROS. Gehl Brothers Co., Inc.
GELENKWEL. . . . . Gelenkwellenbau, GmbH
GEN. EQUIP . . . . General Equipment Mfg. Co.
GERLINGER . . . . Gerlinger Carrier Corp., Gerlinger
Industries (Also see Towmotor)
GILSON . . . . Gilson Bros.
GKN GROUP . . . . GKN Group
GLAENZER . . . . Glaenzer Spicer S.A.
GMB. . . . . GMB Corp.
GMC . . . . General Motors Corp.
GMC-AUSTRALIA . . . . See GMC & HOLDEN
GMC-BRITISH . . . . See GMC & BEDFORD
GMC-GERMANY . . . . See GMC & OPEL
G.M. DIESEL . . . . See GMC

GOODMAN	Goodman Wabco Mining Div. (Also see WABCO)
GOODYEAR	Goodyear Tire & Rubber Co.
GORMAN RUP	Gorman Rupp Co.
GRAIGG CO.	C.V. Graigg Co.
GRAY PRIOR	Gray & Prior Machine Co.
GREEN BALL	The Green Ball Bearing Co.
GROVE CR.	Grove Cranes Ltd., Grove Mfg. Co.
GURRIES	Gurries, Mfg. Corp.
HABAN	Haban Mfg. Co.
HABERLE	Haberle Machine Co., Inc.
HAHN	Hahn Machinery, Inc.
HALLBURTON	Halliburton Co.
HARDIE	Hardie-Tynes Mfg. Co.
HARDY MFG.	Hardy Mfg. Co.
HARDY SPCR	Hardy Spicer Ltd.
HARNISCH.	Harnischfeger Corp.
HAYES	Hayes-Dana Corp. (Also see SPICER, DANA, BORG WARNR, ROCKWELL & DETROIT)
HEIL	Heil Co., Inc.
HENNEY	Henney Mfg. Co.
HERCULES	Hercules Engineering & Mfg.
HESSTON	Hesston Corp. (Also see DANUSER)
HIGHLAND	Highland Tractor Co.
HINO.	Hino Motors, Ltd. (Japan)
HIWAY CHIP	Hi Way Chipper Div.-Hiway Truck Equipment, Inc.
HILLMAN	See CHRYSLER
HOLDEN	Holden Equipment Co. Ltd.
HONDA.	Honda Motors, Ltd. (Japan)
HORWOOD	Horwood Bayshaw Ltd.
HOUGH	Frank G. Hough Co. (Also see IHC)
HOWE	Howe Corp.
HUB CITY	Hub Cty Iron, Safeguard Power Tech Systems
HUBER	Huber-Warco Mfg. Co.
HUDSON	See AMER. MOTOR
HUMBER	See CHRYSLER
HUME, H.D.	H.D. Hume Co.
HUSCO.	Husco International Ltd.
HY-DYNAM	Hy-Dynamic-Div.-Bucyrus-Erie Co. (Also see BUCRY-ERIE)
HYSTER	Hyster Co.
IHC	International Harvester Co. (Also see HOUGH)
IMCO	Interstate Machine Co.
INBRA	Inbra Mfg. Co.
INGER. RAND	Ingersoll Rand Corp.
INNOCENTI.	Innocenti, Ltd.
IOWA MFG.	Iowa Manufacturing Co. (Also see CEDAR RAP)
ISUZU	Isuzu Motors, Ltd. (Japan)
ITALCARDAN	Italian Cardano Corp.
JACOBSEN	Jacobsen Mfg. Co.
JAEGER	Jaeger Machine Co.
JAGUAR.	Jaguar-Daimler Cars. Ltd.
JAMCO	Jamco Western, Inc.
JEBCO	Jebco International Corp.
JEEP.	Willys-Jeep Mfg. Co. (Also see AMER. MOTOR, KAISER, and WILLYS)
JEFFREY	Jeffrey Mfg. Co.
J.G. BRILL	See BRILL, J.G.
JOHN BEAN	See BEAN, JOHN
JOY MFG.	Joy Mfg. Co.
KAISER	Kaiser Corp. (See also AMER. MOTOR, JEEP & WILLYS)
KALAMAZOO	Kalamazoo Manufacturing Co.
KATAOKA	Kataoka Crane Mfg. Co.
KENBAR	Kenbar Industries, Inc.
KENWORTH	Kenworth Motor Truck Co.
KERSHAW	Kershaw Mfg. Co.
KEWANEE	Kewanee Machine & Conveyor Co.
KIEKHAFER	Kiekhaefer Aeromarine, Inc.
KOEHRING	Koehring Co., Inc. (Also see BUFFALO-SP & PARSONS & SCHIELD)
KOMATSU	Komatsu Ltd., (Japan)
KOYO	Koyo Seiko Co., Ltd. (Japan)
L&S BRG.	L&S Bearing Co. (Subsidiary of LSB Industries)
LAKESHORE	Lakeshore, Inc.
LAKEWOOD	Lakewood Engineering, & Mfg. Co.
LANDROVER	Specialist Division of BMC (Also see BMC & ROVER)
LA PORTA	La Porta Lift Corp.
LEADER.	Leader Mfg. Co.
LEMPCO	Lempco, Inc.
LE TOURN.	Le Tourneau-Westinghouse Construction Div. (See WABCO)
LETZ	Letz Mfg. Co.
LEVRATTO	Levratto, Ltd.
LEYLAND.	British-Leyland Motor Corp. (Also see BMC)
LIFT TRUCK	Lift Truck Motor Co.
LILLISTON	Lilliston Implement Mfg. Co.
LINCOLN	See FORD
LINK BELT	Link Belt Mfg. Co.-Div. of FMC Corp.
LÖBRO	Löhr & Bromkamp GmbH
LONG MFG.	Long Manufacturing Co.
LOVEJOY.	Lovejoy Flexible Coupling Co.
LUDWIG	Ludwig Corp.
LUNDELL	Lundell Mfg., Co.
MACK	Mack Trucks, Inc.
MAGIRUS	Magirus Corp.
M.A.N.	M.A.N. Unternenmensbereich Nuffahr-Zuege (W. Germany)
MARATHON	Marathon-Le Tourneau Co. (Also see LE TOURN.)
MASS. FERG.	Massey-Ferguson, Inc.
MASTERS	Masters Mfg. Co. (Also see D&D MACH.)
MATHEWS	Mathews International Corp.
MATRA.	Matra Sports Cars, Ltd. (France)
MATSUBA	Matsuba International, Ltd.
MATSUI	Matsui Mfg. Co.
MAZDA	Mazda Motors, Co., Toyo-Kogyo Co. Ltd.
MCQUAY-NOR	McQuay-Norris, Inc.-An SKF Industries Co.
MECHANICS.	Saginaw-Mechanics (Also see BORG-WARNR)
MERC. KIEK.	Mercury Kiekhafer, Corp.
MERCEDES	Mercedes-Benz, Daimler-Benz AG (W. Germany) (Also see DAIMLER)
MERCURY	Mercury Marine-Brunswick Corp.
MERCURY PASS. CAR	See FORD
MG	See BMC
MICHIGAN	Michigan Shovel Co. (Also see CLARK EQU.)
MIDAS	Midas International Corp.
MIPER	Miper Mfg. Co.
MITSUBUSHI	Mitsubushi Motors Corp. (Japan)

MIXERMOBL . . . Mixermobile Mfg. Co., Inc. (Also see SCOOPMOB)
MOOG . . . . Moog Automotive, Inc.
MOPAR . . . . Mopar Corp. (Also see CHRYSLER)
MORRIS . . . . Morris Motors Ltd. (Great Britain) (Also see BMC)
MOTIVE . . . . Motive Equipment Mfg. Co. (Also see PILOT)
MOTOR MAST . . . . Motor Master Products Corp. (Also see ZELLER)
MPL. MOLINE . . . . Minneapolis Moline, Inc. (Also see
WHITE, WHITE FARM and OLIVER)
M.R.S. . . . . M.R.S. International, Inc.
MUIR-HILL . . . . Muir-Hill (Great Britain) (Also see FORD)
MUNCIE . . . . Muncie Gear Works, Inc.
MURR. TREG. . . . . Murray & Tregurtha, Ltd.
MYERS ELEV . . . . Myers Elevator, Inc.
NASH . . . . Nash-Rambler (See AMER. MOTOR)
NATL. MINE . . . . National Mine Service Co.
NEAPCO . . . . Neapco Corp.-A Division of Berwind Corp.
NELSON . . . . N.P Nelson Mfg. Co.
NEW HOLL. . . . . New Holland Machine-Div. of
Sperry Rand Corp. (CLAYSON)
NEW IDEA . . . . New Idea Mfg. Co.-Div. of Avco Corp.
NISSAN . . . . Nissan Motor Co. (Japan) (Also see DATSUN)
NORTH AMER . . . . North American Eng. & Mfg. Co.
(Also see ALMETAL)
NORTH AMERICAN ROCKWELL. . . . . See ROCKWELL
NSU PRINZ . . . . Prinz, NSU Motorenwerke AG (West Germany)
NTN . . . . NTN Bearing Corp. of America,
Toyo Bearing Mfg. Co., Ltd. (Japan)
NW MOTORS . . . . Northwestern Motors, Inc.
OLDING . . . . Olding Hoist Mfg. Co.
OLDSMOBILE . . . . See GMC
OLIVER . . . . Oliver Corp. (Also see WHITE)
OPEL . . . . Adam Opel AG (W. Germany)
OSHKOSH . . . . Oshkosh Motor Truck, Inc.
OWATONNA . . . . Owatonna Manufacturing Co.
P.M. . . . . P.M. Inc.
PACIF CAR . . . . Pacific Car & Foundry Corp.
PACKARD . . . . See STUDEBAKER
PAPEC MACH . . . . Papec Machine Co.
PAPEX . . . . Papex International Corp.
PARSONS . . . . Parsons Construction Machinery Co., Inc.
(Also see BUFFALO-SP, KOEHRING & SCHIELD)
PAXTON-MIT . . . . Paxton-Mitchell Co.
PEARSON . . . . Ben Pearson Co.
PENNEY, JC . . . . J.C. Penney Power Equipment Co.
PERFECT-CI . . . . Perfect Circle Div.-Dana Corp.
PERFECTION . . . . Perfection Steel Body Co.
PERRY . . . . Perry Company
PETTI-MULL . . . . Pettibone-Mulliken Corp.
PETTIBONE . . . . Pettibone-Mercury Corp.
PILOT . . . . Pilot Mfg. Co. (Also see MOTIVE)
PIONEER . . . . Pioneer Products Co.
PITT. VIOL. . . . . Pitteri Violini, SpA (Italy)
PLYMOUTH . . . . See CHRYSLER
POCLAIN . . . . Poclain S.A. (France)
PONTIAC . . . . See GMC
PORSCHE . . . . Dr. Ing. h.c.F Porsche K.G. (West Germany)
PREC. TOR. . . . . Precision Universal Joint Corp.
(Also see PRECISION & TORQUE)

PRECISION . . . . Precision Universal Joint Corp.
(Also see PREC.TOR & TORQUE)
PULLMAN . . . . Pullman-Standard Corp.
QUINT. HAZ. . . . . Quinton Hazell Ltd.
QUINTON . . . . Quinton Corp.
RAMBLER . . . . See AMER. MOTOR
RAY GO . . . . Ray Go Inc.
RENAULT . . . . Regie Nationale des Usines Renault (France)
REO MOTORS . . . . Reo Motors Inc. (Also see WHITE)
REPCO . . . . Repco Universal Drivelines-Division
of Repco Ltd., Inc. (Australia)
REPUBLIC . . . . Republic Parts, Inc.
REX CHAIN . . . . Rex Chain Belt Co., Inc.
REXNORD . . . . Rexnord, Inc.
REYNOLDS . . . . Reynolds Products, Inc.
RILEY . . . . Specialist Car Division-British Leyland
Motor Corp. Ltd. (Also see BMC)
RITEWAY . . . . Riteway Mfg. Corp.
ROCKWELL . . . . Rockwell Standard Trans. & Axle Co.,
Rockwell International
ROGER IRON . . . . Rogers Iron Co., Inc.
ROKA . . . . Roka Corp.
ROOTES . . . . See CHRYSLER
ROSS GEAR . . . . Ross Steering Gear Mfg. Corp.
ROVER . . . . British Leyland Motor Corp., Ltd.,
The Rover Co. (Also see LAND-ROVER and BMC)
RUEDARSA . . . . Ruedarsa, Ltd.
RUPP . . . . Gorman Rupp Co.
RUST . . . . Rust Cotton Pickers Corp.
SAFE ELEC. . . . . Safety Electric Co., Inc.
SAGINAW-MECHANICS . . . . See BORG-WARNR
SANFOR-DAY . . . . Sanford-Day Mining Equipment Div. of
The Marmon Group, Inc.
SAVIEM . . . . Saviem Ets. (France)
SCAN. VABIS . . . . Scania Vabis Ltd.
SCHIELD BANTAM . . . . Koehring Corp. (Also see BUFFALO-SP,
KOEHRING, & PARSONS)
SCHULTZ . . . . L.H. Schultz Mfg. Co.
SCOOP MOB . . . . Scoop Mobile, Inc. (Also see MIXERMOBL)
SEAGRAVE . . . . Seagrave Corporation
SEARS-ROE . . . . Sears Roebuck & Co.
SERVIS . . . . Servis Equipment Co.
SHOV. SUPP. . . . . Shovel Supply Corp.
SIDEWINDER . . . . Sidewinder Textile Machinery, Enterprise
Machine & Development Co.
SIL. HOIST . . . . Silent Hoist & Crane Co.
SIMCA . . . . . Simca Motors, S&O Poissy Ltd. (Also see CHRYSLER)
SINGER . . . . See CHRYSLER
SKF . . . . SKF-Aktiebolaget Svenska Kullagerfabriken (Sweden)
SMITH, T.L. . . . . T.L. Smith Co.
SPEED-SPR. . . . . Speed Sprayer Plant Machinery Co.
(Also see BEAN, JOHN)
SPICER . . . . Spicer Dana Corp., Dana Corp. (Also see DANA)
STANDARD . . . . . Standard-Triumph Motor Co. Ltd. (Also see BMC)
STEIGER. . . . . Steiger Tractor Co.
STUDEBAKER . . . . Studebaker-Packard Corp.
SUBARU . . . . Fuji Heavy Industries, Ltd. (Japan)
SUNBEAM . . . . See CHRYSLER

SUTTON . . . . Sutton Engineering Co., Inc.
TAMPER. . . . . Tamper Inc.
TAYLOR . . . . Taylor Machine Co.
TEDDY TORQ . . . . Torque Mfg. Co. (Also see PRECISION,
PREC. TOR., and TORQUE)
TEL-E-LECT. . . . . Tel-e-lect Products, Inc.
TEREX . . . . Division of GMC-See GMC
TERRAIN . . . . Terrain King Corp.
THEW SHOV. . . . . Thew Shovel Co.
THOMPSON. . . . . Thompson Grinder Products Co.
TIMBERJACK . . . . Timberjack Inc. (Canada)
TIMBERLAND . . . . Timberland Ellicott, Inc.
TORO . . . . Toro Mfg. Co.
TORQUE . . . . Torque Manufacturing Corp., Precision
Universal Joint Corp. (Also see TEDDY TORQ,
& PRECISION & PREC. TOR.)
TOWMOTOR. . . . . Towmotor Corp. (Also see GERLINGER)
TOYOTA . . . . Toyota Cars & Trucks (Japan)
TRACTOMOT. . . . . See ALLIS-CHALMERS
TRANSMISS. . . . . Transmissions Ltd. (U.K.)
TRIUMPH . . . . Standard Triumph International Ltd.
(Gr. Britain) (Also see BMC)
TROJAN . . . . See EATON
TRU-CROSS . . . . Tru-Cross Corp.
TRW. . . . . TRW, Inc.
TULSA WIN. . . . . Tulsa Winch, Inc.
TUMPER . . . . Tumper Inc.
TWIN DISC . . . . Twin Disc, Inc.
UJS . . . . UJS International, Ltd.
ULRICH. . . . . Ulrich Products Corp.
ULS. . . . . ULS Manufacturing Corp.
UNIC . . . . Unic Truck Division of Fiat (France)
UNION PAC. . . . . Union Pacific Corp.
UNIPART. . . . . British Leyland Motor Corp., Ltd. (Also see BMC)
UNIT PAC . . . . Unit Packaging Machinery Co.

UNIT RIG . . . . Unit Rig Equipment Co.
VANDEN PLAS . . . . See BLMC
VAN DOORNE . . . . See DAF
VAUXHALL . . . . See BEDFORD
VERSATILE. . . . . Versatile Machine Co. (Gear Works)
VOLKSWAGEN . . . . Volkwagen Motors, A.G.
VOLVO . . . . Aktiebolaget Volvo, Ltd. (Sweden)
WABCO. . . . . . . Westinghouse Air Brake Co. (Also see ADAMS-LET.)
WAIN-ROY . . . . Wain-Roy, Inc.
WALTERSCHD . . . . Walterscheid, Inc. (West Germany)
WAYNE . . . . Wayne Sweeper Mfg. Co.
WEASLER . . . . Weasler Engineering, Inc.
WESCO. . . . . Western Universal Joint Co. (Also see
ALCO, ALLOY & CROY)
WEST. LAND . . . . Western Land Roller Co.
WESTINGHSE . . . . Westinghouse Electric Co.
WHITE . . . . White Motor Corp. (Also see DIAMOND T)
WHITE FARM . . . . White Farm Equipment Co. (Also see
OLIVER & MPL. MOLINE) (Subsidiary of
T.I.C. Investment Corp.)
WHITING. . . . . Whiting Corp.
WILK RIP. . . . . Wilk Rippers, Inc.
WILLYS . . Willys-Jeep Corp. (Also see AMER. MOTOR, JEEP KAISER)
WOLSELEY . . . . See BMC
WOOD BROS.. . . . . Wood Brothers Mfg. Co.
WOOLDRIDGE. . . . . Wooldridge Coal Machinery Co.
WORLDPARTS . . . . World Parts Dist. Corp.
WYLIE. . . . . Wylie Mfg. Co.
XLO-GWB . . . . XLO-GLB Cardan Shafts Ltd.
YALE & TOWNE . . . . See EATON
YOUNG FIRE. . . . . Young Fire Apparatus Corp.
Z-WAY . . . . Z-Way Spreader Mfg. Co.
ZELLER. . . . . . . Zeller Corp. (Also see MOTOR MAST & ANDREWS)
ZOOM . . . . Zoom, Inc.

YOUR ATTENTION PLEASE

Since this unique guide is unlike any other cross–reference ever published, and inasmuch as drive line components are complex and confusing, we urge you to read the next few pages carefully. Knowing how to use the International Drive Line Interchange Guide properly will greatly enhance its value to you.

Also...

While we have not recommended the use of our group numbers for use as a basic numbering system for inventory, purchasing or other similar reasons, several firms have elected to do this on their own volition. Should you decide to follow a similar procedure we suggest that you register this fact with us so that we can notify you if any extreme changes are made in the future on our group numbering system. At this time we have no plans or intentions of altering group numbers in any of our Interchange Guides.

HOW TO USE

THE INTERNATIONAL DRIVE LINE INTERCHANGE GUIDE

THE MOST COMPLETE AND EASY-TO-USE SINGLE VOLUME CROSS REFERENCE IN THE WORLD FOR LOCATING ALTERNATIVES FOR DRIVE LINES AND COMPONENTS INCLUDING UNIVERSAL JOINTS, ETC.

This book is designed to serve industry throughout the world, enabling users of drive lines and drive line components to quickly identify options. Interchange, Incorporated researched hundreds of sources to create this world-wide computerized information. Engineers, maintenance and sales personnel, purchasing agents, students and designers have expressed a need for global data in a concise format to assist them in their search for manufacturers and suppliers of universal joints, yokes, shafts, and other drive line components. The I.D.L.I. is the worlds' first labor-saving tool that can take the place of extensive catalog files of cross-reference information. It provides the most complete data on world-wide *producer* and *user* data ever available from a single source.

THE I.D.L.I. GUIDE FEATURES THE EXCLUSIVE RAPID REFERENCE SYSTEM ™

Combining all known numbers of foreign and domestic drive line component manufacturers with the original equipment manufacturers' (O.E.M.) numbers the I.D.L.I. Guide directs the reader into an easy-to-read alphabetized and dimensionally sequenced group. Using the whole part number, including prefixes and suffixes, the index (white pages) is sequenced from left to right giving preference to numerals, directing the reader to the proper group in the back (gold pages) portion of the guide. The group portion has been separated into several sections to facilitate the dimensional progression.

IMPORTANT NOTES

The I.D.L.I. is not intended to be an engineering manual. Its prime function is to provide basic comparisons and interchanges, and to support blueprints and specification data. Competing manufacturers often publish cross-references that may be workable in many instances although they are not 100% equals. While the complete unit or assembly may be acceptable as replacements, the component parts could be quite different from one another. Material content, tolerances, grease fittings, and/or overall design may differ considerably although the complete unit is recommended as an equivalent. This is particularly true when dealing with original equipment or replacement parts from different countries. Detailed deviations are not shown in this guide although they may exist and be acceptable as cross-reference numbers in most instances.

Different manufacturers often use different words to describe driveline components. For example: DRIVE LINE, DRIVE TRAIN and DRIVE SHAFT are a few of the ways used to describe components; CROSSES and TRUNNIONS are the same; ASSEMBLIES and KITS are often the same; etcetera. The configurations, or shapes, of many parts vary from one producers to another. Therefore, we have endeavored to use some of the most commonly accepted terminology and generic drawings to assist the reader as much as possible.

Because of space limitations and computerized compilation of our guide we have restricted dimensional data to only the basic information. The factory manuals should be consulted for more detailed engineering statistics. Where the descriptions in a group "slight variation from above group" there may be differences in dimensions other than those shown, or there may be a keyway or set screw or lube fitting not indicated in our specifications. Some of these details have been intentionally omitted because producers often take liberties in stating cross-references to their items, which, in most cases may be acceptable.

Interchanges are often listed in parts catalogs with very broad liberties having been taken. Complete assemblies are sometimes shown as an alternative for individual parts, and vice versa. Bearings, sometimes referred to as "caps", are not always the same size although the cross, or trunnion, may have been made to offset the differences so that the complete unit is interchangeable with corresponding part numbers. The same applies to the length and/or diameter of the cross, or trunnion, which may vary from one manufacturer to another although they do match up with the bearings designed for them. Lube fittings and center-to-center dimensions frequently vary from one manufacturer to another, although they are said to be interchangeable.

Rectangular bores can be used to replace square bores in most instances according to replacement catalogs.

The information compiled in this guide is accurate to the best of our knowledge. We urge you to advise us of any additions, corrections or suggestions you may have for our future editions. The data contained herein has been compiled from the most reliable sources available to us. Every effort has been made to insure that the data listed in this catalog is correct. However, we cannot assume responsibility for possible errors.

CONTENTS

I.D.L.I. Group Numbers 86600 and Up
Miscellaneous–Clamp Yokes, Tube & Weld Yoke Assemblies,
Double Center Yokes, Yoke Shafts, Etcetera
(Sequenced by inside or mounting recess of ears)

I.D.L.I. Group Numbers 87000 and Up
Miscellaneous–Clamp Yokes, Tube & Weld Yoke Assemblies,
Double Center Yokes, Yoke Shafts, Etcetera
(Sequenced by outside diameter of ears)

I.D.L.I. Group Numbers 87200 and Up
Miscellaneous–Champion Flanges, Center Bearings, Spline Plugs,
Splined Bearing Stubs, Tubing, Sleeves, Shafting, Etcetera
(Sequenced by the spline size or inside bore)

I.D.L.I. Group Numbers 87800 and Up
Miscellaneous–Unwelded Center Assemblies, Etcetera
(Sequenced by end-to-end dimension)

I.D.L.I. Group Numbers 87900 and Up
Miscellaneous–Slip Assemblies, Etcetera
(Sequenced by center-to-center dimension)

I.D.L.I. Group Numbers 88000 and Up
Miscellaneous items from all sections without dimensions or proper
descriptions. (We invite your input on this section, as well as any other
data you may have available that should be considered for future
editions of this guide)

NOTES ABOUT THE

ILLUSTRATIONS & DESCRIPTIONS

Most groups are described to give a general explanation of the item.

Figure numbers also show the illustration on the following pages.

These drawings may or may not look exactly like the part you are replacing. While they may differ in appearance, the alternative cross-references are purported to be workable options to the best of our knowledge.

Descriptions also differ from one producer to another. We have endeavored to identify these drive line components with commonly used terms.

WE INVITE YOUR COMMENTS ON
CHANGES AND ADDITIONS THAT
MIGHT BE MADE IN THE FUTURE
I.D.L.I. GUIDES

DRAWINGS

SHOWN ON THE FOLLOWING PAGES

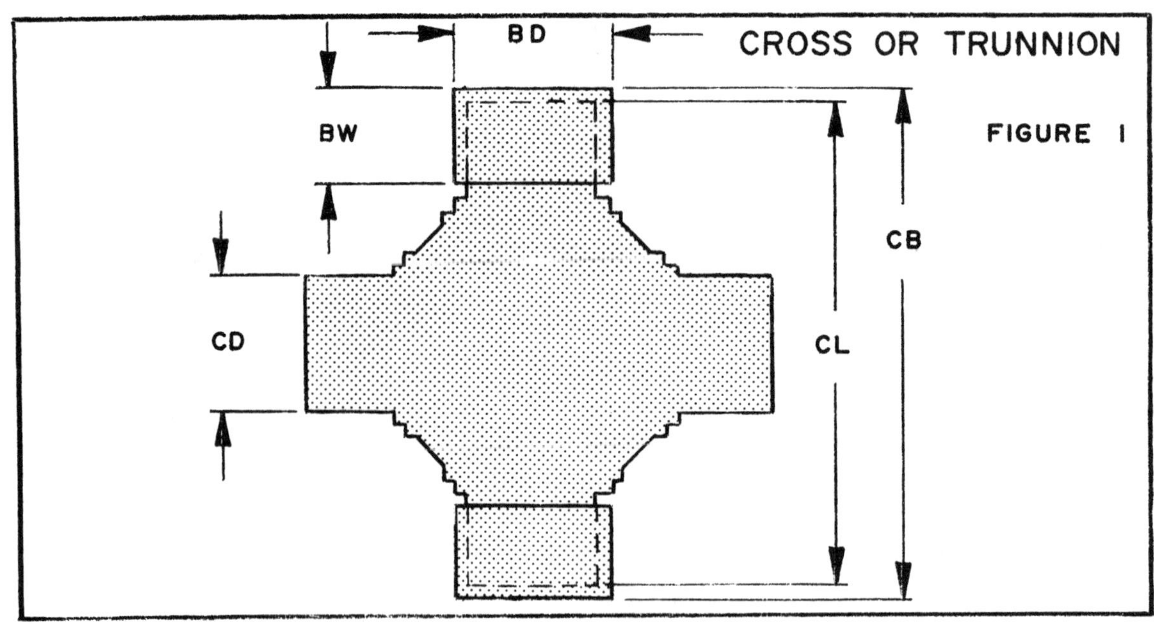

CROSS OR TRUNNION
FIGURE 1
BD
BW
CD
CB
CL

A
BD
BW
PLAIN ROUND
B
BD
BW
ROUND WITH GROOVE
C
BD
BW
SERRATED
D
BD
BW
SLOTTED
E
BD
BW
WITH BACKPLATE
F
BD
BW
FLANGED
G
CC
LOW WING DRILLED
H
CC
LOW WING THREADED
I
CC
HIGH WING DRILLED
J
CC
HIGH WING THREADED
K
CC
MIDWING OR DELTA
L
CC
BLOCK

INSIDE SNAP RING
OUTSIDE SNAP RING
BEARING PLATE
THRUST PLATE
WING BEARING
CAP & BOLT
U-BOLT
STRAP
CAP SCREWS
LOCK PLATE
LOCK PLATE
LOCK PLATE

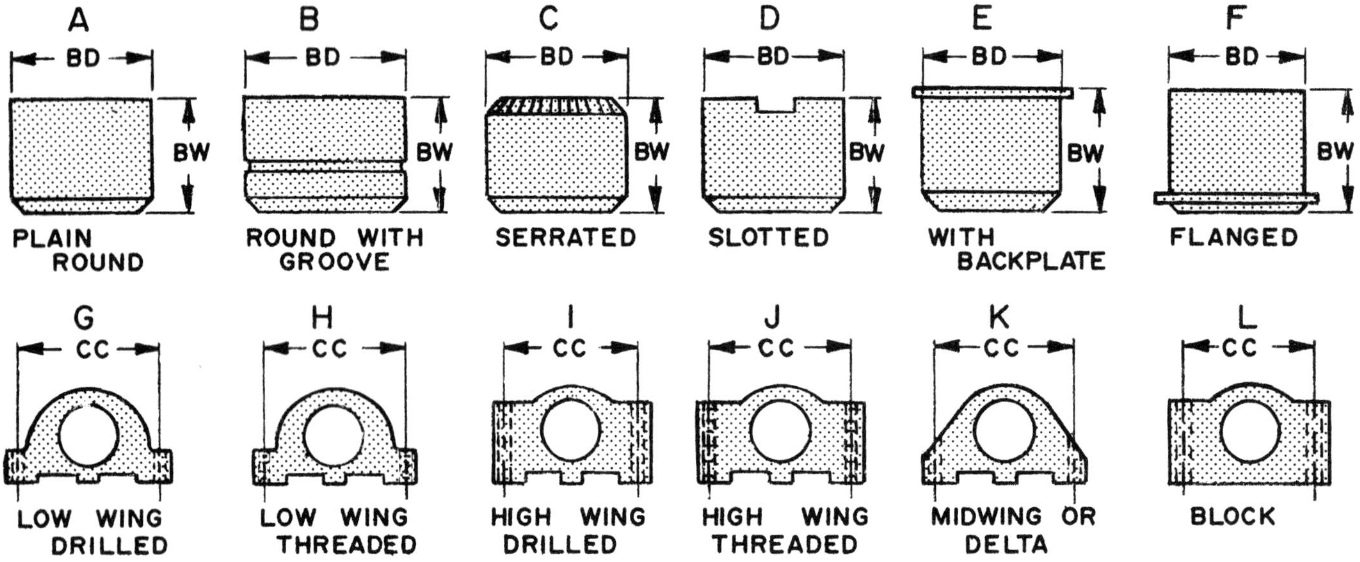

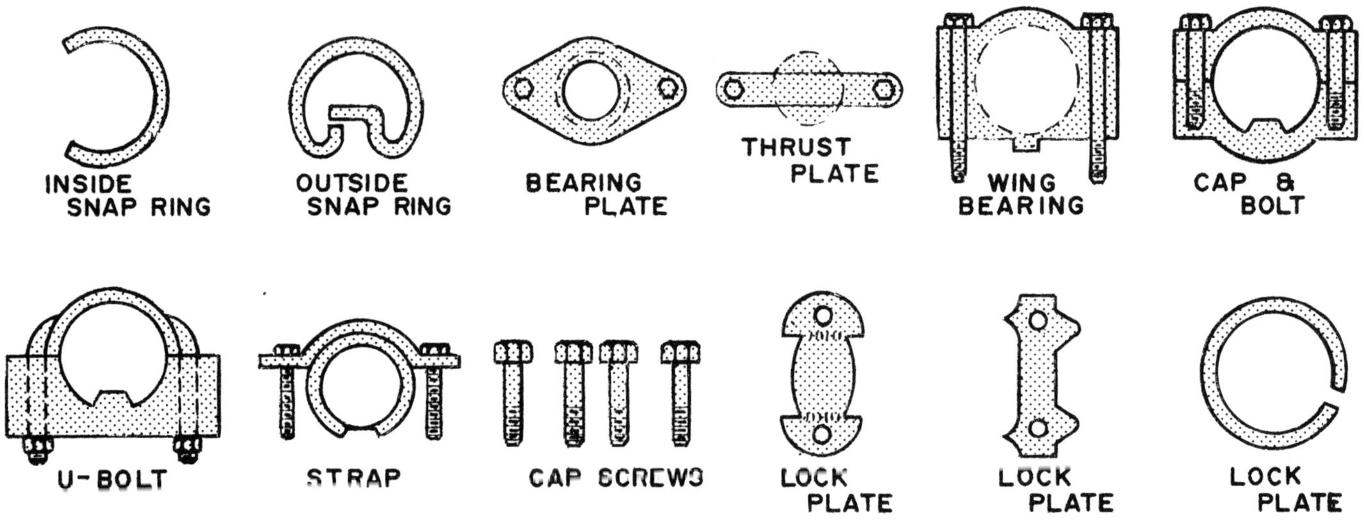

DETROIT POT STYLE REPLACEMENT KITS

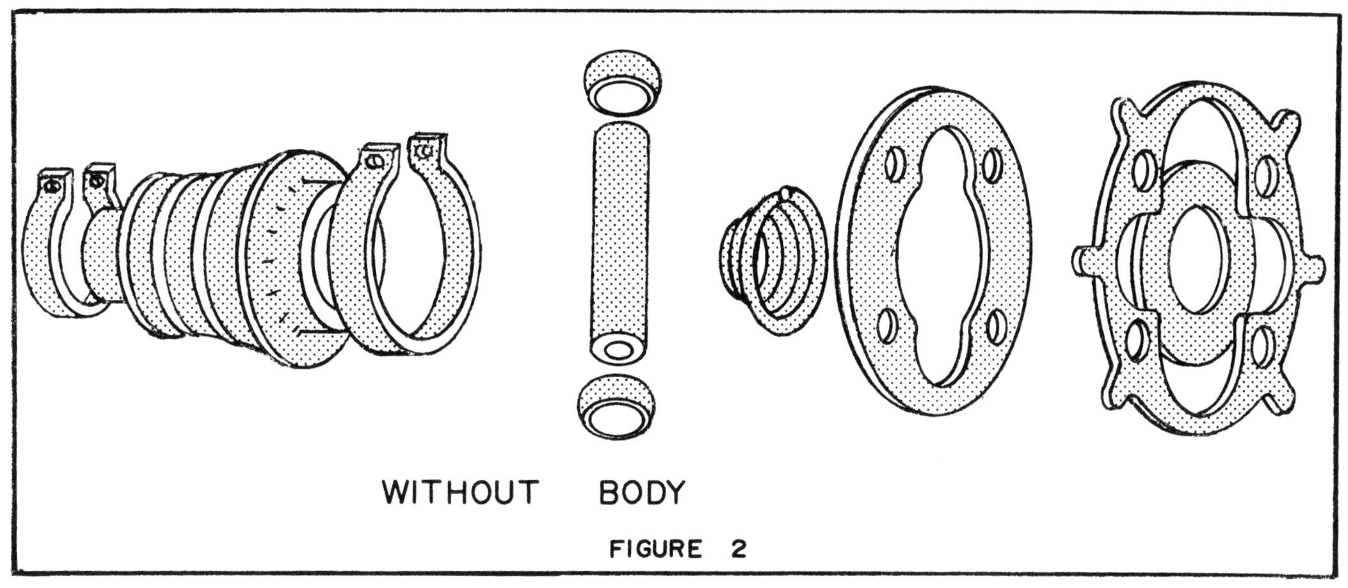

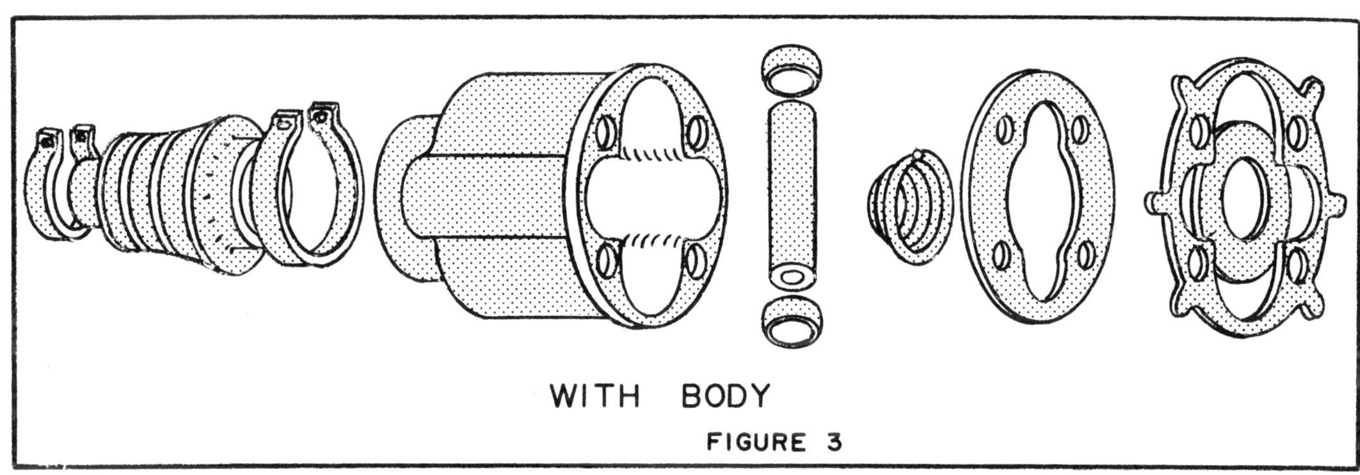

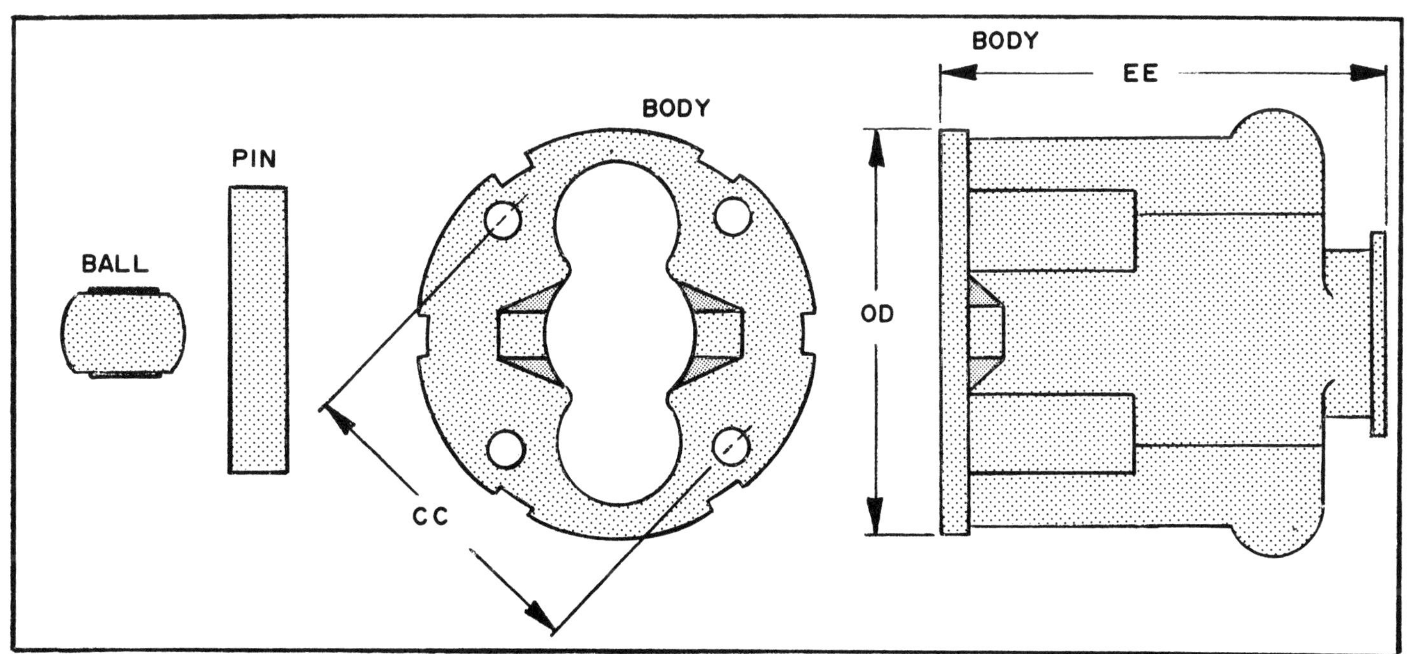

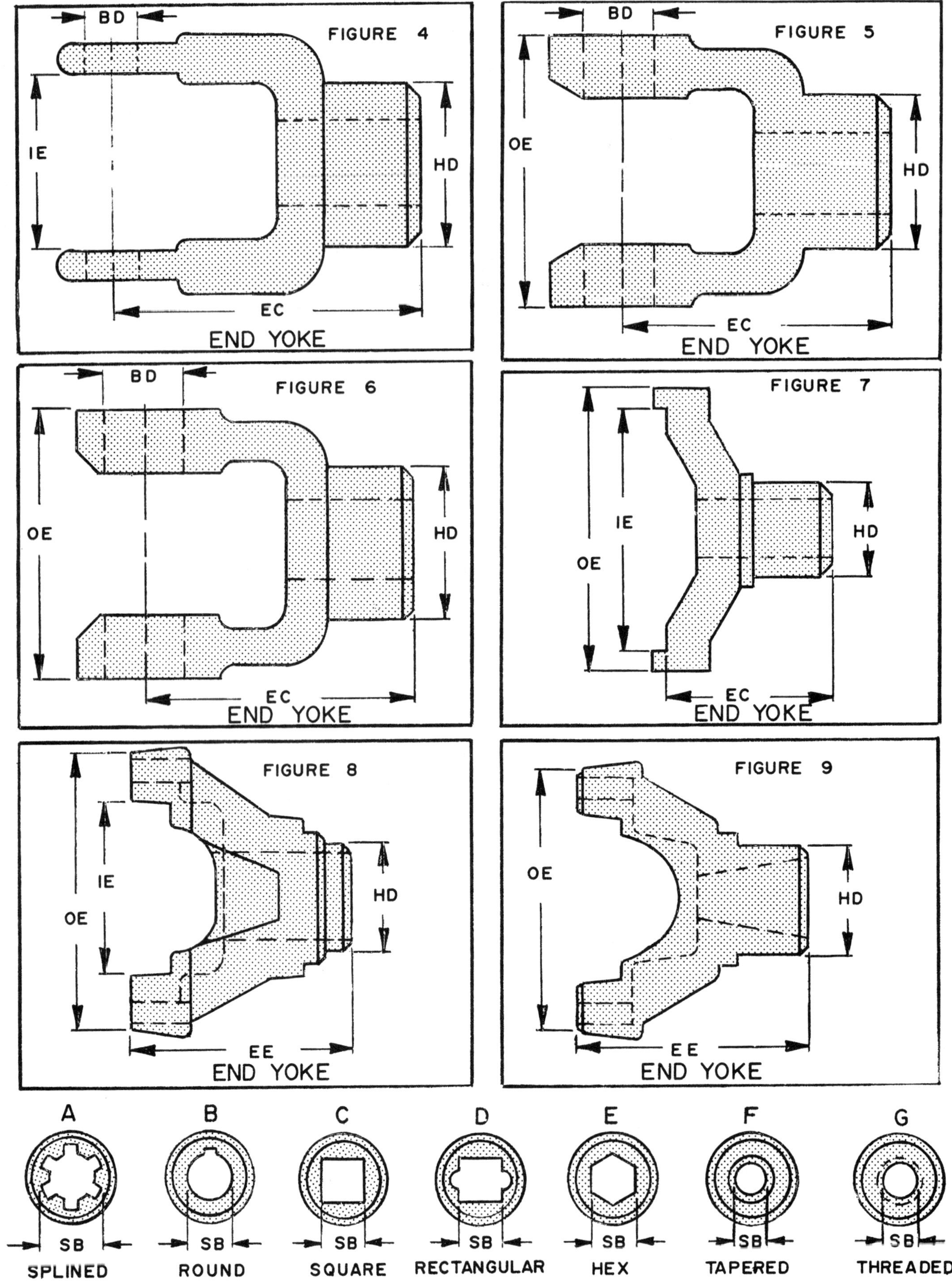

BD
FIGURE 4
IE
HD
EC
END YOKE
BD
FIGURE 5
OE
HD
EC
END YOKE
BD
FIGURE 6
OE
HD
EC
END YOKE
FIGURE 7
IE
OE
HD
EC
END YOKE
FIGURE 8
IE
OE
HD
EE
END YOKE
FIGURE 9
OE
HD
EE
END YOKE
A
SB
SPLINED
B
SB
ROUND
C
SB
SQUARE
D
SB
RECTANGULAR
E
SB
HEX
F
SB
TAPERED
G
SB
THREADED

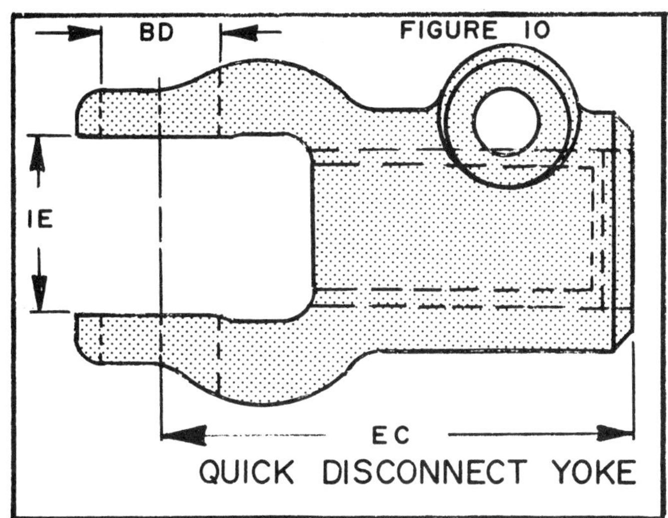

BD
FIGURE 10
IE
EC
QUICK DISCONNECT YOKE

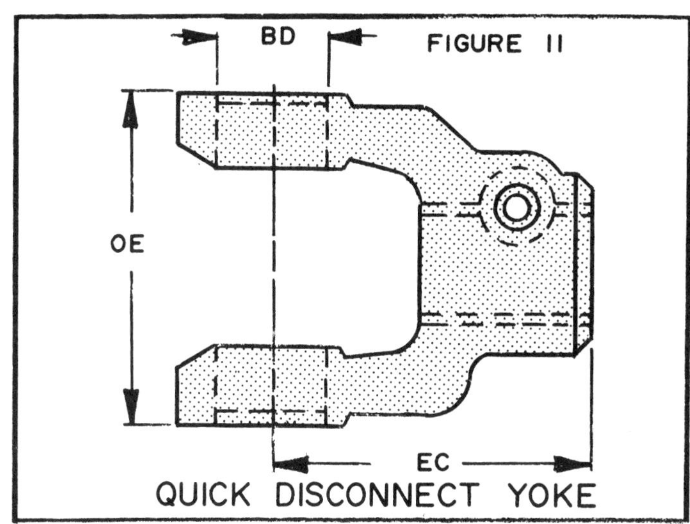

BD
FIGURE 11
OE
EC
QUICK DISCONNECT YOKE

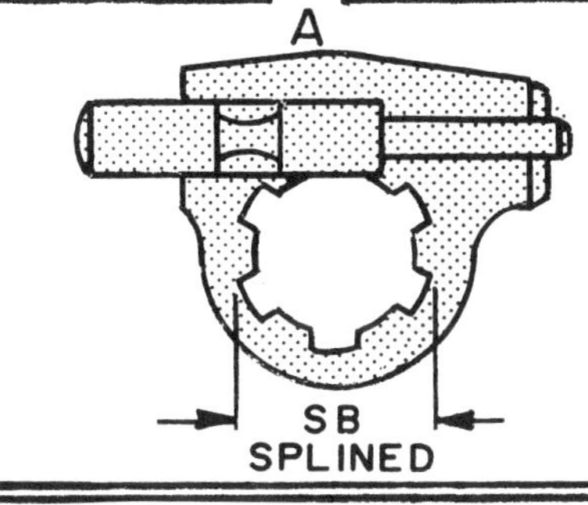

A
SB
SPLINED

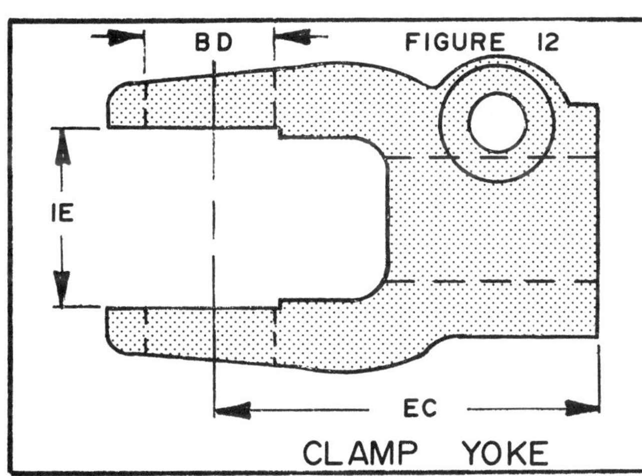

BD
FIGURE 12
IE
EC
CLAMP YOKE

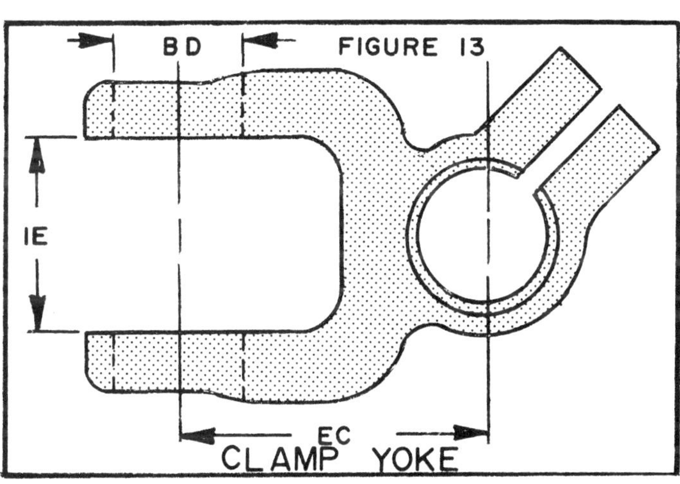

BD
FIGURE 13
IE
EC
CLAMP YOKE

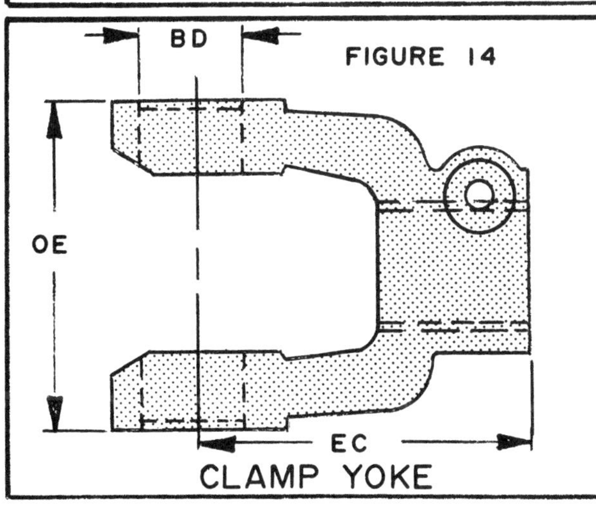

BD
FIGURE 14
OE
EC
CLAMP YOKE

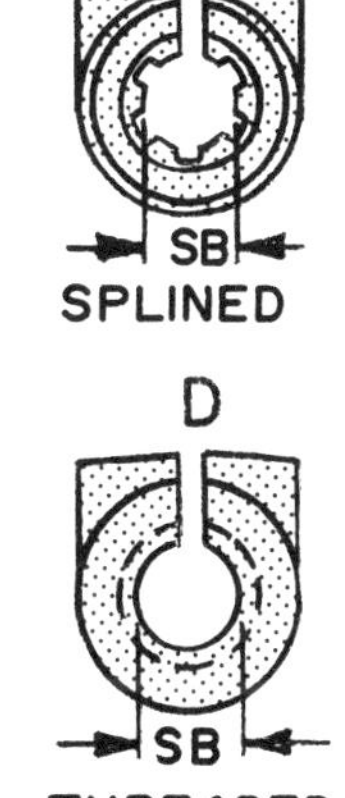

A
SB
SPLINED
D
SB
THREADED

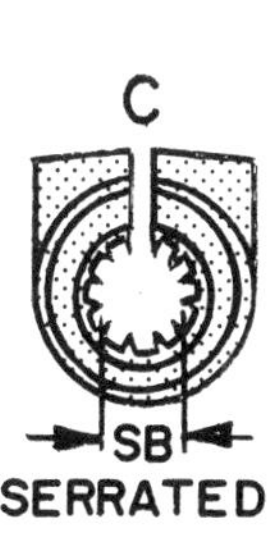

C
SB
SERRATED

B
SB
ROUND
E
SB
HEX

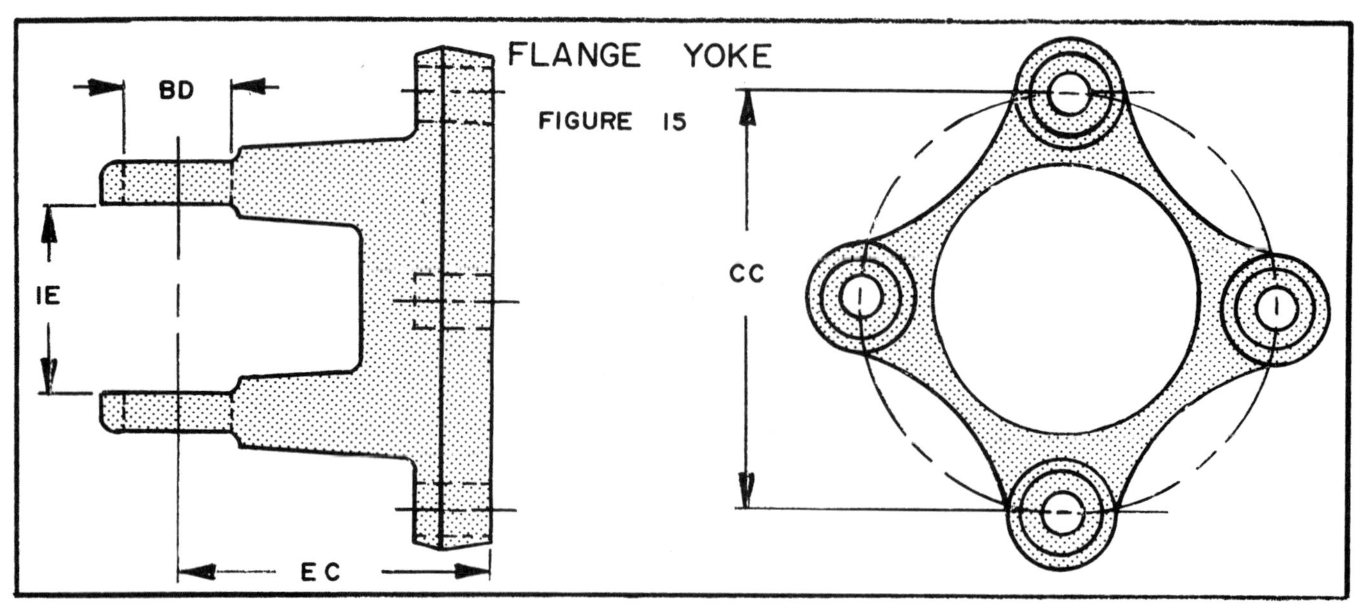

FLANGE YOKE
FIGURE 15
BD
IE
EC
CC

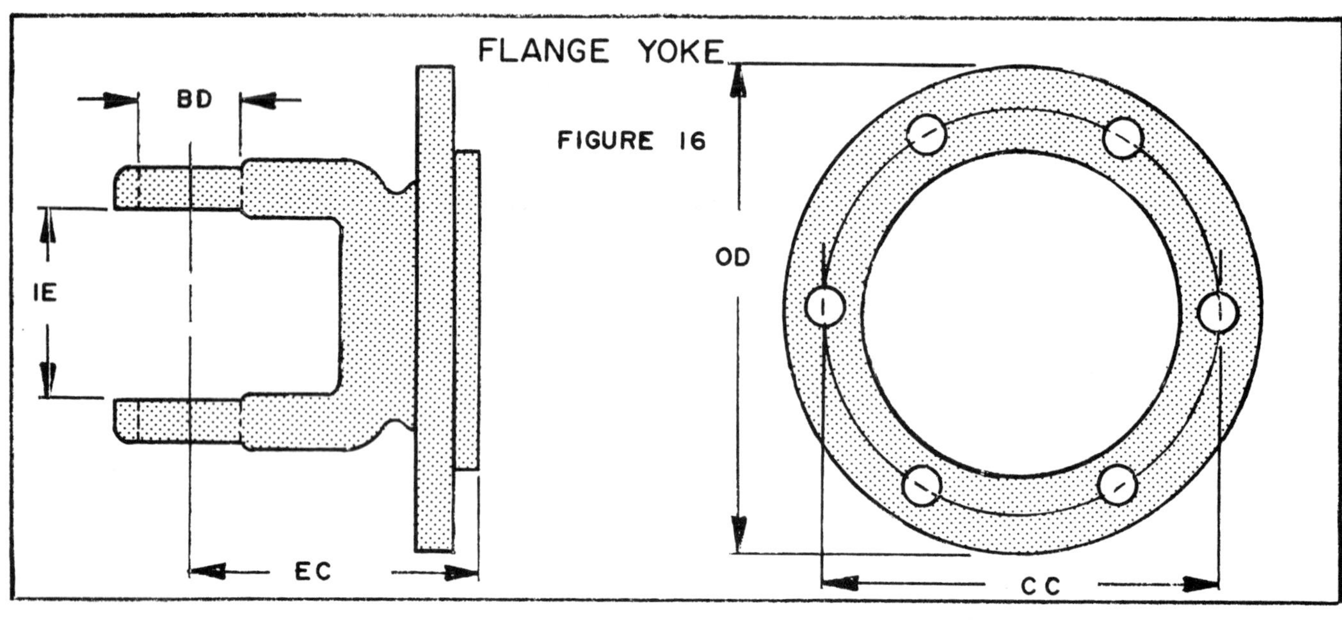

FLANGE YOKE
FIGURE 16
BD
IE
EC
OD
CC

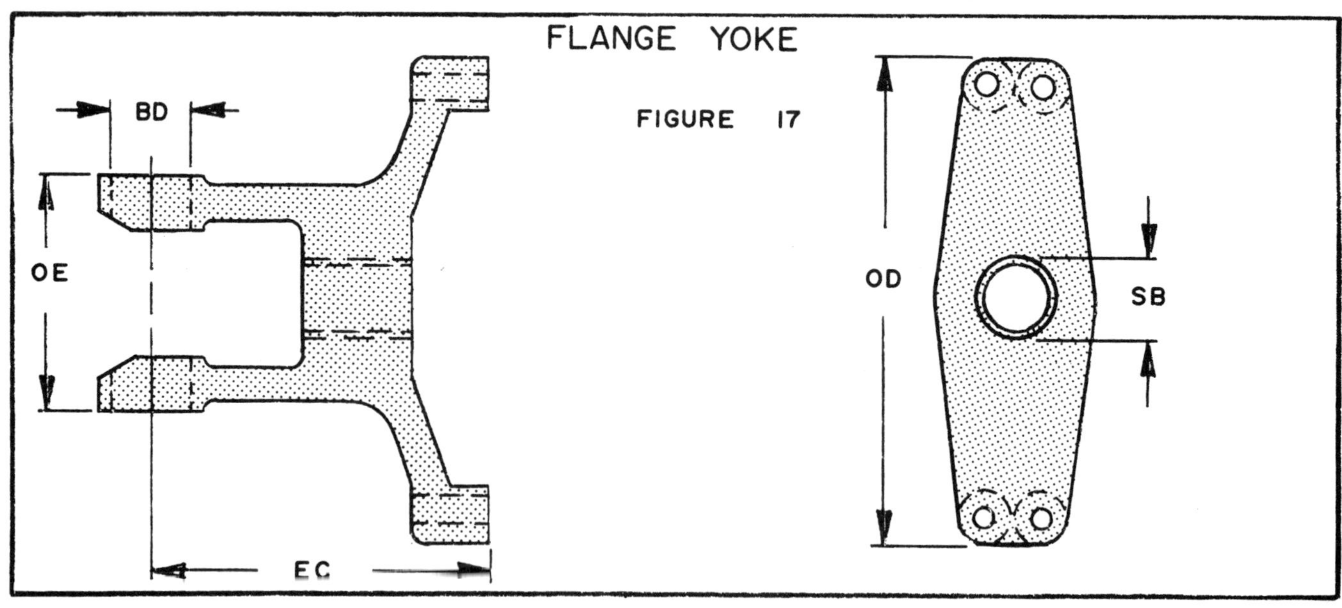

FLANGE YOKE
FIGURE 17
BD
OE
EC
OD
SB

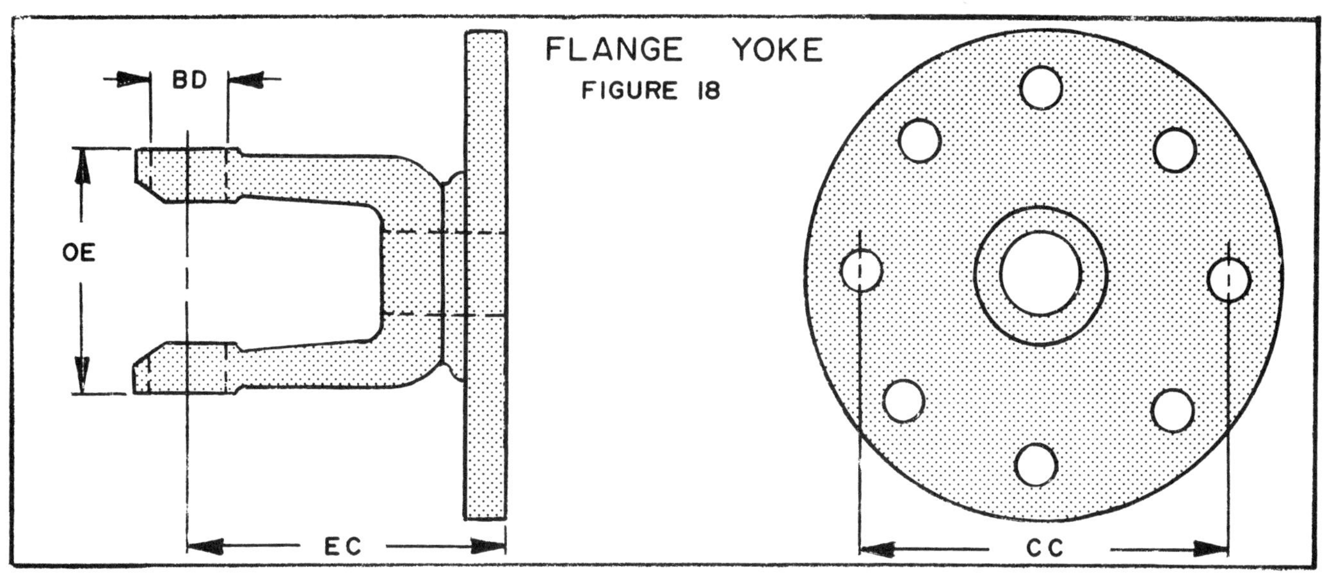

FLANGE YOKE
FIGURE 18
BD
OE
EC
CC

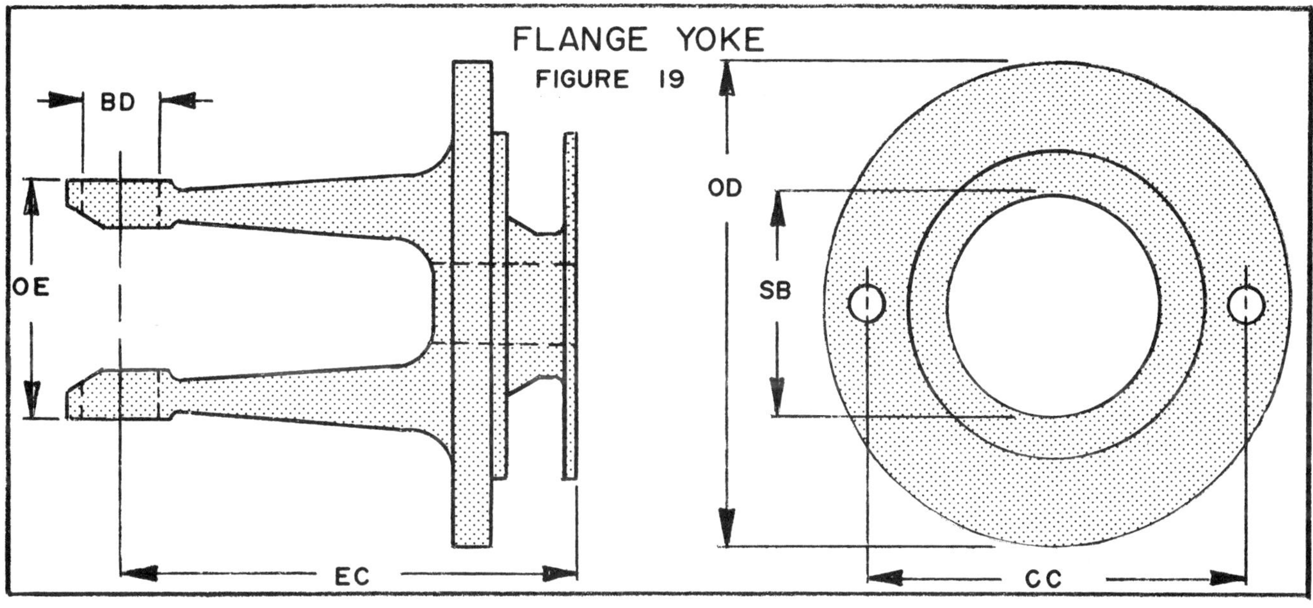

FLANGE YOKE
FIGURE 19
BD
OE
EC
OD
SB
CC

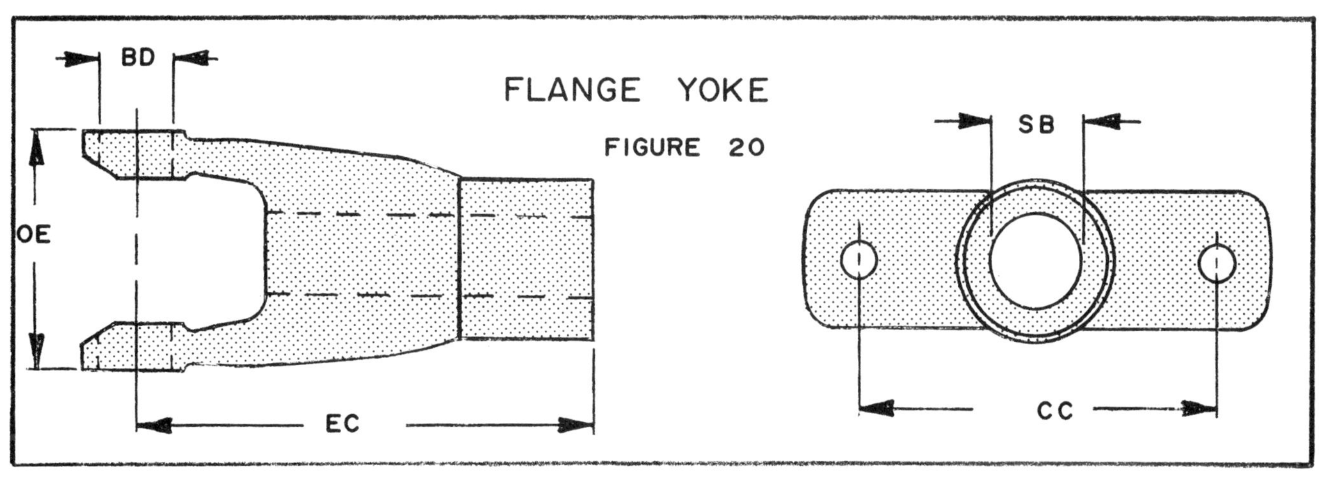

FLANGE YOKE
FIGURE 20
BD
OE
EC
SB
CC

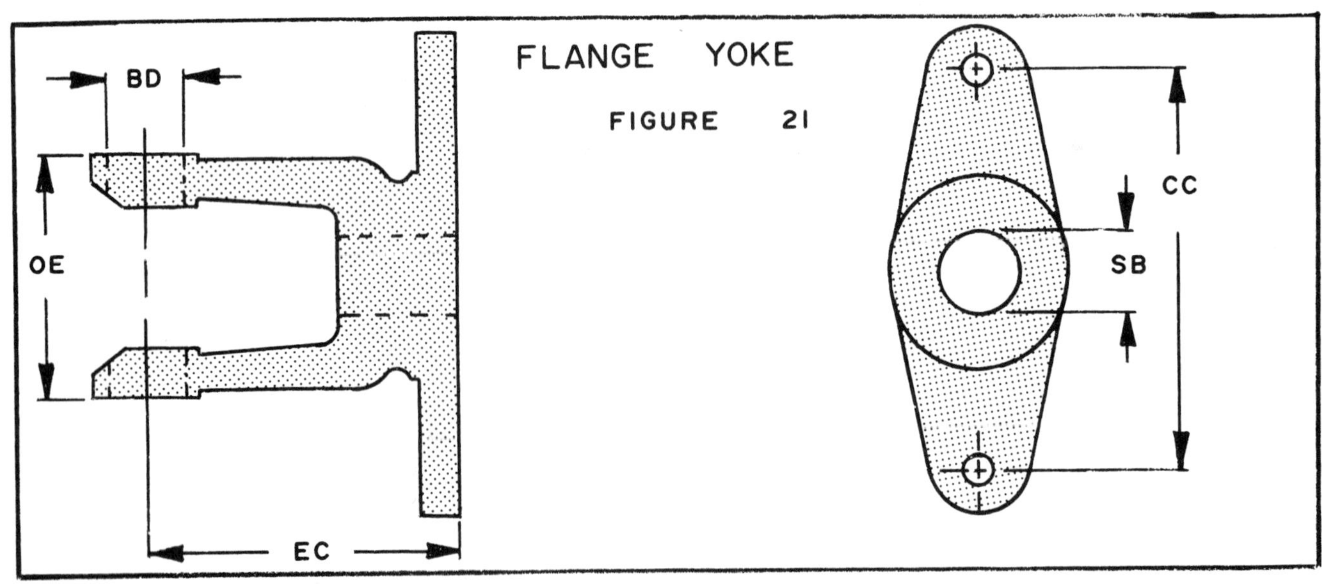

BD
FLANGE YOKE
FIGURE 21
OE
CC
SB
EC

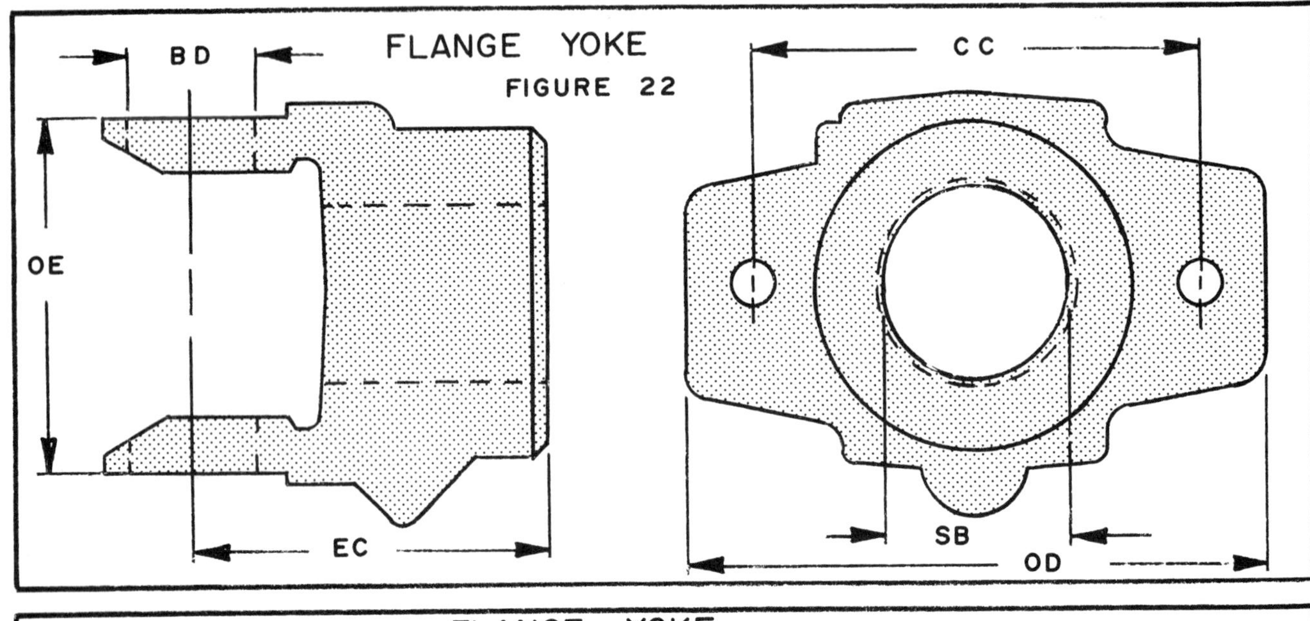

BD
FLANGE YOKE
FIGURE 22
CC
OE
SB
EC
OD

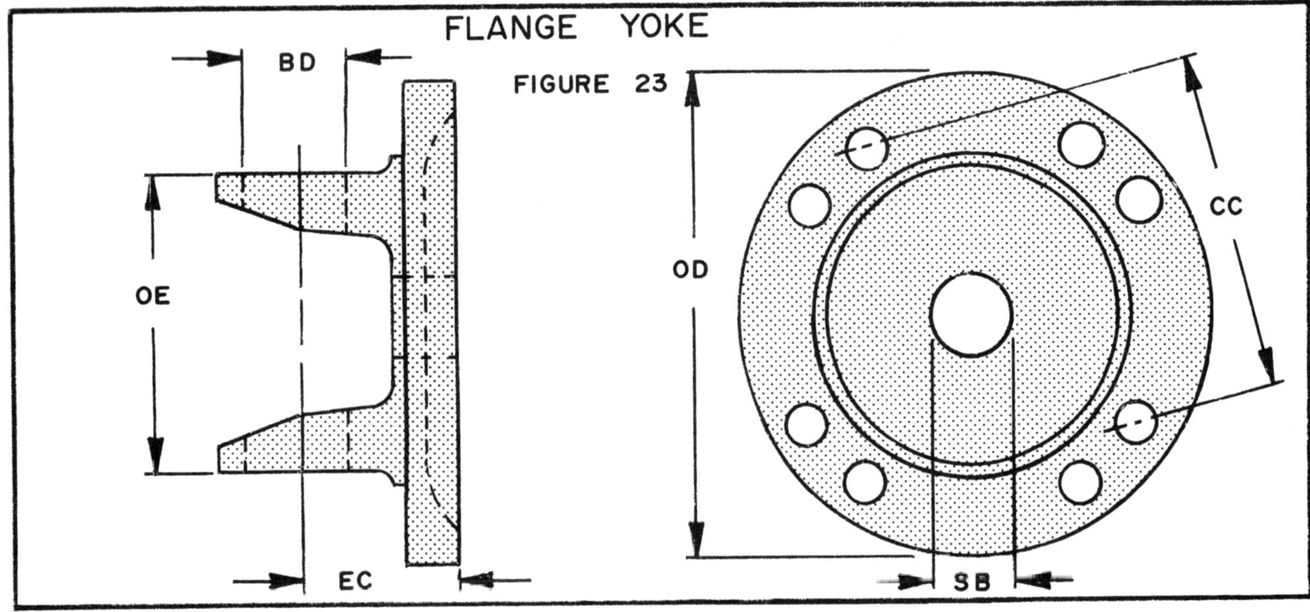

FLANGE YOKE
BD
FIGURE 23
CC
OD
OE
SB
EC

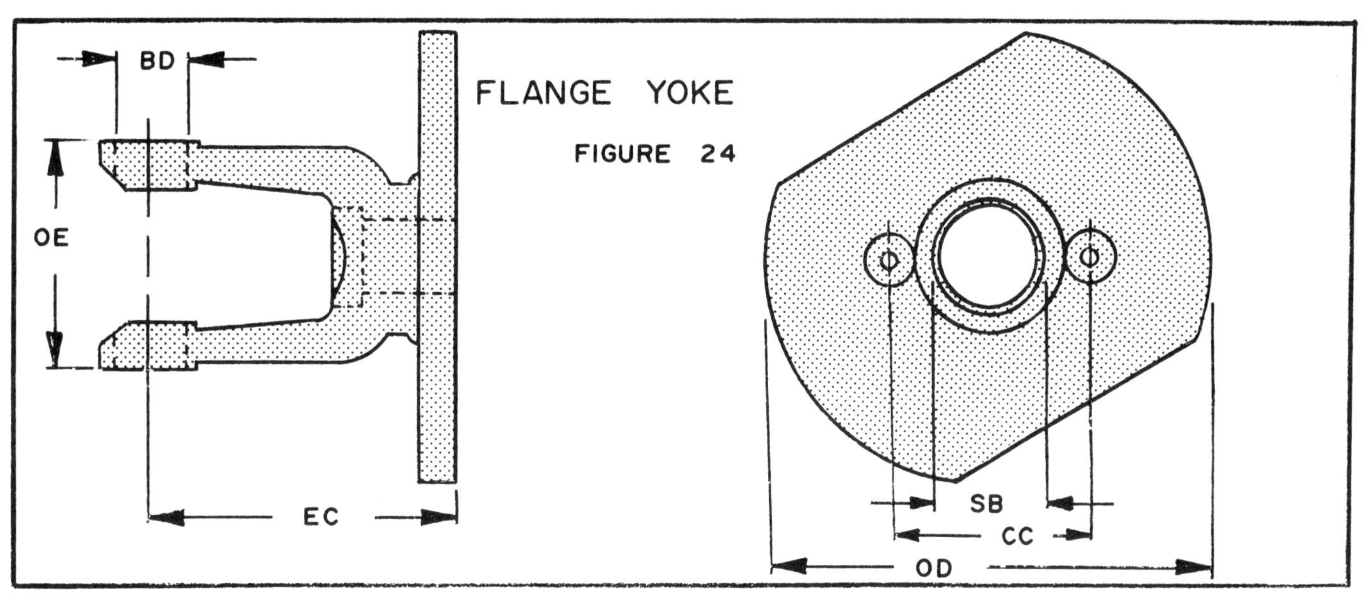

BD
OE
EC
FLANGE YOKE
FIGURE 24
SB
CC
OD

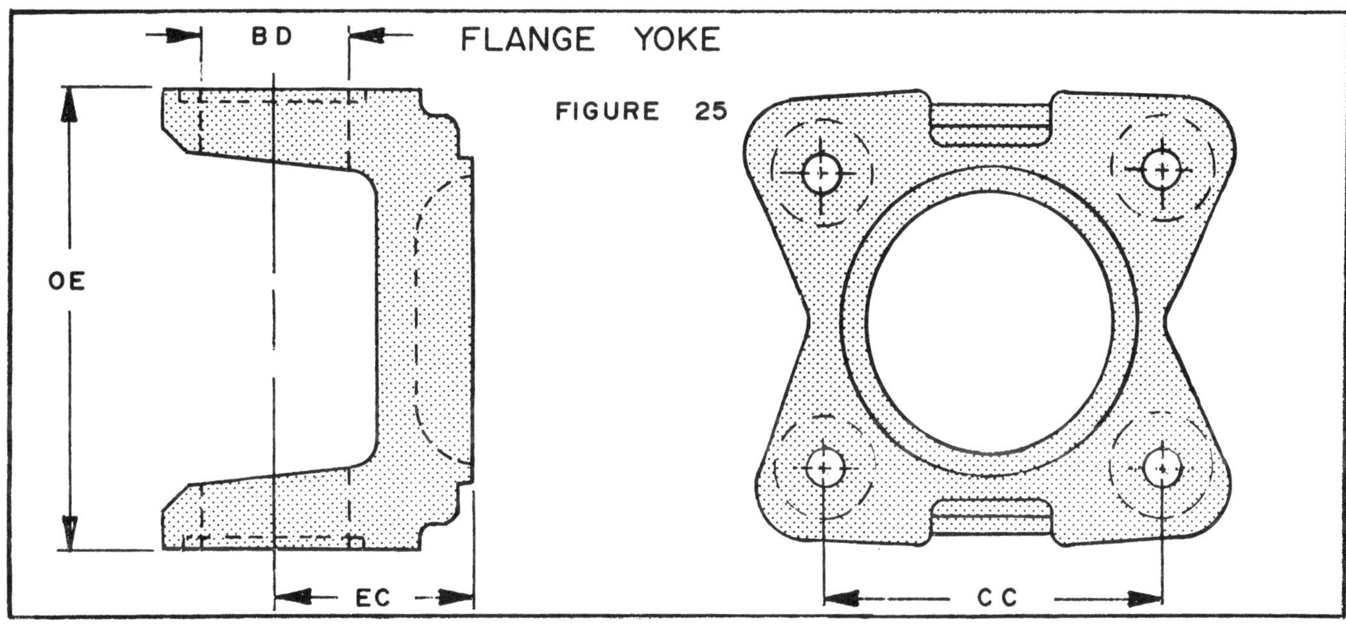

BD
OE
EC
FLANGE YOKE
FIGURE 25
CC

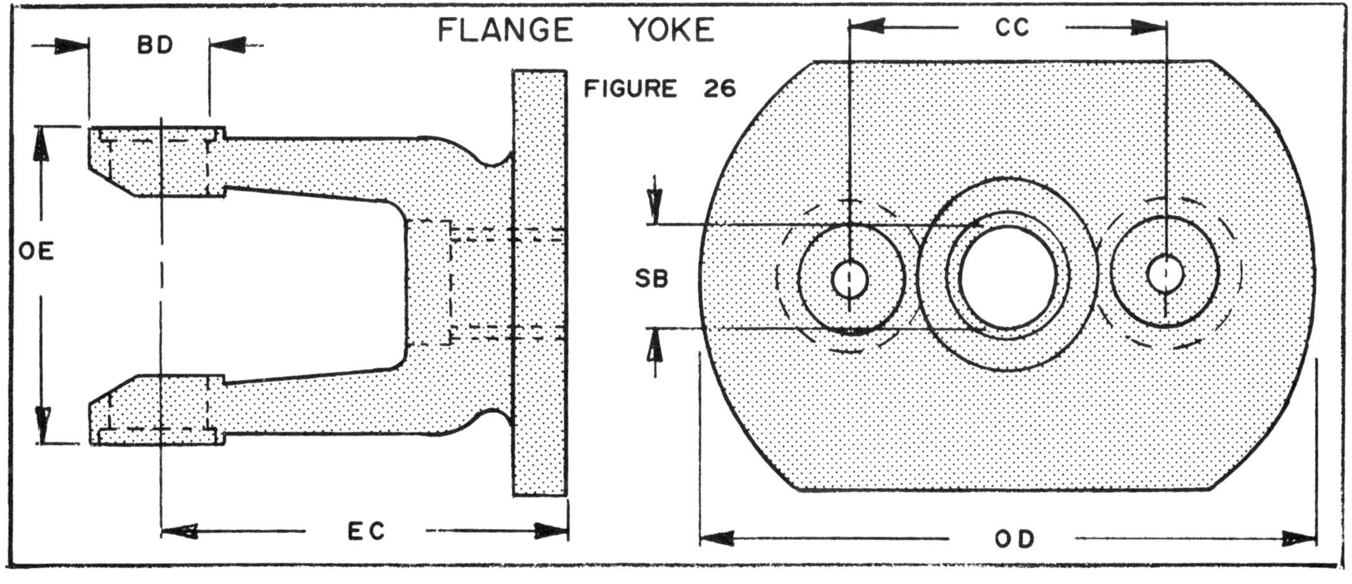

BD
OE
EC
FLANGE YOKE
FIGURE 26
CC
SB
OD

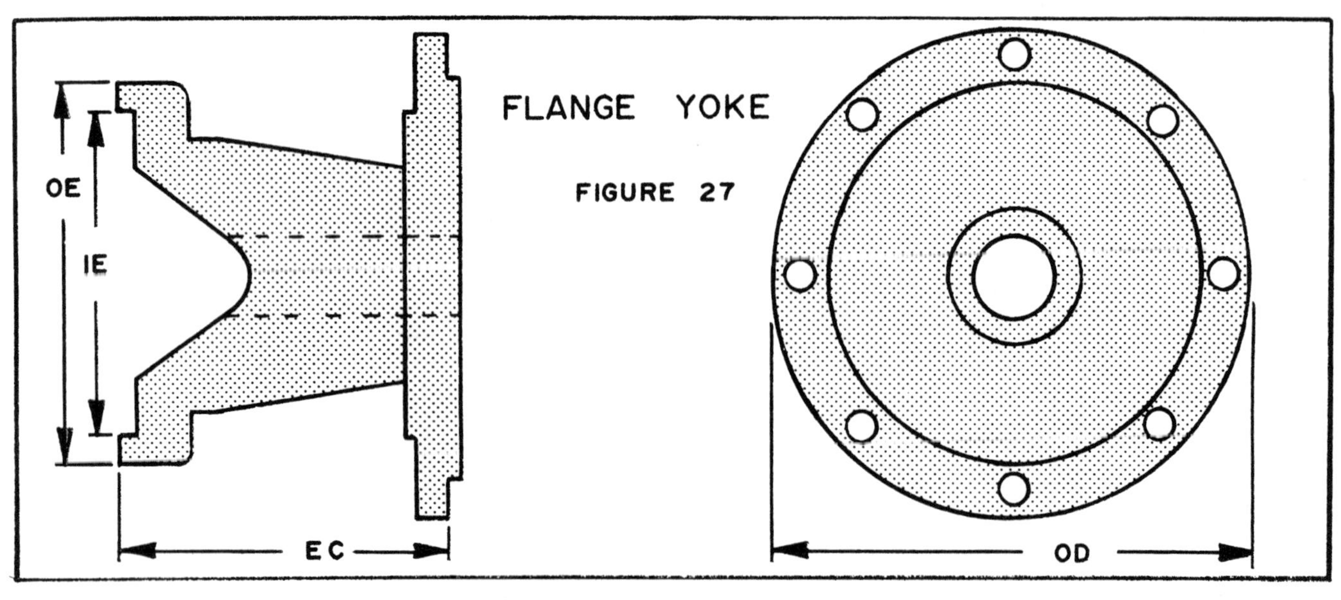

FLANGE YOKE
FIGURE 27
OE
IE
EC
OD

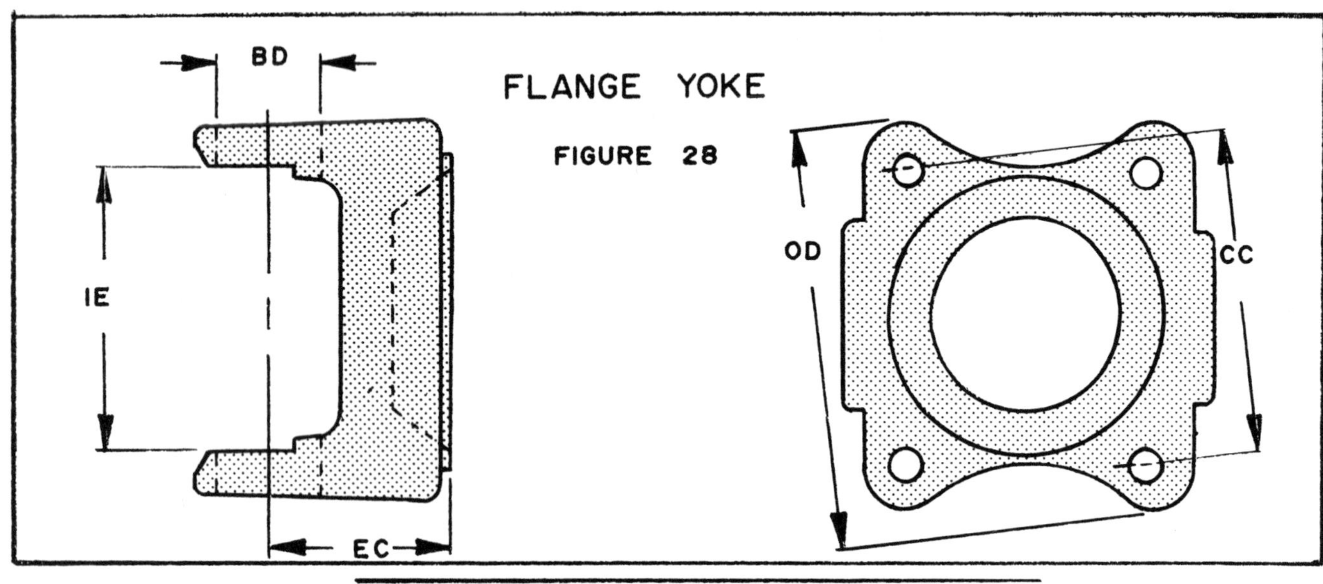

BD
FLANGE YOKE
FIGURE 28
IE
EC
OD
CC

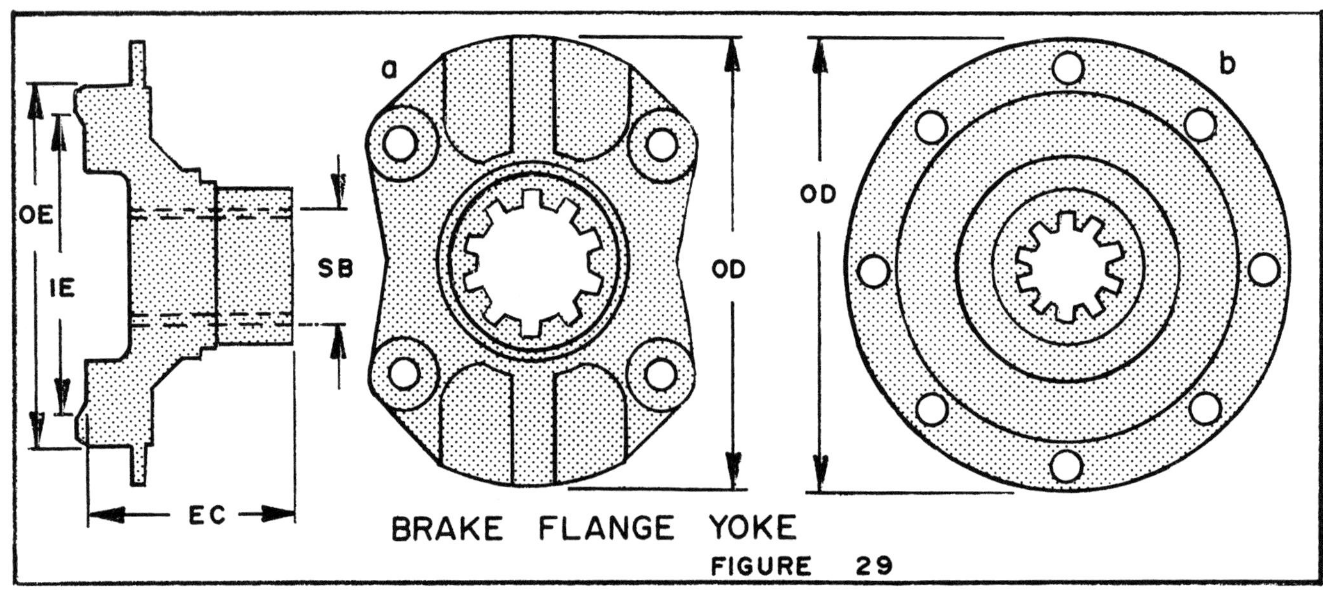

a
b
OE
IE
SB
OD
OD
OD
EC
BRAKE FLANGE YOKE
FIGURE 29

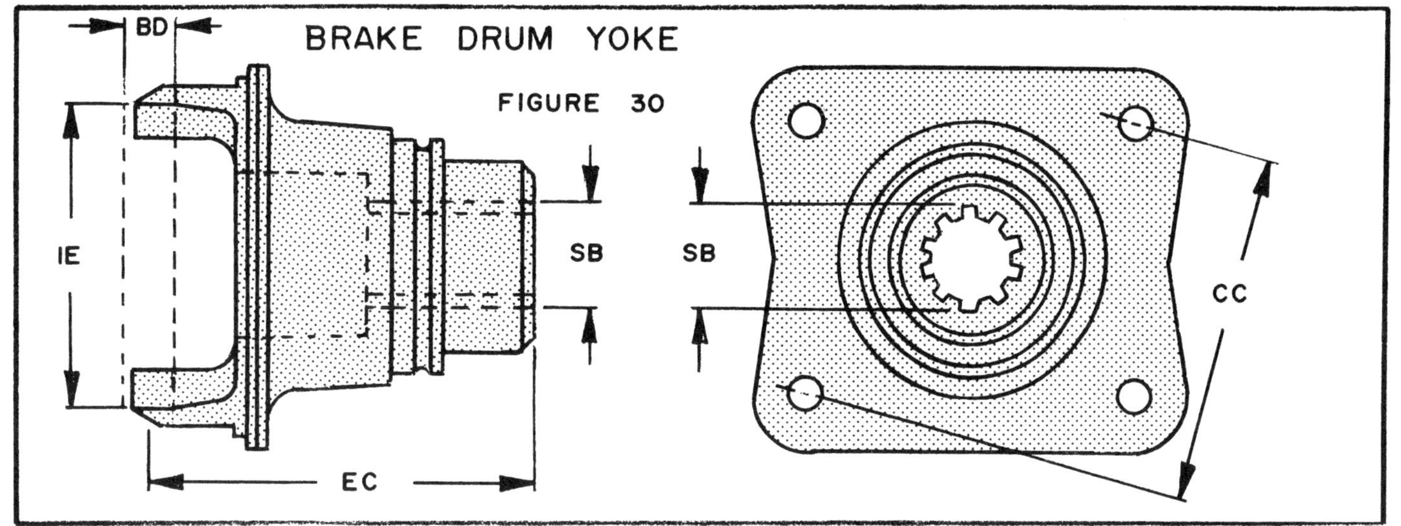

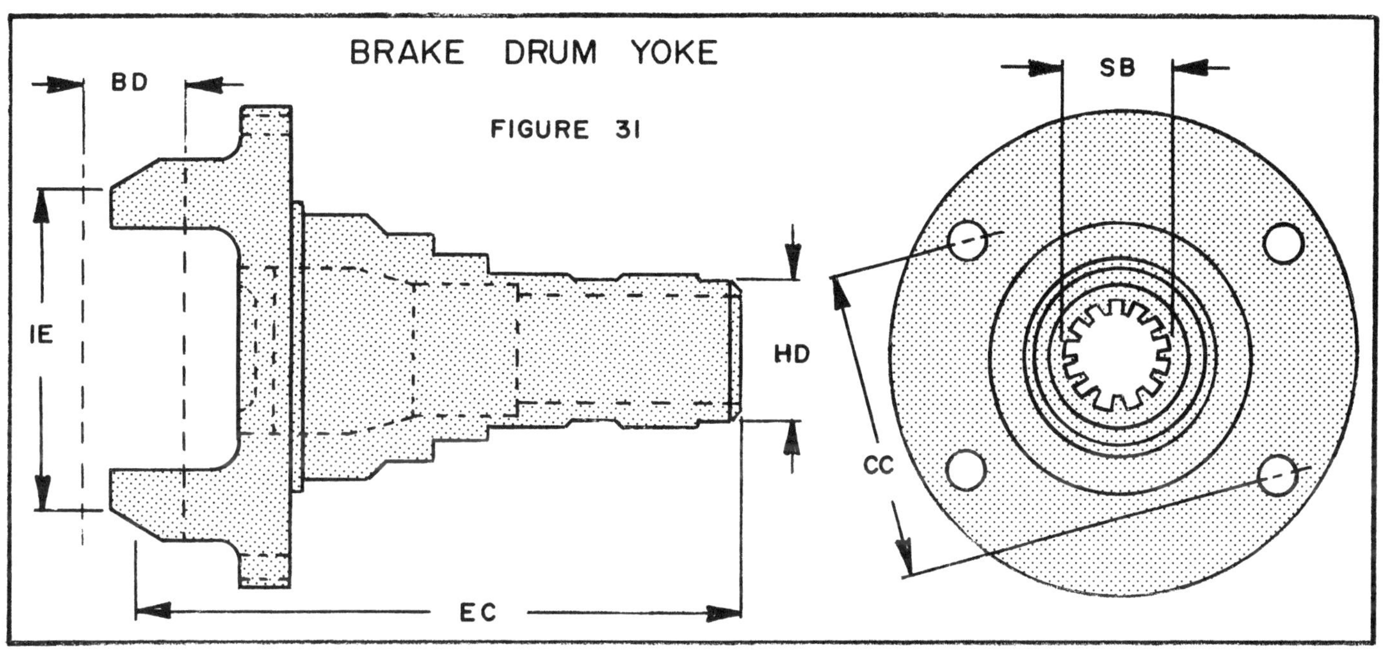

ILLUSTRATIONS & DESCRIPTIONS

These drawings may or may not look exactly like the part you are replacing. While they may differ in appearance, the alternative cross-references are purported to be workable options to the best of our knowledge.

Descriptions also differ from one producer to another. We have endeavored to identify these drive line components with commonly used terms.

COMPANION FLANGES

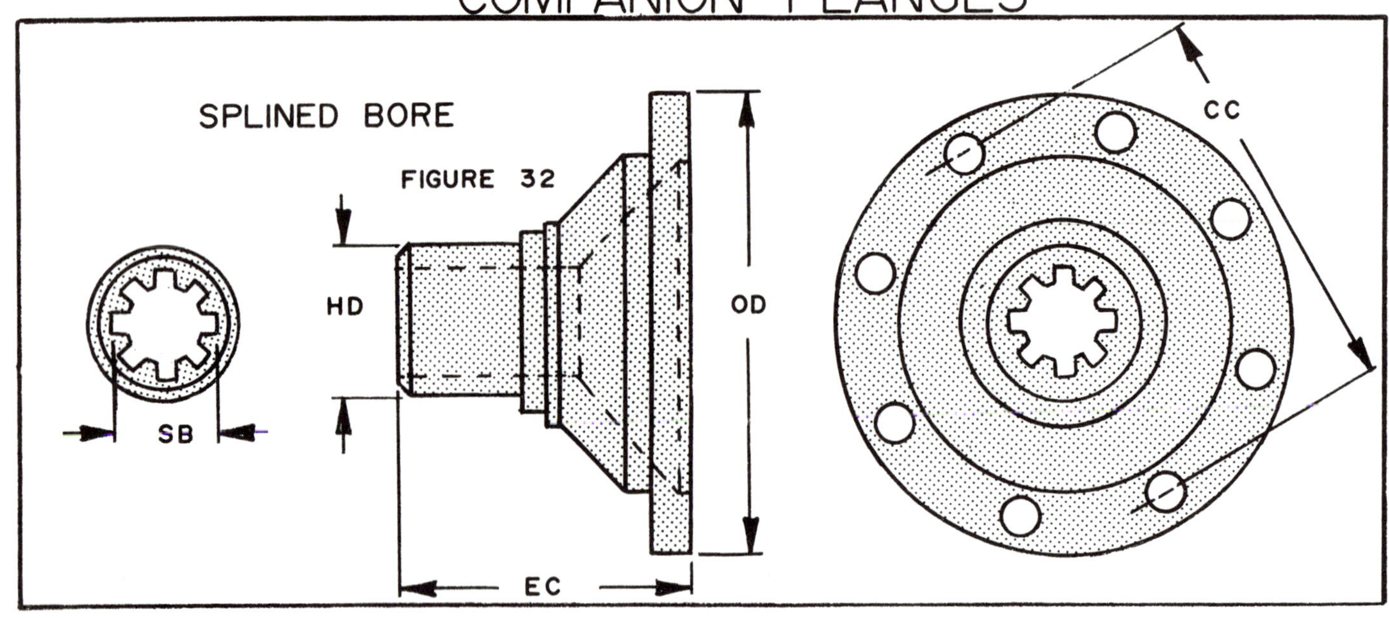

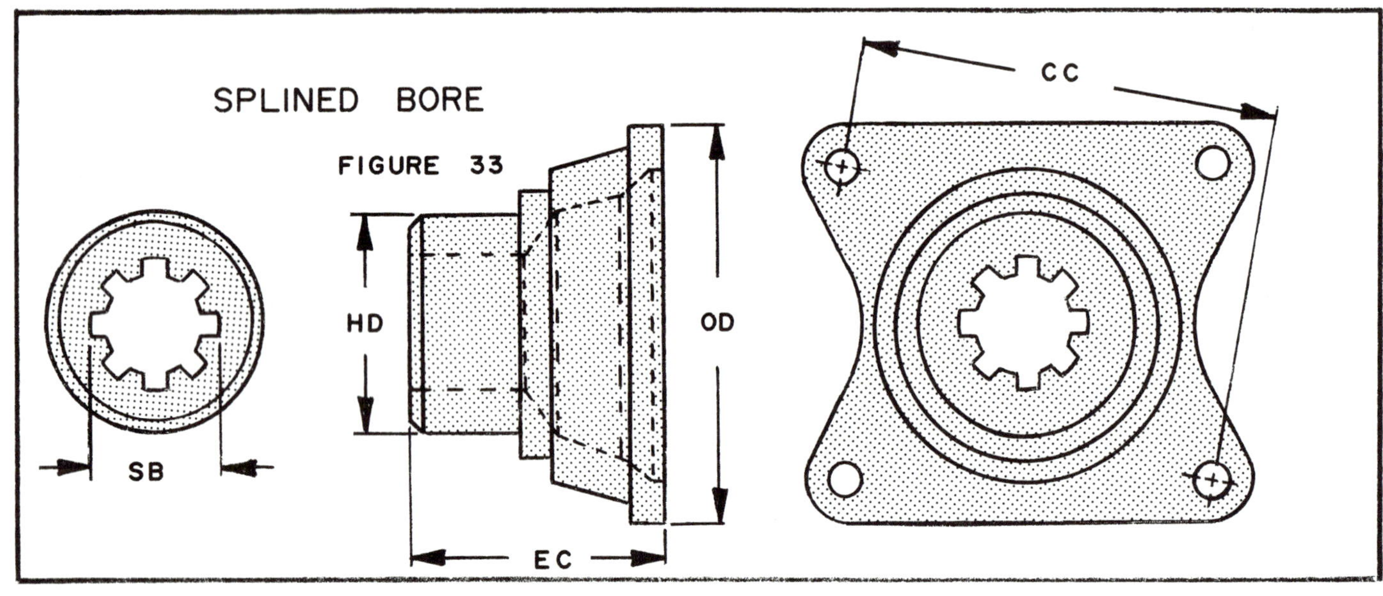

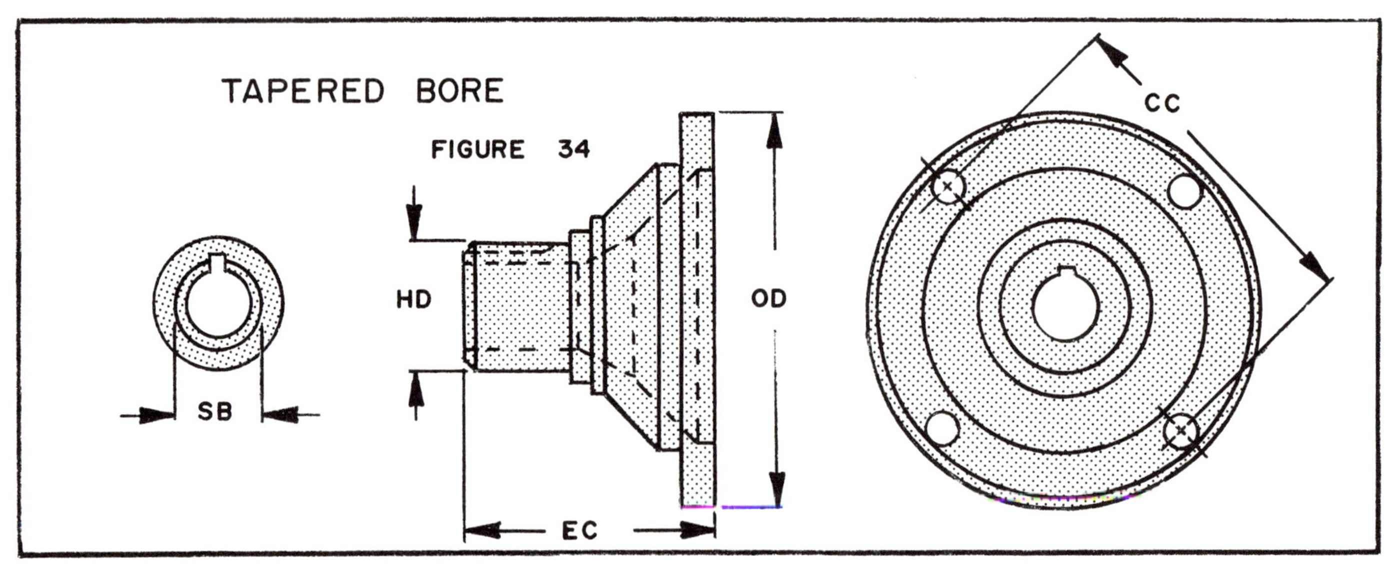

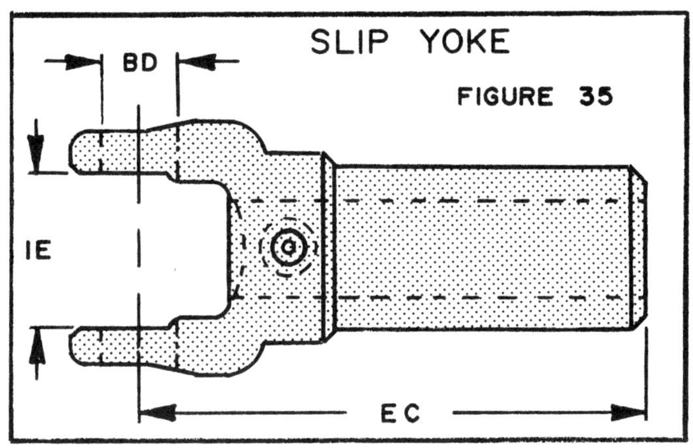

SLIP YOKE
FIGURE 35
BD
IE
EC

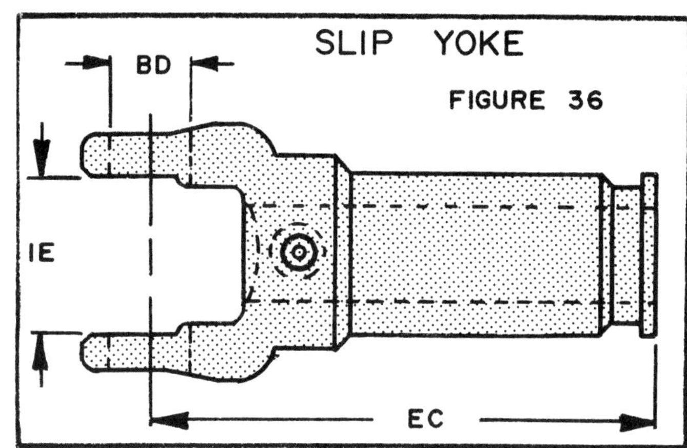

SLIP YOKE
FIGURE 36
BD
IE
EC

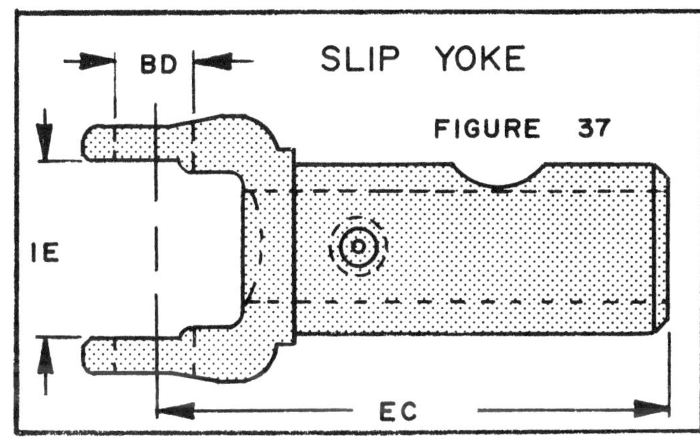

SLIP YOKE
FIGURE 37
BD
IE
EC

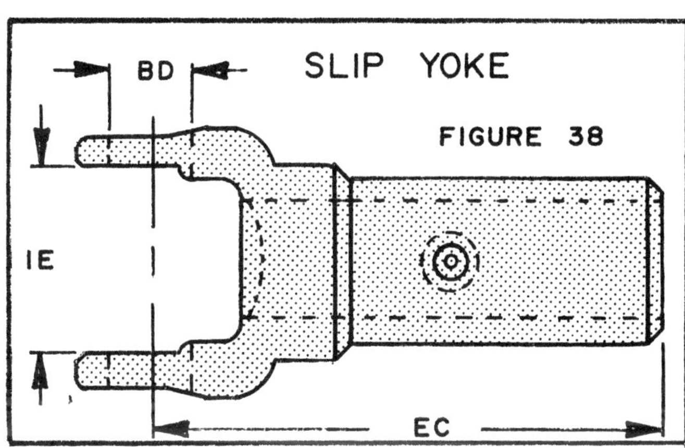

SLIP YOKE
FIGURE 38
BD
IE
EC

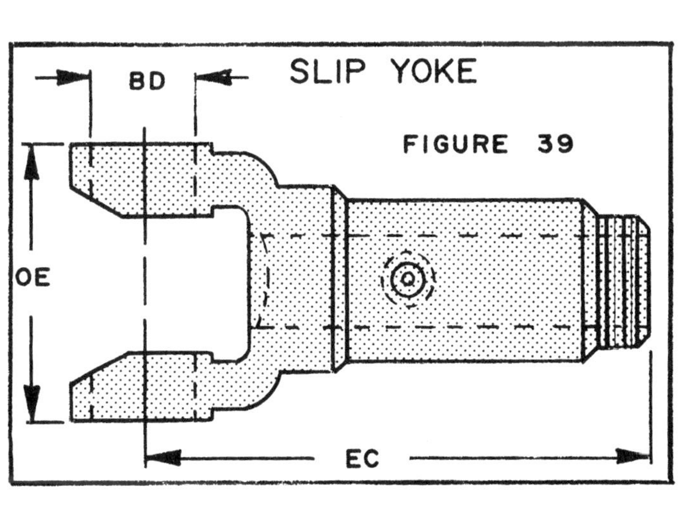

SLIP YOKE
FIGURE 39
BD
OE
EC

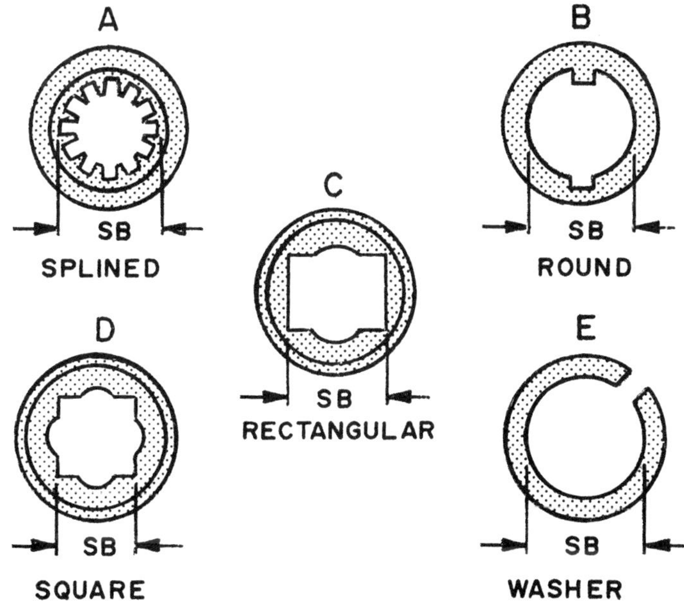

A
SB
SPLINED
B
SB
ROUND
C
SB
RECTANGULAR
D
SB
SQUARE
E
SB
WASHER

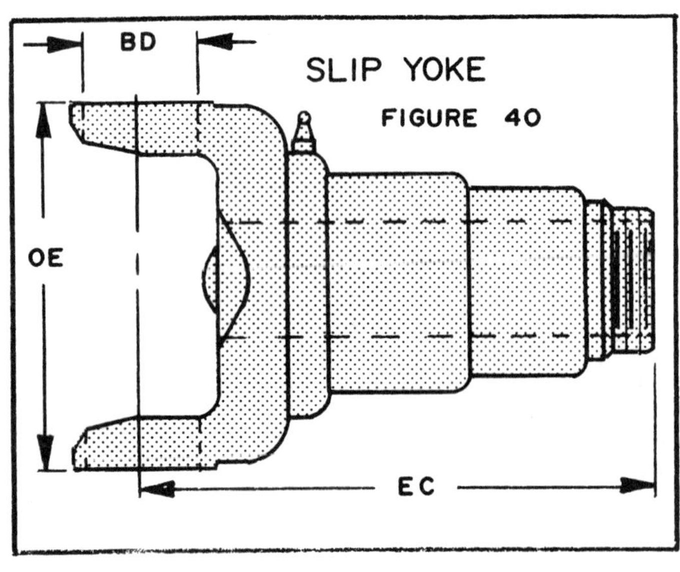

BD
SLIP YOKE
FIGURE 40
OE
EC

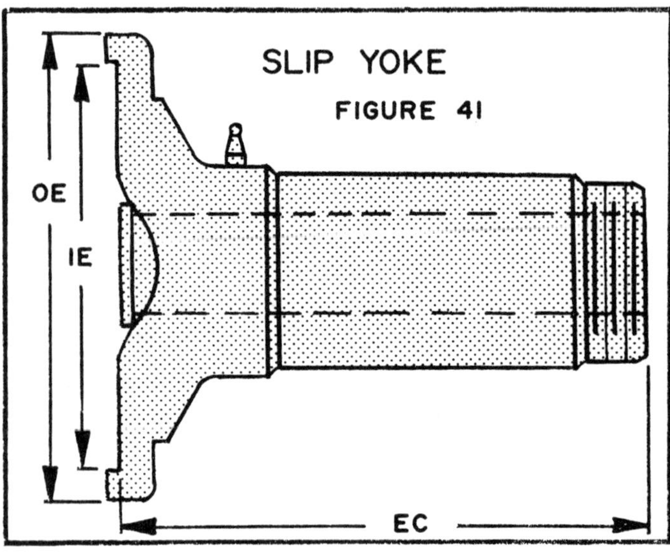

SLIP YOKE
FIGURE 41
OE
IE
EC

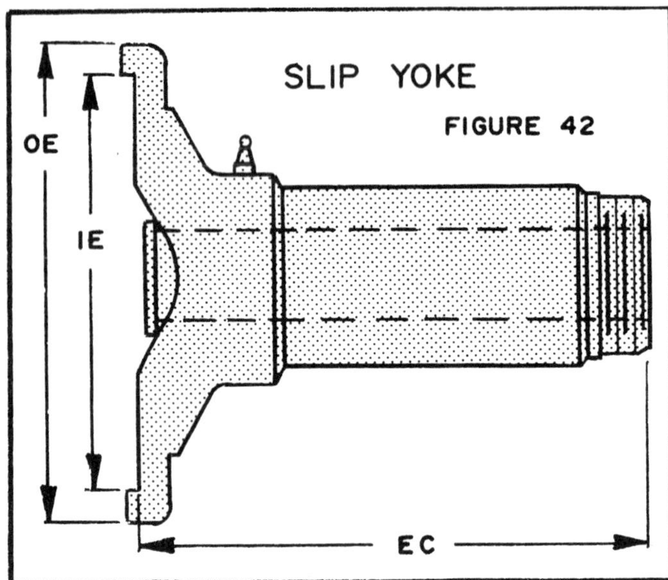

SLIP YOKE
FIGURE 42
OE
IE
BD
EC

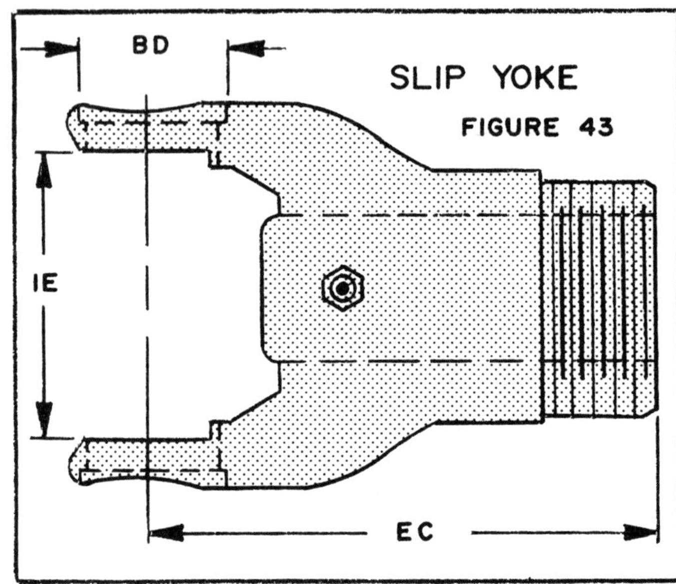

BD
SLIP YOKE
FIGURE 43
IE
EC

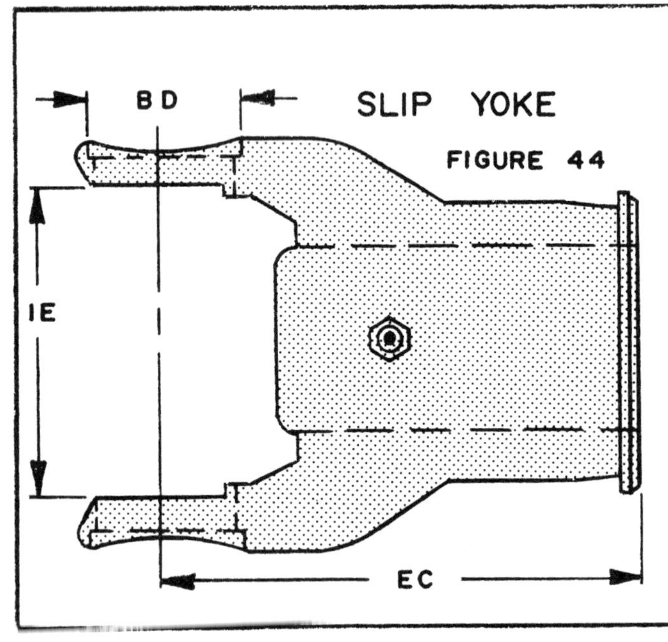

BD
SLIP YOKE
FIGURE 44
IE
EC

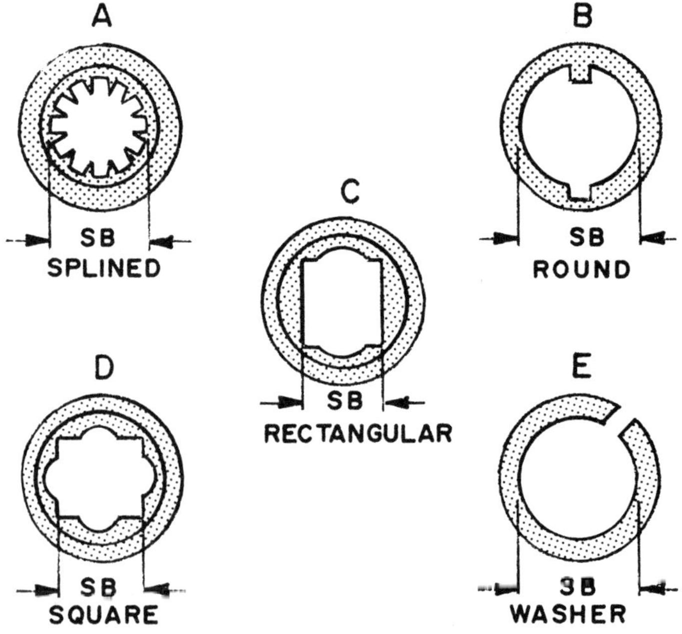

A
SB
SPLINED
B
SB
ROUND
C
SB
RECTANGULAR
D
SB
SQUARE
E
SB
WASHER

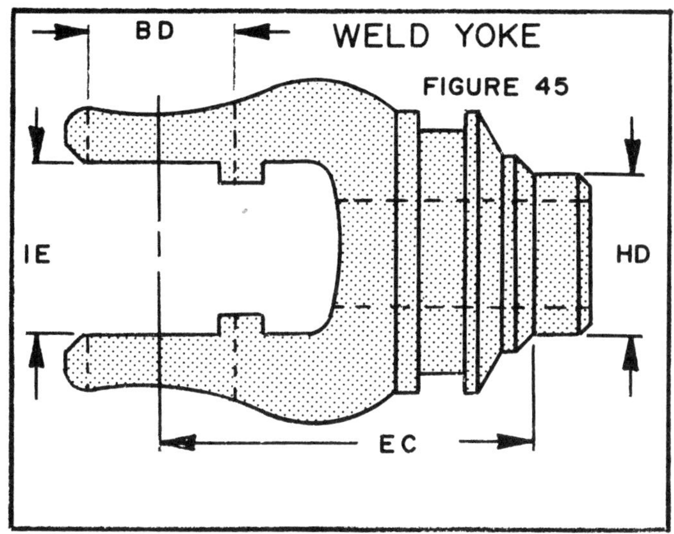

WELD YOKE
FIGURE 45
BD
IE
HD
EC

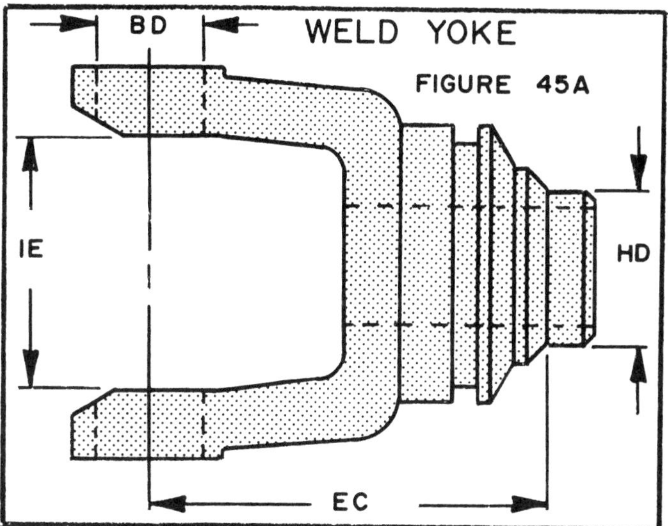

WELD YOKE
FIGURE 45A
BD
IE
HD
EC

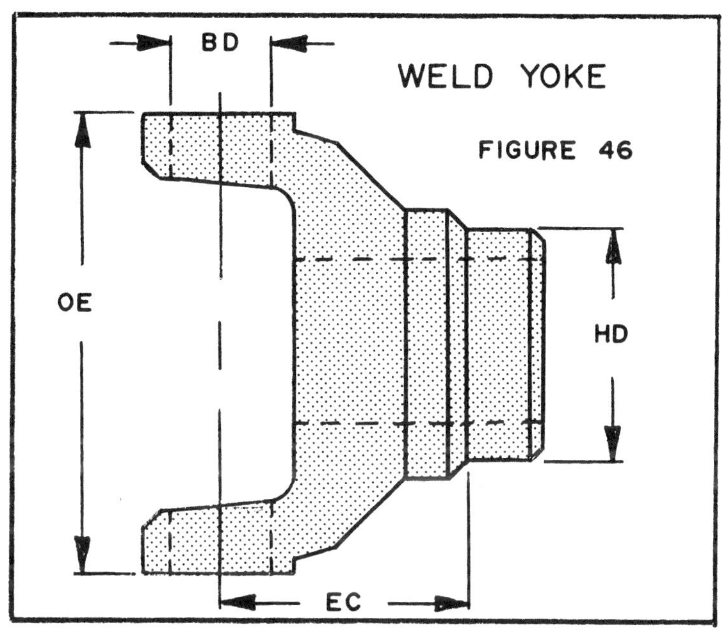

BD
WELD YOKE
FIGURE 46
OE
HD
EC

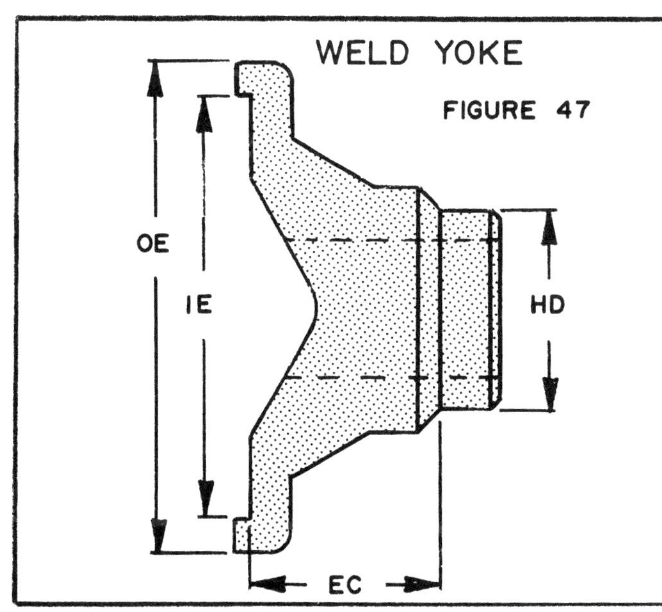

WELD YOKE
FIGURE 47
OE
IE
HD
EC

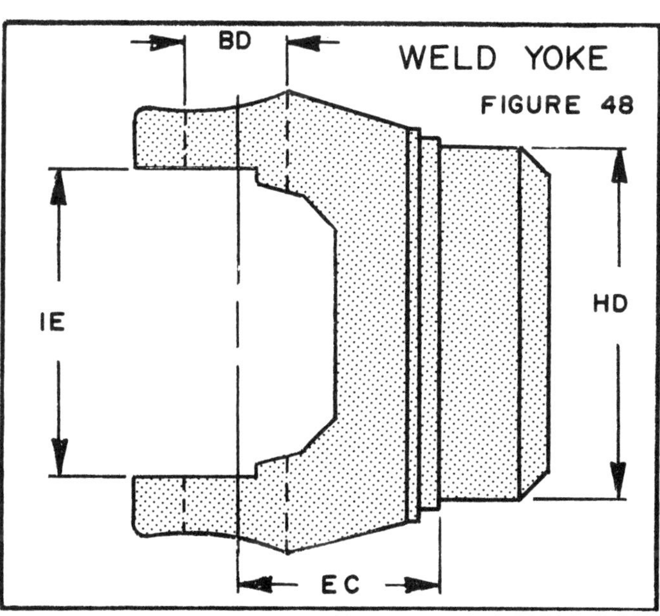

BD
WELD YOKE
FIGURE 48
IE
HD
EC

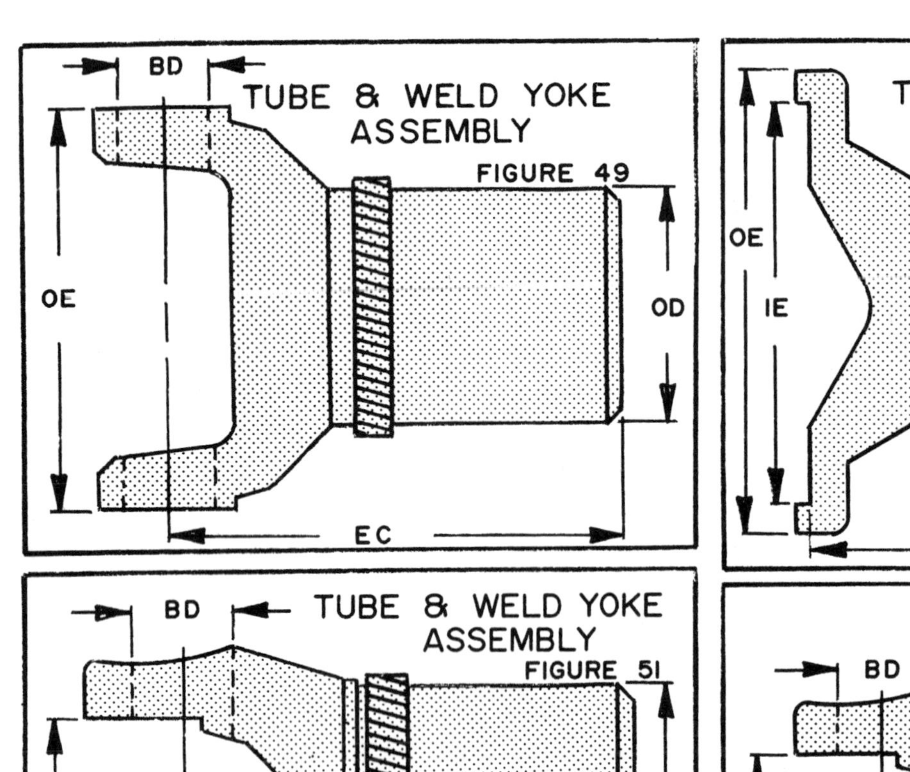

BD
TUBE & WELD YOKE ASSEMBLY
FIGURE 49
OE
OD
EC

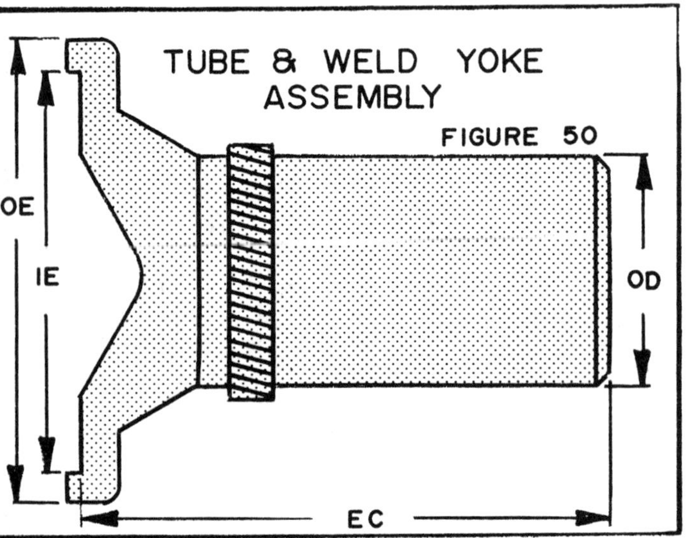

TUBE & WELD YOKE ASSEMBLY
FIGURE 50
OE
IE
OD
EC

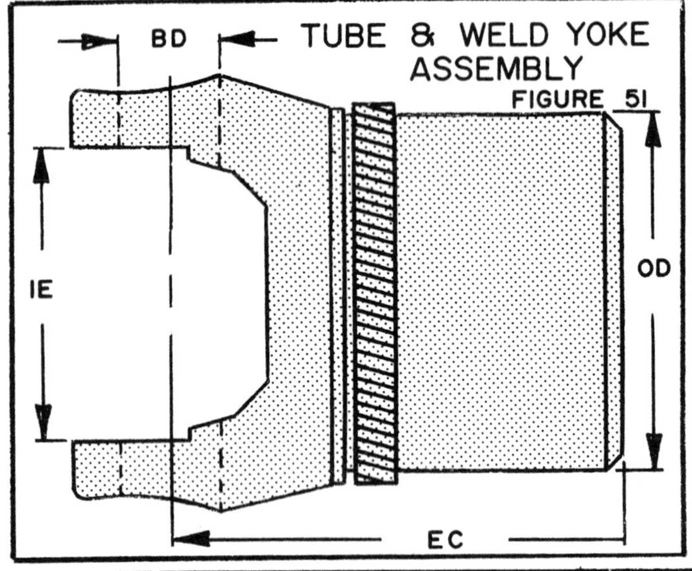

BD
TUBE & WELD YOKE ASSEMBLY
FIGURE 51
IE
OD
EC

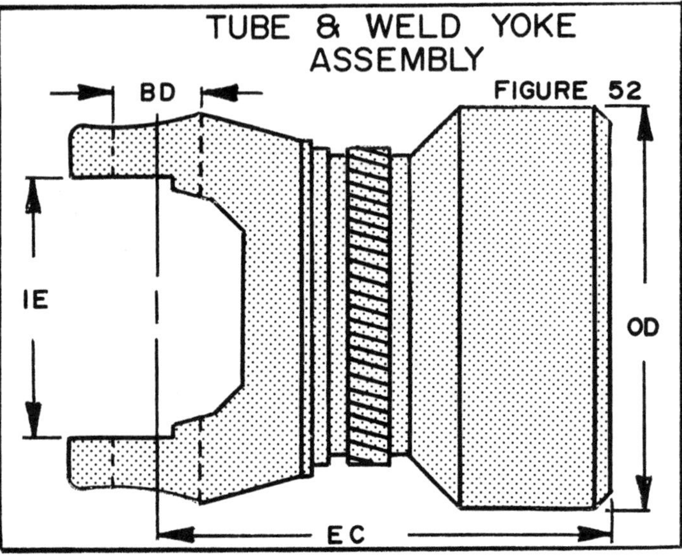

TUBE & WELD YOKE ASSEMBLY
BD
FIGURE 52
IE
OD
EC

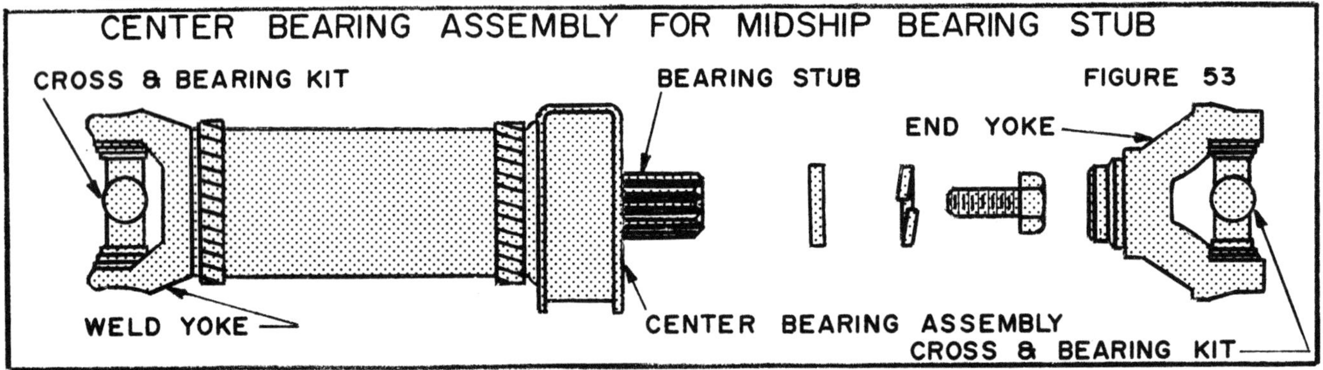

CENTER BEARING ASSEMBLY FOR MIDSHIP BEARING STUB
CROSS & BEARING KIT
BEARING STUB
FIGURE 53
END YOKE
WELD YOKE
CENTER BEARING ASSEMBLY
CROSS & BEARING KIT

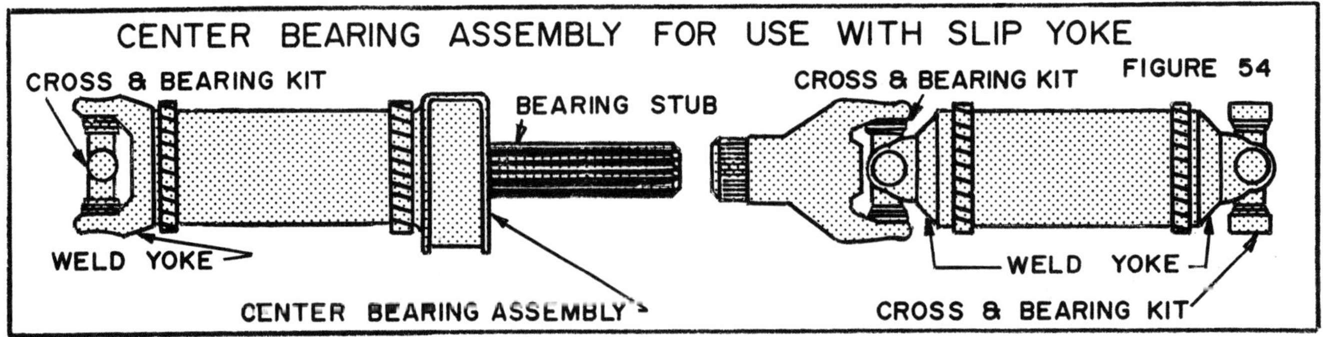

CENTER BEARING ASSEMBLY FOR USE WITH SLIP YOKE
CROSS & BEARING KIT
CROSS & BEARING KIT
FIGURE 54
BEARING STUB
WELD YOKE
CENTER BEARING ASSEMBLY
WELD YOKE
CROSS & BEARING KIT

CENTER BEARINGS

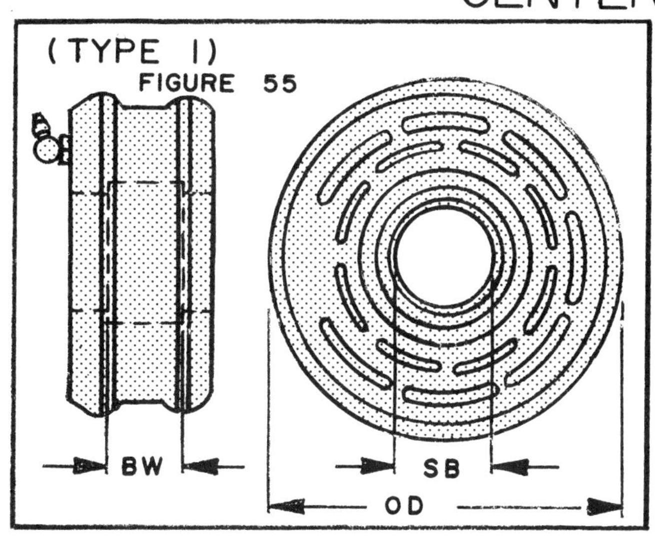

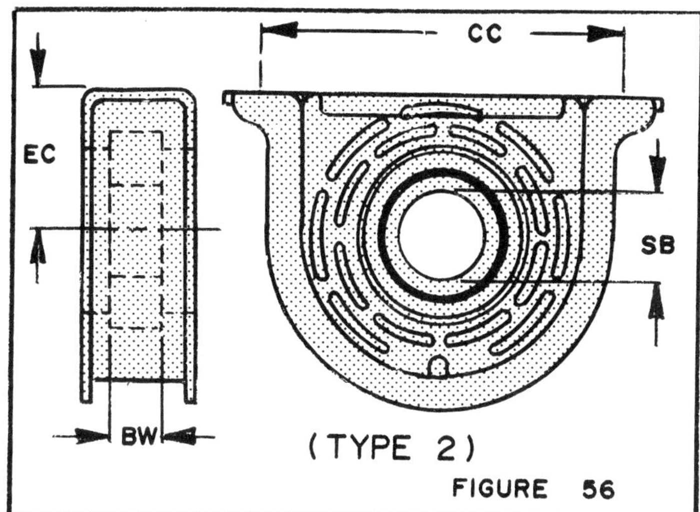

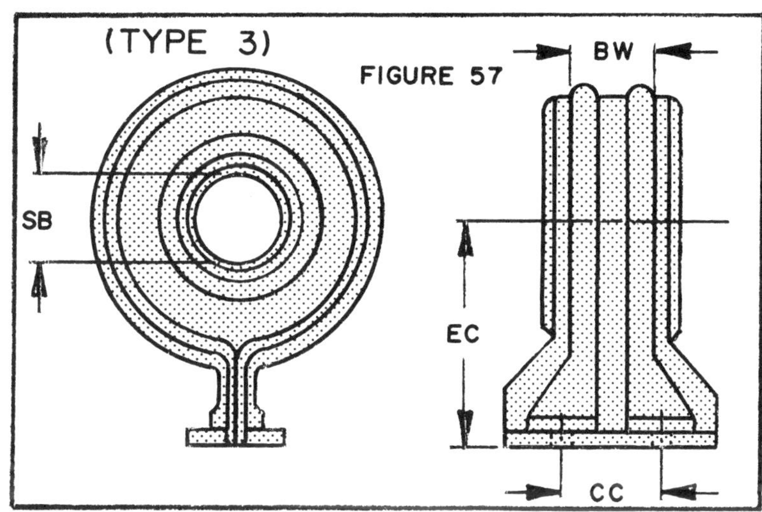

PARTS FOR CENTER BEARINGS

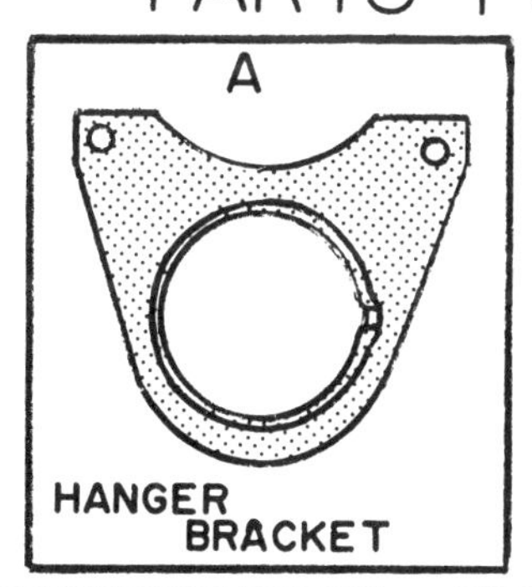

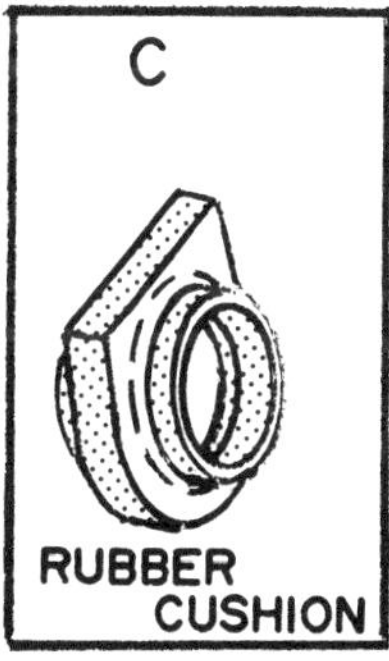

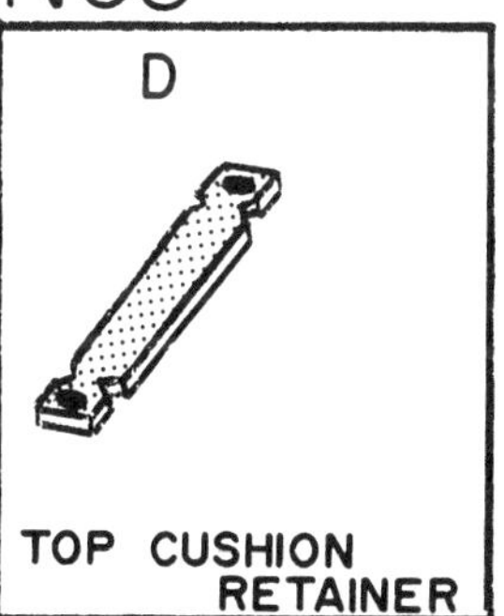

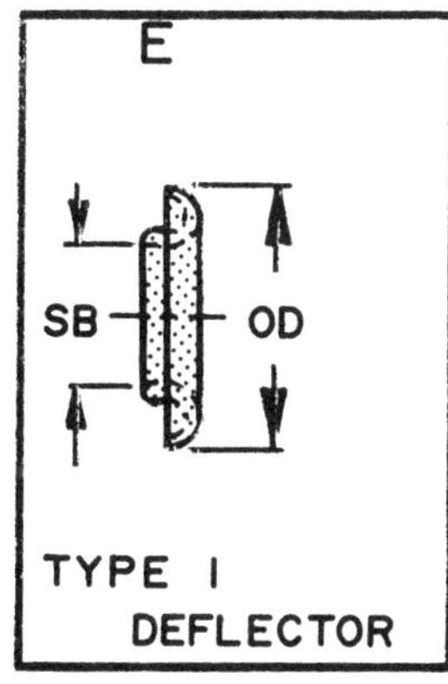

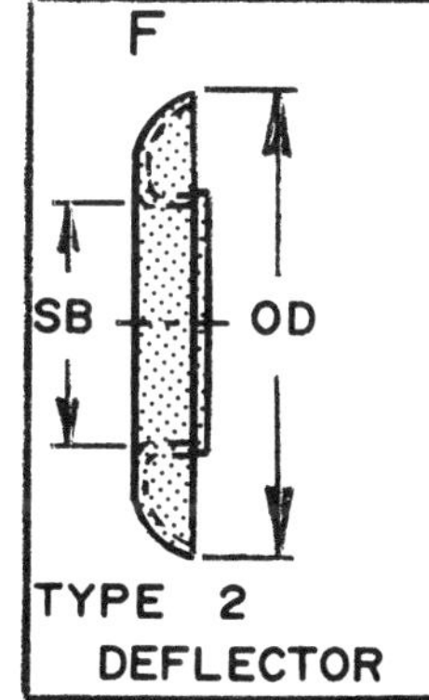

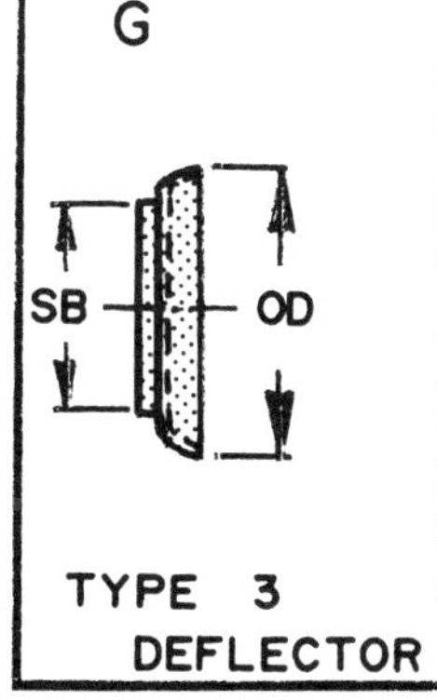

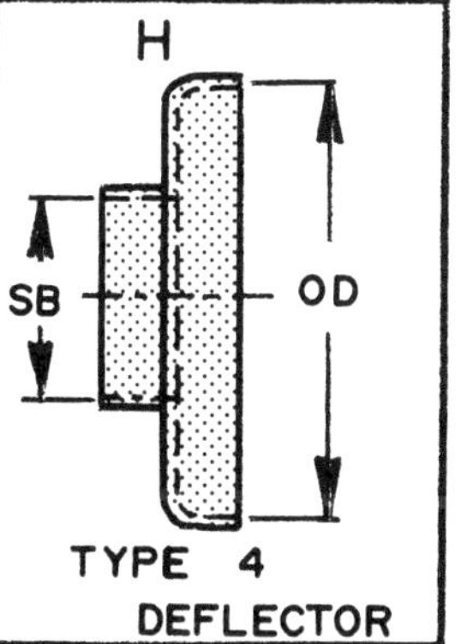

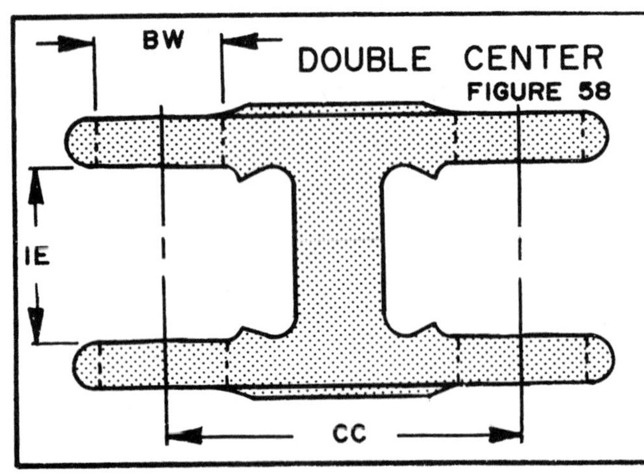

DOUBLE CENTER
FIGURE 58
BW
IE
CC

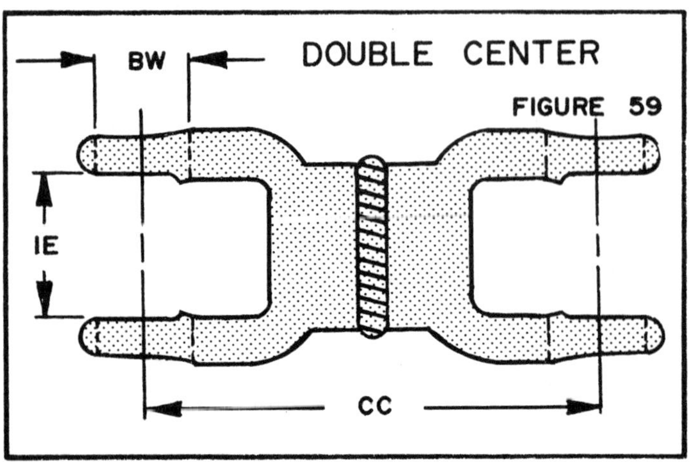

DOUBLE CENTER
FIGURE 59
BW
IE
CC

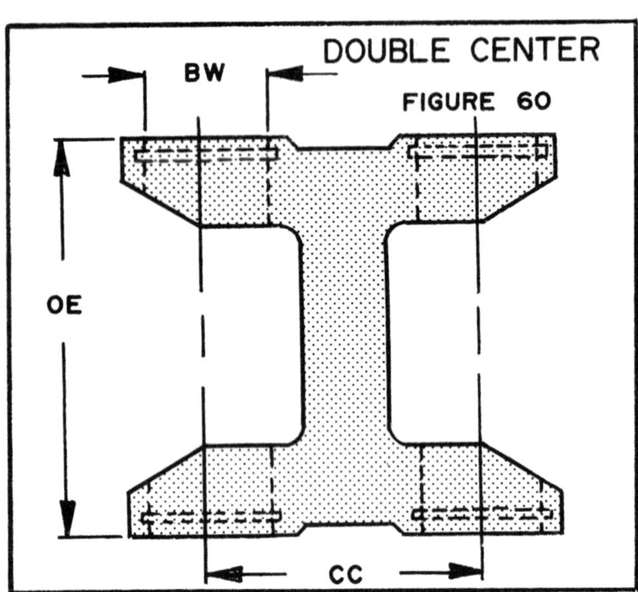

DOUBLE CENTER
FIGURE 60
BW
OE
CC

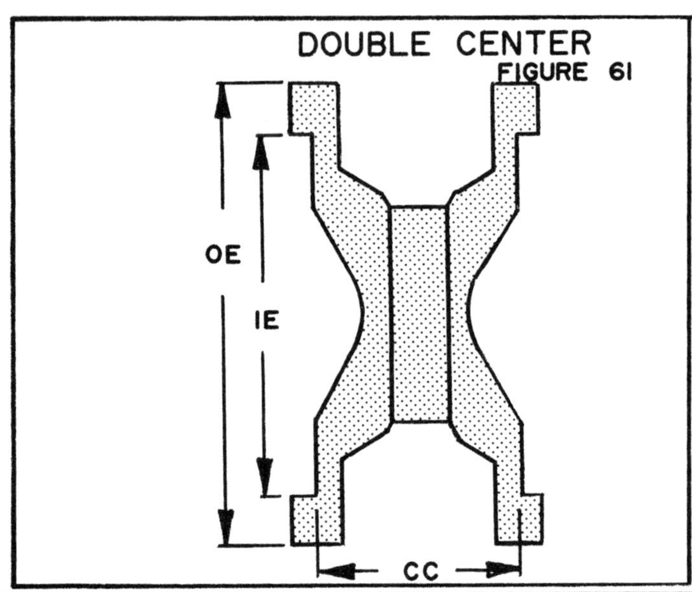

DOUBLE CENTER
FIGURE 61
OE
IE
CC

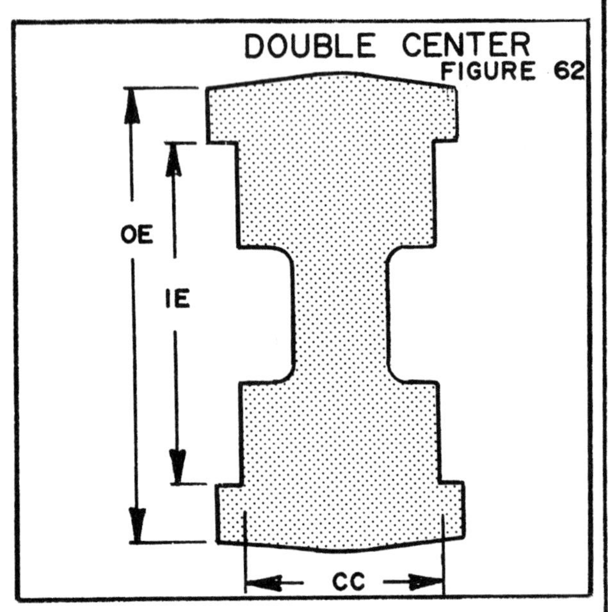

DOUBLE CENTER
FIGURE 62
OE
IE
CC

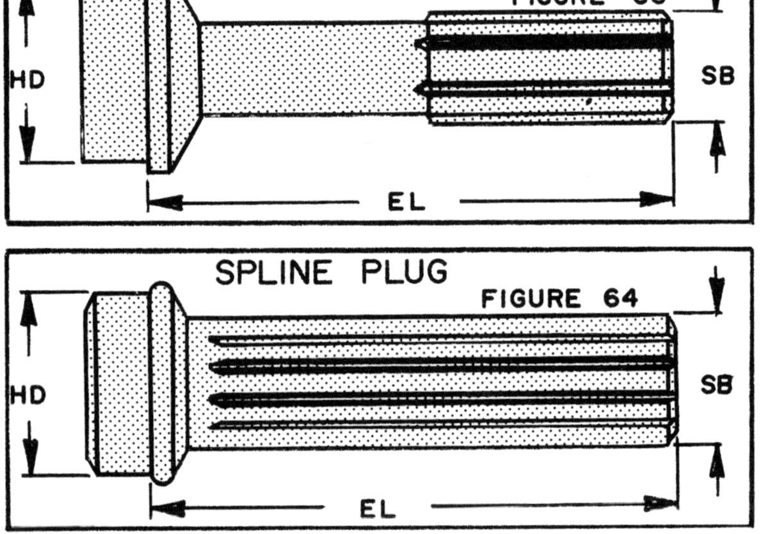

SPLINE PLUG
FIGURE 63
HD
SB
EL
SPLINE PLUG
FIGURE 64
HD
SB
EL

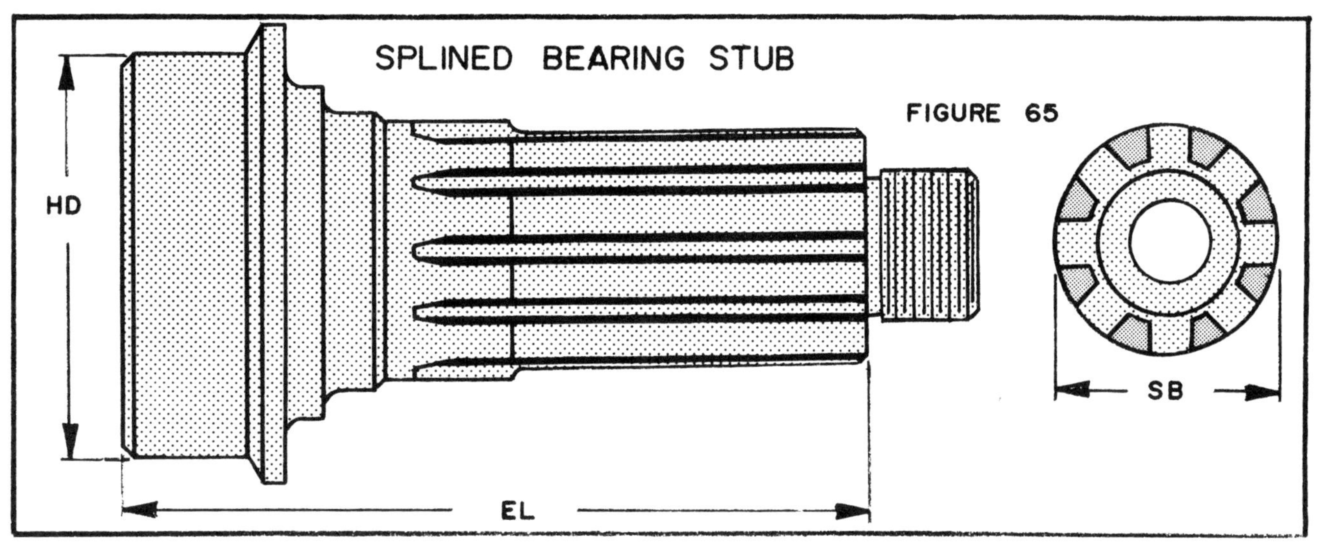

SPLINED BEARING STUB
FIGURE 65
HD
EL
SB

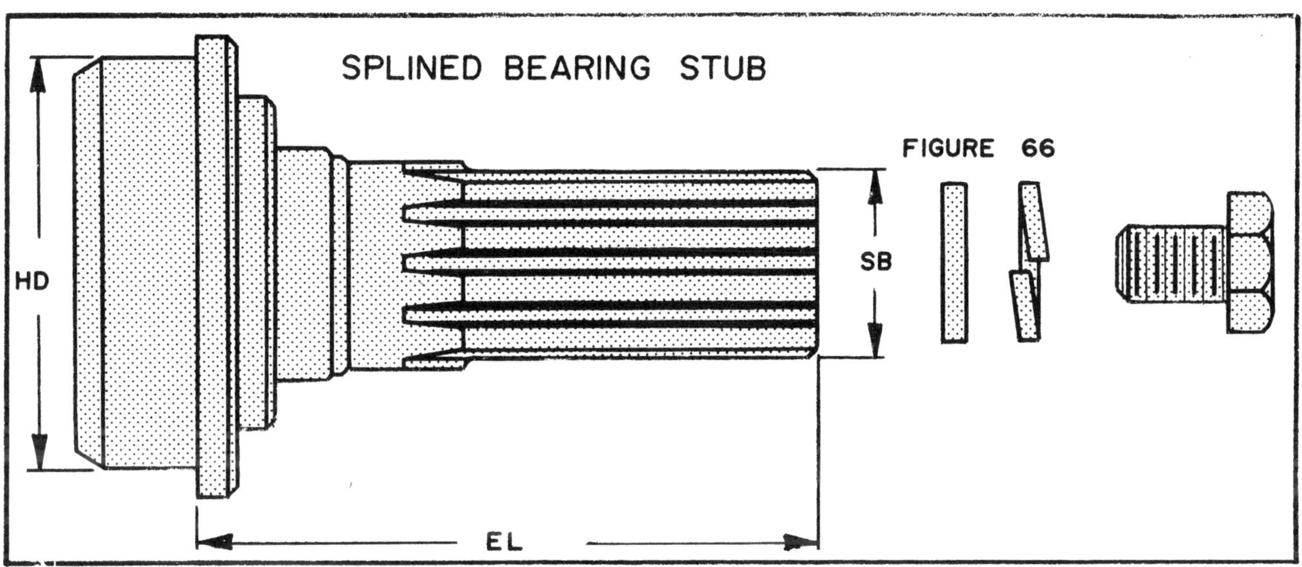

SPLINED BEARING STUB
FIGURE 66
HD
EL
SB

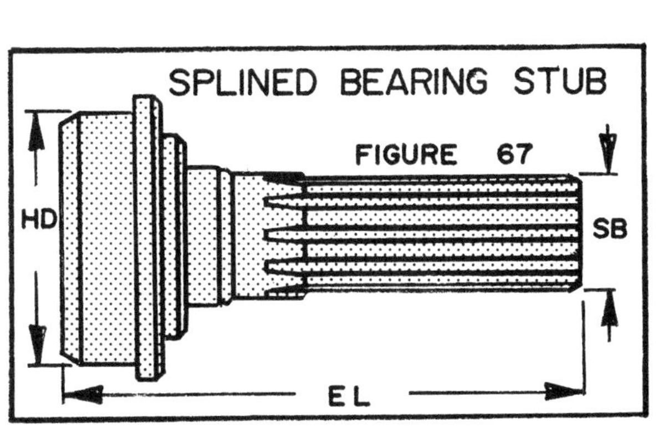

SPLINED BEARING STUB
FIGURE 67
HD
EL
SB

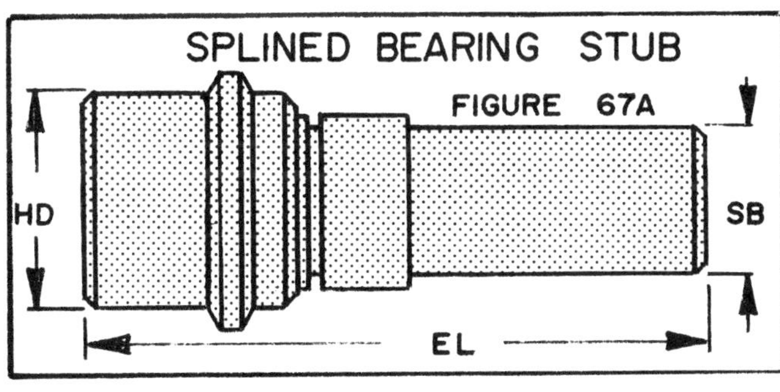

SPLINED BEARING STUB
FIGURE 67A
HD
EL
SB

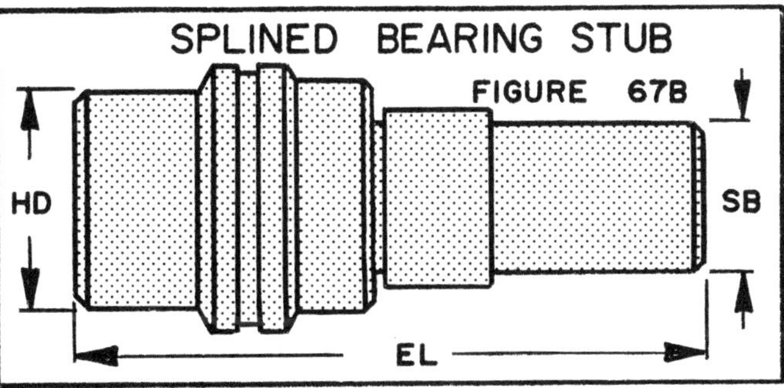

SPLINED BEARING STUB
FIGURE 67B
HD
EL
SB

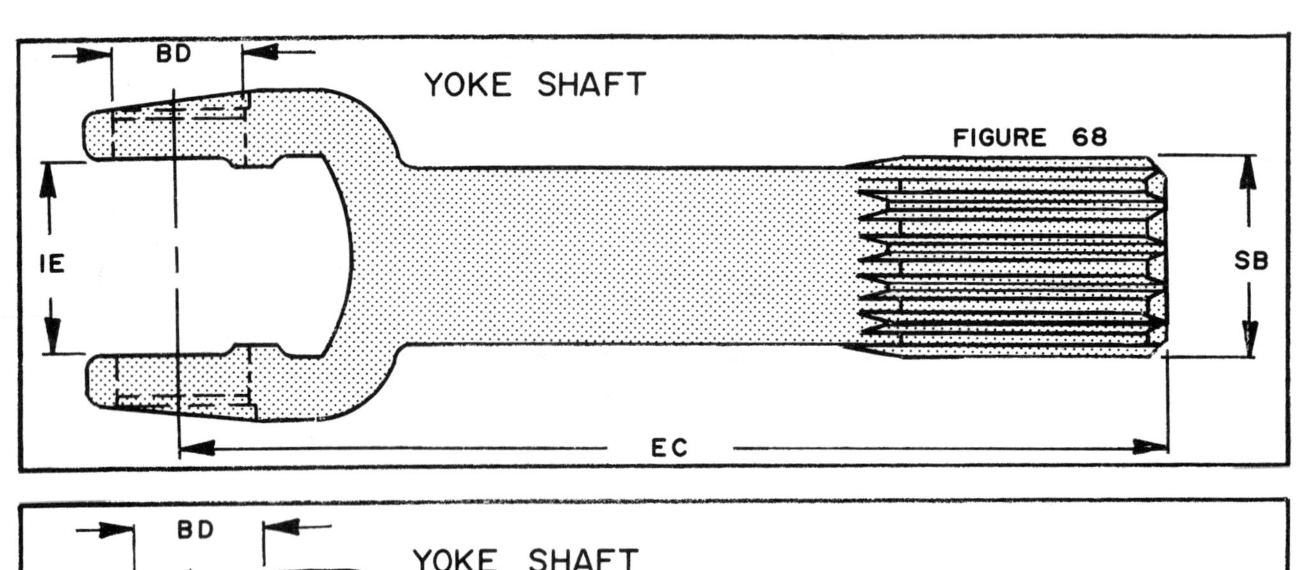

BD
YOKE SHAFT
FIGURE 68
IE
SB
EC

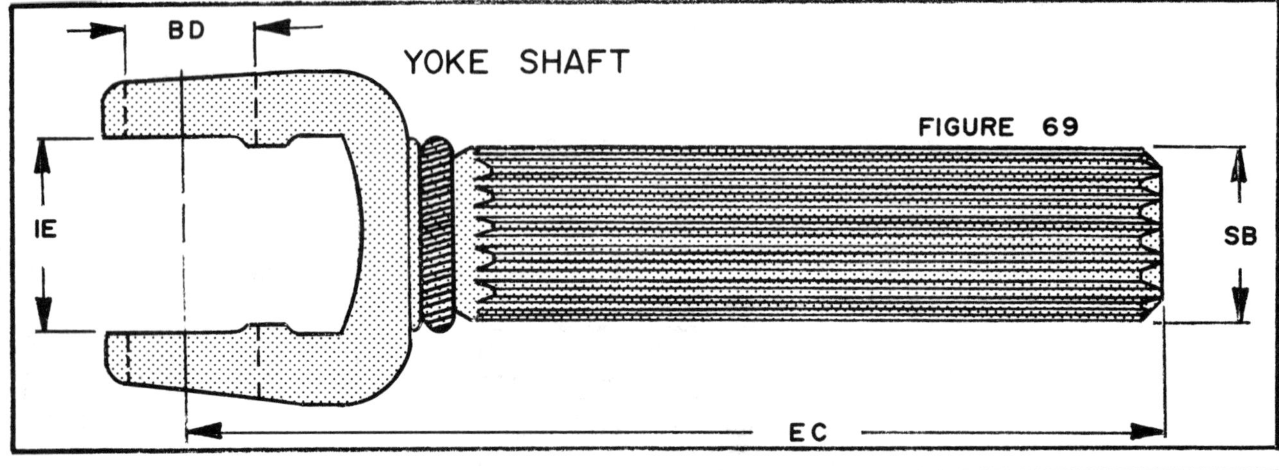

BD
YOKE SHAFT
FIGURE 69
IE
SB
EC

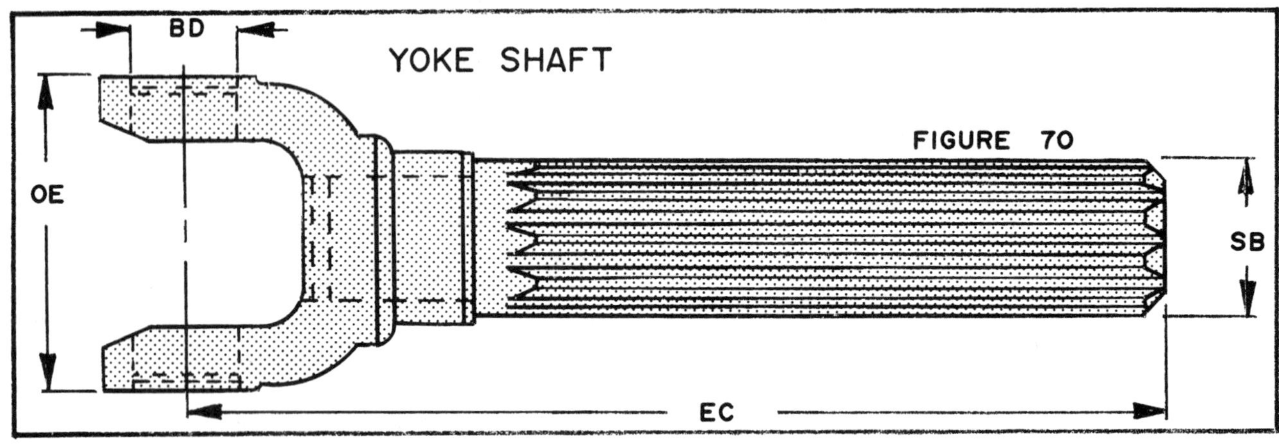

BD
YOKE SHAFT
FIGURE 70
OE
SB
EC

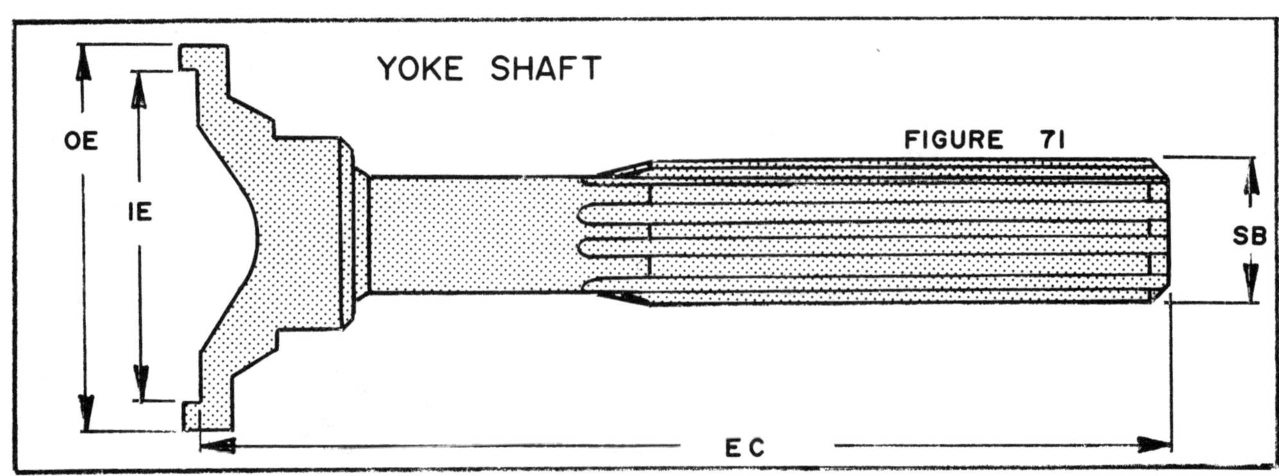

YOKE SHAFT
FIGURE 71
OE
IE
SB
EC

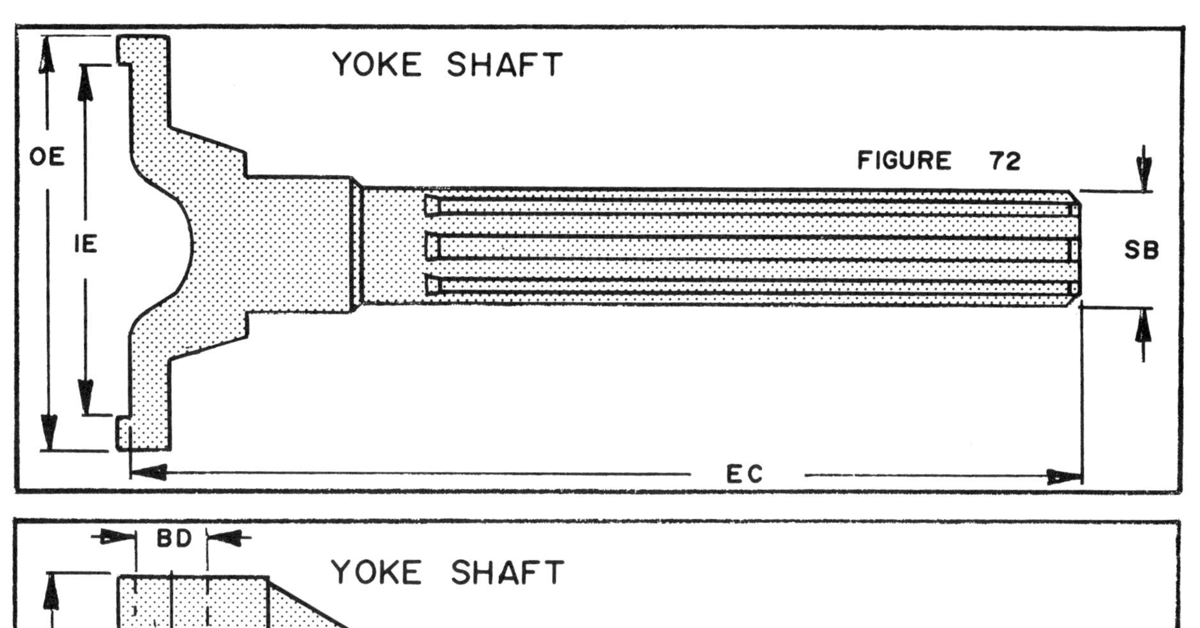
YOKE SHAFT
FIGURE 72
OE
IE
SB
EC

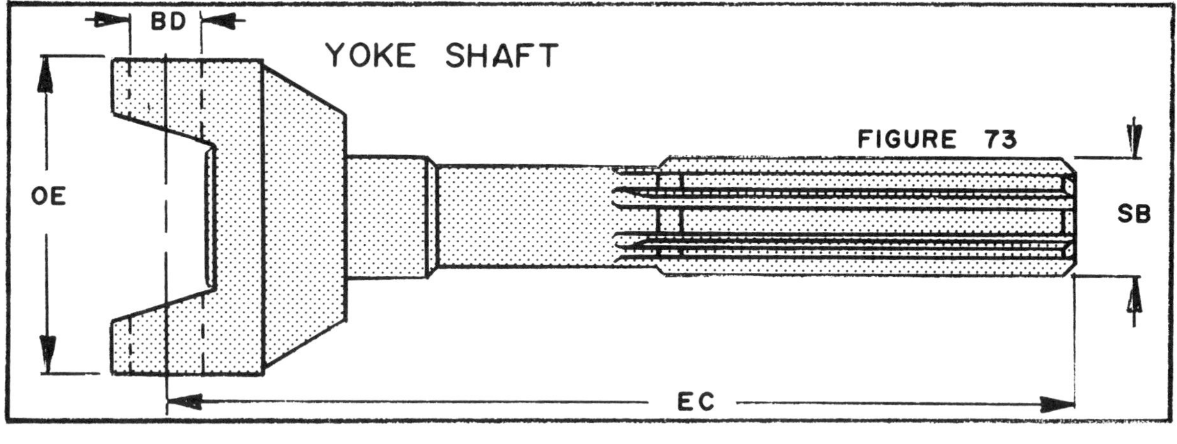
BD
YOKE SHAFT
FIGURE 73
OE
SB
EC

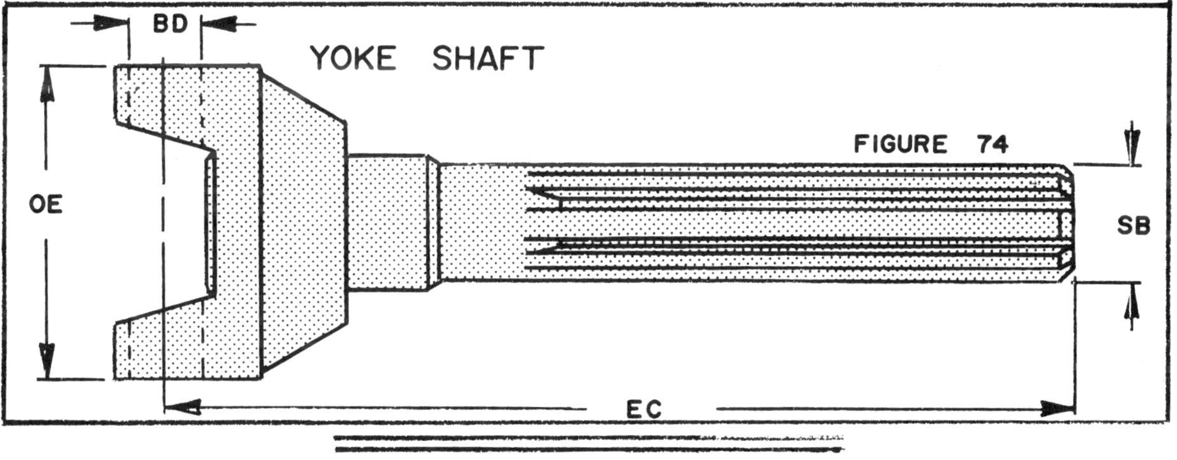
BD
YOKE SHAFT
FIGURE 74
OE
SB
EC

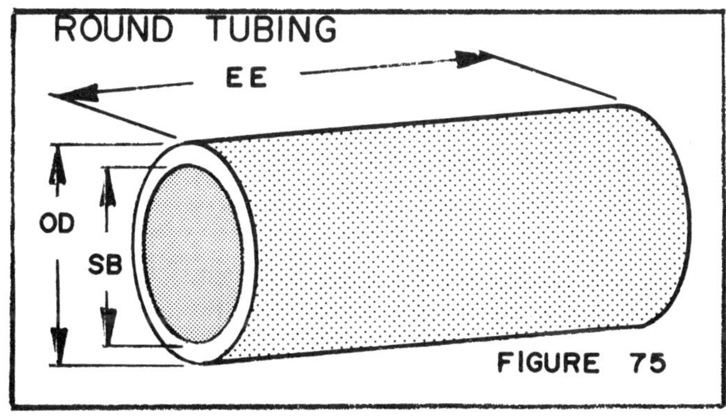
ROUND TUBING
EE
OD
SB
FIGURE 75

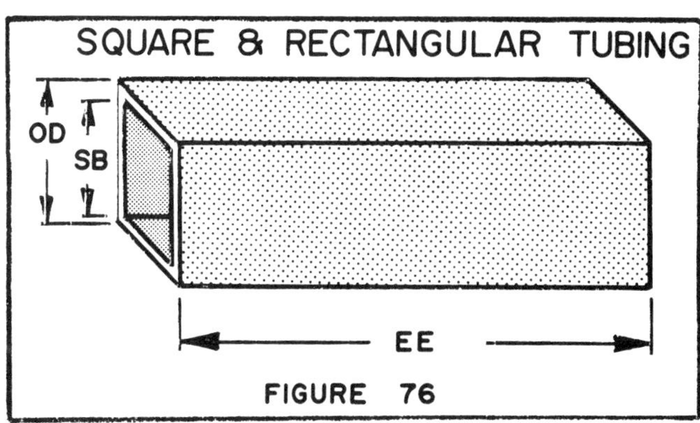
SQUARE & RECTANGULAR TUBING
OD
SB
EE
FIGURE 76

SLIP ASSEMBLIES

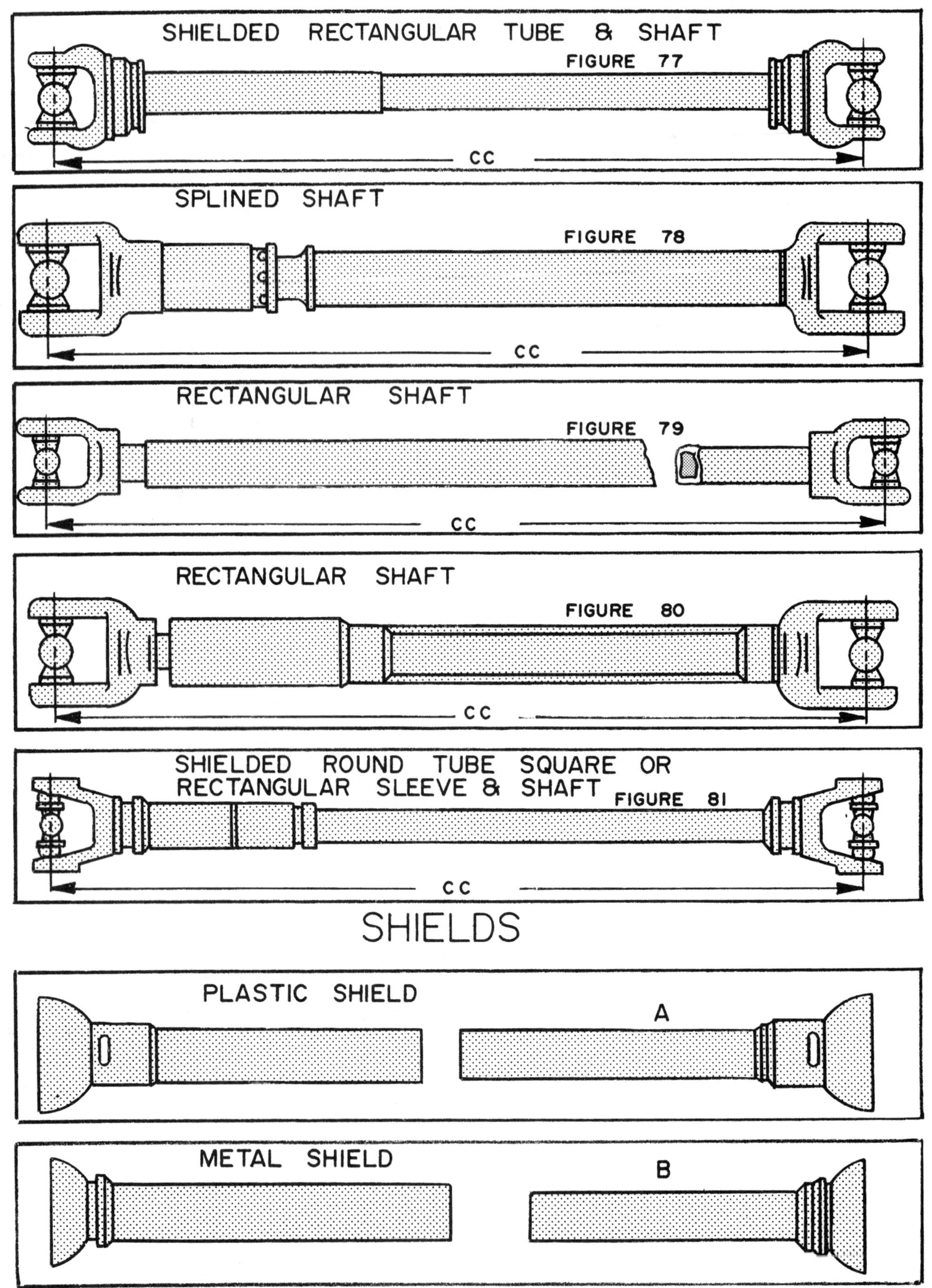

SLEEVES

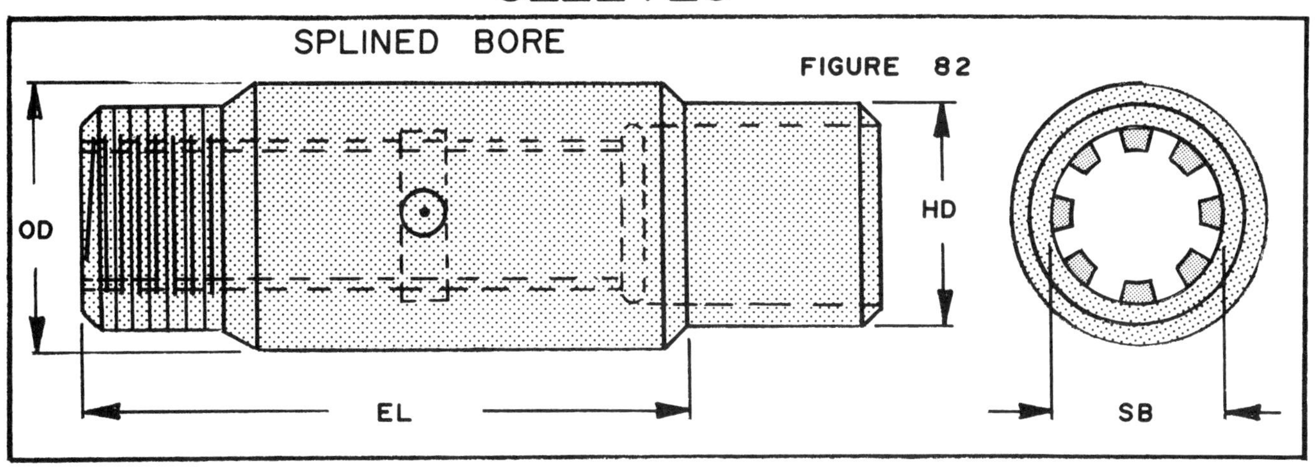

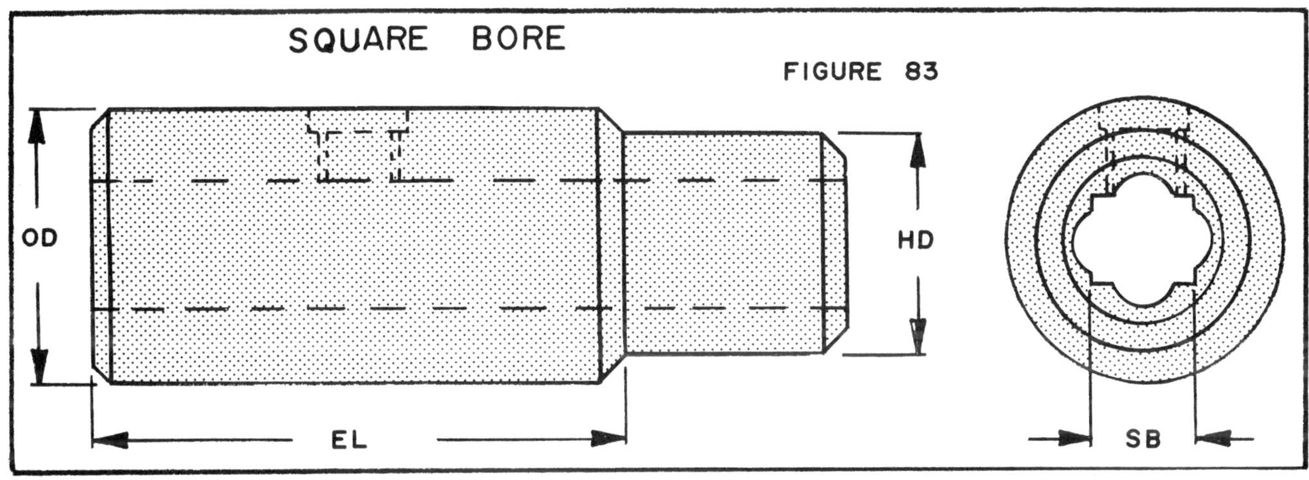

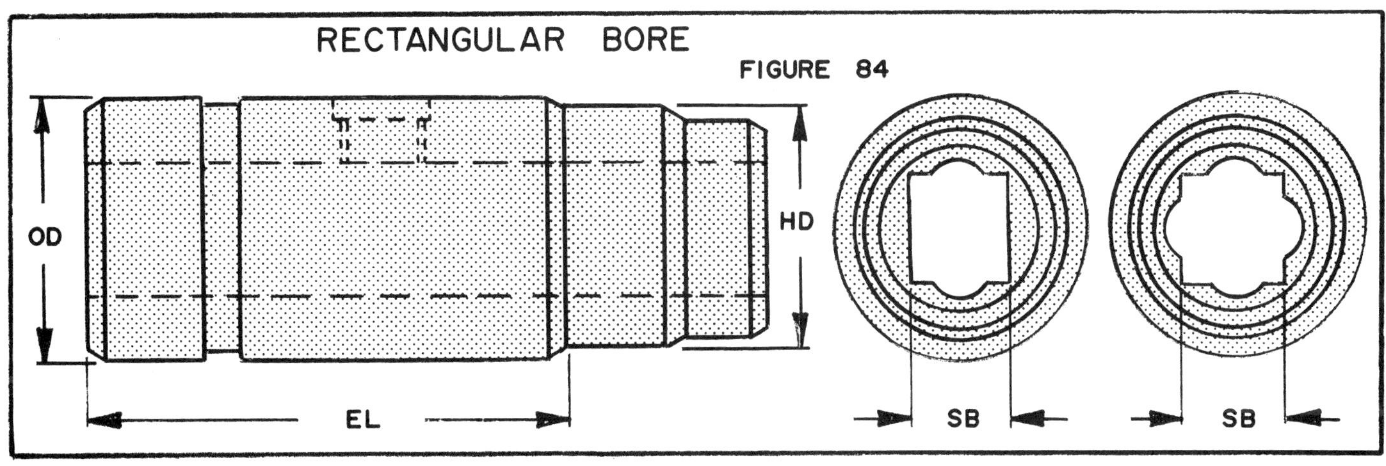

SPLINED BORE
FIGURE 82
OD
HD
EL
SB

SQUARE BORE
FIGURE 83
OD
HD
EL
SB

RECTANGULAR BORE
FIGURE 84
OD
HD
EL
SB
SB

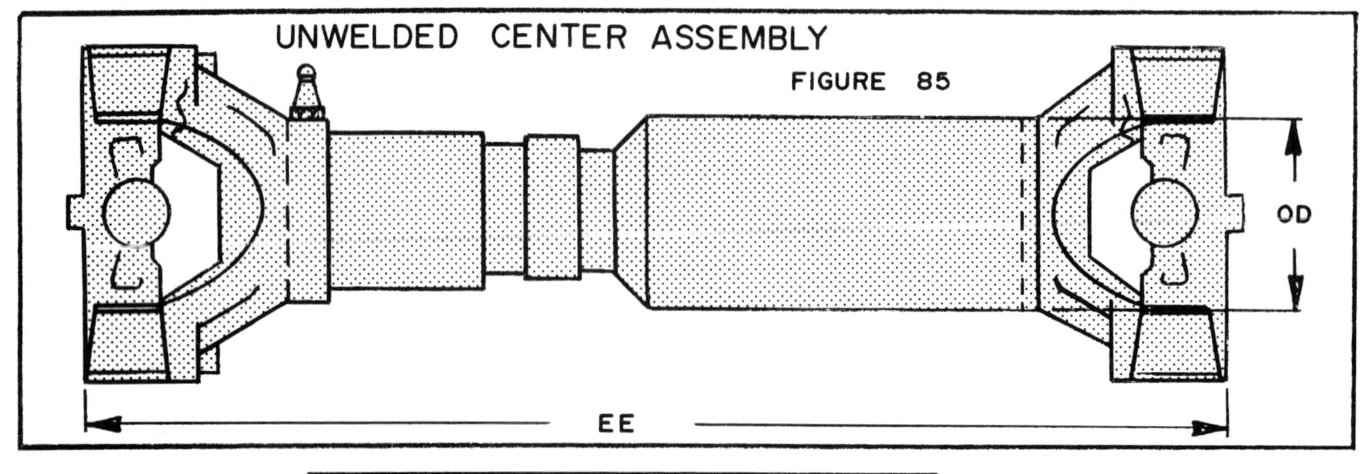

UNWELDED CENTER ASSEMBLY
FIGURE 85
OD
EE

FULLY SPLINED
FIGURE 86
EE
SB

SPLINED & KEYED
FIGURE 87
EE
SB

SQUARE OR RECTANGULAR
FIGURE 88
OD
EE

SHAFTING

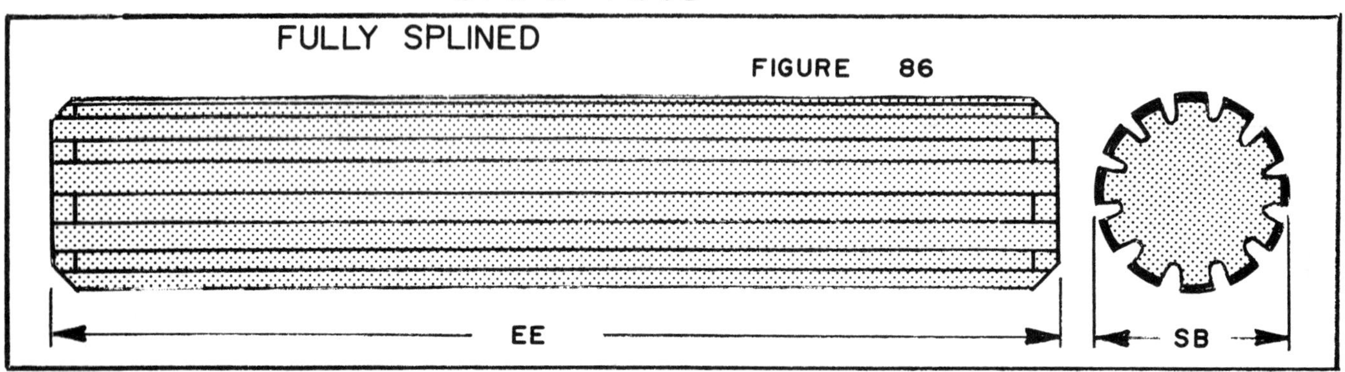

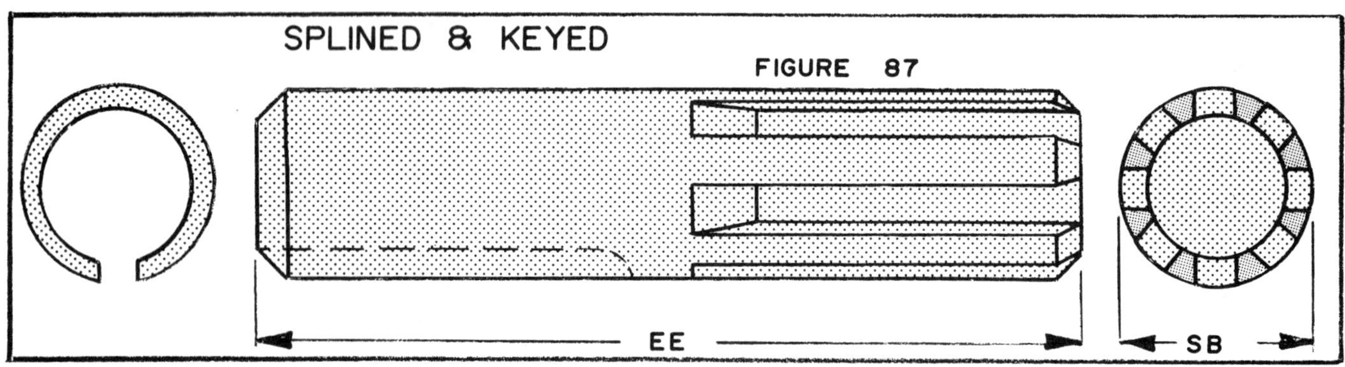

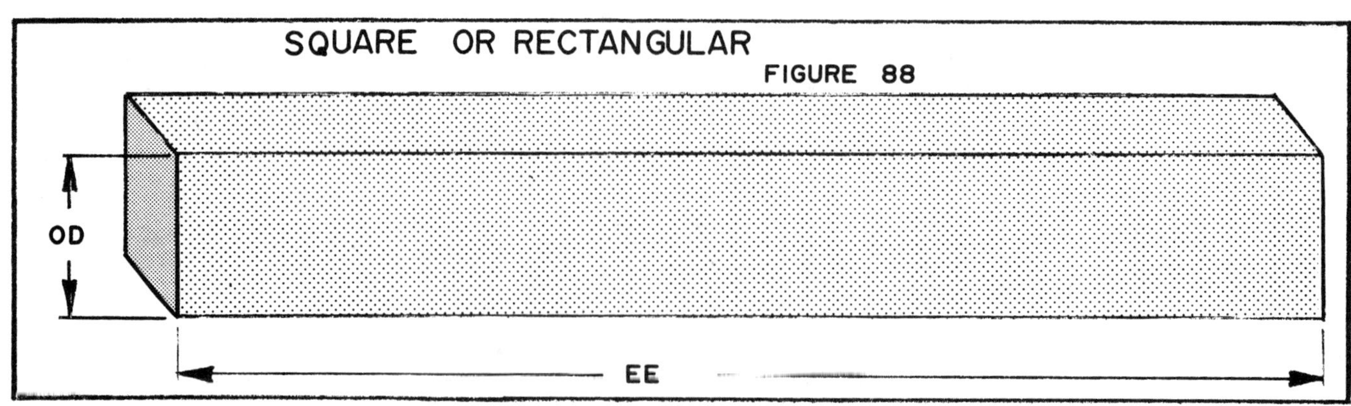

Part No.	Brand	IDLI Group	Part No.	Brand	IDLI Group	Part No.	Brand	IDLI Group	Part No.	Brand	IDLI Group
			01788323-2	FIAT-ALLIS	80242	02-7074113	WHITE	80001	0335-02120	HUB CITY	83349
			0179571	FIAT-ALLIS	80103	02-7074122	WHITE	80140	0335-02123	HUB CITY	83317
FOR NUMBERS STARTING WITH ZERO			0179571-5	FIAT-ALLIS	80103	02-7074361	WHITE	80189	0335-02126	HUB CITY	83316
ALSO CHECK THE ALPHABETICAL O.			0180132	FIAT-ALLIS	80213	02-7074362	WHITE	80221	0335-02132	HUB CITY	83296
NUMBER MAY ALSO BE LISTED WITH			0180132-3	FIAT-ALLIS	80213	02-7074363	WHITE	80227	0335-02133	HUB CITY	83362
NO CHARACTER IN FIRST POSITION.			0180140	FIAT-ALLIS	80224	02-7087816	WHITE	80235	0335-02134	HUB CITY	83357
00-08-00	WALTERSCHD	88004.27	0180140-6	FIAT-ALLIS	80224	02-7091891	WHITE	80219	0335-02135	HUB CITY	83314
00-32-30	HARDY SPCR	88003.61	0185263	FIAT-ALLIS	80234	02-7091918	WHITE	80234	0335-02136	HUB CITY	84529
00-32-40	HARDY SPCR	88003.62	0185263-1	FIAT-ALLIS	80234	0-33	AEC	88000.22	0335-02153	HUB CITY	83334
00-32-50	HARDY SPCR	88003.63	0187-89-251	MAZDA	80012	0335-00002	HUB CITY	88000.67	03-35-900067	HUB CITY	80140
00-32-60	HARDY SPCR	88003.64	019655	PILOT	80138	0335-00003	HUB CITY	88000.64	03-35-90068	HUB CITY	80007
00400-16	POCLAIN	80142	01-9994	STEIGER	80237	03-35-00005	HUB CITY	88000.45	03-35-90069	HUB CITY	80007
00400-34	POCLAIN	80023	01T7039	FORD	80124	03-35-00006	HUB CITY	88000.31	03-35-90070	HUB CITY	80008
00400-38	POCLAIN	80180	01Y7039	FORD	80075	03-35-00007	HUB CITY	88000.52	03-35-90071	HUB CITY	80001
00400-54	POCLAIN	80198	02010006	G & G MFG	88013.84	03-35-00008	HUB CITY	88000.21	03-35-90073	HUB CITY	80044
0053120	FIAT-ALLIS	80001	02010021	G & G MFG	88013.85	0335-00012	HUB CITY	88000.58	03-35-90074	HUB CITY	80161
0053120-2	FIAT-ALLIS	80001	02010221	G & G MFG	88013.85	03-35-00013	HUB CITY	88000.28	03-35-90075	HUB CITY	80035
0069962(2)	FIAT-ALLIS	80223	02-1197	STEIGER	80221	0335-00014	HUB CITY	88000.37	03-35-90075	HUB CITY	80037
0069962-9(2)	FIAT-ALLIS	80223	0-22	AEC	88000.43	03-35-00015	HUB CITY	88000.49	0335-90076	HUB CITY	80069
00-80-00	WALTERSCHD	88004.25	0224890	FIAT-ALLIS	80001	03-35-00016	HUB CITY	88000.55	03-35-90076	HUB CITY	80070
0080470	FIAT-ALLIS	80124	0224890-4	FIAT-ALLIS	80001	03-35-00017	HUB CITY	88000.43	03-501346	BRD	80041
0080470-8	FIAT-ALLIS	80124	0-23	AEC	88000.60	03-35-00019	HUB CITY	88000.44	03-501523	BRD	80040
0081058	FIAT-ALLIS	80001	02324	NEW IDEA	80003	03-35-00025	HUB CITY	88000.36	03-503941	BRD	80041
0081058-0	FIAT-ALLIS	80001	02-501363	BRD	80023	03-35-00026	HUB CITY	88000.35	03521	CHECKER	80077
0081559	FIAT-ALLIS	80142	02-502835	BRD	80023	0335-0004	HUB CITY	88000.51	03-530402	BRD	80041
0081559-7	FIAT-ALLIS	80142	02-530401	BRD	80024	0335-00042	HUB CITY	88000.60	03-531346	BRD	80041
0083616	FIAT-ALLIS	80215	02-532835	BRD	80024	0335-00045	HUB CITY	88000.19	0355-02154	HUB CITY	88013.82
0083616-3	FIAT-ALLIS	80215	0259-25-060	MAZDA	80011	03-35-00046	HUB CITY	88000.25	03568	CHECKER	87331
0084328	FIAT-ALLIS	80195	0259-89-251	FORD	80011	03-35-00046	HUB CITY	88000.27	03568	CHECKER	88008.39
0084328-4	FIAT-ALLIS	80195	0259-89-251	MAZDA	80011	03-35-00061	HUB CITY	88000.38	0-43	AEC	88000.22
0084435	FIAT-ALLIS	80202	02-7010593	WHITE	87570	03-35-00061	HUB CITY	88000.39	04371-10010	TOYOTA	80006
0084435-7	FIAT-ALLIS	80202	02-7011571	WHITE	80221	0335-00063	HUB CITY	88000.53	04371-10011	DAIHATSU	80006
0084537	FIAT-ALLIS	80162	02-7011692	WHITE	80221	0335-00116	HUB CITY	88000.83	04371-10011	TOYOTA	80006
0084537-0	FIAT-ALLIS	80162	02-7011870	WHITE	88008.32	03-35-0063	HUB CITY	88000.53	04371-20010	TOYOTA	80009
0084545	FIAT-ALLIS	80174	02-7011949	WHITE	87382	0335-01202	HUB CITY	84510	04371-20011	TOYOTA	80009
0084545-3	FIAT-ALLIS	80174	02-7012321	WHITE	80189	0335-01204	HUB CITY	81342	04371-22010	TOYOTA	80009
0086284	FIAT-ALLIS	80149	02-7074021	WHITE	80023	0335-01205	HUB CITY	81246	04371-22010C	AUST.MOT	88002.77
0086284-7	FIAT-ALLIS	80149	02-7074022	WHITE	80024	0335-01206	HUB CITY	81237	04371-22010C	TOYOTA	88002.77
0086747	FIAT-ALLIS	80165	02-7074023	WHITE	80072	0335-01213	HUB CITY	81222	04371-22011	TOYOTA	80009
0086747-3	FIAT-ALLIS	80165	02-7074024	WHITE	80077	0335-01215	HUB CITY	81275	04371-25010	TOYOTA	80018
0086752	FIAT-ALLIS	80164	02-7074025	WHITE	80124	0335-01216	HUB CITY	81301	04371-30010	TOYOTA	80018
0086752-3	FIAT-ALLIS	80164	02-7074026	WHITE	80142	0335-01216	HUB CITY	88014.05	04371-30011	TOYOTA	80018
0090311(2)	FIAT-ALLIS	80223	02-7074027	WHITE	80178	0335-01217	HUB CITY	81333	04371-30020	TOYOTA	80016
0090311-2(2)	FIAT-ALLIS	80223	02-7074028	WHITE	80182	0335-01218	HUB CITY	81276	04371-35020	TOYOTA	80016
0-103	AEC	88000.36	02-7074029	WHITE	80189	0335-01226	HUB CITY	88013.90	04371-36010	TOYOTA	80017
010752	CHECKER	80117	02-7074030	WHITE	80190	0335-01227	HUB CITY	81274	04371-36011	TOYOTA	80017
010761	CHECKER	80077	02-7074031	WHITE	80221	0335-01246	HUB CITY	81213	04371-36020	TOYOTA	80115
011073	NEW IDEA	80035	02-7074032	WHITE	80225	0335-01251	HUB CITY	81273	04371-36021	TOYOTA	80115
0-113	AEC	88000.35	02-7074059	WHITE	80197	0335-01254	HUB CITY	88013.82	04371-36036	TOYOTA	80017
0114-26-180	MAZDA	80012	02-7074060	WHITE	80198	0335-01402	HUB CITY	88014.34	04371-40010	TOYOTA	80017
0-115	AEC	88000.55	02-7074061	WHITE	80193	0335-01404	HUB CITY	83116	04371-60010	TOYOTA	80017
0-12	AEC	88000.45	02-7074062	WHITE	80195	0335-01405	HUB CITY	83015	04371-60020	TOYOTA	80017
01200	WHITING	80233	02-7074063	WHITE	80199	0335-01406	HUB CITY	83018	04371-87101	DAIHATSU	80020
0-123	AEC	88000.38	02-7074064	WHITE	80201	0335-01408	HUB CITY	83020	04371-87301	DAIHATSU	80020
012324	NEW IDEA	80003	02-7074065	WHITE	80213	0335-01415	HUB CITY	83031	04371-87303	DAIHATSU	80020
0-124	AEC	88000.37	02-7074066	WHITE	80210	0335-01415	HUB CITY	83057	044-410-06-31	MERCEDES	80017
01288(4)	WHITING	80193	02-7074067	WHITE	80211	0335-01416	HUB CITY	83084	04-501371	BRD	80077
0-13	AEC	88000.20	02-7074068	WHITE	80212	0335-01417	HUB CITY	83102	04-511371	BRD	80077
0136-25-060	MAZDA	80012	02-7074069	WHITE	80238	0335-01417	HUB CITY	83108	045-501470	BRD	80124
015555	PILOT	80141	02-7074070	WHITE	80231	0335-01420	HUB CITY	83127	045-505038	BRD	80124
0-15D	AEC	88000.46	02-7074071	WHITE	80232	0335-01420	HUB CITY	83129	045-510423	BRD	80124
01-61-021	LILLISTON	80037	02-7074072	WHITE	80237	0335-01426	HUB CITY	83082	04-9047101	WHITE	80227
0178183	FIAT-ALLIS	80240	02-7074073	WHITE	80236	0335-01427	HUB CITY	83054	0500	ALLOY	88000.67
0178183-0	FIAT-ALLIS	80240	02-7074074	WHITE	80238	0335-01453	HUB CITY	83125	0501	ALCO	88000.90
0178323	FIAT-ALLIS	80242	02-7074107	WHITE	80168	0335-01481	HUB CITY	88014.28	0502	ALCO	88000.22
0178368	FIAT-ALLIS	80238	02-7074108	WHITE	80101	0335-02100	HUB CITY	83262	0502	ALCO	88000.24
0178368-7	FIAT-ALLIS	80238	02-7074109	WHITE	80185	0335-02102	HUB CITY	84531	0503	ALCO	88000.25
0178461	FIAT-ALLIS	80048	02-7074110	WHITE	80186	0335-02106	HUB CITY	83240	0504	ALCO	88000.41
0178461-0	FIAT-ALLIS	80048	02-7074111	WHITE	80184	0335-02116	HUB CITY	83251	0505	ALCO	88000.29
0178544-3	FIAT-ALLIS	88003.26	02-7074112	WHITE	80007	0335-02117	HUB CITY	83316	0505X	NEAPCO	88000.14
0178554	FIAT-ALLIS	80102									

Part No.	Brand	IDLI Group	Part No.	Brand	IDLI Group	Part No.	Brand	IDLI Group	Part No.	Brand	IDLI Group
0505X1	NEAPCO	80007	05167	NEAPCO	88000.26	051G2	NEAPCO	88000.22	055-55-2	BORG-WARNR	80141
0505X1	NEAPCO	88000.15	05168	NEAPCO	88001	051G2	NEAPCO	88000.70	055-55-2	CLEVELAND	80141
0506	ALCO	88000.19	05169	NEAPCO	88000.94	051G3	NEAPCO	88000.71	0556	ALLOY	88000.60
0507	ALCO	88000.68	0516A	NEAPCO	88000.35	051G4	NEAPCO	88000.72	0557	ALLOY	88000.43
0508	ALCO	88000.70	0517	ALCO	88000.28	051G5	NEAPCO	88000.73	0558	ALLOY	88000.33
0509	ALCO	88000.71	05171	NEAPCO	88000.43	051G6	NEAPCO	88000.68	0559	ALLOY	88000.59
0510	ALCO	88000.72	05172	NEAPCO	88000.47	051G6	NEAPCO	88000.69	0560	ALLOY	88000.61
05101X	NEAPCO	88000.67	05173	NEAPCO	88000.95	051GJ	NEAPCO	88000.33	0561	ALLOY	88000.62
05102X	NEAPCO	88014.06	05175	NEAPCO	88000.92	051GV	NEAPCO	88000.25	0562	ALLOY	88000.56
05103	NEAPCO	88000.20	05175	NEAPCO	88000.93	051H4	NEAPCO	88000.43	0563	ALLOY	88001.01
05105	NEAPCO	88000.52	05175X	NEAPCO	88000.92	051H4	NEAPCO	88000.74	057614	ALLIS-CHLM	80109
0511	ALCO	88000.73	05175X	NEAPCO	88000.93	051H5	NEAPCO	88000.45	057614	ALLIS-CHLM	80149
05111	NEAPCO	88000.20	05175X5	NEAPCO	85087	051H5	NEAPCO	88000.75	057614	FIAT-ALLIS	80149
05111	NEAPCO	88000.21	05175XS	NEAPCO	88000.92	051H6	NEAPCO	88000.74	058735	ALLIS-CHLM	80210
05112	NEAPCO	88000.22	05175XS	NEAPCO	88000.93	051HG	NEAPCO	88000.25	058735	ALLIS-CHLM	80211
05113	NEAPCO	88000.25	05175XS	NEAPCO	88001.02	051J8	NEAPCO	88000.71	058735	FIAT-ALLIS	80215
05113	NEAPCO	88000.27	05176	NEAPCO	88000.86	051K1	NEAPCO	88000.76	05E4	NEAPCO	88000.88
05114	NEAPCO	88000.28	05176	NEAPCO	88000.88	051K2	NEAPCO	88000.77	0600	ALLOY	84510
05115	NEAPCO	88000.30	05177	NEAPCO	88000.63	051KZ	NEAPCO	88000.22	0601	ALLOY	88000.01
05115	NEAPCO	88000.31	05178	NEAPCO	88000.89	051L2	NEAPCO	88000.45	0602	ALLOY	81237
05116	NEAPCO	88000.35	0517A	NEAPCO	88000.37	051P2	NEAPCO	88000.96	0603	ALLOY	88000.04
05117	NEAPCO	88000.37	0518	ALCO	88000.30	051R4	NEAPCO	88000.21	0604	ALLOY	88000.05
05118	NEAPCO	88000.38	05180	NEAPCO	88000.66	051R8	NEAPCO	88000.62	0604-25-060	MAZDA	80020
05118	NEAPCO	88000.39	05182	NEAPCO	88000.44	051R9	NEAPCO	88000.56	0604-89-251	MAZDA	80020
0512X	ALCO	88000.68	05183	NEAPCO	88000.79	051RP	NEAPCO	88000.33	0605	ALLOY	88013.90
0512	ALCO	88000.69	05184	NEAPCO	88000.81	051S8	NEAPCO	88001.01	0605X	NEAPCO	88000.15
05121	NEAPCO	88000.58	05185	NEAPCO	88000.82	051T3	NEAPCO	88000.53	0605X1	NEAPCO	88000.16
05123	NEAPCO	88000.21	05186	NEAPCO	88000.83	0520	ALCO	88000.37	0606	ALLOY	81333
05124	NEAPCO	88000.31	05187	NEAPCO	88000.51	0521	ALCO	88000.38	0607	ALLOY	81277
05125	NEAPCO	88000.39	05189	NEAPCO	88000.45	0522	ALCO	88000.43	060774	MIXERMOBL	80171
05129	NEAPCO	88000.99	0518A	NEAPCO	88000.38	0525	ALCO	88000.21	060774	MIXERMOBL	80197
0513	ALCO	88000.74	0519	ALCO	88000.35	0526	ALCO	88000.49	0608	ALLOY	81271
05132	NEAPCO	88000.22	05191	NEAPCO	88000.34	0527	ALLOY	88000.94	060868	MIXERMOBL	80231
05134	NEAPCO	88000.49	05191	NEAPCO	88000.35	0528	ALCO	88000.23	060868	MIXERMOBL	80237
05136	NEAPCO	88000.52	05193	NEAPCO	88000.46	0529	ALCO	88000.86	060868	MIXERMOBL	80240
05138	NEAPCO	88000.60	05195	NEAPCO	88000.46	0529	ALCO	88000.88	0609	NEAPCO	81246
05139	NEAPCO	88000.46	05195	NEAPCO	88000.59	0529	ALLOY	88000.88	0610	ALLOY	81218
05139	NEAPCO	88000.53	05196	NEAPCO	88000.36	0-53	AEC	88000.52	0611	ALLOY	81342
0513A	NEAPCO	88000.25	05197	NEAPCO	88000.35	0530	ALCO	88000.87	061111	NEAPCO	88000.21
0514	ALCO	88000.75	05199	NEAPCO	88000.84	0530538-8	FIAT-ALLIS	80069	061111	NEAPCO	88000.39
05141	NEAPCO	88000.94	051A1	NEAPCO	88000.52	0531	ALCO	88000.32	061113	NEAPCO	88000.21
05142	NEAPCO	88000.22	051A2	NEAPCO	88000.22	053120	ALLIS-CHLM	80001	061113	NEAPCO	88000.25
05143	NEAPCO	88000.53	051A7	NEAPCO	88000.90	053120	ALLIS-CHLM	80003	061113	NEAPCO	88000.27
05145	NEAPCO	88000.51	051A9	NEAPCO	88000.22	053120	FIAT-ALLIS	80001	061113	NEAPCO	88000.39
05146	NEAPCO	88000.23	051B1	NEAPCO	88000.22	0532	ALCO	88000.80	061184	NEAPCO	88000.39
05147	NEAPCO	88000.86	051B2	NEAPCO	88000.22	0533	ALCO	88000.48	061184	NEAPCO	88000.81
05147	NEAPCO	88000.88	051B3	NEAPCO	88000.25	0534	ALCO	88000.92	0612	ALLOY	81350
05148	NEAPCO	88000.87	051B3	NEAPCO	88000.27	0535	ALCO	88000.91	061212	NEAPCO	88000.22
05148	NEAPCO	88000.88	051B4	NEAPCO	88000.52	0536	ALCO	88000.40	061213	NEAPCO	88000.22
05149	NEAPCO	88000.43	051B5	NEAPCO	88000.25	0537	ALCO	88000.94	061213	NEAPCO	88000.25
05149	NEAPCO	88000.57	051B6	NEAPCO	88000.25	0538	ALCO	88000.47	061213	NEAPCO	88000.27
0515	ALCO	88000.85	051B8	NEAPCO	88000.44	0538	ALCO	88000.60	061214	NEAPCO	88000.22
05150	NEAPCO	88000.23	051BH	NEAPCO	88000.21	0539	ALCO	88000.95	061214	NEAPCO	88000.28
05154	NEAPCO	88000.32	051C8	NEAPCO	88000.29	0540	ALCO	88000.93	061215	NEAPCO	88000.22
05156	NEAPCO	88000.55	051CJ	NEAPCO	88000.27	0541	ALCO	88000.89	061215	NEAPCO	88000.31
05157	NEAPCO	88000.80	051CL	NEAPCO	88000.22	0543	ALCO	88000.79	061230	MIXERMOBL	80171
05158	NEAPCO	88000.48	051CM	NEAPCO	88000.21	0544	ALCO	88000.81	061230	MIXERMOBL	80197
05159	NEAPCO	88000.92	051D2	NEAPCO	88000.55	0545	ALCO	88000.82	0612621	FIAT-ALLIS	80067
05159X	NEAPCO	88000.92	051D4	NEAPCO	88000.19	0546	ALCO	88000.83	0612621-3	FIAT-ALLIS	80067
05159X	NEAPCO	88000.93	051D7	NEAPCO	88000.85	0547	ALCO	88000.51	06-127	NEAPCO	88006.36
0515A	NEAPCO	88000.30	051D8	NEAPCO	88000.36	0549	ALLOY	88000.45	061272	MIXERMOBL	80168
0516	ALCO	88000.20	051E3	NEAPCO	88000.54	0550	ALCO	88000.34	061272	MIXERMOBL	80171
05160	NEAPCO	88000.53	051E4	NEAPCO	88000.98	05-501471	BRD	80142	061282	MIXERMOBL	80168
05162	NEAPCO	88000.22	051E5	NEAPCO	88000.43	05-503303	BRD	80178	061282	MIXERMOBL	80171
05163X1	NEAPCO	88000.48	051E5	NEAPCO	88000.99	0551	ALLOY	88000.46	0613	ALLOY	81268
05163X1	NEAPCO	88000.91	051E6	NEAPCO	88000.61	0552	ALCO	88000.36	061313	NEAPCO	88000.25
05164	NEAPCO	88000.33	051E7	NEAPCO	88000.41	0553	ALCO	88000.84	061313	NEAPCO	88000.27
05165	NEAPCO	88000.40	051EZ	NEAPCO	88000.25	05-540017	BRD	80142	06-139	NEAPCO	88006.36
05166	NEAPCO	88000.64	051G1	NEAPCO	88000.21	05-541471	BRD	80142	0614	ALLOY	81333
05167	NEAPCO	88000.22	051G1	NEAPCO	88000.68	05-551471	BRD	80142	061414	NEAPCO	88000.28

Part No.	Brand	IDLI Group
061515	NEAPCO	88000.31
061576	NEAPCO	88000.31
061576	NEAPCO	88000.88
0616	ALLOY	81346
061616	NEAPCO	88000.35
061617	MIXERMOBL	80196
061617	MIXERMOBL	80216
061618	NEAPCO	88000.35
061618	NEAPCO	88000.38
0617512	FIAT-ALLIS	80077
0617512-9	FIAT-ALLIS	80077
0617909	FIAT-ALLIS	80007
0617909-7	FIAT-ALLIS	80007
061818	NEAPCO	88000.38
0618805	FIAT-ALLIS	80070
0618805-6	FIAT-ALLIS	80070
0619703	FIAT-ALLIS	80100
0619703-2	FIAT-ALLIS	80100
0619915-2	FIAT-ALLIS	80150
062323	NEAPCO	88000.21
062424	NEAPCO	88000.31
06-251	NEAPCO	85102
062525	NEAPCO	88000.38
0625X	NEAPCO	88000.18
0629943	FIAT-ALLIS	80207
0629943-2	FIAT-ALLIS	80207
0-63	AEC	88000.25
0631383	FIAT-ALLIS	80021
0631383-7	FIAT-ALLIS	80021
06-351	NEAPCO	85114
06-351	NEAPCO	88001.14
063939	NEAPCO	88000.53
064053-0	FIAT-ALLIS	80007
064110-8	FIAT-ALLIS	80007
0645053	FIAT-ALLIS	80007
06-451	NEAPCO	88001.12
0648366	FIAT-ALLIS	80035
06483663	ALLIS-CHLM	80035
06483663	FIAT-ALLIS	80035
065-000-0	GLAENZER	80135
065-001-0	GLAENZER	80042
06-501822	BRD	80180
06-510002	BRD	80180
065-100-8	GLAENZER	80205
065-101-8	GLAENZER	80213
065-102-8	GLAENZER	80112
065-103-8	GLAENZER	80235
065-104-8	GLAENZER	80234
065-105-8	GLAENZER	80198
065-106-8	GLAENZER	80214
065-107-8	GLAENZER	80207
065-108-8	GLAENZER	80215
065-109-8	GLAENZER	80239
065-110-8	GLAENZER	80240
065-111-8	GLAENZER	80234
065-112-8	GLAENZER	80242
065-113-8	GLAENZER	80178
065-114-8	GLAENZER	80189
065-115-8	GLAENZER	80221
065-116-8	GLAENZER	80225
065-117-8	GLAENZER	80216
065-118-8	GLAENZER	80149
065-119-8	GLAENZER	80101
065-120-8	GLAENZER	80198
065-121-8	GLAENZER	80213
065-122-8	GLAENZER	80238
065-123-8	GLAENZER	80155
065-124-8	GLAENZER	80150
065-125-8	GLAENZER	80230
065-126-8	GLAENZER	80171
065-127-8	GLAENZER	80070
065-128-8	GLAENZER	80196
065-129-8	GLAENZER	80156
065-131-8	GLAENZER	80238
065-132-8	GLAENZER	80213
065-133-8	GLAENZER	80048
065-134-8	GLAENZER	80047
065-135-8	GLAENZER	80124
065-136-8	GLAENZER	80142
065-137-8	GLAENZER	80195
065-138-8	GLAENZER	80202
065-139-8	GLAENZER	80193
065-140-8	GLAENZER	80049
065-141-8	GLAENZER	80223
065-142-8	GLAENZER	80224
065-143-8	GLAENZER	80069
065-144-8	GLAENZER	80103
065-145-8	GLAENZER	80172
065-146-8	GLAENZER	80221
065-147-8	GLAENZER	80070
065-148-8	GLAENZER	80035
065-149-8	GLAENZER	88003.33
065-150-8	GLAENZER	80182
065-151-8	GLAENZER	80190
065-152-8	GLAENZER	80001
065-153-8	GLAENZER	80021
065-154-8	GLAENZER	88003.39
065-155-8	GLAENZER	80067
065-156-8	GLAENZER	80245
065-157-8	GLAENZER	80154
065-158-8	GLAENZER	80175
065-159-8	GLAENZER	80228
065-160-8	GLAENZER	80233
065-161-8	GLAENZER	80164
065-162-8	GLAENZER	80165
065-164-8	GLAENZER	88003.32
065-165-8	GLAENZER	80140
065-166-8	GLAENZER	80174
065-168-8	GLAENZER	80046
065-169-8	GLAENZER	80166
065-170-8	GLAENZER	80149
065-171-8	GLAENZER	80206
065-172-8	GLAENZER	88003.40
065-173-8	GLAENZER	80104
065-174-8	GLAENZER	80164
065-200-8	GLAENZER	80135
065-201-8	GLAENZER	80064
065-203-8	GLAENZER	80061
065-208-8	GLAENZER	80119
065-209-8	GLAENZER	80140
065-210-8	GLAENZER	80118
065-211-8	GLAENZER	80117
065-212-8	GLAENZER	80090
065-213-8	GLAENZER	80073
065-216-8	GLAENZER	80075
065-217-8	GLAENZER	80138
065-218-8	GLAENZER	80160
065-219-8	GLAENZER	80181
065-220-8	GLAENZER	80123
065-221-8	GLAENZER	80071
065-222-8	GLAENZER	80074
065-223-8	GLAENZER	80141
065-224-8	GLAENZER	80145
065-225-8	GLAENZER	80081
065-226-8	GLAENZER	80091
065-227-8	GLAENZER	80183
065-228-8	GLAENZER	80129
065-229-8	GLAENZER	80131
065-230-8	GLAENZER	80055
065-231-8	GLAENZER	80043
065-232-8	GLAENZER	80049
065-233-8	GLAENZER	80030
065-234-8	GLAENZER	80031
065-235-8	GLAENZER	80051
065-239-8	GLAENZER	80104
065-240-8	GLAENZER	80158
065-241-8	GLAENZER	80148
065-242-8	GLAENZER	80159
065-244-8	GLAENZER	80174
065-245-8	GLAENZER	80173
065-249-8	GLAENZER	80199
065-250-8	GLAENZER	80201
065-256-8	GLAENZER	80032
065-257-8	GLAENZER	88002.77
065-258-8	GLAENZER	80188
065-259-8	GLAENZER	80220
065-260-8	GLAENZER	80186
065-261-8	GLAENZER	80179
065-262-8	GLAENZER	80153
065-263-8	GLAENZER	80161
065-264-8	GLAENZER	80054
065-265-8	GLAENZER	80132
065-266-8	GLAENZER	80121
065-267-8	GLAENZER	80025
065-268-8	GLAENZER	80105
065-269-8	GLAENZER	80066
065-270-8	GLAENZER	80058
065-271-8	GLAENZER	80001
065-272-8	GLAENZER	80033
065-273-8	GLAENZER	80037
065-274-8	GLAENZER	80008
065-275-8	GLAENZER	80133
065-276-8	GLAENZER	80080
065-277-8	GLAENZER	80185
065-280-8	GLAENZER	80151
065-281-8	GLAENZER	80124
065-304-6	GLAENZER	80135
065-351-6	GLAENZER	80140
0653811	FIAT-ALLIS	80001
065454	NEAPCO	88000.31
065-544-6	GLAENZER	80158
065-545-6	GLAENZER	80146
065757	NEAPCO	81020
065-892-6	GLAENZER	80188
065-893-6	GLAENZER	80220
065-906-8	GLAENZER	80135
065-907-8	GLAENZER	80042
065-923-8	GLAENZER	80023
065-935-8	GLAENZER	88002.77
065-936-8	GLAENZER	80032
065-959-8	GLAENZER	88002.77
065-960-8	GLAENZER	88002.77
065-980-8	GLAENZER	88002.77
0665611	FIAT-ALLIS	80070
0665611-0	FIAT-ALLIS	80070
066678	FIAT-ALLIS	80001
0675X	NEAPCO	88000.15
0675X1	NEAPCO	88000.16
0675X1A	NEAPCO	88000.15
0680970	FIAT-ALLIS	80193
0680970-1	FIAT-ALLIS	80193
06809982	ALLIS-CHLM	80167
0680998-2	FIAT-ALLIS	80167
0681014-7	FIAT-ALLIS	80168
0682636-6	FIAT-ALLIS	80109
069953	FIAT-ALLIS	80223
069962	ALLIS-CHLM	80224
069962(2)	FIAT-ALLIS	80223
069962-9(2)	FIAT-ALLIS	80223
06B8B8	NEAPCO	81020
06B8B8	NEAPCO	88000.21
06DYDY	NEAPCO	88000.25
06G1G1	LEMPCO	88000.69
06G2G2	NEAPCO	88000.70
06G3G3	NEAPCO	88000.71
06G4G4	NEAPCO	88000.72
06L6	NEAPCO	88000.15
06N0103	WESCO	80061
06N0123	WESCO	88014.52
06N0134	WESCO	81034
06N0140-30	WESCO	88001.10
06N0140-32	WESCO	81060
06N0154	WESCO	81089
06N0180-21	WESCO	81143
06N0180-26	WESCO	81156
06N048	WESCO	88001.13
06N048F	WESCO	88001.14
06N048M	WESCO	88001.12
06N1434-37	WESCO	81054
06N1634-37	WESCO	81109
06N1845-37	WESCO	81151
06N1945-37	WESCO	88001.09
06N2045-37	WESCO	81175
06NNB1	WESCO	88001.15
06NSSM48	WESCO	85102
0706-89-251	FORD	80020
0706-89-251	MAZDA	80020
0715	ALLOY	83129
0-73	AEC	88000.28
073372	FIAT-ALLIS	80142
0742HSA	NEW IDEA	80033
0750	ALLOY	88014.32
07-501823	BRD	80189
0751	ALLOY	84531
07-510112	BRD	80189
0752	ALLOY	84529
0753	ALLOY	83281
0754	ALLOY	83316
0755	ALLOY	83349
0756	ALLOY	88014.38
075614	ALLIS-CHLM	80109
075614	FIAT-ALLIS	80109
075614	FIAT-ALLIS	80149
0757	ALLOY	83252
0758	ALLOY	83241
0759	ALLOY	83362
0-76	AEC	88001
0760	ALLOY	83283
0761	ALLOY	83262
0762	ALLOY	83357
0763	ALLOY	83240
0764	ALLOY	83251
076456	FIAT-ALLIS	80216
0765	ALLOY	83250
0766	ALLOY	83320
0767	ALLOY	87037
0769	ALLOY	83263
0770	ALLOY	83317
0771	ALLOY	83331
0777HSA	NEW IDEA	80001
077HSA	NEW IDEA	80001
08-000-504-465	ALLIS-CHLM	80210
08-000-504-465	FIAT-ALLIS	80210
08-000-505-01-398	ALLIS-CHLM	80100
08-000-505-01-398	FIAT-ALLIS	80100
08-000-505-01-409	ALLIS-CHLM	80211

Part No.	Brand	IDLI Group
08-000-505-01-409 .... FIAT-ALLIS	...	80211
08-000-505-01-443 .... ALLIS-CHLM	...	80167
08-000-505-01-443 .... FIAT-ALLIS	...	80167
08-000-505-437	ALLIS-CHLM	80235
08-000-505-437	FIAT-ALLIS	80235
08-000-505-61-443 .... ALLIS-CHLM	...	80167
0801	ALLOY	87032
080132	FIAT-ALLIS	80213
080470	ALLIS-CHLM	80124
080470	FIAT-ALLIS	80124
080470-8	ALLIS-CHLM	80124
080470-8	FIAT-ALLIS	80124
08-048-885-63-001 .... ALLIS-CHLM	...	80193
08-048-885-63-001 .... FIAT-ALLIS	...	80193
0807855	EATON	80195
0807856	EATON	80150
0-81	AEC	88000.49
081058	ALLIS-CHLM	80001
081058	ALLIS-CHLM	80003
081058	FIAT-ALLIS	80001
081559	ALLIS-CHLM	80142
081559	ALLIS-CHLM	80143
081559	ALLIS-CHLM	80178
081559	FIAT-ALLIS	80142
081559-7	ALLIS-CHLM	80142
081559-7	FIAT-ALLIS	80142
082073	ALLIS-CHLM	80049
082073	FIAT-ALLIS	80049
082073	FIAT-ALLIS	88003.26
082744	FIAT-ALLIS	80044
0-83	AEC	88000.53
083541	ALLIS-CHLM	80210
083541	FIAT-ALLIS	80215
083616	ALLIS-CHLM	80210
083616	ALLIS-CHLM	80211
083616	FIAT-ALLIS	80215
083616-3	ALLIS-CHLM	80210
083616-3	FIAT-ALLIS	80215
084328	ALLIS-CHLM	80195
084328	FIAT-ALLIS	80195
084395	ALLIS-CHLM	80202
084395	FIAT-ALLIS	80202
084435	ALLIS-CHLM	80185
084435	ALLIS-CHLM	80202
084435	FIAT-ALLIS	80202
084435-7	ALLIS-CHLM	80185
084435-7	FIAT-ALLIS	80202
084537	ALLIS-CHLM	80162
084537	FIAT-ALLIS	80162
084545	ALLIS-CHLM	80174
084545	ALLIS-CHLM	80185
084545	FIAT-ALLIS	80174
084545-3	ALLIS-CHLM	80185
084545-3	FIAT-ALLIS	80174
08-510106	BRD	80221
085918	ALLIS-CHLM	80193
085918	ALLIS-CHLM	80198
085918	FIAT-ALLIS	80193
085918-1	ALLIS-CHLM	80193
085918-1	FIAT-ALLIS	80193
086284	ALLIS-CHLM	80109
086284	ALLIS-CHLM	80158
086284	FIAT-ALLIS	80149
086284-7	ALLIS-CHLM	80109
086284-7	FIAT-ALLIS	80149
08-630-150-00-502 .... ALLIS-CHLM	...	80193
08-630-150-00-502 .... FIAT-ALLIS	...	80193
08-630-403-008	ALLIS-CHLM	80100
08-630-403-008	FIAT-ALLIS	80100
086747	ALLIS-CHLM	80168
086747	FIAT-ALLIS	80165
086747-3	ALLIS-CHLM	80168
086747-3	FIAT-ALLIS	80165
086752	ALLIS-CHLM	80023
086752	ALLIS-CHLM	80167
086752	ALLIS-CHLM	80168
086752	FIAT-ALLIS	80164
086752-3	ALLIS-CHLM	80168
086752-3	FIAT-ALLIS	80164
08.70.290.900	RENAULT	80221
08.70.360.900	RENAULT	80142
08.70.550.300	RENAULT	80180
08744	ALLIS-CHLM	80044
08744	FIAT-ALLIS	80044
090311	ALLIS-CHLM	80224
090311(2)	FIAT-ALLIS	80223
090311-2(2)	FIAT-ALLIS	80223
0-93	AEC	88000.31
0-95	AEC	88000.34
09-520756	BRD	80225
096-55	BORG-WARNR	80138
096-55	CLEVELAND	80138
0A4841	FORD	88002.87

FOR NUMBERS STARTING WITH ONE, ALSO CHECK THE ALPHABETICAL I.

Part No.	Brand	IDLI Group
10	AEC	88008.35
100	TRU-CROSS	80030
100	TRU-CROSS	80031
1000	SPICER	80007
10000J	BORG-WARNR	88015.16
10000J10800	BORG-WARNR	87658
10000J7200	BORG-WARNR	87657
1-0001	NEAPCO	80008
10001P2	DIAMOND R	80182
10001P2	WHITE	80182
10003	AEC	80245
1-0003	NEAPCO	80133
1-0005	NEAPCO	80041
10007	AEC	80245
10008	DIAMOND R	80075
100083	GARD.DENV.	80155
10008P2	WHITE	80077
1000K24	KENWORTH	80220
1000K24A	KENWORTH	80188
1001	ALCO	80072
1001	ALLOY	80072
10-01-00	WALTERSCHD	88004.29
100-103	BORG-WARNR	88000.36
10011	AEC	80245
1-0011	NEAPCO	80006
100-113	BORG-WARNR	88000.35
100-115	BORG-WARNR	88000.55
10011J	BORG-WARNR	87511
100-12	BORG-WARNR	88000.45
1-0012	NEAPCO	80018
100-123	BORG-WARNR	88000.38
100-124	BORG-WARNR	88000.37
10012J	BORG-WARNR	85561
10013	ALLOY	86193
100-13	BORG-WARNR	88000.20
1-0013	NEAPCO	80009
10013	WESCO	86158
10014	ALLOY	86193
1-0014	NEAPCO	80017
10014	TRW	86193
10014	WESCO	86193
10015	ALLOY	86143
10015	TRW	86143
10015	WESCO	86143
10016	ALLOY	86180
10016	TRW	86180
10016	WESCO	86180
10017	ALLOY	86182
10017	ALLOY	86299
10017	WESCO	86182
10018	ALLOY	86149
10018	ALLOY	86295
10018	WESCO	86149
10019	ALLOY	86075
10019	ALLOY	86148
10019	WESCO	86148
1002	ALLOY	80085
10020	ALLOY	86179
10020	ALLOY	86251
10020	ALLOY	86252
10020	ALLOY	86254
10020	ALLOY	86255
10020	ALLOY	86256
10020	ALLOY	86257
10020	ALLOY	86282
10020	WESCO	86179
10021	ALLOY	86147
10021	ALLOY	86236
10021	WESCO	86147
10022	ALLOY	86157
100-22	BORG-WARNR	88000.43
10022	WESCO	86157
10023	ALLOY	86178
10023	ALLOY	86246
10023	ALLOY	86247
1-0023	NEAPCO	88001.70
10023	WESCO	86178
1-0026	NEAPCO	80024
1-0027	NEAPCO	80010
1-0028	NEAPCO	80004
1-0029	NEAPCO	80019
10-0312	NEAPCO	88000.88
100315	GMC	80140
10-0322	NEAPCO	88000.89
10-0323	NEAPCO	88000.80
100-33	BORG-WARNR	88000.26
10-0332	NEAPCO	88000.95
10-0333	NEAPCO	88000.81
100336K1	DIAMOND R	88000.14
10033J	BORG-WARNR	86810
10-0381	NEAPCO	88000.92
10-0383	NEAPCO	88000.83
1004	ALCO	80077
1004	ALLOY	80077
100 410 01 31	MERCEDES	88002.77
10-0412	NEAPCO	88000.45
10-0413	NEAPCO	88000.21
10-0422	NEAPCO	88000.43
10-0423	NEAPCO	88000.44
100-43	BORG-WARNR	88000.22
10-0432	NEAPCO	88000.47
10-0433	NEAPCO	88000.26
10-0443	NEAPCO	88000.22
10-0452	NEAPCO	88000.46
10-0453	NEAPCO	88000.52
10045P2	DIAMOND R	80189
10-0463	NEAPCO	88000.27
10046P2	DIAMOND R	80221
10-0473	NEAPCO	88000.28
10047P2	DIAMOND R	80227
10-0483	NEAPCO	88000.53
10048P2	DIAMOND R	80023
10048P2	DIAMOND R	80140
10048P2	WHITE	80023
10-0493	NEAPCO	88000.31
10049P2	DIAMOND R	80023
10049P2	DIAMOND R	80140
10049P2	WHITE	80024
1004P	ALLOY	80072
1005	ALCO	80075
1005	ALLOY	80072
1005	ALLOY	80075
1-0050	NEAPCO	80092
10050P2	WHITE	80178
10051P2	DIAMOND R	80023
10051P2	DIAMOND R	80140
10051P2	REO MOTORS	80023
10051P2	WHITE	80023
100524060	TRIUMPH	80041
100-53	BORG-WARNR	88000.52
10053P2	DIAMOND R	80023
10053P2	DIAMOND R	80140
10053P2	REO MOTORS	80023
10053P2	WHITE	80024
10054	WESCO	86156
10054P2	DIAMOND R	80072
10054P2	DIAMOND R	80075
10054P2	DIAMOND R	80076
10054P2	REO MOTORS	80077
10054P2	WHITE	80075
10055	WESCO	86155
10055P2	DIAMOND R	80077
10055P2	REO MOTORS	80077
10055P2	WHITE	80077
10056	ALLOY	86184
10056	WESCO	86184
10056P2	DIAMOND R	80124
10056P2	DIAMOND R	80125
10056P2	REO MOTORS	80124
10056P2	WHITE	80124
10057	WESCO	86154
10057P2	DIAMOND R	80142
10057P2	REO MOTORS	88008.97
10057P2	WHITE	80142
10058	WESCO	86153
10058P2	DIAMOND R	80178
10058P2	REO MOTORS	80178
10058P2	WHITE	80178
10059P2	DIAMOND R	80182
10059P2	REO MOTORS	80182
10059P2	WHITE	80182
1006	ALCO	80036
1006	ALLOY	80036
10060P2	DIAMOND R	80189
1-0061	NEAPCO	80021
10061	WESCO	86151
10061P2	DIAMOND R	80190
10061P2	REO MOTORS	80190
10061P2	WHITE	80190
10062	WESCO	86152
10062P2	DIAMOND R	80221
100-63	BORG-WARNR	88000.27
10063P2	DIAMOND R	80225
10064P2	DIAMOND R	80197
10064P2	REO MOTORS	80197
10064P2	WHITE	80196
10065	JAGUAR	80077

Part No.	Brand	IDLI Group
10065P2	DIAMOND R	80198
10065P2	WHITE	80198
10066	ALLOY	86190
10066	WESCO	86190
10067	ALLOY	86163
10067	WESCO	86163
10069P2	DIAMOND R	80193
10069P2	DIAMOND R	80198
10069P2	REO MOTORS	80193
10069P2	WHITE	80193
1007	ALCO	80025
1007	ALLOY	80026
1007023	ALLIS-CHLM	80142
1007023	ALLIS-CHLM	80143
1007023	FIAT-ALLIS	80142
10070613	ALLIS-CHLM	80077
10070613	FIAT-ALLIS	80077
10070P2	DIAMOND R	80195
10071P2	DIAMOND R	80199
10071P2	WHITE	80199
10071PW	REO MOTORS	80199
10072P	DIAMOND R	80201
10072P2	WHITE	80201
100-73	BORG-WARNR	88000.28
10073J	BORG-WARNR	82087
10073P2	DIAMOND R	80213
10073P2	WHITE	80215
10074P2	DIAMOND R	80210
10074P2	REO MOTORS	80210
10074P2	WHITE	80210
10075P2	DIAMOND R	80211
10075P2	REO MOTORS	80211
10075P2	WHITE	80207
100-76	BORG-WARNR	88000.29
10076P2	DIAMOND R	80212
10076P2	REO MOTORS	80218
10076P2	WHITE	80218
10077P2	DIAMOND R	80238
10077P2	WHITE	80238
10078P2	DIAMOND R	80231
10078P2	REO MOTORS	80231
10078P2	WHITE	80240
10079P2	DIAMOND R	80232
10079P2	REO MOTORS	80232
10079P2	WHITE	80239
1008	ALLOY	80013
10080P2	DIAMOND R	80237
10080P2	REO MOTORS	80237
10080P2	WHITE	80237
100-81	BORG-WARNR	88000.49
10081P2	DIAMOND R	80236
10081P2	REO MOTORS	80236
10082P2	DIAMOND R	80238
100-83	BORG-WARNR	88000.53
1-0089	NEAPCO	88013.86
10090P2	DIAMOND R	80221
10090P2	WHITE	80221
100-93	BORG-WARNR	88000.31
100-95	BORG-WARNR	88000.34
1-0099	NEAPCO	88013.87
100E7039	FORD	80024
101	TRU-CROSS	80023
1010	ALCO	80081
1010	ALLOY	80081
10-10-00	WALTERSCHD	88003.73
10104	WESCO	86177
10105	ALLOY	86268
1-0105	NEAPCO	80030
1-0105	NEAPCO	80031
10105	WESCO	86268
1010P	ALLOY	80081
1011	ALCO	80091
1011	ALLOY	80091
10110	ALLOY	86191
10110	WESCO	86191
101-10-00070	KOMATSU	80047
101-10-00080	KOMATSU	80046
101-1010	REPCO	80036
101.10.17100	KOMATSU	80047
101.10.17200	KOMATSU	80046
101-103	BORG-WARNR	88000.94
101-1030	REPCO	80012
101-1031	REPCO	80015
101-1050	REPCO	80010
101-1060	REPCO	80004
101-1070	REPCO	80019
10111	ALLOY	86175
10111	WESCO	86176
101-1100	REPCO	80081
101-1-101	MUNCIE	88000.21
101-1-101	MUNCIE	88000.39
101-1105	REPCO	80091
101-1110	REPCO	80020
101-1-120	MUNCIE	88000.21
101-1-120	MUNCIE	88000.25
101-1-120	MUNCIE	88000.27
101-1-120	MUNCIE	88000.39
101-1120	REPCO	80142
101-1130	REPCO	80011
101-1132	REPCO	80005
101-1140	REPCO	80013
101-12	BORG-WARNR	88000.88
101-1210	REPCO	80024
101-1230	REPCO	80009
101-1240	REPCO	80006
101-1250	REPCO	80018
101-1260	REPCO	80017
101-1270	REPCO	80115
101-1280	REPCO	80041
101-1290	REPCO	80077
101-13	BORG-WARNR	88000.79
101-1-510	MUNCIE	88000.39
101-1-510	MUNCIE	88000.81
1011P	ALLOY	80091
1012	ALCO	80129
1012	ALLOY	80129
1-0121	NEAPCO	80072
10121	WESCO	86175
101-2-101	MUNCIE	88000.21
101-2-101	MUNCIE	88000.39
101-2-120	MUNCIE	88000.21
101-2-120	MUNCIE	88000.25
101-2-120	MUNCIE	88000.27
101-2-120	MUNCIE	88000.39
101-2-510	MUNCIE	88000.39
101-2-510	MUNCIE	88000.81
101281330	TRIUMPH	80024
1012P	ALLOY	80129
1013	ALCO	80131
1013	ALLOY	80131
101-32	BORG-WARNR	88000.95
101-33	BORG-WARNR	88000.81
1-0133	NEAPCO	80116
1-0134	NEAPCO	80118
1013P	ALLOY	80131
1014	ALCO	80107
1014	ALLOY	80104
1014	TEDDY TORQ	80030
1014G	TEDDY TORQ	80031
1014P	ALLOY	80107
1015	ALCO	80104
1015	ALLOY	80104
1015	ALLOY	80107
1015	ALLOY	80112
1-0150	NEAPCO	80119
1-0150	NEAPCO	80120
101-53	BORG-WARNR	88000.83
1-0153	NEAPCO	80077
1016	ALLOY	80015
1016-245M91	MASS.FERG.	80104
101-63	BORG-WARNR	88000.96
1017	ALLOY	80016
1-0170	NEAPCO	80007
1-0170	NEAPCO	88000.15
1-0174	NEAPCO	80078
1-0174	NEAPCO	80090
1018	ALLOY	80115
1018	TEDDY TORQ	80055
101-81	BORG-WARNR	88000.93
10182-127	SUTTON	80196
101-83	BORG-WARNR	88000.92
1018Z203	HARNISCH.	80188
1018Z204	HARNISCH.	80220
1019	ALCO	80061
1019	ALLOY	80061
101-93	BORG-WARNR	88000.84
1019P	ALLOY	80061
102	AEC	88013.86
102	AEC	88013.87
1020	ALCO	80132
1020	ALLOY	80132
1-0200	NEAPCO	80072
1-0200	NEAPCO	80075
1020P	ALLOY	80132
1021	ALCO	80041
1021	ALLOY	80041
10212J	BORG-WARNR	81882
102-13	BORG-WARNR	88000.68
10-217	HY-DYNAM	80149
1021944M91	MASS.FERG.	80174
1022	ALLOY	80005
1022Z487	HARNISCH.	80168
1023	AEC	80024
1023	ALCO	80023
1023	ALLOY	80024
1024	ALCO	80090
10242-16SF	SPICER	88000.25
10242-16SF	SPICER	88000.27
10242-17SF	SPICER	88000.21
10242-17SF	SPICER	88000.25
10242-17SF	SPICER	88000.27
10242-17SF	SPICER	88000.39
10242-18SF	SPICER	88000.21
10242-18SF	SPICER	88000.39
10242-1SF	SPICER	88000.53
10242-21SF	SPICER	88000.22
10242-21SF	SPICER	88000.31
10242-23SF	SPICER	88000.35
10242-23SF	SPICER	88000.38
10242-2SF	SPICER	88000.52
10242-30SF	SPICER	88000.22
10242-32SF	SPICER	88000.38
10242-3SF	SPICER	88000.22
10242-40SF	SPICER	88000.22
10242-40SF	SPICER	88000.28
10242-45SF	SPICER	88000.22
10242-45SF	SPICER	88000.25
10242-45SF	SPICER	88000.27
10242-4SF	SPICER	81020
10242-57SF	SPICER	88000.25
10242-5SF	SPICER	88000.21
10242-5SF	SPICER	88000.39
10242-6SF	SPICER	88000.28
10242-7SF	SPICER	88000.31
10242-8SF	SPICER	88000.38
10242-9SF	SPICER	88000.35
102-43	BORG-WARNR	88000.70
10243-3SF	SPICER	88000.39
10243-3SF	SPICER	88000.81
10243-8SF	SPICER	88000.31
10243-8SF	SPICER	88000.88
1-0248	NEAPCO	80025
1025	ALCO	80090
1025	ALLOY	80090
1026	ALCO	80093
1026	ALLOY	80093
102-63	BORG-WARNR	88000.71
102-684R91	IHC	80210
1026P2	DIAMOND R	80142
1026P2	REO MOTORS	80142
1026P2	WHITE	80142
1027	ALCO	80023
1027	ALCO	80024
1027	ALLOY	80023
1027	TEDDY TORQ	80044
102-73	BORG-WARNR	88000.72
102-76	BORG-WARNR	88000.72
1-0278	NEAPCO	80085
1-0278	NEAPCO	80086
1028	ALCO	80075
1028	ALCO	80089
1028	ALLOY	80089
1028	TEDDY TORQ	80049
10280-1SF	SPICER	88000.71
10280-2SF	SPICER	88000.70
10280-3SF	SPICER	88000.69
10280-7SF	SPICER	88000.72
1029	ALLOY	80020
1029-144M91	MASS.FERG.	80174
1029-148M91	MASS.FERG.	80175
1029-201M91	MASS.FERG.	80149
102-93	BORG-WARNR	88000.73
1-0297	NEAPCO	80079
103	AEC	88013.86
103	AEC	88013.87
1-0300	NEAPCO	80023
1-0300	NEAPCO	80024
10-3-0012X	SPICER	88000.88
10-3-0013X	SPICER	88000.79
10-3-0022X	SPICER	88000.89
10-3-0023X	SPICER	88000.80
10-3-0032	SPICER	88000.95
10-3-0033X	SPICER	88000.81
10-3-0051X	SPICER	88000.85
10-3-0081X	SPICER	88000.92
10-3-0083X	SPICER	88000.83
1-0301	NEAPCO	80023
10-3-0122X	SPICER	88000.37
10-3-0183X	SPICER	88000.84
1031	ALCO	80024
1031	ALCO	80140
1031	ALLOY	80023
1031	WILK RIP.	80072
10-3-103X	SPICER	88000.82
103113	AEC	88000.35
10-3122	NEAPCO	88000.90
10-3-122X	MUNCIE	88000.90
10-3-122X	SPICER	88000.90

Part No.	Brand	IDLI Group
10-3-12X	MUNCIE	88000.88
10-3-12X	SPICER	88000.86
10-3-12X	SPICER	88000.88
10-3-13X	SPICER	88000.79
1-0315	NEAPCO	80026
10-3162	NEAPCO	88001.01
10-3-162X	SPICER	88001.01
10-3163	NEAPCO	88000.85
10-3-163X	MUNCIE	88000.85
10-3-163X	SPICER	88000.84
10-3183	NEAPCO	88000.84
10-3-183X	MUNCIE	88000.84
10-3-183X	SPICER	88000.85
1032	ALCO	80075
1032	ALLOY	80077
10-3-203X	SPICER	88000.82
10-3-22X	MUNCIE	88000.89
10-3-22X	SPICER	88000.89
10-3-23X	MUNCIE	88000.80
10-3-23X	SPICER	88000.80
10324J	BORG-WARNR	81863
10-32-92	SPICER	88014.95
1033	ALCO	80116
1033	ALLOY	80116
10-3-32X	MUNCIE	88000.95
10-3-32X	SPICER	88000.95
10-3-33X	MUNCIE	88000.81
10-3-33X	SPICER	88000.81
1034	ALCO	80118
1034	ALLOY	80118
10-3-41X	SPICER	88001
10345702	HALLBURTON	80215
1-0350	NEAPCO	80036
1-0350	NEAPCO	80044
10-3-51X	SPICER	88000.94
10357764	HALLBURTON	80193
10-3-81X	MUNCIE	88000.92
10-3-81X	SPICER	88000.92
10-3-81X	SPICER	88000.93
10-3-83X	MUNCIE	88000.83
10-3-83X	SPICER	88000.83
1039	ALCO	80076
1039	ALLOY	80076
1039	TEDDY TORQ	80049
1039P2	DIAMOND R	80079
1039P2	DIAMOND R	80189
1039P2	REO MOTORS	80189
1039P2	WHITE	80189
103M520	EIMCO	80238
103M530	EIMCO	80163
103M8226	EIMCO	80210
103M8826	EIMCO	80215
1040	ALCO	80117
1040	ALLOY	80117
10-4-0011	SPICER	88000.49
10-4-0012	SPICER	88000.45
10-4-0013	SPICER	88000.21
10-4-0022	SPICER	88000.43
10-4-0023	SPICER	81020
10-4-0031	SPICER	88000.51
10-4-0032	SPICER	88000.47
10-4-0033	SPICER	88000.22
10-4-0033	SPICER	88000.26
10-4-0043	SPICER	88000.22
10-4-0052	SPICER	88000.46
10-4-0053	SPICER	88000.52
10-4-0062	SPICER	88000.45
10-4-0063	SPICER	88000.27
10-4-0073	SPICER	88000.27
10-4-0083	SPICER	88000.31

Part No.	Brand	IDLI Group
10-4-0093	SPICER	88000.53
10-4-0103	SPICER	88000.36
10-4-0113	SPICER	88000.35
10-4-0123	SPICER	88000.39
10-4-0133	SPICER	88000.21
10-4-0143	SPICER	88000.36
10-4-0153	SPICER	88000.37
10-4-0173	SPICER	88000.22
10-4-0183	SPICER	88000.38
10-4-0183	SPICER	88000.55
10-4-0193	SPICER	88000.39
10-4-0373	SPICER	88000.19
10-4-0683	SPICER	88000.72
10-4-0693	SPICER	88000.69
10-4-0703	SPICER	88000.71
10-4-0723	SPICER	88000.73
10-4-0853	SPICER	88000.31
10-4-0853	SPICER	88000.53
1041	ALCO	80025
1041	ALLOY	80025
10-4-101	SPICER	88000.48
10-4103	NEAPCO	88000.36
10-4-103	SPICER	88000.36
10-4-1043	SPICER	88000.81
10-4-11	SPICER	88000.49
10-4113	NEAPCO	88000.35
10-4-113	SPICER	88000.35
10-4-1193	SPICER	88000.59
10-4-12	MUNCIE	88000.45
10-4-12	SPICER	88000.45
10-4-1203	SPICER	88000.33
10-4-12253	SPICER	88000.54
10-4123	NEAPCO	88000.39
10-4-123	SPICER	88000.39
10-4-13	MUNCIE	88000.21
10-4-13	SPICER	88000.21
10-4133	NEAPCO	88000.20
10-4-133	SPICER	88000.20
10-4-133	SPICER	88000.21
10-4-1363	SPICER	88000.40
10-4141	NEAPCO	88000.67
10-4-141X	SPICER	88000.67
10-4143	NEAPCO	88000.36
10-4-143	SPICER	88000.36
10-4-1473	SPICER	88000.35
10-4-152	SPICER	88000.58
10-4-153	SPICER	88000.37
10-4-163	SPICER	88000.34
10-4-171	SPICER	88000.49
10-4-172	SPICER	88000.58
10-4173	NEAPCO	88000.22
10-4-173	SPICER	88000.22
10-4-182	SPICER	88000.43
10-4-182	SPICER	88000.98
10-4183	NEAPCO	88000.55
10-4-183	SPICER	88000.55
10-4-192	SPICER	88000.45
10-4193	NEAPCO	88000.38
10-4-193	SPICER	88000.38
1042	ALCO	80118
1042	ALCO	80121
1042	ALLOY	80121
1042	ALLOY	80131
1042	TEDDY TORQ	80044
10420	JAGUAR	80142
10-4-203	SPICER	88000.52
10-4-21	SPICER	88000.50
10-4-211	SPICER	88000.49
10-4-22	MUNCIE	88000.43
10-4-22	SPICER	88000.43

Part No.	Brand	IDLI Group
10-4-23	MUNCIE	88000.44
10-4-23	SPICER	88000.44
10-4-253	SPICER	88000.25
10-4282	NEAPCO	88000.62
10-4-282	SPICER	88000.62
10-4-293	SPICER	88000.56
10-4-303	SPICER	88000.20
10-4-303	SPICER	88000.57
10-4-31	SPICER	88000.51
10-4-32	MUNCIE	88000.47
10-4-32	SPICER	88000.47
10432K1	WHITE	80003
10-4-33	MUNCIE	88000.26
10-4-33	SPICER	88000.26
10-4-331X	SPICER	88014.06
10-4363	NEAPCO	88000.40
10-4-373	SPICER	88000.19
104397	HESSTON	80069
104397	HOUGH	80069
104397	IHC	88003.10
104398	HESSTON	80069
104398	HOUGH	80069
104398	IHC	88003.10
104401	HESSTON	80069
104401	HOUGH	80069
104401	IHC	88003.10
104401(5)	IHC	88003.10
104402	HESSTON	80069
104402	HOUGH	80069
104402	IHC	88003.10
10-4-43	MUNCIE	88000.22
10-4-43	SPICER	88000.22
10-4-44	SPICER	88000.81
10-4-443	NEAPCO	88000.32
10-4-443	SPICER	88000.32
10-4-453	NEAPCO	88000.22
10-4-453	SPICER	88000.22
10-4-453	SPICER	88000.25
10-4-473	NEAPCO	88000.35
10-4-483	SPICER	88000.20
10-4-493	SPICER	88000.30
10-4-493	SPICER	88000.31
1045	TEDDY TORQ	80055
10-4-52	MUNCIE	88000.46
10-4-52	SPICER	88000.46
10-4-523	SPICER	88000.31
10-4-53	MUNCIE	88000.52
10-4-53	SPICER	88000.52
10-4-573	NEAPCO	88000.30
10-4-573	SPICER	88000.30
1046	ALCO	80119
1046	ALLOY	80118
1046	ALLOY	80120
10-4-62	SPICER	88000.45
10-463	MUNCIE	88000.27
10-4-63	SPICER	88000.27
10-4-643	SPICER	88000.29
10-4-683	SPICER	88000.72
10-4-693	SPICER	88000.68
1047	ALCO	80130
1047	ALLOY	80130
10-4703	NEAPCO	88000.71
10-4-703	SPICER	88000.71
10-4-713	SPICER	88000.70
10-4-72	SPICER	88000.46
10-4-723	SPICER	88000.73
10-4-73	MUNCIE	88000.28
10-473	NEAPCO	88000.28
10-4-73	SPICER	88000.28
10473J	BORG-WARNR	85598

Part No.	Brand	IDLI Group
104748	IHC	80190
104750	IHC	80193
1048	ALCO	80078
1048	ALLOY	80078
10-48-00	WALTERSCHD	88003.90
10-4-83	MUNCIE	88000.53
10-4-83	SPICER	88000.53
10-4-853	SPICER	88000.30
10-4-853	SPICER	88000.33
1049	ALLOY	80004
104-93	BORG-WARNR	88000.73
10-4-93	MUNCIE	88000.31
10-4-93	SPICER	88000.30
10-4-93	SPICER	88000.31
104M1441	EIMCO	80158
104M472	EIMCO	80069
105	AEC	88002.13
105	TRU-CROSS	80030
105	TRU-CROSS	80031
1050	ALCO	80006
1050	ALLOY	80004
1050	ALLOY	80006
10503J	BORG-WARNR	87364
10503J	BORG-WARNR	88007.90
10504	CHAMPION	80240
10504J	BORG-WARNR	87444
10504J	BORG-WARNR	88007.99
1-0505	NEAPCO	88000.14
1-05-07	BYPY	88003.73
10507J	BORG-WARNR	81661
10508J	BORG-WARNR	81744
1051	ALCO	80019
1051	ALLOY	80019
105-10752-1	BORG-WARNR	86718
105112	IHC	80149
105-11905-1	BORG-WARNR	86759
105-12399-1	BORG-WARNR	86749
105-12768-2	BORG-WARNR	86717
105-16197-2	BORG-WARNR	86719
105-16511-1	BORG-WARNR	88006.70
105-16512-1	BORG-WARNR	88006.67
105-16513-1	BORG-WARNR	88006.66
105-16514-1	BORG-WARNR	88006.68
105-16717-1	BORG-WARNR	88006.65
105-16717-1	BORG-WARNR	88006.69
1052	ALCO	80010
1052	ALLOY	80010
105-21424J	BORG-WARNR	86801
105-2304	BORG-WARNR	88006.71
105-23569J	BORG-WARNR	86801
105-2510	BORG-WARNR	86751
1052614M91	MASS.FERG.	80213
1052619M91	MASS.FERG.	80198
105-2752	BORG-WARNR	86713
1053	ALCO	80009
1053	ALLOY	80009
105-3730	BORG-WARNR	86746
105-4494	BORG-WARNR	86854
105-4499	BORG-WARNR	86747
1055	ALCO	80032
1055	ALLOY	80032
1056	ALCO	80031
1056	ALLOY	80030
1057	ALLOY	80012
1057-1	DANUSER	80035
1058	ALCO	80018
1058	ALLOY	80018
105823	UNIC	80178
1059	ALCO	80011
1059	ALLOY	80011

Part No.	Brand	IDLI Group	Part No.	Brand	IDLI Group	Part No.	Brand	IDLI Group	Part No.	Brand	IDLI Group
105E7039	FORD	80041	1-08-25	BYPY	88004.26	110-2-120	MUNCIE	88000.25	111-10-16110(5)	KOMATSU	80156
105E7039A	FORD	80024	1-08-25-01	BYPY	88004.28	110-2-120	MUNCIE	88000.27	111-10-16120	KOMATSU	80156
105M2383	EIMCO	80207	10834J	BORG-WARNR	87453	110-2-130	MUNCIE	88000.22	11112	NEAPCO	88000
105M2883	EIMCO	80240	10834J	BORG-WARNR	88007.92	110-2-130	MUNCIE	88000.28	11113	NEAPCO	88000.01
105M671	EIMCO	80189	1084	ALCO	80114	110-2-140	MUNCIE	88000.22	11114	NEAPCO	88000.02
105M672	EIMCO	80225	1084	ALLOY	80113	110-2-140	MUNCIE	88000.31	11115	NEAPCO	81237
105M673	EIMCO	80215	1085	ALCO	80107	1-103	AEC	88000.85	11116	NEAPCO	88000.03
106	AEC	88001.70	1085	ALLOY	80105	1103	ALCO	80061	111162	MIXERMOBL	80235
1060	ALCO	80017	1087-314M91	MASS.FERG.	80196	1103	ALCO	80062	111162	MIXERMOBL	80240
1060	ALLOY	80017	1089-666M91	MASS.FERG.	80069	1103	ALLOY	80062	11117	NEAPCO	88000.04
106010	UNIC	80221	1089795	CHRYSLER	80135	1-103	LEMPCO	88000.36	11118	NEAPCO	88000.05
10-60-19-3625X	SPICER	88001.12	108991K92	IHC	80194	11-03-00	WALTERSCHD	88004.30	11119	NEAPCO	88013.90
1-0605	NEAPCO	88000.15	108-991R91	IHC	80194	11031	PILOT	88002.48	11-12	LEMPCO	88000.86
1-0606	NEAPCO	88000.16	108991R92	IHC	80194	110310120	CITROEN	80041	11-12	LEMPCO	88000.88
1061	ALCO	80031	108991R92	IHC	80197	11031J	BORG-WARNR	81937	11120	NEAPCO	81333
1061	ALLOY	80031	1090	ALCO	80042	11032	PILOT	88002.44	11121	NEAPCO	81333
1061	ALLOY	88002.77	1090	ALLOY	80042	11042	PILOT	88002.42	1-113	LEMPCO	88000.35
10-61-033	LILLISTON	80069	1090	RUEDARSA	80041	11046	PILOT	88002.51	11-13	LEMPCO	88000.79
1-0616	NEAPCO	88000.15	1090-2	RUEDARSA	80041	1105	ALCO	80060	11132	NEAPCO	81208
106-16511-1	BORG-WARNR	88006.70	1090-5	RUEDARSA	80041	1105	ALCO	80063	11133	NEAPCO	81213
10617J	BORG-WARNR	86708	10908J	BORG-WARNR	85598	1105	ALLOY	80063	11134	NEAPCO	81222
1062	ALCO	80092	1091-060M91	MASS.FERG.	80178	1-105	NEAPCO	88002.77	11135	NEAPCO	81237
1062	ALLOY	80092	1091-075M91	MASS.FERG.	80210	11050	PILOT	80021	11136	NEAPCO	81249
106244	HUBER	80069	10925J	BORG-WARNR	86808	11051	PILOT	80021	11137	NEAPCO	81261
1063	ALCO	80047	109263	IHC	80239	11051	PILOT	80023	11138	NEAPCO	81286
1063	ALLOY	80047	109265	IHC	80239	11056	PILOT	80021	11139	NEAPCO	81302
106396	UNIC	80142	109266	IHC	80239	11057	PILOT	80021	11140	NEAPCO	81333
1064	ALCO	80049	1092-939M91	MASS.FERG.	80161	1105X	NEAPCO	80035	111 410 00 31	MERCEDES	88002.77
1064	ALLOY	80049	10945J	BORG-WARNR	81760	1105XR	NEAPCO	80035	11144	NEAPCO	81246
1064846	CHRYSLER	80135	10946J	BORG-WARNR	86742	11062	PILOT	88002.58	11148	NEAPCO	81277
1066	ALCO	80055	1094-925M91	MASS.FERG.	80171	11066	PILOT	88002.50	11149	NEAPCO	81271
1066	ALLOY	80055	10A18829	WHITE FARM	80124	110724A1	NEW HOLL.	80003	1-115	LEMPCO	88000.55
1067	ALCO	80058	10H	AEC	80033	11074	PILOT	88002.56	1-115-01	AYRA DUREX	80024
1067	ALLOY	80058	10N	AEC	80033	11078	BAR-GREENE	80224	11152	NEAPCO	81208
1-0675	NEAPCO	88000.15	10P715	MPL.MOLINE	80001	11079	PILOT	88002.42	1-1153	NEAPCO	80088
1-0676	NEAPCO	88000.16	10P715	WHITE FARM	80001	11079	PILOT	88002.43	11154	NEAPCO	81222
1-0677	NEAPCO	88000.15	11	AEC	87373	1107P2	DIAMOND R	80075	11155	NEAPCO	81237
1068	ALCO	80058	11	AEC	87376	1107P2	DIAMOND R	80078	11156	NEAPCO	81249
1068	ALLOY	80058	11	AEC	88007.65	1107P2	REO MOTORS	80072	111562	IHC	80239
106883	HUBER	80178	11	AEC	88008.43	1107P2	WHITE	80075	11157	NEAPCO	81261
106936	UNIC	80142	110	TRU-CROSS	80007	1108402C91	IHC	80245	11158	NEAPCO	81286
1070	ALCO	80007	110	TRU-CROSS	88000.15	1108403C91	IHC	80245	11159	NEAPCO	81239
1070	ALLOY	88000.15	110	WOOD BROS.	80069	110-859R91	IHC	80044	11161	NEAPCO	81350
1070	RUEDARSA	80178	1100	SPICER	80024	110-860R91	IHC	80049	11162	NEAPCO	81268
1070584	DURA-DET.	80136	11003	PILOT	80001	110-861R91	IHC	80054	11163	NEAPCO	81246
1070E	RUEDARSA	80180	11004	PILOT	80001	11088	PILOT	88002.52	11166	NEAPCO	88000.08
1071	ALCO	80044	11007	PILOT	80001	11089	PILOT	88002.57	11168	NEAPCO	81259
1071	ALCO	88001.98	11009	PILOT	80001	11090	PILOT	88002.55	11169	NEAPCO	81302
1071	ALLOY	80044	1101	ALCO	80034	11092	PILOT	88002.44	11170	NEAPCO	81218
1073	ALLOY	80059	1101	ALLOY	80034	110960R91	IHC	80049	11171	NEAPCO	88014.19
10738J	BORG-WARNR	82133	110-1-110	MUNCIE	88000.22	1-11	NEAPCO	80004	11172	NEAPCO	81342
1077	ALCO	80059	110-1-120	MUNCIE	88000.22	1-11	NEAPCO	80006	11173	NEAPCO	85058
1077	ALCO	88002.04	110-1-120	MUNCIE	88000.25	111	TRU-CROSS	80023	11173J	BORG-WARNR	81825
1077	ALLOY	80059	110-1-120	MUNCIE	88000.27	111	TRU-CROSS	80024	11174	NEAPCO	85063
1078	ALCO	80124	110-1-130	MUNCIE	88000.22	1110	SPICER	80024	11175	NEAPCO	81350
1078	ALLOY	80124	110-1-130	MUNCIE	88000.28	11-10-00	WALTERSCHD	88003.75	11176	NEAPCO	88013.90
107-854	FORD	80001	110-1-140	MUNCIE	88000.22	11101	NEAPCO	84510	11177	NEAPCO	81268
107854	FORD	80003	110-1-140	MUNCIE	88000.31	11101X	NEAPCO	84510	11179	NEAPCO	81333
107897	GARD.DENV.	80149	110136	COOK,J.B.	80216	11-103	LEMPCO	88000.94	11181	NEAPCO	81274
107898	GARD.DENV.	80168	110137	COOK,J.B.	80193	11103J	BORG-WARNR	81854	11183	NEAPCO	84510
1078P	ALLOY	80124	110138	COOK,J.B.	80210	111048	HESSTON	80069	11184X	NEAPCO	84508
1079	ALCO	80066	11018	PILOT	88002.60	111048	HOUGH	80069	11185	NEAPCO	88014
1079	ALLOY	80066	1102	ALCO	80064	111048	IHC	88003.10	1118A	NEAPCO	81301
108	AEC	88001.94	1102	ALCO	80065	111050	HESSTON	80069	1118A	NEAPCO	88013.90
10803J	BORG-WARNR	87535	1102	ALLOY	80064	111050	HOUGH	80069	11190	NEAPCO	84510
10804J	BORG-WARNR	85569	11020	ALLIS-CHLM	80069	111050	IHC	88003.10	11197	NEAPCO	81273
1080E	RUEDARSA	80142	11020	ALLIS-CHLM	80204	111051	HESSTON	80069	111A3	NEAPCO	88014.05
1080N	RUEDARSA	80140	11020	FIAT-ALLIS	80070	111051	HOUGH	80069	111A4	NEAPCO	88000.04
1082	RUEDARSA	80140	110-2-110	MUNCIE	88000.22	111051(5)	IHC	88003.10	111AF	NEAPCO	85064
10823	DANLEY	80195	110-2-120	MUNCIE	88000.22	11106	NEAPCO	81237	111AZ	NEAPCO	81246

Part No.	Brand	IDLI Group	Part No.	Brand	IDLI Group	Part No.	Brand	IDLI Group	Part No.	Brand	IDLI Group
111B1	NEAPCO	88014.06	1130	ALLOY	80038	114-10007	BORG-WARNR	80245	114-1504	BORG-WARNR	80021
111B2	NEAPCO	81315	11-3003	NEAPCO	88000.39	114-10011	BORG-WARNR	80245	114-1505	BORG-WARNR	80021
111B8	NEAPCO	85514	11-3008	NEAPCO	88000.88	114-10011A	REPCO	80245	114-1506	BORG-WARNR	80021
111B9	NEAPCO	88000.06	11305	PILOT	80049	114-1002	BORG-WARNR	80001	114-1507	BORG-WARNR	80021
111C9	NEAPCO	88000.04	1131	ALCO	80081	114-1002	BORG-WARNR	88003.38	114-1509	BORG-WARNR	80021
111D6	NEAPCO	81237	1131	ALLOY	80081	114-1003	BORG-WARNR	80001	114-151	BORG-WARNR	80136
111D8	NEAPCO	81203	11-32	LEMPCO	88000.95	114-1003	BORG-WARNR	88003.38	114-1510	BORG-WARNR	80021
111E2	NEAPCO	85055	11-33	LEMPCO	88000.81	114-1003	REPCO	80001	114-1512	BORG-WARNR	80021
111E4	NEAPCO	85063	113362	MIXERMOBL	80216	114-1004	BORG-WARNR	80001	114-1513	BORG-WARNR	80021
111E5	NEAPCO	85054	1134	ALCO	80071	114-1004	BORG-WARNR	80003	114-1514	BORG-WARNR	80021
111F7	NEAPCO	81206	1134	ALLOY	80071	114-1004	REPCO	80001	114-1521	BORG-WARNR	80021
111E8	NEAPCO	81346	1-134	NEAPCO	80118	114-1005	BORG-WARNR	80001	114-1521	REPCO	80021
111E9	NEAPCO	88000.06	11344J	BORG-WARNR	88006.34	114-1005	BORG-WARNR	80003	114-1522	BORG-WARNR	80021
111F2	NEAPCO	88000.06	1134579	CHRYSLER	80022	114-1007	BORG-WARNR	80001	114-1523	BORG-WARNR	80021
111F4	NEAPCO	88000.06	1134579	GMC	80022	114-1007	BORG-WARNR	88003.38	114-154	BORG-WARNR	88001.60
111F8	NEAPCO	81346	1134584	CHRYSLER	88002.86	114-1009	BORG-WARNR	80001	114-156	BORG-WARNR	88001.60
111G2	NEAPCO	81237	1134986	ALLIS-CHLM	80210	114-1009	BORG-WARNR	88003.38	11416	PILOT	80044
111G3	NEAPCO	80189	1134986	FIAT-ALLIS	80215	114-101	BORG-WARNR	80134	11416	PILOT	80047
111H5	NEAPCO	81237	1135	ALCO	80074	114-102	BORG-WARNR	80188	114-1602	BORG-WARNR	80197
111H8	NEAPCO	81342	1135	ALLOY	80074	114-103	BORG-WARNR	80035	114-162	BORG-WARNR	80042
111J2	NEAPCO	81237	113501149	AUDI	86072	114-104	BORG-WARNR	80037	114-162	REPCO	80042
111J2	NEAPCO	88014.01	113501149	AUDI	86143	114-105	BORG-WARNR	80069	11417	PILOT	80092
111K1	NEAPCO	81277	113501149	VOLKSWAGEN	86143	114-105	BORG-WARNR	80070	114-1750	BORG-WARNR	80030
111P1	NEAPCO	81271	113501331	VOLKSWAGEN	86069	114-106	BORG-WARNR	80044	114-1750	BORG-WARNR	88002.77
111P6	NEAPCO	81237	113501331D	PORSCHE	86071	114-106	BORG-WARNR	80179	114-1750	REPCO	88002.77
111P7	NEAPCO	88000.06	113501331D	VOLKSWAGEN	86071	114-107	BORG-WARNR	80185	114-1755	BORG-WARNR	80032
111Z4	NEAPCO	81274	11352	PILOT	80909	114-108	BORG-WARNR	80186	114-1755	REPCO	80032
111Z7	NEAPCO	88000.07	11352R	PILOT	80909	114-109	BORG-WARNR	80184	114-1757	BORG-WARNR	80030
1-12	AEC	88000.86	11354	PILOT	80958	114-10HD	BORG-WARNR	80124	114-1757	BORG-WARNR	80031
1-12	LEMPCO	88000.45	11354R	PILOT	80958	114-10HP	BORG-WARNR	80124	114-1759	BORG-WARNR	80030
1120	RUEDARSA	80189	11355R	PILOT	80956	11411	PILOT	80044	114-1761	BORG-WARNR	80032
1-1200	NEAPCO	80072	1135R	PILOT	88000.14	11411	PILOT	80049	114-1762	BORG-WARNR	88001.95
1-1200	NEAPCO	80075	1135R	PILOT	88000.15	11412	PILOT	80030	114-1762	BORG-WARNR	88002.77
1-1200	NEAPCO	80076	1136	ALCO	80080	11412	PILOT	80044	114-1762	REPCO	88002.77
1-1200	NEAPCO	80089	1136	ALCO	80081	114-120	BORG-WARNR	80038	114-1763	BORG-WARNR	80031
11-2001	NEAPCO	88000.53	1136	ALLOY	80081	114-120	BORG-WARNR	80039	114-1765	BORG-WARNR	80031
11-2002	NEAPCO	88000.52	11360R	PILOT	80903	114-12000	REPCO	80234	114-1767	BORG-WARNR	80031
11-2003	NEAPCO	88000.22	11360R	PILOT	80958	114-12002	BORG-WARNR	80234	114-1767	BWE	80030
11-2004	NEAPCO	81020	11362	PILOT	88001.66	114-12002A	REPCO	80234	114-1767	REPCO	88002.77
11-2005	NEAPCO	88000.21	11362R	PILOT	80901	114-121	BORG-WARNR	80141	114-1771	BORG-WARNR	80031
11-2006	NEAPCO	88000.28	11362R	PILOT	80954	114-122	BORG-WARNR	80117	114-1772	BORG-WARNR	80030
11-2007	NEAPCO	88000.31	11364	PILOT	88001.74	114-122	BORG-WARNR	80129	114-1773	BORG-WARNR	80031
11-2008	NEAPCO	88000.38	11364R	PILOT	88001.74	114-123	BORG-WARNR	80131	11418	PILOT	80053
11-2009	NEAPCO	88000.35	1137	ALCO	80123	114-124	BORG-WARNR	80123	114-180	BORG-WARNR	80186
11-2016	NEAPCO	88000.25	1137	ALLOY	80123	114-125	BORG-WARNR	80081	114-1HD	BORG-WARNR	80132
11-2017	NEAPCO	88000.21	113-7201	BORG-WARNR	80207	114-125	BORG-WARNR	80123	114-1HP	BORG-WARNR	80132
11-2018	NEAPCO	88000.39	11-38-00	WALTERSCHD	88003.93	114-126	BORG-WARNR	80091	114-2000	BORG-WARNR	80049
11-2021	NEAPCO	88000.31	11-39-00	WALTERSCHD	88004.03	114-127	BORG-WARNR	80071	114-2000	BORG-WARNR	80050
11-2023	NEAPCO	88000.35	11398J	BORG-WARNR	81801	114-128	BORG-WARNR	80074	114-2000	BORG-WARNR	88002.09
11-2030	NEAPCO	88000.22	1140	RUEDARSA	80099	114-129	BORG-WARNR	80080	114-2002	BORG-WARNR	80048
11-2032	NEAPCO	88000.38	1140	SPICER	80024	114-130	BORG-WARNR	80145	114-2002	BORG-WARNR	80051
11-2045	NEAPCO	88000.27	11400	PILOT	80049	114-131	BORG-WARNR	80183	114-2002	BORG-WARNR	88002.10
1120-5	RUEDARSA	80189	11400	PILOT	80050	114-132	BORG-WARNR	80138	114-2003	BORG-WARNR	80048
11-22	LEMPCO	88000.89	11401	PILOT	80021	114-1328	BORG-WARNR	88002.61	114-2005	BORG-WARNR	80048
11221J	BORG-WARNR	86811	11402	PILOT	80048	114-133	BORG-WARNR	80068	114-2005	BORG-WARNR	80049
1-123	LEMPCO	88000.38	11402	PILOT	88002.10	114-134	BORG-WARNR	80118	114-2005AB	BORG-WARNR	80049
1-123	LEMPCO	88000.39	11403	PILOT	88002.10	114-134	BORG-WARNR	80130	114-2006	BORG-WARNR	88001.97
11-23	LEMPCO	88000.80	11405	PILOT	80049	114-135	BORG-WARNR	80160	114-2007	BORG-WARNR	80043
1-124	LEMPCO	88000.37	11406	PILOT	88001.97	114-136	BORG-WARNR	80181	114-2008	BORG-WARNR	80048
1124	PILOT	80225	11407	PILOT	80043	11414	PILOT	80045	114-2008	BORG-WARNR	88002.10
1125463	ALLIS-CHLM	80210	11408J	BORG-WARNR	85866	114-145	BORG-WARNR	80001	114-2010	BORG-WARNR	80049
1126463	ALLIS-CHLM	80210	11410	PILOT	80049	114-145	BORG-WARNR	80003	114-2010	BORG-WARNR	88002.09
1126463	FIAT-ALLIS	80215	11410	PILOT	88002.01	114-145	BORG-WARNR	88000.15	114-2011	BORG-WARNR	80044
11270	CASE,J.I.	80077	114-10000	BORG-WARNR	88003.40	114-146	BORG-WARNR	88000.14	114-2012	BORG-WARNR	80030
1-1275	NEAPCO	80035	114-10001	BORG-WARNR	80245	114-148	BORG-WARNR	88000.15	114-2014	BORG-WARNR	80045
1-13	AEC	88000.79	114-10003	BORG-WARNR	80245	114-148	BORG-WARNR	88000.16	114-2016	BORG-WARNR	80044
1-13	LEMPCO	88000.20	114-10003	REPCO	80245	114-150	BORG-WARNR	80161	114-2017	BORG-WARNR	80092
1-13	LEMPCO	88000.21	114-10003A	REPCO	80245	114-1501	BORG-WARNR	80021	114-2017	BORG-WARNR	88001.76
1-13	NEAPCO	80009	114-10006	BORG-WARNR	80245	114-1501	BORG-WARNR	88001.97	114-2018	BORG-WARNR	80053
1130	ALCO	80039	114-10006	REPCO	80245	114-1501	REPCO	80021	114-2024	BORG-WARNR	80055

Part No.	Brand	IDLI Group
114-2025	BORG-WARNR	80001
114-2026	BORG-WARNR	80030
114-2031	BORG-WARNR	80049
114-2031	REPCO	80092
114-2032	BORG-WARNR	80092
114-2032	REPCO	80092
114-2032	REPCO	88001.76
114-2033	BORG-WARNR	80054
114-2036	BORG-WARNR	80049
114-2039	BORG-WARNR	80044
114-2040	BORG-WARNR	80049
114-2040AB	BORG-WARNR	80049
114-2041	BORG-WARNR	80054
114-2044	BORG-WARNR	80047
114-2045	BORG-WARNR	80049
114-2046	BORG-WARNR	80044
114-2047	BORG-WARNR	80046
114-2047	BORG-WARNR	80092
114-2048	BORG-WARNR	80049
114-2049	BORG-WARNR	80054
114-2050	BORG-WARNR	80044
114-2053	BORG-WARNR	80030
114-2062	BORG-WARNR	80049
114-2062	BORG-WARNR	80050
114-2063	BORG-WARNR	80043
114-2064	BORG-WARNR	80044
114-2066	BORG-WARNR	80045
114-2067	BORG-WARNR	80048
114-2068	BORG-WARNR	80049
114-2068	BORG-WARNR	80050
114-2076	BORG-WARNR	80049
114-2077	BORG-WARNR	80044
114-2077	BORG-WARNR	80049
114-2084	BORG-WARNR	80021
114-2093	BORG-WARNR	80104
114-2100	BORG-WARNR	80049
114-2100	BORG-WARNR	88001.96
114-2101	BORG-WARNR	80044
114-2101	BORG-WARNR	88001.98
114-2101	REPCO	80049
114-2102	BORG-WARNR	80054
114-2102	REPCO	80073
114-2102	REPCO	80091
114-2103	BORG-WARNR	80048
114-2103	BORG-WARNR	88002.10
114-2103	REPCO	80048
114-2104	BORG-WARNR	80047
114-2104	REPCO	80047
114-2105	BORG-WARNR	80050
114-2105	REPCO	80050
114-2106	BORG-WARNR	80046
114-2107	BORG-WARNR	80054
114-2107	REPCO	80054
114-2108	BORG-WARNR	80043
114-2108	REPCO	80043
114-2109	BORG-WARNR	80049
114-2109	BORG-WARNR	80050
114-2109	REPCO	80050
114-2110	BORG-WARNR	80048
114-2110	REPCO	80047
114-2111	BORG-WARNR	80049
114-2111	REPCO	80049
114-2111AB	BORG-WARNR	80049
114-2112	BORG-WARNR	80044
114-2112	REPCO	80044
114-2113	BORG-WARNR	80053
114-2113	REPCO	80044
114-2113AB	BORG-WARNR	80053
114-2114	BORG-WARNR	80054
114-2114	REPCO	80054
114-2114AB	BORG-WARNR	80054
114-2115	BORG-WARNR	80047
114-2116	BORG-WARNR	80047
114-2116	REPCO	80047
114-2116A	BORG-WARNR	80047
114-2117	BORG-WARNR	80046
114-2117	REPCO	80046
114-2117A	BORG-WARNR	80046
114-2118	BORG-WARNR	80049
114-2118	REPCO	80049
114-2119	BORG-WARNR	80044
114-2119	REPCO	80044
114-2120	BORG-WARNR	80046
114-2121	BORG-WARNR	80043
114-2121	REPCO	80043
114-2122	BORG-WARNR	80049
114-2122	REPCO	80049
114-2123	BORG-WARNR	80045
114-2123	REPCO	80045
114-2124	BORG-WARNR	80043
114-2124	REPCO	80048
114-2125	BORG-WARNR	80047
114-2126	BORG-WARNR	80049
114-2126	REPCO	80049
114-2127	BORG-WARNR	80044
114-2127	REPCO	80044
114-2128	BORG-WARNR	80045
114-2129	BORG-WARNR	80054
114-213	BORG-WARNR	80117
114-2130	BORG-WARNR	80047
114-2131	BORG-WARNR	80044
114-2131	REPCO	80044
114-2132	BORG-WARNR	80044
114-2133	BORG-WARNR	80044
114-2133	BORG-WARNR	80049
114-2134	BORG-WARNR	80049
114-2134	REPCO	80049
114-2135	BORG-WARNR	80048
114-2135	REPCO	80048
114-2136	BORG-WARNR	80048
114-2137	BORG-WARNR	80044
114-2140	BORG-WARNR	80055
114-2140	REPCO	80055
114-2142	BORG-WARNR	80054
114-2143	BORG-WARNR	80047
114-2143	REPCO	80047
114-2144	BORG-WARNR	80044
114-2145	BORG-WARNR	80044
114-2147	BORG-WARNR	80054
114-2147	BORG-WARNR	80167
114-2148	BORG-WARNR	80049
114-2150	BORG-WARNR	80048
114-2150	REPCO	80048
114-2151	BORG-WARNR	80044
114-2151	BORG-WARNR	80055
114-2153	BORG-WARNR	80055
114-2153	REPCO	80055
114-2155	BORG-WARNR	80049
114-2156	BORG-WARNR	80047
114-2157	BORG-WARNR	80049
114-2158	ALCO	80049
114-2158	BORG-WARNR	80049
114-2160	BORG-WARNR	80044
114-2161	BORG-WARNR	80044
114-2162	BORG-WARNR	80054
114-2163	ALCO	80044
114-2163	ALCO	80049
114-2163	BORG-WARNR	80049
114-2164	BORG-WARNR	80044
114-2165	BORG-WARNR	80058
114-2166	BORG-WARNR	80048
114-2167	BORG-WARNR	80054
114-2168	BORG-WARNR	80048
114-2169	BORG-WARNR	80054
114-2170	BORG-WARNR	80047
114-2171	BORG-WARNR	80048
114-2172	BORG-WARNR	80047
114-2172	REPCO	80047
114-2173	BORG-WARNR	80058
114-2173	REPCO	80058
114-2174	BORG-WARNR	80044
114-2175	BORG-WARNR	80058
114-2175	REPCO	80058
114-2177	BORG-WARNR	80043
114-2178	BORG-WARNR	80058
114-2179	BORG-WARNR	80044
114-2179	BORG-WARNR	80075
114-2180	BORG-WARNR	80058
114-2181	BORG-WARNR	80058
114-2182	BORG-WARNR	80043
114-2183	BORG-WARNR	80054
114-2183	BORG-WARNR	80167
114-2184	BORG-WARNR	80044
114-2185	BORG-WARNR	80047
114-2185A	BORG-WARNR	80047
114-2186	BORG-WARNR	80046
114-2186	BORG-WARNR	80092
114-2186A	BORG-WARNR	80046
114-2187	BORG-WARNR	80044
114-2188	ALCO	80049
114-2188	BORG-WARNR	80049
114-2198	BORG-WARNR	80051
114-2198	REPCO	80051
114-2206	BORG-WARNR	80047
114-2214	BORG-WARNR	80048
114-225	BORG-WARNR	80064
114-225	BORG-WARNR	80065
114-225	BORG-WARNR	88001.63
114-226	BORG-WARNR	80064
114-226	BORG-WARNR	80065
114-231	BORG-WARNR	80132
114-232	BORG-WARNR	80132
114-232	BWE	80121
114-232	BWE	80132
114-232	REPCO	80132
114-233	BORG-WARNR	80061
114-233	BORG-WARNR	80062
114-233	REPCO	80062
114-235	BORG-WARNR	80022
114-2500	BORG-WARNR	80059
114-2500	REPCO	80059
114-2501	BORG-WARNR	80059
114-2501	REPCO	80059
114-2503	BORG-WARNR	80030
114-2505	BORG-WARNR	80059
114-2505	REPCO	80059
114-251	BORG-WARNR	80056
114-251	REPCO	80056
114-2518	BORG-WARNR	80066
114-2518	REPCO	80066
114-252	BORG-WARNR	80034
114-252	BORG-WARNR	88001.58
114-252	REPCO	80034
114-2530	BORG-WARNR	80066
114-2531	BORG-WARNR	80059
114-2531	BORG-WARNR	80066
114-2531	REPCO	80059
114-2531	REPCO	80066
114-2532	BORG-WARNR	80059
114-2532	BORG-WARNR	80066
114-2532	REPCO	80059
114-2532	REPCO	80066
114-2532	REPCO	80067
114-2533	BORG-WARNR	80059
114-254	BORG-WARNR	80057
114-254	BORG-WARNR	88001.58
114-255	BORG-WARNR	80060
114-255	BORG-WARNR	80061
114-255	BORG-WARNR	88001.59
114-255	REPCO	80060
114-256	BORG-WARNR	80056
114-256	BORG-WARNR	80057
114-256	REPCO	80057
114-257	AEC	80095
114-257	BORG-WARNR	80095
114-257	REPCO	80095
114-258	BORG-WARNR	80095
114-258	REPCO	80095
114-259	BORG-WARNR	80096
114-259	REPCO	80062
114-259	REPCO	80096
11426	PILOT	80030
114-261	BORG-WARNR	80064
114-261	BORG-WARNR	88001.68
114-261	REPCO	80064
114-262	BORG-WARNR	80096
114-262	BORG-WARNR	88001.69
114-263	BORG-WARNR	80094
114-263	REPCO	80094
114-264	BORG-WARNR	80094
114-264	BORG-WARNR	80096
114-265	BORG-WARNR	80064
114-265	BORG-WARNR	80065
114-265	BORG-WARNR	88001.63
114-265	REPCO	80064
114-266	BORG-WARNR	80060
114-266	BORG-WARNR	80061
114-266	REPCO	80060
114-267	BORG-WARNR	80060
114-267	BORG-WARNR	80063
114-267	REPCO	80061
114-2HD	BORG-WARNR	80061
114-2HP	BORG-WARNR	80061
114-3000	BORG-WARNR	80099
114-3000	BORG-WARNR	80101
114-3001	BORG-WARNR	80101
114-3003	BORG-WARNR	80101
114-3005	BORG-WARNR	80099
114-3005	BORG-WARNR	80101
114-3008	BORG-WARNR	80107
114-301	BORG-WARNR	80024
114-3010	BORG-WARNR	80100
114-3010	BORG-WARNR	80101
114-3011	BORG-WARNR	80104
114-3011	BORG-WARNR	80107
114-3011	REPCO	80107
114-3012	BORG-WARNR	80110
114-3012	BORG-WARNR	80112
114-3013	BORG-WARNR	80106
114-3014	BORG-WARNR	80114
114-3016	BORG-WARNR	80099
114-3019	BORG-WARNR	80112
114-302	BORG-WARNR	80010
114-3020	BORG-WARNR	80110
114-3020	BORG-WARNR	80112
114-3021	BORG-WARNR	80111
114-3022	BORG-WARNR	80112
114-303	BORG-WARNR	80004
114-3033	BORG-WARNR	80104
114-3036	BORG-WARNR	80101

Part No.	Brand	IDLI Group	Part No.	Brand	IDLI Group	Part No.	Brand	IDLI Group	Part No.	Brand	IDLI Group
114-304	BORG-WARNR	80019	114-3123	BORG-WARNR	80102	114-354	BORG-WARNR	88001.60	114-4109	BORG-WARNR	80157
114-3042	BORG-WARNR	80100	114-3123	REPCO	80102	114-354	BORG-WARNR	88001.62	114-4109	REPCO	80157
114-3043	BORG-WARNR	80110	114-3124	BORG-WARNR	80100	11436	PILOT	80049	114-4110	BORG-WARNR	80155
114-3043	BORG-WARNR	80112	114-3124	REPCO	80100	114-362	BORG-WARNR	80135	114-4110	REPCO	80155
114-3045	BORG-WARNR	80099	114-3125	BORG-WARNR	80103	114-362	BORG-WARNR	88001.65	114-4111	BORG-WARNR	80155
114-3045	BORG-WARNR	80101	114-3125	REPCO	80103	114-362	REPCO	80135	114-4112	BORG-WARNR	80109
114-3047	REPCO	80093	114-3126	BORG-WARNR	80112	114-372	BORG-WARNR	80135	114-4113	BORG-WARNR	80156
114-305	BORG-WARNR	80018	114-3126	REPCO	80112	114-373	BORG-WARNR	80135	114-4113	REPCO	80156
114-306	BORG-WARNR	80009	114-3127	BORG-WARNR	80104	114-373	REPCO	80135	114-4114	BORG-WARNR	80109
114-307	BORG-WARNR	80006	114-3128	BORG-WARNR	80101	11439	PILOT	80044	114-4114	REPCO	80109
114-307	BWE	80004	114-3128	REPCO	80101	114-3HD	BORG-WARNR	80077	114-4115	BORG-WARNR	80150
114-308	BORG-WARNR	80017	114-3128A	BORG-WARNR	80101	114-3HP	BORG-WARNR	80077	114-4115	BORG-WARNR	80156
114-309	BORG-WARNR	80011	114-3129	BORG-WARNR	80100	11440	PILOT	80049	114-4116	BORG-WARNR	80109
11431	PILOT	80049	114-3129	REPCO	80100	114-4000	BORG-WARNR	80109	114-4116	BORG-WARNR	80156
114-310	BORG-WARNR	80012	114-3130	BORG-WARNR	80102	114-4002	BORG-WARNR	80109	114-4116	REPCO	80109
114-3100	BORG-WARNR	80101	114-3131	BORG-WARNR	80101	114-4006	BORG-WARNR	80153	114-4117	BORG-WARNR	80109
114-3100	REPCO	80101	114-3132	BORG-WARNR	80100	114-4006	BORG-WARNR	80156	114-4117	BORG-WARNR	80156
114-3100A	BORG-WARNR	80101	114-3133	BORG-WARNR	80104	114-4006	BORG-WARNR	88003.14	114-4118	BORG-WARNR	80156
114-3101	BORG-WARNR	80100	114-3134	BORG-WARNR	80112	114-4006	REPCO	80150	114-4118	REPCO	80156
114-3101	REPCO	80100	114-3135	BORG-WARNR	80114	114-4007	BORG-WARNR	80147	114-4119	BORG-WARNR	80149
114-3102	BORG-WARNR	80112	114-3136	BORG-WARNR	80111	114-4007	REPCO	80147	114-4120	BORG-WARNR	80156
114-3102	REPCO	80112	114-3137	BORG-WARNR	80101	114-4008	BORG-WARNR	80146	114-4120	REPCO	80156
114-3103	BORG-WARNR	80098	114-3138	BORG-WARNR	80100	114-4008	REPCO	80146	114-4121	BORG-WARNR	80156
114-3103	REPCO	80098	114-3139	BORG-WARNR	80102	114-4011	BORG-WARNR	80147	114-4122	BORG-WARNR	80109
114-3104	BORG-WARNR	80111	114-3141	BORG-WARNR	80100	114-4011	REPCO	80147	114-4122	REPCO	80109
114-3104	REPCO	80111	114-3142	BORG-WARNR	80100	114-4011AB	BORG-WARNR	80147	114-4123	BORG-WARNR	80153
114-3105	BORG-WARNR	80104	114-3143	BORG-WARNR	80103	114-4014	BORG-WARNR	80158	114-4123	REPCO	80153
114-3105	REPCO	80104	114-3144	BORG-WARNR	80101	114-4015	BORG-WARNR	80148	114-4124	BORG-WARNR	80153
114-3106	BORG-WARNR	80099	114-3145	BORG-WARNR	80101	114-4016	BORG-WARNR	80159	114-4125	BORG-WARNR	80155
114-3106	BORG-WARNR	80101	114-3146	BORG-WARNR	80100	114-4017	BORG-WARNR	80158	114-4126	BORG-WARNR	80150
114-3106	REPCO	80099	114-3147	BORG-WARNR	80104	114-4019	BORG-WARNR	80151	114-4126	BORG-WARNR	80156
114-3106AB	BORG-WARNR	80101	114-3148	BORG-WARNR	80102	114-4021	BORG-WARNR	80158	114-4127	BORG-WARNR	80158
114-3107	BORG-WARNR	80100	114-3149	BORG-WARNR	80101	114-4021	BORG-WARNR	88002.16	114-4128	BORG-WARNR	80153
114-3107	REPCO	80100	114-315	BORG-WARNR	80026	114-4022	BORG-WARNR	80109	114-4129	BORG-WARNR	80158
114-3108	BORG-WARNR	80110	114-3150	BORG-WARNR	80100	114-4023	BORG-WARNR	80109	114-4130	BORG-WARNR	80155
114-3108	BORG-WARNR	80112	114-3150	REPCO	80100	114-4024	BORG-WARNR	80153	114-4130	REPCO	80155
114-3108	BORG-WARNR	88002.98	114-3151	BORG-WARNR	80102	114-4029	BORG-WARNR	80109	114-4131	BORG-WARNR	80154
114-3108	REPCO	80110	114-3152	BORG-WARNR	80101	114-4029	REPCO	80109	114-4131	REPCO	80154
114-3109	BORG-WARNR	80098	114-3152	REPCO	80099	114-4032	BORG-WARNR	80159	114-4132	BORG-WARNR	80149
114-311	BORG-WARNR	80020	114-3152	REPCO	80101	114-4033	BORG-WARNR	80148	114-4133	BORG-WARNR	80154
114-3110	BORG-WARNR	80111	114-3152	REPCO	80104	114-4034	BORG-WARNR	80158	114-4134	BORG-WARNR	80158
114-3111	BORG-WARNR	80107	114-3154	BORG-WARNR	80113	114-4036	BORG-WARNR	80109	114-4135	BORG-WARNR	80158
114-3111	REPCO	80107	114-3155	BORG-WARNR	80103	114-407	FORD	80226	114-4136	BORG-WARNR	80149
114-3112	BORG-WARNR	80101	114-3155	REPCO	80103	11441	PILOT	80054	114-4138	BORG-WARNR	80109
114-3112	REPCO	80101	114-3156	BORG-WARNR	80112	114-4100	BORG-WARNR	80109	114-4138	BORG-WARNR	80149
114-3113	BORG-WARNR	80112	114-3157	BORG-WARNR	80099	114-4100	BORG-WARNR	88003.15	114-4138	REPCO	80109
114-3113	BORG-WARNR	88002.14	114-3158	BORG-WARNR	80102	114-4100	REPCO	80109	114-4138	REPCO	80149
114-3113	REPCO	80112	114-3159	BORG-WARNR	80099	114-4100A	BORG-WARNR	80149	114-4139	BORG-WARNR	80153
114-3113AB	BORG-WARNR	80112	114-3159	BORG-WARNR	80101	114-4100AB	BORG-WARNR	80109	114-414	BORG-WARNR	80109
114-3114	BORG-WARNR	80104	114-316	BORG-WARNR	80083	114-4101	BORG-WARNR	80150	114-4140	BORG-WARNR	80155
114-3114	REPCO	80104	114-3160	BORG-WARNR	80100	114-4101	BORG-WARNR	80153	114-4141	BORG-WARNR	80154
114-3115	BORG-WARNR	80112	114-3161	BORG-WARNR	80098	114-4101	REPCO	80153	114-4142	BORG-WARNR	80149
114-3115	BORG-WARNR	80114	114-3162	BORG-WARNR	80100	114-4102	BORG-WARNR	80158	114-4143	BORG-WARNR	80156
114-3115	REPCO	80114	114-3165	BORG-WARNR	80101	114-4102	BORG-WARNR	88003.13	114-4143	BORG-WARNR	88003.17
114-3115AB	BORG-WARNR	80114	114-3166	BORG-WARNR	80102	114-4102	REPCO	80158	114-4146	BORG-WARNR	80155
114-3116	BORG-WARNR	80111	114-3167	BORG-WARNR	80112	114-4102A	BORG-WARNR	80158	114-4147	BORG-WARNR	80149
114-3116	REPCO	80111	114-317	BORG-WARNR	80015	114-4102AB	BORG-WARNR	80158	114-4148	BORG-WARNR	80149
114-3116AB	BORG-WARNR	80111	114-3170	BORG-WARNR	80100	114-4103	BORG-WARNR	80153	114-4149	BORG-WARNR	80156
114-3117	BORG-WARNR	80100	114-3195	BORG-WARNR	80101	114-4103	REPCO	88003.16	114-4150	BORG-WARNR	80153
114-3118	BORG-WARNR	80101	11432	PILOT	80092	114-4104	BORG-WARNR	80151	114-4151	BORG-WARNR	80149
114-3118	REPCO	80101	114-320	BORG-WARNR	80115	114-4104	REPCO	80151	114-4151	REPCO	80149
114-3118AB	BORG-WARNR	80101	114-324	BORG-WARNR	80095	114-4105	BORG-WARNR	80151	114-4152	BORG-WARNR	80158
114-3119	BORG-WARNR	80100	114-324	BORG-WARNR	88001.61	114-4105	REPCO	80151	114-4152	REPCO	80158
114-3119	REPCO	80100	114-325	BORG-WARNR	80096	114-4106	BORG-WARNR	80148	114-4155	BORG-WARNR	80156
114-3119AB	BORG-WARNR	80149	114-325	BORG-WARNR	88001.62	114-4106	REPCO	80148	114-4156	BORG-WARNR	80157
114-3120	BORG-WARNR	80113	114-325	BORG-WARNR	88001.64	114-4107	BORG-WARNR	80159	114-4156	REPCO	80157
114-3120	REPCO	80113	11433	PILOT	80054	114-4107	REPCO	80159	114-4158	BORG-WARNR	80156
114-3121	BORG-WARNR	80102	114-341	BORG-WARNR	80024	114-4107AB	BORG-WARNR	80159	114-4159	BORG-WARNR	80158
114-3121	BORG-WARNR	80113	114-350	BORG-WARNR	80036	114-4108	BORG-WARNR	80109	114-4162	BORG-WARNR	80155
114-3121	REPCO	80102	114-352	BORG-WARNR	88001.62	114-4108	REPCO	80109	114-4163	BORG-WARNR	80109

Part No.	Brand	IDLI Group
114-4179	BORG-WARNR	80155
114-418UB	BORG-WARNR	88013.86
114-418UB	BORG-WARNR	88013.87
114-418UB	BWE	88013.87
114-428	BORG-WARNR	88013.86
114-428UB	BORG-WARNR	88013.86
114-428UB	BORG-WARNR	88013.87
114-438UB	BORG-WARNR	88002.13
114-448SK	BORG-WARNR	88001.70
11445	PILOT	80049
114-458SK	BORG-WARNR	88001.94
11446	PILOT	80044
11446	PILOT	88001.98
114477	IHC	80150
11448	PILOT	80049
11449	PILOT	80054
114-4HD	BORG-WARNR	80104
114-4HP	BORG-WARNR	80104
11450	PILOT	80044
114-500	BORG-WARNR	80072
114-500	BORG-WARNR	80075
114-5000	BORG-WARNR	80168
114-5003	BORG-WARNR	80168
114-5004	BORG-WARNR	80176
114-5006	BORG-WARNR	80167
114-5007	BORG-WARNR	80177
114-5008	BORG-WARNR	80174
114-501	BORG-WARNR	80023
114-5011	BORG-WARNR	80168
114-5012	BORG-WARNR	80177
114-5012	BORG-WARNR	88002.18
114-5014	BORG-WARNR	80168
114-5015	BORG-WARNR	80173
114-502	BORG-WARNR	80041
114-502	BORG-WARNR	88001.85
114-5020	BORG-WARNR	80167
114-5024	BORG-WARNR	80168
114-503	BORG-WARNR	88000.16
114-5031	BORG-WARNR	80171
114-5031	REPCO	80166
114-5032	BORG-WARNR	80174
114-5033	BORG-WARNR	80173
114-5036	BORG-WARNR	80174
114-5037	BORG-WARNR	80177
114-504	BORG-WARNR	80041
114-5043	BORG-WARNR	80167
114-5053	BORG-WARNR	80173
114-5055	BORG-WARNR	80168
114-50A	BORG-WARNR	80153
114-5100	BORG-WARNR	80168
114-5100	REPCO	80168
114-5100A	BORG-WARNR	80168
114-5100A	BORG-WARNR	80216
114-5100AB	BORG-WARNR	80216
114-5101	BORG-WARNR	80167
114-5101	REPCO	80167
114-5102	BORG-WARNR	80171
114-5102	REPCO	80171
114-5103	BORG-WARNR	80174
114-5103	REPCO	80174
114-5103A	BORG-WARNR	80174
114-5105	BORG-WARNR	80169
114-5105	REPCO	80169
114-5108	BORG-WARNR	80172
114-5108	REPCO	80172
114-5109	BORG-WARNR	80162
114-5109	REPCO	80162
114-510X	BORG-WARNR	80142
114-511	MICHIGAN	80077
114-5110	BORG-WARNR	80173
114-5110	REPCO	80173
114-5110A	BORG-WARNR	80173
114-5110AB	BORG-WARNR	80173
114-5111	BORG-WARNR	80175
114-5111	REPCO	80175
114-5112	BORG-WARNR	80177
114-5112	REPCO	80177
114-5112AB	BORG-WARNR	80177
114-5114	BORG-WARNR	80176
114-5114	REPCO	80176
114-5115	BORG-WARNR	80178
114-5117	BORG-WARNR	80172
114-5118	BORG-WARNR	80168
114-5119	BORG-WARNR	80167
114-5120	BORG-WARNR	80168
114-5121	BORG-WARNR	80167
114-5121	REPCO	80167
114-5122	BORG-WARNR	80168
114-5122	REPCO	80168
114-5123	BORG-WARNR	80168
114-5123	REPCO	80168
114-5124	BORG-WARNR	80168
114-5124	REPCO	80168
114-5125	BORG-WARNR	80170
114-5125	BORG-WARNR	80171
114-5125	REPCO	80171
114-5125A	BORG-WARNR	80171
114-5126	BORG-WARNR	80172
114-5127	BORG-WARNR	80167
114-5127	REPCO	80167
114-5127A	BORG-WARNR	80163
114-5128	BORG-WARNR	80168
114-5128	REPCO	80168
114-5129	BORG-WARNR	80172
114-513	BORG-WARNR	80072
114-513	BORG-WARNR	80077
114-513	BWE	80085
114-5130	BORG-WARNR	80170
114-5130	BORG-WARNR	80171
114-5130	REPCO	80171
114-5130A	BORG-WARNR	80171
114-5131	BORG-WARNR	80175
114-5131	REPCO	80175
114-5131A	BORG-WARNR	80175
114-5132	BORG-WARNR	80174
114-5132	REPCO	80174
114-5132A	BORG-WARNR	80174
114-5133	BORG-WARNR	80173
114-5134	BORG-WARNR	80174
114-5135	BORG-WARNR	80174
114-5136	BORG-WARNR	80173
114-5137	BORG-WARNR	80168
114-5138	BORG-WARNR	80177
114-5139	BORG-WARNR	80170
114-5139A	BORG-WARNR	80170
114-513R	BORG-WARNR	80077
114-513R	BORG-WARNR	80081
114-513X	BORG-WARNR	80072
114-513X	BORG-WARNR	80075
114-513X	BORG-WARNR	80077
114-513X	BORG-WARNR	80081
114-514	BORG-WARNR	80023
114-514	BORG-WARNR	80024
114-5140	BORG-WARNR	80175
114-5141	BORG-WARNR	80162
114-5142	BORG-WARNR	80167
114-5143	BORG-WARNR	80175
114-5144	BORG-WARNR	80162
114-5145	BORG-WARNR	80216
114-5146	BORG-WARNR	80172
114-5147	BORG-WARNR	80167
114-5148	BORG-WARNR	80168
114-5148	BORG-WARNR	80169
114-5148	BORG-WARNR	80216
114-5148	REPCO	80169
114-5148A	BORG-WARNR	80169
114-5149	BORG-WARNR	80170
114-5149	BORG-WARNR	80171
114-515	BORG-WARNR	80178
114-5152	BORG-WARNR	80175
114-5154	BORG-WARNR	80175
114-5154	REPCO	80175
114-515R	BORG-WARNR	80178
114-515X	BORG-WARNR	80178
114-516	BORG-WARNR	80189
114-5161	BORG-WARNR	80175
114-5163	BORG-WARNR	80172
114-516R	BORG-WARNR	80189
114-516X	BORG-WARNR	80189
114-517	BORG-WARNR	80221
114-5173	BORG-WARNR	80170
114-5173	BORG-WARNR	80171
114-5173	REPCO	80171
114-5174	BORG-WARNR	80172
114-5175	BORG-WARNR	80175
114-5176	BORG-WARNR	80170
114-5176	BORG-WARNR	80171
114-5177	BORG-WARNR	80172
114-5177	REPCO	80172
114-5178	BORG-WARNR	80168
114-517R	BORG-WARNR	80221
114-517X	BORG-WARNR	80221
114-5180	BORG-WARNR	80172
114-5184	BORG-WARNR	80162
114-518R	BORG-WARNR	80124
114-518X	BORG-WARNR	80124
11451J	BORG-WARNR	87536
114-52	BORG-WARNR	80958
114-52	BORG-WARNR	88001.73
114-52	REPCO	80958
114-520	BORG-WARNR	80075
114-520	BORG-WARNR	80077
114-520U	BORG-WARNR	80076
114-520U	BORG-WARNR	80088
114-521	BORG-WARNR	80072
114-521HD	BORG-WARNR	80075
114-521HD	BORG-WARNR	80077
114-521R	BORG-WARNR	80072
114-521RU	BORG-WARNR	80089
114-521U	BORG-WARNR	80089
114-521X	BORG-WARNR	80072
114-521X	BORG-WARNR	80075
114-524	BORG-WARNR	80225
114-5243	BORG-WARNR	80170
114-52A	BORG-WARNR	80909
114-52A	BORG-WARNR	88001.72
114-52A	REPCO	80909
114-52AR	BORG-WARNR	80909
114-52R	BORG-WARNR	80958
114-53	BORG-WARNR	88001.75
11453	PILOT	80030
114-530	BORG-WARNR	80117
114-531	BORG-WARNR	80121
114-531	BORG-WARNR	80131
114-533	BORG-WARNR	80116
114-533R	BORG-WARNR	80116
114-53A	BORG-WARNR	88001.67
114-54	BORG-WARNR	88001.74
114-54	REPCO	88001.75
114-547	BORG-WARNR	80090
114-548	BORG-WARNR	80025
114-548	HABERLE	80025
114-549	BORG-WARNR	80075
114-54A	BORG-WARNR	80901
114-54A	BORG-WARNR	88001.66
114-54AR	BORG-WARNR	80901
11454J	BORG-WARNR	87434
114-54R	BORG-WARNR	88001.74
114-55	BORG-WARNR	80956
114-55	REPCO	80909
114-550U	BORG-WARNR	80119
114-550	BORG-WARNR	80120
114-550	BWE	80120
114-550R	BORG-WARNR	80119
114-550R	BORG-WARNR	80120
114-552	BORG-WARNR	80118
114-552R	BORG-WARNR	80118
114-553	BORG-WARNR	80072
114-553	BORG-WARNR	80081
114-553R	BORG-WARNR	80081
114-555	BORG-WARNR	80182
114-558	BORG-WARNR	80170
114-55A	BORG-WARNR	80906
114-55A	BORG-WARNR	80909
114-56	BORG-WARNR	80905
114-56	BORG-WARNR	80953
114-560	BORG-WARNR	80142
114-560	BORG-WARNR	80143
114-560R	BORG-WARNR	80142
114-560U	BORG-WARNR	80143
114-565	BORG-WARNR	80190
114-56AR	BORG-WARNR	80907
114-56R	BORG-WARNR	80954
114-57	REPCO	80955
114-570	BORG-WARNR	80007
114-570	BORG-WARNR	88000.15
114-570	BWE	88000.15
114-570P	BORG-WARNR	80007
114-572	BORG-WARNR	80007
114-572	BORG-WARNR	80023
114-573	BORG-WARNR	80078
114-573	BORG-WARNR	88001.71
114-574	BORG-WARNR	80078
114-574	BORG-WARNR	80090
114-575	BORG-WARNR	80007
114-575	BORG-WARNR	88000.15
114-577	BORG-WARNR	80221
114-578	BORG-WARNR	80124
114-578R	BORG-WARNR	80124
114-578U	BORG-WARNR	80124
114-578U	BORG-WARNR	80125
114-579	BORG-WARNR	80073
114-57AR	BORG-WARNR	80908
114-57R	BORG-WARNR	80955
114-581	BORG-WARNR	80023
114-581	BORG-WARNR	80024
114-581	BORG-WARNR	80140
114-585	BORG-WARNR	80228
114-588	BORG-WARNR	80140
114-588	BORG-WARNR	80182
114-588R	BORG-WARNR	80140
114-589	BORG-WARNR	80205
114-590	BORG-WARNR	80093
114-591	BORG-WARNR	80105
114-596	BORG-WARNR	80105
114-596	REPCO	80105
114-596	REPCO	80227
114-5HD	BORG-WARNR	80129
114-5HP	BORG-WARNR	80129
114-60	REPCO	80958

Part No.	Brand	IDLI Group	Part No.	Brand	IDLI Group	Part No.	Brand	IDLI Group	Part No.	Brand	IDLI Group
114-6000	BORG-WARNR	80193	114-6128	BORG-WARNR	80198	114-65	BORG-WARNR	80951	114-7116	BORG-WARNR	80210
114-6003	BORG-WARNR	80193	114-6128	REPCO	80198	114-65	REPCO	80951	114-7116	REPCO	80210
114-6004	BORG-WARNR	80195	114-6129	BORG-WARNR	80193	114-650	BORG-WARNR	80109	114-7117	BORG-WARNR	80213
114-6006	BORG-WARNR	80193	114-6130	BORG-WARNR	80198	114-650	REPCO	80109	114-7118	BORG-WARNR	80210
114-6012	BORG-WARNR	80193	114-6132	BORG-WARNR	80198	114-654	BORG-WARNR	80079	114-7119	BORG-WARNR	80213
114-6012	BORG-WARNR	88010.42	114-6133	BORG-WARNR	80193	114-65AR	BORG-WARNR	80905	114-7119	BORG-WARNR	88002.97
114-6016	BORG-WARNR	80193	114-6134	BORG-WARNR	80194	114-65AR	REPCO	80905	114-7119	BORG-WARNR	88003.21
114-6022	BORG-WARNR	80197	114-6135	BORG-WARNR	80191	114-65R	BORG-WARNR	80951	114-7119	REPCO	80213
114-6023	BORG-WARNR	80199	114-6135	REPCO	80191	114-65R	REPCO	80951	114-7120	BORG-WARNR	80210
114-6024	BORG-WARNR	80201	114-6136	BORG-WARNR	80194	11466	PILOT	80045	114-7121	BORG-WARNR	80213
114-6028	BORG-WARNR	80193	114-6137	BORG-WARNR	80193	114-66	REPCO	80953	114-7122	BORG-WARNR	80213
114-6028	BORG-WARNR	88010.42	114-6139	BORG-WARNR	80203	114-66R	BORG-WARNR	80952	114-7122	BORG-WARNR	88002.97
114-6029	BORG-WARNR	80194	114-6140	BORG-WARNR	80191	114-66R	BORG-WARNR	80953	114-7122	REPCO	80112
114-6029	BORG-WARNR	80195	114-6141	BORG-WARNR	80193	11467	PILOT	80031	114-7122	REPCO	80213
114602A	HYSTER	80207	114-6142	BORG-WARNR	80197	114-67	REPCO	80953	114-7123	BORG-WARNR	80213
114602A	HYSTER	80218	114-6143	BORG-WARNR	80193	114-67AR	BORG-WARNR	80906	114-7124	BORG-WARNR	80213
114-6045	BORG-WARNR	80194	114-6143	BORG-WARNR	80198	114-67AR	REPCO	80906	114-7125	BORG-WARNR	80213
114-60R	BORG-WARNR	80903	114-6143	REPCO	80193	114-67AR	REPCO	80909	114-7126	BORG-WARNR	80213
114-60R	BORG-WARNR	80958	114-6143A	BORG-WARNR	80193	114-67AR	REPCO	80953	114-7126	BWE	80213
114-6100	BORG-WARNR	80193	114-6144	BORG-WARNR	80197	114-67R	BORG-WARNR	80953	114-7126	REPCO	80213
114-6100	BORG-WARNR	88003.11	114-6145	BORG-WARNR	80194	114-67R	REPCO	80906	114-7127	BORG-WARNR	80213
114-6100	REPCO	80193	114-6145	BORG-WARNR	80195	114-67R	REPCO	80909	114-7129	BORG-WARNR	80213
114-6100A	BORG-WARNR	80193	114-6146	BORG-WARNR	80198	114-67R	REPCO	80958	114-7130	BORG-WARNR	80213
114-6100AB	BORG-WARNR	80193	114-6147	BORG-WARNR	80203	114-6HD	BORG-WARNR	80131	114-7131	BORG-WARNR	80213
114-6101	BORG-WARNR	80195	114-6148	BORG-WARNR	80191	114-6HP	BORG-WARNR	80131	11472	PILOT	80030
114-6101	BORG-WARNR	88003.12	114-6149	BORG-WARNR	80198	114-7000	BORG-WARNR	80210	114-7200	BORG-WARNR	80210
114-6101	REPCO	80195	114-6149	BORG-WARNR	88010.38	114-7001	BORG-WARNR	80211	114-7200	BORG-WARNR	80215
114-6102	BORG-WARNR	80196	114-6150	BORG-WARNR	80203	114-7001	BORG-WARNR	80214	114-7200	REPCO	80210
114-6102	BORG-WARNR	80197	114-6150	REPCO	80203	114-7001	BORG-WARNR	88011.14	114-7200A	BORG-WARNR	80210
114-6102	REPCO	80197	114-6151	BORG-WARNR	80198	114-7003	BORG-WARNR	80210	114-7200A	BORG-WARNR	80215
114-6102A	BORG-WARNR	80197	114-6152	BORG-WARNR	80195	114-7004	BORG-WARNR	80211	114-7200AB	BORG-WARNR	80210
114-6103	BORG-WARNR	80194	114-6152	REPCO	80197	114-7006	BORG-WARNR	88001.57	114-7201	BORG-WARNR	80207
114-6103	REPCO	80194	114-6153	REPCO	80193	114-7007	BORG-WARNR	80210	114-7201	BORG-WARNR	80211
114-6103AB	BORG-WARNR	80194	114-6155	BORG-WARNR	80198	114-7010	BORG-WARNR	80210	114-7201	REPCO	80211
114-6104	BORG-WARNR	80192	114-6161	BORG-WARNR	80194	114-7011	BORG-WARNR	80210	114-7202	BORG-WARNR	80212
114-6104	REPCO	80192	114-6161	BORG-WARNR	80195	114-7011	BORG-WARNR	80217	114-7202	BORG-WARNR	80214
114-6105	BORG-WARNR	80196	114-6162	BORG-WARNR	80199	114-7011	BORG-WARNR	88011.12	114-7202	REPCO	80212
114-6106	BORG-WARNR	80198	114-6162	REPCO	80199	114-7012	BORG-WARNR	80210	114-7202A	BORG-WARNR	80212
114-6106	BORG-WARNR	88003.18	114-6162A	BORG-WARNR	80199	114-7012	BORG-WARNR	80217	114-7203	BORG-WARNR	80208
114-6106	REPCO	80198	114-6163	BORG-WARNR	80201	114-7012	BORG-WARNR	88011.12	114-7203	REPCO	80208
114-6107	BORG-WARNR	80202	114-6163	REPCO	80201	114-7013	BORG-WARNR	80210	114-7204	BORG-WARNR	80209
114-6107	REPCO	80202	114-6163A	BORG-WARNR	80201	114-7013	BORG-WARNR	80213	114-7204	REPCO	80209
114-6108	BORG-WARNR	80199	114-6166	BORG-WARNR	80193	114-7013	BORG-WARNR	88011.11	114-7205	BORG-WARNR	80210
114-6108	REPCO	80199	114-6171	BORG-WARNR	80192	114-7015	BORG-WARNR	80210	114-7205	REPCO	80210
114-6108A	BORG-WARNR	80199	114-6172	BORG-WARNR	80193	114-7015	BORG-WARNR	80217	114-7205A	BORG-WARNR	80210
114-6109	BORG-WARNR	80201	114-6173	BORG-WARNR	80198	114-7015	BORG-WARNR	88011.12	114-7206	BORG-WARNR	80211
114-6109	REPCO	80201	114-6179	BORG-WARNR	80198	114-7033	BORG-WARNR	80210	114-7206	REPCO	80211
114-6109A	BORG-WARNR	80201	114-6179	BORG-WARNR	80200	114-7034	BORG-WARNR	80212	114-7207	BORG-WARNR	80212
114-6110	BORG-WARNR	80198	114-6179A	BORG-WARNR	80200	114-7035	BORG-WARNR	80210	114-7207	REPCO	80212
114-6110	BORG-WARNR	88003.18	114-6181	BORG-WARNR	80198	114-7042	BORG-WARNR	80211	114-7207A	BORG-WARNR	80212
114-6111	BORG-WARNR	80198	11462	PILOT	80050	114-7100	BORG-WARNR	80210	114-7208	BORG-WARNR	80208
114-6112	BORG-WARNR	80198	114-62	REPCO	80954	114-7101	BORG-WARNR	80211	114-7208	REPCO	80208
114-6113	BORG-WARNR	80193	114-6228	BORG-WARNR	80199	114-7101	REPCO	80211	114-7209	BORG-WARNR	80209
114-6114	BORG-WARNR	80197	114-6229	BORG-WARNR	80193	114-7102	BORG-WARNR	80212	114-7209	REPCO	80209
114-6114	REPCO	80197	114-62AR	BORG-WARNR	80907	114-7103	BORG-WARNR	80208	114-7211	BORG-WARNR	80210
114-6114A	BORG-WARNR	80197	11462G	PILOT	80049	114-7103	REPCO	80208	114-7212	BORG-WARNR	80210
114-6115	BORG-WARNR	80197	114-62R	BORG-WARNR	80954	114-7104	BORG-WARNR	80209	114-7213	BORG-WARNR	80210
114-6116	BORG-WARNR	80202	11463	PILOT	80043	114-7104	REPCO	80209	114-7214	BORG-WARNR	80211
114-6117	BORG-WARNR	80193	11463	PILOT	88001.96	114-7105	BORG-WARNR	80213	114-7216	BORG-WARNR	80210
114-6118	BORG-WARNR	80198	114-63	REPCO	80957	114-7105	BORG-WARNR	88002.97	114-7217	BORG-WARNR	80214
114-6119	BORG-WARNR	80198	114-635	BORG-WARNR	80124	114-7105	REPCO	80213	114-7218	BORG-WARNR	80210
114-6120	BORG-WARNR	80193	114-636	BORG-WARNR	80139	114-7105A	BORG-WARNR	80213	114-7219	BORG-WARNR	80211
114-6121	BORG-WARNR	80194	11463G	PILOT	80044	114-7107	BORG-WARNR	80210	114-7219AB	BORG-WARNR	80211
114-6121	BORG-WARNR	80195	114-63R	BORG-WARNR	80957	114-7110	BORG-WARNR	80213	114-7220	BORG-WARNR	80212
114-6122	BORG-WARNR	80197	11464	PILOT	80045	114-7111	BORG-WARNR	80213	114-7221	BORG-WARNR	80210
114-6123	BORG-WARNR	80192	114-640	BORG-WARNR	80227	114-7112	BORG-WARNR	80213	114-7222	BORG-WARNR	80210
114-6123	REPCO	80191	114-641	BORG-WARNR	80228	114-7113	BORG-WARNR	80213	114-7223	BORG-WARNR	80212
114-6125	BORG-WARNR	80197	114-642	BORG-WARNR	80229	114-7113	BORG-WARNR	88002.97	114-7224	BORG-WARNR	80211
114-6126	BORG-WARNR	80198	114-644	BORG-WARNR	87481	114-7114	BORG-WARNR	80212	114-7225	BORG-WARNR	80210
114-6127	BORG-WARNR	80198	114-645	BORG-WARNR	80079	114-7115	BORG-WARNR	80211	114-7226	BORG-WARNR	80212

Part No.	Brand	IDLI Group	Part No.	Brand	IDLI Group	Part No.	Brand	IDLI Group	Part No.	Brand	IDLI Group
114-7227	BORG-WARNR	80212	114-8206	REPCO	80232	114-9013	BORG-WARNR	80234	11520	PILOT	80075
114-7228	BORG-WARNR	80210	114-8207	BORG-WARNR	80237	114-9014	BORG-WARNR	80235	115200	PILOT	80075
114-7229	BORG-WARNR	80211	114-8207	REPCO	80237	114-9014	REPCO	80235	115200B	PILOT	80076
114-7230	BORG-WARNR	80210	114-8208	BORG-WARNR	80236	114-9015	BORG-WARNR	80234	115200UB	PILOT	80076
114-7231	BORG-WARNR	80212	114-8208	REPCO	80236	114-9015	REPCO	80234	11521	PILOT	80072
114-7234	BORG-WARNR	80210	114-8209	BORG-WARNR	80238	114-9016	BORG-WARNR	80234	11521HD	PILOT	80075
114-7239	BORG-WARNR	80211	114-8210	BORG-WARNR	80231	114-9016	REPCO	80234	11521HD	PILOT	80077
114-7242	BORG-WARNR	80210	114-8211	BORG-WARNR	80232	114-9016A	BORG-WARNR	80234	11521HD	PILOT	80078
114-7243	BORG-WARNR	80210	114-8212	BORG-WARNR	80231	114-9017	BORG-WARNR	80233	11521HD	PILOT	80085
114-7244	BORG-WARNR	80218	114-8212	REPCO	80231	114-9017	BORG-WARNR	80241	11521UB	PILOT	80075
114-7245	BORG-WARNR	80212	114-8213	BORG-WARNR	80237	114-9017	REPCO	80233	11521UB	PILOT	80089
114-7246	BORG-WARNR	80212	114-8214	BORG-WARNR	80238	114-9018	BORG-WARNR	80235	11522	PILOT	88001.82
114-7248	BORG-WARNR	80210	114-8214A	BORG-WARNR	80238	114-9019	BORG-WARNR	80234	11524	PILOT	80225
114-7249	BORG-WARNR	80212	114-8215	BORG-WARNR	80239	114-9020	BORG-WARNR	80234	11527	PILOT	80072
114-7270	BORG-WARNR	80212	114-8216	BORG-WARNR	80237	114-9021	BORG-WARNR	80233	1153	ALLOY	80083
114-7294	BORG-WARNR	80212	114-8217	BORG-WARNR	80231	114-9022	BORG-WARNR	80234	11-53	LEMPCO	88000.83
1-1475	NEAPCO	80001	114-8218	BORG-WARNR	80231	114-9023	BORG-WARNR	80234	1-153	NEAPCO	80075
1-1475	NEAPCO	80003	114-8219	BORG-WARNR	80237	114-9024	BORG-WARNR	80234	1-153	NEAPCO	80077
114-7HD	BORG-WARNR	80081	114-8221	BORG-WARNR	80232	114-9025	BORG-WARNR	80234	1-153	NEAPCO	80078
114-7HP	BORG-WARNR	80081	114-8222	BORG-WARNR	80231	114-9026	BORG-WARNR	80233	1-153	NEAPCO	80085
114-8000	BORG-WARNR	80231	114-8223	BORG-WARNR	80231	114-9026	BORG-WARNR	88003.09	11530	PILOT	80117
114-8002	BORG-WARNR	80231	114-8226	BORG-WARNR	80231	114-9026	REPCO	80233	11530G	PILOT	80117
114-8003	BORG-WARNR	80232	114-8500	BORG-WARNR	80223	114-9027	BORG-WARNR	80234	11531	PILOT	80121
114-8014	BORG-WARNR	80230	114-8500	BORG-WARNR	88003.28	114-9028	BORG-WARNR	80234	11531	PILOT	80131
114-8014	REPCO	80230	114-8500	REPCO	80223	114-9029	BORG-WARNR	80234	11533	PILOT	80116
114-8018	BORG-WARNR	80231	114-8502	BORG-WARNR	80223	114-9030	BORG-WARNR	80233	11534	PILOT	80118
114-8023	BORG-WARNR	80232	114-8502	BORG-WARNR	80224	114-9031	BORG-WARNR	80234	11545	PILOT	80092
114-8025	BORG-WARNR	80231	114-8502	REPCO	80224	114-9031A	BORG-WARNR	80234	11548	PILOT	80025
114-8100	BORG-WARNR	80231	114-8503	BORG-WARNR	80223	114-9033	BORG-WARNR	80234	11549	PILOT	80075
114-8101	BORG-WARNR	80232	114-8503	BORG-WARNR	80224	114-9034	BORG-WARNR	80234	11549	PILOT	80077
114-8102	BORG-WARNR	80237	114-8504	BORG-WARNR	80223	114-9035	BORG-WARNR	80235	11549UB	PILOT	81213
114-8103	BORG-WARNR	80236	114-8504	BORG-WARNR	88003.28	114-9036	BORG-WARNR	80235	11550	PILOT	80119
114-8105	BORG-WARNR	80238	114-8506	BORG-WARNR	80223	114-9037	BORG-WARNR	80235	11551	PILOT	80119
114-8105	BORG-WARNR	88003.20	114-8506	BORG-WARNR	80224	114-9038	BORG-WARNR	80234	11552	PILOT	80118
114-8105	REPCO	80238	114-8507	BORG-WARNR	80223	114-9039	BORG-WARNR	80234	11553	PILOT	80072
114-8105A	BORG-WARNR	80238	114-8507	BORG-WARNR	80224	114-9040	BORG-WARNR	80234	11553	PILOT	80077
114-8107	BORG-WARNR	80238	114-8507	REPCO	80224	114-9041	BORG-WARNR	80234	1-1554	NEAPCO	80032
114-8111	BORG-WARNR	80238	114-8508	BORG-WARNR	80222	114-9043	BORG-WARNR	80234	1-1555	NEAPCO	88002.77
114-8111	BORG-WARNR	88003.20	114-8511	BORG-WARNR	80223	114-9044	BORG-WARNR	80234	11555	PILOT	80182
114-8112	BORG-WARNR	80238	114-8511	BORG-WARNR	80224	114-9045	BORG-WARNR	80234	1-1556	NEAPCO	88000.16
114-8113	BORG-WARNR	80238	114-8513	BORG-WARNR	80223	114-9046	BORG-WARNR	80235	1-1557	NEAPCO	80012
114-8113	REPCO	80238	114-8513	BORG-WARNR	80224	114-9047	BORG-WARNR	80233	11560	PILOT	80142
114-8115	BORG-WARNR	80238	114-8514	BORG-WARNR	80223	114-9048	BORG-WARNR	80235	11560UB	PILOT	80143
114-8116	BORG-WARNR	80238	114-8515	BORG-WARNR	80222	114-9050	BORG-WARNR	80234	11562	HESSTON	80033
114-8116	REPCO	80238	114-8516	BORG-WARNR	80224	114-9058	BORG-WARNR	80235	11562	IHC	80033
114-8117	BORG-WARNR	80238	114-8516	BORG-WARNR	80236	114-9106	BORG-WARNR	80234	11565	PILOT	80190
114-8118	BORG-WARNR	80238	114-8516	BWE	80236	114-9107	BORG-WARNR	80233	11570	PILOT	80007
114-8119	BORG-WARNR	80238	114-8516	REPCO	80224	11492	PILOT	80048	11571	PILOT	80072
114-8120	BORG-WARNR	80238	114-8HD	BORG-WARNR	80091	11498	PILOT	80051	11572	PILOT	80007
114-8121	BORG-WARNR	80238	114-8HP	BORG-WARNR	80091	11498	PILOT	80054	11573	PILOT	80078
114-8200	BORG-WARNR	80231	114-9000	BORG-WARNR	80235	114-9HD	BORG-WARNR	80097	11573	PILOT	80090
114-8200	BORG-WARNR	80240	114-9001	BORG-WARNR	80234	114-9HP	BORG-WARNR	80097	11573	PILOT	88001.71
114-8200	REPCO	80231	114-9002	BORG-WARNR	80233	115	TRU-CROSS	80178	11574	PILOT	80090
114-8200A	BORG-WARNR	80240	114-9002	BORG-WARNR	88003.09	1-150	AEC	88000.90	11575	PILOT	88000.15
114-8200AB	BORG-WARNR	80231	114-9002	REPCO	80233	1-150	NEAPCO	80120	11577	PILOT	80221
114-8200AB	BORG-WARNR	80240	114-9003	BORG-WARNR	80234	1150	RUEDARSA	80109	11578	PILOT	80124
114-8201	BORG-WARNR	80232	114-9003	REPCO	80234	11500	PILOT	80075	11578UB	PILOT	80125
114-8201	BORG-WARNR	80242	114-9004	BORG-WARNR	80234	11500	PILOT	80077	11579	PILOT	80073
114-8201	REPCO	80232	114-9005	BORG-WARNR	80234	11501	PILOT	80023	11581	PILOT	80023
114-8202	BORG-WARNR	80237	114-9005	REPCO	80234	11502	PILOT	80041	115828	WILLYS	88001.66
114-8202	BORG-WARNR	80239	114-9006	BORG-WARNR	80235	11504	PILOT	80041	11585	PILOT	80228
114-8202	REPCO	80237	114-9007	BORG-WARNR	80235	1150809	CHRYSLER	80077	11586	PILOT	80124
114-8202A	BORG-WARNR	80239	114-9008	BORG-WARNR	80235	11511	PILOT	80023	1-15862	CARDWELL	80215
114-8202AB	BORG-WARNR	80237	114-9008AB	BORG-WARNR	80235	11511	PILOT	80024	11588	PILOT	80140
114-8203	BORG-WARNR	80236	114-9009	BORG-WARNR	80234	115123	AEC	88000.38	11588	PILOT	80182
114-8203	REPCO	80236	114-9010	BORG-WARNR	80233	11513	PILOT	80072	115893R91	IHC	80149
114-8204	BORG-WARNR	80238	114-9011	BORG-WARNR	80235	11514	PILOT	80024	11590	PILOT	80093
114-8205	BORG-WARNR	80231	114-9011	REPCO	80235	11515	PILOT	80178	11598J	BORG-WARNR	85856
114-8205	REPCO	80231	114-9011	REPCO	80233	11516	PILOT	80189	116035	AMER.MOTOR	88003.47
114-8206	BORG-WARNR	80232	114-9012	BORG-WARNR	80233	11517	PILOT	80221	116035	WILLYS	88003.47
			114-9012	REPCO	80233						

Part No.	Brand	IDLI Group
1-1606	NEAPCO	80005
1-1606	NEAPCO	80011
1-1608	NEAPCO	80013
1-1610	NEAPCO	80020
11614	SIMCA	80024
11614L	CHRYSLER	80024
116294	IHC	80048
116-294H1	IHC	80048
116294HI	HOUGH	80048
11-63	LEMPCO	88000.96
11655J	BORG-WARNR	86731
11663	SCHULTZ	80069
1-16648	CARDWELL	80234
1-16649	CARDWELL	80238
1-16655	CARDWELL	80234
11681	JACOBSEN	80063
11682	JAGUAR	80124
116-893R91	IHC	80109
116983A	HYSTER	80210
116E4635A	FORD	80023
116E4635B	FORD	80023
1-170	NEAPCO	80007
1-170	NEAPCO	88000.15
117-110	BORG-WARNR	88002.34
117-110	REPCO	88002.34
117-111	BORG-WARNR	88002.34
117-111	BORG-WARNR	88002.35
117-111	REPCO	88002.34
117-111	REPCO	88002.35
117-112	BORG-WARNR	88002.36
117-112	REPCO	88002.36
117-113	BORG-WARNR	88002.33
117-113	REPCO	88002.33
117-114	BORG-WARNR	88002.33
117-114	REPCO	88002.33
117-114	WESCO	88002.33
117-115	BORG-WARNR	88002.33
117-115A	BORG-WARNR	88002.33
117-115A	REPCO	88002.33
117-116	BORG-WARNR	88002.33
117-116	REPCO	88002.33
117-14346J	BORG-WARNR	81652
117164	AMER.MOTOR	88001.74
117164	WILLYS	88001.66
117-1782J	BORG-WARNR	88002.37
1173J	BORG-WARNR	81825
1175	ALCO	88002.34
1175	ALLOY	88002.34
11756J	BORG-WARNR	85605
11757J	BORG-WARNR	87641
1175FHA	CASE,J.I.	80069
1176	ALCO	88002.33
1176	ALLOY	88002.33
11762	SIDEWINDER	80161
11767	SCHULTZ	80136
1177	ALCO	88002.33
1177	ALLOY	88002.33
1-17785	CARDWELL	80240
1178	ALCO	88002.36
1178	ALLOY	88002.34
1179	ALLOY	88002.33
1180	ALCO	88002.33
1180	ALLOY	88002.33
1180	Z-WAY	80035
1181	ALCO	88002.34
1181	ALLOY	88002.36
11-81	LEMPCO	88000.92
11-81	LEMPCO	88000.93
118122	IHC	80215
118-122H1	IHC	80215
118122HI	HOUGH	80210
118123	IHC	80239
118-123H1	IHC	80239
118123HI	HOUGH	80237
1181657	CHRYSLER	80135
1181735	GMC	80090
11-83	LEMPCO	88000.82
118365	MIXERMOBL	80171
11853J	BORG-WARNR	85605
11854J	BORG-WARNR	87641
11855J	BORG WARNR	88015.28
11855J10800	BORG-WARNR	87680
11855J7200	BORG-WARNR	87679
1186	ALCO	88002.33
1-1875	NEAPCO	80044
1-1876	NEAPCO	80044
11876P11	M.R.S.	80233
119	BUSH HOG	80069
1190	ALLOY	88000.18
1190	Z-WAY	80037
11905	PILOT	88002.70
1191	ALLOY	88013.85
119-11339-1	BORG-WARNR	88006.34
119-13961-1	BORG-WARNR	86737
119-15562-1	BORG-WARNR	86704
119-16985-1	BORG-WARNR	86702
119-16985-10	BORG-WARNR	86704
11917	PILOT	88002.63
119-17900-1	BORG-WARNR	86738
11918	PILOT	88002.64
11918	PILOT	88002.71
119-19369-1	BORG-WARNR	86705
1192	ALLOY	88000.17
11-93	LEMPCO	88000.84
11935	BORG-WARNR	88006.29
11935J	BORG-WARNR	88006.29
11938J	BORG-WARNR	85861
11948	GMC	80142
1195	Z-WAY	80001
11964	PILOT	88002.71
11965J	BORG-WARNR	88006.59
11966J	BORG-WARNR	81850
11975	DATSUN	80023
11985	PILOT	88002.67
11995J	BORG-WARNR	81862
11996J	BORG-WARNR	82085
11998J	BORG-WARNR	88006.30
11B5490	AMER.LAFR.	80193
11DS31	OSHKOSH	80023
11X15873A	DEERE,JOHN	80195
11X15874A	DEERE,JOHN	80193
12	AEC	87326
12	AEC	87330
120	CHAIN BELT	80035
120	REX CHAIN	80035
1200	AEC	80081
1-200	NEAPCO	80076
1200	PRECISION	81203
1200	PRECISION	81204
1200	TRU-CROSS	80081
12-00-00	WALTERSCHD	88004.31
12003	GMB	80234
1201	AEC	80091
1201	PRECISION	81207
1201	PRECISION	81208
1201	TRU-CROSS	80091
120-102-0	ALFA ROMEO	80024
120-1-120	MUNCIE	88000.25
120-1-120	MUNCIE	88000.27
1202	AEC	80118
1202	AEC	80130
1202	PRECISION	88000
120-2-120	MUNCIE	88000.25
120-2-120	MUNCIE	88000.27
1203	AEC	80129
1203	PRECISION	81213
1203	PRECISION	81215
1203	TRU-CROSS	80117
1203	TRU-CROSS	80129
1204	AEC	80131
1204	PRECISION	81212
1204	PRECISION	81214
1204	TRU-CROSS	80121
1204	TRU-CROSS	80131
1205	AEC	80123
1205	PRECISION	88000.01
1205	RUEDARSA	80174
1205	TRU-CROSS	80123
1205	TRU-CROSS	80124
1205X	NEAPCO	80035
1205XR	NEAPCO	80035
1206	AEC	80141
1206	PRECISION	81220
1206	PRECISION	81222
1206	TRU-CROSS	80141
1206	TRU-CROSS	80142
1207	AEC	80071
1207	PRECISION	81219
1207	PRECISION	81221
12074J	BORG-WARNR	85850
1-2075	NEAPCO	80037
1-2076	NEAPCO	80037
1208	AEC	80074
1208	PRECISION	88000.02
12089J	BORG-WARNR	87451
12089J	BORG-WARNR	88007.96
1209	PRECISION	81234
1209	PRECISION	81237
1209	PRECISION	88014.02
1210	AEC	80183
1210	PRECISION	81235
1210	PRECISION	81236
1210	RUEDARSA	80024
1210	SPICER	80027
12-10-00	WALTERSCHD	88003.80
12-103	LEMPCO	88000.94
1210-3	RUEDARSA	80024
1211	AEC	80138
1211	PRECISION	81237
1211	PRECISION	81245
1211	TRU-CROSS	80138
12-1101	NEAPCO	84510
12-1112	NEAPCO	88000
12-1113	NEAPCO	88000.01
12-1114	BORG-WARNR	88000.02
12-1115	NEAPCO	81237
12-1116	NEAPCO	88000.03
12-1117	NEAPCO	88000.04
12-1118	NEAPCO	88000.05
12-1119	NEAPCO	88013.90
12-1120	NEAPCO	81333
12-1121	NEAPCO	81333
12-1132	NEAPCO	81208
12-1133	NEAPCO	81213
12-1134	NEAPCO	81222
12-1135	NEAPCO	81237
12-1137	NEAPCO	81261
12-1138	NEAPCO	81286
12-1139	NEAPCO	81302
12-1140	NEAPCO	81333
12-1148	NEAPCO	81277
12-1149	NEAPCO	81271
12-1153	NEAPCO	81213
12-1155	NEAPCO	81237
12-1156	NEAPCO	81249
12-1157	NEAPCO	81261
12-1158	NEAPCO	81286
12-1159	NEAPCO	81239
12-1161	NEAPCO	81350
12-1162	NEAPCO	81268
12-1163	NEAPCO	81246
12-1166	NEAPCO	88000.08
12-1168	NEAPCO	81259
12-1169	NEAPCO	81302
12-1170	NEAPCO	81218
12-1171	NEAPCO	88014.19
12-1172	NEAPCO	81342
12-1173	NEAPCO	85058
12-1174	NEAPCO	85063
12-1175	NEAPCO	81350
12-1176	NEAPCO	88013.90
12-1177	NEAPCO	81268
12-1181	NEAPCO	81274
12-1184	NEAPCO	84508
12-1185	NEAPCO	88014
12-1197	NEAPCO	81273
1212	PRECISION	88014
12-1218	NEAPCO	81301
12-1218	NEAPCO	88013.90
12-127	NEAPCO	85107
12-13	LEMPCO	88000.21
12-13	LEMPCO	88000.68
12-13	LEMPCO	88000.69
1213	PRECISION	88014.01
12-1301	NEAPCO	81246
12-1304	NEAPCO	88014.05
12-1307	NEAPCO	81315
12-1308	NEAPCO	85514
12-1313	NEAPCO	81203
12-1314	BORG-WARNR	85055
12-1316	NEAPCO	85054
12-1317	BORG-WARNR	81206
12-1318	NEAPCO	81346
12-1319	NEAPCO	88000.06
12-1324	NEAPCO	81346
12-1326	NEAPCO	81268
12-1331	NEAPCO	88014.01
12-1335	NEAPCO	81277
1-2134	NEAPCO	80050
12-1340	NEAPCO	81271
121-345R91	IHC	80124
12-139	NEAPCO	85102
1214	AEC	80145
1214	PRECISION	81249
1214	PRECISION	81250
1214	TRU-CROSS	80145
12140	PILOT	80055
121-406R91	IHC	80072
121406R92	IHC	80075
12141	PILOT	80073
12141	PILOT	80091
1215	PRECISION	88000.03
12153	PILOT	80055
12-158-160	FLXIBLE	80210
1216	PRECISION	81268
1216	PRECISION	81269
1216	PRECISION	81275
121684R91	IHC	80189
121-684R92	IHC	80189
1217	PRECISION	81261

Part No.	Brand	IDLI Group
1217	PRECISION	81264
121-7056	BORG-WARNR	81884
1-2171	NEAPCO	80048
1-2173	NEAPCO	80058
12173	PILOT	80058
12175	PILOT	80058
1217606	VOLVO	80023
121762R91	IHC	80221
121-762R92	IHC	80221
1218	PRECISION	81257
1218	PRECISION	81262
121-8025	BORG-WARNR	82130
1-2185	NEAPCO	80047
1-2186	NEAPCO	80046
1219	PRECISION	88000.04
12190	WHITING	80171
121-9018	BORG-WARNR	82323
121-9028	BORG-WARNR	82368
12197	WHITING	80197
12197(4)	WHITING	80196
12197(4)	WHITING	80215
121PA293	BAL-LM-HAM	80101
1-22	AEC	88000.89
1-22	LEMPCO	88000.43
1221	PRECISION	81285
1221	PRECISION	81286
122123	IHC	80237
122123H1	IHC	80237
122188	IHC	80213
122-188H1	IHC	80213
1222	PRECISION	88000.05
12-227	NEAPCO	88006.97
1222781	BEAN,JOHN	80069
1222781	SPEED-SPR.	80069
122284	HYSTER	80193
122284	HYSTER	80196
1223	PRECISION	81301
1223	PRECISION	81305
1223	PRECISION	81319
12-239	NEAPCO	88006.97
1224	PRECISION	81302
1224	PRECISION	81304
1224706	BEAN,JOHN	80134
1224706	SPEED-SPR.	80134
1225	PRECISION	81303
1225	PRECISION	88014.04
1225X	NEAPCO	88000.17
1226	PRECISION	81308
1226	PRECISION	81317
1226	PRECISION	88013.89
1226	PRECISION	88013.90
1226	PRECISION	88013.91
1226	PRECISION	88014.03
122652	TRIUMPH	80024
12266J	BORG-WARNR	81662
12267J	BORG-WARNR	81666
1227	PRECISION	88014.05
12-275353	BUCYR-ERIE	80101
1228	PRECISION	88000.06
122819(4)	HYSTER	80024
1229	PRECISION	81333
1229	PRECISION	81341
12/2C16	DIGIO-LIFT	80046
12-2C16	LIFT TRUCK	80046
1-23	AEC	88000.80
1-23	LEMPCO	88000.44
1-23	LEMPCO	88000.80
1230	PRECISION	81349
1230	PRECISION	81350
1230	PRECISION	81351
12308J	BORG-WARNR	88006:57
12309J	BORG-WARNR	88006.47
12-327	NEAPCO	88006.99
12-339	NEAPCO	88006.99
12-351	NEAPCO	85126
12-38-00	WALTERSCHD	88003.98
1238415	CHRYSLER	80068
1238416	CHRYSLER	80138
1238571	CHRYSLER	88003.50
12-39-00	WALTERSCHD	88004.08
1239290	BEAN,JOHN	80037
1239290	SPEED-SPR.	80037
1-24	LEMPCO	88000.37
124	TRU-CROSS	80225
1240	PRECISION	81205
1240	PRECISION	81206
1240169	BEAN,JOHN	80140
1240169	SPEED-SPR.	80134
1241	PRECISION	81209
1241	PRECISION	88006.98
1241	PRECISION	88013.95
124125	AEC	88000.37
124166	TRIUMPH	80023
1242	PRECISION	81217
1242	PRECISION	81218
12-43	LEMPCO	88000.22
12-43	LEMPCO	88000.70
1243	PRECISION	81239
1243	PRECISION	81240
1243615	CHRYSLER	80160
12-439	NEAPCO	81209
1244	PRECISION	81246
1244743	BEAN,JOHN	80069
1244743	SPEED-SPR.	80069
1245	PRECISION	81265
1245	PRECISION	81266
1245163	BEAN,JOHN	80140
1245163	SPEED-SPR.	80134
1246	PRECISION	81277
12461	KOYO	80023
124-645R91	IHC	80210
1247	PRECISION	81259
1247	PRECISION	81260
1-248	NEAPCO	80025
1248	PRECISION	81218
12487J	BORG-WARNR	81751
1249	PRECISION	88006.99
124963	IHC	80150
124964	IHC	80109
124967	IHC	80109
1250	CATERPILLR	80048
1250	PRECISION	81274
12-50-00	WALTERSCHD	88003.87
12505	PILOT	80059
1251	PRECISION	81256
1251	PRECISION	88014.14
12513	PILOT	80066
12518	PILOT	80066
1252	PRECISION	81251
1252	PRECISION	81252
1253	AEC	80038
1253	PRECISION	81291
1253	PRECISION	81292
12530	PILOT	80066
12531J	PILOT	80059
12531	PILOT	80066
12532	PILOT	80066
1254	AEC	80038
1254	PRECISION	81300
1254	PRECISION	81314
1254	PRECISION	81315
1254	PRECISION	81316
1255	PRECISION	81300
1255	PRECISION	88014.20
12-551	NEAPCO	88007.06
1256	PRECISION	81331
1256	PRECISION	81332
1257	PRECISION	81331
1257	PRECISION	81342
1258413	BEAN,JOHN	80184
1258413	SPEED-SPR.	80184
125875	HYSTER	80101
125876	HYSTER	80100
1259	PRECISION	88000.07
12595J	BORG-WARNR	88006.90
126	WOOD BROS.	80133
1260	PRECISION	85048
1260	PRECISION	88014.09
126078HI	HOUGH	80158
1261	PRECISION	85050
1261	PRECISION	88014.10
1262	PRECISION	85052
1262	PRECISION	88014.11
12-627	NEAPCO	88006.97
12-63	LEMPCO	88000.71
1263	PRECISION	81247
1263	PRECISION	85054
1263	PRECISION	88014.12
12637J	BORG-WARNR	81728
1264	PRECISION	81282
1264	PRECISION	85059
1264	PRECISION	88014.16
1265	PRECISION	81325
1265	PRECISION	81326
1265	PRECISION	85063
12-651	NEAPCO	88006.97
1265204	BEAN,JOHN	80133
1265204	SPEED-SPR.	80133
1265587	CLARK EQU.	88010.54
12657J	BORG-WARNR	82066
12678J	BORG-WARNR	81109
126812	IHC	80149
126-812H1	IHC	80149
126812HI	HOUGH	80109
126815	IHC	80196
126815H1	IHC	80196
126815HI	HOUGH	80195
126816	IHC	80194
126816	IHC	80195
126-816H1	IHC	80195
126816HI	HOUGH	80198
126819	IHC	80193
126-819H1	IHC	80193
126819HI	HOUGH	80193
12685H1	IHC	80197
12688J	BORG-WARNR	81089
126891H1	IHC	80193
126978	IHC	80158
126-978H1	IHC	80158
1269826	BEAN,JOHN	80186
126NLS18DC	ROCKWELL	88004.37
1-27	NEAPCO	80010
1270	PRECISION	85049
1270	PRECISION	88013.94
1270	SPICER	80072
127002	IHC	80240
127-002H1	IHC	80240
127002HI	HOUGH	80231
127019	IHC	80242
127-019H1	IHC	80242
127019HI	HOUGH	80232
127019HI	HOUGH	80237
1270961	BEAN,JOHN	80001
1270961	FMC	80001
1270UB	REPUBLIC	80076
1271	PRECISION	85051
1271	PRECISION	88013.96
12710J	BORG-WARNR	85596
12711J	BORG-WARNR	87585
127129	HYSTER	80193
12714J	BORG-WARNR	85605
127168	IHC	80153
127188	HYSTER	80193
127188	HYSTER	80196
1272	PRECISION	85055
1272	PRECISION	88013.97
127268	IHC	80150
127268H1	IHC	80150
127269	IHC	80149
127-269H1	IHC	80149
12-727	NEAPCO	88006.95
127290	IHC	88003.16
127290H1	IHC	88003.16
127290HI	HOUGH	80109
12-73	LEMPCO	88000.72
12-739	NEAPCO	88006.99
1275	PRECISION	81281
1275	PRECISION	85058
12750J	BORG-WARNR	85864
12-751	NEAPCO	88006.99
127-573R91	IHC	80124
127-574R91	IHC	80142
1275X	NEAPCO	80035
1275X1	NEAPCO	80035
1275XR	NEAPCO	80035
12-76	LEMPCO	88000.72
1276	PRECISION	85060
1276	PRECISION	88013.93
1276700	BEAN,JOHN	80069
1276704	BEAN,JOHN	80037
1276749	BEAN,JOHN	80035
1277466	BEAN,JOHN	80219
127973R91	IHC	80124
1-28	NEAPCO	80004
1-28	NEAPCO	80006
1280	PRECISION	85053
1280	PRECISION	88013.99
1280	SPICER	80075
1280-0	RUEDARSA	80049
12802J	BORG-WARNR	88007.98
12802J	BORG-WARNR	88008.27
1280-4	RUEDARSA	80045
128-0600	G & G MFG	88001.15
1281	PRECISION	81280
1281	PRECISION	85057
1281	PRECISION	88014.13
128-1200	G & G MFG	88000.09
12812J	BORG-WARNR	85599
128133	TRIUMPH	80023
1282	PRECISION	85061
1282	PRECISION	88014.19
12823E	HIGHLAND	80069
12-827	NEAPCO	88006.94
1-28-277	SPICER	85514
1283	PRECISION	85062
1283	PRECISION	88014.18
12-839	NEAPCO	88006.98
1284	PRECISION	81344
1284	PRECISION	85064
12847J	BORG-WARNR	81751

Part No.	Brand	IDLI Group
1285	RUEDARSA	80049
12-851	NEAPCO	81209
1288571	CHRYSLER	88003.50
1288574	GMC	88003.50
1288621	GMC	88002.43
129	PERFECT-CI	80037
129	TRU-CROSS	80037
1290	PRECISION	88000.08
1290	RUEDARSA	80032
1291	PRECISION	84510
1291	PRECISION	88014.06
1291	PRECISION	88014.07
1291	RUEDARSA	88002.77
1292	PRECISION	84508
1292	PRECISION	88014.06
12927J	BORG-WARNR	81753
12928J	BORG-WARNR	81752
12-93	LEMPCO	88000.73
1293	PRECISION	81346
1294	PRECISION	81271
129437	NEW HOLL.	80037
12949J	BORG-WARNR	81734
1295	PRECISION	86650
12956	BORG-WARNR	81697
12956J	BORG-WARNR	81697
1296	AEC	80068
1-297	NEAPCO	80002
1-297	NEAPCO	80079
12972J	BORG-WARNR	85101
12973J	BORG-WARNR	81828
12976	KOYO	80014
1299	PRECISION	88000.09
129919	TRIUMPH	80023
12DS4	OSHKOSH	80072
12DS4	OSHKOSH	80075
12DS4	OSHKOSH	80225
12DS8	OSHKOSH	80225
12N	AEC	80035
12N	WESCO	80035
12N0140-30	WESCO	81246
12N0144	WESCO	81213
12N0144	WESCO	88000.01
12N0164	WESCO	81237
12N0180-21	WESCO	81274
12N0180-26	WESCO	81277
12N0184	WESCO	81268
12N0220-20	WESCO	84510
12N0220-21	WESCO	81342
12N0225	WESCO	81333
12N048F	WESCO	88006.99
12N048M	WESCO	81209
12N120SH	WESCO	81203
12N120SQ	WESCO	81206
12N123SH	WESCO	81203
12N130SH	WESCO	81208
12N134	WESCO	88000
12N134SH	WESCO	81208
12N140-30	WESCO	81246
12N140SH	WESCO	81213
12N140SQ	WESCO	81218
12N144	WESCO	88000.01
12N144SH	WESCO	81213
12N150SH	WESCO	81222
12N154	WESCO	88000.02
12N154SH	WESCO	81222
12N160LQ	WESCO	85055
12N160SH	WESCO	81237
12N160SL	WESCO	85054
12N160SQ	WESCO	81239
12N164	WESCO	81237
12N164	WESCO	88014.01
12N164L	WESCO	85054
12N164SH	WESCO	81237
12N170SH	WESCO	81249
12N174SH	WESCO	81249
12N180-21	WESCO	81274
12N180-22	WESCO	88000.08
12N180-26	WESCO	81277
12N180-7	WESCO	81257
12N180H	WESCO	81259
12N180HSL	WESCO	85058
12N180LH	WESCO	85058
12N180SH	WESCO	81259
12N184	WESCO	81268
12N184SH	WESCO	81268
12N185SH	WESCO	81261
12N190SH	WESCO	81286
12N195SH	WESCO	81286
12N200-7	WESCO	81303
12N200SH	WESCO	81301
12N200SL	WESCO	85063
12N200SP	WESCO	88014.19
12N2045-37	WESCO	88013.90
12N205L	WESCO	85063
12N205SH	WESCO	81302
12N220-20	WESCO	84510
12N220-21	WESCO	81342
12N220TQB	WESCO	88014.06
12N220TQD	WESCO	84508
12N220TQD	WESCO	88014.06
12N222-0	WESCO	88014.06
12N225	WESCO	81333
12N240SH	WESCO	81350
12N48	WESCO	81218
12N48F	WESCO	88006.99
12N48M	WESCO	81209
12N72	WESCO	88006.99
12N72M	WESCO	81209
12NNB2	WESCO	88000.09
12N-NB2	WESCO	88001.08
12N-SSF48	WESCO	88006.97
12NSSF48	WESCO	88006.97
12N-SSF72	WESCO	88006.97
12NSSM48	WESCO	88007.06
12N-SSM48	WESCO	88007.10
12N-SSM72	WESCO	88007.10
12NW4-1618	WESCO	81271
12NW4-2022	WESCO	81346
12NWY1618	WESCO	81271
12NWY1834	WESCO	85514
12NYR20-58	ROCKWELL	81313
12R	G & G MFG	80035
12R		80035
1300	PRECISION	83010
1300	PRECISION	83018
1300	TRU-CROSS	80042
130007	REX CHAIN	80035
130008	REX CHAIN	80037
130009	REX CHAIN	80069
130062	REX CHAIN	80035
130065	REX CHAIN	80035
130070	REX CHAIN	80037
130071	REX CHAIN	80070
1301	PRECISION	83021
1301	TRU-CROSS	80135
130-1-130	MUNCIE	88000.28
1302	PRECISION	83024
130-2-130	MUNCIE	88000.28
130268	HYSTER	80193
1303	PRECISION	83042
1303	TRU-CROSS	80034
13033J	BORG-WARNR	81772
1304	PRECISION	83031
1304	PRECISION	83044
1304	PRECISION	88014.29
1304	TRU-CROSS	80064
1304	TRU-CROSS	80065
1305	PRECISION	83057
1-305-01	AYRA DUREX	80041
1305151	DENNIS	80189
1306	PRECISION	83047
1306	PRECISION	88014.28
1306	TRU-CROSS	80061
1306-1	TRU-CROSS	80062
13062J	BORG-WARNR	86713
1307	PRECISION	83062
1307	PRECISION	83064
1308	PRECISION	83086
13084J	BORG-WARNR	88006.25
13085J	BORG-WARNR	82096
1309	PRECISION	83076
1309	PRECISION	83082
1309	TRU-CROSS	80132
1309468	GMC	88002.10
13099J	BORG-WARNR	82135
1310	BUCKEYE	80077
1310	PRECISION	83084
1310	PRECISION	83085
1310	RUEDARSA	80221
1310	SPICER	80077
13-102-4	FLXIBLE	80190
1311	PRECISION	83081
1311	PRECISION	88014.30
1311	PRECISION	88014.31
131-10-46110	KOMATSU	80172
13113	ALLIS-CHLM	80069
13113	FIAT-ALLIS	80069
1312	PRECISION	83102
1312	PRECISION	83108
1313	PRECISION	83127
1313	PRECISION	83129
131-32	FWD	80224
13133P11	M.R.S.	80210
13133P11	M.R.S.	80215
13133P11	M.R.S.	80217
1314	PRECISION	83125
1-315-01	AYRA DUREX	80077
13-158-159	FLXIBLE	80218
13-158-160	FLXIBLE	80210
13-1837	KERSHAW	80149
13-1959	KERSHAW	80207
131NF3	ROCKWELL	88004.38
131NF6	ROCKWELL	88004.39
131NLS20-3DC	ROCKWELL	88004.40
131NLS20-6DC	ROCKWELL	88004.41
131NLS20DC	ROCKWELL	88004.42
131NLS24DC	ROCKWELL	88004.43
131NY30	ROCKWELL	88004.44
131NY30-1	ROCKWELL	88004.45
131NY46	ROCKWELL	88004.46
131NY46-1	ROCKWELL	88004.47
131NY46-2	ROCKWELL	88004.48
131NYR12	ROCKWELL	88004.49
131NYR12-1	ROCKWELL	88004.50
131NYR20	ROCKWELL	88004.51
131NYS18-2	ROCKWELL	88004.52
131NYS19	ROCKWELL	88004.53
131NYS19-1	ROCKWELL	88004.54
131NYS19-2	ROCKWELL	88004.55
131NYS20	ROCKWELL	88004.56
131NYS20-1	ROCKWELL	88004.57
131NYS20-2	ROCKWELL	88004.58
131NYS22-4	ROCKWELL	88004.59
131NYS22-5	ROCKWELL	88004.60
1-32	AEC	88000.95
1320	PRECISION	83008
132-083R91	IHC	80021
1321	PRECISION	83004
1321	PRECISION	83015
1321	PRECISION	88014.24
1321115	AUTOCAR	80072
1321115	CHRYSLER	80072
1321134	CHRYSLER	80204
1322	PRECISION	83092
1322	PRECISION	88014.27
13221J	BORG-WARNR	82163
13223J	BORG-WARNR	82193
13225J	BORG-WARNR	82193
1323	PRECISION	83040
1323	PRECISION	88014.22
1323	PRECISION	88014.23
1324	PRECISION	88014.26
1325	PRECISION	83049
1325989	CHRYSLER	80008
1329	PRECISION	88014.25
13295	ALLIS-CHLM	80189
1-33	AEC	88000.81
1-33	LEMPCO	88000.22
1-33	LEMPCO	88000.26
1330	RUEDARSA	80072
1330	SPICER	80117
13301J	BORG-WARNR	85546
1332	RUEDARSA	80117
13323J	BORG-WARNR	85876
1333	RUEDARSA	80121
1334	RUEDARSA	80116
1335	RUEDARSA	80119
1335006	GMC	80225
133507HA	IHC	88001.57
133508HB	IHC	80174
133508HC	IHC	80174
1336	RUEDARSA	80090
13368J	BORG-WARNR	88006.80
134	TRU-CROSS	80118
13411P11	M.R.S.	80168
13411P11	M.R.S.	80216
1342133-02	EATON	80217
13434J	BORG-WARNR	85567
13435J	BORG-WARNR	87531
13446J	BORG-WARNR	85843
13451J	BORG-WARNR	87529
13455J	BORG-WARNR	88006.75
13457J	BORG-WARNR	87452
13457J	BORG-WARNR	88007.97
13458J	BORG-WARNR	85851
1347	PRECISION	88007.08
134778	TRIUMPH	80041
1348	PRECISION	87938
13481J	BORG-WARNR	82154
13482J	BORG-WARNR	82059
13483J	BORG-WARNR	82059
13489J	BORG-WARNR	82044
1349	PRECISION	88007.09
1350	PRECISION	83054
1350	PRECISION	83056
1350	PRECISION	88014.21
1350	SPICER	80124
1-3500	NEAPCO	80049
1350UB	REPUBLIC	88013.87
1351	PRECISION	83090

Part No.	Brand	IDLI Group	Part No.	Brand	IDLI Group	Part No.	Brand	IDLI Group	Part No.	Brand	IDLI Group
1351	RUEDARSA	80044	1370	PRECISION	88014.35	140	REX CHAIN	80037	141-10-00012(4)	KOMATSU	88002.83
1351	TEL-E-LECT	80133	1371	PRECISION	87013	14000	PILOT	80109	141132	FORD	80033
135-14-00030	KOMATSU	80198	137100H	IHC	80193	14002	PILOT	80109	1-4-1141X	HAYES	84508
1351-7	RUEDARSA	80043	137109H	IHC	80147	14006	PILOT	80153	1-4-1141X	SPICER	84508
1352	PRECISION	83116	137109HA	IHC	80147	14007	PILOT	80147	14-114J	BORG-WARNR	87515
1352	PRECISION	88014.33	137109HA	IHC	80210	14008	PILOT	80146	1412	PRECISION	81375
1353	PRECISION	83105	137110H	IHC	80114	1401	RUEDARSA	80124	1412	TEDDY TORQ	80030
1353	PRECISION	83106	137110HA	IHC	80114	140-1-140	MUNCIE	88000.31	1412G	PILOT	80112
1-355-01	AYRA DUREX	80124	137110HB	IHC	80114	14014	PILOT	80158	14138	PILOT	80109
1356	RUEDARSA	80030	137-1900	G & G MFG	88015.41	14015	PILOT	80148	14138	PILOT	80149
13561	TEL-E-LECT	80133	1372	PRECISION	83116	140-1-550	MUNCIE	88000.31	14140	PILOT	80155
13562	TEL-E-LECT	80008	1372	PRECISION	84519	140-1-550	MUNCIE	88000.88	14141	PILOT	80154
13562	TEL-E-LECT	80133	1372	PRECISION	84521	14016	PILOT	80159	14143	PILOT	80156
135669H	IHC	80177	1372	PRECISION	84522	14017	PILOT	80158	14143	PILOT	88003.17
13568J	BORG-WARNR	81725	1372	PRECISION	88014.32	1402	TEDDY TORQ	80109	14144J	BORG-WARNR	87515
135699H	IHC	80177	1372	PRECISION	88014.34	14020	PILOT	80158	1415	TEDDY TORQ	80148
13569J	BORG-WARNR	81723	137-2100	G & G MFG	88015.42	14021	PILOT	80158	1-415-01	AYRA DUREX	80142
1357	RUEDARSA	80031	1373	PRECISION	84517	140-2-140	MUNCIE	88000.31	14151	PILOT	80109
135764	IHC	80156	1374	PRECISION	83038	1-4-0222	SPICER	81246	14152	PILOT	80158
135764H1	IHC	80156	13749	M.R.S.	80163	140-2-550	MUNCIE	88000.31	14153	PILOT	80149
135765	IHC	80198	13749	M.R.S.	80167	140-2-550	MUNCIE	88000.88	14156	PILOT	80157
135-765H1	IHC	80198	1375	PRECISION	85705	14029P11	M.R.S.	80168	14-16	PILOT	80159
135765HI	HOUGH	80198	13780J	BORG-WARNR	85548	1403	TEDDY TORQ	80054	1416	TEDDY TORQ	80159
135766	IHC	80238	13781J	BORG-WARNR	87417	1403764	CHRYSLER	88001.60	1-4-162	HAYES	81271
135-766H1	IHC	80238	13782J	BORG-WARNR	87363	1404	TEDDY TORQ	80059	1-4-162	SPICER	81271
135-766H2	IHC	80238	13782J	BORG-WARNR	88007.89	1404	TRU-CROSS	80951	1417	TEDDY TORQ	80092
135766HI	HOUGH	80238	138	BUSH HOG	80037	14045J	BORG-WARNR	87632	1418	TEDDY TORQ	80053
1358	RUEDARSA	80055	1380	PRECISION	87001	14-04-813	HARDY SPCR	88003.74	1-4-182	SPICER	81218
1359601	EATON	80217	13804J	BORG-WARNR	82207	1405	TRU-CROSS	80905	141859	HYSTER	80198
135N170	ROCKWELL	88004.61	13809	ALLIS-CHLM	80178	1405632	BEAN,JOHN	80185	1419	TEDDY TORQ	80045
135NF2	ROCKWELL	88004.62	13809	FIAT-ALLIS	80178	1405632	FMC	80185	1419	TEDDY TORQ	80053
135NF3	ROCKWELL	88004.63	13818J	BORG-WARNR	87443	1405632	SPEED-SPR.	80186	141N4-1841	ROCKWELL	81385
135NF4	ROCKWELL	88004.64	13818J	BORG-WARNR	88007.93	14058287	FMC	80185	141N4-2241	ROCKWELL	81383
135NLS22-3DC	ROCKWELL	88004.65	13819J	BORG-WARNR	87512	14059J	BORG-WARNR	87448	141N4-2511	ROCKWELL	81386
135NLS22DC	ROCKWELL	88004.66	1388249	GMC	80117	14059J	BORG-WARNR	88007.95	141N4-3941	ROCKWELL	81379
135NY46	ROCKWELL	88004.67	13882R91	IHC	80182	1405X	NEAPCO	80037	141N4-6561	ROCKWELL	81381
135NY46-1	ROCKWELL	88004.68	138839H	IHC	80177	1406	PRECISION	81365	141N4-6631X	ROCKWELL	81382
135NYR14	ROCKWELL	88004.69	138839HC	IHC	80177	1406	TEDDY TORQ	80153	141N53-1031	ROCKWELL	87353
135NYR26	ROCKWELL	88004.70	1389604	GMC	80121	1406	TRU-CROSS	80953	141N53-1071	ROCKWELL	87352
135NYR28	ROCKWELL	88004.71	1389837	GMC	80121	14-0601	NEAPCO	88000.38	141N53-161	ROCKWELL	87290
135NYS22	ROCKWELL	88004.72	1389839	GMC	80117	140601R91	IHC	80104	141NF1	ROCKWELL	84614
135NYS22-1	ROCKWELL	88004.73	1389864	GMC	88002.33	140604K92	IHC	80212	141NF2	ROCKWELL	84615
135NYS22-4	ROCKWELL	88004.74	1389865	GMC	80117	140604R91	IHC	80212	141NLS24-3DC	ROCKWELL	85065
135NYS22-5	ROCKWELL	88004.75	139	BUSH HOG	80037	140-604R92	IHC	80212	141NLS24-6DC	ROCKWELL	85067
135NYS24	ROCKWELL	88004.76	1390	PRECISION	88000.10	14067677	GMC	80128	141NLS24DC	ROCKWELL	85066
135NYS24-1	ROCKWELL	88004.77	1391164	GMC	88002.71	1407	TEDDY TORQ	80147	141NY42	ROCKWELL	85517
135NYS24-2	ROCKWELL	88004.78	1391262	GMC	88002.67	1407	TRU-CROSS	80906	141NY45	ROCKWELL	85521
135NYS24-3	ROCKWELL	88004.79	1391262	GMC	88002.72	1408	PRECISION	81370	141NY45-1	ROCKWELL	85518
135NYS24-4	ROCKWELL	88004.80	13922J	BORG-WARNR	85562	1408	PRECISION	88006.05	141NY45-2	ROCKWELL	88004.83
135NYS26	ROCKWELL	88004.81	13928J	BORG-WARNR	86699	1408	PRECISION	88006.06	141NY45-3	ROCKWELL	85522
135NYS28	ROCKWELL	88004.82	1392987	GMC	80116	1408	TEDDY TORQ	80146	141NY54	ROCKWELL	85519
1-3600	NEAPCO	80044	1393716	GMC	88002.65	14082J	BORG-WARNR	88006.23	141NY54-2	ROCKWELL	85520
136142	IHC	80213	1393717	GMC	88002.63	1-4-0881X	SPICER	84510	141NYS22-1	ROCKWELL	88004.84
136-142H1	IHC	80213	1393718	GMC	88002.64	14089J	BORG-WARNR	85581	141NYS22-1A	ROCKWELL	88004.85
136142HI	HOUGH	80213	1393718	GMC	88002.67	1-4-0921	SPICER	81274	141NYS22-3	ROCKWELL	88004.86
136152HA	IHC	80210	1393719	GMC	88002.67	1409233	BEAN,JOHN	80186	141NYS22-3A	ROCKWELL	88004.87
136-1900	G & G MFG	88015.41	1393719	GMC	88002.71	1409233	SPEED-SPR.	80186	141NYS22-5	ROCKWELL	88004.88
136-2100	G & G MFG	88015.42	13937J	BORG-WARNR	82056	1409290	BEAN,JOHN	80186	141NYS22-6	ROCKWELL	88004.89
136488H	IHC	80168	13937J	BORG-WARNR	82059	1409290	SPEED-SPR.	80186	141NYS22-8	ROCKWELL	88004.90
136-488HA	IHC	80216	13947J	BORG-WARNR	81784	1-4-0981	SPICER	81342	141NYS24-1	ROCKWELL	88004.91
1-3650	NEAPCO	80055	1396838	GMC	88002.33	1410	BUCKEYE	80142	141NYS24-2	ROCKWELL	88004.92
136512H	IHC	80210	139764	IHC	80001	1410	EIMCO	80142	141NYS24-4	ROCKWELL	88004.93
136512HA	IHC	80210	139764H1	IHC	80001	1410	SPICER	80142	141NYS24-4A	ROCKWELL	88004.94
136512HB	IHC	80210	139842	NEW HOLL.	88000.15	1-4-1011X	SPICER	84508	141NYS24A	ROCKWELL	81384
136-512HBK	IHC	80210	139956	BROCKWAY	80189	1-4-1016X	SPICER	84510	141NYS28-2	ROCKWELL	88004.95
136512HBX	IHC	80210	139957	BROCKWAY	80221	1-4-1031	SPICER	81339	141NYS28-2A	ROCKWELL	88004.96
1365-24805	ALFA ROMEO	80024	139959	BROCKWAY	80178	14104J	BORG-WARNR	85854	141NYS28-4	ROCKWELL	88004.97
13670J	BORG-WARNR	81797	13B	AEC	88007.58	1-4-1061X	HAYES	84510	141NYS29	ROCKWELL	88004.98
136908	NEW HOLL.	80037	14	AEC	87331	1-4-1061X	SPICER	84510	141NYS29A	ROCKWELL	88004.99
1370	PRECISION	87016	140	CHAIN BELT	80037	1-4-1091	SPICER	81274	141NYS29A1	ROCKWELL	88005

Part No.	Brand	IDLI Group
141NYS29A2	ROCKWELL	88005.01
141NYSB24-2	ROCKWELL	85802
141NYSB28-2	ROCKWELL	85803
141NYSB29	ROCKWELL	85804
141NYT22	ROCKWELL	88005.02
141NYT22-1	ROCKWELL	81380
142133C1	IHC	80172
1421-352M1	MASS.FERG.	80198
142136C1	IHC	80198
14-2147	BORG-WARNR	80054
14214J	BORG-WARNR	82047
14217J	BORG-WARNR	85847
1422	TEDDY TORQ	80048
1-4-222	HAYES	81246
1-4-222	SPICER	81240
1422427	JOY MFG.	80048
1423	TEDDY TORQ	80044
14240J	BORG-WARNR	81949
1-4-242	SPICER	81267
1-4-242	SPICER	81273
1424820	GMC	80099
1-4-2493	SPICER	88000.71
1425	TEDDY TORQ	80044
14250	PILOT	80059
14252	PILOT	80093
1-4-2543	SPICER	88000.68
1-4-2553	SPICER	88000.70
1426	TEDDY TORQ	80049
14-26-000	HARDY SPCR	88003.92
1-4-262	HAYES	81346
1-4-262	SPICER	81346
1426418K3	JOY MFG.	80189
1427476	GMC	80109
1-4-2783	SPICER	88000.72
1428	TEDDY TORQ	80051
1-4-282	HAYES	81246
1-4-282	SPICER	81246
142-823R91	IHC	80104
142-824R91	IHC	80112
1428G	TEDDY TORQ	80051
1-4-2923	SPICER	88000
142-923R91	IHC	80104
1-4-2953	SPICER	81208
1-4-2953	SPICER	88000
1-4-2973	SPICER	81215
1-4-2973	SPICER	88000.01
142-973R91	IHC	80104
1-4-2983	SPICER	81222
1-4-2983	SPICER	88000.02
1-4-2993	SPICER	81246
142A300	HARDIE	80134
142A370	AMER.PULL.	80178
142A370	HARDIE	80178
142A487	ALLIS-CHLM	80124
142A487	FIAT-ALLIS	80124
1-43	LEMPCO	88000.22
14300	PILOT	80099
14300	PILOT	80101
1-4-3003	HAYES	81237
1-4-3003	SPICER	81245
14301	PILOT	80101
14-30-12	SPICER	88014.97
1-4-3013	SPICER	81237
1-4-3023	SPICER	88000.03
1430287	BEAN,JOHN	80134
1-4-3033	SPICER	81237
1-4-3043	HAYES	88000.04
1-4-3043	SPICER	81275
14-30-52	SPICER	88014.96
1-4-3063	HAYES	88000.04
1-4-3063	SPICER	81264
1-4-3063	SPICER	88000.04
14308	PILOT	80104
1-4-3083	SPICER	81286
1-4-3083	SPICER	88000.05
14308G	PILOT	80104
1-4-3093	HAYES	88000.06
1-4-3093	SPICER	81301
14310	PILOT	80100
1-4-3103	SPICER	81323
14311	PILOT	80107
1-4-3113	HAYES	88000.06
1-4-3113	SPICER	81304
14311J	BORG-WARNR	81848
14312	PILOT	80110
14312	PILOT	80112
1-4-3123	SPICER	81307
14312G	PILOT	80112
1-4-3133	HAYES	81333
1-4-3133	SPICER	81341
14314	PILOT	80114
1-4-3143	SPICER	81333
1-4-3153	SPICER	81351
14318	PILOT	80101
14318J	BORG-WARNR	81673
14319	PILOT	80054
14319	PILOT	80112
14320	PILOT	80113
1-4-3203	SPICER	81257
14321	PILOT	80111
1-4-3213	SPICER	81303
14322	PILOT	80112
1-4-3253	SPICER	88014.01
1-4-3273	HAYES	81333
1433	TEDDY TORQ	80112
1-4-3373	SPICER	88014.05
1434	TEDDY TORQ	80114
14346J	BORG-WARNR	81652
14355	PILOT	80103
14358	PILOT	80102
1437	PREC.TOR.	88007.23
1437	TEDDY TORQ	80111
1438	TEDDY TORQ	80100
14388	PILOT	80105
1438G	TEDDY TORQ	80100
1439	PREC.TOR.	88007.22
1440602	CHRYSLER	80906
1442	PRECISION	81362
1447	PRECISION	88007.24
1448	PRECISION	88000.11
144823	TRIUMPH	80077
144859	TRIUMPH	80041
1449	PRECISION	88000.12
1450249	CHRYSLER	80064
1450445	CHRYSLER	80095
1450446	CHRYSLER	80095
1450446	CHRYSLER	80908
1450448	CHRYSLER	88003.51
1450498	CHRYSLER	80095
1450504	CHRYSLER	80160
1450505	CHRYSLER	80181
1450602	CHRYSLER	80909
1450703	CHRYSLER	80045
1450723	CHRYSLER	80045
1450727	CHRYSLER	80045
1450727	CHRYSLER	80054
1451	RUEDARSA	80193
1451-150N	HARDY SPCR	80023
1451-185R	HARDY SPCR	80024
1451-194T	HARDY SPCR	80024
1451-194W	HARDY SPCR	80024
1451-277N	HARDY SPCR	80023
1451-279W	HARDY SPCR	80024
145-14-00030	KOMATSU	80198
145-14-00190	KOMATSU	80198
145-14-35110	KOMATSU	80198
1452	PRECISION	81362
1452	RUEDARSA	80197
14526J	BORG-WARNR	82133
1453	PRECISION	84512
1453	PRECISION	84515
1454	PRECISION	81356
145-421	FORD	80069
1455	PRECISION	85516
1456503	GMC	80110
1456503	GMC	80112
1456525	GMC	80107
1457	TEDDY TORQ	80177
14577J	BORG-WARNR	82090
14587J	BORG-WARNR	82210
1460	WILK RIP.	80069
14625	DATSUN	80024
1463259	GMC	80110
1463491	CHRYSLER	80049
1463499	CHRYSLER	80055
1-4635	NEAPCO	80091
1463513	CHRYSLER	80146
1463514	CHRYSLER	80158
1463546	GMC	80110
1463547	GMC	80099
1463548	GMC	80107
1464	TEDDY TORQ	80195
14640J	BORG-WARNR	82354
146701H1	IHC	80212
146703H1	IHC	80209
14671	CHAMPION	80198
14678	CHAMPION	80238
14679	CHAMPION	80224
1469	TEDDY TORQ	80201
1470	PREC.TOR.	80217
1470	TEDDY TORQ	88002.82
1-4700	NEAPCO	80034
1470373	GMC	80117
1470374	GMC	80121
1471287	CLARK EQU.	80008
1472	PREC.TOR.	80218
1-4730	NEAPCO	80093
14742J	BORG-WARNR	86743
147511	IHC	80003
147511H1	IHC	80001
147512	IHC	80003
147512H1	IHC	80001
1475X	NEAPCO	80001
1477	PREC.TOR.	80217
14785J	BORG-WARNR	87513
1480	BUCKEYE	80140
1480	PREC.TOR.	80243
1480	SPICER	80140
1482	PREC.TOR.	80243
1482672	GMC	80117
148-357	FORD	80035
1483701	CHRYSLER	80057
1484	PREC.TOR.	80244
1-48402	BONNE-TERR	80109
14841	PILOT	88002.86
1485	TEDDY TORQ	80238
14851	PILOT	88002.87
148-53-291	ROCKWELL	88005.03
1-4-861	SPICER	81315
1-4-871X	SPICER	81315
1-4-881X	SPICER	84510
148959	GARD.DENV.	80100
148N40-971	ROCKWELL	87370
148N4-3061	ROCKWELL	81396
148N4-3081	ROCKWELL	81390
148N4-3091	ROCKWELL	81388
148N4-3231	ROCKWELL	81387
148N4-3541	ROCKWELL	81395
148N4-3571	ROCKWELL	81394
148N4-3781	ROCKWELL	81391
148N4-3861	ROCKWELL	81400
148N4-4051	ROCKWELL	81393
148N4-4421X	ROCKWELL	81389
148N4-5791X	ROCKWELL	81397
148N4-7541	ROCKWELL	81399
148N53-241	ROCKWELL	87360
148N53-291	ROCKWELL	87372
148N53-431	ROCKWELL	87357
148N53-451	ROCKWELL	87371
148NF2	ROCKWELL	84616
148NF3	ROCKWELL	88005.04
148NF4	ROCKWELL	84617
148NLS25DC	ROCKWELL	85068
148NLS25DC1	ROCKWELL	85069
148NLS26-3DC	ROCKWELL	85071
148NLS26DC	ROCKWELL	85070
148NLS32DC	ROCKWELL	85072
148NY54	ROCKWELL	85525
148NY54-2	ROCKWELL	85523
148NY62-1	ROCKWELL	85524
148NY62-2	ROCKWELL	88005.05
148NYS24-4	ROCKWELL	88005.06
148NYS24-5	ROCKWELL	88005.07
148NYS24-7	ROCKWELL	88005.08
148NYS25	ROCKWELL	88005.09
148NYS26-2A	ROCKWELL	81392
148NYS26-3	ROCKWELL	88005.10
148NYS26-3A	ROCKWELL	88005.11
148NYS28-11	ROCKWELL	81398
148NYS28-12	ROCKWELL	88005.12
148NYS28-12A	ROCKWELL	88005.13
148NYS28-14	ROCKWELL	88005.14
148NYS28-15	ROCKWELL	88005.15
148NYS28-17	ROCKWELL	88005.16
148NYS29	ROCKWELL	88005.17
148NYS29-1	ROCKWELL	88005.18
148NYS29-1A	ROCKWELL	88005.19
148NYS29-1A1	ROCKWELL	88005.20
148NYS29A	ROCKWELL	88005.21
148NYS31-1	ROCKWELL	88005.22
148NYS31-1A	ROCKWELL	88005.23
148NYS31-3	ROCKWELL	88005.24
148NYS31-3A	ROCKWELL	88005.25
148NYS31-5	ROCKWELL	88005.26
148NYS34	ROCKWELL	81401
148NYSB24-10	ROCKWELL	85805
148NYSB24-7	ROCKWELL	85806
148NYSB28-17	ROCKWELL	85808
148NYSB28-26	ROCKWELL	85807
148NYSB29	ROCKWELL	85809
1490	PRECISION	87276
1490	PREC.TOR.	80235
1490	TEDDY TORQ	80235
14909J	BORG-WARNR	81877
1-4-921	SPICER	88014.17
1-4-931	SPICER	88014.15
149318	IHC	80047
149-318H1	IHC	80047
149318HI	HOUGH	80047
149320	IHC	80048

Part No.	Brand	IDLI Group	Part No.	Brand	IDLI Group	Part No.	Brand	IDLI Group	Part No.	Brand	IDLI Group
149-320H1	IHC	80048	1501	TRU-CROSS	80010	15148	NEAPCO	88000.87	151B6	NEAPCO	88000.27
149320HI	HOUGH	80048	15010	TRW	86143	15149	NEAPCO	88000.57	151B8	NEAPCO	81020
1-4-951	SPICER	81315	15011	PILOT	80168	1515	ALCO	80178	151C7	NEAPCO	88000.28
149542	JOY MFG.	80124	150-11-12360	KOMATSU	88002.84	1515	PRECISION	88000.24	151C8	NEAPCO	88000.29
14977J	BORG-WARNR	81701	15012	PILOT	80177	15150	NEAPCO	88000.23	151D2	NEAPCO	88000.55
1-4-981	HAYES	88000.07	15014	PILOT	80168	1-515-01	AYRA DUREX	80180	151D4	NEAPCO	88000.19
1-4-981	SPICER	81332	15015	PILOT	80173	15153	NEAPCO	88000.29	151D7	NEAPCO	88000.79
1499	PRECISION	88000.13	15016J	BORG-WARNR	88006.26	15154	NEAPCO	88000.32	151D7	NEAPCO	88000.85
14N	AEC	80037	1502	PRECISION	88000.16	15154	PILOT	80175	151D8	NEAPCO	88000.36
14N	WESCO	80037	1502	TRU-CROSS	80009	15156	NEAPCO	88000.55	151D9	NEAPCO	88000.36
14N0140-30	WESCO	83015	1503	PRECISION	88000.17	15157	NEAPCO	88000.80	151E4	NEAPCO	88000.98
14N0164	WESCO	83018	1503	TRU-CROSS	80012	15158	NEAPCO	88000.48	151E5	NEAPCO	88000.75
14N0200-26	WESCO	88014.27	1504	PRECISION	88000.18	15158	NEAPCO	88000.89	151E6	NEAPCO	88000.61
14N0220-20	WESCO	88014.32	1504	TRU-CROSS	80011	15159	NEAPCO	88000.92	151E7	NEAPCO	88000.41
14N0220-21	WESCO	83116	15049J	BORG-WARNR	85873	1516	ALCO	80189	151G1	NEAPCO	88000.68
14N0220TQD	WESCO	84517	1505	ALCO	80031	1516	PRECISION	88000.25	151G2	NEAPCO	88000.70
14N0225	WESCO	83102	1505	TRU-CROSS	80019	15160	NEAPCO	88000.32	151G3	NEAPCO	88000.71
14N0236	WESCO	83125	15052X	NEAPCO	80007	15162	NEAPCO	88000.26	151G4	NEAPCO	88000.72
14N0246	WESCO	83129	15052X	NEAPCO	88000.14	15163	NEAPCO	88000.91	151G5	NEAPCO	88000.73
14N048F	WESCO	88007.09	1505X	NEAPCO	80007	15164	NEAPCO	88000.30	151G6	NEAPCO	88000.69
14N140-30	WESCO	83015	1506	AEC	80017	15164	NEAPCO	88000.33	151G6	NEAPCO	88000.79
14N164	WESCO	83018	1506100M91	MASS.FERG.	80069	15165	NEAPCO	88000.40	151G7	NEAPCO	88000.70
14N180-21	WESCO	88014.21	150635R91	IHC	80001	15166	NEAPCO	88000.64	151G8	NEAPCO	88000.71
14N1845-37	WESCO	83047	150-635R92	IHC	80001	15168	NEAPCO	88001	151G9	NEAPCO	88000.73
14N1945-37	WESCO	83062	15065J	BORG-WARNR	81865	1516853	CHRYSLER	80142	151H2	NEAPCO	88000.70
14N200-21	WESCO	83090	1507	TRU-CROSS	80020	1516854	CHRYSLER	80124	151H4	NEAPCO	88000.74
14N200-26	WESCO	88014.27	150822	UNIC	80189	15169	NEAPCO	88000.94	151H4	NEAPCO	88000.78
14N2045-37	WESCO	83084	1508762M91	MASS.FERG.	80069	1517	ALCO	80221	151H5	NEAPCO	88000.75
14N220-20	WESCO	88014.32	150965R91	IHC	80001	1517	PRECISION	88000.26	151H6	NEAPCO	88000.74
14N220-21	WESCO	83116	15096J	BORG-WARNR	82335	15170	NEAPCO	88000.53	151H8	NEAPCO	88000.72
14N220T	WESCO	83105	1510	PRECISION	88000.19	15171	NEAPCO	88000.43	151H9	NEAPCO	88000.72
14N220TQD	WESCO	84517	15101X	NEAPCO	88000.67	15172	NEAPCO	88000.47	151J2	NEAPCO	88000.69
14N225	WESCO	83102	151028	WESCO	88000.49	15173	NEAPCO	88000.95	151J3	NEAPCO	88000.71
14N236	WESCO	83125	15102X	NEAPCO	88000.67	15173	PILOT	80171	151J4	NEAPCO	88000.71
14N246	WESCO	83129	15109	PILOT	80162	15174	NEAPCO	88000.55	151J5	NEAPCO	88000.69
14N48F	WESCO	88007.09	1511	PRECISION	88000.20	15175	NEAPCO	88000.93	151J6	NEAPCO	88000.69
14N48FS	WESCO	88007.09	15111	NEAPCO	88000.20	15176	NEAPCO	88000.88	151J9	NEAPCO	88000.71
14N48M	WESCO	88007.08	15112	NEAPCO	88000.22	15177	NEAPCO	88000.63	151K1	NEAPCO	88000.76
14N48TB	WESCO	87938	15113	NEAPCO	88000.25	15177	PILOT	80172	151K2	NEAPCO	88000.77
14N72F	WESCO	88007.09	15113	NEAPCO	88000.27	15177	PILOT	80221	151K5	NEAPCO	88000.30
14N72M	WESCO	88007.08	15114	NEAPCO	88000.28	15178	NEAPCO	88000.89	151L6	NEAPCO	88000.32
14N72SF	WESCO	88007.09	15115	NEAPCO	88000.30	1518	PRECISION	88000.27	151M5	NEAPCO	88000.96
14N-S13	WESCO	88000.10	15116	NEAPCO	88000.35	15180	NEAPCO	88000.66	151M8	NEAPCO	88000.54
14NSA	WESCO	87227	15117	NEAPCO	88000.37	15-181	ALLOY	88008.68	151M9	NEAPCO	88000.28
14NW4-1618	WESCO	83038	15118	NEAPCO	88000.38	15183	NEAPCO	88000.79	151P1	NEAPCO	88000.19
14NWY1618	WESCO	83038	15119	NEAPCO	88000.43	15184	NEAPCO	88000.81	151P2	NEAPCO	88000.96
14NWY1834	WESCO	85704	1512	PRECISION	88000.21	15185	NEAPCO	88000.82	151P8	NEAPCO	88000.58
14NYSB22-8	ROCKWELL	85801	15120	NEAPCO	88000.45	15186	NEAPCO	88000.83	151P9	NEAPCO	88000.66
14P623	MPL.MOLINE	80001	15122	NEAPCO	88000.46	15187	NEAPCO	88000.51	151R2	NEAPCO	88000.56
14R	G & G MFG	80037	15122	PILOT	80168	15188	NEAPCO	88000.49	151R3	NEAPCO	88000.81
14R	WESCO	80037	15122	PILOT	80216	15189	NEAPCO	88000.45	151R5	NEAPCO	88000.80
14S	AEC	87331	15123	NEAPCO	88000.21	1519	PRECISION	88000.28	151R8	NEAPCO	88000.62
15	AEC	87331	15124	NEAPCO	88000.31	15190	NEAPCO	88000.25	151R9	NEAPCO	88000.56
15	AEC	88008.39	15125	NEAPCO	88000.39	15191	NEAPCO	88000.34	151S6	NEAPCO	88000.61
1-5	NEAPCO	80041	15127	PILOT	80163	15193	NEAPCO	88000.46	151S8	NEAPCO	88001.01
1-50	NEAPCO	80092	15127	PILOT	80167	15195	NEAPCO	88000.46	151U7	NEAPCO	88000.65
1500	PRECISION	88000.14	15129	NEAPCO	88000.43	15196	NEAPCO	88000.36	151Z1	NEAPCO	88000.21
1500	SPICER	80178	1513	PRECISION	88000.22	15197	NEAPCO	88000.35	151Z2	NEAPCO	88000.20
1500	TRU-CROSS	80004	15132	PILOT	80174	15198	NEAPCO	88000.36	151Z5	NEAPCO	88000.33
1500	TRU-CROSS	80006	15134	NEAPCO	88000.49	15199	NEAPCO	88000.84	151Z6	NEAPCO	88000.52
15000	PILOT	80168	15136	NEAPCO	88000.52	151A1	NEAPCO	88000.52	151Z9	NEAPCO	88000.43
15003	PILOT	80168	15138	NEAPCO	88000.60	151A6	NEAPCO	88000.30	151Z9	NEAPCO	88000.99
15006	PILOT	80167	15139	NEAPCO	88000.53	151A6	NEAPCO	88000.33	1520	PRECISION	88000.29
15007	PILOT	80177	1514	ALCO	80023	151A7	NEAPCO	88000.90	1521	ALCO	80072
1500765H1	CLARK EQU.	80102	1514	ALCO	80024	151A9	NEAPCO	88000.36	1521	PRECISION	88000.30
15008	PILOT	80174	1514	PRECISION	88000.23	151B2	NEAPCO	88000.22	15214	DURA-DET.	80033
1500 SERIES	COLES	80178	15141	NEAPCO	88000.50	151B3	NEAPCO	88000.25	15214	SIDEWINDER	80069
1501	AEC	80021	15145	NEAPCO	88000.51	151B3	NEAPCO	88000.27	1521-90	WESCO	88013.86
1501	ALCO	88001.97	15146	NEAPCO	88000.23	151B4	NEAPCO	88000.52	1521HD	PILOT	80077
1501	PRECISION	88000.15	15147	NEAPCO	88000.86	151B5	NEAPCO	88000.27	1522	ALCO	80073

Part No.	Brand	IDLI Group
1522	PRECISION	88000.31
152-227R92	IHC	80237
152-275R91	IHC	80240
152276K91	IHC	80237
152276R91	IHC	80237
152-276R92	IHC	80237
15227R92	IHC	80237
15-22875	CURTISS-WR	80124
15-22875	GEN.EQUIP.	80101
15-22875	WOOLDRIDGE	80101
1523	ALCO	80081
1523	ALCO	88001.96
1523	PRECISION	88000.32
15238J	BORG-WARNR	85546
1524	ALCO	80091
1524	PRECISION	88000.33
1524T	ALCO	80225
1525	PRECISION	88000.34
1526	PRECISION	88000.35
15262J	BORG-WARNR	87641
15263J	BORG-WARNR	88006.76
1527	PRECISION	88000.36
15270	DIVCO	80958
1528	PRECISION	88000.37
1529	PRECISION	88000.38
1-53	AEC	88000.83
1-53	LEMPCO	88000.52
153	TRU-CROSS	80077
1530	ALCO	80117
1530	PRECISION	88000.39
1531	ALCO	80121
1531	PRECISION	88000.40
1532	PRECISION	88000.41
1532889	CHRYSLER	80955
1532907	CHRYSLER	88003.44
1532929	CHRYSLER	80095
1532930	CHRYSLER	88001.60
1532937	CHRYSLER	80095
1533	ALCO	80116
1533	PRECISION	88000.42
1533	PREC.TOR.	88000.43
1534	ALCO	80118
1534	PRECISION	88000.43
1534402	CHRYSLER	88003.56
1534428	CHRYSLER	88003.56
1535	PRECISION	88000.44
1536	PREC.TOR.	88000.45
15365J	BORG-WARNR	85549
15367J	BORG-WARNR	81833
1537	PRECISION	88000.45
15-38287	GEN.EQUIP.	80001
1539	PRECISION	88000.46
153915	IHC	80077
153-915H1	IHC	80077
153915HI	HOUGH	80075
15394(4)	SIL.HOIST	80149
15394(4)U	SIL.HOIST	80155
154	ALCO	80041
154	ALCO	88001.85
154	WOOD BROS.	80037
1540	PRECISION	88000.47
154073	IHC	80140
154-073H1	IHC	80140
154073HI	HOUGH	80140
154081	IHC	80124
154-081H1	IHC	80124
154081HI	HOUGH	80124
1541	PRECISION	88000.48
15-4141	NEAPCO	88000.50
1541448	STUDEBAKER	80072
1541K1	WHITE	80001
1542	PRECISION	88000.49
1544	PRECISION	88000.50
1545	PRECISION	88000.51
1546	PRECISION	88000.52
15460	TOWMOTOR	80220
154-66-12100	KOMATSU	80172
1548	PRECISION	88000.53
154-926R91	IHC	80090
154929R91	IHC	80090
155	TRU CROSS	80182
1550	ALCO	80119
1550	ALCO	80120
1550	BAY CITY	80182
1550	PRECISION	88000.54
1550	SPICER	80182
1550426	STUDEBAKER	80072
1550456	CHRYSLER	88003.56
1550466	CHRYSLER	88003.43
1550551	CLARK EQU.	80069
1550551	COLES	80069
1550682	CLARK EQU.	80001
1550682	COLES	80001
1550819	CLARK EQU.	80219
1550819	COLES	80161
15-5101	NEAPCO	88000.67
15-5103	NEAPCO	88000.20
15-5105	NEAPCO	88000.52
15-5111	NEAPCO	88000.20
15-5112	NEAPCO	88000.22
15-5113	NEAPCO	88000.25
15-5114	NEAPCO	88000.28
15-5115	NEAPCO	88000.30
15-5116	NEAPCO	88000.35
15-5117	NEAPCO	88000.37
15-5118	NEAPCO	88000.38
15-5121	NEAPCO	88000.58
15-5123	NEAPCO	88000.21
15-5124	NEAPCO	88000.31
15-5125	NEAPCO	88000.39
15-5129	NEAPCO	88000.99
15-5132	NEAPCO	88000.22
15-5134	NEAPCO	88000.49
15-5136	NEAPCO	88000.52
15-5138	NEAPCO	88000.60
15-5139	NEAPCO	88000.53
15-5145	NEAPCO	88000.51
15-5146	NEAPCO	88000.23
15-5147	NEAPCO	88000.86
15-5147	NEAPCO	88000.88
15-5148	NEAPCO	88000.87
15-5148	NEAPCO	88000.88
15-5149	NEAPCO	88000.43
15-5149	NEAPCO	88000.57
15-5150	NEAPCO	88000.23
15-5154	NEAPCO	88000.32
15-5156	NEAPCO	88000.55
15-5157	NEAPCO	88000.80
15-5158	NEAPCO	88000.48
15-5159	NEAPCO	88000.92
15-5159	NEAPCO	88000.93
15-5162	NEAPCO	88000.22
15-5163	NEAPCO	88000.91
15-5164	NEAPCO	88000.33
15-5165	NEAPCO	88000.40
15-5166	NEAPCO	88000.64
15-5167	NEAPCO	88000.22
15-5167	NEAPCO	88000.26
15-5168	NEAPCO	88001
15-5169	NEAPCO	88000.94
15-5171	NEAPCO	88000.43
15-5172	NEAPCO	88000.47
15-5173	NEAPCO	88000.95
15-5175	NEAPCO	88000.93
15-5176	NEAPCO	88000.86
15-5177	NEAPCO	88000.63
15-5178	NEAPCO	88000.89
15-5180	NEAPCO	88000.66
15-5182	NEAPCO	88000.44
15-5183	NEAPCO	88000.79
15-5184	NEAPCO	88000.81
15-5185	NEAPCO	88000.82
15-5186	NEAPCO	88000.83
15-5187	NEAPCO	88000.51
15-5189	NEAPCO	88000.45
15-5191	NEAPCO	88000.34
15-5193	NEAPCO	88000.46
15-5195	NEAPCO	88000.59
15-5196	NEAPCO	88000.36
15-5197	NEAPCO	88000.35
15-5199	NEAPCO	88000.84
1551P2	DIAMOND R	80178
1551P2	REO MOTORS	80178
1551P2	WHITE	80178
1552	PRECISION	88000.55
15-5213	NEAPCO	88000.25
15-5215	NEAPCO	88000.30
15-5216	NEAPCO	88000.35
15-5217	NEAPCO	88000.37
15-5218	NEAPCO	88000.38
15-5275	NEAPCO	88001.02
155277AS	OLIVER	80216
155-277AS	WHITE FARM	80168
15528J	BORG-WARNR	88006.53
1553	ALCO	80072
1553	ALCO	80077
1553	PRECISION	88000.56
15-5305	NEAPCO	88000.90
15-5306	NEAPCO	88000.22
15-5313	NEAPCO	88000.22
15-5314	NEAPCO	88000.22
15-5315	NEAPCO	88000.27
15-5316	NEAPCO	88000.52
15-5319	NEAPCO	88000.44
15-5324	NEAPCO	88000.29
15-5326	NEAPCO	88000.27
15-5335	NEAPCO	88000.55
15-5336	NEAPCO	88000.19
15-5338	NEAPCO	88000.85
15-5339	NEAPCO	88000.36
15533J	BORG-WARNR	81806
15-5340	NEAPCO	88000.54
15-5341	NEAPCO	88000.98
15-5342	NEAPCO	88000.43
15-5342	NEAPCO	88000.99
15-5343	NEAPCO	88000.61
15-5344	NEAPCO	88000.41
15-5345	NEAPCO	88000.68
15-5346	NEAPCO	88000.70
15-5347	NEAPCO	88000.71
15-5348	NEAPCO	88000.72
15-5349	NEAPCO	88000.73
15534J	BORG-WARNR	88006.92
15-5350	NEAPCO	88000.68
15-5350	NEAPCO	88000.69
15-5354	NEAPCO	88000.74
15-5355	NEAPCO	88000.75
15-5356	NEAPCO	88000.74
15-5359	NEAPCO	88000.71
15-5361	NEAPCO	88000.76
15-5362	NEAPCO	88000.77
15-5363	NEAPCO	88000.45
15-5369	NEAPCO	88000.96
15-5373	NEAPCO	88000.62
15-5374	NEAPCO	88000.56
15-5377	NEAPCO	88001.01
1554	PRECISION	88000.57
1555	ALCO	80182
1555	PRECISION	88000.58
1555	RUEDARSA	80213
1556	PRECISION	88000.59
1556	RUEDARSA	80210
155608	ALLIS-CHLM	80140
1557	PRECISION	88000.60
1557	RUEDARSA	80212
15589J	BORG-WARNR	84702
1559	PRECISION	88000.61
1559481	STUDEBAKER	80044
1559482	STUDEBAKER	80055
15596J	BORG-WARNR	82118
155DS3	OSHKOSH	80182
155N40-461	ROCKWELL	87420
155N4-1891	ROCKWELL	81408
155N4-1901	ROCKWELL	81409
155N4-2041	ROCKWELL	81406
155N4-2051	ROCKWELL	81402
155N4-2431	ROCKWELL	81411
155N4-4101	ROCKWELL	81410
155N4-4211	ROCKWELL	81404
155N4-4241	ROCKWELL	81412
155N53-261	ROCKWELL	87356
155N53-271	ROCKWELL	87361
155N53-61	ROCKWELL	87424
155N53-71	ROCKWELL	87355
155NF2	ROCKWELL	84618
155NF3	ROCKWELL	88005.27
155NLS28-10DC	ROCKWELL	88005.28
155NLS28-3DC	ROCKWELL	85076
155NLS28-6DC	ROCKWELL	85075
155NLS28DC	ROCKWELL	85073
155NLS28DC1	ROCKWELL	85074
155NY53-4	ROCKWELL	85526
155NY53-5	ROCKWELL	85528
155NY62	ROCKWELL	85527
155NY62-1	ROCKWELL	88005.29
155NYS24-1	ROCKWELL	88005.30
155NYS24-1A	ROCKWELL	81403
155NYS24-4A	ROCKWELL	81405
155NYS26	ROCKWELL	88005.31
155NYS26A	ROCKWELL	88005.32
155NYS28-2	ROCKWELL	88005.33
155NYS28-3	ROCKWELL	88005.34
155NYS28-4	ROCKWELL	81407
1560	ALCO	80142
1560	PRECISION	88000.62
15607J	BORG-WARNR	82164
1560U	ALCO	80143
1561999	STUDEBAKER	80031
1562	PRECISION	88000.63
1562019	STUDEBAKER	80075
1562K1	REO MOTORS	80001
1562K1	WHITE	80001
1563	PRECISION	88000.64
1563060	STUDEBAKER	80075
1564	PRECISION	88000.65
1565	ALCO	80190
1565	PRECISION	88000.66
1565240	CLARK EQU.	80007
1565573	CLARK EQU.	80193
1565585	AUST.WEST.	80215

Part No.	Brand	IDLI Group
1565585	CLARK EQU.	80215
1565587	CLARK EQU.	80193
15657J	BORG-WARNR	81691
15658J	BORG-WARNR	81679
1566	PRECISION	88000.67
1566421	AUST.WEST.	80214
1566421	CLARK EQU.	80214
15678	CATERPILLR	80001
15678	CATERPILLR	80003
1570	ALCO	80007
15-70	LEMPCO	80007
15706J	BORG-WARNR	86805
1570944	CLARK EQU.	80224
1570944	COLES	80224
15-70X	LEMPCO	80007
15-70X	LEMPCO	88000.15
15718J	BORG-WARNR	88006.63
15-72X	LEMPCO	80007
1574	ALCO	80090
15753J	BORG-WARNR	88006.89
157714R91	IHC	80044
157734	AMER.MOTOR	80072
157734	AMER.MOTOR	80075
1577604	CLARK EQU.	80214
15777	PILOT	80172
157-794R91	IHC	80044
1578	ALCO	80124
1578-90	WESCO	88013.87
1578U	ALCO	80125
1-5800	NEAPCO	80059
1580982	CLARK EQU.	80224
1581	ALCO	88008.68
158352	HYSTER	80142
15847J	BORG-WARNR	82112
15866J	BORG-WARNR	84649
158746	IHC	80182
158-746H1	IHC	80182
1587A	MOTOR MAST	80001
1587A	ZELLER	80001
1587A	ZELLER	80003
1588	ALCO	80140
158900	IHC	80216
158-900H1	IHC	80216
1-5900	NEAPCO	80081
15927J	BORG-WARNR	82071
159670	CATERPILLR	80198
159707	HYSTER	80172
15973J	BORG-WARNR	82129
15974J	BORG-WARNR	82126
159900	IHC	80168
159900	IHC	80216
159-900H1	IHC	80216
15DS68	OSHKOSH	80178
15P1008	MPL.MOLINE	80001
15P1008	WHITE FARM	80001
15P405	MPL.MOLINE	80037
15P405	WHITE FARM	80037
15P431	MPL.MOLINE	80069
15P431	WHITE FARM	80069
15P623	MPL.MOLINE	80035
15P623	WHITE FARM	80035
15P645	MPL.MOLINE	80104
15P846	MPL.MOLINE	80035
15P846	WHITE FARM	80035
15R108	MPL.MOLINE	80001
15R108	WHITE FARM	80001
15R92	MPL.MOLINE	80133
15R92	WHITE FARM	80133
15R93	MPL.MOLINE	80008
15R93	WHITE FARM	80008
16	AEC	87377
16	AEC	87382
16	AEC	88008.41
160	TRU-CROSS	80142
1600	BAY CITY	80189
1600	BUCKEYE	80189
1600	EIMCO	80189
1600	SPICER	80189
16000	PILOT	80193
16001	PILOT	80194
16001	PILOT	80195
16003	PILOT	80193
16004	PILOT	80194
16006	PILOT	80193
16008J	BORG-WARNR	85556
1600 SERIES	COLES	80189
1601	AEC	80012
1601	JEFFREY	80189
16016	PILOT	80193
1602	AEC	80019
16022	PILOT	80197
16023	PILOT	80199
16024	PILOT	80201
1603	AEC	80010
1603230	CHRYSLER	80094
1603233	CHRYSLER	80096
1603397	CHRYSLER	88003.50
1603398	CHRYSLER	80909
1-60-34-3513X	SPICER	88007.03
1-60-34-3515X	SPICER	88006.99
1-60-34-3519X	BORG-WARNR	88006.99
1-60-34-3519X	HAYES	88006.99
1-60-34-3519X	SPICER	88006.99
1603998	CHRYSLER	80906
1603998	CHRYSLER	80953
1604	AEC	80004
1604	AEC	80006
1604073	GMC	80031
1604073	OPEL	88002.77
1604074	OPEL	80032
1605	AEC	80009
1-605	NEAPCO	88000.14
1-605	NEAPCO	88000.15
1-605-01	AYRA DUREX	80189
1605X	MYERS ELEV	80007
1605X	NEAPCO	80007
1605X	NEAPCO	88000.15
1605X1	NEAPCO	88000.16
1605XBA	HUSCO	88000.16
1606	AEC	80005
1606	AEC	80011
1607	AEC	80225
1608	JEFFREY	80189
1610	AEC	80020
1610	SPICER	80189
16100	PILOT	80193
16101	PILOT	80194
16102	PILOT	80197
16103	PILOT	80194
16104	PILOT	80192
16106	PILOT	80198
16107	PILOT	80202
16108	PILOT	80199
16109	PILOT	80201
161111	NEAPCO	88000.21
161213	NEAPCO	88000.22
161213	NEAPCO	88000.25
161213	NEAPCO	88000.27
161214	NEAPCO	88000.22
161214	NEAPCO	88000.28
161215	NEAPCO	88000.22
16128	PILOT	80198
16128J	BORG-WARNR	88007.20
161294	IHC	80048
161-294H1	IHC	80048
161294HI	HOUGH	80048
161313	NEAPCO	88000.25
16135	PILOT	80191
16143	PILOT	80193
16143	PILOT	80198
16144	PILOT	80197
16150	PILOT	80203
161515	NEAPCO	88000.31
16152	PILOT	80195
161576	NEAPCO	88000.31
161576	NEAPCO	88000.88
1616185	CHRYSLER	88003.45
161818	NEAPCO	88000.38
161850	IHC	80156
161-850H1	IHC	80155
162	AEC	80042
16203	SIDEWINDER	80037
16-20487	CURTISS-WR	80124
16206J	BORG-WARNR	85550
16209J	BORG-WARNR	82077
16235	NW MOTORS	80101
16252	DIVCO	88003.49
16253	DIVCO	88003.45
16254	DIVCO	88003.56
16257J	BORG-WARNR	85864
16258	DIVCO	88003.50
16259	DIVCO	88003.46
16260	DIVCO	88003.42
16262	DIVCO	88003.44
16264	DIVCO	88003.56
16269	DIVCO	80909
16270	DIVCO	80958
16271	DIVCO	88003.44
1-63	AEC	88000.96
1-63	LEMPCO	88000.25
1-63	LEMPCO	88000.27
1-6300	NEAPCO	80062
16301	DIVCO	80056
1-6301	NEAPCO	80061
16-30-32	SPICER	88015
16-30-62	SPICER	88014.99
16316J	BORG-WARNR	81767
16319	DIVCO	80056
163-392	FORD	80133
16343	DIVCO	80077
1635099	CHRYSLER	80957
16362	DIVCO	80124
16364	DIVCO	80072
163966	HYSTER	80216
16417J	BORG-WARNR	82068
16417J	BORG-WARNR	82081
16418J	BORG-WARNR	87477
16418J	BORG-WARNR	88008.01
16430J	BORG-WARNR	81658
1643172	CHRYSLER	80064
1643208	CHRYSLER	80908
1643228	CHRYSLER	80908
1643242	CHRYSLER	80072
1643287	CHRYSLER	80907
1643288	CHRYSLER	80908
164369	AMER.MOTOR	80072
164369	AMER.MOTOR	80075
16453J	BORG-WARNR	81879
16454	DIVCO	80123
165	TRU-CROSS	80190
1650	BAY CITY	80190
1650J	SPICER	80190
165-004-0	GLAENZER	88001.86
1651	JEFFREY	80190
1651	TEL-E-LECT	80133
16515J	BORG-WARNR	88006.41
1652	TEL-E-LECT	80008
1652	TEL-E-LECT	80133
16523J	BORG-WARNR	82170
1653469	CHRYSLER	80061
1653469	CHRYSLER	88001.55
16542J	BORG-WARNR	85542
16544	DIVCO	80081
16545J	BORG-WARNR	87408
1657	BRADY	80037
1657	BRANDY	80037
165757	NEAPCO	81020
1658	JEFFREY	80190
165927	HYSTER	80195
165DS4	OSHKOSH	80190
1661	ALCO	88002.87
16-6101	NEAPCO	84504
16-6111	NEAPCO	88014.52
16-6112	NEAPCO	81034
16-6113	NEAPCO	81054
16-6114	NEAPCO	81089
16-6115	NEAPCO	81109
16-6116	NEAPCO	81153
16-6118	NEAPCO	81175
16-6129	NEAPCO	81268
16-6131	NEAPCO	88001.11
16-6134	NEAPCO	81143
16-6138	NEAPCO	81156
16-6171	NEAPCO	81025
16-6191	NEAPCO	81136
16-6193	NEAPCO	81125
16-6196	NEAPCO	81151
1662	ALCO	88002.86
16-6213	NEAPCO	81070
16-6215	NEAPCO	81115
16-6216	NEAPCO	81151
16-6216	NEAPCO	81153
16-6217	NEAPCO	81169
16-6217	NEAPCO	88001.09
16-6218	NEAPCO	81179
16-6301	NEAPCO	80061
16-6302	NEAPCO	88001.10
16-6303	NEAPCO	81060
16635J	BORG-WARNR	81836
16637J	BORG-WARNR	81846
1663975	CHRYSLER	87382
1663979	CHRYSLER	87382
1663979	CHRYSLER	88008.64
1663981	CHRYSLER	87382
16657J	BORG-WARNR	82076
166686	HYSTER	80225
1666X	NEAPCO	88001.15
1667743	CHRYSLER	87458
1667747	CHRYSLER	87382
1667747	CHRYSLER	88008.65
1667748	CHRYSLER	87461
1667749	CHRYSLER	87461
16720J	BORG-WARNR	88006.44
16728J	BORG-WARNR	88007.94
16728J	BORG-WARNR	88008.23
16745J	BORG-WARNR	88015.25
16745J10800	BORG-WARNR	87666
16745J7200	BORG-WARNR	87665
1-675	NEAPCO	88000.14
1-675	NEAPCO	88000.15

Part No.	Brand	IDLI Group	Part No.	Brand	IDLI Group	Part No.	Brand	IDLI Group	Part No.	Brand	IDLI Group
16752J	BORG-WARNR	85596	16N4-3611X	ROCKWELL	83526	1700537	SKF	88008.39	1710	ALLOY	80159
16757J	BORG-WARNR	87475	16N4-3621	ROCKWELL	83527	1700537A	SKF	87331	1710	SPICER	80221
16757J	BORG-WARNR	88008.02	16N4-3681	ROCKWELL	83544	1700538	SKF	88008.41	17100	PILOT	80210
1675X	NEAPCO	88000.15	16N4-3861	ROCKWELL	83531	1700539	SKF	88008.31	17101	PILOT	80211
1675X1	NEAPCO	88000.16	16N4-3901	ROCKWELL	83529	1700539	SKF	88008.44	17102	PILOT	80212
1675X1A	NEAPCO	88000.15	16N4-4401	ROCKWELL	83549	1700547	SKF	87326	17105	PILOT	80213
1-676	NEAPCO	88000.16	16N4-4441X	ROCKWELL	83531	1700547A	SKF	87326	1711	ALCO	80101
16776J	BORG-WARNR	82383	16N4-4451X	ROCKWELL	83532	1700548	SKF	88008.42	1711	ALLOY	80101
16787J	BORG-WARNR	85583	16N4-4471	ROCKWELL	83545	1700549	SKF	87459	1711	PRECISION	88000.77
1680	BRADY	80069	16N4-4891X	ROCKWELL	83540	1700549	SKF	88008.45	1712	ALCO	80112
1680	BRANDY	80069	16N4-4901X	ROCKWELL	83541	1700810	SKF	87481	1712	ALLOY	80112
16821J	BORG-WARNR	87584	16N4-5031	ROCKWELL	83565	1700810	SKF	88007.77	17123J	BORG-WARNR	85877
16822J	BORG-WARNR	87585	16N4-5061	ROCKWELL	88005.39	17-009	STEIGER	80221	17126	PILOT	80213
16835J	BORG-WARNR	81713	16N4-5071X	ROCKWELL	83536	1700912	SKF	88007.79	171333	FORD	80001
16855J	BORG-WARNR	81718	16N4-5081X	ROCKWELL	83537	1700945	SKF	88007.78	1714-010	TAYLOR	80037
1687069	STUDEBAKER	80081	16N4-5111X	ROCKWELL	83525	1700946	SKF	87373	171407271	VOLKSWAGEN	86004
1687083	STUDEBAKER	80123	16N4-5131X	ROCKWELL	83547	1700946	SKF	88008.43	171407271G	VOLKSWAGEN	86006
1687251	STUDEBAKER	80124	16N4-5281X	ROCKWELL	83544	1700947	SKF	87325	171407271R	VOLKSWAGEN	86004
1687253	STUDEBAKER	80142	16N4-5401	ROCKWELL	83571	1700947	SKF	88007.59	171407272	VOLKSWAGEN	86005
1687692	STUDEBAKER	80077	16N4-5411X	ROCKWELL	83571	1700P2	DIAMOND R	80142	171407272G	VOLKSWAGEN	86007
168778	FORD	80008	16N4-5421	ROCKWELL	83551	1700P2	WHITE	80142	171407272M	VOLKSWAGEN	86005
1690409	STUDEBAKER	80075	16N4-5431X	ROCKWELL	83552	1700 SERIES	COLES	80221	171407282B	VOLKSWAGEN	86073
1691761	STUDEBAKER	80081	16N4-5441	ROCKWELL	83557	1701	PRECISION	88000.69	171407282B	VOLKSWAGEN	86146
1691762	STUDEBAKER	80123	16N4-5451X	ROCKWELL	83558	1701-101X	SPICER	88005.42	171407311	VOLKSWAGEN	86123
1692691	CHRYSLER	80096	16N4-5571	ROCKWELL	83533	1701-104X	SPICER	88005.43	171407311D	AUDI	86122
16946J	BORG-WARNR	88006.47	16N4-5631	ROCKWELL	83570	1701-108X	SPICER	88005.44	171-407-311D	VOLKSWAGEN	86134
16953J	BORG-WARNR	82054	16N4-5741	ROCKWELL	83572	1701-113X	SPICER	88005.45	171407331C	VOLKSWAGEN	86073
16954J	BORG-WARNR	84687	16N4-6241	ROCKWELL	83561	1701-116X	SPICER	88005.46	171407331D	VOLKSWAGEN	86122
1695947	STUDEBAKER	80140	16N4-6251	ROCKWELL	83538	1701-118X	SPICER	88005.47	171407331F	VOLKSWAGEN	86072
1695P2	DIAMOND R	80178	16N4-6261	ROCKWELL	83535	17015J	BORG-WARNR	87525	1714205	FMC	80001
1695P2	REO MOTORS	80178	16N4-6271	ROCKWELL	83555	17016J	BORG-WARNR	87530	1714207	BEAN,JOHN	80001
1695P2	WHITE	80178	16N4-6281X	ROCKWELL	83556	1701-9101X	SPICER	88005.55	1714207	FMC	80001
16974J	BORG-WARNR	85123	16N4-6291X	ROCKWELL	83562	1702	ALCO	80146	1714207	SPEED-SPR.	80001
1698	ALLOY	80156	16N4-6351	ROCKWELL	88005.40	1702	ALLOY	80146	171-498-099B	VOLKSWAGEN	86181
1698021	STUDEBAKER	80075	16N4-781	ROCKWELL	83524	1702	PRECISION	88000.70	171-498-103	VOLKSWAGEN	86072
1699	ALLOY	80097	16N53-141	ROCKWELL	87416	17-020	STEIGER	80182	171-498-103	VOLKSWAGEN	86143
16991J	BORG-WARNR	85093	16NYS28-11	ROCKWELL	83530	1703	ALCO	80109	171-498-103C	VOLKSWAGEN	86099
16A15015	EIMCO	80198	16NYS28-12	ROCKWELL	83528	1703	ALLOY	80109	171-498-103C	VOLKSWAGEN	86143
16B6050	FEDERAL	80008	16NYS28-13A	ROCKWELL	83534	1703	ALLOY	80149	171-498-203	AUDI	86181
16B6050	FEDERAL	88002.88	16NYS28-34A	ROCKWELL	83539	1703	PRECISION	88000.71	171-498-203	VOLKSWAGEN	86181
16DS150	OSHKOSH	80189	16NYS31-3	ROCKWELL	83546	1703	RUEDARSA	80092	171-498-99B	VOLKSWAGEN	86125
16J53(4)	LINK BELT	80214	16NYS32-13	ROCKWELL	83563	17034	PILOT	80212	17150J	BORG-WARNR	88006.40
16N14-19	ROCKWELL	88005.35	16NYS32-15	ROCKWELL	83548	17035	PILOT	80210	171517	WHITE	80007
16N15-23	ROCKWELL	88005.36	16NYS32-31A1	ROCKWELL	83554	1704	ALCO	80153	17-156	STEIGER	80237
16N16-23	ROCKWELL	88005.37	16NYS32-32A	ROCKWELL	83553	1704	ALLOY	80153	17164P11	M.R.S.	80218
16N28-167	ROCKWELL	85723	16NYS32-33A	ROCKWELL	83559	1704	PRECISION	88000.72	17-175	STEIGER	80197
16N28-207	ROCKWELL	85721	16NYS32-33A1	ROCKWELL	83560	1705	ALCO	80158	1717517	WHITE	80007
16N28-627	ROCKWELL	85722	16NYS32-34	ROCKWELL	83550	1705	ALLOY	80158	1717518	WHITE	80007
16N3-108X	ROCKWELL	85312	16NYS34	ROCKWELL	83564	1705	PRECISION	88000.73	1717518	WHITE	88000.15
16N3-128X	ROCKWELL	85314	16NYS36-1	ROCKWELL	83567	1-705-01	AYRA DUREX	80221	17184J	BORG-WARNR	81786
16N3-1411X	ROCKWELL	85315	16NYS36-2	ROCKWELL	83566	170521	FORD	80033	17188	NEAPCO	81315
16N3-2261X	ROCKWELL	85316	16NYS36-3	ROCKWELL	83568	1705339	JOY MFG.	80198	17195C	DIAMOND T	80178
16N3-288X	ROCKWELL	85313	16NYS38-2	ROCKWELL	83572	17-056	STEIGER	80190	17199J	DIAMOND T	80221
16N40-1011	ROCKWELL	87505	16NYS38-3	ROCKWELL	83570	1705-91	OSHKOSH	80221	17199W	DIAMOND T	80142
16N40-471	ROCKWELL	87523	17	AEC	87383	1706	ALLOY	80156	171A6	NEAPCO	83090
16N40-481	ROCKWELL	87497	17	AEC	87384	1706	PRECISION	88000.74	171B8	NEAPCO	83049
16N40-501	ROCKWELL	87526	17	AEC	88008.42	1707	PRECISION	88000.75	171C2	NEAPCO	83082
16N40-511	ROCKWELL	87499	170	TRU-CROSS	80007	17070J	BORG-WARNR	85094	171C8	NEAPCO	83129
16N40-671	ROCKWELL	87503	170	TRU-CROSS	88000.15	170724A1	NEW HOLL.	80003	1720	ALCO	80168
16N40-741	ROCKWELL	87504	1700	ALCO	80051	17072J	BORG-WARNR	87428	1720	ALLOY	80168
16N4-1371	ROCKWELL	83542	1700	ALLOY	80051	17072J	BORG-WARNR	88006.22	17-201	STEIGER	80212
16N4-1631	ROCKWELL	83547	1700	BAY CITY	80221	1708	ALCO	80151	17205	PILOT	80210
16N4-1721	ROCKWELL	83523	1700	BUCKEYE	80221	1708	ALLOY	80151	17206	PILOT	80211
16N4-1931	ROCKWELL	83526	1700	EIMCO	80196	1708	PRECISION	88000.76	17206	PILOT	80212
16N4-2081	ROCKWELL	83525	1700	PRECISION	88000.68	1709	ALCO	80148	17207	PILOT	80212
16N4-2401	ROCKWELL	83522	1700	SPICER	80221	1709	ALLOY	80148	17208	PILOT	80208
16N4-2401	ROCKWELL	88005.38	17000	PILOT	80210	1709P2	DIAMOND R	80142	17208	PILOT	80210
16N4-2501	ROCKWELL	83543	17004	PILOT	80211	1709P2	REO MOTORS	80142	17209	PILOT	80196
16N4-3321	ROCKWELL	83569	1700536	SKF	88007.58	1709P2	WHITE	80142	17209	PILOT	80209
16N4-3601X	ROCKWELL	83526	1700537	SKF	87331	1710	ALCO	80159	1721	ALCO	80167

Part No.	Brand	IDLI Group	Part No.	Brand	IDLI Group	Part No.	Brand	IDLI Group	Part No.	Brand	IDLI Group
1721	ALLOY	80167	1740	ALLOY	80135	1762	MOTOR MAST	80030	17817J	BORG-WARNR	82169
17218J	BORG-WARNR	88006.39	1740	PRECISION	88000.78	1762	PRECISION	88000.91	178183	ALLIS-CHLM	80231
1722	ALCO	80171	17422	PILOT	80060	176297	IHC	80075	178183	ALLIS-CHLM	80237
1722	ALLOY	80171	17422	PILOT	80061	176-297H1	IHC	80124	178183	FIAT-ALLIS	80240
17220	PILOT	80022	17423	PILOT	80094	176297HI	HOUGH	80124	1782	ALCO	88000.14
17221	PILOT	80034	17423	PILOT	80096	1763	ALCO	80141	1782	ALLOY	88000.14
17222	PILOT	80057	17424	PILOT	80094	1763	ALLOY	80141	1783	ALCO	88000.15
17223	PILOT	80060	1742624	CHRYSLER	80061	1763	PRECISION	88000.92	1783	ALLOY	88000.15
17223	PILOT	80061	1743159	CHRYSLER	80953	1764	ALCO	80141	178323	ALLIS-CHLM	80232
17224	PILOT	80064	17441J	BORG-WARNR	85553	1764	ALLOY	80141	178323	ALLIS-CHLM	80237
17225	PILOT	80064	1747	ALLOY	88001.28	1764	PRECISION	88000.93	178323	FIAT-ALLIS	80242
17226	PILOT	80056	174980C91	IHC	80069	1765	PRECISION	88000.94	178323-2	ALLIS-CHLM	80232
17226	PILOT	80057	1750	AEC	80030	1766	PRECISION	88000.95	178323-2	ALLIS-CHLM	80237
17227	PILOT	80056	1750	ALCO	80030	1767	PRECISION	88000.96	178323-2	FIAT-ALLIS	80242
17228	PILOT	80064	1750	ALCO	80142	17673J	BORG-WARNR	81654	178363	ALLIS-CHLM	80238
17228	PILOT	80065	1750	ALCO	88001.95	176771	WABCO	80168	178363	FIAT-ALLIS	80238
17229	PILOT	80060	1750	ALCO	88001.99	1768	PRECISION	88000.97	178368	ALLIS-CHLM	80238
17229	PILOT	80061	1750	ALLOY	80142	1769	PRECISION	88000.98	178368	FIAT-ALLIS	80238
1723	ALCO	80174	1750	PRECISION	88000.79	17691	GERLINGER	80047	178368-7	ALLIS-CHLM	80238
1723	ALLOY	80174	175045	IHC	80213	176N28-17	ROCKWELL	85734	178368-7	FIAT-ALLIS	80238
17230	PILOT	80060	175-045H1	IHC	80213	176N3-21X	ROCKWELL	85340	1784	ALCO	80044
17230	PILOT	80063	175045HI	HOUGH	80213	176N4-11	ROCKWELL	83715	1784	ALLOY	80044
17231	PILOT	80132	1750G	AEC	80031	176N4-121	ROCKWELL	83706	178461	ALLIS-CHLM	80048
17232	PILOT	80132	1751	ALCO	80125	176N4-151	ROCKWELL	83714	178461	ALLIS-CHLM	80049
17233	PILOT	80062	1751	ALLOY	80125	176N4-161	ROCKWELL	83711	178461	ALLIS-CHLM	88001.96
17233J	BORG-WARNR	87423	1751	PRECISION	88000.80	176N4-231X	ROCKWELL	88005.41	178461	FIAT-ALLIS	80048
172-3522	G & G MFG	83312	175-10-00191(4)	KOMATSU	88002.85	176N4-271	ROCKWELL	83713	1785	ALCO	80161
1724	ALCO	80171	1752	ALCO	80143	176N4-321	ROCKWELL	83707	1785	ALLOY	80161
1724	ALLOY	80171	1752	ALLOY	80143	176N4-391	ROCKWELL	83712	178544	ALLIS-CHLM	80049
17-261	STEIGER	80212	1752	PRECISION	88000.81	176N4-411	ROCKWELL	83708	178544	ALLIS-CHLM	88001.96
17-267	STEIGER	80197	175-20-00060	KOMATSU	80224	176N4-421	ROCKWELL	83703	178544	FIAT-ALLIS	88003.26
17272J	BORG-WARNR	88006.56	17522J	BORG-WARNR	82219	176N4-501	ROCKWELL	83709	178544-3	ALLIS-CHLM	80049
1728	ALCO	80172	1752307	CHRYSLER	80140	176N4-531	ROCKWELL	83704	178544-3	FIAT-ALLIS	88003.26
1728	ALLOY	80172	1752317	CHRYSLER	80189	176N4-541X	ROCKWELL	83705	178554	ALLIS-CHLM	80049
17280J	BORG-WARNR	85874	1752325	CHRYSLER	80178	176N4-91	ROCKWELL	83710	178554	ALLIS-CHLM	88001.96
1729	ALCO	80162	1752412	CHRYSLER	80142	1770	PRECISION	88000.99	178554	FIAT-ALLIS	80102
1729	ALLOY	80162	1752559	CHRYSLER	80123	1771	PRECISION	88001	1786	ALLOY	80007
17298	SIDEWINDER	80133	1752623	CHRYSLER	80064	177-1200	G & G MFG	81271	1786	ALLOY	88000.15
1-73	LEMPCO	88000.28	1752624	CHRYSLER	80060	177-1400	G & G MFG	83038	178746	IHC	80182
1730	ALCO	80173	1752680	CHRYSLER	80905	1772	ALLOY	80008	178823	FIAT-ALLIS	80232
1730	ALLOY	80173	1753	PRECISION	88000.82	1772	PRECISION	88001.01	1788287	GMC	80134
17300J	BORG-WARNR	81713	1754	PRECISION	88000.83	1773	ALLOY	80133	17898J	BORG-WARNR	85567
17-30-12	SPICER	88014.98	175419	JOY MFG.	80237	1773	PRECISION	88001.02	1790K1	WHITE	80001
1730J	BORG-WARNR	85605	175419	JOY MFG.	80239	177-3500	G & G MFG	83258	179151	FIAT-ALLIS	80224
1731	ALCO	80175	175-423	FORD	80001	1774	ALLOY	80136	1792216	FORD	80023
1731	ALLOY	80175	175423	FORD	80037	1774	PRECISION	88001.03	1792216	FORD	80024
17310J	BORG-WARNR	81748	1755	AEC	80032	177428	IHC	80231	179502	ALLIS-CHLM	80168
1732	ALCO	80175	1755	ALCO	80134	177428	IHC	80240	179502	FIAT-ALLIS	80216
1732	ALLOY	80175	1755	PRECISION	88000.84	177-428H1	IHC	80240	179504	ALLIS-CHLM	80224
17320J	BORG-WARNR	85605	1755G	AEC	80032	177-4400	G & G MFG	81356	179504	FIAT-ALLIS	80224
17321	PILOT	80034	1756	ALCO	80140	177-482H1	IHC	80243	179507	ALLIS-CHLM	80224
17324	PILOT	80095	1756	ALCO	80182	1775	ALLOY	80134	179507	FIAT-ALLIS	80224
17325	PILOT	80096	1756	ALLOY	80140	1775	PRECISION	88001.04	179571	ALLIS-CHLM	80103
17326	PILOT	80095	1756	PRECISION	88000.85	1776	ALCO	80188	179571	FIAT-ALLIS	80103
17327	PILOT	80135	175-60-11520	KOMATSU	80198	1776	ALLOY	80188	1795P2	REO MOTORS	80072
17328	PILOT	80042	1757	ALCO	80178	1776	PRECISION	88001.05	1795P2	WHITE	80075
17329	PILOT	80096	1757	ALLOY	80178	1777	ALCO	80001	1795PU	DIAMOND R	80076
17329	PILOT	80135	1757	PRECISION	88000.86	1777	ALLOY	80001	1796	ENGLER	80037
17330	PILOT	80135	1758	ALCO	80079	1777	ALLOY	80003	1796	TERRAIN	80037
1734158	CHRYSLER	80953	1758	ALLOY	80079	1778	ALCO	80124	179704	ALLIS-CHLM	80172
1734159	CHRYSLER	88003.58	1758	PRECISION	88000.87	1778	ALLOY	80001	179704	ALLIS-CHLM	80224
17343J	BORG-WARNR	82355	1759	PRECISION	88000.88	1778	ALLOY	80003	179704	FIAT-ALLIS	80224
1735-24704	ALFA ROMEO	80180	1-76	LEMPCO	88000.29	1779	ALCO	80035	179-867H1	IHC	80161
1735-25704	ALFA ROMEO	80189	1760	PRECISION	88000.89	1779	ALLOY	80035	179904	ALLIS-CHLM	80224
1739	ALCO	80135	1760	SPICER	80226	177S	ALCO	80134	179904	ALLIS-CHLM	80231
1739	ALLOY	80135	1760487	COLES	80001	178	TRU-CROSS	80124	179904	FIAT-ALLIS	80224
1739920	CHRYSLER	88003.57	1761	PRECISION	88000.90	1780	ALCO	80037	179909	ALLIS-CHLM	80172
1739927	CHRYSLER	88003.43	17-61-004	LILLISTON	80035	1780	ALLOY	80037	179909	ALLIS-CHLM	80224
173BFA	CASE,J.I.	80133	1762	ALCO	80145	1781	ALCO	80069	179909	FIAT-ALLIS	80224
1740	ALCO	80135	1762	ALLOY	80145	1781	ALLOY	80069	17A2250(4)	LINK BELT	80174

Part No.	Brand	IDLI Group
17A2251(4)	LINK BELT	80174
17A2252(4)	LINK BELT	80174
17A2253(4)	LINK BELT	80174
17A2371(4)	LINK BELT	80149
17DS91	OSHKOSH	80221
17H3835	BMC	80180
17N1-101X	ROCKWELL	88005.42
17N1-104X	ROCKWELL	88005.43
17N1-108X	ROCKWELL	88005.44
17N1-113X	ROCKWELL	88005.45
17N1-116X	ROCKWELL	88005.46
17N1-118X	ROCKWELL	88005.47
17N14-29	ROCKWELL	88005.48
17N15-33	ROCKWELL	88005.49
17N15-73	ROCKWELL	88005.50
17N15-93	ROCKWELL	88005.51
17N16-123	ROCKWELL	88005.52
17N16-53	ROCKWELL	88005.53
17N16-93	ROCKWELL	88005.54
17N1-9101X	ROCKWELL	88005.55
17N1971	ROCKWELL	88005.56
17N28-137	ROCKWELL	85724
17N28-277	ROCKWELL	85729
17N28-347	ROCKWELL	85727
17N28-367	ROCKWELL	85733
17N3-1461X	ROCKWELL	85321
17N3-1501	ROCKWELL	88005.57
17N3-1501X	ROCKWELL	85333
17N3-1521	ROCKWELL	88005.58
17N3-1521X	ROCKWELL	85317
17N3-1541X	ROCKWELL	85326
17N3-2241	ROCKWELL	85322
17N3-2241X	ROCKWELL	85322
17N3-2261	ROCKWELL	88005.59
17N3-2261X	ROCKWELL	85334
17N3-2341	ROCKWELL	88005.60
17N3-2341X	ROCKWELL	85327
17N3-2401X	ROCKWELL	85339
17N3-2651X	ROCKWELL	85335
17N3-2671X	ROCKWELL	85328
17N3-2761X	ROCKWELL	85318
17N3-3021X	ROCKWELL	85330
17N40-161	ROCKWELL	87580
17N40-181	ROCKWELL	87656
17N40-311	ROCKWELL	87601
17N40-401	ROCKWELL	87591
17N40-451	ROCKWELL	87596
17N40-461	ROCKWELL	88005.61
17N40-51	ROCKWELL	87579
17N40-521	ROCKWELL	87598
17N40-541	ROCKWELL	87592
17N40-61	ROCKWELL	87581
17N4-1971	ROCKWELL	83668
17N4-1981	ROCKWELL	83607
17N4-1991	ROCKWELL	83673
17N4-2021	ROCKWELL	83612
17N4-2081	ROCKWELL	83573
17N42-108	ROCKWELL	88005.62
17N42-1081	ROCKWELL	87575
17N42-1081	ROCKWELL	87578
17N4-2131	ROCKWELL	83621
17N4-2141	ROCKWELL	83670
17N4-2151	ROCKWELL	83667
17N4-2161	ROCKWELL	83605
17N4-2181	ROCKWELL	83661
17N4-2191	ROCKWELL	83632
17N4-2211	ROCKWELL	83602
17N4-2241	ROCKWELL	83618
17N4-2271	ROCKWELL	83597
17N4-2361	ROCKWELL	83629
17N4-2371	ROCKWELL	83623
17N4-2381	ROCKWELL	83576
17N4-2391	ROCKWELL	83609
17N4-2441	ROCKWELL	83666
17N4-2501	ROCKWELL	83669
17N4-2511	ROCKWELL	83665
17N4-3031	ROCKWELL	83671
17N4-3121	ROCKWELL	83603
17N4-3141	ROCKWELL	83701
17N4-3171	ROCKWELL	83622
17N4-3181	ROCKWELL	83601
17N4-3191	ROCKWELL	83610
17N4-3221	ROCKWELL	83617
17N4-3231	ROCKWELL	88005.63
17N4-3241	ROCKWELL	83626
17N4-3271	ROCKWELL	83624
17N4-3291	ROCKWELL	83604
17N4-3321	ROCKWELL	88005.64
17N4-3361	ROCKWELL	83598
17N4-3371	ROCKWELL	83600
17N4-3391	ROCKWELL	83577
17N4-3481	ROCKWELL	83614
17N4-3501	ROCKWELL	83677
17N4-3611	ROCKWELL	83695
17N4-3621	ROCKWELL	88005.65
17N4-3721	ROCKWELL	83608
17N4-3791	ROCKWELL	83638
17N4-3901	ROCKWELL	88005.66
17N4-4091	ROCKWELL	83583
17N4-4131	ROCKWELL	83620
17N4-4141	ROCKWELL	83582
17N4-4151	ROCKWELL	83619
17N4-4181X	ROCKWELL	83606
17N4-4191X	ROCKWELL	83674
17N4-4291X	ROCKWELL	83627
17N4-4311	ROCKWELL	83578
17N4-4321	ROCKWELL	83615
17N4-4341	ROCKWELL	83664
17N4-4361X	ROCKWELL	83616
17N4-4391X	ROCKWELL	83625
17N4-4411	ROCKWELL	88005.67
17N4-4411X	ROCKWELL	83675
17N4-4421	ROCKWELL	83693
17N4-4551	ROCKWELL	83700
17N4-4561	ROCKWELL	83682
17N4-4581	ROCKWELL	83691
17N4-4591	ROCKWELL	83663
17N4-4601	ROCKWELL	83678
17N4-4721	ROCKWELL	83599
17N4-4771	ROCKWELL	83683
17N4-4881	ROCKWELL	88005.68
17N4-4911X	ROCKWELL	83593
17N4-4961	ROCKWELL	83660
17N4-4971	ROCKWELL	88005.69
17N4-4981X	ROCKWELL	83653
17N4-5031	ROCKWELL	83595
17N4-5041	ROCKWELL	83596
17N4-5041X	ROCKWELL	88005.70
17N4-5051X	ROCKWELL	83643
17N4-5101X	ROCKWELL	83644
17N4-5111X	ROCKWELL	83594
17N4-5121X	ROCKWELL	83587
17N4-5131X	ROCKWELL	83588
17N4-5141	ROCKWELL	83611
17N4-5351	ROCKWELL	83698
17N4-5451	ROCKWELL	83681
17N4-5461X	ROCKWELL	88005.71
17N4-5501	ROCKWELL	83654
17N4-5511X	ROCKWELL	83655
17N4-5521	ROCKWELL	83646
17N4-5531X	ROCKWELL	83647
17N4-5541	ROCKWELL	83637
17N4-5551X	ROCKWELL	83636
17N4-5591	ROCKWELL	83699
17N4-5661	ROCKWELL	88005.72
17N4-5661X	ROCKWELL	83589
17N4-5711	ROCKWELL	83680
17N4-5791	ROCKWELL	83658
17N4-5821	ROCKWELL	83657
17N4-6321	ROCKWELL	83591
17N4-6331	ROCKWELL	83585
17N4-6371	ROCKWELL	83650
17N4-6381X	ROCKWELL	83651
17N4-6391X	ROCKWELL	83652
17N4-6401	ROCKWELL	83640
17N4-6411X	ROCKWELL	83641
17N4-6421X	ROCKWELL	83642
17N4-6451	ROCKWELL	88005.73
17N4-6481	ROCKWELL	88005.74
17N53-151	ROCKWELL	87482
17N53-201	ROCKWELL	87483
17N53-311	ROCKWELL	87564
17N82-1061-3	ROCKWELL	88005.75
17N82-1091-1	ROCKWELL	87087
17N82-1091-4	ROCKWELL	87095
17N82-1171-12	ROCKWELL	87080
17N82-1171-5	ROCKWELL	88005.76
17N82-1281	ROCKWELL	87092
17N82-971-1X	ROCKWELL	87108
17NLS40-40	ROCKWELL	85323
17NLS40-47	ROCKWELL	85324
17NLS40-49	ROCKWELL	85325
17NLS40-50	ROCKWELL	85319
17NLS40-56	ROCKWELL	85337
17NLS40-72	ROCKWELL	85338
17NLS40-82	ROCKWELL	85332
17NLS40-85	ROCKWELL	85320
17NLS40-86	ROCKWELL	85336
17NLS40-90	ROCKWELL	85329
17NLS40-91	ROCKWELL	85331
17NPS40	ROCKWELL	87589
17NY52	ROCKWELL	85725
17NY58	ROCKWELL	85726
17NY60-3	ROCKWELL	85731
17NY60-7	ROCKWELL	85728
17NY60-8	ROCKWELL	85732
17NY68	ROCKWELL	85730
17NYS28-11	ROCKWELL	83575
17NYS28-19A	ROCKWELL	83584
17NYS28-21	ROCKWELL	83579
17NYS28-21	ROCKWELL	83581
17NYS28-21A1	ROCKWELL	83580
17NYS28-23	ROCKWELL	83574
17NYS28-43A	ROCKWELL	83590
17NYS28-49A	ROCKWELL	83592
17NYS28-5	ROCKWELL	88005.77
17NYS28-50A	ROCKWELL	83586
17NYS31-13A	ROCKWELL	83613
17NYS32-20	ROCKWELL	83628
17NYS32-27A2	ROCKWELL	83645
17NYS32-28	ROCKWELL	83631
17NYS32-30	ROCKWELL	83630
17NYS32-40A1	ROCKWELL	83656
17NYS32-41A1	ROCKWELL	83648
17NYS32-41A2	ROCKWELL	83649
17NYS32-43	ROCKWELL	83635
17NYS32-48	ROCKWELL	83633
17NYS32-52	ROCKWELL	83639
17NYS32-53	ROCKWELL	83634
17NYS34-5	ROCKWELL	83662
17NYS34-9A	ROCKWELL	83659
17NYS36-37	ROCKWELL	83672
17NYS36A1	ROCKWELL	83676
17NYS38-16	ROCKWELL	83687
17NYS38-17	ROCKWELL	83685
17NYS38-17A	ROCKWELL	83686
17NYS38-18	ROCKWELL	83688
17NYS38-18A	ROCKWELL	83689
17NYS38-24	ROCKWELL	83679
17NYS38-9	ROCKWELL	83684
17NYS39-1	ROCKWELL	83690
17NYS39-2	ROCKWELL	83692
17NYS40-106A	ROCKWELL	83696
17NYS40-106A1	ROCKWELL	83697
17NYS40-107	ROCKWELL	83694
17NYS51	ROCKWELL	83702
17NYSM36-28	ROCKWELL	87115
17NYSM40-102	ROCKWELL	87099
17NYSM40-104	ROCKWELL	87098
17NYSM40-33	ROCKWELL	87110
17NYSM40-40	ROCKWELL	87077
17NYSM40-41	ROCKWELL	87085
17NYSM40-42	ROCKWELL	87096
17NYSM40-46	ROCKWELL	87089
17NYSM40-47	ROCKWELL	87101
17NYSM40-50	ROCKWELL	87079
17NYSM40-50	ROCKWELL	87084
17NYSM40-54	ROCKWELL	87100
17NYSM40-55	ROCKWELL	87081
17NYSM40-56	ROCKWELL	87109
17NYSM40-58	ROCKWELL	87086
17NYSM40-59	ROCKWELL	87088
17NYSM40-60	ROCKWELL	87093
17NYSM40-61	ROCKWELL	87097
17NYSM40-62	ROCKWELL	87102
17NYSM40-63	ROCKWELL	87106
17NYSM40-64	ROCKWELL	87078
17NYSM40-68	ROCKWELL	87107
17NYSM40-71	ROCKWELL	87090
17NYSM40-80	ROCKWELL	87104
17NYSM40-81	ROCKWELL	87083
17NYSM40-82	ROCKWELL	87113
17NYSM40-85	ROCKWELL	87082
17NYSM40-86	ROCKWELL	87094
17NYSM40-88	ROCKWELL	87091
17NYSM40-89	ROCKWELL	87076
17NYSM40-91	ROCKWELL	87112
17NYSM40-92	ROCKWELL	87111
17NYSM40-94	ROCKWELL	87114
17NYSM40-95	ROCKWELL	87105
17NYSM40-96	ROCKWELL	87103
18	AEC	87457
18	AEC	87461
18	AEC	88007.76
18	AEC	88008.31
18	AEC	88008.44
18	WYLIE	80024
1800	EIMCO	80225
1800	PRECISION	83240
1800	SPICER	80225
18000	PILOT	80231
18000	PILOT	80240
18002	PILOT	80231
18003	PILOT	80232
1801	ALCO	80190
1801	ALLOY	80190
1801	JEFFREY	80225
1801	PRECISION	83251
180-1200	G & G MFG	81346
180131	ALLIS-CHLM	80213

Part No.	Brand	IDLI Group
180132	ALLIS-CHLM	80213
180132	FIAT-ALLIS	80213
180132-3	ALLIS-CHLM	80213
180132-3	FIAT-ALLIS	80213
180140	ALLIS-CHLM	80224
180140	FIAT-ALLIS	80224
180140-6	FIAT-ALLIS	80224
1802	ALCO	80221
1802	ALLOY	80221
1802	PRECISION	83249
180216	IHC	80223
180-216H2	IHC	80224
18025	PILOT	80231
1803	ALCO	80221
1803	ALLOY	80221
1803	PRECISION	83262
1804	ALCO	80225
1804	ALLOY	80225
1804	PRECISION	83264
180 410 03 31	MERCEDES	88002.77
1805	ALCO	80182
1805	ALLOY	80182
1805	PRECISION	83280
1-805-01	AYRA DUREX	80225
18059J	BORG-WARNR	81864
1805X	NEAPCO	80044
1805X1	NEAPCO	80043
1805X2	NEAPCO	80044
1806	ALCO	80189
1806	ALLOY	80189
1806	PRECISION	83279
1806	PRECISION	88005.99
1807	ALCO	80189
1807	ALLOY	80189
1807	PRECISION	83281
1808	JEFFREY	80225
1808	PRECISION	88014.40
1809	PRECISION	83316
1-81	AEC	88001
1-81	LEMPCO	88000.49
1810	ALCO	80227
1810	ALLOY	80227
1810	PRECISION	83317
1810	PRECISION	88014.41
1810	RUEDARSA	80107
1810	SPICER	80227
18100	PILOT	80231
18101	PILOT	80231
18101	PILOT	80232
18102	PILOT	80237
18103	PILOT	80236
18105	PILOT	80238
1811	PRECISION	83331
1811	PRECISION	83334
1811	RUEDARSA	80147
18113	PILOT	80238
18114J	BORG-WARNR	85596
18114J	BORG-WARNR	88006.84
1812	PRECISION	83349
1812	PRECISION	83353
1812	PRECISION	83354
1812	RUEDARSA	80146
1813	PRECISION	83364
1813	RUEDARSA	80158
1814	RUEDARSA	80148
181-4400	G & G MFG	85516
1815	RUEDARSA	80159
181-5500	G & G MFG	85720
1816	RUEDARSA	80177
18160J	BORG-WARNR	85543
1818779	CHRYSLER	80094
1818920	CHRYSLER	80072
1818921	CHRYSLER	80072
181SJ18	MACK	80140
1820	ALCO	80193
1820	ALLOY	80193
1820	PRECISION	83241
18205	PILOT	80231
18206	PILOT	80232
182-0606	G & G MFG	88000.67
182-0621	G & G MFG	88014.06
18207	PILOT	80237
18207	PILOT	80239
18208	PILOT	80236
18208	PILOT	88002.64
18209	PILOT	80238
1821	ALCO	80195
1821	ALLOY	80195
182-1206	G & G MFG	84510
182-1221	G & G MFG	84508
182-1222	G & G MFG	81342
182-1406	G & G MFG	88014.32
182-1421	G & G MFG	84517
1822	ALCO	80197
1822	ALLOY	80197
1-82-22-3515X	SPICER	88007.02
1-82-22-3715X	BORG-WARNR	88006.98
1-82-22-3715X	HAYES	81209
1-82-22-3715X	SPICER	81209
18-228	NEAPCO	85114
1823	ALCO	80194
1823	ALLOY	80194
182-3506	G & G MFG	84531
182-3521	G & G MFG	84529
1824	ALCO	80192
1824	ALLOY	80192
18-240	NEAPCO	85114
182-4406	G & G MFG	81362
182-4421	G & G MFG	84515
1825	PRECISION	83252
182-5506	G & G MFG	84541
182-5521	G & G MFG	84538
1826	ALCO	80198
1826	ALLOY	80198
1826	PRECISION	83261
1826	PRECISION	83263
1827	ALCO	80202
1827	ALLOY	80202
1827	PRECISION	83275
1827	PRECISION	83283
1828	ALCO	80199
1828	ALLOY	80199
1828	PRECISION	83296
1-82-82-3715X	SPICER	88006.98
1829	ALCO	80201
1829	ALLOY	80201
1829	PRECISION	83357
1-83	AEC	88000.82
1-83	LEMPCO	88000.53
1830	PRECISION	83362
183-0616	G & G MFG	81125
183-1216	G & G MFG	81246
18-32-22	SPICER	88015.01
18-328	NEAPCO	85102
1834-3A	NEAPCO	88015.39
1835	PRECISION	83250
1835	PRECISION	88014.36
183515	MARATHON	80240
183516	MARATHON	80239
18-352	NEAPCO	88001.12
1836	PRECISION	88014.38
18386J	BORG-WARNR	81800
18395J	BORG-WARNR	81733
1840	ALCO	80213
1840	ALLOY	80213
1840	PRECISION	83248
184-0512	G & G MFG	88000.21
184-0513	G & G MFG	88000.22
184-0514	G & G MFG	88000.25
184-0514	G & G MFG	88000.27
184-0514	G & G MFG	88000.52
184-0516	G & G MFG	88000.30
184-0516	G & G MFG	88000.31
184-0516	G & G MFG	88000.32
184-0516	G & G MFG	88000.53
184-0518	G & G MFG	88000.34
184-0518	G & G MFG	88000.35
184-0518	G & G MFG	88000.36
184-0520	G & G MFG	88000.39
184-0520	G & G MFG	88000.56
184-0521	G & G MFG	88000.55
184-0612	G & G MFG	88000.20
184-0612	G & G MFG	88000.21
184-0614	G & G MFG	88000.52
184-0615	G & G MFG	88000.28
184-0616	G & G MFG	88000.30
184-0616	G & G MFG	88000.31
184-0618	G & G MFG	88000.36
184-0620	G & G MFG	88000.55
1841	ALCO	80210
1841	ALLOY	80210
1841	PRECISION	83277
184-1216	G & G MFG	81237
184-1216	G & G MFG	88014.01
184-1218	G & G MFG	88000.04
184-1220	G & G MFG	88000.06
184-1222	G & G MFG	81333
184-1406	G & G MFG	88014.34
184-1416	G & G MFG	83018
184-1418	G & G MFG	83031
184-1418	G & G MFG	83047
184-1420	G & G MFG	83081
184-1420	G & G MFG	83082
184-1420	G & G MFG	83084
184-1420	G & G MFG	83086
184-1421	G & G MFG	84517
184-1422	G & G MFG	83102
184-1422	G & G MFG	83110
184-1423	G & G MFG	83125
184-1424	G & G MFG	83129
1842	ALCO	80211
1842	ALLOY	80211
1842	PRECISION	83320
1842	PRECISION	88014.44
1843	ALCO	80212
1843	ALLOY	80212
1843	PRECISION	83314
1843	PRECISION	83322
1843030	CHRYSLER	80007
1843143	CHRYSLER	80090
1843347	CHRYSLER	80081
184-3516	G & G MFG	83240
184-3518	G & G MFG	83251
184-3519	G & G MFG	83262
184-3520	G & G MFG	83280
184-3520	G & G MFG	83281
184-3522	G & G MFG	80132
184-3522	G & G MFG	83316
184-3522	G & G MFG	83317
184-3523	G & G MFG	83334
184-3524	G & G MFG	83349
184-3524	G & G MFG	88014.40
18-440	NEAPCO	85112
184-4422	G & G MFG	81365
184-4424	G & G MFG	81370
184-4428	G & G MFG	81375
184-5516	G & G MFG	83466
184-5524	G & G MFG	83488
1847	PRECISION	88007.16
1848	PRECISION	87913
1848	PRECISION	87932
1848	PRECISION	87933
184843	IHC	80234
184843H3	HOUGH	80234
184-843H3	IHC	80234
1849	PRECISION	88001.06
185	TRU-CROSS	80228
1850	ALCO	80238
1850	ALLOY	80238
1850	PRECISION	87037
1850	SPICER	80228
18500	PILOT	80223
1850-1	RUEDARSA	80048
18506	PILOT	80224
1851	ALCO	80231
1851	ALLOY	80231
1851	PRECISION	87031
1851	PRECISION	87032
1851	PRECISION	88014.45
185-1400	G & G MFG	88007.09
1852	ALCO	80232
1852	ALLOY	80232
1852	PRECISION	84531
1852	PRECISION	84533
185263	ALLIS-CHLM	80234
185263	FIAT-ALLIS	80234
185263-1	ALLIS-CHLM	80234
185263-1	FIAT-ALLIS	80234
1853	ALCO	80237
1853	ALLOY	80237
1853	PRECISION	84529
185-3500	G & G MFG	88001.06
1854	ALCO	80236
1854	ALLOY	80236
1854	PRECISION	83258
185-4400	G & G MFG	88007.09
1855	PRECISION	85712
1856X	NEAPCO	88000.17
1856X	NEAPCO	88013.85
1859X	NEAPCO	88000.09
1860	PRECISION	88005.97
186-0600	G & G MFG	88000.12
1861	PRECISION	88005.98
186-1200	G & G MFG	88006.99
18613J	BORG-WARNR	87495
18613J	BORG-WARNR	87527
18639J	BORG-WARNR	86805
1864	ALLOY	80235
1865	ALLOY	80233
1866	ALCO	80234
1866	ALLOY	80234
1867	ALCO	80233
1867	ALLOY	80233
186794	WABCO	80168
1868	ALCO	80234
1868	ALLOY	80234
186-874R91	IHC	80081
1868774R91	IHC	80145
186-894R91	IHC	80081
18-7304	NEAPCO	83090

Part No.	Brand	IDLI Group	Part No.	Brand	IDLI Group	Part No.	Brand	IDLI Group	Part No.	Brand	IDLI Group
18-7309	NEAPCO	83049	18N4-2101	ROCKWELL	83742	1903	PRECISION	88014.48	19176	NEAPCO	83102
18-7310	NEAPCO	83082	18N4-2171	ROCKWELL	83746	1903	PRECISION	88014.53	19177	NEAPCO	83018
18-7313	NEAPCO	83129	18N4-2271X	ROCKWELL	83729	1904	PRECISION	81086	1917A	NEAPCO	83062
1874	ALLOY	80185	18N4-2291	ROCKWELL	83728	1904	PRECISION	81089	19183	NEAPCO	83049
1875	ALCO	80220	18N4-2331X	ROCKWELL	83730	1905	PRECISION	81109	19186X	NEAPCO	84517
1875	ALLOY	80220	18N4-2341X	ROCKWELL	83718	1905	PRECISION	88014.54	19187	NEAPCO	88014.21
18750J	BORG-WARNR	88006.73	18N4-2351X	ROCKWELL	83717	1906	PRECISION	81135	19189	NEAPCO	83102
1875X	NEAPCO	80044	18N4-2361X	ROCKWELL	83741	1906	PRECISION	81136	1918A	NEAPCO	83084
1875X1	NEAPCO	80044	18N4-2401X	ROCKWELL	83728	1907	PRECISION	81151	19190	NEAPCO	83129
1876	ALCO	80186	18N4-2441	ROCKWELL	83727	1907	PRECISION	81152	19194	NEAPCO	83110
1876	ALLOY	80186	18N4-2451X	ROCKWELL	83727	190739	HYSTER	80044	191A2	NEAPCO	83031
1877	ALCO	80184	18N4-2471	ROCKWELL	83720	1908	PRECISION	81150	191A2	NEAPCO	83047
1877	ALLOY	80184	18N4-2481X	ROCKWELL	83720	1908	PRECISION	81153	191A4	NEAPCO	83082
1878	ALLOY	80185	18N4-2491X	ROCKWELL	83721	1909	PRECISION	81175	191A4	NEAPCO	83086
187908	IHC	80238	18N4-2531X	ROCKWELL	83731	1909	PRECISION	81178	191A6	NEAPCO	83090
187-908H1	IHC	80238	18N4-2551X	ROCKWELL	83730	1910	ALCO	80953	191B1	NEAPCO	83110
187908HI	HOUGH	80238	18N4-2651	ROCKWELL	83738	1910	ALLOY	80953	191C8	NEAPCO	83129
187964	FIAT-ALLIS	80234	18N4-2701	ROCKWELL	83736	1910	PRECISION	81025	191J3	NEAPCO	83086
188	PERFECT-CI	80140	18N4-2751	ROCKWELL	83748	1910	PRECISION	81026	191L2	NEAPCO	83015
188	TRU-CROSS	80140	18N4-2761X	ROCKWELL	83748	1910	PRECISION	88014.47	1920	ALCO	88002.42
1880	SPICER	80229	18N4-2781	ROCKWELL	83724	19100	PILOT	80235	1920	ALLOY	88002.42
188-0600	G & G MFG	88001.12	18N4-2851X	ROCKWELL	83718	19101	NEAPCO	88014.32	1920	PRECISION	84502
188-1200	G & G MFG	81209	18N4-2931X	ROCKWELL	83747	19101X	NEAPCO	88014.32	1921	ALCO	88002.44
188123H1	HOUGH	80237	18N4-3021	ROCKWELL	83745	19102X	NEAPCO	84517	1921	ALLOY	88002.44
188123H1	IHC	80237	18N4-3171X	ROCKWELL	83725	19105	NEAPCO	83020	1921	PRECISION	84504
188-1272	G & G MFG	81209	18N4-3371	ROCKWELL	83722	1911	ALCO	80951	19210	SCHULTZ	80069
188-1400	G & G MFG	88007.08	18N4-3381X	ROCKWELL	83723	1911	ALLOY	80951	192-1218	G & G MFG	81274
188-1472	G & G MFG	88007.08	18N53-91	ROCKWELL	87567	1911	PRECISION	81125	192-1222	G & G MFG	88000.07
18816557	CHRYSLER	80135	18N82-461-1	ROCKWELL	88005.80	19115	NEAPCO	83018	192-1422	G & G MFG	83116
188-3500	G & G MFG	88007.16	18N82-461-2	ROCKWELL	88005.81	19116	NEAPCO	83024	1922465	CHRYSLER	80135
188-3572	G & G MFG	88007.15	18N82-461-4	ROCKWELL	88005.82	19117	NEAPCO	83047	192-3522	G & G MFG	83320
18870J	BORG-WARNR	87584	18N82-461-8	ROCKWELL	88005.83	19118	NEAPCO	83062	19244J	BORG-WARNR	81840
18872J	BORG-WARNR	87410	18NLS48-10	ROCKWELL	85344	19119	NEAPCO	83082	1925	PRECISION	86601
1889	GORMAN RUP	80070	18NY64-1	ROCKWELL	85735	1912	PRECISION	81060	1925252	CHRYSLER	80081
1890	PRECISION	88001.07	18NYS32-11	ROCKWELL	83719	1912	PRECISION	88014.50	1925552	CHRYSLER	80081
18903J	BORG-WARNR	82037	18NYS32-3A1	ROCKWELL	83726	19120	NEAPCO	83102	1926	PRECISION	86602
18915	SIDEWINDER	80136	18NYS32-4	ROCKWELL	83720	19121	NEAPCO	83082	1927-100R	TAYLOR	80196
1891966	CHRYSLER	80049	18NYS38-13	ROCKWELL	83733	19122	NEAPCO	83082	1927-100R2(4)	TAYLOR	80196
18928J	BORG-WARNR	81842	18NYS38-6	ROCKWELL	83734	19123	NEAPCO	83082	1927-126(4)	TAYLOR	80158
1897	ENGLER	80069	18NYS38-6A	ROCKWELL	83735	19124	NEAPCO	83129	1927-1509R1	TAYLOR	80197
1897	TERRAIN	80069	18NYS38-8A	ROCKWELL	83739	19125	NEAPCO	83102	1927-150R1	TAYLOR	80196
1899	PRECISION	88001.08	18NYS38A	ROCKWELL	83737	19125J	BORG-WARNR	82337	1927R1(4)	TAYLOR	80171
18A1100(4)	LINK BELT	80235	18NYS48-4	ROCKWELL	85341	19125J	BORG-WARNR	82366	1927R1(4)	TAYLOR	80172
18A31(4)	LINK BELT	80235	19	AEC	87459	19126	NEAPCO	83038	192-909R91	IHC	80187
18DS5	OSHKOSH	80225	19	AEC	88008.45	19127	NEAPCO	88014.27	1-93	AEC	88000.84
18DS8	OSHKOSH	80225	19	WYLIE	80024	19127J	BORG-WARNR	81693	1-93	LEMPCO	88000.30
18N1-211	ROCKWELL	88005.78	1900	ALCO	80906	19129	NEAPCO	88014.27	1-93	LEMPCO	88000.31
18N27-9-3813X	ROCKWELL	87116	1900	ALLOY	80906	19129J	BORG-WARNR	82155	1930-030	TAYLOR	80237
18N27-9-5013X	ROCKWELL	87117	1900	PRECISION	81018	1913	PRECISION	81157	1930-110	TAYLOR	80233
18N27-9-6213X	ROCKWELL	87118	1900	PRECISION	81021	1913	PRECISION	81158	1932710	CHRYSLER	87481
18N28-117	ROCKWELL	85736	1900	PRECISION	88014.52	19131	NEAPCO	85705	1935	PRECISION	88001.11
18N3-1351X	ROCKWELL	85343	1900	RUEDARSA	80110	19132	NEAPCO	88014.21	1936	PRECISION	81268
18N3-1371X	ROCKWELL	85345	1900-5	RUEDARSA	80106	19132	NEAPCO	88014.32	1940	PRECISION	87019
18N3-1431X	ROCKWELL	85342	1901	ALCO	80909	19133	NEAPCO	83090	19402J	BORG-WARNR	87518
18N40-151	ROCKWELL	87644	1901	ALLOY	80909	19134	NEAPCO	83116	19403J	BORG-WARNR	85126
18N40-161	ROCKWELL	87648	1901	PRECISION	81020	19135	NEAPCO	83105	1941	PRECISION	84526
18N40-191	ROCKWELL	87649	1901	PRECISION	88014.51	19137	NEAPCO	88014.32	194-1400	G & G MFG	88015.39
18N40-201	ROCKWELL	87645	190-10049	OWATONNA	80037	19137X	NEAPCO	88014.34	194-3500	G & G MFG	88001.07
18N4-1891	ROCKWELL	83731	19015	PILOT	80234	19138	NEAPCO	88014.32	194-4400	G & G MFG	87276
18N4-1901	ROCKWELL	83747	19016	PILOT	80234	19139	NEAPCO	84517	19467J	BORG-WARNR	82206
18N4-1911	ROCKWELL	83730	19017	PILOT	80233	1914	PRECISION	88001.09	1947	PRECISION	88001.12
18N4-1921	ROCKWELL	83718	19017	PILOT	80241	19142	NEAPCO	83062	1947083	COCKSHUTT	80037
18N4-1961	ROCKWELL	83743	1902	ALCO	80905	19147	NEAPCO	83014	1948	PRECISION	88001.13
18N4-1971	ROCKWELL	83729	1902	ALLOY	80905	19149	NEAPCO	83038	1949	PRECISION	88001.14
18N4-1981	ROCKWELL	83732	1902	PRECISION	81034	19151	NEAPCO	83047	19496J	BORG-WARNR	82096
18N4-1991	ROCKWELL	83740	1902	PRECISION	81037	19154	NEAPCO	83125	1-95	LEMPCO	88000.34
18N4-2001	ROCKWELL	83717	19026	PILOT	80233	1916	PRECISION	88001.10	1950	PRECISION	87050
18N4-2011	ROCKWELL	83744	19026	PILOT	88003.09	19166	NEAPCO	83082	1951	PRECISION	84537
18N4-2011	ROCKWELL	88005.79	19028	PILOT	80233	1916A	NEAPCO	88014.29	1951	THOMPSON	80021
18N4-2051	ROCKWELL	83716	1903	PRECISION	81054	19175	NEAPCO	83086	195-20-11100	KOMATSU	80234

Part No.	Brand	IDLI Group	Part No.	Brand	IDLI Group	Part No.	Brand	IDLI Group	Part No.	Brand	IDLI Group
195-221M91	MASS.FERG.	80008	1A140-26	WESCO	88000.45	1A2209-6B	WESCO	88000.51	1C134-43	WESCO	88000.70
195-379M91	MASS.FERG.	80161	1A140-32	WESCO	88000.62	1A3096(4)	LINK BELT	80235	1C140-29	WESCO	88000.75
19543J	BORG-WARNR	82316	1A141-1A	WESCO	88000.52	1A3098(4)	LINK BELT	80240	1C140-44	WESCO	88000.75
19550J	BORG-WARNR	88006.78	1A143	WESCO	88000.52	1A353X	EATON	80069	1C144	WESCO	88000.71
19551J	BORG-WARNR	81699	1A143-1	WESCO	88000.52	1AH140	WESCO	88000.62	1C1442	WESCO	88000.71
1960	RUEDARSA	80213	1A143-1B	WESCO	88000.52	1AH180	WESCO	88000.47	1C150-10	WESCO	88000.72
19605J	BORG-WARNR	88006.42	1A143-22	WESCO	88000.52	1AS120	WESCO	88000.43	1C15045	WESCO	88000.72
196-1218	G & G MFG	81257	1A143-23	WESCO	88000.52	1AS140	WESCO	88000.45	1C154	WESCO	88000.72
196-1220	G & G MFG	81303	1A143-24	WESCO	88000.52	1AS160	WESCO	88000.46	1C1542	WESCO	88000.72
19624J	BORG-WARNR	81981	1A144	WESCO	88000.25	1B120-10	WESCO	88000.86	1C160-46	WESCO	88000.73
196343	IHC	80234	1A144-1	WESCO	88000.27	1B120-24	WESCO	88000.87	1C160-7	WESCO	88000.73
196-343H1	IHC	80234	1A144-12	WESCO	88000.27	1B120-25	WESCO	88000.86	1C164	WESCO	88000.73
196343HI	HOUGH	80234	1A144-13	WESCO	88000.27	1B120-26	WESCO	88000.86	1C1642	WESCO	88000.73
1963CHA	CASE,J.I.	80069	1A144-2	WESCO	88000.25	1B120-34	WESCO	88000.98	1CS140	WESCO	88000.75
197-032R91	IHC	80003	1A144-24	WESCO	88000.25	1B120-47	WESCO	88000.86	1FR	AEC	80008
197-035R91	IHC	80003	1A144-4	WESCO	88000.25	1B120-48	WESCO	88000.87	1FR	WESCO	80008
197603	IHC	80198	1A150-26	WESCO	88000.29	1B120-49	WESCO	88000.98	1FRYD16-1	ROCKWELL	83162
197-603H1	IHC	80198	1A150-9	WESCO	88000.29	1B123-1	WESCO	88000.69	1FRYP12-10	ROCKWELL	83144
197603HI	HOUGH	80198	1A154	WESCO	88000.28	1B123-1	WESCO	88000.79	1FRYQ12-11	ROCKWELL	83134
19767J	BORG-WARNR	85830	1A15401	WESCO	88000.28	1B134-1	WESCO	88000.80	1FRYQ12-13	ROCKWELL	83136
19784J	BORG-WARNR	81710	1A154-1	WESCO	88000.28	1B140-11	WESCO	88001.01	1FRYQ14-3	ROCKWELL	83142
19812J	BORG-WARNR	82145	1A154-4	WESCO	88000.28	1B140-26	WESCO	88000.89	1FRYQ14-9	ROCKWELL	83143
19813J	BORG-WARNR	81948	1A15926	WESCO	88000.29	1B140-28	WESCO	88000.89	1FRYQ15-5	ROCKWELL	83149
19816J	BORG-WARNR	82078	1A160	WESCO	88000.46	1B140-32	WESCO	88001.01	1FRYQ15-8	ROCKWELL	83152
1983	CHAMPION	80007	1A160-10	WESCO	88000.46	1B140-50	WESCO	88000.95	1FRYQ15-9	ROCKWELL	83148
198507	IHC	80198	1A160-24	WESCO	88000.90	1B144-1	WESCO	88000.81	1FRYQ16-10	ROCKWELL	83164
198-507H1	IHC	80198	1A160-26	WESCO	88000.46	1B154-1	WESCO	88000.82	1FRYQ16-24	ROCKWELL	83172
19884J	BORG-WARNR	85838	1A160-27	WESCO	88000.59	1B154-19	WESCO	88000.63	1FRYQ16-4	ROCKWELL	83159
198883AS	OLIVER	80001	1A160-30	WESCO	88000.45	1B154-19	WESCO	88000.82	1FRYQ18	ROCKWELL	83184
198-883AS	WHITE FARM	80001	1A160-32	WESCO	88000.59	1B154-51	WESCO	88000.82	1FRYQ18-1	ROCKWELL	83196
199	BUSH HOG	80069	1A160-33	WESCO	88000.45	1B160-24	WESCO	88000.90	1FRYQ18-3	ROCKWELL	83188
1990418	CLARK EQU.	80170	1A163	WESCO	88000.53	1B160-52	WESCO	88000.90	1FRYQ19-1	ROCKWELL	83200
1990609	CLARK EQU.	80234	1A163-1	WESCO	88000.53	1B160-63	WESCO	88000.91	1FRYQ20-1	ROCKWELL	83212
1990953	CLARK EQU.	80172	1A164	WESCO	88000.30	1B164-1	WESCO	88000.83	1FRYQ21	ROCKWELL	83219
199459	HYSTER	80196	1A164-1	WESCO	88000.31	1B180-21	WESCO	88001	1FRYQ24	ROCKWELL	83230
19951J	BORG-WARNR	85842	1A164-11	WESCO	88000.33	1B180-22	WESCO	88000.64	1FRYQ26	ROCKWELL	83233
19961	PETTI-MULL	80215	1A164-14	WESCO	88000.54	1B180-23	WESCO	88000.93	1FRYR12-12	ROCKWELL	83132
199862	HYSTER	80198	1A164-18	WESCO	88000.54	1B180-23A	WESCO	88000.93	1FRYR12-18	ROCKWELL	83135
1999	PRECISION	88001.15	1A164-2	WESCO	88000.30	1B180-32	WESCO	88000.95	1FRYR12-8	ROCKWELL	83133
1A103	WESCO	88000.19	1A164-28	WESCO	88000.33	1B18054	WESCO	88000.92	1FRYR13-2	ROCKWELL	83138
1A103-1	WESCO	88000.19	1A164-29	WESCO	88000.54	1B18054-10C	WESCO	88000.93	1FRYR13-7	ROCKWELL	83137
1A120-10	WESCO	88000.43	1A164-30	WESCO	88000.54	1B18055-6B	WESCO	88001	1FRYR13-9	ROCKWELL	83139
1A120-14	WESCO	88000.57	1A164-31	WESCO	88000.54	1B18056-10C	WESCO	88000.93	1FRYR14-22	ROCKWELL	83140
1A120-25	WESCO	88000.57	1A164-4	WESCO	88000.30	1B180-57	WESCO	88000.95	1FRYR14-23	ROCKWELL	83141
1A120-26	WESCO	88000.43	1A164-7	WESCO	88000.54	1B184-1	WESCO	88000.96	1FRYR14-26	ROCKWELL	83146
1A120-33	WESCO	88000.99	1A174	WESCO	88000.34	1B185-1	WESCO	88000.84	1FRYR14-29	ROCKWELL	83145
1A123	WESCO	88000.20	1A174-1	WESCO	88000.34	1B1890-6B	WESCO	88001	1FRYR15-15	ROCKWELL	83150
1A123-1	WESCO	88000.21	1A180-11	WESCO	88000.47	1B200-21	WESCO	88000.94	1FRYR15-16	ROCKWELL	83151
1A123-13	WESCO	88000.20	1A180-21	SPICER	88000.49	1B20058-6B	WESCO	88000.94	1FRYR15-7	ROCKWELL	83147
1A123-2	WESCO	88000.20	1A180-21	WESCO	88000.49	1B205-1	WESCO	88000.85	1FRYR16-11	ROCKWELL	83171
1A123-4	WESCO	88000.20	1A180-26	WESCO	88000.60	1B220-20	WESCO	88000.67	1FRYR16-2	ROCKWELL	83169
1A124	WESCO	88000.44	1A180-32	WESCO	88000.47	1B220-21	WESCO	88000.51	1FRYR16-22	ROCKWELL	83156
1A124-1	WESCO	81020	1A184	WESCO	88000.36	1B220-22	WESCO	88000.66	1FRYR16-23	ROCKWELL	83165
1A130-17	WESCO	88000.23	1A184-1	WESCO	88000.36	1BH140	WESCO	88001.01	1FRYR16-26	ROCKWELL	83167
1A130-7	WESCO	88000.23	1A185	WESCO	88000.35	1BH180	WESCO	88001.05	1FRYR16-3	ROCKWELL	83158
1A133	WESCO	88000.26	1A185-1	WESCO	88000.35	1BP180	WESCO	88001	1FRYR16-33	ROCKWELL	83170
1A133-1	WESCO	88000.26	1A195	WESCO	88000.37	1BP200	WESCO	88000.94	1FRYR16-4	ROCKWELL	83155
1A134	WESCO	88000.22	1A195-1	WESCO	88000.37	1BS120	WESCO	88000.86	1FRYR16-48	ROCKWELL	83161
1A134-1	WESCO	88000.22	1A200-21	WESCO	88000.48	1BS140	WESCO	88000.95	1FRYR16-57	ROCKWELL	87017
1A134-15	WESCO	88000.23	1A200-35	WESCO	88000.40	1C120	WESCO	88000.74	1FRYR16-66	ROCKWELL	83168
1A134-16	WESCO	88000.22	1A200-7	WESCO	88000.40	1C120-27	WESCO	88000.74	1FRYR16-74	ROCKWELL	83153
1A134-17	WESCO	88000.22	1A204	WESCO	88000.55	1C120-29	WESCO	88000.74	1FRYR16-80	ROCKWELL	83166
1A134-18	WESCO	88000.23	1A204-1	WESCO	88000.55	1C120-40	WESCO	88000.74	1FRYR16-9	ROCKWELL	83157
1A134-2	WESCO	88000.22	1A204-11	WESCO	88000.55	1C120-41	WESCO	88000.74	1FRYR16-94	ROCKWELL	83163
1A134-20	WESCO	88000.22	1A204-34	WESCO	88000.55	1C123	WESCO	88000.68	1FRYR17	ROCKWELL	83175
1A134-21	WESCO	88000.22	1A204-37	WESCO	88000.55	1C123-1	WESCO	88000.69	1FRYR17-11	ROCKWELL	83174
1A134-4	WESCO	88000.22	1A204-7	WESCO	88000.55	1C1232	WESCO	88000.68	1FRYR17-17	ROCKWELL	83177
1A140-10	WESCO	88000.45	1A205	WESCO	88000.38	1C134	EATON	88000.70	1FRYR17-4	ROCKWELL	83178
1A140-11	WESCO	88000.62	1A205-1	WESCO	88000.39	1C134-13	WESCO	88000.70	1FRYR17-8	ROCKWELL	83173
1A140-25	WESCO	88000.45	1A220-21	WESCO	88000.51	1C1342	WESCO	88000.70	1FRYR17-9	ROCKWELL	83176

Part No.	Brand	IDLI Group
1FRYR18-10	ROCKWELL	83192
1FRYR18-11	ROCKWELL	83179
1FRYR18-32	ROCKWELL	83183
1FRYR18-33	ROCKWELL	83187
1FRYR18-56	ROCKWELL	83182
1FRYR18-6	ROCKWELL	83190
1FRYR18-69	ROCKWELL	87020
1FRYR18-7	ROCKWELL	83193
1FRYR18-79	ROCKWELL	83185
1FRYR18-8	ROCKWELL	83194
1FRYR18-85	ROCKWELL	83195
1FRYR19-12	ROCKWELL	83202
1FRYR19-13	ROCKWELL	83198
1FRYR19-17	ROCKWELL	83197
1FRYR19-19	ROCKWELL	83201
1FRYR19-2	ROCKWELL	83206
1FRYR19-23	ROCKWELL	83199
1FRYR19-25	ROCKWELL	83203
1FRYR19-27	ROCKWELL	83204
1FRYR20-1	ROCKWELL	83218
1FRYR20-10	ROCKWELL	83211
1FRYR20-11	ROCKWELL	83216
1FRYR20-2	ROCKWELL	83217
1FRYR20-21	ROCKWELL	83209
1FRYR20-34	ROCKWELL	83210
1FRYR20-58	ROCKWELL	83213
1FRYR20-64	ROCKWELL	87021
1FRYR20-65	ROCKWELL	83215
1FRYR20-82	ROCKWELL	87022
1FRYR22	ROCKWELL	83222
1FRYR22-1	ROCKWELL	83228
1FRYR22-11	ROCKWELL	83225
1FRYR22-13	ROCKWELL	83221
1FRYR22-2	ROCKWELL	83227
1FRYR22-5	ROCKWELL	83226
1FRYR24	ROCKWELL	83229
1FRYR24-1	ROCKWELL	83232
1FRYR24-2	ROCKWELL	83231
1FRYS16-2	ROCKWELL	83160
1FRYS16-3	ROCKWELL	83154
1FRYS16-8	ROCKWELL	83234
1FRYS18-1	ROCKWELL	87018
1FRYS18-12A	ROCKWELL	84523
1FRYS18-18	ROCKWELL	83181
1FRYS18-2	ROCKWELL	83189
1FRYS18-27	ROCKWELL	83180
1FRYS18-29	ROCKWELL	83186
1FRYS18-3	ROCKWELL	83191
1FRYS19	ROCKWELL	83205
1FRYS20-10	ROCKWELL	83207
1FRYS20-15	ROCKWELL	83214
1FRYS20-21A	ROCKWELL	84524
1FRYS20-25	ROCKWELL	83208
1FRYS22	ROCKWELL	83223
1FRYS22	ROCKWELL	87024
1FRYS22-15A	ROCKWELL	84526
1FRYS22-17	ROCKWELL	83220
1FRYS22-2	ROCKWELL	87019
1FRYS22-28	ROCKWELL	83224
1FRYS22-31A	ROCKWELL	84525
1FRYS22-47	ROCKWELL	87023
1FRYS28-1	ROCKWELL	87025
1FSP20-1	ROCKWELL	87255
1H8022(5)	CATERPILLR	88003.27
1H8024(5)	CATERPILLR	88003.27
1H8025(5)	CATERPILLR	88003.27
1H8499	CATERPILLR	88001.16
1J843	CATERPILLR	80172
1J8584	CATERPILLR	88001.32
1K2741	CATERPILLR	80149
1K9467	CATERPILLR	80109
1K9467	CATERPILLR	80149
1M4410(2)	CATERPILLR	88001.22
1M4411(2)	CATERPILLR	88001.24
1M4841	FORD	88002.86
1M4841A	FORD	88002.86
1M7039	FORD	80049
1M7039A	FORD	80049
1M7042	FORD	80044
1M7042A	FORD	80044
1M7042A	FORD	88001.98
1M8110	CATERPILLR	88001.22
1M8804	CATERPILLR	80238
1N4841	FORD	88002.86
1O3-37	MCQUAY-NOR	80235
1P1569	CATERPILLR	88001.52
1QDK	WESCO	88000.18
1R198	DAFFIN	80035
1R258	DAFFIN	80001
1R368	DAFFIN	80001
1R369	DAFFIN	80001
1R402	FARMHAND	80037
1R505	DAFFIN	80008
1S678	CATERPILLR	80001
1S687	CATERPILLR	80003
1S8753(2)	CATERPILLR	88001.36
1S8754	CATERPILLR	88001.36
1S9163	CATERPILLR	80033
1S9670	CATERPILLR	80198
1T168	DAFFIN	80133
1T170	DAFFIN	80037
1T7039	FORD	80124
1V3639	CATERPILLR	80223
1X8769	CATERPILLR	80195
2	AEC	88013.86
20	AEC	87481
20	AEC	88007.77
20	WYLIE	80077
200	AEC	80075
200	BUSH HOG	80133
200	TRU-CROSS	80072
200	TRU-CROSS	80075
20-00-00	WALTERSCHD	80024
20001	TRW	80001
20001	TRW	80003
20001D	TRW	80001
20001D	TRW	80003
20001D	TRW	80037
20002	TRW	80021
20002D	TRW	80021
20002D	TRW	80023
20002D	TRW	88001.97
20003	TRW	80024
20003	TRW	80030
20003	TRW	80031
20003	TRW	88002.77
20003D	TRW	80024
20003D	TRW	80030
20003D	TRW	80031
20003D	TRW	80032
20003D	TRW	80049
20003D	TRW	88002.77
20004	TRW	80032
20006	TRW	80049
20006	TRW	88001.96
20007	TRW	80044
20007	TRW	88001.98
20010	TRW	80054
20011	TRW	80059
20012	TRW	80059
20012	TRW	80066
20013	TRW	80055
20014	TRW	80058
20015	TRW	80046
20016	TRW	80047
200166A	HYSTER	80149
200167A	HYSTER	80150
20017	TRW	80007
20017	TRW	88000.15
20018	TRW	88000.14
20018	TRW	88000.15
20019	TRW	80007
20019	TRW	88000.16
2002	AEC	80049
2002	ALCO	88002.10
2002	TRU-CROSS	80048
2-0020	NEAPCO	80115
20020	TRW	80023
20020	TRW	80024
20021	TRW	80023
20021	TRW	80140
20022	TRW	80025
20023	TRW	80041
20024	TRW	80072
20024	TRW	80089
20025	TRW	80073
20025	TRW	80091
20026	TRW	80072
20026	TRW	80075
20026	TRW	80076
200260	WESCO	80062
20027	TRW	80034
200-278	WESCO	80085
20028	TRW	80042
200285	WESCO	80132
20029	TRW	80064
20029	TRW	80065
200297	WESCO	80079
2002E4635A	FORD	80189
2002E4635B	FORD	80221
20030	TRW	80060
20030	TRW	80061
200308	WESCO	80104
20031	TRW	80060
20031	TRW	80063
200-311	WESCO	80104
20032	TRW	80039
20033	TRW	80038
20034	TRW	80068
20035	TRW	80071
200353	WESCO	80083
20036	TRW	80035
20037	TRW	80037
20038	TRW	80069
20038	TRW	80070
20039	TRW	80092
20040	TRW	80104
20040	TRW	80107
20040	TRW	80112
20042	TRW	80112
20043	TRW	80101
20044	TRW	80100
20045	TRW	80102
20046	TRW	80113
2004611	EATON	80193
2004613	EATON	80195
200-464C91	IHC	80037
20047	TRW	80105
20048	TRW	80093
20049	TRW	80072
20049	TRW	80075
20049	TRW	80077
20049	TRW	80078
20049	TRW	80085
20050	TRW	80090
20051	TRW	80117
20052	TRW	80116
200520	WESCO	80077
200-522	WESCO	80086
2-0053	NEAPCO	80124
20053	TRW	80118
20053	TRW	80121
20053	TRW	80131
200530	WESCO	80129
200-532	WESCO	80131
2-0054	NEAPCO	80142
20054	TRW	80124
20054	TRW	80125
20055	TRW	80119
20055	TRW	80120
200552	WESCO	80118
200554	WESCO	80081
20056	TRW	80118
20057	TRW	80140
20057	TRW	80182
200574	WESCO	80078
200578	WESCO	80124
20058	TRW	80142
20058	TRW	80143
20059	TRW	80132
2006	ALCO	88001.97
20060	TRW	80081
20060	TRW	80091
20061	TRW	80091
20062	TRW	80123
20063	TRW	80131
20064	TRW	80121
20064	TRW	80129
20065	TRW	80134
20065	TRW	80138
20066	TRW	80141
20067	TRW	80145
2006731	EATON	80212
2006731	EATON	80214
2006739	EATON	80210
2006739	EATON	80215
2006741	EATON	80237
2006741	EATON	80239
2006749	EATON	80195
2006749	EATON	80231
2006749	EATON	80240
20068	TRW	80130
20068	TRW	80134
20068D	TRW	80129
20068D	TRW	80130
20068D	TRW	80134
20068D	TRW	80163
20068D	TRW	80178
20069	TRW	80109
20069	TRW	80149
20070	TRW	80148
20071	TRW	80153
20073	TRW	80156
20073	TRW	88003.17
20074	TRW	80158
20075	TRW	80155
20076	TRW	80157
20077	TRW	80159
20078	TRW	80151
20079	TRW	80154

Part No.	Brand	IDLI Group	Part No.	Brand	IDLI Group	Part No.	Brand	IDLI Group	Part No.	Brand	IDLI Group
20080	TRW	80168	201111	WESCO	80006	201-248	WESCO	80044	20140	TRW	80124
20081	TRW	80172	201112	WESCO	80018	201249	WESCO	80054	20141	TRW	80026
20082	TRW	80174	201113	WESCO	80009	201249	WESCO	88002	20141D	TRW	80026
20082	TRW	80202	201114	WESCO	80017	201249	WESCO	88002.03	20141D	TRW	80083
200822X	SPICER	80225	20112	TRW	80237	20124D	TRW	80061	20142	TRW	80186
20083	TRW	80171	201122	WESCO	80115	20124D	TRW	80062	20142D	TRW	80186
20084	TRW	80167	201126	TRW	80020	20125	TRW	80104	20142D	TRW	88010.16
20085	TRW	80162	201-126	WESCO	80020	20125	TRW	80109	20143	TRW	80184
20086	TRW	80169	201127	TRW	80011	201250	WESCO	80059	20144	TRW	80185
20086	TRW	80216	201127	WESCO	80011	201-251	WESCO	80032	20144D	TRW	80185
20087	TRW	80173	201-128	WESCO	88000.16	201251	WESCO	88002.81	20147	TRW	80018
20088	TRW	80178	201-129	WESCO	80013	201252	WESCO	80034	20150	TRW	86069
20088D	TRW	80044	20113	TRW	80215	201253	WESCO	80031	201504	WESCO	80041
20088D	TRW	80162	20113	TRW	80235	201253	WESCO	88001.99	201504	WESCO	88001.85
20088D	TRW	80168	201-130	WESCO	80005	201254	WESCO	80030	20151	TRW	86071
20088D	TRW	80178	201-131	WESCO	80016	201255	WESCO	80061	20152	TRW	86070
20088D	TRW	80179	201-133	WESCO	80115	201255	WESCO	88001.59	201-520	WESCO	80059
20088D	TRW	80216	201-134	WESCO	80015	201256	WESCO	80057	201-520	WESCO	80075
20088D	TRW	88002.96	20114	TRW	80233	201256	WESCO	88001.58	201520	WESCO	80077
20089	TRW	80182	20114	TRW	88003.09	201257	WESCO	80095	201-521	WESCO	80072
20089D	TRW	80182	20115	TRW	80234	201257	WESCO	88001.61	201521HD	WESCO	80075
20090	TRW	80135	20116	TRW	80234	201258	WESCO	88001.60	201-521HD	WESCO	80077
20091	TRW	80188	201165	WESCO	80091	201259	WESCO	80043	201-523	WESCO	80031
20092	TRW	80193	201167A	HYSTER	80156	201-259	WESCO	88001.98	20153	TRW	86072
20092	TRW	80198	20117	TRW	80909	20125D	TRW	80109	201-530	WESCO	80117
20093	TRW	80198	201177	WESCO	80012	20126	TRW	80078	201530	WESCO	80129
20094	TRW	80194	20118	TRW	80903	201-260	WESCO	80062	201-531	WESCO	80118
20095	TRW	80195	20118	TRW	80909	201263	WESCO	80094	201531	WESCO	80121
20096	TRW	80197	20118	TRW	80958	201266	WESCO	80045	201-531	WESCO	80131
20097	TRW	80201	201180	WESCO	80012	201-269	WESCO	80061	201532	WESCO	80131
20098	TRW	80199	2011875	GMC	80001	20127	TRW	80074	201534	WESCO	80116
20099	TRW	80077	20119	TRW	80951	201273	WESCO	80058	201535	WESCO	80039
20099	TRW	80210	20120	TRW	80061	201-278	WESCO	80085	20154	TRW	86072
200U	AEC	80076	20120	TRW	80905	20128	TRW	80004	20155	TRW	80048
20100	TRW	80212	20-12-00	WALTERSCHD	88004.02	20128	TRW	80006	201-550	WESCO	80119
20-10-00	WALTERSCHD	88003.74	201201	WESCO	80047	201-285	WESCO	80041	201550	WESCO	80120
201-008	WESCO	80001	201202	WESCO	80048	201285	WESCO	80132	201552	WESCO	80118
20100D	TRW	80198	201203	WESCO	80048	20129	TRW	80010	201-552	WESCO	80130
20100D	TRW	80212	201204	WESCO	80066	201297	WESCO	80079	201-553	WESCO	80036
20100D	TRW	80214	201205	WESCO	88000.18	20129D	TRW	80004	201-554	WESCO	80075
20100D	TRW	80215	20120D	TRW	80905	20129D	TRW	80010	201554	WESCO	80081
20100D	TRW	80218	20121	TRW	80953	20129D	TRW	80011	201-554	WESCO	80091
20101	TRW	80213	20-121-157-023	ALLIS-CHLM	80184	20129D	TRW	80012	201555	WESCO	80123
201012	WESCO	80035	201217	WESCO	80092	20129D	TRW	80018	201556	WESCO	80071
20102	TRW	80211	201217	WESCO	88001.76	20129D	TRW	80019	201557	WESCO	80074
20102D	TRW	80207	201217	WESCO	88002.07	2012B	MOTOR MAST	80021	201558	WESCO	80042
20102D	TRW	80211	201218	WESCO	80053	2012B	ZELLER	80021	20156	TRW	80054
20103	TRW	80208	201219	WESCO	80046	20130	TRW	80009	201560	WESCO	80142
201033	KAISER	80049	20121D	TRW	80906	201300	WESCO	80133	201-565	WESCO	80190
201033	WILLYS	80049	20121D	TRW	80909	201300	WESCO	88002.89	20157	TRW	80049
20104	TRW	80189	20121D	TRW	80951	201305	WESCO	80093	201570	WESCO	80007
20105	TRW	80190	20121D	TRW	80953	201308	WESCO	80104	201-570	WESCO	88000.15
20106	TRW	80221	20122	TRW	80906	201309	WESCO	80105	201571	WESCO	80007
20107	TRW	80224	20122	TRW	80909	20131	TRW	80012	201-571	WESCO	88000.14
20108	TRW	80225	20123	TRW	80043	201-311	WESCO	80104	201574	WESCO	80078
20109	TRW	80231	20123	TRW	80044	201311	WESCO	80107	201-574	WESCO	80090
2011	PERFECT-CI	80044	20124	TRW	80060	201-312	WESCO	80112	201575	WESCO	80078
2011	PERFECT-CI	88001.98	20124	TRW	80061	20132	TRW	80227	201576	WESCO	80078
2011	TRU-CROSS	80044	20124	TRW	80062	20133	JEEP	80049	201578	WESCO	80124
2011	TRW	80059	201240	WESCO	80055	20133	TRW	80019	201579	WESCO	80124
20110	TRW	80238	201241	WESCO	80025	20134	TRW	80189	20158	TRW	80044
20-11-00	WALTERSCHD	88003.92	201242	WESCO	80026	20135	TRW	80221	201-581	WESCO	80023
201101	WESCO	80008	201245	WESCO	80049	201-353	WESCO	80083	201581	WESCO	80024
201102	WESCO	88000.17	201245	WESCO	88002.01	20136	TRW	80225	201-581	WESCO	80140
201106	WESCO	80024	201245	WESCO	88002.02	20137	TRW	80227	20159	TRW	80033
201107	WESCO	80010	201245	WESCO	88002.06	201-372	WESCO	80135	20160	TRW	80008
201108	WESCO	80004	201245	WESCO	88002.09	2013785	EATON	80070	201603	WESCO	88002.20
201109	WESCO	80019	201246	WESCO	80044	20138	TRW	80228	201604	WESCO	88002.21
201111	TRW	80232	201246	WESCO	88002.08	2013822	EATON	80070	201606	WESCO	88002.21
201110	WESCO	80033	201247	WESCO	80058	20139	TRW	80079	201607	WESCO	88002.22

Part No.	Brand	IDLI Group	Part No.	Brand	IDLI Group	Part No.	Brand	IDLI Group	Part No.	Brand	IDLI Group
20161	TRW	80100	20189D	TRW	80083	201ST46	MACK	88001.82	2031	TRU-CROSS	80049
20161D	TRW	80100	20190	TRW	80124	202	PRECISION	80153	203117H1	IHC	80155
20162	TRW	80111	20191	TRW	80090	20203	TRW	80136	20319J	BORG-WARNR	85552
20163	TRW	80110	20192	TRW	80078	20204	TRW	80161	2032	ALCO	80092
20164	TRW	80082	20193	TRW	80002	20205	TRW	80061	2032	ALCO	88001.76
20165	TRW	80097	20193	TRW	80079	20205	TRW	80062	2032	ALCO	88002.07
20167	TRW	80144	20194	TRW	80124	20205J	BORG-WARNR	82071	20-32-00	WALTERSCHD	88003.91
20168	TRW	80144	20194	TRW	80126	20206	TRW	80132	203252	WESCO	80019
20169	TRW	80171	20195	TRW	86122	20207	TRW	80104	203260	WESCO	80062
2017	AEC	80092	201965	WESCO	80068	20208	TRW	80077	20-328	NEAPCO	88007.05
20170	TRW	80175	201966	WESCO	80122	20209	TRW	80129	2033	TRU-CROSS	80054
201703	WESCO	88002.24	201967	WESCO	80038	20210	TRW	80118	20-33-00	WALTERSCHD	88004.01
201704	WESCO	88002.26	20198	TRW	88000.16	20210	TRW	80134	203308	WESCO	80104
201705	WESCO	88002.24	20199	TRW	80017	20211	TRW	80081	20-340	NEAPCO	88007.09
201709	WESCO	88002.25	201R	PRECISION	80901	202-13X	SPICER	88002.64	2034362	EATON	80213
20171	TRW	80183	201SJ11	MACK	80124	202-13X	SPICER	88002.71	2034881	EATON	80198
20172	TRW	80209	201SJ11A	MACK	80124	20215	TRW	88002.38	2034A	NEAPCO	88000.10
201-725	WESCO	80064	201SJ12	MACK	80142	202-15X	SPICER	88002.71	2034A	NEAPCO	88015.39
201725	WESCO	80065	201SJ12A	MACK	80142	2021805	GMC	87384	2-0350	NEAPCO	80043
201725	WESCO	88001.68	201SJ13	MACK	80178	2021B	MOTOR MAST	80023	203520	WESCO	80077
20172J	BORG-WARNR	81941	201SJ14	MACK	80189	202-246	WESCO	80044	203765	CATERPILLR	80049
20173	TRW	80223	201SJ14A	MACK	80189	202-253	WESCO	80031	203-787H1	IHC	80023
201735	WESCO	80096	201SJ15	MACK	80221	202-254	WESCO	80031	203787HI	HOUGH	80023
201735	WESCO	88001.64	201SJ15A	MACK	80221	202-255	WESCO	80063	203787HI	HOUGH	80024
201735	WESCO	88001.69	201SJ15B	MACK	80221	202-259	WESCO	80044	203791	IHC	80142
20174	TRW	80233	201SJ16	MACK	80225	202-273	WESCO	80058	203-791H1	IHC	80142
20175	TRW	80245	201SJ16A	MACK	80225	202-285	WESCO	80058	203791HI	HOUGH	80142
201755	WESCO	80032	201SJ16B	MACK	80225	202-308	WESCO	80104	203KC	KENBAR	87285
20176	TRW	80245	201SJ17	MACK	80023	202-311	WESCO	80104	203KC	KENBAR	88008.48
201765	KAISER	88003.50	201SJ19	MACK	80024	202339	IHC	80186	203SJ13	MACK	87570
201766	KAISER	88003.56	201SJ19A	MACK	80024	202-339H1	IHC	80185	2-04	BYPY	88003.66
201767	KAISER	88003.42	201SJ-20	MACK	80188	202-347C91	IHC	80001	204	PRECISION	80906
201768	KAISER	88003.46	201SJ21	MACK	80178	202-348C91	IHC	80001	204	PRECISION	80909
20177	TRW	88003.40	201SJ22	MACK	80072	20239J	BORG-WARNR	82075	2040210	GMC	80226
201772	KAISER	88003.56	201SJ23	MACK	80221	20-240	NEAPCO	88006.97	2040231	EATON	80156
201773	KAISER	88003.45	201SJ25	MACK	80189	202411	HYSTER	80197	2040856	EATON	80187
2017732	EATON	80210	201SJ26	MACK	80124	202411A	HYSTER	80196	20411	PETTI-MULL	80240
2017732	EATON	80212	201SJ27	MACK	80024	202449-1	GAR WOOD	80033	2041234	GMC	80186
2017732(5)	EATON	88003.01	201SJ28	MACK	80077	20-252	NEAPCO	88006.97	2041235	GMC	80184
2017732(5)	EATON	88003.03	201SJ29	MACK	80077	202-520	WESCO	80077	2041472	GMC	80218
201774	KAISER	88003.49	201SJ30	MACK	80077	2025-2B	MOTOR MAST	80032	2041593	WABCO	80101
2017744	EATON	80210	201SJ31	MACK	80182	202-530	WESCO	80129	2041859	GMC	80217
2017744	EATON	80212	201SJ32	MACK	80228	202-531	WESCO	80131	2041860	GMC	80186
2017744	EATON	88003.01	201SJ33	MACK	80190	202-532	WESCO	80131	204192-1	SPICER	88001.13
2017744	EATON	88003.03	201SJ33	MACK	80228	202-550	WESCO	80119	20423J	BORG-WARNR	82021
201775	KAISER	80909	201SJ35	MACK	80225	202-552	WESCO	80118	20-428	NEAPCO	85126
201775	WILLYS	80909	201SJ36	MACK	80007	202-554	WESCO	80081	204301	FWD	80239
2017758	EATON	80212	201SJ37	MACK	80228	202-578	WESCO	80124	204301	MIXERMOBL	80240
2017758	EATON	88003.01	201SJ38	MACK	80140	2025A	MOTOR MAST	80031	204302	FWD	80239
201776	WILLYS	88003.52	201SJ39	MACK	80240	2025A	MOTOR MAST	88002.77	204305	FWD	80235
20178	TRW	88001.93	201SJ40	MACK	80215	2025A	ZELLER	80031	204331J	BORG-WARNR	84651
2018	ALCO	80053	201SJ41	MACK	80228	2025B	MOTOR MAST	80032	204362	FWD	80216
20180	TRW	80229	201SJ46	MACK	88001.82	2025B	ZELLER	80032	204363	FWD	80171
201800	WESCO	88002.41	201SJ47	MACK	80221	202-5X	SPICER	88002.63	20-440	NEAPCO	88007.08
201803	WESCO	88002.29	201SJ48	MACK	80077	20268	ALLIS-CHLM	80044	2045	ALCO	88002.06
201805	WESCO	88002.29	201SJ48	MACK	80210	20268	ALLIS-CHLM	88001.98	2045	ALCO	88002.09
201809	WESCO	88002.30	201SJ48	MACK	80217	20268	FIAT-ALLIS	80044	2045	ALCO	88002.11
20182	TRW	87481	201SJ49	MACK	80212	20268	GMC	80044	2045	ALCO	88002.12
20183	TRW	80019	201SJ49A	MACK	80214	2026823	GAR WOOD	80008	204572-1	SPICER	88007.29
20184	TRW	80036	201SJ50	MACK	80234	202-6X	SPICER	88002.55	204572-2	SPICER	88007.31
20184D	TRW	80036	201SJ51	MACK	80184	202797	FWD	80233	204574-1	SPICER	88007.28
20185	SPICER	80039	201SJ52	MACK	80231	202-7X	SPICER	88002.54	2046	ALCO	88001.98
20185	TRW	80038	201SJ53	MACK	80215	202-8X	SPICER	88002.53	2046	ALCO	88002.08
20185D	TRW	80038	201SJ53	MACK	80234	202931	HYSTER	80240	2046453	WABCO	80109
20185D	TRW	80039	201SJ53	MACK	80235	202938	HYSTER	80239	2046974	WABCO	80213
20186	TRW	88002.37	201SJ54	MACK	80237	202-947C91	IHC	80001	2047681	EATON	80215
20186D	TRW	88002.37	201SJ55	MACK	80048	2-03	BYPY	88003.65	20496J	BORG-WARNR	81729
20187	TRW	80020	201SJ55	MACK	80233	20-30-12	SPICER	88015.06	204989-1	SPICER	88007.30
20188	TRW	80139	201SJ55	MACK	80234	20-30-22	SPICER	88015.05	204E7039	FORD	80023
20189	TRW	80083	201SJ73	MACK	80178	20-30-52	SPICER	88015.04	204E7042	FORD	80023

Part No.	Brand	IDLI Group
204R	PRECISION	80902
20501	TRW	88000.21
20501	TRW	88000.39
20502	TRW	81020
20502	TRW	88000.21
20503	TRW	88000.22
20504	TRW	88000.25
20504	TRW	88000.27
20505	TRW	88000.31
20506	TRW	88000.35
2-05-07	BYPY	88003.74
20507	TRW	88000.38
20508	TRW	88000.71
20508J	BORG-WARNR	82187
20509	TRW	88000.21
20509	TRW	88000.25
20509	TRW	88000.27
20509	TRW	88000.39
20510	TRW	88000.22
20510	TRW	88000.25
20510	TRW	88000.27
20511	TRW	88000.22
20511	TRW	88000.28
20512	TRW	88000.22
20512	TRW	88000.31
20513	TRW	88000.53
20514	TRW	88000.35
20514	TRW	88000.38
2051472	GMC	80212
20515	TRW	88000.72
20516	TRW	88000.39
20516	TRW	88000.81
20517	TRW	88000.31
20517	TRW	88000.88
20523J	BORG-WARNR	85875
20524J	BORG-WARNR	81972
20534J	BORG-WARNR	81965
2053551	WABCO	80167
2053903	WABCO	80188
2053956	BEND.WEST.	80168
2053956	WABCO	80216
2055473	WABCO	80167
2055552	GMC	80227
2055743	WABCO	80021
2055743	WABCO	88001.97
2057273	GMC	80124
2057289	GMC	80140
2058300	GMC	80221
205R	PRECISION	80951
2-06	BYPY	88004.17
20601	TRW	88000.45
20602	TRW	88000.20
20602	TRW	88000.21
20603	TRW	88000.43
20604	TRW	88000.44
20605	TRW	88000.22
20605	TRW	88000.26
20606	TRW	88000.22
20607	TRW	88000.52
20608	TRW	88000.25
20608	TRW	88000.27
20609	TRW	88000.27
20609	TRW	88000.28
2060A	MOTOR MAST	80048
2060A	ZELLER	80049
2060AA	MOTOR MAST	80048
2060AA	ZELLER	80048
2060C	MOTOR MAST	80049
2060C	ZELLER	88001.96
2060CA	MOTOR MAST	80054
2060D	MOTOR MAST	80044
2060D	ZELLER	80044
2060D	ZELLER	88001.98
2060F	MOTOR MAST	80045
2060G	MOTOR MAST	80051
2060G	ZELLER	80051
2060HA	MOTOR MAST	80049
2060J	MOTOR MAST	80044
2060K	MOTOR MAST	80049
2060LA	MOTOR MAST	80044
2060LA	MOTOR MAST	80049
2060LA	MOTOR MAST	88001.98
2060MA	MOTOR MAST	80054
2060MA	ZELLER	80054
2060UJ	ZELLER	80066
2060V	MOTOR MAST	80059
2060V	ZELLER	80059
2060VF	MOTOR MAST	80066
2060VF	ZELLER	80066
2060VJ	MOTOR MAST	80059
2060VJ	MOTOR MAST	80066
2060W	MOTOR MAST	80055
2060W	ZELLER	80055
2060WF	MOTOR MAST	80058
2060-WF	ZELLER	80058
2060X	MOTOR MAST	80046
2060X	ZELLER	80046
2060XA	MOTOR MAST	80046
2060XB	MOTOR MAST	80046
2060XB	MOTOR MAST	80048
2060XC	MOTOR MAST	80046
2060Y	MOTOR MAST	80047
2060Y	ZELLER	80047
2060YA	MOTOR MAST	80047
2060YB	MOTOR MAST	80047
2060YC	MOTOR MAST	80047
2060Z	MOTOR MAST	80046
20611	TRW	88000.31
20611	TRW	88000.53
206113	FWD	80196
20612	TRW	88000.30
20612	TRW	88000.31
20612	TRW	88000.53
20613	TRW	88000.36
20614	TRW	88000.35
20615	TRW	88000.38
20615	TRW	88000.55
20616	TRW	88000.39
20617	TRW	88000.37
20618	TRW	88000.46
2062	AEC	80050
2062	ALCO	80050
20627J	BORG-WARNR	85870
20-628	NEAPCO	88006.97
2063	AEC	80043
2063	ALCO	80043
20630	TRW	88000.86
20630	TRW	88000.88
20630J	BORG-WARNR	85882
20631	TRW	88000.79
20632	TRW	88000.89
20633	TRW	88000.80
20634	TRW	88000.95
20635	TRW	88000.81
20636	TRW	88000.83
20637	TRW	88000.92
20637	TRW	88001
20638	TRW	88000.85
2063824	WABCO	80171
20639	TRW	88000.90
20-640	NEAPCO	88006.97
206421	WESCO	88003.43
206422	WESCO	88003.42
206423	WESCO	88003.47
206426	WESCO	88003.49
206427	WESCO	88003.45
206428	WESCO	88003.48
206429	WESCO	88003.44
2064313	CHRYSLER	80182
20643J	BORG-WARNR	86730
206443	WESCO	88001.66
20645	TRW	88000.68
20645	TRW	88000.69
206453	WESCO	88003.43
20646	TRW	88000.70
20647	TRW	88000.71
20648	TRW	88000.72
20648	TRW	88000.73
20648J	BORG-WARNR	81750
20648J	BORG-WARNR	81763
20649	TRW	88000.72
2065059	GMC	80228
20650J	BORG-WARNR	82094
20651J	BORG-WARNR	82067
206-521	WESCO	88013.86
20652J	BORG-WARNR	81874
20653J	BORG-WARNR	81765
20655J	BORG-WARNR	84701
20656J	BORG-WARNR	81895
20657J	BORG-WARNR	88006.58
20658J	BORG-WARNR	88006.43
20660J	BORG-WARNR	82332
20661J	BORG-WARNR	82333
20662J	BORG-WARNR	82319
20664J	BORG-WARNR	88006.48
20668	DURA-DET.	80033
2067321	WABCO	80215
2069817	WABCO	80193
206R	PRECISION	80952
206SL11	MACK	80033
206SL11A	MACK	80033
206SL12	MACK	80001
206SL13	MACK	80003
206SL13A	MACK	80003
206SL14	MACK	80007
206SL15	MACK	80003
206SL16	MACK	80003
206SL19	MACK	80178
206SL21	MACK	80001
206SL21	MACK	80178
2-07	BYPY	88004.10
20701	TRW	88013.86
20702	TRW	88013.87
20703	TRW	88002.13
20704	TRW	88001.70
20705	TRW	88002.34
207056-1	SPICER	88000.71
207056-2	SPICER	88000.70
207056-3	SPICER	88000.69
207056-7	SPICER	88000.72
20710	TRW	88002.42
20711	TRW	88002.48
20712	TRW	88002.50
20713	IHC	80023
20713	TRW	88002.55
20725	TRW	88002.34
20726	TRW	88002.33
20727	TRW	88002.33
20-728	NEAPCO	88007.05
20728	TRW	88002.36
20729	TRW	88002.34
20730	TRW	88002.33
20731	TRW	88002.33
2073822	BEAN,JOHN	80003
20-740	NEAPCO	88007.09
20-74-11	SPICER	88014.59
20744	UJS	80077
20744	VERSATILE	80075
20744	VERSATILE	80077
20-752	NEAPCO	88007.09
2075X	NEAPCO	80037
2075X1A	NEAPCO	80037
20768J	BORG-WARNR	81916
207R	PRECISION	80953
20801	TRW	87238
20802	TRW	87331
20803	TRW	87326
20804	TRW	87331
20805	TRW	87285
20806	TRW	87377
20807	TRW	87383
208075	JEEP	80072
208075	KAISER	80072
208075	WILLYS	80072
20808	TRW	87337
20809	TRW	87457
20810	TRW	87459
20811	TRW	87399
20812	TRW	87481
20813	TRW	87570
20814	TRW	88008.63
20815	TRW	88008.35
208-156	WESCO	88002.34
208-157	WESCO	88002.34
208-157	WESCO	88002.36
20816	TRW	87373
20816	TRW	88008.43
20817	TRW	87326
20817	TRW	87331
20817	TRW	88008.39
20818	TRW	87326
20818	TRW	87331
208260014	KOEHRING	80210
208260014(4)	BUFFALO-SP	80214
208260014(4)	BUFFALO-SP	80215
208260015(4)	BUFFALO-SP	80215
208260018(4)	BUFFALO-SP	80215
208260020	KOEHRING	80210
208260020	KOEHRING	80212
208260020(4)	BUFFALO-SP	80214
208260020(4)	BUFFALO-SP	80215
208260021(4)	BUFFALO-SP	80215
20826014	KOEHRING	80210
20826014	KOEHRING	80212
20-828	NEAPCO	88007.04
208308	WESCO	88002.34
208308	WESCO	88002.35
208-308	WESCO	88002.36
208309	WESCO	88002.34
20-840	NEAPCO	88007.08
2084313	CHRYSLER	80182
20846J	BORG-WARNR	81667
2084800	CHRYSLER	80123
208555-1	SPICER	87917
208555-2	SPICER	87917
208555-2	SPICER	87943
208555-4	SPICER	87945
208565-4	BORG-WARNR	87944
208565-4	SPICER	81218
208565-4	SPICER	87918

Part No.	Brand	IDLI Group
208565-4	SPICER	87944
20865	DURA-DET.	80033
208655-1	SPICER	87911
208655-2	SPICER	87932
208663-4	SPICER	87933
208683	FALK	80101
208805-1	SPICER	87920
208805-2	SPICER	87946
208838	WESCO	88002.33
208839	WESCO	88002.33
208-864	WESCO	88002.33
208-865	WESCO	88002.33
20886J	BORG-WARNR	82095
208955-1	SPICER	87914
208955-2	SPICER	87914
208955-2	SPICER	87937
208-959R91	IHC	80007
208963-4	SPICER	87938
208R	PRECISION	80954
20901	TRW	80061
20902	TRW	80104
20903	TRW	80132
20904	TRW	80081
20905J	TRW	80091
20906	TRW	80131
20907	TRW	80129
20-9101	NEAPCO	88014.32
20-9102	NEAPCO	84517
20-9115	NEAPCO	83018
20-9116	NEAPCO	83024
20-9117	NEAPCO	83047
20-9118	NEAPCO	83062
20-9119	NEAPCO	83082
20-9120	NEAPCO	83102
20-9126	NEAPCO	83038
20-9127	NEAPCO	88014.27
20-9131	NEAPCO	85705
20-9132	NEAPCO	88014.32
20-9137	NEAPCO	88014.32
20-9147	NEAPCO	83014
20-9149	NEAPCO	83038
20-9154	NEAPCO	83125
20-9166	NEAPCO	83082
20-9175	NEAPCO	83086
20-9176	NEAPCO	83102
20-9177	NEAPCO	83018
20-9183	NEAPCO	83049
20-9187	NEAPCO	88014.21
20-9189	NEAPCO	83102
20-9190	NEAPCO	83129
20-9194	NEAPCO	83110
20-9216	NEAPCO	83031
20-9216	NEAPCO	83047
20-9217	NEAPCO	83062
20-9218	NEAPCO	83082
20-9218	NEAPCO	83084
20-9301	NEAPCO	83031
20-9301	NEAPCO	83047
20-9302	NEAPCO	83086
20-9303	NEAPCO	83090
20-9307	NEAPCO	83129
20-9310	NEAPCO	83015
20940	DATSUN	80023
20956J	BORG-WARNR	86733
20958J	BORG-WARNR	81754
2098536	GMC	80225
209R	PRECISION	80955
20P1380	MPL.MOLINE	80001
20P1380	WHITE FARM	80001
20P727	MPL.MOLINE	80035
20P727	WHITE FARM	80035
20R0447	GAR WOOD	80008
20R0448	GAR WOOD	80008
20R0449-1	GAR WOOD	80033
21	AEC	87570
21	AEC	88007.79
21	WYLIE	80041
2100	ALCO	80049
2100	PRECISION	88001.16
21000	TRW	88000.19
21-00-00	WALTERSCHD	80041
21001	TRW	88000.22
2100-1140-0-0	RUEDARSA	80099
2100-1150-0-0	RUEDARSA	80109
2100-1205-0-0	RUEDARSA	80174
2100-1280-1-0	RUEDARSA	80049
2100-1280-4-0	RUEDARSA	80045
2100-1285-0-0	RUEDARSA	80049
2100-1290-0-0	RUEDARSA	80032
2100-1356-0-0	RUEDARSA	80030
2100-1357-0-0	RUEDARSA	80031
2100-1451-0-0	RUEDARSA	80193
2100-1556-0-0	RUEDARSA	80210
2100-1703-0-0	RUEDARSA	80092
2100-1813-0-0	RUEDARSA	80158
2100-1815-0-0	RUEDARSA	80159
2100-1850-1-0	RUEDARSA	80048
2100-1900-0-0	RUEDARSA	80110
2100-1900-5-0	RUEDARSA	80106
21002	TRW	88000.25
21003	TRW	88000.45
21004	TRW	88000.62
21005	TRW	88001.01
21005J	BORG-WARNR	81871
21006	TRW	88000.72
21007	TRW	88000.30
21008	TRW	88000.87
210083-1X	SPICER	87377
210083-1X	SPICER	88008.41
210084-2X	SPICER	87457
210084-2X	SPICER	87458
210084-2X	SPICER	87461
210085-1X	SPICER	88008.53
210086-1X	SPICER	88008.47
210086X	SPICER	88008.47
210087-1X	SPICER	87326
210087-1X	SPICER	88008.50
210088-1X	SPICER	87327
210088-1X	SPICER	87331
210088-1X	SPICER	87332
210089-1X	SPICER	87326
21009	TRW	88000.46
210090-1X	SPICER	87328
210090-1X	SPICER	87331
210090-1X	SPICER	87332
210091-1X	SPICER	87326
210092-1X	SPICER	88008.56
21010	TRW	88000.90
21011	TRW	88000.34
2101-1452-0-0	RUEDARSA	80197
2101-1555-0-0	RUEDARSA	80213
2101-1557-0-0	RUEDARSA	80212
210119-1X	SPICER	88008.54
2101-1960-0-0	RUEDARSA	80213
21012	TRW	88000.35
210120-1X	SPICER	88008.46
210120-1X	SPICER	88008.60
210121-1X	SPICER	87481
210128-1X	SPICER	87383
210128-1X	SPICER	88007.79
210128-1X	SPICER	88008.42
210129-1X	SPICER	88008.57
21013	TRW	88000.84
210130-1X	SPICER	87459
210137-1X	SPICER	87331
210137-1X	SPICER	88008.39
210139-1X	SPICER	88008.41
21014	TRW	88000.47
210140-1X	SPICER	87378
210140-1X	SPICER	87383
210140-1X	SPICER	87384
210144-1X	SPICER	87379
210144-1X	SPICER	87383
210144-1X	SPICER	87384
21015	TRW	88000.49
21016	TRW	88000.37
21017	TRW	88000.38
21018	TRW	88000.51
21019	TRW	88000.67
2102	AEC	80054
2102	PRECISION	88001.17
21020	TRW	88014.06
210207-1X	SPICER	87457
210207-1X	SPICER	87460
210207-1X	SPICER	87461
210209-1X	SPICER	87459
210209-1X	SPICER	88008.45
210228-1X	SPICER	88008.50
210228X	SPICER	88008.50
21029	TRW	80010
2103	PRECISION	88001.18
21030J	BORG-WARNR	81667
2103-1070-0-0	RUEDARSA	80180
2103-1080-0-0	RUEDARSA	80142
2103-1082-0-0	RUEDARSA	80140
2103-1090-0-0	RUEDARSA	80041
2103-1090-2-0	RUEDARSA	80041
2103-1090-5-0	RUEDARSA	80041
2103-1210-0-0	RUEDARSA	80024
2103-1210-3-0	RUEDARSA	80024
2103-1330-0-0	RUEDARSA	80072
2103-1332-0-0	RUEDARSA	80117
2103-1401-0-0	RUEDARSA	80124
21033J	BORG-WARNR	81820
210367-1X	SPICER	87326
210367-1X	SPICER	87329
210370-1X	SPICER	87326
210370-1X	SPICER	87330
210391-1X	SPICER	87377
210391-1X	SPICER	87380
210391-1X	SPICER	87382
21039J	BORG-WARNR	81935
2104	PRECISION	88001.19
210400-3900X	SPICER	88006.36
210405-3800X	SPICER	88006.97
210405-3800X	SPICER	88007.11
210406-3700X	SPICER	88007.06
210406-3700X	SPICER	88007.10
210410-1X	SPICER	87383
2104-1291-0-0	RUEDARSA	88002.77
2104-1333-0-0	RUEDARSA	80121
2104-1334-0-0	RUEDARSA	80116
2104-1335-0-0	RUEDARSA	80119
2104-1336-0-0	RUEDARSA	80090
2104-1351-0-0	RUEDARSA	80044
2104-1351-7-0	RUEDARSA	80043
2104-1358-0-0	RUEDARSA	80055
2104-1810-0-0	RUEDARSA	80107
2104-1811-0-0	RUEDARSA	80147
2104-1812-0-0	RUEDARSA	80146
2104-1814-0-0	RUEDARSA	80148
2104-1816-0-0	RUEDARSA	80177
210433-1X	SPICER	87377
210433-1X	SPICER	87381
210433-1X	SPICER	87382
21045115X	MACK	80178
21045124X	MACK	80225
21045153X	MACK	80077
21045155Y	MACK	80182
21045160X	MACK	80142
2104-5165X	MACK	80190
21045170X	MACK	80007
21045178X	MACK	80124
21045185X	MACK	80228
21045188X	MACK	80140
21045200X	MACK	80075
21045213X	MACK	80117
2104-5279X	MACK	80189
2104-5280X	MACK	80221
2104-5281X	MACK	80227
21045299X	MACK	80190
2104-5308X	MACK	80229
2104-5407X	MACK	80226
210471-X	SPICER	88008.35
2105	PRECISION	88001.20
21050	TRW	80061
21051	TRW	88014.52
2105-1070-0-0	RUEDARSA	80178
2105-1120-0-0	RUEDARSA	80189
2105-1120-5-0	RUEDARSA	80189
2105-1310-0-0	RUEDARSA	80221
21052	TRW	81034
210527-1X	SPICER	87238
210527X	SPICER	87238
21053	TRW	81054
2105-3500-0-0	RUEDARSA	80225
21054	TRW	81060
21055	TRW	81089
21056	TRW	81109
21057	TRW	88001.10
21058	TRW	81151
210580X	SPICER	88008.35
21059	TRW	81156
2106	PRECISION	88001.21
21060	TRW	81143
21061	TRW	88001.09
21062	TRW	81175
2106-2165-0-0	RUEDARSA	80181
21065	TRW	84504
210661-1X	SPICER	87570
2107	PRECISION	88001.22
210782X	SPICER	88002.37
210786-1X	SPICER	88007.79
210786R91	IHC	80001
210797-2X	SPICER	87332
210797-2X	SPICER	88007.64
2107-A	MOTOR MAST	88000.15
2107A	MOTOR MAST	88000.15
2107A	MOTOR MAST	88000.15
2107-A	MOTOR MAST	88000.15
2107A	ZELLER	88000.15
2108	PRECISION	88001.23
210851-1X	SPICER	87326
210851-1X	SPICER	87330
210865-1X	SPICER	88008.55
210866-1X	SPICER	87373
210866-1X	SPICER	87374
210866-1X	SPICER	87376
210873-1X	SPICER	87373
210873-1X	SPICER	87375

Part No.	Brand	IDLI Group
210873-1X	SPICER	87376
210874-1X	SPICER	87570
210874-1X	SPICER	88007.79
210875-1X	SPICER	88008.36
2109	PRECISION	88001.24
210921X	SPICER	88008.47
210939CX	SPICER	87328
210939X	SPICER	87328
210939X	SPICER	87331
210939X	SPICER	87570
210939-X	SPICER	88008.39
210969X	SPICER	87457
210969X	SPICER	88008.31
210969X	SPICER	88008.44
210R	PRECISION	80903
210SJ20	MACK	80188
210SJ28	MACK	80077
2110	AEC	80048
2110	PRECISION	88001.25
21100	TRW	88000.01
21-10-00	WALTERSCHD	88003.76
21-1004	LUDWIG	86070
211007	SPICER	88002.34
211007	SPICER	88002.37
211007X	SPICER	88002.34
211008	SPICER	88002.34
211008X	SPICER	88002.34
211009	SPICER	88002.36
211009X	SPICER	88002.36
21100D	TRW	88000.01
21101	TRW	81237
21-1010	LUDWIG	86069
211010X	SPICER	88002.33
21-1011	LUDWIG	86071
211011X	SPICER	88002.33
211012	SPICER	88002.33
211012X	SPICER	88002.33
211013	SPICER	88002.33
211013X	SPICER	88002.33
211016	SPICER	87373
211016X	SPICER	87373
21101J	BORG-WARNR	81888
21101X	NEAPCO	84530
21101X	NEAPCO	84531
21102	TRW	81246
21102D	TRW	81246
21102X	NEAPCO	84529
21103	TRW	81246
211036-2X	SPICER	87331
211036-2X	SPICER	88007.64
211036-2X	SPICER	88008.39
211037-1X	SPICER	87382
21104	NEAPCO	83374
21104	TRW	81268
21105	NEAPCO	87032
21105	TRW	81277
21106	TRW	81274
21107	TRW	88000.05
21107J	BORG-WARNR	81857
21108	TRW	85514
21109	TRW	81301
211098-1X	SPICER	87377
211098-1X	SPICER	88008.41
2111	AEC	80049
2111	PRECISION	88001.26
21110	TRW	81333
21-11-00	WALTERSCHD	88003.94
21111	TRW	81342
21112	TRW	84510
21113	TRW	81259
211139X	SPICER	87331
211149	NEAPCO	83258
21115P11	M.R.S.	80224
21115P11	M.R.S.	80237
21116	NEAPCO	83281
211172-1X	SPICER	87461
211172-1X	SPICER	88007.76
211175	NEAPCO	83317
211177X	SPICER	88002.37
211179X	SPICER	88002.38
21118	NEAPCO	83316
211187X	SPICER	87331
21119	NEAPCO	83334
2112	AEC	80044
2112	PRECISION	88001.27
21120	NEAPCO	83349
21-12-00	WALTERSCHD	88004.04
21121	NEAPCO	83362
21122	NEAPCO	88014.38
21123	NEAPCO	83296
211238	FMC	80077
21124	NEAPCO	83252
21125	NEAPCO	83241
211253	FWD	80001
21126	NEAPCO	83362
21128	NEAPCO	83283
21128J	BORG-WARNR	82092
21129	NEAPCO	83262
2113	PRECISION	88001.28
21130	NEAPCO	83357
211302X	SPICER	87285
211303X	SPICER	87337
211304X	SPICER	87399
21131	NEAPCO	83240
21131	NEAPCO	83262
21132	NEAPCO	83251
21133	NEAPCO	83280
21134	NEAPCO	83320
211342X	SPICER	88002.39
21135	NEAPCO	83250
211355X	SPICER	88002.37
21138	NEAPCO	83320
2114	AEC	80054
2114	ALCO	80054
2114	ALCO	88002.03
2114	PRECISION	88001.29
21140	NEAPCO	87037
21142	NEAPCO	83248
21148	NEAPCO	83263
21149	NEAPCO	83263
2115	PRECISION	88001.30
211-501-331B	VOLKSWAGEN	86070
2115-34	TORO	80109
21154	NEAPCO	83349
21155	NEAPCO	83317
21158	NEAPCO	83316
21158	NEAPCO	83317
211-598-101	VOLKSWAGEN	86070
211-598-101	VOLKSWAGEN	86143
211-598-201	VOLKSWAGEN	86144
2115A	MOTOR MAST	80031
2116	AEC	80047
2116	PRECISION	88001.31
2116	TRU-CROSS	80047
21162X	SPICER	87375
21162X	SPICER	88007.66
21169	NEAPCO	83334
2116A	AEC	80047
2117	AEC	80046
2117	PRECISION	88001.32
21173	NEAPCO	83320
2-1175	NEAPCO	80132
21178	NEAPCO	83331
2118	PRECISION	88001.33
2118A	NEAPCO	83282
2119	PRECISION	88001.34
21190	NEAPCO	83252
21191	NEAPCO	83264
21192	NEAPCO	83349
21199J	BORG-WARNR	82195
2119A	NEAPCO	83315
211A2	NEAPCO	88000
211A6	NEAPCO	83312
211A8	NEAPCO	83314
211B9	NEAPCO	83258
211C11-18397	FORD	80072
211D4	NEAPCO	83374
211D6	NEAPCO	83334
211D6	NEAPCO	83335
211D7X	NEAPCO	84530
211D7X	NEAPCO	84531
211F2	NEAPCO	85712
211G5X19-10	NEAPCO	84923
211P9	NEAPCO	83262
211Z3	NEAPCO	85712
211Z4	NEAPCO	83258
212	TRU-CROSS	80121
212	TRU-CROSS	80131
2120	PRECISION	88001.35
21200	TRW	83015
21201	TRW	83018
21202	TRW	83031
21203	TRW	83062
21204	TRW	83084
21205	TRW	88014.27
21206	TRW	83102
21207	TRW	83116
21207	TRW	88014.32
21208	TRW	84517
21-20877	CURTISS-WR	80140
21209	TRW	88014.32
21209J	BORG-WARNR	82157
2121	PRECISION	88001.36
21210	TRW	83125
21211	TRW	83129
212-1X	SPICER	88002.43
2122	PRECISION	88001.37
21-22-100	WALTERSCHD	88003.95
21-22-101	WALTERSCHD	88004.05
21222	GERLINGER	80001
212238	FMC	80077
212247	FMC	80077
2123	AEC	80045
2123	ALCO	80045
2123	PRECISION	88001.38
212-3X	SPICER	88002.42
2124	PRECISION	88001.39
212-4X	SPICER	88002.44
2125	PRECISION	88001.40
21250	TRW	83240
21251	TRW	83241
21252	TRW	83251
21253	TRW	83252
21254	TRW	83248
21255	TRW	83262
21256	TRW	83280
21257	TRW	83283
21258	TRW	83316
21259	TRW	84529
212-5X	SPICER	88002.44
2126	AEC	80049
2126	ALCO	80049
2126	PRECISION	88001.41
21260	TRW	83320
21261	TRW	83320
21262	TRW	84531
21263	TRW	83331
21264	TRW	83334
21265	TRW	83349
21266	TRW	83374
2127	AEC	80043
2127	AEC	80044
2127	ALCO	80044
2127	PRECISION	88001.42
21279J	BORG-WARNR	82064
2128	PRECISION	88001.43
21280J	BORG-WARNR	81927
21282J	BORG-WARNR	82319
21283J	BORG-WARNR	82120
21283J	BORG-WARNR	85867
2129	PRECISION	88001.44
21292J	BORG-WARNR	81757
21296	GMC	80189
2-13	AEC	88000.68
213	TRU-CROSS	80117
213	TRU-CROSS	80131
2130	PRECISION	88001.45
2131	PRECISION	88001.46
2132	PRECISION	88001.47
21-32-22	SPICER	88015.07
21322P11	M.R.S.	80233
21322P11	M.R.S.	80241
21324J	BORG-WARNR	88006.27
2133	PRECISION	88001.48
21333J	BORG-WARNR	81753
2134	PRECISION	88001.49
2135	PRECISION	88001.50
21350	TRW	88000.18
213501	KAISER	80953
213501	WILLYS	80901
21351	TRW	88000.17
21351	TRW	88000.18
21352	TRW	88000.17
2136	PRECISION	88001.51
2137	PRECISION	88001.52
2138	PRECISION	88001.53
213-882R91	IHC	80182
2139	PRECISION	88001.54
213E4635A	FORD	80041
2140	AEC	80055
2140	ALCO	80055
2140	TRU-CROSS	80055
2-1400	NEAPCO	80042
21400	TRW	88001.13
21401	TRW	88001.14
21402	TRW	88001.12
21403	TRW	85102
21415	TRW	81218
21416	TRW	88006.99
21417	TRW	81209
21418	TRW	88007.06
21419	TRW	88006.97
21431	TRW	88001.12
21446	TRW	88007.05
21447	TRW	88007.05
21448	TRW	88007.04
21449	TRW	88006.97
21451	TRW	87938
21452	TRW	88007.09
21453	TRW	88007.08

Part No.	Brand	IDLI Group	Part No.	Brand	IDLI Group	Part No.	Brand	IDLI Group	Part No.	Brand	IDLI Group
21454	TRW	88006.97	2185	AEC	80047	22008R	TRW	86035	22068R	TRW	86023
21456	TRW	87938	2185901	GMC	80142	22009R	TRW	86036	22069R	TRW	86024
21457	TRW	88007.09	2186	AEC	80046	2201	PRECISION	86002	22070R	TRW	86013
21460	TRW	88007.15	21862	ALLIS-CHLM	80188	22010R	TRW	86058	22071R	TRW	86014
21461	TRW	88001.06	21862	FIAT-ALLIS	80188	22011R	TRW	86059	22072R	TRW	86011
21462	TRW	88007.14	2186972	GMC	80124	22012R	TRW	86056	22073R	TRW	86012
21465	TRW	87933	2186973	GMC	80124	22013R	TRW	86057	22074R	TRW	86062
21466	TRW	88001.06	21869J	BORG-WARNR	81762	22014R	PRECISION	86009	22075R	TRW	86063
21467	TRW	88007.16	219-193R91	IHC	80141	22014R	PRECISION	86009	22076R	TRW	86067
21469	TRW	81848	219-2504	GOODYEAR	80031	22015R	TRW	86008	22077R	TRW	86068
21470	TRW	87933	219-2505	GOODYEAR	80072	22016R	TRW	86031	2207A	MOTOR MAST	80007
21471	TRW	88001.06	219-2506	GOODYEAR	80076	22017R	TRW	86030	2207A	ZELLER	80007
21472	TRW	81848	219-2508	GOODYEAR	80104	22018R	TRW	86044	2207A	ZELLER	88000.15
21479	TRW	88006.37	219-2511	GOODYEAR	80061	220-196R91	IHC	80071	2207B	MOTOR MAST	88000.15
21483	TRW	85101	219-2512	GOODYEAR	80081	220-197R91	IHC	80074	2207B	ZELLER	88000.15
21499	TRW	88006.97	219-2513	GOODYEAR	80091	220197R92	IHC	80074	2207C	MOPAR	88000.16
21500	TRW	88001.15	219-2514	GOODYEAR	80129	22019R	TRW	86042	2207C	MOTOR MAST	80007
21501	TRW	88000.09	219-2515	GOODYEAR	80131	2202	PRECISION	86003	2207C	MOTOR MAST	80023
21502	TRW	88001.08	219-2517	GOODYEAR	80105	22020R	TRW	86043	2207C	MOTOR MAST	88000.15
21502	TRW	88013.82	219-2518	GOODYEAR	80132	22021R	TRW	86041	2207C	MOTOR MAST	88000.16
21535J	BORG-WARNR	87514	219-2519	GOODYEAR	80077	22022R	TRW	86039	2207C	ZELLER	80007
2156475	GMC	80158	219-2520	GOODYEAR	80075	22023R	TRW	86038	2207E	MOTOR MAST	88000.15
215649	FMC	80077	219-2521	GOODYEAR	88002.34	22025R	TRW	86019	2207E	MOTOR MAST	88000.16
2-1569	NEAPCO	80105	219-2523	GOODYEAR	88002.33	22026	TRW	80072	2207G	MOTOR MAST	80007
21584R92	IHC	80189	219-2551	GOODYEAR	80061	22026R	TRW	86020	2207P	MOTOR MAST	80007
2158-89-251	MAZDA	80011	219-2552	GOODYEAR	80075	22027R	TRW	86021	2207-P	MOTOR MAST	88000.14
215R	PRECISION	80904	219-2553	GOODYEAR	80077	22028R	TRW	86022	2207P	ZELLER	88000.14
215R	PRECISION	80905	219-2554	GOODYEAR	80132	22029R	TRW	86015	2207PE	MOTOR MAST	88000.14
216035	WHITE	80178	219-2555	GOODYEAR	80081	2203	PRECISION	86004	22081R	TRW	86064
216035A	WHITE	80178	219-2556	GOODYEAR	80091	22030R	TRW	86016	22081R	TRW	86066
216036	WHITE	80189	219-2556	GOODYEAR	80129	22031R	TRW	86017	22082R	TRW	86065
216036A	WHITE	80189	219-2557	GOODYEAR	80129	22032R	TRW	86018	22083R	TRW	86286
216036A	WHITE	80225	219-2558	GOODYEAR	80131	22033R	TRW	86023	22084R	TRW	86284
216038	WHITE	80225	219-2559	GOODYEAR	80107	22034R	TRW	86024	22085R	TRW	86287
216038A	WHITE	80225	219-2560	GOODYEAR	80104	22035R	TRW	86025	22086R	TRW	86285
216038T2	DIAMOND R	80225	219-303R91	IHC	80123	22036R	TRW	86026	22087R	TRW	86288
216038T2	WHITE	80225	2196041	CHRYSLER	80221	22037R	TRW	86013	22089R	TRW	86291
216343	WHITE	80142	2198	ALCO	80051	22038R	TRW	86014	2209	PRECISION	86008
216343A	WHITE	80142	2198	ALCO	80052	22039R	TRW	86011	22090R	TRW	86292
216448	IHC	80234	2198041	CHRYSLER	80221	2204	PRECISION	86005	22091R	TRW	86294
216-448H1	IHC	80234	2198G	AEC	80051	22040R	TRW	86012	22092R	TRW	86293
216448HI	HOUGH	80234	2198G	AEC	80052	22041R	TRW	86033	22093J	BORG-WARNR	86775
216-477H1	IHC	80219	21C18397	FORD	80072	22042R	TRW	86034	22093R	TRW	86289
2165	RUEDARSA	80181	21C18397B	FORD	80072	22043R	TRW	86010	22094R	TRW	86290
21660J	BORG-WARNR	85825	21C7039	FORD	80072	22044R	TRW	86049	22095R	TRW	86197
21665J	BORG-WARNR	87518	21C7039A	FORD	80072	22045R	TRW	86048	22096R	TRW	86196
21665J	BORG-WARNR	87532	21S	AEC	88007.78	22046R	TRW	86005	22097R	TRW	86198
21667J	BORG-WARNR	87586	21X20910A	DEERE,JOHN	80198	22046R	TRW	86007	22098R	TRW	86199
21682	GMC	80215	21X20910C	DEERE,JOHN	80198	22047R	TRW	86004	22099R	TRW	86200
21683	GMC	80215	22	AEC	87285	22048R	TRW	86005	2210	PRECISION	86009
21684R92	IHC	80189	22	AEC	88008.48	22049R	TRW	86006	22-10-00	WALTERSCHD	88003.79
21686	GMC	80240	22	WYLIE	80040	2205	PRECISION	86006	22100J	BORG-WARNR	85849
217	DIAMOND R	80142	2200	PRECISION	86001	2205-2	NEAPCO	80070	22100R	TRW	86201
217	DIAMOND T	80142	22-00-00	WALTERSCHD	80124	220532R91	IHC	80109	22101R	TRW	86222
217	WHITE	80142	22000R	TRW	86051	220-532R91	IHC	80149	22101R	TRW	86224
217-193R91	IHC	80123	22001R	TRW	86051	22054R	TRW	86027	22102R	TRW	86223
2173	AEC	80058	22001R	TRW	86052	22055R	TRW	86032	22102R	TRW	86225
2173	TRU-CROSS	80058	220-020H1	IHC	80037	22057R	TRW	86053	22103R	TRW	86206
2173956	GMC	80189	2200211	GMC	80003	2205X	NEAPCO	80069	22103R	TRW	86220
2173960	GMC	80221	22002R	TRW	86029	2205X2	NEAPCO	80069	22104R	TRW	86207
2175	AEC	80058	22003R	TRW	86028	2205X2	NEAPCO	80070	22104R	TRW	86227
21762R92	IHC	80221	22-0040(4)	WHITING	80214	2205X-Z	NEAPCO	80070	22105R	TRW	86222
2176J	BORG-WARNR	87488	22004R	AMER.MOTOR	86001	2206	PRECISION	86007	22105R	TRW	86228
217-938H1	IHC	80003	22004R	TRW	86001	22061R	TRW	86045	22106R	TRW	86223
217R	PRECISION	80906	22004R	TRW	86019	22062R	TRW	86046	22106R	TRW	86229
218	DIAMOND R	80079	22005R	AMER.MOTOR	86003	22063R	TRW	86047	22107R	TRW	86226
218	DIAMOND T	80189	22005R	TRW	86003	22064R	TRW	86055	22107R	TRW	86228
218	WHITE	80189	22006R	AMER.MOTOR	86002	22065R	TRW	86054	22108R	TRW	86227
218353	NEW HOLL.	80001	22006R	TRW	86002	22066R	TRW	86060	22108R	TRW	86229
21845J	BORG-WARNR	82113	22007R	TRW	86037	22067R	TRW	86061	22109R	TRW	86224

Part No.	Brand	IDLI Group
2211	PRECISION	86010
22-11-00	WALTERSCHD	88003.97
22-1101	NEAPCO	84531
22-1102	NEAPCO	84529
22-1105	NEAPCO	87032
22110R	TRW	86225
22-1116	NEAPCO	83281
22-1118	NEAPCO	83316
22-1119	NEAPCO	83334
22111R	TRW	86208
22-1120	NEAPCO	83349
22-1121	NEAPCO	83362
22-1122	NEAPCO	88014.38
22-1123	NEAPCO	83296
22-1124	NEAPCO	83252
22-1125	NEAPCO	83241
22-1126	NEAPCO	83362
22-1128	NEAPCO	83283
22-1129	NEAPCO	83262
22112R	TRW	86209
22-1130	NEAPCO	83357
22-1131	NEAPCO	83240
22-1132	NEAPCO	83251
22-1133	NEAPCO	83280
22-1134	NEAPCO	83320
22-1135	NEAPCO	83250
22-1138	NEAPCO	83320
22113R	TRW	86210
22-1140	NEAPCO	87037
22-1142	NEAPCO	83248
22-1148	NEAPCO	83263
22-1149	NEAPCO	83263
22114R	TRW	86202
22-1154	NEAPCO	83349
22-1155	NEAPCO	83317
22-1158	NEAPCO	83317
22115R	TRW	86203
22116R	TRW	86204
22-1178	NEAPCO	83331
22117R	TRW	86205
22118R	TRW	86211
22-1190	NEAPCO	83252
22-1191	NEAPCO	83264
22119R	TRW	86212
2212	PRECISION	86011
22120R	TRW	86213
22-1218	NEAPCO	83280
22-1218	NEAPCO	83281
22-1219	NEAPCO	83316
22121R	TRW	86214
22122R	TRW	86215
22123R	TRW	86218
22124R	TRW	86219
22125R	TRW	86221
22126R	TRW	86230
22127R	TRW	86231
22128R	TRW	86233
22-129	NEAPCO	81848
22129R	TRW	86232
2213	PRECISION	86012
22-1301	NEAPCO	88000
22-1303	NEAPCO	83312
22-1305	NEAPCO	83314
22-1308	NEAPCO	83258
22130R	TRW	86195
22-1316	NEAPCO	83374
22-1319	NEAPCO	84531
22131R	TRW	86194
22-1325	NEAPCO	85712
22-1343	NEAPCO	85712
22-1344	NEAPCO	83258
221-350	FORD	80008
22-139	NEAPCO	81848
221396H1	IHC	80212
2214	PRECISION	86013
2215	PRECISION	86014
221501149	VOLKSWAGEN	86070
221501149	VOLKSWAGEN	86144
221501331B	VOLKSWAGEN	86070
22158J	BORG-WARNR	81680
2216	PRECISION	86015
221-613	FORD	80037
22167J	BORG-WARNR	87496
22168J	BORG-WARNR	85560
2217	PRECISION	86016
2218	PRECISION	86017
2219	PRECISION	86018
22-19-00	WALTERSCHD	88003.86
221R	PRECISION	88001.66
221R	PRECISION	88001.74
2220	PRECISION	86019
2221	PRECISION	86020
2222	PRECISION	86021
22223567	GMC	80178
2222567	GMC	80178
2222584	GMC	80189
2222853	GMC	80221
2222854	GMC	80189
2223	PRECISION	86022
2223547	GMC	80178
2223567	GMC	80178
2223568	GMC	80142
2223569	GMC	80142
222361	NEW HOLL.	80001
22-239	NEAPCO	81848
2224	PRECISION	86023
2224005	GMC	80225
2225	MOTOR MAST	80025
2225	PRECISION	86024
2225B	MOTOR MAST	80024
2225B	ZELLER	80024
2225C	MOTOR MAST	80023
2225C	ZELLER	80023
2225C	ZELLER	80024
2225D	MOTOR MAST	80024
2225E	MOTOR MAST	80026
2225E	ZELLER	80026
2225W	MOTOR MAST	80025
2225-W	ZELLER	80025
2-225X(8)	SPICER	86114
2226	PRECISION	86025
22263J	BORG-WARNR	88008.29
2226A	MOTOR MAST	80031
2226B	MOTOR MAST	80030
2226B	ZELLER	80030
2227	PRECISION	86026
2-22746	CURTISS-WR	80211
2-22746	CURTISS-WR	80212
2-22746	FMC	80207
2-22746	GEN.EQUIP.	80215
2-22746	WAYNE	80207
2-22746	WOOLDRIDGE	80210
2-2275	NEAPCO	80069
2-2276	CURTISS-WR	80207
2-2276	NEAPCO	80070
2228	PRECISION	86027
2229	PRECISION	86028
2-22991H1	FMC	80235
2-22991	WAYNE	80235
2230	PRECISION	86029
22300	TRW	86139
22300	TRW	86191
22301	TRW	86175
22302	TRW	86114
22302	TRW	86176
22-30-22	SPICER	88015.08
22303	TRW	86192
22304	TRW	86187
22304	TRW	86188
22305	TRW	86161
22306	TRW	86162
22307	TRW	86189
2230781	GMC	80124
22308	TRW	86159
22309	TRW	86160
2231	PRECISION	86030
22310	TRW	86180
22311	TRW	86069
22311	TRW	86143
22311	TRW	86146
22312	TRW	86181
22313	TRW	86184
22313	TRW	86185
22314	TRW	86081
22314	TRW	86153
22315	TRW	86084
22315	TRW	86156
22316	TRW	86082
22316	TRW	86154
22317	TRW	86151
22318	TRW	86152
22319	TRW	86186
2232	PRECISION	86031
22320	TRW	86234
22321	TRW	86183
2232136	GMC	80189
22322	TRW	86234
22323	TRW	86083
22323	TRW	86155
22324	TRW	86190
22325	TRW	86163
22326	TRW	86302
22327	TRW	86301
22328	FORD	86126
22328	TRW	86182
22-329	NEAPCO	88001.06
22329	TRW	86150
2233	PRECISION	86032
22330	TRW	86183
22331	TRW	86149
22332	TRW	86279
22333	TRW	86269
22334	TRW	86270
22335	TRW	86271
22336	TRW	86272
22337	TRW	86282
22338	TRW	86273
22-339	NEAPCO	85114
22-339	NEAPCO	88001.06
22339	TRW	86274
2234	PRECISION	86033
22340	TRW	86275
22341	TRW	86276
22342	TRW	86280
22343	TRW	86179
22344	TRW	86148
22345	TRW	86147
22346	TRW	86179
22347	TRW	86283
22348	TRW	86281
22349	TRW	86150
22349	TRW	86277
2235	PRECISION	86034
2235	TRW	86193
22350	TRW	86157
22-351	NEAPCO	85114
22351	TRW	86268
22352	TRW	86178
22353	TRW	86177
22353	TRW	86278
22353J	BORG-WARNR	88006.85
22353J	BORG-WARNR	88006.86
22354	TRW	86193
22355	TRW	86158
22356	TRW	86174
223572R93	IHC	87285
223572R93	IHC	88008.48
223573R93	IHC	87337
22358	TRW	86173
2236	PRECISION	86035
2236223	GMC	88008.67
2237	PRECISION	86036
2238	PRECISION	86037
2239	PRECISION	86038
2-23991	CURTISS-WR	80235
2-23991	FMC	80235
2-23991	WAYNE	80235
2-23991	WOOLDRIDGE	80235
224	PRECISION	80956
2240	PRECISION	86039
2-24006	CURTISS-WR	80240
2-24006	FMC	80240
2-24006	WAYNE	80240
2-24006	WOOLDRIDGE	80231
2240231	CHRYSLER	80007
2241	PRECISION	86040
2242	PRECISION	86041
224226	GALION	80198
22-429	NEAPCO	88007.15
2243	PRECISION	86042
22-439	NEAPCO	88007.15
2244	PRECISION	86043
224446(7)	GALION	80198
2245	PRECISION	86044
2246	PRECISION	86045
2247	PRECISION	86046
2248	PRECISION	86047
224890	ALLIS-CHLM	80001
224890	ALLIS-CHLM	80003
224890	FIAT-ALLIS	80001
2249	PRECISION	86048
224R	PRECISION	80956
225	HERCULES	80122
2250	PRECISION	86049
22500	TRW	86114
22500	TRW	86139
22501	TRW	86140
22502	TRW	86114
22503	TRW	86115
22504	TRW	86114
22504	TRW	86116
22505	TRW	86115
22505	TRW	86117
22506	TRW	86134
22506	TRW	86135
22506	TRW	86188
22507	TRW	86095
22508	TRW	86096
22508	TRW	86097
22509	TRW	86136

Part No.	Brand	IDLI Group	Part No.	Brand	IDLI Group	Part No.	Brand	IDLI Group	Part No.	Brand	IDLI Group
2251	DANUSER	80033	22564	TRW	86239	2275B	ZELLER	80041	2298621	GMC	80142
2251	PRECISION	86050	22564	TRW	86253	2275X	NEAPCO	80069	2298823	CHRYSLER	80124
22510	TRW	86133	22564	TRW	86254	2275X3	NEAPCO	80070	2298908	CHRYSLER	80061
22511	TRW	86122	22565	TRW	86241	2276	PRECISION	86067	2298909	CHRYSLER	80061
22511	TRW	86180	22566	TRW	86243	2277	PRECISION	86068	2298940	CHRYSLER	80906
22511	VOLKSWAGEN	86180	22567B	TRW	86265	22794	TULSA WIN.	80008	2298941	CHRYSLER	80905
22512	TRW	86069	22568	TRW	86234	2-2800	NEAPCO	80159	22GB	AEC	80180
22512	TRW	86072	22569	TRW	86235	22802J	BORG-WARNR	81844	22UJ12	OSHKOSH	80001
22513	TRW	86124	2257	PRECISION	86054	22806	GMC	80110	22UJ12	OSHKOSH	80003
22514	TRW	86137	22570	TRW	86259	22806	GMC	80112	22UJ28	OSHKOSH	80213
22515	TRW	86072	22571	TRW	86258	22807	FIAT-ALLIS	80111	22UJ48	OSHKOSH	80238
22515	TRW	86099	22572	TRW	86074	22807	GMC	80111	22UJ63	OSHKOSH	80001
22516	TRW	86128	22573	TRW	86075	2-28-1127	SPICER	85704	23	AEC	87337
22517	TRW	86081	22574	TRW	86236	228161X	SPICER	88001.15	23	AEC	88008.51
2-2518	NEAPCO	80066	22576	TRW	86142	228162X	HAYES	88000.09	23	AEC	88008.52
22518	TRW	86089	22577	TRW	86141	228163X	SPICER	88000.18	23	WYLIE	80124
22518	TRW	86118	22578	TRW	86113	228217-3316X	HAYES	88001.06	2300	PRECISION	86069
22519	NEAPCO	86119	2258	PRECISION	86055	228217-3316X	SPICER	88001.06	23-00-00	WALTERSCHD	88004.32
22519	TRW	86090	22580	TRW	86123	228217-3316X	SPICER	88007.16	2300-2303	PRECISION	86143
22520	TRW	86088	22580	VOLKSWAGEN	86180	228226-2502X	SPICER	88007.05	2301	PRECISION	86070
22521	TRW	86129	22581	TRW	86125	228226-35	SPICER	88007.09	23010	LAKEWOOD	80061
22522	TRW	86130	22582	TRW	86249	228226-3502X	HAYES	88007.09	230104	SPICER	88013.64
22523	TRW	86091	22582	TRW	86251	228226-3502X	SPICER	88007.09	23011	LAKEWOOD	80104
22524	TRW	86092	22582	TRW	86253	228247-3416X	SPICER	81848	2-3011	NEAPCO	80104
22525	TRW	86131	22582	TRW	86254	228305X	SPICER	88000.09	230118	SPICER	88013.73
22526	TRW	86078	22583	TRW	86120	228305X	SPICER	88001.08	230119-1	SPICER	88008.63
22527	TRW	86086	22583	TRW	86121	228311X	NEAPCO	88000.17	230119-1X	SPICER	88008.63
22528	TRW	86087	22583	TRW	86250	228311X	SPICER	88013.84	23012	LAKEWOOD	80132
22-529	NEAPCO	88001.06	22583	TRW	86251	228317X	SPICER	88000.17	2-3012	NEAPCO	80108
22529	TRW	86093	22584	CHRYSLER	86082	228317X	SPICER	88000.18	230121	SPICER	88013.65
2253	PRECISION	86051	22585	TRW	86083	228317X	SPICER	88013.85	230122	SPICER	88007.55
22530	TRW	86094	22586	TRW	86084	228397X	HAYES	88000.09	230124	SPICER	88013.76
22531	TRW	86132	22587	TRW	86079	228397X	SPICER	88000.09	230126-1	SPICER	88008.64
22532	TRW	86085	22588	TRW	86080	228417X	SPICER	88006.97	23013	LAKEWOOD	80081
22533	TRW	86184	22589	TRW	86238	228418X	SPICER	81848	23014	LAKEWOOD	80091
22534	TRW	86234	22589	TRW	86263	2284935	GMC	80178	230141	SPICER	88013.70
22536	TRW	86183	22590	TRW	86240	22855J	BORG-WARNR	81665	230147-3	SPICER	88008.65
22538	TRW	86098	22591	TRW	86242	2285935	GMC	80178	230148	SPICER	88007.56
22538	TRW	86138	22592	TRW	86244	228596X	SPICER	88006.37	230151	SPICER	88013.78
22-539	NEAPCO	88001.06	22593	TRW	86252	228606-3000X	SPICER	85101	2301520	GMC	80221
22539	TRW	86098	22594	TRW	86245	228626-3010X	SPICER	85102	2301520	GMC	88001.79
2254	PRECISION	86052	22595	TRW	86120	22863J	BORG-WARNR	82338	2301544	GMC	88001.78
22540	TRW	86296	22596	TRW	86234	22898J	BORG-WARNR	87514	23016	LAKEWOOD	80129
22541	TRW	86300	2260	PRECISION	86056	22902J	BORG-WARNR	85582	230164	SPICER	88007.57
22542	TRW	86298	2-2600	NEAPCO	80148	229091-3012X	SPICER	88007.23	230166	SPICER	88013.75
22543	NEAPCO	86297	2261	PRECISION	86057	229092-3020X	SPICER	88007.22	230167-1	SPICER	88008.66
22543	TRW	86301	2261160	BEAN,JOHN	80243	2290A	MOTOR MAST	80072	2301775	GMC	88001.81
22544	TRW	86126	2262	PRECISION	86058	2290A	ZELLER	80072	23018	LAKEWOOD	80097
22545	TRW	86077	22623	GMC	80215	2290B	MOTOR MAST	80089	2302	PRECISION	86071
22547	TRW	86127	22-629	NEAPCO	88007.14	2290B	ZELLER	80076	230207	SPICER	88013.77
22548	TRW	86299	2263	PRECISION	86059	2290B	ZELLER	80089	230208	SPICER	88013.67
22549	NEAPCO	86076	2263882	GMC	80142	2290D	MOTOR MAST	80073	230209	SPICER	88013.66
22549	TRW	86076	22-639	NEAPCO	88007.15	2290D	MOTOR MAST	80091	23021	LAKEWOOD	80124
2255	PRECISION	86053	2264	PRECISION	86060	2290D	ZELLER	80073	23025	DEERE,JOHN	80043
22550	TRW	86295	2265	PRECISION	86061	2290UBK2T	MOTOR MAST	88013.86	2303	PRECISION	86072
22-551	NEAPCO	88001.06	226-800R91	IHC	80145	22910C91	IHC	80227	230363	SPICER	88013.69
22551	TRW	86255	2270	PRECISION	86062	229116-3227	SPICER	88007.25	2304	PRECISION	86073
22552	TRW	86248	2-2700	NEAPCO	80054	229116-3227X	SPICER	88000.12	230565	SPICER	88015.35
22553	TRW	86257	22708J	BORG-WARNR	81694	229116-5923X	SPICER	88007.27	230566	SPICER	88015.40
225538	HYSTER	80163	22708J	BORG-WARNR	88006.49	229152X	SPICER	88000.13	230590	SPICER	88015.43
22554	TRW	86246	2271	PRECISION	86063	229153X	SPICER	88013.85	23063J	BORG-WARNR	81830
22555	TRW	86247	22712J	BORG-WARNR	81843	2294607	GMC	80007	2307	PRECISION	86074
22556	TRW	86260	22719J	BORG-WARNR	81694	2295043	CHRYSLER	80190	230746	SPICER	88013.80
22557	TRW	86261	22724J	BORG-WARNR	81684	2296407	GMC	80007	230749-1	SPICER	88008.67
22558	TRW	86262	2273	PRECISION	86064	2296407	GMC	88000.15	230750	SPICER	88007.51
22559	TRW	86263	2274	PRECISION	86065	2297A	MOTOR MAST	80075	2308	PRECISION	86075
22560	TRW	86264	2275	PRECISION	86066	2297A	ZELLER	80075	2308019	BEAN,JOHN	80185
22561	TRW	86266	2275514	CHRYSLER	88003.51	2297AU	MOTOR MAST	80076	2308820	BEAN,JOHN	80001
22562	TRW	86267	2275515	CHRYSLER	80905	2297AV	MOTOR MAST	80076	230909	SPICER	88013.79
22563	TRW	86237	2275B	MOTOR MAST	80041	229840	CHRYSLER	80906	2309277	GMC	87325

Part No.	Brand	IDLI Group
2309277	GMC	88007.59
2309287	FIAT-ALLIS	80124
231	AEC	80132
231	PRECISION	80128
2310	PRECISION	86076
2-3100	NEAPCO	80083
231000	SPICER	88013.68
231004	SPICER	88013.74
231009	VOLVO	80221
2-3102	NEAPCO	80112
231047	SPICER	88013.81
2311	PRECISION	86077
2-3111	NEAPCO	80107
231136	VOLVO	80077
231179	VOLVO	80077
231189	VOLVO	80077
2312	PRECISION	86078
2313	PRECISION	86079
231311	VOLVO	80077
2315	PRECISION	86080
2316	PRECISION	86081
2317	PRECISION	86082
2318	PRECISION	86083
2319	PRECISION	86084
2-3190	NEAPCO	80144
232	DIAMOND R	80075
232	PRECISION	80126
232	WHITE	80075
2320	PRECISION	86085
2321	PRECISION	86086
232-1X	SPICER	88002.57
2322	PRECISION	86087
2322134	GMC	80178
2322135	GMC	80178
2322136	GMC	80189
2322137	GMC	80221
2322138	GMC	80221
2323	PRECISION	86088
23232051	SAVIEM	80041
23232053	SAVIEM	80142
23.232.055	RENAULT	80180
23232055	SAVIEM	80178
23232057	SAVIEM	80189
23.232.059	RENAULT	80221
23232059	SAVIEM	80221
2324	PRECISION	86089
2325	PRECISION	86090
2325029	GMC	80072
2325829	GMC	80072
2326	PRECISION	86091
2327	PRECISION	86092
2328	PRECISION	86093
2329	PRECISION	86094
233	AEC	80062
2330	PRECISION	86160
2330	PRECISION	86168
2330781	GMC	80124
2331	PRECISION	86095
2331-2B	MOTOR MAST	80034
2331-2B	ZELLER	80034
2332	PRECISION	86096
23327J	BORG-WARNR	85603
23328J	BORG-WARNR	87594
2333	PRECISION	86097
23330J	BORG-WARNR	87600
23332J	BORG-WARNR	87517
23-337-041	BARREIROS	80225
2333914	GMC	88008.54
2333918	GMC	88014.86
2333919	GMC	88008.64
2334	PRECISION	86098
2335005	GMC	80225
2335006	GMC	80225
233572R93	IHC	87285
2336424	GMC	88008.65
2336727	GMC	80225
2337	PRECISION	86099
2338	PRECISION	86100
2338-2339	PRECISION	86164
2338929	GMC	88008.56
2339	PRECISION	86101
23391J	BORG-WARNR	82138
23392J	BORG-WARNR	82138
23393J	BORG-WARNR	84657
23394J	BORG-WARNR	84684
23395J	BORG-WARNR	84664
23396J	BORG-WARNR	84668
23397J	BORG-WARNR	84655
23398J	BORG-WARNR	88006.50
23399J	BORG-WARNR	81782
2340	PRECISION	86102
23400J	BORG-WARNR	81940
23401J	BORG-WARNR	82102
23402J	BORG-WARNR	81719
23404J	BORG-WARNR	82079
2341	PRECISION	86103
234-1009-5	BEAN,JOHN	80001
2341009-5	FMC	80001
2341648	GMC	80140
2342	PRECISION	86104
23423(4)	TULSA WIN.	80215
2343	PRECISION	86105
2343325	GMC	87326
2344	PRECISION	86106
2344-2349	PRECISION	86170
2345	PRECISION	86107
2345726	GMC	80140
2346	PRECISION	86108
2347	PRECISION	86109
2348	PRECISION	86110
2349	PRECISION	86111
2-350	NEAPCO	80044
2-350	NEAPCO	88001.98
2350	PRECISION	86112
2-3500	NEAPCO	80049
23505J	BORG-WARNR	82061
2351	PRECISION	86113
23-51-00	WALTERSCHD	88003.81
2351948	GMC	87383
2351948	GMC	87384
2351948	GMC	88008.42
2351949	GMC	87459
23526J	BORG-WARNR	87583
2353	PRECISION	86114
2354	PRECISION	86115
2354320	GMC	80072
2354320	GMC	80077
2354324	GMC	80178
2355	PRECISION	86116
2356	PRECISION	86117
2358	PRECISION	86118
2359	PRECISION	86119
23590J	BORG-WARNR	82093
2360	PRECISION	86120
236092R1	IHC	88013.72
2361	PRECISION	86121
23-61-007	LILLISTON	80069
2362830	GMC	80077
2362830	GMC	80078
2362830	GMC	80085
23635J	BORG-WARNR	84688
23647J	BORG-WARNR	82154
23651J	BORG-WARNR	81797
2366104	GMC	80221
2366104	GMC	88001.79
2366108	ALLIS-CHLM	80221
2366108	FIAT-ALLIS	80221
2366108	GMC	80221
2368954	GMC	80179
236G	LEMPCO	80031
2370	PRECISION	86122
2370-2371	PRECISION	86180
2370728	GMC	80075
2370733	GMC	80221
2371	PRECISION	86123
2372	PRECISION	86124
2372959	FIAT-ALLIS	80101
2373	PRECISION	86125
23749J	BORG-WARNR	81702
2375D	MOTOR MAST	80042
2375D	ZELLER	80042
2375D	ZELLER	80182
2376	PRECISION	86126
237697R91	IHC	80199
2377	PRECISION	86127
2378764	GMC	80162
2378764	GMC	80179
2378954	GMC	80179
237-897R91	IHC	80199
237-899R91	IHC	80215
2379	PRECISION	86128
2380	PRECISION	86129
238-075R91	IHC	80193
238-077R91	IHC	80215
238-079R91	IHC	80240
2381	PRECISION	86130
2382	PRECISION	86131
2383	PRECISION	86132
238-374R91	IHC	80199
238-386R91	IHC	80239
2384	PRECISION	86133
238-448R91	IHC	80193
238-449R91	IHC	80215
238-450R91	IHC	80240
2385	PRECISION	86134
23850J	BORG-WARNR	85557
2386	PRECISION	86135
2-3861	AMER.SNOW	80233
2-3861(4)	AMER.SNOW	80233
2386331	GMC	80003
2387	PRECISION	86136
238-739R91	IHC	80158
238-740R91	IHC	80158
238-796R91	IHC	80216
238-797R91	IHC	80174
238-798R91	IHC	80174
2387A	MOTOR MAST	80065
2387A	ZELLER	80064
2387A	ZELLER	80065
2387BCK	MOTOR MAST	88001.70
2387BCK2	MOTOR MAST	88001.70
2387BCK2	ZELLER	88001.70
2387D	MOTOR MAST	80057
2387E	MOTOR MAST	80060
2387E	MOTOR MAST	80061
2387E	ZELLER	80060
2387E	ZELLER	80061
2387E	ZELLER	80062
2387EA	MOTOR MAST	80062
2387EA	ZELLER	80062
2387EHP	MOTOR MAST	80061
2387F	MOTOR MAST	80063
2387F	ZELLER	80060
2387F	ZELLER	80063
2387FT	MOTOR MAST	80061
2387UBK2T	MOTOR MAST	88001.70
2388	PRECISION	86137
238-800R91	IHC	80173
238-801R91	IHC	80173
23885J	BORG-WARNR	81878
2389	PRECISION	86138
239004	BROCKWAY	80142
239040	BROCKWAY	80124
239044	BROCKWAY	80142
239046	BROCKWAY	80178
239049	BROCKWAY	80189
239052	BROCKWAY	80221
239068	BROCKWAY	80221
239131	BROCKWAY	80190
239220	BROCKWAY	80225
239225	BROCKWAY	80189
2393	PRECISION	86139
239325	BROCKWAY	80189
239326	BROCKWAY	80189
239327	BROCKWAY	80221
2394	PRECISION	86140
23940J	BORG-WARNR	82051
23942C1	BORG-WARNR	81813
23942CI	BORG-WARNR	81813
239-551R91	IHC	80228
2396407	GMC	80007
239-723R91	IHC	80221
239-737R91	IHC	80207
239-738R91	IHC	80207
239-739R91	IHC	80211
2398	PRECISION	86141
2398294	GMC	80186
2398295	GMC	80184
2399	PRECISION	86142
23999J	BORG-WARNR	82037
24	AEC	87399
24	AEC	88008.59
24	WYLIE	80142
2400	PRECISION	86143
24-00-00	WALTERSCHD	80140
2400938	GMC	80182
2400E4635A	FORD	80041
2401	PRECISION	86144
24-01-00	WALTERSCHD	88004.35
2-4-0172	SPICER	83349
240-178R91	IHC	80182
240-190R91	IHC	80182
2402	PRECISION	86145
2402703	GMC	88014.87
2402704	GMC	88008.60
2402705	GMC	88007.57
2402706	GMC	88013.75
2402707	GMC	88008.66
24027J	BORG-WARNR	82366
2402E4635A	FORD	80077
24030J	BORG-WARNR	81671
2403229	ALLIS-CHLM	80190
2403229	FIAT-ALLIS	80190
2403229	GMC	80190
2403700	GMC	80090
2404	PRECISION	86146
24047(4)	TULSA WIN.	80149
24048(4)	TULSA WIN.	80149
2404877	GMC	80007
2405P2	DIAMOND R	80125

Part No.	Brand	IDLI Group
2405P2	WHITE	80124
2406	PRECISION	86147
24063J	BORG-WARNR	81741
24064J	BORG-WARNR	88006.52
2406989	GMC	87481
2407	PRECISION	86148
2409783	CHRYSLER	80090
2409783	GMC	80090
2410	PRECISION	86149
24-10-00	WALTERSCHD	88003.82
241006	WESCO	80001
241007	WESCO	80003
241008	WESCO	88001.15
2-4-102	HAYES	83038
2-4-102	SPICER	83038
2411	PRECISION	86150
241133	WHITING	80047
2-4-1163	SPICER	83110
2-4-1173	SPICER	83102
2412	PRECISION	86151
2-4-122	SPICER	83040
2-4-1263	SPICER	83018
2413	PRECISION	86152
2-4-1333	SPICER	83021
2-4-1343	HAYES	83018
2-4-1343	SPICER	83010
2-4-1353	SPICER	83047
2-4-1363	HAYES	83047
2-4-1363	SPICER	83031
2-4-1373	SPICER	83057
2-4-1383	SPICER	83057
2-4-1393	SPICER	83047
2-4-1403	SPICER	83086
2-4-1413	HAYES	83086
2-4-1413	SPICER	83085
2-4-142	SPICER	83008
2-4-1423	HAYES	83086
2-4-1423	SPICER	83082
2-4-1433	SPICER	83082
2-4-1443	SPICER	83102
2-4-1453	SPICER	83108
2-4-1463	HAYES	83129
2-4-1463	SPICER	83129
2-41485	CURTISS-WR	88001.80
2-41485	FMC	80221
2-41485	WAYNE	80221
2-41485(5)	CURTISS-WR	88001.80
2-4-152	SPICER	83015
2-4-153	SPICER	83015
2-4-1563	SPICER	83086
2-4-1593	SPICER	83057
2415CHA	CASE,J.I.	80033
2416	PRECISION	86153
2-4-1603	SPICER	83057
2-4-1613	HAYES	83129
2-4-162	SPICER	83040
2-4-1623	SPICER	83125
2-4-1633	SPICER	83102
2417	PRECISION	86154
2-4-1703	SPICER	83102
2-4-172	SPICER	83092
2417789	GMC	80190
2417CHA	CASE,J.I.	80069
2418	PRECISION	86155
24-18-00	WALTERSCHD	88003.88
2-4-182	SPICER	83038
2-4-1843	SPICER	83102
2-4185	CURTISS-WR	80221
2-4185	WAYNE	80221
2-4-1853	SPICER	83082
2-4-1863	HAYES	81333
2-4-1873	HAYES	83110
2-4-1873	SPICER	83102
2419	PRECISION	86156
2-4-192	SPICER	83015
2-4-1943	SPICER	83062
242	PERFECT-CI	80035
242	TRU-CROSS	80035
2420606	GMC	80003
2422	PRECISION	86268
2423	PRECISION	86157
2-4-232	SPICER	83014
242521	ROVER	80040
2-4-2731	SPICER	87012
2428	PRECISION	86158
2429	PRECISION	86159
2-4-2971	SPICER	88014.21
2-43	AEC	88000.70
243	PERFECT-CI	80069
243	TRU-CROSS	80069
2430	PRECISION	86160
24-30-12	SPICER	88015.10
243023	ROVER	80041
24-30-32	SPICER	88015.13
2-4-3041X	SPICER	88014.32
24-30-42	SPICER	88015.11
2-4-3061	SPICER	87013
2-4-3061X	SPICER	87013
2-4-3071	SPICER	83054
2-4-3081	SPICER	83090
2-4-3091X	SPICER	83090
243-1	PERFECT-CI	80070
2431	PRECISION	86161
243-1	TRU-CROSS	80070
2-4-3101	SPICER	83116
2-4-3111	SPICER	83105
2432	PRECISION	86162
243203	ROVER	80041
24-32-142	SPICER	88015.09
243-250	FORD	80037
24-32-62	SPICER	88015.10
2434	PRECISION	86163
2434-13A	NEAPCO	88001.07
243-424	FORD	80069
2-4-3481X	SPICER	88014.32
24351-1(4)	SIL.HOIST	80214
24351-1(4)	SIL.HOIST	80240
24351-2(4)	SIL.HOIST	80150
24351-2(4)	SIL.HOIST	80214
24351-2(4)	SIL.HOIST	80240
24351(4)	SIL.HOIST	80196
24351(4)	SIL.HOIST	80240
2-4-352	SPICER	83038
2-4-3681	SPICER	88014.21
2437	WOOD BROS.	80035
2438	PRECISION	86164
2-4-3981	SPICER	83106
2440	PRECISION	86165
2441	PRECISION	86166
2442	PRECISION	86167
2-4-4211	SPICER	84517
2-4-4211X	HAYES	84517
2-4-4211X	SPICER	84517
2443	PRECISION	86168
2444	PRECISION	86169
2445	PRECISION	86170
2447	PRECISION	86171
24475J	BORG-WARNR	81692
2448	PRECISION	86172
2448100	GMC	80072
2448382	GMC	80075
245	CRANE CAR.	80235
2450	PRECISION	86173
2450093	GMC	80124
24508	ALLIS-CHLM	80178
24508	FIAT-ALLIS	80178
2451	PRECISION	86174
24513J	BORG-WARNR	86843
24525J	BORG-WARNR	81834
2453	PRECISION	86175
24539J	BORG-WARNR	81933
2455	COCKSHUTT	80008
2455	PRECISION	86176
2456	PRECISION	86177
2457	PRECISION	86178
245-8	ALLIS-CHLM	80178
245-8	FIAT-ALLIS	80178
2459X	NEAPCO	88001.08
2460	PRECISION	86179
2460455	GMC	80075
2464A	MOTOR MAST	80039
2464A	ZELLER	80039
2464AB	MOTOR MAST	80038
2464AB	MOTOR MAST	80039
2464AB	ZELLER	80038
2464B	MOTOR MAST	80038
24657J	BORG-WARNR	81745
24658J	BORG-WARNR	82069
2466726	GMC	80003
247	CRANE CAR.	80216
2470	PRECISION	86180
2470889	GMC	88008.50
2471	PRECISION	86181
24710J	BORG-WARNR	88015.15
2-4-72	SPICER	83040
2473119	GMC	87330
2476	PRECISION	86182
2477	PRECISION	86183
2479	PRECISION	86184
248	AEC	80025
248	CRANE CAR.	80193
248	TRU-CROSS	80025
2-4800	NEAPCO	80117
2-4800	NEAPCO	80129
2482	PRECISION	86185
2483	PRECISION	86186
2485	PRECISION	86187
2486	PRECISION	86188
2487	PRECISION	86189
2487A	MOTOR MAST	80068
2487A	ZELLER	80068
2489	PRECISION	86190
249	CRANE CAR.	80215
2-4900	NEAPCO	80131
2493	PRECISION	86191
2494	PRECISION	86192
2495043	CHRYSLER	80190
2496A	MOTOR MAST	80071
2496A	ZELLER	80071
2496B	MOTOR MAST	80074
2496B	ZELLER	80074
2496C	MOTOR MAST	80080
2498	PRECISION	86145
2498	PRECISION	86158
2498	PRECISION	86193
2499	PRECISION	86193
25	WYLIE	80180
250	CRANE CAR.	80240
2500	AEC	80059
2500	ALCO	80059
250033	JOY MFG.	80216
2505767	CHRYSLER	80187
2505767	GMC	80187
2508	FIAT-ALLIS	80178
251	DANUSER	80033
251	TEDDY TORQ	80056
251012	WESCO	80035
251015	WESCO	88000.09
25102X	NEAPCO	83349
25105	NEAPCO	87032
251-072	WESCO	80035
2513	CARDWELL	80037
2514402	CHRYSLER	87331
25167X	NEAPCO	83349
2518	AEC	80066
252	AEC	80034
252	TEDDY TORQ	80056
252-015	WESCO	88001.08
2520A	MOTOR MAST	80033
25228J	BORG-WARNR	86739
252382	MUIR-HILL	80172
2525A	MOTOR MAST	80035
2525A	ZELLER	80035
25281N(4)	SIL.HOIST	80149
2528D	MOTOR MAST	80011
2-53	NEAPCO	80124
2-53	NEAPCO	80125
2530	TRU-CROSS	80066
2531	AEC	80066
2533202	CHRYSLER	80117
2537A	MOTOR MAST	80037
2537A	ZELLER	80037
2-54	NEAPCO	80142
2-54	NEAPCO	80143
25420	HYSTER	80023
2-55-102	SPICER	87227
2-55-202	SPICER	88015.39
255-341R91	IHC	80215
255608	ALLIS-CHLM	80140
255608	ALLIS-CHLM	80182
255608	FIAT-ALLIS	80140
2-55-62	HAYES	88015.39
2-55-62	SPICER	88015.39
25586	NEAPCO	80142
255-90	WESCO	88001.70
2-55-92	SPICER	88000.10
255923	MUIR-HILL	80210
2-5700	NEAPCO	80064
2-5700	NEAPCO	80065
2571A	MOTOR MAST	80008
25759(4)	SIL.HOIST	80214
25759(4)	SIL.HOIST	80215
25776-11(4)	SIL.HOIST	80214
25776-11(4)	SIL.HOIST	80215
25776-1(4)	SIL.HOIST	80214
25776-1(4)	SIL.HOIST	80215
25776(4)	SIL.HOIST	80214
25776(4)	SIL.HOIST	80215
25806J	BORG-WARNR	87637
2585WB2	NEAPCO	80184
25896(4)	SIL.HOIST	80214
25896(4)	SIL.HOIST	80215
25897A	HYSTER	80124
259	CRANE CAR.	80215
2-5900	NEAPCO	80121
25933	SCHULTZ	80008
2593A	MOTOR MAST	80069
2593A	MOTOR MAST	80070
2593A	ZELLER	80069
2593B	MOTOR MAST	80069

Part No.	Brand	IDLI Group
259668	FIAT-ALLIS	80198
259770R91	IHC	80213
259770R92	IHC	80213
259-770R93	IHC	80213
25991-1(4)	SIL.HOIST	80171
25991-24	SIL.HOIST	80168
25991-3(4)	SIL.HOIST	80215
26	WYLIE	80178
260	AEC	80078
260	TRU-CROSS	80078
260	TRU-CROSS	80090
26001	AEC	86075
26001	AEC	86148
26002	AEC	86179
26002	AEC	86251
26002	AEC	86252
26002	AEC	86254
26002	AEC	86255
26002	AEC	86256
26002	AEC	86257
26002	AEC	86282
26003	AEC	86149
26003	AEC	86295
26004	AEC	86183
26005	AEC	86143
26006	AEC	86180
26007	AEC	86147
26007	AEC	86236
26008	AEC	86157
26009	AEC	86178
26009	AEC	86246
26009	AEC	86247
26010	AEC	86193
26011	AEC	86184
2-6023	AMER.PARTS	87331
2-6025	AMER.PARTS	88008.64
2-6026	AMER.PARTS	88008.65
2-6028	AMER.PARTS	88008.35
2-6035	AMER.PARTS	87238
2-6035	AMER.PARTS	88007.58
2-6036	AMER.PARTS	87326
2-6037	AMER.PARTS	87331
2-6037	AMER.PARTS	88008.39
2-6038	AMER.PARTS	87377
2-6038	AMER.PARTS	88008.41
2-6039	AMER.PARTS	87383
2-6039	AMER.PARTS	88008.42
2-6040	AMER.PARTS	87457
2-6040	AMER.PARTS	88008.31
2-6040	AMER.PARTS	88008.44
2-6041	AMER.PARTS	87459
2-6041	AMER.PARTS	88008.45
2605X	NEAPCO	80140
261	CRANE CAR.	80235
26-100	MOTOR MAST	88002.43
26101	AEC	86191
26-101	MOTOR MAST	88002.42
26-101	ZELLER	88002.42
261014	WESCO	80037
26102	AEC	86189
26-102	MOTOR MAST	88002.44
26-102	ZELLER	88002.44
26103	AEC	86162
26104	AEC	86151
26105	AEC	86152
26-105	MOTOR MAST	88002.48
26-106	MOTOR MAST	88002.50
26108	AEC	86155
26109	AEC	86156
26110	AEC	86153
261103	MUIR-HILL	80221
26111	AEC	86080
26111	AEC	86085
26111	AEC	86154
261118	MUIR-HILL	80142
26112	AEC	86177
26113	AEC	86268
26114	AEC	86190
26115	AEC	86163
26116	AEC	86150
26117	AEC	86174
26121	AEC	86166
26123	AEC	86160
26124	AEC	86165
26125	AEC	86166
26125	SUBARU	86166
26126	AEC	86169
26126A	HYSTER	80023
26127	AEC	86161
26128	AEC	86191
26-128	MOTOR MAST	88002.55
26129	AEC	86166
26-129	NEAPCO	88006.37
26130	AEC	86166
26-139	NEAPCO	88006.37
26181J	BORG-WARNR	88006.32
2-6200	NEAPCO	80123
26-229	NEAPCO	85101
26-239	NEAPCO	85101
26-251	NEAPCO	85101
262781R92	IHC	80007
2629713	CHRYSLER	80189
2-63	AEC	88000.71
2-6302	NEAPCO	80063
2630272	CHRYSLER	88002.77
26-329	NEAPCO	85102
264	TRU-CROSS	80008
2-6400	NEAPCO	80071
26-429	NEAPCO	88007.14
26-439	NEAPCO	88007.16
26487J	BORG-WARNR	81950
26496MK-D	BLISS	80215
265	AEC	80064
2-6500	NEAPCO	80074
265-005-0	GLAENZER	80024
265-007-0	GLAENZER	80024
265-008-0	GLAENZER	80023
265-014-1	GLAENZER	80023
265-032-0	GLAENZER	80023
265-032-1	GLAENZER	80023
265-033-1	GLAENZER	80023
265-034-0	GLAENZER	80023
265-037-0	GLAENZER	80024
265-038-0	GLAENZER	80024
265-040-0	GLAENZER	80023
265-049-0	GLAENZER	80024
2-6508	AMER.PARTS	87285
2-6508	AMER.PARTS	88008.48
2-6509	AMER.PARTS	87373
2-6509	AMER.PARTS	88008.43
2-6510	AMER.PARTS	87325
2-6510	AMER.PARTS	88007.59
26-5105	NEAPCO	87032
2-6511	AMER.PARTS	87337
2-6511	AMER.PARTS	88008.51
2-6511	AMER.PARTS	88008.52
26512	AMER.PARTS	87399
26512	AMER.PARTS	87399
2-6512	AMER.PARTS	87399
2-6512	AMER.PARTS	87399
2-6512	AMER.PARTS	88008.59
2-6513	AMER.PARTS	87481
2-6513	AMER.PARTS	88007.77
2-6514	AMER.PARTS	87570
2-6514	AMER.PARTS	88007.79
266	AEC	80061
2-6600	NEAPCO	80141
26601	HYSTER	80101
26612	HYSTER	80021
266-1310	AEC	80083
266-1X	SPICER	88002.50
267	AEC	80063
2670487	GMC	80178
26726	ALLIS-CHLM	80069
26726	FIAT-ALLIS	80069
26727	ALLIS-CHLM	80069
26727	FIAT-ALLIS	80069
26728	FIAT-ALLIS	80069
2675X	NEAPCO	80140
2698416	GMC	80142
27	WYLIE	80189
2-70-28X	SPICER	88001.94
2-70-38X	SPICER	88001.70
271035	WESCO	80069
271038	WESCO	88001.08
27106	BRADY	80161
271240	FIAT-ALLIS	80198
27-128	MOTOR MAST	88002.55
271-421R91	IHC	80238
271421R92	IHC	80238
27199W	DIAMOND T	80140
2-7-29	SPICER	88014.90
2-73	AEC	88000.72
27322	HYSTER	80100
27323	HYSTER	80101
27418	HYSTER	80124
27418A	HYSTER	80124
27424	KENBAR	88008.35
27460	DROTT	80109
27460	DROTT	80155
27471	DROTT	80109
27471	DROTT	80155
2750	CARDWELL	80069
2765513	GMC	87326
278	AEC	80085
278	TRU-CROSS	80085
278	TRU-CROSS	80086
27814J	BORG-WARNR	81662
2781A	PETTI-MULL	80109
278269	SPICER	88015.41
27837	HYSTER	80147
278635	SPICER	88015.42
278814	SPICER	88015.45
27889	SCHULTZ	80035
278D225A(4)	PARSONS	80149
278D225A(4)	PARSONS	80150
278D255A	KOEHRING	80109
278D255A	KOEHRING	80150
279	AEC	80189
279	TRU-CROSS	80189
2790-100M	REPUBLIC	88000.21
2790-100M	REPUBLIC	88000.39
2790-105M	REPUBLIC	88000.39
2790-127M	REPUBLIC	88000.22
2790-129M	REPUBLIC	88000.27
2790-131M	REPUBLIC	88000.28
2790-133M	REPUBLIC	88000.31
2790-150M	REPUBLIC	88000.25
2790-169M	REPUBLIC	88000.28
2790-184M	REPUBLIC	88000.53
2790-190M	REPUBLIC	88000.31
2790-195M	REPUBLIC	88000.35
2790-197M	REPUBLIC	88000.38
2790-202M	REPUBLIC	88000.38
27-93-27-10800	SPICER	87554
27-93-27-5400	SPICER	87553
279470R91	IHC	80198
279-470R92	IHC	80198
27969J	BORG-WARNR	81955
27993	NEW HOLL.	80001
279-986R91	IHC	80171
279J	BORG-WARNR	81955
28	WYLIE	80221
280	AEC	80221
280	TRU-CROSS	80221
2800	NEAPCO	80205
2800A	MOTOR MAST	80004
2800A	MOTOR MAST	80006
2800A	ZELLER	80004
2800A	ZELLER	80006
2801	NEAPCO	80215
280100	NEAPCO	80049
280100	NEAPCO	80050
280150	NEAPCO	80050
2802	NEAPCO	80112
280200	NEAPCO	80048
280200	NEAPCO	80051
280200	NEAPCO	88002.10
280200X	NEAPCO	80075
280200XB	NEAPCO	80076
2803	NEAPCO	80235
280300	NEAPCO	80043
280300	NEAPCO	80049
280350	NEAPCO	80043
2804	NEAPCO	80234
280400	NEAPCO	80044
280450	NEAPCO	80044
2805	NEAPCO	80198
280500	NEAPCO	80049
28050X	NEAPCO	80092
28051X	NEAPCO	80072
28051X	NEAPCO	80075
28051XB	NEAPCO	80072
28051XB	NEAPCO	80075
28051XB	NEAPCO	80089
28052X	NEAPCO	80092
28052X	NEAPCO	88001.76
28053X	NEAPCO	80124
28053XB	NEAPCO	80125
28054X	NEAPCO	80142
28054XB	NEAPCO	80134
28054XB	NEAPCO	80143
280550	NEAPCO	80049
28055X	NEAPCO	80178
28055X	NEAPCO	80225
28056	NEAPCO	80135
280564	NEAPCO	80135
280568X	NEAPCO	80135
28056X	NEAPCO	80135
28057X	NEAPCO	80122
28058X	NEAPCO	80189
28059X	NEAPCO	80221
2805X	NEAPCO	80041
2806	NEAPCO	80212
280600	NEAPCO	80099
280600G	NEAPCO	80101
2807	NEAPCO	80211
280700	NEAPCO	80110
28075A	AMER.MOTOR	80041
28075A	AMER.MOTOR	88001.85

Part No.	Brand	IDLI Group
2808	NEAPCO	80210
280800	NEAPCO	80109
28089	NEAPCO	88013.86
28089	NEAPCO	88013.87
2809	NEAPCO	80237
280900	NEAPCO	80045
28099	NEAPCO	88013.86
28099	NEAPCO	88013.87
281	AEC	80227
281	TRU-CROSS	80227
2810	AUTO BIAN.	80142
2810	NEAPCO	80231
28100	NEAPCO	80030
281000	NEAPCO	80096
28100X	NEAPCO	80030
28101X	NEAPCO	80023
281045	WESCO	88000.13
28105X	NEAPCO	80031
28108X	NEAPCO	80043
2811	AUTO BIAN.	80142
2811	NEAPCO	80234
281100	NEAPCO	88001.60
28111X	NEAPCO	80024
281175	NEAPCO	80132
2812	AUTO BIAN.	80142
2812	NEAPCO	80233
281200	NEAPCO	80064
281202X	NEAPCO	80130
28124X	NEAPCO	80225
2813	NEAPCO	80232
281300	NEAPCO	80135
28133X	NEAPCO	80116
28134X	NEAPCO	80118
2814	NEAPCO	80238
281400	NEAPCO	80042
2815	NEAPCO	80178
281500	NEAPCO	80030
281501	NEAPCO	80021
28150X	NEAPCO	80030
28150X	NEAPCO	80031
28150X	NEAPCO	80119
281521	NEAPCO	80021
28152X	NEAPCO	80118
28153X	NEAPCO	80072
28153X	NEAPCO	80075
28153X	NEAPCO	80077
28153XB	NEAPCO	80088
28154X	NEAPCO	80142
281550	NEAPCO	80030
281550G	NEAPCO	80031
281551	NEAPCO	80031
281552	NEAPCO	80030
281553	NEAPCO	80031
281554	NEAPCO	80032
281555	NEAPCO	88002.77
281556	NEAPCO	88000.16
281557	NEAPCO	80012
28155G	NEAPCO	80182
28155X	NEAPCO	80182
281569	NEAPCO	80105
2816	NEAPCO	80189
281600	NEAPCO	80138
281606	NEAPCO	80005
281606	NEAPCO	80011
281608	NEAPCO	80013
281610	NEAPCO	80020
28165X	NEAPCO	80190
2817	NEAPCO	80221
28170	NEAPCO	80007
281700	NEAPCO	80106
28170X	NEAPCO	80007
28174X	NEAPCO	80078
28174X	NEAPCO	80090
2817X	NEAPCO	80007
2818	NEAPCO	80225
281800	NEAPCO	88001.62
28181X	NEAPCO	88008.68
28185X	NEAPCO	80228
281869C91	IHC	88008.56
28186X	NEAPCO	88001.82
28187X	NEAPCO	88001.83
28188X	NEAPCO	80140
28188X	NEAPCO	80182
281900	NEAPCO	80022
281FR	NEAPCO	80008
2820	NEAPCO	80168
282000	NEAPCO	80068
28200X	NEAPCO	80072
28200X	NEAPCO	80075
28200XB	NEAPCO	80076
2821	NEAPCO	80109
282100	NEAPCO	80053
282116	NEAPCO	80047
282117	NEAPCO	80046
282134A	NEAPCO	80050
28213X	NEAPCO	80117
282171	NEAPCO	80048
282172	NEAPCO	80047
282173	NEAPCO	80058
282175	NEAPCO	80058
282185	NEAPCO	80047
282185A	NEAPCO	80047
282186	NEAPCO	80046
282186A	NEAPCO	80046
2822	NEAPCO	80101
282200	NEAPCO	80114
282202X	NEAPCO	80130
2-82-22-2706X	SPICER	88007.04
28226X	NEAPCO	80225
282300	NEAPCO	80147
28231X	NEAPCO	80121
2824	NEAPCO	80198
282400	NEAPCO	80146
28244X	NEAPCO	80075
28244XB	NEAPCO	80076
28248X	NEAPCO	80025
2825	NEAPCO	80198
28250	MOTOR MAST	80011
282500	NEAPCO	80158
282518	NEAPCO	80066
2825A	MOTOR MAST	80010
2825A	ZELLER	80010
2825B	MOTOR MAST	80009
2825B	ZELLER	80009
2825C	MOTOR MAST	80012
2825C	ZELLER	80012
2825D	MOTOR MAST	80011
2825D	ZELLER	80011
282600	NEAPCO	80148
2827	NEAPCO	80213
282700	NEAPCO	80054
28278X	NEAPCO	80086
28278X	NEAPCO	80189
28279X	NEAPCO	80189
2828	NEAPCO	80238
282800	NEAPCO	80159
28280X	NEAPCO	80221
28280X4	NEAPCO	80221
28281X	NEAPCO	80227
2-82-82-2706X	SPICER	85126
2-82-82-3714X	HAYES	88007.08
2-82-82-3714X	SPICER	88007.08
2829	NEAPCO	80155
282900	NEAPCO	80111
28297	NEAPCO	80079
28297X	NEAPCO	80002
28297X	NEAPCO	80079
2830	NEAPCO	80150
28300	NEAPCO	80044
283000	NEAPCO	80168
28300X	NEAPCO	80023
28300X	NEAPCO	80024
28300X	NEAPCO	80140
28300XS	NEAPCO	80024
283011	NEAPCO	80104
283011	NEAPCO	80107
283011C	NEAPCO	80108
28301X	NEAPCO	80023
28-30-22	SPICER	88015.17
283031	NEAPCO	80109
28-30-42	SPICER	88015.20
28-30-52	SPICER	88015.19
28-30-62	SPICER	88015.18
283066C1	IHC	88013.78
283067	IHC	87461
283067C91	IHC	87457
283067C92	IHC	88008.31
283067C92	IHC	88008.44
283068C1	IHC	88013.75
283069C91	IHC	87481
283069C92	IHC	87481
283069C92	IHC	88007.77
283071C91	IHC	88007.79
283071C92	IHC	88007.79
28308X	NEAPCO	80229
28-30-92	SPICER	88015.15
2831	NEAPCO	80232
283100	NEAPCO	80083
283100	NEAPCO	80177
283101	NEAPCO	80100
283102	NEAPCO	80112
283103	NEAPCO	80098
283105	NEAPCO	80104
283105B	NEAPCO	80104
283111	NEAPCO	80107
283125	NEAPCO	80103
283130	NEAPCO	80097
283133	NEAPCO	80144
283140	NEAPCO	80127
283152	NEAPCO	80101
283154	NEAPCO	80113
283154A	NEAPCO	80113
28315J	BORG-WARNR	88006.31
28315X	NEAPCO	80026
283160	NEAPCO	80112
283190	NEAPCO	80077
283190	NEAPCO	80144
283200	NEAPCO	80174
28-32-22	SPICER	88015.17
28-32-42	SPICER	88015.16
2833	NEAPCO	80170
283300	NEAPCO	80094
283300	NEAPCO	80096
28332X	NEAPCO	80139
2834	NEAPCO	80069
283400	NEAPCO	80173
2835	NEAPCO	80168
283500	NEAPCO	80049
283500	NEAPCO	88001.96
283500P	NEAPCO	80049
28350X	NEAPCO	80036
2836	NEAPCO	80197
283600	NEAPCO	80044
283600	NEAPCO	88001.98
283600P	NEAPCO	80044
283650	NEAPCO	80055
2837	NEAPCO	80197
283700	NEAPCO	80051
283700	NEAPCO	80052
2838	NEAPCO	80158
283800	NEAPCO	80095
2839	NEAPCO	80153
283900	NEAPCO	80045
283DN3	NEAPCO	80133
283DR	NEAPCO	80133
283DRN	NEAPCO	80133
2840	NEAPCO	80109
284000	NEAPCO	80193
28407X	NEAPCO	80226
2841	NEAPCO	80153
284103	NEAPCO	80156
284103	NEAPCO	88003.16
284105	NEAPCO	80150
284123	NEAPCO	80150
284123	NEAPCO	80153
284132	NEAPCO	80153
284138	NEAPCO	80109
284138	NEAPCO	80149
284140	NEAPCO	80155
284140X	NEAPCO	80155
284141	NEAPCO	80154
284143	NEAPCO	80156
284143	NEAPCO	88003.17
284152	NEAPCO	80158
284156	NEAPCO	80157
2842	NEAPCO	80156
2843	NEAPCO	80198
284300	NEAPCO	80095
284300	NEAPCO	88001.60
284400	NEAPCO	80057
2844N	NEAPCO	80136
2844R	NEAPCO	80136
2844X	NEAPCO	80136
2845	NEAPCO	80069
284500	NEAPCO	80060
2845N	NEAPCO	80134
2846	NEAPCO	80238
284600	NEAPCO	80056
284600	NEAPCO	80057
284635A	NEAPCO	80073
284635C	NEAPCO	80081
284635D	NEAPCO	80073
284635D	NEAPCO	80091
2847	NEAPCO	80213
284700	NEAPCO	80034
284700G	NEAPCO	80064
284730	NEAPCO	80093
284730X	NEAPCO	80093
2848	NEAPCO	80140
284800	NEAPCO	80117
284800	NEAPCO	80129
28481	NEAPCO	80023
2849	NEAPCO	80124
284900	NEAPCO	80118
284900	NEAPCO	80131
285	LOBRO	80122
285000	NEAPCO	80210
285031	NEAPCO	80171
2851	NEAPCO	80048
285100	NEAPCO	80107

Part No.	Brand	IDLI Group	Part No.	Brand	IDLI Group	Part No.	Brand	IDLI Group	Part No.	Brand	IDLI Group
285101	NEAPCO	80167	286114	NEAPCO	80196	288105	NEAPCO	80238	28T13	NEAPCO	80009
285102	NEAPCO	80170	286114	NEAPCO	80197	288113	NEAPCO	80238	28T14	NEAPCO	80017
285102	NEAPCO	80171	286123	NEAPCO	80191	288-1X	SPICER	88002.50	28T14	NEAPCO	80018
285108	NEAPCO	80172	286128	NEAPCO	80198	288-1X	SPICER	88002.52	28T20	NEAPCO	80115
285109	NEAPCO	80162	286143	NEAPCO	80193	288200	NEAPCO	80231	290-312C91	IHC	80090
285110	NEAPCO	80173	286143	NEAPCO	80198	288201	NEAPCO	80232	29047J	BORG-WARNR	86757
285111	NEAPCO	80175	286152	NEAPCO	80195	288202	NEAPCO	80237	29056J	BORG-WARNR	86755
285112	NEAPCO	80168	286152	NEAPCO	80197	288203	NEAPCO	80236	29057	NEW HOLL.	80163
285112	NEAPCO	80177	286162	NEAPCO	80199	288204	NEAPCO	80238	290728R91	IHC	80172
285122	NEAPCO	80168	286163	NEAPCO	80201	288205	NEAPCO	80231	2908812	CHRYSLER	87331
285127	NEAPCO	80163	2862	NEAPCO	80223	288205	NEAPCO	80240	290-928R91	IHC	80216
285127	NEAPCO	80167	286200	NEAPCO	80123	288206	NEAPCO	80232	291	PRECISION	80095
285132	NEAPCO	80174	286200	TRW	80141	288206	NEAPCO	80242	291-297	FORD	80003
285148	NEAPCO	80169	28626J	BORG-WARNR	82376	288207	NEAPCO	80237	2912B	MOTOR MAST	80023
285148	NEAPCO	80172	2862830	GMC	80077	288207	NEAPCO	80239	2912B	MOTOR MAST	80024
285148	NEAPCO	80216	2863	NEAPCO	80223	288208	NEAPCO	80236	29145J	BORG-WARNR	81786
28514X	NEAPCO	80024	286300	NEAPCO	80060	288209	MOTOR MAST	80238	291-736R91	IHC	80101
285152	NEAPCO	80195	286300	NEAPCO	80062	288209	NEAPCO	80238	291866C1	IHC	88008.65
285154	NEAPCO	80175	2863006	NEAPCO	80061	2882264	CHRYSLER	80063	291867C1	IHC	88008.66
285173	NEAPCO	80171	286300G	NEAPCO	80060	2882487	CHRYSLER	80062	291869C91	IHC	88008.33
285177	NEAPCO	80172	286300G	NEAPCO	80061	288500	NEAPCO	80223	291870C91	IHC	88008.60
285200	NEAPCO	80160	286300GB	NEAPCO	80061	288516	NEAPCO	80223	291872C1	IHC	88007.56
2852175	CHRYSLER	80132	286300GB	NEAPCO	80063	288516	NEAPCO	80224	291873C1	IHC	88007.57
2852175	DETROIT	80132	2-86-348	SPICER	88002.41	288-6013-91	CHAIN BELT	80134	291875C1	IHC	88014.88
2853	NEAPCO	80178	2864	NEAPCO	80069	28862	ALLIS-CHLM	80178	291876C1	IHC	88014.87
285300	NEAPCO	80181	286400	NEAPCO	80071	28862	FIAT-ALLIS	80178	291-971	FORD	80001
28530X	NEAPCO	80117	2865	NEAPCO	80188	2887A	MOTOR MAST	80019	291971	FORD	80003
28531X	NEAPCO	80121	286500	NEAPCO	80071	2887A	ZELLER	80019	292	PRECISION	80057
2854	NEAPCO	80125	286500	NEAPCO	80074	2887D	MOTOR MAST	80020	292000	NEAPCO	80075
285400	NEAPCO	80038	2866	NEAPCO	80102	2888WB10	NEAPCO	80233	2-9201	NEAPCO	88002.34
2855	NEAPCO	80143	286600	NEAPCO	80141	2888WB9	NEAPCO	80235	2920-100M	REPUBLIC	88000.69
285500	NEAPCO	80204	286600	TRW	80141	2889	NEAPCO	80104	2920-139M	REPUBLIC	88000.70
285554	NEAPCO	80081	2867	NEAPCO	80103	2890	NEAPCO	80167	2920-160M	REPUBLIC	88000.71
2855N	NEAPCO	80161	286700	NEAPCO	80145	289000	NEAPCO	80235	2920-169M	REPUBLIC	88000.72
285600	NEAPCO	80039	2868	NEAPCO	80172	289014	NEAPCO	80235	2-9203	NEAPCO	88002.33
2857	NEAPCO	80174	286800	NEAPCO	80183	289015	NEAPCO	80234	2-9205	NEAPCO	88002.33
285700	NEAPCO	80064	2868WB10	NEAPCO	80218	289016	NEAPCO	80234	29214	NEW HOLL.	80069
2857006	NEAPCO	80065	2868WB9	NEAPCO	80217	289017	NEAPCO	80233	2924	NEW HOLL.	80069
285700G	NEAPCO	80064	2869	NEAPCO	80156	289017	NEAPCO	80241	29258	NEW HOLL.	80107
285700G	NEAPCO	80065	2871	WILK RIP.	80035	289026	NEAPCO	80233	2-93	AEC	88000.73
28573	ALLIS-CHLM	80178	287101	NEAPCO	80211	289026	NEAPCO	88003.09	293	PRECISION	80034
28573	FIAT-ALLIS	80178	287102	NEAPCO	80212	28-904-361	BARREIROS	80234	2-9301	NEAPCO	88002.34
2858	NEAPCO	80162	287105	NEAPCO	80213	2891	NEAPCO	80168	2-9302	NEAPCO	88002.36
285800	NEAPCO	80059	287126	NEAPCO	80213	2892	NEAPCO	80168	2-9303	NEAPCO	88002.33
28581	NEAPCO	80023	2872	NEAPCO	80035	2893	NEAPCO	80171	2-9305	NEAPCO	88002.33
28581X	NEAPCO	80023	287200	NEAPCO	80210	2894	NEAPCO	80092	2934	DANUSER	80001
28581X	NEAPCO	80024	287201	NEAPCO	80211	2895	NEAPCO	80024	293-551C91	IHC	80228
28589X	NEAPCO	80205	287202	NEAPCO	80212	2897	NEAPCO	80175	293D608(4)	PARSONS	80216
2858N	NEAPCO	80044	287203	NEAPCO	80208	2899	NEAPCO	80172	294	PRECISION	88001.58
2858N	NEAPCO	80162	287203	NEAPCO	80213	28CP45N	NEAPCO	80134	2-94-18X	SPICER	88013.87
2858N	NEAPCO	80179	287204	NEAPCO	80209	28CP4N	NEAPCO	80134	2-94-28X	SPICER	88013.86
2858WB	NEAPCO	80185	287205	NEAPCO	80210	28CP55N	NEAPCO	80161	2-94-58X	SPICER	88002.13
2858WB1	NEAPCO	80186	287205	NEAPCO	80215	28CP5N	NEAPCO	80178	294-769R91	IHC	80149
2858WB2	NEAPCO	80184	287206	NEAPCO	80207	28CP6N	NEAPCO	80188	295	PRECISION	88001.59
2858WN1	NEAPCO	80186	287206	NEAPCO	80211	28CP7N	NEAPCO	80220	2950025	EATON	80172
2858WN2	NEAPCO	80184	287207	NEAPCO	80212	28CP8N	NEAPCO	88012.23	2950199	EATON	80234
2859	NEAPCO	80174	287207	NEAPCO	80214	28CV82	NEAPCO	88002.37	2950350	EATON	80037
285900	NEAPCO	80081	287208	NEAPCO	80208	28L10N2	NEAPCO	80033	2950390	EATON	80037
2860	NEAPCO	80193	287209	NEAPCO	80209	28L10S	NEAPCO	80033	2950409	EATON	80219
2861	NEAPCO	80049	28737	HYSTER	80147	28L14N	NEAPCO	80037	29537	NEW HOLL.	80033
286100	NEAPCO	80193	2876	MURR.TREG.	80216	28L14S	NEAPCO	80037	29538	NEW HOLL.	80008
286101	NEAPCO	80195	28783J	BORG-WARNR	81925	28L16S	NEAPCO	80069	295-578R91	IHC	80101
286101	NEAPCO	80211	2878WB13	NEAPCO	80243	28L6N8	NEAPCO	80001	296	PRECISION	80056
286102	NEAPCO	80197	2878WB14	NEAPCO	80244	28N26	NEAPCO	80024	29-61-002	LILLISTON	80037
286103	NEAPCO	80194	2879	NEAPCO	80182	28N27	NEAPCO	80010	296-595R91	IHC	80099
286104	NEAPCO	80192	2880	NEAPCO	80190	28N28	NEAPCO	80004	296596R91	IHC	80168
286106	NEAPCO	80198	288014	NEAPCO	80230	28N29	NEAPCO	80019	296-596R91	IHC	80216
286107	NEAPCO	80202	288100	NEAPCO	80231	28N6	NEAPCO	80024	2968	YOUNG FIRE	80210
286108	NEAPCO	80199	288101	NEAPCO	80232	28T11	NEAPCO	80006	297	AEC	80079
286109	NEAPCO	80201	288102	NEAPCO	80237	28T12	NEAPCO	80018	297	PRECISION	88001.60

Part No.	Brand	IDLI Group	Part No.	Brand	IDLI Group	Part No.	Brand	IDLI Group	Part No.	Brand	IDLI Group
297	TRU-CROSS	80079	2L2	LEMPCO	81020	3000443	GAR WOOD	80109	301326	WESCO	80113
2970	YOUNG FIRE	80212	2L22	LEMPCO	88000.28	3000840	ADAMS-LET.	80212	301-327	WESCO	80097
298	PRECISION	88001.61	2L26	LEMPCO	88000.53	3000840	WABCO	80214	301354	WESCO	88001.62
298-6012-91	CHAIN BELT	80178	2L27	LEMPCO	88000.31	3000842	WABCO	80171	301372	WESCO	80135
298-6024-91	CHAIN BELT	80220	2L30	LEMPCO	88000.35	3000844	WABCO	80191	3013809	ALLIS-CHLM	80178
298-6035-91	CHAIN BELT	80238	2L31	LEMPCO	88000.35	3000845	WABCO	80198	3013809	FIAT-ALLIS	80178
298-6035-91	REX CHAIN	80238	2L31	LEMPCO	88000.38	3000845	WABCO	80203	3013809-3	ALLIS-CHLM	80178
298-6044-91	CHAIN BELT	80215	2L32	LEMPCO	88000.38	30011020	FIAT-ALLIS	80069	3013809-3	FIAT-ALLIS	80178
298-6044-91	REX CHAIN	80213	2L4	LEMPCO	88000.21	3001449-3	GAR WOOD	80003	3014	ALCO	80114
298-6056-91	CHAIN BELT	80172	2L4	LEMPCO	88000.25	3001845	GAR WOOD	80133	301402	WESCO	80109
298-6056-91	REX CHAIN	80172	2L4	LEMPCO	88000.27	3002018	GAR WOOD	80133	301404	WESCO	80151
299	PRECISION	88001.62	2L4	LEMPCO	88000.39	300332	WESCO	80139	301407	WESCO	80147
2990674	EATON	80213	2L45	LEMPCO	88000.70	3-0044	NEAPCO	80136	301408	WESCO	80146
29982J	BORG-WARNR	88006.91	2L49	LEMPCO	88000.71	3-0045	NEAPCO	80161	301410	WESCO	80153
29983J	BORG-WARNR	88006.46	2L52	LEMPCO	88000.39	3004542	CHRYSLER	87327	301413	WESCO	80156
29984J	BORG-WARNR	88006.60	2L52	LEMPCO	88000.72	3004690-8	FIAT-ALLIS	80101	301414	WESCO	80158
2A16588-1	BORG-WARNR	80196	2L72	LEMPCO	88000.81	30-04-813	HARDY SPCR	88003.76	301414	WESCO	88002.15
2A5-116X	FWD	80189	2L76	LEMPCO	88000.31	3004900	AEC	80061	301414	WESCO	88002.16
2A5-160X	FWD	80003	2L76	LEMPCO	88000.88	3004900	AEC	80062	301415	WESCO	80148
2A858	CATERPILLR	80215	2M215	CATERPILLR	88001.24	3004900	CHRYSLER	80062	301416	WESCO	80159
2B0001	BORG-WARNR	88000.21	2M2378(2)	CATERPILLR	88001.23	3004900	DETROIT	80061	301417	WESCO	80154
2B0017	BORG-WARNR	88000.25	2M5001	CATERPILLR	88001.23	3004900	DETROIT	80062	301419	WESCO	80157
2B0027	BORG-WARNR	88000.31	2MU4	OSHKOSH	80047	3004900	REPUBLIC	80062	301421	WESCO	80909
2B0032	BORG-WARNR	88000.38	2QDK	WESCO	88000.17	3-0055	NEAPCO	80178	301430	WESCO	80155
2B1	BORG-WARNR	88000.21	2R801	CATERPILLR	80234	3-0056	NEAPCO	80135	301-430	WESCO	80156
2B1	MOTOR MAST	88000.21	2R960	CATERPILLR	80140	30061X	NEAPCO	80021	301441	WESCO	88001.66
2B10	MOTOR MAST	88000.22	2S471	CATERPILLR	88001.40	3006545	FIAT-ALLIS	80048	301482	BMC	80041
2B13	BORG-WARNR	88000.22	2S476(2)	CATERPILLR	88001.40	3006546	ALLIS-CHLM	80048	301482	LEYLAND	80041
2B17	BORG-WARNR	88000.25	2T2910	BORG-WARNR	88002.43	3006546	ALLIS-CHLM	80049	301501	WESCO	80167
2B17	MOTOR MAST	88000.25	2T2910	BORG-WARNR	88002.59	3006546	ALLIS-CHLM	88001.96	301502	WESCO	80170
2B2	MOTOR MAST	88000.21	2T2910	BORG-WARNR	88002.61	3006546	FIAT-ALLIS	80048	301-502	WESCO	80171
2B27	BORG-WARNR	88000.31	2T2910	BORG-WARNR	88002.62	3006546-0	FIAT-ALLIS	80048	301507	WESCO	80177
2B27	MOTOR MAST	88000.31	2T2911	AEC	88002.42	3006595-7	FIAT-ALLIS	80109	301507	WESCO	88002.18
2B30	MOTOR MAST	88000.35	2T2911	BORG-WARNR	88002.42	3008	ALCO	80104	301508	WESCO	80174
2B3053	BORG-WARNR	88002.68	2T2911	GMB	88002.42	3008392	FIAT-ALLIS	80047	301509	WESCO	80162
2B32	BORG-WARNR	88000.38	2T2911	GMB	88002.49	3008392-7	ALLIS-CHLM	80047	301-510	WESCO	80167
2B32	MOTOR MAST	88000.38	2T2912	AEC	88002.44	3008392-7	FIAT-ALLIS	80047	30151-186X	HARDY SPCR	80041
2CP92N5	ROCKWELL	88005.84	2T2912	BORG-WARNR	88002.44	3008E4635A	FORD	80023	301514	WESCO	80176
2D1844	CATERPILLR	80156	2T2912	GMB	88002.44	3009	TRU-CROSS	80105	301515	WESCO	80173
2D2978	CATERPILLR	80021	2T2914	BORG-WARNR	88002.44	301	PILOT	80024	301518	WESCO	80172
2D3765	CATERPILLR	88003.26	2T4295	BORG-WARNR	88002.43	301	PRECISION	88001.64	301-519	WESCO	80168
2D9	SMITH,T.L.	80134	2T95	CATERPILLR	80198	301004	WESCO	80134	301519	WESCO	80169
2DP3281	MPL.MOLINE	80163	2T96	CATERPILLR	80224	3010A	MOTOR MAST	80092	301531	WESCO	80171
2G4157	CATERPILLR	80245	2U2064	ULRICH	80134	3010A	ZELLER	80092	301532	WESCO	80175
2H798(2)	CATERPILLR	88001.26	2U6365	ULRICH	80210	3010E4635A	FORD	80041	301539	WESCO	80170
2H801	CATERPILLR	80230	2V1164	CATERPILLR	80100	3011020	FIAT-ALLIS	80070	3-0155	NEAPCO	80182
2H8257	CATERPILLR	80001	2V7153	CATERPILLR	80224	3011022-5	FIAT-ALLIS	80070	301-55	WESCO	80141
2H8257	CATERPILLR	80003	2WCS22-19	ROCKWELL	88005.85	301-198	WESCO	80051	301553	WESCO	80135
2H8357	CATERPILLR	80213	2WCS22-23	ROCKWELL	88005.86	301198	WESCO	80052	301555	WESCO	80141
2H858	CATERPILLR	80155	2WCS22-75	ROCKWELL	88005.87	3012	ALCO	80110	301556	WESCO	80145
2H858	CATERPILLR	80213	2WCS22-76	ROCKWELL	88005.88	3012	ALCO	80112	301557	WESCO	80183
2H861	CATERPILLR	80001	2WCS24-41	ROCKWELL	88005.89	3012	ALCO	88002.14	301-558	WESCO	80182
2K3631	CATERPILLR	80172	2WCS24-48	ROCKWELL	88005.90	3012	TRU-CROSS	80112	301578R92	IHC	80234
2K7276	CATERPILLR	80213	2WCS24-52	ROCKWELL	88005.91	3012E4635A	FORD	80023	301-587R91	IHC	80234
2K8780	CATERPILLR	88001.56	2WCS24-76	ROCKWELL	88005.92	3013	AEC	80106	301587R92	IHC	80234
2L	AEC	88001.28	2WCS26-13	ROCKWELL	87396	3013	AEC	80112	301588	WESCO	80140
2L1	LEMPCO	88000.21	2WCS26-19	ROCKWELL	88005.93	301-300	WESCO	80101	301-592R91	IHC	80238
2L1	LEMPCO	88000.39	2WCS26-20	ROCKWELL	88005.94	301301	WESCO	80100	301592R92	IHC	80238
2L10	LEMPCO	88000.22	2WCS28-73	ROCKWELL	88005.95	301-308	WESCO	80107	301601	WESCO	88002.19
2L102B	MOTOR MAST	88001.28	2WCS28-78	ROCKWELL	87439	30131130	ALLIS-CHLM	80070	301-680C	OLIVER	80069
2L102B	ZELLER	88001.28	2WCT28-15	ROCKWELL	87440	301312	WESCO	80112	301680E	OLIVER	80070
2L11	LEMPCO	88000.22	3	AEC	88013.87	301312	WESCO	88002.14	301-680E	WHITE FARM	80069
2L11	LEMPCO	88000.25	300	PRECISION	88001.63	301313	WESCO	80106	301689E	OLIVER	80069
2L11	LEMPCO	88000.27	3000	ALCO	80099	301314	WESCO	80114	301701	WESCO	88002.23
2L12	LEMPCO	88000.22	3000	ALCO	80101	301321	WESCO	80111	301725R91	IHC	80231
2L12	LEMPCO	88000.28	3000	TRU-CROSS	80099	301322	WESCO	80103	3-0188	NEAPCO	80140
2L13	LEMPCO	88000.22	3000	TRU-CROSS	80101	301-323	WESCO	80049	301965	WESCO	80160
2L13	LEMPCO	88000.31	3000-100	GAR WOOD	80008	301323	WESCO	80102	301966	WESCO	80138
2L17	LEMPCO	88000.25	3000-100	GAR WOOD	88002.88	301324	WESCO	80098	302	PILOT	80010
2L17	LEMPCO	88000.27	3000297	WABCO	80171	301325	WESCO	80096	302	PRECISION	80022

Part No.	Brand	IDLI Group	Part No.	Brand	IDLI Group	Part No.	Brand	IDLI Group	Part No.	Brand	IDLI Group
302-0300	G & G MFG	80133	3031D	MOTOR MAST	80112	3046102M91	MASS.FERG.	88003.88	3109949	WHITE	80225
302-0600	G & G MFG	80001	3031D	ZELLER	80112	304626	GAR WOOD	80008	311	PILOT	88002.34
302-0675	G & G MFG	88000.15	3031G	MOTOR MAST	80104	304626	GAR WOOD	88002.88	311	PRECISION	80096
3021	ALCO	80111	3031GM	MOTOR MAST	80108	305	PILOT	80018	3111	AEC	80104
302-10X	SPICER	88002.46	3031H	MOTOR MAST	80101	305	PRECISION	80042	3111	PRECISION	88001.55
302-11X	SPICER	88002.60	3031H	ZELLER	80101	3-05-07	BYPY	88003.75	3112002	AMER.MOTOR	80044
302-1200	G & G MFG	80035	3031HB	MOTOR MAST	80099	3051-035N	HARDY SPCR	80040	3112479	AMER.MOTOR	80158
302-13X	SPICER	88002.51	3031HB	MOTOR MAST	80100	3051-161N	HARDY SPCR	80041	3112579	AMER.MOTOR	80158
302-1400	G & G MFG	80037	3031HB	MOTOR MAST	80101	3051-195W	HARDY SPCR	80041	3112942	AMER.MOTOR	80031
302-14X	SPICER	88002.67	3031HC	MOTOR MAST	80100	305-852R91	IHC	80238	3113	AEC	80112
3021862	ALLIS-CHLM	80188	3031L	MOTOR MAST	80111	305-952R91	IHC	80238	3113002	AMER.MOTOR	88002.05
3021862	FIAT-ALLIS	80188	3031N	MOTOR MAST	80112	3-06	BYPY	88004.18	31138E	SCHULTZ	80136
3021862-2	FIAT-ALLIS	80188	3031P	MOTOR MAST	80100	306	PILOT	80009	3113942	AMER.MOTOR	80030
3022	TRU-CROSS	80097	3031P	ZELLER	80100	306	PRECISION	88001.66	3113942	AMER.MOTOR	80031
302-3500	G & G MFG	80069	3031Q	MOTOR MAST	80103	306-114R91	IHC	80101	3114	AEC	80104
302-3X	SPICER	88002.65	3031R	MOTOR MAST	80102	306255	WESCO	88001.70	3114707	AMER.MOTOR	80034
302-4400	G & G MFG	80136	3031S	MOTOR MAST	80106	306265	WESCO	88001.70	3114708	AMER.MOTOR	80060
3024508	ALLIS-CHLM	80178	3031T	MOTOR MAST	80113	306517	WESCO	88001.79	3114708	AMER.MOTOR	80061
3024508	FIAT-ALLIS	80178	3031T	ZELLER	80113	306521	WESCO	88013.86	31-1565100	COCKSHUTT	80033
3024508-8	ALLIS-CHLM	80178	3031V	MOTOR MAST	80105	306-578	WESCO	88013.87	31-1565100	WHITE FARM	80033
3024508-8	FIAT-ALLIS	80178	3031V	ZELLER	80105	3-07	BYPY	88004.11	3115912	AMER.MOTOR	80044
302-5500	G & G MFG	80161	3031W	MOTOR MAST	80098	307	PILOT	80006	3115912	AMER.MOTOR	88002.05
302592R92	IHC	80238	303-255	WESCO	80060	307	PRECISION	88001.67	3116	AEC	80111
30-26-000	HARDY SPCR	88003.94	303255	WESCO	80063	307-949R91	IHC	80070	3116717	AMER.MOTOR	80030
30-26-001	HARDY SPCR	88004.04	303-294R91	IHC	80234	308	AEC	80229	3116717	AMER.MOTOR	80031
30-26-009	HARDY SPCR	88003.95	3032PA	NEW IDEA	80133	308	PILOT	80017	3117373	AMER.MOTOR	80072
30-26-010	HARDY SPCR	88004.05	3-0332	NEAPCO	80139	308R	PRECISION	80907	3117373	AMER.MOTOR	80075
3026726	FIAT-ALLIS	80069	303323404	WHITE	80145	309	PILOT	80011	31-17471510	WHITE	80001
3026726-4	ALLIS-CHLM	80069	3033RA	NEW IDEA	80133	309260005(4)	BUFFALO-SP	80216	3118	AEC	80101
3026726-4	FIAT-ALLIS	80069	3034121	FIAT-ALLIS	80101	3093A	MOTOR MAST	80093	3119	AEC	80100
3026727	FIAT-ALLIS	80069	303520	WESCO	80076	3093A	ZELLER	80093	3119	AEC	80112
3026727-2	ALLIS-CHLM	80069	303-520	WESCO	80088	309-575	FORD	80001	3119116	AMER.MOTOR	80034
3026727-2	FIAT-ALLIS	80069	303520	WESCO	80959	309575	FORD	80003	3119275	AMER.MOTOR	80049
302-725C91	IHC	80240	303-521	WESCO	80076	309R	PRECISION	80908	3119646	AMER.MOTOR	80044
302725R91	IHC	80231	303521	WESCO	80089	30A313-1	HEIL	80037	3119649	AMER.MOTOR	80044
302-726C91	IHC	80239	303554	WESCO	80081	30B301-8	HEIL	80069	3119649	AMER.MOTOR	88002.05
302726R91	IHC	80237	303560	WESCO	80143	3-0M22	DIGIO-LIFT	88001.28	3119731	AEC	80059
302-737C91	IHC	80224	303578	WESCO	80125	3-0M22	LIFT TRUCK	88001.28	3119731	AMER.MOTOR	80059
3028573	ALLIS-CHLM	80178	303710780	CITROEN	80142	3-0R85	DIGIO-LIFT	80124	3119752	AMER.MOTOR	80023
3028573	FIAT-ALLIS	80178	303710810	CITROEN	80142	3-0R85	LIFT TRUCK	80124	3119752	AMER.MOTOR	80024
3028573-8	ALLIS-CHLM	80178	303-852R91	IHC	80238	31	WYLIE	88002.81	312	PILOT	88002.34
3028573-8	FIAT-ALLIS	80178	3038633	FIAT-ALLIS	80220	310	PILOT	80012	312	PRECISION	88001.69
3028862	ALLIS-CHLM	80178	3038633-8	FIAT-ALLIS	80220	310	PRECISION	88001.68	3120	AEC	80113
3028862	FIAT-ALLIS	80178	3039559	FIAT-ALLIS	80187	3100758R91	IHC	80041	3121	AEC	80044
3028862-5	ALLIS-CHLM	80178	3039559-4	FIAT-ALLIS	80187	3100758R91	IHC	80221	3121	AEC	80102
3028862-5	FIAT-ALLIS	80178	303HB	MOTOR MAST	80101	3100A	MOTOR MAST	80082	3125	AEC	80103
302-8X	SPICER	88002.45	303KC	KENBAR	88008.51	3100A	MOTOR MAST	80083	3127013	AMER.MOTOR	88002.69
302927-10	DAIMLER	80041	303KC	KENBAR	88008.52	3100B	MOTOR MAST	80083	313	PILOT	88002.36
302927/10	LEYLAND	80041	304	PILOT	80019	3100B	ZELLER	80083	313	PRECISION	80094
302-9X	SPICER	88002.47	304	PRECISION	80135	3101	AEC	80100	3130A	MOTOR MAST	80097
302-9X	SPICER	88002.48	3040372	FIAT-ALLIS	80235	3102	AEC	80112	3130A	MOTOR MAST	80119
3-03	BYPY	88003.67	3040372-9	FIAT-ALLIS	80235	3103	AEC	80098	3130A	ZELLER	80097
303	PILOT	80004	3040377	FIAT-ALLIS	80235	3-1045	NEAPCO	80134	3130B	MOTOR MAST	80097
303	PILOT	80006	3040377-8	FIAT-ALLIS	80235	3104703	WHITE	80001	3131428	AMER.MOTOR	80075
303	PRECISION	88001.65	304104R91	IHC	80234	3104703	WHITE	80003	3131428	AMER.MOTOR	88002.68
3031	TRU-CROSS	80109	304-105R91	IHC	80234	3104709	WHITE	80003	3131449	AMER.MOTOR	88002.69
3031-28	ZELLER	80107	304106R91	IHC	80238	3105	AEC	80104	3131517	AMER.MOTOR	88002.69
3031-2D	MOTOR MAST	80110	304-106R92	IHC	80238	3106	AEC	80101	3133A	MOTOR MAST	80144
3031-2H	MOTOR MAST	80099	3044128	ALLIS-CHLM	80188	310604005	BMC	80024	3-1-3521	SPICER	88005.89
3031-2J	MOTOR MAST	80104	3044128	FIAT-ALLIS	80188	3106307	WHITE	80217	3136	PRECISION	88001.56
303131580	WHITE	80142	3044128-1	FIAT-ALLIS	80188	3108170	AMER.MOTOR	80030	3138316	AEC	80058
303131629	WHITE	80140	3044617	FIAT-ALLIS	80219	3108170	AMER.MOTOR	80031	3138316	AMER.MOTOR	80058
3031B	MOTOR MAST	80104	3044617-3	FIAT-ALLIS	80219	3108171	AMER.MOTOR	80049	3138316	AMER.MOTOR	88002.68
3031B	ZELLER	80104	304562	GAR WOOD	80008	3108181	AMER.MOTOR	80110	314	PILOT	88002.33
3031BCK2	MOTOR MAST	88001.94	304-608C91	IHC	80235	3108181	AMER.MOTOR	80112	314	PRECISION	80064
3031BCK2T	MOTOR MAST	88002.34	304608C92	IHC	80235	3109409	AMER.MOTOR	80049	3140B	MOTOR MAST	80127
3031BHP	MOTOR MAST	80104	304608R91	IHC	80235	3109413	AMER.MOTOR	80099	3144950	AMER.MOTOR	88002.69
3031BT	MOTOR MAST	80104	304-609C91	IHC	80233	3109413A	AMER.MOTOR	80099	3145293	AMER.MOTOR	80077
3031C	MOTOR MAST	80050	304-609C92	IHC	80233				3145473	AMER.MOTOR	80072
3031C	MOTOR MAST	80109	304609R91	IHC	80233				3145689	AMER.MOTOR	88002.69

Part No.	Brand	IDLI Group
3-14-59	SPICER	88013.54
3147	TRU-CROSS	80104
3147	TRU-CROSS	80107
314G	PRECISION	80065
315	AEC	80026
315	PILOT	88002.33
315	PRECISION	80060
3150425(4)	PARSONS	80021
3151-162N	HARDY SPCR	80077
3151-166W	HARDY SPCR	80077
3151-166Y	HARDY SPCR	80077
3152	AEC	80099
3152	AEC	80101
3-15-41	SPICER	88015.59
315442C91	FORD	87377
315442C91	IHC	87380
3-155	NEAPCO	80182
3-15-51	SPICER	88015.58
3156293	AMER.MOTOR	80077
3-15-63	SPICER	88015.56
315-726M91	MASS.FERG.	80001
315D131(4)	PARSONS	80193
315G	PRECISION	80061
316	PILOT	88002.33
316	PRECISION	80132
31604103	INNOCENTI	80024
3-16-103	SPICER	88015.30
316116	CATERPILLR	80047
316116	TOWMOTOR	80047
316117	CATERPILLR	80046
316117	GERLINGER	80046
316117	TOWMOTOR	80046
316340	DAF	80178
3163514	AMER.MOTOR	80030
3-16-41	SPICER	88015.33
3165080	AMER.MOTOR	80023
3-16-51	SPICER	88015.32
316616	GERLINGER	80047
3166462	AMER.MOTOR	80030
316-726M91	MASS.FERG.	80001
317	PILOT	88002.33
317	PRECISION	80062
3171348	AMER.MOTOR	80030
3172840	AMER.MOTOR	80044
3172840	AMER.MOTOR	88002.05
3172850	AMER.MOTOR	80044
3177	HUBER	80178
317-712R92	IHC	80171
317-713R92	IHC	80163
317-721R91	IHC	80175
317721R92	IHC	80175
317-722R91	IHC	80174
318	PRECISION	80063
3180170	AMER.MOTOR	80030
318-10	PRECISION	88001.70
3188	AEC	80105
3-188	NEAPCO	80140
3-188	NEAPCO	80182
3188632	AMER.MOTOR	80077
319	PRECISION	80083
3190A	MOTOR MAST	80077
3190A	MOTOR MAST	80144
319116	AMER.MOTOR	80034
3195054	AMER.MOTOR	80077
319-711C91	IHC	80080
319711R91	IHC	80074
31G	PRECISION	80132
31X12715A	DEERE,JOHN	80210
31X12KL1715A	DEERE,JOHN	80210
320	PRECISION	80082
3200332	AMER.MOTOR	80044
3200332	AMER.MOTOR	88002.05
3200537	AMER.MOTOR	80030
3200537	AMER.MOTOR	80031
3200538	AMER.MOTOR	80044
3200538	AMER.MOTOR	88002.05
3200539	AMER.MOTOR	80059
320085	AMER.MOTOR	88002.05
3200854	AMER.MOTOR	80030
3200855	AMER.MOTOR	80044
3200855	AMER.MOTOR	80049
3200856	AMER.MOTOR	80059
3200930	CLARK EQU.	80193
3200A	MOTOR MAST	80077
3200A	ZELLER	80072
3200A	ZELLER	80077
3200AHP	MOTOR MAST	80077
3200AT	MOTOR MAST	80081
3200B	MOTOR MAST	80077
3200B	MOTOR MAST	80088
3200C	MOTOR MAST	80090
3200C	ZELLER	80090
3200D	MOTOR MAST	80078
3200D	ZELLER	80078
3200D	ZELLER	88001.71
3200E	MOTOR MAST	80036
3200E	ZELLER	80036
3200F	MOTOR MAST	80085
3200-F	MOTOR MAST	80086
3200F	ZELLER	80085
3200LS	WABCO	80037
3200UBK2	MOTOR MAST	88013.86
3200UBK2	MOTOR MAST	88013.87
3202655	AMER.MOTOR	80072
3202657	AMER.MOTOR	80077
3203655	AMER.MOTOR	80077
3203655	AMER.MOTOR	80081
3203992	AMER.MOTOR	80081
3203992	AMER.MOTOR	88002.05
32046J	BORG-WARNR	88006.88
3204918	AMER.MOTOR	80030
32053	DROTT	80109
32053	DROTT	80155
3205371	AMER.MOTOR	80044
3205371	AMER.MOTOR	88002.05
3206121	AMER.MOTOR	80044
3206122	AMER.MOTOR	80058
3206122	AMER.MOTOR	80092
3206247	AMER.MOTOR	80031
3206980	AMER.MOTOR	80058
3206981	AMER.MOTOR	80058
3207077	AMER.MOTOR	80058
3207078	AMER.MOTOR	88002.05
3207079	AMER.MOTOR	80058
3207174	AMER.MOTOR	80030
32075201	DEERE,JOHN	80189
3207885	AMER.MOTOR	80075
3208473	AMER.MOTOR	80077
3209076	AMER.MOTOR	80030
3210	MOTOR MAST	80079
3210A	MOTOR MAST	80079
3210A	ZELLER	80079
32-1154	PAPEC MACH	80001
3-2-119	SPICER	88004.62
321-291R91	IHC	80225
321407271AE	VOLKSWAGEN	86002
321407271AG	VOLKSWAGEN	86001
321407271AJ	AUDI	86027
321407271AJ	VOLKSWAGEN	86027
321407271AK	AUDI	86046
321407271AK	VOLKSWAGEN	86046
321407271AN	AUDI	86057
321407271AP	AUDI	86059
321407271C	AUDI	86002
321407271C	VOLKSWAGEN	86002
321407271D	AUDI	86001
321407271D	VOLKSWAGEN	86001
321407271K	AUDI	86036
321407271P	AUDI	86037
321407272AA	AUDI	86056
321407272B	AUDI	86003
321407272B	VOLKSWAGEN	86003
321407272D	AMER.MOTOR	86035
321407272P	VOLKSWAGEN	86003
321407272R	AUDI	86045
321407272R	VOLKSWAGEN	86045
321407272S	AUDI	86058
321407285B	AUDI	86122
321407285B	AUDI	86180
321407285B	VOLKSWAGEN	86180
321-407-311B	AUDI	86099
321-407-311L	AUDI	86124
321-407-311L	VOLKSWAGEN	86124
321-407-311M	AUDI	86137
321407331	AUDI	86072
321407331	VOLKSWAGEN	86072
321-485R91	IHC	80044
321-498-103	VOLKSWAGEN	86072
321-498-103	VOLKSWAGEN	86143
321498201	AUDI	86072
321498201	AUDI	86143
321498201	VOLKSWAGEN	86143
321-498-201A	VOLKSWAGEN	86143
321498203	AUDI	86122
321498203	AUDI	86180
321498203	VOLKSWAGEN	86143
321498203	VOLKSWAGEN	86180
3-2-159	SPICER	84614
3218363	CHRYSLER	80180
3218621	GMC	80142
321921	IHC	80225
321-921R91	IHC	80225
3219885	AMER.MOTOR	80026
3219885	AMER.MOTOR	80049
321R	PRECISION	80957
321SC171	CLARK EQU.	80221
321SD171	CLARK EQU.	80225
322	PRECISION	88001.71
3-22746	CURTISS-WR	80215
3-22746	GEN.EQUIP.	80218
3-22746	WAYNE	80215
3-22746	WOOLDRIDGE	80211
322910CN117	DUPONT	80113
323	PRECISION	88001.72
32-30-22	SPICER	88015.24
32-30-52	SPICER	87668
32-30-52	SPICER	88015.22
32-30-72	SPICER	88015.21
3233715	AMER.MOTOR	80083
3233715X	AMER.MOTOR	80083
32360	GERLINGER	80047
32360	TOWMOTOR	80047
32361	GERLINGER	80046
32361	TOWMOTOR	80046
3237089	AEC	86051
3237089	AMER.MOTOR	86051
3237A	MOTOR MAST	80117
3237A	ZELLER	80117
3237B	MOTOR MAST	80116
3237B	ZELLER	80116
3237C	MOTOR MAST	80121
3237C	ZELLER	80121
3238675	AEC	86051
3238675	AMER.MOTOR	86051
3238675	AMER.MOTOR	86052
3-23883	CURTISS-WR	80109
3-23883	FMC	80109
3-23883	WAYNE	80149
3-23883	WOOLDRIDGE	80149
3239621	AMER.MOTOR	86051
323R	PRECISION	80909
324	PRECISION	88001.73
3240A	MOTOR MAST	80124
3240A	ZELLER	80124
3240AHP	MOTOR MAST	80124
3240B	MOTOR MAST	80125
3240B	ZELLER	80125
3240C	MOTOR MAST	80124
324-550R91	IHC	80234
324-667R91	IHC	80238
3-2-479	SPICER	84616
324D258(4)	PARSONS	80193
324D259(4)	PARSONS	80101
324R	PRECISION	80958
325	PRECISION	88001.74
32-50-52	SPICER	87667
325-326R91	IHC	80196
3-2-549	SPICER	84615
325E1574(4)	PARSONS	80215
325E754(4)	PARSONS	80233
325E813(4)	PARSONS	80215
326	AEC	87481
326	PRECISION	88001.75
32603	DEERE,JOHN	80158
3-26-317	SPICER	88004.61
326610	M.R.S.	80219
326610AP	M.R.S.	80219
3-2-667	SPICER	85712
327	PRECISION	88001.76
3-27-16-7120X	SPICER	86670
327204R91	IHC	80197
327-204R92	IHC	80196
3-27-20-7311X	SPICER	86673
32720R91	IHC	80196
3-27-22-6221X	SPICER	86672
3-27-3-6414X	SPICER	86668
3275A	MOTOR MAST	80120
3275A	ZELLER	80119
3275A	ZELLER	80120
3275AE	MOTOR MAST	80118
3275AE	ZELLER	80118
3275B	MOTOR MAST	80119
3276694	CHRYSLER	80189
3-27-8-7317X	SPICER	86669
328	PRECISION	88008.97
328384C91	IHC	88007.58
3-28-507	SPICER	85524
3-28-57	SPICER	88004.67
328-592C91	IHC	80077
328-592C91	IHC	80156
3-28-667	SPICER	85718
3287A	MOTOR MAST	80140
3287A	ZELLER	80140
3287C	MOTOR MAST	80139
3287C	MOTOR MAST	80140
3287C	ZELLER	80139
3287UBK2	ZELLER	88013.86
3-28-97	SPICER	85522
3289A	ZELLER	80140
3289A	ZELLER	80182

Part No.	Brand	IDLI Group
329	HUSCO	80072
329	PRECISION	80072
3290A	MOTOR MAST	80142
3290A	ZELLER	80142
3290B	MOTOR MAST	80143
3290B	ZELLER	80143
3290UBK2	MOTOR MAST	88013.86
3290UBK2	MOTOR MAST	88013.87
3290UBK2	ZELLER	88013.87
3290UBK2T	MOTOR MAST	88013.87
329-10	PRECISION	88013.86
329871	CATERPILLR	80001
32A	WYLIE	80906
330	PRECISION	80142
330-10	PRECISION	88013.87
3-3031	NEAPCO	80109
3306813	CHRYSLER	80023
3308J	BORG-WARNR	87410
33096	VERSATILE	80140
330U	PRECISION	80143
331	PRECISION	80124
3-3100	NEAPCO	80177
3312A	MOTOR MAST	80095
3312B	MOTOR MAST	80096
3312D	MOTOR MAST	80094
3-3130	NEAPCO	80097
3-3133	NEAPCO	80144
3-3140	NEAPCO	80127
3-3152	NEAPCO	80101
3-3154	NEAPCO	80113
331(6)	PRECISION	88001.77
3-3-1601KX	SPICER	85068
3-3-1601X	SPICER	85068
331U	PRECISION	80125
332	AEC	80139
332	PRECISION	80178
332	TRU-CROSS	80139
3-3-2041KX	SPICER	85065
3-3-2041X	SPICER	85065
33-211-105-446	BMW	86164
33211205745	BMW	86100
33211205746	BMW	86101
33211205748	BMW	86102
33211205748	BMW	86165
33211205816	BMW	86164
33211205840	BMW	86102
33211207037	BMW	86102
33211207037	BMW	86165
33-211-207-722	BMW	86103
33-211-208-439	BMW	86104
33-211-208-439	BMW	86264
33-211-208-443	BMW	86167
33-211-209-551	BMW	86105
33-211-209-552	BMW	86105
332-1X	SPICER	88002.56
332-1X	SPICER	88002.58
332-2X	SPICER	88002.58
332-3X	SPICER	88002.48
3-32883	CURTISS-WR	80109
3-32883	CURTISS-WR	80149
332J	SPICER	88014.58
333	PRECISION	80189
3331A	MOTOR MAST	80132
3331A	ZELLER	80132
3331AHP	MOTOR MAST	80132
3331AT	MOTOR MAST	80132
3332	DIAMOND R	80178
333-2	PRECISION	88001.78
3332P2	DIAMOND R	80179
3333P2	DIAMOND R	80178
3333P2	REO MOTORS	80178
3333P2	WHITE	80178
3334	AEC	80012
333-832R91	IHC	80170
334	PRECISION	80221
33410	TORO	80069
334-2	PRECISION	88001.79
334-5	PRECISION	88001.80
3345P2	DIAMOND R	80142
3345P2	REO MOTORS	80142
3345P2	WHITE	80142
334B25384	BUFFALO-SP	80149
334B2538-4	KOEHRING	80109
335	PRECISION	80225
335-2	PRECISION	88001.81
336	PRECISION	88001.82
337	PRECISION	88001.83
337-439R91	IHC	80001
3379RC	JOY MFG.	80240
338	PRECISION	80007
3382446	ELEC.MTR.	80101
3382466	GMC	80101
3382RC	JOY MFG.	80232
3382RC	JOY MFG.	80242
33832R91	IHC	80170
338-412R91	IHC	80213
338421R91	IHC	80213
338-591R91	IHC	80238
338-799R92	IHC	80225
338-923R92	IHC	80169
339981	FIAT-ALLIS	80035
34	WYLIE	80024
340	PRECISION	80023
3400021	BERLIET	80180
3400040	BERLIET	80180
34-004-022	BARREIROS	80228
3-4-0052	SPICER	83241
3-4-00672	SPICER	83283
3400A	MOTOR MAST	80081
3400A	ZELLER	80081
3400AHP	MOTOR MAST	80081
3400AT	MOTOR MAST	80081
3400B	MOTOR MAST	80091
3400B	ZELLER	80091
3400BA	ZELLER	80091
3400BHP	MOTOR MAST	80091
3400BT	MOTOR MAST	80091
3-40-1381	SPICER	88014.75
3-40-1571	SPICER	87370
34022	VERSATILE	80124
3406010M91	MASS.FERG.	88004.29
3406011M91	MASS.FERG.	88003.73
3406015M1	MASS.FERG.	88003.90
3406018M1	MASS.FERG.	88004
3406020M91	MASS.FERG.	80024
3406021M91	MASS.FERG.	88003.74
3406023M1	MASS.FERG.	88004.10
3406025M1	MASS.FERG.	88003.92
3406026M1	MASS.FERG.	88004.17
3406028M1	MASS.FERG.	88004.02
3406030M91	MASS.FERG.	88004.30
3406031M91	MASS.FERG.	88003.75
3406033M1	MASS.FERG.	88004.11
3406035M1	MASS.FERG.	88003.93
3406036M1	MASS.FERG.	88004.18
3406038M1	MASS.FERG.	88004.03
3406040M91	MASS.FERG.	80041
3406041M91	MASS.FERG.	88003.76
3406043M1	MASS.FERG.	88004.12
3406045M1	MASS.FERG.	88003.94
3406046M1	MASS.FERG.	88004.19
3406048M1	MASS.FERG.	88004.04
3406050M91	MASS.FERG.	88004.34
3406051M91	MASS.FERG.	88003.78
3406052M91	MASS.FERG.	88003.85
3406055M1	MASS.FERG.	88003.96
3406058M1	MASS.FERG.	88004.06
3406060M91	MASS.FERG.	88004.33
3406061M91	MASS.FERG.	88003.77
3406062M91	MASS.FERG.	88003.84
3406063M1	MASS.FERG.	88004.13
3406066M1	MASS.FERG.	88004.20
3406070M91	MASS.FERG.	88004.31
3406071M91	MASS.FERG.	88003.80
3406072M91	MASS.FERG.	88003.87
3406075M1	MASS.FERG.	88003.98
3406078M1	MASS.FERG.	88004.08
3406080M91	MASS.FERG.	80124
3406081M91	MASS.FERG.	88003.79
3406082M91	MASS.FERG.	88003.86
3406083M1	MASS.FERG.	88004.14
3406085M1	MASS.FERG.	88003.97
3406086M1	MASS.FERG.	88004.21
3406088M1	MASS.FERG.	88004.07
3406090M91	MASS.FERG.	88004.32
3406091M91	MASS.FERG.	88003.81
3406093M1	MASS.FERG.	88004.15
3406096M1	MASS.FERG.	88004.22
3406100M91	MASS.FERG.	88004.35
3406101M91	MASS.FERG.	88003.82
3406103M1	MASS.FERG.	88004.16
3406106M1	MASS.FERG.	88004.23
3406110M91	MASS.FERG.	88004.36
3406111M91	MASS.FERG.	88003.83
3406112M91	MASS.FERG.	88003.89
3406115M1	MASS.FERG.	88003.99
3406118M1	MASS.FERG.	88004.09
3406211M1	MASS.FERG.	88003.66
3406212M1	MASS.FERG.	88003.65
3406213M1	MASS.FERG.	88003.68
3406214M1	MASS.FERG.	88003.70
3406215M1	MASS.FERG.	88003.72
3406221M1	MASS.FERG.	88003.67
3406222M1	MASS.FERG.	88003.69
3406223M1	MASS.FERG.	88003.71
3406270M1	MASS.FERG.	88003.59
3406271M1	MASS.FERG.	88003.61
340627M1	MASS.FERG.	88003.63
3406280M1	MASS.FERG.	88003.60
3406281M1	MASS.FERG.	88003.62
3406282M1	MASS.FERG.	88003.64
3406451M91	MASS.FERG.	88004.26
3406453M91	MASS.FERG.	88004.25
3406454M91	MASS.FERG.	88004.27
3-4-0743	SPICER	83251
3-40-881	SPICER	88014.75
3-40-971	SPICER	87370
3-41	BYPY	88004.30
341	PRECISION	80024
3-4-1003	SPICER	83334
3-4-1003	SPICER	83335
3-4-12	HAYES	83258
3-4-12	SPICER	83258
34121	ALLIS-CHLM	80099
34121	ALLIS-CHLM	80101
34121	FIAT-ALLIS	80101
3-4123	NEAPCO	80150
3-4123	NEAPCO	80153
3-4-1353	HAYES	83316
3-4-1353	SPICER	83316
3-4-1353	SPICER	83317
3-4138	NEAPCO	80109
3-4138	NEAPCO	80149
3-4140	NEAPCO	80155
3-4143	NEAPCO	80156
3-4152	NEAPCO	80158
3-4-153	SPICER	83316
3.4-16-33	SPICER	88013.83
3-4-1841	SPICER	81383
3-4-1841	SPICER	81385
341G	PRECISION	88001.84
3-4-22	SPICER	83263
34224	CHRYSLER	80024
3-4-2241	SPICER	81383
34244Y	CHRYSLER	80023
3-4-2511	SPICER	81386
34254S	CHRYSLER	80023
3.4-26-11	SPICER	85516
3-4-26-17	SPICER	85516
343	PRECISION	88001.85
3-4-3061	SPICER	81396
3-4-3081	SPICER	81390
3-4-3091	SPICER	81388
3-4-3091	SPICER	88005.06
3-4-3191	SPICER	81394
3-4-3231	SPICER	81387
3-4-3331X	SPICER	81390
3-4-3481X	SPICER	88014.32
3-4-3541	SPICER	81395
3-4-3561	SPICER	88005.26
3-4-3571	SPICER	81394
34371	NEW HOLL.	80044
3-4-3781	SPICER	81391
3437A	MOTOR MAST	80123
3437A	ZELLER	80123
3437B	MOTOR MAST	80122
3437C	MOTOR MAST	80131
3437C	ZELLER	80131
3437CHP	MOTOR MAST	80131
3437CT	MOTOR MAST	80131
3437D	MOTOR MAST	80129
3437D	ZELLER	80117
3437D	ZELLER	80129
3437DHP	MOTOR MAST	80129
3437DT	MOTOR MAST	80129
3437UBK2	MOTOR MAST	88002.13
3437UBK2T	MOTOR MAST	88002.13
3-4-3861	SPICER	81400
3-4-3941	SPICER	81379
3-44	NEAPCO	80136
344	PRECISION	80041
3-4-4051	SPICER	81393
3-4-4-12	SPICER	81356
3-4-4-131X	SPICER	81362
344147	VOLVO	80003
3-4-4161X	SPICER	81385
3-4-4-163	SPICER	81355
3-4-4-173	SPICER	81365
3-4-4-183	SPICER	81370
3-4-4-193	SPICER	81375
3-4-4-221	SPICER	88006.08
3-4-4-23	SPICER	81372
3-4-4-361	SPICER	81376
3-4-4-41X	SPICER	84515
3-4-4421X	SPICER	81389
3-4-4-61X	SPICER	86665
3-4-4-82	SPICER	81356
3-4-4-83	SPICER	81369
3-4-4-91	SPICER	81362
3-45	NEAPCO	80161

Part No.	Brand	IDLI Group
345	PRECISION	88009.06
3-4-5071	SPICER	83248
3-4-5081	SPICER	83277
3-4-5091X	SPICER	83277
3-4-5101	HAYES	83320
3-4-5101	SPICER	83320
3-4-5111	SPICER	83314
3-4-5131X	SPICER	87032
3-4-5181	SPICER	84531
3-4-5181X	SPICER	84531
3-4-52	SPICER	83241
3-4-5221	SPICER	83314
3-4-5231X	SPICER	83314
3-4-5251X	BORG-WARNR	84531
3-4-5251X	HAYES	84531
3-4-5251X	SPICER	84530
3-4-5251X	SPICER	84531
3-4-5251X	SPICER	84532
34532	FMC	80138
3-4-5341	SPICER	83312
345-410-7031	MERCEDES	80017
3-4-55-12	SPICER	87276
3456-510	TAYLOR	80171
3-4-5791X	SPICER	81397
345802	BROCKWAY	80003
3-4-5821X	SPICER	81400
345823	BROCKWAY	80003
345827	BROCKWAY	80001
345884	BROCKWAY	80001
3459095	TAYLOR	80198
346	PRECISION	80089
3-4-6051X	SPICER	81391
346096C91	IHC	80081
3461	TAYLOR	80158
3-4-62	SPICER	83250
3462	TAYLOR	80212
3-4-6221X	SPICER	81391
3-4-6291X	SPICER	84529
3-4-6511X	HAYES	84529
3-4-6511X	SPICER	84528
3-4-6511X	SPICER	84529
3-4-6561	SPICER	81381
3-4-6561	SPICER	88004.91
3-4-6631X	SPICER	81382
3-4-6771X	SPICER	83349
347	PRECISION	80137
3-4-7031	SPICER	81386
3-4-72	SPICER	83283
3-4-723	SPICER	83240
347-385R91	IHC	80221
3-4-743	SPICER	83251
3-4-7541	SPICER	81399
3475A	MOTOR MAST	80130
3475A	ZELLER	80118
3475A	ZELLER	80130
3475A	ZELLER	80131
3475AHP	MOTOR MAST	80118
3475AHP	MOTOR MAST	80130
34761	NEW HOLL.	80037
3-4-763	SPICER	83249
3-4-773	SPICER	83279
3-4-783	HAYES	83280
3-4-783	SPICER	83280
3-4-793	HAYES	83280
3-4-793	SPICER	83281
347-960R91	IHC	80069
348	PRECISION	80127
3-4-803	SPICER	88014.40
3-4-8051X	SPICER	81399
3480934	CHRYSLER	80083
3480936	CHRYSLER	80082
3-4-813	HAYES	83316
3-4-813	SPICER	83316
3481A	MOTOR MAST	80138
3-4-823	SPICER	88014.40
3.4-82-78-3227X	SPICER	88007.24
3.4-82-78-5923X	SPICER	88007.26
3-4-833	HAYES	83349
3-4-833	SPICER	83349
3-4-863	HAYES	83349
34870	NEW HOLL.	80069
3-4-873	SPICER	88014.40
3-4-883	SPICER	88014.40
3-4-893	SPICER	83317
349	PRECISION	88008.68
3-4-913	SPICER	83331
3-4-923	SPICER	83331
3493	MOTOR MAST	80141
3493A	MOTOR MAST	80141
3493A	ZELLER	80141
3493B	MOTOR MAST	80145
3493B	ZELLER	80145
3-4-943	SPICER	88014.40
3-4-993	SPICER	83364
350	AEC	80036
350	CHAIN BELT	80069
350	PRECISION	80119
350	REX CHAIN	80069
3500	RUEDARSA	80225
35-00-00	WALTERSCHD	88004.34
35023X	NEAPCO	88001.70
350-254R91	IHC	80001
35-04-813	HARDY SPCR	88003.79
3506	PERFECTION	88002.72
350G	PRECISION	80120
351	PRECISION	80140
35-10-00	WALTERSCHD	88003.78
35-10-02	WALTERSCHD	88003.85
35-13-00	WALTERSCHD	88003.96
3-5132	NEAPCO	80174
35-14-00	WALTERSCHD	88004.06
3.5-14-39	SPICER	88013.53
352	PRECISION	80116
352	TEDDY TORQ	80095
35-26-000	HARDY SPCR	88003.97
35-26-001	HARDY SPCR	88004.07
353	AEC	80082
353	PRECISION	80118
353	TRU-CROSS	80083
3-53-1031	SPICER	87353
3-53-1031	SPICER	88008.18
3-53-1071	SPICER	87352
3-53-1071	SPICER	88008.17
3-53-151	SPICER	87371
3-53-161	SPICER	87290
3-53-161	SPICER	88008.09
3-53-241	SPICER	87360
3-53-241	SPICER	88008.15
3-53-291	SPICER	87372
3-53-291	SPICER	88005.03
3-53-291	SPICER	88008.20
3-53-341	SPICER	87371
3-53-431	SPICER	87357
3-53-431	SPICER	88008.14
3-53-451	SPICER	87371
3-53-451	SPICER	88008.19
3-53-831	SPICER	88001.07
353E	AEC	80083
354	PRECISION	80117
3-54-611	SPICER	88008.13
3546576	CHRYSLER	80182
354670	HENNEY	80099
354670	STUDEBAKER	80099
354672	STUDEBAKER	80049
354672	STUDEBAKER	80050
355	AEC	80144
3-55	NEAPCO	80179
355	PRECISION	80121
3551-182W	HARDY SPCR	80124
3551-182Y	HARDY SPCR	80124
3551-192N	HARDY SPCR	80124
355-415C91	IHC	80201
3-55-42	HAYES	88001.07
3-55-42	SPICER	88001.07
3555J	BORG-WARNR	85114
355-687C91	IHC	80200
355688C91	IHC	80199
35577	FMC	80081
35577	WAYNE	80081
356	AEC	80097
3-56	NEAPCO	80135
356	PRECISION	80090
3-5600	NEAPCO	80038
3-5600	NEAPCO	80039
35-6024	NEAPCO	87331
35-6025	LEMPCO	87481
35-6025	LEMPCO	88007.77
35-6025B	LEMPCO	88008.60
35-6025R	LEMPCO	88008.66
35-6026	LEMPCO	88007.79
35-6026R	LEMPCO	88008.67
35-6027	ANCHORDOAN	88008.63
35-6027	BORG-WARNR	88008.63
35-6027	REPUBLIC	88008.63
35-6028	ANCHOR	87238
35-6028	ANCHORDOAN	88008.35
35-6028	BORG-WARNR	88008.35
35-6028	LEMPCO	88008.35
35-6028	REPUBLIC	88008.35
35-6035	ANCHOR	87238
35-6035	ANCHORDOAN	87238
35-6035	ANCHORDOAN	88007.58
35-6035	BORG-WARNR	87238
35-6035	LEMPCO	88007.58
35-6035	REPUBLIC	87238
35-6036	ANCHOR	87326
35-6036	ANCHORDOAN	87326
35-6036	BORG-WARNR	87326
35-6036	LEMPCO	87326
35-6036	REPUBLIC	87326
35-6036A	BORG-WARNR	87326
35-6036B	LEMPCO	88008.50
35-6037	ANCHOR	87331
35-6037	ANCHORDOAN	87331
35-6037	ANCHORDOAN	88008.39
35-6037	BORG-WARNR	87331
35-6037	LEMPCO	87331
35-6037	LEMPCO	88008.39
35-6037	REPUBLIC	87331
35-6037B	LEMPCO	88008.47
35-6037R	LEMPCO	88008.63
35-6038	ANCHOR	87377
35-6038	ANCHORDOAN	87377
35-6038	ANCHORDOAN	88008.41
35-6038	BORG-WARNR	87377
35-6038	LEMPCO	88008.41
35-6038B	LEMPCO	88008.53
35-6038R	LEMPCO	88008.64
35-6039	ANCHOR	87459
35-6039	ANCHORDOAN	87383
35-6039	ANCHORDOAN	88008.42
35-6039	BORG-WARNR	87383
35-6039	LEMPCO	88008.42
35-6039B	LEMPCO	88008.54
35-6040	ANCHOR	87457
35-6040	ANCHORDOAN	87457
35-6040	ANCHORDOAN	88008.31
35-6040	ANCHORDOAN	88008.44
35-6040	BORG-WARNR	87457
35-6040	LEMPCO	88008.31
35-6040	LEMPCO	88008.44
35-6040B	LEMPCO	88008.56
35-6040R	LEMPCO	88008.65
35-6041	ANCHOR	87459
35-6041	ANCHORDOAN	87459
35-6041	ANCHORDOAN	88008.45
35-6041	BORG-WARNR	87459
35-6041	LEMPCO	87459
35-6041	LEMPCO	88008.45
35-6041B	LEMPCO	88008.57
35-6045	REPUBLIC	87383
35-6046	REPUBLIC	87459
35-6047	BORG-WARNR	87331
35-6047	REPUBLIC	87377
35-6048	ANCHORDOAN	87481
35-6048	ANCHORDOAN	88007.77
35-6048	BORG-WARNR	87481
35-6048	BORG-WARNR	88007.77
35-6049	ANCHORDOAN	87570
35-6049	ANCHORDOAN	88007.79
35-6049	BORG-WARNR	87570
35-6049	BORG-WARNR	88007.79
35-6049	REPUBLIC	87457
35-6050	ANCHORDOAN	87285
35-6050	ANCHORDOAN	88008.48
35-6050	BORG-WARNR	87285
35-6050	REPUBLIC	87481
35-6051	ANCHORDOAN	87337
35-6051	ANCHORDOAN	88008.51
35-6051	ANCHORDOAN	88008.52
35-6051	BORG-WARNR	87337
35-6051	REPUBLIC	87570
35-6052	ANCHORDOAN	87399
35-6052	ANCHORDOAN	88008.59
35-6052	BORG-WARNR	87399
35-6052	REPUBLIC	87331
35-6053	ANCHORDOAN	87373
35-6053	ANCHORDOAN	88008.43
35-6053	BORG-WARNR	87373
35-6053	REPUBLIC	87285
35-6054	ANCHORDOAN	87325
35-6054	ANCHORDOAN	88007.59
35-6054	BORG-WARNR	87325

Part No.	Brand	IDLI Group	Part No.	Brand	IDLI Group	Part No.	Brand	IDLI Group	Part No.	Brand	IDLI Group
35-6054	REPUBLIC	87337	35N0194	WESCO	83262	36311	KIEKHAFER	80034	3707891	GMC	80124
35-6055	REPUBLIC	87399	35N0200-21	WESCO	83320	36312	KIEKHAFER	80061	3708921	GMC	80124
35-6056	ANCHORDOAN	87326	35N0200-26	WESCO	83283	3631P2	DIAMOND R	80048	3708931	GMC	80124
35-6057	ANCHORDOAN	87331	35N0220-20	WESCO	84531	3631P2	WHITE	80140	3708961	GMC	80124
35-6058	REPUBLIC	88008.43	35N0220TDQ	WESCO	84529	36-32-12	SPICER	88015.28	371	PRECISION	80079
3568A	MOTOR MAST	80133	35N0236	WESCO	83331	3632989	CHRYSLER	87327	37-10-00	WALTERSCHD	88003.83
357	AEC	80144	35N0246	WESCO	83349	3632990	CHRYSLER	87327	3711153	CHRYSLER	80060
357	PRECISION	80073	35N048F	WESCO	88001.06	3632993	CHRYSLER	88008.63	37125	DATSUN	80004
35708(4)	BUFFALO-SP	80149	35N160-26	WESCO	83241	363-308R91	IHC	80001	37125-11975	DATSUN	80023
35708(4)	BUFFALO-SP	80155	35N164	WESCO	83240	363-312R91	IHC	80003	37125-14600	DATSUN	80010
357227	WHITE	80124	35N180-21	WESCO	83248	363-313R91	IHC	80003	37125-14625	DATSUN	80010
357-439R91	IHC	80001	35N180-26	WESCO	83252	363394	CHRYSLER	80078	37125-14626	DATSUN	80010
357439R92	IHC	80001	35N184	WESCO	83251	3633944	CHRYSLER	80078	37125-14627	DATSUN	80010
357439R92	IHC	80003	35N190-26	WESCO	83263	3634563	CHRYSLER	80132	37125-146275	DATSUN	80010
3575A	MOTOR MAST	80134	35N194	WESCO	83262	3634563	DETROIT	80132	37125-18000	DATSUN	80004
3575A	MOTOR MAST	80163	35N200-21	WESCO	83277	3637710	CHRYSLER	88007.63	37125-18025	DATSUN	80004
3575A	ZELLER	80134	35N200-26	WESCO	83283	364	PREC.TOR.	80221	37125-85425	DATSUN	80124
3575B	MOTOR MAST	80136	35N2045-37	WESCO	83281	36-4030	LEMPCO	88008.59	37125-85460	DATSUN	80125
3575D	MOTOR MAST	80042	35N220-20	WESCO	84531	364669	STUDEBAKER	88003.42	37125-85461	DATSUN	80124
357-606R91	IHC	80001	35N220-21	WESCO	83320	364674	STUDEBAKER	88003.46	37125-85461	DATSUN	80125
357-607R91	IHC	80001	35N220T	WESCO	83314	365	PRECISION	80078	37125H1025	DATSUN	80004
3578J	BORG-WARNR	85165	35N220TQD	WESCO	84529	365-000-0	GLAENZER	80041	37125H1025A	DATSUN	88002.77
358	AEC	80127	35N2256-37	WESCO	83316	365-001-0	GLAENZER	80072	37125H1026	DATSUN	80010
358	PRECISION	80182	35N236	WESCO	83331	365-002-0	GLAENZER	80041	37125H3000	DATSUN	80010
35822	GERLINGER	80046	35N246	WESCO	83349	365-007-0	GLAENZER	80040	37125H5000	DATSUN	80006
35823	GERLINGER	80047	35N48	WESCO	87933	365-023-0	GLAENZER	80077	37125P3000	DATSUN	80010
358-412R91	IHC	80213	35N48F	WESCO	88001.06	365-030-0	GLAENZER	80077	37126-14626	DATSUN	80010
35845	DEERE,JOHN	80148	35N48M	WESCO	88007.16	365-037-0	GLAENZER	80077	37132	CHAIN BELT	80220
359	PRECISION	80190	35N60F	WESCO	88001.06	365-040-0	GLAENZER	80041	3713926	GMC	80072
3593J	BORG-WARNR	85087	35N72	WESCO	88001.06	365-047-0	GLAENZER	80041	3714168	GMC	88013.70
35MU210P1	MACK	80240	35N72M	WESCO	88007.15	365-050-0	GLAENZER	80077	3714195	GMC	88014.85
35MU210P2	MACK	80237	35NNB3	WESCO	88001.08	365-051-0	GLAENZER	80077	3717179	GMC	80001
35MU211P1	MACK	80186	35N-NB3	WESCO	88001.08	3653379	GMC	88002.48	372	HERCULES	80092
35MU211P2	MACK	80184	35NNB3	WESCO	88001.08	3653380	GMC	88002.60	372	PRECISION	80097
35MU212	MACK	80234	35N-NB3	WESCO	88001.08	36590	VOLVO	80180	3723609	GMC	88013.77
35MU213	MACK	80235	35N-SE138	WESCO	88001.07	366	PRECISION	80040	3723615	GMC	88013.67
35MU214P1	MACK	80240	35NSSF48	WESCO	81848	36-6050	BORG-WARNR	87285	373	AEC	80135
35MU214P2	MACK	80240	35N-SSF72	WESCO	81848	366-312R91	IHC	80001	373	HERCULES	80092
35MU215P1	MACK	80237	35NW4-1900	WESCO	83258	366-313R91	IHC	80001	373	PRECISION	80026
35MU215P2	MACK	80237	35NWY1900	WESCO	83258	3669J	BORG-WARNR	85545	37300-036	ISUZU	80014
35MU216	MACK	80233	35R	G & G MFG	80069	367	PRECISION	88001.86	37300-065	ISUZU	80014
35MU29P1	MACK	80210	35R	WESCO	80069	3-6700	NEAPCO	80145	373-000-92	ISUZU	80023
35MU29P2	MACK	80218	35R	WESCO	80070	3670J	BORG-WARNR	87413	37300-601	ISUZU	80014
35MU31	MACK	87481	360	PRECISION	80075	3677-6	DANUSER	80037	37300-603	ISUZU	80014
35MU32P1	MACK	87481	36-00-00	WALTERSCHD	88004.36	368	PRECISION	80036	3730827	GMC	87373
35MU32P3	MACK	87570	3-6036	ANCHOR	87326	3-6800	NEAPCO	80183	3730827	GMC	88008.63
35MU32P4	MACK	88007.78	36037(4)	BUFFALO-SP	80149	368390	CHRYSLER	80078	3730828	GMC	88008.64
35MU32PC	MACK	88007.79	36037(4)	BUFFALO-SP	80155	3683980	CHRYSLER	80078	373-1410	AEC	80077
35MUA210P1	MACK	80243	3609288	GMC	88002.52	3685958	GMC	88014.79	373-1410	AEC	80144
35MUA210P2	MACK	80244	360R	PRECISION	80959	368-730R91	IHC	80001	3734D	MOTOR MAST	80129
35MUA211P1	MACK	80186	360U	PRECISION	80076	3689266	GMC	88002.50	3736380	CHRYSLER	80035
35MUA211P2	MACK	80184	361	PRECISION	80025	3689288	FORD	88002.52	3737508	GMC	80140
35MUA212	MACK	80234	36-10-02	WALTERSCHD	88003.89	3689288	GMC	88002.52	3737589	GMC	80140
35MUA214P1	MACK	80243	36-13-00	WALTERSCHD	88003.99	369	PRECISION	80077	3-7-39	SPICER	88014.91
35MUA214P2	MACK	80243	36-14-00	WALTERSCHD	88004.09	3693692	GMC	88002.44	374	HERCULES	80049
35MUA215P1	MACK	80244	36141	GERLINGER	80047	369764	GERLINGER	80047	374	PRECISION	80139
35MUA215P2	MACK	80244	36142	GERLINGER	80046	369764	TOWMOTOR	80047	3741635	GMC	80072
35MUA216	MACK	80233	362	PRECISION	80144	369765	GERLINGER	80046	3741653	GMC	80077
35MUA216A	MACK	80233	36-2030	LEMPCO	87285	369765	TOWMOTOR	80046	374246	GMC	80124
35MUA29P1	MACK	80217	36-2030	LEMPCO	88008.48	369U	PRECISION	80088	3744918	CHRYSLER	80078
35MUA29P2	MACK	80218	3620P2	DIAMOND R	80189	370	PRECISION	80180	3744918	DETROIT	80079
35MUA31	MACK	87481	3620P2	WHITE	80189	3704232	GMC	88002.44	3747J	BORG-WARNR	85559
35N	AEC	80069	3621457	CHRYSLER	87327	3705198	GMC	88013.64	37499R11	IHC	86718
35N	AEC	80070	362668	STUDEBAKER	80109	3-7052(4)	AMER.SNOW	80215	375	HERCULES	80043
35N	WESCO	80069	363	PREC.TOR.	80189	3-7053(4)	AMER.SNOW	80215	375	HERCULES	80044
35N0160-26	WESCO	83241	36-30-12	SPICER	88015.27	3705572	FORD	88002.50	375	PRECISION	80108
35N0164	WESCO	83240	36-30-22	SPICER	88015.29	3705572	GMC	88002.50	3750686	GMC	80075
35N0180-21	WESCO	83248	36-3030	LEMPCO	88008.51	3707163	GMC	80124	376	HERCULES	80048
35N0180-26	WESCO	83252	36-3030	LEMPCO	88008.52	3707508	GMC	80140	376	PRECISION	80226
35N0184	WESCO	83251	36-30-62	SPICER	88015.26	3707830	GMC	80142	3765513	GMC	87326

Part No.	Brand	IDLI Group
3765513	GMC	87330
3766233	GMC	87383
376-892C91	IHC	80215
377	HERCULES	80049
377	PRECISION	80002
378	HERCULES	80045
378	PRECISION	80084
3780158	GMC	80090
3780248	CHRYSLER	80061
3780250	CHRYSLER	80061
3780250	DETROIT	80061
3780272	CHRYSLER	80135
3780272	DETROIT	80135
3780280	CHRYSLER	80083
3780299	CHRYSLER	86128
37830	KIEKHAFER	80061
3785	HOWE	80190
3-7-89	SPICER	88014.92
37-8901	LUDWIG	80009
37-8902	LUDWIG	80006
37-8903	LUDWIG	80018
37-8904	LUDWIG	80017
379	HERCULES	80044
379	PRECISION	80085
380	HERCULES	88002.55
380	PRECISION	80086
38025HX	IHC	85165
381	HERCULES	80099
381	PRECISION	80227
3810-367	TAYLOR	80134
381044	WESCO	80136
381-12-4149	KOMATSU	80213
38161	FMC	80123
38161	WAYNE	80123
3817179	GMC	80003
382	HERCULES	80110
382	HERCULES	80112
3820254	CHRYSLER	80221
3820255	CHRYSLER	80189
3822292	CHRYSLER	80227
3823101	GMC	80124
3823102	GMC	80124
3-82-338-3512X	HAYES	88007.16
3-82-338-3512X	SPICER	88007.15
3-82-448-3230X	SPICER	88007.14
382668	STUDEBAKER	80149
382688	STUDEBAKER	80109
3827046	CHRYSLER	80077
3827050	CHRYSLER	88002.37
382741	STUDEBAKER	80953
382742	STUDEBAKER	80110
382742	STUDEBAKER	80112
3828469	GMC	80072
382852	STUDEBAKER	88001.60
383	HERCULES	80065
383	PRECISION	80014
3830580	GMC	80075
3831798	CHRYSLER	80186
383182C91	IHC	87328
383.472	VOLVO	80041
3837585	CHRYSLER	80139
3837585	CHRYSLER	80140
3837585	CHRYSLER	80182
3837585	DETROIT	80139
384	PRECISION	80020
38403	VOLVO	80041
3840500	GMC	80077
384450R91	IHC	80234
38450R91	IHC	80234
384550R91	IHC	80238
38464	VOLVO	80142
384651	GMC	80077
3848500	GMC	80076
3849500	GMC	80075
385	PRECISION	80005
3851520	GMC	80124
3858A	MOTOR MAST	80017
386	PRECISION	80013
3860801	GMC	88008.35
38633	FIAT-ALLIS	80220
386417C1	IHC	88013.73
386418C91	IHC	87331
386451	GMC	80077
386452	GMC	80119
3869039	GMC	80003
387	PRECISION	80016
387-097R91	IHC	80140
3877040	GMC	80124
38777HX	IHC	85087
38780H	IHC	85545
38781H	IHC	87413
388	PRECISION	80229
3881-1	A.E.LEE	88010.64
3881-1	NATL.MINE	88010.64
3881-2	NATL.MINE	80001
3881-3	A.E.LEE	80069
3881-3	NATL.MINE	80069
3881-4	A.E.LEE	80037
3881-4	NATL.MINE	80037
3881-5	A.E.LEE	80161
3881-5	NATL.MINE	80219
3881-6	NATL.MINE	80070
3881-7	NATL.MINE	80134
388451	GMC	80077
3886579	CHRYSLER	80012
3886579	MITSUBISHI	80012
3889690	GMC	80075
3889696	GMC	80124
389	PRECISION	80115
38UM1040	HAHN	80198
38UM141	HAHN	80215
38UM144	HAHN	80193
38UM175	HAHN	80218
38UM189	HAHN	80201
390	PRECISION	80018
3-901	BALKAMP	80213
390113	GMC	80117
390133	GMC	80117
390134	GMC	80121
3-902	BALKAMP	80112
3-903	BALKAMP	80234
3-903	BALKAMP	80235
3-904	BALKAMP	80234
3-905	BALKAMP	80198
3-906	BALKAMP	80212
3-906	BALKAMP	80214
3-906	BALKAMP	80218
3-907	BALKAMP	80211
3-907	BALKAMP	80214
3-908	BALKAMP	80210
3-908	BALKAMP	80217
39087	DAVEY COM.	80167
3-909	BALKAMP	80237
3-909	BALKAMP	80244
391	PRECISION	80010
3-910	BALKAMP	80231
3-910	BALKAMP	80243
3-910	BALKAMP	88011.37
3910235	GMC	80075
391055	WESCO	80161
391056	WESCO	80134
3-911	BALKAMP	80234
3-912	BALKAMP	80233
3-913	BALKAMP	80232
3-913	BALKAMP	80244
3-914	BALKAMP	80232
3-914	BALKAMP	80238
3-914	BALKAMP	80244
391437	CHRYSLER	88003.47
391437	CHRYSLER	88003.50
3918866	GMC	80003
392	PRECISION	80004
3-920	BALKAMP	80168
3-921	BALKAMP	80109
3-922	BALKAMP	80099
3-922	BALKAMP	80101
3-924	BALKAMP	80201
39241-11M25	DATSUN	86193
3-925	BALKAMP	80198
3926	ALCO	80089
3-927	BALKAMP	80212
3-927	BALKAMP	80214
3-927	BALKAMP	80218
3-927	BALKAMP	80238
3-928	BALKAMP	80109
3-929	BALKAMP	80109
392951C1	IHC	88014.85
392952C1	IHC	88007.55
392953	IHC	88008.38
393	PRECISION	80019
3-934	BALKAMP	80070
3-935	BALKAMP	80193
39360	NEW HOLL.	80035
393-671R91	IHC	80219
3-937	BALKAMP	80193
3-937	BALKAMP	80197
3937988	GMC	80075
3937988	GMC	80117
3-938	BALKAMP	80158
39383	KIEKHAFER	80061
394	PRECISION	80009
3-940	BALKAMP	80109
39414115	UNION PAC.	80109
3-94-18X	SPICER	88013.87
3-94-28X	SPICER	88013.88
3-943	BALKAMP	80198
3-945	BALKAMP	80069
3-94-58X	SPICER	88002.13
3-946	BALKAMP	80238
3-947	BALKAMP	80213
39474	ALLIS-CHLM	80191
395	PRECISION	80006
3-951	BALKAMP	80048
3-952	BALKAMP	80047
3-953	BALKAMP	80178
3-953	BALKAMP	80179
3953659	GMC	80077
3-954	BALKAMP	80124
3-955	BALKAMP	80142
3955571	GMC	80075
39559	FIAT-ALLIS	80187
3-95-59	SPICER	88014.85
3-95-69	SPICER	88014.86
3-959	BALKAMP	80185
3-95-96	SPICER	88014.86
396	PRECISION	80017
3-960	BALKAMP	80185
3-960	BALKAMP	80193
39602101	INNOCENTI	80023
39602102	INNOCENTI	80023
3-961	BALKAMP	80049
3-962	BALKAMP	80223
3-962	BALKAMP	80231
39625-21001	DATSUN	80019
39625-21004	DATSUN	80019
39625-21005	DATSUN	80019
39625-21025	DATSUN	80019
39625-21026	DATSUN	80019
39625-21027	DATSUN	80019
39625-43825	DATSUN	80019
39625H1025	DATSUN	80019
39625N21026	DATSUN	80019
39625N3725	DATSUN	80019
39625N4125	DATSUN	80019
39625N4126	DATSUN	80019
39625U0100	DATSUN	80019
39625U0125	DATSUN	80019
39625U0127	DATSUN	80019
39625U3825	DATSUN	80019
396451	GMC	80077
39674	FIAT-ALLIS	80191
397	PRECISION	80011
3-972	BALKAMP	80035
39741-11M25	DATSUN	86158
3975	CHAMPION	80239
3975	DOMINION	80232
3977	CHAMPION	80210
3979616	GMC	80124
3979616	GMC	80125
3979617	GMC	80124
398	PRECISION	80012
3981590	GMC	80078
3-989	BALKAMP	80049
39-8900	LUDWIG	80019
39-8905	LUDWIG	80010
39-8906	LUDWIG	80006
399	PRECISION	80015
3990291	GMC	87326
3990291	GMC	87331
3990292	GMC	88007.55
399-040C91	IHC	80003
3-991	BALKAMP	80168
3-992	BALKAMP	80168
3A300	BEAN,JOHN	80186
3AA	AUTOCAR	87481
3AA01172	AUTOCAR	80072
3AA01172	AUTOCAR	80075
3AA0970	AUTOCAR	87481
3AA0970	AUTOCAR	88007.77
3AA1172	AUTOCAR	80075
3AB01172	AUTOCAR	80023
3AB0970	AUTOCAR	87570
3AB0970	AUTOCAR	88007.79
3AB1172	AUTOCAR	80023
3AB1172	AUTOCAR	80024
3AD0970	AUTOCAR	87457
3AD0970	AUTOCAR	88008.31
3AD0970	AUTOCAR	88008.44
3AL01150	AUTOCAR	80221
3AL0150	AUTOCAR	80221
3AL150	AUTOCAR	80221
3C37	JEBCO	80104
3C37	PAXTON-MIT	80100
3CA01172	AUTOCAR	80142
3CA1172	AUTOCAR	80142
3CAR29699R	DEERE,JOHN	80099
3CARR29699R	DEERE,JOHN	80099
3DA01172	AUTOCAR	80178
3DR	AEC	80133
3DR	WESCO	80133
3DR	WESCO	88002.89

Part No.	Brand	IDLI Group
3DRYD16	ROCKWELL	83383
3DRYD16-1	ROCKWELL	83387
3DRYD24	ROCKWELL	83452
3DRYP16-7	ROCKWELL	83402
3DRYQ14	ROCKWELL	83379
3DRYQ15-2	ROCKWELL	83380
3DRYQ15-4	ROCKWELL	83381
3DRYQ16	ROCKWELL	83382
3DRYQ16-2	ROCKWELL	83386
3DRYQ17	ROCKWELL	83391
3DRYQ18	ROCKWELL	83403
3DRYQ18-2	ROCKWELL	83401
3DRYQ19	ROCKWELL	83406
3DRYQ20	ROCKWELL	83417
3DRYQ20-2	ROCKWELL	83418
3DRYQ21	ROCKWELL	83426
3DRYQ21-5	ROCKWELL	83429
3DRYQ22	ROCKWELL	83438
3DRYQ24-1	ROCKWELL	83450
3DRYQ26-1	ROCKWELL	83455
3DRYQ28-1	ROCKWELL	83461
3DRYQ30	ROCKWELL	83463
3DRYQ32-1	ROCKWELL	83464
3DRYR12	ROCKWELL	83377
3DRYR14	ROCKWELL	83378
3DRYR16-1	ROCKWELL	83385
3DRYR16-6	ROCKWELL	83384
3DRYR17-1	ROCKWELL	83390
3DRYR17-2	ROCKWELL	83388
3DRYR17-3	ROCKWELL	83389
3DRYR18	ROCKWELL	83398
3DRYR18-1	ROCKWELL	83399
3DRYR18-11	ROCKWELL	83400
3DRYR18-2	ROCKWELL	83392
3DRYR18-20	ROCKWELL	83397
3DRYR18-26	ROCKWELL	83396
3DRYR19-1	ROCKWELL	83409
3DRYR19-5	ROCKWELL	83408
3DRYR19-6	ROCKWELL	83410
3DRYR20-10	ROCKWELL	83419
3DRYR20-2	ROCKWELL	83420
3DRYR20-23	ROCKWELL	83413
3DRYR20-3	ROCKWELL	83421
3DRYR20-41	ROCKWELL	83412
3DRYR20-45	ROCKWELL	87049
3DRYR20-46	ROCKWELL	83416
3DRYR20-9	ROCKWELL	83411
3DRYR21	ROCKWELL	83428
3DRYR21-2	ROCKWELL	83425
3DRYR21-3	ROCKWELL	83423
3DRYR21-4	ROCKWELL	83424
3DRYR21-6	ROCKWELL	83427
3DRYR22	ROCKWELL	83437
3DRYR22-11	ROCKWELL	83430
3DRYR22-12	ROCKWELL	83434
3DRYR22-2	ROCKWELL	83435
3DRYR22-5	ROCKWELL	83436
3DRYR23-10	ROCKWELL	83441
3DRYR23-14	ROCKWELL	83442
3DRYR23-5	ROCKWELL	83443
3DRYR23-9	ROCKWELL	83444
3DRYR24-13	ROCKWELL	83446
3DRYR24-15	ROCKWELL	83454
3DRYR24-19	ROCKWELL	83445
3DRYR24-2	ROCKWELL	83449
3DRYR24-28	ROCKWELL	83451
3DRYR24-9	ROCKWELL	83448
3DRYR26	ROCKWELL	83456
3DRYR26-2	ROCKWELL	83457
3DRYR28-2	ROCKWELL	83460
3DRYS18-1	ROCKWELL	83405
3DRYS18-13A	ROCKWELL	84535
3DRYS18-17	ROCKWELL	83393
3DRYS18-2	ROCKWELL	83395
3DRYS18-6	ROCKWELL	87047
3DRYS18-8	ROCKWELL	83404
3DRYS18-9	ROCKWELL	87046
3DRYS19	ROCKWELL	83407
3DRYS20-2	ROCKWELL	87048
3DRYS20-4	ROCKWELL	83422
3DRYS20-7	ROCKWELL	83415
3DRYS22-1	ROCKWELL	83440
3DRYS22-14	ROCKWELL	83433
3DRYS22-21	ROCKWELL	83431
3DRYS22-22A	ROCKWELL	84536
3DRYS22-2A	ROCKWELL	84537
3DRYS22-4	ROCKWELL	83439
3DRYS22-5	ROCKWELL	87050
3DRYS22-5	ROCKWELL	88005.96
3DRYS23-12	ROCKWELL	87051
3DRYS24-1	ROCKWELL	83447
3DRYS24-30	ROCKWELL	87052
3DRYS24-5	ROCKWELL	83453
3DRYS26-1	ROCKWELL	83458
3DRYS28	ROCKWELL	87053
3DRYS28-2	ROCKWELL	83462
3DRYS28-4	ROCKWELL	83459
3DRYT18-4	ROCKWELL	83394
3DRYT20	ROCKWELL	83414
3DRYT22	ROCKWELL	83432
3EA01172	AUTOCAR	80189
3EA1172	AUTOCAR	80189
3F1518	CATERPILLR	88001.30
3FA01172	AUTOCAR	80221
3GA01172	AUTOCAR	80225
3GC01172	AUTOCAR	80225
3H1945	CATERPILLR	88001.25
3H2579	CATERPILLR	80238
3H2852	CATERPILLR	88001.39
3H2972	CATERPILLR	88001.16
3H301	CATERPILLR	80198
3H301	CATERPILLR	88001.16
3H6104	CATERPILLR	80213
3H6263	CATERPILLR	80001
3H6449	CATERPILLR	80234
3H7364	CATERPILLR	80198
3H8141(2)	CATERPILLR	88001.17
3H8257(5)	CATERPILLR	88002.90
3H8262(5)	CATERPILLR	88002.90
3H8263(5)	CATERPILLR	88002.90
3H9099	CATERPILLR	88001.25
3K7276	CATERPILLR	80213
3K8807	CATERPILLR	88001.38
3K9532(5)	CATERPILLR	88002.92
3L7363	CATERPILLR	80070
3N170-1	ROCKWELL	87027
3N171-1	ROCKWELL	87026
3N173-1	ROCKWELL	88005.97
3NDCA2	ROCKWELL	88005.98
3NF20	ROCKWELL	84906
3NF33	ROCKWELL	84907
3NF5	ROCKWELL	84908
3NF6	ROCKWELL	84912
3NLS24-10DC	ROCKWELL	85308
3NLS24-6DC	ROCKWELL	85305
3NLS24-7A	ROCKWELL	85306
3NLS24-8DC	ROCKWELL	85304
3NLS24DC	ROCKWELL	85307
3NPS24	ROCKWELL	87341
3NPS24-3	ROCKWELL	87344
3N-SH15	ETNYRE	80001
3NY32	ROCKWELL	85709
3NY32-2	ROCKWELL	85708
3NY32-9	ROCKWELL	85710
3NYQ19-5	ROCKWELL	83259
3NYR12-1	ROCKWELL	83235
3NYR14-1	ROCKWELL	83236
3NYR16-4	ROCKWELL	83240
3NYR16-7	ROCKWELL	83238
3NYR18-4	ROCKWELL	83246
3NYR19-1	ROCKWELL	83256
3NYR20-10	ROCKWELL	88005.99
3NYR20-14	ROCKWELL	83272
3NYR20-25	ROCKWELL	83270
3NYR20-5	ROCKWELL	83271
3NYR21-1	ROCKWELL	83293
3NYR22-22	ROCKWELL	83299
3NYR22-25	ROCKWELL	87034
3NYR22-34	ROCKWELL	83303
3NYR23-1	ROCKWELL	83330
3NYR24-1	ROCKWELL	83343
3NYR24-17	ROCKWELL	83356
3NYR24-19	ROCKWELL	83342
3NYR24-37	ROCKWELL	83337
3NYR24-4	ROCKWELL	83341
3NYR27	ROCKWELL	83365
3NYS20-12A	ROCKWELL	85302
3NYS20-13	ROCKWELL	83284
3NYS20-4	ROCKWELL	83286
3NYS20-6	ROCKWELL	83287
3NYS22-10	ROCKWELL	87033
3NYS22-53	ROCKWELL	83304
3NYS22-60	ROCKWELL	83302
3NYS22-61	ROCKWELL	83300
3NYS22-8	ROCKWELL	87037
3NYS24-1	ROCKWELL	83339
3NYS24-17	ROCKWELL	83338
3NYS24-28	ROCKWELL	83340
3NYS24-29	ROCKWELL	83358
3NYS24-5	ROCKWELL	83336
3NYS28-13	ROCKWELL	83370
3NYS28-14	ROCKWELL	83368
3NYS28-8	ROCKWELL	83369
3NYS32	ROCKWELL	83376
3NYSM24-35	ROCKWELL	87039
3NYSM24-37	ROCKWELL	87040
3NYSM24-39	ROCKWELL	87041
3NYSM24-42	ROCKWELL	87042
3NYSM24-43	ROCKWELL	87043
3NYT22-4	ROCKWELL	83324
3P15	JEBCO	80101
3P15	PAXTON-MIT	80101
3P15A	JEBCO	80216
3P43	JEBCO	80112
3R1310	AEC	80097
3R-231	NEAPCO	80144
3R231	NEAPCO	80144
3RCV	AEC	80109
3T1328	AEC	88002.55
3T1328	BORG-WARNR	88002.55
3T-1328	MASTERS	88002.55
3T1602	BORG-WARNR	88002.56
3T2007	BORG-WARNR	88002.57
3T2221	BORG-WARNR	88002.58
3T2222	BORG-WARNR	88002.45
3T2266	BORG-WARNR	88002.54
3T2531	BORG-WARNR	88002.60
3T2595	BORG-WARNR	88002.47
3T2893	BORG-WARNR	88002.48
3T2894	BORG-WARNR	88002.46
3T2895	AEC	88002.48
3T2895	BORG-WARNR	88002.48
3T2896	AEC	88002.50
3T2896	BORG-WARNR	88002.50
3T2897	AEC	88002.51
3T2897	BORG-WARNR	88002.51
3T2898	AEC	88002.52
3T2898	BORG-WARNR	88002.52
3T2899	BORG-WARNR	88002.53
3T3501	BORG-WARNR	88002.63
3T3502	BORG-WARNR	88002.64
3T3503	BORG-WARNR	88002.66
3T3503	BORG-WARNR	88002.67
3T3504	BORG-WARNR	88002.64
3T3505	BORG-WARNR	88002.65
3T3506	BORG-WARNR	88002.66
3T3506	BORG-WARNR	88002.67
3T3507	BORG-WARNR	88002.71
3T3508	BORG-WARNR	88002.72
3T3509	BORG-WARNR	88002.70
3T3510	BORG-WARNR	88002.73
3T3511	BORG-WARNR	88002.74
3UU001172	AUTOCAR	80178
3UU001173	AUTOCAR	80189
3W00-01172	AUTOCAR	80142
3W01172	WHITE	88008.97
3WCS28-93	ROCKWELL	88006
3WCS30-4	ROCKWELL	88006.01
3WCS31-15	ROCKWELL	88006.02
3WCS32-58	ROCKWELL	88006.03
3WCS36-21	ROCKWELL	88006.04
3W001172	AUTOCAR	80142
3W01172	AUTOCAR	80142
4	TRU-CROSS	80041
4000-5	ZELLER	80158
40006	AEC	84504
4000A	MOTOR MAST	80109
4000A	ZELLER	80109
4000A	ZELLER	80149
4000B	MOTOR MAST	80148
4000B	ZELLER	80148
4000DA	MOTOR MAST	80153
4000DA	ZELLER	80153
4000E	MOTOR MAST	80156
4000E	ZELLER	80156
4000E	ZELLER	88003.17
4000J	MOTOR MAST	80158
4000J	ZELLER	80158
4000P	MOTOR MAST	80155
4000P	ZELLER	80155
4000Q	MOTOR MAST	80157
4000Q	ZELLER	80157
4000S	MOTOR MAST	80159
4000S	ZELLER	80159
4000T	MOTOR MAST	80153
4000V	MOTOR MAST	80151
4000V	ZELLER	80151
4000X	MOTOR MAST	80154
4000X	ZELLER	80146
4001	ALLOY	87331
4001	ALLOY	88008.39
40012	AEC	88000.45
40013	AEC	88000.20
4002	ALLOY	87326
4002	TRU-CROSS	80109
4002	TRU-CROSS	80149
40022	AEC	88000.43
40023	AEC	88000.44
40023	AEC	88000.60
4002631	GAR WOOD	80008

Part No.	Brand	IDLI Group
4003	ALLOY	87383
4003	ALLOY	88008.42
40033	AEC	88000.26
400-3-400	MUNCIE	88000.69
4003746	GAR WOOD	80109
4003768	GAR WOOD	80150
4004	ALLOY	87377
4004	ALLOY	88008.41
40043	AEC	88000.22
4005	ALLOY	87459
4005	ALLOY	88008.45
40053	AEC	88000.52
4-0058	NEAPCO	80189
4006	ALLOY	87457
4006	ALLOY	88008.31
4006	ALLOY	88008.44
4-0060	NEAPCO	80188
40063	AEC	88000.25
4007	ALLOY	87481
4007	ALLOY	88007.77
40073	AEC	88000.28
400.755.000	PITT.VIOL.	80221
40076	AEC	88000.29
40076	AEC	88001
4008	ALCO	80146
4008	ALLOY	87570
4008	ALLOY	88007.79
40081	AEC	88000.49
40083	AEC	88000.53
40088M0820	DATSUN	86193
40088M0825	DATSUN	86193
40088M0826	DATSUN	86193
40089M0820	DATSUN	86158
40089M0826	DATSUN	86158
40089M0826	DATSUN	86193
40089M0827	DATSUN	86158
40093	AEC	88000.30
40093	AEC	88000.31
400934	WHITE	80072
400934	WHITE	80075
400934A	WHITE	80072
400934A	WHITE	80075
40095	AEC	88000.34
400E4635	FORD	80023
400E4635A	FORD	80024
401	TEDDY TORQ	80901
4010	ALLOY	87325
4010	ALLOY	88007.59
40103	AEC	88000.36
401062	WESCO	80907
401065	WESCO	80905
401-065	WESCO	80951
401067	WESCO	80953
4011	ALCO	80147
4011	ALLOY	87373
4011	ALLOY	88008.43
401-113	WESCO	88001.28
40113	AEC	88000.35
40115	AEC	88000.55
401159R91	IHC	80171
4012	ALLOY	88008.35
40123	AEC	88000.38
40124	AEC	88000.37
4013	ALLOY	88007.58
4014	TRU-CROSS	80158
4014113	GAR WOOD	80069
4014113	GAR WOOD	80070
401420	WESCO	80958
401421	WESCO	80902
401421	WESCO	80906
401-421	WESCO	80909
401422	WESCO	80958
401423	WESCO	88003.52
401423	WESCO	88003.56
401424	WESCO	88002.44
401424	WESCO	88002.49
401440	WESCO	88001.74
401450	WESCO	80956
401471	WESCO	80908
401484	WESCO	88002.86
4015	TRU-CROSS	80148
40150	AEC	88000.46
4016	TRU-CROSS	80159
401642	WESCO	88002.42
401642	WESCO	88002.59
401642	WESCO	88002.62
401646	WESCO	88002.51
4016484	GMC	80221
4-0165	NEAPCO	80190
4016929	GMC	80190
401709	WESCO	88002.55
401761	TIMBERJACK	80142
40179	ALLOY	88002.38
401794	TIMBERLAND	80215
401953	WESCO	88002.47
401953	WESCO	88002.48
401C	TEDDY TORQ	88001.74
4020	FIAT-ALLIS	80069
4021288	FIAT	80225
402146DS	NATL.MINE	80035
402146DS	OLIVER	80035
402-146DS	WHITE FARM	80035
4023-4120	DAIHATSU	80010
4023541	GMC	80228
4024300	FIAT	80023
4025348	GMC	80227
4025-5	ZELLER	80173
4025A	MOTOR MAST	80168
4025A	ZELLER	80168
4025AA	MOTOR MAST	80168
4025B	MOTOR MAST	80172
4025B	ZELLER	80172
4025D	MOTOR MAST	80174
4025D	ZELLER	80174
4025DH	MOTOR MAST	80174
4025DH	ZELLER	80174
4025E	MOTOR MAST	80171
4025E	ZELLER	80171
4025EX	MOTOR MAST	80171
4025EX	ZELLER	80171
4025FA	MOTOR MAST	80167
4025FA	ZELLER	80167
4025G	MOTOR MAST	80175
4025G	ZELLER	80175
4025GH	MOTOR MAST	80175
4025GH	ZELLER	80175
4025H	MOTOR MAST	80162
4025H	ZELLER	80162
4025HD	MOTOR MAST	80174
4025J	MOTOR MAST	80169
4025J	ZELLER	80216
4025S	MOTOR MAST	80173
4025S	ZELLER	80173
4025T	MOTOR MAST	80177
4025U	MOTOR MAST	80170
4026301	GMC	80225
402631	WABCO	80188
4-0279	NEAPCO	80189
402803DS	COCKSHUTT	80037
402803DS	NATL.MINE	80037
402803DS	OLIVER	80037
402-803DS	WHITE FARM	80037
40292	PETTI-MULL	80047
40293	PETTI-MULL	80048
4030848	CHRYSLER	80107
4030848	DETROIT	80104
4030849	CHRYSLER	88002.36
40309A1	ROSS GEAR	80001
403186	EATON	80142
403186	TIMBERJACK	80142
403414	WHITE	80142
403423A1	ROSS GEAR	80001
40372	ALLIS-CHLM	80235
40372	FIAT-ALLIS	80235
403759	EATON	80140
403759	TIMBERJACK	80140
40377	ALLIS-CHLM	80235
40377	FIAT-ALLIS	80235
403772A1	ROSS GEAR	80001
403KC	KENBAR	88008.59
4-04	BYPY	88003.68
404645	ADAMS-LET.	80035
404645	WABCO	80035
40491	HIWAY CHIP	80124
40495	HIWAY CHIP	80124
4049697	CHRYSLER	88002.36
404E4635A	FORD	80041
405	HERCULES	80041
405	HERCULES	88001.85
4050350	FORD	80077
4-05-07	BYPY	88003.76
405-3-405	MUNCIE	88000.70
405485	TIMBERJACK	80182
4057032	CHRYSLER	86081
4057033	CHRYSLER	86082
4-06	BYPY	88004.19
4-0614	NEAPCO	80196
4-0628	NEAPCO	80198
406428	WESCO	88003.51
4-0643	NEAPCO	80193
406451	WESCO	88003.56
406461	WESCO	88003.57
406606	WHITE	80221
406606A	WHITE	80221
406606B	WHITE	80221
406607	WHITE	80189
407	AEC	80226
4-07	BYPY	88004.12
40.701.000	PITT.VIOL.	80023
40.701.000	PITT.VIOL.	80024
40.711.000	PITT.VIOL.	80180
40.712.000	PITT.VIOL.	80189
40.719.000	PITT.VIOL.	80142
40.744.000	PITT.VIOL.	80042
40.745.000	PITT.VIOL.	80135
40.746.000	PITT.VIOL.	88002.77
40.752.000	PITT.VIOL.	88002.77
40.755.000	PITT.VIOL.	80221
40.756.000	PITT.VIOL.	80225
40.757.000	PITT.VIOL.	80124
40.759.000	PITT.VIOL.	80077
40.760.000	PITT.VIOL.	80210
40782	ALLOY	88002.37
40.811.000	PITT.VIOL.	80178
40.812.000	PITT.VIOL.	80047
40.813.000	PITT.VIOL.	80049
40.814.000	PITT.VIOL.	88002.77
40.815.000	PITT.VIOL.	80045
40.816.000	PITT.VIOL.	80049
408-308	WESCO	80104
408-308	WESCO	80107
408-308	WESCO	80112
408-308	WESCO	88002.34
408-309	WESCO	80104
408-309	WESCO	80107
408-309	WESCO	80112
408-309	WESCO	88002.36
408414	OLDING	80035
40.843.000	PITT.VIOL.	80109
40.851.000	PITT.VIOL.	80023
40.854.000	PITT.VIOL.	80023
40.855.000	PITT.VIOL.	80041
408-838	WESCO	80104
408-838	WESCO	80107
408-838	WESCO	80112
408-838	WESCO	88002.33
408-864	WESCO	80104
408-864	WESCO	80107
408-864	WESCO	80112
40.920.000	PITT.VIOL.	80023
4092P11	M.R.S.	80235
40961	PETTI-MULL	80193
41	DETROIT	88003.55
4100	AEC	80109
4100	AEC	80149
4100	ALCO	80109
4-10010	LEVRATTO	80001
41002	ALLOY	80085
41004	ALCO	80077
41004	ALLOY	80077
4101	AEC	80153
4101	ALCO	80153
41010	ALLOY	80081
41011	ALLOY	80091
41012	ALLOY	80117
41012	ALLOY	80129
41013	ALLOY	80131
41014	ALLOY	80104
4102	AEC	80158
4102	ALCO	80158
4102	ALCO	88002.15
41020	ALLOY	80132
41025	ALCO	80090
41025	ALLOY	80078
41034	ALCO	80118
41034	ALLOY	80118
410-3-410	MUNCIE	88000.71
41048	ALCO	80078
41048	ALLOY	80078
4105	AEC	80151
410529	STUDEBAKER	80072
410529	STUDEBAKER	80077
41058	ALLOY	80018
4-1058	NEAPCO	80044
4-1058	NEAPCO	80179
4106	AEC	80148
4107	AEC	80159
4107	ALCO	80159
41078	ALLOY	80124
41083	BORG-WARNR	88003.47
41085-1	DETROIT	88003.50
4109	AEC	80157
4110	AEC	80155
41103	AEC	88000.85
41103	ALCO	80061
41103	ALLOY	80062
411062	WESCO	80954
411063	WESCO	80957
411065	WESCO	80951
41112	AEC	88000.86

Part No.	Brand	IDLI Group	Part No.	Brand	IDLI Group	Part No.	Brand	IDLI Group	Part No.	Brand	IDLI Group
41112	AEC	88000.88	42012	DETROIT	88003.56	42591	BORG-WARNR	88003.49	431-407-331G	AUDI	86096
41113	AEC	88000.79	42012	HERCULES	88003.56	42591	HERCULES	88003.49	4-3-1431KX	SPICER	85076
41122	AEC	88000.89	42012C	HERCULES	88003.56	4259-25-060	MAZDA	80011	4314783	FIAT	86098
41123	AEC	88000.80	42013	AEC	88000.68	425944	WABCO	80048	4314783	FIAT-ALLIS	86098
4113	AEC	80156	42017	DETROIT	88003.56	426	PRECISION	80074	431498201A	AUDI	86095
41132	AEC	88000.95	42019	DETROIT	88003.56	42-6001	BORG-WARNR	86724	431498201A	AUDI	86161
41133	AEC	88000.81	42021	BORG-WARNR	88003.42	42-6003	BORG-WARNR	86726	431-498-201D	AUDI	86162
411470	WESCO	80955	42021	HERCULES	88003.42	42-6005	BORG-WARNR	86727	431498203C	AUDI	86135
41150	AEC	88000.90	4203-1	BORG-WARNR	88003.43	42-6006	BORG-WARNR	86725	431498203C	AUDI	86188
411501149A	PORSCHE	86071	4203-1	HERCULES	88003.43	42-6007	BORG-WARNR	86728	431-498-203C	VOLKSWAGEN	86187
411501149A	PORSCHE	86145	4204	BORG-WARNR	88003.45	4261	BORG-WARNR	88003.46	431-498-203E	AUDI	86189
411501149A	VOLKSWAGEN	86071	4204	HERCULES	88003.45	4261	HERCULES	88003.46	432	PERFECT-CI	80161
411501149A	VOLKSWAGEN	86145	4204117	FIAT	86098	4265A	MOTOR MAST	80182	432	PRECISION	80183
41153	AEC	88000.83	4204117	FIAT	86163	4265A	ZELLER	80182	432	TRU-CROSS	80161
41154	ALLOY	80086	42043	AEC	88000.70	4-2-669	SPICER	84618	43203	ROVER	80041
41155	ALLOY	80139	42051	BORG-WARNR	88003.48	427	PRECISION	80141	432377C91	IHC	87330
411-598-201	VOLKSWAGEN	86145	42051	HERCULES	88003.48	42-7000	BORG-WARNR	86800	432-433C91	IHC	80235
41163	AEC	88000.96	4205733	CHRYSLER	86080	42-7001	BORG-WARNR	86803	432-473C91	IHC	80003
4117147	CHRYSLER	87331	4205733	CHRYSLER	86082	42-7003	BORG-WARNR	86804	4328468	FIAT	86138
41174	ALLOY	80085	4205852	CHRYSLER	86080	42-7004	BORG-WARNR	86802	4328468	FIAT-ALLIS	86138
41181	AEC	88000.92	4205852	CHRYSLER	86081	4-27-16-6200X	SPICER	86676	433	HERCULES	80072
41181	AEC	88001	4205852	CHRYSLER	86085	4-27-18-6116X	SPICER	86677	433	HERCULES	80077
41183	AEC	88000.82	4205852	CHRYSLER	86089	427271	WABCO	80001	433	PRECISION	80129
411926	WESCO	88002.50	4205852	CHRYSLER	86154	427349C91	IHC	87328	4330 88	FIAT	86163
411928	WESCO	88002.52	4205853	CHRYSLER	86128	427362	WHITE	80178	433-1	TRU-CROSS	80085
41193	AEC	88000.84	4205853	CHRYSLER	86129	427804C91	IHC	80078	433-1	TRU-CROSS	80086
411956R91	IHC	80077	4205853	CHRYSLER	86130	4-279	NEAPCO	80189	4333088	FIAT	86098
4123-4120	DAIHATSU	80010	4205853	CHRYSLER	86131	428	PRECISION	80145	4333088	FIAT	86163
4-1-2761	SPICER	88005.91	4205853	CHRYSLER	86184	42-8001	BORG-WARNR	88007.21	4333088	FIAT-ALLIS	86163
414	PRECISION	80122	42063	AEC	88000.71	42-8002	BORG-WARNR	86830	433951	JEFFREY	80182
4140	TRU-CROSS	80155	42073	AEC	88000.72	42-8003	BORG-WARNR	86829	434	PRECISION	80131
4141	AEC	80154	420819	DETROIT	88003.50	4-28-277	SPICER	85527	43447	NEW HOLL.	80037
4-14-19	SPICER	88005.35	420822-1	BORG-WARNR	88003.50	4287138	FIAT	86138	43449	NEW HOLL.	80069
4143	TRU-CROSS	80156	420822-1	DETROIT	88003.50	4287138	FIAT-ALLIS	86138	43453	NEW HOLL.	80033
4-14-69	SPICER	88013.55	420825-1	DETROIT	88003.50	429	PRECISION	80081	43454	NEW HOLL.	80008
4-1-4791	SPICER	88005.95	42083	DETROIT	88003.47	430	PERFECT-CI	88000.15	435	PRECISION	80080
414-946C91	IHC	80217	42083	DETROIT	88003.50	430	PRECISION	80091	436	PRECISION	80028
415	PRECISION	80138	42083	HERCULES	88003.47	430	TRU-CROSS	88000.15	436087	STUDEBAKER	80065
4151-179W	HARDY SPCR	80142	42093	AEC	88000.73	4300A	MOTOR MAST	80135	436101	STUDEBAKER	80056
4151-179Y	HARDY SPCR	80142	421	PRECISION	80204	4300A	ZELLER	80135	436101	STUDEBAKER	80057
4151-213W	HARDY SPCR	80142	4213200	CHRYSLER	80135	4301	DETROIT	88003.57	436144	STUDEBAKER	80072
415124	JOY MFG.	80142	4215015	FIAT	86138	43023	ROVER	80041	436144	STUDEBAKER	80075
4-15-13	SPICER	88015.57	4215015	FIAT	86190	43063	AEC	88000.25	436144	STUDEBAKER	80077
4151J	BORG-WARNR	87413	4215J	BORG-WARNR	81749	4308412	FIAT	86138	436175	STUDEBAKER	80096
415-3-415	MUNCIE	88000.72	4215J	BORG-WARNR	81750	4308650	FIAT	86138	436179	STUDEBAKER	80955
4156466	CHRYSLER	88002.77	422	HERCULES	80124	4308650	FIAT-ALLIS	86138	436188	STUDEBAKER	80095
4156466	MITSUBISHI	88002.77	422	PRECISION	80039	43093	AEC	88000.30	436198	STUDEBAKER	80954
4157050	CHRYSLER	80060	4221J	BORG-WARNR	81654	431	PRECISION	80130	436199	STUDEBAKER	80955
416	PRECISION	80068	4222J	BORG-WARNR	84662	431	TRU-CROSS	80136	436270	STUDEBAKER	80057
41603	PETTI-MULL	80196	4225A	MOTOR MAST	80178	43113	AEC	88000.35	436277	STUDEBAKER	80954
4-16-143	SPICER	88015.31	4225A	ZELLER	80178	4311356	FIAT	86098	436279	STUDEBAKER	80056
4-165	NEAPCO	80190	4225B	MOTOR MAST	80180	4311356	FIAT	86163	436292	STUDEBAKER	80954
4165J	BORG-WARNR	85559	4227J	BORG-WARNR	85543	43123	AEC	88000.38	436305	STUDEBAKER	80095
416-946C91	IHC	80217	422-910C91	IHC	80227	43123	LEYLAND	80041	436426	STUDEBAKER	88003.48
417	PRECISION	80160	423	PRECISION	80123	4-3-1241KX	SPICER	85073	436426	STUDEBAKER	88003.51
41758	ALCO	80079	42366	PETTI-MULL	80149	4-3-1241X	SPICER	85073	4365	BORG-WARNR	87464
41758	ALLOY	80079	424	PRECISION	88001.87	43125	AEC	88000.37	436552	WHITE	80178
418	PRECISION	80181	424270	WABCO	80188	4313087	FIAT	86138	4365J	BORG-WARNR	88014.99
418-367C91	IHC	80227	4247393	FIAT	86098	4313087	FIAT-ALLIS	86098	4365J10800	BORG-WARNR	87465
4185	BRADY	80140	4247393	FIAT-ALLIS	86098	431407271B	AUDI	86032	4365J-7200	BORG-WARNR	87464
4186890	CHRYSLER	86130	4247399	FIAT	86138	431407271C	AUDI	86030	4367J	BORG-WARNR	88015
4186891	CHRYSLER	86131	4247399	FIAT-ALLIS	86138	431407272A	AUDI	86031	4367J10800	BORG-WARNR	87467
419	PRECISION	80038	4247400	FIAT	86138	431407272L	AUDI	86042	4367J7200	BORG-WARNR	87466
419976	GMC	80043	4247400	FIAT-ALLIS	86138	431407275	AUDI	86135	4368J	BORG-WARNR	88015.03
419977	GMC	80050	425	PRECISION	80071	431407276F	AUDI	86044	4369J	BORG-WARNR	88015.04
42	WYLIE	80135	4251	BORG-WARNR	80951	431407276G	AUDI	86043	4369J10800	BORG-WARNR	87563
420111	DETROIT	88003.50	4251	DETROIT	80951	431407276H	AUDI	86041	4369J7200	BORG-WARNR	87562
420111	DETROIT	88003.58	4257	HERCULES	88003.44	431-407-311C	AUDI	86136	437	PRECISION	80027
420112	DETROIT	88003.50	4257-1	BORG-WARNR	88003.44	431-407-331C	AUDI	86095	437407271E	AUDI	86038
420112	DETROIT	88003.58	4258	PETTI-MULL	80149	431407331G	AUDI	86095	437407272A	AUDI	86039

Part No.	Brand	IDLI Group	Part No.	Brand	IDLI Group	Part No.	Brand	IDLI Group	Part No.	Brand	IDLI Group
43741A	HYSTER	80158	4-4-4211	SPICER	81404	44RYS22-2	ROCKWELL	81368	4551551	FIAT	80023
4375J	BORG-WARNR	88015.09	4-4-4241	SPICER	81412	44RYS22-21A	ROCKWELL	84512	4552008	FIAT	80041
4375J10800	BORG-WARNR	87615	444460	JEFFREY	80225	44RYS22-24A	ROCKWELL	84514	4552009	FIAT	80041
4375J7200	BORG-WARNR	87614	444-8902	LUDWIG	80041	44RYS22-28	ROCKWELL	88006.08	4556515	FIAT	80024
4376J	BORG-WARNR	88015.10	445	HERCULES	80189	44RYS22-36	ROCKWELL	86662	4556519	FIAT	80041
4376J10800	BORG-WARNR	87617	445	PRECISION	80087	44RYS22-3A	ROCKWELL	84511	4556728	FIAT	80225
4376J7200	BORG-WARNR	87616	4450A	MOTOR MAST	80181	44RYS22-4	ROCKWELL	81363	4557	NEW HOLL.	80035
4378J	BORG-WARNR	88015.12	445581	CATERPILLR	80047	44RYS22-5	ROCKWELL	81362	45678	GMC	80112
4378J10800	BORG-WARNR	87611	4455R	LEMPCO	80906	44RYS22-9	ROCKWELL	81364	4575A	MOTOR MAST	80188
4378J7200	BORG-WARNR	87610	445855D	COCKSHUTT	84510	44RYS28-18	ROCKWELL	86666	4575A	ZELLER	80188
4379J	BORG-WARNR	88015.14	446	HERCULES	80221	44RYS28-2	ROCKWELL	81377	4575B	MOTOR MAST	80186
4379J10800	BORG-WARNR	87609	44617	FIAT-ALLIS	80219	44RYS28-21	ROCKWELL	86664	4575B	ZELLER	80186
4379J7200	BORG-WARNR	87608	4465A	MOTOR MAST	80183	44RYS28-26	ROCKWELL	81376	4575C	MOTOR MAST	80184
438	PRECISION	80029	4465A	ZELLER	80183	44RYS28-3	ROCKWELL	86667	4575C	ZELLER	80184
4380J	BORG-WARNR	88015.17	447797C91	IHC	80007	44RYS28-4	ROCKWELL	81374	4575D	MOTOR MAST	80185
4383758	FIAT	86138	4485624	AMER.MOTOR	80077	44RYS28-7	ROCKWELL	81378	4575D	ZELLER	80185
4383758	FIAT	86190	4485624	WILLYS	80077	44RYS28-8	ROCKWELL	81373	4575E	MOTOR MAST	80187
438395	STUDEBAKER	80064	4487436	AMER.MOTOR	80075	44RYS28-9	ROCKWELL	86665	4575E	ZELLER	80161
4387	GILSON	80001	4487436	FORD	80077	44RYT22	ROCKWELL	81366	4-5800	NEAPCO	80185
4-40-141	SPICER	87418	4487847	AMER.MOTOR	80077	44RYT22-1	ROCKWELL	81367	4-5801	NEAPCO	80186
4403A	MOTOR MAST	80160	4488016	AMER.MOTOR	80077	45	TRU-CROSS	80092	458011	OPEL	88002.77
4-40-431	SPICER	87431	4488106	AMER.MOTOR	80077	450-0233	SANFOR-DAY	80231	45802	GMC	80031
4-40-431	SPICER	88014.76	4498505	FIAT-ALLIS	80092	4500416	JOY MFG.	80236	4-5802	NEAPCO	80184
4-40-461	SPICER	87418	4498505	FIAT-ALLIS	80224	4502A	MOTOR MAST	80161	45802	OPEL	88002.77
4-40-461	SPICER	87420	449A42	HARDIE	80069	4510015	JOY MFG.	80178	45803	OPEL	80032
4-40-481	SPICER	87425	44N2100	WESCO	85516	4510193	JOY MFG.	80210	45806	OPEL	80032
4-40-761	SPICER	87420	44N2200	WESCO	81356	4510193	JOY MFG.	80215	45807	GMC	80031
4-40-781	SPICER	88014.76	44N220-20	WESCO	81362	451072	JOY MFG.	80182	45807	OPEL	88002.77
440811-1	DETROIT	88003.51	44N220-21	WESCO	81362	451105	CATERPILLR	80048	4582853	FIAT	80221
44086	BORG-WARNR	88003.47	44N220TQD	WESCO	84515	4-5122	NEAPCO	80168	458395	STUDEBAKER	80065
44128	ALLIS-CHLM	80188	44N2256-37	WESCO	81365	451220	WESCO	88014.06	458396	STUDEBAKER	80096
44128	FIAT-ALLIS	80188	44N240-7	WESCO	81365	451221	WESCO	84510	458420	STUDEBAKER	80954
44132	GMC	80112	44N246	WESCO	81370	4-5127	NEAPCO	80163	4593225	FIAT	80023
44138	GMC	80234	44N48	WESCO	88000.11	4-5127	NEAPCO	80167	459395C91	IHC	80079
441-8901	LUDWIG	80023	44N48F	WESCO	88000.12	4513-2	KENBAR	87326	45N	WESCO	80140
4-4-1891	SPICER	81408	44N48M	WESCO	88007.24	45149	FWD	80178	46	LEYLAND	80041
4-4-1901	SPICER	81409	44N-NB4	WESCO	88000.13	4-5173	NEAPCO	80171	46	WYLIE	80009
4-4-2041	SPICER	81406	44NS06	NEAPCO	87276	4-5177	NEAPCO	80172	4-60	NEAPCO	80188
4-4-2041	SPICER	88005.33	44N-SO6	WESCO	87276	4520128	JOY MFG.	80212	46001	AEC	88000.67
4-4-2051	SPICER	81402	44NW4-1200	WESCO	85516	4520128	JOY MFG.	80214	46002	AEC	88014.06
4-4-2051	SPICER	88005.30	44NW4-2100	WESCO	81356	4520199	JOY MFG.	80196	46012	AEC	84510
4-4-2091	SPICER	81408	44R	WESCO	80136	4520233	JOY MFG.	80231	46014	AEC	88014.32
4-4-2091	SPICER	88005.33	44RDCA1	ROCKWELL	86659	4520233	JOY MFG.	80240	46035	AEC	84531
44-2100	NEAPCO	81356	44RF12	ROCKWELL	84613	4525B	MOTOR MAST	80179	4605734	FIAT	80024
44-2200	NEAPCO	85516	44RF22	ROCKWELL	84612	4525B	ZELLER	80179	4606993	FIAT	80023
44-2201	NEAPCO	84515	44RY24-2	ROCKWELL	88006.05	4528	PETTI-MULL	80109	4-6103	NEAPCO	80194
44-2220	NEAPCO	81362	44RY36-5	ROCKWELL	85516	4-53-111	SPICER	87430	4-6104	NEAPCO	80192
44-2256	NEAPCO	81365	44RYH20	ROCKWELL	86660	4-53-261	SPICER	87356	4610-4120	DAIHATSU	80014
4-4-2431	SPICER	81411	44RYQ18-1	ROCKWELL	81352	4-53-271	SPICER	87361	4610-4120	DAIHATSU	80020
44-2460	NEAPCO	81370	44RYQ21-1	ROCKWELL	81356	4-53-61	SPICER	87424	4-6107	NEAPCO	80202
44-2610	NEAPCO	81372	44RYQ21-2	ROCKWELL	81360	4-53-61	SPICER	88008.25	4-6114	NEAPCO	80197
44-2820	NEAPCO	81375	44RYQ21-4	ROCKWELL	81358	4-53-71	SPICER	87355	461200	WESCO	83090
443	HERCULES	80142	44RYQ21-5	ROCKWELL	81359	4-53-71	SPICER	88008.16	461220	WESCO	84517
44315-634-003	HONDA	86158	44RYQ21-6	ROCKWELL	81357	453750003	IOWA MFG.	80188	461221	WESCO	88014.32
44315-634-013	HONDA	86193	44RYR19	ROCKWELL	81353	45375-017-02	CEDAR RAP.	80101	4-6123	NEAPCO	80191
44315SA2-300	HONDA	86158	44RYR19-1	ROCKWELL	81354	45375-017-02	IOWA MFG.	80220	4-6128	NEAPCO	80198
44315SA2-310	HONDA	86193	44RYR20-2	ROCKWELL	81355	45375-017-03	IOWA MFG.	80220	46135	FWD	80240
44333-634-013	HONDA	86193	44RYR21	ROCKWELL	81361	4537501706	IOWA MFG.	88012.23	46139	FWD	80215
44333SA2-310	HONDA	86193	44RYR22-1	ROCKWELL	81365	45375-017-07	IOWA MFG.	80235	46140	FWD	80235
4-4-3581X	SPICER	81409	44RYR23-1	ROCKWELL	81369	45375-500-03	IOWA MFG.	88010.35	4-6143	NEAPCO	80193
4-4-3601X	SPICER	81408	44RYR24-2	ROCKWELL	88006.06	453-999R91	IHC	80104	4-6152	NEAPCO	80195
4-4-3611X	SPICER	81411	44RYR24-4	ROCKWELL	81370	4542487	JOY MFG.	80193	4-6152	NEAPCO	80197
443-8901	LUDWIG	80041	44RYR24-5	ROCKWELL	86663	4-54-381	SPICER	87449	4-6162	NEAPCO	80199
444	HERCULES	80178	44RYR25	ROCKWELL	81371	4-54-381	SPICER	88007.91	4-6163	NEAPCO	80201
4-4-4071X	SPICER	81411	44RYR26-1	ROCKWELL	81372	4-54-481	SPICER	87472	46188	FWD	80214
4-4-4101	SPICER	81410	44RYR28-2	ROCKWELL	81375	4-54-481	SPICER	88008	46-1910	NEAPCO	84923
444153	WHITE	80023	44RYR28-2	ROCKWELL	88006.07	454-740R91	IHC	80008	4620151	FIAT	80221
444153	WHITE	80024	44RYS22-14	ROCKWELL	86661	4548769	FIAT	80024	46214	FWD	80179
444153A	WHITE	80024	44RYS22-16A	ROCKWELL	84515	455004C1	IHC	88007.53	4622384	FIAT	80225
4441J	BORG-WARNR	85545	44RYS22-1A	ROCKWELL	81362	455005C1	IHC	88007.54	46240	VOLVO	80189

Part No.	Brand	IDLI Group	Part No.	Brand	IDLI Group	Part No.	Brand	IDLI Group	Part No.	Brand	IDLI Group
4625636	FIAT	80041	465208	BAL-LM-HAM	80216	471190	WESCO	83263	485	PRECISION	88001.92
4627-3124	AMER.PULL.	80133	465209	AUST.WEST.	80164	471194	WESCO	83262	4851	ALCO	88002.86
4627-3124	AMER.PULL.	80178	465209	BAL-LM-HAM	80216	471201	WESCO	83277	4851-183Y	HARDY SPCR	80140
462809	GMC	80139	465217	AUST.WEST.	80216	471202	WESCO	83283	4871J	BORG-WARNR	88015.02
463236C91	IHC	87381	465217	BAL-LM-HAM	80216	471220	WESCO	83314	4871J10800	BORG-WARNR	87547
46408	FWD	80213	465225	BAL-LM-HAM	80044	471221	WESCO	84529	4871J7200	BORG-WARNR	87546
46409	FWD	80238	465250	AUST.WEST.	80100	471223	WESCO	84531	48821	TAYLOR	80008
4641105	FIAT	80225	465250	BAL-LM-HAM	80100	471224	WESCO	83320	48827	TAYLOR	80133
46424	FWD	80215	465254	AUST.WEST.	80100	471236	WESCO	83331	489	AEC	80104
4642J	BORG-WARNR	87413	465254	BAL-LM-HAM	80100	471246	WESCO	83349	49	WYLIE	80055
46438	FWD	80238	465286	GMC	88008.35	471256	WESCO	83316	490	AEC	80062
46452	NEW HOLL.	80035	465298	AUST.WEST.	80021	471885	WESCO	85712	490	PRECISION	88001.93
465-002-0	GLAENZER	80142	465298	BAL-LM-HAM	80021	471900	WESCO	83258	4905112	FIAT-ALLIS	80151
465-006-0	GLAENZER	80124	465424	AUST.WEST.	80100	47203	ADAMS-LET.	80049	4905112-1	FIAT-ALLIS	80151
465019	AUST.WEST.	80048	465424	BAL-LM-HAM	80100	47203	WABCO	80077	4906478	FIAT-ALLIS	80046
465019	BAL-LM-HAM	80048	4655161	FIAT	80023	472196	ADAMS-LET.	80049	4906479	FIAT-ALLIS	80047
465-019-0	GLAENZER	80142	4655182	FIAT	80142	472196	WABCO	88003.26	490801	AMER.MOTOR	88003.50
465-019-4	GLAENZER	80142	4658J	BORG-WARNR	85158	472197	ADAMS-LET.	80049	4908839	FIAT-ALLIS	80212
465020	AUST.WEST.	80047	466008	AUST.WEST.	80069	472197	WABCO	88003.26	491	AEC	80132
465020	BAL-LM-HAM	80100	466008	BAL-LM-HAM	80069	472200	ADAMS-LET.	80049	49-100	WORLDPARTS	80006
465-021-1	GLAENZER	80124	466505C91	IHC	87460	472200	WABCO	88003.26	49-101	WORLDPARTS	80009
465022	AUST.WEST.	80047	46653	FWD	80213	473609	ADAMS-LET.	80168	49-102	WORLDPARTS	80017
465028	AUST.WEST.	80104	4667	CARDWELL	80136	473609	WABCO	80168	49-103	WORLDPARTS	80023
465028	BAL-LM-HAM	80104	4675621	FIAT	80142	473609	WABCO	80216	49-103	WORLDPARTS	80024
465029	AUST.WEST.	80104	4675624	FIAT	80142	474041	ADAMS-LET.	80049	49-104	WORLDPARTS	80010
465029	BAL-LM-HAM	80104	46-810-765(4)	BUCYR-ERIE	80240	474041	WABCO	88003.26	49-105	WORLDPARTS	80006
465034	AUST.WEST.	80101	46-810-780(4)	BUCYR-ERIE	80044	474339C1	IHC	88013.64	49-107	WORLDPARTS	80019
465034	BAL-LM-HAM	80099	4-681-3500	BUCYR-ERIE	80101	474340C1	IHC	88013.70	49-108	WORLDPARTS	80018
465035	AUST.WEST.	80101	4-681-3550	BUCYR-ERIE	80197	474524	BEND.WEST.	80171	49-115	WORLDPARTS	80041
465035	BAL-LM-HAM	80099	4-681-3555	BUCYR-ERIE	80215	474524	WABCO	80166	49-117	WORLDPARTS	80023
465036	AUST.WEST.	80104	46-813-675(4)	BUCYR-ERIE	80149	474524	WABCO	80171	49-117	WORLDPARTS	80024
465036	BAL-LM-HAM	80104	46-813-675(4)	BUCYR-ERIE	80214	474-792R91	IHC	80044	49-118	WORLDPARTS	80077
465066	AUST.WEST.	80216	46-813-675(4)	BUCYR-ERIE	80215	474-793R91	IHC	80044	49-120	WORLDPARTS	80077
465066	BAL-LM-HAM	80216	4-681-393(4)	BUCYR-ERIE	80149	474804	ADAMS-LET.	80037	49-129	WORLDPARTS	80012
465067	AUST.WEST.	80216	4-681-397(4)	BUCYR-ERIE	80149	474804	WABCO	80037	49-130	WORLDPARTS	80023
465067	BAL-LM-HAM	80216	46-814-060	BUCYR-ERIE	80150	47483	DANUSER	80035	49-130	WORLDPARTS	80024
465068	AUST.WEST.	80216	46-814-061(4)	BUCYR-ERIE	80149	475	PRECISION	88001.89	49-131	WORLDPARTS	80020
465068	BAL-LM-HAM	80216	46-814-0614	BUCYR-ERIE	80150	475110	WABCO	80100	4914098	FIAT-ALLIS	80046
465071	AUST.WEST.	80216	46847	AUTO BIAN.	80142	475363	ADAMS-LET.	80112	4914099	FIAT-ALLIS	80047
465071	BAL-LM-HAM	80216	4685J	BORG-WARNR	88006.35	475363	WABCO	80112	492	AEC	80139
465072	AUST.WEST.	80163	468656C91	IHC	87374	475370	WABCO	80104	492	PRECISION	80107
465072	BAL-LM-HAM	80163	468666C91	IHC	87375	475397	WABCO	80171	492-10	PRECISION	88001.94
465073	AUST.WEST.	80216	4687J	BORG-WARNR	84656	475471	ADAMS-LET.	80099	493	AEC	80129
465073	BAL-LM-HAM	80216	4690	ADAMS-LET.	80099	475471	ADAMS-LET.	80101	493	PRECISION	80200
465076	AUST.WEST.	80216	4690	ADAMS-LET.	80101	475471	WABCO	80099	4938-487	BUCYR-ERIE	80149
465076	BAL-LM-HAM	80216	4690	ALLIS-CHLM	80048	475471	WABCO	80101	493-849R91	IHC	80136
465077	AUST.WEST.	80163	4690	ALLIS-CHLM	80101	476	PRECISION	88001.90	494	AEC	80131
465077	BAL-LM-HAM	80163	4690	FIAT-ALLIS	80101	47621	NEW HOLL.	80061	494	PRECISION	80157
465111	AUST.WEST.	80193	4690	WABCO	80101	4762487	ALLIS-CHLM	80047	495	AEC	80086
465111	BAL-LM-HAM	80193	4699475	FIAT-ALLIS	80180	4762487	FIAT-ALLIS	80047	495	PRECISION	80098
465123	AUST.WEST.	80215	4699481	FIAT-ALLIS	80189	476-2487-9	FIAT-ALLIS	80047	495	PREC.TOR.	80186
465123	BAL-LM-HAM	80210	47	WYLIE	80044	476-674R91	IHC	80001	4954	HOWE	80215
465124	AUST.WEST.	80215	470-077R91	IHC	80161	47804	SCANIA	80180	496	AEC	80084
465124	BAL-LM-HAM	80210	47041	WABCO	80049	47903	GMC	80225	496	PRECISION	80105
465132	AUST.WEST.	80215	470682C91	IHC	88008.55	47985	GMC	80221	496	PREC.TOR.	80184
465132	BAL-LM-HAM	80210	470683C1	IHC	88013.68	4803J	BORG-WARNR	82036	497	AEC	80002
465136	AUST.WEST.	80215	470684C1	IHC	88013.74	481218	WESCO	85516	497	PRECISION	80066
465136	BAL-LM-HAM	80210	4707800	FIAT-ALLIS	88001.28	481219	WESCO	81356	498	AEC	80126
465137	AUST.WEST.	80215	47079	GMC	80215	481221	WESCO	84515	498	ALLOY	88000.14
465137	BAL-LM-HAM	80215	4-709-11	ATHEY PROD	80149	481228	WESCO	81362	498	PRECISION	80058
465153	AUST.WEST.	80101	47093	GMC	80225	481229	WESCO	81362	4984504	ALLIS-CHLM	80049
4651538	WILLYS	80072	47097	GMC	80210	481246	WESCO	81370	4984504	ALLIS-CHLM	88001.96
465158	AUST.WEST.	80048	471	PRECISION	88001.88	481256	WESCO	81365	4984504	FIAT-ALLIS	80047
465158	BAL-LM-HAM	80048	471045	WESCO	83281	482	PRECISION	88001.91	4984504-3	FIAT-ALLIS	80047
465159	AUST.WEST.	80047	471160	WESCO	83241	48263	NEW HOLL.	80001	4984505	ALLIS-CHLM	80046
465159	BAL-LM-HAM	80047	471-164	WESCO	83240	48263	NEW HOLL.	80003	4984505	FIAT-ALLIS	80046
46520	AUST.WEST.	80100	471180	WESCO	83248	48-39-143	BLACKWELDR	80100	4984505-0	FIAT-ALLIS	80046
46520	BAL-LM-HAM	80100	471182	WESCO	83252	48-39-144	BLACKWELDR	80100	498504	ALLIS-CHLM	80048
465205	BAL-LM-HAM	80216	471184	WESCO	83251	484002C91	IHC	88008.41	498504	FIAT-ALLIS	80047
465208	AUST.WEST.	80165	471-190	WESCO	83258	4841	ALCO	88002.87	498505	FIAT-ALLIS	80046

Part No.	Brand	IDLI Group
4987664	ALLIS-CHLM	80047
4987664	FIAT-ALLIS	80047
499	ALCO	88000.15
499	ALLOY	80007
499	PRECISION	80092
4991328	ALLIS-CHLM	80153
4991328	ALLIS-CHLM	80155
4991328	FIAT-ALLIS	80150
4991328-8	FIAT-ALLIS	80150
4992J	BORG-WARNR	81762
4993479	ALLIS-CHLM	80153
4993479	ALLIS-CHLM	80155
4993479	FIAT-ALLIS	80155
4993479-7	FIAT-ALLIS	80155
4996	ALCO	88013.86
4996	ALLOY	88013.86
4999	ALCO	88013.87
4999	ALLOY	88013.87
499A26	AMER.PULL.	80133
499A26	HARDIE	80133
499A35	AMER.PULL.	80008
499A35	HARDIE	80008
499A42	AMER.PULL.	80069
499A42	HARDIE	80069
4C96-55	ALCO	80204
4D8212	CATERPILLR	88003.26
4D8664	CATERPILLR	80049
4F3056(5)	CATERPILLR	88002.92
4F3057(5)	CATERPILLR	88002.92
4F3058(5)	CATERPILLR	88002.92
4F7024	CATERPILLR	88001.33
4H7808	CATERPILLR	88003.26
4H8461	CATERPILLR	88001.18
4K1917(5)	CATERPILLR	88002.92
4L5920	CATERPILLR	80228
4L6325	CATERPILLR	80214
4L6929	CATERPILLR	80238
4M117	ALLOY	80213
4M5961(2)	CATERPILLR	88001.27
4M5962(2)	CATERPILLR	88001.27
4N	AEC	80134
4NPS24	ROCKWELL	87346
4NPS24-23	ROCKWELL	87349
4NPS24-26	ROCKWELL	87345
4NPS24-29	ROCKWELL	87343
4R7972	CATERPILLR	80238
4RYS22-1A	ROCKWELL	84513
50	WYLIE	88002.77
500	ALLOY	88000.67
500	PRECISION	88001.95
5000	TRU-CROSS	80168
50.00.241.261	RENAULT	80142
500024-16	SPICER	88014.62
50.00.242.492	RENAULT	80180
50.00.243.236	RENAULT	80041
50.00.559.846	RENAULT	80140
5000DA	MOTOR MAST	80213
500267	TRIUMPH	80024
500357	JACOBSEN	80037
500415-9	SPICER	88013.32
500479	JACOBSEN	80044
5004J	BORG-WARNR	87410
5-0059	NEAPCO	80221
500-665E	OLIVER	80035
500-665E	WHITE FARM	80035
5-0070	NEAPCO	80220
5008	TRU-CROSS	80174
5-0080	NEAPCO	88012.23
501	AEC	80023
501	ALLOY	88000.90
501	PRECISION	88001.96
501-0350	ROKA	80024
501-1350	ROKA	80012
501-2351	ROKA	80019
5015	TRU-CROSS	80173
501-5350	ROKA	80011
501-7350	ROKA	80009
501-7353	ROKA	80017
502	ALCO	88000.22
502	ALLOY	88000.22
502	PRECISION	88001.97
50200X	SPICER	80075
5023218	EATON	80210
5025-1J	MOTOR MAST	80197
5025-1K	MOTOR MAST	80201
5025-1M	MOTOR MAST	80199
502529E	OLIVER	80008
502-529E	WHITE FARM	80008
502-579E	OLIVER	80001
502-579E	WHITE FARM	80001
5025A	MOTOR MAST	80193
5025A	ZELLER	80193
5025A	ZELLER	80198
5025AA	MOTOR MAST	80193
5025AF	MOTOR MAST	80193
5025B	MOTOR MAST	80198
5025B	ZELLER	80198
5025FA	MOTOR MAST	80194
5025FA	ZELLER	80194
5025G	MOTOR MAST	80195
5025G	ZELLER	80195
5025G	ZELLER	80199
5025H	MOTOR MAST	80196
5025H	MOTOR MAST	80202
5025J	MOTOR MAST	80197
5025J	ZELLER	80197
5025K	MOTOR MAST	80201
5025K	ZELLER	80201
5025M	MOTOR MAST	80199
5025M	ZELLER	80199
5025MF	MOTOR MAST	80199
5025N	MOTOR MAST	80202
5025N	ZELLER	80202
5025P	MOTOR MAST	80192
5025P	ZELLER	80192
5025R	MOTOR MAST	80203
5025S	MOTOR MAST	80191
5-0280	NEAPCO	80221
5-0284	NEAPCO	80221
50292	PETTI-MULL	80047
503	AEC	80007
503	ALCO	88000.27
503	ALLOY	88000.25
503	ALLOY	88000.27
5-03	BYPY	88003.69
503	PRECISION	88001.98
503012A	HYSTER	80158
5031	AEC	80171
503757	TRIUMPH	80041
50377-2	NATL.MINE	80234
5037918	CHRYSLER	80023
5038495	CHRYSLER	80023
5038930	CHRYSLER	80024
5039570	CHRYSLER	80041
504	AEC	80041
504	ALLOY	88000.41
5-04	BYPY	88003.70
504	PRECISION	80053
5040692	CHRYSLER	80142
5040693	CHRYSLER	80180
5041525	CHRYSLER	80189
5042J	BORG-WARNR	81826
5043828	EATON	80048
504439	ROVER	80077
5044943	CHRYSLER	80041
505	AEC	80031
505	ALLOY	88000.29
505	PRECISION	80055
5-05-07	BYPY	88003.77
5-05-08	BYPY	88003.84
5050A	MOTOR MAST	80210
5050A	ZELLER	80210
5050AA	MOTOR MAST	80210
5050CA	MOTOR MAST	80212
5050CA	ZELLER	80212
5050DA	MOTOR MAST	80213
5050DA	ZELLER	80213
5050F	MOTOR MAST	80211
5050F	ZELLER	80211
5050H	MOTOR MAST	80208
5050J	MOTOR MAST	80209
5050K	MOTOR MAST	80206
5050K	MOTOR MAST	80212
50500A	ZELLER	80213
5051J	BORG-WARNR	81748
5057939	CHRYSLER	80077
5-0589	NEAPCO	80205
506	ALLOY	88000.19
5-06	BYPY	88004.20
506	PRECISION	88001.99
506140	ROSS GEAR	80001
506-602E	OLIVER	80069
506-602E	WHITE FARM	80069
507	ALLOY	88000.68
5-07	BYPY	88004.13
507	PRECISION	80044
50-701-000	ITALCARDAN	80024
50-711000	ITALCARDAN	80180
50-712-000	ITALCARDAN	80189
50-719-000	ITALCARDAN	80142
5-0725	NEAPCO	80215
5-0727	NEAPCO	80214
50-732-000	ITALCARDAN	80225
50743	DATSUN	80024
50-744-000	ITALCARDAN	80042
50-745-000	ITALCARDAN	80135
50-746-000	ITALCARDAN	80031
50-751-000	ITALCARDAN	80032
50-755-000	ITALCARDAN	80221
50-757-000	ITALCARDAN	80124
50-759-000	ITALCARDAN	80072
50-760-000	ITALCARDAN	80210
508	ALLOY	88000.70
508	PRECISION	80049
50-809-000	ITALCARDAN	80003
50-810000	ITALCARDAN	80138
50-811-000	ITALCARDAN	80178
50-812-000	ITALCARDAN	88002.10
50-813-000	ITALCARDAN	80049
50-815000	ITALCARDAN	80048
50-816-000	ITALCARDAN	80044
50-843-000	ITALCARDAN	80109
508536	TRIUMPH	80023
50-854-000	ITALCARDAN	80023
50-855-000	ITALCARDAN	80041
5-086	AEC	80124
5086258	EATON	80213
5086258	EATON	80214
508E7039	FORD	80180
509	ALLOY	88000.71
509	PRECISION	88002
509-047R91	IHC	80037
509206E	OLIVER	80001
509-260E	OLIVER	80001
509-260E	WHITE FARM	80001
509295	GMC	88002.10
509296	GMC	80043
509297	GMC	80050
509456(4)	BUCYR-ERIE	80149
509457(4)	BUCYR-ERIE	80149
5096J	BORG-WARNR	85093
50A	DETROIT	80153
50NPS28-25	ROCKWELL	87405
50NPS28-26	ROCKWELL	87412
50NPS28-27	ROCKWELL	87403
50NPS28-28	ROCKWELL	87414
50NPS28-31	ROCKWELL	87409
50NPS28-38	ROCKWELL	87422
50NPS28-43	ROCKWELL	87430
50NPS28-43	ROCKWELL	87455
50NPS28-44	ROCKWELL	87418
50NPS28-45	ROCKWELL	87425
50NPS28-47	ROCKWELL	87431
50NPS28-48	ROCKWELL	87435
50NPS28-51	ROCKWELL	87433
50NPS28-57	ROCKWELL	87437
50NPS28-59	ROCKWELL	87419
50NPS28-60	ROCKWELL	87427
50Y103	PILOT	88000.36
50Y113	PILOT	88000.35
50Y115	PILOT	88000.55
50Y12	PILOT	88000.45
50Y123	PILOT	88000.38
50Y124	PILOT	88000.37
50Y13	PILOT	88000.20
50Y22	PILOT	88000.43
50Y23	PILOT	88000.44
50Y23	PILOT	88000.80
50Y33	PILOT	88000.26
50Y43	PILOT	88000.22
50Y53	PILOT	88000.52
50Y63	PILOT	88000.25
50Y63	PILOT	88000.27
50Y73	PILOT	88000.28
50Y76	PILOT	88000.29
50Y81	PILOT	88000.93
50Y83	PILOT	88000.53
50Y93	PILOT	88000.31
50Y95	PILOT	88000.34
510	AEC	80007
510	AEC	88000.15
510	AEC	88000.16
510	ALLOY	88000.72
510	PRECISION	88002.01
5-10	SPICER	80142
5100	AEC	80168
5100	ALCO	80168
5100	SEARS-ROE.	80107
5-100	SPICER	80031
510001	WESCO	88000.19
510002	WESCO	88000.43
510002	WESCO	88000.57
510003	WESCO	88000.43
510004	GEHL BROS.	80003
510004	WESCO	88000.43
510004	WESCO	88000.99
510005	WESCO	88000.20
510006	GEHL BROS.	80033
510006	WESCO	88000.21
510007	GEHL BROS.	80037

Part No.	Brand	IDLI Group
510007	WESCO	88000.44
510008	WESCO	88000.23
510009	WESCO	88000.26
510010	WESCO	88000.22
510011	WESCO	88000.22
510012	WESCO	88000.23
510013	WESCO	88000.45
510014	WESCO	88000.62
510015	WESCO	88000.52
510016	WESCO	88000.25
510017	WESCO	88000.27
510018	WESCO	88000.25
510019	WESCO	88000.29
510020	WESCO	88000.28
510021	WESCO	88000.46
510022	WESCO	88000.59
510023	WESCO	88000.53
510024	WESCO	88000.30
510025	WESCO	88000.31
510026	WESCO	88000.54
510027	WESCO	88000.34
510-028	WESCO	88000.49
510029	WESCO	88000.47
510030	WESCO	88000.36
510031	WESCO	88000.35
510032	WESCO	88000.35
510033	WESCO	88000.37
510034	WESCO	88000.40
510035	WESCO	88000.55
510036	GEHL BROS.	80069
510036	WESCO	88000.38
510037	WESCO	88000.39
510038	WESCO	88000.51
510040	GEHL BROS.	80134
510084-2X	BUTLER	87458
510084-2X	BUTLER	87461
510084-2X	BUTLER	88007.74
510084-2X3	BUTLER	87460
510084-2X3	BUTLER	87461
510084-2X3	BUTLER	88007.76
510084-2XWF	BUTLER	87460
510084-2XWF	BUTLER	87461
510084-2XWF	BUTLER	88008.31
510084-2XWF	BUTLER	88008.44
510085-1X	BUTLER	88008.53
510086-1X	BUTLER	88008.47
510088-1X	BUTLER	87328
510088-1X	BUTLER	87331
510088-1X2	BUTLER	87331
510088-1X2	BUTLER	88007.64
510088-1XWF	BUTLER	87331
510088-1XWF	BUTLER	88008.39
510090-1X	BUTLER	87331
510090-1X	BUTLER	88007.63
510092-1X	BUTLER	88008.56
5-100X	SPICER	80030
5-100X	SPICER	80031
5101	AEC	80163
5101	AEC	80167
5-101	SPICER	80023
5-101	SPICER	80024
510-105	WESCO	88000.99
510-106	WESCO	88000.22
510114	GEHL BROS.	80037
510119-1X	BUTLER	88008.54
510120-1X	BUTLER	88008.60
510121-1X	BUTLER	87481
510121-1X	BUTLER	88007.77
5101270	WHITE	80003
510129-1X	BUTLER	88008.57
510129-1X	NEAPCO	88008.57
510140-1X	BUTLER	87378
510140-1X	BUTLER	87384
510140-1X3	BUTLER	87384
510140-1XWF	BUTLER	87384
510140-1XWF	BUTLER	88008.42
510144-1X	BUTLER	87384
510144-1X	BUTLER	88007.71
5101X	NEAPCO	88000.67
5-101X	SPICER	80021
5-101X	SPICER	80023
5-101X	SPICER	80140
5102	AEC	80171
5102	SEARS-ROE.	80031
5-102	SPICER	80024
510207-1X	BUTLER	87460
510207-1X	BUTLER	88007.75
510228X	BUTLER	88008.50
510230-1X	BUTLER	87459
510230-1X	BUTLER	88008.45
5-102X	SPICER	80024
5103	AEC	80174
5103	ALCO	80174
5103	SEARS-ROE.	80061
510367-1X	BUTLER	87326
510367-1X	BUTLER	88007.61
510367-1XWF	BUTLER	87326
510367-1XWF	BUTLER	87330
510370-1X	BUTLER	87326
510370-1X	BUTLER	87330
510391-1X	BUTLER	87380
510391-1X	BUTLER	87382
510391-1X3	BUTLER	87377
510391-1X3	BUTLER	87382
510391-1XWF	BUTLER	87377
510391-1XWF	BUTLER	87382
510391-1XWF	BUTLER	88008.41
5-103X	REPCO	80007
5-103X	SPICER	80007
510433-1X	BUTLER	87377
510433-1X	BUTLER	88007.70
510463	NEW HOLL.	80142
5105	AEC	80169
5105	SEARS-ROE.	80132
5-105	SPICER	80031
510527X	BUTLER	87238
510527X	BUTLER	88007.58
510580X	BUTLER	88008.35
5-105X	SPICER	80024
5-105X	SPICER	80031
510661-1X	BUTLER	87570
510661-1X	BUTLER	88007.79
5-107X	SPICER	80030
5108	AEC	80172
51083	GEHL BROS.	80109
51084	GEHL BROS.	80155
51085X	BUTLER	87325
51085X	BUTLER	88007.59
510866-1X	BUTLER	87373
510866-1X	BUTLER	87374
510866-1X	BUTLER	88007.67
510873-1X	BUTLER	87375
510873X	BUTLER	87375
510873X	BUTLER	88007.66
5-108X	SPICER	80043
5109	AEC	80162
5-10X	SPICER	80142
511	ALLOY	88000.73
511	PRECISION	88002.02
5110	AEC	80173
5110	ALCO	80173
511016X	BUTLER	87373
511016X	BUTLER	87375
511016X	BUTLER	88008.43
511016X3	BUTLER	87373
511016X3	BUTLER	87375
511016X3	BUTLER	87376
511016X3	BUTLER	88007.65
5-110X	SPICER	80007
5-110XBA	HUSCO	88000.16
5111	AEC	80175
5111	NEAPCO	88000.20
5-111	SPICER	80024
5-111X	SPICER	80023
5-111X	SPICER	80024
5-111XBA	HUSCO	80023
5112	ALCO	80177
5112	NEAPCO	88000.22
5112	SEARS-ROE.	80104
51-120-26	WESCO	88000.57
51-1230	WESCO	88000.20
51-1230	WESCO	88000.21
5113	NEAPCO	88000.25
5113	SEARS-ROE.	80091
5-113	SPICER	80077
51-1340	WESCO	88000.22
5-113X	SPICER	80072
5114	NEAPCO	88000.28
5114	SEARS-ROE.	80081
51-140-26	WESCO	88000.45
51-1440	WESCO	88000.27
5-114X	SPICER	80024
5-114X	SPICER	80031
5115	ALCO	80178
5115	NEAPCO	88000.30
51-1540	WESCO	88000.28
5-115X	JEFFREY	80178
5-115X	SPICER	80178
5116	NEAPCO	88000.35
51-160-26	WESCO	88000.46
51-1640	WESCO	88000.30
51-1640	WESCO	88000.31
511-644R91	IHC	80001
5-116X	SPICER	80189
5117	NEAPCO	88000.37
511-705R91	IHC	80021
5-117X	SPICER	80221
5118	NEAPCO	88000.38
51-1850	WESCO	88000.47
5-118X	SPICER	80044
5-11X	SPICER	80072
512	ALLOY	88000.68
512	PRECISION	88002.03
5-1200	AEC	80081
5-1200	SPICER	80081
5-1200X	REPCO	80081
5-1200X	SPICER	80081
5-1201	AEC	80091
5-1201	SPICER	80091
5-1201X	REPCO	80091
5-1201X	SPICER	80073
5-1201X	SPICER	80091
5-1202X	REPCO	80118
5-1202X	REPCO	80130
5-1202X	SPICER	80130
5-1203	AEC	80117
5-1203	AEC	80129
5-1203X	REPCO	80129
5-1203X	SPICER	80129
5-1204	AEC	80131
5-1204X	REPCO	80131
5-1204X	SPICER	80131
5-1205	AEC	80123
51-2050	WESCO	88000.39
5-1205X	REPCO	80123
5-1205X	SPICER	80123
5-1206	AEC	80141
5-1206X	REPCO	80141
5-1206X	SPICER	80141
5-1207X	REPCO	80071
5-1207X	REPCO	80163
5-1207X	SPICER	80071
5-1207X	SPICER	80074
5-1208X	REPCO	80071
5-1208X	REPCO	80074
5-1208X	REPCO	80080
5-1208X	SPICER	80071
5-1208X	SPICER	80074
5-1209X	REPCO	80077
5-1209X	SPICER	80080
5-120X	SPICER	80072
5-121	NEAPCO	80072
5121	TRU-CROSS	80054
5121	TRU-CROSS	80163
5-1210X	REPCO	80175
5-1210X	REPCO	80183
5-1210X	SPICER	80183
5-1211X	REPCO	80118
5-1211X	REPCO	80134
5-1211X	REPCO	80138
5-1211X	REPCO	80181
5-1211X	SPICER	80138
5-1-2121	SPICER	88006.02
5-1214X	REPCO	80145
5-1214X	REPCO	80163
5-1214X	REPCO	80171
5-1214X	SPICER	80145
5-121X	SPICER	80072
5-121XBA	HUSCO	80072
5-121XU	SPICER	80089
5-122X	SPICER	80073
5123	NEAPCO	88000.21
5123	SEARS-ROE.	80075
5-123	SPICER	80091
5-123X	SPICER	80225
5124	NEAPCO	88000.31
5-124	SPICER	80081
5-124X	REPCO	80225
5-124X	SPICER	80225
5125	NEAPCO	88000.39
512544	ROVER	80040
5-125X	SPICER	80075
512629	NEW HOLL.	80003
5126J	BORG-WARNR	85867
5128	SEARS-ROE.	80129
5-1-2841	SPICER	88006.01
5128J	BORG-WARNR	82057
5129	SEARS-ROE.	80131
5-129-1X	SPICER	80037
5-129-2X	SPICER	80037
5-129X	SPICER	80037
5-12X	ALBURUS-SP	80024
5-12X	SPICER	80072
5-12X	SPICER	80077
513	ALLOY	88000.74
513	PRECISION	80059
5-130	SPICER	80117
5-1300	SPICER	80042
5-1300X	SPICER	80042
5-1301X	SPICER	80135

Part No.	Brand	IDLI Group
5-1302X	SPICER	80135
5-1303X	SPICER	80034
5-1304X	SPICER	80064
5-1304X	SPICER	80065
5-1306-1X	SPICER	80061
5-1306-1X	SPICER	80062
5-1306X	SPICER	80060
5-1306X	SPICER	80061
5-1306X	SPICER	80062
5-1306X	SPICER	80063
5-1306X5	SPICER	80061
5-1306XS	SPICER	80060
5-1306XS	SPICER	80063
5-1307X	SPICER	80061
5-1308X	SPICER	80061
5-1309X	SPICER	80132
5-130X	REPCO	80117
5-130X	SPICER	80117
5131	AEC	80175
5-131	SPICER	80121
5-1-3161	SPICER	88006.03
5-131X	REPCO	80121
5-131X	SPICER	80121
5132	AEC	80174
5-132X	REPCO	80075
5-132X	SPICER	80077
5-133X	REPCO	80116
5-133X	SPICER	80116
5134	NEAPCO	88000.49
5-134X	REPCO	80118
5-134X	SPICER	80118
5136	NEAPCO	88000.52
5138	NEAPCO	88000.60
513800	AUST.MOT	80023
513800	AUST.MOT	80024
513800	TRIUMPH	80023
5-138X	SPICER	80118
5139	AEC	80170
5139	NEAPCO	88000.53
513A	NEAPCO	88000.25
513G	PRECISION	88002.04
5-13X	SPICER	80072
514	AEC	80023
514	AEC	80024
514	AEC	80140
514	ALLOY	88000.75
514	PRECISION	80030
5-1400X	SPICER	80909
5-1401X	SPICER	88001.74
5-1402X	SPICER	80901
5-1402X	SPICER	88001.66
5-1403X	SPICER	80903
5-1403X	SPICER	80958
5-1404X	SPICER	80951
5-1405X	SPICER	80905
5-1406X	SPICER	80953
5-1407	SPICER	80909
5-1407X	SPICER	80906
5-1407X	SPICER	80909
5140J	BORG-WARNR	85543
5-143X	SPICER	80124
5-144X	SPICER	80142
5-144X	SPICER	80143
5-1-4511	SPICER	88006.04
5-146X	SPICER	80075
5147J	BORG-WARNR	85107
5148	NEAPCO	88000.87
5-149X	SPICER	80075
514G	PRECISION	80031
5-14X	SPICER	80024
515	AEC	80178
515	ALLOY	88000.85
515	PRECISION	80043
5-15	SPICER	80178
515	TEDDY TORQ	80095
5-1500X	SPICER	80004
5-1500X	SPICER	80006
5-1501X	SPICER	80010
5-1502X	SPICER	80009
5-1503X	SPICER	80012
5 1504X	SPICER	80011
5-1504X	TRW	80011
5-1505X	SPICER	80019
5-1506X	SPICER	80017
5-1507X	SPICER	80020
5-1508X	SPICER	80018
5-1509X	SPICER	80017
5-1509X	SPICER	80115
5150CA	ZELLER	80212
5-150X	REPCO	80119
5-150X	SPICER	80119
5-150X	SPICER	80120
5-1510X	SPICER	80016
5151-176Y	HARDY SPCR	80180
5-1511X	SPICER	80115
5-1513X	SPICER	80013
5-1514X	SPICER	80005
5-1515X	SPICER	80014
5-1516X	SPICER	80015
5-15-23	SPICER	88005.36
515251-2	DETROIT	88001.60
51525-2	DETROIT	88001.60
515253-2	DETROIT	88001.60
5-152X	SPICER	80118
515(3)	PRECISION	88002.05
5-153X	ALBURUS-SP	80077
5-153X	JAEGER	80077
5-153X	MUNCIE	80077
5-153X	SPICER	80072
5-153X	SPICER	80075
5-153X	SPICER	80077
5-153XU	SPICER	80088
5154	NEAPCO	88000.32
5-154X	SPICER	80178
5-155X	REPCO	80182
5-155X	SPICER	80182
5156	NEAPCO	88000.55
515693	WHITE	87481
515697	WHITE	87481
5157	NEAPCO	88000.80
515723	ROVER	80072
5158	NEAPCO	88000.48
5-158X	SPICER	80182
515A	NEAPCO	88000.30
5-15X	BLAW KNOX	80178
5-15X	SPICER	80178
516	AEC	80189
516	ALCO	88000.21
516	ALLOY	88000.20
516	PRECISION	80050
5-16	SPICER	80189
5-1601	ALCO	80010
5-1602	ALCO	80019
5-1603	ALCO	80010
5-1604	ALCO	80009
516046	GMC	80044
5-1605	ALCO	80012
516059	STUDEBAKER	80072
516059	STUDEBAKER	80077
5-1609X	SPICER	80201
5-160X	ALBURUS-SP	80142
5-160X	JEFFREY	80142
5-160X	JOY MFG.	80142
5-160X	RAY GO	80142
5-160X	SPICER	80142
5-160X	SPICER	80143
5-160XU	SPICER	80143
5-1-6121	SPICER	88006
51615-2	DETROIT	80042
5-16-23	SPICER	88005.37
51625-2	ALCO	80042
51625-2	DETROIT	80042
5-163X	SPICER	80072
5-163X	SPICER	80075
5164	NEAPCO	88000.33
516405	GMC	80043
516406	GMC	80050
5-165-4CX	SPICER	80190
5-165X	ALBURUS-SP	80190
5-165X	REPCO	80190
5-165X	SPICER	80190
5168	NEAPCO	88000.92
5-168X	SPICER	80190
5169	NEAPCO	88000.94
516A	NEAPCO	88000.35
5-16X	SPICER	80189
517	ALCO	88000.27
517	ALLOY	88000.28
517	PRECISION	80051
5-17	SPICER	80221
5-1701X	SPICER	80007
517038	FIAT-ALLIS	80133
517039	FIAT-ALLIS	80133
5-170X	MUNCIE	80007
5-170X	REPCO	80007
5-170X	SPICER	80007
5-170X	SPICER	88000.15
5-170XBA	HUSCO	80007
5171	NEAPCO	88000.43
5-171X	SPICER	80221
5172	NEAPCO	88000.47
5-172X	SPICER	80023
5173	NEAPCO	88000.95
517308	FIAT-ALLIS	80033
517309	ALLIS-CHLM	80008
517309	FIAT-ALLIS	80008
517310	FIAT-ALLIS	80008
517311	ALLIS-CHLM	80037
517311	FIAT-ALLIS	80037
517312	ALLIS-CHLM	80069
517312	FIAT-ALLIS	80069
5-174X	SPICER	80078
5175	NEAPCO	88000.93
517594	NEW HOLL.	80140
5176	NEAPCO	88000.86
5177	TRU-CROSS	80172
5-177X	SPICER	80221
5178	NEAPCO	88000.89
5-1784	SPICER	80124
5-178X	ALBURUS-SP	80124
5-178X	BLAW KNOX	80124
5-178X	JEFFREY	80124
5-178X	JOY MFG.	80124
5-178X	REPCO	80124
5-178X	SPICER	80124
5-178XU	SPICER	80125
517A	NEAPCO	88000.37
517G	PRECISION	80052
5-17X	SPICER	80221
518	ALCO	88000.53
518	ALLOY	88000.30
518	PRECISION	80093
5-18	SPICER	80124
518038	FIAT-ALLIS	80133
518039	FIAT-ALLIS	80133
518040	FIAT-ALLIS	80008
5-181X	REPCO	80023
5-181X	SPICER	80023
5183	NEAPCO	88000.79
5184	NEAPCO	88000.81
5184A	MOTOR MAST	80205
5184A	ZELLER	80205
5185	NEAPCO	88000.82
5185J	BORG-WARNR	82033
5-185X	REPCO	80228
5-185X	SPICER	80228
5186	NEAPCO	88000.83
518645	GMC	80072
518648	GMC	80072
5-186X	REPCO	88001.82
5-186X	SPICER	88001.82
5187	NEAPCO	88000.51
5-187X	REPCO	80124
5-187X	REPCO	88001.83
5-187X	SPICER	88001.83
5-188X	REPCO	80140
5-188X	REPCO	80182
5-188X	SPICER	80140
5-188X	SPICER	80182
5189	NEAPCO	88000.45
518A	NEAPCO	88000.38
5-18X	SPICER	80124
519	ALCO	88000.35
519	ALLOY	88000.35
519	PRECISION	80045
5191	NEAPCO	88000.34
5193	NEAPCO	88000.46
5196	NEAPCO	88000.36
5197	NEAPCO	88000.35
5199	NEAPCO	88000.84
51A7	NEAPCO	88000.90
51B1	NEAPCO	88000.22
51B2	NEAPCO	88000.22
51B3	NEAPCO	88000.27
51B8	NEAPCO	88000.44
51D2	NEAPCO	88000.55
51D4	NEAPCO	88000.19
51D7	NEAPCO	88000.85
51D8	NEAPCO	88000.36
51E4	NEAPCO	88000.98
51G1	NEAPCO	88000.68
51G2	NEAPCO	88000.70
51G3	NEAPCO	88000.71
51G4	NEAPCO	88000.72
51G5	NEAPCO	88000.73
51H4	NEAPCO	88000.74
51H5	NEAPCO	88000.75
51R8	NEAPCO	88000.62
51R9	NEAPCO	88000.56
51S8	NEAPCO	88001.01
5-1X	SPICER	80023
5-1X	SPICER	80024
51Y103	PILOT	88000.94
51Y12	PILOT	88000.86
51Y12	PILOT	88000.88
51Y13	PILOT	88000.79
51Y22	PILOT	88000.89
51Y23	PILOT	88000.80
51Y32	PILOT	88000.95
51Y33	PILOT	88000.81

Part No.	Brand	IDLI Group
51Y53	PILOT	88000.83
51Y63	PILOT	88000.96
51Y81	PILOT	88000.92
51Y81	PILOT	88000.93
51Y93	PILOT	88000.84
52	DETROIT	80903
52	DETROIT	80958
52	SPICER	88001.73
52	WYLIE	80077
520	ALCO	88000.37
520	ALLOY	88000.37
520	PRECISION	88002.06
5-200	AEC	80072
5-200	AEC	80075
5-200	SPICER	80075
5-2000X	SPICER	80050
5-2000X	SPICER	80101
5200182	CLARK EQU.	80142
5-2002X	SPICER	80046
5-2002X	SPICER	80048
5-2007X	SPICER	80043
5-2009X	SPICER	80107
5200A	MOTOR MAST	80189
5200A	ZELLER	80189
5200AG	MOTOR MAST	80189
5200AG	ZELLER	80189
5-200U	AEC	80076
5-200X	MUNCIE	80072
5-200X	MUNCIE	80075
5-200X	REPCO	80075
5-200X	SPICER	80072
5-200X	SPICER	80075
5-200X	SPICER	80076
5-200X	SPICER	80089
5-200XU	SPICER	80076
5-2010X	SPICER	80049
5-2011X	SPICER	80043
5-2011X	SPICER	80044
5-2011X	SPICER	88001.98
5-2012X	SPICER	80030
5-2014	SPICER	80045
5-2014X	SPICER	80045
5-2018X	SPICER	80053
5-201X	SPICER	80075
5202074	CLARK EQU.	80104
5202079	CLARK EQU.	80101
5202080	CLARK EQU.	80104
5202096	CLARK EQU.	80216
5202099	CLARK EQU.	80216
5202100	CLARK EQU.	80163
5202127	CLARK EQU.	80215
5202144	CLARK EQU.	80048
5202145	CLARK EQU.	80047
5202168	CLARK EQU.	80165
5202169	CLARK EQU.	80164
5202172	CLARK EQU.	80216
5202193	CLARK EQU.	80100
5202196	CLARK EQU.	80100
5202199	CLARK EQU.	80186
5202224	CLARK EQU.	80163
520243	OLIVER	80044
5-2026X	SPICER	80030
5203	PRECISION	86194
5-2031X	SPICER	80049
5-2031X	SPICER	88001.96
5-2032X	SPICER	80092
5-2033X	SPICER	80054
5204	PRECISION	86195
52043	OLIVER	80044
52043	WHITE FARM	80044
5-2048X	SPICER	80049
5-2049X	SPICER	80054
5-204X	ALBURUS-SP	80077
5205127	CLARK EQU.	80215
5-2053	SPICER	88002.77
5-2053X	SPICER	80030
5206151	AUST.WEST.	80198
5206151	CLARK EQU.	80198
5206151	CLARK EQU.	80213
5-2064X	SPICER	80044
5-2066X	SPICER	80045
5206861	CLARK EQU.	80101
5206990	CLARK EQU.	80102
5207	PRECISION	86196
5207368	CLARK EQU.	80101
520782	WESCO	88002.37
5207973	CLARK EQU.	80214
5-207X	SPICER	80073
5208	PRECISION	86197
5208124	CLARK EQU.	80193
5208189	CLARK EQU.	80195
5208350	AUST.WEST.	80213
5208350	CLARK EQU.	80213
5208351	AUST.WEST.	80214
5208351	CLARK EQU.	80214
5209	PRECISION	86198
5209902	CLARK EQU.	80192
5209J	BORG-WARNR	82101
5-20X	ALBURUS-SP	80090
521	AEC	80072
521	ALCO	88000.39
521	ALLOY	88000.38
521	PRECISION	88002.07
5-21	SPICER	80072
5210	PRECISION	86199
5211	PRECISION	86200
5-2116X	SPICER	80047
5-211X	SPICER	80140
5212	PRECISION	86201
5212552	CHRYSLER	86019
5212553	CHRYSLER	86020
5212554	CHRYSLER	86021
5212555	CHRYSLER	86022
5212558	CHRYSLER	86128
5212567	CHRYSLER	86081
5212567	CHRYSLER	86082
5212567	CHRYSLER	86083
5212567	CHRYSLER	86084
5212568	CHRYSLER	86083
5212569	CHRYSLER	86084
5212572	CHRYSLER	86088
5212574	CHRYSLER	86081
5212575	CHRYSLER	86082
5212576	CHRYSLER	86089
5212577	CHRYSLER	86090
5212591	CHRYSLER	86129
5212592	CHRYSLER	86015
5212593	CHRYSLER	86016
5212594	CHRYSLER	86017
5212595	CHRYSLER	86018
5212602	CHRYSLER	86013
5212603	CHRYSLER	86014
5212604	CHRYSLER	86011
5212605	CHRYSLER	86012
5212606	CHRYSLER	86131
5212607	CHRYSLER	86128
5212607	CHRYSLER	86129
5212607	CHRYSLER	86130
5212607	CHRYSLER	86131
5212608	CHRYSLER	86078
5212608	CHRYSLER	86079
5212612	CHRYSLER	86078
5212612	CHRYSLER	86151
5212613	CHRYSLER	86079
5212613	CHRYSLER	86152
5212614	CHRYSLER	86086
5212615	CHRYSLER	86087
5212621	CHRYSLER	86130
5212622	CHRYSLER	86023
5212623	CHRYSLER	86024
5212624	CHRYSLER	86025
5212625	CHRYSLER	86026
5212627	CHRYSLER	86081
5212640	CHRYSLER	86023
5212641	CHRYSLER	86024
5212644	CHRYSLER	86091
5212644	CHRYSLER	86118
5212645	CHRYSLER	86092
5212652	CHRYSLER	86013
5212653	CHRYSLER	86014
5212654	CHRYSLER	86011
5212655	CHRYSLER	86012
5212656	CHRYSLER	86093
5212657	CHRYSLER	86094
5212670	CHRYSLER	86062
5212671	CHRYSLER	86063
5212809	CHRYSLER	86183
5212810	CHRYSLER	86234
5212811	CHRYSLER	86234
5212830	CHRYSLER	86183
5212852	CHRYSLER	86065
5212853	CHRYSLER	86064
5212854	CHRYSLER	86067
5212855	CHRYSLER	86068
5212856	CHRYSLER	86132
5212858	CHRYSLER	86132
5212858	CHRYSLER	86184
5212859	CHRYSLER	86080
5212859	CHRYSLER	86085
5212859	CHRYSLER	86154
5212869	CHRYSLER	86184
5212877	CHRYSLER	86066
5212896	CHRYSLER	86118
5212897	CHRYSLER	86092
5212897	CHRYSLER	86119
5212898	CHRYSLER	86093
5212899	CHRYSLER	86094
5-212X	REPCO	80121
5-212X	REPCO	88002.43
5-212X	SPICER	80118
5-212X	SPICER	80121
5-212X	SPICER	80131
5213	PRECISION	86202
5-213	SPICER	80117
5-213X	REPCO	80117
5-213X	SPICER	80117
5-213X	SPICER	80131
5214	PRECISION	86203
5-2140	SPICER	80055
5-2140X	SPICER	80055
5-214X	SPICER	80142
5215	PRECISION	86204
5-215X	SPICER	80178
5216	PRECISION	86205
5-216X	SPICER	80189
5217	PRECISION	86206
521720	ALLIS-CHLM	80037
521720	FIAT-ALLIS	80037
5-2173X	SPICER	80058
5-2175X	SPICER	80058
5-217X	SPICER	80221
5218	PRECISION	86207
5218A	MOTOR MAST	80190
5218A	ZELLER	80190
5-218X	SPICER	80225
5-2198X	SPICER	80051
5-2198X	SPICER	80052
521HD	AEC	80077
521HDU	AEC	80088
521U	AEC	80089
5-21X	SPICER	80072
522	ALCO	88000.43
522	ALLOY	88000.43
522	PRECISION	80054
522163	GMC	80107
5-222X	SPICER	80194
5-222X	SPICER	80197
522-311ES	WHITE FARM	80035
522312ES	OLIVER	80037
522-312ES	WHITE FARM	80037
522-313ES	WHITE FARM	80069
5-223X	SPICER	80198
5-224X	SPICER	80194
5-224X	SPICER	80197
5225	PRECISION	86208
5226	PRECISION	86209
5-226X	SPICER	80225
5227	PRECISION	86210
5-227X	SPICER	80228
5228	PRECISION	86211
5-228X	SPICER	80117
5229	PRECISION	86212
5-229X	SPICER	80117
523	ALCO	88000.45
5230	PRECISION	86213
5-230	SPICER	80117
5-230X	SPICER	80117
5231	PRECISION	86214
5-231	SPICER	80121
5-231X	REPCO	80121
5-231X	SPICER	80121
5232	PRECISION	86215
5233	PRECISION	86216
5-233X	SPICER	80221
5234	PRECISION	86217
523425	ROGER IRON	80101
523435	ADAMS-LET.	80099
523435	ADAMS-LET.	80101
523435	BEND.WEST.	80101
523435	WABCO	80101
52345	ROGER IRON	80101
5234700-9	FIRESTONE	80061
5234711-4	FIRESTONE	80132
5234769-6	FIRESTONE	80077
5234830-7	FIRESTONE	80081
5234831-5	FIRESTONE	80091
5234835-8	FIRESTONE	80129
5234836-6	FIRESTONE	80104
5-234X	SPICER	88001.82
5235	PRECISION	86218
5236	PRECISION	86219
5-236X	SPICER	80225
5237	PRECISION	86220
5-237X	SPICER	80140
5238	PRECISION	86221
524	AEC	80225
524	PRECISION	88002.08
52405	STANDARD	80024
52405	TRIUMPH	80024
52406	STANDARD	80023

Part No.	Brand	IDLI Group
52406	TRIUMPH	80024
5-241X	SPICER	80190
5-242-1X	SPICER	80035
5-242X	SPICER	80035
5-243-1X	REPCO	80070
5-243-1X	SPICER	80070
5-243X	REPCO	80069
5-243X	SPICER	80069
5-243X1	SPICER	80070
5-244	SPICER	80075
5-244X	SPICER	80077
524521	ALLIS-CHLM	80008
524521	FIAT-ALLIS	80008
524617	MATRA	80023
5-246X	SPICER	80025
5-246X	SPICER	80072
5-246X	SPICER	80075
5-247X	SPICER	80075
5-248	AEC	80025
5-248X	REPCO	80025
5-248X	SPICER	80025
52495	NEW HOLL.	80001
5-249X	SPICER	80025
5-24X	BORG-WARNR	80225
5-24X	HARDY SPCR	80225
5-24X	REPCO	80225
5-24X	SPICER	80225
5-24X	SPICER	80227
524Z	AEC	80225
525	ALLOY	88000.21
525	PRECISION	80043
525	PRECISION	80044
525	PRECISION	88001.98
525	PRECISION	88002.08
5251	PRECISION	86222
5-251X	SPICER	80077
5252	PRECISION	86223
5252382	CLARK EQU.	80193
5-252X	SPICER	80077
5-252X	SPICER	80182
5253	PRECISION	86224
5-2530X	SPICER	80066
5-2532X	SPICER	80059
5254	PRECISION	86225
5-254X	SPICER	80121
5255	PRECISION	86226
5255753	CLARK EQU.	80215
5-255X	SPICER	80077
5256	PRECISION	86227
5257	PRECISION	86228
5-257X	REPCO	80228
5-257X	SPICER	80228
5258	PRECISION	86229
5-258X	SPICER	80025
5-259X	SPICER	80025
526	ALLOY	88000.49
526	PRECISION	88002.09
5-260	AEC	80078
5-260	AEC	80090
5-260-1X	SPICER	80084
5260862	CLARK EQU.	80101
5-260X	REPCO	80078
5-260X	REPCO	80081
5-260X	REPUBLIC	88001.71
5-260X	SPICER	80078
526299	ALLIS-CHLM	80037
526299	FIAT-ALLIS	80037
5-262X	SPICER	80142
5-263X	ALBURUS-SP	80140
5-263X	SPICER	80140
5-263X	SPICER	80182
5-264X	REPCO	80008
5-264X	SPICER	80008
5-265X	REPCO	80133
5-265X	SPICER	80133
527	ALCO	88000.50
527	PRECISION	88002.10
5271	PRECISION	86230
5272	PRECISION	86231
5-272X	SPICER	80225
5273	PRECISION	86232
5-27-348X	SPICER	87061
5-273X	ALBURUS-SP	80117
5-273X	SPICER	80117
5274	PRECISION	86233
5-27-48X	SPICER	87062
5-275X	ALBURUS-SP	80182
5-275X	SPICER	80182
5-276X	SPICER	88001.82
5-27-7-2726X	SPICER	87062
5-27-7-6722X	SPICER	87062
5-278	AEC	80085
5-278	NEAPCO	80086
5278A	MOTOR MAST	80221
5278A	ZELLER	80221
5278AB	MOTOR MAST	80221
5278AG	MOTOR MAST	80221
5278AG	ZELLER	80221
5278AG4	MOTOR MAST	80221
5-278X	SPICER	80085
5-279-2X	SPICER	80189
5-27-9-6100X	SPICER	87061
5-279X	REPCO	80189
5-279X	SPICER	80189
528	ALLOY	88000.23
528	PRECISION	88002.11
5-280	NEAPCO	80221
5-280-4X	SPICER	80221
5-280X	REPCO	80221
5-280X	REPCO	88003.30
5-280X	REPCO	88009.11
5-280X	SPICER	80221
5-281-4X	SPICER	80227
5-28-167	SPICER	85723
5-281X	CHRYSLER	80227
5-281X	REPCO	80227
5-281X	SPICER	80227
5-281Z	AEC	80227
5-28-207	SPICER	85721
5-28-217	SPICER	85723
5-282X	SPICER	80124
5-28-407	SPICER	85721
5-28-627	SPICER	85722
5-286X	SPICER	80225
5288J	BORG-WARNR	88006.79
528E7039	FORD	80142
529	ALLOY	88000.86
529	PRECISION	88002.12
5290945	CLARK EQU.	80198
5290J	BORG-WARNR	88006.82
5291101	CLARK EQU.	80172
5291326	CLARK EQU.	80216
5291327	CLARK EQU.	80216
5292J	BORG-WARNR	82038
5-297	AEC	80079
5-297-1X	SPICER	80002
5-297X	REPCO	80079
5-297X	REPCO	80139
5-297X	SPICER	80079
5-299X	SPICER	80190
52A	DETROIT	80909
52A	SPICER	88001.72
5-2X	ALBURUS-SP	80023
5-2X	SPICER	80041
52Y13	PILOT	88000.68
52Y43	PILOT	88000.70
52Y63	PILOT	88000.71
52Y73	PILOT	88000.72
52Y93	PILOT	88000.73
53	DETROIT	88001.75
53	WYLIE	80091
530	AEC	80117
530	ALCO	88000.88
530	ALLOY	88000.87
530	PRECISION	80099
5-3000X	SPICER	80099
5-3000X	SPICER	80101
5-3000X	SPICER	80107
530-039	WESCO	88000.74
530040	WESCO	88000.68
530041	WESCO	88000.69
530042	WESCO	88000.70
530-043	WESCO	88000.75
530-044	WESCO	88000.71
530-045	WESCO	88000.71
530046	WESCO	88000.72
530047	WESCO	88000.72
530-048	WESCO	88000.73
5-3008X	SPICER	80104
5-3009X	SPICER	80105
5-300X	SPICER	80024
5-300XBA	HUSCO	80023
5301	PRECISION	86234
530119-1	BUTLER	88008.63
530126-1	BUTLER	88008.64
53012A	HYSTER	80158
5-3012X	SPICER	80104
5-3012X	SPICER	80107
5-3012X	SPICER	80112
5-3013X	SPICER	80106
530147-3	BUTLER	88008.65
5-3014X	SPICER	80114
530167-1	BUTLER	88008.66
530179	WESCO	88002.38
5302	PRECISION	86235
5-3021X	SPICER	80111
5-3022X	SPICER	80097
5-3022X	SPICER	80119
5-3031X	SPICER	80109
53034A	HYSTER	80216
53038	GERLINGER	80047
530538	ALLIS-CHLM	80069
530538	FIAT-ALLIS	80069
530549	BUTLER	88008.67
5-305X	SPICER	80141
53066A	HYSTER	80077
5-306X	SPICER	80221
5307	PRECISION	86074
5307	PRECISION	86075
5307	PRECISION	86236
53075A	HYSTER	80164
53076	HYSTER	80158
5-308	AEC	80229
5308	PRECISION	86236
5-308X	CHRYSLER	80229
5-308X	REPCO	80228
5-308X	REPCO	80229
5-308X	SPICER	80229
5-30X	SPICER	80178
5-30X	SPICER	80189
531	AEC	80121
531	AEC	80131
531	ALLOY	88000.32
531	PRECISION	80110
5-3-108KX	SPICER	85312
5-3-108X	SPICER	85312
5-310X	ALBURUS-SP	80036
5311	PRECISION	86237
5-3111X	SPICER	80107
5312	PRECISION	86238
531220	ALLIS-CHLM	80037
531220	FIAT-ALLIS	80037
5-3-128K-X	SPICER	85314
5-3-128X	SPICER	85314
5-312XR	ALBURUS-SP	80036
5-312XR	ALBURUS-SP	80075
5-3-1411KX	SPICER	85315
5-3-1411X	SPICER	85315
5-3147X	SPICER	80104
5-3147X	SPICER	80107
5-315	AEC	80026
5315	PRECISION	86239
5-315X	REPCO	80026
5-315X	SPICER	80026
5316	PRECISION	86240
531676-1	JOY MFG.	80171
53-17	CRANE CAR.	87461
53-17	CRANE CAR.	88007.79
53-18	CRANE CAR.	87481
53-18	CRANE CAR.	88007.77
53-1834	NEAPCO	88015.39
5319	PRECISION	86241
531G	PRECISION	80112
5-31X	SPICER	80189
532	ALCO	88000.80
532	ALLOY	88000.80
532	PRECISION	80106
5320	PRECISION	86242
53-2034	NEAPCO	88000.10
5-3-2261KX	SPICER	85316
5323	PRECISION	86243
5324	PRECISION	86244
53-2434	NEAPCO	88001.07
5-324X	SPICER	88009.17
5325	PRECISION	86245
5-326X	REPCO	87481
5-326X	SPICER	87481
5-3-288KX	SPICER	85313
5-3-288X	SPICER	85313
533	AEC	80116
533	ALLOY	88000.48
533	PRECISION	80114
5-332X	SPICER	80139
533371	JOY MFG.	80235
534	AEC	80118
534	ALCO	88000.92
534	ALLOY	88000.92
534	PRECISION	80107
534	PRECISION	88009.49
5341	PRECISION	86246
534-10	PRECISION	88002.13
5342	PRECISION	86247
534235	ALLIS-CHLM	80001
534235	ALLIS-CHLM	80003
534235	FIAT-ALLIS	80003
5348	PRECISION	86248
5349	PRECISION	86249
534G	PRECISION	80104
535	ALLOY	88000.91

Part No.	Brand	IDLI Group
535	PRECISION	80100
5-350	AEC	80036
5-350	AEC	80044
5-350X	REPCO	80036
5-350X	SPICER	80036
5352	PRECISION	86250
535251-2	DETROIT	88001.60
535251-2	DETROIT	88001.62
53525-2	DETROIT	80096
53525-2	DETROIT	88001.62
5353	PRECISION	86251
5-353X	SPICER	80083
5354	PRECISION	86252
535554	GMC	80107
5-355X	SPICER	80144
5-356X	SPICER	80097
5-357	SPICER	80144
5-357X	SPICER	80144
5358	PRECISION	86253
5359	PRECISION	86254
536	ALLOY	88000.40
536	PRECISION	80101
536199	STUDEBAKER	80955
5-361X	CHRYSLER	80234
5-361X	SPICER	80234
5362	PRECISION	86255
53625	PILOT	80135
53625-2	ALCO	88001.65
53625-2	DETROIT	80135
53625-2	JAEGER	80135
53625-2	SPICER	80135
53625-2	SPICER	88001.65
5363	AEC	88000.25
5366	PRECISION	86256
5367	PRECISION	86257
537	ALCO	88000.85
537	ALLOY	88000.94
537	PRECISION	88002.14
537	TEDDY TORQ	80135
53725-2	DETROIT	80135
53725-2	JAEGER	80135
53727-2	DETROIT	80135
538	ALLOY	88000.47
538	JOY MFG.	80193
538	PRECISION	80111
5381	PRECISION	86258
5382	PRECISION	86259
53825-2	ALCO	80135
53825-2	DETROIT	80135
53825-3	DETROIT	80135
5383	PRECISION	86260
538-395R91	IHC	80187
5384	PRECISION	86261
538451	LEYLAND	80124
5385	PRECISION	86262
5386	PRECISION	86263
5387	PRECISION	86264
5389	PRECISION	86265
539	ALCO	88000.95
539	ALLOY	88000.95
5390	PRECISION	86266
5391	PRECISION	86267
53A	DETROIT	88001.67
54	DETROIT	88001.74
54	SPICER	88001.74
540	ALLOY	88000.93
540	PRECISION	80109
540	PRECISION	80152
5-4002X	SPICER	80109
5-4002X	SPICER	80149
5-4002X	SPICER	80153
540-049	WESCO	80061
540051	WESCO	81025
540052	WESCO	81268
540053	WESCO	88014.52
540054	WESCO	88014.52
540055	WESCO	81020
540056	WESCO	81034
540058	WESCO	81034
540059	WESCO	81034
540060	WESCO	81034
540061	WESCO	88001.10
540062	WESCO	81125
540063	WESCO	81060
540064	WESCO	81054
540065	WESCO	81054
540066	WESCO	81089
540067	WESCO	81089
540068	WESCO	81109
540069	WESCO	81125
5-4007	SPICER	80147
540071	WESCO	81109
540073	WESCO	81136
540-075	WESCO	81143
540-077	WESCO	81156
540078	WESCO	81157
540079	WESCO	81153
5-4007X	SPICER	80147
540080	WESCO	81175
540081	WESCO	81175
540083	WESCO	81268
540085	WESCO	88001.11
540086	WESCO	81054
540087	WESCO	81109
540088	WESCO	81153
540089	WESCO	88001.09
5-4008X	SPICER	80146
540090	WESCO	81175
5401	PRECISION	86268
5-40-1011	SPICER	87504
5-40-1031	SPICER	87526
5-40-1121	SPICER	87497
5-40-1151	SPICER	87503
5-4014X	SPICER	80158
5-4015X	SPICER	80148
5-4016X	SPICER	80159
5-40-471	SPICER	87523
5-40-481	SPICER	87497
5-40-501	SPICER	87526
5-40-511	SPICER	87499
5-40-671	SPICER	87503
5-40-701	SPICER	87497
5407-1	KENBAR	87331
5407-1	KENBAR	88008.39
5-40-741	SPICER	87504
5407J	BORG-WARNR	81823
5-407X	SPICER	80226
5-407X	SPICER	88009.04
541	ALCO	88000.89
541	ALLOY	88000.89
5-41	BYPY	88004.33
541	PRECISION	80153
5410-1000	ARMSTRONG	80072
5410-1100	ARMSTRONG	80041
5410-150	ARMSTRONG	80023
5410-161	ARMSTRONG	80041
5410-162	ARMSTRONG	80072
5410-164	ARMSTRONG	80142
5410-165	ARMSTRONG	80180
5410-1720	ARMSTRONG	80023
5410-192	ARMSTRONG	80124
5410-2050	ARMSTRONG	80142
5410-3500	ARMSTRONG	80180
5410-400	ARMSTRONG	80023
5410-4000	ARMSTRONG	80189
5410-500	ARMSTRONG	80023
5410-5000	ARMSTRONG	80221
5410-510	ARMSTRONG	80024
5-4106X	SPICER	80159
5410-7010	ARMSTRONG	88002.77
5410-7100	ARMSTRONG	80032
5410-7150	ARMSTRONG	88002.77
5410-7280	ARMSTRONG	88002.77
5411	PRECISION	86269
5412	PRECISION	86270
5-4-1371	SPICER	83542
5-4140X	SPICER	80155
5-4143X	SPICER	80156
5415	PRECISION	86271
5415-1000	ARMSTRONG	80077
5416	PRECISION	86272
5-4-1631	SPICER	83547
5-4-1721	SPICER	83523
541858	LAND ROVER	80072
541858	ROVER	80041
5419	PRECISION	86273
5-4-1931	SPICER	83526
5-419X	SPICER	80075
5-419X	SPICER	80077
5-419X	SPICER	80078
5-419X	SPICER	80085
542	ALLOY	88000.44
542	PRECISION	80147
5420	PRECISION	86274
5-4-2081	SPICER	83525
5-420X	SPICER	80077
5-420X	SPICER	80085
5-420X	SPICER	80086
5423	PRECISION	86275
5424	PRECISION	86276
5-4-2401	SPICER	83522
5425	PRECISION	86277
5-4-2501	SPICER	83543
542544	ROVER	80077
542862	ALLIS-CHLM	80035
542862	FIAT-ALLIS	80035
543	ALCO	88000.79
543	ALLOY	88000.79
543	PRECISION	80146
5-430X	SPICER	88000.15
5-431X	SPICER	80136
5-432X	SPICER	80161
5-4-3311X	SPICER	83569
5-433-1X	SPICER	80086
5-4-3321	SPICER	83569
5-4-3601X	SPICER	83526
5-4-3611X	SPICER	83526
5-4-3621	SPICER	83527
5-4-3681	SPICER	83531
5-4-3861	SPICER	83531
5-4-3901	SPICER	83529
544	ALCO	88000.81
544	ALLOY	88000.81
544	PRECISION	80158
5441	PRECISION	86278
5-4-4271	SPICER	83522
5-443X	SPICER	80027
5-4-4401	SPICER	83549
5-4-4441X	SPICER	83531
5-4-4451X	SPICER	83532
5-4-4471	SPICER	83545
5447	PRECISION	86179
5-4-4761X	SPICER	83529
5-4-4781	SPICER	83540
5-447X	SPICER	80128
5448	PRECISION	86279
5-4-4891X	SPICER	83540
5449	PRECISION	86280
5-4-4901X	SPICER	83541
5-44X	SPICER	88002.61
545	ALCO	80148
545	PRECISION	80148
5-4-5031	SPICER	83565
5-4-5061	SPICER	88005.39
5-4-5071X	SPICER	83536
5-4-5081X	SPICER	83537
5-4-5111X	SPICER	83525
5-4-5131X	SPICER	83547
5-4-5161X	SPICER	83565
5451A	MOTOR MAST	80204
5451A	ZELLER	80138
5-4-5281X	SPICER	83531
5454	PRECISION	86281
5-4-5401	SPICER	83571
5-4-5411X	SPICER	83571
5-4-5421	SPICER	83551
5-4-5431X	SPICER	83552
5-4-5441X	SPICER	83557
5-4-5451X	SPICER	83558
5-4-5571	SPICER	83533
5-4-5631	SPICER	83570
5-456X	SPICER	80028
5-4-5741	SPICER	83572
5-458X	SPICER	80029
5-45X	SPICER	80092
546	ALCO	88000.83
546	ALLOY	88000.83
546	PRECISION	80159
5-460X	SPICER	80127
5461J	BORG-WARNR	86808
5462	PRECISION	86282
5-462X	SPICER	80142
5-4-6351	SPICER	88005.40
5467	PRECISION	86283
546945	ALLIS-CHLM	80035
546945	FIAT-ALLIS	80035
546959-3	WHITING	80001
547	ALLOY	88000.51
547	PRECISION	88002.15
5-4-781	SPICER	83524
5-47X-JA	SPICER	88002.63
5-47X-JA	SPICER	88002.64
548	PRECISION	88002.16
5481J	BORG-WARNR	86808
549	ALCO	88000.45
549	PRECISION	80151
54A	DETROIT	88001.66
54A	SPICER	88001.66
5-4X	SPICER	80041
55	DETROIT	80956
55	WYLIE	80060
550	AEC	80119
550	AEC	80120
550	ALLOY	88000.34
550	PRECISION	80168
5-5000X	SPICER	80168
5-5007X	SPICER	80177
5-5008X	SPICER	80117
5-5008X	SPICER	80174
5-5015X	SPICER	80173

Part No.	Brand	IDLI Group
5-5015X	SPICER	80175
5-5031	NEAPCO	80171
55-055	UJS	80141
5505J	BORG-WARNR	86744
5506J	BORG-WARNR	85102
551	ALCO	88000.46
551	PRECISION	80176
55-1	UJS	80023
55-111	UJS	80024
55-115	MOTIVE	80178
55-115	UJS	80178
55-116	UJS	80189
55-121	UJS	80075
5-5121X	SPICER	80163
5-5121X	SPICER	80167
5-5121X	SPICER	80171
55-124	UJS	80225
55-12L	UJS	80035
55-130	UJS	80117
55-131	UJS	80121
5513-2	KENBAR	87326
55-14LN	UJS	80037
55-150	UJS	80097
55-152	UJS	80118
55-153	UJS	80077
55-155	UJS	80182
55-160	UJS	80142
55-165	UJS	80190
55-1762	UJS	80030
55-1767	UJS	80031
55-177	UJS	80221
5-5177X	SPICER	80172
55-178	UJS	80124
55-188	UJS	80140
55-1FR	UJS	80008
552	ALCO	88000.36
552	ALLOY	88000.36
552	HERCULES	80953
552	HERCULES	80958
552	PRECISION	80167
55-200	UJS	80075
55-2002	UJS	80048
55-2017	UJS	80092
55-2063	UJS	80044
5520J	BORG-WARNR	88006.28
55-2100	UJS	80049
55-2114	UJS	80054
55-212-1	UJS	88002.43
55-2123	UJS	80054
55-212-3	UJS	88002.42
55-2124	UJS	80054
55-2140	UJS	80055
55-2173	UJS	80058
55-2175	UJS	80058
55-248	UJS	80025
55-2518	UJS	80066
552C	HERCULES	80958
553	ALLOY	88000.84
553	PRECISION	80177
55-3000	UJS	80099
55-3008	UJS	80104
55-3008C	UJS	80107
55-3012	UJS	80112
5-53-141	SPICER	87416
55-3188	UJS	80105
55-35N3	IMCO	80069
55-35N3	UJS	80069
55-35N4	UJS	80069
5-537X	SPICER	80144
55-3DR	UJS	80133

Part No.	Brand	IDLI Group
554	PRECISION	80174
55-4	UJS	80041
55-4100	UJS	80109
55-4102	UJS	80158
55-4106	UJS	80148
55-4107	UJS	80159
5545-3042	AUST.MOT	80101
55-4N	IMCO	80134
55-4N	MOTIVE	80134
555	AEC	80182
555	HERCULES	80909
555	PRECISION	88002.17
55506	FMC	80123
55-5100	UJS	80168
55-5101	UJS	80163
55-5103	UJS	80174
55-5110	UJS	80173
55-51625-2	UJS	80042
555354	CATERPILLR	80188
55-53825-2	UJS	80135
55-55	UJS	80141
5556A	MOTOR MAST	80220
5556A	ZELLER	80220
55571	FMC	80081
55-5N	IMCO	80178
556	ALLOY	88000.60
556	PRECISION	88002.18
55-6100	HYSTER	80193
55-6100	IMCO	80193
55-6100	MOTIVE	80193
55-6100	UJS	80193
55-6102	HYSTER	80197
55-6102	IMCO	80197
55-6102	MOTIVE	80197
55-6102	UJS	80197
55-6106	HYSTER	80198
55-6106	IMCO	80198
55-6106	UJS	80198
55-6108	HYSTER	80199
55-6108	IMCO	80199
55-6108	MOTIVE	80199
55-6108	UJS	80199
55-6109	HYSTER	80201
55-6109	IMCO	80201
55-6109	MOTIVE	80201
55-6109	UJS	80201
5562A	MOTOR MAST	80219
55-65AR	UJS	80905
55-65R	UJS	80951
55-67AR	UJS	80906
5567J	BORG-WARNR	86700
55-67R	UJS	80953
55-6LN	UJS	80001
55-6N	IMCO	80188
55-6N	MOTIVE	80188
5-56X	SPICER	80072
557	ALLOY	88000.57
557	PRECISION	80173
55-70	UJS	80007
55-7000	IMCO	80217
55-7035	IMCO	80217
55-7100	HYSTER	80215
55-7100	IMCO	80210
55-7100	MOTIVE	80210
55-7100	UJS	80210
55-7101	HYSTER	80207
55-7101	IMCO	80211
55-7101	MOTIVE	80211
55-7101	UJS	80212
55-7102	HYSTER	80214

Part No.	Brand	IDLI Group
55-7102	IMCO	80212
55-7102	MOTIVE	80212
55-7102	UJS	80212
55-7105	HYSTER	80213
55-7105	IMCO	80213
55-7105	MOTIVE	80213
55-7105	UJS	80213
55-7119	HYSTER	80213
55-7119	IMCO	80213
55-71425-2	UJS	80034
5571J	BORG-WARNR	86707
55-72	UJS	80007
557221	BRILL,J.G.	80109
55-72625-2	UJS	80065
55-72625-3	UJS	80061
55-72925-2	UJS	80132
55-74	UJS	80090
55-7N	IMCO	80220
55-7N	MOTIVE	80220
558	ALLOY	88000.33
558	PRECISION	80170
55-8002	IMCO	80243
5-5800X	SPICER	80185
5-5801X	SPICER	80186
5-5802X	SPICER	80170
5-5802X	SPICER	80184
55-8100	HYSTER	80240
55-8100	IMCO	80231
55-8100	MOTIVE	80231
55-8100	UJS	80231
55-8101	HYSTER	80242
55-8101	IMCO	80232
55-8101	UJS	80232
55-8102	HYSTER	80239
55-8102	IMCO	80237
55-8102	MOTIVE	80237
55-8102	UJS	80237
55-8105	HYSTER	80238
55-8105	IMCO	80238
55-8113	UJS	80238
55-86	UJS	80124
5586J	BORG-WARNR	86734
55-88	UJS	80140
55888	UJS	80081
558947	ACF BRILL	80221
559	ALLOY	88000.59
55-9002	HYSTER	80233
55-9002	IMCO	80233
55-9015	HYSTER	80234
55-9015	IMCO	80234
55-9015	UJS	80234
55-9016	HYSTER	80234
55-9016	IMCO	80234
55-9017	UJS	80233
55-9100	HYSTER	80235
55-9100	IMCO	80235
55-9100	UJS	80235
55-96	UJS	80072
55-96	UJS	80075
559P2	WHITE	80221
55A	DETROIT	80906
55A	DETROIT	80909
55D56	UJS	80145
55N	AEC	80161
55N	WESCO	80161
55N	WESCO	80179
55N170-1	ROCKWELL	87054
55NF29	ROCKWELL	84925
55NF36	ROCKWELL	84926
55NF44	ROCKWELL	84927

Part No.	Brand	IDLI Group
55NY36-4	ROCKWELL	85720
55NYH18	ROCKWELL	83470
55NYH26	ROCKWELL	83501
55NYQ18-2	ROCKWELL	83468
55NYQ19-2	ROCKWELL	83471
55NYQ19-3	ROCKWELL	83472
55NYQ21	ROCKWELL	83475
55NYQ21-3	ROCKWELL	83474
55NYQ26	ROCKWELL	83499
55NYR16-1	ROCKWELL	83465
55NYR16-2	ROCKWELL	83466
55NYR17	ROCKWELL	83467
55NYR18-3	ROCKWELL	83469
55NYR20-7	ROCKWELL	83473
55NYR22	ROCKWELL	83477
55NYR22-4	ROCKWELL	83482
55NYR24	ROCKWELL	83485
55NYR24-1	ROCKWELL	83484
55NYR24-13	ROCKWELL	83488
55NYR24-13	ROCKWELL	88006.09
55NYR24-15	ROCKWELL	87058
55NYR24-3	ROCKWELL	83486
55NYR24-9	ROCKWELL	83487
55NYR25	ROCKWELL	83492
55NYR25-1	ROCKWELL	83489
55NYR25-4	ROCKWELL	83491
55NYR26	ROCKWELL	83495
55NYR26-9	ROCKWELL	83496
55NYR27-3	ROCKWELL	83502
55NYR28	ROCKWELL	83508
55NYR28-1	ROCKWELL	83510
55NYR28-11	ROCKWELL	83514
55NYR28-6	ROCKWELL	83513
55NYR28-7	ROCKWELL	83507
55NYR28-7	ROCKWELL	83509
55NYR28-9	ROCKWELL	83515
55NYR31	ROCKWELL	83517
55NYR31-3	ROCKWELL	83516
55NYR32	ROCKWELL	83520
55NYR32-4	ROCKWELL	83521
55NYS22-17	ROCKWELL	83478
55NYS22-19A	ROCKWELL	84539
55NYS22-25A	ROCKWELL	84540
55NYS22-32	ROCKWELL	83479
55NYS22-33	ROCKWELL	87056
55NYS22-37	ROCKWELL	87055
55NYS22-4	ROCKWELL	83476
55NYS22-56A	ROCKWELL	84538
55NYS22-6	ROCKWELL	87057
55NYS22-7A	ROCKWELL	84541
55NYS22-9A	ROCKWELL	84542
55NYS24	ROCKWELL	83481
55NYS24-1	ROCKWELL	83483
55NYS24-14	ROCKWELL	83480
55NYS25	ROCKWELL	83490
55NYS26-1	ROCKWELL	83494
55NYS26-10	ROCKWELL	83498
55NYS26-19	ROCKWELL	83500
55NYS26-22	ROCKWELL	83493
55NYS26-23	ROCKWELL	83497
55NYS28	ROCKWELL	83505
55NYS28-10	ROCKWELL	83506
55NYS28-19	ROCKWELL	83503
55NYS28-1A	ROCKWELL	84543
55NYS28-21	ROCKWELL	83511
55NYS28-23	ROCKWELL	83504
55NYS28-3	ROCKWELL	87060
55NYS28-36	ROCKWELL	87059
55NYS28-47	ROCKWELL	83512
55NYS28-48A	ROCKWELL	84545

Part No.	Brand	IDLI Group
55NYS28-50A	ROCKWELL	84544
55NYS32	ROCKWELL	83519
55NYS32-7	ROCKWELL	83518
55096	UJS	80138
55P1	DIAMOND R	80041
55P1	REO MOTORS	80041
55P1	WHITE	80041
55P55	UJS	80129
55P675	UJS	80131
55RST55	UJS	80074
55S55	UJS	80081
55S675	UJS	80091
55U56	UJS	80183
5-5X	SPICER	80072
56	WYLIE	88001.71
560	AEC	80142
560	ALLOY	88000.61
560	PRECISION	80193
5-60	SPICER	80142
5-6000X	SPICER	80193
5-6000X	SPICER	80198
560-1	PRECISION	88002.19
560-101	WESCO	88000.10
560-101	WESCO	88015.39
560-102	WESCO	88000.10
560-102	WESCO	88015.39
5-6016X	SPICER	80193
560-3	PRECISION	88002.20
56-0625	NEAPCO	88000.18
56-0666	NEAPCO	88001.15
560U	AEC	80143
5-60X	BLAW KNOX	80142
5-60X	SPICER	80142
561	ALLOY	88000.62
561	PRECISION	80195
5-6102	SPICER	80197
5-6102X	SPICER	80194
5-6102X	SPICER	80195
5-6102X	SPICER	80197
5-6106	SPICER	80198
5-6106X	SPICER	80198
5-6108X	SPICER	80199
5-6109X	SPICER	80201
5-6110X	SPICER	80198
56-1225	NEAPCO	88000.17
561-452C91	IHC	80003
56-1856	NEAPCO	88000.17
56-1859	NEAPCO	88000.09
5-61X	SPICER	80024
562	ALLOY	88000.56
562	PRECISION	80193
562	PRECISION	80198
5620-4	KENBAR	88008.42
56-2299	NEAPCO	88000.09
56-2300	NEAPCO	88000.09
56-2300	NEAPCO	88001.08
56-2459	NEAPCO	88001.08
56289	VERSATILE	80077
563	ALLOY	88001.01
563	PRECISION	80197
563-3B	PRECISION	88002.21
5-63X	SPICER	80072
564	ALLOY	88000.99
564	PRECISION	80194
564274	GMC	80051
5-64X	SPICER	80090
565	AEC	80190
565	PRECISION	80203
565-001-0	GLAENZER	80180
565-019-0	GLAENZER	80178
565-019-1	GLAENZER	80178
565-020-1	GLAENZER	80180
565-023-0	GLAENZER	80180
565-026-1	GLAENZER	80140
565-030-0	GLAENZER	80180
565-040-0	GLAENZER	80140
565225R91	IHC	80037
56564	VERSATILE	80085
565987	GMC	80107
565988	GMC	80110
566	PRECISION	80192
5661385	GMC	88002.67
5665605	GMC	88002.70
5665608	GMC	88002.71
5665939	GMC	88002.67
5665939	GMC	88002.72
56695F11	M.R.S.	80047
56695F11	M.R.S.	80048
567	PRECISION	80191
5670233	GMC	88002.74
5670292	GMC	88002.73
5670293	GMC	88002.74
5670615	GMC	88002.75
5671108	GMC	88002.76
5671712	GMC	80107
5672333	GMC	88002.76
5672334	GMC	88002.75
5674730	GMC	80093
5674739	GMC	80093
5675412	GMC	80044
5675414	GMC	80043
5675415	GMC	80044
5676478	GMC	80093
5677659	GMC	80104
5677659-2	GMC	80104
568	PRECISION	80199
56807	OPEL	80024
56807	OPEL	80030
56807	OPEL	80031
568-3	PRECISION	88002.22
568E7039	FORD	80180
569	PRECISION	80201
569-047R91	IHC	80037
569-085R91	IHC	80101
569262R91	IHC	80035
5698221	GMC	80105
56N-220-20	WESCO	84541
56N220TQD	WESCO	84538
5-6X	SPICER	80072
5-6X	SPICER	80077
57	WYLIE	80060
570	AEC	80007
570	AEC	88000.15
570	PRECISION	80210
570	PREC.TOR.	80217
5-70	SPICER	80007
5-7000	SPICER	80210
5-7000X	SPICER	80210
570-1	PRECISION	88002.23
5702158	AMER.MOTOR	80024
5702158	JEEP	80024
5702158	WILLYS	80024
570-3	PRECISION	88002.24
5703383	JEEP	80081
570-6N8	AEC	88000.15
570816	GMC	80119
570817	GMC	80075
570P	AEC	80007
570P	AEC	88000.15
5-70X	BLAW KNOX	80007
5-70X	SPICER	80007
571	AEC	80007
571	AEC	88000.14
571	PREC.TOR.	80217
5-7105	SPICER	80213
5-7105X	SPICER	80213
5-7109X	SPICER	80201
5-7126	NEAPCO	80213
5-71X	REPCO	80072
5-71X	SPICER	80072
572	PRECISION	80211
5-7200X	SPICER	80210
5-7202	SPICER	80212
5-7202X	SPICER	80211
5-7202X	SPICER	80212
5-7205	NEAPCO	80210
5-7206	NEAPCO	80211
5-7206X	SPICER	80211
5-7207	NEAPCO	80212
5-7208	NEAPCO	80208
5-7209	NEAPCO	80209
572-3A	PRECISION	88002.25
572489	CHRYSLER	88001.60
5732-3	KENBAR	88008.41
574	AEC	80090
574	PRECISION	80212
574	PREC.TOR.	80218
574-3B	PRECISION	88002.26
5-74X	REPCO	80090
5-74X	SPICER	80078
5-74X	SPICER	80090
575	PRECISION	80208
575471	WABCO	80099
575784	CHRYSLER	80135
575787	CHRYSLER	88001.60
57584	CHRYSLER	80135
576	AEC	88000.15
576	ALLOY	88000.15
576	PRECISION	80209
57606	BATUYONG	80023
57606	BATUYONG	80032
576207	EATON	80168
576207	EATON	80171
57689M11	M.R.S.	80196
57689M11	M.R.S.	80197
577	AEC	80221
57735	WABCO	80150
57735	WABCO	80153
577Z	AEC	80221
578	AEC	80124
5-78	SPICER	80124
578-1310	AEC	80127
578805	NEW HOLL.	80136
578U	AEC	80125
5-78X	SPICER	80124
5792J	BORG-WARNR	88015.13
5792J10800	BORG-WARNR	87622
5792J7200	BORG-WARNR	87621
579304	EATON	80168
579304	EATON	80171
5794345	CHRYSLER	88003.42
579445	CHRYSLER	88003.42
57P1	DIAMOND R	80072
57P1	DIAMOND R	80075
57P1	DIAMOND R	80076
57P1	DIAMOND R	80077
57P1	REO MOTORS	80072
57P1	WHITE	80077
5-7X	SPICER	80124
580	PREC.TOR.	80243
5800	TRU-CROSS	80185
58000002	RAY GO	80142
58-000-405	FIAT-ALLIS	80232
58-000-505	FIAT-ALLIS	80232
58-000-505-01-414	ALLIS-CHLM	80167
58-000-505-01-414	FIAT-ALLIS	80167
58-000-505-405	ALLIS-CHLM	80232
58-000-505-405	ALLIS-CHLM	80244
58-000-505-409	ALLIS-CHLM	80213
5-8002	SPICER	80231
5-8002X	SPICER	80231
581	PRECISION	80231
581	PRECISION	88002.27
581	PREC.TOR.	80243
5-81	SPICER	80024
5-8105X	SPICER	80236
5-8105X	SPICER	80238
581-1	PRECISION	88002.28
58122	FWD	80178
581-3	PRECISION	88002.29
5814J	BORG-WARNR	87449
5814J	BORG-WARNR	88007.91
5-8150X	SPICER	80238
58155	FWD	80178
58157	FWD	80178
58159	FWD	80188
58162	FWD	80220
581-954R91	IHC	80044
5-81X	SPICER	80023
582	PRECISION	80232
5-8200	SPICER	80231
5-8200X	SPICER	80231
5-8201	SPICER	80232
5-8201X	SPICER	80232
5-8202	SPICER	80237
5-8202X	SPICER	80232
5-8202X	SPICER	80237
5821	ALLOY	88001.70
582-390(4)	PARSONS	80233
582-3A	PRECISION	88002.30
582-3B	PRECISION	88002.31
58259	WHITE	80077
5-82X	SPICER	80072
583	ALCO	80237
583	PRECISION	80237
583	PREC.TOR.	80244
58359	WHITE	80077
584	PRECISION	80236
5841-5	KENBAR	87459
5841-5	KENBAR	88008.45
585	AEC	80228
5852	AEC	80228
58533F11	M.R.S.	80196
58533F11	M.R.S.	80197
586	AEC	88001.82
5-86X	REPCO	80124
5-86X	SPICER	80124
587	AEC	88001.83
587579	CATERPILLR	80001
588	AEC	80140
588572	CHRYSLER	88001.60
58897F11	M.R.S.	80047
58897F11	M.R.S.	80048
5-88X	REPCO	80139
5-88X	REPCO	80140
5-88X	SPICER	80140
58WB	AEC	80185
58WB1	AEC	80186

Part No.	Brand	IDLI Group
58WB1	PILOT	80186
58WB170	ROCKWELL	86679
58WB170-1	ROCKWELL	86680
58WB2	AEC	80184
58WB2	PILOT	80184
58WBDCA4	ROCKWELL	86681
58WBF1	ROCKWELL	84628
58WBF10	ROCKWELL	84637
58WBF11	ROCKWELL	84638
58WBF12	ROCKWELL	84627
58WBF14	ROCKWELL	84639
58WBF16	ROCKWELL	84643
58WBF19	ROCKWELL	84635
58WBF2	ROCKWELL	84629
58WBF20	ROCKWELL	84631
58WBF21	ROCKWELL	84642
58WBF25	ROCKWELL	84640
58WBF26	ROCKWELL	84648
58WBF28	ROCKWELL	84625
58WBF29	ROCKWELL	84622
58WBF3	ROCKWELL	84630
58WBF31	ROCKWELL	84644
58WBF32	ROCKWELL	84636
58WBF34	ROCKWELL	84620
58WBF35	ROCKWELL	84624
58WBF36	ROCKWELL	84626
58WBF38	ROCKWELL	84632
58WBF4	ROCKWELL	84633
58WBF40	ROCKWELL	84623
58WBF41	ROCKWELL	84641
58WBF42	ROCKWELL	84619
58WBF43	ROCKWELL	84647
58WBF5	ROCKWELL	84645
58WBF6	ROCKWELL	84621
58WBF7	ROCKWELL	84646
58WBF8	ROCKWELL	84634
58WBLS28-1	ROCKWELL	85078
58WBLS28-10	ROCKWELL	85083
58WBLS28-11	ROCKWELL	85084
58WBLS28-12	ROCKWELL	85079
58WBLS28-15	ROCKWELL	85080
58WBLS28-2	ROCKWELL	85081
58WBLS28-21	ROCKWELL	85082
58WBLS28-8	ROCKWELL	85077
58WBLS40-1	ROCKWELL	85085
58WBY53-1	ROCKWELL	85531
58WBY53-4	ROCKWELL	85532
58WBY53-5	ROCKWELL	85538
58WBY53-6	ROCKWELL	85537
58WBY53-8	ROCKWELL	85530
58WBY53-9	ROCKWELL	85536
58WBY59	ROCKWELL	85533
58WBY62	ROCKWELL	85535
58WBY62-3	ROCKWELL	85534
58WBY69-1	ROCKWELL	85539
58WBY94	ROCKWELL	85529
58WBYR16	ROCKWELL	81413
58WBYR16-1	ROCKWELL	81415
58WBYR16-2	ROCKWELL	81414
58WBYR16-3	ROCKWELL	81416
58WBYR17	ROCKWELL	81417
58WBYR18	ROCKWELL	81418
58WBYR18-2	ROCKWELL	81419
58WBYR18-3	ROCKWELL	81421
58WBYR18-4	ROCKWELL	81420
58WBYR20-1	ROCKWELL	81423
58WBYR20-3	ROCKWELL	81424
58WBYR20-4	ROCKWELL	81425
58WBYR20-6	ROCKWELL	81426
58WBYR20-7	ROCKWELL	81422
58WBYR22	ROCKWELL	81427
58WBYR22-1	ROCKWELL	81432
58WBYR22-2	ROCKWELL	81434
58WBYR22-3	ROCKWELL	81433
58WBYR22-4	ROCKWELL	81435
58WBYR22-5	ROCKWELL	81438
58WBYR22-6	ROCKWELL	81436
58WBYR23	ROCKWELL	81442
58WBYR23-1	ROCKWELL	81443
58WBYR24	ROCKWELL	81456
58WBYR24-10	ROCKWELL	81460
58WBYR24-12	ROCKWELL	81650
58WBYR24-13	ROCKWELL	81462
58WBYR24-2	ROCKWELL	81452
58WBYR24-3	ROCKWELL	81458
58WBYR24-4	ROCKWELL	81459
58WBYR24-6	ROCKWELL	81461
58WBYR24-7	ROCKWELL	81451
58WBYR24-8	ROCKWELL	81444
58WBYR24-9	ROCKWELL	81445
58WBYR25	ROCKWELL	81466
58WBYR25-1	ROCKWELL	81465
58WBYR25-2	ROCKWELL	81463
58WBYR25-3	ROCKWELL	81464
58WBYR26	ROCKWELL	81469
58WBYR26-1	ROCKWELL	81467
58WBYR26-2	ROCKWELL	81472
58WBYR26-3	ROCKWELL	81473
58WBYR27	ROCKWELL	81474
58WBYR27-1	ROCKWELL	81475
58WBYR28	ROCKWELL	81503
58WBYR28-11	ROCKWELL	81504
58WBYR28-2	ROCKWELL	81499
58WBYR28-3	ROCKWELL	81500
58WBYR28-4	ROCKWELL	81501
58WBYR28-5	ROCKWELL	81502
58WBYR28-7	ROCKWELL	81481
58WBYR28-8	ROCKWELL	81476
58WBYR30	ROCKWELL	81512
58WBYR30-1	ROCKWELL	81507
58WBYR30-2	ROCKWELL	81511
58WBYR30-4	ROCKWELL	81513
58WBYR30-5	ROCKWELL	81510
58WBYR30-6	ROCKWELL	81514
58WBYR31	ROCKWELL	81516
58WBYR31-1	ROCKWELL	81519
58WBYR31-2	ROCKWELL	81521
58WBYR31-3	ROCKWELL	81518
58WBYR31-4	ROCKWELL	81520
58WBYR31-5	ROCKWELL	81517
58WBYR31-6	ROCKWELL	81522
58WBYR31-7	ROCKWELL	81523
58WBYR32	ROCKWELL	81546
58WBYR32-1	ROCKWELL	81553
58WBYR32-10	ROCKWELL	81543
58WBYR32-11	ROCKWELL	81536
58WBYR32-12	ROCKWELL	81547
58WBYR32-13	ROCKWELL	81535
58WBYR32-16	ROCKWELL	81550
58WBYR32-17	ROCKWELL	81554
58WBYR32-2	ROCKWELL	81542
58WBYR32-21	ROCKWELL	81552
58WBYR32-22	ROCKWELL	81531
58WBYR32-3	ROCKWELL	81544
58WBYR32-4	ROCKWELL	81548
58WBYR32-5	ROCKWELL	81538
58WBYR32-7	ROCKWELL	81539
58WBYR32-9	ROCKWELL	81551
58WBYR34-1	ROCKWELL	81558
58WBYR34-2	ROCKWELL	81556
58WBYR34-4	ROCKWELL	81560
58WBYR34-5	ROCKWELL	81557
58WBYR34-8	ROCKWELL	81559
58WBYR35	ROCKWELL	81562
58WBYR35-1	ROCKWELL	81561
58WBYR35-2	ROCKWELL	81564
58WBYR35-3	ROCKWELL	81565
58WBYR35-4	ROCKWELL	81563
58WBYR35-5	ROCKWELL	81566
58WBYR36-1	ROCKWELL	81578
58WBYR36-10	ROCKWELL	81572
58WBYR36-11	ROCKWELL	81577
58WBYR36-12	ROCKWELL	81576
58WBYR36-14	ROCKWELL	81574
58WBYR36-16	ROCKWELL	81580
58WBYR36-3	ROCKWELL	81570
58WBYR36-4	ROCKWELL	81573
58WBYR36-5	ROCKWELL	81575
58WBYR36-6	ROCKWELL	81571
58WBYR36-7	ROCKWELL	81569
58WBYR36-9	ROCKWELL	81579
58WBYR37	ROCKWELL	81581
58WBYR38-1	ROCKWELL	81582
58WBYR38-4	ROCKWELL	81583
58WBYR38-5	ROCKWELL	81585
58WBYR38-6	ROCKWELL	81587
58WBYR38-8	ROCKWELL	81584
58WBYR38-9	ROCKWELL	81586
58WBYR39	ROCKWELL	81592
58WBYR39-1	ROCKWELL	81591
58WBYR39-2	ROCKWELL	81589
58WBYR39-3	ROCKWELL	81588
58WBYR39-4	ROCKWELL	81593
58WBYR39-5	ROCKWELL	81590
58WBYR40	ROCKWELL	81604
58WBYR40-1	ROCKWELL	81602
58WBYR40-11	ROCKWELL	81594
58WBYR40-13	ROCKWELL	81603
58WBYR40-15	ROCKWELL	81600
58WBYR40-3	ROCKWELL	81599
58WBYR40-4	ROCKWELL	81601
58WBYR40-5	ROCKWELL	81597
58WBYR40-6	ROCKWELL	81595
58WBYR40-9	ROCKWELL	81598
58WBYR41-1	ROCKWELL	81607
58WBYR41-2	ROCKWELL	81605
58WBYR41-3	ROCKWELL	81606
58WBYR42	ROCKWELL	81614
58WBYR42-1	ROCKWELL	81609
58WBYR42-2	ROCKWELL	81615
58WBYR42-3	ROCKWELL	81613
58WBYR42-4	ROCKWELL	81612
58WBYR42-5	ROCKWELL	81611
58WBYR42-6	ROCKWELL	81610
58WBYR42-7	ROCKWELL	81608
58WBYR43	ROCKWELL	81616
58WBYR44-1	ROCKWELL	81619
58WBYR44-10	ROCKWELL	81617
58WBYR44-11	ROCKWELL	81621
58WBYR44-12	ROCKWELL	81622
58WBYR44-2	ROCKWELL	81620
58WBYR44-3	ROCKWELL	81624
58WBYR44-4	ROCKWELL	81625
58WBYR44-5	ROCKWELL	81627
58WBYR44-6	ROCKWELL	81623
58WBYR44-8	ROCKWELL	81626
58WBYR44-9	ROCKWELL	81618
58WBYR46	ROCKWELL	81630
58WBYR46-2	ROCKWELL	81629
58WBYR46-4	ROCKWELL	81628
58WBYR47	ROCKWELL	81632
58WBYR47-1	ROCKWELL	81631
58WBYR48	ROCKWELL	81636
58WBYR48-3	ROCKWELL	81633
58WBYR48-6	ROCKWELL	81634
58WBYR48-7	ROCKWELL	81635
58WBYR50	ROCKWELL	81637
58WBYR51	ROCKWELL	81638
58WBYR52	ROCKWELL	81639
58WBYR54	ROCKWELL	81640
58WBYR54-1	ROCKWELL	81642
58WBYR54-2	ROCKWELL	81641
58WBYR55	ROCKWELL	81643
58WBYR56-2	ROCKWELL	81645
58WBYR56-3	ROCKWELL	81644
58WBYR56-4	ROCKWELL	81646
58WBYR58	ROCKWELL	81647
58WBYR61	ROCKWELL	81648
58WBYR62	ROCKWELL	81649
58WBYS22	ROCKWELL	81428
58WBYS22-1	ROCKWELL	85811
58WBYS22-2	ROCKWELL	85810
58WBYS22-3	ROCKWELL	81440
58WBYS22-4	ROCKWELL	81439
58WBYS22-6	ROCKWELL	81441
58WBYS22-7	ROCKWELL	81437
58WBYS22-8	ROCKWELL	81430
58WBYS22-9	ROCKWELL	81429
58WBYS24	ROCKWELL	85812
58WBYS24-1	ROCKWELL	81454
58WBYS24-13	ROCKWELL	81450
58WBYS24-2	ROCKWELL	81453
58WBYS24-3	ROCKWELL	81447
58WBYS24-4	ROCKWELL	81448
58WBYS24-5	ROCKWELL	81455
58WBYS24-7	ROCKWELL	81449
58WBYS24-9	ROCKWELL	81446
58WBYS26-1	ROCKWELL	81468
58WBYS26-4	ROCKWELL	81470
58WBYS28	ROCKWELL	81484
58WBYS28-1	ROCKWELL	81490
58WBYS28-10	ROCKWELL	81487
58WBYS28-14	ROCKWELL	81491
58WBYS28-2	ROCKWELL	81485
58WBYS28-21	ROCKWELL	81479
58WBYS28-27	ROCKWELL	81506
58WBYS28-3	ROCKWELL	81494
58WBYS28-33	ROCKWELL	81495
58WBYS28-5	ROCKWELL	81651
58WBYS28-51	ROCKWELL	81492
58WBYS28-54	ROCKWELL	85815
58WBYS28-59	ROCKWELL	81478
58WBYS28-6	ROCKWELL	81486
58WBYS28-61	ROCKWELL	81489
58WBYS28-62	ROCKWELL	81477
58WBYS28-67	ROCKWELL	81496
58WBYS28-7	ROCKWELL	81482
58WBYS28-71	ROCKWELL	81488
58WBYS28-72	ROCKWELL	81480
58WBYS28-79	ROCKWELL	81497
58WBYS28-8	ROCKWELL	85813
58WBYS28-81	ROCKWELL	85814
58WBYS28-82	ROCKWELL	81498
58WBYS28-9	ROCKWELL	81483
58WBYS30	ROCKWELL	81509
58WBYS30-2	ROCKWELL	81508
58WBYS31-13	ROCKWELL	81529
58WBYS31-15	ROCKWELL	81533
58WBYS31-17	ROCKWELL	81532
58WBYS31-19	ROCKWELL	81524

Part No.	Brand	IDLI Group	Part No.	Brand	IDLI Group	Part No.	Brand	IDLI Group	Part No.	Brand	IDLI Group
58WBYS31-20	ROCKWELL	81530	5996	CASE,J.I.	80101	5P-4	PILOT	88000.25	6-0124	NEAPCO	80225
58WBYS31-21	ROCKWELL	81526	5-9X	SPICER	80124	5P-45	PILOT	88000.70	601500	WESCO	80168
58WBYS31-3	ROCKWELL	81527	59X11467A	DEERE,JOHN	80069	5P-49	PILOT	88000.71	601500	WESCO	88002.17
58WBYS31-32	ROCKWELL	81525	59X11468A	DEERE,JOHN	80219	5P-52	PILOT	88000.72	601515	WESCO	80178
58WBYS31-4	ROCKWELL	81528	5AB100-18-1	PETTI-MULL	80109	5P-72	PILOT	88000.81	601515	WESCO	88002.96
58WBYS32	ROCKWELL	81549	5B423	CATERPILLR	80221	5P-76	PILOT	88000.31	601-565	WESCO	80190
58WBYS32-11	ROCKWELL	81534	5B426	CATERPILLR	80221	5R56	CATERPILLR	88003.26	601577	LEYLAND	80077
58WBYS32-13	ROCKWELL	81541	5B427	CATERPILLR	80221	5R6109	WABCO	80001	601580	WESCO	80186
58WBYS32-21	ROCKWELL	81540	5B428	CATERPILLR	80221	5S5680	CATERPILLR	80234	601581	WESCO	80185
58WBYS32-25	ROCKWELL	81545	5C327	BRILLION	80069	5S5684	CATERPILLR	88001.36	601582	WESCO	80184
58WBYS32-8	ROCKWELL	81537	5D2167	CATERPILLR	80054	5S7216	CATERPILLR	80238	601600	WESCO	80193
58WBYS32-9	ROCKWELL	85816	5D3248	CATERPILLR	80101	5S7219	CATERPILLR	80238	601602	WESCO	80197
58WBYS33	ROCKWELL	81555	5D4091	CATERPILLR	80113	5S723	CATERPILLR	88001.53	601603	WESCO	80199
58WBYS36	ROCKWELL	81567	5D8293	CATERPILLR	80054	5S7579	CATERPILLR	80001	601604	WESCO	80201
58WBYS40-1	ROCKWELL	85817	5D9153	CATERPILLR	80101	5SR55	UJS	80123	601612	WESCO	80191
58WBYS40-2	ROCKWELL	81596	5F1083	CATERPILLR	88001.29	5T1362	CATERPILLR	80172	601613	WESCO	80194
58WBYSM28-11	ROCKWELL	86683	5F2426	CATERPILLR	80070	5WCS30-2	ROCKWELL	88006.10	601-613	WESCO	80195
58WBYSM28-22	ROCKWELL	86682	5F2476	CATERPILLR	80069	5WCS31-7	ROCKWELL	88006.11	601614	WESCO	80192
58WBYSM28-38	ROCKWELL	86685	5F2478	CATERPILLR	80070	5WCS32-37	ROCKWELL	87537	601616	WESCO	80198
58WBYSM28-4	ROCKWELL	86684	5F2551	CATERPILLR	88001.30	5WCS36-76	ROCKWELL	88006.12	601-617	WESCO	80202
58WBYSM40	ROCKWELL	86691	5F2552	CATERPILLR	88001.33	5WCS38-4	ROCKWELL	88006.13	601617	WESCO	80203
58WBYSM40-10	ROCKWELL	86692	5F6332	CATERPILLR	80069	5Y108	FEDERAL	80220	601619	WESCO	80198
58WBYSM40-12	ROCKWELL	86690	5F6332	CATERPILLR	88001.31	5Y114	FEDERAL	80188	601-619	WESCO	80199
58WBYSM40-21	ROCKWELL	86687	5H1816	CATERPILLR	88001.17	5Y115	FEDERAL	80178	601639	WESCO	80203
58WBYSM40-3	ROCKWELL	86686	5H2308	CATERPILLR	80238	5Y294	FEDERAL	80124	601-647	WESCO	80174
58WBYSM40-5	ROCKWELL	86688	5H2312(2)	CATERPILLR	88001.18	5Y339	FEDERAL	80225	601-649	WESCO	80198
58WBYSM40-6	ROCKWELL	86693	5H3754(2)	CATERPILLR	88001.19	5Y500	FEDERAL	80124	601675	WESCO	80187
58WBYSM40-7	ROCKWELL	86695	5H3756	CATERPILLR	80234	5Y528	FEDERAL	80225	601-699	WESCO	80181
58WBYSM40-8	ROCKWELL	86694	5H445	CATERPILLR	80150	5Y529	FEDERAL	80221	601801	WESCO	88002.28
58WBYSM40-9	ROCKWELL	86689	5H625	CATERPILLR	88001.19	5Y530	FEDERAL	80189	6-0185	NEAPCO	80228
58WBYT22-1	ROCKWELL	81431	5H6609(2)	CATERPILLR	88001.20	5Y531	FEDERAL	80178	6-0186	NEAPCO	88001.82
58WBYT24	ROCKWELL	81457	5H6610	CATERPILLR	88001.20	5Y532	FEDERAL	80142	6-0187	NEAPCO	88001.83
58WBYT26	ROCKWELL	81471	5H7701(2)	CATERPILLR	88001.17	5Y5322	FEDERAL	80124	601889	GMC	88002.57
58WBYT28	ROCKWELL	81493	5H800	CATERPILLR	80155	5Y533	FEDERAL	80124	601955	WESCO	80204
58WBYT28-1	ROCKWELL	81505	5J06	BAR-GREENE	80069	5Y7126	FEDERAL	80178	601965	WESCO	80204
58WBYT31-1	ROCKWELL	81515	5J06	BAR-GREENE	80195	5Y7127	FEDERAL	80189	601966	WESCO	80181
58WBYT37-1	ROCKWELL	81568	5K2005	CATERPILLR	88001.37	5Y7172	FEDERAL	80221	602	ALLOY	81237
5-8X	SPICER	80142	5K2006	CATERPILLR	80213	5Y7218	FEDERAL	80142	602	MOPAR	87330
59	WYLIE	80017	5K3171(2)	CATERPILLR	88001.37	5Y7302	FEDERAL	80124	60-2028	NEAPCO	87916
590	AEC	80093	5K5012	CATERPILLR	80156	5Y7302	FEDERAL	80125	60-2040	NEAPCO	87939
590	PRECISION	88012.34	5L1017	OLIVER	80037	6	AEC	88001.70	6021	LAKE SHORE	80104
5-9000	SPICER	80235	5LP9017	OLIVER	80037	60	WYLIE	80023	602-108R91	IHC	80008
5-9000X	SPICER	80235	5LT017	OLIVER	80037	600	ALLOY	84510	60-2229	NEAPCO	87913
5-9001	SPICER	80234	5LT90-17	OLIVER	80037	600	MOPAR	87238	60-2239	NEAPCO	87934
5-9001X	SPICER	80234	5LT90-17	WHITE FARM	80037	6000	ALLIS-CHLM	80234	6-0226	NEAPCO	80225
5-9002	SPICER	88003.09	5M5072(2)	CATERPILLR	88001.34	6000	TRU-CROSS	80193	602462	GMC	88002.58
5-9002X	SPICER	80233	5M5073	CATERPILLR	88001.34	6000A	MOTOR MAST	80223	602795	GMC	88002.47
5-9002X	SPICER	80241	5M7882	CATERPILLR	88001.35	6000B	MOTOR MAST	80224	602795	GMC	88002.48
5-9002X	SPICER	88003.09	5M7962	CATERPILLR	88001.36	6000B	ZELLER	80224	602796	GMC	80092
5-9016X	SPICER	80234	5M800	CATERPILLR	80155	600174	GMC	88002.56	602796	GMC	88001.76
5-90X	REPCO	80093	5NPS28	ROCKWELL	87404	600448	LEYLAND	80124	6-0281	NEAPCO	80227
5-90X	SPICER	80093	5NPS28-1	ROCKWELL	87407	600490	LEYLAND	80124	602813	CHRYSLER	88003.56
591	PRECISION	88002.32	5NPS28-13	ROCKWELL	87415	6005J	BORG-WARNR	82151	602819	CHRYSLER	88003.45
59127J	BORG-WARNR	81948	5NPS28-24	ROCKWELL	87411	60-0627	NEAPCO	87920	602820	CHRYSLER	88003.49
591642	GMC	88002.42	5NPS28-36	ROCKWELL	87406	600628	ROVER	80024	602-900-022	CLARK EQU.	80219
591646	GMC	88002.51	5NPS28-40	ROCKWELL	87426	60-0639	NEAPCO	87946	602-900-112	CLARK EQU.	80001
5916J	BORG-WARNR	88008.24	5NPS32-12	ROCKWELL	87486	60078	BORG-WARNR	88006.36	603	ALLOY	88000.04
5-92X	SPICER	80001	5P-1	PILOT	88000.39	600865	GMC	88002.68	6-03	BYPY	88003.71
5939	ALCO	88002.66	5P-10	PILOT	88000.22	601	ALLOY	88000.01	603	PRECISION	88002.33
5939	ALCO	88002.67	5P-11	PILOT	88000.27	601	MOPAR	87331	6-0308	NEAPCO	80229
5939	ALCO	88002.72	5P-12	NEAPCO	88000.22	60100	BORG-WARNR	88001.15	6030B	KENBAR	88007.58
5-94X	SPICER	80041	5P-13	PILOT	88000.31	601005	WESCO	80178	6036	ANCHOR	87326
5-95-29	SPICER	88014.88	5P-17	PILOT	88000.27	60101	BORG-WARNR	88000.09	6037	ANCHOR	87331
5954-6	KENBAR	88008.31	5P-2	PILOT	81020	60102	BORG-WARNR	88000.09	60370	BORG-WARNR	87919
5954-6	KENBAR	88008.44	5P-22	PILOT	88000.28	60103	BORG-WARNR	88001.08	60371	BORG-WARNR	87945
598A277-1	BEND.WEST.	80100	5P-26	PILOT	88000.53	601-055	WESCO	80178	603710810	CITROEN	80180
598A277-1	WESTINGHSE	80100	5P-27	PILOT	88000.31	6011557	ROVER	80077	6038	ANCHOR	87377
598A277-2	BEND.WEST.	80098	5P-30	PILOT	88000.35	60-1227	NEAPCO	87919	6039	ANCHOR	87383
598A277-2	WESTINGHSE	80098	5P-31	PILOT	88000.38	60-1239	NEAPCO	81218	603D26A(4)	PARSONS	80155
599250	CHRYSLER	88001.60	5P-32	PILOT	88000.38	60-1239	NEAPCO	87945	603D26A(4)	PARSONS	80216

Part No.	Brand	IDLI Group	Part No.	Brand	IDLI Group	Part No.	Brand	IDLI Group	Part No.	Brand	IDLI Group
603D565(4)	PARSONS	80171	6101	ALCO	80195	61148	BORG-WARNR	81277	6-1251X	SPICER	86159
603D565(4)	PARSONS	80216	61011	BORG-WARNR	84510	61149	BORG-WARNR	81271	61252	BORG-WARNR	85055
603D701(4)	PARSONS	80171	6102	AEC	80197	61151	BORG-WARNR	81264	61255	BORG-WARNR	81247
603D701(4)	PARSONS	80216	6102	ALCO	80197	611516	WESCO	80087	61257	BORG-WARNR	81206
604	ALLOY	88000.05	6102	TRU-CROSS	80194	611516	WESCO	80189	612621	FIAT-ALLIS	80067
604	PRECISION	88002.34	6102	TRU-CROSS	80197	611517	WESCO	80221	61268	BORG-WARNR	81346
6040	ANCHOR	87457	6-1021X	SPICER	86187	61154	BORG-WARNR	88014.06	6-126X(8)	SPICER	86125
6-0407	NEAPCO	80226	6-102X(8)	SPICER	86134	61155	BORG-WARNR	81237	6-1271X	SPICER	86184
6041	ANCHOR	87459	6103	AEC	80194	61157	BORG-WARNR	81261	6-127X(8)	SPICER	86130
604373	GMC	88002.46	6103	AEC	80195	61159	BORG-WARNR	81239	6-128X(8)	SPICER	86131
60448	BORG-WARNR	87923	6103	ALCO	80194	61161	BORG-WARNR	81350	613	ALLOY	81268
605	ALLOY	88013.90	6-1031X	SPICER	86184	6-1161X	SPICER	86145	613	MOPAR	87373
605	MOPAR	87382	6-103X(8)	SPICER	86128	6-1161X	SPICER	86193	61312	BORG-WARNR	88014.01
605	PRECISION	88002.35	6104	AEC	80192	61162	BORG-WARNR	81268	61321	BORG-WARNR	81277
605004	CHRYSLER	88003.46	6-104X(8)	SPICER	86129	61163	BORG-WARNR	81246	613333	WHITE	80124
6-05-07	BYPY	88003.79	6-1051X	SPICER	86182	6-1-1641	SPICER	88006.10	613-471R91	IHC	80069
6-05-08	BYPY	88003.86	6-1051X	SPICER	86183	61166	BORG-WARNR	88000.08	613-630C91	IHC	80197
6051-160Y	HARDY SPCR	80142	6-105X(8)	SPICER	86126	61168	BORG-WARNR	81259	613-878C91	IHC	80196
605118	GMC	88002.60	6106	AEC	80198	61169	BORG-WARNR	81304	6-13X	SPICER	80072
6058776	FORD	86077	6106	TRU-CROSS	80198	61170	BORG-WARNR	81218	6-140X	SPICER	86182
605X	NEAPCO	88000.14	610632	EATON	80212	611700	WESCO	80210	6-140X	SPICER	86299
605X	NEAPCO	88000.15	6-106X(8)	SPICER	86135	611704	WESCO	80211	6-1411X	SPICER	86300
606	ALLOY	81333	6107	AEC	80200	611-705	WESCO	80213	6-1411X	SPICER	86302
6-06	BYPY	88004.21	6107	AEC	80202	611708	WESCO	80208	6-141X	SPICER	86300
606	MOPAR	87384	6-1071X	SPICER	86190	611709	WESCO	80209	6-141X	SPICER	86302
606	PRECISION	88002.36	6-107X(8)	SPICER	86138	61171	BORG-WARNR	88014.19	6-14-29	SPICER	88005.48
6060-10	KENBAR	87481	6108	AEC	80199	611-715	WESCO	80213	6-142X	SPICER	86184
6060-10	KENBAR	88007.77	6108	ALCO	80199	6-1171X	SPICER	86145	6-143X	SPICER	86183
60611	BORG-WARNR	88002.42	6108	TRU-CROSS	80199	6-1171X	SPICER	86166	6-144X	SPICER	86132
606678	LEYLAND	80077	6-1081X	SPICER	86174	61172	BORG-WARNR	81342	6-144X	SPICER	86184
606779	GMC	88002.42	6-108X(8)	SPICER	86113	61173	BORG-WARNR	85058	6-1451X	SPICER	86140
606779	GMC	88002.43	6109	AEC	80201	611734	WESCO	80212	6-1451X	SPICER	86192
607	ALLOY	81277	6109	ALCO	80201	61174	BORG-WARNR	81325	6-145X	SPICER	86140
6-07	BYPY	88004.14	6109	TRU-CROSS	80201	61175	BORG-WARNR	81350	61471	BORG-WARNR	81246
607	PRECISION	88002.37	6-1091X	SPICER	86179	611750	WESCO	80205	6-148X	SPICER	86178
60756	BATUYONG	80077	6-109X(8)	SPICER	86120	61176	BORG-WARNR	88013.90	6-148X	SPICER	86246
607-917R91	IHC	80037	6-109X(8)	SPICER	86121	61176	BORG-WARNR	88014.05	6-148X	SPICER	86247
608	ALLOY	81271	611	ALLOY	81342	61177	BORG-WARNR	81268	6149	AEC	80202
608	MOPAR	87461	611	MOPAR	87481	61179	BORG-WARNR	81333	6-149X	SPICER	86179
608	PRECISION	88002.38	611	PRECISION	88002.41	6118	ALCO	88002.60	6-149X	SPICER	86251
608194	GMC	80178	611006	WESCO	80188	61181	BORG-WARNR	81274	6-149X	SPICER	86252
6-0825	NEAPCO	80240	611007	WESCO	80220	61184	BORG-WARNR	84508	6-149X	SPICER	86254
609	ALLOY	81246	6-1101X	SPICER	86191	61185	BORG-WARNR	88014	6-149X	SPICER	86255
609	PRECISION	88002.39	6-110X(8)	SPICER	86139	61190	BORG-WARNR	84510	6-149X	SPICER	86256
609-144R91	IHC	80035	6-1111X	SPICER	86182	611905	GMC	80221	6-149X	SPICER	86257
6-0916	NEAPCO	80234	61112	BORG-WARNR	88000	6-1191X	SPICER	86193	6-149X	SPICER	86282
6-0917	NEAPCO	80241	61113	BORG-WARNR	88000.01	611934	GMC	80210	614A539-1	BEND.WEST.	80153
609515	GMC	80124	61114	BORG-WARNR	88000.02	61197	BORG-WARNR	81273	614A539-1	WESTINGHSE	80153
609531	GMC	88002.48	61115	BORG-WARNR	81237	6-11X	SPICER	80072	6150	AEC	80203
609944	GMC	80092	61116	BORG-WARNR	88000.03	612	ALLOY	81351	6-1501X	SPICER	86251
609948	GMC	80124	61117	BORG-WARNR	88000.04	6-1201X	SPICER	86178	6-1501X	SPICER	86252
60NPS32-24	ROCKWELL	87510	61118	BORG-WARNR	88000.05	612072	GMC	80007	6-1501X	SPICER	86254
60NPS32-25	ROCKWELL	87489	61119	BORG-WARNR	88013.90	612072	GMC	88000.15	6-1501X	SPICER	86255
60NPS32-26	ROCKWELL	87487	6-111X(8)	SPICER	86127	612094	GMC	80221	6-1501X	SPICER	86256
60NPS32-27	ROCKWELL	87521	61120	BORG-WARNR	81333	6-120X	SPICER	86178	6-1501X	SPICER	86257
60NPS32-30	ROCKWELL	87501	6-1121X	SPICER	86145	6-120X	SPICER	86246	6-1501X	SPICER	86282
60NPS32-35	ROCKWELL	87534	6-1121X	SPICER	86166	6-120X	SPICER	86247	6-1-5081	SPICER	88006.13
60R	AEC	80958	61132	BORG-WARNR	81208	61212	BORG-WARNR	88000	6-150X	SPICER	86179
61	WYLIE	80018	61133	BORG-WARNR	81213	61213	BORG-WARNR	88014.05	6-150X	SPICER	86250
610	ALLOY	81218	61134	BORG-WARNR	81222	6-1-2141	SPICER	87537	6-150X	SPICER	86251
610	PRECISION	88002.40	61135	BORG-WARNR	81237	61216	BORG-WARNR	83312	6-150X	SPICER	86252
6100	AEC	80193	61136	BORG-WARNR	81250	61221	BORG-WARNR	88014.06	6-150X	SPICER	86254
6100	ALCO	80193	61137	BORG-WARNR	81261	6-1221X	SPICER	86189	6-150X	SPICER	86255
6-1001X	SPICER	86122	61138	BORG-WARNR	81286	61222	BORG-WARNR	81315	6-150X	SPICER	86256
6-1001X	SPICER	86180	61139	BORG-WARNR	81302	61229	BORG-WARNR	83258	6-150X	SPICER	86257
6-1005X	SPICER	86177	61140	BORG-WARNR	81333	6-122X(8)	SPICER	86136	6-150X	SPICER	86282
6-100X	SPICER	86122	6-1141X	SPICER	86145	6-1231X	SPICER	86181	6151-001Y	HARDY SPCR	80189
6-100X	SPICER	86134	6-1141X	SPICER	86166	6-123X(8)	SPICER	86124	615-155C91	IHC	80196
6-100X(8)	SPICER	86122	6-1-1421	SPICER	88006.12	61248	BORG-WARNR	81203	6-152X	SPICER	86179
6101	AEC	80195	6-1-1461	SPICER	88006.11	6-124X(8)	SPICER	86137	6-152X	SPICER	86251

Part No.	Brand	IDLI Group
6-152X	SPICER	86252
6-152X	SPICER	86254
6-152X	SPICER	86255
6-152X	SPICER	86256
6-152X	SPICER	86257
6-152X	SPICER	86282
6-1531X	SPICER	86179
6-1531X	SPICER	86251
6-1531X	SPICER	86252
6-1531X	SPICER	86254
6-1531X	SPICER	86255
6-1531X	SPICER	86256
6-1531X	SPICER	86257
6-1531X	SPICER	86282
6-15-33	SPICER	88005.49
6-15-73	SPICER	88005.50
6-15-93	SPICER	88005.51
616	ALLOY	81346
6161-12	KENBAR	88007.79
6161-12S	KENBAR	88007.78
6-16-123	SPICER	88005.52
6164	ALCO	88002.64
6164	ALCO	88002.71
616413	GMC	80047
6-16-53	SPICER	88005.53
6-16-93	SPICER	88005.54
61709	ALLIS-CHLM	88000.15
6174	ALCO	88002.56
617512	ALLIS-CHLM	80075
617512	ALLIS-CHLM	80077
617512	FIAT-ALLIS	80077
6175129	FIAT-ALLIS	80077
61768	BORG-WARNR	87920
617703	FIAT-ALLIS	80101
61774	BORG-WARNR	87946
617861	GERLINGER	80047
617862	GERLINGER	80047
6179	ALCO	80200
617909	ALLIS-CHLM	80007
617909	FIAT-ALLIS	80007
6179097	ALLIS-CHLM	80007
6179097	ALLIS-CHLM	88000.15
6179097	FIAT-ALLIS	80007
61805	FIAT-ALLIS	80069
618805	ALLIS-CHLM	80069
618805	FIAT-ALLIS	80070
618805-06	ALLIS-CHLM	80069
618805-6	FIAT-ALLIS	80070
618D	MATHEWS	80037
619-283-114-1	DAIM.RUBE.	80124
619703	ALLIS-CHLM	80100
619703	FIAT-ALLIS	80100
619915	ALLIS-CHLM	80153
619915	ALLIS-CHLM	80155
619915	FIAT-ALLIS	80150
62	WYLIE	80115
6-2001X	SPICER	86069
6-2001X	SPICER	86143
6-200X	MOOG	86069
6-200X	SPICER	86069
6-200X(8)	SPICER	86069
6-2011X	SPICER	86144
6201J	BORG-WARNR	85165
6-201X	SPICER	86070
6-201X(8)	SPICER	86070
6-2021X	SPICER	86145
6-202X	SPICER	86071
6-202X(8)	SPICER	86071
6-203X	SPICER	86072
6-203X(8)	SPICER	86072
6-2041X	SPICER	86160
6-2046	BORG-WARNR	88002.86
6-204X	SPICER	86160
6-204X(8)	SPICER	86160
6-205-1X	SPICER	86088
6-2051X	SPICER	86156
6-2052X	SPICER	86155
6-205X1	SPICER	86088
6-205X(8)	SPICER	86083
6-205X(8)	SPICER	86084
6-206-1X	SPICER	86089
6-2061X	SPICER	86153
62-0651	NEAPCO	88001.13
6-206X1	SPICER	86089
6-206X1	SPICER	86090
6-206X(8)	SPICER	86081
6-207-1X	SPICER	86090
6-2071X	SPICER	86154
6-2074	BORG-WARNR	88002.86
6-207X1	SPICER	86090
6-207X(8)	SPICER	86082
6-2081X	SPICER	86150
6-208X(8)	SPICER	86077
62101	BORG-WARNR	84531
62102	BORG-WARNR	84529
62105	BORG-WARNR	87032
62112	BORG-WARNR	88013.82
62116	BORG-WARNR	83281
62118	BORG-WARNR	83316
62119	BORG-WARNR	83334
62120	BORG-WARNR	83349
62121	BORG-WARNR	83362
6-2121X	SPICER	86161
62122	BORG-WARNR	88014.38
62123	BORG-WARNR	83296
62124	BORG-WARNR	83252
62125	BORG-WARNR	83241
62-1251	NEAPCO	81218
62126	BORG-WARNR	83362
62128	BORG-WARNR	83283
62129	BORG-WARNR	83262
6-212X(8)	SPICER	86095
62130	BORG-WARNR	83357
62131	BORG-WARNR	83240
6-2131X	SPICER	86164
62132	BORG-WARNR	83251
62133	BORG-WARNR	83280
62134	BORG-WARNR	88014.44
62135	BORG-WARNR	83250
62138	BORG-WARNR	83320
6-213X(8)	SPICER	86101
62140	BORG-WARNR	87037
6-2141X	SPICER	86165
62142	BORG-WARNR	83248
62148	BORG-WARNR	83263
62149	BORG-WARNR	83263
6-214X(8)	SPICER	86102
6-2151X	SPICER	86163
62155	BORG-WARNR	83317
62158	BORG-WARNR	88014.41
6-215X(8)	SPICER	86098
6-2161X	SPICER	86170
62-1651	NEAPCO	88001.13
62167	BORG-WARNR	83349
6-216X(8)	SPICER	86107
62173	BORG-WARNR	83320
62175	BORG-WARNR	88014.41
62178	BORG-WARNR	83331
6-217X(8)	SPICER	86108
6-2181X	SPICER	86171
6-218X(8)	SPICER	86109
62190	BORG-WARNR	83252
62191	BORG-WARNR	83264
6-2191X	SPICER	86172
6-219X(8)	SPICER	86110
6-21X	SPICER	80072
6-2201X	SPICER	86169
62-2040	NEAPCO	87938
62-2040	WESCO	87938
62-2052	NEAPCO	87938
6-220X(8)	SPICER	86111
6-2211X	SPICER	86148
62218	BORG-WARNR	83322
6-221X	SPICER	86075
6-221X	SPICER	86148
6-2221X	SPICER	86147
62-2229	NEAPCO	88007.15
62-2239	NEAPCO	87933
62-2251	NEAPCO	87933
62229	BORG-WARNR	83258
622296	WHITE	87461
6-222X	SPICER	86147
6-222X	SPICER	86236
6-2231X	SPICER	86176
6-223X(8)	SPICER	86116
6-224X	SPICER	86117
62251	BORG-WARNR	88013.84
6-2251X	SPICER	86175
6-225X	SPICER	86114
6-225X8	SPICER	86114
6-226X	SPICER	86115
6-2271X	SPICER	86076
6-2271X	SPICER	86149
6-2271X	SPICER	86295
6-2271X	SPICER	86296
6-227X(8)	SPICER	86076
6228	AEC	80199
6228A	MOTOR MAST	80225
6228A	ZELLER	80225
6228AG	MOTOR MAST	80225
6228AG	ZELLER	80225
6229	AEC	80193
6-2301X	SPICER	86145
6-2301X	SPICER	86166
6-2311X	SPICER	86145
6-2311X	SPICER	86166
6232	ALCO	88002.44
6232	ALCO	88002.49
6232-07050	SUBARU	86166
6232-07051	SUBARU	86166
6232-07052	SUBARU	86166
6232-07091	SUBARU	86166
6232-07190	SUBARU	86166
6-2321X	SPICER	86145
6-2321X	SPICER	86166
6-2341X	SPICER	86145
6-2341X	SPICER	86158
6-2341X	SPICER	86193
6-2351X	SPICER	86145
6-2351X	SPICER	86166
6-2351X	SPICER	86193
6-2361X	SPICER	86157
6-2381X	SPICER	86166
6-238X(8)	SPICER	86103
6-2391X	SPICER	86167
6-239X(8)	SPICER	86104
6-2401X	MOOG	86160
6-2401X	SPICER	86105
6-2401X	SPICER	86168
6-240X	SPICER	86105
6-2411X	SPICER	86162
6-241X(8)	SPICER	86096
6-242X(8)	SPICER	86099
6-245X(8)	SPICER	86079
6-2461X	SPICER	86151
6-246X	SPICER	86086
6-2471X	SPICER	86152
6-247X	SPICER	86087
6-2481X	SPICER	86268
625-002	HORWOOD	80198
625-0046	HORWOOD	80238
625-0061	HORWOOD	80213
625-0066	HORWOOD	80198
6-251	NEAPCO	85102
6-254	SPICER	86091
6-254X	SPICER	86091
6-254X	SPICER	86118
6-254X	SPICER	86119
6-255X	SPICER	86092
6-256X	SPICER	86093
6-257X	SPICER	86094
625808C1	IHC	80111
625810C1	IHC	80098
625811C1	IHC	80100
6258549	GMC	88002.37
6-258X	SPICER	86112
625X	NEAPCO	88000.18
6-2601X	SPICER	86237
6-2601X	SPICER	86238
6-2601X	SPICER	86269
6-2601X	SPICER	86270
6-2611X	SPICER	86237
6-2611X	SPICER	86238
6-2611X	SPICER	86269
6-2611X	SPICER	86270
6-262X	SPICER	86149
6-262X	SPICER	86295
6-263X	SPICER	86296
6-2641X	SPICER	86297
6-2641X	SPICER	86298
6-2641X	SPICER	86301
6-264X	SPICER	86297
6-264X	SPICER	86298
6-264X	SPICER	86301
6-265X	SPICER	86298
6-2671X	SPICER	86234
6268535	GMC	80140
6-268X	SPICER	86080
6-268X	SPICER	86085
6-268X	SPICER	86154
6269	MOTOR MAST	80226
6269A	MOTOR MAST	80226
6-269X	SPICER	86234
6-270X	SPICER	86241
6-270X	SPICER	86242
6-270X	SPICER	86273
6-270X	SPICER	86274
6-27-17-2602X	SPICER	87064
6-27-17-5002X	SPICER	87068
6-27-17-6202X	SPICER	87071
6-27-17-7401X	SPICER	87075
6-27-18X	SPICER	87072
6-271X	SPICER	86237
6-271X	SPICER	86238
6-271X	SPICER	86269
6-271X	SPICER	86270
627211C1	IHC	80001
627-253C1	IHC	80098
627254C1	IHC	80100
6-27-278X	SPICER	87063

Part No.	Brand	IDLI Group
6-272X	SPICER	86239
6-272X	SPICER	86240
6-272X	SPICER	86271
6-272X	SPICER	86272
6-27-38X	SPICER	87072
6273940	GMC	80079
6-273X	SPICER	86243
6-273X	SPICER	86244
6-273X	SPICER	86245
6-273X	SPICER	86275
6-273X	SPICER	86276
6-27-48X	SPICER	87066
6-274X	SPICER	86157
6-274X	SPICER	86235
62-751	WORLDPARTS	86143
62-752	WORLDPARTS	86144
62-753	WORLDPARTS	86145
6275AE	MOTOR MAST	80118
6-27-608X	SPICER	87070
6-27-618X	SPICER	87072
6-2761X	SPICER	86242
6-2761X	SPICER	86274
6-276X	SPICER	86261
6-27-708X	SPICER	87066
6-2771X	SPICER	86241
6-2771X	SPICER	86273
6-2771X	SPICER	86274
6-27-7-2602X	SPICER	87063
6-27-7-3604X	SPICER	87066
6-27-7-5511X	SPICER	87070
6-27-7-6331X	SPICER	87072
6-277X	SPICER	86260
6-2781X	SPICER	86239
6-2781X	SPICER	86240
6-2781X	SPICER	86272
627-879C1	IHC	80100
6-278X	SPICER	86263
6-2791X	SPICER	86239
6-2791X	SPICER	86240
6-2791X	SPICER	86271
6-2791X	SPICER	86272
6-27-948X	SPICER	87063
6-279X	SPICER	86104
6-279X	SPICER	86264
6-280X	SPICER	86259
6-281	NEAPCO	80227
6-2811X	SPICER	86245
6-2811X	SPICER	86277
6-28-1281	SPICER	87092
6-28-137	SPICER	85724
6-28-17	SPICER	85724
6-28-197	SPICER	85724
6-281X	SPICER	86262
6-2821X	SPICER	86244
6-2821X	SPICER	86245
6-2821X	SPICER	86276
62824A	HIGHLAND	80142
6-28-277	SPICER	85729
6-282X	SPICER	86266
6-2831X	SPICER	86243
6-2831X	SPICER	86244
6-2831X	SPICER	86245
6-2831X	SPICER	86267
6-2831X	SPICER	86275
6-2831X	SPICER	86276
6-28-347	SPICER	85727
6-28-367	SPICER	85733
6-283X	SPICER	86267
6-284X	SPICER	86243
6-284X	SPICER	86265
6-284X	SPICER	86267
6-284X	SPICER	86275
6-285X	SPICER	86075
6-285X	SPICER	86147
6-285X	SPICER	86236
6-286X	SPICER	86157
6-286X	SPICER	86234
6-28-87	SPICER	85724
629768	CHRYSLER	88003.44
629770	CHRYSLER	88003.43
629815	CATERPILLR	80210
629815	GERLINGER	80215
629815	TOWMOTOR	80210
629816	GERLINGER	80207
629816	TOWMOTOR	80207
629943	ALLIS-CHLM	80210
629943	ALLIS-CHLM	80211
629943	FIAT-ALLIS	80207
629944	ALLIS-CHLM	80210
629944	FIAT-ALLIS	80215
62N170-11	ROCKWELL	86700
62NDCA1	ROCKWELL	86702
62NDCA10	ROCKWELL	86704
62NDCA11	ROCKWELL	86703
62NDCA2	ROCKWELL	86701
62NDCA3	ROCKWELL	86705
62NF10	ROCKWELL	81656
62NF13	ROCKWELL	81654
62NF14	ROCKWELL	81653
62NF17	ROCKWELL	81657
62NF5	ROCKWELL	81652
62NF9	ROCKWELL	81655
62NLS28	ROCKWELL	85087
62NLS28	ROCKWELL	88006.14
62NLS28-1	ROCKWELL	85089
62NLS28-1	ROCKWELL	88006.15
62NLS28-12	ROCKWELL	81683
62NLS28-12	ROCKWELL	88006.16
62NLS28-2	ROCKWELL	85094
62NLS28-22	ROCKWELL	81677
62NLS28-22	ROCKWELL	88006.17
62NLS28-25	ROCKWELL	81678
62NLS28-25	ROCKWELL	88006.18
62NLS28-27	ROCKWELL	81682
62NLS28-27	ROCKWELL	88006.19
62NLS28-28	ROCKWELL	85092
62NLS28-3	ROCKWELL	85093
62NLS28-35	ROCKWELL	85091
62NLS28-4	ROCKWELL	85095
62NLS28-41	ROCKWELL	85090
62NLS28-43	ROCKWELL	85088
62NLS28-8	ROCKWELL	81676
62NLS28-8	ROCKWELL	88006.20
62NLS28-9	ROCKWELL	81675
62NLS28-9	ROCKWELL	88006.21
62NLS32-2	ROCKWELL	85098
62NLS32-3	ROCKWELL	85096
62NLS32-5	ROCKWELL	85097
62NLS32-7	ROCKWELL	85086
62NPA28-4	ROCKWELL	88006.22
62NPS28-1	ROCKWELL	87432
62NPS28-11	ROCKWELL	88006.23
62NPS28-12	ROCKWELL	87434
62NPS28-13	ROCKWELL	87410
62NPS28-13	ROCKWELL	88006.24
62NPS28-14	ROCKWELL	87413
62NPS28-15	ROCKWELL	87417
62NPS28-19	ROCKWELL	88006.25
62NPS28-32	ROCKWELL	87408
62NPS28-4	ROCKWELL	87428
62NPS28-5	ROCKWELL	87423
62NPS28-8	ROCKWELL	87421
62NY42	ROCKWELL	85542
62NY42-1	ROCKWELL	85552
62NY44	ROCKWELL	85543
62NY44-1	ROCKWELL	85544
62NY44-2	ROCKWELL	88006.26
62NY45	ROCKWELL	85546
62NY45-2	ROCKWELL	85545
62NY45-3	ROCKWELL	85540
62NY45-4	ROCKWELL	85553
62NY45-6	ROCKWELL	88006.27
62NY48	ROCKWELL	85555
62NY52	ROCKWELL	85548
62NY52-3	ROCKWELL	85547
62NY52-5	ROCKWELL	85554
62NY52-7	ROCKWELL	85541
62NY53	ROCKWELL	85556
62NY53-1	ROCKWELL	85549
62NY61	ROCKWELL	85551
62NY61-1	ROCKWELL	85550
62NYS22-10	ROCKWELL	81658
62NYS22-2	ROCKWELL	81660
62NYS22-8	ROCKWELL	81659
62NYS24-1	ROCKWELL	81664
62NYS24-14	ROCKWELL	88006.28
62NYS24-15	ROCKWELL	81663
62NYS24-16	ROCKWELL	88006.29
62NYS24-17	ROCKWELL	81662
62NYS24-21	ROCKWELL	85818
62NYS24-5	ROCKWELL	88006.30
62NYS24-8	ROCKWELL	81661
62NYS26	ROCKWELL	81668
62NYS26-13	ROCKWELL	81666
62NYS26-4	ROCKWELL	81665
62NYS26-5	ROCKWELL	81667
62NYS26-7	ROCKWELL	81669
62NYS28-129	ROCKWELL	81674
62NYS28-14	ROCKWELL	81673
62NYS28-190	ROCKWELL	88006.31
62NYS28-193	ROCKWELL	88006.32
62NYS28-21	ROCKWELL	81672
62NYS28-31	ROCKWELL	81679
62NYS28-32	ROCKWELL	81681
62NYS28-59	ROCKWELL	81684
62NYS28-69	ROCKWELL	81680
62NYS28-71	ROCKWELL	81671
62NYS28-91	ROCKWELL	81670
62NYS30-13	ROCKWELL	81685
62NYS30-7	ROCKWELL	81686
62NYS31	ROCKWELL	81693
62NYS31-1	ROCKWELL	81689
62NYS31-10	ROCKWELL	81694
62NYS31-17	ROCKWELL	81692
62NYS31-2	ROCKWELL	81691
62NYS31-20	ROCKWELL	88006.33
62NYS31-22	ROCKWELL	81690
62NYS31-23	ROCKWELL	81687
62NYS31-5	ROCKWELL	81688
62NYS32-10	ROCKWELL	81695
62NYS32-16	ROCKWELL	81699
62NYS32-23	ROCKWELL	81696
62NYS32-3	ROCKWELL	81698
62NYS32-4	ROCKWELL	81697
62NYS32-7	ROCKWELL	81700
62NYS34-9	ROCKWELL	81701
62NYS36-2	ROCKWELL	81702
62NYS40-3	ROCKWELL	81703
62NYS40-5	ROCKWELL	81704
62NYS40-6	ROCKWELL	81705
62NYS48-1	ROCKWELL	81706
62NYSM28-109	ROCKWELL	86712
62NYSM28-13	ROCKWELL	86707
62NYSM28-16	ROCKWELL	86706
62NYSM28-18	ROCKWELL	86717
62NYSM28-2	ROCKWELL	86718
62NYSM28-23	ROCKWELL	86714
62NYSM28-30	ROCKWELL	86719
62NYSM28-37	ROCKWELL	86708
62NYSM28-51	ROCKWELL	86716
62NYSM28-63	ROCKWELL	86699
62NYSM28-64	ROCKWELL	86715
62NYSM28-7	ROCKWELL	86713
62NYSM28-72	ROCKWELL	86709
62NYSM28-89	ROCKWELL	86710
62NYSM28-9	ROCKWELL	86711
62NYSM30-8	ROCKWELL	86720
62NYSM32-13	ROCKWELL	86721
62NYSM32-17	ROCKWELL	86722
62NYSM34	ROCKWELL	86723
63	WYLIE	80020
630-385C1	IHC	80195
630385C91	IHC	80195
630-386C91	IHC	80192
630-389C91	IHC	80213
631-042C91	IHC	80198
63-1254	NEAPCO	87923
63-1266	NEAPCO	87935
631382	ALLIS-CHLM	80044
631383	ALLIS-CHLM	80044
631383	ALLIS-CHLM	88001.98
631383	FIAT-ALLIS	80021
6-3-1461KX	SPICER	85321
6-3-1461X	SPICER	85321
6-3-1491	SPICER	85333
6-3-1491KX	SPICER	85333
6-3-1501	SPICER	88005.57
6-3-1501KX	SPICER	85333
6-3-1501X	SPICER	85333
6-3-1521	SPICER	88005.58
6-3-1521KX	SPICER	85317
6-3-1521X	SPICER	85317
6-3-1531	SPICER	85326
6-3-1531KX	SPICER	85326
6-3-1541KX	SPICER	85326
6-3-1541X	SPICER	85326
6-3-158X	SPICER	85321
631-934C91	IHC	80214
6-3-2171	SPICER	85317
6-3-2171KX	SPICER	85317
6-3-2241	SPICER	85322
6-3-2241KX	SPICER	85322
6-3-2241X	SPICER	85322
6-3-2251KX	SPICER	85334
6-3-2251X	SPICER	85334
6-3-2261	SPICER	88005.59
6-3-2261KX	SPICER	85334
6-3-2261X	SPICER	85334
6-3-2331KX	SPICER	85327
6-3-2331X	SPICER	85327
6-3-2341	SPICER	88005.60
6-3-2341KX	SPICER	85327
6-3-2341X	SPICER	85327
6-3-2401KX	SPICER	85339
6-3-2401X	SPICER	85339
6-3-2551KX	SPICER	85339
6-3-2571KX	SPICER	85336
6-3-2591KX	SPICER	85337
6-3-2631KX	SPICER	85322
6-3-2651KX	SPICER	85335

Part No.	Brand	IDLI Group	Part No.	Brand	IDLI Group	Part No.	Brand	IDLI Group	Part No.	Brand	IDLI Group
6-3-2651X	SPICER	85335	64-0651	NEAPCO	88001.14	6-4-2501	SPICER	83669	6-4-4481X	SPICER	83620
6-3-2671KX	SPICER	85328	6-40-821	SPICER	87579	6-4-2511	SPICER	83665	6-4-4551	SPICER	83700
6-3-2671X	SPICER	85328	6-4-1061	SPICER	83621	64-2529	NEAPCO	88001.06	6-4-4561	SPICER	83682
6-3-268X	SPICER	85326	6-4-1081	SPICER	83670	64-2539	NEAPCO	88001.06	6-4-4581	SPICER	83691
6-3-2761KX	SPICER	85318	6-4-1101	SPICER	83573	64-2551	NEAPCO	88001.06	6-4-4591	SPICER	83663
6-3-28-17	SPICER	85734	6-4-1151	SPICER	83667	64-2629	NEAPCO	85102	6-4-4601	SPICER	83678
6-3-3021KX	SPICER	85330	641152	CATERPILLR	80193	64-2728	NEAPCO	88007.05	6-4-4671X	SPICER	83664
6-3-3-21KX	SPICER	85340	641152	GERLINGER	80193	64-2739	NEAPCO	88006.99	6-4-4681X	SPICER	83669
6-3-368X	SPICER	85333	641152	TOWMOTOR	80193	64-2740	NEAPCO	88007.09	644683	CATERPILLR	80048
6-3-398X	SPICER	85317	6-4-1211	SPICER	83673	64-2751	NEAPCO	88006.99	644683	GERLINGER	80048
6-3-408X	SPICER	85317	6-4-1211	SPICER	88005.63	64-2752	NEAPCO	88007.09	644683	TOWMOTOR	80048
6.3-4-11	SPICER	83715	64-1251	NEAPCO	85126	6-4-3031	SPICER	83671	6-4-4721	SPICER	83599
6.3-4-121	SPICER	83706	6-4-1391	SPICER	83602	6-4-3041X	SPICER	83671	6-4-4731X	SPICER	83599
6.3-4-151	SPICER	83714	6-4-1411	SPICER	83617	6-4-3121	SPICER	83603	6-4-4741X	SPICER	83626
6.3-4-161	SPICER	83711	6-4-1441	SPICER	83618	6-4-3121	SPICER	83604	6-4-4751X	SPICER	83691
6.3-4-231X	SPICER	88005.41	6-4-1471	SPICER	83597	6-4-3141	SPICER	83701	6-4-4761X	SPICER	88005.66
6.3-4-271	SPICER	83713	6-4-1551	SPICER	83626	6-4-3171	SPICER	83622	6-4-4771	SPICER	83683
6342814	BEDFORD	80024	6-4-1611	SPICER	83598	6-4-3181	SPICER	83601	6-4-4831X	SPICER	83583
6342814	GMC	80024	64-1651	NEAPCO	88001.14	6-4-3191	SPICER	83610	6-4-4841X	SPICER	83583
6.3-4-321	SPICER	83707	6-4-1661	SPICER	83612	6-4-3221	SPICER	83617	6-4-4851X	SPICER	83619
634-362C91	IHC	80225	6417	AEC	88001.93	6-4-3231	SPICER	83673	6-4-4861X	SPICER	83582
6.3-4-391	SPICER	83712	6-4-1731	SPICER	83623	6-4-3231	SPICER	88005.63	6-4-4881	SPICER	83593
6.3-4-411	SPICER	83708	6-4-1741	SPICER	83607	6-4-3241	SPICER	83626	6-4-4881	SPICER	88005.68
6.3-4-421	SPICER	83703	6-4-1751	SPICER	83576	6-4-3271	SPICER	83624	6-4-4891	SPICER	83587
6344J	BORG-WARNR	88006.37	6-4-1771	SPICER	83609	6-4-3291	SPICER	83604	6-4-4901	SPICER	83643
6.3-4-501	SPICER	83709	6-4-1781	SPICER	83624	6-4-3311X	SPICER	88005.64	6-4-4911X	SPICER	83593
6.3-4-531	SPICER	83704	64-1828	NEAPCO	85102	6-4-3361	SPICER	83598	6-4-4961	SPICER	83660
6.3-4-541X	SPICER	83705	64-1852	NEAPCO	88001.12	6-4-3371	SPICER	83600	6-4-4971	SPICER	88005.69
6.3-4-91	SPICER	83710	6-4-1911	SPICER	83669	6-4-3391	SPICER	83577	6-4-4981X	SPICER	83653
63502	BORG-WARNR	81218	6-4-1971	SPICER	83668	6-4-3461X	SPICER	83671	64500	BORG-WARNR	87913
635090	GERLINGER	80235	6-4-1971	SPICER	88005.56	6-4-3481	SPICER	83614	6-4-5031	SPICER	83595
635090	TOWMOTOR	80235	6-4-1981	SPICER	83607	6-4-3501	SPICER	83677	6-4-5041X	SPICER	83595
6-351	NEAPCO	88001.14	6-4-1991	SPICER	83673	64353	BORG-WARNR	88001.07	6-4-5041X	SPICER	88005.70
63569	CHECKER	80072	6-4-2011	SPICER	83673	6-4-3581X	SPICER	83610	6-4-5051X	SPICER	83643
63569	CHECKER	80077	6-4-2011	SPICER	88005.63	6436100	EATON	80001	645053	ALLIS-CHLM	80007
635873	FWD	80001	6-4-2021	SPICER	83612	6436101	EATON	80001	645053	ALLIS-CHLM	88000.15
635900	FWD	80220	6-4-2021	SPICER	83614	6-4-3611	SPICER	83695	645053	FIAT-ALLIS	80007
63619	ELG.SWEEP.	80037	64-2040	NEAPCO	88007.09	6-4-3621	SPICER	88005.65	6-4-5061X	SPICER	83660
63624	ELG.SWEEP.	80069	6-4-2081	SPICER	83573	643633	CATERPILLR	80196	6-4-5071X	SPICER	83678
63625	ELG.SWEEP.	80069	6-42-1081	SPICER	87575	643633	GERLINGER	80196	6-451	NEAPCO	88001.12
63626	ELG.SWEEP.	80161	6-42-1081	SPICER	88005.62	643633	TOWMOTOR	80196	6-4-5101X	SPICER	83644
63635	ELG.SWEEP.	80070	6-42-1101	SPICER	87575	6-4-3721	SPICER	83608	645110	ALLIS-CHLM	80007
636362	CHRYSLER	88003.56	6-4-2131	SPICER	83621	6-4-3791	SPICER	83638	645110	ALLIS-CHLM	88000.15
6371722	BEDFORD	80023	6-4-2141	SPICER	83670	6440025	ETNYRE	80049	645110	FIAT-ALLIS	80007
6373796	WHITE	80133	6-4-2151	SPICER	83667	644007	ETNYRE	80163	6-4-5111X	SPICER	83594
6374032	WHITE	80008	6421591	EATON	88013.90	6440077	ETNYRE	80134	6-4-5121X	SPICER	83587
639511	FIAT-ALLIS	80223	6-4-2161	SPICER	83605	6440103	ETNYRE	80069	6-4-5131X	SPICER	83588
6397J	BORG-WARNR	84652	6-4-2171	SPICER	83671	644055	GERLINGER	80215	6-4-5141	SPICER	83611
639937	CHRYSLER	88003.48	6421711	EATON	80035	644055	TOWMOTOR	80210	6-4-5151X	SPICER	83611
639973	CHRYSLER	88003.48	6-4-2181	SPICER	83661	644056	GERLINGER	80214	6-4-5161X	SPICER	83595
639973	GMC	88003.56	6-4-2191	SPICER	83632	644056	TOWMOTOR	80218	6-4-5251X	SPICER	83700
63RU141	MACK	80221	6-4-2211	SPICER	83602	6-4-4091	SPICER	83583	6-4-5351	SPICER	83698
64005	BORG-WARNR	84923	64-2229	NEAPCO	88001.06	6-4-4131	SPICER	83620	6-4-5451	SPICER	83681
6-40-101	SPICER	87581	6-4-2231	SPICER	83617	644-137C91	IHC	80190	6-4-5461X	SPICER	88005.71
6-40-141	SPICER	87579	64-2239	NEAPCO	85114	6-4-4141	SPICER	83582	6-4-5481X	SPICER	83663
6-40-161	SPICER	87580	6-4-2241	SPICER	83618	6-4-4151	SPICER	83619	6-4-5501	SPICER	83654
6-40-181	SPICER	87656	642249	CATERPILLR	80218	6-4-4171X	SPICER	83661	6-4-5511X	SPICER	83655
6-40-311	SPICER	87601	642249	GERLINGER	80214	6-4-4181X	SPICER	83606	6-4-5521	SPICER	83646
6-40-38X	SPICER	87656	642249	TOWMOTOR	80218	644189C	GERLINGER	80047	6-4-5531X	SPICER	83647
6-40-401	SPICER	87591	64-2251	NEAPCO	85114	6-4-4191X	SPICER	83674	6-4-5541	SPICER	83637
6-40-451	SPICER	87596	6-4-2261	SPICER	88005.77	6-4-4291X	SPICER	83627	6-4-5551X	SPICER	83636
6-40-461	SPICER	87591	6-4-2271	SPICER	83597	6-4-4301X	SPICER	83577	645-573R91	IHC	80077
6-40-461	SPICER	88005.61	6-4-2291	SPICER	83626	6-4-4311	SPICER	83578	645574R91	IHC	80124
6-40-51	SPICER	87579	6-4-2311	SPICER	83598	6-4-4341	SPICER	83664	645-574R92	IHC	80124
6-40-521	SPICER	87596	6-4-2361	SPICER	83629	6-4-4351X	SPICER	83617	645-578R91	IHC	80077
6-40-521	SPICER	87598	6-4-2371	SPICER	83623	6-4-4361X	SPICER	83616	6-4-5581X	SPICER	83573
6-40-541	SPICER	87591	6-4-2381	SPICER	83576	644363C92	IHC	80182	6-4-5591	SPICER	83699
6-40-541	SPICER	87592	6-4-2391	SPICER	83609	6-4-4391X	SPICER	83625	6-4-5601X	SPICER	83699
6-40-61	SPICER	87581	6-4-2411	SPICER	83624	6-4-4411	SPICER	88005.67	6-4-5661	SPICER	88005.72
64-0651	NEAPCO	85114	6-4-2441	SPICER	83666	6-4-4421	SPICER	83693	6-4-5661X	SPICER	83590

Part No.	Brand	IDLI Group
6-4-5701X	SPICER	83667
6-4-5711	SPICER	83680
64575	CHECKER	87331
64575	CHECKER	88008.39
6-4-5791	SPICER	83658
64580	CHECKER	80077
6-4-5821	SPICER	83657
64600	BORG-WARNR	87934
64602	BORG-WARNR	87935
6462	ALCO	88002.58
6462C28	PACIF.CAR	80100
6-4-6451	SPICER	88005.73
6-4-6481	SPICER	88005.74
646649	GERLINGER	80212
647098	ALLIS-CHLM	80100
647098	FIAT-ALLIS	80100
647869R92	IHC	80178
647-869R93	IHC	80178
648366	ALLIS-CHLM	80035
648366	FIAT-ALLIS	80035
648366-3	ALLIS-CHLM	80035
648366-3	FIAT-ALLIS	80035
648397	EATON	80155
648460	STUDEBAKER	80112
6484649	STUDEBAKER	80112
6-4-851	SPICER	83573
649052	GERLINGER	80037
649502	GERLINGER	80037
649502	TOWMOTOR	80037
65	WYLIE	80225
650079	GERLINGER	80035
6-1/2-53-11	SPICER	87557
65-0651	NEAPCO	88001.12
65101	BORG-WARNR	88000.67
65103	BORG-WARNR	88000.57
65107	BORG-WARNR	88000.85
65111	BORG-WARNR	88000.20
65111	NEAPCO	88000.21
65112	BORG-WARNR	88000.22
65113	BORG-WARNR	88000.25
65114	BORG-WARNR	88000.28
65115	BORG-WARNR	88000.30
65116	BORG-WARNR	88000.35
65117	BORG-WARNR	88000.37
65118	BORG-WARNR	88000.38
65119	BORG-WARNR	88000.43
65120	BORG-WARNR	88000.45
65121	BORG-WARNR	88000.58
6.5-1-211	SPICER	88005.78
65123	BORG-WARNR	88000.21
65124	BORG-WARNR	88000.31
65125	BORG-WARNR	88000.39
65129	BORG-WARNR	88000.99
65132	BORG-WARNR	88000.22
65134	BORG-WARNR	88000.49
65136	BORG-WARNR	88000.52
65138	BORG-WARNR	88000.60
65139	BORG-WARNR	88000.53
65141	BORG-WARNR	88000.50
65145	BORG-WARNR	88000.51
65146	BORG-WARNR	88000.23
65147	BORG-WARNR	88000.86
65147	BORG-WARNR	88000.88
65148	BORG-WARNR	88000.87
65149	BORG-WARNR	88000.57
65150	BORG-WARNR	88000.23
65150	BORG-WARNR	88000.57
65154	BORG-WARNR	88000.32
65156	BORG-WARNR	88000.55
65157	BORG-WARNR	88000.80

Part No.	Brand	IDLI Group
65158	BORG-WARNR	88000.48
65159	BORG-WARNR	88000.92
65162	BORG-WARNR	88000.22
65163	BORG-WARNR	88000.91
65164	BORG-WARNR	88000.33
65165	BORG-WARNR	88000.40
65-1651	NEAPCO	88001.12
65166	BORG-WARNR	88000.64
65167	BORG-WARNR	88000.22
65167	BORG-WARNR	88000.26
65168	BORG-WARNR	88001
65169	BORG-WARNR	88000.94
65171	BORG-WARNR	88000.43
65172	BORG-WARNR	88000.47
65173	BORG-WARNR	88000.95
65175	BORG-WARNR	88000.93
65176	BORG-WARNR	88000.86
65177	BORG-WARNR	88000.63
65178	BORG-WARNR	88000.89
65180	BORG-WARNR	88000.66
65182	BORG-WARNR	88000.44
65183	BORG-WARNR	88000.79
65184	BORG-WARNR	88000.81
65-1840	NEAPCO	85112
65185	BORG-WARNR	88000.82
65186	BORG-WARNR	88000.83
65187	BORG-WARNR	88000.51
65189	BORG-WARNR	88000.45
65191	BORG-WARNR	88000.34
65193	BORG-WARNR	88000.46
65195	BORG-WARNR	88000.59
65196	BORG-WARNR	88000.36
65197	BORG-WARNR	88000.35
65199	BORG-WARNR	88000.84
65-2028	NEAPCO	85126
65-2040	NEAPCO	88007.08
65217	BORG-WARNR	88000.90
65219	BORG-WARNR	88000.22
65-2229	NEAPCO	88007.15
65-2239	NEAPCO	88007.15
65261	BORG-WARNR	88000.22
65262	BORG-WARNR	88000.22
65-2629	NEAPCO	88007.14
65263	BORG-WARNR	88000.27
65-2639	NEAPCO	88007.16
65264	BORG-WARNR	88000.52
65264	BORG-WARNR	88000.81
65268	BORG-WARNR	88000.44
65-2751	NEAPCO	88006.99
6.5-27-9-3813X	SPICER	87116
6.5-27-9-5013X	SPICER	87117
6.5-27-9-6213X	SPICER	87118
6.5-28-117	SPICER	85736
65-2828	NEAPCO	88007.04
65-2840	WESCO	88007.08
65-2851	NEAPCO	81209
6531	ALCO	88002.48
6.5-3-1351KX	SPICER	85343
6.5-3-1351X	SPICER	85343
6.5-3-1371KX	SPICER	85345
6.5-3-1371X	SPICER	85345
6.5-3-1431KX	SPICER	85342
6.5-3-1431X	SPICER	85342
6-53-151	SPICER	87482
6-53-201	SPICER	87483
6-53-311	SPICER	87564
65337	BORG-WARNR	88000.28
65338	BORG-WARNR	88000.29
65340R92	IHC	80158
65342	BORG-WARNR	88000.55

Part No.	Brand	IDLI Group
65-343R91	IHC	80049
65343R92	IHC	80049
65344	BORG-WARNR	88000.19
65344R91	IHC	80053
65344R92	IHC	80044
65344R92	IHC	80053
65-345R91	IHC	80049
65345R92	IHC	80049
65346R91	IHC	80099
65346R91	IHC	80101
65-346R92	IHC	80101
65347	BORG-WARNR	88000.85
65-347R91	IHC	80112
65347R92	IHC	80110
65348	BORG-WARNR	88000.36
65-348R91	IHC	80109
65348R92	IHC	80109
65349R91	IHC	80146
65350R91	IHC	80158
65-350R92	IHC	80158
65351K93	IHC	80193
65351R91	IHC	80193
65-351R93	IHC	80193
65354	BORG-WARNR	88000.98
65355	BORG-WARNR	88000.43
65355	BORG-WARNR	88000.99
65356	BORG-WARNR	88000.61
65357	BORG-WARNR	88000.41
653650	UNIC	80225
65371	BORG-WARNR	88000.68
65372	BORG-WARNR	88000.70
65373	BORG-WARNR	88000.71
65374	BORG-WARNR	88000.72
65375	BORG-WARNR	88000.73
65376	BORG-WARNR	88000.68
65376	BORG-WARNR	88000.69
653811	FIAT-ALLIS	80001
65384	BORG-WARNR	88000.74
65385	BORG-WARNR	88000.75
65386	BORG-WARNR	88000.74
65387	BORG-WARNR	88000.73
65388	BORG-WARNR	88000.72
65389	BORG-WARNR	88000.72
6.5-40-151	SPICER	87644
6.5-40-161	SPICER	87648
6.5-40-191	SPICER	87649
6.5-40-201	SPICER	87645
65418	BORG-WARNR	88000.71
6.5-4-1891	SPICER	83731
6.5-4-1901	SPICER	83747
6.5-4-1911	SPICER	83730
6.5-4-1921	SPICER	83718
6.5-4-1961	SPICER	83743
6.5-4-1971	SPICER	83729
6.5-4-1981	SPICER	83732
6.5-4-1991	SPICER	83740
6.5-4-1991	SPICER	83741
6.5-4-2001	SPICER	83717
6.5-4-2011	SPICER	83744
6.5-4-2051	SPICER	83716
65421	BORG-WARNR	88000.76
6.5-4-2171	SPICER	83746
65422	BORG-WARNR	88000.77
6.5-4-2271	SPICER	83729
6.5-4-2271X	SPICER	83729
6.5-4-2331X	SPICER	83730
6.5-4-2341X	SPICER	83718
6.5-4-2351X	SPICER	83717
6.5-4-2361X	SPICER	83741
6.5-4-2401X	SPICER	83728

Part No.	Brand	IDLI Group
6.5-4-2441	SPICER	83727
6.5-4-2451X	SPICER	83727
6.5-4-2471	SPICER	83720
6.5-4-2481X	SPICER	83720
6.5-4-2491X	SPICER	83720
6.5-4-2491X	SPICER	83721
6.5-4-2531X	SPICER	83731
6.5-4-2551X	SPICER	83730
6.5-4-2701	SPICER	83736
6.5-4-2711X	SPICER	83736
6.5-4-2751	SPICER	83748
6.5-4-2761X	SPICER	83748
6.5-4-2781	SPICER	83724
6.5-4-2851X	SPICER	83718
6.5-4-2931X	SPICER	83747
6.5-4-3011	SPICER	83732
6.5-4-3021	SPICER	83745
6.5-4-3171X	SPICER	83725
65432	BORG-WARNR	88000.45
65448	BORG-WARNR	88000.54
6546	ALLIS-CHLM	80048
6546	ALLIS-CHLM	80158
6546	FIAT-ALLIS	80048
65472	BORG-WARNR	88000.96
65478	BORG-WARNR	88000.58
65499	BORG-WARNR	88000.56
655105	GMC	80221
65528	BORG-WARNR	88001.01
6.5-53-91	SPICER	87567
656222R92	IHC	80104
656223R91	IHC	80044
656223R91	IHC	80049
656-223R92	IHC	80044
656224R91	IHC	80070
656225R91	IHC	80037
656-226R91	IHC	80021
656762	VOLVO	80024
65706R91	IHC	80148
65-706R92	IHC	80148
6.5-82-461-1	SPICER	88005.80
6.5-82-461-2	SPICER	88005.81
6.5-82-461-4	SPICER	88005.82
6.5-82-461-8	SPICER	88005.83
6582753	ALLIS-CHLM	80023
6582753	ALLIS-CHLM	80024
6582753	FIAT-ALLIS	80023
65848R91	IHC	80148
658559	GMC	80179
65919	BORG-WARNR	88000.62
659-262R91	IHC	80035
6595	FIAT-ALLIS	80109
659773	VOLVO	80023
659973	GMC	80217
65AR	AEC	80905
65K	WESCO	80003
65MU26	MACK	80077
65R	AEC	80951
6-5X	SPICER	80072
66	WYLIE	80227
660	PRECISION	88002.42
66000	BORG-WARNR	85087
6-6007	BORG-WARNR	85093
6-6009	BORG-WARNR	81675
6-6010	BORG-WARNR	88006.20
660-116R91	IHC	80069
6-6019	BORG-WARNR	88006.20
660195	STUDEBAKER	80122
660196	STUDEBAKER	80138
6-6023	BORG-WARNR	88006.20
6-6025	BORG-WARNR	85095

Part No.	Brand	IDLI Group	Part No.	Brand	IDLI Group	Part No.	Brand	IDLI Group	Part No.	Brand	IDLI Group
6-6026	BORG-WARNR	88006.15	66233	BORG-WARNR	88000.37	67037	BORG-WARNR	88006.99	67R	AEC	80953
6-6032	BORG-WARNR	88006.15	66251	BORG-WARNR	88000.17	67038	BORG-WARNR	88006.99	68	WYLIE	80006
6-6035	BORG-WARNR	88006.20	66251	BORG-WARNR	88000.18	67039	BORG-WARNR	85126	680	PRECISION	88002.60
6-6036	BORG-WARNR	88006.15	66256	BORG-WARNR	88000.38	67040	BORG-WARNR	85102	6-8002	BORG-WARNR	85165
6-6037	BORG-WARNR	81677	66-2629	NEAPCO	88006.37	67042	BORG-WARNR	88001.12	6-8003	BORG-WARNR	85157
6-6038	BORG-WARNR	88006.16	66-2651	NEAPCO	88006.97	67043	BORG-WARNR	88007.05	6-8003	BORG-WARNR	85158
6-6039	BORG-WARNR	81677	663	PRECISION	88002.45	67044	BORG-WARNR	88007.09	6-8012	BORG-WARNR	85168
6-6047	BORG-WARNR	85094	663-120R91	IHC	80001	67046	BORG-WARNR	88001.06	6-8015	BORG-WARNR	85165
6605	ALCO	88002.70	6.631.371	VOLVO	80171	6-7047	BORG-WARNR	85114	6-8020	BORG-WARNR	85159
6-6057	BORG-WARNR	85095	6.631.390	VOLVO	80197	67047	BORG-WARNR	85114	6-8021	BORG-WARNR	85164
66-0639	NEAPCO	88006.36	6.631.405	VOLVO	80212	6-7048	BORG-WARNR	85114	6-8023	BORG-WARNR	85164
6-6066	BORG-WARNR	81675	6.631.406	VOLVO	80210	67048	BORG-WARNR	85114	6-8026	BORG-WARNR	85159
6-6067	BORG-WARNR	85093	663349	CHRYSLER	88001.60	67049	BORG-WARNR	85102	6-8036	BORG-WARNR	85165
6-6068	BORG-WARNR	85094	663366	CHRYSLER	88001.62	67054	BORG-WARNR	88001.12	6-8041	BORG-WARNR	85167
6-6069	BORG-WARNR	85095	6.638.385	VOLVO	80180	67056	BORG-WARNR	81209	6-8045	BORG-WARNR	85169
6-6072	BORG-WARNR	88006.20	664	PRECISION	88002.46	67059	BORG-WARNR	85112	6-8054	BORG-WARNR	85169
6-6082	BORG-WARNR	88006.20	6642	ALCO	88002.42	67061	BORG-WARNR	85126	6805J	BORG-WARNR	85543
6-6084	BORG-WARNR	88006.18	6646	ALCO	88002.51	67062	BORG-WARNR	88007.08	68-0651	NEAPCO	85102
661	PRECISION	88002.43	6646	ALCO	88002.58	67063	BORG-WARNR	88007.15	681	PRECISION	88002.61
66-1227	NEAPCO	85107	665	PRECISION	88002.47	67064	BORG-WARNR	88007.15	681093	GMC	80024
66-1239	NEAPCO	85102	665-000-0	GLAENZER	80189	67065	BORG-WARNR	88007.14	6-8113	NEAPCO	80238
66128	BORG-WARNR	87916	665-002-0	GLAENZER	80189	67066	BORG-WARNR	88007.16	681156	CATERPILLR	80210
661372R91	IHC	80001	665-015-2	GLAENZER	80189	6-7073	BORG-WARNR	85107	681157	CATERPILLR	80218
661-372R91	IHC	80003	665-019-0	GLAENZER	80189	6-7075	BORG-WARNR	85123	681157	GERLINGER	80212
661372R91	IHC	80188	665-137M91	MASS.FERG.	80037	6-7077	BORG-WARNR	85101	68-1828	NEAPCO	85114
66140	BORG-WARNR	87939	665611	FIAT-ALLIS	80070	670774	STUDEBAKER	80142	68-1840	NEAPCO	85114
66141	BORG-WARNR	88000.25	666	PRECISION	88002.48	6707744	STUDEBAKER	80142	682	PRECISION	88002.62
661677	GERLINGER	80195	666184	VOLVO	80023	6-7079	BORG-WARNR	85114	682027	STUDEBAKER	80124
661-677R91	IHC	80001	66642	BORG-WARNR	88001.13	6-7080	BORG-WARNR	85124	6-8205	NEAPCO	80231
661805	GMC	80218	666878	FIAT-ALLIS	80001	6-7086	BORG-WARNR	85108	6-8206	NEAPCO	80232
66187	BORG-WARNR	88000.30	667	PRECISION	88002.49	6-7093	BORG-WARNR	85131	6-8207	NEAPCO	80237
662	PRECISION	88002.44	667467	ALLIS-CHLM	80172	6-7095	BORG-WARNR	85141	6-8208	NEAPCO	80236
662-010	DORMAN	88008.35	667-496R91	IHC	80037	6-7096	BORG-WARNR	85124	6-8209	NEAPCO	80238
662-012	DORMAN	87326	667501	GMC	80186	6-7099	BORG-WARNR	85132	6-82-1061-3	SPICER	88005.75
662-013	DORMAN	87238	667502	GMC	80184	6-7107	BORG-WARNR	88006.64	6-82-1091-1	SPICER	87087
662-013	DORMAN	88007.58	667-558R91	IHC	80037	67131	GERLINGER	80168	6-82-1091-4	SPICER	87095
662-014	DORMAN	87331	668	PRECISION	88002.50	67131	GERLINGER	80171	6-82-1171-12	SPICER	87080
662-015	DORMAN	87331	6680J	BORG-WARNR	82036	6717	ALCO	88002.63	6-82-1171-5	SPICER	88005.76
662-015	DORMAN	88008.39	668-172M91	MASS.FERG.	80033	6-7177	BORG-WARNR	88006.38	68-2239	NEAPCO	81848
662-016	DORMAN	87377	668504	CHRYSLER	88003.56	672037	VOLVO	80023	6824381	ALLIS-CHLM	80190
662-016	DORMAN	88008.41	668504	GMC	88003.56	672037-9	VOLVO	80024	6824381	FIAT-ALLIS	80190
662-017	DORMAN	87383	668526	STUDEBAKER	80138	67216	BORG-WARNR	83090	68-2551	NEAPCO	88007.06
662-017	DORMAN	88008.42	668594	CHRYSLER	88003.56	67228	BORG-WARNR	83049	682580	ALLIS-CHLM	80172
662-018	DORMAN	87457	669	PRECISION	88002.51	67232	BORG-WARNR	83082	682-597R91	IHC	80145
662-018	DORMAN	88008.31	6691	TERRAIN	80136	67238	BORG-WARNR	83129	68-2639	NEAPCO	85101
662-018	DORMAN	88008.44	669-564R91	IHC	80035	673	PRECISION	88002.53	6828A	MOTOR MAST	80225
662-019	DORMAN	87459	6-6X	SPICER	80072	6-73-109	SPICER	88013.36	6-82-971-1X	SPICER	87108
662-019	DORMAN	88008.45	67	WYLIE	80012	673626	CATERPILLR	80049	683	PRECISION	88002.63
662-020	DORMAN	87481	670	PRECISION	88002.52	673627	CATERPILLR	80048	683385	BMC	80024
662-020	DORMAN	88007.77	6-7000	BORG-WARNR	85114	674	PRECISION	88002.54	68343	BORG-WARNR	88015.39
662-021	DORMAN	87570	67002	BORG-WARNR	88006.36	674156	STUDEBAKER	80124	68385	BMC	80024
662-021	DORMAN	88007.79	6700311	EATON	80238	675	AEC	80091	68385	LEYLAND	80024
662-022	DORMAN	87285	67004	BORG-WARNR	85107	675	PRECISION	88002.55	684	PRECISION	88002.64
662-022	DORMAN	88008.48	67005	BORG-WARNR	85102	675082	GMC	80221	685	PRECISION	88002.65
662-023	DORMAN	87337	67013	BORG-WARNR	81848	675X	NEAPCO	88000.14	6-8500	NEAPCO	80223
662-023	DORMAN	88008.51	6-7014	BORG-WARNR	88006.37	675X	NEAPCO	88000.15	68516	BORG-WARNR	88000.17
662-023	DORMAN	88008.52	67018	BORG-WARNR	85102	675X1	NEAPCO	88000.16	6.852.586	VOLVO	80142
662-024	DORMAN	87399	67019	BORG-WARNR	88006.97	676	PRECISION	88002.56	6.852.612	VOLVO	80180
662-024	DORMAN	88008.59	67020	BORG-WARNR	88006.97	677	PRECISION	88002.57	68561	BORG-WARNR	88013.85
662-025	DORMAN	87373	67022	BORG-WARNR	85114	6779	ALCO	88002.43	686	PRECISION	88002.66
662-025	DORMAN	88008.43	6-7022	BORG-WARNR	85114	6779	ALCO	88002.59	6861043	VOLVO	80217
662-026	DORMAN	87325	67023	BORG-WARNR	85114	6779	ALCO	88002.62	686343	CATERPILLR	80238
662-026	DORMAN	88007.59	6-7023	BORG-WARNR	85114	678	PRECISION	88002.58	686704	DENNIS	80180
662-027	DORMAN	87326	67026	BORG-WARNR	88006.97	678510	GMC	80225	687	PRECISION	88002.67
662-027	DORMAN	87331	67026	TRW	88006.97	678510B	GMC	80225	688	PRECISION	88002.68
662044	GMC	88007.79	67027	BORG-WARNR	88006.97	678512	GMC	80225	688016	GMC	80221
66210	BORG-WARNR	88000.35	67027	TRW	88006.97	679	PRECISION	88002.59	68810	BORG-WARNR	88015.43
6.621.656	VOLVO	80180	67032	BORG-WARNR	85101	679937	STUDEBAKER	80138	6889	ALCO	88002.57
6621J	BORG-WARNR	81759	67036	BORG-WARNR	85114	67AR	AEC	80906	689	PRECISION	88002.69
66-2239	NEAPCO	81848	6-7036	BORG-WARNR	85114	67AR	AEC	80909	6892J	BORG-WARNR	88015.19

Part No.	Brand	IDLI Group	Part No.	Brand	IDLI Group	Part No.	Brand	IDLI Group	Part No.	Brand	IDLI Group
689434	WHITE	87382	6951149	EATON	80213	6A7013	BORG-WARNR	81848	6H9294	CATERPILLR	88001.50
69	WYLIE	80004	695149	EATON	80213	6A7014	BORG-WARNR	88006.37	6K316	CATERPILLR	80156
6-9001	BORG-WARNR	85180	6951586	EATON	80234	6A7018	BORG-WARNR	85102	6K855	CATERPILLR	80234
6-9001	BORG-WARNR	85182	6951594	EATON	80224	6A7022	BORG-WARNR	85114	6K885	CATERPILLR	80234
6-9005	BORG-WARNR	85179	6951976	EATON	80198	6A7023	BORG-WARNR	85114	6L1516	CATERPILLR	80101
6-9005	BORG-WARNR	85181	6952210	EATON	80234	6A7032	BORG-WARNR	85101	6L6449	CATERPILLR	80054
6-9008	BORG-WARNR	85173	6952215	EATON	80198	6A7036	BORG-WARNR	85114	6M9762	CATERPILLR	80234
6-9013	BORG-WARNR	85180	695230	GMC	80221	6A7039	BORG-WARNR	85126	6N	AEC	80188
6-9013	BORG-WARNR	85182	695232	GMC	80221	6A7040	BORG-WARNR	85102	6N	WESCO	80001
6-9014	NEAPCO	80235	6952664	EATON	80163	6A7047	BORG-WARNR	85114	6N120-26	WESCO	81025
6-9015	NEAPCO	80234	6952684	EATON	80156	6A7048	BORG-WARNR	85114	6N123	WESCO	88014.52
6-9015	NEAPCO	88002.32	6-95-29	SPICER	88014.87	6A7049	BORG-WARNR	85102	6N123-5	WESCO	88014.52
6-9016	NEAPCO	80234	6953995	EATON	80238	6A7059	BORG-WARNR	85112	6N124	WESCO	81020
6-9017	NEAPCO	80233	6954207	EATON	80215	6A7061	BORG-WARNR	85126	6N134	WESCO	81034
690251	DENNIS	80142	6954995	EATON	80238	6A7073	BORG-WARNR	85107	6N134-5	WESCO	81034
6-9026	NEAPCO	80233	69-550R91	IHC	88009.51	6A7075	BORG-WARNR	85123	6N140-30	WESCO	81125
6-9033	BORG-WARNR	85184	6955338	EATON	80213	6A7077	BORG-WARNR	85101	6N140-32	WESCO	81060
6-9033	BORG-WARNR	88007.17	69557K93	IHC	80231	6A7079	BORG-WARNR	85114	6N1434-37	WESCO	81054
6-9034	BORG-WARNR	85185	69557R91	IHC	80231	6A7080	BORG-WARNR	85124	6N144-2	WESCO	81054
69101	BORG-WARNR	88014.34	69557R92	IHC	80231	6A7086	BORG-WARNR	85108	6N144-5	WESCO	81054
69102	BORG-WARNR	84517	69-557R93	IHC	80231	6A7093	BORG-WARNR	85131	6N154	WESCO	81089
69105	BORG-WARNR	83020	69-558R91	IHC	80173	6A7095	BORG-WARNR	85141	6N160-26	WESCO	81125
69115	BORG-WARNR	83018	69558R92	IHC	80173	6A7096	BORG-WARNR	85124	6N160-27	WESCO	81125
69116	BORG-WARNR	83024	69-559R91	IHC	80159	6A7099	BORG-WARNR	85132	6N1634-37	WESCO	81109
69117	BORG-WARNR	88014.28	69559R92	IHC	80159	6A7107	BORG-WARNR	88006.64	6N164-5	WESCO	88000.31
69118	BORG-WARNR	83062	69560R91	IHC	80111	6A8002	BORG-WARNR	85165	6N164-7	WESCO	88000.31
69119	BORG-WARNR	83082	69-561R91	IHC	80054	6A8003	BORG-WARNR	85157	6N174	WESCO	81136
69120	BORG-WARNR	83108	69573R93	IHC	80240	6A8003	BORG-WARNR	85158	6N1845-37	WESCO	81153
69127	BORG-WARNR	88014.27	6958321	GMC	80003	6A8012	BORG-WARNR	85168	6N185-5	WESCO	88000.36
69130	BORG-WARNR	88014.26	695909	GMC	80189	6A8015	BORG-WARNR	85165	6N200-7	WESCO	88000.55
69132	BORG-WARNR	83116	695990	GMC	80189	6A8020	BORG-WARNR	85159	6N2045-37	WESCO	81175
69147	BORG-WARNR	83014	696	PRECISION	88002.73	6A8021	BORG-WARNR	85159	6N205-5	WESCO	88000.55
69149	BORG-WARNR	83038	697	PRECISION	88002.74	6A8021	BORG-WARNR	85164	6N220-20	WESCO	84504
69154	BORG-WARNR	83125	697320	CHRYSLER	88003.50	6A8023	BORG-WARNR	85164	6N220TQD	WESCO	88014.06
69175	BORG-WARNR	83086	697-689R91	IHC	80189	6A8026	BORG-WARNR	85159	6N48	WESCO	88001.13
69187	BORG-WARNR	83056	697-690R91	IHC	80221	6A8036	BORG-WARNR	85165	6N48F	WESCO	88001.14
692	PRECISION	88002.70	697769R91	IHC	80221	6A8041	BORG-WARNR	85167	6N48M	WESCO	88001.12
69211	BORG-WARNR	83031	698	PRECISION	88002.75	6A8045	BORG-WARNR	85169	6N8	AEC	80001
69211	BORG-WARNR	83047	698-997R91	IHC	80235	6A8054	BORG-WARNR	85169	6N8	AEC	80003
69211	BORG-WARNR	83057	699	PRECISION	88002.76	6A9001	BORG-WARNR	85180	6N-NB1	WESCO	88001.15
69214	BORG-WARNR	83084	6A2046	BORG-WARNR	88002.86	6A9001	BORG-WARNR	85182	6NPS32	ROCKWELL	87499
69214	BORG-WARNR	83086	6A2046	BORG-WARNR	88002.87	6A9005	BORG-WARNR	85179	6NPS32-10	ROCKWELL	87495
69216	BORG-WARNR	83090	6A6000	BORG-WARNR	85087	6A9005	BORG-WARNR	85181	6NPS32-15	ROCKWELL	87491
692296	WHITE	87458	6A6007	BORG-WARNR	85093	6A9008	BORG-WARNR	85173	6NPS32-29	ROCKWELL	87493
692297	WHITE	87458	6A6009	BORG-WARNR	81675	6A9013	BORG-WARNR	85180	6NPS32-31	ROCKWELL	87492
692313	WHITE	87382	6A6010	BORG-WARNR	88006.20	6A9013	BORG-WARNR	85182	6NPS32-32	ROCKWELL	87494
69236	BORG-WARNR	83127	6A6019	BORG-WARNR	88006.20	6A9033	BORG-WARNR	85184	6NPS32-38	ROCKWELL	87516
69240	GERLINGER	80149	6A6023	BORG-WARNR	88006.20	6A9034	BORG-WARNR	85185	6NS	WESCO	80003
69240	GERLINGER	80156	6A6025	BORG-WARNR	85095	6C4	CRANE CAR.	80215	6R	G & G MFG	80001
69302	BORG-WARNR	83015	6A6026	BORG-WARNR	88006.15	6C5	CRANE CAR.	80240	6R	WESCO	80001
6932287	EATON	80198	6A6032	BORG-WARNR	88006.15	6D1-613	WESCO	80195	6S6902	CATERPILLR	80047
6933772	GMC	80235	6A6034	BORG-WARNR	81682	6D2529	CATERPILLR	80054	6S706	CATERPILLR	88001.93
6933776	GMC	80234	6A6035	BORG-WARNR	88006.20	6F2426(5)	CATERPILLR	88003.10	6Y1452KK	GAR WOOD	80134
693864	DENNIS	80124	6A6036	BORG-WARNR	88006.15	6F4339	CATERPILLR	88001.29	6Y1452YY	GAR WOOD	80140
694	PRECISION	88002.71	6A6037	BORG-WARNR	81677	6F4341(5)	CATERPILLR	88003.10	6Y1920	GAR WOOD	80035
694297	WHITE	87327	6A6038	BORG-WARNR	88006.16	6F4342(5)	CATERPILLR	88003.10	6Y53K	GAR WOOD	80133
694968	GMC	80142	6A6039	BORG-WARNR	81677	6F4343(5)	CATERPILLR	88003.10	70	WYLIE	80010
694968	GMC	80221	6A6047	BORG-WARNR	85094	6F5321	CATERPILLR	88001.30	7000	TRU-CROSS	80210
694969	GMC	80141	6A6057	BORG-WARNR	85095	6F5322	CATERPILLR	88001.33	70004020	FIAT-ALLIS	80069
694973	GMC	88013.74	6A6066	BORG-WARNR	81675	6F7160	CATERPILLR	80149	70004690	FIAT-ALLIS	80101
694981	GMC	88013.68	6A6067	BORG-WARNR	85093	6F9190	CATERPILLR	80149	70006546	FIAT-ALLIS	80048
694987	GMC	87373	6A6068	BORG-WARNR	85094	6F9192	CATERPILLR	88003.54	70006595	FIAT-ALLIS	80109
695	PRECISION	88002.72	6A6069	BORG-WARNR	85095	6F9741	CATERPILLR	80069	70007209	FIAT-ALLIS	80101
6950025	EATON	80172	6A6072	BORG-WARNR	88006.20	6G0976	CATERPILLR	80021	70007211	FIAT-ALLIS	80100
6950058	EATON	80198	6A6082	BORG-WARNR	88006.20	6G976	CATERPILLR	80021	70008392	FIAT-ALLIS	80047
6950172	EATON	80213	6A6084	BORG-WARNR	88006.18	6H1261	CATERPILLR	80109	70011020	FIAT-ALLIS	80070
6950199	EATON	80234	6A7000	BORG-WARNR	85114	6H1262	CATERPILLR	80150	70013113	FIAT-ALLIS	80069
6950390	EATON	80037	6A7002	BORG-WARNR	88006.36	6H2570	CATERPILLR	80238	70013809	FIAT-ALLIS	80178
6950840	EATON	80161	6A7004	BORG-WARNR	85107	6H2577	CATERPILLR	80213	70020268	FIAT-ALLIS	80044
6951148	EATON	80213	6A7005	BORG-WARNR	85102	6H2579	CATERPILLR	80238	70021862	FIAT-ALLIS	80188

Part No.	Brand	IDLI Group
70022807	FIAT-ALLIS	80111
70024508	FIAT-ALLIS	80178
70026726	FIAT-ALLIS	80069
70026727	FIAT-ALLIS	80069
70026728	FIAT-ALLIS	80069
70028573	FIAT-ALLIS	80178
70028862	FIAT-ALLIS	80178
70034121	FIAT-ALLIS	80101
70038633	FIAT-ALLIS	80220
70039559	FIAT-ALLIS	80220
70040372	FIAT-ALLIS	80235
70040377	FIAT-ALLIS	80235
70044128	FIAT-ALLIS	80188
70044617	FIAT-ALLIS	80219
70053120	FIAT-ALLIS	80001
70053120-2	FIAT-ALLIS	80001
70057614	FIAT-ALLIS	80149
70057736	FIAT-ALLIS	80210
70058735	FIAT-ALLIS	80215
70058736	FIAT-ALLIS	80210
70069953	FIAT-ALLIS	80223
70069962(2)	FIAT-ALLIS	80223
70069962-9(2)	FIAT-ALLIS	80223
70073372	FIAT-ALLIS	80142
70075614	FIAT-ALLIS	80149
70076456	FIAT-ALLIS	80168
700799-11	JOY MFG.	80069
70080132	FIAT-ALLIS	80213
70080470	FIAT-ALLIS	80124
70080470-8	FIAT-ALLIS	80124
70081058	FIAT-ALLIS	80001
70081058-0	FIAT-ALLIS	80001
70081559	FIAT-ALLIS	80142
70081559-7	FIAT-ALLIS	80142
7-0082	NEAPCO	88002.37
70082073	FIAT-ALLIS	88003.26
70082744	FIAT-ALLIS	80044
70083541	FIAT-ALLIS	80215
70083616	FIAT-ALLIS	80215
70083616-3	FIAT-ALLIS	80215
70084328	FIAT-ALLIS	80195
70084328-4	FIAT-ALLIS	80195
70084395	FIAT-ALLIS	80202
70084435	FIAT-ALLIS	80202
70084435-7	FIAT-ALLIS	80202
70084537	FIAT-ALLIS	80162
70084545	FIAT-ALLIS	80174
70084545-3	FIAT-ALLIS	80174
70085918	FIAT-ALLIS	80193
70085918-1	FIAT-ALLIS	80193
70086284	FIAT-ALLIS	80149
70086284-7	FIAT-ALLIS	80149
70086747	FIAT-ALLIS	80165
70086747-3	FIAT-ALLIS	80165
70086752	FIAT-ALLIS	80164
70086752-3	FIAT-ALLIS	80164
70088665	FIAT-ALLIS	80210
70090311(2)	FIAT-ALLIS	80223
70090311-2(2)	FIAT-ALLIS	80223
7009J	BORG-WARNR	81151
7012	ALCO	88003.56
70148	DROTT	80048
70148M1	DROTT	80048
701639	GMC	80227
70178183	FIAT-ALLIS	80240
70178183-0	FIAT-ALLIS	80240
70178323	FIAT-ALLIS	80242
70178323-2	FIAT-ALLIS	80242
70178368	FIAT-ALLIS	80238
70178368-7	FIAT-ALLIS	80238
70178461	FIAT-ALLIS	80048
70178461-0	FIAT-ALLIS	80048
70178544	FIAT-ALLIS	88003.26
70178544-3	FIAT-ALLIS	88003.26
70178554	FIAT-ALLIS	80102
70178823	FIAT-ALLIS	80232
70179151	FIAT-ALLIS	80224
70179502	FIAT-ALLIS	80216
70179507	FIAT-ALLIS	80224
70179571	FIAT-ALLIS	80103
70179571-5	FIAT-ALLIS	80103
70179704	FIAT-ALLIS	80224
70179904	FIAT-ALLIS	80224
70179909	FIAT-ALLIS	80224
70180132	FIAT-ALLIS	80213
70180132-3	FIAT-ALLIS	80213
70180140	FIAT-ALLIS	80224
70180140-6	FIAT-ALLIS	80224
70185263	FIAT-ALLIS	80234
70185263-1	FIAT-ALLIS	80234
70187964	FIAT-ALLIS	80234
702129	GMC	80221
70224890	FIAT-ALLIS	80001
70224890-4	FIAT-ALLIS	80001
702375C91	IHC	80142
70255608	FIAT-ALLIS	80140
702952C91	IHC	80234
703325	GMC	80007
70339981	FIAT-ALLIS	80035
7-04	BYPY	88003.72
7048	CHAIN BELT	80134
70498504	FIAT-ALLIS	80047
70498505	FIAT-ALLIS	80046
7050	ALCO	88003.42
7-05-07	BYPY	88003.81
7051-017Y	HARDY SPCR	80221
70517038	FIAT-ALLIS	80133
70517039	FIAT-ALLIS	80133
70517308	FIAT-ALLIS	80033
70517309	FIAT-ALLIS	80008
70517310	FIAT-ALLIS	80008
70517311	FIAT-ALLIS	80037
70517312	FIAT-ALLIS	80069
70518038	FIAT-ALLIS	80133
70518039	FIAT-ALLIS	80133
70518040	FIAT-ALLIS	80008
70521720	FIAT-ALLIS	80037
70524521	FIAT-ALLIS	80008
70526299	FIAT-ALLIS	80037
70530538	FIAT-ALLIS	80069
70530538-8	FIAT-ALLIS	80069
70531220	FIAT-ALLIS	80037
70534235	FIAT-ALLIS	80003
70542862	FIAT-ALLIS	80035
70546945	FIAT-ALLIS	80035
7-06	BYPY	88004.22
7060	ALCO	88003.45
70612621	FIAT-ALLIS	80067
70612621-3	FIAT-ALLIS	80067
70617512	FIAT-ALLIS	80077
70617512-9	FIAT-ALLIS	80077
70617703	FIAT-ALLIS	80101
70617909	FIAT-ALLIS	80007
70617909-7	FIAT-ALLIS	80007
70618805	FIAT-ALLIS	80070
70618805-6	FIAT-ALLIS	80070
70619703	FIAT-ALLIS	80100
70619703-2	FIAT-ALLIS	80100
70619915	FIAT-ALLIS	80150
70619915-2	FIAT-ALLIS	80150
70629943	FIAT-ALLIS	80207
70629943-2	FIAT-ALLIS	80207
70629944	FIAT-ALLIS	80215
70631383	FIAT-ALLIS	80021
70631383-7	FIAT-ALLIS	80021
70639511	FIAT-ALLIS	80223
70645053	FIAT-ALLIS	80007
70645053-0	FIAT-ALLIS	80007
70645110	FIAT-ALLIS	80007
70645110-8	FIAT-ALLIS	80007
70647098	FIAT-ALLIS	80100
70648366	FIAT-ALLIS	80035
70648366-3	FIAT-ALLIS	80035
70653811	FIAT-ALLIS	80001
70658275	FIAT-ALLIS	80023
70665314	FIAT-ALLIS	80234
70665611	FIAT-ALLIS	80070
70665611-0	FIAT-ALLIS	80070
70666878	FIAT-ALLIS	80001
70667467	FIAT-ALLIS	80172
70680958	FIAT-ALLIS	80223
70680968	FIAT-ALLIS	80195
70680970	FIAT-ALLIS	80193
70680970-1	FIAT-ALLIS	80193
70680996	FIAT-ALLIS	80168
70680996	FIAT-ALLIS	80216
70680998	FIAT-ALLIS	80167
70681008	FIAT-ALLIS	80215
70681014	FIAT-ALLIS	80168
70681016	FIAT-ALLIS	80109
70681016	FIAT-ALLIS	80150
70681016	FIAT-ALLIS	80156
70681024	FIAT-ALLIS	80207
70681049	FIAT-ALLIS	80222
70682580	FIAT-ALLIS	80172
70682636	FIAT-ALLIS	80149
70682636-6	FIAT-ALLIS	80109
7068345	BEDFORD	80023
7068345	HOLDEN	80023
70687517	FIAT-ALLIS	80224
70689077	FIAT-ALLIS	80196
7-069-000158	GROVE CR.	80212
7-069-000182	GROVE CR.	80070
70692644	FIAT-ALLIS	80214
7-07	BYPY	88004.15
7075	ALCO	88003.46
7075A	MOTOR MAST	80231
7075A	ZELLER	80231
7075B	MOTOR MAST	80238
7075B	ZELLER	80238
7075E	MOTOR MAST	80230
7075F	MOTOR MAST	80232
7075F	ZELLER	80232
7075G	MOTOR MAST	80237
7075G	ZELLER	80237
7075H	MOTOR MAST	80235
7075H	ZELLER	80235
7075HA	MOTOR MAST	80233
7075HA	ZELLER	80233
7075HA	ZELLER	88003.09
7075HB	MOTOR MAST	80234
7075HB	ZELLER	80234
7075HBG	MOTOR MAST	80234
7075HC	MOTOR MAST	80234
7075HC	ZELLER	80234
7075HD	MOTOR MAST	80233
7075HD	ZELLER	80233
7075HD	ZELLER	80241
7075J	MOTOR MAST	80236
7075J	ZELLER	80236
7075K	MOTOR MAST	80238
707680	MIXERMOBL	80233
707747	GMC	80178
7080	ALCO	88003.49
7080421	CHRYSLER	80023
7080428	CHRYSLER	80023
70823456	FIAT-ALLIS	80037
7083616	FIAT-ALLIS	80210
70836163	FIAT-ALLIS	80210
708390	MIXERMOBL	80149
708390	SCOOPMOBLE	80109
708390	SCOOPMOBLE	80149
708392	MIXERMOBL	80196
708394	MIXERMOBL	80193
708395	MIXERMOBL	80215
708398	MIXERMOBL	80196
7084449	BEDFORD	80023
7084449	BEDFORD	80024
7084449	GMC	80024
7084449	HOLDEN	80023
708656	MIXERMOBL	80214
708686	MIXERMOBL	80235
708686	SCOOPMOBLE	88012.34
7087A	MOTOR MAST	80245
7087B	MOTOR MAST	80245
7087C	MOTOR MAST	88003.40
7087D	MOTOR MAST	80245
7087T	MOTOR MAST	88001.93
7088368	BEDFORD	80180
7088368	GMC	80180
7088368	HOLDEN	80180
70883687	BEDFORD	80180
7088370	BEDFORD	80180
7088979	BMC	80023
7090700	BEDFORD	80142
7090700	GMC	80142
7090700	HOLDEN	80142
709715	CATERPILLR	88000.15
70NPS36-18	ROCKWELL	87556
70NPS40-10	ROCKWELL	87577
71	WYLIE	80011
710	AEC	80205
710	PILOT	80032
710	PRECISION	80032
7100	AEC	80210
7100	ALCO	80210
7100-1A	DETROIT	88002.69
71007023	FIAT-ALLIS	80142
7101	AEC	80211
7101	ALCO	80211
7102	AEC	80212
7102	ALCO	80212
7103	AEC	80208
7104	AEC	80209
7105	AEC	80213
7105	TRU-CROSS	80213
710A	AEC	80205
710G	PILOT	80032
7111J	BORG-WARNR	81832
71126463	FIAT-ALLIS	80215
71134986	FIAT-ALLIS	80215
7123	CHAIN BELT	80069
7125-2	DETROIT	80034
7125-2	SPICER	80034
7132J	BORG-WARNR	81816
7133J	BORG-WARNR	88006.55
7140-2A	DETROIT	88002.69
7140J	BORG-WARNR	88006.92
7141J	BORG-WARNR	88006.45
714251-2	DETROIT	80034

Part No.	Brand	IDLI Group
71425-2	ALCO	80034
71425-2	DETROIT	80034
71425-2	SPICER	80034
7142J	BORG-WARNR	82115
7145J	BORG-WARNR	81934
714743	MIXERMOBL	80233
715	ALLOY	83129
715	PILOT	80030
715	PILOT	88002.77
715	PRECISION	88002.77
715	TEDDY TORQ	80034
7151-001Y	HARDY SPCR	80221
7152	DAIMLER	80124
7152	JAGUAR	80124
7156132	BEDFORD	80023
7156132	HOLDEN	80023
7156340	BEDFORD	80023
7156340	HOLDEN	80023
7156385	GMC	80180
7156385	HOLDEN	80180
715G	PILOT	80031
715G	PILOT	88002.77
716	PRECISION	88002.78
7161004	LILLISTON	80035
7169	TAYLOR	80008
7169	TAYLOR	80133
7169J	BORG-WARNR	85823
717	PRECISION	88002.79
7171J	BORG-WARNR	82133
7174-010	TAYLOR	80037
7175-003	TAYLOR	80069
7.175.30.06.13.060	GELENKWEL.	88002.77
7.175.30.06-1R	REPCO	88002.77
7176J	BORG-WARNR	88007.18
71783C1	IHC	80212
7178684	BEDFORD	80180
7178684	GMC	80180
7178684	HOLDEN	80180
7178934	BEDFORD	80142
7178934	GMC	80142
7178934	HOLDEN	80142
7178938	GMC	80142
718	PRECISION	88002.80
7181J	BORG-WARNR	82065
7184J	BORG-WARNR	81668
7.185.10.06.17.060	GELENKWEL.	88002.77
7.185.10.06-1R	REPCO	88002.77
7.185.10.06-2R	REPCO	88002.77
719	PRECISION	88002.81
7190326	BEDFORD	80041
7190326	GMC	80041
7190326	HOLDEN	80041
719413	GMC	87373
719413	GMC	88008.43
7199C	DIAMOND T	80178
7199C	WHITE	80178
7199F	DIAMOND R	80079
7199F	DIAMOND T	80189
7199F	WHITE	80189
7199H	DIAMOND R	80124
7199H	DIAMOND T	80124
7199H	WHITE	80124
7199J	DIAMOND R	80221
7199J	DIAMOND T	80221
7199J	WHITE	80221
7199M	DIAMOND R	80225
7199M	DIAMOND T	80225
7199M	WHITE	80225
7199W	DIAMOND T	80142
7199X	DIAMOND R	80178
7200	ALCO	80210
720001	WESCO	88000.87
720002	WESCO	88000.86
720-002	WESCO	88000.88
720-003	WESCO	88000.86
720004	WESCO	88000.98
720005	WESCO	88000.69
720005	WESCO	88000.79
720006	WESCO	88000.80
720007	WESCO	88000.89
720008	WESCO	88001.01
720009	WESCO	88000.81
720010	WESCO	88000.82
720011	WESCO	88000.90
720012	WESCO	88000.83
720013	WESCO	88001
720-014	WESCO	88001
720015	WESCO	88000.93
720016	WESCO	88000.93
720017	WESCO	88000.95
720018	WESCO	88000.96
720019	WESCO	88000.84
7200-20	DETROIT	88002.68
720020	WESCO	88000.94
720021	WESCO	88000.85
720-022	WESCO	88000.51
720023	WESCO	88000.66
720161	GMC	87373
7202	TRU-CROSS	80212
720424	LONG MFG.	80069
7206	ALCO	88003.44
7206809	BEDFORD	80189
7206809	GMC	80189
7206809	HOLDEN	80189
72-0750	NEAPCO	88015.43
7209	ALLIS-CHLM	80099
7209	ALLIS-CHLM	80101
7209	FIAT-ALLIS	80101
7211	ALLIS-CHLM	80100
7211	FIAT-ALLIS	80100
72-120-126	WESCO	88000.86
72-120-26	WESCO	88000.88
72-1231	WESCO	88000.79
72-140-28	WESCO	88000.89
72-1441	WESCO	88000.80
72-1441	WESCO	88000.81
7-21453	CURTISS-WR	88001.80
7-21453	FMC	80221
7-21453	WAYNE	80221
7-21453(5)	CURTISS-WR	88001.80
72-1541	WESCO	88000.82
72-160-24	WESCO	88000.90
72-1641	WESCO	88000.83
72-180-23	WESCO	88000.92
72251-2	DETROIT	80056
72251-3	DETROIT	80064
7225-2	DETROIT	80064
7225-4	DETROIT	80056
7225-4	DETROIT	80057
7225-5	DETROIT	80060
7225-5	DETROIT	80063
7225-5	SPICER	80060
7225-5	SPICER	88001.59
7225A	MOTOR MAST	80227
7225AG	MOTOR MAST	80227
7225AG	ZELLER	80227
7226	CHAIN BELT	80178
723	TEDDY TORQ	80064
723133	WABCO	80100
7232-22020	SUBARU	86166
72372959	FIAT-ALLIS	80101
723741	ADAMS-LET.	80035
723741	WABCO	80035
7238J	BORG-WARNR	88014.52
7244A	MOTOR MAST	80228
7244AG	MOTOR MAST	80228
7244B	MOTOR MAST	88001.82
7244C	MOTOR MAST	88001.83
7244DA	MOTOR MAST	80229
7244DC	MOTOR MAST	87481
7244G	MOTOR MAST	80228
7245	AEC	80206
7245	AEC	80212
725	TEDDY TORQ	80064
72521-2	DETROIT	80057
725251-2	ALCO	80056
725251-2	DETROIT	80056
725251-2	DETROIT	80057
725251-3	DETROIT	80064
725251-3	DETROIT	80065
72525-2	ALCO	80057
72525-2	ALCO	88001.58
72525-2	DETROIT	80056
72525-2	DETROIT	80057
72525-2	SPICER	80056
72525-3	DETROIT	80064
72525-3	SPICER	80060
725944	WABCO	80048
726	TEDDY TORQ	80064
72620	CHRYSLER	88001.70
726251-12	DETROIT	80061
726251-2	DETROIT	80060
726251-2	DETROIT	80061
726251-2	DETROIT	80063
726251-3	DETROIT	80061
72625-2	ALCO	80064
72625-2	ALCO	88001.63
72625-2	ALCO	88001.68
72625-2	DETROIT	80064
72625-2	DETROIT	80065
72625-2	DETROIT	88001.63
72625-2	SPICER	80064
726252-2	DETROIT	80060
726252-2	DETROIT	80061
726252-2	DETROIT	80063
72625-3	ALCO	80060
72625-3	ALCO	88001.59
72625-3	DETROIT	80060
72625-3	DETROIT	80061
72625-3	SPICER	80060
72625-3	SPICER	88001.59
72625-4	DETROIT	80063
72625-4T	DETROIT	80061
72625-4T	DETROIT	80062
72625-5	DETROIT	80132
72625-6	DETROIT	80061
72652-2	DETROIT	80063
727	HERCULES	88002.43
728	HERCULES	88002.43
729	HERCULES	88002.42
7299J	BORG-WARNR	81838
72DB4635AIA	FORD	80023
72DB4635BIA	FORD	80180
72N170	ROCKWELL	86732
72N170-14	ROCKWELL	86740
72N170-9	ROCKWELL	86729
72N171-1	ROCKWELL	86734
72N172-1	ROCKWELL	86735
72N173-1	ROCKWELL	86736
72N174-1	ROCKWELL	86731
72N175-1	ROCKWELL	86730
72N176-1	ROCKWELL	86733
72N177-1	ROCKWELL	86739
72NDCA1	ROCKWELL	86737
72NDCA2	ROCKWELL	86738
72NDCA3	ROCKWELL	88006.34
72NF1	ROCKWELL	84649
72NF11	ROCKWELL	84663
72NF13	ROCKWELL	84651
72NF14	ROCKWELL	84671
72NF15	ROCKWELL	88006.35
72NF18	ROCKWELL	84662
72NF2	ROCKWELL	84672
72NF24	ROCKWELL	84653
72NF25	ROCKWELL	84668
72NF26	ROCKWELL	84655
72NF27	ROCKWELL	84664
72NF28	ROCKWELL	84661
72NF3	ROCKWELL	84654
72NF30	ROCKWELL	84657
72NF31	ROCKWELL	84670
72NF32	ROCKWELL	84669
72NF34	ROCKWELL	84665
72NF4	ROCKWELL	84666
72NF41	ROCKWELL	84673
72NF50	ROCKWELL	84654
72NF51	ROCKWELL	84659
72NF52	ROCKWELL	84658
72NF53	ROCKWELL	84660
72NF56	ROCKWELL	84650
72NF6	ROCKWELL	84652
72NF8	ROCKWELL	84656
72NF9	ROCKWELL	84667
72NLS31-1	ROCKWELL	85099
72NLS32	ROCKWELL	85114
72NLS32-1	ROCKWELL	85112
72NLS32-113	ROCKWELL	85103
72NLS32-124	ROCKWELL	85122
72NLS32-129	ROCKWELL	85134
72NLS32-13	ROCKWELL	85102
72NLS32-15	ROCKWELL	85126
72NLS32-18	ROCKWELL	85114
72NLS32-19	ROCKWELL	85129
72NLS32-2	ROCKWELL	85123
72NLS32-20	ROCKWELL	85131
72NLS32-21	ROCKWELL	85127
72NLS32-22	ROCKWELL	85133
72NLS32-24	ROCKWELL	85107
72NLS32-25	ROCKWELL	88006.36
72NLS32-27	ROCKWELL	85121
72NLS32-29	ROCKWELL	85101
72NLS32-33	ROCKWELL	85111
72NLS32-34	ROCKWELL	85124
72NLS32-35	ROCKWELL	85132
72NLS32-4	ROCKWELL	85104
72NLS32-50	ROCKWELL	88006.37
72NLS32-51	ROCKWELL	85128
72NLS32-52	ROCKWELL	85110
72NLS32-55	ROCKWELL	85115
72NLS32-6	ROCKWELL	85130
72NLS32-61	ROCKWELL	85108
72NLS32-7	ROCKWELL	85109
72NLS32-71	ROCKWELL	85125
72NLS32-72	ROCKWELL	85120
72NLS32-74	ROCKWELL	85119
72NLS32-77	ROCKWELL	85113
72NLS32-78	ROCKWELL	85106

Part No.	Brand	IDLI Group	Part No.	Brand	IDLI Group	Part No.	Brand	IDLI Group	Part No.	Brand	IDLI Group
72NLS32-8	ROCKWELL	85105	72NY53-3	ROCKWELL	85575	72NYS24-16	ROCKWELL	88006.42	72NYS28-62	ROCKWELL	88006.50
72NLS32-80	ROCKWELL	85116	72NY53-7	ROCKWELL	85557	72NYS24-22	ROCKWELL	81716	72NYS28-63	ROCKWELL	81771
72NLS32-86	ROCKWELL	85118	72NY60-3	ROCKWELL	85566	72NYS24-23	ROCKWELL	81717	72NYS28-65	ROCKWELL	81751
72NLS32-86DC	ROCKWELL	88006.38	72NY61	ROCKWELL	85567	72NYS24-24	ROCKWELL	85819	72NYS28-7	ROCKWELL	81772
72NLS32-89	ROCKWELL	85100	72NY61-1	ROCKWELL	85569	72NYS24-3	ROCKWELL	81710	72NYS28-71	ROCKWELL	81764
72NLS32-90	ROCKWELL	85117	72NY61-10	ROCKWELL	85570	72NYS24-5	ROCKWELL	81714	72NYS28-79	ROCKWELL	81797
72NLS40-1	ROCKWELL	85144	72NY61-3	ROCKWELL	85584	72NYS24-7	ROCKWELL	81718	72NYS28-79	ROCKWELL	88006.51
72NLS40-10	ROCKWELL	85143	72NY61-4	ROCKWELL	85568	72NYS26	ROCKWELL	81725	72NYS28-8	ROCKWELL	81750
72NLS40-12	ROCKWELL	85138	72NYR24	ROCKWELL	81712	72NYS26-1	ROCKWELL	81723	72NYS28-80	ROCKWELL	81778
72NLS40-14	ROCKWELL	85139	72NYR24-2	ROCKWELL	81721	72NYS26-11	ROCKWELL	81726	72NYS28-81	ROCKWELL	88006.52
72NLS40-16	ROCKWELL	85142	72NYR24-4	ROCKWELL	81720	72NYS26-16	ROCKWELL	81727	72NYS28-84	ROCKWELL	81741
72NLS40-3	ROCKWELL	85135	72NYR24-5	ROCKWELL	81711	72NYS26-2	ROCKWELL	81734	72NYS28-85	ROCKWELL	81746
72NLS40-5	ROCKWELL	85140	72NYR25	ROCKWELL	81722	72NYS26-26	ROCKWELL	81732	72NYS28-9	ROCKWELL	81739
72NLS40-6	ROCKWELL	85137	72NYR26	ROCKWELL	81731	72NYS26-3	ROCKWELL	81728	72NYS30	ROCKWELL	88006.53
72NLS40-7	ROCKWELL	85141	72NYR27	ROCKWELL	81736	72NYS26-4	ROCKWELL	81733	72NYS30-1	ROCKWELL	81808
72NLS40-8	ROCKWELL	85136	72NYR27-2	ROCKWELL	81735	72NYS26-5	ROCKWELL	88006.43	72NYS30-11	ROCKWELL	81804
72NLS48-4	ROCKWELL	85145	72NYR28-14	ROCKWELL	81789	72NYS26-6	ROCKWELL	88006.44	72NYS30-16	ROCKWELL	81805
72NPS32	ROCKWELL	87522	72NYR28-2	ROCKWELL	81788	72NYS26-7	ROCKWELL	81729	72NYS30-18	ROCKWELL	85822
72NPS32-10	ROCKWELL	87527	72NYR28-3	ROCKWELL	81737	72NYS26-8	ROCKWELL	81730	72NYS30-19	ROCKWELL	81810
72NPS32-100	ROCKWELL	87496	72NYR28-4	ROCKWELL	81738	72NYS28	ROCKWELL	81749	72NYS30-2	ROCKWELL	81801
72NPS32-13	ROCKWELL	87511	72NYR28-7	ROCKWELL	81787	72NYS28-1	ROCKWELL	81740	72NYS30-21	ROCKWELL	81809
72NPS32-14	ROCKWELL	87509	72NYR30	ROCKWELL	81799	72NYS28-10	ROCKWELL	81786	72NYS30-3	ROCKWELL	81802
72NPS32-17	ROCKWELL	87512	72NYR31	ROCKWELL	81812	72NYS28-105	ROCKWELL	85821	72NYS30-4	ROCKWELL	85823
72NPS32-19	ROCKWELL	87531	72NYR32	ROCKWELL	81858	72NYS28-107	ROCKWELL	81745	72NYS30-5	ROCKWELL	81800
72NPS32-21	ROCKWELL	87535	72NYR32-1	ROCKWELL	81898	72NYS28-11	ROCKWELL	81752	72NYS30-6	ROCKWELL	81807
72NPS32-22	ROCKWELL	87529	72NYR32-10	ROCKWELL	81859	72NYS28-12	ROCKWELL	88006.45	72NYS30-7	ROCKWELL	81806
72NPS32-25	ROCKWELL	87520	72NYR32-11	ROCKWELL	81861	72NYS28-122	ROCKWELL	81775	72NYS30-9	ROCKWELL	81803
72NPS32-26	ROCKWELL	87536	72NYR32-13	ROCKWELL	81817	72NYS28-127	ROCKWELL	81779	72NYS31	ROCKWELL	81836
72NPS32-28	ROCKWELL	87533	72NYR32-2	ROCKWELL	81847	72NYS28-141A	ROCKWELL	81798	72NYS31-1	ROCKWELL	81846
72NPS32-29	ROCKWELL	87518	72NYR32-3	ROCKWELL	81890	72NYS28-145	ROCKWELL	81776	72NYS31-12	ROCKWELL	81848
72NPS32-35	ROCKWELL	87513	72NYR32-5	ROCKWELL	81900	72NYS28-15	ROCKWELL	81758	72NYS31-14	ROCKWELL	81825
72NPS32-44	ROCKWELL	88006.39	72NYR34	ROCKWELL	81914	72NYS28-151A	ROCKWELL	81791	72NYS31-171A	ROCKWELL	85832
72NPS32-45	ROCKWELL	88006.40	72NYR35-1	ROCKWELL	81921	72NYS28-153	ROCKWELL	81766	72NYS31-2	ROCKWELL	81820
72NPS32-5	ROCKWELL	87525	72NYR36	ROCKWELL	81926	72NYS28-154	ROCKWELL	81785	72NYS31-22	ROCKWELL	81818
72NPS32-53	ROCKWELL	87488	72NYR36-1	ROCKWELL	81945	72NYS28-157A	ROCKWELL	81795	72NYS31-26	ROCKWELL	81823
72NPS32-6	ROCKWELL	87530	72NYR36-2	ROCKWELL	81944	72NYS28-16	ROCKWELL	81762	72NYS31-27	ROCKWELL	81842
72NPS32-64	ROCKWELL	87515	72NYR36-6	ROCKWELL	81943	72NYS28-161	ROCKWELL	81773	72NYS31-28	ROCKWELL	81816
72NPS32-65	ROCKWELL	87532	72NYR38	ROCKWELL	81960	72NYS28-163	ROCKWELL	85820	72NYS31-29	ROCKWELL	81838
72NPS32-73	ROCKWELL	87514	72NYR38-1	ROCKWELL	81959	72NYS28-166	ROCKWELL	81793	72NYS31-3	ROCKWELL	81833
72NPS32-74	ROCKWELL	87517	72NYR38-3	ROCKWELL	81961	72NYS28-167	ROCKWELL	81777	72NYS31-33	ROCKWELL	85825
72NPS32-83	ROCKWELL	87498	72NYR39	ROCKWELL	81962	72NYS28-170	ROCKWELL	81769	72NYS31-33	ROCKWELL	88006.54
72NPS32-84	ROCKWELL	87500	72NYR40-11	ROCKWELL	81979	72NYS28-174	ROCKWELL	81754	72NYS31-4	ROCKWELL	81832
72NPS32-87	ROCKWELL	87502	72NYR40-2	ROCKWELL	81978	72NYS28-176	ROCKWELL	81790	72NYS31-41	ROCKWELL	81843
72NPS32-87DC	ROCKWELL	87507	72NYR41	ROCKWELL	81982	72NYS28-176A	ROCKWELL	81792	72NYS31-42	ROCKWELL	81844
72NPS32-9	ROCKWELL	87524	72NYR42	ROCKWELL	81985	72NYS28-177	ROCKWELL	81794	72NYS31-43	ROCKWELL	81827
72NPS32-96	ROCKWELL	87506	72NYR42-1	ROCKWELL	81983	72NYS28-177A	ROCKWELL	81796	72NYS31-46	ROCKWELL	81819
72NPS40-2	ROCKWELL	87583	72NYR42-2	ROCKWELL	81986	72NYS28-180	ROCKWELL	88006.46	72NYS31-47	ROCKWELL	88006.55
72NPS40-5	ROCKWELL	87582	72NYR42-5	ROCKWELL	81987	72NYS28-2	ROCKWELL	81748	72NYS31-48	ROCKWELL	81830
72NPS40-9	ROCKWELL	87588	72NYR44-2	ROCKWELL	81984	72NYS28-20	ROCKWELL	88006.47	72NYS31-49	ROCKWELL	81841
72NY40	ROCKWELL	85558	72NYR44-3	ROCKWELL	81989	72NYS28-21	ROCKWELL	81756	72NYS31-5	ROCKWELL	81815
72NY43	ROCKWELL	85559	72NYR44-4	ROCKWELL	81988	72NYS28-23	ROCKWELL	81770	72NYS31-50	ROCKWELL	81834
72NY43-1	ROCKWELL	85572	72NYR44-6	ROCKWELL	81991	72NYS28-25	ROCKWELL	81781	72NYS31-51	ROCKWELL	81831
72NY43-2	ROCKWELL	85578	72NYR44-7	ROCKWELL	81990	72NYS28-27	ROCKWELL	81761	72NYS31-53	ROCKWELL	81839
72NY43-3	ROCKWELL	85573	72NYR46-1	ROCKWELL	81992	72NYS28-28	ROCKWELL	81763	72NYS31-54	ROCKWELL	81845
72NY43-5	ROCKWELL	85574	72NYR48-1	ROCKWELL	81994	72NYS28-3	ROCKWELL	81774	72NYS31-55	ROCKWELL	81813
72NY44	ROCKWELL	85577	72NYR48-3	ROCKWELL	81993	72NYS28-30	ROCKWELL	81765	72NYS31-6	ROCKWELL	81826
72NY44-2	ROCKWELL	85576	72NYR48-4	ROCKWELL	81995	72NYS28-31	ROCKWELL	88006.48	72NYS31-60	ROCKWELL	81835
72NY50-3	ROCKWELL	85560	72NYR48-5	ROCKWELL	81996	72NYS28-32	ROCKWELL	81744	72NYS31-64	ROCKWELL	85828
72NY51	ROCKWELL	88006.41	72NYR50	ROCKWELL	81997	72NYS28-33	ROCKWELL	81753	72NYS31-65	ROCKWELL	81837
72NY51-1	ROCKWELL	85579	72NYR50-1	ROCKWELL	81998	72NYS28-37	ROCKWELL	81767	72NYS31-69	ROCKWELL	85826
72NY52	ROCKWELL	85581	72NYR52	ROCKWELL	82000	72NYS28-39	ROCKWELL	81784	72NYS31-7	ROCKWELL	81828
72NY52-1	ROCKWELL	85562	72NYR52-2	ROCKWELL	81999	72NYS28-4	ROCKWELL	81759	72NYS31-70	ROCKWELL	81821
72NY52-2	ROCKWELL	85561	72NYR52-3	ROCKWELL	82001	72NYS28-42	ROCKWELL	81757	72NYS31-79	ROCKWELL	85827
72NY52-23	ROCKWELL	85571	72NYR52-6	ROCKWELL	82002	72NYS28-47	ROCKWELL	81760	72NYS31-8	ROCKWELL	81840
72NY52-4	ROCKWELL	85563	72NYS20	ROCKWELL	81707	72NYS28-49	ROCKWELL	81742	72NYS31-83	ROCKWELL	81811
72NY52-5	ROCKWELL	85582	72NYS22	ROCKWELL	81709	72NYS28-5	ROCKWELL	81743	72NYS31-85	ROCKWELL	81822
72NY52-9	ROCKWELL	85580	72NYS22-4	ROCKWELL	81708	72NYS28-51	ROCKWELL	81747	72NYS31-89	ROCKWELL	85824
72NY53	ROCKWELL	85564	72NYS24	ROCKWELL	81713	72NYS28-55	ROCKWELL	88006.49	72NYS31-9	ROCKWELL	81849
72NY53-1	ROCKWELL	85565	72NYS24-12	ROCKWELL	81719	72NYS28-57	ROCKWELL	81782	72NYS31-91	ROCKWELL	81829
72NY53-2	ROCKWELL	85583	72NYS24-13	ROCKWELL	81715	72NYS28-6	ROCKWELL	81755	72NYS31-93	ROCKWELL	81824

Part No.	Brand	IDLI Group
72NYS32	ROCKWELL	81852
72NYS32-100	ROCKWELL	81885
72NYS32-104	ROCKWELL	81877
72NYS32-105	ROCKWELL	81875
72NYS32-106	ROCKWELL	81899
72NYS32-108	ROCKWELL	81868
72NYS32-109	ROCKWELL	81873
72NYS32-110	ROCKWELL	81866
72NYS32-111	ROCKWELL	81867
72NYS32-12	ROCKWELL	81878
72NYS32-123	ROCKWELL	81850
72NYS32-129	ROCKWELL	81860
72NYS32-135	ROCKWELL	81869
72NYS32-14	ROCKWELL	81897
72NYS32-151	ROCKWELL	81881
72NYS32-152	ROCKWELL	81854
72NYS32-155	ROCKWELL	81883
72NYS32-157	ROCKWELL	81892
72NYS32-16	ROCKWELL	81886
72NYS32-167	ROCKWELL	85829
72NYS32-17	ROCKWELL	81896
72NYS32-172	ROCKWELL	81876
72NYS32-173	ROCKWELL	81891
72NYS32-174	ROCKWELL	85831
72NYS32-18	ROCKWELL	81855
72NYS32-186	ROCKWELL	88006.56
72NYS32-217	ROCKWELL	81856
72NYS32-218	ROCKWELL	81905
72NYS32-219	ROCKWELL	81904
72NYS32-220	ROCKWELL	81906
72NYS32-221	ROCKWELL	81909
72NYS32-222	ROCKWELL	81908
72NYS32-239	ROCKWELL	81907
72NYS32-244	ROCKWELL	81901
72NYS32-244A	ROCKWELL	81902
72NYS32-245	ROCKWELL	81910
72NYS32-245A	ROCKWELL	81911
72NYS32-34	ROCKWELL	81853
72NYS32-35	ROCKWELL	85830
72NYS32-38	ROCKWELL	88006.57
72NYS32-39	ROCKWELL	81879
72NYS32-40	ROCKWELL	81889
72NYS32-43	ROCKWELL	81871
72NYS32-44	ROCKWELL	81865
72NYS32-45	ROCKWELL	81874
72NYS32-47	ROCKWELL	81895
72NYS32-48	ROCKWELL	81880
72NYS32-5	ROCKWELL	81870
72NYS32-50	ROCKWELL	81893
72NYS32-55	ROCKWELL	81887
72NYS32-6	ROCKWELL	81862
72NYS32-66	ROCKWELL	88006.58
72NYS32-7	ROCKWELL	81863
72NYS32-70	ROCKWELL	81882
72NYS32-76	ROCKWELL	81857
72NYS32-77	ROCKWELL	81888
72NYS32-8	ROCKWELL	81851
72NYS32-80	ROCKWELL	88006.59
72NYS32-87	ROCKWELL	81872
72NYS32-9	ROCKWELL	81864
72NYS32-98A	ROCKWELL	81884
72NYS33	ROCKWELL	81903
72NYS33-3	ROCKWELL	88006.60
72NYS34-2	ROCKWELL	88006.61
72NYS34-23	ROCKWELL	81920
72NYS34-24	ROCKWELL	81916
72NYS34-28	ROCKWELL	81919
72NYS34-31	ROCKWELL	81917
72NYS34-32	ROCKWELL	81915
72NYS34-33	ROCKWELL	81918
72NYS34-34	ROCKWELL	81913
72NYS34-36	ROCKWELL	81912
72NYS36	ROCKWELL	81929
72NYS36-1	ROCKWELL	81949
72NYS36-13	ROCKWELL	81936
72NYS36-14	ROCKWELL	81938
72NYS36-16	ROCKWELL	81948
72NYS36-17	ROCKWELL	81946
72NYS36-19	ROCKWELL	81939
72NYS36-2	ROCKWELL	81942
72NYS36-21	ROCKWELL	81927
72NYS36-21	ROCKWELL	88006.62
72NYS36-22	ROCKWELL	81922
72NYS36-23	ROCKWELL	81950
72NYS36-36	ROCKWELL	81928
72NYS36-37	ROCKWELL	81947
72NYS36-38	ROCKWELL	81940
72NYS36-4	ROCKWELL	85836
72NYS36-40	ROCKWELL	81934
72NYS36-41	ROCKWELL	81931
72NYS36-42	ROCKWELL	81930
72NYS36-45	ROCKWELL	81932
72NYS36-5	ROCKWELL	81937
72NYS36-51	ROCKWELL	81923
72NYS36-54	ROCKWELL	81935
72NYS36-6	ROCKWELL	81941
72NYS36-61	ROCKWELL	81933
72NYS36-63	ROCKWELL	85833
72NYS36-69	ROCKWELL	81924
72NYS36-7	ROCKWELL	85835
72NYS36-78	ROCKWELL	85834
72NYS36-84	ROCKWELL	81946
72NYS36-88	ROCKWELL	81925
72NYS38	ROCKWELL	81953
72NYS38-1	ROCKWELL	81952
72NYS38-10	ROCKWELL	85837
72NYS38-14	ROCKWELL	81955
72NYS38-15	ROCKWELL	81958
72NYS38-16	ROCKWELL	81956
72NYS38-17	ROCKWELL	81957
72NYS38-2	ROCKWELL	81954
72NYS38-3	ROCKWELL	88006.63
72NYS38-4	ROCKWELL	81951
72NYS39-1	ROCKWELL	85838
72NYS40	ROCKWELL	81965
72NYS40-1	ROCKWELL	81963
72NYS40-19	ROCKWELL	81981
72NYS40-2	ROCKWELL	81969
72NYS40-21	ROCKWELL	85839
72NYS40-22	ROCKWELL	81972
72NYS40-26	ROCKWELL	81971
72NYS40-3	ROCKWELL	81964
72NYS40-31	ROCKWELL	81970
72NYS40-4	ROCKWELL	81966
72NYS40-41	ROCKWELL	81967
72NYS40-5	ROCKWELL	81977
72NYS40-58	ROCKWELL	81968
72NYS40-59	ROCKWELL	81974
72NYS40-59A	ROCKWELL	88006.64
72NYS40-72	ROCKWELL	81975
72NYS40-80	ROCKWELL	81976
72NYS40-90	ROCKWELL	81973
72NYSM30-8	ROCKWELL	86741
72NYSM32-1	ROCKWELL	86746
72NYSM32-11	ROCKWELL	86744
72NYSM32-118	ROCKWELL	86753
72NYSM32-120	ROCKWELL	86769
72NYSM32-121	ROCKWELL	86766
72NYSM32-122	ROCKWELL	86750
72NYSM32-128	ROCKWELL	86763
72NYSM32-131	ROCKWELL	86762
72NYSM32-137	ROCKWELL	86764
72NYSM32-144	ROCKWELL	86755
72NYSM32-161	ROCKWELL	86745
72NYSM32-161	ROCKWELL	86756
72NYSM32-164	ROCKWELL	86748
72NYSM32-21	ROCKWELL	88006.65
72NYSM32-216	ROCKWELL	86760
72NYSM32-23	ROCKWELL	88006.66
72NYSM32-24	ROCKWELL	88006.67
72NYSM32-240	ROCKWELL	86761
72NYSM32-27	ROCKWELL	88006.68
72NYSM32-3	ROCKWELL	86743
72NYSM32-30	ROCKWELL	86765
72NYSM32-31	ROCKWELL	88006.69
72NYSM32-32	ROCKWELL	86767
72NYSM32-33	ROCKWELL	88006.70
72NYSM32-42	ROCKWELL	86757
72NYSM32-54	ROCKWELL	86758
72NYSM32-57	ROCKWELL	86759
72NYSM32-58	ROCKWELL	86751
72NYSM32-59	ROCKWELL	86747
72NYSM32-61	ROCKWELL	88006.71
72NYSM32-65	ROCKWELL	86742
72NYSM32-67	ROCKWELL	86752
72NYSM32-73	ROCKWELL	86754
72NYSM32-74	ROCKWELL	86749
72NYSM32-88	ROCKWELL	86768
72NYSM34-15	ROCKWELL	86771
72NYSM34-19	ROCKWELL	86772
72NYSM34-20	ROCKWELL	86770
72NYSM40-12	ROCKWELL	86774
72NYSM40-13	ROCKWELL	86792
72NYSM40-14	ROCKWELL	86777
72NYSM40-15	ROCKWELL	86796
72NYSM40-16	ROCKWELL	86789
72NYSM40-17	ROCKWELL	86786
72NYSM40-18	ROCKWELL	86799
72NYSM40-34	ROCKWELL	86793
72NYSM40-35	ROCKWELL	86778
72NYSM40-36	ROCKWELL	86787
72NYSM40-37	ROCKWELL	86790
72NYSM40-39	ROCKWELL	86797
72NYSM40-40	ROCKWELL	86775
72NYSM40-49	ROCKWELL	86773
72NYSM40-53	ROCKWELL	86776
72NYSM40-54	ROCKWELL	86779
72NYSM40-55	ROCKWELL	86781
72NYSM40-56	ROCKWELL	86783
72NYSM40-57	ROCKWELL	86788
72NYSM40-66	ROCKWELL	86780
72NYSM40-69	ROCKWELL	86798
72NYSM40-71	ROCKWELL	86795
72NYSM40-76	ROCKWELL	86782
72NYSM40-84	ROCKWELL	86794
72NYSM40-87	ROCKWELL	86791
72NYSM40-88	ROCKWELL	86785
72NYSM40-93	ROCKWELL	86784
72NYSM48-4	ROCKWELL	86801
72NYSM48-4	ROCKWELL	88006.72
72NYT26-1	ROCKWELL	88006.73
72NYT26-2	ROCKWELL	81724
72NYT28-2	ROCKWELL	81783
72NYT28-6	ROCKWELL	81780
72NYT28-8	ROCKWELL	81768
72NYT31-1	ROCKWELL	81814
72NYT32	ROCKWELL	81894
72NYT40	ROCKWELL	81980
73004690-8	FIAT-ALLIS	80101
73006546	FIAT-ALLIS	80048
73006546-0	FIAT-ALLIS	80048
73006595-7	FIAT-ALLIS	80109
73008392-7	FIAT-ALLIS	80047
73011020	FIAT-ALLIS	80070
73011022	FIAT-ALLIS	80070
73013809	FIAT-ALLIS	80178
73013809-3	FIAT-ALLIS	80178
73021862	FIAT-ALLIS	80188
73021862-2	FIAT-ALLIS	80188
73024508	FIAT-ALLIS	80178
73024508-8	FIAT-ALLIS	80178
73026726	FIAT-ALLIS	80069
73026726-4	FIAT-ALLIS	80069
73026727	FIAT-ALLIS	80069
73026727-2	FIAT-ALLIS	80069
73028573	FIAT-ALLIS	80178
73028573-8	FIAT-ALLIS	80178
73028862	FIAT-ALLIS	80178
73028862-5	FIAT-ALLIS	80178
73034121	FIAT-ALLIS	80101
73038633	FIAT-ALLIS	80220
73038633-8	FIAT-ALLIS	80220
73039559	FIAT-ALLIS	80187
73039559-4	FIAT-ALLIS	80187
73040372	FIAT-ALLIS	80235
73040372-9	FIAT-ALLIS	80235
73040377	FIAT-ALLIS	80235
73040377-8	FIAT-ALLIS	80235
73044128	FIAT-ALLIS	80188
73044128-1	FIAT-ALLIS	80188
73044617	FIAT-ALLIS	80219
73044617-3	FIAT-ALLIS	80219
73060000	FIAT-ALLIS	80234
73066875	FIAT-ALLIS	80216
7307J	BORG-WARNR	86831
731	HERCULES	88002.56
73107487	FIAT-ALLIS	80224
73111401	FIAT-ALLIS	80198
73122595	FIAT-ALLIS	80224
73134821	FIAT-ALLIS	80198
73139	DROTT	80021
73140006	FIAT-ALLIS	80213
732	TEDDY TORQ	80061
732	TEDDY TORQ	80096
732324	LONG MFG.	80001
73251-2	DETROIT	80095
7325-2	DETROIT	80096
73252-2	ALCO	88001.64
73252-2	ALCO	88001.69
73252-2	DETROIT	80065
73252-2	DETROIT	80096
7325-4	DETROIT	80095
7325-5	DETROIT	80094
7327J	BORG-WARNR	81883
7329-016	IOWA MFG.	80099
7329-016	IOWA MFG.	80101
7329-021	IOWA MFG.	80109
7329-026	IOWA MFG.	80168
7329-041	IOWA MFG.	80231
7329-102	IOWA MFG.	80109
733	TEDDY TORQ	80096
734	TEDDY TORQ	80094
734-57	BLAW KNOX	80124
734-61	BLAW KNOX	80178
734751	WHITE	80198
7347665	JEEP	80189
7347665	WILLYS	80189
734769	WHITE	80225
734769P2	DIAMOND R	80225
734769P2	WHITE	80225

Part No.	Brand	IDLI Group
735	HERCULES	88002.57
735	TEDDY TORQ	80061
735	TEDDY TORQ	80096
735-19	BLAW KNOX	80007
735-198	BLAW KNOX	80007
7352	ALCO	88001.73
735251-2	ALCO	80095
735251-2	ALCO	88001.61
735251-2	DETROIT	80095
735251-2	PRECISION	88001.60
73525-2	DETROIT	80056
73525-2	DETROIT	80095
73525-2	SPICER	80057
73525-2	SPICER	80095
73525-3	ALCO	80096
73525-3	DETROIT	80096
73525-3	SPICER	80096
73525-3	SPICER	88001.59
73525-4	ALCO	80094
73525-4	DETROIT	80094
7352A	ALCO	88001.72
7354	ALCO	88001.74
7354A	ALCO	88001.66
73-551R93	IHC	80193
735-72	BLAW KNOX	80077
736	HERCULES	88002.58
736	TEDDY TORQ	80060
736268	LONG MFG.	80035
7-364-000029	GROVE CR.	80197
7-364-000330	GROVE	80069
7-364-000331	GROVE	80069
7-364-00330	GROVE	80070
737	HERCULES	88002.60
737100	LONG MFG.	80001
7372959	ALLIS-CHLM	80099
7372959	ALLIS-CHLM	80101
7372959	FIAT-ALLIS	80099
7372959	FIAT-ALLIS	80101
7398J	BORG-WARNR	86732
739XX1	MACK	80001
74	TRU-CROSS	80078
74	TRU-CROSS	80090
74	WYLIE	80081
740	PRECISION	88002.82
7400	CHAIN BELT	80220
7404J	BORG-WARNR	81740
74050092	IHC	80041
7405065	GMC	88002.77
7405065	HOLDEN	88002.77
7405066	GMC	80032
7405066	HOLDEN	80032
7406001	GMC	88002.81
7406001	HOLDEN	88002.81
7407J	BORG-WARNR	82320
7-41	BYPY	88004.32
74123	NEW HOLL.	80001
74180R93	IHC	80225
7421879	GMC	88002.77
7421879	HOLDEN	88002.77
7421880	GMC	80032
7421880	HOLDEN	80032
7421881	GMC	88002.81
7421881	HOLDEN	88002.81
7423286	GMC	88002.77
7423286	HOLDEN	88002.77
7423287	GMC	88002.77
7423287	HOLDEN	88002.77
7424	KENBAR	88008.35
7425515	GMC	88002.81
7425515	HOLDEN	80032
7-4265	NEAPCO	80951
7-4266	NEAPCO	80905
7-4267	NEAPCO	80953
7-4268	NEAPCO	80906
74-289R91	IHC	80142
7428J	BORG-WARNR	85559
74498505	FIAT-ALLIS	80224
74650	DROTT	80047
74650M	DROTT	80047
74-7090	FORD	88002.54
747556	CHRYSLER	80142
74762487	FIAT-ALLIS	80047
74762487-9	FIAT-ALLIS	80047
74905112	FIAT-ALLIS	80151
74905112-1	FIAT-ALLIS	80151
74906478	FIAT-ALLIS	80046
74906478	FIAT-ALLIS	80047
74906478-5	FIAT-ALLIS	80046
74906479	ALLIS-CHLM	80047
74906479-3	FIAT-ALLIS	80047
74984504	FIAT-ALLIS	80047
74984504-3	FIAT-ALLIS	80047
74984505	FIAT-ALLIS	80046
74984505-0	FIAT-ALLIS	80046
74987664	FIAT-ALLIS	80047
74991328	FIAT-ALLIS	80150
74991328-8	FIAT-ALLIS	80150
74993479	FIAT-ALLIS	80154
74993479-7	FIAT-ALLIS	80154
75	WYLIE	80019
750	AEC	80187
750	ALLOY	88014.34
7500055	GMC	80221
750048	WESCO	85055
750049	WESCO	85054
750050	WESCO	85058
750052	WESCO	85063
7501016	GMC	80124
75038	HYSTER	80225
75060444	CHRYSLER	80024
750 610	SCANIA	80180
75-06-60	WALTERSCHD	88003.59
75109A	HYSTER	80210
75109A	HYSTER	80218
75-11-00	WALTERSCHD	88003.61
7512	ALCO	88003.50
75-12-00	WALTERSCHD	88003.60
75-15-00	WALTERSCHD	88003.62
751-762C91	IHC	80190
751762R91	IHC	80190
7520	MATHEWS	80069
7522J	BORG-WARNR	88014.52
75-25-00	WALTERSCHD	88003.63
753	ALLOY	83281
7530J	BORG-WARNR	81820
753111	HESSTON	80035
75-35-00	WALTERSCHD	88003.64
75-351R93	IHC	80193
75351T93	IHC	80193
75371R93	IHC	80193
753772	HESSTON	80037
753889	HESSTON	80069
753-931R91	IHC	80104
7545016	GMC	86075
7545016	GMC	86148
754622	HESSTON	80069
755	ALLOY	83349
755	TEDDY TORQ	80060
756	ALLOY	88014.38
756924	HESSTON	80037
757	ALLOY	83252
757	HERCULES	88002.55
757443	HESSTON	80035
758	ALLOY	83241
758	HERCULES	88002.63
75832	MERCURY	80062
759	HERCULES	88002.64
7590	DANUSER	80035
75N	WESCO	80187
76	WYLIE	80131
760	ALLOY	83283
760004	REX CHAIN	88014.06
760005	REX CHAIN	84517
76002601	FIAT-ALLIS	80198
76002604	FIAT-ALLIS	80213
760029	REX CHAIN	83057
760032	REX CHAIN	88014.05
760035	REX CHAIN	88000.04
760038	REX CHAIN	88013.90
760040	REX CHAIN	88013.90
760041	REX CHAIN	84510
760044	REX CHAIN	81342
760049	REX CHAIN	88014.35
760053	REX CHAIN	83105
760-053	WESCO	87227
760054	REX CHAIN	83084
760054	WESCO	88000.10
760055	REX CHAIN	83129
760058	REX CHAIN	88014.35
760068	REX CHAIN	81342
760086	REX CHAIN	84531
760091	REX CHAIN	83086
760092	REX CHAIN	83129
7601	ALCO	88003.56
7601171	NATL.MINE	88010.64
76011915	NATL.MINE	80069
76012012	NATL.MINE	80037
76012095	NATL.MINE	80070
76012111	NATL.MINE	80161
76012137	NATL.MINE	80134
760-147-341-1	DAIM.RUBE.	80077
7604	ALCO	88003.45
7605	ALCO	88003.56
7607	ALCO	88003.49
7608	ALCO	88003.46
760969	REXNORD	81362
760972	REXNORD	81362
761	ALLOY	83262
76101618	NATL.MINE	80234
7612	ALCO	88003.50
761685	REXNORD	81369
76179097	FIAT-ALLIS	80007
76186	DROTT	80155
76187	DROTT	80109
762	ALLOY	83357
7622	ALCO	88003.42
762349	REXNORD	88006.08
7623J	BORG-WARNR	85594
762451	REXNORD	81376
7627	ALCO	88003.43
763	ALLOY	83240
764	ALLOY	83251
765	ALLOY	83250
765-000-0	GLAENZER	80221
765-002-0	GLAENZER	80221
765-010-2	GLAENZER	80221
765-014-0	GLAENZER	80221
765-016-0	GLAENZER	80221
7654J	BORG-WARNR	81025
7675J	BORG-WARNR	82098
76761	HYSTER	80001
76-8	ALCO	88003.46
7686	RAY GO	80124
7686	REX CHAIN	80124
769	ALLOY	83263
77	WYLIE	80129
770	PRECISION	80217
770058	WESCO	88001.07
77.01.005.432	RENAULT	80180
7701005432	SAVIEM	80180
77.01.008.330	RENAULT	80180
77.01.010.700	RENAULT	80140
7701200922	RENAULT	86174
7701201045	RENAULT	86113
7701348159	RENAULT	86049
7701348759	RENAULT	86049
7701348968	RENAULT	86048
7701450617	RENAULT	86173
7701457538	RENAULT	86174
7701460779	RENAULT	86113
771	ALLOY	83334
7710-1SF	SPICER	88007.31
7710-SF	SPICER	88007.29
77187	SCANIA	80180
773215	WHITE	80225
77372959	FIAT-ALLIS	80101
774	PRECISION	80218
7748998	BLAW KNOX	80104
7765	GURRIES	80221
7765	RAY GO	80221
7765	REX CHAIN	80221
7774J	BORG-WARNR	85856
7781J	BORG-WARNR	81880
7783J	BORG-WARNR	82153
7783J	BORG-WARNR	82156
77-979R91	IHC	80044
7797J	BORG-WARNR	82360
77TT3K187AA	FORD	86126
78	WYLIE	80035
7800	DANUSER	80037
780-059	WESCO	87276
7800666	GMC	86177
78013656-7	FIAT-ALLIS	80149
78013657-5	FIAT-ALLIS	80156
78-019R91	IHC	80054
7802092	GMC	86268
78-020R91	IHC	80049
78-022R91	IHC	80049
78060	SCANIA	80189
7806140	GMC	80107
7808308	GMC	80003
7808979	GMC	88002.33
7809156	GMC	88002.34
781	PRECISION	80243
7814308	GMC	88002.35
7815848	GMC	88002.36
7816006	GMC	88002.34
78-2	MUNCIE	88001.13
7825819	GMC	80036
7826391	FIAT-ALLIS	80149
7828953	GMC	88002.36
7829456	GMC	88002.36
782B3	MACK	80069
782U	MACK	80234
782V	MACK	80037
783	PRECISION	80244
7830112	GMC	86259
7830355	GMC	87331
7831588	GMC	86259
7831589	GMC	86178

Part No.	Brand	IDLI Group	Part No.	Brand	IDLI Group	Part No.	Brand	IDLI Group	Part No.	Brand	IDLI Group
7831589	GMC	86246	7837828	GMC	86236	7842037	GMC	86282	7843917	GMC	86276
7831589	GMC	86247	7838288	GMC	86238	7842039	GMC	86241	7843919	GMC	86179
7831590	GMC	86234	7838298	GMC	86237	7842039	GMC	86242	7843919	GMC	86251
7831591	GMC	86234	7838298	GMC	86238	7842039	GMC	86273	7843919	GMC	86252
7831592	GMC	86246	7838298	GMC	86269	7842039	GMC	86274	7843919	GMC	86254
7832251	GMC	86199	7838298	GMC	86270	7842040	GMC	86241	7843919	GMC	86255
7832252	GMC	86198	7838299	GMC	86237	7842040	GMC	86242	7843919	GMC	86256
7832253	GMC	86197	7838299	GMC	86238	7842040	GMC	86273	7843919	GMC	86257
7832254	GMC	86196	7838299	GMC	86269	7842040	GMC	86274	7843919	GMC	86282
7832255	GMC	86080	7838299	GMC	86270	7842129	GMC	86241	7843920	GMC	86266
7832255	GMC	86085	7838304	GMC	86237	7842129	GMC	86242	7843921	GMC	86243
7832255	GMC	86154	7838304	GMC	86238	7842129	GMC	86273	7843921	GMC	86244
7832257	GMC	86148	7838304	GMC	86269	7842129	GMC	86274	7843921	GMC	86245
7832258	GMC	86120	7838304	GMC	86270	7842152	GMC	86237	7843921	GMC	86275
7832258	GMC	86179	7838306	GMC	86179	7842152	GMC	86238	7843921	GMC	86276
7832258	GMC	86251	7838306	GMC	86248	7842152	GMC	86269	7843922	GMC	86267
7832258	GMC	86252	7838306	GMC	86251	7842152	GMC	86270	7843923	GMC	86243
7832258	GMC	86254	7838306	GMC	86252	7842256	GMC	86215	7843923	GMC	86265
7832258	GMC	86255	7838306	GMC	86254	7842258	GMC	86214	7843923	GMC	86267
7832258	GMC	86256	7838306	GMC	86255	7842304	CLARK EQU.	80193	7843923	GMC	86275
7832258	GMC	86257	7838306	GMC	86256	7842331	GMC	86213	7843924	GMC	86243
7832258	GMC	86282	7838306	GMC	86257	7842332	GMC	86212	7843924	GMC	86244
7832259	GMC	86120	7838306	GMC	86282	7842333	GMC	86215	7843924	GMC	86245
7832259	GMC	86121	7840070	GMC	86075	7842334	GMC	86214	7843924	GMC	86267
7832259	GMC	86248	7840070	GMC	86148	7842335	GMC	86201	7843924	GMC	86275
7832259	GMC	86249	7840071	GMC	86147	7842336	GMC	86200	7843924	GMC	86276
7833098	GMC	86075	7840071	GMC	86236	7842340	GMC	86233	784401R1	IHC	80104
7833098	GMC	86148	7840072	GMC	86074	7842344	GMC	86213	784409R1	IHC	80100
7834082	GMC	86258	7840072	GMC	86075	7842345	GMC	86212	7844536	GMC	86217
7834238	GMC	86157	7840072	GMC	86147	7842346	GMC	86201	7844538	GMC	86245
7834238	GMC	86235	7840072	GMC	86236	7842347	GMC	86200	7844538	GMC	86277
7834239	GMC	86178	7840073	GMC	86197	7842351	GMC	86233	7844539	GMC	86216
7834239	GMC	86246	7840074	GMC	86196	7842378	GMC	86263	7844541	GMC	86231
7834239	GMC	86247	7840075	GMC	86198	7842379	GMC	86104	7844542	GMC	86230
7834241	GMC	86157	7840076	GMC	86199	7842379	GMC	86264	7844671	GMC	86230
7834241	GMC	86235	7840149	GMC	86199	7842380	GMC	86243	7844672	GMC	86231
7834244	GMC	86195	7840150	GMC	86198	7842380	GMC	86262	7844745	GMC	86203
7834245	GMC	86194	7840151	GMC	86197	7842380	GMC	86265	7844746	GMC	86203
7834303	GMC	86194	7840152	GMC	86196	7842380	GMC	86267	7844747	GMC	86179
7834304	GMC	86195	7840364	GMC	86260	7842382	GMC	86239	7844747	GMC	86250
7835070	GMC	86178	7840950	GMC	86179	7842382	GMC	86240	7844747	GMC	86251
7835070	GMC	86246	7840950	GMC	86249	7842382	GMC	86271	7844747	GMC	86252
7835070	GMC	86247	7840950	GMC	86251	7842382	GMC	86272	7844747	GMC	86254
7836405	GMC	86121	7840950	GMC	86252	7842779	ALLIS-CHLM	80001	7844747	GMC	86255
7836405	GMC	86179	7840950	GMC	86254	7842779	FIAT-ALLIS	80001	7844747	GMC	86256
7836405	GMC	86251	7840950	GMC	86255	78433	BMC	80024	7844747	GMC	86257
7836405	GMC	86252	7840950	GMC	86256	78433	LEYLAND	80024	7844747	GMC	86282
7836405	GMC	86254	7840950	GMC	86257	7843576	GMC	86239	7844748	GMC	86202
7836405	GMC	86255	7840950	GMC	86282	7843576	GMC	86240	7844749	GMC	86202
7836405	GMC	86256	7841256	GMC	86212	7843576	GMC	86271	7844750	GMC	86207
7836405	GMC	86257	7841257	GMC	86213	7843576	GMC	86272	7844751	GMC	86207
7836405	GMC	86282	7842034	GMC	86179	7843577	GMC	86239	7844752	GMC	86206
7836409	GMC	86211	7842034	GMC	86251	7843577	GMC	86240	7844752	GMC	86220
7836411	GMC	86211	7842034	GMC	86252	7843577	GMC	86271	7844753	GMC	86220
7836413	GMC	86208	7842034	GMC	86254	7843577	GMC	86272	7844754	GMC	86205
7836414	GMC	86208	7842034	GMC	86255	7843886	GMC	86225	7844755	GMC	86205
7836440	GMC	86209	7842034	GMC	86256	7843887	GMC	86224	7844756	GMC	86204
7836441	GMC	86209	7842034	GMC	86257	7843888	GMC	86227	7844757	GMC	86204
7836462	GMC	86210	7842034	GMC	86282	7843891	GMC	86228	7844758	GMC	86229
7836463	GMC	86210	7842036	GMC	86241	7843892	GMC	86223	7844759	GMC	86223
7836928	GMC	86147	7842036	GMC	86242	7843911	GMC	86225	7844760	GMC	86228
7836928	GMC	86236	7842036	GMC	86273	7843912	GMC	86224	7844761	GMC	86228
7837071	GMC	86259	7842036	GMC	86274	7843913	GMC	86227	7844762	GMC	86221
7837071	GMC	86261	7842037	GMC	86179	7843914	GMC	86226	7844763	GMC	86221
7837135	GMC	86260	7842037	GMC	86251	7843915	GMC	86223	7844764	GMC	86219
7837827	GMC	86074	7842037	GMC	86252	7843916	GMC	86222	7844765	GMC	86219
7837827	GMC	86075	7842037	GMC	86254	7843917	GMC	86243	7844766	GMC	86218
7837827	GMC	86148	7842037	GMC	86255	7843917	GMC	86244	7844767	GMC	86218
7837828	GMC	86075	7842037	GMC	86256	7843917	GMC	86245	7844785	GMC	86179
7837828	GMC	86148	7842037	GMC	86257	7843917	GMC	86275	7844785	GMC	86250

Part No.	Brand	IDLI Group	Part No.	Brand	IDLI Group	Part No.	Brand	IDLI Group	Part No.	Brand	IDLI Group
7844785	GMC	86251	7845020	GMC	86255	7845030	GMC	86282	7K442	CATERPILLR	80223
7844785	GMC	86252	7845020	GMC	86256	7845031	GMC	86179	7M2695	CATERPILLR	80154
7844785	GMC	86254	7845020	GMC	86257	7845031	GMC	86251	7N	AEC	80220
7844785	GMC	86255	7845020	GMC	86282	7845031	GMC	86252	7NPS36	ROCKWELL	87555
7844785	GMC	86256	7845021	GMC	86179	7845031	GMC	86254	7NPS40-17	ROCKWELL	87593
7844785	GMC	86257	7845021	GMC	86251	7845031	GMC	86255	7NPS40-9	ROCKWELL	87587
7844785	GMC	86282	7845021	GMC	86252	7845031	GMC	86256	7S990	CATERPILLR	80155
7844786	GMC	86179	7845021	GMC	86254	7845031	GMC	86257	7V3842	CATERPILLR	88001.90
7844786	GMC	86250	7845021	GMC	86255	7845031	GMC	86282	7V4077	CATERPILLR	88001.89
7844786	GMC	86251	7845021	GMC	86256	7845171	GMC	86074	80	WYLIE	80182
7844786	GMC	86252	7845021	GMC	86257	7845171	GMC	86080	800001	ZOOM	80081
7844786	GMC	86254	7845021	GMC	86282	7845171	GMC	86085	800002	ZOOM	80091
7844786	GMC	86255	7845022	GMC	86239	7845171	GMC	86154	800003	ZOOM	80104
7844786	GMC	86256	7845022	GMC	86240	7845915	GMC	86179	800004	ZOOM	80061
7844786	GMC	86257	7845022	GMC	86271	7845915	GMC	86251	800005	ZOOM	80129
7844786	GMC	86282	7845022	GMC	86272	7845915	GMC	86252	800006	ZOOM	80131
7844804	GMC	86232	7845024	GMC	86179	7845915	GMC	86254	800047	EATON	80958
7844805	GMC	86232	7845024	GMC	86251	7845915	GMC	86255	8004487	GAR WOOD	80008
784-489M1	MASS.FERG.	80001	7845024	GMC	86252	7845915	GMC	86256	801	ALLOY	87032
784489M91	MASS.FERG.	80001	7845024	GMC	86254	7845915	GMC	86257	8012J	BORG-WARNR	85608
7844997	GMC	86120	7845024	GMC	86255	7845915	GMC	86282	8013656-7	FIAT-ALLIS	80149
7844997	GMC	86179	7845024	GMC	86256	785	PRECISION	80223	8013657-5	FIAT-ALLIS	80156
7844997	GMC	86251	7845024	GMC	86257	78719	BMC	80041	8014	AEC	80230
7844997	GMC	86252	7845024	GMC	86282	78TT3A425AA	FORD	86010	801508	CATERPILLR	88000.15
7844997	GMC	86254	7845025	GMC	86243	79	WYLIE	80140	801578	KAISER	80908
7844997	GMC	86255	7845025	GMC	86244	7901J	BORG-WARNR	88006.33	801578	WILLYS	80901
7844997	GMC	86256	7845025	GMC	86245	7-9061	NEAPCO	88002.42	8019J	BORG-WARNR	87647
7844997	GMC	86257	7845025	GMC	86275	791-216-007-0	DAIM.RUBE.	80142	8021J	BORG-WARNR	88006.76
7844997	GMC	86282	7845025	GMC	86276	792	PRECISION	80233	8022J	BORG-WARNR	85598
7845005	GMC	86178	7845026	GMC	86179	7924J	BORG-WARNR	81755	802389	HESSTON	80136
7845005	GMC	86246	7845026	GMC	86251	793	PRECISION	80235	802496	HESSTON	80136
7845005	GMC	86247	7845026	GMC	86252	7936J	BORG-WARNR	86809	8031-2D	MOTOR MAST	80110
7845006	GMC	86234	7845026	GMC	86254	7939J	BORG-WARNR	81020	803407271	AUDI	86029
7845007	GMC	86234	7845026	GMC	86255	793XX1	MACK	88003.38	803407271A	AUDI	86028
7845008	GMC	86246	7845026	GMC	86256	794	PRECISION	88002.83	803407271D	AUDI	86047
7845009	GMC	86120	7845026	GMC	86257	795	PRECISION	88002.84	803407271E	TRW	86054
7845009	GMC	86121	7845026	GMC	86282	79-557R92	IHC	80231	803407272A	AUDI	86055
7845009	GMC	86248	7845027	GMC	86251	796	PRECISION	88002.85	803-407-275	AUDI	86134
7845009	GMC	86249	7845027	GMC	86252	7981J	BORG-WARNR	86732	803-407-275	VOLKSWAGEN	86134
7845012	GMC	86235	7845027	GMC	86254	7985J	BORG-WARNR	86831	803407275A	AUDI	86134
7845013	GMC	86247	7845027	GMC	86255	7990J	BORG-WARNR	85114	803-407-275A	VOLKSWAGEN	86134
7845014	GMC	86157	7845027	GMC	86256	799275R91	IHC	80060	803407283B	AUDI	86160
7845014	GMC	86235	7845027	GMC	86257	799275R91	IHC	80061	803407283C	AUDI	86095
7845015	GMC	86121	7845027	GMC	86282	799281R91	IHC	80073	803407283C	AUDI	86161
7845015	GMC	86179	7845028	GMC	86179	799281R91	IHC	80091	803407331	AUDI	86160
7845015	GMC	86251	7845028	GMC	86250	799282R91	IHC	80172	803498201A	AUDI	86160
7845015	GMC	86252	7845028	GMC	86251	799293R91	IHC	80023	803498203A	AUDI	86134
7845015	GMC	86254	7845028	GMC	86252	799293R91	IHC	80140	803498203A	AUDI	86187
7845015	GMC	86255	7845028	GMC	86254	799295R91	IHC	80030	803-498-203B	AUDI	86159
7845015	GMC	86256	7845028	GMC	86255	79LGB80	HARDY SPCR	80024	804003	AMER.MOTOR	80038
7845015	GMC	86257	7845028	GMC	86256	7C270	BRILLION	80037	804003	KAISER	80038
7845015	GMC	86282	7845028	GMC	86257	7D2059	CATERPILLR	80007	804003	WILLYS	80038
7845016	GMC	86074	7845028	GMC	86282	7D9715	CATERPILLR	80007	804708	ALLIS-CHLM	80124
7845016	GMC	86075	7845029	GMC	86179	7DS15	OSHKOSH	80217	804708	FIAT-ALLIS	80124
7845018	GMC	86179	7845029	GMC	86250	7DS19	OSHKOSH	80168	804735	EATON	80216
7845018	GMC	86248	7845029	GMC	86251	7DS21	OSHKOSH	80211	804785	EATON	80168
7845018	GMC	86251	7845029	GMC	86252	7EO7039	FORD	80178	804989	WABCO	80215
7845018	GMC	86252	7845029	GMC	86254	7EQ7039	FORD	80178	8-05-07	BYPY	88003.82
7845018	GMC	86254	7845029	GMC	86255	7F3679	CATERPILLR	80070	805077	HESSTON	80037
7845018	GMC	86255	7845029	GMC	86256	7H3870	BMC	80024	8-05-08	BYPY	88003.88
7845018	GMC	86256	7845029	GMC	86257	7H3958	CATERPILLR	80234	8051-024Y	HARDY SPCR	80225
7845018	GMC	86257	7845029	GMC	86282	7H4545	CATERPILLR	80198	8-06	BYPY	88004.23
7845018	GMC	86282	7845030	GMC	86179	7H8101	CATERPILLR	80238	8-07	BYPY	88004.16
7845019	GMC	86075	7845030	GMC	86250	7J5241	CATERPILLR	80168	807-034M91	MASS.FERG.	80037
7845019	GMC	86148	7845030	GMC	86251	7J5242	CATERPILLR	80214	807420	HESSTON	80069
7845020	GMC	86179	7845030	GMC	86252	7J5245	CATERPILLR	80163	807806	AMER.MOTOR	80072
7845020	GMC	86249	7845030	GMC	86254	7J5251	CATERPILLR	80216	807806	WHITE	80072
7845020	GMC	86251	7845030	GMC	86255	7J5251	TOWMOTOR	80216	807806	WILLYS	80072
7845020	GMC	86252	7845030	GMC	86256	7K2220	CATERPILLR	80003	807855	EATON	80195
7845020	GMC	86254	7845030	GMC	86257	7K3526	CATERPILLR	80198	807856	EATON	80150

Part No.	Brand	IDLI Group	Part No.	Brand	IDLI Group	Part No.	Brand	IDLI Group	Part No.	Brand	IDLI Group
807856	EATON	80153	81-2122	NEAPCO	88007.62	8133519	AMER.MOTOR	80023	8217	PRECISION	86293
809332	EATON	80044	81-2123	NEAPCO	87326	813-3581	AMER.MOTOR	86114	8218	PRECISION	86294
8100	ALCO	80231	81-2123	NEAPCO	87330	8133581	AMER.MOTOR	86115	82-2010	NEAPCO	88008.47
8101	ALCO	80232	8122993	AMER.MOTOR	80078	813-3582	AMER.MOTOR	86114	82-2011	NEAPCO	88008.50
8103	ALCO	80231	8122993	WILLYS	80077	8133582	AMER.MOTOR	86175	82-2040	NEAPCO	88008.55
8104	AEC	80238	8124589	AMER.MOTOR	80079	813-3583	AMER.MOTOR	86115	82-2110	NEAPCO	88008.63
8105	AEC	80238	8124589	WILLYS	80079	813-4305	DARLINGTON	80047	82-3010	NEAPCO	88008.53
8105	TRU-CROSS	80238	8126603	AMER.MOTOR	80038	813-744	DARLINGTON	88001.28	82-3011	NEAPCO	88008.54
81050D	HIGHLAND	80072	8126605	AMER.MOTOR	80031	813-850	DARLINGTON	80189	82-3110	NEAPCO	88008.64
810580	ALLIS-CHLM	80001	8126606	AMER.MOTOR	80044	81-4010	NEAPCO	88008.31	823456	ALLIS-CHLM	80037
810770	HESSTON	80001	8126606	AMER.MOTOR	88001.98	81-4010	NEAPCO	88008.44	823456	FIAT-ALLIS	80037
810969	AMER.MOTOR	80038	8126607	AMER.MOTOR	80058	81-4011	NEAPCO	87458	8235	MACK	80233
810969	WHITE	80072	8126608	AMER.MOTOR	80090	81-4011	NEAPCO	88007.74	823979R92	IHC	80049
810969	WILLYS	80039	8126610	AMER.MOTOR	80025	81-4012	NEAPCO	87327	82401	TOWMOTOR	80216
811008	WESCO	88012.23	8126611	AMER.MOTOR	80059	81-4012	NEAPCO	87460	82-4010	NEAPCO	88008.56
81-1010	NEAPCO	88008.35	8126611	AMER.MOTOR	80066	81-4012	NEAPCO	88007.75	82-4011	NEAPCO	88008.57
81-1011	NEAPCO	88008.35	8126612	AMER.MOTOR	80059	81-4013	NEAPCO	87461	82-4110	NEAPCO	88008.65
81-1012	NEAPCO	88008.35	8126613	AMER.MOTOR	80072	81-4013	NEAPCO	88007.76	82495	NEW HOLL.	80001
811018	HESSTON	80134	8126613	AMER.MOTOR	80075	81-4020	NEAPCO	87459	82-5010	NEAPCO	87481
81-1020	NEAPCO	87238	8126614	AMER.MOTOR	80077	81-4020	NEAPCO	88008.45	82-5010	NEAPCO	88008.60
81-1020	NEAPCO	88007.58	8126615	AMER.MOTOR	80077	81-4021	NEAPCO	87459	82-5110	NEAPCO	88008.66
81-1022	NEAPCO	87238	8126635	AMER.MOTOR	80026	81-4030	NEAPCO	87399	825X	TEDDY TORQ	80022
811585	WESCO	80228	8126636	AMER.MOTOR	80060	81-4030	NEAPCO	88008.58	82-6110	NEAPCO	88008.67
811800	WESCO	80231	8126636	AMER.MOTOR	80061	81-4030	NEAPCO	88008.59	826461	WHITE	80023
811803	WESCO	80236	8126636	AMER.MOTOR	80077	8146J	BORG-WARNR	82098	826461A	WHITE	80023
811809	WESCO	80238	8126636	BORG-WARNR	80061	81-5010	NEAPCO	87481	8269030	FIAT	80024
811-809	WESCO	80238	8126637	AMER.MOTOR	80078	81-5010	NEAPCO	88007.77	8280920	AMER.MOTOR	80079
811809	WESCO	80238	8126638	AMER.MOTOR	80079	8151-001Y	HARDY SPCR	80225	82-8209	NEAPCO	86226
811-809	WESCO	80238	8126765	AMER.MOTOR	80023	815277R91	IHC	80041	82853	FIAT	80221
811811	WESCO	80232	8127332	AMER.MOTOR	80077	815597	ALLIS-CHLM	80142	8296	DANUSER	80001
811814	WESCO	80230	8127332	WILLYS	80077	815597	FIAT-ALLIS	80142	82N170	ROCKWELL	86809
811815	WESCO	80238	8128J	BORG-WARNR	88015.22	8156J	BORG-WARNR	81109	82N170	ROCKWELL	88006.74
811822	WESCO	80237	8128J	BORG-WARNR	88015.23	815814	WHITE	80178	82N170-1	ROCKWELL	86808
811850	WESCO	80223	8128J10800	BORG-WARNR	87674	815914	WHITE	80178	82N174-1	ROCKWELL	86810
811857	WESCO	80223	8128J7200	BORG-WARNR	87673	815915	REO MOTORS	80188	82NDCA1	ROCKWELL	86811
811858	WESCO	80228	81-3010	NEAPCO	87377	815915	WESCO	80188	82NF1	ROCKWELL	88006.75
81-2010	NEAPCO	87327	81-3010	NEAPCO	88008.41	815915	WHITE	80188	82NF13	ROCKWELL	84682
81-2010	NEAPCO	87331	81-3011	NEAPCO	87380	81-6010	NEAPCO	80146	82NF14	ROCKWELL	84684
81-2011	NEAPCO	87331	81-3011	NEAPCO	88007.69	81-6010	NEAPCO	87570	82NF15	ROCKWELL	84686
81-2012	NEAPCO	87328	81-3012	NEAPCO	87381	81-6010	NEAPCO	88007.79	82NF16	ROCKWELL	84685
81-2012	NEAPCO	88007.60	81-3012	NEAPCO	88007.70	81-6011	NEAPCO	88007.78	82NF18	ROCKWELL	84692
81-2013	NEAPCO	88007.63	81-3013	NEAPCO	87382	817070	CHECKER	80189	82NF19	ROCKWELL	84688
81-2014	NEAPCO	87332	81-3013	NEAPCO	88007.72	8171J	BORG-WARNR	81688	82NF2	ROCKWELL	84687
81-2014	NEAPCO	88007.64	81-3020	NEAPCO	87383	817359	WHITE	80075	82NF22	ROCKWELL	84680
81-2020	NEAPCO	87331	81-3020	NEAPCO	88008.42	817359	WHITE	80077	82NF24	ROCKWELL	84681
81-2021	NEAPCO	87331	81-3021	NEAPCO	87378	8180J	BORG-WARNR	82061	82NF25	ROCKWELL	84690
81-2030	NEAPCO	87285	81-3021	NEAPCO	88007.68	8180J	BORG-WARNR	82063	82NF3	ROCKWELL	84691
81-2030	NEAPCO	88008.48	81-3022	NEAPCO	87379	81C7090	FORD	88002.53	82NF4	ROCKWELL	84689
81-2030D	NEAPCO	88008.49	81-3022	NEAPCO	88007.71	81DB4635BA	FORD	80124	82NF6	ROCKWELL	84683
81-2040	NEAPCO	80147	81-3023	NEAPCO	87384	81DB4635DA	FORD	80180	82NF7	ROCKWELL	84693
81-2040	NEAPCO	87373	81-3023	NEAPCO	88007.73	8200	AEC	80231	82NLS40	ROCKWELL	85159
81-2040	NEAPCO	88008.43	8130297	AMER.MOTOR	80077	8200	TRU-CROSS	80231	82NLS40-1	ROCKWELL	85164
81-2041	NEAPCO	87374	8130298	AMER.MOTOR	80085	820001	WESCO	88000.67	82NLS40-12	ROCKWELL	85152
81-2041	NEAPCO	88007.67	81-3030	NEAPCO	87337	820-003	WESCO	88000.25	82NLS40-16	ROCKWELL	85167
81-2042	NEAPCO	87375	81-3030	NEAPCO	88008.51	820-004	WESCO	88000.31	82NLS40-17	ROCKWELL	85154
81-2042	NEAPCO	88007.66	81-3030	NEAPCO	88008.52	8201	AEC	80232	82NLS40-2	ROCKWELL	85165
81-2043	NEAPCO	87376	8130303	AMER.MOTOR	88002.37	8202	AEC	80237	82NLS40-21	ROCKWELL	85163
81-2043	NEAPCO	88007.65	8130750	AMER.MOTOR	80086	8202	TRU-CROSS	80237	82NLS40-25	ROCKWELL	85162
81-2050	NEAPCO	87325	81-310	NEAPCO	87377	8203	AEC	80236	82NLS40-26	ROCKWELL	85169
81-2050	NEAPCO	88007.59	813-2114	DARLINGTON	80046	8207	PRECISION	86284	82NLS40-3	ROCKWELL	85158
81-2060	NEAPCO	87331	813-223-603-2	DAIM.RUBE.	80221	8209	PRECISION	86285	82NLS40-30	ROCKWELL	85153
81-2061	NEAPCO	87331	8132433	AMER.MOTOR	86139	8210	PRECISION	86286	82NLS40-31	ROCKWELL	85155
81-2062	NEAPCO	87331	8132433	AMER.MOTOR	86191	8211	PRECISION	86287	82NLS40-32	ROCKWELL	85157
81-2070	NEAPCO	87326	8132434	AMER.MOTOR	86114	8212	PRECISION	86288	82NLS40-35	ROCKWELL	85170
81-2071	NEAPCO	87331	813-2434	AMER.MOTOR	86116	8213	PRECISION	86289	82NLS40-36	ROCKWELL	85160
81-2120	NEAPCO	87326	8132435	AMER.MOTOR	86114	8214	PRECISION	86290	82NLS40-44	ROCKWELL	85166
81-2120	NEAPCO	88008.40	813-2435	AMER.MOTOR	86116	8215	PRECISION	86291	82NLS40-52	ROCKWELL	85161
81-2121	NEAPCO	87329	8132435	AMER.MOTOR	86175	8216	PRECISION	86292	82NLS40-54	ROCKWELL	85156
81-2121	NEAPCO	88007.61	813-2437	AMER.MOTOR	86115	82-16-108	CLEVE.STL.	87467	82NLS40-6	ROCKWELL	85168
81-2122	NEAPCO	87330	8132441	AMER.MOTOR	86139	82-16-74	CLEVE.STL.	87466	82NPS40	ROCKWELL	88006.76

Part No.	Brand	IDLI Group
82NPS40-1	ROCKWELL	88006.77
82NPS40-16	ROCKWELL	87584
82NPS40-17	ROCKWELL	87594
82NPS40-19	ROCKWELL	88006.78
82NPS40-21	ROCKWELL	88006.79
82NPS40-22	ROCKWELL	87602
82NPS40-23	ROCKWELL	87586
82NPS40-27	ROCKWELL	87576
82NPS40-3	ROCKWELL	88006.80
82NPS40-48	ROCKWELL	87600
82NPS40-5	ROCKWELL	87585
82NPS40-5	ROCKWELL	88006.81
82NPS40-59	ROCKWELL	87595
82NPS40-60	ROCKWELL	87599
82NPS40-70	ROCKWELL	87597
82NPS40-9	ROCKWELL	88006.82
82NY42	ROCKWELL	85594
82NY50	ROCKWELL	85595
82NY58	ROCKWELL	85596
82NY58	ROCKWELL	88006.83
82NY58-2	ROCKWELL	85597
82NY58-2	ROCKWELL	88006.84
82NY58-3	ROCKWELL	88006.85
82NY58-4	ROCKWELL	88006.86
82NY60	ROCKWELL	85599
82NY60-1	ROCKWELL	85598
82NY60-16	ROCKWELL	85603
82NY60-19	ROCKWELL	85604
82NY60-20	ROCKWELL	85602
82NY60-3	ROCKWELL	85592
82NY60-5	ROCKWELL	85601
82NY60-6	ROCKWELL	85593
82NY66-1	ROCKWELL	85600
82NYR24	ROCKWELL	82031
82NYR24-1	ROCKWELL	82032
82NYR28	ROCKWELL	82035
82NYR28-1	ROCKWELL	82034
82NYR28-3	ROCKWELL	82033
82NYR30	ROCKWELL	82050
82NYR31	ROCKWELL	82053
82NYR31-1	ROCKWELL	82052
82NYR32	ROCKWELL	82107
82NYR34	ROCKWELL	82110
82NYR34-2	ROCKWELL	82111
82NYR35	ROCKWELL	82114
82NYR36-1	ROCKWELL	82140
82NYR36-2	ROCKWELL	82160
82NYR36-3	ROCKWELL	82149
82NYR36-4	ROCKWELL	82159
82NYR38	ROCKWELL	82168
82NYR38-1	ROCKWELL	82167
82NYR38-2	ROCKWELL	82166
82NYR39	ROCKWELL	82172
82NYR39-1	ROCKWELL	82171
82NYR40	ROCKWELL	82183
82NYR40-1	ROCKWELL	82202
82NYR40-10	ROCKWELL	82199
82NYR40-2	ROCKWELL	82197
82NYR40-3	ROCKWELL	82203
82NYR40-4	ROCKWELL	82200
82NYR40-7	ROCKWELL	82194
82NYR40-8	ROCKWELL	82198
82NYR40-9	ROCKWELL	82201
82NYR42	ROCKWELL	82212
82NYR43	ROCKWELL	82213
82NYR44	ROCKWELL	82214
82NYR44-1	ROCKWELL	82217
82NYR44-2	ROCKWELL	82216
82NYR44-3	ROCKWELL	82215
82NYR44-5	ROCKWELL	82218
82NYR46	ROCKWELL	82224
82NYR46-1	ROCKWELL	82223
82NYR46-2	ROCKWELL	82222
82NYR47	ROCKWELL	82226
82NYR47-1	ROCKWELL	82225
82NYR48	ROCKWELL	82242
82NYR48-1	ROCKWELL	82229
82NYR48-10	ROCKWELL	82237
82NYR48-11	ROCKWELL	82239
82NYR48-13	ROCKWELL	82236
82NYR48-2	ROCKWELL	82240
82NYR48-3	ROCKWELL	82238
82NYR48-4	ROCKWELL	82245
82NYR48-5	ROCKWELL	82241
82NYR48-6	ROCKWELL	82246
82NYR48-7	ROCKWELL	82243
82NYR48-8	ROCKWELL	82244
82NYR50	ROCKWELL	82247
82NYR50-1	ROCKWELL	82249
82NYR50-2	ROCKWELL	82248
82NYR51	ROCKWELL	82251
82NYR51-1	ROCKWELL	82250
82NYR52	ROCKWELL	82253
82NYR52-1	ROCKWELL	82252
82NYR52-2	ROCKWELL	82257
82NYR52-4	ROCKWELL	82256
82NYR52-5	ROCKWELL	82254
82NYR52-6	ROCKWELL	82255
82NYR54	ROCKWELL	82259
82NYR54-1	ROCKWELL	82258
82NYR54-2	ROCKWELL	82260
82NYR55-1	ROCKWELL	82262
82NYR55-3	ROCKWELL	82264
82NYR55-4	ROCKWELL	82263
82NYR55-5	ROCKWELL	82261
82NYR56	ROCKWELL	82275
82NYR56-10	ROCKWELL	82265
82NYR56-12	ROCKWELL	82270
82NYR56-13	ROCKWELL	82272
82NYR56-14	ROCKWELL	82271
82NYR56-2	ROCKWELL	82266
82NYR56-4	ROCKWELL	82269
82NYR56-5	ROCKWELL	82276
82NYR56-6	ROCKWELL	82267
82NYR56-7	ROCKWELL	82268
82NYR56-8	ROCKWELL	82273
82NYR56-9	ROCKWELL	82274
82NYR58	ROCKWELL	82277
82NYR59	ROCKWELL	82278
82NYR60	ROCKWELL	82281
82NYR60-1	ROCKWELL	82280
82NYR60-3	ROCKWELL	82279
82NYR62	ROCKWELL	82282
82NYR63	ROCKWELL	82284
82NYR63-1	ROCKWELL	82283
82NYR64	ROCKWELL	82285
82NYR64-1	ROCKWELL	82287
82NYR64-2	ROCKWELL	82286
82NYR68-3	ROCKWELL	82288
82NYR68-4	ROCKWELL	82289
82NYR71	ROCKWELL	82290
82NYR72	ROCKWELL	82293
82NYR72-1	ROCKWELL	82292
82NYR72-2	ROCKWELL	82291
82NYR79	ROCKWELL	82294
82NYR80	ROCKWELL	82295
82NYR80-1	ROCKWELL	82296
82NYR87	ROCKWELL	82297
82NYR88	ROCKWELL	82299
82NYR88-1	ROCKWELL	82298
82NYS28	ROCKWELL	82045
82NYS28-1	ROCKWELL	82044
82NYS28-12	ROCKWELL	82042
82NYS28-14	ROCKWELL	82040
82NYS28-2	ROCKWELL	82036
82NYS28-21	ROCKWELL	82039
82NYS28-3	ROCKWELL	85843
82NYS28-4	ROCKWELL	82043
82NYS28-6	ROCKWELL	82041
82NYS28-8	ROCKWELL	82037
82NYS30	ROCKWELL	85844
82NYS30-1	ROCKWELL	82048
82NYS30-2	ROCKWELL	82046
82NYS30-3	ROCKWELL	82047
82NYS30-5	ROCKWELL	82049
82NYS31	ROCKWELL	82057
82NYS31-1	ROCKWELL	85846
82NYS31-12	ROCKWELL	82058
82NYS31-15	ROCKWELL	82067
82NYS31-16	ROCKWELL	82061
82NYS31-17	ROCKWELL	82068
82NYS31-19	ROCKWELL	82064
82NYS31-2	ROCKWELL	85847
82NYS31-25	ROCKWELL	82062
82NYS31-28	ROCKWELL	82060
82NYS31-29	ROCKWELL	82069
82NYS31-3	ROCKWELL	82059
82NYS31-30	ROCKWELL	82073
82NYS31-31	ROCKWELL	82072
82NYS31-32	ROCKWELL	82051
82NYS31-37	ROCKWELL	82070
82NYS31-4	ROCKWELL	82071
82NYS31-42	ROCKWELL	85845
82NYS31-5	ROCKWELL	82065
82NYS31-6	ROCKWELL	82066
82NYS31-7	ROCKWELL	82063
82NYS31-8	ROCKWELL	82056
82NYS32	ROCKWELL	82090
82NYS32-1	ROCKWELL	82091
82NYS32-10	ROCKWELL	82096
82NYS32-11	ROCKWELL	82087
82NYS32-12	ROCKWELL	85854
82NYS32-13	ROCKWELL	82085
82NYS32-14	ROCKWELL	82099
82NYS32-15	ROCKWELL	85849
82NYS32-17	ROCKWELL	82081
82NYS32-2	ROCKWELL	85850
82NYS32-20	ROCKWELL	82097
82NYS32-21	ROCKWELL	85852
82NYS32-23	ROCKWELL	82103
82NYS32-24	ROCKWELL	82083
82NYS32-25	ROCKWELL	85853
82NYS32-26	ROCKWELL	82078
82NYS32-28	ROCKWELL	82094
82NYS32-29	ROCKWELL	82075
82NYS32-3	ROCKWELL	82088
82NYS32-30	ROCKWELL	82092
82NYS32-31	ROCKWELL	82095
82NYS32-33	ROCKWELL	82082
82NYS32-38	ROCKWELL	82086
82NYS32-39	ROCKWELL	82084
82NYS32-4	ROCKWELL	82098
82NYS32-40	ROCKWELL	82080
82NYS32-41	ROCKWELL	82102
82NYS32-43	ROCKWELL	82079
82NYS32-45	ROCKWELL	85848
82NYS32-46	ROCKWELL	82105
82NYS32-47	ROCKWELL	82093
82NYS32-47	ROCKWELL	88006.87
82NYS32-48	ROCKWELL	82074
82NYS32-49	ROCKWELL	82100
82NYS32-5	ROCKWELL	85851
82NYS32-50	ROCKWELL	82104
82NYS32-51	ROCKWELL	82089
82NYS32-56	ROCKWELL	82106
82NYS32-6	ROCKWELL	82076
82NYS32-9	ROCKWELL	82077
82NYS33-4	ROCKWELL	82108
82NYS33-5	ROCKWELL	88006.88
82NYS34	ROCKWELL	82112
82NYS34-1	ROCKWELL	88006.89
82NYS34-13	ROCKWELL	82109
82NYS34-6	ROCKWELL	88006.90
82NYS34-7	ROCKWELL	82113
82NYS36	ROCKWELL	82163
82NYS36-1	ROCKWELL	85865
82NYS36-10	ROCKWELL	82133
82NYS36-105	ROCKWELL	82144
82NYS36-11	ROCKWELL	85860
82NYS36-12	ROCKWELL	85873
82NYS36-122	ROCKWELL	88006.91
82NYS36-13	ROCKWELL	82139
82NYS36-14	ROCKWELL	85867
82NYS36-15	ROCKWELL	82131
82NYS36-16	ROCKWELL	85862
82NYS36-17	ROCKWELL	82142
82NYS36-18	ROCKWELL	88006.92
82NYS36-19	ROCKWELL	82165
82NYS36-2	ROCKWELL	85856
82NYS36-20	ROCKWELL	82129
82NYS36-21	ROCKWELL	82126
82NYS36-22	ROCKWELL	85864
82NYS36-23	ROCKWELL	82130
82NYS36-24	ROCKWELL	82132
82NYS36-25	ROCKWELL	82134
82NYS36-26	ROCKWELL	82155
82NYS36-27	ROCKWELL	85870
82NYS36-28	ROCKWELL	82153
82NYS36-29	ROCKWELL	85868
82NYS36-3	ROCKWELL	82115
82NYS36-30	ROCKWELL	82154
82NYS36-31	ROCKWELL	85859
82NYS36-32	ROCKWELL	82124
82NYS36-36	ROCKWELL	85869
82NYS36-37	ROCKWELL	82143
82NYS36-4	ROCKWELL	82164
82NYS36-45	ROCKWELL	82147
82NYS36-47	ROCKWELL	82156
82NYS36-49	ROCKWELL	82120
82NYS36-5	ROCKWELL	82150
82NYS36-50	ROCKWELL	82145
82NYS36-50	ROCKWELL	88006.93
82NYS36-51	ROCKWELL	85855
82NYS36-53	ROCKWELL	82162
82NYS36-54	ROCKWELL	82125
82NYS36-56	ROCKWELL	82157
82NYS36-58	ROCKWELL	82117
82NYS36-6	ROCKWELL	85863
82NYS36-62	ROCKWELL	82118
82NYS36-63	ROCKWELL	82116
82NYS36-64	ROCKWELL	82127
82NYS36-65	ROCKWELL	82137
82NYS36-66	ROCKWELL	82141
82NYS36-68	ROCKWELL	82152
82NYS36-69	ROCKWELL	82128
82NYS36-7	ROCKWELL	82151
82NYS36-70	ROCKWELL	82148
82NYS36-71	ROCKWELL	82138
82NYS36-72	ROCKWELL	82146
82NYS36-73	ROCKWELL	82158

Part No.	Brand	IDLI Group
82NYS36-76	ROCKWELL	82121
82NYS36-77	ROCKWELL	85857
82NYS36-78	ROCKWELL	85861
82NYS36-79	ROCKWELL	82161
82NYS36-8	ROCKWELL	85866
82NYS36-80	ROCKWELL	82122
82NYS36-81	ROCKWELL	82136
82NYS36-82	ROCKWELL	85871
82NYS36-83	ROCKWELL	85872
82NYS36-84	ROCKWELL	82119
82NYS36-87	ROCKWELL	82123
82NYS36-9	ROCKWELL	82135
82NYS36-97	ROCKWELL	85858
82NYS38	ROCKWELL	82169
82NYS39	ROCKWELL	82170
82NYS39-1	ROCKWELL	85874
82NYS40-1	ROCKWELL	82193
82NYS40-11	ROCKWELL	82181
82NYS40-12	ROCKWELL	82178
82NYS40-13	ROCKWELL	82186
82NYS40-15	ROCKWELL	82176
82NYS40-16	ROCKWELL	82173
82NYS40-17	ROCKWELL	82188
82NYS40-18	ROCKWELL	82205
82NYS40-2	ROCKWELL	82207
82NYS40-20	ROCKWELL	82190
82NYS40-24	ROCKWELL	82204
82NYS40-26	ROCKWELL	82206
82NYS40-27	ROCKWELL	82189
82NYS40-28	ROCKWELL	82195
82NYS40-3	ROCKWELL	82182
82NYS40-30	ROCKWELL	82208
82NYS40-31	ROCKWELL	82177
82NYS40-32	ROCKWELL	82187
82NYS40-34	ROCKWELL	85875
82NYS40-36	ROCKWELL	82175
82NYS40-42	ROCKWELL	82196
82NYS40-43	ROCKWELL	82185
82NYS40-44	ROCKWELL	82179
82NYS40-47	ROCKWELL	82174
82NYS40-64	ROCKWELL	82180
82NYS40-65	ROCKWELL	82184
82NYS40-67	ROCKWELL	82191
82NYS40-71	ROCKWELL	82192
82NYS42	ROCKWELL	82210
82NYS42-1	ROCKWELL	85876
82NYS42-2	ROCKWELL	82209
82NYS42-5	ROCKWELL	82211
82NYS44	ROCKWELL	85877
82NYS44-1	ROCKWELL	82221
82NYS44-2	ROCKWELL	82219
82NYS46-1	ROCKWELL	82220
82NYS48	ROCKWELL	82234
82NYS48-1	ROCKWELL	82228
82NYS48-11	ROCKWELL	82232
82NYS48-2	ROCKWELL	82233
82NYS48-3	ROCKWELL	82227
82NYS48-6	ROCKWELL	82231
82NYS48-7	ROCKWELL	82230
82NYS48-9	ROCKWELL	82235
82NYSM34-9	ROCKWELL	86812
82NYSM40	ROCKWELL	86818
82NYSM40-14	ROCKWELL	86815
82NYSM40-19	ROCKWELL	86813
82NYSM40-21	ROCKWELL	86824
82NYSM40-4	ROCKWELL	86817
82NYSM40-45	ROCKWELL	86822
82NYSM40-48	ROCKWELL	86814
82NYSM40-49	ROCKWELL	86820
82NYSM40-52	ROCKWELL	86827
82NYSM40-53	ROCKWELL	86825
82NYSM40-54	ROCKWELL	86816
82NYSM40-6	ROCKWELL	86821
82NYSM40-7	ROCKWELL	86826
82NYSM40-75	ROCKWELL	86828
82NYSM40-8	ROCKWELL	86819
82NYSM40-9	ROCKWELL	86823
82NYT28	ROCKWELL	82038
82NYT31	ROCKWELL	82054
82NYT31-1	ROCKWELL	82055
82NYT32	ROCKWELL	82101
8306	PRECISION	86295
8307	PRECISION	86296
8308	PRECISION	86297
8309	PRECISION	86298
830-962M91	MASS.FERG.	80035
8310J	BORG-WARNR	81685
8311J	BORG-WARNR	81802
83-14-108	CLEVE.STL.	87619
83-14-74	CLEVE.STL.	87618
83-16-108	CLEVE.STL.	87622
83-16-74	CLEVE.STL.	87621
83254-2	DETROIT	80022
83254-3	ALCO	80022
83254-3	DETROIT	80022
83254-3	SPICER	80022
833618R91	IHC	80159
8338J	BORG-WARNR	88014.52
83453	VOLVO	80178
8346	PRECISION	86299
8347	PRECISION	86300
835-501M91	MASS.FERG.	80069
835-501M91-8K	MASS.FERG.	80069
835-512M91	MASS.FERG.	80104
835-727M91	MASS.FERG.	80142
835-771M91	MASS.FERG.	80007
835-790M91	MASS.FERG.	80037
835-799M91	MASS.FERG.	80037
835846	MICHIGAN	88008.97
835-991M91	MASS.FERG.	80069
836163	FIAT-ALLIS	80210
836-366M91	MASS.FERG.	80035
836-413M91	MASS.FERG.	80198
836-454M91	MASS.FERG.	80136
836-456M91	MASS.FERG.	80136
836514M91	MASS.FERG.	80001
836-616M91	MASS.FERG.	80198
836-664M91	MASS.FERG.	80197
836-666M91	MASS.FERG.	80214
836-667M91	MASS.FERG.	80213
836-668M91	MASS.FERG.	80213
836-669M91	MASS.FERG.	80238
836-775M91	MASS.FERG.	80212
836-776M91	MASS.FERG.	80210
836-891M91	MASS.FERG.	80001
836-947M91	MASS.FERG.	80198
836-948M91	MASS.FERG.	80198
837-037M91	MASS.FERG.	80037
8390J	BORG-WARNR	81054
8392	ALLIS-CHLM	80204
8392	FIAT-ALLIS	80047
839272	WHITE	80189
8393	AEC	88000.30
840007	WESCO	84504
8401J	BORG-WARNR	81743
8402J	BORG-WARNR	85598
840-302M92	MASS.FERG.	88003.26
840-303M92	MASS.FERG.	80044
840361	GAR WOOD	80150
840363	GAR WOOD	80109
8405	JAGUAR	80077
8406	PRECISION	86301
840-655M91	MASS.FERG.	80133
840-665M91	MASS.FERG.	80133
840676M91	MASS.FERG.	80069
840680M91	MASS.FERG.	80008
840-824M91	MASS.FERG.	80003
840824M92	MASS.FERG.	80001
840-824M92	MASS.FERG.	80003
840830M91	MASS.FERG.	80008
840869M91	MASS.FERG.	80133
840-899M91	MASS.FERG.	80069
840-899M92	MASS.FERG.	80069
8408J	BORG-WARNR	85856
8-41	BYPY	88004.35
841	PRECISION	88002.86
841	TEDDY TORQ	88002.86
84-10-108	CLEVE.STL.	87668
84-10-74	CLEVE.STL.	87667
841-193M91	MASS.FERG.	80001
841-193M92	MASS.FERG.	80001
841-235M91	MASS.FERG.	80161
841-236M91	MASS.FERG.	80035
841-237M91	MASS.FERG.	80037
84-14-108	CLEVE.STL.	87676
84-14-74	CLEVE.STL.	87675
841-603M91	MASS.FERG.	80035
841-604M91	MASS.FERG.	80037
842227	WHITE	80220
8424J	BORG-WARNR	87478
8425J	BORG-WARNR	88006.77
843150	WHITE	80124
843150A	WHITE	80124
8433J	BORG-WARNR	85865
8437J	BORG-WARNR	81747
843942	WHITE	80188
843942A	WHITE	80001
8446	PRECISION	86302
845032	HESSTON	80035
845198	HESSTON	80069
84559	GMC	80225
848036506	KOEHRING	80109
848036509	BUFFALO-SP	80207
848036509	KOEHRING	80211
848036526	KOEHRING	80210
848036529	KOEHRING	80243
848036530	KOEHRING	80243
848036711(4)	BUFFALO-SP	80149
848036711(4)	BUFFALO-SP	80155
848036714(4)	BUFFALO-SP	80149
848036722(4)	BUFFALO-SP	80235
848036724(4)	BUFFALO-SP	80215
848037721(4)	BUFFALO-SP	80214
848037721(4)	BUFFALO-SP	80215
850	AEC	80219
8500	AEC	80223
850009	WESCO	81206
850010	WESCO	81203
850011	WESCO	88000
850012	WESCO	81208
850013	WESCO	81246
850015	WESCO	81218
850-016	WESCO	88000.01
850017	WESCO	81213
850-018	WESCO	88000.02
850021	WESCO	81239
850022	WESCO	81237
850022	WESCO	88014.01
850023	WESCO	81237
850024	WESCO	81249
850025	WESCO	81257
850025	WESCO	81277
850026	WESCO	81274
850027	WESCO	88000.08
850028	WESCO	81277
850029	WESCO	81259
850030	WESCO	81268
850031	WESCO	81268
850032	WESCO	81261
850034	WESCO	81286
850035	WESCO	81303
850037	WESCO	81302
850038	WESCO	81342
850039	WESCO	81333
850040	WESCO	81350
850042	WESCO	88013.90
850043	WESCO	81271
850-044	WESCO	85514
850045	WESCO	81346
8502	AEC	80224
85-0901	NEAPCO	86148
85-0902	NEAPCO	86179
85-0903	NEAPCO	86149
85-0904	NEAPCO	86183
85-0905	NEAPCO	86143
85-0906	NEAPCO	86180
85-0907	NEAPCO	86147
85-0908	NEAPCO	86157
85-0909	NEAPCO	86178
85-0910	NEAPCO	86158
85-0911	NEAPCO	86193
85-0912	NEAPCO	86150
85-0913	NEAPCO	86144
85-0914	NEAPCO	86182
85-0915	NEAPCO	86191
85-0916	NEAPCO	86175
85-0917	NEAPCO	86176
85-0918	NEAPCO	86282
85-0919	NEAPCO	86273
85-0920	NEAPCO	86184
85-0921	NEAPCO	86153
85-0922	NEAPCO	86154
85-0923	NEAPCO	86155
85-0924	NEAPCO	86156
85-0925	NEAPCO	86151
85-0926	NEAPCO	86152
85-0927	NEAPCO	86275
85-0928	NEAPCO	86274
85-0929	NEAPCO	86280
85-0930	NEAPCO	86187
85-0931	NEAPCO	86161
85-0932	NEAPCO	86159
85-0933	NEAPCO	86160
85-0934	NEAPCO	86162
85-0935	NEAPCO	86189
85-0936	NEAPCO	86162
85-0937	NEAPCO	86181
85-0938	NEAPCO	86161
85-0939	NEAPCO	86145
85-0940	NEAPCO	86276
85-0941	NEAPCO	86302
85-0942	NEAPCO	86301
85-0943	NEAPCO	86186
85-0944	NEAPCO	86183
85-0951	NEAPCO	86174
85-0954	NEAPCO	86174
85-0955	NEAPCO	86173
85-0956	NEAPCO	86190
85-0957	NEAPCO	86163
85-0970	NEAPCO	86177

Part No.	Brand	IDLI Group	Part No.	Brand	IDLI Group	Part No.	Brand	IDLI Group	Part No.	Brand	IDLI Group
85-0971	NEAPCO	86268	859107T2	DIAMOND R	88000.15	860054	WESCO	83105	86-0970	NEAPCO	86091
85-0972	NEAPCO	86279	859107T2	WHITE	80007	860-054	WESCO	84517	86-0971	NEAPCO	86092
85-0973	NEAPCO	86270	859568	CLARK EQU.	80135	860056	WESCO	83116	86-0972	NEAPCO	86131
85-0974	NEAPCO	86269	859P2	DIAMOND R	80221	860-056	WESCO	88014.32	86-0973	NEAPCO	86078
85-0975	NEAPCO	86179	859P2	REO MOTORS	80221	860057	WESCO	83102	86-0974	NEAPCO	86079
85-0976	NEAPCO	86271	859P2	WHITE	80221	860058	WESCO	83125	86-0975	NEAPCO	86086
85-0977	NEAPCO	86272	85WB170	ROCKWELL	86805	860060	WESCO	83129	86-0976	NEAPCO	86087
85-0979	NEAPCO	86192	85WBF1	ROCKWELL	84676	860061	WESCO	83038	86-0977	NEAPCO	86093
851	PRECISION	88002.87	85WBF2	ROCKWELL	84674	860062	WESCO	85705	86-0978	NEAPCO	86094
851	TEDDY TORQ	88002.86	85WBF3	ROCKWELL	84675	860065	WESCO	83047	86-0990	NEAPCO	86120
85-1010	NEAPCO	86164	85WBF4	ROCKWELL	84677	860066	WESCO	83062	86-0991	NEAPCO	86075
85-1012	NEAPCO	86165	85WBF5	ROCKWELL	84678	860067	WESCO	83082	86-0992	NEAPCO	86236
85-1013	NEAPCO	86166	85WBF6	ROCKWELL	84679	860070	WESCO	88014.25	86-0993	NEAPCO	86236
85-1014	NEAPCO	86167	85WBLS40-1	ROCKWELL	85148	860665M92	MASS.FERG.	80133	86-0994	NEAPCO	86248
85-1015	NEAPCO	86168	85WBLS40-2	ROCKWELL	85146	86-0901	NEAPCO	86123	86-0995	NEAPCO	86237
85-1020	NEAPCO	86169	85WBLS40-3	ROCKWELL	85147	86-0902	NEAPCO	86072	86-0996	NEAPCO	86260
85-1021	NEAPCO	86170	85WBLS48	ROCKWELL	85150	86-0903	NEAPCO	86125	86-0997	NEAPCO	86261
85-1023	NEAPCO	86172	85WBLS48-1	ROCKWELL	85149	86-0905	NEAPCO	86122	86-0998	NEAPCO	86238
851629R1	IHC	80159	85WBLS48-4	ROCKWELL	85151	86-0908	NEAPCO	86069	86-0999	NEAPCO	86250
851-631R91	IHC	80148	85WBY52-2	ROCKWELL	85585	86-0909	NEAPCO	86070	860P2	DIAMOND R	80189
851-632R91	IHC	80158	85WBY53	ROCKWELL	85586	86-0910	NEAPCO	86133	860P2	REO MOTORS	80189
85-1640	WESCO	81237	85WBY53-1	ROCKWELL	85587	86-0911	NEAPCO	86134	860P2	WHITE	80189
85-2045-37	WESCO	81301	85WBY60	ROCKWELL	85589	86-0912	NEAPCO	86095	861	PRECISION	80001
853	PRECISION	80008	85WBY60-3	ROCKWELL	85591	86-0915	NEAPCO	86096	86-1000	NEAPCO	86262
8-53-41	SPICER	87558	85WBY60-6	ROCKWELL	85588	86-0916	NEAPCO	86099	86-1001	NEAPCO	86239
853411	CLARK EQU.	80070	85WBY66	ROCKWELL	85590	86-0917	NEAPCO	86136	86-1002	NEAPCO	86263
853411	MICHIGAN	80069	85WBYR38	ROCKWELL	82016	86-0918	NEAPCO	86124	86-1003	NEAPCO	86240
85-351R91	IHC	80193	85WBYR38-1	ROCKWELL	82017	86-0919	NEAPCO	86137	86-1004	NEAPCO	86255
854	PRECISION	88002.88	85WBYR42	ROCKWELL	82023	86-0920	NEAPCO	86138	86-1005	NEAPCO	86265
85431	CATERPILLR	80150	85WBYR44	ROCKWELL	82026	86-0921	NEAPCO	86098	86-1006	NEAPCO	86241
85431	TOWMOTOR	80150	85WBYR44-1	ROCKWELL	82025	86-0922	NEAPCO	86118	86-1007	NEAPCO	86243
85432	CATERPILLR	80109	85WBYR44-2	ROCKWELL	82024	86-0924	NEAPCO	86119	86-1008	NEAPCO	86264
85432	TOWMOTOR	80151	85WBYR46	ROCKWELL	82027	86-0927	NEAPCO	86297	86-1009	NEAPCO	86266
854369	MICHIGAN	80069	85WBYR48	ROCKWELL	82029	86-0928	NEAPCO	86300	86-1010	NEAPCO	86100
854639	CLARK EQU.	80069	85WBYR58	ROCKWELL	82030	86-0929	NEAPCO	86298	86-1011	NEAPCO	86101
854639	MICHIGAN	88009.54	85WBYS28	ROCKWELL	82003	86-0930	NEAPCO	86126	86-1012	NEAPCO	86102
855016	WESCO	88000.01	85WBYS30	ROCKWELL	82004	86-0931	NEAPCO	86077	86-1013	NEAPCO	86103
855017	WESCO	81213	85WBYS31	ROCKWELL	82006	86-0933	NEAPCO	86127	86-1014	NEAPCO	86104
855018	WESCO	88000.02	85WBYS31-1	ROCKWELL	82005	86-0934	NEAPCO	86299	86-1015	NEAPCO	86105
855019	WESCO	81222	85WBYS32	ROCKWELL	82007	86-0935	NEAPCO	86076	86-1020	NEAPCO	86106
855047	CLARK EQU.	80124	85WBYS32-1	ROCKWELL	82009	86-0936	NEAPCO	86295	86-1021	NEAPCO	86107
855047	MICHIGAN	80124	85WBYS32-3	ROCKWELL	82008	86-0937	NEAPCO	86296	86-1022	NEAPCO	86108
85525A	HYSTER	80109	85WBYS34	ROCKWELL	82010	86-0939	NEAPCO	86071	86-1023	NEAPCO	86110
85525A	HYSTER	80150	85WBYS36	ROCKWELL	85841	86-0940	NEAPCO	86174	86-1024	NEAPCO	86111
8558-2805	ISUZU	80023	85WBYS36-2	ROCKWELL	82014	86-0940	SPICER	86113	86-1025	NEAPCO	86112
855846	CLARK EQU.	80142	85WBYS36-3	ROCKWELL	82015	86-0943	NEAPCO	86090	86-1030	NEAPCO	86244
855846	MICHIGAN	80142	85WBYS36-4	ROCKWELL	85840	86-0945	NEAPCO	86141	86-1031	NEAPCO	86242
855846	MICHIGAN	80143	85WBYS36-6	ROCKWELL	82011	86-0946	NEAPCO	86142	86-1032	NEAPCO	86254
856	PRECISION	80133	85WBYS36-7	ROCKWELL	82013	86-0947	NEAPCO	86132	86-1033	NEAPCO	86253
85-6035	REPUBLIC	87238	85WBYS36-8	ROCKWELL	82012	86-0948	NEAPCO	86184	86-1034	NEAPCO	86249
856665	CLARK EQU.	80077	85WBYS39-1	ROCKWELL	85842	86-0949	NEAPCO	86183	86-1035	NEAPCO	86251
857	PRECISION	88002.89	85WBYS40	ROCKWELL	82020	86-0950	NEAPCO	86139	86-1037	NEAPCO	86267
857	PREC.TOR.	80133	85WBYS40-2	ROCKWELL	82018	86-0951	NEAPCO	86114	86-1038	NEAPCO	86140
8572J	BORG-WARNR	88006.61	85WBYS40-3	ROCKWELL	82021	86-0952	NEAPCO	86115	86-1039	NEAPCO	86267
857315	GAR WOOD	80240	85WBYS40-4	ROCKWELL	82019	86-0953	NEAPCO	86116	86-1050	NEAPCO	86121
857407271A	AUDI	86061	85WBYS42	ROCKWELL	82022	86-0954	NEAPCO	86117	861452	CLARK EQU.	80142
857407272A	AUDI	86060	85WBYS48-1	ROCKWELL	82028	86-0955	NEAPCO	86085	861452	MICHIGAN	80142
857616	CLARK EQU.	80070	85WBYSM40-5	ROCKWELL	86806	86-0956	NEAPCO	86080	861452	MICHIGAN	80143
857616	MICHIGAN	80069	85WBYSM48-7	ROCKWELL	86807	86-0957	NEAPCO	86234	861453	CLARK EQU.	80142
858001	CHRYSLER	80061	86	AEC	80124	86-0958	NEAPCO	86234	861(5)	PRECISION	88002.90
858002	CHRYSLER	80065	86	TRU-CROSS	80124	86-0960	NEAPCO	86128	861939	WHITE	80140
858017	CHRYSLER	88003.47	860	PRECISION	80003	86-0961	NEAPCO	86081	862	PRECISION	88002.91
858017	CHRYSLER	88003.50	860-041	WESCO	83018	86-0962	NEAPCO	86082	862	PREC.TOR.	80001
858036732	KOEHRING	80109	860-043	WESCO	81271	86-0963	NEAPCO	86083	862	PREC.TOR.	80003
858036732	KOEHRING	80150	860046	WESCO	83015	86-0964	NEAPCO	86084	8627J	BORG-WARNR	81936
858190	CLARK EQU.	80178	860047	WESCO	83018	86-0965	NEAPCO	86089	862837	CLARK EQU.	80069
858190	MICHIGAN	80178	860048	WESCO	88014.21	86-0966	NEAPCO	86090	862837	MICHIGAN	80069
858846	CLARK EQU.	80142	860050	WESCO	83049	86-0967	NEAPCO	86088	863	PREC.TOR.	80001
858P2	WHITE	80221	860052	WESCO	83090	86-0968	NEAPCO	86129	863	PREC.TOR.	80003
859107	WHITE	80007	860053	WESCO	88014.27	86-0969	NEAPCO	86130	863324	CLARK EQU.	80037

Part No.	Brand	IDLI Group	Part No.	Brand	IDLI Group	Part No.	Brand	IDLI Group	Part No.	Brand	IDLI Group
863324	MICHIGAN	80037	875952HI	HOUGH	80198	888282	AUST.WEST.	80196	89-8027	NEAPCO	86020
8645J	BORG-WARNR	84504	875-953C91	IHC	80206	888282	CLARK EQU.	80196	89-8028	NEAPCO	86021
865-000-0	GLAENZER	80225	876044	FIAT	80221	888282	MICHIGAN	80197	89-8029	NEAPCO	86022
865-002-0	GLAENZER	80225	876045	FIAT	80221	8884814	WHITE	80001	89-8030	NEAPCO	86015
865-006-0	GLAENZER	80225	876347	WHITE	88002.95	889	PRECISION	80161	89-8031	NEAPCO	86016
865-011-0	GLAENZER	80225	876347A	WHITE	80069	890	PRECISION	80134	89-8032	NEAPCO	86017
865335	CLARK EQU.	80124	876579	CLARK EQU.	80037	891	PRECISION	88002.96	89-8033	NEAPCO	86018
865335	MICHIGAN	80124	876579	MICHIGAN	80037	891	PREC.TOR.	80178	89-8034	NEAPCO	86023
8655J	BORG-WARNR	85564	877079	CLARK EQU.	80133	891046	CLARK EQU.	80196	89-8035	NEAPCO	86024
865665	CLARK EQU.	80077	877383	CLARK EQU.	80155	891046	MICHIGAN	80197	89-8036	NEAPCO	86025
865731	MICHIGAN	80042	877383	MICHIGAN	80155	892	PRECISION	80188	89-8037	NEAPCO	86026
865751	CLARK EQU.	80042	878268	WHITE	80001	892187	CLARK EQU.	80185	89-8038	TRW	86013
865775	MICHIGAN	80077	878624C91	HOUGH	80225	89-2510	AEC	80005	89-8039	NEAPCO	86014
8657J	BORG-WARNR	85846	878-624C91	IHC	80142	89-2510	AEC	80011	89-8040	NEAPCO	86011
866-785R91	IHC	80211	878626C91	IHC	80225	892837	CLARK EQU.	80155	89-8041	NEAPCO	86012
866887	CLARK EQU.	80149	878771C91	IHC	80245	892837	MICHIGAN	80155	89-8042	NEAPCO	86033
866887	MICHIGAN	80109	878916	WHITE	80182	892850	CLARK EQU.	80240	89-8043	NEAPCO	86034
866887	MICHIGAN	80149	87977	VOLVO	80024	892969	CLARK EQU.	80008	89-8044	NEAPCO	86010
866888	CLARK EQU.	80151	879829R92	IHC	87399	892969	CLARK EQU.	80133	89-8045	NEAPCO	86049
866888	MICHIGAN	80151	87984	HYSTER	80190	892969	MICHIGAN	80008	89-8046	NEAPCO	86048
867294	CLARK EQU.	80033	87984	HYSTER	80196	892989	CLARK EQU.	80008	89-8047	NEAPCO	86005
86751	CLARK EQU.	80042	879849R91	IHC	87399	893	PRECISION	80220	89-8048	NEAPCO	86004
867-635C91	IHC	80182	879849R92	IHC	88008.59	8933001363	AMER.MOTOR	86053	89-8049	NEAPCO	86214
868218	CLARK EQU.	80003	879946	CLARK EQU.	80213	893815	CLARK EQU.	80214	89-8050	NEAPCO	86215
868218	CLARK EQU.	80142	879946	MICHIGAN	80213	893815	MICHIGAN	80237	89-8051	NEAPCO	86212
8687J	BORG-WARNR	81089	88	TRU-CROSS	80140	893998	CLARK EQU.	80157	89-8052	NEAPCO	86213
8694	JAGUAR	80142	880	PRECISION	80037	893998	MICHIGAN	80157	89-8053	NEAPCO	86210
869-620R91	IHC	80140	880798	WHITE	80077	894	PRECISION	80185	89-8054	NEAPCO	86211
870	PRECISION	80033	8-809-11	ATHEY PROD	80240	894362	WHITE	80190	89-8055	NEAPCO	86027
870290900	SAVIEM	80221	881	PRECISION	88002.94	894655	CLARK EQU.	80001	89-8056	NEAPCO	86032
870311	CHRYSLER	88003.50	881	PREC.TOR.	80037	895	PRECISION	80186	89-8060	NEAPCO	86206
870360900	SAVIEM	80142	8812	KENBAR	87331	8-95-49	SPICER	88014.89	89-8060	NEAPCO	86220
870(5)	PRECISION	88002.92	881-908C91	IHC	80172	895600	CLARK EQU.	80037	89-8061	NEAPCO	86221
870538	CHRYSLER	88003.50	882320	FIAT	80041	896	PRECISION	80184	89-8062	NEAPCO	86218
870550300	SAVIEM	80180	882321	FIAT	80041	896-057C1	IHC	80235	89-8063	NEAPCO	86219
870-638R91	IHC	80239	882577R91	IHC	80077	896084	CLARK EQU.	80001	89-8064	NEAPCO	86212
870688	CLARK EQU.	80069	882802	CLARK EQU.	80077	896085	CLARK EQU.	80001	89-8065	NEAPCO	86213
870688	MICHIGAN	80069	88285A	HYSTER	80044	896173	WESCO	80228	89-8066	NEAPCO	86023
871	PRECISION	88002.93	883048	WHITE	80190	896173	WHITE	80228	89-8067	NEAPCO	86024
871	PREC.TOR.	80033	883048P2	DIAMOND R	80190	897	PRECISION	80179	89-8068	NEAPCO	86013
871-280R91	IHC	80055	883048P2	WHITE	80190	897707	VOLVO	80061	89-8069	NEAPCO	86014
871280R92	IHC	80044	883235	FIAT	80024	897732	WHITE	80142	89-8070	NEAPCO	86011
871888	WHITE	80001	883235	NSU-PRINZ	80024	898	PRECISION	80187	89-8071	NEAPCO	86012
871888	WHITE	80003	883236	FIAT	80024	89-8001	NEAPCO	86051	89-8072	NEAPCO	86045
871888	WHITE	80023	883333A	WABCO	80112	89-8002	NEAPCO	86052	89-8073	NEAPCO	86046
871888A	WHITE	80023	883-33A	WABCO	80100	89-8003	NEAPCO	86029	89-8074	NEAPCO	86289
87188A	WHITE	80024	884518	CLARK EQU.	80134	89-8004	NEAPCO	86028	89-8075	NEAPCO	86290
872273	CLARK EQU.	80196	884518	MICHIGAN	80134	89-8005	NEAPCO	86001	89-8076	NEAPCO	86031
872273	MICHIGAN	80197	8845370	BEDFORD	80072	89-8006	NEAPCO	86003	89-8077	NEAPCO	86008
872454	CLARK EQU.	80133	8845370	GMC	80077	89-8007	NEAPCO	86002	89-8078	NEAPCO	86009
872773	CLARK EQU.	80196	8845379	BEDFORD	80024	89-8008	NEAPCO	86037	89-8079	NEAPCO	86047
872773	MICHIGAN	80197	8845379	GMC	80142	89-8009	NEAPCO	86035	89-8080	NEAPCO	86207
873721	CLARK EQU.	80008	8845636	BEDFORD	80023	89-8009	NEAPCO	86037	89-8081	NEAPCO	86055
873726	CLARK EQU.	80008	885	PRECISION	88002.95	89-8010	NEAPCO	86036	89-8082	NEAPCO	86054
873726	CLARK EQU.	88002.88	885	PREC.TOR.	80069	89-8011	NEAPCO	86058	89-8083	NEAPCO	86045
873726	MICHIGAN	80124	885386	CLARK EQU.	88002.88	89-8012	NEAPCO	86059	89-8085	NEAPCO	86060
874026	WHITE	80072	886518	CLARK EQU.	80134	89-8013	NEAPCO	86056	89-8086	NEAPCO	86061
874026	WHITE	80077	886518	MICHIGAN	80134	89-8014	NEAPCO	86057	89-8087	NEAPCO	86062
874789	CLARK EQU.	80155	886676	CLARK EQU.	80133	89-8015	NEAPCO	86208	89-8088	NEAPCO	86063
874789	MICHIGAN	80155	886-759R91	IHC	80077	89-8016	NEAPCO	86209	89-8090	NEAPCO	86286
874845	CLARK EQU.	80033	886-890R91	IHC	80124	89-8017	NEAPCO	86031	89-8091	NEAPCO	86284
874910	CLARK EQU.	80238	886-919R91	IHC	80142	89-8018	NEAPCO	86030	89-8092	NEAPCO	86287
874910	MICHIGAN	80238	8871795	GMC	80210	89-8019	NEAPCO	86044	89-8093	NEAPCO	86285
874969	CLARK EQU.	80238	887737	CLARK EQU.	80156	89-8020	NEAPCO	86042	89-8094	NEAPCO	86288
874969	MICHIGAN	80238	887737	MICHIGAN	80156	89-8021	NEAPCO	86043	89-8095	NEAPCO	86291
87592H1	HOUGH	80198	888	PRECISION	80136	89-8022	NEAPCO	86041	89-8096	NEAPCO	86292
875952	IHC	80198	8882002	KOEHRING	80188	89-8023	NEAPCO	86039	89-8097	NEAPCO	86294
8759529	HOUGH	80198	888280	CLARK EQU.	80192	89-8024	NEAPCO	86038	89-8098	NEAPCO	86293
875-952C91	IHC	80198	888280	CLARK EQU.	80198	89-8025	NEAPCO	86040	89-8100	NEAPCO	86199
875-952H1	IHC	80198	888280	MICHIGAN	80192	89-8026	NEAPCO	86019	89-8101	NEAPCO	86198

Part No.	Brand	IDLI Group	Part No.	Brand	IDLI Group	Part No.	Brand	IDLI Group	Part No.	Brand	IDLI Group
89-8102	NEAPCO	86197	9	WYLIE	88002.77	9022807	GMC	80111	9029107	GMC	80233
89-8103	NEAPCO	86196	900	PRECISION	80205	9023360	GMC	80001	902T110	TIMBERJACK	80182
89-8104	NEAPCO	86053	9000	ALCO	88012.34	9023656	GMC	80178	902T110V11	VERSATILE	80182
89-8105	NEAPCO	86067	9000	TRU-CROSS	80235	9023672	GMC	80238	902T111	TIMBERJACK	80189
89-8106	NEAPCO	86068	9000445	GAR WOOD	80001	9023673	GMC	80077	902T111V1	VERSATILE	80189
89-8113	NEAPCO	86064	9000447	GAR WOOD	80008	9023940	GMC	80213	903	PRECISION	88012.34
89-8114	NEAPCO	86066	9001	TRU-CROSS	80234	9023941	GMC	80213	903	PREC.TOR.	80235
89-8115	NEAPCO	86065	9002	TRU-CROSS	80233	9024660	GMC	80215	9030138	GMC	80228
89-8205	NEAPCO	86200	900314	STUDEBAKER	80075	9024661	GMC	88003	903019	UNIT RIG	80070
89-8206	NEAPCO	86201	9003400	BUFFALO-SP	80214	9024673	GMC	80240	9030220	GMC	80213
89-8207	NEAPCO	86224	9003400	KOEHRING	80212	9024678	GMC	80239	9030469	GMC	80008
89-8208	NEAPCO	86225	900344	UNIT RIG	80070	9024705	GMC	80213	9030836	GMC	80225
89-8209	TRW	86220	9003889	KOEHRING	80186	9024706	GMC	80213	9031	AEC	80234
89-8210	NEAPCO	86227	9004605	BUFFALO-SP	80109	9024786	GMC	80214	9031	NEAPCO	88002.56
89-8211	NEAPCO	86222	9004605	KOEHRING	80210	9024787	GMC	88003	903-10-1	ROCKWELL	87926
89-8212	NEAPCO	86223	9004605	KOEHRING	80213	9024788	GMC	80207	9031109	GMC	80228
89-8213	NEAPCO	86228	9005	NEAPCO	88002.55	9024789	GMC	88003.02	903-11-1	ROCKWELL	87936
89-8214	NEAPCO	86229	9005X	NEAPCO	88002.55	9024790	GMC	80215	903-12-1	ROCKWELL	87927
89-8220	NEAPCO	86202	90072	HUB CITY	80037	9024791	GMC	80239	903-13-1	ROCKWELL	87942
89-8221	NEAPCO	86203	900731	UNIT RIG	80001	9024792	GMC	88003.04	9031316	GMC	80225
89-8222	NEAPCO	86204	9009877	KOEHRING	80069	9024793	GMC	88003.05	903-14-1	ROCKWELL	87951
89-8224	NEAPCO	86205	901	PRECISION	88002.97	9024794	GMC	80240	903-15-1	ROCKWELL	87907
8983500495	AMER.MOTOR	86140	901	PREC.TOR.	80213	9024795	GMC	80242	903-16-1	ROCKWELL	87954
8983500497	AMER.MOTOR	86140	90133203110	PORSCHE	86106	9024796	GMC	88003.07	903-17-1	ROCKWELL	87910
8983500976	AMER.MOTOR	86139	90133203200	PORSCHE	86107	9024891	GMC	80198	9032007	GMC	80245
8983500976	AMER.MOTOR	86191	90133229310	PORSCHE	86169	9024892	GMC	80001	9032043	GMC	80213
899	PRECISION	80219	90133229310	PORSCHE	86170	9024893	GMC	80001	9032043	GMC	80238
899-392R91	IHC	80190	90133229312	PORSCHE	86107	9024945	GMC	80235	9032699	GMC	80245
899599	CLARK EQU.	80198	90133229312	PORSCHE	86108	9024959	GMC	80235	9033306	GMC	80150
899602	CLARK EQU.	80238	90133229312	PORSCHE	86110	9024992	GMC	80001	9033424	GMC	80154
899603	CLARK EQU.	80224	90133229312	PORSCHE	86172	9025	NEAPCO	88002.55	9033740	GMC	80140
8A1822(4)	LINK BELT	80240	9014	AEC	80235	9025191	GMC	80112	9033903	GMC	80193
8A2966(4)	LINK BELT	80215	9015	AEC	80234	9025207	GMC	80240	9033904	GMC	80196
8A3337(4)	LINK BELT	80215	9015	PILOT	80234	9025322	GMC	80235	9034128	GMC	80228
8A3338(4)	LINK BELT	80215	9016	PILOT	80234	9025434	GMC	80233	9034424	GMC	80154
8A3613(4)	LINK BELT	80215	9017	AEC	80233	9025543	GMC	80234	9034661	GMC	80234
8D0537	CATERPILLR	80172	902	PRECISION	88002.98	9025544	GMC	80234	9034950	GMC	80210
8D1505	CATERPILLR	80007	9020268	GMC	88003.26	9025557	GMC	80234	9034965	GMC	80220
8D1508	CATERPILLR	80007	9020304	GMC	80001	9025557	GMC	88003.02	9034965	GMC	80234
8D1566	CATERPILLR	80007	902-10-1	ROCKWELL	87924	9025558	GMC	80213	9035	NEAPCO	88002.57
8D2144	CATERPILLR	80234	902-11-1	ROCKWELL	87940	9025559	GMC	88003.04	903-50-1	ROCKWELL	87920
8D2909	CATERPILLR	80003	902-12-1	ROCKWELL	87925	9025560	GMC	88003.05	903-50-4	ROCKWELL	87921
8D3144	CATERPILLR	80234	9021250	GMC	88003.04	9025561	GMC	88003.26	903-50-5	ROCKWELL	87922
8D537	CATERPILLR	80172	9021251	GMC	80240	9025578	GMC	80216	903-51-1	ROCKWELL	87946
8D7719	CATERPILLR	80213	902-13-1	ROCKWELL	87941	9025579	GMC	80214	903-51-4	ROCKWELL	87947
8F7719	CATERPILLR	80210	902-14-1	ROCKWELL	87950	9025580	GMC	80238	903-51-5	ROCKWELL	87948
8G3000	BMC	80024	902-15-1	ROCKWELL	87906	9025633	GMC	80234	903-54-1	ROCKWELL	87968
8G3001	BMC	80024	9021527	GMC	80123	9025634	GMC	80235	903-54-4	ROCKWELL	87969
8G3002	BMC	80142	9021527	GMC	88003.54	9025635	GMC	80234	903-54-5	ROCKWELL	87970
8G3003	BMC	80124	902-16-1	ROCKWELL	87953	9025903	GMC	80198	9036	NEAPCO	88002.58
8G3008	BMC	80180	9021682	GMC	80207	9025904	GMC	88002.99	9036203	GMC	80182
8G8018	BMC	80077	9021683	GMC	80215	9026	AEC	80233	9036325	GMC	80213
8G8183	BMC	80024	9021686	GMC	80240	9026434	GMC	80233	9036593	GMC	80149
8G8967	LEYLAND	80024	9021691	GMC	88003.05	9026499	GMC	88003	9037278	GMC	80221
8G8979	BMC	80023	902-17-1	ROCKWELL	87909	9026500	GMC	88003.05	9037580	GMC	80196
8G8981	BMC	80041	9021875	GMC	80235	9026501	GMC	80214	9038235	GMC	80215
8G8981	LEYLAND	80041	9022353	GMC	88003.05	9026502	GMC	80207	9038445	GMC	80193
8H3853	CATERPILLR	80149	9022503	GMC	80240	9026503	GMC	80215	9038448	GMC	80001
8H6473	CATERPILLR	88001.38	9022625	GMC	80240	9026504	GMC	80239	9038734	GMC	80171
8K6042	CATERPILLR	80213	9022626	GMC	80214	9026505	GMC	80240	9039278	GMC	80221
8K7002	CATERPILLR	80198	9022733	GMC	80239	9026508	GMC	80231	904	PRECISION	88012.31
8L6301	CATERPILLR	80163	9022734	GMC	80215	9026534	GMC	88003.02	904	PREC.TOR.	80234
8M4841	FORD	88002.86	9022735	GMC	80234	9026536	GMC	80242	9041260	GMC	80220
8M7039	FORD	80049	9022736	GMC	80112	9026541	GMC	88003.06	9041296	GMC	80008
8M7042	FORD	80044	9022753	GMC	80213	9026892	GMC	80238	9041321	GMC	80008
8MTH7039	FORD	80142	9022755	GMC	80207	9026919	GMC	80155	9041334	GMC	80188
8P8132	CATERPILLR	88001.92	9022756	GMC	80242	9027735	GMC	80234	9042141	GMC	80210
8V1540	CATERPILLR	88001.92	9022763	GMC	80240	9027758	GMC	80234	904236	CLARK EQU.	80238
9	AEC	87325	9022766	GMC	80235	9027855	GMC	80213	9042578	GMC	80220
9	AEC	88007.59	9022806	GMC	80112	9027856	GMC	80238	9042907	GMC	80069

Part No.	Brand	IDLI Group	Part No.	Brand	IDLI Group	Part No.	Brand	IDLI Group	Part No.	Brand	IDLI Group
9042934	GMC	80067	906929	AMER.MOTOR	80075	91021384	BEDFORD	80180	9201	NEAPCO	88002.35
9044	NEAPCO	88002.46	906929	WHITE	80072	91021384	GMC	80142	920100	WESCO	88001.13
9044138	GMC	80234	906929	WILLYS	80072	911	PRECISION	80234	920102	WESCO	88001.14
9044140	GMC	80234	907	PRECISION	80207	911284	STUDEBAKER	80142	920103	WESCO	88001.12
9044805	GMC	80205	907	PREC.TOR.	80211	91133203202	PORSCHE	86108	920110	WESCO	81218
9044896	GMC	80049	9071	NEAPCO	88002.74	911-3B	PRECISION	88003.06	920111	WESCO	88006.99
9044898	GMC	80221	907-10-1	ROCKWELL	87923	9113J	BORG-WARNR	88015.24	920112	WESCO	81209
9044923	GMC	88002.77	907-11-1	ROCKWELL	87935	9113J10800	BORG-WARNR	87676	920-113	WESCO	87943
9044943	GMC	80067	907-12-1	ROCKWELL	87949	9113J7200	BORG-WARNR	87675	920-114	WESCO	88007.03
9044961	GMC	80221	907-13-1	ROCKWELL	87901	911524	WESCO	80225	920-115	WESCO	88007.02
9045104	GMC	80205	907-14-1	ROCKWELL	87952	911525	WESCO	80227	920-118	WESCO	87938
904570	GMC	88008.41	907-15-1	ROCKWELL	87908	9119	TRU-CROSS	88002.36	920119	WESCO	87938
9046	NEAPCO	88002.48	9072	NEAPCO	88002.73	911900	WESCO	80235	920-122	WESCO	88007.09
9046303	GMC	88003.04	907-3A	PRECISION	88003.02	911902	WESCO	80233	920123	WESCO	88007.08
9046306	GMC	80240	907-50-1	ROCKWELL	87917	911-902	WESCO	80234	920-126	WESCO	88007.09
9046312	GMC	88003.02	907-50-3	ROCKWELL	88006.94	911915	WESCO	80234	920-127	WESCO	88007.08
9046315	GMC	80215	907-50-4	ROCKWELL	87918	911916	WESCO	80234	920-130	WESCO	88007.09
904802	AMER.MOTOR	80060	907-50-5	ROCKWELL	87919	911917	WESCO	80233	920-131	WESCO	88007.12
904802	AMER.MOTOR	80061	907-50-7	ROCKWELL	88006.95	911931	WESCO	80234	920-134	WESCO	88007.13
9049686	GMC	80225	907-51-1	ROCKWELL	87943	911934	WESCO	80234	920-138	WESCO	88001.06
9049775	GMC	80235	907-51-10	ROCKWELL	88006.96	911A15010	EIMCO	80069	920140	WESCO	87933
905	PRECISION	80198	907-51-11	ROCKWELL	88006.97	912	PRECISION	80241	920141	WESCO	88001.06
9050	NEAPCO	88002.71	907-51-3	ROCKWELL	88006.98	912	PREC.TOR.	80233	920142	WESCO	88007.16
9050449	GMC	80198	907-51-4	ROCKWELL	87944	912-2949455	EIMCO	80217	920-145	WESCO	88001.06
9051	NEAPCO	88002.67	907-51-5	ROCKWELL	87945	912A15020	EIMCO	80193	920-146	WESCO	88001.06
9051148	GMC	80228	907-51-7	ROCKWELL	88006.99	912A15023	EIMCO	80210	920-147	WESCO	88007.15
9051383	GMC	80197	907-52-1	ROCKWELL	87962	912A15034	EIMCO	88010.60	920151	WESCO	88000.11
9051384	GMC	80198	907-52-10	ROCKWELL	88007	912A45084	EIMCO	80001	920152	WESCO	88000.12
9052	NEAPCO	88002.64	907-52-11	ROCKWELL	88007.01	912A45087	EIMCO	80003	920153	WESCO	88007.24
9052137	GMC	80213	907-52-3	ROCKWELL	88007.02	912P15053	EIMCO	80213	9201B	NEAPCO	88002.34
9052139	GMC	80171	907-52-4	ROCKWELL	87963	913	PRECISION	80242	9202B	NEAPCO	88002.36
9052140	GMC	80214	907-52-5	ROCKWELL	87964	913-3A	PRECISION	88003.07	9203	NEAPCO	88002.33
9052141	GMC	80214	907-52-7	ROCKWELL	88007.03	9135J	BORG-WARNR	85551	920-302	WESCO	85102
9052141	GMC	80215	907794	GMC	87238	913-774R91	IHC	80140	920-306	WESCO	88006.97
9052166	GMC	80149	908	PRECISION	80215	9138J	BORG-WARNR	85549	920-306	WESCO	88007.06
9052499	GMC	80001	908	PREC.TOR.	80217	913P150010	EIMCO	80213	920-307	WESCO	88006.97
9052692	GMC	80227	908-5	PRECISION	88003.03	914	PRECISION	88003.08	920-307	WESCO	88007.07
9052699	GMC	80245	909	PRECISION	80239	914236R1	IHC	88007.55	920-308	WESCO	88002.34
9053	NEAPCO	88002.67	909-3	PRECISION	88003.04	914237R1	IHC	88014.85	920-309	WESCO	88002.36
9053683	GMC	80233	909-3B	PRECISION	88003.05	914238R1	IHC	88008.63	920-310	WESCO	88007
9053744	GMC	80182	909476	AMER.MOTOR	80090	914243R1	IHC	88013.73	920-310	WESCO	88007.10
9053746	GMC	80221	909476	JEEP	80090	91433202903	PORSCHE	86111	920-311	WESCO	88007.01
9053747	GMC	80188	909476	WHITE	80090	914381	STUDEBAKER	80178	920-311	WESCO	88007.11
9053747	GMC	80233	909476	WILLYS	80090	914382	STUDEBAKER	80142	920-312	WESCO	81848
9053828	GMC	80001	909-50-1	ROCKWELL	87914	9146J	BORG-WARNR	87421	920-313	WESCO	81848
905-3B	PRECISION	88002.99	909-50-3	ROCKWELL	88007.04	9147J	BORG-WARNR	87432	920-314	WESCO	81848
9054	NEAPCO	88002.60	909-50-4	ROCKWELL	87915	915	PREC.TOR.	80077	920-316	WESCO	81848
9055	NEAPCO	88002.65	909-50-5	ROCKWELL	87916	915	PREC.TOR.	80178	9203B	NEAPCO	88002.33
9056	NEAPCO	88002.43	909-50-6	ROCKWELL	88007.05	915	PREC.TOR.	80210	9205	NEAPCO	88002.33
9057	NEAPCO	88002.60	909-51-1	ROCKWELL	87937	915141	CHRYSLER	88001.60	9205B	NEAPCO	88002.33
9058	NEAPCO	88002.45	909-51-10	ROCKWELL	88007.06	915757	CHRYSLER	88003.51	920-825	WESCO	88006.96
9059	NEAPCO	88002.47	909-51-11	ROCKWELL	88007.07	916	PREC.TOR.	80189	920-825	WESCO	88007.06
9059	NEAPCO	88002.48	909-51-3	ROCKWELL	88007.08	916650R91	IHC	80189	920-838	WESCO	88002.33
906	PRECISION	80214	909-51-4	ROCKWELL	87938	916767	CHRYSLER	88003.47	920A120015	EIMCO	80224
906	PREC.TOR.	80212	909-51-5	ROCKWELL	87939	917	PREC.TOR.	80221	920M940043	EIMCO	88011.35
906	PREC.TOR.	80218	909-51-6	ROCKWELL	88007.09	918	PREC.TOR.	80225	921	PRECISION	80149
9060	NEAPCO	88002.43	909-52-1	ROCKWELL	87965	9182	TRU-CROSS	88002.37	92-1514	PAPEC MACH	80069
9060	NEAPCO	88002.59	909-52-10	ROCKWELL	88007.10	918834R1	IHC	88013.76	921-918C1	IHC	80224
9061	NEAPCO	88002.42	909-52-11	ROCKWELL	88007.11	918835R1	IHC	88014.86	922	PREC.TOR.	80099
9062	NEAPCO	88002.67	909-52-3	ROCKWELL	88007.12	918836R91	IHC	88008.64	922	PREC.TOR.	80101
9063	NEAPCO	88002.44	909-52-4	ROCKWELL	87966	918837R91	IHC	88008.53	922-013	WESCO	88015.43
906-3	PRECISION	88003	909-52-5	ROCKWELL	87967	918842R1	IHC	88013.66	922-014	WESCO	88015.43
906344	AMER.MOTOR	80024	909-52-6	ROCKWELL	88007.13	919	PRECISION	88003.09	922-015	WESCO	88015.43
906344	JEEP	80023	9096F23	M.R.S.	80153	9190690	GMC	80021	922-016	WESCO	88015.43
906344	WILLYS	80023	9096F33	M.R.S.	80150	92	PERFECT-CI	80001	922-017	WESCO	88015.43
9065	NEAPCO	88002.50	9096F33	M.R.S.	80153	92	TRU-CROSS	80001	922-018	WESCO	88015.43
906(5)	PRECISION	88003.01	9-0S12	DIGIO-LIFT	80047	92	TRU-CROSS	80003	922078R1	IHC	88013.69
9066	NEAPCO	88002.52	9-0S12	LIFT TRUCK	80047	920	PRECISION	80216	922-50-1	ROCKWELL	87911
9067	NEAPCO	88002.51	910	PRECISION	80240	920018	WESCO	88002.87	922-50-3	ROCKWELL	88007.14
9069	NEAPCO	88002.70	910	PREC.TOR.	80243	9201	NEAPCO	88002.34	922-50-4	ROCKWELL	87912

Part No.	Brand	IDLI Group
922-50-5	ROCKWELL	87913
922-51-1	ROCKWELL	87932
922-51-3	ROCKWELL	88007.15
922-51-4	ROCKWELL	87933
922-51-5	ROCKWELL	87934
922-51-6	ROCKWELL	88007.16
922-52-1	ROCKWELL	87955
922-52-4	ROCKWELL	87956
922-52-5	ROCKWELL	87957
923	PRECISION	80206
92333203200	PORSCHE	86109
92333203700	PORSCHE	86109
92333203700	PORSCHE	86171
923811	JEEP	80055
924	PREC.TOR.	80198
924	PREC.TOR.	80201
924189	AMER.MOTOR	80074
924189	WILLYS	80074
924190	AMER.MOTOR	80071
924190	JEEP	80163
924190	WILLYS	80071
9244286	GMC	80234
925	PRECISION	88010.38
925	PRECISION	88010.44
9251287	GMC	80233
925-395C1	IHC	80207
925397C1	IHC	80210
925-397C91	IHC	80215
926	PREC.TOR.	80199
9266	ALCO	88002.50
927	PRECISION	80213
927	PREC.TOR.	80212
927-50-1	ROCKWELL	87904
927-50-5	ROCKWELL	87905
927-51-1	ROCKWELL	87930
927-51-5	ROCKWELL	87931
927-52-1	ROCKWELL	87960
927-52-5	ROCKWELL	87961
9276J	BORG-WARNR	87524
9277-2	KENBAR	87325
9277-2	KENBAR	88007.59
9279J	BORG-WARNR	82046
928	PRECISION	80238
9281J	BORG-WARNR	87522
92833202901	PORSCHE	86112
9283J	BORG-WARNR	87450
92850	MICHIGAN	80231
9288	ALCO	88002.52
928P2	DIAMOND R	80178
928P2	REO MOTORS	80178
928P2	WHITE	80178
929	PRECISION	80155
929369C91	IHC	80238
929370C91	IHC	80238
9296SF	SPICER	88007.30
92N170	ROCKWELL	86831
92NDCA1	ROCKWELL	86837
92NDCA2	ROCKWELL	86833
92NDCA4	ROCKWELL	86832
92NDCA5	ROCKWELL	86836
92NDCA6	ROCKWELL	86835
92NDCA8	ROCKWELL	86834
92NF1	ROCKWELL	84713
92NF10	ROCKWELL	84710
92NF11	ROCKWELL	84708
92NF12	ROCKWELL	84694
92NF13	ROCKWELL	84714
92NF14	ROCKWELL	84695
92NF15	ROCKWELL	84712
92NF16	ROCKWELL	84699
92NF17	ROCKWELL	84700
92NF18	ROCKWELL	84705
92NF2	ROCKWELL	84698
92NF20	ROCKWELL	84711
92NF22	ROCKWELL	84706
92NF23	ROCKWELL	84709
92NF24	ROCKWELL	85885
92NF26	ROCKWELL	84704
92NF3	ROCKWELL	84707
92NF4	ROCKWELL	84701
92NF5	ROCKWELL	84702
92NF6	ROCKWELL	84703
92NF7	ROCKWELL	84696
92NF8	ROCKWELL	84697
92NLS48	ROCKWELL	85180
92NLS48-1	ROCKWELL	85179
92NLS48-10	ROCKWELL	85175
92NLS48-12	ROCKWELL	85187
92NLS48-12DC	ROCKWELL	85188
92NLS48-17	ROCKWELL	85177
92NLS48-17DC	ROCKWELL	85178
92NLS48-1DC1	ROCKWELL	85181
92NLS48-2	ROCKWELL	85171
92NLS48-2DC1	ROCKWELL	85172
92NLS48-3	ROCKWELL	85173
92NLS48-37DC	ROCKWELL	85189
92NLS48-47DC	ROCKWELL	85184
92NLS48-49DC	ROCKWELL	85185
92NLS48-52DC	ROCKWELL	85186
92NLS48-55DC	ROCKWELL	88007.17
92NLS48-7	ROCKWELL	85174
92NLS48-9	ROCKWELL	85176
92NLS48DC	ROCKWELL	85182
92NLS48DC1	ROCKWELL	85183
92NPS48-1	ROCKWELL	87641
92NPS48-10	ROCKWELL	87640
92NPS48-12	ROCKWELL	87632
92NPS48-14	ROCKWELL	87639
92NPS48-2	ROCKWELL	87642
92NPS48-21	ROCKWELL	87629
92NPS48-25	ROCKWELL	87634
92NPS48-8	ROCKWELL	87647
92NY51	ROCKWELL	85613
92NY51-2	ROCKWELL	85614
92NY60-6	ROCKWELL	85618
92NY64	ROCKWELL	85615
92NY66	ROCKWELL	85608
92NY66-1	ROCKWELL	85606
92NY66-10	ROCKWELL	85612
92NY66-11	ROCKWELL	85616
92NY66-12	ROCKWELL	85609
92NY66-13	ROCKWELL	85620
92NY66-14	ROCKWELL	85619
92NY66-5	ROCKWELL	85605
92NY66-6	ROCKWELL	85607
92NY66-8	ROCKWELL	85611
92NY68-1	ROCKWELL	85617
92NY68-3	ROCKWELL	85610
92NYR28-3	ROCKWELL	82300
92NYR36	ROCKWELL	82339
92NYR38	ROCKWELL	82349
92NYR40	ROCKWELL	82367
92NYR40-3	ROCKWELL	82365
92NYR40-7	ROCKWELL	82359
92NYR42	ROCKWELL	82373
92NYR42-1	ROCKWELL	82374
92NYR42-2	ROCKWELL	82372
92NYR42-3	ROCKWELL	82371
92NYR44	ROCKWELL	82379
92NYR44-6	ROCKWELL	82380
92NYR46-1	ROCKWELL	82390
92NYR46-2	ROCKWELL	82388
92NYR46-3	ROCKWELL	82389
92NYR47	ROCKWELL	82391
92NYR47-1	ROCKWELL	82392
92NYR48-2	ROCKWELL	82413
92NYR48-4	ROCKWELL	82405
92NYR48-5	ROCKWELL	82407
92NYR48-7	ROCKWELL	82409
92NYR48-8	ROCKWELL	82406
92NYR48-9	ROCKWELL	82410
92NYR50-1	ROCKWELL	82415
92NYR50-2	ROCKWELL	82414
92NYR50-3	ROCKWELL	82416
92NYR50-4	ROCKWELL	82417
92NYR52-1	ROCKWELL	82419
92NYR52-3	ROCKWELL	82420
92NYR52-5	ROCKWELL	82422
92NYR54	ROCKWELL	82427
92NYR54-1	ROCKWELL	82424
92NYR54-3	ROCKWELL	82423
92NYR54-5	ROCKWELL	82425
92NYR54-6	ROCKWELL	82426
92NYR55	ROCKWELL	82429
92NYR55-1	ROCKWELL	82428
92NYR55-2	ROCKWELL	82430
92NYR55-3	ROCKWELL	82431
92NYR55-4	ROCKWELL	82432
92NYR55-5	ROCKWELL	82433
92NYR56	ROCKWELL	82443
92NYR56-1	ROCKWELL	82434
92NYR56-10	ROCKWELL	82441
92NYR56-11	ROCKWELL	82436
92NYR56-12	ROCKWELL	82445
92NYR56-13	ROCKWELL	82437
92NYR56-15	ROCKWELL	82442
92NYR56-2	ROCKWELL	82439
92NYR56-3	ROCKWELL	82440
92NYR56-4	ROCKWELL	82438
92NYR56-5	ROCKWELL	82435
92NYR56-6	ROCKWELL	82444
92NYR58	ROCKWELL	82446
92NYR58-1	ROCKWELL	82448
92NYR58-3	ROCKWELL	82447
92NYR60	ROCKWELL	82450
92NYR60-1	ROCKWELL	82449
92NYR60-2	ROCKWELL	82451
92NYR62-1	ROCKWELL	82453
92NYR62-3	ROCKWELL	82457
92NYR62-4	ROCKWELL	82455
92NYR62-5	ROCKWELL	82456
92NYR62-6	ROCKWELL	82452
92NYR62-7	ROCKWELL	82454
92NYR63-1	ROCKWELL	82459
92NYR63-2	ROCKWELL	82458
92NYR63-3	ROCKWELL	82460
92NYR64-10	ROCKWELL	82471
92NYR64-11	ROCKWELL	82469
92NYR64-12	ROCKWELL	82463
92NYR64-13	ROCKWELL	82470
92NYR64-14	ROCKWELL	82468
92NYR64-15	ROCKWELL	82461
92NYR64-2	ROCKWELL	82462
92NYR64-3	ROCKWELL	82466
92NYR64-4	ROCKWELL	82465
92NYR64-5	ROCKWELL	82467
92NYR64-6	ROCKWELL	82464
92NYR68	ROCKWELL	82472
92NYR70	ROCKWELL	82473
92NYR70-1	ROCKWELL	82474
92NYR70-2	ROCKWELL	82475
92NYR71	ROCKWELL	82477
92NYR71-1	ROCKWELL	82476
92NYR72	ROCKWELL	82483
92NYR72-1	ROCKWELL	82480
92NYR72-4	ROCKWELL	82481
92NYR72-5	ROCKWELL	82482
92NYR72-6	ROCKWELL	82479
92NYR72-7	ROCKWELL	82478
92NYR75	ROCKWELL	82484
92NYR76-1	ROCKWELL	82485
92NYR79	ROCKWELL	82486
92NYR80	ROCKWELL	82488
92NYR80-1	ROCKWELL	82491
92NYR80-2	ROCKWELL	82492
92NYR80-3	ROCKWELL	82493
92NYR80-4	ROCKWELL	82487
92NYR80-5	ROCKWELL	82490
92NYR83	ROCKWELL	82494
92NYR86	ROCKWELL	82495
92NYR88	ROCKWELL	82496
92NYR88-1	ROCKWELL	82499
92NYR88-2	ROCKWELL	82497
92NYR88-3	ROCKWELL	82498
92NYS28-5	ROCKWELL	82301
92NYS28-5A	ROCKWELL	82302
92NYS31	ROCKWELL	82304
92NYS31-2	ROCKWELL	82303
92NYS31-3	ROCKWELL	88007.18
92NYS32-1	ROCKWELL	82306
92NYS32-2	ROCKWELL	82305
92NYS32-25	ROCKWELL	82312
92NYS32-25A	ROCKWELL	82313
92NYS32-26	ROCKWELL	82314
92NYS32-26A	ROCKWELL	82315
92NYS32-27	ROCKWELL	82311
92NYS32-3	ROCKWELL	82307
92NYS32-5	ROCKWELL	82309
92NYS32-7	ROCKWELL	82308
92NYS33	ROCKWELL	82310
92NYS36	ROCKWELL	82335
92NYS36-1	ROCKWELL	82330
92NYS36-10	ROCKWELL	85878
92NYS36-12	ROCKWELL	85879
92NYS36-15	ROCKWELL	82317
92NYS36-16	ROCKWELL	82319
92NYS36-17	ROCKWELL	82338
92NYS36-2	ROCKWELL	82323
92NYS36-2	ROCKWELL	88007.19
92NYS36-22	ROCKWELL	82332
92NYS36-24	ROCKWELL	82333
92NYS36-26	ROCKWELL	82331
92NYS36-27	ROCKWELL	82334
92NYS36-3	ROCKWELL	82337
92NYS36-31	ROCKWELL	82336
92NYS36-32	ROCKWELL	82325
92NYS36-38	ROCKWELL	82328
92NYS36-39	ROCKWELL	85881
92NYS36-4	ROCKWELL	85882
92NYS36-40	ROCKWELL	82326
92NYS36-41	ROCKWELL	82318
92NYS36-47	ROCKWELL	82322
92NYS36-48	ROCKWELL	82320
92NYS36-5	ROCKWELL	82316
92NYS36-6	ROCKWELL	82324
92NYS36-60	ROCKWELL	82327
92NYS36-64	ROCKWELL	82329
92NYS36-7	ROCKWELL	85880
92NYS38	ROCKWELL	82354
92NYS38-1	ROCKWELL	82355

Part No.	Brand	IDLI Group	Part No.	Brand	IDLI Group	Part No.	Brand	IDLI Group	Part No.	Brand	IDLI Group
92NYS38-2	ROCKWELL	82351	92NYSM48-4	ROCKWELL	86848	9385452	GMC	80189	942735	AUST.WEST.	80213
92NYS38-33	ROCKWELL	82340	92NYSM48-55	ROCKWELL	86855	9388999	GMC	80189	942735	CLARK EQU.	80213
92NYS38-43	ROCKWELL	82343	92NYSM48-60	ROCKWELL	86852	9389358	GMC	80225	942735	MICHIGAN	80213
92NYS38-45	ROCKWELL	82341	92NYT36	ROCKWELL	82321	938P2	DIAMOND R	80178	942737	CATERPILLR	80168
92NYS38-46A	ROCKWELL	82347	92NYT38	ROCKWELL	82350	938P2	REO MOTORS	80178	942737	CATERPILLR	80216
92NYS38-52	ROCKWELL	82344	92NYT52	ROCKWELL	82421	938P2	WHITE	80178	942737	CLARK EQU.	80172
92NYS38-52A	ROCKWELL	82345	930	PRECISION	80150	939	PRECISION	88003.14	942737	MICHIGAN	80172
92NYS38-52A1	ROCKWELL	82346	93033203400	PORSCHE	86110	9392534	GMC	80225	942745	CLARK EQU.	80213
92NYS38-55	ROCKWELL	82348	931	PRECISION	80230	9395415	GMC	80149	942758	AMER.MOTOR	80031
92NYS38-6	ROCKWELL	82342	931477	AMER.MOTOR	80072	939700	CHRYSLER	80909	942758	WILLYS	80031
92NYS38A	ROCKWELL	82353	931477	JEEP	80072	940	PRECISION	88003.15	9428	TRU-CROSS	88013.86
92NYS38A1	ROCKWELL	82352	931477	WHITE	80072	940034	AMER.MOTOR	88001.70	942910	AMER.MOTOR	80031
92NYS38A2	ROCKWELL	82356	931477	WILLYS	80072	940036	CLARK EQU.	80149	942910	JEEP	80031
92NYS38A3	ROCKWELL	82357	9317	PRECISION	88001.57	940036	MICHIGAN	80109	942910	WHITE	80031
92NYS39-1	ROCKWELL	82358	932	PRECISION	80228	940036	MICHIGAN	80149	942910	WILLYS	80031
92NYS40-16	ROCKWELL	82366	933	PRECISION	80171	940039	CLARK EQU.	80149	943	PRECISION	88003.18
92NYS40-17	ROCKWELL	82363	934	PRECISION	80070	940039	MICHIGAN	80109	943	PREC.TOR.	80198
92NYS40-18	ROCKWELL	82368	934-5	PRECISION	88003.10	940039	MICHIGAN	80149	943150	WHITE	80124
92NYS40-20	ROCKWELL	82361	9348020	JEEP	80049	940083	CLARK EQU.	80149	943150	WHITE	80125
92NYS40-32	ROCKWELL	82364	934811	AMER.MOTOR	80055	940083	MICHIGAN	80109	943221	CATERPILLR	80109
92NYS40-4	ROCKWELL	82360	934811	JEEP	80055	940083	MICHIGAN	80149	943221	CATERPILLR	80149
92NYS40-7	ROCKWELL	82362	934811	WHITE	80091	940086	CLARK EQU.	80109	943221	CATERPILLR	80153
92NYS42-16	ROCKWELL	82369	934811	WILLYS	80055	940094	CLARK EQU.	80198	943221	CLARK EQU.	80149
92NYS42-4	ROCKWELL	82370	934820	AMER.MOTOR	80049	940094	MICHIGAN	80198	943221	MICHIGAN	80109
92NYS44	ROCKWELL	82385	934820	JEEP	80044	94020606	GMC	80023	943221	MICHIGAN	80149
92NYS44-11	ROCKWELL	82378	934820	WHITE	80044	94021-281	GMC	80023	943235	CATERPILLR	80216
92NYS44-14	ROCKWELL	82381	934820	WILLYS	80044	94021281	GMC	80024	943235	CLARK EQU.	80216
92NYS44-18	ROCKWELL	82375	934920	KAISER	80044	940229	CLARK EQU.	80213	943235	MICHIGAN	80168
92NYS44-19	ROCKWELL	82382	935	PRECISION	88003.11	940235	CLARK EQU.	80238	943424	CLARK EQU.	88003.40
92NYS44-2	ROCKWELL	82387	935	PREC.TOR.	80193	940235	MICHIGAN	80238	943441	CLARK EQU.	80234
92NYS44-20	ROCKWELL	82377	93521	CHECKER	80077	940252	CLARK EQU.	80198	943441	MICHIGAN	80234
92NYS44-21	ROCKWELL	82376	935517	JEEP	80025	940252	MICHIGAN	80198	943517	AMER.MOTOR	80025
92NYS44-7	ROCKWELL	85883	936	PRECISION	88003.12	940258	CLARK EQU.	80238	943517	WHITE	80025
92NYS44-8	ROCKWELL	82386	9-364-100055	GROVE	80197	940258	MICHIGAN	80238	943517	WILLYS	80025
92NYS46	ROCKWELL	82383	9-364-100059	GROVE CR.	80197	94027724	GMC	80024	943540918	BERLIET	80221
92NYS46-1	ROCKWELL	82384	9-364-100059	GROVE CR.	80212	94027724	ISUZU	80023	943541018	BEDFORD	80225
92NYS47-1	ROCKWELL	85884	9364269	GMC	80221	94027768	GMC	86166	943541018	BERLIET	80225
92NYS48	ROCKWELL	82411	936686	AMER.MOTOR	80075	94027769	GMC	86166	943541418	BERLIET	80142
92NYS48-1	ROCKWELL	82399	936686	JEEP	80075	94027819	GMC	80014	943541500	BERLIET	80178
92NYS48-2	ROCKWELL	82397	936686	WHITE	80072	940350	CLARK EQU.	80238	943541518	BERLIET	80142
92NYS48-20	ROCKWELL	82404	936686	WILLYS	80075	940350	MICHIGAN	80238	943541600	BERLIET	80189
92NYS48-22	ROCKWELL	82402	937	PRECISION	80196	9403J	BORG-WARNR	86735	943561	CLARK EQU.	80234
92NYS48-26	ROCKWELL	82398	937	PREC.TOR.	80197	940800	AMER.MOTOR	80906	943561	MICHIGAN	80234
92NYS48-3	ROCKWELL	82395	9-37300-017	GMC	80023	940800	CHRYSLER	80906	943648	AMER.MOTOR	80075
92NYS48-36	ROCKWELL	82403	9-37300-017	ISUZU	80023	940800	WILLYS	80953	943648	JEEP	80075
92NYS48-39	ROCKWELL	82401	9-37300-031	GMC	80023	940802	AMER.MOTOR	80061	943648	WILLYS	80075
92NYS48-41	ROCKWELL	82396	9-37300-031	ISUZU	80023	940802	WILLYS	80061	943724	CLARK EQU.	80245
92NYS48-5	ROCKWELL	82393	9-37300-036	GMC	80014	9408-1	JAGUAR	80142	943767	CLARK EQU.	80234
92NYS48-6	ROCKWELL	82394	9-37300-036	ISUZU	80014	940813	CLARK EQU.	80156	944057	CLARK EQU.	80234
92NYS48-7	ROCKWELL	82408	9-37300-065	GMC	80014	940813	MICHIGAN	80156	944200	WILLYS	80007
92NYS48-8	ROCKWELL	82412	9-37300-065	ISUZU	80014	9409	JAGUAR	80077	94433190300	PORSCHE	86145
92NYS48-9	ROCKWELL	82400	9-37300-091	GMC	80023	941	PRECISION	88003.16	944562	AMER.MOTOR	80077
92NYS51	ROCKWELL	82418	9-37300-092	GMC	80023	9413-1	KENBAR	87373	944562	WILLYS	80077
92NYS80	ROCKWELL	82489	9-37300-601	GMC	80014	9413-1	KENBAR	88008.43	944686	CLARK EQU.	80224
92NYSM34-3	ROCKWELL	86838	9-37300-601	ISUZU	80014	941572	CLARK EQU.	80198	944686	MICHIGAN	80223
92NYSM38-13	ROCKWELL	86841	9-37300-603	GMC	80014	941572	MICHIGAN	80198	944686	MICHIGAN	80224
92NYSM38-23	ROCKWELL	86840	9-37300-603	ISUZU	80014	941572	MICHIGAN	80199	944686	MICHIGAN	88003.28
92NYSM38-8	ROCKWELL	86842	9-37300-605	GMC	80023	941573	CLARK EQU.	80238	94468613	CLARK EQU.	80224
92NYSM38-9	ROCKWELL	86839	9-37300-605	ISUZU	80023	941573	MICHIGAN	80238	944715	CLARK EQU.	80234
92NYSM40-19	ROCKWELL	86843	9-37300-609	GMC	80023	9418	TRU-CROSS	88013.87	944715	MICHIGAN	80234
92NYSM42-8	ROCKWELL	86844	9-37300-609	ISUZU	80023	941862	CLARK EQU.	80198	944842	DAIMLER	80041
92NYSM44-1	ROCKWELL	88007.20	93730-06090	GMC	80023	942	PRECISION	88003.17	944842	LEYLAND	80041
92NYSM44-6	ROCKWELL	86845	9-37300-91	ISUZU	80023	942	PREC.TOR.	80156	944843	DAIMLER	80124
92NYSM48-11	ROCKWELL	86851	9-37300-92	ISUZU	80023	942221	CLARK EQU.	80109	944862	AUST.WEST.	80198
92NYSM48-15	ROCKWELL	86850	937340	AMER.MOTOR	80025	9422980	GMC	88008.54	944862	CLARK EQU.	80198
92NYSM48-19	ROCKWELL	86854	937340	JEEP	80025	94238930	GMC	80014	944864	CLARK EQU.	80245
92NYSM48-21	ROCKWELL	86853	937340	WHITE	80025	94238930	ISUZU	80014	944865	CLARK EQU.	80149
92NYSM48-23	ROCKWELL	86847	937340	WILLYS	80025	942573	MICHIGAN	80232	944865	MICHIGAN	80109
92NYSM48-27	ROCKWELL	86849	938	PRECISION	88003.13	942661	CLARK EQU.	80198	944865	MICHIGAN	80149
92NYSM48-30	ROCKWELL	86846	9385348	GMC	80225	942661	MICHIGAN	80198	944866	CLARK EQU.	80156

Part No.	Brand	IDLI Group	Part No.	Brand	IDLI Group	Part No.	Brand	IDLI Group	Part No.	Brand	IDLI Group
944866	MICHIGAN	80156	948977	WILLYS	80077	960	PREC.TOR.	80193	970163X	SPICER	86187
944882	DAIMLER	80180	948978	AMER.MOTOR	80059	960-001	WESCO	87327	970164X	SPICER	86122
944882	LEYLAND	80142	948978	JEEP	80066	960-001	WESCO	87328	970164X	SPICER	86180
944887	DAIMLER	80189	948978	WILLYS	80059	960-001	WESCO	87331	970167X	SPICER	86143
944887	LEYLAND	80189	949	PREC.TOR.	80124	960-001	WESCO	87332	970168X	SPICER	86070
944966	CLARK EQU.	80156	949	PREC.TOR.	80125	960-002	GMC	87326	970168X	SPICER	86144
945	PRECISION	88003.19	94-90-6475	REYNOLDS	80155	960-002	WESCO	87326	970169X	SPICER	86071
945	PREC.TOR.	80069	949234	CLARK EQU.	80245	960-002	WESCO	87330	970169X	SPICER	86145
945208	AUST.WEST.	80172	949238-1	SPICER	87929	960-003	WESCO	87383	971	PRECISION	88003.31
945208	CLARK EQU.	80172	949238-1	SPICER	88000.11	960-003	WESCO	87384	972	PRECISION	88012.68
945208	MICHIGAN	80172	949248	CLARK EQU.	80245	960-004	WESCO	87377	972	PRECISION	88012.69
945308	CLARK EQU.	80172	949250	CLARK EQU.	80245	960-004	WESCO	87380	9728	TRU-CROSS	88001.70
945387	CLARK EQU.	80174	950	PRECISION	80222	960-004	WESCO	87382	9728	TRU-CROSS	88001.94
945387	MICHIGAN	80174	950003	WESCO	88002.40	960-005	WESCO	87459	9728J	BORG-WARNR	85599
945388	CLARK EQU.	80175	9506SF	SPICER	88007.28	960-006	WESCO	87457	973	PRECISION	88003.32
945388	MICHIGAN	80175	951	PRECISION	80048	960-006	WESCO	87458	9736J	BORG-WARNR	82088
945581	CATERPILLR	80047	9513J	BORG-WARNR	82131	960-006	WESCO	87460	9738	TRU-CROSS	88001.70
945581	CLARK EQU.	80047	9516J	BORG-WARNR	81681	960-006	WESCO	87461	97385A	HYSTER	80197
945581	MICHIGAN	80047	9519J	BORG-WARNR	85095	960-007	WESCO	88007.77	974	PRECISION	88003.33
945741	CLARK EQU.	80213	952	PRECISION	80047	960-008	WESCO	88007.79	974-108R93	IHC	80225
945741	MICHIGAN	80213	952368	CLARK EQU.	80008	960-010	WESCO	87325	974120R91	IHC	80225
9458	TRU-CROSS	88002.13	952369	CLARK EQU.	80134	960-011	WESCO	87373	974180R91	IHC	80225
946	PRECISION	88003.20	952369	CLARK EQU.	80163	960-011	WESCO	87374	974180R93	IHC	80225
946	PREC.TOR.	80238	952369	MICHIGAN	80134	960-011	WESCO	87375	9741J	BORG-WARNR	82091
946020	CLARK EQU.	80003	952371	CLARK EQU.	80178	960-012	WESCO	87238	974-446R91	IHC	80124
946067	CLARK EQU.	80149	952371	MICHIGAN	80178	960-012	WESCO	88008.35	974-449R91	IHC	80023
946067	MICHIGAN	80109	952372	CLARK EQU.	80188	960-013	WESCO	87238	9745J	BORG-WARNR	82045
946067	MICHIGAN	80149	952372	MICHIGAN	80188	960-10-1	ROCKWELL	87818	975	PRECISION	88003.34
946189	CLARK EQU.	80224	952373	CLARK EQU.	80188	960-10-2	ROCKWELL	86857	975-663R91	IHC	80007
946189	MICHIGAN	80223	952373	MICHIGAN	80188	960110	CLARK EQU.	80216	975663R92	IHC	80007
946189	MICHIGAN	80224	952374	CLARK EQU.	80220	960111	CLARK EQU.	80224	975663R92	IHC	88000.15
946322	CLARK EQU.	80196	952375	CLARK EQU.	80008	960-11-1	ROCKWELL	87820	976	PRECISION	88003.35
94650	DROTT	80047	952376	CLARK EQU.	80008	960-11-2	ROCKWELL	86856	976456	FIAT-ALLIS	80168
946568	CLARK EQU.	80103	952377	CLARK EQU.	80008	960699	CLARK EQU.	80224	977	PRECISION	88003.36
946640	CLARK EQU.	80224	952378	CLARK EQU.	80069	960704	CLARK EQU.	80224	978	PRECISION	88003.37
946641	CLARK EQU.	80213	952378	MICHIGAN	80069	961	PRECISION	88003.26	97895A	HYSTER	80150
946641	MICHIGAN	80213	952379	CLARK EQU.	80133	961(5)	PRECISION	88003.27	979	PREC.TOR.	80182
946740	CLARK EQU.	80234	952545	CHRYSLER	88003.56	962	PRECISION	88003.28	980	PREC.TOR.	80190
946840	CLARK EQU.	80184	953	PRECISION	88003.22	962	PREC.TOR.	80223	980136	CLARK EQU.	80186
947	PRECISION	88003.21	953	PREC.TOR.	80178	963	PRECISION	80224	980613	CLARK EQU.	80215
947	PREC.TOR.	80213	95311	NEW HOLL.	80037	964	PRECISION	80069	980613	MICHIGAN	80210
947318	CLARK EQU.	80212	954	PREC.TOR.	80124	965	PRECISION	88003.29	980615	CLARK EQU.	80213
947322	CLARK EQU.	80196	954	PREC.TOR.	80125	965	PREC.TOR.	80188	980615	MICHIGAN	80213
947547	CHRYSLER	80041	954-10-1	ROCKWELL	87816	965-10-1	ROCKWELL	87802	980736	CLARK EQU.	80186
947548	CHRYSLER	88001.60	954-10-2	ROCKWELL	86803	965-10-2	ROCKWELL	86724	981	PRECISION	88003.38
947549	CHRYSLER	80096	954-11-1	ROCKWELL	87814	965-11-1	ROCKWELL	87821	9-810-100006	GROVE CR.	80193
947549	CHRYSLER	88001.62	954-11-2	ROCKWELL	86804	965-11-2	ROCKWELL	86726	9-810-100010	GROVE CR.	80210
94755	CHRYSLER	80064	954-12-1	ROCKWELL	87813	965-12-1	ROCKWELL	87810	9-810-100015	GROVE CR.	80212
94755	CHRYSLER	80065	954-12-2	ROCKWELL	86802	965-12-2	ROCKWELL	86725	9-810-100018	GROVE CR.	80197
947550	CHRYSLER	80064	954-13-1	ROCKWELL	87803	965-13-1	ROCKWELL	87815	9-810-100020	GROVE CR.	80193
947551	CHRYSLER	80096	954-13-2	ROCKWELL	86800	965-13-2	ROCKWELL	86728	9-810-100033	GROVE CR.	80212
947552	CHRYSLER	80122	954796	CLARK EQU.	80220	965-14-2	ROCKWELL	86727	9812055	RENAULT	88002.78
947553	CHRYSLER	80135	955	PREC.TOR.	80142	965-620R91	IHC	80145	9813973	ALFA ROMEO	80041
947555	CHRYSLER	80181	955	PREC.TOR.	80143	965-621R91	IHC	80183	9813973	SAVIEM	80041
947556	CHRYSLER	80042	955-10-1	ROCKWELL	87817	966	PRECISION	80102	981397300	SAVIEM	80041
947977	WILLYS	80066	955-10-2	ROCKWELL	86830	9668J	BORG-WARNR	85543	98-1473	NEAPCO	88014.79
947978	WILLYS	80066	955-11-1	ROCKWELL	87819	9668J	BORG-WARNR	85546	98-1473	SPICER	88014.79
948	PREC.TOR.	80140	955-11-2	ROCKWELL	86829	966932	CHRYSLER	80142	9816522	RENAULT	88002.79
948	PREC.TOR.	80182	955-12-2	ROCKWELL	88007.21	967	PRECISION	80103	9816566	RENAULT	88002.80
948213	CLARK EQU.	80003	956	PRECISION	88003.23	967265	FIAT	80041	98197A	HYSTER	80154
948322	CLARK EQU.	80133	957	PRECISION	80202	967-379R91	IHC	80124	982	PRECISION	88012.89
948361	CLARK EQU.	80218	958	PRECISION	80162	968	PRECISION	80172	982	PRECISION	88012.91
948508	CLARK EQU.	80224	959	PRECISION	88003.24	968294R91	IHC	80161	982	PRECISION	88012.92
9487J	BORG-WARNR	88014.52	959-10-1	ROCKWELL	87801	968-894R91	IHC	80219	982	PRECISION	88012.95
948904	CLARK EQU.	80234	959-10-2	ROCKWELL	86696	969	PRECISION	80156	982	PRECISION	88012.98
948976	AMER.MOTOR	80055	959-10-2	ROCKWELL	86698	9699J	BORG-WARNR	86809	982	PRECISION	88013.01
948976	JEEP	80055	959-11-1	ROCKWELL	87811	96E105	CHAMPLAIN	80044	982	PRECISION	88013.04
948976	WILLYS	80055	959-11-2	ROCKWELL	86697	970	PRECISION	88003.30	982	PRECISION	88013.11
948977	AMER.MOTOR	80066	959-12-1	ROCKWELL	87812	970162X	SPICER	86160	982	PRECISION	88013.12
948977	JEEP	80066	960	PRECISION	88003.25	970163X	SPICER	86134	982	PRECISION	88013.20

Part No.	Brand	IDLI Group	Part No.	Brand	IDLI Group	Part No.	Brand	IDLI Group	Part No.	Brand	IDLI Group
982	PRECISION	88013.26	991790A	WHITE	88007.79	A10182-127	SUTTON	80197	A20499	JOY MFG.	80142
982	PRECISION	88013.27	992	PRECISION	80165	A1103	AMER.MOTOR	80178	A20509-1	PETTI-MULL	80153
982	PRECISION	88013.28	992-10-2	ROCKWELL	86671	A1103	WHITE	80178	A205-1	WESCO	88000.35
982	PRECISION	88013.29	992-11-2	ROCKWELL	88007.35	A1103	WILLYS	80178	A205-1	WESCO	88000.38
982-50-1	ROCKWELL	87902	992-12-2	ROCKWELL	86672	A1177741	ATHEY PROD	80231	A205A205	WESCO	88000.38
982-50-5	ROCKWELL	87903	992-13-2	ROCKWELL	86673	A1228	DANUSER	80044	A20684	CASE,J.I.	80001
982-51-1	ROCKWELL	87928	992-14-2	ROCKWELL	86674	A123-1	WESCO	88000.21	A20684	CASE,J.I.	80003
982-51-10	ROCKWELL	88007.22	992489	CLARK EQU.	80243	A123-1	WESCO	88000.25	A209-0-846(4)	PARSONS	80193
982-51-11	ROCKWELL	88007.23	992624	CLARK EQU.	80001	A123-1	WESCO	88000.27	A22509	PETTI-MULL	80109
982-51-3	ROCKWELL	88007.24	992656	CLARK EQU.	80233	A123-1	WESCO	88000.39	A22509-1	PETTI-MULL	80153
982-51-5	ROCKWELL	87929	992657	CLARK EQU.	80234	A123-1	WESCO	88000.81	A22710	DAVEY COM.	80063
982-51-6	ROCKWELL	88007.25	9929086	FIAT-ALLIS	80180	A123A123	WESCO	88000.21	A23260	DAVEY COM.	80193
982-52-1	ROCKWELL	87958	993	PRECISION	80166	A124-1	WESCO	81020	A23697	DAVEY COM.	80193
982-52-3	ROCKWELL	88007.26	993-10-2	ROCKWELL	88007.36	A126012	BMC	80024	A24209	DAVEY COM.	80124
982-52-5	ROCKWELL	87959	993-11-2	ROCKWELL	88007.37	A126102	BMC	80024	A24520	DAVEY COM.	80142
982-52-6	ROCKWELL	88007.27	993-12-2	ROCKWELL	88007.38	A126433	CHRYSLER	80012	A24615	DAVEY COM.	80215
983	PRECISION	80021	993-13-2	ROCKWELL	88007.39	A12877P	CASE,J.I.	80215	A24616	DAVEY COM.	80216
983-10-2	ROCKWELL	87062	993142	CLARK EQU.	80213	A12878	CASE,J.I.	80214	A25151	CASE,J.I.	80219
983-13-2	ROCKWELL	87061	993-14-2	ROCKWELL	88007.40	A134-1	WESCO	88000.22	A253000	JOY MFG.	80077
983-268R91	IHC	80001	993-15-2	ROCKWELL	88007.41	A134-1	WESCO	88000.25	A259525	JOY MFG.	80075
983-922R91	IHC	80024	993-16-2	ROCKWELL	88007.42	A134-1	WESCO	88000.27	A262457	JOY MFG.	80189
984	PRECISION	88003.39	993-17-2	ROCKWELL	88007.43	A134-1	WESCO	88000.28	A262467	JOY MFG.	80189
984-10-1	ROCKWELL	88007.28	993-18-2	ROCKWELL	88007.44	A134-1	WESCO	88000.31	A26294A	JOY MFG.	80168
984-10-2	ROCKWELL	87066	994	PRECISION	80046	A134A164	WESCO	88000.22	A26294A	JOY MFG.	80216
984-11-1	ROCKWELL	88007.29	994063	CLARK EQU.	80233	A134A205	WESCO	88000.22	A27232-1	P.M.	80196
984-11-2	ROCKWELL	87070	994085	CLARK EQU.	80213	A138977	HYSTER	80196	A27232-1	P.M.	80197
984-12-1	ROCKWELL	88007.30	994-10-2	ROCKWELL	88007.45	A138977	HYSTER	80198	A277742	ATHEY PROD	80231
984-13-1	ROCKWELL	88007.31	994-11-2	ROCKWELL	88007.46	A1433	AMER.MOTOR	80072	A281868	FORD	80178
984-13-2	ROCKWELL	87072	994-12-2	ROCKWELL	88007.47	A1433	WHITE	80072	A28451-1	PETTI-MULL	80163
984-16-2	ROCKWELL	87073	994-13-2	ROCKWELL	88007.48	A1433	WILLYS	80072	A-2900-846	KOEHRING	80193
984-19-2	ROCKWELL	87067	994-14-2	ROCKWELL	88007.49	A144-1	WESCO	88000.21	A302	MUNCIE	85114
984-21-2	ROCKWELL	87063	994-15-2	ROCKWELL	88007.50	A144-1	WESCO	88000.22	A-302	MUNCIE	88001.14
984-22-2	ROCKWELL	87064	994173	AMER.MOTOR	80077	A144-1	WESCO	88000.25	A30408	CASE,J.I.	80001
984-22-2	ROCKWELL	88007.32	994173	JEEP	80077	A144-1	WESCO	88000.27	A30425	CASE,J.I.	80001
984-23-2	ROCKWELL	87068	994173	WILLYS	80077	A144-1	WESCO	88000.39	A30428	CASE,J.I.	80001
984-24-2	ROCKWELL	87071	994823	AMER.MOTOR	88002.37	A144385	HYSTER	80195	A30435	CASE,J.I.	80001
984-25-2	ROCKWELL	87075	994827	AMER.MOTOR	80075	A144A144	WESCO	88000.25	A31364	DAVEY COM.	80007
984-26-2	ROCKWELL	87065	994827	WILLYS	80077	A14725	CASE,J.I.	80047	A33628	DAVEY COM.	80218
984-27-2	ROCKWELL	87069	994839	CLARK EQU.	80243	A154-1	WESCO	88000.22	A337-21	CRANE CAR.	80240
984-28-2	ROCKWELL	87074	995	PRECISION	80035	A154-1	WESCO	88000.28	A34800	SCHIELD	80207
984899A	HYSTER	80109	995800	ACF BRILL	80189	A163-1	WESCO	88000.53	A353	EATON	80069
984908A	HYSTER	80193	996	PRECISION	80154	A164-1	WESCO	88000.22	A353	EATON	80070
985	PRECISION	80067	9961728	GMC	80142	A164-1	WESCO	88000.31	A353X	EATON	80069
986	PRECISION	88003.40	9961730	GMC	80180	A164-1	WESCO	88000.88	A353X	EATON	80070
986493	CHRYSLER	80045	997	PRECISION	80175	A164A164	WESCO	88000.31	A354	EATON	80069
986956	CHRYSLER	88001.67	998	PRECISION	80113	A17330H	DEERE,JOHN	80001	A354	EATON	80070
986963	UNIT RIG	80240	999	PRECISION	80169	A17330H	DEERE,JOHN	80003	A354-1X	EATON	80069
986980	UNIT RIG	80070	999791	ACF BRILL	80178	A177741	ATHEY PROD	80231	A354-1X	EATON	80070
987967	UNIT RIG	80134	9F3761	CATERPILLR	88001.26	A177741	ATHEY PROD	80240	A354X	EATON	80069
988	PRECISION	80245	9H2477	CATERPILLR	80198	A177741	JOY MFG.	80231	A354X	EATON	80070
988175	UNIT RIG	80001	9H2478	CATERPILLR	80198	A177741	JOY MFG.	80240	A357	EATON	80069
988188	UNIT RIG	80239	9H3766(2)	CATERPILLR	88002.93	A177744	JOY MFG.	80232	A357	EATON	80070
989	PRECISION	88003.41	9H3772(5)	CATERPILLR	88002.92	A177744	JOY MFG.	80242	A365789	JOY MFG.	80189
989717	UNIT RIG	80240	9H5107(2)	CATERPILLR	88001.21	A1809-6B	WESCO	88000.49	A369158	JOY MFG.	80210
990	PRECISION	80163	9H5108	CATERPILLR	88001.21	A183515	JOY MFG.	80231	A369158	JOY MFG.	80215
990-10-2	ROCKWELL	88007.33	9H6423	CATERPILLR	80234	A183515	JOY MFG.	80240	A369518	JOY MFG.	80210
990-11-2	ROCKWELL	86675	9H9491	CATERPILLR	80235	A183516	JOY MFG.	80232	A370374-1	JOY MFG.	80007
990-12-2	ROCKWELL	86676	9K1971	CATERPILLR	80101	A183516	JOY MFG.	80242	A381868	FORD	80178
990-13-2	ROCKWELL	86678	9K1976	CATERPILLR	80100	A18382	CASE,J.I.	80171	A386002	ALLIS-CHLM	80192
990-14-2	ROCKWELL	86677	9K3969	CATERPILLR	80150	A18383R	CASE,J.I.	80216	A386002	FIAT-ALLIS	80192
990192	AMER.MOTOR	80001	9K3970	CATERPILLR	80216	A18398	CASE,J.I.	80187	A39087	DAVEY COM.	80163
990192	AMER.MOTOR	80003	9M2036(2)	CATERPILLR	88001.35	A185-1	WESCO	88000.35	A40106	CASE,J.I.	80069
990192	AMER.MOTOR	80076	9P356	CATERPILLR	80213	A185-1	WESCO	88000.38	A-402	MUNCIE	88001.12
990937	AMER.MOTOR	80076	9P4809	CATERPILLR	80238	A19522	PETTI-MULL	80193	A402	MUNCIE	88001.12
991	PRECISION	80164	9P604	CATERPILLR	80234	A19920	CASE,J.I.	80193	A40200	CASE,J.I.	80171
991-10-2	ROCKWELL	86668	9S7120	CATERPILLR	80235	A19921	CASE,J.I.	80196	A40435	CASE,J.I.	80001
991-11-2	ROCKWELL	86669	9S7184	CATERPILLR	88001.54	A19961	PETTI-MULL	80210	A44176	CASE,J.I.	80210
991-12-2	ROCKWELL	86670	A03-500660	BRD	80041	A20411	P.M.	80231	A44844	JOY MFG.	80124
991-13-2	ROCKWELL	88007.34	A0T7039A	FORD	80124	A20411	P.M.	80240	A45040	ALLIS-CHLM	80047
991790	WHITE	87570	A101082-112	SUTTON	80216	A20435	CASE,J.I.	80001	A45040	CASE,J.I.	80238

Part No.	Brand	IDLI Group	Part No.	Brand	IDLI Group	Part No.	Brand	IDLI Group	Part No.	Brand	IDLI Group
A45040	FIAT-ALLIS	80047	AD5200	REPUBLIC	80909	AE11033W	DEERE,JOHN	80037	AE2185	AEC	80047
A45050	ALLIS-CHLM	80046	AD5200	REPUBLIC	80958	AE111	AEC	80175	AE2186	AEC	80046
A45050	FIAT-ALLIS	80046	AD5200	REPUBLIC	88001.73	AE112	AEC	88000.86	AE2198	AEC	80051
A45060	ALLIS-CHLM	80167	AD5200A	REPUBLIC	80909	AE112	AEC	88000.88	AE2198G	AEC	80051
A45060	FIAT-ALLIS	80163	AD5200A	REPUBLIC	88001.72	AE113	AEC	88000.79	AE2198G	AEC	80052
A45070	ALLIS-CHLM	80168	AD5200AD	REPUBLIC	80909	AE1193	AEC	88000.85	AE22	AEC	87285
A45070	FIAT-ALLIS	80168	AD5200AR	REPUBLIC	80909	AE12	AEC	87326	AE225	AEC	80064
A45595	CASE,J.I.	80172	AD5200R	REPUBLIC	80903	AE1205	AEC	80123	AE22753	DEERE,JOHN	80037
A45596	CASE,J.I.	80212	AD5200R	REPUBLIC	80958	AE122	AEC	88000.89	AE23	AEC	87337
A45634	CASE,J.I.	80238	AD5299R	REPUBLIC	80958	AE123	AEC	88000.80	AE231	AEC	80132
A48402	JOY MFG.	80149	AD5300	REPUBLIC	88001.75	AE123716	DEERE,JOHN	80035	AE232	AEC	80132
A48709	JOY MFG.	80142	AD5300A	REPUBLIC	88001.67	AE12N000	AEC	80035	AE233	AEC	80061
A49524	JOY MFG.	80124	AD5400	REPUBLIC	80951	AE133	AEC	88000.81	AE233	AEC	80062
A49542	JOY MFG.	80124	AD5400	REPUBLIC	88001.74	AE13B	AEC	88008.34	AE23516B	DEERE,JOHN	80069
A49562	JOY MFG.	80124	AD5400A	REPUBLIC	88001.66	AE14978	DEERE,JOHN	80001	AE23516E	DEERE,JOHN	80069
A53116(7)	GALION	80077	AD5400AR	REPUBLIC	80901	AE14978E	DEERE,JOHN	80001	AE23716	DEERE,JOHN	80035
A59654	JOY MFG.	80210	AD5400AR	REPUBLIC	80909	AE14S	AEC	87331	AE24	AEC	87399
A59654	JOY MFG.	80215	AD5400R	REPUBLIC	88001.74	AE15	AEC	87331	AE243	AEC	88000.22
A649502	KOEHRING	80037	AD5500	REPUBLIC	80956	AE1501	AEC	80021	AE248	AEC	80025
A650079	GERLINGER	80035	AD5500A	REPUBLIC	80906	AE153	AEC	88000.83	AE2500	AEC	80059
A650079	TOWMOTOR	80035	AD5500A	REPUBLIC	80909	AE16	AEC	87377	AE2505	AEC	80059
A661677	GERLINGER	80195	AD5500AR	REPUBLIC	80909	AE1601	AEC	80012	AE251	AEC	80056
A661678	GERLINGER	80196	AD5500R	REPUBLIC	80956	AE1602	AEC	80019	AE2518	AEC	80066
A661678	TOWMOTOR	80196	AD5600AR	REPUBLIC	80907	AE1603	AEC	80012	AE252	AEC	80034
A66178	GERLINGER	80197	AD5600R	REPUBLIC	80954	AE1604	AEC	80004	AE2530	AEC	80237
A7H4	PETTI-MULL	80099	AD5700	REPUBLIC	80955	AE1605	AEC	80009	AE2531	AEC	80066
A7HA	PETTI-MULL	80101	AD5700AR	REPUBLIC	80908	AE1606	AEC	80011	AE255	AEC	80060
A80-7039A	FORD	80178	AD5700R	REPUBLIC	80955	AE162	AEC	80042	AE256	AEC	80056
A826	DANUSER	80008	AD6000R	REPUBLIC	80958	AE17	AEC	87383	AE257	AEC	80095
A891	DANUSER	80008	AD6200AR	REPUBLIC	80907	AE1750	AEC	80030	AE259	AEC	80096
A8Q7039A	FORD	80178	AD6200R	REPUBLIC	80954	AE1750	AEC	80031	AE261	AEC	80064
A90678	JOY MFG.	80124	AD6300R	REPUBLIC	80957	AE1750G	AEC	80031	AE262	AEC	80096
A94633	JOY MFG.	80077	AD6500AR	REPUBLIC	80905	AE1755	AEC	80032	AE26238	DEERE,JOHN	80001
A96078	JOY MFG.	80124	AD6500R	REPUBLIC	80951	AE1755G	AEC	80032	AE263	AEC	80094
A9LY4635	FORD	80072	AD6600R	REPUBLIC	80953	AE1762	AEC	80030	AE263	AEC	88000.71
A9LY4635A	FORD	80072	AD6700AR	REPUBLIC	80906	AE1767	AEC	80031	AE265	AEC	80064
AAK8264	BMC	80024	AD6700R	REPUBLIC	80953	AE18	AEC	87457	AE265	AEC	80065
AB15225B	DEERE,JOHN	80186	AD9353	ADAMS-LET.	80212	AE181	AEC	88000.92	AE266	AEC	80060
AB4851	FORD	88002.87	AD9353	WABCO	80214	AE183	AEC	88000.82	AE266	AEC	80061
AB4851A	FORD	88002.87	ADG5600AR	REPUBLIC	80907	AE19	AEC	87459	AE266-1310	AEC	80083
AB5A7039	FORD	80049	AE0103	AEC	88000.36	AE20	AEC	87481	AE266BCK2	AEC	88001.70
AB5A7042A	FORD	80049	AE01103	AEC	88000.94	AE2017	AEC	80092	AE267	AEC	80060
AB6T7039A	FORD	80142	AE0113	AEC	88000.35	AE2062	AEC	80050	AE267	AEC	80063
AC15878	DEERE,JOHN	80136	AE0115	AEC	88000.55	AE2063	AEC	80043	AE26884	DEERE,JOHN	80037
AC1718	OLIVER	80133	AE012	AEC	88000.45	AE21	AEC	87570	AE2727	AEC	88001.98
AC1719	OLIVER	80133	AE0123	AEC	88000.38	AE2100	AEC	80049	AE273	AEC	88000.72
AC17-20	OLIVER	80133	AE0124	AEC	88000.37	AE2101	AEC	80044	AE281	AEC	80227
AC17-20	WHITE FARM	80133	AE013	AEC	88000.20	AE2102	AEC	80054	AE293	AEC	88000.73
AC1721	OLIVER	80008	AE022	AEC	88000.43	AE2103	AEC	80048	AE2-94-28X2	AEC	88013.86
AC1722	OLIVER	80008	AE023	AEC	88000.44	AE2109	AEC	80049	AE297	AEC	80079
AC1722	WHITE FARM	80008	AE033	AEC	88000.26	AE2110	AEC	80048	AE2B17	AEC	88000.25
AC17-77	OLIVER	80008	AE043	AEC	88000.22	AE2111	AEC	80049	AE2B49	AEC	88000.25
AC1777	OLIVER	80037	AE053	AEC	88000.52	AE2112	AEC	80044	AE2B72	AEC	88000.39
AC17-77	OLIVER	88002.94	AE055-55-2	AEC	80141	AE21122	DEERE,JOHN	80069	AE2L102B	AEC	88001.28
AC17-77	WHITE FARM	80037	AE063	AEC	88000.27	AE2114	AEC	80054	AE2T2910	AEC	88002.43
AC17-81	OLIVER	80134	AE073	AEC	88000.28	AE2116	AEC	80047	AE2T2911	AEC	88002.42
AC1781	OLIVER	80140	AE076	AEC	88000.29	AE2116A	AEC	80047	AE2T2912	AEC	88002.44
AC17-81	WHITE FARM	80134	AE081	AEC	88000.49	AE2117	AEC	80046	AE3000	AEC	80099
AC4393(4)	LINK BELT	80240	AE083	AEC	88000.53	AE2117A	AEC	80234	AE3000	AEC	80101
AC4503(4)	LINK BELT	80193	AE093	AEC	88000.30	AE2123	AEC	80045	AE3008	AEC	80107
AC4504(4)	LINK BELT	80216	AE095	AEC	88000.34	AE2126	AEC	80049	AE3011	AEC	80107
AC4640(4)	LINK BELT	80171	AE096-55	AEC	80138	AE2126	AEC	88001.96	AE3012	AEC	80110
AC4640(4)	LINK BELT	80216	AE10	AEC	88008.35	AE2127	AEC	80044	AE3012	AEC	80112
AC4998(4)	LINK BELT	80240	AE10003	AEC	80245	AE2127	AEC	88001.98	AE3013	AEC	80106
AC5422(4)	LINK BELT	80240	AE1004	AEC	80001	AE213	AEC	88000.68	AE3019	AEC	80112
AC5424(4)	LINK BELT	80215	AE10091E	DEERE,JOHN	80021	AE2133	AEC	80045	AE3021	AEC	80111
AD10091E	DEERE,JOHN	80023	AE1025	AEC	80019	AE2140	AEC	80055	AE3100	AEC	80101
AD4251	REPUBLIC	80905	AE11	AEC	87373	AE2153	AEC	80055	AE3101	AEC	80100
AD4251	REPUBLIC	80953	AE11027	DEERE,JOHN	80044	AE2173	AEC	80058	AE3102	AEC	80112
AD5000A	REPUBLIC	80153	AE11027E	DEERE,JOHN	80044	AE2175	AEC	80058	AE3103	AEC	80098

Part No.	Brand	IDLI Group
AE3105	AEC	80104
AE3105BCK2	AEC	88001.94
AE3106	AEC	80099
AE3108	AEC	80110
AE3108	AEC	80112
AE3111	AEC	80107
AE3113	AEC	80112
AE3114	AEC	80104
AE3116	AEC	80111
AE3118	AEC	80101
AE3119	AEC	80100
AE3120	AEC	80113
AE3121	AEC	80102
AE3125	AEC	80103
AE3152	AEC	80101
AE3188	AEC	80105
AE32413	DEERE,JOHN	80035
AE3334	AEC	80012
AE35N3	AEC	80069
AE35N4	AEC	80069
AE361	AEC	80135
AE362	AEC	80135
AE372	AEC	80135
AE373	AEC	80135
AE373-1410	AEC	80144
AE374	AEC	80135
AE3-94-18X2	AEC	88013.87
AE3R1310	AEC	80097
AE3RCV	AEC	80109
AE3T1328	AEC	88002.55
AE3T2895	AEC	88002.48
AE3T2896	AEC	88002.50
AE3T2897	AEC	88002.51
AE3T2898	AEC	88002.52
AE3T3501	AEC	88002.63
AE3T3501	AEC	88002.67
AE3T3507	AEC	88002.71
AE40013	AEC	88000.21
AE40023	AEC	88000.44
AE40043	AEC	88000.44
AE40073	AEC	88000.28
AE40093	AEC	88000.31
AE40103	AEC	88000.36
AE40123	AEC	88000.39
AE4100	AEC	80109
AE4100	AEC	80149
AE4101	AEC	80153
AE4102	AEC	80158
AE4103	AEC	80153
AE4104	AEC	80151
AE4105	AEC	80151
AE4106	AEC	80148
AE4107	AEC	80159
AE4109	AEC	80157
AE4110	AEC	80155
AE41103	AEC	88000.85
AE4113	AEC	80156
AE4113	AEC	88003.17
AE4141	AEC	80154
AE4143	AEC	80156
AE500	AEC	80030
AE500U	AEC	80076
AE501	AEC	80023
AE501200	AEC	80081
AE502	AEC	80041
AE503	AEC	80007
AE5031	AEC	80171
AE504	AEC	80041
AE505	AEC	80031
AE507	AEC	80030
AE508	AEC	80043
AE5-086	AEC	80124
AE5-088	AEC	80140
AE50U	AEC	80143
AE510	AEC	80007
AE5100	AEC	80168
AE5101	AEC	80167
AE5102	AEC	80171
AE5103	AEC	80174
AE5105	AEC	80169
AE5105	AEC	80216
AE5108	AEC	80172
AE5109	AEC	80162
AE511	AEC	80024
AE5110	AEC	80173
AE5111	AEC	80175
AE5-1200	AEC	80081
AE5-1201	AEC	80073
AE5-1201	AEC	80091
AE5-1202	AEC	80130
AE5-1203	AEC	80129
AE5-1204	AEC	80131
AE5-1204UBK2	AEC	88002.13
AE5-1205	AEC	80123
AE5-1206	AEC	80141
AE5121	AEC	80163
AE5127	AEC	80167
AE512U	AEC	80075
AE512U	AEC	80076
AE513	AEC	80077
AE5131	AEC	80175
AE5132	AEC	80174
AE5139	AEC	80170
AE514	AEC	80024
AE515	AEC	80178
AE515R	AEC	80178
AE516	AEC	80189
AE516Z	AEC	80189
AE520	AEC	80075
AE5-200	AEC	80075
AE5-200U	AEC	80076
AE520U	AEC	80076
AE521	AEC	80072
AE521H0U	AEC	80088
AE521HD	AEC	80072
AE521HD	AEC	80075
AE521HD	AEC	80077
AE521HDU	AEC	80077
AE521U	AEC	80089
AE524	AEC	80225
AE5-248	AEC	80025
AE5-249	AEC	80025
AE524Z	AEC	80225
AE5-260	AEC	80078
AE5-260	AEC	80081
AE5-278	AEC	80085
AE5-281	AEC	80227
AE5-281Z	AEC	80227
AE5-297	AEC	80079
AE52AR	AEC	80909
AE52R	AEC	80958
AE530	AEC	80117
AE5-308	AEC	80229
AE531	AEC	80121
AE531	BORG-WARNR	80118
AE5-315	AEC	80026
AE532	AEC	80075
AE533	AEC	80116
AE534	AEC	80118
AE5-350	AEC	80036
AE5-353	AEC	80083
AE54AR	AEC	80901
AE54R	AEC	88001.74
AE550	AEC	80119
AE553	AEC	80077
AE555	AEC	80182
AE560	AEC	80142
AE560U	AEC	80142
AE560U	AEC	80143
AE565	AEC	80190
AE56E	AEC	80190
AE570	AEC	80007
AE572	AEC	80007
AE572	AEC	88000.16
AE574	AEC	80090
AE575	AEC	80007
AE575	AEC	88000.15
AE577	AEC	80221
AE577Z	AEC	80221
AE578	AEC	80124
AE578U	AEC	80125
AE579	AEC	80091
AE579	AEC	80955
AE57AR	AEC	80908
AE57R	AEC	80955
AE581	AEC	80023
AE581	AEC	80140
AE585	AEC	80228
AE585Z	AEC	80228
AE586	AEC	88001.82
AE587	AEC	88001.83
AE588	AEC	80140
AE588	AEC	80182
AE589	AEC	80229
AE590	AEC	80093
AE5S55-55-4	AEC	80081
AE60R	AEC	80903
AE60R	AEC	80958
AE6100	AEC	80193
AE6100	AEC	80198
AE6101	AEC	80195
AE6102	AEC	80197
AE6103	AEC	80194
AE6104	AEC	80192
AE6106	AEC	80198
AE6107	AEC	80202
AE6108	AEC	80199
AE6109	AEC	80201
AE6110	AEC	80198
AE6111	AEC	80198
AE6128	AEC	80198
AE6149	AEC	80198
AE6150	AEC	80203
AE62AR	AEC	80907
AE62R	AEC	80954
AE63R	AEC	80957
AE65AR	AEC	80905
AE65R	AEC	80951
AE675	AEC	80091
AE679E	DEERE,JOHN	80001
AE67AR	AEC	80906
AE67HR	AEC	80906
AE67R	AEC	80953
AE6A2046	AEC	88002.86
AE6A2046	AEC	88002.87
AE6N8	AEC	80001
AE7100	AEC	80210
AE7101	AEC	80211
AE7102	AEC	80212
AE7103	AEC	80208
AE7104	AEC	80209
AE7105	AEC	80213
AE7105	AEC	88002.97
AE710A	AEC	80205
AE7200	AEC	80210
AE7202	AEC	80214
AE7202	AEC	80218
AE7245	AEC	80212
AE7939E	DEERE,JOHN	80001
AE793E	DEERE,JOHN	80001
AE8000	AEC	80231
AE8000	AEC	80243
AE8014	AEC	80230
AE8101	AEC	80232
AE8101	AEC	80244
AE8102	AEC	80237
AE8102	AEC	80244
AE8105	AEC	80238
AE8105	AEC	88003.20
AE8113	AEC	80238
AE8200	AEC	80231
AE8201	AEC	80232
AE8202	AEC	80237
AE8203	AEC	80230
AE8203	AEC	80236
AE8214	AEC	80238
AE8500	AEC	80223
AE8500	AEC	88003.28
AE8502	AEC	80224
AE89-2510	AEC	80011
AE9	AEC	87325
AE9000	AEC	80235
AE9001	AEC	80234
AE9002	AEC	80233
AE9002	AEC	88003.09
AE9013	AEC	80234
AE9014	AEC	80235
AE9015	AEC	80234
AE9016	AEC	80234
AE9017	AEC	80233
AE9017	AEC	80241
AE9026	AEC	80233
AE9026	AEC	80241
AE9026	AEC	88003.09
AE9031	AEC	80234
AE9100	AEC	80235
AE-C96-55	AEC	80204
AE-CP1FR	AEC	80008
AE-CP35N3	AEC	80069
AECP35N384	AEC	80069
AE-CP35N4	AEC	80070
AE-CP3DR	AEC	80133
AE-CP4N	AEC	80134
AE-CP4N	AEC	80178
AE-CP55N	AEC	80161
AECP58N	AEC	80161
AE-CP58N4	AEC	80179
AE-CP58WB	AEC	80185
AE-CP58WB1	AEC	80186
AE-CP58WB2	AEC	80184
AE-CP5N	AEC	80178
AE-CP6N	AEC	80001
AE-CP6N	AEC	80188
AE-CP750N1	AEC	80187
AE-CP7N	AEC	80220
AE-CP850N4	AEC	80219
AE-CPL10N2	AEC	80033
AE-CPL10N2	AEC	88002.93
AECPL106	AEC	80033
AE-CPL12N	AEC	80035

Part No.	Brand	IDLI Group	Part No.	Brand	IDLI Group	Part No.	Brand	IDLI Group	Part No.	Brand	IDLI Group
AE-CPL14N	AEC	80037	AE-T2010	AEC	80009	AOX33323	DEERE,JOHN	80021	AV12520V	DEERE,JOHN	80238
AECPL6N	AEC	80001	AE-T55-55-2	AEC	80071	AP21678	DEERE,JOHN	80021	AV12521	DEERE,JOHN	80242
AE-CPL6N8	AEC	80001	AE-T555-55-3	AEC	80080	AP22718H	DEERE,JOHN	80207	AV21521V	DEERE,JOHN	80232
AE-CVK1	AEC	88002.34	AE-TS55-55-2	AEC	80074	AP23260	DAVEY COM.	80193	AW13174	DEERE,JOHN	80001
AE-CVK1	AEC	88002.35	AE-TS55-55-3	AEC	80081	AP23697	DAVEY COM.	80193	AW371052	AUST.WEST.	80142
AE-CVK2	AEC	88002.34	AE-TS55-55-4	AEC	80081	AP24580H	DEERE,JOHN	80037	AW465019	AUST.WEST.	80048
AE-CVK2A	AEC	88002.36	AE-U56-55-2	AEC	80183	AP24615	DAVEY COM.	80210	AW465020	AUST.WEST.	80047
AECVK3	AEC	88002.33	AE-U56-55-Z	AEC	80183	APF4X5	FOX RIVER	80198	AW465028	AUST.WEST.	80104
AE-CVK4	AEC	88002.33	AE-U96-55	AEC	80181	AQX33323	DEERE,JOHN	80021	AW465029	AUST.WEST.	80104
AECVK4	AEC	88002.33	AEX55-55	AEC	80007	AR21101	DEERE,JOHN	80001	AW465034	AUST.WEST.	80099
AECVK5	AEC	88002.33	AE-Y1001	AEC	88002.86	AR21101R	DEERE,JOHN	80001	AW465035	AUST.WEST.	80099
AE-CVK6	AEC	88002.33	AE-Y1001	AEC	88002.87	AR29054R	DEERE,JOHN	80231	AW465036	AUST.WEST.	80104
AE-D1025	AEC	80019	AEZ13	AEC	88000.68	AR29584	DEERE,JOHN	80237	AW465066	AUST.WEST.	80168
AE-D4627	AEC	80010	AEZ43	AEC	88000.70	AR29584R	DEERE,JOHN	80239	AW465067	AUST.WEST.	80168
AE-D56-55-2	AEC	80145	AEZ63	AEC	88000.71	AR29699	DEERE,JOHN	80101	AW465068	AUST.WEST.	80168
AE-D8025	AEC	80004	AEZ73	AEC	88000.72	AR29699	DEERE,JOHN	80104	AW465071	AUST.WEST.	80168
AE-D96-55-4	AEC	80160	AEZ76	AEC	88000.72	AR29699R	DEERE,JOHN	80099	AW465072	AUST.WEST.	80163
AE-DS56-55-2	AEC	80145	AEZ93	AEC	88000.73	AR3008E4635A	FORD	80023	AW465073	AUST.WEST.	80168
AEI103	AEC	88000.85	AF4841A	FORD	88002.86	AR37096	DEERE,JOHN	80185	AW465076	AUST.WEST.	80168
AEI12	AEC	88000.88	AGRICULT.	MCQUAY-NOR	80008	AR37097	DEERE,JOHN	80185	AW465077	AUST.WEST.	80163
AEI122	AEC	88000.89	AGRICULT.	MCQUAY-NOR	80136	AR84046	DEERE,JOHN	80212	AW465111	AUST.WEST.	80193
AEI13	AEC	88000.79	AH12461H	DEERE,JOHN	80215	AR87394	DEERE,JOHN	80184	AW465123	AUST.WEST.	80210
AEI22	AEC	88000.89	AH1261H	DEERE,JOHN	80210	AR89AF7039A	FORD	80077	AW465124	AUST.WEST.	80210
AEI23	AEC	88000.80	AH18327	DEERE,JOHN	80001	AR90543	DEERE,JOHN	80213	AW465132	AUST.WEST.	80210
AEI32	AEC	88000.95	AH18327H	DEERE,JOHN	80003	AR95974	DEERE,JOHN	88001.91	AW465136	AUST.WEST.	80210
AEI33	AEC	88000.81	AH79074	DEERE,JOHN	80001	ARC4DZ4635A	FORD	88002.77	AW465137	AUST.WEST.	80210
AEI53	AEC	88000.83	AH81534	DEERE,JOHN	80190	ARC6DZ4635A	FORD	80077	AW465158	AUST.WEST.	80048
AEI63	AEC	88000.96	AH8153A	DEERE,JOHN	80142	ARC6DZ4635B	FORD	80091	AW465159	AUST.WEST.	80047
AEI81	AEC	88001	AH82541	DEERE,JOHN	80001	AS60190P	WEST.LAND	88002.94	AW46520	AUST.WEST.	80100
AEI83	AEC	88000.82	AH83541	DEERE,JOHN	80001	AS6779	WESCO	88002.43	AW46520	BAL-LM-HAM	80100
AEI93	AEC	88000.84	AJ12823E	DEERE,JOHN	80044	AS72C19D	WEST.LAND	80035	AW465205	BAL-LM-HAM	80168
AE-K5GB22	AEC	80180	AJ3294H	DEERE,JOHN	80021	AS72C19P	WEST.LAND	80035	AW465208	AUST.WEST.	80168
AEO103	AEC	88000.36	AJ5729H	DEERE,JOHN	80007	ASF5762	AUST.WEST.	80101	AW465209	AUST.WEST.	80163
AEO113	AEC	88000.35	AJ7329	DEERE,JOHN	80007	ASF5762	BAL-LM-HAM	80101	AW465217	AUST.WEST.	80168
AEO115	AEC	88000.55	AJ7329H	DEERE,JOHN	80021	ASF5762-55A	AUST.WEST.	80101	AW465225	AUST.WEST.	80049
AEO12	AEC	88000.45	AJ7329HN	DEERE,JOHN	80007	ASF5762-55A	BAL-LM-HAM	80101	AW465225	BAL-LM-HAM	80044
AEO123	AEC	88000.38	AJ7330H	DEERE,JOHN	80001	AS-US10AP	WEST.LAND	80069	AW465250	AUST.WEST.	80100
AEO124	AEC	88000.37	AJ7330HN	DEERE,JOHN	80044	ASUS130P	WEST.LAND	80136	AW4652505	BAL-LM-HAM	80100
AEO13	AEC	88000.20	AJ7331H	DEERE,JOHN	88003.38	AS-US21P	WEST.LAND	80001	AW465254	AUST.WEST.	80100
AEO150	AEC	88000.46	AJ8977	DEERE,JOHN	80133	AS-US90P	WEST.LAND	80161	AW465298	AUST.WEST.	80163
AEO22	AEC	88000.43	AJ9007N	DEERE,JOHN	80021	AT23096	DEERE,JOHN	80184	AW465345	AUST.WEST.	80198
AEO23	AEC	88000.60	AJ9907N	DEERE,JOHN	80021	AT25144	DEERE,JOHN	80198	AW465346	AUST.WEST.	80192
AEO-33	AEC	88000.26	AL1564N	DEERE,JOHN	80001	AT26446	DEERE,JOHN	80172	AW465424	AUST.WEST.	80100
AEO43	AEC	88000.22	AL430N	DEERE,JOHN	80054	AT27353	DEERE,JOHN	80213	AW466008	AUST.WEST.	80069
AEO53	AEC	88000.52	AL9907N	DEERE,JOHN	80021	AT27787	DEERE,JOHN	80182	AXC5351E	DEERE,JOHN	80104
AEO63	AEC	88000.25	AM1117E	DEERE,JOHN	80104	AT29361	DEERE,JOHN	80066	AXMT19725	CASE,J.I.	80008
AEO63	AEC	88000.27	AM1793E	DEERE,JOHN	80001	AT29362	DEERE,JOHN	80059	AXMT1972S	CASE,J.I.	80008
AEO73	AEC	88000.28	AM1861	DEERE,JOHN	80110	AT33511	DEERE,JOHN	80172	A-XP427	GERLINGER	80195
AEO76	AEC	88000.29	AM1861E	DEERE,JOHN	80100	AT3427	DEERE,JOHN	88001.88	AZ3304	DEERE,JOHN	80023
AEO81	AEC	88000.49	AM1862E	DEERE,JOHN	80101	AT36353	DEERE,JOHN	80218	B1-101K	BEAR.INC.	87238
AEO83	AEC	88000.53	AM2568E	DEERE,JOHN	80069	AT36356	DEERE,JOHN	80214	B115952A	HYSTER	80237
AEO95	AEC	88000.34	AM2854E	DEERE,JOHN	80044	AT36360	DEERE,JOHN	80216	B115952A	HYSTER	80240
AEO96-55	AEC	80134	AM2871E	DEERE,JOHN	80035	AT44365	DEERE,JOHN	80196	B1200N	HUSCO	80035
AE-P55-55-140	AEC	80081	AM3014	DEERE,JOHN	80104	AT49181	DEERE,JOHN	80141	B120-26	WESCO	88000.31
AE-P55-55-140	AEC	80129	AM3014E	DEERE,JOHN	80107	AT49181	DEERE,JOHN	80196	B120-26	WESCO	88000.88
AE-P55-55-145	AEC	80091	AM3025E	DEERE,JOHN	80099	AT54901	DEERE,JOHN	80085	B1-202KD	BEAR.INC.	87331
AE-P55-55-145	AEC	80131	AM3062E	DEERE,JOHN	80100	AT55470	DEERE,JOHN	80238	B1-203KC	BEAR.INC.	87285
AE-P96-55-1	AEC	80039	AM793E	DEERE,JOHN	80001	AT57090	DEERE,JOHN	80212	B1-303KC	BEAR.INC.	87337
AE-P96-55-1	AEC	80068	AMX46117	ALLIS-CHLM	80109	AT59126	DEERE,JOHN	80184	B13270	CASE,J.I.	80037
AE-R55-55-2	AEC	80123	AMX46117	ALLIS-CHLM	80149	AT59127	DEERE,JOHN	80185	B13272	CASE,J.I.	88014.32
AE-R96-55-1	AEC	80122	AMX46117	FIAT-ALLIS	80149	AT64485	DEERE,JOHN	88001.88	B13275	CASE,J.I.	80035
AER96-55-1	CLEVE.MOT.	80123	AN15327N	DEERE,JOHN	80104	AT73775	DEERE,JOHN	80198	B139472	HYSTER	80109
AE-S1200	AEC	80081	AN30706	DEERE,JOHN	80048	AT74268	DEERE,JOHN	80184	B139472	HYSTER	80150
AE-S53-55	AEC	80039	AN30706N	DEERE,JOHN	80048	AT74269	DEERE,JOHN	80185	B1400N	BPC	80037
AE-S55-55-1330	AEC	80129	AN30893N	DEERE,JOHN	80049	AU12520V	DEERE,JOHN	80231	B1403KC	BEAR.INC.	87399
AE-S55-55-1330	AEC	80130	AN31873N	DEERE,JOHN	80048	AU13093U	DEERE,JOHN	80069	B144-1	WESCO	88000.39
AE-S55-55-2	AEC	80081	AN32800	DEERE,JOHN	80077	AU21521U	DEERE,JOHN	80232	B144-1	WESCO	88000.81
AE-S55-55-675	AEC	80091	AN32803	DEERE,JOHN	80048	AU2152U	DEERE,JOHN	80232	B1-501KF	BEAR.INC.	87481
AE-S96-55	AEC	80038	AN32803N	DEERE,JOHN	80048	AU42404ED	DEERE,JOHN	80182	B15275	CASE,J.I.	80035
AE-T1010	AEC	80006	AOT7039A	FORD	80124	AU5729H	DEERE,JOHN	80007	B1-601KF	BEAR.INC.	88007.79

Part No.	Brand	IDLI Group	Part No.	Brand	IDLI Group	Part No.	Brand	IDLI Group	Part No.	Brand	IDLI Group
B1729	GERLINGER	80046	B4TZ4635E	FORD	80212	B7AZ4635A	FORD	80072	BB7199W	WHITE	80142
B1729	TOWMOTOR	88001.28	B4TZ4635F	FORD	80118	B7N	WESCO	80220	BF88510	FED.BEAR.	87481
B1769	GERLINGER	80047	B4TZ4635G	FORD	80201	B7S7039A	FORD	80072	BG3000	BMC	80024
B17691	GERLINGER	80047	B4TZ4635H	FORD	80197	B7T3815A	FORD	80024	BHR110-1	ROCKWELL	88007.51
B18132	CASE,J.I.	80001	B4TZ4635J	FORD	80210	B80H4635A	FORD	80231	BHR56-1	ROCKWELL	88007.52
B18132	CASE,J.I.	80003	B4TZ4635K	FORD	80218	B80H4635B	FORD	80237	BHR56-2	ROCKWELL	88007.53
B1FR	PILOT	80008	B4TZ4635N	FORD	80212	B87045	REO MOTORS	80221	BHR62-3	ROCKWELL	88007.54
B20-7039A	FORD	80178	B50N	PILOT	80178	B8N	WESCO	80225	BHR72-1	ROCKWELL	88007.55
B262	TEDDY TORQ	88002.67	B5556	MOTOR MAST	80220	B8Q4816C	FORD	86713	BHR86-1	ROCKWELL	88007.56
B2Q7039	FORD	80178	B55N	PILOT	80161	B8QH4635A	FORD	80231	BHR90-1	ROCKWELL	88007.57
B2Q7039A	FORD	80178	B58749	MERCURY	80062	B8QH4635B	FORD	80237	BI101	BEAR.INC.	87238
B2TZ4635A	FORD	80178	B58N	PILOT	80179	B8QH4635C	FORD	80240	BI101K	BEAR.INC.	87238
B300N	HUSCO	80023	B58N4	PILOT	80179	B8QH4635D	FORD	80237	BI202KD	BEAR.INC.	87332
B34NR	PILOT	80069	B58WB	PILOT	80185	B8QH4841C	FORD	85165	BI501KF	BEAR.INC.	88007.77
B3500N	BPC	80069	B58WB1	PILOT	80186	B8QH4841D	FORD	85126	BI601KF	BEAR.INC.	88007.79
B3575	MOTOR MAST	80134	B58WB2	PILOT	80184	B8QZ4635A	FORD	80243	BL10N	PILOT	80003
B35N3	PILOT	80069	B58WN	PILOT	80185	B8TZ4385B	FORD	86713	BL12N	PILOT	80035
B35N4	PILOT	80070	B58WN1	PILOT	80186	B92318	TOWMOTOR	80049	BL14N	PILOT	80037
B36311	MERCURY	80034	B58WN2	PILOT	80184	B9A7039A	FORD	80055	BL6N	PILOT	80001
B36312	MERCURY	80062	B5A4841A	FORD	88002.86	B9AZ4635A	FORD	80055	BL6N	PILOT	80003
B374	TEDDY TORQ	88002.63	B5A7039	FORD	80049	B9T2-3249A	FORD	80078	BL6N8	PILOT	80001
B385	TEDDY TORQ	88002.67	B5A7039A	FORD	80049	B9T4635A	FORD	80090	BR102-57	WAIN-ROY	80001
B3DR	PILOT	80133	B5A7042A	FORD	80044	B9T4724A	FORD	80118	BRAR19-2	ROCKWELL	88007.58
B3TZ4635A	FORD	80193	B5AZ4635A	FORD	80049	B9TF4724A	FORD	80118	BRAR22-2	ROCKWELL	88007.59
B3TZ4635B	FORD	80193	B5AZ4635B	FORD	80044	B9TT4724	FORD	80118	BRAR22-4	ROCKWELL	88007.60
B40N	PILOT	80134	B5AZ7039A	FORD	80049	B9TT4724A	FORD	80118	BRAR22-6	ROCKWELL	88007.61
B41083	BORG-WARNR	88003.47	B5N	WESCO	80178	B9TT4800A	FORD	87457	BRAR22-7	ROCKWELL	88007.62
B420111	BORG-WARNR	88003.58	B5Q4805B	FORD	88013.76	B9TT4826A	FORD	88008.65	BRAR22-8	ROCKWELL	88007.63
B420112	BORG-WARNR	80953	B5Q4865A	FORD	81744	B9TZ3249A	FORD	80090	BRAR22-9	ROCKWELL	88007.64
B42012	BORG-WARNR	88003.56	B5Q4865B	FORD	81661	B9TZ4635A	FORD	80090	BRAR25-12	ROCKWELL	88007.65
B42013	BORG-WARNR	88003.56	B5QH4818B	FORD	88014.86	B9TZ4840A	FORD	88008.47	BRAR25-13	ROCKWELL	88007.66
B42017	REPUBLIC	88003.53	B5QH4826B	FORD	88008.64	BB114-8515A	BAR-GREENE	80185	BRAR25-14	ROCKWELL	88007.67
B42019	BORG-WARNR	88003.56	B5T4805A	FORD	88013.73	BB17041	DIAMOND T	80178	BRAR25-3	ROCKWELL	88007.68
B42021	BORG-WARNR	88003.42	B5T4818A	FORD	88014.85	BB17041	WHITE	80178	BRAR25-4	ROCKWELL	88007.69
B42031	BORG-WARNR	88003.43	B5T4835A	FORD	88013.65	BB17042	DIAMOND R	80079	BRAR25-5	ROCKWELL	88007.70
B4204	BORG-WARNR	88003.45	B5T7039	FORD	80124	BB17042	DIAMOND T	80189	BRAR25-6	ROCKWELL	88007.71
B42042	BORG-WARNR	88003.45	B5T7039A	FORD	80124	BB17042	WHITE	80189	BRAR25-7	ROCKWELL	88007.72
B42051	BORG-WARNR	88003.48	B5TZ3249A	FORD	80124	BB17044	DIAMOND R	80221	BRAR25-8	ROCKWELL	88007.73
B420821	BORG-WARNR	88003.50	B5TZ4635A	FORD	80124	BB17044	DIAMOND T	80221	BRAR28-1	ROCKWELL	88007.74
B420822-1	BORG-WARNR	88003.50	B5TZ4840A	FORD	88008.56	BB17044	WHITE	80221	BRAR28-2	ROCKWELL	88007.75
B420825-1	BORG-WARNR	88003.50	B5TZ4840B	FORD	88008.53	BB17199J	DIAMOND R	80221	BRAR28-5	ROCKWELL	88007.76
B42083	BORG-WARNR	88003.47	B600N	BPC	88003.38	BB17199J	WHITE	80221	BRAR32-1	ROCKWELL	88007.77
B42083	BORG-WARNR	88003.50	B600N	HUSCO	80001	BB17199W	DIAMOND R	80142	BRAR38-1	ROCKWELL	88007.78
B42083	NEAPCO	88003.47	B60N	PILOT	80188	BB17199W	WHITE	80142	BRAR38-2	ROCKWELL	88007.79
B4257-1	BORG-WARNR	88003.44	B62	CARDWELL	80001	BB1742	DIAMOND T	80189	BRBAR19-1	ROCKWELL	88007.80
B42591	BORG-WARNR	88003.49	B663	TEDDY TORQ	88002.71	BB176041	DIAMOND R	80178	BRBAR22-6	ROCKWELL	88007.81
B4261	BORG-WARNR	88003.46	B69240	TOWMOTOR	80109	BB1799F	DIAMOND T	80189	BRBAR25-4	ROCKWELL	88007.82
B42703(7)	GALION	80240	B69240	TOWMOTOR	80150	BB27199W	DIAMOND R	80140	BRBAR25-5	ROCKWELL	88007.83
B4301	BORG-WARNR	88003.57	B6A7039A	FORD	80077	BB27199W	WHITE	80140	BRBAR28-3	ROCKWELL	88007.84
B4303-1	BORG-WARNR	88003.43	B6D7039	FORD	80077	BB7041	DIAMOND R	80178	BRBAR32-6	ROCKWELL	88007.85
B432464	NELSON	80219	B6D7039A	FORD	80072	BB7041	DIAMOND T	80178	BRBAR38-2	ROCKWELL	88007.86
B44016	BORG-WARNR	80957	B6N	WESCO	80188	BB7041	WHITE	80178	BRNR24-2	ROCKWELL	88007.87
B44023	BORG-WARNR	88003.51	B6S7039A	FORD	80072	BB7042	DIAMOND R	80079	BRNR28-2	ROCKWELL	88007.88
B440811-1	BORG-WARNR	88003.51	B6T7039A	FORD	80142	BB7042	DIAMOND T	80189	BS21S	MOTOR MAST	88007.78
B44086	BORG-WARNR	88003.47	B6T7039A	FORD	80143	BB7042	WHITE	80189	BS524-11	ROCKWELL	88007.89
B445014	BORG-WARNR	80955	B6TZ039A	FORD	80142	BB7043	DIAMOND T	80189	BS524-8	ROCKWELL	88007.90
B4575	MOTOR MAST	80188	B6TZ4635	FORD	80142	BB7044	DIAMOND R	80221	BS528-10	ROCKWELL	88007.91
B4634B	FORD	80142	B6TZ4635A	FORD	80075	BB7044	DIAMOND T	80221	BS528-11	ROCKWELL	88007.92
B4635A	FORD	80142	B6TZ4635B	FORD	80142	BB7044	WHITE	80221	BS528-12	ROCKWELL	88007.93
B4635B	FORD	80142	B6TZ4635C	FORD	80055	BB7045	DIAMOND R	80221	BS528-16	ROCKWELL	88007.94
B48153	MERCURY	80104	B6TZ4635D	FORD	80031	BB7199C	DIAMOND T	80178	BS528-28	ROCKWELL	88007.95
B4HZ4635	FORD	80199	B7084	FORD	88002.61	BB7199C	WHITE	80178	BS528-4	ROCKWELL	88007.96
B4HZ4635B	FORD	80194	B7090	ALCO	88002.55	BB7199F	DIAMOND T	80189	BS528-6	ROCKWELL	88007.97
B4N	WESCO	80134	B7090	FORD	88002.55	BB7199F	WHITE	80189	BS528-7	ROCKWELL	88007.98
B4QH4841B	FORD	85114	B70N	PILOT	80220	BB7199H	DIAMOND T	80124	BS528-9	ROCKWELL	88007.99
B4QH4841C	FORD	85093	B7100-1A	BORG-WARNR	88002.69	BB7199H	WHITE	80124	BS531-14	ROCKWELL	88008
B4TZ4635A	FORD	80199	B7140-2A	BORG-WARNR	88002.69	BB7199J	DIAMOND T	80221	BS531-3	ROCKWELL	88008.01
B4TZ4635B	FORD	80201	B7200-20	BORG-WARNR	88002.68	BB7199J	WHITE	80221	BS531-4	ROCKWELL	88008.02
B4TZ4635C	FORD	80197	B7A4635A	FORD	80072	BB7199W	DIAMOND T	80142	BSA-R12-1	ROCKWELL	87201
B4TZ4635D	FORD	80210	B7A7039A	FORD	80072	BB7199W	DIAMOND T	80143	BSA-R12-2	ROCKWELL	87202

Part No.	Brand	IDLI Group
BSA-R14-1	ROCKWELL	87206
BSA-R16-1	ROCKWELL	87212
BSA-R16-3	ROCKWELL	87211
BSA-R18-6	ROCKWELL	87232
BSA-R18-7	ROCKWELL	87236
BSA-R19-4	ROCKWELL	87243
BSA-R20-1	ROCKWELL	87244
BSA-R20-10	ROCKWELL	87269
BSA-R20-19	ROCKWELL	87270
BSA-R20-2	ROCKWELL	87267
BSA-R20-3	ROCKWELL	87271
BSA-R20-6	ROCKWELL	87268
BSA-R23-5	ROCKWELL	87333
BSA-R23-6	ROCKWELL	87334
BSA-S18-1	ROCKWELL	87230
BSA-S18-2	ROCKWELL	87231
BSA-S28-2	ROCKWELL	87456
BSA-S28-3	ROCKWELL	87446
BSA-S28-5	ROCKWELL	87441
BS-R16-2	ROCKWELL	87210
BS-R18-1	ROCKWELL	87228
BS-R18-20	ROCKWELL	87235
BS-R18-23	ROCKWELL	87234
BS-R18-25	ROCKWELL	87233
BS-R18-32	ROCKWELL	87237
BS-R18-5	ROCKWELL	87229
BS-R21-1	ROCKWELL	87279
BS-R21-14	ROCKWELL	87281
BS-R21-19	ROCKWELL	87282
BS-R21-2	ROCKWELL	87280
BS-R21-22	ROCKWELL	87283
BS-R21-23	ROCKWELL	87284
BS-R21-3	ROCKWELL	87278
BS-R22-1	ROCKWELL	87321
BS-R22-14	ROCKWELL	87304
BS-R22-15	ROCKWELL	87302
BS-R24-5	ROCKWELL	87365
BS-S20-1	ROCKWELL	87258
BS-S20-11	ROCKWELL	87259
BS-S20-13	ROCKWELL	87260
BS-S20-14	ROCKWELL	87264
BS-S20-15	ROCKWELL	87262
BS-S20-16	ROCKWELL	87266
BS-S20-17	ROCKWELL	87263
BS-S20-23	ROCKWELL	87261
BS-S20-24	ROCKWELL	87265
BS-S20-3	ROCKWELL	87257
BS-S20-9	ROCKWELL	87256
BS-S22-1	ROCKWELL	87295
BS-S22-10	ROCKWELL	87316
BS-S22-11	ROCKWELL	87309
BS-S22-114	ROCKWELL	88008.03
BS-S22-115	ROCKWELL	88008.04
BS-S22-116	ROCKWELL	88008.05
BS-S22-117	ROCKWELL	88008.06
BS-S22-118	ROCKWELL	88008.07
BS-S22-119	ROCKWELL	88008.08
BS-S22-12	ROCKWELL	87310
BS-S22-120	ROCKWELL	88008.09
BS-S22-121	ROCKWELL	88008.10
BS-S22-122	ROCKWELL	88008.11
BS-S22-123	ROCKWELL	88008.12
BS-S22-124	ROCKWELL	88008.13
BS-S22-12A	ROCKWELL	87311
BS-S22-13	ROCKWELL	87303
BS-S22-16	ROCKWELL	87292
BS-S22-17	ROCKWELL	87308
BS-S22-25	ROCKWELL	87306
BS-S22-29	ROCKWELL	87307
BS-S22-34	ROCKWELL	87301
BS-S22-36	ROCKWELL	87317
BS-S22-38	ROCKWELL	87296
BS-S22-39	ROCKWELL	87294
BS-S22-4	ROCKWELL	87300
BS-S22-40	ROCKWELL	87293
BS-S22-41	ROCKWELL	87305
BS-S22-43	ROCKWELL	87315
BS-S22-44	ROCKWELL	87291
BS-S22-49	ROCKWELL	87299
BS-S22-50	ROCKWELL	87312
BS-S22-51	ROCKWELL	87320
BS-S22-53	ROCKWELL	87318
BS-S22-57	ROCKWELL	87313
BS-S22-60	ROCKWELL	87297
BS-S22-64	ROCKWELL	87324
BS-S22-65	ROCKWELL	87298
BS-S22-72	ROCKWELL	87322
BS-S22-82	ROCKWELL	87319
BS-S22-9	ROCKWELL	87323
BS-S24-11	ROCKWELL	87363
BS-S24-30	ROCKWELL	88008.14
BS-S24-31	ROCKWELL	88008.15
BS-S24-32	ROCKWELL	88008.16
BS-S24-36	ROCKWELL	87350
BS-S24-37	ROCKWELL	87351
BS-S24-38	ROCKWELL	87359
BS-S24-39	ROCKWELL	87354
BS-S24-40	ROCKWELL	87358
BS-S24-40A	ROCKWELL	87362
BS-S24-41	ROCKWELL	88008.17
BS-S24-42	ROCKWELL	88008.18
BS-S24-8	ROCKWELL	87364
BS-S25-3	ROCKWELL	88008.19
BS-S25-4	ROCKWELL	88008.20
BS-S26-12	ROCKWELL	87398
BS-S26-14	ROCKWELL	88008.21
BS-S26-15	ROCKWELL	88008.22
BS-S26-16	ROCKWELL	87391
BS-S26-16A	ROCKWELL	87394
BS-S26-3	ROCKWELL	87395
BS-S26-6	ROCKWELL	87397
BS-S28-10	ROCKWELL	87449
BS-S28-11	ROCKWELL	87453
BS-S28-12	ROCKWELL	87314
BS-S28-12	ROCKWELL	87443
BS-S28-16	ROCKWELL	88008.23
BS-S28-17	ROCKWELL	87447
BS-S28-28	ROCKWELL	87448
BS-S28-33	ROCKWELL	87445
BS-S28-36	ROCKWELL	87442
BS-S28-4	ROCKWELL	87451
BS-S28-42	ROCKWELL	87454
BS-S28-49	ROCKWELL	88008.24
BS-S28-6	ROCKWELL	87452
BS-S28-61	ROCKWELL	88008.25
BS-S28-62	ROCKWELL	87436
BS-S28-67	ROCKWELL	87429
BS-S28-67A	ROCKWELL	87438
BS-S28-68	ROCKWELL	88008.26
BS-S28-7	ROCKWELL	88008.27
BS-S28-8	ROCKWELL	87450
BS-S28-9	ROCKWELL	87444
BS-S31-1	ROCKWELL	87478
BS-S31-12	ROCKWELL	87471
BS-S31-13	ROCKWELL	87473
BS-S31-14	ROCKWELL	87472
BS-S31-18	ROCKWELL	88008.28
BS-S31-19	ROCKWELL	87470
BS-S31-21	ROCKWELL	87476
BS-S31-3	ROCKWELL	87477
BS-S31-4	ROCKWELL	87475
BS-S31-9	ROCKWELL	87474
BS-S32-12	ROCKWELL	87541
BS-S32-15	ROCKWELL	87543
BS-S32-16	ROCKWELL	87542
BS-S32-19	ROCKWELL	87540
BS-S32-7A	ROCKWELL	88008.29
BS-S32-8	ROCKWELL	87539
BS-S32-9	ROCKWELL	87538
BS-S33-1	ROCKWELL	87548
BS-S36-10	ROCKWELL	87558
BS-S36-12	ROCKWELL	87557
BS-S36-3	ROCKWELL	87559
BS-S38-1	ROCKWELL	87565
BS-S38-4	ROCKWELL	87566
BS-S38-5	ROCKWELL	87569
BS-S38-6	ROCKWELL	87568
BS-S40-4	ROCKWELL	87603
BS-S46-1	ROCKWELL	87620
BU200	REPUBLIC	88002.63
BU201	REPUBLIC	88002.64
BU202	REPUBLIC	88002.67
BU203	REPUBLIC	88002.65
BU204	REPUBLIC	88002.67
BU205	REPUBLIC	88002.71
BU206	REPUBLIC	88002.70
BU207	REPUBLIC	88002.67
BU207	REPUBLIC	88002.72
BU208	REPUBLIC	88002.73
BU209	REPUBLIC	88002.74
BU210	REPUBLIC	88002.75
BU211	REPUBLIC	88002.76
C0-9655	BALKAMP	80138
C0D2-7039A	FORD	80031
C0DZ4635A	FORD	80030
C1009	REPUBLIC	88002.43
C1009	REPUBLIC	88002.62
C1109	REPUBLIC	88002.42
C1110	REPUBLIC	88002.47
C1209	REPUBLIC	88002.44
C1210	REPUBLIC	88002.60
C123-1	WESCO	88000.69
C1309	REPUBLIC	88002.44
C134	WESCO	88000.70
C1400	BEAN,JOHN	80161
C1410	REPUBLIC	88002.46
C144	WESCO	88000.71
C154	WESCO	88000.72
C1552	REPUBLIC	80118
C1610	REPUBLIC	88002.48
C1709	REPUBLIC	88002.51
C1710	REPUBLIC	88002.51
C1810	REPUBLIC	88002.50
C1910	REPUBLIC	88002.50
C1910	REPUBLIC	88002.52
C19655	PILOT	80068
C19655	PILOT	80204
C1AA4724A	FORD	80081
C1AA4724B	FORD	80091
C1AZ4635A	FORD	80073
C1AZ4635A	FORD	80091
C1AZ4635A	FORD	80140
C1AZ4635C	FORD	80081
C1AZ4635D	FORD	80091
C1T2-3249A	FORD	80124
C1TT4724A	FORD	80140
C1TT4826A	FORD	88008.66
C1TT4835A	FORD	88013.66
C1TT4840A	FORD	87481
C1TZ3249A	FORD	80128
C1TZ3249A	FORD	88001.77
C1TZ4557C	FORD	88006.66
C1TZ4557F	FORD	88006.70
C1TZ4635A	FORD	80140
C1TZ4635B	FORD	80118
C1TZ4841C	FORD	85102
C1UZ4635A	FORD	80031
C1VV4497A	FORD	80077
C1VV4997	FORD	80072
C1VV4997A	FORD	80072
C1VV4997A	FORD	80077
C21222	TOWMOTOR	80001
C25A4724A	FORD	80118
C2665	ALVIS	80041
C2AZ4635D	FORD	80091
C2SA4724A	FORD	80130
C2SZ4635A	FORD	80130
C2SZ4724A	FORD	80130
C302	MOTOR MAST	88006.99
C-302	MOTOR MAST	88006.99
C3106	MIPER	80023
C3107	MIPER	80023
C3108	MIPER	80024
C3109	MIPER	80023
C3112	MIPER	80023
C3116	MIPER	88001.86
C3120	MIPER	80041
C3121	MIPER	80041
C3122	MIPER	80041
C3123	MIPER	80041
C3127	MIPER	80072
C3128	MIPER	80077
C3135	MIPER	80124
C3136	MIPER	80124
C3146	MIPER	80142
C3147	MIPER	80142
C3148	MIPER	80040
C3163	MIPER	80140
C3169	MIPER	80180
C3171	MIPER	80178
C3176	MIPER	80189
C3190	MIPER	80188
C3195	MIPER	80221
C3197	MIPER	80221
C3200	MIPER	80221
C3218	MIPER	80182
C3234	MIPER	80037
C3265	MIPER	80225
C3318-6	MIPER	88002.77
C3320	MIPER	80044
C3352	MIPER	80090
C3360	MIPER	80042
C3368	MIPER	80135
C3452	MIPER	80133
C3460	MIPER	80039
C3461	MIPER	80038
C3462	MIPER	80122
C3464	MIPER	80138
C3466	MIPER	80160
C3468	MIPER	80181
C3520	MIPER	80210
C3670-6	MIPER	80032
C3674	MIPER	80049
C3A4635A	FORD	80081
C3AA4635A	FORD	80081
C3AA4635B	FORD	80091
C3AZ4635A	FORD	80081
C3AZ4635B	FORD	80091
C3AZ4635C	FORD	80072
C3AZ4635D	FORD	80073

Part No.	Brand	IDLI Group
C3AZ4635E	FORD	80081
C3AZ4635F	FORD	80091
C3AZ4635G	FORD	80129
C3AZ4635H	FORD	80131
C3DZ4635A	FORD	80030
C3SZ4635A	FORD	80081
C3T23815A	FORD	80003
C3TZ3815A	FORD	80003
C3TZ4841E	FORD	85114
C3TZ4841G	FORD	85094
C3TZ4841H	FORD	88007.15
C3UZ4635A	FORD	80031
C3UZ4635B	FORD	80031
C3VY4497A	FORD	80117
C3VY4997A	FORD	80117
C402	MUNCIE	81209
C-402	MUNCIE	81209
C409	REPUBLIC	88002.56
C420825-1	PILOT	88003.50
C42083	PILOT	88003.50
C4996	ALCO	88013.86
C4999	ALCO	88013.87
C4AZ4635A	FORD	80231
C4GW-4635A	FORD	80129
C4SZ4635A	FORD	80117
C4SZ4635B	FORD	80117
C4TA4635A	FORD	80075
C4TA4635B	FORD	80210
C4TA4635E	FORD	80197
C4TA4635H	FORD	80212
C4TA4635J	FORD	80237
C4TA4724A	FORD	80075
C4TZ4635A	FORD	80075
C4TZ4724A	FORD	80075
C4VW4635A	FORD	80121
C5000	WESCO	80168
C5700AR	PILOT	80908
C5700R	PILOT	80955
C5DZ4635A	FORD	80058
C5DZ4635B	FORD	80044
C5TZ23815A	FORD	80003
C5TZ3815A	FORD	80003
C5TZ4635A	FORD	80123
C5TZ4635B	FORD	80141
C5TZ4635C	FORD	80075
C5TZ4635D	FORD	80075
C5TZ4635E	FORD	80124
C5TZ4635F	FORD	80118
C5TZ4635G	FORD	80141
C5TZ4805A	FORD	88013.71
C5UZ4635A	FORD	80031
C5UZ4635B	FORD	80031
C5ZZ4841A	FORD	88014.41
C609	REPUBLIC	88002.57
C6200AR	PILOT	80907
C6200R	PILOT	80954
C6300	PILOT	80957
C6300R	PILOT	80957
C6500AR	PILOT	80905
C6500ARS	PILOT	80905
C6500R	PILOT	80951
C6600R	PILOT	80952
C661-106	CLEVE.STL.	83527
C6699R	PILOT	80953
C6700AR	PILOT	80906
C6700ARS	PILOT	80906
C6700R	PILOT	80953
C6700RS	PILOT	80953
C67-110	CLEVE.STL.	88008.28
C67-1-100	CLEVE.STL.	83695
C67-1-122	CLEVE.STL.	83583
C67-1-123	CLEVE.STL.	83607
C67-12	CLEVE.STL.	88014.55
C67-13	CLEVE.STL.	88013.36
C67-55-2	CLEVE.STL.	80221
C67-7-4-10	CLEVE.STL.	87596
C67-7-4-105	CLEVE.STL.	88005.61
C6TA4635C	FORD	80210
C6TA4635D	FORD	80212
C6TA4635E	FORD	80197
C6TA4635F	FORD	80193
C6TA4635G	FORD	80199
C6TA4635H	FORD	80201
C6TTZ3249A	FORD	80078
C6TTZ3249A	FORD	80090
C6TZ3249A	FORD	80078
C6TZ3815A	FORD	80001
C6TZ3815B	FORD	80003
C6TZ4635	FORD	80189
C6TZ4635A	FORD	80066
C6TZ4635C	FORD	80075
C6TZ4635D	FORD	80104
C6TZ4635E	FORD	80190
C6TZ4635F	FORD	80221
C6TZ4635G	FORD	80189
C6TZ4635H	FORD	80059
C6TZ4635J	FORD	80190
C6TZ4635K	FORD	80221
C6TZ4635L	FORD	80189
C709	REPUBLIC	88002.58
C7TZ4635A	FORD	80225
C809	REPUBLIC	88002.43
C809	REPUBLIC	88002.59
C8HA4800A	FORD	88007.79
C8HA4800B	FORD	87481
C8HA4800B	FORD	88007.77
C8SZ4635A	FORD	80104
C8SZ4635A	FORD	80107
C8SZ4A031	FORD	88002.34
C8SZ4A031A	FORD	80023
C8SZ4A031A	FORD	80073
C8SZ4A031A	FORD	88002.34
C8TZ24635A	FORD	80066
C8TZ4635A	FORD	80066
C8TZ4826A	FORD	88008.67
C901	REPUBLIC	88002.45
C910	REPUBLIC	88002.45
C96	TEDDY TORQ	80204
C96-55	AEC	80204
C96-55	ALCO	80204
C96-55	BORG-WARNR	80204
C96-55	CLEVELAND	80068
C96-55	CLEVE.STL.	80204
C96-55	REPCO	80204
C96-55	SPICER	80204
C9SZ4635A	FORD	80107
C9TZ4635A	FORD	80072
C9TZ4635B	FORD	80077
C9TZ4782A	FORD	88002.37
CA205	INBRA	80142
CA2100	INBRA	80049
CA2101	INBRA	80044
CA220	INBRA	80124
CA266	INBRA	80061
CA372	INBRA	80135
CA375	INBRA	80042
CA504	INBRA	80041
CA505	INBRA	80031
CA506	INBRA	80032
CA508	INBRA	80023
CA514	INBRA	80024
CA521	INBRA	80072
CA521HD	INBRA	80077
CA8-42	JOY MFG.	80069
CA8U2	JOY MFG.	80069
CAD.	MCQUAY-NOR	80099
CAD.	MCQUAY-NOR	80110
CASR30-1	ROCKWELL	88008.30
CAT401794	JEFFREY	80142
CAT410151	JEFFREY	80178
CAT415325	JEFFREY	80124
CAW8055A	DEERE,JOHN	80210
CB	REPUBLIC	80136
CB055	REPUBLIC	80141
CB096	REPUBLIC	80138
CB1100	REPUBLIC	80023
CB1100	REPUBLIC	80024
CB1109	REPUBLIC	88002.42
CB1209	REPUBLIC	88002.49
CB1248	REPUBLIC	80025
CB1249	REPUBLIC	80066
CB1250B	REPUBLIC	80041
CB1251	REPUBLIC	80023
CB1251	REPUBLIC	80024
CB1270	REPUBLIC	80077
CB1270B	REPUBLIC	80072
CB1270B	REPUBLIC	80075
CB1270B	REPUBLIC	80077
CB1270BHD	REPUBLIC	80072
CB1270C	REPUBLIC	80072
CB1270HD	REPUBLIC	80072
CB1270HD	REPUBLIC	80075
CB1270HD	REPUBLIC	80077
CB1270HDC	REPUBLIC	80077
CB1270L	REPUBLIC	80072
CB1270L	REPUBLIC	80075
CB1270U	REPUBLIC	80089
CB1270UB	REPUBLIC	88013.86
CB1271	REPUBLIC	80073
CB1271	REPUBLIC	80091
CB1271HDC	REPUBLIC	80073
CB1272	REPUBLIC	80036
CB1279BHD	REPUBLIC	80077
CB1300	REPUBLIC	80117
CB1301	REPUBLIC	80121
CB1301	REPUBLIC	80131
CB1301X	REPUBLIC	80131
CB13-1	REPUBLIC	80121
CB1310	REPUBLIC	80079
CB1332	REPUBLIC	80139
CB1333	REPUBLIC	80116
CB1350A	REPUBLIC	80124
CB1350U	REPUBLIC	80125
CB1350UB	REPUBLIC	88013.87
CB1350W	REPUBLIC	80124
CB1372	REPUBLIC	80105
CB1372	REPUBLIC	80227
CB1399	REPUBLIC	80083
CB140P	REPUBLIC	80007
CB140P	REPUBLIC	88000.14
CB140R	REPUBLIC	80007
CB140R	REPUBLIC	88000.15
CB1410	REPUBLIC	80142
CB1410A	REPUBLIC	80142
CB1410U	REPUBLIC	80143
CB14DR	REPUBLIC	80007
CB14DR	REPUBLIC	88000.15
CB14OP	REPUBLIC	80007
CB14OR	REPUBLIC	80007
CB1500	REPUBLIC	80178
CB1501	REPUBLIC	80021
CB1501	REPUBLIC	80023
CB1505	REPUBLIC	80119
CB1552	REPUBLIC	80118
CB1555	REPUBLIC	80182
CB1574	REPUBLIC	80090
CB1574S	REPUBLIC	80078
CB1575	REPUBLIC	80090
CB1600	REPUBLIC	80189
CB1600AG	REPUBLIC	80189
CB1600HD	REPUBLIC	80189
CB1650HD	REPUBLIC	80190
CB1655	REPUBLIC	80190
CB1700	REPUBLIC	80221
CB1700-4	REPUBLIC	80221
CB1700AG	REPUBLIC	80221
CB1700HD	REPUBLIC	80221
CB1720C	REPUBLIC	80075
CB1720L	REPUBLIC	80077
CB1755	REPUBLIC	80032
CB1755G	REPUBLIC	80032
CB1755G	REPUBLIC	88002.81
CB1800	REPUBLIC	80225
CB1800AG	REPUBLIC	80225
CB1800HD	REPUBLIC	80225
CB180D	REPUBLIC	80225
CB1810	REPUBLIC	80227
CB1810AG	REPUBLIC	80227
CB1810HD	REPUBLIC	80227
CB1855	REPUBLIC	80228
CB1855AG	REPUBLIC	80228
CB1855B	REPUBLIC	88001.82
CB1855HD	REPUBLIC	80228
CB1860	REPUBLIC	80225
CB1880	REPUBLIC	80140
CB1880A	REPUBLIC	80007
CB1880A	REPUBLIC	80229
CB1880C	REPUBLIC	87481
CB1888	REPUBLIC	80140
CB1888	REPUBLIC	80182
CB1FR	REPUBLIC	80008
CB1FR4	REPUBLIC	80008
CB1FRN4	REPUBLIC	80008
CB1FRN4	REPUBLIC	88002.88
CB2000	REPUBLIC	80049
CB2000	REPUBLIC	80050
CB2000	REPUBLIC	88002.09
CB2002	REPUBLIC	80048
CB2002	REPUBLIC	80051
CB2002	REPUBLIC	88002.10
CB2005	REPUBLIC	80049
CB2006	REPUBLIC	80044
CB2007	REPUBLIC	80043
CB2007	REPUBLIC	80044
CB2007C	REPUBLIC	80044
CB2010	REPUBLIC	80049
CB2010	REPUBLIC	88002.09
CB2010C	REPUBLIC	80049
CB2011	REPUBLIC	80044
CB2011C	REPUBLIC	80044
CB2012	REPUBLIC	80030
CB2014	REPUBLIC	80045
CB2014	REPUBLIC	88002.09
CB2016	REPUBLIC	80044
CB2017	ALMETAL	80092
CB2017	ALMETAL	88001.76
CB2017	REPUBLIC	80092
CB2017	REPUBLIC	88001.76
CB2018	REPUBLIC	80053

Part No.	Brand	IDLI Group	Part No.	Brand	IDLI Group	Part No.	Brand	IDLI Group	Part No.	Brand	IDLI Group
CB2026	REPUBLIC	80030	CB2887	REPUBLIC	80020	CB51525	REPUBLIC	88001.60	CB719	REPUBLIC	80233
CB2031	REPUBLIC	80049	CB3000	REPUBLIC	80099	CB515251	REPUBLIC	88001.60	CB719	REPUBLIC	88003.09
CB2032	REPUBLIC	80092	CB3000	REPUBLIC	80101	CB515252	REPUBLIC	88001.60	CB720	REPUBLIC	80216
CB2033	REPUBLIC	80054	CB3001	REPUBLIC	80099	CB515253	REPUBLIC	88001.60	CB721	REPUBLIC	80149
CB2039	REPUBLIC	80044	CB3001	REPUBLIC	80101	CB51625	REPUBLIC	80042	CB722	REPUBLIC	80101
CB2040	REPUBLIC	80049	CB3008	REPUBLIC	80107	CB51626	REPUBLIC	80042	CB7225	REPUBLIC	80064
CB2040	REPUBLIC	80053	CB3008G	REPUBLIC	80104	CB53525	REPUBLIC	88001.62	CB7225	REPUBLIC	88001.63
CB2041	REPUBLIC	80054	CB3008HDC	REPUBLIC	80104	CB535251	REPUBLIC	88001.62	CB72251	REPUBLIC	80064
CB2045	REPUBLIC	80049	CB3008L	REPUBLIC	80104	CB53625	REPUBLIC	80135	CB72251-3	REPUBLIC	80064
CB2046	REPUBLIC	80044	CB3010	REPUBLIC	80100	CB53725	REPUBLIC	80135	CB72251-3	REPUBLIC	80065
CB2046	REPUBLIC	88001.98	CB3011	REPUBLIC	80104	CB53825-2	REPUBLIC	80135	CB7225-4	REPUBLIC	80056
CB2048	REPUBLIC	80049	CB3011	REPUBLIC	80107	CB553	REPUBLIC	80039	CB7225-4	REPUBLIC	80057
CB2049	REPUBLIC	80054	CB3011C	REPUBLIC	80109	CB55N	REPUBLIC	80161	CB7225-5	REPUBLIC	80060
CB2053	REPUBLIC	80030	CB3011HDC	REPUBLIC	80104	CB565-6	REPUBLIC	80073	CB723	REPUBLIC	80206
CB2053G	REPUBLIC	80031	CB3012	REPUBLIC	80110	CB58N4	REPUBLIC	80178	CB723	REPUBLIC	80212
CB2062	REPUBLIC	80050	CB3012	REPUBLIC	80112	CB58N4	REPUBLIC	80179	CB724	REPUBLIC	80198
CB2063	REPUBLIC	80043	CB3012G	REPUBLIC	80112	CB58N4	REPUBLIC	80186	CB725	REPUBLIC	80198
CB2063	REPUBLIC	80044	CB3013	REPUBLIC	80106	CB58NB2	REPUBLIC	80184	CB72525	REPUBLIC	80056
CB2063	REPUBLIC	88001.96	CB3014	REPUBLIC	80114	CB58WB	REPUBLIC	80185	CB72525	REPUBLIC	80057
CB2064	REPUBLIC	80044	CB3019	REPUBLIC	80110	CB58WB1	REPCO	80186	CB725251	REPUBLIC	80056
CB2066	REPUBLIC	80045	CB3019	REPUBLIC	80112	CB58WB1	REPUBLIC	80186	CB72525-1	REPUBLIC	80057
CB210	REPUBLIC	80059	CB3021	REPUBLIC	80111	CB58WB17	ROCKWELL	88008.37	CB72525-3	REPUBLIC	80064
CB2100	REPUBLIC	80026	CB3031BHP	REPUBLIC	80104	CB58WB2	REPUBLIC	80184	CB72525-3	REPUBLIC	80065
CB210084-1X	ROCKWELL	88008.32	CB3047	REPUBLIC	80093	CB5N	REPUBLIC	80178	CB72575-3	REPUBLIC	80065
CB210084-2X	ROCKWELL	87458	CB3050A	REPUBLIC	80124	CB5R40-1	ROCKWELL	88008.38	CB726	REPUBLIC	80199
CB210088-1X	ROCKWELL	87327	CB310001-1X	ROCKWELL	87377	CB6000	REPUBLIC	80193	CB7262-2	REPUBLIC	80065
CB210090-1X	ROCKWELL	87328	CB310001-2X	ROCKWELL	87382	CB6000	REPUBLIC	80198	CB72625-2	REPUBLIC	80064
CB210094-2X	ROCKWELL	88008.33	CB310002-1X	ROCKWELL	87383	CB6016	REPUBLIC	80193	CB72625-2	REPUBLIC	80065
CB210121-1X	ROCKWELL	87481	CB310002-2X	ROCKWELL	87384	CB6022	REPUBLIC	80196	CB72625-3	REPUBLIC	80061
CB210130-1X	ROCKWELL	87459	CB310003-1X	ROCKWELL	87376	CB6022	REPUBLIC	80197	CB72625-4	REPUBLIC	80061
CB210140-1X	ROCKWELL	87378	CB310005-1X	ROCKWELL	87326	CB6023	REPUBLIC	80199	CB72625-4	REPUBLIC	80063
CB210144-1X	ROCKWELL	87379	CB310006-1X	ROCKWELL	87331	CB6024	REPUBLIC	80201	CB72625-4C	REPUBLIC	80061
CB210207-1X	ROCKWELL	87460	CB310006-2X	ROCKWELL	87332	CB6103	REPUBLIC	80194	CB72625-9	REPUBLIC	80062
CB210367-1X	ROCKWELL	87329	CB310007-1X	ROCKWELL	87457	CB6N	REPUBLIC	80188	CB727	REPUBLIC	80213
CB210370-1X	ROCKWELL	87330	CB310007-2X	ROCKWELL	87461	CB700	REPUBLIC	80205	CB728	REPUBLIC	80238
CB210391-1X	ROCKWELL	87380	CB310008-1X	ROCKWELL	87285	CB7000	REPUBLIC	80210	CB729	REPUBLIC	80155
CB210433-1X	ROCKWELL	87381	CB310008-2X	ROCKWELL	87286	CB7004	REPUBLIC	80207	CB72N57	REPUBLIC	80209
CB210527X	ROCKWELL	87238	CB310009-1X	ROCKWELL	87399	CB7004	REPUBLIC	80211	CB730	REPUBLIC	80150
CB21057X	ROCKWELL	88008.34	CB310009-2X	ROCKWELL	87400	CB700A	REPUBLIC	80205	CB730	REPUBLIC	80153
CB210580X	ROCKWELL	88008.35	CB310010-1X	ROCKWELL	87337	CB701	REPUBLIC	80213	CB731	REPUBLIC	80230
CB210661-1X	ROCKWELL	87570	CB310010-2X	ROCKWELL	87338	CB702	REPUBLIC	80112	CB731	REPUBLIC	80232
CB210866-1X	ROCKWELL	87374	CB310011-1X	ROCKWELL	87325	CB703	REPUBLIC	80235	CB732	REPUBLIC	80228
CB210873-1X	ROCKWELL	87375	CB3100AX	REPUBLIC	80082	CB7034	REPUBLIC	80212	CB7325	REPUBLIC	80096
CB210875-1X	ROCKWELL	88008.36	CB3100X	REPUBLIC	80083	CB7034	REPUBLIC	80218	CB7325	REPUBLIC	88001.64
CB2110	REPUBLIC	80026	CB3100X	REPUBLIC	80177	CB7039	REPUBLIC	80049	CB73252-2	REPUBLIC	80096
CB2110	REPUBLIC	80049	CB3110X	REPUBLIC	80097	CB704	REPUBLIC	80234	CB73252-2	REPUBLIC	88001.64
CB211016X	ROCKWELL	87373	CB3110X	REPUBLIC	80120	CB705	REPUBLIC	80198	CB73252-3	REPUBLIC	80096
CB2140	REPUBLIC	80055	CB3140X	REPUBLIC	80144	CB706	REPUBLIC	80214	CB7325-4	REPUBLIC	80056
CB2173	REPUBLIC	80058	CB3150X	REPUBLIC	80144	CB707	REPUBLIC	80207	CB7325-4	REPUBLIC	80095
CB2175	REPUBLIC	80058	CB3278	REPUBLIC	80077	CB708	REPUBLIC	80215	CB7325-5	REPUBLIC	80094
CB2177	REPUBLIC	80024	CB3278	REPUBLIC	80085	CB709	REPUBLIC	80237	CB733	REPUBLIC	80171
CB2178	REPUBLIC	80010	CB3400AHP	REPUBLIC	80081	CB709	REPUBLIC	80239	CB734	REPUBLIC	80070
CB2179	REPUBLIC	80004	CB3DR	REPUBLIC	80133	CB710	REPUBLIC	80231	CB735	REPUBLIC	80193
CB2180	REPUBLIC	80019	CB4000	REPUBLIC	80109	CB710	REPUBLIC	80240	CB73525	REPUBLIC	80095
CB2181	REPUBLIC	80018	CB4002	REPUBLIC	80109	CB7100	REPUBLIC	80210	CB735251	REPUBLIC	80095
CB2182	REPUBLIC	80009	CB4007	REPUBLIC	80147	CB711	REPUBLIC	80234	CB73525-2	REPUBLIC	80095
CB2183	REPUBLIC	80004	CB4008	REPUBLIC	80146	CB7111	REPUBLIC	80234	CB73525-3	REPUBLIC	80096
CB2183	REPUBLIC	80006	CB4010	REPUBLIC	80153	CB712	REPUBLIC	80233	CB735254	REPUBLIC	80060
CB2184	REPUBLIC	80017	CB4010	REPUBLIC	80156	CB712	REPUBLIC	80241	CB73525-4	REPUBLIC	80094
CB2185	REPUBLIC	80011	CB4014	REPUBLIC	80158	CB7125-2	REPUBLIC	80034	CB736	REPUBLIC	80195
CB2186	REPUBLIC	80012	CB4015	REPUBLIC	80148	CB7125-2	REPUBLIC	80057	CB737	REPUBLIC	80196
CB2198	REPUBLIC	80051	CB4016	REPUBLIC	80159	CB713	REPUBLIC	80232	CB738	REPUBLIC	80158
CB2198	REPUBLIC	80052	CB4017	REPUBLIC	80158	CB713	REPUBLIC	80237	CB739	REPUBLIC	80150
CB2500	REPUBLIC	80059	CB4N	REPUBLIC	80134	CB713	REPUBLIC	80242	CB740	REPUBLIC	80149
CB2501	REPUBLIC	80023	CB5000	REPUBLIC	80168	CB714	REPUBLIC	80238	CB741	REPUBLIC	80153
CB2503	REPUBLIC	80030	CB5000	REPUBLIC	80216	CB71425-2	REPUBLIC	80034	CB741	REPUBLIC	88003.16
CB2503G	REPUBLIC	80024	CB5007	REPUBLIC	80177	CB715	REPUBLIC	80178	CB742	REPUBLIC	80156
CB2503G	REPUBLIC	80031	CB5008	REPUBLIC	80174	CB716	REPUBLIC	80189	CB742	REPUBLIC	88003.17
CB2582	REPUBLIC	80132	CB5014	REPUBLIC	80168	CB717	REPUBLIC	80221	CB743	REPUBLIC	80198
CB2852	REPUBLIC	80132	CB5015	REPUBLIC	80173	CB718	REPUBLIC	80225	CB744	REPUBLIC	80224

Part No.	Brand	IDLI Group
CB745	REPUBLIC	80070
CB746	REPUBLIC	80238
CB747	REPUBLIC	80213
CB748	REPUBLIC	80140
CB749	REPUBLIC	80124
CB750	REPUBLIC	80222
CB750N1	REPUBLIC	80187
CB751	REPUBLIC	80048
CB752	REPUBLIC	80047
CB753	REPUBLIC	80178
CB753	REPUBLIC	88003.22
CB754	REPUBLIC	80124
CB755	REPUBLIC	80142
CB756	REPUBLIC	80195
CB757	REPUBLIC	80202
CB758	REPUBLIC	80162
CB759	REPUBLIC	80174
CB760	REPUBLIC	80193
CB761	REPUBLIC	88003.26
CB762	REPUBLIC	80223
CB7625-2	REPUBLIC	80064
CB763	REPUBLIC	80224
CB764	REPUBLIC	80069
CB765	REPUBLIC	80188
CB765	REPUBLIC	88003.29
CB766	REPUBLIC	80049
CB766	REPUBLIC	80102
CB767	REPUBLIC	80103
CB768	REPUBLIC	80172
CB769	REPUBLIC	80156
CB770	REPUBLIC	88003.30
CB771	REPUBLIC	88003.31
CB772	REPUBLIC	80035
CB773	REPUBLIC	88003.32
CB774	REPUBLIC	88003.33
CB775	REPUBLIC	88003.34
CB776	REPUBLIC	88003.35
CB777	REPUBLIC	88003.36
CB778	REPUBLIC	88003.37
CB779	REPUBLIC	80182
CB780	REPUBLIC	80190
CB781	REPUBLIC	80001
CB782	REPUBLIC	80001
CB783	REPUBLIC	80021
CB785	REPUBLIC	80067
CB786	REPUBLIC	88003.40
CB788	REPUBLIC	80245
CB789	REPUBLIC	80104
CB790	REPUBLIC	80163
CB790	REPUBLIC	80167
CB791	REPUBLIC	80164
CB791	REPUBLIC	80168
CB792	REPUBLIC	80165
CB792	REPUBLIC	80168
CB793	REPUBLIC	80166
CB793	REPUBLIC	80171
CB794	REPUBLIC	80046
CB795	REPUBLIC	80035
CB796	REPUBLIC	80154
CB797	REPUBLIC	80175
CB798	REPUBLIC	80113
CB799	REPUBLIC	80169
CB799	REPUBLIC	80216
CB7N	REPUBLIC	80220
CB8002	REPUBLIC	80231
CB8002	REPUBLIC	80240
CB8101	REPUBLIC	80232
CB8101	REPUBLIC	80239
CB8102	REPUBLIC	80237
CB8103	REPUBLIC	80236
CB83254	ALMETAL	80022
CB83254	REPUBLIC	80022
CB83254A	REPUBLIC	80022
CB835254A	REPUBLIC	80022
CB850N4	REPUBLIC	80219
CB85WB7	REPUBLIC	80222
CB-8N	REPUBLIC	88012.23
CB9000	REPUBLIC	80235
CB96	REPUBLIC	80138
CBAR22-1	ROCKWELL	88008.39
CBAR22-2	ROCKWELL	88008.40
CBAR25-1	ROCKWELL	88008.41
CBAR25-2	ROCKWELL	88008.42
CBAR25-3	ROCKWELL	88008.43
CBAR28-1	ROCKWELL	88008.44
CBAR28-2	ROCKWELL	88008.45
CBAR32-1	ROCKWELL	88008.46
CB-C96	REPUBLIC	80204
CB-D96	REPUBLIC	80160
CB-L10N2	REPUBLIC	80003
CB-L10N2	REPUBLIC	80033
CB-L10S	REPUBLIC	80033
CB-L14N	REPUBLIC	80037
CB-L14S	REPUBLIC	80037
CB-L16S	REPUBLIC	88002.95
CB-L69-8	REPUBLIC	80001
CB-L6N11	REPUBLIC	80001
CB-L6N5	REPUBLIC	88002.91
CB-L6N8	REPUBLIC	80001
CBL6NS9	REPUBLIC	80003
CB-L6S5	REPUBLIC	80003
CB-L6S9	REPUBLIC	80003
CBN3550A	FORD	80001
CBN3550B	FORD	80001
CB-O55	REPUBLIC	80141
CB-O96	REPUBLIC	80138
CB-P55	REPUBLIC	80129
CB-P55-5	REPUBLIC	80131
CB-P55-5UB	REPUBLIC	88002.13
CB-P55-6	REPUBLIC	80131
CB-P96-1	REPUBLIC	80068
CB-P96G1	REPUBLIC	80068
CBPN3550A	FORD	80001
CBPN3550A	FORD	80003
CBPN3550B	FORD	80001
CBPN3550B	FORD	80003
CB-R55	REPUBLIC	80123
CB-R96	REPUBLIC	80122
CBRR22-1	ROCKWELL	88008.47
CBRR22-2	ROCKWELL	88008.48
CBRR22-2A	ROCKWELL	88008.49
CBRR22-5	ROCKWELL	88008.50
CBRR24-1	ROCKWELL	88008.51
CBRR24-1A	ROCKWELL	88008.52
CBRR25-1	ROCKWELL	88008.53
CBRR25-2	ROCKWELL	88008.54
CBRR25-3	ROCKWELL	88008.55
CBRR28-1	ROCKWELL	88008.56
CBRR28-2	ROCKWELL	88008.57
CBRR28-4	ROCKWELL	88008.58
CBRR28-4A	ROCKWELL	88008.59
CBRR32-1	ROCKWELL	88008.60
CB-S53	REPUBLIC	80039
CB-S55	REPUBLIC	80081
CBS55-13	ALEPUBLIC	80118
CB-S55-13	REPUBLIC	80130
CB-S55-6	REPUBLIC	80091
CB-S55-6HDC	REPUBLIC	80091
CB-S55-8	REPUBLIC	80130
CB-S55HD	REPUBLIC	80081
CB-S55HDC	REPUBLIC	80081
CBS55L	REPUBLIC	80081
CBS55L	REPUBLIC	80091
CB-S56	REPUBLIC	80145
CB-S96	REPUBLIC	80038
CBSR30-1	ROCKWELL	88008.61
CBSR38-1R	ROCKWELL	88008.62
CBSR40-1	ROCKWELL	88008.63
CBSR45-1	ROCKWELL	88008.64
CBSR48-1	ROCKWELL	88008.65
CB3R51-1	ROCKWELL	88008.66
CBSR64-3	ROCKWELL	88008.67
CB-T55	REPUBLIC	80071
CB-T55L	REPUBLIC	80080
CB-T55L	REPUBLIC	80081
CB-TS55	REPUBLIC	80074
CB-U56	REPUBLIC	80183
CB-U96	REPUBLIC	80181
CBW114-4100	INGER.RAND	80109
CBW114-5100	INGER.RAND	80216
CBW114-6100	INGER.RAND	80193
CBW114-7100	INGER.RAND	80210
CC372	HYSTER	80135
CC375	INBRA	80042
CC4002	INBRA	80109
CC5000	INBRA	80168
CC5008	INBRA	80174
CC5015	INBRA	80173
CC515	INBRA	80178
CC516	INBRA	80189
CC517	INBRA	80221
CC588	INBRA	80140
CC6000	INBRA	80193
CC61	CLEVE.STL.	88007.85
CC7000	INBRA	80210
CC8200	INBRA	80190
CC996426	CHRYSLER	80008
CCB100	CLEVE.STL.	88008.60
CCB101	CLEVE.STL.	87481
CCB1-4	CLEVE.STL.	88007.57
CCB1-4A	CLEVE.STL.	88014.87
CCB2	CLEVE.STL.	88013.75
CCB50-1	CLEVE.STL.	88008.66
CCB61	CLEVE.STL.	88007.85
CD33-79	FORD	80008
CD5AZ4635A	FORD	80182
CD5AZ4635B	FORD	80121
CD5AZ4635C	FORD	80135
CD5AZ4635D	FORD	80061
CD5AZ4635E	FORD	80063
CD5AZ4635F	FORD	80132
CD5AZ4635G	FORD	80951
CD5AZ4635H	FORD	80905
CD5AZ4635J	FORD	80953
CD5AZ4635K	FORD	80906
CD5AZ4635L	FORD	80227
CD5AZ4635M	FORD	80109
CD9655	BALKAMP	80160
CF28-24	OLIVER	80003
CF28-24	WHITE FARM	80003
CF4635A	FORD	88002.77
CH4635A	FORD	88002.77
CIAZ4635A	FORD	80091
CIAZ4635C	FORD	80081
CIAZ4635D	FORD	80091
CITT4385A	FORD	86744
CITT4841E	FORD	85114
CITT4865A	FORD	82081
CITZ4635A	FORD	80140
CITZ4635B	FORD	80118
CIUZ4635A	FORD	88002.77
CJ15-37N	NTN	80010
CLARK	MCQUAY-NOR	80170
CM10002	CLEVE.MOT.	88003.40
CM10003	CLEVE.MOT.	80245
CM10011	CLEVE.MOT.	80245
CM1003	CLEVE.MOT.	80001
CM1003	CLEVE.MOT.	80003
CM103SL	CLEVE.MOT.	88000.96
CM103ST	CLEVE.MOT.	88000.36
CM103ST	CMP	88000.36
CM111X	CLEVE.MOT.	80023
CM111X	CLEVE.MOT.	80024
CM113SL	CLEVE.MOT.	88000.84
CM113ST	CLEVE.MOT.	88000.35
CM113ST	CMP	88000.35
CM115ST	CLEVE.MOT.	88000.37
CM115ST	CMP	88000.37
CM123ST	CLEVE.MOT.	88000.55
CM123ST	CMP	88000.55
CM124SL	CLEVE.MOT.	88000.94
CM124SL	CMP	88000.94
CM124ST	CLEVE.MOT.	88000.38
CM124ST	CMP	88000.38
CM124ST	CMP	88000.39
CM12SL	CLEVE.MOT.	88000.86
CM12SL	CMP	88000.88
CM12ST	CLEVE.MOT.	88000.43
CM12ST	CMP	88000.43
CM13SH	CMP	88000.21
CM13SL	CMP	88000.79
CM13ST	CLEVE.MOT.	88000.20
CM13ST	CMP	88000.21
CM1501	CLEVE.MOT.	80021
CM1501	CLEVE.MOT.	88001.97
CM1750	CLEVE.MOT.	80030
CM1750	CLEVE.MOT.	80031
CM1750	CMP	80030
CM1750G	CLEVE.MOT.	80031
CM1750G	CMP	80031
CM1750GS	CLEVE.MOT.	80031
CM1752	CLEVE.MOT.	80034
CM1755	CLEVE.MOT.	80032
CM1755	CMP	80032
CM1759	CLEVE.MOT.	80030
CM1759	CLEVE.MOT.	80031
CM1759	CMP	80030
CM1759G	CLEVE.MOT.	80031
CM1762	CLEVE.MOT.	80030
CM1762	CLEVE.MOT.	80031
CM1762	CMP	80030
CM1767	CLEVE.MOT.	80031
CM1767	CMP	80031
CM1810	CLEVE.MOT.	80227
CM1810	CMP	80227
CM2002	CLEVE.MOT.	80048
CM2002	CLEVE.MOT.	80049
CM2002	CLEVE.MOT.	88001.96
CM2002	CMP	80048
CM2006	CLEVE.MOT.	80045
CM2006	CLEVE.MOT.	80054
CM2006	CMP	80054
CM2017	CLEVE.MOT.	80092
CM2017	CLEVE.MOT.	88001.76
CM2017	CMP	80092
CM2031	CLEVE.MOT.	80092
CM2032	CLEVE.MOT.	80092
CM2032	CMP	80092
CM2049	CLEVE.MOT.	80054
CM2049	CMP	80054

Part No.	Brand	IDLI Group	Part No.	Brand	IDLI Group	Part No.	Brand	IDLI Group	Part No.	Brand	IDLI Group
CM2062	CLEVE.MOT.	80049	CM3000	CMP	80099	CM4110	CMP	80155	CM5-111X	CLEVE.MOT.	80023
CM2062	CLEVE.MOT.	88001.96	CM3000J	CMP	80006	CM4113	CLEVE.MOT.	80156	CM5-111X	CLEVE.MOT.	80024
CM2062	CMP	80049	CM3004900	CLEVE.MOT.	80062	CM4113	CMP	80156	CM5-111X	CLEVE.MOT.	80140
CM2063	CLEVE.MOT.	80044	CM3004900	CMP	80061	CM4138	CLEVE.MOT.	80109	CM5-111X	CMP	80024
CM2063	CLEVE.MOT.	88001.98	CM3008	CLEVE.MOT.	80104	CM4138	CMP	80109	CM5112	CLEVE.MOT.	80173
CM2063	CMP	80044	CM3008	CMP	80104	CM4143	CLEVE.MOT.	80156	CM5-115X	CLEVE.MOT.	80178
CM2063	CMP	88001.98	CM300J	CLEVE.MOT.	80004	CM4143	CMP	80156	CM5-115X	CMP	80178
CM206417	CLEVE.MOT.	88001.93	CM3010	CLEVE.MOT.	80100	CM4152	CLEVE.MOT.	80158	CM5-116X	CLEVE.MOT.	80189
CM2100	CLEVE.MOT.	80049	CM3011CV	CLEVE.MOT.	80104	CM4152	CMP	80158	CM5-116X	CMP	80189
CM2100	CLEVE.MOT.	88001.96	CM3011CV	CMP	80107	CM4200	CLEVE.MOT.	80109	CM5-11X	CLEVE.MOT.	80024
CM2100	CMP	80049	CM3011CVX	CLEVE.MOT.	80107	CM4200	CMP	80109	CM5-1202X	CLEVE.MOT.	80130
CM2101	CLEVE.MOT.	80044	CM3012	CLEVE.MOT.	80112	CM43SH	CLEVE.MOT.	88000.70	CM5-1202X	CMP	80130
CM2101	CLEVE.MOT.	88001.98	CM3012	CMP	80112	CM43SH	CMP	88000.22	CM5-121X	CLEVE.MOT.	80072
CM2101	CMP	80044	CM3013	CLEVE.MOT.	80106	CM43SL	CLEVE.MOT.	88000.80	CM5-121X	CMP	80072
CM2103	CLEVE.MOT.	80048	CM3015	CMP	80104	CM43SL	CMP	88000.80	CM5-121X	CMP	80075
CM2103	CMP	80048	CM3019	CLEVE.MOT.	80110	CM43ST	CLEVE.MOT.	88000.22	CM5-121XUB	CLEVE.MOT.	80072
CM2109	CLEVE.MOT.	80049	CM3021	CLEVE.MOT.	80111	CM43ST	CMP	88000.22	CM5-121XUB	CLEVE.MOT.	80075
CM2109	CMP	80049	CM3031CV	CLEVE.MOT.	80109	CM5000	CLEVE.MOT.	80168	CM5-121XUB	CMP	80075
CM2110	CLEVE.MOT.	80048	CM3031CV	CMP	80109	CM5000	CMP	80168	CM5-121XUB	CMP	80089
CM2110	CMP	80048	CM3100	CLEVE.MOT.	80099	CM5006	CLEVE.MOT.	80047	CM5122	CLEVE.MOT.	80216
CM2112	CLEVE.MOT.	80044	CM3100	CMP	80099	CM5006	CLEVE.MOT.	80163	CM5122	CMP	80168
CM2114	CLEVE.MOT.	80054	CM3100	CMP	80101	CM5006	CMP	80163	CM5-124X	CLEVE.MOT.	80225
CM2114	CMP	80054	CM3101	CLEVE.MOT.	80100	CM5-007	CLEVE.MOT.	88013.86	CM5-124X	CMP	80225
CM2116	CLEVE.MOT.	80047	CM3102	CLEVE.MOT.	80112	CM5-008	CLEVE.MOT.	88013.87	CM5-124XGC	CLEVE.MOT.	80225
CM2116	CMP	80047	CM3102	CMP	80112	CM5008	CMP	80174	CM5-125X	CMP	80077
CM2117	CLEVE.MOT.	80046	CM3103	CLEVE.MOT.	80098	CM500J	CLEVE.MOT.	80009	CM5-126X	CLEVE.MOT.	80037
CM2121	CLEVE.MOT.	80044	CM3105	CLEVE.MOT.	80104	CM500J	CMP	80009	CM5127	CLEVE.MOT.	80163
CM2123	CLEVE.MOT.	80045	CM3105	CMP	80104	CM5-013	CLEVE.MOT.	88001.94	CM5127	CMP	80163
CM2123	CMP	80054	CM3106	CLEVE.MOT.	80099	CM5015	CLEVE.MOT.	80173	CM5-129X	CLEVE.MOT.	80037
CM2126	CLEVE.MOT.	80049	CM3106	CLEVE.MOT.	80101	CM5015	CMP	80173	CM5132	CLEVE.MOT.	80174
CM2126	CMP	80049	CM3106	CMP	80099	CM5031	CLEVE.MOT.	80171	CM5-133X	CLEVE.MOT.	80116
CM2127	CLEVE.MOT.	80044	CM3108	CLEVE.MOT.	80112	CM5100	CLEVE.MOT.	80168	CM5-133X	CMP	80116
CM2127	CMP	80044	CM3108	CMP	80112	CM5100	CLEVE.MOT.	80216	CM5-134X	CLEVE.MOT.	80118
CM2140	CLEVE.MOT.	80055	CM3113	CLEVE.MOT.	80110	CM5100	CMP	80168	CM5-134X	CMP	80118
CM2140	CMP	80055	CM3116	CLEVE.MOT.	80111	CM5-100X	CLEVE.MOT.	80030	CM5-145X	CLEVE.MOT.	80178
CM2141	CLEVE.MOT.	80078	CM3120	CLEVE.MOT.	80113	CM5-100X	CLEVE.MOT.	80031	CM5-14X	CLEVE.MOT.	80024
CM2141	CLEVE.MOT.	80091	CM3121	CLEVE.MOT.	80102	CM5-100X	CMP	80030	CM5-150X	CLEVE.MOT.	80119
CM2141	CMP	80091	CM3125	CLEVE.MOT.	80103	CM5-100X	CMP	80031	CM5-150X	CMP	80097
CM2153	CLEVE.MOT.	80055	CM3158	CLEVE.MOT.	80102	CM5101	CLEVE.MOT.	80163	CM5-150X	CMP	80119
CM2153	CMP	80055	CM3188	CLEVE.MOT.	80105	CM5101	CLEVE.MOT.	80167	CM5-153X	CLEVE.MOT.	80077
CM2173	CLEVE.MOT.	80058	CM3188	CLEVE.MOT.	80227	CM5101	CMP	80163	CM5-153X	CMP	80075
CM2173	CMP	80058	CM3188	CMP	80105	CM5101	CMP	80167	CM5-153X	CMP	80077
CM2175	CLEVE.MOT.	80058	CM32SL	CLEVE.MOT.	88000.95	CM5-101X	CLEVE.MOT.	80023	CM5-153XUB	CLEVE.MOT.	80088
CM2175	CMP	80058	CM32SL	CMP	88000.95	CM5-101X	CLEVE.MOT.	80024	CM5-154X	CMP	80178
CM2179	CLEVE.MOT.	80055	CM-33ST	CLEVE.MOT.	88000.22	CM5-101X	CLEVE.MOT.	80140	CM5-155X	CLEVE.MOT.	80182
CM2179	CMP	80055	CM33ST	CMP	88000.26	CM5-101X	CMP	80023	CM5-155X	CMP	80182
CM2198	CLEVE.MOT.	80051	CM3689266	CLEVE.MOT.	88002.50	CM5102	CLEVE.MOT.	80170	CM5-158X	CLEVE.MOT.	80182
CM229X	CLEVE.MOT.	80117	CM4000J	CMP	80010	CM5102	CLEVE.MOT.	80171	CM5-158X	CMP	80182
CM22SL	CLEVE.MOT.	88000.89	CM400J	CLEVE.MOT.	80010	CM5103	CLEVE.MOT.	80174	CM5-160X	CLEVE.MOT.	80142
CM22SL	CMP	88000.89	CM4015	CLEVE.MOT.	80148	CM5103	CMP	80174	CM5-160X	CLEVE.MOT.	80143
CM22ST	CLEVE.MOT.	88000.45	CM4015	CMP	80148	CM5-103X	CLEVE.MOT.	88000.15	CM5-160X	CMP	80142
CM22ST	CMP	88000.45	CM4016	CLEVE.MOT.	80159	CM5105	CLEVE.MOT.	80168	CM5-160XUB	CLEVE.MOT.	80142
CM23ST	CLEVE.MOT.	88000.44	CM4016	CMP	80159	CM5105	CLEVE.MOT.	80169	CM5-160XUB	CLEVE.MOT.	80143
CM23ST	CMP	88000.44	CM4019	CLEVE.MOT.	80158	CM5105	CLEVE.MOT.	80216	CM5-160XUB	CMP	80142
CM2500	CLEVE.MOT.	80059	CM4019	CMP	80158	CM5-105X	CLEVE.MOT.	80031	CM5-160XUB	CMP	80143
CM2500	CMP	80059	CM4100	CLEVE.MOT.	80109	CM5-105X	CMP	80031	CM51625-2	CLEVE.MOT.	80042
CM2505	CLEVE.MOT.	80059	CM4100	CLEVE.MOT.	80149	CM5-107X	CLEVE.MOT.	80030	CM51625-2	CMP	80042
CM2518	CLEVE.MOT.	80059	CM4100	CMP	80109	CM5-107X	CLEVE.MOT.	80031	CM5-165X	CLEVE.MOT.	80190
CM2518	CLEVE.MOT.	80066	CM4101	CLEVE.MOT.	80150	CM5-107X	CMP	80030	CM5-165X	CMP	80190
CM2518	CMP	80066	CM4101	CLEVE.MOT.	80153	CM5108	CLEVE.MOT.	80172	CM5-170PB	CLEVE.MOT.	80007
CM2520	CLEVE.MOT.	80093	CM4102	CLEVE.MOT.	80158	CM5108	CMP	80172	CM5-170X	CLEVE.MOT.	80007
CM2530	CLEVE.MOT.	80059	CM4102	CMP	80158	CM5-108X	CLEVE.MOT.	80044	CM5-170X	CMP	80007
CM2530	CLEVE.MOT.	80066	CM4104	CLEVE.MOT.	80150	CM5-108X	CLEVE.MOT.	88001.98	CM5-171X	CLEVE.MOT.	80072
CM2530	CMP	80066	CM4105	CLEVE.MOT.	80150	CM5-108X	CMP	80044	CM5-174X	CLEVE.MOT.	80078
CM2531	CLEVE.MOT.	80066	CM4106	CLEVE.MOT.	80148	CM5109	CLEVE.MOT.	80162	CM5-174X	CLEVE.MOT.	80090
CM281175	CMP	80132	CM4106	CMP	80148	CM5110	CLEVE.MOT.	80173	CM5-174X	CMP	80078
CM285175	CLEVE.MOT.	80132	CM4107	CLEVE.MOT.	80159	CM5110	CLEVE.MOT.	80177	CM5-174X	CMP	80090
CM2852175	CLEVE.MOT.	80132	CM4107	CMP	80159	CM5110	CMP	80173	CM5-175X	CLEVE.MOT.	80007
CM2852175	CMP	80132	CM4109	CLEVE.MOT.	80157	CM5111	CLEVE.MOT.	80175	CM5177	CMP	80172
CM3000	CLEVE.MOT.	80099	CM4110	CLEVE.MOT.	80155	CM5111	CMP	80173	CM5-177X	CLEVE.MOT.	80221

Part No.	Brand	IDLI Group
CM5-177X	CMP	80221
CM5-178X	CLEVE.MOT.	80124
CM5-178X	CMP	80124
CM5-178XUB	CLEVE.MOT.	80125
CM5-178XUB	CMP	80124
CM5-181X	CLEVE.MOT.	80023
CM5-181X	CLEVE.MOT.	80024
CM5-181X	CLEVE.MOT.	80140
CM5-181X	CMP	80140
CM5-185X	CLEVE.MOT.	80228
CM5-185X	CMP	80228
CM5-185XGC	CLEVE.MOT.	80228
CM5-186X	CLEVE.MOT.	88001.82
CM5-186X	CMP	88001.82
CM5-187X	CLEVE.MOT.	88001.83
CM5-187X	CMP	88001.83
CM5-188X	CLEVE.MOT.	80140
CM5-188X	CLEVE.MOT.	80182
CM5-188X	CMP	80140
CM5-189X	CLEVE.MOT.	80205
CM5189X	CMP	80229
CM5-200X	CLEVE.MOT.	80075
CM5-200X	CMP	80075
CM5-200XUB	CLEVE.MOT.	80075
CM5-200XUB	CLEVE.MOT.	80076
CM5-200XUB	CMP	80076
CM5-212X	CLEVE.MOT.	80121
CM5-213X	CLEVE.MOT.	80117
CM5-213X	CMP	80117
CM5-229X	CMP	80117
CM5-231X	CLEVE.MOT.	80121
CM5-231X	CMP	80118
CM5-231X	CMP	80121
CM5-233X	CLEVE.MOT.	80221
CM5-233X	CMP	80221
CM5-243X	CLEVE.MOT.	80069
CM5-248X	CLEVE.MOT.	80025
CM5-248X	CMP	80025
CM5-249X	CLEVE.MOT.	80076
CM5-249X	CMP	80025
CM5-260X	CLEVE.MOT.	80078
CM5-260X	CMP	80078
CM5-272X	CLEVE.MOT.	80225
CM5-272X	CMP	80225
CM5-278X	CLEVE.MOT.	80086
CM5279XGC	CLEVE.MOT.	80189
CM5-280XGC	CLEVE.MOT.	80221
CM5-281X	CLEVE.MOT.	80227
CM5-281X	CMP	80227
CM5-281XGC	CLEVE.MOT.	80227
CM5-286XGC	CLEVE.MOT.	80225
CM5-297X	CLEVE.MOT.	80079
CM5-297X	CMP	80079
CM52AR	CLEVE.MOT.	80909
CM52AR	CMP	80909
CM5-308X	CLEVE.MOT.	80229
CM5-315X	CLEVE.MOT.	80026
CM5-350X	CLEVE.MOT.	80036
CM5-353X	CLEVE.MOT.	80083
CM5355S	CLEVE.MOT.	80038
CM5355S	CLEVE.MOT.	80039
CM536252	CLEVE.MOT.	80135
CM53625-2	CMP	80135
CM537252	CLEVE.MOT.	80135
CM53725-2	CMP	80135
CM53825-2	CLEVE.MOT.	80135
CM53825-2	CMP	80135
CM53ST	CLEVE.MOT.	88000.52
CM53ST	CMP	88000.52
CM5-4X	CLEVE.MOT.	80041
CM5-4X	CMP	80041
CM5555-0	CLEVE.MOT.	80141
CM55550	CMP	80141
CM55550	CLEVE.MOT.	80141
CM5555P	CLEVE.MOT.	80129
CM5555P	CMP	80129
CM5555R	CLEVE.MOT.	80123
CM5555R	CLEVE.MOT.	80131
CM5555R	CMP	80123
CM5555S	CLEVE.MOT.	80081
CM5555S	CLEVE.MOT.	80123
CM5555S	CMP	80081
CM5555T	CLEVE.MOT.	80071
CM5555T	CMP	80071
CM5555TS	CLEVE.MOT.	80074
CM5555TS	CMP	80074
CM5555X	CLEVE.MOT.	80007
CM5555X	CLEVE.MOT.	88000.15
CM5555X	CMP	80007
CM5655-2B	CLEVE.MOT.	80145
CM5655-2D	CLEVE.MOT.	80141
CM5655-2D	CLEVE.MOT.	80145
CM5655-2D	CMP	80145
CM5655D	CLEVE.MOT.	80141
CM5655U	CLEVE.MOT.	80183
CM5655U	CMP	80183
CM5-7	CMP	88013.86
CM5-8	CMP	88013.87
CM5-86X	CLEVE.MOT.	80124
CM5-90X	CLEVE.MOT.	80093
CM591642	CLEVE.MOT.	88002.42
CM591642	CMP	88002.42
CM5-92X	CLEVE.MOT.	80001
CM600J	CLEVE.MOT.	80012
CM600J	CMP	80012
CM6016	CLEVE.MOT.	80193
CM6016	CMP	80193
CM6023	CLEVE.MOT.	80199
CM6023	CMP	80199
CM6024	CLEVE.MOT.	80201
CM6024	CMP	80201
CM6044	CLEVE.MOT.	80195
CM6044	CMP	80197
CM606779	CMP	88002.43
CM609531	CLEVE.MOT.	88002.48
CM60R	CLEVE.MOT.	80903
CM60R	CLEVE.MOT.	80953
CM60R	CLEVE.MOT.	80958
CM6100	CLEVE.MOT.	80193
CM6100	CMP	80193
CM6101	CLEVE.MOT.	80195
CM6101	CMP	80197
CM6102	CLEVE.MOT.	80196
CM6102	CLEVE.MOT.	80197
CM6102	CMP	80197
CM6103	CLEVE.MOT.	80194
CM6103	CMP	80197
CM6104	CLEVE.MOT.	80192
CM6106	CLEVE.MOT.	80198
CM6106	CMP	80198
CM6107	CLEVE.MOT.	80202
CM6108	CLEVE.MOT.	80199
CM6108	CMP	80199
CM6109	CLEVE.MOT.	80199
CM6109	CLEVE.MOT.	80201
CM6109	CMP	80201
CM6114	CLEVE.MOT.	80196
CM6139	CLEVE.MOT.	80203
CM6143P	CLEVE.MOT.	80193
CM6143	CMP	80193
CM6162	CLEVE.MOT.	80199
CM6162	CMP	80199
CM6163	CLEVE.MOT.	80201
CM6163	CMP	80201
CM63SH	CLEVE.MOT.	88000.71
CM63SH	CMP	88000.71
CM-63SL	CLEVE.MOT.	88000.52
CM63SL	CMP	88000.81
CM63ST	CLEVE.MOT.	88000.25
CM63ST	CMP	88000.25
CM63ST	CMP	88000.27
CM65AR1	CLEVE.MOT.	80905
CM65AR1	CMP	80905
CM65R1	CLEVE.MOT.	80951
CM65R1	CMP	80951
CM674-1	CMP	80953
CM67AR1	CLEVE.MOT.	80906
CM67AR1	CMP	80906
CM67R1	CLEVE.MOT.	80953
CM67R1	CMP	80953
CM7000	CLEVE.MOT.	80210
CM7000	CMP	80210
CM700J	CLEVE.MOT.	80019
CM700J	CMP	80011
CM7100	CLEVE.MOT.	80210
CM7100	CMP	80210
CM7101	CLEVE.MOT.	80207
CM7101	CLEVE.MOT.	80211
CM7101	CMP	80212
CM7102	CLEVE.MOT.	80212
CM7102	CLEVE.MOT.	80218
CM7102	CMP	80212
CM7103	CLEVE.MOT.	80208
CM7104	CLEVE.MOT.	80209
CM7105	CLEVE.MOT.	80213
CM7105	CLEVE.MOT.	80215
CM7105	CMP	80213
CM7125-2	CMP	80034
CM714251-2	CLEVE.MOT.	80034
CM714251-2	CMP	80034
CM71425-2	CLEVE.MOT.	80034
CM71425-2	CMP	80034
CM7200	CLEVE.MOT.	80210
CM7200	CMP	80210
CM7202	CLEVE.MOT.	80218
CM7202	CMP	80212
CM7205	CLEVE.MOT.	80215
CM7205	CMP	80213
CM7207	CMP	80212
CM725253	CLEVE.MOT.	80064
CM725253	CLEVE.MOT.	80065
CM72525-3	CMP	80065
CM726251-2	CLEVE.MOT.	80063
CM726251-2	CMP	80061
CM726251-3	CLEVE.MOT.	80061
CM726251-3	CMP	80061
CM72625-2	CLEVE.MOT.	80064
CM72625-2	CLEVE.MOT.	80065
CM72625-2	CLEVE.MOT.	88001.63
CM72625-2	CMP	80064
CM726252-2	CLEVE.MOT.	80061
CM726252-2	CMP	80061
CM72625-3	CLEVE.MOT.	80060
CM72625-3	CLEVE.MOT.	80061
CM72625-3	CMP	80061
CM72625-4	CLEVE.MOT.	80060
CM72625-4	CLEVE.MOT.	80063
CM72625-4	CMP	80060
CM72625-4	CMP	80061
CM732522	CLEVE.MOT.	80042
CM73252-2	CMP	80042
CM735253	CLEVE.MOT.	80042
CM73525-3	CMP	80042
CM73SH	CMP	88000.72
CM73ST	CLEVE.MOT.	88000.28
CM73ST	CMP	88000.28
CM76ST	CMP	88000.29
CM7853X	CLEVE.MOT.	80127
CM8	PETTI-MULL	80109
CM8	PETTI-MULL	80149
CM8003	CLEVE.MOT.	80232
CM8003	CLEVE.MOT.	80237
CM8003	CMP	80232
CM800J	CLEVE.MOT.	80011
CM800J	CMP	80019
CM8014	CLEVE.MOT.	80238
CM8100	CLEVE.MOT.	80231
CM8100	CLEVE.MOT.	80240
CM8100	CMP	80231
CM8101	CLEVE.MOT.	80232
CM8101	CLEVE.MOT.	80239
CM8101	CMP	80232
CM8102	CLEVE.MOT.	80237
CM8102	CMP	80237
CM8103	CLEVE.MOT.	80236
CM8104	CLEVE.MOT.	80238
CM8105	CLEVE.MOT.	80238
CM8105	CMP	80238
CM8111	CLEVE.MOT.	80238
CM8111	CMP	80238
CM8113	CLEVE.MOT.	80238
CM8113	CMP	80238
CM81SL	CLEVE.MOT.	88000.93
CM81SL	CMP	88000.92
CM81ST	CMP	88000.49
CM8200	CLEVE.MOT.	80240
CM8200	CMP	80231
CM8202	CLEVE.MOT.	80237
CM8202	CMP	80237
CM8205	CMP	80231
CM8206	CLEVE.MOT.	80232
CM8206	CLEVE.MOT.	80237
CM8206	CMP	80232
CM8207	CLEVE.MOT.	80237
CM8207	CMP	80237
CM83ST	CLEVE.MOT.	88000.53
CM83ST	CMP	88000.53
CM8500	CLEVE.MOT.	80223
CM8502	CLEVE.MOT.	80224
CM8516	CLEVE.MOT.	80224
CM9000	CLEVE.MOT.	80235
CM9000	CMP	80235
CM9001	CLEVE.MOT.	80234
CM9001	CMP	80234
CM9002	CMP	80233
CM9012	CLEVE.MOT.	80233
CM9012	CMP	80233
CM9013	CLEVE.MOT.	80234
CM9013	CMP	80234
CM9014	CLEVE.MOT.	80235
CM9014	CMP	80235
CM9015	CLEVE.MOT.	80234
CM9015	CMP	80234
CM9016	CLEVE.MOT.	80234
CM9016	CMP	80234
CM9017	CLEVE.MOT.	80233
CM9017	CLEVE.MOT.	80241
CM9017	CMP	80233
CM9026	CLEVE.MOT.	80233
CM9026	CMP	80233

Part No.	Brand	IDLI Group	Part No.	Brand	IDLI Group	Part No.	Brand	IDLI Group	Part No.	Brand	IDLI Group
CM902HP	CLEVE.MOT.	80082	CM-CP72N33	CMP	80212	CP1205X	ROCKWELL	88008.77	CP2002	ROCKWELL	88009.39
CM9031	CLEVE.MOT.	80234	CM-CP72N34	CLEVE.MOT.	80207	CP1206X	ROCKWELL	88008.78	CP200X	ROCKWELL	88009.40
CM9031	CMP	80234	CM-CP72N34	CMP	80212	CP1207X	ROCKWELL	88008.79	CP2011	ROCKWELL	88009.41
CM903HP	CLEVE.MOT.	80144	CMCP72N55	CLEVE.MOT.	80208	CP1208X	ROCKWELL	88008.80	CP2017	ROCKWELL	80092
CM904HP	CLEVE.MOT.	80144	CM-CP78WB12	CLEVE.MOT.	80238	CP1209X	ROCKWELL	88008.81	CP2017	ROCKWELL	88001.76
CM9200	CLEVE.MOT.	80233	CM-CP78WB13	CLEVE.MOT.	80240	CP1210X	ROCKWELL	88008.82	CP2031	ROCKWELL	88009.42
CM93SH	CLEVE.MOT.	88000.73	CM-CP78WB14	CLEVE.MOT.	80237	CP1211X	ROCKWELL	88008.83	CP2033	ROCKWELL	88009.43
CM93SH	CMP	88000.73	CM-CP78WB2	CLEVE.MOT.	80238	CP1214X	ROCKWELL	88008.84	CP2116	ROCKWELL	88009.44
CM93SL	CLEVE.MOT.	88000.83	CMCP7N	CLEVE.MOT.	80220	CP121X	ROCKWELL	88008.85	CP213X	ROCKWELL	88009.45
CM93SL	CMP	88000.83	CM-CP82N27	CLEVE.MOT.	80238	CP124X	ROCKWELL	88008.86	CP2140	ROCKWELL	88009.46
CM93ST	CLEVE.MOT.	88000.30	CM-CP82N27	CMP	80238	CP12N	ALCO	80035	CP248X	ROCKWELL	88009.47
CM93ST	CMP	88000.31	CM-CP82N28	CLEVE.MOT.	80240	CP131N	ROCKWELL	88008.87	CP2711	PEARSON	80008
CM95ST	CMP	88000.34	CM-CP82N28	CMP	80231	CP131N1	ROCKWELL	88008.88	CP2711	RUST	80008
CM9655-0	CLEVE.MOT.	80138	CM-CP82N29	CLEVE.MOT.	80237	CP133N	ROCKWELL	88008.89	CP3000	ROCKWELL	88009.48
CM9655C	CLEVE.MOT.	80068	CM-CP82N29	CMP	80237	CP133N1	ROCKWELL	88008.90	CP3008	ROCKWELL	88009.49
CM9655O	CLEVE.MOT.	80138	CM-CP82N30	CLEVE.MOT.	80239	CP134X	ROCKWELL	88008.91	CP3012	ROCKWELL	88009.50
CM9655O	CMP	80138	CM-CP82N30	CMP	80232	CP135N	ROCKWELL	88008.92	CP3021	ROCKWELL	88009.51
CM9655P	CLEVE.MOT.	80068	CM-CP85WB12	CLEVE.MOT.	80224	CP141N	ROCKWELL	88008.93	CP3101	ROCKWELL	80100
CM9675P	CLEVE.MOT.	80131	CM-CP85WB20	CLEVE.MOT.	80224	CP148N	ROCKWELL	88008.94	CP3120	ROCKWELL	88009.52
CM9675P	CMP	80131	CMCP88WB10	CLEVE.MOT.	80233	CP14N	ALCO	80037	CP3121	ROCKWELL	88009.53
CM9675S	CLEVE.MOT.	80091	CM-CP88WB6	CLEVE.MOT.	80234	CP153X	ROCKWELL	88008.95	CP3339	FORD	80008
CM9675S	CLEVE.MOT.	80131	CM-CP88WB9	CLEVE.MOT.	80235	CP155N	ROCKWELL	88008.96	CP33-39	FORD	80133
CM9675S	CMP	80091	CMCP92N14	CLEVE.MOT.	80233	CP155X	ROCKWELL	80182	CP33-46	FORD	80133
CMA1333	LETZ	80069	CM-CP92N14	CMP	80233	CP160X	ROCKWELL	88008.97	CP33-47	FORD	80008
CM-B7090	CLEVE.MOT.	88002.55	CM-CP92N15	CLEVE.MOT.	80234	CP165X	ROCKWELL	88008.98	CP3379	FORD	80008
CM-CP1FR	CLEVE.MOT.	80008	CM-CP92N15	CMP	80234	CP16N	ROCKWELL	88008.99	CP35N	ALCO	80069
CM-CP1FR1	CLEVE.MOT.	80008	CM-CP92N17	CLEVE.MOT.	80234	CP16N1	ROCKWELL	88009	CP35N	ALCO	80070
CM-CP1FR1	CMP	80008	CM-CP92N17	CMP	80234	CP16N2	ROCKWELL	88009.01	CP35N	ALLOY	80069
CM-CP35N	CLEVE.MOT.	80069	CM-CP92N18	CLEVE.MOT.	80235	CP16N3	ROCKWELL	88009.02	CP35N	REPCO	80069
CM-CP35N3	CLEVE.MOT.	80069	CM-CP92N18	CMP	80235	CP170X	ROCKWELL	88009.03	CP35N	REPCO	80070
CM-CP35N3	CMP	80069	CMCP92N19	CLEVE.MOT.	80233	CP176N	ROCKWELL	88009.04	CP35N	REPCO	88003.19
CM-CP35N3-4	CLEVE.MOT.	80069	CM-CP92N19	CMP	80233	CP177X	ROCKWELL	88009.05	CP35N	REPCO	88003.31
CM-CP35N4	CLEVE.MOT.	80069	CM-CPK10	CLEVE.MOT.	88002.34	CP178X	ROCKWELL	88009.06	CP35N	REPCO	88009.54
CM-CP35R	CLEVE.MOT.	80069	CM-CPL10N2	CLEVE.MOT.	80003	CP17N	ROCKWELL	88009.07	CP35N	ROCKWELL	88009.54
CM-CP35R	CMP	80069	CM-CPL12N	CLEVE.MOT.	80035	CP17N12	ROCKWELL	88009.08	CP35N1	ALCO	80070
CM-CP3DR	CLEVE.MOT.	80133	CM-CPL12N	MOTIVE	80035	CP17N20	ROCKWELL	88009.09	CP35N1	ALLOY	80070
CM-CP3DR	CMP	80133	CM-CPL12N1	CMP	80035	CP17N22	ROCKWELL	88009.10	CP35N10	ROCKWELL	88009.55
CM-CP4N	CLEVE.MOT.	80134	CM-CPL12R	CLEVE.MOT.	80035	CP17N3	ROCKWELL	88009.11	CP35N2	ROCKWELL	88009.56
CM-CP4N	CMP	80134	CM-CPL12R	CMP	80035	CP17N7	ROCKWELL	88009.12	CP35N20	ROCKWELL	88009.57
CM-CP55N	CLEVE.MOT.	80161	CM-CPL14N	CLEVE.MOT.	80037	CP188X	ROCKWELL	88009.13	CP35N3	AEC	80069
CM-CP55N	CMP	80161	CM-CPL14N	CMP	80037	CP18N1	ROCKWELL	88009.14	CP35N3	ALCO	80069
CM-CP58N	CLEVE.MOT.	80179	CM-CPL14N1	CLEVE.MOT.	80037	CP18N3	REPCO	80227	CP35N3	ALLOY	80069
CM-CP58WB	CLEVE.MOT.	80185	CM-CPL14N1	CMP	80037	CP18N3	ROCKWELL	88009.15	CP35N3	BALKAMP	80069
CM-CP58WB	CMP	80185	CM-CPL14R	CLEVE.MOT.	80037	CP18N8	ROCKWELL	88009.16	CP35N3	BORG-WARNR	80069
CM-CP58WB1	CLEVE.MOT.	80186	CM-CPL6N	CLEVE.MOT.	80001	CP18N95	ROCKWELL	88009.17	CP35N3	PILOT	80069
CM-CP-58WB2	CLEVE.MOT.	80184	CM-CPL6N	CLEVE.MOT.	80003	CP1F6	ROCKWELL	88009.18	CP35N3	REPCO	80069
CM-CP58WB20	CLEVE.MOT.	80186	CM-CPL6N8	CLEVE.MOT.	80001	CP1FN12	ROCKWELL	88009.19	CP35N3	ROCKWELL	88009.58
CMCP5N	CLEVE.MOT.	80178	CM-CPL6N8	CLEVE.MOT.	80003	CP1FN7	ROCKWELL	88009.20	CP35N4	AEC	80069
CMCP5N1	CLEVE.MOT.	80178	CM-CPL6N8	CMP	80001	CP1FN8	ROCKWELL	88009.21	CP35N4	ALCO	80070
CM-CP5N1	CMP	80178	CM-CVK10	CLEVE.MOT.	88002.34	CP1FN9	ROCKWELL	88009.22	CP35N4	BORG-WARNR	80069
CM-CP62N12	CLEVE.MOT.	80193	CM-CVK10A	CLEVE.MOT.	88002.34	CP1FR	AEC	80008	CP35N4	BORG-WARNR	80070
CM-CP62N12	CMP	80193	CM-CVK10B	CLEVE.MOT.	88002.36	CP1FR	HUSCO	80008	CP35N4	HUSCO	80070
CM-CP62N13	CLEVE.MOT.	80196	CMCVK20	CLEVE.MOT.	88002.33	CP1FR	REPCO	80008	CP35N4	REPCO	80070
CM-CP62N13	CMP	80197	CMCVK20A	CLEVE.MOT.	88002.33	CP1FR	ROCKWELL	88009.23	CP35N4	ROCKWELL	88009.59
CM-CP62N14	CLEVE.MOT.	80198	CM-CVK30	CLEVE.MOT.	88002.33	CP1FR1	ROCKWELL	88009.24	CP35N5	ROCKWELL	88009.60
CM-CP62N14	CMP	80198	CM-CVK30A	CLEVE.MOT.	88002.33	CP1FR2	ROCKWELL	88009.25	CP35N8	ROCKWELL	88009.61
CM-CP62N15	CLEVE.MOT.	80199	CMS124X	CLEVE.MOT.	80225	CP1FR4	ROCKWELL	88009.26	CP35N9	ROCKWELL	88009.62
CM-CP62N15	CMP	80199	CMS179XUB	CLEVE.MOT.	80124	CP1FR5	ROCKWELL	88009.27	CP35R	HUSCO	80070
CM-CP62N16	CLEVE.MOT.	80201	CODZ4635	FORD	80030	CP1FR7	ROCKWELL	88009.28	CP35R	ROCKWELL	88009.63
CM-CP62N16	CMP	80201	CODZ7039A	FORD	80030	CP1FRN	ROCKWELL	88009.29	CP35R17	ROCKWELL	88009.64
CM-CP68WB1	CLEVE.MOT.	80215	CODZ7039A	FORD	88002.77	CP1FRN2	ROCKWELL	88009.30	CP35R3	ROCKWELL	88009.65
CM-CP68WB10	CLEVE.MOT.	80218	CP101X	ROCKWELL	88008.68	CP1FRN4	REPCO	88002.88	CP3D	ROCKWELL	88009.66
CM-CP68WB9	CLEVE.MOT.	80193	CP105X	ROCKWELL	88008.69	CP1FRN4	ROCKWELL	88009.31	CP3D1	ROCKWELL	88009.67
CM-CP6N	CLEVE.MOT.	80188	CP111X	ROCKWELL	88008.70	CP1FRN5	ROCKWELL	88009.32	CP3DN3	ROCKWELL	88009.68
CM-CP6N	CMP	80188	CP115X	ROCKWELL	88008.71	CP1FRN5	ROCKWELL	88009.33	CP3DN4	ROCKWELL	88009.69
CM-CP72N31	CLEVE.MOT.	80215	CP116X	ROCKWELL	88008.72	CP1FS1	ROCKWELL	88009.34	CP3DR	AEC	80133
CM-CP72N31	CMP	80213	CP1200X	ROCKWELL	88008.73	CP1FS2	ROCKWELL	88009.35	CP3DR	REPCO	80133
CM-CP72N32	CLEVE.MOT.	80210	CP1201X	ROCKWELL	88008.74	CP1FS5	ROCKWELL	88009.36	CP3DR	ROCKWELL	88009.70
CM-CP72N32	CMP	80210	CP1203X	ROCKWELL	88008.75	CP1FS8	ROCKWELL	88009.37	CP3DRN	ROCKWELL	88009.71
CM-CP72N33	CLEVE.MOT.	80218	CP1204X	ROCKWELL	88008.76	CP1FS9	ROCKWELL	88009.38	CP3DRN3	ROCKWELL	88009.72

Part No.	Brand	IDLI Group
CP3DS1	ROCKWELL	88009.73
CP3DS4	ROCKWELL	88009.74
CP4002	ROCKWELL	88009.75
CP4007	ROCKWELL	88009.76
CP4008	ROCKWELL	88009.77
CP4012	ROCKWELL	88009.78
CP4015	ROCKWELL	88009.79
CP4016	ROCKWELL	88009.80
CP4101	ROCKWELL	88009.81
CP4102	ROCKWELL	88009.82
CP4113	ROCKWELL	88009.83
CP4130	ROCKWELL	88009.84
CP4156	ROCKWELL	88009.85
CP44N	ALLOY	80136
CP44R	ROCKWELL	88009.86
CP45N	REPCO	80134
CP45N	ROCKWELL	88009.87
CP45N1	ROCKWELL	88009.88
CP4N	AEC	80134
CP4N	ALCO	80134
CP4N	ALLOY	80134
CP4N	BALKAMP	80134
CP4N	BORG-WARNR	80134
CP4N	PILOT	80134
CP4N	REPCO	80134
CP4N	ROCKWELL	88009.89
CP4N	SPICER	80134
CP4N1	ROCKWELL	88009.90
CP4N5	ROCKWELL	88009.91
CP4N7	ROCKWELL	88009.92
CP5000	ROCKWELL	88009.93
CP5007	ROCKWELL	88009.94
CP5008	ROCKWELL	88009.95
CP5015	ROCKWELL	88009.96
CP5101	ROCKWELL	88009.97
CP5109	ROCKWELL	88009.98
CP5122	ROCKWELL	88009.99
CP5154	ROCKWELL	88010
CP51625-2	ROCKWELL	88010.01
CP5177	ROCKWELL	88010.02
CP52625-2	ROCKWELL	88010.03
CP53625-2	ROCKWELL	88010.04
CP55N	AEC	80161
CP55N	ALLOY	80161
CP55N	REPCO	80161
CP55N	REPCO	80219
CP55N	ROCKWELL	88010.05
CP55N1	ROCKWELL	88010.06
CP55N2	ROCKWELL	88010.07
CP55N3	ROCKWELL	88010.08
CP55N4	ROCKWELL	88010.09
CP58N	ROCKWELL	88010.10
CP58N3	ALCO	80179
CP58N3	ROCKWELL	88010.11
CP58N3	SPICER	80069
CP58N3	SPICER	80179
CP58N4	ALLOY	80179
CP58N4	BORG-WARNR	80179
CP58N4	REPCO	80179
CP58N4	ROCKWELL	88010.12
CP58NBW-4	ALCO	88010.13
CP58NBW4	ROCKWELL	88010.13
CP58NWB4	ALLOY	80179
CP58WB	AEC	80185
CP58WB	ALCO	80185
CP58WB	ALLOY	80185
CP58WB	BORG-WARNR	80185
CP58WB	REPCO	80185
CP58WB	ROCKWELL	88010.14
CP58WB	SPICER	80105
CP58WB1	AEC	80186
CP58WB1	ALCO	80186
CP58WB1	ALLOY	80186
CP58WB1	BORG-WARNR	80186
CP58WB1	REPCO	80186
CP58WB1	REPUBLIC	80186
CP58WB1	ROCKWELL	88010.15
CP58WB17	ROCKWELL	88010.16
CP58WB18	ROCKWELL	88010.17
CP58WB2	AEC	80184
CP58WB2	ALCO	80184
CP58WB2	ALLOY	80184
CP58WB2	BORG-WARNR	80184
CP58WB2	REPCO	80184
CP58WB2	REPUBLIC	80184
CP58WB2	ROCKWELL	88010.18
CP58WB20	ROCKWELL	88010.19
CP58WB21	ROCKWELL	88010.20
CP58WB29	REPCO	80186
CP58WB29	ROCKWELL	88010.21
CP58WB3	ROCKWELL	88010.22
CP58WB37	REPCO	80184
CP58WB37	ROCKWELL	88010.23
CP58WB38	REPCO	80185
CP58WB38	REPUBLIC	80185
CP58WB38	ROCKWELL	88010.24
CP58WB38	SPICER	80185
CP58WB4	ALCO	80184
CP58WB4	ROCKWELL	88010.25
CP58WB5	ROCKWELL	88010.26
CP58WB6	ROCKWELL	88010.27
CP58WB7	ROCKWELL	88010.28
CP58WB9	ROCKWELL	88010.29
CP58WN	SPICER	80185
CP5N	BORG-WARNR	80178
CP5N	BORG-WARNR	88002.96
CP5N	PILOT	80178
CP5N	REPCO	80178
CP5N	ROCKWELL	88010.30
CP5N1	ROCKWELL	88010.31
CP5N14	ROCKWELL	88010.32
CP5N4	ROCKWELL	88010.33
CP5N9	ROCKWELL	88010.34
CP6	ROCKWELL	88010.35
CP6107	ROCKWELL	88010.36
CP6135	ROCKWELL	88010.37
CP62N	ROCKWELL	88010.38
CP62N10	ROCKWELL	88010.39
CP62N11	ROCKWELL	88010.40
CP62N12	ROCKWELL	88010.41
CP62N12	ROCKWELL	88010.42
CP62N13	ROCKWELL	88010.43
CP62N14	ROCKWELL	88010.44
CP62N15	ROCKWELL	88010.45
CP62N16	ROCKWELL	88010.46
CP62N17	ROCKWELL	88010.47
CP62N18	ROCKWELL	88010.48
CP62N19	REPUBLIC	80198
CP62N23	ROCKWELL	88010.49
CP62N25	ROCKWELL	88010.50
CP62N26	ROCKWELL	88010.51
CP62N3	ROCKWELL	88010.52
CP62N4	ROCKWELL	88010.53
CP62N46	ROCKWELL	88010.54
CP62N47	ROCKWELL	88010.55
CP62N48	ROCKWELL	88010.56
CP62N49	ROCKWELL	88010.57
CP62N50	ROCKWELL	88010.58
CP62N51	ROCKWELL	88010.59
CP62N56	ROCKWELL	88010.60
CP62N6	ROCKWELL	88010.61
CP62N8	ROCKWELL	88010.62
CP62N9	ROCKWELL	88010.63
CP65N	ROCKWELL	88010.64
CP68WB	ROCKWELL	88010.65
CP68WB1	ROCKWELL	88010.66
CP68WB10	ROCKWELL	88010.67
CP68WB2	ROCKWELL	88010.68
CP68WB6	ROCKWELL	88010.69
CP68WB64	ROCKWELL	88010.70
CP68WB7	ROCKWELL	88010.71
CP68WB8	ROCKWELL	88010.72
CP68WB9	ALCO	80217
CP68WB9	ROCKWELL	88010.73
CP6N	AEC	80188
CP6N	ALCO	80188
CP6N	ALLOY	80188
CP6N	BALKAMP	80188
CP6N	BORG-WARNR	80001
CP6N	BORG-WARNR	80188
CP6N	PILOT	80188
CP6N	REPCO	80188
CP6N	ROCKWELL	88010.74
CP6N	SPICER	80188
CP6N1	ROCKWELL	88010.75
CP6N10	ROCKWELL	88010.76
CP6N5	ROCKWELL	88010.77
CP6N8	ROCKWELL	88010.78
CP6N9	ROCKWELL	88010.79
CP713	REPUBLIC	80242
CP715	REPUBLIC	80178
CP726251-2	ROCKWELL	88010.80
CP72625-2	ROCKWELL	88010.81
CP72625-3	ROCKWELL	88010.82
CP72625-4	ROCKWELL	88010.83
CP72N1	ROCKWELL	88010.84
CP72N10	ROCKWELL	88010.85
CP72N103	ROCKWELL	88010.86
CP72N104	ROCKWELL	88010.87
CP72N11	ROCKWELL	88010.88
CP72N112	ROCKWELL	88010.89
CP72N113	ROCKWELL	88010.90
CP72N12	ROCKWELL	88010.91
CP72N13	ROCKWELL	88010.92
CP72N14	ROCKWELL	88010.93
CP72N14A	ROCKWELL	88010.94
CP72N15	ROCKWELL	88010.95
CP72N16	ROCKWELL	88010.96
CP72N17	ROCKWELL	88010.97
CP72N18	ROCKWELL	88010.98
CP72N19	ROCKWELL	88010.99
CP72N2	ROCKWELL	88011
CP72N20	ROCKWELL	88011.01
CP72N21	ROCKWELL	88011.02
CP72N22	ROCKWELL	88011.03
CP72N23	ROCKWELL	88011.04
CP72N24	ROCKWELL	88011.05
CP72N25	ROCKWELL	88011.06
CP72N26	ROCKWELL	88011.07
CP72N27	ROCKWELL	88011.08
CP72N28	ROCKWELL	88011.09
CP72N3	ROCKWELL	88011.10
CP72N31	ROCKWELL	88011.11
CP72N32	ROCKWELL	88011.12
CP72N33	ROCKWELL	88011.13
CP72N34	ROCKWELL	88011.14
CP72N35	ROCKWELL	88011.15
CP72N36	ROCKWELL	88011.16
CP72N37	ROCKWELL	88011.17
CP72N38	ROCKWELL	88011.18
CP72N39	ROCKWELL	88011.19
CP72N4	ROCKWELL	88011.20
CP72N40	ROCKWELL	88011.21
CP72N42	ROCKWELL	88011.22
CP72N43	ROCKWELL	88011.23
CP72N44	ROCKWELL	88011.24
CP72N45	ROCKWELL	88011.25
CP72N47	ROCKWELL	88011.26
CP72N48	ROCKWELL	88011.27
CP72N49	ROCKWELL	88011.28
CP72N5	ROCKWELL	88011.29
CP72N54	REPUBLIC	80215
CP72N55	ROCKWELL	88011.30
CP72N57	ROCKWELL	88011.31
CP72N6	ROCKWELL	88011.32
CP72N68	ROCKWELL	88011.33
CP72N7	ROCKWELL	88011.34
CP72N72	ROCKWELL	88011.35
CP72N73	ROCKWELL	88011.36
CP72N76	ROCKWELL	88011.37
CP72N8	ROCKWELL	88011.38
CP72N9	ROCKWELL	88011.39
CP74X	ROCKWELL	88011.40
CP750N	ROCKWELL	88011.41
CP750N1	REPCO	80187
CP750N1	REPUBLIC	80187
CP750N1	ROCKWELL	88011.42
CP75N	ROCKWELL	88011.43
CP75N1	ROCKWELL	88011.44
CP75N2	ROCKWELL	88011.45
CP75N4	ROCKWELL	88011.46
CP75WB13	ROCKWELL	88011.47
CP78WB12	ROCKWELL	88011.48
CP78WB13	ROCKWELL	88011.49
CP78WB14	ALCO	80244
CP78WB14	ROCKWELL	88011.50
CP78WB15	ROCKWELL	88011.51
CP78WB16	ROCKWELL	88011.52
CP78WB2	ROCKWELL	88011.53
CP78WB5	ROCKWELL	88011.54
CP78WB6	ROCKWELL	88011.55
CP78X	ROCKWELL	88011.56
CP7N	AEC	80220
CP7N	ALCO	80220
CP7N	ALLOY	80220
CP7N	BALKAMP	80220
CP7N	BORG-WARNR	80220
CP7N	PILOT	80220
CP7N	REPCO	80220
CP7N	ROCKWELL	88011.57
CP7N1	ROCKWELL	88011.58
CP7N10	ROCKWELL	88011.59
CP7N2	ROCKWELL	88011.60
CP7N5	ROCKWELL	88011.61
CP7N6	ROCKWELL	88011.62
CP7N7	ROCKWELL	88011.63
CP7N8	ROCKWELL	88011.64
CP7N9	ROCKWELL	88011.65
CP82N1	ROCKWELL	88011.66
CP82N10	ROCKWELL	88011.67
CP82N11	ROCKWELL	88011.68
CP82N12	ROCKWELL	88011.69
CP82N13	ROCKWELL	88011.70
CP82N14	ROCKWELL	88011.71
CP82N14A	ROCKWELL	88011.72
CP82N15	ROCKWELL	88011.73
CP82N16	ROCKWELL	88011.74
CP82N17	ROCKWELL	88011.75
CP82N18	ROCKWELL	88011.76
CP82N19	ROCKWELL	88011.77

Part No.	Brand	IDLI Group
CP82N2	ROCKWELL	88011.78
CP82N20	ROCKWELL	88011.79
CP82N22	ROCKWELL	88011.80
CP82N24	ROCKWELL	88011.81
CP82N25	ROCKWELL	88011.82
CP82N27	ROCKWELL	88011.83
CP82N28	ROCKWELL	88011.84
CP82N29	ROCKWELL	88011.85
CP82N3	ROCKWELL	88011.86
CP82N30	ROCKWELL	88011.87
CP82N35	ROCKWELL	88011.88
CP82N36	ROCKWELL	88011.89
CP82N37	ROCKWELL	88011.90
CP82N4	ROCKWELL	88011.91
CP82N47	ROCKWELL	88011.92
CP82N5	ROCKWELL	88011.93
CP82N6	ROCKWELL	88011.94
CP82N72	ROCKWELL	88011.95
CP82N75	ROCKWELL	88011.96
CP82N76	ROCKWELL	88011.97
CP82N9	ROCKWELL	88011.98
CP83N75	ROCKWELL	88011.99
CP850N	ROCKWELL	88012
CP850N4	REPCO	80219
CP850N4	ROCKWELL	88012.01
CP85WB1	ROCKWELL	88012.02
CP85WB10	ROCKWELL	88012.03
CP85WB12	ROCKWELL	88012.04
CP85WB13	ROCKWELL	88012.05
CP85WB14	ROCKWELL	88012.06
CP85WB19	ROCKWELL	88012.07
CP85WB2	ROCKWELL	88012.08
CP85WB20	ROCKWELL	88012.09
CP85WB22	ROCKWELL	88012.10
CP85WB23	ROCKWELL	88012.11
CP85WB3	ROCKWELL	88012.12
CP85WB46	ROCKWELL	88012.13
CP85WB47	ROCKWELL	88012.14
CP85WB48	ROCKWELL	88012.15
CP85WB5	ROCKWELL	88012.16
CP85WB6	ROCKWELL	88012.17
CP85WB8	ROCKWELL	88012.18
CP88WB	ROCKWELL	88012.19
CP88WB10	ALCO	80218
CP88WB10	ALCO	80233
CP88WB10	ROCKWELL	88012.20
CP88WB6	ROCKWELL	88012.21
CP88WB9	ALCO	80235
CP88WB9	ROCKWELL	88012.22
CP8N	ROCKWELL	88012.23
CP92N	ROCKWELL	88012.24
CP92N10	ROCKWELL	88012.25
CP92N103	ROCKWELL	88012.26
CP92N11	ROCKWELL	88012.27
CP92N12	ROCKWELL	88012.28
CP92N13	ROCKWELL	88012.29
CP92N14	ROCKWELL	88012.30
CP92N15	ROCKWELL	88012.31
CP92N16	ROCKWELL	88012.32
CP92N17	ROCKWELL	88012.33
CP92N18	ROCKWELL	88012.34
CP92N19	ROCKWELL	88012.35
CP92N20	ROCKWELL	88012.36
CP92N21	ROCKWELL	88012.37
CP92N22	ROCKWELL	88012.38
CP92N24	ROCKWELL	88012.39
CP92N25	ROCKWELL	88012.40
CP92N26	ROCKWELL	88012.41
CP92N3	ROCKWELL	88012.42
CP92N4	ROCKWELL	88012.43
CP92N49	ROCKWELL	88012.44
CP92N5	ROCKWELL	88012.45
CP92N50	ROCKWELL	88012.46
CP92N6	ROCKWELL	88012.47
CP92N7	ROCKWELL	88012.48
CP92N71	ROCKWELL	88012.49
CP92N72	ROCKWELL	88012.50
CP92N73	ROCKWELL	88012.51
CP92N8	ROCKWELL	88012.52
CP92N9	ROCKWELL	88012.53
CP92N93	ROCKWELL	88012.54
CPF11964	BAL-LM-HAM	80235
CPL105	ROCKWELL	88012.55
CPL10N1	ROCKWELL	88012.56
CPL10N2	REPCO	80033
CPL10N2	ROCKWELL	88012.57
CPL10N3	ROCKWELL	88012.58
CPL10N4	ROCKWELL	88012.59
CPL10S	REPCO	80033
CPL10S	ROCKWELL	88012.60
CPL10S1	ROCKWELL	88012.61
CPL10S10	ROCKWELL	88012.62
CPL10S3	ROCKWELL	88012.63
CPL10S5	ROCKWELL	88012.64
CPL10S7	ROCKWELL	88012.65
CPL10S8	ROCKWELL	88012.66
CPL10S9	ROCKWELL	88012.67
CPL12N	AEC	80035
CPL12N	ALCO	80035
CPL12N	ALLOY	80035
CPL12N	BORG-WARNR	80035
CPL12N	REPCO	80035
CPL12N	ROCKWELL	88012.68
CPL12N1	ROCKWELL	88012.69
CPL12N2	ROCKWELL	88012.70
CPL12N3	ROCKWELL	88012.71
CPL12N4	ROCKWELL	88012.72
CPL12N5	ROCKWELL	88012.73
CPL12N7	ROCKWELL	88012.74
CPL12N8	ROCKWELL	88012.75
CPL12R	REPCO	80035
CPL12R	ROCKWELL	88012.76
CPL12R12	ROCKWELL	88012.77
CPL14N	AEC	80037
CPL14N	ALCO	80037
CPL14N	ALLOY	80037
CPL14N	BMB	80037
CPL14N	BORG-WARNR	80037
CPL14N	REPCO	80037
CPL14N	ROCKWELL	88012.78
CPL14N1	ALLOY	80037
CPL14N1	ROCKWELL	88012.79
CPL14N2	ROCKWELL	88012.80
CPL14N4	ROCKWELL	88012.81
CPL14R	HUSCO	80037
CPL14R	REPCO	80037
CPL14R	ROCKWELL	88012.82
CPL14R1	REPCO	80037
CPL14R1	ROCKWELL	88012.83
CPL14S	REPCO	80037
CPL14S	ROCKWELL	88012.84
CPL14S1	ROCKWELL	88012.85
CPL14S2	ROCKWELL	88012.86
CPL16S	REPCO	80069
CPL16S	ROCKWELL	88012.87
CPL16S1	ROCKWELL	88012.88
CPL35N	BEAR CAT	80069
CPL35N	BMB	80069
CPL35N3	GEHL BROS.	80069
CPL35N3	PAPEX	80069
CPL659	ALLOY	80001
CPL6N	ALCO	80001
CPL6N	ALLOY	80001
CPL6N	BORG-WARNR	80001
CPL6N	ROCKWELL	88012.89
CPL6N1	ROCKWELL	88012.90
CPL6N10	ROCKWELL	88012.91
CPL6N11	REPCO	80001
CPL6N11	ROCKWELL	88012.92
CPL6N12	ROCKWELL	88012.93
CPL6N13	ROCKWELL	88012.94
CPL6N14	ROCKWELL	88012.95
CPL6N17	ROCKWELL	88012.96
CPL6N19	ROCKWELL	88012.97
CPL6N2	ROCKWELL	88012.98
CPL6N20	ROCKWELL	88012.99
CPL6N4	ROCKWELL	88013
CPL6N5	ROCKWELL	88013.01
CPL6N6	ROCKWELL	88013.02
CPL6N7	ROCKWELL	88013.03
CPL6N8	AEC	80001
CPL6N8	AEC	80003
CPL6N8	ALCO	80001
CPL6N8	BORG-WARNR	80001
CPL6N8	REPCO	80001
CPL6N8	ROCKWELL	88013.04
CPL6N9	ROCKWELL	88013.05
CPL6R	ROCKWELL	88013.06
CPL6R30	ROCKWELL	88013.07
CPL6R34	ROCKWELL	88013.08
CPL6R39	ROCKWELL	88013.09
CPL6R40	ROCKWELL	88013.10
CPL6S10	ROCKWELL	88013.11
CPL6S11	ROCKWELL	88013.12
CPL6S12	ROCKWELL	88013.13
CPL6S13	ROCKWELL	88013.14
CPL6S14	ROCKWELL	88013.15
CPL6S15	ROCKWELL	88013.16
CPL6S16	ROCKWELL	88013.17
CPL6S19	ROCKWELL	88013.18
CPL6S20	ROCKWELL	88013.19
CPL6S22	REPCO	80003
CPL6S22	REPCO	80142
CPL6S22	ROCKWELL	88013.20
CPL6S26	ROCKWELL	88013.21
CPL6S3	ROCKWELL	88013.22
CPL6S36	ROCKWELL	88013.23
CPL6S4	ROCKWELL	88013.24
CPL6S48	ROCKWELL	88013.25
CPL6S5	REPCO	80003
CPL6S5	ROCKWELL	88013.26
CPL6S7	ROCKWELL	88013.27
CPL6S8	HUSCO	80001
CPL6S8	ROCKWELL	88013.28
CPL6S9	ALCO	80003
CPL6S9	ALLOY	80003
CPL6S9	REPCO	80003
CPL6S9	ROCKWELL	88013.29
CPL7S48	ROCKWELL	88013.30
CPLSN10	ROCKWELL	88013.31
CP-P55	REPUBLIC	80129
CPR109690	CHRYSLER	80234
CPR109691	CHRYSLER	80243
CPS96	REPUBLIC	80038
CR1210	REPUBLIC	88002.60
CR9655	BALKAMP	80122
CS-H12-16-6	ROCKWELL	88013.32
CS-H4-20-13	ROCKWELL	88013.33
CS-H5-18-27	ROCKWELL	88013.34
CS-H5-18-28	ROCKWELL	88013.35
CS-H6-24-49	ROCKWELL	88013.36
CT10-000	INBRA	80031
CT10-005	INBRA	80049
CT10-010	INBRA	80171
CT10-015	INBRA	80048
CT11-000	INBRA	80070
CT11-005	INBRA	80100
CT11-010	INBRA	80188
CT11015	INBRA	80178
CT11020	INBRA	80001
CT11-025	INBRA	80215
CT11-026	INBRA	80211
CT11-030	INBRA	80124
CT11-035	INBRA	80001
CT11-040	INBRA	80142
CT11-045	INBRA	88003.23
CT11-050	INBRA	80202
CT11-055	INBRA	80162
CT11-060	INBRA	80216
CT11-065	INBRA	80193
CT11-070	INBRA	80149
CT11-075	INBRA	80164
CT11-081	INBRA	80223
CT11-085	INBRA	80240
CT11-090	INBRA	80238
CT11-095	INBRA	80213
CT11-100	INBRA	80224
CT13-005	INBRA	80198
CT13-010	INBRA	80021
CT13-015	INBRA	88003.26
CT13-020	INBRA	80230
CT13-025	INBRA	80156
CT13-030	INBRA	80213
CT13-035	INBRA	80198
CT13-040	INBRA	80234
CT13-045	INBRA	80238
CT13-050	INBRA	80172
CT13-070	INBRA	80001
CT13-075	INBRA	80033
CT13-080	INBRA	80223
CT16-000	INBRA	80156
CT16-010	INBRA	80198
CT16-020	INBRA	80172
CT-16-030	INBRA	80193
CT17-000	INBRA	80189
CT17-005	INBRA	80221
CT17-010	INBRA	80112
CT17-030	INBRA	80225
CT17-035	INBRA	80182
CT17-040	INBRA	80190
CT17-050	INBRA	80239
CT17-055	INBRA	80210
CT17-065	INBRA	80234
CT17-090	INBRA	80234
CT17-100	INBRA	80214
CT17-110	INBRA	80233
CT17-116	INBRA	80196
CT17-130	INBRA	80021
CT18-005	INBRA	80041
CT19-005	INBRA	80101
CT19-010	INBRA	80021
CT19-015	INBRA	80100
CT19-020	INBRA	80099
CT20-031	INBRA	80175
CT20-032	INBRA	80174
CT20-507	INBRA	80047
CT20-510	INBRA	80171
CT20-515	INBRA	80168
CT20-520	INBRA	80187
CT20-545	INBRA	80197

Part No.	Brand	IDLI Group
CT20-570	INBRA	80219
CT20-580	INBRA	80142
CT21-000	INBRA	80193
CT22-000	INBRA	80234
CT22-005	INBRA	80213
CT23-000	INBRA	80171
CT23-020	INBRA	80213
CT24-000	INBRA	80238
CT24-005	INBRA	80213
CT24-020	INBRA	80198
CT24-025	INBRA	80109
CT24-030	INBRA	80156
CT24-035	INBRA	80172
CT24-040	INBRA	80001
CT25-000	INBRA	80198
CT55L	REPUBLIC	80081
CT55L	REPUBLIC	80091
CTBA18310	FORD	80023
CTD145518	CHRYSLER	80186
CTD145519	CHRYSLER	80184
CTZ4724A	FORD	80075
CU9655	BALKAMP	80181
CV1000	MIDAS	86001
CV1001	MIDAS	86002
CV1002	MIDAS	86003
CV1003	MIDAS	86004
CV1004	MIDAS	86005
CV1005	MIDAS	86006
CV1006	MIDAS	86007
CV1007	MIDAS	86008
CV1008	MIDAS	86009
CV1009	MIDAS	86010
CV1010	MIDAS	86011
CV1011	MIDAS	86012
CV1012	MIDAS	86013
CV1013	MIDAS	86014
CV1014	MIDAS	86015
CV1015	MIDAS	86016
CV1016	MIDAS	86017
CV1017	MIDAS	86018
CV1018	MIDAS	86019
CV1019	MIDAS	86020
CV1020	MIDAS	86021
CV1021	MIDAS	86022
CV1022	MIDAS	86023
CV1023	MIDAS	86024
CV1024	MIDAS	86025
CV1025	MIDAS	86026
CV1026	MIDAS	86027
CV1027	MIDAS	86028
CV1028	MIDAS	86029
CV1029	MIDAS	86030
CV1030	MIDAS	86031
CV1031	MIDAS	86032
CV1032	MIDAS	86033
CV1033	MIDAS	86034
CV1034	MIDAS	86035
CV1035	MIDAS	86036
CV1036	MIDAS	86037
CV1037	MIDAS	86038
CV1038	MIDAS	86039
CV1039	MIDAS	86040
CV1040	MIDAS	86041
CV1041	MIDAS	86042
CV1042	MIDAS	86043
CV1043	MIDAS	86044
CV1044	MIDAS	86045
CV1045	MIDAS	86046
CV1046	MIDAS	86047
CV1047	MIDAS	86048
CV1048	MIDAS	86049
CV1049	MIDAS	86050
CV1050	MIDAS	86051
CV1051	MIDAS	86052
CV1052	MIDAS	86056
CV1053	MIDAS	86057
CV1054	MIDAS	86058
CV1055	MIDAS	86059
CV1056	MIDAS	86060
CV1057	MIDAS	86061
CV1058	MIDAS	86069
CV1059	MIDAS	86070
CV1060	MIDAS	86071
CV1061	MIDAS	86072
CV1062	MIDAS	86073
CV1063	MIDAS	86076
CV1064	MIDAS	86077
CV1065	MIDAS	86078
CV1066	MIDAS	86079
CV1067	MIDAS	86081
CV1068	MIDAS	86082
CV1069	MIDAS	86083
CV1070	MIDAS	86084
CV1071	MIDAS	86086
CV1072	MIDAS	86087
CV1073	MIDAS	86088
CV1074	MIDAS	86089
CV1075	MIDAS	86090
CV1076	MIDAS	86091
CV1077	MIDAS	86092
CV1078	MIDAS	86093
CV1079	MIDAS	86094
CV1080	MIDAS	86095
CV1081	MIDAS	86096
CV1082	MIDAS	86097
CV1083	MIDAS	86098
CV1084	MIDAS	86099
CV1085	MIDAS	86100
CV1086	MIDAS	86101
CV1087	MIDAS	86102
CV1088	MIDAS	86103
CV1089	MIDAS	86104
CV1090	MIDAS	86105
CV1091	MIDAS	86106
CV1092	MIDAS	86107
CV1093	MIDAS	86108
CV1094	MIDAS	86109
CV1095	MIDAS	86110
CV1096	MIDAS	86111
CV1097	MIDAS	86112
CV1098	MIDAS	86114
CV1099	MIDAS	86115
CV1100	MIDAS	86116
CV1101	MIDAS	86117
CV1102	MIDAS	86122
CV1103	MIDAS	86123
CV1104	MIDAS	86124
CV1105	MIDAS	86125
CV1106	MIDAS	86126
CV1107	MIDAS	86127
CV1108	MIDAS	86128
CV1109	MIDAS	86129
CV1110	MIDAS	86130
CV1111	MIDAS	86131
CV1112	MIDAS	86133
CV1113	MIDAS	86134
CV1114	MIDAS	86135
CV1115	MIDAS	86136
CV1116	MIDAS	86137
CV1117	MIDAS	86138
CV1118	MIDAS	86139
CV1119	MIDAS	86143
CV1120	MIDAS	86144
CV1121	MIDAS	86145
CV1122	MIDAS	86146
CV1123	MIDAS	86147
CV1124	MIDAS	86148
CV1125	MIDAS	86149
CV1126	MIDAS	86150
CV1127	MIDAS	86151
CV1128	MIDAS	86152
CV1129	MIDAS	86153
CV1130	MIDAS	86154
CV1131	MIDAS	86155
CV1132	MIDAS	86156
CV1133	MIDAS	86158
CV1134	MIDAS	86159
CV1135	MIDAS	86160
CV1136	MIDAS	86161
CV1137	MIDAS	86162
CV1138	MIDAS	86163
CV1139	MIDAS	86164
CV1140	MIDAS	86165
CV1141	MIDAS	86166
CV1142	MIDAS	86167
CV1143	MIDAS	86168
CV1144	MIDAS	86169
CV1145	MIDAS	86170
CV1146	MIDAS	86171
CV1147	MIDAS	86172
CV1148	MIDAS	86173
CV1149	MIDAS	86174
CV1150	MIDAS	86175
CV1151	MIDAS	86176
CV1152	MIDAS	86179
CV1153	MIDAS	86180
CV1154	MIDAS	86181
CV1155	MIDAS	86182
CV1156	MIDAS	86183
CV1157	MIDAS	86184
CV1158	MIDAS	86185
CV1159	MIDAS	86187
CV1160	MIDAS	86188
CV1161	MIDAS	86189
CV1162	MIDAS	86190
CV1163	MIDAS	86191
CV1164	MIDAS	86192
CV1165	MIDAS	86193
CV2200	MOOG	86001
CV2201	MOOG	86002
CV2202	MOOG	86003
CV2203	MOOG	86004
CV2204	MOOG	86005
CV2205	MOOG	86006
CV2206	MOOG	86007
CV2209	MOOG	86008
CV2210	MOOG	86009
CV2211	MOOG	86010
CV2212	MOOG	86011
CV2212	MOOG	86012
CV2213	MOOG	86012
CV2214	MOOG	86013
CV2215	MOOG	86014
CV2216	MOOG	86015
CV2217	MOOG	86016
CV2218	MOOG	86017
CV2219	MOOG	86018
CV2220	MOOG	86019
CV2221	MOOG	86020
CV2222	MOOG	86021
CV2223	MOOG	86022
CV2224	MOOG	86023
CV2225	MOOG	86024
CV2226	MOOG	86025
CV2227	MOOG	86026
CV2228	MOOG	86027
CV2229	MOOG	86028
CV2230	MOOG	86029
CV2231	MOOG	86030
CV2232	MOOG	86031
CV2233	MOOG	86032
CV2234	MOOG	86033
CV2235	MOOG	86034
CV2236	MOOG	86035
CV2237	MOOG	86036
CV2238	MOOG	86037
CV2239	MOOG	86038
CV2240	MOOG	86039
CV2242	MOOG	86041
CV2243	MOOG	86042
CV2244	MOOG	86043
CV2246	MOOG	86045
CV2247	MOOG	86046
CV2248	MOOG	86047
CV2249	MOOG	86048
CV2250	MOOG	86049
CV2253	MOOG	86051
CV2254	MOOG	86051
CV2255	MOOG	86053
CV2257	MOOG	86054
CV2258	MOOG	86055
CV2260	MOOG	86056
CV2261	MOOG	86057
CV2262	MOOG	86058
CV2263	MOOG	86059
CV2264	MOOG	86060
CV2265	MOOG	86061
CV2270	MOOG	86062
CV2271	MOOG	86063
CV2273	MOOG	86064
CV2274	MOOG	86065
CV2275	MOOG	86064
CV2276	MOOG	86067
CV2277	CHRYSLER	86068
CV2300	MOOG	86069
CV2300-2303	MOOG	86143
CV2301	MOOG	86070
CV2302	MOOG	86071
CV2303	MOOG	86072
CV2303	VOLKSWAGEN	86072
CV2304	MOOG	86122
CV2307	MOOG	86075
CV2307	MOOG	86148
CV2308	MOOG	86075
CV2308	MOOG	86147
CV2308	MOOG	86236
CV2310	MOOG	86076
CV2311	MOOG	86077
CV2312	MOOG	86079
CV2312	MOOG	86151
CV2313	MOOG	86079
CV2313	MOOG	86152
CV2315	MOOG	86080
CV2316	MOOG	86081
CV2317	MOOG	86082
CV2318	MOOG	86083
CV2319	MOOG	86084
CV2320	MOOG	86153
CV2321	MOOG	86086
CV2322	MOOG	86087

Part No.	Brand	IDLI Group	Part No.	Brand	IDLI Group	Part No.	Brand	IDLI Group	Part No.	Brand	IDLI Group
CV2323	MOOG	86088	CV2377	MOOG	86182	CV2460	MOOG	86256	CV5315	MOOG	86240
CV2324	MOOG	86089	CV2377	MOOG	86299	CV2460	MOOG	86257	CV5315	MOOG	86271
CV2325	MOOG	86090	CV2379	MOOG	86128	CV2460	MOOG	86282	CV5315	MOOG	86272
CV2326	MOOG	86118	CV2380	MOOG	86129	CV2470	MOOG	86180	CV5316	MOOG	86239
CV2327	MOOG	86092	CV2381	MOOG	86130	CV2471	MOOG	86181	CV5316	MOOG	86240
CV2328	MOOG	86093	CV2381	MOOG	86184	CV2476	MOOG	86182	CV5316	MOOG	86271
CV2329	MOOG	86094	CV2382	MOOG	86131	CV2477	MOOG	86182	CV5316	MOOG	86272
CV2330	MOOG	86160	CV2382	MOOG	86184	CV2477	MOOG	86299	CV5319	MOOG	86241
CV2331	MOOG	86095	CV2383	MOOG	86184	CV2479	MOOG	86184	CV5319	MOOG	86242
CV2332	MOOG	86096	CV2384	MOOG	86134	CV2482	MOOG	86184	CV5319	MOOG	86273
CV2332	MOOG	86162	CV2384	MOOG	86187	CV2483	MOOG	86184	CV5319	MOOG	86274
CV2333	MOOG	86096	CV2385	MOOG	86134	CV2485	MOOG	86187	CV5320	MOOG	86241
CV2333	MOOG	86162	CV2386	MOOG	86135	CV2486	MOOG	86188	CV5320	MOOG	86242
CV2334	MOOG	86098	CV2386	MOOG	86187	CV2487	MOOG	86189	CV5320	MOOG	86274
CV2337	MOOG	86099	CV2387	MOOG	86136	CV2489	MOOG	86190	CV5323	MOOG	86243
CV2337	MOOG	86143	CV2388	MOOG	86137	CV2493	MOOG	86191	CV5323	MOOG	86244
CV2338	MOOG	86100	CV2389	MOOG	86138	CV2494	MOOG	86140	CV5323	MOOG	86245
CV2338-2339	MOOG	86164	CV2389	MOOG	86190	CV2499	MOOG	86193	CV5323	MOOG	86267
CV2339	MOOG	86101	CV2393	MOOG	86139	CV5203	MOOG	86194	CV5323	MOOG	86275
CV2340	MOOG	86102	CV2394	MOOG	86140	CV5204	MOOG	86195	CV5323	MOOG	86276
CV2341	MOOG	86103	CV2398	MOOG	86141	CV5207	MOOG	86196	CV5324	MOOG	86243
CV2341	MOOG	86166	CV2399	MOOG	86142	CV5208	MOOG	86197	CV5324	MOOG	86244
CV2342	MOOG	86104	CV2400	MOOG	86143	CV5209	MOOG	86198	CV5324	MOOG	86245
CV2342	MOOG	86167	CV2401	MOOG	86144	CV5210	MOOG	86199	CV5324	MOOG	86275
CV2342	MOOG	86264	CV2402	MOOG	86145	CV5211	MOOG	86200	CV5324	MOOG	86276
CV2343	MOOG	86105	CV2404	MOOG	86146	CV5212	MOOG	86201	CV5325	MOOG	86243
CV2344-2349	MOOG	86170	CV2406	MOOG	86147	CV5213	MOOG	86202	CV5325	MOOG	86244
CV2345	MOOG	86107	CV2406	MOOG	86236	CV5214	MOOG	86203	CV5325	MOOG	86245
CV2346	MOOG	86108	CV2407	MOOG	86075	CV5215	MOOG	86204	CV5325	MOOG	86275
CV2347	MOOG	86109	CV2407	MOOG	86148	CV5216	MOOG	86205	CV5325	MOOG	86276
CV2348	MOOG	86110	CV2410	MOOG	86149	CV5217	MOOG	86220	CV5325	MOOG	86277
CV2349	MOOG	86111	CV2410	MOOG	86295	CV5218	MOOG	86207	CV5341	MOOG	86178
CV2349	MOOG	86169	CV2411	MOOG	86150	CV5225	MOOG	86208	CV5341	MOOG	86246
CV2350	MOOG	86172	CV2412	MOOG	86151	CV5226	MOOG	86209	CV5341	MOOG	86247
CV2351	MOOG	86113	CV2413	MOOG	86152	CV5227	MOOG	86210	CV5342	MOOG	86178
CV2351	MOOG	86174	CV2416	MOOG	86153	CV5228	MOOG	86211	CV5342	MOOG	86246
CV2353	MOOG	86114	CV2417	MOOG	86080	CV5229	MOOG	86212	CV5342	MOOG	86247
CV2354	MOOG	86115	CV2417	MOOG	86085	CV5230	MOOG	86213	CV5347	MOOG	86120
CV2355	MOOG	86116	CV2417	MOOG	86154	CV5231	MOOG	86214	CV5347	MOOG	86179
CV2355	MOOG	86176	CV2418	MOOG	86155	CV5232	MOOG	86215	CV5347	MOOG	86251
CV2356	MOOG	86117	CV2419	MOOG	86156	CV5235	MOOG	86218	CV5347	MOOG	86252
CV2358	MOOG	86118	CV2422	MOOG	86268	CV5236	MOOG	86219	CV5347	MOOG	86254
CV2359	MOOG	86118	CV2423	MOOG	86157	CV5237	MOOG	86220	CV5347	MOOG	86255
CV2360	MOOG	86121	CV2428	MOOG	86158	CV5238	MOOG	86221	CV5347	MOOG	86256
CV2360	MOOG	86179	CV2429	MOOG	86159	CV5251	GMC	86222	CV5347	MOOG	86257
CV2360	MOOG	86251	CV2430	MOOG	86160	CV5251	MOOG	86222	CV5347	MOOG	86282
CV2360	MOOG	86252	CV2431	MOOG	86161	CV5252	MOOG	86223	CV5348	MOOG	86120
CV2360	MOOG	86254	CV2432	MOOG	86162	CV5253	MOOG	86224	CV5348	MOOG	86179
CV2360	MOOG	86255	CV2434	MOOG	86163	CV5254	MOOG	86225	CV5348	MOOG	86251
CV2360	MOOG	86256	CV2438	MOOG	86164	CV5255	MOOG	86228	CV5348	MOOG	86252
CV2360	MOOG	86257	CV2440	MOOG	86165	CV5256	MOOG	86227	CV5348	MOOG	86254
CV2360	MOOG	86282	CV2441	MOOG	86166	CV5257	MOOG	86228	CV5348	MOOG	86255
CV2361	MOOG	86121	CV2442	MOOG	86167	CV5258	MOOG	86223	CV5348	MOOG	86256
CV2361	MOOG	86179	CV2444	MOOG	86169	CV5271	MOOG	86230	CV5348	MOOG	86257
CV2361	MOOG	86251	CV2445	MOOG	86170	CV5272	MOOG	86231	CV5348	MOOG	86282
CV2361	MOOG	86252	CV2447	MOOG	86171	CV5273	GMC	86232	CV5349	MOOG	86120
CV2361	MOOG	86254	CV2448	MOOG	86172	CV5274	MOOG	86233	CV5349	MOOG	86179
CV2361	MOOG	86255	CV2450	MOOG	86173	CV5301	MOOG	86157	CV5349	MOOG	86251
CV2361	MOOG	86256	CV2451	MOOG	86174	CV5302	MOOG	86157	CV5349	MOOG	86252
CV2361	MOOG	86257	CV2453	MOOG	86175	CV5308	MOOG	86147	CV5349	MOOG	86254
CV2361	MOOG	86282	CV2455	MOOG	86176	CV5308	MOOG	86236	CV5349	MOOG	86255
CV2370	MOOG	86122	CV2456	MOOG	86177	CV5311	MOOG	86237	CV5349	MOOG	86256
CV2370-2371	MOOG	86180	CV2457	MOOG	86178	CV5311	MOOG	86238	CV5349	MOOG	86257
CV2371	MOOG	86123	CV2457	MOOG	86246	CV5311	MOOG	86269	CV5349	MOOG	86282
CV2372	MOOG	86124	CV2457	MOOG	86247	CV5311	MOOG	86270	CV5352	MOOG	86179
CV2372	MOOG	86181	CV2460	MOOG	86179	CV5312	MOOG	86237	CV5352	MOOG	86251
CV2373	MOOG	86125	CV2460	MOOG	86251	CV5312	MOOG	86238	CV5352	MOOG	86252
CV2373	MOOG	86181	CV2460	MOOG	86252	CV5312	MOOG	86269	CV5352	MOOG	86254
CV2376	MOOG	86126	CV2460	MOOG	86254	CV5312	MOOG	86270	CV5352	MOOG	86255
CV2377	MOOG	86127	CV2460	MOOG	86255	CV5315	MOOG	86239	CV5352	MOOG	86256

Part No.	Brand	IDLI Group	Part No.	Brand	IDLI Group	Part No.	Brand	IDLI Group	Part No.	Brand	IDLI Group
CV5352	MOOG	86257	CV5391	MOOG	86244	CV5462	MOOG	86254	CVK5	MOTOR MAST	88002.33
CV5352	MOOG	86282	CV5391	MOOG	86245	CV5462	MOOG	86256	CVK5	REPUBLIC	88002.33
CV5353	MOOG	86179	CV5391	MOOG	86267	CV5462	MOOG	86257	CVK5	ZELLER	88002.33
CV5353	MOOG	86251	CV5391	MOOG	86275	CV5462	MOOG	86282	CVK5-115A	ZELLER	88002.33
CV5353	MOOG	86252	CV5391	MOOG	86276	CV5467	GMC	86251	CVK5A	MOTOR MAST	88002.33
CV5353	MOOG	86254	CV5401	MOOG	86157	CV5467	MOOG	86179	CVK5A	REPUBLIC	88002.33
CV5353	MOOG	86255	CV5411	MOOG	86237	CV5467	MOOG	86252	CVK6	AEC	88002.33
CV5353	MOOG	86256	CV5411	MOOG	86238	CV5467	MOOG	86254	CVK6	MOTOR MAST	88002.33
CV5353	MOOG	86257	CV5411	MOOG	86269	CV5467	MOOG	86256	CVK6	ZELLER	88002.33
CV5353	MOOG	86282	CV5411	MOOG	86270	CV5467	MOOG	86257	CVK6A	REPUBLIC	88002.33
CV5354	MOOG	86179	CV5412	MOOG	86237	CV5467	MOOG	86282	CVK8	REPUBLIC	88002.37
CV5354	MOOG	86251	CV5412	MOOG	86238	CV8207	MOOG	86284	CX1TZ3249A	FORD	80124
CV5354	MOOG	86252	CV5412	MOOG	86269	CV8209	MOOG	86285	CY123CY123	CHELSEA	88000.69
CV5354	MOOG	86254	CV5412	MOOG	86270	CV8210	MOOG	86286	CY134CY134	CHELSEA	88000.70
CV5354	MOOG	86255	CV5415	MOOG	86239	CV8211	MOOG	86287	CY144CY144	CHELSEA	88000.71
CV5354	MOOG	86256	CV5415	MOOG	86240	CV8212	MOOG	86288	CY154CY154	CHELSEA	88000.72
CV5354	MOOG	86257	CV5415	MOOG	86271	CV8213	MOOG	86289	D0RY4635A	FORD	80012
CV5354	MOOG	86282	CV5415	MOOG	86272	CV8214	MOOG	86290	D0TZ4808A	FORD	88007.53
CV5358	MOOG	86179	CV5416	MOOG	86239	CV8215	MOOG	86291	D0TZ4808B	FORD	88007.55
CV5358	MOOG	86251	CV5416	MOOG	86240	CV8216	MOOG	86292	D0TZ4840A	FORD	87337
CV5358	MOOG	86252	CV5416	MOOG	86271	CV8217	MOOG	86293	D0TZ4840B	FORD	87285
CV5358	MOOG	86254	CV5416	MOOG	86272	CV8218	MOOG	86294	D1	MOPAR	80021
CV5358	MOOG	86255	CV5419	MOOG	86241	CV8306	MOOG	86149	D1	MOPAR	88001.97
CV5358	MOOG	86256	CV5419	MOOG	86242	CV8306	MOOG	86295	D100	MOPAR	80068
CV5358	MOOG	86257	CV5419	MOOG	86273	CV8307	MOOG	86296	D101341	GALION	80001
CV5358	MOOG	86282	CV5419	MOOG	86274	CV8308	MOOG	86297	D1025	AEC	80019
CV5359	MOOG	86251	CV5420	MOOG	86241	CV8308	MOOG	86298	D103	MOPAR	80183
CV5359	MOOG	86252	CV5420	MOOG	86242	CV8308	MOOG	86301	D104	MOPAR	80141
CV5359	MOOG	86254	CV5420	MOOG	86273	CV8309	MOOG	86297	D105	MOPAR	80138
CV5359	MOOG	86255	CV5420	MOOG	86274	CV8309	MOOG	86298	D106	MOPAR	80068
CV5359	MOOG	86256	CV5423	MOOG	86243	CV8309	MOOG	86301	D106251	GALION	80142
CV5359	MOOG	86257	CV5423	MOOG	86244	CV8346	MOOG	86182	D107	MOPAR	80123
CV5359	MOOG	86282	CV5423	MOOG	86245	CV8346	MOOG	86299	D108	MOPAR	80122
CV5362	GMC	86251	CV5423	MOOG	86267	CV8347	MOOG	86300	D109	MOPAR	80038
CV5362	MOOG	86179	CV5423	MOOG	86275	CV8347	MOOG	86302	D109	MOPAR	80039
CV5362	MOOG	86251	CV5423	MOOG	86276	CV8406	MOOG	86297	D10A	SPICER	88013.52
CV5362	MOOG	86252	CV5424	MOOG	86244	CV8406	MOOG	86298	D11	MOPAR	80045
CV5362	MOOG	86254	CV5424	MOOG	86276	CV8406	MOOG	86301	D110	MOPAR	80081
CV5362	MOOG	86255	CV5425	MOOG	86277	CV8446	MOOG	86300	D110676	GALION	80182
CV5362	MOOG	86256	CV5441	MOOG	86178	CV8446	MOOG	86302	D111	MOPAR	80038
CV5362	MOOG	86257	CV5441	MOOG	86246	CVK001	AEC	88002.34	D112	MOPAR	80071
CV5362	MOOG	86282	CV5441	MOOG	86247	CVK002A	AEC	88002.36	D113	MOPAR	80074
CV5366	MOOG	86179	CV5447	GMC	86251	CVK003	AEC	88002.33	D115952A	HYSTER	80237
CV5366	MOOG	86251	CV5447	MOOG	86179	CVK005	AEC	88002.33	D12	MOPAR	80049
CV5366	MOOG	86252	CV5447	MOOG	86252	CVK1	AEC	88002.34	D12	MOPAR	88001.96
CV5366	MOOG	86254	CV5447	MOOG	86254	CVK1	MOTOR MAST	88002.34	D122813	GALION	80054
CV5366	MOOG	86255	CV5447	MOOG	86255	CVK1	MOTOR MAST	88002.35	D123329	GALION	80182
CV5366	MOOG	86256	CV5447	MOOG	86256	CVK1	REPUBLIC	88002.34	D126	MOPAR	80042
CV5366	MOOG	86257	CV5447	MOOG	86257	CVK1	ZELLER	88002.34	D127	MOPAR	80135
CV5366	MOOG	86282	CV5447	MOOG	86282	CVK1-111	ZELLER	88002.34	D127	MOPAR	88001.65
CV5367	MOOG	86179	CV5448	GMC	86251	CVK1-111	ZELLER	88002.35	D128	MOPAR	80135
CV5367	MOOG	86251	CV5448	MOOG	86179	CVK2	AEC	88002.34	D13	MOPAR	80044
CV5367	MOOG	86252	CV5448	MOOG	86252	CVK2	MOTOR MAST	88002.34	D13	MOPAR	88001.98
CV5367	MOOG	86254	CV5448	MOOG	86254	CVK2	REPUBLIC	88002.34	D130	MOPAR	80034
CV5367	MOOG	86255	CV5448	MOOG	86255	CVK2	ZELLER	88002.34	D132	MOPAR	80064
CV5367	MOOG	86256	CV5448	MOOG	86256	CVK2A	AEC	88002.36	D132	MOPAR	80065
CV5367	MOOG	86257	CV5448	MOOG	86257	CVK2A	MOTOR MAST	88002.36	D132	MOPAR	88001.63
CV5367	MOOG	86282	CV5448	MOOG	86282	CVK2A	REPUBLIC	88002.36	D133	MOPAR	80060
CV5381	MOOG	86258	CV5449	GMC	86251	CVK2A	ZELLER	88002.36	D134	MOPAR	80060
CV5382	MOOG	86259	CV5449	MOOG	86179	CVK2A-112	ZELLER	88002.36	D134	MOPAR	80061
CV5383	MOOG	86260	CV5449	MOOG	86252	CVK3	AEC	88002.33	D134	MOPAR	80063
CV5384	MOOG	86261	CV5449	MOOG	86254	CVK3	MOTOR MAST	88002.33	D134A	MOPAR	80061
CV5385	MOOG	86262	CV5449	MOOG	86255	CVK3	REPUBLIC	88002.33	D134A	MOPAR	80063
CV5386	MOOG	86263	CV5449	MOOG	86256	CVK-3	ZELLER	88002.33	D134A	MOPAR	88001.70
CV5387	MOOG	86104	CV5449	MOOG	86257	CVK4	AEC	88002.33	D135	MOPAR	80096
CV5387	MOOG	86264	CV5449	MOOG	86282	CVK 4	AEC	88002.33	D136	MOPAR	80096
CV5389	MOOG	86265	CV5454	MOOG	86281	CVK4	MOTOR MAST	88002.33	D137	MOPAR	80094
CV5389	MOOG	86267	CV5462	GMC	86251	CVK-4	REPUBLIC	88002.33	D138	MOPAR	80056
CV5390	MOOG	86266	CV5462	MOOG	86179	CVK4	REPUBLIC	88002.33	D139	MOPAR	80095
CV5391	MOOG	86243	CV5462	MOOG	86252	CVK5	AEC	88002.33	D14	MOPAR	80055

Part No.	Brand	IDLI Group
D15	MOPAR	80051
D150	MOPAR	80909
D151	MOPAR	80909
D154	MOPAR	80908
D154	WESCO	88001.60
D155	MOPAR	80907
D156	MOPAR	80905
D156	WESCO	80042
D156	WESCO	88001.60
D15655	PILOT	80145
D157	MOPAR	80906
D158	MOPAR	80903
D158	MOPAR	80958
D16	MOPAR	80059
D1601	MOPAR	80004
D1602	MOPAR	80019
D1603	MOPAR	80010
D1604	MOPAR	80009
D1605	MOPAR	80012
D1608	MOPAR	80017
D162	MOPAR	80955
D163	MOPAR	80903
D163	MOPAR	80958
D164	MOPAR	80954
D165	MOPAR	80957
D166	MOPAR	80951
D168	MOPAR	80953
D169	MOPAR	80132
D17	MOPAR	80101
D18	MOPAR	80110
D18	MOPAR	80112
D18	MOPAR	88002.98
D180	MOPAR	88002.43
D18049	CHRYSLER	80142
D181	MOPAR	88002.42
D182	MOPAR	88002.44
D183	MOPAR	88002.55
D18363	CHRYSLER	80180
D184	MOPAR	88002.60
D185	MOPAR	88002.48
D186	MOPAR	88002.50
D187	MOPAR	88002.51
D188	MOPAR	88002.52
D18888-5	CHRYSLER	80041
D19	MOPAR	80104
D19655	PILOT	80160
D19834-7	CHRYSLER	80189
D1HZ4635A	FORD	80225
D1TZ3249A	FORD	80186
D2	MOPAR	80030
D2	MOPAR	80031
D20	MOPAR	80099
D20	MOPAR	80101
D200	MOPAR	80032
D201	MOPAR	80066
D2012X	NEAPCO	88003.56
D202	MOPAR	80058
D202	MOPAR	88002.77
D2022	NEAPCO	88003.42
D203	MOPAR	80058
D20311	NEAPCO	88003.43
D204	MOPAR	80105
D204	NEAPCO	88003.45
D205	MOPAR	80066
D206	MOPAR	80104
D2060C	MOTOR MAST	80050
D2060D	MOTOR MAST	80043
D2060D	MOTOR MAST	80044
D2060D	ZELLER	80043
D2060G	MOTOR MAST	80051
D207	MOPAR	80060
D207	MOPAR	80061
D207	MOPAR	80063
D208	MOPAR	88002.34
D209	MOPAR	88002.33
D210	MOPAR	88002.33
D211	MOPAR	80036
D212	MOPAR	80079
D213	MOPAR	80104
D226	WESCO	80064
D23	MOPAR	80146
D24	MOPAR	80147
D240811X	NEAPCO	88003.51
D25	MOPAR	80109
D25	MOPAR	80149
D251	WESCO	80056
D251	WESCO	80061
D252	WESCO	80034
D254	WESCO	80056
D255	WESCO	80060
D255	WESCO	80061
D255	WESCO	88001.59
D255-1	WESCO	80060
D255-1	WESCO	80061
D255-1	WESCO	80062
D255C	WESCO	80063
D256	WESCO	80056
D256	WESCO	80057
D256	WESCO	88001.58
D257	WESCO	80095
D257	WESCO	88001.61
D258	WESCO	80095
D258	WESCO	88001.60
D259	WESCO	80096
D26	MOPAR	80153
D261	WESCO	80064
D262	WESCO	80096
D26204	CASE,J.I.	80077
D263	WESCO	80094
D27	MOPAR	80158
D27Z4758A	FORD	80020
D28	MOPAR	80148
D285	WESCO	80132
D2880	KOYO	80020
D2880A	KOYO	80020
D29	MOPAR	80159
D2A	MOPAR	80031
D2A	SPICER	88000.92
D2HA4635AA	FORD	80231
D2HA4635BA	FORD	80212
D2HA4635CA	FORD	80210
D2HA4635DA	FORD	80237
D2HZ4635A	FORD	80227
D2HZ4635B	FORD	80225
D2TZ4782A	FORD	88002.37
D2TZ4782Z	FORD	88002.37
D2TZ4840A	FORD	88008.47
D2TZ4840B	FORD	87327
D2TZ4840B	FORD	87331
D2X2202	CATERPILLR	88003.26
D2X2203	CATERPILLR	80047
D3	MOPAR	80092
D3	MOPAR	88001.76
D30	MOPAR	80168
D300	CHRYSLER	88000.14
D300	MOPAR	88000.14
D302	CHRYSLER	88000.90
D302	MOPAR	88000.90
D303	CHRYSLER	88000.22
D303	MOPAR	88000.22
D304	CHRYSLER	88000.25
D304	MOPAR	88000.25
D305	MOPAR	88000.41
D306	MOPAR	88000.29
D307	MOPAR	88000.19
D308	MOPAR	88000.68
D309	MOPAR	88000.70
D31	MOPAR	80174
D310	CHRYSLER	88000.71
D310	MOPAR	88000.71
D311	MOPAR	88000.72
D312	CHRYSLER	88000.73
D312	MOPAR	88000.73
D-313	MOPAR	88000.69
D313	MOPAR	88000.69
D313	MOPAR	88000.69
D-313	MOPAR	88000.69
D314	MOPAR	88000.74
D31437	CASE,J.I.	80174
D315	MOPAR	88000.75
D317	CHRYSLER	88000.85
D317	MOPAR	88000.85
D318	CHRYSLER	88000.20
D318	MOPAR	88000.20
D319	CHRYSLER	88000.28
D319	MOPAR	88000.28
D32	MOPAR	80162
D320	CHRYSLER	88000.30
D320	MOPAR	88000.30
D321	CHRYSLER	88000.35
D321	MOPAR	88000.35
D322	CHRYSLER	88000.37
D322	MOPAR	88000.37
D323	CHRYSLER	88000.38
D323	MOPAR	88000.38
D324	CHRYSLER	88000.43
D324	MOPAR	88000.43
D324	WESCO	80095
D-325	MOPAR	88000.58
D-325	MOPAR	88000.58
D325	MOPAR	88000.58
D325	MOPAR	88000.58
D325	WESCO	80096
D326	MOPAR	88000.21
D327	MOPAR	88000.49
D329	MOPAR	88000.23
D33	MOPAR	80173
D331	MOPAR	88000.87
D332	MOPAR	88000.32
D333	CHRYSLER	88000.80
D333	MOPAR	88000.80
D334	MOPAR	88000.48
D335	MOPAR	88000.92
D336	MOPAR	88000.48
D336	MOPAR	88000.91
D336	MOPAR	88000.91
D-336	MOPAR	88000.91
D-336	MOPAR	88000.91
D337	MOPAR	88000.40
D338	MOPAR	88000.94
D339	MOPAR	88000.47
D34	MOPAR	80177
D340	CHRYSLER	88000.95
D340	MOPAR	88000.95
D341	MOPAR	88000.93
D342	CHRYSLER	88000.89
D342	MOPAR	88000.89
D-343	MOPAR	88000.44
D-343	MOPAR	88000.44
D343	MOPAR	88000.44
D343	MOPAR	88000.44
D344	MOPAR	88000.79
D345	CHRYSLER	88000.81
D345	MOPAR	88000.81
D346	MOPAR	88000.82
D347	CHRYSLER	88000.83
D347	MOPAR	88000.83
D348	MOPAR	88000.51
D34948	CASE,J.I.	80047
D35	MOPAR	80193
D350	CHRYSLER	88000.45
D350	MOPAR	88000.45
D351	MOPAR	88000.34
D352	CHRYSLER	88000.46
D352	MOPAR	88000.46
D352	WESCO	88001.62
D353	CHRYSLER	88000.36
D353	MOPAR	88000.36
D354	MOPAR	88000.84
D354	WESCO	80095
D354	WESCO	88001.62
D36	MOPAR	80197
D37	MOPAR	80194
D37040	CASE,J.I.	80112
D372	WESCO	80135
D38	MOPAR	80198
D39	MOPAR	80199
D3H	SPICER	88013.44
D3HZ4635A	FORD	80210
D3HZ4635B	FORD	80212
D3HZ4635C	FORD	80231
D3HZ4635D	FORD	80237
D3J	SPICER	88013.45
D3K	SPICER	88013.42
D3SZ4635A	FORD	80081
D3SZ4635A	FORD	80091
D3TZ3249A	FORD	80079
D3TZ4635A	FORD	80189
D3TZ4635B	FORD	80221
D4	MOPAR	80049
D4	MOPAR	88001.96
D40	MOPAR	80201
D4010	WESCO	80153
D4015	REPUBLIC	88003.56
D41	MOPAR	80210
D41083	REPUBLIC	88003.47
D4141R	NEAPCO	80901
D4141X	NEAPCO	80901
D4141X	NEAPCO	88001.66
D4141XR	NEAPCO	80901
D4190	NEAPCO	88001.74
D41904	NEAPCO	88001.74
D4190R	NEAPCO	88001.74
D42	MOPAR	80213
D4200	WESCO	80958
D4200HD	WESCO	80958
D420111	REPUBLIC	88003.58
D420112	REPUBLIC	80953
D42012	REPUBLIC	88003.56
D42013	ALMETAL	88003.56
D-42013	REPUBLIC	88003.56
D42017	REPUBLIC	88003.52
D42017	REPUBLIC	88003.56
D42019	REPUBLIC	88003.56
D4201HD	WESCO	80958
D4202	WESCO	80958
D42021	REPUBLIC	88003.42
D4202AD	WESCO	80958
D4202HD	WESCO	80903
D4202HD	WESCO	80958

Part No.	Brand	IDLI Group	Part No.	Brand	IDLI Group	Part No.	Brand	IDLI Group	Part No.	Brand	IDLI Group
D42031	REPUBLIC	88003.43	D440811-1	REPUBLIC	88003.48	D56	MOPAR	80041	D-614	MOPAR	88007.59
D4204	REPUBLIC	88003.45	D440811-1	REPUBLIC	88003.51	D56	TEDDY TORQ	80145	D62	MOPAR	80189
D42042	REPUBLIC	88003.45	D44086	REPUBLIC	88003.47	D56-10	CLEVE.STL.	88015.52	D62	WESCO	80954
D42051	NEAPCO	88003.48	D4410	WESCO	80901	D56-104	CLEVE.STL.	88008.21	D625	WESCO	80064
D42051	REPUBLIC	88003.48	D4410	WESCO	88001.66	D56-1-1075	CLEVE.STL.	88005.09	D626	WESCO	80060
D420819	ALMETAL	88003.50	D4441X	NEAPCO	88001.67	D56-1-1215	CLEVE.STL.	88005.08	D62A	WESCO	80907
D-420819	REPUBLIC	88003.50	D445014	REPUBLIC	80955	D56-1-1220	CLEVE.STL.	88005.16	D63	MOPAR	80072
D420822-1	REPUBLIC	88003.50	D4490	NEAPCO	88001.75	D56-1-135-1	CLEVE.STL.	88005.10	D63	WESCO	80957
D420825-1	REPUBLIC	88003.50	D45	MOPAR	80238	D56-1-135-1A	CLEVE.STL.	88005.11	D63084	GALION	88001.28
D42083	ALMETAL	88003.47	D4500	WESCO	80956	D56-1-139-1	CLEVE.STL.	88005.22	D64	MOPAR	80089
D42083	NEAPCO	88003.47	D4510	WESCO	80909	D56-1-139-1A	CLEVE.STL.	88005.23	D65	MOPAR	80073
D42083	NEAPCO	88003.50	D45162	CASE,J.I.	80189	D56-1-141-1	CLEVE.STL.	88005.12	D65	MOPAR	80091
D42083	REPUBLIC	88003.47	D45166	CASE,J.I.	80182	D56-1-141-1A	CLEVE.STL.	88005.13	D65	WESCO	80951
D42086	REPUBLIC	88003.47	D4551XR	NEAPCO	80908	D56-1-142-1	CLEVE.STL.	88005.24	D65A	WESCO	80905
D42086	REPUBLIC	88003.50	D4556R	NEAPCO	80954	D56-1-142-1A	CLEVE.STL.	88005.25	D66	MOPAR	80225
D4210	WESCO	80909	D4557R	NEAPCO	80955	D56-1-1642	CLEVE.STL.	88005.15	D66	WESCO	80953
D4211	WESCO	80909	D4579	WESCO	80022	D56-120	CLEVE.STL.	87391	D67	WESCO	80953
D4212	WESCO	80902	D4579C	WESCO	80022	D56-125	CLEVE.STL.	88008.22	D67A	WESCO	80906
D4212	WESCO	80909	D45982	CASE,J.I.	80186	D56-12S	CLEVE.STL.	88014.92	D69	MOPAR	80077
D4220-1	CHRYSLER	80142	D46	MOPAR	80231	D56-13	CLEVE.STL.	88013.34	D6A	SPICER	88013.50
D4220HD	WESCO	88003.52	D4600	WESCO	80954	D56-1-631	CLEVE.STL.	88005.93	D6B	SPICER	88013.50
D4220HD	WESCO	88003.56	D46065	CASE,J.I.	80134	D56-1-642	CLEVE.STL.	88005.15	D6F	SPICER	88013.51
D4221	WESCO	88003.43	D4610	WESCO	80907	D56-1-645	CLEVE.STL.	88005.92	D6TJ4800CA	FORD	87458
D4222	WESCO	88003.42	D4625R	MCQUAY-NOR	80951	D56-17	CLEVE.STL.	88013.43	D70	MOPAR	80116
D4223	WESCO	88003.47	D4627	AEC	80010	D56-1-713	CLEVE.STL.	88005.07	D7004	WESCO	80195
D4223	WESCO	88003.50	D47	MOPAR	80232	D56-1-743	CLEVE.STL.	88005.14	D70496	GALION	80101
D4225	WESCO	88003.50	D4700	WESCO	80955	D56-1-785	CLEVE.STL.	88005.17	D70629	CASE,J.I.	80077
D4226	WESCO	88003.49	D4710	WESCO	80908	D56-1-887	CLEVE.STL.	88005.18	D70654(2)	GALION	80182
D4227	WESCO	88003.45	D48	MOPAR	80235	D56-2-3	CLEVE.STL.	88005.04	D71	MOPAR	80118
D4228	WESCO	88003.48	D48	MOPAR	80237	D56-43-102	CLEVE.STL.	86671	D72	MOPAR	80119
D4229	WESCO	88003.44	D49	MOPAR	80215	D56-43-321	CLEVE.STL.	88007.35	D72	MOPAR	80120
D4229-30	WESCO	88003.44	D4HZ4635A	FORD	80199	D56-43-926	CLEVE.STL.	86674	D7225	WESCO	80065
D4230	WESCO	88003.46	D4HZ4635B	FORD	80193	D56-54	CLEVE.STL.	85070	D7225	WESCO	88001.68
D4241X	NEAPCO	80909	D4J	SPICER	88013.47	D56-54-140	CLEVE.STL.	85071	D73	MOPAR	80077
D4241XP	NEAPCO	80909	D4NN4N167A	FORD	80142	D56-55-2	AEC	80145	D7325	WESCO	80095
D4241XR	NEAPCO	80909	D4TA4800AA	FORD	87375	D56-55-2	BORG-WARNR	80145	D7325	WESCO	80096
D4252	NEAPCO	80958	D4TA4800BA	FORD	87374	D56-55-2	CLEVELAND	80145	D7325	WESCO	88001.64
D4252R	NEAPCO	80903	D4TJ4800AA	FORD	87377	D56-55-2	CLEVE.STL.	80145	D7325	WESCO	88001.69
D4252R	NEAPCO	80958	D4TJ4800BA	FORD	87332	D56-55-2	REPCO	80145	D73644	CASE,J.I.	80171
D4252XR	NEAPCO	80958	D4TZ4800A	FORD	87331	D56-6	CLEVE.STL.	88014.81	D73645	CASE,J.I.	80100
D4255R	NEAPCO	80956	D4TZ4800B	FORD	87331	D56-7-138	CLEVE.STL.	87390	D73646	CASE,J.I.	80101
D4257-1	NEAPCO	88003.44	D4TZ4800C	FORD	87382	D56-884	CLEVE.STL.	88005.21	D73649	CASE,J.I.	80077
D4257-1	REPUBLIC	88003.44	D4TZ4800D	FORD	87331	D56-888	CLEVE.STL.	88005.19	D73653	CASE,J.I.	80198
D4257-2	NEAPCO	88003.44	D4TZ4800D	FORD	87382	D56-888A	CLEVE.STL.	88005.20	D74	MOPAR	80182
D4257R	NEAPCO	80955	D4TZ4800E	FORD	87458	D56-9	CLEVE.STL.	88013.61	D74601	GALION	80182
D4257XR	NEAPCO	80908	D4TZ4800F	FORD	87457	D57	MOPAR	80023	D75	MOPAR	80142
D4259-1	NEAPCO	88003.49	D4TZ4800G	FORD	87481	D57	MOPAR	80024	D75	MOPAR	80143
D42591	REPUBLIC	88003.49	D4TZ4800G	FORD	88007.77	D5A	SPICER	88013.49	D75441	GALION	80156
D4261	NEAPCO	88003.46	D4TZ4800H	FORD	87481	D5TZ4800A	FORD	87376	D75442	GALION	80198
D4261	REPUBLIC	88003.46	D4TZ4800H	FORD	88007.77	D6	MOPAR	80048	D76	MOPAR	80143
D4262R	NEAPCO	80954	D4TZ4800J	FORD	88007.79	D6	MOPAR	88002.10	D76-10	CLEVE.STL.	88015.30
D4262XR	NEAPCO	80907	D4TZ4800K	FORD	87570	D60	MOPAR	80031	D76-101	CLEVE.STL.	88008.14
D4263R	NEAPCO	80957	D4TZ4800K	FORD	88007.79	D600	MOPAR	87238	D76-10-1	CLEVE.STL.	88015.33
D4265R	NEAPCO	80951	D5	MOPAR	80044	D-601	MOPAR	87331	D76-102	CLEVE.STL.	88008.15
D4265XP	NEAPCO	80905	D5	MOPAR	88001.98	D601	MOPAR	88008.39	D76-11	CLEVE.STL.	88015.56
D4265XR	NEAPCO	80905	D50	MOPAR	80078	D602	MOPAR	87326	D76-1-100	CLEVE.STL.	88005.06
D4267R	NEAPCO	80953	D50	MOPAR	80090	D603	MOPAR	87331	D76-11-1	CLEVE.STL.	88015.59
D4267XR	NEAPCO	80906	D50866	CASE,J.I.	80215	D604	MOPAR	87285	D76-1-129	CLEVE.STL.	88005.26
D428	NEAPCO	88003.50	D51	MOPAR	80093	D605	MOPAR	87377	D76-125	CLEVE.STL.	88008.19
D43	MOPAR	80211	D51104	GALION	80001	D606	MOPAR	87383	D76-126	CLEVE.STL.	88008.20
D4300	WESCO	88001.75	D52	MOPAR	80023	D607	MOPAR	87337	D76-17	CLEVE.STL.	88013.44
D4301	REPUBLIC	88003.57	D5200HD	WESCO	88001.73	D608	MOPAR	87457	D76-17-1	CLEVE.STL.	88013.45
D-4303-1	REPUBLIC	88003.43	D525	WESCO	80096	D609	MOPAR	87459	D76-2	CLEVE.STL.	84616
D4310	WESCO	88001.67	D526	WESCO	80064	D61	MOPAR	80178	D76-3-4-14	CLEVE.STL.	85524
D44	MOPAR	80212	D52A	DETROIT	80906	D610	MOPAR	87399	D76-3-4-14D	CLEVE.STL.	88005.05
D4400	WESCO	88001.74	D53	MOPAR	80234	D611	MOPAR	87481	D76-43-158	CLEVE.STL.	86672
D44015	REPUBLIC	80954	D54	MOPAR	80233	D612	MOPAR	87570	D76-43-159	CLEVE.STL.	86673
D44016	REPUBLIC	80957	D54	MOPAR	80241	D613	MOPAR	87373	D76462	GALION	80193
D44023	ALMETAL	88003.51	D55	MOPAR	80234	D-613	MOPAR	88008.43	D76-54	CLEVE.STL.	85068
D-44023	REPUBLIC	88003.51	D55145	GALION	80001	D614	MOPAR	87325	D76-7-4-14N	CLEVE.STL.	88014.75

Part No.	Brand	IDLI Group	Part No.	Brand	IDLI Group	Part No.	Brand	IDLI Group	Part No.	Brand	IDLI Group
D76-9	CLEVE.STL.	88013.54	D9TZ4800A	FORD	87331	DS101	NEAPCO	88008.35	DS203KC	CLEVE.MOT.	87285
D76949	CASE,J.I.	80174	DC126N	ROCKWELL	88013.37	DS101-1	NEAPCO	88008.35	DS203KC	NEAPCO	87285
D76950	CASE,J.I.	80198	DC131N	ROCKWELL	88013.38	DS101-2	NEAPCO	88008.35	DS203KC	NEAPCO	88008.48
D77	MOPAR	80190	DC131N1	ROCKWELL	88013.39	DS101K	CLEVE.MOT.	87238	DS203KC	NEAPCO	88008.49
D77092	GALION	80046	DC135N	ROCKWELL	88013.40	DS101K	CMP	87238	DS204BF	NEAPCO	88008.55
D77092	GALION	80092	DC141N	ROCKWELL	88013.41	DS101K	NEAPCO	87238	DS204KF	CLEVE.MOT.	87373
D77093	GALION	80047	DC141N1	ROCKWELL	88013.42	DS101K	NEAPCO	88007.58	DS204KF	CMP	87373
D78	MOPAR	80007	DC148N	ROCKWELL	88013.43	DS101K2	CMP	87238	DS204KF	CMP	87375
D78942	GALION	80190	DC148N1	ROCKWELL	88013.44	DS101K2	NEAPCO	87238	DS204KF	NEAPCO	80147
D79	MOPAR	80221	DC148N2	ROCKWELL	88013.45	DS11	MOTOR MAST	87373	DS204KF	NEAPCO	87373
D7TZ4800A	FORD	87331	DC155N	ROCKWELL	88013.46	DS12	DORMAN	87326	DS204KF	NEAPCO	88008.43
D80	MOPAR	80124	DC155N1	ROCKWELL	88013.47	DS12	MOTOR MAST	87326	DS204KF1	CMP	87373
D8025	AEC	80004	DC155N2	ROCKWELL	88013.48	DS13B	DORMAN	87238	DS204KF1	NEAPCO	87374
D80321(7)	GALION	80198	DC16N	ROCKWELL	88013.49	DS13B	MOTOR MAST	87238	DS204KF1	NEAPCO	88007.67
D80329	CASE,J.I.	80213	DC17N	ROCKWELL	88013.50	DS13B	WESCO	87238	DS204KF2	CMP	87375
D80481	GALION	80189	DC17N2	ROCKWELL	88013.51	DS14	DORMAN	87332	DS204KF2	NEAPCO	87375
D80485(2)	GALION	80189	DC17N3	ROCKWELL	88013.52	DS14	MOTOR MAST	87331	DS204KF2	NEAPCO	88007.66
D81	MOPAR	80125	DCB100	CLEVE.STL.	88008.53	DS145	MOTOR MAST	87331	DS204KF3	CMP	87373
D82	MOPAR	80023	DCB101	CLEVE.STL.	87377	DS14S	MOTOR MAST	87331	DS204KF3	CMP	87375
D82	MOPAR	80024	DCB102	CLEVE.STL.	87380	DS15	DORMAN	87331	DS204KF3	NEAPCO	87376
D82	MOPAR	80140	DCB103	CLEVE.STL.	87381	DS15	MOTOR MAST	87331	DS204KF3	NEAPCO	88007.65
D82809	GALION	80101	DCB104	CLEVE.STL.	88008.54	DS15655	PILOT	80145	DS205KG	CLEVE.MOT.	87325
D83	MOPAR	80140	DCB105	CLEVE.STL.	87383	DS16	DORMAN	87377	DS205KG	CMP	87325
D83	MOPAR	80182	DCB106	CLEVE.STL.	87378	DS16	MOTOR MAST	87377	DS205KG	NEAPCO	87325
D84	MOPAR	80076	DCB107	CLEVE.STL.	87379	DS17	DORMAN	87383	DS205KG	NEAPCO	88007.59
D84A	MOPAR	80075	DCB108	CLEVE.STL.	87382	DS17	MOTOR MAST	87383	DS206KF	NEAPCO	87331
D85	MOPAR	80117	DCB109	CLEVE.STL.	87384	DS18	DORMAN	87457	DS206KF1	NEAPCO	87331
D87	MOPAR	80121	DCB1-4A	CLEVE.STL.	88014.86	DS18	MOTOR MAST	87457	DS206KF2	NEAPCO	87331
D87263	CASE,J.I.	80101	DCB2	CLEVE.STL.	88013.76	DS19	DORMAN	87459	DS207KF	NEAPCO	87326
D88	MOPAR	80081	DCB2-1	CLEVE.STL.	88013.66	DS19	MOTOR MAST	87459	DS207KF1	NEAPCO	87331
D88	MOPAR	80123	DCB2-2	CLEVE.MOT.	88013.77	DS20	DORMAN	87481	DS21	DORMAN	88007.79
D89	MOPAR	80091	DCB2-3	CLEVE.MOT.	88013.67	DS20	DORMAN	88007.77	DS21	MOTOR MAST	87570
D8QH4635C	FORD	80238	DCB50-1	CLEVE.STL.	88008.64	DS20	MOTOR MAST	87481	DS21S	MOTOR MAST	88007.78
D8RZ3249B	FORD	86077	DCB61	CLEVE.STL.	88007.82	DS201BF	NEAPCO	87331	DS22	DORMAN	87285
D8RZ3A331A	FORD	86126	DCD24-1	ROCKWELL	88013.53	DS201BF	NEAPCO	88008.47	DS22	MOTOR MAST	87285
D8RZ3A331A	FORD	86182	DCD25-1	ROCKWELL	88013.54	DS201BG	NEAPCO	88008.50	DS23	DORMAN	87337
D8RZ3A331B	FORD	86077	DCD28-1	ROCKWELL	88013.55	DS201KDS	NEAPCO	87327	DS23	MOTOR MAST	87337
D8RZ3A331B	FORD	86150	DCR20-2	ROCKWELL	88013.56	DS201KDS	NEAPCO	87331	DS24	DORMAN	87399
D8RZ3A426B	FORD	86010	DCS22-1	ROCKWELL	88013.57	DS201KF	CLEVE.MOT.	87331	DS24	MOTOR MAST	87399
D8RZ3K187A	FORD	86126	DCS22-2	ROCKWELL	88013.58	DS201KF	CMP	87331	DS240KF	CMP	87373
D8TZ3249A	FORD	80139	DCS24-1	ROCKWELL	88013.59	DS201KF	NEAPCO	87327	DS25	DORMAN	87373
D8TZ4800A	FORD	87331	DCS24-2	ROCKWELL	88013.60	DS201KF	NEAPCO	87331	DS26	DORMAN	87325
D8Z	MOPAR	88008.68	DCS26-1	ROCKWELL	88013.61	DS201KF	NEAPCO	88008.39	DS301BF	NEAPCO	88008.53
D9	MOPAR	80054	DCS28-1	ROCKWELL	88013.62	DS201KF1	CMP	87331	DS301BG	NEAPCO	88008.54
D90	MOPAR	80025	DCS30-1	ROCKWELL	88013.63	DS201KF1	NEAPCO	87328	DS301KF	CLEVE.MOT.	87377
D91	MOPAR	80130	DEFR22-4	ROCKWELL	88013.64	DS201KF1	NEAPCO	88007.60	DS301KF	CMP	87377
D91128	GALION	80001	DEFR22-5	ROCKWELL	88013.65	DS201KF2	CMP	87331	DS301KF	CMP	87382
D92	MOPAR	80129	DEFR25-1	ROCKWELL	88013.66	DS201KF2	NEAPCO	88007.63	DS301KF	NEAPCO	87377
D93	MOPAR	80131	DEFR25-2	ROCKWELL	88013.67	DS201KF4	CMP	87331	DS301KF	NEAPCO	87382
D94	MOPAR	80036	DEFR25-4	ROCKWELL	88013.68	DS201KF4	NEAPCO	87332	DS301KF	NEAPCO	88008.41
D94	MOPAR	80044	DEFR28-3	ROCKWELL	88013.69	DS201KF4	NEAPCO	88007.64	DS301KF1	CMP	87382
D95	MOPAR	80079	DEFR28-4	ROCKWELL	88013.70	DS201KG	CLEVE.MOT.	87326	DS301KF1	NEAPCO	87380
D96	TEDDY TORQ	80160	DEFR30-1	ROCKWELL	88013.71	DS201KG	CMP	87326	DS301KF1	NEAPCO	88007.69
D96-10	CLEVE.STL.	88015.50	DEFR32-3	ROCKWELL	88013.72	DS201KG	CMP	87330	DS301KF2	CMP	87377
D96-13	CLEVE.STL.	88013.33	DEFR33-1	ROCKWELL	88013.73	DS201KG	NEAPCO	87326	DS301KF2	NEAPCO	87381
D96-23	CLEVE.STL.	88014.83	DEFR34-1	ROCKWELL	88013.74	DS201KG	NEAPCO	87330	DS301KF2	NEAPCO	88007.70
D96-55	AEC	80160	DEFR35-1	ROCKWELL	88013.75	DS201KG	NEAPCO	88008.40	DS301KF3	CMP	87377
D96-55	ALCO	80160	DEFR37-1	ROCKWELL	88013.76	DS201KG1	CMP	87326	DS301KF3	CMP	87382
D96-55	BORG-WARNR	80160	DEFR40-5	ROCKWELL	88013.77	DS201KG1	NEAPCO	87329	DS301KF3	NEAPCO	87382
D96-55	CLEVE.STL.	80160	DEFR41-1	ROCKWELL	88013.78	DS201KG1	NEAPCO	88007.61	DS301KF3	NEAPCO	88007.72
D96-55	REPCO	80160	DEFR42-1	ROCKWELL	88013.79	DS201KG2	CMP	87330	DS301KG	CLEVE.MOT.	87383
D96-55	SPICER	80160	DEFR55-1	ROCKWELL	88013.80	DS201KG2	NEAPCO	87330	DS301KG	CMP	87384
D96-55-4	CLEVE.STL.	80160	DEFR56-2	ROCKWELL	88013.81	DS201KG2	NEAPCO	88007.62	DS301KG	NEAPCO	87383
D96-9	CLEVE.STL.	88013.59	DH4635A	FORD	80182	DS201KG3	CMP	87326	DS301KG	NEAPCO	87384
D9HZ4635A	FORD	80182	DIA., REO, WHITE	MCQUAY-NOR	80195	DS201KG3	CMP	87330	DS301KG	NEAPCO	88008.42
D9HZ4635B	FORD	80226	DN	WEASLER	80069	DS201R	NEAPCO	88008.63	DS301KG1	CMP	87384
D9HZ4800A	FORD	87461	DS10	DORMAN	87238	DS202KD	CLEVE.MOT.	87331	DS301KG1	NEAPCO	87378
D9HZ4800A	FORD	88007.76	DS10	MOTOR MAST	88008.35	DS202KD	CMP	87332	DS301KG1	NEAPCO	88007.68
D9TZ4782A	FORD	88002.38	DS101	CLEVE.MOT.	88008.35	DS202KD	NEAPCO	87331	DS301KG2	CMP	87384
D9TZ4782Z	FORD	88002.38	DS101	CMP	87238	DS202KD1	NEAPCO	87331	DS301KG2	NEAPCO	87379

Part No.	Brand	IDLI Group
DS301KG2	NEAPCO	88007.71
DS301KG3	CMP	87384
DS301R	NEAPCO	88008.64
DS303KC	CLEVE.MOT.	87337
DS303KC	NEAPCO	87337
DS303KC	NEAPCO	88008.51
DS303KC	NEAPCO	88008.52
DS401BF	NEAPCO	88008.56
DS401BG	NEAPCO	88008.57
DS401KF	CLEVE.MOT.	87457
DS401KF	CMP	87460
DS401KF	CMP	87461
DS401KF	NEAPCO	87458
DS401KF	NEAPCO	88008.31
DS401KF	NEAPCO	88008.44
DS401KF1	CMP	87461
DS401KF1	NEAPCO	87458
DS401KF1	NEAPCO	88007.74
DS401KF2	CMP	87460
DS401KF2	NEAPCO	87327
DS401KF2	NEAPCO	87460
DS401KF2	NEAPCO	88007.75
DS401KF3	CMP	87460
DS401KF3	CMP	87461
DS401KF3	NEAPCO	87461
DS401KF3	NEAPCO	88007.76
DS401KG	CLEVE.MOT.	87459
DS401KG	CMP	87459
DS401KG	NEAPCO	87459
DS401KG	NEAPCO	88008.45
DS401KG1	NEAPCO	87459
DS401R	NEAPCO	88008.65
DS403KC	CLEVE.MOT.	87399
DS403KC	NEAPCO	87399
DS403KC	NEAPCO	88008.58
DS403KC	NEAPCO	88008.59
DS501BF	NEAPCO	88008.60
DS501KF	CLEVE.MOT.	87481
DS501KF	CMP	87481
DS501KF	CMP	88007.77
DS501KF	NEAPCO	87481
DS501KF	NEAPCO	88007.77
DS501R	NEAPCO	88008.66
DS55-55-2	ALCO	80145
DS55-55-2	BORG-WARNR	80145
DS55-55-2	CLEVE.STL.	80145
DS56-55-2	AEC	80145
DS56-55-2	BORG-WARNR	80145
DS56-55-2	CLEVE.STL.	80145
DS56-55-2	REPCO	80145
DS601KF	CLEVE.MOT.	87570
DS601KF	CMP	87570
DS601KF	CMP	88007.79
DS601KF	NEAPCO	80146
DS601KF	NEAPCO	87570
DS601KF	NEAPCO	88007.79
DS601KFS	CMP	88008.36
DS601KFS	NEAPCO	88007.78
DS601R	NEAPCO	88008.67
DS701KF	CMP	88007.79
DS9	MOTOR MAST	87325
DSTZ4800A	FORD	87373
DSVZ4800A	FORD	87331
E0A7039	FORD	80041
E0TZ3249A	FORD	80128
E1F2-3A331B	FORD	86076
E1FZ3A331A	FORD	86127
E1FZ3A331A	FORD	86182
E1FZ3A331A	FORD	86183
E1FZ3A331A	FORD	86299
E1FZ3A331B	FORD	86149
E1FZ3A331B	FORD	86295
E1FZ3A331B	FORD	86296
E25435	EATON	80031
E2GZ3B413A	FORD	86127
E2GZ3B413B	FORD	86127
E2GZ3B414A	FORD	86076
E2GZ3B414B	FORD	86149
E2GZ3B414B	FORD	86295
E2GZ3B436A	FORD	86008
E2GZ3B436B	FORD	86009
E3FZ3B413B	FORD	86182
E3FZ3B413B	FORD	86299
E3FZ3B414C	FORD	86296
E3FZ3B436C	FORD	86288
E3FZ3B436D	FORD	86284
E3FZ3B436E	FORD	86288
E3FZ3B436F	FORD	86286
E3FZ3B436G	FORD	86285
E3FZ3B436H	FORD	86287
E3TZ3249A	FORD	80028
E3TZ4635B	FORD	80027
E3TZ4782A	FORD	88002.39
E3TZ7A399B	FORD	80029
E43Z3A331A	FORD	86300
E43Z3A331A	FORD	86302
E43Z3A331B	FORD	86297
E43Z3A331B	FORD	86298
E43Z3A331B	FORD	86301
E43Z3B413B	FORD	86300
E43Z3B413B	FORD	86302
E43Z3B414B	FORD	86297
E43Z3B414B	FORD	86301
E43Z3B414C	FORD	86298
E43Z3B436A	FORD	86290
E43Z3B436B	FORD	86289
E43Z3B436C	FORD	86292
E43Z3B436D	FORD	86291
E4GZ3B436C	FORD	86294
E4GZ3B436D	FORD	86293
E540131	AMER.HOIST	80001
E56241	GMC	80021
ECN2-4392	CHAIN BELT	80238
ECN2-4392	REX CHAIN	80238
EOA7039	FORD	80041
EY100	MUNCIE	88000.19
EY101	MUNCIE	88000.20
EY102	MUNCIE	88000.21
EY102	MUNCIE	88000.22
EY103	MUNCIE	88000.44
EY110	MUNCIE	88000.22
EY111	MUNCIE	88000.23
EY112	MUNCIE	88000.22
EY12	MUNCIE	88000.25
EY12	MUNCIE	88000.27
EY120	MUNCIE	88000.25
EY121	MUNCIE	88000.27
EY122	MUNCIE	88000.27
EY123	MUNCIE	88000.52
EY130	MUNCIE	88000.28
EY131	MUNCIE	88000.29
EY131	MUNCIE	88001
EY132	MUNCIE	88000.95
EY140	MUNCIE	88000.30
EY140	MUNCIE	88000.31
EY141	MUNCIE	88000.31
EY142	MUNCIE	88000.32
EY143	MUNCIE	88000.53
EY150	MUNCIE	88000.34
EY160	MUNCIE	88000.35
EY161	MUNCIE	88000.36
EY162	MUNCIE	88000.35
EY163	MUNCIE	88000.22
EY170	MUNCIE	88000.37
EY180	MUNCIE	88000.38
EY181	MUNCIE	88000.39
EY183	MUNCIE	88000.55
EY184	MUNCIE	88000.40
EY190	MUNCIE	88000.41
EY202	MUNCIE	88000.43
EY203	MUNCIE	88000.57
EY210	MUNCIE	88000.44
EY-210	MUNCIE	88000.44
EY-210	MUNCIE	88000.44
EY210	MUNCIE	88000.44
EY220	MUNCIE	88000.45
EY221	MUNCIE	88000.45
EY230	MUNCIE	88000.58
EY240	MUNCIE	88000.46
EY250	MUNCIE	88000.60
EY325	MUNCIE	88000.47
EY330	MUNCIE	88000.62
EY340	MUNCIE	88000.43
EY340	MUNCIE	88000.43
EY-340	MUNCIE	88000.43
EY-340	MUNCIE	88000.43
EY340	MUNCIE	88000.99
EY345	MUNCIE	88000.61
EY370	MUNCIE	88000.48
EY380	MUNCIE	88000.49
EY390	MUNCIE	88000.50
EY395	MUNCIE	88000.51
EY400	MUNCIE	88000.68
EY405	MUNCIE	88000.70
EY410	MUNCIE	88000.71
EY415	MUNCIE	88000.72
EY416	MUNCIE	88000.72
EY420	MUNCIE	88000.73
EY421	MUNCIE	88000.73
EY430	MUNCIE	88000.77
EY450	MUNCIE	88000.74
EY451	MUNCIE	88000.74
EY460	MUNCIE	88000.75
EY470	MUNCIE	88000.71
EY550	MUNCIE	88000.87
EY720	MUNCIE	88000.67
EY721	MUNCIE	88000.66
EY-722	MUNCIE	88000.51
EY725	MUNCIE	88000.64
F00200	NEAPCO	88001.28
F1	TEDDY TORQ	88002.55
F100	REPUBLIC	88002.55
F111	ALMETAL	88002.54
F111	REPUBLIC	88002.54
F115	REPUBLIC	88002.53
F13270	CASE,J.I.	80037
F164369	AMER.MOTOR	80072
F42A7039A	FORD	80024
F4CX10	FOX RIVER	80134
F4CX5	FOX RIVER	80069
F4CX5	FOX RIVER	88002.95
F4CX8	FOX RIVER	80069
F4CX8B	FOX RIVER	80161
F4CX8BW	FOX RIVER	80161
F4CX8H	FOX RIVER	80069
F50537	CASE,J.I.	80001
F6527-2A	PETTIBONE	80104
F85470	CASE,J.I.	80037
F9E7039A	FORD	80077
FB13278	CASE,J.I.	84510
FEX1707	ALLIS-CHLM	80210
FEX1707	FIAT-ALLIS	80215
FFH788	COCKSHUTT	80001
FH1131	COCKSHUTT	88002.95
FH756924	CASE,J.I.	80037
FH845032	CASE,J.I.	80035
FHH788	COCKSHUTT	80001
FHY116	COCKSHUTT	80069
FHY169	COCKSHUTT	80001
FHY289	COCKSHUTT	80008
FHY342	COCKSHUTT	80035
FL4635A	FORD	80077
FM4635A	FORD	80077
FM4635B	FORD	80124
G1000	GMB	80072
G1003	GMB	80001
G1003	GMB	80044
G1003	GMB	80054
G1003	GMB	88003.38
G1004	GMB	80001
G1004	GMB	80044
G1007	GMB	80001
G1009	GMB	80001
G12000	GMB	80234
G12002	GMB	80234
G12100	GMB	80049
G1328	PERFECTION	88002.55
G1501	GMB	80021
G1501	GMB	80023
G1501	GMB	80035
G1501	GMB	80044
G1501	GMB	80163
G1501	GMB	80167
G1504	GMB	80021
G1506	GMB	80021
G1509	GMB	80021
G1514	GMB	80021
G1516	WESCO	80189
G1517	WESCO	80221
G1520R	WESCO	80075
G1520RU	WESCO	80076
G1521R	WESCO	80072
G1530	WESCO	80129
G1531	WESCO	80121
G1531-1	WESCO	80131
G1550	WESCO	80119
G1552	WESCO	80118
G1554	WESCO	80081
G1554RU	WESCO	80081
G1560R	WESCO	80142
G1578R	WESCO	80124
G1578RU	WESCO	80125
G1579R	WESCO	80073
G1602	GMB	80197
G1602	PERFECTION	88002.56
G1661	GMB	80072
G1661	GMB	88002.87
G1661	WESCO	88002.87
G1662	GMB	88002.86
G1750	GMB	80030
G1750	TEDDY TORQ	80030
G1750G	GMB	80030
G1750G	GMB	80031
G1750G	GMB	88002.77
G1755	GMB	80050
G1755G	GMB	88002.81
G1762	GMB	80031
G1767	GMB	80031
G2007	PERFECTION	88002.57
G2008G	WESCO	80104

Part No.	Brand	IDLI Group
G2013	GMB	80049
G20139	NORTH AMER	80035
G2017	GMB	80058
G2017	GMB	80092
G2019	GMB	80049
G2045G	WESCO	88002.06
G2046-1	WESCO	80043
G2046G	WESCO	80044
G2053	WESCO	80031
G2053-1	WESCO	80030
G2100	GMB	80049
G2100	TEDDY TORQ	80049
G2101	GMB	80044
G2101	TEDDY TORQ	80044
G2102	GMB	80044
G2103	GMB	80048
G2103	GMB	88002.10
G2103	TEDDY TORQ	80048
G2104	GMB	80047
G2104	GMB	80100
G210782X	GMB	88002.37
G2109	GMB	80049
G2109	TEDDY TORQ	80049
G2110	GMB	80047
G2110	GMB	80048
G2110	GMB	80049
G2110	GMB	80218
G2111	GMB	80049
G2112	GMB	80044
G2113	GMB	80049
G2113	GMB	80053
G2114	GMB	80054
G2114	TEDDY TORQ	80054
G2116	GMB	80047
G2117	GMB	80046
G2121	GMB	80043
G2121	TEDDY TORQ	80044
G2123	GMB	80045
G2125	GMB	80047
G2126	GMB	80049
G2126	GMB	80054
G2127	GMB	80044
G2127	WESCO	80047
G2131	GMB	80044
G2134	GMB	80044
G2140	GMB	80055
G2140	TEDDY TORQ	80055
G2140	WESCO	80055
G2150	GMB	80049
G2153	GMB	80055
G21625-3	GMB	80061
G217	GMB	80046
G2173	GMB	80058
G2173	GMB	80066
G2173	WESCO	80058
G2175	GMB	80058
G2198	GMB	80051
G2198	TEDDY TORQ	80051
G2216	NORTH AMER	80008
G2221	PERFECTION	88002.58
G2500	GMB	80059
G2503	GMB	80030
G2505	GMB	80059
G2505	GMB	80067
G2518	GMB	80066
G252	WESCO	80034
G2530	GMB	80066
G2531	GMB	80066
G2531	PERFECTION	88002.60
G255	WESCO	80061

Part No.	Brand	IDLI Group
G2801	GMB	80232
G285	WESCO	80132
G2895	PERFECTION	88002.48
G2896	PERFECTION	88002.50
G2897	PERFECTION	88002.51
G2898	PERFECTION	88002.52
G2910	PERFECTION	88002.43
G2911	PERFECTION	88002.42
G2912	PERFECTION	88002.44
G3000G	WESCO	80099
G3008	GMB	80104
G3008G	WESCO	80104
G3009	WESCO	80105
G3011	GMB	80107
G3011	WESCO	80107
G3012G	WESCO	80110
G3013	GMB	80106
G3013	GMB	80112
G3100	GMB	80101
G3101	GMB	80100
G31011	GMB	80100
G3102	GMB	80112
G3103	GMB	80098
G3103	GMB	80112
G3105	GMB	80100
G3105	GMB	80104
G3105	GMB	80107
G3105	GMB	80112
G3106	GMB	80099
G3108	GMB	80112
G3111	GMB	80104
G3111	GMB	88003.41
G3113	GMB	80112
G3114	GMB	80104
G3115	GMB	80104
G3116	GMB	80106
G3116	GMB	80111
G3118	GMB	80099
G3118	GMB	80101
G3118	GMB	80104
G3119	GMB	80100
G3119	GMB	80101
G3120	GMB	80113
G3-1204UB	GMB	88002.13
G3121	GMB	80102
G3-121UB	GMB	80076
G3-121XU	GMB	88013.86
G3123	GMB	80102
G3125	GMB	80103
G3155	GMB	80103
G3-178UB	GMB	88013.87
G3-178XU	GMB	88013.87
G3188	GMB	80107
G3-266S	GMB	88001.70
G3-3105S	GMB	88001.94
G3501	PERFECTION	88002.63
G3502	PERFECTION	88002.63
G3506	PERFECTION	88002.67
G3507	PERFECTION	88002.71
G3508	PERFECTION	88002.72
G3509	PERFECTION	88002.70
G35N	G & G MFG	80069
G4007	GMB	80147
G4008	GMB	80146
G4008	TEDDY TORQ	80146
G4011	GMB	80147
G4029	GMB	80109
G4100	GMB	80109
G4100	TEDDY TORQ	80109
G4101	GMB	80153

Part No.	Brand	IDLI Group
G4101	TEDDY TORQ	80153
G4102	GMB	80158
G4103	GMB	80150
G4104	GMB	80151
G4104	GMB	80156
G4105	GMB	80150
G4106	GMB	80148
G4106	TEDDY TORQ	80148
G4107	GMB	80159
G4107	TEDDY TORQ	80159
G4109	GMB	80157
G4109	GMB	80159
G4110	GMB	80109
G4110	GMB	80155
G4113	GMB	80155
G4113	GMB	80156
G4113	GMB	80158
G4113	GMB	88003.16
G4115	GMB	80150
G4115	GMB	80153
G4118	GMB	80104
G4120	GMB	80109
G4120	GMB	80156
G4122	GMB	80109
G4123	GMB	88003.14
G4130	GMB	80155
G4133	GMB	80154
G4138	GMB	80109
G4138	GMB	88003.15
G4151	GMB	80109
G4151	GMB	80149
G4152	GMB	80158
G4152	GMB	88003.13
G4156	GMB	80157
G4158	GMB	80156
G4196	GMB	80148
G4295	PERFECTION	88002.43
G4295	PERFECTION	88002.62
G499	GMB	88000.15
G4C96-55	GMB	80204
G5006	GMB	80163
G5006	GMB	80167
G5007	WESCO	80177
G5008	WESCO	80174
G5015	WESCO	80173
G5031	GMB	80171
G5031	GMB	80234
G5100	GMB	80168
G5-10003	GMB	80245
G5-1004	GMB	80001
G5-100X	GMB	80030
G5-101	GMB	80023
G5101	GMB	80167
G5-101X	GMB	80023
G5-102	GMB	80024
G5102	GMB	80171
G5102	GMB	80238
G5103	GMB	80174
G5-103X	GMB	88000.14
G5-103X	GMB	88000.15
G5-104	GMB	80037
G5-105	GMB	80031
G5105	GMB	80167
G5-105X	CENTRAL	80030
G5-105X	GMB	80030
G5-105X	GMB	80031
G5-107	GMB	80185
G5108	GMB	80172
G5109	GMB	80162
G5110	GMB	80173

Part No.	Brand	IDLI Group
G5-110X	GMB	88000.16
G5111	GMB	80173
G5111	GMB	80175
G5-111X	GMB	80024
G5112	GMB	80021
G5112	GMB	80113
G5112	GMB	80173
G5112	GMB	80177
G5114	GMB	80176
G5-115	GMB	80178
G5-115X	GMB	80178
G5-116	GMB	80189
G5-116	TEDDY TORQ	80189
G5-116X	CENTRAL	80189
G5-116X	CENTRAL	80189
G5-117	GMB	80221
G5118	GMB	80216
G5-1200	GMB	80081
G5-1200X	CENTRAL	80081
G5-1200X	CENTRAL	80081
G5-1201	GMB	80091
G5-1201	TEDDY TORQ	80091
G5-1201X	CENTRAL	80091
G5-1201X	GMB	80091
G5-1202	GMB	80130
G5-1202	TEDDY TORQ	80130
G5-1202X	GMB	80130
G5-1203	GMB	80129
G5-1203	TEDDY TORQ	80129
G5-1203X	CENTRAL	80129
G5-1203X	GMB	80129
G5-1204	GMB	80131
G5-1204	TEDDY TORQ	80131
G5-1204X	CENTRAL	80131
G5-1204X	GMB	80131
G5-1205X	GMB	80123
G5-1206X	GMB	80141
G5-1207X	GMB	80071
G5-1207X	GMB	80074
G5-1208X	GMB	80071
G5-1208X	GMB	80074
G5-121	GMB	80072
G5121	GMB	80168
G5121	GMB	80216
G5-121	TEDDY TORQ	80072
G5-1210	GMB	80091
G5-1210X	GMB	80183
G5-1211X	GMB	80138
G5-1214X	GMB	80068
G5-1214X	GMB	80145
G5-121B	GMB	80075
G5-121B	GMB	80089
G5-121B	GMB	88001.82
G5-121UB	GMB	80088
G5-121X	CENTRAL	80075
G5-121X	GMB	80072
G5-121X	GMB	80075
G5-121XU	GMB	80075
G5-121XU	GMB	80089
G5-122	GMB	80073
G5122	GMB	80164
G5122	GMB	80168
G5-122	TEDDY TORQ	80073
G5-1226X	GMB	80047
G5-123	GMB	80131
G5-1230X	GMB	80122
G5-1231X	GMB	80160
G5-1232	GMB	80181
G5-1232X	GMB	80181
G5-1233X	GMB	80038

Part No.	Brand	IDLI Group	Part No.	Brand	IDLI Group	Part No.	Brand	IDLI Group	Part No.	Brand	IDLI Group
G5-1239X	GMB	80038	G5146	GMB	80172	G5-188	TEDDY TORQ	80140	G5-266X	GMB	80225
G5-124	GMB	80225	G5148	GMB	80168	G5-188X	CENTRAL	80140	G5-267	GMB	80060
G5-124	TEDDY TORQ	80225	G5148	GMB	80169	G5-188X	GMB	80140	G5-278X	GMB	80085
G5-1241X	GMB	80145	G5148	GMB	80216	G52	PERFECTION	88001.73	G5-279X	GMB	80189
G5-124T	GMB	80225	G5149	GMB	80171	G5-200	GMB	80075	G5-280X	GMB	80221
G5-124X	GMB	80225	G5-14X	GMB	80024	G5-200	TEDDY TORQ	80075	G5-281X	GMB	80227
G5125	GMB	80238	G5-150	GMB	80119	G5-2001X	GMB	88003.26	G5-2907X	GMB	80079
G5-1253X	GMB	80038	G5-150	TEDDY TORQ	80119	G5-2002X	GMB	80048	G5-297X	GMB	80079
G5-1254X	GMB	80038	G5-1501	GMB	80021	G5-2007	GMB	80049	G52R	PERFECTION	80958
G5-1254X	GMB	80039	G5-150X	GMB	80030	G5-200B	GMB	80075	G5-3000X	GMB	80099
G5-126	GMB	80091	G5-150X	GMB	80031	G5-200B	GMB	80076	G5-3000X	GMB	80101
G5127	GMB	80163	G5-150X	GMB	80097	G5-200X	CENTRAL	80075	G5-3006X	GMB	80104
G5-128	GMB	80074	G5-150X	GMB	80120	G5-200X	GMB	80075	G5-3008X	CENTRAL	80104
G5-129	GMB	80081	G51525-2	GMB	88001.60	G5-200XU	GMB	80075	G5-3008X	GMB	80104
G5-129-1X	GMB	80037	G5-153	GMB	80077	G5-200XU	GMB	80076	G5-3009X	GMB	80105
G5-1296X	GMB	80039	G5-153B	GMB	80077	G5-2011X	CENTRAL	80044	G5-300X	GMB	80024
G5-1296X	GMB	80068	G5-153X	CENTRAL	80077	G5-2011X	GMB	80044	G5-301	GMB	80024
G5-1296X	GMB	80204	G5-153X	GMB	80077	G5-2011XA	GMB	80044	G5-3011	GMB	80104
G5-129X	GMB	80037	G5-153XU	GMB	80077	G5-2017	GMB	80092	G5-3011X	CENTRAL	80107
G5130	GMB	80163	G5-153XU	GMB	80088	G5-2022X	GMB	80044	G5-3011X	GMB	80104
G5130	GMB	80171	G5-154X	GMB	80075	G5-2031X	CENTRAL	80049	G5-3011X	GMB	80107
G5-1300X	GMB	80042	G5-154X	GMB	80077	G5-2031X	GMB	80049	G5-3012X	GMB	80112
G5-1301X	CENTRAL	80135	G5-154X	GMB	80078	G5-2031XA	GMB	80049	G5-3013X	GMB	80106
G5-1301X	GMB	80135	G5-154X	GMB	80085	G5-2033X	GMB	80054	G5-3014X	GMB	80114
G5-1303X	GMB	80034	G5-155	GMB	80182	G5-2100	GMB	80049	G5-3015X	GMB	80104
G5-1304X	GMB	80064	G5-155	TEDDY TORQ	80182	G5-2101	GMB	80049	G5-302	GMB	80010
G5-1304X	GMB	80065	G5-155X	CENTRAL	80178	G5-2110	GMB	80048	G5-3022X	GMB	80097
G5-1306-1X	GMB	80062	G5-155X	GMB	80182	G5-2111	GMB	80049	G5-303	GMB	80004
G51306X	CENTRAL	80061	G5-160	GMB	80072	G5-2114	GMB	80054	G5-3031X	GMB	80109
G5-1306X	GMB	80061	G5-160	GMB	80142	G5-2116X	GMB	80047	G5-304	GMB	80019
G5-1306XS	GMB	80061	G5-160	TEDDY TORQ	80142	G5-2117	GMB	80046	G5-305	GMB	80018
G5-1306XS	GMB	80063	G5-160B	GMB	80142	G5-213	GMB	80117	G5-306	GMB	80009
G5-1307XS	GMB	80063	G5-160B	GMB	80143	G5-213	TEDDY TORQ	80117	G5-307	GMB	80004
G51309X	CENTRAL	80061	G5-160X	CENTRAL	80142	G5-213X	CENTRAL	80117	G5-308	GMB	80017
G5-1309X	GMB	80132	G5-160X	GMB	80142	G5-213X	GMB	80117	G5-308X	GMB	80205
G5-130X	GMB	80117	G5-160XU	GMB	80142	G5-213X	GMB	80131	G5-309	GMB	80011
G5131	GMB	80173	G5-160XU	GMB	80143	G5-2140X	GMB	80055	G5-310	GMB	80012
G5131	GMB	80175	G5-162	GMB	80042	G5-2153	GMB	80055	G5-3101X	GMB	80100
G5-1310X	GMB	80095	G51625-2	GMB	80042	G5-2172	GMB	80047	G5-3102	GMB	80112
G5-1310X	GMB	88001.61	G5-165	GMB	80190	G5-2173	GMB	80058	G5-3103X	GMB	80098
G5-1311X	GMB	80056	G5-165	TEDDY TORQ	80190	G5-2173X	GMB	80058	G5-3105	GMB	80104
G5-1311X	GMB	80057	G5-165X	GMB	80190	G5-2175X	GMB	80058	G5-3105-1X	GMB	80104
G5-1311X	GMB	88001.58	G5-170	GMB	80007	G5-2198X	GMB	80051	G5-3105-1X	GMB	80107
G5-1312X	GMB	80056	G5-170	TEDDY TORQ	80007	G5-231	GMB	80121	G5-3105-1X	GMB	80112
G5-1312X	GMB	80061	G5-170-1X	GMB	80007	G5-231	TEDDY TORQ	80121	G5-3105S	GMB	88001.94
G5-1313X	GMB	80095	G5-170X	CENTRAL	80007	G5-231X	GMB	80121	G5-3105X	GMB	80104
G5-1313X	GMB	80096	G5-170X	GMB	80007	G5-232	GMB	80132	G5-3111	GMB	80104
G5-1313X	GMB	88001.62	G5-170X	GMB	88000.15	G5-233	GMB	80060	G5-3116X	GMB	80111
G5-1313X	GMB	88001.64	G5-174	GMB	80090	G5-242X	GMB	80035	G5-3120X	GMB	80113
G5-1314X	GMB	80095	G5-1767	GMB	88002.77	G5-243-1X	GMB	80070	G5-3120X	GMB	80114
G5-1315X	GMB	80022	G5-177	GMB	80221	G5-243X	GMB	80069	G5-3121X	GMB	80102
G5-132	GMB	80077	G5-177	TEDDY TORQ	80221	G5-248	GMB	80025	G5-3121X	GMB	80113
G5132	GMB	80174	G5-177X	CENTRAL	80221	G5-248	TEDDY TORQ	80025	G5-3125X	GMB	80103
G5132	GMB	88003.24	G5-177X	GMB	80221	G5-248X	GMB	80025	G5-3147X	GMB	80104
G5-133	GMB	80116	G5-178	GMB	80124	G5-249	GMB	80025	G5-3150	GMB	80100
G5-133	TEDDY TORQ	80116	G5-178B	GMB	80124	G5-2500X	GMB	80059	G5-3152	GMB	80101
G5-133X	GMB	80116	G5-178B	GMB	80125	G5-2530X	GMB	80066	G5-3158	GMB	80102
G5-134	GMB	80118	G5-178UB	GMB	80124	G5-2531	GMB	80059	G5-315X	GMB	80026
G5-134	TEDDY TORQ	80118	G5-178X	CENTRAL	80124	G5-2532	GMB	80067	G5-324X	GMB	88009.17
G5-134X	GMB	80118	G5-178X	GMB	80124	G5-260A	GMB	80090	G5-332X	GMB	80139
G5139	GMB	80170	G5-178XU	GMB	80124	G5-260X	GMB	80078	G5-350X	GMB	80036
G5139	GMB	80172	G5-178XU	GMB	80125	G5-260X	GMB	80090	G5-351X	GMB	80082
G5-14	GMB	80041	G5-181	GMB	80023	G5-262X	GMB	80142	G53525-2	GMB	88001.62
G5-1403	GMB	80958	G5-181	GMB	80024	G5-263X	GMB	80140	G5-352X	GMB	80130
G5-1403X	GMB	80958	G5-185	GMB	80228	G5-264X	GMB	80008	G5-353X	GMB	80083
G5-1404X	GMB	80951	G5-185X	GMB	80228	G5-264X	GMB	88002.88	G5-355X	GMB	80144
G5-1405X	GMB	80905	G5-186X	GMB	88001.82	G5-265	GMB	80064	G5-356X	GMB	80097
G5-1406X	TEDDY TORQ	80953	G5-187X	GMB	80124	G5-265X	GMB	80133	G5-357X	GMB	80144
G5-1407X	GMB	80906	G5-187X	GMB	88001.83	G5-265X	GMB	88002.09	G5-358X	GMB	80127
G5-1407X	GMB	80909	G5-188	GMB	80140	G5-266	GMB	80060	G53625-2	GMB	80135

Part No.	Brand	IDLI Group	Part No.	Brand	IDLI Group	Part No.	Brand	IDLI Group	Part No.	Brand	IDLI Group
G5-369X	GMB	80137	G5-552	GMB	80118	G5-8202X	GMB	80237	G66R	PERFECTION	80953
G5-373	GMB	80135	G5-555	GMB	80182	G5-8203X	GMB	80236	G67	PERFECTION	80953
G53825-2	GMB	80135	G5-560	GMB	80142	G5-8205	GMB	80231	G67R	PERFECTION	80953
G53825-2	TEDDY TORQ	80135	G5-565	GMB	80190	G5-8206	GMB	80232	G67R	PERFECTION	80957
G5-4	GMB	80023	G5-570	GMB	80007	G5-8207	GMB	80237	G7000	WESCO	80210
G54	PERFECTION	88001.74	G5-572	GMB	80007	G5-8500X	GMB	80223	G7004	WESCO	80212
G5-4002X	GMB	80109	G5-573	GMB	80078	G5-8500X	GMB	88003.28	G7033	GMB	80210
G5-4014X	GMB	80158	G5-574	GMB	80090	G5-8502X	GMB	80224	G7034	WESCO	80212
G5-4015X	GMB	80148	G5-575	GMB	80007	G5-850X	GMB	80219	G7105	GMB	80213
G5-4016X	GMB	80159	G5-578	GMB	80124	G5-8516	GMB	80224	G7107	GMB	80210
G5-4017X	GMB	80159	G5-579	GMB	80091	G5-86R	GMB	80124	G7112	GMB	80213
G5-407X	GMB	80226	G5-5800X	GMB	80185	G5-86X	GMB	80124	G7113	GMB	80213
G5-4104X	GMB	80151	G5-5801X	GMB	80186	G5-88X	GMB	80140	G7113	GMB	80215
G5-4109X	GMB	80157	G5-5802X	GMB	80184	G5-90	GMB	80093	G7117	GMB	80213
G5-4110X	GMB	80155	G5-581	GMB	80023	G5-9000X	GMB	80235	G7117	GMB	80215
G5-4123	GMB	80153	G5-588	GMB	80140	G5-9001X	GMB	80234	G7122	GMB	80213
G5-4138	GMB	80109	G5-596	GMB	80105	G5-9002X	GMB	80233	G7126	GMB	80213
G5-4140	GMB	80155	G5-596-55	GMB	80038	G5-9015	GMB	80234	G7126	GMB	88003.21
G5-4140X	GMB	80153	G5-6000X	GMB	80193	G5-9015X	GMB	80234	G71425-2	GMB	80034
G5-4140X	GMB	80155	G5-6101X	GMB	80197	G5-9016	GMB	80234	G7150G	GMB	80031
G5-4141X	GMB	80146	G5-6102X	GMB	80194	G5-9017	GMB	80233	G7200	GMB	80210
G5-4141X	GMB	80154	G5-6102X	GMB	80195	G5-9017X	GMB	80233	G7200	GMB	80213
G5-4143	GMB	80156	G5-6102X	GMB	80197	G5-9017X	GMB	80241	G7200	GMB	80215
G5-4143X	GMB	80156	G5-6103	GMB	80194	G5-9026	GMB	80233	G7200	GMB	88002.82
G5-4152	GMB	80158	G5-6104X	GMB	80192	G5-90X	GMB	80093	G7201	GMB	80211
G5-4156	GMB	80157	G5-6106X	GMB	80198	G5-92X	GMB	80001	G7202	GMB	80212
G5-4200X	GMB	80150	G5-6108X	GMB	80199	G5S53-55	GMB	80039	G7203	GMB	80208
G5-45X	GMB	80092	G5-6109X	GMB	80201	G6000	WESCO	80193	G7203	GMB	80213
G5-4X	GMB	80041	G5-6114	GMB	80197	G6004	GMB	80195	G7204	GMB	80209
G5-5000X	GMB	80168	G5-6128	GMB	80198	G6022	WESCO	80197	G7204	GMB	80213
G5-5008X	GMB	80174	G5-6132X	GMB	80202	G6023	WESCO	80199	G7205	GMB	80210
G5-5015X	GMB	80173	G5-6139X	GMB	80203	G6024	WESCO	80201	G7205	GMB	80215
G5-5031	GMB	80166	G5-6143	GMB	80193	G60R	PERFECTION	80903	G7206	GMB	80207
G5-5031X	GMB	80050	G56152	GMB	80197	G60R	PERFECTION	80953	G7206	GMB	80211
G5-504	GMB	80041	G5-6162	GMB	80199	G60R	PERFECTION	80958	G7206	GMB	80212
G5-5108X	GMB	80172	G5-6163	GMB	80201	G6100	GMB	80193	G7208	GMB	80210
G5-5109	GMB	80162	G5-635	GMB	80124	G6101	GMB	80195	G7225-2	GMB	88001.84
G5-5109X	GMB	80162	G5-640	GMB	80227	G6101	GMB	80196	G7242	GMB	80044
G5-5110	GMB	80173	G5-645	GMB	80139	G6101	GMB	80197	G7246	GMB	80212
G5-5121X	GMB	80163	G5-650	GMB	80109	G6101	GMB	80198	G725251-2	GMB	80056
G5-5121X	GMB	80167	G5-65AR	GMB	80905	G6102	GMB	80197	G725251-2	GMB	80061
G5-5122	GMB	80168	G5-65R	GMB	80951	G6103	GMB	80194	G72525-2	GMB	80031
G5-5127	GMB	80167	G5-67AR	GMB	80906	G6103	WESCO	80194	G72525-4	GMB	80061
G5-513	GMB	80077	G5-67R	GMB	80953	G6104	GMB	80192	G7256	GMB	80218
G5-5132	GMB	80174	G5-70	GMB	80007	G6104	GMB	80198	G72625-2	GMB	80064
G5-5139X	GMB	80170	G5-7000X	GMB	80210	G6105	GMB	80215	G72625-2	GMB	80065
G5-514	GMB	80024	G5-7105X	GMB	80213	G6106	GMB	80198	G72625-3	GMB	80061
G5-5148	GMB	80169	G5-710X	GMB	80205	G6107	GMB	80174	G72625-4	GMB	80061
G5-5148X	GMB	80168	G5-7126	GMB	80213	G6107	GMB	80202	G72625-4	GMB	80063
G5-5148X	GMB	80169	G5-7202X	GMB	80212	G6108	GMB	80199	G72625-5	GMB	80132
G5-5148X	GMB	80216	G5-7203X	GMB	80208	G6109	GMB	80201	G72625-2/4	GMB	80065
G5-515	GMB	80178	G5-7204X	GMB	80209	G6114	GMB	80196	G72625-6	GMB	80062
G5-5152	GMB	80174	G5-7205	GMB	80210	G6114	GMB	80197	G72625-3/4	GMB	80061
G5-5154	GMB	80175	G5-7206	GMB	80211	G6123	GMB	80191	G7325-2	GMB	80096
G5-5154X	GMB	80175	G5-7206X	GMB	80211	G6139	GMB	80194	G73252-2	GMB	80061
G5-516	GMB	80189	G5-7207	GMB	80212	G6139	GMB	80203	G735251-2	GMB	80095
G5-517	GMB	80221	G5-7245X	GMB	80206	G6140	GMB	80191	G73525-2	GMB	80061
G5-5173	GMB	80171	G5-7245X	GMB	80212	G6140	GMB	80193	G73525-3	GMB	80061
G5-5173X	GMB	80171	G5-74	GMB	80090	G6140	GMB	88003.25	G73525-3	GMB	80096
G5-5177	GMB	80172	G5-74X	CENTRAL	80090	G6141	GMB	80193	G73525-4	GMB	80061
G5-5177X	GMB	80172	G5-74X	GMB	80090	G6142	GMB	80197	G73525-4	GMB	80094
G5-520	GMB	80077	G5-750X	GMB	80187	G6143	GMB	88003.11	G73525-3/4	GMB	80096
G5-5201X	GMB	80186	G57R	PERFECTION	80955	G6152	GMB	80197	G7601	GMB	88003.56
G5-521	GMB	80077	G5-8014X	GMB	80230	G6152	GMB	88003.12	G7604	GMB	88003.45
G5-524	GMB	80225	G5-8104X	GMB	80238	G6152	GMB	88003.23	G7605	GMB	88003.48
G5-530	GMB	80117	G5-8105X	GMB	80238	G6163	GMB	80201	G7605	GMB	88003.56
G5-531	GMB	80121	G5-8113	GMB	80238	G6179	GMB	80200	G7607	GMB	88003.49
G5-548	GMB	80025	G5-8200X	GMB	80231	G62R	PERFECTION	80954	G7608	GMB	88003.46
G5-549	GMB	80077	G5-8201X	GMB	80232	G63R	PERFECTION	80957	G7609	GMB	88003.50
G5-550	GMB	80119	G5-8201X	GMB	80237	G65R	PERFECTION	80951	G7612	GMB	88003.50

Part No.	Brand	IDLI Group	Part No.	Brand	IDLI Group	Part No.	Brand	IDLI Group	Part No.	Brand	IDLI Group
G7622	GMB	88003.42	GCPU2N	GMB	80035	GR60	GMB	88001.73	GU3000	GMB	80178
G7627	GMB	88003.43	GCVK1	GMB	88002.34	GR62A	GMB	80907	GU350	GMB	80180
G8014	GMB	80230	GCVK2	GMB	88002.34	GR62A	TEDDY TORQ	80907	GU3500	GMB	80180
G8105	GMB	80238	GCVK2	GMB	88002.36	GR65A	GMB	80905	GU3T1328	GMB	80208
G8107	GMB	80238	GCVK2A	GMB	88002.36	GR65A	TEDDY TORQ	80905	GU3T1328	GMB	88002.55
G8113	GMB	80238	GCVK3	GMB	88002.33	GR67A	GMB	80906	GU3T1328	GMB	88002.61
G8113	GMB	88003.08	GCVK4	GMB	88002.33	GR67A	TEDDY TORQ	80906	GU3T1595	GMB	88002.47
G8113	GMB	88003.20	GCVK4	GMB	88002.33	GR96-55	GMB	80122	GU3T1602	GMB	80104
G8200	GMB	80231	GCVK-4	GMB	88002.33	GR96-55	TEDDY TORQ	80122	GU3T1602	GMB	88002.56
G8201	GMB	80232	GCVK5	GMB	88002.33	GS53-55	GMB	80039	GU3T2007	GMB	88002.57
G8202	GMB	80237	GCVK6	GMB	88002.33	GS53-55	TEDDY TORQ	80039	GU3T2221	GMB	88002.58
G8203	GMB	80236	GCVK7	GMB	88002.33	GS55-55	GMB	80081	GU3T2222	GMB	88002.45
G8203	GMB	80238	GD115X	CATERPILLR	80178	GS55-55	TEDDY TORQ	80081	GU3T2266	GMB	88002.54
G8205	GMB	80231	GD96-55	GMB	80160	GS56-55	GMB	80145	GU3T2501	GMB	88002.63
G8205	GMB	80232	GD96-55	TEDDY TORQ	80160	GS96-55	GMB	80038	GU3T2531	GMB	88002.60
G8205	GMB	80240	GD96-55-4	GMB	80160	GS96-55	TEDDY TORQ	80038	GU3T2538	GMB	88002.60
G8206	GMB	80184	GDS55-55-2	GMB	80145	GT55-55-2	GMB	80071	GU3T2894	GMB	88002.46
G8206	GMB	80242	GDS56-55	GMB	80145	GT55-55-2	TEDDY TORQ	80071	GU3T2895	GMB	88002.48
G8207	GMB	80237	GDS56-55-2	GMB	80145	GTBA18310	FORD	80024	GU3T2896	GMB	88002.50
G8215	GMB	80184	GF9190	CATERPILLR	80109	GTS55-55-2	GMB	80074	GU3T2897	GMB	88002.51
G8215	GMB	80237	GGPL10H	GMB	80033	GTS55-55-2	TEDDY TORQ	80074	GU3T2898	GMB	88002.52
G8215	GMB	80239	GGPL10N	GMB	80033	GU055-55-2	GMB	80141	GU3T2899	GMB	88002.53
G83254-3	GMB	80022	GGPL10N	GMB	88002.93	GU096-55	GMB	80138	GU3T3407	GMB	88002.71
G8500	GMB	80224	GGPL12R	GMB	80035	GU100	CENTRAL	80041	GU3T3501	GMB	88002.63
G8500	GMB	88003.28	GGPL14R	GMB	80037	GU1000	GMB	80072	GU3T3502	GMB	88002.64
G8502	GMB	80224	GGPL14R1X	GMB	80037	GU1000	GMB	80075	GU3T3505	GMB	88002.65
G8506	GMB	80224	GGPL34R	GMB	80069	GU1000A	GMB	80072	GU3T3506	GMB	88002.67
G8507	GMB	80224	GGPL35N	GMB	80161	GU1000SI	GMB	80072	GU3T3507	GMB	88002.71
G8515	GMB	80222	GGPL35R	GMB	80069	GU1000SI	GMB	80077	GU3T3508	GMB	88002.72
G8515	GMB	80224	GGPL35R1X	GMB	80070	GU1000SL	GMB	80077	GU3T3509	GMB	88002.70
G8T18397	FORD	80124	GGPL44R	GMB	80136	GU100A	GMB	80077	GU3T3510	GMB	88002.73
G9000	GMB	80234	GGPL4N	GMB	80134	GU1030	GMB	80077	GU3T3511	GMB	88002.74
G9001	GMB	80234	GGPL4N	GMB	80178	GU1100	GMB	80041	GU400	GMB	80023
G9002	GMB	80233	GGPL55N	GMB	80161	GU1100SL	GMB	80041	GU4000	GMB	80189
G9002	GMB	88003.09	GGPL55R	GMB	80161	GU114-252	GMB	80034	GU500	CENTRAL	80024
G9010	GMB	80233	GGPL5N	GMB	80178	GU114-266	GMB	80060	GU500	GMB	80023
G9010	GMB	80235	GGPL6N	GMB	80188	GU114-362	GMB	80135	GU500	GMB	80024
G9013	GMB	80234	GGPL6R	GMB	80001	GU1170	GMB	80040	GU5000	GMB	80221
G9014	GMB	80234	GGPL6R	GMB	80003	GU1S51	GMB	80023	GU5000SL	GMB	80221
G9014	GMB	80235	GGPL7N	GMB	80220	GU1S52	GMB	80014	GU500SL	GMB	80023
G9015	GMB	80234	GGPL9R	GMB	80001	GU1S56	GMB	80024	GU500SL	GMB	80140
G9016	GMB	80234	GH	WEASLER	80037	GU2000	GMB	80078	GU510	GMB	80023
G9016	GMB	88012.36	GH2577	CATERPILLR	80155	GU2050	GMB	80142	GU55-55-2	GMB	80183
G9017	GMB	80233	GH2579	CATERPILLR	80238	GU2050SL	GMB	80142	GU56-55-2	GMB	80183
G9026	GMB	80233	GMZ7	CENTRAL	80011	GU2100	GMB	80003	GU5800	GMB	80188
G9030	GMB	80233	GN	WEASLER	80037	GU2103	GMB	80048	GU5900	GMB	80220
G9031	GMB	80234	GN100W	GEHL BROS.	80037	GU2109	GMB	80049	GU5950	GMB	80030
GC96-55	GMB	80204	GO55-55	GMB	80141	GU2114	GMB	80047	GU5950	GMB	80031
GC96-55	TEDDY TORQ	80204	GO55-55	TEDDY TORQ	80141	GU2114	GMB	80054	GU5960	GMB	80032
GCP12N	GMB	80035	GO55-55-2	GMB	80141	GU2116	GMB	80047	GU5970	GMB	80030
GCP14N	GMB	80037	GO96-55	GMB	80138	GU2200	GMB	80124	GU6010	GMB	80049
GCP14N1	GMB	80037	GO96-55	TEDDY TORQ	80138	GU2210	GMB	80124	GU6050	GMB	80049
GCP35N	GMB	80069	GP18397	FORD	80072	GU2210A	GMB	80123	GU6050	GMB	80050
GCP35N1	GMB	80069	GP55-55-140	GMB	80129	GU2210SL	GMB	80124	GU6100	GMB	80049
GCP35N3	GMB	80047	GP55-55-145	GMB	80131	GU2300	GMB	80140	GU6110	GMB	80049
GCP35N3	GMB	80069	GP96-55	GMB	80068	GU2500	GMB	80042	GU6120	GMB	80049
GCP35N3	GMB	80070	GP96-55	TEDDY TORQ	80068	GU2518	GMB	80066	GU6120	GMB	88002.77
GCP35N3	GMB	80178	GPW18397	FORD	80072	GU2570	GMB	80064	GU6200	GMB	80044
GCP35N3	GMB	88003.19	GR5165	WESCO	80091	GU2600	GMB	80135	GU6300	GMB	80099
GCP35N4	GMB	80069	GR52A	GMB	80906	GU2650	GMB	80135	GU6320	GMB	80112
GCP35N4	GMB	80070	GR52A	GMB	80909	GU2760	GMB	80039	GU6370	GMB	80104
GCP4N	GMB	80134	GR52A	TEDDY TORQ	80909	GU2770	GMB	80038	GU6410	GMB	80159
GCP58N3	GMB	80179	GR54	GMB	80951	GU2800	GMB	80138	GU6420	GMB	80146
GCP58WB	GMB	80185	GR54	GMB	88001.74	GU2850	GMB	80122	GU6450	GMB	80158
GCP58WB1	GMB	80186	GR54A	GMB	80901	GU2870	GMB	80160	GU6470	GMB	80148
GCP58WB2	GMB	80184	GR54A	TEDDY TORQ	80901	GU2T2910	GMB	88002.43	GU6500	GMB	80168
GCP6N	GMB	80188	GR55-55	GMB	80129	GU2T2911	GMB	88002.42	GU6520	GMB	80177
GCP7N	GMB	80220	GR55-55	TEDDY TORQ	80123	GU2T2912	GMB	88002.44	GU6550	GMB	80174
GCPL14N	GMB	80037	GR55-55-2	GMB	80117	GU300	GMB	80178	GU6600	GMB	80193
GCPL6N	GMB	80001	GR57A	GMB	80908	GU3000	GMB	80119	GU670	GMB	80210

Part No.	Brand	IDLI Group
GU6700	GMB	80210
GU700	GMB	80041
GU7010	GMB	88000.16
GU7100	GMB	80032
GU7150	GMB	88002.77
GU730	GMB	88001.86
GU96-55	GMB	80181
GU96-55	TEDDY TORQ	80181
GU96-55-4	GMB	80181
GUD81	GMB	80010
GUD83	GMB	80010
GUD86	GMB	80010
GUIS56	GMB	80023
GUIS61	GMB	80024
GUJ1001	BMC	80041
GUJ1001	UNIPART	80023
GUJ1002	BMC	80041
GUJ1002	UNIPART	80041
GUJ101	BMC	80023
GUJ101	UNIPART	80023
GUJ102	BMC	80041
GUJ102	UNIPART	80041
GUJ103	BMC	80041
GUJ103	UNIPART	80041
GUJ104	BMC	80077
GUJ104	UNIPART	80077
GUJ105	BMC	80077
GUJ105	UNIPART	80077
GUJ106	BMC	80023
GUJ106	UNIPART	80023
GUJ107	BMC	80041
GUJ108	BMC	80077
GUJ108	UNIPART	80077
GUJ110	BMC	80142
GUJ110	UNIPART	80142
GUJ111	BMC	80142
GUJ111	UNIPART	80142
GUJ112	BMC	80124
GUJ112	UNIPART	80124
GUJ114	BMC	80024
GUJ114	UNIPART	80024
GUJ115	BMC	80024
GUJ116	BMC	80041
GUJ117	BMC	80077
GUK01	GMB	80047
GUK013	GMB	80172
GUK02	GMB	80046
GUK04	GMB	80156
GUK05	GMB	80172
GUK06	GMB	80198
GUK08	GMB	80224
GUM79	GMB	80012
GUM81	GMB	80012
GUM88	GMB	80015
GUMZ1	GMB	80010
GUMZ10	GMB	80005
GUMZ2	GMB	80010
GUMZ6	GMB	80020
GUMZ7	GMB	80011
GUMZ9	GMB	80013
GUN11	GMB	80006
GUN26	GMB	80004
GUN26	GMB	80006
GUN26	GMB	80023
GUN26	GMB	80024
GUN27	CENTRAL	80010
GUN27	GMB	80010
GUN28	CENTRAL	80006
GUN28	GMB	80004
GUN29	CENTRAL	80019
GUN29	GMB	80019
GUN30	GMB	80124
GU-P55-55-140	GMB	80129
GU-P55-55-145	GMB	80131
GUPA1	GMB	80156
GUPA2	GMB	80213
GUPA3	GMB	80140
GUPA4	GMB	80124
GUPA6	GMB	80172
GUR52	GMB	80958
GUR54	GMB	88001.74
GUR54	TEDDY TORQ	88001.74
GU-R55-55-2	GMB	80123
GUR57	GMB	80955
GUR57	TEDDY TORQ	80955
GUR60	GMB	80958
GUR60	TEDDY TORQ	80958
GUR62	GMB	80954
GUR62	GMB	80958
GUR62	TEDDY TORQ	80954
GUR63	GMB	80957
GUR63	TEDDY TORQ	80957
GUR65	GMB	80951
GUR65	TEDDY TORQ	80951
GUR66	GMB	80953
GUR67	GMB	80953
GUR67	TEDDY TORQ	80953
GU-S53-55	GMB	80039
GU-S55-55-2	GMB	80081
GU-S55-55-675	GMB	80091
GU-S96-55	GMB	80039
GUSR4	GMB	80010
GUT11	CENTRAL	80006
GUT11	GMB	80006
GUT12	CENTRAL	80018
GUT12	GMB	80018
GUT13	CENTRAL	80009
GUT13	GMB	80009
GUT14	CENTRAL	80017
GUT14	GMB	80017
GUT17	GMB	80016
GUT19	GMB	80006
GUT20	GMB	80115
GV11312	GMB	86069
GV11312	GMB	86071
GV12310	GMB	86070
GY2455	COCKSHUTT	80008
H09655	HABERLE	80138
H1000	PILOT	88008.35
H1001	PILOT	88007.58
H-1001	PILOT	88007.58
H1001	PILOT	88007.58
H-1001	PILOT	88007.58
H-1002	PILOT	87331
H1002	PILOT	88008.39
H1003	PILOT	87326
H-1003	PILOT	87326
H1003	PILOT	87326
H-1003	PILOT	87326
H1004	PILOT	87331
H-1004	PILOT	87331
H-1005	PILOT	87285
H1005	PILOT	88008.48
H1007	PILOT	87325
H-1007	PILOT	88007.59
H-1008	PILOT	88008.41
H-1008	PILOT	88008.41
H1008	PILOT	88008.41
H1008	PILOT	88008.41
H1009	PILOT	88008.42
H1009	PILOT	88008.42
H-1009	PILOT	88008.42
H-1009	PILOT	88008.42
H-1010	PILOT	88008.51
H1010	PILOT	88008.52
H-1011	PILOT	88008.31
H1011	PILOT	88008.44
H-1012	PILOT	87459
H1012	PILOT	88008.45
H-1013	PILOT	88008.59
H-1013	PILOT	88008.59
H-1013	PILOT	88008.59
H1013	PILOT	88008.59
H-1014	PILOT	87481
H1014	PILOT	88007.77
H1016	PILOT	88007.79
H1016	PILOT	88007.79
H-1016	PILOT	88007.79
H-1016	PILOT	88007.79
H12EY-1	BPC	81218
H12EY-1-1/16R-1/4KW	BPC	81249
H12EY-1-1/8	BPC	81259
H12EY-1-1/8	BPC	81266
H12EY-1-1/8	BPC	81274
H12EY-1-1/8R-1/4KW	BPC	81268
H12EY-1-1/8R-5/16KW	BPC	81261
H12EY-1-3/16R-5/16KW	BPC	81286
H12EY-1-1/4	BPC	81300
H12EY-1-1/4R-1/4KW	BPC	81301
H12EY-1-1/4R-5/16KW	BPC	81302
H12EY-1-3/8	BPC	81331
H12EY-1-3/8	BPC	84510
H12EY-1-3/8R-5/16KW	BPC	81333
H12EY-1-1/2R-5/16KW	BPC	81350
H12EY-1R-1/4KW	BPC	81237
H12EY-1R-5/16KW	BPC	81235
H12EY-3/4	BPC	81206
H12EY-3/4R-3/16KW	BPC	81203
H12EY-13/16	BPC	81209
H12EY-13/16R-1/4KW	BPC	81208
H12EY-7/8	BPC	81218
H12EY-7/8R-3/16KW	BPC	81212
H12EY-7/8R-1/4KW	BPC	81213
H12EY-15/16R	BPC	81221
H12EY-15/16R-1/4KW	BPC	81222
H12SY-1	BPC	85055
H12SY-1-1/8	BPC	85057
H12SY-1-1/8	BPC	85058
H12SY-1-1/8R-5/16KW	BPC	81282
H12SY-1-1/4	BPC	85062
H12SY-1-1/4	BPC	88014.19
H12SY-1-1/4R-5/16KW	BPC	85063
H12SY-1-3/8	BPC	85064
H12SY-1R	BPC	88013.99
H12SY-1R-1/4KW	BPC	85054
H12SY-3/4	BPC	85049
H12SY-3/4R-3/16KW	BPC	88014.09
H12SY-7/8	BPC	88013.96
H12SY-7/8R-1/4KW	BPC	85050
H12SY-15/16R-1/4KW	BPC	88014.11
H1510X	HABERLE	80142
H1513X	HABERLE	80077
H1515X	HABERLE	80178
H1516X	HABERLE	80189
H1517X	HABERLE	80221
H1518X	HABERLE	80124
H1521X	HABERLE	80072
H1521X	HABERLE	80075
H1521X	HABERLE	80077
H1524X	HABERLE	80023
H1524X	HABERLE	80024
H1548X	HABERLE	80025
H154X	HABERLE	80041
H1560	HABERLE	80142
H1560	HABERLE	80143
H1578X	HABERLE	80124
H2000	HABERLE	80049
H2000	HABERLE	88001.96
H2007	HABERLE	80055
H2010	HABERLE	80049
H2010	HABERLE	88001.96
H2011	HABERLE	80044
H2011	HABERLE	88001.98
H2012	HABERLE	80031
H2014	HABERLE	80045
H2016	HABERLE	80044
H2016	HABERLE	88001.98
H2017	HABERLE	80092
H2017	HABERLE	88001.76
H2026	HABERLE	80031
H2032	HABERLE	80092
H2032	HABERLE	88001.76
H2033	HABERLE	80054
H2036	HABERLE	80044
H2036	HABERLE	88001.98
H2039	HABERLE	80044
H2039	HABERLE	88001.98
H2041	HABERLE	80054
H2045	HABERLE	80044
H2045	HABERLE	88001.98
H2046	HABERLE	80049
H2046	HABERLE	88001.96
H2049	HABERLE	80054
H2053	HABERLE	80031
H2062	HABERLE	80049
H2062	HABERLE	88001.96
H2063	HABERLE	80055
H2064	HABERLE	80044
H2064	HABERLE	88001.98
H2066	HABERLE	80045
H2198	HABERLE	80051
H3000	HABERLE	80101
H3008	HABERLE	80104
H3010	HABERLE	80100
H3012	HABERLE	80112
H4002	HABERLE	80109
H4002	HABERLE	80149
H4007	HABERLE	80147
H4008	HABERLE	80146

Part No.	Brand	IDLI Group
H4014	HABERLE	80158
H4015	HABERLE	80148
H4016	HABERLE	80159
H5000	HABERLE	80168
H5007	HABERLE	80177
H5008	HABERLE	80174
H5015	HABERLE	80173
H6000	HABERLE	80193
H6000	HARERLE	80193
H6016	HABERLE	80193
H6016	HARERLE	80193
H6022	HARERLE	80197
H6022	HARERLE	80197
H6023	HABERLE	80199
H6024	HABERLE	80201
H62152(7)	GALION	80003
H625124	GALION	80003
H6531	HABERLE	88002.48
H6642	HABERLE	88002.42
H66746(2)	GALION	80182
H66773(7)	GALION	80182
H69215(7)	GALION	80001
H7000	HABERLE	80210
H7000	HARERLE	80210
H7034	HABERLE	80212
H7034	HARERLE	80212
H7090	HABERLE	88002.55
H7125	HABERLE	80042
H7254	HABERLE	80095
H751EY-1R-1/4KW	BPC	88000.73
H751EY-3/4	BPC	88000.74
H751EY-3/4R-3/16KW	BPC	88000.68
H751EY-13/16R-1/4KW	BPC	88000.70
H751EY-7/8	BPC	88000.75
H751EY-7/8R-1/4KW	BPC	88000.71
H751EY-15/16R-1/4KW	BPC	88000.72
H7523	HABERLE	80096
H7525	HABERLE	80095
H7753	HABERLE	80064
H7753	HABERLE	80065
H7753	HABERLE	88001.63
H7EY-1	BPC	88000.46
H7EY-1-1/16R-1/4KW	BPC	88000.34
H7EY-1-1/8	BPC	88000.47
H7EY-1-1/8	BPC	88000.49
H7EY-1-1/8R-1/4KW	BPC	88000.36
H7EY-1-1/8R-5/16KW	BPC	88000.35
H7EY-1-3/16R-5/16KW	BPC	88000.37
H7EY-1-1/4	BPC	88000.50
H7EY-1-1/4R-1/4KW	BPC	88000.55
H7EY-1-1/4R-5/16KW	BPC	88000.38
H7EY-1-3/8	BPC	88000.51
H7EY-1-3/8	BPC	88000.67
H7EY-1R-3/16KW	BPC	88000.53
H7EY-1R-1/4KW	BPC	88000.30
H7EY-1R-3/8PH	BPC	88000.32
H7EY-5/8R-3/16KW	BPC	88000.19
H7EY-3/4	BPC	88000.43
H7EY-3/4R-3/16KW	BPC	88000.20
H7EY-13/16R-1/4KW	BPC	88000.22
H7EY7/8	BPC	88000.45
H7EY-7/8	BPC	88000.62
H7EY-7/8R-3/16KW	BPC	88000.52
H7EY-7/8R-1/4KW	BPC	88000.25
H7EY15/16R	BPC	88000.29
H7EY-15/16R-1/4KW	BPC	88000.28
H7SY-1	BPC	88000.90
H7SY-1-1/8	BPC	88000.92
H7SY-1-1/8	BPC	88000.95
H7SY-1-1/8	BPC	88001
H7SY-1-1/8R-1/4KW	BPC	88000.96
H7SY-1-1/8R-5/16KW	BPC	88000.84
H7SY-1-1/4	BPC	88000.94
H7SY-1-1/4R-5/16KW	BPC	88000.85
H7SY-1R	BPC	88000.91
H7SY-1R-1/4KW	BPC	88000.83
H7SY-3/4	BPC	88000.86
H7SY-3/4R-3/16KW	BPC	88000.79
H7SY-13/16R-1/4KW	BPC	88000.80
H7SY-7/8	BPC	88000.89
H7SY-7/8	BPC	88001.01
H7SY-7/8R-1/4KW	BPC	88000.81
H7SY-15/16R-1/4KW	BPC	88000.82
H8002	HABERLE	80231
H8002	HARERLE	80231
H81402	GALION	80001
H9000	HABERLE	80235
H9266	HABERLE	88002.50
HAR1405	THOMPSON	80149
HB1	FED.MOGUL	87285
HB101	FED.BEAR.	87285
HB102	FED.BEAR.	87337
HB103	FED.BEAR.	87399
HB1206LL	FED.BEAR.	87238
HB1206LL	FEDERAL	87238
HB2	FED.MOGUL	87337
HB20600	FED.MOGUL	87238
HB206DD	FED.MOGUL	87238
HB206FF	FED.MOGUL	87238
HB206FF	FED.MOGUL	88007.58
HB3	FED.MOGUL	87399
HB5	FED.MOGUL	87325
HB8805	FED.MOGUL	87377
HB88107	FED.BEAR.	87326
HB88107	FEDERAL	87326
HB88107	FED.MOGUL	87331
HB88107A	FED.MOGUL	87326
HB88508	FED.BEAR.	87383
HB88508	FEDERAL	87383
HB88508	FED.MOGUL	87377
HB88508A	FED.MOGUL	87373
HB88509	FED.BEAR.	87459
HB88509	FEDERAL	87459
HB88509	FED.MOGUL	87457
HB88510	FED.MOGUL	87481
HB88512A	FED.MOGUL	87570
HBD88107	FED.BEAR.	87331
HBD88107	FEDERAL	87332
HBD88107	FED.MOGUL	87331
HBF88107	FED.BEAR.	87331
HBF88107	FEDERAL	87332
HBF88507A	FED.BEAR.	87373
HBF88507A	FEDERAL	87373
HBF88508	FED.BEAR.	87377
HBF88508	FEDERAL	87377
HBF88509	FED.BEAR.	87457
HBF88509	FEDERAL	87457
HBF88510	FED.BEAR.	87481
HBF88510	FEDERAL	88007.77
HBF88512A	FED.BEAR.	87570
HBF88512A	FEDERAL	88007.79
HBG88107	FEDERAL	87325
HCB101	CLEVE.STL.	87570
HCB1-4	CLEVE.STL.	88007.51
HCB-1-4A	CLEVE.STL.	88014.89
HCB2	CLEVE.STL.	88013.80
HCB50-1	CLEVE.STL.	88008.67
HCB61	CLEVE.STL.	88007.86
HCF15790	AUST.WEST.	80101
HCF16144	AUST.WEST.	80102
HCF16263	AUST.WEST.	80196
HCF19058	AUST.WEST.	80101
HCF2098-57	AUST.WEST.	80215
HCF2098-57	BAL-LM-HAM	80210
HCF6641A	AUST.WEST.	80048
HCF6641A	BAL-LM-HAM	80048
HD TRK.	MCQUAY-NOR	80229
HGB88107	FED.BEAR.	87325
HI-PERF.CONVERSION KIT	MCQUAY-NOR	80097
HO9655	HABERLE	80138
HP555	HABERLE	80129
HP9655	HABERLE	80068
HP9675	HABERLE	80131
HR5555	HABERLE	80123
HR7360	HABERLE	80903
HR7360	HABERLE	80953
HR7360	HABERLE	80958
HS5355	HABERLE	80038
HS5355	HABERLE	80039
HS555	HABERLE	80081
HS555	HABERLE	80123
HS9675	HABERLE	80091
HY284-10M	REPUBLIC	88000.73
HY284-1M	REPUBLIC	88000.68
HY284-4M	REPUBLIC	88000.70
HY284-6M	REPUBLIC	88000.71
HY284-8M	REPUBLIC	88000.72
HY311	REPUBLIC	88002.86
HY311	REPUBLIC	88002.87
HY362-1	REPUBLIC	88000.20
HY362-1	REPUBLIC	88000.21
HY362-10	REPUBLIC	88000.30
HY362-10	REPUBLIC	88000.31
HY362-1A	REPUBLIC	88000.44
HY362-2	REPUBLIC	88000.20
HY362-2	REPUBLIC	88000.21
HY362-3	REPUBLIC	88000.26
HY362-3A	REPUBLIC	88000.22
HY362-4	REPUBLIC	88000.22
HY362-5	REPUBLIC	88000.52
HY362-6	REPUBLIC	88000.25
HY362-6	REPUBLIC	88000.27
HY362-6A	REPUBLIC	88000.25
HY362-7	REPUBLIC	88000.26
HY362-7	REPUBLIC	88000.28
HY362-9	REPUBLIC	88000.53
HY363-1	REPUBLIC	88000.34
HY363-1	REPUBLIC	88000.35
HY363-1A	REPUBLIC	88000.34
HY363-2	REPUBLIC	88000.36
HY363-3	REPUBLIC	88000.39
HY363-3A	REPUBLIC	88000.36
HY363-5	REPUBLIC	88000.37
HY363-6	REPUBLIC	88000.55
HY363-6A	REPUBLIC	88000.38
HY363-6A	REPUBLIC	88000.39
HY363-7	REPUBLIC	88000.56
HY363-8	REPUBLIC	88000.20
HY363-9	REPUBLIC	88000.52
HY364-1	REPUBLIC	88000.43
HY364-3	REPUBLIC	88000.45
HY364-4	REPUBLIC	88000.46
HY365	REPUBLIC	88002.86
HY365	REPUBLIC	88002.87
HYL1944	BMC	80142
HYL2306	BMC	80030
HYL2307	BMC	80032
HYL3090	BMC	88002.77
HYL3091	BMC	80032
HYL3826	LEYLAND	88002.77
HYL3873	BMC	80189
HYL4007	BMC	80043
HYL4674	LEYLAND	80044
HYL4675	LEYLAND	80060

FOR NUMBERS THAT START WITH I, ALSO CHECK FOR THE NUMERAL ONE.

Part No.	Brand	IDLI Group
IHC	MCQUAY-NOR	80173
IHC	MCQUAY-NOR	80208
IM7039	FORD	80049
IND.	MCQUAY-NOR	80176
IND.	MCQUAY-NOR	80191
IND.	MCQUAY-NOR	80192
IND.	MCQUAY-NOR	80203
IND.	MCQUAY-NOR	80209
IND.	MCQUAY-NOR	80236
IT7039	FORD	80124
J055-55	ALCO	80141
J055-55	ALLOY	80141
J055-55-2	ALLOY	80141
J096-55	ALCO	80138
J1003	ALCO	80001
J1004	ALCO	80001
J1004	ALLOY	88013.28
J1007	ALCO	80001
J1009	ALCO	80001
J101	ALLOY	80024
J12000	ALCO	80234
J12002	ALCO	80234
J12003	ALCO	80234
J1501	ALCO	80021
J1501	ALCO	88001.97
J1506	ALCO	80021
J1509	ALCO	80021
J1514	ALCO	80021
J1750	ALCO	80030
J1750	TEDDY TORQ	80030
J1750G	ALCO	80031
J1750G	ALLOY	80031
J1755	ALCO	80032
J1755	ALLOY	80032
J1755G	ALCO	88002.81
J1755L	ALCO	80032
J1762	ALCO	80031
J2000	BORG-WARNR	80049

Part No.	Brand	IDLI Group	Part No.	Brand	IDLI Group	Part No.	Brand	IDLI Group	Part No.	Brand	IDLI Group
J2002	BORG-WARNR	80048	J2198	ALLOY	80051	J4102	TEDDY TORQ	80158	J5-1200	ALCO	80081
J2005	BORG-WARNR	80049	J22-257F	MIXERMOBL	80196	J4103	ALCO	80153	J5-1200	ALLOY	80081
J2006	BORG-WARNR	80021	J22-257F	MIXERMOBL	80199	J4105	ALCO	80151	J5-1200	TEDDY TORQ	80081
J2007	BORG-WARNR	80044	J2500	BORG-WARNR	80059	J4105	ALLOY	80151	J5-1201	ALCO	80091
J2010	BORG-WARNR	80049	J2503	ALCO	80030	J4106	ALCO	80148	J5-1201	ALLOY	80091
J2011	BORG-WARNR	80044	J2503	ALLOY	80030	J4106	ALLOY	80148	J5-1202	ALCO	80130
J2012	BORG-WARNR	80030	J2505	ALCO	80059	J4107	ALCO	80159	J5-1202	ALLOY	80130
J2014	BORG-WARNR	80045	J2505	ALLOY	80059	J4107	ALLOY	80159	J5-1202	SPICER	80130
J2016	BORG-WARNR	80044	J2518	ALCO	80066	J4109	ALCO	80157	J5-1203	ALCO	80129
J2017	ALCO	80092	J2518	ALLOY	80066	J4109	ALLOY	80157	J5-1203	ALLOY	80129
J2017	ALLOY	80092	J2530	ALCO	80066	J4110	ALCO	80155	J5-1204	ALCO	80131
J2018	BORG-WARNR	80053	J2801	ALCO	80232	J4113	ALCO	80109	J5-1204	ALLOY	80131
J2026	BORG-WARNR	80030	J3000	BORG-WARNR	80101	J4113	ALLOY	80156	J5-121	ALCO	80072
J2032	BORG-WARNR	80092	J3001	ALLOY	80104	J4118	ALCO	80156	J5121	ALCO	80167
J2033	BORG-WARNR	80054	J3008	BORG-WARNR	80104	J4118	ALLOY	80156	J5-121	ALLOY	80072
J2039	BORG-WARNR	80044	J3011	ALCO	80107	J4120	ALCO	80109	J5-121B	ALCO	80089
J2040	BORG-WARNR	80049	J3011	ALLOY	80104	J4130	ALCO	80155	J5-121B	ALLOY	80089
J2041	BORG-WARNR	80054	J3011	ALLOY	80107	J4133	ALCO	80154	J5-122	ALCO	80073
J2045	BORG-WARNR	80049	J3012	BORG-WARNR	80112	J4151	ALCO	80109	J5-122	ALLOY	80073
J2046	BORG-WARNR	80044	J3013	ALCO	80106	J4151	ALLOY	80149	J5-123	SPICER	80091
J2048	BORG-WARNR	80049	J3014	BORG-WARNR	80114	J4156	ALCO	80157	J5-124	ALCO	80225
J2053	BORG-WARNR	80030	J3018	ALCO	80110	J4158	ALCO	80109	J5-124	ALLOY	80225
J2062	BORG-WARNR	80049	J3022	ALLOY	80097	J4-1601	ALLOY	80006	J5-124	SPICER	80081
J2063	BORG-WARNR	80044	J3100	ALCO	80101	J419	ALCO	80148	J5-124T	ALCO	80225
J2064	BORG-WARNR	80044	J3100	ALLOY	80101	J4-2	ALLOY	88002.34	J5-124T	ALLOY	80225
J2066	BORG-WARNR	80053	J3101	ALCO	80100	J427B	LEMPCO	80908	J5125	ALCO	80171
J2100	ALCO	80049	J3101	ALLOY	80100	J427R	LEMPCO	80955	J5130	ALCO	80171
J2100	ALLOY	88001.96	J3102	ALCO	80112	J4-3	ALCO	88002.33	J5130	ALLOY	80171
J2100	REPUBLIC	80049	J3102	ALLOY	80112	J4-3	ALLOY	88002.33	J5-130	SPICER	80117
J2101	ALCO	80044	J3102	GEN.EQUIP.	80112	J4-4	ALLOY	88002.33	J5131	ALCO	80175
J2101	ALLOY	88001.98	J3103	ALCO	80098	J4-5	ALCO	88002.33	J5131	ALLOY	80175
J2103	ALCO	80048	J3105	ALCO	80104	J4-5	ALLOY	88002.33	J5-131	SPICER	80121
J2103	ALLOY	80048	J3105	ALLOY	80104	J4-6	ALLOY	88002.33	J5-132	ALCO	80075
J2104	ALCO	80047	J3105	ALLOY	80107	J4-7	ALLOY	88002.36	J5-132	ALLOY	80077
J2104	ALLOY	80047	J3105	ALLOY	80112	J5000	BORG-WARNR	80168	J5-133	ALCO	80116
J2108	ALCO	80044	J3106	ALCO	80101	J5007	BORG-WARNR	80177	J5-133	ALLOY	80116
J2109	ALCO	80049	J3106	ALLOY	80101	J5008	BORG-WARNR	80174	J5-133	REPUBLIC	80116
J2110	ALCO	80048	J3108	ALCO	80112	J5015	ALCO	80178	J5-134	ALCO	80118
J2111	ALCO	80049	J3111	ALCO	80104	J5015	BORG-WARNR	80173	J5-134	ALLOY	80118
J2111	BORG-WARNR	80049	J3113	ALCO	80112	J5031	ALCO	80171	J5-138	ALCO	80072
J2112	ALCO	80044	J3114	ALCO	80104	J5031	ALLOY	80171	J5139	ALCO	80170
J2113	ALCO	80053	J3115	ALCO	80114	J5-10	SPICER	80142	J5-14	ALCO	80024
J2113	SPICER	80053	J3116	ALCO	80111	J5100	ALCO	80168	J5-14	ALLOY	80024
J2114	ALCO	80054	J3118	ALCO	80101	J5100	ALLOY	80168	J5146	ALCO	80172
J2114	ALLOY	80054	J3119	ALCO	80100	J5-101	ALCO	80023	J5148	ALCO	80168
J2116	ALCO	80047	J3120	ALCO	80113	J5101	ALCO	80167	J5148	ALLOY	80216
J2117	ALCO	80046	J3120	ALLOY	80113	J5-101	ALLOY	80023	J5149	ALCO	80171
J2117	ALLOY	80046	J3121	ALCO	80102	J5101	ALLOY	80167	J5-15	ALCO	80119
J2121	ALCO	80044	J3123	ALCO	80102	J5102	ALCO	80171	J5-15	SPICER	80178
J2123	ALCO	80045	J3125	ALCO	80103	J5102	ALLOY	80171	J5-150	ALCO	80119
J2124	ALCO	80047	J3155	ALCO	80103	J5103	ALCO	80174	J5-150	ALLOY	80119
J2125	ALCO	80047	J3188	ALCO	80105	J5103	ALLOY	80174	J5-150	ALLOY	80120
J2126	ALCO	80049	J3188	ALLOY	80105	J5-105	ALCO	80031	J51525-2	ALCO	88001.60
J2126	ALLOY	80049	J4002	BORG-WARNR	80109	J5108	ALCO	80172	J5-153	ALCO	80077
J2127	ALCO	80044	J4007	BORG-WARNR	80147	J5108	ALLOY	80172	J5-153	ALLOY	80072
J2127	ALLOY	80044	J4008	ALCO	80146	J5109	ALCO	80162	J5-153	ALLOY	80077
J2131	ALCO	80044	J4008	ALLOY	80146	J5109	ALLOY	80162	J5153B	ALCO	80077
J2134	ALCO	80049	J4011	ALCO	80147	J5110	ALCO	80173	J5-155	ALCO	80182
J2140	ALCO	80055	J4014	BORG-WARNR	80158	J5110	ALLOY	80173	J5-155	ALLOY	80182
J2140	ALLOY	80055	J4015	BORG-WARNR	80148	J5111	ALCO	80175	J5-16	SPICER	80189
J2150	ALCO	80048	J4016	BORG-WARNR	80159	J5111	ALLOY	80175	J5-160	ALCO	80142
J2153	ALCO	80055	J4029	ALCO	80109	J5112	ALCO	80177	J5-160	ALLOY	80142
J2153	ALLOY	80055	J4-1	ALCO	88002.34	J5-1124	SPICER	80225	J5-1601	ALCO	80004
J2160	ALCO	80055	J4-1	ALLOY	88002.34	J5-113	SPICER	80077	J5-1601	ALLOY	80006
J2173	ALCO	80058	J4100	ALCO	80109	J5114	TEDDY TORQ	80176	J5-16015	ALCO	80004
J2173	ALLOY	80058	J4100	ALLOY	80109	J5-115	ALCO	80178	J5-1601A	ALLOY	80004
J2175	ALCO	80058	J4101	ALCO	80153	J5-115	ALLOY	80178	J5-1601S	ALLOY	80004
J2175	ALLOY	80058	J4101	ALLOY	80153	J5-116	ALCO	80189	J5-1602	ALCO	80019
J2198	ALCO	80051	J4102	ALCO	80158	J5-116	ALLOY	80189	J5-1602	ALLOY	80019
J2198	ALCO	80052	J4102	ALLOY	80158	J5118	ALCO	80168	J5-1603	ALCO	80010

Part No.	Brand	IDLI Group
J5-1603	ALLOY	80010
J5-1604	ALCO	80009
J5-1604	ALLOY	80009
J5-1605	ALLOY	80012
J5-1606	ALLOY	80018
J5-1607	ALLOY	80011
J5-1608	ALLOY	80017
J5-160B	ALCO	80143
J5-160B	ALLOY	80143
J51625-2	ALCO	80042
J51625-2	ALLOY	80042
J5-165	ALCO	80190
J5-165	ALLOY	80190
J5-17	SPICER	80221
J5-170	ALCO	80007
J5-170	ALLOY	80007
J5-170	ALLOY	88000.15
J5-177	ALCO	80221
J5-177	ALLOY	80221
J5-178	ALCO	80124
J5-178	ALLOY	80124
J5-178	TEDDY TORQ	80124
J5-178B	ALCO	80125
J5-178B	ALLOY	80125
J5-178UB	ALCO	80125
J5-18	SPICER	80124
J5-181	ALCO	80024
J5-181	ALCO	80140
J5-181	ALLOY	80023
J5-185	ALCO	80228
J5-188	ALCO	80140
J5-188	ALCO	80182
J5-188	ALLOY	80140
J5-188	ALLOY	80182
J5-200	ALCO	80075
J5-200	ALLOY	80075
J5-200B	ALCO	80076
J5-200B	ALLOY	80076
J5-21	SPICER	80072
J5-213	ALCO	80117
J5-213	ALLOY	80117
J5-230	SPICER	80117
J5-231	ALCO	80121
J5-231	ALLOY	80121
J5-248	ALCO	80025
J5-248	ALLOY	80025
J5-260	ALLOY	80078
J5-278	ALLOY	80085
J5-279	ALLOY	80189
J5-280	ALLOY	80221
J5-281	ALLOY	80227
J5-297	ALLOY	80079
J52A	ALCO	88001.66
J53025-2	ALLOY	80135
J5-315	ALLOY	80026
J5-350	ALLOY	80036
J53525-2	ALCO	88001.62
J5-353	ALCO	80083
J5-353	ALLOY	80083
J53625-2	ALCO	80135
J53625-2	ALLOY	80135
J53725-2	DETROIT	80135
J53825-2	ALCO	80135
J53825-2	ALLOY	80135
J5-4	ALCO	80041
J5-4	ALCO	88001.85
J5-4	ALLOY	80041
J54A	ALCO	80901
J54A	ALCO	88001.66
J5-60	SPICER	80142
J5-70	SPICER	80007
J5-74	ALCO	80090
J5-74	ALLOY	80090
J5-75	ALCO	80078
J5-75	ALCO	80090
J5-75	ALLOY	80090
J5-78	SPICER	80124
J5-81	SPICER	80024
J5-90	ALCO	80093
J5-90	ALLOY	80093
J5-74/75	ALLOY	80090
J6000	ALLOY	80193
J6000	BORG-WARNR	80193
J6100	ALCO	80193
J6100	ALLOY	80193
J6100	ALLOY	80198
J6101	ALCO	80195
J6101	ALLOY	80195
J6102	ALCO	80197
J6102	ALLOY	80197
J6103	ALCO	80194
J6103	ALLOY	80194
J6104	ALCO	80192
J6104	ALLOY	80192
J6105	FWD	80213
J6106	ALCO	80198
J6106	ALLOY	80198
J6107	ALCO	80202
J6107	ALLOY	80202
J6108	ALCO	80172
J6108	ALCO	80199
J6108	ALLOY	80199
J6109	ALCO	80201
J6109	ALLOY	80201
J6123	ALCO	80191
J6139	ALCO	80203
J6140	ALCO	80191
J6141	ALCO	80193
J6142	ALCO	80197
J6179	ALCO	80200
J7000	ALLOY	80217
J7000	BORG-WARNR	80210
J7033	ALCO	80210
J7033	ALLOY	80217
J7100	ALLOY	80210
J7100	ALLOY	80217
J7100	BORG-WARNR	80210
J7101	ALLOY	80214
J7101	BORG-WARNR	80211
J7102	ALLOY	80212
J7102	ALLOY	80214
J7102	ALLOY	80218
J7102	BORG-WARNR	80212
J7105	ALCO	80213
J7105	ALCO	88002.97
J7105	ALLOY	80213
J7112	ALCO	80213
J7112	ALCO	88002.97
J7113	ALCO	80213
J7113	ALCO	88002.97
J7117	ALCO	80213
J7117	ALCO	88002.97
J7122	ALCO	80213
J7122	ALCO	88002.97
J7125-2	DETROIT	80034
J71425-2	ALCO	80034
J71425-2	ALLOY	80034
J7200	ALCO	80210
J7200	ALCO	88002.82
J7200	ALLOY	80210
J7201	ALCO	80211
J7201	ALLOY	80211
J7202	ALCO	80212
J7202	ALLOY	80212
J7203	ALCO	80208
J7204	ALCO	80209
J7205	ALCO	80210
J7206	ALCO	80211
J7208	ALCO	80208
J7210	ALCO	80211
J72251-2	DETROIT	80056
J7225-2	ALCO	80064
J7225-2	DETROIT	88001.63
J7242	ALCO	80210
J72425-2	ALCO	80034
J725251-2	ALCO	80056
J72525-2	ALCO	80056
J72525-3	DETROIT	88001.63
J7256	ALCO	80214
J72625-2	ALCO	80064
J72625-2	ALLOY	80064
J72625-2	ALLOY	80065
J72625-3	ALCO	80061
J72625-3	DETROIT	80060
J72625-4	ALCO	80060
J72625-4	ALLOY	80060
J72625-4	ALLOY	80061
J72625-5	ALCO	80132
J72625-5	ALLOY	80132
J72625-6	ALCO	80061
J72625-6	ALLOY	80062
J72625-7	ALCO	80063
J72625-7	ALLOY	80063
J7325-2	ALCO	80096
J7325-2	DETROIT	80135
J73252-2	ALCO	80096
J7325-5	ALCO	80096
J7352-4	ALCO	80094
J735251-2	ALCO	80095
J73525-2	ALCO	80095
J73525-2	DETROIT	80056
J73525-3	ALCO	80096
J73525-4	ALCO	80094
J7604	ALCO	88003.45
J7605	ALCO	88003.56
J7607	ALCO	88003.49
J7608	ALCO	88003.46
J7612	ALCO	88003.50
J7622	ALCO	88003.42
J8002	ALLOY	80243
J8002	BORG-WARNR	80231
J8014	ALCO	80230
J8100	ALLOY	80231
J8100	ALLOY	80243
J8100	BORG-WARNR	80231
J8100	GEN.EQUIP.	80240
J8101	ALLOY	80244
J8101	BORG-WARNR	80232
J8102	ALLOY	80237
J8102	ALLOY	80244
J8102	BORG-WARNR	80237
J8102	GEN.EQUIP.	80237
J8104	ALCO	80230
J8105	ALCO	80238
J8105	ALCO	88003.20
J8105	ALLOY	80238
J8107	ALCO	80238
J8107	ALCO	88003.20
J8107	ALLOY	80238
J8113	ALCO	80238
J8113	ALCO	88003.20
J8200	ALCO	80231
J8200	ALLOY	80231
J8201	ALCO	80232
J8201	ALLOY	80232
J8202	ALCO	80237
J8202	ALLOY	80237
J8203	ALCO	80236
J8203	ALLOY	80236
J8205	ALCO	80240
J8205	ALLOY	80231
J8207	ALCO	80239
J8215	ALCO	80237
J83254-3	ALCO	80022
J8500	ALCO	88003.28
J8500	ALLOY	80224
J8502	ALCO	80224
J8506	ALCO	80237
J8507	ALCO	80237
J8515	ALCO	80222
J9000	BORG-WARNR	80235
J9001	BORG-WARNR	80234
J9002	ALCO	88003.09
J9010	ALCO	80233
J9010	ALCO	80241
J9014	ALCO	80235
J9014	ALLOY	80235
J9015	ALCO	80234
J9015	ALLOY	80234
J9016	ALCO	80234
J9016	ALCO	88012.36
J9016	ALLOY	80234
J9017	ALCO	80233
J9017	ALCO	80241
J9017	ALLOY	80233
J9017	ALLOY	80241
J9017	ALLOY	88003.09
J9026	ALLOY	80233
J9030	ALCO	80233
J9031	ALCO	80234
J9031	ALLOY	80234
J9038	ALCO	80234
J906926	CHRYSLER	80077
JC96-55	ALCO	80204
JD1002	BMC	80024
JD96-55	ALCO	80160
JD96-55-4	ALCO	80160
JDS55-55-2	ALCO	80145
JDS55-55-2	ALLOY	80145
JDS56-55-2	ALCO	80145
JDS56-55-2	ALLOY	80145
JJ6109	ALLOY	80201
J055-55	ALCO	80141
J055-55	ALLOY	80141
J055-55-2	ALCO	80141
J096-55	ALCO	80138
J096-55	ALLOY	80138
J0S56-55-2	ALLOY	80145
JP55-55-140	CLEVE.STL.	80129
JP55-55-145	CLEVE.STL.	80131
JP96-55	ALCO	80068
JP96-55	ALLOY	80068
JR5100	ALCO	80168
JR52A	ALCO	80909
JR52A	ALLOY	80909
JR54	ALCO	88001.74
JR54A	ALCO	80901
JR54A	ALCO	88001.66
JR55-55	ALCO	80123
JR55-55	ALLOY	80123

Part No.	Brand	IDLI Group	Part No.	Brand	IDLI Group	Part No.	Brand	IDLI Group	Part No.	Brand	IDLI Group
JR56A	DETROIT	80907	K3591A	NEW IDEA	80037	K5GB127	HARDY SPCR	80178	K5LGB194	HARDY SPCR	80023
JR57A	ALCO	80908	K3595	NEW IDEA	80037	K5GB15	HARDY SPCR	80178	K5LGB194	HARDY SPCR	80024
JR60	ALCO	88001.73	K3595A	NEW IDEA	80037	K5GB150	HARDY SPCR	80023	K5LGB195	HARDY SPCR	80024
JR62A	ALCO	80907	K3803A	NEW IDEA	80044	K5GB150	REPCO	80023	K5LGB195	HARDY SPCR	80041
JR65A	ALCO	80905	K4	TEDDY TORQ	80041	K5GB152	HARDY SPCR	80024	K5LGB2	HARDY SPCR	80024
JR65A	ALLOY	80905	K4464	PULLMAN	80193	K5GB15R	REPCO	80178	K5LGB21	HARDY SPCR	80072
JR67A	ALCO	80906	K4464	UNION PAC.	80193	K5GB16	HARDY SPCR	80189	K5LGB213	HARDY SPCR	80142
JR67A	ALCO	80953	K50	TEDDY TORQ	80119	K5GB160	HARDY SPCR	80189	K5LGB274	HARDY SPCR	80024
JR67A	ALLOY	80906	K50G	TEDDY TORQ	80119	K5GB161	HARDY SPCR	80041	K5LGB279	HARDY SPCR	80024
JR96-55	ALCO	80122	K5-1	MUNCIE	80023	K5GB162	HARDY SPCR	80077	K5LGB33	HARDY SPCR	80077
JS101	ALLOY	80023	K5-10X	DIAMOND T	80142	K5GB163	HARDY SPCR	80124	K5LGB36	HARDY SPCR	80041
JS53-55	ALCO	80039	K510X	REO MOTORS	80142	K5GB164	HARDY SPCR	80142	K5LGB36R	REPCO	80041
JS53-55	ALLOY	80038	K5-13XR	GMB	80077	K5GB165	HARDY SPCR	80180	K5LGB41	HARDY SPCR	80024
JS53-55	ALLOY	80039	K5-13XR	HARDY SPCR	80077	K5GB166	HARDY SPCR	80077	K5LGB50	HARDY SPCR	80023
JS55-55	ALCO	80081	K513XR	REPCO	80077	K5GB17	HARDY SPCR	80221	K5LGB50R	REPCO	80023
JS55-55	ALCO	88001.87	K5-16	HARDY SPCR	80189	K5GB179	HARDY SPCR	80142	K5LGB50R	REPCO	80024
JS55-55	ALLOY	80081	K5-16R	REPCO	80189	K5GB18	HARDY SPCR	80124	K5LGB50R	REPCO	80140
JS55-55-1330	CLEVE.STL.	80130	K5-17	HARDY SPCR	80221	K5GB192	HARDY SPCR	80124	K5LGB64	HARDY SPCR	80023
JS56-55	ALCO	80145	K5-17R	REPCO	80221	K5GB206	HARDY SPCR	88001.82	K5LGB64	HARDY SPCR	80024
JS96-55	ALCO	80038	K5-18X	DIAMOND T	80124	K5GB22	HARDY SPCR	80180	K5LGB74	HARDY SPCR	80077
JT55-55-2	ALCO	80071	K5-21	HARDY SPCR	80023	K5GB22R	REPCO	80180	K5LGB74	REPCO	80075
JT55-55-2	ALLOY	80071	K5-21XR	REPCO	80072	K5GB26	HARDY SPCR	80023	K5LGB74	REPCO	80077
JTS55-35-2	ALCO	80074	K5-22	HARDY SPCR	80024	K5GB277	HARDY SPCR	80023	K5LGB74	REPCO	80078
JTS55-55-2	ALCO	80074	K5-24	HARDY SPCR	80041	K5GB35	HARDY SPCR	80040	K5LGB74	REPCO	80085
JTS55-55-2	ALLOY	80074	K5279X	REPCO	80189	K5GB35	REPCO	80040	K5LGB80	HARDY SPCR	80024
JTS55-55-4	ALCO	80081	K5280X	REPCO	80221	K5GB50	HARDY SPCR	80023	K5LGB80R	REPCO	80024
JTS55-55-4	ALLOY	80081	K5281X	REPCO	80227	K5GB50	HARDY SPCR	80024	K5LGB81	HARDY SPCR	80023
JU55-55-2	ALCO	80183	K53	TEDDY TORQ	80118	K5GB55	HARDY SPCR	80221	K5LGB81	HARDY SPCR	80024
JU56-55-2	ALCO	80183	K5-4	HARDY SPCR	80041	K5GB65	HARDY SPCR	80189	K5LGB82	HARDY SPCR	80077
JU96-55	ALCO	80181	K54	TEDDY TORQ	80073	K5GB65	REPCO	80189	K5LGB83	HARDY SPCR	80041
JU96-55-4	ALCO	80181	K5-4R	REPCO	80041	K5GB79	HARDY SPCR	80024	K5LGB87	HARDY SPCR	80041
K1	TEDDY TORQ	80023	K55	TEDDY TORQ	80182	K5GB79	REPCO	80023	K5LGB99	HARDY SPCR	80023
K10	TEDDY TORQ	80142	K5-70	MUNCIE	80007	K5GB79	REPCO	80024	K5LGB99	HARDY SPCR	80024
K10U	TEDDY TORQ	80143	K5A13XR	REPCO	80077	K5GB79	REPCO	80140	K5PGB55	HARDY SPCR	80221
K11513	PULLMAN	80210	K5A502	REPCO	80081	K5GB91	HARDY SPCR	80124	K5PL10	HARDY SPCR	80142
K14	TEDDY TORQ	80023	K5A507	REPCO	80091	K5GB96	HARDY SPCR	80023	K5PL39	HARDY SPCR	80041
K14	TEDDY TORQ	80024	K5A510	REPCO	80010	K5L1	HARDY SPCR	80024	K5SR13XR	REPCO	80077
K15	TEDDY TORQ	80178	K5A517	REPCO	80006	K5L1	REPCO	80024	K5SR178X	REPCO	80124
K16	TEDDY TORQ	80189	K5A519	REPCO	80017	K5L12	HARDY SPCR	80023	K60	TEDDY TORQ	80075
K17	TEDDY TORQ	80221	K5A64R	REPCO	80024	K5L4	HARDY SPCR	80041	K60452	DEERE,JOHN	80134
K18244	CHRYSLER	80023	K5EF118	REPCO	80124	K5L4	REPCO	80041	K60U	TEDDY TORQ	80076
K18246	CHRYSLER	80041	K5EF118	REPCO	80125	K5L49	REPCO	80041	K61	TEDDY TORQ	80025
K18U	TEDDY TORQ	80125	K5G81	REPCO	80023	K5L4R	REPCO	80041	K65	TEDDY TORQ	80190
K1911A	NEW IDEA	80008	K5G81	REPCO	80140	K5LBG176	BORG-WARNR	80180	K67709	DEERE,JOHN	80185
K1912A	NEW IDEA	80008	K5GB	HARDY SPCR	80024	K5LGB100	HARDY SPCR	80023	K6GB106	HARDY SPCR	80072
K1965A	NEW IDEA	80133	K5GB1	HARDY SPCR	80023	K5LGB102	HARDY SPCR	80023	K70	TEDDY TORQ	80007
K2	TEDDY TORQ	80041	K5GB1	REPCO	80023	K5LGB102	HARDY SPCR	80024	K74	TEDDY TORQ	80090
K21	TEDDY TORQ	80072	K5GB10	HARDY SPCR	80142	K5LGB102R	REPCO	80023	K81	TEDDY TORQ	80024
K22173	CHRYSLER	80180	K5GB103	HARDY SPCR	80142	K5LGB102R	REPCO	80024	K88	TEDDY TORQ	80140
K22318	CHRYSLER	80124	K5GB104	HARDY SPCR	80077	K5LGB102R	REPCO	80140	K9568	MICHIGAN	80135
K23645	CHRYSLER	80142	K5GB106	HARDY SPCR	80072	K5LGB103	HARDY SPCR	80142	KAR758	OLIVER	80001
K23703	CHRYSLER	80023	K5GB106	REPCO	80075	K5LGB104	HARDY SPCR	80024	KAR758	WHITE FARM	80001
K23703	CHRYSLER	80024	K5GB106	REPCO	80077	K5LGB104R	REPCO	80075	KC2766	WHITE	80109
K23704	CHRYSLER	80024	K5GB106	REPCO	80078	K5LGB104R	REPCO	80077	KC7012	DIAMOND R	80210
K23958	CHRYSLER	80024	K5GB107	HARDY SPCR	80124	K5LGB104R	REPCO	80078	KC7012	DIAMOND T	80210
K24	TEDDY TORQ	80225	K5GB10R	REPCO	80142	K5LGB104R	REPCO	80085	KC7012	WHITE	80210
K2550A	NEW IDEA	80001	K5GB113	HARDY SPCR	80023	K5LGB106	REPCO	80085	KC7012	WHITE	80217
K26110	CHRYSLER	80189	K5GB115	HARDY SPCR	80023	K5LGB128	HARDY SPCR	80077	KC7013	WHITE	80243
K26345	CHRYSLER	80142	K5GB115	REPCO	80023	K5LGB130	HARDY SPCR	80041	KD10X	WHITE	80142
K2701	AEC BRIT.	80189	K5GB115	REPCO	80024	K5LGB133	HARDY SPCR	80077	KF10X	DIAMOND R	80142
K2703	AEC BRIT.	80221	K5GB115	REPCO	80140	K5LGB166	HARDY SPCR	80077	KF10X	DIAMOND R	80143
K2708	AEC BRIT.	80225	K5GB116	HARDY SPCR	80041	K5LGB174	HARDY SPCR	80023	KIT ONLY	CHRYSLER	80905
K2964A	NEW IDEA	84531	K5GB117	HARDY SPCR	80077	K5LGB174	HARDY SPCR	80024	KIT ONLY	CHRYSLER	80906
K2980A	NEW IDEA	80069	K5GB117	REPCO	80075	K5LGB176	HARDY SPCR	80178	KS4GB3458	HARDY SPCR	88003.81
K30	TEDDY TORQ	80117	K5GB117	REPCO	80077	K5LGB176	HARDY SPCR	80180	KS-BG166	REPCO	80118
K31	TEDDY TORQ	80121	K5GB117	REPCO	80078	K5LGB179	HARDY SPCR	80142	KS-GB179	HARDY SPCR	88004.32
K33	TEDDY TORQ	80116	K5GB117	REPCO	80085	K5LGB182	HARDY SPCR	80124	KT105	ROCKWELL	88013.82
K3339A	NEW IDEA	84510	K5GB117R	REPCO	80077	K5LGB183	HARDY SPCR	80140	KT109	ROCKWELL	88013.83
K3347A	NEW IDEA	80035	K5GB118	HARDY SPCR	80124	K5LGB185	HARDY SPCR	80023	KT110	ROCKWELL	88013.84
K3410A	NEW IDEA	80037	K5GB119	HARDY SPCR	80142	K5LGB186	HARDY SPCR	80041	KT111	ROCKWELL	88013.85

Part No.	Brand	IDLI Group
KT116	ROCKWELL	88013.86
KT117	ROCKWELL	88013.87
KT118	ROCKWELL	88013.88
L12-10	ROCKWELL	81203
L12-11	ROCKWELL	81207
L12-12	ROCKWELL	81213
L12-13	ROCKWELL	81214
L12-14	ROCKWELL	81220
L12-15	ROCKWELL	81221
L12-16	ROCKWELL	81234
L12-17	ROCKWELL	81235
L12-18	ROCKWELL	81249
L12-19	ROCKWELL	81268
L12-20	ROCKWELL	81261
L12-21	ROCKWELL	81262
L12-22	ROCKWELL	81285
L12-23	ROCKWELL	81301
L12-24	ROCKWELL	81302
L12-25	ROCKWELL	81303
L12-26	ROCKWELL	81349
L12-27	ROCKWELL	81205
L12-28	ROCKWELL	81209
L12-29	ROCKWELL	81217
L12-30	ROCKWELL	81239
L12-31	ROCKWELL	81265
L12-32	ROCKWELL	81256
L12-33	ROCKWELL	81251
L12-34	ROCKWELL	81291
L12-35	ROCKWELL	81314
L12-36	ROCKWELL	81300
L12-37	ROCKWELL	81331
L12-38	ROCKWELL	81259
L12-39	ROCKWELL	85048
L12-40	ROCKWELL	85050
L12-41	ROCKWELL	85052
L12-42	ROCKWELL	85054
L12-43	ROCKWELL	85059
L12-44	ROCKWELL	85063
L12-45	ROCKWELL	85049
L12-46	ROCKWELL	85051
L12-47	ROCKWELL	85055
L12-48	ROCKWELL	85053
L12-49	ROCKWELL	85057
L12-50	ROCKWELL	85061
L12-51	ROCKWELL	85062
L12-52	ROCKWELL	85064
L12-53	ROCKWELL	85058
L12-54	ROCKWELL	85060
L12N170-1	ROCKWELL	86650
L12N2424	ROCKWELL	88013.89
L12N2438	ROCKWELL	88013.90
L12N2453	ROCKWELL	88013.91
L12NF100	ROCKWELL	84607
L12NF123A	ROCKWELL	84610
L12NF130	ROCKWELL	84611
L12NF138	ROCKWELL	84609
L12NF81	ROCKWELL	84608
L12NLS18DC	ROCKWELL	85056
L12NPS18	ROCKWELL	87223
L12NY20	ROCKWELL	85509
L12NY30-14	ROCKWELL	85515
L12NY30-23	ROCKWELL	85514
L12NY30-23	ROCKWELL	88013.92
L12NY30-26	ROCKWELL	85512
L12NY30-3	ROCKWELL	85510
L12NY30-5	ROCKWELL	85513
L12NY30-8	ROCKWELL	85511
L12NYD16-3	ROCKWELL	81228
L12NYH16-3	ROCKWELL	81224
L12NYH16-4	ROCKWELL	81230
L12NYH16-5	ROCKWELL	81243
L12NYH18	ROCKWELL	81260
L12NYH18-2	ROCKWELL	81281
L12NYH20	ROCKWELL	88013.93
L12NYP12	ROCKWELL	81211
L12NYP16-11	ROCKWELL	81267
L12NYP16-13	ROCKWELL	81253
L12NYP16-20	ROCKWELL	81271
L12NYP16-27	ROCKWELL	81273
L12NYP16-39	ROCKWELL	81255
L12NYP20-1	ROCKWELL	81345
L12NYP20-13	ROCKWELL	81347
L12NYP20-8	ROCKWELL	81330
L12NYP20-9	ROCKWELL	81346
L12NYQ12-15	ROCKWELL	88013.94
L12NYQ12-2	ROCKWELL	81206
L12NYQ13-2	ROCKWELL	88013.95
L12NYQ14	ROCKWELL	86651
L12NYQ14-2	ROCKWELL	81218
L12NYQ14-4	ROCKWELL	88013.96
L12NYQ16-16	ROCKWELL	88013.97
L12NYQ16-17A	ROCKWELL	84505
L12NYQ16-26	ROCKWELL	81246
L12NYQ16-3	ROCKWELL	81240
L12NYQ16-4	ROCKWELL	81227
L12NYQ16-6	ROCKWELL	81229
L12NYQ18-3	ROCKWELL	81277
L12NYQ18-4	ROCKWELL	81266
L12NYQ18-5	ROCKWELL	81276
L12NYQ18-6	ROCKWELL	81272
L12NYR12-3	ROCKWELL	81204
L12NYR12-3	ROCKWELL	88013.98
L12NYR13	ROCKWELL	81208
L12NYR14-18	ROCKWELL	81210
L12NYR14-4	ROCKWELL	81215
L12NYR14-7	ROCKWELL	81216
L12NYR14-8	ROCKWELL	81212
L12NYR15-1	ROCKWELL	81222
L12NYR15-10	ROCKWELL	81223
L12NYR15-2	ROCKWELL	81219
L12NYR16	ROCKWELL	86654
L12NYR16-105	ROCKWELL	81241
L12NYR16-106	ROCKWELL	81248
L12NYR16-13	ROCKWELL	81247
L12NYR16-14	ROCKWELL	81236
L12NYR16-15	ROCKWELL	88013.99
L12NYR16-23	ROCKWELL	81237
L12NYR16-27	ROCKWELL	81238
L12NYR16-29	ROCKWELL	81242
L12NYR16-32	ROCKWELL	81244
L12NYR16-36	ROCKWELL	88014
L12NYR16-5	ROCKWELL	88014.01
L12NYR16-55	ROCKWELL	86653
L12NYR16-8	ROCKWELL	81245
L12NYR16-9	ROCKWELL	88014.02
L12NYR17-1	ROCKWELL	81250
L12NYR18	ROCKWELL	81201
L12NYR18-10	ROCKWELL	81202
L12NYR18-12	ROCKWELL	81270
L12NYR18-16	ROCKWELL	81275
L12NYR18-25	ROCKWELL	81282
L12NYR18-30	ROCKWELL	81269
L12NYR18-4	ROCKWELL	81264
L12NYR18-5	ROCKWELL	81257
L12NYR18-53	ROCKWELL	81263
L12NYR18-57	ROCKWELL	81200
L12NYR18-60	ROCKWELL	86655
L12NYR18-73	ROCKWELL	86656
L12NYR18-93	ROCKWELL	81278
L12NYR18-98	ROCKWELL	81258
L12NYR19-14	ROCKWELL	81288
L12NYR19-29	ROCKWELL	81289
L12NYR19-3	ROCKWELL	81286
L12NYR19-7	ROCKWELL	81287
L12NYR20	ROCKWELL	81317
L12NYR20-10	ROCKWELL	81298
L12NYR20-102	ROCKWELL	81309
L12NYR20-15	ROCKWELL	81304
L12NYR20-17	ROCKWELL	81296
L12NYR20-173	ROCKWELL	81308
L12NYR20-18	ROCKWELL	81310
L12NYR20-19	ROCKWELL	81319
L12NYR20-20	ROCKWELL	81320
L12NYR20-25	ROCKWELL	81305
L12NYR20-3	ROCKWELL	81325
L12NYR20-30	ROCKWELL	81306
L12NYR20-37	ROCKWELL	88014.03
L12NYR20-40	ROCKWELL	88014.04
L12NYR20-42	ROCKWELL	88014.05
L12NYR20-43	ROCKWELL	81321
L12NYR20-45	ROCKWELL	81293
L12NYR20-48	ROCKWELL	81322
L12NYR20-51	ROCKWELL	81323
L12NYR20-52	ROCKWELL	81297
L12NYR20-61	ROCKWELL	86658
L12NYR20-62	ROCKWELL	81312
L12NYR20-85	ROCKWELL	81307
L12NYR20-9	ROCKWELL	81318
L12NYR22	ROCKWELL	81341
L12NYR22-12	ROCKWELL	81340
L12NYR22-16	ROCKWELL	88014.06
L12NYR22-20	ROCKWELL	81335
L12NYR22-27	ROCKWELL	81333
L12NYR22A	ROCKWELL	88014.07
L12NYR23-9	ROCKWELL	81348
L12NYR24-4	ROCKWELL	81350
L12NYR24-7	ROCKWELL	81351
L12NYR24-7	ROCKWELL	88014.08
L12NYS12-2	ROCKWELL	88014.09
L12NYS14-11	ROCKWELL	88014.10
L12NYS15-4	ROCKWELL	88014.11
L12NYS16	ROCKWELL	86652
L12NYS16-11	ROCKWELL	81225
L12NYS16-12	ROCKWELL	81231
L12NYS16-13	ROCKWELL	88014.12
L12NYS16-14	ROCKWELL	81232
L12NYS16-22	ROCKWELL	81226
L12NYS16-8	ROCKWELL	81233
L12NYS18	ROCKWELL	81279
L12NYS18-1	ROCKWELL	81274
L12NYS18-10	ROCKWELL	88014.13
L12NYS18-11	ROCKWELL	81252
L12NYS18-15	ROCKWELL	88014.14
L12NYS18-17	ROCKWELL	88014.15
L12NYS18-18	ROCKWELL	81254
L12NYS18-25	ROCKWELL	88014.16
L12NYS18-54	ROCKWELL	81280
L12NYS18-8	ROCKWELL	88014.17
L12NYS19-10	ROCKWELL	81283
L12NYS19-3	ROCKWELL	81290
L12NYS19-4	ROCKWELL	81284
L12NYS20	ROCKWELL	81294
L12NYS20-1	ROCKWELL	81292
L12NYS20-10A	ROCKWELL	84507
L12NYS20-15	ROCKWELL	88014.18
L12NYS20-17	ROCKWELL	81315
L12NYS20-21	ROCKWELL	81316
L12NYS20-23	ROCKWELL	81329
L12NYS20-3	ROCKWELL	81326
L12NYS20-34	ROCKWELL	86657
L12NYS20-46	ROCKWELL	81311
L12NYS20-47A	ROCKWELL	84506
L12NYS20-52	ROCKWELL	81295
L12NYS20-55	ROCKWELL	81299
L12NYS20-58	ROCKWELL	81327
L12NYS20-65	ROCKWELL	81324
L12NYS20-65	ROCKWELL	81328
L12NYS20-7	ROCKWELL	88014.19
L12NYS20-9	ROCKWELL	88014.20
L12NYS22-1	ROCKWELL	81338
L12NYS22-10	ROCKWELL	81332
L12NYS22-12	ROCKWELL	81342
L12NYS22-16A	ROCKWELL	84508
L12NYS22-21	ROCKWELL	81339
L12NYS22-23	ROCKWELL	81344
L12NYS22-29	ROCKWELL	81334
L12NYS22-30	ROCKWELL	81337
L12NYS22-50	ROCKWELL	81343
L12NYS22-60	ROCKWELL	81336
L12NYS22-6A	ROCKWELL	84509
L12NYS22A	ROCKWELL	84510
L14S170-1	ROCKWELL	87001
L14SDCA8	ROCKWELL	87002
L14SF113	ROCKWELL	84903
L14SF159	ROCKWELL	84904
L14SF168A	ROCKWELL	84905
L14SF184	ROCKWELL	84901
L14SF89	ROCKWELL	84902
L14SLS20-4DC	ROCKWELL	85301
L14SPS20-2	ROCKWELL	87253
L14SY18-5	ROCKWELL	88014.21
L14SY29-1	ROCKWELL	85702
L14SY29-10	ROCKWELL	85701
L14SY29-18	ROCKWELL	85704
L14SY30-39	ROCKWELL	85706
L14SY30-43	ROCKWELL	85705
L14SY32-17	ROCKWELL	85707
L14SY32-6	ROCKWELL	85703
L14SYH18-1	ROCKWELL	83046
L14SYH18-2	ROCKWELL	83058
L14SYH18-3	ROCKWELL	83051
L14SYH18-5	ROCKWELL	87004
L14SYH18-7	ROCKWELL	83045
L14SYH18-8	ROCKWELL	83029
L14SYH20	ROCKWELL	87009
L14SYP16-1	ROCKWELL	83036
L14SYP16-11	ROCKWELL	83040
L14SYP16-18	ROCKWELL	83050
L14SYP16-33	ROCKWELL	83038
L14SYP16-40	ROCKWELL	83037
L14SYP16-55	ROCKWELL	83041
L14SYP16-6	ROCKWELL	83043
L14SYP16-71	ROCKWELL	83039
L14SYP20-1	ROCKWELL	83124
L14SYP20-11	ROCKWELL	83123
L14SYQ14-2	ROCKWELL	83004
L14SYQ14-3	ROCKWELL	83005
L14SYQ15-3	ROCKWELL	83008
L14SYQ15-4	ROCKWELL	83009
L14SYQ16-11	ROCKWELL	88014.22
L14SYQ16-15	ROCKWELL	83013
L14SYQ16-18	ROCKWELL	88014.23
L14SYQ16-30	ROCKWELL	83017
L14SYQ16-34	ROCKWELL	83012
L14SYQ16-4	ROCKWELL	88014.24
L14SYQ16-41	ROCKWELL	83014
L14SYQ16-52	ROCKWELL	83016
L14SYQ16-8	ROCKWELL	83015
L14SYQ18	ROCKWELL	83049
L14SYQ19-15	ROCKWELL	88014.25

Part No.	Brand	IDLI Group	Part No.	Brand	IDLI Group	Part No.	Brand	IDLI Group	Part No.	Brand	IDLI Group
L14SYQ19-17	ROCKWELL	83060	L14SYR22-11	ROCKWELL	83112	L16SF195A	ROCKWELL	84918	L16SYR21-11	ROCKWELL	83298
L14SYQ19-5	ROCKWELL	83065	L14SYR22-12	ROCKWELL	83110	L16SF20	ROCKWELL	84909	L16SYR21-4	ROCKWELL	83295
L14SYQ20	ROCKWELL	83092	L14SYR22-14	ROCKWELL	83101	L16SF201	ROCKWELL	84919	L16SYR22	ROCKWELL	83306
L14SYQ20-1	ROCKWELL	88014.26	L14SYR22-1A	ROCKWELL	88014.32	L16SF206	ROCKWELL	84924	L16SYR22-10	ROCKWELL	83316
L14SYQ20-5	ROCKWELL	88014.27	L14SYR22-21	ROCKWELL	83102	L16SF220	ROCKWELL	84920	L16SYR22-106	ROCKWELL	83315
L14SYQ21	ROCKWELL	83099	L14SYR22-31	ROCKWELL	83113	L16SF227	ROCKWELL	84922	L16SYR22-115	ROCKWELL	88014.41
L14SYR13-1	ROCKWELL	83001	L14SYR22-37	ROCKWELL	83117	L16SF238A	ROCKWELL	84913	L16SYR22-15	ROCKWELL	83317
L14SYR14-9	ROCKWELL	83003	L14SYR22-43	ROCKWELL	83114	L16SF4	ROCKWELL	84910	L16SYR22-36	ROCKWELL	83323
L14SYR15	ROCKWELL	83007	L14SYR22-44	ROCKWELL	87014	L16SF47	ROCKWELL	84915	L16SYR22-57	ROCKWELL	83305
L14SYR15-1	ROCKWELL	83006	L14SYR22-47	ROCKWELL	83111	L16SF59A	ROCKWELL	84917	L16SYR22-83	ROCKWELL	83309
L14SYR16	ROCKWELL	83010	L14SYR22-6	ROCKWELL	83118	L16SF61A	ROCKWELL	84911	L16SYR23	ROCKWELL	83331
L14SYR16-11	ROCKWELL	83020	L14SYR22-60	ROCKWELL	83119	L16SF88	ROCKWELL	84923	L16SYR23-10	ROCKWELL	83333
L14SYR16-15	ROCKWELL	83011	L14SYR22-63	ROCKWELL	83120	L16SF98	ROCKWELL	84916	L16SYR23-11	ROCKWELL	83334
L14SYR16-19	ROCKWELL	83019	L14SYR22-65	ROCKWELL	87011	L16SLS24-6DC	ROCKWELL	85309	L16SYR23-13	ROCKWELL	83335
L14SYR16-30	ROCKWELL	83022	L14SYR22-7	ROCKWELL	83108	L16SLS28A	ROCKWELL	85311	L16SYR23-18	ROCKWELL	87038
L14SYR16-6	ROCKWELL	83018	L14SYR22-8	ROCKWELL	83109	L16SY30-19	ROCKWELL	85716	L16SYR23-2	ROCKWELL	83332
L14SYR16-9	ROCKWELL	83021	L14SYR22-93	ROCKWELL	83103	L16SY30-3	ROCKWELL	85712	L16SYR24	ROCKWELL	83345
L14SYR17-10	ROCKWELL	83026	L14SYR23-4	ROCKWELL	83125	L16SY30-31	ROCKWELL	85717	L16SYR24-11	ROCKWELL	83349
L14SYR17-3	ROCKWELL	83025	L14SYR23-8	ROCKWELL	83126	L16SY32	ROCKWELL	85713	L16SYR24-12	ROCKWELL	83353
L14SYR17-5	ROCKWELL	83023	L14SYR24-10	ROCKWELL	83129	L16SY32-10	ROCKWELL	85714	L16SYR24-19	ROCKWELL	83350
L14SYR17-7	ROCKWELL	83024	L14SYR24-24	ROCKWELL	83128	L16SY32-35	ROCKWELL	85718	L16SYR24-20	ROCKWELL	83354
L14SYR17-8	ROCKWELL	83028	L14SYR24-3	ROCKWELL	83127	L16SY32-52	ROCKWELL	85715	L16SYR24-22	ROCKWELL	83347
L14SYR17-9	ROCKWELL	83027	L14SYS14-1	ROCKWELL	83002	L16SY36-3	ROCKWELL	85711	L16SYR24-4	ROCKWELL	83352
L14SYR18	ROCKWELL	83053	L14SYS16	ROCKWELL	88014.33	L16SY36-4	ROCKWELL	85719	L16SYR24-41	ROCKWELL	83348
L14SYR18-10	ROCKWELL	83042	L14SYS18-11	ROCKWELL	87007	L16SYH18	ROCKWELL	83245	L16SYR25-2	ROCKWELL	83360
L14SYR18-11	ROCKWELL	83035	L14SYS18-20	ROCKWELL	87006	L16SYH18-1	ROCKWELL	83243	L16SYR26-19	ROCKWELL	83361
L14SYR18-16	ROCKWELL	83047	L14SYS18-23	ROCKWELL	87005	L16SYH20	ROCKWELL	87029	L16SYR26-3	ROCKWELL	83364
L14SYR18-27	ROCKWELL	83055	L14SYS18-3	ROCKWELL	83052	L16SYH22	ROCKWELL	87035	L16SYR26-9	ROCKWELL	83363
L14SYR18-34	ROCKWELL	83033	L14SYS18-5	ROCKWELL	83054	L16SYP16-3	ROCKWELL	83250	L16SYR27	ROCKWELL	83366
L14SYR18-37	ROCKWELL	83044	L14SYS18-6	ROCKWELL	83030	L16SYP20-7	ROCKWELL	83311	L16SYR27-2	ROCKWELL	83367
L14SYR18-39	ROCKWELL	83057	L14SYS18-9A	ROCKWELL	84516	L16SYQ15-1	ROCKWELL	83237	L16SYR28-6	ROCKWELL	83375
L14SYR18-41	ROCKWELL	88014.28	L14SYS20	ROCKWELL	83090	L16SYQ16-1	ROCKWELL	83241	L16SYR28-8	ROCKWELL	83374
L14SYR18-44	ROCKWELL	83031	L14SYS20-1	ROCKWELL	83071	L16SYQ16-3	ROCKWELL	88014.36	L16SYR28-8	ROCKWELL	88014.42
L14SYR18-47	ROCKWELL	83034	L14SYS20-10	ROCKWELL	83073	L16SYQ18-1	ROCKWELL	83252	L16SYS18	ROCKWELL	83253
L14SYR18-5	ROCKWELL	83056	L14SYS20-11	ROCKWELL	83078	L16SYQ19	ROCKWELL	83255	L16SYS18-1	ROCKWELL	83248
L14SYR18-52	ROCKWELL	83032	L14SYS20-2	ROCKWELL	83077	L16SYQ19-13	ROCKWELL	83261	L16SYS18-3	ROCKWELL	87028
L14SYR18-72	ROCKWELL	87003	L14SYS20-36	ROCKWELL	83093	L16SYQ19-19	ROCKWELL	83258	L16SYS20-1	ROCKWELL	83276
L14SYR18-82	ROCKWELL	83048	L14SYS20-39	ROCKWELL	83089	L16SYQ19-2	ROCKWELL	83263	L16SYS20-2	ROCKWELL	83277
L14SYR18-82	ROCKWELL	88014.29	L14SYS20-40	ROCKWELL	83094	L16SYQ19-25	ROCKWELL	83265	L16SYS20-24	ROCKWELL	83288
L14SYR19-10	ROCKWELL	83066	L14SYS20-41	ROCKWELL	83069	L16SYQ19-5	ROCKWELL	83266	L16SYS20-30	ROCKWELL	83268
L14SYR19-11	ROCKWELL	83061	L14SYS20-5	ROCKWELL	83072	L16SYQ19-53	ROCKWELL	83257	L16SYS20-34	ROCKWELL	83291
L14SYR19-17	ROCKWELL	83059	L14SYS22-12	ROCKWELL	83107	L16SYQ19-54	ROCKWELL	83254	L16SYS20-35	ROCKWELL	83290
L14SYR19-2	ROCKWELL	83063	L14SYS22-127A	ROCKWELL	88014.34	L16SYQ19-55	ROCKWELL	83260	L16SYS20-36	ROCKWELL	83273
L14SYR19-3	ROCKWELL	83064	L14SYS22-14A	ROCKWELL	84519	L16SYQ20	ROCKWELL	83275	L16SYS20-37	ROCKWELL	83269
L14SYR19-31	ROCKWELL	83062	L14SYS22-16	ROCKWELL	83116	L16SYQ20-3	ROCKWELL	83283	L16SYS20-38	ROCKWELL	83285
L14SYR19-6	ROCKWELL	83067	L14SYS22-1A	ROCKWELL	84522	L16SYQ20-3	ROCKWELL	88014.37	L16SYS20-47	ROCKWELL	83292
L14SYR20-1	ROCKWELL	83079	L14SYS22-2	ROCKWELL	87016	L16SYQ20-7	ROCKWELL	88014.38	L16SYS20-9	ROCKWELL	83289
L14SYR20-10	ROCKWELL	83086	L14SYS22-24A	ROCKWELL	84521	L16SYQ21	ROCKWELL	83294	L16SYS22-182	ROCKWELL	83307
L14SYR20-102	ROCKWELL	83080	L14SYS22-26	ROCKWELL	87012	L16SYQ21-5	ROCKWELL	83296	L16SYS22-187	ROCKWELL	87031
L14SYR20-114	ROCKWELL	87008	L14SYS22-28	ROCKWELL	87015	L16SYQ24	ROCKWELL	83357	L16SYS22-188	ROCKWELL	83301
L14SYR20-17	ROCKWELL	83082	L14SYS22-31A	ROCKWELL	84517	L16SYQ26	ROCKWELL	83362	L16SYS22-19	ROCKWELL	83308
L14SYR20-174	ROCKWELL	83084	L14SYS22-33	ROCKWELL	83106	L16SYQ26	ROCKWELL	88014.39	L16SYS22-194	ROCKWELL	85303
L14SYR20-2	ROCKWELL	83081	L14SYS22-36	ROCKWELL	83105	L16SYR16-1	ROCKWELL	83239	L16SYS22-20	ROCKWELL	83312
L14SYR20-25	ROCKWELL	83096	L14SYS22-4	ROCKWELL	88014.35	L16SYR17-2	ROCKWELL	83242	L16SYS22-204A	ROCKWELL	84529
L14SYR20-29	ROCKWELL	87010	L14SYS22-40	ROCKWELL	83104	L16SYR18	ROCKWELL	83244	L16SYS22-205A	ROCKWELL	84531
L14SYR20-30	ROCKWELL	83095	L14SYS22-41	ROCKWELL	83121	L16SYR18-11	ROCKWELL	83249	L16SYS22-21	ROCKWELL	83326
L14SYR20-34	ROCKWELL	83074	L14SYS22-43	ROCKWELL	87013	L16SYR18-12	ROCKWELL	83251	L16SYS22-25	ROCKWELL	83320
L14SYR20-38	ROCKWELL	83075	L14SYS22-54	ROCKWELL	83070	L16SYR18-2	ROCKWELL	83247	L16SYS22-30A	ROCKWELL	84532
L14SYR20-43	ROCKWELL	88014.30	L14SYS22-77	ROCKWELL	83122	L16SYR19	ROCKWELL	83264	L16SYS22-31	ROCKWELL	83328
L14SYR20-44	ROCKWELL	83083	L14SYS22-89A	ROCKWELL	84520	L16SYR19-2	ROCKWELL	83262	L16SYS2236	ROCKWELL	84529
L14SYR20-47	ROCKWELL	83085	L14SYS22-93	ROCKWELL	83115	L16SYR20	ROCKWELL	83267	L16SYS22-36A	ROCKWELL	84528
L14SYR20-51	ROCKWELL	88014.31	L14SYS24-6	ROCKWELL	83130	L16SYR20-10	ROCKWELL	83281	L16SYS22-37A	ROCKWELL	84530
L14SYR20-56	ROCKWELL	83068	L14SYS28-4	ROCKWELL	83131	L16SYR20-25	ROCKWELL	88014.40	L16SYS22-39	ROCKWELL	83322
L14SYR20-58	ROCKWELL	83088	L15460	GERLINGER	80220	L16SYR20-33	ROCKWELL	83274	L16SYS22-3A	ROCKWELL	84533
L14SYR20-84	ROCKWELL	83091	L15460	TOWMOTOR	80188	L16SYR20-35	ROCKWELL	83279	L16SYS223A	ROCKWELL	88014.43
L14SYR20-9	ROCKWELL	83076	L1663-20	PETTI-MULL	80193	L16SYR20-40	ROCKWELL	83278	L16SYS22-40	ROCKWELL	88014.44
L14SYR20-99	ROCKWELL	83087	L1663-6	PETTI-MULL	80168	L16SYR20-52	ROCKWELL	87030	L16SYS22-41	ROCKWELL	83314
L14SYR21-10	ROCKWELL	83097	L1663-6	PETTI-MULL	80216	L16SYR20-6	ROCKWELL	83280	L16SYS22-46	ROCKWELL	87032
L14SYR21-2	ROCKWELL	83098	L16SF102A	ROCKWELL	84914	L16SYR20-81	ROCKWELL	83282	L16SYS22-46A	ROCKWELL	88014.45
L14SYR21-8	ROCKWELL	83100	L16SF181A	ROCKWELL	84921	L16SYR21	ROCKWELL	83297	L16SYS22-48	ROCKWELL	87036

Part No.	Brand	IDLI Group
L16SYS22-50	ROCKWELL	83313
L16SYS22-51	ROCKWELL	83318
L16SYS22-52	ROCKWELL	83327
L16SYS22-53A	ROCKWELL	84527
L16SYS22-59	ROCKWELL	83321
L16SYS22-67	ROCKWELL	83329
L16SYS22-77	ROCKWELL	83319
L16SYS22-95	ROCKWELL	83310
L16SYS24-12	ROCKWELL	83351
L16SYS24-13	ROCKWELL	83346
L16SYS24-15	ROCKWELL	83359
L16SYS24-26	ROCKWELL	83344
L16SYS24-3	ROCKWELL	83355
L16SYS28	ROCKWELL	87045
L16SYS28-11	ROCKWELL	83373
L16SYS28-13	ROCKWELL	83371
L16SYS28-25	ROCKWELL	85310
L16SYS28-2A	ROCKWELL	84534
L16SYS28-31	ROCKWELL	87044
L16SYS28-5	ROCKWELL	83372
L16SYT22-1	ROCKWELL	83325
L17600	GERLINGER	80220
L1773	DEERE,JOHN	80149
L1773C	DEERE,JOHN	80149
L1827C	DEERE,JOHN	80044
L18383	CASE,J.I.	80168
L2596-14	PETTI-MULL	80101
L2596-2	PETTI-MULL	80048
L2596-2	PETTI-MULL	80049
L2596-20	PETTI-MULL	80240
L2618-1	PETTI-MULL	80124
L2618-18	PETTI-MULL	80178
L2893-3	PETTIBONE	80231
L2893-3	PETTIBONE	80240
L30492	CASE,J.I.	80047
L30634	CASE,J.I.	80046
L31173	CASE,J.I.	80214
L31776	CASE,J.I.	80198
L31782	CASE,J.I.	80213
L32423	CASE,J.I.	80197
L33742	CASE,J.I.	80134
L33945	CASE,J.I.	80198
L35511	CASE,J.I.	80185
L36151	CASE,J.I.	80198
L36152	CASE,J.I.	80216
L41666	TAMPER	80142
L48344	CASE,J.I.	80198
L5-105	ALLOY	80031
L5-115	ALLOY	80178
L6-10	ROCKWELL	81018
L6-100	ROCKWELL	81016
L6-101	ROCKWELL	81052
L6101X	NEAPCO	84504
L6-102	ROCKWELL	81103
L6-103	ROCKWELL	81142
L6-104	ROCKWELL	81148
L6-105	ROCKWELL	81173
L6-106	ROCKWELL	81189
L6-107	ROCKWELL	81060
L6-107	ROCKWELL	88014.46
L6-108	ROCKWELL	81157
L6-109	ROCKWELL	85001
L6-110	ROCKWELL	85004
L6111	NEAPCO	88014.52
L6-111	ROCKWELL	85009
L6112	NEAPCO	81034
L6-112	ROCKWELL	85011
L6113	NEAPCO	81054
L6-113	ROCKWELL	85019
L6114	NEAPCO	81089
L6-114	ROCKWELL	85023
L6115	NEAPCO	81109
L6-115	ROCKWELL	85024
L6116	NEAPCO	81153
L6-116	ROCKWELL	85032
L6-117	ROCKWELL	85002
L6118	NEAPCO	81175
L6-118	ROCKWELL	85012
L6-119	ROCKWELL	85017
L6-120	ROCKWELL	85021
L6-121	ROCKWELL	85035
L6-122	ROCKWELL	85034
L6-123	ROCKWELL	85047
L6-124	ROCKWELL	85005
L6-125	ROCKWELL	85027
L6129	NEAPCO	81268
L6131	NEAPCO	88001.11
L6134	NEAPCO	81143
L6138	NEAPCO	81156
L613A	NEAPCO	81070
L6-14	ROCKWELL	81033
L6-15	ROCKWELL	81034
L615A	NEAPCO	81115
L616A	NEAPCO	81151
L616A	NEAPCO	81153
L6-17	ROCKWELL	81035
L6171	NEAPCO	81025
L617A	NEAPCO	81169
L617A	NEAPCO	88001.09
L618A	NEAPCO	81179
L6-19	ROCKWELL	81053
L6191	NEAPCO	81136
L6193	NEAPCO	81125
L6196	NEAPCO	81151
L61D4	NEAPCO	80061
L61L2	NEAPCO	88001.10
L61R8	NEAPCO	81060
L6-23	ROCKWELL	88014.47
L6-26	ROCKWELL	81150
L6-27	ROCKWELL	81151
L6-28	ROCKWELL	88014.48
L6-29	ROCKWELL	81175
L6-30	ROCKWELL	81176
L6-31	ROCKWELL	81024
L6-32	ROCKWELL	81025
L6-34	ROCKWELL	81056
L6-35	ROCKWELL	81123
L6-351	NEAPCO	88001.14
L6-36	ROCKWELL	81124
L6-37	ROCKWELL	81084
L6-38	ROCKWELL	81054
L6-39	ROCKWELL	81004
L6-40	ROCKWELL	81005
L6-41	ROCKWELL	81019
L6-42	ROCKWELL	81020
L6-43	ROCKWELL	81055
L6-44	ROCKWELL	81085
L6-45	ROCKWELL	81086
L6-451	NEAPCO	88001.12
L6-46	ROCKWELL	81108
L6-47	ROCKWELL	81109
L6-48	ROCKWELL	81135
L6-49	ROCKWELL	81133
L6-56	ROCKWELL	85014
L6-58	ROCKWELL	85010
L6-61	ROCKWELL	85026
L6-69	ROCKWELL	81144
L6-70	ROCKWELL	81007
L6-72	ROCKWELL	81030
L6-76	ROCKWELL	81040
L6-79	ROCKWELL	81081
L6-80	ROCKWELL	81082
L6-82	ROCKWELL	81006
L6-83	ROCKWELL	81044
L6-84	ROCKWELL	81145
L6-85	ROCKWELL	81165
L6-86	ROCKWELL	81167
L6-87	ROCKWELL	81168
L6-88	ROCKWELL	81177
L6-89	ROCKWELL	81190
L6-90	ROCKWELL	81194
L6-91	ROCKWELL	81196
L6-92	ROCKWELL	81001
L6-93	ROCKWELL	81041
L6-94	ROCKWELL	81096
L6-95	ROCKWELL	81009
L6-96	ROCKWELL	81038
L6-97	ROCKWELL	81091
L6-98	ROCKWELL	81092
L6-99	ROCKWELL	81155
L6N15-15	ROCKWELL	88014.49
L6N170-1	ROCKWELL	86601
L6N171-1	ROCKWELL	86602
L6NDCA22	ROCKWELL	86603
L6NDCA3	ROCKWELL	86604
L6NF115	ROCKWELL	84603
L6NF123	ROCKWELL	84601
L6NF134	ROCKWELL	84602
L6NF20	ROCKWELL	84606
L6NF43	ROCKWELL	84604
L6NF53	ROCKWELL	84605
L6NLS18-1DC	ROCKWELL	85028
L6NLS18-3	ROCKWELL	85020
L6NLS18DC	ROCKWELL	85022
L6NLS20-1DC2	ROCKWELL	85036
L6NLS20-22DC	ROCKWELL	85030
L6NLS20-25DC	ROCKWELL	85045
L6NLS20-31DC	ROCKWELL	85029
L6NLS20-35DC	ROCKWELL	85041
L6NLS20-50DC	ROCKWELL	85046
L6NLS20-55DC	ROCKWELL	85039
L6NLS20-63DC1	ROCKWELL	85042
L6NLS20-65DC	ROCKWELL	85038
L6NLS20-65DC1	ROCKWELL	85037
L6NLS20-6DC	ROCKWELL	85043
L6NLS20-71DC	ROCKWELL	85033
L6NLS20-77DC	ROCKWELL	85040
L6NLS20-94DC	ROCKWELL	85044
L6NPS18	ROCKWELL	87218
L6NPS18-1	ROCKWELL	87219
L6NPS18-3	ROCKWELL	87217
L6NPS18-8	ROCKWELL	87224
L6NY16-1	ROCKWELL	85502
L6NY20	ROCKWELL	85503
L6NY21	ROCKWELL	85504
L6NY21-1	ROCKWELL	85501
L6NY21-8	ROCKWELL	85508
L6NY24-6	ROCKWELL	85507
L6NY25	ROCKWELL	85505
L6NY26	ROCKWELL	85506
L6NYD14-7	ROCKWELL	81051
L6NYD16-5	ROCKWELL	86643
L6NYH14	ROCKWELL	81076
L6NYH14-17	ROCKWELL	81077
L6NYH14-3	ROCKWELL	88014.50
L6NYH16-1	ROCKWELL	81121
L6NYH18	ROCKWELL	81158
L6NYP12-1	ROCKWELL	81079
L6NYP12-12	ROCKWELL	81047
L6NYP12-14	ROCKWELL	81074
L6NYP12-23	ROCKWELL	81058
L6NYP12-28A	ROCKWELL	85007
L6NYP12-32	ROCKWELL	81075
L6NYP12-35	ROCKWELL	81061
L6NYP12-48	ROCKWELL	81078
L6NYP12-5	ROCKWELL	81057
L6NYP12-66	ROCKWELL	81059
L6NYP16-17	ROCKWELL	81187
L6NYP16-19	ROCKWELL	81188
L6NYP16-5	ROCKWELL	81138
L6NYP16-9	ROCKWELL	81140
L6NYQ12	ROCKWELL	81010
L6NYQ12-3	ROCKWELL	81026
L6NYQ12-67	ROCKWELL	81015
L6NYQ13	ROCKWELL	81039
L6NYQ14-14	ROCKWELL	81063
L6NYQ14-2A	ROCKWELL	85008
L6NYQ15	ROCKWELL	81093
L6NYQ15-8	ROCKWELL	81083
L6NYQ16	ROCKWELL	81122
L6NYQ16-1A	ROCKWELL	85016
L6NYQ16-2	ROCKWELL	81125
L6NYQ16-23	ROCKWELL	81098
L6NYQ16-35	ROCKWELL	81130
L6NYQ16-4	ROCKWELL	81126
L6NYQ17-1	ROCKWELL	81137
L6NYQ18	ROCKWELL	81156
L6NYQ20-1	ROCKWELL	81172
L6NYR10-12	ROCKWELL	81003
L6NYR10-6	ROCKWELL	81002
L6NYR12-1	ROCKWELL	81008
L6NYR12-10	ROCKWELL	88014.51
L6NYR12-140	ROCKWELL	86610
L6NYR12-146	ROCKWELL	86609
L6NYR12-162	ROCKWELL	81028
L6NYR12-18	ROCKWELL	81022
L6NYR12-25	ROCKWELL	81013
L6NYR12-30	ROCKWELL	86611
L6NYR12-33	ROCKWELL	81023
L6NYR12-4	ROCKWELL	81021
L6NYR12-44	ROCKWELL	81012
L6NYR12-9	ROCKWELL	88014.52
L6NYR12-92	ROCKWELL	81014
L6NYR13-17A	ROCKWELL	85003
L6NYR13-19	ROCKWELL	81031
L6NYR13-19	ROCKWELL	81032
L6NYR13-5	ROCKWELL	81036
L6NYR13-6	ROCKWELL	81037
L6NYR14	ROCKWELL	81049
L6NYR14	ROCKWELL	81050
L6NYR14-10	ROCKWELL	88014.53
L6NYR14-132	ROCKWELL	81071
L6NYR14-149	ROCKWELL	86628
L6NYR14-16	ROCKWELL	86629
L6NYR14-17	ROCKWELL	81072
L6NYR14-19	ROCKWELL	81073
L6NYR14-21	ROCKWELL	81070
L6NYR14-22A	ROCKWELL	85006
L6NYR14-3	ROCKWELL	81042
L6NYR14-47	ROCKWELL	81043
L6NYR14-51	ROCKWELL	81069
L6NYR14-7	ROCKWELL	86626
L6NYR14-8	ROCKWELL	81046
L6NYR14-93	ROCKWELL	81068
L6NYR14-96	ROCKWELL	86625
L6NYR15-12	ROCKWELL	81089
L6NYR15-6	ROCKWELL	81087
L6NYR15-7	ROCKWELL	81088
L6NYR16	ROCKWELL	81110
L6NYR16-1	ROCKWELL	81111

Part No.	Brand	IDLI Group	Part No.	Brand	IDLI Group	Part No.	Brand	IDLI Group	Part No.	Brand	IDLI Group
L6NYR16-114	ROCKWELL	86631	L6NYS13-31	ROCKWELL	86614	LA425-3	PETTI-MULL	80109	M2032	BALKAMP	88001.76
L6NYR16-146	ROCKWELL	81118	L6NYS13-6	ROCKWELL	86616	LA425-4	PETTI-MULL	80168	M2033	BALKAMP	80054
L6NYR16-15	ROCKWELL	81112	L6NYS14-100	ROCKWELL	86621	LA425-5	PETTI-MULL	80193	M2035	BALKAMP	80054
L6NYR16-151	ROCKWELL	81100	L6NYS14-101	ROCKWELL	81048	LA425-6	PETTI-MULL	80210	M2036	BALKAMP	80049
L6NYR16-188	ROCKWELL	81119	L6NYS14-109	ROCKWELL	86627	LA425-7	PETTI-MULL	80231	M2036	BALKAMP	88001.96
L6NYR16-198	ROCKWELL	81104	L6NYS14-122	ROCKWELL	86623	LA425-8	PETTI-MULL	80235	M2039	BALKAMP	80044
L6NYR16-201	ROCKWELL	85015	L6NYS14-123	ROCKWELL	86620	LA452-2	PETTI-MULL	80099	M2039	BALKAMP	88001.98
L6NYR16-219	ROCKWELL	81105	L6NYS14-16	ROCKWELL	86618	LA518-10	PETTI-MULL	80100	M2040	BALKAMP	80049
L6NYR16-242	ROCKWELL	81129	L6NYS14-17	ROCKWELL	86622	LA5415	LA PORTA	80216	M2040	BALKAMP	88001.96
L6NYR16-245	ROCKWELL	81097	L6NYS14-18	ROCKWELL	81065	LA6301	LA PORTA	80150	M2041	BALKAMP	80044
L6NYR16-259	ROCKWELL	81120	L6NYS14-18	ROCKWELL	81066	LA737	PETTI-MULL	80214	M2041	BALKAMP	88001.98
L6NYR16-287	ROCKWELL	86638	L6NYS14-19	ROCKWELL	81067	LA7957	LA PORTA	88001.28	M2044	BALKAMP	80047
L6NYR16-37	ROCKWELL	81094	L6NYS14-20	ROCKWELL	81080	LA854N	GRAIGG CO.	80213	M2045	BALKAMP	80049
L6NYR16-38A	ROCKWELL	85013	L6NYS14-22	ROCKWELL	81045	LA927	GRAIGG CO.	80234	M2045	BALKAMP	88001.96
L6NYR16-44A	ROCKWELL	85018	L6NYS14-27	ROCKWELL	86613	LANDCRUISER	MCQUAY-NOR	80017	M2046	BALKAMP	80044
L6NYR16-51	ROCKWELL	81113	L6NYS14-31	ROCKWELL	81062	LB32548	CASE,J.I.	80047	M2046	BALKAMP	88001.98
L6NYR16-54	ROCKWELL	86634	L6NYS14-39	ROCKWELL	86615	LD44610A	FORD	80072	M2048	BALKAMP	80049
L6NYR16-59	ROCKWELL	81095	L6NYS14-42	ROCKWELL	81064	LL240-14	PETTI-MULL	80189	M2048	BALKAMP	88001.96
L6NYR16-59	ROCKWELL	81132	L6NYS14-65	ROCKWELL	86624	LL240-22	PETTI-MULL	80189	M2049	BALKAMP	80054
L6NYR16-64	ROCKWELL	81114	L6NYS14A	ROCKWELL	84501	LL240-23	PETTI-MULL	80189	M2050	BALKAMP	80044
L6NYR16-66	ROCKWELL	81115	L6NYS16-1	ROCKWELL	81128	LL240-8	PETTI-MULL	80189	M2053	BALKAMP	80031
L6NYR16-78	ROCKWELL	86641	L6NYS16-14	ROCKWELL	86639	LL242-64	PETTIBONE	80023	M2062	BALKAMP	80049
L6NYR16-8	ROCKWELL	88014.54	L6NYS16-18	ROCKWELL	86633	LL242-8	PETTI-MULL	80142	M2062	BALKAMP	80050
L6NYR16-85	ROCKWELL	81116	L6NYS16-2	ROCKWELL	81090	LL273-2	PETTI-MULL	80210	M2063	BALKAMP	80044
L6NYR16-86	ROCKWELL	86642	L6NYS16-27	ROCKWELL	81131	LL3323-1	PETTI-MULL	80210	M2063	BALKAMP	88001.98
L6NYR16-96	ROCKWELL	81117	L6NYS16-34	ROCKWELL	86635	LL773-111	PETTI-MULL	80210	M2064	BALKAMP	80044
L6NYR17	ROCKWELL	81136	L6NYS16-46	ROCKWELL	86636	LL773-12	PETTI-MULL	80193	M2064	BALKAMP	88001.98
L6NYR17-7	ROCKWELL	81134	L6NYS16-54	ROCKWELL	81101	LL773-123	PETTI-MULL	80153	M2066	BALKAMP	80045
L6NYR18-104	ROCKWELL	86645	L6NYS16-54	ROCKWELL	81102	LL773-124	PETTI-MULL	80109	M2067	BALKAMP	80048
L6NYR18-105	ROCKWELL	81163	L6NYS16-59	ROCKWELL	86630	LL773-183	PETTI-MULL	80101	M2077	BALKAMP	80044
L6NYR18-109	ROCKWELL	81164	L6NYS16-60	ROCKWELL	81107	LL773-186	PETTI-MULL	80149	M2100	BALKAMP	80049
L6NYR18-21	ROCKWELL	81141	L6NYS16-64	ROCKWELL	86637	LL773-188	PETTIBONE	80194	M21005	CHRYSLER	80189
L6NYR18-26	ROCKWELL	81139	L6NYS16-65	ROCKWELL	81106	LL773-2	PETTI-MULL	80215	M2101	BALKAMP	80044
L6NYR18-42	ROCKWELL	81160	L6NYS16-66	ROCKWELL	86640	LL773-246	PETTIBONE	80235	M2103	BALKAMP	80048
L6NYR18-62	ROCKWELL	86644	L6NYS16-77	ROCKWELL	86632	LL773-270	PETTIBONE	80196	M2104	BALKAMP	80047
L6NYR18-66	ROCKWELL	81146	L6NYS16-78	ROCKWELL	81099	LL773-270	PETTIBONE	80197	M2105	BALKAMP	80049
L6NYR18-79	ROCKWELL	81154	L6NYS18	ROCKWELL	81143	LL773-272	PETTIBONE	80197	M2108	BALKAMP	80044
L6NYR18-8	ROCKWELL	81152	L6NYS18-15	ROCKWELL	81149	LL773-45	PETTI-MULL	80214	M2109	BALKAMP	80049
L6NYR18-84	ROCKWELL	81161	L6NYS18-19	ROCKWELL	81147	LL773-62	PETTI-MULL	80099	M2110	BALKAMP	80048
L6NYR18-9	ROCKWELL	81153	L6NYS18-23A	ROCKWELL	84502	LL773-63	PETTI-MULL	80101	M2111	BALKAMP	80044
L6NYR19-1	ROCKWELL	81169	L6NYS18-25	ROCKWELL	81159	LL773-77	PETTI-MULL	80167	M2112	BALKAMP	80044
L6NYR19-12	ROCKWELL	81170	L6NYS18-6A	ROCKWELL	85025	LL773-87	PETTI-MULL	80168	M2114	BALKAMP	80054
L6NYR19-3	ROCKWELL	86646	L6NYS18-8	ROCKWELL	81162	LL773-88	PETTI-MULL	80168	M2116	BALKAMP	80047
L6NYR19-4	ROCKWELL	81166	L6NYS20	ROCKWELL	81174	LP-HC39-3	ROCKWELL	88014.55	M2118	BALKAMP	80049
L6NYR19-9	ROCKWELL	81171	L6NYS20-11	ROCKWELL	81182	LYSYS22-127A	ROCKWELL	84518	M2119	BALKAMP	80044
L6NYR20-1	ROCKWELL	81183	L6NYS20-5	ROCKWELL	81185	M1750	BALKAMP	80030	M2121	BALKAMP	80044
L6NYR20-1	ROCKWELL	81199	L6NYS20-5A	ROCKWELL	81186	M1750	BALKAMP	80031	M2123	BALKAMP	80054
L6NYR20-26	ROCKWELL	81181	L6NYS22-1A	ROCKWELL	84504	M1759	BALKAMP	80030	M2126	BALKAMP	80049
L6NYR20-38	ROCKWELL	81179	L6NYS22-20A	ROCKWELL	84503	M1759	BALKAMP	80031	M2127	BALKAMP	80044
L6NYR20-39	ROCKWELL	81180	L6NYSM20-18	ROCKWELL	86647	M1762	BALKAMP	80030	M2137	BALKAMP	80049
L6NYR20-4	ROCKWELL	81178	L6NYSM20-61	ROCKWELL	86648	M1762	BALKAMP	80031	M2140	BALKAMP	80055
L6NYR20-60	ROCKWELL	81184	L6NYSM22-5	ROCKWELL	86649	M2000	BALKAMP	80049	M2153	BALKAMP	80055
L6NYR20-9A	ROCKWELL	85031	L6NYT16-2	ROCKWELL	81127	M2000	BALKAMP	80050	M2173	BALKAMP	80058
L6NYR22-1	ROCKWELL	81191	LA123-12	PETTI-MULL	80149	M2002	BALKAMP	80048	M2198	BALKAMP	80051
L6NYR22-11	ROCKWELL	81193	LA1516	LA PORTA	80047	M2003	BALKAMP	80048	M24616	DAVEY COM.	80168
L6NYR22-2	ROCKWELL	81192	LA17205	PETTI-MULL	80210	M2007	BALKAMP	80045	M2564	KOYO	80012
L6NYR23-1	ROCKWELL	81195	LA205	GRAIGG CO.	80168	M2010	BALKAMP	80049	M2564B-E	KOYO	80020
L6NYR24-1	ROCKWELL	81197	LA205B	GRAIGG CO.	80212	M2010	BALKAMP	88001.96	M3000	BALKAMP	80099
L6NYR24-2	ROCKWELL	81198	LA301-1	PETTI-MULL	80189	M2011	BALKAMP	80044	M3000100	GAR WOOD	80008
L6NYS12-12	ROCKWELL	81029	LA3200	LA PORTA	80046	M2011	BALKAMP	88001.98	M3001	BALKAMP	80099
L6NYS12-13	ROCKWELL	81027	LA337	GRAIGG CO.	80210	M2012	BALKAMP	80031	M3003	BALKAMP	80099
L6NYS12-27	ROCKWELL	86606	LA348-3	PETTIBONE	80198	M2014	BALKAMP	80045	M3005	BALKAMP	80099
L6NYS12-3	ROCKWELL	81017	LA384-3	PETTI-MULL	80198	M2016	BALKAMP	80044	M3008	BALKAMP	80104
L6NYS12-39	ROCKWELL	86605	LA384-6	PETTI-MULL	80210	M2017	BALKAMP	80092	M3012	BALKAMP	80112
L6NYS12-4	ROCKWELL	81011	LA384-7	PETTI-MULL	80197	M2017	BALKAMP	88001.76	M3014	BALKAMP	80114
L6NYS12-42	ROCKWELL	86607	LA384-8	PETTI-MULL	80172	M2018	BALKAMP	80053	M3019	BALKAMP	80112
L6NYS12-48	ROCKWELL	86608	LA425-1	PETTI-MULL	80048	M2026	BALKAMP	80031	M3021	BALKAMP	80111
L6NYS13-20	ROCKWELL	86619	LA425-1	PETTI-MULL	80049	M2031	BALKAMP	80049	M3022	BALKAMP	80112
L6NYS13-21	ROCKWELL	86617	LA425-2	PETTI-MULL	80099	M2031	BALKAMP	88001.96	M3033	BALKAMP	80104
L6NYS13-25	ROCKWELL	86612	LA425-2	PETTI-MULL	80101	M2032	BALKAMP	80092	M3036	BALKAMP	80099

Part No.	Brand	IDLI Group	Part No.	Brand	IDLI Group	Part No.	Brand	IDLI Group	Part No.	Brand	IDLI Group
M3036	BALKAMP	80101	M6023	BALKAMP	80199	M8101	BALKAMP	80232	MB109184	MITSUBISHI	86158
M3043	BALKAMP	80112	M6024	BALKAMP	80201	M8101	BALKAMP	80237	MB176162	CHRYSLER	86193
M3045	BALKAMP	80099	M6028	BALKAMP	80193	M8102	BALKAMP	80237	MB176162	MITSUBISHI	86193
M3045	BALKAMP	80101	M6100	BALKAMP	80193	M8102	BALKAMP	80244	MB202	DETROIT	87332
M306012	GAR WOOD	80109	M6102	BALKAMP	80197	M8105	BALKAMP	80238	MB202	MOPAR	87327
M307712	GAR WOOD	80001	M6103	BALKAMP	80197	M8107	BALKAMP	80238	MB202	MOPAR	87331
M3100	BALKAMP	80099	M6106	BALKAMP	80198	M8200	BALKAMP	80231	MB600	DETROIT	87238
M3100	BALKAMP	80101	M6108	BALKAMP	80199	M8201	BALKAMP	80232	MB601	DETROIT	87332
M3105	BALKAMP	80099	M6109	BALKAMP	80201	M8201	BALKAMP	80237	MB602	DETROIT	87326
M3105	BALKAMP	80101	M6110	BALKAMP	80198	M8202	BALKAMP	80237	MB603	DETROIT	87332
M3106	BALKAMP	80099	M6111	BALKAMP	80198	M8205	BALKAMP	80231	MB605	DETROIT	87377
M3106	BALKAMP	80101	M6112	BALKAMP	80198	M8206	BALKAMP	80232	MB606	DETROIT	87383
M3108	BALKAMP	80112	M6113	BALKAMP	80193	M8206	BALKAMP	80237	MB608	DETROIT	87457
M3111	BALKAMP	80104	M6114	BALKAMP	80197	M8212	BALKAMP	80231	MB609	DETROIT	87459
M3112	BALKAMP	80099	M6118	BALKAMP	80198	M8218	BALKAMP	80232	MB611	DETROIT	88007.77
M3112	BALKAMP	80101	M6119	BALKAMP	80198	M8218	BALKAMP	80237	MB612	DETROIT	88007.79
M3113	BALKAMP	80112	M6127	BALKAMP	80197	M9000	BALKAMP	80235	MBCP4N	D&D MACH.	80134
M3114	BALKAMP	80099	M6129	BALKAMP	80193	M9001	BALKAMP	80234	MBCP58	D&D MACH.	80185
M3114	BALKAMP	80101	M7000	BALKAMP	80210	M9001	BALKAMP	80235	MBCP58	MASTERS	80178
M3117	BALKAMP	80099	M7001	BALKAMP	80211	M9002	BALKAMP	80233	MBCP58	MASTERS	80185
M3117	BALKAMP	80101	M7003	BALKAMP	80210	M9003	BALKAMP	80234	MBCP581	D&D MACH.	80186
M3128	BALKAMP	80099	M7004	BALKAMP	80211	M9004	BALKAMP	80233	MBCP581	MASTERS	80186
M3128	BALKAMP	80101	M7010	BALKAMP	80210	M9005	BALKAMP	80234	MBCP582	D&D MACH.	80184
M4000	BALKAMP	80109	M7011	BALKAMP	80210	M9006	BALKAMP	80235	MBCP582	MASTERS	80184
M4002	BALKAMP	80109	M7012	BALKAMP	80210	M9007	BALKAMP	80235	MBCP6N	D&D MACH.	80188
M4007	BALKAMP	80147	M7013	BALKAMP	80213	M9008	BALKAMP	80235	MC675	D&D MACH.	80091
M4008	BALKAMP	80146	M7033	BALKAMP	80210	M9009	BALKAMP	80234	MC675	MASTERS	80091
M4014	BALKAMP	80158	M7034	BALKAMP	80210	M9010	BALKAMP	80233	MCD96	D&D MACH.	80160
M4014	BALKAMP	88003.13	M7034	BALKAMP	80212	M9012	BALKAMP	80233	MCD96	MASTERS	80068
M4015	BALKAMP	80148	M7042	BALKAMP	80211	M9013	BALKAMP	80234	MCDS56	D&D MACH.	80145
M4016	BALKAMP	80159	M7100	BALKAMP	80210	M9014	BALKAMP	80235	MCO55	D&D MACH.	80141
M4017	BALKAMP	80158	M7100	BALKAMP	80217	M9015	BALKAMP	80234	MCO55	MASTERS	80141
M4034	BALKAMP	80158	M7101	BALKAMP	80212	M9016	BALKAMP	80234	MCO96	D&D MACH.	80138
M4036	BALKAMP	80109	M7102	BALKAMP	80212	M9017A	BALKAMP	80233	MCO96	MASTERS	80138
M4100	BALKAMP	80109	M7102	BALKAMP	80214	M9018	BALKAMP	80235	MCP55	D&D MACH.	80129
M4102	BALKAMP	80158	M7102	BALKAMP	80218	M9020	BALKAMP	80234	MCP55	MASTERS	80129
M4106	BALKAMP	80148	M7103	BALKAMP	80210	M9022	BALKAMP	80234	MCP56	D&D MACH.	80131
M4108	BALKAMP	80109	M7105	BALKAMP	80210	M9023	BALKAMP	80234	MCP56	MASTERS	80091
M4130	BALKAMP	80109	M7105	BALKAMP	80213	M9024	BALKAMP	80234	MCP96	D&D MACH.	80068
M5000	BALKAMP	80168	M7107	BALKAMP	80210	M9026	BALKAMP	80233	MCP96	MASTERS	80068
M5003	BALKAMP	80168	M7112	BALKAMP	80213	M9028	BALKAMP	80234	MCR55	D&D MACH.	80123
M5006	BALKAMP	80163	M7114	BALKAMP	80212	M9029	BALKAMP	80234	MCR55	MASTERS	80123
M5006	BALKAMP	80167	M7115	BALKAMP	80211	M9031	BALKAMP	80234	MCR96	D&D MACH.	80122
M5007	BALKAMP	80177	M7116	BALKAMP	80210	MA0126433	CHRYSLER	80012	MCS53	MASTERS	80038
M5008	BALKAMP	80174	M7119	BALKAMP	80213	MA126433	CHRYSLER	80012	MCS53	MASTERS	80039
M5011	BALKAMP	80168	M7120	BALKAMP	80210	MA126433	MITSUBISHI	80012	MCS55	D&D MACH.	80081
M5014	BALKAMP	80168	M7122	BALKAMP	80213	MA126434	CHRYSLER	80012	MCS55	MASTERS	80081
M5015	BALKAMP	80173	M7126	BALKAMP	80213	MA54X	MASTERS	80041	MCS55	MASTERS	80123
M5024	BALKAMP	80168	M7200	BALKAMP	80210	MACK	MCQUAY-NOR	80188	MCS96	D&D MACH.	80038
M5032	BALKAMP	80174	M7201	BALKAMP	80211	MACK	MCQUAY-NOR	80220	MCS96	MASTERS	80038
M5033	BALKAMP	80173	M7201A	BALKAMP	80211	MACK	MCQUAY-NOR	80233	MCT55	MASTERS	80071
M5036	BALKAMP	80174	M7202	BALKAMP	80212	MAZDA	MCQUAY-NOR	80005	MCTS55	MASTERS	80074
M5037	BALKAMP	80168	M7203	BALKAMP	80210	MAZDA	MCQUAY-NOR	80013	MCU96	D&D MACH.	80181
M5043	BALKAMP	80163	M7205	BALKAMP	80210	MB000345	CHRYSLER	80012	MCVK1	D&D MACH.	88002.34
M5043	BALKAMP	80167	M7206	BALKAMP	80211	MB000345	MITSUBISHI	80012	MCVK2	D&D MACH.	88002.34
M5100	BALKAMP	80168	M7207	BALKAMP	80212	MB000391	CHRYSLER	80012	MCVK2A	D&D MACH.	88002.36
M5101	BALKAMP	80163	M7208	BALKAMP	80210	MB000391	MITSUBISHI	80012	MD000392	CHRYSLER	80012
M5101	BALKAMP	80167	M7213	BALKAMP	80210	MB000392	MITSUBISHI	80012	MD2255	D&D MACH.	80063
M5103	BALKAMP	80174	M7214	BALKAMP	80211	MB000823	CHRYSLER	80015	MD2255	MASTERS	80060
M5110	BALKAMP	80173	M8000	BALKAMP	80231	MB000823	MITSUBISHI	80015	MD2255	MASTERS	80063
M5120	BALKAMP	80168	M8002	BALKAMP	80231	MB000826	CHRYSLER	80015	MD2255	MASTERS	80064
M5122	BALKAMP	80168	M8003	BALKAMP	80232	MB000826	MITSUBISHI	80015	MD2255	MASTERS	80065
M5124	BALKAMP	80168	M8003	BALKAMP	80237	MB000827	CHRYSLER	80012	MD2526	D&D MACH.	80042
M5272	CONS.MACH.	80101	M8018	BALKAMP	80231	MB000827	MITSUBISHI	80012	MD2526	MASTERS	80042
M6000	BALKAMP	80193	M8023	BALKAMP	80232	MB00391	MITSUBISHI	80012	MD2625	D&D MACH.	80064
M6003	BALKAMP	80193	M8023	BALKAMP	80237	MB109012	CHRYSLER	86158	MD2625	MASTERS	80064
M6006	BALKAMP	80193	M8025	BALKAMP	80231	MB109012	MITSUBISHI	86158	MD2625	MASTERS	80065
M6012	BALKAMP	80193	M8027	BALKAMP	80231	MB109183	CHRYSLER	86193	MD2625	MASTERS	88001.63
M6016	BALKAMP	80193	M8100	BALKAMP	80231	MB109183	MITSUBISHI	86193	MD2683L	DEERE,JOHN	80077
M6022	BALKAMP	80197	M8100	BALKAMP	80243	MB109184	CHRYSLER	86158	MD4252	D&D MACH.	80034

Part No.	Brand	IDLI Group
MD4252	MASTERS	80034
MD4252	MASTERS	88001.62
MD5253	MASTERS	80064
MD5253	MASTERS	80065
MD5254	D&D MACH.	80094
MD5254	MASTERS	80064
MD5254	MASTERS	80065
MD6252	MASTERS	80135
MD6253	D&D MACH.	80061
MD6253	D&D MACH.	80094
MD6253	MASTERS	80060
MD6253	MASTERS	80061
MD6255	D&D MACH.	80132
MD6255	MASTERS	80132
MD6256	D&D MACH.	80062
MD7252	D&D MACH.	80135
MD7252	MASTERS	80135
MDH414	CASE,J.I.	80072
MDR52	MASTERS	80903
MDR52	MASTERS	80953
MDR52	MASTERS	80958
MDR54A	D&D MACH.	80901
MDR57	D&D MACH.	80955
MDR57A	D&D MACH.	80908
MDR60	D&D MACH.	80958
MDR60	MASTERS	80903
MDR60	MASTERS	80958
MDR60A	D&D MACH.	80909
MDR62	D&D MACH.	80954
MDR62A	D&D MACH.	80907
MDR63	D&D MACH.	80957
MDR63A	D&D MACH.	80907
MDR65	D&D MACH.	80951
MDR65	MASTERS	80951
MDR65A	D&D MACH.	80905
MDR65A	MASTERS	80905
MDR67	D&D MACH.	80953
MDR67	MASTERS	80953
MDR67A	D&D MACH.	80906
MDR67A	MASTERS	80906
MEK7039	FORD	80049
MEK7039A	FORD	80049
MEK7039B	FORD	80072
MEK7042	FORD	80044
MEK7042A	FORD	80044
MF25989	D&D MACH.	80011
MF26434	D&D MACH.	80012
MF37125	D&D MACH.	80010
MF39625	D&D MACH.	80019
MF43712	D&D MACH.	80009
MF43715	D&D MACH.	80006
MFK7039	FORD	80072
MFK7039A	FORD	80072
MFK7039B	FORD	80072
MI85-1194	KATAOKA	80142
MI85-1961	KATAOKA	80225
MI85-3211	KATAOKA	80140
MI85-5680	KATAOKA	80077
MIK7039A	FORD	80049
MJY10	MUNCIE	81208
MJY14	MUNCIE	81268
MJY14A	MUNCIE	81261
MJY15	MUNCIE	81302
MJY15A	MUNCIE	81301
MJY15A	MUNCIE	88013.90
MJY17	MUNCIE	81286
MJY18	MUNCIE	81350
MJY18A	MUNCIE	81350
MJY18A	MUNCIE	81351
MJY22	MUNCIE	81206
MJY23	MUNCIE	81218
MJY24	MUNCIE	81239
MJY24A	MUNCIE	81246
MJY25	MUNCIE	81259
MJY26	MUNCIE	84510
MJY27	MUNCIE	88000.08
MJY29	MUNCIE	81333
MJY-30	MUNCIE	81277
MJY6	MUNCIE	81203
MJY7	MUNCIE	81213
MJY8	MUNCIE	81222
MJY9	MUNCIE	81237
MK1X	MUNCIE	88000.14
MK2X	MUNCIE	80007
MK2X	MUNCIE	88000.15
MK35X	MUNCIE	80069
MK3X	MUNCIE	88000.15
MK3X	MUNCIE	88000.16
MK3XA	MUNCIE	80007
MK5X	MUNCIE	80035
MM1200	MASTERS	80081
MM1201	MASTERS	80091
MM1750	D&D MACH.	80030
MM1750	MASTERS	80030
MM1750	MASTERS	80031
MM1750	MASTERS	80061
MM1755	D&D MACH.	80032
MM1759	D&D MACH.	80030
MM1759	MASTERS	80030
MM1759	MASTERS	80031
MM1759	MASTERS	80061
MM1762	D&D MACH.	80031
MM1762	MASTERS	80030
MM1762	MASTERS	80031
MM1762	MASTERS	80061
MM1767	D&D MACH.	80031
MM1767	MASTERS	80030
MM1767	MASTERS	80031
MM2002	D&D MACH.	80048
MM2002	MASTERS	80048
MM2002	MASTERS	80049
MM2018	MASTERS	80049
MM2032	D&D MACH.	80092
MM2032	MASTERS	80092
MM2032	MASTERS	88001.76
MM2062	D&D MACH.	80050
MM2062	MASTERS	80049
MM2063	D&D MACH.	80043
MM2063	MASTERS	80044
MM2063	MASTERS	88001.98
MM2100	D&D MACH.	80049
MM2100	MASTERS	80049
MM2101	D&D MACH.	80044
MM2101	MASTERS	80044
MM2101	MASTERS	88001.98
MM2114	D&D MACH.	80054
MM2114	MASTERS	80054
MM2116	D&D MACH.	80047
MM2116	MASTERS	80047
MM2117	D&D MACH.	80046
MM2117	MASTERS	80092
MM2123	MASTERS	80054
MM2126	D&D MACH.	80049
MM2126	MASTERS	80049
MM2127	D&D MACH.	80044
MM2127	MASTERS	80058
MM2140	D&D MACH.	80055
MM2140	MASTERS	80055
MM2173	D&D MACH.	80058
MM2173	MASTERS	80058
MM2175	D&D MACH.	80058
MM2175	MASTERS	80058
MM2179	MASTERS	80055
MM2198	D&D MACH.	80051
MM2198	MASTERS	80051
MM2500	D&D MACH.	80059
MM2500	MASTERS	80059
MM2518	D&D MACH.	80066
MM2518	MASTERS	80066
MM3000	D&D MACH.	80099
MM3000	MASTERS	80099
MM3000	MASTERS	80101
MM3008	D&D MACH.	80107
MM3008	MASTERS	80104
MM3010	MASTERS	80100
MM3011	D&D MACH.	80104
MM3011	MASTERS	80104
MM3012	D&D MACH.	80110
MM3012	MASTERS	80112
MM3019	MASTERS	80112
MM3101	D&D MACH.	80100
MM3105	D&D MACH.	80104
MM3105	MASTERS	80104
MM3106	MASTERS	80099
MM3106	MASTERS	80101
MM3111	D&D MACH.	80107
MM3111	MASTERS	80104
MM3118	MASTERS	80101
MM3121	D&D MACH.	80102
MM3125	MASTERS	80097
MM3188	MASTERS	80105
MM4006	D&D MACH.	80153
MM4006	MASTERS	80153
MM4008	MASTERS	80146
MM4100	D&D MACH.	80109
MM4100	MASTERS	80149
MM4101	D&D MACH.	80153
MM4102	D&D MACH.	80158
MM4102	MASTERS	80158
MM4106	D&D MACH.	80148
MM4106	MASTERS	80148
MM4107	D&D MACH.	80159
MM4107	MASTERS	80159
MM4110	D&D MACH.	80155
MM4110	MASTERS	80153
MM4113	D&D MACH.	80156
MM4113	MASTERS	80156
MM5100	D&D MACH.	80168
MM5100	MASTERS	80168
MM5101	D&D MACH.	80167
MM5101	MASTERS	80167
MM5102	D&D MACH.	80171
MM5103	D&D MACH.	80174
MM5103	MASTERS	80174
MM5108	D&D MACH.	80172
MM5108	MASTERS	80172
MM5110	D&D MACH.	80173
MM5110	MASTERS	80173
MM5111	D&D MACH.	80175
MM5111	MASTERS	80175
MM5112	D&D MACH.	80177
MM5139	MASTERS	80170
MM6000	MASTERS	80193
MM6027	MCQUAY-NOR	88008.63
MM6028	MCQUAY-NOR	88008.35
MM6035	EATON	87238
MM6035	MCQUAY-NOR	87238
MM6035	MCQUAY-NOR	88007.58
MM6036	EATON	87326
MM6036	MCQUAY-NOR	87326
MM6037	EATON	87331
MM6037	MCQUAY-NOR	87331
MM6037	MCQUAY-NOR	88008.39
MM6038	EATON	87377
MM6038	MCQUAY-NOR	87377
MM6038	MCQUAY-NOR	88008.41
MM6039	EATON	87383
MM6039	MCQUAY-NOR	87383
MM6039	MCQUAY-NOR	88008.42
MM6040	EATON	87457
MM6040	MCQUAY-NOR	87457
MM6040	MCQUAY-NOR	88008.31
MM6040	MCQUAY-NOR	88008.44
MM6041	EATON	87459
MM6041	MCQUAY-NOR	87459
MM6041	MCQUAY-NOR	88008.45
MM6042	MCQUAY-NOR	87326
MM6050	EATON	87481
MM6050	MCQUAY-NOR	87481
MM6050	MCQUAY-NOR	88007.77
MM6051	EATON	87570
MM6051	MCQUAY-NOR	87570
MM6051	MCQUAY-NOR	88007.79
MM6053	EATON	87285
MM6053	MCQUAY-NOR	87285
MM6053	MCQUAY-NOR	88008.48
MM6054	EATON	87337
MM6054	MCQUAY-NOR	87285
MM6054	MCQUAY-NOR	87337
MM6054	MCQUAY-NOR	88008.48
MM6055	EATON	87399
MM6055	MCQUAY-NOR	87399
MM6055	MCQUAY-NOR	88008.59
MM6100	D&D MACH.	80193
MM6100	MASTERS	80193
MM6101	D&D MACH.	80195
MM6101	MASTERS	80194
MM6101	MASTERS	80195
MM6101	MASTERS	80197
MM6102	D&D MACH.	80197
MM6102	MASTERS	80194
MM6102	MASTERS	80195
MM6102	MASTERS	80197
MM6103	D&D MACH.	80194
MM6103	MASTERS	80194
MM6103	MASTERS	80195
MM6103	MASTERS	80197
MM6104	MASTERS	80192
MM6106	D&D MACH.	80198
MM6106	MASTERS	80198
MM6107	D&D MACH.	80202
MM6108	D&D MACH.	80199
MM6108	MASTERS	80199
MM6109	D&D MACH.	80201
MM6109	MASTERS	80201
MM6139	MASTERS	80203
MM7100	D&D MACH.	80210
MM7100	MASTERS	80210
MM7101	D&D MACH.	80211
MM7101	MASTERS	80211
MM7102	D&D MACH.	80212
MM7102	MASTERS	80211
MM7102	MASTERS	80212
MM7105	D&D MACH.	80213
MM7105	MASTERS	80213
MM8100	D&D MACH.	80231
MM8100	MASTERS	80231
MM8101	D&D MACH.	80232
MM8101	MASTERS	80232
MM8101	MASTERS	80237

Part No.	Brand	IDLI Group
MM8102	D&D MACH.	80237
MM8102	MASTERS	80237
MM8103	MASTERS	80236
MM8105	D&D MACH.	80238
MM9000	MASTERS	80235
MM9000	MOPAR	88012.34
MM9014	D&D MACH.	80235
MM9015	D&D MACH.	80234
MM9015	MASTERS	80234
MM9016	D&D MACH.	80234
MM9016	MASTERS	80234
MM9017	D&D MACH.	80233
MM9017	MASTERS	80233
MR52A	MOTOR MAST	80909
MR52A	ZELLER	80909
MR54	MOTOR MAST	88001.74
MR54A	MOTOR MAST	80901
MR57A	MOTOR MAST	80908
MR60	MOTOR MAST	80958
MR60	ZELLER	80903
MR62	MOTOR MAST	80954
MR62A	MOTOR MAST	80907
MR63	MOTOR MAST	80957
MR65	MOTOR MAST	80951
MR65	ZELLER	80951
MR65A	MOTOR MAST	80905
MR65A	ZELLER	80905
MR67	MOTOR MAST	80953
MR67	ZELLER	80953
MR67A	MOTOR MAST	80906
MR67A	ZELLER	80906
MR67A	ZELLER	80909
MS100X	MASTERS	80031
MS100X	MASTERS	80061
MS101X	D&D MACH.	80023
MS101X	MASTERS	80023
MS101X	MASTERS	80024
MS105X	D&D MACH.	80031
MS105X	MASTERS	80030
MS105X	MASTERS	80031
MS105X	MASTERS	80061
MS111X	MASTERS	80023
MS111X	MASTERS	80024
MS1202X	D&D MACH.	80130
MS1202X	MASTERS	80130
MS1210	MASTERS	88013.86
MS121U	D&D MACH.	80089
MS121U	MASTERS	80072
MS121U	MASTERS	80075
MS121U	MASTERS	80089
MS121X	D&D MACH.	80072
MS121X	MASTERS	80072
MS121X	MASTERS	80075
MS122X	MASTERS	80091
MS130X	D&D MACH.	80117
MS130X	MASTERS	80117
MS130X	MASTERS	80129
MS131X	D&D MACH.	80121
MS131X	MASTERS	80121
MS131X	MASTERS	80131
MS133X	D&D MACH.	80116
MS133X	MASTERS	80116
MS134X	D&D MACH.	80118
MS134X	MASTERS	80118
MS150X	D&D MACH.	80120
MS150X	MASTERS	80097
MS150X	MASTERS	80119
MS150X	MASTERS	80120
MS153X	D&D MACH.	80077
MS153X	MASTERS	80077
MS155X	D&D MACH.	80182
MS155X	MASTERS	80182
MS165X	D&D MACH.	80190
MS165X	MASTERS	80190
MS200U	D&D MACH.	80076
MS200U	MASTERS	80075
MS200X	D&D MACH.	80075
MS200X	MASTERS	80075
MS260X	D&D MACH.	80078
MS260X	MASTERS	80078
MS260X	MASTERS	80090
MS280X	D&D MACH.	80221
MS297X	D&D MACH.	80079
MS297X	MASTERS	80079
MS514X	D&D MACH.	80024
MS514X	MASTERS	80023
MS514X	MASTERS	80024
MS515X	MASTERS	80178
MS516X	D&D MACH.	80189
MS516X	MASTERS	80189
MS517X	D&D MACH.	80221
MS517X	MASTERS	80221
MS524X	D&D MACH.	80225
MS524X	MASTERS	80225
MS548X	D&D MACH.	80025
MS548X	MASTERS	80025
MS54X	MASTERS	80092
MS560U	D&D MACH.	80143
MS560U	MASTERS	80142
MS560U	MASTERS	80143
MS560X	D&D MACH.	80142
MS560X	MASTERS	80142
MS560X	MASTERS	80143
MS570X	D&D MACH.	80007
MS570X	MASTERS	80007
MS570X	MASTERS	88000.15
MS574X	D&D MACH.	80090
MS574X	MASTERS	80078
MS574X	MASTERS	80090
MS577X	MASTERS	80221
MS578U	D&D MACH.	80125
MS578U	MASTERS	80125
MS578X	D&D MACH.	80124
MS578X	MASTERS	80124
MS581X	MASTERS	80023
MS581X	MASTERS	80024
MS581X	MASTERS	80140
MS588X	D&D MACH.	80140
MS588X	MASTERS	80140
MS588X	MASTERS	80182
MST00024	MITSUBISHI	80012
MST05012	MITSUBISHI	80012
MSY54	MUNCIE	85054
MSY57	MUNCIE	85063
MSY60	MUNCIE	85058
MSY62	MUNCIE	88014.19
MSY69	MUNCIE	85055
MT1642	MASTERS	88002.42
MT202157	MITSUBISHI	80012
MT3692	MASTERS	88002.44
MT372505	CHRYSLER	80012
MT372505	MITSUBISHI	80012
MT7090	MASTERS	88002.55
MT9531	MASTERS	88002.48
MXR3020	COCKSHUTT	80133
MYO-103	MOTOR MAST	88000.36
MYO-103113	MOTOR MAST	88000.35
MYO-113	MOTOR MAST	88000.35
MYO-115	MOTOR MAST	88000.38
MYO-115	MOTOR MAST	88000.55
MYO-115123	MOTOR MAST	88000.38
MYO-12	MOTOR MAST	88000.45
MYO-123	MOTOR MAST	88000.38
MYO-123	MOTOR MAST	88000.39
MYO-124	MOTOR MAST	88000.37
MYO-124125	MOTOR MAST	88000.37
MYO-13	MOTOR MAST	88000.20
MYO-13	MOTOR MAST	88000.21
MYO-13	MOTOR MAST	88000.39
MYO-150	MOTOR MAST	88000.46
MYO-22	MOTOR MAST	88000.43
MYO-23	MOTOR MAST	88000.60
MYO-33	MOTOR MAST	88000.22
MYO-33	MOTOR MAST	88000.26
MYO-43	MOTOR MAST	88000.22
MYO-53	MOTOR MAST	88000.52
MYO-5363	MOTOR MAST	88000.25
MYO-63	MOTOR MAST	88000.25
MYO-63	MOTOR MAST	88000.27
MYO-73	MOTOR MAST	88000.28
MYO-76	MOTOR MAST	88000.29
MYO-81	MOTOR MAST	88000.49
MYO-83	MOTOR MAST	88000.53
MYO-8393	MOTOR MAST	88000.30
MYO-93	MOTOR MAST	88000.31
MYO-95	MOTOR MAST	88000.34
MY1-103	MOTOR MAST	88000.85
MY1-12	MOTOR MAST	88000.86
MY1-12	MOTOR MAST	88000.88
MY1-13	MOTOR MAST	88000.79
MY1-150	MOTOR MAST	88000.90
MY1-22	MOTOR MAST	88000.89
MY1-23	MOTOR MAST	88000.80
MY1-32	MOTOR MAST	88000.95
MY1-33	MOTOR MAST	88000.81
MY1-53	MOTOR MAST	88000.83
MY1-63	MOTOR MAST	88000.96
MY1-81	MOTOR MAST	88000.92
MY1-81	MOTOR MAST	88001
MY1-83	MOTOR MAST	88000.82
MY1-93	MOTOR MAST	88000.84
MY2-13	MOTOR MAST	88000.68
MY2-13	MOTOR MAST	88000.69
MY2-43	MOTOR MAST	88000.70
MY2-63	MOTOR MAST	88000.71
MY2-73	MOTOR MAST	88000.72
MY2-76	MOTOR MAST	88000.72
MY2-93	MOTOR MAST	88000.73
MYI12	MUNCIE	88000.86
MYO12	MOTOR MAST	88000.45
MYO13	MOTOR MAST	88000.21
N0555	WESCO	80141
N-0666X	NEAPCO	88001.15
N0965	WESCO	80138
N10242-1	NEAPCO	88000.53
N10242-16	NEAPCO	88000.25
N10242-17	NEAPCO	88000.21
N10242-18	NEAPCO	88000.39
N10242-2	NEAPCO	88000.52
N10242-21	NEAPCO	88000.31
N10242-23	NEAPCO	88000.35
N10242-3	NEAPCO	88000.22
N10242-30	NEAPCO	88000.22
N10242-32	NEAPCO	88000.38
N10242-4	NEAPCO	81020
N10242-45	NEAPCO	88000.27
N10242-5	NEAPCO	88000.21
N10242-6	NEAPCO	88000.28
N10242-7	NEAPCO	88000.31
N10242-8	NEAPCO	88000.38
N10242-9	NEAPCO	88000.35
N10243-3	NEAPCO	88000.39
N10243-8	NEAPCO	88000.88
N103122X	NEAPCO	88000.90
N10312X	NEAPCO	88000.88
N103162X	NEAPCO	88001.01
N103163X	NEAPCO	88000.85
N103183X	NEAPCO	88000.84
N10322X	NEAPCO	88000.89
N10323X	NEAPCO	88000.80
N10332X	NEAPCO	88000.95
N10333X	NEAPCO	88000.81
N10381X	NEAPCO	88000.92
N10383X	NEAPCO	88000.83
N104103	NEAPCO	88000.36
N104113	NEAPCO	88000.35
N10412	NEAPCO	88000.45
N104123	NEAPCO	88000.39
N10413	NEAPCO	88000.21
N104133	NEAPCO	88000.20
N1041363	NEAPCO	88000.40
N104141X	NEAPCO	88000.67
N104143	NEAPCO	88000.36
N1041473	NEAPCO	88000.35
N104173	NEAPCO	88000.22
N104183	NEAPCO	88000.55
N104193	NEAPCO	88000.38
N10422	NEAPCO	88000.43
N10423	NEAPCO	88000.44
N104282	NEAPCO	88000.62
N10432	NEAPCO	88000.47
N10433	NEAPCO	88000.26
N10443	NEAPCO	88000.22
N104443	NEAPCO	88000.32
N104453	NEAPCO	88000.22
N10452	NEAPCO	88000.46
N10453	NEAPCO	88000.52
N104573	NEAPCO	88000.30
N10463	NEAPCO	88000.27
N104703	NEAPCO	88000.71
N10473	NEAPCO	88000.28
N10483	NEAPCO	88000.53
N10493	NEAPCO	88000.31
N1052A16	BMC	80041
N12	PRECISION	88003.42
N12	WESCO	80035
N13	PRECISION	88003.43
N14	PRECISION	88003.44
N14	WESCO	80037
N15	PRECISION	88003.45
N1501	WESCO	80023
N1501	WESCO	80135
N1502	WESCO	80041
N1504	WESCO	80041
N1504	WESCO	88001.85
N1513	WESCO	80072
N1515	WESCO	80178
N1515	WESCO	88002.96
N1516	WESCO	80087
N1516	WESCO	80189
N1517	WESCO	80221
N1518	WESCO	80124
N1520R	WESCO	80075
N1520RU	WESCO	80076
N1520RU	WESCO	80959
N1521	WESCO	80072
N1521HD	WESCO	80077
N1521R	WESCO	80072
N1521R	WESCO	80075
N1521RHD	WESCO	80072

Part No.	Brand	IDLI Group	Part No.	Brand	IDLI Group	Part No.	Brand	IDLI Group	Part No.	Brand	IDLI Group
N1521RU	WESCO	80089	N2045G	WESCO	80049	N3120	WESCO	80113	N6104	WESCO	80192
N1524	WESCO	80225	N2045G	WESCO	88002.01	N3-40-971	NEAPCO	87370	N6106	WESCO	80198
N1525	WESCO	80227	N2045G	WESCO	88002.02	N35	WESCO	80069	N6107	WESCO	80202
N1530	WESCO	80117	N2045G	WESCO	88002.06	N3-53-431	NEAPCO	87357	N6119	WESCO	88010.38
N1530	WESCO	80129	N2045G	WESCO	88002.09	N4002	WESCO	80109	N6123	WESCO	80191
N1530-1	WESCO	80129	N2046	WESCO	80044	N4002	WESCO	80149	N6132	WESCO	80199
N1531	WESCO	80121	N2046-1	WESCO	80043	N4007	WESCO	80147	N6139	WESCO	80203
N153-1	WESCO	80131	N2046G	WESCO	80043	N4008	WESCO	80146	N6149	WESCO	80198
N1531-1	WESCO	80131	N2046G	WESCO	80044	N4010	WESCO	80153	N6168	CASE,J.I.	80172
N1533	WESCO	80116	N2046G	WESCO	88001.98	N4014	WESCO	80158	N666X	NEAPCO	88001.15
N1550	WESCO	80119	N2046G	WESCO	88002.08	N4014	WESCO	88002.15	N6K	WESCO	80001
N1552	WESCO	80118	N2047	WESCO	80058	N4014	WESCO	88002.16	N6S	WESCO	80001
N1552-1	WESCO	80130	N2048	WESCO	80049	N4015	WESCO	80148	N7000	WESCO	80210
N1553	WESCO	80077	N2049	WESCO	80054	N4016	WESCO	80159	N7004	WESCO	80211
N1553RB	WESCO	80076	N2049	WESCO	88002	N4103	WESCO	80150	N7008	WESCO	80208
N1553RV	WESCO	80076	N2049	WESCO	88002.03	N4104	WESCO	80151	N7009	WESCO	80209
N1554	WESCO	80072	N2053	WESCO	80031	N4109	WESCO	80151	N7034	WESCO	80212
N1554	WESCO	80081	N2053	WESCO	88001.99	N4113	WESCO	80156	N7100-1	REPUBLIC	88002.69
N1554RU	WESCO	80081	N2053-1	WESCO	80030	N4113	WESCO	88003.17	N7105	WESCO	80213
N1555	WESCO	80182	N2053G	WESCO	80031	N4130	WESCO	80155	N7140-2A	REPUBLIC	88002.69
N1560	WESCO	80142	N2057	KOYO	80004	N4141	WESCO	80154	N7200-20	REPUBLIC	88002.68
N1560R	WESCO	80142	N2062	WESCO	80050	N4156	WESCO	80157	N7325	WESCO	80096
N1560RU	WESCO	80143	N2063	WESCO	80043	N42	PRECISION	88003.52	N73321	DIAMOND T	80124
N1565	WESCO	80190	N2064	WESCO	80044	N44N220-20	NEAPCO	81362	N7500	WESCO	80205
N1570	WESCO	80007	N2066	WESCO	80045	N44N220-21	NEAPCO	81368	N8000	WESCO	80231
N1570	WESCO	88000.15	N2117	WESCO	80046	N44N220TQD	NEAPCO	84511	N8003	WESCO	80236
N1570P	WESCO	80007	N2140	WESCO	80055	N44N2256-37	NEAPCO	81365	N8014	WESCO	80230
N1574	WESCO	80078	N2150	WESCO	80025	N44N246	NEAPCO	81370	N8101	WESCO	80232
N1574	WESCO	80090	N2155	WESCO	80083	N44NWY2100	NEAPCO	81356	N8102	WESCO	80237
N1574-1	WESCO	80090	N2173	ALCO	80058	N44NWY2200	NEAPCO	85516	N8105	WESCO	80232
N1574G	WESCO	80078	N2173	WESCO	80058	N44RYR26-1	NEAPCO	81372	N8105	WESCO	80238
N1578	WESCO	80124	N2198	WESCO	80051	N44RYR28-2	NEAPCO	81375	N8202	WESCO	80237
N15789RU	WESCO	80124	N2299-1X	NEAPCO	88000.09	N4-53-71	NEAPCO	88008.16	N8209	WESCO	80238
N15789RU	WESCO	80125	N2299X	NEAPCO	88000.09	N47	PRECISION	88003.53	N8500	WESCO	80223
N1578R	WESCO	80124	N-2299X-1	NEAPCO	88000.09	N5	PRECISION	88003.54	N8507	WESCO	80223
N1578RU	WESCO	80125	N-2299X-1	NEAPCO	88001.08	N5000	WESCO	80168	N8811	CASE,J.I.	80213
N1578S	WESCO	80124	N2299X1	NEAPCO	88013.82	N5000	WESCO	88002.17	N9000	WESCO	80235
N1579	WESCO	80073	N230590	NEAPCO	88015.43	N5007	WESCO	80177	N9002	WESCO	80233
N1579R	WESCO	80073	N24	PRECISION	88003.51	N5007	WESCO	88002.18	N9002	WESCO	88003.09
N1579RU	WESCO	80091	N2459X	NEAPCO	88001.08	N5008	WESCO	80174	N9015	WESCO	80234
N1581	WESCO	80024	N2461	KOYO	80024	N5015	WESCO	80173	N9016	WESCO	80234
N1585	WESCO	80228	N2500	WESCO	80059	N5031	WESCO	80171	N9017	WESCO	80233
N1588	WESCO	80140	N2503	WESCO	80032	N5101	WESCO	80167	N9017	WESCO	80241
N1588	WESCO	80182	N2540	WESCO	80066	N5102	WESCO	80170	N9031	WESCO	80234
N1597	WESCO	80079	N2563	KOYO	80010	N5102	WESCO	80171	N9034	WESCO	80234
N16	PRECISION	88003.46	N2604X	NEAPCO	88006.37	N5108	WESCO	80172	NB10A	HUSCO	80031
N1666X	NEAPCO	88001.15	N2691X	NEAPCO	88006.37	N5109	WESCO	80162	NB12A	BPC	80035
N17	PRECISION	88003.47	N2880RA	KOYO	80019	N5114	WESCO	80176	NB7	HUSCO	88000.16
N1755	WESCO	80032	N3000	WESCO	80099	N5131	WESCO	80175	NB751	HUSCO	88000.16
N18	PRECISION	88003.48	N3000	WESCO	80101	N5139	WESCO	80170	NB751A	HUSCO	88000.16
N1859X	NEAPCO	88000.09	N3000G	WESCO	80101	N5148	WESCO	80169	NB7A	BPC	88000.15
N19	PRECISION	88003.49	N3001	WESCO	80101	N5155	WESCO	80081	NB7A	HUSCO	80007
N20	PRECISION	88003.50	N3005	WESCO	80093	N5165	WESCO	80091	NB7A2	HUSCO	80007
N2000	WESCO	80049	N3008	WESCO	80107	N5-177	ALCO	80221	NC965	WESCO	80204
N2001	WESCO	80047	N3008G	WESCO	80104	N5-200	ALCO	80075	ND1915D	DEERE,JOHN	80001
N2002	WESCO	80048	N3008G	WESCO	80107	N55	WESCO	80161	ND965	WESCO	80160
N2002-1	WESCO	88002.10	N3009	WESCO	80105	N553	WESCO	80135	ND965-4	WESCO	80160
N2004X	NEAPCO	88000.09	N3010	WESCO	80100	N556	WESCO	80042	NDS555	WESCO	80145
N2012	WESCO	80030	N3011	WESCO	80107	N5800	WESCO	80185	NF4841A	NEAPCO	88002.86
N2014	WESCO	80045	N3012	WESCO	80112	N5800	WESCO	80186	NF4851A	NEAPCO	88002.87
N2017	WESCO	80092	N3012G	WESCO	80112	N5801	WESCO	80185	NH47621	IHC	80044
N2018	WESCO	80053	N3012G	WESCO	88002.14	N5801	WESCO	80186	NO555	WESCO	80141
N2026	WESCO	80030	N3013	WESCO	80106	N5802	WESCO	80184	NO5555	WESCO	80141
N2031	WESCO	80049	N3014	WESCO	80114	N6	WESCO	80001	NO965	WESCO	80138
N2033	WESCO	80054	N3019	WESCO	80112	N6000	WESCO	80193	NOS555	WESCO	80183
N2039	WESCO	80049	N3021	WESCO	80111	N6000	WESCO	80198	NP965	WESCO	80068
N2040	WESCO	80049	N3022	WESCO	80103	N6022	WESCO	80197	NR1565	WESCO	80091
N2041	WESCO	80054	N3023	WESCO	80102	N6023	WESCO	80199	NR5165	WESCO	80073
N2045	WESCO	80049	N3092	KOYO	80124	N6024	WESCO	80201	NR5165	WESCO	80091
N20456	WESCO	88001.96	N3103	WESCO	80098	N6103	WESCO	80194	NR555	WESCO	80123

Part No.	Brand	IDLI Group	Part No.	Brand	IDLI Group	Part No.	Brand	IDLI Group	Part No.	Brand	IDLI Group
NR965	WESCO	80122	O55-3-3-14D	CLEVE.STL.	88004.83	P32193	THEW SHOV.	80221	PB7A	BPC	88000.14
NRF2885	AUST.WEST.	80216	O55-3-3-14S	CLEVE.STL.	85522	P32225	ATHEY PROD.	80172	PB7A	HUSCO	80007
NRF2885	BAL-LM-HAM	80216	O55-43-169	CLEVE.STL.	86669	P33408	THEW SHOV.	80215	PBS1	L & S BRG.	87285
NS535	WESCO	80039	O55-43-171	CLEVE.STL.	88007.34	P33423	THEW SHOV.	80214	PBS10	L & S BRG.	87377
NS555	WESCO	80081	O55-43-175	CLEVE.STL.	86668	P33433	THEW SHOV.	80215	PBS10	SKF	87377
NS555-1	WESCO	80081	O55-43-176	CLEVE.STL.	86670	P33438	THEW SHOV.	80213	PBS11	L & S BRG.	87457
NS555-1	WESCO	80038	O55-54-1	CLEVE.STL.	85066	P33443	THEW SHOV.	80238	PBS11	SKF	87457
NS965	WESCO	80071	O55-54-240	CLEVE.STL.	85067	P33444	THEW SHOV.	80239	PBS12	L & S BRG.	87481
NT555	WESCO	80074	O55-54-5	CLEVE.STL.	85065	P33446	THEW SHOV.	80224	PBS13	L & S BRG.	87570
NTS555	WESCO	80181	O55-55	ALCO	80141	P33459	THEW SHOV	80239	PBS2	L & S BRG.	87337
NU965	WESCO	80181	O55-55	CLEVE.STL.	80141	P33460	THEW SHOV.	80213	PBS3	L & S BRG.	87399
NU965-4	WESCO	80181	O55-55-2	BORG-WARNR	80141	P33466	THEW SHOV.	80224	PBS6	L & S BRG.	87238
NUS555	WESCO	80183	O55-55-2	CLEVE.STL.	80141	P335	THEW SHOV.	80188	PBS6	SKF	87238
NUSL16-20-1	ROCKWELL	88014.56	O55-55-2	REPCO	80141	P33502	THEW SHOV.	80238	PBS7	L & S BRG.	87331
NUSL16-20-2	ROCKWELL	88014.57	O55-6	CLEVE.STL.	88014.80	P33508	THEW SHOV.	80234	PBS7	SKF	87326
NU-SL20-18-1	ROCKWELL	88014.58	O55-7-241	CLEVE.STL.	88014.72	P33516	KOEHRING	80069	PBS8	SKF	87331
NU-SL20-18-3	ROCKWELL	88014.59	O55-7-242	CLEVE.STL.	88014.73	P33516	THEW SHOV.	80069	PBS9	L & S BRG.	87331
NYRYR26-1	NEAPCO	81372	O55-7-3-14	CLEVE.STL.	88014.71	P33526	KOEHRING	80001	PGF10797-56A	AUST.WEST.	80215
FOR NUMBERS THAT START WITH O, ALSO CHECK THE NUMBER ZERO.			O55-9	CLEVE.STL.	88013.60	P33526	THEW SHOV.	80001	PGF10797-56A	BAL-LM-HAM	80215
O10530	NEW IDEA	80035	O55-9-3	CLEVE.STL.	88013.53	P33536	KOEHRING	80001	PGF10797-58A	AUST.WEST.	80213
O11070	NEW IDEA	80069	O96	TEDDY TORQ	80138	P33559	THEW SHOV.	80198	PGF10798-56A	AUST.WEST.	80210
O12014	NEW IDEA	80037	O96-55	ALCO	80138	P33576	MASS.FERG.	80219	PGF10798-56A	BAL-LM-HAM	80210
O12324	NEW IDEA	80003	O96-55	BORG-WARNR	80138	P33576	THEW SHOV.	80219	PGF1891	AUST.WEST.	80212
O12716	NEW IDEA	80035	O96-55	CLEVE.STL.	80138	P33588	THEW SHOV.	80197	PGF18913	AUST.WEST.	80214
O15555	PILOT	80141	O96-55	REPCO	80138	P33596	THEW SHOV.	80217	PGF18913	BAL-LM-HAM	80218
O19655	PILOT	80138	O96-55	SPICER	80138	P353	THEW SHOV.	80220	PGF6732	AUST.WEST.	80193
O1Y18397	FORD	80077	O96-9	CLEVE.STL.	88013.57	P371	THEW SHOV.	80220	PGF6732-56A	AUST.WEST.	80193
O3TZ3249A	FORD	80079	OA4851	FORD	88002.87	P38776	ATHEY PROD.	80101	PGF6732-56A	BAL-LM-HAM	80193
O4NN4N167A	FORD	80142	OCB100L	CLEVE.STL.	87337	P398	THEW SHOV.	80188	PGF7652	BAL-LM-HAM	80215
O55	TEDDY TORQ	80141	OCB100LS	CLEVE.STL.	87338	P399	THEW SHOV.	80220	PGF7653	BAL-LM-HAM	80215
O55-10	CLEVE.STL.	88015.51	OCB103	CLEVE.STL.	87373	P41	PRECISION	88003.55	PGF8508-56	AUST.WEST.	80193
O55-10-4	CLEVE.STL.	88015.32	OCB105	CLEVE.STL.	88008.55	P42	PRECISION	88003.56	PGF8508-56	BAL-LM-HAM	80193
O55-11-1	CLEVE.STL.	88015.58	OCB106	CLEVE.STL.	87376	P43	PRECISION	88003.57	PGF8537-56	AUST.WEST.	80215
O55-1-1186	CLEVE.STL.	88004.88	OCB107	CLEVE.STL.	87375	P443	KOEHRING	80178	PGF8537-56	BAL-LM-HAM	80210
O55-1-1188	CLEVE.STL.	88004.98	OCB108	CLEVE.STL.	87374	P443	THEW SHOV.	80178	PGF8585-56	AUST.WEST.	80193
O55-1-1188A	CLEVE.STL.	88004.99	OCB109	CLEVE.STL.	87374	P466	KOEHRING	80134	PGF8585-56	BAL-LM-HAM	80193
O55-1-1255	CLEVE.STL.	88004.89	OCB2	CLEVE.STL.	88013.74	P466	THEW SHOV.	80134	PGF8807-57	AUST.WEST.	80215
O55-1-1305	CLEVE.STL.	88004.86	OCB2-1	CLEVE.STL.	88013.68	P48776	ATHEY PROD.	80099	PGF8807-57	BAL-LM-HAM	80210
O55-1-1306	CLEVE.STL.	88004.87	OCB51-2	CLEVE.STL.	88008.30	P48776	ATHEY PROD.	80101	PGH7652	AUST.WEST.	80215
O55-1-1409	CLEVE.STL.	88004.91	OCB51-2	CLEVE.STL.	88008.61	P55	TEDDY TORQ	80129	PGH7653	AUST.WEST.	80215
O55-1-1548	CLEVE.STL.	88004.90	OCB61	CLEVE.STL.	88007.87	P55-55-1330	CLEVE.STL.	80129	PK96-55-279	CLEVE.STL.	80062
O55-1-1694	CLEVE.WEST.	88005.02	OCB61-1	CLEVE.STL.	88007.83	P55-55-140	CLEVE.STL.	80129	PK96-55-302	CLEVE.STL.	80062
O55-1-1705	CLEVE.STL.	88004.92	OFF-HWY.	MCQUAY-NOR	80134	P55-55-145	BORG-WARNR	80131	PL6R40	ROCKWELL	88014.60
O55-1-1714	CLEVE.STL.	88004.97	OFF-HWY.	MCQUAY-NOR	80223	P55-55-145	CLEVE.STL.	80131	PNCO2-32-6	ROCKWELL	88014.61
O55-12	CLEVE.STL.	88014.94	OFF RD. EQUIP.	MCQUAY-NOR	80098	P55-55-150	CLEVELAND	80129	PN-CO2-40-4	ROCKWELL	88014.62
O55-1259	CLEVE.STL.	88005	OIY7039	FORD	80077	P55-55-2	BORG-WARNR	80129	PNS11	SKF	87457
O55-1259A	CLEVE.STL.	88005.01	P0009	REO MOTORS	80124	P55-55-2	CLEVELAND	80129	PS18-24-2	ROCKWELL	88014.63
O55-129	CLEVE.STL.	87354	P1005	NATL.MINE	80161	P55-55-2	CLEVE.STL.	80129	PS20-10-1	ROCKWELL	87254
O55-1295	CLEVE.STL.	88008.09	P1006	NATL.MINE	80069	P55-55-675	BORG-WARNR	80131	PS20-32-1	ROCKWELL	88014.64
O55-12S	CLEVE.STL.	88014.91	P12169	ATHEY PROD	80104	P55-55-675	CLEVELAND	80131	PS20-32-13	ROCKWELL	88014.65
O55-130	CLEVE.STL.	88008.03	P12558	ATHEY PROD	80163	P55-55-675	CLEVE.STL.	80131	PS20-32-2	ROCKWELL	88014.66
O55-1303	CLEVE.STL.	88008.13	P12559	ATHEY PROD	80176	P55-55-675	CLEVE.STL.	88008.90	PS20-32-5	ROCKWELL	88014.67
O55-131	CLEVE.STL.	88008.04	P12560	ATHEY PROD	80048	P57	PRECISION	88003.58	PS20-32-6	ROCKWELL	88014.68
O55-132	CLEVE.STL.	87358	P14368	ATHEY PROD	80216	P75	TEDDY TORQ	80131	PS22-32-3	ROCKWELL	88014.69
O55-135	CLEVE.STL.	88008.05	P15555	PILOT	80129	P8808	THEW SHOV.	80070	PS22-32-4	ROCKWELL	88014.70
O55-130	CLEVE.STL.	88008.03	P19655	PILOT	80068	P8809	THEW SHOV.	80070	PS24-32-1	ROCKWELL	88014.71
O55-1408	CLEVE.STL.	88008.18	P19675	PILOT	80131	P8814	THEW SHOV.	80213	PS24-32-5	ROCKWELL	88014.72
O55-1410	CLEVE.STL.	88008.17	P1FR	WESCO	80008	P8900	KOEHRING	80069	PS24-32-6	ROCKWELL	88014.73
O55-17	CLEVE.STL.	88013.41	P20261	ATHEY PROD	88003.26	P8900	THEW SHOV.	80070	PS24-32-7	ROCKWELL	88014.74
O55-17-1	CLEVE.STL.	88013.42	P20305	ATHEY PROD	80048	P8908	KOEHRING	80070	PS25-16-2	ROCKWELL	88014.75
O55-1-812	CLEVE.STL.	88004.95	P20322	ATHEY PROD	80048	P96	TEDDY TORQ	80068	PS26-32-1	ROCKWELL	87389
O55-1-812A	CLEVE.STL.	88004.96	P20331	ATHEY PROD	88003.26	P96-55	ALCO	80068	PS26-32-1A	ROCKWELL	87392
O55-1-877	CLEVE.STL.	88004.93	P22172	ATHEY PROD	80037	P96-55-1	AEC	80068	PS26-32-2	ROCKWELL	87390
O55-1-877A	CLEVE.STL.	88004.94	P22181	ATHEY PROD	80104	P96-55-1	BORG-WARNR	80068	PS26-32-2A	ROCKWELL	87393
O55-1-961	WESCO	88004.84	P22239	ATHEY PROD	80044	P96-55-1	CLEVELAND	80068	PS28-16-6	ROCKWELL	88014.76
O55-1-961A	AUST.WEST.	88004.85	P27423	THEW SHOV.	80178	P96-55-1	CLEVE.STL.	80068	PS28-32-4	ROCKWELL	88014.77
O55-2	CLEVE.STL.	84614	P27525	THEW SHOV.	80221	P96-55-1	REPCO	80068	PS28-32-5	ROCKWELL	88014.78
O55-2-2	CLEVE.STL.	84615	P301	BMC	80041	P96-55-1	SPICER	80068	PS32-16-16	ROCKWELL	87519
O55-245	CLEVE.STL.	87359	P304	BMC	80124	PA679	PIONEER	80134	PS32-16-31	ROCKWELL	87508
O55-3-3-14	CLEVE.STL.	85518	P305	BMC	80124	PB418558	NELSON	80198	PS32-16-43	ROCKWELL	87528

Part No.	Brand	IDLI Group
PS40-16-18	ROCKWELL	87590
PS48-10-13	ROCKWELL	87646
PS48-16-1	ROCKWELL	87636
PS48-16-1	ROCKWELL	87638
PS48-16-10	ROCKWELL	87637
PS48-16-17	ROCKWELL	87643
PS48-16-19	ROCKWELL	87633
PS48-16-9	ROCKWELL	87630
PS48-23-3	ROCKWELL	87635
PS48-23-4	ROCKWELL	87631
QL1005	QUINT.HAZ.	80142
QL102	QUINT.HAZ.	80024
QL11007	QUINT.HAZ.	80006
QL11007	QUINT.HAZ.	88002.77
QL11007	QUINTON	80004
QL11009	QUINT.HAZ.	80010
QL11009	QUINTON	80010
QL11010	QUINT.HAZ.	80011
QL12	QUINT.HAZ.	80018
QL14	QUINT.HAZ.	80017
QL15002	QUINT.HAZ.	80023
QL155	QUINT.HAZ.	80182
QL16007	QUINT.HAZ.	80189
QL1607	QUINT.HAZ.	80189
QL16103	QUINT.HAZ.	80041
QL16204	QUINT.HAZ.	80072
QL16405	QUINT.HAZ.	80142
QL16506	QUINT.HAZ.	80180
QL1708	QUINT.HAZ.	80221
QL1758	QUINT.HAZ.	88002.81
QL1761	QUINT.HAZ.	88002.77
QL1762	QUINT.HAZ.	88002.77
QL1763	QUINT.HAZ.	80032
QL1764	QUINT.HAZ.	80032
QL1765	QUINT.HAZ.	88002.77
QL1776	QUINT.HAZ.	80009
QL1780	QUINT.HAZ.	80004
QL1783	QUINT.HAZ.	80012
QL1786	QUINT.HAZ.	88002.77
QL188X	QUINT.HAZ.	80140
QL19245	QUINT.HAZ.	80124
QL203	QUINT.HAZ.	88002.77
QL2030	QUINT.HAZ.	80044
QL2038	QUINT.HAZ.	80060
QL2039	QUINT.HAZ.	80091
QL204	QUINT.HAZ.	88002.77
QL205	QUINT.HAZ.	80032
QL206	QUINT.HAZ.	80032
QL2104	QUINT.HAZ.	80072
QL211	QUINT.HAZ.	80023
QL212	QUINT.HAZ.	88002.77
QL213	QUINT.HAZ.	80023
QL214	QUINT.HAZ.	80023
QL216	QUINT.HAZ.	80006
QL216	QUINTON	80004
QL217	QUINT.HAZ.	80009
QL217	QUINTON	80009
QL218	QUINT.HAZ.	80024
QL219	QUINT.HAZ.	80012
QL220	QUINT.HAZ.	80047
QL2206	QUINT.HAZ.	80180
QL2409	QUINT.HAZ.	80225
QL26	QUINT.HAZ.	80024
QL27	QUINT.HAZ.	80010
QL291	QUINT.HAZ.	80041
QL291	QUINTON	80041
QL300	QUINT.HAZ.	88002.77
QL312	QUINT.HAZ.	88002.77
QL403	QUINT.HAZ.	80024
QL502	QUINT.HAZ.	80041

Part No.	Brand	IDLI Group
QL6	QUINT.HAZ.	80020
QL6200	QUINT.HAZ.	80049
QL6700	QUINT.HAZ.	80210
R004-01-29	AUST.MOT	80044
R055-55-2	PERFECTION	80081
R101	TEDDY TORQ	80909
R101C	TEDDY TORQ	80958
R1025	HUME,H.D.	80008
R1028	HUME,H.D.	80008
R12	WESCO	80035
R12N	CHAIN BELT	80035
R12N	REX CHAIN	80035
R14	WESCO	80037
R14744	CASE,J.I.	80069
R14N	CHAIN BELT	80037
R14N	REX CHAIN	80037
R15555	PILOT	80123
R19655	PILOT	80122
R1D1116	TAYLOR	80214
R1D8952	TAYLOR	80205
R2713	CASE,J.I.	80124
R2X1972	CATERPILLR	80104
R30068	CASE,J.I.	80174
R30673	CASE,J.I.	80193
R3092	CASE,J.I.	80124
R31N	CHAIN BELT	80035
R31N	REX CHAIN	80035
R35	WESCO	80069
R35N	CHAIN BELT	80069
R35N	REX CHAIN	80069
R401	TEDDY TORQ	80901
R401C	TEDDY TORQ	88001.74
R4177795	FMC	80222
R424	PERFECTION	88002.86
R424	PERFECTION	88002.87
R425	PERFECTION	88002.86
R425	PERFECTION	88002.87
R514X	PERFECTION	80023
R514X	PERFECTION	80024
R52	DETROIT	80903
R52	DETROIT	80958
R52	SPICER	80958
R52A	DETROIT	80909
R52A	SPICER	80909
R54	DETROIT	88001.66
R54A	DETROIT	80901
R54A	DETROIT	88001.66
R55	DETROIT	80956
R55	TEDDY TORQ	80123
R55-10	CLEVE.STL.	88015.49
R55-1-101	CLEVE.STL.	88004.76
R55-1-104	CLEVE.STL.	88004.77
R55-1-107	CLEVE.STL.	88004.80
R55-1-112	CLEVE.STL.	88004.78
R55-1-1226	CLEVE.STL.	88004.74
R55-1-1434	CLEVE.STL.	88004.69
R55-1-172	CLEVE.STL.	88004.79
R55-1-179	CLEVE.STL.	88004.75
R55-1-188	CLEVE.STL.	88004.82
R55-1233	CLEVE.STL.	88008.12
R55-1-240	CLEVE.STL.	88004.81
R55-1-324	CLEVE.STL.	88004.72
R55-1-446	CLEVE.STL.	88004.73
R55-1-544	CLEVE.STL.	88004.70
R55-1-669	CLEVE.STL.	88004.71
R55-17	CLEVE.STL.	88013.40
R55-19	CLEVE.STL.	88004.61
R55-2	CLEVE.STL.	88004.62
R55-2-1	CLEVE.STL.	88004.63
R55-2-4	CLEVE.STL.	88004.64

Part No.	Brand	IDLI Group
R55-3-3-14	CLEVE.STL.	88004.67
R55-3-3-14D	CLEVE.STL.	88004.68
R55-43-311	CLEVE.STL.	88007.45
R55-43-313	CLEVE.STL.	88007.46
R55-43-314	CLEVE.STL.	88007.47
R55-43-315	CLEVE.STL.	88007.49
R55-43-316	CLEVE.STL.	88007.50
R55-43-593	CLEVE.STL.	88007.48
R55-449	CLEVE.STL.	88008.06
R55-450	CLEVE.STL.	88008.07
R55-452	CLEVE.STL.	88008.08
R55-54-1	CLEVE.STL.	88004.66
R55-54-16	CLEVE.STL.	88004.65
R55-55	ALCO	80123
R55-55	BORG-WARNR	80123
R55-55	CLEVE.STL.	80123
R55-55	SPICER	80123
R55-55-2	BORG-WARNR	80123
R55-55-2	CLEVELAND	80123
R55-55-2	CLEVE.STL.	80123
R55-55-2	REPCO	80123
R55-7-3-14	CLEVE.STL.	88014.69
R55-7-3-14L	CLEVE.STL.	88014.70
R55-9	CLEVE.STL.	88013.58
R56	DETROIT	80954
R56A	DETROIT	80907
R57	DETROIT	80955
R57	SPICER	80955
R57A	DETROIT	80908
R57A	SPICER	80908
R60	DETROIT	80903
R60	DETROIT	80958
R60	SPICER	80958
R601	TEDDY TORQ	80907
R601C	TEDDY TORQ	80954
R603C	TEDDY TORQ	80957
R605	TEDDY TORQ	80905
R605C	TEDDY TORQ	80951
R606C	TEDDY TORQ	80953
R607	TEDDY TORQ	80906
R607C	TEDDY TORQ	80953
R62	DETROIT	80954
R62	SPICER	80954
R62A	DETROIT	80907
R63	DETROIT	80957
R63	SPICER	80957
R65	DETROIT	80951
R65	SPICER	80951
R65A	DETROIT	80905
R65A	SPICER	80905
R66	DETROIT	80953
R661C	TEDDY TORQ	80958
R67	DETROIT	80953
R67	SPICER	80953
R67	SPICER	80957
R67A	DETROIT	80906
R67A	SPICER	80906
R701	TEDDY TORQ	80908
R701C	TEDDY TORQ	80955
R7352A	ALCO	80909
R7354	ALCO	88001.74
R7354A	ALCO	80901
R7356	DETROIT	80954
R7356A	DETROIT	80907
R7357	ALCO	80955
R7357A	ALCO	80908
R7360	ALCO	80903
R7362	ALCO	80954
R7362A	ALCO	80907
R7363	ALCO	80957

Part No.	Brand	IDLI Group
R7365	ALCO	80951
R7365A	ALCO	80905
R7366	ALCO	80952
R7367	ALCO	80953
R7367	ALCO	80957
R7367A	ALCO	80906
R7X7352	CATERPILLR	80112
R7X7355	CATERPILLR	80104
R7X7356	CATERPILLR	80112
R96	TEDDY TORQ	80122
R96-1-792	CLEVE.STL.	88005.86
R96-1-793	CLEVE.STL.	88005.90
R96-1-809	CLEVE.STL.	88005.94
R96-1-811	CLEVE.STL.	88005.88
R96-1-812	CLEVE.STL.	88005.87
R96-1-814	CLEVE.STL.	88005.85
R96-55	ALCO	80122
R96-55	BORG-WARNR	80122
R96-55	CLEVE.STL.	80122
R96-55	REPCO	80122
R96-55	SPICER	80122
R96-55-1	AEC	80122
RASB46248(7)	GALION	80193
RB51477	GALION	80101
RBR28-1	ROCKWELL	88014.79
RCB100L	CLEVE.STL.	87285
RCB100LS	CLEVE.STL.	87286
RCB1-10	CLEVE.STL.	88007.53
RCB1-2	CLEVE.STL.	88007.52
RCB2-1	CLEVE.STL.	88013.71
RD51477(7)	GALION	80101
RD96-55	PERFECTION	80160
REBC141	ROCKWELL	88014.80
REBC148	ROCKWELL	88014.81
RECB131	ROCKWELL	88014.82
RECP148	ROCKWELL	88014.83
RECP155	ROCKWELL	88014.84
REDI-MIX TRUCK	RITEWAY	80102
REDI-MIX TRUCK	RITEWAY	80210
REP26-2	ROCKWELL	88014.85
REP29-1	ROCKWELL	88014.86
REP31-1	ROCKWELL	88014.87
REP31-2	ROCKWELL	88014.88
REP35-4	ROCKWELL	88014.89
RG096-55	PERFECTION	80138
RG154	PERFECTION	80056
RG154	PERFECTION	80057
RG162	PERFECTION	80042
RG2000	PERFECTION	80049
RG2002	PERFECTION	80048
RG2002	PERFECTION	88002.10
RG2006	PERFECTION	80021
RG2006	PERFECTION	88001.97
RG2012	PERFECTION	80030
RG2014	PERFECTION	80054
RG2016	PERFECTION	80044
RG2017	PERFECTION	80092
RG2017	PERFECTION	88001.76
RG2018	PERFECTION	80053
RG2023	PERFECTION	80092
RG2023	PERFECTION	88001.76
RG2026	PERFECTION	80030
RG2026	PERFECTION	88001.99
RG2031	PERFECTION	80049
RG2032	PERFECTION	80092
RG2035	PERFECTION	80030
RG2036	PERFECTION	80049
RG2039	PERFECTION	80044
RG2040	PERFECTION	80049
RG2041	PERFECTION	80054

Part No.	Brand	IDLI Group
RG2045	PERFECTION	80049
RG2045	PERFECTION	88001.96
RG2045	PERFECTION	88002.06
RG2046	PERFECTION	80044
RG2046	PERFECTION	88001.98
RG2049	PERFECTION	80054
RG2049	PERFECTION	88002.03
RG2053	PERFECTION	80030
RG2053	PERFECTION	80031
RG2053	PERFECTION	88001.95
RG2053G	PERFECTION	80031
RG2062	PERFECTION	80031
RG2062	PERFECTION	80049
RG2062	PERFECTION	80050
RG2063	PERFECTION	80043
RG2063	PERFECTION	80044
RG2063	PERFECTION	80055
RG2064	PERFECTION	80044
RG2066	PERFECTION	80045
RG2066	PERFECTION	80054
RG2140	PERFECTION	80055
RG2153	PERFECTION	80055
RG2162	PERFECTION	80049
RG2163	PERFECTION	80044
RG2198	PERFECTION	80051
RG225	PERFECTION	80064
RG225	PERFECTION	80065
RG225	PERFECTION	88001.63
RG235	PERFECTION	80022
RG2500	PERFECTION	80059
RG252	PERFECTION	80034
RG254	PERFECTION	80056
RG254	PERFECTION	80057
RG254X	PERFECTION	80225
RG255	PERFECTION	80060
RG255	PERFECTION	88001.59
RG256	PERFECTION	80056
RG256	PERFECTION	80057
RG257	PERFECTION	80095
RG259	PERFECTION	80096
RG261	PERFECTION	80064
RG261	PERFECTION	88001.63
RG262	PERFECTION	88001.64
RG263	PERFECTION	80094
RG265	PERFECTION	80064
RG265	PERFECTION	80065
RG265	PERFECTION	88001.63
RG266	PERFECTION	80060
RG266	PERFECTION	80061
RG266	PERFECTION	88001.59
RG3000	PERFECTION	80099
RG3000	PERFECTION	80101
RG3000G	PERFECTION	80099
RG3000G	PERFECTION	80101
RG3008	PERFECTION	80104
RG3008G	PERFECTION	80104
RG3011	PERFECTION	80099
RG3011	PERFECTION	80101
RG3012	PERFECTION	80110
RG3012	PERFECTION	80112
RG3012G	PERFECTION	80110
RG3012G	PERFECTION	80112
RG3013	PERFECTION	80106
RG3014	PERFECTION	80114
RG3019	PERFECTION	80110
RG3019	PERFECTION	80112
RG3021	PERFECTION	80111
RG324	PERFECTION	80056
RG324	PERFECTION	80095
RG325	PERFECTION	80061
RG325	PERFECTION	88001.64
RG350R	PERFECTION	80117
RG362	PERFECTION	80135
RG362	PERFECTION	88001.65
RG372	PERFECTION	80135
RG373	PERFECTION	80135
RG373	PERFECTION	88001.65
RG4002	PERFECTION	80109
RG4002	PERFECTION	80149
RG4007	PERFECTION	80147
RG4008	PERFECTION	80146
RG4014	PERFECTION	80158
RG4015	PERFECTION	80148
RG4016	PERFECTION	80159
RG4N	PERFECTION	80134
RG5000	PERFECTION	80168
RG5007	PERFECTION	80177
RG5008	PERFECTION	80174
RG5015	PERFECTION	80173
RG511X	PERFECTION	80075
RG514X	PERFECTION	80023
RG514X	PERFECTION	80024
RG515X	PERFECTION	80178
RG516X	PERFECTION	80189
RG517X	PERFECTION	80221
RG518X	PERFECTION	80124
RG5-1X	PERFECTION	80023
RG5-1X	PERFECTION	80024
RG521B	PERFECTION	80075
RG521B	PERFECTION	80089
RG521X	PERFECTION	80072
RG521X	PERFECTION	80077
RG524X	PERFECTION	80225
RG52A	PERFECTION	80909
RG52A	REPUBLIC	88001.66
RG52AR	PERFECTION	80909
RG5-2X	PERFECTION	80041
RG5-2X	PERFECTION	88001.85
RG530R	PERFECTION	80117
RG531R	PERFECTION	80121
RG533R	PERFECTION	80116
RG533R	PERFECTION	80117
RG548X	PERFECTION	80025
RG549X	PERFECTION	80075
RG549X	PERFECTION	80077
RG54A	PERFECTION	88001.66
RG5-4X	PERFECTION	80041
RG5-4X	PERFECTION	88001.85
RG550R	PERFECTION	80097
RG550R	PERFECTION	80119
RG552	PERFECTION	80118
RG552R	PERFECTION	80118
RG555	PERFECTION	80182
RG55-55	PERFECTION	80123
RG55-55-2	PERFECTION	80081
RG55-55-2	PERFECTION	80123
RG555R	PERFECTION	80182
RG560X	PERFECTION	80142
RG560X	PERFECTION	80143
RG570K	PERFECTION	80007
RG570K	PERFECTION	88000.15
RG570X	PERFECTION	80007
RG574	PERFECTION	80078
RG574	PERFECTION	80090
RG574R	PERFECTION	80090
RG579	PERFECTION	80073
RG579	PERFECTION	80091
RG579R	PERFECTION	80073
RG579R	PERFECTION	80091
RG57AR	PERFECTION	80908
RG581X	ALCO	88008.68
RG581X	PERFECTION	80023
RG581X	PERFECTION	80024
RG581X	PERFECTION	80140
RG588R	PERFECTION	80140
RG588R	PERFECTION	80182
RG6000	PERFECTION	80193
RG6003	PERFECTION	80193
RG6005	PERFECTION	80193
RG6015	PERFECTION	80193
RG6016	PERFECTION	80193
RG6102	PERFECTION	80197
RG6103	PERFECTION	80194
RG6108	PERFECTION	80197
RG6108	PERFECTION	80199
RG6109	PERFECTION	80201
RG62AR	PERFECTION	80907
RG65AR	PERFECTION	80905
RG65AR	PERFECTION	80951
RG67AR	PERFECTION	80906
RG6N	PERFECTION	80188
RG7000	PERFECTION	80210
RG7004	PERFECTION	80211
RG7102	PERFECTION	80212
RG7N	PERFECTION	80220
RG8002	PERFECTION	80231
RG8002	PERFECTION	80243
RG8003	PERFECTION	80232
RG8102	PERFECTION	80244
RG8202	PERFECTION	80231
RG9000	PERFECTION	80235
RG96-55	PERFECTION	80122
RG096-55	PERFECTION	80138
RO55-55-2	PERFECTION	80141
RP55-55-165	PERFECTION	80131
RP55-55-2	PERFECTION	80129
RP96-55-1	PERFECTION	80068
RR1R19-6	ROCKWELL	88014.90
RR1R22-6	ROCKWELL	88014.91
RR1R24-2	ROCKWELL	88014.92
RRER16-5	ROCKWELL	88014.93
RRER17-3	ROCKWELL	88014.94
RRWA8-1E	WHITING	80104
RS2	GREEN BALL	88008.35
RS53-55	PERFECTION	80038
RS53-55	PERFECTION	80039
RS55-165	PERFECTION	80091
RS55-55-1	PERFECTION	80081
RS55-55-1	PERFECTION	88001.87
RS55-55-2	PERFECTION	80081
RS55-55-2	PERFECTION	80123
RS96-55	ALCO	80038
RS96-55	PERFECTION	80038
RSB10	GREEN BALL	87481
RSB10	GREEN BALL	88007.77
RSB11	GREEN BALL	88007.79
RSB12	GREEN BALL	87337
RSB12	GREEN BALL	88008.48
RSB13	GREEN BALL	87285
RSB13	GREEN BALL	88008.51
RSB13	GREEN BALL	88008.52
RSB14	GREEN BALL	87399
RSB14	GREEN BALL	88008.59
RSB2	GREEN BALL	88007.58
RSB3	GREEN BALL	87326
RSB4	GREEN BALL	87331
RSB4	GREEN BALL	88008.39
RSB5	GREEN BALL	87331
RSB6	GREEN BALL	87383
RSB6	GREEN BALL	88008.42
RSB7	GREEN BALL	87377
RSB7	GREEN BALL	88008.41
RSB8	GREEN BALL	87459
RSB9	GREEN BALL	87457
RSB9	GREEN BALL	88008.31
RSB9	GREEN BALL	88008.44
RT125-11	ROCKWELL	88014.95
RT125-11-108	ROCKWELL	87214
RT125-11-72	ROCKWELL	87213
RT15-11-108	ROCKWELL	87273
RT15-11-72	ROCKWELL	87272
RT15-13-108	ROCKWELL	87275
RT15-13-72	ROCKWELL	87274
RT1625-14-108	ROCKWELL	87336
RT1625-14-72	ROCKWELL	87335
RT175-11-108	ROCKWELL	87367
RT175-11-72	ROCKWELL	87366
RT175-13-108	ROCKWELL	87369
RT175-13-72	ROCKWELL	87368
RT175-14	ROCKWELL	88014.96
RT175-14-108	ROCKWELL	87386
RT175-14-72	ROCKWELL	87385
RT175-16	ROCKWELL	88014.97
RT175-16-108	ROCKWELL	87388
RT175-16-72	ROCKWELL	87387
RT1937-16-108	ROCKWELL	87463
RT1937-16-72	ROCKWELL	87462
RT2125-11-108	ROCKWELL	87469
RT2125-11-72	ROCKWELL	87468
RT2125-14	ROCKWELL	88014.98
RT2125-14-108	ROCKWELL	87480
RT2125-14-72	ROCKWELL	87479
RT2125-16-108	ROCKWELL	87485
RT2125-16-72	ROCKWELL	87484
RT2-14	ROCKWELL	88014.99
RT2-14-108	ROCKWELL	87465
RT2-14-72	ROCKWELL	87464
RT2-16	ROCKWELL	88015
RT2-16-108	ROCKWELL	87467
RT2-16-72	ROCKWELL	87466
RT225-11	ROCKWELL	88015.01
RT225-11-108	ROCKWELL	87545
RT225-11-72	ROCKWELL	87544
RT225-14	ROCKWELL	88015.02
RT225-14-108	ROCKWELL	87547
RT225-14-72	ROCKWELL	87546
RT225-18-108	ROCKWELL	87550
RT225-18-72	ROCKWELL	87549
RT2375-14-108	ROCKWELL	87552
RT2375-14-72	ROCKWELL	87551
RT25-10-108	ROCKWELL	87554
RT25-10-72	ROCKWELL	87553
RT25-12	ROCKWELL	88015.03
RT25-12-108	ROCKWELL	87561
RT25-12-72	ROCKWELL	87560
RT25-13	ROCKWELL	88015.04
RT25-13-108	ROCKWELL	87563
RT25-13-72	ROCKWELL	87562
RT25-14	ROCKWELL	88015.05
RT25-16	ROCKWELL	88015.06
RT25-18-108	ROCKWELL	87574
RT25-18-72	ROCKWELL	87573
RT2625-11	ROCKWELL	88015.07
RT2625-11-108	ROCKWELL	87572
RT2625-11-72	ROCKWELL	87571
RT275-12-108	ROCKWELL	87605
RT275-12-72	ROCKWELL	87604
RT275-13-108	ROCKWELL	87607
RT275-13-72	ROCKWELL	87606
RT275-14	ROCKWELL	88015.08

Part No.	Brand	IDLI Group	Part No.	Brand	IDLI Group	Part No.	Brand	IDLI Group	Part No.	Brand	IDLI Group
RT2875-16-108	ROCKWELL	87613	RT45-3-72	ROCKWELL	87677	RUJ2039SL	REPCO	80148	S53-55	ALCO	80039
RT2875-16-72	ROCKWELL	87612	RT45-375-108	ROCKWELL	87672	RUJ2041	REPCO	80115	S53-55	BORG-WARNR	80039
RT3-11	ROCKWELL	88015.09	RT45-375-72	ROCKWELL	87671	RUJ2045	REPCO	80015	S53-55	CLEVELAND	80038
RT3-11-108	ROCKWELL	87615	RT55-55-2	PERFECTION	80071	RUJ2100	REPCO	80016	S53-55	CLEVELAND	80039
RT3-11-72	ROCKWELL	87614	RT6-14-108	ROCKWELL	87686	RUJ3000	REPCO	80131	S53-55	CLEVE.STL.	80039
RT3125-13-108	ROCKWELL	87626	RT6-14-72	ROCKWELL	87685	RUJ3001	REPCO	80123	S53-55	REPCO	80039
RT3125-13-72	ROCKWELL	87625	RTS55-55-2	PERFECTION	80074	RUJ3002	REPCO	80129	S53-55	SPICER	80039
RT3-13	ROCKWELL	88015.10	RU96-55	PERFECTION	80181	RUJ3142	REPCO	80100	S55	TEDDY TORQ	80081
RT3-13-108	ROCKWELL	87617	RUJ1750	REPCO	80032	RUJ4000	REPCO	80146	S55-10	CLEVE.STL.	88015.48
RT3-13-72	ROCKWELL	87616	RUJ1755	REPCO	80031	RUJ4001	REPCO	80158	S55-10-1	CLEVE.STL.	88015.46
RT3-14	ROCKWELL	88015.11	RUJ1756	REPCO	80032	RUJ4002	REPCO	80159	S55-11	CLEVE.STL.	88015.55
RT3-14-108	ROCKWELL	87619	RUJ1757	REPCO	80032	RUJ4003	REPCO	80153	S55-1-1170	CLEVE.STL.	88004.60
RT3-14-72	ROCKWELL	87618	RUJ1758	REPCO	88002.81	RUJ4004	REPCO	80151	S55-1-129	CLEVE.STL.	88004.51
RT3-156	ROCKWELL	88015.12	RUJ1759	REPCO	80032	RUJ4005	REPCO	80158	S55-1-130	CLEVE.STL.	88004.56
RT3-156-108	ROCKWELL	87611	RUJ1760	REPCO	88002.77	RUJ4006	REPCO	80158	S55-1-131	CLEVE.STL.	88004.57
RT3-156-72	ROCKWELL	87610	RUJ1761	REPCO	88002.77	RUJ4007	REPCO	80135	S55-1-132	CLEVE.STL.	88004.52
RT3-16	ROCKWELL	88015.13	RUJ1762	REPCO	88002.77	RUJ4008	REPCO	80141	S55-1-133	CLEVE.STL.	88004.53
RT3-16-108	ROCKWELL	87622	RUJ1763	REPCO	80032	RUJ6000	REPCO	80198	S55-1-1431	CLEVE.STL.	88004.50
RT3-16-72	ROCKWELL	87621	RUJ1764	REPCO	80032	RUJ6001	REPCO	80196	S55-1-153	CLEVE.STL.	88004.54
RT3-187	ROCKWELL	88015.14	RUJ1765	REPCO	88002.77	RUJ6002	REPCO	80198	S55-1-1551	CLEVE.STL.	88004.49
RT3-187-108	ROCKWELL	87609	RUJ1768	REPCO	80032	RUJ6003	REPCO	80198	S55-1168	CLEVE.STL.	88008.10
RT3-187-72	ROCKWELL	87608	RUJ1769	REPCO	80032	S10004	GEHL BROS.	80003	S55-1169	CLEVE.STL.	88008.11
RT325-11-108	ROCKWELL	87651	RUJ1773	REPCO	88002.77	S10092	BELLE CITY	80037	S55-1-218	CLEVE.STL.	88004.55
RT325-11-72	ROCKWELL	87650	RUJ1776	REPCO	80009	S10093	BELLE CITY	80008	S55-1-225	CLEVE.STL.	88004.59
RT325-13-108	ROCKWELL	87653	RUJ1776SL	REPCO	80009	S1018	WESCO	88002.60	S55-1-229	CLEVE.STL.	88004.58
RT325-13-72	ROCKWELL	87652	RUJ1777	REPCO	80031	S1164	WESCO	88002.64	S55-12S	CLEVE.STL.	88014.90
RT325-187-108	ROCKWELL	87624	RUJ1779	REPCO	80031	S119	LUNDELL	80133	S55-17	CLEVE.STL.	88013.38
RT325-187-72	ROCKWELL	87623	RUJ1780	REPCO	80004	S1210AR	COLES	80178	S55-17-1	CLEVE.STL.	88013.39
RT35-10	ROCKWELL	88015.15	RUJ1781	REPCO	80010	S1210AR	COLES	80189	S55-2	CLEVE.STL.	88004.38
RT35-10-108	ROCKWELL	87662	RUJ1781SL	REPCO	80010	S1210AR	COLES	80221	S55-2-1	CLEVE.STL.	88004.39
RT35-10-72	ROCKWELL	87661	RUJ1781SL	REPCO	80012	S1385	WESCO	88002.66	S55-3-2-16	CLEVE.STL.	88004.44
RT35-11	ROCKWELL	88015.16	RUJ1782	REPCO	80010	S1385	WESCO	88002.67	S55-3-2-16D	CLEVE.STL.	88004.45
RT35-11-108	ROCKWELL	87658	RUJ1783	REPCO	80012	S144Y123	CHELSEA	88000.39	S55-3-3-14	CLEVE.STL.	88004.46
RT35-11-72	ROCKWELL	87657	RUJ1783SL	REPCO	80012	S144Y123	CHELSEA	88000.81	S55-3-3-16	CLEVE.STL.	88004.47
RT35-13	ROCKWELL	88015.17	RUJ1784	REPCO	80011	S15355	PILOT	80039	S55-3-3-16D	CLEVE.STL.	88004.48
RT35-13-108	ROCKWELL	87660	RUJ1784SL	REPCO	80011	S15513	PILOT	80118	S55-43-201	CLEVE.STL.	88007.38
RT35-13-72	ROCKWELL	87659	RUJ1786	REPCO	88002.77	S15513	PILOT	80130	S55-43-202	CLEVE.STL.	88007.39
RT35-14	ROCKWELL	88015.18	RUJ1788	REPCO	80035	S15555	PILOT	80081	S55-43-203	CLEVE.STL.	88007.40
RT35-14-108	ROCKWELL	87664	RUJ2000	REPCO	80055	S1642	WESCO	88002.42	S55-43-204	CLEVE.STL.	88007.41
RT35-14-72	ROCKWELL	87663	RUJ2001	REPCO	80049	S1642	WESCO	88002.59	S55-43-209	CLEVE.STL.	88007.43
RT35-156	ROCKWELL	88015.19	RUJ2002	REPCO	80055	S1642	WESCO	88002.62	S55-43-210	CLEVE.STL.	88007.44
RT35-156-108	ROCKWELL	87655	RUJ2003	REPCO	80048	S1646	WESCO	88002.51	S55-43-469	CLEVE.STL.	88007.42
RT35-156-72	ROCKWELL	87654	RUJ2004	REPCO	80046	S166	LUNDELL	80037	S55-54-1	CLEVE.STL.	88004.42
RT35-16	ROCKWELL	88015.20	RUJ2005	REPCO	80044	S188	LUNDELL	80069	S55-54-1167	CLEVE.STL.	88004.43
RT4095-7	ROCKWELL	88015.21	RUJ2006	REPCO	80049	S1889	RUPP	80069	S55-54-1L	CLEVE.STL.	88004.40
RT4095-7-108	ROCKWELL	87670	RUJ2008	REPCO	80063	S1889	WESCO	88002.57	S55-54-960	CLEVE.STL.	88004.41
RT4095-7-72	ROCKWELL	87669	RUJ2009	REPCO	80058	S1890	GORMAN RUP	80037	S55-55	ALCO	88001.87
RT4-10	ROCKWELL	88015.22	RUJ2010	REPCO	80043	S1890	RUPP	80037	S55-5503	CLEVE.STL.	80081
RT4-10-108	ROCKWELL	87668	RUJ2011	REPCO	80055	S1891	GORMAN RUP	80161	S55-55-1	BORG-WARNR	80081
RT4-10-72	ROCKWELL	87667	RUJ2012	REPCO	80063	S1891	RUPP	80219	S55-55-1	BORG-WARNR	88001.87
RT4-12	ROCKWELL	88015.23	RUJ2016	REPCO	80018	S19655	PILOT	80038	S55-55-1	CLEVE.STL.	80081
RT4-12-108	ROCKWELL	87674	RUJ2016SL	REPCO	80018	S19675	PILOT	80091	S55-55-1330	BORG-WARNR	80130
RT4-12-72	ROCKWELL	87673	RUJ2017	REPCO	80092	S201007	CASE,J.I.	80077	S55-55-1330	CLEVELAND	80118
RT4-14	ROCKWELL	88015.24	RUJ2022	REPCO	80060	S2017	WESCO	80092	S55-55-1330	CLEVELAND	80130
RT4-14-108	ROCKWELL	87676	RUJ2023	REPCO	80020	S2017	WESCO	88001.76	S55-55-1330	CLEVE.STL.	80130
RT4-14-72	ROCKWELL	87675	RUJ2026	REPCO	80049	S2017	WESCO	88002.07	S55-55-2	BORG-WARNR	80081
RT4-187	ROCKWELL	88015.25	RUJ2027	REPCO	80081	S2462	WESCO	88002.58	S55-55-2	CLEVELAND	80081
RT4-187-108	ROCKWELL	87666	RUJ2027SL	REPCO	80080	S2765	KOYO	80012	S55-55-2	CLEVELAND	80123
RT4-187-72	ROCKWELL	87665	RUJ2027SL	REPCO	80081	S2782	KOYO	80077	S55-55-2	CLEVE.STL.	80081
RT45-10	ROCKWELL	88015.26	RUJ2028	REPCO	80019	S31	TEDDY TORQ	80130	S55-55-2	REPCO	80081
RT45-10-108	ROCKWELL	87682	RUJ2028SL	REPCO	80019	S3692	WESCO	88002.44	S55-55-675	BORG-WARNR	80091
RT45-10-72	ROCKWELL	87681	RUJ2029	REPCO	88001.71	S3717	WESCO	88002.63	S55-55-675	CLEVELAND	80091
RT45-13	ROCKWELL	88015.27	RUJ2030	REPCO	80044	S40	TEDDY TORQ	80091	S55-55-675	CLEVE.STL.	80091
RT45-13-108	ROCKWELL	87684	RUJ2032	REPCO	80055	S4232	WESCO	88002.44	S55-55-675	REPCO	80091
RT45-13-72	ROCKWELL	87683	RUJ2033	REPCO	80014	S4232	WESCO	88002.49	S55-55-735	CLEVE.STL.	80081
RT45-187	ROCKWELL	88015.28	RUJ2034	REPCO	80058	S437	HABAN	80031	S55-55-740	CLEVE.STL.	80091
RT45-187-108	ROCKWELL	87680	RUJ2038	REPCO	80060	S46959-3	WHITING	80001	S55-55-870	CLEVE.STL.	80081
RT45-187-72	ROCKWELL	87679	RUJ2039	REPCO	80091	S53	TEDDY TORQ	80039	S55-55-875	CLEVE.STL.	80091
RT45-3	ROCKWELL	88015.29	RUJ2039SL	REPCO	80073	S53-55	AEC	80038	S55-6	CLEVE.STL.	88014.82
RT45-3-108	ROCKWELL	87678	RUJ2039SL	REPCO	80091	S53-55	AEC	80039	S55-7-1166	CLEVE.STL.	88014.74

Part No.	Brand	IDLI Group	Part No.	Brand	IDLI Group	Part No.	Brand	IDLI Group	Part No.	Brand	IDLI Group
S55-7-2-16	CLEVE.STL.	88014.64	SH-P12-11-108	ROCKWELL	87823	SQS100-108	ROCKWELL	87824	SY550	MUNCIE	88000.86
S55-7-2-16L	CLEVE.STL.	88014.66	SH-P12-11-72	ROCKWELL	87805	SQS100-72	ROCKWELL	87806	SY550	MUNCIE	88000.87
S55-7-3-16	CLEVE.STL.	88014.67	SH-P16-7	ROCKWELL	88015.35	SQS1187	ROCKWELL	88015.41	SY560	MUNCIE	88000.89
S55-7-3-16L	CLEVE.STL.	88014.68	SH-P16-7-108	ROCKWELL	87825	SQS1187-108	ROCKWELL	87826	SY570	MUNCIE	88000.90
S55-7-3-16LN	CLEVE.STL.	88014.65	SH-P16-7-72	ROCKWELL	87807	SQS1187-72	ROCKWELL	87808	SY630	MUNCIE	88000.95
S55-9-1	CLEVE.STL.	88013.56	SH-S20-26	ROCKWELL	87250	SQS13125	ROCKWELL	88015.42	SY635	MUNCIE	88001.01
S5605	WESCO	88002.70	SH-S20-27	ROCKWELL	87251	SQS13125-108	ROCKWELL	87827	SY670	MUNCIE	88000.91
S5R55	UJS	80123	SH-S20-88-12	ROCKWELL	87247	SQS13125-72	ROCKWELL	87809	SY675	MUNCIE	88000.93
S7090	WESCO	88002.55	SH-S20-90-12	ROCKWELL	87248	SQS75	ROCKWELL	88015.43	SY676	MUNCIE	88000.93
S7409	WESCO	88002.54	SH-S20-90-24	ROCKWELL	87249	SQS75-108	ROCKWELL	87822	SY677	MUNCIE	88001
S8571	MIXERMOBL	80149	SH-S22-88-12	ROCKWELL	87287	SQS75-72	ROCKWELL	87804	SY680	MUNCIE	88000.94
S8571	SCOOPMOBLE	80109	SH-S24-53-12	ROCKWELL	87339	SQT13125-108	ROCKWELL	87216	T10403	LEADER	80221
S9266	WESCO	88002.50	SH-S24-53-24	ROCKWELL	87340	SQT13125-72	ROCKWELL	87215	T10404	LEADER	80189
S9288	WESCO	88002.52	SH-S28-48-12	ROCKWELL	87401	SQT1625-108	ROCKWELL	87289	T106-01-39	AUST.MOT	88002.77
S9531	WESCO	88002.47	SH-S28-48-24	ROCKWELL	87402	SQT1625-72	ROCKWELL	87288	T11774	CASE,J.I.	80021
S9531	WESCO	88002.48	SHS28-91	ROCKWELL	88015.36	SR6109	ADAMS-LET.	80001	T15555	PILOT	80071
S96	TEDDY TORQ	80038	SHS28-92	ROCKWELL	88015.37	SR6109	ADAMS-LET.	80003	T2010	AEC	80009
S96-55	ALCO	80038	SH-S48-21-12	ROCKWELL	87627	SR6109	WABCO	80001	T2057	KOYO	80006
S96-55	BORG-WARNR	80038	SH-S48-21-24	ROCKWELL	87628	SR6779	WESCO	88002.43	T23616	CASE,J.I.	80008
S96-55	CLEVE.STL.	80038	SK3281E	HARDY SPCR	88004.24	SR8778-3	ADAMS-LET.	80109	T23626	CASE,J.I.	80008
S96-55	REPCO	80038	SK413(4)	AMER.SNOW	80215	SR8778-3	WABCO	80156	T2667	KOYO	80009
S96-55	SPICER	80038	SK414(4)	AMER.SNOW	80215	SRR8778	ADAMS-LET.	80109	T2680C	KOYO	80018
SA35043	CLARK EQU.	80168	SK518(4)	AMER.SNOW	80215	SR-R8778	WABCO	80156	T2731	SAFE ELEC.	80193
SA35043	CLARK EQU.	80216	SK523(4)	AMER.SNOW	80215	SS12Y164	CHELSEA	88000.31	T27790	CASE,J.I.	80069
SA35043	MICHIGAN	80168	SKA11653	REPCO	88002.80	SS12Y164	CHELSEA	88000.88	T2978	KOYO	80017
SA5043	CLARK EQU.	80168	SKA11654	REPCO	88002.78	SS55-12	CLEVE.STL.	88014.93	T30746	CASE,J.I.	80136
SAB100-18-1	PETTI-MULL	80150	SKA11655	REPCO	88002.79	SSFL5-18-3	ROCKWELL	88015.44	T32970	HARNISCH.	80104
SAP200-24	PETTI-MULL	80155	SL8673-1	ADAMS-LET.	80099	ST0024	MITSUBISHI	80012	T34905	HARNISCH.	80237
SB145	CHAMPION	80104	SL8673-1	ADAMS-LET.	80101	ST12-002700	NEAPCO	87919	T36575	CASE,J.I.	80001
SB145	DOMINION	80104	SL8673-1	WABCO	80101	ST12-003900	NEAPCO	87945	T37173	CASE,J.I.	80001
SB171	CHAMPION	80037	SL8673-3	ADAMS-LET.	80099	ST20-002800	NEAPCO	87916	T37895	CASE,J.I.	80069
SB539-8695	RENAULT	88002.77	SL8673-3	ADAMS-LET.	80101	ST20-002800P	NEAPCO	87915	T40106	CASE,J.I.	80069
SCB101	CLEVE.STL.	87331	SL8673-3	WABCO	80101	ST20-004000	NEAPCO	87939	T41672	CASE,J.I.	80069
SCB102	CLEVE.STL.	87327	SLA-P16-14	ROCKWELL	87227	ST20-004000P	NEAPCO	87938	T41672	CASE,J.I.	80070
SCB103	CLEVE.STL.	87326	SLA-P16-14	ROCKWELL	88015.38	ST22-002900	NEAPCO	87913	T462-255	HYSTER	80216
SCB104	CLEVE.STL.	87329	SLA-Q21-7	ROCKWELL	87277	ST22-002900P	NEAPCO	87912	T462-295	HYSTER	80168
SCB106	CLEVE.STL.	87330	SL-P12-1	ROCKWELL	87222	ST22-003900	NEAPCO	87934	T468	HYSTER	80193
SCB107	CLEVE.STL.	87328	SL-P12-31	ROCKWELL	87220	ST22-003900P	NEAPCO	87933	T52	TEDDY TORQ	80074
SCB108	CLEVE.STL.	88008.50	SL-P12-32	ROCKWELL	87205	STR06-002700	NEAPCO	87920	T55	TEDDY TORQ	80071
SCB109	CLEVE.STL.	88008.47	SL-P12-38	ROCKWELL	87221	STR06-003900	NEAPCO	87946	T55-10	CLEVE.STL.	88015.47
SCB110	CLEVE.STL.	87332	SL-P16-40	ROCKWELL	87225	STR12-002700	NEAPCO	87919	T55-17	CLEVE.STL.	88013.37
SCB1-4	CLEVE.STL.	88007.55	SL-P16-47	ROCKWELL	87226	STR12-002700P	NEAPCO	87918	T55-54-170	CLEVE.STL.	88004.37
SCB1-4A	CLEVE.STL.	88014.85	SL-P16-47	ROCKWELL	88015.39	STR12-003900	NEAPCO	87945	T55-55-2	AEC	80071
SCB2-2	CLEVE.STL.	88013.73	SL-Q16-64	ROCKWELL	87209	STR12-003900P	NEAPCO	87944	T55-55-2	ALCO	80071
SCB2-3	CLEVE.STL.	88013.70	SL-Q19-102	ROCKWELL	87241	SU8485	WABCO	80235	T55-55-2	BORG-WARNR	80071
SCB2-4	CLEVE.STL.	88014.79	SL-Q19-53	ROCKWELL	87240	SV8485	ADAMS-LET.	80235	T55-55-2	CLEVELAND	80071
SCB2-5	CLEVE.STL.	88013.64	SL-Q19-82	ROCKWELL	87242	SV8485	WABCO	80235	T55-55-2	CLEVE.STL.	80071
SCB2-6	CLEVE.STL.	88013.65	SL-Q19-83	ROCKWELL	87239	SV8490	ADAMS-LET.	80234	T55-55-2	REPCO	80071
SCB50-1	CLEVE.STL.	88008.63	SL-Q21-8	ROCKWELL	87276	SV8490	WABCO	80234	T55-7-2-16	CLEVE.STL.	88014.63
SCB61	CLEVE.STL.	88007.81	SL-S20-15	ROCKWELL	87252	SY360-1	REPUBLIC	88000.79	T61-01-85	AUST.MOT	88002.77
SCB61-1	CLEVE.STL.	88007.80	SL-S24-3	ROCKWELL	87342	SY360-11	REPUBLIC	88000.86	T698-1	EATON	80210
SCB75	CLEVE.STL.	88008.35	SL-S24-30	ROCKWELL	87347	SY360-11	REPUBLIC	88000.88	T698-1	EATON	80212
SCB76	CLEVE.STL.	87238	SL-S24-32	ROCKWELL	87348	SY360-13	REPUBLIC	88000.89	T698-1	EATON	88003.01
SCB85	CLEVE.STL.	87325	SN1157	CHAMPION	80024	SY360-15	REPUBLIC	88000.90	T698-1	EATON	88003.03
SCP1FR	REPCO	80008	SN6779	WESCO	88002.43	SY360-2	REPUBLIC	88000.80	T698-7	EATON	80210
SCP1FR	REPCO	88002.88	SO174	WESCO	88002.73	SY360-21	REPUBLIC	88001	T698-7	EATON	80212
SCV082	AEC	88002.37	SO292	WESCO	88002.73	SY360-22	REPUBLIC	88001.02	T698-7(5)	EATON	88003.01
SCV79	MOTOR MAST	88002.38	SO293	WESCO	88002.74	SY360-23	REPUBLIC	88000.85	T698-7(5)	EATON	88003.03
SCV79	ZELLER	88002.38	SP112529(4)	SHOV.SUPP.	80149	SY360-23	REPUBLIC	88000.94	T699-1	EATON	80212
SCV82	AEC	88002.37	SP112540	MASS.FERG.	80109	SY360-3	REPUBLIC	88000.81	T699-1	EATON	88003.01
SCV82	MOTOR MAST	88002.37	SP112635	MASS.FERG.	80109	SY360-31	REPUBLIC	88000.95	T6AA4635A	FORD	80075
SCV82	ZELLER	88002.37	SP120804(4)	SHOV.SUPP.	80193	SY360-5	REPUBLIC	88000.83	T8767	GERLINGER	80178
SD2167	CATERPILLR	80054	SP65142(4)	SHOV.SUPP.	80216	SY370-22	REPUBLIC	88000.92	T9502	GERLINGER	80178
SECR20-7	ROCKWELL	88015.30	SP65143(4)	SHOV.SUPP.	80149	SY500	MUNCIE	88000.79	TA78-2	MUNCIE	88001.13
SECR22-8	ROCKWELL	88015.31	SP730B	BROS	80178	SY505	MUNCIE	88000.80	TA78-5	MUNCIE	81218
SECS24-7	ROCKWELL	88015.32	SPF4CX5	FOX RIVER	80197	SY510	MUNCIE	88000.81	TB94402	GMC	80031
SECS25-3	ROCKWELL	88015.33	SPF4CXS	FOX RIVER	80198	SY515	MUNCIE	88000.82	TB94402	OPEL	88002.77
SECS28-2	ROCKWELL	88015.34	SPF4X5	FOX RIVER	80198	SY520	MUNCIE	88000.83	TB94403	OPEL	80032
SG612	WHITE	80023	SPFCXS	FOX	80198	SY530	MUNCIE	88000.84	TC4412-1	ADAMS-LET.	80213
SH1481-129	ANTHONY	80047	SQS100	ROCKWELL	88015.40	SY540	MUNCIE	88000.85	TC4412-1	WABCO	80213

Part No.	Brand	IDLI Group
TD5261	ADAMS-LET.	80231
TD5261	WABCO	80240
TD9791	WABCO	80048
TF1592-56	AUST.WEST.	80193
TF1592-56	BAL-LM-HAM	80193
TF9192	WABCO	80047
TF9791	ADAMS-LET.	80047
TF9791	WABCO	80048
TF9792	ADAMS-LET.	80047
TF9792	WABCO	80047
TG4614	ADAMS-LET.	80001
TG4614	ADAMS-LET.	80003
TG4614	WABCO	80001
TG5575	WABCO	80077
TG6102	PERFECTION	80197
TG9695	ADAMS-LET.	80077
TG9695	WABCO	80077
TGAA4635A	FORD	80193
TGAA4841A	FORD	85087
TGAY4865A	FORD	81661
TGAY4865A	FORD	81744
TH7348	WABCO	80102
TH7436	ADAMS-LET.	80225
TH7436	WABCO	80225
TH7986	ADAMS-LET.	80213
TH7986	WABCO	80213
TK4635A	FORD	80081
TL4948	WABCO	80228
TL7604	ADAMS-LET.	80077
TL7604	WABCO	80077
TM2055	KOYO	80010
TM2564A	KOYO	80012
TMD1027-1	JOY MFG.	80243
TOYOTA	MCQUAY-NOR	80016
TOYOTA	MCQUAY-NOR	80018
TOYOTA	MCQUAY-NOR	80115
TP4635A	FORD	80044
TP4635B	FORD	80060
TRK. & IND.	MCQUAY-NOR	80157
TS15553	PILOT	80080
TS15553	PILOT	80081
TS15554	PILOT	80081
TS15555	PILOT	80074
TS4CS2	FOX	80001
TS4CX2	FOX RIVER	80001
TS55-43-181	CLEVE.SIL.	88007.36
TS55-43-182	CLEVE.STL.	88007.37
TS55-55-2	AEC	80074
TS55-55-2	ALCO	80074
TS55-55-2	BORG-WARNR	80074
TS55-55-2	CLEVELAND	80074
TS55-55-2	CLEVE.STL.	80074
TS55-55-2	REPCO	80074
TS55-55-3	AEC	80081
TS55-55-3	BORG-WARNR	80080
TS55-55-3	CLEVE.STL.	80080
TS55-55-4	AEC	80081
TS55-55-4	CLEVE.STL.	80080
TSB11	GREEN BALL	88007.79
TSB12	GREEN BALL	87285
TSB8	GREEN BALL	88008.45
TU-P16-5-108	ROCKWELL	87208
TU-P16-5-72	ROCKWELL	87207
TU-P20-11	ROCKWELL	88015.45
TU-P20-11-108	ROCKWELL	87246
TU-P20-11-72	ROCKWELL	87245
TU-Q17-1-108	ROCKWELL	87204
TU-Q17-1-72	ROCKWELL	87203
TV1499	ADAMS-LET.	80234
TV1499	WABCO	80234
TV2971	ADAMS-LET.	80234
TV2971	WABCO	80234
TV3259	ADAMS-LET.	80198
TV3259	WABCO	80198
TVAA4635	FORD	80201
TVAA4635A	FORD	80201
TVBV44627A	FORD	86713
TWAA4635A	FORD	80197
TWAA4635B	FORD	80212
TWAA4635C	FORD	80199
TWAA4635D	FORD	80210
TWAA4841A	FORD	85114
TWAA4841B	FORD	85093
TWAM4865A	FORD	81744
TX55-55-2	AUST.WEST.	80074
U01-1	MCQUAY-NOR	80024
U01-1	MCQUAY-NOR	80030
U01-1	MCQUAY-NOR	80031
U01-1	MCQUAY-NOR	88002.77
U01-10	MCQUAY-NOR	80023
U01-10	MCQUAY-NOR	80024
U01-11	MCQUAY-NOR	80025
U01-12	MCQUAY-NOR	80030
U01-12	MCQUAY-NOR	80031
U01-12	MCQUAY-NOR	88002.77
U01-13	MCQUAY-NOR	80041
U01-14	MCQUAY-NOR	80072
U01-14	MCQUAY-NOR	80075
U01-14	MCQUAY-NOR	80077
U01-15	MCQUAY-NOR	80072
U01-15	MCQUAY-NOR	80075
U01-15	MCQUAY-NOR	80076
U01-15	MCQUAY-NOR	80077
U01-15	MCQUAY-NOR	80089
U01-16	MCQUAY-NOR	80073
U01-16	MCQUAY-NOR	80091
U01-17	MCQUAY-NOR	80072
U01-17	MCQUAY-NOR	80075
U01-17	MCQUAY-NOR	80076
U01-17	MCQUAY-NOR	80077
U01-18	MCQUAY-NOR	80064
U01-18	MCQUAY-NOR	80065
U01-19	MCQUAY-NOR	80060
U01-19	MCQUAY-NOR	80061
U01-2	MCQUAY-NOR	80032
U01-20	MCQUAY NOR	80061
U01-20	MCQUAY-NOR	80063
U01-21	MCQUAY-NOR	80104
U01-21	MCQUAY-NOR	80107
U01-22	MCQUAY-NOR	80104
U01-22	MCQUAY-NOR	80107
U01-22	MCQUAY-NOR	80112
U01-23	MCQUAY-NOR	80109
U01-24	MCQUAY-NOR	80112
U01-25	MCQUAY-NOR	80101
U01-26	MCQUAY-NOR	80100
U01-26	MCQUAY-NOR	80101
U01-27	MCQUAY-NOR	80110
U01-27	MCQUAY-NOR	80112
U01-28	MCQUAY-NOR	80105
U01-29	MCQUAY-NOR	80026
U01-29	MCQUAY-NOR	80077
U01-29	MCQUAY-NOR	80085
U01-29	MCQUAY-NOR	80086
U01-3	MCQUAY-NOR	88003.26
U01-30	MCQUAY-NOR	80090
U01-31	MCQUAY-NOR	80117
U01-31	MCQUAY-NOR	80129
U01-31	MCQUAY-NOR	80130
U01-32	MCQUAY-NOR	80116
U01-33	MCQUAY-NOR	80118
U01-33	MCQUAY-NOR	80121
U01-33	MCQUAY-NOR	80131
U01-34	MCQUAY-NOR	80120
U01-35	MCQUAY-NOR	80118
U01-36	MCQUAY-NOR	80132
U01-37	MCQUAY-NOR	80091
U01-38	MCQUAY-NOR	80108
U01-38	MCQUAY-NOR	80131
U01-39	MCQUAY-NOR	80121
U01-39	MCQUAY-NOR	80129
U01-4	MCQUAY-NOR	80043
U01-40	MCQUAY-NOR	80004
U01-40	MCQUAY-NOR	80006
U01-41	MCQUAY-NOR	80010
U01-42	MCQUAY-NOR	80009
U01-43	MCQUAY-NOR	80012
U01-44	MCQUAY-NOR	80011
U01-45	MCQUAY-NOR	80019
U01-46	MCQUAY-NOR	88002.34
U01-46	MCQUAY-NOR	88002.35
U01-47	MCQUAY-NOR	88002.36
U01-48	MCQUAY-NOR	88002.33
U01-49	MCQUAY-NOR	88002.33
U01-5	MCQUAY-NOR	80043
U01-5	MCQUAY-NOR	80044
U01-50	MCQUAY-NOR	80072
U01-50	MCQUAY-NOR	80075
U01-50	MCQUAY-NOR	80076
U01-50	MCQUAY-NOR	80077
U01-51	MCQUAY-NOR	80142
U01-51	MCQUAY-NOR	80143
U01-52	MCQUAY-NOR	80092
U01-53	MCQUAY-NOR	88002.33
U01-54	MCQUAY-NOR	88002.33
U01-55	MCQUAY-NOR	80083
U01-6	MCQUAY-NOR	80067
U01-60	MCQUAY-NOR	80034
U01-61	MCQUAY-NOR	88002.42
U01-64	MCQUAY-NOR	80036
U01-65	MCQUAY-NOR	88002.34
U01-66	MCQUAY-NOR	80080
U01-66	MCQUAY-NOR	80081
U01-67	MCQUAY-NOR	80088
U01-68	MCQUAY-NOR	86070
U01-7	MCQUAY-NOR	80066
U01-8	MCQUAY-NOR	80058
U01-9	MCQUAY-NOR	80023
U01-9	MCQUAY-NOR	80024
U01-9	MCQUAY-NOR	80140
U02-1	MCQUAY-NOR	80055
U02-10	MCQUAY-NOR	80142
U02-10	MCQUAY-NOR	80143
U02-11	MCQUAY-NOR	80124
U02-11	MCQUAY-NOR	80125
U02-12	MCQUAY-NOR	80123
U02-13	MCQUAY-NOR	80141
U02-14	MCQUAY-NOR	80077
U02-14	MCQUAY-NOR	80145
U02-15	MCQUAY-NOR	80148
U02-16	MCQUAY-NOR	80149
U02-16	MCQUAY-NOR	80159
U02-16	MCQUAY-NOR	80192
U02-17	MCQUAY-NOR	80178
U02-17	MCQUAY-NOR	80179
U02-18	MCQUAY-NOR	80182
U02-19	MCQUAY-NOR	80135
U02-2	MCQUAY-NOR	80042
U02-20	MCQUAY-NOR	80183
U02-21	MCQUAY-NOR	80186
U02-22	MCQUAY-NOR	80184
U02-23	MCQUAY-NOR	80185
U02-24	MCQUAY-NOR	80193
U02-25	MCQUAY-NOR	80194
U02-26	MCQUAY-NOR	80201
U02-27	MCQUAY-NOR	80199
U02-28	MCQUAY-NOR	80105
U02-28	MCQUAY-NOR	80227
U02-29	MCQUAY-NOR	80190
U02-3	MCQUAY-NOR	80060
U02-3	MCQUAY-NOR	80061
U02-3	MCQUAY-NOR	80062
U02-30	MCQUAY-NOR	80221
U02-31	MCQUAY-NOR	80225
U02-32	MCQUAY-NOR	80048
U02-34	MCQUAY-NOR	80020
U02-35	MCQUAY-NOR	88002.37
U02-4	MCQUAY-NOR	80106
U02-5	MCQUAY-NOR	80078
U02-6	MCQUAY-NOR	80002
U02-6	MCQUAY-NOR	80079
U02-6	MCQUAY-NOR	80139
U02-7	MCQUAY-NOR	80124
U02-7	MCQUAY-NOR	80125
U02-8	MCQUAY-NOR	80124
U02-9	MCQUAY-NOR	80140
U02-9	MCQUAY-NOR	80182
U03-1	MCQUAY-NOR	80001
U03-10	MCQUAY-NOR	80033
U03-101	MCQUAY-NOR	83349
U03-103	MCQUAY-NOR	81025
U03-104	MCQUAY-NOR	81268
U03-105	MCQUAY-NOR	81268
U03-106	MCQUAY-NOR	83018
U03-107	MCQUAY-NOR	81237
U03-109	MCQUAY-NOR	88001.10
U03-11	MCQUAY-NOR	80035
U03-110	MCQUAY-NOR	88000.18
U03-111	MCQUAY-NOR	81218
U03-112	MCQUAY-NOR	81125
U03-12	MCQUAY-NOR	80069
U03-12	MCQUAY-NOR	80070
U03-13	MCQUAY-NOR	80100
U03-14	MCQUAY-NOR	80102
U03-15	MCQUAY-NOR	80113
U03-16	MCQUAY-NOR	80165
U03-17	MCQUAY-NOR	80153
U03-17	MCQUAY-NOR	88003.16
U03-18	MCQUAY-NOR	80156
U03-19	MCQUAY-NOR	80158
U03-2	MCQUAY-NOR	80021
U03-20	MCQUAY-NOR	80155
U03-21	MCQUAY-NOR	80151
U03-22	MCQUAY-NOR	80154
U03-22	MCQUAY-NOR	80158
U03-23	MCQUAY-NOR	80168
U03-24	MCQUAY-NOR	80172
U03-25	MCQUAY-NOR	80174
U03-26	MCQUAY-NOR	80174
U03-27	MCQUAY-NOR	80163
U03-27	MCQUAY-NOR	80164
U03-27	MCQUAY-NOR	80167
U03-28	MCQUAY-NOR	80175
U03-29	MCQUAY-NOR	80198
U03-3	MCQUAY-NOR	80047
U03-3	MCQUAY-NOR	80048
U03-30	MCQUAY-NOR	80197
U03-31	MCQUAY-NOR	80210
U03-31	MCQUAY-NOR	80215
U03-31	MCQUAY-NOR	80217

Part No.	Brand	IDLI Group	Part No.	Brand	IDLI Group	Part No.	Brand	IDLI Group	Part No.	Brand	IDLI Group
U03-32	MCQUAY-NOR	80212	U03-86	MCQUAY-NOR	88014.32	U112	LEMPCO	80225	U142	LEMPCO	80036
U03-32	MCQUAY-NOR	80214	U03-87	MCQUAY-NOR	83116	U112	LOBRO	80041	U142	LEMPCO	80044
U03-32	MCQUAY-NOR	80218	U03-87	MCQUAY-NOR	88014.32	U113	EXCEL	80089	U143	LEMPCO	80227
U03-33	MCQUAY-NOR	80213	U03-88	MCQUAY-NOR	83102	U113	LEMPCO	80089	U15	LEMPCO	88002.67
U03-34	MCQUAY-NOR	80228	U03-89	MCQUAY-NOR	81089	U113	LOBRO	80063	U15616	ERF	80180
U03-35	MCQUAY-NOR	80234	U03-9	MCQUAY-NOR	88000.14	U114	LEMPCO	80124	U15655	PILOT	80183
U03-36	MCQUAY-NOR	80234	U03-90	MCQUAY-NOR	81143	U114	LOBRO	80062	U17	LEMPCO	88002.55
U03-37	MCQUAY-NOR	80215	U03-92	MCQUAY-NOR	83281	U115	EXCEL	80024	U172	LOBRO	80023
U03-37	MCQUAY-NOR	80235	U03-94	MCQUAY-NOR	88006.99	U115	LEMPCO	80023	U18	LEMPCO	88002.67
U03-38	MCQUAY-NOR	80239	U03-95	MCQUAY-NOR	81333	U115	LEMPCO	80024	U18	LEMPCO	88002.72
U03-38	MCQUAY-NOR	80244	U03-98	MCQUAY-NOR	83082	U115	LOBRO	80041	U180	LOBRO	80006
U03-39	MCQUAY-NOR	80232	U04-1	MCQUAY-NOR	80905	U116	LEMPCO	80072	U181	LOBRO	80004
U03-39	MCQUAY-NOR	80242	U04-17	MCQUAY-NOR	80951	U117	EXCEL	80023	U182	LOBRO	80010
U03-4	MCQUAY-NOR	80049	U04-18	MCQUAY-NOR	80953	U117	EXCEL	80140	U183	LOBRO	80012
U03-40	MCQUAY-NOR	80238	U04-19	MCQUAY-NOR	80906	U117	LEMPCO	80023	U184	LOBRO	80011
U03-41	MCQUAY-NOR	80231	U04-19	MCQUAY-NOR	80909	U117	LEMPCO	80024	U185	LOBRO	80009
U03-41	MCQUAY-NOR	80243	U04-19	MCQUAY-NOR	88001.72	U117	LOBRO	80041	U186	LOBRO	80019
U03-42	MCQUAY-NOR	80189	U04-2	MCQUAY-NOR	80906	U118	EXCEL	80116	U19	LEMPCO	88002.71
U03-43	MCQUAY-NOR	80211	U04-2	MCQUAY-NOR	80909	U118	LEMPCO	80116	U190	LOBRO	80017
U03-44	MCQUAY-NOR	80037	U04-23	MCQUAY-NOR	80088	U11803	ERF	80189	U193	LOBRO	80018
U03-45	MCQUAY-NOR	80161	U04-24	MCQUAY-NOR	88013.87	U11805	ERF	80221	U19397	ERF	80225
U03-46	MCQUAY-NOR	88000.43	U04-26	MCQUAY-NOR	88002.13	U11806	ERF	80024	U195	LOBRO	80012
U03-47	MCQUAY-NOR	88000.20	U04-27	MCQUAY-NOR	88001.70	U119	EXCEL	80119	U19655	PILOT	80181
U03-47	MCQUAY-NOR	88000.21	U04-29	MCQUAY-NOR	88001.94	U119	LEMPCO	80119	U2	LEMPCO	88002.55
U03-48	MCQUAY-NOR	88000.60	U1	LEMPCO	88002.61	U119	LEMPCO	80120	U20	LEMPCO	88002.67
U03-49	MCQUAY-NOR	88000.22	U10	LEMPCO	88002.42	U12	LEMPCO	88002.44	U20	LEMPCO	88002.70
U03-5	MCQUAY-NOR	80044	U10	LEMPCO	88002.43	U120	EXCEL	80182	U200	EXCEL	80092
U03-50	MCQUAY-NOR	88000.45	U100	EXCEL	80041	U120	LEMPCO	80140	U200	LEMPCO	80092
U03-51	MCQUAY-NOR	88000.52	U100	LEMPCO	80041	U121	EXCEL	80118	U200	LOBRO	80142
U03-52	MCQUAY-NOR	88000.25	U100	LOBRO	80072	U121	LEMPCO	80121	U201	LEMPCO	80092
U03-52	MCQUAY-NOR	88000.27	U100	PERRY	80133	U121	LEMPCO	80131	U20138	ERF	80225
U03-53	MCQUAY-NOR	88000.28	U101	EXCEL	80072	U122	LEMPCO	80140	U202	LEMPCO	80049
U03-54	MCQUAY-NOR	88000.46	U101	LEMPCO	80072	U122	LEMPCO	80182	U202	LEMPCO	80050
U03-55	MCQUAY-NOR	88000.46	U101	LOBRO	80077	U123	LEMPCO	80072	U203	LEMPCO	80048
U03-55	MCQUAY-NOR	88000.53	U101C	LEMPCO	80072	U125	LEMPCO	80090	U203	LEMPCO	88002.10
U03-56	MCQUAY-NOR	88000.30	U101F	LEMPCO	80072	U12521U	DEERE,JOHN	80231	U204	LEMPCO	80043
U03-56	MCQUAY-NOR	88000.31	U101HD	EXCEL	80075	U1252U	DEERE,JOHN	80232	U205	LEMPCO	80044
U03-57	MCQUAY-NOR	88000.34	U101HD	EXCEL	80077	U126	LEMPCO	80135	U205	LOBRO	80142
U03-58	MCQUAY-NOR	88000.36	U101HD	LEMPCO	80075	U127	LEMPCO	80178	U206	LEMPCO	80049
U03-59	MCQUAY-NOR	88000.34	U101HD	LEMPCO	80077	U128	EXCEL	80117	U206	LEMPCO	88002.06
U03-59	MCQUAY-NOR	88000.35	U101HDU	LEMPCO	80088	U128	LEMPCO	80117	U207	EXCEL	80101
U03-6	MCQUAY-NOR	80054	U101HP	LEMPCO	80077	U129	EXCEL	80091	U207	LEMPCO	80099
U03-60	MCQUAY-NOR	88000.37	U101HU	LEMPCO	80077	U129	LEMPCO	80073	U207	LEMPCO	80101
U03-61	MCQUAY-NOR	88000.55	U102	EXCEL	80124	U129	LEMPCO	80091	U208	LEMPCO	80110
U03-62	MCQUAY-NOR	88000.38	U102	LEMPCO	80124	U13	LEMPCO	88002.69	U208	LEMPCO	80112
U03-62	MCQUAY-NOR	88000.39	U102	LOBRO	80090	U130	EXCEL	80078	U2086	EXCEL	80112
U03-63	MCQUAY-NOR	88000.79	U102U	LEMPCO	80125	U130	EXCEL	80090	U208G	LEMPCO	80112
U03-64	MCQUAY-NOR	88000.80	U103	EXCEL	80142	U130	LEMPCO	80090	U209	EXCEL	80109
U03-65	MCQUAY-NOR	88000.89	U103	LEMPCO	80142	U131	EXCEL	80118	U209	LEMPCO	80109
U03-66	MCQUAY-NOR	88000.81	U103	LOBRO	80077	U131	LEMPCO	80118	U209	LEMPCO	80149
U03-67	MCQUAY-NOR	88000.83	U103U	EXCEL	80143	U132	EXCEL	80182	U21	LEMPCO	88002.44
U03-68	MCQUAY-NOR	88000.95	U103U	LEMPCO	80143	U132	LEMPCO	80182	U210	LEMPCO	80045
U03-69	MCQUAY-NOR	88000.85	U104	EXCEL	80178	U133	EXCEL	80190	U210	LOBRO	80003
U03-7	MCQUAY-NOR	80046	U104	LEMPCO	80178	U133	LEMPCO	80190	U211	LEMPCO	80030
U03-70	MCQUAY-NOR	88000.73	U105	EXCEL	80189	U134	EXCEL	80075	U212	LEMPCO	80106
U03-72	MCQUAY-NOR	84531	U105	LEMPCO	80189	U134	LEMPCO	80075	U213	LEMPCO	80053
U03-73	MCQUAY-NOR	83316	U105	LOBRO	80077	U134B	LEMPCO	80076	U214	LEMPCO	80114
U03-74	MCQUAY-NOR	80133	U106	EXCEL	80221	U134U	EXCEL	80076	U215	LEMPCO	80147
U03-75	MCQUAY-NOR	88001.13	U106	LEMPCO	80221	U134U	LEMPCO	80076	U216	LEMPCO	80146
U03-76	MCQUAY-NOR	84504	U107	EXCEL	80225	U135	LEMPCO	80044	U217	EXCEL	80158
U03-77	MCQUAY-NOR	81054	U107	LEMPCO	80225	U136	EXCEL	80025	U217	LEMPCO	80158
U03-78	MCQUAY-NOR	81109	U108	LEMPCO	88001.82	U136	LEMPCO	80025	U218	EXCEL	88001.98
U03-79	MCQUAY-NOR	81153	U109	LEMPCO	88001.83	U137	LEMPCO	80228	U218	LEMPCO	80044
U03-8	MCQUAY-NOR	80047	U109	PERRY	80069	U138	EXCEL	80124	U218	LEMPCO	88001.98
U03-80	MCQUAY-NOR	81175	U11	LEMPCO	88002.42	U138	LEMPCO	80124	U219	EXCEL	80049
U03-81	MCQUAY-NOR	81218	U11	LOBRO	80041	U14	LEMPCO	88002.68	U219	LEMPCO	80049
U03-82	MCQUAY-NOR	81246	U110	LEMPCO	88001.82	U140	LEMPCO	80024	U22	LEMPCO	88002.50
U03-83	MCQUAY-NOR	81237	U110	LOBRO	80041	U140	LEMPCO	80036	U220	EXCEL	80148
U03-84	MCQUAY-NOR	84510	U111	LEMPCO	80023	U140	LEMPCO	80079	U220	LLMPCO	80148
U03-85	MCQUAY-NOR	88013.90	U111	LOBRO	80041	U141	LEMPCO	80026	U220	LOBRO	80124

Part No.	Brand	IDLI Group
U221	LEMPCO	80054
U221	LOBRO	80124
U222	EXCEL	80159
U222	EXCEL	80167
U222	LEMPCO	80159
U222	LEMPCO	80167
U223	LEMPCO	80111
U223	LOBRO	80125
U224	EXCEL	80168
U224	LEMPCO	80168
U225	LEMPCO	80177
U225	LOBRO	80124
U226	EXCEL	80174
U226	LEMPCO	80174
U227	EXCEL	80173
U227	LEMPCO	80173
U228	LEMPCO	80104
U228	LEMPCO	80107
U228G	EXCEL	80104
U228G	EXCEL	80107
U228G	LEMPCO	80104
U229	EXCEL	80193
U229	LEMPCO	80193
U229	LEMPCO	80198
U23	LEMPCO	88002.52
U230	EXCEL	80210
U230	LEMPCO	80210
U230	LOBRO	80140
U231	EXCEL	80231
U231	LEMPCO	80231
U232	LEMPCO	80235
U233	LEMPCO	80043
U234	LEMPCO	80104
U235	LEMPCO	80045
U236	LEMPCO	80030
U236	LEMPCO	80031
U236G	EXCEL	80031
U236G	LEMPCO	80031
U237	LEMPCO	80030
U238	EXCEL	80049
U238	LEMPCO	80049
U238	LEMPCO	80050
U238	LOBRO	80119
U238G	LEMPCO	80049
U239	EXCEL	88001.98
U239	LEMPCO	80043
U239	LEMPCO	88001.96
U239G	LEMPCO	80044
U24	LEMPCO	88002.51
U240	LEMPCO	80051
U240	LOBRO	80118
U241	LEMPCO	80193
U24125	ERF	80041
U242	LEMPCO	80100
U243	LEMPCO	80101
U244	LEMPCO	80021
U244	LEMPCO	88001.97
U245	LEMPCO	80054
U246	LEMPCO	80049
U247	LEMPCO	80054
U248	EXCEL	80197
U248	LEMPCO	80197
U249	EXCEL	80199
U249	LEMPCO	80199
U25	LEMPCO	88002.73
U250	EXCEL	80201
U250	LEMPCO	80201
U250	LOBRO	80042
U251	EXCEL	80212
U251	LEMPCO	80212
U2518	ALCO	80066
U252	EXCEL	80059
U252	LEMPCO	80059
U253	LEMPCO	80194
U253	LEMPCO	80195
U254	LEMPCO	80211
U255	EXCEL	80031
U255	LEMPCO	80030
U256	LEMPCO	80023
U257	EXCEL	80055
U257	LEMPCO	80055
U257	LOBRO	80065
U258	EXCEL	80093
U258	LEMPCO	80093
U258	LOBRO	80061
U259	EXCEL	80237
U259	LEMPCO	80237
U26	LEMPCO	88002.74
U260	LEMPCO	80058
U260	LOBRO	80135
U261	LEMPCO	80032
U261G	EXCEL	80032
U261G	LEMPCO	80032
U262	LEMPCO	88002.77
U262G	LEMPCO	88002.77
U263	EXCEL	80066
U263	LEMPCO	80066
U263G	LEMPCO	80066
U264	EXCEL	80058
U264	LEMPCO	80058
U265	LOBRO	80135
U266	EXCEL	80107
U266	LEMPCO	80105
U267	LEMPCO	80234
U27	LEMPCO	88002.76
U270	LOBRO	80181
U276	LOBRO	80039
U277	LOBRO	80038
U28	LEMPCO	88002.75
U280	LOBRO	80138
U285	LOBRO	80122
U287	LOBRO	80160
U29	LEMPCO	88002.75
U290	LOBRO	80181
U2T2910	ALCO	88002.43
U2T2911	ALCO	88002.42
U2T2911	ALLOY	88002.42
U2T2912	ALCO	88002.44
U2T2912	ALLOY	88002.44
U2T2914	TORQUE	88002.44
U3	LEMPCO	88002.56
U30	LEMPCO	88002.76
U300	LEMPCO	80122
U300	LOBRO	80178
U301	LEMPCO	80138
U302	LEMPCO	80068
U303	LEMPCO	80160
U3031C	LEMPCO	80109
U304	LEMPCO	80038
U305	LEMPCO	80181
U306	LEMPCO	80204
U307	LEMPCO	80038
U307	LEMPCO	80039
U308	EXCEL	80123
U308	LEMPCO	80123
U309	LEMPCO	80081
U310	EXCEL	80081
U310	LEMPCO	80081
U311	EXCEL	80091
U311	LEMPCO	80091
U312	EXCEL	80071
U312	LEMPCO	80071
U313	EXCEL	80074
U313	LEMPCO	80074
U3130A	LEMPCO	80097
U314	EXCEL	80141
U314	LEMPCO	80141
U315	EXCEL	80129
U315	LEMPCO	80129
U316	EXCEL	80131
U316	LEMPCO	80131
U317	LEMPCO	80131
U317	LEMPCO	80145
U318	EXCEL	80130
U318	LEMPCO	80130
U319	LEMPCO	80007
U350	LOBRO	80182
U3T1328	ALCO	88002.55
U3T1328	ALCO	88002.61
U3T1595	ALCO	88002.47
U3T1602	ALCO	88002.56
U3T2007	ALCO	88002.57
U3T2221	ALCO	88002.58
U3T2222	ALCO	88002.45
U3T2266	ALCO	88002.54
U3T2501	ALCO	88002.63
U3T2531	ALCO	88002.60
U3T2538	ALCO	88002.60
U3T2893	TORQUE	88002.48
U3T2894	ALCO	88002.46
U3T2895	ALCO	88002.48
U3T2896	ALCO	88002.50
U3T2897	ALCO	88002.51
U3T2898	ALCO	88002.52
U3T2899	ALCO	88002.53
U3T3407	ALCO	88002.71
U3T3501	ALCO	88002.63
U3T3501	ALCO	88002.64
U3T3501	ALCO	88002.67
U3T3502	ALCO	88002.64
U3T3502	TORQUE	88002.63
U3T3502	TORQUE	88002.67
U3T3503	TORQUE	88002.72
U3T3505	ALCO	88002.65
U3T3506	ALCO	88002.67
U3T3507	ALCO	88002.71
U3T3508	ALCO	88002.72
U3T3508	TORQUE	88002.67
U3T3508	TORQUE	88002.72
U3T3509	ALCO	88002.70
U3T3510	ALCO	88002.73
U3T3511	ALCO	88002.74
U4	LEMPCO	88002.57
U40	LOBRO	80023
U400	LOBRO	80189
U401	LEMPCO	88003.50
U401B	LEMPCO	88003.51
U404	LEMPCO	80909
U404R	EXCEL	80906
U404R	LEMPCO	80906
U404R	LEMPCO	80909
U405	LEMPCO	80958
U405	LOBRO	80189
U405R	LEMPCO	80958
U409	LEMPCO	80135
U411	LEMPCO	88003.50
U413	LEMPCO	80064
U414	LEMPCO	80135
U415	EXCEL	80042
U415	EXCEL	80135
U415	LEMPCO	80042
U415	LEMPCO	80135
U416	LEMPCO	80096
U417	LEMPCO	88001.60
U417R	LEMPCO	80951
U418	LEMPCO	88001.62
U419	LEMPCO	80022
U420	LEMPCO	88001.74
U421	LEMPCO	80901
U421	LEMPCO	88001.66
U422	LEMPCO	88001.75
U422R	LEMPCO	80901
U423	LEMPCO	88001.67
U423R	LEMPCO	88001.67
U424	EXCEL	80034
U424	LEMPCO	80034
U425	LEMPCO	80060
U426B	LEMPCO	80901
U426B	LEMPCO	80909
U426R	LEMPCO	88001.74
U427	LEMPCO	80955
U427A	LEMPCO	80908
U427B	LEMPCO	80908
U427R	LEMPCO	80955
U428	LEMPCO	80095
U429	LEMPCO	80064
U430	LEMPCO	80956
U430A	LEMPCO	80953
U430B	LEMPCO	80905
U430B	LEMPCO	80906
U430B	LEMPCO	80956
U431R	LEMPCO	80954
U432	LEMPCO	80056
U432	LEMPCO	80095
U433	LEMPCO	80056
U433	LEMPCO	80057
U434	LEMPCO	80056
U434	LEMPCO	80057
U435	LEMPCO	80956
U435R	LEMPCO	80903
U435R	LEMPCO	80956
U436	LEMPCO	80954
U436A	LEMPCO	80907
U436R	LEMPCO	80954
U437	LEMPCO	80056
U438	LEMPCO	80061
U438	LEMPCO	80095
U439	LEMPCO	80096
U440	LEMPCO	80064
U441	LEMPCO	80096
U442	LEMPCO	80094
U443	LEMPCO	80096
U444R	LEMPCO	80907
U445	LEMPCO	80957
U445A	LEMPCO	80907
U445R	LEMPCO	80957
U446	LEMPCO	80951
U446A	LEMPCO	80905
U446B	LEMPCO	80905
U446R	LEMPCO	80951
U447	EXCEL	80064
U447	LEMPCO	80064
U447	LEMPCO	80065
U448	EXCEL	80060
U448	LEMPCO	80060
U448	LEMPCO	80061
U448A	EXCEL	80061
U448A	LEMPCO	80062
U449	LEMPCO	80953
U449B	LEMPCO	80906

Part No.	Brand	IDLI Group	Part No.	Brand	IDLI Group	Part No.	Brand	IDLI Group	Part No.	Brand	IDLI Group
U449R	EXCEL	80953	U592	LOBRO	88002.77	U76-1-133	CLEVE.STL.	88005.34	UJ1003	MIDAS	80081
U449R	LEMPCO	80953	U595	LOBRO	88002.77	U76-1-135	CLEVE.STL.	81407	UJ1004	MIDAS	80091
U450	LEMPCO	80953	U596	LOBRO	80030	U76-17	CLEVE.STL.	88013.47	UJ1005	MIDAS	80129
U450R	LEMPCO	80953	U597	LOBRO	88002.77	U76-17-1	CLEVE.STL.	88013.48	UJ1006	MIDAS	80131
U451	EXCEL	80132	U5A	LEMPCO	88002.46	U76-2	CLEVE.STL.	84618	UJ1007	MIDAS	80107
U451	LEMPCO	80132	U5B	LEMPCO	88002.48	U76-3-4-14	CLEVE.STL.	85527	UJ1008	MIDAS	80031
U452	EXCEL	80061	U6	LEMPCO	88002.63	U76-3-4-14D	CLEVE.STL.	88005.29	UJ1009	MIDAS	80104
U452	LEMPCO	80060	U600	LOBRO	80048	U76-43-198	CLEVE.STL.	86677	UJ105	JAMCO	80031
U452	LEMPCO	80061	U601	LOBRO	80048	U76-43-199	CLEVE.STL.	86676	UJ110	MATSUBA	80024
U452	LEMPCO	80063	U601K	LEMPCO	88002.34	U76-43-200	CLEVE.STL.	86678	UJ111	MATSUI	80010
U453	EXCEL	80135	U602A	LEMPCO	88002.36	U76-54	CLEVE.STL.	85073	UJ112	MATSUI	80004
U453	LEMPCO	80135	U602AK	LEMPCO	88002.36	U76-54L	CLEVE.STL.	85076	UJ114	MATSUI	80019
U454	EXCEL	80951	U602K	LEMPCO	88002.34	U76-7-4-14N	CLEVE.STL.	88014.76	UJ115	MATSUBA	80124
U454	LEMPCO	80951	U603K	LEMPCO	88002.33	U76-9	CLEVE.STL.	88013.55	UJ1200	JAMCO	80081
U454R	EXCEL	80905	U603K	LEMPCO	88002.33	U7A	LEMPCO	88002.67	UJ1201	JAMCO	80091
U454R	LEMPCO	80905	U-603K	LEMPCO	88002.33	U7B	LEMPCO	88002.65	UJ1203	JAMCO	80129
U455	LEMPCO	80909	U604K	LEMPCO	88002.33	U8	LEMPCO	88002.43	UJ1204	JAMCO	80131
U455	LEMPCO	80953	U605	LOBRO	80050	U8A	LEMPCO	88002.71	UJ1205	JAMCO	80123
U455R	EXCEL	80909	U605K	LEMPCO	88002.33	U9	LEMPCO	88002.60	UJ1275	MATSUI	80035
U455R	LEMPCO	80905	U606K	LEMPCO	88002.33	U96	TEDDY TORQ	80181	UJ1306	JAMCO	80061
U455R	LEMPCO	80906	U607	LOBRO	80044	U96-10	CLEVE.STL.	88015.54	UJ1309	JAMCO	80132
U455R	LEMPCO	80909	U607T	LEMPCO	88002.37	U96-13	CLEVE.STL.	88013.35	UJ134	JAMCO	80118
U5	LEMPCO	88002.58	U610	LOBRO	80049	U96-55	ALCO	80181	UJ200	JAMCO	80075
U50	LOBRO	80024	U611	LOBRO	80049	U96-55	BORG-WARNR	80181	UJ210	MATSUI	80018
U500	LOBRO	80221	U612	LOBRO	80030	U96-55	CLEVE.STL.	80181	UJ211	MATSUI	80009
U506	EXCEL	80134	U614	LOBRO	80045	U96-55	REPCO	80181	UJ212	MATSUI	80006
U506	LEMPCO	80134	U620	LOBRO	80044	U96-55	SPICER	80181	UJ214	MATSUI	80017
U507	LEMPCO	80178	U630	LOBRO	80099	U96-55-4	CLEVE.STL.	80181	UJ218	MATSUBA	80010
U508	EXCEL	80188	U632	LOBRO	80112	U96-55-4	REPCO	80181	UJ219	MATSUBA	80010
U508	LEMPCO	80188	U635	LOBRO	80106	U96-9	CLEVE.STL.	88013.63	UJ220	MATSUBA	80020
U509	LEMPCO	80220	U636	LOBRO	80104	U9A	LEMPCO	88002.60	UJ221	MATSUBA	80009
U51	LOBRO	80023	U637	LOBRO	80104	U9B	LEMPCO	88002.45	UJ222	MATSUBA	80006
U510	EXCEL	80069	U638	LOBRO	80107	U9C	LEMPCO	88002.47	UJ230	MATSUI	80016
U510	LEMPCO	80069	U640	LOBRO	80109	UB200K	D&D MACH.	88013.86	UJ231	MATSUI	80009
U511	LEMPCO	80186	U641	LOBRO	80147	UB253K	D&D MACH.	88001.70	UJ234	MATSUI	80115
U512	ALLOY	80184	U642	LOBRO	80146	UB560K	D&D MACH.	88013.87	UJ2387	JAMCO	88001.70
U512	LEMPCO	80184	U645	LOBRO	80158	UCB100L	CLEVE.STL.	87399	UJ279	JAMCO	80189
U513	LEMPCO	80179	U647	LOBRO	80148	UCB100LS	CLEVE.STL.	87400	UJ280	JAMCO	80131
U514	LEMPCO	80227	U650	LOBRO	80168	UCB103	CLEVE.STL.	87457	UJ280	JAMCO	80221
U52	ALCO	80909	U651	LOBRO	80173	UCB104	CLEVE.STL.	87458	UJ2844	MATSUI	80136
U52	ALCO	80958	U652	LOBRO	80177	UCB105	CLEVE.STL.	87460	UJ2855	MATSUI	80161
U52	ALCO	88001.73	U655	LOBRO	80174	UCB106	CLEVE.STL.	88008.56	UJ3009	JAMCO	80105
U54	ALCO	88001.74	U660	LOBRO	80193	UCB107	CLEVE.STL.	87459	UJ313	MATSUBA	80023
U550	LOBRO	80178	U670	LOBRO	80210	UCB108	CLEVE.STL.	88008.57	UJ314	MATSUBA	80014
U56	TEDDY TORQ	80181	U672	LOBRO	88001.57	UCB109	CLEVE.STL.	87461	UJ3147	JAMCO	80104
U56-10	CLEVE.STL.	88015.53	U674	LOBRO	80215	UCB1-4	CLEVE.STL.	88007.56	UJ323	MATSUI	80024
U56-102	CLEVE.STL.	87429	U675	LOBRO	80212	UCB1-4A	CLEVE.STL.	88014.88	UJ324	MATSUI	80014
U56-106	CLEVE.STL.	88008.26	U7	LEMPCO	88002.64	UCB1-5	CLEVE.STL.	88007.54	UJ411	MATSUBA	80012
U56-1-108-1	CLEVE.STL.	88005.31	U700	LEMPCO	80004	UCB2-1	CLEVE.STL.	88013.72	UJ412	MATSUBA	80010
U56-1-108-1A	CLEVE.STL.	88005.32	U700	LEMPCO	80006	UCB2-2	CLEVE.STL.	88013.78	UJ413	MATSUI	80011
U56-17	CLEVE.STL.	88013.46	U701	LEMPCO	80010	UCB2-3	CLEVE.STL.	88013.69	UJ415	MATSUI	80020
U56-2-1	CLEVE.STL.	88005.27	U701	LEMPCO	80018	UCB50-1	CLEVE.STL.	88008.65	UJ417	MATSUI	80013
U56-23	CLEVE.STL.	88014.84	U701	LOBRO	88002.77	UCB51-2	CLEVE.STL.	88008.62	UJ418	MATSUI	80005
U56-38	CLEVELAND	80183	U702	LEMPCO	80009	UCB61	CLEVE.STL.	88007.88	UJ507	LILLISTON	80133
U56-43-131	CLEVE.STL.	88007.33	U703	LEMPCO	80012	UCB61-1	CLEVE.STL.	88007.84	UJ520	LILLISTON	80021
U56-43-154	CLEVE.STL.	86675	U704	LEMPCO	80019	UF1	LEMPCO	88002.87	UJ520	LILLISTON	80035
U56-54	CLEVE.STL.	85075	U704	LOBRO	80035	UJ011	MATSUBA	80032	UJ616	MATSUBA	80012
U56-54-109	CLEVE.STL.	88005.28	U705	LEMPCO	80011	UJ015	MATSUBA	80041	UJ620	MATSUBA	80077
U56-55-2	AEC	80183	U710	LOBRO	80032	UJ016	MATSUBA	80077	UJ625	MATSUI	80012
U56-55-2	BORG-WARNR	80183	U715	LOBRO	88002.77	UJ017	MATSUBA	80124	UJLC	BOST.GEAR	80035
U56-55-2	CLEVELAND	80183	U728	LOBRO	88002.77	UJ020	MATSUBA	80049	UJ-M2564	KOYO	80012
U56-55-2	CLEVE.STL.	80183	U728/3	LOBRO	88002.77	UJ021	MATSUBA	80049	UJ-N2057	KOYO	80004
U56-7-110	CLEVE.STL.	88014.77	U76-10	CLEVE.STL.	88015.31	UJ022	MATSUBA	80142	UJ-N2563	KOYO	80010
U56-7-111	CLEVE.STL.	88014.78	U76-101	CLEVE.STL.	88008.16	UJ023	MATSUBA	80024	UJ-N2880B	KOYO	80019
U56-7-137	CLEVE.STL.	88015.37	U76-10-1	CLEVE.STL.	88015.34	UJ030	MATSUBA	88002.77	UJSC	BOST.GEAR	80001
U56-7-140	CLEVE.STL.	88015.36	U76-102	CLEVE.STL.	88008.25	UJ031	MATSUBA	88002.81	UJ-T2057	KOYO	80006
U56-9	CLEVE.STL.	88013.62	U76-103	CLEVE.STL.	87436	UJ055	MATSUI	80041	UJ-T2667A	KOYO	80009
U580	LOBRO	80188	U76-11	CLEVE.STL.	88015.57	UJ1000	MIDAS	80061	UJ-T2978A	KOYO	80017
U590	LOBRO	80220	U76-1-100	CLEVE.STL.	88005.30	UJ1001	MIDAS	80132	UJ-TM2546B	KOYO	80011
U591	LOBRO	88002.81	U76-1-132	CLEVE.STL.	88005.33	UJ1002	MIDAS	80075	UJ-TM2564B	KOYO	80011

Part No.	Brand	IDLI Group	Part No.	Brand	IDLI Group	Part No.	Brand	IDLI Group	Part No.	Brand	IDLI Group
UJ-TM2564E	KOYO	80011	UO1-33	MCQUAY-NOR	80121	UO3-22	MCQUAY-NOR	80154	UT22-3900	NEAPCO	87933
UK108	TRANSMISS.	80024	UO1-34	MCQUAY-NOR	80119	UO3-23	MCQUAY-NOR	80216	UT22-5100	NEAPCO	87933
UK120	TRANSMISS.	80041	UO1-35	MCQUAY-NOR	80118	UO3-24	MCQUAY-NOR	80172	UT26-002900	NEAPCO	85102
UK121	TRANSMISS.	80041	UO1-36	MCQUAY-NOR	80132	UO3-25	MCQUAY-NOR	80174	UT2912	ALCO	88002.44
UK123	TRANSMISS.	80041	UO1-37	MCQUAY-NOR	80091	UO3-26	MCQUAY-NOR	80174	UTP06-005100	NEAPCO	88001.13
UK127	TRANSMISS.	80075	UO1-38	MCQUAY-NOR	80131	UO3-27	MCQUAY-NOR	80163	UTP12-005400	NEAPCO	87923
UK135/6	TRANSMISS.	80124	UO1-39	MCQUAY-NOR	80129	UO3-28	MCQUAY-NOR	80175	UTP12-006600	NEAPCO	87935
UK136	TRANSMISS.	80124	UO1-4	MCQUAY-NOR	80044	UO3-29	MCQUAY-NOR	80198	UTR06-002700	NEAPCO	88001.13
UK146	TRANSMISS.	80142	UO1-40	MCQUAY-NOR	80004	UO3-3	MCQUAY-NOR	80048	UTR06-003900	NEAPCO	88001.13
UK147	TRANSMISS.	80142	UO1-41	MCQUAY-NOR	80010	UO3-30	MCQUAY-NOR	80196	UTR06-005100	NEAPCO	88001.13
UK163	TRANSMISS.	80140	UO1-42	MCQUAY-NOR	80009	UO3-31	MCQUAY-NOR	80210	UTR06-5100	NEAPCO	88001.13
UK169	TRANSMISS.	80178	UO1-44	MCQUAY-NOR	80011	UO3-32	MCQUAY-NOR	80212	UTR12-002700	NEAPCO	81209
UK169/6	TRANSMISS.	80180	UO1-45	MCQUAY-NOR	80019	UO3-33	MCQUAY-NOR	80215	UTR12-002700	NEAPCO	88006.99
UK171	TRANSMISS.	80180	UO1-46	MCQUAY-NOR	88002.34	UO3-34	MCQUAY-NOR	80228	UTR12-003900	NEAPCO	81209
UK176	TRANSMISS.	80189	UO1-47	MCQUAY-NOR	88002.36	UO3-36	MCQUAY-NOR	80234	UTR12-003900	NEAPCO	88006.99
UK190	TRANSMISS.	80188	UO1-49	MCQUAY-NOR	88002.33	UO3-38	MCQUAY-NOR	80237	UTR12-005100	NEAPCO	81218
UK200	TRANSMISS.	80221	UO1-50	MCQUAY-NOR	80076	UO3-39	MCQUAY-NOR	80239	UTR12-5100	NEAPCO	81218
UK202	TRANSMISS.	80220	UO1-51	MCQUAY-NOR	80143	UO3-4	MCQUAY-NOR	80049	UTRL6-005100	NEAPCO	88001.13
UK265	TRANSMISS.	80225	UO1-52	MCQUAY-NOR	80092	UO3-40	MCQUAY-NOR	80238	UY1	LEMPCO	88002.86
UK275	TRANSMISS.	80031	UO1-6	MCQUAY-NOR	80059	UO3-41	MCQUAY-NOR	80240	V1082	BEGE MFG.	80001
UK318	TRANSMISS.	80031	UO1-60	MCQUAY-NOR	80034	UO3-42	MCQUAY-NOR	80189	V8385	BMC	80142
UK318/6	TRANSMISS.	80031	UO1-61	MCQUAY-NOR	88002.42	UO3-43	MCQUAY-NOR	80207	V8385	LEYLAND	80142
UK320	TRANSMISS.	80044	UO1-64	MCQUAY-NOR	80036	UO3-44	MCQUAY-NOR	80037	V9009	WABCO	80234
UK320/6	TRANSMISS.	80044	UO1-66	MCQUAY-NOR	80081	UO3-5	MCQUAY-NOR	80044	VA1775	WABCO	80001
UK351	TRANSMISS.	80104	UO1-7	MCQUAY-NOR	80059	UO3-6	MCQUAY-NOR	80054	VCB100L	CLEVE.STL.	87399
UK352	TRANSMISS.	80090	UO1-8	MCQUAY-NOR	80058	UO3-74	MCQUAY-NOR	80133	VD0222	WABCO	80156
UK360	TRANSMISS.	80042	UO1-9	MCQUAY-NOR	80024	UO3-9	MCQUAY-NOR	80007	VD1557	WABCO	80234
UK368	TRANSMISS.	80135	UO2-1	MCQUAY-NOR	80055	UO4-1	MCQUAY-NOR	80905	VG8035	WABCO	80234
UK405	TRANSMISS.	80151	UO2-10	MCQUAY-NOR	80142	UO4-17	MCQUAY-NOR	80951	VG8302	WABCO	80234
UK460	TRANSMISS.	80039	UO2-11	MCQUAY-NOR	80125	UO4-18	MCQUAY-NOR	80953	VH1311	WABCO	80245
UK461	TRANSMISS.	80039	UO2-12	MCQUAY-NOR	80123	UO4-2	MCQUAY-NOR	80906	VI9523	WABCO	80077
UK464	TRANSMISS.	80138	UO2-13	MCQUAY-NOR	80141	UO4-23	MCQUAY-NOR	88013.86	VJ0635	WABCO	80048
UK503	TRANSMISS.	80048	UO2-14	MCQUAY-NOR	80145	UO4-24	MCQUAY-NOR	88013.87	VJ0636	WABCO	80047
UK506	TRANSMISS.	80101	UO2-15	MCQUAY-NOR	80148	UO4-29	MCQUAY-NOR	88001.94	VJ0684	WABCO	80196
UK511	TRANSMISS.	80168	UO2-16	MCQUAY-NOR	80159	UR12010	HUB CITY	80035	VJ0763	WABCO	80213
UK516	TRANSMISS.	80193	UO2-17	MCQUAY-NOR	80178	UR14N	HUB CITY	80037	VJ3051	WABCO	80215
UK520S	TRANSMISS.	80210	UO2-18	MCQUAY-NOR	80182	UR17110	HUB CITY	80044	VJ3948	WABCO	80213
UK536	TRANSMISS.	80109	UO2-19	MCQUAY-NOR	80135	UR1FR	HUB CITY	80008	VJ4057	WABCO	88003.54
UK585S	TRANSMISS.	80213	UO2-2	MCQUAY-NOR	80042	UR20	BEAN,JOHN	80001	VJ4059	WABCO	80240
UK670	TRANSMISS.	80032	UO2-20	MCQUAY-NOR	80183	UR21110	HUB CITY	80069	VJ4060	WABCO	80240
UK671	TRANSMISS.	80032	UO2-21	MCQUAY-NOR	80186	UR52	ALCO	80958	VJ4180	WABCO	88003.54
UK674	TRANSMISS.	80049	UO2-22	MCQUAY-NOR	80184	UR52	DETROIT	80958	VJ5827	WABCO	80174
UK674/6	TRANSMISS.	80050	UO2-24	MCQUAY-NOR	80193	UR52A	ALCO	80906	VJ5949	WABCO	80174
UK676	TRANSMISS.	80045	UO2-26	MCQUAY-NOR	80201	UR54	ALCO	88001.74	VJ6789	WABCO	88003.54
UK680	TRANSMISS.	80112	UO2-27	MCQUAY-NOR	80199	UR56	DETROIT	80954	VJ6880	WABCO	80216
UK682	TRANSMISS.	80158	UO2-29	MCQUAY-NOR	80190	UR57	ALCO	80955	VJ6881	WABCO	80163
UK686	TRANSMISS.	80174	UO2-3	MCQUAY-NOR	80062	UR60	ALCO	80958	VJ763	WABCO	80215
UO1-1	MCQUAY-NOR	80031	UO2-30	MCQUAY-NOR	80221	UR60	ALLOY	80903	VJ8125	WABCO	80240
UO1-10	MCQUAY-NOR	80023	UO2-31	MCQUAY-NOR	80225	UR62	ALCO	80954	VK107	TRANSMISS.	80023
UO1-11	MCQUAY-NOR	80025	UO2-32	MCQUAY-NOR	88002.10	UR63	ALCO	80957	VK109	TRANSMISS.	80023
UO1-12	MCQUAY-NOR	80030	UO2-4	MCQUAY-NOR	80106	UR65	ALCO	80951	VK112	TRANSMISS.	80023
UO1-13	MCQUAY-NOR	80041	UO2-5	MCQUAY-NOR	80078	UR65	ALLOY	80951	VK128	TRANSMISS.	80077
UO1-14	MCQUAY-NOR	80072	UO2-6	MCQUAY-NOR	80079	UR66	ALCO	80953	VK130	TRANSMISS.	80077
UO1-15	MCQUAY-NOR	80089	UO2-7	MCQUAY-NOR	80124	UR66	DETROIT	80953	VL8642	ADAMS-LET.	80234
UO1-16	MCQUAY-NOR	80091	UO2-8	MCQUAY-NOR	80124	UR67	ALCO	80953	VL8642	WABCO	80234
UO1-17	MCQUAY-NOR	80075	UO2-9	MCQUAY-NOR	80140	UR67	ALLOY	80953	VL8643	ADAMS-LET.	80234
UO1-18	MCQUAY-NOR	80065	UO3-1	MCQUAY-NOR	80003	UR6N	HUB CITY	80001	VL8643	WABCO	80234
UO1-19	MCQUAY-NOR	80060	UO3-10	MCQUAY-NOR	80003	US56-55-2	CLEVE.STL.	80183	VL9009	ADAMS-LET.	80234
UO1-19	MCQUAY-NOR	80061	UO3-11	MCQUAY-NOR	80035	UT12-005100	NEAPCO	81218	VL9009	WABCO	80234
UO1-2	MCQUAY-NOR	80032	UO3-12	MCQUAY-NOR	80069	UT20-002800	NEAPCO	88007.05	VL9523	ADAMS-LET.	80077
UO1-21	MCQUAY-NOR	80104	UO3-13	MCQUAY-NOR	80100	UT20-004000	NEAPCO	88007.08	VL9523	WABCO	80077
UO1-24	MCQUAY-NOR	80112	UO3-14	MCQUAY-NOR	80102	UT20-004000	NEAPCO	88007.09	VN2793	ADAMS-LET.	80077
UO1-25	MCQUAY-NOR	80099	UO3-15	MCQUAY-NOR	80113	UT20-005200	NEAPCO	87938	VN2793	WABCO	80077
UO1-26	MCQUAY-NOR	80101	UO3-16	MCQUAY-NOR	80109	UT20-4000	NEAPCO	87938	VP1493	PEARSON	80101
UO1-27	MCQUAY-NOR	80110	UO3-17	MCQUAY-NOR	80150	UT22-002900	NEAPCO	88001.06	VP1494	PEARSON	80100
UO1-29	MCQUAY-NOR	80077	UO3-18	MCQUAY-NOR	80156	UT22-002900	NEAPCO	88007.15	VP14949	RUST	80101
UO1-3	MCQUAY-NOR	80049	UO3-19	MCQUAY-NOR	80158	UT22-003900	NEAPCO	88001.06	VP1779	PEARSON	80037
UO1-30	MCQUAY-NOR	80090	UO3-2	MCQUAY-NOR	80021	UT22-003900	NEAPCO	88007.15	VS0297	WABCO	80174
UO1-31	MCQUAY-NOR	80117	UO3-20	MCQUAY-NOR	80155	UT22-005100	NEAPCO	87933	VS0372	WABCO	80007
UO1-32	MCQUAY-NOR	80116	UO3-21	MCQUAY-NOR	80150	UT22-2900	NEAPCO	88007.15	VS10604	GMC	88002.77

Part No.	Brand	IDLI Group	Part No.	Brand	IDLI Group	Part No.	Brand	IDLI Group	Part No.	Brand	IDLI Group
VS10605	GMC	88002.77	VTE3212	VAPORMATIC	88004.22	W62-746	WORLDPARTS	86145	WAFS28-5	ROCKWELL	88015.53
VS10606	GMC	80044	VTE3213	VAPORMATIC	88004.23	W62-747	WORLDPARTS	86193	WAFS30-1	ROCKWELL	88015.54
VS10607	GMC	80044	VTE3900	VAPORMATIC	88004.24	W62-751	WORLDPARTS	86069	WAR18-15	ROCKWELL	88015.55
VS10798	GMC	80055	VTE3901	VAPORMATIC	88004.25	W62-751	WORLDPARTS	86143	WAR21-4	ROCKWELL	88015.56
VS11996	GMC	80221	VTE3902	VAPORMATIC	88004.26	W62-752	WORLDPARTS	86070	WAR23-4	ROCKWELL	88015.57
VTA3127	CASE,J.I.	80001	VTE3903	VAPORMATIC	88004.27	W62-752	WORLDPARTS	86144	WARS24-5	ROCKWELL	88015.58
VTA3227	CASE,J.I.	80001	VTE3904	VAPORMATIC	88004.28	W62-753	WORLDPARTS	86071	WASR25-4	ROCKWELL	88015.59
VTE2000	VAPORMATIC	88003.59	VTE5000	VAPORMATIC	88004.29	W62-753	WORLDPARTS	86145	WFR2885	AUST.WEST.	80168
VTE2001	VAPORMATIC	88003.60	VTE5001	VAPORMATIC	88004.30	W62-754	WORLDPARTS	86143	WM8-1538-3	SMITH,T.L.	80100
VTE2002	VAPORMATIC	88003.61	VTE5002	VAPORMATIC	88004.31	W62-756	WORLDPARTS	86144	WM8-1538-4	SMITH,T.L.	80101
VTE2003	VAPORMATIC	88003.62	VTE5003	VAPORMATIC	80024	W62-756	WORLDPARTS	86145	WRF1064-57	AUST.WEST.	80195
VTE2004	VAPORMATIC	88003.63	VTE5004	VAPORMATIC	80041	W62-764	WORLDPARTS	86070	WRF1604-57	BAL-LM-HAM	80195
VTE2005	VAPORMATIC	88003.64	VTE5005	VAPORMATIC	80124	W62-766	WORLDPARTS	86069	WRF1679	AUST.WEST.	80213
VTE2010	VAPORMATIC	88003.65	VTE5006	VAPORMATIC	88004.32	W62-769	WORLDPARTS	86072	WRF1679-56	AUST.WEST.	80213
VTE2011	VAPORMATIC	88003.66	VTE5007	VAPORMATIC	88004.33	W62-771	WORLDPARTS	86122	WRF1679-56	BAL-LM-HAM	80215
VTE2012	VAPORMATIC	88003.67	VTE5008	VAPORMATIC	88004.34	W62-771	WORLDPARTS	86180	WRF2882	AUST.WEST.	80214
VTE2013	VAPORMATIC	88003.68	VTE5009	VAPORMATIC	88004.35	W62-772	WORLDPARTS	86122	WRF2882	BAL-LM-HAM	80218
VTE2014	VAPORMATIC	88003.69	VTE5010	VAPORMATIC	88004.36	W62-772	WORLDPARTS	86181	WRF2885	AUST.WEST.	80168
VTE2015	VAPORMATIC	88003.70	W279	SERVIS	80069	W62-773	WORLDPARTS	86160	WRF2885	AUST.WEST.	80222
VTE2016	VAPORMATIC	88003.71	W279R	SERVIS	80069	W62-774	WORLDPARTS	86135	WRF2885E	BAL-LM-HAM	80216
VTE2017	VAPORMATIC	88003.72	W292A29	KENWORTH	80188	W62-774	WORLDPARTS	86187	X11-39	WHITE	80220
VTE3000	VAPORMATIC	88003.73	W298	SERVIS	80037	W62-775	WORLDPARTS	86096	X11999	SEAGRAVE	80186
VTE3001	VAPORMATIC	88003.74	W298A	SERVIS	80037	W62-775	WORLDPARTS	86162	X12861	SEAGRAVE	80178
VTE3002	VAPORMATIC	88003.75	W42SR	SERVIS	80133	W62-776	WORLDPARTS	86135	X1345	CLEVE.STL.	88015.44
VTE3003	VAPORMATIC	88003.76	W49-100	WORLDPARTS	80006	W62-776	WORLDPARTS	86187	X13754	SEAGRAVE	80008
VTE3004	VAPORMATIC	88003.77	W49-101	WORLDPARTS	80009	W62-777	WORLDPARTS	86110	X1391	CLEVE.STL.	88014.61
VTE3005	VAPORMATIC	88003.78	W49-102	WORLDPARTS	80115	W62-777	WORLDPARTS	86172	X1394	CLEVE.STL.	88014.57
VTE3006	VAPORMATIC	88003.79	W49-103	WORLDPARTS	80024	W62-778	WORLDPARTS	86109	X1405	CLEVE.STL.	88013.32
VTE3007	VAPORMATIC	88003.80	W49-104	WORLDPARTS	80010	W62-778	WORLDPARTS	86171	X1510X	HABERLE	80142
VTE3008	VAPORMATIC	88003.81	W49-105	WORLDPARTS	80004	W62-779	WORLDPARTS	86108	X1510X	HABERLE	80143
VTE3009	VAPORMATIC	88003.82	W49-107	WORLDPARTS	80019	W62-779	WORLDPARTS	86170	X15555	PILOT	80007
VTE3010	VAPORMATIC	88003.83	W49-113	WORLDPARTS	80024	W62-780	WORLDPARTS	86126	X200	CASE,J.I.	80008
VTE3020	VAPORMATIC	88003.84	W49-115	WORLDPARTS	80041	W62-780	WORLDPARTS	86183	X206262	MASS.FERG.	80234
VTE3021	VAPORMATIC	88003.85	W49-116	WORLDPARTS	80024	W62-781	WORLDPARTS	86077	X206262	TWIN DISC	80234
VTE3022	VAPORMATIC	88003.86	W49-117	WORLDPARTS	80024	W62-781	WORLDPARTS	86150	X206340	TWIN DISC	80224
VTE3023	VAPORMATIC	88003.87	W49-118	WORLDPARTS	80142	W62-782	WORLDPARTS	86098	X206417	TWIN DISC	88001.93
VTE3025	VAPORMATIC	88003.88	W49-120	WORLDPARTS	80041	W62-782	WORLDPARTS	86163	X210-570	ATHEY PROD	80100
VTE3026	VAPORMATIC	88003.89	W49-121	WORLDPARTS	80041	W62-783	WORLDPARTS	86138	X213	ULRICH	80069
VTE3100	VAPORMATIC	88003.90	W49-129	WORLDPARTS	80012	W62-783	WORLDPARTS	86190	X214	ULRICH	80104
VTE3101	VAPORMATIC	88003.91	W49-130	WORLDPARTS	80024	W62-785	WORLDPARTS	86101	X23	CLEVE.STL.	88014.56
VTE3102	VAPORMATIC	88003.92	W49-131	WORLDPARTS	80013	W62-785	WORLDPARTS	86164	X23192	ALLIS-CHLM	80153
VTE3103	VAPORMATIC	88003.93	W49-132	WORLDPARTS	80020	W62-786	WORLDPARTS	86102	X23192	FIAT-ALLIS	80150
VTE3104	VAPORMATIC	88003.94	W620744	WORLDPARTS	86166	W62-788	WORLDPARTS	86107	X34000	ALLIS-CHLM	80102
VTE3105	VAPORMATIC	88003.95	W62-701	WORLDPARTS	86180	W62-788	WORLDPARTS	86165	X34000	FIAT-ALLIS	80102
VTE3106	VAPORMATIC	88003.96	W62-702	WORLDPARTS	86160	W62-788	WORLDPARTS	86170	X47259	GMC	80221
VTE3107	VAPORMATIC	88003.97	W62-703	WORLDPARTS	86187	W62-790	WORLDPARTS	86111	X55-55	AEC	80007
VTE3108	VAPORMATIC	88003.98	W62-704	WORLDPARTS	86161	W62-790	WORLDPARTS	86169	X55-55	BORG-WARNR	80007
VTE3109	VAPORMATIC	88003.99	W62-705	WORLDPARTS	86187	W62-801	WORLDPARTS	86090	X55-55	CLEVE.STL.	80007
VTE3110	VAPORMATIC	88004	W62-707	WORLDPARTS	86172	W62-802	WORLDPARTS	86090	X55-55-1	CLEVE.STL.	80007
VTE3111	VAPORMATIC	88004.01	W62-709	WORLDPARTS	86183	W62-803	WORLDPARTS	86088	X55-55-675	AUST.WEST.	80091
VTE3112	VAPORMATIC	88004.02	W62-710	WORLDPARTS	86150	W62-804	WORLDPARTS	86128	X629-29	SCHIELD	80216
VTE3113	VAPORMATIC	88004.03	W62-711	WORLDPARTS	86163	W62-804	WORLDPARTS	86129	X629-30	SCHIELD	80193
VTE3114	VAPORMATIC	88004.04	W62-712	WORLDPARTS	86190	W62-804	WORLDPARTS	86130	X7080-428	CHRYSLER	80061
VTE3115	VAPORMATIC	88004.05	W62-713	WORLDPARTS	86164	W62-804	WORLDPARTS	86131	X850-203	KOEHRING	80193
VTE3116	VAPORMATIC	88004.06	W62-714	WORLDPARTS	86165	W62-804	WORLDPARTS	86184	X8569	JAEGER	80178
VTE3117	VAPORMATIC	88004.07	W62-715	WORLDPARTS	86169	W62-805	WORLDPARTS	86084	X9357	KEWANEE	80001
VTE3118	VAPORMATIC	88004.08	W62-716	WORLDPARTS	86145	W62-850	WORLDPARTS	86193	XA4635AA	FORD	80060
VTE3119	VAPORMATIC	88004.09	W62-717	WORLDPARTS	86143	W62-851	WORLDPARTS	86166	XAC5351E	DEERE,JOHN	80104
VTE3200	VAPORMATIC	88004.10	W62-718	WORLDPARTS	86144	W62-898	WORLDPARTS	86144	XAF205A	COCKSHUTT	80037
VTE3201	VAPORMATIC	88004.11	W62-733	WORLDPARTS	86193	W62-898	WORLDPARTS	86145	XAG2230E	DEERE,JOHN	80044
VTE3202	VAPORMATIC	88004.12	W62-734	WORLDPARTS	86193	W62-898	WORLDPARTS	86166	XAG2513E	DEERE,JOHN	80044
VTE3203	VAPORMATIC	88004.13	W62-735	WORLDPARTS	86166	W62-898	WORLDPARTS	86193	XAG2530E	DEERE,JOHN	80044
VTE3204	VAPORMATIC	88004.14	W62-736	WORLDPARTS	86145	W840	SERVIS	80136	XAG3796A	DEERE,JOHN	80001
VTE3205	VAPORMATIC	88004.15	W62-737	WORLDPARTS	86166	WAFR17-7	ROCKWELL	88015.46	XAG63796A	DEERE,JOHN	80001
VTE3206	VAPORMATIC	88004.16	W62-738	WORLDPARTS	86166	WAFS18-2	ROCKWELL	88015.47	XAG63796H	DEERE,JOHN	80001
VTE3207	VAPORMATIC	88004.17	W62-739	WORLDPARTS	86166	WAFS20-6	ROCKWELL	88015.48	XAG6379GH	DEERE,JOHN	80001
VTE3208	VAPORMATIC	88004.18	W62-740	WORLDPARTS	86145	WAFS22-1	ROCKWELL	88015.49	XAG6379GH	DEERE,JOHN	80003
VTE3209	VAPORMATIC	88004.19	W62-741	WORLDPARTS	86145	WAFS24-3	ROCKWELL	88015.50	XAL5101N	DEERE,JOHN	80049
VTE3210	VAPORMATIC	88004.20	W62-742	WORLDPARTS	86193	WAFS24-4	ROCKWELL	88015.51	XAW70210A	DEERE,JOHN	80099
VTE3211	VAPORMATIC	88004.21	W62-743	WORLDPARTS	86166	WAFS26-4	ROCKWELL	88015.52	XAW70220A	DEERE,JOHN	80238

Part No.	Brand	IDLI Group	Part No.	Brand	IDLI Group	Part No.	Brand	IDLI Group	Part No.	Brand	IDLI Group
XAW70226A	DEERE,JOHN	80237	YBS140	WESCO	88000.89	ZAD11581	FLEETRITE	80023	ZAD302	FLEETRITE	80010
XAW70227H	DEERE,JOHN	80239	YBX120	WESCO	88000.88	ZAD11586	FLEETRITE	80124	ZAD303	FLEETRITE	80004
XAW7022A	DEERE,JOHN	80243	YC1232	WESCO	88000.68	ZAD11588	FLEETRITE	80140	ZAD304	FLEETRITE	80019
XAW8055A	DEERE,JOHN	80210	YC1342	WESCO	88000.70	ZAD12140	FLEETRITE	80055	ZAD305	FLEETRITE	80018
XAW9153A	DEERE,JOHN	80099	YC1442	WESCO	88000.96	ZAD12141	FLEETRITE	80091	ZAD306	FLEETRITE	80009
XAW9525A	DEERE,JOHN	80231	YC1542	WESCO	88000.72	ZAD12173	FLEETRITE	80058	ZAD307	FLEETRITE	80006
XB4635A	FORD	80077	YC1642	WESCO	88000.84	ZAD12505	FLEETRITE	80059	ZAD308	FLEETRITE	80017
XB4635B	FORD	80091	Z1020	COCKSHUTT	80037	ZAD12518	FLEETRITE	80066	ZAD310	FLEETRITE	80012
XB4635C	FORD	80060	Z10424	LEADER	80024	ZAD14002	FLEETRITE	80109	ZAD311	FLEETRITE	88002.34
XB4635DA	FORD	88001.71	Z11C11-18397	FORD	80072	ZAD14014	FLEETRITE	80158	ZAD313	FLEETRITE	88002.36
XD500818	FORD	80178	Z1C18397	FORD	80072	ZAD14015	FLEETRITE	80148	ZAD314	FLEETRITE	88002.33
XL512825	FORD	80077	Z1C18397B	FORD	80072	ZAD14016	FLEETRITE	80159	ZAD316	FLEETRITE	88002.33
XM767	EIMCO	80193	Z2595	KALAMAZOO	80069	ZAD14138	FLEETRITE	80109	ZAD50Y103	FLEETRITE	88000.36
XM853	EIMCO	80195	Z2598	KALAMAZOO	80186	ZAD14140	FLEETRITE	80155	ZAD50Y113	FLEETRITE	88000.35
XW4635A	FORD	80060	Z2782	KALAMAZOO	80193	ZAD14143	FLEETRITE	80156	ZAD50Y115	FLEETRITE	88000.55
XW4635B	FORD	88001.71	Z2838	KALAMAZOO	80069	ZAD14151	FLEETRITE	80109	ZAD50Y12	FLEETRITE	88000.45
XW4635C	FORD	80077	Z862	COCKSHUTT	80003	ZAD14300	FLEETRITE	80099	ZAD50Y123	FLEETRITE	88000.38
XW4635D	FORD	80091	ZAD11003	FLEETRITE	80001	ZAD14300	FLEETRITE	80101	ZAD50Y124	FLEETRITE	88000.37
XX1804-6	CATERPILLR	80048	ZAD11042	FLEETRITE	88002.42	ZAD14308	FLEETRITE	80104	ZAD50Y13	FLEETRITE	88000.20
XX1894-6	CATERPILLR	80048	ZAD11402	FLEETRITE	80048	ZAD14311	FLEETRITE	80107	ZAD50Y22	FLEETRITE	88000.43
XX1894-7	CATERPILLR	80216	ZAD11416	FLEETRITE	80047	ZAD14320	FLEETRITE	80113	ZAD50Y23	FLEETRITE	88000.44
XX7146	CATERPILLR	80240	ZAD11417	FLEETRITE	80046	ZAD14388	FLEETRITE	80105	ZAD50Y43	FLEETRITE	88000.22
XY501497	FORD	80077	ZAD11432	FLEETRITE	80092	ZAD15000	FLEETRITE	80168	ZAD50Y52	FLEETRITE	88000.46
Y00601312	FODENS	80189	ZAD11445	FLEETRITE	80049	ZAD15006	FLEETRITE	80163	ZAD50Y53	FLEETRITE	88000.52
Y00601508	FODENS	80221	ZAD11446	FLEETRITE	80044	ZAD15008	FLEETRITE	80174	ZAD50Y63	FLEETRITE	88000.25
Y00601802	FODENS	80225	ZAD11449	FLEETRITE	80054	ZAD15015	FLEETRITE	80173	ZAD50Y73	FLEETRITE	88000.28
Y1001	BORG-WARNR	88002.87	ZAD11453	FLEETRITE	80030	ZAD15122	FLEETRITE	80168	ZAD50Y83	FLEETRITE	88000.53
Y123Y123	CHELSEA	88000.21	ZAD11463	FLEETRITE	80044	ZAD15127	FLEETRITE	80163	ZAD50Y93	FLEETRITE	88000.30
Y123Y123	CHELSEA	88000.39	ZAD11466	FLEETRITE	80054	ZAD15127	FLEETRITE	80167	ZAD50Y95	FLEETRITE	88000.34
Y124Y124	CHELSEA	81020	ZAD11467	FLEETRITE	80031	ZAD15132	FLEETRITE	80174	ZAD51Y103	FLEETRITE	88000.94
Y134Y134	CHELSEA	88000.22	ZAD11472	FLEETRITE	80030	ZAD15173	FLEETRITE	80171	ZAD51Y12	FLEETRITE	88000.86
Y134Y144	CHELSEA	88000.22	ZAD11501	FLEETRITE	80023	ZAD15177	FLEETRITE	80172	ZAD51Y13	FLEETRITE	88000.79
Y134Y144	CHELSEA	88000.25	ZAD11504	FLEETRITE	80041	ZAD16000	FLEETRITE	80193	ZAD51Y22	FLEETRITE	88000.89
Y134Y144	CHELSEA	88000.27	ZAD11511	FLEETRITE	80024	ZAD16102	FLEETRITE	80197	ZAD51Y23	FLEETRITE	88000.80
Y134Y154	CHELSEA	88000.22	ZAD11514	FLEETRITE	80024	ZAD16103	FLEETRITE	80194	ZAD51Y32	FLEETRITE	88000.95
Y134Y154	CHELSEA	88000.28	ZAD11515	FLEETRITE	80178	ZAD16106	FLEETRITE	80198	ZAD51Y33	FLEETRITE	88000.81
Y134Y164	CHELSEA	88000.22	ZAD11516	FLEETRITE	80189	ZAD16108	FLEETRITE	80047	ZAD51Y52	FLEETRITE	88000.90
Y134Y164	CHELSEA	88000.31	ZAD11517	FLEETRITE	80221	ZAD16108	FLEETRITE	80199	ZAD51Y53	FLEETRITE	88000.83
Y144Y144	CHELSEA	88000.25	ZAD11520	FLEETRITE	80072	ZAD16109	FLEETRITE	80201	ZAD51Y83	FLEETRITE	88000.82
Y144Y144	CHELSEA	88000.27	ZAD11520	FLEETRITE	80075	ZAD16128	FLEETRITE	80198	ZAD51Y93	FLEETRITE	88000.84
Y154Y154	CHELSEA	88000.28	ZAD115200	FLEETRITE	80075	ZAD16135	FLEETRITE	80191	ZAD52Y13	FLEETRITE	88000.68
Y163Y163	CHELSEA	88000.53	ZAD115200	FLEETRITE	80232	ZAD16144	FLEETRITE	80197	ZAD52Y43	FLEETRITE	88000.70
Y164Y164	CHELSEA	88000.31	ZAD11520UB	FLEETRITE	80076	ZAD17000	FLEETRITE	80210	ZAD52Y63	FLEETRITE	88000.71
Y185Y185	CHELSEA	88000.35	ZAD11520UB	FLEETRITE	80089	ZAD17101	FLEETRITE	80212	ZAD55N	FLEETRITE	80161
Y185Y205	CHELSEA	88000.35	ZAD11521	FLEETRITE	80075	ZAD17102	FLEETRITE	80212	ZAD5P1	FLEETRITE	88000.39
Y185Y205	CHELSEA	88000.38	ZAD11521HD	FLEETRITE	80077	ZAD17105	FLEETRITE	80213	ZAD5P10	FLEETRITE	88000.22
Y205Y205	CHELSEA	88000.38	ZAD11521UB	FLEETRITE	80075	ZAD17126	FLEETRITE	80213	ZAD5P11	FLEETRITE	88000.27
Y2724-4	SMITH,T.L.	80193	ZAD11524	FLEETRITE	80225	ZAD17206	FLEETRITE	80211	ZAD5P12	FLEETRITE	88000.22
Y4841	WESCO	88002.86	ZAD11530	FLEETRITE	80117	ZAD17207	FLEETRITE	80212	ZAD5P13	FLEETRITE	88000.31
Y4851	WESCO	88002.87	ZAD11531	FLEETRITE	80121	ZAD17208	FLEETRITE	80208	ZAD5P17	FLEETRITE	88000.27
YA123	WESCO	88000.20	ZAD11534	FLEETRITE	80118	ZAD17228	FLEETRITE	80065	ZAD5P2	FLEETRITE	88000.21
YA133	WESCO	88000.26	ZAD11548	FLEETRITE	80025	ZAD17229	FLEETRITE	80061	ZAD5P22	FLEETRITE	88000.28
YA134	WESCO	88000.22	ZAD11549	FLEETRITE	80075	ZAD17230	FLEETRITE	80061	ZAD5P26	FLEETRITE	88000.53
YA143	WESCO	88000.52	ZAD11550	FLEETRITE	80097	ZAD17231	FLEETRITE	80132	ZAD5P27	FLEETRITE	88000.31
YA144	WESCO	88000.27	ZAD11552	FLEETRITE	80118	ZAD17232	FLEETRITE	80132	ZAD5P30	FLEETRITE	88000.35
YA154	WESCO	88000.28	ZAD11553	FLEETRITE	80077	ZAD17321	FLEETRITE	80034	ZAD5P31	FLEETRITE	88000.38
YA163	WESCO	88000.53	ZAD11553UB	FLEETRITE	80088	ZAD17328	FLEETRITE	80042	ZAD5P32	FLEETRITE	88000.38
YA164	WESCO	88000.31	ZAD11555	FLEETRITE	80182	ZAD17330	FLEETRITE	80135	ZAD5P4	FLEETRITE	88000.25
YA184	WESCO	88000.36	ZAD11560	FLEETRITE	80142	ZAD18101	FLEETRITE	80232	ZAD5P45	FLEETRITE	88000.70
YA185	WESCO	88000.35	ZAD11560UB	FLEETRITE	80134	ZAD18102	FLEETRITE	80237	ZAD5P49	FLEETRITE	88000.71
YA204	WESCO	88000.55	ZAD11560UB	FLEETRITE	80142	ZAD18105	FLEETRITE	80238	ZAD5P52	FLEETRITE	88000.72
YA205	WESCO	88000.38	ZAD11565	FLEETRITE	80190	ZAD18113	FLEETRITE	80238	ZAD5P72	FLEETRITE	88000.81
YAS120	WESCO	88000.43	ZAD11570	FLEETRITE	80007	ZAD18205	FLEETRITE	80231	ZAD5P76	FLEETRITE	88000.31
YAS140	WESCO	88000.45	ZAD11572	FLEETRITE	80007	ZAD18206	FLEETRITE	80232	ZAD6700AR	FLEETRITE	80906
YB123	WESCO	88000.79	ZAD11574	FLEETRITE	80090	ZAD18207	FLEETRITE	80237	ZAD6700R	FLEETRITE	80953
YB134	WESCO	88000.80	ZAD11577	FLEETRITE	80221	ZAD19015	FLEETRITE	80234	ZAD710	FLEETRITE	80032
YB144	WESCO	88000.81	ZAD11578	FLEETRITE	80124	ZAD19016	FLEETRITE	80234	ZAD710G	FLEETRITE	80032
YB154	WESCO	88000.92	ZAD11578UB	FLEETRITE	80124	ZAD19017	FLEETRITE	80233	ZAD715	FLEETRITE	80030
YB164	WESCO	88000.83	ZAD11578UB	FLEETRITE	80125	ZAD19100	FLEETRITE	80235	ZAD715	FLEETRITE	88002.77
YBH180	WESCO	88000.95	ZAD1158	FLEETRITE	80140	ZAD301	BORG-WARNR	80024	ZAD715G	FLEETRITE	80031

Part No.	Brand	IDLI Group
ZAD715G	FLEETRITE	88002.77
ZAD81FR	FLEETRITE	80008
ZAD83DR	FLEETRITE	88002.89
ZAD-B1FR	FLEETRITE	80008
ZAD-B35N3	FLEETRITE	80069
ZAD-B35N4	FLEETRITE	80070
ZAD-B3DR	FLEETRITE	80133
ZADB40N	FLEETRITE	80134
ZADB58N	FLEETRITE	80044
ZADB58N	FLEETRITE	80179
ZAD-B58WB	FLEETRITE	80185
ZADB58WB1	FLEETRITE	80186
ZADB58WB2	FLEETRITE	80184
ZADB60N	FLEETRITE	80188
ZADB70N	FLEETRITE	80220
ZAD-BL12N	FLEETRITE	80035
ZAD-BL14N	FLEETRITE	80037
ZAD-BL6N	FLEETRITE	80001
ZADBLN12N	FLEETRITE	80035
ZADBLN14N	FLEETRITE	80037
ZADBLN6N	FLEETRITE	80001
ZADBLN6N	FLEETRITE	80003
ZADD15655	FLEETRITE	80145
ZAD-O15555	FLEETRITE	80141
ZAD-O15655	FLEETRITE	80145
ZAD-P15555	FLEETRITE	80129
ZAD-P19675	FLEETRITE	80131
ZAD-R15555	FLEETRITE	80123
ZAD-S10675	FLEETRITE	80091
ZADS15355	FLEETRITE	80038
ZADS15355	FLEETRITE	80039
ZAD-S15513	FLEETRITE	80130
ZAD-S15555	FLEETRITE	80081
ZAD-S19675	FLEETRITE	80091
ZADT15555	FLEETRITE	80071
ZAD-TS15555	FLEETRITE	80074
ZAD-U15655	FLEETRITE	80183

Other Valuable Interchange Publications

The International Bearing Interchange Guide

New Ninth Edition

The I.B.I. Guide has over 365,000 listings of 25,326 ball and roller bearings available worldwide and dating back to 1918. It includes manufacturers, original equipment users and government part numbers, with THOUSANDS of new listings not included in earlier editions!

I.S.B.N. #916966-17-8

The International Seal Interchange Guide

New Sixth Edition

The I.S.I. Guide has over 150,000 listings of 12,632 different seals and "O" rings. It includes government numbers as well as producers and users throughout the world—over 700 páges of information to save you time and money!

I.S.B.N. #916966-16-X

The International Drive Belt Interchange Guide

New Fourth Edition

The I.D.B.I. Guide has more than 115,000 listings for nearly 5,863 types and sizes of drive and V-belts. Listings include manufacturers, original equipment and government users from all over the world. 760 pages.

I.S.B.N. #916966-15-1

LETTERS FROM USERS

One of the satisfactions of our business is helping people solve problems by using our Guides. Many of them write to tell us of the time and money they've saved, to order additional copies and to offer to send us published data for future editions. After all, it's in their own self-interest. We'd like to hear from you, too. Just write to: Interchange, Incorporated, P. O. Box 16012, St. Louis Park, MN, U.S.A. Following are excerpts from some letters we've received:

CALIFORNIA VEHICLE PARTS COMPANY

"Your interchange books have been the most helpful cross-reference catalogs in our library consisting of well over 2,000 individual books plus the unknown numbers in our micro fiche system. Thanks for such a helpful tool." (President)

NEW ZEALAND ENGINEERING FIRM

"We have found your I.B.I. book on bearings to be worth its weight in gold! I'm interested in obtaining a copy of your I.S.I. guide to seals. Finally, if you like, I can send you copies of our lists of bearing part numbers and equivalents for future editions." (Purchasing Mgr.)

MINNESOTA PHOTOGRAPHIC PROCESSOR

"A recent incident demonstrates the value of your guides. Local parts distributors were out of stock on an item we urgently needed. Using your guide, we located alternative bearings to complete machines with a dollar value of $500,000. It took only three days compared to a wait of two weeks from the factory." (Purchasing Agent)

SWEDISH BEARING DISTRIBUTOR

We now use both your I.B.I and I.S.I Interchange Guides and find them of incalculable value. We have also introduced them to our business contacts and understand that many have ordered." (Purchasing)

NEW YORK GAS & ELECTRIC COMPANY

"After ordering a set of your bearing and seals interchange books, and reviewing them, I decided to order sets for all five of our power plants." (Buyer)

LONDON BEARING MANUFACTURER

"Many of our distributors and end-users throughout the world make use of your I.B.I. Guide.. We would like to cooperate fully with your company in having our complete product line listed in your next edition." (Vice President)

OIL PRODUCER IN DUBAI

"I have used your I.B.I. and I.S.I. manuals for over a year and I am very impressed. They have saved much valuable time and a considerable amount of money." (Engineer)

BOMBAY ENGINEERING, EXPORTING CO.

"We are interested in receiving latest editions of your I.B.I. and I.S.I. Guides by air. We must say that we find your guides to be immensely useful and, truly, we cannot afford to be without them in our business." (Partner)

ILLINOIS FARM IMPLEMENT DEALER

"We were surprised, recently, to see a competitive dealer stop in to use our I.B.I. Guide to find a bearing substitute for a customer." (Owner)

IDLI GROUP 80001

CROSS + 4 GROOVED BEARINGS
+ INSIDE SNAP RINGS
SEE FIGURE 1B

.9688 in	BD	24.61 mm
.5625 in	BW	14.29 mm
1.8750 in	CL	47.63 mm
.5 0 in	CD	12.70 mm

Manufacturer	Part No.	Manufacturer	Part No.	Manufacturer	Part No.	Manufacturer	Part No.
ADAMS-LET.	SR6109	CASE,J.I.	T36575	FIAT-ALLIS	053120	GMC	9024893
ADAMS-LET.	TG4614	CASE,J.I.	T37173	FIAT-ALLIS	0653811	GMC	9024992
AEC	6N8	CASE,J.I.	VTA3127	FIAT-ALLIS	066678	GMC	9038448
AEC	AE1004	CASE,J.I.	VTA3227	FIAT-ALLIS	081058	GMC	9052499
AEC	AE6N8	CATERPILLR	15678	FIAT-ALLIS	224890	GMC	9053828
AEC	AE-CP6N	CATERPILLR	1S678	FIAT-ALLIS	653811	HESSTON	810770
AEC	AECPL6N	CATERPILLR	2H8257	FIAT-ALLIS	666878	HUB CITY	03-35-90071
AEC	AE-CPL6N8	CATERPILLR	2H861	FIAT-ALLIS	70053120	HUB CITY	UR6N
AEC	CPL6N8	CATERPILLR	329871	FIAT-ALLIS	70053120-2	HUSCO	B600N
ALCO	1777	CATERPILLR	3H6263	FIAT-ALLIS	70081058	HUSCO	CPL6S8
ALCO	CPL6N	CATERPILLR	587579	FIAT-ALLIS	70081058-0	HYSTER	76761
ALCO	CPL6N8	CATERPILLR	5S7579	FIAT-ALLIS	70224890	IHC	139764
ALCO	J1003	CLARK EQU.	1550682	FIAT-ALLIS	70224890-4	IHC	139764H1
ALCO	J1004	CLARK EQU.	602-900-112	FIAT-ALLIS	70653811	IHC	147511H1
ALCO	J1007	CLARK EQU.	894655	FIAT-ALLIS	70666878	IHC	147512H1
ALCO	J1009	CLARK EQU.	896084	FIAT-ALLIS	7842779	IHC	150635R91
ALLIS-CHLM	053120	CLARK EQU.	896085	FLEETRITE	ZAD11003	IHC	150-635R92
ALLIS-CHLM	081058	CLARK EQU.	992624	FLEETRITE	ZAD-BL6N	IHC	150965R91
ALLIS-CHLM	224890	CLEVE.MOT.	CM1003	FLEETRITE	ZADBLN6N	IHC	202-347C91
ALLIS-CHLM	534235	CLEVE.MOT.	CM5-92X	FMC	1270961	IHC	202-348C91
ALLIS-CHLM	7842779	CLEVE.MOT.	CM-CPL6N	FMC	1714205	IHC	202-947C91
ALLIS-CHLM	810580	CLEVE.MOT.	CM-CPL6N8	FMC	1714207	IHC	210786R91
ALLOY	1777	CMP	CM-CPL6N8	FMC	2341009-5	IHC	337-439R91
ALLOY	1778	COCKSHUTT	FFH788	FORD	107-854	IHC	350-254R91
ALLOY	CPL659	COCKSHUTT	FHH788	FORD	171333	IHC	357-439R91
ALLOY	CPL6N	COCKSHUTT	FHY169	FORD	175-423	IHC	357439R92
AMER.HOIST	E540131	COLES	1550682	FORD	291-971	IHC	357-606R91
AMER.MOTOR	990192	COLES	1760487	FORD	309-575	IHC	357-607R91
BEAN,JOHN	1270961	DAFFIN	1R258	FORD	C6TZ3815A	IHC	363-308R91
BEAN,JOHN	1714207	DAFFIN	1R368	FORD	CBN3550A	IHC	366-312R91
BEAN,JOHN	2308820	DAFFIN	1R369	FORD	CBN3550B	IHC	366-313R91
BEAN,JOHN	234-1009-5	DANUSER	2934	FORD	CBPN3550A	IHC	368-730R91
BEAN,JOHN	UR20	DANUSER	8296	FORD	CBPN3550B	IHC	476-674R91
BEGE MFG.	V1082	DEERE,JOHN	A17330H	FOX	TS4CS2	IHC	511-644R91
BORG-WARNR	114-1002	DEERE,JOHN	AE14978	FOX RIVER	TS4CX2	IHC	627211C1
BORG-WARNR	114-1003	DEERE,JOHN	AE14978E	FWD	211253	IHC	661372R91
BORG-WARNR	114-1004	DEERE,JOHN	AE26238	FWD	635873	IHC	661-677R91
BORG-WARNR	114-1005	DEERE,JOHN	AE679E	G & G MFG	302-0600	IHC	663-120R91
BORG-WARNR	114-1007	DEERE,JOHN	AE7939E	G & G MFG	6R	IHC	983-268R91
BORG-WARNR	114-1009	DEERE,JOHN	AE793E	GALION	D101341	INBRA	CT11020
BORG-WARNR	114-145	DEERE,JOHN	AH18327	GALION	D51104	INBRA	CT11-035
BORG-WARNR	114-2025	DEERE,JOHN	AH79074	GALION	D55145	INBRA	CT13-070
BORG-WARNR	CP6N	DEERE,JOHN	AH82541	GALION	D91128	INBRA	CT24-040
BORG-WARNR	CPL6N	DEERE,JOHN	AH83541	GALION	H69215(7)	KEWANEE	X9357
BORG-WARNR	CPL6N8	DEERE,JOHN	AJ7330H	GALION	H81402	KOEHRING	P33526
BOST.GEAR	UJSC	DEERE,JOHN	AL1564N	GAR WOOD	9000445	KOEHRING	P33536
BROCKWAY	345827	DEERE,JOHN	AM1793E	GAR WOOD	M307712	LEVRATTO	4-10010
BROCKWAY	345884	DEERE,JOHN	AM793E	GEN.EQUIP.	15-38287	LONG MFG.	732324
CARDWELL	B62	DEERE,JOHN	AR21101	GERLINGER	21222	LONG MFG.	737100
CASE,J.I.	A20435	DEERE,JOHN	AR21101R	GILSON	4387	MACK	206SL12
CASE,J.I.	A20684	DEERE,JOHN	AW13174	GLAENZER	065-152-8	MACK	206SL21
CASE,J.I.	A30408	DEERE,JOHN	ND1915D	GLAENZER	065-271-8	MACK	739XX1
CASE,J.I.	A30425	DEERE,JOHN	XAG3796A	GMB	G1003	MASS.FERG.	315-726M91
CASE,J.I.	A30428	DEERE,JOHN	XAG63796A	GMB	G1004	MASS.FERG.	316-726M91
CASE,J.I.	A30435	DEERE,JOHN	XAG63796H	GMB	G1007	MASS.FERG.	784-489M1
CASE,J.I.	A40435	DEERE,JOHN	XAG6379GH	GMB	G1009	MASS.FERG.	784489M91
CASE,J.I.	B18132	EATON	6436100	GMB	G5-1004	MASS.FERG.	836514M91
CASE,J.I.	F50537	EATON	6436101	GMB	G5-92X	MASS.FERG.	836-891M91
		EIMCO	912A45084	GMB	GCPL6N	MASS.FERG.	840824M92
		ETNYRE	3N-SH15	GMB	GGPL6R	MASS.FERG.	841-193M91
		FIAT-ALLIS	0053120	GMB	GGPL9R	MASS.FERG.	841-193M92
		FIAT-ALLIS	0053120-2	GMC	2011875	MCQUAY-NOR	U03-1
		FIAT-ALLIS	0081058	GMC	3717179	MOTOR MAST	1587A
		FIAT-ALLIS	0081058-0	GMC	9020304	MPL.MOLINE	10P715
		FIAT-ALLIS	0224890	GMC	9023360	MPL.MOLINE	14P623
		FIAT-ALLIS	0224890-4	GMC	9024892	MPL.MOLINE	15P1008

IDLI GUIDE   COPYRIGHT © INTERCHANGE, INC.   ST. LOUIS PARK, MN. 55416 USA

CAUTION: BE SURE TO REFER TO ENGINEERING CATALOGS FOR SPECIAL APPLICATIONS THAT REQUIRE SPECIFIC MATERIAL CONTENT, TOLERANCES, ETC. SEE FOOTNOTE.

MPL.MOLINE	15R108
MPL.MOLINE	20P1380
NATL.MINE	3881-2
NFAPCO	1-1475
NEAPCO	1475X
NEAPCO	28L6N8
NEW HOLL.	218353
NEW HOLL.	222361
NEW HOLL.	27993
NEW HOLL.	48263
NEW HOLL.	52495
NEW HOLL.	74123
NEW HOLL.	82495
NEW IDEA	0777HSA
NEW IDEA	077HSA
NEW IDEA	K2550A
OLIVER	198883AS
OLIVER	502-579E
OLIVER	509206E
OLIVER	509-260E
OLIVER	KAR758
OSHKOSH	22UJ12
OSHKOSH	22UJ63
PAPEC MACH	32-1154
PERFECT-CI	92
PILOT	11003
PILOT	11004
PILOT	11007
PILOT	11009
PILOT	BL6N
PILOT	BL6N8
PRECISION	861
PREC.TOR.	862
PREC.TOR.	863
REO MOTORS	1562K1
REPCO	114-1003
REPCO	114-1004
REPCO	CPL6N11
REPCO	CPL6N8
REPUBLIC	CB781
REPUBLIC	CB782
REPUBLIC	CB-L69-8
REPUBLIC	CB-L6N11
REPUBLIC	CB-L6N8
ROSS GEAR	40309A1
ROSS GEAR	403423A1
ROSS GEAR	403772A1
ROSS GEAR	506140
SPEED-SPR.	1714207
SPICER	5-92X
THEW SHOV.	P33526
TOWMOTOR	C21222
TRU-CROSS	92
TRW	20001
TRW	20001D
UJS	55-6LN
UNIT RIG	900731
UNIT RIG	988175
WABCO	427271
WABCO	5R6109
WABCO	SR6109

WABCO	TG4614
WABCO	VA1775
WAIN-ROY	BR102-57
WESCO	201-008
WESCO	241006
WESCO	6N
WESCO	6R
WESCO	N6
WESCO	N6K
WESCO	N6S
WEST.LAND	AS-US21P
WHITE	02-7074113
WHITE	1541K1
WHITE	1562K1
WHITE	1790K1
WHITE	3104703
WHITE	31-17471510
WHITE	843942A
WHITE	871888
WHITE	878268
WHITE	8884814
WHITE FARM	10P715
WHITE FARM	15P1008
WHITE FARM	15R108
WHITE FARM	198-883AS
WHITE FARM	20P1380
WHITE FARM	502-579E
WHITE FARM	509-260E
WHITE FARM	KAR758
WHITING	546959-3
WHITING	S46959-3
ZELLER	1587A
Z-WAY	1195

IDLI GROUP 80002

**CROSS + 2 ROUND BEARINGS
2 MIDWING DRILLED BEARINGS
+ INSIDE AND OUTSIDE SNAP RINGS
SEE FIGURE 1AK**

1.1875 in	BD	30.16 mm
.7656 in	BW	19.45 mm
1.9688 in	CL	50.01 mm
.750 in	CD	19.05 mm

AEC	497
MCQUAY-NOR	U02-6
NEAPCO	1-297
NEAPCO	28297X
PRECISION	377
SPICER	5-297-1X
TRW	20193

IDLI GROUP 80003

**CROSS + 4 GROOVED BEARINGS
+ INSIDE SNAP RINGS
SEE FIGURE 1B**

.9688 in	BD	24.61 mm
.6250 in	BW	15.88 mm
2.0 in	CL	50.80 mm
.6250 in	CD	15.88 mm

ADAMS-LET.	SR6109
ADAMS-LET.	TG4614
AEC	6N8
AEC	CPL6N8
ALCO	CPL6S9
ALLIS-CHLM	053120
ALLIS-CHLM	081058

ALLIS-CHLM	224890
ALLIS-CHLM	534235
ALLOY	1777
ALLOY	1778
ALLOY	CPL6S9
AMER.MOTOR	990192
BEAN,JOHN	2073822
BORG-WARNR	114-1004
BORG-WARNR	114-1005
BORG-WARNR	114-145
BROCKWAY	345802
BROCKWAY	345823
CASE,J.I.	A20684
CASE,J.I.	B18132
CATERPILLR	15678
CATERPILLR	1S687
CATERPILLR	2H8257
CATERPILLR	7K2220
CATERPILLR	8D2909
CLARK EQU.	868218
CLARK EQU.	946020
CLARK EQU.	948213
CLEVE.MOT.	CM1003
CLEVE.MOT.	CM-CPL10N2
CLEVE.MOT.	CM-CPL6N
CLEVE.MOT.	CM-CPL6N8
COCKSHUTT	Z862
DEERE,JOHN	A17330H
DEERE,JOHN	AH18327H
DEERE,JOHN	XAG6379GH
EIMCO	912A45087
FIAT-ALLIS	534235
FIAT-ALLIS	70534235
FLEETRITE	ZADBLN6N
FORD	107854
FORD	291-297
FORD	291971
FORD	309575
FORD	C3T23815A
FORD	C3TZ3815A
FORD	C5TZ23815A
FORD	C5TZ3815A
FORD	C6TZ3815B
FORD	CBPN3550A
FORD	CBPN3550B
FWD	2A5-160X
GALION	H62152(7)
GALION	H625124
GAR WOOD	3001449-3
GEHL BROS.	510004
GEHL BROS.	S10004
GMB	GGPL6R
GMB	GU2100
GMC	2200211
GMC	2386331
GMC	2420606
GMC	2466726
GMC	3817179
GMC	3869039
GMC	3918866
GMC	6958321
GMC	7808308
IHC	147511
IHC	147512
IHC	197-032R91
IHC	197-035R91

IHC	217-938H1
IHC	357439R92
IHC	363-312R91
IHC	363-313R91
IHC	399-040C91
IHC	432-473C91
IHC	561-452C91
IHC	661-372R91
ITALCARDAN	50-809-000
LOBRO	U210
MACK	206SL13
MACK	206SL13A
MACK	206SL15
MACK	206SL16
MASS.FERG.	840-824M91
MASS.FERG.	840-824M92
MCQUAY-NOR	UO3-1
MCQUAY-NOR	UO3-10
NEAPCO	1-1475
NEW HOLL.	110724A1
NEW HOLL.	170724A1
NEW HOLL.	48263
NEW HOLL.	512629
NEW IDEA	012324
NEW IDEA	02324
NEW IDEA	O12324
OLIVER	CF28-24
OSHKOSH	22UJ12
PILOT	BL10N
PILOT	BL6N
PRECISION	860
PREC.TOR.	862
PREC.TOR.	863
REPCO	CPL6S22
REPCO	CPL6S5
REPCO	CPL6S9
REPUBLIC	CB-L10N2
REPUBLIC	CBL6NS9
REPUBLIC	CB-L6S5
REPUBLIC	CB-L6S9
TRU-CROSS	92
TRW	20001
TRW	20001D
VOLVO	344147
WESCO	241007
WESCO	65K
WESCO	6NS
WHITE	10432K1
WHITE	3104703
WHITE	3104709
WHITE	5101270
WHITE	871888
WHITE FARM	CF28-24
ZELLER	1587A

IDLI GROUP 80004

**CROSS + 4 GROOVED BEARINGS
+ INSIDE SNAP RINGS
SEE FIGURE 1B**

1.4960 in	IE	38.00 mm
.7880 in	BD	20.02 mm
2.0080 in	CL	51.00 mm

AEC	1604
AEC	AE1604
AEC	AE-D8025
AEC	D8025

BD = BEARING DIAMETER (outside)   BW = BEARING WIDTH   CB = CROSS LENGTH WITH BEARINGS   CC = CENTER TO CENTER   CD = CROSS DIAMETER
CL = CROSS LENGTH WITHOUT BEARINGS   EC = END TO CENTER or FACE TO FACE   EE = END TO END   EL = EFFECTIVE LENGTH
HD = HUB DIAMETER (or insert)   IE = INSIDE OF EARS (or recess)   OD = OUTSIDE DIAMETER   OE = OUTSIDE OF EARS   SB = SPLINE OR BORE SIZE

Column 1

ALCO	J5-1601
ALCO	J5-16015
ALLOY	1049
ALLOY	1050
ALLOY	J5-1601A
ALLOY	J5-1601S
BORG-WARNR	114-303
BWE	114-307
CLEVE.MOT.	CM300J
DATSUN	37125
DATSUN	37125-18000
DATSUN	37125-18025
DATSUN	37125H1025
FLEETRITE	ZAD303
GMB	G5-303
GMB	G5-307
GMB	GUN26
GMB	GUN28
KOYO	N2057
KOYO	UJ-N2057
LEMPCO	U700
LOBRO	U181
MATSUI	UJ112
MCQUAY-NOR	U01-40
MCQUAY-NOR	UO1-40
MOPAR	D1601
MOTOR MAST	2800A
NEAPCO	1-0028
NEAPCO	1-11
NEAPCO	1-28
NEAPCO	28N28
PILOT	303
PRECISION	392
QUINT.HAZ.	QL1780
QUINTON	QL11007
QUINTON	QL216
REPCO	101-1060
REPCO	RUJ1780
REPUBLIC	CB2179
REPUBLIC	CB2183
SPICER	5-1500X
TRU-CROSS	1500
TRW	20128
TRW	20129D
WESCO	201108
WORLDPARTS	W49-105
WYLIE	69
ZELLER	2800A

IDLI GROUP 80005

**CROSS + 4 GROOVED BEARINGS
+ INSIDE SNAP RINGS
SEE FIGURE 1B**

1.4960 in	IE	38.00 mm
.8860 in	BD	22.50 mm
2.0150 in	CL	51.18 mm

AEC	1606
AEC	89-2510
ALLOY	1022
GMB	GUMZ10
MATSUI	UJ418
MCQUAY-NOR	MAZDA
NEAPCO	1-1606
NEAPCO	281606
PRECISION	385
REPCO	101-1132

Column 2

SPICER	5-1514X
WESCO	201-130

IDLI GROUP 80006

**CROSS + 4 GROOVED BEARINGS
+ INSIDE SNAP RINGS
SEE FIGURE 1A**

2.2440 in	OE	57.00 mm
.7880 in	BD	20.02 mm
2.0470 in	CL	51.99 mm

AEC	1604
AEC	AE-T1010
ALCO	1050
ALLOY	1050
ALLOY	J4-1601
ALLOY	J5-1601
BORG-WARNR	114-307
CENTRAL	GUN28
CENTRAL	GUT11
CMP	CM3000J
D&D MACH.	MF43715
DAIHATSU	04371-10011
DATSUN	37125H5000
FLEETRITE	ZAD307
GMB	GUN11
GMB	GUN26
GMB	GUT11
GMB	GUT19
KOYO	T2057
KOYO	UJ-T2057
LEMPCO	U700
LOBRO	U180
LUDWIG	37-8902
LUDWIG	39-8906
MATSUBA	UJ222
MATSUI	UJ212
MCQUAY-NOR	U01-40
MOTOR MAST	2800A
NEAPCO	1-0011
NEAPCO	1-11
NEAPCO	1-28
NEAPCO	28T11
PILOT	303
PILOT	307
PRECISION	395
QUINT.HAZ.	QL11007
QUINT.HAZ.	QL216
REPCO	101-1240
REPCO	K5A517
REPUBLIC	CB2183
SPICER	5-1500X
TOYOTA	04371-10010
TOYOTA	04371-10011
TRU-CROSS	1500
TRW	20128
WESCO	201111
WORLDPARTS	49-100
WORLDPARTS	49-105
WORLDPARTS	W49-100
WYLIE	68
ZELLER	2800A

Column 3

IDLI GROUP 80007

**CROSS + 4 GROOVED BEARINGS
+ INSIDE SNAP RINGS
SEE FIGURE 1B**

.9375 in	BD	23.81 mm
.6250 in	BW	15.88 mm
2.0781 in	CL	52.78 mm
.50 in	CD	12.70 mm

AEC	503
AEC	510
AEC	570
AEC	570P
AEC	571
AEC	AE503
AEC	AE510
AEC	AE570
AEC	AE572
AEC	AE575
AEC	AEX55-55
AEC	X55-55
ALCO	1070
ALCO	1570
ALCO	J5-170
ALLIS-CHLM	617909
ALLIS-CHLM	6179097
ALLIS-CHLM	645053
ALLIS-CHLM	645110
ALLOY	1786
ALLOY	499
ALLOY	J5-170
BLAW KNOX	5-70X
BLAW KNOX	735-19
BLAW KNOX	735-198
BORG-WARNR	114-570
BORG-WARNR	114-570P
BORG-WARNR	114-572
BORG-WARNR	114-575
BORG-WARNR	X55-55
CATERPILLR	7D2059
CATERPILLR	7D9715
CATERPILLR	8D1505
CATERPILLR	8D1508
CATERPILLR	8D1566
CENTRAL	G5-170X
CHAMPION	1983
CHRYSLER	1843030
CHRYSLER	2240231
CLARK EQU.	1565240
CLEVE.MOT.	CM5-170PB
CLEVE.MOT.	CM5-170X
CLEVE.MOT.	CM5-175X
CLEVE.MOT.	CM5555X
CLEVE.STL.	X55-55
CLEVE.STL.	X55-55-1
CMP	CM5-170X
CMP	CM5555X
D&D MACH.	MS570X
DAVEY COM.	A31364
DEERE,JOHN	AJ5729H
DEERE,JOHN	AJ7329
DEERE,JOHN	AJ7329HN
DEERE,JOHN	AU5729H
FIAT-ALLIS	0617909
FIAT-ALLIS	0617909-7
FIAT-ALLIS	064053-0
FIAT-ALLIS	064110-8

Column 4

FIAT-ALLIS	0645053
FIAT-ALLIS	617909
FIAT-ALLIS	6179097
FIAT-ALLIS	645053
FIAT-ALLIS	645110
FIAT-ALLIS	70617909
FIAT-ALLIS	70617909-7
FIAT-ALLIS	70645053
FIAT-ALLIS	70645053-0
FIAT-ALLIS	70645110
FIAT-ALLIS	70645110-8
FIAT-ALLIS	76179097
FLEETRITE	ZAD11570
FLEETRITE	ZAD11572
GMB	G5-170
GMB	G5-170-1X
GMB	G5-170X
GMB	G5-570
GMB	G5-572
GMB	G5-575
GMB	G5-70
GMC	2294607
GMC	2296407
GMC	2396407
GMC	2404877
GMC	612072
GMC	703325
HUB CITY	03-35-90068
HUB CITY	03-35-90069
HUSCO	5-170XBA
HUSCO	NB7A
HUSCO	NB7A2
HUSCO	PB7A
IHC	208-959R91
IHC	262781R92
IHC	447797C91
IHC	975-663R91
IHC	975663R92
JOY MFG.	A370374-1
LEMPCO	15-70
LEMPCO	15-70X
LEMPCO	15-72X
LEMPCO	U319
MACK	201SJ36
MACK	206SL14
MACK	21045170X
MASS.FERG.	835-771M91
MASTERS	MS570X
MCQUAY-NOR	UO3-9
MOPAR	D78
MOTOR MAST	2207A
MOTOR MAST	2207C
MOTOR MAST	2207G
MOTOR MAST	2207P
MUNCIE	5-170X
MUNCIE	K5-70
MUNCIE	MK2X
MUNCIE	MK3XA
MYERS ELEV	1605X
NEAPCO	0505X1
NEAPCO	1-0170
NEAPCO	1-170
NEAPCO	15052X
NEAPCO	1505X
NEAPCO	1605X
NEAPCO	28170

CAUTION: BE SURE TO REFER TO ENGINEERING CATALOGS FOR SPECIAL APPLICATIONS THAT REQUIRE SPECIFIC MATERIAL CONTENT, TOLERANCES, ETC. SEE FOOTNOTE.

NEAPCO	28170X
NEAPCO	2817X
PERFECTION	RG570K
PERFECTION	RG570X
PILOT	11570
PILOT	11572
PILOT	X15555
PRECISION	338
REPCO	5-103X
REPCO	5-170X
REPUBLIC	CB140P
REPUBLIC	CB140R
REPUBLIC	CB14DR
REPUBLIC	CB14OP
REPUBLIC	CB14OR
REPUBLIC	CB1880A
SPICER	1000
SPICER	5-103X
SPICER	5-110X
SPICER	5-1701X
SPICER	5-170X
SPICER	5-70
SPICER	5-70X
SPICER	J5-70
TEDDY TORQ	G5-170
TEDDY TORQ	K70
TRU-CROSS	110
TRU-CROSS	170
TRW	20017
TRW	20019
UJS	55-70
UJS	55-72
WABCO	VS0372
WESCO	201570
WESCO	201571
WESCO	N1570
WESCO	N1570P
WHITE	02-7074112
WHITE	171517
WHITE	1717517
WHITE	1717518
WHITE	859107
WHITE	859107T2
WILLYS	944200
ZELLER	2207A
ZELLER	2207C

IDLI GROUP 80008

CROSS + 4 ROUND BEARINGS + OUTSIDE SNAP RINGS
SEE FIGURE 1A

.8750 in	BD	22.23 mm
.7188 in	BW	18.26 mm
2.0938 in	CL	53.18 mm
.6563 in	CD	16.67 mm

AEC	1FR
AEC	AE-CP1FR
AEC	CP1FR
ALLIS-CHLM	517309
ALLIS-CHLM	524521
ALLOY	1772
AMER.PULL.	499A35
BELLE CITY	S10093
CASE,J.I.	AXMT19725
CASE,J.I.	AXMT1972S
CASE,J.I.	T23616
CASE,J.I.	T23626
CASE,J.I.	X200
CHRYSLER	1325989
CHRYSLER	CC996426
CLARK EQU.	1471287
CLARK EQU.	873721
CLARK EQU.	873726
CLARK EQU.	892969
CLARK EQU.	892989
CLARK EQU.	952368
CLARK EQU.	952375
CLARK EQU.	952376
CLARK EQU.	952377
CLEVE.MOT.	CM-CP1FR
CLEVE.MOT.	CM-CP1FR1
CMP	CM-CP1FR1
COCKSHUTT	2455
COCKSHUTT	FHY289
COCKSHUTT	GY2455
DAFFIN	1R505
DANUSER	A826
DANUSER	A891
FEDERAL	16B6050
FIAT-ALLIS	517309
FIAT-ALLIS	517310
FIAT-ALLIS	518040
FIAT-ALLIS	524521
FIAT-ALLIS	70517309
FIAT-ALLIS	70517310
FIAT-ALLIS	70518040
FIAT-ALLIS	70524521
FLEETRITE	ZAD81FR
FLEETRITE	ZAD-B1FR
FORD	168778
FORD	221-350
FORD	CD33-79
FORD	CP3339
FORD	CP33-47
FORD	CP3379
GAR WOOD	2026823
GAR WOOD	20R0447
GAR WOOD	20R0448
GAR WOOD	3000-100
GAR WOOD	304562
GAR WOOD	304626
GAR WOOD	4002631
GAR WOOD	8004487
GAR WOOD	9000447
GAR WOOD	M3000100
GLAENZER	065-274-8
GMB	G5-264X
GMC	9030469
GMC	9041296
GMC	9041321
HARDIE	499A35
HUB CITY	03-35-90070
HUB CITY	UR1FR
HUME,H.D.	R1025
HUME,H.D.	R1028
HUSCO	CP1FR
IHC	454-740R91
IHC	602-108R91
MASS.FERG.	195-221M91
MASS.FERG.	840680M91
MASS.FERG.	840830M91
MCQUAY-NOR	AGRICULT.
MICHIGAN	892969
MOTOR MAST	2571A
MPL.MOLINE	15R93
NEAPCO	1-0001
NEAPCO	281FR
NEW HOLL.	29538
NEW HOLL.	43454
NEW IDEA	K1911A
NEW IDEA	K1912A
NORTH AMER	G2216
OLIVER	502529E
OLIVER	AC1721
OLIVER	AC1722
OLIVER	AC17-77
PEARSON	CP2711
PILOT	B1FR
PRECISION	853
REPCO	5-264X
REPCO	CP1FR
REPCO	SCP1FR
REPUBLIC	CB1FR
REPUBLIC	CB1FR4
REPUBLIC	CB1FRN4
RUST	CP2711
SCHULTZ	25933
SEAGRAVE	X13754
SPICER	5-264X
TAYLOR	48821
TAYLOR	7169
TEL-E-LECT	13562
TEL-E-LECT	1652
TRU-CROSS	264
TRW	20160
TULSA WIN.	22794
UJS	55-1FR
WESCO	1FR
WESCO	201101
WESCO	P1FR
WHITE	6374032
WHITE FARM	15R93
WHITE FARM	502-529E
WHITE FARM	AC1722

IDLI GROUP 80009

CROSS + 4 GROOVED BEARINGS + INSIDE SNAP RINGS
SEE FIGURE 1B

1.8430 in	IE	46.81 mm
1.0240 in	BD	26.01 mm
2.40 in	CL	60.96 mm

AEC	1605
AEC	AE1605
AEC	AE-T2010
AEC	T2010
ALCO	1053
ALCO	5-1604
ALCO	J5-1604
ALLOY	1053
ALLOY	J5-1604
BORG-WARNR	114-306
CENTRAL	GUT13
CLEVE.MOT.	CM500J
CMP	CM500J
D&D MACH.	MF43712
FLEETRITE	ZAD306
GMB	G5-306
GMB	GUT13
KOYO	T2667
KOYO	UJ-T2667A
LEMPCO	U702
LOBRO	U185
LUDWIG	37-8901
MATSUBA	UJ221
MATSUI	UJ211
MATSUI	UJ231
MCQUAY-NOR	U01-42
MCQUAY-NOR	UO1-42
MOPAR	D1604
MOTOR MAST	2825B
NEAPCO	1-0013
NEAPCO	1-13
NEAPCO	28T13
PILOT	306
PRECISION	394
QUINT.HAZ.	QL1776
QUINT.HAZ.	QL217
QUINTON	QL217
REPCO	101-1230
REPCO	RUJ1776
REPCO	RUJ1776SL
REPUBLIC	CB2182
ROKA	501-7350
SPICER	5-1502X
TOYOTA	04371-20010
TOYOTA	04371-20011
TOYOTA	04371-22010
TOYOTA	04371-22011
TRU-CROSS	1502
TRW	20130
WESCO	201113
WORLDPARTS	49-101
WORLDPARTS	W49-101
WYLIE	46
ZELLER	2825B

IDLI GROUP 80010

CROSS + 4 GROOVED BEARINGS + INSIDE SNAP RINGS
SEE FIGURE 1B

1.7320 in	IE	43.99 mm
.9850 in	BD	25.02 mm
2.2440 in	CL	57.00 mm

AEC	1603
AEC	AE-D4627
AEC	D4627
ALCO	1052
ALCO	5-1601
ALCO	5-1603
ALCO	J5-1603
ALLOY	1052
ALLOY	J5-1603
BORG-WARNR	114-302
CENTRAL	GUN27
CLEVE.MOT.	CM400J
CMP	CM4000J
D&D MACH.	MF37125
DAIHATSU	4023-4120

BD=BEARING DIAMETER (outside) BW=BEARING WIDTH CB=CROSS LENGTH WITH BEARINGS CC=CENTER TO CENTER CD=CROSS DIAMETER
CL=CROSS LENGTH WITHOUT BEARINGS EC=END TO CENTER or FACE TO FACE EE=END TO END EL=EFFECTIVE LENGTH
HD=HUB DIAMETER (or insert) IE=INSIDE OF EARS (or recess) OD=OUTSIDE DIAMETER OE=OUTSIDE OF EARS SB=SPLINE OR BORE SIZE

Column 1

DAIHATSU	4123-4120
DATSUN	37125-14600
DATSUN	37125-14625
DATSUN	37125-14626
DATSUN	37125-14627
DATSUN	37125-146275
DATSUN	37125H1026
DATSUN	37125H3000
DATSUN	37125P3000
DATSUN	37126-14626
FLEETRITE	ZAD302
GMB	G5-302
GMB	GUD81
GMB	GUD83
GMB	GUD86
GMB	GUMZ1
GMB	GUMZ2
GMB	GUN27
GMB	GUSR4
KOYO	N2563
KOYO	TM2055
KOYO	UJ-N2563
LEMPCO	U701
LOBRO	U182
LUDWIG	39-8905
MATSUBA	UJ218
MATSUBA	UJ219
MATSUBA	UJ412
MATSUI	UJ111
MCQUAY-NOR	U01-41
MCQUAY-NOR	UO1-41
MOPAR	D1603
MOTOR MAST	2825A
NEAPCO	1-0027
NEAPCO	1-27
NEAPCO	28N27
NTN	CJ15-37N
PILOT	302
PRECISION	391
QUINT.HAZ.	QL11009
QUINT.HAZ.	QL27
QUINTON	QL11009
REPCO	101-1050
REPCO	K5A510
REPCO	RUJ1781
REPCO	RUJ1781SL
REPCO	RUJ1782
REPUBLIC	CB2178
SPICER	5-1501X
TRU-CROSS	1501
TRW	20129
TRW	20129D
TRW	21029
WESCO	201107
WORLDPARTS	49-104
WORLDPARTS	W49-104
WYLIE	70
ZELLER	2825A

IDLI GROUP 80011

CROSS + 4 GROOVED BEARINGS + INSIDE SNAP RINGS
SEE FIGURE 1B

1.7717 in	IE	45.00 mm
.9850 in	BD	25.02 mm
2.2440 in	CL	57.00 mm

AEC	1606
AEC	89-2510
AEC	AE1606
AEC	AE89-2510
ALCO	1059
ALLOY	1059
ALLOY	J5-1607
BORG-WARNR	114-309
CENTRAL	GMZ7
CLEVE.MOT.	CM800J
CMP	CM700J
D&D MACH.	MF25989
FORD	0259-89-251
GMB	G5-309
GMB	GUMZ7
KOYO	UJ-TM2546B
KOYO	UJ-TM2564B
KOYO	UJ-TM2564E
LEMPCO	U705
LOBRO	U184
MATSUI	UJ413
MAZDA	0259-25-060
MAZDA	0259-89-251
MAZDA	2158-89-251
MAZDA	4259-25-060
MCQUAY-NOR	U01-44
MCQUAY-NOR	UO1-44
MOTOR MAST	2528D
MOTOR MAST	28250
MOTOR MAST	2825D
NEAPCO	1-1606
NEAPCO	281606
PILOT	309
PRECISION	397
QUINT.HAZ.	QL11010
REPCO	101-1130
REPCO	RUJ1784
REPCO	RUJ1784SL
REPUBLIC	CB2185
ROKA	501-5350
SPICER	5-1504X
TRU-CROSS	1504
TRW	201127
TRW	20129D
TRW	5-1504X
WESCO	201127
WYLIE	71
ZELLER	2825D

IDLI GROUP 80012

CROSS + 4 ROUND BEARINGS + OUTSIDE SNAP RINGS
SEE FIGURE 1A

2.5260 in	OE	64.16 mm
.9850 in	BD	25.02 mm
2.2580 in	CL	57.35 mm

AEC	1601
AEC	3334
AEC	AE1601

Column 3

AEC	AE1603
AEC	AE3334
ALCO	5-1605
ALLOY	1057
ALLOY	J5-1605
BORG-WARNR	114-310
CHRYSLER	3886579
CHRYSLER	A126433
CHRYSLER	MA0126433
CHRYSLER	MA126433
CHRYSLER	MA126434
CHRYSLER	MB000345
CHRYSLER	MB000391
CHRYSLER	MB000827
CHRYSLER	MD000392
CHRYSLER	MT372505
CLEVE.MOT.	CM600J
CMP	CM600J
D&D MACH.	MF26434
FLEETRITE	ZAD310
FORD	D0RY4635A
GMB	G5-310
GMB	GUM79
GMB	GUM81
KOYO	M2564
KOYO	S2765
KOYO	TM2564A
KOYO	UJ-M2564
LEMPCO	U703
LOBRO	U183
LOBRO	U195
MATSUBA	UJ411
MATSUBA	UJ616
MATSUI	UJ625
MAZDA	0114-26-180
MAZDA	0136-25-060
MAZDA	0187-89-251
MCQUAY-NOR	U01-43
MITSUBISHI	3886579
MITSUBISHI	MA126433
MITSUBISHI	MB000345
MITSUBISHI	MB000391
MITSUBISHI	MB000392
MITSUBISHI	MB000827
MITSUBISHI	MB00391
MITSUBISHI	MST00024
MITSUBISHI	MST05012
MITSUBISHI	MT202157
MITSUBISHI	MT372505
MITSUBISHI	ST0024
MOPAR	D1605
MOTOR MAST	2825C
NEAPCO	1-1557
NEAPCO	281557
PILOT	310
PRECISION	398
QUINT.HAZ.	QL1783
QUINT.HAZ.	QL219
REPCO	101-1030
REPCO	RUJ1781SL
REPCO	RUJ1783
REPCO	RUJ1783SL
REPUBLIC	CB2186
ROKA	501-1350
SPICER	5-1503X
TRU-CROSS	1503

Column 4

TRW	20129D
TRW	20131
WESCO	201177
WESCO	201180
WORLDPARTS	49-129
WORLDPARTS	W49-129
WYLIE	67
ZELLER	2825C

IDLI GROUP 80013

CROSS + 4 GROOVED BEARINGS + INSIDE SNAP RINGS
SEE FIGURE 1B

2.0080 in	IE	51.00 mm
1.0440 in	BD	26.52 mm
2.5180 in	CL	63.96 mm

ALLOY	1008
GMB	GUMZ9
MATSUI	UJ417
MCQUAY-NOR	MAZDA
NEAPCO	1-1608
NEAPCO	281608
PRECISION	386
REPCO	101-1140
SPICER	5-1513X
WESCO	201-129
WORLDPARTS	W49-131

IDLI GROUP 80014

CROSS + 4 GROOVED BEARINGS + INSIDE SNAP RINGS
SEE FIGURE 1B

2.0870 in	IE	53.01 mm
1.1420 in	BD	29.01 mm
2.7170 in	CL	69.01 mm

DAIHATSU	4610-4120
GMB	GU1S52
GMC	9-37300-036
GMC	9-37300-065
GMC	9-37300-601
GMC	9-37300-603
GMC	94027819
GMC	94238930
ISUZU	37300-036
ISUZU	37300-065
ISUZU	37300-601
ISUZU	37300-603
ISUZU	9-37300-036
ISUZU	9-37300-065
ISUZU	9-37300-601
ISUZU	9-37300-603
ISUZU	94238930
KOYO	12976
MATSUBA	UJ314
MATSUI	UJ324
PRECISION	383
REPCO	RUJ2033
SPICER	5-1515X

CAUTION: BE SURE TO REFER TO ENGINEERING CATALOGS FOR SPECIAL APPLICATIONS THAT REQUIRE SPECIFIC MATERIAL CONTENT, TOLERANCES, ETC. SEE FOOTNOTE.

IDLI GROUP 80015

CROSS + 4 ROUND BEARINGS + OUTSIDE SNAP RINGS
SEE FIGURE 1A

3.0120 in	OE	76.50 mm
0.9850 in	BD	25.02 mm
2.7510 in	CL	69.88 mm

ALLOY	1016
BORG-WARNR	114-317
CHRYSLER	MB000823
CHRYSLER	MB000826
GMB	GUM88
MITSUBISHI	MB000823
MITSUBISHI	MB000826
PRECISION	399
REPCO	101-1031
REPCO	RUJ2045
SPICER	5-1516X
WESCO	201-134

IDLI GROUP 80016

CROSS + 4 GROOVED BEARINGS + INSIDE SNAP RINGS
SEE FIGURE 1B

2.0470 in	IE	51.99 mm
1.1420 in	BD	29.01 mm
2.0470 in	BW	51.99 mm
2.7540 in	CL	69.95 mm

ALLOY	1017
GMB	GUT17
MATSUI	UJ230
MCQUAY-NOR	TOYOTA
PRECISION	387
REPCO	RUJ2100
SPICER	5-1510X
TOYOTA	04371-30020
TOYOTA	04371-35020
WESCO	201-131

IDLI GROUP 80017

CROSS + 4 ROUND BEARINGS + OUTSIDE SNAP RINGS
SEE FIGURE 1A

3.0710 in	OE	78.00 mm
1.1230 in	BD	28.52 mm
2.8340 in	CL	71.98 mm

AEC	1506
ALCO	1060
ALLOY	1060
ALLOY	J5-1608
BORG-WARNR	114-308
CENTRAL	GUT14
FLEETRITE	ZAD308
GMB	G5-308
GMB	GUT14
KOYO	T2978
KOYO	UJ-T2978A
LOBRO	U190
LUDWIG	37-8904
MATSUI	UJ214
MCQUAY-NOR	LANDCRUISER
MERCEDES	044-410-06-31
MERCEDES	345-410-7031
MOPAR	D1608
MOTOR MAST	3858A
NEAPCO	1-0014
NEAPCO	28T14
PILOT	308
PRECISION	396
QUINT.HAZ.	QL14
REPCO	101-1260
REPCO	K5A519
REPUBLIC	CB2184
ROKA	501-7353
SPICER	5-1506X
SPICER	5-1509X
TOYOTA	04371-36010
TOYOTA	04371-36011
TOYOTA	04371-36036
TOYOTA	04371-40010
TOYOTA	04371-60010
TOYOTA	04371-60020
TRW	20199
WESCO	201114
WORLDPARTS	49-102
WYLIE	59

IDLI GROUP 80018

CROSS + 4 GROOVED BEARINGS + INSIDE SNAP RINGS
SEE FIGURE 1B

2.2050 in	IE	56.01 mm
1.0240 in	BD	26.01 mm
2.2050 in	BW	56.01 mm
2.8730 in	CL	72.97 mm

ALCO	1058
ALLOY	1058
ALLOY	41058
ALLOY	J5-1606
BORG-WARNR	114-305
CENTRAL	GUT12
FLEETRITE	ZAD305
GMB	G5-305
GMB	GUT12
KOYO	T2680C
LEMPCO	U701
LOBRO	U193
LUDWIG	37-8903
MATSUI	UJ210
MCQUAY-NOR	TOYOTA
NEAPCO	1-0012
NEAPCO	28T12
NEAPCO	28T14
PILOT	305
PRECISION	390
QUINT.HAZ.	QL12
REPCO	101-1250
REPCO	RUJ2016
REPCO	RUJ2016SL
REPUBLIC	CB2181
SPICER	5-1508X
TOYOTA	04371-25010
TOYOTA	04371-30010
TOYOTA	04371-30011
TRW	20129D
TRW	20147
WESCO	201112
WORLDPARTS	49-108
WYLIE	61

IDLI GROUP 80019

CROSS + 4 GROOVED BEARINGS + INSIDE SNAP RINGS
SEE FIGURE 1B

2.2050 in	IE	56.01 mm
1.1030 in	BD	28.02 mm
2.2050 in	BW	56.01 mm
2.8730 in	CL	72.97 mm

AEC	1602
AEC	AE1025
AEC	AE1602
AEC	AE-D1025
AEC	D1025
ALCO	1051
ALCO	5-1602
ALCO	J5-1602
ALLOY	1051
ALLOY	J5-1602
BORG-WARNR	114-304
CENTRAL	GUN29
CLEVE.MOT.	CM700J
CMP	CM800J
D&D MACH.	MF39625
DATSUN	39625-21001
DATSUN	39625-21004
DATSUN	39625-21005
DATSUN	39625-21025
DATSUN	39625-21026
DATSUN	39625-21027
DATSUN	39625-43825
DATSUN	39625H1025
DATSUN	39625N21026
DATSUN	39625N3725
DATSUN	39625N4125
DATSUN	39625N4126
DATSUN	39625U0100
DATSUN	39625U0125
DATSUN	39625U0127
DATSUN	39625U3825
FLEETRITE	ZAD304
GMB	G5-304
GMB	GUN29
KOYO	N2880RA
KOYO	UJ-N2880B
LEMPCO	U704
LOBRO	U186
LUDWIG	39-8900
MATSUI	UJ114
MCQUAY-NOR	U01-45
MCQUAY-NOR	UO1-45
MOPAR	D1602
MOTOR MAST	2887A
NEAPCO	1-0029
NEAPCO	28N29
PILOT	304
PRECISION	393
REPCO	101-1070
REPCO	RUJ2028
REPCO	RUJ2028SL
REPUBLIC	CB2180
ROKA	501-2351
SPICER	5-1505X
TRU-CROSS	1505
TRW	20129D
TRW	20133
TRW	20183
WESCO	201109
WESCO	203252
WORLDPARTS	49-107
WORLDPARTS	W49-107
WYLIE	75
ZELLER	2887A

IDLI GROUP 80020

CROSS + 4 GROOVED BEARINGS + INSIDE SNAP RINGS
SEE FIGURE 1B

2.3230 in	IE	59.00 mm
1.1030 in	BD	28.02 mm
2.3230 in	BW	59.00 mm
2.8730 in	CL	72.97 mm

AEC	1610
ALLOY	1029
BORG-WARNR	114-311
DAIHATSU	04371-87101
DAIHATSU	04371-87301
DAIHATSU	04371-87303
DAIHATSU	4610-4120
FORD	0706-89-251
FORD	D27Z4758A
GMB	GUMZ6
KOYO	D2880
KOYO	D2880A
KOYO	M2564B-E
MATSUBA	UJ220
MATSUI	UJ415
MAZDA	0604-25-060
MAZDA	0604-89-251
MAZDA	0706-89-251
MCQUAY-NOR	U02-34
MOTOR MAST	2887D
NEAPCO	1-1610
NEAPCO	281610
PRECISION	384
QUINT.HAZ.	QL6
REPCO	101-1110
REPCO	RUJ2023
REPUBLIC	CB2887
SPICER	5-1507X
TRU-CROSS	1507
TRW	201126
TRW	20187
WESCO	201-126
WORLDPARTS	49-131
WORLDPARTS	W49-132
WYLIE	63

IDLI GROUP 80021

CROSS + 4 GROOVED BEARINGS + INSIDE SNAP RINGS
SEE FIGURE 1B

0.9688 in	BD	24.61 mm
0.5625 in	BW	14.29 mm
2.1250 in	CL	53.98 mm
0.4844 in	CD	12.30 mm

AEC	1501
AEC	AE1501
ALCO	J1501
ALCO	J1506
ALCO	J1509

BD = BEARING DIAMETER (outside) **BW** = BEARING WIDTH **CB** = CROSS LENGTH WITH BEARINGS **CC** = CENTER TO CENTER **CD** = CROSS DIAMETER
CL = CROSS LENGTH WITHOUT BEARINGS **EC** = END TO CENTER or FACE TO FACE **EE** = END TO END **EL** = EFFECTIVE LENGTH
HD = HUB DIAMETER (or insert) **IE** = INSIDE OF EARS (or recess) **OD** = OUTSIDE DIAMETER **OE** = OUTSIDE OF EARS **SB** = SPLINE OR BORE SIZE

Column 1

ALCO	J1514
AUST.WEST.	465298
BAL-LM-HAM	465298
BORG-WARNR	114-1501
BORG-WARNR	114-1504
BORG-WARNR	114-1505
BORG-WARNR	114-1506
BORG-WARNR	114-1507
BORG-WARNR	114-1509
BORG-WARNR	114-1510
BORG-WARNR	114-1512
BORG-WARNR	114-1513
BORG-WARNR	114-1514
BORG-WARNR	114-1521
BORG-WARNR	114-1522
BORG-WARNR	114-1523
BORG-WARNR	114-2084
BORG-WARNR	J2006
CASE,J.I.	T11774
CATERPILLR	2D2978
CATERPILLR	6G0976
CATERPILLR	6G976
CLEVE.MOT.	CM1501
DEERE,JOHN	AE10091E
DEERE,JOHN	AJ3294H
DEERE,JOHN	AJ7329H
DEERE,JOHN	AJ9007N
DEERE,JOHN	AJ9907N
DEERE,JOHN	AL9907N
DEERE,JOHN	AOX33323
DEERE,JOHN	AP21678
DEERE,JOHN	AQX33323
DROTT	73139
FIAT-ALLIS	0631383
FIAT-ALLIS	0631383-7
FIAT-ALLIS	631383
FIAT-ALLIS	70631383
FIAT-ALLIS	70631383-7
GLAENZER	065-153-8
GMB	G1501
GMB	G1504
GMB	G1506
GMB	G1509
GMB	G1514
GMB	G5112
GMB	G5-1501
GMC	9190690
GMC	E56241
HYSTER	26612
IHC	132-083R91
IHC	511-705R91
IHC	656-226R91
INBRA	CT13-010
INBRA	CT17-130
INBRA	CT19-010
LEMPCO	U244
LILLISTON	UJ520
MCQUAY-NOR	U03-2
MCQUAY-NOR	UO3-2
MOPAR	D1
MOTOR MAST	2012B
NEAPCO	1-0061
NEAPCO	281501
NEAPCO	281521
NEAPCO	30061X
PARSONS	3150425(4)

Column 2

PERFECTION	RG2006
PILOT	11050
PILOT	11051
PILOT	11056
PILOT	11057
PILOT	11401
PRECISION	983
REPCO	114-1501
REPCO	114-1521
REPUBLIC	CB1501
REPUBLIC	CB783
SPICER	5-101X
THOMPSON	1951
TRW	20002
TRW	20002D
WABCO	2055743
ZELLER	2012B

IDLI GROUP 80022

CROSS + 4 ROUND BEARINGS + OUTSIDE SNAP RINGS

.8906 in	BD	22.62 mm
.6719 in	BW	17.07 mm
2.1875 in	CL	55.56 mm
.4688 in	CD	11.91 mm

ALCO	83254-3
ALCO	J83254-3
ALMETAL	CB83254
BORG-WARNR	114-235
CHRYSLER	1134579
DETROIT	83254-2
DETROIT	83254-3
GMB	G5-1315X
GMB	G83254-3
GMC	1134579
LEMPCO	U419
NEAPCO	281900
PERFECTION	RG235
PILOT	17220
PRECISION	302
REPUBLIC	CB83254
REPUBLIC	CB83254A
REPUBLIC	CB835254A
SPICER	83254-3
TEDDY TORQ	825X
WESCO	D4579
WESCO	D4579C

IDLI GROUP 80023

CROSS + 4 ROUND BEARINGS + OUTSIDE SNAP RINGS

SEE FIGURE 1A

.9375 in	BD	23.81 mm
.6250 in	BW	15.88 mm
2.2031 in	CL	55.96 mm
.5 0 in	CD	12.70 mm

AEC	501
AEC	514
AEC	AE501
AEC	AE581
ALBURUS-SP	5-2X
ALCO	1023
ALCO	1027
ALCO	1514
ALCO	J5-101

Column 3

ALLIS-CHLM	086752
ALLIS-CHLM	6582753
ALLOY	1027
ALLOY	1031
ALLOY	J5-101
ALLOY	J5-181
ALLOY	JS101
AMER.MOTOR	3119752
AMER.MOTOR	3165080
AMER.MOTOR	8126765
AMER.MOTOR	8133519
ARMSTRONG	5410-150
ARMSTRONG	5410-1720
ARMSTRONG	5410-400
ARMSTRONG	5410-500
AUST.MOT	513800
AUTOCAR	3AB01172
AUTOCAR	3AB1172
BATUYONG	57606
BEDFORD	6371722
BEDFORD	7068345
BEDFORD	7084449
BEDFORD	7156132
BEDFORD	7156340
BEDFORD	8845636
BMC	7088979
BMC	8G8979
BMC	GUJ101
BMC	GUJ106
BORG-WARNR	114-501
BORG-WARNR	114-514
BORG-WARNR	114-572
BORG-WARNR	114-581
BRD	02-501363
BRD	02-502835
CHRYSLER	3306813
CHRYSLER	34244Y
CHRYSLER	34254S
CHRYSLER	5037918
CHRYSLER	5038495
CHRYSLER	7080421
CHRYSLER	7080428
CHRYSLER	K18244
CHRYSLER	K23703
CLEVE.MOT.	CM111X
CLEVE.MOT.	CM5-101X
CLEVE.MOT.	CM5-111X
CLEVE.MOT.	CM5-181X
CMP	CM5-101X
D&D MACH.	MS101X
DATSUN	11975
DATSUN	20940
DATSUN	37125-11975
DEERE,JOHN	AD10091E
DEERE,JOHN	AZ3304
DIAMOND R	10048P2
DIAMOND R	10049P2
DIAMOND R	10051P2
DIAMOND R	10053P2
EXCEL	U117
FIAT	4024300
FIAT	4551551
FIAT	4593225
FIAT	4606993
FIAT	4655161
FIAT-ALLIS	6582753

Column 4

FIAT-ALLIS	70658275
FLEETRITE	ZAD11501
FLEETRITE	ZAD11581
FORD	116E4635A
FORD	116E4635B
FORD	1792216
FORD	204E7039
FORD	204E7042
FORD	3008E4635A
FORD	3012E4635A
FORD	400E4635
FORD	72DB4635AIA
FORD	AR3008E4635A
FORD	C8SZ4A031A
FORD	CTBA18310
GLAENZER	065-923-8
GLAENZER	265-008-0
GLAENZER	265-014-1
GLAENZER	265-032-0
GLAENZER	265-032-1
GLAENZER	265-033-1
GLAENZER	265-034-0
GLAENZER	265-040-0
GMB	G1501
GMB	G5-101
GMB	G5-101X
GMB	G5-181
GMB	G5-4
GMB	G5-581
GMB	GU1S51
GMB	GU400
GMB	GU500
GMB	GU500SL
GMB	GU510
GMB	GUIS56
GMB	GUN26
GMC	9-37300-017
GMC	9-37300-031
GMC	9-37300-091
GMC	9-37300-092
GMC	9-37300-605
GMC	9-37300-609
GMC	93730-06090
GMC	94020606
GMC	94021-281
HABERLE	H1524X
HARDY SPCR	1451-150N
HARDY SPCR	1451-277N
HARDY SPCR	K5-21
HARDY SPCR	K5GB1
HARDY SPCR	K5GB113
HARDY SPCR	K5GB115
HARDY SPCR	K5GB150
HARDY SPCR	K5GB26
HARDY SPCR	K5GB277
HARDY SPCR	K5GB50
HARDY SPCR	K5GB96
HARDY SPCR	K5L12
HARDY SPCR	K5LGB100
HARDY SPCR	K5LGB102
HARDY SPCR	K5LGB174
HARDY SPCR	K5LGB185
HARDY SPCR	K5LGB194
HARDY SPCR	K5LGB50
HARDY SPCR	K5LGB64
HARDY SPCR	K5LGB81

CAUTION: BE SURE TO REFER TO ENGINEERING CATALOGS FOR SPECIAL APPLICATIONS THAT REQUIRE SPECIFIC MATERIAL CONTENT, TOLERANCES, ETC. SEE FOOTNOTE.

HARDY SPCR	K5LGB99
HOLDEN	7068345
HOLDEN	7084449
HOLDEN	7156132
HOLDEN	7156340
HOUGH	203787HI
HUSCO	5-111XBA
HUSCO	5-300XBA
HUSCO	B300N
HYSTER	25420
HYSTER	26126A
IHC	203-787H1
IHC	20713
IHC	799293R91
IHC	974-449R91
INBRA	CA508
INNOCENTI	39602101
INNOCENTI	39602102
ISUZU	373-000-92
ISUZU	8558-2805
ISUZU	9-37300-017
ISUZU	9-37300-031
ISUZU	9-37300-605
ISUZU	9-37300-609
ISUZU	9-37300-91
ISUZU	9-37300-92
ISUZU	94027724
ITALCARDAN	50-854-000
JEEP	906344
KOYO	12461
LEMPCO	U111
LEMPCO	U115
LEMPCO	U117
LEMPCO	U256
LOBRO	U172
LOBRO	U40
LOBRO	U51
LUDWIG	441-8901
MACK	201SJ17
MASTERS	MS101X
MASTERS	MS111X
MASTERS	MS514X
MASTERS	MS581X
MATRA	524617
MATSUBA	UJ313
MCQUAY-NOR	U01-10
MCQUAY-NOR	U01-9
MCQUAY-NOR	UO1-10
MIPER	C3106
MIPER	C3107
MIPER	C3109
MIPER	C3112
MOPAR	D52
MOPAR	D57
MOPAR	D82
MOTOR MAST	2021B
MOTOR MAST	2207C
MOTOR MAST	2225C
MOTOR MAST	2912B
MUNCIE	K5-1
NEAPCO	1-0300

NEAPCO	1-0301
NEAPCO	28101X
NEAPCO	28300X
NEAPCO	28301X
NEAPCO	28481
NEAPCO	28581
NEAPCO	28581X
OSHKOSH	11DS31
PERFECTION	R514X
PERFECTION	RG514X
PERFECTION	RG5-1X
PERFECTION	RG581X
PETTIBONE	LL242-64
PILOT	11051
PILOT	11501
PILOT	11511
PILOT	11581
PITT.VIOL.	40.701.000
PITT.VIOL.	40.851.000
PITT.VIOL.	40.854.000
PITT.VIOL.	40.920.000
POCLAIN	00400-34
PRECISION	340
QUINT.HAZ.	QL15002
QUINT.HAZ.	QL211
QUINT.HAZ.	QL213
QUINT.HAZ.	QL214
REO MOTORS	10051P2
REO MOTORS	10053P2
REPCO	5-181X
REPCO	K5G81
REPCO	K5GB1
REPCO	K5GB115
REPCO	K5GB150
REPCO	K5GB79
REPCO	K5LGB102R
REPCO	K5LGB50R
REPUBLIC	CB1100
REPUBLIC	CB1251
REPUBLIC	CB1501
REPUBLIC	CB2501
SPICER	5-101
SPICER	5-101X
SPICER	5-111X
SPICER	5-172X
SPICER	5-181X
SPICER	5-1X
SPICER	5-81X
STANDARD	52406
TEDDY TORQ	K1
TEDDY TORQ	K14
TRANSMISS.	VK107
TRANSMISS.	VK109
TRANSMISS.	VK112
TRIUMPH	124166
TRIUMPH	128133
TRIUMPH	129919
TRIUMPH	508536
TRIUMPH	513800
TRU-CROSS	101
TRU-CROSS	111
TRW	20002D
TRW	20020
TRW	20021
UJS	55-1
UNIPART	GUJ1001

UNIPART	GUJ101
UNIPART	GUJ106
VOLVO	1217606
VOLVO	659773
VOLVO	666184
VOLVO	672037
WESCO	201-581
WESCO	N1501
WHITE	02-7074021
WHITE	10048P2
WHITE	10051P2
WHITE	444153
WHITE	826461
WHITE	826461A
WHITE	871888
WHITE	871888A
WHITE	SG612
WILLYS	906344
WORLDPARTS	49-103
WORLDPARTS	49-117
WORLDPARTS	49-130
WYLIE	60
ZELLER	2225C

IDLI GROUP 80024

CROSS + 4 ROUND BEARINGS + OUTSIDE SNAP RINGS
slight variation from above group
SEE FIGURE 1A

.9375 in	BD	23.81 mm
.6250 in	BW	15.88 mm
2.2031 in	CL	55.96 mm
.50 in	CD	12.70 mm

AEC	1023
AEC	514
AEC	AE511
AEC	AE514
ALBURUS-SP	5-12X
ALCO	1027
ALCO	1031
ALCO	1514
ALCO	J5-14
ALCO	J5-181
ALFA ROMEO	120-102-0
ALFA ROMEO	1365-24805
ALLIS-CHLM	6582753
ALLOY	1023
ALLOY	J101
ALLOY	J5-14
AMER.MOTOR	3119752
AMER.MOTOR	5702158
AMER.MOTOR	906344
ARMSTRONG	5410-510
AUST.MOT	513800
AUTOCAR	3AB1172
AYRA DUREX	1-115-01
BEDFORD	6342814
BEDFORD	7084449
BEDFORD	8845379
BMC	310604005
BMC	683385
BMC	68385
BMC	78433
BMC	7H3870
BMC	8G3000
BMC	8G3001

BMC	8G8183
BMC	A126012
BMC	A126102
BMC	AAK8264
BMC	BG3000
BMC	GUJ114
BMC	GUJ115
BMC	JD1002
BORG-WARNR	114-301
BORG-WARNR	114-341
BORG-WARNR	114-514
BORG-WARNR	114-581
BORG-WARNR	ZAD301
BRD	02-530401
BRD	02-532835
CENTRAL	GU500
CHAMPION	SN1157
CHRYSLER	11614L
CHRYSLER	34224
CHRYSLER	5038930
CHRYSLER	75060444
CHRYSLER	K23703
CHRYSLER	K23704
CHRYSLER	K23958
CLEVE.MOT.	CM111X
CLEVE.MOT.	CM5-101X
CLEVE.MOT.	CM5-111X
CLEVE.MOT.	CM5-11X
CLEVE.MOT.	CM5-14X
CLEVE.MOT.	CM5-181X
CMP	CM5-111X
D&D MACH.	MS514X
DATSUN	14625
DATSUN	50743
ERF	U11806
EXCEL	U115
FIAT	4548769
FIAT	4556515
FIAT	4605734
FIAT	8269030
FIAT	883235
FIAT	883236
FLEETRITE	ZAD11511
FLEETRITE	ZAD11514
FORD	100E7039
FORD	105E7039A
FORD	1792216
FORD	400E4635A
FORD	B7T3815A
FORD	F42A7039A
FORD	GTBA18310
GLAENZER	265-005-0
GLAENZER	265-007-0
GLAENZER	265-037-0
GLAENZER	265-038-0
GLAENZER	265-049-0
GMB	G5-102
GMB	G5-111X
GMB	G5-14X
GMB	G5-181
GMB	G5-300X
GMB	G5-301
GMB	G5-514
GMB	GU1S56
GMB	GU500
GMB	GUIS61

BD=BEARING DIAMETER (outside)   **BW**=BEARING WIDTH   **CB**=CROSS LENGTH WITH BEARINGS   **CC**=CENTER TO CENTER   **CD**=CROSS DIAMETER
CL=CROSS LENGTH WITHOUT BEARINGS   **EC**=END TO CENTER or FACE TO FACE   **EE**=END TO END   **EL**=EFFECTIVE LENGTH
HD=HUB DIAMETER (or insert)   **IE**=INSIDE OF EARS (or recess)   **OD**=OUTSIDE DIAMETER   **OE**=OUTSIDE OF EARS   **SB**=SPLINE OR BORE SIZE

Make	Number
GMB	GUN26
GMC	6342814
GMC	681093
GMC	7084449
GMC	94021281
GMC	94027724
HABERLE	H1524X
HARDY SPCR	1451-185R
HARDY SPCR	1451-194T
HARDY SPCR	1451-194W
HARDY SPCR	1451-279W
HARDY SPCR	79LGB80
HARDY SPCR	K5-22
HARDY SPCR	K5GB
HARDY SPCR	K5GB152
HARDY SPCR	K5GB50
HARDY SPCR	K5GB79
HARDY SPCR	K5L1
HARDY SPCR	K5LGB102
HARDY SPCR	K5LGB104
HARDY SPCR	K5LGB174
HARDY SPCR	K5LGB194
HARDY SPCR	K5LGB195
HARDY SPCR	K5LGB2
HARDY SPCR	K5LGB274
HARDY SPCR	K5LGB279
HARDY SPCR	K5LGB41
HARDY SPCR	K5LGB64
HARDY SPCR	K5LGB80
HARDY SPCR	K5LGB81
HARDY SPCR	K5LGB99
HOUGH	203787HI
HYSTER	122819(4)
IHC	983-922R91
INBRA	CA514
INNOCENTI	31604103
ITALCARDAN	50-701-000
JEEP	5702158
KOYO	N2461
LEADER	Z10424
LEMPCO	U115
LEMPCO	U117
LEMPCO	U140
LEYLAND	68385
LEYLAND	78433
LEYLAND	8G8967
LOBRO	U50
MACK	201SJ19
MACK	201SJ19A
MACK	201SJ27
MASS.FERG.	3406020M91
MASTERS	MS101X
MASTERS	MS111X
MASTERS	MS514X
MASTERS	MS581X
MATSUBA	UJ023
MATSUBA	UJ110
MATSUI	UJ323
MCQUAY-NOR	U01-1
MCQUAY-NOR	U01-10
MCQUAY-NOR	UO1-9
MCQUAY-NOR	UO1-9
MIPER	C3108
MOPAR	D57
MOPAR	D82
MOTOR MAST	2225B
MOTOR MAST	2225D
MOTOR MAST	2912B
NEAPCO	1-0026
NEAPCO	1-0300
NEAPCO	28111X
NEAPCO	28300X
NEAPCO	28300XS
NEAPCO	28514X
NEAPCO	28581X
NEAPCO	2895
NEAPCO	28N26
NEAPCO	28N6
NSU-PRINZ	883235
OPEL	56807
PERFECTION	R514X
PERFECTION	RG514X
PERFECTION	RG5-1X
PERFECTION	RG581X
PILOT	11511
PILOT	11514
PILOT	301
PITT.VIOL.	40.701.000
PRECISION	341
QUINT.HAZ.	QL102
QUINT.HAZ.	QL218
QUINT.HAZ.	QL26
QUINT.HAZ.	QL403
REPCO	101-1210
REPCO	K5A64R
REPCO	K5GB115
REPCO	K5GB79
REPCO	K5L1
REPCO	K5LGB102R
REPCO	K5LGB50R
REPCO	K5LGB80R
REPUBLIC	CB1100
REPUBLIC	CB1251
REPUBLIC	CB2177
REPUBLIC	CB2503G
ROKA	501-0350
ROVER	600628
RUEDARSA	1210
RUEDARSA	1210-3
RUEDARSA	2103-1210-0-0
RUEDARSA	2103-1210-3-0
SIMCA	11614
SPICER	1100
SPICER	1110
SPICER	1140
SPICER	5-101
SPICER	5-102
SPICER	5-102X
SPICER	5-105X
SPICER	5-111
SPICER	5-111X
SPICER	5-114X
SPICER	5-14X
SPICER	5-1X
SPICER	5-300X
SPICER	5-61X
SPICER	5-81
SPICER	J5-81
STANDARD	52405
TEDDY TORQ	K14
TEDDY TORQ	K81
TRANSMISS.	UK108
TRIUMPH	101281330
TRIUMPH	122652
TRIUMPH	500267
TRIUMPH	52405
TRIUMPH	52406
TRU-CROSS	111
TRW	20003
TRW	20003D
TRW	20020
UJS	55-111
UNIPART	GUJ114
VAPORMATIC	VTE5003
VOLVO	656762
VOLVO	672037-9
VOLVO	87977
WALTERSCHD	20-00-00
WESCO	201106
WESCO	201581
WESCO	N1581
WHITE	02-7074022
WHITE	10049P2
WHITE	10053P2
WHITE	444153
WHITE	444153A
WHITE	87188A
WILLYS	5702158
WORLDPARTS	49-103
WORLDPARTS	49-117
WORLDPARTS	49-130
WORLDPARTS	W49-103
WORLDPARTS	W49-113
WORLDPARTS	W49-116
WORLDPARTS	W49-117
WORLDPARTS	W49-130
WYLIE	18
WYLIE	19
WYLIE	34
ZELLER	2225B
ZELLER	2225C

IDLI GROUP 80025

CROSS + 4 GROOVED BEARINGS
+ INSIDE SNAP RINGS
SEE FIGURE 1AB

1.0625 in	BD	26.99 mm
.6563 in	BW	16.67 mm
2.2031 in	CL	55.96 mm
.5469 in	CD	13.89 mm

Make	Number
AEC	248
AEC	5-248
AEC	AE248
AEC	AE5-248
AEC	AE5-249
ALCO	1007
ALCO	1041
ALCO	J5-248
ALLOY	1041
ALLOY	J5-248
AMER.MOTOR	8126610
AMER.MOTOR	937340
AMER.MOTOR	943517
BORG-WARNR	114-548
CLEVE.MOT.	CM5-248X
CMP	CM5-248X
CMP	CM5-249X
D&D MACH.	MS548X

Make	Number
EXCEL	U136
FLEETRITE	ZAD11548
GLAENZER	065-267-8
GMB	G5-248
GMB	G5-248X
GMB	G5-249
GMB	G5-548
HABERLE	114-548
HABERLE	H1548X
JEEP	935517
JEEP	937340
LEMPCO	U136
MASTERS	MS548X
MCQUAY-NOR	U01-11
MCQUAY-NOR	UO1-11
MOPAR	D90
MOTOR MAST	2225
MOTOR MAST	2225W
NEAPCO	1-0248
NEAPCO	1-248
NEAPCO	28248X
PERFECTION	RG548X
PILOT	11548
PRECISION	361
REPCO	5-248X
REPUBLIC	CB1248
SPICER	5-246X
SPICER	5-248X
SPICER	5-249X
SPICER	5-258X
SPICER	5-259X
TEDDY TORQ	G5-248
TEDDY TORQ	K61
TRU-CROSS	248
TRW	20022
UJS	55-248
WESCO	201241
WESCO	N2150
WHITE	937340
WHITE	943517
WILLYS	937340
WILLYS	943517
ZELLER	2225-W

IDLI GROUP 80026

CROSS + 4 ROUND BEARINGS
+ OUTSIDE SNAP RINGS
SEE FIGURE 1A

1.0625 in	BD	26.99 mm
.6563 in	BW	16.67 mm
2.2031 in	CL	55.96 mm
.6250 in	CD	15.88 mm

Make	Number
AEC	315
AEC	5-315
AEC	AE5-315
ALLOY	1007
ALLOY	J5-315
AMER.MOTOR	3219885
AMER.MOTOR	8126635
BORG-WARNR	114-315
CLEVE.MOT.	CM5-315X
GMB	G5-315X
LEMPCO	U141
MCQUAY-NOR	U01-29
MOTOR MAST	2225E
NEAPCO	1-0315

ENGINEERING CATALOGS MUST BE CONSULTED FOR DETAILS NOT INCLUDED
IN THIS GUIDE. SPECIFIC DESIGNS, MATERIAL CONTENT, TOLERANCES
LUBE FITTINGS AND OTHER DIMENSIONS ARE INTENTIONALLY OMITTED HERE.

CAUTION: BE SURE TO REFER TO ENGINEERING CATALOGS FOR SPECIAL APPLICATIONS THAT REQUIRE SPECIFIC MATERIAL CONTENT, TOLERANCES, ETC. SEE FOOTNOTE.

NEAPCO	28315X
PRECISION	373
REPCO	5-315X
REPUBLIC	CB2100
REPUBLIC	CB2110
SPICER	5-315X
TRW	20141
TRW	20141D
WESCO	201242
ZELLER	2225E

IDLI GROUP 80027
SEE FIGURE 1

1.0625 in	BD	26.99 mm
.6406 in	BW	16.27 mm
2.2031 in	CL	55.96 mm
.6875 in	CD	17.46 mm

FORD	E3TZ4635B
PRECISION	437
SPICER	1210
SPICER	5-443X

IDLI GROUP 80028
SEE FIGURE 1

1.0625 in	BD	26.99 mm
.6406 in	BW	16.27 mm
2.2031 in	CL	55.96 mm
.6875 in	CD	17.46 mm

FORD	E3TZ3249A
PRECISION	436
SPICER	5-456X

IDLI GROUP 80029
SEE FIGURE 1

1.0625 in	BD	26.99 mm
.6406 in	BW	16.27 mm
2.2031 in	CL	55.96 mm
.6875 in	CD	17.46 mm

FORD	E3TZ7A399B
PRECISION	438
SPICER	5-458X

IDLI GROUP 80030
CROSS + 4 GROOVED BEARINGS + INSIDE SNAP RINGS
SEE FIGURE 1B

1.0000 in	BD	25.40 mm
.6719 in	BW	17.07 mm
2.250 in	CL	57.15 mm
.5781 in	CD	14.68 mm

AEC	1750
AEC	AE1750
AEC	AE1762
AEC	AE500
AEC	AE507
ALCO	1750
ALCO	J1750
ALCO	J2503
ALLOY	1056
ALLOY	J2503
AMER.MOTOR	3108170
AMER.MOTOR	3113942
AMER.MOTOR	3116717
AMER.MOTOR	3163514
AMER.MOTOR	3166462
AMER.MOTOR	3171348
AMER.MOTOR	3180170
AMER.MOTOR	3200537
AMER.MOTOR	3200854
AMER.MOTOR	3204918
AMER.MOTOR	3207174
AMER.MOTOR	3209076
BALKAMP	M1750
BALKAMP	M1759
BALKAMP	M1762
BMC	HYL2306
BORG-WARNR	114-1750
BORG-WARNR	114-1757
BORG-WARNR	114-1759
BORG-WARNR	114-1772
BORG-WARNR	114-2012
BORG-WARNR	114-2026
BORG-WARNR	114-2053
BORG-WARNR	114-2503
BORG-WARNR	J2012
BORG-WARNR	J2026
BORG-WARNR	J2053
BWE	114-1767
CENTRAL	G5-105X
CLEVE.MOT.	CM1750
CLEVE.MOT.	CM1759
CLEVE.MOT.	CM1762
CLEVE.MOT.	CM5-100X
CLEVE.MOT.	CM5-107X
CMP	CM1750
CMP	CM1759
CMP	CM1762
CMP	CM5-100X
CMP	CM5-107X
D&D MACH.	MM1750
D&D MACH.	MM1759
FLEETRITE	ZAD11453
FLEETRITE	ZAD11472
FLEETRITE	ZAD715
FORD	CODZ4635A
FORD	C3DZ4635A
FORD	CODZ4635
FORD	CODZ7039A
GLAENZER	065-233-8
GMB	G1750
GMB	G1750G
GMB	G2503
GMB	G5-100X
GMB	G5-105X
GMB	G5-150X
GMB	GU5950
GMB	GU5970
IHC	799295R91
LEMPCO	U211
LEMPCO	U236
LEMPCO	U237
LEMPCO	U255
LOBRO	U596
LOBRO	U612
MASTERS	MM1750
MASTERS	MM1759
MASTERS	MM1762
MASTERS	MM1767
MASTERS	MS105X
MCQUAY-NOR	U01-1
MCQUAY-NOR	U01-12
MCQUAY-NOR	UO1-12
MOPAR	D2
MOTOR MAST	1762
MOTOR MAST	2226B
NEAPCO	1-0105
NEAPCO	28100
NEAPCO	28100X
NEAPCO	281500
NEAPCO	28150X
NEAPCO	281550
NEAPCO	281552
OPEL	56807
PERFECTION	RG2012
PERFECTION	RG2026
PERFECTION	RG2035
PERFECTION	RG2053
PILOT	11412
PILOT	11426
PILOT	11453
PILOT	11472
PILOT	715
PRECISION	514
REPUBLIC	CB2012
REPUBLIC	CB2026
REPUBLIC	CB2053
REPUBLIC	CB2503
RUEDARSA	1356
RUEDARSA	2100-1356-0-0
SPICER	5-100X
SPICER	5-107X
SPICER	5-2012X
SPICER	5-2026X
SPICER	5-2053X
TEDDY TORQ	1014
TEDDY TORQ	1412
TEDDY TORQ	G1750
TEDDY TORQ	J1750
TRU-CROSS	100
TRU-CROSS	105
TRW	20003
TRW	20003D
UJS	55-1762
WESCO	201254
WESCO	G2053-1
WESCO	N2012
WESCO	N2026
WESCO	N2053-1
ZELLER	2226B

IDLI GROUP 80031
CROSS + 4 GROOVED BEARINGS + INSIDE SNAP RINGS
slight variation from above group
SEE FIGURE 1B

1.0000 in	BD	25.40 mm
.6719 in	BW	17.07 mm
2.250 in	CL	57.15 mm
.5781 in	CD	14.68 mm

AEC	1750G
AEC	505
AEC	AE1750
AEC	AE1750G
AEC	AE1767
AEC	AE505
ALCO	1056
ALCO	1061
ALCO	1505
ALCO	J1750G
ALCO	J1762
ALCO	J5-105
ALLOY	1061
ALLOY	J1750G
ALLOY	L5-105
AMER.MOTOR	3108170
AMER.MOTOR	3112942
AMER.MOTOR	3113942
AMER.MOTOR	3116717
AMER.MOTOR	3200537
AMER.MOTOR	3206247
AMER.MOTOR	8126605
AMER.MOTOR	942758
AMER.MOTOR	942910
BALKAMP	M1750
BALKAMP	M1759
BALKAMP	M1762
BALKAMP	M2012
BALKAMP	M2026
BALKAMP	M2053
BORG-WARNR	114-1757
BORG-WARNR	114-1763
BORG-WARNR	114-1765
BORG-WARNR	114-1767
BORG-WARNR	114-1771
BORG-WARNR	114-1773
CLEVE.MOT.	CM1750
CLEVE.MOT.	CM1750G
CLEVE.MOT.	CM1750GS
CLEVE.MOT.	CM1759
CLEVE.MOT.	CM1759G
CLEVE.MOT.	CM1762
CLEVE.MOT.	CM1767
CLEVE.MOT.	CM5-100X
CLEVE.MOT.	CM5-105X
CLEVE.MOT.	CM5-107X
CMP	CM1750G
CMP	CM1767
CMP	CM5-100X
CMP	CM5-105X
D&D MACH.	MM1762
D&D MACH.	MM1767
D&D MACH.	MS105X
EATON	E25435
EXCEL	U236G
EXCEL	U255
FLEETRITE	ZAD11467
FLEETRITE	ZAD715G
FORD	B6TZ4635D
FORD	C0D2-7039A
FORD	C1UZ4635A
FORD	C3UZ4635A
FORD	C3UZ4635B
FORD	C5UZ4635A
FORD	C5UZ4635B
GLAENZER	065-234-8
GMB	G1750G
GMB	G1762
GMB	G1767

BD = BEARING DIAMETER (outside)   **BW** = BEARING WIDTH   **CB** = CROSS LENGTH WITH BEARINGS   **CC** = CENTER TO CENTER   **CD** = CROSS DIAMETER
CL = CROSS LENGTH WITHOUT BEARINGS   **EC** = END TO CENTER or FACE TO FACE   **EE** = END TO END   **EL** = EFFECTIVE LENGTH
HD = HUB DIAMETER (or insert)   **IE** = INSIDE OF EARS (or recess)   **OD** = OUTSIDE DIAMETER   **OE** = OUTSIDE OF EARS   **SB** = SPLINE OR BORE SIZE

GMB	G5-105
GMB	G5-105X
GMB	G5-150X
GMB	G7150G
GMB	G72525-2
GMB	GU5950
GMC	1604073
GMC	45802
GMC	45807
GMC	TB94402
GOODYEAR	219-2504
HABAN	S437
HABERLE	H2012
HABERLE	H2026
HABERLE	H2053
HUSCO	NB10A
INBRA	CA505
INBRA	CT10-000
ITALCARDAN	50-746-000
JAMCO	UJ105
JEEP	942910
LEMPCO	236G
LEMPCO	U236
LEMPCO	U236G
MASTERS	MM1750
MASTERS	MM1759
MASTERS	MM1762
MASTERS	MM1767
MASTERS	MS100X
MASTERS	MS105X
MCQUAY-NOR	U01-1
MCQUAY-NOR	U01-12
MCQUAY-NOR	U01-1
MIDAS	UJ1008
MOPAR	D2
MOPAR	D2A
MOPAR	D60
MOTOR MAST	2025A
MOTOR MAST	2115A
MOTOR MAST	2226A
NEAPCO	1-0105
NEAPCO	28105X
NEAPCO	28150X
NEAPCO	281550G
NEAPCO	281551
NEAPCO	281553
OPEL	56807
PERFECTION	RG2053
PERFECTION	RG2053G
PERFECTION	RG2062
PILOT	11467
PILOT	715G
PRECISION	514G
REPCO	RUJ1755
REPCO	RUJ1777
REPCO	RUJ1779
REPUBLIC	CB2053G
REPUBLIC	CB2503G
RUEDARSA	1357
RUEDARSA	2100-1357-0-0
SEARS-ROE	5102
SPICER	5-100
SPICER	5-100X
SPICER	5-105
SPICER	5-105X
SPICER	5-114X
STUDEBAKER	1561999
TEDDY TORQ	1014G
TRANSMISS.	UK275
TRANSMISS.	UK318
TRANSMISS.	UK318/6
TRU-CROSS	100
TRU-CROSS	105
TRW	20003
TRW	20003D
UJS	55-1767
WESCO	201253
WESCO	201-523
WESCO	202-253
WESCO	202-254
WESCO	G2053
WESCO	N2053
WESCO	N2053G
WHITE	942910
WILLYS	942758
WILLYS	942910
ZELLER	2025A

IDLI GROUP 80032

CROSS + 2 WING TYPE BEARINGS + 2 ROUND GROOVED BEARINGS
SEE FIGURE 1BJ

1.0000 in	BD	25.40 mm	
1.3125 in	CC	33.34 mm	
.6719 in	BW	17.07 mm	
2.250 in	CL	57.15 mm	
.5781 in	CD	14.68 mm	

AEC	1755
AEC	1755G
AEC	AE1755
AEC	AE1755G
ALCO	1055
ALCO	J1755
ALCO	J1755L
ALLOY	1055
ALLOY	J1755
ARMSTRONG	5410-7100
BATUYONG	57606
BMC	HYL2307
BMC	HYL3091
BORG-WARNR	114-1755
BORG-WARNR	114-1761
CLEVE.MOT.	CM1755
CMP	CM1755
D&D MACH.	MM1755
EXCEL	U261G
FLEETRITE	ZAD710
FLEETRITE	ZAD710G
GLAENZER	065-256-8
GLAENZER	065-936-8
GMB	GU5960
GMB	GU7100
GMC	7405066
GMC	7421880
HOLDEN	7405066
HOLDEN	7421880
HOLDEN	7425515
INBRA	CA506
ITALCARDAN	50-751-000
LEMPCO	U261
LEMPCO	U261G
LOBRO	U710
MATSUBA	UJ011
MCQUAY-NOR	U01-2
MCQUAY-NOR	UO1-2
MIPER	C3670-6
MOPAR	D200
MOTOR MAST	2025-2B
MOTOR MAST	2025B
NEAPCO	1-1554
NEAPCO	281554
OPEL	1604074
OPEL	45803
OPEL	45806
OPEL	TB94403
PILOT	710
PILOT	710G
PRECISION	710
QUINT.HAZ.	QL1763
QUINT.HAZ.	QL1764
QUINT.HAZ.	QL205
QUINT.HAZ.	QL206
REPCO	114-1755
REPCO	RUJ1750
REPCO	RUJ1756
REPCO	RUJ1757
REPCO	RUJ1759
REPCO	RUJ1763
REPCO	RUJ1764
REPCO	RUJ1768
REPCO	RUJ1769
REPUBLIC	CB1755
REPUBLIC	CB1755G
RUEDARSA	1290
RUEDARSA	2100-1290-0-0
TRANSMISS.	UK670
TRANSMISS.	UK671
TRW	20003D
TRW	20004
WESCO	201-251
WESCO	201755
WESCO	N1755
WESCO	N2503
ZELLER	2025B

IDLI GROUP 80033

CROSS + 4 ROUND BEARINGS + OUTSIDE SNAP RINGS
SEE FIGURE 1A

.8750 in	BD	22.23 mm	
.7031 in	BW	17.86 mm	
2.250 in	CL	57.15 mm	
.6250 in	CD	15.88 mm	

AEC	10H
AEC	10N
AEC	AE-CPL10N2
AEC	AECPL10S
CASE,J.I.	2415CHA
CATERPILLR	1S9163
CLARK EQU.	867294
CLARK EQU.	874845
COCKSHUTT	31-1565100
DANUSER	2251
DANUSER	251
DURA-DET.	15214
DURA-DET.	20668
DURA-DET.	20865
FIAT-ALLIS	517308

FIAT-ALLIS	70517308
FORD	141132
FORD	170521
GAR WOOD	202449-1
GAR WOOD	20RO449-1
GEHL BROS.	510006
GLAENZER	065-272-8
GMB	GGPL10H
GMB	GGPL10N
HESSTON	11562
IHC	11562
INBRA	CT13-075
MACK	206SL11
MACK	206SL11A
MASS.FERG.	668-172M91
MCQUAY-NOR	U03-10
MOTOR MAST	2520A
NEAPCO	28L10N2
NEAPCO	28L10S
NEW HOLL.	29537
NEW HOLL.	43453
NEW IDEA	0742HSA
PRECISION	870
PREC.TOR.	871
REPCO	CPL10N2
REPCO	CPL10S
REPUBLIC	CB-L10N2
REPUBLIC	CB-L10S
TRW	20159
WESCO	201110
WHITE FARM	31-1565100

IDLI GROUP 80034

CROSS + 4 GROOVED BEARINGS + INSIDE SNAP RINGS
SEE FIGURE 1B

1.0100 in	BD	25.65 mm	
.6250 in	BW	15.88 mm	
2.3125 in	CL	58.74 mm	
.5469 in	CD	13.89 mm	

AEC	252
AEC	AE252
ALCO	1101
ALCO	71425-2
ALCO	J71425-2
ALCO	J72425-2
ALLOY	1101
ALLOY	J71425-2
AMER.MOTOR	3114707
AMER.MOTOR	3119116
AMER.MOTOR	319116
BORG-WARNR	114-252
CLEVE.MOT.	CM1752
CLEVE.MOT.	CM714251-2
CLEVE.MOT.	CM71425-2
CMP	CM7125-2
CMP	CM714251-2
CMP	CM71425-2
D&D MACH.	MD4252
DETROIT	7125-2
DETROIT	714251-2
DETROIT	71425-2
DETROIT	J7125-2
EXCEL	U424
FLEETRITE	ZAD17321
GMB	G5-1303X

CAUTION: BE SURE TO REFER TO ENGINEERING CATALOGS FOR SPECIAL APPLICATIONS THAT REQUIRE SPECIFIC MATERIAL CONTENT, TOLERANCES, ETC. SEE FOOTNOTE.

GMB G71425-2
GMB GU114-252
KIEKHAFER 36311
LEMPCO U424
MASTERS MD4252
MCQUAY-NOR U01-60
MCQUAY-NOR UO1-60
MERCURY B36311
MOPAR D130
MOTOR MAST 2331-2B
NEAPCO 1-4700
NEAPCO 284700
PERFECTION RG252
PILOT 17221
PILOT 17321
PRECISION 293
REPCO 114-252
REPUBLIC CB7125-2
REPUBLIC CB71425-2
SPICER 5-1303X
SPICER 7125-2
SPICER 71425-2
TEDDY TORQ 715
TRU-CROSS 1303
TRW 20027
UJS 55-71425-2
WESCO 201252
WESCO D252
WESCO G252
ZELLER 2331-2B

IDLI GROUP 80035

CROSS + 4 GROOVED BEARINGS + INSIDE SNAP RINGS
SEE FIGURE 1B

1.0625 in	BD	26.99 mm
.6250 in	BW	15.88 mm
2.3125 in	CL	58.74 mm
.6406 in	CD	16.27 mm

ADAMS-LET. 404645
ADAMS-LET. 723741
AEC 12N
AEC AE12N000
AEC AE-CPL12N
AEC CPL12N
ALCO 1779
ALCO CP12N
ALCO CPL12N
ALLIS-CHLM 06483663
ALLIS-CHLM 542862
ALLIS-CHLM 546945
ALLIS-CHLM 648366
ALLIS-CHLM 648366-3
ALLOY 1779
ALLOY CPL12N
BALKAMP 3-972
BEAN,JOHN 1276749
BORG-WARNR 114-103
BORG-WARNR CPL12N
BOST.GEAR .... UJLC
BPC NB12A

CASE,J.I. B13275
CASE,J.I. B15275
CASE,J.I. FH845032
CHAIN BELT 120
CHAIN BELT R12N
CHAIN BELT R31N
CHRYSLER 3736380
CLEVE.MOT. .... CM CPL12N
CLEVE.MOT. CM-CPL12R
CMP CM-CPL12N1
CMP CM-CPL12R
COCKSHUTT FHY342
DAFFIN 1R198
DANUSER 1057-1
DANUSER 47483
DANUSER 7590
DEERE,JOHN AE123716
DEERE,JOHN AE23716
DEERE,JOHN AE32413
DEERE,JOHN AM2871E
EATON 6421711
FIAT-ALLIS 0648366
FIAT-ALLIS 06483663
FIAT-ALLIS 339981
FIAT-ALLIS 542862
FIAT-ALLIS 546945
FIAT-ALLIS 648366
FIAT-ALLIS 648366-3
FIAT-ALLIS 70339981
FIAT-ALLIS 70542862
FIAT-ALLIS 70546945
FIAT-ALLIS 70648366
FIAT-ALLIS 70648366-3
FLEETRITE ZAD-BL12N
FLEETRITE ZADBLN12N
FORD 148-357
G & G MFG 12R
G & G MFG 302-1200
GAR WOOD 6Y1920
GERLINGER 650079
GERLINGER A650079
GLAENZER 065-148-8
GMB G1501
GMB G5-242X
GMB GCP12N
GMB GCPU2N
GMB GGPL12R
HESSTON 753111
HESSTON 757443
HESSTON 845032
HUB CITY 03-35-90075
HUB CITY UR12010
HUSCO B1200N
IHC 569262R91
IHC 609-144R91
IHC 659-262R91
IHC 669-564R91
LILLISTON 17-61-004
LILLISTON 7161004
LILLISTON UJ520
LOBRO U704
LONG MFG. 736268
MASS.FERG. 830-962M91
MASS.FERG. 836-366M91
MASS.FERG. 841-236M91
MASS.FERG. 841-603M91

MATSUI UJ1275
MCQUAY-NOR U03-11
MCQUAY-NOR UO3-11
MOTIVE CM-CPL12N
MOTOR MAST 2525A
MPL.MOLINE 15P623
MPL.MOLINE 15P846
MPL.MOLINE 20P727
MUNCIE MK5X
NATL.MINE 402146DS
NEAPCO 1105X
NEAPCO 1105XR
NEAPCO 1-1275
NEAPCO 1205X
NEAPCO 1205XR
NEAPCO 1275X
NEAPCO 1275X1
NEAPCO 1275XR
NEAPCO 2872
NEW HOLL. 39360
NEW HOLL. 4557
NEW HOLL. 46452
NEW IDEA 011073
NEW IDEA K3347A
NEW IDEA O10530
NEW IDEA O12716
NORTH AMER G20139
OLDING 408414
OLIVER 402146DS
OLIVER 500-665E
PERFECT-CI 242
PILOT BL12N
PRECISION 995
REPCO CPL12N
REPCO CPL12R
REPCO RUJ1788
REPUBLIC CB772
REPUBLIC CB795
REX CHAIN 120
REX CHAIN 130007
REX CHAIN 130062
REX CHAIN 130065
REX CHAIN R12N
REX CHAIN R31N
SCHULTZ 27889
SPICER 5-242-1X
SPICER 5-242X
TOWMOTOR A650079
TRU-CROSS 242
TRW 20036
UJS 55-12L
WABCO 404645
WABCO 723741
WESCO 12N
WESCO 12R
WESCO 201012
WESCO 251012
WESCO 251-072
WESCO N12
WESCO R12
WEST.LAND AS72C19D
WEST.LAND AS72C19P
WHITE FARM 15P623
WHITE FARM 15P846
WHITE FARM 20P727
WHITE FARM 402-146DS

WHITE FARM 500-665E
WHITE FARM 522-311ES
WILK RIP. 2871
WOOD BROS. 2437
WYLIE 78
ZELLER 2525A
Z-WAY 1180

IDLI GROUP 80036

CROSS + 4 GROOVED BEARINGS + INSIDE SNAP RINGS
SEE FIGURE 1B

1.0625 in	BD	26.99 mm
.6875 in	BW	17.46 mm
2.3438 in	CL	59.53 mm
.6563 in	CD	16.67 mm

AEC 350
AEC 5-350
AEC AE5-350
ALBURUS-SP 5-310X
ALBURUS-SP 5-312XR
ALCO 1006
ALLOY 1006
ALLOY J5-350
BORG-WARNR 114-350
CLEVE.MOT. CM5-350X
GMB G5-350X
GMC 7825819
LEMPCO U140
LEMPCO U142
MCQUAY-NOR U01-64
MCQUAY-NOR UO1-64
MOPAR D211
MOPAR D94
MOTOR MAST 3200E
NEAPCO 1-0350
NEAPCO 28350X
PRECISION 368
REPCO 101-1010
REPCO 5-350X
REPUBLIC CB1272
SPICER 5-350X
TRW 20184
TRW 20184D
WESCO 201-553
ZELLER 3200E

IDLI GROUP 80037

CROSS + 4 ROUND BEARINGS + OUTSIDE SNAP RINGS
SEE FIGURE 1A

1.1250 in	BD	28.58 mm
.5938 in	BW	15.08 mm
2.3750 in	CL	60.33 mm
.6875 in	CD	17.46 mm

ADAMS-LET. 474804
AEC 14N
AEC AE-CPL14N
AEC CPL14N
A.E.LEE 3881-4
ALCO 1780
ALCO CP14N
ALCO CPL14N
ALLIS-CHLM 517311
ALLIS-CHLM 521720
ALLIS-CHLM 520299

BD = BEARING DIAMETER (outside)   BW = BEARING WIDTH   CB = CROSS LENGTH WITH BEARINGS   CC = CENTER TO CENTER   CD = CROSS DIAMETER
CL = CROSS LENGTH WITHOUT BEARINGS   EC = END TO CENTER or FACE TO FACE   EE = END TO END   EL = EFFECTIVE LENGTH
HD = HUB DIAMETER (or insert)   IE = INSIDE OF EARS (or recess)   OD = OUTSIDE DIAMETER   OE = OUTSIDE OF EARS   SB = SPLINE OR BORE SIZE

IDLI GUIDE — COPYRIGHT © INTERCHANGE, INC. — ST. LOUIS PARK, MN. 55416 USA

ALLIS-CHLM	531220
ALLIS-CHLM	823456
ALLOY	1780
ALLOY	CPL14N
ALLOY	CPL14N1
ATHEY PROD	P22172
BEAN,JOHN	1239290
BEAN,JOHN	1276704
BELLE CITY	S10092
BMB	CPL14N
BORG-WARNR	114-104
BORG-WARNR	CPL14N
BPC	B1400N
BRADY	1657
BRANDY	1657
BRILLION	7C270
BUSH HOG	138
BUSH HOG	139
CARDWELL	2513
CASE,J.I.	B13270
CASE,J.I.	F13270
CASE,J.I.	F85470
CASE,J.I.	FH756924
CHAIN BELT	140
CHAIN BELT	R14N
CHAMPION	SB171
CLARK EQU.	863324
CLARK EQU.	876579
CLARK EQU.	895600
CLEVE.MOT.	CM5-126X
CLEVE.MOT.	CM5-129X
CLEVE.MOT.	CM-CPL14N
CLEVE.MOT.	CM-CPL14N1
CLEVE.MOT.	CM-CPL14R
CMP	CM-CPL14N
CMP	CM-CPL14N1
COCKSHUTT	1947083
COCKSHUTT	402803DS
COCKSHUTT	XAF205A
COCKSHUTT	Z1020
DAFFIN	1T170
DANUSER	3677-6
DANUSER	7800
DEERE,JOHN	AE11033W
DEERE,JOHN	AE22753
DEERE,JOHN	AE26884
DEERE,JOHN	AP24580H
EATON	2950350
EATON	2950390
EATON	6950390
ELG.SWEEP.	63619
ENGLER	1796
FARMHAND	1R402
FIAT-ALLIS	517311
FIAT-ALLIS	521720
FIAT-ALLIS	526299
FIAT-ALLIS	531220
FIAT-ALLIS	70517311
FIAT-ALLIS	70521720
FIAT-ALLIS	70526299
FIAT-ALLIS	70531220
FIAT-ALLIS	70823456
FIAT-ALLIS	823456
FLEETRITE	ZAD-BL14N
FLEETRITE	ZADBLN14N
FORD	175423
FORD	221-613
FORD	243-250
G & G MFG	14R
G & G MFG	302-1400
GEHL BROS.	510007
GEHL BROS.	510114
GEHL BROS.	GN100W
GERLINGER	649052
GERLINGER	649502
GLAENZER	065-273-8
GMB	G5-104
GMB	G5-129-1X
GMB	G5-129X
GMB	GCP14N
GMB	GCP14N1
GMB	GCPL14N
GMB	GGPL14R
GMB	GGPL14R1X
GORMAN RUP	S1890
HEIL	30A313-1
HESSTON	753772
HESSTON	756924
HESSTON	805077
HUB CITY	03-35-90075
HUB CITY	90072
HUB CITY	UR14N
HUSCO	CPL14R
IHC	200-464C91
IHC	220-020H1
IHC	509-047R91
IHC	565225R91
IHC	569-047R91
IHC	607-917R91
IHC	656225R91
IHC	667-496R91
IHC	667-558R91
JACOBSEN	500357
KOEHRING	A649502
LILLISTON	01-61-021
LILLISTON	29-61-002
LUNDELL	S166
MACK	782V
MASS.FERG.	665-137M91
MASS.FERG.	807-034M91
MASS.FERG.	835-790M91
MASS.FERG.	835-799M91
MASS.FERG.	837-037M91
MASS.FERG.	841-237M91
MASS.FERG.	841-604M91
MATHEWS	618D
MCQUAY-NOR	U03-44
MCQUAY-NOR	UO3-44
MICHIGAN	863324
MICHIGAN	876579
MIPER	C3234
MOTOR MAST	2537A
MPL.MOLINE	15P405
NATL.MINE	3881-4
NATL.MINE	402803DS
NATL.MINE	76012012
NEAPCO	1-2075
NEAPCO	1-2076
NEAPCO	1405X
NEAPCO	2075X
NEAPCO	2075X1A
NEAPCO	28L14N
NEAPCO	28L14S
NEW HOLL.	129437
NEW HOLL.	136908
NEW HOLL.	34761
NEW HOLL.	43447
NEW HOLL.	95311
NEW IDEA	K3410A
NEW IDEA	K3591A
NEW IDEA	K3595
NEW IDEA	K3595A
NEW IDEA	O12014
OLIVER	402803DS
OLIVER	522312ES
OLIVER	5L1017
OLIVER	5LP9017
OLIVER	5LT017
OLIVER	5LT90-17
OLIVER	AC1777
OWATONNA	190-10049
PEARSON	VP1779
PERFECT-CI	129
PILOT	BL14N
PRECISION	880
PREC.TOR.	881
REPCO	CPL14N
REPCO	CPL14R
REPCO	CPL14R1
REPCO	CPL14S
REPUBLIC	CB-L14N
REPUBLIC	CB-L14S
REX CHAIN	130008
REX CHAIN	130070
REX CHAIN	140
REX CHAIN	R14N
RUPP	S1890
SERVIS	W298
SERVIS	W298A
SIDEWINDER	16203
SPEED-SPR.	1239290
SPICER	5-129-1X
SPICER	5-129-2X
SPICER	5-129X
TAYLOR	1714-010
TAYLOR	7174-010
TERRAIN	1796
TOWMOTOR	649502
TRU-CROSS	129
TRW	20001D
TRW	20037
UJS	55-14LN
WABCO	3200LS
WABCO	474804
WEASLER	GH
WEASLER	GN
WESCO	14N
WESCO	14R
WESCO	261014
WESCO	N14
WESCO	R14
WHITE FARM	15P405
WHITE FARM	402-803DS
WHITE FARM	522-312ES
WHITE FARM	5LT90-17
WHITE FARM	AC17-77
WOOD BROS.	154
ZELLER	2537A
Z-WAY	1190

IDLI GROUP 80038

CROSS + 4 SERRATED BEARINGS + THRUST AND LOCK PLATES
SEE FIGURE 1C

1.1406 in	BD	28.97 mm	
.7656 in	BW	19.45 mm	
2.6406 in	CL	67.07 mm	
.6563 in	CD	16.67 mm	

AEC	1253
AEC	1254
AEC	AE-S96-55
AEC	S53-55
ALCO	JS96-55
ALCO	RS96-55
ALCO	S96-55
ALLOY	1130
ALLOY	JS53-55
AMER.MOTOR	804003
AMER.MOTOR	810969
AMER.MOTOR	8126603
BORG-WARNR	114-120
BORG-WARNR	S96-55
CLEVELAND	S53-55
CLEVE.MOT.	CM5355S
CLEVE.STL.	S96-55
D&D MACH.	MCS96
FLEETRITE	ZADS15355
GMB	G5-1233X
GMB	G5-1239X
GMB	G5-1253X
GMB	G5-1254X
GMB	G5-596-55
GMB	GS96-55
GMB	GU2770
HABERLE	HS5355
KAISER	804003
LEMPCO	U304
LEMPCO	U307
LOBRO	U277
MASTERS	MCS53
MASTERS	MCS96
MIPER	C3461
MOPAR	D109
MOPAR	D111
MOTOR MAST	2464AB
MOTOR MAST	2464B
NEAPCO	285400
NEAPCO	3-5600
PERFECTION	RS53-55
PERFECTION	RS96-55
PILOT	S19655
PRECISION	419
REPCO	S96-55
REPUBLIC	CB-S96
REPUBLIC	CPS96
SPICER	S96-55
TEDDY TORQ	GS96-55
TEDDY TORQ	S96
TRW	20033
TRW	20185
TRW	20185D
WESCO	201967
WESCO	NS965
WILLYS	804003

ENGINEERING CATALOGS MUST BE CONSULTED FOR DETAILS NOT INCLUDED
IN THIS GUIDE. SPECIFIC DESIGNS, MATERIAL CONTENT, TOLERANCES
LUBE FITTINGS AND OTHER DIMENSIONS ARE INTENTIONALLY OMITTED HERE.

CAUTION: BE SURE TO REFER TO ENGINEERING CATALOGS FOR SPECIAL APPLICATIONS THAT REQUIRE SPECIFIC MATERIAL CONTENT, TOLERANCES, ETC. SEE FOOTNOTE.

ZELLER	2464AB

IDLI GROUP 80039

CROSS + 4 SLOTTED BEARINGS + THRUST AND LOCK PLATES
SEE FIGURE 1D

1.1406 in	BD	28.97 mm
.8125 in	BW	20.64 mm
2.6406 in	CL	67.07 mm
.6563 in	CD	16.67 mm

AEC	AE-P96-55-1
AEC	AE-S53-55
AEC	S53-55
ALCO	1130
ALCO	JS53-55
ALCO	S53-55
ALLOY	JS53-55
BORG-WARNR	114-120
BORG-WARNR	S53-55
CLEVELAND	S53-55
CLEVE.MOT.	CM5355S
CLEVE.STL.	S53-55
FLEETRITE	ZADS15355
GMB	G5-1254X
GMB	G5-1296X
GMB	G5S53-55
GMB	GS53-55
GMB	GU2760
GMB	GU-S53-55
GMB	GU-S96-55
HABERLE	HS5355
LEMPCO	U307
LOBRO	U276
MASTERS	MCS53
MIPER	C3460
MOPAR	D109
MOTOR MAST	2464A
MOTOR MAST	2464AB
NEAPCO	285600
NEAPCO	3-5600
PERFECTION	RS53-55
PILOT	S15355
PRECISION	422
REPCO	S53-55
REPUBLIC	CB553
REPUBLIC	CB-S53
SPICER	20185
SPICER	S53-55
TEDDY TORQ	GS53-55
TEDDY TORQ	S53
TRANSMISS.	UK460
TRANSMISS.	UK461
TRW	20032
TRW	20185D
WESCO	201535
WESCO	NS535
WILLYS	810969
ZELLER	2464A

IDLI GROUP 80040

CROSS + 4 ROUND BEARINGS + OUTSIDE SNAP RINGS
SEE FIGURE 1A

1.0625 in	BD	26.99 mm
.5313 in	BW	13.50 mm
2.750 in	CL	69.85 mm
.6563 in	CD	16.67 mm

BRD	03-501523
GLAENZER	365-007-0
GMB	GU1170
HARDY SPCR	3051-035N
HARDY SPCR	K5GB35
MIPER	C3148
PRECISION	366
REPCO	K5GB35
ROVER	242521
ROVER	512544
WYLIE	22

IDLI GROUP 80041

CROSS + 4 ROUND BEARINGS + OUTSIDE SNAP RINGS
slight variation from above group
SEE FIGURE 1A

1.0625 in	BD	26.99 mm
.5313 in	BW	13.50 mm
2.750 in	CL	69.85 mm
.6563 in	CD	16.67 mm

AEC	504
AEC	AE502
AEC	AE504
ALCO	1021
ALCO	154
ALCO	J5-4
ALFA ROMEO	9813973
ALLOY	1021
ALLOY	J5-4
ALVIS	C2665
AMER.MOTOR	28075A
ARMSTRONG	5410-1100
ARMSTRONG	5410-161
AYRA DUREX	1-305-01
BEDFORD	7190326
BMC	301482
BMC	78719
BMC	8G8981
BMC	GUJ1001
BMC	GUJ1002
BMC	GUJ102
BMC	GUJ103
BMC	GUJ107
BMC	GUJ116
BMC	N1052A16
BMC	P301
BORG-WARNR	114-502
BORG-WARNR	114-504
BRD	03-501346
BRD	03-503941
BRD	03-530402
BRD	03-531346
BRD	A03-500660
CENTRAL	GU100
CHRYSLER	5039570
CHRYSLER	5044943
CHRYSLER	947547
CHRYSLER	D18888-5
CHRYSLER	K18246
CITROEN	110310120
CLEVE.MOT.	CM5-4X
CMP	CM5-4X
DAIMLER	302927-10
DAIMLER	944842
DIAMOND R	55P1
ERF	U24125
EXCEL	U100
FIAT	4552008
FIAT	4552009
FIAT	4556519
FIAT	4625636
FIAT	882320
FIAT	882321
FIAT	967265
FLEETRITE	ZAD11504
FORD	105E7039
FORD	213E4635A
FORD	2400E4635A
FORD	3010E4635A
FORD	404E4635A
FORD	E0A7039
FORD	E0A7039
GLAENZER	365-000-0
GLAENZER	365-002-0
GLAENZER	365-040-0
GLAENZER	365-047-0
GMB	G5-14
GMB	G5-4X
GMB	G5-504
GMB	GU1100
GMB	GU1100SL
GMB	GU700
GMC	7190326
HABERLE	H154X
HARDY SPCR	30151-186X
HARDY SPCR	3051-161N
HARDY SPCR	3051-195W
HARDY SPCR	K5-24
HARDY SPCR	K5-4
HARDY SPCR	K5GB116
HARDY SPCR	K5GB161
HARDY SPCR	K5L4
HARDY SPCR	K5LGB130
HARDY SPCR	K5LGB186
HARDY SPCR	K5LGB195
HARDY SPCR	K5LGB36
HARDY SPCR	K5LGB83
HARDY SPCR	K5LGB87
HARDY SPCR	K5PL39
HERCULES	405
HOLDEN	7190326
IHC	3100758R91
IHC	74050092
IHC	815277R91
INBRA	CA504
INBRA	CT18-005
ITALCARDAN	50-855-000
LEMPCO	U100
LEYLAND	301482
LEYLAND	302927/10
LEYLAND	43123
LEYLAND	46
LEYLAND	8G8981
LEYLAND	944842
LOBRO	U11
LOBRO	U110
LOBRO	U111
LOBRO	U112
LOBRO	U115
LOBRO	U117
LUDWIG	443-8901
LUDWIG	444-8902
MASS.FERG.	3406040M91
MASTERS	MA54X
MATSUBA	UJ015
MATSUI	UJ055
MCQUAY-NOR	U01-13
MCQUAY-NOR	U01-13
MIPER	C3120
MIPER	C3121
MIPER	C3122
MIPER	C3123
MOPAR	D56
MOTOR MAST	2275B
NEAPCO	1-0005
NEAPCO	1-5
NEAPCO	2805X
PERFECTION	RG5-2X
PERFECTION	RG5-4X
PILOT	11502
PILOT	11504
PITT.VIOL.	40.855.000
PRECISION	344
QUINT.HAZ.	QL16103
QUINT.HAZ.	QL291
QUINT.HAZ.	QL502
QUINTON	QL291
RENAULT	50.00.243.236
REO MOTORS	55P1
REPCO	101-1280
REPCO	K5-4R
REPCO	K5L4
REPCO	K5L49
REPCO	K5L4R
REPCO	K5LGB36R
REPUBLIC	CB1250B
ROVER	243023
ROVER	243203
ROVER	43023
ROVER	43203
ROVER	541858
RUEDARSA	1090
RUEDARSA	1090-2
RUEDARSA	1090-5
RUEDARSA	2103-1090-0-0
RUEDARSA	2103-1090-2-0
RUEDARSA	2103-1090-5-0
SAVIEM	23232051
SAVIEM	9813973
SAVIEM	981397300
SPICER	5-2X
SPICER	5-4X
SPICER	5-94X
TEDDY TORQ	K2
TEDDY TORQ	K4
TRANSMISS.	UK120
TRANSMISS.	UK121
TRANSMISS.	UK123
TRIUMPH	100524060

BD = BEARING DIAMETER (outside)   **BW** = BEARING WIDTH   **CB** = CROSS LENGTH WITH BEARINGS   **CC** = CENTER TO CENTER   **CD** = CROSS DIAMETER
CL = CROSS LENGTH WITHOUT BEARINGS   **EC** = END TO CENTER or FACE TO FACE   **EE** = END TO END   **EL** = EFFECTIVE LENGTH
HD = HUB DIAMETER (or insert)   **IE** = INSIDE OF EARS (or recess)   **OD** = OUTSIDE DIAMETER   **OE** = OUTSIDE OF EARS   **SB** = SPLINE OR BORE SIZE

TRIUMPH	134778
TRIUMPH	144859
TRIUMPH	503757
TRU-CROSS	4
TRW	20023
UJS	55-4
UNIPART	GUJ1002
UNIPART	GUJ102
UNIPART	GUJ103
VAPORMATIC	VTE5004
VOLVO	383.472
VOLVO	38403
WALTERSCHD	21-00-00
WESCO	201-285
WESCO	201504
WESCO	N1502
WESCO	N1504
WHITE	55P1
WORLDPARTS	49-115
WORLDPARTS	W49-115
WORLDPARTS	W49-120
WORLDPARTS	W49-121
WYLIE	21
ZELLER	2275B

IDLI GROUP 80042

CROSS + 4 GROOVED BEARINGS + INSIDE SNAP RINGS
SEE FIGURE 1B

1.2344 in	BD	31.35 mm
.8125 in	BW	20.64 mm
2.750 in	CL	69.85 mm
.7188 in	CD	18.26 mm

AEC	162
AEC	AE162
ALCO	1090
ALCO	51625-2
ALCO	J51625-2
ALLOY	1090
ALLOY	J51625-2
BORG-WARNR	114-162
CHRYSLER	947556
CLARK EQU	865751
CLARK EQU.	86751
CLEVE.MOT.	CM51625-2
CLEVE.MOT.	CM732522
CLEVE.MOT.	CM735253
CMP	CM51625-2
CMP	CM73252-2
CMP	CM73525-3
D&D MACH.	MD2526
DETROIT	51615-2
DETROIT	51625-2
EXCEL	U415
FLEETRITE	ZAD17328
GLAENZER	065-001-0
GLAENZER	065-907-8
GMB	G5-1300X
GMB	G5-162
GMB	G51625-2
GMB	GU2500
HABERLE	H7125
INBRA	CA375
INBRA	CC375
ITALCARDAN	50-744-000
LEMPCO	U415

LOBRO	U250
MASTERS	MD2526
MCQUAY-NOR	U02-2
MCQUAY-NOR	UO2-2
MICHIGAN	865731
MIPER	C3360
MOPAR	D126
MOTOR MAST	2375D
MOTOR MAST	3575D
NEAPCO	2-1400
NEAPCO	281400
PERFECTION	RG162
PILOT	17328
PITT.VIOL.	40.744.000
PRECISION	305
REPCO	114-162
REPUBLIC	CB51625
REPUBLIC	CB51626
SPICER	5-1300
SPICER	5-1300X
TRANSMISS.	UK360
TRU-CROSS	1300
TRW	20028
UJS	55-51625-2
WESCO	201558
WESCO	D156
WESCO	N556
ZELLER	2375D

IDLI GROUP 80043

CROSS + 4 GROOVED BEARINGS + INSIDE SNAP RINGS
SEE FIGURE 1B

1.0000 in	BD	25.40 mm
.6719 in	BW	17.07 mm
2.8750 in	CL	73.03 mm
.5781 in	CD	14.68 mm

AEC	2063
AEC	2127
AEC	AE2063
AEC	AE508
ALCO	2063
BMC	HYL4007
BORG-WARNR	114-2007
BORG-WARNR	114-2063
BORG-WARNR	114-2108
BORG-WARNR	114-2121
BORG-WARNR	114-2124
BORG-WARNR	114-2177
BORG-WARNR	114-2182
D&D MACH.	MM2063
DEERE,JOHN	23025
GLAENZER	065-231-8
GMB	G2121
GMC	419976
GMC	509296
GMC	516405
GMC	5675414
HERCULES	375
LEMPCO	U204
LEMPCO	U233
LEMPCO	U239
MCQUAY-NOR	U01-4
MCQUAY-NOR	U01-5
MOTOR MAST	D2060D
NEAPCO	1805X1

NEAPCO	2-0350
NEAPCO	280300
NEAPCO	280350
NEAPCO	28108X
PERFECTION	RG2063
PILOT	11407
PILOT	11463
PRECISION	515
PRECISION	525
REPCO	114-2108
REPCO	114-2121
REPCO	RUJ2010
REPUBLIC	CB2007
REPUBLIC	CB2063
RUEDARSA	1351-7
RUEDARSA	2104-1351-7-0
SPICER	5-108X
SPICER	5-2007X
SPICER	5-2011X
TRW	20123
WESCO	201259
WESCO	G2046-1
WESCO	N2046-1
WESCO	N2046G
WESCO	N2063
ZELLER	D2060D

IDLI GROUP 80044

CROSS + 4 GROOVED BEARINGS + INSIDE SNAP RINGS
slight variation from above group
SEE FIGURE 1B

1.0000 in	BD	25.40 mm
.6719 in	BW	17.07 mm
2.8750 in	CL	73.03 mm
.5781 in	CD	14.68 mm

AEC	2112
AEC	2127
AEC	3121
AEC	5-350
AEC	AE2101
AEC	AE2112
AEC	AE2127
ALCO	1071
ALCO	114-2163
ALCO	1784
ALCO	2127
ALCO	J2101
ALCO	J2108
ALCO	J2112
ALCO	J2121
ALCO	J2127
ALCO	J2131
ALLIS-CHLM	08744
ALLIS-CHLM	20268
ALLIS-CHLM	631382
ALLIS-CHLM	631383
ALLOY	1071
ALLOY	1784
ALLOY	J2127
AMER.MOTOR	3112002
AMER.MOTOR	3115912
AMER.MOTOR	3119646
AMER.MOTOR	3119649
AMER.MOTOR	3172840
AMER.MOTOR	3172850

AMER.MOTOR	3200332
AMER.MOTOR	3200538
AMER.MOTOR	3200855
AMER.MOTOR	3205371
AMER.MOTOR	3206121
AMER.MOTOR	8126606
ATHEY PROD	P22239
AUST.MOT	R004-01-29
BALKAMP	M2011
BALKAMP	M2016
BALKAMP	M2039
BALKAMP	M2041
BALKAMP	M2046
BALKAMP	M2050
BALKAMP	M2063
BALKAMP	M2064
BALKAMP	M2077
BALKAMP	M2101
BALKAMP	M2108
BALKAMP	M2111
BALKAMP	M2112
BALKAMP	M2119
BALKAMP	M2121
BALKAMP	M2127
BAL-LM-HAM	465225
BAL-LM-HAM	AW465225
BORG-WARNR	114-106
BORG-WARNR	114-2011
BORG-WARNR	114-2016
BORG-WARNR	114-2039
BORG-WARNR	114-2046
BORG-WARNR	114-2050
BORG-WARNR	114-2064
BORG-WARNR	114-2077
BORG-WARNR	114-2101
BORG-WARNR	114-2112
BORG-WARNR	114-2119
BORG-WARNR	114-2127
BORG-WARNR	114-2131
BORG-WARNR	114-2132
BORG-WARNR	114-2133
BORG-WARNR	114-2137
BORG-WARNR	114-2144
BORG-WARNR	114-2145
BORG-WARNR	114-2151
BORG-WARNR	114-2160
BORG-WARNR	114-2161
BORG-WARNR	114-2164
BORG-WARNR	114-2174
BORG-WARNR	114-2179
BORG-WARNR	114-2184
BORG-WARNR	114-2187
BORG-WARNR	J2007
BORG-WARNR	J2011
BORG-WARNR	J2016
BORG-WARNR	J2039
BORG-WARNR	J2046
BORG-WARNR	J2063
BORG-WARNR	J2064
BUCYR-ERIE	46-810-780(4)
CENTRAL	G5-2011X
CHAMPLAIN	96E105
CLEVE.MOT.	CM2063
CLEVE.MOT.	CM2101
CLEVE.MOT.	CM2112
CLEVE.MOT.	CM2121

ENGINEERING CATALOGS MUST BE CONSULTED FOR DETAILS NOT INCLUDED
IN THIS GUIDE. SPECIFIC DESIGNS, MATERIAL CONTENT, TOLERANCES
LUBE FITTINGS AND OTHER DIMENSIONS ARE INTENTIONALLY OMITTED HERE.

IDLI GUIDE COPYRIGHT © INTERCHANGE, INC. ST. LOUIS PARK, MN. 55416 USA

CAUTION: BE SURE TO REFER TO ENGINEERING CATALOGS FOR SPECIAL APPLICATIONS THAT REQUIRE SPECIFIC MATERIAL CONTENT, TOLERANCES, ETC. SEE FOOTNOTE.

Manufacturer	Part Number
CLEVE.MOT.	CM2127
CLEVE.MOT.	CM5-108X
CMP	CM2063
CMP	CM2101
CMP	CM2127
CMP	CM5-108X
D&D MACH.	MM2101
D&D MACH.	MM2127
DANUSER	A1228
DEERE,JOHN	AE11027
DEERE,JOHN	AE11027E
DEERE,JOHN	AJ12823E
DEERE,JOHN	AJ7330HN
DEERE,JOHN	AM2854E
DEERE,JOHN	L1827C
DEERE,JOHN	XAG2230E
DEERE,JOHN	XAG2513E
DEERE,JOHN	XAG2530E
EATON	809332
FIAT-ALLIS	082744
FIAT-ALLIS	08744
FIAT-ALLIS	20268
FIAT-ALLIS	70020268
FIAT-ALLIS	70082744
FLEETRITE	ZAD11446
FLEETRITE	ZAD11463
FLEETRITE	ZADB58N
FORD	1M7042
FORD	1M7042A
FORD	8M7042
FORD	B5A7042A
FORD	B5AZ4635B
FORD	C5DZ4635B
FORD	MEK7042
FORD	MEK7042A
FORD	TP4635A
GMB	G1003
GMB	G1004
GMB	G1501
GMB	G2101
GMB	G2102
GMB	G2112
GMB	G2127
GMB	G2131
GMB	G2134
GMB	G5-2011X
GMB	G5-2011XA
GMB	G5-2022X
GMB	G7242
GMB	GU6200
GMC	20268
GMC	516046
GMC	5675412
GMC	5675415
GMC	VS10606
GMC	VS10607
HABERLE	H2011
HABERLE	H2016
HABERLE	H2036
HABERLE	H2039
HABERLE	H2045
HABERLE	H2064
HERCULES	375
HERCULES	379
HUB CITY	03-35-90073
HUB CITY	UR17110
HYSTER	190739
HYSTER	88285A
IHC	110-859R91
IHC	157714R91
IHC	157-794R91
IHC	321-485R91
IHC	474-792R91
IHC	474-793R91
IHC	581-954R91
IHC	65344R92
IHC	656223R91
IHC	656-223R92
IHC	77-979R91
IHC	871280R92
IHC	NH47621
INBRA	CA2101
ITALCARDAN	50-816-000
JACOBSEN	500479
JEEP	934820
KAISER	934920
LEMPCO	U135
LEMPCO	U142
LEMPCO	U205
LEMPCO	U218
LEMPCO	U239G
LEYLAND	HYL4674
LOBRO	U607
LOBRO	U620
MASS.FERG.	840-303M92
MASTERS	MM2063
MASTERS	MM2101
MCQUAY-NOR	U01-5
MCQUAY-NOR	U03-5
MCQUAY-NOR	UO1-4
MCQUAY-NOR	UO3-5
MIPER	C3320
MOPAR	D13
MOPAR	D5
MOPAR	D94
MOTOR MAST	2060D
MOTOR MAST	2060J
MOTOR MAST	2060LA
MOTOR MAST	D2060D
NEAPCO	1-0350
NEAPCO	1-1875
NEAPCO	1-1876
NEAPCO	1-3600
NEAPCO	1805X
NEAPCO	1805X2
NEAPCO	1875X
NEAPCO	1875X1
NEAPCO	2-350
NEAPCO	280400
NEAPCO	280450
NEAPCO	28300
NEAPCO	283600
NEAPCO	283600P
NEAPCO	2858N
NEAPCO	4-1058
NEW HOLL.	34371
NEW IDEA	K3803A
OLIVER	520243
OLIVER	52043
PERFECT-CI	2011
PERFECTION	RG2016
PERFECTION	RG2039
PERFECTION	RG2046
PERFECTION	RG2063
PERFECTION	RG2064
PERFECTION	RG2163
PILOT	11411
PILOT	11412
PILOT	11416
PILOT	11439
PILOT	11446
PILOT	11450
PILOT	11463G
PRECISION	507
PRECISION	525
QUINT.HAZ.	QL2030
REPCO	114-2112
REPCO	114-2113
REPCO	114-2119
REPCO	114-2127
REPCO	114-2131
REPCO	RUJ2005
REPCO	RUJ2030
REPUBLIC	CB2006
REPUBLIC	CB2007
REPUBLIC	CB2007C
REPUBLIC	CB2011
REPUBLIC	CB2011C
REPUBLIC	CB2016
REPUBLIC	CB2039
REPUBLIC	CB2046
REPUBLIC	CB2063
REPUBLIC	CB2064
RUEDARSA	1351
RUEDARSA	2104-1351-0-0
SPICER	5-118X
SPICER	5-2011X
SPICER	5-2064X
STUDEBAKER	1559481
TEDDY TORQ	1027
TEDDY TORQ	1042
TEDDY TORQ	1423
TEDDY TORQ	1425
TEDDY TORQ	G2101
TEDDY TORQ	G2121
TRANSMISS.	UK320
TRANSMISS.	UK320/6
TRU-CROSS	2011
TRW	20007
TRW	20088D
TRW	20123
TRW	20158
UJS	55-2063
WESCO	201246
WESCO	201-248
WESCO	202-246
WESCO	202-259
WESCO	G2046G
WESCO	N2046
WESCO	N2046G
WESCO	N2064
WHITE	934820
WHITE FARM	52043
WILLYS	934820
WYLIE	47
ZELLER	2060D

IDLI GROUP 80045

CROSS + 2 WING TYPE BEARINGS + 2 ROUND GROOVED BEARINGS
SEE FIGURE 1BI

1.0000 in	BD	25.40 mm
.6719 in	BW	17.07 mm
2.8750 in	CL	73.03 mm
.5781 in	CD	14.68 mm

Manufacturer	Part Number
AEC	2123
AEC	AE2123
AEC	AE2133
ALCO	2123
ALCO	J2123
BALKAMP	M2007
BALKAMP	M2014
BALKAMP	M2066
BORG-WARNR	114-2014
BORG-WARNR	114-2066
BORG-WARNR	114-2123
BORG-WARNR	114-2128
BORG-WARNR	J2014
CHRYSLER	1450703
CHRYSLER	1450723
CHRYSLER	1450727
CHRYSLER	986493
CLEVE.MOT.	CM2006
CLEVE.MOT.	CM2123
GMB	G2123
HABERLE	H2014
HABERLE	H2066
HERCULES	378
LEMPCO	U210
LEMPCO	U235
LOBRO	U614
MOPAR	D11
MOTOR MAST	2060F
NEAPCO	280900
NEAPCO	283900
PERFECTION	RG2066
PILOT	11414
PILOT	11464
PILOT	11466
PITT.VIOL.	40.815.000
PRECISION	519
REPCO	114-2123
REPUBLIC	CB2014
REPUBLIC	CB2066
RUEDARSA	1280-4
RUEDARSA	2100-1280-4-0
SPICER	5-2014
SPICER	5-2014X
SPICER	5-2066X
TEDDY TORQ	1419
TRANSMISS.	UK676
WESCO	201266
WESCO	N2014
WESCO	N2066

BD=BEARING DIAMETER (outside) BW=BEARING WIDTH CB=CROSS LENGTH WITH BEARINGS CC=CENTER TO CENTER CD=CROSS DIAMETER
CL=CROSS LENGTH WITHOUT BEARINGS EC=END TO CENTER or FACE TO FACE EE=END TO END EL=EFFECTIVE LENGTH
HD=HUB DIAMETER (or insert) IE=INSIDE OF EARS (or recess) OD=OUTSIDE DIAMETER OE=OUTSIDE OF EARS SB=SPLINE OR BORE SIZE

IDLI GROUP 80046

**CROSS AND LOW OR HIGH WING TYPE
DRILLED OR THREADED BEARINGS
SEE FIGURE 1I**

1.3125 in	CC	33.34 mm
2.8750 in	CL	73.03 mm
.5781 in	CD	14.68 mm

AEC	2117
AEC	2186
AEC	AE2117
AEC	AE2186
ALCO	J2117
ALLIS-CHLM	4984505
ALLIS-CHLM	A45050
ALLOY	J2117
BORG-WARNR	114-2047
BORG-WARNR	114-2106
BORG-WARNR	114-2117
BORG-WARNR	114-2117A
BORG-WARNR	114-2120
BORG-WARNR	114-2186
BORG-WARNR	114-2186A
CASE,J.I.	L30634
CATERPILLR	316117
CLEVE.MOT.	CM2117
D&D MACH.	MM2117
DARLINGTON	813-2114
DIGIO-LIFT	12/2C16
FIAT-ALLIS	4906478
FIAT-ALLIS	4914098
FIAT-ALLIS	4984505
FIAT-ALLIS	4984505-0
FIAT-ALLIS	498505
FIAT-ALLIS	70498505
FIAT-ALLIS	74906478
FIAT-ALLIS	74906478-5
FIAT-ALLIS	74984505
FIAT-ALLIS	74984505-0
FIAT-ALLIS	A45050
FLEETRITE	ZAD11417
GALION	D77092
GERLINGER	316117
GERLINGER	32361
GERLINGER	35822
GERLINGER	36142
GERLINGER	369765
GERLINGER	B1729
GLAENZER	065-168-8
GMB	G2117
GMB	G217
GMB	G5-2117
GMB	GUK02
KOMATSU	101-10-00080
KOMATSU	101.10.17200
LA PORTA	LA3200
LIFT TRUCK	12-2C16
MCQUAY-NOR	U03-7
MOTOR MAST	2060X
MOTOR MAST	2060XA
MOTOR MAST	2060XB
MOTOR MAST	2060XC
MOTOR MAST	2060Z
NEAPCO	1-2186
NEAPCO	282117
NEAPCO	282186
NEAPCO	282186A
PRECISION	994
REPCO	114-2117
REPCO	RUJ2004
REPUBLIC	CB794
SPICER	5-2002X
TOWMOTOR	316117
TOWMOTOR	32361
TOWMOTOR	369765
TRW	20015
WESCO	201219
WESCO	N2117
ZELLER	2060X

IDLI GROUP 80047

**CROSS + 2 WING TYPE BEARINGS
+ 2 ROUND GROOVED BEARINGS
SEE FIGURE 1AJ**

1.3125 in	CC	33.34 mm
2.8750 in	CL	73.03 mm
.5781 in	CD	14.68 mm

ADAMS-LET.	TF9791
ADAMS-LET.	TF9792
AEC	2116
AEC	2116A
AEC	2185
AEC	AE2116
AEC	AE2116A
AEC	AE2185
ALCO	1063
ALCO	J2104
ALCO	J2116
ALCO	J2124
ALCO	J2125
ALLIS-CHLM	3008392-7
ALLIS-CHLM	4762487
ALLIS-CHLM	4987664
ALLIS-CHLM	74906479
ALLIS-CHLM	A45040
ALLOY	1063
ALLOY	J2104
ANTHONY	SH1481-129
AUST.WEST.	465020
AUST.WEST.	465022
AUST.WEST.	465159
AUST.WEST.	AW465020
AUST.WEST.	AW465159
BALKAMP	3-952
BALKAMP	M2044
BALKAMP	M2104
BALKAMP	M2116
BAL-LM-HAM	465159
BORG-WARNR	114-2044
BORG-WARNR	114-2104
BORG-WARNR	114-2115
BORG-WARNR	114-2116
BORG-WARNR	114-2116A
BORG-WARNR	114-2125
BORG-WARNR	114-2130
BORG-WARNR	114-2143
BORG-WARNR	114-2156
BORG-WARNR	114-2170
BORG-WARNR	114-2172
BORG-WARNR	114-2185
BORG-WARNR	114-2185A
BORG-WARNR	114-2206
CASE,J.I.	A14725
CASE,J.I.	D34948
CASE,J.I.	L30492
CASE,J.I.	LB32548
CATERPILLR	316116
CATERPILLR	445581
CATERPILLR	6S6902
CATERPILLR	945581
CATERPILLR	D2X2203
CLARK EQU.	5202145
CLARK EQU.	945581
CLEVE.MOT.	CM2116
CLEVE.MOT.	CM5006
CMP	CM2116
D&D MACH.	MM2116
DARLINGTON	813-4305
DIGIO-LIFT	9-0S12
DROTT	74650
DROTT	74650M
DROTT	94650
FIAT-ALLIS	3008392
FIAT-ALLIS	3008392-7
FIAT-ALLIS	4762487
FIAT-ALLIS	476-2487-9
FIAT-ALLIS	4906479
FIAT-ALLIS	4914099
FIAT-ALLIS	4984504
FIAT-ALLIS	4984504-3
FIAT-ALLIS	498504
FIAT-ALLIS	4987664
FIAT-ALLIS	70008392
FIAT-ALLIS	70498504
FIAT-ALLIS	73008392-7
FIAT-ALLIS	74762487
FIAT-ALLIS	74762487-9
FIAT-ALLIS	74906478
FIAT-ALLIS	74906479-3
FIAT-ALLIS	74984504
FIAT-ALLIS	74984504-3
FIAT-ALLIS	74987664
FIAT-ALLIS	8392
FIAT-ALLIS	A45040
FLEETRITE	ZAD11416
FLEETRITE	ZAD16108
GALION	D77093
GERLINGER	17691
GERLINGER	316616
GERLINGER	32360
GERLINGER	35823
GERLINGER	36141
GERLINGER	369764
GERLINGER	53038
GERLINGER	617861
GERLINGER	617862
GERLINGER	644189C
GERLINGER	B1769
GERLINGER	B17691
GLAENZER	065-134-8
GMB	G2104
GMB	G2110
GMB	G2116
GMB	G2125
GMB	G5-1226X
GMB	G5-2116X
GMB	G5-2172
GMB	GCP35N3
GMB	GU2114
GMB	GU2116
GMB	GUK01
GMC	616413
HOUGH	149318HI
IHC	149318
IHC	149-318H1
INBRA	CT20-507
KOMATSU	101-10-00070
KOMATSU	101.10.17100
LA PORTA	LA1516
LIFT TRUCK	9-0S12
MASTERS	MM2116
MCQUAY-NOR	U03-3
MCQUAY-NOR	U03-8
MICHIGAN	945581
MOTOR MAST	2060Y
MOTOR MAST	2060YA
MOTOR MAST	2060YB
MOTOR MAST	2060YC
M.R.S.	56695F11
M.R.S.	58897F11
NEAPCO	1-2185
NEAPCO	282116
NEAPCO	282172
NEAPCO	282185
NEAPCO	282185A
OSHKOSH	2MU4
PETTI-MULL	40292
PETTI-MULL	50292
PILOT	11416
PITT.VIOL.	40.812.000
PRECISION	952
QUINT.HAZ.	QL220
REPCO	114-2104
REPCO	114-2110
REPCO	114-2116
REPCO	114-2143
REPCO	114-2172
REPUBLIC	CB752
SPICER	5-2116X
TOWMOTOR	316116
TOWMOTOR	32360
TOWMOTOR	369764
TRU-CROSS	2116
TRW	20016
WABCO	TF9192
WABCO	TF9792
WABCO	VJ0636
WESCO	201201
WESCO	G2127
WESCO	N2001
WHITING	241133
ZELLER	2060Y

IDLI GROUP 80048

**CROSS + 2 WING TYPE BEARINGS
+ 2 ROUND GROOVED BEARINGS
slight variation from above group
SEE FIGURE 1AJ**

1.3125 in	CC	33.34 mm
2.8750 in	CL	73.03 mm
.5781 in	CD	14.68 mm

AEC	2110
AEC	AE2103
AEC	AE2110
ALCO	J2103

ENGINEERING CATALOGS MUST BE CONSULTED FOR DETAILS NOT INCLUDED
IN THIS GUIDE. SPECIFIC DESIGNS, MATERIAL CONTENT, TOLERANCES
LUBE FITTINGS AND OTHER DIMENSIONS ARE INTENTIONALLY OMITTED HERE.

CAUTION: BE SURE TO REFER TO ENGINEERING CATALOGS FOR SPECIAL APPLICATIONS THAT REQUIRE SPECIFIC MATERIAL CONTENT, TOLERANCES, ETC. SEE FOOTNOTE.

Manufacturer	Part Number
ALCO	J2110
ALCO	J2150
ALLIS-CHLM	178461
ALLIS-CHLM	3006546
ALLIS-CHLM	4690
ALLIS-CHLM	498504
ALLIS-CHLM	6546
ALLOY	J2103
ATHEY PROD	P12560
ATHEY PROD	P20305
ATHEY PROD	P20322
AUST.WEST.	465019
AUST.WEST.	465158
AUST.WEST.	AW465019
AUST.WEST.	AW465158
AUST.WEST.	HCF6641A
BALKAMP	3-951
BALKAMP	M2002
BALKAMP	M2003
BALKAMP	M2067
BALKAMP	M2103
BALKAMP	M2110
BAL-LM-HAM	465019
BAL-LM-HAM	465158
BAL-LM-HAM	HCF6641A
BORG-WARNR	114-2002
BORG-WARNR	114-2003
BORG-WARNR	114-2005
BORG-WARNR	114-2008
BORG-WARNR	114-2067
BORG-WARNR	114-2103
BORG-WARNR	114-2110
BORG-WARNR	114-2135
BORG-WARNR	114-2136
BORG-WARNR	114-2150
BORG-WARNR	114-2166
BORG-WARNR	114-2168
BORG-WARNR	114-2171
BORG-WARNR	114-2214
BORG-WARNR	J2002
CATERPILLR	1250
CATERPILLR	451105
CATERPILLR	644683
CATERPILLR	673627
CATERPILLR	XX1804-6
CATERPILLR	XX1894-6
CLARK EQU.	5202144
CLEVE.MOT.	CM2002
CLEVE.MOT.	CM2103
CLEVE.MOT.	CM2110
CMP	CM2002
CMP	CM2103
CMP	CM2110
D&D MACH.	MM2002
DEERE,JOHN	AN30706
DEERE,JOHN	AN30706N
DEERE,JOHN	AN31873N
DEERE,JOHN	AN32803
DEERE,JOHN	AN32803N
DIAMOND R	3631P2
DROTT	70148
DROTT	70148M1
EATON	5043828
FIAT-ALLIS	0178461
FIAT-ALLIS	0178461-0
FIAT-ALLIS	178461
FIAT-ALLIS	3006545
FIAT-ALLIS	3006546
FIAT-ALLIS	3006546-0
FIAT-ALLIS	6546
FIAT-ALLIS	70006546
FIAT-ALLIS	70178461
FIAT-ALLIS	70178461-0
FIAT-ALLIS	73006546
FIAT-ALLIS	73006546-0
FLEETRITE	ZAD11402
GERLINGER	644683
GLAENZER	065-133-8
GMB	G2103
GMB	G2110
GMB	G5-2002X
GMB	G5-2110
GMB	GU2103
HERCULES	376
HOUGH	116294HI
HOUGH	149320HI
HOUGH	161294HI
IHC	116294
IHC	116-294H1
IHC	149320
IHC	149-320H1
IHC	161294
IHC	161-294H1
INBRA	CT10-015
ITALCARDAN	50-815000
JOY MFG.	1422427
LEMPCO	U203
LOBRO	U600
LOBRO	U601
MACK	201SJ55
MASTERS	MM2002
MCQUAY-NOR	U02-32
MCQUAY-NOR	U03-3
MCQUAY-NOR	UO3-3
MOPAR	D6
MOTOR MAST	2060A
MOTOR MAST	2060AA
MOTOR MAST	2060XB
M.R.S.	56695F11
M.R.S.	58897F11
NEAPCO	1-2171
NEAPCO	280200
NEAPCO	282171
NEAPCO	2851
PERFECTION	RG2002
PETTI-MULL	40293
PETTI-MULL	L2596-2
PETTI-MULL	LA425-1
PILOT	11402
PILOT	11492
PRECISION	951
REPCO	114-2103
REPCO	114-2124
REPCO	114-2135
REPCO	114-2150
REPCO	RUJ2003
REPUBLIC	CB2002
REPUBLIC	CB751
RUEDARSA	1850-1
RUEDARSA	2100-1850-1-0
SPICER	5-2002X
TEDDY TORQ	1422
TEDDY TORQ	G2103
TOWMOTOR	644683
TRANSMISS	UK503
TRU-CROSS	2002
TRW	20155
UJS	55-2002
WABCO	425944
WABCO	725944
WABCO	TD9791
WABCO	TF9791
WABCO	VJ0635
WESCO	201202
WESCO	201203
WESCO	N2002
ZELLER	2060AA

IDLI GROUP 80049

CROSS + 2 WING TYPE BEARINGS + 2 ROUND GROOVED BEARINGS
SEE FIGURE 1BJ

1.0000 in	BD	25.40 mm
1.3125 in	CC	33.34 mm
.6719 in	BW	17.07 mm
2.8750 in	CL	73.03 mm
.5781 in	CD	14.68 mm

Manufacturer	Part Number
ADAMS-LET.	47203
ADAMS-LET.	472196
ADAMS-LET.	472197
ADAMS-LET.	472200
ADAMS-LET.	474041
AEC	2002
AEC	2111
AEC	2126
AEC	AE2100
AEC	AE2109
AEC	AE2111
AEC	AE2126
ALCO	1064
ALCO	114-2158
ALCO	114-2163
ALCO	114-2188
ALCO	2100
ALCO	2126
ALCO	J2100
ALCO	J2109
ALCO	J2111
ALCO	J2126
ALCO	J2134
ALLIS-CHLM	082073
ALLIS-CHLM	178461
ALLIS-CHLM	178544
ALLIS-CHLM	178544-3
ALLIS-CHLM	178554
ALLIS-CHLM	3006546
ALLIS-CHLM	4984504
ALLOY	1064
ALLOY	J2126
AMER.MOTOR	3108171
AMER.MOTOR	3109409
AMER.MOTOR	3119275
AMER.MOTOR	3200855
AMER.MOTOR	3219885
AMER.MOTOR	934820
AUST.WEST.	AW465225
BALKAMP	3-961
BALKAMP	3-989
BALKAMP	M2000
BALKAMP	M2010
BALKAMP	M2031
BALKAMP	M2036
BALKAMP	M2040
BALKAMP	M2045
BALKAMP	M2048
BALKAMP	M2062
BALKAMP	M2100
BALKAMP	M2105
BALKAMP	M2109
BALKAMP	M2118
BALKAMP	M2126
BALKAMP	M2137
BORG-WARNR	114-2000
BORG-WARNR	114-2005
BORG-WARNR	114-2005AB
BORG-WARNR	114-2010
BORG-WARNR	114-2031
BORG-WARNR	114-2036
BORG-WARNR	114-2040
BORG-WARNR	114-2040AB
BORG-WARNR	114-2045
BORG-WARNR	114-2048
BORG-WARNR	114-2062
BORG-WARNR	114-2068
BORG-WARNR	114-2076
BORG-WARNR	114-2077
BORG-WARNR	114-2100
BORG-WARNR	114-2109
BORG-WARNR	114-2111
BORG-WARNR	114-2111AB
BORG-WARNR	114-2118
BORG-WARNR	114-2122
BORG-WARNR	114-2126
BORG-WARNR	114-2133
BORG-WARNR	114-2134
BORG-WARNR	114-2148
BORG-WARNR	114-2155
BORG-WARNR	114-2157
BORG-WARNR	114-2158
BORG-WARNR	114-2163
BORG-WARNR	114-2188
BORG-WARNR	J2000
BORG-WARNR	J2005
BORG-WARNR	J2010
BORG-WARNR	J2040
BORG-WARNR	J2045
BORG-WARNR	J2048
BORG-WARNR	J2062
BORG-WARNR	J2111
CATERPILLR	203765
CATERPILLR	4D8664
CATERPILLR	673626
CENTRAL	G5-2031X
CHRYSLER	1463491
CHRYSLER	1891966
CLEVE.MOT.	CM2002
CLEVE.MOT.	CM2062
CLEVE.MOT.	CM2100
CLEVE.MOT.	CM2109

BD=BEARING DIAMETER (outside)   BW=BEARING WIDTH   CB=CROSS LENGTH WITH BEARINGS   CC=CENTER TO CENTER   CD=CROSS DIAMETER
CL=CROSS LENGTH WITHOUT BEARINGS   EC=END TO CENTER or FACE TO FACE   EE=END TO END   EL=EFFECTIVE LENGTH
HD=HUB DIAMETER (or insert)   IE=INSIDE OF EARS (or recess)   OD=OUTSIDE DIAMETER   OE=OUTSIDE OF EARS   SB=SPLINE OR BORE SIZE

Make	Number	Make	Number	Make	Number
CLEVE.MOT.	CM2126	IHC	78-022R91	REPCO	114-2101
CMP	CM2062	IHC	823979R92	REPCO	114-2111
CMP	CM2100	INBRA	CA2100	REPCO	114-2118
CMP	CM2109	INBRA	CT10-005	REPCO	114-2122
CMP	CM2126	ITALCARDAN	50-813-000	REPCO	114-2126
D&D MACH.	MM2100	JEEP	20133	REPCO	114-2134
D&D MACH.	MM2126	JEEP	9348020	REPCO	RUJ2001
DEERE,JOHN	AN30893N	KAISER	201033	REPCO	RUJ2006
DEERE,JOHN	XAL5101N	LEMPCO	U202	REPCO	RUJ2026
ETNYRE	6440025	LEMPCO	U206	REPUBLIC	CB2000
EXCEL	U219	LEMPCO	U219	REPUBLIC	CB2005
EXCEL	U238	LEMPCO	U238	REPUBLIC	CB2010
FIAT-ALLIS	082073	LEMPCO	U238G	REPUBLIC	CB2010C
FLEETRITE	ZAD11445	LEMPCO	U246	REPUBLIC	CB2031
FORD	1M7039	LOBRO	U610	REPUBLIC	CB2040
FORD	1M7039A	LOBRO	U611	REPUBLIC	CB2045
FORD	8M7039	MASTERS	MM2002	REPUBLIC	CB2048
FORD	AB5A7039	MASTERS	MM2018	REPUBLIC	CB2110
FORD	AB5A7042A	MASTERS	MM2062	REPUBLIC	CB7039
FORD	B5A7039	MASTERS	MM2100	REPUBLIC	CB766
FORD	B5A7039A	MASTERS	MM2126	REPUBLIC	J2100
FORD	B5AZ4635A	MATSUBA	UJ020	RUEDARSA	1280-0
FORD	B5AZ7039A	MATSUBA	UJ021	RUEDARSA	1285
FORD	IM7039	MCQUAY-NOR	U03-4	RUEDARSA	2100-1280-1-0
FORD	MEK7039	MCQUAY-NOR	UO1-3	RUEDARSA	2100-1285-0-0
FORD	MEK7039A	MCQUAY-NOR	UO3-4	SPICER	5-2010X
FORD	MIK7039A	MIPER	C3674	SPICER	5-2031X
GLAENZER	065-140-8	MOPAR	D12	SPICER	5-2048X
GLAENZER	065-232-8	MOPAR	D4	STUDEBAKER	354672
GMB	G12100	MOTOR MAST	2060C	TEDDY TORQ	1028
GMB	G2013	MOTOR MAST	2060HA	TEDDY TORQ	1039
GMB	G2019	MOTOR MAST	2060K	TEDDY TORQ	1426
GMB	G2100	MOTOR MAST	2060LA	TEDDY TORQ	G2100
GMB	G2109	NEAPCO	1-3500	TEDDY TORQ	G2109
GMB	G2110	NEAPCO	2-3500	TOWMOTOR	B92318
GMB	G2111	NEAPCO	280100	TRANSMISS.	UK674
GMB	G2113	NEAPCO	280300	TRU-CROSS	2031
GMB	G2126	NEAPCO	280500	TRW	20003D
GMB	G2150	NEAPCO	280550	TRW	20006
GMB	G5-2007	NEAPCO	283500	TRW	20157
GMB	G5-2031X	NEAPCO	283500P	UJS	55-2100
GMB	G5-2031XA	NEAPCO	2861	WABCO	47041
GMB	G5-2100	PERFECTION	RG2000	WESCO	201245
GMB	G5-2101	PERFECTION	RG2031	WESCO	301-323
GMB	G5-2111	PERFECTION	RG2036	WESCO	N2000
GMB	GU2109	PERFECTION	RG2040	WESCO	N2031
GMB	GU6010	PERFECTION	RG2045	WESCO	N2039
GMB	GU6050	PERFECTION	RG2062	WESCO	N2040
GMB	GU6100	PERFECTION	RG2162	WESCO	N2045
GMB	GU6110	PETTI-MULL	L2596-2	WESCO	N2045G
GMB	GU6120	PETTI-MULL	LA425-1	WESCO	N2048
GMC	9044896	PILOT	11305	WILLYS	201033
HABERLE	H2000	PILOT	11400	ZELLER	2060A
HABERLE	H2010	PILOT	11405		
HABERLE	H2046	PILOT	11410		
HABERLE	H2062	PILOT	11411		
HERCULES	374	PILOT	11431		
HERCULES	377	PILOT	11436		
IHC	110-860R91	PILOT	11440		
IHC	110960R91	PILOT	11445		
IHC	65-343R91	PILOT	11448		
IHC	65343R92	PILOT	11462G		
IHC	65-345R91	PITT.VIOL.	40.813.000		
IHC	65345R92	PITT.VIOL.	40.816.000		
IHC	656223R91	PRECISION	508		
IHC	78-020R91	QUINT.HAZ.	QL6200		

IDLI GROUP 80050

CROSS + 2 WING TYPE BEARINGS
+ 2 ROUND GROOVED BEARINGS
slight variation from above group
SEE FIGURE 1BJ

1.0000 in	BD	25.40 mm
1.3125 in	CC	33.34 mm
.6719 in	BW	17.07 mm
2.8750 in	CL	73.03 mm
.5781 in	CD	14.68 mm

Make	Number
AEC	2062
AEC	AE2062

IDLI GROUP 80051

CROSS + 2 WING TYPE BEARINGS
+ 2 ROUND GROOVED BEARINGS
SEE FIGURE 1BJ

1.0000 in	BD	25.40 mm
1.4375 in	CC	36.51 mm
.6719 in	BW	17.07 mm
2.8750 in	CL	73.03 mm
.5781 in	CD	14.68 mm

Make	Number	Make	Number
ALCO	2062	AEC	2198G
BALKAMP	M2000	AEC	AE2198
BALKAMP	M2062	AEC	AE2198G
BORG-WARNR	114-2000	ALCO	1700
BORG-WARNR	114-2062	ALCO	2198
BORG-WARNR	114-2068	ALCO	J2198
BORG-WARNR	114-2105	ALLOY	1700
BORG-WARNR	114-2109	ALLOY	J2198
D&D MACH.	MM2062	BALKAMP	M2198
GMB	G1755	BORG-WARNR	114-2002
GMB	G5-5031X	BORG-WARNR	114-2198
GMB	GU6050	CLEVE.MOT.	CM2198
GMC	419977	D&D MACH.	MM2198
GMC	509297	GLAENZER	065-235-8
GMC	516406	GMB	G2198
LEMPCO	U202	GMB	G5-2198X
LEMPCO	U238	GMC	564274
LOBRO	U605	HABERLE	H2198
MOTOR MAST	3031C	LEMPCO	U240
MOTOR MAST	D2060C	MASTERS	MM2198
NEAPCO	1-2134		
NEAPCO	280100		
NEAPCO	280150		
NEAPCO	282134A		
PERFECTION	RG2062		
PILOT	11400		
PILOT	11462		
PRECISION	516		
REPCO	114-2105		
REPCO	114-2109		
REPUBLIC	CB2000		
REPUBLIC	CB2062		
SPICER	5-2000X		
STUDEBAKER	354672		
TRANSMISS.	UK674/6		
WESCO	N2062		

ENGINEERING CATALOGS MUST BE CONSULTED FOR DETAILS NOT INCLUDED
IN THIS GUIDE. SPECIFIC DESIGNS, MATERIAL CONTENT, TOLERANCES
LUBE FITTINGS AND OTHER DIMENSIONS ARE INTENTIONALLY OMITTED HERE.

CAUTION: BE SURE TO REFER TO ENGINEERING CATALOGS FOR SPECIAL APPLICATIONS THAT REQUIRE SPECIFIC MATERIAL CONTENT, TOLERANCES, ETC. SEE FOOTNOTE.

MOPAR	D15
MOTOR MAST	2060G
MOTOR MAST	D2060G
NEAPCO	280200
NEAPCO	283700
PERFECTION	RG2198
PILOT	11498
PRECISION	517
REPCO	114-2198
REPUBLIC	CB2002
REPUBLIC	CB2198
SPICER	5-2198X
TEDDY TORQ	1428
TEDDY TORQ	1428G
TEDDY TORQ	G2198
WESCO	301-198
WESCO	N2198
ZELLER	2060G

IDLI GROUP 80052

CROSS + 2 WING TYPE BEARINGS + 2 ROUND GROOVED BEARINGS
slight variation from above group
SEE FIGURE 1BJ

1.0000 in	BD	25.40 mm
1.4375 in	CC	36.51 mm
.6719 in	BW	17.07 mm
2.8750 in	CL	73.03 mm
.5781 in	CD	14.68 mm

AEC	2198G
AEC	AE2198G
ALCO	2198
ALCO	J2198
NEAPCO	283700
PRECISION	517G
REPUBLIC	CB2198
SPICER	5-2198X
WESCO	301198

IDLI GROUP 80053

CROSS + 2 ROUND BEARINGS 2 MIDWING DRILLED BEARINGS + INSIDE AND OUTSIDE SNAP RINGS
SEE FIGURE 1AK

1.0000 in	BD	25.40 mm
1.5000 in	CC	38.10 mm
.6719 in	BW	17.07 mm
2.8750 in	CL	73.03 mm
.5781 in	CD	14.68 mm

ALCO	2018
ALCO	J2113
BALKAMP	M2018
BORG-WARNR	114-2018
BORG-WARNR	114-2113
BORG-WARNR	114-2113AB
BORG-WARNR	J2018
BORG-WARNR	J2066
GMB	G2113
IHC	65344R91
IHC	65344R92
LEMPCO	U213
NEAPCO	282100

PERFECTION	RG2018
PILOT	11418
PRECISION	504
REPUBLIC	CB2018
REPUBLIC	CB2040
SPICER	5-2018X
SPICER	J2113
TEDDY TORQ	1418
TEDDY TORQ	1419
WESCO	201218
WESCO	N2018

IDLI GROUP 80054

CROSS + 2 WING TYPE BEARINGS + 2 ROUND GROOVED BEARINGS
SEE FIGURE 1AI

1.0000 in	BD	25.40 mm
1.9375 in	CC	49.21 mm
.6719 in	BW	17.07 mm
2.8750 in	CL	73.03 mm
.5781 in	CD	14.68 mm

AEC	2102
AEC	2114
AEC	AE2102
AEC	AE2114
ALCO	2114
ALCO	J2114
ALLOY	J2114
BALKAMP	M2033
BALKAMP	M2035
BALKAMP	M2049
BALKAMP	M2114
BALKAMP	M2123
BORG-WARNR	114-2033
BORG-WARNR	114-2041
BORG-WARNR	114-2049
BORG-WARNR	114-2102
BORG-WARNR	114-2107
BORG-WARNR	114-2114
BORG-WARNR	114-2114AB
BORG-WARNR	114-2129
BORG-WARNR	114-2142
BORG-WARNR	114-2147
BORG-WARNR	114-2162
BORG-WARNR	114-2167
BORG-WARNR	114-2169
BORG-WARNR	114-2183
BORG-WARNR	14-2147
BORG-WARNR	J2033
BORG-WARNR	J2041
CATERPILLR	5D2167
CATERPILLR	5D8293
CATERPILLR	6D2529
CATERPILLR	6L6449
CATERPILLR	SD2167
CHRYSLER	1450727
CLEVE.MOT.	CM2006
CLEVE.MOT.	CM2049
CLEVE.MOT.	CM2114
CMP	CM2006
CMP	CM2049
CMP	CM2114
CMP	CM2123
D&D MACH.	MM2114
DEERE,JOHN	AL430N
FLEETRITE	ZAD11449

FLEETRITE	ZAD11466
GALION	D122813
GLAENZER	065-264-8
GMB	G1003
GMB	G2114
GMB	G2126
GMB	G5-2033X
GMB	G5-2114
GMB	GU2114
HABERLE	H2033
HABERLE	H2041
HABERLE	H2049
IHC	110-861R91
IHC	69-561R91
IHC	78-019R91
LEMPCO	U221
LEMPCO	U245
LEMPCO	U247
MASTERS	MM2114
MASTERS	MM2123
MCQUAY-NOR	U03-6
MCQUAY-NOR	U03-6
MOPAR	D9
MOTOR MAST	2060CA
MOTOR MAST	2060MA
NEAPCO	2-2700
NEAPCO	282700
PERFECTION	RG2014
PERFECTION	RG2041
PERFECTION	RG2049
PERFECTION	RG2066
PILOT	11433
PILOT	11441
PILOT	11449
PILOT	11498
PILOT	14319
PRECISION	522
REPCO	114-2107
REPCO	114-2114
REPUBLIC	CB2033
REPUBLIC	CB2041
REPUBLIC	CB2049
SPICER	5-2033X
SPICER	5-2049X
TEDDY TORQ	1403
TEDDY TORQ	G2114
TRU-CROSS	2033
TRU-CROSS	5121
TRW	20010
TRW	20156
UJS	55-2114
UJS	55-2123
UJS	55-2124
WESCO	201249
WESCO	N2033
WESCO	N2041
WESCO	N2049
ZELLER	2060MA

IDLI GROUP 80055

CROSS + 4 GROOVED BEARINGS + INSIDE SNAP RINGS
largest BD dimension is shown
SEE FIGURE 1B

1.0625 in	BD	26.99 mm
.7031 in	BW	17.86 mm
2.8750 in	CL	73.03 mm
.5781 in	CD	14.68 mm

AEC	2140
AEC	AE2140
AEC	AE2153
ALCO	1066
ALCO	2140
ALCO	J2140
ALCO	J2153
ALCO	J2160
ALLOY	1066
ALLOY	J2140
ALLOY	J2153
AMER.MOTOR	934811
AMER.MOTOR	948976
BALKAMP	M2140
BALKAMP	M2153
BORG-WARNR	114-2024
BORG-WARNR	114-2140
BORG-WARNR	114-2151
BORG-WARNR	114-2153
CHRYSLER	1463499
CLEVE.MOT.	CM2140
CLEVE.MOT.	CM2153
CLEVE.MOT.	CM2179
CMP	CM2140
CMP	CM2153
CMP	CM2179
D&D MACH.	MM2140
EXCEL	U257
FLEETRITE	ZAD12140
FORD	B6TZ4635C
FORD	B9A7039A
FORD	B9AZ4635A
GLAENZER	065-230-8
GMB	G2140
GMB	G2153
GMB	G5-2140X
GMB	G5-2153
GMC	VS10798
HABERLE	H2007
HABERLE	H2063
IHC	871-280R91
JEEP	923811
JEEP	934811
JEEP	948976
LEMPCO	U257
MASTERS	MM2140
MASTERS	MM2179
MCQUAY-NOR	U02-1
MCQUAY-NOR	UO2-1
MOPAR	D14
MOTOR MAST	2060W
NEAPCO	1-3650
NEAPCO	283650
PERFECTION	RG2063
PERFECTION	RG2140
PERFECTION	RG2153
PILOT	12140

BD = BEARING DIAMETER (outside)   BW = BEARING WIDTH   CD = CROSS LENGTH WITH BEARINGS   CC = CENTER TO CENTER   CD = CROSS DIAMETER
CL = CROSS LENGTH WITHOUT BEARINGS   EC = END TO CENTER or FACE TO FACE   EE = END TO END   EL = EFFECTIVE LENGTH
HD = HUB DIAMETER (or insert)   IE = INSIDE OF EARS (or recess)   OD = OUTSIDE DIAMETER   OE = OUTSIDE OF EARS   SB = SPLINE OR BORE SIZE

PILOT 12153
PRECISION 505
REPCO 114-2140
REPCO 114-2153
REPCO RUJ2000
REPCO RUJ2002
REPCO RUJ2011
REPCO RUJ2032
REPUBLIC CB2140
RUEDARSA 1358
RUEDARSA 2104-1358-0-0
SPICER 5-2140
SPICER 5-2140X
STUDEBAKER 1559482
TEDDY TORQ 1018
TEDDY TORQ 1045
TEDDY TORQ G2140
TRU-CROSS 2140
TRW 20013
UJS 55-2140
WESCO 201240
WESCO G2140
WESCO N2140
WILLYS 934811
WILLYS 948976
WYLIE 49
ZELLER 2060W

IDLI GROUP 80056

CROSS + 2 SLOTTED BEARINGS + 2 ROUND BEARINGS
SEE FIGURE 1BD

1.0781 in	BD	27.38 mm
.8281 in	BW	21.03 mm
2.8750 in	CL	73.03 mm
.5781 in	CD	14.68 mm

AEC AE251
AEC AE256
ALCO 725251-2
ALCO J725251-2
ALCO J72525-2
BORG-WARNR 114-251
BORG-WARNR 114-256
DETROIT 72251-2
DETROIT 7225-4
DETROIT 725251-2
DETROIT 72525-2
DETROIT 73525-2
DETROIT J72251-2
DETROIT J73525-2
DIVCO 16301
DIVCO 16319
GMB G5-1311X
GMB G5-1312X
GMB G725251-2
LEMPCO U432
LEMPCO U433
LEMPCO U434
LEMPCO U437
MOPAR D138
NEAPCO 284600
PERFECTION RG154
PERFECTION RG254
PERFECTION RG256
PERFECTION RG324
PILOT 17226

PILOT 17227
PRECISION 296
REPCO 114-251
REPUBLIC CB7225-4
REPUBLIC CB72525
REPUBLIC CB725251
REPUBLIC CB7325-4
SPICER 72525-2
STUDEBAKER 436101
STUDEBAKER 436279
TEDDY TORQ 251
TEDDY TORQ 252
WESCO D251
WESCO D254
WESCO D256

IDLI GROUP 80057

CROSS + 2 SLOTTED BEARINGS + 2 ROUND BEARINGS
slight variation from above group
SEE FIGURE 1BD

1.0781 in	BD	27.38 mm
.8281 in	BW	21.03 mm
2.8750 in	CL	73.03 mm
.5781 in	CD	14.68 mm

ALCO 72525-2
BORG-WARNR 114-254
BORG-WARNR 114-256
CHRYSLER 1483701
DETROIT 7225-4
DETROIT 72521-2
DETROIT 725251-2
DETROIT 72525-2
GMB G5-1311X
LEMPCO U433
LEMPCO U434
MOTOR MAST 2387D
NEAPCO 284400
NEAPCO 284600
PERFECTION RG154
PERFECTION RG254
PERFECTION RG256
PILOT 17222
PILOT 17226
PRECISION 292
REPCO 114-256
REPUBLIC CB7125-2
REPUBLIC CB7225-4
REPUBLIC CB72525
REPUBLIC CB72525-1
SPICER 73525-2
STUDEBAKER 436101
STUDEBAKER 436270
WESCO 201256
WESCO D256

IDLI GROUP 80058

CROSS + 4 GROOVED BEARINGS + INSIDE SNAP RINGS
largest BD dimension is shown
SEE FIGURE 1B

1.1250 in	BD	28.58 mm
.6719 in	BW	17.07 mm
2.8750 in	CL	73.03 mm
.5781 in	CD	14.68 mm

AEC 2173

AEC 2175
AEC 3138316
AEC AE2173
AEC AE2175
ALCO 1067
ALCO 1068
ALCO J2173
ALCO J2175
ALCO N2173
ALLOY 1067
ALLOY 1068
ALLOY J2173
ALLOY J2175
AMER.MOTOR 3138316
AMER.MOTOR 3206122
AMER.MOTOR 3206980
AMER.MOTOR 3206981
AMER.MOTOR 3207077
AMER.MOTOR 3207079
AMER.MOTOR 8126607
BALKAMP M2173
BORG-WARNR 114-2165
BORG-WARNR 114-2173
BORG-WARNR 114-2175
BORG-WARNR 114-2178
BORG-WARNR 114-2180
BORG-WARNR 114-2181
CLEVE.MOT. CM2173
CLEVE.MOT. CM2175
CMP CM2173
CMP CM2175
D&D MACH. MM2173
D&D MACH. MM2175
EXCEL U264
FLEETRITE ZAD12173
FORD C5DZ4635A
GLAENZER 065-270-8
GMB G2017
GMB G2173
GMB G2175
GMB G5-2173
GMB G5-2173X
GMB G5-2175X
LEMPCO U260
LEMPCO U264
MASTERS MM2127
MASTERS MM2173
MASTERS MM2175
MCQUAY-NOR U01-8
MCQUAY-NOR UO1-8
MOPAR D202
MOPAR D203
MOTOR MAST 2060WF
NEAPCO 1-2173
NEAPCO 282173
NEAPCO 282175
PILOT 12173
PILOT 12175
PRECISION 498
REPCO 114-2173
REPCO 114-2175
REPCO RUJ2009
REPCO RUJ2034
REPUBLIC CB2173
REPUBLIC CB2175
SPICER 5-2173X

SPICER 5-2175X
TRU-CROSS 2173
TRW 20014
UJS 55-2173
UJS 55-2175
WESCO 201247
WESCO 201273
WESCO 202-273
WESCO 202-285
WESCO G2173
WESCO N2047
WESCO N2173
ZELLER 2060-WF

IDLI GROUP 80059

CROSS + 4 GROOVED BEARINGS + INSIDE SNAP RINGS
SEE FIGURE 1B

1.1250 in	BD	28.58 mm
.7656 in	BW	19.45 mm
2.8750 in	CL	73.03 mm
.6094 in	CD	15.48 mm

AEC 2500
AEC 3119731
AEC AE2500
AEC AE2505
ALCO 1077
ALCO 2500
ALCO J2505
ALLOY 1073
ALLOY 1077
ALLOY J2505
AMER.MOTOR 3119731
AMER.MOTOR 3200539
AMER.MOTOR 3200856
AMER.MOTOR 8126611
AMER.MOTOR 8126612
AMER.MOTOR 948978
BORG-WARNR 114-2500
BORG-WARNR 114-2501
BORG-WARNR 114-2505
BORG-WARNR 114-2531
BORG-WARNR 114-2532
BORG-WARNR 114-2533
BORG-WARNR J2500
CLEVE.MOT. CM2500
CLEVE.MOT. CM2505
CLEVE.MOT. CM2518
CLEVE.MOT. CM2530
CMP CM2500
D&D MACH. MM2500
DEERE,JOHN AT29362
EXCEL U252
FLEETRITE ZAD12505
FORD C6TZ4635H
GMB G2500
GMB G2505
GMB G5-2500X
GMB G5-2531
LEMPCO U252
MASTERS MM2500
MCQUAY-NOR UO1-6
MCQUAY-NOR UO1-7
MOPAR D16
MOTOR MAST 2060V
MOTOR MAST 2060VJ

ENGINEERING CATALOGS MUST BE CONSULTED FOR DETAILS NOT INCLUDED
IN THIS GUIDE. SPECIFIC DESIGNS, MATERIAL CONTENT, TOLERANCES
LUBE FITTINGS AND OTHER DIMENSIONS ARE INTENTIONALLY OMITTED HERE.

IDLI GUIDE   COPYRIGHT ® INTERCHANGE, INC.   ST. LOUIS PARK, MN. 55416 USA

CAUTION: BE SURE TO REFER TO ENGINEERING CATALOGS FOR SPECIAL APPLICATIONS THAT REQUIRE SPECIFIC MATERIAL CONTENT, TOLERANCES, ETC. SEE FOOTNOTE.

NEAPCO	1-5800
NEAPCO	285800
PERFECTION	RG2500
PILOT	12505
PILOT	12531
PILOT	14250
PRECISION	513
REPCO	114-2500
REPCO	114-2501
REPCO	114-2505
REPCO	114-2531
REPCO	114-2532
REPUBLIC	CB210
REPUBLIC	CB2500
SPICER	5-2532X
TEDDY TORQ	1404
TRW	20011
TRW	20012
TRW	2011
WESCO	201250
WESCO	201-520
WESCO	N2500
WILLYS	948978
ZELLER	2060V

IDLI GROUP 80060

CROSS + 4 GROOVED BEARINGS + INSIDE SNAP RINGS
SEE FIGURE 1B

1.0781 in	BD	27.38 mm
.7031 in	BW	17.86 mm
2.8750 in	CL	73.03 mm
.6406 in	CD	16.27 mm

AEC	AE255
AEC	AE266
AEC	AE267
ALCO	1105
ALCO	72625-3
ALCO	J72625-4
ALLOY	J72625-4
AMER.MOTOR	3114708
AMER.MOTOR	8126636
AMER.MOTOR	904802
BORG-WARNR	114-255
BORG-WARNR	114-266
BORG-WARNR	114-267
CHRYSLER	1752624
CHRYSLER	3711153
CHRYSLER	4157050
CLEVE.MOT.	CM72625-3
CLEVE.MOT.	CM72625-4
CMP	CM72625-4
DETROIT	7225-5
DETROIT	726251-2
DETROIT	726252-2
DETROIT	72625-3
DETROIT	J72625-3
EXCEL	U448
FORD	TP4635B
FORD	XA4635AA
FORD	XB4635C
FORD	XW4635A
GMB	G5-233
GMB	G5-266
GMB	G5-267
GMB	GU114-266
IHC	799275R91
LEMPCO	U425
LEMPCO	U448
LEMPCO	U452
LEYLAND	HYL4675
MASTERS	MD2255
MASTERS	MD6253
MCQUAY-NOR	U01-19
MCQUAY-NOR	U02-3
MCQUAY-NOR	UO1-19
MOPAR	D133
MOPAR	D134
MOPAR	D207
MOTOR MAST	2387E
NEAPCO	284500
NEAPCO	286300
NEAPCO	286300G
PERFECTION	RG255
PERFECTION	RG266
PILOT	17223
PILOT	17229
PILOT	17230
PILOT	17422
PRECISION	315
QUINT.HAZ.	QL2038
REPCO	114-255
REPCO	114-266
REPCO	RUJ2022
REPCO	RUJ2038
REPUBLIC	CB7225-5
REPUBLIC	CB735254
SPICER	5-1306X
SPICER	5-1306XS
SPICER	7225-5
SPICER	72525-3
SPICER	72625-3
TEDDY TORQ	736
TEDDY TORQ	755
TRW	20030
TRW	20031
TRW	20124
WESCO	303-255
WESCO	D255
WESCO	D255-1
WESCO	D626
WYLIE	55
WYLIE	57
ZELLER	2387E
ZELLER	2387F

IDLI GROUP 80061

CROSS + 4 GROOVED BEARINGS + INSIDE SNAP RINGS
slight variation from above group
SEE FIGURE 1B

1.0781 in	BD	27.38 mm
.7031 in	BW	17.86 mm
2.8750 in	CL	73.03 mm
.6406 in	CD	16.27 mm

AEC	266
AEC	3004900
AEC	AE233
AEC	AE266
ALCO	1019
ALCO	1103
ALCO	41103
ALCO	J72625-3
ALCO	J72625-6
ALLOY	1019
ALLOY	1019P
ALLOY	J72625-4
AMER.MOTOR	3114708
AMER.MOTOR	8126636
AMER.MOTOR	904802
AMER.MOTOR	940802
BORG-WARNR	114-233
BORG-WARNR	114-255
BORG-WARNR	114-266
BORG-WARNR	114-2HD
BORG-WARNR	114-2HP
BORG-WARNR	8126636
CENTRAL	G51306X
CENTRAL	G51309X
CHRYSLER	1653469
CHRYSLER	1742624
CHRYSLER	2298908
CHRYSLER	2298909
CHRYSLER	3780248
CHRYSLER	3780250
CHRYSLER	858001
CHRYSLER	X7080-428
CLEVE.MOT.	CM726251-3
CLEVE.MOT.	CM726252-2
CLEVE.MOT.	CM72625-3
CMP	CM3004900
CMP	CM726251-2
CMP	CM726251-3
CMP	CM726252-2
CMP	CM72625-3
CMP	CM72625-4
D&D MACH.	MD6253
DETROIT	3004900
DETROIT	3780250
DETROIT	726251-12
DETROIT	726251-2
DETROIT	726251-3
DETROIT	726252-2
DETROIT	72625-3
DETROIT	72625-4T
DETROIT	72625-6
EXCEL	U448A
EXCEL	U452
FIRESTONE	5234700-9
FLEETRITE	ZAD17229
FLEETRITE	ZAD17230
FORD	CD5AZ4635D
GLAENZER	065-203-8
GMB	G21625-3
GMB	G5-1306X
GMB	G5-1306XS
GMB	G5-1312X
GMB	G725251-2
GMB	G72525-4
GMB	G72625-3
GMB	G72625-4
GMB	G72625-3/4
GMB	G73252-2
GMB	G73525-2
GMB	G73525-3
GMB	G73525-4
GOODYEAR	219-2511
GOODYEAR	219-2551
IHC	799275R91
INBRA	CA266
JAMCO	UJ1306
KIEKHAFER	36312
KIEKHAFER	37830
KIEKHAFER	39383
LAKEWOOD	23010
LEMPCO	U438
LEMPCO	U448
LEMPCO	U452
LOBRO	U258
MASTERS	MD6253
MASTERS	MM1750
MASTERS	MM1759
MASTERS	MM1762
MASTERS	MS100X
MASTERS	MS105X
MCQUAY-NOR	U01-19
MCQUAY-NOR	U01-20
MCQUAY-NOR	U02-3
MCQUAY-NOR	UO1-19
MIDAS	UJ1000
MOPAR	D134
MOPAR	D134A
MOPAR	D207
MOTOR MAST	2387E
MOTOR MAST	2387EHP
MOTOR MAST	2387FT
NEAPCO	1-6301
NEAPCO	16-6301
NEAPCO	2863006
NEAPCO	286300G
NEAPCO	286300GB
NEAPCO	L61D4
NEW HOLL.	47621
PERFECTION	RG266
PERFECTION	RG325
PILOT	17223
PILOT	17229
PILOT	17422
PRECISION	315G
REPCO	114-267
REPUBLIC	CB72625-3
REPUBLIC	CB72625-4
REPUBLIC	CB72625-4C
SEARS-ROE.	5103
SPICER	5-1306-1X
SPICER	5-1306X
SPICER	5-1306X5
SPICER	5-1307X
SPICER	5-1308X
TEDDY TORQ	732
TEDDY TORQ	735
TRU-CROSS	1306
TRW	20030
TRW	20120
TRW	20124
TRW	20124D
TRW	20205
TRW	20901
TRW	21050

BD=BEARING DIAMETER (outside)   **BW**=BEARING WIDTH   **CB**=CROSS LENGTH WITH BEARINGS   **CC**=CENTER TO CENTER   **CD**=CROSS DIAMETER
CL=CROSS LENGTH WITHOUT BEARINGS   **EC**=END TO CENTER or FACE TO FACE   **EE**=END TO END   **EL**=EFFECTIVE LENGTH
HD=HUB DIAMETER (or insert)   **IE**=INSIDE OF EARS (or recess)   **OD**=OUTSIDE DIAMETER   **OE**=OUTSIDE OF EARS   **SB**=SPLINE OR BORE SIZE

Column 1

UJS	55-72625-3
VOLVO	897707
WESCO	06N0103
WESCO	201255
WESCO	201-269
WESCO	540-049
WESCO	D251
WESCO	D255
WESCO	D255-1
WESCO	G255
WILLYS	940802
ZELLER	2387E
ZOOM	800004

IDLI GROUP 80062

CROSS + 4 GROOVED BEARINGS + INSIDE SNAP RINGS
slight variation from above group
SEE FIGURE 1B

1.0781 in	BD	27.38 mm
.7031 in	BW	17.86 mm
2.8750 in	CL	73.03 mm
.6406 in	CD	16.27 mm

AEC	233
AEC	3004900
AEC	490
AEC	AE233
ALCO	1103
ALLOY	1103
ALLOY	41103
ALLOY	J72625-6
BORG-WARNR	114-233
CHRYSLER	2882487
CHRYSLER	3004900
CLEVE.MOT.	CM3004900
CLEVE.STL.	PK96-55-279
CLEVE.STL.	PK96-55-302
D&D MACH.	MD6256
DETROIT	3004900
DETROIT	72625-4T
GMB	G5-1306-1X
GMB	G72625-6
LEMPCO	U448A
LOBRO	U114
MCQUAY-NOR	U02-3
MCQUAY-NOR	UO2-3
MERCURY	75832
MERCURY	B36312
MERCURY	B58749
MOTOR MAST	2387EA
NEAPCO	1-6300
NEAPCO	286300
PILOT	17233
PRECISION	317
REPCO	114-233
REPCO	114-259
REPUBLIC	3004900
REPUBLIC	CB72625-9
SPICER	5-1306-1X
SPICER	5-1306X
TRU-CROSS	1306-1
TRW	20124
TRW	20124D
TRW	20205
WESCO	200260
WESCO	201-260

Column 2

WESCO	203260
WESCO	D255-1
ZELLER	2387E
ZELLER	2387EA

IDLI GROUP 80063

CROSS + 4 GROOVED BEARINGS + INSIDE SNAP RINGS
slight variation from above group
SEE FIGURE 1B

1.0781 in	BD	27.38 mm
.7031 in	BW	17.86 mm
2.8750 in	CL	73.03 mm
.6406 in	CD	16.27 mm

AEC	267
AEC	AE267
ALCO	1105
ALCO	J72625-7
ALLOY	1105
ALLOY	J72625-7
BORG-WARNR	114-267
CHRYSLER	2882264
CLEVE.MOT.	CM726251-2
CLEVE.MOT.	CM72625-4
D&D MACH.	MD2255
DAVEY COM.	A22710
DETROIT	7225-5
DETROIT	726251-2
DETROIT	726252-2
DETROIT	72625-4
DETROIT	72652-2
FORD	CD5AZ4635E
GMB	G5-1306XS
GMB	G5-1307XS
GMB	G72625-4
JACOBSEN	11681
LEMPCO	U452
LOBRO	U113
MASTERS	MD2255
MCQUAY-NOR	U01-20
MOPAR	D134
MOPAR	D134A
MOPAR	D207
MOTOR MAST	2387F
NEAPCO	2-6302
NEAPCO	286300GB
PILOT	17230
PRECISION	318
REPCO	RUJ2008
REPCO	RUJ2012
REPUBLIC	CB72625-4
SPICER	5-1306X
SPICER	5-1306XS
TRW	20031
WESCO	202-255
WESCO	303255
WESCO	D255C
ZELLER	2387F

Column 3

IDLI GROUP 80064

CROSS + 2 GROOVED BEARINGS + 2 LOW WING DRILLED BEARINGS + INSIDE SNAP RINGS
SEE FIGURE 1BG

1.0781 in	BD	27.38 mm
1.750 in	BW	44.45 mm
2.8750 in	CL	73.03 mm
.6406 in	CD	16.27 mm

AEC	265
AEC	AE225
AEC	AE261
AEC	AE265
ALCO	1102
ALCO	72625-2
ALCO	J7225-2
ALCO	J72625-2
ALLOY	1102
ALLOY	J72625-2
BORG-WARNR	114-225
BORG-WARNR	114-226
BORG-WARNR	114-261
BORG-WARNR	114-265
CHRYSLER	1450249
CHRYSLER	1643172
CHRYSLER	1752623
CHRYSLER	94755
CHRYSLER	947550
CLEVE.MOT.	CM725253
CLEVE.MOT.	CM72625-2
CMP	CM72625-2
D&D MACH.	MD2625
DETROIT	72251-3
DETROIT	7225-2
DETROIT	725251-3
DETROIT	72525-3
DETROIT	72625-2
EXCEL	U447
GLAENZER	065-201-8
GMB	G5-1304X
GMB	G5-265
GMB	G72625-2
GMB	GU2570
HABERLE	H7753
LEMPCO	U413
LEMPCO	U429
LEMPCO	U440
LEMPCO	U447
MASTERS	MD2255
MASTERS	MD2625
MASTERS	MD5253
MASTERS	MD5254
MCQUAY-NOR	U01-18
MOPAR	D132
NEAPCO	2-5700
NEAPCO	281200
NEAPCO	284700G
NEAPCO	285700
NEAPCO	285700G
PERFECTION	RG225
PERFECTION	RG261
PERFECTION	RG265
PILOT	17224
PILOT	17225
PILOT	17228
PRECISION	314

Column 4

REPCO	114-261
REPCO	114-265
REPUBLIC	CB7225
REPUBLIC	CB72251
REPUBLIC	CB72251-3
REPUBLIC	CB72525-3
REPUBLIC	CB72625-2
REPUBLIC	CB7625-2
SPICER	5-1304X
SPICER	72625-2
STUDEBAKER	438395
TEDDY TORQ	723
TEDDY TORQ	725
TEDDY TORQ	726
TRU-CROSS	1304
TRW	20029
WESCO	201-725
WESCO	D226
WESCO	D261
WESCO	D526
WESCO	D625
ZELLER	2387A

IDLI GROUP 80065

CROSS + 2 GROOVED BEARINGS + 2 LOW WING DRILLED BEARINGS + INSIDE SNAP RINGS
slight variation from above group
SEE FIGURE 1BG

1.0781 in	BD	27.38 mm
1.750 in	BW	44.45 mm
2.8750 in	CL	73.03 mm
.6406 in	CD	16.27 mm

AEC	AE265
ALCO	1102
ALLOY	J72625-2
BORG-WARNR	114-225
BORG-WARNR	114-226
BORG-WARNR	114-265
CHRYSLER	858002
CHRYSLER	94755
CLEVE.MOT.	CM725253
CLEVE.MOT.	CM72625-2
CMP	CM72525-3
DETROIT	725251-3
DETROIT	72625-2
DETROIT	73252-2
FLEETRITE	ZAD17228
GMB	G5-1304X
GMB	G72625-2
GMB	G72625-2/4
HABERLE	H7753
HERCULES	383
LEMPCO	U447
LOBRO	U257
MASTERS	MD2255
MASTERS	MD2625
MASTERS	MD5253
MASTERS	MD5254
MCQUAY-NOR	U01-18
MCQUAY-NOR	UO1-18
MOPAR	D132
MOTOR MAST	2387A
NEAPCO	2-5700
NEAPCO	2857006
NEAPCO	285700G

CAUTION: BE SURE TO REFER TO ENGINEERING CATALOGS FOR SPECIAL APPLICATIONS THAT REQUIRE SPECIFIC MATERIAL CONTENT, TOLERANCES, ETC. SEE FOOTNOTE.

PERFECTION	RG225
PERFECTION	RG265
PILOT	17228
PRECISION	314G
REPUBLIC	CB72251-3
REPUBLIC	CB72525-3
REPUBLIC	CB72575-3
REPUBLIC	CB7262-2
REPUBLIC	CB72625-2
SPICER	5-1304X
STUDEBAKER	436087
STUDEBAKER	458395
TRU-CROSS	1304
TRW	20029
UJS	55-72625-2
WESCO	201725
WESCO	D7225
ZELLER	2387A

IDLI GROUP 80066

CROSS + 4 GROOVED BEARINGS + INSIDE SNAP RINGS
slight variation from above group
SEE FIGURE 1B

1.1250 in	BD	28.58 mm
.7031 in	BW	17.86 mm
2.8750 in	CL	73.03 mm
.6719 in	CD	17.07 mm

AEC	2518
AEC	2531
AEC	AE2518
AEC	AE2531
ALCO	1079
ALCO	J2518
ALCO	J2530
ALCO	U2518
ALLOY	1079
ALLOY	J2518
AMER.MOTOR	8126611
AMER.MOTOR	948977
BORG-WARNR	114-2518
BORG-WARNR	114-2530
BORG-WARNR	114-2531
BORG-WARNR	114-2532
CLEVE.MOT.	CM2518
CLEVE.MOT.	CM2530
CLEVE.MOT.	CM2531
CMP	CM2518
CMP	CM2530
D&D MACH.	MM2518
DEERE,JOHN	AT29361
EXCEL	U263
FLEETRITE	ZAD12518
FORD	C6TZ4635A
FORD	C8TZ24635A
FORD	C8TZ4635A
GLAENZER	065-269-8
GMB	G2173
GMB	G2518
GMB	G2530
GMB	G2531

GMB	G5-2530X
GMB	GU2518
JEEP	948977
JEEP	948978
LEMPCO	U263
LEMPCO	U263G
MASTERS	MM2518
MCQUAY-NOR	U01-7
MOPAR	D201
MOPAR	D205
MOTOR MAST	2060VF
MOTOR MAST	2060VJ
NEAPCO	2-2518
NEAPCO	282518
PILOT	12513
PILOT	12518
PILOT	12530
PILOT	12531
PILOT	12532
PRECISION	497
REPCO	114-2518
REPCO	114-2531
REPCO	114-2532
REPUBLIC	CB1249
SPICER	5-2530X
TRU-CROSS	2530
TRW	20012
UJS	55-2518
WESCO	201204
WESCO	N2540
WILLYS	947977
WILLYS	947978
ZELLER	2060UJ
ZELLER	2060VF

IDLI GROUP 80067

CROSS + 4 GROOVED BEARINGS + INSIDE SNAP RINGS
SEE FIGURE 1B

1.1250 in	BD	28.58 mm
.8281 in	BW	21.03 mm
2.8750 in	CL	73.03 mm
.6719 in	CD	17.07 mm

FIAT-ALLIS	0612621
FIAT-ALLIS	0612621-3
FIAT-ALLIS	612621
FIAT-ALLIS	70612621
FIAT-ALLIS	70612621-3
GLAENZER	065-155-8
GMB	G2505
GMB	G5-2532
GMC	9042934
GMC	9044943
MCQUAY-NOR	U01-6
PRECISION	985
REPCO	114-2532
REPUBLIC	CB785

IDLI GROUP 80068

CROSS + 4 SERRATED BEARINGS + THRUST AND LOCK PLATES
SEE FIGURE 1C

1.1406 in	BD	28.97 mm
.8750 in	BW	22.23 mm
2.8750 in	CL	73.03 mm
.7188 in	CD	18.26 mm

AEC	1296
AEC	AE-P96-55-1
AEC	P96-55-1
ALCO	JP96-55
ALCO	P96-55
ALLOY	JP96-55
BORG-WARNR	114-133
BORG-WARNR	P96-55-1
CHRYSLER	1238415
CLEVELAND	C96-55
CLEVELAND	P96-55-1
CLEVE.MOT.	CM9655C
CLEVE.MOT.	CM9655P
CLEVE.STL.	P96-55-1
D&D MACH.	MCP96
GMB	G5-1214X
GMB	G5-1296X
GMB	GP96-55
HABERLE	HP9655
LEMPCO	U302
MASTERS	MCD96
MASTERS	MCP96
MOPAR	D100
MOPAR	D106
MOTOR MAST	2487A
NEAPCO	282000
PERFECTION	RP96-55-1
PILOT	C19655
PILOT	P19655
PRECISION	416
REPCO	P96-55-1
REPUBLIC	CB-P96-1
REPUBLIC	CB-P96G1
SPICER	P96-55-1
TEDDY TORQ	GP96-55
TEDDY TORQ	P96
TRW	20034
WESCO	201965
WESCO	NP965
ZELLER	2487A

IDLI GROUP 80069

CROSS + 4 ROUND BEARINGS + OUTSIDE SNAP RINGS
SEE FIGURE 1A

1.2500 in	BD	31.75 mm
.7344 in	BW	18.65 mm
2.9375 in	CL	74.61 mm
.750 in	CD	19.05 mm

AEC	35N
AEC	AE35N3
AEC	AE35N4
AEC	AE-CP35N3
AEC	AECP35N384
AEC	CP35N3
AEC	CP35N4
A.E.LEE	3881-3
ALCO	1781

ALCO	CP35N
ALCO	CP35N3
ALLIS-CHLM	11020
ALLIS-CHLM	13113
ALLIS-CHLM	26726
ALLIS-CHLM	26727
ALLIS-CHLM	3026726-4
ALLIS-CHLM	3026727-2
ALLIS-CHLM	517312
ALLIS-CHLM	530538
ALLIS-CHLM	618805
ALLIS-CHLM	618805-06
ALLOY	1781
ALLOY	CP35N
ALLOY	CP35N3
AMER.PULL.	499A42
AUST.WEST.	466008
AUST.WEST.	AW466008
BALKAMP	3-945
BALKAMP	CP35N3
BAL-LM-HAM	466008
BAR-GREENE	5J06
BEAN,JOHN	1222781
BEAN,JOHN	1244743
BEAN,JOHN	1276700
BEAR CAT	CPL35N
BMB	CPL35N
BORG-WARNR	114-105
BORG-WARNR	CP35N3
BORG-WARNR	CP35N4
BPC	B3500N
BRADY	1680
BRANDY	1680
BRILLION	5C327
BUSH HOG	119
BUSH HOG	199
CARDWELL	2750
CASE,J.I.	1175FHA
CASE,J.I.	1963CHA
CASE,J.I.	2417CHA
CASE,J.I.	A40106
CASE,J.I.	R14744
CASE,J.I.	T27790
CASE,J.I.	T37895
CASE,J.I.	T40106
CASE,J.I.	T41672
CATERPILLR	5F2476
CATERPILLR	5F6332
CATERPILLR	6F9741
CHAIN BELT	350
CHAIN BELT	7123
CHAIN BELT	R35N
CLARK EQU.	1550551
CLARK EQU.	854639
CLARK EQU.	862837
CLARK EQU.	870688
CLARK EQU.	952378
CLEVE.MOT.	CM5-243X
CLEVE.MOT.	CM-CP35N
CLEVE.MOT.	CM-CP35N3
CLEVE.MOT.	CM-CP35N3-4
CLEVE.MOT.	CM-CP35N4
CLEVE.MOT.	CM-CP35R
CMP	CM-CP35N3
CMP	CM-CP35R
COCKSHUTT	FHY116

BD = BEARING DIAMETER (outside)   **BW** = BEARING WIDTH   **CB** = CROSS LENGTH WITH BEARINGS   **CC** = CENTER TO CENTER   **CD** = CROSS DIAMETER
CL = CROSS LENGTH WITHOUT BEARINGS   **EC** = END TO CENTER or FACE TO FACE   **EE** = END TO END   **EL** = EFFECTIVE LENGTH
HD = HUB DIAMETER (or insert)   **IE** = INSIDE OF EARS (or recess)   **OD** = OUTSIDE DIAMETER   **OE** = OUTSIDE OF EARS   **SB** = SPLINE OR BORE SIZE

COLES	1550551	GMB	GGPL35R
DEERE,JOHN	59X11467A	GMC	9042907
DEERE,JOHN	AE21122	GROVE	7-364-000330
DEERE,JOHN	AE23516	GROVE	7-364-000331
DEERE,JOHN	AE23516E	HARDIE	449A42
DEERE,JOHN	AM2568E	HARDIE	499A42
DEERE,JOHN	AU13093U	HEIL	30B301-8
EATON	1A353X	HESSTON	104397
EATON	A353	HESSTON	104398
EATON	A353X	HESSTON	104401
EATON	A354	HESSTON	104402
EATON	A354-1X	HESSTON	111048
EATON	A354X	HESSTON	111050
EATON	A357	HESSTON	111051
EIMCO	104M472	HESSTON	753889
EIMCO	911A15010	HESSTON	754622
ELG.SWEEP.	63624	HESSTON	807420
ELG.SWEEP.	63625	HESSTON	845198
ENGLER	1897	HIGHLAND	12823E
ETNYRE	6440103	HOUGH	104397
EXCEL	U510	HOUGH	104398
FIAT-ALLIS	0530538-8	HOUGH	104401
FIAT-ALLIS	13113	HOUGH	104402
FIAT-ALLIS	26726	HOUGH	111048
FIAT-ALLIS	26727	HOUGH	111050
FIAT-ALLIS	26728	HOUGH	111051
FIAT-ALLIS	30011020	HUB CITY	0335-90076
FIAT-ALLIS	3026726	HUB CITY	UR21110
FIAT-ALLIS	3026726-4	HUBER	106244
FIAT-ALLIS	3026727	IHC	174980C91
FIAT-ALLIS	3026727-2	IHC	347-960R91
FIAT-ALLIS	4020	IHC	613-471R91
FIAT-ALLIS	517312	IHC	660-116R91
FIAT-ALLIS	530538	IMCO	55-35N3
FIAT-ALLIS	61805	JOY MFG.	700799-11
FIAT-ALLIS	70004020	JOY MFG.	CA8-42
FIAT-ALLIS	70013113	JOY MFG.	CA8U2
FIAT-ALLIS	70026726	KALAMAZOO	Z2595
FIAT-ALLIS	70026727	KALAMAZOO	Z2838
FIAT-ALLIS	70026728	KOEHRING	9009877
FIAT-ALLIS	70517312	KOEHRING	P33516
FIAT-ALLIS	70530538	KOEHRING	P8900
FIAT-ALLIS	70530538-8	LEMPCO	U510
FIAT-ALLIS	73026726	LETZ	CMA1333
FIAT-ALLIS	73026726-4	LILLISTON	10-61-033
FIAT-ALLIS	73026727	LILLISTON	23-61-007
FIAT-ALLIS	73026727-2	LONG MFG.	720424
FLEETRITE	ZAD-B35N3	LUNDELL	S188
FORD	145-421	MACK	782B3
FORD	243-424	MASS.FERG.	1089-666M91
FOX RIVER	F4CX5	MASS.FERG.	1506100M91
FOX RIVER	F4CX8	MASS.FERG.	1508762M91
FOX RIVER	F4CX8H	MASS.FERG.	835-501M91
G & G MFG	302-3500	MASS.FERG.	835-501M91-8K
G & G MFG	35R	MASS.FERG.	835-991M91
G & G MFG	G35N	MASS.FERG.	840676M91
GAR WOOD	4014113	MASS.FERG.	840-899M91
GEHL BROS.	510036	MASS.FERG.	840-899M92
GEHL BROS.	CPL35N3	MATHEWS	7520
GLAENZER	065-143-8	MCQUAY-NOR	UO3-12
GMB	G5-243X	MCQUAY-NOR	UO3-12
GMB	GCP35N	MICHIGAN	853411
GMB	GCP35N1	MICHIGAN	854369
GMB	GCP35N3	MICHIGAN	857616
GMB	GCP35N4	MICHIGAN	862837
GMB	GGPL34R	MICHIGAN	870688

MICHIGAN	952378	WESCO	N35
MOTOR MAST	2593A	WESCO	R35
MOTOR MAST	2593B	WEST.LAND	AS-US10AP
MPL.MOLINE	15P431	WHITE	876347A
MUNCIE	MK35X	WHITE FARM	15P431
NATL.MINE	3881-3	WHITE FARM	301-680E
NATL.MINE	76011915	WHITE FARM	506-602E
NATL.MINE	P1006	WHITE FARM	522-313ES
NEAPCO	2205X	WILK RIP.	1460
NEAPCO	2205X2	WOOD BROS.	110
NEAPCO	2-2275	ZELLER	2593A
NEAPCO	2275X		
NEAPCO	2834		
NEAPCO	2845		

IDLI GROUP 80070

**CROSS + 4 ROUND BEARINGS
+ OUTSIDE SNAP RINGS**
slight variation from above group
SEE FIGURE 1A

1.2500 in	BD	31.75 mm
.7344 in	BW	18.65 mm
2.9375 in	CL	74.61 mm
.750 in	CD	19.05 mm

AEC	35N
AEC	AE-CP35N4
ALCO	CP35N
ALCO	CP35N1
ALCO	CP35N4
ALLIS-CHLM	30131130
ALLOY	CP35N1
BALKAMP	3-934
BORG-WARNR	114-105
BORG-WARNR	CP35N4
CASE,J.I.	T41672
CATERPILLR	3L7363
CATERPILLR	5F2426
CATERPILLR	5F2478
CATERPILLR	7F3679
CLARK EQU.	853411
CLARK EQU.	857616
EATON	2013785
EATON	2013822
EATON	A353
EATON	A353X
EATON	A354
EATON	A354-1X
EATON	A354X
EATON	A357
ELG.SWEEP.	63635
FIAT-ALLIS	0618805
FIAT-ALLIS	0618805-6
FIAT-ALLIS	0665611
FIAT-ALLIS	0665611-0
FIAT-ALLIS	11020
FIAT-ALLIS	3011020
FIAT-ALLIS	3011022-5
FIAT-ALLIS	618805
FIAT-ALLIS	618805-6
FIAT-ALLIS	665611
FIAT-ALLIS	70011020
FIAT-ALLIS	70618805
FIAT-ALLIS	70618805-6
FIAT-ALLIS	70665611
FIAT-ALLIS	70665611-0
FIAT-ALLIS	73011020
FIAT-ALLIS	73011022
FLEETRITE	ZAD-B35N4
GAR WOOD	4014113

The following continue the third/fourth columns:

NEAPCO	2864		
NEAPCO	28L16S		
NEW HOLL.	29214		
NEW HOLL.	2924		
NEW HOLL.	34870		
NEW HOLL.	43449		
NEW IDEA	K2980A		
NEW IDEA	O11070		
OLIVER	301-680C		
OLIVER	301689E		
OLIVER	506-602E		
PAPEC MACH	92-1514		
PAPEX	CPL35N3		
PERFECT-CI	243		
PERRY	U109		
PILOT	B34NR		
PILOT	B35N3		
PILOT	CP35N3		
PRECISION	964		
PREC.TOR.	885		
PREC.TOR.	945		
REPCO	5-243X		
REPCO	CP35N		
REPCO	CP35N3		
REPCO	CPL16S		
REPUBLIC	CB764		
REX CHAIN	130009		
REX CHAIN	350		
REX CHAIN	R35N		
RUPP	S1889		
SCHULTZ	11663		
SCHULTZ	19210		
SERVIS	W279		
SERVIS	W279R		
SIDEWINDER	15214		
SPEED-SPR.	1222781		
SPEED-SPR.	1244743		
SPICER	5-243X		
SPICER	CP58N3		
TAYLOR	7175-003		
TERRAIN	1897		
THEW SHOV.	P33516		
TORO	33410		
TRU-CROSS	243		
TRW	20038		
UJS	55-35N3		
UJS	55-35N4		
ULRICH	X213		
WEASLER	DN		
WESCO	271035		
WESCO	35N		
WESCO	35R		

CAUTION: BE SURE TO REFER TO ENGINEERING CATALOGS FOR SPECIAL APPLICATIONS THAT REQUIRE SPECIFIC MATERIAL CONTENT, TOLERANCES, ETC. SEE FOOTNOTE.

Make	Part Number
GLAENZER	065-127-8
GLAENZER	065-147-8
GMB	G5-243-1X
GMB	GCP35N3
GMB	GCP35N4
GMB	GGPL35R1X
GORMAN RUP	1889
GROVE	7-364-00330
GROVE CR.	7-069-000182
HUB CITY	03-35-90076
HUSCO	CP35N4
HUSCO	CP35R
IHC	307-949R91
IHC	656224R91
INBRA	CT11-000
KOEHRING	P8908
MCQUAY-NOR	U03-12
MOTOR MAST	2593A
NATL.MINE	3881-6
NATL.MINE	76012095
NEAPCO	2205-2
NEAPCO	2205X2
NEAPCO	2205X-Z
NEAPCO	2-2276
NEAPCO	2275X3
OLIVER	301680E
PERFECT-CI	243-1
PILOT	B35N4
PRECISION	934
REPCO	5-243-1X
REPCO	CP35N
REPCO	CP35N4
REPUBLIC	CB734
REPUBLIC	CB745
REX CHAIN	130071
SPICER	5-243-1X
SPICER	5-243X1
THEW SHOV.	P8808
THEW SHOV.	P8809
THEW SHOV.	P8900
TRU-CROSS	243-1
TRW	20038
UNIT RIG	900344
UNIT RIG	903019
UNIT RIG	986980
WESCO	35R

IDLI GROUP 80071

**CROSS + 4 GROOVED BEARINGS + INSIDE SNAP RINGS
SEE FIGURE 1B**

.9375 in	BD	23.81 mm
.6094 in	BW	15.48 mm
2.9688 in	CL	75.41 mm
.5469 in	CD	13.89 mm

Make	Part Number
AEC	1207
AEC	AE-T55-55-2
AEC	T55-55-2
ALCO	1134
ALCO	JT55-55-2
ALCO	T55-55-2
ALLOY	1134
ALLOY	JT55-55-2
AMER.MOTOR	924190
BORG-WARNR	114-127
BORG-WARNR	T55-55-2
CLEVELAND	T55-55-2
CLEVE.MOT.	CM5555T
CLEVE.STL.	T55-55-2
CMP	CM5555T
EXCEL	U312
FLEETRITE	ZADT15555
GLAENZER	065-221-8
GMB	G5-1207X
GMB	G5-1208X
GMB	GT55-55-2
IHC	220-196R91
LEMPCO	U312
MASTERS	MCT55
MOPAR	D112
MOTOR MAST	2496A
NEAPCO	2-6400
NEAPCO	286400
NEAPCO	286500
PERFECTION	RT55-55-2
PILOT	T15555
PRECISION	425
REPCO	5-1207X
REPCO	5-1208X
REPCO	T55-55-2
REPUBLIC	CB-T55
SPICER	5-1207X
SPICER	5-1208X
TEDDY TORQ	GT55-55-2
TEDDY TORQ	T55
TRW	20035
WESCO	201556
WESCO	NT555
WILLYS	924190
ZELLER	2496A

IDLI GROUP 80072

**CROSS + 4 ROUND BEARINGS + OUTSIDE SNAP RINGS
SEE FIGURE 1A**

1.0625 in	BD	26.99 mm
.5625 in	BW	14.29 mm
2.9688 in	CL	75.41 mm
.5938 in	CD	15.08 mm

Make	Part Number
AEC	5-200
AEC	521
AEC	AE521
AEC	AE521HD
ALCO	1001
ALCO	1521
ALCO	1553
ALCO	J5-121
ALCO	J5-138
ALLOY	1001
ALLOY	1004P
ALLOY	1005
ALLOY	J5-121
ALLOY	J5-153
AMER.MOTOR	157734
AMER.MOTOR	164369
AMER.MOTOR	3117373
AMER.MOTOR	3145473
AMER.MOTOR	3202655
AMER.MOTOR	807806
AMER.MOTOR	8126613
AMER.MOTOR	931477
AMER.MOTOR	A1433
AMER.MOTOR	F164369
ARMSTRONG	5410-1000
ARMSTRONG	5410-162
AUTOCAR	1321115
AUTOCAR	3AA01172
BEDFORD	8845370
BORG-WARNR	114-500
BORG-WARNR	114-513
BORG-WARNR	114-513X
BORG-WARNR	114-521
BORG-WARNR	114-521R
BORG-WARNR	114-521X
BORG-WARNR	114-553
CASE,J.I.	MDH414
CHECKER	63569
CHRYSLER	1321115
CHRYSLER	1643242
CHRYSLER	1818920
CHRYSLER	1818921
CLEVE.MOT.	CM5-121X
CLEVE.MOT.	CM5-121XUB
CLEVE.MOT.	CM5-171X
CMP	CM5-121X
D&D MACH.	MS121X
DIAMOND R	10054P2
DIAMOND R	57P1
DIVCO	16364
EXCEL	U101
FLEETRITE	ZAD11520
FORD	211C11-18397
FORD	21C18397
FORD	21C18397B
FORD	21C7039
FORD	21C7039A
FORD	A9LY4635
FORD	A9LY4635A
FORD	B6D7039A
FORD	B6S7039A
FORD	B7A4635A
FORD	B7A7039A
FORD	B7AZ4635A
FORD	B7S7039A
FORD	C1VV4997
FORD	C1VV4997A
FORD	C3AZ4635C
FORD	C9TZ4635A
FORD	GP18397
FORD	GPW18397
FORD	LD44610A
FORD	MEK7039B
FORD	MFK7039
FORD	MFK7039A
FORD	MFK7039B
FORD	Z11C11-18397
FORD	Z1C18397
FORD	Z1C18397B
GLAENZER	365-001-0
GMB	G1000
GMB	G1661
GMB	G5-121
GMB	G5-121X
GMB	G5-160
GMB	GU1000
GMB	GU1000A
GMB	GU1000SI
GMC	2325029
GMC	2325829
GMC	2354320
GMC	2448100
GMC	3713926
GMC	3741635
GMC	3828469
GMC	518645
GMC	518648
GOODYEAR	219-2505
HABERLE	H1521X
HARDY SPCR	K5GB106
HARDY SPCR	K5LGB21
HARDY SPCR	K6GB106
HERCULES	433
HIGHLAND	81050D
HUSCO	329
HUSCO	5-121XBA
IHC	121-406R91
INBRA	CA521
ITALCARDAN	50-759-000
JEEP	208075
JEEP	931477
KAISER	208075
LAND ROVER	541858
LEMPCO	U101
LEMPCO	U101C
LEMPCO	U101F
LEMPCO	U116
LEMPCO	U123
LOBRO	U100
MACK	201SJ22
MASTERS	MS121U
MASTERS	MS121X
MCQUAY-NOR	U01-14
MCQUAY-NOR	U01-15
MCQUAY-NOR	U01-17
MCQUAY-NOR	U01-50
MCQUAY-NOR	UO1-14
MIPER	C3127
MOPAR	D63
MOTOR MAST	2290A
MUNCIE	5-200X
NEAPCO	1-0121
NEAPCO	1-0200
NEAPCO	1-1200
NEAPCO	28051X
NEAPCO	28051XB
NEAPCO	28153X
NEAPCO	28200X
NEAPCO	5-121
OSHKOSH	12DS4
PERFECTION	RG521X
PILOT	11513
PILOT	11521
PILOT	11527
PILOT	11553
PILOT	11571
PRECISION	329
QUINT.HAZ.	QL16204
QUINT.HAZ.	QL2104
REO MOTORS	1107P2

REO MOTORS	1795P2
REO MOTORS	57P1
REPCO	5-71X
REPCO	K5-21XR
REPUBLIC	CB1270B
REPUBLIC	CB1270BHD
REPUBLIC	CB1270C
REPUBLIC	CB1270HD
REPUBLIC	CB1270L
ROVER	515723
RUEDARSA	1330
RUEDARSA	2103-1330-0-0
SPICER	1270
SPICER	5-113X
SPICER	5-11X
SPICER	5-120X
SPICER	5-121X
SPICER	5-12X
SPICER	5-13X
SPICER	5-153X
SPICER	5-163X
SPICER	5-200X
SPICER	5-21
SPICER	5-21X
SPICER	5-246X
SPICER	5-56X
SPICER	5-5X
SPICER	5-63X
SPICER	5-6X
SPICER	5-71X
SPICER	5-82X
SPICER	6-11X
SPICER	6-13X
SPICER	6-21X
SPICER	6-5X
SPICER	6-6X
SPICER	J5-21
STUDEBAKER	1541448
STUDEBAKER	1550426
STUDEBAKER	410529
STUDEBAKER	436144
STUDEBAKER	516059
TEDDY TORQ	G5-121
TEDDY TORQ	K21
TRU-CROSS	200
TRW	20024
TRW	20026
TRW	20049
TRW	22026
UJS	55-96
WESCO	201-521
WESCO	G1521R
WESCO	N1513
WESCO	N1521
WESCO	N1521R
WESCO	N1521RHD
WESCO	N1554
WHITE	02-7074023
WHITE	400934
WHITE	400934A
WHITE	807806
WHITE	810969
WHITE	874026
WHITE	906929
WHITE	931477
WHITE	936686

WHITE	A1433
WILK RIP.	1031
WILLYS	208075
WILLYS	4651538
WILLYS	807806
WILLYS	906929
WILLYS	931477
WILLYS	A1433
ZELLER	2290A
ZELLER	3200A

IDLI GROUP 80073

CROSS + 4 ROUND BEARINGS + OUTSIDE SNAP RINGS
slight variation from above group
SEE FIGURE 1A

1.1250 in	BD	28.58 mm
.7188 in	BW	18.26 mm
2.9688 in	CL	75.41 mm
.5938 in	CD	15.08 mm

AEC	AE5-1201
ALCO	1522
ALCO	J5-122
ALLOY	J5-122
BORG-WARNR	114-579
FORD	C1AZ4635A
FORD	C3AZ4635D
FORD	C8SZ4A031A
GLAENZER	065-213-8
GMB	G5-122
IHC	799281R91
LEMPCO	U129
MCQUAY-NOR	U01-16
MOPAR	D65
MOTOR MAST	2290D
NEAPCO	284635A
NEAPCO	284635D
PERFECTION	RG579
PERFECTION	RG579R
PILOT	11579
PILOT	12141
PRECISION	357
REPCO	114-2102
REPCO	RUJ2039SL
REPUBLIC	CB1271
REPUBLIC	CB1271HDC
REPUBLIC	CB565-6
SPICER	5-1201X
SPICER	5-122X
SPICER	5-207X
TEDDY TORQ	G5-122
TEDDY TORQ	K54
TRW	20025
WESCO	G1579R
WESCO	N1579
WESCO	N1579R
WESCO	NR5165
ZELLER	2290D

IDLI GROUP 80074

CROSS + 4 GROOVED BEARINGS + INSIDE SNAP RINGS
largest BD dimension is shown
SEE FIGURE 1B

1.0625 in	BD	26.99 mm
.6094 in	BW	15.48 mm
2.9688 in	CL	75.41 mm
.5469 in	CD	13.89 mm

AEC	1208
AEC	AE-TS55-55-2
AEC	TS55-55-2
ALCO	1135
ALCO	JTS55-35-2
ALCO	JTS55-55-2
ALCO	TS55-55-2
ALLOY	1135
ALLOY	JTS55-55-2
AMER.MOTOR	924189
AUST.WEST.	TX55-55-2
BORG-WARNR	114-128
BORG-WARNR	TS55-55-2
CLEVELAND	TS55-55-2
CLEVE.MOT.	CM5555TS
CLEVE.STL.	TS55-55-2
CMP	CM5555TS
EXCEL	U313
FLEETRITE	ZAD-TS15555
GLAENZER	065-222-8
GMB	G5-1207X
GMB	G5-1208X
GMB	G5-128
GMB	GTS55-55-2
IHC	220-197R91
IHC	220197R92
IHC	319711R91
LEMPCO	U313
MASTERS	MCTS55
MOPAR	D113
MOTOR MAST	2496B
NEAPCO	2-6500
NEAPCO	286500
PERFECTION	RTS55-55-2
PILOT	TS15555
PRECISION	426
REPCO	5-1208X
REPCO	TS55-55-2
REPUBLIC	CB-TS55
SPICER	5-1207X
SPICER	5-1208X
TEDDY TORQ	GTS55-55-2
TEDDY TORQ	T52
TRW	20127
UJS	55RST55
WESCO	201557
WESCO	NTS555
WILLYS	924189
ZELLER	2496B

IDLI GROUP 80075

CROSS + 4 ROUND BEARINGS + OUTSIDE SNAP RINGS
SEE FIGURE 1A

1.0625 in	BD	26.99 mm
.6563 in	BW	16.67 mm
2.9688 in	CL	75.41 mm
.6094 in	CD	15.48 mm

AEC	200
AEC	5-200
AEC	AE512U
AEC	AE520
AEC	AE5-200
AEC	AE521HD
AEC	AE532
ALBURUS-SP	5-312XR
ALCO	1005
ALCO	1028
ALCO	1032
ALCO	J5-132
ALCO	J5-200
ALCO	N5-200
ALLIS-CHLM	617512
ALLOY	1005
ALLOY	J5-200
AMER.MOTOR	157734
AMER.MOTOR	164369
AMER.MOTOR	3117373
AMER.MOTOR	3131428
AMER.MOTOR	3207885
AMER.MOTOR	4487436
AMER.MOTOR	8126613
AMER.MOTOR	906929
AMER.MOTOR	936686
AMER.MOTOR	943648
AMER.MOTOR	994827
AUTOCAR	3AA01172
AUTOCAR	3AA1172
BORG-WARNR	114-2179
BORG-WARNR	114-500
BORG-WARNR	114-513X
BORG-WARNR	114-520
BORG-WARNR	114-521HD
BORG-WARNR	114-521X
BORG-WARNR	114-549
CENTRAL	G5-121X
CENTRAL	G5-200X
CLEVE.MOT.	CM5-121XUB
CLEVE.MOT.	CM5-200X
CLEVE.MOT.	CM5-200XUB
CMP	CM5-121X
CMP	CM5-121XUB
CMP	CM5-153X
CMP	CM5-200X
D&D MACH.	MS200X
DIAMOND R	10008
DIAMOND R	10054P2
DIAMOND R	1107P2
DIAMOND R	232
DIAMOND R	57P1
EXCEL	U101HD
EXCEL	U134
FLEETRITE	ZAD11520
FLEETRITE	ZAD115200
FLEETRITE	ZAD11521
FLEETRITE	ZAD11521UB

ENGINEERING CATALOGS MUST BE CONSULTED FOR DETAILS NOT INCLUDED
IN THIS GUIDE. SPECIFIC DESIGNS, MATERIAL CONTENT, TOLERANCES
LUBE FITTINGS AND OTHER DIMENSIONS ARE INTENTIONALLY OMITTED HERE.

CAUTION: BE SURE TO REFER TO ENGINEERING CATALOGS FOR SPECIAL APPLICATIONS THAT REQUIRE SPECIFIC MATERIAL CONTENT, TOLERANCES, ETC. SEE FOOTNOTE.

FLEETRITE	ZAD11549
FORD	01Y7039
FORD	B6TZ4635A
FORD	C4TA4635A
FORD	C4TA4724A
FORD	C4TZ4635A
FORD	C4TZ4724A
FORD	C5TZ4635C
FORD	C5TZ4635D
FORD	C6TZ4635C
FORD	CTZ4724A
FORD	T6AA4635A
GLAENZER	065-216-8
GMB	G5-121B
GMB	G5-121X
GMB	G5-121XU
GMB	G5-154X
GMB	G5-200
GMB	G5-200B
GMB	G5-200X
GMB	G5-200XU
GMB	GU1000
GMC	2370728
GMC	2448382
GMC	2460455
GMC	3750686
GMC	3830580
GMC	3849500
GMC	3889690
GMC	3910235
GMC	3937988
GMC	3955571
GMC	570817
GOODYEAR	219-2520
GOODYEAR	219-2552
HABERLE	H1521X
HOUGH	153915HI
IHC	121406R92
IHC	176297
JAMCO	UJ200
JEEP	936686
JEEP	943648
JOY MFG.	A259525
LEMPCO	U101HD
LEMPCO	U134
MACK	21045200X
MASTERS	MS121U
MASTERS	MS121X
MASTERS	MS200U
MASTERS	MS200X
MCQUAY-NOR	U01-14
MCQUAY-NOR	U01-15
MCQUAY-NOR	U01-17
MCQUAY-NOR	U01-50
MCQUAY-NOR	UO1-17
MIDAS	UJ1002
MOPAR	D84A
MOTOR MAST	2297A
MUNCIE	5-200X
NEAPCO	1-0200
NEAPCO	1-1200

NEAPCO	1-153
NEAPCO	280200X
NEAPCO	28051X
NEAPCO	28051XB
NEAPCO	28153X
NEAPCO	28200X
NEAPCO	28244X
NEAPCO	292000
OSHKOSH	12DS4
PERFECTION	RG511X
PERFECTION	RG521B
PERFECTION	RG549X
PILOT	11500
PILOT	11520
PILOT	115200
PILOT	11521HD
PILOT	11521UB
PILOT	11549
PRECISION	360
REPCO	5-132X
REPCO	5-200X
REPCO	K5GB106
REPCO	K5GB117
REPCO	K5LGB104R
REPCO	K5LGB74
REPUBLIC	CB1270B
REPUBLIC	CB1270HD
REPUBLIC	CB1270L
REPUBLIC	CB1720C
SEARS-ROE.	5123
SPICER	1280
SPICER	50200X
SPICER	5-125X
SPICER	5-146X
SPICER	5-149X
SPICER	5-153X
SPICER	5-163X
SPICER	5-200
SPICER	5-200X
SPICER	5-201X
SPICER	5-244
SPICER	5-246X
SPICER	5-247X
SPICER	5-419X
STUDEBAKER	1562019
STUDEBAKER	1563060
STUDEBAKER	1690409
STUDEBAKER	1698021
STUDEBAKER	436144
STUDEBAKER	900314
TEDDY TORQ	G5-200
TEDDY TORQ	K60
TRANSMISS.	UK127
TRU-CROSS	200
TRW	20026
TRW	20049
UJS	55-121
UJS	55-200
UJS	55-96
VERSATILE	20744
WESCO	201-520
WESCO	201521HD
WESCO	201-554
WESCO	G1520R
WESCO	N1520R
WESCO	N1521R

WHITE	10054P2
WHITE	1107P2
WHITE	1795P2
WHITE	232
WHITE	400934
WHITE	400934A
WHITE	817359
WILLYS	936686
WILLYS	943648
ZELLER	2297A

IDLI GROUP 80076

CROSS + 4 ROUND BEARINGS + OUTSIDE SNAP RINGS
slight variation from above group
SEE FIGURE 1A

1.0625 in	BD	26.99 mm
.6563 in	BW	16.67 mm
2.9688 in	CL	75.41 mm
.6094 in	CD	15.48 mm

AEC	200U
AEC	5-200U
AEC	AE500U
AEC	AE512U
AEC	AE5-200U
AEC	AE520U
ALCO	1039
ALCO	J5-200B
ALLOY	1039
ALLOY	J5-200B
AMER.MOTOR	990192
AMER.MOTOR	990937
BORG-WARNR	114-520U
CLEVE.MOT.	CM5-200XUB
CLEVE.MOT.	CM5-249X
CMP	CM5-200XUB
D&D MACH.	MS200U
DIAMOND R	10054P2
DIAMOND R	1795PU
DIAMOND R	57P1
EXCEL	U134U
FLEETRITE	ZAD11520UB
GMB	G3-121UB
GMB	G5-200B
GMB	G5-200XU
GMC	3848500
GOODYEAR	219-2506
LEMPCO	U134B
LEMPCO	U134U
MCQUAY-NOR	U01-15
MCQUAY-NOR	U01-17
MCQUAY-NOR	U01-50
MCQUAY-NOR	UO1-50
MOPAR	D84
MOTOR MAST	2297AU
MOTOR MAST	2297AV
NEAPCO	1-1200
NEAPCO	1-200
NEAPCO	280200XB
NEAPCO	28200XB
NEAPCO	28244XB
PILOT	115200B
PILOT	115200UB
PRECISION	360U
REPUBLIC	1270UB
SPICER	5-200X

SPICER	5-200XU
TEDDY TORQ	K60U
TRW	20026
WESCO	303520
WESCO	303-521
WESCO	G1520RU
WESCO	N1520RU
WESCO	N1553RB
WESCO	N1553RV
ZELLER	2290B

IDLI GROUP 80077

CROSS + 4 ROUND BEARINGS + OUTSIDE SNAP RINGS
SEE FIGURE 1A

1.0625 in	BD	26.99 mm
.6875 in	BW	17.46 mm
2.9688 in	CL	75.41 mm
.6563 in	CD	16.67 mm

ADAMS-LET.	TG9695
ADAMS-LET.	TL7604
ADAMS-LET.	VL9523
ADAMS-LET.	VN2793
AEC	373-1410
AEC	521HD
AEC	AE513
AEC	AE521HD
AEC	AE521HDU
AEC	AE553
ALBURUS-SP	5-153X
ALBURUS-SP	5-204X
ALCO	1004
ALCO	1553
ALCO	41004
ALCO	J5-153
ALCO	J5153B
ALLIS-CHLM	10070613
ALLIS-CHLM	617512
ALLOY	1004
ALLOY	1032
ALLOY	41004
ALLOY	J5-132
ALLOY	J5-153
AMER.MOTOR	3145293
AMER.MOTOR	3156293
AMER.MOTOR	3188632
AMER.MOTOR	3195054
AMER.MOTOR	3202657
AMER.MOTOR	3203655
AMER.MOTOR	3208473
AMER.MOTOR	4485624
AMER.MOTOR	4487847
AMER.MOTOR	4488016
AMER.MOTOR	4488106
AMER.MOTOR	8126614
AMER.MOTOR	8126615
AMER.MOTOR	8126636
AMER.MOTOR	8127332
AMER.MOTOR	8130297
AMER.MOTOR	944562
AMER.MOTOR	994173
ARMSTRONG	5415-1000
AYRA DUREX	1-315-01
BATUYONG	60756
BLAW KNOX	735-72
BMC	8G8018

IDLI GUIDE   COPYRIGHT © INTERCHANGE, INC.   ST. LOUIS PARK, MN. 55416 USA

Brand	Part	Brand	Part	Brand	Part	Brand	Part
BMC	GUJ104	FORD	C9TZ4635B	IHC	411956R91	REPCO	101-1290
BMC	GUJ105	FORD	F9E7039A	IHC	645-573R91	REPCO	5-1209X
BMC	GUJ108	FORD	FL4635A	IHC	645-578R91	REPCO	K513XR
BMC	GUJ117	FORD	FM4635A	IHC	882577R91	REPCO	K5A13XR
BORG-WARNR	114-3HD	FORD	O1Y18397	IHC	886-759R91	REPCO	K5GB106
BORG-WARNR	114-3HP	FORD	OIY7039	INBRA	CA521HD	REPCO	K5GB117
BORG-WARNR	114-513	FORD	XB4635A	JAEGER	5-153X	REPCO	K5GB117R
BORG-WARNR	114-513R	FORD	XL512825	JAGUAR	10065	REPCO	K5LGB104R
BORG-WARNR	114-513X	FORD	XW4635C	JAGUAR	8405	REPCO	K5LGB74
BORG-WARNR	114-520	FORD	XY501497	JAGUAR	9409	REPCO	K5SR13XR
BORG-WARNR	114-521HD	GALION	A53116(7)	JEEP	994173	REPUBLIC	CB1270
BRD	04-501371	GLAENZER	365-023-0	JOY MFG.	A253000	REPUBLIC	CB1270B
BRD	04-511371	GLAENZER	365-030-0	JOY MFG.	A94633	REPUBLIC	CB1270HD
BUCKEYE	1310	GLAENZER	365-037-0	KATAOKA	MI85-5680	REPUBLIC	CB1270HDC
CASE,J.I.	11270	GLAENZER	365-050-0	KOYO	S2782	REPUBLIC	CB1279BHD
CASE,J.I.	D26204	GLAENZER	365-051-0	LEMPCO	U101HD	REPUBLIC	CB1720L
CASE,J.I.	D70629	GMB	G5-132	LEMPCO	U101HP	REPUBLIC	CB3278
CASE,J.I.	D73649	GMB	G5-153	LEMPCO	U101HU	ROVER	504439
CASE,J.I.	S201007	GMB	G5-153B	LEYLAND	601577	ROVER	542544
CENTRAL	G5-153X	GMB	G5-153X	LEYLAND	606678	ROVER	6011557
CHECKER	010761	GMB	G5-153XU	LOBRO	U101	SPICER	1310
CHECKER	03521	GMB	G5-154X	LOBRO	U103	SPICER	5-113
CHECKER	63569	GMB	G5-513	LOBRO	U105	SPICER	5-12X
CHECKER	64580	GMB	G5-520	MACK	201SJ28	SPICER	5-132X
CHECKER	93521	GMB	G5-521	MACK	201SJ29	SPICER	5-153X
CHRYSLER	1150809	GMB	G5-549	MACK	201SJ30	SPICER	5-244X
CHRYSLER	3827046	GMB	GU1000SI	MACK	201SJ48	SPICER	5-251X
CHRYSLER	5057939	GMB	GU1000SL	MACK	21045153X	SPICER	5-252X
CHRYSLER	J906926	GMB	GU100A	MACK	210SJ28	SPICER	5-255X
CLARK EQU.	856665	GMB	GU1030	MACK	65MU26	SPICER	5-419X
CLARK EQU.	865665	GMB	K5-13XR	MASTERS	MS153X	SPICER	5-420X
CLARK EQU.	882802	GMC	2354320	MATSUBA	UJ016	SPICER	5-6X
CLEVE.MOT.	CM5-153X	GMC	2362830	MATSUBA	UJ620	SPICER	J5-113
CMP	CM5-125X	GMC	2862830	MCQUAY-NOR	U01-14	STUDEBAKER	1687692
CMP	CM5-153X	GMC	3741653	MCQUAY-NOR	U01-15	STUDEBAKER	410529
D&D MACH.	MS153X	GMC	3840500	MCQUAY-NOR	U01-17	STUDEBAKER	436144
DAIM.RUBE.	760-147-341-1	GMC	384651	MCQUAY-NOR	U01-29	STUDEBAKER	516059
DEERE,JOHN	AN32800	GMC	386451	MCQUAY-NOR	U01-50	TRANSMISS.	VK128
DEERE,JOHN	MD2683L	GMC	388451	MCQUAY-NOR	U02-14	TRANSMISS.	VK130
DIAMOND R	10055P2	GMC	3953659	MCQUAY-NOR	UO1-29	TRIUMPH	144823
DIAMOND R	57P1	GMC	396451	MICHIGAN	114-511	TRU-CROSS	153
DIVCO	16343	GMC	8845370	MICHIGAN	865775	TRW	20049
EXCEL	U101HD	GMC	9023673	MIPER	C3128	TRW	20099
FIAT-ALLIS	0617512	GOODYEAR	219-2519	MOPAR	D69	TRW	20208
FIAT-ALLIS	0617512-9	GOODYEAR	219-2553	MOPAR	D73	UJS	20744
FIAT-ALLIS	10070613	HABERLE	H1513X	MOTOR MAST	3190A	UJS	55-153
FIAT-ALLIS	617512	HABERLE	H1521X	MOTOR MAST	3200A	UNIPART	GUJ104
FIAT-ALLIS	6175129	HARDY SPCR	3151-162N	MOTOR MAST	3200AHP	UNIPART	GUJ105
FIAT-ALLIS	70617512	HARDY SPCR	3151-166W	MOTOR MAST	3200B	UNIPART	GUJ108
FIAT-ALLIS	70617512-9	HARDY SPCR	3151-166Y	MUNCIE	5-153X	VERSATILE	20744
FIRESTONE	5234769-6	HARDY SPCR	K5-13XR	NEAPCO	1-0153	VERSATILE	56289
FLEETRITE	ZAD11521HD	HARDY SPCR	K5GB104	NEAPCO	1-153	VOLVO	231136
FLEETRITE	ZAD11553	HARDY SPCR	K5GB117	NEAPCO	28153X	VOLVO	231179
FMC	211238	HARDY SPCR	K5GB162	NEAPCO	283190	VOLVO	231189
FMC	212238	HARDY SPCR	K5GB166	PERFECTION	RG521X	VOLVO	231311
FMC	212247	HARDY SPCR	K5LGB128	PERFECTION	RG549X	WABCO	47203
FMC	215649	HARDY SPCR	K5LGB133	PILOT	11500	WABCO	TG5575
FORD	2402E4635A	HARDY SPCR	K5LGB166	PILOT	11521HD	WABCO	TG9695
FORD	4050350	HARDY SPCR	K5LGB33	PILOT	11549	WABCO	TL7604
FORD	4487436	HARDY SPCR	K5LGB74	PILOT	11553	WABCO	VI9523
FORD	AR89AF7039A	HARDY SPCR	K5LGB82	PILOT	1521HD	WABCO	VL9523
FORD	ARC6DZ4635A	HERCULES	433	PITT.VIOL.	40.759.000	WABCO	VN2793
FORD	B6A7039A	HYSTER	53066A	PRECISION	369	WESCO	200520
FORD	B6D7039	IHC	153915	PREC.TOR.	915	WESCO	201520
FORD	C1VV4497A	IHC	153-915H1	REO MOTORS	10054P2	WESCO	201-521HD
FORD	C1VV4997A	IHC	328-592C91	REO MOTORS	10055P2	WESCO	202-520

ENGINEERING CATALOGS MUST BE CONSULTED FOR DETAILS NOT INCLUDED
IN THIS GUIDE. SPECIFIC DESIGNS, MATERIAL CONTENT, TOLERANCES
LUBE FITTINGS AND OTHER DIMENSIONS ARE INTENTIONALLY OMITTED HERE.

CAUTION: BE SURE TO REFER TO ENGINEERING CATALOGS FOR SPECIAL APPLICATIONS THAT REQUIRE SPECIFIC MATERIAL CONTENT, TOLERANCES, ETC. SEE FOOTNOTE.

WESCO	203520
WESCO	N1521HD
WESCO	N1553
WHITE	02-7074024
WHITE	10008P2
WHITE	10055P2
WHITE	57P1
WHITE	58259
WHITE	58359
WHITE	817359
WHITE	874026
WHITE	880798
WILLYS	4485624
WILLYS	8122993
WILLYS	8127332
WILLYS	944562
WILLYS	948977
WILLYS	994173
WILLYS	994827
WORLDPARTS	49-118
WORLDPARTS	49-120
WYLIE	20
WYLIE	52
ZELLER	3200A

IDLI GROUP 80078

CROSS + 4 GROOVED BEARINGS + INSIDE SNAP RINGS
SEE FIGURE 1B

1.0625 in	BD	26.99 mm
.750 in	BW	19.05 mm
2.9688 in	CL	75.41 mm
.6563 in	CD	16.67 mm

AEC	260
AEC	.5-260
AEC	AE5-260
ALCO	1048
ALCO	41048
ALCO	J5-75
ALLOY	1048
ALLOY	41025
ALLOY	41048
ALLOY	J5-260
AMER.MOTOR	8122993
AMER.MOTOR	8126637
BORG-WARNR	114-573
BORG-WARNR	114-574
CHRYSLER	363394
CHRYSLER	3633944
CHRYSLER	368390
CHRYSLER	3683980
CHRYSLER	3744918
CLEVE.MOT.	CM2141
CLEVE.MOT.	CM5-174X
CLEVE.MOT.	CM5-260X
CMP	CM5-174X
CMP	CM5-260X
D&D MACH.	MS260X
DIAMOND R	1107P2
EXCEL	U130
FORD	B9T2-3249A

FORD	C6TTZ3249A
FORD	C6TZ3249A
GMB	G5-154X
GMB	G5-260X
GMB	G5-573
GMB	GU2000
GMC	2362830
GMC	3981590
IHC	427804C91
MASTERS	MS260X
MASTERS	MS574X
MCQUAY-NOR	U02-5
MCQUAY-NOR	UO2-5
MOPAR	D50
MOTOR MAST	3200D
NEAPCO	1-0174
NEAPCO	1-153
NEAPCO	28174X
PERFECTION	RG574
PILOT	11521HD
PILOT	11573
PRECISION	365
REPCO	5-260X
REPCO	K5GB106
REPCO	K5GB117
REPCO	K5LGB104R
REPCO	K5LGB74
REPUBLIC	CB1574S
SPICER	5-174X
SPICER	5-260X
SPICER	5-419X
SPICER	5-74X
TRU-CROSS	260
TRU-CROSS	74
TRW	20049
TRW	20126
TRW	20192
WESCO	200574
WESCO	201574
WESCO	201575
WESCO	201576
WESCO	N1574
WESCO	N1574G
ZELLER	3200D

IDLI GROUP 80079

CROSS + 4 GROOVED BEARINGS + INSIDE SNAP RINGS
SEE FIGURE 1B

1.1875 in	BD	30.16 mm
.7656 in	BW	19.45 mm
2.9688 in	CL	75.41 mm
.750 in	CD	19.05 mm

AEC	297
AEC	.5-297
AEC	AE297
AEC	AE5-297
ALCO	1758
ALCO	41758
ALLOY	1758
ALLOY	41758
ALLOY	J5-297
AMER.MOTOR	8124589
AMER.MOTOR	8126638
AMER.MOTOR	8280920
BORG-WARNR	114-645

BORG-WARNR	114-654
CLEVE.MOT.	CM5-297X
CMP	CM5-297X
D&D MACH.	MS297X
DETROIT	3744918
DIAMOND R	1039P2
DIAMOND R	218
DIAMOND R	7199F
DIAMOND R	BB17042
DIAMOND R	BB7042
FORD	D3TZ3249A
FORD	O3TZ3249A
GMB	G5-2907X
GMB	G5-297X
GMC	6273940
IHC	459395C91
LEMPCO	U140
MASTERS	MS297X
MCQUAY-NOR	U02-6
MCQUAY-NOR	UO2-6
MOPAR	D212
MOPAR	D95
MOTOR MAST	3210
MOTOR MAST	3210A
NEAPCO	1-0297
NEAPCO	1-297
NEAPCO	28297
NEAPCO	28297X
PRECISION	371
REPCO	5-297X
REPUBLIC	CB1310
SPICER	5-297X
TRU-CROSS	297
TRW	20139
TRW	20193
WESCO	200297
WESCO	201297
WESCO	N1597
WILLYS	8124589
ZELLER	3210A

IDLI GROUP 80080

CROSS + 4 GROOVED BEARINGS + INSIDE SNAP RINGS
SEE FIGURE 1B

1.0625 in	BD	26.99 mm
.3594 in	BW	9.13 mm
3.0 in	CL	76.20 mm
.6875 in	CD	17.46 mm

AEC	AE-T555-55-3
ALCO	1136
BORG-WARNR	114-129
BORG-WARNR	TS55-55-3
CLEVE.STL.	TS55-55-3
CLEVE.STL.	TS55-55-4
GLAENZER	065-276-8
IHC	319-711C91
MCQUAY-NOR	U01-66
MOTOR MAST	2496C
PILOT	TS15553
PRECISION	435
REPCO	5-1208X
REPCO	RUJ2027SL
REPUBLIC	CB-T55L
SPICER	5-1209X

IDLI GROUP 80081

CROSS + 4 GROOVED BEARINGS + INSIDE SNAP RINGS
SEE FIGURE 1B

1.0625 in	BD	26.99 mm
.6406 in	BW	16.27 mm
3.0 in	CL	76.20 mm
.6875 in	CD	17.46 mm

AEC	1200
AEC	.5-1200
AEC	AE501200
AEC	AE5-1200
AEC	AE5-260
AEC	AE5S55-55-4
AEC	AE-P55-55-140
AEC	AE-S1200
AEC	AE-S55-55-2
AEC	AE-TS55-55-3
AEC	AE-TS55-55-4
AEC	TS55-55-3
AEC	TS55-55-4
ALCO	1010
ALCO	1131
ALCO	1136
ALCO	1523
ALCO	J5-1200
ALCO	JS55-55
ALCO	JTS55-55-4
ALLOY	1010
ALLOY	1010P
ALLOY	1131
ALLOY	1136
ALLOY	41010
ALLOY	J5-1200
ALLOY	JS55-55
ALLOY	JTS55-55-4
AMER.MOTOR	3203655
AMER.MOTOR	3203992
BORG-WARNR	114-125
BORG-WARNR	114-513R
BORG-WARNR	114-513X
BORG-WARNR	114-553
BORG-WARNR	114-553R
BORG-WARNR	114-7HD
BORG-WARNR	114-7HP
BORG-WARNR	S55-55-1
BORG-WARNR	S55-55-2
CENTRAL	G5-1200X
CHRYSLER	1843347
CHRYSLER	1925252
CHRYSLER	1925552
CLEVELAND	S55-55-2
CLEVE.MOT.	CM5555S
CLEVE.STL.	S55-5503
CLEVE.STL.	S55-55-1
CLEVE.STL.	S55-55-2
CLEVE.STL.	S55-55-735
CLEVE.STL.	S55-55-870
CMP	CM5555S
D&D MACH.	MCS55
DIVCO	16544
EXCEL	U310
FIRESTONE	5234830-7
FLEETRITE	ZAD-S15555
FMC	35577
FMC	55571

FORD	C1AA4724A
FORD	C1AZ4635C
FORD	C3A4635A
FORD	C3AA4635A
FORD	C3AZ4635A
FORD	C3AZ4635E
FORD	C3SZ4635A
FORD	CIAZ4635C
FORD	D3SZ4635A
FORD	TK4635A
GLAENZER	065-225-8
GMB	G5-1200
GMB	G5-1200X
GMB	G5-129
GMB	GS55-55
GMB	GU-S55-55-2
GOODYEAR	219-2512
GOODYEAR	219-2555
HABERLE	HS555
IHC	186-874R91
IHC	186-894R91
IHC	346096C91
JAMCO	UJ1200
JEEP	5703383
LAKEWOOD	23013
LEMPCO	U309
LEMPCO	U310
MASTERS	MCS55
MASTERS	MM1200
MCQUAY-NOR	U01-66
MCQUAY-NOR	UO1-66
MIDAS	UJ1003
MOPAR	D110
MOPAR	D88
MOTOR MAST	3200AT
MOTOR MAST	3400A
MOTOR MAST	3400AHP
MOTOR MAST	3400AT
NEAPCO	1-5900
NEAPCO	284635C
NEAPCO	285554
NEAPCO	285900
PERFECTION	R055-55-2
PERFECTION	RG55-55-2
PERFECTION	RS55-55-1
PERFECTION	RS55-55-2
PILOT	S15555
PILOT	TS15553
PILOT	TS15554
PRECISION	429
REPCO	101-1100
REPCO	5-1200X
REPCO	5-260X
REPCO	K5A502
REPCO	RUJ2027
REPCO	RUJ2027SL
REPCO	S55-55-2
REPUBLIC	CB3400AHP
REPUBLIC	CB-S55
REPUBLIC	CB-S55HD
REPUBLIC	CB-S55HDC
REPUBLIC	CBS55L
REPUBLIC	CB-T55L
REPUBLIC	CT55L
SEARS-ROE.	5114
SPICER	5-1200

SPICER	5-1200X
SPICER	5-124
SPICER	J5-124
STUDEBAKER	1687069
STUDEBAKER	1691761
TEDDY TORQ	GS55-55
TEDDY TORQ	J5-1200
TEDDY TORQ	S55
TRU-CROSS	1200
TRW	20060
TRW	20211
TRW	20904
UJS	55888
UJS	55S55
WAYNE	35577
WESCO	200554
WESCO	201554
WESCO	202-554
WESCO	303554
WESCO	G1554
WESCO	G1554RU
WESCO	N1554
WESCO	N1554RU
WESCO	N5155
WESCO	NS555
WESCO	NS555-1
WYLIE	74
ZELLER	3400A
ZOOM	800001

IDLI GROUP 80082

CROSS + 4 ROUND BEARINGS
2 INSIDE + 2 OUTSIDE SNAP RINGS
SEE FIGURE 1AB

1.0625 in	BD	26.99 mm
.6406 in	BW	16.27 mm
3.0 in	CL	76.20 mm
.6875 in	CD	17.46 mm

AEC	353
CHRYSLER	3480936
CLEVE.MOT.	CM902HP
GMB	G5-351X
MOTOR MAST	3100A
PRECISION	320
REPUBLIC	CB3100AX
TRW	20164

IDLI GROUP 80083

CROSS + 4 ROUND BEARINGS
2 INSIDE + 2 OUTSIDE SNAP RINGS
SEE FIGURE 1AB

1.0625 in	BD	26.99 mm
.6406 in	BW	16.27 mm
3.0 in	CL	76.20 mm
3.2188 in	CB	81.76 mm
.6875 in	CD	17.46 mm

AEC	266-1310
AEC	353E
AEC	AE266-1310
AEC	AE5-353
ALCO	J5-353
ALLOY	1153
ALLOY	J5-353
AMER.MOTOR	3233715
AMER.MOTOR	3233715X
BORG-WARNR	114-316

CHRYSLER	3480934
CHRYSLER	3780280
CLEVE.MOT.	CM5-353X
GMB	G5-353X
MCQUAY-NOR	U01-55
MOTOR MAST	3100A
MOTOR MAST	3100B
NEAPCO	2-3100
NEAPCO	283100
PRECISION	319
REPUBLIC	CB1399
REPUBLIC	CB3100X
SPICER	5-353X
TRU-CROSS	353
TRW	20141D
TRW	20189
TRW	20189D
WESCO	200353
WESCO	201-353
WESCO	N2155
ZELLER	3100B

IDLI GROUP 80084

CROSS + 2 ROUND BEARINGS
2 MIDWING DRILLED BEARINGS
+ INSIDE AND OUTSIDE SNAP RINGS
SEE FIGURE 1AK

1.0625 in	BD	26.99 mm
.6406 in	BW	16.27 mm
3.0 in	CL	76.20 mm
.6875 in	CD	17.46 mm

AEC	496
PRECISION	378
SPICER	5-260-1X

IDLI GROUP 80085

CROSS + 2 ROUND BEARINGS
2 MIDWING DRILLED BEARINGS
+ INSIDE AND OUTSIDE SNAP RINGS
slight variation from above group
SEE FIGURE 1AK

1.0625 in	BD	26.99 mm
.6406 in	BW	16.27 mm
3.0 in	CL	76.20 mm
.6875 in	CD	17.46 mm

AEC	278
AEC	5-278
AEC	AE5-278
ALLOY	1002
ALLOY	41002
ALLOY	41174
ALLOY	J5-278
AMER.MOTOR	8130298
BWE	114-513
DEERE,JOHN	AT54901
GMB	G5-154X
GMB	G5-278X
GMC	2362830
MCQUAY-NOR	U01-29
MOTOR MAST	3200F
NEAPCO	1-0278
NEAPCO	1-153
PILOT	11521HD
PRECISION	379
REPCO	K5GB117
REPCO	K5LGB104R

REPCO	K5LGB106
REPCO	K5LGB74
REPUBLIC	CB3278
SPICER	5-278X
SPICER	5-419X
SPICER	5-420X
TRU-CROSS	278
TRU-CROSS	433-1
TRW	20049
VERSATILE	56564
WESCO	200-278
WESCO	201-278
ZELLER	3200F

IDLI GROUP 80086

CROSS + 2 ROUND BEARINGS
2 MIDWING DRILLED BEARINGS
+ INSIDE AND OUTSIDE SNAP RINGS
slight variation from above group
SEE FIGURE 1AK

1.0625 in	BD	26.99 mm
.6406 in	BW	16.27 mm
3.0 in	CL	76.20 mm
.6875 in	CD	17.46 mm

AEC	495
ALLOY	41154
AMER.MOTOR	8130750
CLEVE.MOT.	CM5-278X
MCQUAY-NOR	U01-29
MOTOR MAST	3200-F
NEAPCO	1-0278
NEAPCO	28278X
NEAPCO	5-278
PRECISION	380
SPICER	5-420X
SPICER	5-433-1X
TRU-CROSS	278
TRU-CROSS	433-1
WESCO	200-522

IDLI GROUP 80087

CROSS + 2 ROUND BEARINGS
2 MIDWING DRILLED BEARINGS
+ INSIDE AND OUTSIDE SNAP RINGS
slight variation from above group
SEE FIGURE 1AK

1.0625 in	BD	26.99 mm
.6406 in	BW	16.27 mm
3.0 in	CL	76.20 mm
.6875 in	CD	17.46 mm

PRECISION	445
WESCO	611516
WESCO	N1516

IDLI GROUP 80088

CROSS + 2 ROUND BEARINGS
2 MIDWING DRILLED BEARINGS
+ INSIDE AND OUTSIDE SNAP RINGS
slight variation from above group
SEE FIGURE 1AK

1.0625 in	BD	26.99 mm
.6406 in	BW	16.27 mm
3.0 in	CL	76.20 mm
.6875 in	CD	17.46 mm

AEC	521HDU
AEC	AE521H0U

CAUTION: BE SURE TO REFER TO ENGINEERING CATALOGS FOR SPECIAL APPLICATIONS THAT REQUIRE SPECIFIC MATERIAL CONTENT, TOLERANCES, ETC. SEE FOOTNOTE.

BORG-WARNR	114-520U
CLEVE.MOT.	CM5-153XUB
FLEETRITE	ZAD11553UB
GMD	G5-121UB
GMB	G5-153XU
LEMPCO	U101HDU
MCQUAY-NOR	U01-67
MCQUAY-NOR	U04-23
MOTOR MAST	3200B
NEAPCO	1-1153
NEAPCO	28153XB
PRECISION	369U
SPICER	5-153XU
WESCO	303-520

IDLI GROUP 80089

CROSS + 2 ROUND BEARINGS 2 MIDWING DRILLED BEARINGS + INSIDE AND OUTSIDE SNAP RINGS SEE FIGURE 1GI

1.0625 in	BD	26.99 mm
.6406 in	BW	16.27 mm
3.0 in	CL	76.20 mm
.6875 in	CD	17.46 mm

AEC	521U
AEC	AE521U
ALCO	1028
ALCO	3926
ALCO	J5-121B
ALLOY	1028
ALLOY	J5-121B
BORG-WARNR	114-521RU
BORG-WARNR	114-521U
CMP	CM5-121XUB
D&D MACH.	MS121U
EXCEL	U113
FLEETRITE	ZAD11520UB
GMB	G5-121B
GMB	G5-121XU
LEMPCO	U113
MASTERS	MS121U
MCQUAY-NOR	U01-15
MCQUAY-NOR	UO1-15
MOPAR	D64
MOTOR MAST	2290B
NEAPCO	1-1200
NEAPCO	28051XB
PERFECTION	RG521B
PILOT	11521UB
PRECISION	346
REPUBLIC	CB1270U
SPICER	5-121XU
SPICER	5-200X
TRW	20024
WESCO	303521
WESCO	N1521RU
ZELLER	2290B

IDLI GROUP 80090

CROSS + 4 GROOVED BEARINGS + INSIDE SNAP RINGS SEE FIGURE 1B

1.0625 in	BD	26.99 mm
.6719 in	BW	17.07 mm
3.0 in	CL	76.20 mm
.6563 in	CD	16.67 mm

AEC	5-260
AEC	574
AEC	AE574
ALBURUS-SP	5-20X
ALCO	1024
ALCO	1025
ALCO	1574
ALCO	41025
ALCO	J5-74
ALCO	J5-75
ALLOY	1025
ALLOY	J5-74
ALLOY	J5-75
ALLOY	J5-74/75
AMER.MOTOR	8126608
AMER.MOTOR	909476
BORG-WARNR	114-547
BORG-WARNR	114-574
CENTRAL	G5-74X
CHRYSLER	1843143
CHRYSLER	2409783
CLEVE.MOT.	CM5-174X
CMP	CM5-174X
D&D MACH.	MS574X
EXCEL	U130
FLEETRITE	ZAD11574
FORD	B9T4635A
FORD	B9TZ3249A
FORD	B9TZ4635A
FORD	C6TTZ3249A
GLAENZER	065-212-8
GMB	G5-174
GMB	G5-260A
GMB	G5-260X
GMB	G5-574
GMB	G5-74
GMB	G5-74X
GMC	1181735
GMC	2403700
GMC	2409783
GMC	3780158
IHC	154-926R91
IHC	154929R91
IHC	290-312C91
JEEP	909476
LEMPCO	U125
LEMPCO	U130
LOBRO	U102
MASTERS	MS260X
MASTERS	MS574X
MCQUAY-NOR	U01-30
MCQUAY-NOR	UO1-30
MIPER	C3352
MOPAR	D50
MOTOR MAST	3200C
NEAPCO	1-0174
NEAPCO	28174X
PERFECTION	RG574
PERFECTION	RG574R
PILOT	11573
PILOT	11574
PRECISION	356
REPCO	5-74X
REPUBLIC	CB1574
REPUBLIC	CB1575
RUFDARSA	1336
RUEDARSA	2104-1336-0-0
SPICER	5-64X
SPICER	5-74X
TEDDY TORQ	K74
TRANSMISS.	UK352
TRU-CROSS	260
TRU-CROSS	74
TRW	20050
TRW	20191
UJS	55-74
WESCO	201-574
WESCO	N1574
WESCO	N1574-1
WHITE	909476
WILLYS	909476
ZELLER	3200C

IDLI GROUP 80091

CROSS + 4 GROOVED BEARINGS + INSIDE SNAP RINGS largest BD dimension is shown SEE FIGURE 1B

1.1250 in	BD	28.58 mm
.6406 in	BW	16.27 mm
3.0 in	CL	76.20 mm
.6875 in	CD	17.46 mm

AEC	1201
AEC	5-1201
AEC	675
AEC	AE5-1201
AEC	AE579
AEC	AE675
AEC	AE-P55-55-145
AEC	AE-S55-55-675
ALCO	1011
ALCO	1524
ALCO	J5-1201
ALLOY	1011
ALLOY	1011P
ALLOY	41011
ALLOY	J5-1201
AUST.WEST.	X55-55-675
BORG-WARNR	114-126
BORG-WARNR	114-8HD
BORG-WARNR	114-8HP
BORG-WARNR	S55-55-675
CENTRAL	G5-1201X
CLEVELAND	S55-55-675
CLEVE.MOT.	CM2141
CLEVE.MOT.	CM9675S
CLEVE.STL.	S55-55-675
CLEVE.STL.	S55-55-740
CLEVE.STL.	S55-55-875
CMP	CM2141
CMP	CM9675S
D&D MACH.	MC675
EXCEL	U129
EXCEL	U311
FIRESTONE	5234831-5
FLEETRITE	ZAD12141
FLEETRITE	ZAD-S10675
FLEETRITE	ZAD-S19675
FORD	ARC6DZ4635B
FORD	C1AA4724B
FORD	C1AZ4635A
FORD	C1AZ4635D
FORD	C2AZ4635D
FORD	C3AA4635B
FORD	C3AZ4635B
FORD	C3AZ4635F
FORD	CIAZ4635A
FORD	CIAZ4635D
FORD	D3SZ4635A
FORD	XB4635B
FORD	XW4635D
GLAENZER	065-226-8
GMB	G5-1201
GMB	G5-1201X
GMB	G5-1210
GMB	G5-126
GMB	G5-579
GMB	GU-S55-55-675
GOODYEAR	219-2513
GOODYEAR	219-2556
HABERLE	HS9675
IHC	799281R91
JAMCO	UJ1201
LAKEWOOD	23014
LEMPCO	U129
LEMPCO	U311
MASTERS	MC675
MASTERS	MCP56
MASTERS	MM1201
MASTERS	MS122X
MCQUAY-NOR	U01-16
MCQUAY-NOR	U01-37
MCQUAY-NOR	UO1-16
MCQUAY-NOR	UO1-37
MIDAS	UJ1004
MOPAR	D65
MOPAR	D89
MOTOR MAST	2290D
MOTOR MAST	3400B
MOTOR MAST	3400BHP
MOTOR MAST	3400BT
NEAPCO	1-4635
NEAPCO	284635D
PERFECTION	RG579
PERFECTION	RG579R
PERFECTION	RS55-165
PILOT	12141
PILOT	S19675
PRECISION	430
QUINT.HAZ.	QL2039
REPCO	101-1105
REPCO	114-2102
REPCO	5-1201X
REPCO	K5A507
REPCO	RUJ2039
REPCO	RUJ2039SL
REPCO	S55-55-675
REPUBLIC	CB1271
REPUBLIC	CB-S55-6
REPUBLIC	CB-S55-6HDC

REPUBLIC CBS55L
REPUBLIC CT55L
SEARS-ROE. 5113
SPICER 5-1201
SPICER 5-1201X
SPICER 5-123
SPICER J5-123
TEDDY TORQ G5-1201
TEDDY TORQ S40
TRU-CROSS 1201
TRW 20025
TRW 20060
TRW 20061
TRW 20905
UJS 55S675
WESCO 201165
WESCO 201-554
WESCO GR5165
WESCO N1579RU
WESCO N5165
WESCO NR1565
WESCO NR5165
WHITE 934811
WYLIE 53
ZELLER 3400B
ZELLER 3400BA
ZOOM 800002

IDLI GROUP 80092

**CROSS + 2 WING TYPE BEARINGS
+ 2 ROUND GROOVED BEARINGS
SEE FIGURE 1BI**

1.0000 in	BD	25.40 mm
.6719 in	BW	17.07 mm
3.0625 in	CL	77.79 mm
.6875 in	CD	17.46 mm

AEC 2017
AEC AE2017
ALCO 1062
ALCO 2032
ALCO J2017
ALLOY 1062
ALLOY J2017
ALMETAL CB2017
AMER.MOTOR 3206122
BALKAMP M2017
BALKAMP M2032
BORG-WARNR 114-2017
BORG-WARNR 114-2032
BORG-WARNR 114-2047
BORG-WARNR 114-2186
BORG-WARNR J2032
CLEVE.MOT. CM2017
CLEVE.MOT. CM2031
CLEVE.MOT. CM2032
CMP CM2017
CMP CM2032
D&D MACH. MM2032
EXCEL U200
FIAT-ALLIS 4498505
FLEETRITE ZAD11432
GALION D77092
GMB G2017
GMB G5-2017
GMB G5-45X
GMC 602796

GMC 609944
HABERLE H2017
HABERLE H2032
HERCULES 372
HERCULES 373
LEMPCO U200
LEMPCO U201
MASTERS MM2032
MASTERS MM2117
MASTERS MS54X
MCQUAY-NOR U01-52
MCQUAY-NOR UO1-52
MOPAR D3
MOTOR MAST 3010A
NEAPCO 1-0050
NEAPCO 1-50
NEAPCO 28050X
NEAPCO 28052X
NEAPCO 2894
PERFECTION RG2017
PERFECTION RG2023
PERFECTION RG2032
PILOT 11417
PILOT 11432
PILOT 11545
PRECISION 499
REPCO 114-2031
REPCO 114-2032
REPCO RUJ2017
REPUBLIC CB2017
REPUBLIC CB2032
ROCKWELL CP2017
RUEDARSA 1703
RUEDARSA 2100-1703-0-0
SPICER 5-2032X
SPICER 5-45X
TEDDY TORQ 1417
TRU-CROSS 45
TRW 20039
UJS 55-2017
WESCO 201217
WESCO N2017
WESCO S2017
ZELLER 3010A

IDLI GROUP 80093

**CROSS + 4 GROOVED BEARINGS
+ INSIDE SNAP RINGS
SEE FIGURE 1B**

1.1250 in	BD	28.58 mm
.7656 in	BW	19.45 mm
3.0938 in	CL	78.58 mm
.6719 in	CD	17.07 mm

AEC 590
AEC AE590
ALCO 1026
ALCO J5-90
ALLOY 1026
ALLOY J5-90
BORG-WARNR 114-590
CLEVE.MOT. CM2520
CLEVE.MOT. CM5-90X
EXCEL U258
GMB G5-90
GMB G5-90X
GMC 5674730

GMC 5674739
GMC 5676478
LEMPCO U258
MOPAR D51
MOTOR MAST 3093A
NEAPCO 1-4730
NEAPCO 284730
NEAPCO 284730X
PILOT 11590
PILOT 14252
PRECISION 518
REPCO 114-3047
REPCO 5-90X
REPUBLIC CB3047
SPICER 5-90X
TRW 20048
WESCO 201305
WESCO N3005
ZELLER 3093A

IDLI GROUP 80094

**CROSS + 4 GROOVED BEARINGS
+ INSIDE SNAP RINGS
SEE FIGURE 1B**

1.1719 in	BD	29.77 mm
.8750 in	BW	22.23 mm
3.1250 in	CL	79.38 mm
.6406 in	CD	16.27 mm

AEC AE263
ALCO 73525-4
ALCO J7352-4
ALCO J73525-4
BORG-WARNR 114-263
BORG-WARNR 114-264
CHRYSLER 1603230
CHRYSLER 1818779
D&D MACH. MD5254
D&D MACH. MD6253
DETROIT 7325-5
DETROIT 73525-4
GMB G73525-4
LEMPCO U442
MOPAR D137
MOTOR MAST 3312D
NEAPCO 283300
PERFECTION RG263
PILOT 17423
PILOT 17424
PRECISION 313
REPCO 114-263
REPUBLIC CB7325-5
REPUBLIC CB73525-4
TEDDY TORQ 734
WESCO 201263
WESCO D263

IDLI GROUP 80095

**CROSS + 2 SLOTTED BEARINGS
+ 2 ROUND BEARINGS
SEE FIGURE 1BD**

1.1719 in	BD	29.77 mm
.9531 in	BW	24.21 mm
3.1250 in	CL	79.38 mm
.6406 in	CD	16.27 mm

AEC 114-257
AEC AE257

ALCO 735251-2
ALCO J735251-2
ALCO J73525-2
BORG-WARNR 114-257
BORG-WARNR 114-258
BORG-WARNR 114-324
CHRYSLER 1450445
CHRYSLER 1450446
CHRYSLER 1450498
CHRYSLER 1532929
CHRYSLER 1532937
DETROIT 73251-2
DETROIT 7325-4
DETROIT 735251-2
DETROIT 73525-2
GMB G5-1310X
GMB G5-1313X
GMB G5-1314X
GMB G735251-2
HABERLE H7254
HABERLE H7525
LEMPCO U428
LEMPCO U432
LEMPCO U438
MOPAR D139
MOTOR MAST 3312A
NEAPCO 283800
NEAPCO 284300
PERFECTION RG257
PERFECTION RG324
PILOT 17324
PILOT 17326
PRECISION 291
REPCO 114-257
REPCO 114-258
REPUBLIC CB7325-4
REPUBLIC CB73525
REPUBLIC CB735251
REPUBLIC CB73525-2
SPICER 73525-2
STUDEBAKER 436188
STUDEBAKER 436305
TEDDY TORQ 352
TEDDY TORQ 515
WESCO 201257
WESCO D257
WESCO D258
WESCO D324
WESCO D354
WESCO D7325

IDLI GROUP 80096

**CROSS + 2 GROOVED BEARINGS
+ 2 LOW WING DRILLED BEARINGS
+ INSIDE SNAP RINGS
SEE FIGURE 1BH**

1.1719 in	BD	29.77 mm
1.8125 in	BW	46.04 mm
3.1250 in	CL	79.38 mm
.6406 in	CD	16.27 mm

AEC AE259
AEC AE262
ALCO 73525-3
ALCO J7325-2
ALCO J73252-2
ALCO J7325-5

ENGINEERING CATALOGS MUST BE CONSULTED FOR DETAILS NOT INCLUDED
IN THIS GUIDE. SPECIFIC DESIGNS, MATERIAL CONTENT, TOLERANCES
LUBE FITTINGS AND OTHER DIMENSIONS ARE INTENTIONALLY OMITTED HERE.

CAUTION: BE SURE TO REFER TO ENGINEERING CATALOGS FOR SPECIAL APPLICATIONS THAT REQUIRE SPECIFIC MATERIAL CONTENT, TOLERANCES, ETC. SEE FOOTNOTE.

ALCO	J73525-3
BORG-WARNR	114-259
BORG-WARNR	114-262
BORG-WARNR	114-264
BORG-WARNR	114-325
CHRYSLER	1603233
CHRYSLER	1692691
CHRYSLER	947549
CHRYSLER	947551
DETROIT	53525-2
DETROIT	7325-2
DETROIT	73252-2
DETROIT	73525-3
GMB	G5-1313X
GMB	G7325-2
GMB	G73525-3
GMB	G73525-3/4
HABERLE	H7523
LEMPCO	U416
LEMPCO	U439
LEMPCO	U441
LEMPCO	U443
MOPAR	D135
MOPAR	D136
MOTOR MAST	3312B
NEAPCO	281000
NEAPCO	283300
PERFECTION	RG259
PILOT	17325
PILOT	17329
PILOT	17423
PRECISION	311
REPCO	114-259
REPUBLIC	CB7325
REPUBLIC	CB73252-2
REPUBLIC	CB73252-3
REPUBLIC	CB73525-3
SPICER	73525-3
STUDEBAKER	436175
STUDEBAKER	458396
TEDDY TORQ	732
TEDDY TORQ	733
TEDDY TORQ	735
WESCO	201735
WESCO	301325
WESCO	D259
WESCO	D262
WESCO	D325
WESCO	D525
WESCO	D7325
WESCO	N7325

IDLI GROUP 80097

CROSS + 2 ROUND BEARINGS 2 MIDWING DRILLED BEARINGS + INSIDE AND OUTSIDE SNAP RINGS
SEE FIGURE 1AK

1.0625 in	BD	26.99 mm
.6875 in	BW	17.46 mm
3.2188 in	CL	81.76 mm
3.3750 in	CB	85.73 mm
.6563 in	CD	16.67 mm

AEC	356
AEC	3R1310
AEC	AE3R1310
ALLOY	1699
ALLOY	J3022
BORG-WARNR	114-9HD
BORG-WARNR	114-9HP
CMP	CM5-150X
FLEETRITE	ZAD11550
GMB	G5-150X
GMB	G5-3022X
GMB	G5-356X
LAKEWOOD	23018
LEMPCO	U3130A
MASTERS	MM3125
MASTERS	MS150X
MCQUAY-NOR	HI-PERF.CONVERSION KIT
MOTOR MAST	3130A
MOTOR MAST	3130B
NEAPCO	283130
NEAPCO	3-3130
PERFECTION	RG550R
PRECISION	372
REPUBLIC	CB3110X
SPICER	5-3022X
SPICER	5-356X
TRU-CROSS	3022
TRW	20165
UJS	55-150
WESCO	301-327
ZELLER	3130A

IDLI GROUP 80098

CROSS AND LOW OR HIGH WING TYPE DRILLED OR THREADED BEARINGS
SEE FIGURE 1I

1.4375 in	CC	36.51 mm
3.3125 in	CL	84.14 mm
.6719 in	CD	17.07 mm

AEC	3103
AEC	AE3103
ALCO	J3103
BEND.WEST.	598A277-2
BORG-WARNR	114-3103
BORG-WARNR	114-3109
BORG-WARNR	114-3161
CLEVE.MOT.	CM3103
GMB	G3103
GMB	G5-3103X
IHC	625810C1
IHC	627-253C1
MCQUAY-NOR	OFF RD. EQUIP.
MOTOR MAST	3031W
NEAPCO	283103
PRECISION	495
REPCO	114-3103
WESCO	301324
WESCO	N3103
WESTINGHSE	598A277-2

IDLI GROUP 80099

CROSS AND LOW OR HIGH WING TYPE DRILLED OR THREADED BEARINGS
SEE FIGURE 1IJ

1.4375 in	CC	36.51 mm
3.3125 in	CL	84.14 mm
.6719 in	CD	17.07 mm

ADAMS-LET.	4690
ADAMS-LET.	475471
ADAMS-LET.	523435
ADAMS-LET.	SL8673-1
ADAMS-LET.	SL8673-3
AEC	3152
AEC	AE3000
AEC	AE3106
ALCO	3000
ALLIS-CHLM	34121
ALLIS-CHLM	7209
ALLIS-CHLM	7372959
AMER.MOTOR	3109413
AMER.MOTOR	3109413A
ATHEY PROD	P48776
AUST.WEST.	AW465034
AUST.WEST.	AW465035
BALKAMP	3-922
BALKAMP	M3000
BALKAMP	M3001
BALKAMP	M3003
BALKAMP	M3005
BALKAMP	M3036
BALKAMP	M3045
BALKAMP	M3100
BALKAMP	M3105
BALKAMP	M3106
BALKAMP	M3112
BALKAMP	M3114
BALKAMP	M3117
BALKAMP	M3128
BAL-LM-HAM	465034
BAL-LM-HAM	465035
BORG-WARNR	114-3000
BORG-WARNR	114-3005
BORG-WARNR	114-3016
BORG-WARNR	114-3045
BORG-WARNR	114-3106
BORG-WARNR	114-3157
BORG-WARNR	114-3159
CLEVE.MOT.	CM3000
CLEVE.MOT.	CM3100
CLEVE.MOT.	CM3106
CMP	CM3000
CMP	CM3100
CMP	CM3106
D&D MACH.	MM3000
DEERE,JOHN	3CAR29699R
DEERE,JOHN	3CARR29699R
DEERE,JOHN	AM3025E
DEERE,JOHN	AR29699R
DEERE,JOHN	XAW70210A
DEERE,JOHN	XAW9153A
FIAT-ALLIS	7372959
FLEETRITE	ZAD14300
GMB	G3106
GMB	G3118
GMB	G5-3000X
GMB	GU6300
GMC	1424820
GMC	1463547
HENNEY	354670
HERCULES	381
IHC	296-595R91
IHC	65346R91
INBRA	CT19-020
IOWA MFG.	7329-016
LEMPCO	U207
LOBRO	U630
MASTERS	MM3000
MASTERS	MM3106
MCQUAY-NOR	CAD.
MCQUAY-NOR	UO1-25
MOPAR	D20
MOTOR MAST	3031-2H
MOTOR MAST	3031HB
NEAPCO	280600
PERFECTION	RG3000
PERFECTION	RG3000G
PERFECTION	RG3011
PETTI-MULL	A7H4
PETTI-MULL	LA425-2
PETTI-MULL	LA452-2
PETTI-MULL	LL773-62
PILOT	14300
PRECISION	530
PREC.TOR.	922
REPCO	114-3106
REPCO	114-3152
REPUBLIC	CB3000
REPUBLIC	CB3001
RUEDARSA	1140
RUEDARSA	2100-1140-0-0
SPICER	5-3000X
STUDEBAKER	354670
TRU-CROSS	3000
UJS	55-3000
WABCO	475471
WABCO	575471
WESCO	G3000G
WESCO	N3000

IDLI GROUP 80100

CROSS AND LOW OR HIGH WING TYPE DRILLED OR THREADED BEARINGS
slight variation from above group
SEE FIGURE 1IJ

1.4375 in	CC	36.51 mm
3.3125 in	CL	84.14 mm
.6719 in	CD	17.07 mm

AEC	3101
AEC	3119
AEC	AE3101
AEC	AE3119
ALCO	J3101
ALCO	J3119
ALLIS-CHLM	08-000-505-01-398
ALLIS-CHLM	08-630-403-008
ALLIS-CHLM	619703
ALLIS-CHLM	647098
ALLIS-CHLM	7211
ALLOY	J3101
ATHEY PROD	X210-570
AUST.WEST.	46520
AUST.WEST.	465250

BD = BEARING DIAMETER (outside)   BW = BEARING WIDTH   CB = CROSS LENGTH WITH BEARINGS   CC = CENTER TO CENTER   CD = CROSS DIAMETER
CL = CROSS LENGTH WITHOUT BEARINGS   EC = END TO CENTER or FACE TO FACE   EE = END TO END   EL = EFFECTIVE LENGTH
HD = HUB DIAMETER (or insert)   IE = INSIDE OF EARS (or recess)   OD = OUTSIDE DIAMETER   OE = OUTSIDE OF EARS   SB = SPLINE OR BORE SIZE

AUST.WEST. 465254	IHC 627254C1	ALCO J3106	CATERPILLR 6L1516
AUST.WEST. 465424	IHC 627-879C1	ALCO J3118	CATERPILLR 9K1971
AUST.WEST. AW46520	IHC 784409R1	ALLIS-CHLM 34121	CEDAR RAP. 45375-017-02
AUST.WEST. AW465250	INBRA CT11-005	ALLIS-CHLM 4690	CLARK EQU. 5202079
AUST.WEST. AW465254	INBRA CT19-015	ALLIS-CHLM 7209	CLARK EQU. 5206861
AUST.WEST. AW465424	LEMPCO U242	ALLIS-CHLM 7372959	CLARK EQU. 5207368
BAL-LM-HAM 465020	MASTERS MM3010	ALLOY 1711	CLARK EQU. 5260862
BAL-LM-HAM 46520	MCQUAY-NOR U01-26	ALLOY J3100	CLEVE.MOT. CM3106
BAL-LM-HAM 465250	MCQUAY-NOR U03-13	ALLOY J3106	CMP CM3100
BAL-LM-HAM 465254	MCQUAY-NOR UO3-13	ATHEY PROD P38776	CONS.MACH. M5272
BAL-LM-HAM 465424	MOTOR MAST 3031HB	ATHEY PROD P48776	DEERE,JOHN AM1862E
BAL-LM-HAM AW46520	MOTOR MAST 3031HC	AUST.MOT 5545-3042	DEERE,JOHN AR29699
BAL-LM-HAM AW4652505	MOTOR MAST 3031P	AUST.WEST. 465034	ELEC.MTR. 3382446
BEND.WEST. 598A277-1	NEAPCO 283101	AUST.WEST. 465035	EXCEL U207
BLACKWELDR 48-39-143	PACIF.CAR 6462C28	AUST.WEST. 465153	FALK 208683
BLACKWELDR 48-39-144	PAXTON-MIT 3C37	AUST.WEST. ASF5762	FIAT-ALLIS 2372959
BORG-WARNR 114-3010	PEARSON VP1494	AUST.WEST. ASF5762-55A	FIAT-ALLIS 3004690-8
BORG-WARNR 114-3042	PETTI-MULL LA518-10	AUST.WEST. HCF15790	FIAT-ALLIS 3034121
BORG-WARNR 114-3101	PILOT 14310	AUST.WEST. HCF19058	FIAT-ALLIS 34121
BORG-WARNR 114-3107	PRECISION 535	BALKAMP 3-922	FIAT-ALLIS 4690
BORG-WARNR 114-3117	REPCO 114-3101	BALKAMP M3036	FIAT-ALLIS 617703
BORG-WARNR 114-3119	REPCO 114-3107	BALKAMP M3045	FIAT-ALLIS 70004690
BORG-WARNR 114-3124	REPCO 114-3119	BALKAMP M3100	FIAT-ALLIS 70007209
BORG-WARNR 114-3129	REPCO 114-3124	BALKAMP M3105	FIAT-ALLIS 70034121
BORG-WARNR 114-3132	REPCO 114-3129	BALKAMP M3106	FIAT-ALLIS 70617703
BORG-WARNR 114-3138	REPCO 114-3150	BALKAMP M3112	FIAT-ALLIS 7209
BORG-WARNR 114-3141	REPCO RUJ3142	BALKAMP M3114	FIAT-ALLIS 72372959
BORG-WARNR 114-3142	REPUBLIC CB3010	BALKAMP M3117	FIAT-ALLIS 73004690-8
BORG-WARNR 114-3146	ROCKWELL CP3101	BALKAMP M3128	FIAT-ALLIS 73034121
BORG-WARNR 114-3150	SMITH,T.L. WM8-1538-3	BAL-LM-HAM 121PA293	FIAT-ALLIS 7372959
BORG-WARNR 114-3160	TEDDY TORQ 1438	BAL-LM-HAM ASF5762	FIAT-ALLIS 77372959
BORG-WARNR 114-3162	TEDDY TORQ 1438G	BAL-LM-HAM ASF5762-55A	FLEETRITE ZAD14300
BORG-WARNR 114-3170	TRW 20044	BEND.WEST. 523435	GALION D70496
CASE,J.I. D73645	TRW 20161	BORG-WARNR 114-3000	GALION D82809
CATERPILLR 2V1164	TRW 20161D	BORG-WARNR 114-3001	GALION RB51477
CATERPILLR 9K1976	WABCO 475110	BORG-WARNR 114-3003	GALION .... RD51477(7)
CLARK EQU. 5202193	WABCO 723133	BORG-WARNR 114-3005	GEN.EQUIP. 15-22875
CLARK EQU. 5202196	WABCO 883-33A	BORG-WARNR 114-3010	GLAENZER 065-119-8
CLEVE.MOT. CM3010	WESCO 301301	BORG-WARNR 114-3036	GMB G3100
CLEVE.MOT. CM3101	WESCO N3010	BORG-WARNR 114-3045	GMB G3118
D&D MACH. MM3101	WESTINGHSE 598A277-1	BORG-WARNR 114-3100	GMB G3119
DEERE,JOHN AM1861E	ZELLER 3031P	BORG-WARNR 114-3100A	GMB G5-3000X
DEERE,JOHN AM3062E		BORG-WARNR 114-3106	GMB G5-3152
FIAT-ALLIS 0619703		BORG-WARNR 114-3106AB	GMC 3382466
FIAT-ALLIS 0619703-2	IDLI GROUP 80101	BORG-WARNR 114-3112	HABERLE H3000
FIAT-ALLIS 08-000-505-01-398	CROSS AND LOW OR HIGH WING TYPE	BORG-WARNR 114-3118	HYSTER 125875
FIAT-ALLIS 08-630-403-008	DRILLED OR THREADED BEARINGS	BORG-WARNR 114-3118AB	HYSTER 26601
FIAT-ALLIS 619703	slight variation from above group	BORG-WARNR 114-3128	HYSTER 27323
FIAT-ALLIS 647098	SEE FIGURE 1J	BORG-WARNR 114-3128A	IHC 291-736R91
FIAT-ALLIS 70007211	1.4375 in CC 36.51 mm	BORG-WARNR 114-3131	IHC 295-578R91
FIAT-ALLIS 70619703	3.3125 in CL 84.14 mm	BORG-WARNR 114-3137	IHC 306-114R91
FIAT-ALLIS 70619703-2	.6719 in CD 17.07 mm	BORG-WARNR 114-3144	IHC 569-085R91
FIAT-ALLIS 70647098	ADAMS-LET. 4690	BORG-WARNR 114-3145	IHC 65346R91
FIAT-ALLIS 7211	ADAMS-LET. 475471	BORG-WARNR 114-3149	IHC 65-346R92
GARD.DENV. 148959	ADAMS-LET. 523435	BORG-WARNR 114-3152	INBRA CT19-005
GMB G2104	ADAMS-LET. SL8673-1	BORG-WARNR 114-3159	IOWA MFG. 7329-016
GMB G3101	ADAMS-LET. SL8673-3	BORG-WARNR 114-3165	JEBCO 3P15
GMB G31011	AEC 3106	BORG-WARNR 114-3195	LEMPCO U207
GMB G3105	AEC 3118	BORG-WARNR J3000	LEMPCO U243
GMB G3119	AEC 3152	BUCYR-ERIE 12-275353	MASTERS MM3000
GMB G5-3101X	AEC AE3000	BUCYR-ERIE 4-681-3500	MASTERS MM3106
GMB G5-3150	AEC AE3100	CASE,J.I. 5996	MASTERS MM3118
HABERLE H3010	AEC AE3118	CASE,J.I. D73646	MCQUAY-NOR U01-25
HYSTER 125876	AEC AE3152	CASE,J.I. D87263	MCQUAY-NOR U01-26
HYSTER 27322	ALCO 1711	CATERPILLR 5D3248	MCQUAY-NOR UO1-26
IHC 625811C1	ALCO 3000	CATERPILLR 5D9153	MOPAR D17
	ALCO J3100		

CAUTION: BE SURE TO REFER TO ENGINEERING CATALOGS FOR SPECIAL APPLICATIONS THAT REQUIRE SPECIFIC MATERIAL CONTENT, TOLERANCES, ETC. SEE FOOTNOTE.

Manufacturer	Part No.
MOPAR	D20
MOTOR MAST	3031H
MOTOR MAST	3031HB
MOTOR MAST	303HR
NEAPCO	280600G
NEAPCO	2822
NEAPCO	283152
NEAPCO	3-3152
NW MOTORS	16235
PARSONS	324D259(4)
PAXTON-MIT	3P15
PEARSON	VP1493
PERFECTION	RG3000
PERFECTION	RG3000G
PERFECTION	RG3011
PETTI-MULL	A7HA
PETTI-MULL	L2596-14
PETTI-MULL	LA425-2
PETTI-MULL	LL773-183
PETTI-MULL	LL773-63
PILOT	14300
PILOT	14301
PILOT	14318
PRECISION	536
PREC.TOR.	922
REPCO	114-3100
REPCO	114-3112
REPCO	114-3118
REPCO	114-3128
REPCO	114-3152
REPUBLIC	CB3000
REPUBLIC	CB3001
REPUBLIC	CB722
ROGER IRON	523425
ROGER IRON	52345
RUST	VP14949
SMITH,T.L.	WM8-1538-4
SPICER	5-2000X
SPICER	5-3000X
TRANSMISS.	UK506
TRU-CROSS	3000
TRW	20043
WABCO	2041593
WABCO	4690
WABCO	475471
WABCO	523435
WABCO	SL8673-1
WABCO	SL8673-3
WESCO	301-300
WESCO	N3000
WESCO	N3000G
WESCO	N3001
WHITE	02-7074108
WOOLDRIDGE	15-22875
ZELLER	3031H

IDLI GROUP 80102

CROSS + 2 LOW WING THREADED + 2 BLOCK BEARINGS
SEE FIGURE 1JL

1.4375 in	CC	36.51 mm
3.3125 in	CL	84.14 mm
.6719 in	CD	17.07 mm

Manufacturer	Part No.
AEC	3121
AEC	AE3121
ALCO	J3121
ALCO	J3123
ALLIS-CHLM	X34000
AUST.WEST.	HCF16144
BORG-WARNR	114-3121
BORG-WARNR	114-3123
BORG-WARNR	114-3130
BORG-WARNR	114-3139
BORG-WARNR	114-3148
BORG-WARNR	114-3151
BORG-WARNR	114-3158
BORG-WARNR	114-3166
CLARK EQU.	1500765
CLARK EQU.	5206990
CLEVE.MOT.	CM3121
CLEVE.MOT.	CM3158
D&D MACH.	MM3121
FIAT-ALLIS	0178554
FIAT-ALLIS	178554
FIAT-ALLIS	70178554
FIAT-ALLIS	X34000
GMB	G3121
GMB	G3123
GMB	G5-3121X
GMB	G5-3158
MCQUAY-NOR	U03-14
MCQUAY-NOR	U03-14
MOTOR MAST	3031R
NEAPCO	2866
PILOT	14358
PRECISION	966
REPCO	114-3121
REPCO	114-3123
REPUBLIC	CB766
RITEWAY	REDI-MIX TRUCK
TRW	20045
WABCO	TH7348
WESCO	301323
WESCO	N3023

IDLI GROUP 80103

CROSS WITH HOLE IN THE CENTER + 2 HIGH WING DRILLED + 2 BLOCK BRGS.
SEE FIGURE 1L

1.4375 in	CC	36.51 mm
3.3125 in	CL	84.14 mm
.6719 in	CD	17.07 mm

Manufacturer	Part No.
AEC	3125
AEC	AE3125
ALCO	J3125
ALCO	J3155
ALLIS-CHLM	179571
BORG-WARNR	114-3125
BORG-WARNR	114-3143
BORG-WARNR	114-3155
CLARK EQU.	946568
CLEVE.MOT.	CM3125
FIAT-ALLIS	0179571
FIAT-ALLIS	0179571-5
FIAT-ALLIS	179571
FIAT-ALLIS	70179571
FIAT-ALLIS	70179571-5
GLAENZER	065-144-8
GMB	G3125
GMB	G3155
GMB	G5-3125X
MOTOR MAST	3031Q
NEAPCO	283125
NEAPCO	2867
PILOT	14355
PRECISION	967
REPCO	114-3125
REPCO	114-3155
REPUBLIC	CB767
WESCO	301322
WESCO	N3022

IDLI GROUP 80104

CROSS + 4 GROOVED BEARINGS + INSIDE SNAP RINGS
SEE FIGURE 1B

1.1250 in	BD	28.58 mm
.7969 in	BW	20.24 mm
3.3125 in	CL	84.14 mm
.6719 in	CD	17.07 mm

Manufacturer	Part No.
AEC	3105
AEC	3111
AEC	3114
AEC	489
AEC	AE3105
AEC	AE3114
ALCO	1015
ALCO	3008
ALCO	J3105
ALCO	J3111
ALCO	J3114
ALLOY	1014
ALLOY	1015
ALLOY	41014
ALLOY	J3001
ALLOY	J3011
ALLOY	J3105
ATHEY PROD	P12169
ATHEY PROD	P22181
AUST.WEST.	465028
AUST.WEST.	465029
AUST.WEST.	465036
AUST.WEST.	AW465028
AUST.WEST.	AW465029
AUST.WEST.	AW465036
BALKAMP	M3008
BALKAMP	M3033
BALKAMP	M3111
BAL-LM-HAM	465028
BAL-LM-HAM	465029
BAL-LM-HAM	465036
BLAW KNOX	7748998
BORG-WARNR	114-2093
BORG-WARNR	114-3011
BORG-WARNR	114-3033
BORG-WARNR	114-3105
BORG-WARNR	114-3114
BORG-WARNR	114-3127
BORG-WARNR	114-3133
BORG-WARNR	114-3147
BORG-WARNR	114-4HD
BORG-WARNR	114-4HP
BORG-WARNR	J3008
CATERPILLR	R2■1972
CATERPILLR	R7X7355
CENTRAL	G5-3008X
CHAMPION	SB145
CLARK EQU.	5202074
CLARK EQU.	5202080
CLEVE.MOT.	CM3008
CLEVE.MOT.	CM3011CV
CLEVE.MOT.	CM3105
CMP	CM3008
CMP	CM3015
CMP	CM3105
D&D MACH.	MM3011
D&D MACH.	MM3105
DEERE,JOHN	AM1117E
DEERE,JOHN	AM3014
DEERE,JOHN	AN15327N
DEERE,JOHN	AR29699
DEERE,JOHN	AXC5351E
DEERE,JOHN	XAC5351E
DETROIT	4030848
DOMINION	SB145
EXCEL	U228G
FIRESTONE	5234836-6
FLEETRITE	ZAD14308
FORD	C6TZ4635D
FORD	C8SZ4635A
GLAENZER	065-173-8
GLAENZER	065-239-8
GMB	G3008
GMB	G3105
GMB	G3111
GMB	G3114
GMB	G3115
GMB	G3118
GMB	G4118
GMB	G5-3006X
GMB	G5-3008X
GMB	G5-3011
GMB	G5-3011X
GMB	G5-3015X
GMB	G5-3105
GMB	G5-3105-1X
GMB	G5-3105X
GMB	G5-3111
GMB	G5-3147X
GMB	GU3T1602
GMB	GU6370
GMC	5677659
GMC	5677659-2
GOODYEAR	219-2508
GOODYEAR	219-2560
HABERLE	H3008
HARNISCH.	T32970
IHC	140601R91
IHC	142-823R91
IHC	142-923R91
IHC	142-973R91
IHC	453-999R91
IHC	656222R92

Column 1

IHC	753-931R91
IHC	784401R1
JAMCO	UJ3147
JEBCO	3C37
LAKE SHORE	6021
LAKEWOOD	23011
LEMPCO	U228
LEMPCO	U228G
LEMPCO	U234
LOBRO	U636
LOBRO	U637
MASS.FERG.	1016-245M91
MASS.FERG.	835-512M91
MASTERS	MM3008
MASTERS	MM3011
MASTERS	MM3105
MASTERS	MM3111
MCQUAY-NOR	U01-21
MCQUAY-NOR	U01-22
MCQUAY-NOR	UO1-21
MERCURY	B48153
MIDAS	UJ1009
MOPAR	D19
MOPAR	D206
MOPAR	D213
MOTOR MAST	3031-2J
MOTOR MAST	3031B
MOTOR MAST	3031BHP
MOTOR MAST	3031BT
MOTOR MAST	3031G
MPL.MOLINE	15P645
NEAPCO	2-3011
NEAPCO	283011
NEAPCO	283105
NEAPCO	283105B
NEAPCO	2889
PERFECTION	RG3008
PERFECTION	RG3008G
PETTIBONE	F6527-2A
PILOT	14308
PILOT	14308G
PRECISION	534G
REPCO	114-3105
REPCO	114-3114
REPCO	114-3152
REPUBLIC	CB3008G
REPUBLIC	CB3008HDC
REPUBLIC	CB3008L
REPUBLIC	CB3011
REPUBLIC	CB3011HDC
REPUBLIC	CB3031BHP
REPUBLIC	CB789
SEARS-ROE.	5112
SPICER	5-3008X
SPICER	5-3012X
SPICER	5-3147X
TRANSMISS.	UK351
TRU-CROSS	3147
TRW	20040
TRW	20125
TRW	20207
TRW	20902
UJS	55-3008
ULRICH	X214
WABCO	475370
WESCO	200308

Column 2

WESCO	200-311
WESCO	201308
WESCO	201-311
WESCO	202-308
WESCO	202-311
WESCO	203308
WESCO	408-308
WESCO	408-309
WESCO	408-838
WESCO	408-864
WESCO	G2008G
WESCO	G3008G
WESCO	N3008G
WHITING	RRWA8-1E
ZELLER	3031B
ZOOM	800003

IDLI GROUP 80105

CROSS + 4 GROOVED BEARINGS + INSIDE SNAP RINGS
largest BD dimension is shown
SEE FIGURE 1B

1.1250 in	BD	28.58 mm
.7969 in	BW	20.24 mm
3.3125 in	CL	84.14 mm
.6719 in	CD	17.07 mm

AEC	3188
AEC	AE3188
ALCO	J3188
ALLOY	1085
ALLOY	J3188
BORG-WARNR	114-591
BORG-WARNR	114-596
CLEVE.MOT.	CM3188
CMP	CM3188
FLEETRITE	ZAD14388
GLAENZER	065-268-8
GMB	G5-3009X
GMB	G5-596
GMC	5698221
GOODYEAR	219-2517
JAMCO	UJ3009
LEMPCO	U266
MASTERS	MM3188
MCQUAY-NOR	U01-28
MCQUAY-NOR	U02-28
MOPAR	D204
MOTOR MAST	3031V
NEAPCO	2-1569
NEAPCO	281569
PILOT	14388
PRECISION	496
REPCO	114-596
REPUBLIC	CB1372
SPICER	5-3009X
TRU-CROSS	3009
TRW	20047
UJS	55-3188
WESCO	201309
WESCO	G3009
WESCO	N3009
ZELLER	3031V

Column 3

IDLI GROUP 80106

CROSS + 2 GROOVED BEARINGS + 2 LOW WING DRILLED BEARINGS + INSIDE SNAP RINGS
SEE FIGURE 1BG

1.1250 in	BD	28.58 mm
.7969 in	BW	20.24 mm
3.3125 in	CL	84.14 mm
.6719 in	CD	17.07 mm

AEC	3013
AEC	AE3013
ALCO	J3013
BORG-WARNR	114-3013
CLEVE.MOT.	CM3013
GMB	G3013
GMB	G3116
GMB	G5-3013X
LEMPCO	U212
LOBRO	U635
MCQUAY-NOR	U02-4
MCQUAY-NOR	UO2-4
MOTOR MAST	3031S
NEAPCO	281700
PERFECTION	RG3013
PRECISION	532
REPUBLIC	CB3013
RUEDARSA	1900-5
RUEDARSA	2100-1900-5-0
SPICER	5-3013X
WESCO	301313
WESCO	N3013

IDLI GROUP 80107

CROSS + 4 GROOVED BEARINGS + INSIDE SNAP RINGS
SEE FIGURE 1B

1.1250 in	BD	28.58 mm
.8281 in	BW	21.03 mm
3.3125 in	CL	84.14 mm
.6719 in	CD	17.07 mm

AEC	AE3008
AEC	AE3011
AEC	AE3111
ALCO	1014
ALCO	1085
ALCO	J3011
ALLOY	1014P
ALLOY	1015
ALLOY	J3011
ALLOY	J3105
BORG-WARNR	114-3008
BORG-WARNR	114-3011
BORG-WARNR	114-3111
CENTRAL	G5-3011X
CHRYSLER	4030848
CLEVE.MOT.	CM3011CVX
CMP	CM3011CV
D&D MACH.	MM3008
D&D MACH.	MM3111
DEERE,JOHN	AM3014E
EXCEL	U228G
EXCEL	U266
FLEETRITE	ZAD14311
FORD	C8SZ4635A
FORD	C9SZ4635A
GMB	G3011

Column 4

GMB	G3105
GMB	G3188
GMB	G5-3011X
GMB	G5-3105-1X
GMC	1456525
GMC	1463548
GMC	522163
GMC	535554
GMC	565987
GMC	5671712
GMC	7806140
GOODYEAR	219-2559
LEMPCO	U228
LOBRO	U638
MCQUAY-NOR	U01-21
MCQUAY-NOR	U01-22
MIDAS	UJ1007
NEAPCO	2-3111
NEAPCO	283011
NEAPCO	283111
NEAPCO	285100
NEW HOLL.	29258
PILOT	14311
PRECISION	492
PRECISION	534
REPCO	114-3011
REPCO	114-3111
REPUBLIC	CB3008
REPUBLIC	CB3011
RUEDARSA	1810
RUEDARSA	2104-1810-0-0
SEARS-ROE.	5100
SPICER	5-2009X
SPICER	5-3000X
SPICER	5-3011X
SPICER	5-3012X
SPICER	5-3111X
SPICER	5-3147X
TRU-CROSS	3147
TRW	20040
UJS	55-3008C
WESCO	201311
WESCO	301-308
WESCO	408-308
WESCO	408-309
WESCO	408-838
WESCO	408-864
WESCO	G3011
WESCO	N3008
WESCO	N3008G
WESCO	N3011
ZELLER	3031-28

IDLI GROUP 80108

CROSS + 2 ROUND BEARINGS 2 MIDWING DRILLED BEARINGS + INSIDE AND OUTSIDE SNAP RINGS
SEE FIGURE 1AK

1.1250 in	BD	28.58 mm
.8281 in	BW	21.03 mm
3.3125 in	CL	84.14 mm
.6719 in	CD	17.07 mm

MCQUAY-NOR	U01-38
MOTOR MAST	3031GM
NEAPCO	2-3012
NEAPCO	283011C

ENGINEERING CATALOGS MUST BE CONSULTED FOR DETAILS NOT INCLUDED
IN THIS GUIDE. SPECIFIC DESIGNS, MATERIAL CONTENT, TOLERANCES
LUBE FITTINGS AND OTHER DIMENSIONS ARE INTENTIONALLY OMITTED HERE.

CAUTION: BE SURE TO REFER TO ENGINEERING CATALOGS FOR SPECIAL APPLICATIONS THAT REQUIRE SPECIFIC MATERIAL CONTENT, TOLERANCES, ETC. SEE FOOTNOTE.

PRECISION 375

IDLI GROUP 80109

**CROSS + 2 ROUND BEARINGS
2 MIDWING DRILLED BEARINGS
+ INSIDE AND OUTSIDE SNAP RINGS
SEE FIGURE 1K**

1.1250 in	BD	28.58 mm
.8281 in	BW	21.03 mm
3.3125 in	CL	84.14 mm
.6719 in	CD	17.07 mm

ADAMS-LET.	SR8778-3
ADAMS-LET.	SRR8778
AEC	3RCV
AEC	4100
AEC	AE3RCV
AEC	AE4100
ALCO	1703
ALCO	4100
ALCO	J4029
ALCO	J4100
ALCO	J4113
ALCO	J4120
ALCO	J4151
ALCO	J4158
ALLIS-CHLM	057614
ALLIS-CHLM	075614
ALLIS-CHLM	086284
ALLIS-CHLM	086284-7
ALLIS-CHLM	AMX46117
ALLOY	1703
ALLOY	J4100
BALKAMP	3-921
BALKAMP	3-929
BALKAMP	3-940
BALKAMP	M4000
BALKAMP	M4002
BALKAMP	M4036
BALKAMP	M4100
BALKAMP	M4108
BALKAMP	M4130
BONNE-TERR	1-48402
BORG-WARNR	114-4000
BORG-WARNR	114-4002
BORG-WARNR	114-4022
BORG-WARNR	114-4023
BORG-WARNR	114-4029
BORG-WARNR	114-4036
BORG-WARNR	114-4100
BORG-WARNR	114-4100AB
BORG-WARNR	114-4108
BORG-WARNR	114-4112
BORG-WARNR	114-4114
BORG-WARNR	114-4116
BORG-WARNR	114-4117
BORG-WARNR	114-4122
BORG-WARNR	114-4138
BORG-WARNR	114-414
BORG-WARNR	114-4163
BORG-WARNR	114-650
BORG-WARNR	J4002
BRILL,J.G.	557221
BUFFALO-SP	9004605
CATERPILLR	1K9467
CATERPILLR	6H1261
CATERPILLR	85432
CATERPILLR	943221
CATERPILLR	GF9190
CLARK EQU.	940086
CLARK EQU.	942221
CLEVE.MOT.	CM3031CV
CLEVE.MOT.	CM4100
CLEVE.MOT.	CM4138
CLEVE.MOT.	CM4200
CMP	CM3031CV
CMP	CM4100
CMP	CM4138
CMP	CM4200
CURTISS-WR	3-23883
CURTISS-WR	3-32883
D&D MACH.	MM4100
DROTT	27460
DROTT	27471
DROTT	32053
DROTT	76187
EXCEL	U209
FIAT-ALLIS	0682636-6
FIAT-ALLIS	075614
FIAT-ALLIS	3006595-7
FIAT-ALLIS	6595
FIAT-ALLIS	70006595
FIAT-ALLIS	70681016
FIAT-ALLIS	70682636-6
FIAT-ALLIS	73006595-7
FLEETRITE	ZAD14002
FLEETRITE	ZAD14138
FLEETRITE	ZAD14151
FMC	3-23883
FORD	CD5AZ4635M
GAR WOOD	3000443
GAR WOOD	4003746
GAR WOOD	840363
GAR WOOD	M306012
GEHL BROS.	51083
GMB	G4029
GMB	G4100
GMB	G4110
GMB	G4120
GMB	G4122
GMB	G4138
GMB	G4151
GMB	G5-3031X
GMB	G5-4002X
GMB	G5-4138
GMB	G5-650
GMC	1427476
HABERLE	H4002
HOUGH	126812HI
HOUGH	127290HI
HYSTER	85525A
HYSTER	984899A
HYSTER	B139472
IHC	116-893R91
IHC	124964
IHC	124967
IHC	220532R91
IHC	65-348R91
IHC	65348R92
INBRA	CC4002
INBRA	CT24-025
INGER.RAND	CBW114-4100
IOWA MFG.	7329-021
IOWA MFG.	7329-102
ITALCARDAN	50-843-000
KOEHRING	278D255A
KOEHRING	334B2538-4
KOEHRING	848036506
KOEHRING	858036732
LEMPCO	U209
LEMPCO	U3031C
LOBRO	U640
MASS.FERG.	SP112540
MASS.FERG.	SP112635
MCQUAY-NOR	U01-23
MCQUAY-NOR	UO3-16
MICHIGAN	866887
MICHIGAN	940036
MICHIGAN	940039
MICHIGAN	940083
MICHIGAN	943221
MICHIGAN	944865
MICHIGAN	946067
MOPAR	D25
MOTOR MAST	3031C
MOTOR MAST	4000A
NEAPCO	280800
NEAPCO	2821
NEAPCO	283031
NEAPCO	2840
NEAPCO	284138
NEAPCO	3-3031
NEAPCO	3-4138
PERFECTION	RG4002
PETTI-MULL	2781A
PETTI-MULL	4528
PETTI-MULL	5AB100-18-1
PETTI-MULL	A22509
PETTI-MULL	CM8
PETTI-MULL	LA425-3
PETTI-MULL	LL773-124
PILOT	14000
PILOT	14002
PILOT	14138
PILOT	14151
PITT.VIOL.	40.843.000
PRECISION	540
REPCO	114-4029
REPCO	114-4100
REPCO	114-4108
REPCO	114-4114
REPCO	114-4116
REPCO	114-4122
REPCO	114-4138
REPCO	114-650
REPUBLIC	CB3011C
REPUBLIC	CB4000
REPUBLIC	CB4002
RUEDARSA	1150
RUEDARSA	2100-1150-0-0
SCOOPMOBLE	708390
SCOOPMOBLE	S8571
SPICER	5-3031X
SPICER	5-4002X
STUDEBAKER	362668
STUDEBAKER	382688
TEDDY TORQ	1402
TEDDY TORQ	G4100
TORO	2115-34
TOWMOTOR	B69240
TRANSMISS.	UK536
TRU-CROSS	3031
TRU-CROSS	4002
TRW	20069
TRW	20125
TRW	20125D
UJS	55-4100
UNION PAC.	39414115
WABCO	2046453
WESCO	301402
WESCO	N4002
WHITE	KC2766
ZELLER	4000A

IDLI GROUP 80110

**CROSS + 2 WING TYPE BEARINGS
+ 2 ROUND GROOVED BEARINGS
SEE FIGURE 1BJ**

1.1250 in	BD	28.58 mm
1.4375 in	CC	36.51 mm
3.3125 in	CL	84.14 mm
.6719 in	CD	17.07 mm

AEC	AE3012
AEC	AE3108
ALCO	3012
ALCO	J3018
AMER.MOTOR	3108181
BORG-WARNR	114-3012
BORG-WARNR	114-3020
BORG-WARNR	114-3043
BORG-WARNR	114-3108
CLEVE.MOT.	CM3019
CLEVE.MOT.	CM3113
D&D MACH.	MM3012
DEERE,JOHN	AM1861
GMC	1456503
GMC	1463259
GMC	1463546
GMC	22806
GMC	565988
HERCULES	382
IHC	65347R92
LEMPCO	U208
MCQUAY-NOR	CAD.
MCQUAY-NOR	U01-27
MCQUAY-NOR	UO1-27
MOPAR	D18
MOTOR MAST	3031-2D
MOTOR MAST	8031-2D
NEAPCO	280700
PERFECTION	RG3012
PERFECTION	RG3012G
PERFECTION	RG3019
PILOT	14312
PRECISION	531
REPCO	114-3108
REPUBLIC	CB3012
REPUBLIC	CB3019
RUEDARSA	1900
RUEDARSA	2100-1900-0-0

BD=BEARING DIAMETER (outside)   BW=BEARING WIDTH   CB=CROSS LENGTH WITH BEARINGS   CC=CENTER TO CENTER   CD=CROSS DIAMETER
CL=CROSS LENGTH WITHOUT BEARINGS   EC=END TO CENTER or FACE TO FACE   EE=END TO END   EL=EFFECTIVE LENGTH
HD=HUB DIAMETER (or insert)   IE=INSIDE OF EARS (or recess)   OD=OUTSIDE DIAMETER   OE=OUTSIDE OF EARS   SB=SPLINE OR BORE SIZE

STUDEBAKER 382742
TRW 20163
WESCO G3012G

IDLI GROUP 80111

**CROSS + 2 WING TYPE BEARINGS
+ 2 ROUND GROOVED BEARINGS
SEE FIGURE 1BI**

1.1250 in	BD	28.58 mm
1.4375 in	CC	36.51 mm
.7969 in	BW	20.24 mm
3.3125 in	CL	84.14 mm
.6719 in	CD	17.07 mm

AEC 3116
AEC AE3021
AEC AE3116
ALCO 3021
ALCO J3116
BALKAMP M3021
BORG-WARNR 114-3021
BORG-WARNR 114-3104
BORG-WARNR 114-3110
BORG-WARNR 114-3116
BORG-WARNR 114-3116AB
BORG-WARNR 114-3136
CLEVE.MOT. CM3021
CLEVE.MOT. CM3116
FIAT-ALLIS 22807
FIAT-ALLIS 70022807
GMB G3116
GMB G5-3116X
GMC 22807
GMC 9022807
IHC 625808C1
IHC 69560R91
LEMPCO U223
MOTOR MAST 3031L
NEAPCO 282900
PERFECTION RG3021
PILOT 14321
PRECISION 538
REPCO 114-3104
REPCO 114-3116
REPUBLIC CB3021
SPICER 5-3021X
TEDDY TORQ 1437
TRW 20162
WESCO 301321
WESCO N3021

IDLI GROUP 80112

**CROSS + 2 WING TYPE BEARINGS
+ 2 ROUND GROOVED BEARINGS
slight variation from above group
SEE FIGURE 1BJ**

1.1250 in	BD	28.58 mm
1.4375 in	CC	36.51 mm
.7969 in	BW	20.24 mm
3.3125 in	CL	84.14 mm
.6719 in	CD	17.07 mm

ADAMS-LET. 475363
AEC 3013
AEC 3102
AEC 3113
AEC 3119
AEC AE3012

AEC AE3019
AEC AE3102
AEC AE3108
AEC AE3113
ALCO 1712
ALCO 3012
ALCO J3102
ALCO J3108
ALCO J3113
ALLOY 1015
ALLOY 1712
ALLOY J3102
ALLOY J3105
AMER.MOTOR 3108181
BALKAMP 3-902
BALKAMP M3012
BALKAMP M3019
BALKAMP M3022
BALKAMP M3043
BALKAMP M3108
BALKAMP M3113
BORG-WARNR 114-3012
BORG-WARNR 114-3019
BORG-WARNR 114-3020
BORG-WARNR 114-3022
BORG-WARNR 114-3043
BORG-WARNR 114-3102
BORG-WARNR 114-3108
BORG-WARNR 114-3113
BORG-WARNR 114-3113AB
BORG-WARNR 114-3115
BORG-WARNR 114-3126
BORG-WARNR 114-3134
BORG-WARNR 114-3156
BORG-WARNR 114-3167
BORG-WARNR J3012
CASE,J.I. D37040
CATERPILLR R7X7352
CATERPILLR R7X7356
CLEVE.MOT. CM3012
CLEVE.MOT. CM3102
CLEVE.MOT. CM3108
CMP CM3012
CMP CM3102
CMP CM3108
EXCEL U2086
GEN.EQUIP. J3102
GLAENZER 065-102-8
GMB G3013
GMB G3102
GMB G3103
GMB G3105
GMB G3108
GMB G3113
GMB G5-3012X
GMB G5-3102
GMB G5-3105-1X
GMB GU6320
GMC 1456503
GMC 22806
GMC 44132
GMC 45678
GMC 9022736
GMC 9022806
GMC 9025191
HABERLE H3012

HERCULES 382
IHC 142-824R91
IHC 65-347R91
INBRA CT17-010
JEBCO 3P43
LEMPCO U208
LEMPCO U208G
LOBRO U632
MASTERS MM3012
MASTERS MM3019
MCQUAY-NOR U01-22
MCQUAY-NOR U01-24
MCQUAY-NOR U01-27
MCQUAY-NOR U01-24
MOPAR D18
MOTOR MAST 3031D
MOTOR MAST 3031N
NEAPCO 2-3102
NEAPCO 2802
NEAPCO 283102
NEAPCO 283160
PERFECTION RG3012
PERFECTION RG3012G
PERFECTION RG3019
PILOT 1412G
PILOT 14312
PILOT 14312G
PILOT 14319
PILOT 14322
PRECISION 531G
REPCO 114-3102
REPCO 114-3113
REPCO 114-3126
REPCO 114-7122
REPUBLIC CB3012
REPUBLIC CB3012G
REPUBLIC CB3019
REPUBLIC CB702
SPICER 5-3012X
STUDEBAKER 382742
STUDEBAKER 648460
STUDEBAKER 6484649
TEDDY TORQ 1433
TRANSMISS. UK680
TRU-CROSS 3012
TRW 20040
TRW 20042
UJS 55-3012
WABCO 475363
WABCO 883333A
WESCO 201-312
WESCO 301312
WESCO 408-308
WESCO 408-309
WESCO 408-838
WESCO 408-864
WESCO N3012
WESCO N3012G
WESCO N3019
ZELLER 3031D

IDLI GROUP 80113

**CROSS + 2 WING TYPE BEARINGS
+ 2 ROUND GROOVED BEARINGS
slight variation from above group
SEE FIGURE 1BL**

1.1250 in	BD	28.58 mm
1.4375 in	CC	36.51 mm
.7969 in	BW	20.24 mm
3.3125 in	CL	84.14 mm
.6719 in	CD	17.07 mm

AEC 3120
AEC AE3120
ALCO J3120
ALLOY 1084
ALLOY J3120
BORG-WARNR 114-3120
BORG-WARNR 114-3121
BORG-WARNR 114-3154
CATERPILLR 5D4091
CLEVE.MOT. CM3120
DUPONT 322910CN117
FLEETRITE ZAD14320
GMB G3120
GMB G5112
GMB G5-3120X
GMB G5-3121X
MCQUAY-NOR U03-15
MCQUAY-NOR UO3-15
MOTOR MAST 3031T
NEAPCO 283154
NEAPCO 283154A
NEAPCO 3-3154
PILOT 14320
PRECISION 998
REPCO 114-3120
REPUBLIC CB798
TRW 20046
WESCO 301326
WESCO N3120
ZELLER 3031T

IDLI GROUP 80114

**CROSS + 2 ROUND BEARINGS
2 MIDWING DRILLED BEARINGS
+ INSIDE AND OUTSIDE SNAP RINGS
SEE FIGURE 1BK**

1.1250 in	BD	28.58 mm
1.6250 in	CC	41.28 mm
.7969 in	BW	20.24 mm
3.3125 in	CL	84.14 mm
.6719 in	CD	17.07 mm

ALCO 1084
ALCO 3014
ALCO J3115
BALKAMP M3014
BORG-WARNR 114-3014
BORG-WARNR 114-3115
BORG-WARNR 114-3115AB
BORG-WARNR 114-3135
BORG-WARNR J3014
GMB G5-3014X
GMB G5-3120X
IHC 137110H
IHC 137110HA
IHC 137110HB
LEMPCO U214
NEAPCO 282200

CAUTION: BE SURE TO REFER TO ENGINEERING CATALOGS FOR SPECIAL APPLICATIONS THAT REQUIRE SPECIFIC MATERIAL CONTENT, TOLERANCES, ETC. SEE FOOTNOTE.

PERFECTION RG3014
PILOT 14314
PRECISION 533
REPCO 114-3115
REPUBLIC CB3014
SPICER 5-3014X
TEDDY TORQ 1434
WESCO 301314
WESCO N3014

IDLI GROUP 80115

CROSS + 4 GROOVED BEARINGS + INSIDE SNAP RINGS
SEE FIGURE 1B

2.5200 in	IE	64.01 mm
1.2600 in	BD	32.00 mm
3.3450 in	CL	84.96 mm

ALLOY 1018
BORG-WARNR 114-320
GMB GUT20
MATSUI UJ234
MCQUAY-NOR .... TOYOTA
NEAPCO 2-0020
NEAPCO 28T20
PRECISION 389
REPCO 101-1270
REPCO RUJ2041
SPICER 5-1509X
SPICER 5-1511X
TOYOTA 04371-36020
TOYOTA 04371-36021
WESCO 201122
WESCO 201-133
WORLDPARTS W49-102
WYLIE 62

IDLI GROUP 80116

CROSS + 4 GROOVED BEARINGS + INSIDE SNAP RINGS
SEE FIGURE 1B

1.0625 in	BD	26.99 mm
.7188 in	BW	18.26 mm
3.3750 in	CL	85.73 mm
.6719 in	CD	17.07 mm

AEC 533
AEC AE533
ALCO 1033
ALCO 1533
ALCO J5-133
ALLOY 1033
ALLOY J5-133
BORG-WARNR 114-533
BORG-WARNR 114-533R
CLEVE.MOT. CM5-133X
CMP CM5-133X
D&D MACH. MS133X
EXCEL U118
GMB G5-133
GMB G5-133X
GMC 1392987
LEMPCO U118

MASTERS MS133X
MCQUAY-NOR U01-32
MCQUAY-NOR UO1-32
MOPAR D70
MOTOR MAST 3237B
NEAPCO 1-0133
NEAPCO 28133X
PERFECTION RG533R
PILOT 11533
PRECISION 352
REPCO 5-133X
REPUBLIC CB1333
REPUBLIC J5-133
RUEDARSA 1334
RUEDARSA 2104-1334-0-0
SPICER 5-133X
TEDDY TORQ G5-133
TEDDY TORQ K33
TRW 20052
WESCO 201534
WESCO N1533
ZELLER 3237B

IDLI GROUP 80117

CROSS + 4 ROUND BEARINGS + OUTSIDE SNAP RINGS
SEE FIGURE 1A

1.0625 in	BD	26.99 mm
.6875 in	BW	17.46 mm
3.3750 in	CL	85.73 mm
.6563 in	CD	16.67 mm

AEC 5-1203
AEC 530
AEC AE530
ALBURUS-SP 5-273X
ALCO 1040
ALCO 1530
ALCO J5-213
ALLOY 1040
ALLOY 41012
ALLOY J5-213
BORG-WARNR 114-122
BORG-WARNR 114-213
BORG-WARNR 114-530
CENTRAL G5-213X
CHECKER 010752
CHRYSLER 2533202
CLEVE.MOT. CM229X
CLEVE.MOT. CM5-213X
CMP CM5-213X
CMP CM5-229X
D&D MACH. MS130X
EXCEL U128
FLEETRITE ZAD11530
FORD C3VY4497A
FORD C3VY4997A
FORD C4SZ4635A
FORD C4SZ4635B
GLAENZER 065-211-8
GMB G5-130X
GMB G5-213
GMB G5-213X
GMB G5-530
GMB GR55-55-2
GMC 1388249
GMC 1389839

GMC 1389865
GMC 1470373
GMC 1482672
GMC 390113
GMC 390133
GMC 3937988
LEMPCO U128
MACK 21045213X
MASTERS MS130X
MCQUAY-NOR U01-31
MCQUAY-NOR UO1-31
MOPAR D85
MOTOR MAST 3237A
NEAPCO 2-4800
NEAPCO 28213X
NEAPCO 284800
NEAPCO 28530X
PERFECTION RG350R
PERFECTION RG530R
PERFECTION RG533R
PILOT 11530
PILOT 11530G
PRECISION 354
REPCO 5-130X
REPCO 5-213X
REPUBLIC CB1300
RUEDARSA 1332
RUEDARSA 2103-1332-0-0
SPICER 1330
SPICER 5-130
SPICER 5-130X
SPICER 5-213
SPICER 5-213X
SPICER 5-228X
SPICER 5-229X
SPICER 5-230
SPICER 5-230X
SPICER 5-273X
SPICER 5-5008X
SPICER J5-130
SPICER J5-230
TEDDY TORQ G5-213
TEDDY TORQ K30
TRU-CROSS 1203
TRU-CROSS 213
TRW 20051
UJS 55-130
WESCO 201-530
WESCO N1530
ZELLER 3237A
ZELLER 3437D

IDLI GROUP 80118

CROSS + 4 ROUND BEARINGS + OUTSIDE SNAP RINGS
slight variation from above group
SEE FIGURE 1A

1.0625 in	BD	26.99 mm
.6875 in	BW	17.46 mm
3.3750 in	CL	85.73 mm
.6563 in	CD	16.67 mm

AEC 1202
AEC 534
AEC AE534
ALCO 1034
ALCO 1042

ALCO 1534
ALCO 41034
ALCO J5-134
ALLOY 1034
ALLOY 1046
ALLOY 41034
ALLOY J5-134
BORG-WARNR 114-134
BORG-WARNR 114-552
BORG-WARNR 114-552R
BORG-WARNR AE531
CLEVELAND S55-55-1330
CLEVE.MOT. CM5-134X
CMP CM5-134X
CMP CM5-231X
D&D MACH. MS134X
EXCEL U121
EXCEL U131
FLEETRITE ZAD11534
FLEETRITE ZAD11552
FORD B4TZ4635F
FORD B9T4724A
FORD B9TF4724A
FORD B9TT4724
FORD B9TT4724A
FORD C1TZ4635B
FORD C25A4724A
FORD C5TZ4635F
FORD CITZ4635B
GLAENZER 065-210-8
GMB G5-134
GMB G5-134X
GMB G5-552
JAMCO UJ134
LEMPCO U131
LOBRO U240
MASTERS MS134X
MCQUAY-NOR U01-33
MCQUAY-NOR UO1-35
MCQUAY-NOR UO1-35
MOPAR D71
MOTOR MAST 3275AE
MOTOR MAST 3475AHP
MOTOR MAST 6275AE
NEAPCO 1-0134
NEAPCO 1-134
NEAPCO 28134X
NEAPCO 28152X
NEAPCO 284900
PERFECTION RG552
PERFECTION RG552R
PILOT 11534
PILOT 11552
PILOT S15513
PRECISION 353
REPCO 5-1202X
REPCO 5-1211X
REPCO 5-134X
REPCO KS-BG166
REPUBLIC C1552
REPUBLIC CB1552
REPUBLIC CBS55-13
SPICER 5-134X
SPICER 5-138X
SPICER 5-152X
SPICER 5-212X

BD = BEARING DIAMETER (outside)　　BW = BEARING WIDTH　　CB = CROSS LENGTH WITH BEARINGS　　CC = CENTER TO CENTER　　CD = CROSS DIAMETER
CL = CROSS LENGTH WITHOUT BEARINGS　　EC = END TO CENTER or FACE TO FACE　　EE = END TO END　　EL = EFFECTIVE LENGTH
HD = HUB DIAMETER (or insert)　　IE = INSIDE OF EARS (or recess)　　OD = OUTSIDE DIAMETER　　OE = OUTSIDE OF EARS　　SB = SPLINE OR BORE SIZE

TEDDY TORQ	G5-134
TEDDY TORQ	K53
TRU-CROSS	134
TRW	20053
TRW	20056
TRW	20210
UJS	55-152
WESCO	200552
WESCO	201-531
WESCO	201552
WESCO	202-552
WESCO	G1552
WESCO	N1552
ZELLER	3275AE
ZELLER	3475A

IDLI GROUP 80119

CROSS + 4 ROUND BEARINGS + OUTSIDE SNAP RINGS
largest BD dimension is shown
SEE FIGURE 1AF

1.1250 in	BD	28.58 mm
.7188 in	BW	18.26 mm
3.3750 in	CL	85.73 mm
.6563 in	CD	16.67 mm

AEC	550
AEC	AE550
ALCO	1046
ALCO	1550
ALCO	J5-15
ALCO	J5-150
ALLOY	J5-150
BORG-WARNR	114-550
BORG-WARNR	114-550R
CLEVE.MOT.	CM5-150X
CMP	CM5-150X
EXCEL	U119
GLAENZER	065-208-8
GMB	G5-150
GMB	G5-550
GMB	GU3000
GMC	386452
GMC	570816
LEMPCO	U119
LOBRO	U238
MASTERS	MS150X
MCQUAY-NOR	UO1-34
MOPAR	D72
MOTOR MAST	3130A
MOTOR MAST	3275B
NEAPCO	1-0150
NEAPCO	28150X
PERFECTION	RG550R
PILOT	11550
PILOT	11551
PRECISION	350
REPCO	5-150X
REPUBLIC	CB1505
RUEDARSA	1335
RUEDARSA	2104-1335-0-0
SPICER	5-150X
SPICER	5-3022X
TEDDY TORQ	G5-150
TEDDY TORQ	K50
TEDDY TORQ	K50G
TRW	20055

WESCO	201-550
WESCO	202-550
WESCO	G1550
WESCO	N1550
ZELLER	3275A

IDLI GROUP 80120

CROSS + 4 ROUND BEARINGS + OUTSIDE SNAP RINGS
slight variation from above group
SEE FIGURE 1AF

1.1250 in	BD	28.58 mm
.7188 in	BW	18.26 mm
3.3750 in	CL	85.73 mm
.6563 in	CD	16.67 mm

AEC	550
ALCO	1550
ALLOY	1046
ALLOY	J5-150
BORG-WARNR	114-550
BORG-WARNR	114-550R
BWE	114-550
D&D MACH.	MS150X
GMB	G5-150X
LEMPCO	U119
MASTERS	MS150X
MCQUAY-NOR	UO1-34
MOPAR	D72
MOTOR MAST	3275A
NEAPCO	1-0150
NEAPCO	1-150
PRECISION	350G
REPUBLIC	CB3110X
SPICER	5-150X
TRW	20055
WESCO	201550
ZELLER	3275A

IDLI GROUP 80121

CROSS + 4 ROUND BEARINGS + OUTSIDE SNAP RINGS
slight variation from above group
SEE FIGURE 1AF

1.1250 in	BD	28.58 mm
.7188 in	BW	18.26 mm
3.3750 in	CL	85.73 mm
.6563 in	CD	16.67 mm

AEC	531
AEC	AE531
ALCO	1042
ALCO	1531
ALCO	J5-231
ALLOY	1042
ALLOY	J5-231
BORG-WARNR	114-531
BWE	114-232
CLEVE.MOT.	CM5-212X
CLEVE.MOT.	CM5-231X
CMP	CM5-231X
D&D MACH.	MS131X
FLEETRITE	ZAD11531
FORD	C4VW4635A
FORD	CD5AZ4635B
GLAENZER	065-266-8
GMB	G5-231
GMB	G5-231X

GMB	G5-531
GMC	1389604
GMC	1389837
GMC	1470374
GMC	390134
LEMPCO	U121
MASTERS	MS131X
MCQUAY-NOR	U01-33
MCQUAY-NOR	U01-39
MCQUAY-NOR	UO1-33
MOPAR	D87
MOTOR MAST	3237C
NEAPCO	2-5900
NEAPCO	28231X
NEAPCO	28531X
PERFECTION	RG531R
PILOT	11531
PRECISION	355
REPCO	5-131X
REPCO	5-212X
REPCO	5-231X
REPUBLIC	CB1301
REPUBLIC	CB13-1
RUEDARSA	1333
RUEDARSA	2104-1333-0-0
SPICER	5-131
SPICER	5-131X
SPICER	5-212X
SPICER	5-231
SPICER	5-231X
SPICER	5-254X
SPICER	J5-131
TEDDY TORQ	G5-231
TEDDY TORQ	K31
TRU-CROSS	1204
TRU-CROSS	212
TRW	20053
TRW	20064
UJS	55-131
WESCO	201531
WESCO	G1531
WESCO	N1531
ZELLER	3237C

IDLI GROUP 80122

**CROSS + 4 SERRATED BEARINGS + THRUST AND LOCK PLATES
SEE FIGURE 1C**

1.1250 in	BD	28.58 mm
.8750 in	BW	22.23 mm
3.3750 in	CL	85.73 mm
.7188 in	CD	18.26 mm

AEC	AE-R96-55-1
AEC	R96-55-1
ALCO	JR96-55
ALCO	R96-55
BALKAMP	CR9655
BORG-WARNR	R96-55
CHRYSLER	947552
CLEVE.STL.	R96-55
D&D MACH.	MCR96
GMB	G5-1230X
GMB	GR96-55
GMB	GU2850
HERCULES	225
LEMPCO	U300

GMB	G5-531
LOBRO	285
LOBRO	U285
MIPER	C3462
MOPAR	D108
MOTOR MAST	3437B
NEAPCO	28057X
PERFECTION	RG96-55
PILOT	R19655
PRECISION	414
REPCO	R96-55
REPUBLIC	CB-R96
SPICER	R96-55
STUDEBAKER	660195
TEDDY TORQ	GR96-55
TEDDY TORQ	R96
WESCO	201966
WESCO	NR965

IDLI GROUP 80123

CROSS + 4 GROOVED BEARINGS + INSIDE SNAP RINGS
largest BD dimension is shown
SEE FIGURE 1B

1.1875 in	BD	30.16 mm
.7813 in	BW	19.85 mm
3.3750 in	CL	85.73 mm
.7656 in	CD	19.45 mm

AEC	1205
AEC	5-1205
AEC	AE1205
AEC	AE5-1205
AEC	AE-R55-55-2
ALCO	1137
ALCO	JR55-55
ALCO	R55-55
ALLOY	1137
ALLOY	JR55-55
BORG-WARNR	114-124
BORG-WARNR	114-125
BORG-WARNR	R55-55
BORG-WARNR	R55-55-2
CHRYSLER	1752559
CHRYSLER	2084800
CLEVELAND	R55-55-2
CLEVELAND	S55-55-2
CLEVE.MOT.	AER96-55-1
CLEVE.MOT.	CM5555R
CLEVE.MOT.	CM5555S
CLEVE.STL.	R55-55
CLEVE.STL.	R55-55-2
CMP	CM5555R
D&D MACH.	MCR55
DIVCO	16454
EXCEL	U308
FLEETRITE	ZAD-R15555
FMC	38161
FMC	55506
FORD	C5TZ4635A
GLAENZER	065-220-8
GMB	G5-1205X
GMB	GU2210A
GMB	GU-R55-55-2
GMC	9021527
HABERLE	HR5555
HABERLE	HS555
IHC	217-193R91

ENGINEERING CATALOGS MUST BE CONSULTED FOR DETAILS NOT INCLUDED
IN THIS GUIDE. SPECIFIC DESIGNS, MATERIAL CONTENT, TOLERANCES
LUBE FITTINGS AND OTHER DIMENSIONS ARE INTENTIONALLY OMITTED HERE.

CAUTION: BE SURE TO REFER TO ENGINEERING CATALOGS FOR SPECIAL APPLICATIONS THAT REQUIRE SPECIFIC MATERIAL CONTENT, TOLERANCES, ETC. SEE FOOTNOTE.

IHC	219-303R91
JAMCO	UJ1205
LEMPCO	U308
MASTERS	MCR55
MASTERS	MCS55
MCQUAY-NOR	U02-12
MCQUAY-NOR	UO2-12
MOPAR	D107
MOPAR	D88
MOTOR MAST	3437A
NEAPCO	2-6200
NEAPCO	286200
PERFECTION	RG55-55
PERFECTION	RG55-55-2
PERFECTION	RS55-55-2
PILOT	R15555
PRECISION	423
REPCO	5-1205X
REPCO	R55-55-2
REPCO	RUJ3001
REPUBLIC	CB-R55
SPICER	5-1205X
SPICER	R55-55
STUDEBAKER	1687083
STUDEBAKER	1691762
TEDDY TORQ	GR55-55
TEDDY TORQ	R55
TRU-CROSS	1205
TRW	20062
UJS	5SR55
UJS	S5R55
WAYNE	38161
WESCO	201555
WESCO	NR555
ZELLER	3437A

IDLI GROUP 80124

CROSS + 4 ROUND BEARINGS + OUTSIDE SNAP RINGS SEE FIGURE 1A

1.1875 in	BD	30.16 mm
.7813 in	BW	19.85 mm
3.3750 in	CL	85.73 mm
.7656 in	CD	19.45 mm

AEC	5-086
AEC	578
AEC	86
AEC	AE5-086
AEC	AE578
ALBURUS-SP	5-178X
ALCO	1078
ALCO	1578
ALCO	1778
ALCO	J5-178
ALLIS-CHLM	080470
ALLIS-CHLM	080470-8
ALLIS-CHLM	142A487
ALLIS-CHLM	804708
ALLOY	1078
ALLOY	1078P
ALLOY	41078

ALLOY	J5-178
ARMSTRONG	5410-192
AYRA DUREX	1-355-01
BALKAMP	3-954
BLAW KNOX	5-178X
BLAW KNOX	734-57
BMC	8G3003
BMC	GUJ112
BMC	P304
BMC	P305
BORG-WARNR	114-10HD
BORG-WARNR	114-10HP
BORG-WARNR	114-518R
BORG-WARNR	114-518X
BORG-WARNR	114-578
BORG-WARNR	114-578R
BORG-WARNR	114-578U
BORG-WARNR	114-635
BRD	045-501470
BRD	045-505038
BRD	045-510423
BROCKWAY	239040
CASE,J.I.	R2713
CASE,J.I.	R3092
CENTRAL	G5-178X
CHRYSLER	1516854
CHRYSLER	2298823
CHRYSLER	K22318
CLARK EQU.	855047
CLARK EQU.	865335
CLEVE.MOT.	CM5-178X
CLEVE.MOT.	CM5-86X
CLEVE.MOT.	CMS179XUB
CMP	CM5-178X
CMP	CM5-178XUB
CURTISS-WR	15-22875
CURTISS-WR	16-20487
D&D MACH.	MS578X
DAIMLER	7152
DAIMLER	944843
DAIM.RUBE.	619-283-114-1
DATSUN	37125-85425
DATSUN	37125-85461
DAVEY COM.	A24209
DENNIS	693864
DIAMOND R	10056P2
DIAMOND R	7199H
DIAMOND T	7199H
DIAMOND T	BB7199H
DIAMOND T	K5-18X
DIAMOND T	N73321
DIGIO-LIFT	3-0R85
DIVCO	16362
EXCEL	U102
EXCEL	U138
FEDERAL	5Y294
FEDERAL	5Y500
FEDERAL	5Y5322
FEDERAL	5Y533
FEDERAL	5Y7302
FIAT-ALLIS	0080470
FIAT-ALLIS	0080470-8
FIAT-ALLIS	080470
FIAT-ALLIS	080470-8
FIAT-ALLIS	142A487
FIAT-ALLIS	2309287

FIAT-ALLIS	70080470
FIAT-ALLIS	70080470-8
FIAT-ALLIS	804708
FLEETRITE	ZAD11578
FLEETRITE	ZAD11578UB
FLEETRITE	ZAD11586
FORD	01T7039
FORD	1T7039
FORD	81DB4635BA
FORD	A0T7039A
FORD	AOT7039A
FORD	B5T7039
FORD	B5T7039A
FORD	B5TZ3249A
FORD	B5TZ4635A
FORD	C1T2-3249A
FORD	C5TZ4635E
FORD	CX1TZ3249A
FORD	FM4635B
FORD	G8T18397
FORD	IT7039
GLAENZER	065-135-8
GLAENZER	065-281-8
GLAENZER	465-006-0
GLAENZER	465-021-1
GMB	G5-178
GMB	G5-178B
GMB	G5-178UB
GMB	G5-178X
GMB	G5-178XU
GMB	G5-187X
GMB	G5-578
GMB	G5-635
GMB	G5-86R
GMB	G5-86X
GMB	GU2200
GMB	GU2210
GMB	GU2210SL
GMB	GUN30
GMB	GUPA4
GMC	2057273
GMC	2186972
GMC	2186973
GMC	2230781
GMC	2330781
GMC	2450093
GMC	3707163
GMC	3707891
GMC	3708921
GMC	3708931
GMC	3708961
GMC	374246
GMC	3823101
GMC	3823102
GMC	3851520
GMC	3877040
GMC	3889696
GMC	3979616
GMC	3979617
GMC	609515
GMC	609948
GMC	7501016
HABERLE	H1518X
HABERLE	H1578X
HARDY SPCR	3551-182W
HARDY SPCR	3551-182Y

HARDY SPCR	3551-192N
HARDY SPCR	K5GB107
HARDY SPCR	K5GB118
HARDY SPCR	K5GB163
HARDY SPCR	K5GB18
HARDY SPCR	K5GB192
HARDY SPCR	K5GB91
HARDY SPCR	K5LGB182
HERCULES	422
HIWAY CHIP	40491
HIWAY CHIP	40495
HOUGH	154081HI
HOUGH	176297HI
HYSTER	25897A
HYSTER	27418
HYSTER	27418A
IHC	121-345R91
IHC	127-573R91
IHC	127973R91
IHC	154081
IHC	154-081H1
IHC	176-297H1
IHC	645574R91
IHC	645-574R92
IHC	886-890R91
IHC	967-379R91
IHC	974-446R91
INBRA	CA220
INBRA	CT11-030
ITALCARDAN	50-757-000
JAGUAR	11682
JAGUAR	7152
JEFFREY	5-178X
JEFFREY	CAT415325
JOY MFG.	149542
JOY MFG.	5-178X
JOY MFG.	A44844
JOY MFG.	A49524
JOY MFG.	A49542
JOY MFG.	A49562
JOY MFG.	A90678
JOY MFG.	A96078
KOYO	N3092
LAKEWOOD	23021
LEMPCO	U102
LEMPCO	U114
LEMPCO	U138
LEYLAND	538451
LEYLAND	600448
LEYLAND	600490
LIFT TRUCK	3-0R85
LOBRO	U220
LOBRO	U221
LOBRO	U225
MACK	201SJ11
MACK	201SJ11A
MACK	201SJ26
MACK	21045178X
MASS.FERG.	3406080M91
MASTERS	MS578X
MATSUBA	UJ017
MATSUBA	UJ115
MCQUAY-NOR	U02-11
MCQUAY-NOR	U02-7
MCQUAY-NOR	U02-8
MCQUAY-NOR	UO2-7

MCQUAY-NOR	UO2-8
MICHIGAN	855047
MICHIGAN	865335
MICHIGAN	873726
MIPER	C3135
MIPER	C3136
MOPAR	D80
MOTOR MAST	3240A
MOTOR MAST	3240AHP
MOTOR MAST	3240C
NEAPCO	2-0053
NEAPCO	2-53
NEAPCO	28053X
NEAPCO	2849
PERFECTION	RG518X
PETTI-MULL	L2618-1
PILOT	11578
PILOT	11586
PITT.VIOL.	40.757.000
PRECISION	331
PREC.TOR.	949
PREC.TOR.	954
QUINT.HAZ.	QL19245
RAY GO	7686
REO MOTORS	10056P2
REO MOTORS	P0009
REPCO	5-178X
REPCO	5-187X
REPCO	5-86X
REPCO	K5EF118
REPCO	K5SR178X
REPUBLIC	CB1350A
REPUBLIC	CB1350W
REPUBLIC	CB3050A
REPUBLIC	CB749
REPUBLIC	CB754
REX CHAIN	7686
RUEDARSA	1401
RUEDARSA	2103-1401-0-0
SPICER	1350
SPICER	5-143X
SPICER	5-1784
SPICER	5-178X
SPICER	5-18
SPICER	5-18X
SPICER	5-282X
SPICER	5-78
SPICER	5-78X
SPICER	5-7X
SPICER	5-86X
SPICER	5-9X
SPICER	J5-18
SPICER	J5-78
STUDEBAKER	1687251
STUDEBAKER	674156
STUDEBAKER	682027
TEDDY TORQ	J5-178
TRANSMISS.	UK135/6
TRANSMISS.	UK136
TRU-CROSS	1205
TRU-CROSS	178
TRU-CROSS	86
TRW	20054
TRW	20140
TRW	20190
TRW	20194

UJS	55-178
UJS	55-86
UNIPART	GUJ112
VAPORMATIC	VTE5005
VERSATILE	34022
WALTERSCHD	22-00-00
WESCO	200578
WESCO	201578
WESCO	201579
WESCO	202-578
WESCO	G1578R
WESCO	N1518
WESCO	N1578
WESCO	N15789RU
WESCO	N1578R
WESCO	N1578S
WHITE	02-7074025
WHITE	10056P2
WHITE	2405P2
WHITE	357227
WHITE	613333
WHITE	7199H
WHITE	843150
WHITE	843150A
WHITE	943150
WHITE	BB7199H
WHITE FARM	10A18829
WYLIE	23
ZELLER	3240A

IDLI GROUP 80125

CROSS + 4 ROUND BEARINGS
+ OUTSIDE SNAP RINGS
largest BD dimension is shown
SEE FIGURE 1A

1.1875 in	BD	30.16 mm
.7813 in	BW	19.85 mm
3.3750 in	CL	85.73 mm
.7656 in	CD	19.45 mm

AEC	578U
AEC	AE578U
ALCO	1578U
ALCO	1751
ALCO	J5-178B
ALCO	J5-178UB
ALLOY	1751
ALLOY	J5-178B
BORG-WARNR	114-578U
CLEVE.MOT.	CM5-178XUB
D&D MACH.	MS578U
DATSUN	37125-85460
DATSUN	37125-85461
DIAMOND R	10056P2
DIAMOND R	2405P2
FEDERAL	5Y7302
FLEETRITE	ZAD11578UB
GMB	G5-178B
GMB	G5-178XU
GMC	3979616
LEMPCO	U102U
LOBRO	U223
MASTERS	MS578U
MCQUAY-NOR	U02-11
MCQUAY-NOR	U02-7
MCQUAY-NOR	UO2-11
MOPAR	D81

MOTOR MAST	3240B
NEAPCO	2-53
NEAPCO	28053XB
NEAPCO	2854
NEAPCO	3-1053
PILOT	11578UB
PRECISION	331U
PREC.TOR.	949
PREC.TOR.	954
REPCO	K5EF118
REPUBLIC	CB1350U
SPICER	5-178XU
TEDDY TORQ	K18U
TRW	20054
WESCO	303578
WESCO	G1578RU
WESCO	N15789RU
WESCO	N1578RU
WHITE	943150
ZELLER	3240B

IDLI GROUP 80126

CROSS + 4 ROUND BEARINGS
+ OUTSIDE SNAP RINGS
slight variation from above group

1.1875 in	BD	30.16 mm
.7813 in	BW	19.85 mm
3.3750 in	CL	85.73 mm
.7656 in	CD	19.45 mm

AEC	498
PRECISION	232
TRW	20194

IDLI GROUP 80127

CROSS + 4 ROUND BEARINGS
2 INSIDE + 2 OUTSIDE SNAP RINGS
SEE FIGURE 1AB

1.1875 in	BD	30.16 mm
.7813 in	BW	19.85 mm
3.3750 in	CL	85.73 mm
3.6250 in	CB	92.08 mm
.7656 in	CD	19.45 mm

AEC	358
AEC	578-1310
CLEVE.MOT.	CM7853X
GMB	G5-358X
MOTOR MAST	3140B
NEAPCO	283140
NEAPCO	3-3140
PRECISION	348
SPICER	5-460X

IDLI GROUP 80128

CROSS + 2 ROUND BEARINGS
2 MIDWING DRILLED BEARINGS
+ INSIDE AND OUTSIDE SNAP RINGS
SEE FIGURE 1AK

1.1875 in	BD	30.16 mm
.7813 in	BW	19.85 mm
3.3750 in	CL	85.73 mm
.7656 in	CD	19.45 mm

FORD	C1TZ3249A
FORD	E0TZ3249A
GMC	14067677
PRECISION	231
SPICER	5-447X

IDLI GROUP 80129

CROSS + 4 GROOVED BEARINGS
+ INSIDE SNAP RINGS
SEE FIGURE 1B

1.0625 in	BD	26.99 mm
.6406 in	BW	16.27 mm
3.4063 in	CL	86.52 mm
.6875 in	CD	17.46 mm

AEC	1203
AEC	493
AEC	5-1203
AEC	AE5-1203
AEC	AE-P55-55-140
AEC	AE-S55-55-1330
ALCO	1012
ALCO	J5-1203
ALLOY	1012
ALLOY	1012P
ALLOY	41012
ALLOY	J5-1203
BORG-WARNR	114-122
BORG-WARNR	114-5HD
BORG-WARNR	114-5HP
BORG-WARNR	P55-55-2
CENTRAL	G5-1203X
CLEVELAND	P55-55-150
CLEVELAND	P55-55-2
CLEVE.MOT.	CM5555P
CLEVE.STL.	JP55-55-140
CLEVE.STL.	P55-55-1330
CLEVE.STL.	P55-55-140
CLEVE.STL.	P55-55-2
CMP	CM5555P
D&D MACH.	MCP55
EXCEL	U315
FIRESTONE	5234835-8
FLEETRITE	ZAD-P15555
FORD	C3AZ4635G
FORD	C4GW-4635A
GLAENZER	065-228-8
GMB	G5-1203
GMB	G5-1203X
GMB	GP55-55-140
GMB	GR55-55
GMB	GU-P55-55-140
GOODYEAR	219-2514
GOODYEAR	219-2556
GOODYEAR	219-2557
HABERLE	HP555
JAMCO	UJ1203
LAKEWOOD	23016
LEMPCO	U315
MASTERS	MCP55
MASTERS	MS130X
MCQUAY-NOR	U01-31
MCQUAY-NOR	U01-39
MCQUAY-NOR	UO1-39
MIDAS	UJ1005
MOPAR	D92
MOTOR MAST	3437D
MOTOR MAST	3437DHP
MOTOR MAST	3437DT
MOTOR MAST	3734D
NEAPCO	2-4800
NEAPCO	284800
PERFECTION	RP55-55-2

CAUTION: BE SURE TO REFER TO ENGINEERING CATALOGS FOR SPECIAL APPLICATIONS THAT REQUIRE SPECIFIC MATERIAL CONTENT, TOLERANCES, ETC. SEE FOOTNOTE.

PILOT	P15555
PRECISION	433
REPCO	5-1203X
REPCO	RUJ3002
REPUBLIC	CB-P55
REPUBLIC	CP-P55
SEARS-ROE.	5128
SPICER	5-1203X
TEDDY TORQ	G5-1203
TEDDY TORQ	P55
TRU-CROSS	1203
TRW	20064
TRW	20068D
TRW	20209
TRW	20907
UJS	55P55
WESCO	200530
WESCO	201530
WESCO	202-530
WESCO	G1530
WESCO	N1530
WESCO	N1530-1
WYLIE	77
ZELLER	3437D
ZOOM	800005

IDLI GROUP 80130

CROSS + 4 GROOVED BEARINGS + INSIDE SNAP RINGS
slight variation from above group
SEE FIGURE 1B

1.0625 in	BD	26.99 mm
.6406 in	BW	16.27 mm
3.4063 in	CL	86.52 mm
.6875 in	CD	17.46 mm

AEC	1202
AEC	AE5-1202
AEC	AE-S55-55-1330
ALCO	1047
ALCO	J5-1202
ALLOY	1047
ALLOY	J5-1202
BORG-WARNR	114-134
BORG-WARNR	S55-55-1330
CLEVELAND	S55-55-1330
CLEVE.MOT.	CM5-1202X
CLEVE.STL.	JS55-55-1330
CLEVE.STL.	S55-55-1330
CMP	CM5-1202X
D&D MACH.	MS1202X
EXCEL	U318
FLEETRITE	ZAD-S15513
FORD	C2SA4724A
FORD	C2SZ4635A
FORD	C2SZ4724A
GMB	G5-1202
GMB	G5-1202X
GMB	G5-352X
LEMPCO	U318
MASTERS	MS1202X
MCQUAY-NOR	U01-31
MOPAR	D91
MOTOR MAST	3475A
MOTOR MAST	3475AHP
NEAPCO	281202X
NEAPCO	282202X
PILOT	S15513
PRECISION	431
REPCO	5-1202X
REPUBLIC	CB-S55-13
REPUBLIC	CB-S55-8
SPICER	5-1202X
SPICER	J5-1202
TEDDY TORQ	G5-1202
TEDDY TORQ	S31
TRW	20068
TRW	20068D
WESCO	201-552
WESCO	N1552-1
ZELLER	3475A

IDLI GROUP 80131

CROSS + 4 GROOVED BEARINGS + INSIDE SNAP RINGS
largest BD dimension is shown
SEE FIGURE 1B

1.1250 in	BD	28.58 mm
.6406 in	BW	16.27 mm
3.4063 in	CL	86.52 mm
.6875 in	CD	17.46 mm

AEC	1204
AEC	494
AEC	5-1204
AEC	531
AEC	AE5-1204
AEC	AE-P55-55-145
ALCO	1013
ALCO	J5-1204
ALLOY	1013
ALLOY	1013P
ALLOY	1042
ALLOY	41013
ALLOY	J5-1204
BORG-WARNR	114-123
BORG-WARNR	114-531
BORG-WARNR	114-6HD
BORG-WARNR	114-6HP
BORG-WARNR	P55-55-145
BORG-WARNR	P55-55-675
CENTRAL	G5-1204X
CLEVELAND	P55-55-675
CLEVE.MOT.	CM5555R
CLEVE.MOT.	CM9675P
CLEVE.MOT.	CM9675S
CLEVE.STL.	JP55-55-145
CLEVE.STL.	P55-55-145
CLEVE.STL.	P55-55-675
CMP	CM9675P
D&D MACH.	MCP56
EXCEL	U316
FLEETRITE	ZAD-P19675
FORD	C3AZ4635H
GLAENZER	065-229-8
GMB	G5-1204
GMB	G5-1204X
GMB	G5-123
GMB	G5-213X
GMB	GP55-55-145
GMB	GU-P55-55-145
GOODYEAR	219-2515
GOODYEAR	219-2558
HABERLE	HP9675
JAMCO	UJ1204
JAMCO	UJ280
LEMPCO	U121
LEMPCO	U316
LEMPCO	U317
MASTERS	MS131X
MCQUAY-NOR	U01-33
MCQUAY-NOR	U01-38
MCQUAY-NOR	UO1-38
MIDAS	UJ1006
MOPAR	D93
MOTOR MAST	3437C
MOTOR MAST	3437CHP
MOTOR MAST	3437CT
NEAPCO	2-4900
NEAPCO	284900
PERFECTION	RP55-55-165
PILOT	11531
PILOT	P19675
PRECISION	434
REPCO	5-1204X
REPCO	RUJ3000
REPUBLIC	CB1301
REPUBLIC	CB1301X
REPUBLIC	CB-P55-5
REPUBLIC	CB-P55-6
SEARS-ROE.	5129
SPICER	5-1204X
SPICER	5-212X
SPICER	5-213X
TEDDY TORQ	G5-1204
TEDDY TORQ	P75
TRU-CROSS	1204
TRU-CROSS	212
TRU-CROSS	213
TRW	20053
TRW	20063
TRW	20906
UJS	55P675
WESCO	200-532
WESCO	201-531
WESCO	201532
WESCO	202-531
WESCO	202-532
WESCO	G1531-1
WESCO	N153-1
WESCO	N1531-1
WYLIE	76
ZELLER	3437C
ZELLER	3475A
ZOOM	800006

IDLI GROUP 80132

CROSS + 4 GROOVED BEARINGS + INSIDE SNAP RINGS
SEE FIGURE 1B

1.1250 in	BD	28.58 mm
.7344 in	BW	18.65 mm
3.4375 in	CL	87.31 mm
.6875 in	CD	17.46 mm

AEC	231
AEC	491
AEC	AE231
AEC	AE232
ALCO	1020
ALCO	J72625-5
ALLOY	1020
ALLOY	1020P
ALLOY	41020
ALLOY	J72625-5
BORG-WARNR	114-1HD
BORG-WARNR	114-1HP
BORG-WARNR	114-231
BORG-WARNR	114-232
BWE	114-232
CHRYSLER	2852175
CHRYSLER	3634563
CLEVE.MOT.	CM285175
CLEVE.MOT.	CM2852175
CMP	CM281175
CMP	CM2852175
D&D MACH.	MD6255
DETROIT	2852175
DETROIT	3634563
DETROIT	72625-5
EXCEL	U451
FIRESTONE	5234711-4
FLEETRITE	ZAD17231
FLEETRITE	ZAD17232
FORD	CD5AZ4635F
G & G MFG	184-3522
GLAENZER	065-265-8
GMB	G5-1309X
GMB	G5-232
GMB	G72625-5
GOODYEAR	219-2518
GOODYEAR	219-2554
JAMCO	UJ1309
LAKEWOOD	23012
LEMPCO	U451
MASTERS	MD6255
MCQUAY-NOR	U01-36
MCQUAY-NOR	UO1-36
MIDAS	UJ1001
MOPAR	D169
MOTOR MAST	3331A
MOTOR MAST	3331AHP
MOTOR MAST	3331AT
NEAPCO	2-1175
NEAPCO	281175
PILOT	17231
PILOT	17232
PRECISION	316
PRECISION	31G
REPCO	114-232
REPUBLIC	CB2582
REPUBLIC	CB2852
SEARS-ROE.	5106

BD=BEARING DIAMETER (outside)   BW=BEARING WIDTH   CB=CROSS LENGTH WITH BEARINGS   CC=CENTER TO CENTER   CD=CROSS DIAMETER
CL=CROSS LENGTH WITHOUT BEARINGS   EC=END TO CENTER or FACE TO FACE   EE=END TO END   EL=EFFECTIVE LENGTH
HD=HUB DIAMETER (or insert)   IE=INSIDE OF EARS (or recess)   OD=OUTSIDE DIAMETER   OE=OUTSIDE OF EARS   SB=SPLINE OR BORE SIZE

SPICER	5-1309X
TRU-CROSS	1309
TRW	20059
TRW	20206
TRW	20903
UJS	55-72925-2
WESCO	200285
WESCO	201285
WESCO	D285
WESCO	G285
ZELLER	3331A

IDLI GROUP 80133

CROSS + 4 ROUND BEARINGS + OUTSIDE SNAP RINGS
SEE FIGURE 1A

1.0000 in	BD	25.40 mm
.9844 in	BW	25.00 mm
3.6875 in	CL	93.66 mm
.750 in	CD	19.05 mm

AEC	3DR
AEC	AE-CP3DR
AEC	CP3DR
ALLOY	1773
AMER.PULL.	4627-3124
AMER.PULL.	499A26
BEAN,JOHN	1265204
BUSH HOG	200
CASE,J.I.	173BFA
CLARK EQU.	872454
CLARK EQU.	877079
CLARK EQU.	886676
CLARK EQU.	892969
CLARK EQU.	948322
CLARK EQU.	952379
CLEVE.MOT.	CM-CP3DR
CMP	CM-CP3DR
COCKSHUTT	MXR3020
DAFFIN	1T168
DEERE,JOHN	AJ8977
FIAT-ALLIS	517038
FIAT-ALLIS	517039
FIAT-ALLIS	518038
FIAT-ALLIS	518039
FIAT-ALLIS	70517038
FIAT-ALLIS	70517039
FIAT-ALLIS	70518038
FIAT-ALLIS	70518039
FLEETRITE	ZAD-B3DR
FORD	163-392
FORD	CP33-39
FORD	CP33-46
G & G MFG	302-0300
GAR WOOD	3001845
GAR WOOD	3002018
GAR WOOD	6Y53K
GLAENZER	065-275-8
GMB	G5-265X
HARDIE	499A26
LILLISTON	UJ507
LUNDELL	S119
MASS.FERG.	840-655M91
MASS.FERG.	840-665M91
MASS.FERG.	840869M91
MASS.FERG.	860665M92
MCQUAY-NOR	U03-74

MCQUAY-NOR	U03-74
MIPER	C3452
MOTOR MAST	3568A
MPL.MOLINE	15R92
NEAPCO	1-0003
NEAPCO	283DN3
NEAPCO	283DR
NEAPCO	283DRN
NEW IDEA	3032PA
NEW IDEA	3033RA
NEW IDEA	K1965A
OLIVER	AC1718
OLIVER	AC1719
OLIVER	AC17-20
PERRY	U100
PILOT	B3DR
PRECISION	856
PREC.TOR.	857
REPCO	5-265X
REPCO	CP3DR
REPUBLIC	CB3DR
SERVIS	W42SR
SIDEWINDER	17298
SPEED-SPR.	1265204
SPICER	5-265X
TAYLOR	48827
TAYLOR	7169
TEL-E-LECT	1351
TEL-E-LECT	13561
TEL-E-LECT	13562
TEL-E-LECT	1651
TEL-E-LECT	1652
UJS	55-3DR
WESCO	201300
WESCO	3DR
WHITE	6373796
WHITE FARM	15R92
WHITE FARM	AC17-20
WOOD BROS.	126

IDLI GROUP 80134

CROSS + 4 SLOTTED BEARINGS + THRUST AND LOCK PLATES
SEE FIGURE 1D

1.2500 in	BD	31.75 mm
.9375 in	BW	23.81 mm
3.750 in	CL	95.25 mm
.750 in	CD	19.05 mm

AEC	4N
AEC	AE-CP4N
AEC	AEO96-55
AEC	CP4N
ALCO	1755
ALCO	177S
ALCO	CP4N
ALLOY	1775
ALLOY	CP4N
BALKAMP	CP4N
BEAN,JOHN	1224706
BEAN,JOHN	1430287
BORG-WARNR	114-101
BORG-WARNR	CP4N
CASE,J.I.	D46065
CASE,J.I.	L33742
CHAIN BELT	288-6013-91
CHAIN BELT	7048

CLARK EQU.	884518
CLARK EQU.	886518
CLARK EQU.	952369
CLEVE.MOT.	CM-CP4N
CMP	CM-CP4N
D&D MACH.	MBCP4N
DEERE,JOHN	K60452
ETNYRE	6440077
EXCEL	U506
FLEETRITE	ZAD11560UB
FLEETRITE	ZADB40N
FOX RIVER	F4CX10
GAR WOOD	6Y1452KK
GEHL BROS.	510040
GMB	GCP4N
GMB	GGPL4N
GMC	1788287
HARDIE	142A300
HESSTON	811018
IMCO	55-4N
KOEHRING	P466
LEMPCO	U506
MCQUAY-NOR	OFF-HWY.
MICHIGAN	884518
MICHIGAN	886518
MICHIGAN	952369
MOTIVE	55-4N
MOTOR MAST	3575A
MOTOR MAST	B3575
NATL.MINE	3881-7
NATL.MINE	76012137
NEAPCO	28054XB
NEAPCO	2845N
NEAPCO	28CP45N
NEAPCO	28CP4N
NEAPCO	3-1045
OLIVER	AC17-81
PERFECTION	RG4N
PILOT	B40N
PILOT	CP4N
PIONEER	PA679
PRECISION	890
REPCO	5-1211X
REPCO	CP45N
REPCO	CP4N
REPUBLIC	CB4N
SMITH,T.L.	2D9
SPEED-SPR.	1224706
SPEED-SPR.	1240169
SPEED-SPR.	1245163
SPICER	CP4N
TAYLOR	3810-367
THEW SHOV.	P466
TRW	20065
TRW	20068
TRW	20068D
TRW	20210
ULRICH	2U2064
UNIT RIG	987967
WESCO	301004
WESCO	391056
WESCO	B4N
WHITE FARM	AC17-81
ZELLER	3575A

IDLI GROUP 80135

CROSS + 4 GROOVED BEARINGS + INSIDE SNAP RINGS
SEE FIGURE 1B

1.2344 in	BD	31.35 mm
.8125 in	BW	20.64 mm
3.750 in	CL	95.25 mm
.7656 in	CD	19.45 mm

AEC	373
AEC	AE361
AEC	AE362
AEC	AE372
AEC	AE373
AEC	AE374
ALCO	1739
ALCO	1740
ALCO	53825-2
ALCO	J53625-2
ALCO	J53825-2
ALLOY	1739
ALLOY	1740
ALLOY	J53025-2
ALLOY	J53625-2
ALLOY	J53825-2
BORG-WARNR	114-362
BORG-WARNR	114-372
BORG-WARNR	114-373
CENTRAL	G5-1301X
CHRYSLER	1064846
CHRYSLER	1089795
CHRYSLER	1181657
CHRYSLER	18816557
CHRYSLER	1922465
CHRYSLER	3780272
CHRYSLER	4213200
CHRYSLER	575784
CHRYSLER	57584
CHRYSLER	947553
CLARK EQU.	859568
CLEVE.MOT.	CM536252
CLEVE.MOT.	CM537252
CLEVE.MOT.	CM53825-2
CMP	CM53625-2
CMP	CM53725-2
CMP	CM53825-2
D&D MACH.	MD7252
DETROIT	3780272
DETROIT	53625-2
DETROIT	53725-2
DETROIT	53727-2
DETROIT	53825-2
DETROIT	53825-3
DETROIT	J53725-2
DETROIT	J7325-2
EXCEL	U415
EXCEL	U453
FLEETRITE	ZAD17330
FORD	CD5AZ4635C
GLAENZER	065-000-0
GLAENZER	065-200-8
GLAENZER	065-304-6
GLAENZER	065-906-8
GMB	G5-1301X
GMB	G53625-2
GMB	G5-373
GMB	G53825-2

ENGINEERING CATALOGS MUST BE CONSULTED FOR DETAILS NOT INCLUDED
IN THIS GUIDE. SPECIFIC DESIGNS, MATERIAL CONTENT, TOLERANCES
LUBE FITTINGS AND OTHER DIMENSIONS ARE INTENTIONALLY OMITTED HERE.

CAUTION: BE SURE TO REFER TO ENGINEERING CATALOGS FOR SPECIAL APPLICATIONS THAT REQUIRE SPECIFIC MATERIAL CONTENT, TOLERANCES, ETC. SEE FOOTNOTE.

GMB	GU114-362
GMB	GU2600
GMB	GU2650
HYSTER	CC372
INBRA	CA372
ITALCARDAN	50-745-000
JAEGER	53625-2
JAEGER	53725-2
LEMPCO	U126
LEMPCO	U409
LEMPCO	U414
LEMPCO	U415
LEMPCO	U453
LOBRO	U260
LOBRO	U265
MASTERS	MD6252
MASTERS	MD7252
MCQUAY-NOR	U02-19
MCQUAY-NOR	UO2-19
MICHIGAN	K9568
MIPER	C3368
MOPAR	D127
MOPAR	D128
MOTOR MAST	4300A
NEAPCO	28056
NEAPCO	280564
NEAPCO	280568X
NEAPCO	28056X
NEAPCO	281300
NEAPCO	3-0056
NEAPCO	3-56
PERFECTION	RG362
PERFECTION	RG372
PERFECTION	RG373
PILOT	17327
PILOT	17329
PILOT	17330
PILOT	53625
PITT.VIOL.	40.745.000
PRECISION	304
REPCO	114-362
REPCO	114-373
REPCO	RUJ4007
REPUBLIC	CB53625
REPUBLIC	CB53725
REPUBLIC	CB53825-2
SPICER	5-1301X
SPICER	5-1302X
SPICER	53625-2
TEDDY TORQ	537
TEDDY TORQ	G53825-2
TRANSMISS.	UK368
TRU-CROSS	1301
TRW	20090
UJS	55-53825-2
WESCO	201-372
WESCO	301372
WESCO	301553
WESCO	D372
WESCO	N1501
WESCO	N553

WYLIE	42
ZELLER	4300A

IDLI GROUP 80136

CROSS + 4 GROOVED BEARINGS + INSIDE SNAP RINGS
SEE FIGURE 1B

1.3125 in	BD	33.34 mm
.8594 in	BW	21.83 mm
3.75 0 in	CL	95.25 mm
.8438 in	CD	21.43 mm

ALLOY	1774
ALLOY	CP44N
BORG-WARNR	114-151
CARDWELL	4667
CASE,J.I.	T30746
DEERE,JOHN	AC15878
DURA-DET.	1070584
G & G MFG	302-4400
GMB	GGPL44R
HESSTON	802389
HESSTON	802496
IHC	493-849R91
MASS.FERG.	836-454M91
MASS.FERG.	836-456M91
MATSUI	UJ2844
MCQUAY-NOR	AGRICULT.
MOTOR MAST	3575B
NEAPCO	2844N
NEAPCO	2844R
NEAPCO	2844X
NEAPCO	3-0044
NEAPCO	3-44
NEW HOLL.	578805
PRECISION	888
REPUBLIC	CB
SCHULTZ	11767
SCHULTZ	31138E
SERVIS	W840
SIDEWINDER	18915
SPICER	5-431X
TERRAIN	6691
TRU-CROSS	431
TRW	20203
WESCO	381044
WESCO	44R
WEST.LAND	ASUS130P

IDLI GROUP 80137

CROSS + 4 ROUND BEARINGS 2 INSIDE + 2 OUTSIDE SNAP RINGS
SEE FIGURE 1AB

1.0625 in	BD	26.99 mm
.6406 in	BW	16.27 mm
3.8125 in	CL	96.84 mm
.6875 in	CD	17.46 mm

GMB	G5-369X
PRECISION	347

IDLI GROUP 80138

CROSS + 4 SERRATED BEARINGS + THRUST AND LOCK PLATES
SEE FIGURE 1C

1.1250 in	BD	28.58 mm
.8750 in	BW	22.23 mm
3.8125 in	CL	96.84 mm
.7188 in	CD	18.26 mm

AEC	1211
AEC	AE096-55
ALCO	JO96-55
ALCO	JO96-55
ALCO	O96-55
ALLOY	JO96-55
BALKAMP	C0-9655
BORG-WARNR	096-55
BORG-WARNR	114-132
BORG-WARNR	O96-55
CHRYSLER	1238416
CLEVELAND	096-55
CLEVE.MOT.	CM9655-0
CLEVE.MOT.	CM9655O
CLEVE.STL.	O96-55
CMP	CM9655O
D&D MACH.	MCO96
FMC	34532
GLAENZER	065-217-8
GMB	G5-1211X
GMB	GO96-55
GMB	GU096-55
GMB	GU2800
HABERLE	H09655
HABERLE	HO9655
ITALCARDAN	50-810000
LEMPCO	U301
LOBRO	U280
MASTERS	MCO96
MIPER	C3464
MOPAR	D105
MOTOR MAST	3481A
NEAPCO	281600
PERFECTION	RG096-55
PERFECTION	RGO96-55
PILOT	O19655
PILOT	O19655
PRECISION	415
REPCO	5-1211X
REPCO	O96-55
REPUBLIC	CB096
REPUBLIC	CB96
REPUBLIC	CB-O96
SPICER	5-1211X
SPICER	O96-55
STUDEBAKER	660196
STUDEBAKER	668526
STUDEBAKER	679937
TEDDY TORQ	GO96-55
TEDDY TORQ	O96
TRANSMISS.	UK464
TRU-CROSS	1211
TRW	20065
UJS	55O96
WESCO	301966
WESCO	NO965
WESCO	NO965
ZELLER	5451A

IDLI GROUP 80139

CROSS + 4 GROOVED BEARINGS + INSIDE SNAP RINGS
SEE FIGURE 1B

1.3750 in	BD	34.93 mm
.9375 in	BW	23.81 mm
3.8750 in	CL	98.43 mm
.8906 in	CD	22.62 mm

AEC	332
AEC	492
ALLOY	41155
BORG-WARNR	114-636
CHRYSLER	3837585
DETROIT	3837585
FORD	D8TZ3249A
GMB	G5-332X
GMB	G5-645
GMC	462809
MCQUAY-NOR	U02-6
MOTOR MAST	3287C
NEAPCO	28332X
NEAPCO	3-0332
PRECISION	374
REPCO	5-297X
REPCO	5-88X
REPUBLIC	CB1332
SPICER	5-332X
TRU-CROSS	332
TRW	20188
WESCO	300332
ZELLER	3287C

IDLI GROUP 80140

CROSS + 4 ROUND BEARINGS + OUTSIDE SNAP RINGS
SEE FIGURE 1A

1.3750 in	BD	34.93 mm
.9375 in	BW	23.81 mm
3.8750 in	CL	98.43 mm
.8906 in	CD	22.62 mm

AEC	514
AEC	588
AEC	AE5-088
AEC	AE581
AEC	AE588
ALBURUS-SP	5-263X
ALCO	1031
ALCO	1588
ALCO	1756
ALCO	J5-181
ALCO	J5-188
ALLIS-CHLM	155608
ALLIS-CHLM	255608
ALLOY	1756
ALLOY	J5-188
BEAN,JOHN	1240169
BEAN,JOHN	1245163
BORG-WARNR	114-581
BORG-WARNR	114-588
BORG-WARNR	114-588R
BRADY	4185
BUCKEYE	1480
CATERPILLR	2R960
CENTRAL	G5-188X
CHRYSLER	1752307
CHRYSLER	3837585

BD = BEARING DIAMETER (outside)   BW = BEARING WIDTH   CB = CROSS LENGTH WITH BEARINGS   CC = CENTER TO CENTER   CD = CROSS DIAMETER
CL = CROSS LENGTH WITHOUT BEARINGS   EC = END TO CENTER or FACE TO FACE   EE = END TO END   EL = EFFECTIVE LENGTH
HD = HUB DIAMETER (or insert)   IE = INSIDE OF EARS (or recess)   OD = OUTSIDE DIAMETER   OE = OUTSIDE OF EARS   SB = SPLINE OR BORE SIZE

CLEVE.MOT.	CM5-101X
CLEVE.MOT.	CM5-111X
CLEVE.MOT.	CM5-181X
CLEVE.MOT.	CM5-188X
CMP	CM5-181X
CMP	CM5-188X
CURTISS-WR	21-20877
D&D MACH.	MS588X
DIAMOND R	10048P2
DIAMOND R	10049P2
DIAMOND R	10051P2
DIAMOND R	10053P2
DIAMOND R	BB27199W
DIAMOND T	27199W
EATON	403759
EXCEL	U117
FIAT-ALLIS	255608
FIAT-ALLIS	70255608
FLEETRITE	ZAD1158
FLEETRITE	ZAD11588
FORD	C1AZ4635A
FORD	C1TT4724A
FORD	C1TZ4635A
FORD	CITZ4635A
GAR WOOD	6Y1452YY
GLAENZER	065-165-8
GLAENZER	065-209-8
GLAENZER	065-351-6
GLAENZER	565-026-1
GLAENZER	565-040-0
GMB	G5-188
GMB	G5-188X
GMB	G5-263X
GMB	G5-588
GMB	G5-88X
GMB	GU2300
GMB	GU500SL
GMB	GUPA3
GMC	100315
GMC	2057289
GMC	2341648
GMC	2345726
GMC	3707508
GMC	3737508
GMC	3737589
GMC	6268535
GMC	9033740
HARDY SPCR	4851-183Y
HARDY SPCR	K5LGB183
HOUGH	154073HI
HUB CITY	03-35-900067
IHC	154073
IHC	154-073H1
IHC	387-097R91
IHC	799293R91
IHC	869-620R91
IHC	913-774R91
INBRA	CC588
KATAOKA	MI85-3211
LEMPCO	U120
LEMPCO	U122
LOBRO	U230
MACK	181SJ18
MACK	201SJ38
MACK	21045188X
MASTERS	MS581X
MASTERS	MS588X
MCQUAY-NOR	U01-9
MCQUAY-NOR	U02-9
MCQUAY-NOR	UO2-9
MIPER	C3163
MOPAR	D82
MOPAR	D83
MOTOR MAST	3287A
MOTOR MAST	3287C
NEAPCO	2605X
NEAPCO	2675X
NEAPCO	28188X
NEAPCO	28300X
NEAPCO	2848
NEAPCO	3-0188
NEAPCO	3-188
NEW HOLL.	517594
OLIVER	AC1781
PERFECT-CI	188
PERFECTION	RG581X
PERFECTION	RG588R
PILOT	11588
PRECISION	351
PREC.TOR.	948
QUINT.HAZ.	QL188X
RENAULT	50.00.559.846
RENAULT	77.01.010.700
REPCO	5-188X
REPCO	5-88X
REPCO	K5G81
REPCO	K5GB115
REPCO	K5GB79
REPCO	K5LGB102R
REPCO	K5LGB50R
REPUBLIC	CB1880
REPUBLIC	CB1888
REPUBLIC	CB748
RUEDARSA	1080N
RUEDARSA	1082
RUEDARSA	2103-1082-0-0
SPICER	1480
SPICER	5-101X
SPICER	5-188X
SPICER	5-211X
SPICER	5-237X
SPICER	5-263X
SPICER	5-88X
STUDEBAKER	1695947
TEDDY TORQ	G5-188
TEDDY TORQ	K88
TIMBERJACK	403759
TRANSMISS.	UK163
TRU-CROSS	188
TRU-CROSS	88
TRW	20021
TRW	20057
UJS	55-188
UJS	55-88
VERSATILE	33096
WALTERSCHD	24-00-00
WESCO	201-581
WESCO	301588
WESCO	45N
WESCO	N1588
WHITE	02-7074122
WHITE	303131629
WHITE	3631P2
WHITE	861939
WHITE	BB27199W
WYLIE	79
ZELLER	3287A
ZELLER	3289A

IDLI GROUP 80141

CROSS + 4 GROOVED BEARINGS + INSIDE SNAP RINGS
SEE FIGURE 1B

1.1875 in	BD	30.16 mm
.7813 in	BW	19.85 mm
3.9375 in	CL	100.01 mm
.7656 in	CD	19.45 mm

AEC	1206
AEC	5-1206
AEC	AE055-55-2
AEC	AE5-1206
ALCO	1763
ALCO	1764
ALCO	J055-55
ALCO	JO55-55
ALCO	JO55-55-2
ALCO	O55-55
ALLOY	1763
ALLOY	1764
ALLOY	J055-55
ALLOY	J055-55-2
ALLOY	J055-55
BORG-WARNR	055-55-2
BORG-WARNR	114-121
BORG-WARNR	O55-55-2
CLEVELAND	055-55-2
CLEVE.MOT.	CM5555-0
CLEVE.MOT.	CM55550
CLEVE.MOT.	CM5655-2D
CLEVE.MOT.	CM5655D
CLEVE.STL.	O55-55
CLEVE.STL.	O55-55-2
CMP	CM55550
D&D MACH.	MCO55
DEERE,JOHN	AT49181
EXCEL	U314
FLEETRITE	ZAD-O15555
FORD	C5TZ4635B
FORD	C5TZ4635G
GLAENZER	065-223-8
GMB	G5-1206X
GMB	GO55-55
GMB	GO55-55-2
GMB	GU055-55-2
GMC	694969
IHC	219-193R91
LEMPCO	U314
MASTERS	MCO55
MCQUAY-NOR	U02-13
MCQUAY-NOR	UO2-13
MOPAR	D104
MOTOR MAST	3493
MOTOR MAST	3493A
NEAPCO	2-6600
NEAPCO	286600
PERFECTION	RO55-55-2
PILOT	015555
PILOT	015555
PRECISION	427
REPCO	5-1206X
REPCO	O55-55-2
REPCO	RUJ4008
REPUBLIC	CB055
REPUBLIC	CB-O55
SPICER	5-1206X
SPICER	5-305X
TEDDY TORQ	GO55-55
TEDDY TORQ	O55
TRU-CROSS	1206
TRW	20066
TRW	286200
TRW	286600
UJS	55-055
UJS	55-55
WESCO	301-55
WESCO	301555
WESCO	NO555
WESCO	NO555
WESCO	NO5555
ZELLER	3493A

IDLI GROUP 80142

CROSS + 4 ROUND BEARINGS + OUTSIDE SNAP RINGS
SEE FIGURE 1A

1.1875 in	BD	30.16 mm
.7813 in	BW	19.85 mm
3.9375 in	CL	100.01 mm
.7656 in	CD	19.45 mm

AEC	560
AEC	AE560
AEC	AE560U
ALBURUS-SP	5-160X
ALCO	1560
ALCO	1750
ALCO	J5-160
ALLIS-CHLM	081559
ALLIS-CHLM	081559-7
ALLIS-CHLM	1007023
ALLIS-CHLM	815597
ALLOY	1750
ALLOY	J5-160
ARMSTRONG	5410-164
ARMSTRONG	5410-2050
AUST.WEST.	AW371052
AUTO BIAN.	2810
AUTO BIAN.	2811
AUTO BIAN.	2812
AUTO BIAN.	46847
AUTOCAR	3CA01172
AUTOCAR	3CA1172
AUTOCAR	3W00-01172
AUTOCAR	3WO01172
AUTOCAR	3WO1172
AYRA DUREX	1-415-01
BALKAMP	3-955
BEDFORD	7090700
BEDFORD	7178934
BERLIET	943541418
BERLIET	943541518
BLAW KNOX	5-60X
BMC	8G3002
BMC	GUJ110
BMC	GUJ111

ENGINEERING CATALOGS MUST BE CONSULTED FOR DETAILS NOT INCLUDED
IN THIS GUIDE. SPECIFIC DESIGNS, MATERIAL CONTENT, TOLERANCES
LUBE FITTINGS AND OTHER DIMENSIONS ARE INTENTIONALLY OMITTED HERE.

IDLI GUIDE   COPYRIGHT © INTERCHANGE, INC.   ST. LOUIS PARK, MN. 55416 USA

CAUTION: BE SURE TO REFER TO ENGINEERING CATALOGS FOR SPECIAL APPLICATIONS THAT REQUIRE SPECIFIC MATERIAL CONTENT, TOLERANCES, ETC. SEE FOOTNOTE.

Make	Number	Make	Number
BMC	HYL1944	HARDY SPCR	K5LGB179
BMC	V8385	HARDY SPCR	K5LGB213
BORG-WARNR	114-510X	HARDY SPCR	K5PL10
BORG-WARNR	114-560	HERCULES	443
BORG-WARNR	114-560R	HIGHLAND	62824A
BRD	05-501471	HOLDEN	7090700
BRD	05-540017	HOLDEN	7178934
BRD	05-541471	HOUGH	203791HI
BRD	05-551471	HYSTER	158352
BROCKWAY	239004	IHC	127-574R91
BROCKWAY	239044	IHC	203791
BUCKEYE	1410	IHC	203-791H1
CENTRAL	G5-160X	IHC	702375C91
CHRYSLER	1516853	IHC	74-289R91
CHRYSLER	1752412	IHC	878-624C91
CHRYSLER	5040692	IHC	886-919R91
CHRYSLER	747556	INBRA	CA205
CHRYSLER	966932	INBRA	CT11-040
CHRYSLER	D18049	INBRA	CT20-580
CHRYSLER	D4220-1	ITALCARDAN	50-719-000
CHRYSLER	K23645	JAGUAR	10420
CHRYSLER	K26345	JAGUAR	8694
CITROEN	303710780	JAGUAR	9408-1
CITROEN	303710810	JEFFREY	5-160X
CLARK EQU.	5200182	JEFFREY	CAT401794
CLARK EQU.	855846	JOY MFG.	415124
CLARK EQU.	858846	JOY MFG.	5-160X
CLARK EQU.	861452	JOY MFG.	A20499
CLARK EQU.	861453	JOY MFG.	A48709
CLARK EQU.	868218	KATAOKA	MI85-1194
CLEVE.MOT.	CM5-160X	LEMPCO	U103
CLEVE.MOT.	CM5-160XUB	LEYLAND	944882
CMP	CM5-160X	LEYLAND	V8385
CMP	CM5-160XUB	LOBRO	U200
D&D MACH.	MS560X	LOBRO	U205
DAIM.RUBE.	791-216-007-0	MACK	201SJ12
DAVEY COM.	A24520	MACK	201SJ12A
DEERE,JOHN	AH8153A	MACK	21045160X
DENNIS	690251	MASS.FERG.	835-727M91
DIAMOND R	10057P2	MASTERS	MS560U
DIAMOND R	1026P2	MASTERS	MS560X
DIAMOND R	1700P2	MATSUBA	UJ022
DIAMOND R	1709P2	MCQUAY-NOR	U01-51
DIAMOND R	217	MCQUAY-NOR	U02-10
DIAMOND R	3345P2	MCQUAY-NOR	UO2-10
DIAMOND R	BB17199W	MICHIGAN	855846
DIAMOND R	KF10X	MICHIGAN	861452
DIAMOND T	17199W	MIPER	C3146
DIAMOND T	217	MIPER	C3147
DIAMOND T	7199W	MOPAR	D75
DIAMOND T	BB7199W	MOTOR MAST	3290A
DIAMOND T	K5-10X	MUIR-HILL	261118
EATON	403186	NEAPCO	2-0054
EIMCO	1410	NEAPCO	2-54
EXCEL	U103	NEAPCO	25586
FEDERAL	5Y532	NEAPCO	28054X
FEDERAL	5Y7218	NEAPCO	28154X
FIAT	4655182	NEW HOLL.	510463
FIAT	4675621	PERFECTION	RG560X
FIAT	4675624	PETTI-MULL	LL242-8
FIAT-ALLIS	0081559	PILOT	11560
FIAT-ALLIS	0081559-7	PITT.VIOL.	40.719.000
FIAT-ALLIS	073372	POCLAIN	00400-16
FIAT-ALLIS	081559	PRECISION	330
FIAT-ALLIS	081559-7	PREC.TOR.	955
FIAT-ALLIS	1007023	QUINT.HAZ.	QL1005
FIAT-ALLIS	70073372	QUINT.HAZ.	QL16405
FIAT-ALLIS	70081559	RAY GO	5-160X
FIAT-ALLIS	70081559-7	RAY GO	58000002
FIAT-ALLIS	71007023	RENAULT	08.70.360.900
FIAT-ALLIS	815597	RENAULT	50.00.241.261
FLEETRITE	ZAD11560	REO MOTORS	1026P2
FLEETRITE	ZAD11560UB	REO MOTORS	1709P2
FORD	528E7039	REO MOTORS	3345P2
FORD	8MTH7039	REO MOTORS	K510X
FORD	AB6T7039A	REPCO	101-1120
FORD	B4634B	REPCO	CPL6S22
FORD	B4635A	REPCO	K5GB10R
FORD	B4635B	REPUBLIC	CB1410
FORD	B6T7039A	REPUBLIC	CB1410A
FORD	B6TZ039A	REPUBLIC	CB755
FORD	B6TZ4635	RUEDARSA	1080E
FORD	B6TZ4635B	RUEDARSA	2103-1080-0-0
FORD	D4NN4N167A	SAVIEM	23232053
FORD	O4NN4N167A	SAVIEM	870360900
GALION	D106251	SPICER	1410
GLAENZER	065-136-8	SPICER	5-10
GLAENZER	465-002-0	SPICER	5-10X
GLAENZER	465-019-0	SPICER	5-144X
GLAENZER	465-019-4	SPICER	5-160X
GMB	G5-160	SPICER	5-214X
GMB	G5-160B	SPICER	5-262X
GMB	G5-160X	SPICER	5-462X
GMB	G5-160XU	SPICER	5-60
GMB	G5-262X	SPICER	5-60X
GMB	G5-560	SPICER	5-8X
GMB	GU2050	SPICER	J5-10
GMB	GU2050SL	SPICER	J5-60
GMC	11948	STUDEBAKER	1687253
GMC	2185901	STUDEBAKER	670774
GMC	2223568	STUDEBAKER	6707744
GMC	2223569	STUDEBAKER	911284
GMC	2263882	STUDEBAKER	914382
GMC	2298621	TAMPER	L41666
GMC	2698416	TEDDY TORQ	G5-160
GMC	3218621	TEDDY TORQ	K10
GMC	3707830	TIMBERJACK	401761
GMC	694968	TIMBERJACK	403186
GMC	7090700	TRANSMISS.	UK146
GMC	7178934	TRANSMISS.	UK147
GMC	7178938	TRU-CROSS	1206
GMC	8845379	TRU-CROSS	160
GMC	91021384	TRW	20058
GMC	9961728	UJS	55-160
HABERLE	H1510X	UNIC	106396
HABERLE	H1560	UNIC	106936
HABERLE	X1510X	UNIPART	GUJ110
HARDY SPCR	4151-179W	UNIPART	GUJ111
HARDY SPCR	4151-179Y	VOLVO	38464
HARDY SPCR	4151-213W	VOLVO	6.852.586
HARDY SPCR	6051-160Y	WESCO	201560
HARDY SPCR	K5GB10	WESCO	G1560R
HARDY SPCR	K5GB103	WESCO	N1560
HARDY SPCR	K5GB119	WESCO	N1560R
HARDY SPCR	K5GB164	WHITE	02-7074026
HARDY SPCR	K5GB179	WHITE	10057P2
HARDY SPCR	K5LGB103	WHITE	1026P2
		WHITE	1700P2
		WHITE	1709P2
		WHITE	216343
		WHITE	216343A
		WHITE	217

BD = BEARING DIAMETER (outside)   BW = BEARING WIDTH   CB = CROSS LENGTH WITH BEARINGS   CC = CENTER TO CENTER   CD = CROSS DIAMETER
CL = CROSS LENGTH WITHOUT BEARINGS   EC = END TO CENTER or FACE TO FACE   EE = END TO END   EL = EFFECTIVE LENGTH
HD = HUB DIAMETER (or insert)   IE = INSIDE OF EARS (or recess)   OD = OUTSIDE DIAMETER   OE = OUTSIDE OF EARS   SB = SPLINE OR BORE SIZE

IDLI GUIDE   COPYRIGHT ® INTERCHANGE, INC.   ST. LOUIS PARK, MN. 55416 USA

WHITE	303131580
WHITE	3345P2
WHITE	403414
WHITE	897732
WHITE	BB17199W
WHITE	BB7199W
WHITE	KD10X
WORLDPARTS	W49-118
WYLIE	24
ZELLER	3290A

IDLI GROUP 80143

CROSS + 4 ROUND BEARINGS + OUTSIDE SNAP RINGS
slight variation from above group
SEE FIGURE 1A

1.1875 in	BD	30.16 mm
.7813 in	BW	19.85 mm
3.9375 in	CL	100.01 mm
.7656 in	CD	19.45 mm

AEC	560U
AEC	AE50U
AEC	AE560U
ALCO	1560U
ALCO	1752
ALCO	J5-160B
ALLIS-CHLM	081559
ALLIS-CHLM	1007023
ALLOY	1752
ALLOY	J5-160B
BORG-WARNR	114-560
BORG-WARNR	114-560U
CLEVE.MOT.	CM5-160X
CLEVE.MOT.	CM5-160XUB
CMP	CM5-160XUB
D&D MACH.	MS560U
DIAMOND R	KF10X
DIAMOND T	BB7199W
EXCEL	U103U
FORD	B6T7039A
GMB	G5-160B
GMB	G5-160XU
HABERLE	H1560
HABERLE	X1510X
LEMPCO	U103U
MASTERS	MS560U
MASTERS	MS560X
MCQUAY-NOR	U01-51
MCQUAY-NOR	U02-10
MCQUAY-NOR	UO1-51
MICHIGAN	855846
MICHIGAN	861452
MOPAR	D75
MOPAR	D76
MOTOR MAST	3290B
NEAPCO	2-54
NEAPCO	28054XB
NEAPCO	2855
NEAPCO	3-1054
PERFECTION	RG560X
PILOT	11560UB
PRECISION	330U
PREC.TOR.	955
REPUBLIC	CB1410U
SPICER	5-144X
SPICER	5-160X

SPICER	5-160XU
TEDDY TORQ	K10U
TRW	20058
WESCO	303560
WESCO	N1560RU
ZELLER	3290B

IDLI GROUP 80144

**CROSS + 4 ROUND BEARINGS
2 INSIDE + 2 OUTSIDE SNAP RINGS
SEE FIGURE 1AB**

1.1875 in	BD	30.16 mm
.7813 in	BW	19.85 mm
3.9375 in	CL	100.01 mm
4.1875 in	CB	106.36 mm

AEC	355
AEC	357
AEC	373-1410
AEC	AE373-1410
CLEVE.MOT.	CM903HP
CLEVE.MOT.	CM904HP
GMB	G5-355X
GMB	G5-357X
MOTOR MAST	3133A
MOTOR MAST	3190A
NEAPCO	2-3190
NEAPCO	283133
NEAPCO	283190
NEAPCO	3-3133
NEAPCO	3R231
NEAPCO	3R-231
PRECISION	362
REPUBLIC	CB3140X
REPUBLIC	CB3150X
SPICER	5-355X
SPICER	5-357
SPICER	5-357X
SPICER	5-537X
TRW	20167
TRW	20168

IDLI GROUP 80145

**CROSS + 4 SERRATED BEARINGS
+ THRUST AND LOCK PLATES
SEE FIGURE 1C**

1.3750 in	BD	34.93 mm
.8750 in	BW	22.23 mm
3.9375 in	CL	100.01 mm
.8750 in	CD	22.23 mm

AEC	1214
AEC	AE-D56-55-2
AEC	AE-DS56-55-2
AEC	D56-55-2
AEC	DS56-55-2
ALCO	1762
ALCO	DS55-55-2
ALCO	JDS55-55-2
ALCO	JDS56-55-2
ALCO	JS56-55
ALLOY	1762
ALLOY	JDS55-55-2
ALLOY	JDS56-55-2
ALLOY	JOS56-55-2
BORG-WARNR	114-130
BORG-WARNR	D56-55-2
BORG-WARNR	DS55-55-2

BORG-WARNR	DS56-55-2
CLEVELAND	D56-55-2
CLEVE.MOT.	CM5655-2B
CLEVE.MOT.	CM5655-2D
CLEVE.STL.	D56-55-2
CLEVE.STL.	DS55-55-2
CLEVE.STL.	DS56-55-2
CMP	CM5655-2D
D&D MACH.	MCDS56
FLEETRITE	ZADD15655
FLEETRITE	ZAD-O15655
GLAENZER	065-224-8
GMB	G5-1214X
GMB	G5-1241X
GMB	GDS55-55-2
GMB	GDS56-55
GMB	GDS56-55-2
GMB	GS56-55
IHC	1868774R91
IHC	226-800R91
IHC	682-597R91
IHC	965-620R91
LEMPCO	U317
MCQUAY-NOR	U02-14
MCQUAY-NOR	UO2-14
MOTOR MAST	3493B
NEAPCO	286700
NEAPCO	3-6700
PILOT	D15655
PILOT	DS15655
PRECISION	428
REPCO	5-1214X
REPCO	D56-55-2
REPCO	DS56-55-2
REPUBLIC	CB-S56
SPICER	5-1214X
TEDDY TORQ	D56
TRU-CROSS	1214
TRW	20067
UJS	55D56
WESCO	301556
WESCO	NDS555
WHITE	303323404
ZELLER	3493B

IDLI GROUP 80146

**CROSS + 4 GROOVED BEARINGS
+ INSIDE SNAP RINGS
SEE FIGURE 1B**

1.1406 in	BD	28.97 mm
.7969 in	BW	20.24 mm
4.0 in	CL	101.60 mm
.6563 in	CD	16.67 mm

ALCO	1702
ALCO	4008
ALCO	J4008
ALLOY	1702
ALLOY	J4008
BALKAMP	M4008
BORG-WARNR	114-4008
CHRYSLER	1463513
GLAENZER	065-545-6
GMB	G4008
GMB	G5-4141X
GMB	GU6420
HABERLE	H4008

IHC	65349R91
LEMPCO	U216
LOBRO	U642
MASTERS	MM4008
MOPAR	D23
NEAPCO	282400
NEAPCO	81-6010
NEAPCO	DS601KF
PERFECTION	RG4008
PILOT	14008
PRECISION	543
REPCO	114-4008
REPCO	RUJ4000
REPUBLIC	CB4008
RUEDARSA	1812
RUEDARSA	2104-1812-0-0
SPICER	5-4008X
TEDDY TORQ	1408
TEDDY TORQ	G4008
WESCO	301408
WESCO	N4008
ZELLER	4000X

IDLI GROUP 80147

**CROSS + 2 ROUND BEARINGS
2 MIDWING DRILLED BEARINGS
+ INSIDE AND OUTSIDE SNAP RINGS
SEE FIGURE 1BI**

1.1406 in	BD	28.97 mm
.7969 in	BW	20.24 mm
4.0 in	CL	101.60 mm
.6563 in	CD	16.67 mm

ALCO	4011
ALCO	J4011
BALKAMP	M4007
BORG-WARNR	114-4007
BORG-WARNR	114-4011
BORG-WARNR	114-4011AB
BORG-WARNR	J4007
GMB	G4007
GMB	G4011
HABERLE	H4007
HYSTER	27837
HYSTER	28737
IHC	137109H
IHC	137109HA
LEMPCO	U215
LOBRO	U641
MOPAR	D24
NEAPCO	282300
NEAPCO	81-2040
NEAPCO	DS204KF
PERFECTION	RG4007
PILOT	14007
PRECISION	542
REPCO	114-4007
REPCO	114-4011
REPUBLIC	CB4007
RUEDARSA	1811
RUEDARSA	2104-1811-0-0
SPICER	5-4007
SPICER	5-4007X
TEDDY TORQ	1407
WESCO	301407
WESCO	N4007

ENGINEERING CATALOGS MUST BE CONSULTED FOR DETAILS NOT INCLUDED
IN THIS GUIDE. SPECIFIC DESIGNS, MATERIAL CONTENT, TOLERANCES
LUBE FITTINGS AND OTHER DIMENSIONS ARE INTENTIONALLY OMITTED HERE.

CAUTION: BE SURE TO REFER TO ENGINEERING CATALOGS FOR SPECIAL APPLICATIONS THAT REQUIRE SPECIFIC MATERIAL CONTENT, TOLERANCES, ETC. SEE FOOTNOTE.

IDLI GROUP 80148

CROSS + 4 GROOVED BEARINGS + INSIDE SNAP RINGS
SEE FIGURE 1B

1.1250 in	BD	28.58 mm
.7969 in	BW	20.24 mm
4.0 in	CL	101.60 mm
.6719 in	CD	17.07 mm

AEC	4106
AEC	AE4106
ALCO	1709
ALCO	545
ALCO	J4106
ALCO	J419
ALLOY	1709
ALLOY	J4106
BALKAMP	M4015
BALKAMP	M4106
BORG-WARNR	114-4015
BORG-WARNR	114-4033
BORG-WARNR	114-4106
BORG-WARNR	J4015
CLEVE.MOT.	CM4015
CLEVE.MOT.	CM4106
CMP	CM4015
CMP	CM4106
D&D MACH.	MM4106
DEERE,JOHN	35845
EXCEL	U220
FLEETRITE	ZAD14015
GLAENZER	065-241-8
GMB	G4106
GMB	G4196
GMB	G5-4015X
GMB	GU6470
HABERLE	H4015
IHC	65706R91
IHC	65-706R92
IHC	65848R91
IHC	851-631R91
LEMPCO	U220
LOBRO	U647
MASTERS	MM4106
MCQUAY-NOR	U02-15
MCQUAY-NOR	UO2-15
MOPAR	D28
MOTOR MAST	4000B
NEAPCO	2-2600
NEAPCO	282600
PERFECTION	RG4015
PILOT	14015
PRECISION	545
REPCO	114-4106
REPCO	RUJ2039SL
REPUBLIC	CB4015
RUEDARSA	1814
RUEDARSA	2104-1814-0-0
SPICER	5-4015X
TEDDY TORQ	1415
TEDDY TORQ	G4106
TRU-CROSS	4015
TRW	20070

UJS	55-4106
WESCO	301415
WESCO	N4015
ZELLER	4000B

IDLI GROUP 80149

CROSS + 4 LOW WING DRILLED BEARINGS
SEE FIGURE 1G

1.4375 in	CC	36.51 mm
4.0 in	CL	101.60 mm
.6719 in	CD	17.07 mm

AEC	4100
AEC	AE4100
ALLIS-CHLM	057614
ALLIS-CHLM	AMX46117
ALLOY	1703
ALLOY	J4151
ATHEY PROD	4-709-11
BORG-WARNR	114-3119AB
BORG-WARNR	114-4100A
BORG-WARNR	114-4119
BORG-WARNR	114-4132
BORG-WARNR	114-4136
BORG-WARNR	114-4138
BORG-WARNR	114-4142
BORG-WARNR	114-4147
BORG-WARNR	114-4148
BORG-WARNR	114-4151
BUCYR-ERIE	46-813-675(4)
BUCYR-ERIE	4-681-393(4)
BUCYR-ERIE	4-681-397(4)
BUCYR-ERIE	46-814-061(4)
BUCYR-ERIE	4938-487
BUCYR-ERIE	509456(4)
BUCYR-ERIE	509457(4)
BUFFALO-SP	334B25384
BUFFALO-SP	35708(4)
BUFFALO-SP	36037(4)
BUFFALO-SP	848036711(4)
BUFFALO-SP	848036714(4)
CATERPILLR	1K2741
CATERPILLR	1K9467
CATERPILLR	6F7160
CATERPILLR	6F9190
CATERPILLR	8H3853
CATERPILLR	943221
CLARK EQU.	866887
CLARK EQU.	940036
CLARK EQU.	940039
CLARK EQU.	940083
CLARK EQU.	943221
CLARK EQU.	944865
CLARK EQU.	946067
CLEVE.MOT.	CM4100
CURTISS-WR	3-32883
DEERE,JOHN	L1773
DEERE,JOHN	L1773C
FIAT-ALLIS	0086284
FIAT-ALLIS	0086284-7
FIAT-ALLIS	057614
FIAT-ALLIS	075614
FIAT-ALLIS	086284
FIAT-ALLIS	086284-7
FIAT-ALLIS	70057614
FIAT-ALLIS	70075614

FIAT-ALLIS	70086284
FIAT-ALLIS	70086284-7
FIAT-ALLIS	70682636
FIAT-ALLIS	78013656-7
FIAT-ALLIS	7826391
FIAT-ALLIS	8013656-7
FIAT-ALLIS	AMX46117
GARD.DENV.	107897
GERLINGER	69240
GLAENZER	065-118-8
GLAENZER	065-170-8
GMB	G4151
GMC	9036593
GMC	9052166
GMC	9395415
HABERLE	H4002
HY-DYNAM	10-217
HYSTER	200166A
IHC	105112
IHC	115893R91
IHC	126812
IHC	126-812H1
IHC	127269
IHC	127-269H1
IHC	220-532R91
IHC	294-769R91
INBRA	CT11-070
JOY MFG.	A48402
KERSHAW	13-1837
LEMPCO	U209
LINK BELT	17A2371(4)
MASS.FERG.	1029-201M91
MASTERS	MM4100
MCQUAY-NOR	U02-16
MICHIGAN	866887
MICHIGAN	940036
MICHIGAN	940039
MICHIGAN	940083
MICHIGAN	943221
MICHIGAN	944865
MICHIGAN	946067
MIXERMOBL	708390
MIXERMOBL	S8571
MOPAR	D25
NEAPCO	284138
NEAPCO	3-4138
PARSONS	278D225A(4)
PERFECTION	RG4002
PETTI-MULL	42366
PETTI-MULL	4258
PETTI-MULL	CM8
PETTI-MULL	LA123-12
PETTI-MULL	LL773-186
PILOT	14138
PILOT	14153
PRECISION	921
REPCO	114-4138
REPCO	114-4151
REPUBLIC	CB721
REPUBLIC	CB740
SCOOPMOBLE	708390
SHOV.SUPP.	SP112529(4)
SHOV.SUPP.	SP65143(4)
SIL.HOIST	15394(4)
SIL.HOIST	25281N(4)
SPICER	5-4002X

STUDEBAKER	382668
THOMPSON	HAR1405
TRU-CROSS	4002
TRW	20069
TULSA WIN.	24047(4)
TULSA WIN.	24048(4)
WAYNE	3-23883
WESCO	N4002
WOOLDRIDGE	3-23883
ZELLER	4000A

IDLI GROUP 80150

CROSS + 4 LOW WING DRILLED BEARINGS
slight variation from above group
SEE FIGURE 1G

1.4375 in	CC	36.51 mm
4.0 in	CL	101.60 mm
.6719 in	CD	17.07 mm

BORG-WARNR	114-4101
BORG-WARNR	114-4115
BORG-WARNR	114-4126
BUCYR-ERIE	46-814-060
BUCYR-ERIE	46-814-0614
CATERPILLR	5H445
CATERPILLR	6H1262
CATERPILLR	85431
CATERPILLR	9K3969
CLEVE.MOT.	CM4101
CLEVE.MOT.	CM4104
CLEVE.MOT.	CM4105
EATON	0807856
EATON	807856
FIAT-ALLIS	0619915-2
FIAT-ALLIS	4991328
FIAT-ALLIS	4991328-8
FIAT-ALLIS	619915
FIAT-ALLIS	70619915
FIAT-ALLIS	70619915-2
FIAT-ALLIS	70681016
FIAT-ALLIS	74991328
FIAT-ALLIS	74991328-8
FIAT-ALLIS	X23192
GAR WOOD	4003768
GAR WOOD	840361
GLAENZER	065-124-8
GMB	G4103
GMB	G4105
GMB	G4115
GMB	G5-4200X
GMC	9033306
HYSTER	200167A
HYSTER	85525A
HYSTER	97895A
HYSTER	B139472
IHC	114477
IHC	124963
IHC	127268
IHC	127268H1
KOEHRING	278D255A
KOEHRING	858036732
LA PORTA	LA6301
MCQUAY-NOR	UO3-17
MCQUAY-NOR	UO3-21
M.R.S.	9096F33
NEAPCO	2830

BD=BEARING DIAMETER (outside)   **BW**=BEARING WIDTH   **CB**=CROSS LENGTH WITH BEARINGS   **CC**=CENTER TO CENTER   **CD**=CROSS DIAMETER
CL=CROSS LENGTH WITHOUT BEARINGS   **EC**=END TO CENTER or FACE TO FACE   **EE**=END TO END   **EL**=EFFECTIVE LENGTH
HD=HUB DIAMETER (or insert)   **IE**=INSIDE OF EARS (or recess)   **OD**=OUTSIDE DIAMETER   **OE**=OUTSIDE OF EARS   **SB**=SPLINE OR BORE SIZE

Column 1

NEAPCO	284105
NEAPCO	284123
NEAPCO	3-4123
PARSONS	278D225A(4)
PETTI-MULL	SAB100-18-1
PRECISION	930
REPCO	114-4006
REPUBLIC	CB730
REPUBLIC	CB739
SIL.HOIST	24351-2(4)
TOWMOTOR	85431
TOWMOTOR	B69240
WABCO	57735
WESCO	N4103

IDLI GROUP 80151

CROSS + 4 ROUND BEARINGS
+ OUTSIDE SNAP RINGS
SEE FIGURE 1I

1.4375 in	CC	36.51 mm
4. 0 in	CL	101.60 mm
.6719 in	CD	17.07 mm

AEC	4105
AEC	AE4104
AEC	AE4105
ALCO	1708
ALCO	J4105
ALLOY	1708
ALLOY	J4105
BORG-WARNR	114-4019
BORG-WARNR	114-4104
BORG-WARNR	114-4105
CLARK EQU.	866888
FIAT-ALLIS	4905112
FIAT-ALLIS	4905112-1
FIAT-ALLIS	74905112
FIAT-ALLIS	74905112-1
GLAENZER	065-280-8
GMB	G4104
GMB	G5-4104X
MCQUAY-NOR	U03-21
MICHIGAN	866888
MOTOR MAST	4000V
PRECISION	549
REPCO	114-4104
REPCO	114-4105
REPCO	RUJ4004
TOWMOTOR	85432
TRANSMISS.	UK405
TRW	20078
WESCO	301404
WESCO	N4104
WESCO	N4109
ZELLER	4000V

IDLI GROUP 80152

CROSS AND LOW OR HIGH WING TYPE
DRILLED OR THREADED BEARINGS
SEE FIGURE 1J

1.4375 in	CC	36.51 mm
4. 0 in	CL	101.60 mm
.6719 in	CD	17.07 mm

PRECISION	540

Column 2

IDLI GROUP 80153

CROSS AND LOW OR HIGH WING TYPE
DRILLED OR THREADED BEARINGS
SEE FIGURE 1IJ

1.4375 in	CC	36.51 mm
4. 0 in	CL	101.60 mm
.6719 in	CD	17.07 mm

AEC	4101
AEC	AE4101
AEC	AE4103
ALCO	1704
ALCO	4101
ALCO	J4101
ALCO	J4103
ALLIS-CHLM	4991328
ALLIS-CHLM	4993479
ALLIS-CHLM	619915
ALLIS-CHLM	X23192
ALLOY	1704
ALLOY	J4101
BEND.WEST.	614A539-1
BORG-WARNR	114-4006
BORG-WARNR	114-4024
BORG-WARNR	114-4101
BORG-WARNR	114-4103
BORG-WARNR	114-4123
BORG-WARNR	114-4124
BORG-WARNR	114-4128
BORG-WARNR	114-4139
BORG-WARNR	114-4150
BORG-WARNR	114-50A
CATERPILLR	943221
CLEVE.MOT.	CM4101
D&D MACH.	MM4006
D&D MACH.	MM4101
DETROIT	50A
EATON	807856
GLAENZER	065-262-8
GMB	G4101
GMB	G4115
GMB	G5-4123
GMB	G5-4140X
IHC	127168
MASTERS	MM4006
MASTERS	MM4110
MCQUAY-NOR	U03-17
MOPAR	D26
MOTOR MAST	4000DA
MOTOR MAST	4000T
M.R.S.	9096F23
M.R.S.	9096F33
NEAPCO	2839
NEAPCO	2841
NEAPCO	284123
NEAPCO	284132
NEAPCO	3-4123
PETTI-MULL	A20509-1
PETTI-MULL	A22509-1
PETTI-MULL	LL773-123
PILOT	14006
PRECISION	202
PRECISION	541
REPCO	114-4101
REPCO	114-4123
REPCO	RUJ4003
REPUBLIC	AD5000A

Column 3

REPUBLIC	CB4010
REPUBLIC	CB730
REPUBLIC	CB741
SPICER	5-4002X
TEDDY TORQ	1406
TEDDY TORQ	G4101
TRW	20071
WABCO	57735
WESCO	301410
WESCO	D4010
WESCO	N4010
WESTINGHSE	614A539-1
ZELLER	4000DA

IDLI GROUP 80154

CROSS + 2 LOW WING THREADED
+ 2 BLOCK BEARINGS
SEE FIGURE 1IL

1.4375 in	CC	36.51 mm
4. 0 in	CL	101.60 mm
.6719 in	CD	17.07 mm

AEC	4141
AEC	AE4141
ALCO	J4133
BORG-WARNR	114-4131
BORG-WARNR	114-4133
BORG-WARNR	114-4141
CATERPILLR	7M2695
FIAT-ALLIS	74993479
FIAT-ALLIS	74993479-7
GLAENZER	065-157-8
GMB	G4133
GMB	G5-4141X
GMC	9033424
GMC	9034424
HYSTER	98197A
MCQUAY-NOR	U03-22
MCQUAY-NOR	UO3-22
MOTOR MAST	4000X
NEAPCO	284141
PILOT	14141
PRECISION	996
REPCO	114-4131
REPUBLIC	CB796
TRW	20079
WESCO	301417
WESCO	N4141

IDLI GROUP 80155

CROSS + 2 LOW WING THREADED
+ 2 BLOCK BEARINGS
slight variation from above group
SEE FIGURE 1IL

1.4375 in	CC	36.51 mm
4. 0 in	CL	101.60 mm
.6719 in	CD	17.07 mm

AEC	4110
AEC	AE4110
ALCO	J4110
ALCO	J4130
ALLIS-CHLM	4991328
ALLIS-CHLM	4993479
ALLIS-CHLM	619915
BORG-WARNR	114-4110
BORG-WARNR	114-4111

Column 4

BORG-WARNR	114-4125
BORG-WARNR	114-4130
BORG-WARNR	114-4140
BORG-WARNR	114-4146
BORG-WARNR	114-4162
BORG-WARNR	114-4179
BUFFALO-SP	35708(4)
BUFFALO-SP	36037(4)
BUFFALO-SP	848036711(4)
CATERPILLR	2H858
CATERPILLR	5H800
CATERPILLR	5M800
CATERPILLR	7S990
CATERPILLR	GH2577
CLARK EQU.	874789
CLARK EQU.	877383
CLARK EQU.	892837
CLEVE.MOT.	CM4110
CMP	CM4110
D&D MACH.	MM4110
DROTT	27460
DROTT	27471
DROTT	32053
DROTT	76186
EATON	648397
FIAT-ALLIS	4993479
FIAT-ALLIS	4993479-7
FLEETRITE	ZAD14140
GARD.DENV.	100083
GEHL BROS.	51084
GLAENZER	065-123-8
GMB	G4110
GMB	G4113
GMB	G4130
GMB	G5-4110X
GMB	G5-4140
GMB	G5-4140X
GMC	9026919
IHC	161-850H1
IHC	203117H1
MCQUAY-NOR	U03-20
MCQUAY-NOR	UO3-20
MICHIGAN	874789
MICHIGAN	877383
MICHIGAN	892837
MOTOR MAST	4000P
NEAPCO	2829
NEAPCO	284140
NEAPCO	284140X
NEAPCO	3-4140
PARSONS	603D26A(4)
PETTI-MULL	SAP200-24
PILOT	14140
PRECISION	929
REPCO	114-4110
REPCO	114-4130
REPUBLIC	CB729
REYNOLDS	94-90-6475
SIL.HOIST	15394(4)
SPICER	5-4140X
TRU-CROSS	4140
TRW	20075
WESCO	301430
WESCO	N4130
ZELLER	4000P

ENGINEERING CATALOGS MUST BE CONSULTED FOR DETAILS NOT INCLUDED
IN THIS GUIDE. SPECIFIC DESIGNS, MATERIAL CONTENT, TOLERANCES
LUBE FITTINGS AND OTHER DIMENSIONS ARE INTENTIONALLY OMITTED HERE.

CAUTION: BE SURE TO REFER TO ENGINEERING CATALOGS FOR SPECIAL APPLICATIONS THAT REQUIRE SPECIFIC MATERIAL CONTENT, TOLERANCES, ETC. SEE FOOTNOTE.

IDLI GROUP 80156

CROSS WITH HOLE IN THE CENTER + 2 HIGH WING DRILLED + 2 BLOCK BRGS.

SEE FIGURE 1L

1.4375 in	CC	36.51 mm
4. 0 in	CL	101.60 mm
.6719 in	CD	17.07 mm

AEC	4113
AEC	AE4113
AEC	AE4143
ALCO	J4118
ALLOY	1698
ALLOY	1706
ALLOY	J4113
ALLOY	J4118
BORG-WARNR	114-4006
BORG-WARNR	114-4113
BORG-WARNR	114-4115
BORG-WARNR	114-4116
BORG-WARNR	114-4117
BORG-WARNR	114-4118
BORG-WARNR	114-4120
BORG-WARNR	114-4121
BORG-WARNR	114-4126
BORG-WARNR	114-4143
BORG-WARNR	114-4149
BORG-WARNR	114-4155
BORG-WARNR	114-4158
CATERPILLR	2D1844
CATERPILLR	5K5012
CATERPILLR	6K316
CLARK EQU.	887737
CLARK EQU.	940813
CLARK EQU.	944866
CLARK EQU.	944966
CLEVE.MOT.	CM4113
CLEVE.MOT.	CM4143
CMP	CM4113
CMP	CM4143
D&D MACH.	MM4113
EATON	2040231
EATON	6952684
FIAT-ALLIS	70681016
FIAT-ALLIS	78013657-5
FIAT-ALLIS	8013657-5
FLEETRITE	ZAD14143
GALION	D75441
GERLINGER	69240
GLAENZER	065-129-8
GMB	G4104
GMB	G4113
GMB	G4120
GMB	G4158
GMB	G5-4143
GMB	G5-4143X
GMB	GUK04
GMB	GUPA1
HYSTER	201167A
IHC	135764
IHC	135764H1
IHC	161850

IHC	328-592C91
INBRA	CT13-025
INBRA	CT16-000
INBRA	CT24-030
KOMATSU	111-10-16110(5)
KOMATSU	111-10-16120
MASTERS	MM4113
MCQUAY-NOR	U03-18
MCQUAY-NOR	UO3-18
MICHIGAN	887737
MICHIGAN	940813
MICHIGAN	944866
MOTOR MAST	4000E
NEAPCO	284103
NEAPCO	284143
NEAPCO	2842
NEAPCO	2869
NEAPCO	3-4143
PILOT	14143
PRECISION	969
PREC.TOR.	942
REPCO	114-4113
REPCO	114-4118
REPCO	114-4120
REPUBLIC	CB4010
REPUBLIC	CB742
REPUBLIC	CB769
SPICER	5-4143X
TRU-CROSS	4143
TRW	20073
WABCO	SR8778-3
WABCO	SR-R8778
WABCO	VD0222
WESCO	301413
WESCO	301-430
WESCO	N4113
ZELLER	4000E

IDLI GROUP 80157

CROSS + 2 WING TYPE BEARINGS + 2 ROUND GROOVED BEARINGS

SEE FIGURE 1BL

1.1250 in	BD	28.58 mm
1.4375 in	CC	36.51 mm
.7969 in	BW	20.24 mm
4. 0 in	CL	101.60 mm
.6719 in	CD	17.07 mm

AEC	4109
AEC	AE4109
ALCO	J4109
ALCO	J4156
ALLOY	J4109
BORG-WARNR	114-4109
BORG-WARNR	114-4156
CLARK EQU.	893998
CLEVE.MOT.	CM4109
GMB	G4109
GMB	G4156
GMB	G5-4109X
GMB	G5-4156
MCQUAY-NOR	TRK. & IND.
MICHIGAN	893998
MOTOR MAST	4000Q
NEAPCO	284156
PILOT	14156
PRECISION	494

REPCO	114-4109
REPCO	114-4156
TRW	20076
WESCO	301419
WESCO	N4156
ZELLER	4000Q

IDLI GROUP 80158

CROSS + 2 WING TYPE BEARINGS + 2 ROUND GROOVED BEARINGS

SEE FIGURE 1BJ

1.1250 in	BD	28.58 mm
1.4375 in	CC	36.51 mm
.7969 in	BW	20.24 mm
4. 0 in	CL	101.60 mm
.6719 in	CD	17.07 mm

AEC	4102
AEC	AE4102
ALCO	1705
ALCO	4102
ALCO	J4102
ALLIS-CHLM	086284
ALLIS-CHLM	6546
ALLOY	1705
ALLOY	J4102
AMER.MOTOR	3112479
AMER.MOTOR	3112579
BALKAMP	3-938
BALKAMP	M4014
BALKAMP	M4017
BALKAMP	M4034
BALKAMP	M4102
BORG-WARNR	114-4014
BORG-WARNR	114-4017
BORG-WARNR	114-4021
BORG-WARNR	114-4034
BORG-WARNR	114-4102
BORG-WARNR	114-4102A
BORG-WARNR	114-4102AB
BORG-WARNR	114-4127
BORG-WARNR	114-4129
BORG-WARNR	114-4134
BORG-WARNR	114-4135
BORG-WARNR	114-4152
BORG-WARNR	114-4159
BORG-WARNR	J4014
CHRYSLER	1463514
CLEVE.MOT.	CM4019
CLEVE.MOT.	CM4102
CLEVE.MOT.	CM4152
CMP	CM4019
CMP	CM4102
CMP	CM4152
D&D MACH.	MM4102
DEERE,JOHN	32603
EIMCO	104M1441
EXCEL	U217
FLEETRITE	ZAD14014
GLAENZER	065-240-8
GLAENZER	065-544-6
GMB	G4102
GMB	G4113
GMB	G4152
GMB	G5-4014X
GMB	G5-4152
GMB	GU6450

GMC	2156475
HABERLE	H4014
HOUGH	126078HI
HYSTER	43741A
HYSTER	503012A
HYSTER	53012A
HYSTER	53076
IHC	126978
IHC	126-978H1
IHC	238-739R91
IHC	238-740R91
IHC	65340R92
IHC	65350R91
IHC	65-350R92
IHC	851-632R91
LEMPCO	U217
LOBRO	U645
MASTERS	MM4102
MCQUAY-NOR	U03-19
MCQUAY-NOR	U03-22
MCQUAY-NOR	UO3-19
MOPAR	D27
MOTOR MAST	4000J
NEAPCO	282500
NEAPCO	2838
NEAPCO	284152
NEAPCO	3-4152
PERFECTION	RG4014
PILOT	14014
PILOT	14017
PILOT	14020
PILOT	14021
PILOT	14152
PRECISION	544
REPCO	114-4102
REPCO	114-4152
REPCO	RUJ4001
REPCO	RUJ4005
REPCO	RUJ4006
REPUBLIC	CB4014
REPUBLIC	CB4017
REPUBLIC	CB738
RUEDARSA	1813
RUEDARSA	2100-1813-0-0
SPICER	5-4014X
TAYLOR	1927-126(4)
TAYLOR	3461
TEDDY TORQ	J4102
TRANSMISS.	UK682
TRU-CROSS	4014
TRW	20074
UJS	55-4102
WESCO	301414
WESCO	N4014
ZELLER	4000-5
ZELLER	4000J

BD = BEARING DIAMETER (outside)   **BW** = BEARING WIDTH   **CB** = CROSS LENGTH WITH BEARINGS   **CC** = CENTER TO CENTER   **CD** = CROSS DIAMETER
CL = CROSS LENGTH WITHOUT BEARINGS   **EC** = END TO CENTER or FACE TO FACE   **EE** = END TO END   **EL** = EFFECTIVE LENGTH
HD = HUB DIAMETER (or insert)   **IE** = INSIDE OF EARS (or recess)   **OD** = OUTSIDE DIAMETER   **OE** = OUTSIDE OF EARS   **SB** = SPLINE OR BORE SIZE

IDLI GROUP 80159

CROSS + 2 WING TYPE BEARINGS + 2 ROUND GROOVED BEARINGS
SEE FIGURE 1BI

1.1250 in	BD	28.58 mm
1.4375 in	CC	36.51 mm
.7969 in	BW	20.24 mm
4.0 in	CL	101.60 mm
.6719 in	CD	17.07 mm

AEC	4107
AEC	AE4107
ALCO	1710
ALCO	4107
ALCO	J4107
ALLOY	1710
ALLOY	J4107
BALKAMP	M4016
BORG-WARNR	114-4016
BORG-WARNR	114-4032
BORG-WARNR	114-4107
BORG-WARNR	114-4107AB
BORG-WARNR	J4016
CLEVE.MOT.	CM4016
CLEVE.MOT.	CM4107
CMP	CM4016
CMP	CM4107
D&D MACH.	MM4107
EXCEL	U222
FLEETRITE	ZAD14016
GLAENZER	065-242-8
GMB	G4107
GMB	G4109
GMB	G5-4016X
GMB	G5-4017X
GMB	GU6410
HABERLE	H4016
IHC	69-559R91
IHC	69559R92
IHC	833618R91
IHC	851629R1
LEMPCO	U222
MASTERS	MM4107
MCQUAY-NOR	U02-16
MCQUAY-NOR	U02-16
MOPAR	D29
MOTOR MAST	4000S
NEAPCO	2-2800
NEAPCO	282800
PERFECTION	RG4016
PILOT	14016
PILOT	14-16
PRECISION	546
REPCO	114-4107
REPCO	RUJ4002
REPUBLIC	CB4016
RUEDARSA	1815
RUEDARSA	2100-1815-0-0
SPICER	5-4016X
SPICER	5-4106X
TEDDY TORQ	1416
TEDDY TORQ	G4107
TRU-CROSS	4016
TRW	20077
UJS	55-4107
WESCO	301416
WESCO	N4016

ZELLER	4000S

IDLI GROUP 80160

CROSS + 4 SERRATED BEARINGS + THRUST AND LOCK PLATES
SEE FIGURE 1C

1.2656 in	BD	32.15 mm
.8906 in	BW	22.62 mm
4.0313 in	CL	102.40 mm
.8281 in	CD	21.03 mm

AEC	AE-D96-55-4
AEC	D96-55
ALCO	D96-55
ALCO	JD96-55
ALCO	JD96-55-4
BALKAMP	CD9655
BORG-WARNR	114-135
BORG-WARNR	D96-55
CHRYSLER	1243615
CHRYSLER	1450504
CLEVE.STL.	D96-55
CLEVE.STL.	D96-55-4
D&D MACH.	MCD96
GLAENZER	065-218-8
GMB	G5-1231X
GMB	GD96-55
GMB	GD96-55-4
GMB	GU2870
LEMPCO	U303
LOBRO	U287
MIPER	C3466
MOTOR MAST	4403A
NEAPCO	285200
PERFECTION	RD96-55
PILOT	D19655
PRECISION	417
REPCO	D96-55
REPUBLIC	CB-D96
SPICER	D96-55
TEDDY TORQ	D96
TEDDY TORQ	GD96-55
WESCO	301965
WESCO	ND965
WESCO	ND965-4

IDLI GROUP 80161

CROSS + 4 ROUND BEARINGS + OUTSIDE SNAP RINGS
SEE FIGURE 1A

1.5313 in	BD	38.90 mm
.8125 in	BW	20.64 mm
4.0313 in	CL	102.40 mm
1.0781 in	CD	27.38 mm

AEC	55N
AEC	AE-CP55N
AEC	AECP58N
AEC	CP55N
A.E.LEE	3881-5
ALCO	1785
ALLOY	1785
ALLOY	CP55N
BEAN,JOHN	C1400
BORG-WARNR	114-150
BRADY	27106
CLEVE.MOT.	CM-CP55N
CMP	CM-CP55N

IDLI GROUP (continued)

COLES	1550819
EATON	6950840
ELG.SWEEP.	63626
FLEETRITE	ZAD55N
FOX RIVER	F4CX8B
FOX RIVER	F4CX8BW
G & G MFG	302-5500
GLAENZER	065-263-8
GMB	GGPL35N
GMB	GGPL55N
GMB	GGPL55R
GORMAN RUP	S1891
HUB CITY	03-35-90074
IHC	179-867H1
IHC	470-077R91
IHC	968294R91
MASS.FERG.	1092-939M91
MASS.FERG.	195-379M91
MASS.FERG.	841-235M91
MATSUI	UJ2855
MCQUAY-NOR	U03-45
MOTOR MAST	4502A
NATL.MINE	76012111
NATL.MINE	P1005
NEAPCO	2855N
NEAPCO	28CP55N
NEAPCO	3-0045
NEAPCO	3-45
PERFECT-CI	432
PILOT	B55N
PRECISION	889
REPCO	CP55N
REPUBLIC	CB55N
SIDEWINDER	11762
SPICER	5-432X
TRU-CROSS	432
TRW	20204
WESCO	391055
WESCO	55N
WESCO	N55
WEST.LAND	AS-US90P
ZELLER	4575E

IDLI GROUP 80162

CROSS + 4 GROOVED BEARINGS + INSIDE SNAP RINGS
SEE FIGURE 1B

1.3125 in	BD	33.34 mm
1.0 in	BW	25.40 mm
4.25 0 in	CL	107.95 mm
.8125 in	CD	20.64 mm

AEC	5109
AEC	AE5109
ALCO	1729
ALCO	J5109
ALLIS-CHLM	084537
ALLOY	1729
ALLOY	J5109
BORG-WARNR	114-5109
BORG-WARNR	114-5141
BORG-WARNR	114-5144
BORG-WARNR	114-5184
CLEVE.MOT.	CM5109
FIAT-ALLIS	0084537
FIAT-ALLIS	0084537-0
FIAT-ALLIS	084537
FIAT-ALLIS	70084537
GMB	G5109
GMB	G5-5109
GMB	G5-5109X
GMC	2378764
INBRA	CT11-055
MOPAR	D32
MOTOR MAST	4025H
NEAPCO	285109
NEAPCO	2858
NEAPCO	2858N
PILOT	15109
PRECISION	958
REPCO	114-5109
REPUBLIC	CB758
TRW	20085
TRW	20088D
WESCO	301509
WESCO	N5109
ZELLER	4025H

IDLI GROUP 80163

CROSS + 4 LOW WING DRILLED BEARINGS
SEE FIGURE 1G

1.6875 in	CC	42.86 mm
4.25 0 in	CL	107.95 mm
.8125 in	CD	20.64 mm

AEC	5101
AEC	AE5121
ATHEY PROD	P12558
AUST.WEST.	465072
AUST.WEST.	465077
AUST.WEST.	AW465072
AUST.WEST.	AW465077
AUST.WEST.	AW465209
AUST.WEST.	AW465298
BALKAMP	M5006
BALKAMP	M5043
BALKAMP	M5101
BAL-LM-HAM	465072
BAL-LM-HAM	465077
BORG-WARNR	114-5127A
CATERPILLR	7J5245
CATERPILLR	8L6301
CLARK EQU.	5202100
CLARK EQU.	5202224
CLARK EQU.	952369
CLEVE.MOT.	CM5006
CLEVE.MOT.	CM5101
CLEVE.MOT.	CM5127
CMP	CM5006
CMP	CM5101
CMP	CM5127
DAVEY COM.	A39087
EATON	6952664
EIMCO	103M530
ETNYRE	644007
FIAT-ALLIS	A45060
FLEETRITE	ZAD15006
FLEETRITE	ZAD15127
GMB	G1501
GMB	G5006
GMB	G5127
GMB	G5130
GMB	G5-5121X

CAUTION: BE SURE TO REFER TO ENGINEERING CATALOGS FOR SPECIAL APPLICATIONS THAT REQUIRE SPECIFIC MATERIAL CONTENT, TOLERANCES, ETC. SEE FOOTNOTE.

HYSTER	225538
IHC	317-713R92
JEEP	924190
MCQUAY-NOR	U03-27
MCQUAY-NOR	UO3-27
MOTOR MAST	3575A
MPL.MOLINE	2DP3281
M.R.S.	13749
NEAPCO	285127
NEAPCO	4-5127
NEW HOLL.	29057
PETTI-MULL	A28451-1
PILOT	15127
PRECISION	990
REPCO	5-1207X
REPCO	5-1214X
REPUBLIC	CB790
SPICER	5-5121X
TRU-CROSS	5121
TRW	20068D
UJS	55-5101
WABCO	VJ6881

IDLI GROUP 80164

CROSS + 4 LOW WING DRILLED BEARINGS
slight variation from above group
SEE FIGURE 1G

1.6875 in	CC	42.86 mm
4.25 0 in	CL	107.95 mm
.8125 in	CD	20.64 mm

AUST.WEST.	465209
CLARK EQU.	5202169
FIAT-ALLIS	0086752
FIAT-ALLIS	0086752-3
FIAT-ALLIS	086752
FIAT-ALLIS	086752-3
FIAT-ALLIS	70086752
FIAT-ALLIS	70086752-3
GLAENZER	065-161-8
GLAENZER	065-174-8
GMB	G5122
HYSTER	53075A
INBRA	CT11-075
MCQUAY-NOR	U03-27
PRECISION	991
REPUBLIC	CB791

IDLI GROUP 80165

CROSS + 4 LOW WING DRILLED BEARINGS
slight variation from above group
SEE FIGURE 1G

1.6875 in	CC	42.86 mm
4.25 0 in	CL	107.95 mm
.8125 in	CD	20.64 mm

AUST.WEST.	465208
CLARK EQU.	5202168
FIAT-ALLIS	0086747
FIAT-ALLIS	0086747-3
FIAT-ALLIS	086747

FIAT-ALLIS	086747-3
FIAT-ALLIS	70086747
FIAT-ALLIS	70086747-3
GLAENZER	065-162-8
MCQUAY-NOR	U03-16
PRECISION	992
REPUBLIC	CB792

IDLI GROUP 80166

CROSS + 4 LOW WING DRILLED BEARINGS
slight variation from above group
SEE FIGURE 1G

1.6875 in	CC	42.86 mm
4.25 0 in	CL	107.95 mm
.8125 in	CD	20.64 mm

GLAENZER	065-169-8
GMB	G5-5031
PRECISION	993
REPCO	114-5031
REPUBLIC	CB793
WABCO	474524

IDLI GROUP 80167

CROSS AND LOW OR HIGH WING TYPE DRILLED OR THREADED BEARINGS
SEE FIGURE 1IJ

1.6875 in	CC	42.86 mm
4.25 0 in	CL	107.95 mm
.8125 in	CD	20.64 mm

AEC	5101
AEC	AE5101
AEC	AE5127
ALCO	1721
ALCO	J5101
ALCO	J5121
ALLIS-CHLM	06809982
ALLIS-CHLM	08-000-505-01-443
ALLIS-CHLM	08-000-505-61-443
ALLIS-CHLM	086752
ALLIS-CHLM	58-000-505-01-414
ALLIS-CHLM	A45060
ALLOY	1721
ALLOY	J5101
BALKAMP	M5006
BALKAMP	M5043
BALKAMP	M5101
BORG-WARNR	114-2147
BORG-WARNR	114-2183
BORG-WARNR	114-5006
BORG-WARNR	114-5020
BORG-WARNR	114-5043
BORG-WARNR	114-5101
BORG-WARNR	114-5119
BORG-WARNR	114-5121
BORG-WARNR	114-5127
BORG-WARNR	114-5142
BORG-WARNR	114-5147
CLEVE.MOT.	CM5101
CMP	CM5101
D&D MACH.	MM5101
DAVEY COM.	39087
EXCEL	U222
FIAT-ALLIS	0680998-2
FIAT-ALLIS	08-000-505-01-443

FIAT-ALLIS	58-000-505-01-414
FIAT-ALLIS	70680998
FLEETRITE	ZAD15127
GMB	G1501
GMB	G5006
GMB	G5101
GMB	G5105
GMB	G5-5121X
GMB	G5-5127
LEMPCO	U222
MASTERS	MM5101
MCQUAY-NOR	U03-27
MOTOR MAST	4025FA
M.R.S.	13749
NEAPCO	285101
NEAPCO	285127
NEAPCO	2890
NEAPCO	4-5127
PETTI-MULL	LL773-77
PILOT	15006
PILOT	15127
PRECISION	552
REPCO	114-5101
REPCO	114-5121
REPCO	114-5127
REPUBLIC	CB790
SPICER	5-5121X
TRW	20084
WABCO	2053551
WABCO	2055473
WESCO	301501
WESCO	301-510
WESCO	N5101
ZELLER	4025FA

IDLI GROUP 80168

CROSS AND LOW OR HIGH WING TYPE DRILLED OR THREADED BEARINGS
SEE FIGURE 1J

1.6875 in	CC	42.86 mm
4.25 0 in	CL	107.95 mm
.8125 in	CD	20.64 mm

ADAMS-LET.	473609
AEC	5100
AEC	AE5100
ALCO	1720
ALCO	5100
ALCO	J5100
ALCO	J5118
ALCO	J5148
ALCO	JR5100
ALLIS-CHLM	086747
ALLIS-CHLM	086747-3
ALLIS-CHLM	086752
ALLIS-CHLM	086752-3
ALLIS-CHLM	179502
ALLIS-CHLM	A45070
ALLOY	1720
ALLOY	J5100
AUST.WEST.	AW465066
AUST.WEST.	AW465067
AUST.WEST.	AW465068
AUST.WEST.	AW465071
AUST.WEST.	AW465073
AUST.WEST.	AW465076
AUST.WEST.	AW465208

AUST.WEST.	AW465217
AUST.WEST.	WFR2885
AUST.WEST.	WRF2885
BALKAMP	3-920
BALKAMP	3-991
BALKAMP	3-992
BALKAMP	M5000
BALKAMP	M5003
BALKAMP	M5011
BALKAMP	M5014
BALKAMP	M5024
BALKAMP	M5037
BALKAMP	M5100
BALKAMP	M5120
BALKAMP	M5122
BALKAMP	M5124
BAL-LM-HAM	AW465205
BEND.WEST.	2053956
BORG-WARNR	114-5000
BORG-WARNR	114-5003
BORG-WARNR	114-5011
BORG-WARNR	114-5014
BORG-WARNR	114-5024
BORG-WARNR	114-5055
BORG-WARNR	114-5100
BORG-WARNR	114-5100A
BORG-WARNR	114-5118
BORG-WARNR	114-5120
BORG-WARNR	114-5122
BORG-WARNR	114-5123
BORG-WARNR	114-5124
BORG-WARNR	114-5128
BORG-WARNR	114-5137
BORG-WARNR	114-5148
BORG-WARNR	114-5178
BORG-WARNR	J5000
CASE,J.I.	L18383
CATERPILLR	7J5241
CATERPILLR	942737
CLARK EQU.	SA35043
CLARK EQU.	SA5043
CLEVE.MOT.	CM5000
CLEVE.MOT.	CM5100
CLEVE.MOT.	CM5105
CMP	CM5000
CMP	CM5100
CMP	CM5122
D&D MACH.	MM5100
DAVEY COM.	M24616
EATON	576207
EATON	579304
EATON	804785
EXCEL	U224
FIAT-ALLIS	0681014-7
FIAT-ALLIS	70076456
FIAT-ALLIS	70680996
FIAT-ALLIS	70681014
FIAT-ALLIS	976456
FIAT-ALLIS	A45070
FLEETRITE	ZAD15000
FLEETRITE	ZAD15122
GARD.DENV.	107898
GERLINGER	67131
GMB	G5100
GMB	G5121
GMB	G5122

BD = BEARING DIAMETER (outside)   BW = BEARING WIDTH   CB = CROSS LENGTH WITH BEARINGS   CC = CENTER TO CENTER   CD = CROSS DIAMETER
CL = CROSS LENGTH WITHOUT BEARINGS   EC = END TO CENTER or FACE TO FACE   EE = END TO END   EL = EFFECTIVE LENGTH
HD = HUB DIAMETER (or insert)   IE = INSIDE OF EARS (or recess)   OD = OUTSIDE DIAMETER   OE = OUTSIDE OF EARS   SB = SPLINE OR BORE SIZE

Column 1

GMB	G5148
GMB	G5-5000X
GMB	G5-5122
GMB	G5-5148X
GMB	GU6500
GRAIGG CO.	LA205
HABERLE	H5000
HARNISCH.	1022Z487
HYSTER	T462-295
IHC	136488H
IHC	159900
IHC	296596R91
INBRA	CC5000
INBRA	CT20-515
IOWA MFG.	7329-026
JOY MFG.	A26294A
LEMPCO	U224
LOBRO	U650
MASTERS	MM5100
MCQUAY-NOR	U03-23
MICHIGAN	943235
MICHIGAN	SA35043
MIXERMOBL	061272
MIXERMOBL	061282
MOPAR	D30
MOTOR MAST	4025A
MOTOR MAST	4025AA
M.R.S.	13411P11
M.R.S.	14029P11
NEAPCO	2820
NEAPCO	283000
NEAPCO	2835
NEAPCO	285112
NEAPCO	285122
NEAPCO	2891
NEAPCO	2892
NEAPCO	4-5122
OSHKOSH	7DS19
PERFECTION	RG5000
PETTI-MULL	L1663-6
PETTI-MULL	LA425-4
PETTI-MULL	LL773-87
PETTI-MULL	LL773-88
PILOT	15000
PILOT	15003
PILOT	15011
PILOT	15014
PILOT	15122
PRECISION	550
REPCO	114-5100
REPCO	114-5122
REPCO	114-5123
REPCO	114-5124
REPCO	114-5128
REPUBLIC	CB5000
REPUBLIC	CB5014
REPUBLIC	CB791
REPUBLIC	CB792
SIL.HOIST	25991-24
SPICER	5-5000X
TRANSMISS.	UK511
TRU-CROSS	5000
TRW	20080
TRW	20088D
UJS	55-5100
WABCO	176771

Column 2

WABCO	186794
WABCO	473609
WESCO	301-519
WESCO	601500
WESCO	C5000
WESCO	N5000
WHITE	02-7074107
WHITE FARM	155-277AS
ZELLER	4025A

IDLI GROUP 80169

CROSS AND LOW OR HIGH WING TYPE DRILLED OR THREADED BEARINGS
SEE FIGURE 1I

1.6875 in	CC	42.86 mm
4.25 0 in	CL	107.95 mm
.8125 in	CD	20.64 mm

AEC	5105
AEC	AE5105
BORG-WARNR	114-5105
BORG-WARNR	114-5148
BORG-WARNR	114-5148A
CLEVE.MOT.	CM5105
GMB	G5148
GMB	G5-5148
GMB	G5-5148X
IHC	338-923R92
MOTOR MAST	4025J
NEAPCO	285148
PRECISION	999
REPCO	114-5105
REPCO	114-5148
REPUBLIC	CB799
TRW	20086
WESCO	301519
WESCO	N5148

IDLI GROUP 80170

CROSS AND LOW OR HIGH WING TYPE DRILLED OR THREADED BEARINGS
SEE FIGURE 1IL

1.6875 in	CC	42.86 mm
4.25 0 in	CL	107.95 mm
.8125 in	CD	20.64 mm

AEC	5139
AEC	AE5139
ALCO	J5139
BORG-WARNR	114-5125
BORG-WARNR	114-5130
BORG-WARNR	114-5139
BORG-WARNR	114-5139A
BORG-WARNR	114-5149
BORG-WARNR	114-5173
BORG-WARNR	114-5176
BORG-WARNR	114-5243
BORG-WARNR	114-558
CLARK EQU.	1990418
CLEVE.MOT.	CM5102
GMB	G5139
GMB	G5-5139X
IHC	333-832R91
IHC	33832R91
MASTERS	MM5139
MCQUAY-NOR	CLARK
MOTOR MAST	4025U

Column 3

NEAPCO	2833
NEAPCO	285102
PRECISION	558
SPICER	5-5802X
WESCO	301502
WESCO	301539
WESCO	N5102
WESCO	N5139

IDLI GROUP 80171

CROSS + 2 LOW WING THREADED + 2 BLOCK BEARINGS
SEE FIGURE 1JL

1.6875 in	CC	42.86 mm
4.25 0 in	CL	107.95 mm
.8125 in	CD	20.64 mm

AEC	5031
AEC	5102
AEC	AE5031
AEC	AE5102
ALCO	1722
ALCO	1724
ALCO	J5031
ALCO	J5102
ALCO	J5125
ALCO	J5130
ALCO	J5149
ALLOY	1722
ALLOY	1724
ALLOY	J5031
ALLOY	J5102
ALLOY	J5130
BEND.WEST.	474524
BORG-WARNR	114-5031
BORG-WARNR	114-5102
BORG-WARNR	114-5125
BORG-WARNR	114-5125A
BORG-WARNR	114-5130
BORG-WARNR	114-5130A
BORG-WARNR	114-5149
BORG-WARNR	114-5173
BORG-WARNR	114-5176
CASE,J.I.	A18382
CASE,J.I.	A40200
CASE,J.I.	D73644
CLEVE.MOT.	CM5031
CLEVE.MOT.	CM5102
D&D MACH.	MM5102
EATON	576207
EATON	579304
FLEETRITE	ZAD15173
FWD	204363
GERLINGER	67131
GLAENZER	065-126-8
GMB	G5031
GMB	G5102
GMB	G5130
GMB	G5149
GMB	G5-5173
GMB	G5-5173X
GMC	9038734
GMC	9052139
IHC	279-986R91
IHC	317-712R92
IHC	401159R91
INBRA	CT10-010

Column 4

INBRA	CT20-510
INBRA	CT23-000
JOY MFG.	531676-1
LINK BELT	AC4640(4)
MASS.FERG.	1094-925M91
MIXERMOBL	060774
MIXERMOBL	061230
MIXERMOBL	061272
MIXERMOBL	061282
MIXERMOBL	118365
MOTOR MAST	4025E
MOTOR MAST	4025EX
NEAPCO	285031
NEAPCO	285102
NEAPCO	285173
NEAPCO	2893
NEAPCO	4-5173
NEAPCO	5-5031
PARSONS	603D565(4)
PARSONS	603D701(4)
PILOT	15173
PRECISION	933
REPCO	114-5102
REPCO	114-5125
REPCO	114-5130
REPCO	114-5173
REPCO	5-1214X
REPUBLIC	CB733
REPUBLIC	CB793
SIL.HOIST	25991-1(4)
SPICER	5-5121X
TAYLOR	1927R1(4)
TAYLOR	3456-510
TRW	20083
TRW	20169
VOLVO	6.631.371
WABCO	2063824
WABCO	3000297
WABCO	3000842
WABCO	474524
WABCO	475397
WESCO	301-502
WESCO	301531
WESCO	N5031
WESCO	N5102
WHITING	12190
ZELLER	4025E
ZELLER	4025EX

IDLI GROUP 80172

CROSS WITH HOLE IN THE CENTER + 2 HIGH WING DRILLED + 2 BLOCK BRGS.
SEE FIGURE 1L

1.6875 in	CC	42.86 mm
4.25 0 in	CL	107.95 mm
.8125 in	CD	20.64 mm

AEC	5108
AEC	AE5108
ALCO	1728
ALCO	J5108
ALCO	J5146
ALCO	J6108
ALLIS-CHLM	179704
ALLIS-CHLM	179909
ALLIS-CHLM	667467

ENGINEERING CATALOGS MUST BE CONSULTED FOR DETAILS NOT INCLUDED
IN THIS GUIDE. SPECIFIC DESIGNS, MATERIAL CONTENT, TOLERANCES
LUBE FITTINGS AND OTHER DIMENSIONS ARE INTENTIONALLY OMITTED HERE.

IDLI GUIDE　　COPYRIGHT ® INTERCHANGE, INC.　　ST. LOUIS PARK, MN. 55416 USA

CAUTION: BE SURE TO REFER TO ENGINEERING CATALOGS FOR SPECIAL APPLICATIONS THAT REQUIRE SPECIFIC MATERIAL CONTENT, TOLERANCES, ETC. SEE FOOTNOTE.

Manufacturer	Part Number
ALLIS-CHLM	682580
ALLOY	1728
ALLOY	J5108
ATHEY PROD	P32225
AUST.WEST.	945208
BORG-WARNR	114-5108
BORG-WARNR	114-5117
BORG-WARNR	114-5126
BORG-WARNR	114-5129
BORG-WARNR	114-5146
BORG-WARNR	114-5163
BORG-WARNR	114-5174
BORG-WARNR	114-5177
BORG-WARNR	114-5180
CASE,J.I.	A45595
CASE,J.I.	N6168
CATERPILLR	1J843
CATERPILLR	2K3631
CATERPILLR	5T1362
CATERPILLR	8D0537
CATERPILLR	8D537
CHAIN BELT	298-6056-91
CLARK EQU.	1990953
CLARK EQU.	5291101
CLARK EQU.	942737
CLARK EQU.	945208
CLARK EQU.	945308
CLEVE.MOT.	CM5108
CMP	CM5108
CMP	CM5177
D&D MACH.	MM5108
DEERE,JOHN	AT26446
DEERE,JOHN	AT33511
EATON	2950025
EATON	6950025
FIAT-ALLIS	70667467
FIAT-ALLIS	70682580
FLEETRITE	ZAD15177
GLAENZER	065-145-8
GMB	G5108
GMB	G5139
GMB	G5146
GMB	G5-5108X
GMB	G5-5177
GMB	G5-5177X
GMB	GUK013
GMB	GUK05
GMB	GUPA6
HYSTER	159707
IHC	142133C1
IHC	290728R91
IHC	799282R91
IHC	881-908C91
INBRA	CT13-050
INBRA	CT16-020
INBRA	CT24-035
KOMATSU	131-10-46110
KOMATSU	154-66-12100
MASTERS	MM5108
MCQUAY-NOR	U03-24
MCQUAY-NOR	UO3-24
MICHIGAN	942737
MICHIGAN	945208
MOTOR MAST	4025B
MUIR-HILL	252382
NEAPCO	285108
NEAPCO	285148
NEAPCO	285177
NEAPCO	2868
NEAPCO	2899
NEAPCO	4-5177
PETTI-MULL	LA384-8
PILOT	15177
PILOT	15777
PRECISION	968
REPCO	114-5108
REPCO	114-5177
REPUBLIC	CB768
REX CHAIN	298-6056-91
SPICER	5-5177X
TAYLOR	1927R1(4)
TRU-CROSS	5177
TRW	20081
WESCO	301518
WESCO	N5108
ZELLER	4025B

IDLI GROUP 80173

CROSS + 2 WING TYPE BEARINGS + 2 ROUND GROOVED BEARINGS
SEE FIGURE 1BI

1.3125 in	BD	33.34 mm
1.6875 in	CC	42.86 mm
1.0 in	BW	25.40 mm
4.250 in	CL	107.95 mm
.8125 in	CD	20.64 mm

Manufacturer	Part Number
AEC	5110
AEC	AE5110
ALCO	1730
ALCO	5110
ALCO	J5110
ALLOY	1730
ALLOY	J5110
BALKAMP	M5015
BALKAMP	M5033
BALKAMP	M5110
BORG-WARNR	114-5015
BORG-WARNR	114-5033
BORG-WARNR	114-5053
BORG-WARNR	114-5110
BORG-WARNR	114-5110A
BORG-WARNR	114-5110AB
BORG-WARNR	114-5133
BORG-WARNR	114-5136
BORG-WARNR	J5015
CLEVE.MOT.	CM5015
CLEVE.MOT.	CM5110
CLEVE.MOT.	CM5112
CMP	CM5015
CMP	CM5110
CMP	CM5111
D&D MACH.	MM5110
EXCEL	U227
FLEETRITE	ZAD15015
GLAENZER	065-245-8
GMB	G5110
GMB	G5111
GMB	G5112
GMB	G5131
GMB	G5-5015X
GMB	G5-5110
HABERLE	H5015
IHC	238-800R91
IHC	238-801R91
IHC	69-558R91
IHC	69558R92
INBRA	CC5015
LEMPCO	U227
LOBRO	U651
MASTERS	MM5110
MCQUAY-NOR	IHC
MOPAR	D33
MOTOR MAST	4025S
NEAPCO	283400
NEAPCO	285110
PERFECTION	RG5015
PILOT	15015
PRECISION	557
REPCO	114-5110
REPUBLIC	CB5015
SPICER	5-5015X
TRU-CROSS	5015
TRW	20087
UJS	55-5110
WESCO	301515
WESCO	G5015
WESCO	N5015
ZELLER	4025-5
ZELLER	4025S

IDLI GROUP 80174

CROSS + 2 WING TYPE BEARINGS + 2 ROUND GROOVED BEARINGS
SEE FIGURE 1BJ

1.3125 in	BD	33.34 mm
1.6875 in	CC	42.86 mm
1.0 in	BW	25.40 mm
4.250 in	CL	107.95 mm
.8125 in	CD	20.64 mm

Manufacturer	Part Number
AEC	5103
AEC	5132
AEC	AE5103
AEC	AE5132
ALCO	1723
ALCO	5103
ALCO	J5103
ALLIS-CHLM	084545
ALLOY	1723
ALLOY	J5103
BALKAMP	M5008
BALKAMP	M5032
BALKAMP	M5036
BALKAMP	M5103
BORG-WARNR	114-5008
BORG-WARNR	114-5032
BORG-WARNR	114-5036
BORG-WARNR	114-5103
BORG-WARNR	114-5103A
BORG-WARNR	114-5132
BORG-WARNR	114-5132A
BORG-WARNR	114-5134
BORG-WARNR	114-5135
BORG-WARNR	J5008
CASE,J.I.	D31437
CASE,J.I.	D76949
CASE,J.I.	R30068
CLARK EQU.	945387
CLEVE.MOT.	CM5103
CLEVE.MOT.	CM5132
CMP	CM5008
CMP	CM5103
D&D MACH.	MM5103
EXCEL	U226
FIAT-ALLIS	0084545
FIAT-ALLIS	0084545-3
FIAT-ALLIS	084545
FIAT-ALLIS	084545-3
FIAT-ALLIS	70084545
FIAT-ALLIS	70084545-3
FLEETRITE	ZAD15008
FLEETRITE	ZAD15132
GLAENZER	065-166-8
GLAENZER	065-244-8
GMB	G5103
GMB	G5132
GMB	G5-5008X
GMB	G5-5132
GMB	G5-5152
GMB	G6107
GMB	GU6550
HABERLE	H5008
IHC	133508HB
IHC	133508HC
IHC	238-797R91
IHC	238-798R91
IHC	317-722R91
INBRA	CC5008
INBRA	CT20-032
LEMPCO	U226
LINK BELT	17A2250(4)
LINK BELT	17A2251(4)
LINK BELT	17A2252(4)
LINK BELT	17A2253(4)
LOBRO	U655
MASS.FERG.	1021944M91
MASS.FERG.	1029-144M91
MASTERS	MM5103
MCQUAY-NOR	U03-25
MCQUAY-NOR	U03-26
MCQUAY-NOR	UO3-25
MCQUAY-NOR	UO3-26
MICHIGAN	945387
MOPAR	D31
MOTOR MAST	4025D
MOTOR MAST	4025DH
MOTOR MAST	4025HD
NEAPCO	283200
NEAPCO	285132
NEAPCO	2857
NEAPCO	2859
NEAPCO	3-5132
PERFECTION	RG5008
PILOT	15008
PILOT	15132
PRECISION	554
REPCO	114-5103
REPCO	114-5132
REPUBLIC	CB5008
REPUBLIC	CB759

BD = BEARING DIAMETER (outside)　**BW** = BEARING WIDTH　**CB** = CROSS LENGTH WITH BEARINGS　**CC** = CENTER TO CENTER　**CD** = CROSS DIAMETER
CL = CROSS LENGTH WITHOUT BEARINGS　**EC** = END TO CENTER or FACE TO FACE　**EE** = END TO END　**EL** = EFFECTIVE LENGTH
HD = HUB DIAMETER (or insert)　**IE** = INSIDE OF EARS (or recess)　**OD** = OUTSIDE DIAMETER　**OE** = OUTSIDE OF EARS　**SB** = SPLINE OR BORE SIZE

RUEDARSA	1205
RUEDARSA	2100-1205-0-0
SPICER	5-5008X
TRANSMISS.	UK686
TRU-CROSS	5008
TRW	20082
UJS	55-5103
WABCO	VJ5827
WABCO	VJ5949
WABCO	VS0297
WESCO	301508
WESCO	601-647
WESCO	G5008
WESCO	N5008
ZELLER	4025D
ZELLER	4025DH

IDLI GROUP 80175

CROSS + 2 WING TYPE BEARINGS + 2 ROUND GROOVED BEARINGS
SEE FIGURE 1BL

1.3125 in	BD	33.34 mm
1.6875 in	CC	42.86 mm
1.0 in	BW	25.40 mm
4.25 0 in	CL	107.95 mm
.8125 in	CD	20.64 mm

AEC	5111
AEC	5131
AEC	AE111
AEC	AE5111
AEC	AE5131
ALCO	1731
ALCO	1732
ALCO	J5111
ALCO	J5131
ALLOY	1731
ALLOY	1732
ALLOY	J5111
ALLOY	J5131
BORG-WARNR	114-5111
BORG-WARNR	114-5131
BORG-WARNR	114-5131A
BORG-WARNR	114-5140
BORG-WARNR	114-5143
BORG-WARNR	114-5152
BORG-WARNR	114-5154
BORG-WARNR	114-5161
BORG-WARNR	114-5175
CLARK EQU.	945388
CLEVE.MOT.	CM5111
D&D MACH.	MM5111
GLAENZER	065-158-8
GMB	G5111
GMB	G5131
GMB	G5-5154
GMB	G5-5154X
IHC	317-721R91
IHC	317721R92
INBRA	CT20-031
MASS.FERG.	1029-148M91
MASTERS	MM5111
MCQUAY-NOR	U03-28
MCQUAY-NOR	UO3-28
MICHIGAN	945388
MOTOR MAST	4025G
MOTOR MAST	4025GH

NEAPCO	285111
NEAPCO	285154
NEAPCO	2897
PILOT	15154
PRECISION	997
REPCO	114-5111
REPCO	114-5131
REPCO	114-5154
REPCO	5-1210X
REPUBLIC	CB797
SPICER	5-5015X
TRW	20170
WESCO	301532
WESCO	N5131
ZELLER	4025G
ZELLER	4025GH

IDLI GROUP 80176

CROSS + 2 ROUND BEARINGS 2 MIDWING DRILLED BEARINGS + INSIDE AND OUTSIDE SNAP RINGS
SEE FIGURE 1IJ

1.8125 in	CC	46.04 mm
4.25 0 in	CL	107.95 mm
.8125 in	CD	20.64 mm

ALCO	J5114
ATHEY PROD	P12559
BORG-WARNR	114-5004
BORG-WARNR	114-5114
GMB	G5114
MCQUAY-NOR	IND.
PRECISION	551
REPCO	114-5114
WESCO	301514
WESCO	N5114

IDLI GROUP 80177

CROSS + 2 ROUND BEARINGS 2 MIDWING DRILLED BEARINGS + INSIDE AND OUTSIDE SNAP RINGS
SEE FIGURE 1BI

1.3125 in	BD	33.34 mm
1.8125 in	CC	46.04 mm
1.0 in	BW	25.40 mm
4.25 0 in	CL	107.95 mm
.8125 in	CD	20.64 mm

ALCO	5112
ALCO	J5112
BALKAMP	M5007
BORG-WARNR	114-5007
BORG-WARNR	114-5012
BORG-WARNR	114-5037
BORG-WARNR	114-5112
BORG-WARNR	114-5112AB
BORG-WARNR	114-5138
BORG-WARNR	J5007
CLEVE.MOT.	CM5110
D&D MACH.	MM5112
GMB	G5112
GMB	GU6520
HABERLE	H5007
IHC	135669H
IHC	135699H
IHC	138839H
IHC	138839HC
LEMPCO	U225

LOBRO	U652
MOPAR	D34
MOTOR MAST	4025T
NEAPCO	283100
NEAPCO	285112
NEAPCO	3-3100
PERFECTION	RG5007
PILOT	15007
PILOT	15012
PRECISION	553
REPCO	114-5112
REPUBLIC	CB3100X
REPUBLIC	CB5007
RUEDARSA	1816
RUEDARSA	2104-1816-0-0
SPICER	5-5007X
TEDDY TORQ	1457
WESCO	301507
WESCO	G5007
WESCO	N5007

IDLI GROUP 80178

CROSS + 4 BACKPLATE BEARINGS
SEE FIGURE 1E

1.5625 in	BD	39.69 mm
.8906 in	BW	22.62 mm
4.25 0 in	CL	107.95 mm
1.0 in	CD	25.40 mm

ACF BRILL	999791
AEC	515
AEC	AE515
AEC	AE515R
AEC	AE-CP4N
AEC	AE-CP5N
ALCO	1515
ALCO	1757
ALCO	5115
ALCO	J5015
ALCO	J5-115
ALLIS-CHLM	081559
ALLIS-CHLM	13809
ALLIS-CHLM	24508
ALLIS-CHLM	245-8
ALLIS-CHLM	28573
ALLIS-CHLM	28862
ALLIS-CHLM	3013809
ALLIS-CHLM	3013809-3
ALLIS-CHLM	3024508
ALLIS-CHLM	3024508-8
ALLIS-CHLM	3028573
ALLIS-CHLM	3028573-8
ALLIS-CHLM	3028862
ALLIS-CHLM	3028862-5
ALLOY	1757
ALLOY	J5-115
ALLOY	L5-115
AMER.MOTOR	A1103
AMER.PULL.	142A370
AMER.PULL.	4627-3124
AUTOCAR	3DA01172
AUTOCAR	3UU001172
BALKAMP	3-953
BERLIET	943541500
BLAW KNOX	5-15X
BLAW KNOX	734-61
BORG-WARNR	114-5115

BORG-WARNR	114-515
BORG-WARNR	114-515R
BORG-WARNR	114-515X
BORG-WARNR	CP5N
BRD	05-503303
BROCKWAY	139959
BROCKWAY	239046
BROS	SP730B
CATERPILLR	GD115X
CENTRAL	G5-155X
CHAIN BELT	298-6012-91
CHAIN BELT	7226
CHRYSLER	1752325
CLARK EQU.	858190
CLARK EQU.	952371
CLEVE.MOT.	CM5-115X
CLEVE.MOT.	CM5-145X
CLEVE.MOT.	CMCP5N
CLEVE.MOT.	CMCP5N1
CMP	CM5-115X
CMP	CM5-154X
CMP	CM-CP5N1
COLES	1500 SERIES
COLES	S1210AR
DAF	316340
DIAMOND R	10058P2
DIAMOND R	1551P2
DIAMOND R	1695P2
DIAMOND R	3332
DIAMOND R	3333P2
DIAMOND R	7199X
DIAMOND R	928P2
DIAMOND R	938P2
DIAMOND R	BB176041
DIAMOND R	BB7041
DIAMOND T	17195C
DIAMOND T	7199C
DIAMOND T	BB17041
DIAMOND T	BB7041
DIAMOND T	BB7199C
EXCEL	U104
FEDERAL	5Y115
FEDERAL	5Y531
FEDERAL	5Y7126
FIAT-ALLIS	13809
FIAT-ALLIS	24508
FIAT-ALLIS	245-8
FIAT-ALLIS	2508
FIAT-ALLIS	28573
FIAT-ALLIS	28862
FIAT-ALLIS	3013809
FIAT-ALLIS	3013809-3
FIAT-ALLIS	3024508
FIAT-ALLIS	3024508-8
FIAT-ALLIS	3028573
FIAT-ALLIS	3028573-8
FIAT-ALLIS	3028862
FIAT-ALLIS	3028862-5
FIAT-ALLIS	70013809
FIAT-ALLIS	70024508
FIAT-ALLIS	70028573
FIAT-ALLIS	70028862
FIAT-ALLIS	73013809
FIAT-ALLIS	73013809-3
FIAT-ALLIS	73024508
FIAT-ALLIS	73024508-8

CAUTION: BE SURE TO REFER TO ENGINEERING CATALOGS FOR SPECIAL APPLICATIONS THAT REQUIRE SPECIFIC MATERIAL CONTENT, TOLERANCES, ETC. SEE FOOTNOTE.

FIAT-ALLIS ... 73028573
FIAT-ALLIS ... 73028573-8
FIAT-ALLIS ... 73028862
FIAT-ALLIS ... 73028862-5
FLEETRITE ... ZAD11515
FORD ... 7EO7039
FORD ... 7EQ7039
FORD ... A281868
FORD ... A381868
FORD ... A80-7039A
FORD ... A8Q7039A
FORD ... B20-7039A
FORD ... B2Q7039
FORD ... B2Q7039A
FORD ... B2TZ4635A
FORD ... XD500818
FWD ... 45149
FWD ... 58122
FWD ... 58155
FWD ... 58157
GERLINGER ... T8767
GERLINGER ... T9502
GLAENZER ... 065-113-8
GLAENZER ... 565-019-0
GLAENZER ... 565-019-1
GMB ... G5-115
GMB ... G5-115X
GMB ... G5-515
GMB ... GCP35N3
GMB ... GGPL4N
GMB ... GGPL5N
GMB ... GU300
GMB ... GU3000
GMC ... 22223567
GMC ... 2222567
GMC ... 2223547
GMC ... 2223567
GMC ... 2284935
GMC ... 2285935
GMC ... 2322134
GMC ... 2322135
GMC ... 2354324
GMC ... 2670487
GMC ... 608194
GMC ... 707747
GMC ... 9023656
HABERLE ... H1515X
HARDIE ... 142A370
HARDY SPCR ... K5GB127
HARDY SPCR ... K5GB15
HARDY SPCR ... K5LGB176
HERCULES ... 444
HUBER ... 106883
HUBER ... 3177
IHC ... 647869R92
IHC ... 647-869R93
IMCO ... 55-5N
INBRA ... CC515
INBRA ... CT11015
ITALCARDAN ... 50-811-000
JAEGER ... X8569

JEFFREY ... 5-115X
JEFFREY ... CAT410151
JOY MFG. ... 4510015
KOEHRING ... P443
LEMPCO ... U104
LEMPCO ... U127
LEMPCO ... U507
LOBRO ... U300
LOBRO ... U550
MACK ... 201SJ13
MACK ... 201SJ21
MACK ... 201SJ73
MACK ... 206SL19
MACK ... 206SL21
MACK ... 21045115X
MASS.FERG. ... 1091-060M91
MASTERS ... MBCP58
MASTERS ... MS515X
MCQUAY-NOR ... U02-17
MCQUAY-NOR ... UO2-17
MICHIGAN ... 858190
MICHIGAN ... 952371
MIPER ... C3171
MOPAR ... D61
MOTIVE ... 55-115
MOTOR MAST ... 4225A
NEAPCO ... 28055X
NEAPCO ... 2815
NEAPCO ... 2853
NEAPCO ... 28CP5N
NEAPCO ... 3-0055
OSHKOSH ... 15DS68
PERFECTION ... RG515X
PETTI-MULL ... L2618-18
PILOT ... 11515
PILOT ... B50N
PILOT ... CP5N
PITT.VIOL. ... 40.811.000
PRECISION ... 332
PREC.TOR. ... 891
PREC.TOR. ... 915
PREC.TOR. ... 953
REO MOTORS ... 10058P2
REO MOTORS ... 1551P2
REO MOTORS ... 1695P2
REO MOTORS ... 3333P2
REO MOTORS ... 928P2
REO MOTORS ... 938P2
REPCO ... CP5N
REPCO ... K5GB15R
REPUBLIC ... CB1500
REPUBLIC ... CB58N4
REPUBLIC ... CB5N
REPUBLIC ... CB715
REPUBLIC ... CB753
REPUBLIC ... CP715
RUEDARSA ... 1070
RUEDARSA ... 2105-1070-0-0
SAVIEM ... 23232055
SEAGRAVE ... X12861
SPICER ... 1500
SPICER ... 5-115X
SPICER ... 5-15
SPICER ... 5-154X
SPICER ... 5-15X
SPICER ... 5-215X

SPICER ... 5-30X
SPICER ... J5-15
STUDEBAKER ... 914381
TEDDY TORQ ... K15
THEW SHOV. ... P27423
THEW SHOV. ... P443
TRANSMISS. ... UK169
TRU-CROSS ... 115
TRW ... 20068D
TRW ... 20088
TRW ... 20088D
UJS ... 55-115
UNIC ... 105823
VOLVO ... 83453
WESCO ... 601005
WESCO ... 601-055
WESCO ... 601515
WESCO ... B5N
WESCO ... N1515
WHITE ... 02-7074027
WHITE ... 10050P2
WHITE ... 10058P2
WHITE ... 1551P2
WHITE ... 1695P2
WHITE ... 216035
WHITE ... 216035A
WHITE ... 3333P2
WHITE ... 427362
WHITE ... 436552
WHITE ... 7199C
WHITE ... 815814
WHITE ... 815914
WHITE ... 928P2
WHITE ... 938P2
WHITE ... A1103
WHITE ... BB17041
WHITE ... BB7041
WHITE ... BB7199C
WILLYS ... A1103
WYLIE ... 26
ZELLER ... 4225A

IDLI GROUP 80179

CROSS + 4 SLOTTED BEARINGS + THRUST AND LOCK PLATES
SEE FIGURE 1D

1.5625 in	BD	39.69 mm
.8281 in	BW	21.03 mm
4.25 0 in	CL	107.95 mm
1.0625 in	CD	26.99 mm

AEC ... AE-CP58N4
ALCO ... CP58N3
ALLOY ... CP58N4
ALLOY ... CP58NWB4
BALKAMP ... 3-953
BORG-WARNR ... 114-106
BORG-WARNR ... CP58N4
CLEVE.MOT. ... CM-CP58N
DIAMOND R ... 3332P2
FLEETRITE ... ZADB58N
FWD ... 46214
GLAENZER ... 065-261-8
GMB ... GCP58N3
GMC ... 2368954
GMC ... 2378764
GMC ... 2378954

GMC ... 658559
LEMPCO ... U513
MCQUAY-NOR ... U02-17
MOTOR MAST ... 4525B
NEAPCO ... 2858N
NEAPCO ... 3-55
NEAPCO ... 4-1058
PILOT ... B58N
PILOT ... B58N4
PRECISION ... 897
REPCO ... CP58N4
REPUBLIC ... CB58N4
SPICER ... CP58N3
TRW ... 20088D
WESCO ... 55N
ZELLER ... 4525B

IDLI GROUP 80180

CROSS + 4 ROUND BEARINGS + OUTSIDE SNAP RINGS
SEE FIGURE 1A

1.5625 in	BD	39.69 mm
.8906 in	BW	22.62 mm
4.25 0 in	CL	107.95 mm
1.0625 in	CD	26.99 mm

AEC ... 22GB
AEC ... AE-K5GB22
ALFA ROMEO ... 1735-24704
ARMSTRONG ... 5410-165
ARMSTRONG ... 5410-3500
AYRA DUREX ... 1-515-01
BEDFORD ... 7088368
BEDFORD ... 70883687
BEDFORD ... 7088370
BEDFORD ... 7178684
BEDFORD ... 91021384
BERLIET ... 3400021
BERLIET ... 3400040
BMC ... 17H3835
BMC ... 8G3008
BORG-WARNR ... K5LBG176
BRD ... 06-501822
BRD ... 06-510002
CHRYSLER ... 3218363
CHRYSLER ... 5040693
CHRYSLER ... D18363
CHRYSLER ... K22173
CITROEN ... 603710810
DAIMLER ... 944882
DENNIS ... 686704
ERF ... U15616
FIAT-ALLIS ... 4699475
FIAT-ALLIS ... 9929086
FORD ... 508E7039
FORD ... 568E7039
FORD ... 72DB4635BIA
FORD ... 81DB4635DA
GLAENZER ... 565-001-0
GLAENZER ... 565-020-1
GLAENZER ... 565-023-0
GLAENZER ... 565-030-0
GMB ... GU350
GMB ... GU3500
GMC ... 7088368
GMC ... 7156385
GMC ... 7178684

GMC 9961730
HARDY SPCR 5151-176Y
HARDY SPCR K5GB165
HARDY SPCR K5GB22
HARDY SPCR K5LGB176
HOLDEN 7088368
HOLDEN 7156385
HOLDEN 7178684
ITALCARDAN 50-711000
MIPER C3169
MOTOR MAST 4225B
PITT.VIOL. 40.711.000
POCLAIN 00400-38
PRECISION 370
QUINT.HAZ. QL16506
QUINT.HAZ. QL2206
RENAULT 08.70.550.300
RENAULT 23.232.055
RENAULT 50.00.242.492
RENAULT 77.01.005.432
RENAULT 77.01.008.330
REPCO K5GB22R
RUEDARSA 1070E
RUEDARSA 2103-1070-0-0
SAVIEM 7701005432
SAVIEM 870550300
SCANIA 47804
SCANIA 750 610
SCANIA 77187
TRANSMISS. UK169/6
TRANSMISS. UK171
VOLVO 36590
VOLVO 6.621.656
VOLVO 6.638.385
VOLVO 6.852.612
WYLIE 25

IDLI GROUP 80181

CROSS + 4 SERRATED BEARINGS + THRUST AND LOCK PLATES
SEE FIGURE 1C

1.6406 in	BD	41.67 mm
.9219 in	BW	23.42 mm
4.50 in	CL	114.30 mm
1.1406 in	CD	28.97 mm

AEC AE-U96-55
ALCO JU96-55
ALCO JU96-55-4
ALCO U96-55
BALKAMP CU9655
BORG-WARNR 114-136
BORG-WARNR U96-55
CHRYSLER 1450505
CHRYSLER 947555
CLEVE.STL. U96-55
CLEVE.STL. U96-55-4
D&D MACH. MCU96
GLAENZER 065-219-8
GMB G5-1232
GMB G5-1232X
GMB GU96-55
GMB GU96-55-4
LEMPCO U305
LOBRO U270
LOBRO U290
MIPER C3468

MOTOR MAST 4450A
NEAPCO 285300
PERFECTION RU96-55
PILOT U19655
PRECISION 418
REPCO 5-1211X
REPCO U96-55
REPCO U96-55-4
REPUBLIC CB-U96
RUEDARSA 2106-2165-0-0
RUEDARSA 2165
SPICER U96-55
TEDDY TORQ GU96-55
TEDDY TORQ U56
TEDDY TORQ U96
WESCO 601-699
WESCO 601966
WESCO NU965
WESCO NU965-4

IDLI GROUP 80182

CROSS + 4 ROUND BEARINGS + OUTSIDE SNAP RINGS
SEE FIGURE 1A

1.3750 in	BD	34.93 mm
.9375 in	BW	23.81 mm
4.6563 in	CL	118.27 mm
.8906 in	CD	22.62 mm

AEC 555
AEC AE555
AEC AE588
ALBURUS-SP 5-275X
ALCO 1555
ALCO 1756
ALCO 1805
ALCO J5-155
ALCO J5-188
ALLIS-CHLM 255608
ALLOY 1805
ALLOY J5-155
ALLOY J5-188
BAY CITY 1550
BORG-WARNR 114-555
BORG-WARNR 114-588
CASE,J.I. D45166
CHRYSLER 2064313
CHRYSLER 2084313
CHRYSLER 3546576
CHRYSLER 3837585
CLEVE.MOT. CM5-155X
CLEVE.MOT. CM5-158X
CLEVE.MOT. CM5-188X
CMP CM5-155X
CMP CM5-158X
D&D MACH. MS155X
DEERE,JOHN AT27787
DEERE,JOHN AU42404ED
DIAMOND R 10001P2
DIAMOND R 10059P2
EXCEL U120
EXCEL U132
FLEETRITE ZAD11555
FORD CD5AZ4635A
FORD D9HZ4635A
FORD DH4635A
GALION D110676

GALION D123329
GALION . . . . D70654(2)
GALION D74601
GALION . . . . H66746(2)
GALION . . . . H66773(7)
GLAENZER 065-150-8
GMB G5-155
GMB G5-155X
GMB G5-555
GMC 2400938
GMC 9036203
GMC 9053744
IHC 13882R91
IHC 158746
IHC 158-746H1
IHC 178746
IHC 213-882R91
IHC 240-178R91
IHC 240-190R91
IHC 644363C92
IHC 867-635C91
INBRA CT17-035
JEFFREY 433951
JOY MFG. 451072
LEMPCO U122
LEMPCO U132
LOBRO U350
MACK 201SJ31
MACK 21045155Y
MASTERS MS155X
MASTERS MS588X
MCQUAY-NOR U02-18
MCQUAY-NOR U02-9
MCQUAY-NOR U02-18
MIPER C3218
MOPAR D74
MOPAR D83
MOTOR MAST 4265A
NEAPCO 28155G
NEAPCO 28155X
NEAPCO 28188X
NEAPCO 2879
NEAPCO 3-0155
NEAPCO 3 155
NEAPCO 3-188
OSHKOSH 155DS3
PERFECTION RG555
PERFECTION RG555R
PERFECTION RG588R
PILOT 11555
PILOT 11588
PRECISION 358
PREC.TOR. 948
PREC.TOR. 979
QUINT.HAZ. QL155
REO MOTORS 10059P2
REPCO 5-155X
REPCO 5-188X
REPUBLIC CB1555
REPUBLIC CB1888
REPUBLIC CB779
ROCKWELL CP155X
SPICER 1550
SPICER 5-155X
SPICER 5-158X
SPICER 5-188X

SPICER 5-252X
SPICER 5-263X
SPICER 5-275X
STEIGER 17-020
TEDDY TORQ G5-155
TEDDY TORQ K55
TIMBERJACK 405485
TIMBERJACK 902T110
TRU-CROSS 155
TRW 20057
TRW 20089
TRW 20089D
UJS 55-155
VERSATILE 902T110V11
WESCO 301-558
WESCO N1555
WESCO N1588
WHITE 02-7074028
WHITE 10001P2
WHITE 10059P2
WHITE 878916
WYLIE 80
ZELLER 2375D
ZELLER 3289A
ZELLER 4265A

IDLI GROUP 80183

CROSS + 4 SERRATED BEARINGS + THRUST AND LOCK PLATES
SEE FIGURE 1C

1.3750 in	BD	34.93 mm
.8750 in	BW	22.23 mm
4.7188 in	CL	119.86 mm
.8750 in	CD	22.23 mm

AEC 1210
AEC AE-U56-55-2
AEC AE-U56-55-Z
AEC U56-55-2
ALCO JU55-55-2
ALCO JU56-55-2
BORG-WARNR 114-131
BORG-WARNR U56-55-2
CLEVELAND U56-38
CLEVELAND U56-55-2
CLEVE.MOT. CM5655U
CLEVE.STL. U56-55-2
CLEVE.STL. US56-55-2
CMP CM5655U
FLEETRITE ZAD-U15655
GLAENZER 065-227-8
GMB G5-1210X
GMB GU55-55-2
GMB GU56-55-2
IHC 965-621R91
MCQUAY-NOR U02-20
MCQUAY-NOR UO2-20
MOPAR D103
MOTOR MAST 4465A
NEAPCO 286800
NEAPCO 3-6800
PILOT U15655
PRECISION 432
REPCO 5-1210X
REPUBLIC CB-U56
SPICER 5-1210X
TRW 20171

CAUTION: BE SURE TO REFER TO ENGINEERING CATALOGS FOR SPECIAL APPLICATIONS THAT REQUIRE SPECIFIC MATERIAL CONTENT, TOLERANCES, ETC. SEE FOOTNOTE.

UJS	55U56
WESCO	301557
WESCO	NOS555
WESCO	NUS555
ZELLER	4465A

IDLI GROUP 80184

CROSS + 2 LOW WING THREADED + 2 BLOCK BEARINGS
SEE FIGURE 1KL

1.9375 in	BW	49.21 mm
4.75 0 in	CL	120.65 mm
1.0625 in	CD	26.99 mm

AEC	58WB2
AEC	AE-CP58WB2
AEC	CP58WB2
ALCO	1877
ALCO	CP58WB2
ALCO	CP58WB4
ALLIS-CHLM	20-121-157-023
ALLOY	1877
ALLOY	CP58WB2
ALLOY	U512
BEAN,JOHN	1258413
BORG-WARNR	114-109
BORG-WARNR	CP58WB2
CHRYSLER	CTD145519
CLARK EQU.	946840
CLEVE.MOT.	CM-CP-58WB2
D&D MACH.	MBCP582
DEERE,JOHN	AR87394
DEERE,JOHN	AT23096
DEERE,JOHN	AT59126
DEERE,JOHN	AT74268
FLEETRITE	ZADB58WB2
GMB	G5-5802X
GMB	G8206
GMB	G8215
GMB	GCP58WB2
GMC	2041235
GMC	2398295
GMC	667502
LEMPCO	U512
MACK	201SJ51
MACK	35MU211P2
MACK	35MUA211P2
MASTERS	MBCP582
MCQUAY-NOR	U02-22
MCQUAY-NOR	UO2-22
MOTOR MAST	4575C
NEAPCO	2585WB2
NEAPCO	2858WB2
NEAPCO	2858WN2
NEAPCO	4-5802
PILOT	58WB2
PILOT	B58WB2
PILOT	B58WN2
PRECISION	896
PREC.TOR.	496
REPCO	CP58WB2
REPCO	CP58WB37

REPUBLIC	CB58NB2
REPUBLIC	CB58WB2
REPUBLIC	CP58WB2
SPEED-SPR.	1258413
SPICER	5-5802X
TRW	20143
WESCO	601582
WESCO	N5802
WHITE	02-7074111
ZELLER	4575C

IDLI GROUP 80185

CROSS + 4 BLOCK BEARINGS
SEE FIGURE 1L

1.9375 in	BW	49.21 mm
4.75 0 in	CL	120.65 mm
1.0625 in	CD	26.99 mm

AEC	58WB
AEC	AE-CP58WB
AEC	CP58WB
ALCO	CP58WB
ALLIS-CHLM	084435
ALLIS-CHLM	084435-7
ALLIS-CHLM	084545
ALLIS-CHLM	084545-3
ALLOY	1874
ALLOY	1878
ALLOY	CP58WB
BALKAMP	3-959
BALKAMP	3-960
BAR-GREENE	BB114-8515A
BEAN,JOHN	1405632
BEAN,JOHN	2308019
BORG-WARNR	114-107
BORG-WARNR	CP58WB
CASE,J.I.	L35511
CLARK EQU.	892187
CLEVE.MOT.	CM-CP58WB
CMP	CM-CP58WB
D&D MACH.	MBCP58
DEERE,JOHN	AR37096
DEERE,JOHN	AR37097
DEERE,JOHN	AT59127
DEERE,JOHN	AT74269
DEERE,JOHN	K67709
FLEETRITE	ZAD-B58WB
FMC	1405632
FMC	14058287
GLAENZER	065-277-8
GMB	G5-107
GMB	G5-5800X
GMB	GCP58WB
IHC	202-339H1
MASTERS	MBCP58
MCQUAY-NOR	U02-23
MOTOR MAST	4575D
NEAPCO	2858WB
NEAPCO	4-5800
PILOT	B58WB
PILOT	B58WN
PRECISION	894
REPCO	CP58WB
REPCO	CP58WB38
REPUBLIC	CB58WB
REPUBLIC	CP58WB38

SPICER	5-5800X
SPICER	CP58WB
SPICER	CP58WB38
SPICER	CP58WN
TRU-CROSS	5800
TRW	20144
TRW	20144D
WESCO	601581
WESCO	N5800
WESCO	N5801
WHITE	02-7074109
ZELLER	4575D

IDLI GROUP 80186

CROSS + 4 DELTA WING THREADED BOLT HOLE BEARINGS
SEE FIGURE 1K

1.9375 in	BW	49.21 mm
4.75 0 in	CL	120.65 mm
1.0625 in	CD	26.99 mm

AEC	58WB1
AEC	AE-CP58WB1
AEC	CP58WB1
ALCO	1876
ALCO	CP58WB1
ALLOY	1876
ALLOY	CP58WB1
BEAN,JOHN	1269826
BEAN,JOHN	1409233
BEAN,JOHN	1409290
BEAN,JOHN	3A300
BORG-WARNR	114-108
BORG-WARNR	114-180
BORG-WARNR	CP58WB1
CASE,J.I.	D45982
CHRYSLER	3831798
CHRYSLER	CTD145518
CLARK EQU.	5202199
CLARK EQU.	980136
CLARK EQU.	980736
CLEVE.MOT.	CM-CP58WB1
CLEVE.MOT.	CM-CP58WB20
D&D MACH.	MBCP581
DEERE,JOHN	AB15225B
FLEETRITE	ZADB58WB1
FORD	D1TZ3249A
GLAENZER	065-260-8
GMB	G5-5201X
GMB	G5-5801X
GMB	GCP58WB1
GMC	2041234
GMC	2041860
GMC	2398294
GMC	667501
IHC	202339
KALAMAZOO	Z2598
KOEHRING	9003889
LEMPCO	U511
MACK	35MU211P1
MACK	35MUA211P1
MASTERS	MBCP581
MCQUAY-NOR	U02-21
MCQUAY-NOR	UO2-21
MOTOR MAST	4575B
NEAPCO	2858WB1
NEAPCO	2858WN1

NEAPCO	4-5801
PILOT	58WB1
PILOT	B58WB1
PILOT	B58WN1
PRECISION	895
PREC.TOR.	495
REPCO	CB58WB1
REPCO	CP58WB1
REPCO	CP58WB29
REPUBLIC	CB58N4
REPUBLIC	CB58WB1
REPUBLIC	CP58WB1
SEAGRAVE	X11999
SPEED-SPR.	1405632
SPEED-SPR.	1409233
SPEED-SPR.	1409290
SPICER	5-5801X
TRW	20142
TRW	20142D
WESCO	601580
WESCO	N5800
WESCO	N5801
WHITE	02-7074110
ZELLER	4575B

IDLI GROUP 80187

CROSS + 4 ROUND BEARINGS + OUTSIDE SNAP RINGS
SEE FIGURE 1A

1.6250 in	BD	41.28 mm
.9688 in	BW	24.61 mm
4.75 0 in	CL	120.65 mm
1.1563 in	CD	29.37 mm

AEC	750
AEC	AE-CP750N1
CASE,J.I.	A18398
CHRYSLER	2505767
EATON	2040856
FIAT-ALLIS	3039559
FIAT-ALLIS	3039559-4
FIAT-ALLIS	39559
FIAT-ALLIS	73039559
FIAT-ALLIS	73039559-4
GMB	G5-750X
GMC	2505767
IHC	192-909R91
IHC	538-395R91
INBRA	CT20-520
MOTOR MAST	4575E
PRECISION	898
REPCO	CP750N1
REPUBLIC	CB750N1
REPUBLIC	CP750N1
WESCO	601675
WESCO	75N

IDLI GROUP 80188

CROSS + 4 SLOTTED BEARINGS + THRUST AND LOCK PLATES
SEE FIGURE 1D

1.9375 in	BD	49.21 mm
1.1094 in	BW	28.18 mm
4.75 0 in	CL	120.65 mm
1.3281 in	CD	33.73 mm

AEC	6N
AEC	AE-CP6N

BD=BEARING DIAMETER (outside) **BW**=BEARING WIDTH **CB**=CROSS LENGTH WITH BEARINGS **CC**=CENTER TO CENTER **CD**=CROSS DIAMETER **CL**=CROSS LENGTH WITHOUT BEARINGS **EC**=END TO CENTER or FACE TO FACE **EE**=END TO END **EL**=EFFECTIVE LENGTH **HD**=HUB DIAMETER (or insert) **IE**=INSIDE OF EARS (or recess) **OD**=OUTSIDE DIAMETER **OE**=OUTSIDE OF EARS **SB**=SPLINE OR BORE SIZE

AEC	CP6N
ALCO	1776
ALCO	CP6N
ALLIS-CHLM	21862
ALLIS-CHLM	3021862
ALLIS-CHLM	3044128
ALLIS-CHLM	44128
ALLOY	1776
ALLOY	CP6N
BALKAMP	CP6N
BORG-WARNR	114-102
BORG-WARNR	CP6N
CATERPILLR	555354
CLARK EQU.	952372
CLARK EQU.	952373
CLEVE.MOT.	CM-CP6N
CMP	CM-CP6N
D&D MACH.	MBCP6N
EXCEL	U508
FEDERAL	5Y114
FIAT-ALLIS	21862
FIAT-ALLIS	3021862
FIAT-ALLIS	3021862-2
FIAT-ALLIS	3044128
FIAT-ALLIS	3044128-1
FIAT-ALLIS	44128
FIAT-ALLIS	70021862
FIAT-ALLIS	70044128
FIAT-ALLIS	73021862
FIAT-ALLIS	73021862-2
FIAT-ALLIS	73044128
FIAT-ALLIS	73044128-1
FLEETRITE	ZADB60N
FWD	58159
GLAENZER	065-258-8
GLAENZER	065-892-6
GMB	GCP6N
GMB	GGPL6N
GMB	GU5800
GMC	9041334
GMC	9053747
HARNISCH.	1018Z203
IHC	661372R91
IMCO	55-6N
INBRA	CT11-010
IOWA MFG.	453750003
KENWORTH	1000K24A
KENWORTH	W292A29
KOEHRING	8882002
LEMPCO	U508
LOBRO	U580
MACK	201SJ-20
MACK	210SJ20
MCQUAY-NOR	MACK
MICHIGAN	952372
MICHIGAN	952373
MIPER	C3190
MOTIVE	55-6N
MOTOR MAST	4575A
MOTOR MAST	B4575
NEAPCO	2865
NEAPCO	28CP6N
NEAPCO	4-0060
NEAPCO	4-60
PERFECTION	RG6N
PILOT	B60N
PILOT	CP6N
PRECISION	892
PREC.TOR.	965
REO MOTORS	815915
REPCO	CP6N
REPUBLIC	CB6N
REPUBLIC	CB765
SPICER	CP6N
THEW SHOV.	P335
THEW SHOV.	P398
TOWMOTOR	L15460
TRANSMISS.	UK190
TRW	20091
WABCO	2053903
WABCO	402631
WABCO	424270
WESCO	611006
WESCO	815915
WESCO	B6N
WHITE	815915
WHITE	843942
ZELLER	4575A

IDLI GROUP 80189

CROSS + 4 BACKPLATE BEARINGS
SEE FIGURE 1E

1.8750 in	BD	47.63 mm	
.9219 in	BW	23.42 mm	
5.0 in	CL	127.00 mm	
1.2813 in	CD	32.55 mm	

ACF BRILL	995800
AEC	279
AEC	516
AEC	AE516
AEC	AE516Z
AEC BRIT.	K2701
ALCO	1516
ALCO	1806
ALCO	1807
ALCO	J5-116
ALFA ROMEO	1735-25704
ALLIS-CHLM	13295
ALLOY	1806
ALLOY	1807
ALLOY	J5-116
ALLOY	J5-279
ARMSTRONG	5410-4000
AUTOCAR	3EA01172
AUTOCAR	3EA1172
AUTOCAR	3UUO01173
AYRA DUREX	1-605-01
BAY CITY	1600
BEDFORD	7206809
BERLIET	943541600
BMC	HYL3873
BORG-WARNR	114-516
BORG-WARNR	114-516R
BORG-WARNR	114-516X
BRD	07-501823
BRD	07-510112
BROCKWAY	139956
BROCKWAY	239049
BROCKWAY	239225
BROCKWAY	239325
BROCKWAY	239326
BUCKEYE	1600
CASE,J.I.	D45162
CENTRAL	G5-116X
CHECKER	817070
CHRYSLER	1752317
CHRYSLER	2629713
CHRYSLER	3276694
CHRYSLER	3820255
CHRYSLER	5041525
CHRYSLER	D19834-7
CHRYSLER	K26110
CHRYSLER	M21005
CLEVE.MOT.	CM5-116X
CLEVE.MOT.	CM5279XGC
CMP	CM5-116X
COLES	1600 SERIES
COLES	S1210AR
D&D MACH.	MS516X
DAIMLER	944887
DARLINGTON	813-850
DEERE,JOHN	32075201
DENNIS	1305151
DIAMOND R	10045P2
DIAMOND R	10060P2
DIAMOND R	1039P2
DIAMOND R	3620P2
DIAMOND R	860P2
DIAMOND T	218
DIAMOND T	7199F
DIAMOND T	BB17042
DIAMOND T	BB1742
DIAMOND T	BB1799F
DIAMOND T	BB7042
DIAMOND T	BB7043
DIAMOND T	BB7199F
EIMCO	105M671
EIMCO	1600
ERF	U11803
EXCEL	U105
FEDERAL	5Y530
FEDERAL	5Y7127
FIAT-ALLIS	4699481
FLEETRITE	ZAD11516
FODENS	Y00601312
FORD	2002F4635A
FORD	C6TZ4635
FORD	C6TZ4635G
FORD	C6TZ4635L
FORD	D3TZ4635A
FWD	2A5-116X
GALION	D80481
GALION	D80485(2)
GLAENZER	065-114-8
GLAENZER	665-000-0
GLAENZER	665-002-0
GLAENZER	665-015-2
GLAENZER	665-019-0
GMB	G5-116
GMB	G5-116X
GMB	G5-279X
GMB	G5-516
GMB	GU4000
GMC	21296
GMC	2173956
GMC	2222584
GMC	2222854
GMC	2232136
GMC	2322136
GMC	695909
GMC	695990
GMC	7206809
GMC	9385452
GMC	9388999
HABERLE	H1516X
HARDY SPCR	6151-001Y
HARDY SPCR	K5-16
HARDY SPCR	K5GB16
HARDY SPCR	K5GB160
HARDY SPCR	K5GB65
HERCULES	445
HOLDEN	7206809
IHC	121684R91
IHC	121-684R92
IHC	21584R92
IHC	21684R92
IHC	697-689R91
IHC	916650R91
INBRA	CC516
INBRA	CT17-000
ITALCARDAN	50-712-000
JAMCO	UJ279
JEEP	7347665
JEFFREY	1601
JEFFREY	1608
JOY MFG.	1426418K3
JOY MFG.	A262457
JOY MFG.	A262467
JOY MFG.	A365789
LEADER	T10404
LEMPCO	U105
LEYLAND	944887
LOBRO	U400
LOBRO	U405
MACK	201SJ14
MACK	201SJ14A
MACK	201SJ25
MACK	2104-5279X
MASTERS	MS516X
MCQUAY-NOR	U03-42
MCQUAY-NOR	UO3-42
MIPER	C3176
MOPAR	D62
MOTOR MAST	5200A
MOTOR MAST	5200AG
NEAPCO	111G3
NEAPCO	28058X
NEAPCO	2816
NEAPCO	28278X
NEAPCO	28279X
NEAPCO	4-0058
NEAPCO	4-0279
NEAPCO	4-279
OSHKOSH	16DS150
PERFECTION	RG516X
PETTI-MULL	LA301-1
PETTI-MULL	LL240-14
PETTI-MULL	LL240-22
PETTI-MULL	LL240-23
PETTI-MULL	LL240-8
PILOT	11516
PITT.VIOL.	40.712.000
PRECISION	333
PREC.TOR.	363

CAUTION: BE SURE TO REFER TO ENGINEERING CATALOGS FOR SPECIAL APPLICATIONS THAT REQUIRE SPECIFIC MATERIAL CONTENT, TOLERANCES, ETC. SEE FOOTNOTE.

PREC.TOR.	916
QUINT.HAZ.	QL16007
QUINT.HAZ.	QL1607
REO MOTORS	1039P2
REO MOTORS	860P2
REPCO	5-279X
REPCO	K5-16R
REPCO	K5279X
REPCO	K5GB65
REPUBLIC	CB1600
REPUBLIC	CB1600AG
REPUBLIC	CB1600HD
REPUBLIC	CB716
RUEDARSA	1120
RUEDARSA	1120-5
RUEDARSA	2105-1120-0-0
RUEDARSA	2105-1120-5-0
SAVIEM	23232057
SCANIA	78060
SPICER	1600
SPICER	1610
SPICER	5-116X
SPICER	5-16
SPICER	5-16X
SPICER	5-216X
SPICER	5-279-2X
SPICER	5-279X
SPICER	5-30X
SPICER	5-31X
SPICER	J5-16
TEDDY TORQ	G5-116
TEDDY TORQ	K16
TIMBERJACK	902T111
TRANSMISS.	UK176
TRU-CROSS	279
TRW	20104
TRW	20134
UJS	55-116
UNIC	150822
VERSATILE	902T111V1
VOLVO	46240
WESCO	611516
WESCO	G1516
WESCO	N1516
WHITE	02-7012321
WHITE	02-7074029
WHITE	02-7074361
WHITE	1039P2
WHITE	216036
WHITE	216036A
WHITE	218
WHITE	3620P2
WHITE	406607
WHITE	7199F
WHITE	839272
WHITE	860P2
WHITE	BB17042
WHITE	BB7042
WHITE	BB7199F
WILLYS	7347665
WYLIE	27

ZELLER	5200A
ZELLER	5200AG

IDLI GROUP 80190
CROSS + 4 ROUND BEARINGS + OUTSIDE SNAP RINGS
SEE FIGURE 1A

1.6250 in	BD	41.28 mm
1.1094 in	BW	28.18 mm
5.1875 in	CL	131.76 mm
1.0469 in	CD	26.59 mm

AEC	565
AEC	AE565
AEC	AE56E
ALBURUS-SP	5-165X
ALCO	1565
ALCO	1801
ALCO	J5-165
ALLIS-CHLM	2403229
ALLIS-CHLM	6824381
ALLOY	1801
ALLOY	J5-165
BAY CITY	1650
BORG-WARNR	114-565
BROCKWAY	239131
CHRYSLER	2295043
CHRYSLER	2495043
CLEVE.MOT.	CM5-165X
CMP	CM5-165X
D&D MACH.	MS165X
DEERE,JOHN	AH81534
DIAMOND R	10061P2
DIAMOND R	883048P2
EXCEL	U133
FIAT-ALLIS	2403229
FIAT-ALLIS	6824381
FLEETRITE	ZAD11565
FLXIBLE	13-102-4
FORD	C6TZ4635E
FORD	C6TZ4635J
GALION	D78942
GLAENZER	065-151-8
GMB	G5-165
GMB	G5-165X
GMB	G5-565
GMC	2403229
GMC	2417789
GMC	4016929
HOWE	3785
HYSTER	87984
IHC	104748
IHC	644-137C91
IHC	751-762C91
IHC	751762R91
IHC	899-392R91
INBRA	CC8200
INBRA	CT17-040
JEFFREY	1651
JEFFREY	1658
LEMPCO	U133
MACK	201SJ33
MACK	2104-5165X
MACK	21045299X
MASTERS	MS165X
MCQUAY-NOR	U02-29
MCQUAY-NOR	U02-29

MOPAR	D77
MOTOR MAST	5218A
NEAPCO	28165X
NEAPCO	2880
NEAPCO	4-0165
NEAPCO	4-165
OSHKOSH	165DS4
PILOT	11565
PRECISION	359
PREC.TOR.	980
REO MOTORS	10061P2
REPCO	5-165X
REPUBLIC	CB1650HD
REPUBLIC	CB1655
REPUBLIC	CB780
SPICER	1650
SPICER	5-165-4CX
SPICER	5-165X
SPICER	5-168X
SPICER	5-241X
SPICER	5-299X
STEIGER	17-056
TEDDY TORQ	G5-165
TEDDY TORQ	K65
TRU-CROSS	165
TRW	20105
UJS	55-165
WESCO	201-565
WESCO	601-565
WESCO	N1565
WHITE	02-7074030
WHITE	10061P2
WHITE	883048
WHITE	883048P2
WHITE	894362
ZELLER	5218A

IDLI GROUP 80191
CROSS AND LOW OR HIGH WING TYPE DRILLED OR THREADED BEARINGS
SEE FIGURE 1B

1.3125 in	BD	33.34 mm
1.0 in	BW	25.40 mm
5.250 in	CL	133.35 mm
.8125 in	CD	20.64 mm

ALCO	J6123
ALCO	J6140
ALLIS-CHLM	39474
BORG-WARNR	114-6135
BORG-WARNR	114-6140
BORG-WARNR	114-6148
FIAT-ALLIS	39674
FLEETRITE	ZAD16135
GMB	G6123
GMB	G6140
MCQUAY-NOR	IND.
MOTOR MAST	5025S
NEAPCO	286123
NEAPCO	4-6123
PILOT	16135
PRECISION	567
REPCO	114-6123
REPCO	114-6135
WABCO	3000844
WESCO	601612
WESCO	N6123

IDLI GROUP 80192
CROSS AND LOW OR HIGH WING TYPE DRILLED OR THREADED BEARINGS
SEE FIGURE 1I

1.6875 in	CC	42.86 mm
5.250 in	CL	133.35 mm
.8125 in	CD	20.64 mm

AEC	6104
AEC	AE6104
ALCO	1824
ALCO	J6104
ALLIS-CHLM	A386002
ALLOY	1824
ALLOY	J6104
AUST.WEST.	AW465346
BORG-WARNR	114-6104
BORG-WARNR	114-6123
BORG-WARNR	114-6171
CLARK EQU.	5209902
CLARK EQU.	888280
CLEVE.MOT.	CM6104
FIAT-ALLIS	A386002
GMB	G5-6104X
GMB	G6104
IHC	630-386C91
MASTERS	MM6104
MCQUAY-NOR	IND.
MCQUAY-NOR	U02-16
MICHIGAN	888280
MOTOR MAST	5025P
NEAPCO	286104
NEAPCO	4-6104
PILOT	16104
PRECISION	566
REPCO	114-6104
WESCO	601614
WESCO	N6104
ZELLER	5025P

IDLI GROUP 80193
CROSS AND LOW OR HIGH WING TYPE DRILLED OR THREADED BEARINGS
SEE FIGURE 1J

1.6875 in	CC	42.86 mm
5.250 in	CL	133.35 mm
.8125 in	CD	20.64 mm

AEC	6100
AEC	6229
AEC	AE6100
ALCO	1820
ALCO	6100
ALCO	J6100
ALCO	J6141
ALLIS-CHLM	08-048-885-63-001
ALLIS-CHLM	085918
ALLIS-CHLM	085918-1
ALLIS-CHLM	08-630-150-00-502
ALLOY	1820
ALLOY	J6000
ALLOY	J6100
AMER.LAFR.	11B5490
AUST.WEST.	465111
AUST.WEST.	AW465111
AUST.WEST.	PGF6732
AUST.WEST.	PGF6732-56A
AUST.WEST.	PGF8508-56

AUST.WEST.	PGF8585-56
AUST.WEST.	TF1592-56
BALKAMP	3-935
BALKAMP	3-937
BALKAMP	3-960
BALKAMP	M6000
BALKAMP	M6003
BALKAMP	M6006
BALKAMP	M6012
BALKAMP	M6016
BALKAMP	M6028
BALKAMP	M6100
BALKAMP	M6113
BALKAMP	M6129
BAL-LM-HAM	465111
BAL-LM-HAM	PGF6732-56A
BAL-LM-HAM	PGF8508-56
BAL-LM-HAM	PGF8585-56
BAL-LM-HAM	TF1592-56
BORG-WARNR	114-6000
BORG-WARNR	114-6003
BORG-WARNR	114-6006
BORG-WARNR	114-6012
BORG-WARNR	114-6016
BORG-WARNR	114-6028
BORG-WARNR	114-6100
BORG-WARNR	114-6100A
BORG-WARNR	114-6100AB
BORG-WARNR	114-6113
BORG-WARNR	114-6117
BORG-WARNR	114-6120
BORG-WARNR	114-6129
BORG-WARNR	114-6133
BORG-WARNR	114-6137
BORG-WARNR	114-6141
BORG-WARNR	114-6143
BORG-WARNR	114-6143A
BORG-WARNR	114-6166
BORG-WARNR	114-6172
BORG-WARNR	114-6229
BORG-WARNR	J6000
CASE,J.I.	A19920
CASE,J.I.	R30673
CATERPILLR	641152
CLARK EQU.	1565573
CLARK EQU.	1565587
CLARK EQU.	3200930
CLARK EQU.	5208124
CLARK EQU.	5252382
CLARK EQU.	7842304
CLEVE.MOT.	CM6016
CLEVE.MOT.	CM6100
CLEVE.MOT.	CM6143
CLEVE.MOT.	CM-CP62N12
CLEVE.MOT.	CM-CP68WB9
CMP	CM6016
CMP	CM6100
CMP	CM6143
CMP	CM-CP62N12
COOK,J.B.	110137
CRANE CAR.	248
D&D MACH.	MM6100
DAVEY COM.	A23260
DAVEY COM.	A23697
DAVEY COM.	AP23260
DAVEY COM.	AP23697
DEERE,JOHN	11X15874A
DIAMOND R	10069P2
EATON	2004611
EIMCO	912A15020
EIMCO	XM767
EXCEL	U229
FIAT-ALLIS	0680970
FIAT-ALLIS	0680970-1
FIAT-ALLIS	08-048-885-63-001
FIAT-ALLIS	085918
FIAT-ALLIS	085918-1
FIAT-ALLIS	08-630-150-00-502
FIAT-ALLIS	70085918
FIAT-ALLIS	70085918-1
FIAT-ALLIS	70680970
FIAT-ALLIS	70680970-1
FLEETRITE	ZAD16000
FORD	B3TZ4635A
FORD	B3TZ4635B
FORD	C6TA4635F
FORD	D4HZ4635B
FORD	TGAA4635A
GALION	D76462
GALION	RASB46248(7)
GERLINGER	641152
GLAENZER	065-139-8
GMB	G5-6000X
GMB	G5-6143
GMB	G6100
GMB	G6140
GMB	G6141
GMB	GU6600
GMC	9033903
GMC	9038445
GROVE CR.	9-810-100006
GROVE CR.	9-810-100020
HABERLE	H6000
HABERLE	H6016
HAHN	38UM144
HALLBURTON	10357764
HARERLE	H6000
HARERLE	H6016
HOUGH	126819HI
HYSTER	122284
HYSTER	127129
HYSTER	127188
HYSTER	130268
HYSTER	55-6100
HYSTER	984908A
HYSTER	T468
IHC	104750
IHC	126819
IHC	126-819H1
IHC	126891H1
IHC	137100H
IHC	238-075R91
IHC	238-448R91
IHC	65351K93
IHC	65351R91
IHC	65-351R93
IHC	73-551R93
IHC	75-351R93
IHC	75351T93
IHC	75371R93
IHC	85-351R91
IMCO	55-6100
INBRA	CC6000
INBRA	CT11-065
INBRA	CT-16-030
INBRA	CT21-000
INGER.RAND	CBW114-6100
JOY MFG.	4542487
JOY MFG.	538
KALAMAZOO	Z2782
KOEHRING	A-2900-846
KOEHRING	X850-203
LEMPCO	U229
LEMPCO	U241
LINK BELT	AC4503(4)
LOBRO	U660
MASTERS	MM6000
MASTERS	MM6100
MCQUAY-NOR	U02-24
MCQUAY-NOR	U02-24
MIXERMOBL	708394
MOPAR	D35
MOTIVE	55-6100
MOTOR MAST	5025A
MOTOR MAST	5025AA
MOTOR MAST	5025AF
NEAPCO	284000
NEAPCO	2860
NEAPCO	286100
NEAPCO	286143
NEAPCO	4-0643
NEAPCO	4-6143
PARSONS	315D131(4)
PARSONS	324D258(4)
PARSONS	A209-0-846(4)
PERFECTION	RG6000
PERFECTION	RG6003
PERFECTION	RG6005
PERFECTION	RG6015
PERFECTION	RG6016
PETTI-MULL	40961
PETTI-MULL	A19522
PETTI-MULL	L1663-20
PETTI-MULL	LA425-5
PETTI-MULL	LL773-12
PILOT	16000
PILOT	16003
PILOT	16006
PILOT	16016
PILOT	16100
PILOT	16143
PRECISION	560
PRECISION	562
PREC.TOR.	935
PREC.TOR.	960
PULLMAN	K4464
REO MOTORS	10069P2
REPCO	114-6100
REPCO	114-6143
REPCO	114-6153
REPUBLIC	CB6000
REPUBLIC	CB6016
REPUBLIC	CB735
REPUBLIC	CB760
RUEDARSA	1451
RUEDARSA	2100-1451-0-0
SAFE ELEC.	T2731
SCHIELD	X629-30
SHOV.SUPP.	SP120804(4)
SMITH,T.L.	Y2724-4
SPICER	5-6000X
SPICER	5-6016X
TOWMOTOR	641152
TRANSMISS.	UK516
TRU-CROSS	6000
TRW	20092
UJS	55-6100
UNION PAC.	K4464
WABCO	2069817
WESCO	601600
WESCO	G6000
WESCO	N6000
WHITE	02-7074061
WHITE	10069P2
WHITING	01288(4)
ZELLER	5025A

IDLI GROUP 80194

CROSS AND LOW OR HIGH WING TYPE DRILLED OR THREADED BEARINGS SEE FIGURE 1IJ

1.6875 in	**CC**	**42.86 mm**
5.25 0 in	**CL**	**133.35 mm**
.8125 in	**CD**	**20.64 mm**

AEC	6103
AEC	AE6103
ALCO	1823
ALCO	6103
ALCO	J6103
ALLOY	1823
ALLOY	J6103
BORG-WARNR	114-6029
BORG-WARNR	114-6045
BORG-WARNR	114-6103
BORG-WARNR	114-6103AB
BORG-WARNR	114-6121
BORG-WARNR	114-6134
BORG-WARNR	114-6136
BORG-WARNR	114-6145
BORG-WARNR	114-6161
CLEVE.MOT.	CM6103
D&D MACH.	MM6103
FLEETRITE	ZAD16103
FORD	B4HZ4635B
GMB	G5-6102X
GMB	G5-6103
GMB	G6103
GMB	G6139
IHC	108991K92
IHC	108-991R91
IHC	108991R92
IHC	126816
LEMPCO	U253
MASTERS	MM6101
MASTERS	MM6102
MASTERS	MM6103
MCQUAY-NOR	U02-25
MOPAR	D37
MOTOR MAST	5025FA
NEAPCO	286103
NEAPCO	4-6103
PERFECTION	RG6103
PETTIBONE	LL773-188
PILOT	16001

ENGINEERING CATALOGS MUST BE CONSULTED FOR DETAILS NOT INCLUDED IN THIS GUIDE. SPECIFIC DESIGNS, MATERIAL CONTENT, TOLERANCES LUBE FITTINGS AND OTHER DIMENSIONS ARE INTENTIONALLY OMITTED HERE.

CAUTION: BE SURE TO REFER TO ENGINEERING CATALOGS FOR SPECIAL APPLICATIONS THAT REQUIRE SPECIFIC MATERIAL CONTENT, TOLERANCES, ETC. SEE FOOTNOTE.

PILOT	16004
PILOT	16101
PILOT	16103
PRECISION	564
REPCO	114-6103
REPUBLIC	CB6103
SPICER	5-222X
SPICER	5-224X
SPICER	5-6102X
TRU-CROSS	6102
TRW	20094
WESCO	601613
WESCO	G6103
WESCO	N6103
ZELLER	5025FA

IDLI GROUP 80195

CROSS AND LOW OR HIGH WING TYPE DRILLED OR THREADED BEARINGS
slight variation from above group
SEE FIGURE 1IJ

1.6875 in	CC	42.86 mm
5.250 in	CL	133.35 mm
.8125 in	CD	20.64 mm

AEC	6101
AEC	6103
AEC	AE6101
ALCO	1821
ALCO	6101
ALCO	J6101
ALLIS-CHLM	084328
ALLOY	1821
ALLOY	J6101
AUST.WEST.	WRF1064-57
BAL-LM-HAM	WRF1604-57
BAR-GREENE	5J06
BORG-WARNR	114-6004
BORG-WARNR	114-6029
BORG-WARNR	114-6101
BORG-WARNR	114-6121
BORG-WARNR	114-6145
BORG-WARNR	114-6152
BORG-WARNR	114-6161
CATERPILLR	1X8769
CLARK EQU.	5208189
CLEVE.MOT.	CM6044
CLEVE.MOT.	CM6101
D&D MACH.	MM6101
DANLEY	10823
DEERE,JOHN	11X15873A
DIAMOND R	10070P2
EATON	0807855
EATON	2004613
EATON	2006749
EATON	807855
EIMCO	XM853
FIAT-ALLIS	0084328
FIAT-ALLIS	0084328-4
FIAT-ALLIS	084328
FIAT-ALLIS	70084328
FIAT-ALLIS	70084328-4

FIAT-ALLIS	70680968
GERLINGER	661677
GERLINGER	A661677
GERLINGER	A-XP427
GLAENZER	065-137-8
GMB	G5-6102X
GMB	G6004
GMB	G6101
HOUGH	126815HI
HYSTER	165927
HYSTER	A144385
IHC	126816
IHC	126-816H1
IHC	630-385C1
IHC	630385C91
LEMPCO	U253
MASTERS	MM6101
MASTERS	MM6102
MASTERS	MM6103
MCQUAY-NOR	DIA., REO, WHITE
MOTOR MAST	5025G
NEAPCO	285152
NEAPCO	286101
NEAPCO	286152
NEAPCO	4-6152
PILOT	16001
PILOT	16152
PRECISION	561
REPCO	114-6101
REPUBLIC	CB736
REPUBLIC	CB756
SPICER	5-6102X
TEDDY TORQ	1464
TRW	20095
WESCO	601-613
WESCO	6D1-613
WESCO	D7004
WHITE	02-7074062
ZELLER	5025G

IDLI GROUP 80196

CROSS + 4 LOW WING DRILLED BEARINGS
SEE FIGURE 1G

1.6875 in	CC	42.86 mm
5.250 in	CL	133.35 mm
.8125 in	CD	20.64 mm

AUST.WEST.	888282
AUST.WEST.	HCF16263
BORG-WARNR	114-6102
BORG-WARNR	114-6105
BORG-WARNR	2A16588-1
CASE,J.I.	A19921
CATERPILLR	643633
CLARK EQU.	872273
CLARK EQU.	872773
CLARK EQU.	888282
CLARK EQU.	891046
CLARK EQU.	946322
CLARK EQU.	947322
CLEVE.MOT.	CM6102
CLEVE.MOT.	CM6114
CLEVE.MOT.	CM-CP62N13
DEERE,JOHN	AT44365
DEERE,JOHN	AT49181
EIMCO	1700

FIAT-ALLIS	70689077
FWD	206113
GERLINGER	643633
GERLINGER	A661678
GLAENZER	065-128-8
GMB	G6101
GMB	G6114
GMC	9033904
GMC	9037580
HYSTER	122284
HYSTER	127188
HYSTER	199459
HYSTER	202411A
HYSTER	87984
HYSTER	A138977
IHC	126815
IHC	126815H1
IHC	325-326R91
IHC	327-204R92
IHC	32720R91
IHC	613-878C91
IHC	615-155C91
INBRA	CT17-116
JOY MFG.	4520199
MASS.FERG.	1087-314M91
MCQUAY-NOR	UO3-30
MIXERMOBL	061617
MIXERMOBL	708392
MIXERMOBL	708398
MIXERMOBL	J22-257F
MOTOR MAST	5025H
M.R.S.	57689M11
M.R.S.	58533F11
NEAPCO	286114
NEAPCO	4-0614
PETTIBONE	LL773-270
PETTI-MULL	41603
PILOT	17209
P.M.	A27232-1
PRECISION	937
REPCO	RUJ6001
REPUBLIC	CB6022
REPUBLIC	CB737
SIL.HOIST	24351(4)
SUTTON	10182-127
TAYLOR	1927-100R
TAYLOR	1927-100R2(4)
TAYLOR	1927-150R1
TOWMOTOR	643633
TOWMOTOR	A661678
WABCO	VJ0684
WHITE	10064P2
WHITING	12197(4)

IDLI GROUP 80197

CROSS + 2 LOW WING THREADED + 2 BLOCK BEARINGS
SEE FIGURE 1JL

1.6875 in	CC	42.86 mm
5.250 in	CL	133.35 mm
.8125 in	CD	20.64 mm

AEC	6102
AEC	AE6102
ALCO	1822
ALCO	6102
ALCO	J6102

ALCO	J6142
ALLOY	1822
ALLOY	J6102
BALKAMP	3-937
BALKAMP	M6022
BALKAMP	M6102
BALKAMP	M6103
BALKAMP	M6114
BALKAMP	M6127
BORG-WARNR	114-1602
BORG-WARNR	114-6022
BORG-WARNR	114-6102
BORG-WARNR	114-6102A
BORG-WARNR	114-6114
BORG-WARNR	114-6114A
BORG-WARNR	114-6115
BORG-WARNR	114-6122
BORG-WARNR	114-6125
BORG-WARNR	114-6142
BORG-WARNR	114-6144
BUCYR-ERIE	4-681-3550
CASE,J.I.	L32423
CLEVE.MOT.	CM6102
CMP	CM6044
CMP	CM6101
CMP	CM6102
CMP	CM6103
CMP	CM-CP62N13
D&D MACH.	MM6102
DIAMOND R	10064P2
EXCEL	U248
FLEETRITE	ZAD16102
FLEETRITE	ZAD16144
FORD	B4TZ4635C
FORD	B4TZ4635H
FORD	C4TA4635E
FORD	C6TA4635E
FORD	TWAA4635A
FOX RIVER	SPF4CX5
GERLINGER	A66178
GMB	G1602
GMB	G5-6101X
GMB	G5-6102X
GMB	G5-6114
GMB	G56152
GMB	G6101
GMB	G6102
GMB	G6114
GMB	G6142
GMB	G6152
GMC	9051383
GROVE	9-364-100055
GROVE CR.	7-364-000029
GROVE CR.	9-364-100059
GROVE CR.	9-810-100018
HABERLE	H6022
HARERLE	H6022
HYSTER	202411
HYSTER	55-6102
HYSTER	97385A
IHC	108991R92
IHC	12685H1
IHC	327204R91
IHC	613-630C91
IMCO	55-6102
INBRA	CT20-545

BD=BEARING DIAMETER (outside)   BW=BEARING WIDTH   CB=CROSS LENGTH WITH BEARINGS   CC=CENTER TO CENTER   CD=CROSS DIAMETER
CL=CROSS LENGTH WITHOUT BEARINGS   EC=END TO CENTER or FACE TO FACE   EE=END TO END   EL=EFFECTIVE LENGTH
HD=HUB DIAMETER (or insert)   IE=INSIDE OF EARS (or recess)   OD=OUTSIDE DIAMETER   OE=OUTSIDE OF EARS   SB=SPLINE OR BORE SIZE

Column 1

LEMPCO	U248
MASS.FERG.	836-664M91
MASTERS	MM6101
MASTERS	MM6102
MASTERS	MM6103
MCQUAY-NOR	U03-30
MICHIGAN	872273
MICHIGAN	872773
MICHIGAN	888282
MICHIGAN	891046
MIXERMOBL	060774
MIXERMOBL	061230
MOPAR	D36
MOTIVE	55-6102
MOTOR MAST	5025-1J
MOTOR MAST	5025J
M.R.S.	57689M11
M.R.S.	58533F11
NEAPCO	2836
NEAPCO	2837
NEAPCO	286102
NEAPCO	286114
NEAPCO	286152
NEAPCO	4-6114
NEAPCO	4-6152
PERFECTION	RG6102
PERFECTION	RG6108
PERFECTION	TG6102
PETTIBONE	LL773-270
PETTIBONE	LL773-272
PETTI-MULL	LA384-7
PILOT	16022
PILOT	16102
PILOT	16144
P.M.	A27232-1
PRECISION	563
PREC.TOR.	937
REO MOTORS	10064P2
REPCO	114-6102
REPCO	114-6114
REPCO	114-6152
REPUBLIC	CB6022
RUEDARSA	1452
RUEDARSA	2101-1452-0-0
SPICER	5-222X
SPICER	5-224X
SPICER	5-6102
SPICER	5-6102X
STEIGER	17-175
STEIGER	17-267
SUTTON	A10182-127
TAYLOR	1927-1509R1
THEW SHOV.	P33588
TRU-CROSS	6102
TRW	20096
UJS	55-6102
VOLVO	6.631.390
WESCO	601602
WESCO	G6022
WESCO	N6022
WHITE	02-7074059
WHITING	12197
ZELLER	5025J

Column 2

IDLI GROUP 80198

CROSS WITH HOLE IN THE CENTER + 2 HIGH WING DRILLED + 2 BLOCK BRGS.
SEE FIGURE 1L

1.6875 in	CC	42.86 mm
5.250 in	CL	133.35 mm
.8125 in	CD	20.64 mm

ADAMS-LET.	TV3259
AEC	6106
AEC	AE6100
AEC	AE6106
AEC	AE6110
AEC	AE6111
AEC	AE6128
AEC	AE6149
ALCO	1826
ALCO	J6106
ALLIS-CHLM	085918
ALLOY	1826
ALLOY	J6100
ALLOY	J6106
AUST.WEST.	5206151
AUST.WEST.	944862
AUST.WEST.	AW465345
BALKAMP	3-905
BALKAMP	3-925
BALKAMP	3-943
BALKAMP	M6106
BALKAMP	M6110
BALKAMP	M6111
BALKAMP	M6112
BALKAMP	M6118
BALKAMP	M6119
BORG-WARNR	114-6106
BORG-WARNR	114-6110
BORG-WARNR	114-6111
BORG-WARNR	114-6112
BORG-WARNR	114-6118
BORG-WARNR	114-6119
BORG-WARNR	114-6126
BORG-WARNR	114-6127
BORG-WARNR	114-6128
BORG-WARNR	114-6130
BORG-WARNR	114-6132
BORG-WARNR	114-6143
BORG-WARNR	114-6146
BORG-WARNR	114-6149
BORG-WARNR	114-6151
BORG-WARNR	114-6155
BORG-WARNR	114-6173
BORG-WARNR	114-6179
BORG-WARNR	114-6181
CASE,J.I.	D73653
CASE,J.I.	D76950
CASE,J.I.	L31776
CASE,J.I.	L33945
CASE,J.I.	L36151
CASE,J.I.	L48344
CATERPILLR	159670
CATERPILLR	1S9670
CATERPILLR	2T95
CATERPILLR	3H301
CATERPILLR	3H7364
CATERPILLR	7H4545
CATERPILLR	7K3526

Column 3

CATERPILLR	8K7002
CATERPILLR	9H2477
CATERPILLR	9H2478
CHAMPION	14671
CLARK EQU.	5206151
CLARK EQU.	5290945
CLARK EQU.	888280
CLARK EQU.	899599
CLARK EQU.	940094
CLARK EQU.	940252
CLARK EQU.	941572
CLARK EQU.	941862
CLARK EQU.	942661
CLARK EQU.	944862
CLEVE.MOT.	CM6106
CLEVE.MOT.	CM-CP62N14
CMP	CM6106
CMP	CM-CP62N14
D&D MACH.	MM6106
DEERE,JOHN	21X20910A
DEERE,JOHN	21X20910C
DEERE,JOHN	AT25144
DEERE,JOHN	AT73775
DIAMOND R	10065P2
DIAMOND R	10069P2
EATON	2034881
EATON	6932287
EATON	6950058
EATON	6951976
EATON	6952215
EIMCO	16A15015
FIAT-ALLIS	259668
FIAT-ALLIS	271240
FIAT-ALLIS	73111401
FIAT-ALLIS	73134821
FIAT-ALLIS	76002601
FLEETRITE	ZAD16106
FLEETRITE	ZAD16128
FOX	SPFCXS
FOX RIVER	APF4X5
FOX RIVER	SPF4CXS
FOX RIVER	SPF4X5
GALION	224226
GALION	224446(7)
GALION	D75442
GALION	D80321(7)
GLAENZER	065-105-8
GLAENZER	065-120-8
GMB	G5-6106X
GMB	G5-6128
GMB	G6101
GMB	G6104
GMB	G6106
GMB	GUK06
GMC	9024891
GMC	9025903
GMC	9050449
GMC	9051384
HAHN	38UM1040
HORWOOD	625-002
HORWOOD	625-0066
HOUGH	126816HI
HOUGH	135765HI
HOUGH	197603HI
HOUGH	87592H1
HOUGH	8759529

Column 4

HOUGH	875952HI
HYSTER	141859
HYSTER	199862
HYSTER	55-6106
HYSTER	A138977
IHC	135765
IHC	135-765H1
IHC	142136C1
IHC	197603
IHC	197-603H1
IHC	198507
IHC	198-507H1
IHC	279470R91
IHC	279-470R92
IHC	631-042C91
IHC	875952
IHC	875-952C91
IHC	875-952H1
IMCO	55-6106
INBRA	CT13-005
INBRA	CT13-035
INBRA	CT16-010
INBRA	CT24-020
INBRA	CT25-000
JOY MFG.	1705339
KOMATSU	135-14-00030
KOMATSU	145-14-00030
KOMATSU	145-14-00190
KOMATSU	145-14-35110
KOMATSU	175-60-11520
LEMPCO	U229
MASS.FERG.	1052619M91
MASS.FERG.	1421-352M1
MASS.FERG.	836-413M91
MASS.FERG.	836-616M91
MASS.FERG.	836-947M91
MASS.FERG.	836-948M91
MASTERS	MM6106
MCQUAY-NOR	U03-29
MCQUAY-NOR	UO3-29
MICHIGAN	940094
MICHIGAN	940252
MICHIGAN	941572
MICHIGAN	942661
MOPAR	D38
MOTOR MAST	5025B
NEAPCO	2805
NEAPCO	2824
NEAPCO	2825
NEAPCO	2843
NEAPCO	286106
NEAPCO	286128
NEAPCO	286143
NEAPCO	4-0628
NEAPCO	4-6128
NELSON	PB418558
PETTIBONE	LA348-3
PETTI-MULL	LA384-3
PILOT	16106
PILOT	16128
PILOT	16143
POCLAIN	00400-54
PRECISION	562
PRECISION	905
PREC.TOR.	924
PREC.TOR.	943

CAUTION: BE SURE TO REFER TO ENGINEERING CATALOGS FOR SPECIAL APPLICATIONS THAT REQUIRE SPECIFIC MATERIAL CONTENT, TOLERANCES, ETC. SEE FOOTNOTE.

REPCO	114-6106
REPCO	114-6128
REPCO	RUJ6000
REPCO	RUJ6002
REPCO	RUJ6003
REPUBLIC	CB6000
REPUBLIC	CB705
REPUBLIC	CB724
REPUBLIC	CB725
REPUBLIC	CB743
REPUBLIC	CP62N19
SPICER	5-223X
SPICER	5-6000X
SPICER	5-6106
SPICER	5-6106X
SPICER	5-6110X
TAYLOR	3459095
THEW SHOV.	P33559
TRU-CROSS	6106
TRW	20092
TRW	20093
TRW	20100D
UJS	55-6106
WABCO	3000845
WABCO	TV3259
WESCO	601616
WESCO	601619
WESCO	601-649
WESCO	N6000
WESCO	N6106
WESCO	N6149
WHITE	02-7074060
WHITE	10065P2
WHITE	734751
ZELLER	5025A
ZELLER	5025B

IDLI GROUP 80199

CROSS AND LOW OR HIGH WING TYPE DRILLED OR THREADED BEARINGS
SEE FIGURE 1J

1.9375 in	CC	49.21 mm
5.25 0 in	CL	133.35 mm
.8125 in	CD	20.64 mm

AEC	6108
AEC	6228
AEC	AE6108
ALCO	1828
ALCO	6108
ALCO	J6108
ALLOY	1828
ALLOY	J6108
BALKAMP	M6023
BALKAMP	M6108
BORG-WARNR	114-6023
BORG-WARNR	114-6108
BORG-WARNR	114-6108A
BORG-WARNR	114-6162
BORG-WARNR	114-6162A
BORG-WARNR	114-6228
CLEVE.MOT.	CM6023
CLEVE.MOT.	CM6108
CLEVE.MOT.	CM6109
CLEVE.MOT.	CM6162
CLEVE.MOT.	CM-CP62N15
CMP	CM6023
CMP	CM6108
CMP	CM6162
CMP	CM-CP62N15
D&D MACH.	MM6108
DIAMOND R	10071P2
EXCEL	U249
FLEETRITE	ZAD16108
FORD	B4HZ4635
FORD	B4TZ4635A
FORD	C6TA4635G
FORD	D4HZ4635A
FORD	TWAA4635C
GLAENZER	065-249-8
GMB	G5-6108X
GMB	G5-6162
GMB	G6108
HABERLE	H6023
HYSTER	55-6108
IHC	237697R91
IHC	237-897R91
IHC	238-374R91
IHC	355688C91
IMCO	55-6108
LEMPCO	U249
MASTERS	MM6108
MCQUAY-NOR	U02-27
MCQUAY-NOR	UO2-27
MICHIGAN	941572
MIXERMOBL	J22-257F
MOPAR	D39
MOTIVE	55-6108
MOTOR MAST	5025-1M
MOTOR MAST	5025M
MOTOR MAST	5025MF
NEAPCO	286108
NEAPCO	286162
NEAPCO	4-6162
PERFECTION	RG6108
PILOT	16023
PILOT	16108
PRECISION	568
PREC.TOR.	926
REO MOTORS	10071PW
REPCO	114-6108
REPCO	114-6162
REPUBLIC	CB6023
REPUBLIC	CB726
SPICER	5-6108X
TRU-CROSS	6108
TRW	20098
UJS	55-6108
WESCO	601603
WESCO	601-619
WESCO	G6023
WESCO	N6023
WESCO	N6132
WHITE	02-7074063
WHITE	10071P2
ZELLER	5025G
ZELLER	5025M

IDLI GROUP 80200

CROSS AND LOW OR HIGH WING TYPE DRILLED OR THREADED BEARINGS
SEE FIGURE 1IJ

1.9375 in	CC	49.21 mm
5.25 0 in	CL	133.35 mm
.8125 in	CD	20.64 mm

AEC	6107
ALCO	G179
ALCO	J6179
BORG-WARNR	114-6179
BORG-WARNR	114-6179A
GMB	G6179
IHC	355-687C91
PRECISION	493

IDLI GROUP 80201

CROSS + 2 LOW WING THREADED + 2 BLOCK BEARINGS
SEE FIGURE 1JL

1.9375 in	CC	49.21 mm
5.25 0 in	CL	133.35 mm
.8125 in	CD	20.64 mm

AEC	6109
AEC	AE6109
ALCO	1829
ALCO	6109
ALCO	J6109
ALLOY	1829
ALLOY	J6109
ALLOY	JJ6109
BALKAMP	3-924
BALKAMP	M6024
BALKAMP	M6109
BORG-WARNR	114-6024
BORG-WARNR	114-6109
BORG-WARNR	114-6109A
BORG-WARNR	114-6163
BORG-WARNR	114-6163A
CLEVE.MOT.	CM6024
CLEVE.MOT.	CM6109
CLEVE.MOT.	CM6163
CLEVE.MOT.	CM-CP62N16
CMP	CM6024
CMP	CM6109
CMP	CM6163
CMP	CM-CP62N16
D&D MACH.	MM6109
DIAMOND R	10072P
EXCEL	U250
FLEETRITE	ZAD16109
FORD	B4TZ4635B
FORD	B4TZ4635G
FORD	C6TA4635H
FORD	TVAA4635
FORD	TVAA4635A
GLAENZER	065-250-8
GMB	G5-6109X
GMB	G5-6163
GMB	G6109
GMB	G6163
HABERLE	H6024
HAHN	38UM189
HYSTER	55-6109
IHC	355-415C91
IMCO	55-6109

LEMPCO	U250
MASTERS	MM6109
MCQUAY-NOR	U02-26
MCQUAY-NOR	UO2-26
MOPAR	D40
MOTIVE	55-6109
MOTOR MAST	5025-1K
MOTOR MAST	5025K
NEAPCO	286109
NEAPCO	286163
NEAPCO	4-6163
PERFECTION	RG6109
PILOT	16024
PILOT	16109
PRECISION	569
PREC.TOR.	924
REPCO	114-6109
REPCO	114-6163
REPUBLIC	CB6024
SPICER	5-1609X
SPICER	5-6109X
SPICER	5-7109X
TEDDY TORQ	1469
TRU-CROSS	6109
TRW	20097
UJS	55-6109
WESCO	601604
WESCO	G6024
WESCO	N6024
WHITE	02-7074064
WHITE	10072P2
ZELLER	5025K

IDLI GROUP 80202

CROSS + 2 WING TYPE BEARINGS + 2 ROUND GROOVED BEARINGS
SEE FIGURE 1BJ

1.3125 in	BD	33.34 mm
1.6875 in	CC	42.86 mm
1.0 in	BW	25.40 mm
5.25 0 in	CL	133.35 mm
.8125 in	CD	20.64 mm

AEC	6107
AEC	6149
AEC	AE6107
ALCO	1827
ALCO	J6107
ALLIS-CHLM	084395
ALLIS-CHLM	084435
ALLOY	1827
ALLOY	J6107
BORG-WARNR	114-6107
BORG-WARNR	114-6116
CLEVE.MOT.	CM6107
D&D MACH.	MM6107
FIAT-ALLIS	0084435
FIAT-ALLIS	0084435-7
FIAT-ALLIS	084395
FIAT-ALLIS	084435
FIAT-ALLIS	084435-7
FIAT-ALLIS	70084395
FIAT-ALLIS	70084435
FIAT-ALLIS	70084435-7
GLAENZER	065-138-8
GMB	G5-6132X
GMB	G6107

BD = BEARING DIAMETER (outside)   BW = BEARING WIDTH   CB = CROSS LENGTH WITH BEARINGS   CC = CENTER TO CENTER   CD = CROSS DIAMETER
CL = CROSS LENGTH WITHOUT BEARINGS   EC = END TO CENTER or FACE TO FACE   EE = END TO END   EL = EFFECTIVE LENGTH
HD = HUB DIAMETER (or insert)   IE = INSIDE OF EARS (or recess)   OD = OUTSIDE DIAMETER   OE = OUTSIDE OF EARS   SB = SPLINE OR BORE SIZE

INBRA	CT11-050
MOTOR MAST	5025H
MOTOR MAST	5025N
NEAPCO	286107
NEAPCO	4-6107
PILOT	16107
PRECISION	957
REPCO	114-6107
REPUBLIC	CB757
TRW	20082
WESCO	601-617
WESCO	N6107
ZELLER	5025N

IDLI GROUP 80203

CROSS AND LOW OR HIGH WING TYPE DRILLED OR THREADED BEARINGS
SEE FIGURE 1BL

1.3125 in	BD	33.34 mm
1.6875 in	CC	42.86 mm
1.0 in	BW	25.40 mm
5.25 0 in	CL	133.35 mm
.8125 in	CD	20.64 mm

AEC	6150
AEC	AE6150
ALCO	J6139
BORG-WARNR	114-6139
BORG-WARNR	114-6147
BORG-WARNR	114-6150
CLEVE.MOT.	CM6139
GMB	G5-6139X
GMB	G6139
MASTERS	MM6139
MCQUAY-NOR	IND.
MOTOR MAST	5025R
PILOT	16150
PRECISION	565
REPCO	114-6150
WABCO	3000845
WESCO	601617
WESCO	601639
WESCO	N6139

IDLI GROUP 80204

CROSS + 4 SERRATED BEARINGS + THRUST AND LOCK PLATES
SEE FIGURE 1C

1.9375 in	BD	49.21 mm
1.0 in	BW	25.40 mm
5.4688 in	CL	138.91 mm
1.4063 in	CD	35.72 mm

AEC	AE-C96-55
AEC	C96-55
ALCO	4C96-55
ALCO	C96-55
ALCO	JC96-55
ALLIS-CHLM	11020
ALLIS-CHLM	8392
BORG-WARNR	C96-55
CHRYSLER	1321134
CLEVE.STL.	C96-55
GMB	G4C96-55
GMB	G5-1296X
GMB	GC96-55
LEMPCO	U306
MOTOR MAST	5451A

NEAPCO	285500
PILOT	C19655
PRECISION	421
REPCO	C96-55
REPUBLIC	CB-C96
SPICER	C96-55
TEDDY TORQ	C96
TEDDY TORQ	GC96-55
WESCO	601955
WESCO	601965
WESCO	NC965

IDLI GROUP 80205

SEE FIGURE 1

1.8120 in	BD	46.02 mm
5.5 0 in	CL	139.70 mm
1.0 in	CD	25.40 mm

AEC	710
AEC	710A
AEC	AE710A
BORG-WARNR	114-589
CLEVE.MOT.	CM5-189X
GLAENZER	065-100-8
GMB	G5-308X
GMB	G5-710X
GMC	9044805
GMC	9045104
MOTOR MAST	5184A
NEAPCO	2800
NEAPCO	28589X
NEAPCO	5-0589
PRECISION	900
REPUBLIC	CB700
REPUBLIC	CB700A
TAYLOR	R1D8952
WESCO	611750
WESCO	N7500
ZELLER	5184A

IDLI GROUP 80206

CROSS + 4 GROOVED BEARINGS + INSIDE SNAP RINGS
SEE FIGURE 1B

1.6250 in	BD	41.28 mm
1.2969 in	BW	32.94 mm
5.5 0 in	CL	139.70 mm
1.0 in	CD	25.40 mm

AEC	7245
GLAENZER	065-171-8
GMB	G5-7245X
IHC	875-953C91
MOTOR MAST	5050K
PRECISION	923
REPUBLIC	CB723

IDLI GROUP 80207

CROSS + 4 LOW WING DRILLED BEARINGS
SEE FIGURE 1G

1.9375 in	CC	49.21 mm
5.5 0 in	CL	139.70 mm
1.0 in	CD	25.40 mm

BORG-WARNR	113-7201
BORG-WARNR	114-7201
BUFFALO-SP	848036509

CLEVE.MOT.	CM7101
CLEVE.MOT.	CM-CP72N34
CURTISS-WR	2-2276
DEERE,JOHN	AP22718H
EIMCO	105M2383
FIAT-ALLIS	0629943
FIAT-ALLIS	0629943-2
FIAT-ALLIS	629943
FIAT-ALLIS	70629943
FIAT-ALLIS	70629943-2
FIAT-ALLIS	70681024
FMC	2-22746
GERLINGER	629816
GLAENZER	065-107-8
GMB	G7206
GMC	9021682
GMC	9022755
GMC	9024788
GMC	9026502
HYSTER	114602A
HYSTER	55-7101
IHC	239-737R91
IHC	239-738R91
IHC	925-395C1
KERSHAW	13-1959
MCQUAY-NOR	UO3-43
NEAPCO	287206
PRECISION	907
REPUBLIC	CB7004
REPUBLIC	CB707
SCHIELD	A34800
TOWMOTOR	629816
TRW	20102D
WAYNE	2-22746
WHITE	10075P2

IDLI GROUP 80208

CROSS AND LOW OR HIGH WING TYPE DRILLED OR THREADED BEARINGS
SEE FIGURE 1I

1.9375 in	CC	49.21 mm
5.5 0 in	CL	139.70 mm
1.0 in	CD	25.40 mm

AEC	7103
AEC	AE7103
ALCO	J7203
ALCO	J7208
BORG-WARNR	114-7103
BORG-WARNR	114-7203
BORG-WARNR	114-7208
CLEVE.MOT.	CM7103
CLEVE.MOT.	CMCP72N55
FLEETRITE	ZAD17208
GMB	G5-7203X
GMB	G7203
GMB	GU3T1328
MCQUAY-NOR	IHC
MOTOR MAST	5050H
NEAPCO	287203
NEAPCO	287208
NEAPCO	5-7208
PILOT	17208
PRECISION	575
REPCO	114-7103
REPCO	114-7203
REPCO	114-7208

TRW	20103
WESCO	611708
WESCO	N7008

IDLI GROUP 80209

CROSS AND LOW OR HIGH WING TYPE DRILLED OR THREADED BEARINGS
SEE FIGURE 1IL

1.9375 in	CC	49.21 mm
5.5 0 in	CL	139.70 mm
1.0 in	CD	25.40 mm

AEC	7104
AEC	AE7104
ALCO	J7204
BORG-WARNR	114-7104
BORG-WARNR	114-7204
BORG-WARNR	114-7209
CLEVE.MOT.	CM7104
GMB	G5-7204X
GMB	G7204
IHC	146703H1
MCQUAY-NOR	IND.
MOTOR MAST	5050J
NEAPCO	287204
NEAPCO	287209
NEAPCO	5-7209
PILOT	17209
PRECISION	576
REPCO	114-7104
REPCO	114-7204
REPCO	114-7209
REPUBLIC	CB72N57
TRW	20172
WESCO	611709
WESCO	N7009

IDLI GROUP 80210

CROSS AND LOW OR HIGH WING TYPE DRILLED OR THREADED BEARINGS
SEE FIGURE 1J

1.9375 in	CC	49.21 mm
5.5 0 in	CL	139.70 mm
1.0 in	CD	25.40 mm

AEC	7100
AEC	AE7100
AEC	AE7200
ALCO	1841
ALCO	7100
ALCO	7200
ALCO	J7033
ALCO	J7200
ALCO	J7205
ALCO	J7242
ALLIS-CHLM	058735
ALLIS-CHLM	08-000-504-465
ALLIS-CHLM	083541
ALLIS-CHLM	083616
ALLIS-CHLM	083616-3
ALLIS-CHLM	1125463
ALLIS-CHLM	1126463
ALLIS-CHLM	1134986
ALLIS-CHLM	629943
ALLIS-CHLM	629944
ALLIS-CHLM	FEX1707
ALLOY	1841

CAUTION: BE SURE TO REFER TO ENGINEERING CATALOGS FOR SPECIAL APPLICATIONS THAT REQUIRE SPECIFIC MATERIAL CONTENT, TOLERANCES, ETC. SEE FOOTNOTE.

ALLOY	J7100
ALLOY	J7200
AUST.WEST.	AW465123
AUST.WEST.	AW465124
AUST.WEST.	AW465132
AUST.WEST.	AW465136
AUST.WEST.	AW465137
AUST.WEST.	PGF10798-56A
BALKAMP	3-908
BALKAMP	M7000
BALKAMP	M7003
BALKAMP	M7010
BALKAMP	M7011
BALKAMP	M7012
BALKAMP	M7033
BALKAMP	M7034
BALKAMP	M7100
BALKAMP	M7103
BALKAMP	M7105
BALKAMP	M7107
BALKAMP	M7116
BALKAMP	M7120
BALKAMP	M7200
BALKAMP	M7203
BALKAMP	M7205
BALKAMP	M7208
BALKAMP	M7213
BAL-LM-HAM	465123
BAL-LM-HAM	465124
BAL-LM-HAM	465132
BAL-LM-HAM	465136
BAL-LM-HAM	HCF2098-57
BAL-LM-HAM	PGF10798-56A
BAL-LM-HAM	PGF8537-56
BAL-LM-HAM	PGF8807-57
BORG-WARNR	114-7000
BORG-WARNR	114-7003
BORG-WARNR	114-7007
BORG-WARNR	114-7010
BORG-WARNR	114-7011
BORG-WARNR	114-7012
BORG-WARNR	114-7013
BORG-WARNR	114-7015
BORG-WARNR	114-7033
BORG-WARNR	114-7035
BORG-WARNR	114-7100
BORG-WARNR	114-7107
BORG-WARNR	114-7116
BORG-WARNR	114-7118
BORG-WARNR	114-7120
BORG-WARNR	114-7200
BORG-WARNR	114-7200A
BORG-WARNR	114-7200AB
BORG-WARNR	114-7205
BORG-WARNR	114-7205A
BORG-WARNR	114-7211
BORG-WARNR	114-7212
BORG-WARNR	114-7213
BORG-WARNR	114-7216
BORG-WARNR	114-7218
BORG-WARNR	114-7221

BORG-WARNR	114-7222
BORG-WARNR	114-7225
BORG-WARNR	114-7228
BORG-WARNR	114-7230
BORG-WARNR	114-7234
BORG-WARNR	114-7242
BORG-WARNR	114-7243
BORG-WARNR	114-7248
BORG-WARNR	J7000
BORG-WARNR	J7100
CASE,J.I.	A44176
CATERPILLR	629815
CATERPILLR	681156
CATERPILLR	8F7719
CHAMPION	3977
CLEVE.MOT.	CM7000
CLEVE.MOT.	CM7100
CLEVE.MOT.	CM7200
CLEVE.MOT.	CM-CP72N32
CMP	CM7000
CMP	CM7100
CMP	CM7200
CMP	CM-CP72N32
COOK,J.B.	110138
D&D MACH.	MM7100
DAVEY COM.	AP24615
DEERE,JOHN	31X12715A
DEERE,JOHN	31X12KL1715A
DEERE,JOHN	AH1261H
DEERE,JOHN	CAW8055A
DEERE,JOHN	XAW8055A
DIAMOND R	10074P2
DIAMOND R	KC7012
DIAMOND T	KC7012
EATON	2006739
EATON	2017732
EATON	2017744
EATON	5023218
EATON	T698-1
EATON	T698-7
EIMCO	103M8226
EIMCO	912A15023
EXCEL	U230
FIAT-ALLIS	08-000-504-465
FIAT-ALLIS	70057736
FIAT-ALLIS	70058736
FIAT-ALLIS	70088665
FIAT-ALLIS	7083616
FIAT-ALLIS	70836163
FIAT-ALLIS	836163
FLEETRITE	ZAD17000
FLXIBLE	12-158-160
FLXIBLE	13-158-160
FORD	B4TZ4635D
FORD	B4TZ4635J
FORD	C4TA4635B
FORD	C6TA4635C
FORD	D2HA4635CA
FORD	D3HZ4635A
FORD	TWAA4635D
GMB	G5-7000X
GMB	G5-7205
GMB	G7033
GMB	G7107
GMB	G7200
GMB	G7205

GMB	G7208
GMB	GU670
GMB	GU6700
GMC	47097
GMC	611934
GMC	8871795
GMC	9034950
GMC	9042141
GRAIGG CO.	LA337
GROVE CR.	9-810-100010
HABERLE	H7000
HARERLE	H7000
HOUGH	118122HI
HYSTER	116983A
HYSTER	75109A
IHC	102-684R91
IHC	124-645R91
IHC	136152HA
IHC	136512H
IHC	136512HA
IHC	136512HB
IHC	136-512HBK
IHC	136512HBX
IHC	137109HA
IHC	925397C1
IMCO	55-7100
INBRA	CC7000
INBRA	CT17-055
INGER.RAND	CBW114-7100
ITALCARDAN	50-760-000
JOY MFG.	4510193
JOY MFG.	A369158
JOY MFG.	A369518
JOY MFG.	A59654
KOEHRING	208260014
KOEHRING	208260020
KOEHRING	20826014
KOEHRING	848036526
KOEHRING	9004605
LEMPCO	U230
LOBRO	U670
MACK	201SJ48
MACK	35MU29P1
MASS.FERG.	1091-075M91
MASS.FERG.	836-776M91
MASTERS	MM7100
MCQUAY-NOR	U03-31
MCQUAY-NOR	UO3-31
MICHIGAN	980613
MIPER	C3520
MOPAR	D41
MOTIVE	55-7100
MOTOR MAST	5050A
MOTOR MAST	5050AA
M.R.S.	13133P11
MUIR-HILL	255923
NEAPCO	2808
NEAPCO	285000
NEAPCO	287200
NEAPCO	287205
NEAPCO	5-7205
PERFECTION	RG7000
PETTI-MULL	A19961
PETTI-MULL	LA17205
PETTI-MULL	LA384-6
PETTI-MULL	LA425-6

PETTI-MULL	LL273-2
PETTI-MULL	LL3323-1
PETTI-MULL	LL773-111
PILOT	17000
PILOT	17035
PILOT	17100
PILOT	17205
PILOT	17208
PITT.VIOL.	40.760.000
PRECISION	570
PREC.TOR.	915
PULLMAN	K11513
QUINT.HAZ.	QL6700
REO MOTORS	10074P2
REPCO	114-7116
REPCO	114-7200
REPCO	114-7205
REPUBLIC	CB7000
REPUBLIC	CB7100
RITEWAY	REDI-MIX TRUCK
RUEDARSA	1556
RUEDARSA	2100-1556-0-0
SPICER	5-7000
SPICER	5-7000X
SPICER	5-7200X
TOWMOTOR	629815
TOWMOTOR	644055
TRANSMISS.	UK520S
TRU-CROSS	7000
TRW	20099
UJS	55-7100
ULRICH	2U6365
VOLVO	6.631.406
WESCO	611700
WESCO	G7000
WESCO	N7000
WHITE	02-7074066
WHITE	10074P2
WHITE	KC7012
WOOLDRIDGE	2-22746
YOUNG FIRE	2968
ZELLER	5050A

IDLI GROUP 80211

CROSS AND LOW OR HIGH WING TYPE DRILLED OR THREADED BEARINGS
SEE FIGURE 1IJ

1.9375 in	CC	49.21 mm
5.50 in	CL	139.70 mm
1.0 in	CD	25.40 mm

AEC	7101
AEC	AE7101
ALCO	1842
ALCO	7101
ALCO	J7201
ALCO	J7206
ALCO	J7210
ALLIS-CHLM	058735
ALLIS-CHLM	08-000-505-01-409
ALLIS-CHLM	083616
ALLIS-CHLM	629943
ALLOY	1842
ALLOY	J7201
BALKAMP	3-907
BALKAMP	M7001
BALKAMP	M7004

BD = BEARING DIAMETER (outside)   BW = BEARING WIDTH   CD = CROSS LENGTH WITH BEARINGS   CC = CENTER TO CENTER   CD = CROSS DIAMETER
CL = CROSS LENGTH WITHOUT BEARINGS   EC = END TO CENTER or FACE TO FACE   EE = END TO END   EL = EFFECTIVE LENGTH
HD = HUB DIAMETER (or insert)   IE = INSIDE OF EARS (or recess)   OD = OUTSIDE DIAMETER   OE = OUTSIDE OF EARS   SB = SPLINE OR BORE SIZE

BALKAMP	M7042
BALKAMP	M7115
BALKAMP	M7201
BALKAMP	M7201A
BALKAMP	M7206
BALKAMP	M7214
BORG-WARNR	114-7001
BORG-WARNR	114-7004
BORG-WARNR	114-7042
BORG-WARNR	114-7101
BORG-WARNR	114-7115
BORG-WARNR	114-7201
BORG-WARNR	114-7206
BORG-WARNR	114-7214
BORG-WARNR	114-7219
BORG-WARNR	114-7219AB
BORG-WARNR	114-7224
BORG-WARNR	114-7229
BORG-WARNR	114-7239
BORG-WARNR	J7101
CLEVE.MOT.	CM7101
CURTISS-WR	2-22746
D&D MACH.	MM7101
DIAMOND R	10075P2
FIAT-ALLIS	08-000-505-01-409
FLEETRITE	ZAD17206
GMB	G5-7206
GMB	G5-7206X
GMB	G7201
GMB	G7206
IHC	239-739R91
IHC	866-785R91
IMCO	55-7101
INBRA	CT11-026
KOEHRING	848036509
LEMPCO	U254
MASTERS	MM7101
MASTERS	MM7102
MCQUAY-NOR	U03-43
MOPAR	D43
MOTIVE	55-7101
MOTOR MAST	5050F
NEAPCO	2807
NEAPCO	286101
NEAPCO	287101
NEAPCO	287201
NEAPCO	287206
NEAPCO	.5-7206
OSHKOSH	7DS21
PERFECTION	RG7004
PILOT	17004
PILOT	17101
PILOT	17206
PRECISION	572
PREC.TOR.	907
REO MOTORS	10075P2
REPCO	114-7101
REPCO	114-7201
REPCO	114-7206
REPUBLIC	CB7004
SPICER	5-7202X
SPICER	5-7206X
TRW	20102
TRW	20102D
WESCO	611704
WESCO	N7004
WHITE	02-7074067
WOOLDRIDGE	3-22746
ZELLER	5050F

IDLI GROUP 80212

CROSS + 2 LOW WING THREADED + 2 BLOCK BEARINGS

SEE FIGURE 1JL

1.9375 in	CC	49.21 mm
5.50 in	CL	139.70 mm
1.0 in	CD	25.40 mm

ADAMS-LET.	3000840
ADAMS-LET.	AD9353
AEC	7102
AEC	7245
AEC	AE7102
AEC	AE7245
ALCO	1843
ALCO	7102
ALCO	J7202
ALLOY	1843
ALLOY	J7102
ALLOY	J7202
AUST.WEST.	PGF1891
BALKAMP	3-906
BALKAMP	3-927
BALKAMP	M7034
BALKAMP	M7101
BALKAMP	M7102
BALKAMP	M7114
BALKAMP	M7202
BALKAMP	M7207
BORG-WARNR	114-7034
BORG-WARNR	114-7102
BORG-WARNR	114-7114
BORG-WARNR	114-7202
BORG-WARNR	114-7202A
BORG-WARNR	114-7207
BORG-WARNR	114-7207A
BORG-WARNR	114-7220
BORG-WARNR	114-7223
BORG-WARNR	114-7226
BORG-WARNR	114-7227
BORG-WARNR	114-7231
BORG-WARNR	114-7245
BORG-WARNR	114-7246
BORG-WARNR	114-7249
BORG-WARNR	114-7270
BORG-WARNR	114-7294
BORG-WARNR	J7102
CASE,J.I.	A45596
CLARK EQU.	947318
CLEVE.MOT.	CM7102
CMP	CM7101
CMP	CM7102
CMP	CM7202
CMP	CM7207
CMP	CM-CP72N33
CMP	CM-CP72N34
CURTISS-WR	2-22746
D&D MACH.	MM7102
DEERE,JOHN	AR84046
DEERE,JOHN	AT57090
DIAMOND R	10076P2
EATON	2006731
EATON	2017732
EATON	2017744
EATON	2017758
EATON	610632
EATON	T698-1
EATON	T698-7
EATON	T699-1
EXCEL	U251
FIAT-ALLIS	4908839
FLEETRITE	ZAD17101
FLEETRITE	ZAD17102
FLEETRITE	ZAD17207
FORD	B4TZ4635E
FORD	B4TZ4635N
FORD	C4TA4635H
FORD	C6TA4635D
FORD	D2HA4635BA
FORD	D3HZ4635B
FORD	TWAA4635B
GERLINGER	646649
GERLINGER	681157
GMB	G5-7202X
GMB	G5-7207
GMB	G5-7245X
GMB	G7202
GMB	G7206
GMB	G7246
GMC	2051472
GRAIGG CO.	LA205B
GROVE CR.	7-069-000158
GROVE CR.	9-364-100059
GROVE CR.	9-810-100015
GROVE CR.	9-810-100033
HABERLE	H7034
HARERLE	H7034
IHC	140604K92
IHC	140604R91
IHC	140-604R92
IHC	146701H1
IHC	221396H1
IHC	71783C1
IMCO	55-7102
JOY MFG.	4520128
KOEHRING	208260020
KOEHRING	20826014
KOEHRING	9003400
LEMPCO	U251
LOBRO	U675
MACK	201SJ49
MASS.FERG.	836-775M91
MASTERS	MM7102
MCQUAY-NOR	U03-32
MCQUAY-NOR	UO3-32
MOPAR	D44
MOTIVE	55-7102
MOTOR MAST	5050CA
MOTOR MAST	5050K
NEAPCO	2806
NEAPCO	287102
NEAPCO	287202
NEAPCO	287207
NEAPCO	.5-7207
PERFECTION	RG7102
PILOT	17034
PILOT	17102
PILOT	17206
PILOT	17207
PRECISION	574
PREC.TOR.	906
PREC.TOR.	927
REPCO	114-7202
REPCO	114-7207
REPUBLIC	CB7034
REPUBLIC	CB723
RUEDARSA	1557
RUEDARSA	2101-1557-0-0
SPICER	.5-7202
SPICER	5-7202X
STEIGER	17-201
STEIGER	17-261
TAYLOR	3462
TRU-CROSS	7202
TRW	20100
TRW	20100D
UJS	55-7101
UJS	55-7102
VOLVO	6.631.405
WESCO	611734
WESCO	G7004
WESCO	G7034
WESCO	N7034
WHITE	02-7074068
YOUNG FIRE	2970
ZELLER	5050CA
ZELLER	5150CA

IDLI GROUP 80213

CROSS WITH HOLE IN THE CENTER + 2 HIGH WING DRILLED + 2 BLOCK BRGS.

SEE FIGURE 1L

1.9375 in	CC	49.21 mm
5.50 in	CL	139.70 mm
1.0 in	CD	25.40 mm

ADAMS-LET.	TC4412-1
ADAMS-LET.	TH7986
AEC	7105
AEC	AE7105
ALCO	1840
ALCO	J7105
ALCO	J7112
ALCO	J7113
ALCO	J7117
ALCO	J7122
ALLIS-CHLM	180131
ALLIS-CHLM	180132
ALLIS-CHLM	180132-3
ALLIS-CHLM	58-000-505-409
ALLOY	1840
ALLOY	4M117
ALLOY	J7105
AUST.WEST.	5208350
AUST.WEST.	942735
AUST.WEST.	PGF10797-58A
AUST.WEST.	WRF1679
AUST.WEST.	WRF1679-56
BALKAMP	3-901
BALKAMP	3-947
BALKAMP	M7013
BALKAMP	M7105
BALKAMP	M7112
BALKAMP	M7119
BALKAMP	M7122

CAUTION: BE SURE TO REFER TO ENGINEERING CATALOGS FOR SPECIAL APPLICATIONS THAT REQUIRE SPECIFIC MATERIAL CONTENT, TOLERANCES, ETC. SEE FOOTNOTE.

BALKAMP	M7126
BORG-WARNR	114-7013
BORG-WARNR	114-7105
BORG-WARNR	114-7105A
BORG-WARNR	114-7110
BORG-WARNR	114-7111
BORG-WARNR	114-7112
BORG-WARNR	114-7113
BORG-WARNR	114-7117
BORG-WARNR	114-7119
BORG-WARNR	114-7121
BORG-WARNR	114-7122
BORG-WARNR	114-7123
BORG-WARNR	114-7124
BORG-WARNR	114-7125
BORG-WARNR	114-7126
BORG-WARNR	114-7127
BORG-WARNR	114-7129
BORG-WARNR	114-7130
BORG-WARNR	114-7131
BWE	114-7126
CASE,J.I.	D80329
CASE,J.I.	L31782
CASE,J.I.	N8811
CATERPILLR	2H8357
CATERPILLR	2H858
CATERPILLR	2K7276
CATERPILLR	3H6104
CATERPILLR	3K7276
CATERPILLR	5K2006
CATERPILLR	6H2577
CATERPILLR	8D7719
CATERPILLR	8K6042
CATERPILLR	9P356
CLARK EQU.	5206151
CLARK EQU.	5208350
CLARK EQU.	879946
CLARK EQU.	940229
CLARK EQU.	942735
CLARK EQU.	942745
CLARK EQU.	945741
CLARK EQU.	946641
CLARK EQU.	980615
CLARK EQU.	993142
CLARK EQU.	994085
CLEVE.MOT.	CM7105
CMP	CM7105
CMP	CM7205
CMP	CM-CP72N31
D&D MACH.	MM7105
DEERE,JOHN	AR90543
DEERE,JOHN	AT27353
DIAMOND R	10073P2
EATON	2034362
EATON	2990674
EATON	5086258
EATON	6950172
EATON	6951148
EATON	6951149
EATON	695149
EATON	6955338

EIMCO	912P15053
EIMCO	913P150010
FIAT-ALLIS	0180132
FIAT-ALLIS	0180132-3
FIAT-ALLIS	080132
FIAT-ALLIS	180132
FIAT-ALLIS	180132-3
FIAT-ALLIS	70080132
FIAT-ALLIS	70180132
FIAT-ALLIS	70180132-3
FIAT-ALLIS	73140006
FIAT-ALLIS	76002604
FLEETRITE	ZAD17105
FLEETRITE	ZAD17126
FWD	46408
FWD	46653
FWD	J6105
GLAENZER	065-101-8
GLAENZER	065-121-8
GLAENZER	065-132-8
GMB	G5-7105X
GMB	G5-7126
GMB	G7105
GMB	G7112
GMB	G7113
GMB	G7117
GMB	G7122
GMB	G7126
GMB	G7200
GMB	G7203
GMB	G7204
GMB	GUPA2
GMC	9022753
GMC	9023940
GMC	9023941
GMC	9024705
GMC	9024706
GMC	9025558
GMC	9027855
GMC	9030220
GMC	9032043
GMC	9036325
GMC	9052137
GRAIGG CO.	LA854N
HORWOOD	625-0061
HOUGH	136142HI
HOUGH	175045HI
HYSTER	55-7105
HYSTER	55-7119
IHC	122188
IHC	122-188H1
IHC	136142
IHC	136-142H1
IHC	175045
IHC	175-045H1
IHC	259770R91
IHC	259770R92
IHC	259-770R93
IHC	338-412R91
IHC	338421R91
IHC	358-412R91
IHC	630-389C91
IMCO	55-7105
IMCO	55-7119
INBRA	CT11-095
INBRA	CT13-030

INBRA	CT22-005
INBRA	CT23-020
INBRA	CT24-005
KOEHRING	9004605
KOMATSU	381-12-4149
MASS.FERG.	1052614M91
MASS.FERG.	836-667M91
MASS.FERG.	836-668M91
MASTERS	MM7105
MCQUAY-NOR	U03-33
MICHIGAN	879946
MICHIGAN	942735
MICHIGAN	945741
MICHIGAN	946641
MICHIGAN	980615
MOPAR	D42
MOTIVE	55-7105
MOTOR MAST	5000DA
MOTOR MAST	5050DA
NEAPCO	2827
NEAPCO	2847
NEAPCO	287105
NEAPCO	287126
NEAPCO	287203
NEAPCO	5-7126
OSHKOSH	22UJ28
PILOT	17105
PILOT	17126
PRECISION	927
PREC.TOR.	901
PREC.TOR.	947
REPCO	114-7105
REPCO	114-7119
REPCO	114-7122
REPCO	114-7126
REPUBLIC	CB701
REPUBLIC	CB727
REPUBLIC	CB747
REX CHAIN	298-6044-91
RUEDARSA	1555
RUEDARSA	1960
RUEDARSA	2101-1555-0-0
RUEDARSA	2101-1960-0-0
SPICER	5-7105
SPICER	5-7105X
THEW SHOV.	P33438
THEW SHOV.	P33460
THEW SHOV.	P8814
TRANSMISS.	UK585S
TRU-CROSS	7105
TRW	20101
UJS	55-7105
WABCO	2046974
WABCO	TC4412-1
WABCO	TH7986
WABCO	VJ0763
WABCO	VJ3948
WESCO	611-705
WESCO	611-715
WESCO	N7105
WHITE	02-7074065
ZELLER	5050DA
ZELLER	5050OA

IDLI GROUP 80214

CROSS + 4 LOW WING DRILLED BEARINGS

SEE FIGURE 1G

1.9375 in	CC	49.21 mm
5.50 in	CL	139.70 mm
1.0 in	CD	25.40 mm

AEC	AE7202
ALCO	J7256
ALLOY	J7101
ALLOY	J7102
AUST.WEST.	1566421
AUST.WEST.	5208351
AUST.WEST.	PGF18913
AUST.WEST.	WRF2882
BALKAMP	3-906
BALKAMP	3-907
BALKAMP	3-927
BALKAMP	M7102
BORG-WARNR	114-7001
BORG-WARNR	114-7202
BORG-WARNR	114-7217
BUCYR-ERIE	46-813-675(4)
BUFFALO-SP	208260014(4)
BUFFALO-SP	208260020(4)
BUFFALO-SP	848037721(4)
BUFFALO-SP	9003400
CASE,J.I.	A12878
CASE,J.I.	L31173
CATERPILLR	4L6325
CATERPILLR	7J5242
CLARK EQU.	1566421
CLARK EQU.	1577604
CLARK EQU.	5207973
CLARK EQU.	5208351
CLARK EQU.	893815
DEERE,JOHN	AT36356
EATON	2006731
EATON	5086258
FIAT-ALLIS	70692644
FWD	46188
GERLINGER	642249
GERLINGER	644056
GLAENZER	065-106-8
GMC	9022626
GMC	9024786
GMC	9025579
GMC	9026501
GMC	9052140
GMC	9052141
HYSTER	55-7102
IHC	631-934C91
INBRA	CT17-100
JOY MFG.	4520128
LINK BELT	16J53(4)
MACK	201SJ49A
MASS.FERG.	836-666M91
MCQUAY-NOR	U03-32
MIXERMOBL	708656
NEAPCO	287207
NEAPCO	5-0727
PETTI-MULL	LA737
PETTI-MULL	LL773-45
PRECISION	906
REPUBLIC	CB706
SIL.HOIST	24351-1(4)

SIL.HOIST	24351-2(4)
SIL.HOIST	25759(4)
SIL.HOIST	25776-11(4)
SIL.HOIST	25776-1(4)
SIL.HOIST	25776(4)
SIL.HOIST	25896(4)
TAYLOR	R1D1116
THEW SHOV.	P33423
TRW	20100D
WABCO	3000840
WABCO	AD9353
WHITING	22-0040(4)

IDLI GROUP 80215

CROSS + 4 LOW WING DRILLED BEARINGS
slight variation from above group
SEE FIGURE 1G

1.9375 in	CC	49.21 mm
5.50 in	CL	139.70 mm
1.0 in	CD	25.40 mm

AMER.SNOW	3-7052(4)
AMER.SNOW	3-7053(4)
AMER.SNOW	SK413(4)
AMER.SNOW	SK414(4)
AMER.SNOW	SK518(4)
AMER.SNOW	SK523(4)
AUST.WEST.	1565585
AUST.WEST.	465123
AUST.WEST.	465124
AUST.WEST.	465132
AUST.WEST.	465136
AUST.WEST.	465137
AUST.WEST.	HCF2098-57
AUST.WEST.	PGF10797-56A
AUST.WEST.	PGF8537-56
AUST.WEST.	PGF8807-57
AUST.WEST.	PGH7652
AUST.WEST.	PGH7653
BAL-LM-HAM	465137
BAL-LM-HAM	PGF10797-56A
BAL-LM-HAM	PGF7652
BAL-LM-HAM	PGF7653
BAL-LM-HAM	WRF1679-56
BLISS	26496MK-D
BORG-WARNR	114-7200
BORG-WARNR	114-7200A
BUCYR-ERIE	4-681-3555
BUCYR-ERIE	46-813-675(4)
BUFFALO-SP	208260014(4)
BUFFALO-SP	208260015(4)
BUFFALO-SP	208260018(4)
BUFFALO-SP	208260020(4)
BUFFALO-SP	208260021(4)
BUFFALO-SP	848036724(4)
BUFFALO-SP	848037721(4)
CARDWELL	1-15862
CASE,J.I.	A12877P
CASE,J.I.	D50866
CATERPILLR	2A858
CHAIN BELT	298-6044-91
CLARK EQU.	1565585
CLARK EQU.	5202127
CLARK EQU.	5205127
CLARK EQU.	5255753
CLARK EQU.	980613

CLEVE.MOT.	CM7105
CLEVE.MOT.	CM7205
CLEVE.MOT.	CM-CP68WB1
CLEVE.MOT.	CM-CP72N31
CRANE CAR.	249
CRANE CAR.	259
CRANE CAR.	6C4
CURTISS-WR	3-22746
DAVEY COM.	A24615
DEERE,JOHN	AH12461H
EATON	2006739
EATON	2047681
EATON	6954207
EIMCO	103M8826
EIMCO	105M673
FIAT-ALLIS	0083616
FIAT-ALLIS	0083616-3
FIAT-ALLIS	058735
FIAT-ALLIS	083541
FIAT-ALLIS	083616
FIAT-ALLIS	083616-3
FIAT-ALLIS	1126463
FIAT-ALLIS	1134986
FIAT-ALLIS	629944
FIAT-ALLIS	70058735
FIAT-ALLIS	70083541
FIAT-ALLIS	70083616
FIAT-ALLIS	70083616-3
FIAT-ALLIS	70629944
FIAT-ALLIS	70681008
FIAT-ALLIS	71126463
FIAT-ALLIS	71134986
FIAT-ALLIS	FEX1707
FWD	46139
FWD	46424
GEN.EQUIP.	2-22746
GERLINGER	629815
GERLINGER	644055
GLAENZER	065-108-8
GMB	G6105
GMB	G7113
GMB	G7117
GMB	G7200
GMB	G7205
GMC	21682
GMC	21683
GMC	22623
GMC	47079
GMC	9021683
GMC	9022734
GMC	9024660
GMC	9024790
GMC	9026503
GMC	9038235
GMC	9046315
GMC	9052141
HAHN	38UM141
HALLBURTON	10345702
HOWE	4954
HYSTER	55-7100
IHC	118122
IHC	118-122H1
IHC	237-899R91
IHC	238-077R91
IHC	238-449R91
IHC	255-341R91

IHC	376-892C91
IHC	925-397C91
INBRA	CT11-025
JOY MFG.	4510193
JOY MFG.	A369158
JOY MFG.	A59654
LINK BELT	8A2966(4)
LINK BELT	8A3337(4)
LINK BELT	8A3338(4)
LINK BELT	8A3613(4)
LINK BELT	AC5424(4)
LOBRO	U674
MACK	201SJ40
MACK	201SJ53
MCQUAY-NOR	U03-31
MCQUAY-NOR	U03-37
MCQUAY-NOR	U03-33
MIXERMOBL	708395
MOPAR	D49
M.R.S.	13133P11
NEAPCO	2801
NEAPCO	287205
NEAPCO	5-0725
PARSONS	325E1574(4)
PARSONS	325E813(4)
PETTI-MULL	19961
PETTI-MULL	LL773-2
PRECISION	908
REPUBLIC	CB708
REPUBLIC	CP72N54
SIL.HOIST	25759(4)
SIL.HOIST	25776-11(4)
SIL.HOIST	25776-1(4)
SIL.HOIST	25776(4)
SIL.HOIST	25896(4)
SIL.HOIST	25991-3(4)
THEW SHOV.	P33408
THEW SHOV.	P33433
TIMBERLAND	401794
TRW	20100D
TRW	20113
TULSA WIN.	23423(4)
WABCO	2067321
WABCO	804989
WABCO	VJ3051
WABCO	VJ763
WAYNE	3-22746
WHITE	10073P2
WHITING	12197(4)

IDLI GROUP 80216

CROSS + 4 LOW WING DRILLED BEARINGS
slight variation from above group
SEE FIGURE 1G

1.9375 in	CC	49.21 mm
5.50 in	CL	139.70 mm
1.0 in	CD	25.40 mm

AEC	AE5105
ALLOY	J5148
ATHEY PROD	P14368
AUST.WEST.	465066
AUST.WEST.	465067
AUST.WEST.	465068
AUST.WEST.	465071
AUST.WEST.	465073

AUST.WEST.	465076
AUST.WEST.	465217
AUST.WEST.	NRF2885
BAL-LM-HAM	465066
BAL-LM-HAM	465067
BAL-LM-HAM	465068
BAL-LM-HAM	465071
BAL-LM-HAM	465073
BAL-LM-HAM	465076
BAL-LM-HAM	465205
BAL-LM-HAM	465208
BAL-LM-HAM	465209
BAL-LM-HAM	465217
BAL-LM-HAM	NRF2885
BAL-LM-HAM	WRF2885
BORG-WARNR	114-5100A
BORG-WARNR	114-5100AB
BORG-WARNR	114-5145
BORG-WARNR	114-5148
BUFFALO-SP	309260005(4)
CASE,J.I.	A18383R
CASE,J.I.	L36152
CATERPILLR	7J5251
CATERPILLR	942737
CATERPILLR	943235
CATERPILLR	9K3970
CATERPILLR	XX1894-7
CLARK EQU.	5202096
CLARK EQU.	5202099
CLARK EQU.	5202172
CLARK EQU.	5291326
CLARK EQU.	5291327
CLARK EQU.	943235
CLARK EQU.	960110
CLARK EQU.	SA35043
CLEVE.MOT.	CM5100
CLEVE.MOT.	CM5105
CLEVE.MOT.	CM5122
COOK,J.B.	110136
CRANE CAR.	247
DAVEY COM.	A24616
DEERE,JOHN	AT36360
EATON	804735
FIAT-ALLIS	076456
FIAT-ALLIS	179502
FIAT-ALLIS	70179502
FIAT-ALLIS	70680996
FIAT-ALLIS	73066875
FWD	204362
GLAENZER	065-117-8
GMB	G5118
GMB	G5121
GMB	G5148
GMB	G5-5148X
GMC	9025578
HYSTER	163966
HYSTER	53034A
HYSTER	T462-255
IHC	136-488HA
IHC	158900
IHC	158-900H1
IHC	159900
IHC	159-900H1
IHC	238-796R91
IHC	290-928R91
IHC	296-596R91

CAUTION: BE SURE TO REFER TO ENGINEERING CATALOGS FOR SPECIAL APPLICATIONS THAT REQUIRE SPECIFIC MATERIAL CONTENT, TOLERANCES, ETC. SEE FOOTNOTE.

INBRA	CT11-060
INGER.RAND	CBW114-5100
JEBCO	3P15A
JOY MFG.	250033
JOY MFG.	A26294A
LA PORTA	LA5415
LINK BELT	AC4504(4)
LINK BELT	AC4640(4)
MCQUAY-NOR	UO3-23
MIXERMOBL	061617
MIXERMOBL	113362
M.R.S.	13411P11
MURR.TREG.	2876
NEAPCO	285148
OLIVER	155277AS
PARSONS	293D608(4)
PARSONS	603D26A(4)
PARSONS	603D565(4)
PARSONS	603D701(4)
PETTI-MULL	L1663-6
PILOT	15122
PRECISION	920
REPUBLIC	CB5000
REPUBLIC	CB720
REPUBLIC	CB799
SCHIELD	X629-29
SHOV.SUPP.	SP65142(4)
SUTTON	A101082-112
TOWMOTOR	7J5251
TOWMOTOR	82401
TRW	20086
TRW	20088D
WABCO	2053956
WABCO	473609
WABCO	VJ6880
ZELLER	4025J

IDLI GROUP 80217

CROSS + 4 DELTA WING THREADED BOLT HOLE BEARINGS
SEE FIGURE 1K

1.9375 in	BW	49.21 mm
5.50 in	CL	139.70 mm
1.0625 in	CD	26.99 mm

ALCO	CP68WB9
ALLOY	J7000
ALLOY	J7033
ALLOY	J7100
BALKAMP	3-908
BALKAMP	M7100
BORG-WARNR	114-7011
BORG-WARNR	114-7012
BORG-WARNR	114-7015
EATON	1342133-02
EATON	1359601
EIMCO	912-2949455
GMC	2041859
GMC	659973
IHC	414-946C91
IHC	416-946C91
IMCO	55-7000

IMCO	55-7035
MACK	201SJ48
MACK	35MUA29P1
MCQUAY-NOR	U03-31
M.R.S.	13133P11
NEAPCO	2868WB9
OSHKOSH	7DS15
PRECISION	770
PREC.TOR.	1470
PREC.TOR.	1477
PREC.TOR.	570
PREC.TOR.	571
PREC.TOR.	908
THEW SHOV.	P33596
VOLVO	6861043
WHITE	3106307
WHITE	KC7012

IDLI GROUP 80218

CROSS + 2 LOW WING THREADED + 2 BLOCK BEARINGS
SEE FIGURE 1KL

1.9375 in	BW	49.21 mm
5.50 in	CL	139.70 mm
1.0625 in	CD	26.99 mm

AEC	AE7202
ALCO	CP88WB10
ALLOY	J7102
BALKAMP	3-906
BALKAMP	3-927
BALKAMP	M7102
BAL-LM-HAM	PGF18913
BAL-LM-HAM	WRF2882
BORG-WARNR	114-7244
CATERPILLR	642249
CATERPILLR	681157
CLARK EQU.	948361
CLEVE.MOT.	CM7102
CLEVE.MOT.	CM7202
CLEVE.MOT.	CM-CP68WB10
CLEVE.MOT.	CM-CP72N33
DAVEY COM.	A33628
DEERE,JOHN	AT36353
FLXIBLE	13-158-159
FORD	B4TZ4635K
GEN.EQUIP.	3-22746
GMB	G2110
GMB	G7256
GMC	2041472
GMC	661805
HAHN	38UM175
HYSTER	114602A
HYSTER	75109A
MACK	35MU29P2
MACK	35MUA29P2
MCQUAY-NOR	U03-32
M.R.S.	17164P11
NEAPCO	2868WB10
PRECISION	774
PREC.TOR.	1472
PREC.TOR.	574
PREC.TOR.	906
REO MOTORS	10076P2
REPUBLIC	CB7034
TOWMOTOR	642249
TOWMOTOR	644056

TRW	20100D
WHITE	10076P2

IDLI GROUP 80219

CROSS + 4 ROUND BEARINGS + OUTSIDE SNAP RINGS
SEE FIGURE 1A

1.9375 in	BD	49.21 mm
1.1406 in	BW	28.97 mm
5.6250 in	CL	142.88 mm
1.3438 in	CD	34.13 mm

AEC	850
AEC	AE-CP850N4
BEAN,JOHN	1277466
CASE,J.I.	A25151
CLARK EQU.	1550819
CLARK EQU.	602-900-022
DEERE,JOHN	59X11468A
EATON	2950409
FIAT-ALLIS	3044617
FIAT-ALLIS	3044617-3
FIAT-ALLIS	44617
FIAT-ALLIS	70044617
FIAT-ALLIS	73044617
FIAT-ALLIS	73044617-3
GMB	G5-850X
IHC	216-477H1
IHC	393-671R91
IHC	968-894R91
INBRA	CT20-570
MASS.FERG.	P33576
MOTOR MAST	5562A
M.R.S.	326610
M.R.S.	326610AP
NATL.MINE	3881-5
NELSON	B432464
PRECISION	899
REPCO	CP55N
REPCO	CP850N4
REPUBLIC	CB850N4
RUPP	S1891
THEW SHOV.	P33576
WHITE	02-7091891

IDLI GROUP 80220

CROSS + 4 SLOTTED BEARINGS + THRUST AND LOCK PLATES
SEE FIGURE 1D

2.0000 in	BD	50.80 mm
1.2188 in	BW	30.96 mm
5.5625 in	CL	141.29 mm
1.3906 in	CD	35.32 mm

AEC	7N
AEC	AE-CP7N
AEC	CP7N
ALCO	1875
ALCO	CP7N
ALLOY	1875
ALLOY	CP7N
BALKAMP	CP7N
BORG-WARNR	CP7N
CHAIN BELT	298-6024-91
CHAIN BELT	37132
CHAIN BELT	7400
CLARK EQU.	952374
CLARK EQU.	954796

CLEVE.MOT.	CMCP7N
FEDERAL	5Y108
FIAT-ALLIS	3038633
FIAT-ALLIS	3038633-8
FIAT-ALLIS	38633
FIAT-ALLIS	70038633
FIAT-ALLIS	70039559
FIAT-ALLIS	73038633
FIAT-ALLIS	73038633-8
FLEETRITE	ZADB70N
FWD	58162
FWD	635900
GERLINGER	L15460
GERLINGER	L17600
GLAENZER	065-259-8
GLAENZER	065-893-6
GMB	GCP7N
GMB	GGPL7N
GMB	GU5900
GMC	9034965
GMC	9041260
GMC	9042578
HARNISCH.	1018Z204
IMCO	55-7N
IOWA MFG.	45375-017-02
IOWA MFG.	45375-017-03
KENWORTH	1000K24
LEMPCO	U509
LOBRO	U590
MCQUAY-NOR	MACK
MOTIVE	55-7N
MOTOR MAST	5556A
MOTOR MAST	B5556
NEAPCO	28CP7N
NEAPCO	5-0070
PERFECTION	RG7N
PILOT	B70N
PILOT	CP7N
PRECISION	893
REPCO	CP7N
REPUBLIC	CB7N
THEW SHOV.	P353
THEW SHOV.	P371
THEW SHOV.	P399
TOWMOTOR	15460
TRANSMISS.	UK202
WESCO	611007
WESCO	B7N
WHITE	842227
WHITE	X11-39
ZELLER	5556A

IDLI GROUP 80221

CROSS + 4 BACKPLATE BEARINGS
SEE FIGURE 1E

1.9375 in	BD	49.21 mm
1.1875 in	BW	30.16 mm
5.7813 in	CL	146.85 mm
1.2969 in	CD	32.94 mm

ACF BRILL	558947
AEC	280
AEC	577
AEC	577Z
AEC	AE577
AEC	AE577Z
AEC BRIT.	K2703

BD = BEARING DIAMETER (outside)   BW = BEARING WIDTH   CB = CROSS LENGTH WITH BEARINGS   CC = CENTER TO CENTER   CD = CROSS DIAMETER
CL = CROSS LENGTH WITHOUT BEARINGS   EC = END TO CENTER or FACE TO FACE   EE = END TO END   EL = EFFECTIVE LENGTH
HD = HUB DIAMETER (or insert)   IE = INSIDE OF EARS (or recess)   OD = OUTSIDE DIAMETER   OE = OUTSIDE OF EARS   SB = SPLINE OR BORE SIZE

Make	Part	Make	Part	Make	Part	Make	Part
ALCO	1517	FEDERAL	5Y7172	HARDY SPCR	K5PGB55	REPUBLIC	CB1700-4
ALCO	1802	FIAT	4582853	HERCULES	446	REPUBLIC	CB1700AG
ALCO	1803	FIAT	4620151	IHC	121762R91	REPUBLIC	CB1700HD
ALCO	J5-177	FIAT	82853	IHC	121-762R92	REPUBLIC	CB717
ALCO	N5-177	FIAT	876044	IHC	21762R92	REX CHAIN	7765
ALLIS-CHLM	2366108	FIAT	876045	IHC	239-723R91	RUEDARSA	1310
ALLOY	1802	FIAT-ALLIS	2366108	IHC	3100758R91	RUEDARSA	2105-1310-0-0
ALLOY	1803	FLEETRITE	ZAD11517	IHC	347-385R91	SAVIEM	23232059
ALLOY	J5-177	FLEETRITE	ZAD11577	IHC	697-690R91	SAVIEM	870290900
ALLOY	J5-280	FMC	2-41485	IHC	697769R91	SPICER	1700
ARMSTRONG	5410-5000	FMC	7-21453	INBRA	CC517	SPICER	1710
AUTOCAR	3AL01150	FODENS	Y00601508	INBRA	CT17-005	SPICER	5-117X
AUTOCAR	3AL0150	FORD	2002E4635B	ITALCARDAN	50-755-000	SPICER	5-17
AUTOCAR	3AL150	FORD	C6TZ4635F	JAMCO	UJ280	SPICER	5-171X
AUTOCAR	3FA01172	FORD	C6TZ4635K	LEADER	T10403	SPICER	5-177X
AYRA DUREX	1-705-01	FORD	D3TZ4635B	LEMPCO	U106	SPICER	5-17X
BAY CITY	1700	GLAENZER	065-115-8	LOBRO	U500	SPICER	5-217X
BERLIET	943540918	GLAENZER	065-146-8	MACK	201SJ15	SPICER	5-233X
BORG-WARNR	114-517	GLAENZER	765-000-0	MACK	201SJ15A	SPICER	5-280-4X
BORG-WARNR	114-517R	GLAENZER	765-002-0	MACK	201SJ15B	SPICER	5-280X
BORG-WARNR	114-517X	GLAENZER	765-010-2	MACK	201SJ23	SPICER	5-306X
BORG-WARNR	114-577	GLAENZER	765-014-0	MACK	201SJ47	SPICER	J5-17
BRD	08-510106	GLAENZER	765-016-0	MACK	2104-5280X	STEIGER	02-1197
BROCKWAY	139957	GMB	G5-117	MACK	63RU141	STEIGER	17-009
BROCKWAY	239052	GMB	G5-177	MASTERS	MS517X	TEDDY TORQ	G5-177
BROCKWAY	239068	GMB	G5-177X	MASTERS	MS577X	TEDDY TORQ	K17
BROCKWAY	239327	GMB	G5-280X	MCQUAY-NOR	U02-30	THEW SHOV.	P27525
BUCKEYE	1700	GMB	G5-517	MCQUAY-NOR	UO2-30	THEW SHOV.	P32193
CATERPILLR	5B423	GMB	GU5000	MIPER	C3195	TRANSMISS.	UK200
CATERPILLR	5B426	GMB	GU5000SL	MIPER	C3197	TRU-CROSS	280
CATERPILLR	5B427	GMC	2058300	MIPER	C3200	TRW	20106
CATERPILLR	5B428	GMC	2173960	MOPAR	D79	TRW	20135
CENTRAL	G5-177X	GMC	2222853	MOTOR MAST	5278A	UJS	55-177
CHRYSLER	2196041	GMC	2301520	MOTOR MAST	5278AB	UNIC	106010
CHRYSLER	2198041	GMC	2322137	MOTOR MAST	5278AG	VOLVO	231009
CHRYSLER	3820254	GMC	2322138	MOTOR MAST	5278AG4	WAYNE	2-41485
CLARK EQU.	321SC171	GMC	2366104	MUIR-HILL	261103	WAYNE	2-4185
CLEVE.MOT.	CM5-177X	GMC	2366108	NEAPCO	28059X	WAYNE	7-21453
CLEVE.MOT.	CM5-233X	GMC	2370733	NEAPCO	2817	WESCO	611517
CLEVE.MOT.	CM5-280XGC	GMC	4016484	NEAPCO	28280X	WESCO	G1517
CLEVE.STL.	C67-55-2	GMC	47985	NEAPCO	28280X4	WESCO	N1517
CMP	CM5-177X	GMC	611905	NEAPCO	5-0059	WHITE	02-7011571
CMP	CM5-233X	GMC	612094	NEAPCO	5-0280	WHITE	02-7011692
COLES	1700 SERIES	GMC	655105	NEAPCO	5-0284	WHITE	02-7074031
COLES	S1210AR	GMC	675082	NEAPCO	5-280	WHITE	02-7074362
CURTISS-WR	2-4185	GMC	688016	OSHKOSH	1705-91	WHITE	10090P2
D&D MACH.	MS280X	GMC	694968	OSHKOSH	17DS91	WHITE	406606
D&D MACH.	MS517X	GMC	695230	PERFECTION	RG517X	WHITE	406606A
DAIM.RUBE.	813-223-603-2	GMC	695232	PILOT	11517	WHITE	406606B
DIAMOND R	10046P2	GMC	702129	PILOT	11577	WHITE	559P2
DIAMOND R	10062P2	GMC	7500055	PILOT	15177	WHITE	7199J
DIAMOND R	10090P2	GMC	9037278	PITT.VIOL.	400.755.000	WHITE	858P2
DIAMOND R	7199J	GMC	9039278	PITT.VIOL.	40.755.000	WHITE	859P2
DIAMOND R	859P2	GMC	9044898	PRECISION	334	WHITE	BB17044
DIAMOND R	BB17044	GMC	9044961	PREC.TOR.	364	WHITE	BB17199J
DIAMOND R	BB17199J	GMC	9053746	PREC.TOR.	917	WHITE	BB7044
DIAMOND R	BB7044	GMC	9364269	QUINT.HAZ.	QL1708	WHITE	BB7199J
DIAMOND R	BB7045	GMC	VS11996	RAY GO	7765	WYLIE	28
DIAMOND T	17199J	GMC	X47259	RENAULT	08.70.290.900	ZELLER	5278A
DIAMOND T	7199J	GURRIES	7765	RENAULT	23.232.059	ZELLER	5278AG
DIAMOND T	BB17044	HABERLE	H1517X	REO MOTORS	859P2		
DIAMOND T	BB7044	HARDY SPCR	7051-017Y	REO MOTORS	B87045		
DIAMOND T	BB7199J	HARDY SPCR	7151-001Y	REPCO	5-280X		
ERF	U11805	HARDY SPCR	K5-17	REPCO	K5-17R		
EXCEL	U106	HARDY SPCR	K5GB17	REPCO	K5280X		
FEDERAL	5Y529	HARDY SPCR	K5GB55	REPUBLIC	CB1700		

CAUTION: BE SURE TO REFER TO ENGINEERING CATALOGS FOR SPECIAL APPLICATIONS THAT REQUIRE SPECIFIC MATERIAL CONTENT, TOLERANCES, ETC. SEE FOOTNOTE.

IDLI GROUP 80221

CROSS + 2 LOW WING THREADED + 2 BLOCK BEARINGS
SEE FIGURE 1JL

2.8125 in	CC	71.44 mm
5.8438 in	CL	148.43 mm
1.50 in	CD	38.10 mm

ALCO	J8515
AUST.WEST.	WRF2885
BORG-WARNR	114-8508
BORG-WARNR	114-8515
FIAT-ALLIS	70681049
FMC	R4177795
GMB	G8515
PRECISION	950
REPUBLIC	CB750
REPUBLIC	CB85WB7

IDLI GROUP 80223

CROSS + 4 DELTA WING THREADED BOLT HOLE BEARINGS
SEE FIGURE 1K

2.8125 in	CC	71.44 mm
5.8438 in	CL	148.43 mm
1.50 in	CD	38.10 mm

AEC	8500
AEC	AE8500
BALKAMP	3-962
BORG-WARNR	114-8500
BORG-WARNR	114-8502
BORG-WARNR	114-8503
BORG-WARNR	114-8504
BORG-WARNR	114-8506
BORG-WARNR	114-8507
BORG-WARNR	114-8511
BORG-WARNR	114-8513
BORG-WARNR	114-8514
CATERPILLR	1V3639
CATERPILLR	7K442
CLEVE.MOT.	CM8500
FIAT-ALLIS	0069962(2)
FIAT-ALLIS	0069962-9(2)
FIAT-ALLIS	0090311(2)
FIAT-ALLIS	0090311-2(2)
FIAT-ALLIS	069953
FIAT-ALLIS	069962(2)
FIAT-ALLIS	069962-9(2)
FIAT-ALLIS	090311(2)
FIAT-ALLIS	090311-2(2)
FIAT-ALLIS	639511
FIAT-ALLIS	70069953
FIAT-ALLIS	70069962(2)
FIAT-ALLIS	70069962-9(2)
FIAT-ALLIS	70090311(2)
FIAT-ALLIS	70090311-2(2)
FIAT-ALLIS	70639511
FIAT-ALLIS	70680958
GLAENZER	065-141-8
GMB	G5-8500X
IHC	180216
INBRA	CT11-081
INBRA	CT13-080

MCQUAY-NOR	OFF-HWY.
MICHIGAN	944686
MICHIGAN	946189
MOTOR MAST	6000A
NEAPCO	2862
NEAPCO	2863
NEAPCO	288500
NEAPCO	288516
NEAPCO	6-8500
PILOT	18500
PRECISION	785
PREC.TOR.	962
REPCO	114-8500
REPUBLIC	CB762
TRW	20173
WESCO	811850
WESCO	811857
WESCO	N8500
WESCO	N8507

IDLI GROUP 80224

CROSS WITH HOLE IN THE CENTER + 2 HIGH WING DRILLED + 2 BLOCK BRGS.
SEE FIGURE 1L

2.8125 in	CC	71.44 mm
6.0 in	CL	152.40 mm
1.3281 in	CD	33.73 mm

AEC	8502
AEC	AE8502
ALCO	J8502
ALLIS-CHLM	069962
ALLIS-CHLM	090311
ALLIS-CHLM	179504
ALLIS-CHLM	179507
ALLIS-CHLM	179704
ALLIS-CHLM	179904
ALLIS-CHLM	179909
ALLIS-CHLM	180140
ALLOY	J8500
BAR-GREENE	11078
BORG-WARNR	114-8502
BORG-WARNR	114-8503
BORG-WARNR	114-8506
BORG-WARNR	114-8507
BORG-WARNR	114-8511
BORG-WARNR	114-8513
BORG-WARNR	114-8516
CATERPILLR	2T96
CATERPILLR	2V7153
CHAMPION	14679
CLARK EQU.	1570944
CLARK EQU.	1580982
CLARK EQU.	899603
CLARK EQU.	944686
CLARK EQU.	94468613
CLARK EQU.	946189
CLARK EQU.	946640
CLARK EQU.	948508
CLARK EQU.	960111
CLARK EQU.	960699
CLARK EQU.	960704
CLEVE.MOT.	CM8502
CLEVE.MOT.	CM8516
CLEVE.MOT.	CM-CP85WB12
CLEVE.MOT.	CM-CP85WB20

COLES	1570944
EATON	6951594
EIMCO	920A120015
FIAT-ALLIS	0180140
FIAT-ALLIS	0180140-6
FIAT-ALLIS	179151
FIAT-ALLIS	179504
FIAT-ALLIS	179507
FIAT-ALLIS	179704
FIAT-ALLIS	179904
FIAT-ALLIS	179909
FIAT-ALLIS	180140
FIAT-ALLIS	180140-6
FIAT-ALLIS	4498505
FIAT-ALLIS	70179151
FIAT-ALLIS	70179507
FIAT-ALLIS	70179704
FIAT-ALLIS	70179904
FIAT-ALLIS	70179909
FIAT-ALLIS	70180140
FIAT-ALLIS	70180140-6
FIAT-ALLIS	70687517
FIAT-ALLIS	73107487
FIAT-ALLIS	73122595
FIAT-ALLIS	74498505
FWD	131-32
GLAENZER	065-142-8
GMB	G5-8502X
GMB	G5-8516
GMB	G8500
GMB	G8502
GMB	G8506
GMB	G8507
GMB	G8515
GMB	GUK08
IHC	180-216H2
IHC	302-737C91
IHC	921-918C1
INBRA	CT11-100
KOMATSU	175-20-00060
MICHIGAN	944686
MICHIGAN	946189
MOTOR MAST	6000B
M.R.S.	21115P11
NEAPCO	288516
PILOT	18506
PRECISION	963
REPCO	114-8502
REPCO	114-8507
REPCO	114-8516
REPUBLIC	CB744
REPUBLIC	CB763
THEW SHOV.	P33446
THEW SHOV.	P33466
TRW	20107
TWIN DISC	X206340
ZELLER	6000B

IDLI GROUP 80225

CROSS + 4 BACKPLATE BEARINGS
SEE FIGURE 1E

2.3125 in	BD	58.74 mm
1.1875 in	BW	30.16 mm
6.2813 in	CL	159.55 mm
1.6875 in	CD	42.86 mm

ADAMS-LET.	TH7436
AEC	1607
AEC	524
AEC	524Z
AEC	AE524
AEC	AE524Z
AEC BRIT.	K2708
ALCO	1524T
ALCO	1804
ALCO	J5-124
ALCO	J5-124T
ALLOY	1804
ALLOY	J5-124
ALLOY	J5-124T
AUTOCAR	3GA01172
AUTOCAR	3GC01172
AYRA DUREX	1-805-01
BARREIROS	23-337-041
BEDFORD	943541018
BERLIET	943541018
BORG-WARNR	114-524
BORG-WARNR	5-24X
BRD	09-520756
BROCKWAY	239220
CLARK EQU.	321SD171
CLEVE.MOT.	CM5-124X
CLEVE.MOT.	CM5-124XGC
CLEVE.MOT.	CM5-272X
CLEVE.MOT.	CM5-286XGC
CLEVE.MOT.	CMS124X
CMP	CM5-124X
CMP	CM5-272X
D&D MACH.	MS524X
DIAMOND R	10063P2
DIAMOND R	216038T2
DIAMOND R	7199M
DIAMOND R	734769P2
DIAMOND T	7199M
EIMCO	105M672
EIMCO	1800
ERF	U19397
ERF	U20138
EXCEL	U107
FEDERAL	5Y339
FEDERAL	5Y528
FIAT	4021288
FIAT	4556728
FIAT	4622384
FIAT	4641105
FLEETRITE	ZAD11524
FODENS	Y00601802
FORD	C7TZ4635A
FORD	D1HZ4635A
FORD	D2HZ4635B
GLAENZER	065-116-8
GLAENZER	865-000-0
GLAENZER	865-002-0
GLAENZER	865-006-0
GLAENZER	865-011-0

BD = BEARING DIAMETER (outside)   DW = BEARING WIDTH   CB = CROSS LENGTH WITH BEARINGS   CC = CENTER TO CENTER   CD = CROSS DIAMETER
CL = CROSS LENGTH WITHOUT BEARINGS   EC = END TO CENTER or FACE TO FACE   EE = END TO END   EL = EFFECTIVE LENGTH
HD = HUB DIAMETER (or insert)   IE = INSIDE OF EARS (or recess)   OD = OUTSIDE DIAMETER   OE = OUTSIDE OF EARS   SB = SPLINE OR BORE SIZE

GMB	G5-124
GMB	G5-124T
GMB	G5-124X
GMB	G5-266X
GMB	G5-524
GMC	1335006
GMC	2098536
GMC	2224005
GMC	2335005
GMC	2335006
GMC	2336727
GMC	4026301
GMC	47093
GMC	47903
GMC	678510
GMC	678510B
GMC	678512
GMC	84559
GMC	9030836
GMC	9031316
GMC	9049686
GMC	9385348
GMC	9389358
GMC	9392534
HARDY SPCR	5-24X
HARDY SPCR	8051-024Y
HARDY SPCR	8151-001Y
HOUGH	878624C91
HYSTER	166686
HYSTER	75038
IHC	321-291R91
IHC	321921
IHC	321-921R91
IHC	338-799R92
IHC	634-362C91
IHC	74180R93
IHC	878626C91
IHC	974-108R93
IHC	974120R91
IHC	974180R91
IHC	974180R93
INBRA	CT17-030
ITALCARDAN	50-732-000
JEFFREY	1801
JEFFREY	1808
JEFFREY	444460
KATAOKA	MI85-1961
LEMPCO	U107
LEMPCO	U112
MACK	201SJ16
MACK	201SJ16A
MACK	201SJ16B
MACK	201SJ35
MACK	21045124X
MASTERS	MS524X
MCQUAY-NOR	U02-31
MCQUAY-NOR	UO2-31
MIPER	C3265
MOPAR	D66
MOTOR MAST	6228A
MOTOR MAST	6228AG
MOTOR MAST	6828A
NEAPCO	28055X
NEAPCO	28124X
NEAPCO	2818
NEAPCO	28226X
NEAPCO	6-0124
NEAPCO	6-0226
OSHKOSH	12DS4
OSHKOSH	12DS8
OSHKOSH	18DS5
OSHKOSH	18DS8
PERFECTION	RG254X
PERFECTION	RG524X
PILOT	1124
PILOT	11524
PITT.VIOL.	40.756.000
PRECISION	335
PREC.TOR.	918
QUINT.HAZ.	QL2409
REPCO	5-124X
REPCO	5-24X
REPUBLIC	CB1800
REPUBLIC	CB1800AG
REPUBLIC	CB1800HD
REPUBLIC	CB180D
REPUBLIC	CB1860
REPUBLIC	CB718
RUEDARSA	2105-3500-0-0
RUEDARSA	3500
SPICER	1800
SPICER	200822X
SPICER	5-123X
SPICER	5-124X
SPICER	5-218X
SPICER	5-226X
SPICER	5-236X
SPICER	5-24X
SPICER	5-272X
SPICER	5-286X
SPICER	J5-1124
TEDDY TORQ	G5-124
TEDDY TORQ	K24
TRANSMISS.	UK265
TRU-CROSS	124
TRW	20108
TRW	20136
UJS	55-124
UNIC	653650
WABCO	TH7436
WESCO	911524
WESCO	B8N
WESCO	N1524
WHITE	02-7074032
WHITE	216036A
WHITE	216038
WHITE	216038A
WHITE	216038T2
WHITE	3109949
WHITE	7199M
WHITE	734769
WHITE	734769P2
WHITE	773215
WYLIE	65
ZELLER	6228A
ZELLER	6228AG

IDLI GROUP 80226

CROSS + 4 BACKPLATE BEARINGS
SEE FIGURE 1E

1.9375 in	BD	49.21 mm
1.25 0 in	BW	31.75 mm
6.6875 in	CL	169.86 mm
1.2969 in	CD	32.94 mm

AEC	407
FORD	114-407
FORD	D9HZ4635B
GMB	G5-407X
GMC	2040210
MACK	2104-5407X
MOTOR MAST	6269
MOTOR MAST	6269A
NEAPCO	28407X
NEAPCO	6-0407
PRECISION	376
SPICER	1760
SPICER	5-407X

IDLI GROUP 80227

CROSS + 4 BACKPLATE BEARINGS
SEE FIGURE 1E

1.9375 in	BD	49.21 mm
1.25 0 in	BW	31.75 mm
7.25 0 in	CL	184.15 mm
1.2969 in	CD	32.94 mm

AEC	281
AEC	5-281Z
AEC	AE281
AEC	AE5-281
AEC	AE5-281Z
ALCO	1810
ALLOY	1810
ALLOY	J5-281
BORG-WARNR	114-640
CHRYSLER	3822292
CHRYSLER	5-281X
CLEVE.MOT.	CM1810
CLEVE.MOT.	CM3188
CLEVE.MOT.	CM5-281X
CLEVE.MOT.	CM5-281XGC
CMP	CM1810
CMP	CM5-281X
DIAMOND R	10047P2
FORD	CD5AZ4635L
FORD	D2HZ4635A
GMB	G5-281X
GMB	G5-640
GMC	2055552
GMC	4025348
GMC	701639
GMC	9052692
IHC	22910C91
IHC	418-367C91
IHC	422-910C91
LEMPCO	U143
LEMPCO	U514
MACK	2104-5281X
MCQUAY-NOR	U02-28
MOTOR MAST	7225A
MOTOR MAST	7225AG
NEAPCO	28281X
NEAPCO	6-0281
NEAPCO	6-281

PRECISION	381
REPCO	114-596
REPCO	5-281X
REPCO	CP18N3
REPCO	K5281X
REPUBLIC	CB1372
REPUBLIC	CB1810
REPUBLIC	CB1810AG
REPUBLIC	CB1810HD
SPICER	1810
SPICER	5-24X
SPICER	5-281-4X
SPICER	5-281X
TRU-CROSS	281
TRW	20132
TRW	20137
WESCO	911525
WESCO	N1525
WHITE	02-7074363
WHITE	04-9047101
WYLIE	66
ZELLER	7225AG

IDLI GROUP 80228

CROSS + 4 ROUND BEARINGS
+ OUTSIDE SNAP RINGS
SEE FIGURE 1A

2.1875 in	BD	55.56 mm
1.4375 in	BW	36.51 mm
7.5938 in	CL	192.88 mm
1.3750 in	CD	34.93 mm

AEC	585
AEC	5852
AEC	AE585
AEC	AE585Z
ALCO	J5-185
BARREIROS	34-004-022
BORG-WARNR	114-585
BORG-WARNR	114-641
CATERPILLR	4L5920
CLEVE.MOT.	CM5-185X
CLEVE.MOT.	CM5-185XGC
CMP	CM5-185X
GLAENZER	065-159 8
GMB	G5-185
GMB	G5-185X
GMC	2065059
GMC	4023541
GMC	9030138
GMC	9031109
GMC	9034128
GMC	9051148
IHC	239-551R91
IHC	293-551C91
LEMPCO	U137
MACK	201SJ32
MACK	201SJ33
MACK	201SJ37
MACK	201SJ41
MACK	21045185X
MCQUAY-NOR	U03-34
MCQUAY-NOR	UO3-34
MOTOR MAST	7244A
MOTOR MAST	7244AG
MOTOR MAST	7244G
NEAPCO	28185X

CAUTION: BE SURE TO REFER TO ENGINEERING CATALOGS FOR SPECIAL APPLICATIONS THAT REQUIRE SPECIFIC MATERIAL CONTENT, TOLERANCES, ETC. SEE FOOTNOTE.

NEAPCO	6-0185
PILOT	11585
PRECISION	932
REPCO	5-185X
REPCO	5-257X
REPCO	5-308X
REPUBLIC	CB1855
REPUBLIC	CB1855AG
REPUBLIC	CB1855HD
REPUBLIC	CB732
SPICER	1850
SPICER	5-185X
SPICER	5-227X
SPICER	5-257X
TRU-CROSS	185
TRW	20138
WABCO	TL4948
WESCO	811585
WESCO	811858
WESCO	896173
WESCO	N1585
WHITE	896173

IDLI GROUP 80229

CROSS + 4 BACKPLATE BEARINGS
SEE FIGURE 1E

2.1875 in	BD	55.56 mm
1.75 0 in	BW	44.45 mm
7.5938 in	CL	192.88 mm
1.3750 in	CD	34.93 mm

AEC	308
AEC	5-308
AEC	AE5-308
AEC	AE589
BORG-WARNR	114-642
CHRYSLER	5-308X
CLEVE.MOT.	CM5-308X
CMP	CM5189X
MACK	2104-5308X
MCQUAY-NOR	HD TRK.
MOTOR MAST	7244DA
NEAPCO	28308X
NEAPCO	6-0308
PRECISION	388
REPCO	5-308X
REPUBLIC	CB1880A
SPICER	1880
SPICER	5-308X
TRW	20180

IDLI GROUP 80230

CROSS + 2 ROUND BEARINGS
2 MIDWING DRILLED BEARINGS
+ INSIDE AND OUTSIDE SNAP RINGS
SEE FIGURE 1AK

1.6250 in	BD	41.28 mm
1.9375 in	CC	49.21 mm
7.75 0 in	CL	196.85 mm
1.0 in	CD	25.40 mm

AEC	8014
AEC	AE8014

AEC	AE8203
ALCO	J8014
ALCO	J8104
BORG-WARNR	114-8014
CATERPILLR	2H801
GLAENZER	065-125-8
GMB	G5-8014X
GMB	G8014
INBRA	CT13-020
MOTOR MAST	7075E
NEAPCO	288014
PRECISION	931
REPCO	114-8014
REPUBLIC	CB731
WESCO	811814
WESCO	N8014

IDLI GROUP 80231

CROSS AND LOW OR HIGH WING TYPE
DRILLED OR THREADED BEARINGS
SEE FIGURE 1J

1.9375 in	CC	49.21 mm
7.75 0 in	CL	196.85 mm
1.0 in	CD	25.40 mm

ADAMS-LET.	TD5261
AEC	8200
AEC	AE8000
AEC	AE8200
ALCO	1851
ALCO	8100
ALCO	8103
ALCO	J8200
ALLIS-CHLM	178183
ALLIS-CHLM	179904
ALLOY	1851
ALLOY	J8100
ALLOY	J8200
ALLOY	J8205
ATHEY PROD	A1177741
ATHEY PROD	A177741
ATHEY PROD	A277742
BALKAMP	3-910
BALKAMP	3-962
BALKAMP	M8000
BALKAMP	M8002
BALKAMP	M8018
BALKAMP	M8025
BALKAMP	M8027
BALKAMP	M8100
BALKAMP	M8200
BALKAMP	M8205
BALKAMP	M8212
BORG-WARNR	114-8000
BORG-WARNR	114-8002
BORG-WARNR	114-8018
BORG-WARNR	114-8025
BORG-WARNR	114-8100
BORG-WARNR	114-8200
BORG-WARNR	114-8200AB
BORG-WARNR	114-8205
BORG-WARNR	114-8210
BORG-WARNR	114-8212
BORG-WARNR	114-8217
BORG-WARNR	114-8218
BORG-WARNR	114-8222
BORG-WARNR	114-8223

BORG-WARNR	114-8226
BORG-WARNR	J8002
BORG-WARNR	J8100
CLEVE.MOT.	CM8100
CMP	CM8100
CMP	CM8200
CMP	CM8205
CMP	CM-CP82N28
D&D MACH.	MM8100
DEERE,JOHN	AR29054R
DEERE,JOHN	AU12520V
DEERE,JOHN	U12521U
DEERE,JOHN	XAW9525A
DIAMOND R	10078P2
EATON	2006749
EXCEL	U231
FLEETRITE	ZAD18205
FORD	B80H4635A
FORD	B8QH4635A
FORD	C4AZ4635A
FORD	D2HA4635AA
FORD	D3HZ4635C
GMB	G5-8200X
GMB	G5-8205
GMB	G8200
GMB	G8205
GMC	9026508
HABERLE	H8002
HARERLE	H8002
HOUGH	127002HI
IHC	177428
IHC	301725R91
IHC	302725R91
IHC	69557K93
IHC	69557R91
IHC	69557R92
IHC	69-557R93
IHC	79-557R92
IMCO	55-8100
IOWA MFG.	7329-041
JOY MFG.	4520233
JOY MFG.	A177741
JOY MFG.	A183515
LEMPCO	U231
MACK	201SJ52
MASTERS	MM8100
MCQUAY-NOR	U03-41
MICHIGAN	92850
MIXERMOBL	060868
MOPAR	D46
MOTIVE	55-8100
MOTOR MAST	7075A
NEAPCO	2810
NEAPCO	288100
NEAPCO	288200
NEAPCO	288205
NEAPCO	6-8205
PERFECTION	RG8002
PERFECTION	RG8202
PETTIBONE	L2893-3
PETTI-MULL	LA425-7
PILOT	18000
PILOT	18002
PILOT	18025
PILOT	18100
PILOT	18101

PILOT	18205
P.M.	A20411
PRECISION	581
REO MOTORS	10078P2
REPCO	114-8200
REPCO	114-8205
REPCO	114-8212
REPUBLIC	CB710
REPUBLIC	CB8002
SANFOR-DAY	450-0233
SPICER	5-8002
SPICER	5-8002X
SPICER	5-8200
SPICER	5-8200X
TRU-CROSS	8200
TRW	20109
UJS	55-8100
WESCO	811800
WESCO	N8000
WHITE	02-7074070
WOOLDRIDGE	2-24006
ZELLER	7075A

IDLI GROUP 80232

CROSS AND LOW OR HIGH WING TYPE
DRILLED OR THREADED BEARINGS
SEE FIGURE 1IJ

1.9375 in	CC	49.21 mm
7.75 0 in	CL	196.85 mm
1.0 in	CD	25.40 mm

AEC	8201
AEC	AE8101
AEC	AE8201
ALCO	1852
ALCO	8101
ALCO	J2801
ALCO	J8201
ALLIS-CHLM	178323
ALLIS-CHLM	178323-2
ALLIS-CHLM	58-000-505-405
ALLOY	1852
ALLOY	J8201
BALKAMP	3-913
BALKAMP	3-914
BALKAMP	M8003
BALKAMP	M8023
BALKAMP	M8101
BALKAMP	M8201
BALKAMP	M8206
BALKAMP	M8218
BORG-WARNR	114-8003
BORG-WARNR	114-8023
BORG-WARNR	114-8101
BORG-WARNR	114-8201
BORG-WARNR	114-8206
BORG-WARNR	114-8211
BORG-WARNR	114-8221
BORG-WARNR	J8101
CLEVE.MOT.	CM8003
CLEVE.MOT.	CM8101
CLEVE.MOT.	CM8206
CMP	CM8003
CMP	CM8101
CMP	CM8206
CMP	CM-CP82N30
D&D MACH.	MM8101

BD = BEARING DIAMETER (outside)   BW = BEARING WIDTH   CB = CROSS LENGTH WITH BEARINGS   CC = CENTER TO CENTER   CD = CROSS DIAMETER
CL = CROSS LENGTH WITHOUT BEARINGS   EC = END TO CENTER or FACE TO FACE   EE = END TO END   EL = EFFECTIVE LENGTH
HD = HUB DIAMETER (or insert)   IE = INSIDE OF EARS (or recess)   OD = OUTSIDE DIAMETER   OE = OUTSIDE OF EARS   SB = SPLINE OR BORE SIZE

IDLI GUIDE   COPYRIGHT ® INTERCHANGE, INC.   ST. LOUIS PARK, MN. 55416 USA

DEERE,JOHN	AU21521U
DEERE,JOHN	AU2152U
DEERE,JOHN	AV21521V
DEERE,JOHN	U1252U
DIAMOND R	10079P2
DOMINION	3975
FIAT-ALLIS	178823
FIAT-ALLIS	58-000-405
FIAT-ALLIS	58-000-505
FIAT-ALLIS	70178823
FLEETRITE	ZAD115200
FLEETRITE	ZAD18101
FLEETRITE	ZAD18206
GMB	G2801
GMB	G5-8201X
GMB	G5-8206
GMB	G8201
GMB	G8205
HOUGH	127019HI
IMCO	55-8101
JOY MFG.	3382RC
JOY MFG.	A177744
JOY MFG.	A183516
MASTERS	MM8101
MCQUAY-NOR	U03-39
MICHIGAN	942573
MOPAR	D47
MOTOR MAST	7075F
NEAPCO	2813
NEAPCO	2831
NEAPCO	288101
NEAPCO	288201
NEAPCO	288206
NEAPCO	.6-8206
PERFECTION	RG8003
PILOT	18003
PILOT	18101
PILOT	18206
PRECISION	582
REO MOTORS	10079P2
REPCO	114-8201
REPCO	114-8206
REPUBLIC	CB713
REPUBLIC	CB731
REPUBLIC	CB8101
SPICER	5-8201
SPICER	5-8201X
SPICER	5-8202X
TRW	20111
UJS	55-8101
WESCO	811811
WESCO	N8101
WESCO	N8105
WHITE	02-7074071
ZELLER	7075F

IDLI GROUP 80233

CROSS + 2 LOW WING THREADED + 2 BLOCK BEARINGS
SEE FIGURE 1JL

2.8125 in	CC	71.44 mm
7.75 0 in	CL	196.85 mm
1.3281 in	CD	33.73 mm

AEC	9017
AEC	9026
AEC	AE9002
AEC	AE9017
AEC	AE9026
ALCO	1867
ALCO	CP88WB10
ALCO	J9010
ALCO	J9017
ALCO	J9030
ALLOY	1865
ALLOY	1867
ALLOY	J9017
ALLOY	J9026
AMER.SNOW	.2-3861
AMER.SNOW	2-3861(4)
BALKAMP	.3-912
BALKAMP	M9002
BALKAMP	M9004
BALKAMP	M9010
BALKAMP	M9012
BALKAMP	M9017A
BALKAMP	M9026
BORG-WARNR	114-9002
BORG-WARNR	114-9010
BORG-WARNR	114-9012
BORG-WARNR	114-9017
BORG-WARNR	114-9021
BORG-WARNR	114-9026
BORG-WARNR	114-9030
BORG-WARNR	114-9047
BORG-WARNR	114-9107
CLARK EQU.	992656
CLARK EQU.	994063
CLEVE.MOT.	CM9012
CLEVE.MOT.	CM9017
CLEVE.MOT.	CM9026
CLEVE.MOT.	CM9200
CLEVE.MOT.	CMCP88WB10
CLEVE.MOT.	CMCP92N14
CLEVE.MOT.	CMCP92N19
CMP	CM9002
CMP	CM9012
CMP	CM9017
CMP	CM9026
CMP	CM-CP92N14
CMP	CM-CP92N19
D&D MACH.	MM9017
FLEETRITE	ZAD19017
FWD	202797
GLAENZER	065-160-8
GMB	G5-9002X
GMB	G5-9017
GMB	G5-9017X
GMB	G5-9026
GMB	G9002
GMB	G9010
GMB	G9017
GMB	G9026
GMB	G9030
GMC	9025434
GMC	9026434
GMC	9029107
GMC	9053683
GMC	9053747
GMC	9251287
HYSTER	55-9002
IHC	304-609C91
IHC	304-609C92
IHC	304609R91
IMCO	55-9002
INBRA	CT17-110
MACK	201SJ55
MACK	35MU216
MACK	35MUA216
MACK	35MUA216A
MACK	8235
MASTERS	MM9017
MCQUAY-NOR	MACK
MIXERMOBL	707680
MIXERMOBL	714743
MOPAR	D54
MOTOR MAST	7075HA
MOTOR MAST	7075HD
M.R.S.	11876P11
M.R.S.	21322P11
NEAPCO	2812
NEAPCO	2888WB10
NEAPCO	289017
NEAPCO	289026
NEAPCO	.6-9017
NEAPCO	.6-9026
PARSONS	325E754(4)
PARSONS	582-390(4)
PILOT	19017
PILOT	19026
PILOT	19028
PRECISION	792
PREC.TOR.	912
REPCO	114-9002
REPCO	114-9012
REPCO	114-9017
REPCO	114-9026
REPUBLIC	CB712
REPUBLIC	CB719
SPICER	5-9002X
TAYLOR	1930-110
TRU-CROSS	9002
TRW	20114
TRW	20174
UJS	55-9017
WESCO	911902
WESCO	911917
WESCO	N9002
WESCO	N9017
WHITING	01200
ZELLER	7075HA
ZELLER	7075HD

IDLI GROUP 80234

CROSS WITH HOLE IN THE CENTER + 2 HIGH WING DRILLED + 2 BLOCK BRGS.
SEE FIGURE 1L

2.8125 in	CC	71.44 mm
7.75 0 in	CL	196.85 mm
1.3281 in	CD	33.73 mm

ADAMS-LET.	SV8490
ADAMS-LET.	TV1499
ADAMS-LET.	TV2971
ADAMS-LET.	VL8642
ADAMS-LET.	VL8643
ADAMS-LET.	VL9009
AEC	9015
AEC	9031
AEC	AE2117A
AEC	AE9001
AEC	AE9013
AEC	AE9015
AEC	AE9016
AEC	AE9031
ALCO	1866
ALCO	1868
ALCO	J12000
ALCO	J12002
ALCO	J12003
ALCO	J9015
ALCO	J9016
ALCO	J9031
ALCO	J9038
ALLIS-CHLM	185263
ALLIS-CHLM	185263-1
ALLIS-CHLM	6000
ALLOY	1866
ALLOY	1868
ALLOY	J9015
ALLOY	J9016
ALLOY	J9031
BALKAMP	.3-903
BALKAMP	.3-904
BALKAMP	.3-911
BALKAMP	M9001
BALKAMP	M9003
BALKAMP	M9005
BALKAMP	M9009
BALKAMP	M9013
BALKAMP	M9015
BALKAMP	M9016
BALKAMP	M9020
BALKAMP	M9022
BALKAMP	M9023
BALKAMP	M9024
BALKAMP	M9028
BALKAMP	M9029
BALKAMP	M9031
BARREIROS	28-904-361
BORG-WARNR	114-12002
BORG-WARNR	114-9001
BORG-WARNR	114-9003
BORG-WARNR	114-9004
BORG-WARNR	114-9005
BORG-WARNR	114-9009
BORG-WARNR	114-9013
BORG-WARNR	114-9015
BORG-WARNR	114-9016
BORG-WARNR	114-9016A
BORG-WARNR	114-9019
BORG-WARNR	114-9020
BORG-WARNR	114-9022
BORG-WARNR	114-9023
BORG-WARNR	114-9024
BORG-WARNR	114-9025
BORG-WARNR	114-9027
BORG-WARNR	114-9028
BORG-WARNR	114-9029
BORG-WARNR	114-9031
BORG-WARNR	114-9031A
BORG-WARNR	114-9033
BORG-WARNR	114-9034
BORG-WARNR	114-9038
BORG-WARNR	114-9039

ENGINEERING CATALOGS MUST BE CONSULTED FOR DETAILS NOT INCLUDED IN THIS GUIDE. SPECIFIC DESIGNS, MATERIAL CONTENT, TOLERANCES LUBE FITTINGS AND OTHER DIMENSIONS ARE INTENTIONALLY OMITTED HERE.

CAUTION: BE SURE TO REFER TO ENGINEERING CATALOGS FOR SPECIAL APPLICATIONS THAT REQUIRE SPECIFIC MATERIAL CONTENT, TOLERANCES, ETC. SEE FOOTNOTE.

BORG-WARNR	114-9040
BORG-WARNR	114-9041
BORG-WARNR	114-9043
BORG-WARNR	114-9044
BORG-WARNR	114-9045
BORG-WARNR	114-9050
BORG-WARNR	114-9106
BORG-WARNR	J9001
CARDWELL	1-16648
CARDWELL	1-16655
CATERPILLR	2R801
CATERPILLR	3H6449
CATERPILLR	5H3756
CATERPILLR	5S5680
CATERPILLR	6K855
CATERPILLR	6K885
CATERPILLR	6M9762
CATERPILLR	7H3958
CATERPILLR	8D2144
CATERPILLR	8D3144
CATERPILLR	9H6423
CATERPILLR	9P604
CHRYSLER	5-361X
CHRYSLER	CPR109690
CLARK EQU.	1990609
CLARK EQU.	943441
CLARK EQU.	943561
CLARK EQU.	943767
CLARK EQU.	944057
CLARK EQU.	944715
CLARK EQU.	946740
CLARK EQU.	948904
CLARK EQU.	992657
CLEVE.MOT.	CM9001
CLEVE.MOT.	CM9013
CLEVE.MOT.	CM9015
CLEVE.MOT.	CM9016
CLEVE.MOT.	CM9031
CLEVE.MOT.	CM-CP88WB6
CLEVE.MOT.	CM-CP92N15
CLEVE.MOT.	CM-CP92N17
CMP	CM9001
CMP	CM9013
CMP	CM9015
CMP	CM9016
CMP	CM9031
CMP	CM-CP92N15
CMP	CM-CP92N17
D&D MACH.	MM9015
D&D MACH.	MM9016
EATON	2950199
EATON	6950199
EATON	6951586
EATON	6952210
FIAT-ALLIS	0185263
FIAT-ALLIS	0185263-1
FIAT-ALLIS	185263
FIAT-ALLIS	185263-1
FIAT-ALLIS	187964
FIAT-ALLIS	70185263
FIAT-ALLIS	70185263-1
FIAT-ALLIS	70187964
FIAT-ALLIS	70665314
FIAT-ALLIS	73060000
FLEETRITE	ZAD19015
FLEETRITE	ZAD19016
GLAENZER	065-104-8
GLAENZER	065-111-8
GMB	12003
GMB	G12000
GMB	G12002
GMB	G5031
GMB	G5-9001X
GMB	G5-9015
GMB	G5-9015X
GMB	G5-9016
GMB	G9000
GMB	G9001
GMB	G9013
GMB	G9014
GMB	G9015
GMB	G9016
GMB	G9031
GMC	44138
GMC	6933776
GMC	9022735
GMC	9025543
GMC	9025544
GMC	9025557
GMC	9025633
GMC	9025635
GMC	9027735
GMC	9027758
GMC	9034661
GMC	9034965
GMC	9044138
GMC	9044140
GMC	9244286
GRAIGG CO.	LA927
HOUGH	184843H3
HOUGH	196343HI
HOUGH	216448HI
HYSTER	55-9015
HYSTER	55-9016
IHC	184843
IHC	184-843H3
IHC	196343
IHC	196-343H1
IHC	216448
IHC	216-448H1
IHC	301578R92
IHC	301-587R91
IHC	301587R92
IHC	303-294R91
IHC	304104R91
IHC	304-105R91
IHC	324-550R91
IHC	384450R91
IHC	38450R91
IHC	702952C91
IMCO	55-9015
IMCO	55-9016
INBRA	CT13-040
INBRA	CT17-065
INBRA	CT17-090
INBRA	CT22-000
KOMATSU	195-20-11100
LEMPCO	U267
MACK	201SJ50
MACK	201SJ53
MACK	201SJ55
MACK	35MU212
MACK	35MUA212
MACK	782U
MASS.FERG.	X206262
MASTERS	MM9015
MASTERS	MM9016
MCQUAY-NOR	U03-35
MCQUAY-NOR	U03-36
MCQUAY-NOR	UO3-36
MICHIGAN	943441
MICHIGAN	943561
MICHIGAN	944715
MOPAR	D53
MOPAR	D55
MOTOR MAST	7075HB
MOTOR MAST	7075HBG
MOTOR MAST	7075HC
NATL.MINE	50377-2
NATL.MINE	76101618
NEAPCO	2804
NEAPCO	2811
NEAPCO	289015
NEAPCO	289016
NEAPCO	6-0916
NEAPCO	6-9015
NEAPCO	6-9016
PILOT	19015
PILOT	19016
PILOT	9015
PILOT	9016
PRECISION	911
PREC.TOR.	904
REPCO	114-12000
REPCO	114-12002A
REPCO	114-9003
REPCO	114-9005
REPCO	114-9015
REPCO	114-9016
REPUBLIC	CB704
REPUBLIC	CB711
REPUBLIC	CB7111
SPICER	5-361X
SPICER	5-9001
SPICER	5-9001X
SPICER	5-9016X
THEW SHOV.	P33508
TRU-CROSS	9001
TRW	20115
TRW	20116
TWIN DISC	X206262
UJS	55-9015
WABCO	SV8490
WABCO	TV1499
WABCO	TV2971
WABCO	V9009
WABCO	VD1557
WABCO	VG8035
WABCO	VG8302
WABCO	VL8642
WABCO	VL8643
WABCO	VL9009
WESCO	911-902
WESCO	911915
WESCO	911916
WESCO	911931
WESCO	911934
WESCO	N9015
WESCO	N9016
WESCO	N9031
WESCO	N9034
WHITE	02-7091918
ZELLER	7075HB
ZELLER	7075HC

IDLI GROUP 80235

CROSS + 4 DELTA WING THREADED BOLT HOLE BEARINGS SEE FIGURE 1K

2.8125 in	CC	71.44 mm
7.750 in	CL	196.85 mm
1.3281 in	CD	33.73 mm

ADAMS-LET.	SV8485
AEC	9014
AEC	AE9000
AEC	AE9014
AEC	AE9100
ALCO	CP88WB9
ALCO	J9014
ALLIS-CHLM	08-000-505-437
ALLIS-CHLM	40372
ALLIS-CHLM	40377
ALLOY	1864
ALLOY	J9014
BALKAMP	3-903
BALKAMP	M9000
BALKAMP	M9001
BALKAMP	M9006
BALKAMP	M9007
BALKAMP	M9008
BALKAMP	M9014
BALKAMP	M9018
BAL-LM-HAM	CPF11964
BORG-WARNR	114-9000
BORG-WARNR	114-9006
BORG-WARNR	114-9007
BORG-WARNR	114-9008
BORG-WARNR	114-9008AB
BORG-WARNR	114-9011
BORG-WARNR	114-9014
BORG-WARNR	114-9018
BORG-WARNR	114-9035
BORG-WARNR	114-9036
BORG-WARNR	114-9037
BORG-WARNR	114-9046
BORG-WARNR	114-9048
BORG-WARNR	114-9058
BORG-WARNR	J9000
BUFFALO-SP	848036722(4)
CATERPILLR	9H9491
CATERPILLR	9S7120
CLEVE.MOT.	CM9000
CLEVE.MOT.	CM9014
CLEVE.MOT.	CM-CP88WB9
CLEVE.MOT.	CM-CP92N18
CMP	CM9000
CMP	CM9014
CMP	CM-CP92N18
CRANE CAR.	245

RD = BEARING DIAMETER (outside) BW = BEARING WIDTH CB = CROSS LENGTH WITH BEARINGS CC = CENTER TO CENTER CD = CROSS DIAMETER
CL = CROSS LENGTH WITHOUT BEARINGS EC = END TO CENTER or FACE TO FACE EE = END TO END EL = EFFECTIVE LENGTH
HD = HUB DIAMETER (or insert) IE = INSIDE OF EARS (or recess) OD = OUTSIDE DIAMETER OE = OUTSIDE OF EARS SB = SPLINE OR BORE SIZE

CRANE CAR.	261
CURTISS-WR	2-23991
D&D MACH.	MM9014
FIAT-ALLIS	08-000-505-437
FIAT-ALLIS	3040372
FIAT-ALLIS	3040372-9
FIAT-ALLIS	3040377
FIAT-ALLIS	3040377-8
FIAT-ALLIS	40372
FIAT-ALLIS	40377
FIAT-ALLIS	70040372
FIAT-ALLIS	70040377
FIAT-ALLIS	73040372
FIAT-ALLIS	73040372-9
FIAT-ALLIS	73040377
FIAT-ALLIS	73040377-8
FLEETRITE	ZAD19100
FMC	2-22991
FMC	2-23991
FWD	204305
FWD	46140
GERLINGER	635090
GLAENZER	065-103-8
GMB	G5-9000X
GMB	G9010
GMB	G9014
GMC	6933772
GMC	9021875
GMC	9022766
GMC	9024945
GMC	9024959
GMC	9025322
GMC	9025634
GMC	9049775
HABERLE	H9000
HYSTER	55-9100
IHC	304-608C91
IHC	304608C92
IHC	304608R91
IHC	432-433C91
IHC	698-997R91
IHC	896-057C1
IMCO	55-9100
IOWA MFG.	45375-017-07
JOY MFG.	533371
LEMPCO	U232
LINK BELT	18A1100(4)
LINK BELT	18A31(4)
LINK BELT	1A3096(4)
MACK	201SJ53
MACK	35MU213
MASTERS	MM9000
MCQUAY-NOR	1O3-37
MCQUAY-NOR	U03-37
MIXERMOBL	111162
MIXERMOBL	708686
MOPAR	D48
MOTOR MAST	7075H
M.R.S.	4092P11
NEAPCO	2803
NEAPCO	2888WB9
NEAPCO	289000
NEAPCO	289014
NEAPCO	6-9014
PERFECTION	RG9000
PETTIBONE	LL773-246
PETTI-MULL	LA425-8
PILOT	19100
PRECISION	793
PREC.TOR.	1490
PREC.TOR.	903
REPCO	114-9011
REPCO	114-9014
REPUBLIC	CB703
REPUBLIC	CB9000
SPICER	5-9000
SPICER	5-9000X
TEDDY TORQ	1490
TOWMOTOR	635090
TRU-CROSS	9000
TRW	20113
UJS	55-9100
WABCO	SU8485
WABCO	SV8485
WAYNE	2-22991
WAYNE	2-23991
WESCO	911900
WESCO	N9000
WHITE	02-7087816
WOOLDRIDGE	2-23991
ZELLER	7075H

IDLI GROUP 80236

CROSS AND LOW OR HIGH WING TYPE DRILLED OR THREADED BEARINGS
SEE FIGURE 1I

1.9375 in	CC	49.21 mm
7.75 0 in	CL	196.85 mm
1.0 in	CD	25.40 mm

AEC	8203
AEC	AE8203
ALCO	1854
ALCO	J8203
ALLOY	1854
ALLOY	J8203
BORG-WARNR	114-8103
BORG-WARNR	114-8203
BORG-WARNR	114-8208
BORG-WARNR	114-8516
BWE	114-8516
CLEVE.MOT.	CM8103
DIAMOND R	10081P2
GMB	G5-8203X
GMB	G8203
JOY MFG.	4500416
MASTERS	MM8103
MCQUAY-NOR	IND.
MOTOR MAST	7075J
NEAPCO	288203
NEAPCO	288208
NEAPCO	6-8208
PILOT	18103
PILOT	18208
PRECISION	584
REO MOTORS	10081P2
REPCO	114-8203
REPCO	114-8208
REPUBLIC	CB8103
SPICER	5-8105X
WESCO	811803
WESCO	N8003
WHITE	02-7074073
ZELLER	7075J

IDLI GROUP 80237

CROSS AND LOW OR HIGH WING TYPE DRILLED OR THREADED BEARINGS
SEE FIGURE 1JL

1.9375 in	CC	49.21 mm
7.75 0 in	CL	196.85 mm
1.0 in	CD	25.40 mm

AEC	8202
AEC	AE2530
AEC	AE8102
AEC	AE8202
ALCO	1853
ALCO	583
ALCO	J8202
ALCO	J8215
ALCO	J8506
ALCO	J8507
ALLIS-CHLM	178183
ALLIS-CHLM	178323
ALLIS-CHLM	178323-2
ALLOY	1853
ALLOY	J8102
ALLOY	J8202
BALKAMP	3-909
BALKAMP	M8003
BALKAMP	M8023
BALKAMP	M8101
BALKAMP	M8102
BALKAMP	M8201
BALKAMP	M8202
BALKAMP	M8206
BALKAMP	M8218
BORG-WARNR	114-8102
BORG-WARNR	114-8202
BORG-WARNR	114-8202AB
BORG-WARNR	114-8207
BORG-WARNR	114-8213
BORG-WARNR	114-8216
BORG-WARNR	114-8219
BORG-WARNR	J8102
CLEVE.MOT.	CM8003
CLEVE.MOT.	CM8102
CLEVE.MOT.	CM8202
CLEVE.MOT.	CM8206
CLEVE.MOT.	CM8207
CLEVE.MOT.	CM-CP78WB14
CLEVE.MOT.	CM-CP82N29
CMP	CM8102
CMP	CM8202
CMP	CM8207
CMP	CM-CP82N29
D&D MACH.	MM8102
DEERE,JOHN	AR29584
DEERE,JOHN	XAW70226A
DIAMOND R	10080P2
EATON	2006741
EXCEL	U259
FLEETRITE	ZAD18102
FLEETRITE	ZAD18207
FORD	B80H4635B
FORD	B8QH4635B
FORD	B8QH4635D
FORD	C4TA4635J
FORD	D2HA4635DA
FORD	D3HZ4635D
GEN.EQUIP.	J8102
GMB	G5-8201X
GMB	G5-8202X
GMB	G5-8207
GMB	G8202
GMB	G8207
GMB	G8215
HARNISCH.	T34905
HOUGH	118123HI
HOUGH	127019HI
HOUGH	188123H1
HYSTER	B115952A
HYSTER	D115952A
IHC	122123
IHC	122123H1
IHC	152-227R92
IHC	152276K91
IHC	152276R91
IHC	152-276R92
IHC	15227R92
IHC	188123H1
IHC	302726R91
IMCO	55-8102
JOY MFG.	175419
LEMPCO	U259
MACK	201SJ54
MACK	35MU210P2
MACK	35MU215P1
MACK	35MU215P2
MASTERS	MM8101
MASTERS	MM8102
MCQUAY-NOR	UO3-38
MICHIGAN	893815
MIXERMOBL	060868
MOPAR	D48
MOTIVE	55-8102
MOTOR MAST	7075G
M.R.S.	21115P11
NEAPCO	2809
NEAPCO	288102
NEAPCO	288202
NEAPCO	288207
NEAPCO	6-8207
PILOT	18102
PILOT	18207
PRECISION	583
REO MOTORS	10080P2
REPCO	114-8202
REPCO	114-8207
REPUBLIC	CB709
REPUBLIC	CB713
REPUBLIC	CB8102
SPICER	5-8202
SPICER	5-8202X
STEIGER	01-9994
STEIGER	17-156
TAYLOR	1930-030
TRU-CROSS	8202
TRW	20112
UJS	55-8102
WESCO	811822
WESCO	N8102
WESCO	N8202
WHITE	02-7074072
WHITE	10080P2

CAUTION: BE SURE TO REFER TO ENGINEERING CATALOGS FOR SPECIAL APPLICATIONS THAT REQUIRE SPECIFIC MATERIAL CONTENT, TOLERANCES, ETC. SEE FOOTNOTE.

ZELLER 7075G

IDLI GROUP 80238

CROSS AND LOW OR HIGH WING TYPE DRILLED OR THREADED BEARINGS
SEE FIGURE 1L

1.9375 in	CC	49.21 mm
7.75 0 in	CL	196.85 mm
1.0 in	CD	25.40 mm

AEC	8104
AEC	8105
AEC	AE8105
AEC	AE8113
AEC	AE8214
ALCO	1850
ALCO	J8105
ALCO	J8107
ALCO	J8113
ALLIS-CHLM	178363
ALLIS-CHLM	178368
ALLIS-CHLM	178368-7
ALLOY	1850
ALLOY	J8105
ALLOY	J8107
BALKAMP	3-914
BALKAMP	3-928
BALKAMP	3-946
BALKAMP	M8105
BALKAMP	M8107
BORG-WARNR	114-8105
BORG-WARNR	114-8105A
BORG-WARNR	114-8107
BORG-WARNR	114-8111
BORG-WARNR	114-8112
BORG-WARNR	114-8113
BORG-WARNR	114-8115
BORG-WARNR	114-8116
BORG-WARNR	114-8117
BORG-WARNR	114-8118
BORG-WARNR	114-8119
BORG-WARNR	114-8120
BORG-WARNR	114-8121
BORG-WARNR	114-8204
BORG-WARNR	114-8209
BORG-WARNR	114-8214
BORG-WARNR	114-8214A
CARDWELL	1-16649
CASE,J.I.	A45040
CASE,J.I.	A45634
CATERPILLR	1M8804
CATERPILLR	3H2579
CATERPILLR	4L6929
CATERPILLR	4R7972
CATERPILLR	5H2308
CATERPILLR	5S7216
CATERPILLR	5S7219
CATERPILLR	686343
CATERPILLR	6H2570
CATERPILLR	6H2579
CATERPILLR	7H8101
CATERPILLR	9P4809
CATERPILLR	GH2579
CHAIN BELT	298-6035-91
CHAIN BELT	ECN2-4392
CHAMPION	14678
CLARK EQU.	874910
CLARK EQU.	874969
CLARK EQU.	899602
CLARK EQU.	904236
CLARK EQU.	940235
CLARK EQU.	940258
CLARK EQU.	940350
CLARK EQU.	941573
CLEVE.MOT.	CM8014
CLEVE.MOT.	CM8104
CLEVE.MOT.	CM8105
CLEVE.MOT.	CM8111
CLEVE.MOT.	CM8113
CLEVE.MOT.	CM-CP78WB12
CLEVE.MOT.	CM-CP78WB2
CLEVE.MOT.	CM-CP82N27
CMP	CM8105
CMP	CM8111
CMP	CM8113
CMP	CM-CP82N27
D&D MACH.	MM8105
DEERE,JOHN	AT55470
DEERE,JOHN	AV12520V
DEERE,JOHN	XAW70220A
DIAMOND R	10077P2
DIAMOND R	10082P2
EATON	6700311
EATON	6953995
EATON	6954995
EIMCO	103M520
FIAT-ALLIS	0178368
FIAT-ALLIS	0178368-7
FIAT-ALLIS	178363
FIAT-ALLIS	178368
FIAT-ALLIS	178368-7
FIAT-ALLIS	70178368
FIAT-ALLIS	70178368-7
FLEETRITE	ZAD18105
FLEETRITE	ZAD18113
FORD	D8QH4635C
FWD	46409
FWD	46438
GLAENZER	065-122-8
GLAENZER	065-131-8
GMB	G5102
GMB	G5125
GMB	G5-8104X
GMB	G5-8105X
GMB	G5-8113
GMB	G8105
GMB	G8107
GMB	G8113
GMB	G8203
GMC	9023672
GMC	9025580
GMC	9026892
GMC	9027856
GMC	9032043
HORWOOD	625-0046
HOUGH	135766HI
HOUGH	187908HI
HYSTER	55-8105
IHC	135766
IHC	135-766H1
IHC	135-766H2
IHC	187908
IHC	187-908H1
IHC	271-421R91
IHC	271421R92
IHC	301-592R91
IHC	301592R92
IHC	302592R92
IHC	303-852R91
IHC	304106R91
IHC	304-106R92
IHC	305-852R91
IHC	305-952R91
IHC	324-667R91
IHC	338-591R91
IHC	384550R91
IHC	929369C91
IHC	929370C91
IMCO	55-8105
INBRA	CT11-090
INBRA	CT13-045
INBRA	CT24-000
MASS.FERG.	836-669M91
MCQUAY-NOR	U03-40
MCQUAY-NOR	UO3-40
MICHIGAN	874910
MICHIGAN	874969
MICHIGAN	940235
MICHIGAN	940258
MICHIGAN	940350
MICHIGAN	941573
MOPAR	D45
MOTOR MAST	288209
MOTOR MAST	7075B
MOTOR MAST	7075K
NEAPCO	2814
NEAPCO	2828
NEAPCO	2846
NEAPCO	288105
NEAPCO	288113
NEAPCO	288204
NEAPCO	288209
NEAPCO	6-8113
NEAPCO	6-8209
OSHKOSH	22UJ48
PILOT	18105
PILOT	18113
PILOT	18209
PRECISION	928
PREC.TOR.	946
REPCO	114-8105
REPCO	114-8113
REPCO	114-8116
REPUBLIC	CB714
REPUBLIC	CB728
REPUBLIC	CB746
REX CHAIN	298-6035-91
REX CHAIN	ECN2-4392
SPICER	5-8105X
SPICER	5-8150X
TEDDY TORQ	1485
THEW SHOV.	P33443
THEW SHOV.	P33502
TRU-CROSS	8105
TRW	20110
UJS	55-8113
WESCO	811-809
WESCO	811809
WESCO	811-809
WESCO	811809
WESCO	811815
WESCO	N8105
WESCO	N8209
WHITE	02-7074069
WHITE	02-7074074
WHITE	10077P2
ZELLER	7075B

IDLI GROUP 80239

CROSS + 4 LOW WING DRILLED BEARINGS
SEE FIGURE 1G

1.9375 in	CC	49.21 mm
7.75 0 in	CL	196.85 mm
1.0 in	CD	25.40 mm

ALCO	J8207
BORG-WARNR	114-8202
BORG-WARNR	114-8202A
BORG-WARNR	114-8215
CHAMPION	3975
CLEVE.MOT.	CM8101
CLEVE.MOT.	CM-CP82N30
DEERE,JOHN	AR29584R
DEERE,JOHN	XAW70227H
EATON	2006741
FWD	204301
FWD	204302
GLAENZER	065-109-8
GMB	G8215
GMC	9022733
GMC	9024678
GMC	9024791
GMC	9026504
HYSTER	202938
HYSTER	55-8102
IHC	109263
IHC	109265
IHC	109266
IHC	111562
IHC	118123
IHC	118-123H1
IHC	238-386R91
IHC	302-726C91
IHC	870-638R91
INBRA	CT17-050
JOY MFG.	175419
MARATHON	183516
MCQUAY-NOR	U03-38
MCQUAY-NOR	UO3-39
NEAPCO	288207
PILOT	18207
PRECISION	909
REPUBLIC	CB709
REPUBLIC	CB8101
THEW SHOV.	P33444
THEW SHOV.	P33459
UNIT RIG	988188
WHITE	10079P2

BD = BEARING DIAMETER (outside)   BW = BEARING WIDTH   CB = CROSS LENGTH WITH BEARINGS   CC = CENTER TO CENTER   CD = CROSS DIAMETER
CL = CROSS LENGTH WITHOUT BEARINGS   EC = END TO CENTER or FACE TO FACE   EE = END TO END   EL = EFFECTIVE LENGTH
HD = HUB DIAMETER (or insert)   IE = INSIDE OF EARS (or recess)   OD = OUTSIDE DIAMETER   OE = OUTSIDE OF EARS   SB = SPLINE OR BORE SIZE

IDLI GUIDE   COPYRIGHT ® INTERCHANGE, INC.   ST. LOUIS PARK, MN. 55416 USA

IDLI GROUP 80240

CROSS + 4 LOW WING DRILLED BEARINGS
slight variation from above group
SEE FIGURE 1G

1.9375 in	CC	49.21 mm
7.75 0 in	CL	196.85 mm
1. 0 in	CD	25.40 mm

ALCO	J8205
ATHEY PROD	8-809-11
ATHEY PROD	A177741
BORG-WARNR	114-8200
BORG-WARNR	114-8200A
BORG-WARNR	114-8200AB
BUCYR-ERIE	46-810-765(4)
CARDWELL	1-17785
CATERPILLR	XX7146
CHAMPION	10504
CLARK EQU.	892850
CLEVE.MOT.	CM8100
CLEVE.MOT.	CM8200
CLEVE.MOT.	CM-CP78WB13
CLEVE.MOT.	CM-CP82N28
CRANE CAR.	250
CRANE CAR.	6C5
CRANE CAR.	A337-21
CURTISS-WR	2-24006
EATON	2006749
EIMCO	105M2883
FIAT-ALLIS	0178183
FIAT-ALLIS	0178183-0
FIAT-ALLIS	178183
FIAT-ALLIS	70178183
FIAT-ALLIS	70178183-0
FMC	2-24006
FORD	B8QH4635C
FWD	46135
GALION	B42703(7)
GAR WOOD	857315
GEN.EQUIP.	J8100
GLAENZER	065-110-8
GMB	G8205
GMC	21686
GMC	9021251
GMC	9021686
GMC	9022503
GMC	9022625
GMC	9022763
GMC	9024673
GMC	9024794
GMC	9025207
GMC	9026505
GMC	9046306
HYSTER	202931
HYSTER	55-8100
HYSTER	B115952A
IHC	127002
IHC	127-002H1
IHC	152-275R91
IHC	177428
IHC	177-428H1
IHC	238-079R91
IHC	238-450R91
IHC	302-725C91
IHC	69573R93
INBRA	CT11-085

JOY MFG.	3379RC
JOY MFG.	4520233
JOY MFG.	A177741
JOY MFG.	A183515
LINK BELT	1A3098(4)
LINK BELT	8A1822(4)
LINK BELT	AC4393(4)
LINK BELT	AC4998(4)
LINK BELT	AC5422(4)
MACK	201SJ39
MACK	35MU210P1
MACK	35MU214P1
MACK	35MU214P2
MARATHON	183515
MCQUAY-NOR	UO3-41
MIXERMOBL	060868
MIXERMOBL	111162
MIXERMOBL	204301
NEAPCO	288205
NEAPCO	6-0825
PETTIBONE	L2893-3
PETTI-MULL	20411
PETTI-MULL	L2596-20
PILOT	18000
P.M.	A20411
PRECISION	910
REPUBLIC	CB710
REPUBLIC	CB8002
SIL.HOIST	24351-1(4)
SIL.HOIST	24351-2(4)
SIL.HOIST	24351(4)
UNIT RIG	986963
UNIT RIG	989717
WABCO	TD5261
WABCO	VJ4059
WABCO	VJ4060
WABCO	VJ8125
WAYNE	2-24006
WHITE	10078P2

IDLI GROUP 80241

CROSS + 4 LOW WING DRILLED BEARINGS
slight variation from above group
SEE FIGURE 1G

1.9375 in	CC	49.21 mm
7.75 0 in	CL	196.85 mm
1. 0 in	CD	25.40 mm

AEC	AE9017
AEC	AE9026
ALCO	J9010
ALCO	J9017
ALLOY	J9017
BORG-WARNR	114-9017
CLEVE.MOT.	CM9017
GMB	G5-9017X
MOPAR	D54
M.R.S.	21322P11
NEAPCO	289017
NEAPCO	6-0917
PILOT	19017
PRECISION	912
REPUBLIC	CB712
SPICER	5-9002X
WESCO	N9017
ZELLER	7075HD

IDLI GROUP 80242

CROSS + 4 LOW WING DRILLED BEARINGS
slight variation from above group
SEE FIGURE 1G

1.9375 in	CC	49.21 mm
7.75 0 in	CL	196.85 mm
1. 0 in	CD	25.40 mm

BORG-WARNR	114-8201
DEERE,JOHN	AV12521
FIAT-ALLIS	0178323
FIAT-ALLIS	01788323-2
FIAT-ALLIS	178323
FIAT-ALLIS	178323-2
FIAT-ALLIS	70178323
FIAT-ALLIS	70178323-2
GLAENZER	065-112-8
GMB	G8206
GMC	9022756
GMC	9024795
GMC	9026536
HYSTER	55-8101
IHC	127019
IHC	127-019H1
JOY MFG.	3382RC
JOY MFG.	A177744
JOY MFG.	A183516
MCQUAY-NOR	U03-39
NEAPCO	288206
PRECISION	913
REPUBLIC	CB713
REPUBLIC	CP713

IDLI GROUP 80243

CROSS + 4 DELTA WING THREADED BOLT HOLE BEARINGS
SEE FIGURE 1K

1.9375 in	CC	49.21 mm
7.75 0 in	CL	196.85 mm
1.0625 in	CD	26.99 mm

AEC	AE8000
ALLOY	J8002
ALLOY	J8100
BALKAMP	3-910
BALKAMP	M8100
BEAN,JOHN	2261160
CHRYSLER	CPR109691
CLARK EQU.	992489
CLARK EQU.	994839
DEERE,JOHN	XAW7022A
FORD	B8QZ4635A
IHC	177-482H1
IMCO	55-8002
JOY MFG.	TMD1027-1
KOEHRING	848036529
KOEHRING	848036530
MACK	35MUA210P1
MACK	35MUA214P1
MACK	35MUA214P2
MCQUAY-NOR	U03-41
NEAPCO	2878WB13
PERFECTION	RG8002
PRECISION	781
PREC.TOR.	1480
PREC.TOR.	1482
PREC.TOR.	580

PREC.TOR.	581
PREC.TOR.	910
WHITE	KC7013

IDLI GROUP 80244

CROSS + 2 LOW WING THREADED + 2 BLOCK BEARINGS
SEE FIGURE 1KL

1.9375 in	CC	49.21 mm
7.75 0 in	CL	196.85 mm
1.0625 in	CD	26.99 mm

AEC	AE8101
AEC	AE8102
ALCO	CP78WB14
ALLIS-CHLM	58-000-505-405
ALLOY	J8101
ALLOY	J8102
BALKAMP	3-909
BALKAMP	3-913
BALKAMP	3-914
BALKAMP	M8102
MACK	35MUA210P2
MACK	35MUA215P1
MACK	35MUA215P2
MCQUAY-NOR	U03-38
NEAPCO	2878WB14
PERFECTION	RG8102
PRECISION	783
PREC.TOR.	1484
PREC.TOR.	583

IDLI GROUP 80245

CROSS WITH HOLE IN THE CENTER + 2 HIGH WING DRILLED + 2 BLOCK BRGS.
SEE FIGURE 1L

3.6250 in	CC	92.08 mm
7.8750 in	CL	200.03 mm
1.6719 in	CD	42.47 mm

AEC	10003
AEC	10007
AEC	10011
AEC	AE10003
BORG-WARNR	114-10001
BORG-WARNR	114-10003
BORG-WARNR	114-10006
BORG-WARNR	114-10007
BORG-WARNR	114-10011
CATERPILLR	2G4157
CLARK EQU.	943724
CLARK EQU.	944864
CLARK EQU.	949234
CLARK EQU.	949248
CLARK EQU.	949250
CLEVE.MOT.	CM10003
CLEVE.MOT.	CM10011
GLAENZER	065-156-8
GMB	G5-10003
GMC	9032007
GMC	9032699
GMC	9052699
IHC	1108402C91
IHC	1108403C91
IHC	878771C91
MOTOR MAST	7087A

CAUTION: BE SURE TO REFER TO ENGINEERING CATALOGS FOR SPECIAL APPLICATIONS THAT REQUIRE SPECIFIC MATERIAL CONTENT, TOLERANCES, ETC. SEE FOOTNOTE.

MOTOR MAST	7087B
MOTOR MAST	7087D
PRECISION	988
REPCO	114-10003
REPCO	114-10003A
REPCO	114-10006
REPCO	114-10011A
REPUBLIC	CB788
TRW	20175
TRW	20176
WABCO	VH1311

NAME MISSPELLED?
SOMETHING OMITTED?

PLEASE TELL US IF THERE ARE CORRECTIONS OR ADDITIONS THAT SHOULD BE MADE IN ANY PART OF THIS GUIDE.

WE ARE CONSTANTLY STRIVING TO MAKE THIS MORE COMPLETE, MORE CORRECT AND MORE VALUABLE TO YOU AND ALL OF THE INTERCHANGE CLIENTS.

YOUR HELP WILL BE GREATLY APPRECIATED.

BD=BEARING DIAMETER (outside)   **BW**=BEARING WIDTH   **CB**=CROSS LENGTH WITH BEARINGS   **CC**=CENTER TO CENTER   **CD**=CROSS DIAMETER
CL=CROSS LENGTH WITHOUT BEARINGS   **EC**=END TO CENTER or FACE TO FACE   **EE**=END TO END   **EL**=EFFECTIVE LENGTH
HD=HUB DIAMETER (or insert)   **IE**=INSIDE OF EARS (or recess)   **OD**=OUTSIDE DIAMETER   **OE**=OUTSIDE OF EARS   **SB**=SPLINE OR BORE SIZE

IDLI GROUP 80901

**DETROIT POT STYLE KIT WITHOUT BODY
SEE FIGURE 2**

AEC	AE54AR
ALCO	J54A
ALCO	JR54A
ALCO	R7354A
BORG-WARNR	114-54A
BORG-WARNR	114-54AR
D&D MACH.	MDR54A
DETROIT	R54A
GMB	GR54A
LEMPCO	U421
LEMPCO	U422R
LEMPCO	U426B
MOTOR MAST	MR54A
NEAPCO	D4141R
NEAPCO	D4141X
NEAPCO	D4141XR
PILOT	11362R
PRECISION	201R
REPUBLIC	AD5400AR
SPICER	5-1402X
TEDDY TORQ	401
TEDDY TORQ	GR54A
TEDDY TORQ	R401
WESCO	D4410
WILLYS	213501
WILLYS	801578

IDLI GROUP 80902

**DETROIT POT STYLE KIT WITHOUT BODY
SEE FIGURE 2**

PRECISION	204R
WESCO	401421
WESCO	D4212

IDLI GROUP 80903

**DETROIT POT STYLE KIT WITH BODY
SEE FIGURE 2**

AEC	AE60R
ALCO	R7360
ALLOY	UR60
BORG-WARNR	114-60R
CLEVE.MOT.	CM60R
DETROIT	52
DETROIT	R52
DETROIT	R60
HABERLE	HR7360
LEMPCO	U435R
MASTERS	MDR52
MASTERS	MDR60
MOPAR	D158
MOPAR	D163
NEAPCO	D4252R
PERFECTION	G60R
PILOT	11360R
PRECISION	210R
REPUBLIC	AD5200R
SPICER	5-1403X
TRW	20118
WESCO	D4202HD
ZELLER	MR60

IDLI GROUP 80904

**DETROIT POT STYLE KIT WITHOUT BODY
SEE FIGURE 2**

PRECISION	215R

IDLI GROUP 80905

**DETROIT POT STYLE KIT WITHOUT BODY
SEE FIGURE 2**

AEC	65AR
AEC	AE65AR
ALCO	1902
ALCO	JR65A
ALCO	R7365A
ALLOY	1902
ALLOY	JR65A
BORG-WARNR	114-56
BORG-WARNR	114-65AR
CHRYSLER	1752680
CHRYSLER	2275515
CHRYSLER	2298941
CHRYSLER	KIT ONLY
CLEVE.MOT.	CM65AR1
CMP	CM65AR1
D&D MACH.	MDR65A
DETROIT	R65A
EXCEL	U454R
FORD	CD5AZ4635H
GMB	G5-1405X
GMB	G5-65AR
GMB	GR65A
LEMPCO	U430B
LEMPCO	U446A
LEMPCO	U446B
LEMPCO	U454R
LEMPCO	U455R
MASTERS	MDR65A
MCQUAY-NOR	U04-1
MCQUAY-NOR	UO4-1
MOPAR	D156
MOTOR MAST	MR65A
NEAPCO	7-4266
NEAPCO	D4265XP
NEAPCO	D4265XR
PERFECTION	RG65AR
PILOT	C6500AR
PILOT	C6500ARS
PRECISION	215R
REPCO	114-65AR
REPUBLIC	AD4251
REPUBLIC	AD6500AR
SPICER	5-1405X
SPICER	R65A
TEDDY TORQ	GR65A
TEDDY TORQ	R605
TRU-CROSS	1405
TRW	20120
TRW	20120D
UJS	55-65AR
WESCO	401065
WESCO	D65A
ZELLER	MR65A

IDLI GROUP 80906

**DETROIT POT STYLE KIT WITHOUT BODY
SEE FIGURE 2**

AEC	67AR
AEC	AE67AR
AEC	AE67HR
ALCO	1900
ALCO	JR67A
ALCO	R7367A
ALCO	UR52A
ALLOY	1900
ALLOY	JR67A
AMER.MOTOR	940800
BORG-WARNR	114-55A
BORG-WARNR	114-67AR
CHRYSLER	1440602
CHRYSLER	1603998
CHRYSLER	229840
CHRYSLER	2298940
CHRYSLER	940800
CHRYSLER	KIT ONLY
CLEVE.MOT.	CM67AR1
CMP	CM67AR1
D&D MACH.	MDR67A
DETROIT	55A
DETROIT	D52A
DETROIT	R67A
EXCEL	U404R
FLEETRITE	ZAD6700AR
FORD	CD5AZ4635K
GMB	G5-1407X
GMB	G5-67AR
GMB	GR52A
GMB	GR67A
LEMPCO	4455R
LEMPCO	U404R
LEMPCO	U430B
LEMPCO	U449B
LEMPCO	U455R
MASTERS	MDR67A
MCQUAY-NOR	U04-19
MCQUAY-NOR	U04-2
MCQUAY-NOR	UO4-2
MOPAR	D157
MOTOR MAST	MR67A
NEAPCO	7-4268
NEAPCO	D4267XR
PERFECTION	RG67AR
PILOT	C6700AR
PILOT	C6700ARS
PRECISION	204
PRECISION	217R
REPCO	114-67AR
REPCO	114-67R
REPUBLIC	AD5500A
REPUBLIC	AD6700AR
SPICER	5-1407X
SPICER	R67A
TEDDY TORQ	GR67A
TEDDY TORQ	R607
TRU-CROSS	1407
TRW	20121D
TRW	20122
UJS	55-67AR
WESCO	401421
WESCO	D67A

WYLIE	32A
ZELLER	MR67A

IDLI GROUP 80907

**DETROIT POT STYLE KIT WITHOUT BODY
SEE FIGURE 2**

AEC	AE62AR
ALCO	JR62A
ALCO	R7362A
BORG-WARNR	114-56AR
BORG-WARNR	114-62AR
CHRYSLER	1643287
D&D MACH.	MDR62A
D&D MACH.	MDR63A
DETROIT	JR56A
DETROIT	R56A
DETROIT	R62A
DETROIT	R7356A
GMB	GR62A
LEMPCO	U436A
LEMPCO	U444R
LEMPCO	U445A
MOPAR	D155
MOTOR MAST	MR62A
NEAPCO	D4262XR
PERFECTION	RG62AR
PILOT	C6200AR
PRECISION	308R
REPUBLIC	AD5600AR
REPUBLIC	AD6200AR
REPUBLIC	ADG5600AR
TEDDY TORQ	GR62A
TEDDY TORQ	R601
WESCO	401062
WESCO	D4610
WESCO	D62A

IDLI GROUP 80908

**DETROIT POT STYLE KIT WITHOUT BODY
SEE FIGURE 2**

AEC	AE57AR
ALCO	JR57A
ALCO	R7357A
BORG-WARNR	114-57AR
CHRYSLER	1450446
CHRYSLER	1643208
CHRYSLER	1643228
CHRYSLER	1643288
D&D MACH.	MDR57A
DETROIT	R57A
GMB	GR57A
KAISER	801578
LEMPCO	J427B
LEMPCO	U427A
LEMPCO	U427B
MOPAR	D154
MOTOR MAST	MR57A
NEAPCO	D4257XR
NEAPCO	D4551XR
PERFECTION	RG57AR
PILOT	C5700AR
PRECISION	309R
REPUBLIC	AD5700AR
SPICER	R57A
TEDDY TORQ	R701

ENGINEERING CATALOGS MUST BE CONSULTED FOR DETAILS NOT INCLUDED
IN THIS GUIDE. SPECIFIC DESIGNS, MATERIAL CONTENT, TOLERANCES
LUBE FITTINGS AND OTHER DIMENSIONS ARE INTENTIONALLY OMITTED HERE.

CAUTION: BE SURE TO REFER TO ENGINEERING CATALOGS FOR SPECIAL APPLICATIONS THAT REQUIRE SPECIFIC MATERIAL CONTENT, TOLERANCES, ETC. SEE FOOTNOTE.

WESCO	401471
WESCO	D4710

IDLI GROUP 80909

**DETROIT POT STYLE KIT WITHOUT BODY
SEE FIGURE 2**

AEC	67AR
AEC	AE52AR
ALCO	1901
ALCO	JR52A
ALCO	R7352A
ALCO	U52
ALLOY	1901
ALLOY	JR52A
BORG-WARNR	114-52A
BORG-WARNR	114-52AR
BORG-WARNR	114-55A
CHRYSLER	1450602
CHRYSLER	1603398
CHRYSLER	939700
CLEVE.MOT.	CM52AR
CMP	CM52AR
D&D MACH.	MDR60A
DETROIT	52A
DETROIT	55A
DETROIT	R52A
DIVCO	16269
EXCEL	U455R
GMB	G5-1407X
GMB	GR52A
HERCULES	555
KAISER	201775
LEMPCO	U404
LEMPCO	U404R
LEMPCO	U426B
LEMPCO	U455
LEMPCO	U455R
MCQUAY-NOR	U04-19
MCQUAY-NOR	U04-2
MOPAR	D150
MOPAR	D151
MOTOR MAST	MR52A
NEAPCO	D4241X
NEAPCO	D4241XP
NEAPCO	D4241XR
PERFECTION	RG52A
PERFECTION	RG52AR
PILOT	11352
PILOT	11352R
PRECISION	204
PRECISION	323R
REPCO	114-52A
REPCO	114-55
REPCO	114-67AR
REPCO	114-67R
REPUBLIC	AD5200
REPUBLIC	AD5200A
REPUBLIC	AD5200AD
REPUBLIC	AD5200AR
REPUBLIC	AD5400AR
REPUBLIC	AD5500A
REPUBLIC	AD5500AR
SPICER	5-1400X
SPICER	5-1407
SPICER	5-1407X
SPICER	R52A
TEDDY TORQ	GR52A
TEDDY TORQ	R101
TRW	20117
TRW	20118
TRW	20121D
TRW	20122
WESCO	301421
WESCO	401-421
WESCO	D4210
WESCO	D4211
WESCO	D4212
WESCO	D4510
WILLYS	201775
ZELLER	MR52A
ZELLER	MR67A

BD=BEARING DIAMETER (outside)   **BW**=BEARING WIDTH   **CB**=CROSS LENGTH WITH BEARINGS   **CC**=CENTER TO CENTER   **CD**=CROSS DIAMETER
CL=CROSS LENGTH WITHOUT BEARINGS   **EC**=END TO CENTER or FACE TO FACE   **EE**=END TO END   **EL**=EFFECTIVE LENGTH
HD=HUB DIAMETER (or insert)   **IE**=INSIDE OF EARS (or recess)   **OD**=OUTSIDE DIAMETER   **OE**=OUTSIDE OF EARS   **SB**=SPLINE OR BORE SIZE

IDLI GUIDE   COPYRIGHT ® INTERCHANGE, INC.   ST. LOUIS PARK, MN. 55416 USA

IDLI GROUP 80951
DETROIT POT STYLE KIT WITH BODY
SEE FIGURE 3

AEC	65R
AEC	AE65R
ALCO	1911
ALCO	R7365
ALCO	UR65
ALLOY	1911
ALLOY	UR65
BORG-WARNR	114-65
BORG-WARNR	114-65R
BORG-WARNR	4251
CLEVE.MOT.	CM65R1
CMP	CM65R1
D&D MACH.	MDR65
DETROIT	4251
DETROIT	R65
EXCEL	U454
FORD	CD5AZ4635G
GMB	G5-1404X
GMB	G5-65R
GMB	GR54
GMB	GUR65
LEMPCO	U417R
LEMPCO	U446
LEMPCO	U446R
LEMPCO	U454
MASTERS	MDR65
MCQUAY-NOR	D4625R
MCQUAY-NOR	U04-17
MCQUAY-NOR	UO4-17
MOPAR	D166
MOTOR MAST	MR65
NEAPCO	7-4265
NEAPCO	D4265R
PERFECTION	G65R
PERFECTION	RG65AR
PILOT	C6500R
PRECISION	205R
REPCO	114-65
REPCO	114-65R
REPUBLIC	AD5400
REPUBLIC	AD6500R
SPICER	5-1404X
SPICER	R65
TEDDY TORQ	GUR65
TEDDY TORQ	R605C
TRU-CROSS	1404
TRW	20119
TRW	20121D
UJS	55-65R
WESCO	401-065
WESCO	411065
WESCO	D65
ZELLER	MR65

IDLI GROUP 80952
DETROIT POT STYLE KIT WITH BODY
SEE FIGURE 3

ALCO	R7366
BORG-WARNR	114-66R
PILOT	C6600R
PRECISION	206R

IDLI GROUP 80953
DETROIT POT STYLE KIT WITH BODY
SEE FIGURE 3

AEC	67R
AEC	AE67R
ALCO	1910
ALCO	JR67A
ALCO	R7367
ALCO	UR66
ALCO	UR67
ALLOY	1910
ALLOY	UR67
BORG-WARNR	114-56
BORG-WARNR	114-66R
BORG-WARNR	114-67R
BORG-WARNR	B420112
CHRYSLER	1603998
CHRYSLER	1734158
CHRYSLER	1743159
CLEVE.MOT.	CM60R
CLEVE.MOT.	CM67R1
CMP	CM674-1
CMP	CM67R1
D&D MACH.	MDR67
DETROIT	R66
DETROIT	R67
DETROIT	UR66
EXCEL	U449R
FLEETRITE	ZAD6700R
FORD	CD5AZ4635J
GMB	G5-1406X
GMB	G5-67R
GMB	GUR66
GMB	GUR67
HABERLE	HR7360
HERCULES	552
KAISER	213501
LEMPCO	U430A
LEMPCO	U449
LEMPCO	U449R
LEMPCO	U450
LEMPCO	U450R
LEMPCO	U455
MASTERS	MDR52
MASTERS	MDR67
MCQUAY-NOR	U04-18
MCQUAY-NOR	UO4-18
MOPAR	D168
MOTOR MAST	MR67
NEAPCO	7-4267
NEAPCO	D4267R
PERFECTION	G60R
PERFECTION	G66R
PERFECTION	G67
PERFECTION	G67R
PILOT	C6699R
PILOT	C6700R
PILOT	C6700RS
PRECISION	207R
REPCO	114-66
REPCO	114-67
REPCO	114-67AR
REPUBLIC	AD4251
REPUBLIC	AD6600R
REPUBLIC	AD6700R
REPUBLIC	D420112
SPICER	5-1406X
SPICER	R67
STUDEBAKER	382741
TEDDY TORQ	GUR67
TEDDY TORQ	R606C
TEDDY TORQ	R607C
TRU-CROSS	1406
TRW	20121
TRW	20121D
UJS	55-67R
WESCO	401067
WESCO	D66
WESCO	D67
WILLYS	940800
ZELLER	MR67

IDLI GROUP 80954
DETROIT POT STYLE KIT WITH BODY
SEE FIGURE 3

AEC	AE62R
ALCO	R7362
ALCO	UR62
BORG-WARNR	114-56R
BORG-WARNR	114-62R
D&D MACH.	MDR62
DETROIT	R56
DETROIT	R62
DETROIT	R7356
DETROIT	UR56
GMB	GUR62
LEMPCO	U431R
LEMPCO	U436
LEMPCO	U436R
MOPAR	D164
MOTOR MAST	MR62
NEAPCO	D4262R
NEAPCO	D4556R
PERFECTION	G62R
PILOT	11362R
PILOT	C6200R
PRECISION	208R
REPCO	114-62
REPUBLIC	AD5600R
REPUBLIC	AD6200R
REPUBLIC	D44015
SPICER	R62
STUDEBAKER	436198
STUDEBAKER	436277
STUDEBAKER	436292
STUDEBAKER	458420
TEDDY TORQ	GUR62
TEDDY TORQ	R601C
WESCO	411062
WESCO	D4600
WESCO	D62

IDLI GROUP 80955
DETROIT POT STYLE KIT WITH BODY
SEE FIGURE 3

AEC	AE579
AEC	AE57R
ALCO	R7357
ALCO	UR57
BORG-WARNR	114-57R
BORG-WARNR	B445014
CHRYSLER	1532889
D&D MACH.	MDR57
DETROIT	R57
GMB	GUR57
LEMPCO	J427R
LEMPCO	U427
LEMPCO	U427R
MOPAR	D162
NEAPCO	D4257R
NEAPCO	D4557R
PERFECTION	G57R
PILOT	C5700R
PRECISION	209R
REPCO	114-57
REPUBLIC	AD5700
REPUBLIC	AD5700R
REPUBLIC	D445014
SPICER	R57
STUDEBAKER	436179
STUDEBAKER	436199
STUDEBAKER	536199
TEDDY TORQ	GUR57
TEDDY TORQ	R701C
WESCO	411470
WESCO	D4700

IDLI GROUP 80956
DETROIT POT STYLE KIT WITH BODY
SEE FIGURE 3

BORG-WARNR	114-55
DETROIT	55
DETROIT	R55
LEMPCO	U430
LEMPCO	U430B
LEMPCO	U435
LEMPCO	U435R
NEAPCO	D4255R
PILOT	11355R
PRECISION	224
PRECISION	224R
REPUBLIC	AD5500
REPUBLIC	AD5500R
WESCO	401450
WESCO	D4500

IDLI GROUP 80957
DETROIT POT STYLE KIT WITH BODY
SEE FIGURE 3

AEC	AE63R
ALCO	R7363
ALCO	R7367
ALCO	UR63
BORG-WARNR	114-63R
BORG-WARNR	B44016
CHRYSLER	1635099
D&D MACH.	MDR63

CAUTION: BE SURE TO REFER TO ENGINEERING CATALOGS FOR SPECIAL APPLICATIONS THAT REQUIRE SPECIFIC MATERIAL CONTENT, TOLERANCES, ETC. SEE FOOTNOTE.

DETROIT	R63
GMB	GUR63
LEMPCO	U445
LEMPCO	U445R
MOPAR	D165
MOTOR MAST	MR63
NEAPCO	D4263R
PERFECTION	G63R
PERFECTION	G67R
PILOT	C6300
PILOT	C6300R
PRECISION	321R
REPCO	114-63
REPUBLIC	AD6300R
REPUBLIC	D44016
SPICER	R63
SPICER	R67
TEDDY TORQ	GUR63
TEDDY TORQ	R603C
WESCO	411063
WESCO	D63

IDLI GROUP 80958

**DETROIT POT STYLE KIT WITH BODY
SEE FIGURE 3**

AEC	60R
AEC	AE52R
AEC	AE60R
ALCO	U52
ALCO	UR52
ALCO	UR60
BORG-WARNR	114-52
BORG-WARNR	114-52R
BORG-WARNR	114-60R
CLEVE.MOT.	CM60R
D&D MACH.	MDR60
DETROIT	52
DETROIT	R52
DETROIT	R60
DETROIT	UR52
DIVCO	15270
DIVCO	16270
EATON	800047
GMB	G5-1403
GMB	G5-1403X
GMB	GUR52
GMB	GUR60
GMB	GUR62
HABERLE	HR7360
HERCULES	552
HERCULES	552C
LEMPCO	U405
LEMPCO	U405R
MASTERS	MDR52
MASTERS	MDR60
MOPAR	D158
MOPAR	D163
MOTOR MAST	MR60
NEAPCO	D4252
NEAPCO	D4252R
NEAPCO	D4252XR

PERFECTION	G52R
PERFECTION	G60R
PILOT	11354
PILOT	11354R
PILOT	11360R
PRECISION	324R
REPCO	114-52
REPCO	114-60
REPCO	114-67R
REPUBLIC	AD5200
REPUBLIC	AD5200R
REPUBLIC	AD5299R
REPUBLIC	AD6000R
SPICER	5-1403X
SPICER	R52
SPICER	R60
TEDDY TORQ	GUR60
TEDDY TORQ	R101C
TEDDY TORQ	R661C
TRW	20118
WESCO	401420
WESCO	401422
WESCO	D4200
WESCO	D4200HD
WESCO	D4201HD
WESCO	D4202
WESCO	D4202AD
WESCO	D4202HD

IDLI GROUP 80959

**DETROIT POT STYLE KIT WITH BODY
SEE FIGURE 3**

PRECISION	360R
WESCO	303520
WESCO	N1520RU

BD = BEARING DIAMETER (outside)   BW = BEARING WIDTH   CB = CROSS LENGTH WITH BEARINGS   CC = CENTER TO CENTER   CD = CROSS DIAMETER
CL = CROSS LENGTH WITHOUT BEARINGS   EC = END TO CENTER or FACE TO FACE   EE = END TO END   EL = EFFECTIVE LENGTH
HD = HUB DIAMETER (or insert)   IE = INSIDE OF EARS (or recess)   OD = OUTSIDE DIAMETER   OE = OUTSIDE OF EARS   SB = SPLINE OR BORE SIZE

IDLI GROUP 81001

END YOKE WITH ROUND BORE
SEE FIGURE 4B

1.4375 in	IE	36.51 mm
.9688 in	BD	24.61 mm
.6250 in	SB	15.88 mm
1.6875 in	EC	42.86 mm
1.6250 in	HD	41.28 mm

ROCKWELL L6-92

IDLI GROUP 81002

END YOKE WITH ROUND BORE
slight variation from above group
SEE FIGURE 4B

1.4375 in	IE	36.51 mm
.9688 in	BD	24.61 mm
.6250 in	SB	15.88 mm
1.6875 in	EC	42.86 mm
1.6250 in	HD	41.28 mm

ROCKWELL L6NYR10-6

IDLI GROUP 81003

END YOKE WITH ROUND BORE
SEE FIGURE 4B

1.4375 in	IE	36.51 mm
.9688 in	BD	24.61 mm
.6250 in	SB	15.88 mm
2.5000 in	EC	63.50 mm
1.5000 in	HD	38.10 mm

ROCKWELL L6NYR10-12

IDLI GROUP 81004

END YOKE WITH ROUND BORE
SEE FIGURE 4B

1.4375 in	IE	36.51 mm
.9688 in	BD	24.61 mm
.6250 in	SB	15.88 mm
2.5000 in	EC	63.50 mm
1.6250 in	HD	41.28 mm

ROCKWELL L6-39

IDLI GROUP 81005

END YOKE WITH ROUND BORE
slight variation from above group
SEE FIGURE 4B

1.4375 in	IE	36.51 mm
.9688 in	BD	24.61 mm
.6250 in	SB	15.88 mm
2.5000 in	EC	63.50 mm
1.6250 in	HD	41.28 mm

ROCKWELL L6-40

IDLI GROUP 81006

END YOKE WITH SQUARE BORE
SEE FIGURE 4C

1.4375 in	IE	36.51 mm
.9688 in	BD	24.61 mm
.7500 in	SB	19.05 mm
1.0000 in	EC	25.40 mm
1.6250 in	HD	41.28 mm

ROCKWELL L6-82

IDLI GROUP 81007

END YOKE WITH ROUND BORE
SEE FIGURE 4B

1.4375 in	IE	36.51 mm
.9688 in	BD	24.61 mm
.7500 in	SB	19.05 mm
1.6875 in	EC	42.86 mm
1.6250 in	HD	41.28 mm

ROCKWELL L6-70

IDLI GROUP 81008

END YOKE WITH ROUND BORE
slight variation from above group
SEE FIGURE 4B

1.4375 in	IE	36.51 mm
.9688 in	BD	24.61 mm
.7500 in	SB	19.05 mm
1.6875 in	EC	42.86 mm
1.6250 in	HD	41.28 mm

ROCKWELL L6NYR12-1

IDLI GROUP 81009

END YOKE WITH SQUARE BORE
SEE FIGURE 4C

1.4375 in	IE	36.51 mm
.9688 in	BD	24.61 mm
.7500 in	SB	19.05 mm
1.6875 in	EC	42.86 mm
1.6250 in	HD	41.28 mm

ROCKWELL L6-95

IDLI GROUP 81010

END YOKE WITH SQUARE BORE
slight variation from above group
SEE FIGURE 4C

1.4375 in	IE	36.51 mm
.9688 in	BD	24.61 mm
.7500 in	SB	19.05 mm
1.6875 in	EC	42.86 mm
1.6250 in	HD	41.28 mm

ROCKWELL L6NYQ12

IDLI GROUP 81011

END YOKE WITH 6 SPLINES
SEE FIGURE 4A

1.4375 in	IE	36.51 mm
.9688 in	BD	24.61 mm
.7500 in	SB	19.05 mm
2.5000 in	EC	63.50 mm
1.3750 in	HD	34.93 mm

ROCKWELL L6NYS12-4

IDLI GROUP 81012

END YOKE WITH ROUND BORE
SEE FIGURE 4B

1.4375 in	IE	36.51 mm
.9688 in	BD	24.61 mm
.7500 in	SB	19.05 mm
2.5000 in	EC	63.50 mm
1.3750 in	HD	34.93 mm

ROCKWELL L6NYR12-44

IDLI GROUP 81013

END YOKE WITH ROUND BORE
SEE FIGURE 4B

1.4375 in	IE	36.51 mm
.9688 in	BD	24.61 mm
.7500 in	SB	19.05 mm
2.5000 in	EC	63.50 mm
1.5000 in	HD	38.10 mm

ROCKWELL L6NYR12-25

IDLI GROUP 81014

END YOKE WITH ROUND BORE
slight variation from above group
SEE FIGURE 4B

1.4375 in	IE	36.51 mm
.9688 in	BD	24.61 mm
.7500 in	SB	19.05 mm
2.5000 in	EC	63.50 mm
1.5000 in	HD	38.10 mm

ROCKWELL L6NYR12-92

IDLI GROUP 81015

END YOKE WITH SQUARE BORE
SEE FIGURE 4C

1.4375 in	IE	36.51 mm
.9688 in	BD	24.61 mm
.7500 in	SB	19.05 mm
2.5000 in	EC	63.50 mm
1.5000 in	HD	38.10 mm

ROCKWELL L6NYQ12-67

IDLI GROUP 81016

END YOKE WITH 6 SPLINES
SEE FIGURE 4A

1.4375 in	IE	36.51 mm
.9688 in	BD	24.61 mm
.7500 in	SB	19.05 mm
2.5000 in	EC	63.50 mm
1.6250 in	HD	41.28 mm

ROCKWELL L6-100

IDLI GROUP 81017

END YOKE WITH 6 SPLINES
slight variation from above group
SEE FIGURE 4A

1.4375 in	IE	36.51 mm
.9688 in	BD	24.61 mm
.7500 in	SB	19.05 mm
2.5000 in	EC	63.50 mm
1.6250 in	HD	41.28 mm

ROCKWELL L6NYS12-3

IDLI GROUP 81018

END YOKE WITH ROUND BORE
SEE FIGURE 4B

1.4375 in	IE	36.51 mm
.9688 in	BD	24.61 mm
.7500 in	SB	19.05 mm
2.5000 in	EC	63.50 mm
1.6250 in	HD	41.28 mm

PRECISION 1900
ROCKWELL L6-10

IDLI GROUP 81019

END YOKE WITH ROUND BORE
slight variation from above group
SEE FIGURE 4B

1.4375 in	IE	36.51 mm
.9688 in	BD	24.61 mm
.7500 in	SB	19.05 mm
2.5000 in	EC	63.50 mm
1.6250 in	HD	41.28 mm

ROCKWELL L6-41

IDLI GROUP 81020

END YOKE WITH ROUND BORE
slight variation from above group
SEE FIGURE 4B

1.4375 in	IE	36.51 mm
.9688 in	BD	24.61 mm
.7500 in	SB	19.05 mm
2.5000 in	EC	63.50 mm
1.6250 in	HD	41.28 mm

BORG-WARNR	. . . . 7939J
CHELSEA	. . . . Y124Y124
LEMPCO	. . . . 2L2
NEAPCO	 065757
NEAPCO	. . . . 06B8B8
NEAPCO	 11-2004
NEAPCO	. . . . 151B8
NEAPCO	 165757
NEAPCO	. . . . N10242-4
PILOT	. . . . 5P-2
PRECISION	 1901
ROCKWELL	. . . . L6-42
SPICER	. . . . 10242-4SF
SPICER	 10-4-0023
TRW	 20502
WESCO	. . . . 1A124-1
WESCO	 540055
WESCO	. . . . 6N124
WESCO	. . . . A124-1

IDLI GROUP 81021

END YOKE WITH ROUND BORE
slight variation from above group
SEE FIGURE 4B

1.4375 in	IE	36.51 mm
.9688 in	BD	24.61 mm
.7500 in	SB	19.05 mm
2.5000 in	EC	63.50 mm
1.6250 in	HD	41.28 mm

PRECISION 1900
ROCKWELL L6NYR12-4

IDLI GROUP 81022

END YOKE WITH ROUND BORE
slight variation from above group
SEE FIGURE 4B

1.4375 in	IE	36.51 mm
.9688 in	BD	24.61 mm
.7500 in	SB	19.05 mm
2.5000 in	EC	63.50 mm
1.6250 in	HD	41.28 mm

ROCKWELL L6NYR12-18

ENGINEERING CATALOGS MUST BE CONSULTED FOR DETAILS NOT INCLUDED
IN THIS GUIDE. SPECIFIC DESIGNS, MATERIAL CONTENT, TOLERANCES
LUBE FITTINGS AND OTHER DIMENSIONS ARE INTENTIONALLY OMITTED HERE.

CAUTION: BE SURE TO REFER TO ENGINEERING CATALOGS FOR SPECIAL APPLICATIONS THAT REQUIRE SPECIFIC MATERIAL CONTENT, TOLERANCES, ETC. SEE FOOTNOTE.

IDLI GROUP 81023

END YOKE WITH ROUND BORE
slight variation from above group
SEE FIGURE 4B

1.4375 in	IE	36.51 mm
.9688 in	BD	24.61 mm
.7500 in	SB	19.05 mm
2.5000 in	EC	63.50 mm
1.6250 in	HD	41.28 mm

ROCKWELL L6NYR12-33

IDLI GROUP 81024

END YOKE WITH SQUARE BORE
SEE FIGURE 4C

1.4375 in	IE	36.51 mm
.9688 in	BD	24.61 mm
.7500 in	SB	19.05 mm
2.5000 in	EC	63.50 mm
1.6250 in	HD	41.28 mm

ROCKWELL L6-31

IDLI GROUP 81025

END YOKE WITH SQUARE BORE
slight variation from above group
SEE FIGURE 4C

1.4375 in	IE	36.51 mm
.9688 in	BD	24.61 mm
.7500 in	SB	19.05 mm
2.5000 in	EC	63.50 mm
1.6250 in	HD	41.28 mm

BORG-WARNR 7654J
MCQUAY-NOR U03-103
NEAPCO 16-6171
NEAPCO L6171
PRECISION 1910
ROCKWELL L6-32
WESCO 540051
WESCO 6N120-26

IDLI GROUP 81026

END YOKE WITH SQUARE BORE
slight variation from above group
SEE FIGURE 4C

1.4375 in	IE	36.51 mm
.9688 in	BD	24.61 mm
.7500 in	SB	19.05 mm
2.5000 in	EC	63.50 mm
1.6250 in	HD	41.28 mm

PRECISION 1910
ROCKWELL L6NYQ12-3

IDLI GROUP 81027

END YOKE WITH 8 INVOLUTED SPLINES
SEE FIGURE 4A

1.4375 in	IE	36.51 mm
.9688 in	BD	24.61 mm
.7500 in	SB	19.05 mm
3.2500 in	EC	82.55 mm
1.3750 in	HD	34.93 mm

ROCKWELL L6NYS12-13

IDLI GROUP 81028

END YOKE WITH ROUND BORE
SEE FIGURE 4B

1.4375 in	IE	36.51 mm
.9688 in	BD	24.61 mm
.7500 in	SB	19.05 mm
3.5000 in	EC	88.90 mm
2.0000 in	HD	50.80 mm

ROCKWELL L6NYR12-162

IDLI GROUP 81029

END YOKE WITH 8 INVOLUTED SPLINES
SEE FIGURE 4A

1.4375 in	IE	36.51 mm
.9688 in	BD	24.61 mm
.7500 in	SB	19.05 mm
4.3750 in	EC	111.13 mm
1.5000 in	HD	38.10 mm

ROCKWELL L6NYS12-12

IDLI GROUP 81030

END YOKE WITH ROUND BORE
SEE FIGURE 4B

1.4375 in	IE	36.51 mm
.9688 in	BD	24.61 mm
.8125 in	SB	20.64 mm
1.6875 in	EC	42.86 mm
1.6250 in	HD	41.28 mm

ROCKWELL L6-72

IDLI GROUP 81031

END YOKE WITH ROUND BORE
SEE FIGURE 4B

1.4375 in	IE	36.51 mm
.9688 in	BD	24.61 mm
.8125 in	SB	20.64 mm
2.5000 in	EC	63.50 mm
1.5000 in	HD	38.10 mm

ROCKWELL L6NYR13-19

IDLI GROUP 81032

END YOKE WITH ROUND BORE
slight variation from above group
SEE FIGURE 4B

1.4375 in	IE	36.51 mm
.9688 in	BD	24.61 mm
.8125 in	SB	20.64 mm
2.5000 in	EC	63.50 mm
1.5000 in	HD	38.10 mm

ROCKWELL L6NYR13-19

IDLI GROUP 81033

END YOKE WITH ROUND BORE
SEE FIGURE 4B

1.4375 in	IE	36.51 mm
.9688 in	BD	24.61 mm
.8125 in	SB	20.64 mm
2.5000 in	EC	63.50 mm
1.6250 in	HD	41.28 mm

ROCKWELL L6-14

IDLI GROUP 81034

END YOKE WITH ROUND BORE
slight variation from above group
SEE FIGURE 4B

1.4375 in	IE	36.51 mm
.9688 in	BD	24.61 mm
.8125 in	SB	20.64 mm
2.5000 in	EC	63.50 mm
1.6250 in	HD	41.28 mm

NEAPCO 16-6112
NEAPCO L6112
PRECISION 1902
ROCKWELL L6-15
TRW 21052
WESCO 06N0134
WESCO 540056
WESCO 540058
WESCO 540059
WESCO 540060
WESCO 6N134
WESCO 6N134-5

IDLI GROUP 81035

END YOKE WITH ROUND BORE
slight variation from above group
SEE FIGURE 4B

1.4375 in	IE	36.51 mm
.9688 in	BD	24.61 mm
.8125 in	SB	20.64 mm
2.5000 in	EC	63.50 mm
1.6250 in	HD	41.28 mm

ROCKWELL L6-17

IDLI GROUP 81036

END YOKE WITH ROUND BORE
slight variation from above group
SEE FIGURE 4B

1.4375 in	IE	36.51 mm
.9688 in	BD	24.61 mm
.8125 in	SB	20.64 mm
2.5000 in	EC	63.50 mm
1.6250 in	HD	41.28 mm

ROCKWELL L6NYR13-5

IDLI GROUP 81037

END YOKE WITH ROUND BORE
slight variation from above group
SEE FIGURE 4B

1.4375 in	IE	36.51 mm
.9688 in	BD	24.61 mm
.8125 in	SB	20.64 mm
2.5000 in	EC	63.50 mm
1.6250 in	HD	41.28 mm

PRECISION 1902
ROCKWELL L6NYR13-6

IDLI GROUP 81038

END YOKE WITH SQUARE BORE
SEE FIGURE 4C

1.4375 in	IE	36.51 mm
.9688 in	BD	24.61 mm
.8125 in	SB	20.64 mm
2.5000 in	EC	63.50 mm
1.6250 in	HD	41.28 mm

ROCKWELL L6-96

IDLI GROUP 81039

END YOKE WITH SQUARE BORE
slight variation from above group
SEE FIGURE 4C

1.4375 in	IE	36.51 mm
.9688 in	BD	24.61 mm
.8125 in	SB	20.64 mm
2.5000 in	EC	63.50 mm
1.6250 in	HD	41.28 mm

ROCKWELL L6NYQ13

IDLI GROUP 81040

END YOKE WITH ROUND BORE
SEE FIGURE 4B

1.4375 in	IE	36.51 mm
.9688 in	BD	24.61 mm
.8750 in	SB	22.23 mm
1.6875 in	EC	42.86 mm
1.6250 in	HD	41.28 mm

ROCKWELL L6-76

IDLI GROUP 81041

END YOKE WITH ROUND BORE
slight variation from above group
SEE FIGURE 4B

1.4375 in	IE	36.51 mm
.9688 in	BD	24.61 mm
.8750 in	SB	22.23 mm
1.6875 in	EC	42.86 mm
1.6250 in	HD	41.28 mm

ROCKWELL L6-93

IDLI GROUP 81042

END YOKE WITH ROUND BORE
slight variation from above group
SEE FIGURE 4B

1.4375 in	IE	36.51 mm
.9688 in	BD	24.61 mm
.8750 in	SB	22.23 mm
1.6875 in	EC	42.86 mm
1.6250 in	HD	41.28 mm

ROCKWELL L6NYR14-3

IDLI GROUP 81043

END YOKE WITH ROUND BORE
slight variation from above group
SEE FIGURE 4B

1.4375 in	IE	36.51 mm
.9688 in	BD	24.61 mm
.8750 in	SB	22.23 mm
1.6875 in	EC	42.86 mm
1.6250 in	HD	41.28 mm

ROCKWELL L6NYR14-47

IDLI GROUP 81044

END YOKE WITH SQUARE BORE
SEE FIGURE 4C

1.4375 in	IE	36.51 mm
.9688 in	BD	24.61 mm
.8750 in	SB	22.23 mm
1.6875 in	EC	42.86 mm
1.6250 in	HD	41.28 mm

ROCKWELL L6-83

BD = BEARING DIAMETER (outside)   BW = BEARING WIDTH   CB = CROSS LENGTH WITH BEARINGS   CC = CENTER TO CENTER   CD = CROSS DIAMETER
CL = CROSS LENGTH WITHOUT BEARINGS   EC = END TO CENTER or FACE TO FACE   EE = END TO END   EL = EFFECTIVE LENGTH
HD = HUB DIAMETER (or insert)   IE = INSIDE OF EARS (or recess)   OD = OUTSIDE DIAMETER   OE = OUTSIDE OF EARS   SB = SPLINE OR BORE SIZE

IDLI GROUP 81045

END YOKE WITH 13 INVOLUTED SPLINES
SEE FIGURE 4A

1.4375 in	IE	36.51 mm
.9688 in	BD	24.61 mm
.8750 in	SB	22.23 mm
1.8125 in	EC	46.04 mm
1.6250 in	HD	41.28 mm

ROCKWELL L6NYS14-22

IDLI GROUP 81046

END YOKE WITH ROUND BORE
SEE FIGURE 4B

1.4375 in	IE	36.51 mm
.9688 in	BD	24.61 mm
.8750 in	SB	22.23 mm
2.1250 in	EC	53.98 mm
1.6250 in	HD	41.28 mm

ROCKWELL L6NYR14-8

IDLI GROUP 81047

END YOKE WITH RECTANGULAR BORE
maximum SB dimension shown
SEE FIGURE 4D

1.4375 in	IE	36.51 mm
.9688 in	BD	24.61 mm
.8750 in	SB	22.23 mm
2.1875 in	EC	55.56 mm

ROCKWELL L6NYP12-12

IDLI GROUP 81048

END YOKE WITH 20 INVOLUTED SPLINES
SEE FIGURE 4A

1.4375 in	IE	36.51 mm
.9688 in	BD	24.61 mm
.8750 in	SB	22.23 mm
2.3750 in	EC	60.33 mm
1.3750 in	HD	34.93 mm

ROCKWELL L6NYS14-101

IDLI GROUP 81049

END YOKE WITH ROUND BORE
SEE FIGURE 4B

1.4375 in	IE	36.51 mm
.9688 in	BD	24.61 mm
.8750 in	SB	22.23 mm
2.3750 in	EC	60.33 mm
1.3750 in	HD	34.93 mm

ROCKWELL L6NYR14

IDLI GROUP 81050

END YOKE WITH ROUND BORE
slight variation from above group
SEE FIGURE 4B

1.4375 in	IE	36.51 mm
.9688 in	BD	24.61 mm
.8750 in	SB	22.23 mm
2.3750 in	EC	60.33 mm
1.3750 in	HD	34.93 mm

ROCKWELL L6NYR14

IDLI GROUP 81051

END YOKE WITH THREADED BORE
SEE FIGURE 4G

1.4375 in	IE	36.51 mm
.9688 in	BD	24.61 mm
.8750 in	SB	22.23 mm
2.4375 in	EC	61.91 mm
1.5000 in	HD	38.10 mm

ROCKWELL L6NYD14-7

IDLI GROUP 81052

END YOKE WITH 6 SPLINES
SEE FIGURE 4A

1.4375 in	IE	36.51 mm
.9688 in	BD	24.61 mm
.8750 in	SB	22.23 mm
2.5000 in	EC	63.50 mm
1.6250 in	HD	41.28 mm

ROCKWELL L6-101

IDLI GROUP 81053

END YOKE WITH ROUND BORE
SEE FIGURE 4B

1.4375 in	IE	36.51 mm
.9688 in	BD	24.61 mm
.8750 in	SB	22.23 mm
2.5000 in	EC	63.50 mm
1.6250 in	HD	41.28 mm

ROCKWELL L6-19

IDLI GROUP 81054

END YOKE WITH ROUND BORE
slight variation from above group
SEE FIGURE 4B

1.4375 in	IE	36.51 mm
.9688 in	BD	24.61 mm
.8750 in	SB	22.23 mm
2.5000 in	EC	63.50 mm
1.6250 in	HD	41.28 mm

BORG-WARNR	8390J
MCQUAY-NOR	U03-77
NEAPCO	16-6113
NEAPCO	L6113
PRECISION	1903
ROCKWELL	L6-38
TRW	21053
WESCO	06N1434-37
WESCO	540064
WESCO	540065
WESCO	540086
WESCO	6N1434-37
WESCO	6N144-2
WESCO	6N144-5

IDLI GROUP 81055

END YOKE WITH ROUND BORE
slight variation from above group
SEE FIGURE 4B

1.4375 in	IE	36.51 mm
.9688 in	BD	24.61 mm
.8750 in	SB	22.23 mm
2.5000 in	EC	63.50 mm
1.6250 in	HD	41.28 mm

ROCKWELL L6-43

IDLI GROUP 81056

END YOKE WITH SQUARE BORE
SEE FIGURE 4C

1.4375 in	IE	36.51 mm
.9688 in	BD	24.61 mm
.8750 in	SB	22.23 mm
2.5000 in	EC	63.50 mm
1.6250 in	HD	41.28 mm

ROCKWELL L6-34

IDLI GROUP 81057

END YOKE WITH RECTANGULAR BORE
maximum SB dimension shown
slight variation from above group
SEE FIGURE 4D

1.4375 in	IE	36.51 mm
.9688 in	BD	24.61 mm
.8750 in	SB	22.23 mm
2.5000 in	EC	63.50 mm

ROCKWELL L6NYP12-5

IDLI GROUP 81058

END YOKE WITH RECTANGULAR BORE
maximum SB dimension shown
slight variation from above group
SEE FIGURE 4D

1.4375 in	IE	36.51 mm
.9688 in	BD	24.61 mm
.8750 in	SB	22.23 mm
2.5000 in	EC	63.50 mm

ROCKWELL L6NYP12-23

IDLI GROUP 81059

END YOKE WITH RECTANGULAR BORE
maximum SB dimension shown
slight variation from above group
SEE FIGURE 4D

1.4375 in	IE	36.51 mm
.9688 in	BD	24.61 mm
.8750 in	SB	22.23 mm
2.5000 in	EC	63.50 mm

ROCKWELL L6NYP12-66

IDLI GROUP 81060

END YOKE WITH HEX BORE
SEE FIGURE 4E

1.4375 in	IE	36.51 mm
.9688 in	BD	24.61 mm
.8750 in	SB	22.23 mm
2.5000 in	EC	63.50 mm
1.6250 in	HD	41.28 mm

NEAPCO	16-6303
NEAPCO	L61R8
PRECISION	1912
ROCKWELL	L6-107
TRW	21054
WESCO	06N0140-32
WESCO	540063
WESCO	6N140-32

IDLI GROUP 81061

END YOKE WITH RECTANGULAR BORE
maximum SB dimension shown
SEE FIGURE 4D

1.4375 in	IE	36.51 mm
.9688 in	BD	24.61 mm
.8750 in	SB	22.23 mm
2.5000 in	EC	63.50 mm
1.3750 in	HD	34.93 mm

ROCKWELL L6NYP12-35

IDLI GROUP 81062

END YOKE WITH 13 INVOLUTED SPLINES
SEE FIGURE 4A

1.4375 in	IE	36.51 mm
.9688 in	BD	24.61 mm
.8750 in	SB	22.23 mm
2.5000 in	EC	63.50 mm
1.5000 in	HD	38.10 mm

ROCKWELL L6NYS14-31

IDLI GROUP 81063

END YOKE WITH SQUARE BORE
SEE FIGURE 4C

1.4375 in	IE	36.51 mm
.9688 in	BD	24.61 mm
.8750 in	SB	22.23 mm
2.5000 in	EC	63.50 mm
1.5000 in	HD	38.10 mm

ROCKWELL L6NYQ14-14

IDLI GROUP 81064

END YOKE WITH 13 INVOLUTED SPLINES
SEE FIGURE 4A

1.4375 in	IE	36.51 mm
.9688 in	BD	24.61 mm
.8750 in	SB	22.23 mm
2.5000 in	EC	63.50 mm
1.5625 in	HD	39.69 mm

ROCKWELL L6NYS14-42

IDLI GROUP 81065

END YOKE WITH 6 SPLINES
SEE FIGURE 4A

1.4375 in	IE	36.51 mm
.9688 in	BD	24.61 mm
.8750 in	SB	22.23 mm
2.5000 in	EC	63.50 mm
1.6250 in	HD	41.28 mm

ROCKWELL L6NYS14-18

IDLI GROUP 81066

END YOKE WITH 6 SPLINES
slight variation from above group
SEE FIGURE 4A

1.4375 in	IE	36.51 mm
.9688 in	BD	24.61 mm
.8750 in	SB	22.23 mm
2.5000 in	EC	63.50 mm
1.6250 in	HD	41.28 mm

ROCKWELL L6NYS14-18

ENGINEERING CATALOGS MUST BE CONSULTED FOR DETAILS NOT INCLUDED
IN THIS GUIDE. SPECIFIC DESIGNS, MATERIAL CONTENT, TOLERANCES
LUBE FITTINGS AND OTHER DIMENSIONS ARE INTENTIONALLY OMITTED HERE.

CAUTION: BE SURE TO REFER TO ENGINEERING CATALOGS FOR SPECIAL APPLICATIONS THAT REQUIRE SPECIFIC MATERIAL CONTENT, TOLERANCES, ETC. SEE FOOTNOTE.

IDLI GROUP 81067

END YOKE WITH 13 INVOLUTED SPLINES
SEE FIGURE 4A

1.4375 in	IE	36.51 mm
.9688 in	BD	24.61 mm
.8750 in	SB	22.23 mm
2.5000 in	EC	63.50 mm
1.6250 in	HD	41.28 mm

ROCKWELL L6NYS14-19

IDLI GROUP 81068

END YOKE WITH ROUND BORE
SEE FIGURE 4B

1.4375 in	IE	36.51 mm
.9688 in	BD	24.61 mm
.8750 in	SB	22.23 mm
2.5000 in	EC	63.50 mm
1.6250 in	HD	41.28 mm

ROCKWELL L6NYR14-93

IDLI GROUP 81069

END YOKE WITH ROUND BORE
slight variation from above group
SEE FIGURE 4B

1.4375 in	IE	36.51 mm
.9688 in	BD	24.61 mm
.8750 in	SB	22.23 mm
2.5000 in	EC	63.50 mm
1.6250 in	HD	41.28 mm

ROCKWELL L6NYR14-51

IDLI GROUP 81070

END YOKE WITH ROUND BORE
slight variation from above group
SEE FIGURE 4B

1.4375 in	IE	36.51 mm
.9688 in	BD	24.61 mm
.8750 in	SB	22.23 mm
2.5000 in	EC	63.50 mm
1.6250 in	HD	41.28 mm

NEAPCO 16-6213
NEAPCO L613A
ROCKWELL L6NYR14-21

IDLI GROUP 81071

END YOKE WITH ROUND BORE
slight variation from above group
SEE FIGURE 4B

1.4375 in	IE	36.51 mm
.9688 in	BD	24.61 mm
.8750 in	SB	22.23 mm
2.5000 in	EC	63.50 mm
1.6250 in	HD	41.28 mm

ROCKWELL L6NYR14-132

IDLI GROUP 81072

END YOKE WITH ROUND BORE
slight variation from above group
SEE FIGURE 4B

1.4375 in	IE	36.51 mm
.9688 in	BD	24.61 mm
.8750 in	SB	22.23 mm
2.5000 in	EC	63.50 mm
1.6250 in	HD	41.28 mm

ROCKWELL L6NYR14-17

IDLI GROUP 81073

END YOKE WITH ROUND BORE
slight variation from above group
SEE FIGURE 4B

1.4375 in	IE	36.51 mm
.9688 in	BD	24.61 mm
.8750 in	SB	22.23 mm
2.5000 in	EC	63.50 mm
1.6250 in	HD	41.28 mm

ROCKWELL L6NYR14-19

IDLI GROUP 81074

END YOKE WITH RECTANGULAR BORE
maximum SB dimension shown
SEE FIGURE 4D

1.4375 in	IE	36.51 mm
.9688 in	BD	24.61 mm
.8750 in	SB	22.23 mm
2.5000 in	EC	63.50 mm
1.6250 in	HD	41.28 mm

ROCKWELL L6NYP12-14

IDLI GROUP 81075

END YOKE WITH RECTANGULAR BORE
maximum SB dimension shown
slight variation from above group
SEE FIGURE 4D

1.4375 in	IE	36.51 mm
.9688 in	BD	24.61 mm
.8750 in	SB	22.23 mm
2.5000 in	EC	63.50 mm
1.6250 in	HD	41.28 mm

ROCKWELL L6NYP12-32

IDLI GROUP 81076

END YOKE WITH HEX BORE
SEE FIGURE 4E

1.4375 in	IE	36.51 mm
.9688 in	BD	24.61 mm
.8750 in	SB	22.23 mm
2.5000 in	EC	63.50 mm
1.6250 in	HD	41.28 mm

ROCKWELL L6NYH14

IDLI GROUP 81077

END YOKE WITH HEX BORE
SEE FIGURE 4E

1.4375 in	IE	36.51 mm
.9688 in	BD	24.61 mm
.8750 in	SB	22.23 mm
2.7500 in	EC	69.85 mm
1.6250 in	HD	41.28 mm

ROCKWELL L6NYH14-17

IDLI GROUP 81078

END YOKE WITH RECTANGULAR BORE
maximum SB dimension shown
SEE FIGURE 4D

1.4375 in	IE	36.51 mm
.9688 in	BD	24.61 mm
.8750 in	SB	22.23 mm
3.1250 in	EC	79.38 mm
1.3750 in	HD	34.93 mm

ROCKWELL L6NYP12-48

IDLI GROUP 81079

END YOKE WITH RECTANGULAR BORE
maximum SB dimension shown
SEE FIGURE 4D

1.4375 in	IE	36.51 mm
.9688 in	BD	24.61 mm
.8750 in	SB	22.23 mm
3.5625 in	EC	90.49 mm
1.8125 in	HD	46.04 mm

ROCKWELL L6NYP12-1

IDLI GROUP 81080

END YOKE WITH 13 INVOLUTED SPLINES
SEE FIGURE 4A

1.4375 in	IE	36.51 mm
.9688 in	BD	24.61 mm
.8750 in	SB	22.23 mm
4.5000 in	EC	114.30 mm
1.6250 in	HD	41.28 mm

ROCKWELL L6NYS14-20

IDLI GROUP 81081

END YOKE WITH ROUND BORE
SEE FIGURE 4B

1.4375 in	IE	36.51 mm
.9688 in	BD	24.61 mm
.9375 in	SB	23.81 mm
1.6875 in	EC	42.86 mm
1.6250 in	HD	41.28 mm

ROCKWELL L6-79

IDLI GROUP 81082

END YOKE WITH ROUND BORE
slight variation from above group
SEE FIGURE 4B

1.4375 in	IE	36.51 mm
.9688 in	BD	24.61 mm
.9375 in	SB	23.81 mm
1.6875 in	EC	42.86 mm
1.6250 in	HD	41.28 mm

ROCKWELL L6-80

IDLI GROUP 81083

END YOKE WITH SQUARE BORE AND HUB
SEE FIGURE 4C

1.4375 in	IE	36.51 mm
.9688 in	BD	24.61 mm
.9375 in	SB	23.81 mm
2.1875 in	EC	55.56 mm
1.7500 in	HD	44.45 mm

ROCKWELL L6NYQ15-8

IDLI GROUP 81084

END YOKE WITH ROUND BORE
SEE FIGURE 4B

1.4375 in	IE	36.51 mm
.9688 in	BD	24.61 mm
.9375 in	SB	23.81 mm
2.5000 in	EC	63.50 mm
1.6250 in	HD	41.28 mm

ROCKWELL L6-37

IDLI GROUP 81085

END YOKE WITH ROUND BORE
slight variation from above group
SEE FIGURE 4B

1.4375 in	IE	36.51 mm
.9688 in	BD	24.61 mm
.9375 in	SB	23.81 mm
2.5000 in	EC	63.50 mm
1.6250 in	HD	41.28 mm

ROCKWELL L6-44

IDLI GROUP 81086

END YOKE WITH ROUND BORE
slight variation from above group
SEE FIGURE 4B

1.4375 in	IE	36.51 mm
.9688 in	BD	24.61 mm
.9375 in	SB	23.81 mm
2.5000 in	EC	63.50 mm
1.6250 in	HD	41.28 mm

PRECISION 1904
ROCKWELL L6-45

IDLI GROUP 81087

END YOKE WITH ROUND BORE
slight variation from above group
SEE FIGURE 4B

1.4375 in	IE	36.51 mm
.9688 in	BD	24.61 mm
.9375 in	SB	23.81 mm
2.5000 in	EC	63.50 mm
1.6250 in	HD	41.28 mm

ROCKWELL L6NYR15-6

IDLI GROUP 81088

END YOKE WITH ROUND BORE
slight variation from above group
SEE FIGURE 4B

1.4375 in	IE	36.51 mm
.9688 in	BD	24.61 mm
.9375 in	SB	23.81 mm
2.5000 in	EC	63.50 mm
1.6250 in	HD	41.28 mm

ROCKWELL L6NYR15-7

IDLI GROUP 81089

END YOKE WITH ROUND BORE
slight variation from above group
SEE FIGURE 4B

1.4375 in	IE	36.51 mm
.9688 in	BD	24.61 mm
.9375 in	SB	23.81 mm
2.5000 in	EC	63.50 mm
1.6250 in	HD	41.28 mm

BORG-WARNR 12688J
BORG-WARNR 8687J

BD = BEARING DIAMETER (outside)   **BW** = BEARING WIDTH   **CB** = CROSS LENGTH WITH BEARINGS   **CC** = CENTER TO CENTER   **CD** = CROSS DIAMETER
CL = CROSS LENGTH WITHOUT BEARINGS   **EC** = END TO CENTER or FACE TO FACE   **EE** = END TO END   **EL** = EFFECTIVE LENGTH
HD = HUB DIAMETER (or insert)   **IE** = INSIDE OF EARS (or recess)   **OD** = OUTSIDE DIAMETER   **OE** = OUTSIDE OF EARS   **SB** = SPLINE OR BORE SIZE

MCQUAY-NOR U03-89
NEAPCO 16-6114
NEAPCO L6114
PRECISION 1904
ROCKWELL L6NYR15-12
TRW 21055
WESCO 06N0154
WESCO 540066
WESCO 540067
WESCO 6N154

IDLI GROUP 81090
END YOKE WITH 6 SPLINES
SEE FIGURE 4A

1.4375 in	IE	36.51 mm
.9688 in	BD	24.61 mm
.9844 in	SB	25.00 mm
2.5000 in	EC	63.50 mm
1.5000 in	HD	38.10 mm

ROCKWELL L6NYS16-2

IDLI GROUP 81091
END YOKE WITH SQUARE BORE
SEE FIGURE 4C

1.4375 in	IE	36.51 mm
.9688 in	BD	24.61 mm
.9375 in	SB	23.81 mm
2.5000 in	EC	63.50 mm
2.1250 in	HD	53.98 mm

ROCKWELL L6-97

IDLI GROUP 81092
END YOKE WITH SQUARE BORE
slight variation from above group
SEE FIGURE 4C

1.4375 in	IE	36.51 mm
.9688 in	BD	24.61 mm
.9375 in	SB	23.81 mm
2.5000 in	EC	63.50 mm
2.1250 in	HD	53.98 mm

ROCKWELL L6-98

IDLI GROUP 81093
END YOKE WITH SQUARE BORE
slight variation from above group
SEE FIGURE 4C

1.4375 in	IE	36.51 mm
.9688 in	BD	24.61 mm
.9375 in	SB	23.81 mm
2.5000 in	EC	63.50 mm
2.1250 in	HD	53.98 mm

ROCKWELL L6NYQ15

IDLI GROUP 81094
END YOKE WITH ROUND BORE
SEE FIGURE 4B

1.4375 in	IE	36.51 mm
.9688 in	BD	24.61 mm
.9844 in	SB	25.00 mm
2.7500 in	EC	69.85 mm
1.6250 in	HD	41.28 mm

ROCKWELL L6NYR16-37

IDLI GROUP 81095
END YOKE WITH ROUND BORE
SEE FIGURE 4B

1.4375 in	IE	36.51 mm
.9688 in	BD	24.61 mm
.9844 in	SB	25.00 mm
3.3125 in	EC	84.14 mm
2.0000 in	HD	50.80 mm

ROCKWELL L6NYR16-59

IDLI GROUP 81096
END YOKE WITH ROUND BORE
SEE FIGURE 4B

1.4375 in	IE	36.51 mm
.9688 in	BD	24.61 mm
1.0000 in	SB	25.40 mm
1.6875 in	EC	42.86 mm
1.6250 in	HD	41.28 mm

ROCKWELL L6-94

IDLI GROUP 81097
END YOKE WITH ROUND BORE
SEE FIGURE 4B

1.4375 in	IE	36.51 mm
.9688 in	BD	24.61 mm
1.0000 in	SB	25.40 mm
2.0000 in	EC	50.80 mm
1.5156 in	HD	38.50 mm

ROCKWELL L6NYR16-245

IDLI GROUP 81098
END YOKE WITH SQUARE BORE AND HUB
SEE FIGURE 4C

1.4375 in	IE	36.51 mm
.9688 in	BD	24.61 mm
1.0000 in	SB	25.40 mm
2.1875 in	EC	55.56 mm
1.7500 in	HD	44.45 mm

ROCKWELL L6NYQ16-23

IDLI GROUP 81099
END YOKE WITH 15 INVOLUTED SPLINES
SEE FIGURE 4A

1.4375 in	IE	36.51 mm
.9688 in	BD	24.61 mm
1.0000 in	SB	25.40 mm
2.2500 in	EC	57.15 mm
1.6250 in	HD	41.28 mm

ROCKWELL L6NYS16-78

IDLI GROUP 81100
END YOKE WITH ROUND BORE
SEE FIGURE 4B

1.4375 in	IE	36.51 mm
.9688 in	BD	24.61 mm
1.0000 in	SB	25.40 mm
2.3750 in	EC	60.33 mm
1.6250 in	HD	41.28 mm

ROCKWELL L6NYR16-151

IDLI GROUP 81101
END YOKE WITH 15 INVOLUTED SPLINES
SEE FIGURE 4A

1.4375 in	IE	36.51 mm
.9688 in	BD	24.61 mm
1.0000 in	SB	25.40 mm
2.5000 in	EC	63.50 mm
1.3750 in	HD	34.93 mm

ROCKWELL L6NYS16-54

IDLI GROUP 81102
END YOKE WITH 15 INVOLUTED SPLINES
slight variation from above group
SEE FIGURE 4A

1.4375 in	IE	36.51 mm
.9688 in	BD	24.61 mm
1.0000 in	SB	25.40 mm
2.5000 in	EC	63.50 mm
1.3750 in	HD	34.93 mm

ROCKWELL L6NYS16-54

IDLI GROUP 81103
END YOKE WITH 6 SPLINES
SEE FIGURE 4A

1.4375 in	IE	36.51 mm
.9688 in	BD	24.61 mm
1.0000 in	SB	25.40 mm
2.5000 in	EC	63.50 mm
1.5000 in	HD	38.10 mm

ROCKWELL L6-102

IDLI GROUP 81104
END YOKE WITH ROUND BORE
SEE FIGURE 4B

1.4375 in	IE	36.51 mm
.9688 in	BD	24.61 mm
1.0000 in	SB	25.40 mm
2.5000 in	EC	63.50 mm
1.5000 in	HD	38.10 mm

ROCKWELL L6NYR16-198

IDLI GROUP 81105
END YOKE WITH ROUND BORE
slight variation from above group
SEE FIGURE 4B

1.4375 in	IE	36.51 mm
.9688 in	BD	24.61 mm
1.0000 in	SB	25.40 mm
2.5000 in	EC	63.50 mm
1.5000 in	HD	38.10 mm

ROCKWELL L6NYR16-219

IDLI GROUP 81106
END YOKE WITH 19 INVOLUTED SPLINES
SEE FIGURE 4A

1.4375 in	IE	36.51 mm
.9688 in	BD	24.61 mm
1.0000 in	SB	25.40 mm
2.5000 in	EC	63.50 mm
1.5625 in	HD	39.69 mm

ROCKWELL L6NYS16-65

IDLI GROUP 81107
END YOKE WITH 15 INVOLUTED SPLINES
SEE FIGURE 4A

1.4375 in	IE	36.51 mm
.9688 in	BD	24.61 mm
1.0000 in	SB	25.40 mm
2.5000 in	EC	63.50 mm
1.6250 in	HD	41.28 mm

ROCKWELL L6NYS16-60

IDLI GROUP 81108
END YOKE WITH ROUND BORE
SEE FIGURE 4B

1.4375 in	IE	36.51 mm
.9688 in	BD	24.61 mm
1.0000 in	SB	25.40 mm
2.5000 in	EC	63.50 mm
1.6250 in	HD	41.28 mm

ROCKWELL L6-46

IDLI GROUP 81109
END YOKE WITH ROUND BORE
slight variation from above group
SEE FIGURE 4B

1.4375 in	IE	36.51 mm
.9688 in	BD	24.61 mm
1.0000 in	SB	25.40 mm
2.5000 in	EC	63.50 mm
1.6250 in	HD	41.28 mm

BORG-WARNR 12678J
BORG-WARNR 8156J
MCQUAY-NOR U03-78
NEAPCO 16-6115
NEAPCO L6115
PRECISION 1905
ROCKWELL L6-47
TRW 21056
WESCO 06N1634-37
WESCO 540068
WESCO 540071
WESCO 540087
WESCO 6N1634-37

IDLI GROUP 81110
END YOKE WITH ROUND BORE
slight variation from above group
SEE FIGURE 4B

1.4375 in	IE	36.51 mm
.9688 in	BD	24.61 mm
1.0000 in	SB	25.40 mm
2.5000 in	EC	63.50 mm
1.6250 in	HD	41.28 mm

ROCKWELL L6NYR16

IDLI GROUP 81111
END YOKE WITH ROUND BORE
slight variation from above group
SEE FIGURE 4B

1.4375 in	IE	36.51 mm
.9688 in	BD	24.61 mm
1.0000 in	SB	25.40 mm
2.5000 in	EC	63.50 mm
1.6250 in	HD	41.28 mm

ROCKWELL L6NYR16-1

ENGINEERING CATALOGS MUST BE CONSULTED FOR DETAILS NOT INCLUDED
IN THIS GUIDE. SPECIFIC DESIGNS, MATERIAL CONTENT, TOLERANCES
LUBE FITTINGS AND OTHER DIMENSIONS ARE INTENTIONALLY OMITTED HERE.

IDLI GUIDE   COPYRIGHT ® INTERCHANGE, INC.   ST. LOUIS PARK, MN. 55416 USA

CAUTION: BE SURE TO REFER TO ENGINEERING CATALOGS FOR SPECIAL APPLICATIONS THAT REQUIRE SPECIFIC MATERIAL CONTENT, TOLERANCES, ETC. SEE FOOTNOTE.

IDLI GROUP 81112

END YOKE WITH ROUND BORE
slight variation from above group
SEE FIGURE 4B

1.4375 in	IE	36.51 mm
.9688 in	BD	24.61 mm
1.0000 in	SB	25.40 mm
2.5000 in	EC	63.50 mm
1.6250 in	HD	41.28 mm

ROCKWELL L6NYR16-15

IDLI GROUP 81113

END YOKE WITH ROUND BORE
slight variation from above group
SEE FIGURE 4B

1.4375 in	IE	36.51 mm
.9688 in	BD	24.61 mm
1.0000 in	SB	25.40 mm
2.5000 in	EC	63.50 mm
1.6250 in	HD	41.28 mm

ROCKWELL L6NYR16-51

IDLI GROUP 81114

END YOKE WITH ROUND BORE
slight variation from above group
SEE FIGURE 4B

1.4375 in	IE	36.51 mm
.9688 in	BD	24.61 mm
1.0000 in	SB	25.40 mm
2.5000 in	EC	63.50 mm
1.6250 in	HD	41.28 mm

ROCKWELL L6NYR16-64

IDLI GROUP 81115

END YOKE WITH ROUND BORE
slight variation from above group
SEE FIGURE 4B

1.4375 in	IE	36.51 mm
.9688 in	BD	24.61 mm
1.0000 in	SB	25.40 mm
2.5000 in	EC	63.50 mm
1.6250 in	HD	41.28 mm

NEAPCO 16-6215
NEAPCO L615A
ROCKWELL L6NYR16-66

IDLI GROUP 81116

END YOKE WITH ROUND BORE
slight variation from above group
SEE FIGURE 4B

1.4375 in	IE	36.51 mm
.9688 in	BD	24.61 mm
1.0000 in	SB	25.40 mm
2.5000 in	EC	63.50 mm
1.6250 in	HD	41.28 mm

ROCKWELL L6NYR16-85

IDLI GROUP 81117

END YOKE WITH ROUND BORE
slight variation from above group
SEE FIGURE 4B

1.4375 in	IE	36.51 mm
.9688 in	BD	24.61 mm
1.0000 in	SB	25.40 mm
2.5000 in	EC	63.50 mm
1.6250 in	HD	41.28 mm

ROCKWELL L6NYR16-96

IDLI GROUP 81118

END YOKE WITH ROUND BORE
slight variation from above group
SEE FIGURE 4B

1.4375 in	IE	36.51 mm
.9688 in	BD	24.61 mm
1.0000 in	SB	25.40 mm
2.5000 in	EC	63.50 mm
1.6250 in	HD	41.28 mm

ROCKWELL L6NYR16-146

IDLI GROUP 81119

END YOKE WITH ROUND BORE
slight variation from above group
SEE FIGURE 4B

1.4375 in	IE	36.51 mm
.9688 in	BD	24.61 mm
1.0000 in	SB	25.40 mm
2.5000 in	EC	63.50 mm
1.6250 in	HD	41.28 mm

ROCKWELL L6NYR16-188

IDLI GROUP 81120

END YOKE WITH ROUND BORE
slight variation from above group
SEE FIGURE 4B

1.4375 in	IE	36.51 mm
.9688 in	BD	24.61 mm
1.0000 in	SB	25.40 mm
2.5000 in	EC	63.50 mm
1.6250 in	HD	41.28 mm

ROCKWELL L6NYR16-259

IDLI GROUP 81121

END YOKE WITH HEX BORE
SEE FIGURE 4E

1.4375 in	IE	36.51 mm
.9688 in	BD	24.61 mm
1.0000 in	SB	25.40 mm
2.5000 in	EC	63.50 mm
1.6250 in	HD	41.28 mm

ROCKWELL L6NYH16-1

IDLI GROUP 81122

END YOKE WITH SQUARE BORE
SEE FIGURE 4C

1.4375 in	IE	36.51 mm
.9688 in	BD	24.61 mm
1.0000 in	SB	25.40 mm
2.5000 in	EC	63.50 mm
1.8750 in	HD	47.63 mm

ROCKWELL L6NYQ16

IDLI GROUP 81123

END YOKE WITH SQUARE BORE
SEE FIGURE 4C

1.4375 in	IE	36.51 mm
.9688 in	BD	24.61 mm
1.0000 in	SB	25.40 mm
2.5000 in	EC	63.50 mm
2.1250 in	HD	53.98 mm

ROCKWELL L6-35

IDLI GROUP 81124

END YOKE WITH SQUARE BORE
slight variation from above group
SEE FIGURE 4C

1.4375 in	IE	36.51 mm
.9688 in	BD	24.61 mm
1.0000 in	SB	25.40 mm
2.5000 in	EC	63.50 mm
2.1250 in	HD	53.98 mm

ROCKWELL L6-36

IDLI GROUP 81125

END YOKE WITH SQUARE BORE
slight variation from above group
SEE FIGURE 4C

1.4375 in	IE	36.51 mm
.9688 in	BD	24.61 mm
1.0000 in	SB	25.40 mm
2.5000 in	EC	63.50 mm
2.1250 in	HD	53.98 mm

G & G MFG 183-0616
MCQUAY-NOR U03-112
NEAPCO 16-6193
NEAPCO L6193
PRECISION 1911
ROCKWELL L6NYQ16-2
WESCO 540062
WESCO 540069
WESCO 6N140-30
WESCO 6N160-26
WESCO 6N160-27

IDLI GROUP 81126

END YOKE WITH SQUARE BORE
slight variation from above group
SEE FIGURE 4C

1.4375 in	IE	36.51 mm
.9688 in	BD	24.61 mm
1.0000 in	SB	25.40 mm
2.5000 in	EC	63.50 mm
2.1250 in	HD	53.98 mm

ROCKWELL L6NYQ16-4

IDLI GROUP 81127

END YOKE WITH TAPERED BORE
SEE FIGURE 4F

1.4375 in	IE	36.51 mm
.9688 in	BD	24.61 mm
1.0000 in	SB	25.40 mm
2.6250 in	EC	66.68 mm
1.5000 in	HD	38.10 mm

ROCKWELL L6NYT16-2

IDLI GROUP 81128

END YOKE WITH 6 SPLINES
SEE FIGURE 4A

1.4375 in	IE	36.51 mm
.9688 in	BD	24.61 mm
1.0000 in	SB	25.40 mm
2.7500 in	EC	69.85 mm
1.6250 in	HD	41.28 mm

ROCKWELL L6NYS16-1

IDLI GROUP 81129

END YOKE WITH ROUND BORE
SEE FIGURE 4B

1.4375 in	IE	36.51 mm
.9688 in	BD	24.61 mm
1.0000 in	SB	25.40 mm
2.8125 in	EC	71.44 mm
1.6250 in	HD	41.28 mm

ROCKWELL L6NYR16-242

IDLI GROUP 81130

END YOKE WITH SQUARE BORE
SEE FIGURE 4C

1.4375 in	IE	36.51 mm
.9688 in	BD	24.61 mm
1.0000 in	SB	25.40 mm
3.0000 in	EC	76.20 mm
2.0000 in	HD	50.80 mm

ROCKWELL L6NYQ16-35

IDLI GROUP 81131

END YOKE WITH 6 SPLINES
slight variation from above group
SEE FIGURE 4A

1.4375 in	IE	36.51 mm
.9688 in	BD	24.61 mm
1.0000 in	SB	25.40 mm
3.3125 in	EC	84.14 mm
1.9375 in	HD	49.21 mm

ROCKWELL L6NYS16-27

IDLI GROUP 81132

END YOKE WITH ROUND BORE
SEE FIGURE 4B

1.4375 in	IE	36.51 mm
.9688 in	BD	24.61 mm
1.0000 in	SB	25.40 mm
3.3125 in	EC	84.14 mm
2.0000 in	HD	50.80 mm

ROCKWELL L6NYR16-59

IDLI GROUP 81133

END YOKE WITH ROUND BORE
SEE FIGURE 4B

1.4375 in	IE	36.51 mm
.9688 in	BD	24.61 mm
1.0625 in	SB	26.99 mm
2.5000 in	EC	63.50 mm
1.6250 in	HD	41.28 mm

ROCKWELL L6-49

BD=BEARING DIAMETER (outside)   **BW**=BEARING WIDTH   **CB**=CROSS LENGTH WITH BEARINGS   **CC**=CENTER TO CENTER   **CD**=CROSS DIAMETER
CL=CROSS LENGTH WITHOUT BEARINGS   **EC**=END TO CENTER or FACE TO FACE   **EE**=END TO END   **EL**=EFFECTIVE LENGTH
HD=HUB DIAMETER (or insert)   **IE**=INSIDE OF EARS (or recess)   **OD**=OUTSIDE DIAMETER   **OE**=OUTSIDE OF EARS   **SB**=SPLINE OR BORE SIZE

IDLI GROUP 81134

END YOKE WITH ROUND BORE
slight variation from above group
SEE FIGURE 4B

1.4375 in	IE	36.51 mm
.9688 in	BD	24.61 mm
1.0625 in	SB	26.99 mm
2.5000 in	EC	63.50 mm
1.6250 in	HD	41.28 mm

ROCKWELL L6NYR17-7

IDLI GROUP 81135

END YOKE WITH ROUND BORE
SEE FIGURE 4B

1.4375 in	IE	36.51 mm
.9688 in	BD	24.61 mm
1.0625 in	SB	26.99 mm
2.5000 in	EC	63.50 mm
1.8750 in	HD	47.63 mm

PRECISION 1906
ROCKWELL L6-48

IDLI GROUP 81136

END YOKE WITH ROUND BORE
slight variation from above group
SEE FIGURE 4B

1.4375 in	IE	36.51 mm
.9688 in	BD	24.61 mm
1.0625 in	SB	26.99 mm
2.5000 in	EC	63.50 mm
1.8750 in	HD	47.63 mm

NEAPCO 16-6191
NEAPCO L6191
PRECISION 1906
ROCKWELL L6NYR17
WESCO 540073
WESCO 6N174

IDLI GROUP 81137

END YOKE WITH SQUARE BORE AND HUB
SEE FIGURE 4C

1.4375 in	IE	36.51 mm
.9688 in	BD	24.61 mm
1.0938 in	SB	27.78 mm
2.1875 in	EC	55.56 mm
1.7500 in	HD	44.45 mm

ROCKWELL L6NYQ17-1

IDLI GROUP 81138

END YOKE WITH RECTANGULAR BORE
maximum SB dimension shown
SEE FIGURE 4D

1.4375 in	IE	36.51 mm
.9688 in	BD	24.61 mm
1.1250 in	SB	28.58 mm
2.1875 in	EC	55.56 mm

ROCKWELL L6NYP16-5

IDLI GROUP 81139

END YOKE WITH ROUND BORE
SEE FIGURE 4B

1.4375 in	IE	36.51 mm
.9688 in	BD	24.61 mm
1.1250 in	SB	28.58 mm
2.4375 in	EC	61.91 mm
1.6250 in	HD	41.28 mm

ROCKWELL L6NYR18-26

IDLI GROUP 81140

END YOKE WITH RECTANGULAR BORE
maximum SB dimension shown
SEE FIGURE 4D

1.4375 in	IE	36.51 mm
.9688 in	BD	24.61 mm
1.1250 in	SB	28.58 mm
2.5000 in	EC	63.50 mm

ROCKWELL L6NYP16-9

IDLI GROUP 81141

END YOKE WITH ROUND BORE
SEE FIGURE 4B

1.4375 in	IE	36.51 mm
.9688 in	BD	24.61 mm
1.1250 in	SB	28.58 mm
2.5000 in	EC	63.50 mm
1.5000 in	HD	38.10 mm

ROCKWELL L6NYR18-21

IDLI GROUP 81142

END YOKE WITH 6 SPLINES
SEE FIGURE 4A

1.4375 in	IE	36.51 mm
.9688 in	BD	24.61 mm
1.1250 in	SB	28.58 mm
2.5000 in	EC	63.50 mm
1.6250 in	HD	41.28 mm

ROCKWELL L6-103

IDLI GROUP 81143

END YOKE WITH 6 SPLINES
slight variation from above group
SEE FIGURE 4A

1.4375 in	IE	36.51 mm
.9688 in	BD	24.61 mm
1.1250 in	SB	28.58 mm
2.5000 in	EC	63.50 mm
1.6250 in	HD	41.28 mm

MCQUAY-NOR U03-90
NEAPCO 16-6134
NEAPCO L6134
ROCKWELL L6NYS18
TRW 21060
WESCO 06N0180-21
WESCO 540-075

IDLI GROUP 81144

END YOKE WITH ROUND BORE
SEE FIGURE 4B

1.4375 in	IE	36.51 mm
.9688 in	BD	24.61 mm
1.1250 in	SB	28.58 mm
2.5000 in	EC	63.50 mm
1.6250 in	HD	41.28 mm

ROCKWELL L6-69

IDLI GROUP 81145

END YOKE WITH ROUND BORE
slight variation from above group
SEE FIGURE 4B

1.4375 in	IE	36.51 mm
.9688 in	BD	24.61 mm
1.1250 in	SB	28.58 mm
2.5000 in	EC	63.50 mm
1.6250 in	HD	41.28 mm

ROCKWELL L6-84

IDLI GROUP 81146

END YOKE WITH ROUND BORE
SEE FIGURE 4B

1.4375 in	IE	36.51 mm
.9688 in	BD	24.61 mm
1.1250 in	SB	28.58 mm
2.5000 in	EC	63.50 mm
1.6250 in	HD	41.28 mm

ROCKWELL L6NYR18-66

IDLI GROUP 81147

END YOKE WITH 6 SPLINES
SEE FIGURE 4A

1.4375 in	IE	36.51 mm
.9688 in	BD	24.61 mm
1.1250 in	SB	28.58 mm
2.5000 in	EC	63.50 mm
1.7500 in	HD	44.45 mm

ROCKWELL L6NYS18-19

IDLI GROUP 81148

END YOKE WITH 6 SPLINES
SEE FIGURE 4A

1.4375 in	IE	36.51 mm
.9688 in	BD	24.61 mm
1.1250 in	SB	28.58 mm
2.5000 in	EC	63.50 mm
2.1250 in	HD	53.98 mm

ROCKWELL L6-104

IDLI GROUP 81149

END YOKE WITH 6 SPLINES
SEE FIGURE 4A

1.4375 in	IE	36.51 mm
.9688 in	BD	24.61 mm
1.1250 in	SB	28.58 mm
2.5000 in	EC	63.50 mm
2.1250 in	HD	53.98 mm

ROCKWELL L6NYS18-15

IDLI GROUP 81150

END YOKE WITH ROUND BORE
SEE FIGURE 4B

1.4375 in	IE	36.51 mm
.9688 in	BD	24.61 mm
1.1250 in	SB	28.58 mm
2.5000 in	EC	63.50 mm
2.1250 in	HD	53.98 mm

PRECISION 1908
ROCKWELL L6-26

IDLI GROUP 81151

END YOKE WITH ROUND BORE
slight variation from above group
SEE FIGURE 4B

1.4375 in	IE	36.51 mm
.9688 in	BD	24.61 mm
1.1250 in	SB	28.58 mm
2.5000 in	EC	63.50 mm
2.1250 in	HD	53.98 mm

BORG-WARNR 7009J
NEAPCO 16-6196
NEAPCO 16-6216
NEAPCO L616A
NEAPCO L6196
PRECISION 1907
ROCKWELL L6-27
TRW 21058
WESCO 06N1845-37

IDLI GROUP 81152

END YOKE WITH ROUND BORE
slight variation from above group
SEE FIGURE 4B

1.4375 in	IE	36.51 mm
.9688 in	BD	24.61 mm
1.1250 in	SB	28.58 mm
2.5000 in	EC	63.50 mm
2.1250 in	HD	53.98 mm

PRECISION 1907
ROCKWELL L6NYR18-8

IDLI GROUP 81153

END YOKE WITH ROUND BORE
slight variation from above group
SEE FIGURE 4B

1.4375 in	IE	36.51 mm
.9688 in	BD	24.61 mm
1.1250 in	SB	28.58 mm
2.5000 in	EC	63.50 mm
2.1250 in	HD	53.98 mm

MCQUAY-NOR U03-79
NEAPCO 16-6116
NEAPCO 16-6216
NEAPCO L6116
NEAPCO L616A
PRECISION 1908
ROCKWELL L6NYR18-9
WESCO 540079
WESCO 540088
WESCO 6N1845-37

ENGINEERING CATALOGS MUST BE CONSULTED FOR DETAILS NOT INCLUDED
IN THIS GUIDE. SPECIFIC DESIGNS, MATERIAL CONTENT, TOLERANCES
LUBE FITTINGS AND OTHER DIMENSIONS ARE INTENTIONALLY OMITTED HERE.

IDLI GUIDE   COPYRIGHT ® INTERCHANGE, INC.   ST. LOUIS PARK, MN. 55416 USA

CAUTION: BE SURE TO REFER TO ENGINEERING CATALOGS FOR SPECIAL APPLICATIONS THAT REQUIRE SPECIFIC MATERIAL CONTENT, TOLERANCES, ETC. SEE FOOTNOTE.

IDLI GROUP 81154

END YOKE WITH ROUND BORE
slight variation from above group
SEE FIGURE 4B

1.4375 in	IE	36.51 mm
.9688 in	BD	24.61 mm
1.1250 in	SB	28.58 mm
2.5000 in	EC	63.50 mm
2.1250 in	HD	53.98 mm

ROCKWELL L6NYR18-79

IDLI GROUP 81155

END YOKE WITH SQUARE BORE
SEE FIGURE 4C

1.4375 in	IE	36.51 mm
.9688 in	BD	24.61 mm
1.1250 in	SB	28.58 mm
2.5000 in	EC	63.50 mm
2.1250 in	HD	53.98 mm

ROCKWELL L6-99

IDLI GROUP 81156

END YOKE WITH SQUARE BORE
slight variation from above group
SEE FIGURE 4C

1.4375 in	IE	36.51 mm
.9688 in	BD	24.61 mm
1.1250 in	SB	28.58 mm
2.5000 in	EC	63.50 mm
2.1250 in	HD	53.98 mm

NEAPCO 16-6138
NEAPCO L6138
ROCKWELL L6NYQ18
TRW 21059
WESCO 06N0180-26
WESCO 540-077

IDLI GROUP 81157

END YOKE WITH HEX BORE
SEE FIGURE 4E

1.4375 in	IE	36.51 mm
.9688 in	BD	24.61 mm
1.1250 in	SB	28.58 mm
2.5000 in	EC	63.50 mm
2.1250 in	HD	53.98 mm

PRECISION 1913
ROCKWELL L6-108
WESCO 540078

IDLI GROUP 81158

END YOKE WITH HEX BORE
slight variation from above group
SEE FIGURE 4E

1.4375 in	IE	36.51 mm
.9688 in	BD	24.61 mm
1.1250 in	SB	28.58 mm
2.5000 in	EC	63.50 mm
2.1250 in	HD	53.98 mm

PRECISION 1913
ROCKWELL L6NYH18

IDLI GROUP 81159

END YOKE WITH 17 INVOLUTED SPLINES
SEE FIGURE 4A

1.4375 in	IE	36.51 mm
.9688 in	BD	24.61 mm
1.1250 in	SB	28.58 mm
2.7500 in	EC	69.85 mm
1.5000 in	HD	38.10 mm

ROCKWELL L6NYS18-25

IDLI GROUP 81160

END YOKE WITH ROUND BORE
SEE FIGURE 4B

1.4375 in	IE	36.51 mm
.9688 in	BD	24.61 mm
1.1250 in	SB	28.58 mm
2.7500 in	EC	69.85 mm
2.0000 in	HD	50.80 mm

ROCKWELL L6NYR18-42

IDLI GROUP 81161

END YOKE WITH ROUND BORE
SEE FIGURE 4B

1.4375 in	IE	36.51 mm
.9688 in	BD	24.61 mm
1.1250 in	SB	28.58 mm
3.0000 in	EC	76.20 mm
1.7500 in	HD	44.45 mm

ROCKWELL L6NYR18-84

IDLI GROUP 81162

END YOKE WITH 6 SPLINES
SEE FIGURE 4A

1.4375 in	IE	36.51 mm
.9688 in	BD	24.61 mm
1.1250 in	SB	28.58 mm
3.1250 in	EC	79.38 mm
2.0000 in	HD	50.80 mm

ROCKWELL L6NYS18-8

IDLI GROUP 81163

END YOKE WITH ROUND BORE
SEE FIGURE 4B

1.4375 in	IE	36.51 mm
.9688 in	BD	24.61 mm
1.1250 in	SB	28.58 mm
3.1875 in	EC	80.96 mm
1.6250 in	HD	41.28 mm

ROCKWELL L6NYR18-105

IDLI GROUP 81164

END YOKE WITH ROUND BORE
SEE FIGURE 4B

1.4375 in	IE	36.51 mm
.9688 in	BD	24.61 mm
1.1250 in	SB	28.58 mm
3.3125 in	EC	84.14 mm
2.0000 in	HD	50.80 mm

ROCKWELL L6NYR18-109

IDLI GROUP 81165

END YOKE WITH ROUND BORE
SEE FIGURE 4B

1.4375 in	IE	36.51 mm
.9688 in	BD	24.61 mm
1.1875 in	SB	30.16 mm
2.5000 in	EC	63.50 mm
1.6250 in	HD	41.28 mm

ROCKWELL L6-85

IDLI GROUP 81166

END YOKE WITH ROUND BORE
slight variation from above group
SEE FIGURE 4B

1.4375 in	IE	36.51 mm
.9688 in	BD	24.61 mm
1.1875 in	SB	30.16 mm
2.5000 in	EC	63.50 mm
1.6250 in	HD	41.28 mm

ROCKWELL L6NYR19-4

IDLI GROUP 81167

END YOKE WITH ROUND BORE
SEE FIGURE 4B

1.4375 in	IE	36.51 mm
.9688 in	BD	24.61 mm
1.1875 in	SB	30.16 mm
2.5000 in	EC	63.50 mm
2.1250 in	HD	53.98 mm

ROCKWELL L6-86

IDLI GROUP 81168

END YOKE WITH ROUND BORE
slight variation from above group
SEE FIGURE 4B

1.4375 in	IE	36.51 mm
.9688 in	BD	24.61 mm
1.1875 in	SB	30.16 mm
2.5000 in	EC	63.50 mm
2.1250 in	HD	53.98 mm

ROCKWELL L6-87

IDLI GROUP 81169

END YOKE WITH ROUND BORE
slight variation from above group
SEE FIGURE 4B

1.4375 in	IE	36.51 mm
.9688 in	BD	24.61 mm
1.1875 in	SB	30.16 mm
2.5000 in	EC	63.50 mm
2.1250 in	HD	53.98 mm

NEAPCO 16-6217
NEAPCO L617A
ROCKWELL L6NYR19-1

IDLI GROUP 81170

END YOKE WITH ROUND BORE
slight variation from above group
SEE FIGURE 4B

1.4375 in	IE	36.51 mm
.9688 in	BD	24.61 mm
1.1875 in	SB	30.16 mm
2.5000 in	EC	63.50 mm
2.1250 in	HD	53.98 mm

ROCKWELL L6NYR19-12

IDLI GROUP 81171

END YOKE WITH ROUND BORE
SEE FIGURE 4B

1.4375 in	IE	36.51 mm
.9688 in	BD	24.61 mm
1.1875 in	SB	30.16 mm
3.3750 in	EC	85.73 mm
2.0000 in	HD	50.80 mm

ROCKWELL L6NYR19-9

IDLI GROUP 81172

END YOKE WITH SQUARE BORE AND HUB
SEE FIGURE 4C

1.4375 in	IE	36.51 mm
.9688 in	BD	24.61 mm
1.2500 in	SB	31.75 mm
2.1875 in	EC	55.56 mm
1.7500 in	HD	44.45 mm

ROCKWELL L6NYQ20-1

IDLI GROUP 81173

END YOKE WITH 6 SPLINES
SEE FIGURE 4A

1.4375 in	IE	36.51 mm
.9688 in	BD	24.61 mm
1.2500 in	SB	31.75 mm
2.5000 in	EC	63.50 mm
2.1250 in	HD	53.98 mm

ROCKWELL L6-105

IDLI GROUP 81174

END YOKE WITH 6 SPLINES
slight variation from above group
SEE FIGURE 4A

1.4375 in	IE	36.51 mm
.9688 in	BD	24.61 mm
1.2500 in	SB	31.75 mm
2.5000 in	EC	63.50 mm
2.1250 in	HD	53.98 mm

ROCKWELL L6NYS20

IDLI GROUP 81175

END YOKE WITH ROUND BORE
SEE FIGURE 4B

1.4375 in	IE	36.51 mm
.9688 in	BD	24.61 mm
1.2500 in	SB	31.75 mm
2.5000 in	EC	63.50 mm
2.1250 in	HD	53.98 mm

MCQUAY-NOR U03-80
NEAPCO 16-6118
NEAPCO L6118
PRECISION 1909
ROCKWELL L6-29
TRW 21062
WESCO 06N2045-37
WESCO 540080
WESCO 540081
WESCO 540090
WESCO 6N2045-37

BD = BEARING DIAMETER (outside)   **BW** = BEARING WIDTH   **CB** = CROSS LENGTH WITH BEARINGS   **CC** = CENTER TO CENTER   **CD** = CROSS DIAMETER
CL = CROSS LENGTH WITHOUT BEARINGS   **EC** = END TO CENTER or FACE TO FACE   **EE** = END TO END   **EL** = EFFECTIVE LENGTH
HD = HUB DIAMETER (or insert)   **IE** = INSIDE OF EARS (or recess)   **OD** = OUTSIDE DIAMETER   **OE** = OUTSIDE OF EARS   **SB** = SPLINE OR BORE SIZE

IDLI GUIDE　　COPYRIGHT ® INTERCHANGE, INC.　　ST. LOUIS PARK, MN. 55416 USA

IDLI GROUP 81176

END YOKE WITH ROUND BORE
slight variation from above group
SEE FIGURE 4B

1.4375 in	IE	36.51 mm
.9688 in	BD	24.61 mm
1.2500 in	SB	31.75 mm
2.5000 in	EC	63.50 mm
2.1250 in	HD	53.98 mm

ROCKWELL L6-30

IDLI GROUP 81177

END YOKE WITH ROUND BORE
slight variation from above group
SEE FIGURE 4B

1.4375 in	IE	36.51 mm
.9688 in	BD	24.61 mm
1.2500 in	SB	31.75 mm
2.5000 in	EC	63.50 mm
2.1250 in	HD	53.98 mm

ROCKWELL L6-88

IDLI GROUP 81178

END YOKE WITH ROUND BORE
slight variation from above group
SEE FIGURE 4B

1.4375 in	IE	36.51 mm
.9688 in	BD	24.61 mm
1.2500 in	SB	31.75 mm
2.5000 in	EC	63.50 mm
2.1250 in	HD	53.98 mm

PRECISION 1909
ROCKWELL L6NYR20-4

IDLI GROUP 81179

END YOKE WITH ROUND BORE
slight variation from above group
SEE FIGURE 4B

1.4375 in	IE	36.51 mm
.9688 in	BD	24.61 mm
1.2500 in	SB	31.75 mm
2.5000 in	EC	63.50 mm
2.1250 in	HD	53.98 mm

NEAPCO 16-6218
NEAPCO L618A
ROCKWELL L6NYR20-38

IDLI GROUP 81180

END YOKE WITH ROUND BORE
slight variation from above group
SEE FIGURE 4B

1.4375 in	IE	36.51 mm
.9688 in	BD	24.61 mm
1.2500 in	SB	31.75 mm
2.5000 in	EC	63.50 mm
2.1250 in	HD	53.98 mm

ROCKWELL L6NYR20-39

IDLI GROUP 81181

END YOKE WITH ROUND BORE
SEE FIGURE 4B

1.4375 in	IE	36.51 mm
.9688 in	BD	24.61 mm
1.2500 in	SB	31.75 mm
2.7500 in	EC	69.85 mm
2.0000 in	HD	50.80 mm

ROCKWELL L6NYR20-26

IDLI GROUP 81182

END YOKE WITH 19 INVOLUTED SPLINES
SEE FIGURE 4A

1.4375 in	IE	36.51 mm
.9688 in	BD	24.61 mm
1.2500 in	SB	31.75 mm
3.1250 in	EC	79.38 mm
2.0000 in	HD	50.80 mm

ROCKWELL L6NYS20-11

IDLI GROUP 81183

END YOKE WITH ROUND BORE
SEE FIGURE 4B

1.4375 in	IE	36.51 mm
.9688 in	BD	24.61 mm
1.2500 in	SB	31.75 mm
3.2500 in	EC	82.55 mm
2.0000 in	HD	50.80 mm

ROCKWELL L6NYR20-1

IDLI GROUP 81184

END YOKE WITH ROUND BORE
slight variation from above group
SEE FIGURE 4B

1.4375 in	IE	36.51 mm
.9688 in	BD	24.61 mm
1.2500 in	SB	31.75 mm
3.2500 in	EC	82.55 mm
2.0000 in	HD	50.80 mm

ROCKWELL L6NYR20-60

IDLI GROUP 81185

END YOKE WITH 6 SPLINES
SEE FIGURE 4A

1.4375 in	IE	36.51 mm
.9688 in	BD	24.61 mm
1.2500 in	SB	31.75 mm
3.2813 in	EC	83.35 mm
2.0000 in	HD	50.80 mm

ROCKWELL L6NYS20-5

IDLI GROUP 81186

END YOKE WITH 6 SPLINES
slight variation from above group
SEE FIGURE 4A

1.4375 in	IE	36.51 mm
.9688 in	BD	24.61 mm
1.2500 in	SB	31.75 mm
3.2813 in	EC	83.35 mm
2.0000 in	HD	50.80 mm

ROCKWELL L6NYS20-5A

IDLI GROUP 81187

END YOKE WITH RECTANGULAR BORE
maximum SB dimension shown
SEE FIGURE 4D

1.4375 in	IE	36.51 mm
.9688 in	BD	24.61 mm
1.1250 in	SB	28.58 mm
3.4375 in	EC	87.31 mm

ROCKWELL L6NYP16-17

IDLI GROUP 81188

END YOKE WITH RECTANGULAR BORE
maximum SB dimension shown
SEE FIGURE 4D

1.4375 in	IE	36.51 mm
.9688 in	BD	24.61 mm
1.1250 in	SB	28.58 mm
3.5625 in	EC	90.49 mm
2.0000 in	HD	50.80 mm

ROCKWELL L6NYP16-19

IDLI GROUP 81189

END YOKE WITH 6 SPLINES
SEE FIGURE 4A

1.4375 in	IE	36.51 mm
.9688 in	BD	24.61 mm
1.3750 in	SB	34.93 mm
2.5000 in	EC	63.50 mm
2.1250 in	HD	53.98 mm

ROCKWELL L6-106

IDLI GROUP 81190

END YOKE WITH ROUND BORE
SEE FIGURE 4B

1.4375 in	IE	36.51 mm
.9688 in	BD	24.61 mm
1.3750 in	SB	34.93 mm
2.5000 in	EC	63.50 mm
2.1250 in	HD	53.98 mm

ROCKWELL L6-89

IDLI GROUP 81191

END YOKE WITH ROUND BORE
slight variation from above group
SEE FIGURE 4B

1.4375 in	IE	36.51 mm
.9688 in	BD	24.61 mm
1.3750 in	SB	34.93 mm
2.5000 in	EC	63.50 mm
2.1250 in	HD	53.98 mm

ROCKWELL L6NYR22-1

IDLI GROUP 81192

END YOKE WITH ROUND BORE
SEE FIGURE 4B

1.4375 in	IE	36.51 mm
.9688 in	BD	24.61 mm
1.3750 in	SB	34.93 mm
3.1250 in	EC	79.38 mm
2.0000 in	HD	50.80 mm

ROCKWELL L6NYR22-2

IDLI GROUP 81193

END YOKE WITH ROUND BORE
SEE FIGURE 4B

1.4375 in	IE	36.51 mm
.9688 in	BD	24.61 mm
1.3750 in	SB	34.93 mm
3.3750 in	EC	85.73 mm
1.9375 in	HD	49.21 mm

ROCKWELL L6NYR22-11

IDLI GROUP 81194

END YOKE WITH ROUND BORE
SEE FIGURE 4B

1.4375 in	IE	36.51 mm
.9688 in	BD	24.61 mm
1.4375 in	SB	36.51 mm
2.5000 in	EC	63.50 mm
2.1250 in	HD	53.98 mm

ROCKWELL L6-90

IDLI GROUP 81195

END YOKE WITH ROUND BORE
slight variation from above group
SEE FIGURE 4B

1.4375 in	IE	36.51 mm
.9688 in	BD	24.61 mm
1.4375 in	SB	36.51 mm
2.5000 in	EC	63.50 mm
2.1250 in	HD	53.98 mm

ROCKWELL L6NYR23-1

IDLI GROUP 81196

END YOKE WITH ROUND BORE
SEE FIGURE 4B

1.4375 in	IE	36.51 mm
.9688 in	BD	24.61 mm
1.5000 in	SB	38.10 mm
2.5000 in	EC	63.50 mm
2.1250 in	HD	53.98 mm

ROCKWELL L6-91

IDLI GROUP 81197

END YOKE WITH ROUND BORE
slight variation from above group
SEE FIGURE 4B

1.4375 in	IE	36.51 mm
.9688 in	BD	24.61 mm
1.5000 in	SB	38.10 mm
2.5000 in	EC	63.50 mm
2.1250 in	HD	53.98 mm

ROCKWELL L6NYR24-1

IDLI GROUP 81198

END YOKE WITH ROUND BORE
slight variation from above group
SEE FIGURE 4B

1.4375 in	IE	36.51 mm
.9688 in	BD	24.61 mm
1.5000 in	SB	38.10 mm
2.5000 in	EC	63.50 mm
2.1250 in	HD	53.98 mm

ROCKWELL L6NYR24-2

ENGINEERING CATALOGS MUST BE CONSULTED FOR DETAILS NOT INCLUDED
IN THIS GUIDE. SPECIFIC DESIGNS, MATERIAL CONTENT, TOLERANCES
LUBE FITTINGS AND OTHER DIMENSIONS ARE INTENTIONALLY OMITTED HERE.

CAUTION: BE SURE TO REFER TO ENGINEERING CATALOGS FOR SPECIAL APPLICATIONS THAT REQUIRE SPECIFIC MATERIAL CONTENT, TOLERANCES, ETC. SEE FOOTNOTE.

IDLI GROUP 81199

END YOKE WITH ROUND BORE
SEE FIGURE 4B

1.4375 in	IE	36.51 mm
.9688 in	BD	24.61 mm
1.7500 in	SB	44.45 mm
3.2500 in	EC	82.55 mm
2.0000 in	HD	50.80 mm

ROCKWELL L6NYR20-1

IDLI GROUP 81200

END YOKE WITH ROUND BORE
SEE FIGURE 4B

1.8125 in	IE	46.04 mm
1.0362 in	BD	26.32 mm
1.1250 in	SB	28.58 mm
3.4063 in	EC	86.52 mm
2.0000 in	HD	50.80 mm

ROCKWELL L12NYR18-57

IDLI GROUP 81201

END YOKE WITH ROUND BORE
SEE FIGURE 4B

1.8125 in	IE	46.04 mm
1.0362 in	BD	26.32 mm
1.1250 in	SB	28.58 mm
3.5000 in	EC	88.90 mm
1.9375 in	HD	49.21 mm

ROCKWELL L12NYR18

IDLI GROUP 81202

END YOKE WITH ROUND BORE
SEE FIGURE 4B

1.8125 in	IE	46.04 mm
1.0362 in	BD	26.32 mm
1.1250 in	SB	28.58 mm
3.6250 in	EC	92.08 mm
1.7500 in	HD	44.45 mm

ROCKWELL L12NYR18-10

IDLI GROUP 81203

END YOKE WITH ROUND BORE
SEE FIGURE 4B

1.8125 in	IE	46.04 mm
1.0625 in	BD	26.99 mm
.7500 in	SB	19.05 mm
2.7500 in	EC	69.85 mm
2.0000 in	HD	50.80 mm

BORG-WARNR 61248
BPC H12EY-3/4R-3/16KW
MUNCIE MJY6
NEAPCO 111D8
NEAPCO 12-1313
PRECISION 1200
ROCKWELL L12-10
WESCO 12N120SH
WESCO 12N123SH
WESCO 850010

IDLI GROUP 81204

END YOKE WITH ROUND BORE
slight variation from above group
SEE FIGURE 4B

1.8125 in	IE	46.04 mm
1.0625 in	BD	26.99 mm
.7500 in	SB	19.05 mm
2.7500 in	EC	69.85 mm
2.0000 in	HD	50.80 mm

PRECISION 1200
ROCKWELL L12NYR12-3

IDLI GROUP 81205

END YOKE WITH SQUARE BORE
SEE FIGURE 4C

1.8125 in	IE	46.04 mm
1.0625 in	BD	26.99 mm
.7500 in	SB	19.05 mm
2.7500 in	EC	69.85 mm
2.0000 in	HD	50.80 mm

PRECISION 1240
ROCKWELL L12-27

IDLI GROUP 81206

END YOKE WITH SQUARE BORE
slight variation from above group
SEE FIGURE 4C

1.8125 in	IE	46.04 mm
1.0625 in	BD	26.99 mm
.7500 in	SB	19.05 mm
2.7500 in	EC	69.85 mm
2.0000 in	HD	50.80 mm

BORG-WARNR 12-1317
BORG-WARNR 61257
BPC H12EY-3/4
MUNCIE MJY22
NEAPCO 111E7
PRECISION 1240
ROCKWELL L12NYQ12-2
WESCO 12N120SQ
WESCO 850009

IDLI GROUP 81207

END YOKE WITH ROUND BORE
SEE FIGURE 4B

1.8125 in	IE	46.04 mm
1.0625 in	BD	26.99 mm
.8125 in	SB	20.64 mm
2.7500 in	EC	69.85 mm
2.0000 in	HD	50.80 mm

PRECISION 1201
ROCKWELL L12-11

IDLI GROUP 81208

END YOKE WITH ROUND BORE
slight variation from above group
SEE FIGURE 4B

1.8125 in	IE	46.04 mm
1.0625 in	BD	26.99 mm
.8125 in	SB	20.64 mm
2.7500 in	EC	69.85 mm
2.0000 in	HD	50.80 mm

BORG-WARNR 61132
BPC H12EY-13/16R-1/4KW
MUNCIE MJY10
NEAPCO 11132

IDLI GROUP 81209

END YOKE WITH SQUARE BORE
SEE FIGURE 4C

1.8125 in	IE	46.04 mm
1.0625 in	BD	26.99 mm
.8125 in	SB	20.64 mm
2.7500 in	EC	69.85 mm
2.0000 in	HD	50.80 mm

BORG-WARNR 67056
BPC H12EY-13/16
G & G MFG 188-1200
G & G MFG 188-1272
HAYES 1-82-22-3715X
MUNCIE C402
MUNCIE C-402
NEAPCO 12-439
NEAPCO 12-851
NEAPCO 65-2851
NEAPCO UTR12-002700
NEAPCO UTR12-003900
PRECISION 1241
ROCKWELL L12-28
SPICER 1-82-22-3715X
TRW 21417
WESCO 12N048M
WESCO 12N48M
WESCO 12N72M
WESCO 920112

IDLI GROUP 81210

END YOKE WITH ROUND BORE
SEE FIGURE 4B

1.8125 in	IE	46.04 mm
1.0625 in	BD	26.99 mm
.8750 in	SB	22.23 mm
2.1250 in	EC	53.98 mm

ROCKWELL L12NYR14-18

IDLI GROUP 81211

END YOKE WITH RECTANGULAR BORE AND SQUARE HUB
SEE FIGURE 4D

1.8125 in	IE	46.04 mm
1.0625 in	BD	26.99 mm
.8750 in	SB	22.23 mm
2.2500 in	EC	57.15 mm

ROCKWELL L12NYP12

IDLI GROUP 81212

END YOKE WITH ROUND BORE
SEE FIGURE 4B

1.8125 in	IE	46.04 mm
1.0625 in	BD	26.99 mm
.8750 in	SB	22.23 mm
2.7500 in	EC	69.85 mm

BPC H12EY-7/8R-3/16KW
PRECISION 1204
ROCKWELL L12NYR14-8

IDLI GROUP 81213

END YOKE WITH ROUND BORE
slight variation from above group
SEE FIGURE 4B

1.8125 in	IE	46.04 mm
1.0625 in	BD	26.99 mm
.8750 in	SB	22.23 mm
2.7500 in	EC	69.85 mm
2.0000 in	HD	50.80 mm

BORG-WARNR 61133
BPC H12EY-7/8R-1/4KW
HUB CITY 0335-01246
MUNCIE MJY7
NEAPCO 11133
NEAPCO 12-1133
NEAPCO 12-1153
PILOT 11549UB
PRECISION 1203
ROCKWELL L12-12
WESCO 12N0144
WESCO 12N140SH
WESCO 12N144SH
WESCO 850017
WESCO 855017

IDLI GROUP 81214

END YOKE WITH ROUND BORE
slight variation from above group
SEE FIGURE 4B

1.8125 in	IE	46.04 mm
1.0625 in	BD	26.99 mm
.8750 in	SB	22.23 mm
2.7500 in	EC	69.85 mm
2.0000 in	HD	50.80 mm

PRECISION 1204
ROCKWELL L12-13

IDLI GROUP 81215

END YOKE WITH ROUND BORE
slight variation from above group
SEE FIGURE 4B

1.8125 in	IE	46.04 mm
1.0625 in	BD	26.99 mm
.8750 in	SB	22.23 mm
2.7500 in	EC	69.85 mm
2.0000 in	HD	50.80 mm

PRECISION 1203
ROCKWELL L12NYR14-4
SPICER 1-4-2973

The following applications are listed under IDLI GROUP 81204 (NEAPCO column):

NEAPCO 11152
NEAPCO 12-1132
PRECISION 1201
ROCKWELL L12NYR13
SPICER 1-4-2953
WESCO 12N130SH
WESCO 12N134SH
WESCO 850012

BD = BEARING DIAMETER (outside)   **BW** = BEARING WIDTH   **CB** = CROSS LENGTH WITH BEARINGS   **CC** = CENTER TO CENTER   **CD** = CROSS DIAMETER
CL = CROSS LENGTH WITHOUT BEARINGS   **EC** = END TO CENTER or FACE TO FACE   **EE** = END TO END   **EL** = EFFECTIVE LENGTH
HD = HUB DIAMETER (or insert)   **IE** = INSIDE OF EARS (or recess)   **OD** = OUTSIDE DIAMETER   **OE** = OUTSIDE OF EARS   **SB** = SPLINE OR BORE SIZE

IDLI GROUP 81216

END YOKE WITH ROUND BORE
slight variation from above group
SEE FIGURE 4B

1.8125 in	IE	46.04 mm
1.0625 in	BD	26.99 mm
.8750 in	SB	22.23 mm
2.7500 in	EC	69.85 mm
2.0000 in	HD	50.80 mm

ROCKWELL L12NYR14-7

IDLI GROUP 81217

END YOKE WITH SQUARE BORE
SEE FIGURE 4C

1.8125 in	IE	46.04 mm
1.0625 in	BD	26.99 mm
.8750 in	SB	22.23 mm
2.7500 in	EC	69.85 mm
2.0000 in	HD	50.80 mm

PRECISION 1242
ROCKWELL L12-29

IDLI GROUP 81218

END YOKE WITH SQUARE BORE
slight variation from above group
SEE FIGURE 4C

1.8125 in	IE	46.04 mm
1.0625 in	BD	26.99 mm
.8750 in	SB	22.23 mm
2.7500 in	EC	69.85 mm
2.0000 in	HD	50.80 mm

ALLOY 0610
ALLOY 610
BORG-WARNR 61170
BORG-WARNR 63502
BPC H12EY-1
BPC H12EY-7/8
MCQUAY-NOR U03-111
MCQUAY-NOR U03-81
MUNCIE MJY23
MUNCIE TA78-5
NEAPCO 11170
NEAPCO 12-1170
NEAPCO 60-1239
NEAPCO 62-1251
NEAPCO UT12-005100
NEAPCO UTR12-005100
NEAPCO UTR12-5100
PRECISION 1242
PRECISION 1248
ROCKWELL L12NYQ14-2
SPICER 1-4-182
SPICER 208565-4
TRW 21415
WESCO 12N140SQ
WESCO 12N48
WESCO 850015
WESCO 920110

IDLI GROUP 81219

END YOKE WITH ROUND BORE
SEE FIGURE 4B

1.8125 in	IE	46.04 mm
1.0625 in	BD	26.99 mm
.9375 in	SB	23.81 mm
2.6250 in	EC	66.68 mm
2.0000 in	HD	50.80 mm

PRECISION 1207
ROCKWELL L12NYR15-2

IDLI GROUP 81220

END YOKE WITH ROUND BORE
SEE FIGURE 4B

1.8125 in	IE	46.04 mm
1.0625 in	BD	26.99 mm
.9375 in	SB	23.81 mm
2.7500 in	EC	69.85 mm
2.0000 in	HD	50.80 mm

PRECISION 1206
ROCKWELL L12-14

IDLI GROUP 81221

END YOKE WITH ROUND BORE
slight variation from above group
SEE FIGURE 4B

1.8125 in	IE	46.04 mm
1.0625 in	BD	26.99 mm
.9375 in	SB	23.81 mm
2.7500 in	EC	69.85 mm
2.0000 in	HD	50.80 mm

BPC H12EY-15/16R
PRECISION 1207
ROCKWELL L12-15

IDLI GROUP 81222

END YOKE WITH ROUND BORE
slight variation from above group
SEE FIGURE 4B

1.8125 in	IE	46.04 mm
1.0625 in	BD	26.99 mm
.9375 in	SB	23.81 mm
2.7500 in	EC	69.85 mm
2.0000 in	HD	50.80 mm

BORG-WARNR 61134
BPC H12EY-15/16R-1/4KW
HUB CITY 0335-01213
MUNCIE MJY8
NEAPCO 11134
NEAPCO 11154
NEAPCO 12-1134
PRECISION 1206
ROCKWELL L12NYR15-1
SPICER 1-4-2983
WESCO 12N150SH
WESCO 12N154SH
WESCO 855019

IDLI GROUP 81223

END YOKE WITH ROUND BORE
slight variation from above group
SEE FIGURE 4B

1.8125 in	IE	46.04 mm
1.0625 in	BD	26.99 mm
.9375 in	SB	23.81 mm
2.7500 in	EC	69.85 mm
2.0000 in	HD	50.80 mm

ROCKWELL L12NYR15-10

IDLI GROUP 81224

END YOKE WITH HEX BORE
SEE FIGURE 4E

1.8125 in	IE	46.04 mm
1.0625 in	BD	26.99 mm
1.0000 in	SB	25.40 mm
2.1875 in	EC	55.56 mm
2.0625 in	HD	52.39 mm

ROCKWELL L12NYH16-3

IDLI GROUP 81225

END YOKE WITH 15 INVOLUTED SPLINES
SEE FIGURE 4A

1.8125 in	IE	46.04 mm
1.0625 in	BD	26.99 mm
1.0000 in	SB	25.40 mm
2.3750 in	EC	60.33 mm
2.0000 in	HD	50.80 mm

ROCKWELL L12NYS16-11

IDLI GROUP 81226

END YOKE WITH 15 INVOLUTED SPLINES
slight variation from above group
SEE FIGURE 4A

1.8125 in	IE	46.04 mm
1.0625 in	BD	26.99 mm
1.0000 in	SB	25.40 mm
2.3750 in	EC	60.33 mm
2.0000 in	HD	50.80 mm

ROCKWELL L12NYS16-22

IDLI GROUP 81227

END YOKE WITH SQUARE BORE AND HUB
SEE FIGURE 4C

1.8125 in	IE	46.04 mm
1.0625 in	BD	26.99 mm
1.0000 in	SB	25.40 mm
2.3750 in	EC	60.33 mm
2.0000 in	HD	50.80 mm

ROCKWELL L12NYQ16-4

IDLI GROUP 81228

END YOKE WITH THREADED BORE
SEE FIGURE 4G

1.8125 in	IE	46.04 mm
1.0625 in	BD	26.99 mm
1.0000 in	SB	25.40 mm
2.3750 in	EC	60.33 mm
2.2500 in	HD	57.15 mm

ROCKWELL L12NYD16-3

IDLI GROUP 81229

END YOKE WITH SQUARE BORE
SEE FIGURE 4C

1.8125 in	IE	46.04 mm
1.0625 in	BD	26.99 mm
1.0000 in	SB	25.40 mm
2.5000 in	EC	63.50 mm
2.2500 in	HD	57.15 mm

ROCKWELL L12NYQ16-6

IDLI GROUP 81230

END YOKE WITH HEX BORE
SEE FIGURE 4E

1.8125 in	IE	46.04 mm
1.0625 in	BD	26.99 mm
1.0000 in	SB	25.40 mm
2.6250 in	EC	66.68 mm
2.0000 in	HD	50.80 mm

ROCKWELL L12NYH16-4

IDLI GROUP 81231

END YOKE WITH 15 INVOLUTED SPLINES
SEE FIGURE 4A

1.8125 in	IE	46.04 mm
1.0625 in	BD	26.99 mm
1.0000 in	SB	25.40 mm
2.7500 in	EC	69.85 mm
1.5000 in	HD	38.10 mm

ROCKWELL L12NYS16-12

IDLI GROUP 81232

END YOKE WITH 7 INVOLUTED SPLINES
SEE FIGURE 4A

1.8125 in	IE	46.04 mm
1.0625 in	BD	26.99 mm
1.0000 in	SB	25.40 mm
2.7500 in	EC	69.85 mm
2.0000 in	HD	50.80 mm

ROCKWELL L12NYS16-14

IDLI GROUP 81233

END YOKE WITH 19 INVOLUTED SPLINES
SEE FIGURE 4A

1.8125 in	IE	46.04 mm
1.0625 in	BD	26.99 mm
1.0000 in	SB	25.40 mm
2.7500 in	EC	69.85 mm
2.0000 in	HD	50.80 mm

ROCKWELL L12NYS16-8

IDLI GROUP 81234

END YOKE WITH ROUND BORE
SEE FIGURE 4B

1.8125 in	IE	46.04 mm
1.0625 in	BD	26.99 mm
1.0000 in	SB	25.40 mm
2.7500 in	EC	69.85 mm
2.0000 in	HD	50.80 mm

PRECISION 1209
ROCKWELL L12-16

ENGINEERING CATALOGS MUST BE CONSULTED FOR DETAILS NOT INCLUDED
IN THIS GUIDE. SPECIFIC DESIGNS, MATERIAL CONTENT, TOLERANCES
LUBE FITTINGS AND OTHER DIMENSIONS ARE INTENTIONALLY OMITTED HERE.

CAUTION: BE SURE TO REFER TO ENGINEERING CATALOGS FOR SPECIAL APPLICATIONS THAT REQUIRE SPECIFIC MATERIAL CONTENT, TOLERANCES, ETC. SEE FOOTNOTE.

IDLI GROUP 81235

END YOKE WITH ROUND BORE
slight variation from above group
SEE FIGURE 4B

1.8125 in	IE	46.04 mm
1.0625 in	BD	26.99 mm
1.0000 in	SB	25.40 mm
2.7500 in	EC	69.85 mm
2.0000 in	HD	50.80 mm

BPC H12EY-1R-5/16KW
PRECISION 1210
ROCKWELL L12-17

IDLI GROUP 81236

END YOKE WITH ROUND BORE
slight variation from above group
SEE FIGURE 4B

1.8125 in	IE	46.04 mm
1.0625 in	BD	26.99 mm
1.0000 in	SB	25.40 mm
2.7500 in	EC	69.85 mm
2.0000 in	HD	50.80 mm

PRECISION 1210
ROCKWELL L12NYR16-14

IDLI GROUP 81237

END YOKE WITH ROUND BORE
slight variation from above group
SEE FIGURE 4B

1.8125 in	IE	46.04 mm
1.0625 in	BD	26.99 mm
1.0000 in	SB	25.40 mm
2.7500 in	EC	69.85 mm
2.0000 in	HD	50.80 mm

ALLOY 0602
ALLOY 602
BORG-WARNR 61115
BORG-WARNR 61135
BORG-WARNR 61155
BPC H12EY-1R-1/4KW
G & G MFG 184-1216
HAYES 1-4-3003
HUB CITY 0335-01206
MCQUAY-NOR U03-107
MCQUAY-NOR U03-83
MUNCIE MJY9
NEAPCO 11106
NEAPCO 11115
NEAPCO 11135
NEAPCO 11155
NEAPCO 111D6
NEAPCO 111G2
NEAPCO 111H5
NEAPCO 111J2
NEAPCO 111P6
NEAPCO 12-1115
NEAPCO 12-1135
NEAPCO 12-1155
PRECISION 1209
PRECISION 1211
ROCKWELL L12NYR16-23
SPICER 1-4-3013

SPICER 1-4-3033
TRW 21101
WESCO 12N0164
WESCO 12N160SH
WESCO 12N164
WESCO 12N164SH
WESCO 850022
WESCO 850023
WESCO 85-1640

IDLI GROUP 81238

END YOKE WITH ROUND BORE
slight variation from above group
SEE FIGURE 4B

1.8125 in	IE	46.04 mm
1.0625 in	BD	26.99 mm
1.0000 in	SB	25.40 mm
2.7500 in	EC	69.85 mm
2.0000 in	HD	50.80 mm

ROCKWELL L12NYR16-27

IDLI GROUP 81239

END YOKE WITH SQUARE BORE
SEE FIGURE 4C

1.8125 in	IE	46.04 mm
1.0625 in	BD	26.99 mm
1.0000 in	SB	25.40 mm
2.7500 in	EC	69.85 mm
2.0000 in	HD	50.80 mm

BORG-WARNR 61159
MUNCIE MJY24
NEAPCO 11159
NEAPCO 12-1159
PRECISION 1243
ROCKWELL L12-30
WESCO 12N160SQ
WESCO 850021

IDLI GROUP 81240

END YOKE WITH SQUARE BORE
slight variation from above group
SEE FIGURE 4C

1.8125 in	IE	46.04 mm
1.0625 in	BD	26.99 mm
1.0000 in	SB	25.40 mm
2.7500 in	EC	69.85 mm
2.0000 in	HD	50.80 mm

PRECISION 1243
ROCKWELL L12NYQ16-3
SPICER 1-4-222

IDLI GROUP 81241

END YOKE WITH ROUND BORE
SEE FIGURE 4B

1.8125 in	IE	46.04 mm
1.0625 in	BD	26.99 mm
1.0000 in	SB	25.40 mm
3.0000 in	EC	76.20 mm
2.0625 in	HD	52.39 mm

ROCKWELL L12NYR16-105

IDLI GROUP 81242

END YOKE WITH ROUND BORE
SEE FIGURE 4B

1.8125 in	IE	46.04 mm
1.0625 in	BD	26.99 mm
1.0000 in	SB	25.40 mm
3.2500 in	EC	82.55 mm
2.0000 in	HD	50.80 mm

ROCKWELL L12NYR16-29

IDLI GROUP 81243

END YOKE WITH HEX BORE
SEE FIGURE 4E

1.8125 in	IE	46.04 mm
1.0625 in	BD	26.99 mm
1.0000 in	SB	25.40 mm
3.3750 in	EC	85.73 mm
1.9375 in	HD	49.21 mm

ROCKWELL L12NYH16-5

IDLI GROUP 81244

END YOKE WITH ROUND BORE
SEE FIGURE 4B

1.8125 in	IE	46.04 mm
1.0625 in	BD	26.99 mm
1.0000 in	SB	25.40 mm
3.5000 in	EC	88.90 mm
2.0000 in	HD	50.80 mm

ROCKWELL L12NYR16-32

IDLI GROUP 81245

END YOKE WITH ROUND BORE
slight variation from above group
SEE FIGURE 4B

1.8125 in	IE	46.04 mm
1.0625 in	BD	26.99 mm
1.0000 in	SB	25.40 mm
3.5000 in	EC	88.90 mm
2.0000 in	HD	50.80 mm

PRECISION 1211
ROCKWELL L12NYR16-8
SPICER 1-4-3003

IDLI GROUP 81246

END YOKE WITH SQUARE BORE
SEE FIGURE 4C

1.8125 in	IE	46.04 mm
1.0625 in	BD	26.99 mm
1.0000 in	SB	25.40 mm
3.6875 in	EC	93.66 mm
1.9375 in	HD	49.21 mm

ALLOY 609
BORG-WARNR 61163
BORG-WARNR 61471
G & G MFG 183-1216
HAYES 1-4-222
HAYES 1-4-282
HUB CITY 0335-01205
MCQUAY-NOR U03-82
MUNCIE MJY24A
NEAPCO 0609
NEAPCO 11144
NEAPCO 11163
NEAPCO 111AZ
NEAPCO 12-1163
NEAPCO 12-1301

IDLI GROUP 81247

END YOKE WITH ROUND BORE
SEE FIGURE 4B

1.8125 in	IE	46.04 mm
1.0625 in	BD	26.99 mm
1.0000 in	SB	25.40 mm
4.7500 in	EC	120.65 mm
1.7500 in	HD	44.45 mm

BORG-WARNR 61255
PRECISION 1263
ROCKWELL L12NYR16-13

IDLI GROUP 81248

END YOKE WITH ROUND BORE
SEE FIGURE 4B

1.8125 in	IE	46.04 mm
1.0625 in	BD	26.99 mm
1.0156 in	SB	25.80 mm
2.4375 in	EC	61.91 mm

ROCKWELL L12NYR16-106

IDLI GROUP 81249

END YOKE WITH ROUND BORE
SEE FIGURE 4B

1.8125 in	IE	46.04 mm
1.0625 in	BD	26.99 mm
1.0625 in	SB	26.99 mm
2.7500 in	EC	69.85 mm
2.0000 in	HD	50.80 mm

BPC H12EY-1-1/16R-1/4KW
NEAPCO 11136
NEAPCO 11156
NEAPCO 12-1156
PRECISION 1214
ROCKWELL L12-18
WESCO 12N170SH
WESCO 12N174SH
WESCO 850024

IDLI GROUP 81250

END YOKE WITH ROUND BORE
slight variation from above group
SEE FIGURE 4B

1.8125 in	IE	46.04 mm
1.0625 in	BD	26.99 mm
1.0625 in	SB	26.99 mm
2.7500 in	EC	69.85 mm
2.0000 in	HD	50.80 mm

BORG-WARNR 61136
PRECISION 1214
ROCKWELL L12NYR17-1

PRECISION 1244
ROCKWELL L12NYQ16-26
SPICER 1-4-0222
SPICER 1-4-282
SPICER 1-4-2993
TRW 21102
TRW 21102D
TRW 21103
WESCO 12N0140-30
WESCO 12N140-30
WESCO 850013

BD = BEARING DIAMETER (outside)   **DW** = BEARING WIDTH   **CB** = CROSS LENGTH WITH BEARINGS   **CC** = CENTER TO CENTER   **CD** = CROSS DIAMETER
CL = CROSS LENGTH WITHOUT BEARINGS   **EC** = END TO CENTER or FACE TO FACE   **EE** = END TO END   **EL** = EFFECTIVE LENGTH
HD = HUB DIAMETER (or insert)   **IE** = INSIDE OF EARS (or recess)   **OD** = OUTSIDE DIAMETER   **OE** = OUTSIDE OF EARS   **SB** = SPLINE OR BORE SIZE

IDLI GROUP 81251

END YOKE WITH 6 SPLINES
SEE FIGURE 4A

1.8125 in	IE	46.04 mm
1.0625 in	BD	26.99 mm
1.1250 in	SB	28.58 mm
2.3125 in	EC	58.74 mm
2.0000 in	HD	50.80 mm

PRECISION 1252
ROCKWELL L12-33

IDLI GROUP 81252

END YOKE WITH 6 SPLINES
slight variation from above group
SEE FIGURE 4A

1.8125 in	IE	46.04 mm
1.0625 in	BD	26.99 mm
1.1250 in	SB	28.58 mm
2.3125 in	EC	58.74 mm
2.0000 in	HD	50.80 mm

PRECISION 1252
ROCKWELL L12NYS18-11

IDLI GROUP 81253

END YOKE WITH RECTANGULAR BORE
AND SQUARE HUB
SEE FIGURE 4D

1.8125 in	IE	46.04 mm
1.0625 in	BD	26.99 mm
1.1250 in	SB	28.58 mm
2.3750 in	EC	60.33 mm

ROCKWELL L12NYP16-13

IDLI GROUP 81254

END YOKE WITH 6 SPLINES AND SQUARE
HUB
SEE FIGURE 4A

1.8125 in	IE	46.04 mm
1.0625 in	BD	26.99 mm
1.1250 in	SB	28.58 mm
2.3750 in	EC	60.33 mm
2.0000 in	HD	50.80 mm

ROCKWELL L12NYS18-18

IDLI GROUP 81255

END YOKE WITH RECTANGULAR BORE
maximum SB dimension shown
SEE FIGURE 4D

1.8125 in	IE	46.04 mm
1.0625 in	BD	26.99 mm
1.1250 in	SB	28.58 mm
2.5000 in	EC	63.50 mm
2.0000 in	HD	50.80 mm

ROCKWELL L12NYP16-39

IDLI GROUP 81256

END YOKE WITH 6 SPLINES
SEE FIGURE 4A

1.8125 in	IE	46.04 mm
1.0625 in	BD	26.99 mm
1.1250 in	SB	28.58 mm
2.6250 in	EC	66.68 mm
2.0000 in	HD	50.80 mm

PRECISION 1251
ROCKWELL L12-32

IDLI GROUP 81257

END YOKE WITH ROUND BORE
SEE FIGURE 4B

1.8125 in	IE	46.04 mm
1.0625 in	BD	26.99 mm
1.1250 in	SB	28.58 mm
2.6250 in	EC	66.68 mm
2.0000 in	HD	50.80 mm

G & G MFG 196-1218
PRECISION 1218
ROCKWELL L12NYR18-5
SPICER 1-4-3203
WESCO 12N180-7
WESCO 850025

IDLI GROUP 81258

END YOKE WITH ROUND BORE
SEE FIGURE 4B

1.8125 in	IE	46.04 mm
1.0625 in	BD	26.99 mm
1.1250 in	SB	28.58 mm
2.7500 in	EC	69.85 mm
1.8750 in	HD	47.63 mm

ROCKWELL L12NYR18-98

IDLI GROUP 81259

END YOKE WITH HEX BORE
SEE FIGURE 4E

1.8125 in	IE	46.04 mm
1.0625 in	BD	26.99 mm
1.1250 in	SB	28.58 mm
2.7500 in	EC	69.85 mm
1.8750 in	HD	47.63 mm

BORG-WARNR 61168
BPC H12EY-1-1/8
MUNCIE MJY25
NEAPCO 11168
NEAPCO 12-1168
PRECISION 1247
ROCKWELL L12-38
TRW 21113
WESCO 12N180H
WESCO 12N180SH
WESCO 850029

IDLI GROUP 81260

END YOKE WITH HEX BORE
slight variation from above group
SEE FIGURE 4E

1.8125 in	IE	46.04 mm
1.0625 in	BD	26.99 mm
1.1250 in	SB	28.58 mm
2.7500 in	EC	69.85 mm
1.8750 in	HD	47.63 mm

PRECISION 1247
ROCKWELL L12NYH18

IDLI GROUP 81261

END YOKE WITH ROUND BORE
SEE FIGURE 4B

1.8125 in	IE	46.04 mm
1.0625 in	BD	26.99 mm
1.1250 in	SB	28.58 mm
2.7500 in	EC	69.85 mm
2.0000 in	HD	50.80 mm

BORG-WARNR 61137
BORG-WARNR 61157
BPC H12EY-1-1/8R-5/16KW
MUNCIE MJY14A
NEAPCO 11137
NEAPCO 11157
NEAPCO 12-1137
NEAPCO 12-1157
PRECISION 1217
ROCKWELL L12-20
WESCO 12N185SH
WESCO 850032

IDLI GROUP 81262

END YOKE WITH ROUND BORE
slight variation from above group
SEE FIGURE 4B

1.8125 in	IE	46.04 mm
1.0625 in	BD	26.99 mm
1.1250 in	SB	28.58 mm
2.7500 in	EC	69.85 mm
2.0000 in	HD	50.80 mm

PRECISION 1218
ROCKWELL L12-21

IDLI GROUP 81263

END YOKE WITH ROUND BORE
slight variation from above group
SEE FIGURE 4B

1.8125 in	IE	46.04 mm
1.0625 in	BD	26.99 mm
1.1250 in	SB	28.58 mm
2.7500 in	EC	69.85 mm
2.0000 in	HD	50.80 mm

ROCKWELL L12NYR18-53

IDLI GROUP 81264

END YOKE WITH ROUND BORE
slight variation from above group
SEE FIGURE 4B

1.8125 in	IE	46.04 mm
1.0625 in	BD	26.99 mm
1.1250 in	SB	28.58 mm
2.7500 in	EC	69.85 mm
2.0000 in	HD	50.80 mm

BORG-WARNR 61151
PRECISION 1217
ROCKWELL L12NYR18-4
SPICER 1-4-3063

IDLI GROUP 81265

END YOKE WITH SQUARE BORE
SEE FIGURE 4C

1.8125 in	IE	46.04 mm
1.0625 in	BD	26.99 mm
1.1250 in	SB	28.58 mm
2.7500 in	EC	69.85 mm
2.0000 in	HD	50.80 mm

PRECISION 1245
ROCKWELL L12-31

IDLI GROUP 81266

END YOKE WITH SQUARE BORE
slight variation from above group
SEE FIGURE 4C

1.8125 in	IE	46.04 mm
1.0625 in	BD	26.99 mm
1.1250 in	SB	28.58 mm
2.7500 in	EC	69.85 mm
2.0000 in	HD	50.80 mm

BPC H12EY-1-1/8
PRECISION 1245
ROCKWELL L12NYQ18-4

IDLI GROUP 81267

END YOKE WITH RECTANGULAR BORE
maximum SB dimension shown
SEE FIGURE 4D

1.8125 in	IE	46.04 mm
1.0625 in	BD	26.99 mm
1.1250 in	SB	28.58 mm
2.7500 in	EC	69.85 mm
2.0000 in	HD	50.80 mm

ROCKWELL L12NYP16-11
SPICER 1-4-242

IDLI GROUP 81268

END YOKE WITH ROUND BORE
SEE FIGURE 4B

1.8125 in	IE	46.04 mm
1.0625 in	BD	26.99 mm
1.1250 in	SB	28.58 mm
2.7500 in	EC	69.85 mm
2.2500 in	HD	57.15 mm

ALLOY 0613
ALLOY 613
BORG-WARNR 61162
BORG-WARNR 61177
BPC H12EY-1-1/8R-1/4KW
MCQUAY-NOR U03-104
MCQUAY-NOR U03-105
MUNCIE MJY14
NEAPCO 11162
NEAPCO 11177
NEAPCO 12-1162
NEAPCO 12-1177
NEAPCO 12-1326
NEAPCO 16-6129
NEAPCO L6129
PRECISION 1216
PRECISION 1936
ROCKWELL L12-19
TRW 21104
WESCO 12N0184
WESCO 12N184
WESCO 12N184SH

ENGINEERING CATALOGS MUST BE CONSULTED FOR DETAILS NOT INCLUDED
IN THIS GUIDE. SPECIFIC DESIGNS, MATERIAL CONTENT, TOLERANCES
LUBE FITTINGS AND OTHER DIMENSIONS ARE INTENTIONALLY OMITTED HERE.

CAUTION: BE SURE TO REFER TO ENGINEERING CATALOGS FOR SPECIAL APPLICATIONS THAT REQUIRE SPECIFIC MATERIAL CONTENT, TOLERANCES, ETC. SEE FOOTNOTE.

WESCO	540052
WESCO	540083
WESCO	850030
WESCO	850031

IDLI GROUP 81269

END YOKE WITH ROUND BORE
slight variation from above group
SEE FIGURE 4B

1.8125 in	IE	46.04 mm
1.0625 in	BD	26.99 mm
1.1250 in	SB	28.58 mm
2.7500 in	EC	69.85 mm
2.2500 in	HD	57.15 mm

| PRECISION | 1216 |
| ROCKWELL | L12NYR18-30 |

IDLI GROUP 81270

END YOKE WITH ROUND BORE
SEE FIGURE 4B

1.8125 in	IE	46.04 mm
1.0625 in	BD	26.99 mm
1.1250 in	SB	28.58 mm
2.8750 in	EC	73.03 mm
1.6875 in	HD	42.86 mm

| ROCKWELL | L12NYR18-12 |

IDLI GROUP 81271

END YOKE WITH RECTANGULAR BORE
maximum SB dimension shown
SEE FIGURE 4D

1.8125 in	IE	46.04 mm
1.0625 in	BD	26.99 mm
1.1250 in	SB	28.58 mm
2.9375 in	EC	74.61 mm

ALLOY	0608
ALLOY	608
BORG-WARNR	61149
G & G MFG	177-1200
HAYES	1-4-162
NEAPCO	11149
NEAPCO	111P1
NEAPCO	12-1149
NEAPCO	12-1340
PRECISION	1294
ROCKWELL	L12NYP16-20
SPICER	1-4-162
WESCO	12NW4-1618
WESCO	12NWY1618
WESCO	850043
WESCO	860-043

IDLI GROUP 81272

END YOKE WITH SQUARE BORE
SEE FIGURE 4C

1.8125 in	IE	46.04 mm
1.0625 in	BD	26.99 mm
1.1250 in	SB	28.58 mm
2.9375 in	EC	74.61 mm
2.2500 in	HD	57.15 mm

| ROCKWELL | L12NYQ18-6 |

IDLI GROUP 81273

END YOKE WITH RECTANGULAR BORE
maximum SB dimension shown
SEE FIGURE 4D

1.8125 in	IE	46.04 mm
1.0625 in	BD	26.99 mm
1.1250 in	SB	28.58 mm
3.2500 in	EC	82.55 mm
2.0625 in	HD	52.39 mm

BORG-WARNR	61197
HUB CITY	0335-01251
NEAPCO	11197
NEAPCO	12-1197
ROCKWELL	L12NYP16-27
SPICER	1-4-242

IDLI GROUP 81274

END YOKE WITH 6 SPLINES
SEE FIGURE 4A

1.8125 in	IE	46.04 mm
1.0625 in	BD	26.99 mm
1.1250 in	SB	28.58 mm
3.3750 in	EC	85.73 mm
2.0000 in	HD	50.80 mm

BORG-WARNR	61181
BPC	H12EY-1-1/8
G & G MFG	192-1218
HUB CITY	0335-01227
NEAPCO	11181
NEAPCO	111Z4
NEAPCO	12-1181
PRECISION	1250
ROCKWELL	L12NYS18-1
SPICER	1-4-0921
SPICER	1-4-1091
TRW	21106
WESCO	12N0180-21
WESCO	12N180-21
WESCO	850026

IDLI GROUP 81275

END YOKE WITH ROUND BORE
SEE FIGURE 4B

1.8125 in	IE	46.04 mm
1.0625 in	BD	26.99 mm
1.1250 in	SB	28.58 mm
3.5000 in	EC	88.90 mm
2.0000 in	HD	50.80 mm

HUB CITY	0335-01215
PRECISION	1216
ROCKWELL	L12NYR18-16
SPICER	1-4-3043

IDLI GROUP 81276

END YOKE WITH SQUARE BORE
SEE FIGURE 4C

1.8125 in	IE	46.04 mm
1.0625 in	BD	26.99 mm
1.1250 in	SB	28.58 mm
3.5000 in	EC	88.90 mm
2.0000 in	HD	50.80 mm

| HUB CITY | 0335-01218 |
| ROCKWELL | L12NYQ18-5 |

IDLI GROUP 81277

END YOKE WITH SQUARE BORE
slight variation from above group
SEE FIGURE 4C

1.8125 in	IE	46.04 mm
1.0625 in	BD	26.99 mm
1.1250 in	SB	28.58 mm
3.5000 in	EC	88.90 mm
2.0000 in	HD	50.80 mm

ALLOY	0607
ALLOY	607
BORG-WARNR	61148
BORG-WARNR	61321
MUNCIE	MJY-30
NEAPCO	11148
NEAPCO	111K1
NEAPCO	12-1148
NEAPCO	12-1335
PRECISION	1246
ROCKWELL	L12NYQ18-3
TRW	21105
WESCO	12N0180-26
WESCO	12N180-26
WESCO	850025
WESCO	850028

IDLI GROUP 81278

END YOKE WITH ROUND BORE
SEE FIGURE 4B

1.8125 in	IE	46.04 mm
1.0625 in	BD	26.99 mm
1.1250 in	SB	28.58 mm
3.5000 in	EC	88.90 mm
2.0625 in	HD	52.39 mm

| ROCKWELL | L12NYR18-93 |

IDLI GROUP 81279

END YOKE WITH 6 SPLINES
SEE FIGURE 4A

1.8125 in	IE	46.04 mm
1.0625 in	BD	26.99 mm
1.1250 in	SB	28.58 mm
3.9375 in	EC	100.01 mm
1.3750 in	HD	34.93 mm

| ROCKWELL | L12NYS18 |

IDLI GROUP 81280

END YOKE WITH 6 SPLINES
SEE FIGURE 4A

1.8125 in	IE	46.04 mm
1.0625 in	BD	26.99 mm
1.1250 in	SB	28.58 mm
5.2500 in	EC	133.35 mm
1.8750 in	HD	47.63 mm

| PRECISION | 1281 |
| ROCKWELL | L12NYS18-54 |

IDLI GROUP 81281

END YOKE WITH HEX BORE
SEE FIGURE 4E

1.8125 in	IE	46.04 mm
1.0625 in	BD	26.99 mm
1.1250 in	SB	28.58 mm
5.2500 in	EC	133.35 mm
1.8750 in	HD	47.63 mm

| PRECISION | 1275 |
| ROCKWELL | L12NYH18-2 |

IDLI GROUP 81282

END YOKE WITH ROUND BORE
SEE FIGURE 4B

1.8125 in	IE	46.04 mm
1.0625 in	BD	26.99 mm
1.1250 in	SB	28.58 mm
5.2500 in	EC	133.35 mm
2.0000 in	HD	50.80 mm

BPC	H12SY-1-1/8R-5/16KW
PRECISION	1264
ROCKWELL	L12NYR18-25

IDLI GROUP 81283

END YOKE WITH 8 INVOLUTED SPLINES
SEE FIGURE 4A

1.8125 in	IE	46.04 mm
1.0625 in	BD	26.99 mm
1.1563 in	SB	29.37 mm
3.3750 in	EC	85.73 mm
1.9375 in	HD	49.21 mm

| ROCKWELL | L12NYS19-10 |

IDLI GROUP 81284

END YOKE WITH 8 INVOLUTED SPLINES
SEE FIGURE 4A

1.8125 in	IE	46.04 mm
1.0625 in	BD	26.99 mm
1.1563 in	SB	29.37 mm
3.4375 in	EC	87.31 mm
1.7500 in	HD	44.45 mm

| ROCKWELL | L12NYS19-4 |

IDLI GROUP 81285

END YOKE WITH ROUND BORE
SEE FIGURE 4B

1.8125 in	IE	46.04 mm
1.0625 in	BD	26.99 mm
1.1875 in	SB	30.16 mm
2.7500 in	EC	69.85 mm
2.0000 in	HD	50.80 mm

| PRECISION | 1221 |
| ROCKWELL | L12-22 |

IDLI GROUP 81286

END YOKE WITH ROUND BORE
slight variation from above group
SEE FIGURE 4B

1.8125 in	IE	46.04 mm
1.0625 in	BD	26.99 mm
1.1875 in	SB	30.16 mm
2.7500 in	EC	69.85 mm
2.0000 in	HD	50.80 mm

BORG-WARNR	61138
BPC	H12EY-1-3/16R-5/16KW
MUNCIE	MJY17

BD = BEARING DIAMETER (outside)   BW = BEARING WIDTH   CB = CROSS LENGTH WITH BEARINGS   CC = CENTER TO CENTER   CD = CROSS DIAMETER
CL = CROSS LENGTH WITHOUT BEARINGS   EC = END TO CENTER or FACE TO FACE   EE = END TO END   EL = EFFECTIVE LENGTH
HD = HUB DIAMETER (or insert)   IE = INSIDE OF EARS (or recess)   OD = OUTSIDE DIAMETER   OE = OUTSIDE OF EARS   SB = SPLINE OR BORE SIZE

NEAPCO 11138
NEAPCO 11158
NEAPCO 12-1138
NEAPCO 12-1158
PRECISION 1221
ROCKWELL L12NYR19-3
SPICER 1-4-3083
WESCO 12N190SH
WESCO 12N195SH
WESCO 850034

IDLI GROUP 81287

END YOKE WITH ROUND BORE
slight variation from above group
SEE FIGURE 4B

1.8125 in	IE	46.04 mm
1.0625 in	BD	26.99 mm
1.1875 in	SB	30.16 mm
2.7500 in	EC	69.85 mm
2.0000 in	HD	50.80 mm

ROCKWELL L12NYR19-7

IDLI GROUP 81288

END YOKE WITH ROUND BORE
slight variation from above group
SEE FIGURE 4B

1.8125 in	IE	46.04 mm
1.0625 in	BD	26.99 mm
1.1875 in	SB	30.16 mm
2.7500 in	EC	69.85 mm
2.0000 in	HD	50.80 mm

ROCKWELL L12NYR19-14

IDLI GROUP 81289

END YOKE WITH ROUND BORE
slight variation from above group
SEE FIGURE 4B

1.8125 in	IE	46.04 mm
1.0625 in	BD	26.99 mm
1.1875 in	SB	30.16 mm
2.7500 in	EC	69.85 mm
2.0000 in	HD	50.80 mm

ROCKWELL L12NYR19-29

IDLI GROUP 81290

END YOKE WITH 10 SPLINES
SEE FIGURE 4A

1.8125 in	IE	46.04 mm
1.0625 in	BD	26.99 mm
1.1875 in	SB	30.16 mm
3.1875 in	EC	80.96 mm
1.6250 in	HD	41.28 mm

ROCKWELL L12NYS19-3

IDLI GROUP 81291

END YOKE WITH 6 SPLINES
SEE FIGURE 4A

1.8125 in	IE	46.04 mm
1.0625 in	BD	26.99 mm
1.2500 in	SB	31.75 mm
2.2500 in	EC	57.15 mm
2.0000 in	HD	50.80 mm

PRECISION 1253
ROCKWELL L12-34

IDLI GROUP 81292

END YOKE WITH 6 SPLINES
slight variation from above group
SEE FIGURE 4A

1.8125 in	IE	46.04 mm
1.0625 in	BD	26.99 mm
1.2500 in	SB	31.75 mm
2.2500 in	EC	57.15 mm
2.0000 in	HD	50.80 mm

PRECISION 1253
ROCKWELL L12NYS20-1

IDLI GROUP 81293

END YOKE WITH ROUND BORE
SEE FIGURE 4B

1.8125 in	IE	46.04 mm
1.0625 in	BD	26.99 mm
1.2500 in	SB	31.75 mm
2.2500 in	EC	57.15 mm
2.0000 in	HD	50.80 mm

ROCKWELL L12NYR20-45

IDLI GROUP 81294

END YOKE WITH 10 SPLINES
SEE FIGURE 4A

1.8125 in	IE	46.04 mm
1.0625 in	BD	26.99 mm
1.2500 in	SB	31.75 mm
2.5000 in	EC	63.50 mm
2.0000 in	HD	50.80 mm

ROCKWELL L12NYS20

IDLI GROUP 81295

END YOKE WITH 19 INVOLUTED SPLINES
SEE FIGURE 4A

1.8125 in	IE	46.04 mm
1.0625 in	BD	26.99 mm
1.2500 in	SB	31.75 mm
2.5000 in	EC	63.50 mm
2.0000 in	HD	50.80 mm

ROCKWELL L12NYS20-52

IDLI GROUP 81296

END YOKE WITH ROUND BORE
SEE FIGURE 4B

1.8125 in	IE	46.04 mm
1.0625 in	BD	26.99 mm
1.2500 in	SB	31.75 mm
2.6250 in	EC	66.68 mm
2.1250 in	HD	53.98 mm

ROCKWELL L12NYR20-17

IDLI GROUP 81297

END YOKE WITH ROUND BORE
SEE FIGURE 4B

1.8125 in	IE	46.04 mm
1.0625 in	BD	26.99 mm
1.2500 in	SB	31.75 mm
2.6875 in	EC	68.26 mm
1.7500 in	HD	44.45 mm

ROCKWELL L12NYR20-52

IDLI GROUP 81298

END YOKE WITH ROUND BORE
SEE FIGURE 4B

1.8125 in	IE	46.04 mm
1.0625 in	BD	26.99 mm
1.2500 in	SB	31.75 mm
2.7188 in	EC	69.06 mm
2.0000 in	HD	50.80 mm

ROCKWELL L12NYR20-10

IDLI GROUP 81299

END YOKE WITH 19 INVOLUTED SPLINES
SEE FIGURE 4A

1.8125 in	IE	46.04 mm
1.0625 in	BD	26.99 mm
1.2500 in	SB	31.75 mm
2.7500 in	EC	69.85 mm
1.6250 in	HD	41.28 mm

ROCKWELL L12NYS20-55

IDLI GROUP 81300

END YOKE WITH 6 SPLINES
SEE FIGURE 4A

1.8125 in	IE	46.04 mm
1.0625 in	BD	26.99 mm
1.2500 in	SB	31.75 mm
2.7500 in	EC	69.85 mm
2.0000 in	HD	50.80 mm

BPC H12EY-1-1/4
PRECISION 1254
PRECISION 1255
ROCKWELL L12-36

IDLI GROUP 81301

END YOKE WITH ROUND BORE
SEE FIGURE 4B

1.8125 in	IE	46.04 mm
1.0625 in	BD	26.99 mm
1.2500 in	SB	31.75 mm
2.7500 in	EC	69.85 mm
2.0000 in	HD	50.80 mm

BPC H12EY-1-1/4R-1/4KW
HUB CITY 0335-01216
MUNCIE MJY15A
NEAPCO 1118A
NEAPCO 12-1218
PRECISION 1223
ROCKWELL L12-23
SPICER 1-4-3093
TRW 21109
WESCO 12N200SH
WESCO 85-2045-37

IDLI GROUP 81302

END YOKE WITH ROUND BORE
slight variation from above group
SEE FIGURE 4B

1.8125 in	IE	46.04 mm
1.0625 in	BD	26.99 mm
1.2500 in	SB	31.75 mm
2.7500 in	EC	69.85 mm
2.0000 in	HD	50.80 mm

BORG-WARNR 61139
BPC H12EY-1-1/4R-5/16KW
MUNCIE MJY15
NEAPCO 11139

NEAPCO 11169
NEAPCO 12-1139
NEAPCO 12-1169
PRECISION 1224
ROCKWELL L12-24
WESCO 12N205SH
WESCO 850037

IDLI GROUP 81303

END YOKE WITH ROUND BORE
slight variation from above group
SEE FIGURE 4B

1.8125 in	IE	46.04 mm
1.0625 in	BD	26.99 mm
1.2500 in	SB	31.75 mm
2.7500 in	EC	69.85 mm
2.0000 in	HD	50.80 mm

G & G MFG 196-1220
PRECISION 1225
ROCKWELL L12-25
SPICER 1-4-3213
WESCO 12N200-7
WESCO 850035

IDLI GROUP 81304

END YOKE WITH ROUND BORE
slight variation from above group
SEE FIGURE 4B

1.8125 in	IE	46.04 mm
1.0625 in	BD	26.99 mm
1.2500 in	SB	31.75 mm
2.7500 in	EC	69.85 mm
2.0000 in	HD	50.80 mm

BORG-WARNR 61169
PRECISION 1224
ROCKWELL L12NYR20-15
SPICER 1-4-3113

IDLI GROUP 81305

END YOKE WITH ROUND BORE
slight variation from above group
SEE FIGURE 4B

1.8125 in	IE	46.04 mm
1.0625 in	BD	26.99 mm
1.2500 in	SB	31.75 mm
2.7500 in	EC	69.85 mm
2.0000 in	HD	50.80 mm

PRECISION 1223
ROCKWELL L12NYR20-25

IDLI GROUP 81306

END YOKE WITH ROUND BORE
slight variation from above group
SEE FIGURE 4B

1.8125 in	IE	46,04 mm
1.0625 in	BD	26.99 mm
1.2500 in	SB	31.75 mm
2.7500 in	EC	69.85 mm
2.0000 in	HD	50.80 mm

ROCKWELL L12NYR20-30

CAUTION: BE SURE TO REFER TO ENGINEERING CATALOGS FOR SPECIAL APPLICATIONS THAT REQUIRE SPECIFIC MATERIAL CONTENT, TOLERANCES, ETC. SEE FOOTNOTE.

IDLI GROUP 81307

END YOKE WITH ROUND BORE
slight variation from above group
SEE FIGURE 4B

1.8125 in	IE	46.04 mm
1.0625 in	BD	26.99 mm
1.2500 in	SB	31.75 mm
2.7500 in	EC	69.85 mm
2.0000 in	HD	50.80 mm

ROCKWELL L12NYR20-85
SPICER 1-4-3123

IDLI GROUP 81308

END YOKE WITH ROUND BORE
slight variation from above group
SEE FIGURE 4B

1.8125 in	IE	46.04 mm
1.0625 in	BD	26.99 mm
1.2500 in	SB	31.75 mm
2.7500 in	EC	69.85 mm
2.0000 in	HD	50.80 mm

PRECISION 1226
ROCKWELL L12NYR20-173

IDLI GROUP 81309

END YOKE WITH ROUND BORE
SEE FIGURE 4B

1.8125 in	IE	46.04 mm
1.0625 in	BD	26.99 mm
1.2500 in	SB	31.75 mm
2.9375 in	EC	74.61 mm
2.2500 in	HD	57.15 mm

ROCKWELL L12NYR20-102

IDLI GROUP 81310

END YOKE WITH ROUND BORE
SEE FIGURE 4B

1.8125 in	IE	46.04 mm
1.0625 in	BD	26.99 mm
1.2500 in	SB	31.75 mm
3.0000 in	EC	76.20 mm
1.8125 in	HD	46.04 mm

ROCKWELL L12NYR20-18

IDLI GROUP 81311

END YOKE WITH 10 SPLINES
SEE FIGURE 4A

1.8125 in	IE	46.04 mm
1.0625 in	BD	26.99 mm
1.2500 in	SB	31.75 mm
3.4375 in	EC	87.31 mm
2.0000 in	HD	50.80 mm

ROCKWELL L12NYS20-46

IDLI GROUP 81312

END YOKE WITH ROUND BORE
SEE FIGURE 4B

1.8125 in	IE	46.04 mm
1.0625 in	BD	26.99 mm
1.2500 in	SB	31.75 mm
3.5000 in	EC	88.90 mm
1.8750 in	HD	47.63 mm

ROCKWELL L12NYR20-62

IDLI GROUP 81313

END YOKE WITH ROUND BORE
SEE FIGURE 4B

1.8125 in	IE	46.04 mm
1.0625 in	BD	26.99 mm
1.2500 in	SB	31.75 mm
3.5000 in	EC	88.90 mm
1.9375 in	HD	49.21 mm

ROCKWELL 12NYR20-58

IDLI GROUP 81314

END YOKE WITH 6 SPLINES
SEE FIGURE 4A

1.8125 in	IE	46.04 mm
1.0625 in	BD	26.99 mm
1.2500 in	SB	31.75 mm
3.5000 in	EC	88.90 mm
2.0000 in	HD	50.80 mm

PRECISION 1254
ROCKWELL L12-35

IDLI GROUP 81315

END YOKE WITH 6 SPLINES
slight variation from above group
SEE FIGURE 4A

1.8125 in	IE	46.04 mm
1.0625 in	BD	26.99 mm
1.2500 in	SB	31.75 mm
3.5000 in	EC	88.90 mm
2.0000 in	HD	50.80 mm

BORG-WARNR 61222
NEAPCO 111B2
NEAPCO 12-1307
NEAPCO 17188
PRECISION 1254
ROCKWELL L12NYS20-17
SPICER 1-4-861
SPICER 1-4-871X
SPICER 1-4-951

IDLI GROUP 81316

END YOKE WITH 6 SPLINES
slight variation from above group
SEE FIGURE 4A

1.8125 in	IE	46.04 mm
1.0625 in	BD	26.99 mm
1.2500 in	SB	31.75 mm
3.5000 in	EC	88.90 mm
2.0000 in	HD	50.80 mm

PRECISION 1254
ROCKWELL L12NYS20-21

IDLI GROUP 81317

END YOKE WITH ROUND BORE
SEE FIGURE 4B

1.8125 in	IE	46.04 mm
1.0625 in	BD	26.99 mm
1.2500 in	SB	31.75 mm
3.5000 in	EC	88.90 mm
2.0000 in	HD	50.80 mm

PRECISION 1226
ROCKWELL L12NYR20

IDLI GROUP 81318

END YOKE WITH ROUND BORE
slight variation from above group
SEE FIGURE 4B

1.8125 in	IE	46.04 mm
1.0625 in	BD	26.99 mm
1.2500 in	SB	31.75 mm
3.5000 in	EC	88.90 mm
2.0000 in	HD	50.80 mm

ROCKWELL L12NYR20-9

IDLI GROUP 81319

END YOKE WITH ROUND BORE
slight variation from above group
SEE FIGURE 4B

1.8125 in	IE	46.04 mm
1.0625 in	BD	26.99 mm
1.2500 in	SB	31.75 mm
3.5000 in	EC	88.90 mm
2.0000 in	HD	50.80 mm

PRECISION 1223
ROCKWELL L12NYR20-19

IDLI GROUP 81320

END YOKE WITH ROUND BORE
slight variation from above group
SEE FIGURE 4B

1.8125 in	IE	46.04 mm
1.0625 in	BD	26.99 mm
1.2500 in	SB	31.75 mm
3.5000 in	EC	88.90 mm
2.0000 in	HD	50.80 mm

ROCKWELL L12NYR20-20

IDLI GROUP 81321

END YOKE WITH ROUND BORE
slight variation from above group
SEE FIGURE 4B

1.8125 in	IE	46.04 mm
1.0625 in	BD	26.99 mm
1.2500 in	SB	31.75 mm
3.5000 in	EC	88.90 mm
2.0000 in	HD	50.80 mm

ROCKWELL L12NYR20-43

IDLI GROUP 81322

END YOKE WITH ROUND BORE
slight variation from above group
SEE FIGURE 4B

1.8125 in	IE	46.04 mm
1.0625 in	BD	26.99 mm
1.2500 in	SB	31.75 mm
3.5000 in	EC	88.90 mm
2.0000 in	HD	50.80 mm

ROCKWELL L12NYR20-48

IDLI GROUP 81323

END YOKE WITH ROUND BORE
slight variation from above group
SEE FIGURE 4B

1.8125 in	IE	46.04 mm
1.0625 in	BD	26.99 mm
1.2500 in	SB	31.75 mm
3.5000 in	EC	88.90 mm
2.0000 in	HD	50.80 mm

ROCKWELL L12NYR20-51
SPICER 1-4-3103

IDLI GROUP 81324

END YOKE WITH 19 INVOLUTED SPLINES
SEE FIGURE 4A

1.8125 in	IE	46.04 mm
1.0625 in	BD	26.99 mm
1.2500 in	SB	31.75 mm
3.6250 in	EC	92.08 mm
1.6875 in	HD	42.86 mm

ROCKWELL L12NYS20-65

IDLI GROUP 81325

END YOKE WITH ROUND BORE
SEE FIGURE 4B

1.8125 in	IE	46.04 mm
1.0625 in	BD	26.99 mm
1.2500 in	SB	31.75 mm
5.2500 in	EC	133.35 mm
2.0000 in	HD	50.80 mm

BORG-WARNR 61174
PRECISION 1265
ROCKWELL L12NYR20-3

IDLI GROUP 81326

END YOKE WITH 14 INVOLUTED SPLINES
SEE FIGURE 4A

1.8125 in	IE	46.04 mm
1.0625 in	BD	26.99 mm
1.2813 in	SB	32.55 mm
2.6250 in	EC	66.68 mm
2.0000 in	HD	50.80 mm

PRECISION 1265
ROCKWELL L12NYS20-3

IDLI GROUP 81327

END YOKE WITH 9 INVOLUTED SPLINES
SEE FIGURE 4A

1.8125 in	IE	46.04 mm
1.0625 in	BD	26.99 mm
1.2813 in	SB	32.55 mm
3.0625 in	EC	77.79 mm
1.6250 in	HD	41.28 mm

ROCKWELL L12NYS20-58

IDLI GROUP 81328

END YOKE WITH 19 INVOLUTED SPLINES
SEE FIGURE 4A

1.8125 in	IE	46.04 mm
1.0625 in	BD	26.99 mm
1.2813 in	SB	32.55 mm
3.6250 in	EC	92.08 mm
1.6875 in	HD	42.86 mm

ROCKWELL L12NYS20-65

BD=BEARING DIAMETER (outside)   **BW**=BEARING WIDTH   **CB**=CROSS LENGTH WITH BEARINGS   **CC**=CENTER TO CENTER   **CD**=CROSS DIAMETER
CL=CROSS LENGTH WITHOUT BEARINGS   **EC**=END TO CENTER or FACE TO FACE   **EE**=END TO END   **EL**=EFFECTIVE LENGTH
HD=HUB DIAMETER (or insert)   **IE**=INSIDE OF EARS (or recess)   **OD**=OUTSIDE DIAMETER   **OE**=OUTSIDE OF EARS   **SB**=SPLINE OR BORE SIZE

IDLI GROUP 81329

END YOKE WITH 9 INVOLUTED SPLINES
SEE FIGURE 4A

1.8125 in	IE	46.04 mm
1.0625 in	BD	26.99 mm
1.2813 in	SB	32.55 mm
3.6250 in	EC	92.08 mm
1.9375 in	HD	49.21 mm

ROCKWELL L12NYS20-23

IDLI GROUP 81330

END YOKE WITH RECTANGULAR BORE
AND SQUARE HUB
SEE FIGURE 4D

1.8125 in	IE	46.04 mm
1.0625 in	BD	26.99 mm
1.3750 in	SB	34.93 mm
2.3750 in	EC	60.33 mm

ROCKWELL L12NYP20-8

IDLI GROUP 81331

END YOKE WITH 6 SPLINES
SEE FIGURE 4A

1.8125 in	IE	46.04 mm
1.0625 in	BD	26.99 mm
1.3750 in	SB	34.93 mm
2.7500 in	EC	69.85 mm
2.0000 in	HD	50.80 mm

BPC H12EY-1-3/8
PRECISION 1256
PRECISION 1257
ROCKWELL L12-37

IDLI GROUP 81332

END YOKE WITH 6 SPLINES
slight variation from above group
SEE FIGURE 4A

1.8125 in	IE	46.04 mm
1.0625 in	BD	26.99 mm
1.3750 in	SB	34.93 mm
2.7500 in	EC	69.85 mm
2.0000 in	HD	50.80 mm

PRECISION 1256
ROCKWELL L12NYS22-10
SPICER 1-4-981

IDLI GROUP 81333

END YOKE WITH ROUND BORE
SEE FIGURE 4B

1.8125 in	IE	46.04 mm
1.0625 in	BD	26.99 mm
1.3750 in	SB	34.93 mm
2.9375 in	EC	74.61 mm
2.2500 in	HD	57.15 mm

ALLOY 0606
ALLOY 0614
ALLOY 606
BORG-WARNR 61120
BORG-WARNR 61140
BORG-WARNR 61179
BPC H12EY-1-3/8R-5/16KW
G & G MFG 184-1222
HAYES 1-4-3133
HAYES 1-4-3273
HAYES 2-4-1863
HUB CITY 0335-01217

MCQUAY-NOR U03-95
MUNCIE MJY29
NEAPCO 11120
NEAPCO 11121
NEAPCO 11140
NEAPCO 11179
NEAPCO 12-1120
NEAPCO 12-1121
NEAPCO 12-1140
PRECISION 1229
ROCKWELL L12NYR22-27
SPICER 1-4-3143
TRW 21110
WESCO 12N0225
WESCO 12N225
WESCO 850039

IDLI GROUP 81334

END YOKE WITH 21 INVOLUTED SPLINES
SEE FIGURE 4A

1.8125 in	IE	46.04 mm
1.0625 in	BD	26.99 mm
1.3750 in	SB	34.93 mm
3.0000 in	EC	76.20 mm
2.0000 in	HD	50.80 mm

ROCKWELL L12NYS22-29

IDLI GROUP 81335

END YOKE WITH ROUND BORE
SEE FIGURE 4B

1.8125 in	IE	46.04 mm
1.0625 in	BD	26.99 mm
1.3750 in	SB	34.93 mm
3.2500 in	EC	82.55 mm
2.0000 in	HD	50.80 mm

ROCKWELL L12NYR22-20

IDLI GROUP 81336

END YOKE WITH 21 INVOLUTED SPLINES
SEE FIGURE 4A

1.8125 in	IE	46.04 mm
1.0625 in	BD	26.99 mm
1.3750 in	SB	34.93 mm
3.3750 in	EC	85.73 mm
1.8125 in	HD	46.04 mm

ROCKWELL L12NYS22-60

IDLI GROUP 81337

END YOKE WITH 21 INVOLUTED SPLINES
SEE FIGURE 4A

1.8125 in	IE	46.04 mm
1.0625 in	BD	26.99 mm
1.3750 in	SB	34.93 mm
3.5000 in	EC	88.90 mm
1.7500 in	HD	44.45 mm

ROCKWELL L12NYS22-30

IDLI GROUP 81338

END YOKE WITH 6 SPLINES
SEE FIGURE 4A

1.8125 in	IE	46.04 mm
1.0625 in	BD	26.99 mm
1.3750 in	SB	34.93 mm
3.5000 in	EC	88.90 mm
2.0000 in	HD	50.80 mm

ROCKWELL L12NYS22-1

IDLI GROUP 81339

END YOKE WITH 21 INVOLUTED SPLINES
SEE FIGURE 4A

1.8125 in	IE	46.04 mm
1.0625 in	BD	26.99 mm
1.3750 in	SB	34.93 mm
3.5000 in	EC	88.90 mm
2.0000 in	HD	50.80 mm

ROCKWELL L12NYS22-21
SPICER 1-4-1031

IDLI GROUP 81340

END YOKE WITH ROUND BORE
SEE FIGURE 4B

1.8125 in	IE	46.04 mm
1.0625 in	BD	26.99 mm
1.3750 in	SB	34.93 mm
3.5000 in	EC	88.90 mm
2.0000 in	HD	50.80 mm

ROCKWELL L12NYR22-12

IDLI GROUP 81341

END YOKE WITH ROUND BORE
slight variation from above group
SEE FIGURE 4B

1.8125 in	IE	46.04 mm
1.0625 in	BD	26.99 mm
1.3750 in	SB	34.93 mm
3.5000 in	EC	88.90 mm
2.0000 in	HD	50.80 mm

PRECISION 1229
ROCKWELL L12NYR22
SPICER 1-4-3133

IDLI GROUP 81342

END YOKE WITH 6 SPLINES
SEE FIGURE 4A

1.8125 in	IE	46.04 mm
1.0625 in	BD	26.99 mm
1.3750 in	SB	34.93 mm
3.5000 in	EC	88.90 mm
2.0625 in	HD	52.39 mm

ALLOY 0611
ALLOY 611
BORG-WARNR 61172
G & G MFG 182-1222
HUB CITY 0335-01204
NEAPCO 11172
NEAPCO 111H8
NEAPCO 12-1172
PRECISION 1257
REX CHAIN 760044
REX CHAIN 760068
ROCKWELL L12NYS22-12
SPICER 1-4-0981
TRW 21111

WESCO 12N0220-21
WESCO 12N220-21
WESCO 850038

IDLI GROUP 81343

END YOKE WITH 10 INVOLUTED SPLINES
SEE FIGURE 4A

1.8125 in	IE	46.04 mm
1.0625 in	BD	26.99 mm
1.3750 in	SB	34.93 mm
3.9375 in	EC	100.01 mm
1.8125 in	HD	46.04 mm

ROCKWELL L12NYS22-50

IDLI GROUP 81344

END YOKE WITH 6 SPLINES
SEE FIGURE 4A

1.8125 in	IE	46.04 mm
1.0625 in	BD	26.99 mm
1.3750 in	SB	34.93 mm
4.5000 in	EC	114.30 mm
2.0000 in	HD	50.80 mm

PRECISION 1284
ROCKWELL L12NYS22-23

IDLI GROUP 81345

END YOKE WITH RECTANGULAR BORE
AND SQUARE HUB
SEE FIGURE 4D

1.8125 in	IE	46.04 mm
1.0625 in	BD	26.99 mm
1.4063 in	SB	35.72 mm
2.3750 in	EC	60.33 mm

ROCKWELL L12NYP20-1

IDLI GROUP 81346

END YOKE WITH RECTANGULAR BORE
maximum SB dimension shown
SEE FIGURE 4D

1.8125 in	IE	46.04 mm
1.0625 in	BD	26.99 mm
1.4063 in	SB	35.72 mm
2.9375 in	EC	74.61 mm

ALLOY 0616
ALLOY 616
BORG-WARNR 61268
G & G MFG 180-1200
HAYES 1-4-262
NEAPCO 111E8
NEAPCO 111F8
NEAPCO 12-1318
NEAPCO 12-1324
PRECISION 1293
ROCKWELL L12NYP20-9
SPICER 1-4-262
WESCO 12NW4-2022
WESCO 850045

ENGINEERING CATALOGS MUST BE CONSULTED FOR DETAILS NOT INCLUDED
IN THIS GUIDE. SPECIFIC DESIGNS, MATERIAL CONTENT, TOLERANCES
LUBE FITTINGS AND OTHER DIMENSIONS ARE INTENTIONALLY OMITTED HERE.

CAUTION: BE SURE TO REFER TO ENGINEERING CATALOGS FOR SPECIAL APPLICATIONS THAT REQUIRE SPECIFIC MATERIAL CONTENT, TOLERANCES, ETC. SEE FOOTNOTE.

IDLI GROUP 81347

END YOKE WITH RECTANGULAR BORE
maximum SB dimension shown
SEE FIGURE 4D

1.8125 in	IE	46.04 mm
1.0625 in	BD	26.99 mm
1.4063 in	SB	35.72 mm
2.9375 in	EC	74.61 mm
2.3750 in	HD	60.33 mm

ROCKWELL L12NYP20-13

IDLI GROUP 81348

END YOKE WITH ROUND BORE
SEE FIGURE 4B

1.8125 in	IE	46.04 mm
1.0625 in	BD	26.99 mm
1.4844 in	SB	37.70 mm
2.9375 in	EC	74.61 mm
2.2500 in	HD	57.15 mm

ROCKWELL L12NYR23-9

IDLI GROUP 81349

END YOKE WITH ROUND BORE
SEE FIGURE 4B

1.8125 in	IE	46.04 mm
1.0625 in	BD	26.99 mm
1.5000 in	SB	38.10 mm
2.7500 in	EC	69.85 mm
2.2500 in	HD	57.15 mm

PRECISION 1230
ROCKWELL L12-26

IDLI GROUP 81350

END YOKE WITH ROUND BORE
slight variation from above group
SEE FIGURE 4B

1.8125 in	IE	46.04 mm
1.0625 in	BD	26.99 mm
1.5000 in	SB	38.10 mm
2.7500 in	EC	69.85 mm
2.2500 in	HD	57.15 mm

ALLOY 0612
BORG-WARNR 61161
BORG-WARNR 61175
BPC H12EY-1-1/2R-5/16KW
MUNCIE MJY18
MUNCIE MJY18A
NEAPCO 11161
NEAPCO 11175
NEAPCO 12-1161
NEAPCO 12-1175
PRECISION 1230
ROCKWELL L12NYR24-4
WESCO 12N240SH
WESCO 850040

IDLI GROUP 81351

END YOKE WITH ROUND BORE
SEE FIGURE 4B

1.8125 in	IE	46.04 mm
1.0625 in	BD	26.99 mm
1.5000 in	SB	38.10 mm
2.9375 in	EC	74.61 mm
2.2500 in	HD	57.15 mm

ALLOY 612
MUNCIE MJY18A
PRECISION 1230
ROCKWELL L12NYR24-7
SPICER 1-4-3153

IDLI GROUP 81352

END YOKE WITH SQUARE BORE
SEE FIGURE 4C

2.7969 in	IE	71.04 mm
1.3125 in	BD	33.34 mm
1.1250 in	SB	28.58 mm
4.0000 in	EC	101.60 mm
3.0000 in	HD	76.20 mm

ROCKWELL 44RYQ18-1

IDLI GROUP 81353

END YOKE WITH ROUND BORE
SEE FIGURE 4B

2.7969 in	IE	71.04 mm
1.3125 in	BD	33.34 mm
1.1875 in	SB	30.16 mm
4.1250 in	EC	104.78 mm
2.6250 in	HD	66.68 mm

ROCKWELL 44RYR19

IDLI GROUP 81354

END YOKE WITH ROUND BORE
SEE FIGURE 4B

2.7969 in	IE	71.04 mm
1.3125 in	BD	33.34 mm
1.2188 in	SB	30.96 mm
4.0000 in	EC	101.60 mm
3.0000 in	HD	76.20 mm

ROCKWELL 44RYR19-1

IDLI GROUP 81355

END YOKE WITH ROUND BORE
SEE FIGURE 4B

2.7969 in	IE	71.04 mm
1.3125 in	BD	33.34 mm
1.2500 in	SB	31.75 mm
4.1250 in	EC	104.78 mm
3.0000 in	HD	76.20 mm

ROCKWELL 44RYR20-2
SPICER 3-4-4-163

IDLI GROUP 81356

END YOKE WITH SQUARE BORE
SEE FIGURE 4C

2.7969 in	IE	71.04 mm
1.3125 in	BD	33.34 mm
1.3125 in	SB	33.34 mm
4.1250 in	EC	104.78 mm

G & G MFG 177-4400
NEAPCO 44-2100
NEAPCO N44NWY2100
PRECISION 1454

IDLI GROUP 81357

END YOKE WITH SQUARE BORE
SEE FIGURE 4C

2.7969 in	IE	71.04 mm
1.3125 in	BD	33.34 mm
1.3125 in	SB	33.34 mm
4.1250 in	EC	104.78 mm
2.6250 in	HD	66.68 mm

ROCKWELL 44RYQ21-6

IDLI GROUP 81358

END YOKE WITH SQUARE BORE
SEE FIGURE 4C

2.7969 in	IE	71.04 mm
1.3125 in	BD	33.34 mm
1.3125 in	SB	33.34 mm
4.1250 in	EC	104.78 mm
3.0000 in	HD	76.20 mm

ROCKWELL 44RYQ21-4

IDLI GROUP 81359

END YOKE WITH SQUARE BORE
slight variation from above group
SEE FIGURE 4C

2.7969 in	IE	71.04 mm
1.3125 in	BD	33.34 mm
1.3125 in	SB	33.34 mm
4.1250 in	EC	104.78 mm
3.0000 in	HD	76.20 mm

ROCKWELL 44RYQ21-5

IDLI GROUP 81360

END YOKE WITH SQUARE BORE
SEE FIGURE 4C

2.7969 in	IE	71.04 mm
1.3125 in	BD	33.34 mm
1.3125 in	SB	33.34 mm
4.9375 in	EC	125.41 mm

ROCKWELL 44RYQ21-2

IDLI GROUP 81361

END YOKE WITH ROUND BORE
SEE FIGURE 4B

2.7969 in	IE	71.04 mm
1.3125 in	BD	33.34 mm
1.3438 in	SB	34.13 mm
4.0000 in	EC	101.60 mm
3.0000 in	HD	76.20 mm

ROCKWELL 44RYR21

ROCKWELL 44RYQ21-1

SPICER 3-4-4-12
SPICER 3.4-4-82
WESCO 44N2200
WESCO 44NW4-2100
WESCO 481219

(continued)

IDLI GROUP 81362

END YOKE WITH 6 SPLINES
SEE FIGURE 4A

2.7969 in	IE	71.04 mm
1.3125 in	BD	33.34 mm
1.3750 in	SB	34.93 mm
4.1250 in	EC	104.78 mm
3.0000 in	HD	76.20 mm

G & G MFG 182-4406
NEAPCO 44-2220
NEAPCO N44N220-20
PRECISION 1442
PRECISION 1452
REXNORD 760969
REXNORD 760972
ROCKWELL 44RYS22-1A
ROCKWELL 44RYS22-5
SPICER 3-4-4-131X
SPICER 3-4-4-91
WESCO 44N220-20
WESCO 44N220-21
WESCO 481228
WESCO 481229

IDLI GROUP 81363

END YOKE WITH 6 SPLINES
slight variation from above group
SEE FIGURE 4A

2.7969 in	IE	71.04 mm
1.3125 in	BD	33.34 mm
1.3750 in	SB	34.93 mm
4.1250 in	EC	104.78 mm
3.0000 in	HD	76.20 mm

ROCKWELL 44RYS22-4

IDLI GROUP 81364

END YOKE WITH 6 SPLINES
slight variation from above group
SEE FIGURE 4A

2.7969 in	IE	71.04 mm
1.3125 in	BD	33.34 mm
1.3750 in	SB	34.93 mm
4.1250 in	EC	104.78 mm
3.0000 in	HD	76.20 mm

ROCKWELL 44RYS22-9

IDLI GROUP 81365

END YOKE WITH ROUND BORE
SEE FIGURE 4B

2.7969 in	IE	71.04 mm
1.3125 in	BD	33.34 mm
1.3750 in	SB	34.93 mm
4.1250 in	EC	104.78 mm
3.0000 in	HD	76.20 mm

G & G MFG 184-4422
NEAPCO 44-2256
NEAPCO N44N2256-37
PRECISION 1406
ROCKWELL 44RYR22-1
SPICER 3-4-4-173
WESCO 44N2256-37
WESCO 44N240-7
WESCO 481256

BD = BEARING DIAMETER (outside)   BW = BEARING WIDTH   CB = CROSS LENGTH WITH BEARINGS   CC = CENTER TO CENTER   CD = CROSS DIAMETER
CL = CROSS LENGTH WITHOUT BEARINGS   EC = END TO CENTER or FACE TO FACE   EE = END TO END   EL = EFFECTIVE LENGTH
HD = HUB DIAMETER (or insert)   IE = INSIDE OF EARS (or recess)   OD = OUTSIDE DIAMETER   OE = OUTSIDE OF EARS   SB = SPLINE OR BORE SIZE

IDLI GROUP 81366

END YOKE WITH TAPERED BORE
SEE FIGURE 4F

2.7969 in	IE	71.04 mm
1.3125 in	BD	33.34 mm
1.3750 in	SB	34.93 mm
4.5000 in	EC	114.30 mm
3.0000 in	HD	76.20 mm

ROCKWELL 44RYT22

IDLI GROUP 81367

END YOKE WITH TAPERED BORE
SEE FIGURE 4F

2.7969 in	IE	71.04 mm
1.3125 in	BD	33.34 mm
1.3750 in	SB	34.93 mm
5.0000 in	EC	127.00 mm
2.5000 in	HD	63.50 mm

ROCKWELL 44RYT22-1

IDLI GROUP 81368

END YOKE WITH 6 SPLINES
SEE FIGURE 4A

2.7969 in	IE	71.04 mm
1.3125 in	BD	33.34 mm
1.3750 in	SB	34.93 mm
5.0000 in	EC	127.00 mm
3.0000 in	HD	76.20 mm

NEAPCO N44N220-21
ROCKWELL 44RYS22-2

IDLI GROUP 81369

END YOKE WITH ROUND BORE
SEE FIGURE 4B

2.7969 in	IE	71.04 mm
1.3125 in	BD	33.34 mm
1.4375 in	SB	36.51 mm
4.1250 in	EC	104.78 mm
2.6250 in	HD	66.68 mm

REXNORD 761685
ROCKWELL 44RYR23-1
SPICER 3.4-4-83

IDLI GROUP 81370

END YOKE WITH ROUND BORE
SEE FIGURE 4B

2.7969 in	IE	71.04 mm
1.3125 in	BD	33.34 mm
1.5000 in	SB	38.10 mm
4.1250 in	EC	104.78 mm
3.0000 in	HD	76.20 mm

G & G MFG 184-4424
NEAPCO 44-2460
NEAPCO N44N246
PRECISION 1408
ROCKWELL 44RYR24-4
SPICER 3-4-4-183
WESCO 44N246
WESCO 481246

IDLI GROUP 81371

END YOKE WITH ROUND BORE
SEE FIGURE 4B

2.7969 in	IE	71.04 mm
1.3125 in	BD	33.34 mm
1.5781 in	SB	40.08 mm
3.0313 in	EC	77.00 mm
2.7500 in	HD	69.85 mm

ROCKWELL 44RYR25

IDLI GROUP 81372

END YOKE WITH ROUND BORE
SEE FIGURE 4B

2.7969 in	IE	71.04 mm
1.3125 in	BD	33.34 mm
1.6250 in	SB	41.28 mm
4.1250 in	EC	104.78 mm
3.0000 in	HD	76.20 mm

NEAPCO 44-2610
NEAPCO N44RYR26-1
NEAPCO NYRYR26-1
ROCKWELL 44RYR26-1
SPICER 3.4-4-23

IDLI GROUP 81373

END YOKE WITH 27 INVOLUTED SPLINES
SEE FIGURE 4A

2.7969 in	IE	71.04 mm
1.3125 in	BD	33.34 mm
1.7500 in	SB	44.45 mm
4.1250 in	EC	104.78 mm
2.6250 in	HD	66.68 mm

ROCKWELL 44RYS28-8

IDLI GROUP 81374

END YOKE WITH 6 SPLINES
SEE FIGURE 4A

2.7969 in	IE	71.04 mm
1.3125 in	BD	33.34 mm
1.7500 in	SB	44.45 mm
4.1250 in	EC	104.78 mm
3.0000 in	HD	76.20 mm

ROCKWELL 44RYS28-4

IDLI GROUP 81375

END YOKE WITH ROUND BORE
SEE FIGURE 4B

2.7969 in	IE	71.04 mm
1.3125 in	BD	33.34 mm
1.7500 in	SB	44.45 mm
4.1250 in	EC	104.78 mm
3.0000 in	HD	76.20 mm

G & G MFG 184-4428
NEAPCO 44-2820
NEAPCO N44RYR28-2
PRECISION 1412
ROCKWELL 44RYR28-2
SPICER 3-4-4-193

IDLI GROUP 81376

END YOKE WITH 27 INVOLUTED SPLINES
SEE FIGURE 4A

2.7969 in	IE	71.04 mm
1.3125 in	BD	33.34 mm
1.7500 in	SB	44.45 mm
4.3125 in	EC	109.54 mm
3.0000 in	HD	76.20 mm

REXNORD 762451
ROCKWELL 44RYS28-26
SPICER 3.4-4-361

IDLI GROUP 81377

END YOKE WITH 6 SPLINES
SEE FIGURE 4A

2.7969 in	IE	71.04 mm
1.3125 in	BD	33.34 mm
1.7500 in	SB	44.45 mm
4.9375 in	EC	125.41 mm
2.5000 in	HD	63.50 mm

ROCKWELL 44RYS28-2

IDLI GROUP 81378

END YOKE WITH 27 INVOLUTED SPLINES
SEE FIGURE 4A

2.7969 in	IE	71.04 mm
1.3125 in	BD	33.34 mm
1.7500 in	SB	44.45 mm
5.0625 in	EC	128.59 mm
2.8750 in	HD	73.03 mm

ROCKWELL 44RYS28-7

IDLI GROUP 81379

END YOKE WITH 10 SPLINES
SEE FIGURE 8A

3.1250 in	IE	79.38 mm
1.1875 in	BD	30.16 mm
1.3750 in	SB	34.93 mm
3.5000 in	EC	88.90 mm
2.1250 in	HD	53.98 mm

ROCKWELL 141N4-3941
SPICER 3-4-3941

IDLI GROUP 81380

END YOKE WITH TAPERED BORE
SEE FIGURE 8F

3.1250 in	IE	79.38 mm
1.1875 in	BD	30.16 mm
1.3750 in	SB	34.93 mm
4.0469 in	EC	102.79 mm
2.5000 in	HD	63.50 mm

ROCKWELL 141NYT22-1

IDLI GROUP 81381

END YOKE WITH 10 SPLINES
SEE FIGURE 8A

3.1250 in	IE	79.38 mm
1.1875 in	BD	30.16 mm
1.5000 in	SB	38.10 mm
3.6875 in	EC	93.66 mm
1.8125 in	HD	46.04 mm

ROCKWELL 141N4-6561
SPICER 3-4-6561

IDLI GROUP 81382

END YOKE WITH 10 SPLINES
slight variation from above group
SEE FIGURE 8A

3.1250 in	IE	79.38 mm
1.1875 in	BD	30.16 mm
1.5000 in	SB	38.10 mm
3.6875 in	EC	93.66 mm
1.8125 in	HD	46.04 mm

ROCKWELL 141N4-6631X
SPICER 3-4-6631X

IDLI GROUP 81383

END YOKE WITH 10 SPLINES
SEE FIGURE 8A

3.1250 in	IE	79.38 mm
1.1875 in	BD	30.16 mm
1.5000 in	SB	38.10 mm
3.8750 in	EC	98.43 mm
2.2500 in	HD	57.15 mm

ROCKWELL 141N4-2241
SPICER 3-4-1841
SPICER 3-4-2241

IDLI GROUP 81384

END YOKE WITH 10 SPLINES
slight variation from above group
SEE FIGURE 8A

3.1250 in	IE	79.38 mm
1.1875 in	BD	30.16 mm
1.5000 in	SB	38.10 mm
3.8750 in	EC	98.43 mm
2.2500 in	HD	57.15 mm

ROCKWELL 141NYS24A

IDLI GROUP 81385

END YOKE WITH 10 SPLINES
SEE FIGURE 8A

3.1250 in	IE	79.38 mm
1.1875 in	BD	30.16 mm
1.7500 in	SB	44.45 mm
4.1875 in	EC	106.36 mm
2.3750 in	HD	60.33 mm

ROCKWELL 141N4-1841
SPICER 3-4-1841
SPICER 3-4-4161X

IDLI GROUP 81386

END YOKE WITH 10 SPLINES
SEE FIGURE 8A

3.1250 in	IE	79.38 mm
1.1875 in	BD	30.16 mm
1.7500 in	SB	44.45 mm
4.2500 in	EC	107.95 mm
2.6250 in	HD	66.68 mm

ROCKWELL 141N4-2511
SPICER 3-4-2511
SPICER 3-4-7031

ENGINEERING CATALOGS MUST BE CONSULTED FOR DETAILS NOT INCLUDED
IN THIS GUIDE. SPECIFIC DESIGNS, MATERIAL CONTENT, TOLERANCES
LUBE FITTINGS AND OTHER DIMENSIONS ARE INTENTIONALLY OMITTED HERE.

CAUTION: BE SURE TO REFER TO ENGINEERING CATALOGS FOR SPECIAL APPLICATIONS THAT REQUIRE SPECIFIC MATERIAL CONTENT, TOLERANCES, ETC. SEE FOOTNOTE.

IDLI GROUP 81387

END YOKE WITH 10 SPLINES
SEE FIGURE 8A

4.1875 in	IE	106.36 mm
1.3750 in	BD	34.93 mm
1.3750 in	SB	34.93 mm
3.9688 in	EC	100.81 mm
2.1250 in	HD	53.98 mm

ROCKWELL 148N4-3231
SPICER 3-4-3231

IDLI GROUP 81388

END YOKE WITH 10 SPLINES
SEE FIGURE 8A

4.1875 in	IE	106.36 mm
1.3750 in	BD	34.93 mm
1.5000 in	SB	38.10 mm
4.0313 in	EC	102.40 mm
1.9375 in	HD	49.21 mm

ROCKWELL 148N4-3091
SPICER 3-4-3091

IDLI GROUP 81389

END YOKE WITH 10 SPLINES
slight variation from above group
SEE FIGURE 8A

4.1875 in	IE	106.36 mm
1.3750 in	BD	34.93 mm
1.5000 in	SB	38.10 mm
4.0313 in	EC	102.40 mm
1.9375 in	HD	49.21 mm

ROCKWELL 148N4-4421X
SPICER 3-4-4421X

IDLI GROUP 81390

END YOKE WITH 10 SPLINES
SEE FIGURE 8A

4.1875 in	IE	106.36 mm
1.3750 in	BD	34.93 mm
1.5000 in	SB	38.10 mm
4.0313 in	EC	102.40 mm
2.0625 in	HD	52.39 mm

ROCKWELL 148N4-3081
SPICER 3-4-3081
SPICER 3-4-3331X

IDLI GROUP 81391

END YOKE WITH 10 SPLINES
SEE FIGURE 8A

4.1875 in	IE	106.36 mm
1.3750 in	BD	34.93 mm
1.5000 in	SB	38.10 mm
4.1875 in	EC	106.36 mm
2.2500 in	HD	57.15 mm

ROCKWELL 148N4-3781
SPICER 3-4-3781
SPICER 3-4-6051X
SPICER 3-4-6221X

IDLI GROUP 81392

END YOKE WITH 10 SPLINES
SEE FIGURE 8A

4.1875 in	IE	106.36 mm
1.3750 in	BD	34.93 mm
1.6250 in	SB	41.28 mm
4.1875 in	EC	106.36 mm
2.5000 in	HD	63.50 mm

ROCKWELL 148NYS2G-2A

IDLI GROUP 81393

END YOKE WITH 10 SPLINES
SEE FIGURE 8A

4.1875 in	IE	106.36 mm
1.3750 in	BD	34.93 mm
1.7500 in	SB	44.45 mm
3.9688 in	EC	100.81 mm
2.6250 in	HD	66.68 mm

ROCKWELL 148N4-4051
SPICER 3-4-4051

IDLI GROUP 81394

END YOKE WITH 10 SPLINES
SEE FIGURE 8A

4.1875 in	IE	106.36 mm
1.3750 in	BD	34.93 mm
1.7500 in	SB	44.45 mm
4.2188 in	EC	107.16 mm
2.3750 in	HD	60.33 mm

ROCKWELL 148N4-3571
SPICER 3-4-3191
SPICER 3-4-3571

IDLI GROUP 81395

END YOKE WITH 10 SPLINES
SEE FIGURE 8A

4.1875 in	IE	106.36 mm
1.3750 in	BD	34.93 mm
1.7500 in	SB	44.45 mm
4.2813 in	EC	108.75 mm
2.6250 in	HD	66.68 mm

ROCKWELL 148N4-3541
SPICER 3-4-3541

IDLI GROUP 81396

END YOKE WITH 10 SPLINES
slight variation from above group
SEE FIGURE 8A

4.1875 in	IE	106.36 mm
1.3750 in	BD	34.93 mm
1.7500 in	SB	44.45 mm
4.2813 in	EC	108.75 mm
2.6250 in	HD	66.68 mm

ROCKWELL 148N4-3061
SPICER 3-4-3061

IDLI GROUP 81397

END YOKE WITH 10 SPLINES
slight variation from above group
SEE FIGURE 8A

4.1875 in	IE	106.36 mm
1.3750 in	BD	34.93 mm
1.7500 in	SB	44.45 mm
4.2813 in	EC	108.75 mm
2.6250 in	HD	66.68 mm

ROCKWELL 148N4-5791X

SPICER 3-4-5791X

IDLI GROUP 81398

END YOKE WITH 10 SPLINES
SEE FIGURE 8A

4.1875 in	IE	106.36 mm
1.3750 in	BD	34.93 mm
1.7500 in	SB	44.45 mm
4.3125 in	EC	109.54 mm
2.3750 in	HD	60.33 mm

ROCKWELL 148NYS28-11

IDLI GROUP 81399

END YOKE WITH 34 SLANTED SPLINES
SEE FIGURE 8A

4.1875 in	IE	106.36 mm
1.3750 in	BD	34.93 mm
1.7656 in	SB	44.85 mm
4.4688 in	EC	113.51 mm
2.5000 in	HD	63.50 mm

ROCKWELL 148N4-7541
SPICER 3-4-7541
SPICER 3-4-8051X

IDLI GROUP 81400

END YOKE WITH 10 SPLINES
SEE FIGURE 8A

4.1875 in	IE	106.36 mm
1.3750 in	BD	34.93 mm
1.9688 in	SB	50.01 mm
4.4063 in	EC	111.92 mm
2.6250 in	HD	66.68 mm

ROCKWELL 148N4-3861
SPICER 3-4-3861
SPICER 3-4-5821X

IDLI GROUP 81401

END YOKE WITH 32 INVOLUTED SPLINES
SEE FIGURE 8A

4.1875 in	IE	106.36 mm
1.3750 in	BD	34.93 mm
2.1094 in	SB	53.58 mm
4.2813 in	EC	108.75 mm
3.0000 in	HD	76.20 mm

ROCKWELL 148NYS34

IDLI GROUP 81402

END YOKE WITH 10 SPLINES
SEE FIGURE 8A

4.9688 in	IE	126.21 mm
1.3750 in	BD	34.93 mm
1.5000 in	SB	38.10 mm
3.7813 in	EC	96.05 mm
1.9375 in	HD	49.21 mm

ROCKWELL 155N4-2051
SPICER 4-4-2051

IDLI GROUP 81403

END YOKE WITH 10 SPLINES
slight variation from above group
SEE FIGURE 8A

4.9688 in	IE	126.21 mm
1.3750 in	BD	34.93 mm
1.5000 in	SB	38.10 mm
3.7813 in	EC	96.05 mm
1.9375 in	HD	49.21 mm

ROCKWELL 155NYS24-1A

IDLI GROUP 81404

END YOKE WITH 10 SPLINES
SEE FIGURE 8A

4.9688 in	IE	126.21 mm
1.3750 in	BD	34.93 mm
1.5000 in	SB	38.10 mm
3.7969 in	EC	96.44 mm
2.1875 in	HD	55.56 mm

ROCKWELL 155N4-4211
SPICER 4-4-4211

IDLI GROUP 81405

END YOKE WITH 10 SPLINES
slight variation from above group
SEE FIGURE 8A

4.9688 in	IE	126.21 mm
1.3750 in	BD	34.93 mm
1.5000 in	SB	38.10 mm
3.7969 in	EC	96.44 mm
2.1875 in	HD	55.56 mm

ROCKWELL 155NYS24-4A

IDLI GROUP 81406

END YOKE WITH 10 SPLINES
SEE FIGURE 8A

4.9688 in	IE	126.21 mm
1.3750 in	BD	34.93 mm
1.7500 in	SB	44.45 mm
4.0000 in	EC	101.60 mm
2.3750 in	HD	60.33 mm

ROCKWELL 155N4-2041
SPICER 4-4-2041

IDLI GROUP 81407

END YOKE WITH 10 SPLINES
SEE FIGURE 8A

4.9688 in	IE	126.21 mm
1.3750 in	BD	34.93 mm
1.7500 in	SB	44.45 mm
4.0000 in	EC	101.60 mm
2.6250 in	HD	66.68 mm

CLEVE.STL. U76-1-135
ROCKWELL 155NYS28-4

IDLI GROUP 81408

END YOKE WITH 10 SPLINES
SEE FIGURE 8A

4.9688 in	IE	126.21 mm
1.3750 in	BD	34.93 mm
1.7500 in	SB	44.45 mm
4.3125 in	EC	109.54 mm
2.6250 in	HD	66.68 mm

ROCKWELL 155N4-1891
SPICER 4-4-1891
SPICER 4-4-2091

BD=BEARING DIAMETER (outside)   **BW**=BEARING WIDTH   **CB**=CROSS LENGTH WITH BEARINGS   **CC**=CENTER TO CENTER   **CD**=CROSS DIAMETER
CL=CROSS LENGTH WITHOUT BEARINGS   **EC**=END TO CENTER or FACE TO FACE   **EE**=END TO END   **EL**=EFFECTIVE LENGTH
HD=HUB DIAMETER (or insert)   **IE**=INSIDE OF EARS (or recess)   **OD**=OUTSIDE DIAMETER   **OE**=OUTSIDE OF EARS   **SB**=SPLINE OR BORE SIZE

IDLI GUIDE   COPYRIGHT © INTERCHANGE, INC.   ST. LOUIS PARK, MN. 55416 USA

Column 1

SPICER 4-4-3601X

IDLI GROUP 81409

**END YOKE WITH 10 SPLINES
SEE FIGURE 8A**

4.9688 in	IE	126.21 mm
1.3750 in	BD	34.93 mm
1.7500 in	SB	44.45 mm
4.6563 in	EC	118.27 mm
2.6250 in	HD	66.68 mm

ROCKWELL 155N4-1901
SPICER 4-4-1901
SPICER 4-4-3581X

IDLI GROUP 81410

**END YOKE WITH 34 SLANTED SPLINES
SEE FIGURE 8A**

4.9688 in	IE	126.21 mm
1.3750 in	BD	34.93 mm
1.7656 in	SB	44.85 mm
4.6875 in	EC	119.06 mm
2.5000 in	HD	63.50 mm

ROCKWELL 155N4-4101
SPICER 4-4-4101

IDLI GROUP 81411

**END YOKE WITH 10 SPLINES
SEE FIGURE 8A**

4.9688 in	IE	126.21 mm
1.3750 in	BD	34.93 mm
1.9688 in	SB	50.01 mm
4.2813 in	EC	108.75 mm
2.6250 in	HD	66.68 mm

ROCKWELL 155N4-2431
SPICER 4-4-2431
SPICER 4-4-3611X
SPICER 4-4-4071X

IDLI GROUP 81412

**END YOKE WITH 32 INVOLUTED SPLINES
SEE FIGURE 8A**

4.9688 in	IE	126.21 mm
1.3750 in	BD	34.93 mm
2.0938 in	SB	53.18 mm
4.3125 in	EC	109.54 mm
3.0000 in	HD	76.20 mm

ROCKWELL 155N4-4241
SPICER 4-4-4241

IDLI GROUP 81413

**END YOKE WITH ROUND BORE
SEE FIGURE 7B**

5.3125 in	IE	134.94 mm
5.6250 in	OE	142.88 mm
1.0000 in	SB	25.40 mm
3.0000 in	EC	76.20 mm
3.5000 in	HD	88.90 mm

ROCKWELL 58WBYR16

Column 2

IDLI GROUP 81414

**END YOKE WITH ROUND BORE
SEE FIGURE 7B**

5.3125 in	IE	134.94 mm
5.6250 in	OE	142.88 mm
1.0000 in	SB	25.40 mm
3.6875 in	EC	93.66 mm
3.5000 in	HD	88.90 mm

ROCKWELL 58WBYR16-2

IDLI GROUP 81415

**END YOKE WITH ROUND BORE
SEE FIGURE 7B**

5.3125 in	IE	134.94 mm
5.6250 in	OE	142.88 mm
1.0000 in	SB	25.40 mm
3.8125 in	EC	96.84 mm
2.7500 in	HD	69.85 mm

ROCKWELL 58WBYR16-1

IDLI GROUP 81416

**END YOKE WITH ROUND BORE
SEE FIGURE 7B**

5.3125 in	IE	134.94 mm
5.6250 in	OE	142.88 mm
1.0000 in	SB	25.40 mm
4.9375 in	EC	125.41 mm
4.0000 in	HD	101.60 mm

ROCKWELL 58WBYR16-3

IDLI GROUP 81417

**END YOKE WITH ROUND BORE
SEE FIGURE 7B**

5.3125 in	IE	134.94 mm
5.6250 in	OE	142.88 mm
1.0625 in	SB	26.99 mm
3.8750 in	EC	98.43 mm
2.6250 in	HD	66.68 mm

ROCKWELL 58WBYR17

IDLI GROUP 81418

**END YOKE WITH ROUND BORE
SEE FIGURE 7B**

5.3125 in	IE	134.94 mm
5.6250 in	OE	142.88 mm
1.1250 in	SB	28.58 mm
3.0000 in	EC	76.20 mm
3.5000 in	HD	88.90 mm

ROCKWELL 58WBYR18

IDLI GROUP 81419

**END YOKE WITH ROUND BORE
SEE FIGURE 7B**

5.3125 in	IE	134.94 mm
5.6250 in	OE	142.88 mm
1.1250 in	SB	28.58 mm
3.4375 in	EC	87.31 mm
2.6250 in	HD	66.68 mm

ROCKWELL 58WBYR18-2

Column 3

IDLI GROUP 81420

**END YOKE WITH ROUND BORE
SEE FIGURE 7B**

5.3125 in	IE	134.94 mm
5.6250 in	OE	142.88 mm
1.1250 in	SB	28.58 mm
3.6875 in	EC	93.66 mm
2.7500 in	HD	69.85 mm

ROCKWELL 58WBYR18-4

IDLI GROUP 81421

**END YOKE WITH ROUND BORE
SEE FIGURE 7B**

5.3125 in	IE	134.94 mm
5.6250 in	OE	142.88 mm
1.1250 in	SB	28.58 mm
3.8125 in	EC	96.84 mm
2.7500 in	HD	69.85 mm

ROCKWELL 58WBYR18-3

IDLI GROUP 81422

**END YOKE WITH ROUND BORE
SEE FIGURE 7B**

5.3125 in	IE	134.94 mm
5.6250 in	OE	142.88 mm
1.2500 in	SB	31.75 mm
3.8750 in	EC	98.43 mm
2.6250 in	HD	66.68 mm

ROCKWELL 58WBYR20-7

IDLI GROUP 81423

**END YOKE WITH ROUND BORE
slight variation from above group
SEE FIGURE 7B**

5.3125 in	IE	134.94 mm
5.6250 in	OE	142.88 mm
1.2500 in	SB	31.75 mm
3.8750 in	EC	98.43 mm
2.6250 in	HD	66.68 mm

ROCKWELL 58WBYR20-1

IDLI GROUP 81424

**END YOKE WITH ROUND BORE
slight variation from above group
SEE FIGURE 7B**

5.3125 in	IE	134.94 mm
5.6250 in	OE	142.88 mm
1.2500 in	SB	31.75 mm
3.8750 in	EC	98.43 mm
2.6250 in	HD	66.68 mm

ROCKWELL 58WBYR20-3

IDLI GROUP 81425

**END YOKE WITH ROUND BORE
slight variation from above group
SEE FIGURE 7B**

5.3125 in	IE	134.94 mm
5.6250 in	OE	142.88 mm
1.2500 in	SB	31.75 mm
3.8750 in	EC	98.43 mm
2.6250 in	HD	66.68 mm

ROCKWELL 58WBYR20-4

Column 4

IDLI GROUP 81426

**END YOKE WITH ROUND BORE
SEE FIGURE 7B**

5.3125 in	IE	134.94 mm
5.6250 in	OE	142.88 mm
1.2500 in	SB	31.75 mm
4.8750 in	EC	123.83 mm
3.5000 in	HD	88.90 mm

ROCKWELL 58WBYR20-6

IDLI GROUP 81427

**END YOKE WITH ROUND BORE
SEE FIGURE 7B**

5.3125 in	IE	134.94 mm
5.6250 in	OE	142.88 mm
1.3594 in	SB	34.53 mm
3.3750 in	EC	85.73 mm
2.1406 in	HD	54.37 mm

ROCKWELL 58WBYR22

IDLI GROUP 81428

**END YOKE WITH 10 SPLINES
SEE FIGURE 7A**

5.3125 in	IE	134.94 mm
5.6250 in	OE	142.88 mm
1.3750 in	SB	34.93 mm
3.0625 in	EC	77.79 mm
2.1250 in	HD	53.98 mm

ROCKWELL 58WBYS22

IDLI GROUP 81429

**END YOKE WITH 10 SPLINES
SEE FIGURE 7A**

5.3125 in	IE	134.94 mm
5.6250 in	OE	142.88 mm
1.3750 in	SB	34.93 mm
3.4375 in	EC	87.31 mm
2.1250 in	HD	53.98 mm

ROCKWELL 58WBYS22-9

IDLI GROUP 81430

**END YOKE WITH 10 SPLINES
SEE FIGURE 7A**

5.3125 in	IE	134.94 mm
5.6250 in	OE	142.88 mm
1.3750 in	SB	34.93 mm
3.6250 in	EC	92.08 mm
2.1250 in	HD	53.98 mm

ROCKWELL 58WBYS22-8

IDLI GROUP 81431

**END YOKE WITH TAPERED BORE
SEE FIGURE 7F**

5.3125 in	IE	134.94 mm
5.6250 in	OE	142.88 mm
1.3750 in	SB	34.93 mm
3.6250 in	EC	92.08 mm
2.6406 in	HD	67.07 mm

ROCKWELL 58WBYT22-1

ENGINEERING CATALOGS MUST BE CONSULTED FOR DETAILS NOT INCLUDED
IN THIS GUIDE. SPECIFIC DESIGNS, MATERIAL CONTENT, TOLERANCES
LUBE FITTINGS AND OTHER DIMENSIONS ARE INTENTIONALLY OMITTED HERE.

CAUTION: BE SURE TO REFER TO ENGINEERING CATALOGS FOR SPECIAL APPLICATIONS THAT REQUIRE SPECIFIC MATERIAL CONTENT, TOLERANCES, ETC. SEE FOOTNOTE.

IDLI GROUP 81432

END YOKE WITH ROUND BORE
SEE FIGURE 7B

5.3125 in	IE	134.94 mm
5.6250 in	OE	142.88 mm
1.3750 in	SB	34.93 mm
3.6875 in	EC	93.66 mm
3.5000 in	HD	88.90 mm

ROCKWELL 58WBYR22-1

IDLI GROUP 81433

END YOKE WITH ROUND BORE
slight variation from above group
SEE FIGURE 7B

5.3125 in	IE	134.94 mm
5.6250 in	OE	142.88 mm
1.3750 in	SB	34.93 mm
3.6875 in	EC	93.66 mm
3.5000 in	HD	88.90 mm

ROCKWELL 58WBYR22-3

IDLI GROUP 81434

END YOKE WITH ROUND BORE
slight variation from above group
SEE FIGURE 7B

5.3125 in	IE	134.94 mm
5.6250 in	OE	142.88 mm
1.3750 in	SB	34.93 mm
3.6875 in	EC	93.66 mm
3.5000 in	HD	88.90 mm

ROCKWELL 58WBYR22-2

IDLI GROUP 81435

END YOKE WITH ROUND BORE
slight variation from above group
SEE FIGURE 7B

5.3125 in	IE	134.94 mm
5.6250 in	OE	142.88 mm
1.3750 in	SB	34.93 mm
3.6875 in	EC	93.66 mm
3.5000 in	HD	88.90 mm

ROCKWELL 58WBYR22-4

IDLI GROUP 81436

END YOKE WITH ROUND BORE
slight variation from above group
SEE FIGURE 7B

5.3125 in	IE	134.94 mm
5.6250 in	OE	142.88 mm
1.3750 in	SB	34.93 mm
3.6875 in	EC	93.66 mm
3.5000 in	HD	88.90 mm

ROCKWELL 58WBYR22-6

IDLI GROUP 81437

END YOKE WITH 10 SPLINES
SEE FIGURE 7A

5.3125 in	IE	134.94 mm
5.6250 in	OE	142.88 mm
1.3750 in	SB	34.93 mm
3.8750 in	EC	98.43 mm
1.9375 in	HD	49.21 mm

ROCKWELL 58WBYS22-7

IDLI GROUP 81438

END YOKE WITH ROUND BORE
SEE FIGURE 7B

5.3125 in	IE	134.94 mm
5.6250 in	OE	142.88 mm
1.3750 in	SB	34.93 mm
4.8125 in	EC	122.24 mm
4.0000 in	HD	101.60 mm

ROCKWELL 58WBYR22-5

IDLI GROUP 81439

END YOKE WITH 21 INVOLUTED SPLINES
SEE FIGURE 7A

5.3125 in	IE	134.94 mm
5.6250 in	OE	142.88 mm
1.4375 in	SB	36.51 mm
2.6250 in	EC	66.68 mm
2.1250 in	HD	53.98 mm

ROCKWELL 58WBYS22-4

IDLI GROUP 81440

END YOKE WITH 21 INVOLUTED SPLINES
SEE FIGURE 7A

5.3125 in	IE	134.94 mm
5.6250 in	OE	142.88 mm
1.4375 in	SB	36.51 mm
2.9375 in	EC	74.61 mm
2.1250 in	HD	53.98 mm

ROCKWELL 58WBYS22-3

IDLI GROUP 81441

END YOKE WITH 21 INVOLUTED SPLINES
SEE FIGURE 7A

5.3125 in	IE	134.94 mm
5.6250 in	OE	142.88 mm
1.4375 in	SB	36.51 mm
3.3750 in	EC	85.73 mm
2.1250 in	HD	53.98 mm

ROCKWELL 58WBYS22-6

IDLI GROUP 81442

END YOKE WITH ROUND BORE
SEE FIGURE 7B

5.3125 in	IE	134.94 mm
5.6250 in	OE	142.88 mm
1.4375 in	SB	36.51 mm
3.5625 in	EC	90.49 mm
2.5000 in	HD	63.50 mm

ROCKWELL 58WBYR23

IDLI GROUP 81443

END YOKE WITH ROUND BORE
SEE FIGURE 7B

5.3125 in	IE	134.94 mm
5.6250 in	OE	142.88 mm
1.4375 in	SB	36.51 mm
4.5625 in	EC	115.89 mm
4.0000 in	HD	101.60 mm

ROCKWELL 58WBYR23-1

IDLI GROUP 81444

END YOKE WITH ROUND BORE
SEE FIGURE 7B

5.3125 in	IE	134.94 mm
5.6250 in	OE	142.88 mm
1.5000 in	SB	38.10 mm
2.6250 in	EC	66.68 mm
2.6250 in	HD	66.68 mm

ROCKWELL 58WBYR24-8

IDLI GROUP 81445

END YOKE WITH ROUND BORE
SEE FIGURE 7B

5.3125 in	IE	134.94 mm
5.6250 in	OE	142.88 mm
1.5000 in	SB	38.10 mm
3.3750 in	EC	85.73 mm
2.6250 in	HD	66.68 mm

ROCKWELL 58WBYR24-9

IDLI GROUP 81446

END YOKE WITH 10 SPLINES
SEE FIGURE 7A

5.3125 in	IE	134.94 mm
5.6250 in	OE	142.88 mm
1.5000 in	SB	38.10 mm
3.4375 in	EC	87.31 mm
2.2500 in	HD	57.15 mm

ROCKWELL 58WBYS24-9

IDLI GROUP 81447

END YOKE WITH 10 SPLINES
slight variation from above group
SEE FIGURE 7A

5.3125 in	IE	134.94 mm
5.6250 in	OE	142.88 mm
1.5000 in	SB	38.10 mm
3.4375 in	EC	87.31 mm
2.2500 in	HD	57.15 mm

ROCKWELL 58WBYS24-3

IDLI GROUP 81448

END YOKE WITH 10 SPLINES
SEE FIGURE 7A

5.3125 in	IE	134.94 mm
5.6250 in	OE	142.88 mm
1.5000 in	SB	38.10 mm
3.6250 in	EC	92.08 mm
2.5000 in	HD	63.50 mm

ROCKWELL 58WBYS24-4

IDLI GROUP 81449

END YOKE WITH 10 SPLINES
SEE FIGURE 7A

5.3125 in	IE	134.94 mm
5.6250 in	OE	142.88 mm
1.5000 in	SB	38.10 mm
3.6875 in	EC	93.66 mm
2.1250 in	HD	53.98 mm

ROCKWELL 58WBYS24-7

IDLI GROUP 81450

END YOKE WITH 10 SPLINES
SEE FIGURE 7A

5.3125 in	IE	134.94 mm
5.6250 in	OE	142.88 mm
1.5000 in	SB	38.10 mm
3.6875 in	EC	93.66 mm
2.6250 in	HD	66.68 mm

ROCKWELL 58WBYS24-13

IDLI GROUP 81451

END YOKE WITH ROUND BORE
SEE FIGURE 7B

5.3125 in	IE	134.94 mm
5.6250 in	OE	142.88 mm
1.5000 in	SB	38.10 mm
3.7500 in	EC	95.25 mm
2.5000 in	HD	63.50 mm

ROCKWELL 58WBYR24-7

IDLI GROUP 81452

END YOKE WITH ROUND BORE
SEE FIGURE 7B

5.3125 in	IE	134.94 mm
5.6250 in	OE	142.88 mm
1.5000 in	SB	38.10 mm
3.8125 in	EC	96.84 mm
2.6250 in	HD	66.68 mm

ROCKWELL 58WBYR24-2

IDLI GROUP 81453

END YOKE WITH 10 SPLINES
SEE FIGURE 7A

5.3125 in	IE	134.94 mm
5.6250 in	OE	142.88 mm
1.5000 in	SB	38.10 mm
3.8750 in	EC	98.43 mm
2.1250 in	HD	53.98 mm

ROCKWELL 58WBYS24-2

IDLI GROUP 81454

END YOKE WITH 10 SPLINES
SEE FIGURE 7A

5.3125 in	IE	134.94 mm
5.6250 in	OE	142.88 mm
1.5000 in	SB	38.10 mm
3.8750 in	EC	98.43 mm
2.2500 in	HD	57.15 mm

ROCKWELL 58WBYS24-1

BD=BEARING DIAMETER (outside)   **BW**=BEARING WIDTH   **CB**=CROSS LENGTH WITH BEARINGS   **CC**=CENTER TO CENTER   **CD**=CROSS DIAMETER
CL=CROSS LENGTH WITHOUT BEARINGS   **EC**=END TO CENTER or FACE TO FACE   **EE**=END TO END   **EL**=EFFECTIVE LENGTH
HD=HUB DIAMETER (or insert)   **IE**=INSIDE OF EARS (or recess)   **OD**=OUTSIDE DIAMETER   **OE**=OUTSIDE OF EARS   **SB**=SPLINE OR BORE SIZE

IDLI GROUP 81455

END YOKE WITH 10 SPLINES
SEE FIGURE 7A

5.3125 in	IE	134.94 mm
5.6250 in	OE	142.88 mm
1.5000 in	SB	38.10 mm
3.8750 in	EC	98.43 mm
2.6250 in	HD	66.68 mm

ROCKWELL 58WBYS24-5

IDLI GROUP 81461

END YOKE WITH ROUND BORE
SEE FIGURE 7B

5.3125 in	IE	134.94 mm
5.6250 in	OE	142.88 mm
1.5000 in	SB	38.10 mm
4.9375 in	EC	125.41 mm
4.0000 in	HD	101.60 mm

ROCKWELL 58WBYR24-6

IDLI GROUP 81467

END YOKE WITH ROUND BORE
SEE FIGURE 7B

5.3125 in	IE	134.94 mm
5.6250 in	OE	142.88 mm
1.6250 in	SB	41.28 mm
3.6875 in	EC	93.66 mm
2.8750 in	HD	73.03 mm

ROCKWELL 58WBYR26-1

IDLI GROUP 81473

END YOKE WITH ROUND BORE
SEE FIGURE 7B

5.3125 in	IE	134.94 mm
5.6250 in	OE	142.88 mm
1.6250 in	SB	41.28 mm
4.9375 in	EC	125.41 mm
4.0000 in	HD	101.60 mm

ROCKWELL 58WBYR26-3

IDLI GROUP 81456

END YOKE WITH ROUND BORE
SEE FIGURE 7B

5.3125 in	IE	134.94 mm
5.6250 in	OE	142.88 mm
1.5000 in	SB	38.10 mm
3.8750 in	EC	98.43 mm
2.6250 in	HD	66.68 mm

ROCKWELL 58WBYR24

IDLI GROUP 81462

END YOKE WITH ROUND BORE
slight variation from above group
SEE FIGURE 7B

5.3125 in	IE	134.94 mm
5.6250 in	OE	142.88 mm
1.5000 in	SB	38.10 mm
4.9375 in	EC	125.41 mm
4.0000 in	HD	101.60 mm

ROCKWELL 58WBYR24-13

IDLI GROUP 81468

END YOKE WITH 10 SPLINES
SEE FIGURE 7A

5.3125 in	IE	134.94 mm
5.6250 in	OE	142.88 mm
1.6250 in	SB	41.28 mm
3.7500 in	EC	95.25 mm
2.3750 in	HD	60.33 mm

ROCKWELL 58WBYS26-1

IDLI GROUP 81474

END YOKE WITH ROUND BORE
SEE FIGURE 7B

5.3125 in	IE	134.94 mm
5.6250 in	OE	142.88 mm
1.6875 in	SB	42.86 mm
3.6875 in	EC	93.66 mm
2.8750 in	HD	73.03 mm

ROCKWELL 58WBYR27

IDLI GROUP 81457

END YOKE WITH TAPERED BORE
SEE FIGURE 7F

5.3125 in	IE	134.94 mm
5.6250 in	OE	142.88 mm
1.5000 in	SB	38.10 mm
4.1875 in	EC	106.36 mm
2.2500 in	HD	57.15 mm

ROCKWELL 58WBYT24

IDLI GROUP 81463

END YOKE WITH ROUND BORE
SEE FIGURE 7B

5.3125 in	IE	134.94 mm
5.6250 in	OE	142.88 mm
1.5625 in	SB	39.69 mm
3.6875 in	EC	93.66 mm
2.8750 in	HD	73.03 mm

ROCKWELL 58WBYR25-2

IDLI GROUP 81469

END YOKE WITH ROUND BORE
SEE FIGURE 7B

5.3125 in	IE	134.94 mm
5.6250 in	OE	142.88 mm
1.6250 in	SB	41.28 mm
3.8125 in	EC	96.84 mm
2.6250 in	HD	66.68 mm

ROCKWELL 58WBYR26

IDLI GROUP 81475

END YOKE WITH ROUND BORE
SEE FIGURE 7B

5.3125 in	IE	134.94 mm
5.6250 in	OE	142.88 mm
1.6875 in	SB	42.86 mm
4.8125 in	EC	122.24 mm
4.0000 in	HD	101.60 mm

ROCKWELL 58WBYR27-1

IDLI GROUP 81458

END YOKE WITH ROUND BORE
SEE FIGURE 7B

5.3125 in	IE	134.94 mm
5.6250 in	OE	142.88 mm
1.5000 in	SB	38.10 mm
4.1875 in	EC	106.36 mm
4.1250 in	HD	104.78 mm

ROCKWELL 58WBYR24-3

IDLI GROUP 81464

END YOKE WITH ROUND BORE
SEE FIGURE 7B

5.3125 in	IE	134.94 mm
5.6250 in	OE	142.88 mm
1.5625 in	SB	39.69 mm
4.9375 in	EC	125.41 mm
4.0000 in	HD	101.60 mm

ROCKWELL 58WBYR25-3

IDLI GROUP 81470

END YOKE WITH 10 SPLINES
SEE FIGURE 7A

5.3125 in	IE	134.94 mm
5.6250 in	OE	142.88 mm
1.6250 in	SB	41.28 mm
3.8750 in	EC	98.43 mm
2.3750 in	HD	60.33 mm

ROCKWELL 58WBYS26-4

IDLI GROUP 81476

END YOKE WITH ROUND BORE
SEE FIGURE 7B

5.3125 in	IE	134.94 mm
5.6250 in	OE	142.88 mm
1.7188 in	SB	43.66 mm
4.5625 in	EC	115.89 mm
4.0000 in	HD	101.60 mm

ROCKWELL 58WBYR28-8

IDLI GROUP 81459

END YOKE WITH ROUND BORE
SEE FIGURE 7B

5.3125 in	IE	134.94 mm
5.6250 in	OE	142.88 mm
1.5000 in	SB	38.10 mm
4.5625 in	EC	115.89 mm
4.0000 in	HD	101.60 mm

ROCKWELL 58WBYR24-4

IDLI GROUP 81465

END YOKE WITH ROUND BORE
SEE FIGURE 7B

5.3125 in	IE	134.94 mm
5.6250 in	OE	142.88 mm
1.5781 in	SB	40.08 mm
4.1875 in	EC	106.36 mm
4.0625 in	HD	103.19 mm

ROCKWELL 58WBYR25-1

IDLI GROUP 81471

END YOKE WITH TAPERED BORE
SEE FIGURE 7F

5.3125 in	IE	134.94 mm
5.6250 in	OE	142.88 mm
1.6250 in	SB	41.28 mm
3.8750 in	EC	98.43 mm
2.6250 in	HD	66.68 mm

ROCKWELL 58WBYT26

IDLI GROUP 81477

END YOKE WITH 16 SPLINES
SEE FIGURE 7A

5.3125 in	IE	134.94 mm
5.6250 in	OE	142.88 mm
1.7500 in	SB	44.45 mm
1.7500 in	EC	44.45 mm
3.0000 in	HD	76.20 mm

ROCKWELL 58WBYS28-62

IDLI GROUP 81460

END YOKE WITH ROUND BORE
slight variation from above group
SEE FIGURE 7B

5.3125 in	IE	134.94 mm
5.6250 in	OE	142.88 mm
1.5000 in	SB	38.10 mm
4.5625 in	EC	115.89 mm
4.0000 in	HD	101.60 mm

ROCKWELL 58WBYR24-10

IDLI GROUP 81466

END YOKE WITH ROUND BORE
slight variation from above group
SEE FIGURE 7B

5.3125 in	IE	134.94 mm
5.6250 in	OE	142.88 mm
1.5781 in	SB	40.08 mm
4.1875 in	EC	106.36 mm
4.0625 in	HD	103.19 mm

ROCKWELL 58WBYR25

IDLI GROUP 81472

END YOKE WITH ROUND BORE
SEE FIGURE 7B

5.3125 in	IE	134.94 mm
5.6250 in	OE	142.88 mm
1.6250 in	SB	41.28 mm
4.5625 in	EC	115.89 mm
4.0000 in	HD	101.60 mm

ROCKWELL 58WBYR26-2

IDLI GROUP 81478

END YOKE WITH 10 SPLINES
SEE FIGURE 7A

5.3125 in	IE	134.94 mm
5.6250 in	OE	142.88 mm
1.7500 in	SB	44.45 mm
2.7500 in	EC	69.85 mm
2.8125 in	HD	71.44 mm

ROCKWELL 58WBYS28-59

ENGINEERING CATALOGS MUST BE CONSULTED FOR DETAILS NOT INCLUDED
IN THIS GUIDE. SPECIFIC DESIGNS, MATERIAL CONTENT, TOLERANCES
LUBE FITTINGS AND OTHER DIMENSIONS ARE INTENTIONALLY OMITTED HERE.

CAUTION: BE SURE TO REFER TO ENGINEERING CATALOGS FOR SPECIAL APPLICATIONS THAT REQUIRE SPECIFIC MATERIAL CONTENT, TOLERANCES, ETC. SEE FOOTNOTE.

IDLI GROUP 81479

END YOKE WITH 10 SPLINES
slight variation from above group
SEE FIGURE 7A

5.3125 in	IE	134.94 mm
5.6250 in	OE	142.88 mm
1.7500 in	SB	44.45 mm
2.7500 in	EC	69.85 mm
2.8125 in	HD	71.44 mm

ROCKWELL 58WBYS28-21

IDLI GROUP 81480

END YOKE WITH 16 SPLINES
SEE FIGURE 7A

5.3125 in	IE	134.94 mm
5.6250 in	OE	142.88 mm
1.7500 in	SB	44.45 mm
2.7500 in	EC	69.85 mm
2.8125 in	HD	71.44 mm

ROCKWELL 58WBYS28-72

IDLI GROUP 81481

END YOKE WITH ROUND BORE
SEE FIGURE 7B

5.3125 in	IE	134.94 mm
5.6250 in	OE	142.88 mm
1.7500 in	SB	44.45 mm
3.1875 in	EC	80.96 mm
3.0000 in	HD	76.20 mm

ROCKWELL 58WBYR28-7

IDLI GROUP 81482

END YOKE WITH 16 SPLINES
SEE FIGURE 7A

5.3125 in	IE	134.94 mm
5.6250 in	OE	142.88 mm
1.7500 in	SB	44.45 mm
3.5000 in	EC	88.90 mm
3.2500 in	HD	82.55 mm

ROCKWELL 58WBYS28-7

IDLI GROUP 81483

END YOKE WITH 10 SPLINES
SEE FIGURE 7A

5.3125 in	IE	134.94 mm
5.6250 in	OE	142.88 mm
1.7500 in	SB	44.45 mm
3.8125 in	EC	96.84 mm
2.3750 in	HD	60.33 mm

ROCKWELL 58WBYS28-9

IDLI GROUP 81484

END YOKE WITH 10 SPLINES
slight variation from above group
SEE FIGURE 7A

5.3125 in	IE	134.94 mm
5.6250 in	OE	142.88 mm
1.7500 in	SB	44.45 mm
3.8125 in	EC	96.84 mm
2.3750 in	HD	60.33 mm

ROCKWELL 58WBYS28

IDLI GROUP 81485

END YOKE WITH 10 SPLINES
slight variation from above group
SEE FIGURE 7A

5.3125 in	IE	134.94 mm
5.6250 in	OE	142.88 mm
1.7500 in	SB	44.45 mm
3.8125 in	EC	96.84 mm
2.3750 in	HD	60.33 mm

ROCKWELL 58WBYS28-2

IDLI GROUP 81486

END YOKE WITH 10 SPLINES
SEE FIGURE 7A

5.3125 in	IE	134.94 mm
5.6250 in	OE	142.88 mm
1.7500 in	SB	44.45 mm
3.8125 in	EC	96.84 mm
2.4375 in	HD	61.91 mm

ROCKWELL 58WBYS28-6

IDLI GROUP 81487

END YOKE WITH 10 SPLINES
SEE FIGURE 7A

5.3125 in	IE	134.94 mm
5.6250 in	OE	142.88 mm
1.7500 in	SB	44.45 mm
3.8750 in	EC	98.43 mm
2.2500 in	HD	57.15 mm

ROCKWELL 58WBYS28-10

IDLI GROUP 81488

END YOKE WITH 10 SPLINES
SEE FIGURE 7A

5.3125 in	IE	134.94 mm
5.6250 in	OE	142.88 mm
1.7500 in	SB	44.45 mm
3.8750 in	EC	98.43 mm
2.5000 in	HD	63.50 mm

ROCKWELL 58WBYS28-71

IDLI GROUP 81489

END YOKE WITH 10 SPLINES
SEE FIGURE 7A

5.3125 in	IE	134.94 mm
5.6250 in	OE	142.88 mm
1.7500 in	SB	44.45 mm
3.8750 in	EC	98.43 mm
2.6250 in	HD	66.68 mm

ROCKWELL 58WBYS28-61

IDLI GROUP 81490

END YOKE WITH 10 SPLINES
slight variation from above group
SEE FIGURE 7A

5.3125 in	IE	134.94 mm
5.6250 in	OE	142.88 mm
1.7500 in	SB	44.45 mm
3.8750 in	EC	98.43 mm
2.6250 in	HD	66.68 mm

ROCKWELL 58WBYS28-1

IDLI GROUP 81491

END YOKE WITH 10 SPLINES
slight variation from above group
SEE FIGURE 7A

5.3125 in	IE	134.94 mm
5.6250 in	OE	142.88 mm
1.7500 in	SB	44.45 mm
3.8750 in	EC	98.43 mm
2.6250 in	HD	66.68 mm

ROCKWELL 58WBYS28-14

IDLI GROUP 81492

END YOKE WITH 10 SPLINES
slight variation from above group
SEE FIGURE 7A

5.3125 in	IE	134.94 mm
5.6250 in	OE	142.88 mm
1.7500 in	SB	44.45 mm
3.8750 in	EC	98.43 mm
2.6250 in	HD	66.68 mm

ROCKWELL 58WBYS28-51

IDLI GROUP 81493

END YOKE WITH TAPERED BORE
SEE FIGURE 7F

5.3125 in	IE	134.94 mm
5.6250 in	OE	142.88 mm
1.7500 in	SB	44.45 mm
3.8750 in	EC	98.43 mm
2.6250 in	HD	66.68 mm

ROCKWELL 58WBYT28

IDLI GROUP 81494

END YOKE WITH 10 SPLINES
SEE FIGURE 7A

5.3125 in	IE	134.94 mm
5.6250 in	OE	142.88 mm
1.7500 in	SB	44.45 mm
3.8750 in	EC	98.43 mm
2.7500 in	HD	69.85 mm

ROCKWELL 58WBYS28-3

IDLI GROUP 81495

END YOKE WITH 10 SPLINES
slight variation from above group
SEE FIGURE 7A

5.3125 in	IE	134.94 mm
5.6250 in	OE	142.88 mm
1.7500 in	SB	44.45 mm
3.8750 in	EC	98.43 mm
2.7500 in	HD	69.85 mm

ROCKWELL 58WBYS28-33

IDLI GROUP 81496

END YOKE WITH 10 SPLINES
slight variation from above group
SEE FIGURE 7A

5.3125 in	IE	134.94 mm
5.6250 in	OE	142.88 mm
1.7500 in	SB	44.45 mm
3.8750 in	EC	98.43 mm
2.7500 in	HD	69.85 mm

ROCKWELL 58WBYS28-67

IDLI GROUP 81497

END YOKE WITH 10 SPLINES
SEE FIGURE 7A

5.3125 in	IE	134.94 mm
5.6250 in	OE	142.88 mm
1.7500 in	SB	44.45 mm
3.8750 in	EC	98.43 mm
3.0000 in	HD	76.20 mm

ROCKWELL 58WBYS28-79

IDLI GROUP 81498

END YOKE WITH 10 SPLINES
SEE FIGURE 7A

5.3125 in	IE	134.94 mm
5.6250 in	OE	142.88 mm
1.7500 in	SB	44.45 mm
3.9063 in	EC	99.22 mm
2.5000 in	HD	63.50 mm

ROCKWELL 58WBYS28-82

IDLI GROUP 81499

END YOKE WITH ROUND BORE
SEE FIGURE 7B

5.3125 in	IE	134.94 mm
5.6250 in	OE	142.88 mm
1.7500 in	SB	44.45 mm
4.1875 in	EC	106.36 mm
4.0625 in	HD	103.19 mm

ROCKWELL 58WBYR28-2

IDLI GROUP 81500

END YOKE WITH ROUND BORE
slight variation from above group
SEE FIGURE 7B

5.3125 in	IE	134.94 mm
5.6250 in	OE	142.88 mm
1.7500 in	SB	44.45 mm
4.1875 in	EC	106.36 mm
4.0625 in	HD	103.19 mm

ROCKWELL 58WBYR28-3

IDLI GROUP 81501

END YOKE WITH ROUND BORE
slight variation from above group
SEE FIGURE 7B

5.3125 in	IE	134.94 mm
5.6250 in	OE	142.88 mm
1.7500 in	SB	44.45 mm
4.1875 in	EC	106.36 mm
4.0625 in	HD	103.19 mm

ROCKWELL 58WBYR28-4

IDLI GROUP 81502

END YOKE WITH ROUND BORE
slight variation from above group
SEE FIGURE 7B

5.3125 in	IE	134.94 mm
5.6250 in	OE	142.88 mm
1.7500 in	SB	44.45 mm
4.1875 in	EC	106.36 mm
4.0625 in	HD	103.19 mm

ROCKWELL 58WBYR28-5

BD = BEARING DIAMETER (outside)   **BW** = BEARING WIDTH   **CB** = CROSS LENGTH WITH BEARINGS   **CC** = CENTER TO CENTER   **CD** = CROSS DIAMETER
CL = CROSS LENGTH WITHOUT BEARINGS   **EC** = END TO CENTER or FACE TO FACE   **EE** = END TO END   **EL** = EFFECTIVE LENGTH
HD = HUB DIAMETER (or insert)   **IE** = INSIDE OF EARS (or recess)   **OD** = OUTSIDE DIAMETER   **OE** = OUTSIDE OF EARS   **SB** = SPLINE OR BORE SIZE

IDLI GROUP 81503

END YOKE WITH ROUND BORE
SEE FIGURE 7B

5.3125 in	IE	134.94 mm
5.6250 in	OE	142.88 mm
1.7500 in	SB	44.45 mm
4.5625 in	EC	115.89 mm
4.0000 in	HD	101.60 mm

ROCKWELL 58WBYR28

IDLI GROUP 81504

END YOKE WITH ROUND BORE
slight variation from above group
SEE FIGURE 7B

5.3125 in	IE	134.94 mm
5.6250 in	OE	142.88 mm
1.7500 in	SB	44.45 mm
4.5625 in	EC	115.89 mm
4.0000 in	HD	101.60 mm

ROCKWELL 58WBYR28-11

IDLI GROUP 81505

END YOKE WITH TAPERED BORE
SEE FIGURE 7F

5.3125 in	IE	134.94 mm
5.6250 in	OE	142.88 mm
1.7500 in	SB	44.45 mm
6.7500 in	EC	171.45 mm
2.8125 in	HD	71.44 mm

ROCKWELL 58WBYT28-1

IDLI GROUP 81506

END YOKE WITH 16 INVOLUTED SPLINES
SEE FIGURE 7A

5.3125 in	IE	134.94 mm
5.6250 in	OE	142.88 mm
1.7656 in	SB	44.85 mm
3.8750 in	EC	98.43 mm
2.6250 in	HD	66.68 mm

ROCKWELL 58WBYS28-27

IDLI GROUP 81507

END YOKE WITH ROUND BORE
SEE FIGURE 7B

5.3125 in	IE	134.94 mm
5.6250 in	OE	142.88 mm
1.8750 in	SB	47.63 mm
3.7500 in	EC	95.25 mm
3.8750 in	HD	98.43 mm

ROCKWELL 58WBYR30-1

IDLI GROUP 81508

END YOKE WITH 10 SPLINES
SEE FIGURE 7A

5.3125 in	IE	134.94 mm
5.6250 in	OE	142.88 mm
1.8750 in	SB	47.63 mm
4.0000 in	EC	101.60 mm
2.7500 in	HD	69.85 mm

ROCKWELL 58WBYS30-2

IDLI GROUP 81509

END YOKE WITH 10 SPLINES
slight variation from above group
SEE FIGURE 7A

5.3125 in	IE	134.94 mm
5.6250 in	OE	142.88 mm
1.8750 in	SB	47.63 mm
4.0000 in	EC	101.60 mm
2.7500 in	HD	69.85 mm

ROCKWELL 58WBYS30

IDLI GROUP 81510

END YOKE WITH ROUND BORE
SEE FIGURE 7B

5.3125 in	IE	134.94 mm
5.6250 in	OE	142.88 mm
1.8750 in	SB	47.63 mm
4.3125 in	EC	109.54 mm
4.0000 in	HD	101.60 mm

ROCKWELL 58WBYR30-5

IDLI GROUP 81511

END YOKE WITH ROUND BORE
SEE FIGURE 7B

5.3125 in	IE	134.94 mm
5.6250 in	OE	142.88 mm
1.8750 in	SB	47.63 mm
4.3750 in	EC	111.13 mm
4.0000 in	HD	101.60 mm

ROCKWELL 58WBYR30-2

IDLI GROUP 81512

END YOKE WITH ROUND BORE
SEE FIGURE 7B

5.3125 in	IE	134.94 mm
5.6250 in	OE	142.88 mm
1.8750 in	SB	47.63 mm
4.5625 in	EC	115.89 mm
4.0000 in	HD	101.60 mm

ROCKWELL 58WBYR30

IDLI GROUP 81513

END YOKE WITH ROUND BORE
SEE FIGURE 7B

5.3125 in	IE	134.94 mm
5.6250 in	OE	142.88 mm
1.8750 in	SB	47.63 mm
4.8125 in	EC	122.24 mm
4.0000 in	HD	101.60 mm

ROCKWELL 58WBYR30-4

IDLI GROUP 81514

END YOKE WITH ROUND BORE
SEE FIGURE 7B

5.3125 in	IE	134.94 mm
5.6250 in	OE	142.88 mm
1.8750 in	SB	47.63 mm
4.9375 in	EC	125.41 mm
4.0000 in	HD	101.60 mm

ROCKWELL 58WBYR30-6

IDLI GROUP 81515

END YOKE WITH TAPERED BORE
SEE FIGURE 7F

5.3125 in	IE	134.94 mm
5.6250 in	OE	142.88 mm
1.9219 in	SB	48.82 mm
4.0000 in	EC	101.60 mm
2.7500 in	HD	69.85 mm

ROCKWELL 58WBYT31-1

IDLI GROUP 81516

END YOKE WITH ROUND BORE
SEE FIGURE 7B

5.3125 in	IE	134.94 mm
5.6250 in	OE	142.88 mm
1.9375 in	SB	49.21 mm
3.6875 in	EC	93.66 mm
2.8750 in	HD	73.03 mm

ROCKWELL 58WBYR31

IDLI GROUP 81517

END YOKE WITH ROUND BORE
SEE FIGURE 7B

5.3125 in	IE	134.94 mm
5.6250 in	OE	142.88 mm
1.9375 in	SB	49.21 mm
3.8750 in	EC	98.43 mm
2.8125 in	HD	71.44 mm

ROCKWELL 58WBYR31-5

IDLI GROUP 81518

END YOKE WITH ROUND BORE
SEE FIGURE 7B

5.3125 in	IE	134.94 mm
5.6250 in	OE	142.88 mm
1.9375 in	SB	49.21 mm
4.1875 in	EC	106.36 mm
4.0625 in	HD	103.19 mm

ROCKWELL 58WBYR31-3

IDLI GROUP 81519

END YOKE WITH ROUND BORE
slight variation from above group
SEE FIGURE 7B

5.3125 in	IE	134.94 mm
5.6250 in	OE	142.88 mm
1.9375 in	SB	49.21 mm
4.1875 in	EC	106.36 mm
4.0625 in	HD	103.19 mm

ROCKWELL 58WBYR31-1

IDLI GROUP 81520

END YOKE WITH ROUND BORE
slight variation from above group
SEE FIGURE 7B

5.3125 in	IE	134.94 mm
5.6250 in	OE	142.88 mm
1.9375 in	SB	49.21 mm
4.1875 in	EC	106.36 mm
4.0625 in	HD	103.19 mm

ROCKWELL 58WBYR31-4

IDLI GROUP 81521

END YOKE WITH ROUND BORE
SEE FIGURE 7B

5.3125 in	IE	134.94 mm
5.6250 in	OE	142.88 mm
1.9375 in	SB	49.21 mm
4.3750 in	EC	111.13 mm
4.0000 in	HD	101.60 mm

ROCKWELL 58WBYR31-2

IDLI GROUP 81522

END YOKE WITH ROUND BORE
SEE FIGURE 7B

5.3125 in	IE	134.94 mm
5.6250 in	OE	142.88 mm
1.9375 in	SB	49.21 mm
4.5625 in	EC	115.89 mm
4.0000 in	HD	101.60 mm

ROCKWELL 58WBYR31-6

IDLI GROUP 81523

END YOKE WITH ROUND BORE
SEE FIGURE 7B

5.3125 in	IE	134.94 mm
5.6250 in	OE	142.88 mm
1.9375 in	SB	49.21 mm
4.9375 in	EC	125.41 mm
4.0000 in	HD	101.60 mm

ROCKWELL 58WBYR31-7

IDLI GROUP 81524

END YOKE WITH 10 SPLINES
SEE FIGURE 7A

5.3125 in	IE	134.94 mm
5.6250 in	OE	142.88 mm
1.9688 in	SB	50.01 mm
3.8750 in	EC	98.43 mm
2.6250 in	HD	66.68 mm

ROCKWELL 58WBYS31-19

IDLI GROUP 81525

END YOKE WITH 10 SPLINES
slight variation from above group
SEE FIGURE 7A

5.3125 in	IE	134.94 mm
5.6250 in	OE	142.88 mm
1.9688 in	SB	50.01 mm
3.8750 in	EC	98.43 mm
2.6250 in	HD	66.68 mm

ROCKWELL 58WBYS31-32

IDLI GROUP 81526

END YOKE WITH 10 SPLINES
slight variation from above group
SEE FIGURE 7A

5.3125 in	IE	134.94 mm
5.6250 in	OE	142.88 mm
1.9688 in	SB	50.01 mm
3.8750 in	EC	98.43 mm
2.6250 in	HD	66.68 mm

ROCKWELL 58WBYS31-21

ENGINEERING CATALOGS MUST BE CONSULTED FOR DETAILS NOT INCLUDED
IN THIS GUIDE. SPECIFIC DESIGNS, MATERIAL CONTENT, TOLERANCES
LUBE FITTINGS AND OTHER DIMENSIONS ARE INTENTIONALLY OMITTED HERE.

CAUTION: BE SURE TO REFER TO ENGINEERING CATALOGS FOR SPECIAL APPLICATIONS THAT REQUIRE SPECIFIC MATERIAL CONTENT, TOLERANCES, ETC. SEE FOOTNOTE.

IDLI GROUP 81527

END YOKE WITH 10 SPLINES
SEE FIGURE 7A

5.3125 in	IE	134.94 mm
5.6250 in	OE	142.88 mm
1.9688 in	SB	50.01 mm
3.8750 in	EC	98.43 mm
2.8750 in	HD	73.03 mm

ROCKWELL 58WBYS31-3

IDLI GROUP 81528

END YOKE WITH 10 SPLINES
SEE FIGURE 7A

5.3125 in	IE	134.94 mm
5.6250 in	OE	142.88 mm
1.9688 in	SB	50.01 mm
3.8750 in	EC	98.43 mm
2.9375 in	HD	74.61 mm

ROCKWELL 58WBYS31-4

IDLI GROUP 81529

END YOKE WITH 6 SPLINES
SEE FIGURE 7A

5.3125 in	IE	134.94 mm
5.6250 in	OE	142.88 mm
1.9688 in	SB	50.01 mm
4.0000 in	EC	101.60 mm
2.7344 in	HD	69.45 mm

ROCKWELL 58WBYS31-13

IDLI GROUP 81530

END YOKE WITH 10 SPLINES
SEE FIGURE 7A

5.3125 in	IE	134.94 mm
5.6250 in	OE	142.88 mm
1.9688 in	SB	50.01 mm
4.0000 in	EC	101.60 mm
2.7500 in	HD	69.85 mm

ROCKWELL 58WBYS31-20

IDLI GROUP 81531

END YOKE WITH ROUND BORE
SEE FIGURE 7B

5.3125 in	IE	134.94 mm
5.6250 in	OE	142.88 mm
1.9688 in	SB	50.01 mm
4.1875 in	EC	106.36 mm
4.0625 in	HD	103.19 mm

ROCKWELL 58WBYR32-22

IDLI GROUP 81532

END YOKE WITH 10 SPLINES
SEE FIGURE 7A

5.3125 in	IE	134.94 mm
5.6250 in	OE	142.88 mm
1.9688 in	SB	50.01 mm
4.2500 in	EC	107.95 mm
2.7500 in	HD	69.85 mm

ROCKWELL 58WBYS31-17

IDLI GROUP 81533

END YOKE WITH 10 SPLINES
slight variation from above group
SEE FIGURE 7A

5.3125 in	IE	134.94 mm
5.6250 in	OE	142.88 mm
1.9688 in	SB	50.01 mm
4.2500 in	EC	107.95 mm
2.7500 in	HD	69.85 mm

ROCKWELL 58WBYS31-15

IDLI GROUP 81534

END YOKE WITH 15 INVOLUTED SPLINES
SEE FIGURE 7A

5.3125 in	IE	134.94 mm
5.6250 in	OE	142.88 mm
2.0000 in	SB	50.80 mm
3.0000 in	EC	76.20 mm
2.7500 in	HD	69.85 mm

ROCKWELL 58WBYS32-11

IDLI GROUP 81535

END YOKE WITH ROUND BORE
SEE FIGURE 7B

5.3125 in	IE	134.94 mm
5.6250 in	OE	142.88 mm
2.0000 in	SB	50.80 mm
3.3750 in	EC	85.73 mm
2.9375 in	HD	74.61 mm

ROCKWELL 58WBYR32-13

IDLI GROUP 81536

END YOKE WITH ROUND BORE
SEE FIGURE 7B

5.3125 in	IE	134.94 mm
5.6250 in	OE	142.88 mm
2.0000 in	SB	50.80 mm
3.4375 in	EC	87.31 mm
2.9375 in	HD	74.61 mm

ROCKWELL 58WBYR32-11

IDLI GROUP 81537

END YOKE WITH 15 INVOLUTED SPLINES
SEE FIGURE 7A

5.3125 in	IE	134.94 mm
5.6250 in	OE	142.88 mm
2.0000 in	SB	50.80 mm
3.5625 in	EC	90.49 mm
2.6250 in	HD	66.68 mm

ROCKWELL 58WBYS32-8

IDLI GROUP 81538

END YOKE WITH ROUND BORE
SEE FIGURE 7B

5.3125 in	IE	134.94 mm
5.6250 in	OE	142.88 mm
2.0000 in	SB	50.80 mm
3.6250 in	EC	92.08 mm
3.2500 in	HD	82.55 mm

ROCKWELL 58WBYR32-5

IDLI GROUP 81539

END YOKE WITH ROUND BORE
SEE FIGURE 7B

5.3125 in	IE	134.94 mm
5.6250 in	OE	142.88 mm
2.0000 in	SB	50.80 mm
3.8125 in	EC	96.84 mm
3.0000 in	HD	76.20 mm

ROCKWELL 58WBYR32-7

IDLI GROUP 81540

END YOKE WITH 10 SPLINES
SEE FIGURE 7A

5.3125 in	IE	134.94 mm
5.6250 in	OE	142.88 mm
2.0000 in	SB	50.80 mm
3.8750 in	EC	98.43 mm
2.7500 in	HD	69.85 mm

ROCKWELL 58WBYS32-21

IDLI GROUP 81541

END YOKE WITH 10 SPLINES
SEE FIGURE 7A

5.3125 in	IE	134.94 mm
5.6250 in	OE	142.88 mm
2.0000 in	SB	50.80 mm
3.9375 in	EC	100.01 mm
2.9375 in	HD	74.61 mm

ROCKWELL 58WBYS32-13

IDLI GROUP 81542

END YOKE WITH ROUND BORE
slight variation from above group
SEE FIGURE 7B

5.3125 in	IE	134.94 mm
5.6250 in	OE	142.88 mm
2.0000 in	SB	50.80 mm
4.1875 in	EC	106.36 mm
4.0625 in	HD	103.19 mm

ROCKWELL 58WBYR32-2

IDLI GROUP 81543

END YOKE WITH ROUND BORE
SEE FIGURE 7B

5.3125 in	IE	134.94 mm
5.6250 in	OE	142.88 mm
2.0000 in	SB	50.80 mm
4.1875 in	EC	106.36 mm
4.0625 in	HD	103.19 mm

ROCKWELL 58WBYR32-10

IDLI GROUP 81544

END YOKE WITH ROUND BORE
slight variation from above group
SEE FIGURE 7B

5.3125 in	IE	134.94 mm
5.6250 in	OE	142.88 mm
2.0000 in	SB	50.80 mm
4.1875 in	EC	106.36 mm
4.0625 in	HD	103.19 mm

ROCKWELL 58WBYR32-3

IDLI GROUP 81545

END YOKE WITH 10 SPLINES
SEE FIGURE 7A

5.3125 in	IE	134.94 mm
5.6250 in	OE	142.88 mm
2.0000 in	SB	50.80 mm
4.5625 in	EC	115.89 mm
3.2500 in	HD	82.55 mm

ROCKWELL 58WBYS32-25

IDLI GROUP 81546

END YOKE WITH ROUND BORE
SEE FIGURE 7B

5.3125 in	IE	134.94 mm
5.6250 in	OE	142.88 mm
2.0000 in	SB	50.80 mm
4.5625 in	EC	115.89 mm
4.0000 in	HD	101.60 mm

ROCKWELL 58WBYR32

IDLI GROUP 81547

END YOKE WITH ROUND BORE
SEE FIGURE 7B

5.3125 in	IE	134.94 mm
5.6250 in	OE	142.88 mm
2.0000 in	SB	50.80 mm
4.6250 in	EC	117.48 mm
3.2344 in	HD	82.15 mm

ROCKWELL 58WBYR32-12

IDLI GROUP 81548

END YOKE WITH ROUND BORE
SEE FIGURE 7B

5.3125 in	IE	134.94 mm
5.6250 in	OE	142.88 mm
2.0000 in	SB	50.80 mm
4.6250 in	EC	117.48 mm
4.0000 in	HD	101.60 mm

ROCKWELL 58WBYR32-4

IDLI GROUP 81549

END YOKE WITH 10 SPLINES
SEE FIGURE 7A

5.3125 in	IE	134.94 mm
5.6250 in	OE	142.88 mm
2.0000 in	SB	50.80 mm
4.6875 in	EC	119.06 mm
2.9375 in	HD	74.61 mm

ROCKWELL 58WBYS32

IDLI GROUP 81550

END YOKE WITH ROUND BORE
SEE FIGURE 7B

5.3125 in	IE	134.94 mm
5.6250 in	OE	142.88 mm
2.0000 in	SB	50.80 mm
4.8750 in	EC	123.83 mm
4.0000 in	HD	101.60 mm

ROCKWELL 58WBYR32-16

BD=BEARING DIAMETER (outside)   **BW**=BEARING WIDTH   **CB**=CROSS LENGTH WITH BEARINGS   **CC**=CENTER TO CENTER   **CD**=CROSS DIAMETER
CL=CROSS LENGTH WITHOUT BEARINGS   **EC**=END TO CENTER or FACE TO FACE   **EE**=END TO END   **EL**=EFFECTIVE LENGTH
HD=HUB DIAMETER (or insert)   **IE**=INSIDE OF EARS (or recess)   **OD**=OUTSIDE DIAMETER   **OE**=OUTSIDE OF EARS   **SB**=SPLINE OR BORE SIZE

IDLI GROUP 81551

END YOKE WITH ROUND BORE
SEE FIGURE 7B

5.3125 in	IE	134.94 mm
5.6250 in	OE	142.88 mm
2.0000 in	SB	50.80 mm
4.9375 in	EC	125.41 mm
4.0000 in	HD	101.60 mm

ROCKWELL 58WBYR32-9

IDLI GROUP 81552

END YOKE WITH ROUND BORE
slight variation from above group
SEE FIGURE 7B

5.3125 in	IE	134.94 mm
5.6250 in	OE	142.88 mm
2.0000 in	SB	50.80 mm
4.9375 in	EC	125.41 mm
4.0000 in	HD	101.60 mm

ROCKWELL 58WBYR32-21

IDLI GROUP 81553

END YOKE WITH ROUND BORE
slight variation from above group
SEE FIGURE 7B

5.3125 in	IE	134.94 mm
5.6250 in	OE	142.88 mm
2.0000 in	SB	50.80 mm
4.9375 in	EC	125.41 mm
4.0000 in	HD	101.60 mm

ROCKWELL 58WBYR32-1

IDLI GROUP 81554

END YOKE WITH ROUND BORE
slight variation from above group
SEE FIGURE 7B

5.3125 in	IE	134.94 mm
5.6250 in	OE	142.88 mm
2.0000 in	SB	50.80 mm
4.9375 in	EC	125.41 mm
4.0000 in	HD	101.60 mm

ROCKWELL 58WBYR32-17

IDLI GROUP 81555

END YOKE WITH 10 SPLINES
SEE FIGURE 7A

5.3125 in	IE	134.94 mm
5.6250 in	OE	142.88 mm
2.0313 in	SB	51.60 mm
3.8750 in	EC	98.43 mm
2.7500 in	HD	69.85 mm

ROCKWELL 58WBYS33

IDLI GROUP 81556

END YOKE WITH ROUND BORE
SEE FIGURE 7B

5.3125 in	IE	134.94 mm
5.6250 in	OE	142.88 mm
2.1250 in	SB	53.98 mm
4.5625 in	EC	115.89 mm
4.0000 in	HD	101.60 mm

ROCKWELL 58WBYR34-2

IDLI GROUP 81557

END YOKE WITH ROUND BORE
slight variation from above group
SEE FIGURE 7B

5.3125 in	IE	134.94 mm
5.6250 in	OE	142.88 mm
2.1250 in	SB	53.98 mm
4.5625 in	EC	115.89 mm
4.0000 in	HD	101.60 mm

ROCKWELL 58WBYR34-5

IDLI GROUP 81558

END YOKE WITH ROUND BORE
SEE FIGURE 7B

5.3125 in	IE	134.94 mm
5.6250 in	OE	142.88 mm
2.1250 in	SB	53.98 mm
4.8750 in	EC	123.83 mm
4.0000 in	HD	101.60 mm

ROCKWELL 58WBYR34-1

IDLI GROUP 81559

END YOKE WITH ROUND BORE
slight variation from above group
SEE FIGURE 7B

5.3125 in	IE	134.94 mm
5.6250 in	OE	142.88 mm
2.1250 in	SB	53.98 mm
4.8750 in	EC	123.83 mm
4.0000 in	HD	101.60 mm

ROCKWELL 58WBYR34-8

IDLI GROUP 81560

END YOKE WITH ROUND BORE
SEE FIGURE 7B

5.3125 in	IE	134.94 mm
5.6250 in	OE	142.88 mm
2.1250 in	SB	53.98 mm
4.9375 in	EC	125.41 mm
4.0000 in	HD	101.60 mm

ROCKWELL 58WBYR34-4

IDLI GROUP 81561

END YOKE WITH ROUND BORE
SEE FIGURE 7B

5.3125 in	IE	134.94 mm
5.6250 in	OE	142.88 mm
2.1875 in	SB	55.56 mm
4.6250 in	EC	117.48 mm
4.0000 in	HD	101.60 mm

ROCKWELL 58WBYR35-1

IDLI GROUP 81562

END YOKE WITH ROUND BORE
SEE FIGURE 7B

5.3125 in	IE	134.94 mm
5.6250 in	OE	142.88 mm
2.1875 in	SB	55.56 mm
4.8750 in	EC	123.83 mm
4.0000 in	HD	101.60 mm

ROCKWELL 58WBYR35

IDLI GROUP 81563

END YOKE WITH ROUND BORE
slight variation from above group
SEE FIGURE 7B

5.3125 in	IE	134.94 mm
5.6250 in	OE	142.88 mm
2.1875 in	SB	55.56 mm
4.8750 in	EC	123.83 mm
4.0000 in	HD	101.60 mm

ROCKWELL 58WBYR35-4

IDLI GROUP 81564

END YOKE WITH ROUND BORE
slight variation from above group
SEE FIGURE 7B

5.3125 in	IE	134.94 mm
5.6250 in	OE	142.88 mm
2.1875 in	SB	55.56 mm
4.8750 in	EC	123.83 mm
4.0000 in	HD	101.60 mm

ROCKWELL 58WBYR35-2

IDLI GROUP 81565

END YOKE WITH ROUND BORE
SEE FIGURE 7B

5.3125 in	IE	134.94 mm
5.6250 in	OE	142.88 mm
2.1875 in	SB	55.56 mm
4.9375 in	EC	125.41 mm
4.0000 in	HD	101.60 mm

ROCKWELL 58WBYR35-3

IDLI GROUP 81566

END YOKE WITH ROUND BORE
slight variation from above group
SEE FIGURE 7B

5.3125 in	IE	134.94 mm
5.6250 in	OE	142.88 mm
2.1875 in	SB	55.56 mm
4.9375 in	EC	125.41 mm
4.0000 in	HD	101.60 mm

ROCKWELL 58WBYR35-5

IDLI GROUP 81567

END YOKE WITH 10 SPLINES
SEE FIGURE 7A

5.3125 in	IE	134.94 mm
5.6250 in	OE	142.88 mm
2.2500 in	SB	57.15 mm
3.8750 in	EC	98.43 mm
3.1250 in	HD	79.38 mm

ROCKWELL 58WBYS36

IDLI GROUP 81568

END YOKE WITH TAPERED BORE
SEE FIGURE 7F

5.3125 in	IE	134.94 mm
5.6250 in	OE	142.88 mm
2.3438 in	SB	59.53 mm
4.0000 in	EC	101.60 mm
3.2500 in	HD	82.55 mm

ROCKWELL 58WBYT37-1

IDLI GROUP 81569

END YOKE WITH ROUND BORE
SEE FIGURE 7B

5.3125 in	IE	134.94 mm
5.6250 in	OE	142.88 mm
2.2500 in	SB	57.15 mm
4.1250 in	EC	104.78 mm
4.0625 in	HD	103.19 mm

ROCKWELL 58WBYR36-7

IDLI GROUP 81570

END YOKE WITH ROUND BORE
slight variation from above group
SEE FIGURE 7B

5.3125 in	IE	134.94 mm
5.6250 in	OE	142.88 mm
2.2500 in	SB	57.15 mm
4.1875 in	EC	106.36 mm
4.0625 in	HD	103.19 mm

ROCKWELL 58WBYR36-3

IDLI GROUP 81571

END YOKE WITH ROUND BORE
slight variation from above group
SEE FIGURE 7B

5.3125 in	IE	134.94 mm
5.6250 in	OE	142.88 mm
2.2500 in	SB	57.15 mm
4.1875 in	EC	106.36 mm
4.0625 in	HD	103.19 mm

ROCKWELL 58WBYR36-6

IDLI GROUP 81572

END YOKE WITH ROUND BORE
slight variation from above group
SEE FIGURE 7B

5.3125 in	IE	134.94 mm
5.6250 in	OE	142.88 mm
2.2500 in	SB	57.15 mm
4.1875 in	EC	106.36 mm
4.0625 in	HD	103.19 mm

ROCKWELL 58WBYR36-10

IDLI GROUP 81573

END YOKE WITH ROUND BORE
SEE FIGURE 7B

5.3125 in	IE	134.94 mm
5.6250 in	OE	142.88 mm
2.2500 in	SB	57.15 mm
4.3750 in	EC	111.13 mm
4.0000 in	HD	101.60 mm

ROCKWELL 58WBYR36-4

IDLI GROUP 81574

END YOKE WITH ROUND BORE
SEE FIGURE 7B

5.3125 in	IE	134.94 mm
5.6250 in	OE	142.88 mm
2.2500 in	SB	57.15 mm
4.5625 in	EC	115.89 mm
4.0000 in	HD	101.60 mm

ROCKWELL 58WBYR36-14

ENGINEERING CATALOGS MUST BE CONSULTED FOR DETAILS NOT INCLUDED
IN THIS GUIDE. SPECIFIC DESIGNS, MATERIAL CONTENT, TOLERANCES
LUBE FITTINGS AND OTHER DIMENSIONS ARE INTENTIONALLY OMITTED HERE.

IDLI GUIDE   COPYRIGHT © INTERCHANGE, INC.   ST. LOUIS PARK, MN. 55416 USA

CAUTION: BE SURE TO REFER TO ENGINEERING CATALOGS FOR SPECIAL APPLICATIONS THAT REQUIRE SPECIFIC MATERIAL CONTENT, TOLERANCES, ETC. SEE FOOTNOTE.

IDLI GROUP 81575

END YOKE WITH ROUND BORE
slight variation from above group
SEE FIGURE 7B

5.3125 in	IE	134.94 mm
5.6250 in	OE	142.88 mm
2.2500 in	SB	57.15 mm
4.5625 in	EC	115.89 mm
4.0000 in	HD	101.60 mm

ROCKWELL 58WBYR36-5

IDLI GROUP 81576

END YOKE WITH ROUND BORE
SEE FIGURE 7B

5.3125 in	IE	134.94 mm
5.6250 in	OE	142.88 mm
2.2500 in	SB	57.15 mm
4.7500 in	EC	120.65 mm
4.0000 in	HD	101.60 mm

ROCKWELL 58WBYR36-12

IDLI GROUP 81577

END YOKE WITH ROUND BORE
SEE FIGURE 7B

5.3125 in	IE	134.94 mm
5.6250 in	OE	142.88 mm
2.2500 in	SB	57.15 mm
4.8750 in	EC	123.83 mm
4.0000 in	HD	101.60 mm

ROCKWELL 58WBYR36-11

IDLI GROUP 81578

END YOKE WITH ROUND BORE
slight variation from above group
SEE FIGURE 7B

5.3125 in	IE	134.94 mm
5.6250 in	OE	142.88 mm
2.2500 in	SB	57.15 mm
4.8750 in	EC	123.83 mm
4.0000 in	HD	101.60 mm

ROCKWELL 58WBYR36-1

IDLI GROUP 81579

END YOKE WITH ROUND BORE
SEE FIGURE 7B

5.3125 in	IE	134.94 mm
5.6250 in	OE	142.88 mm
2.2500 in	SB	57.15 mm
4.9375 in	EC	125.41 mm
4.0000 in	HD	101.60 mm

ROCKWELL 58WBYR36-9

IDLI GROUP 81580

END YOKE WITH ROUND BORE
slight variation from above group
SEE FIGURE 7B

5.3125 in	IE	134.94 mm
5.6250 in	OE	142.88 mm
2.2500 in	SB	57.15 mm
4.9375 in	EC	125.41 mm
4.0000 in	HD	101.60 mm

ROCKWELL 58WBYR36-16

IDLI GROUP 81581

END YOKE WITH ROUND BORE
SEE FIGURE 7B

5.3125 in	IE	134.94 mm
5.6250 in	OE	142.88 mm
2.3125 in	SB	58.74 mm
4.5625 in	EC	115.89 mm
4.0000 in	HD	101.60 mm

ROCKWELL 58WBYR37

IDLI GROUP 81582

END YOKE WITH ROUND BORE
SEE FIGURE 7B

5.3125 in	IE	134.94 mm
5.6250 in	OE	142.88 mm
2.3750 in	SB	60.33 mm
4.1875 in	EC	106.36 mm
4.0625 in	HD	103.19 mm

ROCKWELL 58WBYR38-1

IDLI GROUP 81583

END YOKE WITH ROUND BORE
slight variation from above group
SEE FIGURE 7B

5.3125 in	IE	134.94 mm
5.6250 in	OE	142.88 mm
2.3750 in	SB	60.33 mm
4.1875 in	EC	106.36 mm
4.0625 in	HD	103.19 mm

ROCKWELL 58WBYR38-4

IDLI GROUP 81584

END YOKE WITH ROUND BORE
SEE FIGURE 7B

5.3125 in	IE	134.94 mm
5.6250 in	OE	142.88 mm
2.3750 in	SB	60.33 mm
4.5625 in	EC	115.89 mm
4.0000 in	HD	101.60 mm

ROCKWELL 58WBYR38-8

IDLI GROUP 81585

END YOKE WITH ROUND BORE
SEE FIGURE 7B

5.3125 in	IE	134.94 mm
5.6250 in	OE	142.88 mm
2.3750 in	SB	60.33 mm
4.6875 in	EC	119.06 mm
4.0000 in	HD	101.60 mm

ROCKWELL 58WBYR38-5

IDLI GROUP 81586

END YOKE WITH ROUND BORE
SEE FIGURE 7B

5.3125 in	IE	134.94 mm
5.6250 in	OE	142.88 mm
2.3750 in	SB	60.33 mm
4.8750 in	EC	123.83 mm
4.0000 in	HD	101.60 mm

ROCKWELL 58WBYR38-9

IDLI GROUP 81587

END YOKE WITH ROUND BORE
slight variation from above group
SEE FIGURE 7B

5.3125 in	IE	134.94 mm
5.6250 in	OE	142.88 mm
2.3750 in	SB	60.33 mm
4.8750 in	EC	123.83 mm
4.0000 in	HD	101.60 mm

ROCKWELL 58WBYR38-6

IDLI GROUP 81588

END YOKE WITH ROUND BORE
SEE FIGURE 7B

5.3125 in	IE	134.94 mm
5.6250 in	OE	142.88 mm
2.4375 in	SB	61.91 mm
4.5625 in	EC	115.89 mm
4.0000 in	HD	101.60 mm

ROCKWELL 58WBYR39-3

IDLI GROUP 81589

END YOKE WITH ROUND BORE
slight variation from above group
SEE FIGURE 7B

5.3125 in	IE	134.94 mm
5.6250 in	OE	142.88 mm
2.4375 in	SB	61.91 mm
4.5625 in	EC	115.89 mm
4.0000 in	HD	101.60 mm

ROCKWELL 58WBYR39-2

IDLI GROUP 81590

END YOKE WITH ROUND BORE
slight variation from above group
SEE FIGURE 7B

5.3125 in	IE	134.94 mm
5.6250 in	OE	142.88 mm
2.4375 in	SB	61.91 mm
4.5625 in	EC	115.89 mm
4.0000 in	HD	101.60 mm

ROCKWELL 58WBYR39-5

IDLI GROUP 81591

END YOKE WITH ROUND BORE
SEE FIGURE 7B

5.3125 in	IE	134.94 mm
5.6250 in	OE	142.88 mm
2.4375 in	SB	61.91 mm
4.8750 in	EC	123.83 mm
4.0000 in	HD	101.60 mm

ROCKWELL 58WBYR39-1

IDLI GROUP 81592

END YOKE WITH ROUND BORE
slight variation from above group
SEE FIGURE 7B

5.3125 in	IE	134.94 mm
5.6250 in	OE	142.88 mm
2.4375 in	SB	61.91 mm
4.8750 in	EC	123.83 mm
4.0000 in	HD	101.60 mm

ROCKWELL 58WBYR39

IDLI GROUP 81593

END YOKE WITH ROUND BORE
SEE FIGURE 7B

5.3125 in	IE	134.94 mm
5.6250 in	OE	142.88 mm
2.4375 in	SB	61.91 mm
4.9375 in	EC	125.41 mm
4.0000 in	HD	101.60 mm

ROCKWELL 58WBYR39-4

IDLI GROUP 81594

END YOKE WITH ROUND BORE
SEE FIGURE 7B

5.3125 in	IE	134.94 mm
5.6250 in	OE	142.88 mm
2.5000 in	SB	63.50 mm
3.3750 in	EC	85.73 mm
4.2500 in	HD	107.95 mm

ROCKWELL 58WBYR40-11

IDLI GROUP 81595

END YOKE WITH ROUND BORE
SEE FIGURE 7B

5.3125 in	IE	134.94 mm
5.6250 in	OE	142.88 mm
2.5000 in	SB	63.50 mm
3.7500 in	EC	95.25 mm
3.8750 in	HD	98.43 mm

ROCKWELL 58WBYR40-6

IDLI GROUP 81596

END YOKE WITH 10 SPLINES
SEE FIGURE 7A

5.3125 in	IE	134.94 mm
5.6250 in	OE	142.88 mm
2.5000 in	SB	63.50 mm
3.8750 in	EC	98.43 mm
2.2500 in	HD	57.15 mm

ROCKWELL 58WBYS40-2

IDLI GROUP 81597

END YOKE WITH ROUND BORE
SEE FIGURE 7B

5.3125 in	IE	134.94 mm
5.6250 in	OE	142.88 mm
2.5000 in	SB	63.50 mm
4.5625 in	EC	115.89 mm
4.0000 in	HD	101.60 mm

ROCKWELL 58WBYR40-5

IDLI GROUP 81598

END YOKE WITH ROUND BORE
slight variation from above group
SEE FIGURE 7B

5.3125 in	IE	134.94 mm
5.6250 in	OE	142.88 mm
2.5000 in	SB	63.50 mm
4.5625 in	EC	115.89 mm
4.0000 in	HD	101.60 mm

ROCKWELL 58WBYR40-9

BD = BEARING DIAMETER (outside)   BW = BEARING WIDTH   CB = CROSS LENGTH WITH BEARINGS   CC = CENTER TO CENTER   CD = CROSS DIAMETER
CL = CROSS LENGTH WITHOUT BEARINGS   EC = END TO CENTER or FACE TO FACE   EE = END TO END   EL = EFFECTIVE LENGTH
HD = HUB DIAMETER (or insert)   IE = INSIDE OF EARS (or recess)   OD = OUTSIDE DIAMETER   OE = OUTSIDE OF EARS   SB = SPLINE OR BORE SIZE

IDLI GROUP 81599

END YOKE WITH ROUND BORE
SEE FIGURE 7B

5.3125 in	IE	134.94 mm
5.6250 in	OE	142.88 mm
2.5000 in	SB	63.50 mm
4.6250 in	EC	117.48 mm
4.0000 in	HD	101.60 mm

ROCKWELL 58WBYR40-3

IDLI GROUP 81600

END YOKE WITH ROUND BORE
SEE FIGURE 7B

5.3125 in	IE	134.94 mm
5.6250 in	OE	142.88 mm
2.5000 in	SB	63.50 mm
4.8750 in	EC	123.83 mm
4.0000 in	HD	101.60 mm

ROCKWELL 58WBYR40-15

IDLI GROUP 81601

END YOKE WITH ROUND BORE
slight variation from above group
SEE FIGURE 7B

5.3125 in	IE	134.94 mm
5.6250 in	OE	142.88 mm
2.5000 in	SB	63.50 mm
4.8750 in	EC	123.83 mm
4.0000 in	HD	101.60 mm

ROCKWELL 58WBYR40-4

IDLI GROUP 81602

END YOKE WITH ROUND BORE
slight variation from above group
SEE FIGURE 7B

5.3125 in	IE	134.94 mm
5.6250 in	OE	142.88 mm
2.5000 in	SB	63.50 mm
4.8750 in	EC	123.83 mm
4.0000 in	HD	101.60 mm

ROCKWELL 58WBYR40-1

IDLI GROUP 81603

END YOKE WITH ROUND BORE
SEE FIGURE 7B

5.3125 in	IE	134.94 mm
5.6250 in	OE	142.88 mm
2.5000 in	SB	63.50 mm
4.9375 in	EC	125.41 mm
4.0000 in	HD	101.60 mm

ROCKWELL 58WBYR40-13

IDLI GROUP 81604

END YOKE WITH ROUND BORE
slight variation from above group
SEE FIGURE 7B

5.3125 in	IE	134.94 mm
5.6250 in	OE	142.88 mm
2.5000 in	SB	63.50 mm
4.9375 in	EC	125.41 mm
4.0000 in	HD	101.60 mm

ROCKWELL 58WBYR40

IDLI GROUP 81605

END YOKE WITH ROUND BORE
SEE FIGURE 7B

5.3125 in	IE	134.94 mm
5.6250 in	OE	142.88 mm
2.5625 in	SB	65.09 mm
4.9375 in	EC	125.41 mm
4.0000 in	HD	101.60 mm

ROCKWELL 58WBYR41-2

IDLI GROUP 81606

END YOKE WITH ROUND BORE
slight variation from above group
SEE FIGURE 7B

5.3125 in	IE	134.94 mm
5.6250 in	OE	142.88 mm
2.5625 in	SB	65.09 mm
4.9375 in	EC	125.41 mm
4.0000 in	HD	101.60 mm

ROCKWELL 58WBYR41-3

IDLI GROUP 81607

END YOKE WITH ROUND BORE
SEE FIGURE 7B

5.3125 in	IE	134.94 mm
5.6250 in	OE	142.88 mm
2.5781 in	SB	65.48 mm
4.9375 in	EC	125.41 mm
4.0000 in	HD	101.60 mm

ROCKWELL 58WBYR41-1

IDLI GROUP 81608

END YOKE WITH ROUND BORE
SEE FIGURE 7B

5.3125 in	IE	134.94 mm
5.6250 in	OE	142.88 mm
2.6250 in	SB	66.68 mm
4.1875 in	EC	106.36 mm
4.0625 in	HD	103.19 mm

ROCKWELL 58WBYR42-7

IDLI GROUP 81609

END YOKE WITH ROUND BORE
slight variation from above group
SEE FIGURE 7B

5.3125 in	IE	134.94 mm
5.6250 in	OE	142.88 mm
2.6250 in	SB	66.68 mm
4.1875 in	EC	106.36 mm
4.0625 in	HD	103.19 mm

ROCKWELL 58WBYR42-1

IDLI GROUP 81610

END YOKE WITH ROUND BORE
SEE FIGURE 7B

5.3125 in	IE	134.94 mm
5.6250 in	OE	142.88 mm
2.6250 in	SB	66.68 mm
4.5625 in	EC	115.89 mm
4.0000 in	HD	101.60 mm

ROCKWELL 58WBYR42-6

IDLI GROUP 81611

END YOKE WITH ROUND BORE
SEE FIGURE 7B

5.3125 in	IE	134.94 mm
5.6250 in	OE	142.88 mm
2.6250 in	SB	66.68 mm
4.7500 in	EC	120.65 mm
4.0000 in	HD	101.60 mm

ROCKWELL 58WBYR42-5

IDLI GROUP 81612

END YOKE WITH ROUND BORE
SEE FIGURE 7B

5.3125 in	IE	134.94 mm
5.6250 in	OE	142.88 mm
2.6250 in	SB	66.68 mm
4.8125 in	EC	122.24 mm
4.0000 in	HD	101.60 mm

ROCKWELL 58WBYR42-4

IDLI GROUP 81613

END YOKE WITH ROUND BORE
SEE FIGURE 7B

5.3125 in	IE	134.94 mm
5.6250 in	OE	142.88 mm
2.6250 in	SB	66.68 mm
4.8750 in	EC	123.83 mm
4.0000 in	HD	101.60 mm

ROCKWELL 58WBYR42-3

IDLI GROUP 81614

END YOKE WITH ROUND BORE
SEE FIGURE 7B

5.3125 in	IE	134.94 mm
5.6250 in	OE	142.88 mm
2.6250 in	SB	66.68 mm
4.9375 in	EC	125.41 mm
4.0000 in	HD	101.60 mm

ROCKWELL 58WBYR42

IDLI GROUP 81615

END YOKE WITH ROUND BORE
slight variation from above group
SEE FIGURE 7B

5.3125 in	IE	134.94 mm
5.6250 in	OE	142.88 mm
2.6250 in	SB	66.68 mm
4.9375 in	EC	125.41 mm
4.0000 in	HD	101.60 mm

ROCKWELL 58WBYR42-2

IDLI GROUP 81616

END YOKE WITH ROUND BORE
SEE FIGURE 7B

5.3125 in	IE	134.94 mm
5.6250 in	OE	142.88 mm
2.6875 in	SB	68.26 mm
4.9375 in	EC	125.41 mm
4.0000 in	HD	101.60 mm

ROCKWELL 58WBYR43

IDLI GROUP 81617

END YOKE WITH ROUND BORE
SEE FIGURE 7B

5.3125 in	IE	134.94 mm
5.6250 in	OE	142.88 mm
2.7500 in	SB	69.85 mm
3.4375 in	EC	87.31 mm
4.2500 in	HD	107.95 mm

ROCKWELL 58WBYR44-10

IDLI GROUP 81618

END YOKE WITH ROUND BORE
SEE FIGURE 7B

5.3125 in	IE	134.94 mm
5.6250 in	OE	142.88 mm
2.7500 in	SB	69.85 mm
4.5625 in	EC	115.89 mm
4.0000 in	HD	101.60 mm

ROCKWELL 58WBYR44-9

IDLI GROUP 81619

END YOKE WITH ROUND BORE
slight variation from above group
SEE FIGURE 7B

5.3125 in	IE	134.94 mm
5.6250 in	OE	142.88 mm
2.7500 in	SB	69.85 mm
4.5625 in	EC	115.89 mm
4.0000 in	HD	101.60 mm

ROCKWELL 58WBYR44-1

IDLI GROUP 81620

END YOKE WITH ROUND BORE
slight variation from above group
SEE FIGURE 7B

5.3125 in	IE	134.94 mm
5.6250 in	OE	142.88 mm
2.7500 in	SB	69.85 mm
4.5625 in	EC	115.89 mm
4.0000 in	HD	101.60 mm

ROCKWELL 58WBYR44-2

IDLI GROUP 81621

END YOKE WITH ROUND BORE
SEE FIGURE 7B

5.3125 in	IE	134.94 mm
5.6250 in	OE	142.88 mm
2.7500 in	SB	69.85 mm
4.6250 in	EC	117.48 mm
4.0000 in	HD	101.60 mm

ROCKWELL 58WBYR44-11

IDLI GROUP 81622

END YOKE WITH ROUND BORE
slight variation from above group
SEE FIGURE 7B

5.3125 in	IE	134.94 mm
5.6250 in	OE	142.88 mm
2.7500 in	SB	69.85 mm
4.6250 in	EC	117.48 mm
4.0000 in	HD	101.60 mm

ROCKWELL 58WBYR44-12

ENGINEERING CATALOGS MUST BE CONSULTED FOR DETAILS NOT INCLUDED
IN THIS GUIDE. SPECIFIC DESIGNS, MATERIAL CONTENT, TOLERANCES
LUBE FITTINGS AND OTHER DIMENSIONS ARE INTENTIONALLY OMITTED HERE.

CAUTION: BE SURE TO REFER TO ENGINEERING CATALOGS FOR SPECIAL APPLICATIONS THAT REQUIRE SPECIFIC MATERIAL CONTENT, TOLERANCES, ETC. SEE FOOTNOTE.

IDLI GROUP 81623

END YOKE WITH ROUND BORE
SEE FIGURE 7B

5.3125 in	IE	134.94 mm
5.6250 in	OE	142.88 mm
2.7500 in	SB	69.85 mm
4.8750 in	EC	123.83 mm
4.0000 in	HD	101.60 mm

ROCKWELL 58WBYR44-6

IDLI GROUP 81624

END YOKE WITH ROUND BORE
slight variation from above group
SEE FIGURE 7B

5.3125 in	IE	134.94 mm
5.6250 in	OE	142.88 mm
2.7500 in	SB	69.85 mm
4.8750 in	EC	123.83 mm
4.0000 in	HD	101.60 mm

ROCKWELL 58WBYR44-3

IDLI GROUP 81625

END YOKE WITH ROUND BORE
slight variation from above group
SEE FIGURE 7B

5.3125 in	IE	134.94 mm
5.6250 in	OE	142.88 mm
2.7500 in	SB	69.85 mm
4.8750 in	EC	123.83 mm
4.0000 in	HD	101.60 mm

ROCKWELL 58WBYR44-4

IDLI GROUP 81626

END YOKE WITH ROUND BORE
slight variation from above group
SEE FIGURE 7B

5.3125 in	IE	134.94 mm
5.6250 in	OE	142.88 mm
2.7500 in	SB	69.85 mm
4.8750 in	EC	123.83 mm
4.0000 in	HD	101.60 mm

ROCKWELL 58WBYR44-8

IDLI GROUP 81627

END YOKE WITH ROUND BORE
SEE FIGURE 7B

5.3125 in	IE	134.94 mm
5.6250 in	OE	142.88 mm
2.7500 in	SB	69.85 mm
4.9375 in	EC	125.41 mm
4.0000 in	HD	101.60 mm

ROCKWELL 58WBYR44-5

IDLI GROUP 81628

END YOKE WITH ROUND BORE
SEE FIGURE 7B

5.3125 in	IE	134.94 mm
5.6250 in	OE	142.88 mm
2.8750 in	SB	73.03 mm
4.8750 in	EC	123.83 mm
4.0000 in	HD	101.60 mm

ROCKWELL 58WBYR46-4

IDLI GROUP 81629

END YOKE WITH ROUND BORE
slight variation from above group
SEE FIGURE 7B

5.3125 in	IE	134.94 mm
5.6250 in	OE	142.88 mm
2.8750 in	SB	73.03 mm
4.8750 in	EC	123.83 mm
4.0000 in	HD	101.60 mm

ROCKWELL 58WBYR46-2

IDLI GROUP 81630

END YOKE WITH ROUND BORE
SEE FIGURE 7B

5.3125 in	IE	134.94 mm
5.6250 in	OE	142.88 mm
2.8750 in	SB	73.03 mm
4.9375 in	EC	125.41 mm
4.0000 in	HD	101.60 mm

ROCKWELL 58WBYR46

IDLI GROUP 81631

END YOKE WITH ROUND BORE
SEE FIGURE 7B

5.3125 in	IE	134.94 mm
5.6250 in	OE	142.88 mm
2.9375 in	SB	74.61 mm
4.8750 in	EC	123.83 mm
4.0000 in	HD	101.60 mm

ROCKWELL 58WBYR47-1

IDLI GROUP 81632

END YOKE WITH ROUND BORE
SEE FIGURE 7B

5.3125 in	IE	134.94 mm
5.6250 in	OE	142.88 mm
2.9375 in	SB	74.61 mm
4.9375 in	EC	125.41 mm
4.0000 in	HD	101.60 mm

ROCKWELL 58WBYR47

IDLI GROUP 81633

END YOKE WITH ROUND BORE
SEE FIGURE 7B

5.3125 in	IE	134.94 mm
5.6250 in	OE	142.88 mm
3.0000 in	SB	76.20 mm
4.1875 in	EC	106.36 mm
4.0625 in	HD	103.19 mm

ROCKWELL 58WBYR48-3

IDLI GROUP 81634

END YOKE WITH ROUND BORE
SEE FIGURE 7B

5.3125 in	IE	134.94 mm
5.6250 in	OE	142.88 mm
3.0000 in	SB	76.20 mm
4.5625 in	EC	115.89 mm
4.0000 in	HD	101.60 mm

ROCKWELL 58WBYR48-6

IDLI GROUP 81635

END YOKE WITH ROUND BORE
SEE FIGURE 7B

5.3125 in	IE	134.94 mm
5.6250 in	OE	142.88 mm
3.0000 in	SB	76.20 mm
4.8750 in	EC	123.83 mm
4.0000 in	HD	101.60 mm

ROCKWELL 58WBYR48-7

IDLI GROUP 81636

END YOKE WITH ROUND BORE
slight variation from above group
SEE FIGURE 7B

5.3125 in	IE	134.94 mm
5.6250 in	OE	142.88 mm
3.0000 in	SB	76.20 mm
4.8750 in	EC	123.83 mm
4.0000 in	HD	101.60 mm

ROCKWELL 58WBYR48

IDLI GROUP 81637

END YOKE WITH ROUND BORE
SEE FIGURE 7B

5.3125 in	IE	134.94 mm
5.6250 in	OE	142.88 mm
3.1250 in	SB	79.38 mm
4.8750 in	EC	123.83 mm
4.0000 in	HD	101.60 mm

ROCKWELL 58WBYR50

IDLI GROUP 81638

END YOKE WITH ROUND BORE
SEE FIGURE 7B

5.3125 in	IE	134.94 mm
5.6250 in	OE	142.88 mm
3.1875 in	SB	80.96 mm
4.8750 in	EC	123.83 mm
4.0000 in	HD	101.60 mm

ROCKWELL 58WBYR51

IDLI GROUP 81639

END YOKE WITH ROUND BORE
SEE FIGURE 7B

5.3125 in	IE	134.94 mm
5.6250 in	OE	142.88 mm
3.2500 in	SB	82.55 mm
5.1875 in	EC	131.76 mm
5.6250 in	HD	142.88 mm

ROCKWELL 58WBYR52

IDLI GROUP 81640

END YOKE WITH ROUND BORE
SEE FIGURE 7B

5.3125 in	IE	134.94 mm
5.6250 in	OE	142.88 mm
3.3438 in	SB	84.93 mm
4.6875 in	EC	119.06 mm
5.6250 in	HD	142.88 mm

ROCKWELL 58WBYR54

IDLI GROUP 81641

END YOKE WITH ROUND BORE
SEE FIGURE 7B

5.3125 in	IE	134.94 mm
5.6250 in	OE	142.88 mm
3.3750 in	SB	85.73 mm
4.9375 in	EC	125.41 mm
5.6250 in	HD	142.88 mm

ROCKWELL 58WBYR54-2

IDLI GROUP 81642

END YOKE WITH ROUND BORE
SEE FIGURE 7B

5.3125 in	IE	134.94 mm
5.6250 in	OE	142.88 mm
3.3750 in	SB	85.73 mm
5.1875 in	EC	131.76 mm
5.6250 in	HD	142.88 mm

ROCKWELL 58WBYR54-1

IDLI GROUP 81643

END YOKE WITH ROUND BORE
SEE FIGURE 7B

5.3125 in	IE	134.94 mm
5.6250 in	OE	142.88 mm
3.4375 in	SB	87.31 mm
4.1875 in	EC	106.36 mm
5.6250 in	HD	142.88 mm

ROCKWELL 58WBYR55

IDLI GROUP 81644

END YOKE WITH ROUND BORE
SEE FIGURE 7B

5.3125 in	IE	134.94 mm
5.6250 in	OE	142.88 mm
3.5000 in	SB	88.90 mm
4.6875 in	EC	119.06 mm
5.6250 in	HD	142.88 mm

ROCKWELL 58WBYR56-3

IDLI GROUP 81645

END YOKE WITH ROUND BORE
slight variation from above group
SEE FIGURE 7B

5.3125 in	IE	134.94 mm
5.6250 in	OE	142.88 mm
3.5000 in	SB	88.90 mm
4.6875 in	EC	119.06 mm
5.6250 in	HD	142.88 mm

ROCKWELL 58WBYR56-2

BD = BEARING DIAMETER (outside)   **BW** = BEARING WIDTH   **CB** = CROSS LENGTH WITH BEARINGS   **CC** = CENTER TO CENTER   **CD** = CROSS DIAMETER
CL = CROSS LENGTH WITHOUT BEARINGS   **EC** = END TO CENTER or FACE TO FACE   **EE** = END TO END   **EL** = EFFECTIVE LENGTH
HD = HUB DIAMETER (or insert)   **IE** = INSIDE OF EARS (or recess)   **OD** = OUTSIDE DIAMETER   **OE** = OUTSIDE OF EARS   **SB** = SPLINE OR BORE SIZE

IDLI GUIDE   COPYRIGHT © INTERCHANGE, INC.   ST. LOUIS PARK, MN. 55416 USA

IDLI GROUP 81646

END YOKE WITH ROUND BORE
slight variation from above group
SEE FIGURE 7B

5.3125 in	IE	134.94 mm
5.6250 in	OE	142.88 mm
3.5000 in	SB	88.90 mm
4.6875 in	EC	119.06 mm
5.6250 in	HD	142.88 mm

ROCKWELL 58WBYR56-4

IDLI GROUP 81647

END YOKE WITH ROUND BORE
SEE FIGURE 7B

5.3125 in	IE	134.94 mm
5.6250 in	OE	142.88 mm
3.6250 in	SB	92.08 mm
4.6875 in	EC	119.06 mm
5.6250 in	HD	142.88 mm

ROCKWELL 58WBYR58

IDLI GROUP 81648

END YOKE WITH ROUND BORE
SEE FIGURE 7B

5.3125 in	IE	134.94 mm
5.6250 in	OE	142.88 mm
3.8125 in	SB	96.84 mm
4.9375 in	EC	125.41 mm
6.0000 in	HD	152.40 mm

ROCKWELL 58WBYR61

IDLI GROUP 81649

END YOKE WITH ROUND BORE
SEE FIGURE 7B

5.3125 in	IE	134.94 mm
5.6250 in	OE	142.88 mm
3.8750 in	SB	98.43 mm
4.9375 in	EC	125.41 mm
6.0000 in	HD	152.40 mm

ROCKWELL 58WBYR62

IDLI GROUP 81650

END YOKE WITH ROUND BORE
slight variation from above group
SEE FIGURE 7B

5.3125 in	IE	134.94 mm
6.2500 in	OE	158.75 mm
1.5000 in	SB	38.10 mm
4.9375 in	EC	125.41 mm
4.0000 in	HD	101.60 mm

ROCKWELL 58WBYR24-12

IDLI GROUP 81651

END YOKE WITH 16 SPLINES
slight variation from above group
SEE FIGURE 7A

5.3162 in	IE	135.03 mm
5.6250 in	OE	142.88 mm
1.7500 in	SB	44.45 mm
2.7500 in	EC	69.85 mm
2.8125 in	HD	71.44 mm

ROCKWELL 58WBYS28-5

IDLI GROUP 81652

FLANGE YOKE
SEE FIGURE 27

5.5313 in	IE	140.50 mm
5.8438 in	OE	148.43 mm
1.7188 in	EC	43.66 mm
11.75 0 in	OD	298.45 mm

BORG-WARNR 117-14346J
BORG-WARNR 14346J
ROCKWELL 62NF5

IDLI GROUP 81653

FLANGE YOKE
SEE FIGURE 27

5.5313 in	IE	140.50 mm
5.8438 in	OE	148.43 mm
1.8750 in	EC	47.63 mm
10.25 0 in	OD	260.35 mm

ROCKWELL 62NF14

IDLI GROUP 81654

FLANGE YOKE
SEE FIGURE 27

5.5313 in	IE	140.50 mm
5.8438 in	OE	148.43 mm
2.2500 in	EC	57.15 mm
6.8750 in	OD	174.63 mm

BORG-WARNR 17673J
BORG-WARNR 4221J
ROCKWELL 62NF13

IDLI GROUP 81655

FLANGE YOKE
SEE FIGURE 27

5.5313 in	IE	140.50 mm
5.8438 in	OE	148.43 mm
2.3125 in	EC	58.74 mm
10.25 0 in	OD	260.35 mm

ROCKWELL 62NF9

IDLI GROUP 81656

FLANGE YOKE
slight variation from above group
SEE FIGURE 27

5.5313 in	IE	140.50 mm
5.8438 in	OE	148.43 mm
2.3125 in	EC	58.74 mm
10.25 0 in	OD	260.35 mm

ROCKWELL 62NF10

IDLI GROUP 81657

FLANGE YOKE
SEE FIGURE 27

5.5313 in	IE	140.50 mm
5.8438 in	OE	148.43 mm
2.3438 in	EC	59.53 mm
5.75 0 in	OD	146.05 mm

ROCKWELL 62NF17

IDLI GROUP 81658

END YOKE WITH 10 SPLINES
SEE FIGURE 7A

5.5313 in	IE	140.50 mm
5.8438 in	OE	148.43 mm
1.3750 in	SB	34.93 mm
2.9375 in	EC	74.61 mm
2.1250 in	HD	53.98 mm

BORG-WARNR 16430J
ROCKWELL 62NYS22-10

IDLI GROUP 81659

END YOKE WITH 10 SPLINES
SEE FIGURE 7A

5.5313 in	IE	140.50 mm
5.8438 in	OE	148.43 mm
1.3750 in	SB	34.93 mm
3.2500 in	EC	82.55 mm
2.1250 in	HD	53.98 mm

ROCKWELL 62NYS22-8

IDLI GROUP 81660

END YOKE WITH 21 INVOLUTED SPLINES
SEE FIGURE 7A

5.5313 in	IE	140.50 mm
5.8438 in	OE	148.43 mm
1.4375 in	SB	36.51 mm
2.9375 in	EC	74.61 mm
2.1250 in	HD	53.98 mm

ROCKWELL 62NYS22-2

IDLI GROUP 81661

END YOKE WITH 10 SPLINES
SEE FIGURE 7A

5.5313 in	IE	140.50 mm
5.8438 in	OE	148.43 mm
1.5000 in	SB	38.10 mm
3.0625 in	EC	77.79 mm
1.9375 in	HD	49.21 mm

BORG-WARNR 10507J
FORD B5Q4865B
FORD TGAY4865A
ROCKWELL 62NYS24-8

IDLI GROUP 81662

END YOKE WITH 10 SPLINES
SEE FIGURE 7A

5.5313 in	IE	140.50 mm
5.8438 in	OE	148.43 mm
1.5000 in	SB	38.10 mm
3.4688 in	EC	88.11 mm
2.2500 in	HD	57.15 mm

BORG-WARNR 12266J
BORG-WARNR 27814J
ROCKWELL 62NYS24-17

IDLI GROUP 81663

END YOKE WITH 10 SPLINES
SEE FIGURE 7A

5.5313 in	IE	140.50 mm
5.8438 in	OE	148.43 mm
1.5000 in	SB	38.10 mm
3.5000 in	EC	88.90 mm
2.2500 in	HD	57.15 mm

ROCKWELL 62NYS24-15

IDLI GROUP 81664

END YOKE WITH 10 SPLINES
SEE FIGURE 7A

5.5313 in	IE	140.50 mm
5.8438 in	OE	148.43 mm
1.5000 in	SB	38.10 mm
3.5000 in	EC	88.90 mm
2.5000 in	HD	63.50 mm

ROCKWELL 62NYS24-1

IDLI GROUP 81665

END YOKE WITH 15 INVOLUTED SPLINES
SEE FIGURE 7A

5.5313 in	IE	140.50 mm
5.8438 in	OE	148.43 mm
1.6094 in	SB	40.88 mm
4.2500 in	EC	107.95 mm
2.1250 in	HD	53.98 mm

BORG-WARNR 22855J
ROCKWELL 62NYS26-4

IDLI GROUP 81666

END YOKE WITH 10 SPLINES
SEE FIGURE 7A

5.5313 in	IE	140.50 mm
5.8438 in	OE	148.43 mm
1.6250 in	SB	41.28 mm
3.1875 in	EC	80.96 mm
2.3750 in	HD	60.33 mm

BORG-WARNR 12267J
ROCKWELL 62NYS26-13

IDLI GROUP 81667

END YOKE WITH 10 SPLINES
slight variation from above group
SEE FIGURE 7A

5.5313 in	IE	140.50 mm
5.8438 in	OE	148.43 mm
1.6250 in	SB	41.28 mm
3.1875 in	EC	80.96 mm
2.3750 in	HD	60.33 mm

BORG-WARNR 20846J
BORG-WARNR 21030J
ROCKWELL 62NYS26 5

IDLI GROUP 81668

END YOKE WITH 10 SPLINES
SEE FIGURE 7A

5.5313 in	IE	140.50 mm
5.8438 in	OE	148.43 mm
1.6250 in	SB	41.28 mm
3.3125 in	EC	84.14 mm
2.5000 in	HD	63.50 mm

BORG-WARNR 7184J
ROCKWELL 62NYS26

IDLI GROUP 81669

END YOKE WITH 10 SPLINES
SEE FIGURE 7A

5.5313 in	IE	140.50 mm
5.8438 in	OE	148.43 mm
1.6250 in	SB	41.28 mm
3.5000 in	EC	88.90 mm
2.5000 in	HD	63.50 mm

ROCKWELL 62NYS26-7

ENGINEERING CATALOGS MUST BE CONSULTED FOR DETAILS NOT INCLUDED
IN THIS GUIDE. SPECIFIC DESIGNS, MATERIAL CONTENT, TOLERANCES
LUBE FITTINGS AND OTHER DIMENSIONS ARE INTENTIONALLY OMITTED HERE.

IDLI GUIDE   COPYRIGHT © INTERCHANGE, INC.   ST. LOUIS PARK, MN. 55416 USA

CAUTION: BE SURE TO REFER TO ENGINEERING CATALOGS FOR SPECIAL APPLICATIONS THAT REQUIRE SPECIFIC MATERIAL CONTENT, TOLERANCES, ETC. SEE FOOTNOTE.

IDLI GROUP 81670

END YOKE WITH 10 SPLINES
SEE FIGURE 7A

5.5313 in	IE	140.50 mm
5.8438 in	OE	148.43 mm
1.7500 in	SB	44.45 mm
3.3125 in	EC	84.14 mm
2.5000 in	HD	63.50 mm

ROCKWELL 62NYS28-91

IDLI GROUP 81671

END YOKE WITH 20 INVOLUTED SPLINES
SEE FIGURE 7A

5.5313 in	IE	140.50 mm
5.8438 in	OE	148.43 mm
1.7500 in	SB	44.45 mm
3.3125 in	EC	84.14 mm
2.8750 in	HD	73.03 mm

BORG-WARNR 24030J
ROCKWELL 62NYS28-71

IDLI GROUP 81672

END YOKE WITH 10 SPLINES
SEE FIGURE 7A

5.5313 in	IE	140.50 mm
5.8438 in	OE	148.43 mm
1.7500 in	SB	44.45 mm
3.3438 in	EC	84.93 mm
2.3750 in	HD	60.33 mm

ROCKWELL 62NYS28-21

IDLI GROUP 81673

END YOKE WITH 10 SPLINES
slight variation from above group
SEE FIGURE 7A

5.5313 in	IE	140.50 mm
5.8438 in	OE	148.43 mm
1.7500 in	SB	44.45 mm
3.3438 in	EC	84.93 mm
2.3750 in	HD	60.33 mm

BORG-WARNR 14318J
ROCKWELL 62NYS28-14

IDLI GROUP 81674

END YOKE WITH 10 SPLINES
SEE FIGURE 7A

5.5313 in	IE	140.50 mm
5.8438 in	OE	148.43 mm
1.7500 in	SB	44.45 mm
3.3438 in	EC	84.93 mm
2.8125 in	HD	71.44 mm

ROCKWELL 62NYS28-129

IDLI GROUP 81675

SLIP YOKE WITH 10 SPLINES
SEE FIGURE 41

5.5313 in	IE	140.50 mm
5.8438 in	OE	148.43 mm
1.7500 in	SB	44.45 mm
3.5000 in	EC	88.90 mm

BORG-WARNR 6-6009
BORG-WARNR 6-6066
BORG-WARNR 6A6009
BORG-WARNR 6A6066
ROCKWELL 62NLS28-9

IDLI GROUP 81676

SLIP YOKE WITH 10 SPLINES
SEE FIGURE 41

5.5313 in	IE	140.50 mm
5.8438 in	OE	148.43 mm
1.7500 in	SB	44.45 mm
3.7500 in	EC	95.25 mm

ROCKWELL 62NLS28-8

IDLI GROUP 81677

SLIP YOKE WITH 10 SPLINES
slight variation from above group
SEE FIGURE 41

5.5313 in	IE	140.50 mm
5.8438 in	OE	148.43 mm
1.7500 in	SB	44.45 mm
3.7500 in	EC	95.25 mm

BORG-WARNR 6-6037
BORG-WARNR 6-6039
BORG-WARNR 6A6037
BORG-WARNR 6A6039
ROCKWELL 62NLS28-22

IDLI GROUP 81678

SLIP YOKE WITH 10 SPLINES
SEE FIGURE 41

5.5313 in	IE	140.50 mm
5.8438 in	OE	148.43 mm
1.7500 in	SB	44.45 mm
3.8125 in	EC	96.84 mm

ROCKWELL 62NLS28-25

IDLI GROUP 81679

END YOKE WITH 10 SPLINES
SEE FIGURE 7A

5.5313 in	IE	140.50 mm
5.8438 in	OE	148.43 mm
1.7500 in	SB	44.45 mm
3.9375 in	EC	100.01 mm
2.3750 in	HD	60.33 mm

BORG-WARNR 15658J
ROCKWELL 62NYS28-31

IDLI GROUP 81680

END YOKE WITH 10 SPLINES
SEE FIGURE 7A

5.5313 in	IE	140.50 mm
5.8438 in	OE	148.43 mm
1.7500 in	SB	44.45 mm
4.1875 in	EC	106.36 mm
2.4375 in	HD	61.91 mm

BORG-WARNR 22158J

IDLI GROUP 81681

END YOKE WITH 10 SPLINES
SEE FIGURE 7A

5.5313 in	IE	140.50 mm
5.8438 in	OE	148.43 mm
1.7500 in	SB	44.45 mm
4.2500 in	EC	107.95 mm
2.6250 in	HD	66.68 mm

BORG-WARNR 9516J
ROCKWELL 62NYS28-32

IDLI GROUP 81682

SLIP YOKE WITH 10 SPLINES
SEE FIGURE 41

5.5313 in	IE	140.50 mm
5.8438 in	OE	148.43 mm
1.7500 in	SB	44.45 mm
4.5625 in	EC	115.89 mm

BORG-WARNR 6A6034
ROCKWELL 62NLS28-27

IDLI GROUP 81683

SLIP YOKE WITH 10 SPLINES
slight variation from above group
SEE FIGURE 41

5.5313 in	IE	140.50 mm
5.8438 in	OE	148.43 mm
1.7500 in	SB	44.45 mm
4.5625 in	EC	115.89 mm

ROCKWELL 62NLS28-12

IDLI GROUP 81684

END YOKE WITH 10 SPLINES
SEE FIGURE 7A

5.5313 in	IE	140.50 mm
5.8438 in	OE	148.43 mm
1.7500 in	SB	44.45 mm
4.5938 in	EC	116.68 mm
2.7500 in	HD	69.85 mm

BORG-WARNR 22724J
ROCKWELL 62NYS28-59

IDLI GROUP 81685

END YOKE WITH 10 SPLINES
SEE FIGURE 7A

5.5313 in	IE	140.50 mm
5.8438 in	OE	148.43 mm
1.8750 in	SB	47.63 mm
3.4375 in	EC	87.31 mm
2.7500 in	HD	69.85 mm

BORG-WARNR 8310J
ROCKWELL 62NYS30-13

IDLI GROUP 81686

END YOKE WITH 14 INVOLUTED SPLINES
SEE FIGURE 7A

5.5313 in	IE	140.50 mm
5.8438 in	OE	148.43 mm
1.9219 in	SB	48.82 mm
2.7500 in	EC	69.85 mm
2.4688 in	HD	62.71 mm

ROCKWELL 62NYS30-7

IDLI GROUP 81687

END YOKE WITH 10 SPLINES
SEE FIGURE 7A

5.5313 in	IE	140.50 mm
5.8438 in	OE	148.43 mm
1.9688 in	SB	50.01 mm
3.1250 in	EC	79.38 mm
2.5938 in	HD	65.88 mm

ROCKWELL 62NYS31-23

IDLI GROUP 81688

END YOKE WITH 10 SPLINES
SEE FIGURE 7A

5.5313 in	IE	140.50 mm
5.8438 in	OE	148.43 mm
1.9688 in	SB	50.01 mm
3.3125 in	EC	84.14 mm
2.6250 in	HD	66.68 mm

BORG-WARNR 8171J
ROCKWELL 62NYS31-5

IDLI GROUP 81689

END YOKE WITH 10 SPLINES
SEE FIGURE 7A

5.5313 in	IE	140.50 mm
5.8438 in	OE	148.43 mm
1.9688 in	SB	50.01 mm
3.3750 in	EC	85.73 mm
2.3750 in	HD	60.33 mm

ROCKWELL 62NYS31-1

IDLI GROUP 81690

END YOKE WITH 10 SPLINES
SEE FIGURE 7A

5.5313 in	IE	140.50 mm
5.8438 in	OE	148.43 mm
1.9688 in	SB	50.01 mm
3.6250 in	EC	92.08 mm
2.8750 in	HD	73.03 mm

ROCKWELL 62NYS31-22

IDLI GROUP 81691

END YOKE WITH 10 SPLINES
SEE FIGURE 7A

5.5313 in	IE	140.50 mm
5.8438 in	OE	148.43 mm
1.9688 in	SB	50.01 mm
3.7500 in	EC	95.25 mm
2.5000 in	HD	63.50 mm

BORG-WARNR 15657J
ROCKWELL 62NYS31-2

IDLI GROUP 81692

END YOKE WITH 10 SPLINES
SEE FIGURE 7A

5.5313 in	IE	140.50 mm
5.8438 in	OE	148.43 mm
1.9688 in	SB	50.01 mm
3.8750 in	EC	98.43 mm
2.7500 in	HD	69.85 mm

BORG-WARNR 24475J
ROCKWELL 62NYS31-17

RD = BEARING DIAMETER (outside)   BW = BEARING WIDTH   CB = CROSS LENGTH WITH BEARINGS   CC = CENTER TO CENTER   CD = CROSS DIAMETER
CL = CROSS LENGTH WITHOUT BEARINGS   EC = END TO CENTER or FACE TO FACE   EE = END TO END   EL = EFFECTIVE LENGTH
HD = HUB DIAMETER (or insert)   IE = INSIDE OF EARS (or recess)   OD = OUTSIDE DIAMETER   OE = OUTSIDE OF EARS   SB = SPLINE OR BORE SIZE

IDLI GROUP 81693

END YOKE WITH 10 SPLINES
SEE FIGURE 7A

5.5313 in	IE	140.50 mm
5.8438 in	OE	148.43 mm
1.9688 in	SB	50.01 mm
4.2500 in	EC	107.95 mm
3.0000 in	HD	76.20 mm

BORG-WARNR 19127J
ROCKWELL 62NYS31

IDLI GROUP 81694

END YOKE WITH 10 SPLINES
SEE FIGURE 7A

5.5313 in	IE	140.50 mm
5.8438 in	OE	148.43 mm
1.9688 in	SB	50.01 mm
4.5625 in	EC	115.89 mm
3.0000 in	HD	76.20 mm

BORG-WARNR 22708J
BORG-WARNR 22719J
ROCKWELL 62NYS31-10

IDLI GROUP 81695

END YOKE WITH 19 INVOLUTED SPLINES
SEE FIGURE 7A

5.5313 in	IE	140.50 mm
5.8438 in	OE	148.43 mm
2.0000 in	SB	50.80 mm
3.5000 in	EC	88.90 mm
2.8750 in	HD	73.03 mm

ROCKWELL 62NYS32-10

IDLI GROUP 81696

END YOKE WITH 10 SPLINES
SEE FIGURE 7A

5.5313 in	IE	140.50 mm
5.8438 in	OE	148.43 mm
2.0000 in	SB	50.80 mm
3.8125 in	EC	96.84 mm
3.3750 in	HD	85.73 mm

ROCKWELL 62NYS32-23

IDLI GROUP 81697

END YOKE WITH 10 SPLINES
SEE FIGURE 7A

5.5313 in	IE	140.50 mm
5.8438 in	OE	148.43 mm
2.0000 in	SB	50.80 mm
3.9375 in	EC	100.01 mm
2.9375 in	HD	74.61 mm

BORG-WARNR 12956
BORG-WARNR 12956J
ROCKWELL 62NYS32-4

IDLI GROUP 81698

END YOKE WITH 10 SPLINES
SEE FIGURE 7A

5.5313 in	IE	140.50 mm
5.8438 in	OE	148.43 mm
2.0000 in	SB	50.80 mm
4.1875 in	EC	106.36 mm
3.2500 in	HD	82.55 mm

ROCKWELL 62NYS32-3

IDLI GROUP 81699

END YOKE WITH 10 SPLINES
SEE FIGURE 7A

5.5313 in	IE	140.50 mm
5.8438 in	OE	148.43 mm
2.0000 in	SB	50.80 mm
4.2500 in	EC	107.95 mm
2.9375 in	HD	74.61 mm

BORG-WARNR 19551J
ROCKWELL 62NYS32-16

IDLI GROUP 81700

END YOKE WITH 16 SPLINES
SEE FIGURE 7A

5.5313 in	IE	140.50 mm
5.8438 in	OE	148.43 mm
2.0000 in	SB	50.80 mm
4.2500 in	EC	107.95 mm
3.0000 in	HD	76.20 mm

ROCKWELL 62NYS32-7

IDLI GROUP 81701

END YOKE WITH 16 INVOLUTED SPLINES
SEE FIGURE 7A

5.5313 in	IE	140.50 mm
5.8438 in	OE	148.43 mm
2.1250 in	SB	53.98 mm
3.5000 in	EC	88.90 mm
2.6250 in	HD	66.68 mm

BORG-WARNR 14977J
ROCKWELL 62NYS34-9

IDLI GROUP 81702

END YOKE WITH 16 INVOLUTED SPLINES
SEE FIGURE 7A

5.5313 in	IE	140.50 mm
5.8438 in	OE	148.43 mm
2.2188 in	SB	56.36 mm
3.3750 in	EC	85.73 mm
2.6250 in	HD	66.68 mm

BORG-WARNR 23749J
ROCKWELL 62NYS36-2

IDLI GROUP 81703

END YOKE WITH 16 SPLINES
SEE FIGURE 7A

5.5313 in	IE	140.50 mm
5.8438 in	OE	148.43 mm
2.5000 in	SB	63.50 mm
3.6250 in	EC	92.08 mm
3.1250 in	HD	79.38 mm

ROCKWELL 62NYS40-3

IDLI GROUP 81704

END YOKE WITH 10 SPLINES
SEE FIGURE 7A

5.5313 in	IE	140.50 mm
5.8438 in	OE	148.43 mm
2.5000 in	SB	63.50 mm
3.8750 in	EC	98.43 mm
3.1250 in	HD	79.38 mm

ROCKWELL 62NYS40-5

IDLI GROUP 81705

END YOKE WITH 10 SPLINES
SEE FIGURE 7A

5.5313 in	IE	140.50 mm
5.8438 in	OE	148.43 mm
2.5000 in	SB	63.50 mm
3.8750 in	EC	98.43 mm
3.2500 in	HD	82.55 mm

ROCKWELL 62NYS40-6

IDLI GROUP 81706

END YOKE WITH 29 INVOLUTED SPLINES
SEE FIGURE 7A

5.5313 in	IE	140.50 mm
5.8438 in	OE	148.43 mm
3.0000 in	SB	76.20 mm
3.4219 in	EC	86.92 mm
4.9375 in	HD	125.41 mm

ROCKWELL 62NYS48-1

IDLI GROUP 81707

END YOKE WITH 10 SPLINES
SEE FIGURE 7A

5.8438 in	IE	148.43 mm
6.2188 in	OE	157.96 mm
1.2500 in	SB	31.75 mm
2.8750 in	EC	73.03 mm
1.8750 in	HD	47.63 mm

ROCKWELL 72NYS20

IDLI GROUP 81708

END YOKE WITH 10 SPLINES
SEE FIGURE 7A

5.8438 in	IE	148.43 mm
6.2188 in	OE	157.96 mm
1.3750 in	SB	34.93 mm
3.2500 in	EC	82.55 mm
2.1250 in	HD	53.98 mm

ROCKWELL 72NYS22-4

IDLI GROUP 81709

END YOKE WITH 10 SPLINES
SEE FIGURE 7A

5.8438 in	IE	148.43 mm
6.2188 in	OE	157.96 mm
1.3750 in	SB	34.93 mm
3.3750 in	EC	85.73 mm
2.0000 in	HD	50.80 mm

ROCKWELL 72NYS22

IDLI GROUP 81710

END YOKE WITH 10 SPLINES
SEE FIGURE 7A

5.8438 in	IE	148.43 mm
6.2188 in	OE	157.96 mm
1.5000 in	SB	38.10 mm
3.0000 in	EC	76.20 mm
2.2500 in	HD	57.15 mm

BORG-WARNR 19784J
ROCKWELL 72NYS24-3

IDLI GROUP 81711

END YOKE WITH ROUND BORE
SEE FIGURE 7B

5.8438 in	IE	148.43 mm
6.2188 in	OE	157.96 mm
1.5000 in	SB	38.10 mm
3.2500 in	EC	82.55 mm
4.0000 in	HD	101.60 mm

ROCKWELL 72NYR24-5

IDLI GROUP 81712

END YOKE WITH ROUND BORE
SEE FIGURE 7B

5.8438 in	IE	148.43 mm
6.2188 in	OE	157.96 mm
1.5000 in	SB	38.10 mm
3.3750 in	EC	85.73 mm
3.5000 in	HD	88.90 mm

ROCKWELL 72NYR24

IDLI GROUP 81713

END YOKE WITH 10 SPLINES
SEE FIGURE 7A

5.8438 in	IE	148.43 mm
6.2188 in	OE	157.96 mm
1.5000 in	SB	38.10 mm
3.4375 in	EC	87.31 mm
2.2500 in	HD	57.15 mm

BORG-WARNR 16835J
BORG-WARNR 17300J
ROCKWELL 72NYS24

IDLI GROUP 81714

END YOKE WITH 10 SPLINES
SEE FIGURE 7A

5.8438 in	IE	148.43 mm
6.2188 in	OE	157.96 mm
1.5000 in	SB	38.10 mm
3.5000 in	EC	88.90 mm
2.2500 in	HD	57.15 mm

ROCKWELL 72NYS24-5

IDLI GROUP 81715

END YOKE WITH 10 SPLINES
slight variation from above group
SEE FIGURE 7A

5.8438 in	IE	148.43 mm
6.2188 in	OE	157.96 mm
1.5000 in	SB	38.10 mm
3.5000 in	EC	88.90 mm
2.2500 in	HD	57.15 mm

ROCKWELL 72NYS24-13

IDLI GROUP 81716

END YOKE WITH 10 SPLINES
slight variation from above group
SEE FIGURE 7A

5.8438 in	IE	148.43 mm
6.2188 in	OE	157.96 mm
1.5000 in	SB	38.10 mm
3.5000 in	EC	88.90 mm
2.2500 in	HD	57.15 mm

ROCKWELL 72NYS24-22

ENGINEERING CATALOGS MUST BE CONSULTED FOR DETAILS NOT INCLUDED
IN THIS GUIDE. SPECIFIC DESIGNS, MATERIAL CONTENT, TOLERANCES
LUBE FITTINGS AND OTHER DIMENSIONS ARE INTENTIONALLY OMITTED HERE.

IDLI GUIDE   COPYRIGHT © INTERCHANGE, INC.   ST. LOUIS PARK, MN. 55416 USA

CAUTION: BE SURE TO REFER TO ENGINEERING CATALOGS FOR SPECIAL APPLICATIONS THAT REQUIRE SPECIFIC MATERIAL CONTENT, TOLERANCES, ETC. SEE FOOTNOTE.

IDLI GROUP 81717

END YOKE WITH 10 SPLINES
SEE FIGURE 7A

5.8438 in	IE	148.43 mm
6.2188 in	OE	157.96 mm
1.5000 in	SB	38.10 mm
3.5625 in	EC	90.49 mm
2.1250 in	HD	53.98 mm

ROCKWELL 72NYS24-23

IDLI GROUP 81718

END YOKE WITH 10 SPLINES
slight variation from above group
SEE FIGURE 7A

5.8438 in	IE	148.43 mm
6.2188 in	OE	157.96 mm
1.5000 in	SB	38.10 mm
3.5625 in	EC	90.49 mm
2.1250 in	HD	53.98 mm

BORG-WARNR 16855J
ROCKWELL 72NYS24-7

IDLI GROUP 81719

END YOKE WITH 10 SPLINES
slight variation from above group
SEE FIGURE 7A

5.8438 in	IE	148.43 mm
6.2188 in	OE	157.96 mm
1.5000 in	SB	38.10 mm
3.5625 in	EC	90.49 mm
2.1250 in	HD	53.98 mm

BORG-WARNR 23402J
ROCKWELL 72NYS24-12

IDLI GROUP 81720

END YOKE WITH ROUND BORE
SEE FIGURE 7B

5.8438 in	IE	148.43 mm
6.2188 in	OE	157.96 mm
1.5000 in	SB	38.10 mm
4.0000 in	EC	101.60 mm
3.0000 in	HD	76.20 mm

ROCKWELL 72NYR24-4

IDLI GROUP 81721

END YOKE WITH ROUND BORE
SEE FIGURE 7B

5.8438 in	IE	148.43 mm
6.2188 in	OE	157.96 mm
1.5000 in	SB	38.10 mm
4.9375 in	EC	125.41 mm
4.0000 in	HD	101.60 mm

ROCKWELL 72NYR24-2

IDLI GROUP 81722

END YOKE WITH ROUND BORE
SEE FIGURE 7B

5.8438 in	IE	148.43 mm
6.2188 in	OE	157.96 mm
1.5781 in	SB	40.08 mm
3.5625 in	EC	90.49 mm
3.3750 in	HD	85.73 mm

ROCKWELL 72NYR25

IDLI GROUP 81723

END YOKE WITH 25 INVOLUTED SPLINES
SEE FIGURE 7A

5.8438 in	IE	148.43 mm
6.2188 in	OE	157.96 mm
1.6250 in	SB	41.28 mm
3.2188 in	EC	81.76 mm
2.2500 in	HD	57.15 mm

BORG-WARNR 13569J
ROCKWELL 72NYS26-1

IDLI GROUP 81724

END YOKE WITH TAPERED BORE
SEE FIGURE 7F

5.8438 in	IE	148.43 mm
6.2188 in	OE	157.96 mm
1.6250 in	SB	41.28 mm
3.2500 in	EC	82.55 mm
3.0000 in	HD	76.20 mm

ROCKWELL 72NYT26-2

IDLI GROUP 81725

END YOKE WITH 25 INVOLUTED SPLINES
SEE FIGURE 7A

5.8438 in	IE	148.43 mm
6.2188 in	OE	157.96 mm
1.6250 in	SB	41.28 mm
3.3125 in	EC	84.14 mm
2.3594 in	HD	59.93 mm

BORG-WARNR 13568J
ROCKWELL 72NYS26

IDLI GROUP 81726

END YOKE WITH 25 INVOLUTED SPLINES
slight variation from above group
SEE FIGURE 7A

5.8438 in	IE	148.43 mm
6.2188 in	OE	157.96 mm
1.6250 in	SB	41.28 mm
3.3125 in	EC	84.14 mm
2.3594 in	HD	59.93 mm

ROCKWELL 72NYS26-11

IDLI GROUP 81727

END YOKE WITH 10 SPLINES
SEE FIGURE 7A

5.8438 in	IE	148.43 mm
6.2188 in	OE	157.96 mm
1.6250 in	SB	41.28 mm
3.3750 in	EC	85.73 mm
2.2500 in	HD	57.15 mm

ROCKWELL 72NYS26-16

IDLI GROUP 81728

END YOKE WITH 10 SPLINES
SEE FIGURE 7A

5.8438 in	IE	148.43 mm
6.2188 in	OE	157.96 mm
1.6250 in	SB	41.28 mm
3.3750 in	EC	85.73 mm
2.3750 in	HD	60.33 mm

BORG-WARNR 12637J
ROCKWELL 72NYS26-3

IDLI GROUP 81729

END YOKE WITH 10 SPLINES
slight variation from above group
SEE FIGURE 7A

5.8438 in	IE	148.43 mm
6.2188 in	OE	157.96 mm
1.6250 in	SB	41.28 mm
3.3750 in	EC	85.73 mm
2.3750 in	HD	60.33 mm

BORG-WARNR 20496J
ROCKWELL 72NYS26-7

IDLI GROUP 81730

END YOKE WITH 10 SPLINES
slight variation from above group
SEE FIGURE 7A

5.8438 in	IE	148.43 mm
6.2188 in	OE	157.96 mm
1.6250 in	SB	41.28 mm
3.3750 in	EC	85.73 mm
2.3750 in	HD	60.33 mm

ROCKWELL 72NYS26-8

IDLI GROUP 81731

END YOKE WITH ROUND BORE
SEE FIGURE 7B

5.8438 in	IE	148.43 mm
6.2188 in	OE	157.96 mm
1.6250 in	SB	41.28 mm
3.5625 in	EC	90.49 mm
2.7500 in	HD	69.85 mm

ROCKWELL 72NYR26

IDLI GROUP 81732

END YOKE WITH 10 SPLINES
SEE FIGURE 7A

5.8438 in	IE	148.43 mm
6.2188 in	OE	157.96 mm
1.6250 in	SB	41.28 mm
3.9375 in	EC	100.01 mm
2.3750 in	HD	60.33 mm

ROCKWELL 72NYS26-26

IDLI GROUP 81733

END YOKE WITH 10 SPLINES
SEE FIGURE 7A

5.8438 in	IE	148.43 mm
6.2188 in	OE	157.96 mm
1.6250 in	SB	41.28 mm
4.2500 in	EC	107.95 mm
2.3750 in	HD	60.33 mm

BORG-WARNR 18395J
ROCKWELL 72NYS26-4

IDLI GROUP 81734

END YOKE WITH 10 SPLINES
SEE FIGURE 7A

5.8438 in	IE	148.43 mm
6.2188 in	OE	157.96 mm
1.6406 in	SB	41.67 mm
3.2500 in	EC	82.55 mm
2.7500 in	HD	69.85 mm

BORG-WARNR 12949J
ROCKWELL 72NYS26-2

IDLI GROUP 81735

END YOKE WITH ROUND BORE
SEE FIGURE 7B

5.8438 in	IE	148.43 mm
6.2188 in	OE	157.96 mm
1.6875 in	SB	42.86 mm
3.5625 in	EC	90.49 mm
3.3750 in	HD	85.73 mm

ROCKWELL 72NYR27-2

IDLI GROUP 81736

END YOKE WITH ROUND BORE
SEE FIGURE 7B

5.8438 in	IE	148.43 mm
6.2188 in	OE	157.96 mm
1.6875 in	SB	42.86 mm
4.0000 in	EC	101.60 mm
3.0000 in	HD	76.20 mm

ROCKWELL 72NYR27

IDLI GROUP 81737

END YOKE WITH ROUND BORE
SEE FIGURE 7B

5.8438 in	IE	148.43 mm
6.2188 in	OE	157.96 mm
1.7500 in	SB	44.45 mm
3.1250 in	EC	79.38 mm
2.7500 in	HD	69.85 mm

ROCKWELL 72NYR28-3

IDLI GROUP 81738

END YOKE WITH ROUND BORE
SEE FIGURE 7B

5.8438 in	IE	148.43 mm
6.2188 in	OE	157.96 mm
1.7500 in	SB	44.45 mm
3.1875 in	EC	80.96 mm
2.7500 in	HD	69.85 mm

ROCKWELL 72NYR28-4

IDLI GROUP 81739

END YOKE WITH 10 SPLINES
SEE FIGURE 7A

5.8438 in	IE	148.43 mm
6.2188 in	OE	157.96 mm
1.7500 in	SB	44.45 mm
3.2500 in	EC	82.55 mm
2.5000 in	HD	63.50 mm

ROCKWELL 72NYS28-9

BD=BEARING DIAMETER (outside)   BW=BEARING WIDTH   CB=CROSS LENGTH WITH BEARINGS   CC=CENTER TO CENTER   CD=CROSS DIAMETER
CL=CROSS LENGTH WITHOUT BEARINGS   EC=END TO CENTER or FACE TO FACE   EE=END TO END   EL=EFFECTIVE LENGTH
HD=HUB DIAMETER (or insert)   IE=INSIDE OF EARS (or recess)   OD=OUTSIDE DIAMETER   OE=OUTSIDE OF EARS   SB=SPLINE OR BORE SIZE

IDLI GROUP 81740

END YOKE WITH 10 SPLINES
SEE FIGURE 7A

5.8438 in	IE	148.43 mm
6.2188 in	OE	157.96 mm
1.7500 in	SB	44.45 mm
3.2500 in	EC	82.55 mm
2.6250 in	HD	66.68 mm

BORG-WARNR 7404J
ROCKWELL 72NYS28-1

IDLI GROUP 81741

END YOKE WITH 27 INVOLUTED SPLINES
SEE FIGURE 7A

5.8438 in	IE	148.43 mm
6.2188 in	OE	157.96 mm
1.7500 in	SB	44.45 mm
3.3125 in	EC	84.14 mm
2.6250 in	HD	66.68 mm

BORG-WARNR 24063J
ROCKWELL 72NYS28-84

IDLI GROUP 81742

END YOKE WITH 10 SPLINES
SEE FIGURE 7A

5.8438 in	IE	148.43 mm
6.2188 in	OE	157.96 mm
1.7500 in	SB	44.45 mm
3.3750 in	EC	85.73 mm
3.0000 in	HD	76.20 mm

ROCKWELL 72NYS28-49

IDLI GROUP 81743

END YOKE WITH 10 SPLINES
SEE FIGURE 7A

5.8438 in	IE	148.43 mm
6.2188 in	OE	157.96 mm
1.7500 in	SB	44.45 mm
3.4375 in	EC	87.31 mm
2.3750 in	HD	60.33 mm

BORG-WARNR 8401J
ROCKWELL 72NYS28-5

IDLI GROUP 81744

END YOKE WITH 10 SPLINES
SEE FIGURE 7A

5.8438 in	IE	148.43 mm
6.2188 in	OE	157.96 mm
1.7500 in	SB	44.45 mm
3.5000 in	EC	88.90 mm
2.1875 in	HD	55.56 mm

BORG-WARNR 10508J
FORD B5Q4865A
FORD TGAY4865A
FORD TWAM4865A
ROCKWELL 72NYS28-32

IDLI GROUP 81745

END YOKE WITH 10 SPLINES
slight variation from above group
SEE FIGURE 7A

5.8438 in	IE	148.43 mm
6.2188 in	OE	157.96 mm
1.7500 in	SB	44.45 mm
3.5000 in	EC	88.90 mm
2.1875 in	HD	55.56 mm

BORG-WARNR 24657J
ROCKWELL 72NYS28-107

IDLI GROUP 81746

END YOKE WITH 10 SPLINES
SEE FIGURE 7A

5.8438 in	IE	148.43 mm
6.2188 in	OE	157.96 mm
1.7500 in	SB	44.45 mm
3.5000 in	EC	88.90 mm
2.5000 in	HD	63.50 mm

ROCKWELL 72NYS28-85

IDLI GROUP 81747

END YOKE WITH 10 SPLINES
slight variation from above group
SEE FIGURE 7A

5.8438 in	IE	148.43 mm
6.2188 in	OE	157.96 mm
1.7500 in	SB	44.45 mm
3.5000 in	EC	88.90 mm
2.5000 in	HD	63.50 mm

BORG-WARNR 8437J
ROCKWELL 72NYS28-51

IDLI GROUP 81748

END YOKE WITH 10 SPLINES
SEE FIGURE 7A

5.8438 in	IE	148.43 mm
6.2188 in	OE	157.96 mm
1.7500 in	SB	44.45 mm
3.5000 in	EC	88.90 mm
2.6250 in	HD	66.68 mm

BORG-WARNR 17310J
BORG-WARNR 5051J
ROCKWELL 72NYS28-2

IDLI GROUP 81749

END YOKE WITH 10 SPLINES
slight variation from above group
SEE FIGURE 7A

5.8438 in	IE	148.43 mm
6.2188 in	OE	157.96 mm
1.7500 in	SB	44.45 mm
3.5000 in	EC	88.90 mm
2.6250 in	HD	66.68 mm

BORG-WARNR 4215J
ROCKWELL 72NYS28

IDLI GROUP 81750

END YOKE WITH 10 SPLINES
slight variation from above group
SEE FIGURE 7A

5.8438 in	IE	148.43 mm
6.2188 in	OE	157.96 mm
1.7500 in	SB	44.45 mm
3.5000 in	EC	88.90 mm
2.6250 in	HD	66.68 mm

BORG-WARNR 20648J
BORG-WARNR 4215J
ROCKWELL 72NYS28-8

IDLI GROUP 81751

END YOKE WITH 10 SPLINES
slight variation from above group
SEE FIGURE 7A

5.8438 in	IE	148.43 mm
6.2188 in	OE	157.96 mm
1.7500 in	SB	44.45 mm
3.5000 in	EC	88.90 mm
2.6250 in	HD	66.68 mm

BORG-WARNR 12487J
BORG-WARNR 12847J
ROCKWELL 72NYS28-65

IDLI GROUP 81752

END YOKE WITH 16 SPLINES
SEE FIGURE 7A

5.8438 in	IE	148.43 mm
6.2188 in	OE	157.96 mm
1.7500 in	SB	44.45 mm
3.5000 in	EC	88.90 mm
2.6250 in	HD	66.68 mm

BORG-WARNR 12928J
ROCKWELL 72NYS28-11

IDLI GROUP 81753

END YOKE WITH 16 SPLINES
SEE FIGURE 7A

5.8438 in	IE	148.43 mm
6.2188 in	OE	157.96 mm
1.7500 in	SB	44.45 mm
3.7500 in	EC	95.25 mm
2.6875 in	HD	68.26 mm

BORG-WARNR 12927J
BORG-WARNR 21333J
ROCKWELL 72NYS28-33

IDLI GROUP 81754

END YOKE WITH 10 SPLINES
SEE FIGURE 7A

5.8438 in	IE	148.43 mm
6.2188 in	OE	157.96 mm
1.7500 in	SB	44.45 mm
3.5000 in	EC	88.90 mm
2.7500 in	HD	69.85 mm

BORG-WARNR 20958J
ROCKWELL 72NYS28-174

IDLI GROUP 81755

END YOKE WITH 10 SPLINES
slight variation from above group
SEE FIGURE 7A

5.8438 in	IE	148.43 mm
6.2188 in	OE	157.96 mm
1.7500 in	SB	44.45 mm
3.5000 in	EC	88.90 mm
2.7500 in	HD	69.85 mm

BORG-WARNR 7924J
ROCKWELL 72NYS28-6

IDLI GROUP 81756

END YOKE WITH 10 SPLINES
slight variation from above group
SEE FIGURE 7A

5.8438 in	IE	148.43 mm
6.2188 in	OE	157.96 mm
1.7500 in	SB	44.45 mm
3.5000 in	EC	88.90 mm
2.7500 in	HD	69.85 mm

ROCKWELL 72NYS28-21

IDLI GROUP 81757

END YOKE WITH 10 SPLINES
slight variation from above group
SEE FIGURE 7A

5.8438 in	IE	148.43 mm
6.2188 in	OE	157.96 mm
1.7500 in	SB	44.45 mm
3.5000 in	EC	88.90 mm
2.7500 in	HD	69.85 mm

BORG-WARNR 21292J
ROCKWELL 72NYS28-42

IDLI GROUP 81758

END YOKE WITH 10 SPLINES
SEE FIGURE 7A

5.8438 in	IE	148.43 mm
6.2188 in	OE	157.96 mm
1.7500 in	SB	44.45 mm
3.5000 in	EC	88.90 mm
3.0000 in	HD	76.20 mm

ROCKWELL 72NYS28-15

IDLI GROUP 81759

END YOKE WITH 10 SPLINES
slight variation from above group
SEE FIGURE 7A

5.8438 in	IE	148.43 mm
6.2188 in	OE	157.96 mm
1.7500 in	SB	44.45 mm
3.5000 in	EC	88.90 mm
3.0000 in	HD	76.20 mm

BORG-WARNR 6621J
ROCKWELL 72NYS28-4

IDLI GROUP 81760

END YOKE WITH 10 SPLINES
SEE FIGURE 7A

5.8438 in	IE	148.43 mm
6.2188 in	OE	157.96 mm
1.7500 in	SB	44.45 mm
3.7500 in	EC	95.25 mm
2.6250 in	HD	66.68 mm

BORG-WARNR 10945J

ENGINEERING CATALOGS MUST BE CONSULTED FOR DETAILS NOT INCLUDED
IN THIS GUIDE. SPECIFIC DESIGNS, MATERIAL CONTENT, TOLERANCES
LUBE FITTINGS AND OTHER DIMENSIONS ARE INTENTIONALLY OMITTED HERE.

CAUTION: BE SURE TO REFER TO ENGINEERING CATALOGS FOR SPECIAL APPLICATIONS THAT REQUIRE SPECIFIC MATERIAL CONTENT, TOLERANCES, ETC. SEE FOOTNOTE.

ROCKWELL 72NYS28-47

IDLI GROUP 81761

END YOKE WITH 10 SPLINES
SEE FIGURE 7A

5.8438 in	IE	148.43 mm
6.2188 in	OE	157.96 mm
1.7500 in	SB	44.45 mm
3.7500 in	EC	95.25 mm
2.7500 in	HD	69.85 mm

ROCKWELL 72NYS28-27

IDLI GROUP 81762

END YOKE WITH 10 SPLINES
SEE FIGURE 7A

5.8438 in	IE	148.43 mm
6.2188 in	OE	157.96 mm
1.7500 in	SB	44.45 mm
3.8125 in	EC	96.84 mm
2.3750 in	HD	60.33 mm

BORG-WARNR 21869J
BORG-WARNR 4992J
ROCKWELL 72NYS28-16

IDLI GROUP 81763

END YOKE WITH 10 SPLINES
SEE FIGURE 7A

5.8438 in	IE	148.43 mm
6.2188 in	OE	157.96 mm
1.7500 in	SB	44.45 mm
3.8125 in	EC	96.84 mm
2.6250 in	HD	66.68 mm

BORG-WARNR 20648J
ROCKWELL 72NYS28-28

IDLI GROUP 81764

END YOKE WITH 10 SPLINES
SEE FIGURE 7A

5.8438 in	IE	148.43 mm
6.2188 in	OE	157.96 mm
1.7500 in	SB	44.45 mm
3.8750 in	EC	98.43 mm
2.5000 in	HD	63.50 mm

ROCKWELL 72NYS28-71

IDLI GROUP 81765

END YOKE WITH 10 SPLINES
SEE FIGURE 7A

5.8438 in	IE	148.43 mm
6.2188 in	OE	157.96 mm
1.7500 in	SB	44.45 mm
3.8750 in	EC	98.43 mm
2.7500 in	HD	69.85 mm

BORG-WARNR 20653J
ROCKWELL 72NYS28-30

IDLI GROUP 81766

END YOKE WITH 10 SPLINES
SEE FIGURE 7A

5.8438 in	IE	148.43 mm
6.2188 in	OE	157.96 mm
1.7500 in	SB	44.45 mm
3.9375 in	EC	100.01 mm
2.3750 in	HD	60.33 mm

ROCKWELL 72NYS28-153

IDLI GROUP 81767

END YOKE WITH 10 SPLINES
slight variation from above group
SEE FIGURE 7A

5.8438 in	IE	148.43 mm
6.2188 in	OE	157.96 mm
1.7500 in	SB	44.45 mm
3.9375 in	EC	100.01 mm
2.3750 in	HD	60.33 mm

BORG-WARNR 16316J
ROCKWELL 72NYS28-37

IDLI GROUP 81768

END YOKE WITH TAPERED BORE
SEE FIGURE 7F

5.8438 in	IE	148.43 mm
6.2188 in	OE	157.96 mm
1.7500 in	SB	44.45 mm
3.9375 in	EC	100.01 mm
3.0000 in	HD	76.20 mm

ROCKWELL 72NYT28-8

IDLI GROUP 81769

END YOKE WITH 10 SPLINES
SEE FIGURE 7A

5.8438 in	IE	148.43 mm
6.2188 in	OE	157.96 mm
1.7500 in	SB	44.45 mm
4.0000 in	EC	101.60 mm
2.5000 in	HD	63.50 mm

ROCKWELL 72NYS28-170

IDLI GROUP 81770

END YOKE WITH 10 SPLINES
SEE FIGURE 7A

5.8438 in	IE	148.43 mm
6.2188 in	OE	157.96 mm
1.7500 in	SB	44.45 mm
4.0000 in	EC	101.60 mm
2.6250 in	HD	66.68 mm

ROCKWELL 72NYS28-23

IDLI GROUP 81771

END YOKE WITH 10 SPLINES
slight variation from above group
SEE FIGURE 7A

5.8438 in	IE	148.43 mm
6.2188 in	OE	157.96 mm
1.7500 in	SB	44.45 mm
4.0000 in	EC	101.60 mm
2.6250 in	HD	66.68 mm

ROCKWELL 72NYS28-63

IDLI GROUP 81772

END YOKE WITH 10 SPLINES
SEE FIGURE 7A

5.8438 in	IE	148.43 mm
6.2188 in	OE	157.96 mm
1.7500 in	SB	44.45 mm
4.0000 in	EC	101.60 mm
2.7500 in	HD	69.85 mm

BORG-WARNR 13033J
ROCKWELL 72NYS28-7

IDLI GROUP 81773

END YOKE WITH 10 SPLINES
SEE FIGURE 7A

5.8438 in	IE	148.43 mm
6.2188 in	OE	157.96 mm
1.7500 in	SB	44.45 mm
4.0000 in	EC	101.60 mm
3.3750 in	HD	85.73 mm

ROCKWELL 72NYS28-161

IDLI GROUP 81774

END YOKE WITH 10 SPLINES
SEE FIGURE 7A

5.8438 in	IE	148.43 mm
6.2188 in	OE	157.96 mm
1.7500 in	SB	44.45 mm
4.1094 in	EC	104.38 mm
2.5000 in	HD	63.50 mm

ROCKWELL 72NYS28-3

IDLI GROUP 81775

END YOKE WITH 27 INVOLUTED SPLINES
SEE FIGURE 7A

5.8438 in	IE	148.43 mm
6.2188 in	OE	157.96 mm
1.7500 in	SB	44.45 mm
4.1250 in	EC	104.78 mm
2.6250 in	HD	66.68 mm

ROCKWELL 72NYS28-122

IDLI GROUP 81776

5.8438 in	IE	148.43 mm
6.2188 in	OE	157.96 mm
1.7500 in	SB	44.45 mm
4.1250 in	EC	104.78 mm
2.6250 in	HD	66.68 mm

ROCKWELL 72NYS28-145

IDLI GROUP 81777

5.8438 in	IE	148.43 mm
6.2188 in	OE	157.96 mm
1.7500 in	SB	44.45 mm
4.1250 in	EC	104.78 mm
2.6250 in	HD	66.68 mm

ROCKWELL 72NYS28-167

IDLI GROUP 81778

END YOKE WITH 10 SPLINES
SEE FIGURE 7A

5.8438 in	IE	148.43 mm
6.2188 in	OE	157.96 mm
1.7500 in	SB	44.45 mm
4.1250 in	EC	104.78 mm
2.7500 in	HD	69.85 mm

ROCKWELL 72NYS28-80

IDLI GROUP 81779

END YOKE WITH 27 INVOLUTED SPLINES
SEE FIGURE 7A

5.8438 in	IE	148.43 mm
6.2188 in	OE	157.96 mm
1.7500 in	SB	44.45 mm
4.1250 in	EC	104.78 mm
3.3750 in	HD	85.73 mm

ROCKWELL 72NYS28-127

IDLI GROUP 81780

END YOKE WITH TAPERED BORE
SEE FIGURE 7F

5.8438 in	IE	148.43 mm
6.2188 in	OE	157.96 mm
1.7500 in	SB	44.45 mm
4.1875 in	EC	106.36 mm
2.7500 in	HD	69.85 mm

ROCKWELL 72NYT28-6

IDLI GROUP 81781

END YOKE WITH 10 SPLINES
SEE FIGURE 7A

5.8438 in	IE	148.43 mm
6.2188 in	OE	157.96 mm
1.7500 in	SB	44.45 mm
4.1875 in	EC	106.36 mm
3.0000 in	HD	76.20 mm

ROCKWELL 72NYS28-25

IDLI GROUP 81782

END YOKE WITH 10 SPLINES
slight variation from above group
SEE FIGURE 7A

5.8438 in	IE	148.43 mm
6.2188 in	OE	157.96 mm
1.7500 in	SB	44.45 mm
4.1875 in	EC	106.36 mm
3.0000 in	HD	76.20 mm

BORG-WARNR 23399J
ROCKWELL 72NYS28-57

IDLI GROUP 81783

END YOKE WITH TAPERED BORE
SEE FIGURE 7F

5.8438 in	IE	148.43 mm
6.2188 in	OE	157.96 mm
1.7500 in	SB	44.45 mm
4.1875 in	EC	106.36 mm
3.0000 in	HD	76.20 mm

ROCKWELL 72NYT28-2

BD = BEARING DIAMETER (outside)   BW = BEARING WIDTH   CD = CROSS LENGTH WITH BEARINGS   CC = CENTER TO CENTER   CD = CROSS DIAMETER
CL = CROSS LENGTH WITHOUT BEARINGS   EC = END TO CENTER or FACE TO FACE   EE = END TO END   EL = EFFECTIVE LENGTH
HD = HUB DIAMETER (or insert)   IE = INSIDE OF EARS (or recess)   OD = OUTSIDE DIAMETER   OE = OUTSIDE OF EARS   SB = SPLINE OR BORE SIZE

IDLI GROUP 81784

END YOKE WITH 10 SPLINES
SEE FIGURE 7A

5.8438 in	IE	148.43 mm
6.2188 in	OE	157.96 mm
1.7500 in	SB	44.45 mm
4.3750 in	EC	111.13 mm
2.6250 in	HD	66.68 mm

BORG-WARNR 13947J
ROCKWELL 72NYS28-39

IDLI GROUP 81785

END YOKE WITH 10 SPLINES
slight variation from above group
SEE FIGURE 7A

5.8438 in	IE	148.43 mm
6.2188 in	OE	157.96 mm
1.7500 in	SB	44.45 mm
4.3750 in	EC	111.13 mm
2.6250 in	HD	66.68 mm

ROCKWELL 72NYS28-154

IDLI GROUP 81786

END YOKE WITH 10 SPLINES
SEE FIGURE 7A

5.8438 in	IE	148.43 mm
6.2188 in	OE	157.96 mm
1.7500 in	SB	44.45 mm
4.5000 in	EC	114.30 mm
2.7500 in	HD	69.85 mm

BORG-WARNR 17184J
BORG-WARNR 29145J
ROCKWELL 72NYS28-10

IDLI GROUP 81787

END YOKE WITH ROUND BORE
SEE FIGURE 7B

5.8438 in	IE	148.43 mm
6.2188 in	OE	157.96 mm
1.7500 in	SB	44.45 mm
4.6250 in	EC	117.48 mm
2.4375 in	HD	61.91 mm

ROCKWELL 72NYR28-7

IDLI GROUP 81788

END YOKE WITH ROUND BORE
SEE FIGURE 7B

5.8438 in	IE	148.43 mm
6.2188 in	OE	157.96 mm
1.7500 in	SB	44.45 mm
4.6250 in	EC	117.48 mm
4.0000 in	HD	101.60 mm

ROCKWELL 72NYR28-2

IDLI GROUP 81789

END YOKE WITH ROUND BORE
SEE FIGURE 7B

5.8438 in	IE	148.43 mm
6.2188 in	OE	157.96 mm
1.7500 in	SB	44.45 mm
4.8125 in	EC	122.24 mm
4.0000 in	HD	101.60 mm

ROCKWELL 72NYR28-14

IDLI GROUP 81790

END YOKE WITH 34 SLANTED SPLINES
SEE FIGURE 7A

5.8438 in	IE	148.43 mm
6.2188 in	OE	157.96 mm
1.7656 in	SB	44.85 mm
3.6250 in	EC	92.08 mm
2.5000 in	HD	63.50 mm

ROCKWELL 72NYS28-176

IDLI GROUP 81791

END YOKE WITH 34 SLANTED SPLINES
slight variation from above group
SEE FIGURE 7A

5.8438 in	IE	148.43 mm
6.2188 in	OE	157.96 mm
1.7656 in	SB	44.85 mm
3.6250 in	EC	92.08 mm
2.5000 in	HD	63.50 mm

ROCKWELL 72NYS28-151A

IDLI GROUP 81792

END YOKE WITH 34 SLANTED SPLINES
slight variation from above group
SEE FIGURE 7A

5.8438 in	IE	148.43 mm
6.2188 in	OE	157.96 mm
1.7656 in	SB	44.85 mm
3.6250 in	EC	92.08 mm
2.5000 in	HD	63.50 mm

ROCKWELL 72NYS28-176A

IDLI GROUP 81793

END YOKE WITH 34 SLANTED SPLINES
SEE FIGURE 7A

5.8438 in	IE	148.43 mm
6.2188 in	OE	157.96 mm
1.7656 in	SB	44.85 mm
3.7500 in	EC	95.25 mm
2.5000 in	HD	63.50 mm

ROCKWELL 72NYS28-166

IDLI GROUP 81794

END YOKE WITH 34 SLANTED SPLINES
SEE FIGURE 7A

5.8438 in	IE	148.43 mm
6.2188 in	OE	157.96 mm
1.7656 in	SB	44.85 mm
4.6875 in	EC	119.06 mm
2.5000 in	HD	63.50 mm

ROCKWELL 72NYS28-177

IDLI GROUP 81795

END YOKE WITH 34 SLANTED SPLINES
slight variation from above group
SEE FIGURE 7A

5.8438 in	IE	148.43 mm
6.2188 in	OE	157.96 mm
1.7656 in	SB	44.85 mm
4.6875 in	EC	119.06 mm
2.5000 in	HD	63.50 mm

ROCKWELL 72NYS28-157A

IDLI GROUP 81796

END YOKE WITH 34 SLANTED SPLINES
slight variation from above group
SEE FIGURE 7A

5.8438 in	IE	148.43 mm
6.2188 in	OE	157.96 mm
1.7656 in	SB	44.85 mm
4.6875 in	EC	119.06 mm
2.5000 in	HD	63.50 mm

ROCKWELL 72NYS28-177A

IDLI GROUP 81797

END YOKE WITH 16 INVOLUTED SPLINES
SEE FIGURE 7A

5.8438 in	IE	148.43 mm
6.2188 in	OE	157.96 mm
1.7813 in	SB	45.25 mm
3.2500 in	EC	82.55 mm
2.3750 in	HD	60.33 mm

BORG-WARNR 13670J
BORG-WARNR 23651J
ROCKWELL 72NYS28-79

IDLI GROUP 81798

END YOKE WITH 34 INVOLUTED SPLINES
SEE FIGURE 7A

5.8438 in	IE	148.43 mm
6.2188 in	OE	157.96 mm
1.7969 in	SB	45.64 mm
4.6875 in	EC	119.06 mm
2.5625 in	HD	65.09 mm

ROCKWELL 72NYS28-141A

IDLI GROUP 81799

END YOKE WITH ROUND BORE
SEE FIGURE 7B

5.8438 in	IE	148.43 mm
6.2188 in	OE	157.96 mm
1.8750 in	SB	47.63 mm
3.5625 in	EC	90.49 mm
3.5000 in	HD	88.90 mm

ROCKWELL 72NYR30

IDLI GROUP 81800

END YOKE WITH 10 SPLINES
SEE FIGURE 7A

5.8438 in	IE	148.43 mm
6.2188 in	OE	157.96 mm
1.8750 in	SB	47.63 mm
3.6250 in	EC	92.08 mm
2.7500 in	HD	69.85 mm

BORG-WARNR 18386J
ROCKWELL 72NYS30-5

IDLI GROUP 81801

END YOKE WITH 10 SPLINES
slight variation from above group
SEE FIGURE 7A

5.8438 in	IE	148.43 mm
6.2188 in	OE	157.96 mm
1.8750 in	SB	47.63 mm
3.6250 in	EC	92.08 mm
2.7500 in	HD	69.85 mm

BORG-WARNR 11398J
ROCKWELL 72NYS30-2

IDLI GROUP 81802

END YOKE WITH 10 SPLINES
slight variation from above group
SEE FIGURE 7A

5.8438 in	IE	148.43 mm
6.2188 in	OE	157.96 mm
1.8750 in	SB	47.63 mm
3.6250 in	EC	92.08 mm
2.7500 in	HD	69.85 mm

BORG-WARNR 8311J
ROCKWELL 72NYS30-3

IDLI GROUP 81803

END YOKE WITH 10 SPLINES
SEE FIGURE 7A

5.8438 in	IE	148.43 mm
6.2188 in	OE	157.96 mm
1.8750 in	SB	47.63 mm
4.0000 in	EC	101.60 mm
2.7500 in	HD	69.85 mm

ROCKWELL 72NYS30-9

IDLI GROUP 81804

END YOKE WITH 10 SPLINES
slight variation from above group
SEE FIGURE 7A

5.8438 in	IE	148.43 mm
6.2188 in	OE	157.96 mm
1.8750 in	SB	47.63 mm
4.0000 in	EC	101.60 mm
2.7500 in	HD	69.85 mm

ROCKWELL 72NYS30-11

IDLI GROUP 81805

END YOKE WITH 10 SPLINES
SEE FIGURE 7A

5.8438 in	IE	148.43 mm
6.2188 in	OE	157.96 mm
1.8750 in	SB	47.63 mm
4.2500 in	EC	107.95 mm
2.7500 in	HD	69.85 mm

ROCKWELL 72NYS30-16

IDLI GROUP 81806

END YOKE WITH 10 SPLINES
SEE FIGURE 7A

5.8438 in	IE	148.43 mm
6.2188 in	OE	157.96 mm
1.8750 in	SB	47.63 mm
4.5000 in	EC	114.30 mm
2.7500 in	HD	69.85 mm

BORG-WARNR 15533J
ROCKWELL 72NYS30-7

IDLI GROUP 81807

END YOKE WITH 14 INVOLUTED SPLINES
SEE FIGURE 7A

5.8438 in	IE	148.43 mm
6.2188 in	OE	157.96 mm
1.9063 in	SB	48.42 mm
3.2500 in	EC	82.55 mm
2.7500 in	HD	69.85 mm

ROCKWELL 72NYS30-6

CAUTION: BE SURE TO REFER TO ENGINEERING CATALOGS FOR SPECIAL APPLICATIONS THAT REQUIRE SPECIFIC MATERIAL CONTENT, TOLERANCES, ETC. SEE FOOTNOTE.

IDLI GROUP 81808
END YOKE WITH 14 INVOLUTED SPLINES
SEE FIGURE 7A

5.8438 in	IE	148.43 mm
6.2188 in	OE	157.96 mm
1.9063 in	SB	48.42 mm
3.7500 in	EC	95.25 mm
2.6250 in	HD	66.68 mm

ROCKWELL 72NYS30-1

IDLI GROUP 81809
END YOKE WITH 14 INVOLUTED SPLINES
SEE FIGURE 7A

5.8438 in	IE	148.43 mm
6.2188 in	OE	157.96 mm
1.9063 in	SB	48.42 mm
4.1250 in	EC	104.78 mm
3.2500 in	HD	82.55 mm

ROCKWELL 72NYS30-21

IDLI GROUP 81810
END YOKE WITH 14 INVOLUTED SPLINES
SEE FIGURE 7A

5.8438 in	IE	148.43 mm
6.2188 in	OE	157.96 mm
1.9219 in	SB	48.82 mm
4.1875 in	EC	106.36 mm
2.7500 in	HD	69.85 mm

ROCKWELL 72NYS30-19

IDLI GROUP 81811
END YOKE WITH 30 INVOLUTED SPLINES
SEE FIGURE 7A

5.8438 in	IE	148.43 mm
6.2188 in	OE	157.96 mm
1.9375 in	SB	49.21 mm
3.3125 in	EC	84.14 mm
3.1250 in	HD	79.38 mm

ROCKWELL 72NYS31-83

IDLI GROUP 81812
END YOKE WITH ROUND BORE
SEE FIGURE 7B

5.8438 in	IE	148.43 mm
6.2188 in	OE	157.96 mm
1.9375 in	SB	49.21 mm
3.6250 in	EC	92.08 mm
3.0000 in	HD	76.20 mm

ROCKWELL 72NYR31

IDLI GROUP 81813
END YOKE WITH 30 INVOLUTED SPLINES
SEE FIGURE 7A

5.8438 in	IE	148.43 mm
6.2188 in	OE	157.96 mm
1.9375 in	SB	49.21 mm
3.6875 in	EC	93.66 mm
3.0000 in	HD	76.20 mm

BORG-WARNR 23942C1
BORG-WARNR 23942CI
ROCKWELL 72NYS31-55

IDLI GROUP 81814
END YOKE WITH TAPERED BORE
SEE FIGURE 7F

5.8438 in	IE	148.43 mm
6.2188 in	OE	157.96 mm
1.9375 in	SB	49.21 mm
4.6250 in	EC	117.48 mm
3.1250 in	HD	79.38 mm

ROCKWELL 72NYT31-1

IDLI GROUP 81815
END YOKE WITH 10 SPLINES
SEE FIGURE 7A

5.8438 in	IE	148.43 mm
6.2188 in	OE	157.96 mm
1.9688 in	SB	50.01 mm
3.1875 in	EC	80.96 mm
2.7500 in	HD	69.85 mm

ROCKWELL 72NYS31-5

IDLI GROUP 81816
END YOKE WITH 10 SPLINES
SEE FIGURE 7A

5.8438 in	IE	148.43 mm
6.2188 in	OE	157.96 mm
1.9688 in	SB	50.01 mm
3.1875 in	EC	80.96 mm
3.0000 in	HD	76.20 mm

BORG-WARNR 7132J
ROCKWELL 72NYS31-28

IDLI GROUP 81817
END YOKE WITH ROUND BORE
SEE FIGURE 7B

5.8438 in	IE	148.43 mm
6.2188 in	OE	157.96 mm
1.9688 in	SB	50.01 mm
3.2500 in	EC	82.55 mm
3.0000 in	HD	76.20 mm

ROCKWELL 72NYR32-13

IDLI GROUP 81818
END YOKE WITH 6 SPLINES
SEE FIGURE 7A

5.8438 in	IE	148.43 mm
6.2188 in	OE	157.96 mm
1.9688 in	SB	50.01 mm
3.2500 in	EC	82.55 mm
3.1250 in	HD	79.38 mm

ROCKWELL 72NYS31-22

IDLI GROUP 81819
END YOKE WITH 10 SPLINES
SEE FIGURE 7A

5.8438 in	IE	148.43 mm
6.2188 in	OE	157.96 mm
1.9688 in	SB	50.01 mm
3.5000 in	EC	88.90 mm
2.6250 in	HD	66.68 mm

ROCKWELL 72NYS31-46

IDLI GROUP 81820
END YOKE WITH 10 SPLINES
slight variation from above group
SEE FIGURE 7A

5.8438 in	IE	148.43 mm
6.2188 in	OE	157.96 mm
1.9688 in	SB	50.01 mm
3.5000 in	EC	88.90 mm
2.6250 in	HD	66.68 mm

BORG-WARNR 21033J
BORG-WARNR 7530J
ROCKWELL 72NYS31-2

IDLI GROUP 81821
END YOKE WITH 10 SPLINES
slight variation from above group
SEE FIGURE 7A

5.8438 in	IE	148.43 mm
6.2188 in	OE	157.96 mm
1.9688 in	SB	50.01 mm
3.5000 in	EC	88.90 mm
2.6250 in	HD	66.68 mm

ROCKWELL 72NYS31-70

IDLI GROUP 81822
END YOKE WITH 10 SPLINES
slight variation from above group
SEE FIGURE 7A

5.8438 in	IE	148.43 mm
6.2188 in	OE	157.96 mm
1.9688 in	SB	50.01 mm
3.5000 in	EC	88.90 mm
2.6250 in	HD	66.68 mm

ROCKWELL 72NYS31-85

IDLI GROUP 81823
END YOKE WITH 10 SPLINES
SEE FIGURE 7A

5.8438 in	IE	148.43 mm
6.2188 in	OE	157.96 mm
1.9688 in	SB	50.01 mm
3.5000 in	EC	88.90 mm
2.8750 in	HD	73.03 mm

BORG-WARNR 5407J
ROCKWELL 72NYS31-26

IDLI GROUP 81824
END YOKE WITH 10 SPLINES
SEE FIGURE 7A

5.8438 in	IE	148.43 mm
6.2188 in	OE	157.96 mm
1.9688 in	SB	50.01 mm
3.5000 in	EC	88.90 mm
2.9375 in	HD	74.61 mm

ROCKWELL 72NYS31-93

IDLI GROUP 81825
END YOKE WITH 10 SPLINES
slight variation from above group
SEE FIGURE 7A

5.8438 in	IE	148.43 mm
6.2188 in	OE	157.96 mm
1.9688 in	SB	50.01 mm
3.5000 in	EC	88.90 mm
2.9375 in	HD	74.61 mm

BORG-WARNR 11173J

IDLI GROUP 81826
END YOKE WITH 10 SPLINES
SEE FIGURE 7A

5.8438 in	IE	148.43 mm
6.2188 in	OE	157.96 mm
1.9688 in	SB	50.01 mm
3.5000 in	EC	88.90 mm
3.0000 in	HD	76.20 mm

BORG-WARNR 5042J
ROCKWELL 72NYS31-6

IDLI GROUP 81827
END YOKE WITH 10 SPLINES
SEE FIGURE 7A

5.8438 in	IE	148.43 mm
6.2188 in	OE	157.96 mm
1.9688 in	SB	50.01 mm
3.6875 in	EC	93.66 mm
2.7500 in	HD	69.85 mm

ROCKWELL 72NYS31-43

IDLI GROUP 81828
END YOKE WITH 10 SPLINES
SEE FIGURE 7A

5.8438 in	IE	148.43 mm
6.2188 in	OE	157.96 mm
1.9688 in	SB	50.01 mm
3.6875 in	EC	93.66 mm
3.0000 in	HD	76.20 mm

BORG-WARNR 12973J
ROCKWELL 72NYS31-7

IDLI GROUP 81829
END YOKE WITH 10 SPLINES
SEE FIGURE 7A

5.8438 in	IE	148.43 mm
6.2188 in	OE	157.96 mm
1.9688 in	SB	50.01 mm
3.7500 in	EC	95.25 mm
2.6875 in	HD	68.26 mm

ROCKWELL 72NYS31-91

IDLI GROUP 81830
END YOKE WITH 10 SPLINES
SEE FIGURE 7A

5.8438 in	IE	148.43 mm
6.2188 in	OE	157.96 mm
1.9688 in	SB	50.01 mm
3.7500 in	EC	95.25 mm
2.7500 in	HD	69.85 mm

BORG-WARNR 23063J
ROCKWELL 72NYS31-48

IDLI GROUP 81831
END YOKE WITH 10 SPLINES
SEE FIGURE 7A

5.8438 in	IE	148.43 mm
6.2188 in	OE	157.96 mm
1.9688 in	SB	50.01 mm
3.8750 in	EC	98.43 mm
2.6250 in	HD	66.68 mm

ROCKWELL 72NYS31-51

BD=BEARING DIAMETER (outside)   **BW**=BEARING WIDTH   **CB**=CROSS LENGTH WITH BEARINGS   **CC**=CENTER TO CENTER   **CD**=CROSS DIAMETER
CL=CROSS LENGTH WITHOUT BEARINGS   **EC**=END TO CENTER or FACE TO FACE   **EE**=END TO END   **EL**=EFFECTIVE LENGTH
HD=HUB DIAMETER (or insert)   **IE**=INSIDE OF EARS (or recess)   **OD**=OUTSIDE DIAMETER   **OE**=OUTSIDE OF EARS   **SB**=SPLINE OR BORE SIZE

IDLI GROUP 81832

END YOKE WITH 10 SPLINES
SEE FIGURE 7A

5.8438 in	IE	148.43 mm
6.2188 in	OE	157.96 mm
1.9688 in	SB	50.01 mm
3.8750 in	EC	98.43 mm
2.7500 in	HD	69.85 mm

BORG-WARNR 7111J
ROCKWELL 72NYS31-4

IDLI GROUP 81833

END YOKE WITH 10 SPLINES
slight variation from above group
SEE FIGURE 7A

5.8438 in	IE	148.43 mm
6.2188 in	OE	157.96 mm
1.9688 in	SB	50.01 mm
3.8750 in	EC	98.43 mm
2.7500 in	HD	69.85 mm

BORG-WARNR 15367J
ROCKWELL 72NYS31-3

IDLI GROUP 81834

END YOKE WITH 10 SPLINES
SEE FIGURE 7A

5.8438 in	IE	148.43 mm
6.2188 in	OE	157.96 mm
1.9688 in	SB	50.01 mm
3.8750 in	EC	98.43 mm
2.8750 in	HD	73.03 mm

BORG-WARNR 24525J
ROCKWELL 72NYS31-50

IDLI GROUP 81835

END YOKE WITH 10 SPLINES
SEE FIGURE 7A

5.8438 in	IE	148.43 mm
6.2188 in	OE	157.96 mm
1.9688 in	SB	50.01 mm
4.2500 in	EC	107.95 mm
2.4375 in	HD	61.91 mm

ROCKWELL 72NYS31-60

IDLI GROUP 81836

END YOKE WITH 10 SPLINES
SEE FIGURE 7A

5.8438 in	IE	148.43 mm
6.2188 in	OE	157.96 mm
1.9688 in	SB	50.01 mm
4.2500 in	EC	107.95 mm
2.5000 in	HD	63.50 mm

BORG-WARNR 16635J
ROCKWELL 72NYS31

IDLI GROUP 81837

END YOKE WITH 6 SPLINES
SEE FIGURE 7A

5.8438 in	IE	148.43 mm
6.2188 in	OE	157.96 mm
1.9688 in	SB	50.01 mm
4.2500 in	EC	107.95 mm
2.7420 in	HD	69.65 mm

ROCKWELL 72NYS31-65

IDLI GROUP 81838

END YOKE WITH 6 SPLINES
SEE FIGURE 7A

5.8438 in	IE	148.43 mm
6.2188 in	OE	157.96 mm
1.9688 in	SB	50.01 mm
4.2500 in	EC	107.95 mm
2.7500 in	HD	69.85 mm

BORG-WARNR 7299J
ROCKWELL 72NYS31-29

IDLI GROUP 81839

END YOKE WITH 10 SPLINES
SEE FIGURE 7A

5.8438 in	IE	148.43 mm
6.2188 in	OE	157.96 mm
1.9688 in	SB	50.01 mm
4.3125 in	EC	109.54 mm
2.7500 in	HD	69.85 mm

ROCKWELL 72NYS31-53

IDLI GROUP 81840

END YOKE WITH 10 SPLINES
slight variation from above group
SEE FIGURE 7A

5.8438 in	IE	148.43 mm
6.2188 in	OE	157.96 mm
1.9688 in	SB	50.01 mm
4.3125 in	EC	109.54 mm
2.7500 in	HD	69.85 mm

BORG-WARNR 19244J
ROCKWELL 72NYS31-8

IDLI GROUP 81841

END YOKE WITH 10 SPLINES
slight variation from above group
SEE FIGURE 7A

5.8438 in	IE	148.43 mm
6.2188 in	OE	157.96 mm
1.9688 in	SB	50.01 mm
4.3125 in	EC	109.54 mm
2.7500 in	HD	69.85 mm

ROCKWELL 72NYS31-49

IDLI GROUP 81842

END YOKE WITH 10 SPLINES
SEE FIGURE 7A

5.8438 in	IE	148.43 mm
6.2188 in	OE	157.96 mm
1.9688 in	SB	50.01 mm
4.3750 in	EC	111.13 mm
2.6250 in	HD	66.68 mm

BORG-WARNR 18928J
ROCKWELL 72NYS31-27

IDLI GROUP 81843

END YOKE WITH 10 SPLINES
SEE FIGURE 7A

5.8438 in	IE	148.43 mm
6.2188 in	OE	157.96 mm
1.9688 in	SB	50.01 mm
4.4375 in	EC	112.71 mm
2.5000 in	HD	63.50 mm

BORG-WARNR 22712J
ROCKWELL 72NYS31-41

IDLI GROUP 81844

END YOKE WITH 10 SPLINES
SEE FIGURE 7A

5.8438 in	IE	148.43 mm
6.2188 in	OE	157.96 mm
1.9688 in	SB	50.01 mm
4.4375 in	EC	112.71 mm
3.0000 in	HD	76.20 mm

BORG-WARNR 22802J
ROCKWELL 72NYS31-42

IDLI GROUP 81845

END YOKE WITH 10 SPLINES
SEE FIGURE 7A

5.8438 in	IE	148.43 mm
6.2188 in	OE	157.96 mm
1.9688 in	SB	50.01 mm
4.5000 in	EC	114.30 mm
2.7500 in	HD	69.85 mm

ROCKWELL 72NYS31-54

IDLI GROUP 81846

END YOKE WITH 10 SPLINES
SEE FIGURE 7A

5.8438 in	IE	148.43 mm
6.2188 in	OE	157.96 mm
1.9688 in	SB	50.01 mm
4.5000 in	EC	114.30 mm
3.0000 in	HD	76.20 mm

BORG-WARNR 16637J
ROCKWELL 72NYS31-1

IDLI GROUP 81847

END YOKE WITH ROUND BORE
SEE FIGURE 7B

5.8438 in	IE	148.43 mm
6.2188 in	OE	157.96 mm
1.9688 in	SB	50.01 mm
4.6250 in	EC	117.48 mm
4.0000 in	HD	101.60 mm

ROCKWELL 72NYR32-2

IDLI GROUP 81848

END YOKE WITH 10 SPLINES
SEE FIGURE 7A

5.8438 in	IE	148.43 mm
6.2188 in	OE	157.96 mm
1.9688 in	SB	50.01 mm
4.7813 in	EC	121.45 mm
2.6250 in	HD	66.68 mm

BORG-WARNR 14311J
BORG-WARNR 67013
BORG-WARNR 6A7013
NEAPCO 22-129
NEAPCO 22-139
NEAPCO 22-239
NEAPCO 66-2239
NEAPCO 68-2239
ROCKWELL 72NYS31-12
SPICER 228247-3416X
SPICER 228418X
TRW 21469
TRW 21472
WESCO 35NSSF48
WESCO 35N-SSF72
WESCO 920-312
WESCO 920-313
WESCO 920-314
WESCO 920-316

IDLI GROUP 81849

END YOKE WITH 10 SPLINES
SEE FIGURE 7A

5.8438 in	IE	148.43 mm
6.2188 in	OE	157.96 mm
1.9688 in	SB	50.01 mm
5.5625 in	EC	141.29 mm
2.7500 in	HD	69.85 mm

ROCKWELL 72NYS31-9

IDLI GROUP 81850

END YOKE WITH 10 SPLINES
SEE FIGURE 7A

5.8438 in	IE	148.43 mm
6.2188 in	OE	157.96 mm
2.0000 in	SB	50.80 mm
2.2500 in	EC	57.15 mm
2.8750 in	HD	73.03 mm

BORG-WARNR 11966J
ROCKWELL 72NYS32-123

IDLI GROUP 81851

END YOKE WITH 10 SPLINES
SEE FIGURE 7A

5.8438 in	IE	148.43 mm
6.2188 in	OE	157.96 mm
2.0000 in	SB	50.80 mm
2.5625 in	EC	65.09 mm
3.0000 in	HD	76.20 mm

ROCKWELL 72NYS32-8

IDLI GROUP 81852

END YOKE WITH 10 SPLINES
SEE FIGURE 7A

5.8438 in	IE	148.43 mm
6.2188 in	OE	157.96 mm
2.0000 in	SB	50.80 mm
3.1250 in	EC	79.38 mm
2.6250 in	HD	66.68 mm

ROCKWELL 72NYS32

IDLI GROUP 81853

END YOKE WITH 10 SPLINES
SEE FIGURE 7A

5.8438 in	IE	148.43 mm
6.2188 in	OE	157.96 mm
2.0000 in	SB	50.80 mm
3.2500 in	EC	82.55 mm
2.7500 in	HD	69.85 mm

ROCKWELL 72NYS32-34

IDLI GROUP 81854

END YOKE WITH 10 SPLINES
SEE FIGURE 7A

5.8438 in	IE	148.43 mm
6.2188 in	OE	157.96 mm
2.0000 in	SB	50.80 mm
3.5000 in	EC	88.90 mm
2.6250 in	HD	66.68 mm

BORG-WARNR 11103J
ROCKWELL 72NYS32-152

ENGINEERING CATALOGS MUST BE CONSULTED FOR DETAILS NOT INCLUDED
IN THIS GUIDE. SPECIFIC DESIGNS, MATERIAL CONTENT, TOLERANCES
LUBE FITTINGS AND OTHER DIMENSIONS ARE INTENTIONALLY OMITTED HERE.

CAUTION: BE SURE TO REFER TO ENGINEERING CATALOGS FOR SPECIAL APPLICATIONS THAT REQUIRE SPECIFIC MATERIAL CONTENT, TOLERANCES, ETC. SEE FOOTNOTE.

IDLI GROUP 81855

END YOKE WITH 10 SPLINES
SEE FIGURE 7A

5.8438 in	IE	148.43 mm
6.2188 in	OE	157.96 mm
2.0000 in	SB	50.80 mm
3.5000 in	EC	88.90 mm
2.7500 in	HD	69.85 mm

ROCKWELL 72NYS32-18

IDLI GROUP 81856

END YOKE WITH 10 SPLINES
SEE FIGURE 7A

5.8438 in	IE	148.43 mm
6.2188 in	OE	157.96 mm
2.0000 in	SB	50.80 mm
3.5000 in	EC	88.90 mm
3.0000 in	HD	76.20 mm

ROCKWELL 72NYS32-217

IDLI GROUP 81857

END YOKE WITH 15 INVOLUTED SPLINES
SEE FIGURE 7A

5.8438 in	IE	148.43 mm
6.2188 in	OE	157.96 mm
2.0000 in	SB	50.80 mm
3.6875 in	EC	93.66 mm
3.0000 in	HD	76.20 mm

BORG-WARNR 21107J
ROCKWELL 72NYS32-76

IDLI GROUP 81858

END YOKE WITH ROUND BORE
SEE FIGURE 7B

5.8438 in	IE	148.43 mm
6.2188 in	OE	157.96 mm
2.0000 in	SB	50.80 mm
3.6875 in	EC	93.66 mm
3.2500 in	HD	82.55 mm

ROCKWELL 72NYR32

IDLI GROUP 81859

END YOKE WITH ROUND BORE
SEE FIGURE 7B

5.8438 in	IE	148.43 mm
6.2188 in	OE	157.96 mm
2.0000 in	SB	50.80 mm
3.7500 in	EC	95.25 mm
3.0000 in	HD	76.20 mm

ROCKWELL 72NYR32-10

IDLI GROUP 81860

END YOKE WITH 10 SPLINES
SEE FIGURE 7A

5.8438 in	IE	148.43 mm
6.2188 in	OE	157.96 mm
2.0000 in	SB	50.80 mm
3.8750 in	EC	98.43 mm
2.7500 in	HD	69.85 mm

ROCKWELL 72NYS32-129

IDLI GROUP 81861

END YOKE WITH ROUND BORE
SEE FIGURE 7B

5.8438 in	IE	148.43 mm
6.2188 in	OE	157.96 mm
2.0000 in	SB	50.80 mm
3.8750 in	EC	98.43 mm
3.5000 in	HD	88.90 mm

ROCKWELL 72NYR32-11

IDLI GROUP 81862

END YOKE WITH 10 SPLINES
SEE FIGURE 7A

5.8438 in	IE	148.43 mm
6.2188 in	OE	157.96 mm
2.0000 in	SB	50.80 mm
3.9375 in	EC	100.01 mm
2.7500 in	HD	69.85 mm

BORG-WARNR 11995J
ROCKWELL 72NYS32-6

IDLI GROUP 81863

END YOKE WITH 10 SPLINES
SEE FIGURE 7A

5.8438 in	IE	148.43 mm
6.2188 in	OE	157.96 mm
2.0000 in	SB	50.80 mm
3.9375 in	EC	100.01 mm
2.9375 in	HD	74.61 mm

BORG-WARNR 10324J
ROCKWELL 72NYS32-7

IDLI GROUP 81864

END YOKE WITH 10 SPLINES
slight variation from above group
SEE FIGURE 7A

5.8438 in	IE	148.43 mm
6.2188 in	OE	157.96 mm
2.0000 in	SB	50.80 mm
3.9375 in	EC	100.01 mm
2.9375 in	HD	74.61 mm

BORG-WARNR 18059J
ROCKWELL 72NYS32-9

IDLI GROUP 81865

END YOKE WITH 10 SPLINES
SEE FIGURE 7A

5.8438 in	IE	148.43 mm
6.2188 in	OE	157.96 mm
2.0000 in	SB	50.80 mm
4.0000 in	EC	101.60 mm
2.7500 in	HD	69.85 mm

BORG-WARNR 15065J
ROCKWELL 72NYS32-44

IDLI GROUP 81866

END YOKE WITH 10 SPLINES
slight variation from above group
SEE FIGURE 7A

5.8438 in	IE	148.43 mm
6.2188 in	OE	157.96 mm
2.0000 in	SB	50.80 mm
4.0000 in	EC	101.60 mm
2.7500 in	HD	69.85 mm

ROCKWELL 72NYS32-110

IDLI GROUP 81867

END YOKE WITH 10 SPLINES
slight variation from above group
SEE FIGURE 7A

5.8438 in	IE	148.43 mm
6.2188 in	OE	157.96 mm
2.0000 in	SB	50.80 mm
4.0000 in	EC	101.60 mm
2.7500 in	HD	69.85 mm

ROCKWELL 72NYS32-111

IDLI GROUP 81868

END YOKE WITH 10 SPLINES
SEE FIGURE 7A

5.8438 in	IE	148.43 mm
6.2188 in	OE	157.96 mm
2.0000 in	SB	50.80 mm
4.0625 in	EC	103.19 mm
2.7500 in	HD	69.85 mm

ROCKWELL 72NYS32-108

IDLI GROUP 81869

END YOKE WITH 10 SPLINES
SEE FIGURE 7A

5.8438 in	IE	148.43 mm
6.2188 in	OE	157.96 mm
2.0000 in	SB	50.80 mm
4.0625 in	EC	103.19 mm
2.9375 in	HD	74.61 mm

ROCKWELL 72NYS32-135

IDLI GROUP 81870

END YOKE WITH 10 SPLINES
SEE FIGURE 7A

5.8438 in	IE	148.43 mm
6.2188 in	OE	157.96 mm
2.0000 in	SB	50.80 mm
4.0938 in	EC	103.98 mm
2.8750 in	HD	73.03 mm

ROCKWELL 72NYS32-5

IDLI GROUP 81871

END YOKE WITH 10 SPLINES
SEE FIGURE 7A

5.8438 in	IE	148.43 mm
6.2188 in	OE	157.96 mm
2.0000 in	SB	50.80 mm
4.1250 in	EC	104.78 mm
2.7500 in	HD	69.85 mm

BORG-WARNR 21005J
ROCKWELL 72NYS32-43

IDLI GROUP 81872

END YOKE WITH 10 SPLINES
SEE FIGURE 7A

5.8438 in	IE	148.43 mm
6.2188 in	OE	157.96 mm
2.0000 in	SB	50.80 mm
4.1250 in	EC	104.78 mm
2.9375 in	HD	74.61 mm

ROCKWELL 72NYS32-87

IDLI GROUP 81873

END YOKE WITH 10 SPLINES
slight variation from above group
SEE FIGURE 7A

5.8438 in	IE	148.43 mm
6.2188 in	OE	157.96 mm
2.0000 in	SB	50.80 mm
4.1250 in	EC	104.78 mm
2.9375 in	HD	74.61 mm

ROCKWELL 72NYS32-109

IDLI GROUP 81874

END YOKE WITH 10 SPLINES
SEE FIGURE 7A

5.8438 in	IE	148.43 mm
6.2188 in	OE	157.96 mm
2.0000 in	SB	50.80 mm
4.1875 in	EC	106.36 mm
3.0000 in	HD	76.20 mm

BORG-WARNR 20652J
ROCKWELL 72NYS32-45

IDLI GROUP 81875

END YOKE WITH 10 SPLINES
slight variation from above group
SEE FIGURE 7A

5.8438 in	IE	148.43 mm
6.2188 in	OE	157.96 mm
2.0000 in	SB	50.80 mm
4.1875 in	EC	106.36 mm
3.0000 in	HD	76.20 mm

ROCKWELL 72NYS32-105

IDLI GROUP 81876

END YOKE WITH 10 SPLINES
SEE FIGURE 7A

5.8438 in	IE	148.43 mm
6.2188 in	OE	157.96 mm
2.0000 in	SB	50.80 mm
4.2500 in	EC	107.95 mm
2.7500 in	HD	69.85 mm

ROCKWELL 72NYS32-172

IDLI GROUP 81877

END YOKE WITH 10 SPLINES
slight variation from above group
SEE FIGURE 7A

5.8438 in	IE	148.43 mm
6.2188 in	OE	157.96 mm
2.0000 in	SB	50.80 mm
4.2500 in	EC	107.95 mm
2.7500 in	HD	69.85 mm

BORG-WARNR 14909J
ROCKWELL 72NYS32-104

IDLI GROUP 81878

END YOKE WITH 10 SPLINES
SEE FIGURE 7A

5.8438 in	IE	148.43 mm
6.2188 in	OE	157.96 mm
2.0000 in	SB	50.80 mm
4.2500 in	EC	107.95 mm
2.9375 in	HD	74.61 mm

BORG-WARNR 23885J
ROCKWELL 72NYS32-12

BD = BEARING DIAMETER (outside)   **BW** = BEARING WIDTH   **CB** = CROSS LENGTH WITH BEARINGS   **CC** = CENTER TO CENTER   **CD** = CROSS DIAMETER
CL = CROSS LENGTH WITHOUT BEARINGS   **EC** = END TO CENTER or FACE TO FACE   **EE** = END TO END   **EL** = EFFECTIVE LENGTH
HD = HUB DIAMETER (or insert)   **IE** = INSIDE OF EARS (or recess)   **OD** = OUTSIDE DIAMETER   **OE** = OUTSIDE OF EARS   **SB** = SPLINE OR BORE SIZE

IDLI GROUP 81879

END YOKE WITH 10 SPLINES
slight variation from above group
SEE FIGURE 7A

5.8438 in	IE	148.43 mm
6.2188 in	OE	157.96 mm
2.0000 in	SB	50.80 mm
4.2500 in	EC	107.95 mm
2.9375 in	HD	74.61 mm

BORG-WARNR 16453J
ROCKWELL 72NYS32-39

IDLI GROUP 81880

END YOKE WITH 10 SPLINES
slight variation from above group
SEE FIGURE 7A

5.8438 in	IE	148.43 mm
6.2188 in	OE	157.96 mm
2.0000 in	SB	50.80 mm
4.2500 in	EC	107.95 mm
2.9375 in	HD	74.61 mm

BORG-WARNR 7781J
ROCKWELL 72NYS32-48

IDLI GROUP 81881

END YOKE WITH 10 SPLINES
slight variation from above group
SEE FIGURE 7A

5.8438 in	IE	148.43 mm
6.2188 in	OE	157.96 mm
2.0000 in	SB	50.80 mm
4.2500 in	EC	107.95 mm
2.9375 in	HD	74.61 mm

ROCKWELL 72NYS32-151

IDLI GROUP 81882

END YOKE WITH 10 SPLINES
SEE FIGURE 7A

5.8438 in	IE	148.43 mm
6.2188 in	OE	157.96 mm
2.0000 in	SB	50.80 mm
4.2500 in	EC	107.95 mm
3.0000 in	HD	76.20 mm

BORG-WARNR 10212J
ROCKWELL 72NYS32-70

IDLI GROUP 81883

END YOKE WITH 10 SPLINES
SEE FIGURE 7A

5.8438 in	IE	148.43 mm
6.2188 in	OE	157.96 mm
2.0000 in	SB	50.80 mm
4.2500 in	EC	107.95 mm
3.2500 in	HD	82.55 mm

BORG-WARNR 7327J
ROCKWELL 72NYS32-155

IDLI GROUP 81884

END YOKE WITH 10 SPLINES
SEE FIGURE 7A

5.8438 in	IE	148.43 mm
6.2188 in	OE	157.96 mm
2.0000 in	SB	50.80 mm
4.3125 in	EC	109.54 mm
3.0000 in	HD	76.20 mm

BORG-WARNR 121-7056

ROCKWELL 72NYS32-98A

IDLI GROUP 81885

END YOKE WITH 10 SPLINES
SEE FIGURE 7A

5.8438 in	IE	148.43 mm
6.2188 in	OE	157.96 mm
2.0000 in	SB	50.80 mm
4.3750 in	EC	111.13 mm
3.0000 in	HD	76.20 mm

ROCKWELL 72NYS32-100

IDLI GROUP 81886

END YOKE WITH 10 SPLINES
SEE FIGURE 7A

5.8438 in	IE	148.43 mm
6.2188 in	OE	157.96 mm
2.0000 in	SB	50.80 mm
4.3750 in	EC	111.13 mm
3.2500 in	HD	82.55 mm

ROCKWELL 72NYS32-16

IDLI GROUP 81887

END YOKE WITH 10 SPLINES
slight variation from above group
SEE FIGURE 7A

5.8438 in	IE	148.43 mm
6.2188 in	OE	157.96 mm
2.0000 in	SB	50.80 mm
4.3750 in	EC	111.13 mm
3.2500 in	HD	82.55 mm

ROCKWELL 72NYS32-55

IDLI GROUP 81888

END YOKE WITH 15 INVOLUTED SPLINES
SEE FIGURE 7A

5.8438 in	IE	148.43 mm
6.2188 in	OE	157.96 mm
2.0000 in	SB	50.80 mm
4.4375 in	EC	112.71 mm
2.6250 in	HD	66.68 mm

BORG-WARNR 21101J
ROCKWELL 72NYS32-77

IDLI GROUP 81889

END YOKE WITH 10 SPLINES
SEE FIGURE 7A

5.8438 in	IE	148.43 mm
6.2188 in	OE	157.96 mm
2.0000 in	SB	50.80 mm
4.5625 in	EC	115.89 mm
3.2500 in	HD	82.55 mm

ROCKWELL 72NYS32-40

IDLI GROUP 81890

END YOKE WITH ROUND BORE
SEE FIGURE 7B

5.8438 in	IE	148.43 mm
6.2188 in	OE	157.96 mm
2.0000 in	SB	50.80 mm
4.5625 in	EC	115.89 mm
4.0000 in	HD	101.60 mm

ROCKWELL 72NYR32-3

IDLI GROUP 81891

END YOKE WITH 10 SPLINES
SEE FIGURE 7A

5.8438 in	IE	148.43 mm
6.2188 in	OE	157.96 mm
2.0000 in	SB	50.80 mm
4.6250 in	EC	117.48 mm
2.5000 in	HD	63.50 mm

ROCKWELL 72NYS32-173

IDLI GROUP 81892

END YOKE WITH 10 SPLINES
SEE FIGURE 7A

5.8438 in	IE	148.43 mm
6.2188 in	OE	157.96 mm
2.0000 in	SB	50.80 mm
4.6250 in	EC	117.48 mm
2.7500 in	HD	69.85 mm

ROCKWELL 72NYS32-157

IDLI GROUP 81893

END YOKE WITH 10 SPLINES
slight variation from above group
SEE FIGURE 7A

5.8438 in	IE	148.43 mm
6.2188 in	OE	157.96 mm
2.0000 in	SB	50.80 mm
4.6250 in	EC	117.48 mm
2.7500 in	HD	69.85 mm

ROCKWELL 72NYS32-50

IDLI GROUP 81894

END YOKE WITH TAPERED BORE
SEE FIGURE 7F

5.8438 in	IE	148.43 mm
6.2188 in	OE	157.96 mm
2.0000 in	SB	50.80 mm
4.6250 in	EC	117.48 mm
3.0000 in	HD	76.20 mm

ROCKWELL 72NYT32

IDLI GROUP 81895

END YOKE WITH 10 SPLINES
SEE FIGURE 7A

5.8438 in	IE	148.43 mm
6.2188 in	OE	157.96 mm
2.0000 in	SB	50.80 mm
4.6250 in	EC	117.48 mm
3.2500 in	HD	82.55 mm

BORG-WARNR 20656J
ROCKWELL 72NYS32-47

IDLI GROUP 81896

END YOKE WITH 10 SPLINES
slight variation from above group
SEE FIGURE 7A

5.8438 in	IE	148.43 mm
6.2188 in	OE	157.96 mm
2.0000 in	SB	50.80 mm
4.6250 in	EC	117.48 mm
3.2500 in	HD	82.55 mm

ROCKWELL 72NYS32-17

IDLI GROUP 81897

END YOKE WITH 10 SPLINES
SEE FIGURE 7A

5.8438 in	IE	148.43 mm
6.2188 in	OE	157.96 mm
2.0000 in	SB	50.80 mm
4.6250 in	EC	117.48 mm
3.5000 in	HD	88.90 mm

ROCKWELL 72NYS32-14

IDLI GROUP 81898

END YOKE WITH ROUND BORE
SEE FIGURE 7B

5.8438 in	IE	148.43 mm
6.2188 in	OE	157.96 mm
2.0000 in	SB	50.80 mm
4.6250 in	EC	117.48 mm
4.0000 in	HD	101.60 mm

ROCKWELL 72NYR32-1

IDLI GROUP 81899

END YOKE WITH 10 SPLINES
SEE FIGURE 7A

5.8438 in	IE	148.43 mm
6.2188 in	OE	157.96 mm
2.0000 in	SB	50.80 mm
4.6875 in	EC	119.06 mm
2.9375 in	HD	74.61 mm

ROCKWELL 72NYS32-106

IDLI GROUP 81900

END YOKE WITH ROUND BORE
SEE FIGURE 7B

5.8438 in	IE	148.43 mm
6.2188 in	OE	157.96 mm
2.0000 in	SB	50.80 mm
4.8125 in	EC	122.24 mm
4.0000 in	HD	101.60 mm

ROCKWELL 72NYR32-5

IDLI GROUP 81901

END YOKE WITH 39 SLANTED SPLINES
SEE FIGURE 7A

5.8438 in	IE	148.43 mm
6.2188 in	OE	157.96 mm
2.0313 in	SB	51.60 mm
3.8125 in	EC	96.84 mm
3.0000 in	HD	76.20 mm

ROCKWELL 72NYS32-244

IDLI GROUP 81902

END YOKE WITH 39 SLANTED SPLINES
slight variation from above group
SEE FIGURE 7A

5.8438 in	IE	148.43 mm
6.2188 in	OE	157.96 mm
2.0313 in	SB	51.60 mm
3.8125 in	EC	96.84 mm
3.0000 in	HD	76.20 mm

ROCKWELL 72NYS32-244A

ENGINEERING CATALOGS MUST BE CONSULTED FOR DETAILS NOT INCLUDED
IN THIS GUIDE. SPECIFIC DESIGNS, MATERIAL CONTENT, TOLERANCES
LUBE FITTINGS AND OTHER DIMENSIONS ARE INTENTIONALLY OMITTED HERE.

CAUTION: BE SURE TO REFER TO ENGINEERING CATALOGS FOR SPECIAL APPLICATIONS THAT REQUIRE SPECIFIC MATERIAL CONTENT, TOLERANCES, ETC. SEE FOOTNOTE.

IDLI GROUP 81903

END YOKE WITH 10 SPLINES
SEE FIGURE 7A

5.8438 in	IE	148.43 mm
6.2188 in	OE	157.96 mm
2.0313 in	SB	51.60 mm
3.8750 in	EC	98.43 mm
2.7500 in	HD	69.85 mm

ROCKWELL 72NYS33

IDLI GROUP 81904

END YOKE WITH 39 SLANTED SPLINES
SEE FIGURE 7A

5.8438 in	IE	148.43 mm
6.2188 in	OE	157.96 mm
2.0313 in	SB	51.60 mm
3.8750 in	EC	98.43 mm
2.7500 in	HD	69.85 mm

ROCKWELL 72NYS32-219

IDLI GROUP 81905

END YOKE WITH 39 SLANTED SPLINES
slight variation from above group
SEE FIGURE 7A

5.8438 in	IE	148.43 mm
6.2188 in	OE	157.96 mm
2.0313 in	SB	51.60 mm
3.8750 in	EC	98.43 mm
2.7500 in	HD	69.85 mm

ROCKWELL 72NYS32-218

IDLI GROUP 81906

END YOKE WITH 39 SLANTED SPLINES
slight variation from above group
SEE FIGURE 7A

5.8438 in	IE	148.43 mm
6.2188 in	OE	157.96 mm
2.0313 in	SB	51.60 mm
3.8750 in	EC	98.43 mm
2.7500 in	HD	69.85 mm

ROCKWELL 72NYS32-220

IDLI GROUP 81907

END YOKE WITH 39 SLANTED SPLINES
slight variation from above group
SEE FIGURE 7A

5.8438 in	IE	148.43 mm
6.2188 in	OE	157.96 mm
2.0313 in	SB	51.60 mm
3.8750 in	EC	98.43 mm
2.7500 in	HD	69.85 mm

ROCKWELL 72NYS32-239

IDLI GROUP 81908

END YOKE WITH 39 SLANTED SPLINES
SEE FIGURE 7A

5.8438 in	IE	148.43 mm
6.2188 in	OE	157.96 mm
2.0313 in	SB	51.60 mm
4.3125 in	EC	109.54 mm
2.7500 in	HD	69.85 mm

ROCKWELL 72NYS32-222

IDLI GROUP 81909

END YOKE WITH 39 SLANTED SPLINES
slight variation from above group
SEE FIGURE 7A

5.8438 in	IE	148.43 mm
6.2188 in	OE	157.96 mm
2.0313 in	SB	51.60 mm
4.3125 in	EC	109.54 mm
2.7500 in	HD	69.85 mm

ROCKWELL 72NYS32-221

IDLI GROUP 81910

END YOKE WITH 39 SLANTED SPLINES
SEE FIGURE 7A

5.8438 in	IE	148.43 mm
6.2188 in	OE	157.96 mm
2.0313 in	SB	51.60 mm
4.8750 in	EC	123.83 mm
3.0000 in	HD	76.20 mm

ROCKWELL 72NYS32-245

IDLI GROUP 81911

END YOKE WITH 39 SLANTED SPLINES
slight variation from above group
SEE FIGURE 7A

5.8438 in	IE	148.43 mm
6.2188 in	OE	157.96 mm
2.0313 in	SB	51.60 mm
4.8750 in	EC	123.83 mm
3.0000 in	HD	76.20 mm

ROCKWELL 72NYS32-245A

IDLI GROUP 81912

END YOKE WITH 32 INVOLUTED SPLINES
SEE FIGURE 7A

5.8438 in	IE	148.43 mm
6.2188 in	OE	157.96 mm
2.1250 in	SB	53.98 mm
3.6250 in	EC	92.08 mm
3.0000 in	HD	76.20 mm

ROCKWELL 72NYS34-36

IDLI GROUP 81913

END YOKE WITH 32 INVOLUTED SPLINES
SEE FIGURE 7A

5.8438 in	IE	148.43 mm
6.2188 in	OE	157.96 mm
2.1250 in	SB	53.98 mm
3.6875 in	EC	93.66 mm
3.0000 in	HD	76.20 mm

ROCKWELL 72NYS34-34

IDLI GROUP 81914

END YOKE WITH ROUND BORE
SEE FIGURE 7B

5.8438 in	IE	148.43 mm
6.2188 in	OE	157.96 mm
2.1250 in	SB	53.98 mm
4.6250 in	EC	117.48 mm
4.0000 in	HD	101.60 mm

ROCKWELL 72NYR34

IDLI GROUP 81915

END YOKE WITH 16 SPLINES
SEE FIGURE 7A

5.8438 in	IE	148.43 mm
6.2188 in	OE	157.96 mm
2.1563 in	SB	54.77 mm
4.2500 in	EC	107.95 mm
2.9375 in	HD	74.61 mm

ROCKWELL 72NYS34-32

IDLI GROUP 81916

END YOKE WITH 16 SPLINES
slight variation from above group
SEE FIGURE 7A

5.8438 in	IE	148.43 mm
6.2188 in	OE	157.96 mm
2.1563 in	SB	54.77 mm
4.2500 in	EC	107.95 mm
2.9375 in	HD	74.61 mm

BORG-WARNR 20768J
ROCKWELL 72NYS34-24

IDLI GROUP 81917

END YOKE WITH 16 SPLINES
slight variation from above group
SEE FIGURE 7A

5.8438 in	IE	148.43 mm
6.2188 in	OE	157.96 mm
2.1563 in	SB	54.77 mm
4.2500 in	EC	107.95 mm
2.9375 in	HD	74.61 mm

ROCKWELL 72NYS34-31

IDLI GROUP 81918

END YOKE WITH 32 INVOLUTED SPLINES
SEE FIGURE 7A

5.8438 in	IE	148.43 mm
6.2188 in	OE	157.96 mm
2.1250 in	SB	53.98 mm
4.5000 in	EC	114.30 mm
3.0000 in	HD	76.20 mm

ROCKWELL 72NYS34-33

IDLI GROUP 81919

END YOKE WITH 16 SPLINES
SEE FIGURE 7A

5.8438 in	IE	148.43 mm
6.2188 in	OE	157.96 mm
2.1250 in	SB	53.98 mm
4.6875 in	EC	119.06 mm
2.8750 in	HD	73.03 mm

ROCKWELL 72NYS34-28

IDLI GROUP 81920

END YOKE WITH 16 SPLINES
SEE FIGURE 7A

5.8438 in	IE	148.43 mm
6.2188 in	OE	157.96 mm
2.1563 in	SB	54.77 mm
4.6875 in	EC	119.06 mm
2.9375 in	HD	74.61 mm

ROCKWELL 72NYS34-23

IDLI GROUP 81921

END YOKE WITH ROUND BORE
SEE FIGURE 7B

5.8438 in	IE	148.43 mm
6.2188 in	OE	157.96 mm
2.1875 in	SB	55.56 mm
4.6250 in	EC	117.48 mm
4.0000 in	HD	101.60 mm

ROCKWELL 72NYR35-1

IDLI GROUP 81922

END YOKE WITH 6 SPLINES
SEE FIGURE 7A

5.8438 in	IE	148.43 mm
6.2188 in	OE	157.96 mm
2.2188 in	SB	56.36 mm
4.6875 in	EC	119.06 mm
3.0000 in	HD	76.20 mm

ROCKWELL 72NYS36-22

IDLI GROUP 81923

END YOKE WITH 6 SPLINES
slight variation from above group
SEE FIGURE 7A

5.8438 in	IE	148.43 mm
6.2188 in	OE	157.96 mm
2.2188 in	SB	56.36 mm
4.6875 in	EC	119.06 mm
3.0000 in	HD	76.20 mm

ROCKWELL 72NYS36-51

IDLI GROUP 81924

END YOKE WITH 10 SPLINES
SEE FIGURE 7A

5.8438 in	IE	148.43 mm
6.2188 in	OE	157.96 mm
2.2500 in	SB	57.15 mm
3.0625 in	EC	77.79 mm
3.0000 in	HD	76.20 mm

ROCKWELL 72NYS36-69

IDLI GROUP 81925

END YOKE WITH 10 SPLINES
SEE FIGURE 7A

5.8438 in	IE	148.43 mm
6.2188 in	OE	157.96 mm
2.2500 in	SB	57.15 mm
3.3750 in	EC	85.73 mm
3.1250 in	HD	79.38 mm

BORG-WARNR 28783J
ROCKWELL 72NYS36-88

BD = BEARING DIAMETER (outside)   BW = BEARING WIDTH   CB = CROSS LENGTH WITH BEARINGS   CC = CENTER TO CENTER   CD = CROSS DIAMETER
CL = CROSS LENGTH WITHOUT BEARINGS   EC = END TO CENTER or FACE TO FACE   EE = END TO END   EL = EFFECTIVE LENGTH
HD = HUB DIAMETER (or insert)   IE = INSIDE OF EARS (or recess)   OD = OUTSIDE DIAMETER   OE = OUTSIDE OF EARS   SB = SPLINE OR BORE SIZE

IDLI GROUP 81926

END YOKE WITH ROUND BORE
SEE FIGURE 7B

5.8438 in	IE	148.43 mm
6.2188 in	OE	157.96 mm
2.2500 in	SB	57.15 mm
3.6250 in	EC	92.08 mm
4.2500 in	HD	107.95 mm

ROCKWELL 72NYR36

IDLI GROUP 81927

END YOKE WITH 10 SPLINES
SEE FIGURE 7A

5.8438 in	IE	148.43 mm
6.2188 in	OE	157.96 mm
2.2500 in	SB	57.15 mm
3.9375 in	EC	100.01 mm
3.2500 in	HD	82.55 mm

BORG-WARNR 21280J
ROCKWELL 72NYS36-21

IDLI GROUP 81928

END YOKE WITH 10 SPLINES
slight variation from above group
SEE FIGURE 7A

5.8438 in	IE	148.43 mm
6.2188 in	OE	157.96 mm
2.2500 in	SB	57.15 mm
3.9375 in	EC	100.01 mm
3.2500 in	HD	82.55 mm

ROCKWELL 72NYS36-36

IDLI GROUP 81929

END YOKE WITH 6 SPLINES
SEE FIGURE 7A

5.8438 in	IE	148.43 mm
6.2188 in	OE	157.96 mm
2.2500 in	SB	57.15 mm
4.0000 in	EC	101.60 mm
3.0000 in	HD	76.20 mm

ROCKWELL 72NYS36

IDLI GROUP 81930

END YOKE WITH 10 SPLINES
SEE FIGURE 7A

5.8438 in	IE	148.43 mm
6.2188 in	OE	157.96 mm
2.2500 in	SB	57.15 mm
4.1250 in	EC	104.78 mm
3.1250 in	HD	79.38 mm

ROCKWELL 72NYS36-42

IDLI GROUP 81931

END YOKE WITH 10 SPLINES
SEE FIGURE 7A

5.8438 in	IE	148.43 mm
6.2188 in	OE	157.96 mm
2.2500 in	SB	57.15 mm
4.1250 in	EC	104.78 mm
3.2500 in	HD	82.55 mm

ROCKWELL 72NYS36-41

IDLI GROUP 81932

END YOKE WITH 6 SPLINES
SEE FIGURE 7A

5.8438 in	IE	148.43 mm
6.2188 in	OE	157.96 mm
2.2656 in	SB	57.55 mm
4.1250 in	EC	104.78 mm
3.2500 in	HD	82.55 mm

ROCKWELL 72NYS36-45

IDLI GROUP 81933

END YOKE WITH 10 SPLINES
SEE FIGURE 7A

5.8438 in	IE	148.43 mm
6.2188 in	OE	157.96 mm
2.2500 in	SB	57.15 mm
4.1875 in	EC	106.36 mm
2.9375 in	HD	74.61 mm

BORG-WARNR 24539J
ROCKWELL 72NYS36-61

IDLI GROUP 81934

END YOKE WITH 10 SPLINES
SEE FIGURE 7A

5.8438 in	IE	148.43 mm
6.2188 in	OE	157.96 mm
2.2500 in	SB	57.15 mm
4.2500 in	EC	107.95 mm
3.0000 in	HD	76.20 mm

BORG-WARNR 7145J
ROCKWELL 72NYS36-40

IDLI GROUP 81935

END YOKE WITH 10 SPLINES
slight variation from above group
SEE FIGURE 7A

5.8438 in	IE	148.43 mm
6.2188 in	OE	157.96 mm
2.2500 in	SB	57.15 mm
4.2500 in	EC	107.95 mm
3.0000 in	HD	76.20 mm

BORG-WARNR 21039J
ROCKWELL 72NYS36-54

IDLI GROUP 81936

END YOKE WITH 10 SPLINES
SEE FIGURE 7A

5.8438 in	IE	148.43 mm
6.2188 in	OE	157.96 mm
2.2500 in	SB	57.15 mm
4.2500 in	EC	107.95 mm
3.5000 in	HD	88.90 mm

BORG-WARNR 8627J
ROCKWELL 72NYS36-13

IDLI GROUP 81937

END YOKE WITH 10 SPLINES
SEE FIGURE 7A

5.8438 in	IE	148.43 mm
6.2188 in	OE	157.96 mm
2.2500 in	SB	57.15 mm
4.3125 in	EC	109.54 mm
3.0000 in	HD	76.20 mm

BORG-WARNR 11031J
ROCKWELL 72NYS36-5

IDLI GROUP 81938

END YOKE WITH 10 SPLINES
SEE FIGURE 7A

5.8438 in	IE	148.43 mm
6.2188 in	OE	157.96 mm
2.2500 in	SB	57.15 mm
4.3750 in	EC	111.13 mm
3.5000 in	HD	88.90 mm

ROCKWELL 72NYS36-14

IDLI GROUP 81939

END YOKE WITH 10 SPLINES
SEE FIGURE 7A

5.8438 in	IE	148.43 mm
6.2188 in	OE	157.96 mm
2.2500 in	SB	57.15 mm
4.5000 in	EC	114.30 mm
3.5000 in	HD	88.90 mm

ROCKWELL 72NYS36-19

IDLI GROUP 81940

END YOKE WITH 10 SPLINES
slight variation from above group
SEE FIGURE 7A

5.8438 in	IE	148.43 mm
6.2188 in	OE	157.96 mm
2.2500 in	SB	57.15 mm
4.5000 in	EC	114.30 mm
3.5000 in	HD	88.90 mm

BORG-WARNR 23400J
ROCKWELL 72NYS36-38

IDLI GROUP 81941

END YOKE WITH 10 SPLINES
SEE FIGURE 7A

5.8438 in	IE	148.43 mm
6.2188 in	OE	157.96 mm
2.2500 in	SB	57.15 mm
4.5625 in	EC	115.89 mm
3.0000 in	HD	76.20 mm

BORG-WARNR 20172J
ROCKWELL 72NYS36-6

IDLI GROUP 81942

END YOKE WITH 10 SPLINES
slight variation from above group
SEE FIGURE 7A

5.8438 in	IE	148.43 mm
6.2188 in	OE	157.96 mm
2.2500 in	SB	57.15 mm
4.5625 in	EC	115.89 mm
3.0000 in	HD	76.20 mm

ROCKWELL 72NYS36-2

IDLI GROUP 81943

END YOKE WITH ROUND BORE
SEE FIGURE 7B

5.8438 in	IE	148.43 mm
6.2188 in	OE	157.96 mm
2.2500 in	SB	57.15 mm
4.6250 in	EC	117.48 mm
3.6250 in	HD	92.08 mm

ROCKWELL 72NYR36-6

IDLI GROUP 81944

END YOKE WITH ROUND BORE
SEE FIGURE 7B

5.8438 in	IE	148.43 mm
6.2188 in	OE	157.96 mm
2.2500 in	SB	57.15 mm
4.6250 in	EC	117.48 mm
4.0000 in	HD	101.60 mm

ROCKWELL 72NYR36-2

IDLI GROUP 81945

END YOKE WITH ROUND BORE
slight variation from above group
SEE FIGURE 7B

5.8438 in	IE	148.43 mm
6.2188 in	OE	157.96 mm
2.2500 in	SB	57.15 mm
4.6250 in	EC	117.48 mm
4.0000 in	HD	101.60 mm

ROCKWELL 72NYR36-1

IDLI GROUP 81946

END YOKE WITH 10 SPLINES
SEE FIGURE 7A

5.8438 in	IE	148.43 mm
6.2188 in	OE	157.96 mm
2.2500 in	SB	57.15 mm
4.6875 in	EC	119.06 mm
3.0000 in	HD	76.20 mm

ROCKWELL 72NYS36-17
ROCKWELL 72NYS36-84

IDLI GROUP 81947

END YOKE WITH 10 SPLINES
SEE FIGURE 7A

5.8438 in	IE	148.43 mm
6.2188 in	OE	157.96 mm
2.2500 in	SB	57.15 mm
4.6875 in	EC	119.06 mm
3.5000 in	HD	88.90 mm

ROCKWELL 72NYS36-37

IDLI GROUP 81948

END YOKE WITH 10 SPLINES
slight variation from above group
SEE FIGURE 7A

5.8438 in	IE	148.43 mm
6.2188 in	OE	157.96 mm
2.2500 in	SB	57.15 mm
4.6875 in	EC	119.06 mm
3.5000 in	HD	88.90 mm

BORG-WARNR 19813J
BORG-WARNR 59127J
ROCKWELL 72NYS36-16

IDLI GROUP 81949

END YOKE WITH 17 INVOLUTED SPLINES
SEE FIGURE 7A

5.8438 in	IE	148.43 mm
6.2188 in	OE	157.96 mm
2.2813 in	SB	57.95 mm
3.5000 in	EC	88.90 mm
3.0000 in	HD	76.20 mm

BORG-WARNR 14240J
ROCKWELL 72NYS36-1

ENGINEERING CATALOGS MUST BE CONSULTED FOR DETAILS NOT INCLUDED
IN THIS GUIDE. SPECIFIC DESIGNS, MATERIAL CONTENT, TOLERANCES
LUBE FITTINGS AND OTHER DIMENSIONS ARE INTENTIONALLY OMITTED HERE.

CAUTION: BE SURE TO REFER TO ENGINEERING CATALOGS FOR SPECIAL APPLICATIONS THAT REQUIRE SPECIFIC MATERIAL CONTENT, TOLERANCES, ETC. SEE FOOTNOTE.

IDLI GROUP 81950

END YOKE WITH 17 INVOLUTED SPLINES
SEE FIGURE 7A

5.8438 in	IE	148.43 mm
6.2188 in	OE	157.96 mm
2.2813 in	SB	57.95 mm
4.3750 in	EC	111.13 mm
3.0000 in	HD	76.20 mm

BORG-WARNR 26487J
ROCKWELL 72NYS36-23

IDLI GROUP 81951

END YOKE WITH 18 INVOLUTED SPLINES
SEE FIGURE 7A

5.8438 in	IE	148.43 mm
6.2188 in	OE	157.96 mm
2.3594 in	SB	59.93 mm
2.9844 in	EC	75.80 mm
3.0000 in	HD	76.20 mm

ROCKWELL 72NYS38-4

IDLI GROUP 81952

END YOKE WITH 18 INVOLUTED SPLINES
SEE FIGURE 7A

5.8438 in	IE	148.43 mm
6.2188 in	OE	157.96 mm
2.3594 in	SB	59.93 mm
3.5781 in	EC	90.88 mm
3.0000 in	HD	76.20 mm

ROCKWELL 72NYS38-1

IDLI GROUP 81953

END YOKE WITH 18 INVOLUTED SPLINES
SEE FIGURE 7A

5.8438 in	IE	148.43 mm
6.2188 in	OE	157.96 mm
2.3594 in	SB	59.93 mm
4.0000 in	EC	101.60 mm
3.0000 in	HD	76.20 mm

ROCKWELL 72NYS38

IDLI GROUP 81954

END YOKE WITH 18 INVOLUTED SPLINES
SEE FIGURE 7A

5.8438 in	IE	148.43 mm
6.2188 in	OE	157.96 mm
2.3594 in	SB	59.93 mm
4.0781 in	EC	103.58 mm
3.0000 in	HD	76.20 mm

ROCKWELL 72NYS38-2

IDLI GROUP 81955

END YOKE WITH 46 SLANTED SPLINES
SEE FIGURE 7A

5.8438 in	IE	148.43 mm
6.2188 in	OE	157.96 mm
2.3750 in	SB	60.33 mm
4.0625 in	EC	103.19 mm
3.2500 in	HD	82.55 mm

BORG-WARNR 27969J
BORG-WARNR 279J

ROCKWELL 72NYS38-14

IDLI GROUP 81956

END YOKE WITH 36 INVOLUTED SPLINES
SEE FIGURE 7A

5.8438 in	IE	148.43 mm
6.2188 in	OE	157.96 mm
2.3750 in	SB	60.33 mm
4.3125 in	EC	109.54 mm
3.0000 in	HD	76.20 mm

ROCKWELL 72NYS38-16

IDLI GROUP 81957

END YOKE WITH 46 SLANTED SPLINES
SEE FIGURE 7A

5.8438 in	IE	148.43 mm
6.2188 in	OE	157.96 mm
2.3750 in	SB	60.33 mm
4.3125 in	EC	109.54 mm
3.5000 in	HD	88.90 mm

ROCKWELL 72NYS38-17

IDLI GROUP 81958

END YOKE WITH 46 SLANTED SPLINES
SEE FIGURE 7A

5.8438 in	IE	148.43 mm
6.2188 in	OE	157.96 mm
2.3750 in	SB	60.33 mm
4.3750 in	EC	111.13 mm
3.2500 in	HD	82.55 mm

ROCKWELL 72NYS38-15

IDLI GROUP 81959

END YOKE WITH ROUND BORE
SEE FIGURE 7B

5.8438 in	IE	148.43 mm
6.2188 in	OE	157.96 mm
2.3750 in	SB	60.33 mm
4.6250 in	EC	117.48 mm
4.0000 in	HD	101.60 mm

ROCKWELL 72NYR38-1

IDLI GROUP 81960

END YOKE WITH ROUND BORE
slight variation from above group
SEE FIGURE 7B

5.8438 in	IE	148.43 mm
6.2188 in	OE	157.96 mm
2.3750 in	SB	60.33 mm
4.6250 in	EC	117.48 mm
4.0000 in	HD	101.60 mm

ROCKWELL 72NYR38

IDLI GROUP 81961

END YOKE WITH ROUND BORE
SEE FIGURE 7B

5.8438 in	IE	148.43 mm
6.2188 in	OE	157.96 mm
2.3750 in	SB	60.33 mm
4.9375 in	EC	125.41 mm
4.0000 in	HD	101.60 mm

ROCKWELL 72NYR38-3

IDLI GROUP 81962

END YOKE WITH ROUND BORE
SEE FIGURE 7B

5.8438 in	IE	148.43 mm
6.2188 in	OE	157.96 mm
2.4375 in	SB	61.91 mm
4.6250 in	EC	117.48 mm
4.0000 in	HD	101.60 mm

ROCKWELL 72NYR39

IDLI GROUP 81963

END YOKE WITH 16 SPLINES
SEE FIGURE 7A

5.8438 in	IE	148.43 mm
6.2188 in	OE	157.96 mm
2.5000 in	SB	63.50 mm
1.6250 in	EC	41.28 mm
3.2656 in	HD	82.95 mm

ROCKWELL 72NYS40-1

IDLI GROUP 81964

END YOKE WITH 16 SPLINES
SEE FIGURE 7A

5.8438 in	IE	148.43 mm
6.2188 in	OE	157.96 mm
2.5000 in	SB	63.50 mm
1.7500 in	EC	44.45 mm
3.2656 in	HD	82.95 mm

ROCKWELL 72NYS40-3

IDLI GROUP 81965

END YOKE WITH 16 SPLINES
SEE FIGURE 7A

5.8438 in	IE	148.43 mm
6.2188 in	OE	157.96 mm
2.5000 in	SB	63.50 mm
2.1250 in	EC	53.98 mm
3.2500 in	HD	82.55 mm

BORG-WARNR 20534J
ROCKWELL 72NYS40

IDLI GROUP 81966

END YOKE WITH 14 INVOLUTED SPLINES
SEE FIGURE 7A

5.8438 in	IE	148.43 mm
6.2188 in	OE	157.96 mm
2.5000 in	SB	63.50 mm
2.6250 in	EC	66.68 mm
3.2500 in	HD	82.55 mm

ROCKWELL 72NYS40-4

IDLI GROUP 81967

END YOKE WITH 16 SPLINES
SEE FIGURE 7A

5.8438 in	IE	148.43 mm
6.2188 in	OE	157.96 mm
2.5000 in	SB	63.50 mm
2.7500 in	EC	69.85 mm
3.0625 in	HD	77.79 mm

ROCKWELL 72NYS40-41

IDLI GROUP 81968

END YOKE WITH 16 SPLINES
SEE FIGURE 7A

5.8438 in	IE	148.43 mm
6.2188 in	OE	157.96 mm
2.5000 in	SB	63.50 mm
2.7500 in	EC	69.85 mm
3.2500 in	HD	82.55 mm

ROCKWELL 72NYS40-58

IDLI GROUP 81969

END YOKE WITH 16 SPLINES
SEE FIGURE 7A

5.8438 in	IE	148.43 mm
6.2188 in	OE	157.96 mm
2.5000 in	SB	63.50 mm
3.0000 in	EC	76.20 mm
3.0625 in	HD	77.79 mm

ROCKWELL 72NYS40-2

IDLI GROUP 81970

END YOKE WITH 16 SPLINES
SEE FIGURE 7A

5.8438 in	IE	148.43 mm
6.2188 in	OE	157.96 mm
2.5000 in	SB	63.50 mm
3.1875 in	EC	80.96 mm
3.3750 in	HD	85.73 mm

ROCKWELL 72NYS40-31

IDLI GROUP 81971

END YOKE WITH 16 SPLINES
SEE FIGURE 7A

5.8438 in	IE	148.43 mm
6.2188 in	OE	157.96 mm
2.5000 in	SB	63.50 mm
3.8750 in	EC	98.43 mm
3.0625 in	HD	77.79 mm

ROCKWELL 72NYS40-26

IDLI GROUP 81972

END YOKE WITH 10 SPLINES
SEE FIGURE 7A

5.8438 in	IE	148.43 mm
6.2188 in	OE	157.96 mm
2.5000 in	SB	63.50 mm
3.8750 in	EC	98.43 mm
3.2500 in	HD	82.55 mm

BORG-WARNR 20524J
ROCKWELL 72NYS40-22

IDLI GROUP 81973

END YOKE WITH 10 SPLINES
slight variation from above group
SEE FIGURE 7A

5.8438 in	IE	148.43 mm
6.2188 in	OE	157.96 mm
2.5000 in	SB	63.50 mm
3.8750 in	EC	98.43 mm
3.2500 in	HD	82.55 mm

ROCKWELL 72NYS40-90

BD = BEARING DIAMETER (outside)   **BW** = BEARING WIDTH   **CB** = CROSS LENGTH WITH BEARINGS   **CC** = CENTER TO CENTER   **CD** = CROSS DIAMETER
CL = CROSS LENGTH WITHOUT BEARINGS   **EC** = END TO CENTER or FACE TO FACE   **EE** = END TO END   **EL** = EFFECTIVE LENGTH
HD = HUB DIAMETER (or insert)   **IE** = INSIDE OF EARS (or recess)   **OD** = OUTSIDE DIAMETER   **OE** = OUTSIDE OF EARS   **SB** = SPLINE OR BORE SIZE

IDLI GROUP 81974

END YOKE WITH 16 SPLINES
SEE FIGURE 7A

5.8438 in	IE	148.43 mm
6.2188 in	OE	157.96 mm
2.5000 in	SB	63.50 mm
4.0625 in	EC	103.19 mm
3.2500 in	HD	82.55 mm

ROCKWELL 72NYS40-59

IDLI GROUP 81975

END YOKE WITH 10 SPLINES
SEE FIGURE 7A

5.8438 in	IE	148.43 mm
6.2188 in	OE	157.96 mm
2.5000 in	SB	63.50 mm
4.3125 in	EC	109.54 mm
3.5000 in	HD	88.90 mm

ROCKWELL 72NYS40-72

IDLI GROUP 81976

END YOKE WITH 10 SPLINES
SEE FIGURE 7A

5.8438 in	IE	148.43 mm
6.2188 in	OE	157.96 mm
2.5000 in	SB	63.50 mm
4.4375 in	EC	112.71 mm
2.9375 in	HD	74.61 mm

ROCKWELL 72NYS40-80

IDLI GROUP 81977

END YOKE WITH 16 SPLINES
SEE FIGURE 7A

5.8438 in	IE	148.43 mm
6.2188 in	OE	157.96 mm
2.5000 in	SB	63.50 mm
4.6250 in	EC	117.48 mm
3.5000 in	HD	88.90 mm

ROCKWELL 72NYS40-5

IDLI GROUP 81978

END YOKE WITH ROUND BORE
SEE FIGURE 7B

5.8438 in	IE	148.43 mm
6.2188 in	OE	157.96 mm
2.5000 in	SB	63.50 mm
4.6250 in	EC	117.48 mm
4.0000 in	HD	101.60 mm

ROCKWELL 72NYR40-2

IDLI GROUP 81979

END YOKE WITH ROUND BORE
SEE FIGURE 7B

5.8438 in	IE	148.43 mm
6.2188 in	OE	157.96 mm
2.5000 in	SB	63.50 mm
4.8125 in	EC	122.24 mm
4.0000 in	HD	101.60 mm

ROCKWELL 72NYR40-11

IDLI GROUP 81980

END YOKE WITH TAPERED BORE
SEE FIGURE 7F

5.8438 in	IE	148.43 mm
6.2188 in	OE	157.96 mm
2.5000 in	SB	63.50 mm
4.8125 in	EC	122.24 mm
4.0000 in	HD	101.60 mm

ROCKWELL 72NYT40

IDLI GROUP 81981

END YOKE WITH 19 INVOLUTED SPLINES
SEE FIGURE 7A

5.8438 in	IE	148.43 mm
6.2188 in	OE	157.96 mm
2.5313 in	SB	64.30 mm
4.5938 in	EC	116.68 mm
3.1250 in	HD	79.38 mm

BORG-WARNR 19624J
ROCKWELL 72NYS40-19

IDLI GROUP 81982

END YOKE WITH ROUND BORE
SEE FIGURE 7B

5.8438 in	IE	148.43 mm
6.2188 in	OE	157.96 mm
2.5625 in	SB	65.09 mm
4.6250 in	EC	117.48 mm
4.0000 in	HD	101.60 mm

ROCKWELL 72NYR41

IDLI GROUP 81983

END YOKE WITH ROUND BORE
SEE FIGURE 7B

5.8438 in	IE	148.43 mm
6.2188 in	OE	157.96 mm
2.6250 in	SB	66.68 mm
4.6250 in	EC	117.48 mm
4.0000 in	HD	101.60 mm

ROCKWELL 72NYR42-1

IDLI GROUP 81984

END YOKE WITH ROUND BORE
slight variation from above group
SEE FIGURE 7B

5.8438 in	IE	148.43 mm
6.2188 in	OE	157.96 mm
2.6250 in	SB	66.68 mm
4.6250 in	EC	117.48 mm
4.0000 in	HD	101.60 mm

ROCKWELL 72NYR44-2

IDLI GROUP 81985

END YOKE WITH ROUND BORE
slight variation from above group
SEE FIGURE 7B

5.8438 in	IE	148.43 mm
6.2188 in	OE	157.96 mm
2.6250 in	SB	66.68 mm
4.6250 in	EC	117.48 mm
4.0000 in	HD	101.60 mm

ROCKWELL 72NYR42

IDLI GROUP 81986

END YOKE WITH ROUND BORE
slight variation from above group
SEE FIGURE 7B

5.8438 in	IE	148.43 mm
6.2188 in	OE	157.96 mm
2.6250 in	SB	66.68 mm
4.6250 in	EC	117.48 mm
4.0000 in	HD	101.60 mm

ROCKWELL 72NYR42-2

IDLI GROUP 81987

END YOKE WITH ROUND BORE
SEE FIGURE 7B

5.8438 in	IE	148.43 mm
6.2188 in	OE	157.96 mm
2.6250 in	SB	66.68 mm
4.8125 in	EC	122.24 mm
4.0000 in	HD	101.60 mm

ROCKWELL 72NYR42-5

IDLI GROUP 81988

END YOKE WITH ROUND BORE
SEE FIGURE 7B

5.8438 in	IE	148.43 mm
6.2188 in	OE	157.96 mm
2.7500 in	SB	69.85 mm
4.6250 in	EC	117.48 mm
4.0000 in	HD	101.60 mm

ROCKWELL 72NYR44-4

IDLI GROUP 81989

END YOKE WITH ROUND BORE
slight variation from above group
SEE FIGURE 7B

5.8438 in	IE	148.43 mm
6.2188 in	OE	157.96 mm
2.7500 in	SB	69.85 mm
4.6250 in	EC	117.48 mm
4.0000 in	HD	101.60 mm

ROCKWELL 72NYR44-3

IDLI GROUP 81990

END YOKE WITH ROUND BORE
slight variation from above group
SEE FIGURE 7B

5.8438 in	IE	148.43 mm
6.2188 in	OE	157.96 mm
2.7500 in	SB	69.85 mm
4.6250 in	EC	117.48 mm
4.0000 in	HD	101.60 mm

ROCKWELL 72NYR44-7

IDLI GROUP 81991

END YOKE WITH ROUND BORE
SEE FIGURE 7B

5.8438 in	IE	148.43 mm
6.2188 in	OE	157.96 mm
2.7500 in	SB	69.85 mm
4.7500 in	EC	120.65 mm
3.8750 in	HD	98.43 mm

ROCKWELL 72NYR44-6

IDLI GROUP 81992

END YOKE WITH ROUND BORE
SEE FIGURE 7B

5.8438 in	IE	148.43 mm
6.2188 in	OE	157.96 mm
2.8750 in	SB	73.03 mm
4.6250 in	EC	117.48 mm
4.0000 in	HD	101.60 mm

ROCKWELL 72NYR46-1

IDLI GROUP 81993

END YOKE WITH ROUND BORE
SEE FIGURE 7B

5.8438 in	IE	148.43 mm
6.2188 in	OE	157.96 mm
3.0000 in	SB	76.20 mm
4.6250 in	EC	117.48 mm
4.0000 in	HD	101.60 mm

ROCKWELL 72NYR48-3

IDLI GROUP 81994

END YOKE WITH ROUND BORE
slight variation from above group
SEE FIGURE 7B

5.8438 in	IE	148.43 mm
6.2188 in	OE	157.96 mm
3.0000 in	SB	76.20 mm
4.6250 in	EC	117.48 mm
4.0000 in	HD	101.60 mm

ROCKWELL 72NYR48-1

IDLI GROUP 81995

END YOKE WITH ROUND BORE
slight variation from above group
SEE FIGURE 7B

5.8438 in	IE	148.43 mm
6.2188 in	OE	157.96 mm
3.0000 in	SB	76.20 mm
4.6250 in	EC	117.48 mm
4.0000 in	HD	101.60 mm

ROCKWELL 72NYR48-4

IDLI GROUP 81996

END YOKE WITH ROUND BORE
SEE FIGURE 7B

5.8438 in	IE	148.43 mm
6.2188 in	OE	157.96 mm
3.0000 in	SB	76.20 mm
4.8125 in	EC	122.24 mm
4.0000 in	HD	101.60 mm

ROCKWELL 72NYR48-5

IDLI GROUP 81997

END YOKE WITH ROUND BORE
SEE FIGURE 7B

5.8438 in	IE	148.43 mm
6.2188 in	OE	157.96 mm
3.1250 in	SB	79.38 mm
4.6250 in	EC	117.48 mm
4.0000 in	HD	101.60 mm

ROCKWELL 72NYR50

ENGINEERING CATALOGS MUST BE CONSULTED FOR DETAILS NOT INCLUDED
IN THIS GUIDE. SPECIFIC DESIGNS, MATERIAL CONTENT, TOLERANCES
LUBE FITTINGS AND OTHER DIMENSIONS ARE INTENTIONALLY OMITTED HERE.

CAUTION: BE SURE TO REFER TO ENGINEERING CATALOGS FOR SPECIAL APPLICATIONS THAT REQUIRE SPECIFIC MATERIAL CONTENT, TOLERANCES, ETC. SEE FOOTNOTE.

IDLI GROUP 81998

END YOKE WITH ROUND BORE
slight variation from above group
SEE FIGURE 7B

5.8438 in	IE	148.43 mm
6.2188 in	OE	157.96 mm
3.1250 in	SB	79.38 mm
4.6250 in	EC	117.48 mm
4.0000 in	HD	101.60 mm

ROCKWELL 72NYR50-1

IDLI GROUP 81999

END YOKE WITH ROUND BORE
SEE FIGURE 7B

5.8438 in	IE	148.43 mm
6.2188 in	OE	157.96 mm
3.2500 in	SB	82.55 mm
4.6250 in	EC	117.48 mm
4.0000 in	HD	101.60 mm

ROCKWELL 72NYR52-2

IDLI GROUP 82000

END YOKE WITH ROUND BORE
slight variation from above group
SEE FIGURE 7B

5.8438 in	IE	148.43 mm
6.2188 in	OE	157.96 mm
3.2500 in	SB	82.55 mm
4.6250 in	EC	117.48 mm
4.0000 in	HD	101.60 mm

ROCKWELL 72NYR52

IDLI GROUP 82001

END YOKE WITH ROUND BORE
slight variation from above group
SEE FIGURE 7B

5.8438 in	IE	148.43 mm
6.2188 in	OE	157.96 mm
3.2500 in	SB	82.55 mm
4.6250 in	EC	117.48 mm
4.0000 in	HD	101.60 mm

ROCKWELL 72NYR52-3

IDLI GROUP 82002

END YOKE WITH ROUND BORE
slight variation from above group
SEE FIGURE 7B

5.8438 in	IE	148.43 mm
6.2188 in	OE	157.96 mm
3.2500 in	SB	82.55 mm
4.6250 in	EC	117.48 mm
4.0000 in	HD	101.60 mm

ROCKWELL 72NYR52-6

IDLI GROUP 82003

END YOKE WITH 10 SPLINES
SEE FIGURE 7A

6.5000 in	IE	165.10 mm
6.8750 in	OE	174.63 mm
1.7500 in	SB	44.45 mm
3.2500 in	EC	82.55 mm
2.2500 in	HD	57.15 mm

ROCKWELL 85WBYS28

IDLI GROUP 82004

END YOKE WITH 10 SPLINES
SEE FIGURE 7A

6.5000 in	IE	165.10 mm
6.8750 in	OE	174.63 mm
1.8750 in	SB	47.63 mm
3.1250 in	EC	79.38 mm
2.7500 in	HD	69.85 mm

ROCKWELL 85WBYS30

IDLI GROUP 82005

END YOKE WITH 10 SPLINES
SEE FIGURE 7A

6.5000 in	IE	165.10 mm
6.8750 in	OE	174.63 mm
1.9688 in	SB	50.01 mm
3.1250 in	EC	79.38 mm
2.7500 in	HD	69.85 mm

ROCKWELL 85WBYS31-1

IDLI GROUP 82006

END YOKE WITH 10 SPLINES
SEE FIGURE 7A

6.5000 in	IE	165.10 mm
6.8750 in	OE	174.63 mm
1.9688 in	SB	50.01 mm
3.5000 in	EC	88.90 mm
2.6250 in	HD	66.68 mm

ROCKWELL 85WBYS31

IDLI GROUP 82007

END YOKE WITH 16 SPLINES
SEE FIGURE 7A

6.5000 in	IE	165.10 mm
6.8750 in	OE	174.63 mm
2.0000 in	SB	50.80 mm
2.5000 in	EC	63.50 mm
2.5000 in	HD	63.50 mm

ROCKWELL 85WBYS32

IDLI GROUP 82008

END YOKE WITH 16 SPLINES
slight variation from above group
SEE FIGURE 7A

6.5000 in	IE	165.10 mm
6.8750 in	OE	174.63 mm
2.0000 in	SB	50.80 mm
2.5000 in	EC	63.50 mm
2.5000 in	HD	63.50 mm

ROCKWELL 85WBYS32-3

IDLI GROUP 82009

END YOKE WITH 10 SPLINES
SEE FIGURE 7A

6.5000 in	IE	165.10 mm
6.8750 in	OE	174.63 mm
2.0000 in	SB	50.80 mm
4.7500 in	EC	120.65 mm
2.9375 in	HD	74.61 mm

ROCKWELL 85WBYS32-1

IDLI GROUP 82010

END YOKE WITH 16 INVOLUTED SPLINES
SEE FIGURE 7A

6.5000 in	IE	165.10 mm
6.8750 in	OE	174.63 mm
2.1250 in	SB	53.98 mm
3.7500 in	EC	95.25 mm
2.8750 in	HD	73.03 mm

ROCKWELL 85WBYS34

IDLI GROUP 82011

END YOKE WITH 10 SPLINES
SEE FIGURE 7A

6.5000 in	IE	165.10 mm
6.8750 in	OE	174.63 mm
2.2500 in	SB	57.15 mm
3.8125 in	EC	96.84 mm
2.8750 in	HD	73.03 mm

ROCKWELL 85WBYS36-6

IDLI GROUP 82012

END YOKE WITH 6 SPLINES
SEE FIGURE 7A

6.5000 in	IE	165.10 mm
6.8750 in	OE	174.63 mm
2.2500 in	SB	57.15 mm
4.8125 in	EC	122.24 mm
3.3750 in	HD	85.73 mm

ROCKWELL 85WBYS36-8

IDLI GROUP 82013

END YOKE WITH 10 SPLINES
SEE FIGURE 7A

6.5000 in	IE	165.10 mm
6.8750 in	OE	174.63 mm
2.2500 in	SB	57.15 mm
5.1250 in	EC	130.18 mm
3.0000 in	HD	76.20 mm

ROCKWELL 85WBYS36-7

IDLI GROUP 82014

END YOKE WITH 17 INVOLUTED SPLINES
SEE FIGURE 7A

6.5000 in	IE	165.10 mm
6.8750 in	OE	174.63 mm
2.2813 in	SB	57.95 mm
3.2188 in	EC	81.76 mm
3.1250 in	HD	79.38 mm

ROCKWELL 85WBYS36-2

IDLI GROUP 82015

END YOKE WITH 17 INVOLUTED SPLINES
SEE FIGURE 7A

6.5000 in	IE	165.10 mm
6.8750 in	OE	174.63 mm
2.2813 in	SB	57.95 mm
3.4375 in	EC	87.31 mm
3.0000 in	HD	76.20 mm

ROCKWELL 85WBYS36-3

IDLI GROUP 82016

END YOKE WITH ROUND BORE
SEE FIGURE 7B

6.5000 in	IE	165.10 mm
6.8750 in	OE	174.63 mm
2.3750 in	SB	60.33 mm
5.0000 in	EC	127.00 mm
4.5000 in	HD	114.30 mm

ROCKWELL 85WBYR38

IDLI GROUP 82017

END YOKE WITH ROUND BORE
SEE FIGURE 7B

6.5000 in	IE	165.10 mm
6.8750 in	OE	174.63 mm
2.3750 in	SB	60.33 mm
5.2500 in	EC	133.35 mm
4.5000 in	HD	114.30 mm

ROCKWELL 85WBYR38-1

IDLI GROUP 82018

END YOKE WITH 16 SPLINES
SEE FIGURE 7A

6.5000 in	IE	165.10 mm
6.8750 in	OE	174.63 mm
2.5000 in	SB	63.50 mm
1.7500 in	EC	44.45 mm
3.3125 in	HD	84.14 mm

ROCKWELL 85WBYS40-2

IDLI GROUP 82019

END YOKE WITH 16 SPLINES
SEE FIGURE 7A

6.5000 in	IE	165.10 mm
6.8750 in	OE	174.63 mm
2.5000 in	SB	63.50 mm
2.6250 in	EC	66.68 mm
3.8750 in	HD	98.43 mm

ROCKWELL 85WBYS40-4

IDLI GROUP 82020

END YOKE WITH 16 SPLINES
SEE FIGURE 7A

6.5000 in	IE	165.10 mm
6.8750 in	OE	174.63 mm
2.5313 in	SB	64.30 mm
3.5000 in	EC	88.90 mm
4.5000 in	HD	114.30 mm

ROCKWELL 85WBYS40

BD = BEARING DIAMETER (outside)　　**DW** = BEARING WIDTH　　**CB** = CROSS LENGTH WITH BEARINGS　　**CC** = CENTER TO CENTER　　**CD** = CROSS DIAMETER
CL = CROSS LENGTH WITHOUT BEARINGS　　**EC** = END TO CENTER or FACE TO FACE　　**EE** = END TO END　　**EL** = EFFECTIVE LENGTH
HD = HUB DIAMETER (or insert)　　**IE** = INSIDE OF EARS (or recess)　　**OD** = OUTSIDE DIAMETER　　**OE** = OUTSIDE OF EARS　　**SB** = SPLINE OR BORE SIZE

IDLI GROUP 82021

END YOKE WITH 19 INVOLUTED SPLINES
SEE FIGURE 7A

6.5000 in	IE	165.10 mm
6.8750 in	OE	174.63 mm
2.5313 in	SB	64.30 mm
4.5938 in	EC	116.68 mm
3.1250 in	HD	79.38 mm

BORG-WARNR 20423J
ROCKWELL 85WBYS40-3

IDLI GROUP 82022

END YOKE WITH 25 INVOLUTED SPLINES
SEE FIGURE 7A

6.5000 in	IE	165.10 mm
6.8750 in	OE	174.63 mm
2.6250 in	SB	66.68 mm
4.7500 in	EC	120.65 mm
3.6250 in	HD	92.08 mm

ROCKWELL 85WBYS42

IDLI GROUP 82023

END YOKE WITH ROUND BORE
SEE FIGURE 7B

6.5000 in	IE	165.10 mm
6.8750 in	OE	174.63 mm
2.6250 in	SB	66.68 mm
5.2500 in	EC	133.35 mm
4.5000 in	HD	114.30 mm

ROCKWELL 85WBYR42

IDLI GROUP 82024

END YOKE WITH ROUND BORE
SEE FIGURE 7B

6.5000 in	IE	165.10 mm
6.8750 in	OE	174.63 mm
2.7500 in	SB	69.85 mm
5.0000 in	EC	127.00 mm
4.5000 in	HD	114.30 mm

ROCKWELL 85WBYR44-2

IDLI GROUP 82025

END YOKE WITH ROUND BORE
slight variation from above group
SEE FIGURE 7B

6.5000 in	IE	165.10 mm
6.8750 in	OE	174.63 mm
2.7500 in	SB	69.85 mm
5.0000 in	EC	127.00 mm
4.5000 in	HD	114.30 mm

ROCKWELL 85WBYR44-1

IDLI GROUP 82026

END YOKE WITH ROUND BORE
SEE FIGURE 7B

6.5000 in	IE	165.10 mm
6.8750 in	OE	174.63 mm
2.7500 in	SB	69.85 mm
5.2500 in	EC	133.35 mm
4.5000 in	HD	114.30 mm

ROCKWELL 85WBYR44

IDLI GROUP 82027

END YOKE WITH ROUND BORE
SEE FIGURE 7B

6.5000 in	IE	165.10 mm
6.8750 in	OE	174.63 mm
2.8750 in	SB	73.03 mm
5.0000 in	EC	127.00 mm
4.5000 in	HD	114.30 mm

ROCKWELL 85WBYR46

IDLI GROUP 82028

END YOKE WITH 10 SPLINES
SEE FIGURE 7A

6.5000 in	IE	165.10 mm
6.8750 in	OE	174.63 mm
3.0000 in	SB	76.20 mm
3.5000 in	EC	88.90 mm
4.3750 in	HD	111.13 mm

ROCKWELL 85WBYS48-1

IDLI GROUP 82029

END YOKE WITH ROUND BORE
SEE FIGURE 7B

6.5000 in	IE	165.10 mm
6.8750 in	OE	174.63 mm
3.0000 in	SB	76.20 mm
5.0000 in	EC	127.00 mm
4.5000 in	HD	114.30 mm

ROCKWELL 85WBYR48

IDLI GROUP 82030

END YOKE WITH ROUND BORE
SEE FIGURE 7B

6.5000 in	IE	165.10 mm
6.8750 in	OE	174.63 mm
3.6250 in	SB	92.08 mm
5.2500 in	EC	133.35 mm
6.9375 in	HD	176.21 mm

ROCKWELL 85WBYR58

IDLI GROUP 82031

END YOKE WITH ROUND BORE
SEE FIGURE 7B

8.1250 in	IE	206.38 mm
8.5000 in	OE	215.90 mm
1.5000 in	SB	38.10 mm
4.8750 in	EC	123.83 mm
6.0000 in	HD	152.40 mm

ROCKWELL 82NYR24

IDLI GROUP 82032

END YOKE WITH ROUND BORE
SEE FIGURE 7B

8.1250 in	IE	206.38 mm
8.5000 in	OE	215.90 mm
1.5000 in	SB	38.10 mm
6.5625 in	EC	166.69 mm
6.0000 in	HD	152.40 mm

ROCKWELL 82NYR24-1

IDLI GROUP 82033

END YOKE WITH ROUND BORE
SEE FIGURE 7B

8.1250 in	IE	206.38 mm
8.5000 in	OE	215.90 mm
1.7500 in	SB	44.45 mm
1.8125 in	EC	46.04 mm
3.0000 in	HD	76.20 mm

BORG-WARNR 5185J
ROCKWELL 82NYR28-3

IDLI GROUP 82034

END YOKE WITH ROUND BORE
SEE FIGURE 7B

8.1250 in	IE	206.38 mm
8.5000 in	OE	215.90 mm
1.7500 in	SB	44.45 mm
2.2500 in	EC	57.15 mm
3.0000 in	HD	76.20 mm

ROCKWELL 82NYR28-1

IDLI GROUP 82035

END YOKE WITH ROUND BORE
slight variation from above group
SEE FIGURE 7B

8.1250 in	IE	206.38 mm
8.5000 in	OE	215.90 mm
1.7500 in	SB	44.45 mm
2.2500 in	EC	57.15 mm
3.0000 in	HD	76.20 mm

ROCKWELL 82NYR28

IDLI GROUP 82036

END YOKE WITH 10 SPLINES
SEE FIGURE 7A

8.1250 in	IE	206.38 mm
8.5000 in	OE	215.90 mm
1.7500 in	SB	44.45 mm
3.5000 in	EC	88.90 mm
2.6250 in	HD	66.68 mm

BORG-WARNR 4803J
BORG-WARNR 6680J
ROCKWELL 82NYS28-2

IDLI GROUP 82037

END YOKE WITH 10 SPLINES
SEE FIGURE 7A

8.1250 in	IE	206.38 mm
8.5000 in	OE	215.90 mm
1.7500 in	SB	44.45 mm
3.5000 in	EC	88.90 mm
2.7500 in	HD	69.85 mm

BORG-WARNR 18903J
BORG-WARNR 23999J
ROCKWELL 82NYS28-8

IDLI GROUP 82038

END YOKE WITH TAPERED BORE
SEE FIGURE 7F

8.1250 in	IE	206.38 mm
8.5000 in	OE	215.90 mm
1.7500 in	SB	44.45 mm
3.5000 in	EC	88.90 mm
3.3750 in	HD	85.73 mm

BORG-WARNR 5292J
ROCKWELL 82NYT28

IDLI GROUP 82039

END YOKE WITH 10 SPLINES
SEE FIGURE 7A

8.1250 in	IE	206.38 mm
8.5000 in	OE	215.90 mm
1.7500 in	SB	44.45 mm
3.6250 in	EC	92.08 mm
2.5000 in	HD	63.50 mm

ROCKWELL 82NYS28-21

IDLI GROUP 82040

END YOKE WITH 10 SPLINES
SEE FIGURE 7A

8.1250 in	IE	206.38 mm
8.5000 in	OE	215.90 mm
1.7500 in	SB	44.45 mm
3.7500 in	EC	95.25 mm
2.3125 in	HD	58.74 mm

ROCKWELL 82NYS28-14

IDLI GROUP 82041

END YOKE WITH 16 SPLINES
SEE FIGURE 7A

8.1250 in	IE	206.38 mm
8.5000 in	OE	215.90 mm
1.7500 in	SB	44.45 mm
3.8750 in	EC	98.43 mm
2.7500 in	HD	69.85 mm

ROCKWELL 82NYS28-6

IDLI GROUP 82042

END YOKE WITH 10 SPLINES
SEE FIGURE 7A

8.1250 in	IE	206.38 mm
8.5000 in	OE	215.90 mm
1.7500 in	SB	44.45 mm
4.0625 in	EC	103.19 mm
2.4375 in	HD	61.91 mm

ROCKWELL 82NYS28-12

IDLI GROUP 82043

END YOKE WITH 24 SLANTED SPLINES
SEE FIGURE 7A

8.1250 in	IE	206.38 mm
8.5000 in	OE	215.90 mm
1.7500 in	SB	44.45 mm
4.0625 in	EC	103.19 mm
2.7500 in	HD	69.85 mm

ROCKWELL 82NYS28-4

IDLI GROUP 82044

END YOKE WITH 16 INVOLUTED SPLINES
SEE FIGURE 7A

8.1250 in	IE	206.38 mm
8.5000 in	OE	215.90 mm
1.7813 in	SB	45.25 mm
3.7500 in	EC	95.25 mm
2.3750 in	HD	60.33 mm

BORG-WARNR 13489J
ROCKWELL 82NYS28-1

ENGINEERING CATALOGS MUST BE CONSULTED FOR DETAILS NOT INCLUDED
IN THIS GUIDE. SPECIFIC DESIGNS, MATERIAL CONTENT, TOLERANCES
LUBE FITTINGS AND OTHER DIMENSIONS ARE INTENTIONALLY OMITTED HERE.

CAUTION: BE SURE TO REFER TO ENGINEERING CATALOGS FOR SPECIAL APPLICATIONS THAT REQUIRE SPECIFIC MATERIAL CONTENT, TOLERANCES, ETC. SEE FOOTNOTE.

IDLI GROUP 82045

END YOKE WITH 16 INVOLUTED SPLINES
SEE FIGURE 7A

8.1250 in	IE	206.38 mm
8.5000 in	OE	215.90 mm
1.7813 in	SB	45.25 mm
3.8750 in	EC	98.43 mm
2.6250 in	HD	66.68 mm

BORG-WARNR 9745J
ROCKWELL 82NYS28

IDLI GROUP 82046

END YOKE WITH 10 SPLINES
SEE FIGURE 7A

8.1250 in	IE	206.38 mm
8.5000 in	OE	215.90 mm
1.8750 in	SB	47.63 mm
3.8125 in	EC	96.84 mm
2.7500 in	HD	69.85 mm

BORG-WARNR 9279J
ROCKWELL 82NYS30-2

IDLI GROUP 82047

END YOKE WITH 10 SPLINES
slight variation from above group
SEE FIGURE 7A

8.1250 in	IE	206.38 mm
8.5000 in	OE	215.90 mm
1.8750 in	SB	47.63 mm
3.8125 in	EC	96.84 mm
2.7500 in	HD	69.85 mm

BORG-WARNR 14214J
ROCKWELL 82NYS30-3

IDLI GROUP 82048

END YOKE WITH 10 SPLINES
SEE FIGURE 7A

8.1250 in	IE	206.38 mm
8.5000 in	OE	215.90 mm
1.8750 in	SB	47.63 mm
3.8750 in	EC	98.43 mm
2.7500 in	HD	69.85 mm

ROCKWELL 82NYS30-1

IDLI GROUP 82049

END YOKE WITH 10 SPLINES
SEE FIGURE 7A

8.1250 in	IE	206.38 mm
8.5000 in	OE	215.90 mm
1.8750 in	SB	47.63 mm
4.0000 in	EC	101.60 mm
2.7500 in	HD	69.85 mm

ROCKWELL 82NYS30-5

IDLI GROUP 82050

END YOKE WITH ROUND BORE
SEE FIGURE 7B

8.1250 in	IE	206.38 mm
8.5000 in	OE	215.90 mm
1.8750 in	SB	47.63 mm
5.2500 in	EC	133.35 mm
6.0000 in	HD	152.40 mm

ROCKWELL 82NYR30

IDLI GROUP 82051

END YOKE WITH 30 INVOLUTED SPLINES
SEE FIGURE 7A

8.1250 in	IE	206.38 mm
8.5000 in	OE	215.90 mm
1.9375 in	SB	49.21 mm
3.6875 in	EC	93.66 mm
3.0000 in	HD	76.20 mm

BORG-WARNR 23940J
ROCKWELL 82NYS31-32

IDLI GROUP 82052

END YOKE WITH ROUND BORE
SEE FIGURE 7B

8.1250 in	IE	206.38 mm
8.5000 in	OE	215.90 mm
1.9375 in	SB	49.21 mm
4.1875 in	EC	106.36 mm
3.5000 in	HD	88.90 mm

ROCKWELL 82NYR31-1

IDLI GROUP 82053

END YOKE WITH ROUND BORE
slight variation from above group
SEE FIGURE 7B

8.1250 in	IE	206.38 mm
8.5000 in	OE	215.90 mm
1.9375 in	SB	49.21 mm
4.1875 in	EC	106.36 mm
3.5000 in	HD	88.90 mm

ROCKWELL 82NYR31

IDLI GROUP 82054

END YOKE WITH TAPERED BORE
SEE FIGURE 7F

8.1250 in	IE	206.38 mm
8.5000 in	OE	215.90 mm
1.9375 in	SB	49.21 mm
4.2500 in	EC	107.95 mm
3.2500 in	HD	82.55 mm

BORG-WARNR 16953J
ROCKWELL 82NYT31

IDLI GROUP 82055

END YOKE WITH TAPERED BORE
slight variation from above group
SEE FIGURE 7F

8.1250 in	IE	206.38 mm
8.5000 in	OE	215.90 mm
1.9375 in	SB	49.21 mm
4.2500 in	EC	107.95 mm
3.2500 in	HD	82.55 mm

ROCKWELL 82NYT31-1

IDLI GROUP 82056

END YOKE WITH 10 SPLINES
SEE FIGURE 7A

8.1250 in	IE	206.38 mm
8.5000 in	OE	215.90 mm
1.9688 in	SB	50.01 mm
3.5000 in	EC	88.90 mm
2.6250 in	HD	66.68 mm

BORG-WARNR 13937J
ROCKWELL 82NYS31-8

IDLI GROUP 82057

END YOKE WITH 10 SPLINES
SEE FIGURE 7A

8.1250 in	IE	206.38 mm
8.5000 in	OE	215.90 mm
1.9688 in	SB	50.01 mm
3.5000 in	EC	88.90 mm
3.0000 in	HD	76.20 mm

BORG-WARNR 5128J
ROCKWELL 82NYS31

IDLI GROUP 82058

END YOKE WITH 10 SPLINES
SEE FIGURE 7A

8.1250 in	IE	206.38 mm
8.5000 in	OE	215.90 mm
1.9688 in	SB	50.01 mm
3.6250 in	EC	92.08 mm
2.6250 in	HD	66.68 mm

ROCKWELL 82NYS31-12

IDLI GROUP 82059

END YOKE WITH 10 SPLINES
SEE FIGURE 7A

8.1250 in	IE	206.38 mm
8.5000 in	OE	215.90 mm
1.9688 in	SB	50.01 mm
3.6875 in	EC	93.66 mm
3.0000 in	HD	76.20 mm

BORG-WARNR 13482J
BORG-WARNR 13483J
BORG-WARNR 13937J
ROCKWELL 82NYS31-3

IDLI GROUP 82060

END YOKE WITH 10 SPLINES
SEE FIGURE 7A

8.1250 in	IE	206.38 mm
8.5000 in	OE	215.90 mm
1.9688 in	SB	50.01 mm
3.7500 in	EC	95.25 mm
2.7500 in	HD	69.85 mm

ROCKWELL 82NYS31-28

IDLI GROUP 82061

END YOKE WITH 10 SPLINES
SEE FIGURE 7A

8.1250 in	IE	206.38 mm
8.5000 in	OE	215.90 mm
1.9688 in	SB	50.01 mm
3.8750 in	EC	98.43 mm
2.6250 in	HD	66.68 mm

BORG-WARNR 23505J
BORG-WARNR 8180J
ROCKWELL 82NYS31-16

IDLI GROUP 82062

END YOKE WITH 10 SPLINES
slight variation from above group
SEE FIGURE 7A

8.1250 in	IE	206.38 mm
8.5000 in	OE	215.90 mm
1.9688 in	SB	50.01 mm
3.8750 in	EC	98.43 mm
2.6250 in	HD	66.68 mm

ROCKWELL 82NYS31-25

IDLI GROUP 82063

END YOKE WITH 10 SPLINES
slight variation from above group
SEE FIGURE 7A

8.1250 in	IE	206.38 mm
8.5000 in	OE	215.90 mm
1.9688 in	SB	50.01 mm
3.8750 in	EC	98.43 mm
2.6250 in	HD	66.68 mm

BORG-WARNR 8180J
ROCKWELL 82NYS31-7

IDLI GROUP 82064

END YOKE WITH 10 SPLINES
slight variation from above group
SEE FIGURE 7A

8.1250 in	IE	206.38 mm
8.5000 in	OE	215.90 mm
1.9688 in	SB	50.01 mm
3.8750 in	EC	98.43 mm
2.6250 in	HD	66.68 mm

BORG-WARNR 21279J
ROCKWELL 82NYS31-19

IDLI GROUP 82065

END YOKE WITH 10 SPLINES
SEE FIGURE 7A

8.1250 in	IE	206.38 mm
8.5000 in	OE	215.90 mm
1.9688 in	SB	50.01 mm
3.8750 in	EC	98.43 mm
2.7500 in	HD	69.85 mm

BORG-WARNR 7181J
ROCKWELL 82NYS31-5

IDLI GROUP 82066

END YOKE WITH 10 SPLINES
slight variation from above group
SEE FIGURE 7A

8.1250 in	IE	206.38 mm
8.5000 in	OE	215.90 mm
1.9688 in	SB	50.01 mm
3.8750 in	EC	98.43 mm
2.7500 in	HD	69.85 mm

BORG-WARNR 12657J
ROCKWELL 82NYS31-6

BD = BEARING DIAMETER (outside)   **BW** = BEARING WIDTH   **CB** = CROSS LENGTH WITH BEARINGS   **CC** = CENTER TO CENTER   **CD** = CROSS DIAMETER
CL = CROSS LENGTH WITHOUT BEARINGS   **EC** = END TO CENTER or FACE TO FACE   **EE** = END TO END   **EL** = EFFECTIVE LENGTH
HD = HUB DIAMETER (or insert)   **IE** = INSIDE OF EARS (or recess)   **OD** = OUTSIDE DIAMETER   **OE** = OUTSIDE OF EARS   **SB** = SPLINE OR BORE SIZE

IDLI GROUP 82067

END YOKE WITH 10 SPLINES
slight variation from above group
SEE FIGURE 7A

8.1250 in	IE	206.38 mm
8.5000 in	OE	215.90 mm
1.9688 in	SB	50.01 mm
3.8750 in	EC	98.43 mm
2.7500 in	HD	69.85 mm

BORG-WARNR 20651J
ROCKWELL 82NYS31-15

IDLI GROUP 82068

END YOKE WITH 10 SPLINES
SEE FIGURE 7A

8.1250 in	IE	206.38 mm
8.5000 in	OE	215.90 mm
1.9688 in	SB	50.01 mm
4.1250 in	EC	104.78 mm
2.3750 in	HD	60.33 mm

BORG-WARNR 16417J
ROCKWELL 82NYS31-17

IDLI GROUP 82069

END YOKE WITH 10 SPLINES
slight variation from above group
SEE FIGURE 7A

8.1250 in	IE	206.38 mm
8.5000 in	OE	215.90 mm
1.9688 in	SB	50.01 mm
4.1250 in	EC	104.78 mm
2.3750 in	HD	60.33 mm

BORG-WARNR 24658J
ROCKWELL 82NYS31-29

IDLI GROUP 82070

END YOKE WITH 10 SPLINES
SEE FIGURE 7A

8.1250 in	IE	206.38 mm
8.5000 in	OE	215.90 mm
1.9688 in	SB	50.01 mm
4.1875 in	EC	106.36 mm
2.4375 in	HD	61.91 mm

ROCKWELL 82NYS31-37

IDLI GROUP 82071

END YOKE WITH 6 SPLINES
SEE FIGURE 7A

8.1250 in	IE	206.38 mm
8.5000 in	OE	215.90 mm
1.9688 in	SB	50.01 mm
4.2500 in	EC	107.95 mm
2.7500 in	HD	69.85 mm

BORG-WARNR 15927J
BORG-WARNR 20205J
ROCKWELL 82NYS31-4

IDLI GROUP 82072

END YOKE WITH 10 SPLINES
SEE FIGURE 7A

8.1250 in	IE	206.38 mm
8.5000 in	OE	215.90 mm
1.9688 in	SB	50.01 mm
4.2500 in	EC	107.95 mm
2.7500 in	HD	69.85 mm

ROCKWELL 82NYS31-31

IDLI GROUP 82073

END YOKE WITH 10 SPLINES
slight variation from above group
SEE FIGURE 7A

8.1250 in	IE	206.38 mm
8.5000 in	OE	215.90 mm
1.9688 in	SB	50.01 mm
4.2500 in	EC	107.95 mm
2.7500 in	HD	69.85 mm

ROCKWELL 82NYS31-30

IDLI GROUP 82074

END YOKE WITH 10 SPLINES
SEE FIGURE 7A

8.1250 in	IE	206.38 mm
8.5000 in	OE	215.90 mm
2.0000 in	SB	50.80 mm
2.2500 in	EC	57.15 mm
4.0000 in	HD	101.60 mm

ROCKWELL 82NYS32-48

IDLI GROUP 82075

END YOKE WITH 10 SPLINES
SEE FIGURE 7A

8.1250 in	IE	206.38 mm
8.5000 in	OE	215.90 mm
2.0000 in	SB	50.80 mm
3.1875 in	EC	80.96 mm
2.8594 in	HD	72.63 mm

BORG-WARNR 20239J
ROCKWELL 82NYS32-29

IDLI GROUP 82076

END YOKE WITH 10 SPLINES
SEE FIGURE 7A

8.1250 in	IE	206.38 mm
8.5000 in	OE	215.90 mm
2.0000 in	SB	50.80 mm
3.1875 in	EC	80.96 mm
2.9375 in	HD	74.61 mm

BORG-WARNR 16657J
ROCKWELL 82NYS32-6

IDLI GROUP 82077

END YOKE WITH 10 SPLINES
SEE FIGURE 7A

8.1250 in	IE	206.38 mm
8.5000 in	OE	215.90 mm
2.0000 in	SB	50.80 mm
3.5000 in	EC	88.90 mm
2.6250 in	HD	66.68 mm

BORG-WARNR 16209J
ROCKWELL 82NYS32-9

IDLI GROUP 82078

END YOKE WITH 10 SPLINES
SEE FIGURE 7A

8.1250 in	IE	206.38 mm
8.5000 in	OE	215.90 mm
2.0000 in	SB	50.80 mm
3.5000 in	EC	88.90 mm
2.7500 in	HD	69.85 mm

BORG-WARNR 19816J
ROCKWELL 82NYS32-26

IDLI GROUP 82079

END YOKE WITH 10 SPLINES
slight variation from above group
SEE FIGURE 7A

8.1250 in	IE	206.38 mm
8.5000 in	OE	215.90 mm
2.0000 in	SB	50.80 mm
3.5000 in	EC	88.90 mm
2.7500 in	HD	69.85 mm

BORG-WARNR 23404J
ROCKWELL 82NYS32-43

IDLI GROUP 82080

END YOKE WITH 6 SPLINES
SEE FIGURE 7A

8.1250 in	IE	206.38 mm
8.5000 in	OE	215.90 mm
2.0000 in	SB	50.80 mm
3.5000 in	EC	88.90 mm
3.0000 in	HD	76.20 mm

ROCKWELL 82NYS32-40

IDLI GROUP 82081

END YOKE WITH 10 SPLINES
SEE FIGURE 7A

8.1250 in	IE	206.38 mm
8.5000 in	OE	215.90 mm
2.0000 in	SB	50.80 mm
3.6250 in	EC	92.08 mm
2.7500 in	HD	69.85 mm

BORG-WARNR 16417J
FORD CITT4865A
ROCKWELL 82NYS32-17

IDLI GROUP 82082

END YOKE WITH 15 INVOLUTED SPLINES
SEE FIGURE 7A

8.1250 in	IE	206.38 mm
8.5000 in	OE	215.90 mm
2.0000 in	SB	50.80 mm
3.6875 in	EC	93.66 mm
3.0000 in	HD	76.20 mm

ROCKWELL 82NYS32-33

IDLI GROUP 82083

END YOKE WITH 10 SPLINES
SEE FIGURE 7A

8.1250 in	IE	206.38 mm
8.5000 in	OE	215.90 mm
2.0000 in	SB	50.80 mm
3.7500 in	EC	95.25 mm
3.0000 in	HD	76.20 mm

ROCKWELL 82NYS32-24

IDLI GROUP 82084

END YOKE WITH 10 SPLINES
SEE FIGURE 7A

8.1250 in	IE	206.38 mm
8.5000 in	OE	215.90 mm
2.0000 in	SB	50.80 mm
3.8125 in	EC	96.84 mm
2.7500 in	HD	69.85 mm

ROCKWELL 82NYS32-39

IDLI GROUP 82085

END YOKE WITH 10 SPLINES
slight variation from above group
SEE FIGURE 7A

8.1250 in	IE	206.38 mm
8.5000 in	OE	215.90 mm
2.0000 in	SB	50.80 mm
3.8125 in	EC	96.84 mm
2.7500 in	HD	69.85 mm

BORG-WARNR 11996J
ROCKWELL 82NYS32-13

IDLI GROUP 82086

END YOKE WITH 10 SPLINES
SEE FIGURE 7A

8.1250 in	IE	206.38 mm
8.5000 in	OE	215.90 mm
2.0000 in	SB	50.80 mm
3.8125 in	EC	96.84 mm
2.9375 in	HD	74.61 mm

ROCKWELL 82NYS32-38

IDLI GROUP 82087

END YOKE WITH 10 SPLINES
slight variation from above group
SEE FIGURE 7A

8.1250 in	IE	206.38 mm
8.5000 in	OE	215.90 mm
2.0000 in	SB	50.80 mm
3.8125 in	EC	96.84 mm
2.9375 in	HD	74.61 mm

BORG-WARNR 10073J
ROCKWELL 82NYS32-11

IDLI GROUP 82088

END YOKE WITH 15 INVOLUTED SPLINES
SEE FIGURE 7A

8.1250 in	IE	206.38 mm
8.5000 in	OE	215.90 mm
2.0000 in	SB	50.80 mm
3.8750 in	EC	98.43 mm
2.5000 in	HD	63.50 mm

BORG-WARNR 9736J
ROCKWELL 82NYS32-3

IDLI GROUP 82089

END YOKE WITH 10 SPLINES
SEE FIGURE 7A

8.1250 in	IE	206.38 mm
8.5000 in	OE	215.90 mm
2.0000 in	SB	50.80 mm
3.8750 in	EC	98.43 mm
2.7500 in	HD	69.85 mm

ROCKWELL 82NYS32-51

CAUTION: BE SURE TO REFER TO ENGINEERING CATALOGS FOR SPECIAL APPLICATIONS THAT REQUIRE SPECIFIC MATERIAL CONTENT, TOLERANCES, ETC. SEE FOOTNOTE.

IDLI GROUP 82090

END YOKE WITH 15 INVOLUTED SPLINES
SEE FIGURE 7A

8.1250 in	IE	206.38 mm
8.5000 in	OE	215.90 mm
2.0000 in	SB	50.80 mm
3.8750 in	EC	98.43 mm
2.8750 in	HD	73.03 mm

BORG-WARNR 14577J
ROCKWELL 82NYS32

IDLI GROUP 82091

END YOKE WITH 15 INVOLUTED SPLINES
slight variation from above group
SEE FIGURE 7A

8.1250 in	IE	206.38 mm
8.5000 in	OE	215.90 mm
2.0000 in	SB	50.80 mm
3.8750 in	EC	98.43 mm
2.8750 in	HD	73.03 mm

BORG-WARNR 9741J
ROCKWELL 82NYS32-1

IDLI GROUP 82092

END YOKE WITH 15 INVOLUTED SPLINES
SEE FIGURE 7A

8.1250 in	IE	206.38 mm
8.5000 in	OE	215.90 mm
2.0000 in	SB	50.80 mm
3.8750 in	EC	98.43 mm
3.0000 in	HD	76.20 mm

BORG-WARNR 21128J
ROCKWELL 82NYS32-30

IDLI GROUP 82093

END YOKE WITH 10 SPLINES
SEE FIGURE 7A

8.1250 in	IE	206.38 mm
8.5000 in	OE	215.90 mm
2.0000 in	SB	50.80 mm
4.0469 in	EC	102.79 mm
2.7500 in	HD	69.85 mm

BORG-WARNR 23590J
ROCKWELL 82NYS32-47

IDLI GROUP 82094

END YOKE WITH 10 SPLINES
SEE FIGURE 7A

8.1250 in	IE	206.38 mm
8.5000 in	OE	215.90 mm
2.0000 in	SB	50.80 mm
4.1250 in	EC	104.78 mm
2.9375 in	HD	74.61 mm

BORG-WARNR 20650J
ROCKWELL 82NYS32-28

IDLI GROUP 82095

END YOKE WITH 15 INVOLUTED SPLINES
SEE FIGURE 7A

8.1250 in	IE	206.38 mm
8.5000 in	OE	215.90 mm
2.0000 in	SB	50.80 mm
4.1875 in	EC	106.36 mm
2.6250 in	HD	66.68 mm

BORG-WARNR 20886J
ROCKWELL 82NYS32-31

IDLI GROUP 82096

END YOKE WITH 10 SPLINES
SEE FIGURE 7A

8.1250 in	IE	206.38 mm
8.5000 in	OE	215.90 mm
2.0000 in	SB	50.80 mm
4.2500 in	EC	107.95 mm
3.0000 in	HD	76.20 mm

BORG-WARNR 13085J
BORG-WARNR 19496J
ROCKWELL 82NYS32-10

IDLI GROUP 82097

END YOKE WITH 10 SPLINES
SEE FIGURE 7A

8.1250 in	IE	206.38 mm
8.5000 in	OE	215.90 mm
2.0000 in	SB	50.80 mm
4.2500 in	EC	107.95 mm
3.2500 in	HD	82.55 mm

ROCKWELL 82NYS32-20

IDLI GROUP 82098

END YOKE WITH 10 SPLINES
slight variation from above group
SEE FIGURE 7A

8.1250 in	IE	206.38 mm
8.5000 in	OE	215.90 mm
2.0000 in	SB	50.80 mm
4.2500 in	EC	107.95 mm
3.2500 in	HD	82.55 mm

BORG-WARNR 7675J
BORG-WARNR 8146J
ROCKWELL 82NYS32-4

IDLI GROUP 82099

END YOKE WITH 10 SPLINES
slight variation from above group
SEE FIGURE 7A

8.1250 in	IE	206.38 mm
8.5000 in	OE	215.90 mm
2.0000 in	SB	50.80 mm
4.2500 in	EC	107.95 mm
3.2500 in	HD	82.55 mm

ROCKWELL 82NYS32-14

IDLI GROUP 82100

END YOKE WITH 10 SPLINES
slight variation from above group
SEE FIGURE 7A

8.1250 in	IE	206.38 mm
8.5000 in	OE	215.90 mm
2.0000 in	SB	50.80 mm
4.2500 in	EC	107.95 mm
3.2500 in	HD	82.55 mm

ROCKWELL 82NYS32-49

IDLI GROUP 82101

END YOKE WITH TAPERED BORE
SEE FIGURE 7F

8.1250 in	IE	206.38 mm
8.5000 in	OE	215.90 mm
2.0000 in	SB	50.80 mm
4.2500 in	EC	107.95 mm
3.3125 in	HD	84.14 mm

BORG-WARNR 5209J
ROCKWELL 82NYT32

IDLI GROUP 82102

END YOKE WITH 10 SPLINES
slight variation from above group
SEE FIGURE 7A

8.1250 in	IE	206.38 mm
8.5000 in	OE	215.90 mm
2.0000 in	SB	50.80 mm
4.5000 in	EC	114.30 mm
2.2500 in	HD	57.15 mm

BORG-WARNR 23401J
ROCKWELL 82NYS32-41

IDLI GROUP 82103

END YOKE WITH 10 SPLINES
SEE FIGURE 7A

8.1250 in	IE	206.38 mm
8.5000 in	OE	215.90 mm
2.0000 in	SB	50.80 mm
4.5000 in	EC	114.30 mm
3.2500 in	HD	82.55 mm

ROCKWELL 82NYS32-23

IDLI GROUP 82104

END YOKE WITH 10 SPLINES
SEE FIGURE 7A

8.1250 in	IE	206.38 mm
8.5000 in	OE	215.90 mm
2.0000 in	SB	50.80 mm
4.6250 in	EC	117.48 mm
2.7500 in	HD	69.85 mm

ROCKWELL 82NYS32-50

IDLI GROUP 82105

END YOKE WITH 10 SPLINES
SEE FIGURE 7A

8.1250 in	IE	206.38 mm
8.5000 in	OE	215.90 mm
2.0000 in	SB	50.80 mm
4.6250 in	EC	117.48 mm
2.9375 in	HD	74.61 mm

ROCKWELL 82NYS32-46

IDLI GROUP 82106

END YOKE WITH 10 SPLINES
SEE FIGURE 7A

8.1250 in	IE	206.38 mm
8.5000 in	OE	215.90 mm
2.0000 in	SB	50.80 mm
4.7500 in	EC	120.65 mm
2.9375 in	HD	74.61 mm

ROCKWELL 82NYS32-56

IDLI GROUP 82107

END YOKE WITH ROUND BORE
SEE FIGURE 7B

8.1250 in	IE	206.38 mm
8.5000 in	OE	215.90 mm
2.0000 in	SB	50.80 mm
5.5000 in	EC	139.70 mm
5.7500 in	HD	146.05 mm

ROCKWELL 82NYR32

IDLI GROUP 82108

END YOKE WITH 10 SPLINES
SEE FIGURE 7A

8.1250 in	IE	206.38 mm
8.5000 in	OE	215.90 mm
2.0313 in	SB	51.60 mm
3.6250 in	EC	92.08 mm
2.7500 in	HD	69.85 mm

ROCKWELL 82NYS33-4

IDLI GROUP 82109

END YOKE WITH 16 INVOLUTED SPLINES
SEE FIGURE 7A

8.1250 in	IE	206.38 mm
8.5000 in	OE	215.90 mm
2.1250 in	SB	53.98 mm
3.7500 in	EC	95.25 mm
2.8750 in	HD	73.03 mm

ROCKWELL 82NYS34-13

IDLI GROUP 82110

END YOKE WITH ROUND BORE
SEE FIGURE 7B

8.1250 in	IE	206.38 mm
8.5000 in	OE	215.90 mm
2.1250 in	SB	53.98 mm
4.1875 in	EC	106.36 mm
3.5000 in	HD	88.90 mm

ROCKWELL 82NYR34

IDLI GROUP 82111

END YOKE WITH ROUND BORE
SEE FIGURE 7B

8.1250 in	IE	206.38 mm
8.5000 in	OE	215.90 mm
2.1250 in	SB	53.98 mm
4.4375 in	EC	112.71 mm
3.8750 in	HD	98.43 mm

ROCKWELL 82NYR34-2

BD = BEARING DIAMETER (outside)   **BW** = BEARING WIDTH   **CB** = CROSS LENGTH WITH BEARINGS   **CC** = CENTER TO CENTER   **CD** = CROSS DIAMETER
CL = CROSS LENGTH WITHOUT BEARINGS   **EC** = END TO CENTER or FACE TO FACE   **EE** = END TO END   **EL** = EFFECTIVE LENGTH
HD = HUB DIAMETER (or insert)   **IE** = INSIDE OF EARS (or recess)   **OD** = OUTSIDE DIAMETER   **OE** = OUTSIDE OF EARS   **SB** = SPLINE OR BORE SIZE

IDLI GROUP 82112

END YOKE WITH 25 INVOLUTED SPLINES
SEE FIGURE 7A

8.1250 in	IE	206.38 mm
8.5000 in	OE	215.90 mm
2.1719 in	SB	55.17 mm
3.8438 in	EC	97.63 mm
2.8750 in	HD	73.03 mm

BORG-WARNR 15847J
ROCKWELL 82NYS34

IDLI GROUP 82113

8.1250 in	IE	206.38 mm
8.5000 in	OE	215.90 mm
2.1719 in	SB	55.17 mm
4.5313 in	EC	115.10 mm
2.8750 in	HD	73.03 mm

BORG-WARNR 21845J
ROCKWELL 82NYS34-7

IDLI GROUP 82114

END YOKE WITH ROUND BORE
SEE FIGURE 7B

8.1250 in	IE	206.38 mm
8.5000 in	OE	215.90 mm
2.1875 in	SB	55.56 mm
4.3750 in	EC	111.13 mm
3.6250 in	HD	92.08 mm

ROCKWELL 82NYR35

IDLI GROUP 82115

END YOKE WITH 6 SPLINES
SEE FIGURE 7A

8.1250 in	IE	206.38 mm
8.5000 in	OE	215.90 mm
2.2188 in	SB	56.36 mm
4.5000 in	EC	114.30 mm
3.0000 in	HD	76.20 mm

BORG-WARNR 7142J
ROCKWELL 82NYS36-3

IDLI GROUP 82116

END YOKE WITH 6 SPLINES
slight variation from above group
SEE FIGURE 7A

8.1250 in	IE	206.38 mm
8.5000 in	OE	215.90 mm
2.2188 in	SB	56.36 mm
4.5000 in	EC	114.30 mm
3.0000 in	HD	76.20 mm

ROCKWELL 82NYS36-63

IDLI GROUP 82117

END YOKE WITH 6 SPLINES
SEE FIGURE 7A

8.1250 in	IE	206.38 mm
8.5000 in	OE	215.90 mm
2.2188 in	SB	56.36 mm
4.7500 in	EC	120.65 mm
3.0000 in	HD	76.20 mm

ROCKWELL 82NYS36-58

IDLI GROUP 82118

END YOKE WITH 10 SPLINES
SEE FIGURE 7A

8.1250 in	IE	206.38 mm
8.5000 in	OE	215.90 mm
2.2500 in	SB	57.15 mm
3.2500 in	EC	82.55 mm
3.5000 in	HD	88.90 mm

BORG-WARNR 15596J
ROCKWELL 82NYS36-62

IDLI GROUP 82119

END YOKE WITH 10 SPLINES
SEE FIGURE 7A

8.1250 in	IE	206.38 mm
8.5000 in	OE	215.90 mm
2.2500 in	SB	57.15 mm
3.5000 in	EC	88.90 mm
3.5000 in	HD	88.90 mm

ROCKWELL 82NYS36-84

IDLI GROUP 82120

END YOKE WITH 10 SPLINES
SEE FIGURE 7A

8.1250 in	IE	206.38 mm
8.5000 in	OE	215.90 mm
2.2500 in	SB	57.15 mm
3.7500 in	EC	95.25 mm
3.2500 in	HD	82.55 mm

BORG-WARNR 21283J
ROCKWELL 82NYS36-49

IDLI GROUP 82121

END YOKE WITH 10 SPLINES
slight variation from above group
SEE FIGURE 7A

8.1250 in	IE	206.38 mm
8.5000 in	OE	215.90 mm
2.2500 in	SB	57.15 mm
3.7500 in	EC	95.25 mm
3.2500 in	HD	82.55 mm

ROCKWELL 82NYS36-76

IDLI GROUP 82122

END YOKE WITH 10 SPLINES
SEE FIGURE 7A

8.1250 in	IE	206.38 mm
8.5000 in	OE	215.90 mm
2.2500 in	SB	57.15 mm
3.8750 in	EC	98.43 mm
3.1250 in	HD	79.38 mm

ROCKWELL 82NYS36-80

IDLI GROUP 82123

END YOKE WITH 10 SPLINES
SEE FIGURE 7A

8.1250 in	IE	206.38 mm
8.5000 in	OE	215.90 mm
2.2500 in	SB	57.15 mm
3.8750 in	EC	98.43 mm
3.2500 in	HD	82.55 mm

ROCKWELL 82NYS36-87

IDLI GROUP 82124

END YOKE WITH 10 SPLINES
slight variation from above group
SEE FIGURE 7A

8.1250 in	IE	206.38 mm
8.5000 in	OE	215.90 mm
2.2500 in	SB	57.15 mm
3.8750 in	EC	98.43 mm
3.2500 in	HD	82.55 mm

ROCKWELL 82NYS36-32

IDLI GROUP 82125

END YOKE WITH 10 SPLINES
SEE FIGURE 7A

8.1250 in	IE	206.38 mm
8.5000 in	OE	215.90 mm
2.2500 in	SB	57.15 mm
4.1875 in	EC	106.36 mm
3.1250 in	HD	79.38 mm

ROCKWELL 82NYS36-54

IDLI GROUP 82126

END YOKE WITH 10 SPLINES
SEE FIGURE 7A

8.1250 in	IE	206.38 mm
8.5000 in	OE	215.90 mm
2.2500 in	SB	57.15 mm
4.2500 in	EC	107.95 mm
2.6875 in	HD	68.26 mm

BORG-WARNR 15974J
ROCKWELL 82NYS36-21

IDLI GROUP 82127

END YOKE WITH 10 SPLINES
SEE FIGURE 7A

8.1250 in	IE	206.38 mm
8.5000 in	OE	215.90 mm
2.2500 in	SB	57.15 mm
4.2500 in	EC	107.95 mm
2.8750 in	HD	73.03 mm

ROCKWELL 82NYS36-64

IDLI GROUP 82128

END YOKE WITH 10 SPLINES
slight variation from above group
SEE FIGURE 7A

8.1250 in	IE	206.38 mm
8.5000 in	OE	215.90 mm
2.2500 in	SB	57.15 mm
4.2500 in	EC	107.95 mm
2.8750 in	HD	73.03 mm

ROCKWELL 82NYS36-69

IDLI GROUP 82129

END YOKE WITH 10 SPLINES
SEE FIGURE 7A

8.1250 in	IE	206.38 mm
8.5000 in	OE	215.90 mm
2.2500 in	SB	57.15 mm
4.2500 in	EC	107.95 mm
3.0000 in	HD	76.20 mm

BORG-WARNR 15973J
ROCKWELL 82NYS36-20

IDLI GROUP 82130

END YOKE WITH 10 SPLINES
SEE FIGURE 7A

8.1250 in	IE	206.38 mm
8.5000 in	OE	215.90 mm
2.2500 in	SB	57.15 mm
4.2500 in	EC	107.95 mm
3.3750 in	HD	85.73 mm

BORG-WARNR 121-8025
ROCKWELL 82NYS36-23

IDLI GROUP 82131

END YOKE WITH 10 SPLINES
SEE FIGURE 7A

8.1250 in	IE	206.38 mm
8.5000 in	OE	215.90 mm
2.2500 in	SB	57.15 mm
4.2500 in	EC	107.95 mm
3.5000 in	HD	88.90 mm

BORG-WARNR 9513J
ROCKWELL 82NYS36-15

IDLI GROUP 82132

END YOKE WITH 10 SPLINES
slight variation from above group
SEE FIGURE 7A

8.1250 in	IE	206.38 mm
8.5000 in	OE	215.90 mm
2.2500 in	SB	57.15 mm
4.2500 in	EC	107.95 mm
3.5000 in	HD	88.90 mm

ROCKWELL 82NYS36-24

IDLI GROUP 82133

END YOKE WITH 10 SPLINES
slight variation from above group
SEE FIGURE 7A

8.1250 in	IE	206.38 mm
8.5000 in	OE	215.90 mm
2.2500 in	SB	57.15 mm
4.2500 in	EC	107.95 mm
3.5000 in	HD	88.90 mm

BORG-WARNR 10738J
BORG-WARNR 14526J
BORG-WARNR 7171J
ROCKWELL 82NYS36-10

IDLI GROUP 82134

END YOKE WITH 10 SPLINES
slight variation from above group
SEE FIGURE 7A

8.1250 in	IE	206.38 mm
8.5000 in	OE	215.90 mm
2.2500 in	SB	57.15 mm
4.2500 in	EC	107.95 mm
3.5000 in	HD	88.90 mm

ROCKWELL 82NYS36-25

CAUTION: BE SURE TO REFER TO ENGINEERING CATALOGS FOR SPECIAL APPLICATIONS THAT REQUIRE SPECIFIC MATERIAL CONTENT, TOLERANCES, ETC. SEE FOOTNOTE.

IDLI GROUP 82135

END YOKE WITH 10 SPLINES
SEE FIGURE 7A

8.1250 in	IE	206.38 mm
8.5000 in	OE	215.90 mm
2.2500 in	SB	57.15 mm
4.3125 in	EC	109.54 mm
3.0000 in	HD	76.20 mm

BORG-WARNR 13099J
ROCKWELL 82NYS36-9

IDLI GROUP 82136

END YOKE WITH 10 SPLINES
slight variation from above group
SEE FIGURE 7A

8.1250 in	IE	206.38 mm
8.5000 in	OE	215.90 mm
2.2500 in	SB	57.15 mm
4.3125 in	EC	109.54 mm
3.0000 in	HD	76.20 mm

ROCKWELL 82NYS36-81

IDLI GROUP 82137

END YOKE WITH 10 SPLINES
slight variation from above group
SEE FIGURE 7A

8.1250 in	IE	206.38 mm
8.5000 in	OE	215.90 mm
2.2500 in	SB	57.15 mm
4.3125 in	EC	109.54 mm
3.0000 in	HD	76.20 mm

ROCKWELL 82NYS36-65

IDLI GROUP 82138

END YOKE WITH 10 SPLINES
slight variation from above group
SEE FIGURE 7A

8.1250 in	IE	206.38 mm
8.5000 in	OE	215.90 mm
2.2500 in	SB	57.15 mm
4.3125 in	EC	109.54 mm
3.0000 in	HD	76.20 mm

BORG-WARNR 23391J
BORG-WARNR 23392J
ROCKWELL 82NYS36-71

IDLI GROUP 82139

END YOKE WITH 10 SPLINES
SEE FIGURE 7A

8.1250 in	IE	206.38 mm
8.5000 in	OE	215.90 mm
2.2500 in	SB	57.15 mm
4.3125 in	EC	109.54 mm
3.5000 in	HD	88.90 mm

ROCKWELL 82NYS36-13

IDLI GROUP 82140

END YOKE WITH ROUND BORE
SEE FIGURE 7B

8.1250 in	IE	206.38 mm
8.5000 in	OE	215.90 mm
2.2500 in	SB	57.15 mm
4.3750 in	EC	111.13 mm
5.7500 in	HD	146.05 mm

ROCKWELL 82NYR36-1

IDLI GROUP 82141

END YOKE WITH 10 SPLINES
SEE FIGURE 7A

8.1250 in	IE	206.38 mm
8.5000 in	OE	215.90 mm
2.2500 in	SB	57.15 mm
4.5000 in	EC	114.30 mm
3.0000 in	HD	76.20 mm

ROCKWELL 82NYS36-66

IDLI GROUP 82142

END YOKE WITH 10 SPLINES
slight variation from above group
SEE FIGURE 7A

8.1250 in	IE	206.38 mm
8.5000 in	OE	215.90 mm
2.2500 in	SB	57.15 mm
4.5000 in	EC	114.30 mm
3.0000 in	HD	76.20 mm

ROCKWELL 82NYS36-17

IDLI GROUP 82143

END YOKE WITH 10 SPLINES
slight variation from above group
SEE FIGURE 7A

8.1250 in	IE	206.38 mm
8.5000 in	OE	215.90 mm
2.2500 in	SB	57.15 mm
4.5000 in	EC	114.30 mm
3.0000 in	HD	76.20 mm

ROCKWELL 82NYS36-37

IDLI GROUP 82144

END YOKE WITH 6 SPLINES
SEE FIGURE 7A

8.1250 in	IE	206.38 mm
8.5000 in	OE	215.90 mm
2.2500 in	SB	57.15 mm
4.5000 in	EC	114.30 mm
3.3750 in	HD	85.73 mm

ROCKWELL 82NYS36-105

IDLI GROUP 82145

END YOKE WITH 10 SPLINES
SEE FIGURE 7A

8.1250 in	IE	206.38 mm
8.5000 in	OE	215.90 mm
2.2500 in	SB	57.15 mm
4.5000 in	EC	114.30 mm
3.5000 in	HD	88.90 mm

BORG-WARNR 19812J
ROCKWELL 82NYS36-50

IDLI GROUP 82146

END YOKE WITH 10 SPLINES
slight variation from above group
SEE FIGURE 7A

8.1250 in	IE	206.38 mm
8.5000 in	OE	215.90 mm
2.2500 in	SB	57.15 mm
4.5000 in	EC	114.30 mm
3.5000 in	HD	88.90 mm

ROCKWELL 82NYS36-72

IDLI GROUP 82147

END YOKE WITH 10 SPLINES
slight variation from above group
SEE FIGURE 7A

8.1250 in	IE	206.38 mm
8.5000 in	OE	215.90 mm
2.2500 in	SB	57.15 mm
4.5000 in	EC	114.30 mm
3.5000 in	HD	88.90 mm

ROCKWELL 82NYS36-45

IDLI GROUP 82148

END YOKE WITH 10 SPLINES
slight variation from above group
SEE FIGURE 7A

8.1250 in	IE	206.38 mm
8.5000 in	OE	215.90 mm
2.2500 in	SB	57.15 mm
4.5000 in	EC	114.30 mm
3.5000 in	HD	88.90 mm

ROCKWELL 82NYS36-70

IDLI GROUP 82149

END YOKE WITH ROUND BORE
SEE FIGURE 7B

8.1250 in	IE	206.38 mm
8.5000 in	OE	215.90 mm
2.2500 in	SB	57.15 mm
4.5625 in	EC	115.89 mm
3.5000 in	HD	88.90 mm

ROCKWELL 82NYR36-3

IDLI GROUP 82150

END YOKE WITH 10 SPLINES
SEE FIGURE 7A

8.1250 in	IE	206.38 mm
8.5000 in	OE	215.90 mm
2.2500 in	SB	57.15 mm
4.6250 in	EC	117.48 mm
3.0000 in	HD	76.20 mm

ROCKWELL 82NYS36-5

IDLI GROUP 82151

END YOKE WITH 10 SPLINES
SEE FIGURE 7A

8.1250 in	IE	206.38 mm
8.5000 in	OE	215.90 mm
2.2500 in	SB	57.15 mm
4.6250 in	EC	117.48 mm
3.5000 in	HD	88.90 mm

BORG-WARNR 6005J
ROCKWELL 82NYS36-7

IDLI GROUP 82152

END YOKE WITH 10 SPLINES
SEE FIGURE 7A

8.1250 in	IE	206.38 mm
8.5000 in	OE	215.90 mm
2.2500 in	SB	57.15 mm
4.7500 in	EC	120.65 mm
2.8750 in	HD	73.03 mm

ROCKWELL 82NYS36-68

IDLI GROUP 82153

END YOKE WITH 10 SPLINES
SEE FIGURE 7A

8.1250 in	IE	206.38 mm
8.5000 in	OE	215.90 mm
2.2500 in	SB	57.15 mm
4.7500 in	EC	120.65 mm
3.0000 in	HD	76.20 mm

BORG-WARNR 7783J
ROCKWELL 82NYS36-28

IDLI GROUP 82154

END YOKE WITH 10 SPLINES
slight variation from above group
SEE FIGURE 7A

8.1250 in	IE	206.38 mm
8.5000 in	OE	215.90 mm
2.2500 in	SB	57.15 mm
4.7500 in	EC	120.65 mm
3.0000 in	HD	76.20 mm

BORG-WARNR 13481J
BORG-WARNR 23647J
ROCKWELL 82NYS36-30

IDLI GROUP 82155

END YOKE WITH 10 SPLINES
slight variation from above group
SEE FIGURE 7A

8.1250 in	IE	206.38 mm
8.5000 in	OE	215.90 mm
2.2500 in	SB	57.15 mm
4.7500 in	EC	120.65 mm
3.0000 in	HD	76.20 mm

BORG-WARNR 19129J
ROCKWELL 82NYS36-26

IDLI GROUP 82156

END YOKE WITH 10 SPLINES
slight variation from above group
SEE FIGURE 7A

8.1250 in	IE	206.38 mm
8.5000 in	OE	215.90 mm
2.2500 in	SB	57.15 mm
4.7500 in	EC	120.65 mm
3.0000 in	HD	76.20 mm

BORG-WARNR 7783J
ROCKWELL 82NYS36-47

RD = BEARING DIAMETER (outside)   **BW** = BEARING WIDTH   **CB** = CROSS LENGTH WITH BEARINGS   **CC** = CENTER TO CENTER   **CD** = CROSS DIAMETER
CL = CROSS LENGTH WITHOUT BEARINGS   **EC** = END TO CENTER or FACE TO FACE   **EE** = END TO END   **EL** = EFFECTIVE LENGTH
HD = HUB DIAMETER (or insert)   **IE** = INSIDE OF EARS (or recess)   **OD** = OUTSIDE DIAMETER   **OE** = OUTSIDE OF EARS   **SB** = SPLINE OR BORE SIZE

IDLI GUIDE COPYRIGHT ® INTERCHANGE, INC. ST. LOUIS PARK, MN. 55416 USA

IDLI GROUP 82157

END YOKE WITH 10 SPLINES
SEE FIGURE 7A

8.1250 in	IE	206.38 mm
8.5000 in	OE	215.90 mm
2.2500 in	SB	57.15 mm
4.7500 in	EC	120.65 mm
3.5000 in	HD	88.90 mm

BORG-WARNR 21209J
ROCKWELL 82NYS36-56

IDLI GROUP 82158

END YOKE WITH 10 SPLINES
slight variation from above group
SEE FIGURE 7A

8.1250 in	IE	206.38 mm
8.5000 in	OE	215.90 mm
2.2500 in	SB	57.15 mm
4.7500 in	EC	120.65 mm
3.5000 in	HD	88.90 mm

ROCKWELL 82NYS36-73

IDLI GROUP 82159

END YOKE WITH ROUND BORE
SEE FIGURE 7B

8.1250 in	IE	206.38 mm
8.5000 in	OE	215.90 mm
2.2500 in	SB	57.15 mm
4.7500 in	EC	120.65 mm
3.6250 in	HD	92.08 mm

ROCKWELL 82NYR36-4

IDLI GROUP 82160

END YOKE WITH ROUND BORE
SEE FIGURE 7B

8.1250 in	IE	206.38 mm
8.5000 in	OE	215.90 mm
2.2500 in	SB	57.15 mm
5.2500 in	EC	133.35 mm
6.0000 in	HD	152.40 mm

ROCKWELL 82NYR36-2

IDLI GROUP 82161

END YOKE WITH 6 SPLINES
SEE FIGURE 7A

8.1250 in	IE	206.38 mm
8.5000 in	OE	215.90 mm
2.2656 in	SB	57.55 mm
3.8750 in	EC	98.43 mm
3.2500 in	HD	82.55 mm

ROCKWELL 82NYS36-79

IDLI GROUP 82162

END YOKE WITH 17 INVOLUTED SPLINES
SEE FIGURE 7A

8.1250 in	IE	206.38 mm
8.5000 in	OE	215.90 mm
2.2813 in	SB	57.95 mm
3.8750 in	EC	98.43 mm
3.0000 in	HD	76.20 mm

ROCKWELL 82NYS36-53

IDLI GROUP 82163

END YOKE WITH 17 INVOLUTED SPLINES
SEE FIGURE 7A

8.1250 in	IE	206.38 mm
8.5000 in	OE	215.90 mm
2.2813 in	SB	57.95 mm
4.2500 in	EC	107.95 mm
3.1094 in	HD	78.98 mm

BORG-WARNR 13221J
ROCKWELL 82NYS36

IDLI GROUP 82164

END YOKE WITH 26 INVOLUTED SPLINES
SEE FIGURE 7A

8.1250 in	IE	206.38 mm
8.5000 in	OE	215.90 mm
2.2813 in	SB	57.95 mm
4.2500 in	EC	107.95 mm
3.1094 in	HD	78.98 mm

BORG-WARNR 15607J
ROCKWELL 82NYS36-4

IDLI GROUP 82165

END YOKE WITH 17 INVOLUTED SPLINES
SEE FIGURE 7A

8.1250 in	IE	206.38 mm
8.5000 in	OE	215.90 mm
2.2813 in	SB	57.95 mm
4.6094 in	EC	117.08 mm
3.5000 in	HD	88.90 mm

ROCKWELL 82NYS36-19

IDLI GROUP 82166

END YOKE WITH ROUND BORE
SEE FIGURE 7B

8.1250 in	IE	206.38 mm
8.5000 in	OE	215.90 mm
2.3750 in	SB	60.33 mm
4.0625 in	EC	103.19 mm
3.5000 in	HD	88.90 mm

ROCKWELL 82NYR38-2

IDLI GROUP 82167

END YOKE WITH ROUND BORE
SEE FIGURE 7B

8.1250 in	IE	206.38 mm
8.5000 in	OE	215.90 mm
2.3750 in	SB	60.33 mm
4.6250 in	EC	117.48 mm
3.6250 in	HD	92.08 mm

ROCKWELL 82NYR38-1

IDLI GROUP 82168

END YOKE WITH ROUND BORE
SEE FIGURE 7B

8.1250 in	IE	206.38 mm
8.5000 in	OE	215.90 mm
2.3750 in	SB	60.33 mm
5.8125 in	EC	147.64 mm
5.5000 in	HD	139.70 mm

ROCKWELL 82NYR38

IDLI GROUP 82169

END YOKE WITH 24 INVOLUTED SPLINES
SEE FIGURE 7A

8.1250 in	IE	206.38 mm
8.5000 in	OE	215.90 mm
2.4063 in	SB	61.12 mm
3.7500 in	EC	95.25 mm
3.1250 in	HD	79.38 mm

BORG-WARNR 17817J
ROCKWELL 82NYS38

IDLI GROUP 82170

END YOKE WITH 28 INVOLUTED SPLINES
SEE FIGURE 7A

8.1250 in	IE	206.38 mm
8.5000 in	OE	215.90 mm
2.4375 in	SB	61.91 mm
3.8750 in	EC	98.43 mm
3.3750 in	HD	85.73 mm

BORG-WARNR 16523J
ROCKWELL 82NYS39

IDLI GROUP 82171

END YOKE WITH ROUND BORE
SEE FIGURE 7B

8.1250 in	IE	206.38 mm
8.5000 in	OE	215.90 mm
2.4375 in	SB	61.91 mm
5.0000 in	EC	127.00 mm
6.0000 in	HD	152.40 mm

ROCKWELL 82NYR39-1

IDLI GROUP 82172

END YOKE WITH ROUND BORE
SEE FIGURE 7B

8.1250 in	IE	206.38 mm
8.5000 in	OE	215.90 mm
2.4375 in	SB	61.91 mm
6.0000 in	EC	152.40 mm
3.3125 in	HD	84.14 mm

ROCKWELL 82NYR39

IDLI GROUP 82173

END YOKE WITH 16 SPLINES
SEE FIGURE 7A

8.1250 in	IE	206.38 mm
8.5000 in	OE	215.90 mm
2.5000 in	SB	63.50 mm
1.6250 in	EC	41.28 mm
4.5625 in	HD	115.89 mm

ROCKWELL 82NYS40-16

IDLI GROUP 82174

END YOKE WITH 16 SPLINES
SEE FIGURE 7A

8.1250 in	IE	206.38 mm
8.5000 in	OE	215.90 mm
2.5000 in	SB	63.50 mm
2.3750 in	EC	60.33 mm
3.5000 in	HD	88.90 mm

ROCKWELL 82NYS40-47

IDLI GROUP 82175

END YOKE WITH 16 SPLINES
slight variation from above group
SEE FIGURE 7A

8.1250 in	IE	206.38 mm
8.5000 in	OE	215.90 mm
2.5000 in	SB	63.50 mm
2.3750 in	EC	60.33 mm
3.5000 in	HD	88.90 mm

ROCKWELL 82NYS40-36

IDLI GROUP 82176

END YOKE WITH 10 SPLINES
SEE FIGURE 7A

8.1250 in	IE	206.38 mm
8.5000 in	OE	215.90 mm
2.5000 in	SB	63.50 mm
2.4375 in	EC	61.91 mm
3.0625 in	HD	77.79 mm

ROCKWELL 82NYS40-15

IDLI GROUP 82177

END YOKE WITH 10 SPLINES
slight variation from above group
SEE FIGURE 7A

8.1250 in	IE	206.38 mm
8.5000 in	OE	215.90 mm
2.5000 in	SB	63.50 mm
2.4375 in	EC	61.91 mm
3.0625 in	HD	77.79 mm

ROCKWELL 82NYS40-31

IDLI GROUP 82178

END YOKE WITH 10 SPLINES
SEE FIGURE 7A

8.1250 in	IE	206.38 mm
8.5000 in	OE	215.90 mm
2.5000 in	SB	63.50 mm
2.5000 in	EC	63.50 mm
3.6250 in	HD	92.08 mm

ROCKWELL 82NYS40-12

IDLI GROUP 82179

END YOKE WITH 10 SPLINES
slight variation from above group
SEE FIGURE 7A

8.1250 in	IE	206.38 mm
8.5000 in	OE	215.90 mm
2.5000 in	SB	63.50 mm
2.5000 in	EC	63.50 mm
3.6250 in	HD	92.08 mm

ROCKWELL 82NYS40-44

IDLI GROUP 82180

END YOKE WITH 10 SPLINES
SEE FIGURE 7A

8.1250 in	IE	206.38 mm
8.5000 in	OE	215.90 mm
2.5000 in	SB	63.50 mm
2.5000 in	EC	63.50 mm
4.3125 in	HD	109.54 mm

ROCKWELL 82NYS40-64

ENGINEERING CATALOGS MUST BE CONSULTED FOR DETAILS NOT INCLUDED
IN THIS GUIDE. SPECIFIC DESIGNS, MATERIAL CONTENT, TOLERANCES
LUBE FITTINGS AND OTHER DIMENSIONS ARE INTENTIONALLY OMITTED HERE.

IDLI GUIDE   COPYRIGHT © INTERCHANGE, INC.   ST. LOUIS PARK, MN. 55416 USA

CAUTION: BE SURE TO REFER TO ENGINEERING CATALOGS FOR SPECIAL APPLICATIONS THAT REQUIRE SPECIFIC MATERIAL CONTENT, TOLERANCES, ETC. SEE FOOTNOTE.

IDLI GROUP 82181

END YOKE WITH 10 SPLINES
SEE FIGURE 7A

8.1250 in	IE	206.38 mm
8.5000 in	OE	215.90 mm
2.5000 in	SB	63.50 mm
2.6250 in	EC	66.68 mm
3.6250 in	HD	92.08 mm

ROCKWELL 82NYS40-11

IDLI GROUP 82182

END YOKE WITH 10 SPLINES
SEE FIGURE 7A

8.1250 in	IE	206.38 mm
8.5000 in	OE	215.90 mm
2.5000 in	SB	63.50 mm
2.8125 in	EC	71.44 mm
3.2500 in	HD	82.55 mm

ROCKWELL 82NYS40-3

IDLI GROUP 82183

END YOKE WITH ROUND BORE
SEE FIGURE 7B

8.1250 in	IE	206.38 mm
8.5000 in	OE	215.90 mm
2.5000 in	SB	63.50 mm
3.1250 in	EC	79.38 mm
6.0000 in	HD	152.40 mm

ROCKWELL 82NYR40

IDLI GROUP 82184

END YOKE WITH 10 SPLINES
SEE FIGURE 7A

8.1250 in	IE	206.38 mm
8.5000 in	OE	215.90 mm
2.5000 in	SB	63.50 mm
3.3750 in	EC	85.73 mm
3.6250 in	HD	92.08 mm

ROCKWELL 82NYS40-65

IDLI GROUP 82185

END YOKE WITH 10 SPLINES
slight variation from above group
SEE FIGURE 7A

8.1250 in	IE	206.38 mm
8.5000 in	OE	215.90 mm
2.5000 in	SB	63.50 mm
3.3750 in	EC	85.73 mm
3.6250 in	HD	92.08 mm

ROCKWELL 82NYS40-43

IDLI GROUP 82186

END YOKE WITH 10 SPLINES
SEE FIGURE 7A

8.1250 in	IE	206.38 mm
8.5000 in	OE	215.90 mm
2.5000 in	SB	63.50 mm
3.7500 in	EC	95.25 mm
3.2500 in	HD	82.55 mm

ROCKWELL 82NYS40-13

IDLI GROUP 82187

END YOKE WITH 10 SPLINES
SEE FIGURE 7A

8.1250 in	IE	206.38 mm
8.5000 in	OE	215.90 mm
2.5000 in	SB	63.50 mm
3.8750 in	EC	98.43 mm
3.2500 in	HD	82.55 mm

BORG-WARNR 20508J
ROCKWELL 82NYS40-32

IDLI GROUP 82188

END YOKE WITH 10 SPLINES
SEE FIGURE 7A

8.1250 in	IE	206.38 mm
8.5000 in	OE	215.90 mm
2.5000 in	SB	63.50 mm
4.0000 in	EC	101.60 mm
3.6250 in	HD	92.08 mm

ROCKWELL 82NYS40-17

IDLI GROUP 82189

END YOKE WITH 16 SPLINES
SEE FIGURE 7A

8.1250 in	IE	206.38 mm
8.5000 in	OE	215.90 mm
2.5000 in	SB	63.50 mm
4.0000 in	EC	101.60 mm
3.6250 in	HD	92.08 mm

ROCKWELL 82NYS40-27

IDLI GROUP 82190

END YOKE WITH 16 SPLINES
SEE FIGURE 7A

8.1250 in	IE	206.38 mm
8.5000 in	OE	215.90 mm
2.5000 in	SB	63.50 mm
4.0625 in	EC	103.19 mm
3.2500 in	HD	82.55 mm

ROCKWELL 82NYS40-20

IDLI GROUP 82191

END YOKE WITH 10 SPLINES
SEE FIGURE 7A

8.1250 in	IE	206.38 mm
8.5000 in	OE	215.90 mm
2.5000 in	SB	63.50 mm
4.0625 in	EC	103.19 mm
3.5000 in	HD	88.90 mm

ROCKWELL 82NYS40-67

IDLI GROUP 82192

END YOKE WITH 10 SPLINES
slight variation from above group
SEE FIGURE 7A

8.1250 in	IE	206.38 mm
8.5000 in	OE	215.90 mm
2.5000 in	SB	63.50 mm
4.0625 in	EC	103.19 mm
3.5000 in	HD	88.90 mm

ROCKWELL 82NYS40-71

IDLI GROUP 82193

END YOKE WITH 10 SPLINES
SEE FIGURE 7A

8.1250 in	IE	206.38 mm
8.5000 in	OE	215.90 mm
2.5000 in	SB	63.50 mm
4.2500 in	EC	107.95 mm
3.2500 in	HD	82.55 mm

BORG-WARNR 13223J
BORG-WARNR 13225J
ROCKWELL 82NYS40-1

IDLI GROUP 82194

END YOKE WITH ROUND BORE
SEE FIGURE 7B

8.1250 in	IE	206.38 mm
8.5000 in	OE	215.90 mm
2.5000 in	SB	63.50 mm
4.4375 in	EC	112.71 mm
3.8750 in	HD	98.43 mm

ROCKWELL 82NYR40-7

IDLI GROUP 82195

END YOKE WITH 10 SPLINES
SEE FIGURE 7A

8.1250 in	IE	206.38 mm
8.5000 in	OE	215.90 mm
2.5000 in	SB	63.50 mm
4.7500 in	EC	120.65 mm
2.9375 in	HD	74.61 mm

BORG-WARNR 21199J
ROCKWELL 82NYS40-28

IDLI GROUP 82196

END YOKE WITH 10 SPLINES
slight variation from above group
SEE FIGURE 7A

8.1250 in	IE	206.38 mm
8.5000 in	OE	215.90 mm
2.5000 in	SB	63.50 mm
4.7500 in	EC	120.65 mm
2.9375 in	HD	74.61 mm

ROCKWELL 82NYS40-42

IDLI GROUP 82197

END YOKE WITH ROUND BORE
SEE FIGURE 7B

8.1250 in	IE	206.38 mm
8.5000 in	OE	215.90 mm
2.5000 in	SB	63.50 mm
4.7500 in	EC	120.65 mm
5.7500 in	HD	146.05 mm

ROCKWELL 82NYR40-2

IDLI GROUP 82198

END YOKE WITH ROUND BORE
SEE FIGURE 7B

8.1250 in	IE	206.38 mm
8.5000 in	OE	215.90 mm
2.5000 in	SB	63.50 mm
5.0000 in	EC	127.00 mm
6.0000 in	HD	152.40 mm

ROCKWELL 82NYR40-8

IDLI GROUP 82199

END YOKE WITH ROUND BORE
SEE FIGURE 7B

8.1250 in	IE	206.38 mm
8.5000 in	OE	215.90 mm
2.5000 in	SB	63.50 mm
5.2500 in	EC	133.35 mm
6.0000 in	HD	152.40 mm

ROCKWELL 82NYR40-10

IDLI GROUP 82200

END YOKE WITH ROUND BORE
slight variation from above group
SEE FIGURE 7B

8.1250 in	IE	206.38 mm
8.5000 in	OE	215.90 mm
2.5000 in	SB	63.50 mm
5.2500 in	EC	133.35 mm
6.0000 in	HD	152.40 mm

ROCKWELL 82NYR40-4

IDLI GROUP 82201

END YOKE WITH ROUND BORE
slight variation from above group
SEE FIGURE 7B

8.1250 in	IE	206.38 mm
8.5000 in	OE	215.90 mm
2.5000 in	SB	63.50 mm
5.2500 in	EC	133.35 mm
6.0000 in	HD	152.40 mm

ROCKWELL 82NYR40-9

IDLI GROUP 82202

END YOKE WITH ROUND BORE
SEE FIGURE 7B

8.1250 in	IE	206.38 mm
8.5000 in	OE	215.90 mm
2.5000 in	SB	63.50 mm
6.3750 in	EC	161.93 mm
5.7500 in	HD	146.05 mm

ROCKWELL 82NYR40-1

IDLI GROUP 82203

END YOKE WITH ROUND BORE
SEE FIGURE 7B

8.1250 in	IE	206.38 mm
8.5000 in	OE	215.90 mm
2.5000 in	SB	63.50 mm
6.5000 in	EC	165.10 mm
6.0000 in	HD	152.40 mm

ROCKWELL 82NYR40-3

IDLI GROUP 82204

END YOKE WITH 19 INVOLUTED SPLINES
SEE FIGURE 7A

8.1250 in	IE	206.38 mm
8.5000 in	OE	215.90 mm
2.5313 in	SB	64.30 mm
4.1250 in	EC	104.78 mm
3.5000 in	HD	88.90 mm

ROCKWELL 82NYS40-24

BD = BEARING DIAMETER (outside)   **BW** = BEARING WIDTH   **CB** = CROSS LENGTH WITH BEARINGS   **CC** = CENTER TO CENTER   **CD** = CROSS DIAMETER
CL = CROSS LENGTH WITHOUT BEARINGS   **EC** = END TO CENTER or FACE TO FACE   **EE** = END TO END   **EL** = EFFECTIVE LENGTH
HD = HUB DIAMETER (or insert)   **IE** = INSIDE OF EARS (or recess)   **OD** = OUTSIDE DIAMETER   **OE** = OUTSIDE OF EARS   **SB** = SPLINE OR BORE SIZE

IDLI GROUP 82205

END YOKE WITH 19 INVOLUTED SPLINES
slight variation from above group
SEE FIGURE 7A

8.1250 in	IE	206.38 mm
8.5000 in	OE	215.90 mm
2.5313 in	SB	64.30 mm
4.1250 in	EC	104.78 mm
3.5000 in	HD	88.90 mm

ROCKWELL 82NYS40-18

IDLI GROUP 82206

END YOKE WITH 19 INVOLUTED SPLINES
SEE FIGURE 7A

8.1250 in	IE	206.38 mm
8.5000 in	OE	215.90 mm
2.5313 in	SB	64.30 mm
4.5938 in	EC	116.68 mm
3.1250 in	HD	79.38 mm

BORG-WARNR 19467J
ROCKWELL 82NYS40-26

IDLI GROUP 82207

END YOKE WITH 19 INVOLUTED SPLINES
SEE FIGURE 7A

8.1250 in	IE	206.38 mm
8.5000 in	OE	215.90 mm
2.5313 in	SB	64.30 mm
4.5938 in	EC	116.68 mm
3.3750 in	HD	85.73 mm

BORG-WARNR 13804J
ROCKWELL 82NYS40-2

IDLI GROUP 82208

END YOKE WITH 16 INVOLUTED SPLINES
SEE FIGURE 7A

8.1250 in	IE	206.38 mm
8.5000 in	OE	215.90 mm
2.5313 in	SB	64.30 mm
4.7500 in	EC	120.65 mm
3.0000 in	HD	76.20 mm

ROCKWELL 82NYS40-30

IDLI GROUP 82209

END YOKE WITH 25 INVOLUTED SPLINES
SEE FIGURE 7A

8.1250 in	IE	206.38 mm
8.5000 in	OE	215.90 mm
2.6250 in	SB	66.68 mm
4.2500 in	EC	107.95 mm
3.6250 in	HD	92.08 mm

ROCKWELL 82NYS42-2

IDLI GROUP 82210

END YOKE WITH 25 INVOLUTED SPLINES
slight variation from above group
SEE FIGURE 7A

8.1250 in	IE	206.38 mm
8.5000 in	OE	215.90 mm
2.6250 in	SB	66.68 mm
4.2500 in	EC	107.95 mm
3.6250 in	HD	92.08 mm

BORG-WARNR 14587J
ROCKWELL 82NYS42

IDLI GROUP 82211

END YOKE WITH 25 INVOLUTED SPLINES
slight variation from above group
SEE FIGURE 7A

8.1250 in	IE	206.38 mm
8.5000 in	OE	215.90 mm
2.6250 in	SB	66.68 mm
4.2500 in	EC	107.95 mm
3.6250 in	HD	92.08 mm

ROCKWELL 82NYS42-5

IDLI GROUP 82212

END YOKE WITH ROUND BORE
SEE FIGURE 7B

8.1250 in	IE	206.38 mm
8.5000 in	OE	215.90 mm
2.6250 in	SB	66.68 mm
5.0000 in	EC	127.00 mm
6.0000 in	HD	152.40 mm

ROCKWELL 82NYR42

IDLI GROUP 82213

END YOKE WITH ROUND BORE
SEE FIGURE 7B

8.1250 in	IE	206.38 mm
8.5000 in	OE	215.90 mm
2.6875 in	SB	68.26 mm
5.0000 in	EC	127.00 mm
6.0000 in	HD	152.40 mm

ROCKWELL 82NYR43

IDLI GROUP 82214

END YOKE WITH ROUND BORE
SEE FIGURE 7B

8.1250 in	IE	206.38 mm
8.5000 in	OE	215.90 mm
2.7500 in	SB	69.85 mm
5.5000 in	EC	139.70 mm
5.7500 in	HD	146.05 mm

ROCKWELL 82NYR44

IDLI GROUP 82215

END YOKE WITH ROUND BORE
SEE FIGURE 7B

8.1250 in	IE	206.38 mm
8.5000 in	OE	215.90 mm
2.7500 in	SB	69.85 mm
5.5000 in	EC	139.70 mm
6.0000 in	HD	152.40 mm

ROCKWELL 82NYR44-3

IDLI GROUP 82216

END YOKE WITH ROUND BORE
SEE FIGURE 7B

8.1250 in	IE	206.38 mm
8.5000 in	OE	215.90 mm
2.7500 in	SB	69.85 mm
5.7500 in	EC	146.05 mm
6.0000 in	HD	152.40 mm

ROCKWELL 82NYR44-2

IDLI GROUP 82217

END YOKE WITH ROUND BORE
SEE FIGURE 7B

8.1250 in	IE	206.38 mm
8.5000 in	OE	215.90 mm
2.7500 in	SB	69.85 mm
6.5000 in	EC	165.10 mm
6.0000 in	HD	152.40 mm

ROCKWELL 82NYR44-1

IDLI GROUP 82218

END YOKE WITH ROUND BORE
slight variation from above group
SEE FIGURE 7B

8.1250 in	IE	206.38 mm
8.5000 in	OE	215.90 mm
2.7500 in	SB	69.85 mm
6.5000 in	EC	165.10 mm
6.0000 in	HD	152.40 mm

ROCKWELL 82NYR44-5

IDLI GROUP 82219

END YOKE WITH 21 INVOLUTED SPLINES
SEE FIGURE 7A

8.1250 in	IE	206.38 mm
8.5000 in	OE	215.90 mm
2.7813 in	SB	70.65 mm
4.5938 in	EC	116.68 mm
3.3750 in	HD	85.73 mm

BORG-WARNR 17522J
ROCKWELL 82NYS44-2

IDLI GROUP 82220

END YOKE WITH 21 INVOLUTED SPLINES
SEE FIGURE 7A

8.1250 in	IE	206.38 mm
8.5000 in	OE	215.90 mm
2.8594 in	SB	72.63 mm
3.9375 in	EC	100.01 mm
3.5000 in	HD	88.90 mm

ROCKWELL 82NYS46-1

IDLI GROUP 82221

END YOKE WITH 21 INVOLUTED SPLINES
SEE FIGURE 7A

8.1250 in	IE	206.38 mm
8.5000 in	OE	215.90 mm
2.8594 in	SB	72.63 mm
4.6250 in	EC	117.48 mm
3.5000 in	HD	88.90 mm

ROCKWELL 82NYS44-1

IDLI GROUP 82222

END YOKE WITH ROUND BORE
SEE FIGURE 7B

8.1250 in	IE	206.38 mm
8.5000 in	OE	215.90 mm
2.8750 in	SB	73.03 mm
5.0000 in	EC	127.00 mm
6.0000 in	HD	152.40 mm

ROCKWELL 82NYR46-2

IDLI GROUP 82223

END YOKE WITH ROUND BORE
SEE FIGURE 7B

8.1250 in	IE	206.38 mm
8.5000 in	OE	215.90 mm
2.8750 in	SB	73.03 mm
5.2500 in	EC	133.35 mm
6.0000 in	HD	152.40 mm

ROCKWELL 82NYR46-1

IDLI GROUP 82224

END YOKE WITH ROUND BORE
slight variation from above group
SEE FIGURE 7B

8.1250 in	IE	206.38 mm
8.5000 in	OE	215.90 mm
2.8750 in	SB	73.03 mm
5.2500 in	EC	133.35 mm
6.0000 in	HD	152.40 mm

ROCKWELL 82NYR46

IDLI GROUP 82225

END YOKE WITH ROUND BORE
SEE FIGURE 7B

8.1250 in	IE	206.38 mm
8.5000 in	OE	215.90 mm
2.9375 in	SB	74.61 mm
5.2500 in	EC	133.35 mm
6.0000 in	HD	152.40 mm

ROCKWELL 82NYR47-1

IDLI GROUP 82226

END YOKE WITH ROUND BORE
SEE FIGURE 7B

8.1250 in	IE	206.38 mm
8.5000 in	OE	215.90 mm
2.9375 in	SB	74.61 mm
5.6250 in	EC	142.88 mm
5.7500 in	HD	146.05 mm

ROCKWELL 82NYR47

IDLI GROUP 82227

END YOKE WITH 10 SPLINES
SEE FIGURE 7A

8.1250 in	IE	206.38 mm
8.5000 in	OE	215.90 mm
3.0000 in	SB	76.20 mm
2.0000 in	EC	50.80 mm
3.7500 in	HD	95.25 mm

ROCKWELL 82NYS48-3

IDLI GROUP 82228

END YOKE WITH 10 SPLINES
slight variation from above group
SEE FIGURE 7A

8.1250 in	IE	206.38 mm
8.5000 in	OE	215.90 mm
3.0000 in	SB	76.20 mm
2.0000 in	EC	50.80 mm
3.7500 in	HD	95.25 mm

ROCKWELL 82NYS48-1

CAUTION: BE SURE TO REFER TO ENGINEERING CATALOGS FOR SPECIAL APPLICATIONS THAT REQUIRE SPECIFIC MATERIAL CONTENT, TOLERANCES, ETC. SEE FOOTNOTE.

IDLI GROUP 82229

END YOKE WITH ROUND BORE
SEE FIGURE 7B

8.1250 in	IE	206.38 mm
8.5000 in	OE	215.90 mm
3.0000 in	SB	76.20 mm
2.6875 in	EC	68.26 mm
3.4375 in	HD	87.31 mm

ROCKWELL 82NYR48-1

IDLI GROUP 82230

END YOKE WITH 10 SPLINES
SEE FIGURE 7A

8.1250 in	IE	206.38 mm
8.5000 in	OE	215.90 mm
3.0000 in	SB	76.20 mm
3.0000 in	EC	76.20 mm
3.5000 in	HD	88.90 mm

ROCKWELL 82NYS48-7

IDLI GROUP 82231

END YOKE WITH 10 SPLINES
slight variation from above group
SEE FIGURE 7A

8.1250 in	IE	206.38 mm
8.5000 in	OE	215.90 mm
3.0000 in	SB	76.20 mm
3.0000 in	EC	76.20 mm
3.5000 in	HD	88.90 mm

ROCKWELL 82NYS48-6

IDLI GROUP 82232

END YOKE WITH 10 SPLINES
SEE FIGURE 7A

8.1250 in	IE	206.38 mm
8.5000 in	OE	215.90 mm
3.0000 in	SB	76.20 mm
3.6875 in	EC	93.66 mm
3.5000 in	HD	88.90 mm

ROCKWELL 82NYS48-11

IDLI GROUP 82233

END YOKE WITH 10 SPLINES
SEE FIGURE 7A

8.1250 in	IE	206.38 mm
8.5000 in	OE	215.90 mm
3.0000 in	SB	76.20 mm
3.8750 in	EC	98.43 mm
3.5000 in	HD	88.90 mm

ROCKWELL 82NYS48-2

IDLI GROUP 82234

END YOKE WITH 10 SPLINES
slight variation from above group
SEE FIGURE 7A

8.1250 in	IE	206.38 mm
8.5000 in	OE	215.90 mm
3.0000 in	SB	76.20 mm
3.8750 in	EC	98.43 mm
3.5000 in	HD	88.90 mm

ROCKWELL 82NYS48

IDLI GROUP 82235

END YOKE WITH 29 INVOLUTED SPLINES
SEE FIGURE 7A

8.1250 in	IE	206.38 mm
8.5000 in	OE	215.90 mm
3.0000 in	SB	76.20 mm
4.7500 in	EC	120.65 mm
4.0000 in	HD	101.60 mm

ROCKWELL 82NYS48-9

IDLI GROUP 82236

END YOKE WITH ROUND BORE
SEE FIGURE 7B

8.1250 in	IE	206.38 mm
8.5000 in	OE	215.90 mm
3.0000 in	SB	76.20 mm
5.0000 in	EC	127.00 mm
6.0000 in	HD	152.40 mm

ROCKWELL 82NYR48-13

IDLI GROUP 82237

END YOKE WITH ROUND BORE
SEE FIGURE 7B

8.1250 in	IE	206.38 mm
8.5000 in	OE	215.90 mm
3.0000 in	SB	76.20 mm
5.2500 in	EC	133.35 mm
6.0000 in	HD	152.40 mm

ROCKWELL 82NYR48-10

IDLI GROUP 82238

END YOKE WITH ROUND BORE
SEE FIGURE 7B

8.1250 in	IE	206.38 mm
8.5000 in	OE	215.90 mm
3.0000 in	SB	76.20 mm
5.5000 in	EC	139.70 mm
6.0000 in	HD	152.40 mm

ROCKWELL 82NYR48-3

IDLI GROUP 82239

END YOKE WITH ROUND BORE
SEE FIGURE 7B

8.1250 in	IE	206.38 mm
8.5000 in	OE	215.90 mm
3.0000 in	SB	76.20 mm
6.1250 in	EC	155.58 mm
6.0000 in	HD	152.40 mm

ROCKWELL 82NYR48-11

IDLI GROUP 82240

END YOKE WITH ROUND BORE
slight variation from above group
SEE FIGURE 7B

8.1250 in	IE	206.38 mm
8.5000 in	OE	215.90 mm
3.0000 in	SB	76.20 mm
6.1250 in	EC	155.58 mm
6.0000 in	HD	152.40 mm

ROCKWELL 82NYR48-2

IDLI GROUP 82241

END YOKE WITH ROUND BORE
SEE FIGURE 7B

8.1250 in	IE	206.38 mm
8.5000 in	OE	215.90 mm
3.0000 in	SB	76.20 mm
6.3125 in	EC	160.34 mm
6.0000 in	HD	152.40 mm

ROCKWELL 82NYR48-5

IDLI GROUP 82242

END YOKE WITH ROUND BORE
SEE FIGURE 7B

8.1250 in	IE	206.38 mm
8.5000 in	OE	215.90 mm
3.0000 in	SB	76.20 mm
6.5000 in	EC	165.10 mm
5.5000 in	HD	139.70 mm

ROCKWELL 82NYR48

IDLI GROUP 82243

END YOKE WITH ROUND BORE
SEE FIGURE 7B

8.1250 in	IE	206.38 mm
8.5000 in	OE	215.90 mm
3.0000 in	SB	76.20 mm
6.5000 in	EC	165.10 mm
6.0000 in	HD	152.40 mm

ROCKWELL 82NYR48-7

IDLI GROUP 82244

END YOKE WITH ROUND BORE
slight variation from above group
SEE FIGURE 7B

8.1250 in	IE	206.38 mm
8.5000 in	OE	215.90 mm
3.0000 in	SB	76.20 mm
6.5000 in	EC	165.10 mm
6.0000 in	HD	152.40 mm

ROCKWELL 82NYR48-8

IDLI GROUP 82245

END YOKE WITH ROUND BORE
SEE FIGURE 7B

8.1250 in	IE	206.38 mm
8.5000 in	OE	215.90 mm
3.0000 in	SB	76.20 mm
6.5625 in	EC	166.69 mm
6.0000 in	HD	152.40 mm

ROCKWELL 82NYR48-4

IDLI GROUP 82246

END YOKE WITH ROUND BORE
slight variation from above group
SEE FIGURE 7B

8.1250 in	IE	206.38 mm
8.5000 in	OE	215.90 mm
3.0000 in	SB	76.20 mm
6.5625 in	EC	166.69 mm
6.0000 in	HD	152.40 mm

ROCKWELL 82NYR48-6

IDLI GROUP 82247

END YOKE WITH ROUND BORE
SEE FIGURE 7B

8.1250 in	IE	206.38 mm
8.5000 in	OE	215.90 mm
3.1250 in	SB	79.38 mm
5.0000 in	EC	127.00 mm
4.8750 in	HD	123.83 mm

ROCKWELL 82NYR50

IDLI GROUP 82248

END YOKE WITH ROUND BORE
slight variation from above group
SEE FIGURE 7B

8.1250 in	IE	206.38 mm
8.5000 in	OE	215.90 mm
3.1250 in	SB	79.38 mm
5.0000 in	EC	127.00 mm
4.8750 in	HD	123.83 mm

ROCKWELL 82NYR50-2

IDLI GROUP 82249

END YOKE WITH ROUND BORE
SEE FIGURE 7B

8.1250 in	IE	206.38 mm
8.5000 in	OE	215.90 mm
3.1250 in	SB	79.38 mm
6.5000 in	EC	165.10 mm
6.0000 in	HD	152.40 mm

ROCKWELL 82NYR50-1

IDLI GROUP 82250

END YOKE WITH ROUND BORE
SEE FIGURE 7B

8.1250 in	IE	206.38 mm
8.5000 in	OE	215.90 mm
3.1875 in	SB	80.96 mm
5.7500 in	EC	146.05 mm
6.0000 in	HD	152.40 mm

ROCKWELL 82NYR51-1

IDLI GROUP 82251

END YOKE WITH ROUND BORE
SEE FIGURE 7B

8.1250 in	IE	206.38 mm
8.5000 in	OE	215.90 mm
3.1875 in	SB	80.96 mm
6.5000 in	EC	165.10 mm
6.0000 in	HD	152.40 mm

ROCKWELL 82NYR51

IDLI GROUP 82252

END YOKE WITH ROUND BORE
SEE FIGURE 7B

8.1250 in	IE	206.38 mm
8.5000 in	OE	215.90 mm
3.2500 in	SB	82.55 mm
5.0000 in	EC	127.00 mm
4.8750 in	HD	123.83 mm

ROCKWELL 82NYR52-1

BD = BEARING DIAMETER (outside)   **BW** = BEARING WIDTH   **CB** = CROSS LENGTH WITH BEARINGS   **CC** = CENTER TO CENTER   **CD** = CROSS DIAMETER
CL = CROSS LENGTH WITHOUT BEARINGS   **EC** = END TO CENTER or FACE TO FACE   **EE** = END TO END   **EL** = EFFECTIVE LENGTH
HD = HUB DIAMETER (or insert)   **IE** = INSIDE OF EARS (or recess)   **OD** = OUTSIDE DIAMETER   **OE** = OUTSIDE OF EARS   **SB** = SPLINE OR BORE SIZE

IDLI GROUP 82253

END YOKE WITH ROUND BORE
slight variation from above group
SEE FIGURE 7B

8.1250 in	IE	206.38 mm
8.5000 in	OE	215.90 mm
3.2500 in	SB	82.55 mm
5.0000 in	EC	127.00 mm
4.8750 in	HD	123.83 mm

ROCKWELL 82NYR52

IDLI GROUP 82254

END YOKE WITH ROUND BORE
SEE FIGURE 7B

8.1250 in	IE	206.38 mm
8.5000 in	OE	215.90 mm
3.2500 in	SB	82.55 mm
6.5000 in	EC	165.10 mm
6.0000 in	HD	152.40 mm

ROCKWELL 82NYR52-5

IDLI GROUP 82255

END YOKE WITH ROUND BORE
slight variation from above group
SEE FIGURE 7B

8.1250 in	IE	206.38 mm
8.5000 in	OE	215.90 mm
3.2500 in	SB	82.55 mm
6.5000 in	EC	165.10 mm
6.0000 in	HD	152.40 mm

ROCKWELL 82NYR52-6

IDLI GROUP 82256

END YOKE WITH ROUND BORE
slight variation from above group
SEE FIGURE 7B

8.1250 in	IE	206.38 mm
8.5000 in	OE	215.90 mm
3.2500 in	SB	82.55 mm
6.5000 in	EC	165.10 mm
6.0000 in	HD	152.40 mm

ROCKWELL 82NYR52-4

IDLI GROUP 82257

END YOKE WITH ROUND BORE
SEE FIGURE 7B

8.1250 in	IE	206.38 mm
8.5000 in	OE	215.90 mm
3.2500 in	SB	82.55 mm
6.5625 in	EC	166.69 mm
6.0000 in	HD	152.40 mm

ROCKWELL 82NYR52-2

IDLI GROUP 82258

END YOKE WITH ROUND BORE
SEE FIGURE 7B

8.1250 in	IE	206.38 mm
8.5000 in	OE	215.90 mm
3.3750 in	SB	85.73 mm
5.8125 in	EC	147.64 mm
6.0000 in	HD	152.40 mm

ROCKWELL 82NYR54-1

IDLI GROUP 82259

END YOKE WITH ROUND BORE
SEE FIGURE 7B

8.1250 in	IE	206.38 mm
8.5000 in	OE	215.90 mm
3.3750 in	SB	85.73 mm
6.5000 in	EC	165.10 mm
5.5000 in	HD	139.70 mm

ROCKWELL 82NYR54

IDLI GROUP 82260

END YOKE WITH ROUND BORE
SEE FIGURE 7B

8.1250 in	IE	206.38 mm
8.5000 in	OE	215.90 mm
3.3750 in	SB	85.73 mm
6.5000 in	EC	165.10 mm
6.0000 in	HD	152.40 mm

ROCKWELL 82NYR54-2

IDLI GROUP 82261

END YOKE WITH ROUND BORE
SEE FIGURE 7B

8.1250 in	IE	206.38 mm
8.5000 in	OE	215.90 mm
3.4375 in	SB	87.31 mm
5.3125 in	EC	134.94 mm
6.0000 in	HD	152.40 mm

ROCKWELL 82NYR55-5

IDLI GROUP 82262

END YOKE WITH ROUND BORE
SEE FIGURE 7B

8.1250 in	IE	206.38 mm
8.5000 in	OE	215.90 mm
3.4375 in	SB	87.31 mm
6.5000 in	EC	165.10 mm
6.0000 in	HD	152.40 mm

ROCKWELL 82NYR55-1

IDLI GROUP 82263

END YOKE WITH ROUND BORE
slight variation from above group
SEE FIGURE 7B

8.1250 in	IE	206.38 mm
8.5000 in	OE	215.90 mm
3.4375 in	SB	87.31 mm
6.5000 in	EC	165.10 mm
6.0000 in	HD	152.40 mm

ROCKWELL 82NYR55-4

IDLI GROUP 82264

END YOKE WITH ROUND BORE
slight variation from above group
SEE FIGURE 7B

8.1250 in	IE	206.38 mm
8.5000 in	OE	215.90 mm
3.4375 in	SB	87.31 mm
6.5000 in	EC	165.10 mm
6.0000 in	HD	152.40 mm

ROCKWELL 82NYR55-3

IDLI GROUP 82265

END YOKE WITH ROUND BORE
SEE FIGURE 7B

8.1250 in	IE	206.38 mm
8.5000 in	OE	215.90 mm
3.5000 in	SB	88.90 mm
4.7500 in	EC	120.65 mm
6.0000 in	HD	152.40 mm

ROCKWELL 82NYR56-10

IDLI GROUP 82266

END YOKE WITH ROUND BORE
SEE FIGURE 7B

8.1250 in	IE	206.38 mm
8.5000 in	OE	215.90 mm
3.5000 in	SB	88.90 mm
5.1250 in	EC	130.18 mm
6.0000 in	HD	152.40 mm

ROCKWELL 82NYR56-2

IDLI GROUP 82267

END YOKE WITH ROUND BORE
SEE FIGURE 7B

8.1250 in	IE	206.38 mm
8.5000 in	OE	215.90 mm
3.5000 in	SB	88.90 mm
5.4375 in	EC	138.11 mm
5.5000 in	HD	139.70 mm

ROCKWELL 82NYR56-6

IDLI GROUP 82268

END YOKE WITH ROUND BORE
SEE FIGURE 7B

8.1250 in	IE	206.38 mm
8.5000 in	OE	215.90 mm
3.5000 in	SB	88.90 mm
5.5625 in	EC	141.29 mm
6.0000 in	HD	152.40 mm

ROCKWELL 82NYR56-7

IDLI GROUP 82269

END YOKE WITH ROUND BORE
SEE FIGURE 7B

8.1250 in	IE	206.38 mm
8.5000 in	OE	215.90 mm
3.5000 in	SB	88.90 mm
6.5000 in	EC	165.10 mm
6.0000 in	HD	152.40 mm

ROCKWELL 82NYR56-4

IDLI GROUP 82270

END YOKE WITH ROUND BORE
slight variation from above group
SEE FIGURE 7B

8.1250 in	IE	206.38 mm
8.5000 in	OE	215.90 mm
3.5000 in	SB	88.90 mm
6.5000 in	EC	165.10 mm
6.0000 in	HD	152.40 mm

ROCKWELL 82NYR56-12

IDLI GROUP 82271

END YOKE WITH ROUND BORE
slight variation from above group
SEE FIGURE 7B

8.1250 in	IE	206.38 mm
8.5000 in	OE	215.90 mm
3.5000 in	SB	88.90 mm
6.5000 in	EC	165.10 mm
6.0000 in	HD	152.40 mm

ROCKWELL 82NYR56-14

IDLI GROUP 82272

END YOKE WITH ROUND BORE
slight variation from above group
SEE FIGURE 7B

8.1250 in	IE	206.38 mm
8.5000 in	OE	215.90 mm
3.5000 in	SB	88.90 mm
6.5000 in	EC	165.10 mm
6.0000 in	HD	152.40 mm

ROCKWELL 82NYR56-13

IDLI GROUP 82273

END YOKE WITH ROUND BORE
slight variation from above group
SEE FIGURE 7B

8.1250 in	IE	206.38 mm
8.5000 in	OE	215.90 mm
3.5000 in	SB	88.90 mm
6.5000 in	EC	165.10 mm
6.0000 in	HD	152.40 mm

ROCKWELL 82NYR56-8

IDLI GROUP 82274

END YOKE WITH ROUND BORE
slight variation from above group
SEE FIGURE 7B

8.1250 in	IE	206.38 mm
8.5000 in	OE	215.90 mm
3.5000 in	SB	88.90 mm
6.5000 in	EC	165.10 mm
6.0000 in	HD	152.40 mm

ROCKWELL 82NYR56-9

IDLI GROUP 82275

END YOKE WITH ROUND BORE
slight variation from above group
SEE FIGURE 7B

8.1250 in	IE	206.38 mm
8.5000 in	OE	215.90 mm
3.5000 in	SB	88.90 mm
6.5000 in	EC	165.10 mm
6.0000 in	HD	152.40 mm

ROCKWELL 82NYR56

IDLI GROUP 82276

END YOKE WITH ROUND BORE
SEE FIGURE 7B

8.1250 in	IE	206.38 mm
8.5000 in	OE	215.90 mm
3.5000 in	SB	88.90 mm
6.5625 in	EC	166.69 mm
6.0000 in	HD	152.40 mm

ROCKWELL 82NYR56-5

ENGINEERING CATALOGS MUST BE CONSULTED FOR DETAILS NOT INCLUDED
IN THIS GUIDE. SPECIFIC DESIGNS, MATERIAL CONTENT, TOLERANCES
LUBE FITTINGS AND OTHER DIMENSIONS ARE INTENTIONALLY OMITTED HERE.

IDLI GUIDE   COPYRIGHT © INTERCHANGE, INC.   ST. LOUIS PARK, MN. 55416 USA

CAUTION: BE SURE TO REFER TO ENGINEERING CATALOGS FOR SPECIAL APPLICATIONS THAT REQUIRE SPECIFIC MATERIAL CONTENT, TOLERANCES, ETC. SEE FOOTNOTE.

IDLI GROUP 82277

END YOKE WITH ROUND BORE
SEE FIGURE 7B

8.1250 in	IE	206.38 mm
8.5000 in	OE	215.90 mm
3.6250 in	SB	92.08 mm
6.5000 in	EC	165.10 mm
6.0000 in	HD	152.40 mm

ROCKWELL 82NYR58

IDLI GROUP 82278

END YOKE WITH ROUND BORE
SEE FIGURE 7B

8.1250 in	IE	206.38 mm
8.5000 in	OE	215.90 mm
3.6875 in	SB	93.66 mm
4.7500 in	EC	120.65 mm
6.0000 in	HD	152.40 mm

ROCKWELL 82NYR59

IDLI GROUP 82279

END YOKE WITH ROUND BORE
SEE FIGURE 7B

8.1250 in	IE	206.38 mm
8.5000 in	OE	215.90 mm
3.7500 in	SB	95.25 mm
6.3750 in	EC	161.93 mm
6.0000 in	HD	152.40 mm

ROCKWELL 82NYR60-3

IDLI GROUP 82280

END YOKE WITH ROUND BORE
SEE FIGURE 7B

8.1250 in	IE	206.38 mm
8.5000 in	OE	215.90 mm
3.7500 in	SB	95.25 mm
6.5000 in	EC	165.10 mm
6.0000 in	HD	152.40 mm

ROCKWELL 82NYR60-1

IDLI GROUP 82281

END YOKE WITH ROUND BORE
SEE FIGURE 7B

8.1250 in	IE	206.38 mm
8.5000 in	OE	215.90 mm
3.7500 in	SB	95.25 mm
6.5625 in	EC	166.69 mm
6.0000 in	HD	152.40 mm

ROCKWELL 82NYR60

IDLI GROUP 82282

END YOKE WITH ROUND BORE
SEE FIGURE 7B

8.1250 in	IE	206.38 mm
8.5000 in	OE	215.90 mm
3.8750 in	SB	98.43 mm
6.5000 in	EC	165.10 mm
6.0000 in	HD	152.40 mm

ROCKWELL 82NYR62

IDLI GROUP 82283

END YOKE WITH ROUND BORE
SEE FIGURE 7B

8.1250 in	IE	206.38 mm
8.5000 in	OE	215.90 mm
3.9375 in	SB	100.01 mm
6.5000 in	EC	165.10 mm
6.0000 in	HD	152.40 mm

ROCKWELL 82NYR63-1

IDLI GROUP 82284

END YOKE WITH ROUND BORE
slight variation from above group
SEE FIGURE 7B

8.1250 in	IE	206.38 mm
8.5000 in	OE	215.90 mm
3.9375 in	SB	100.01 mm
6.5000 in	EC	165.10 mm
6.0000 in	HD	152.40 mm

ROCKWELL 82NYR63

IDLI GROUP 82285

END YOKE WITH ROUND BORE
SEE FIGURE 7B

8.1250 in	IE	206.38 mm
8.5000 in	OE	215.90 mm
4.0000 in	SB	101.60 mm
6.2500 in	EC	158.75 mm
6.0000 in	HD	152.40 mm

ROCKWELL 82NYR64

IDLI GROUP 82286

END YOKE WITH ROUND BORE
slight variation from above group
SEE FIGURE 7B

8.1250 in	IE	206.38 mm
8.5000 in	OE	215.90 mm
4.0000 in	SB	101.60 mm
6.2500 in	EC	158.75 mm
6.0000 in	HD	152.40 mm

ROCKWELL 82NYR64-2

IDLI GROUP 82287

END YOKE WITH ROUND BORE
SEE FIGURE 7B

8.1250 in	IE	206.38 mm
8.5000 in	OE	215.90 mm
4.0000 in	SB	101.60 mm
6.5000 in	EC	165.10 mm
6.0000 in	HD	152.40 mm

ROCKWELL 82NYR64-1

IDLI GROUP 82288

END YOKE WITH ROUND BORE
SEE FIGURE 7B

8.1250 in	IE	206.38 mm
8.5000 in	OE	215.90 mm
4.2500 in	SB	107.95 mm
6.0000 in	EC	152.40 mm
5.7500 in	HD	146.05 mm

ROCKWELL 82NYR68-3

IDLI GROUP 82289

END YOKE WITH ROUND BORE
slight variation from above group
SEE FIGURE 7B

8.1250 in	IE	206.38 mm
8.5000 in	OE	215.90 mm
4.2500 in	SB	107.95 mm
6.0000 in	EC	152.40 mm
5.7500 in	HD	146.05 mm

ROCKWELL 82NYR68-4

IDLI GROUP 82290

END YOKE WITH ROUND BORE
SEE FIGURE 7B

8.1250 in	IE	206.38 mm
8.5000 in	OE	215.90 mm
4.4375 in	SB	112.71 mm
6.0000 in	EC	152.40 mm
5.7500 in	HD	146.05 mm

ROCKWELL 82NYR71

IDLI GROUP 82291

END YOKE WITH ROUND BORE
SEE FIGURE 7B

8.1250 in	IE	206.38 mm
8.5000 in	OE	215.90 mm
4.5000 in	SB	114.30 mm
5.7500 in	EC	146.05 mm
8.5000 in	HD	215.90 mm

ROCKWELL 82NYR72-2

IDLI GROUP 82292

END YOKE WITH ROUND BORE
slight variation from above group
SEE FIGURE 7B

8.1250 in	IE	206.38 mm
8.5000 in	OE	215.90 mm
4.5000 in	SB	114.30 mm
5.7500 in	EC	146.05 mm
8.5000 in	HD	215.90 mm

ROCKWELL 82NYR72-1

IDLI GROUP 82293

END YOKE WITH ROUND BORE
slight variation from above group
SEE FIGURE 7B

8.1250 in	IE	206.38 mm
8.5000 in	OE	215.90 mm
4.5000 in	SB	114.30 mm
5.7500 in	EC	146.05 mm
8.5000 in	HD	215.90 mm

ROCKWELL 82NYR72

IDLI GROUP 82294

END YOKE WITH ROUND BORE
SEE FIGURE 7B

8.1250 in	IE	206.38 mm
8.5000 in	OE	215.90 mm
4.9375 in	SB	125.41 mm
6.2500 in	EC	158.75 mm
8.5000 in	HD	215.90 mm

ROCKWELL 82NYR79

IDLI GROUP 82295

END YOKE WITH ROUND BORE
SEE FIGURE 7B

8.1250 in	IE	206.38 mm
8.5000 in	OE	215.90 mm
5.0000 in	SB	127.00 mm
5.7500 in	EC	146.05 mm
8.5000 in	HD	215.90 mm

ROCKWELL 82NYR80

IDLI GROUP 82296

END YOKE WITH ROUND BORE
SEE FIGURE 7B

8.1250 in	IE	206.38 mm
8.5000 in	OE	215.90 mm
5.0000 in	SB	127.00 mm
6.2500 in	EC	158.75 mm
8.5000 in	HD	215.90 mm

ROCKWELL 82NYR80-1

IDLI GROUP 82297

END YOKE WITH ROUND BORE
SEE FIGURE 7B

8.1250 in	IE	206.38 mm
8.5000 in	OE	215.90 mm
5.4375 in	SB	138.11 mm
6.2500 in	EC	158.75 mm
8.5000 in	HD	215.90 mm

ROCKWELL 82NYR87

IDLI GROUP 82298

END YOKE WITH ROUND BORE
SEE FIGURE 7B

8.1250 in	IE	206.38 mm
8.5000 in	OE	215.90 mm
5.4688 in	SB	138.91 mm
6.2500 in	EC	158.75 mm
8.5000 in	HD	215.90 mm

ROCKWELL 82NYR88-1

IDLI GROUP 82299

END YOKE WITH ROUND BORE
SEE FIGURE 7B

8.1250 in	IE	206.38 mm
8.5000 in	OE	215.90 mm
5.5000 in	SB	139.70 mm
6.2500 in	EC	158.75 mm
8.5000 in	HD	215.90 mm

ROCKWELL 82NYR88

IDLI GROUP 82300

END YOKE WITH ROUND BORE
SEE FIGURE 7B

8.2500 in	IE	209.55 mm
8.6250 in	OE	219.08 mm
1.7500 in	SB	44.45 mm
6.5625 in	EC	166.69 mm
5.7500 in	HD	146.05 mm

ROCKWELL 92NYR28-3

BD = BEARING DIAMETER (outside)   **BW** = BEARING WIDTH   **CB** = CROSS LENGTH WITH BEARINGS   **CC** = CENTER TO CENTER   **CD** = CROSS DIAMETER
CL = CROSS LENGTH WITHOUT BEARINGS   **EC** = END TO CENTER or FACE TO FACE   **EE** = END TO END   **EL** = EFFECTIVE LENGTH
HD = HUB DIAMETER (or insert)   **IE** = INSIDE OF EARS (or recess)   **OD** = OUTSIDE DIAMETER   **OE** = OUTSIDE OF EARS   **SB** = SPLINE OR BORE SIZE

IDLI GROUP 82301

END YOKE WITH 34 SLANTED SPLINES
SEE FIGURE 7A

8.2500 in	IE	209.55 mm
8.6250 in	OE	219.08 mm
1.7656 in	SB	44.85 mm
4.2500 in	EC	107.95 mm
2.5000 in	HD	63.50 mm

ROCKWELL 92NYS28-5

IDLI GROUP 82302

END YOKE WITH 34 SLANTED SPLINES
slight variation from above group
SEE FIGURE 7A

8.2500 in	IE	209.55 mm
8.6250 in	OE	219.08 mm
1.7656 in	SB	44.85 mm
4.2500 in	EC	107.95 mm
2.5000 in	HD	63.50 mm

ROCKWELL 92NYS28-5A

IDLI GROUP 82303

END YOKE WITH 10 SPLINES
SEE FIGURE 7A

8.2500 in	IE	209.55 mm
8.6250 in	OE	219.08 mm
1.9688 in	SB	50.01 mm
3.9375 in	EC	100.01 mm
2.6250 in	HD	66.68 mm

ROCKWELL 92NYS31-2

IDLI GROUP 82304

END YOKE WITH 10 SPLINES
SEE FIGURE 7A

8.2500 in	IE	209.55 mm
8.6250 in	OE	219.08 mm
1.9688 in	SB	50.01 mm
4.2500 in	EC	107.95 mm
2.6250 in	HD	66.68 mm

ROCKWELL 92NYS31

IDLI GROUP 82305

END YOKE WITH 10 SPLINES
SEE FIGURE 7A

8.2500 in	IE	209.55 mm
8.6250 in	OE	219.08 mm
2.0000 in	SB	50.80 mm
3.8750 in	EC	98.43 mm
2.9375 in	HD	74.61 mm

ROCKWELL 92NYS32-2

IDLI GROUP 82306

END YOKE WITH 10 SPLINES
slight variation from above group
SEE FIGURE 7A

8.2500 in	IE	209.55 mm
8.6250 in	OE	219.08 mm
2.0000 in	SB	50.80 mm
3.8750 in	EC	98.43 mm
2.9375 in	HD	74.61 mm

ROCKWELL 92NYS32-1

IDLI GROUP 82307

END YOKE WITH 10 SPLINES
SEE FIGURE 7A

8.2500 in	IE	209.55 mm
8.6250 in	OE	219.08 mm
2.0000 in	SB	50.80 mm
4.1875 in	EC	106.36 mm
2.5625 in	HD	65.09 mm

ROCKWELL 92NYS32-3

IDLI GROUP 82308

END YOKE WITH 10 SPLINES
SEE FIGURE 7A

8.2500 in	IE	209.55 mm
8.6250 in	OE	219.08 mm
2.0000 in	SB	50.80 mm
4.2500 in	EC	107.95 mm
2.7500 in	HD	69.85 mm

ROCKWELL 92NYS32-7

IDLI GROUP 82309

END YOKE WITH 10 SPLINES
SEE FIGURE 7A

8.2500 in	IE	209.55 mm
8.6250 in	OE	219.08 mm
2.0000 in	SB	50.80 mm
4.2500 in	EC	107.95 mm
2.9375 in	HD	74.61 mm

ROCKWELL 92NYS32-5

IDLI GROUP 82310

END YOKE WITH 10 SPLINES
SEE FIGURE 7A

8.2500 in	IE	209.55 mm
8.6250 in	OE	219.08 mm
2.0313 in	SB	51.60 mm
3.9375 in	EC	100.01 mm
2.7500 in	HD	69.85 mm

ROCKWELL 92NYS33

IDLI GROUP 82311

END YOKE WITH 39 SLANTED SPLINES
SEE FIGURE 7A

8.2500 in	IE	209.55 mm
8.6250 in	OE	219.08 mm
2.0313 in	SB	51.60 mm
3.9375 in	EC	100.01 mm
2.7500 in	HD	69.85 mm

ROCKWELL 92NYS32-27

IDLI GROUP 82312

END YOKE WITH 39 SLANTED SPLINES
SEE FIGURE 7A

8.2500 in	IE	209.55 mm
8.6250 in	OE	219.08 mm
2.0313 in	SB	51.60 mm
4.2500 in	EC	107.95 mm
3.0000 in	HD	76.20 mm

ROCKWELL 92NYS32-25

IDLI GROUP 82313

END YOKE WITH 39 SLANTED SPLINES
slight variation from above group
SEE FIGURE 7A

8.2500 in	IE	209.55 mm
8.6250 in	OE	219.08 mm
2.0313 in	SB	51.60 mm
4.2500 in	EC	107.95 mm
3.0000 in	HD	76.20 mm

ROCKWELL 92NYS32-25A

IDLI GROUP 82314

END YOKE WITH 39 SLANTED SPLINES
SEE FIGURE 7A

8.2500 in	IE	209.55 mm
8.6250 in	OE	219.08 mm
2.0313 in	SB	51.60 mm
4.4375 in	EC	112.71 mm
3.0000 in	HD	76.20 mm

ROCKWELL 92NYS32-26

IDLI GROUP 82315

END YOKE WITH 39 SLANTED SPLINES
slight variation from above group
SEE FIGURE 7A

8.2500 in	IE	209.55 mm
8.6250 in	OE	219.08 mm
2.0313 in	SB	51.60 mm
4.4375 in	EC	112.71 mm
3.0000 in	HD	76.20 mm

ROCKWELL 92NYS32-26A

IDLI GROUP 82316

END YOKE WITH 6 SPLINES
SEE FIGURE 7A

8.2500 in	IE	209.55 mm
8.6250 in	OE	219.08 mm
2.2188 in	SB	56.36 mm
4.2500 in	EC	107.95 mm
3.0000 in	HD	76.20 mm

BORG-WARNR 19543J
ROCKWELL 92NYS36-5

IDLI GROUP 82317

END YOKE WITH 6 SPLINES
SEE FIGURE 7A

8.2500 in	IE	209.55 mm
8.6250 in	OE	219.08 mm
2.2188 in	SB	56.36 mm
4.9375 in	EC	125.41 mm
3.0000 in	HD	76.20 mm

ROCKWELL 92NYS36-15

IDLI GROUP 82318

END YOKE WITH 6 SPLINES
SEE FIGURE 7A

8.2500 in	IE	209.55 mm
8.6250 in	OE	219.08 mm
2.2188 in	SB	56.36 mm
5.1875 in	EC	131.76 mm
3.0000 in	HD	76.20 mm

ROCKWELL 92NYS36-41

IDLI GROUP 82319

END YOKE WITH 10 SPLINES
SEE FIGURE 7A

8.2500 in	IE	209.55 mm
8.6250 in	OE	219.08 mm
2.2500 in	SB	57.15 mm
3.9375 in	EC	100.01 mm
3.2500 in	HD	82.55 mm

BORG-WARNR 20662J
BORG-WARNR 21282J
ROCKWELL 92NYS36-16

IDLI GROUP 82320

END YOKE WITH 10 SPLINES
SEE FIGURE 7A

8.2500 in	IE	209.55 mm
8.6250 in	OE	219.08 mm
2.2500 in	SB	57.15 mm
4.1250 in	EC	104.78 mm
3.5000 in	HD	88.90 mm

BORG-WARNR 7407J
ROCKWELL 92NYS36-48

IDLI GROUP 82321

END YOKE WITH TAPERED BORE
SEE FIGURE 7F

8.2500 in	IE	209.55 mm
8.6250 in	OE	219.08 mm
2.2500 in	SB	57.15 mm
4.1250 in	EC	104.78 mm
4.3750 in	HD	111.13 mm

ROCKWELL 92NYT36

IDLI GROUP 82322

END YOKE WITH 10 SPLINES
SEE FIGURE 7A

8.2500 in	IE	209.55 mm
8.6250 in	OE	219.08 mm
2.2500 in	SB	57.15 mm
4.1875 in	EC	106.36 mm
2.8750 in	HD	73.03 mm

ROCKWELL 92NYS36-47

IDLI GROUP 82323

END YOKE WITH 10 SPLINES
SEE FIGURE 7A

8.2500 in	IE	209.55 mm
8.6250 in	OE	219.08 mm
2.2500 in	SB	57.15 mm
4.1875 in	EC	106.36 mm
3.0000 in	HD	76.20 mm

BORG-WARNR 121-9018
ROCKWELL 92NYS36-2

IDLI GROUP 82324

END YOKE WITH 10 SPLINES
slight variation from above group
SEE FIGURE 7A

8.2500 in	IE	209.55 mm
8.6250 in	OE	219.08 mm
2.2500 in	SB	57.15 mm
4.1875 in	EC	106.36 mm
3.0000 in	HD	76.20 mm

ROCKWELL 92NYS36-6

ENGINEERING CATALOGS MUST BE CONSULTED FOR DETAILS NOT INCLUDED
IN THIS GUIDE. SPECIFIC DESIGNS, MATERIAL CONTENT, TOLERANCES
LUBE FITTINGS AND OTHER DIMENSIONS ARE INTENTIONALLY OMITTED HERE.

CAUTION: BE SURE TO REFER TO ENGINEERING CATALOGS FOR SPECIAL APPLICATIONS THAT REQUIRE SPECIFIC MATERIAL CONTENT, TOLERANCES, ETC. SEE FOOTNOTE.

IDLI GROUP 82325

END YOKE WITH 10 SPLINES
slight variation from above group
SEE FIGURE 7A

8.2500 in	IE	209.55 mm
8.6250 in	OE	219.08 mm
2.2500 in	SB	57.15 mm
4.1875 in	EC	106.36 mm
3.0000 in	HD	76.20 mm

ROCKWELL 92NYS36-32

IDLI GROUP 82326

END YOKE WITH 10 SPLINES
SEE FIGURE 7A

8.2500 in	IE	209.55 mm
8.6250 in	OE	219.08 mm
2.2500 in	SB	57.15 mm
4.1875 in	EC	106.36 mm
3.1250 in	HD	79.38 mm

ROCKWELL 92NYS36-40

IDLI GROUP 82327

END YOKE WITH 10 SPLINES
slight variation from above group
SEE FIGURE 7A

8.2500 in	IE	209.55 mm
8.6250 in	OE	219.08 mm
2.2500 in	SB	57.15 mm
4.1875 in	EC	106.36 mm
3.1250 in	HD	79.38 mm

ROCKWELL 92NYS36-60

IDLI GROUP 82328

END YOKE WITH 10 SPLINES
SEE FIGURE 7A

8.2500 in	IE	209.55 mm
8.6250 in	OE	219.08 mm
2.2500 in	SB	57.15 mm
4.1875 in	EC	106.36 mm
3.2500 in	HD	82.55 mm

ROCKWELL 92NYS36-38

IDLI GROUP 82329

END YOKE WITH 10 SPLINES
slight variation from above group
SEE FIGURE 7A

8.2500 in	IE	209.55 mm
8.6250 in	OE	219.08 mm
2.2500 in	SB	57.15 mm
4.1875 in	EC	106.36 mm
3.2500 in	HD	82.55 mm

ROCKWELL 92NYS36-64

IDLI GROUP 82330

END YOKE WITH 10 SPLINES
SEE FIGURE 7A

8.2500 in	IE	209.55 mm
8.6250 in	OE	219.08 mm
2.2500 in	SB	57.15 mm
4.1875 in	EC	106.36 mm
3.5000 in	HD	88.90 mm

ROCKWELL 92NYS36-1

IDLI GROUP 82331

END YOKE WITH 10 SPLINES
SEE FIGURE 7A

8.2500 in	IE	209.55 mm
8.6250 in	OE	219.08 mm
2.2500 in	SB	57.15 mm
4.2500 in	EC	107.95 mm
3.0000 in	HD	76.20 mm

ROCKWELL 92NYS36-26

IDLI GROUP 82332

END YOKE WITH 10 SPLINES
SEE FIGURE 7A

8.2500 in	IE	209.55 mm
8.6250 in	OE	219.08 mm
2.2500 in	SB	57.15 mm
4.2500 in	EC	107.95 mm
3.5000 in	HD	88.90 mm

BORG-WARNR 20660J
ROCKWELL 92NYS36-22

IDLI GROUP 82333

END YOKE WITH 10 SPLINES
slight variation from above group
SEE FIGURE 7A

8.2500 in	IE	209.55 mm
8.6250 in	OE	219.08 mm
2.2500 in	SB	57.15 mm
4.2500 in	EC	107.95 mm
3.5000 in	HD	88.90 mm

BORG-WARNR 20661J
ROCKWELL 92NYS36-24

IDLI GROUP 82334

END YOKE WITH 10 SPLINES
slight variation from above group
SEE FIGURE 7A

8.2500 in	IE	209.55 mm
8.6250 in	OE	219.08 mm
2.2500 in	SB	57.15 mm
4.2500 in	EC	107.95 mm
3.5000 in	HD	88.90 mm

ROCKWELL 92NYS36-27

IDLI GROUP 82335

END YOKE WITH 6 SPLINES
SEE FIGURE 7A

8.2500 in	IE	209.55 mm
8.6250 in	OE	219.08 mm
2.2500 in	SB	57.15 mm
4.7500 in	EC	120.65 mm
3.0000 in	HD	76.20 mm

BORG-WARNR 15096J
ROCKWELL 92NYS36

IDLI GROUP 82336

END YOKE WITH 10 SPLINES
SEE FIGURE 7A

8.2500 in	IE	209.55 mm
8.6250 in	OE	219.08 mm
2.2500 in	SB	57.15 mm
4.8750 in	EC	123.83 mm
2.8750 in	HD	73.03 mm

ROCKWELL 92NYS36-31

IDLI GROUP 82337

END YOKE WITH 10 SPLINES
SEE FIGURE 7A

8.2500 in	IE	209.55 mm
8.6250 in	OE	219.08 mm
2.2500 in	SB	57.15 mm
5.0000 in	EC	127.00 mm
3.5000 in	HD	88.90 mm

BORG-WARNR 19125J
ROCKWELL 92NYS36-3

IDLI GROUP 82338

END YOKE WITH 10 SPLINES
SEE FIGURE 7A

8.2500 in	IE	209.55 mm
8.6250 in	OE	219.08 mm
2.2500 in	SB	57.15 mm
5.7500 in	EC	146.05 mm
3.5000 in	HD	88.90 mm

BORG-WARNR 22863J
ROCKWELL 92NYS36-17

IDLI GROUP 82339

END YOKE WITH ROUND BORE
SEE FIGURE 7B

8.2500 in	IE	209.55 mm
8.6250 in	OE	219.08 mm
2.2500 in	SB	57.15 mm
6.0000 in	EC	152.40 mm
5.7500 in	HD	146.05 mm

ROCKWELL 92NYR36

IDLI GROUP 82340

END YOKE WITH 16 SPLINES
SEE FIGURE 7A

8.2500 in	IE	209.55 mm
8.6250 in	OE	219.08 mm
2.3438 in	SB	59.53 mm
4.1250 in	EC	104.78 mm
2.8750 in	HD	73.03 mm

ROCKWELL 92NYS38-33

IDLI GROUP 82341

END YOKE WITH 16 SPLINES
SEE FIGURE 7A

8.2500 in	IE	209.55 mm
8.6250 in	OE	219.08 mm
2.3438 in	SB	59.53 mm
4.1875 in	EC	106.36 mm
2.8125 in	HD	71.44 mm

ROCKWELL 92NYS38-45

IDLI GROUP 82342

END YOKE WITH 18 INVOLUTED SPLINES
SEE FIGURE 7A

8.2500 in	IE	209.55 mm
8.6250 in	OE	219.08 mm
2.3594 in	SB	59.93 mm
4.4219 in	EC	112.32 mm
3.5000 in	HD	88.90 mm

ROCKWELL 92NYS38-6

IDLI GROUP 82343

END YOKE WITH 18 INVOLUTED SPLINES
SEE FIGURE 7A

8.2500 in	IE	209.55 mm
8.6250 in	OE	219.08 mm
2.3594 in	SB	59.93 mm
5.2188 in	EC	132.56 mm
3.5000 in	HD	88.90 mm

ROCKWELL 92NYS38-43

IDLI GROUP 82344

END YOKE WITH 46 SLANTED SPLINES
SEE FIGURE 7A

8.2500 in	IE	209.55 mm
8.6250 in	OE	219.08 mm
2.3750 in	SB	60.33 mm
4.1875 in	EC	106.36 mm
3.2500 in	HD	82.55 mm

ROCKWELL 92NYS38-52

IDLI GROUP 82345

END YOKE WITH 46 SLANTED SPLINES
slight variation from above group
SEE FIGURE 7A

8.2500 in	IE	209.55 mm
8.6250 in	OE	219.08 mm
2.3750 in	SB	60.33 mm
4.1875 in	EC	106.36 mm
3.2500 in	HD	82.55 mm

ROCKWELL 92NYS38-52A

IDLI GROUP 82346

END YOKE WITH 46 SLANTED SPLINES
slight variation from above group
SEE FIGURE 7A

8.2500 in	IE	209.55 mm
8.6250 in	OE	219.08 mm
2.3750 in	SB	60.33 mm
4.1875 in	EC	106.36 mm
3.2500 in	HD	82.55 mm

ROCKWELL 92NYS38-52A1

IDLI GROUP 82347

END YOKE WITH 46 SLANTED SPLINES
SEE FIGURE 7A

8.2500 in	IE	209.55 mm
8.6250 in	OE	219.08 mm
2.3750 in	SB	60.33 mm
4.2500 in	EC	107.95 mm
3.2500 in	HD	82.55 mm

ROCKWELL 92NYS38-46A

BD = BEARING DIAMETER (outside)　　BW = BEARING WIDTH　　CB = CROSS LENGTH WITH BEARINGS　　CC = CENTER TO CENTER　　CD = CROSS DIAMETER
CL = CROSS LENGTH WITHOUT BEARINGS　　EC = END TO CENTER or FACE TO FACE　　EE = END TO END　　EL = EFFECTIVE LENGTH
HD = HUB DIAMETER (or insert)　　IE = INSIDE OF EARS (or recess)　　OD = OUTSIDE DIAMETER　　OE = OUTSIDE OF EARS　　SB = SPLINE OR BORE SIZE

IDLI GROUP 82348

END YOKE WITH 46 SLANTED SPLINES
SEE FIGURE 7A

8.2500 in	IE	209.55 mm
8.6250 in	OE	219.08 mm
2.3750 in	SB	60.33 mm
4.2500 in	EC	107.95 mm
3.5000 in	HD	88.90 mm

ROCKWELL 92NYS38-55

IDLI GROUP 82349

END YOKE WITH ROUND BORE
SEE FIGURE 7B

8.2500 in	IE	209.55 mm
8.6250 in	OE	219.08 mm
2.3750 in	SB	60.33 mm
6.5000 in	EC	165.10 mm
5.5000 in	HD	139.70 mm

ROCKWELL 92NYR38

IDLI GROUP 82350

END YOKE WITH TAPERED BORE
SEE FIGURE 7F

8.2500 in	IE	209.55 mm
8.6250 in	OE	219.08 mm
2.3906 in	SB	60.72 mm
4.1250 in	EC	104.78 mm
3.5000 in	HD	88.90 mm

ROCKWELL 92NYT38

IDLI GROUP 82351

END YOKE WITH 24 INVOLUTED SPLINES
SEE FIGURE 7A

8.2500 in	IE	209.55 mm
8.6250 in	OE	219.08 mm
2.4063 in	SB	61.12 mm
3.4375 in	EC	87.31 mm
3.1250 in	HD	79.38 mm

ROCKWELL 92NYS38-2

IDLI GROUP 82352

END YOKE WITH 24 SLANTED SPLINES
SEE FIGURE 7A

8.2500 in	IE	209.55 mm
8.6250 in	OE	219.08 mm
2.4063 in	SB	61.12 mm
3.7500 in	EC	95.25 mm
3.1250 in	HD	79.38 mm

ROCKWELL 92NYS38A1

IDLI GROUP 82353

END YOKE WITH 24 SLANTED SPLINES
slight variation from above group
SEE FIGURE 7A

8.2500 in	IE	209.55 mm
8.6250 in	OE	219.08 mm
2.4063 in	SB	61.12 mm
3.7500 in	EC	95.25 mm
3.1250 in	HD	79.38 mm

ROCKWELL 92NYS38A

IDLI GROUP 82354

END YOKE WITH 24 INVOLUTED SPLINES
SEE FIGURE 7A

8.2500 in	IE	209.55 mm
8.6250 in	OE	219.08 mm
2.4063 in	SB	61.12 mm
3.7500 in	EC	95.25 mm
3.1250 in	HD	79.38 mm

BORG-WARNR 14640J
ROCKWELL 92NYS38

IDLI GROUP 82355

END YOKE WITH 24 INVOLUTED SPLINES
slight variation from above group
SEE FIGURE 7A

8.2500 in	IE	209.55 mm
8.6250 in	OE	219.08 mm
2.4063 in	SB	61.12 mm
3.7500 in	EC	95.25 mm
3.1250 in	HD	79.38 mm

BORG-WARNR 17343J
ROCKWELL 92NYS38-1

IDLI GROUP 82356

END YOKE WITH 24 SLANTED SPLINES
SEE FIGURE 7A

8.2500 in	IE	209.55 mm
8.6250 in	OE	219.08 mm
2.4063 in	SB	61.12 mm
3.7500 in	EC	95.25 mm
3.2500 in	HD	82.55 mm

ROCKWELL 92NYS38A2

IDLI GROUP 82357

END YOKE WITH 24 SLANTED SPLINES
slight variation from above group
SEE FIGURE 7A

8.2500 in	IE	209.55 mm
8.6250 in	OE	219.08 mm
2.4063 in	SB	61.12 mm
3.7500 in	EC	95.25 mm
3.2500 in	HD	82.55 mm

ROCKWELL 92NYS38A3

IDLI GROUP 82358

END YOKE WITH 28 INVOLUTED SPLINES
SEE FIGURE 7A

8.2500 in	IE	209.55 mm
8.6250 in	OE	219.08 mm
2.4375 in	SB	61.91 mm
3.8750 in	EC	98.43 mm
3.3750 in	HD	85.73 mm

ROCKWELL 92NYS39-1

IDLI GROUP 82359

END YOKE WITH ROUND BORE
SEE FIGURE 7B

8.2500 in	IE	209.55 mm
8.6250 in	OE	219.08 mm
2.5000 in	SB	63.50 mm
3.8750 in	EC	98.43 mm
4.6250 in	HD	117.48 mm

ROCKWELL 92NYR40-7

IDLI GROUP 82360

END YOKE WITH 10 SPLINES
SEE FIGURE 7A

8.2500 in	IE	209.55 mm
8.6250 in	OE	219.08 mm
2.5000 in	SB	63.50 mm
4.1875 in	EC	106.36 mm
3.7500 in	HD	95.25 mm

BORG-WARNR 7797J
ROCKWELL 92NYS40-4

IDLI GROUP 82361

END YOKE WITH 16 SPLINES
SEE FIGURE 7A

8.2500 in	IE	209.55 mm
8.6250 in	OE	219.08 mm
2.5000 in	SB	63.50 mm
4.2188 in	EC	107.16 mm
3.2500 in	HD	82.55 mm

ROCKWELL 92NYS40-20

IDLI GROUP 82362

END YOKE WITH 10 SPLINES
SEE FIGURE 7A

8.2500 in	IE	209.55 mm
8.6250 in	OE	219.08 mm
2.5000 in	SB	63.50 mm
4.2500 in	EC	107.95 mm
2.9375 in	HD	74.61 mm

ROCKWELL 92NYS40-7

IDLI GROUP 82363

END YOKE WITH 10 SPLINES
SEE FIGURE 7A

8.2500 in	IE	209.55 mm
8.6250 in	OE	219.08 mm
2.5000 in	SB	63.50 mm
4.2500 in	EC	107.95 mm
3.5000 in	HD	88.90 mm

ROCKWELL 92NYS40-17

IDLI GROUP 82364

END YOKE WITH 10 SPLINES
SEE FIGURE 7A

8.2500 in	IE	209.55 mm
8.6250 in	OE	219.08 mm
2.5000 in	SB	63.50 mm
4.2500 in	EC	107.95 mm
3.6250 in	HD	92.08 mm

ROCKWELL 92NYS40-32

IDLI GROUP 82365

END YOKE WITH ROUND BORE
SEE FIGURE 7B

8.2500 in	IE	209.55 mm
8.6250 in	OE	219.08 mm
2.5000 in	SB	63.50 mm
4.3125 in	EC	109.54 mm
4.5625 in	HD	115.89 mm

ROCKWELL 92NYR40-3

IDLI GROUP 82366

END YOKE WITH 10 SPLINES
SEE FIGURE 7A

8.2500 in	IE	209.55 mm
8.6250 in	OE	219.08 mm
2.5000 in	SB	63.50 mm
5.0000 in	EC	127.00 mm
3.5000 in	HD	88.90 mm

BORG-WARNR 19125J
BORG-WARNR 24027J
ROCKWELL 92NYS40-16

IDLI GROUP 82367

END YOKE WITH ROUND BORE
SEE FIGURE 7B

8.2500 in	IE	209.55 mm
8.6250 in	OE	219.08 mm
2.5000 in	SB	63.50 mm
6.5000 in	EC	165.10 mm
5.7500 in	HD	146.05 mm

ROCKWELL 92NYR40

IDLI GROUP 82368

END YOKE WITH 24 INVOLUTED SPLINES
SEE FIGURE 7A

8.2500 in	IE	209.55 mm
8.6250 in	OE	219.08 mm
2.5000 in	SB	63.50 mm
8.5313 in	EC	216.70 mm
3.5469 in	HD	90.09 mm

BORG-WARNR 121-9028
ROCKWELL 92NYS40-18

IDLI GROUP 82369

END YOKE WITH 20 INVOLUTED SPLINES
SEE FIGURE 7A

8.2500 in	IE	209.55 mm
8.6250 in	OE	219.08 mm
2.5625 in	SB	65.09 mm
5.1719 in	EC	131.37 mm
3.5000 in	HD	88.90 mm

ROCKWELL 92NYS42-16

IDLI GROUP 82370

END YOKE WITH 25 INVOLUTED SPLINES
SEE FIGURE 7A

8.2500 in	IE	209.55 mm
8.6250 in	OE	219.08 mm
2.6250 in	SB	66.68 mm
4.1250 in	EC	104.78 mm
3.6250 in	HD	92.08 mm

ROCKWELL 92NYS42-4

IDLI GROUP 82371

END YOKE WITH ROUND BORE
SEE FIGURE 7B

8.2500 in	IE	209.55 mm
8.6250 in	OE	219.08 mm
2.6250 in	SB	66.68 mm
5.3125 in	EC	134.94 mm
5.7500 in	HD	146.05 mm

ROCKWELL 92NYR42-3

ENGINEERING CATALOGS MUST BE CONSULTED FOR DETAILS NOT INCLUDED
IN THIS GUIDE. SPECIFIC DESIGNS, MATERIAL CONTENT, TOLERANCES
LUBE FITTINGS AND OTHER DIMENSIONS ARE INTENTIONALLY OMITTED HERE.

CAUTION: BE SURE TO REFER TO ENGINEERING CATALOGS FOR SPECIAL APPLICATIONS THAT REQUIRE SPECIFIC MATERIAL CONTENT, TOLERANCES, ETC. SEE FOOTNOTE.

IDLI GROUP 82372

END YOKE WITH ROUND BORE
SEE FIGURE 7B

8.2500 in	IE	209.55 mm
8.6250 in	OE	219.08 mm
2.6250 in	SB	66.68 mm
6.0000 in	EC	152.40 mm
5.7500 in	HD	146.05 mm

ROCKWELL 92NYR42-2

IDLI GROUP 82373

END YOKE WITH ROUND BORE
SEE FIGURE 7B

8.2500 in	IE	209.55 mm
8.6250 in	OE	219.08 mm
2.6250 in	SB	66.68 mm
6.1250 in	EC	155.58 mm
5.5000 in	HD	139.70 mm

ROCKWELL 92NYR42

IDLI GROUP 82374

END YOKE WITH ROUND BORE
SEE FIGURE 7B

8.2500 in	IE	209.55 mm
8.6250 in	OE	219.08 mm
2.6250 in	SB	66.68 mm
6.5625 in	EC	166.69 mm
5.7500 in	HD	146.05 mm

ROCKWELL 92NYR42-1

IDLI GROUP 82375

END YOKE WITH 10 SPLINES
SEE FIGURE 7A

8.2500 in	IE	209.55 mm
8.6250 in	OE	219.08 mm
2.7500 in	SB	69.85 mm
4.3750 in	EC	111.13 mm
3.3750 in	HD	85.73 mm

ROCKWELL 92NYS44-18

IDLI GROUP 82376

END YOKE WITH 10 SPLINES
SEE FIGURE 7A

8.2500 in	IE	209.55 mm
8.6250 in	OE	219.08 mm
2.7500 in	SB	69.85 mm
4.5625 in	EC	115.89 mm
3.7500 in	HD	95.25 mm

BORG-WARNR 28626J
ROCKWELL 92NYS44-21

IDLI GROUP 82377

END YOKE WITH 10 SPLINES
SEE FIGURE 7A

8.2500 in	IE	209.55 mm
8.6250 in	OE	219.08 mm
2.7500 in	SB	69.85 mm
4.7188 in	EC	119.86 mm
3.7500 in	HD	95.25 mm

ROCKWELL 92NYS44-20

IDLI GROUP 82378

END YOKE WITH 21 INVOLUTED SPLINES
SEE FIGURE 7A

8.2500 in	IE	209.55 mm
8.6250 in	OE	219.08 mm
2.7500 in	SB	69.85 mm
5.7500 in	EC	146.05 mm
3.5000 in	HD	88.90 mm

ROCKWELL 92NYS44-11

IDLI GROUP 82379

END YOKE WITH ROUND BORE
SEE FIGURE 7B

8.2500 in	IE	209.55 mm
8.6250 in	OE	219.08 mm
2.7500 in	SB	69.85 mm
6.3750 in	EC	161.93 mm
5.6250 in	HD	142.88 mm

ROCKWELL 92NYR44

IDLI GROUP 82380

END YOKE WITH ROUND BORE
SEE FIGURE 7B

8.2500 in	IE	209.55 mm
8.6250 in	OE	219.08 mm
2.7500 in	SB	69.85 mm
6.3750 in	EC	161.93 mm
5.7500 in	HD	146.05 mm

ROCKWELL 92NYR44-6

IDLI GROUP 82381

END YOKE WITH 43 INVOLUTED SPLINES
SEE FIGURE 7A

8.2500 in	IE	209.55 mm
8.6250 in	OE	219.08 mm
2.7500 in	SB	69.85 mm
8.5313 in	EC	216.70 mm
3.5469 in	HD	90.09 mm

ROCKWELL 92NYS44-14

IDLI GROUP 82382

END YOKE WITH 21 INVOLUTED SPLINES
SEE FIGURE 7A

8.2500 in	IE	209.55 mm
8.6250 in	OE	219.08 mm
2.8438 in	SB	72.23 mm
4.1875 in	EC	106.36 mm
4.0000 in	HD	101.60 mm

ROCKWELL 92NYS44-19

IDLI GROUP 82383

END YOKE WITH 21 INVOLUTED SPLINES
SEE FIGURE 7A

8.2500 in	IE	209.55 mm
8.6250 in	OE	219.08 mm
2.8594 in	SB	72.63 mm
3.9375 in	EC	100.01 mm
3.5000 in	HD	88.90 mm

BORG-WARNR 16776J
ROCKWELL 92NYS46

IDLI GROUP 82384

END YOKE WITH 21 INVOLUTED SPLINES
slight variation from above group
SEE FIGURE 7A

8.2500 in	IE	209.55 mm
8.6250 in	OE	219.08 mm
2.8594 in	SB	72.63 mm
3.9375 in	EC	100.01 mm
3.5000 in	HD	88.90 mm

ROCKWELL 92NYS46-1

IDLI GROUP 82385

END YOKE WITH 21 INVOLUTED SPLINES
SEE FIGURE 7A

8.2500 in	IE	209.55 mm
8.6250 in	OE	219.08 mm
2.8594 in	SB	72.63 mm
4.1875 in	EC	106.36 mm
4.0000 in	HD	101.60 mm

ROCKWELL 92NYS44

IDLI GROUP 82386

END YOKE WITH 21 INVOLUTED SPLINES
slight variation from above group
SEE FIGURE 7A

8.2500 in	IE	209.55 mm
8.6250 in	OE	219.08 mm
2.8594 in	SB	72.63 mm
4.1875 in	EC	106.36 mm
4.0000 in	HD	101.60 mm

ROCKWELL 92NYS44-8

IDLI GROUP 82387

END YOKE WITH 21 INVOLUTED SPLINES
SEE FIGURE 7A

8.2500 in	IE	209.55 mm
8.6250 in	OE	219.08 mm
2.8594 in	SB	72.63 mm
4.2500 in	EC	107.95 mm
3.5000 in	HD	88.90 mm

ROCKWELL 92NYS44-2

IDLI GROUP 82388

END YOKE WITH ROUND BORE
SEE FIGURE 7B

8.2500 in	IE	209.55 mm
8.6250 in	OE	219.08 mm
2.8750 in	SB	73.03 mm
4.2500 in	EC	107.95 mm
4.3750 in	HD	111.13 mm

ROCKWELL 92NYR46-2

IDLI GROUP 82389

END YOKE WITH ROUND BORE
SEE FIGURE 7B

8.2500 in	IE	209.55 mm
8.6250 in	OE	219.08 mm
2.8750 in	SB	73.03 mm
6.0000 in	EC	152.40 mm
5.7500 in	HD	146.05 mm

ROCKWELL 92NYR46-3

IDLI GROUP 82390

END YOKE WITH ROUND BORE
SEE FIGURE 7B

8.2500 in	IE	209.55 mm
8.6250 in	OE	219.08 mm
2.8750 in	SB	73.03 mm
6.5000 in	EC	165.10 mm
5.5000 in	HD	139.70 mm

ROCKWELL 92NYR46-1

IDLI GROUP 82391

END YOKE WITH ROUND BORE
SEE FIGURE 7B

8.2500 in	IE	209.55 mm
8.6250 in	OE	219.08 mm
2.9375 in	SB	74.61 mm
6.1250 in	EC	155.58 mm
5.6250 in	HD	142.88 mm

ROCKWELL 92NYR47

IDLI GROUP 82392

END YOKE WITH ROUND BORE
SEE FIGURE 7B

8.2500 in	IE	209.55 mm
8.6250 in	OE	219.08 mm
2.9375 in	SB	74.61 mm
6.5000 in	EC	165.10 mm
5.7500 in	HD	146.05 mm

ROCKWELL 92NYR47-1

IDLI GROUP 82393

END YOKE WITH 10 SPLINES
SEE FIGURE 7A

8.2500 in	IE	209.55 mm
8.6250 in	OE	219.08 mm
3.0000 in	SB	76.20 mm
2.5000 in	EC	63.50 mm
4.6875 in	HD	119.06 mm

ROCKWELL 92NYS48-5

IDLI GROUP 82394

END YOKE WITH 10 SPLINES
SEE FIGURE 7A

8.2500 in	IE	209.55 mm
8.6250 in	OE	219.08 mm
3.0000 in	SB	76.20 mm
3.0625 in	EC	77.79 mm
4.6875 in	HD	119.06 mm

ROCKWELL 92NYS48-6

IDLI GROUP 82395

END YOKE WITH 10 SPLINES
SEE FIGURE 7A

8.2500 in	IE	209.55 mm
8.6250 in	OE	219.08 mm
3.0000 in	SB	76.20 mm
3.2500 in	EC	82.55 mm
4.6250 in	HD	117.48 mm

ROCKWELL 92NYS48-3

BD=BEARING DIAMETER (outside)   **BW**=BEARING WIDTH   **CB**=CROSS LENGTH WITH BEARINGS   **CC**=CENTER TO CENTER   **CD**=CROSS DIAMETER
CL=CROSS LENGTH WITHOUT BEARINGS   **EC**=END TO CENTER or FACE TO FACE   **EE**=END TO END   **EL**=EFFECTIVE LENGTH
HD=HUB DIAMETER (or insert)   **IE**=INSIDE OF EARS (or recess)   **OD**=OUTSIDE DIAMETER   **OE**=OUTSIDE OF EARS   **SB**=SPLINE OR BORE SIZE

IDLI GROUP 82396

END YOKE WITH 10 SPLINES
slight variation from above group
SEE FIGURE 7A

8.2500 in	IE	209.55 mm
8.6250 in	OE	219.08 mm
3.0000 in	SB	76.20 mm
3.2500 in	EC	82.55 mm
4.6250 in	HD	117.48 mm

ROCKWELL 92NYS48-41

IDLI GROUP 82397

END YOKE WITH 29 INVOLUTED SPLINES
SEE FIGURE 7A

8.2500 in	IE	209.55 mm
8.6250 in	OE	219.08 mm
3.0000 in	SB	76.20 mm
3.6875 in	EC	93.66 mm
4.0000 in	HD	101.60 mm

ROCKWELL 92NYS48-2

IDLI GROUP 82398

END YOKE WITH 23 INVOLUTED SPLINES
SEE FIGURE 7A

8.2500 in	IE	209.55 mm
8.6250 in	OE	219.08 mm
3.0000 in	SB	76.20 mm
3.8125 in	EC	96.84 mm
4.5625 in	HD	115.89 mm

ROCKWELL 92NYS48-26

IDLI GROUP 82399

END YOKE WITH 29 INVOLUTED SPLINES
SEE FIGURE 7A

8.2500 in	IE	209.55 mm
8.6250 in	OE	219.08 mm
3.0000 in	SB	76.20 mm
4.1250 in	EC	104.78 mm
4.0000 in	HD	101.60 mm

ROCKWELL 92NYS48-1

IDLI GROUP 82400

END YOKE WITH 6 SPLINES
SEE FIGURE 7A

8.2500 in	IE	209.55 mm
8.6250 in	OE	219.08 mm
3.0000 in	SB	76.20 mm
4.1250 in	EC	104.78 mm
4.5000 in	HD	114.30 mm

ROCKWELL 92NYS48-9

IDLI GROUP 82401

END YOKE WITH 10 SPLINES
SEE FIGURE 7A

8.2500 in	IE	209.55 mm
8.6250 in	OE	219.08 mm
3.0000 in	SB	76.20 mm
4.1250 in	EC	104.78 mm
4.5000 in	HD	114.30 mm

ROCKWELL 92NYS48-39

IDLI GROUP 82402

END YOKE WITH 10 SPLINES
slight variation from above group
SEE FIGURE 7A

8.2500 in	IE	209.55 mm
8.6250 in	OE	219.08 mm
3.0000 in	SB	76.20 mm
4.1250 in	EC	104.78 mm
4.5000 in	HD	114.30 mm

ROCKWELL 92NYS48-22

IDLI GROUP 82403

END YOKE WITH 10 SPLINES
slight variation from above group
SEE FIGURE 7A

8.2500 in	IE	209.55 mm
8.6250 in	OE	219.08 mm
3.0000 in	SB	76.20 mm
4.1250 in	EC	104.78 mm
4.5000 in	HD	114.30 mm

ROCKWELL 92NYS48-36

IDLI GROUP 82404

END YOKE WITH 6 SPLINES
SEE FIGURE 7A

8.2500 in	IE	209.55 mm
8.6250 in	OE	219.08 mm
3.0000 in	SB	76.20 mm
4.2500 in	EC	107.95 mm
4.5000 in	HD	114.30 mm

ROCKWELL 92NYS48-20

IDLI GROUP 82405

END YOKE WITH ROUND BORE
SEE FIGURE 7B

8.2500 in	IE	209.55 mm
8.6250 in	OE	219.08 mm
3.0000 in	SB	76.20 mm
4.8125 in	EC	122.24 mm
5.7500 in	HD	146.05 mm

ROCKWELL 92NYR48-4

IDLI GROUP 82406

END YOKE WITH ROUND BORE
SEE FIGURE 7B

8.2500 in	IE	209.55 mm
8.6250 in	OE	219.08 mm
3.0000 in	SB	76.20 mm
4.8750 in	EC	123.83 mm
5.7500 in	HD	146.05 mm

ROCKWELL 92NYR48-8

IDLI GROUP 82407

END YOKE WITH ROUND BORE
slight variation from above group
SEE FIGURE 7B

8.2500 in	IE	209.55 mm
8.6250 in	OE	219.08 mm
3.0000 in	SB	76.20 mm
4.8750 in	EC	123.83 mm
5.7500 in	HD	146.05 mm

ROCKWELL 92NYR48-5

IDLI GROUP 82408

END YOKE WITH 10 SPLINES
SEE FIGURE 7A

8.2500 in	IE	209.55 mm
8.6250 in	OE	219.08 mm
3.0000 in	SB	76.20 mm
5.1875 in	EC	131.76 mm
4.0000 in	HD	101.60 mm

ROCKWELL 92NYS48-7

IDLI GROUP 82409

END YOKE WITH ROUND BORE
SEE FIGURE 7B

8.2500 in	IE	209.55 mm
8.6250 in	OE	219.08 mm
3.0000 in	SB	76.20 mm
5.5000 in	EC	139.70 mm
5.7500 in	HD	146.05 mm

ROCKWELL 92NYR48-7

IDLI GROUP 82410

END YOKE WITH ROUND BORE
SEE FIGURE 7B

8.2500 in	IE	209.55 mm
8.6250 in	OE	219.08 mm
3.0000 in	SB	76.20 mm
5.8750 in	EC	149.23 mm
5.7500 in	HD	146.05 mm

ROCKWELL 92NYR48-9

IDLI GROUP 82411

END YOKE WITH 10 SPLINES
SEE FIGURE 7A

8.2500 in	IE	209.55 mm
8.6250 in	OE	219.08 mm
3.0000 in	SB	76.20 mm
6.3750 in	EC	161.93 mm
4.0000 in	HD	101.60 mm

ROCKWELL 92NYS48

IDLI GROUP 82412

END YOKE WITH 10 SPLINES
SEE FIGURE 7A

8.2500 in	IE	209.55 mm
8.6250 in	OE	219.08 mm
3.0000 in	SB	76.20 mm
6.5625 in	EC	166.69 mm
4.0000 in	HD	101.60 mm

ROCKWELL 92NYS48-8

IDLI GROUP 82413

END YOKE WITH ROUND BORE
SEE FIGURE 7B

8.2500 in	IE	209.55 mm
8.6250 in	OE	219.08 mm
3.0000 in	SB	76.20 mm
6.5625 in	EC	166.69 mm
5.7500 in	HD	146.05 mm

ROCKWELL 92NYR48-2

IDLI GROUP 82414

END YOKE WITH ROUND BORE
SEE FIGURE 7B

8.2500 in	IE	209.55 mm
8.6250 in	OE	219.08 mm
3.1250 in	SB	79.38 mm
5.0000 in	EC	127.00 mm
4.8750 in	HD	123.83 mm

ROCKWELL 92NYR50-2

IDLI GROUP 82415

END YOKE WITH ROUND BORE
SEE FIGURE 7B

8.2500 in	IE	209.55 mm
8.6250 in	OE	219.08 mm
3.1250 in	SB	79.38 mm
5.0000 in	EC	127.00 mm
5.7500 in	HD	146.05 mm

ROCKWELL 92NYR50-1

IDLI GROUP 82416

END YOKE WITH ROUND BORE
SEE FIGURE 7B

8.2500 in	IE	209.55 mm
8.6250 in	OE	219.08 mm
3.1250 in	SB	79.38 mm
5.5000 in	EC	139.70 mm
5.7500 in	HD	146.05 mm

ROCKWELL 92NYR50-3

IDLI GROUP 82417

END YOKE WITH ROUND BORE
SEE FIGURE 7B

8.2500 in	IE	209.55 mm
8.6250 in	OE	219.08 mm
3.1250 in	SB	79.38 mm
6.5000 in	EC	165.10 mm
5.7500 in	HD	146.05 mm

ROCKWELL 92NYR50-4

IDLI GROUP 82418

END YOKE WITH 50 INVOLUTED SPLINES
SEE FIGURE 7A

8.2500 in	IE	209.55 mm
8.6250 in	OE	219.08 mm
3.1875 in	SB	80.96 mm
4.0000 in	EC	101.60 mm
3.7188 in	HD	94.46 mm

ROCKWELL 92NYS51

IDLI GROUP 82419

END YOKE WITH ROUND BORE
SEE FIGURE 7B

8.2500 in	IE	209.55 mm
8.6250 in	OE	219.08 mm
3.2500 in	SB	82.55 mm
5.0000 in	EC	127.00 mm
4.8750 in	HD	123.83 mm

ROCKWELL 92NYR52-1

ENGINEERING CATALOGS MUST BE CONSULTED FOR DETAILS NOT INCLUDED
IN THIS GUIDE. SPECIFIC DESIGNS, MATERIAL CONTENT, TOLERANCES
LUBE FITTINGS AND OTHER DIMENSIONS ARE INTENTIONALLY OMITTED HERE.

CAUTION: BE SURE TO REFER TO ENGINEERING CATALOGS FOR SPECIAL APPLICATIONS THAT REQUIRE SPECIFIC MATERIAL CONTENT, TOLERANCES, ETC. SEE FOOTNOTE.

IDLI GROUP 82420

END YOKE WITH ROUND BORE
SEE FIGURE 7B

8.2500 in	IE	209.55 mm
8.6250 in	OE	219.08 mm
3.2500 in	SB	82.55 mm
5.5000 in	EC	139.70 mm
5.7500 in	HD	146.05 mm

ROCKWELL 92NYR52-3

IDLI GROUP 82421

END YOKE WITH TAPERED BORE
SEE FIGURE 7F

8.2500 in	IE	209.55 mm
8.6250 in	OE	219.08 mm
3.2500 in	SB	82.55 mm
6.3750 in	EC	161.93 mm
8.6250 in	HD	219.08 mm

ROCKWELL 92NYT52

IDLI GROUP 82422

END YOKE WITH ROUND BORE
SEE FIGURE 7B

8.2500 in	IE	209.55 mm
8.6250 in	OE	219.08 mm
3.2500 in	SB	82.55 mm
6.5000 in	EC	165.10 mm
5.7500 in	HD	146.05 mm

ROCKWELL 92NYR52-5

IDLI GROUP 82423

END YOKE WITH ROUND BORE
SEE FIGURE 7B

8.2500 in	IE	209.55 mm
8.6250 in	OE	219.08 mm
3.3750 in	SB	85.73 mm
5.8125 in	EC	147.64 mm
5.7500 in	HD	146.05 mm

ROCKWELL 92NYR54-3

IDLI GROUP 82424

END YOKE WITH ROUND BORE
SEE FIGURE 7B

8.2500 in	IE	209.55 mm
8.6250 in	OE	219.08 mm
3.3750 in	SB	85.73 mm
6.3750 in	EC	161.93 mm
5.7500 in	HD	146.05 mm

ROCKWELL 92NYR54-1

IDLI GROUP 82425

END YOKE WITH ROUND BORE
SEE FIGURE 7B

8.2500 in	IE	209.55 mm
8.6250 in	OE	219.08 mm
3.3750 in	SB	85.73 mm
6.5000 in	EC	165.10 mm
5.6250 in	HD	142.88 mm

ROCKWELL 92NYR54-5

IDLI GROUP 82426

END YOKE WITH ROUND BORE
SEE FIGURE 7B

8.2500 in	IE	209.55 mm
8.6250 in	OE	219.08 mm
3.3750 in	SB	85.73 mm
6.5000 in	EC	165.10 mm
5.7500 in	HD	146.05 mm

ROCKWELL 92NYR54-6

IDLI GROUP 82427

END YOKE WITH ROUND BORE
SEE FIGURE 7B

8.2500 in	IE	209.55 mm
8.6250 in	OE	219.08 mm
3.3750 in	SB	85.73 mm
6.5625 in	EC	166.69 mm
5.7500 in	HD	146.05 mm

ROCKWELL 92NYR54

IDLI GROUP 82428

END YOKE WITH ROUND BORE
SEE FIGURE 7B

8.2500 in	IE	209.55 mm
8.6250 in	OE	219.08 mm
3.4375 in	SB	87.31 mm
5.5000 in	EC	139.70 mm
5.7500 in	HD	146.05 mm

ROCKWELL 92NYR55-1

IDLI GROUP 82429

END YOKE WITH ROUND BORE
slight variation from above group
SEE FIGURE 7B

8.2500 in	IE	209.55 mm
8.6250 in	OE	219.08 mm
3.4375 in	SB	87.31 mm
5.5000 in	EC	139.70 mm
5.7500 in	HD	146.05 mm

ROCKWELL 92NYR55

IDLI GROUP 82430

END YOKE WITH ROUND BORE
SEE FIGURE 7B

8.2500 in	IE	209.55 mm
8.6250 in	OE	219.08 mm
3.4375 in	SB	87.31 mm
5.8750 in	EC	149.23 mm
5.7500 in	HD	146.05 mm

ROCKWELL 92NYR55-2

IDLI GROUP 82431

END YOKE WITH ROUND BORE
SEE FIGURE 7B

8.2500 in	IE	209.55 mm
8.6250 in	OE	219.08 mm
3.4375 in	SB	87.31 mm
6.3750 in	EC	161.93 mm
5.7500 in	HD	146.05 mm

ROCKWELL 92NYR55-3

IDLI GROUP 82432

END YOKE WITH ROUND BORE
SEE FIGURE 7B

8.2500 in	IE	209.55 mm
8.6250 in	OE	219.08 mm
3.4375 in	SB	87.31 mm
6.5000 in	EC	165.10 mm
5.7500 in	HD	146.05 mm

ROCKWELL 92NYR55-4

IDLI GROUP 82433

END YOKE WITH ROUND BORE
slight variation from above group
SEE FIGURE 7B

8.2500 in	IE	209.55 mm
8.6250 in	OE	219.08 mm
3.4375 in	SB	87.31 mm
6.5000 in	EC	165.10 mm
5.7500 in	HD	146.05 mm

ROCKWELL 92NYR55-5

IDLI GROUP 82434

END YOKE WITH ROUND BORE
SEE FIGURE 7B

8.2500 in	IE	209.55 mm
8.6250 in	OE	219.08 mm
3.5000 in	SB	88.90 mm
5.0000 in	EC	127.00 mm
5.2500 in	HD	133.35 mm

ROCKWELL 92NYR56-1

IDLI GROUP 82435

END YOKE WITH ROUND BORE
SEE FIGURE 7B

8.2500 in	IE	209.55 mm
8.6250 in	OE	219.08 mm
3.5000 in	SB	88.90 mm
5.5000 in	EC	139.70 mm
5.6250 in	HD	142.88 mm

ROCKWELL 92NYR56-5

IDLI GROUP 82436

END YOKE WITH ROUND BORE
SEE FIGURE 7B

8.2500 in	IE	209.55 mm
8.6250 in	OE	219.08 mm
3.5000 in	SB	88.90 mm
5.5000 in	EC	139.70 mm
5.7500 in	HD	146.05 mm

ROCKWELL 92NYR56-11

IDLI GROUP 82437

END YOKE WITH ROUND BORE
slight variation from above group
SEE FIGURE 7B

8.2500 in	IE	209.55 mm
8.6250 in	OE	219.08 mm
3.5000 in	SB	88.90 mm
5.5000 in	EC	139.70 mm
5.7500 in	HD	146.05 mm

ROCKWELL 92NYR56-13

IDLI GROUP 82438

END YOKE WITH ROUND BORE
SEE FIGURE 7B

8.2500 in	IE	209.55 mm
8.6250 in	OE	219.08 mm
3.5000 in	SB	88.90 mm
5.5625 in	EC	141.29 mm
5.7500 in	HD	146.05 mm

ROCKWELL 92NYR56-4

IDLI GROUP 82439

END YOKE WITH ROUND BORE
SEE FIGURE 7B

8.2500 in	IE	209.55 mm
8.6250 in	OE	219.08 mm
3.5000 in	SB	88.90 mm
6.3750 in	EC	161.93 mm
5.7500 in	HD	146.05 mm

ROCKWELL 92NYR56-2

IDLI GROUP 82440

END YOKE WITH ROUND BORE
slight variation from above group
SEE FIGURE 7B

8.2500 in	IE	209.55 mm
8.6250 in	OE	219.08 mm
3.5000 in	SB	88.90 mm
6.3750 in	EC	161.93 mm
5.7500 in	HD	146.05 mm

ROCKWELL 92NYR56-3

IDLI GROUP 82441

END YOKE WITH ROUND BORE
SEE FIGURE 7B

8.2500 in	IE	209.55 mm
8.6250 in	OE	219.08 mm
3.5000 in	SB	88.90 mm
6.5000 in	EC	165.10 mm
5.6250 in	HD	142.88 mm

ROCKWELL 92NYR56-10

IDLI GROUP 82442

END YOKE WITH ROUND BORE
SEE FIGURE 7B

8.2500 in	IE	209.55 mm
8.6250 in	OE	219.08 mm
3.5000 in	SB	88.90 mm
6.5000 in	EC	165.10 mm
5.7500 in	HD	146.05 mm

ROCKWELL 92NYR56-15

IDLI GROUP 82443

END YOKE WITH ROUND BORE
slight variation from above group
SEE FIGURE 7B

8.2500 in	IE	209.55 mm
8.6250 in	OE	219.08 mm
3.5000 in	SB	88.90 mm
6.5000 in	EC	165.10 mm
5.7500 in	HD	146.05 mm

ROCKWELL 92NYR56

BD – BEARING DIAMETER (outside)   **BW** = BEARING WIDTH   **CB** = CROSS LENGTH WITH BEARINGS   **CC** = CENTER TO CENTER   **CD** = CROSS DIAMETER
CL = CROSS LENGTH WITHOUT BEARINGS   **EC** = END TO CENTER or FACE TO FACE   **EE** = END TO END   **EL** = EFFECTIVE LENGTH
HD = HUB DIAMETER (or insert)   **IE** = INSIDE OF EARS (or recess)   **OD** = OUTSIDE DIAMETER   **OE** = OUTSIDE OF EARS   **SB** = SPLINE OR BORE SIZE

IDLI GROUP 82444

END YOKE WITH ROUND BORE
slight variation from above group
SEE FIGURE 7B

8.2500 in	IE	209.55 mm
8.6250 in	OE	219.08 mm
3.5000 in	SB	88.90 mm
6.5000 in	EC	165.10 mm
5.7500 in	HD	146.05 mm

ROCKWELL 92NYR56-6

IDLI GROUP 82445

END YOKE WITH ROUND BORE
SEE FIGURE 7B

8.2500 in	IE	209.55 mm
8.6250 in	OE	219.08 mm
3.5000 in	SB	88.90 mm
6.6250 in	EC	168.28 mm
5.7500 in	HD	146.05 mm

ROCKWELL 92NYR56-12

IDLI GROUP 82446

END YOKE WITH ROUND BORE
SEE FIGURE 7B

8.2500 in	IE	209.55 mm
8.6250 in	OE	219.08 mm
3.6250 in	SB	92.08 mm
5.5625 in	EC	141.29 mm
5.7500 in	HD	146.05 mm

ROCKWELL 92NYR58

IDLI GROUP 82447

END YOKE WITH ROUND BORE
SEE FIGURE 7B

8.2500 in	IE	209.55 mm
8.6250 in	OE	219.08 mm
3.6250 in	SB	92.08 mm
6.5000 in	EC	165.10 mm
5.7500 in	HD	146.05 mm

ROCKWELL 92NYR58-3

IDLI GROUP 82448

END YOKE WITH ROUND BORE
slight variation from above group
SEE FIGURE 7B

8.2500 in	IE	209.55 mm
8.6250 in	OE	219.08 mm
3.6250 in	SB	92.08 mm
6.5000 in	EC	165.10 mm
5.7500 in	HD	146.05 mm

ROCKWELL 92NYR58-1

IDLI GROUP 82449

END YOKE WITH ROUND BORE
SEE FIGURE 7B

8.2500 in	IE	209.55 mm
8.6250 in	OE	219.08 mm
3.7500 in	SB	95.25 mm
6.5000 in	EC	165.10 mm
5.7500 in	HD	146.05 mm

ROCKWELL 92NYR60-1

IDLI GROUP 82450

END YOKE WITH ROUND BORE
slight variation from above group
SEE FIGURE 7B

8.2500 in	IE	209.55 mm
8.6250 in	OE	219.08 mm
3.7500 in	SB	95.25 mm
6.5000 in	EC	165.10 mm
5.7500 in	HD	146.05 mm

ROCKWELL 92NYR60

IDLI GROUP 82451

END YOKE WITH ROUND BORE
slight variation from above group
SEE FIGURE 7B

8.2500 in	IE	209.55 mm
8.6250 in	OE	219.08 mm
3.7500 in	SB	95.25 mm
6.5000 in	EC	165.10 mm
5.7500 in	HD	146.05 mm

ROCKWELL 92NYR60-2

IDLI GROUP 82452

END YOKE WITH ROUND BORE
SEE FIGURE 7B

8.2500 in	IE	209.55 mm
8.6250 in	OE	219.08 mm
3.8750 in	SB	98.43 mm
5.2500 in	EC	133.35 mm
5.7500 in	HD	146.05 mm

ROCKWELL 92NYR62-6

IDLI GROUP 82453

END YOKE WITH ROUND BORE
SEE FIGURE 7B

8.2500 in	IE	209.55 mm
8.6250 in	OE	219.08 mm
3.8750 in	SB	98.43 mm
6.3750 in	EC	161.93 mm
5.7500 in	HD	146.05 mm

ROCKWELL 92NYR62-1

IDLI GROUP 82454

END YOKE WITH ROUND BORE
SEE FIGURE 7B

8.2500 in	IE	209.55 mm
8.6250 in	OE	219.08 mm
3.8750 in	SB	98.43 mm
6.5000 in	EC	165.10 mm
5.7500 in	HD	146.05 mm

ROCKWELL 92NYR62-7

IDLI GROUP 82455

END YOKE WITH ROUND BORE
slight variation from above group
SEE FIGURE 7B

8.2500 in	IE	209.55 mm
8.6250 in	OE	219.08 mm
3.8750 in	SB	98.43 mm
6.5000 in	EC	165.10 mm
5.7500 in	HD	146.05 mm

ROCKWELL 92NYR62-4

IDLI GROUP 82456

END YOKE WITH ROUND BORE
SEE FIGURE 7B

8.2500 in	IE	209.55 mm
8.6250 in	OE	219.08 mm
3.8750 in	SB	98.43 mm
6.5625 in	EC	166.69 mm
5.7500 in	HD	146.05 mm

ROCKWELL 92NYR62-5

IDLI GROUP 82457

END YOKE WITH ROUND BORE
slight variation from above group
SEE FIGURE 7B

8.2500 in	IE	209.55 mm
8.6250 in	OE	219.08 mm
3.8750 in	SB	98.43 mm
6.5625 in	EC	166.69 mm
5.7500 in	HD	146.05 mm

ROCKWELL 92NYR62-3

IDLI GROUP 82458

END YOKE WITH ROUND BORE
SEE FIGURE 7B

8.2500 in	IE	209.55 mm
8.6250 in	OE	219.08 mm
3.9375 in	SB	100.01 mm
5.5000 in	EC	139.70 mm
5.7500 in	HD	146.05 mm

ROCKWELL 92NYR63-2

IDLI GROUP 82459

END YOKE WITH ROUND BORE
slight variation from above group
SEE FIGURE 7B

8.2500 in	IE	209.55 mm
8.6250 in	OE	219.08 mm
3.9375 in	SB	100.01 mm
5.5000 in	EC	139.70 mm
5.7500 in	HD	146.05 mm

ROCKWELL 92NYR63-1

IDLI GROUP 82460

END YOKE WITH ROUND BORE
slight variation from above group
SEE FIGURE 7B

8.2500 in	IE	209.55 mm
8.6250 in	OE	219.08 mm
3.9375 in	SB	100.01 mm
5.5000 in	EC	139.70 mm
5.7500 in	HD	146.05 mm

ROCKWELL 92NYR63-3

IDLI GROUP 82461

END YOKE WITH ROUND BORE
SEE FIGURE 7B

8.2500 in	IE	209.55 mm
8.6250 in	OE	219.08 mm
4.0000 in	SB	101.60 mm
5.5000 in	EC	139.70 mm
5.7500 in	HD	146.05 mm

ROCKWELL 92NYR64-15

IDLI GROUP 82462

END YOKE WITH ROUND BORE
SEE FIGURE 7B

8.2500 in	IE	209.55 mm
8.6250 in	OE	219.08 mm
4.0000 in	SB	101.60 mm
6.0000 in	EC	152.40 mm
5.7500 in	HD	146.05 mm

ROCKWELL 92NYR64-2

IDLI GROUP 82463

END YOKE WITH ROUND BORE
SEE FIGURE 7B

8.2500 in	IE	209.55 mm
8.6250 in	OE	219.08 mm
4.0000 in	SB	101.60 mm
6.1250 in	EC	155.58 mm
5.7500 in	HD	146.05 mm

ROCKWELL 92NYR64-12

IDLI GROUP 82464

END YOKE WITH ROUND BORE
SEE FIGURE 7B

8.2500 in	IE	209.55 mm
8.6250 in	OE	219.08 mm
4.0000 in	SB	101.60 mm
6.5000 in	EC	165.10 mm
5.7500 in	HD	146.05 mm

ROCKWELL 92NYR64-6

IDLI GROUP 82465

END YOKE WITH ROUND BORE
slight variation from above group
SEE FIGURE 7B

8.2500 in	IE	209.55 mm
8.6250 in	OE	219.08 mm
4.0000 in	SB	101.60 mm
6.5000 in	EC	165.10 mm
5.7500 in	HD	146.05 mm

ROCKWELL 92NYR64-4

IDLI GROUP 82466

END YOKE WITH ROUND BORE
slight variation from above group
SEE FIGURE 7B

8.2500 in	IE	209.55 mm
8.6250 in	OE	219.08 mm
4.0000 in	SB	101.60 mm
6.5000 in	EC	165.10 mm
5.7500 in	HD	146.05 mm

ROCKWELL 92NYR64-3

IDLI GROUP 82467

END YOKE WITH ROUND BORE
slight variation from above group
SEE FIGURE 7B

8.2500 in	IE	209.55 mm
8.6250 in	OE	219.08 mm
4.0000 in	SB	101.60 mm
6.5000 in	EC	165.10 mm
5.7500 in	HD	146.05 mm

ROCKWELL 92NYR64-5

ENGINEERING CATALOGS MUST BE CONSULTED FOR DETAILS NOT INCLUDED
IN THIS GUIDE. SPECIFIC DESIGNS, MATERIAL CONTENT, TOLERANCES
LUBE FITTINGS AND OTHER DIMENSIONS ARE INTENTIONALLY OMITTED HERE.

CAUTION: BE SURE TO REFER TO ENGINEERING CATALOGS FOR SPECIFIC APPLICATIONS THAT REQUIRE SPECIFIC MATERIAL CONTENT, TOLERANCES, ETC. SEE FOOTNOTE.

IDLI GROUP 82468

END YOKE WITH ROUND BORE
slight variation from above group
SEE FIGURE 7B

8.2500 in	IE	209.55 mm
8.6250 in	OE	219.08 mm
4.0000 in	SB	101.60 mm
6.5000 in	EC	165.10 mm
5.7500 in	HD	146.05 mm

ROCKWELL 92NYR64-14

IDLI GROUP 82469

END YOKE WITH ROUND BORE
slight variation from above group
SEE FIGURE 7B

8.2500 in	IE	209.55 mm
8.6250 in	OE	219.08 mm
4.0000 in	SB	101.60 mm
6.5000 in	EC	165.10 mm
5.7500 in	HD	146.05 mm

ROCKWELL 92NYR64-11

IDLI GROUP 82470

END YOKE WITH ROUND BORE
slight variation from above group
SEE FIGURE 7B

8.2500 in	IE	209.55 mm
8.6250 in	OE	219.08 mm
4.0000 in	SB	101.60 mm
6.5000 in	EC	165.10 mm
5.7500 in	HD	146.05 mm

ROCKWELL 92NYR64-13

IDLI GROUP 82471

END YOKE WITH ROUND BORE
slight variation from above group
SEE FIGURE 7B

8.2500 in	IE	209.55 mm
8.6250 in	OE	219.08 mm
4.0000 in	SB	101.60 mm
6.5000 in	EC	165.10 mm
5.7500 in	HD	146.05 mm

ROCKWELL 92NYR64-10

IDLI GROUP 82472

END YOKE WITH ROUND BORE
SEE FIGURE 7B

8.2500 in	IE	209.55 mm
8.6250 in	OE	219.08 mm
4.2500 in	SB	107.95 mm
5.7500 in	EC	146.05 mm
8.3750 in	HD	212.73 mm

ROCKWELL 92NYR68

IDLI GROUP 82473

END YOKE WITH ROUND BORE
SEE FIGURE 7B

8.2500 in	IE	209.55 mm
8.6250 in	OE	219.08 mm
4.3750 in	SB	111.13 mm
5.7500 in	EC	146.05 mm
8.3750 in	HD	212.73 mm

ROCKWELL 92NYR70

IDLI GROUP 82474

END YOKE WITH ROUND BORE
SEE FIGURE 7B

8.2500 in	IE	209.55 mm
8.6250 in	OE	219.08 mm
4.3750 in	SB	111.13 mm
6.3750 in	EC	161.93 mm
8.3750 in	HD	212.73 mm

ROCKWELL 92NYR70-1

IDLI GROUP 82475

END YOKE WITH ROUND BORE
slight variation from above group
SEE FIGURE 7B

8.2500 in	IE	209.55 mm
8.6250 in	OE	219.08 mm
4.3750 in	SB	111.13 mm
6.3750 in	EC	161.93 mm
8.3750 in	HD	212.73 mm

ROCKWELL 92NYR70-2

IDLI GROUP 82476

END YOKE WITH ROUND BORE
SEE FIGURE 7B

8.2500 in	IE	209.55 mm
8.6250 in	OE	219.08 mm
4.4375 in	SB	112.71 mm
5.2500 in	EC	133.35 mm
8.3750 in	HD	212.73 mm

ROCKWELL 92NYR71-1

IDLI GROUP 82477

END YOKE WITH ROUND BORE
slight variation from above group
SEE FIGURE 7B

8.2500 in	IE	209.55 mm
8.6250 in	OE	219.08 mm
4.4375 in	SB	112.71 mm
5.2500 in	EC	133.35 mm
8.3750 in	HD	212.73 mm

ROCKWELL 92NYR71

IDLI GROUP 82478

END YOKE WITH ROUND BORE
SEE FIGURE 7B

8.2500 in	IE	209.55 mm
8.6250 in	OE	219.08 mm
4.5000 in	SB	114.30 mm
6.3750 in	EC	161.93 mm
8.3750 in	HD	212.73 mm

ROCKWELL 92NYR72-7

IDLI GROUP 82479

END YOKE WITH ROUND BORE
slight variation from above group
SEE FIGURE 7B

8.2500 in	IE	209.55 mm
8.6250 in	OE	219.08 mm
4.5000 in	SB	114.30 mm
6.3750 in	EC	161.93 mm
8.3750 in	HD	212.73 mm

ROCKWELL 92NYR72-6

IDLI GROUP 82480

END YOKE WITH ROUND BORE
slight variation from above group
SEE FIGURE 7B

8.2500 in	IE	209.55 mm
8.6250 in	OE	219.08 mm
4.5000 in	SB	114.30 mm
6.3750 in	EC	161.93 mm
8.3750 in	HD	212.73 mm

ROCKWELL 92NYR72-1

IDLI GROUP 82481

END YOKE WITH ROUND BORE
slight variation from above group
SEE FIGURE 7B

8.2500 in	IE	209.55 mm
8.6250 in	OE	219.08 mm
4.5000 in	SB	114.30 mm
6.3750 in	EC	161.93 mm
8.3750 in	HD	212.73 mm

ROCKWELL 92NYR72-4

IDLI GROUP 82482

END YOKE WITH ROUND BORE
slight variation from above group
SEE FIGURE 7B

8.2500 in	IE	209.55 mm
8.6250 in	OE	219.08 mm
4.5000 in	SB	114.30 mm
6.3750 in	EC	161.93 mm
8.3750 in	HD	212.73 mm

ROCKWELL 92NYR72-5

IDLI GROUP 82483

END YOKE WITH ROUND BORE
slight variation from above group
SEE FIGURE 7B

8.2500 in	IE	209.55 mm
8.6250 in	OE	219.08 mm
4.5000 in	SB	114.30 mm
6.3750 in	EC	161.93 mm
8.3750 in	HD	212.73 mm

ROCKWELL 92NYR72

IDLI GROUP 82484

END YOKE WITH ROUND BORE
SEE FIGURE 7B

8.2500 in	IE	209.55 mm
8.6250 in	OE	219.08 mm
4.6875 in	SB	119.06 mm
6.3750 in	EC	161.93 mm
8.3750 in	HD	212.73 mm

ROCKWELL 92NYR75

IDLI GROUP 82485

END YOKE WITH ROUND BORE
SEE FIGURE 7B

8.2500 in	IE	209.55 mm
8.6250 in	OE	219.08 mm
4.7500 in	SB	120.65 mm
6.3750 in	EC	161.93 mm
8.3750 in	HD	212.73 mm

ROCKWELL 92NYR76-1

IDLI GROUP 82486

END YOKE WITH ROUND BORE
SEE FIGURE 7B

8.2500 in	IE	209.55 mm
8.6250 in	OE	219.08 mm
4.9375 in	SB	125.41 mm
6.3750 in	EC	161.93 mm
8.3750 in	HD	212.73 mm

ROCKWELL 92NYR79

IDLI GROUP 82487

END YOKE WITH ROUND BORE
SEE FIGURE 7B

8.2500 in	IE	209.55 mm
8.6250 in	OE	219.08 mm
5.0000 in	SB	127.00 mm
5.2500 in	EC	133.35 mm
8.3750 in	HD	212.73 mm

ROCKWELL 92NYR80-4

IDLI GROUP 82488

END YOKE WITH ROUND BORE
SEE FIGURE 7B

8.2500 in	IE	209.55 mm
8.6250 in	OE	219.08 mm
5.0000 in	SB	127.00 mm
5.7500 in	EC	146.05 mm
8.6250 in	HD	219.08 mm

ROCKWELL 92NYR80

IDLI GROUP 82489

END YOKE WITH 10 SPLINES
SEE FIGURE 7A

8.2500 in	IE	209.55 mm
8.6250 in	OE	219.08 mm
5.0000 in	SB	127.00 mm
6.3750 in	EC	161.93 mm
8.3750 in	HD	212.73 mm

ROCKWELL 92NYS80

IDLI GROUP 82490

END YOKE WITH ROUND BORE
SEE FIGURE 7B

8.2500 in	IE	209.55 mm
8.6250 in	OE	219.08 mm
5.0000 in	SB	127.00 mm
6.3750 in	EC	161.93 mm
8.3750 in	HD	212.73 mm

ROCKWELL 92NYR80-5

BD = BEARING DIAMETER (outside)   **BW** = BEARING WIDTH   **CB** = CROSS LENGTH WITH BEARINGS   **CC** = CENTER TO CENTER   **CD** = CROSS DIAMETER
CL = CROSS LENGTH WITHOUT BEARINGS   **EC** = END TO CENTER or FACE TO FACE   **EE** = END TO END   **EL** = EFFECTIVE LENGTH
HD = HUB DIAMETER (or insert)   **IE** = INSIDE OF EARS (or recess)   **OD** = OUTSIDE DIAMETER   **OE** = OUTSIDE OF EARS   **SB** = SPLINE OR BORE SIZE

IDLI GROUP 82491

END YOKE WITH ROUND BORE
slight variation from above group
SEE FIGURE 7B

8.2500 in	IE	209.55 mm
8.6250 in	OE	219.08 mm
5.0000 in	SB	127.00 mm
6.3750 in	EC	161.93 mm
8.3750 in	HD	212.73 mm

ROCKWELL 92NYR80-1

IDLI GROUP 82492

END YOKE WITH ROUND BORE
slight variation from above group
SEE FIGURE 7B

8.2500 in	IE	209.55 mm
8.6250 in	OE	219.08 mm
5.0000 in	SB	127.00 mm
6.3750 in	EC	161.93 mm
8.3750 in	HD	212.73 mm

ROCKWELL 92NYR80-2

IDLI GROUP 82493

END YOKE WITH ROUND BORE
SEE FIGURE 7B

8.2500 in	IE	209.55 mm
8.6250 in	OE	219.08 mm
5.0000 in	SB	127.00 mm
6.3750 in	EC	161.93 mm
8.6250 in	HD	219.08 mm

ROCKWELL 92NYR80-3

IDLI GROUP 82494

END YOKE WITH ROUND BORE
SEE FIGURE 7B

8.2500 in	IE	209.55 mm
8.6250 in	OE	219.08 mm
5.1875 in	SB	131.76 mm
6.3750 in	EC	161.93 mm
8.3750 in	HD	212.73 mm

ROCKWELL 92NYR83

IDLI GROUP 82495

END YOKE WITH ROUND BORE
SEE FIGURE 7B

8.2500 in	IE	209.55 mm
8.6250 in	OE	219.08 mm
5.3750 in	SB	136.53 mm
5.9375 in	EC	150.81 mm
8.3750 in	HD	212.73 mm

ROCKWELL 92NYR86

IDLI GROUP 82496

END YOKE WITH ROUND BORE
SEE FIGURE 7B

8.2500 in	IE	209.55 mm
8.6250 in	OE	219.08 mm
5.5000 in	SB	139.70 mm
6.2500 in	EC	158.75 mm
8.3750 in	HD	212.73 mm

ROCKWELL 92NYR88

IDLI GROUP 82497

END YOKE WITH ROUND BORE
SEE FIGURE 7B

8.2500 in	IE	209.55 mm
8.6250 in	OE	219.08 mm
5.5000 in	SB	139.70 mm
6.2500 in	EC	158.75 mm
8.6250 in	HD	219.08 mm

ROCKWELL 92NYR88-2

IDLI GROUP 82498

END YOKE WITH ROUND BORE
SEE FIGURE 7B

8.2500 in	IE	209.55 mm
8.6250 in	OE	219.08 mm
5.5000 in	SB	139.70 mm
6.3750 in	EC	161.93 mm
8.3750 in	HD	212.73 mm

ROCKWELL 92NYR88-3

IDLI GROUP 82499

END YOKE WITH ROUND BORE
SEE FIGURE 7B

8.2500 in	IE	209.55 mm
8.6250 in	OE	219.08 mm
5.5000 in	SB	139.70 mm
6.5000 in	EC	165.10 mm
8.6250 in	HD	219.08 mm

ROCKWELL 92NYR88-1

ENGINEERING CATALOGS MUST BE CONSULTED FOR DETAILS NOT INCLUDED
IN THIS GUIDE. SPECIFIC DESIGNS, MATERIAL CONTENT, TOLERANCES
LUBE FITTINGS AND OTHER DIMENSIONS ARE INTENTIONALLY OMITTED HERE.

FOR YOUR CONVENIENCE
NOTES ON OTHER INFORMATION

We urge you to send us any data that
you feel should be included with the
future editions of the I.D.L.I. Guide.

IDLI GROUP 83001

END YOKE WITH ROUND BORE
SEE FIGURE 5B

2.9063 in	OE	73.82 mm
1.1250 in	BD	28.58 mm
.8125 in	SB	20.64 mm
3.4375 in	EC	87.31 mm
2.0000 in	HD	50.80 mm

ROCKWELL L14SYR13-1

IDLI GROUP 83002

END YOKE WITH 13 INVOLUTED SPLINES
SEE FIGURE 5A

2.9063 in	OE	73.82 mm
1.1250 in	BD	28.58 mm
.8750 in	SB	22.23 mm
2.1875 in	EC	55.56 mm
1.8750 in	HD	47.63 mm

ROCKWELL L14SYS14-1

IDLI GROUP 83003

END YOKE WITH ROUND BORE
SEE FIGURE 5B

2.9063 in	OE	73.82 mm
1.1250 in	BD	28.58 mm
.8750 in	SB	22.23 mm
3.4375 in	EC	87.31 mm
2.0000 in	HD	50.80 mm

ROCKWELL L14SYR14-9

IDLI GROUP 83004

END YOKE WITH SQUARE BORE
SEE FIGURE 5C

2.9063 in	OE	73.82 mm
1.1250 in	BD	28.58 mm
.8750 in	SB	22.23 mm
3.4375 in	EC	87.31 mm
2.0000 in	HD	50.80 mm

PRECISION 1321
ROCKWELL L14SYQ14-2

IDLI GROUP 83005

END YOKE WITH SQUARE BORE
slight variation from above group
SEE FIGURE 5C

2.9063 in	OE	73.82 mm
1.1250 in	BD	28.58 mm
.8750 in	SB	22.23 mm
3.4375 in	EC	87.31 mm
2.0000 in	HD	50.80 mm

ROCKWELL L14SYQ14-3

IDLI GROUP 83006

END YOKE WITH ROUND BORE
SEE FIGURE 5B

2.9063 in	OE	73.82 mm
1.1250 in	BD	28.58 mm
.9375 in	SB	23.81 mm
2.6250 in	EC	66.68 mm
2.0000 in	HD	50.80 mm

ROCKWELL L14SYR15-1

IDLI GROUP 83007

END YOKE WITH ROUND BORE
SEE FIGURE 5B

2.9063 in	OE	73.82 mm
1.1250 in	BD	28.58 mm
.9375 in	SB	23.81 mm
3.4375 in	EC	87.31 mm
2.0000 in	HD	50.80 mm

ROCKWELL L14SYR15

IDLI GROUP 83008

END YOKE WITH SQUARE BORE
SEE FIGURE 5C

2.9063 in	OE	73.82 mm
1.1250 in	BD	28.58 mm
.9375 in	SB	23.81 mm
3.4375 in	EC	87.31 mm
2.0000 in	HD	50.80 mm

PRECISION 1320
ROCKWELL L14SYQ15-3
SPICER 2-4-142

IDLI GROUP 83009

END YOKE WITH SQUARE BORE
SEE FIGURE 5C

2.9063 in	OE	73.82 mm
1.1250 in	BD	28.58 mm
.9375 in	SB	23.81 mm
3.8750 in	EC	98.43 mm
2.0000 in	HD	50.80 mm

ROCKWELL L14SYQ15-4

IDLI GROUP 83010

END YOKE WITH ROUND BORE
SEE FIGURE 5B

2.9063 in	OE	73.82 mm
1.1250 in	BD	28.58 mm
1.0000 in	SB	25.40 mm
2.7500 in	EC	69.85 mm
2.0000 in	HD	50.80 mm

PRECISION 1300
ROCKWELL L14SYR16
SPICER 2-4-1343

IDLI GROUP 83011

END YOKE WITH ROUND BORE
slight variation from above group
SEE FIGURE 5B

2.9063 in	OE	73.82 mm
1.1250 in	BD	28.58 mm
1.0000 in	SB	25.40 mm
2.7500 in	EC	69.85 mm
2.0000 in	HD	50.80 mm

ROCKWELL L14SYR16-15

IDLI GROUP 83012

END YOKE WITH SQUARE BORE
SEE FIGURE 5C

2.9063 in	OE	73.82 mm
1.1250 in	BD	28.58 mm
1.0000 in	SB	25.40 mm
2.7500 in	EC	69.85 mm
2.0000 in	HD	50.80 mm

ROCKWELL L14SYQ16-34

IDLI GROUP 83013

END YOKE WITH SQUARE BORE
SEE FIGURE 5C

2.9063 in	OE	73.82 mm
1.1250 in	BD	28.58 mm
1.0000 in	SB	25.40 mm
2.7500 in	EC	69.85 mm
2.3750 in	HD	60.33 mm

ROCKWELL L14SYQ16-15

IDLI GROUP 83014

END YOKE WITH SQUARE BORE
SEE FIGURE 5C

2.9063 in	OE	73.82 mm
1.1250 in	BD	28.58 mm
1.0000 in	SB	25.40 mm
3.1875 in	EC	80.96 mm

BORG-WARNR 69147
NEAPCO 19147
NEAPCO 20-9147
ROCKWELL L14SYQ16-41
SPICER 2-4-232

IDLI GROUP 83015

END YOKE WITH SQUARE BORE
SEE FIGURE 5C

2.9063 in	OE	73.82 mm
1.1250 in	BD	28.58 mm
1.0000 in	SB	25.40 mm
3.3750 in	EC	85.73 mm
2.0000 in	HD	50.80 mm

BORG-WARNR 69302
HUB CITY 0335-01405
NEAPCO 191L2
NEAPCO 20-9310
PRECISION 1321
ROCKWELL L14SYQ16-8
SPICER 2-4-152
SPICER 2-4-153
SPICER 2-4-192
TRW 21200
WESCO 14N0140-30
WESCO 14N140-30
WESCO 860046

IDLI GROUP 83016

END YOKE WITH SQUARE BORE
slight variation from above group
SEE FIGURE 5C

2.9063 in	OE	73.82 mm
1.1250 in	BD	28.58 mm
1.0000 in	SB	25.40 mm
3.3750 in	EC	85.73 mm
2.0000 in	HD	50.80 mm

ROCKWELL L14SYQ16-52

IDLI GROUP 83017

END YOKE WITH SQUARE BORE
slight variation from above group
SEE FIGURE 5C

2.9063 in	OE	73.82 mm
1.1250 in	BD	28.58 mm
1.0000 in	SB	25.40 mm
3.3750 in	EC	85.73 mm
2.0000 in	HD	50.80 mm

ROCKWELL L14SYQ16-30

IDLI GROUP 83018

END YOKE WITH ROUND BORE
SEE FIGURE 5B

2.9063 in	OE	73.82 mm
1.1250 in	BD	28.58 mm
1.0000 in	SB	25.40 mm
3.4375 in	EC	87.31 mm
2.0000 in	HD	50.80 mm

BORG-WARNR 69115
G & G MFG 184-1416
HAYES 2-4-1343
HUB CITY 0335-01406
MCQUAY-NOR U03-106
NEAPCO 19115
NEAPCO 19177
NEAPCO 20-9115
NEAPCO 20-9177
PRECISION 1300
ROCKWELL L14SYR16-6
SPICER 2-4-1263
TRW 21201
WESCO 14N0164
WESCO 14N164
WESCO 860-041
WESCO 860047

IDLI GROUP 83019

END YOKE WITH ROUND BORE
SEE FIGURE 5B

2.9063 in	OE	73.82 mm
1.1250 in	BD	28.58 mm
1.0000 in	SB	25.40 mm
3.8125 in	EC	96.84 mm
2.0000 in	HD	50.80 mm

ROCKWELL L14SYR16-19

IDLI GROUP 83020

END YOKE WITH ROUND BORE
SEE FIGURE 5B

2.9063 in	OE	73.82 mm
1.1250 in	BD	28.58 mm
1.0000 in	SB	25.40 mm
3.8750 in	EC	98.43 mm
2.0000 in	HD	50.80 mm

BORG-WARNR 69105
HUB CITY 0335-01408
NEAPCO 19105
ROCKWELL L14SYR16-11

IDLI GROUP 83021

END YOKE WITH ROUND BORE
SEE FIGURE 5B

2.9063 in	OE	73.82 mm
1.1250 in	BD	28.58 mm
1.0000 in	SB	25.40 mm
3.8750 in	EC	98.43 mm
2.0000 in	HD	50.80 mm

PRECISION 1301
ROCKWELL L14SYR16-9
SPICER 2-4-1333

BD=BEARING DIAMETER (outside)   **BW**=BEARING WIDTH   **CB**=CROSS LENGTH WITH BEARINGS   **CC**=CENTER TO CENTER   **CD**=CROSS DIAMETER
CL=CROSS LENGTH WITHOUT BEARINGS   **EC**=END TO CENTER or FACE TO FACE   **EE**=END TO END   **EL**=EFFECTIVE LENGTH
HD=HUB DIAMETER (or insert)   **IE**=INSIDE OF EARS (or recess)   **OD**=OUTSIDE DIAMETER   **OE**=OUTSIDE OF EARS   **SB**=SPLINE OR BORE SIZE

CAUTION: BE SURE TO REFER TO ENGINEERING CATALOGS FOR SPECIAL APPLICATIONS THAT REQUIRE SPECIFIC MATERIAL CONTENT, TOLERANCES, ETC. SEE FOOTNOTE.

IDLI GROUP 83022

END YOKE WITH ROUND BORE
slight variation from above group
SEE FIGURE 5B

2.9063 in	OE	73.82 mm
1.1250 in	BD	28.58 mm
1.0000 in	SB	25.40 mm
3.8750 in	EC	98.43 mm
2.0000 in	HD	50.80 mm

ROCKWELL L14SYR16-30

IDLI GROUP 83023

END YOKE WITH ROUND BORE
SEE FIGURE 5B

2.9063 in	OE	73.82 mm
1.1250 in	BD	28.58 mm
1.0625 in	SB	26.99 mm
2.7500 in	EC	69.85 mm
2.0000 in	HD	50.80 mm

ROCKWELL L14SYR17-5

IDLI GROUP 83024

END YOKE WITH ROUND BORE
SEE FIGURE 5B

2.9063 in	OE	73.82 mm
1.1250 in	BD	28.58 mm
1.0625 in	SB	26.99 mm
3.3750 in	EC	85.73 mm
2.0000 in	HD	50.80 mm

BORG-WARNR 69116
NEAPCO 19116
NEAPCO 20-9116
PRECISION 1302
ROCKWELL L14SYR17-7

IDLI GROUP 83025

END YOKE WITH ROUND BORE
SEE FIGURE 5B

2.9063 in	OE	73.82 mm
1.1250 in	BD	28.58 mm
1.0625 in	SB	26.99 mm
3.4375 in	EC	87.31 mm
2.0000 in	HD	50.80 mm

ROCKWELL L14SYR17-3

IDLI GROUP 83026

END YOKE WITH ROUND BORE
SEE FIGURE 5B

2.9063 in	OE	73.82 mm
1.1250 in	BD	28.58 mm
1.0625 in	SB	26.99 mm
3.8750 in	EC	98.43 mm
1.8906 in	HD	48.02 mm

ROCKWELL L14SYR17-10

IDLI GROUP 83027

END YOKE WITH ROUND BORE
SEE FIGURE 5B

2.9063 in	OE	73.82 mm
1.1250 in	BD	28.58 mm
1.0938 in	SB	27.78 mm
2.7500 in	EC	69.85 mm
1.8750 in	HD	47.63 mm

ROCKWELL L14SYR17-9

IDLI GROUP 83028

END YOKE WITH ROUND BORE
SEE FIGURE 5B

2.9063 in	OE	73.82 mm
1.1250 in	BD	28.58 mm
1.0938 in	SB	27.78 mm
3.5625 in	EC	90.49 mm
1.0000 in	HD	25.40 mm

ROCKWELL L14SYR17-8

IDLI GROUP 83029

END YOKE WITH HEX BORE
SEE FIGURE 5E

2.9063 in	OE	73.82 mm
1.1250 in	BD	28.58 mm
1.1250 in	SB	28.58 mm
2.7500 in	EC	69.85 mm
1.7500 in	HD	44.45 mm

ROCKWELL L14SYH18-8

IDLI GROUP 83030

END YOKE WITH 6 SPLINES
SEE FIGURE 5A

2.9063 in	OE	73.82 mm
1.1250 in	BD	28.58 mm
1.1250 in	SB	28.58 mm
2.7500 in	EC	69.85 mm
2.0000 in	HD	50.80 mm

ROCKWELL L14SYS18-6

IDLI GROUP 83031

END YOKE WITH ROUND BORE
SEE FIGURE 5B

2.9063 in	OE	73.82 mm
1.1250 in	BD	28.58 mm
1.1250 in	SB	28.58 mm
2.7500 in	EC	69.85 mm
2.0000 in	HD	50.80 mm

BORG-WARNR 69211
G & G MFG 184-1418
HUB CITY 0335-01415
NEAPCO 191A2
NEAPCO 20-9216
NEAPCO 20-9301
PRECISION 1304
ROCKWELL L14SYR18-44
SPICER 2-4-1363
TRW 21202

IDLI GROUP 83032

END YOKE WITH ROUND BORE
slight variation from above group
SEE FIGURE 5B

2.9063 in	OE	73.82 mm
1.1250 in	BD	28.58 mm
1.1250 in	SB	28.58 mm
2.7500 in	EC	69.85 mm
2.0000 in	HD	50.80 mm

ROCKWELL L14SYR18-52

IDLI GROUP 83033

END YOKE WITH ROUND BORE
slight variation from above group
SEE FIGURE 5B

2.9063 in	OE	73.82 mm
1.1250 in	BD	28.58 mm
1.1250 in	SB	28.58 mm
2.7500 in	EC	69.85 mm
2.0000 in	HD	50.80 mm

ROCKWELL L14SYR18-34

IDLI GROUP 83034

END YOKE WITH ROUND BORE
slight variation from above group
SEE FIGURE 5B

2.9063 in	OE	73.82 mm
1.1250 in	BD	28.58 mm
1.1250 in	SB	28.58 mm
2.7500 in	EC	69.85 mm
2.0000 in	HD	50.80 mm

ROCKWELL L14SYR18-47

IDLI GROUP 83035

END YOKE WITH ROUND BORE
slight variation from above group
SEE FIGURE 5B

2.9063 in	OE	73.82 mm
1.1250 in	BD	28.58 mm
1.1250 in	SB	28.58 mm
2.7500 in	EC	69.85 mm
2.0000 in	HD	50.80 mm

ROCKWELL L14SYR18-11

IDLI GROUP 83036

END YOKE WITH RECTANGULAR BORE
maximum SB dimension shown
SEE FIGURE 5D

2.9063 in	OE	73.82 mm
1.1250 in	BD	28.58 mm
1.1250 in	SB	28.58 mm
2.7500 in	EC	69.85 mm
2.0000 in	HD	50.80 mm

ROCKWELL L14SYP16-1

IDLI GROUP 83037

END YOKE WITH RECTANGULAR BORE
maximum SB dimension shown
slight variation from above group
SEE FIGURE 5D

2.9063 in	OE	73.82 mm
1.1250 in	BD	28.58 mm
1.1250 in	SB	28.58 mm
2.7500 in	EC	69.85 mm
2.0000 in	HD	50.80 mm

ROCKWELL L14SYP16-40

IDLI GROUP 83038

END YOKE WITH RECTANGULAR BORE
maximum SB dimension shown
SEE FIGURE 5D

2.9063 in	OE	73.82 mm
1.1250 in	BD	28.58 mm
1.1250 in	SB	28.58 mm
3.0625 in	EC	77.79 mm

BORG-WARNR 69149
G & G MFG 177-1400
HAYES 2-4-102
NEAPCO 19126
NEAPCO 19149
NEAPCO 20-9126
NEAPCO 20-9149
PRECISION 1374
ROCKWELL L14SYP16-33
SPICER 2-4-102
SPICER 2-4-182
SPICER 2-4-352
WESCO 14NW4-1618
WESCO 14NWY1618
WESCO 860061

IDLI GROUP 83039

END YOKE WITH RECTANGULAR BORE
maximum SB dimension shown
slight variation from above group
SEE FIGURE 5D

2.9063 in	OE	73.82 mm
1.1250 in	BD	28.58 mm
1.1250 in	SB	28.58 mm
3.0625 in	EC	77.79 mm

ROCKWELL L14SYP16-71

IDLI GROUP 83040

END YOKE WITH RECTANGULAR BORE
maximum SB dimension shown
SEE FIGURE 5D

2.9063 in	OE	73.82 mm
1.1250 in	BD	28.58 mm
1.1250 in	SB	28.58 mm
3.1250 in	EC	79.38 mm
2.1250 in	HD	53.98 mm

PRECISION 1323
ROCKWELL L14SYP16-11
SPICER 2-4-122
SPICER 2-4-162
SPICER 2-4-72

IDLI GROUP 83041

END YOKE WITH RECTANGULAR BORE
maximum SB dimension shown
SEE FIGURE 5D

2.9063 in	OE	73.82 mm
1.1250 in	BD	28.58 mm
1.1250 in	SB	28.58 mm
3.1875 in	EC	80.96 mm

ROCKWELL L14SYP16-55

IDLI GROUP 83042

END YOKE WITH ROUND BORE
SEE FIGURE 5B

2.9063 in	OE	73.82 mm
1.1250 in	BD	28.58 mm
1.1250 in	SB	28.58 mm
3.2500 in	EC	82.55 mm
2.0000 in	HD	50.80 mm

PRECISION 1303
ROCKWELL L14SYR18-10

IDLI GROUP 83043

END YOKE WITH RECTANGULAR BORE
maximum SB dimension shown
SEE FIGURE 5D

2.9063 in	OE	73.82 mm
1.1250 in	BD	28.58 mm
1.1250 in	SB	28.58 mm
3.2500 in	EC	82.55 mm
2.0000 in	HD	50.80 mm

ROCKWELL L14SYP16-6

IDLI GROUP 83044

END YOKE WITH ROUND BORE
SEE FIGURE 5B

2.9063 in	OE	73.82 mm
1.1250 in	BD	28.58 mm
1.1250 in	SB	28.58 mm
3.3750 in	EC	85.73 mm
2.0000 in	HD	50.80 mm

PRECISION 1304
ROCKWELL L14SYR18-37

IDLI GROUP 83045

END YOKE WITH HEX BORE
SEE FIGURE 5E

2.9063 in	OE	73.82 mm
1.1250 in	BD	28.58 mm
1.1250 in	SB	28.58 mm
3.3750 in	EC	85.73 mm
2.0000 in	HD	50.80 mm

ROCKWELL L14SYH18-7

IDLI GROUP 83046

END YOKE WITH HEX BORE
slight variation from above group
SEE FIGURE 5E

2.9063 in	OE	73.82 mm
1.1250 in	BD	28.58 mm
1.1250 in	SB	28.58 mm
3.3750 in	EC	85.73 mm
2.0000 in	HD	50.80 mm

ROCKWELL L14SYH18-1

IDLI GROUP 83047

END YOKE WITH ROUND BORE
SEE FIGURE 5B

2.9063 in	OE	73.82 mm
1.1250 in	BD	28.58 mm
1.1250 in	SB	28.58 mm
3.4375 in	EC	87.31 mm
2.0000 in	HD	50.80 mm

BORG-WARNR 69211
G & G MFG 184-1418
HAYES 2-4-1363
NEAPCO 19117

NEAPCO 19151
NEAPCO 191A2
NEAPCO 20-9117
NEAPCO 20-9216
NEAPCO 20-9301
PRECISION 1306
ROCKWELL L14SYR18-16
SPICER 2-4-1353
SPICER 2-4-1393
WESCO 14N1845-37
WESCO 860065

IDLI GROUP 83048

END YOKE WITH ROUND BORE
slight variation from above group
SEE FIGURE 5B

2.9063 in	OE	73.82 mm
1.1250 in	BD	28.58 mm
1.1250 in	SB	28.58 mm
3.4375 in	EC	87.31 mm
2.0000 in	HD	50.80 mm

ROCKWELL L14SYR18-82

IDLI GROUP 83049

END YOKE WITH SQUARE BORE
SEE FIGURE 5C

2.9063 in	OE	73.82 mm
1.1250 in	BD	28.58 mm
1.1250 in	SB	28.58 mm
3.5625 in	EC	90.49 mm
2.2500 in	HD	57.15 mm

BORG-WARNR 67228
NEAPCO 171B8
NEAPCO 18-7309
NEAPCO 19183
NEAPCO 20-9183
PRECISION 1325
ROCKWELL L14SYQ18
WESCO 860050

IDLI GROUP 83050

END YOKE WITH RECTANGULAR BORE
maximum SB dimension shown
SEE FIGURE 5D

2.9063 in	OE	73.82 mm
1.1250 in	BD	28.58 mm
1.1250 in	SB	28.58 mm
3.5625 in	EC	90.49 mm
2.2500 in	HD	57.15 mm

ROCKWELL L14SYP16-18

IDLI GROUP 83051

END YOKE WITH HEX BORE
SEE FIGURE 5E

2.9063 in	OE	73.82 mm
1.1250 in	BD	28.58 mm
1.1250 in	SB	28.58 mm
3.7500 in	EC	95.25 mm
2.0000 in	HD	50.80 mm

ROCKWELL L14SYH18-3

IDLI GROUP 83052

END YOKE WITH 6 SPLINES
SEE FIGURE 5A

2.9063 in	OE	73.82 mm
1.1250 in	BD	28.58 mm
1.1250 in	SB	28.58 mm
3.8750 in	EC	98.43 mm
1.3750 in	HD	34.93 mm

ROCKWELL L14SYS18-3

IDLI GROUP 83053

END YOKE WITH ROUND BORE
SEE FIGURE 5B

2.9063 in	OE	73.82 mm
1.1250 in	BD	28.58 mm
1.1250 in	SB	28.58 mm
3.8750 in	EC	98.43 mm
1.7500 in	HD	44.45 mm

ROCKWELL L14SYR18

IDLI GROUP 83054

END YOKE WITH 6 SPLINES
SEE FIGURE 5A

2.9063 in	OE	73.82 mm
1.1250 in	BD	28.58 mm
1.1250 in	SB	28.58 mm
3.8750 in	EC	98.43 mm
2.0000 in	HD	50.80 mm

HUB CITY 0335-01427
PRECISION 1350
ROCKWELL L14SYS18-5
SPICER 2-4-3071

IDLI GROUP 83055

END YOKE WITH ROUND BORE
SEE FIGURE 5B

2.9063 in	OE	73.82 mm
1.1250 in	BD	28.58 mm
1.1250 in	SB	28.58 mm
3.8750 in	EC	98.43 mm
2.0000 in	HD	50.80 mm

ROCKWELL L14SYR18-27

IDLI GROUP 83056

END YOKE WITH ROUND BORE
slight variation from above group
SEE FIGURE 5B

2.9063 in	OE	73.82 mm
1.1250 in	BD	28.58 mm
1.1250 in	SB	28.58 mm
3.8750 in	EC	98.43 mm
2.0000 in	HD	50.80 mm

BORG-WARNR 69187
PRECISION 1350
ROCKWELL L14SYR18-5

IDLI GROUP 83057

END YOKE WITH ROUND BORE
slight variation from above group
SEE FIGURE 5B

2.9063 in	OE	73.82 mm
1.1250 in	BD	28.58 mm
1.1250 in	SB	28.58 mm
3.8750 in	EC	98.43 mm
2.0000 in	HD	50.80 mm

BORG-WARNR 69211

HUB CITY 0335-01415
PRECISION 1305
REX CHAIN 760029
ROCKWELL L14SYR18-39
SPICER 2-4-1373
SPICER 2-4-1383
SPICER 2-4-1593
SPICER 2-4-1603

IDLI GROUP 83058

END YOKE WITH HEX BORE
SEE FIGURE 5E

2.9063 in	OE	73.82 mm
1.1250 in	BD	28.58 mm
1.1250 in	SB	28.58 mm
4.6875 in	EC	119.06 mm
1.8125 in	HD	46.04 mm

ROCKWELL L14SYH18-2

IDLI GROUP 83059

END YOKE WITH ROUND BORE
SEE FIGURE 5B

2.9063 in	OE	73.82 mm
1.1250 in	BD	28.58 mm
1.1875 in	SB	30.16 mm
2.7500 in	EC	69.85 mm
1.8750 in	HD	47.63 mm

ROCKWELL L14SYR19-17

IDLI GROUP 83060

END YOKE WITH SQUARE BORE
SEE FIGURE 5C

2.9063 in	OE	73.82 mm
1.1250 in	BD	28.58 mm
1.1875 in	SB	30.16 mm
3.1875 in	EC	80.96 mm

ROCKWELL L14SYQ19-17

IDLI GROUP 83061

END YOKE WITH ROUND BORE
SEE FIGURE 5B

2.9063 in	OE	73.82 mm
1.1250 in	BD	28.58 mm
1.1875 in	SB	30.16 mm
3.4375 in	EC	87.31 mm
2.0000 in	HD	50.80 mm

ROCKWELL L14SYR19-11

IDLI GROUP 83062

END YOKE WITH ROUND BORE
slight variation from above group
SEE FIGURE 5B

2.9063 in	OE	73.82 mm
1.1250 in	BD	28.58 mm
1.1875 in	SB	30.16 mm
3.4375 in	EC	87.31 mm
2.0000 in	HD	50.80 mm

BORG-WARNR 69118
NEAPCO 19118
NEAPCO 19142
NEAPCO 1917A
NEAPCO 20-9118
NEAPCO 20-9217
PRECISION 1307
ROCKWELL L14SYR19-31

BD = BEARING DIAMETER (outside)   **BW** = BEARING WIDTH   **CB** = CROSS LENGTH WITH BEARINGS   **CC** = CENTER TO CENTER   **CD** = CROSS DIAMETER
CL = CROSS LENGTH WITHOUT BEARINGS   **EC** = END TO CENTER or FACE TO FACE   **EE** = END TO END   **EL** = EFFECTIVE LENGTH
HD = HUB DIAMETER (or insert)   **IE** = INSIDE OF EARS (or recess)   **OD** = OUTSIDE DIAMETER   **OE** = OUTSIDE OF EARS   **SB** = SPLINE OR BORE SIZE

IDLI GUIDE   COPYRIGHT ® INTERCHANGE, INC.   ST. LOUIS PARK, MN. 55416 USA

CAUTION: BE SURE TO REFER TO ENGINEERING CATALOGS FOR SPECIAL APPLICATIONS THAT REQUIRE SPECIFIC MATERIAL CONTENT, TOLERANCES, ETC. SEE FOOTNOTE.

SPICER		2-4-1943
TRW		21203
WESCO		14N1945-37
WESCO		860066

IDLI GROUP 83063

END YOKE WITH ROUND BORE
slight variation from above group
SEE FIGURE 5B

2.9063 in	OE	73.82 mm
1.1250 in	BD	28.58 mm
1.1875 in	SB	30.16 mm
3.4375 in	EC	87.31 mm
2.0000 in	HD	50.80 mm

ROCKWELL L14SYR19-2

IDLI GROUP 83064

END YOKE WITH ROUND BORE
slight variation from above group
SEE FIGURE 5B

2.9063 in	OE	73.82 mm
1.1250 in	BD	28.58 mm
1.1875 in	SB	30.16 mm
3.4375 in	EC	87.31 mm
2.0000 in	HD	50.80 mm

PRECISION 1307
ROCKWELL L14SYR19-3

IDLI GROUP 83065

END YOKE WITH SQUARE BORE
SEE FIGURE 5C

2.9063 in	OE	73.82 mm
1.1250 in	BD	28.58 mm
1.1875 in	SB	30.16 mm
3.5000 in	EC	88.90 mm
2.2500 in	HD	57.15 mm

ROCKWELL L14SYQ19-5

IDLI GROUP 83066

END YOKE WITH ROUND BORE
SEE FIGURE 5B

2.9063 in	OE	73.82 mm
1.1250 in	BD	28.58 mm
1.1875 in	SB	30.16 mm
3.8750 in	EC	98.43 mm
2.0000 in	HD	50.80 mm

ROCKWELL L14SYR19-10

IDLI GROUP 83067

END YOKE WITH ROUND BORE
SEE FIGURE 5B

2.9063 in	OE	73.82 mm
1.1250 in	BD	28.58 mm
1.1875 in	SB	30.16 mm
4.1250 in	EC	104.78 mm
2.0000 in	HD	50.80 mm

ROCKWELL L14SYR19-6

IDLI GROUP 83068

END YOKE WITH ROUND BORE
SEE FIGURE 5B

2.9063 in	OE	73.82 mm
1.1250 in	BD	28.58 mm
1.2500 in	SB	31.75 mm
2.3750 in	EC	60.33 mm
2.0000 in	HD	50.80 mm

ROCKWELL L14SYR20-56

IDLI GROUP 83069

END YOKE WITH 19 INVOLUTED SPLINES
SEE FIGURE 5A

2.9063 in	OE	73.82 mm
1.1250 in	BD	28.58 mm
1.2500 in	SB	31.75 mm
2.4063 in	EC	61.12 mm
2.0000 in	HD	50.80 mm

ROCKWELL L14SYS20-41

IDLI GROUP 83070

END YOKE WITH 6 SPLINES
SEE FIGURE 5A

2.9063 in	OE	73.82 mm
1.1250 in	BD	28.58 mm
1.2500 in	SB	31.75 mm
2.7500 in	EC	69.85 mm
2.0000 in	HD	50.80 mm

ROCKWELL L14SYS22-54

IDLI GROUP 83071

END YOKE WITH 10 SPLINES
SEE FIGURE 5A

2.9063 in	OE	73.82 mm
1.1250 in	BD	28.58 mm
1.2500 in	SB	31.75 mm
2.7500 in	EC	69.85 mm
2.0000 in	HD	50.80 mm

ROCKWELL L14SYS20-1

IDLI GROUP 83072

END YOKE WITH 14 INVOLUTED SPLINES
SEE FIGURE 5A

2.9063 in	OE	73.82 mm
1.1250 in	BD	28.58 mm
1.2500 in	SB	31.75 mm
2.7500 in	EC	69.85 mm
2.0000 in	HD	50.80 mm

ROCKWELL L14SYS20-5

IDLI GROUP 83073

END YOKE WITH 19 INVOLUTED SPLINES
SEE FIGURE 5A

2.9063 in	OE	73.82 mm
1.1250 in	BD	28.58 mm
1.2500 in	SB	31.75 mm
2.7500 in	EC	69.85 mm
2.0000 in	HD	50.80 mm

ROCKWELL L14SYS20-10

IDLI GROUP 83074

END YOKE WITH ROUND BORE
SEE FIGURE 5B

2.9063 in	OE	73.82 mm
1.1250 in	BD	28.58 mm
1.2500 in	SB	31.75 mm
2.7500 in	EC	69.85 mm
2.0000 in	HD	50.80 mm

ROCKWELL L14SYR20-34

IDLI GROUP 83075

END YOKE WITH ROUND BORE
slight variation from above group
SEE FIGURE 5B

2.9063 in	OE	73.82 mm
1.1250 in	BD	28.58 mm
1.2500 in	SB	31.75 mm
2.7500 in	EC	69.85 mm
2.0000 in	HD	50.80 mm

ROCKWELL L14SYR20-38

IDLI GROUP 83076

END YOKE WITH ROUND BORE
slight variation from above group
SEE FIGURE 5B

2.9063 in	OE	73.82 mm
1.1250 in	BD	28.58 mm
1.2500 in	SB	31.75 mm
2.7500 in	EC	69.85 mm
2.0000 in	HD	50.80 mm

PRECISION 1309
ROCKWELL L14SYR20-9

IDLI GROUP 83077

END YOKE WITH 10 SPLINES
SEE FIGURE 5A

2.9063 in	OE	73.82 mm
1.1250 in	BD	28.58 mm
1.2500 in	SB	31.75 mm
3.3750 in	EC	85.73 mm
2.0000 in	HD	50.80 mm

ROCKWELL L14SYS20-2

IDLI GROUP 83078

END YOKE WITH 19 INVOLUTED SPLINES
SEE FIGURE 5A

2.9063 in	OE	73.82 mm
1.1250 in	BD	28.58 mm
1.2500 in	SB	31.75 mm
3.4375 in	EC	87.31 mm
2.0000 in	HD	50.80 mm

ROCKWELL L14SYS20-11

IDLI GROUP 83079

END YOKE WITH ROUND BORE
SEE FIGURE 5B

2.9063 in	OE	73.82 mm
1.1250 in	BD	28.58 mm
1.2500 in	SB	31.75 mm
3.4375 in	EC	87.31 mm
2.0000 in	HD	50.80 mm

ROCKWELL L14SYR20-1

IDLI GROUP 83080

END YOKE WITH ROUND BORE
slight variation from above group
SEE FIGURE 5B

2.9063 in	OE	73.82 mm
1.1250 in	BD	28.58 mm
1.2500 in	SB	31.75 mm
3.4375 in	EC	87.31 mm
2.0000 in	HD	50.80 mm

ROCKWELL L14SYR20-102

IDLI GROUP 83081

END YOKE WITH ROUND BORE
slight variation from above group
SEE FIGURE 5B

2.9063 in	OE	73.82 mm
1.1250 in	BD	28.58 mm
1.2500 in	SB	31.75 mm
3.4375 in	EC	87.31 mm
2.0000 in	HD	50.80 mm

G & G MFG 184-1420
PRECISION 1311
ROCKWELL L14SYR20-2

IDLI GROUP 83082

END YOKE WITH ROUND BORE
slight variation from above group
SEE FIGURE 5B

2.9063 in	OE	73.82 mm
1.1250 in	BD	28.58 mm
1.2500 in	SB	31.75 mm
3.4375 in	EC	87.31 mm
2.0000 in	HD	50.80 mm

BORG-WARNR		67232
BORG-WARNR		69119
G & G MFG		184-1420
HUB CITY		0335-01426
MCQUAY-NOR		U03-98
NEAPCO		171C2
NEAPCO		18-7310
NEAPCO		19119
NEAPCO		19121
NEAPCO		19122
NEAPCO		19123
NEAPCO		19166
NEAPCO		191A4
NEAPCO		20-9119
NEAPCO		20-9166
NEAPCO		20-9218
PRECISION		1309
ROCKWELL		L14SYR20-17
SPICER		2-4-1423
SPICER		2-4-1433
SPICER		2-4-1853
WESCO		860067

IDLI GROUP 83083

END YOKE WITH ROUND BORE
slight variation from above group
SEE FIGURE 5B

2.9063 in	OE	73.82 mm
1.1250 in	BD	28.58 mm
1.2500 in	SB	31.75 mm
3.4375 in	EC	87.31 mm
2.0000 in	HD	50.80 mm

ROCKWELL L14SYR20-44

ENGINEERING CATALOGS MUST BE CONSULTED FOR DETAILS NOT INCLUDED
IN THIS GUIDE. SPECIFIC DESIGNS, MATERIAL CONTENT, TOLERANCES
LUBE FITTINGS AND OTHER DIMENSIONS ARE INTENTIONALLY OMITTED HERE.

IDLI GROUP 83084

END YOKE WITH ROUND BORE
slight variation from above group
SEE FIGURE 5B

2.9063 in	OE	73.82 mm
1.1250 in	BD	28.58 mm
1.2500 in	SB	31.75 mm
3.4375 in	EC	87.31 mm
2.0000 in	HD	50.80 mm

BORG-WARNR 69214
G & G MFG 184-1420
HUB CITY 0335-01416
NEAPCO 1918A
NEAPCO 20-9218
PRECISION 1310
REX CHAIN 760054
ROCKWELL L14SYR20-174
TRW 21204
WESCO 14N2045-37

IDLI GROUP 83085

END YOKE WITH ROUND BORE
slight variation from above group
SEE FIGURE 5B

2.9063 in	OE	73.82 mm
1.1250 in	BD	28.58 mm
1.2500 in	SB	31.75 mm
3.4375 in	EC	87.31 mm
2.0000 in	HD	50.80 mm

PRECISION 1310
ROCKWELL L14SYR20-47
SPICER 2-4-1413

IDLI GROUP 83086

END YOKE WITH ROUND BORE
slight variation from above group
SEE FIGURE 5B

2.9063 in	OE	73.82 mm
1.1250 in	BD	28.58 mm
1.2500 in	SB	31.75 mm
3.4375 in	EC	87.31 mm
2.0000 in	HD	50.80 mm

BORG-WARNR 69175
BORG-WARNR 69214
G & G MFG 184-1420
HAYES 2-4-1413
HAYES 2-4-1423
NEAPCO 19175
NEAPCO 191A4
NEAPCO 191J3
NEAPCO 20-9175
NEAPCO 20-9302
PRECISION 1308
REX CHAIN 760091
ROCKWELL L14SYR20-10
SPICER 2-4-1403
SPICER 2-4-1563

IDLI GROUP 83087

END YOKE WITH ROUND BORE
slight variation from above group
SEE FIGURE 5B

2.9063 in	OE	73.82 mm
1.1250 in	BD	28.58 mm
1.2500 in	SB	31.75 mm
3.4375 in	EC	87.31 mm
2.0000 in	HD	50.80 mm

ROCKWELL L14SYR20-99

IDLI GROUP 83088

END YOKE WITH ROUND BORE
slight variation from above group
SEE FIGURE 5B

2.9063 in	OE	73.82 mm
1.1250 in	BD	28.58 mm
1.2500 in	SB	31.75 mm
3.4375 in	EC	87.31 mm
2.0000 in	HD	50.80 mm

ROCKWELL L14SYR20-58

IDLI GROUP 83089

END YOKE WITH 19 INVOLUTED SPLINES
SEE FIGURE 5A

2.9063 in	OE	73.82 mm
1.1250 in	BD	28.58 mm
1.2500 in	SB	31.75 mm
3.5000 in	EC	88.90 mm
2.0000 in	HD	50.80 mm

ROCKWELL L14SYS20-39

IDLI GROUP 83090

END YOKE WITH 6 SPLINES
SEE FIGURE 5A

2.9063 in	OE	73.82 mm
1.1250 in	BD	28.58 mm
1.2500 in	SB	31.75 mm
3.5000 in	EC	88.90 mm
2.2500 in	HD	57.15 mm

BORG-WARNR 67216
BORG-WARNR 69216
NEAPCO 171A6
NEAPCO 18-7304
NEAPCO 19133
NEAPCO 191A6
NEAPCO 20-9303
PRECISION 1351
ROCKWELL L14SYS20
SPICER 2-4-3081
SPICER 2-4-3091X
WESCO 14N200-21
WESCO 461200
WESCO 860052

IDLI GROUP 83091

END YOKE WITH ROUND BORE
SEE FIGURE 5B

2.9063 in	OE	73.82 mm
1.1250 in	BD	28.58 mm
1.2500 in	SB	31.75 mm
3.5625 in	EC	90.49 mm
2.2500 in	HD	57.15 mm

ROCKWELL L14SYR20-84

IDLI GROUP 83092

END YOKE WITH SQUARE BORE
SEE FIGURE 5C

2.9063 in	OE	73.82 mm
1.1250 in	BD	28.58 mm
1.2500 in	SB	31.75 mm
3.5625 in	EC	90.49 mm
2.2500 in	HD	57.15 mm

PRECISION 1322
ROCKWELL L14SYQ20
SPICER 2-4-172

IDLI GROUP 83093

END YOKE WITH 9 INVOLUTED SPLINES
SEE FIGURE 5A

2.9063 in	OE	73.82 mm
1.1250 in	BD	28.58 mm
1.2500 in	SB	31.75 mm
3.8125 in	EC	96.84 mm
2.0000 in	HD	50.80 mm

ROCKWELL L14SYS20-36

IDLI GROUP 83094

END YOKE WITH 19 INVOLUTED SPLINES
SEE FIGURE 5A

2.9063 in	OE	73.82 mm
1.1250 in	BD	28.58 mm
1.2500 in	SB	31.75 mm
3.8750 in	EC	98.43 mm
2.0000 in	HD	50.80 mm

ROCKWELL L14SYS20-40

IDLI GROUP 83095

END YOKE WITH ROUND BORE
SEE FIGURE 5B

2.9063 in	OE	73.82 mm
1.1250 in	BD	28.58 mm
1.2500 in	SB	31.75 mm
3.8750 in	EC	98.43 mm
2.0000 in	HD	50.80 mm

ROCKWELL L14SYR20-30

IDLI GROUP 83096

END YOKE WITH ROUND BORE
slight variation from above group
SEE FIGURE 5B

2.9063 in	OE	73.82 mm
1.1250 in	BD	28.58 mm
1.2500 in	SB	31.75 mm
3.8750 in	EC	98.43 mm
2.0000 in	HD	50.80 mm

ROCKWELL L14SYR20-25

IDLI GROUP 83097

END YOKE WITH ROUND BORE
SEE FIGURE 5B

2.9063 in	OE	73.82 mm
1.1250 in	BD	28.58 mm
1.3125 in	SB	33.34 mm
3.5625 in	EC	90.49 mm
2.1250 in	HD	53.98 mm

ROCKWELL L14SYR21-10

IDLI GROUP 83098

END YOKE WITH ROUND BORE
SEE FIGURE 5B

2.9063 in	OE	73.82 mm
1.1250 in	BD	28.58 mm
1.3125 in	SB	33.34 mm
3.5625 in	EC	90.49 mm
2.2500 in	HD	57.15 mm

ROCKWELL L14SYR21-2

IDLI GROUP 83099

END YOKE WITH SQUARE BORE
SEE FIGURE 5C

2.9063 in	OE	73.82 mm
1.1250 in	BD	28.58 mm
1.3125 in	SB	33.34 mm
3.5625 in	EC	90.49 mm
2.2500 in	HD	57.15 mm

ROCKWELL L14SYQ21

IDLI GROUP 83100

END YOKE WITH ROUND BORE
SEE FIGURE 5B

2.9063 in	OE	73.82 mm
1.1250 in	BD	28.58 mm
1.3125 in	SB	33.34 mm
4.1250 in	EC	104.78 mm
2.0000 in	HD	50.80 mm

ROCKWELL L14SYR21-8

IDLI GROUP 83101

END YOKE WITH ROUND BORE
SEE FIGURE 5B

2.9063 in	OE	73.82 mm
1.1250 in	BD	28.58 mm
1.3750 in	SB	34.93 mm
2.7500 in	EC	69.85 mm
1.8750 in	HD	47.63 mm

ROCKWELL L14SYR22-14

IDLI GROUP 83102

END YOKE WITH ROUND BORE
SEE FIGURE 5B

2.9063 in	OE	73.82 mm
1.1250 in	BD	28.58 mm
1.3750 in	SB	34.93 mm
2.7500 in	EC	69.85 mm
2.5000 in	HD	63.50 mm

G & G MFG 184-1422
HUB CITY 0335-01417
MCQUAY-NOR U03-88
NEAPCO 19120
NEAPCO 19125
NEAPCO 19176
NEAPCO 19189
NEAPCO 20-9120
NEAPCO 20-9176
NEAPCO 20-9189
PRECISION 1312
ROCKWELL L14SYR22-21
SPICER 2-4-1173
SPICER 2-4-1443
SPICER 2-4-1633
SPICER 2-4-1703
SPICER 2-4-1843
SPICER 2-4-1873

BD=BEARING DIAMETER (outside)   **BW**=BEARING WIDTH   **CB**=CROSS LENGTH WITH BEARINGS   **CC**=CENTER TO CENTER   **CD**=CROSS DIAMETER
CL=CROSS LENGTH WITHOUT BEARINGS   **EC**=END TO CENTER or FACE TO FACE   **EE**=END TO END   **EL**=EFFECTIVE LENGTH
HD=HUB DIAMETER (or insert)   **IE**=INSIDE OF EARS (or recess)   **OD**=OUTSIDE DIAMETER   **OE**=OUTSIDE OF EARS   **SB**=SPLINE OR BORE SIZE

CAUTION: BE SURE TO REFER TO ENGINEERING CATALOGS FOR SPECIAL APPLICATIONS THAT REQUIRE SPECIFIC MATERIAL CONTENT, TOLERANCES, ETC. SEE FOOTNOTE.

TRW 21206
WESCO 14N0225
WESCO 14N225
WESCO 860057

IDLI GROUP 83103

END YOKE WITH ROUND BORE
SEE FIGURE 5B

2.9063 in	OE	73.82 mm
1.1250 in	BD	28.58 mm
1.3750 in	SB	34.93 mm
3.5000 in	EC	88.90 mm
2.2500 in	HD	57.15 mm

ROCKWELL L14SYR22-93

IDLI GROUP 83104

END YOKE WITH 6 SPLINES
SEE FIGURE 5A

2.9063 in	OE	73.82 mm
1.1250 in	BD	28.58 mm
1.3750 in	SB	34.93 mm
3.4375 in	EC	87.31 mm
2.0000 in	HD	50.80 mm

ROCKWELL L14SYS22-40

IDLI GROUP 83105

END YOKE WITH 21 INVOLUTED SPLINES
SEE FIGURE 5A

2.9063 in	OE	73.82 mm
1.1250 in	BD	28.58 mm
1.3750 in	SB	34.93 mm
3.4375 in	EC	87.31 mm
2.0000 in	HD	50.80 mm

NEAPCO 19135
PRECISION 1353
REX CHAIN 760053
ROCKWELL L14SYS22-36
SPICER 2-4-3111
WESCO 14N220T
WESCO 860054

IDLI GROUP 83106

END YOKE WITH 21 INVOLUTED SPLINES
SEE FIGURE 5A

2.9063 in	OE	73.82 mm
1.1250 in	BD	28.58 mm
1.3750 in	SB	34.93 mm
3.5000 in	EC	88.90 mm
2.2500 in	HD	57.15 mm

PRECISION 1353
ROCKWELL L14SYS22-33
SPICER 2-4-3981

IDLI GROUP 83107

END YOKE WITH 6 SPLINES
SEE FIGURE 5A

2.9063 in	OE	73.82 mm
1.1250 in	BD	28.58 mm
1.3750 in	SB	34.93 mm
3.5625 in	EC	90.49 mm
2.2500 in	HD	57.15 mm

ROCKWELL L14SYS22-12

IDLI GROUP 83108

END YOKE WITH ROUND BORE
SEE FIGURE 5B

2.9063 in	OE	73.82 mm
1.1250 in	BD	28.58 mm
1.3750 in	SB	34.93 mm
3.5625 in	EC	90.49 mm
2.2500 in	HD	57.15 mm

BORG-WARNR 69120
HUB CITY 0335-01417
PRECISION 1312
ROCKWELL L14SYR22-7
SPICER 2-4-1453

IDLI GROUP 83109

END YOKE WITH ROUND BORE
slight variation from above group
SEE FIGURE 5B

2.9063 in	OE	73.82 mm
1.1250 in	BD	28.58 mm
1.3750 in	SB	34.93 mm
3.5625 in	EC	90.49 mm
2.2500 in	HD	57.15 mm

ROCKWELL L14SYR22-8

IDLI GROUP 83110

END YOKE WITH ROUND BORE
slight variation from above group
SEE FIGURE 5B

2.9063 in	OE	73.82 mm
1.1250 in	BD	28.58 mm
1.3750 in	SB	34.93 mm
3.5625 in	EC	90.49 mm
2.2500 in	HD	57.15 mm

G & G MFG 184-1422
HAYES 2-4-1873
NEAPCO 19194
NEAPCO 191B1
NEAPCO 20-9194
ROCKWELL L14SYR22-12
SPICER 2-4-1163

IDLI GROUP 83111

END YOKE WITH ROUND BORE
slight variation from above group
SEE FIGURE 5B

2.9063 in	OE	73.82 mm
1.1250 in	BD	28.58 mm
1.3750 in	SB	34.93 mm
3.5625 in	EC	90.49 mm
2.2500 in	HD	57.15 mm

ROCKWELL L14SYR22-47

IDLI GROUP 83112

END YOKE WITH ROUND BORE
slight variation from above group
SEE FIGURE 5B

2.9063 in	OE	73.82 mm
1.1250 in	BD	28.58 mm
1.3750 in	SB	34.93 mm
3.5625 in	EC	90.49 mm
2.2500 in	HD	57.15 mm

ROCKWELL L14SYR22-11

IDLI GROUP 83113

END YOKE WITH ROUND BORE
slight variation from above group
SEE FIGURE 5B

2.9063 in	OE	73.82 mm
1.1250 in	BD	28.58 mm
1.3750 in	SB	34.93 mm
3.5625 in	EC	90.49 mm
2.2500 in	HD	57.15 mm

ROCKWELL L14SYR22-31

IDLI GROUP 83114

END YOKE WITH ROUND BORE
slight variation from above group
SEE FIGURE 5B

2.9063 in	OE	73.82 mm
1.1250 in	BD	28.58 mm
1.3750 in	SB	34.93 mm
3.5625 in	EC	90.49 mm
2.2500 in	HD	57.15 mm

ROCKWELL L14SYR22-43

IDLI GROUP 83115

END YOKE WITH 21 INVOLUTED SPLINES
SEE FIGURE 5A

2.9063 in	OE	73.82 mm
1.1250 in	BD	28.58 mm
1.3750 in	SB	34.93 mm
3.8750 in	EC	98.43 mm
1.7500 in	HD	44.45 mm

ROCKWELL L14SYS22-93

IDLI GROUP 83116

END YOKE WITH 6 SPLINES
SEE FIGURE 5A

2.9063 in	OE	73.82 mm
1.1250 in	BD	28.58 mm
1.3750 in	SB	34.93 mm
3.8750 in	EC	98.43 mm
2.0000 in	HD	50.80 mm

BORG-WARNR 69132
G & G MFG 192-1422
HUB CITY 0335-01404
MCQUAY-NOR U03-87
NEAPCO 19134
PRECISION 1352
PRECISION 1372
ROCKWELL L14SYS22-16
SPICER 2-4-3101
TRW 21207
WESCO 14N0220-21
WESCO 14N220-21
WESCO 860056

IDLI GROUP 83117

END YOKE WITH ROUND BORE
SEE FIGURE 5B

2.9063 in	OE	73.82 mm
1.1250 in	BD	28.58 mm
1.3750 in	SB	34.93 mm
3.8750 in	EC	98.43 mm
2.0000 in	HD	50.80 mm

ROCKWELL L14SYR22-37

IDLI GROUP 83118

END YOKE WITH ROUND BORE
slight variation from above group
SEE FIGURE 5B

2.9063 in	OE	73.82 mm
1.1250 in	BD	28.58 mm
1.3750 in	SB	34.93 mm
3.8750 in	EC	98.43 mm
2.0000 in	HD	50.80 mm

ROCKWELL L14SYR22-6

IDLI GROUP 83119

END YOKE WITH ROUND BORE
slight variation from above group
SEE FIGURE 5B

2.9063 in	OE	73.82 mm
1.1250 in	BD	28.58 mm
1.3750 in	SB	34.93 mm
3.8750 in	EC	98.43 mm
2.0000 in	HD	50.80 mm

ROCKWELL L14SYR22-60

IDLI GROUP 83120

END YOKE WITH ROUND BORE
slight variation from above group
SEE FIGURE 5B

2.9063 in	OE	73.82 mm
1.1250 in	BD	28.58 mm
1.3750 in	SB	34.93 mm
3.8750 in	EC	98.43 mm
2.0000 in	HD	50.80 mm

ROCKWELL L14SYR22-63

IDLI GROUP 83121

END YOKE WITH 10 SPLINES
SEE FIGURE 5A

2.9063 in	OE	73.82 mm
1.1250 in	BD	28.58 mm
1.3750 in	SB	34.93 mm
3.9375 in	EC	100.01 mm
1.7500 in	HD	44.45 mm

ROCKWELL L14SYS22-41

IDLI GROUP 83122

END YOKE WITH 6 SPLINES
SEE FIGURE 5A

2.9063 in	OE	73.82 mm
1.1250 in	BD	28.58 mm
1.3750 in	SB	34.93 mm
4.7500 in	EC	120.65 mm
2.0000 in	HD	50.80 mm

ROCKWELL L14SYS22-77

ENGINEERING CATALOGS MUST BE CONSULTED FOR DETAILS NOT INCLUDED
IN THIS GUIDE. SPECIFIC DESIGNS, MATERIAL CONTENT, TOLERANCES
LUBE FITTINGS AND OTHER DIMENSIONS ARE INTENTIONALLY OMITTED HERE.

IDLI GROUP 83123

END YOKE WITH RECTANGULAR BORE
maximum SB dimension shown
SEE FIGURE 5D

2.9063 in	OE	73.82 mm
1.1250 in	BD	28.58 mm
1.4063 in	SB	35.72 mm
3.1875 in	EC	80.96 mm

ROCKWELL L14SYP20-11

IDLI GROUP 83124

END YOKE WITH RECTANGULAR BORE
maximum SB dimension shown
SEE FIGURE 5D

2.9063 in	OE	73.82 mm
1.1250 in	BD	28.58 mm
1.4063 in	SB	35.72 mm
3.1875 in	EC	80.96 mm
2.3750 in	HD	60.33 mm

ROCKWELL L14SYP20-1

IDLI GROUP 83125

END YOKE WITH ROUND BORE
SEE FIGURE 5B

2.9063 in	OE	73.82 mm
1.1250 in	BD	28.58 mm
1.4375 in	SB	36.51 mm
3.5625 in	EC	90.49 mm
2.2500 in	HD	57.15 mm

BORG-WARNR 69154
G & G MFG 184-1423
HUB CITY 0335-01453
NEAPCO 19154
NEAPCO 20-9154
PRECISION 1314
ROCKWELL L14SYR23-4
SPICER 2-4-1623
TRW 21210
WESCO 14N0236
WESCO 14N236
WESCO 860058

IDLI GROUP 83126

END YOKE WITH ROUND BORE
slight variation from above group
SEE FIGURE 5B

2.9063 in	OE	73.82 mm
1.1250 in	BD	28.58 mm
1.4375 in	SB	36.51 mm
3.5625 in	EC	90.49 mm
2.2500 in	HD	57.15 mm

ROCKWELL L14SYR23-8

IDLI GROUP 83127

END YOKE WITH ROUND BORE
SEE FIGURE 5B

2.9063 in	OE	73.82 mm
1.1250 in	BD	28.58 mm
1.5000 in	SB	38.10 mm
3.5625 in	EC	90.49 mm
2.2500 in	HD	57.15 mm

BORG-WARNR 69236
HUB CITY 0335-01420
PRECISION 1313
ROCKWELL L14SYR24-3

IDLI GROUP 83128

END YOKE WITH ROUND BORE
slight variation from above group
SEE FIGURE 5B

2.9063 in	OE	73.82 mm
1.1250 in	BD	28.58 mm
1.5000 in	SB	38.10 mm
3.5625 in	EC	90.49 mm
2.2500 in	HD	57.15 mm

ROCKWELL L14SYR24-24

IDLI GROUP 83129

END YOKE WITH ROUND BORE
slight variation from above group
SEE FIGURE 5B

2.9063 in	OE	73.82 mm
1.1250 in	BD	28.58 mm
1.5000 in	SB	38.10 mm
3.5625 in	EC	90.49 mm
2.2500 in	HD	57.15 mm

ALLOY 0715
ALLOY 715
BORG-WARNR 67238
G & G MFG 184-1424
HAYES 2-4-1463
HAYES 2-4-1613
HUB CITY 0335-01420
NEAPCO 171C8
NEAPCO 18-7313
NEAPCO 19124
NEAPCO 19190
NEAPCO 191C8
NEAPCO 20-9190
NEAPCO 20-9307
PRECISION 1313
REX CHAIN 760055
REX CHAIN 760092
ROCKWELL L14SYR24-10
SPICER 2-4-1463
TRW 21211
WESCO 14N0246
WESCO 14N246
WESCO 860060

IDLI GROUP 83130

END YOKE WITH 11 INVOLUTED SPLINES
SEE FIGURE 5A

2.9063 in	OE	73.82 mm
1.1250 in	BD	28.58 mm
1.5000 in	SB	38.10 mm
3.6250 in	EC	92.08 mm
2.1250 in	HD	53.98 mm

ROCKWELL L14SYS24-6

IDLI GROUP 83131

END YOKE WITH 27 INVOLUTED SPLINES
SEE FIGURE 5A

2.9063 in	OE	73.82 mm
1.1250 in	BD	28.58 mm
1.7500 in	SB	44.45 mm
3.5625 in	EC	90.49 mm
2.2500 in	HD	57.15 mm

ROCKWELL L14SYS28-4

IDLI GROUP 83132

END YOKE WITH ROUND BORE
SEE FIGURE 5B

3.1250 in	OE	79.38 mm
.8750 in	BD	22.23 mm
.7500 in	SB	19.05 mm
2.5000 in	EC	63.50 mm
1.7500 in	HD	44.45 mm

ROCKWELL 1FRYR12-12

IDLI GROUP 83133

END YOKE WITH ROUND BORE
SEE FIGURE 5B

3.1250 in	OE	79.38 mm
.8750 in	BD	22.23 mm
.7500 in	SB	19.05 mm
2.5000 in	EC	63.50 mm
2.0000 in	HD	50.80 mm

ROCKWELL 1FRYR12-8

IDLI GROUP 83134

END YOKE WITH SQUARE BORE
SEE FIGURE 5C

3.1250 in	OE	79.38 mm
.8750 in	BD	22.23 mm
.7500 in	SB	19.05 mm
2.5000 in	EC	63.50 mm
2.0000 in	HD	50.80 mm

ROCKWELL 1FRYQ12-11

IDLI GROUP 83135

END YOKE WITH ROUND BORE
SEE FIGURE 5B

3.1250 in	OE	79.38 mm
.8750 in	BD	22.23 mm
.7500 in	SB	19.05 mm
3.6250 in	EC	92.08 mm
2.0000 in	HD	50.80 mm

ROCKWELL 1FRYR12-18

IDLI GROUP 83136

END YOKE WITH SQUARE BORE
SEE FIGURE 5C

3.1250 in	OE	79.38 mm
.8750 in	BD	22.23 mm
.7500 in	SB	19.05 mm
3.7500 in	EC	95.25 mm
2.0000 in	HD	50.80 mm

ROCKWELL 1FRYQ12-13

IDLI GROUP 83137

END YOKE WITH ROUND BORE
SEE FIGURE 5B

3.1250 in	OE	79.38 mm
.8750 in	BD	22.23 mm
.8125 in	SB	20.64 mm
2.5000 in	EC	63.50 mm
2.0000 in	HD	50.80 mm

ROCKWELL 1FRYR13-7

IDLI GROUP 83138

END YOKE WITH ROUND BORE
slight variation from above group
SEE FIGURE 5B

3.1250 in	OE	79.38 mm
.8750 in	BD	22.23 mm
.8125 in	SB	20.64 mm
2.5000 in	EC	63.50 mm
2.0000 in	HD	50.80 mm

ROCKWELL 1FRYR13-2

IDLI GROUP 83139

END YOKE WITH ROUND BORE
SEE FIGURE 5B

3.1250 in	OE	79.38 mm
.8750 in	BD	22.23 mm
.8125 in	SB	20.64 mm
3.7500 in	EC	95.25 mm
2.0000 in	HD	50.80 mm

ROCKWELL 1FRYR13-9

IDLI GROUP 83140

END YOKE WITH ROUND BORE
SEE FIGURE 5B

3.1250 in	OE	79.38 mm
.8750 in	BD	22.23 mm
.8750 in	SB	22.23 mm
2.5000 in	EC	63.50 mm
2.0000 in	HD	50.80 mm

ROCKWELL 1FRYR14-22

IDLI GROUP 83141

END YOKE WITH ROUND BORE
slight variation from above group
SEE FIGURE 5B

3.1250 in	OE	79.38 mm
.8750 in	BD	22.23 mm
.8750 in	SB	22.23 mm
2.5000 in	EC	63.50 mm
2.0000 in	HD	50.80 mm

ROCKWELL 1FRYR14-23

IDLI GROUP 83142

END YOKE WITH SQUARE BORE
SEE FIGURE 5C

3.1250 in	OE	79.38 mm
.8750 in	BD	22.23 mm
.8750 in	SB	22.23 mm
2.5000 in	EC	63.50 mm
2.0000 in	HD	50.80 mm

ROCKWELL 1FRYQ14-3

IDLI GROUP 83143

END YOKE WITH SQUARE BORE
SEE FIGURE 5C

3.1250 in	OE	79.38 mm
.8750 in	BD	22.23 mm
.8750 in	SB	22.23 mm
3.1875 in	EC	80.96 mm
1.5000 in	HD	38.10 mm

ROCKWELL 1FRYQ14-9

BD = BEARING DIAMETER (outside)   **BW** = BEARING WIDTH   **CB** = CROSS LENGTH WITH BEARINGS   **CC** = CENTER TO CENTER   **CD** = CROSS DIAMETER
CL = CROSS LENGTH WITHOUT BEARINGS   **EC** = END TO CENTER or FACE TO FACE   **EE** = END TO END   **EL** = EFFECTIVE LENGTH
HD = HUB DIAMETER (or insert)   **IE** = INSIDE OF EARS (or recess)   **OD** = OUTSIDE DIAMETER   **OE** = OUTSIDE OF EARS   **SB** = SPLINE OR BORE SIZE

CAUTION: BE SURE TO REFER TO ENGINEERING CATALOGS FOR SPECIAL APPLICATIONS THAT REQUIRE SPECIFIC MATERIAL CONTENT, TOLERANCES, ETC. SEE FOOTNOTE.

IDLI GROUP 83144

END YOKE WITH RECTANGULAR BORE
maximum SB dimension shown
SEE FIGURE 5D

3.1250 in	OE	79.38 mm
.8750 in	BD	22.23 mm
.8750 in	SB	22.23 mm
3.1875 in	EC	80.96 mm
1.5000 in	HD	38.10 mm
ROCKWELL		1FRYP12-10

IDLI GROUP 83145

END YOKE WITH ROUND BORE
SEE FIGURE 5B

3.1250 in	OE	79.38 mm
.8750 in	BD	22.23 mm
.8750 in	SB	22.23 mm
3.7500 in	EC	95.25 mm
2.0000 in	HD	50.80 mm
ROCKWELL		1FRYR14-29

IDLI GROUP 83146

END YOKE WITH ROUND BORE
slight variation from above group
SEE FIGURE 5B

3.1250 in	OE	79.38 mm
.8750 in	BD	22.23 mm
.8750 in	SB	22.23 mm
3.7500 in	EC	95.25 mm
2.0000 in	HD	50.80 mm
ROCKWELL		1FRYR14-26

IDLI GROUP 83147

END YOKE WITH ROUND BORE
SEE FIGURE 5B

3.1250 in	OE	79.38 mm
.8750 in	BD	22.23 mm
.9375 in	SB	23.81 mm
2.5000 in	EC	63.50 mm
1.7500 in	HD	44.45 mm
ROCKWELL		1FRYR15-7

IDLI GROUP 83148

END YOKE WITH SQUARE BORE
SEE FIGURE 5C

3.1250 in	OE	79.38 mm
.8750 in	BD	22.23 mm
.9375 in	SB	23.81 mm
2.5000 in	EC	63.50 mm
2.0000 in	HD	50.80 mm
ROCKWELL		1FRYQ15-9

IDLI GROUP 83149

END YOKE WITH SQUARE BORE
SEE FIGURE 5C

3.1250 in	OE	79.38 mm
.8750 in	BD	22.23 mm
.9375 in	SB	23.81 mm
3.1875 in	EC	80.96 mm
1.5000 in	HD	38.10 mm
ROCKWELL		1FRYQ15-5

IDLI GROUP 83150

END YOKE WITH ROUND BORE
SEE FIGURE 5B

3.1250 in	OE	79.38 mm
.8750 in	BD	22.23 mm
.9375 in	SB	23.81 mm
3.7500 in	EC	95.25 mm
2.0000 in	HD	50.80 mm
ROCKWELL		1FRYR15-15

IDLI GROUP 83151

END YOKE WITH ROUND BORE
slight variation from above group
SEE FIGURE 5B

3.1250 in	OE	79.38 mm
.8750 in	BD	22.23 mm
.9375 in	SB	23.81 mm
3.7500 in	EC	95.25 mm
2.0000 in	HD	50.80 mm
ROCKWELL		1FRYR15-16

IDLI GROUP 83152

END YOKE WITH SQUARE BORE
SEE FIGURE 5C

3.1250 in	OE	79.38 mm
.8750 in	BD	22.23 mm
.9375 in	SB	23.81 mm
3.8750 in	EC	98.43 mm
1.9375 in	HD	49.21 mm
ROCKWELL		1FRYQ15-8

IDLI GROUP 83153

END YOKE WITH ROUND BORE
SEE FIGURE 5B

3.1250 in	OE	79.38 mm
.8750 in	BD	22.23 mm
.9844 in	SB	25.00 mm
3.7500 in	EC	95.25 mm
2.0000 in	HD	50.80 mm
ROCKWELL		1FRYR16-74

IDLI GROUP 83154

END YOKE WITH 10 SPLINES
SEE FIGURE 5A

3.1250 in	OE	79.38 mm
.8750 in	BD	22.23 mm
1.0000 in	SB	25.40 mm
2.2500 in	EC	57.15 mm
1.8750 in	HD	47.63 mm
ROCKWELL		1FRYS16-3

IDLI GROUP 83155

END YOKE WITH ROUND BORE
SEE FIGURE 5B

3.1250 in	OE	79.38 mm
.8750 in	BD	22.23 mm
1.0000 in	SB	25.40 mm
2.5000 in	EC	63.50 mm
1.8750 in	HD	47.63 mm
ROCKWELL		1FRYR16-4

IDLI GROUP 83156

END YOKE WITH ROUND BORE
slight variation from above group
SEE FIGURE 5B

3.1250 in	OE	79.38 mm
.8750 in	BD	22.23 mm
1.0000 in	SB	25.40 mm
2.5000 in	EC	63.50 mm
1.8750 in	HD	47.63 mm
ROCKWELL		1FRYR16-22

IDLI GROUP 83157

END YOKE WITH ROUND BORE
SEE FIGURE 5B

3.1250 in	OE	79.38 mm
.8750 in	BD	22.23 mm
1.0000 in	SB	25.40 mm
2.5000 in	EC	63.50 mm
2.0000 in	HD	50.80 mm
ROCKWELL		1FRYR16-9

IDLI GROUP 83158

END YOKE WITH ROUND BORE
slight variation from above group
SEE FIGURE 5B

3.1250 in	OE	79.38 mm
.8750 in	BD	22.23 mm
1.0000 in	SB	25.40 mm
2.5000 in	EC	63.50 mm
2.0000 in	HD	50.80 mm
ROCKWELL		1FRYR16-3

IDLI GROUP 83159

END YOKE WITH SQUARE BORE
SEE FIGURE 5C

3.1250 in	OE	79.38 mm
.8750 in	BD	22.23 mm
1.0000 in	SB	25.40 mm
2.5000 in	EC	63.50 mm
2.0000 in	HD	50.80 mm
ROCKWELL		1FRYQ16-4

IDLI GROUP 83160

END YOKE WITH 6 SPLINES
SEE FIGURE 5A

3.1250 in	OE	79.38 mm
.8750 in	BD	22.23 mm
1.0000 in	SB	25.40 mm
2.6250 in	EC	66.68 mm
1.8750 in	HD	47.63 mm
ROCKWELL		1FRYS16-2

IDLI GROUP 83161

END YOKE WITH ROUND BORE
SEE FIGURE 5B

3.1250 in	OE	79.38 mm
.8750 in	BD	22.23 mm
1.0000 in	SB	25.40 mm
2.8750 in	EC	73.03 mm
2.0000 in	HD	50.80 mm
ROCKWELL		1FRYR16-48

IDLI GROUP 83162

END YOKE WITH THREADED BORE
SEE FIGURE 5G

3.1250 in	OE	79.38 mm
.8750 in	BD	22.23 mm
1.0000 in	SB	25.40 mm
2.8750 in	EC	73.03 mm
2.0000 in	HD	50.80 mm
ROCKWELL		1FRYD16-1

IDLI GROUP 83163

END YOKE WITH ROUND BORE
SEE FIGURE 5B

3.1250 in	OE	79.38 mm
.8750 in	BD	22.23 mm
1.0000 in	SB	25.40 mm
3.1250 in	EC	79.38 mm
2.0000 in	HD	50.80 mm
ROCKWELL		1FRYR16-94

IDLI GROUP 83164

END YOKE WITH SQUARE BORE
SEE FIGURE 5C

3.1250 in	OE	79.38 mm
.8750 in	BD	22.23 mm
1.0000 in	SB	25.40 mm
3.1875 in	EC	80.96 mm
1.7500 in	HD	44.45 mm
ROCKWELL		1FRYQ16-10

IDLI GROUP 83165

END YOKE WITH ROUND BORE
SEE FIGURE 5B

3.1250 in	OE	79.38 mm
.8750 in	BD	22.23 mm
1.0000 in	SB	25.40 mm
3.2500 in	EC	82.55 mm
2.0000 in	HD	50.80 mm
ROCKWELL		1FRYR16-23

IDLI GROUP 83166

END YOKE WITH ROUND BORE
SEE FIGURE 5B

3.1250 in	OE	79.38 mm
.8750 in	BD	22.23 mm
1.0000 in	SB	25.40 mm
3.7500 in	EC	95.25 mm
1.7500 in	HD	44.45 mm
ROCKWELL		1FRYR16-80

IDLI GROUP 83167

END YOKE WITH ROUND BORE
SEE FIGURE 5B

3.1250 in	OE	79.38 mm
.8750 in	BD	22.23 mm
1.0000 in	SB	25.40 mm
3.7500 in	EC	95.25 mm
2.0000 in	HD	50.80 mm
ROCKWELL		1FRYR16-26

ENGINEERING CATALOGS MUST BE CONSULTED FOR DETAILS NOT INCLUDED IN THIS GUIDE. SPECIFIC DESIGNS, MATERIAL CONTENT, TOLERANCES LUBE FITTINGS AND OTHER DIMENSIONS ARE INTENTIONALLY OMITTED HERE.

IDLI GROUP 83168

END YOKE WITH ROUND BORE
slight variation from above group
SEE FIGURE 5B

3.1250 in	OE	79.38 mm
.8750 in	BD	22.23 mm
1.0000 in	SB	25.40 mm
3.7500 in	EC	95.25 mm
2.0000 in	HD	50.80 mm

ROCKWELL 1FRYR16-66

IDLI GROUP 83169

END YOKE WITH ROUND BORE
slight variation from above group
SEE FIGURE 5B

3.1250 in	OE	79.38 mm
.8750 in	BD	22.23 mm
1.0000 in	SB	25.40 mm
3.7500 in	EC	95.25 mm
2.0000 in	HD	50.80 mm

ROCKWELL 1FRYR16-2

IDLI GROUP 83170

END YOKE WITH ROUND BORE
slight variation from above group
SEE FIGURE 5B

3.1250 in	OE	79.38 mm
.8750 in	BD	22.23 mm
1.0000 in	SB	25.40 mm
3.7500 in	EC	95.25 mm
2.0000 in	HD	50.80 mm

ROCKWELL 1FRYR16-33

IDLI GROUP 83171

END YOKE WITH ROUND BORE
slight variation from above group
SEE FIGURE 5B

3.1250 in	OE	79.38 mm
.8750 in	BD	22.23 mm
1.0000 in	SB	25.40 mm
3.7500 in	EC	95.25 mm
2.0000 in	HD	50.80 mm

ROCKWELL 1FRYR16-11

IDLI GROUP 83172

END YOKE WITH SQUARE BORE
SEE FIGURE 5C

3.1250 in	OE	79.38 mm
.8750 in	BD	22.23 mm
1.0000 in	SB	25.40 mm
3.7500 in	EC	95.25 mm
2.0000 in	HD	50.80 mm

ROCKWELL 1FRYQ16-24

IDLI GROUP 83173

END YOKE WITH ROUND BORE
SEE FIGURE 5B

3.1250 in	OE	79.38 mm
.8750 in	BD	22.23 mm
1.0625 in	SB	26.99 mm
2.5000 in	EC	63.50 mm
2.0000 in	HD	50.80 mm

ROCKWELL 1FRYR17-8

IDLI GROUP 83174

END YOKE WITH ROUND BORE
slight variation from above group
SEE FIGURE 5B

3.1250 in	OE	79.38 mm
.8750 in	BD	22.23 mm
1.0625 in	SB	26.99 mm
2.5000 in	EC	63.50 mm
2.0000 in	HD	50.80 mm

ROCKWELL 1FRYR17-11

IDLI GROUP 83175

END YOKE WITH ROUND BORE
SEE FIGURE 5B

3.1250 in	OE	79.38 mm
.8750 in	BD	22.23 mm
1.0625 in	SB	26.99 mm
2.7500 in	EC	69.85 mm
1.7500 in	HD	44.45 mm

ROCKWELL 1FRYR17

IDLI GROUP 83176

END YOKE WITH ROUND BORE
SEE FIGURE 5B

3.1250 in	OE	79.38 mm
.8750 in	BD	22.23 mm
1.0625 in	SB	26.99 mm
2.8750 in	EC	73.03 mm
2.0000 in	HD	50.80 mm

ROCKWELL 1FRYR17-9

IDLI GROUP 83177

END YOKE WITH ROUND BORE
SEE FIGURE 5B

3.1250 in	OE	79.38 mm
.8750 in	BD	22.23 mm
1.0625 in	SB	26.99 mm
3.7500 in	EC	95.25 mm
2.0000 in	HD	50.80 mm

ROCKWELL 1FRYR17-17

IDLI GROUP 83178

END YOKE WITH ROUND BORE
slight variation from above group
SEE FIGURE 5B

3.1250 in	OE	79.38 mm
.8750 in	BD	22.23 mm
1.0625 in	SB	26.99 mm
3.7500 in	EC	95.25 mm
2.0000 in	HD	50.80 mm

ROCKWELL 1FRYR17-4

IDLI GROUP 83179

END YOKE WITH ROUND BORE
SEE FIGURE 5B

3.1250 in	OE	79.38 mm
.8750 in	BD	22.23 mm
1.1250 in	SB	28.58 mm
2.5000 in	EC	63.50 mm
1.8750 in	HD	47.63 mm

ROCKWELL 1FRYR18-11

IDLI GROUP 83180

END YOKE WITH 6 SPLINES
SEE FIGURE 5A

3.1250 in	OE	79.38 mm
.8750 in	BD	22.23 mm
1.1250 in	SB	28.58 mm
2.5000 in	EC	63.50 mm
2.0000 in	HD	50.80 mm

ROCKWELL 1FRYS18-27

IDLI GROUP 83181

END YOKE WITH 10 SPLINES
SEE FIGURE 5A

3.1250 in	OE	79.38 mm
.8750 in	BD	22.23 mm
1.1250 in	SB	28.58 mm
2.5000 in	EC	63.50 mm
2.0000 in	HD	50.80 mm

ROCKWELL 1FRYS18-18

IDLI GROUP 83182

END YOKE WITH ROUND BORE
SEE FIGURE 5B

3.1250 in	OE	79.38 mm
.8750 in	BD	22.23 mm
1.1250 in	SB	28.58 mm
2.5000 in	EC	63.50 mm
2.0000 in	HD	50.80 mm

ROCKWELL 1FRYR18-56

IDLI GROUP 83183

END YOKE WITH ROUND BORE
slight variation from above group
SEE FIGURE 5B

3.1250 in	OE	79.38 mm
.8750 in	BD	22.23 mm
1.1250 in	SB	28.58 mm
2.5000 in	EC	63.50 mm
2.0000 in	HD	50.80 mm

ROCKWELL 1FRYR18-32

IDLI GROUP 83184

END YOKE WITH SQUARE BORE
SEE FIGURE 5C

3.1250 in	OE	79.38 mm
.8750 in	BD	22.23 mm
1.1250 in	SB	28.58 mm
2.5000 in	EC	63.50 mm
2.0000 in	HD	50.80 mm

ROCKWELL 1FRYQ18

IDLI GROUP 83185

END YOKE WITH ROUND BORE
SEE FIGURE 5B

3.1250 in	OE	79.38 mm
.8750 in	BD	22.23 mm
1.1250 in	SB	28.58 mm
2.5625 in	EC	65.09 mm
2.0000 in	HD	50.80 mm

ROCKWELL 1FRYR18-79

IDLI GROUP 83186

END YOKE WITH 6 SPLINES
SEE FIGURE 5A

3.1250 in	OE	79.38 mm
.8750 in	BD	22.23 mm
1.1250 in	SB	28.58 mm
2.7500 in	EC	69.85 mm
1.7500 in	HD	44.45 mm

ROCKWELL 1FRYS18-29

IDLI GROUP 83187

END YOKE WITH ROUND BORE
SEE FIGURE 5B

3.1250 in	OE	79.38 mm
.8750 in	BD	22.23 mm
1.1250 in	SB	28.58 mm
2.8750 in	EC	73.03 mm
2.0000 in	HD	50.80 mm

ROCKWELL 1FRYR18-33

IDLI GROUP 83188

END YOKE WITH SQUARE BORE
SEE FIGURE 5C

3.1250 in	OE	79.38 mm
.8750 in	BD	22.23 mm
1.1250 in	SB	28.58 mm
3.1875 in	EC	80.96 mm
1.7500 in	HD	44.45 mm

ROCKWELL 1FRYQ18-3

IDLI GROUP 83189

END YOKE WITH 6 SPLINES
SEE FIGURE 5A

3.1250 in	OE	79.38 mm
.8750 in	BD	22.23 mm
1.1250 in	SB	28.58 mm
3.6250 in	EC	92.08 mm
2.0000 in	HD	50.80 mm

ROCKWELL 1FRYS18-2

IDLI GROUP 83190

END YOKE WITH ROUND BORE
SEE FIGURE 5B

3.1250 in	OE	79.38 mm
.8750 in	BD	22.23 mm
1.1250 in	SB	28.58 mm
3.6250 in	EC	92.08 mm
2.0000 in	HD	50.80 mm

ROCKWELL 1FRYR18-6

IDLI GROUP 83191

END YOKE WITH 6 SPLINES
SEE FIGURE 5A

3.1250 in	OE	79.38 mm
.8750 in	BD	22.23 mm
1.1250 in	SB	28.58 mm
3.7500 in	EC	95.25 mm
1.8750 in	HD	47.63 mm

ROCKWELL 1FRYS18-3

BD = BEARING DIAMETER (outside)   **BW** = BEARING WIDTH   **CB** = CROSS LENGTH WITH BEARINGS   **CC** = CENTER TO CENTER   **CD** = CROSS DIAMETER
CL = CROSS LENGTH WITHOUT BEARINGS   **EC** = END TO CENTER or FACE TO FACE   **EE** = END TO END   **EL** = EFFECTIVE LENGTH
HD = HUB DIAMETER (or insert)   **IE** = INSIDE OF EARS (or recess)   **OD** = OUTSIDE DIAMETER   **OE** = OUTSIDE OF EARS   **SB** = SPLINE OR BORE SIZE

CAUTION: BE SURE TO REFER TO ENGINEERING CATALOGS FOR SPECIAL APPLICATIONS THAT REQUIRE SPECIFIC MATERIAL CONTENT, TOLERANCES, ETC. SEE FOOTNOTE.

IDLI GROUP 83192

END YOKE WITH ROUND BORE
SEE FIGURE 5B

3.1250 in	OE	79.38 mm
.8750 in	BD	22.23 mm
1.1250 in	SB	28.58 mm
3.7500 in	EC	95.25 mm
2.0000 in	HD	50.80 mm

ROCKWELL 1FRYR18-10

IDLI GROUP 83193

END YOKE WITH ROUND BORE
slight variation from above group
SEE FIGURE 5B

3.1250 in	OE	79.38 mm
.8750 in	BD	22.23 mm
1.1250 in	SB	28.58 mm
3.7500 in	EC	95.25 mm
2.0000 in	HD	50.80 mm

ROCKWELL 1FRYR18-7

IDLI GROUP 83194

END YOKE WITH ROUND BORE
slight variation from above group
SEE FIGURE 5B

3.1250 in	OE	79.38 mm
.8750 in	BD	22.23 mm
1.1250 in	SB	28.58 mm
3.7500 in	EC	95.25 mm
2.0000 in	HD	50.80 mm

ROCKWELL 1FRYR18-8

IDLI GROUP 83195

END YOKE WITH ROUND BORE
slight variation from above group
SEE FIGURE 5B

3.1250 in	OE	79.38 mm
.8750 in	BD	22.23 mm
1.1250 in	SB	28.58 mm
3.7500 in	EC	95.25 mm
2.0000 in	HD	50.80 mm

ROCKWELL 1FRYR18-85

IDLI GROUP 83196

END YOKE WITH SQUARE BORE
SEE FIGURE 5C

3.1250 in	OE	79.38 mm
.8750 in	BD	22.23 mm
1.1250 in	SB	28.58 mm
3.7500 in	EC	95.25 mm
2.0000 in	HD	50.80 mm

ROCKWELL 1FRYQ18-1

IDLI GROUP 83197

END YOKE WITH ROUND BORE
SEE FIGURE 5B

3.1250 in	OE	79.38 mm
.8750 in	BD	22.23 mm
1.1875 in	SB	30.16 mm
2.5000 in	EC	63.50 mm
1.8750 in	HD	47.63 mm

ROCKWELL 1FRYR19-17

IDLI GROUP 83198

END YOKE WITH ROUND BORE
SEE FIGURE 5B

3.1250 in	OE	79.38 mm
.8750 in	BD	22.23 mm
1.1875 in	SB	30.16 mm
2.5000 in	EC	63.50 mm
2.0000 in	HD	50.80 mm

ROCKWELL 1FRYR19-13

IDLI GROUP 83199

END YOKE WITH ROUND BORE
SEE FIGURE 5B

3.1250 in	OE	79.38 mm
.8750 in	BD	22.23 mm
1.1875 in	SB	30.16 mm
2.5625 in	EC	65.09 mm
2.0000 in	HD	50.80 mm

ROCKWELL 1FRYR19-23

IDLI GROUP 83200

END YOKE WITH SQUARE BORE
SEE FIGURE 5C

3.1250 in	OE	79.38 mm
.8750 in	BD	22.23 mm
1.1875 in	SB	30.16 mm
3.1875 in	EC	80.96 mm
1.7500 in	HD	44.45 mm

ROCKWELL 1FRYQ19-1

IDLI GROUP 83201

END YOKE WITH ROUND BORE
SEE FIGURE 5B

3.1250 in	OE	79.38 mm
.8750 in	BD	22.23 mm
1.1875 in	SB	30.16 mm
3.5000 in	EC	88.90 mm
2.0000 in	HD	50.80 mm

ROCKWELL 1FRYR19-19

IDLI GROUP 83202

END YOKE WITH ROUND BORE
SEE FIGURE 5B

3.1250 in	OE	79.38 mm
.8750 in	BD	22.23 mm
1.1875 in	SB	30.16 mm
3.7500 in	EC	95.25 mm
1.9375 in	HD	49.21 mm

ROCKWELL 1FRYR19-12

IDLI GROUP 83203

END YOKE WITH ROUND BORE
SEE FIGURE 5B

3.1250 in	OE	79.38 mm
.8750 in	BD	22.23 mm
1.1875 in	SB	30.16 mm
3.7500 in	EC	95.25 mm
2.0000 in	HD	50.80 mm

ROCKWELL 1FRYR19-25

IDLI GROUP 83204

END YOKE WITH ROUND BORE
slight variation from above group
SEE FIGURE 5B

3.1250 in	OE	79.38 mm
.8750 in	BD	22.23 mm
1.1875 in	SB	30.16 mm
3.7500 in	EC	95.25 mm
2.0000 in	HD	50.80 mm

ROCKWELL 1FRYR19-27

IDLI GROUP 83205

END YOKE WITH 10 SPLINES
SEE FIGURE 5A

3.1250 in	OE	79.38 mm
.8750 in	BD	22.23 mm
1.1875 in	SB	30.16 mm
3.8750 in	EC	98.43 mm
1.6250 in	HD	41.28 mm

ROCKWELL 1FRYS19

IDLI GROUP 83206

END YOKE WITH ROUND BORE
SEE FIGURE 5B

3.1250 in	OE	79.38 mm
.8750 in	BD	22.23 mm
1.1875 in	SB	30.16 mm
3.8750 in	EC	98.43 mm
1.7500 in	HD	44.45 mm

ROCKWELL 1FRYR19-2

IDLI GROUP 83207

END YOKE WITH 6 SPLINES
SEE FIGURE 5A

3.1250 in	OE	79.38 mm
.8750 in	BD	22.23 mm
1.2500 in	SB	31.75 mm
2.5000 in	EC	63.50 mm
2.0000 in	HD	50.80 mm

ROCKWELL 1FRYS20-10

IDLI GROUP 83208

END YOKE WITH 19 INVOLUTED SPLINES
SEE FIGURE 5A

3.1250 in	OE	79.38 mm
.8750 in	BD	22.23 mm
1.2500 in	SB	31.75 mm
2.5000 in	EC	63.50 mm
2.0000 in	HD	50.80 mm

ROCKWELL 1FRYS20-25

IDLI GROUP 83209

END YOKE WITH ROUND BORE
SEE FIGURE 5B

3.1250 in	OE	79.38 mm
.8750 in	BD	22.23 mm
1.2500 in	SB	31.75 mm
2.5000 in	EC	63.50 mm
2.0000 in	HD	50.80 mm

ROCKWELL 1FRYR20-21

IDLI GROUP 83210

END YOKE WITH ROUND BORE
slight variation from above group
SEE FIGURE 5B

3.1250 in	OE	79.38 mm
.8750 in	BD	22.23 mm
1.2500 in	SB	31.75 mm
2.5000 in	EC	63.50 mm
2.0000 in	HD	50.80 mm

ROCKWELL 1FRYR20-34

IDLI GROUP 83211

END YOKE WITH ROUND BORE
slight variation from above group
SEE FIGURE 5B

3.1250 in	OE	79.38 mm
.8750 in	BD	22.23 mm
1.2500 in	SB	31.75 mm
2.5000 in	EC	63.50 mm
2.0000 in	HD	50.80 mm

ROCKWELL 1FRYR20-10

IDLI GROUP 83212

END YOKE WITH SQUARE BORE
SEE FIGURE 5C

3.1250 in	OE	79.38 mm
.8750 in	BD	22.23 mm
1.2500 in	SB	31.75 mm
3.1875 in	EC	80.96 mm
1.7500 in	HD	44.45 mm

ROCKWELL 1FRYQ20-1

IDLI GROUP 83213

END YOKE WITH ROUND BORE
SEE FIGURE 5B

3.1250 in	OE	79.38 mm
.8750 in	BD	22.23 mm
1.2500 in	SB	31.75 mm
3.5000 in	EC	88.90 mm
2.0000 in	HD	50.80 mm

ROCKWELL 1FRYR20-58

IDLI GROUP 83214

END YOKE WITH 6 SPLINES
SEE FIGURE 5A

3.1250 in	OE	79.38 mm
.8750 in	BD	22.23 mm
1.2500 in	SB	31.75 mm
3.7500 in	EC	95.25 mm
2.0000 in	HD	50.80 mm

ROCKWELL 1FRYS20-15

ENGINEERING CATALOGS MUST BE CONSULTED FOR DETAILS NOT INCLUDED
IN THIS GUIDE. SPECIFIC DESIGNS, MATERIAL CONTENT, TOLERANCES
LUBE FITTINGS AND OTHER DIMENSIONS ARE INTENTIONALLY OMITTED HERE.

IDLI GROUP 83215

END YOKE WITH ROUND BORE
SEE FIGURE 5B

3.1250 in	OE	79.38 mm
.8750 in	BD	22.23 mm
1.2500 in	SB	31.75 mm
3.7500 in	EC	95.25 mm
2.0000 in	HD	50.80 mm

ROCKWELL 1FRYR20-65

IDLI GROUP 83216

END YOKE WITH ROUND BORE
slight variation from above group
SEE FIGURE 5B

3.1250 in	OE	79.38 mm
.8750 in	BD	22.23 mm
1.2500 in	SB	31.75 mm
3.7500 in	EC	95.25 mm
2.0000 in	HD	50.80 mm

ROCKWELL 1FRYR20-11

IDLI GROUP 83217

END YOKE WITH ROUND BORE
slight variation from above group
SEE FIGURE 5B

3.1250 in	OE	79.38 mm
.8750 in	BD	22.23 mm
1.2500 in	SB	31.75 mm
3.7500 in	EC	95.25 mm
2.0000 in	HD	50.80 mm

ROCKWELL 1FRYR20-2

IDLI GROUP 83218

END YOKE WITH ROUND BORE
slight variation from above group
SEE FIGURE 5B

3.1250 in	OE	79.38 mm
.8750 in	BD	22.23 mm
1.2500 in	SB	31.75 mm
3.7500 in	EC	95.25 mm
2.0000 in	HD	50.80 mm

ROCKWELL 1FRYR20-1

IDLI GROUP 83219

END YOKE WITH SQUARE BORE
SEE FIGURE 5C

3.1250 in	OE	79.38 mm
.8750 in	BD	22.23 mm
1.3125 in	SB	33.34 mm
3.1875 in	EC	80.96 mm
2.0000 in	HD	50.80 mm

ROCKWELL 1FRYQ21

IDLI GROUP 83220

END YOKE WITH 6 SPLINES
SEE FIGURE 5A

3.1250 in	OE	79.38 mm
.8750 in	BD	22.23 mm
1.3750 in	SB	34.93 mm
2.5000 in	EC	63.50 mm
1.8750 in	HD	47.63 mm

ROCKWELL 1FRYS22-17

IDLI GROUP 83221

END YOKE WITH ROUND BORE
SEE FIGURE 5B

3.1250 in	OE	79.38 mm
.8750 in	BD	22.23 mm
1.3750 in	SB	34.93 mm
2.5000 in	EC	63.50 mm
2.0000 in	HD	50.80 mm

ROCKWELL 1FRYR22-13

IDLI GROUP 83222

END YOKE WITH ROUND BORE
slight variation from above group
SEE FIGURE 5B

3.1250 in	OE	79.38 mm
.8750 in	BD	22.23 mm
1.3750 in	SB	34.93 mm
2.5000 in	EC	63.50 mm
2.0000 in	HD	50.80 mm

ROCKWELL 1FRYR22

IDLI GROUP 83223

END YOKE WITH 6 SPLINES
SEE FIGURE 5A

3.1250 in	OE	79.38 mm
.8750 in	BD	22.23 mm
1.3750 in	SB	34.93 mm
3.7500 in	EC	95.25 mm
2.0000 in	HD	50.80 mm

ROCKWELL 1FRYS22

IDLI GROUP 83224

END YOKE WITH 10 SPLINES
SEE FIGURE 5A

3.1250 in	OE	79.38 mm
.8750 in	BD	22.23 mm
1.3750 in	SB	34.93 mm
3.7500 in	EC	95.25 mm
2.0000 in	HD	50.80 mm

ROCKWELL 1FRYS22-28

IDLI GROUP 83225

END YOKE WITH ROUND BORE
SEE FIGURE 5B

3.1250 in	OE	79.38 mm
.8750 in	BD	22.23 mm
1.3750 in	SB	34.93 mm
3.7500 in	EC	95.25 mm
2.0000 in	HD	50.80 mm

ROCKWELL 1FRYR22-11

IDLI GROUP 83226

END YOKE WITH ROUND BORE
slight variation from above group
SEE FIGURE 5B

3.1250 in	OE	79.38 mm
.8750 in	BD	22.23 mm
1.3750 in	SB	34.93 mm
3.7500 in	EC	95.25 mm
2.0000 in	HD	50.80 mm

ROCKWELL 1FRYR22-5

IDLI GROUP 83227

END YOKE WITH ROUND BORE
slight variation from above group
SEE FIGURE 5B

3.1250 in	OE	79.38 mm
.8750 in	BD	22.23 mm
1.3750 in	SB	34.93 mm
3.7500 in	EC	95.25 mm
2.0000 in	HD	50.80 mm

ROCKWELL 1FRYR22-2

IDLI GROUP 83228

END YOKE WITH ROUND BORE
slight variation from above group
SEE FIGURE 5B

3.1250 in	OE	79.38 mm
.8750 in	BD	22.23 mm
1.3750 in	SB	34.93 mm
3.7500 in	EC	95.25 mm
2.0000 in	HD	50.80 mm

ROCKWELL 1FRYR22-1

IDLI GROUP 83229

END YOKE WITH ROUND BORE
SEE FIGURE 5B

3.1250 in	OE	79.38 mm
.8750 in	BD	22.23 mm
1.5000 in	SB	38.10 mm
2.5000 in	EC	63.50 mm
2.0000 in	HD	50.80 mm

ROCKWELL 1FRYR24

IDLI GROUP 83230

END YOKE WITH SQUARE BORE
SEE FIGURE 5C

3.1250 in	OE	79.38 mm
.8750 in	BD	22.23 mm
1.5000 in	SB	38.10 mm
3.1875 in	EC	80.96 mm
2.0000 in	HD	50.80 mm

ROCKWELL 1FRYQ24

IDLI GROUP 83231

END YOKE WITH ROUND BORE
SEE FIGURE 5B

3.1250 in	OE	79.38 mm
.8750 in	BD	22.23 mm
1.5000 in	SB	38.10 mm
3.7500 in	EC	95.25 mm
2.0000 in	HD	50.80 mm

ROCKWELL 1FRYR24-2

IDLI GROUP 83232

END YOKE WITH ROUND BORE
SEE FIGURE 5B

3.1250 in	OE	79.38 mm
.8750 in	BD	22.23 mm
1.5000 in	SB	38.10 mm
3.7500 in	EC	95.25 mm
2.2500 in	HD	57.15 mm

ROCKWELL 1FRYR24-1

IDLI GROUP 83233

END YOKE WITH SQUARE BORE
SEE FIGURE 5C

3.1250 in	OE	79.38 mm
.8750 in	BD	22.23 mm
1.6250 in	SB	41.28 mm
3.1875 in	EC	80.96 mm
2.0000 in	HD	50.80 mm

ROCKWELL 1FRYQ26

IDLI GROUP 83234

END YOKE WITH 10 INVOLUTED SPLINES
SEE FIGURE 5A

3.1250 in	OE	79.38 mm
.8750 in	BD	22.23 mm
1.9844 in	SB	50.40 mm
2.5625 in	EC	65.09 mm
2.0000 in	HD	50.80 mm

ROCKWELL 1FRYS16-8

IDLI GROUP 83235

END YOKE WITH ROUND BORE
SEE FIGURE 5B

3.4844 in	OE	88.50 mm
1.2500 in	BD	31.75 mm
.7500 in	SB	19.05 mm
3.6250 in	EC	92.08 mm
2.2500 in	HD	57.15 mm

ROCKWELL 3NYR12-1

IDLI GROUP 83236

END YOKE WITH ROUND BORE
SEE FIGURE 5B

3.4844 in	OE	88.50 mm
1.2500 in	BD	31.75 mm
.8750 in	SB	22.23 mm
3.6250 in	EC	92.08 mm
2.2500 in	HD	57.15 mm

ROCKWELL 3NYR14-1

IDLI GROUP 83237

END YOKE WITH SQUARE BORE
SEE FIGURE 5C

3.4844 in	OE	88.50 mm
1.2500 in	BD	31.75 mm
.9375 in	SB	23.81 mm
3.7500 in	EC	95.25 mm
2.3125 in	HD	58.74 mm

ROCKWELL L16SYQ15-1

IDLI GROUP 83238

END YOKE WITH ROUND BORE
SEE FIGURE 5B

3.4844 in	OE	88.50 mm
1.2500 in	BD	31.75 mm
1.0000 in	SB	25.40 mm
1.9375 in	EC	49.21 mm
2.0000 in	HD	50.80 mm

ROCKWELL 3NYR16-7

BD = BEARING DIAMETER (outside)   **BW** = BEARING WIDTH   **CB** = CROSS LENGTH WITH BEARINGS   **CC** = CENTER TO CENTER   **CD** = CROSS DIAMETER
CL = CROSS LENGTH WITHOUT BEARINGS   **EC** = END TO CENTER or FACE TO FACE   **EE** = END TO END   **EL** = EFFECTIVE LENGTH
HD = HUB DIAMETER (or insert)   **IE** = INSIDE OF EARS (or recess)   **OD** = OUTSIDE DIAMETER   **OE** = OUTSIDE OF EARS   **SB** = SPLINE OR BORE SIZE

CAUTION: BE SURE TO REFER TO ENGINEERING CATALOGS FOR SPECIAL APPLICATIONS THAT REQUIRE SPECIFIC MATERIAL CONTENT, TOLERANCES, ETC. SEE FOOTNOTE.

IDLI GROUP 83239

END YOKE WITH ROUND BORE
SEE FIGURE 5B

3.4044 in	OE	88.50 mm
1.2500 in	BD	31.75 mm
1.0000 in	SB	25.40 mm
3.5000 in	EC	88.90 mm
2.3750 in	HD	60.33 mm

ROCKWELL L16SYR16-1

IDLI GROUP 83240

END YOKE WITH ROUND BORE
SEE FIGURE 5B

3.4844 in	OE	88.50 mm
1.2500 in	BD	31.75 mm
1.0000 in	SB	25.40 mm
4.1250 in	EC	104.78 mm
2.3125 in	HD	58.74 mm

ALLOY 0763
ALLOY 763
BORG-WARNR 62131
G & G MFG 184-3516
HUB CITY 0335-02106
NEAPCO 21131
NEAPCO 22-1131
PRECISION 1800
ROCKWELL 3NYR16-4
SPICER 3-4-723
TRW 21250
WESCO 35N0164
WESCO 35N164
WESCO 471-164

IDLI GROUP 83241

END YOKE WITH SQUARE BORE
SEE FIGURE 5C

3.4844 in	OE	88.50 mm
1.2500 in	BD	31.75 mm
1.0000 in	SB	25.40 mm
4.1875 in	EC	106.36 mm
2.3125 in	HD	58.74 mm

ALLOY 0758
ALLOY 758
BORG-WARNR 62125
NEAPCO 21125
NEAPCO 22-1125
PRECISION 1820
ROCKWELL L16SYQ16-1
SPICER 3-4-0052
SPICER 3-4-52
TRW 21251
WESCO 35N0160-26
WESCO 35N160-26
WESCO 471160

IDLI GROUP 83242

END YOKE WITH ROUND BORE
SEE FIGURE 5B

3.4844 in	OE	88.50 mm
1.2500 in	BD	31.75 mm
1.0625 in	SB	26.99 mm
4.1875 in	EC	106.36 mm
2.3125 in	HD	58.74 mm

ROCKWELL L16SYR17-2

IDLI GROUP 83243

END YOKE WITH HEX BORE
SEE FIGURE 5E

3.4844 in	OE	88.50 mm
1.2500 in	BD	31.75 mm
1.1250 in	SB	28.58 mm
3.0000 in	EC	76.20 mm
5.3125 in	HD	134.94 mm

ROCKWELL L16SYH18-1

IDLI GROUP 83244

END YOKE WITH ROUND BORE
SEE FIGURE 5B

3.4844 in	OE	88.50 mm
1.2500 in	BD	31.75 mm
1.1250 in	SB	28.58 mm
3.1875 in	EC	80.96 mm
2.2500 in	HD	57.15 mm

ROCKWELL L16SYR18

IDLI GROUP 83245

END YOKE WITH HEX BORE
SEE FIGURE 5E

3.4844 in	OE	88.50 mm
1.2500 in	BD	31.75 mm
1.1250 in	SB	28.58 mm
3.0000 in	EC	76.20 mm
4.1875 in	HD	106.36 mm

ROCKWELL L16SYH18

IDLI GROUP 83246

END YOKE WITH ROUND BORE
SEE FIGURE 5B

3.4844 in	OE	88.50 mm
1.2500 in	BD	31.75 mm
1.1250 in	SB	28.58 mm
3.6250 in	EC	92.08 mm
2.5000 in	HD	63.50 mm

ROCKWELL 3NYR18-4

IDLI GROUP 83247

END YOKE WITH ROUND BORE
slight variation from above group
SEE FIGURE 5B

3.4844 in	OE	88.50 mm
1.2500 in	BD	31.75 mm
1.1250 in	SB	28.58 mm
3.6250 in	EC	92.08 mm
2.5000 in	HD	63.50 mm

ROCKWELL L16SYR18-2

IDLI GROUP 83248

END YOKE WITH 6 SPLINES
SEE FIGURE 5A

3.4844 in	OE	88.50 mm
1.2500 in	BD	31.75 mm
1.1250 in	SB	28.58 mm
4.1250 in	EC	104.78 mm
2.3125 in	HD	58.74 mm

BORG WARNR 62142
NEAPCO 21142
NEAPCO 22-1142
PRECISION 1840
ROCKWELL L16SYS18-1
SPICER 3-4-5071
TRW 21254
WESCO 35N0180-21
WESCO 35N180-21
WESCO 471180

IDLI GROUP 83249

END YOKE WITH ROUND BORE
SEE FIGURE 5B

3.4844 in	OE	88.50 mm
1.2500 in	BD	31.75 mm
1.1250 in	SB	28.58 mm
4.1250 in	EC	104.78 mm
2.3125 in	HD	58.74 mm

PRECISION 1802
ROCKWELL L16SYR18-11
SPICER 3-4-763

IDLI GROUP 83250

END YOKE WITH RECTANGULAR BORE
maximum SB dimension shown
SEE FIGURE 5D

3.4844 in	OE	88.50 mm
1.2500 in	BD	31.75 mm
1.1250 in	SB	28.58 mm
4.1250 in	EC	104.78 mm
2.3125 in	HD	58.74 mm

ALLOY 0765
ALLOY 765
BORG-WARNR 62135
NEAPCO 21135
NEAPCO 22-1135
PRECISION 1835
ROCKWELL L16SYP16-3
SPICER 3-4-62

IDLI GROUP 83251

END YOKE WITH ROUND BORE
SEE FIGURE 5B

3.4844 in	OE	88.50 mm
1.2500 in	BD	31.75 mm
1.1250 in	SB	28.58 mm
4.1875 in	EC	106.36 mm
2.3125 in	HD	58.74 mm

ALLOY 0764
ALLOY 764
BORG-WARNR 62132
G & G MFG 184-3518
HUB CITY 0335-02116
NEAPCO 21132
NEAPCO 22-1132
PRECISION 1801
ROCKWELL L16SYR18-12

SPICER 3-4-0743
SPICER 3-4-743
TRW 21252
WESCO 35N0184
WESCO 35N184
WESCO 471184

IDLI GROUP 83252

END YOKE WITH SQUARE BORE
SEE FIGURE 5C

3.4844 in	OE	88.50 mm
1.2500 in	BD	31.75 mm
1.1250 in	SB	28.58 mm
4.1875 in	EC	106.36 mm
2.3125 in	HD	58.74 mm

ALLOY 0757
ALLOY 757
BORG-WARNR 62124
BORG-WARNR 62190
NEAPCO 21124
NEAPCO 21190
NEAPCO 22-1124
NEAPCO 22-1190
PRECISION 1825
ROCKWELL L16SYQ18-1
TRW 21253
WESCO 35N0180-26
WESCO 35N180-26
WESCO 471182

IDLI GROUP 83253

END YOKE WITH 6 SPLINES
SEE FIGURE 5A

3.4844 in	OE	88.50 mm
1.2500 in	BD	31.75 mm
1.1250 in	SB	28.58 mm
5.3125 in	EC	134.94 mm
2.2500 in	HD	57.15 mm

ROCKWELL L16SYS18

IDLI GROUP 83254

END YOKE WITH SQUARE BORE
SEE FIGURE 5C

3.4844 in	OE	88.50 mm
1.2500 in	BD	31.75 mm
1.1875 in	SB	30.16 mm
2.7500 in	EC	69.85 mm
2.5000 in	HD	63.50 mm

ROCKWELL L16SYQ19-54

IDLI GROUP 83255

END YOKE WITH SQUARE BORE
SEE FIGURE 5C

3.4844 in	OE	88.50 mm
1.2500 in	BD	31.75 mm
1.1875 in	SB	30.16 mm
3.0000 in	EC	76.20 mm
2.5000 in	HD	63.50 mm

ROCKWELL L16SYQ19

IDLI GUIDE   COPYRIGHT ® INTERCHANGE, INC.   ST. LOUIS PARK, MN. 55416 USA

IDLI GROUP 83256

END YOKE WITH ROUND BORE
SEE FIGURE 5B

3.4844 in	OE	88.50 mm
1.2500 in	BD	31.75 mm
1.1875 in	SB	30.16 mm
3.6250 in	EC	92.08 mm
2.5000 in	HD	63.50 mm

ROCKWELL 3NYR19-1

IDLI GROUP 83257

END YOKE WITH SQUARE BORE
SEE FIGURE 5C

3.4844 in	OE	88.50 mm
1.2500 in	BD	31.75 mm
1.1875 in	SB	30.16 mm
3.6875 in	EC	93.66 mm
2.0625 in	HD	52.39 mm

ROCKWELL L16SYQ19-53

IDLI GROUP 83258

END YOKE WITH SQUARE BORE
SEE FIGURE 5C

3.4844 in	OE	88.50 mm
1.2500 in	BD	31.75 mm
1.1875 in	SB	30.16 mm
3.7344 in	EC	94.85 mm

BORG-WARNR 61229
BORG-WARNR 62229
G & G MFG 177-3500
HAYES 3-4-12
NEAPCO 211149
NEAPCO 211B9
NEAPCO 211Z4
NEAPCO 22-1308
NEAPCO 22-1344
PRECISION 1854
ROCKWELL L16SYQ19-19
SPICER 3-4-12
WESCO 35NW4-1900
WESCO 35NWY1900
WESCO 471-190
WESCO 471900

IDLI GROUP 83259

END YOKE WITH SQUARE BORE AND HUB
SEE FIGURE 5C

3.4844 in	OE	88.50 mm
1.2500 in	BD	31.75 mm
1.1875 in	SB	30.16 mm
3.7500 in	EC	95.25 mm
2.1400 in	HD	54.36 mm

ROCKWELL 3NYQ19-5

IDLI GROUP 83260

END YOKE WITH SQUARE BORE
SEE FIGURE 5C

3.4844 in	OE	88.50 mm
1.2500 in	BD	31.75 mm
1.1875 in	SB	30.16 mm
3.7500 in	EC	95.25 mm
2.3125 in	HD	58.74 mm

ROCKWELL L16SYQ19-55

IDLI GROUP 83261

END YOKE WITH SQUARE BORE
SEE FIGURE 5C

3.4844 in	OE	88.50 mm
1.2500 in	BD	31.75 mm
1.1875 in	SB	30.16 mm
4.0625 in	EC	103.19 mm
2.3125 in	HD	58.74 mm

PRECISION 1826
ROCKWELL L16SYQ19-13

IDLI GROUP 83262

END YOKE WITH ROUND BORE
SEE FIGURE 5B

3.4844 in	OE	88.50 mm
1.2500 in	BD	31.75 mm
1.1875 in	SB	30.16 mm
4.1250 in	EC	104.78 mm
2.3125 in	HD	58.74 mm

ALLOY 0761
ALLOY 761
BORG-WARNR 62129
G & G MFG 184-3519
HUB CITY 0335-02100
NEAPCO 21129
NEAPCO 21131
NEAPCO 211P9
NEAPCO 22-1129
PRECISION 1803
ROCKWELL L16SYR19-2
TRW 21255
WESCO 35N0194
WESCO 35N194
WESCO 471194

IDLI GROUP 83263

END YOKE WITH SQUARE BORE
SEE FIGURE 5C

3.4844 in	OE	88.50 mm
1.2500 in	BD	31.75 mm
1.1875 in	SB	30.16 mm
4.1875 in	EC	106.36 mm
2.3125 in	HD	58.74 mm

ALLOY 0769
ALLOY 769
BORG-WARNR 62148
BORG-WARNR 62149
NEAPCO 21148
NEAPCO 21149
NEAPCO 22-1148
NEAPCO 22-1149
PRECISION 1826
ROCKWELL L16SYQ19-2
SPICER 3-4-22
WESCO 35N190-26
WESCO 471190

IDLI GROUP 83264

END YOKE WITH ROUND BORE
SEE FIGURE 5B

3.4844 in	OE	88.50 mm
1.2500 in	BD	31.75 mm
1.1875 in	SB	30.16 mm
4.1875 in	EC	106.36 mm
2.3125 in	HD	58.74 mm

BORG-WARNR 62191
NEAPCO 21191
NEAPCO 22-1191
PRECISION 1804
ROCKWELL L16SYR19

IDLI GROUP 83265

END YOKE WITH SQUARE BORE
SEE FIGURE 5C

3.4844 in	OE	88.50 mm
1.2500 in	BD	31.75 mm
1.1875 in	SB	30.16 mm
5.0000 in	EC	127.00 mm

ROCKWELL L16SYQ19-25

IDLI GROUP 83266

END YOKE WITH SQUARE BORE
SEE FIGURE 5C

3.4844 in	OE	88.50 mm
1.2500 in	BD	31.75 mm
1.1875 in	SB	30.16 mm
5.2500 in	EC	133.35 mm
2.3750 in	HD	60.33 mm

ROCKWELL L16SYQ19-5

IDLI GROUP 83267

END YOKE WITH ROUND BORE
SEE FIGURE 5B

3.4844 in	OE	88.50 mm
1.2500 in	BD	31.75 mm
1.2500 in	SB	31.75 mm
3.0000 in	EC	76.20 mm
2.0000 in	HD	50.80 mm

ROCKWELL L16SYR20

IDLI GROUP 83268

END YOKE WITH 14 INVOLUTED SPLINES
SEE FIGURE 5A

3.4844 in	OE	88.50 mm
1.2500 in	BD	31.75 mm
1.2500 in	SB	31.75 mm
3.0000 in	EC	76.20 mm
2.3750 in	HD	60.33 mm

ROCKWELL L16SYS20-30

IDLI GROUP 83269

END YOKE WITH 19 INVOLUTED SPLINES
SEE FIGURE 5A

3.4844 in	OE	88.50 mm
1.2500 in	BD	31.75 mm
1.2500 in	SB	31.75 mm
3.5000 in	EC	88.90 mm
2.3750 in	HD	60.33 mm

ROCKWELL L16SYS20-37

IDLI GROUP 83270

END YOKE WITH ROUND BORE
SEE FIGURE 5B

3.4844 in	OE	88.50 mm
1.2500 in	BD	31.75 mm
1.2500 in	SB	31.75 mm
3.6250 in	EC	92.08 mm
2.3750 in	HD	60.33 mm

ROCKWELL 3NYR20-25

IDLI GROUP 83271

END YOKE WITH ROUND BORE
SEE FIGURE 5B

3.4844 in	OE	88.50 mm
1.2500 in	BD	31.75 mm
1.2500 in	SB	31.75 mm
3.6250 in	EC	92.08 mm
2.2500 in	HD	57.15 mm

ROCKWELL 3NYR20-5

IDLI GROUP 83272

END YOKE WITH ROUND BORE
SEE FIGURE 5B

3.4844 in	OE	88.50 mm
1.2500 in	BD	31.75 mm
1.2500 in	SB	31.75 mm
3.6250 in	EC	92.08 mm
2.5000 in	HD	63.50 mm

ROCKWELL 3NYR20-14

IDLI GROUP 83273

END YOKE WITH 19 INVOLUTED SPLINES
SEE FIGURE 5A

3.4844 in	OE	88.50 mm
1.2500 in	BD	31.75 mm
1.2500 in	SB	31.75 mm
3.8750 in	EC	98.43 mm
2.3750 in	HD	60.33 mm

ROCKWELL L16SYS20-36

IDLI GROUP 83274

END YOKE WITH ROUND BORE
SEE FIGURE 5B

3.4844 in	OE	88.50 mm
1.2500 in	BD	31.75 mm
1.2500 in	SB	31.75 mm
4.0000 in	EC	101.60 mm
2.3125 in	HD	58.74 mm

ROCKWELL L16SYR20-33

IDLI GROUP 83275

END YOKE WITH SQUARE BORE AND HUB
SEE FIGURE 5C

3.4844 in	OE	88.50 mm
1.2500 in	BD	31.75 mm
1.2500 in	SB	31.75 mm
4.1250 in	EC	104.78 mm
2.2500 in	HD	57.15 mm

PRECISION 1827
ROCKWELL L16SYQ20

BD=BEARING DIAMETER (outside)   **BW**=BEARING WIDTH   **CB**=CROSS LENGTH WITH BEARINGS   **CC**=CENTER TO CENTER   **CD**=CROSS DIAMETER
CL=CROSS LENGTH WITHOUT BEARINGS   **EC**=END TO CENTER or FACE TO FACE   **EE**=END TO END   **EL**=EFFECTIVE LENGTH
HD=HUB DIAMETER (or insert)   **IE**=INSIDE OF EARS (or recess)   **OD**=OUTSIDE DIAMETER   **OE**=OUTSIDE OF EARS   **SB**=SPLINE OR BORE SIZE

CAUTION: BE SURE TO REFER TO ENGINEERING CATALOGS FOR SPECIAL APPLICATIONS THAT REQUIRE SPECIFIC MATERIAL CONTENT, TOLERANCES, ETC. SEE FOOTNOTE.

IDLI GROUP 83276

END YOKE WITH 6 SPLINES
SEE FIGURE 5A

3.4844 in	OE	88.50 mm
1.2500 in	BD	31.75 mm
1.2500 in	SB	31.75 mm
4.1250 in	EC	104.78 mm
2.3125 in	HD	58.74 mm

ROCKWELL L16SYS20-1

IDLI GROUP 83277

END YOKE WITH 6 SPLINES
slight variation from above group
SEE FIGURE 5A

3.4844 in	OE	88.50 mm
1.2500 in	BD	31.75 mm
1.2500 in	SB	31.75 mm
4.1250 in	EC	104.78 mm
2.3125 in	HD	58.74 mm

PRECISION 1841
ROCKWELL L16SYS20-2
SPICER 3-4-5081
SPICER 3-4-5091X
WESCO 35N200-21
WESCO 471201

IDLI GROUP 83278

END YOKE WITH ROUND BORE
slight variation from above group
SEE FIGURE 5B

3.4844 in	OE	88.50 mm
1.2500 in	BD	31.75 mm
1.2500 in	SB	31.75 mm
4.1250 in	EC	104.78 mm
2.3125 in	HD	58.74 mm

ROCKWELL L16SYR20-40

IDLI GROUP 83279

END YOKE WITH ROUND BORE
slight variation from above group
SEE FIGURE 5B

3.4844 in	OE	88.50 mm
1.2500 in	BD	31.75 mm
1.2500 in	SB	31.75 mm
4.1250 in	EC	104.78 mm
2.3125 in	HD	58.74 mm

PRECISION 1806
ROCKWELL L16SYR20-35
SPICER 3-4-773

IDLI GROUP 83280

END YOKE WITH ROUND BORE
slight variation from above group
SEE FIGURE 5B

3.4844 in	OE	88.50 mm
1.2500 in	BD	31.75 mm
1.2500 in	SB	31.75 mm
4.1250 in	EC	104.78 mm
2.3125 in	HD	58.74 mm

BORG-WARNR 62133
G & G MFG 184-3520
HAYES 3-4-783

HAYES 3-4-793
NEAPCO 21133
NEAPCO 22-1133
NEAPCO 22-1218
PRECISION 1805
ROCKWELL L16SYR20-6
SPICER 3-4-783
TRW 21256

IDLI GROUP 83281

END YOKE WITH ROUND BORE
slight variation from above group
SEE FIGURE 5B

3.4844 in	OE	88.50 mm
1.2500 in	BD	31.75 mm
1.2500 in	SB	31.75 mm
4.1250 in	EC	104.78 mm
2.3125 in	HD	58.74 mm

ALLOY 0753
ALLOY 753
BORG-WARNR 62116
G & G MFG 184-3520
MCQUAY-NOR U03-92
NEAPCO 21116
NEAPCO 22-1116
NEAPCO 22-1218
PRECISION 1807
ROCKWELL L16SYR20-10
SPICER 3-4-793
WESCO 35N2045-37
WESCO 471045

IDLI GROUP 83282

END YOKE WITH ROUND BORE
slight variation from above group
SEE FIGURE 5B

3.4844 in	OE	88.50 mm
1.2500 in	BD	31.75 mm
1.2500 in	SB	31.75 mm
4.1250 in	EC	104.78 mm
2.3125 in	HD	58.74 mm

NEAPCO 2118A
ROCKWELL L16SYR20-81

IDLI GROUP 83283

END YOKE WITH SQUARE BORE
SEE FIGURE 5C

3.4844 in	OE	88.50 mm
1.2500 in	BD	31.75 mm
1.2500 in	SB	31.75 mm
4.1250 in	EC	104.78 mm
2.3125 in	HD	58.74 mm

ALLOY 0760
ALLOY 760
BORG-WARNR 62128
NEAPCO 21128
NEAPCO 22-1128
PRECISION 1827
ROCKWELL L16SYQ20-3
SPICER 3-4-00672
SPICER 3-4-72
TRW 21257
WESCO 35N0200-26
WESCO 35N200-26
WESCO 471202

IDLI GROUP 83284

END YOKE WITH 14 INVOLUTED SPLINES
SEE FIGURE 5A

3.4844 in	OE	88.50 mm
1.2500 in	BD	31.75 mm
1.2813 in	SB	32.55 mm
2.5000 in	EC	63.50 mm
1.8750 in	HD	47.63 mm

ROCKWELL 3NYS20-13

IDLI GROUP 83285

END YOKE WITH 14 INVOLUTED SPLINES
SEE FIGURE 5A

3.4844 in	OE	88.50 mm
1.2500 in	BD	31.75 mm
1.2813 in	SB	32.55 mm
2.6875 in	EC	68.26 mm
2.0000 in	HD	50.80 mm

ROCKWELL L16SYS20-38

IDLI GROUP 83286

END YOKE WITH 14 INVOLUTED SPLINES
SEE FIGURE 5A

3.4844 in	OE	88.50 mm
1.2500 in	BD	31.75 mm
1.2813 in	SB	32.55 mm
2.7500 in	EC	69.85 mm
2.3750 in	HD	60.33 mm

ROCKWELL 3NYS20-4

IDLI GROUP 83287

END YOKE WITH 14 INVOLUTED SPLINES
SEE FIGURE 5A

3.4844 in	OE	88.50 mm
1.2500 in	BD	31.75 mm
1.2813 in	SB	32.55 mm
3.0000 in	EC	76.20 mm
2.3750 in	HD	60.33 mm

ROCKWELL 3NYS20-6

IDLI GROUP 83288

END YOKE WITH 14 INVOLUTED SPLINES
SEE FIGURE 5A

3.4844 in	OE	88.50 mm
1.2500 in	BD	31.75 mm
1.2813 in	SB	32.55 mm
3.6250 in	EC	92.08 mm
2.3750 in	HD	60.33 mm

ROCKWELL L16SYS20-24

IDLI GROUP 83289

END YOKE WITH 14 INVOLUTED SPLINES
SEE FIGURE 5A

3.4844 in	OE	88.50 mm
1.2500 in	BD	31.75 mm
1.2813 in	SB	32.55 mm
4.0000 in	EC	101.60 mm
2.0000 in	HD	50.80 mm

ROCKWELL L16SYS20-9

IDLI GROUP 83290

END YOKE WITH 19 INVOLUTED SPLINES
SEE FIGURE 5A

3.4844 in	OE	88.50 mm
1.2500 in	BD	31.75 mm
1.2813 in	SB	32.55 mm
4.0625 in	EC	103.19 mm
2.3125 in	HD	58.74 mm

ROCKWELL L16SYS20-35

IDLI GROUP 83291

END YOKE WITH 19 INVOLUTED SPLINES
SEE FIGURE 5A

3.4844 in	OE	88.50 mm
1.2500 in	BD	31.75 mm
1.2813 in	SB	32.55 mm
4.1875 in	EC	106.36 mm
1.9375 in	HD	49.21 mm

ROCKWELL L16SYS20-34

IDLI GROUP 83292

END YOKE WITH 19 INVOLUTED SPLINES
SEE FIGURE 5A

3.4844 in	OE	88.50 mm
1.2500 in	BD	31.75 mm
1.2813 in	SB	32.55 mm
4.1875 in	EC	106.36 mm
2.0000 in	HD	50.80 mm

ROCKWELL L16SYS20-47

IDLI GROUP 83293

END YOKE WITH ROUND BORE
SEE FIGURE 5B

3.4844 in	OE	88.50 mm
1.2500 in	BD	31.75 mm
1.3125 in	SB	33.34 mm
3.1250 in	EC	79.38 mm
2.5000 in	HD	63.50 mm

ROCKWELL 3NYR21-1

IDLI GROUP 83294

END YOKE WITH SQUARE BORE AND HUB
SEE FIGURE 5C

3.4844 in	OE	88.50 mm
1.2500 in	BD	31.75 mm
1.3125 in	SB	33.34 mm
4.1250 in	EC	104.78 mm

ROCKWELL L16SYQ21

IDLI GROUP 83295

END YOKE WITH ROUND BORE
SEE FIGURE 5B

3.4844 in	OE	88.50 mm
1.2500 in	BD	31.75 mm
1.3125 in	SB	33.34 mm
4.1250 in	EC	104.78 mm
2.3125 in	HD	58.74 mm

ROCKWELL L16SYR21-4

ENGINEERING CATALOGS MUST BE CONSULTED FOR DETAILS NOT INCLUDED IN THIS GUIDE. SPECIFIC DESIGNS, MATERIAL CONTENT, TOLERANCES LUBE FITTINGS AND OTHER DIMENSIONS ARE INTENTIONALLY OMITTED HERE.

IDLI GROUP 83296

END YOKE WITH SQUARE BORE
SEE FIGURE 5C

3.4844 in	OE	88.50 mm
1.2500 in	BD	31.75 mm
1.3125 in	SB	33.34 mm
4.1250 in	EC	104.78 mm
2.3125 in	HD	58.74 mm

BORG-WARNR 62123
HUB CITY 0335-02132
NEAPCO 21123
NEAPCO 22-1123
PRECISION 1828
ROCKWELL L16SYQ21-5

IDLI GROUP 83297

END YOKE WITH ROUND BORE
SEE FIGURE 5B

3.4844 in	OE	88.50 mm
1.2500 in	BD	31.75 mm
1.3125 in	SB	33.34 mm
4.1875 in	EC	106.36 mm
2.1250 in	HD	53.98 mm

ROCKWELL L16SYR21

IDLI GROUP 83298

END YOKE WITH ROUND BORE
SEE FIGURE 5B

3.4844 in	OE	88.50 mm
1.2500 in	BD	31.75 mm
1.3125 in	SB	33.34 mm
4.1875 in	EC	106.36 mm
2.1406 in	HD	54.37 mm

ROCKWELL L16SYR21-11

IDLI GROUP 83299

END YOKE WITH ROUND BORE
SEE FIGURE 5B

3.4844 in	OE	88.50 mm
1.2500 in	BD	31.75 mm
1.3750 in	SB	34.93 mm
1.6875 in	EC	42.86 mm
2.0000 in	HD	50.80 mm

ROCKWELL 3NYR22-22

IDLI GROUP 83300

END YOKE WITH 6 SPLINES
SEE FIGURE 5A

3.4844 in	OE	88.50 mm
1.2500 in	BD	31.75 mm
1.3750 in	SB	34.93 mm
2.2500 in	EC	57.15 mm
1.8750 in	HD	47.63 mm

ROCKWELL 3NYS22-61

IDLI GROUP 83301

END YOKE WITH 21 INVOLUTED SPLINES
SEE FIGURE 5A

3.4844 in	OE	88.50 mm
1.2500 in	BD	31.75 mm
1.3750 in	SB	34.93 mm
2.4375 in	EC	61.91 mm
2.3750 in	HD	60.33 mm

ROCKWELL L16SYS22-188

IDLI GROUP 83302

END YOKE WITH 6 SPLINES
SEE FIGURE 5A

3.4844 in	OE	88.50 mm
1.2500 in	BD	31.75 mm
1.3750 in	SB	34.93 mm
2.5000 in	EC	63.50 mm
1.8750 in	HD	47.63 mm

ROCKWELL 3NYS22-60

IDLI GROUP 83303

END YOKE WITH ROUND BORE
SEE FIGURE 5B

3.4844 in	OE	88.50 mm
1.2500 in	BD	31.75 mm
1.3750 in	SB	34.93 mm
2.7500 in	EC	69.85 mm
2.5000 in	HD	63.50 mm

ROCKWELL 3NYR22-34

IDLI GROUP 83304

END YOKE WITH 21 INVOLUTED SPLINES
SEE FIGURE 5A

3.4844 in	OE	88.50 mm
1.2500 in	BD	31.75 mm
1.3750 in	SB	34.93 mm
3.0000 in	EC	76.20 mm
2.3750 in	HD	60.33 mm

ROCKWELL 3NYS22-53

IDLI GROUP 83305

END YOKE WITH ROUND BORE
SEE FIGURE 5B

3.4844 in	OE	88.50 mm
1.2500 in	BD	31.75 mm
1.3750 in	SB	34.93 mm
3.5000 in	EC	88.90 mm
2.3750 in	HD	60.33 mm

ROCKWELL L16SYR22-57

IDLI GROUP 83306

END YOKE WITH ROUND BORE
SEE FIGURE 5B

3.4844 in	OE	88.50 mm
1.2500 in	BD	31.75 mm
1.3750 in	SB	34.93 mm
3.6250 in	EC	92.08 mm
2.3125 in	HD	58.74 mm

ROCKWELL L16SYR22

IDLI GROUP 83307

END YOKE WITH 6 SPLINES
SEE FIGURE 5A

3.4844 in	OE	88.50 mm
1.2500 in	BD	31.75 mm
1.3750 in	SB	34.93 mm
3.6250 in	EC	92.08 mm
2.3750 in	HD	60.33 mm

ROCKWELL L16SYS22-182

IDLI GROUP 83308

END YOKE WITH 6 SPLINES
SEE FIGURE 5A

3.4844 in	OE	88.50 mm
1.2500 in	BD	31.75 mm
1.3750 in	SB	34.93 mm
3.6250 in	EC	92.08 mm
2.5000 in	HD	63.50 mm

ROCKWELL L16SYS22-19

IDLI GROUP 83309

END YOKE WITH ROUND BORE
SEE FIGURE 5B

3.4844 in	OE	88.50 mm
1.2500 in	BD	31.75 mm
1.3750 in	SB	34.93 mm
4.0000 in	EC	101.60 mm
2.3750 in	HD	60.33 mm

ROCKWELL L16SYR22-83

IDLI GROUP 83310

END YOKE WITH 21 INVOLUTED SPLINES
SEE FIGURE 5A

3.4844 in	OE	88.50 mm
1.2500 in	BD	31.75 mm
1.3750 in	SB	34.93 mm
4.1250 in	EC	104.78 mm
2.2500 in	HD	57.15 mm

ROCKWELL L16SYS22-95

IDLI GROUP 83311

END YOKE WITH RECTANGULAR BORE
AND SQUARE HUB
SEE FIGURE 5D

3.4844 in	OE	88.50 mm
1.2500 in	BD	31.75 mm
1.3750 in	SB	34.93 mm
4.1250 in	EC	104.78 mm
2.2500 in	HD	57.15 mm

ROCKWELL L16SYP20-7

IDLI GROUP 83312

END YOKE WITH 6 SPLINES
SEE FIGURE 5A

3.4844 in	OE	88.50 mm
1.2500 in	BD	31.75 mm
1.3750 in	SB	34.93 mm
4.1250 in	EC	104.78 mm
2.3125 in	HD	58.74 mm

BORG-WARNR 61216
G & G MFG 172-3522
NEAPCO 211A6
NEAPCO 22-1303
ROCKWELL L16SYS22-20
SPICER 3-4-5341

IDLI GROUP 83313

END YOKE WITH 10 INVOLUTED SPLINES
SEE FIGURE 5A

3.4844 in	OE	88.50 mm
1.2500 in	BD	31.75 mm
1.3750 in	SB	34.93 mm
4.1250 in	EC	104.78 mm
2.3125 in	HD	58.74 mm

ROCKWELL L16SYS22-50

IDLI GROUP 83314

END YOKE WITH 21 INVOLUTED SPLINES
SEE FIGURE 5A

3.4844 in	OE	88.50 mm
1.2500 in	BD	31.75 mm
1.3750 in	SB	34.93 mm
4.1250 in	EC	104.78 mm
2.3125 in	HD	58.74 mm

HUB CITY 0335-02135
NEAPCO 211A8
NEAPCO 22-1305
PRECISION 1843
ROCKWELL L16SYS22-41
SPICER 3-4-5111
SPICER 3-4-5221
SPICER 3-4-5231X
WESCO 35N220T
WESCO 471220

IDLI GROUP 83315

END YOKE WITH ROUND BORE
SEE FIGURE 5B

3.4844 in	OE	88.50 mm
1.2500 in	BD	31.75 mm
1.3750 in	SB	34.93 mm
4.1250 in	EC	104.78 mm
2.3125 in	HD	58.74 mm

NEAPCO 2119A
ROCKWELL L16SYR22-106

IDLI GROUP 83316

END YOKE WITH ROUND BORE
slight variation from above group
SEE FIGURE 5B

3.4844 in	OE	88.50 mm
1.2500 in	BD	31.75 mm
1.3750 in	SB	34.93 mm
4.1250 in	EC	104.78 mm
2.3125 in	HD	58.74 mm

ALLOY 0754
BORG-WARNR 62118
G & G MFG 184-3522
HAYES 3-4-1353
HAYES 3-4-813
HUB CITY 0335-02117
HUB CITY 0335-02126
MCQUAY-NOR U03-73
NEAPCO 21118
NEAPCO 21158
NEAPCO 22-1118
NEAPCO 22-1219
PRECISION 1809
ROCKWELL L16SYR22-10
SPICER 3-4-1353
SPICER 3-4-153
SPICER 3-4-813
TRW 21258
WESCO 35N2256-37
WESCO 471256

BD=BEARING DIAMETER (outside)   **BW**=BEARING WIDTH   **CB**=CROSS LENGTH WITH BEARINGS   **CC**=CENTER TO CENTER   **CD**=CROSS DIAMETER
CL=CROSS LENGTH WITHOUT BEARINGS   **EC**=END TO CENTER or FACE TO FACE   **EE**=END TO END   **EL**=EFFECTIVE LENGTH
HD=HUB DIAMETER (or insert)   **IE**=INSIDE OF EARS (or recess)   **OD**=OUTSIDE DIAMETER   **OE**=OUTSIDE OF EARS   **SB**=SPLINE OR BORE SIZE

CAUTION: BE SURE TO REFER TO ENGINEERING CATALOGS FOR SPECIAL APPLICATIONS THAT REQUIRE SPECIFIC MATERIAL CONTENT, TOLERANCES, ETC. SEE FOOTNOTE.

IDLI GROUP 83317

END YOKE WITH ROUND BORE
slight variation from above group
SEE FIGURE 5B

3.4844 in	OE	88.50 mm
1.2500 in	BD	31.75 mm
1.3750 in	SB	34.93 mm
4.1250 in	EC	104.78 mm
2.3125 in	HD	58.74 mm

ALLOY 0770
BORG-WARNR 62155
G & G MFG 184-3522
HUB CITY 0335-02123
NEAPCO 211175
NEAPCO 21155
NEAPCO 21158
NEAPCO 22-1155
NEAPCO 22-1158
PRECISION 1810
ROCKWELL L16SYR22-15
SPICER 3-4-1353
SPICER 3-4-893

IDLI GROUP 83318

END YOKE WITH 21 INVOLUTED SPLINES
SEE FIGURE 5A

3.4844 in	OE	88.50 mm
1.2500 in	BD	31.75 mm
1.3750 in	SB	34.93 mm
4.1875 in	EC	106.36 mm
1.6875 in	HD	42.86 mm

ROCKWELL L16SYS22-51

IDLI GROUP 83319

END YOKE WITH 21 INVOLUTED SPLINES
SEE FIGURE 5A

3.4844 in	OE	88.50 mm
1.2500 in	BD	31.75 mm
1.3750 in	SB	34.93 mm
4.1875 in	EC	106.36 mm
2.2500 in	HD	57.15 mm

ROCKWELL L16SYS22-77

IDLI GROUP 83320

END YOKE WITH 6 SPLINES
SEE FIGURE 5A

3.4844 in	OE	88.50 mm
1.2500 in	BD	31.75 mm
1.3750 in	SB	34.93 mm
4.1875 in	EC	106.36 mm
2.3125 in	HD	58.74 mm

ALLOY 0766
BORG-WARNR 62138
BORG-WARNR 62173
G & G MFG 192-3522
HAYES 3-4-5101
NEAPCO 21134
NEAPCO 21138
NEAPCO 21173
NEAPCO 22-1134
NEAPCO 22-1138
PRECISION 1842

ROCKWELL L16SYS22-25
SPICER 3-4-5101
TRW 21260
TRW 21261
WESCO 35N0200-21
WESCO 35N220-21
WESCO 471224

IDLI GROUP 83321

END YOKE WITH 21 INVOLUTED SPLINES
SEE FIGURE 5A

3.4844 in	OE	88.50 mm
1.2500 in	BD	31.75 mm
1.3750 in	SB	34.93 mm
4.1875 in	EC	106.36 mm
2.3125 in	HD	58.74 mm

ROCKWELL L16SYS22-59

IDLI GROUP 83322

END YOKE WITH 21 INVOLUTED SPLINES
slight variation from above group
SEE FIGURE 5A

3.4844 in	OE	88.50 mm
1.2500 in	BD	31.75 mm
1.3750 in	SB	34.93 mm
4.1875 in	EC	106.36 mm
2.3125 in	HD	58.74 mm

BORG-WARNR 62218
PRECISION 1843
ROCKWELL L16SYS22-39

IDLI GROUP 83323

END YOKE WITH ROUND BORE
SEE FIGURE 5B

3.4844 in	OE	88.50 mm
1.2500 in	BD	31.75 mm
1.3750 in	SB	34.93 mm
4.1875 in	EC	106.36 mm
2.3125 in	HD	58.74 mm

ROCKWELL L16SYR22-36

IDLI GROUP 83324

END YOKE WITH TAPERED BORE
SEE FIGURE 5F

3.4844 in	OE	88.50 mm
1.2500 in	BD	31.75 mm
1.3750 in	SB	34.93 mm
4.3750 in	EC	111.13 mm
2.2500 in	HD	57.15 mm

ROCKWELL 3NYT22-4

IDLI GROUP 83325

END YOKE WITH TAPERED BORE
slight variation from above group
SEE FIGURE 5F

3.4844 in	OE	88.50 mm
1.2500 in	BD	31.75 mm
1.3750 in	SB	34.93 mm
4.3750 in	EC	111.13 mm
2.2500 in	HD	57.15 mm

ROCKWELL L16SYT22-1

IDLI GROUP 83326

END YOKE WITH 6 SPLINES
SEE FIGURE 5A

3.4844 in	OE	88.50 mm
1.2500 in	BD	31.75 mm
1.3750 in	SB	34.93 mm
4.3750 in	EC	111.13 mm
2.5000 in	HD	63.50 mm

ROCKWELL L16SYS22-21

IDLI GROUP 83327

END YOKE WITH 10 SPLINES
SEE FIGURE 5A

3.4844 in	OE	88.50 mm
1.2500 in	BD	31.75 mm
1.3750 in	SB	34.93 mm
4.4375 in	EC	112.71 mm
1.8750 in	HD	47.63 mm

ROCKWELL L16SYS22-52

IDLI GROUP 83328

END YOKE WITH 10 SPLINES
SEE FIGURE 5A

3.4844 in	OE	88.50 mm
1.2500 in	BD	31.75 mm
1.3750 in	SB	34.93 mm
4.4375 in	EC	112.71 mm
2.0000 in	HD	50.80 mm

ROCKWELL L16SYS22-31

IDLI GROUP 83329

END YOKE WITH 43 SERRATIONS
SEE FIGURE 5A

3.4844 in	OE	88.50 mm
1.2500 in	BD	31.75 mm
1.3750 in	SB	34.93 mm
5.1250 in	EC	130.18 mm
2.3750 in	HD	60.33 mm

ROCKWELL L16SYS22-67

IDLI GROUP 83330

END YOKE WITH ROUND BORE
SEE FIGURE 5B

3.4844 in	OE	88.50 mm
1.2500 in	BD	31.75 mm
1.4375 in	SB	36.51 mm
3.6250 in	EC	92.08 mm
2.2500 in	HD	57.15 mm

ROCKWELL 3NYR23-1

IDLI GROUP 83331

END YOKE WITH ROUND BORE
SEE FIGURE 5A

3.4844 in	OE	88.50 mm
1.2500 in	BD	31.75 mm
1.4375 in	SB	36.51 mm
3.6250 in	EC	92.08 mm
2.5000 in	HD	63.50 mm

ALLOY 0771
BORG-WARNR 62178
NEAPCO 21178
NEAPCO 22-1178
PRECISION 1811
ROCKWELL L16SYR23
SPICER 3-4-913
SPICER 3-4-923

TRW 21263
WESCO 35N0236
WESCO 35N236
WESCO 471236

IDLI GROUP 83332

END YOKE WITH ROUND BORE
SEE FIGURE 5B

3.4844 in	OE	88.50 mm
1.2500 in	BD	31.75 mm
1.4375 in	SB	36.51 mm
4.0625 in	EC	103.19 mm
2.3125 in	HD	58.74 mm

ROCKWELL L16SYR23-2

IDLI GROUP 83333

END YOKE WITH ROUND BORE
SEE FIGURE 5B

3.4844 in	OE	88.50 mm
1.2500 in	BD	31.75 mm
1.4375 in	SB	36.51 mm
4.1875 in	EC	106.36 mm
2.3125 in	HD	58.74 mm

ROCKWELL L16SYR23-10

IDLI GROUP 83334

END YOKE WITH ROUND BORE
slight variation from above group
SEE FIGURE 5B

3.4844 in	OE	88.50 mm
1.2500 in	BD	31.75 mm
1.4375 in	SB	36.51 mm
4.1875 in	EC	106.36 mm
2.3125 in	HD	58.74 mm

ALLOY 771
BORG-WARNR 62119
G & G MFG 184-3523
HUB CITY 0335-02153
NEAPCO 21119
NEAPCO 21169
NEAPCO 211D6
NEAPCO 22-1119
PRECISION 1811
ROCKWELL L16SYR23-11
SPICER 3-4-1003
TRW 21264

IDLI GROUP 83335

END YOKE WITH ROUND BORE
SEE FIGURE 5B

3.4844 in	OE	88.50 mm
1.2500 in	BD	31.75 mm
1.4375 in	SB	36.51 mm
4.3750 in	EC	111.13 mm
2.6875 in	HD	68.26 mm

NEAPCO 211D6
ROCKWELL L16SYR23-13
SPICER 3-4-1003

ENGINEERING CATALOGS MUST BE CONSULTED FOR DETAILS NOT INCLUDED IN THIS GUIDE. SPECIFIC DESIGNS, MATERIAL CONTENT, TOLERANCES LUBE FITTINGS AND OTHER DIMENSIONS ARE INTENTIONALLY OMITTED HERE.

IDLI GROUP 83336

END YOKE WITH 10 SPLINES
SEE FIGURE 5A

3.4844 in	OE	88.50 mm
1.2500 in	BD	31.75 mm
1.5000 in	SB	38.10 mm
2.5000 in	EC	63.50 mm
2.2500 in	HD	57.15 mm

ROCKWELL 3NYS24-5

IDLI GROUP 83337

END YOKE WITH ROUND BORE
SEE FIGURE 5B

3.4844 in	OE	88.50 mm
1.2500 in	BD	31.75 mm
1.5000 in	SB	38.10 mm
2.7500 in	EC	69.85 mm
2.5000 in	HD	63.50 mm

ROCKWELL 3NYR24-37

IDLI GROUP 83338

END YOKE WITH 10 SPLINES
SEE FIGURE 5A

3.4844 in	OE	88.50 mm
1.2500 in	BD	31.75 mm
1.5000 in	SB	38.10 mm
2.8750 in	EC	73.03 mm
2.2500 in	HD	57.15 mm

ROCKWELL 3NYS24-17

IDLI GROUP 83339

END YOKE WITH 10 SPLINES
SEE FIGURE 5A

3.4844 in	OE	88.50 mm
1.2500 in	BD	31.75 mm
1.5000 in	SB	38.10 mm
3.0000 in	EC	76.20 mm
2.2500 in	HD	57.15 mm

ROCKWELL 3NYS24-1

IDLI GROUP 83340

END YOKE WITH 10 SPLINES
SEE FIGURE 5A

3.4844 in	OE	88.50 mm
1.2500 in	BD	31.75 mm
1.5000 in	SB	38.10 mm
3.1250 in	EC	79.38 mm
2.2500 in	HD	57.15 mm

ROCKWELL 3NYS24-28

IDLI GROUP 83341

END YOKE WITH ROUND BORE
SEE FIGURE 5B

3.4844 in	OE	88.50 mm
1.2500 in	BD	31.75 mm
1.5000 in	SB	38.10 mm
3.6250 in	EC	92.08 mm
2.2500 in	HD	57.15 mm

ROCKWELL 3NYR24-4

IDLI GROUP 83342

END YOKE WITH ROUND BORE
SEE FIGURE 5B

3.4844 in	OE	88.50 mm
1.2500 in	BD	31.75 mm
1.5000 in	SB	38.10 mm
3.6250 in	EC	92.08 mm
2.5000 in	HD	63.50 mm

ROCKWELL 3NYR24-19

IDLI GROUP 83343

END YOKE WITH ROUND BORE
slight variation from above group
SEE FIGURE 5B

3.4844 in	OE	88.50 mm
1.2500 in	BD	31.75 mm
1.5000 in	SB	38.10 mm
3.6250 in	EC	92.08 mm
2.5000 in	HD	63.50 mm

ROCKWELL 3NYR24-1

IDLI GROUP 83344

END YOKE WITH 23 INVOLUTED SPLINES
SEE FIGURE 5A

3.4844 in	OE	88.50 mm
1.2500 in	BD	31.75 mm
1.5000 in	SB	38.10 mm
4.1250 in	EC	104.78 mm
2.3125 in	HD	58.74 mm

ROCKWELL L16SYS24-26

IDLI GROUP 83345

END YOKE WITH ROUND BORE
SEE FIGURE 5B

3.4844 in	OE	88.50 mm
1.2500 in	BD	31.75 mm
1.5000 in	SB	38.10 mm
4.1875 in	EC	106.36 mm
2.2500 in	HD	57.15 mm

ROCKWELL L16SYR24

IDLI GROUP 83346

END YOKE WITH 6 SPLINES
SEE FIGURE 5A

3.4844 in	OE	88.50 mm
1.2500 in	BD	31.75 mm
1.5000 in	SB	38.10 mm
4.1875 in	EC	106.36 mm
2.3125 in	HD	58.74 mm

ROCKWELL L16SYS24-13

IDLI GROUP 83347

END YOKE WITH ROUND BORE
SEE FIGURE 5B

3.4844 in	OE	88.50 mm
1.2500 in	BD	31.75 mm
1.5000 in	SB	38.10 mm
4.1875 in	EC	106.36 mm
2.3125 in	HD	58.74 mm

ROCKWELL L16SYR24-22

IDLI GROUP 83348

END YOKE WITH ROUND BORE
slight variation from above group
SEE FIGURE 5B

3.4844 in	OE	88.50 mm
1.2500 in	BD	31.75 mm
1.5000 in	SB	38.10 mm
4.1875 in	EC	106.36 mm
2.3125 in	HD	58.74 mm

ROCKWELL L16SYR24-41

IDLI GROUP 83349

END YOKE WITH ROUND BORE
slight variation from above group
SEE FIGURE 5B

3.4844 in	OE	88.50 mm
1.2500 in	BD	31.75 mm
1.5000 in	SB	38.10 mm
4.1875 in	EC	106.36 mm
2.3125 in	HD	58.74 mm

ALLOY	 0755
ALLOY	 755
BORG-WARNR	 62120
BORG-WARNR	 62167
G & G MFG	 184-3524
HAYES	 3-4-833
HAYES	 3-4-863
HUB CITY	 0335-02120
MCQUAY-NOR	. . . . U03-101
NEAPCO	 21120
NEAPCO	 21154
NEAPCO	 21192
NEAPCO	 22-1120
NEAPCO	 22-1154
NEAPCO	. . . . 25102X
NEAPCO	. . . . 25167X
PRECISION	 1812
ROCKWELL	. . . . L16SYR24-11
SPICER	 2-4-0172
SPICER	. . . . 3-4-6771X
SPICER	 3-4-833
TRW	 21265
WESCO	. . . . 35N0246
WESCO	. . . . 35N246
WESCO	 471246

IDLI GROUP 83350

END YOKE WITH ROUND BORE
slight variation from above group
SEE FIGURE 5B

3.4844 in	OE	88.50 mm
1.2500 in	BD	31.75 mm
1.5000 in	SB	38.10 mm
4.1875 in	EC	106.36 mm
2.3125 in	HD	58.74 mm

ROCKWELL L16SYR24-19

IDLI GROUP 83351

END YOKE WITH 10 SPLINES
SEE FIGURE 5A

3.4844 in	OE	88.50 mm
1.2500 in	BD	31.75 mm
1.5000 in	SB	38.10 mm
4.1875 in	EC	106.36 mm
2.6875 in	HD	68.26 mm

ROCKWELL L16SYS24-12

IDLI GROUP 83352

END YOKE WITH ROUND BORE
SEE FIGURE 5B

3.4844 in	OE	88.50 mm
1.2500 in	BD	31.75 mm
1.5000 in	SB	38.10 mm
4.2500 in	EC	107.95 mm
2.3125 in	HD	58.74 mm

ROCKWELL L16SYR24-4

IDLI GROUP 83353

END YOKE WITH ROUND BORE
SEE FIGURE 5B

3.4844 in	OE	88.50 mm
1.2500 in	BD	31.75 mm
1.5000 in	SB	38.10 mm
4.3750 in	EC	111.13 mm
2.6875 in	HD	68.26 mm

PRECISION 1812
ROCKWELL L16SYR24-12

IDLI GROUP 83354

END YOKE WITH ROUND BORE
slight variation from above group
SEE FIGURE 5B

3.4844 in	OE	88.50 mm
1.2500 in	BD	31.75 mm
1.5000 in	SB	38.10 mm
4.3750 in	EC	111.13 mm
2.6875 in	HD	68.26 mm

PRECISION 1812
ROCKWELL L16SYR24-20

IDLI GROUP 83355

END YOKE WITH 6 SPLINES
SEE FIGURE 5A

3.4844 in	OE	88.50 mm
1.2500 in	BD	31.75 mm
1.5000 in	SB	38.10 mm
4.5625 in	EC	115.89 mm
2.3750 in	HD	60.33 mm

ROCKWELL L16SYS24-3

IDLI GROUP 83356

END YOKE WITH ROUND BORE
SEE FIGURE 5B

3.4844 in	OE	88.50 mm
1.2500 in	BD	31.75 mm
1.5000 in	SB	38.10 mm
4.6875 in	EC	119.06 mm
2.2500 in	HD	57.15 mm

ROCKWELL 3NYR24-17

IDLI GROUP 83357

END YOKE WITH SQUARE BORE AND
HUB
SEE FIGURE 5C

3.4844 in	OE	88.50 mm
1.2500 in	BD	31.75 mm
1.5156 in	SB	38.50 mm
4.1250 in	EC	104.78 mm

ALLOY	 0762
ALLOY	 762
BORG-WARNR	 62130
HUB CITY	 0335-02134
NEAPCO	 21130

BD = BEARING DIAMETER (outside)　　**BW** = BEARING WIDTH　　**CB** = CROSS LENGTH WITH BEARINGS　　**CC** = CENTER TO CENTER　　**CD** = CROSS DIAMETER
CL = CROSS LENGTH WITHOUT BEARINGS　　**EC** = END TO CENTER or FACE TO FACE　　**EE** = END TO END　　**EL** = EFFECTIVE LENGTH
HD = HUB DIAMETER (or insert)　　**IE** = INSIDE OF EARS (or recess)　　**OD** = OUTSIDE DIAMETER　　**OE** = OUTSIDE OF EARS　　**SB** = SPLINE OR BORE SIZE

CAUTION: BE SURE TO REFER TO ENGINEERING CATALOGS FOR SPECIAL APPLICATIONS THAT REQUIRE SPECIFIC MATERIAL CONTENT, TOLERANCES, ETC. SEE FOOTNOTE.

NEAPCO	22-1130
PRECISION	1829
ROCKWELL	L16SYQ24

IDLI GROUP 83358

END YOKE WITH 14 INVOLUTED SPLINES
SEE FIGURE 5A

3.4844 in	OE	88.50 mm
1.2500 in	BD	31.75 mm
1.5313 in	SB	38.90 mm
3.4375 in	EC	87.31 mm
2.2500 in	HD	57.15 mm

ROCKWELL	3NYS24-29

IDLI GROUP 83359

END YOKE WITH 14 INVOLUTED SPLINES
SEE FIGURE 5A

3.4844 in	OE	88.50 mm
1.2500 in	BD	31.75 mm
1.5313 in	SB	38.90 mm
4.3750 in	EC	111.13 mm
2.2500 in	HD	57.15 mm

ROCKWELL	L16SYS24-15

IDLI GROUP 83360

END YOKE WITH ROUND BORE
SEE FIGURE 5B

3.4844 in	OE	88.50 mm
1.2500 in	BD	31.75 mm
1.5781 in	SB	40.08 mm
4.1250 in	EC	104.78 mm
2.3125 in	HD	58.74 mm

ROCKWELL	L16SYR25-2

IDLI GROUP 83361

END YOKE WITH ROUND BORE
SEE FIGURE 5B

3.4844 in	OE	88.50 mm
1.2500 in	BD	31.75 mm
1.6250 in	SB	41.28 mm
3.6250 in	EC	92.08 mm
2.5000 in	HD	63.50 mm

ROCKWELL	L16SYR26-19

IDLI GROUP 83362

END YOKE WITH SQUARE BORE AND HUB
SEE FIGURE 5C

3.4844 in	OE	88.50 mm
1.2500 in	BD	31.75 mm
1.6250 in	SB	41.28 mm
4.1250 in	EC	104.78 mm

ALLOY	0759
BORG-WARNR	62121
BORG-WARNR	62126
HUB CITY	0335-02133
NEAPCO	21121
NEAPCO	21126
NEAPCO	22-1121
NEAPCO	22-1126
PRECISION	1830

ROCKWELL	L16SYQ26

IDLI GROUP 83363

END YOKE WITH ROUND BORE
SEE FIGURE 5B

3.4844 in	OE	88.50 mm
1.2500 in	BD	31.75 mm
1.6250 in	SB	41.28 mm
4.3750 in	EC	111.13 mm
2.6875 in	HD	68.26 mm

ROCKWELL	L16SYR26-9

IDLI GROUP 83364

END YOKE WITH ROUND BORE
slight variation from above group
SEE FIGURE 5B

3.4844 in	OE	88.50 mm
1.2500 in	BD	31.75 mm
1.6250 in	SB	41.28 mm
4.3750 in	EC	111.13 mm
2.6875 in	HD	68.26 mm

PRECISION	1813
ROCKWELL	L16SYR26-3
SPICER	3-4-993

IDLI GROUP 83365

END YOKE WITH ROUND BORE
SEE FIGURE 5B

3.4844 in	OE	88.50 mm
1.2500 in	BD	31.75 mm
1.6875 in	SB	42.86 mm
3.6250 in	EC	92.08 mm
2.5000 in	HD	63.50 mm

ROCKWELL	3NYR27

IDLI GROUP 83366

END YOKE WITH ROUND BORE
SEE FIGURE 5B

3.4844 in	OE	88.50 mm
1.2500 in	BD	31.75 mm
1.6875 in	SB	42.86 mm
4.3750 in	EC	111.13 mm
2.6875 in	HD	68.26 mm

ROCKWELL	L16SYR27

IDLI GROUP 83367

END YOKE WITH ROUND BORE
slight variation from above group
SEE FIGURE 5B

3.4844 in	OE	88.50 mm
1.2500 in	BD	31.75 mm
1.6875 in	SB	42.86 mm
4.3750 in	EC	111.13 mm
2.6875 in	HD	68.26 mm

ROCKWELL	L16SYR27-2

IDLI GROUP 83368

END YOKE WITH 13 INVOLUTED SPLINES
SEE FIGURE 5A

3.4844 in	OE	88.50 mm
1.2500 in	BD	31.75 mm
1.7500 in	SB	44.45 mm
2.7500 in	EC	69.85 mm
2.5000 in	HD	63.50 mm

ROCKWELL	3NYS28-14

IDLI GROUP 83369

END YOKE WITH 16 SPLINES
SEE FIGURE 5A

3.4844 in	OE	88.50 mm
1.2500 in	BD	31.75 mm
1.7500 in	SB	44.45 mm
3.5000 in	EC	88.90 mm
2.3125 in	HD	58.74 mm

ROCKWELL	3NYS28-8

IDLI GROUP 83370

END YOKE WITH 13 INVOLUTED SPLINES
SEE FIGURE 5A

3.4844 in	OE	88.50 mm
1.2500 in	BD	31.75 mm
1.7500 in	SB	44.45 mm
3.5625 in	EC	90.49 mm
2.3750 in	HD	60.33 mm

ROCKWELL	3NYS28-13

IDLI GROUP 83371

END YOKE WITH 27 INVOLUTED SPLINES
SEE FIGURE 5A

3.4844 in	OE	88.50 mm
1.2500 in	BD	31.75 mm
1.7500 in	SB	44.45 mm
3.6875 in	EC	93.66 mm
2.5000 in	HD	63.50 mm

ROCKWELL	L16SYS28-13

IDLI GROUP 83372

END YOKE WITH 6 SPLINES
SEE FIGURE 5A

3.4844 in	OE	88.50 mm
1.2500 in	BD	31.75 mm
1.7500 in	SB	44.45 mm
4.3750 in	EC	111.13 mm
2.6875 in	HD	68.26 mm

ROCKWELL	L16SYS28-5

IDLI GROUP 83373

END YOKE WITH 27 INVOLUTED SPLINES
SEE FIGURE 5A

3.4844 in	OE	88.50 mm
1.2500 in	BD	31.75 mm
1.7500 in	SB	44.45 mm
4.3750 in	EC	111.13 mm
2.6875 in	HD	68.26 mm

ROCKWELL	L16SYS28-11

IDLI GROUP 83374

END YOKE WITH ROUND BORE
SEE FIGURE 5B

3.4844 in	OE	88.50 mm
1.2500 in	BD	31.75 mm
1.7500 in	SB	44.45 mm
4.3750 in	EC	111.13 mm
2.6875 in	HD	68.26 mm

NEAPCO	21104
NEAPCO	211D4
NEAPCO	22-1316
ROCKWELL	L16SYR28-8
TRW	21266

IDLI GROUP 83375

END YOKE WITH ROUND BORE
slight variation from above group
SEE FIGURE 5B

3.4844 in	OE	88.50 mm
1.2500 in	BD	31.75 mm
1.7500 in	SB	44.45 mm
4.3750 in	EC	111.13 mm
2.6875 in	HD	68.26 mm

ROCKWELL	L16SYR28-6

IDLI GROUP 83376

END YOKE WITH 10 SPLINES
SEE FIGURE 5A

3.4844 in	OE	88.50 mm
1.2500 in	BD	31.75 mm
1.9688 in	SB	50.01 mm
3.5625 in	EC	90.49 mm
2.5000 in	HD	63.50 mm

ROCKWELL	3NYS32

IDLI GROUP 83377

END YOKE WITH ROUND BORE
SEE FIGURE 5B

4.1875 in	OE	106.36 mm
1.0000 in	BD	25.40 mm
.7500 in	SB	19.05 mm
4.1875 in	EC	106.36 mm
2.2500 in	HD	57.15 mm

ROCKWELL	3DRYR12

IDLI GROUP 83378

END YOKE WITH ROUND BORE
SEE FIGURE 5B

4.1875 in	OE	106.36 mm
1.0000 in	BD	25.40 mm
.8750 in	SB	22.23 mm
3.0000 in	EC	76.20 mm
2.0000 in	HD	50.80 mm

ROCKWELL	3DRYR14

IDLI GROUP 83379

END YOKE WITH SQUARE BORE
SEE FIGURE 5C

4.1875 in	OE	106.36 mm
1.0000 in	BD	25.40 mm
.8750 in	SB	22.23 mm
4.1250 in	EC	104.78 mm
2.2500 in	HD	57.15 mm

ROCKWELL	3DRYQ14

IDLI GROUP 83380

END YOKE WITH SQUARE BORE
SEE FIGURE 5C

4.1875 in	OE	106.36 mm
1.0000 in	BD	25.40 mm
.9375 in	SB	23.81 mm
4.1250 in	EC	104.78 mm
2.2500 in	HD	57.15 mm

ROCKWELL	3DRYQ15-2

ENGINEERING CATALOGS MUST BE CONSULTED FOR DETAILS NOT INCLUDED
IN THIS GUIDE. SPECIFIC DESIGNS, MATERIAL CONTENT, TOLERANCES
LUBE FITTINGS AND OTHER DIMENSIONS ARE INTENTIONALLY OMITTED HERE.

IDLI GROUP 83381

END YOKE WITH SQUARE BORE
SEE FIGURE 5C

4.1875 in	OE	106.36 mm
1.0000 in	BD	25.40 mm
.9375 in	SB	23.81 mm
5.3750 in	EC	136.53 mm
2.3750 in	HD	60.33 mm

ROCKWELL 3DRYQ15-4

IDLI GROUP 83382

END YOKE WITH SQUARE BORE
SEE FIGURE 5C

4.1875 in	OE	106.36 mm
1.0000 in	BD	25.40 mm
1.0000 in	SB	25.40 mm
3.0000 in	EC	76.20 mm
2.0000 in	HD	50.80 mm

ROCKWELL 3DRYQ16

IDLI GROUP 83383

END YOKE WITH THREADED BORE
SEE FIGURE 5G

4.1875 in	OE	106.36 mm
1.0000 in	BD	25.40 mm
1.0000 in	SB	25.40 mm
3.2500 in	EC	82.55 mm
2.2500 in	HD	57.15 mm

ROCKWELL 3DRYD16

IDLI GROUP 83384

END YOKE WITH ROUND BORE
SEE FIGURE 5B

4.1875 in	OE	106.36 mm
1.0000 in	BD	25.40 mm
1.0000 in	SB	25.40 mm
4.1250 in	EC	104.78 mm
2.2500 in	HD	57.15 mm

ROCKWELL 3DRYR16-6

IDLI GROUP 83385

END YOKE WITH ROUND BORE
slight variation from above group
SEE FIGURE 5B

4.1875 in	OE	106.36 mm
1.0000 in	BD	25.40 mm
1.0000 in	SB	25.40 mm
4.1250 in	EC	104.78 mm
2.2500 in	HD	57.15 mm

ROCKWELL 3DRYR16-1

IDLI GROUP 83386

END YOKE WITH SQUARE BORE
SEE FIGURE 5C

4.1875 in	OE	106.36 mm
1.0000 in	BD	25.40 mm
1.0000 in	SB	25.40 mm
4.1250 in	EC	104.78 mm
2.2500 in	HD	57.15 mm

ROCKWELL 3DRYQ16-2

IDLI GROUP 83387

END YOKE WITH THREADED BORE
SEE FIGURE 5G

4.1875 in	OE	106.36 mm
1.0000 in	BD	25.40 mm
1.0000 in	SB	25.40 mm
4.1250 in	EC	104.78 mm
2.2500 in	HD	57.15 mm

ROCKWELL 3DRYD16-1

IDLI GROUP 83388

END YOKE WITH ROUND BORE
SEE FIGURE 5B

4.1875 in	OE	106.36 mm
1.0000 in	BD	25.40 mm
1.0625 in	SB	26.99 mm
4.1250 in	EC	104.78 mm
2.2500 in	HD	57.15 mm

ROCKWELL 3DRYR17-2

IDLI GROUP 83389

END YOKE WITH ROUND BORE
slight variation from above group
SEE FIGURE 5B

4.1875 in	OE	106.36 mm
1.0000 in	BD	25.40 mm
1.0625 in	SB	26.99 mm
4.1250 in	EC	104.78 mm
2.2500 in	HD	57.15 mm

ROCKWELL 3DRYR17-3

IDLI GROUP 83390

END YOKE WITH ROUND BORE
slight variation from above group
SEE FIGURE 5B

4.1875 in	OE	106.36 mm
1.0000 in	BD	25.40 mm
1.0625 in	SB	26.99 mm
4.1250 in	EC	104.78 mm
2.2500 in	HD	57.15 mm

ROCKWELL 3DRYR17-1

IDLI GROUP 83391

END YOKE WITH SQUARE BORE
SEE FIGURE 5C

4.1875 in	OE	106.36 mm
1.0000 in	BD	25.40 mm
1.0625 in	SB	26.99 mm
4.1875 in	EC	106.36 mm
2.2500 in	HD	57.15 mm

ROCKWELL 3DRYQ17

IDLI GROUP 83392

END YOKE WITH ROUND BORE
SEE FIGURE 5B

4.1875 in	OE	106.36 mm
1.0000 in	BD	25.40 mm
1.1250 in	SB	28.58 mm
2.6250 in	EC	66.68 mm
2.0000 in	HD	50.80 mm

ROCKWELL 3DRYR18-2

IDLI GROUP 83393

END YOKE WITH 6 SPLINES
SEE FIGURE 5A

4.1875 in	OE	106.36 mm
1.0000 in	BD	25.40 mm
1.1250 in	SB	28.58 mm
3.8750 in	EC	98.43 mm
2.2500 in	HD	57.15 mm

ROCKWELL 3DRYS18-17

IDLI GROUP 83394

END YOKE WITH TAPERED BORE
SEE FIGURE 5F

4.1875 in	OE	106.36 mm
1.0000 in	BD	25.40 mm
1.1250 in	SB	28.58 mm
3.9375 in	EC	100.01 mm
1.8750 in	HD	47.63 mm

ROCKWELL 3DRYT18-4

IDLI GROUP 83395

END YOKE WITH 6 SPLINES
SEE FIGURE 5A

4.1875 in	OE	106.36 mm
1.0000 in	BD	25.40 mm
1.1250 in	SB	28.58 mm
4.1250 in	EC	104.78 mm
2.2500 in	HD	57.15 mm

ROCKWELL 3DRYS18-2

IDLI GROUP 83396

END YOKE WITH ROUND BORE
SEE FIGURE 5B

4.1875 in	OE	106.36 mm
1.0000 in	BD	25.40 mm
1.1250 in	SB	28.58 mm
4.1250 in	EC	104.78 mm
2.2500 in	HD	57.15 mm

ROCKWELL 3DRYR18-26

IDLI GROUP 83397

END YOKE WITH ROUND BORE
slight variation from above group
SEE FIGURE 5B

4.1875 in	OE	106.36 mm
1.0000 in	BD	25.40 mm
1.1250 in	SB	28.58 mm
4.1250 in	EC	104.78 mm
2.2500 in	HD	57.15 mm

ROCKWELL 3DRYR18-20

IDLI GROUP 83398

END YOKE WITH ROUND BORE
slight variation from above group
SEE FIGURE 5B

4.1875 in	OE	106.36 mm
1.0000 in	BD	25.40 mm
1.1250 in	SB	28.58 mm
4.1250 in	EC	104.78 mm
2.2500 in	HD	57.15 mm

ROCKWELL 3DRYR18

IDLI GROUP 83399

END YOKE WITH ROUND BORE
slight variation from above group
SEE FIGURE 5B

4.1875 in	OE	106.36 mm
1.0000 in	BD	25.40 mm
1.1250 in	SB	28.58 mm
4.1250 in	EC	104.78 mm
2.2500 in	HD	57.15 mm

ROCKWELL 3DRYR18-1

IDLI GROUP 83400

END YOKE WITH ROUND BORE
slight variation from above group
SEE FIGURE 5B

4.1875 in	OE	106.36 mm
1.0000 in	BD	25.40 mm
1.1250 in	SB	28.58 mm
4.1250 in	EC	104.78 mm
2.2500 in	HD	57.15 mm

ROCKWELL 3DRYR18-11

IDLI GROUP 83401

END YOKE WITH SQUARE BORE
SEE FIGURE 5C

4.1875 in	OE	106.36 mm
1.0000 in	BD	25.40 mm
1.1250 in	SB	28.58 mm
4.1250 in	EC	104.78 mm
2.2500 in	HD	57.15 mm

ROCKWELL 3DRYQ18-2

IDLI GROUP 83402

END YOKE WITH RECTANGULAR BORE
AND SQUARE HUB
SEE FIGURE 5D

4.1875 in	OE	106.36 mm
1.0000 in	BD	25.40 mm
1.1250 in	SB	28.58 mm
4.1250 in	EC	104.78 mm
2.2500 in	HD	57.15 mm

ROCKWELL 3DRYP16-7

IDLI GROUP 83403

END YOKE WITH SQUARE BORE AND
HUB
SEE FIGURE 5C

4.1875 in	OE	106.36 mm
1.0000 in	BD	25.40 mm
1.1250 in	SB	28.58 mm
4.2500 in	EC	107.95 mm
2.2500 in	HD	57.15 mm

ROCKWELL 3DRYQ18

IDLI GROUP 83404

END YOKE WITH 6 SPLINES
SEE FIGURE 5A

4.1875 in	OE	106.36 mm
1.0000 in	BD	25.40 mm
1.1250 in	SB	28.58 mm
5.0000 in	EC	127.00 mm
2.3750 in	HD	60.33 mm

ROCKWELL 3DRYS18-8

BD = BEARING DIAMETER (outside)   **BW** = BEARING WIDTH   **CB** = CROSS LENGTH WITH BEARINGS   **CC** = CENTER TO CENTER   **CD** = CROSS DIAMETER
CL = CROSS LENGTH WITHOUT BEARINGS   **EC** = END TO CENTER or FACE TO FACE   **EE** = END TO END   **EL** = EFFECTIVE LENGTH
HD = HUB DIAMETER (or insert)   **IE** = INSIDE OF EARS (or recess)   **OD** = OUTSIDE DIAMETER   **OE** = OUTSIDE OF EARS   **SB** = SPLINE OR BORE SIZE

IDLI GUIDE   COPYRIGHT ® INTERCHANGE, INC.   ST. LOUIS PARK, MN. 55416 USA

CAUTION: BE SURE TO REFER TO ENGINEERING CATALOGS FOR SPECIAL APPLICATIONS THAT REQUIRE SPECIFIC MATERIAL CONTENT, TOLERANCES, ETC. SEE FOOTNOTE.

IDLI GROUP 83405

END YOKE WITH 6 SPLINES
SEE FIGURE 5A

4.1875 in	OE	106.36 mm
1.0000 in	BD	25.40 mm
1.1250 in	SB	28.58 mm
5.3125 in	EC	134.94 mm
2.2500 in	HD	57.15 mm

ROCKWELL 3DRYS18-1

IDLI GROUP 83406

END YOKE WITH SQUARE BORE
SEE FIGURE 5C

4.1875 in	OE	106.36 mm
1.0000 in	BD	25.40 mm
1.1875 in	SB	30.16 mm
4.0625 in	EC	103.19 mm
2.2500 in	HD	57.15 mm

ROCKWELL 3DRYQ19

IDLI GROUP 83407

END YOKE WITH 6 SPLINES
SEE FIGURE 5A

4.1875 in	OE	106.36 mm
1.0000 in	BD	25.40 mm
1.1875 in	SB	30.16 mm
4.1250 in	EC	104.78 mm
2.0000 in	HD	50.80 mm

ROCKWELL 3DRYS19

IDLI GROUP 83408

END YOKE WITH ROUND BORE
SEE FIGURE 5B

4.1875 in	OE	106.36 mm
1.0000 in	BD	25.40 mm
1.1875 in	SB	30.16 mm
4.1250 in	EC	104.78 mm
2.2500 in	HD	57.15 mm

ROCKWELL 3DRYR19-5

IDLI GROUP 83409

END YOKE WITH ROUND BORE
slight variation from above group
SEE FIGURE 5B

4.1875 in	OE	106.36 mm
1.0000 in	BD	25.40 mm
1.1875 in	SB	30.16 mm
4.1250 in	EC	104.78 mm
2.2500 in	HD	57.15 mm

ROCKWELL 3DRYR19-1

IDLI GROUP 83410

END YOKE WITH ROUND BORE
SEE FIGURE 5B

4.1875 in	OE	106.36 mm
1.0000 in	BD	25.40 mm
1.1875 in	SB	30.16 mm
4.1875 in	EC	106.36 mm
2.2500 in	HD	57.15 mm

ROCKWELL 3DRYR19-6

IDLI GROUP 83411

END YOKE WITH ROUND BORE
SEE FIGURE 5B

4.1875 in	OE	106.36 mm
1.0000 in	BD	25.40 mm
1.2500 in	SB	31.75 mm
2.7500 in	EC	69.85 mm
2.0000 in	HD	50.80 mm

ROCKWELL 3DRYR20-9

IDLI GROUP 83412

END YOKE WITH ROUND BORE
slight variation from above group
SEE FIGURE 5B

4.1875 in	OE	106.36 mm
1.0000 in	BD	25.40 mm
1.2500 in	SB	31.75 mm
2.7500 in	EC	69.85 mm
2.0000 in	HD	50.80 mm

ROCKWELL 3DRYR20-41

IDLI GROUP 83413

END YOKE WITH ROUND BORE
SEE FIGURE 5B

4.1875 in	OE	106.36 mm
1.0000 in	BD	25.40 mm
1.2500 in	SB	31.75 mm
3.0000 in	EC	76.20 mm
2.0000 in	HD	50.80 mm

ROCKWELL 3DRYR20-23

IDLI GROUP 83414

END YOKE WITH TAPERED BORE
SEE FIGURE 5F

4.1875 in	OE	106.36 mm
1.0000 in	BD	25.40 mm
1.2500 in	SB	31.75 mm
3.3750 in	EC	85.73 mm
2.4375 in	HD	61.91 mm

ROCKWELL 3DRYT20

IDLI GROUP 83415

END YOKE WITH 6 SPLINES
SEE FIGURE 5A

4.1875 in	OE	106.36 mm
1.0000 in	BD	25.40 mm
1.2500 in	SB	31.75 mm
4.0000 in	EC	101.60 mm
2.2500 in	HD	57.15 mm

ROCKWELL 3DRYS20-7

IDLI GROUP 83416

END YOKE WITH ROUND BORE
SEE FIGURE 5B

4.1875 in	OE	106.36 mm
1.0000 in	BD	25.40 mm
1.2500 in	SB	31.75 mm
4.1250 in	EC	104.78 mm
2.2500 in	HD	57.15 mm

ROCKWELL 3DRYR20-46

IDLI GROUP 83417

END YOKE WITH SQUARE BORE
SEE FIGURE 5C

4.1875 in	OE	106.36 mm
1.0000 in	BD	25.40 mm
1.2500 in	SB	31.75 mm
4.1250 in	EC	104.78 mm
2.2500 in	HD	57.15 mm

ROCKWELL 3DRYQ20

IDLI GROUP 83418

END YOKE WITH SQUARE BORE AND HUB
SEE FIGURE 5C

4.1875 in	OE	106.36 mm
1.0000 in	BD	25.40 mm
1.2500 in	SB	31.75 mm
4.1250 in	EC	104.78 mm
2.2500 in	HD	57.15 mm

ROCKWELL 3DRYQ20-2

IDLI GROUP 83419

END YOKE WITH ROUND BORE
SEE FIGURE 5B

4.1875 in	OE	106.36 mm
1.0000 in	BD	25.40 mm
1.2500 in	SB	31.75 mm
4.1875 in	EC	106.36 mm
2.1250 in	HD	53.98 mm

ROCKWELL 3DRYR20-10

IDLI GROUP 83420

END YOKE WITH ROUND BORE
SEE FIGURE 5B

4.1875 in	OE	106.36 mm
1.0000 in	BD	25.40 mm
1.2500 in	SB	31.75 mm
4.1875 in	EC	106.36 mm
2.2500 in	HD	57.15 mm

ROCKWELL 3DRYR20-2

IDLI GROUP 83421

END YOKE WITH ROUND BORE
slight variation from above group
SEE FIGURE 5B

4.1875 in	OE	106.36 mm
1.0000 in	BD	25.40 mm
1.2500 in	SB	31.75 mm
4.1875 in	EC	106.36 mm
2.2500 in	HD	57.15 mm

ROCKWELL 3DRYR20-3

IDLI GROUP 83422

END YOKE WITH 6 SPLINES
SEE FIGURE 5A

4.1875 in	OE	106.36 mm
1.0000 in	BD	25.40 mm
1.2500 in	SB	31.75 mm
5.2500 in	EC	133.35 mm
2.2500 in	HD	57.15 mm

ROCKWELL 3DRYS20-4

IDLI GROUP 83423

END YOKE WITH ROUND BORE
SEE FIGURE 5B

4.1875 in	OE	106.36 mm
1.0000 in	BD	25.40 mm
1.3125 in	SB	33.34 mm
3.0000 in	EC	76.20 mm
2.0000 in	HD	50.80 mm

ROCKWELL 3DRYR21-3

IDLI GROUP 83424

END YOKE WITH ROUND BORE
SEE FIGURE 5B

4.1875 in	OE	106.36 mm
1.0000 in	BD	25.40 mm
1.3125 in	SB	33.34 mm
4.1250 in	EC	104.78 mm
2.2500 in	HD	57.15 mm

ROCKWELL 3DRYR21-4

IDLI GROUP 83425

END YOKE WITH ROUND BORE
slight variation from above group
SEE FIGURE 5B

4.1875 in	OE	106.36 mm
1.0000 in	BD	25.40 mm
1.3125 in	SB	33.34 mm
4.1250 in	EC	104.78 mm
2.2500 in	HD	57.15 mm

ROCKWELL 3DRYR21-2

IDLI GROUP 83426

END YOKE WITH SQUARE BORE
SEE FIGURE 5C

4.1875 in	OE	106.36 mm
1.0000 in	BD	25.40 mm
1.3125 in	SB	33.34 mm
4.1250 in	EC	104.78 mm
2.2500 in	HD	57.15 mm

ROCKWELL 3DRYQ21

IDLI GROUP 83427

END YOKE WITH ROUND BORE
SEE FIGURE 5B

4.1875 in	OE	106.36 mm
1.0000 in	BD	25.40 mm
1.3125 in	SB	33.34 mm
4.1875 in	EC	106.36 mm
1.7500 in	HD	44.45 mm

ROCKWELL 3DRYR21-6

IDLI GROUP 83428

END YOKE WITH ROUND BORE
SEE FIGURE 5B

4.1875 in	OE	106.36 mm
1.0000 in	BD	25.40 mm
1.3125 in	SB	33.34 mm
4.1875 in	EC	106.36 mm
2.1250 in	HD	53.98 mm

ROCKWELL 3DRYR21

ENGINEERING CATALOGS MUST BE CONSULTED FOR DETAILS NOT INCLUDED IN THIS GUIDE. SPECIFIC DESIGNS, MATERIAL CONTENT, TOLERANCES LUBE FITTINGS AND OTHER DIMENSIONS ARE INTENTIONALLY OMITTED HERE.

IDLI GROUP 83429

END YOKE WITH SQUARE BORE AND HUB
SEE FIGURE 5C

4.1875 in	OE	106.36 mm
1.0000 in	BD	25.40 mm
1.3125 in	SB	33.34 mm
4.2500 in	EC	107.95 mm
2.2500 in	HD	57.15 mm

ROCKWELL 3DRYQ21-5

IDLI GROUP 83430

END YOKE WITH ROUND BORE
SEE FIGURE 5B

4.1875 in	OE	106.36 mm
1.0000 in	BD	25.40 mm
1.3750 in	SB	34.93 mm
2.8750 in	EC	73.03 mm
2.0000 in	HD	50.80 mm

ROCKWELL 3DRYR22-11

IDLI GROUP 83431

END YOKE WITH 6 SPLINES
SEE FIGURE 5A

4.1875 in	OE	106.36 mm
1.0000 in	BD	25.40 mm
1.3750 in	SB	34.93 mm
3.8750 in	EC	98.43 mm
2.2500 in	HD	57.15 mm

ROCKWELL 3DRYS22-21

IDLI GROUP 83432

END YOKE WITH TAPERED BORE
SEE FIGURE 5F

4.1875 in	OE	106.36 mm
1.0000 in	BD	25.40 mm
1.3750 in	SB	34.93 mm
3.8750 in	EC	98.43 mm
2.2500 in	HD	57.15 mm

ROCKWELL 3DRYT22

IDLI GROUP 83433

END YOKE WITH 6 SPLINES
SEE FIGURE 5A

4.1875 in	OE	106.36 mm
1.0000 in	BD	25.40 mm
1.3750 in	SB	34.93 mm
4.1250 in	EC	104.78 mm
2.2500 in	HD	57.15 mm

ROCKWELL 3DRYS22-14

IDLI GROUP 83434

END YOKE WITH ROUND BORE
SEE FIGURE 5B

4.1875 in	OE	106.36 mm
1.0000 in	BD	25.40 mm
1.3750 in	SB	34.93 mm
4.1250 in	EC	104.78 mm
2.2500 in	HD	57.15 mm

ROCKWELL 3DRYR22-12

IDLI GROUP 83435

END YOKE WITH ROUND BORE
slight variation from above group
SEE FIGURE 5B

4.1875 in	OE	106.36 mm
1.0000 in	BD	25.40 mm
1.3750 in	SB	34.93 mm
4.1250 in	EC	104.78 mm
2.2500 in	HD	57.15 mm

ROCKWELL 3DRYR22-2

IDLI GROUP 83436

END YOKE WITH ROUND BORE
slight variation from above group
SEE FIGURE 5B

4.1875 in	OE	106.36 mm
1.0000 in	BD	25.40 mm
1.3750 in	SB	34.93 mm
4.1250 in	EC	104.78 mm
2.2500 in	HD	57.15 mm

ROCKWELL 3DRYR22-5

IDLI GROUP 83437

END YOKE WITH ROUND BORE
slight variation from above group
SEE FIGURE 5B

4.1875 in	OE	106.36 mm
1.0000 in	BD	25.40 mm
1.3750 in	SB	34.93 mm
4.1250 in	EC	104.78 mm
2.2500 in	HD	57.15 mm

ROCKWELL 3DRYR22

IDLI GROUP 83438

END YOKE WITH SQUARE BORE AND HUB
SEE FIGURE 5C

4.1875 in	OE	106.36 mm
1.0000 in	BD	25.40 mm
1.3750 in	SB	34.93 mm
4.1250 in	EC	104.78 mm
2.2500 in	HD	57.15 mm

ROCKWELL 3DRYQ22

IDLI GROUP 83439

END YOKE WITH 6 SPLINES
SEE FIGURE 5A

4.1875 in	OE	106.36 mm
1.0000 in	BD	25.40 mm
1.3750 in	SB	34.93 mm
4.8750 in	EC	123.83 mm
1.8750 in	HD	47.63 mm

ROCKWELL 3DRYS22-4

IDLI GROUP 83440

END YOKE WITH 6 SPLINES
SEE FIGURE 5A

4.1875 in	OE	106.36 mm
1.0000 in	BD	25.40 mm
1.3750 in	SB	34.93 mm
5.3125 in	EC	134.94 mm
2.2500 in	HD	57.15 mm

ROCKWELL 3DRYS22-1

IDLI GROUP 83441

END YOKE WITH ROUND BORE
SEE FIGURE 5B

4.1875 in	OE	106.36 mm
1.0000 in	BD	25.40 mm
1.4375 in	SB	36.51 mm
4.0000 in	EC	101.60 mm
2.2500 in	HD	57.15 mm

ROCKWELL 3DRYR23-10

IDLI GROUP 83442

END YOKE WITH ROUND BORE
SEE FIGURE 5B

4.1875 in	OE	106.36 mm
1.0000 in	BD	25.40 mm
1.4375 in	SB	36.51 mm
4.1250 in	EC	104.78 mm
2.2500 in	HD	57.15 mm

ROCKWELL 3DRYR23-14

IDLI GROUP 83443

END YOKE WITH ROUND BORE
slight variation from above group
SEE FIGURE 5B

4.1875 in	OE	106.36 mm
1.0000 in	BD	25.40 mm
1.4375 in	SB	36.51 mm
4.1250 in	EC	104.78 mm
2.2500 in	HD	57.15 mm

ROCKWELL 3DRYR23-5

IDLI GROUP 83444

END YOKE WITH ROUND BORE
slight variation from above group
SEE FIGURE 5B

4.1875 in	OE	106.36 mm
1.0000 in	BD	25.40 mm
1.4375 in	SB	36.51 mm
4.1250 in	EC	104.78 mm
2.2500 in	HD	57.15 mm

ROCKWELL 3DRYR23-9

IDLI GROUP 83445

END YOKE WITH ROUND BORE
SEE FIGURE 5B

4.1875 in	OE	106.36 mm
1.0000 in	BD	25.40 mm
1.5000 in	SB	38.10 mm
3.8750 in	EC	98.43 mm
2.2500 in	HD	57.15 mm

ROCKWELL 3DRYR24-19

IDLI GROUP 83446

END YOKE WITH ROUND BORE
slight variation from above group
SEE FIGURE 5B

4.1875 in	OE	106.36 mm
1.0000 in	BD	25.40 mm
1.5000 in	SB	38.10 mm
3.8750 in	EC	98.43 mm
2.2500 in	HD	57.15 mm

ROCKWELL 3DRYR24-13

IDLI GROUP 83447

END YOKE WITH 10 SPLINES
SEE FIGURE 5A

4.1875 in	OE	106.36 mm
1.0000 in	BD	25.40 mm
1.5000 in	SB	38.10 mm
4.0000 in	EC	101.60 mm
2.4375 in	HD	61.91 mm

ROCKWELL 3DRYS24-1

IDLI GROUP 83448

END YOKE WITH ROUND BORE
SEE FIGURE 5B

4.1875 in	OE	106.36 mm
1.0000 in	BD	25.40 mm
1.5000 in	SB	38.10 mm
4.1250 in	EC	104.78 mm
2.2500 in	HD	57.15 mm

ROCKWELL 3DRYR24-9

IDLI GROUP 83449

END YOKE WITH ROUND BORE
slight variation from above group
SEE FIGURE 5B

4.1875 in	OE	106.36 mm
1.0000 in	BD	25.40 mm
1.5000 in	SB	38.10 mm
4.1250 in	EC	104.78 mm
2.2500 in	HD	57.15 mm

ROCKWELL 3DRYR24-2

IDLI GROUP 83450

END YOKE WITH SQUARE BORE AND HUB
SEE FIGURE 5C

4.1875 in	OE	106.36 mm
1.0000 in	BD	25.40 mm
1.5000 in	SB	38.10 mm
4.1250 in	EC	104.78 mm
2.2500 in	HD	57.15 mm

ROCKWELL 3DRYQ24-1

IDLI GROUP 83451

END YOKE WITH ROUND BORE
SEE FIGURE 5B

4.1875 in	OE	106.36 mm
1.0000 in	BD	25.40 mm
1.5000 in	SB	38.10 mm
4.1875 in	EC	106.36 mm
2.2500 in	HD	57.15 mm

ROCKWELL 3DRYR24-28

IDLI GROUP 83452

END YOKE WITH THREADED BORE
SEE FIGURE 5G

4.1875 in	OE	106.36 mm
1.0000 in	BD	25.40 mm
1.5000 in	SB	38.10 mm
4.7500 in	EC	120.65 mm
1.9375 in	HD	49.21 mm

ROCKWELL 3DRYD24

BD = BEARING DIAMETER (outside)   BW = BEARING WIDTH   CB = CROSS LENGTH WITH BEARINGS   CC = CENTER TO CENTER   CD = CROSS DIAMETER
CL = CROSS LENGTH WITHOUT BEARINGS   EC = END TO CENTER or FACE TO FACE   EE = END TO END   EL = EFFECTIVE LENGTH
HD = HUB DIAMETER (or insert)   IE = INSIDE OF EARS (or recess)   OD = OUTSIDE DIAMETER   OE = OUTSIDE OF EARS   SB = SPLINE OR BORE SIZE

IDLI GUIDE   COPYRIGHT ® INTERCHANGE, INC.   ST. LOUIS PARK, MN. 55416 USA

CAUTION: BE SURE TO REFER TO ENGINEERING CATALOGS FOR SPECIAL APPLICATIONS THAT REQUIRE SPECIFIC MATERIAL CONTENT, TOLERANCES, ETC. SEE FOOTNOTE.

IDLI GROUP 83453

END YOKE WITH 6 SPLINES
SEE FIGURE 5A

4.1875 in	OE	106.36 mm
1.0000 in	BD	25.40 mm
1.5000 in	SB	38.10 mm
5.3125 in	EC	134.94 mm
2.2500 in	HD	57.15 mm

ROCKWELL 3DRYS24-5

IDLI GROUP 83454

END YOKE WITH ROUND BORE
SEE FIGURE 5B

4.1875 in	OE	106.36 mm
1.0000 in	BD	25.40 mm
1.5000 in	SB	38.10 mm
5.3750 in	EC	136.53 mm
2.3750 in	HD	60.33 mm

ROCKWELL 3DRYR24-15

IDLI GROUP 83455

END YOKE WITH SQUARE BORE AND HUB
SEE FIGURE 5C

4.1875 in	OE	106.36 mm
1.0000 in	BD	25.40 mm
1.6250 in	SB	41.28 mm
4.1250 in	EC	104.78 mm
2.2500 in	HD	57.15 mm

ROCKWELL 3DRYQ26-1

IDLI GROUP 83456

END YOKE WITH ROUND BORE
SEE FIGURE 5B

4.1875 in	OE	106.36 mm
1.0000 in	BD	25.40 mm
1.6250 in	SB	41.28 mm
4.1250 in	EC	104.78 mm
2.5000 in	HD	63.50 mm

ROCKWELL 3DRYR26

IDLI GROUP 83457

END YOKE WITH ROUND BORE
SEE FIGURE 5B

4.1875 in	OE	106.36 mm
1.0000 in	BD	25.40 mm
1.6250 in	SB	41.28 mm
4.1250 in	EC	104.78 mm
2.6250 in	HD	66.68 mm

ROCKWELL 3DRYR26-2

IDLI GROUP 83458

END YOKE WITH 6 SPLINES
SEE FIGURE 5A

4.1875 in	OE	106.36 mm
1.0000 in	BD	25.40 mm
1.6250 in	SB	41.28 mm
5.3125 in	EC	134.94 mm
2.3750 in	HD	60.33 mm

ROCKWELL 3DRYS26-1

IDLI GROUP 83459

END YOKE WITH 6 SPLINES
SEE FIGURE 5A

4.1875 in	OE	106.36 mm
1.0000 in	BD	25.40 mm
1.7500 in	SB	44.45 mm
4.0625 in	EC	103.19 mm
2.2500 in	HD	57.15 mm

ROCKWELL 3DRYS28-4

IDLI GROUP 83460

END YOKE WITH ROUND BORE
SEE FIGURE 5B

4.1875 in	OE	106.36 mm
1.0000 in	BD	25.40 mm
1.7500 in	SB	44.45 mm
4.1250 in	EC	104.78 mm
2.6250 in	HD	66.68 mm

ROCKWELL 3DRYR28-2

IDLI GROUP 83461

END YOKE WITH SQUARE BORE AND HUB
SEE FIGURE 5C

4.1875 in	OE	106.36 mm
1.0000 in	BD	25.40 mm
1.7500 in	SB	44.45 mm
4.1250 in	EC	104.78 mm
3.0000 in	HD	76.20 mm

ROCKWELL 3DRYQ28-1

IDLI GROUP 83462

END YOKE WITH 6 SPLINES
SEE FIGURE 5A

4.1875 in	OE	106.36 mm
1.0000 in	BD	25.40 mm
1.7500 in	SB	44.45 mm
5.3125 in	EC	134.94 mm
2.3750 in	HD	60.33 mm

ROCKWELL 3DRYS28-2

IDLI GROUP 83463

END YOKE WITH SQUARE BORE AND HUB
SEE FIGURE 5C

4.1875 in	OE	106.36 mm
1.0000 in	BD	25.40 mm
1.8750 in	SB	47.63 mm
4.1250 in	EC	104.78 mm
3.0000 in	HD	76.20 mm

ROCKWELL 3DRYQ30

IDLI GROUP 83464

END YOKE WITH SQUARE BORE AND HUB
SEE FIGURE 5C

4.1875 in	OE	106.36 mm
1.0000 in	BD	25.40 mm
2.0000 in	SB	50.80 mm
4.1875 in	EC	106.36 mm
3.0000 in	HD	76.20 mm

ROCKWELL 3DRYQ32-1

IDLI GROUP 83465

END YOKE WITH ROUND BORE
SEE FIGURE 5B

4.6250 in	OE	117.48 mm
1.5313 in	BD	38.90 mm
1.0000 in	SB	25.40 mm
4.9375 in	EC	125.41 mm
2.8750 in	HD	73.03 mm

ROCKWELL 55NYR16-1

IDLI GROUP 83466

END YOKE WITH ROUND BORE
SEE FIGURE 5B

4.6250 in	OE	117.48 mm
1.5313 in	BD	38.90 mm
1.0000 in	SB	25.40 mm
5.5625 in	EC	141.29 mm
2.8750 in	HD	73.03 mm

G & G MFG 184-5516
ROCKWELL 55NYR16-2

IDLI GROUP 83467

END YOKE WITH ROUND BORE
SEE FIGURE 5B

4.6250 in	OE	117.48 mm
1.5313 in	BD	38.90 mm
1.0781 in	SB	27.38 mm
4.0000 in	EC	101.60 mm
1.8750 in	HD	47.63 mm

ROCKWELL 55NYR17

IDLI GROUP 83468

END YOKE WITH SQUARE BORE
SEE FIGURE 5C

4.6250 in	OE	117.48 mm
1.5313 in	BD	38.90 mm
1.1250 in	SB	28.58 mm
4.0000 in	EC	101.60 mm
3.0000 in	HD	76.20 mm

ROCKWELL 55NYQ18-2

IDLI GROUP 83469

END YOKE WITH ROUND BORE
SEE FIGURE 5B

4.6250 in	OE	117.48 mm
1.5313 in	BD	38.90 mm
1.1250 in	SB	28.58 mm
4.0313 in	EC	102.40 mm
1.9375 in	HD	49.21 mm

ROCKWELL 55NYR18-3

IDLI GROUP 83470

END YOKE WITH HEX BORE
SEE FIGURE 5E

4.6250 in	OE	117.48 mm
1.5313 in	BD	38.90 mm
1.1563 in	SB	29.37 mm
8.3125 in	EC	211.14 mm
2.3750 in	HD	60.33 mm

ROCKWELL 55NYH18

IDLI GROUP 83471

END YOKE WITH SQUARE BORE
SEE FIGURE 5C

4.6250 in	OE	117.48 mm
1.5313 in	BD	38.90 mm
1.1875 in	SB	30.16 mm
4.0625 in	EC	103.19 mm
3.0000 in	HD	76.20 mm

ROCKWELL 55NYQ19-2

IDLI GROUP 83472

END YOKE WITH SQUARE BORE
SEE FIGURE 5C

4.6250 in	OE	117.48 mm
1.5313 in	BD	38.90 mm
1.1875 in	SB	30.16 mm
5.5000 in	EC	139.70 mm
2.6875 in	HD	68.26 mm

ROCKWELL 55NYQ19-3

IDLI GROUP 83473

END YOKE WITH ROUND BORE
SEE FIGURE 5B

4.6250 in	OE	117.48 mm
1.5313 in	BD	38.90 mm
1.2500 in	SB	31.75 mm
5.5625 in	EC	141.29 mm
2.8750 in	HD	73.03 mm

ROCKWELL 55NYR20-7

IDLI GROUP 83474

END YOKE WITH SQUARE BORE
SEE FIGURE 5C

4.6250 in	OE	117.48 mm
1.5313 in	BD	38.90 mm
1.3125 in	SB	33.34 mm
4.5000 in	EC	114.30 mm

ROCKWELL 55NYQ21-3

IDLI GROUP 83475

END YOKE WITH SQUARE BORE
SEE FIGURE 5C

4.6250 in	OE	117.48 mm
1.5313 in	BD	38.90 mm
1.3125 in	SB	33.34 mm
4.7500 in	EC	120.65 mm
2.6875 in	HD	68.26 mm

ROCKWELL 55NYQ21

IDLI GROUP 83476

END YOKE WITH 6 SPLINES
SEE FIGURE 5A

4.6250 in	OE	117.48 mm
1.5313 in	BD	38.90 mm
1.3750 in	SB	34.93 mm
4.0625 in	EC	103.19 mm
2.0000 in	HD	50.80 mm

ROCKWELL 55NYS22-4

IDLI GROUP 83477

END YOKE WITH ROUND BORE
SEE FIGURE 5B

4.6250 in	OE	117.48 mm
1.5313 in	BD	38.90 mm
1.3750 in	SB	34.93 mm
4.5000 in	EC	114.30 mm
2.6875 in	HD	68.26 mm

ROCKWELL 55NYR22

IDLI GROUP 83478

END YOKE WITH 21 INVOLUTED SPLINES
SEE FIGURE 5A

4.6250 in	OE	117.48 mm
1.5313 in	BD	38.90 mm
1.3750 in	SB	34.93 mm
4.6250 in	EC	117.48 mm
2.5000 in	HD	63.50 mm

ROCKWELL 55NYS22-17

IDLI GROUP 83479

END YOKE WITH 6 SPLINES
SEE FIGURE 5A

4.6250 in	OE	117.48 mm
1.5313 in	BD	38.90 mm
1.3750 in	SB	34.93 mm
4.6250 in	EC	117.48 mm
2.6875 in	HD	68.26 mm

ROCKWELL 55NYS22-32

IDLI GROUP 83480

END YOKE WITH 23 INVOLUTED SPLINES
SEE FIGURE 5A

4.6250 in	OE	117.48 mm
1.5313 in	BD	38.90 mm
1.5000 in	SB	38.10 mm
4.6250 in	EC	117.48 mm
2.5000 in	HD	63.50 mm

ROCKWELL 55NYS24-14

IDLI GROUP 83481

END YOKE WITH 10 SPLINES
SEE FIGURE 5A

4.6250 in	OE	117.48 mm
1.5313 in	BD	38.90 mm
1.5000 in	SB	38.10 mm
4.6250 in	EC	117.48 mm
2.6875 in	HD	68.26 mm

ROCKWELL 55NYS24

IDLI GROUP 83482

END YOKE WITH ROUND BORE
SEE FIGURE 5B

4.6250 in	OE	117.48 mm
1.5313 in	BD	38.90 mm
1.3750 in	SB	34.93 mm
5.5000 in	EC	139.70 mm
2.6875 in	HD	68.26 mm

ROCKWELL 55NYR22-4

IDLI GROUP 83483

END YOKE WITH 23 INVOLUTED SPLINES
SEE FIGURE 5A

4.6250 in	OE	117.48 mm
1.5313 in	BD	38.90 mm
1.5000 in	SB	38.10 mm
2.8750 in	EC	73.03 mm
2.0000 in	HD	50.80 mm

ROCKWELL 55NYS24-1

IDLI GROUP 83484

END YOKE WITH ROUND BORE
SEE FIGURE 5B

4.6250 in	OE	117.48 mm
1.5313 in	BD	38.90 mm
1.5000 in	SB	38.10 mm
4.0625 in	EC	103.19 mm
2.8750 in	HD	73.03 mm

ROCKWELL 55NYR24-1

IDLI GROUP 83485

END YOKE WITH ROUND BORE
SEE FIGURE 5B

4.6250 in	OE	117.48 mm
1.5313 in	BD	38.90 mm
1.5000 in	SB	38.10 mm
4.3750 in	EC	111.13 mm
2.8750 in	HD	73.03 mm

ROCKWELL 55NYR24

IDLI GROUP 83486

END YOKE WITH ROUND BORE
SEE FIGURE 5B

4.6250 in	OE	117.48 mm
1.5313 in	BD	38.90 mm
1.5000 in	SB	38.10 mm
5.5625 in	EC	141.29 mm
2.6875 in	HD	68.26 mm

ROCKWELL 55NYR24-3

IDLI GROUP 83487

END YOKE WITH ROUND BORE
slight variation from above group
SEE FIGURE 5B

4.6250 in	OE	117.48 mm
1.5313 in	BD	38.90 mm
1.5000 in	SB	38.10 mm
5.5625 in	EC	141.29 mm
2.6875 in	HD	68.26 mm

ROCKWELL 55NYR24-9

IDLI GROUP 83488

END YOKE WITH ROUND BORE
SEE FIGURE 5B

4.6250 in	OE	117.48 mm
1.5313 in	BD	38.90 mm
1.5000 in	SB	38.10 mm
5.5625 in	EC	141.29 mm
2.8750 in	HD	73.03 mm

G & G MFG 184-5524
ROCKWELL 55NYR24-13

IDLI GROUP 83489

END YOKE WITH ROUND BORE
SEE FIGURE 5B

4.6250 in	OE	117.48 mm
1.5313 in	BD	38.90 mm
1.5313 in	SB	38.90 mm
5.5625 in	EC	141.29 mm
2.8750 in	HD	73.03 mm

ROCKWELL 55NYR25-1

IDLI GROUP 83490

END YOKE WITH 16 SPLINES
SEE FIGURE 5A

4.6250 in	OE	117.48 mm
1.5313 in	BD	38.90 mm
1.5469 in	SB	39.29 mm
4.6875 in	EC	119.06 mm
2.0000 in	HD	50.80 mm

ROCKWELL 55NYS25

IDLI GROUP 83491

END YOKE WITH ROUND BORE
SEE FIGURE 5B

4.6250 in	OE	117.48 mm
1.5313 in	BD	38.90 mm
1.5781 in	SB	40.08 mm
4.0000 in	EC	101.60 mm
3.0000 in	HD	76.20 mm

ROCKWELL 55NYR25-4

IDLI GROUP 83492

END YOKE WITH ROUND BORE
slight variation from above group
SEE FIGURE 5B

4.6250 in	OE	117.48 mm
1.5313 in	BD	38.90 mm
1.5781 in	SB	40.08 mm
4.0000 in	EC	101.60 mm
3.0000 in	HD	76.20 mm

ROCKWELL 55NYR25

IDLI GROUP 83493

END YOKE WITH 16 INVOLUTED SPLINES
SEE FIGURE 5A

4.6250 in	OE	117.48 mm
1.5313 in	BD	38.90 mm
1.5938 in	SB	40.48 mm
4.5000 in	EC	114.30 mm

ROCKWELL 55NYS26-22

IDLI GROUP 83494

END YOKE WITH 6 SPLINES
SEE FIGURE 5A

4.6250 in	OE	117.48 mm
1.5313 in	BD	38.90 mm
1.6250 in	SB	41.28 mm
4.0625 in	EC	103.19 mm
2.2500 in	HD	57.15 mm

ROCKWELL 55NYS26-1

IDLI GROUP 83495

END YOKE WITH ROUND BORE
SEE FIGURE 5B

4.6250 in	OE	117.48 mm
1.5313 in	BD	38.90 mm
1.6250 in	SB	41.28 mm
4.0625 in	EC	103.19 mm
3.0000 in	HD	76.20 mm

ROCKWELL 55NYR26

IDLI GROUP 83496

END YOKE WITH ROUND BORE
SEE FIGURE 5B

4.6250 in	OE	117.48 mm
1.5313 in	BD	38.90 mm
1.6250 in	SB	41.28 mm
4.4375 in	EC	112.71 mm
2.6875 in	HD	68.26 mm

ROCKWELL 55NYR26-9

IDLI GROUP 83497

END YOKE WITH 10 SPLINES
SEE FIGURE 5A

4.6250 in	OE	117.48 mm
1.5313 in	BD	38.90 mm
1.6250 in	SB	41.28 mm
4.6250 in	EC	117.48 mm
2.6875 in	HD	68.26 mm

ROCKWELL 55NYS26-23

IDLI GROUP 83498

END YOKE WITH 25 INVOLUTED SPLINES
SEE FIGURE 5A

4.6250 in	OE	117.48 mm
1.5313 in	BD	38.90 mm
1.6250 in	SB	41.28 mm
4.6875 in	EC	119.06 mm
2.0000 in	HD	50.80 mm

ROCKWELL 55NYS26-10

IDLI GROUP 83499

END YOKE WITH SQUARE BORE
SEE FIGURE 5C

4.6250 in	OE	117.48 mm
1.5313 in	BD	38.90 mm
1.6250 in	SB	41.28 mm
4.7500 in	EC	120.65 mm
2.6875 in	HD	68.26 mm

ROCKWELL 55NYQ26

IDLI GROUP 83500

END YOKE WITH 10 SPLINES
SEE FIGURE 5A

4.6250 in	OE	117.48 mm
1.5313 in	BD	38.90 mm
1.6250 in	SB	41.28 mm
4.8750 in	EC	123.83 mm
2.5000 in	HD	63.50 mm

ROCKWELL 55NYS26-19

BD = BEARING DIAMETER (outside)   **BW** = BEARING WIDTH   **CB** = CROSS LENGTH WITH BEARINGS   **CC** = CENTER TO CENTER   **CD** = CROSS DIAMETER
CL = CROSS LENGTH WITHOUT BEARINGS   **EC** = END TO CENTER or FACE TO FACE   **EE** = END TO END   **EL** = EFFECTIVE LENGTH
HD = HUB DIAMETER (or insert)   **IE** = INSIDE OF EARS (or recess)   **OD** = OUTSIDE DIAMETER   **OE** = OUTSIDE OF EARS   **SB** = SPLINE OR BORE SIZE

CAUTION: BE SURE TO REFER TO ENGINEERING CATALOGS FOR SPECIAL APPLICATIONS THAT REQUIRE SPECIFIC MATERIAL CONTENT, TOLERANCES, ETC. SEE FOOTNOTE.

IDLI GROUP 83501

END YOKE WITH HEX BORE
SEE FIGURE 5E

4.6250 in	OE	117.48 mm
1.5313 in	BD	38.90 mm
1.6563 in	SB	42.07 mm
8.4375 in	EC	214.31 mm
2.3750 in	HD	60.33 mm

ROCKWELL 55NYH26

IDLI GROUP 83502

END YOKE WITH ROUND BORE
SEE FIGURE 5B

4.6250 in	OE	117.48 mm
1.5313 in	BD	38.90 mm
1.6875 in	SB	42.86 mm
4.8125 in	EC	122.24 mm
3.0000 in	HD	76.20 mm

ROCKWELL 55NYR27-3

IDLI GROUP 83503

END YOKE WITH 27 INVOLUTED SPLINES
SEE FIGURE 5A

4.6250 in	OE	117.48 mm
1.5313 in	BD	38.90 mm
1.7500 in	SB	44.45 mm
3.0000 in	EC	76.20 mm
2.2500 in	HD	57.15 mm

ROCKWELL 55NYS28-19

IDLI GROUP 83504

END YOKE WITH 6 SPLINES
SEE FIGURE 5A

4.6250 in	OE	117.48 mm
1.5313 in	BD	38.90 mm
1.7500 in	SB	44.45 mm
4.0000 in	EC	101.60 mm
3.0000 in	HD	76.20 mm

ROCKWELL 55NYS28-23

IDLI GROUP 83505

END YOKE WITH 16 SPLINES
SEE FIGURE 5A

4.6250 in	OE	117.48 mm
1.5313 in	BD	38.90 mm
1.7500 in	SB	44.45 mm
4.0000 in	EC	101.60 mm
3.0000 in	HD	76.20 mm

ROCKWELL 55NYS28

IDLI GROUP 83506

END YOKE WITH 16 SPLINES
slight variation from above group
SEE FIGURE 5A

4.6250 in	OE	117.48 mm
1.5313 in	BD	38.90 mm
1.7500 in	SB	44.45 mm
4.0000 in	EC	101.60 mm
3.0000 in	HD	76.20 mm

ROCKWELL 55NYS28-10

IDLI GROUP 83507

END YOKE WITH ROUND BORE
SEE FIGURE 5B

4.6250 in	OE	117.48 mm
1.5313 in	BD	38.90 mm
1.7500 in	SB	44.45 mm
4.0625 in	EC	103.19 mm
3.0000 in	HD	76.20 mm

ROCKWELL 55NYR28-7

IDLI GROUP 83508

END YOKE WITH ROUND BORE
slight variation from above group
SEE FIGURE 5B

4.6250 in	OE	117.48 mm
1.5313 in	BD	38.90 mm
1.7500 in	SB	44.45 mm
4.0625 in	EC	103.19 mm
3.0000 in	HD	76.20 mm

ROCKWELL 55NYR28

IDLI GROUP 83509

END YOKE WITH ROUND BORE
slight variation from above group
SEE FIGURE 5B

4.6250 in	OE	117.48 mm
1.5313 in	BD	38.90 mm
1.7500 in	SB	44.45 mm
4.0625 in	EC	103.19 mm
3.0000 in	HD	76.20 mm

ROCKWELL 55NYR28-7

IDLI GROUP 83510

END YOKE WITH ROUND BORE
SEE FIGURE 5B

4.6250 in	OE	117.48 mm
1.5313 in	BD	38.90 mm
1.7500 in	SB	44.45 mm
4.5000 in	EC	114.30 mm
2.6875 in	HD	68.26 mm

ROCKWELL 55NYR28-1

IDLI GROUP 83511

END YOKE WITH 10 SPLINES
SEE FIGURE 5A

4.6250 in	OE	117.48 mm
1.5313 in	BD	38.90 mm
1.7500 in	SB	44.45 mm
4.6250 in	EC	117.48 mm
2.6875 in	HD	68.26 mm

ROCKWELL 55NYS28-21

IDLI GROUP 83512

END YOKE WITH 27 INVOLUTED SPLINES
SEE FIGURE 5A

4.6250 in	OE	117.48 mm
1.5313 in	BD	38.90 mm
1.7500 in	SB	44.45 mm
5.5000 in	EC	139.70 mm
2.6875 in	HD	68.26 mm

ROCKWELL 55NYS28-47

IDLI GROUP 83513

END YOKE WITH ROUND BORE
SEE FIGURE 5B

4.6250 in	OE	117.48 mm
1.5313 in	BD	38.90 mm
1.7500 in	SB	44.45 mm
5.5625 in	EC	141.29 mm
3.0000 in	HD	76.20 mm

ROCKWELL 55NYR28-6

IDLI GROUP 83514

END YOKE WITH ROUND BORE
slight variation from above group
SEE FIGURE 5B

4.6250 in	OE	117.48 mm
1.5313 in	BD	38.90 mm
1.7500 in	SB	44.45 mm
5.5625 in	EC	141.29 mm
3.0000 in	HD	76.20 mm

ROCKWELL 55NYR28-11

IDLI GROUP 83515

END YOKE WITH ROUND BORE
slight variation from above group
SEE FIGURE 5B

4.6250 in	OE	117.48 mm
1.5313 in	BD	38.90 mm
1.7500 in	SB	44.45 mm
5.5625 in	EC	141.29 mm
3.0000 in	HD	76.20 mm

ROCKWELL 55NYR28-9

IDLI GROUP 83516

END YOKE WITH ROUND BORE
SEE FIGURE 5B

4.6250 in	OE	117.48 mm
1.5313 in	BD	38.90 mm
1.9219 in	SB	48.82 mm
4.0625 in	EC	103.19 mm
3.0000 in	HD	76.20 mm

ROCKWELL 55NYR31-3

IDLI GROUP 83517

END YOKE WITH ROUND BORE
SEE FIGURE 5B

4.6250 in	OE	117.48 mm
1.5313 in	BD	38.90 mm
1.9375 in	SB	49.21 mm
4.0625 in	EC	103.19 mm
3.0000 in	HD	76.20 mm

ROCKWELL 55NYR31

IDLI GROUP 83518

END YOKE WITH 10 SPLINES
SEE FIGURE 5A

4.6250 in	OE	117.48 mm
1.5313 in	BD	38.90 mm
1.9688 in	SB	50.01 mm
4.5000 in	EC	114.30 mm
2.5938 in	HD	65.88 mm

ROCKWELL 55NYS32-7

IDLI GROUP 83519

END YOKE WITH 16 SPLINES
SEE FIGURE 5A

4.6250 in	OE	117.48 mm
1.5313 in	BD	38.90 mm
2.0000 in	SB	50.80 mm
4.0000 in	EC	101.60 mm
3.0000 in	HD	76.20 mm

ROCKWELL 55NYS32

IDLI GROUP 83520

END YOKE WITH ROUND BORE
SEE FIGURE 5B

4.6250 in	OE	117.48 mm
1.5313 in	BD	38.90 mm
2.0000 in	SB	50.80 mm
5.5000 in	EC	139.70 mm
2.6875 in	HD	68.26 mm

ROCKWELL 55NYR32

IDLI GROUP 83521

END YOKE WITH ROUND BORE
SEE FIGURE 5B

4.6250 in	OE	117.48 mm
1.5313 in	BD	38.90 mm
2.0000 in	SB	50.80 mm
5.5625 in	EC	141.29 mm
3.0000 in	HD	76.20 mm

ROCKWELL 55NYR32-4

IDLI GROUP 83522

END YOKE WITH 10 SPLINES
SEE FIGURE 6A

5.3125 in	OE	134.94 mm
1.8750 in	BD	47.63 mm
1.6250 in	SB	41.28 mm
4.1250 in	EC	104.78 mm
2.3750 in	HD	60.33 mm

ROCKWELL 16N4-2401
SPICER 5-4-2401
SPICER 5-4-4271

IDLI GROUP 83523

END YOKE WITH 10 SPLINES
SEE FIGURE 6A

5.3125 in	OE	134.94 mm
1.8750 in	BD	47.63 mm
1.7500 in	SB	44.45 mm
4.1250 in	EC	104.78 mm
2.1875 in	HD	55.56 mm

ROCKWELL 16N4-1721
SPICER 5-4-1721

IDLI GROUP 83524

END YOKE WITH 10 SPLINES
SEE FIGURE 6A

5.3125 in	OE	134.94 mm
1.8750 in	BD	47.63 mm
1.7500 in	SB	44.45 mm
4.1250 in	EC	104.78 mm
2.6250 in	HD	66.68 mm

ROCKWELL 16N4-781
SPICER 5-4-781

IDLI GROUP 83525

END YOKE WITH 10 SPLINES
SEE FIGURE 6A

5.3125 in	OE	134.94 mm
1.8750 in	BD	47.63 mm
1.7500 in	SB	44.45 mm
4.7500 in	EC	120.65 mm
2.6250 in	HD	66.68 mm

ROCKWELL 16N4-2081
ROCKWELL 16N4-5111X
SPICER 5-4-2081
SPICER 5-4-5111X

IDLI GROUP 83526

END YOKE WITH 10 SPLINES
SEE FIGURE 6A

5.3125 in	OE	134.94 mm
1.8750 in	BD	47.63 mm
1.7500 in	SB	44.45 mm
4.7500 in	EC	120.65 mm
2.7500 in	HD	69.85 mm

ROCKWELL 16N4-1931
ROCKWELL 16N4-3601X
ROCKWELL 16N4-3611X
SPICER 5-4-1931
SPICER 5-4-3601X
SPICER 5-4-3611X

IDLI GROUP 83527

END YOKE WITH 10 SPLINES
SEE FIGURE 6A

5.3125 in	OE	134.94 mm
1.8750 in	BD	47.63 mm
1.7500 in	SB	44.45 mm
5.0000 in	EC	127.00 mm
2.7500 in	HD	69.85 mm

CLEVE.STL. C661-106
ROCKWELL 16N4-3621
SPICER 5-4-3621

IDLI GROUP 83528

END YOKE WITH 10 SPLINES
SEE FIGURE 6A

5.3125 in	OE	134.94 mm
1.8750 in	BD	47.63 mm
1.7500 in	SB	44.45 mm
5.3438 in	EC	135.73 mm
2.6250 in	HD	66.68 mm

ROCKWELL 16NYS28-12

IDLI GROUP 83529

END YOKE WITH 10 SPLINES
SEE FIGURE 6A

5.3125 in	OE	134.94 mm
1.8750 in	BD	47.63 mm
1.7500 in	SB	44.45 mm
6.0000 in	EC	152.40 mm
2.6250 in	HD	66.68 mm

ROCKWELL 16N4-3901
SPICER 5-4-3901
SPICER 5-4-4761X

IDLI GROUP 83530

END YOKE WITH 10 SPLINES
SEE FIGURE 6A

5.3125 in	OE	134.94 mm
1.8750 in	BD	47.63 mm
1.7500 in	SB	44.45 mm
6.0000 in	EC	152.40 mm
2.7500 in	HD	69.85 mm

ROCKWELL 16NYS28-11

IDLI GROUP 83531

END YOKE WITH 10 SPLINES
slight variation from above group
SEE FIGURE 6A

5.3125 in	OE	134.94 mm
1.8750 in	BD	47.63 mm
1.7500 in	SB	44.45 mm
6.0000 in	EC	152.40 mm
2.7500 in	HD	69.85 mm

ROCKWELL 16N4-3861
ROCKWELL 16N4-4441X
SPICER 5-4-3681
SPICER 5-4-3861
SPICER 5-4-4441X
SPICER 5-4-5281X

IDLI GROUP 83532

END YOKE WITH 10 SPLINES
slight variation from above group
SEE FIGURE 6A

5.3125 in	OE	134.94 mm
1.8750 in	BD	47.63 mm
1.7500 in	SB	44.45 mm
6.0000 in	EC	152.40 mm
2.7500 in	HD	69.85 mm

ROCKWELL 16N4-4451X
SPICER 5-4-4451X

IDLI GROUP 83533

END YOKE WITH 34 SLANTED SPLINES
SEE FIGURE 6A

5.3125 in	OE	134.94 mm
1.8750 in	BD	47.63 mm
1.7656 in	SB	44.85 mm
4.8750 in	EC	123.83 mm
2.5000 in	HD	63.50 mm

ROCKWELL 16N4-5571
SPICER 5-4-5571

IDLI GROUP 83534

END YOKE WITH 34 SLANTED SPLINES
slight variation from above group
SEE FIGURE 6A

5.3125 in	OE	134.94 mm
1.8750 in	BD	47.63 mm
1.7656 in	SB	44.85 mm
4.8750 in	EC	123.83 mm
2.5000 in	HD	63.50 mm

ROCKWELL 16NYS28-13A

IDLI GROUP 83535

END YOKE WITH 34 SLANTED SPLINES
SEE FIGURE 6A

5.3125 in	OE	134.94 mm
1.8750 in	BD	47.63 mm
1.7656 in	SB	44.85 mm
5.4375 in	EC	138.11 mm
2.5000 in	HD	63.50 mm

ROCKWELL 16N4-6261

IDLI GROUP 83536

END YOKE WITH 34 SLANTED SPLINES
slight variation from above group
SEE FIGURE 6A

5.3125 in	OE	134.94 mm
1.8750 in	BD	47.63 mm
1.7656 in	SB	44.85 mm
5.4375 in	EC	138.11 mm
2.5000 in	HD	63.50 mm

ROCKWELL 16N4-5071X
SPICER 5-4-5071X

IDLI GROUP 83537

END YOKE WITH 34 SLANTED SPLINES
slight variation from above group
SEE FIGURE 6A

5.3125 in	OE	134.94 mm
1.8750 in	BD	47.63 mm
1.7656 in	SB	44.85 mm
5.4375 in	EC	138.11 mm
2.5000 in	HD	63.50 mm

ROCKWELL 16N4-5081X
SPICER 5-4-5081X

IDLI GROUP 83538

END YOKE WITH 34 SLANTED SPLINES
SEE FIGURE 6A

5.3125 in	OE	134.94 mm
1.8750 in	BD	47.63 mm
1.7656 in	SB	44.85 mm
6.0313 in	EC	153.20 mm
2.5000 in	HD	63.50 mm

ROCKWELL 16N4-6251

IDLI GROUP 83539

END YOKE WITH 34 SLANTED SPLINES
slight variation from above group
SEE FIGURE 6A

5.3125 in	OE	134.94 mm
1.8750 in	BD	47.63 mm
1.7656 in	SB	44.85 mm
6.0313 in	EC	153.20 mm
2.5000 in	HD	63.50 mm

ROCKWELL 16NYS28-34A

IDLI GROUP 83540

END YOKE WITH 34 SLANTED SPLINES
slight variation from above group
SEE FIGURE 6A

5.3125 in	OE	134.94 mm
1.8750 in	BD	47.63 mm
1.7656 in	SB	44.85 mm
6.0313 in	EC	153.20 mm
2.5000 in	HD	63.50 mm

ROCKWELL 16N4-4891X
SPICER 5-4-4781

IDLI GROUP 83541

END YOKE WITH 34 SLANTED SPLINES
slight variation from above group
SEE FIGURE 6A

SPICER 5-4-4891X

5.3125 in	OE	134.94 mm
1.8750 in	BD	47.63 mm
1.7656 in	SB	44.85 mm
6.0313 in	EC	153.20 mm
2.5000 in	HD	63.50 mm

ROCKWELL 16N4-4901X
SPICER 5-4-4901X

IDLI GROUP 83542

END YOKE WITH 10 SPLINES
SEE FIGURE 6A

5.3125 in	OE	134.94 mm
1.8750 in	BD	47.63 mm
1.8750 in	SB	47.63 mm
4.5625 in	EC	115.89 mm
2.7500 in	HD	69.85 mm

ROCKWELL 16N4-1371
SPICER 5-4-1371

IDLI GROUP 83543

END YOKE WITH 10 SPLINES
SEE FIGURE 6A

5.3125 in	OE	134.94 mm
1.8750 in	BD	47.63 mm
1.8750 in	SB	47.63 mm
5.0000 in	EC	127.00 mm
2.7500 in	HD	69.85 mm

ROCKWELL 16N4-2501
SPICER 5-4-2501

IDLI GROUP 83544

END YOKE WITH 10 SPLINES
SEE FIGURE 6A

5.3125 in	OE	134.94 mm
1.8750 in	BD	47.63 mm
1.9688 in	SB	50.01 mm
5.1250 in	EC	130.18 mm
2.7500 in	HD	69.85 mm

ROCKWELL 16N4-3681
ROCKWELL 16N4-5281X

IDLI GROUP 83545

END YOKE WITH 10 SPLINES
SEE FIGURE 6A

5.3125 in	OE	134.94 mm
1.8750 in	BD	47.63 mm
1.9688 in	SB	50.01 mm
5.3125 in	EC	134.94 mm
2.6250 in	HD	66.68 mm

ROCKWELL 16N4-4471
SPICER 5-4-4471

BD = BEARING DIAMETER (outside)   **BW** = BEARING WIDTH   **CB** = CROSS LENGTH WITH BEARINGS   **CC** = CENTER TO CENTER   **CD** = CROSS DIAMETER
CL = CROSS LENGTH WITHOUT BEARINGS   **EC** = END TO CENTER or FACE TO FACE   **EE** = END TO END   **EL** = EFFECTIVE LENGTH
HD = HUB DIAMETER (or insert)   **IE** = INSIDE OF EARS (or recess)   **OD** = OUTSIDE DIAMETER   **OE** = OUTSIDE OF EARS   **SB** = SPLINE OR BORE SIZE

CAUTION: BE SURE TO REFER TO ENGINEERING CATALOGS FOR SPECIAL APPLICATIONS THAT REQUIRE SPECIFIC MATERIAL CONTENT, TOLERANCES, ETC. SEE FOOTNOTE.

IDLI GROUP 83546

END YOKE WITH 10 SPLINES
SEE FIGURE 6A

5.3125 in	OE	134.94 mm
1.8750 in	BD	47.63 mm
1.9688 in	SB	50.01 mm
5.3750 in	EC	136.53 mm
3.0000 in	HD	76.20 mm

ROCKWELL 16NYS31-3

IDLI GROUP 83547

END YOKE WITH 10 SPLINES
SEE FIGURE 6A

5.3125 in	OE	134.94 mm
1.8750 in	BD	47.63 mm
2.0000 in	SB	50.80 mm
5.1875 in	EC	131.76 mm
2.9375 in	HD	74.61 mm

ROCKWELL 16N4-1631
ROCKWELL 16N4-5131X
SPICER 5-4-1631
SPICER 5-4-5131X

IDLI GROUP 83548

END YOKE WITH 30 INVOLUTED SPLINES
SEE FIGURE 6A

5.3125 in	OE	134.94 mm
1.8750 in	BD	47.63 mm
2.0000 in	SB	50.80 mm
6.0000 in	EC	152.40 mm
3.0000 in	HD	76.20 mm

ROCKWELL 16NYS32-15

IDLI GROUP 83549

END YOKE WITH 10 SPLINES
SEE FIGURE 6A

5.3125 in	OE	134.94 mm
1.8750 in	BD	47.63 mm
2.0313 in	SB	51.60 mm
4.5000 in	EC	114.30 mm
2.7500 in	HD	69.85 mm

ROCKWELL 16N4-4401
SPICER 5-4-4401

IDLI GROUP 83550

END YOKE WITH 39 SLANTED SPLINES
SEE FIGURE 6A

5.3125 in	OE	134.94 mm
1.8750 in	BD	47.63 mm
2.0313 in	SB	51.60 mm
4.5000 in	EC	114.30 mm
2.7500 in	HD	69.85 mm

ROCKWELL 16NYS32-34

IDLI GROUP 83551

END YOKE WITH 39 SLANTED SPLINES
SEE FIGURE 6A

5.3125 in	OE	134.94 mm
1.8750 in	BD	47.63 mm
2.0313 in	SB	51.60 mm
5.2500 in	EC	133.35 mm
2.7500 in	HD	69.85 mm

ROCKWELL 16N4-5421
SPICER 5-4-5421

IDLI GROUP 83552

END YOKE WITH 39 SLANTED SPLINES
slight variation from above group
SEE FIGURE 6A

5.3125 in	OE	134.94 mm
1.8750 in	BD	47.63 mm
2.0313 in	SB	51.60 mm
5.2500 in	EC	133.35 mm
2.7500 in	HD	69.85 mm

ROCKWELL 16N4-5431X
SPICER 5-4-5431X

IDLI GROUP 83553

END YOKE WITH 39 SLANTED SPLINES
SEE FIGURE 6A

5.3125 in	OE	134.94 mm
1.8750 in	BD	47.63 mm
2.0313 in	SB	51.60 mm
5.4063 in	EC	137.32 mm
3.0000 in	HD	76.20 mm

ROCKWELL 16NYS32-32A

IDLI GROUP 83554

END YOKE WITH 39 SLANTED SPLINES
slight variation from above group
SEE FIGURE 6A

5.3125 in	OE	134.94 mm
1.8750 in	BD	47.63 mm
2.0313 in	SB	51.60 mm
5.4063 in	EC	137.32 mm
3.0000 in	HD	76.20 mm

ROCKWELL 16NYS32-31A1

IDLI GROUP 83555

END YOKE WITH 39 SLANTED SPLINES
SEE FIGURE 6A

5.3125 in	OE	134.94 mm
1.8750 in	BD	47.63 mm
2.0313 in	SB	51.60 mm
5.5313 in	EC	140.50 mm
3.0000 in	HD	76.20 mm

ROCKWELL 16N4-6271

IDLI GROUP 83556

END YOKE WITH 39 SLANTED SPLINES
slight variation from above group
SEE FIGURE 6A

5.3125 in	OE	134.94 mm
1.8750 in	BD	47.63 mm
2.0313 in	SB	51.60 mm
5.5313 in	EC	140.50 mm
3.0000 in	HD	76.20 mm

ROCKWELL 16N4-6281X

IDLI GROUP 83557

END YOKE WITH 39 SLANTED SPLINES
SEE FIGURE 6A

5.3125 in	OE	134.94 mm
1.8750 in	BD	47.63 mm
2.0313 in	SB	51.60 mm
5.9375 in	EC	150.81 mm
2.7500 in	HD	69.85 mm

ROCKWELL 16N4-5441
SPICER 5-4-5441X

IDLI GROUP 83558

END YOKE WITH 39 SLANTED SPLINES
slight variation from above group
SEE FIGURE 6A

5.3125 in	OE	134.94 mm
1.8750 in	BD	47.63 mm
2.0313 in	SB	51.60 mm
5.9375 in	EC	150.81 mm
2.7500 in	HD	69.85 mm

ROCKWELL 16N4-5451X
SPICER 5-4-5451X

IDLI GROUP 83559

END YOKE WITH 39 SLANTED SPLINES
SEE FIGURE 6A

5.3125 in	OE	134.94 mm
1.8750 in	BD	47.63 mm
2.0313 in	SB	51.60 mm
6.0313 in	EC	153.20 mm
3.0000 in	HD	76.20 mm

ROCKWELL 16NYS32-33A

IDLI GROUP 83560

END YOKE WITH 39 SLANTED SPLINES
slight variation from above group
SEE FIGURE 6A

5.3125 in	OE	134.94 mm
1.8750 in	BD	47.63 mm
2.0313 in	SB	51.60 mm
6.0313 in	EC	153.20 mm
3.0000 in	HD	76.20 mm

ROCKWELL 16NYS32-33A1

IDLI GROUP 83561

END YOKE WITH 39 SLANTED SPLINES
SEE FIGURE 6A

5.3125 in	OE	134.94 mm
1.8750 in	BD	47.63 mm
2.0313 in	SB	51.60 mm
6.2188 in	EC	157.96 mm
3.0000 in	HD	76.20 mm

ROCKWELL 16N4-6241

IDLI GROUP 83562

END YOKE WITH 39 SLANTED SPLINES
slight variation from above group
SEE FIGURE 6A

5.3125 in	OE	134.94 mm
1.8750 in	BD	47.63 mm
2.0313 in	SB	51.60 mm
6.2188 in	EC	157.96 mm
3.0000 in	HD	76.20 mm

ROCKWELL 16N4-6291X

IDLI GROUP 83563

END YOKE WITH 39 SLANTED SPLINES
SEE FIGURE 6A

5.3125 in	OE	134.94 mm
1.8750 in	BD	47.63 mm
2.0313 in	SB	51.60 mm
6.2500 in	EC	158.75 mm
2.7500 in	HD	69.85 mm

ROCKWELL 16NYS32-13

IDLI GROUP 83564

END YOKE WITH 32 INVOLUTED SPLINES
SEE FIGURE 6A

5.3125 in	OE	134.94 mm
1.8750 in	BD	47.63 mm
2.1250 in	SB	53.98 mm
6.0000 in	EC	152.40 mm
3.0000 in	HD	76.20 mm

ROCKWELL 16NYS34

IDLI GROUP 83565

END YOKE WITH 16 SPLINES
SEE FIGURE 6A

5.3125 in	OE	134.94 mm
1.8750 in	BD	47.63 mm
2.1563 in	SB	54.77 mm
6.2500 in	EC	158.75 mm
2.9375 in	HD	74.61 mm

ROCKWELL 16N4-5031
SPICER 5-4-5031
SPICER 5-4-5161X

IDLI GROUP 83566

END YOKE WITH 10 SPLINES
SEE FIGURE 6A

5.3125 in	OE	134.94 mm
1.8750 in	BD	47.63 mm
2.2500 in	SB	57.15 mm
4.9375 in	EC	125.41 mm
3.1250 in	HD	79.38 mm

ROCKWELL 16NYS36-2

IDLI GROUP 83567

END YOKE WITH 6 SPLINES
SEE FIGURE 6A

5.3125 in	OE	134.94 mm
1.8750 in	BD	47.63 mm
2.2500 in	SB	57.15 mm
4.9375 in	EC	125.41 mm
3.2500 in	HD	82.55 mm

ROCKWELL 16NYS36-1

IDLI GROUP 83568

END YOKE WITH 10 SPLINES
SEE FIGURE 6A

5.3125 in	OE	134.94 mm
1.8750 in	BD	47.63 mm
2.2500 in	SB	57.15 mm
4.9375 in	EC	125.41 mm
3.2500 in	HD	82.55 mm

ROCKWELL 16NYS36-3

ENGINEERING CATALOGS MUST BE CONSULTED FOR DETAILS NOT INCLUDED
IN THIS GUIDE. SPECIFIC DESIGNS, MATERIAL CONTENT, TOLERANCES
LUBE FITTINGS AND OTHER DIMENSIONS ARE INTENTIONALLY OMITTED HERE.

IDLI GROUP 83569

END YOKE WITH 10 SPLINES
SEE FIGURE 6A

5.3125 in	OE	134.94 mm
1.8750 in	BD	47.63 mm
2.2500 in	SB	57.15 mm
5.5000 in	EC	139.70 mm
3.5000 in	HD	88.90 mm

ROCKWELL 16N4-3321
SPICER 5-4-3311X
SPICER 5-4-3321

IDLI GROUP 83570

END YOKE WITH 36 INVOLUTED SPLINES
SEE FIGURE 6A

5.3125 in	OE	134.94 mm
1.8750 in	BD	47.63 mm
2.3750 in	SB	60.33 mm
5.1563 in	EC	130.97 mm
3.0000 in	HD	76.20 mm

ROCKWELL 16N4-5631
ROCKWELL 16NYS38-3
SPICER 5-4-5631

IDLI GROUP 83571

END YOKE WITH 46 SLANTED SPLINES
SEE FIGURE 6A

5.3125 in	OE	134.94 mm
1.8750 in	BD	47.63 mm
2.3750 in	SB	60.33 mm
5.3125 in	EC	134.94 mm
3.2500 in	HD	82.55 mm

ROCKWELL 16N4-5401
ROCKWELL 16N4-5411X
SPICER 5-4-5401
SPICER 5-4-5411X

IDLI GROUP 83572

END YOKE WITH 36 INVOLUTED SPLINES
SEE FIGURE 6A

5.3125 in	OE	134.94 mm
1.8750 in	BD	47.63 mm
2.3750 in	SB	60.33 mm
6.0000 in	EC	152.40 mm
3.0000 in	HD	76.20 mm

ROCKWELL 16N4-5741
ROCKWELL 16NYS38-2
SPICER 5-4-5741

IDLI GROUP 83573

END YOKE WITH 10 SPLINES
SEE FIGURE 6A

6.0938 in	OE	154.78 mm
1.9375 in	BD	49.21 mm
1.7500 in	SB	44.45 mm
4.5000 in	EC	114.30 mm
2.6250 in	HD	66.68 mm

ROCKWELL 17N4-2081
SPICER 6-4-1101
SPICER 6-4-2081
SPICER 6-4-5581X
SPICER 6-4-851

IDLI GROUP 83574

END YOKE WITH 10 SPLINES
SEE FIGURE 6A

6.0938 in	OE	154.78 mm
1.9375 in	BD	49.21 mm
1.7500 in	SB	44.45 mm
4.8750 in	EC	123.83 mm
2.6250 in	HD	66.68 mm

ROCKWELL 17NYS28-23

IDLI GROUP 83575

END YOKE WITH 10 SPLINES
SEE FIGURE 6A

6.0938 in	OE	154.78 mm
1.9375 in	BD	49.21 mm
1.7500 in	SB	44.45 mm
4.8750 in	EC	123.83 mm
2.7500 in	HD	69.85 mm

ROCKWELL 17NYS28-11

IDLI GROUP 83576

END YOKE WITH 10 SPLINES
slight variation from above group
SEE FIGURE 6A

6.0938 in	OE	154.78 mm
1.9375 in	BD	49.21 mm
1.7500 in	SB	44.45 mm
4.8750 in	EC	123.83 mm
2.7500 in	HD	69.85 mm

ROCKWELL 17N4-2381
SPICER 6-4-1751
SPICER 6-4-2381

IDLI GROUP 83577

END YOKE WITH 10 SPLINES
SEE FIGURE 6A

6.0938 in	OE	154.78 mm
1.9375 in	BD	49.21 mm
1.7500 in	SB	44.45 mm
5.2500 in	EC	133.35 mm
2.6250 in	HD	66.68 mm

ROCKWELL 17N4-3391
SPICER 6-4-3391
SPICER 6-4-4301X

IDLI GROUP 83578

END YOKE WITH 10 SPLINES
SEE FIGURE 6A

6.0938 in	OE	154.78 mm
1.9375 in	BD	49.21 mm
1.7500 in	SB	44.45 mm
5.2500 in	EC	133.35 mm
2.7500 in	HD	69.85 mm

ROCKWELL 17N4-4311
SPICER 6-4-4311

IDLI GROUP 83579

END YOKE WITH 10 SPLINES
SEE FIGURE 6A

6.0938 in	OE	154.78 mm
1.9375 in	BD	49.21 mm
1.7500 in	SB	44.45 mm
6.0313 in	EC	153.20 mm
2.7500 in	HD	69.85 mm

ROCKWELL 17NYS28-21

IDLI GROUP 83580

END YOKE WITH 10 SPLINES
slight variation from above group
SEE FIGURE 6A

6.0938 in	OE	154.78 mm
1.9375 in	BD	49.21 mm
1.7500 in	SB	44.45 mm
6.0313 in	EC	153.20 mm
2.7500 in	HD	69.85 mm

ROCKWELL 17NYS28-21A1

IDLI GROUP 83581

END YOKE WITH 10 SPLINES
slight variation from above group
SEE FIGURE 6A

6.0938 in	OE	154.78 mm
1.9375 in	BD	49.21 mm
1.7500 in	SB	44.45 mm
6.0313 in	EC	153.20 mm
2.7500 in	HD	69.85 mm

ROCKWELL 17NYS28-21

IDLI GROUP 83582

END YOKE WITH 10 SPLINES
SEE FIGURE 6A

6.0938 in	OE	154.78 mm
1.9375 in	BD	49.21 mm
1.7500 in	SB	44.45 mm
6.1250 in	EC	155.58 mm
2.6250 in	HD	66.68 mm

ROCKWELL 17N4-4141
SPICER 6-4-4141
SPICER 6-4-4861X

IDLI GROUP 83583

END YOKE WITH 10 SPLINES
SEE FIGURE 6A

6.0938 in	OE	154.78 mm
1.9375 in	BD	49.21 mm
1.7500 in	SB	44.45 mm
6.2500 in	EC	158.75 mm
2.7500 in	HD	69.85 mm

CLEVE.STL. C67-1-122
ROCKWELL 17N4-4091
SPICER 6-4-4091
SPICER 6-4-4831X
SPICER 6-4-4841X

IDLI GROUP 83584

END YOKE WITH 34 SLANTED SPLINES
SEE FIGURE 6A

6.0938 in	OE	154.78 mm
1.9375 in	BD	49.21 mm
1.7656 in	SB	44.85 mm
4.7500 in	EC	120.65 mm
2.5000 in	HD	63.50 mm

ROCKWELL 17NYS28-19A

IDLI GROUP 83585

END YOKE WITH 34 SLANTED SPLINES
SEE FIGURE 6A

6.0938 in	OE	154.78 mm
1.9375 in	BD	49.21 mm
1.7656 in	SB	44.85 mm
4.8750 in	EC	123.83 mm
2.5000 in	HD	63.50 mm

ROCKWELL 17N4-6331

IDLI GROUP 83586

END YOKE WITH 34 SLANTED SPLINES
slight variation from above group
SEE FIGURE 6A

6.0938 in	OE	154.78 mm
1.9375 in	BD	49.21 mm
1.7656 in	SB	44.85 mm
4.8750 in	EC	123.83 mm
2.5000 in	HD	63.50 mm

ROCKWELL 17NYS28-50A

IDLI GROUP 83587

END YOKE WITH 34 SLANTED SPLINES
slight variation from above group
SEE FIGURE 6A

6.0938 in	OE	154.78 mm
1.9375 in	BD	49.21 mm
1.7656 in	SB	44.85 mm
4.8750 in	EC	123.83 mm
2.5000 in	HD	63.50 mm

ROCKWELL 17N4-5121X
SPICER 6-4-4891
SPICER 6-4-5121X

IDLI GROUP 83588

END YOKE WITH 34 SLANTED SPLINES
slight variation from above group
SEE FIGURE 6A

6.0938 in	OE	154.78 mm
1.9375 in	BD	49.21 mm
1.7656 in	SB	44.85 mm
4.8750 in	EC	123.83 mm
2.5000 in	HD	63.50 mm

ROCKWELL 17N4-5131X
SPICER 6-4-5131X

IDLI GROUP 83589

END YOKE WITH 34 SLANTED SPLINES
slight variation from above group
SEE FIGURE 6A

6.0938 in	OE	154.78 mm
1.9375 in	BD	49.21 mm
1.7656 in	SB	44.85 mm
4.8750 in	EC	123.83 mm
2.5000 in	HD	63.50 mm

ROCKWELL 17N4-5661X

IDLI GROUP 83590

END YOKE WITH 34 SLANTED SPLINES
slight variation from above group
SEE FIGURE 6A

6.0938 in	OE	154.78 mm
1.9375 in	BD	49.21 mm
1.7656 in	SB	44.85 mm
4.8750 in	EC	123.83 mm
2.5000 in	HD	63.50 mm

BD=BEARING DIAMETER (outside)   **BW**=BEARING WIDTH   **CB**=CROSS LENGTH WITH BEARINGS   **CC**=CENTER TO CENTER   **CD**=CROSS DIAMETER
CL=CROSS LENGTH WITHOUT BEARINGS   **EC**=END TO CENTER or FACE TO FACE   **EE**=END TO END   **EL**=EFFECTIVE LENGTH
HD=HUB DIAMETER (or insert)   **IE**=INSIDE OF EARS (or recess)   **OD**=OUTSIDE DIAMETER   **OE**=OUTSIDE OF EARS   **SB**=SPLINE OR BORE SIZE

CAUTION: BE SURE TO REFER TO ENGINEERING CATALOGS FOR SPECIAL APPLICATIONS THAT REQUIRE SPECIFIC MATERIAL CONTENT, TOLERANCES, ETC. SEE FOOTNOTE.

ROCKWELL 17NYS28-43A
SPICER 6-4-5661X

IDLI GROUP 83591

END YOKE WITH 34 SLANTED SPLINES
SEE FIGURE 6A

6.0938 in	OE	154.78 mm
1.9375 in	BD	49.21 mm
1.7656 in	SB	44.85 mm
5.9688 in	EC	151.61 mm
2.5000 in	HD	63.50 mm

ROCKWELL 17N4-6321

IDLI GROUP 83592

END YOKE WITH 34 SLANTED SPLINES
slight variation from above group
SEE FIGURE 6A

6.0938 in	OE	154.78 mm
1.9375 in	BD	49.21 mm
1.7656 in	SB	44.85 mm
5.9688 in	EC	151.61 mm
2.5000 in	HD	63.50 mm

ROCKWELL 17NYS28-49A

IDLI GROUP 83593

END YOKE WITH 34 SLANTED SPLINES
slight variation from above group
SEE FIGURE 6A

6.0938 in	OE	154.78 mm
1.9375 in	BD	49.21 mm
1.7656 in	SB	44.85 mm
5.9688 in	EC	151.61 mm
2.5000 in	HD	63.50 mm

ROCKWELL 17N4-4911X
SPICER 6-4-4881
SPICER 6-4-4911X

IDLI GROUP 83594

END YOKE WITH 34 SLANTED SPLINES
slight variation from above group
SEE FIGURE 6A

6.0938 in	OE	154.78 mm
1.9375 in	BD	49.21 mm
1.7656 in	SB	44.85 mm
5.9688 in	EC	151.61 mm
2.5000 in	HD	63.50 mm

ROCKWELL 17N4-5111X
SPICER 6-4-5111X

IDLI GROUP 83595

END YOKE WITH 34 INVOLUTED SPLINES
SEE FIGURE 6A

6.0938 in	OE	154.78 mm
1.9375 in	BD	49.21 mm
1.7656 in	SB	44.85 mm
7.1875 in	EC	182.56 mm
2.5625 in	HD	65.09 mm

ROCKWELL 17N4-5031
SPICER 6-4-5031
SPICER 6-4-5041X
SPICER 6-4-5161X

IDLI GROUP 83596

END YOKE WITH 34 INVOLUTED SPLINES
slight variation from above
SEE FIGURE 6A

6.0938 in	OE	154.78 mm
1.9375 in	BD	49.21 mm
1.7656 in	SB	44.85 mm
7.1875 in	EC	182.56 mm
2.5625 in	HD	65.09 mm

ROCKWELL 17N4-5041

IDLI GROUP 83597

END YOKE WITH 10 SPLINES
SEE FIGURE 6A

6.0938 in	OE	154.78 mm
1.9375 in	BD	49.21 mm
1.8750 in	SB	47.63 mm
4.5000 in	EC	114.30 mm
2.7500 in	HD	69.85 mm

ROCKWELL 17N4-2271
SPICER 6-4-1471
SPICER 6-4-2271

IDLI GROUP 83598

END YOKE WITH 10 SPLINES
SEE FIGURE 6A

6.0938 in	OE	154.78 mm
1.9375 in	BD	49.21 mm
1.8750 in	SB	47.63 mm
5.3750 in	EC	136.53 mm
2.7500 in	HD	69.85 mm

ROCKWELL 17N4-3361
SPICER 6-4-1611
SPICER 6-4-2311
SPICER 6-4-3361

IDLI GROUP 83599

END YOKE WITH 10 SPLINES
SEE FIGURE 6A

6.0938 in	OE	154.78 mm
1.9375 in	BD	49.21 mm
1.9688 in	SB	50.01 mm
4.5000 in	EC	114.30 mm
2.6250 in	HD	66.68 mm

ROCKWELL 17N4-4721
SPICER 6-4-4721
SPICER 6-4-4731X

IDLI GROUP 83600

END YOKE WITH 10 SPLINES
slight variation from above group
SEE FIGURE 6A

6.0938 in	OE	154.78 mm
1.9375 in	BD	49.21 mm
1.9688 in	SB	50.01 mm
4.5000 in	EC	114.30 mm
2.6250 in	HD	66.68 mm

ROCKWELL 17N4-3371
SPICER 6-4-3371

IDLI GROUP 83601

END YOKE WITH 10 SPLINES
slight variation from above group
SEE FIGURE 6A

6.0938 in	OE	154.78 mm
1.9375 in	BD	49.21 mm
1.9688 in	SB	50.01 mm
4.5000 in	EC	114.30 mm
2.6250 in	HD	66.68 mm

ROCKWELL 17N4-3181
SPICER 6-4-3181

IDLI GROUP 83602

END YOKE WITH 10 SPLINES
slight variation from above group
SEE FIGURE 6A

6.0938 in	OE	154.78 mm
1.9375 in	BD	49.21 mm
1.9688 in	SB	50.01 mm
4.5000 in	EC	114.30 mm
2.6250 in	HD	66.68 mm

ROCKWELL 17N4-2211
SPICER 6-4-1391
SPICER 6-4-2211

IDLI GROUP 83603

END YOKE WITH 10 SPLINES
SEE FIGURE 6A

6.0938 in	OE	154.78 mm
1.9375 in	BD	49.21 mm
1.9688 in	SB	50.01 mm
4.5625 in	EC	115.89 mm
3.0000 in	HD	76.20 mm

ROCKWELL 17N4-3121
SPICER 6-4-3121

IDLI GROUP 83604

END YOKE WITH 10 SPLINES
slight variation from above group
SEE FIGURE 6A

6.0938 in	OE	154.78 mm
1.9375 in	BD	49.21 mm
1.9688 in	SB	50.01 mm
4.5625 in	EC	115.89 mm
3.0000 in	HD	76.20 mm

ROCKWELL 17N4-3291
SPICER 6-4-3121
SPICER 6-4-3291

IDLI GROUP 83605

END YOKE WITH 10 SPLINES
SEE FIGURE 6A

6.0938 in	OE	154.78 mm
1.9375 in	BD	49.21 mm
1.9688 in	SB	50.01 mm
4.8125 in	EC	122.24 mm
2.7500 in	HD	69.85 mm

ROCKWELL 17N4-2161
SPICER 6-4-2161

IDLI GROUP 83606

END YOKE WITH 10 SPLINES
slight variation from above group
SEE FIGURE 6A

6.0938 in	OE	154.78 mm
1.9375 in	BD	49.21 mm
1.9688 in	SB	50.01 mm
4.8125 in	EC	122.24 mm
2.7500 in	HD	69.85 mm

ROCKWELL 17N4-4181X
SPICER 6-4-4181X

IDLI GROUP 83607

END YOKE WITH 10 SPLINES
SEE FIGURE 6A

6.0938 in	OE	154.78 mm
1.9375 in	BD	49.21 mm
1.9688 in	SB	50.01 mm
5.0000 in	EC	127.00 mm
2.4375 in	HD	61.91 mm

CLEVE.STL. C67-1-123
ROCKWELL 17N4-1981
SPICER 6-4-1741
SPICER 6-4-1981

IDLI GROUP 83608

END YOKE WITH 10 SPLINES
SEE FIGURE 6A

6.0938 in	OE	154.78 mm
1.9375 in	BD	49.21 mm
1.9688 in	SB	50.01 mm
5.0000 in	EC	127.00 mm
2.5000 in	HD	63.50 mm

ROCKWELL 17N4-3721
SPICER 6-4-3721

IDLI GROUP 83609

END YOKE WITH 10 SPLINES
SEE FIGURE 6A

6.0938 in	OE	154.78 mm
1.9375 in	BD	49.21 mm
1.9688 in	SB	50.01 mm
5.0000 in	EC	127.00 mm
2.6250 in	HD	66.68 mm

ROCKWELL 17N4-2391
SPICER 6-4-1771
SPICER 6-4-2391

IDLI GROUP 83610

END YOKE WITH 6 SPLINES
SEE FIGURE 6A

6.0938 in	OE	154.78 mm
1.9375 in	BD	49.21 mm
1.9688 in	SB	50.01 mm
5.0938 in	EC	129.38 mm
2.7500 in	HD	69.85 mm

ROCKWELL 17N4-3191
SPICER 6-4-3191
SPICER 6-4-3581X

IDLI GROUP 83611

END YOKE WITH 10 SPLINES
SEE FIGURE 6A

6.0938 in	OE	154.78 mm
1.9375 in	BD	49.21 mm
1.9688 in	SB	50.01 mm
5.2500 in	EC	133.35 mm
2.6250 in	HD	66.68 mm

ROCKWELL 17N4-5141
SPICER 6-4-5141
SPICER 6-4-5151X

IDLI GROUP 83612

END YOKE WITH 10 SPLINES
slight variation from above group
SEE FIGURE 6A

6.0938 in	OE	154.78 mm
1.9375 in	BD	49.21 mm
1.9688 in	SB	50.01 mm
5.2500 in	EC	133.35 mm
2.6250 in	HD	66.68 mm

ROCKWELL 17N4-2021
SPICER 6-4-1661
SPICER 6-4-2021

IDLI GROUP 83613

END YOKE WITH 10 SPLINES
slight variation from above group
SEE FIGURE 6A

6.0938 in	OE	154.78 mm
1.9375 in	BD	49.21 mm
1.9688 in	SB	50.01 mm
5.2500 in	EC	133.35 mm
2.6250 in	HD	66.68 mm

ROCKWELL 17NYS31-13A

IDLI GROUP 83614

END YOKE WITH 10 SPLINES
slight variation from above group
SEE FIGURE 6A

6.0938 in	OE	154.78 mm
1.9375 in	BD	49.21 mm
1.9688 in	SB	50.01 mm
5.2500 in	EC	133.35 mm
2.6250 in	HD	66.68 mm

ROCKWELL 17N4-3481
SPICER 6-4-2021
SPICER 6-4-3481

IDLI GROUP 83615

END YOKE WITH 10 SPLINES
SEE FIGURE 6A

6.0938 in	OE	154.78 mm
1.9375 in	BD	49.21 mm
1.9688 in	SB	50.01 mm
5.2500 in	EC	133.35 mm
2.7500 in	HD	69.85 mm

ROCKWELL 17N4-4321

IDLI GROUP 83616

END YOKE WITH 10 SPLINES
slight variation from above group
SEE FIGURE 6A

6.0938 in	OE	154.78 mm
1.9375 in	BD	49.21 mm
1.9688 in	SB	50.01 mm
5.2500 in	EC	133.35 mm
2.7500 in	HD	69.85 mm

ROCKWELL 17N4-4361X
SPICER 6-4-4361X

IDLI GROUP 83617

END YOKE WITH 10 SPLINES
slight variation from above group
SEE FIGURE 6A

6.0938 in	OE	154.78 mm
1.9375 in	BD	49.21 mm
1.9688 in	SB	50.01 mm
5.2500 in	EC	133.35 mm
2.7500 in	HD	69.85 mm

ROCKWELL 17N4-3221
SPICER 6-4-1411
SPICER 6-4-2231
SPICER 6-4-3221
SPICER 6-4-4351X

IDLI GROUP 83618

END YOKE WITH 10 SPLINES
SEE FIGURE 6A

6.0938 in	OE	154.78 mm
1.9375 in	BD	49.21 mm
1.9688 in	SB	50.01 mm
5.8750 in	EC	149.23 mm
2.9375 in	HD	74.61 mm

ROCKWELL 17N4-2241
SPICER 6-4-1441
SPICER 6-4-2241

IDLI GROUP 83619

END YOKE WITH 10 SPLINES
SEE FIGURE 6A

6.0938 in	OE	154.78 mm
1.9375 in	BD	49.21 mm
1.9688 in	SB	50.01 mm
6.1250 in	EC	155.58 mm
2.6250 in	HD	66.68 mm

ROCKWELL 17N4-4151
SPICER 6-4-4151
SPICER 6-4-4851X

IDLI GROUP 83620

END YOKE WITH 10 SPLINES
SEE FIGURE 6A

6.0938 in	OE	154.78 mm
1.9375 in	BD	49.21 mm
1.9688 in	SB	50.01 mm
6.5000 in	EC	165.10 mm
2.7500 in	HD	69.85 mm

ROCKWELL 17N4-4131
SPICER 6-4-4131
SPICER 6-4-4481X

IDLI GROUP 83621

END YOKE WITH 10 SPLINES
SEE FIGURE 6A

6.0938 in	OE	154.78 mm
1.9375 in	BD	49.21 mm
2.0000 in	SB	50.80 mm
4.8125 in	EC	122.24 mm
3.2500 in	HD	82.55 mm

ROCKWELL 17N4-2131
SPICER 6-4-1061
SPICER 6-4-2131

IDLI GROUP 83622

END YOKE WITH 10 SPLINES
slight variation from above group
SEE FIGURE 6A

6.0938 in	OE	154.78 mm
1.9375 in	BD	49.21 mm
2.0000 in	SB	50.80 mm
4.8125 in	EC	122.24 mm
3.2500 in	HD	82.55 mm

ROCKWELL 17N4-3171
SPICER 6-4-3171

IDLI GROUP 83623

END YOKE WITH 10 SPLINES
SEE FIGURE 6A

6.0938 in	OE	154.78 mm
1.9375 in	BD	49.21 mm
2.0000 in	SB	50.80 mm
5.0000 in	EC	127.00 mm
2.7500 in	HD	69.85 mm

ROCKWELL 17N4-2371
SPICER 6-4-1731
SPICER 6-4-2371

IDLI GROUP 83624

END YOKE WITH 10 SPLINES
SEE FIGURE 6A

6.0938 in	OE	154.78 mm
1.9375 in	BD	49.21 mm
2.0000 in	SB	50.80 mm
5.1875 in	EC	131.76 mm
2.7500 in	HD	69.85 mm

ROCKWELL 17N4-3271
SPICER 6-4-1781
SPICER 6-4-2411
SPICER 6-4-3271

IDLI GROUP 83625

END YOKE WITH 10 SPLINES
slight variation from above group
SEE FIGURE 6A

6.0938 in	OE	154.78 mm
1.9375 in	BD	49.21 mm
2.0000 in	SB	50.80 mm
5.1875 in	EC	131.76 mm
2.7500 in	HD	69.85 mm

ROCKWELL 17N4-4391X
SPICER 6-4-4391X

IDLI GROUP 83626

END YOKE WITH 10 SPLINES
SEE FIGURE 6A

6.0938 in	OE	154.78 mm
1.9375 in	BD	49.21 mm
2.0000 in	SB	50.80 mm
5.1875 in	EC	131.76 mm
2.9375 in	HD	74.61 mm

ROCKWELL 17N4-3241
SPICER 6-4-1551
SPICER 6-4-2291
SPICER 6-4-3241
SPICER 6-4-4741X

IDLI GROUP 83627

END YOKE WITH 10 SPLINES
slight variation from above group
SEE FIGURE 6A

6.0938 in	OE	154.78 mm
1.9375 in	BD	49.21 mm
2.0000 in	SB	50.80 mm
5.1875 in	EC	131.76 mm
2.9375 in	HD	74.61 mm

ROCKWELL 17N4-4291X
SPICER 6-4-4291X

IDLI GROUP 83628

END YOKE WITH 10 SPLINES
SEE FIGURE 6A

6.0938 in	OE	154.78 mm
1.9375 in	BD	49.21 mm
2.0000 in	SB	50.80 mm
5.2500 in	EC	133.35 mm
2.7500 in	HD	69.85 mm

ROCKWELL 17NYS32-20

IDLI GROUP 83629

END YOKE WITH 10 SPLINES
slight variation from above group
SEE FIGURE 6A

6.0938 in	OE	154.78 mm
1.9375 in	BD	49.21 mm
2.0000 in	SB	50.80 mm
5.2500 in	EC	133.35 mm
2.7500 in	HD	69.85 mm

ROCKWELL 17N4-2361
SPICER 6-4-2361

IDLI GROUP 83630

END YOKE WITH 10 SPLINES
SEE FIGURE 6A

6.0938 in	OE	154.78 mm
1.9375 in	BD	49.21 mm
2.0000 in	SB	50.80 mm
5.2500 in	EC	133.35 mm
3.0000 in	HD	76.20 mm

ROCKWELL 17NYS32-30

BD=BEARING DIAMETER (outside)   **BW**=BEARING WIDTH   **CB**=CROSS LENGTH WITH BEARINGS   **CC**=CENTER TO CENTER   **CD**=CROSS DIAMETER
CL=CROSS LENGTH WITHOUT BEARINGS   **EC**=END TO CENTER or FACE TO FACE   **EE**=END TO END   **EL**=EFFECTIVE LENGTH
HD=HUB DIAMETER (or insert)   **IE**=INSIDE OF EARS (or recess)   **OD**=OUTSIDE DIAMETER   **OE**=OUTSIDE OF EARS   **SB**=SPLINE OR BORE SIZE

CAUTION: BE SURE TO REFER TO ENGINEERING CATALOGS FOR SPECIAL APPLICATIONS THAT REQUIRE SPECIFIC MATERIAL CONTENT, TOLERANCES, ETC. SEE FOOTNOTE.

IDLI GROUP 83631

END YOKE WITH 10 SPLINES
SEE FIGURE 6A

6.0938 in	OE	154.78 mm
1.9375 in	BD	49.21 mm
2.0000 in	SB	50.80 mm
5.4375 in	EC	138.11 mm
2.9375 in	HD	74.61 mm

ROCKWELL 17NYS32-28

IDLI GROUP 83632

END YOKE WITH 10 SPLINES
SEE FIGURE 6A

6.0938 in	OE	154.78 mm
1.9375 in	BD	49.21 mm
2.0000 in	SB	50.80 mm
5.6250 in	EC	142.88 mm
2.9375 in	HD	74.61 mm

ROCKWELL 17N4-2191
SPICER 6-4-2191

IDLI GROUP 83633

END YOKE WITH 30 INVOLUTED SPLINES
SEE FIGURE 6A

6.0938 in	OE	154.78 mm
1.9375 in	BD	49.21 mm
2.0000 in	SB	50.80 mm
6.0938 in	EC	154.78 mm
3.0000 in	HD	76.20 mm

ROCKWELL 17NYS32-48

IDLI GROUP 83634

END YOKE WITH 30 INVOLUTED SPLINES
SEE FIGURE 6A

6.0938 in	OE	154.78 mm
1.9375 in	BD	49.21 mm
2.0000 in	SB	50.80 mm
6.9375 in	EC	176.21 mm
3.0000 in	HD	76.20 mm

ROCKWELL 17NYS32-53

IDLI GROUP 83635

END YOKE WITH 39 SLANTED SPLINES
SEE FIGURE 6A

6.0938 in	OE	154.78 mm
1.9375 in	BD	49.21 mm
2.0313 in	SB	51.60 mm
4.8125 in	EC	122.24 mm
2.7500 in	HD	69.85 mm

ROCKWELL 17NYS32-43

IDLI GROUP 83636

END YOKE WITH 39 SLANTED SPLINES
slight variation from above group
SEE FIGURE 6A

6.0938 in	OE	154.78 mm
1.9375 in	BD	49.21 mm
2.0313 in	SB	51.60 mm
4.8125 in	EC	122.24 mm
2.7500 in	HD	69.85 mm

ROCKWELL 17N4-5551X
SPICER 6-4-5551X

IDLI GROUP 83637

END YOKE WITH 39 SLANTED SPLINES
slight variation from above group
SEE FIGURE 6A

6.0938 in	OE	154.78 mm
1.9375 in	BD	49.21 mm
2.0313 in	SB	51.60 mm
4.8125 in	EC	122.24 mm
2.7500 in	HD	69.85 mm

ROCKWELL 17N4-5541
SPICER 6-4-5541

IDLI GROUP 83638

END YOKE WITH 10 SPLINES
SEE FIGURE 6A

6.0938 in	OE	154.78 mm
1.9375 in	BD	49.21 mm
2.0313 in	SB	51.60 mm
5.0000 in	EC	127.00 mm
2.7500 in	HD	69.85 mm

ROCKWELL 17N4-3791
SPICER 6-4-3791

IDLI GROUP 83639

END YOKE WITH 39 SLANTED SPLINES
SEE FIGURE 6A

6.0938 in	OE	154.78 mm
1.9375 in	BD	49.21 mm
2.0313 in	SB	51.60 mm
5.0000 in	EC	127.00 mm
2.7500 in	HD	69.85 mm

ROCKWELL 17NYS32-52

IDLI GROUP 83640

END YOKE WITH 39 SLANTED SPLINES
SEE FIGURE 6A

6.0938 in	OE	154.78 mm
1.9375 in	BD	49.21 mm
2.0313 in	SB	51.60 mm
5.1563 in	EC	130.97 mm
3.0000 in	HD	76.20 mm

ROCKWELL 17N4-6401

IDLI GROUP 83641

END YOKE WITH 39 SLANTED SPLINES
slight variation from above group
SEE FIGURE 6A

6.0938 in	OE	154.78 mm
1.9375 in	BD	49.21 mm
2.0313 in	SB	51.60 mm
5.1563 in	EC	130.97 mm
3.0000 in	HD	76.20 mm

ROCKWELL 17N4-6411X

IDLI GROUP 83642

END YOKE WITH 39 SLANTED SPLINES
slight variation from above group
SEE FIGURE 6A

6.0938 in	OE	154.78 mm
1.9375 in	BD	49.21 mm
2.0313 in	SB	51.60 mm
5.1563 in	EC	130.97 mm
3.0000 in	HD	76.20 mm

ROCKWELL 17N4-6421X

IDLI GROUP 83643

END YOKE WITH 39 SLANTED SPLINES
slight variation from above group
SEE FIGURE 6A

6.0938 in	OE	154.78 mm
1.9375 in	BD	49.21 mm
2.0313 in	SB	51.60 mm
5.1563 in	EC	130.97 mm
3.0000 in	IID	76.20 mm

ROCKWELL 17N4-5051X
SPICER 6-4-4901
SPICER 6-4-5051X

IDLI GROUP 83644

END YOKE WITH 39 SLANTED SPLINES
slight variation from above group
SEE FIGURE 6A

6.0938 in	OE	154.78 mm
1.9375 in	BD	49.21 mm
2.0313 in	SB	51.60 mm
5.1563 in	EC	130.97 mm
3.0000 in	HD	76.20 mm

ROCKWELL 17N4-5101X
SPICER 6-4-5101X

IDLI GROUP 83645

END YOKE WITH 39 SLANTED SPLINES
slight variation from above group
SEE FIGURE 6A

6.0938 in	OE	154.78 mm
1.9375 in	BD	49.21 mm
2.0313 in	SB	51.60 mm
5.1563 in	EC	130.97 mm
3.0000 in	HD	76.20 mm

ROCKWELL 17NYS32-27A2

IDLI GROUP 83646

END YOKE WITH 39 SLANTED SPLINES
SEE FIGURE 6A

6.0938 in	OE	154.78 mm
1.9375 in	BD	49.21 mm
2.0313 in	SB	51.60 mm
5.2500 in	EC	133.35 mm
2.7500 in	HD	69.85 mm

ROCKWELL 17N4-5521
SPICER 6-4-5521

IDLI GROUP 83647

END YOKE WITH 39 SLANTED SPLINES
slight variation from above group
SEE FIGURE 6A

6.0938 in	OE	154.78 mm
1.9375 in	BD	49.21 mm
2.0313 in	SB	51.60 mm
5.2500 in	EC	133.35 mm
2.7500 in	HD	69.85 mm

ROCKWELL 17N4-5531X
SPICER 6-4-5531X

IDLI GROUP 83648

END YOKE WITH 39 SLANTED SPLINES
slight variation from above group
SEE FIGURE 6A

6.0938 in	OE	154.78 mm
1.9375 in	BD	49.21 mm
2.0313 in	SB	51.60 mm
5.2500 in	EC	133.35 mm
2.7500 in	HD	69.85 mm

ROCKWELL 17NYS32-41A1

IDLI GROUP 83649

END YOKE WITH 39 SLANTED SPLINES
slight variation from above group
SEE FIGURE 6A

6.0938 in	OE	154.78 mm
1.9375 in	BD	49.21 mm
2.0313 in	SB	51.60 mm
5.2500 in	EC	133.35 mm
2.7500 in	HD	69.85 mm

ROCKWELL 17NYS32-41A2

IDLI GROUP 83650

END YOKE WITH 39 SLANTED SPLINES
SEE FIGURE 6A

6.0938 in	OE	154.78 mm
1.9375 in	BD	49.21 mm
2.0313 in	SB	51.60 mm
6.2500 in	EC	158.75 mm
3.0000 in	HD	76.20 mm

ROCKWELL 17N4-6371

IDLI GROUP 83651

END YOKE WITH 39 SLANTED SPLINES
slight variation from above group
SEE FIGURE 6A

6.0938 in	OE	154.78 mm
1.9375 in	BD	49.21 mm
2.0313 in	SB	51.60 mm
6.2500 in	EC	158.75 mm
3.0000 in	HD	76.20 mm

ROCKWELL 17N4-6381X

IDLI GROUP 83652

END YOKE WITH 39 SLANTED SPLINES
slight variation from above group
SEE FIGURE 6A

6.0938 in	OE	154.78 mm
1.9375 in	BD	49.21 mm
2.0313 in	SB	51.60 mm
6.2500 in	EC	158.75 mm
3.0000 in	HD	76.20 mm

ROCKWELL 17N4-6391X

IDLI GROUP 83653

END YOKE WITH 39 SLANTED SPLINES
SEE FIGURE 6A

6.0938 in	OE	154.78 mm
1.9375 in	BD	49.21 mm
2.0313 in	SB	51.60 mm
6.2813 in	EC	159.55 mm
3.0000 in	HD	76.20 mm

ROCKWELL 17N4-4981X
SPICER 6-4-4981X

IDLI GROUP 83654

END YOKE WITH 39 SLANTED SPLINES
SEE FIGURE 6A

6.0938 in	OE	154.78 mm
1.9375 in	BD	49.21 mm
2.0313 in	SB	51.60 mm
6.5000 in	EC	165.10 mm
2.7500 in	HD	69.85 mm

ROCKWELL 17N4-5501
SPICER 6-4-5501

IDLI GROUP 83655

END YOKE WITH 39 SLANTED SPLINES
slight variation from above group
SEE FIGURE 6A

6.0938 in	OE	154.78 mm
1.9375 in	BD	49.21 mm
2.0313 in	SB	51.60 mm
6.5000 in	EC	165.10 mm
2.7500 in	HD	69.85 mm

ROCKWELL 17N4-5511X
SPICER 6-4-5511X

IDLI GROUP 83656

END YOKE WITH 39 SLANTED SPLINES
slight variation from above group
SEE FIGURE 6A

6.0938 in	OE	154.78 mm
1.9375 in	BD	49.21 mm
2.0313 in	SB	51.60 mm
6.5000 in	EC	165.10 mm
2.7500 in	HD	69.85 mm

ROCKWELL 17NYS32-40A1

IDLI GROUP 83657

END YOKE WITH 32 INVOLUTED SPLINES
SEE FIGURE 6A

6.0938 in	OE	154.78 mm
1.9375 in	BD	49.21 mm
2.1094 in	SB	53.58 mm
5.0000 in	EC	127.00 mm
3.0000 in	HD	76.20 mm

ROCKWELL 17N4-5821
SPICER 6-4-5821

IDLI GROUP 83658

END YOKE WITH 32 INVOLUTED SPLINES
slight variation from above group
SEE FIGURE 6A

6.0938 in	OE	154.78 mm
1.9375 in	BD	49.21 mm
2.1094 in	SB	53.58 mm
5.0000 in	EC	127.00 mm
3.0000 in	HD	76.20 mm

ROCKWELL 17N4-5791
SPICER 6-4-5791

IDLI GROUP 83659

END YOKE WITH 32 INVOLUTED SPLINES
SEE FIGURE 6A

6.0938 in	OE	154.78 mm
1.9375 in	BD	49.21 mm
2.1094 in	SB	53.58 mm
6.2500 in	EC	158.75 mm
3.0000 in	HD	76.20 mm

ROCKWELL 17NYS34-9A

IDLI GROUP 83660

END YOKE WITH 16 SPLINES
SEE FIGURE 6A

6.0938 in	OE	154.78 mm
1.9375 in	BD	49.21 mm
2.1563 in	SB	54.77 mm
5.1875 in	EC	131.76 mm
2.9375 in	HD	74.61 mm

ROCKWELL 17N4-4961
SPICER 6-4-4961
SPICER 6-4-5061X

IDLI GROUP 83661

END YOKE WITH 16 SPLINES
SEE FIGURE 6A

6.0938 in	OE	154.78 mm
1.9375 in	BD	49.21 mm
2.1563 in	SB	54.77 mm
5.5000 in	EC	139.70 mm
2.9375 in	HD	74.61 mm

ROCKWELL 17N4-2181
SPICER 6-4-2181
SPICER 6-4-4171X

IDLI GROUP 83662

END YOKE WITH 16 SPLINES
SEE FIGURE 6A

6.0938 in	OE	154.78 mm
1.9375 in	BD	49.21 mm
2.1563 in	SB	54.77 mm
6.3125 in	EC	160.34 mm
2.9375 in	HD	74.61 mm

ROCKWELL 17NYS34-5

IDLI GROUP 83663

END YOKE WITH 16 SPLINES
SEE FIGURE 6A

6.0938 in	OE	154.78 mm
1.9375 in	BD	49.21 mm
2.1563 in	SB	54.77 mm
6.6250 in	EC	168.28 mm
2.9375 in	HD	74.61 mm

ROCKWELL 17N4-4591
SPICER 6-4-4591
SPICER 6-4-5481X

IDLI GROUP 83664

END YOKE WITH 6 SPLINES
SEE FIGURE 6A

6.0938 in	OE	154.78 mm
1.9375 in	BD	49.21 mm
2.2188 in	SB	56.36 mm
5.6250 in	EC	142.88 mm
3.0000 in	HD	76.20 mm

ROCKWELL 17N4-4341
SPICER 6-4-4341
SPICER 6-4-4671X

IDLI GROUP 83665

END YOKE WITH 10 SPLINES
SEE FIGURE 6A

6.0938 in	OE	154.78 mm
1.9375 in	BD	49.21 mm
2.2500 in	SB	57.15 mm
5.2500 in	EC	133.35 mm
3.0000 in	HD	76.20 mm

ROCKWELL 17N4-2511
SPICER 6-4-2511

IDLI GROUP 83666

END YOKE WITH 10 SPLINES
SEE FIGURE 6A

6.0938 in	OE	154.78 mm
1.9375 in	BD	49.21 mm
2.2500 in	SB	57.15 mm
5.2500 in	EC	133.35 mm
3.1250 in	HD	79.38 mm

ROCKWELL 17N4-2441
SPICER 6-4-2441

IDLI GROUP 83667

END YOKE WITH 10 SPLINES
slight variation from above group
SEE FIGURE 6A

6.0938 in	OE	154.78 mm
1.9375 in	BD	49.21 mm
2.2500 in	SB	57.15 mm
5.2500 in	EC	133.35 mm
3.1250 in	HD	79.38 mm

ROCKWELL 17N4-2151
SPICER 6-4-1151
SPICER 6-4-2151
SPICER 6-4-5701X

IDLI GROUP 83668

END YOKE WITH 10 SPLINES
SEE FIGURE 6A

6.0938 in	OE	154.78 mm
1.9375 in	BD	49.21 mm
2.2500 in	SB	57.15 mm
5.2500 in	EC	133.35 mm
3.2500 in	HD	82.55 mm

ROCKWELL 17N4-1971
SPICER 6-4-1971

IDLI GROUP 83669

END YOKE WITH 10 SPLINES
slight variation from above group
SEE FIGURE 6A

6.0938 in	OE	154.78 mm
1.9375 in	BD	49.21 mm
2.2500 in	SB	57.15 mm
5.2500 in	EC	133.35 mm
3.2500 in	HD	82.55 mm

ROCKWELL 17N4-2501
SPICER 6-4-1911
SPICER 6-4-2501
SPICER 6-4-4681X

IDLI GROUP 83670

END YOKE WITH 10 SPLINES
SEE FIGURE 6A

6.0938 in	OE	154.78 mm
1.9350 in	BD	49.15 mm
2.2500 in	SB	57.15 mm
5.2500 in	EC	133.35 mm
3.5000 in	HD	88.90 mm

ROCKWELL 17N4-2141
SPICER 6-4-1081
SPICER 6-4-2141

IDLI GROUP 83671

END YOKE WITH 6 SPLINES
SEE FIGURE 6A

6.0938 in	OE	154.78 mm
1.9375 in	BD	49.21 mm
2.8438 in	SB	72.23 mm
5.3125 in	EC	134.94 mm
3.0000 in	HD	76.20 mm

ROCKWELL 17N4-3031
SPICER 6-4-2171
SPICER 6-4-3031
SPICER 6-4-3041X
SPICER 6-4-3461X

IDLI GROUP 83672

END YOKE WITH 10 SPLINES
SEE FIGURE 6A

6.0938 in	OE	154.78 mm
1.9375 in	BD	49.21 mm
2.2500 in	SB	57.15 mm
5.8750 in	EC	149.23 mm
3.0000 in	HD	76.20 mm

ROCKWELL 17NYS36-37

IDLI GROUP 83673

END YOKE WITH 10 SPLINES
SEE FIGURE 6A

6.0938 in	OE	154.78 mm
1.9375 in	BD	49.21 mm
2.2500 in	SB	57.15 mm
6.0625 in	EC	153.99 mm
3.0000 in	HD	76.20 mm

ROCKWELL 17N4-1991
SPICER 6-4-1211
SPICER 6-4-1991
SPICER 6-4-2011
SPICER 6-4-3231

IDLI GROUP 83674

END YOKE WITH 10 SPLINES
slight variation from above group
SEE FIGURE 6A

6.0938 in	OE	154.78 mm
1.9375 in	BD	49.21 mm
2.2500 in	SB	57.15 mm
6.0625 in	EC	153.99 mm
3.0000 in	HD	76.20 mm

ROCKWELL 17N4-4191X
SPICER 6-4-4191X

BD = BEARING DIAMETER (outside) BW = BEARING WIDTH CB = CROSS LENGTH WITH BEARINGS CC = CENTER TO CENTER CD = CROSS DIAMETER
CL = CROSS LENGTH WITHOUT BEARINGS EC = END TO CENTER or FACE TO FACE EE = END TO END EL = EFFECTIVE LENGTH
HD = HUB DIAMETER (or insert) IE = INSIDE OF EARS (or recess) OD = OUTSIDE DIAMETER OE = OUTSIDE OF EARS SB = SPLINE OR BORE SIZE

CAUTION: BE SURE TO REFER TO ENGINEERING CATALOGS FOR SPECIAL APPLICATIONS THAT REQUIRE SPECIFIC MATERIAL CONTENT, TOLERANCES, ETC. SEE FOOTNOTE.

IDLI GROUP 83675

END YOKE WITH 10 SPLINES
slight variation from above group
SEE FIGURE 6A

6.0938 in	OE	154.78 mm
1.9375 in	BD	49.21 mm
2.2500 in	SB	57.15 mm
6.0625 in	EC	153.99 mm
3.0000 in	HD	76.20 mm

ROCKWELL 17N4-4411X

IDLI GROUP 83676

END YOKE WITH 10 SPLINES
slight variation from above group
SEE FIGURE 6A

6.0938 in	OE	154.78 mm
1.9375 in	BD	49.21 mm
2.2500 in	SB	57.15 mm
6.0625 in	EC	153.99 mm
3.0000 in	HD	76.20 mm

ROCKWELL 17NYS36A1

IDLI GROUP 83677

END YOKE WITH 6 SPLINES
SEE FIGURE 6A

6.0938 in	OE	154.78 mm
1.9375 in	BD	49.21 mm
2.2656 in	SB	57.55 mm
5.2500 in	EC	133.35 mm
3.2500 in	HD	82.55 mm

ROCKWELL 17N4-3501
SPICER 6-4-3501

IDLI GROUP 83678

END YOKE WITH 16 SPLINES
SEE FIGURE 6A

6.0938 in	OE	154.78 mm
1.9375 in	BD	49.21 mm
2.3438 in	SB	59.53 mm
4.9375 in	EC	125.41 mm
2.8750 in	HD	73.03 mm

ROCKWELL 17N4-4601
SPICER 6-4-4601
SPICER 6-4-5071X

IDLI GROUP 83679

END YOKE WITH 46 SLANTED SPLINES
SEE FIGURE 6A

6.0938 in	OE	154.78 mm
1.9375 in	BD	49.21 mm
2.3750 in	SB	60.33 mm
5.1563 in	EC	130.97 mm
3.5000 in	HD	88.90 mm

ROCKWELL 17NYS38-24

IDLI GROUP 83680

END YOKE WITH 36 INVOLUTED SPLINES
SEE FIGURE 6A

6.0938 in	OE	154.78 mm
1.9375 in	BD	49.21 mm
2.3750 in	SB	60.33 mm
5.4375 in	EC	138.11 mm
3.0000 in	HD	76.20 mm

ROCKWELL 17N4-5711
SPICER 6-4-5711

IDLI GROUP 83681

END YOKE WITH 46 SLANTED SPLINES
SEE FIGURE 6A

6.0938 in	OE	154.78 mm
1.9375 in	BD	49.21 mm
2.3750 in	SB	60.33 mm
5.6875 in	EC	144.46 mm
3.2500 in	HD	82.55 mm

ROCKWELL 17N4-5451
SPICER 6-4-5451

IDLI GROUP 83682

END YOKE WITH 18 INVOLUTED SPLINES
SEE FIGURE 6A

6.0938 in	OE	154.78 mm
1.9375 in	BD	49.21 mm
2.3750 in	SB	60.33 mm
5.7500 in	EC	146.05 mm
3.5000 in	HD	88.90 mm

ROCKWELL 17N4-4561
SPICER 6-4-4561

IDLI GROUP 83683

END YOKE WITH 46 INVOLUTED SPLINES
SEE FIGURE 6A

6.0938 in	OE	154.78 mm
1.9375 in	BD	49.21 mm
2.3750 in	SB	60.33 mm
5.7500 in	EC	146.05 mm
3.5000 in	HD	88.90 mm

ROCKWELL 17N4-4771
SPICER 6-4-4771

IDLI GROUP 83684

END YOKE WITH 46 SLANTED SPLINES
SEE FIGURE 6A

6.0938 in	OE	154.78 mm
1.9375 in	BD	49.21 mm
2.3750 in	SB	60.33 mm
6.0625 in	EC	153.99 mm
3.2500 in	HD	82.55 mm

ROCKWELL 17NYS38-9

IDLI GROUP 83685

END YOKE WITH 46 SLANTED SPLINES
SEE FIGURE 6A

6.0938 in	OE	154.78 mm
1.9375 in	BD	49.21 mm
2.3750 in	SB	60.33 mm
6.1875 in	EC	157.16 mm

ROCKWELL 17NYS38-17

IDLI GROUP 83686

END YOKE WITH 46 SLANTED SPLINES
slight variation from above group
SEE FIGURE 6A

6.0938 in	OE	154.78 mm
1.9375 in	BD	49.21 mm
2.3750 in	SB	60.33 mm
6.1875 in	EC	157.16 mm

ROCKWELL 17NYS38-17A

IDLI GROUP 83687

END YOKE WITH 36 INVOLUTED SPLINES
SEE FIGURE 6A

6.0938 in	OE	154.78 mm
1.9375 in	BD	49.21 mm
2.3750 in	SB	60.33 mm
6.5000 in	EC	165.10 mm
3.0000 in	HD	76.20 mm

ROCKWELL 17NYS38-16

IDLI GROUP 83688

END YOKE WITH 46 SLANTED SPLINES
SEE FIGURE 6A

6.0938 in	OE	154.78 mm
1.9375 in	BD	49.21 mm
2.3750 in	SB	60.33 mm
6.8438 in	EC	173.83 mm
3.2500 in	HD	82.55 mm

ROCKWELL 17NYS38-18

IDLI GROUP 83689

END YOKE WITH 46 SLANTED SPLINES
slight variation from above group
SEE FIGURE 6A

6.0938 in	OE	154.78 mm
1.9375 in	BD	49.21 mm
2.3750 in	SB	60.33 mm
6.8438 in	EC	173.83 mm
3.2500 in	HD	82.55 mm

ROCKWELL 17NYS38-18A

IDLI GROUP 83690

END YOKE WITH 18 INVOLUTED SPLINES
SEE FIGURE 6A

6.0938 in	OE	154.78 mm
1.9375 in	BD	49.21 mm
2.4375 in	SB	61.91 mm
5.2500 in	EC	133.35 mm
3.3750 in	HD	85.73 mm

ROCKWELL 17NYS39-1

IDLI GROUP 83691

END YOKE WITH 18 INVOLUTED SPLINES
SEE FIGURE 6A

6.0938 in	OE	154.78 mm
1.9375 in	BD	49.21 mm
2.4375 in	SB	61.91 mm
5.8125 in	EC	147.64 mm
3.3750 in	HD	85.73 mm

ROCKWELL 17N4-4581
SPICER 6-4-4581
SPICER 6-4-4751X

IDLI GROUP 83692

END YOKE WITH 18 INVOLUTED SPLINES
SEE FIGURE 6A

6.0938 in	OE	154.78 mm
1.9375 in	BD	49.21 mm
2.4375 in	SB	61.91 mm
6.9688 in	EC	177.01 mm
3.3750 in	HD	85.73 mm

ROCKWELL 17NYS39-2

IDLI GROUP 83693

END YOKE WITH 10 SPLINES
SEE FIGURE 6A

6.0938 in	OE	154.78 mm
1.9375 in	BD	49.21 mm
2.5000 in	SB	63.50 mm
5.1563 in	EC	130.97 mm
3.5000 in	HD	88.90 mm

ROCKWELL 17N4-4421
SPICER 6-4-4421

IDLI GROUP 83694

END YOKE WITH 10 SPLINES
SEE FIGURE 6A

6.0938 in	OE	154.78 mm
1.9375 in	BD	49.21 mm
2.5000 in	SB	63.50 mm
5.1563 in	EC	130.97 mm
4.1250 in	HD	104.78 mm

ROCKWELL 17NYS40-107

IDLI GROUP 83695

END YOKE WITH 10 SPLINES
SEE FIGURE 6A

6.0938 in	OE	154.78 mm
1.9375 in	BD	49.21 mm
2.5000 in	SB	63.50 mm
5.6250 in	EC	142.88 mm
2.9375 in	HD	74.61 mm

CLEVE.STL. C67-1-100
ROCKWELL 17N4-3611
SPICER 6-4-3611

IDLI GROUP 83696

END YOKE WITH 16 SLANTED SPLINES
SEE FIGURE 6A

6.0938 in	OE	154.78 mm
1.9375 in	BD	49.21 mm
2.5313 in	SB	64.30 mm
6.4375 in	EC	163.51 mm
3.5000 in	HD	88.90 mm

ROCKWELL 17NYS40-106A

IDLI GROUP 83697

END YOKE WITH 16 SLANTED SPLINES
slight variation from above group
SEE FIGURE 6A

6.0938 in	OE	154.78 mm
1.9375 in	BD	49.21 mm
2.5313 in	SB	64.30 mm
6.4375 in	EC	163.51 mm
3.5000 in	HD	88.90 mm

ROCKWELL 17NYS40-106A1

IDLI GROUP 83698

END YOKE WITH 20 INVOLUTED SPLINES
SEE FIGURE 6A

6.0938 in	OE	154.78 mm
1.9375 in	BD	49.21 mm
2.6250 in	SB	66.68 mm
5.1875 in	EC	131.76 mm
3.0000 in	HD	76.20 mm

ROCKWELL 17N4-5351
SPICER 6-4-5351

IDLI GROUP 83699

END YOKE WITH 10 SPLINES
SEE FIGURE 6A

6.0938 in	OE	154.78 mm
1.9375 in	BD	49.21 mm
2.7500 in	SB	69.85 mm
5.7500 in	EC	146.05 mm
3.3750 in	HD	85.73 mm

ROCKWELL 17N4-5591
SPICER 6-4-5591
SPICER 6-4-5601X

IDLI GROUP 83700

END YOKE WITH 10 SPLINES
SEE FIGURE 6A

6.0938 in	OE	154.78 mm
1.9375 in	BD	49.21 mm
2.7500 in	SB	69.85 mm
5.7500 in	EC	146.05 mm
3.5000 in	HD	88.90 mm

ROCKWELL 17N4-4551
SPICER 6-4-4551
SPICER 6-4-5251X

IDLI GROUP 83701

END YOKE WITH 23 INVOLUTED SPLINES
SEE FIGURE 6A

6.0938 in	OE	154.78 mm
1.9375 in	BD	49.21 mm
3.0000 in	SB	76.20 mm
4.8125 in	EC	122.24 mm
4.0000 in	HD	101.60 mm

ROCKWELL 17N4-3141
SPICER 6-4-3141

IDLI GROUP 83702

END YOKE WITH 50 INVOLUTED SPLINES
SEE FIGURE 6A

6.0938 in	OE	154.78 mm
1.9375 in	BD	49.21 mm
3.1875 in	SB	80.96 mm
5.3750 in	EC	136.53 mm
3.7188 in	HD	94.46 mm

ROCKWELL 17NYS51

IDLI GROUP 83703

END YOKE WITH 10 SPLINES
SEE FIGURE 6A

7.0000 in	OE	177.80 mm
1.9375 in	BD	49.21 mm
2.0313 in	SB	51.60 mm
5.0000 in	EC	127.00 mm
2.7500 in	HD	69.85 mm

ROCKWELL 176N4-421
SPICER 6.3-4-421

IDLI GROUP 83704

END YOKE WITH 39 SLANTED SPLINES
SEE FIGURE 6A

7.0000 in	OE	177.80 mm
1.9375 in	BD	49.21 mm
2.0313 in	SB	51.60 mm
5.1875 in	EC	131.76 mm
3.0000 in	HD	76.20 mm

ROCKWELL 176N4-531
SPICER 6.3-4-531

IDLI GROUP 83705

END YOKE WITH 39 SLANTED SPLINES
slight variation from above group
SEE FIGURE 6A

7.0000 in	OE	177.80 mm
1.9375 in	BD	49.21 mm
2.0313 in	SB	51.60 mm
5.1875 in	EC	131.76 mm
3.0000 in	HD	76.20 mm

ROCKWELL 176N4-541X
SPICER 6.3-4-541X

IDLI GROUP 83706

END YOKE WITH 16 SPLINES
SEE FIGURE 6A

7.0000 in	OE	177.80 mm
1.9375 in	BD	49.21 mm
2.1563 in	SB	54.77 mm
5.2500 in	EC	133.35 mm
2.9375 in	HD	74.61 mm

ROCKWELL 176N4-121
SPICER 6.3-4-121

IDLI GROUP 83707

END YOKE WITH 44 INVOLUTED SPLINES
SEE FIGURE 6A

7.0000 in	OE	177.80 mm
1.9375 in	BD	49.21 mm
2.2813 in	SB	57.95 mm
5.2500 in	EC	133.35 mm
2.9375 in	HD	74.61 mm

ROCKWELL 176N4-321
SPICER 6.3-4-321

IDLI GROUP 83708

END YOKE WITH 10 SPLINES
SEE FIGURE 6A

7.0000 in	OE	177.80 mm
1.9375 in	BD	49.21 mm
2.2500 in	SB	57.15 mm
5.2500 in	EC	133.35 mm
3.1250 in	HD	79.38 mm

ROCKWELL 176N4-411
SPICER 6.3-4-411

IDLI GROUP 83709

END YOKE WITH 10 SPLINES
SEE FIGURE 6A

7.0000 in	OE	177.80 mm
1.9375 in	BD	49.21 mm
2.2500 in	SB	57.15 mm
5.2500 in	EC	133.35 mm
3.2500 in	HD	82.55 mm

ROCKWELL 176N4-501
SPICER 6.3-4-501

IDLI GROUP 83710

END YOKE WITH 10 SPLINES
SEE FIGURE 6A

7.0000 in	OE	177.80 mm
1.9375 in	BD	49.21 mm
2.2500 in	SB	57.15 mm
5.8125 in	EC	147.64 mm
3.0000 in	HD	76.20 mm

ROCKWELL 176N4-91
SPICER 6.3-4-91

IDLI GROUP 83711

END YOKE WITH 16 SPLINES
SEE FIGURE 6A

7.0000 in	OE	177.80 mm
1.9375 in	BD	49.21 mm
2.3438 in	SB	59.53 mm
5.5000 in	EC	139.70 mm
2.8750 in	HD	73.03 mm

ROCKWELL 176N4-161
SPICER 6.3-4-161

IDLI GROUP 83712

END YOKE WITH 46 SLANTED SPLINES
SEE FIGURE 6A

7.0000 in	OE	177.80 mm
1.9375 in	BD	49.21 mm
2.3750 in	SB	60.33 mm
5.3750 in	EC	136.53 mm
3.5000 in	HD	88.90 mm

ROCKWELL 176N4-391
SPICER 6.3-4-391

IDLI GROUP 83713

END YOKE WITH 18 INVOLUTED SPLINES
SEE FIGURE 6A

7.0000 in	OE	177.80 mm
1.9375 in	BD	49.21 mm
2.3750 in	SB	60.33 mm
6.1250 in	EC	155.58 mm
3.5000 in	HD	88.90 mm

ROCKWELL 176N4-271
SPICER 6.3-4-271

IDLI GROUP 83714

END YOKE WITH 10 SPLINES
SEE FIGURE 6A

7.0000 in	OE	177.80 mm
1.9375 in	BD	49.21 mm
2.5000 in	SB	63.50 mm
5.4063 in	EC	137.32 mm
3.5000 in	HD	88.90 mm

ROCKWELL 176N4-151
SPICER 6.3-4-151

IDLI GROUP 83715

END YOKE WITH 10 SPLINES
SEE FIGURE 6A

7.0000 in	OE	177.80 mm
1.9375 in	BD	49.21 mm
2.7500 in	SB	69.85 mm
5.9688 in	EC	151.61 mm
3.7500 in	HD	95.25 mm

ROCKWELL 176N4-11
SPICER 6.3-4-11

IDLI GROUP 83716

END YOKE WITH 30 INVOLUTED SPLINES
SEE FIGURE 6A

7.5469 in	OE	191.69 mm
1.9375 in	BD	49.21 mm
1.9375 in	SB	49.21 mm
5.3750 in	EC	136.53 mm
3.0000 in	HD	76.20 mm

ROCKWELL 18N4-2051
SPICER 6.5-4-2051

IDLI GROUP 83717

END YOKE WITH 10 SPLINES
SEE FIGURE 6A

7.5469 in	OE	191.69 mm
1.9375 in	BD	49.21 mm
1.9688 in	SB	50.01 mm
5.6250 in	EC	142.88 mm
2.6250 in	HD	66.68 mm

ROCKWELL 18N4-2001
ROCKWELL 18N4-2351X
SPICER 6.5-4-2001
SPICER 6.5-4-2351X

IDLI GROUP 83718

END YOKE WITH 10 SPLINES
SEE FIGURE 6A

7.5469 in	OE	191.69 mm
1.9375 in	BD	49.21 mm
2.0000 in	SB	50.80 mm
5.3750 in	EC	136.53 mm
2.9375 in	HD	74.61 mm

ROCKWELL 18N4-1921
ROCKWELL 18N4-2341X
ROCKWELL 18N4-2851X
SPICER 6.5-4-1921
SPICER 6.5-4-2341X
SPICER 6.5-4-2851X

IDLI GROUP 83719

END YOKE WITH 10 SPLINES
slight variation from above group
SEE FIGURE 6A

7.5469 in	OE	191.69 mm
1.9375 in	BD	49.21 mm
2.0000 in	SB	50.80 mm
5.3750 in	EC	136.53 mm
2.9375 in	HD	74.61 mm

ROCKWELL 18NYS32-11

IDLI GROUP 83720

END YOKE WITH 39 SLANTED SPLINES
SEE FIGURE 6A

7.5469 in	OE	191.69 mm
1.9375 in	BD	49.21 mm
2.0313 in	SB	51.60 mm
5.1563 in	EC	130.97 mm
3.0000 in	HD	76.20 mm

ROCKWELL 18N4-2471
ROCKWELL 18N4-2481X
ROCKWELL 18NYS32-4
SPICER 6.5-4-2471
SPICER 6.5-4-2481X
SPICER 6.5-4-2491X

BD=BEARING DIAMETER (outside)   **BW**=BEARING WIDTH   **CB**=CROSS LENGTH WITH BEARINGS   **CC**=CENTER TO CENTER   **CD**=CROSS DIAMETER
CL=CROSS LENGTH WITHOUT BEARINGS   **EC**=END TO CENTER or FACE TO FACE   **EE**=END TO END   **EL**=EFFECTIVE LENGTH
HD=HUB DIAMETER (or insert)   **IE**=INSIDE OF EARS (or recess)   **OD**=OUTSIDE DIAMETER   **OE**=OUTSIDE OF EARS   **SB**=SPLINE OR BORE SIZE

CAUTION: BE SURE TO REFER TO ENGINEERING CATALOGS FOR SPECIAL APPLICATIONS THAT REQUIRE SPECIFIC MATERIAL CONTENT, TOLERANCES, ETC. SEE FOOTNOTE.

IDLI GROUP 83721

END YOKE WITH 39 SLANTED SPLINES
slight variation from above group
SEE FIGURE 6A

7.5469 in	OE	191.69 mm
1.9375 in	BD	49.21 mm
2.0313 in	SB	51.60 mm
5.1563 in	EC	130.97 mm
3.0000 in	HD	76.20 mm

ROCKWELL 18N4-2491X
SPICER 6.5-4-2491X

IDLI GROUP 83722

END YOKE WITH 39 SLANTED SPLINES
SEE FIGURE 6A

7.5469 in	OE	191.69 mm
1.9375 in	BD	49.21 mm
2.0313 in	SB	51.60 mm
5.1875 in	EC	131.76 mm
3.0000 in	HD	76.20 mm

ROCKWELL 18N4-3371

IDLI GROUP 83723

END YOKE WITH 39 SLANTED SPLINES
slight variation from above group

7.5469 in	OE	191.69 mm
1.9375 in	BD	49.21 mm
2.0313 in	SB	51.60 mm
5.1875 in	EC	131.76 mm
3.0000 in	HD	76.20 mm

ROCKWELL 18N4-3381X

IDLI GROUP 83724

END YOKE WITH 39 SLANTED SPLINES
SEE FIGURE 6A

7.5469 in	OE	191.69 mm
1.9375 in	BD	49.21 mm
2.0313 in	SB	51.60 mm
5.5313 in	EC	140.50 mm
2.7500 in	HD	69.85 mm

ROCKWELL 18N4-2781
SPICER 6.5-4-2781

IDLI GROUP 83725

END YOKE WITH 39 SLANTED SPLINES
slight variation from above group
SEE FIGURE 6A

7.5469 in	OE	191.69 mm
1.9375 in	BD	49.21 mm
2.0313 in	SB	51.60 mm
5.5313 in	EC	140.50 mm
2.7500 in	HD	69.85 mm

ROCKWELL 18N4-3171X
SPICER 6.5-4-3171X

IDLI GROUP 83726

END YOKE WITH 39 SLANTED SPLINES
slight variation from above group
SEE FIGURE 6A

7.5469 in	OE	191.69 mm
1.9375 in	BD	49.21 mm
2.0313 in	SB	51.60 mm
5.5313 in	EC	140.50 mm
2.7500 in	HD	69.85 mm

ROCKWELL 18NYS32-3A1

IDLI GROUP 83727

END YOKE WITH 16 SPLINES
SEE FIGURE 6A

7.5469 in	OE	191.69 mm
1.9375 in	BD	49.21 mm
2.1563 in	SB	54.77 mm
5.2500 in	EC	133.35 mm
2.9375 in	HD	74.61 mm

ROCKWELL 18N4-2441
ROCKWELL 18N4-2451X
SPICER 6.5-4-2441
SPICER 6.5-4-2451X

IDLI GROUP 83728

END YOKE WITH 16 SPLINES
SEE FIGURE 6A

7.5469 in	OE	191.69 mm
1.9375 in	BD	49.21 mm
2.1563 in	SB	54.77 mm
5.8125 in	EC	147.64 mm
2.9375 in	HD	74.61 mm

ROCKWELL 18N4-2291
ROCKWELL 18N4-2401X
SPICER 6.5-4-2401X

IDLI GROUP 83729

END YOKE WITH 6 SPLINES
SEE FIGURE 6A

7.5469 in	OE	191.69 mm
1.9375 in	BD	49.21 mm
2.2188 in	SB	56.36 mm
6.0000 in	EC	152.40 mm
3.0000 in	HD	76.20 mm

ROCKWELL 18N4-1971
ROCKWELL 18N4-2271X
SPICER 6.5-4-1971
SPICER 6.5-4-2271
SPICER 6.5-4-2271X

IDLI GROUP 83730

END YOKE WITH 10 SPLINES
SEE FIGURE 6A

7.5469 in	OE	191.69 mm
1.9375 in	BD	49.21 mm
2.2500 in	SB	57.15 mm
5.8438 in	EC	148.43 mm
3.0000 in	HD	76.20 mm

ROCKWELL 18N4-1911
ROCKWELL 18N4-2331X
ROCKWELL 18N4-2551X
SPICER 6.5-4-1911
SPICER 6.5-4-2331X
SPICER 6.5-4-2551X

IDLI GROUP 83731

END YOKE WITH 16 SPLINES
SEE FIGURE 6A

7.5469 in	OE	191.69 mm
1.9375 in	BD	49.21 mm
2.3438 in	SB	59.53 mm
5.5000 in	EC	139.70 mm
2.8750 in	HD	73.03 mm

ROCKWELL 18N4-1891
ROCKWELL 18N4-2531X
SPICER 6.5-4-1891
SPICER 6.5-4-2531X

IDLI GROUP 83732

END YOKE WITH 16 SPLINES
SEE FIGURE 6A

7.5469 in	OE	191.69 mm
1.9375 in	BD	49.21 mm
2.3438 in	SB	59.53 mm
5.6250 in	EC	142.88 mm
2.8125 in	HD	71.44 mm

ROCKWELL 18N4-1981
SPICER 6.5-4-1981
SPICER 6.5-4-3011

IDLI GROUP 83733

END YOKE WITH 46 SLANTED SPLINES
SEE FIGURE 6A

7.5469 in	OE	191.69 mm
1.9375 in	BD	49.21 mm
2.3750 in	SB	60.33 mm
5.4063 in	EC	137.32 mm
3.5000 in	HD	88.90 mm

ROCKWELL 18NYS38-13

IDLI GROUP 83734

END YOKE WITH 36 INVOLUTED SPLINES
SEE FIGURE 6A

7.5469 in	OE	191.69 mm
1.9375 in	BD	49.21 mm
2.3750 in	SB	60.33 mm
5.4375 in	EC	138.11 mm
3.0000 in	HD	76.20 mm

ROCKWELL 18NYS38-6

IDLI GROUP 83735

END YOKE WITH 36 INVOLUTED SPLINES
slight variation from above group
SEE FIGURE 6A

7.5469 in	OE	191.69 mm
1.9375 in	BD	49.21 mm
2.3750 in	SB	60.33 mm
5.4375 in	EC	138.11 mm
3.0000 in	HD	76.20 mm

ROCKWELL 18NYS38-6A

IDLI GROUP 83736

END YOKE WITH 46 SLANTED SPLINES
SEE FIGURE 6A

7.5469 in	OE	191.69 mm
1.9375 in	BD	49.21 mm
2.3750 in	SB	60.33 mm
5.4375 in	EC	138.11 mm
3.2500 in	HD	82.55 mm

ROCKWELL 18N4-2701
SPICER 6.5-4-2701

IDLI GROUP 83737

END YOKE WITH 46 SLANTED SPLINES
slight variation from above group
SEE FIGURE 6A

7.5469 in	OE	191.69 mm
1.9375 in	BD	49.21 mm
2.3750 in	SB	60.33 mm
5.4375 in	EC	138.11 mm
3.2500 in	HD	82.55 mm

ROCKWELL 18NYS38A

IDLI GROUP 83738

END YOKE WITH 46 SLANTED SPLINES
SEE FIGURE 6A

7.5469 in	OE	191.69 mm
1.9375 in	BD	49.21 mm
2.3750 in	SB	60.33 mm
5.6875 in	EC	144.46 mm
3.2500 in	HD	82.55 mm

ROCKWELL 18N4-2651

IDLI GROUP 83739

END YOKE WITH 46 SLANTED SPLINES
slight variation from above group
SEE FIGURE 6A

7.5469 in	OE	191.69 mm
1.9375 in	BD	49.21 mm
2.3750 in	SB	60.33 mm
5.6875 in	EC	144.46 mm
3.2500 in	HD	82.55 mm

ROCKWELL 18NYS38-8A

IDLI GROUP 83740

END YOKE WITH 18 INVOLUTED SPLINES
SEE FIGURE 6A

7.5469 in	OE	191.69 mm
1.9375 in	BD	49.21 mm
2.4219 in	SB	61.52 mm
5.8125 in	EC	147.64 mm
3.3750 in	HD	85.73 mm

ROCKWELL 18N4-1991
SPICER 6.5-4-1991

IDLI GROUP 83741

END YOKE WITH 18 INVOLUTED SPLINES
slight variation from above group
SEE FIGURE 6A

7.5469 in	OE	191.69 mm
1.9375 in	BD	49.21 mm
2.4219 in	SB	61.52 mm
5.8125 in	EC	147.64 mm
3.3750 in	HD	85.73 mm

ROCKWELL 18N4-2361X
SPICER 6.5-4-1991
SPICER 6.5-4-2361X

SPICER 6.5-4-2711X

ENGINEERING CATALOGS MUST BE CONSULTED FOR DETAILS NOT INCLUDED IN THIS GUIDE. SPECIFIC DESIGNS, MATERIAL CONTENT, TOLERANCES LUBE FITTINGS AND OTHER DIMENSIONS ARE INTENTIONALLY OMITTED HERE.

IDLI GROUP 83742

END YOKE WITH 10 SPLINES
SEE FIGURE 6A

7.5469 in	OE	191.69 mm
1.9375 in	BD	49.21 mm
2.5000 in	SB	63.50 mm
5.4063 in	EC	137.32 mm
3.5000 in	HD	88.90 mm

ROCKWELL 18N4-2101

IDLI GROUP 83743

END YOKE WITH 10 SPLINES
SEE FIGURE 6A

7.5469 in	OE	191.69 mm
1.9375 in	BD	49.21 mm
2.5000 in	SB	63.50 mm
5.8750 in	EC	149.23 mm
2.9375 in	HD	74.61 mm

ROCKWELL 18N4-1961
SPICER 6.5-4-1961

IDLI GROUP 83744

END YOKE WITH 20 INVOLUTED SPLINES
SEE FIGURE 6A

7.5469 in	OE	191.69 mm
1.9375 in	BD	49.21 mm
2.6250 in	SB	66.68 mm
5.5000 in	EC	139.70 mm
3.0000 in	HD	76.20 mm

ROCKWELL 18N4-2011
SPICER 6.5-4-2011

IDLI GROUP 83745

END YOKE WITH 20 INVOLUTED SPLINES
SEE FIGURE 6A

7.5469 in	OE	191.69 mm
1.9375 in	BD	49.21 mm
2.6250 in	SB	66.68 mm
5.5625 in	EC	141.29 mm
3.0000 in	HD	76.20 mm

ROCKWELL 18N4-3021
SPICER 6.5-4-3021

IDLI GROUP 83746

END YOKE WITH 10 SPLINES
SEE FIGURE 6A

7.5469 in	OE	191.69 mm
1.9375 in	BD	49.21 mm
2.7500 in	SB	69.85 mm
6.0000 in	EC	152.40 mm
3.3750 in	HD	85.73 mm

ROCKWELL 18N4-2171
SPICER 6.5-4-2171

IDLI GROUP 83747

END YOKE WITH 10 SPLINES
SEE FIGURE 6A

7.5469 in	OE	191.69 mm
1.9375 in	BD	49.21 mm
2.7500 in	SB	69.85 mm
6.0625 in	EC	153.99 mm
3.5000 in	HD	88.90 mm

ROCKWELL 18N4-1901
ROCKWELL 18N4-2931X
SPICER 6.5-4-1901
SPICER 6.5-4-2931X

IDLI GROUP 83748

END YOKE WITH 10 SPLINES
SEE FIGURE 6A

7.5469 in	OE	191.69 mm
1.9375 in	BD	49.21 mm
2.7500 in	SB	69.85 mm
6.0625 in	EC	153.99 mm
3.7500 in	HD	95.25 mm

ROCKWELL 18N4-2751
ROCKWELL 18N4-2761X
SPICER 6.5-4-2751
SPICER 6.5-4-2761X

BD = BEARING DIAMETER (outside)   **BW** = BEARING WIDTH   **CB** = CROSS LENGTH WITH BEARINGS   **CC** = CENTER TO CENTER   **CD** = CROSS DIAMETER
CL = CROSS LENGTH WITHOUT BEARINGS   **EC** = END TO CENTER or FACE TO FACE   **EE** = END TO END   **EL** = EFFECTIVE LENGTH
HD = HUB DIAMETER (or insert)   **IE** = INSIDE OF EARS (or recess)   **OD** = OUTSIDE DIAMETER   **OE** = OUTSIDE OF EARS   **SB** = SPLINE OR BORE SIZE

Other Valuable Interchange Publications

The International Bearing Interchange Guide

New Ninth Edition

The I.B.I. Guide has over 365,000 listings of 25,326 ball and roller bearings available worldwide and dating back to 1918. It includes manufacturers, original equipment users and government part numbers, with THOUSANDS of new listings not included in earlier editions!

I.S.B.N. #916966-17-8

The International Seal Interchange Guide

New Sixth Edition

The I.S.I. Guide has over 150,000 listings of 12,632 different seals and "O" rings. It includes government numbers as well as producers and users throughout the world—over 700 páges of information to save you time and money!

I.S.B.N. #916966-16-X

The International Drive Belt Interchange Guide

New Fourth Edition

The I.D.B.I. Guide has more than 115,000 listings for nearly 5,863 types and sizes of drive and V-belts. Listings include manufacturers, original equipment and government users from all over the world. 760 pages.

I.S.B.N. #916966-15-1

IDLI GROUP 84501

QUICK DISCONNECT YOKE
WITH 6 SPLINES
SEE FIGURE 10A

1.438 in	IE	36.51 mm
.969 in	BD	24.61 mm
.875 in	SB	22.23 mm
3.375 in	EC	85.73 mm

ROCKWELL L6NYS14A

IDLI GROUP 84502

QUICK DISCONNECT YOKE
WITH 6 SPLINES
SEE FIGURE 10A

1.438 in	IE	36.51 mm
.969 in	BD	24.61 mm
1.125 in	SB	28.58 mm
3.813 in	EC	96.84 mm

PRECISION 1920
ROCKWELL L6NYS18-23A

IDLI GROUP 84503

QUICK DISCONNECT YOKE
WITH 21 INVOLUTED SPLINES
SEE FIGURE 10A

1.438 in	IE	36.51 mm
.969 in	BD	24.61 mm
1.375 in	SB	34.93 mm
3.313 in	EC	84.14 mm

ROCKWELL L6NYS22-20A

IDLI GROUP 84504

QUICK DISCONNECT YOKE
WITH 6 SPLINES
SEE FIGURE 10A

1.438 in	IE	36.51 mm
.969 in	BD	24.61 mm
1.375 in	SB	34.93 mm
3.813 in	EC	96.84 mm

AEC 40006
BORG-WARNR 8645J
MCQUAY-NOR U03-76
NEAPCO 16-6101
NEAPCO L6101X
PRECISION 1921
ROCKWELL L6NYS22-1A
TRW 21065
WESCO 6N220-20
WESCO 840007

IDLI GROUP 84505

QUICK DISCONNECT YOKE
WITH SQUARE BORE
SEE FIGURE 10A

1.813 in	IE	46.04 mm
1.063 in	BD	26.99 mm
1.000 in	SB	25.40 mm
4.063 in	EC	103.19 mm

ROCKWELL L12NYQ16-17A

IDLI GROUP 84506

QUICK DISCONNECT YOKE
WITH 9 INVOLUTED SPLINES
SEE FIGURE 10A

1.813 in	IE	46.04 mm
1.063 in	BD	26.99 mm
1.250 in	SB	31.75 mm
3.281 in	EC	83.35 mm

ROCKWELL L12NYS20-47A

IDLI GROUP 84507

QUICK DISCONNECT YOKE
WITH 6 SPLINES
SEE FIGURE 10A

1.813 in	IE	46.04 mm
1.063 in	BD	26.99 mm
1.250 in	SB	31.75 mm
4.063 in	EC	103.19 mm

ROCKWELL L12NYS20-10A

IDLI GROUP 84508

QUICK DISCONNECT YOKE
WITH 21 INVOLUTED SPLINES
SEE FIGURE 10A

1.813 in	IE	46.04 mm
1.063 in	BD	26.99 mm
1.375 in	SB	34.93 mm
3.563 in	EC	90.49 mm

BORG-WARNR 61184
G & G MFG 182-1221
HAYES 1-4-1141X
NEAPCO 11184X
NEAPCO 12-1184
PRECISION 1292
ROCKWELL L12NYS22-16A
SPICER 1-4-1011X
SPICER 1-4-1141X
WESCO 12N220TQD

IDLI GROUP 84509

QUICK DISCONNECT YOKE
WITH 6 SPLINES
SEE FIGURE 10A

1.813 in	IE	46.04 mm
1.063 in	BD	26.99 mm
1.375 in	SB	34.93 mm
4.063 in	EC	103.19 mm

ROCKWELL L12NYS22-6A

IDLI GROUP 84510

QUICK DISCONNECT YOKE
WITH 6 SPLINES
slight variation from above group
SEE FIGURE 10A

1.813 in	IE	46.04 mm
1.063 in	BD	26.99 mm
1.375 in	SB	34.93 mm
4.063 in	EC	103.19 mm

AEC 46012
ALLOY 0600
ALLOY 600
BORG-WARNR 61011
BORG-WARNR 61190
BPC H12EY-1-3/8
CASE,J.I. FB13278
COCKSHUTT 445855D

G & G MFG 182-1206
HAYES 1-4-1061X
HUB CITY 0335-01202
MCQUAY-NOR U03-84
MUNCIE MJY26
NEAPCO 11101
NEAPCO 11101X
NEAPCO 11183
NEAPCO 11190
NEAPCO 12-1101
NEW IDEA K3339A
PRECISION 1291
REX CHAIN 760041
ROCKWELL L12NYS22A
SPICER 1-4-0881X
SPICER 1-4-1016X
SPICER 1-4-1061X
SPICER 1-4-881X
TRW 21112
WESCO 12N0220-20
WESCO 12N220-20
WESCO 451221

IDLI GROUP 84511

QUICK DISCONNECT YOKE
WITH 21 INVOLUTED SPLINES
SEE FIGURE 10A

2.797 in	IE	71.04 mm
1.313 in	BD	33.34 mm
1.375 in	SB	34.93 mm
4.438 in	EC	112.71 mm

NEAPCO N44N220TQD
ROCKWELL 44RYS22-3A

IDLI GROUP 84512

QUICK DISCONNECT YOKE
WITH 21 INVOLUTED SPLINES
slight variation from above group
SEE FIGURE 10A

2.797 in	IE	71.04 mm
1.313 in	BD	33.34 mm
1.375 in	SB	34.93 mm
4.438 in	EC	112.71 mm

PRECISION 1453
ROCKWELL 44RYS22-21A

IDLI GROUP 84513

QUICK DISCONNECT YOKE
WITH 6 SPLINES
SEE FIGURE 10A

2.797 in	IE	71.04 mm
1.313 in	BD	33.34 mm
1.375 in	SB	34.93 mm
4.938 in	EC	125.41 mm

ROCKWELL 4RYS22-1A

IDLI GROUP 84514

QUICK DISCONNECT YOKE
WITH 21 INVOLUTED SPLINES
slight variation from above group
SEE FIGURE 10A

2.797 in	IE	71.04 mm
1.313 in	BD	33.34 mm
1.375 in	SB	34.93 mm
4.938 in	EC	125.41 mm

ROCKWELL 44RYS22-24A

IDLI GROUP 84515

QUICK DISCONNECT YOKE
WITH 21 INVOLUTED SPLINES
SEE FIGURE 10A

2.797 in	IE	71.04 mm
1.438 in	BD	36.51 mm
1.375 in	SB	34.93 mm
4.938 in	EC	125.41 mm

G & G MFG 182-4421
NEAPCO 44-2201
PRECISION 1453
ROCKWELL 44RYS22-16A
SPICER 3-4-4-41X
WESCO 44N220TQD
WESCO 481221

IDLI GROUP 84516

QUICK DISCONNECT YOKE
WITH 6 SPLINES
SEE FIGURE 11A

2.906 in	OE	73.82 mm
1.125 in	BD	28.58 mm
1.125 in	SB	28.58 mm
4.313 in	EC	109.54 mm

ROCKWELL L14SYS18-9A

IDLI GROUP 84517

QUICK DISCONNECT YOKE
WITH 21 INVOLUTED SPLINES
SEE FIGURE 11A

2.906 in	OE	73.82 mm
1.125 in	BD	28.58 mm
1.375 in	SB	34.93 mm
3.688 in	EC	93.66 mm

BORG-WARNR 69102
G & G MFG 182-1421
G & G MFG 184-1421
HAYES 2-4-4211X
NEAPCO 19102X
NEAPCO 19139
NEAPCO 19186X
NEAPCO 20-9102
PRECISION 1373
REX CHAIN 760005
ROCKWELL L14SYS22-31A
SPICER 2-4-4211
SPICER 2-4-4211X
TRW 21208
WESCO 14N0220TQD
WESCO 14N220TQD
WESCO 461220
WESCO 860-054

IDLI GROUP 84518

QUICK DISCONNECT YOKE
WITH 6 SPLINES
SEE FIGURE 11A

2.906 in	OE	73.82 mm
1.125 in	BD	28.58 mm
1.375 in	SB	34.93 mm
4.063 in	EC	103.19 mm

ROCKWELL LYSYS22-127A

ENGINEERING CATALOGS MUST BE CONSULTED FOR DETAILS NOT INCLUDED
IN THIS GUIDE. SPECIFIC DESIGNS, MATERIAL CONTENT, TOLERANCES
LUBE FITTINGS AND OTHER DIMENSIONS ARE INTENTIONALLY OMITTED HERE.

IDLI GUIDE   COPYRIGHT ® INTERCHANGE, INC.   ST. LOUIS PARK, MN. 55416 USA

CAUTION: BE SURE TO REFER TO ENGINEERING CATALOGS FOR SPECIAL APPLICATIONS THAT REQUIRE SPECIFIC MATERIAL CONTENT, TOLERANCES, ETC. SEE FOOTNOTE.

IDLI GROUP 84519

QUICK DISCONNECT YOKE
WITH 6 SPLINES
slight variation from above group
SEE FIGURE 11A

2.906 in	OE	73.82 mm
1.125 in	BD	28.58 mm
1.375 in	SB	34.93 mm
4.063 in	EC	103.19 mm

PRECISION 1372
ROCKWELL L14SYS22-14A

IDLI GROUP 84520

QUICK DISCONNECT YOKE
WITH 5 SPLINES IN 6 SPLINE SPACES
SEE FIGURE 11A

2.906 in	OE	73.82 mm
1.125 in	BD	28.58 mm
1.375 in	SB	34.93 mm
4.313 in	EC	109.54 mm

ROCKWELL L14SYS22-89A

IDLI GROUP 84521

QUICK DISCONNECT YOKE
WITH 6 SPLINES
SEE FIGURE 11A

2.906 in	OE	73.82 mm
1.125 in	BD	28.58 mm
1.375 in	SB	34.93 mm
4.563 in	EC	115.89 mm

PRECISION 1372
ROCKWELL L14SYS22-24A

IDLI GROUP 84522

QUICK DISCONNECT YOKE
WITH 6 SPLINES
SEE FIGURE 11A

2.906 in	OE	73.82 mm
1.125 in	BD	28.58 mm
1.375 in	SB	34.93 mm
4.813 in	EC	122.24 mm

PRECISION 1372
ROCKWELL L14SYS22-1A

IDLI GROUP 84523

QUICK DISCONNECT YOKE
WITH 6 SPLINES
SEE FIGURE 11A

3.125 in	OE	79.38 mm
.875 in	BD	22.23 mm
1.125 in	SB	28.58 mm
4.313 in	EC	109.54 mm

ROCKWELL 1FRYS18-12A

IDLI GROUP 84524

QUICK DISCONNECT YOKE
WITH 6 SPLINES
SEE FIGURE 11A

3.125 in	OE	79.38 mm
.875 in	BD	22.23 mm
1.250 in	SB	31.75 mm
4.313 in	EC	109.54 mm

ROCKWELL 1FRYS20-21A

IDLI GROUP 84525

QUICK DISCONNECT YOKE
WITH 6 SPLINES
SEE FIGURE 11A

3.125 in	OE	79.38 mm
.875 in	BD	22.23 mm
1.375 in	SB	34.93 mm
4.125 in	EC	104.78 mm

ROCKWELL 1FRYS22-31A

IDLI GROUP 84526

QUICK DISCONNECT YOKE
WITH 6 SPLINES
SEE FIGURE 11A

3.125 in	OE	79.38 mm
.875 in	BD	22.23 mm
1.375 in	SB	34.93 mm
4.313 in	EC	109.54 mm

PRECISION 1941
ROCKWELL 1FRYS22-15A

IDLI GROUP 84527

QUICK DISCONNECT YOKE
WITH 6 SPLINES
SEE FIGURE 11A

3.484 in	OE	88.50 mm
1.250 in	BD	31.75 mm
1.375 in	SB	34.93 mm
3.750 in	EC	95.25 mm

ROCKWELL L16SYS22-53A

IDLI GROUP 84528

QUICK DISCONNECT YOKE
WITH 21 INVOLUTED SPLINES
SEE FIGURE 11A

3.484 in	OE	88.50 mm
1.250 in	BD	31.75 mm
1.375 in	SB	34.93 mm
4.000 in	EC	101.60 mm

ROCKWELL L16SYS22-36A
SPICER 3-4-6511X

IDLI GROUP 84529

QUICK DISCONNECT YOKE
WITH 21 INVOLUTED SPLINES
SEE FIGURE 11A

3.484 in	OE	88.50 mm
1.250 in	BD	31.75 mm
1.375 in	SB	34.93 mm
4.000 in	EC	101.60 mm

ALLOY 0752
BORG-WARNR 62102
G & G MFG 182-3521
HAYES 3-4-6511X
HUB CITY 0335-02136
NEAPCO 21102X
NEAPCO 22-1102
PRECISION 1853
ROCKWELL L16SYS22-204A
ROCKWELL L16SYS2236
SPICER 3-4-6291X
SPICER 3-4-6511X
TRW 21259
WESCO 35N0220TDQ
WESCO 35N220TQD
WESCO 471221

IDLI GROUP 84530

QUICK DISCONNECT YOKE
WITH 6 SPLINES
SEE FIGURE 11A

3.484 in	OE	88.50 mm
1.250 in	BD	31.75 mm
1.375 in	SB	34.93 mm
4.563 in	EC	115.89 mm

NEAPCO 21101X
NEAPCO 211D7X
ROCKWELL L16SYS22-37A
SPICER 3-4-5251X

IDLI GROUP 84531

QUICK DISCONNECT YOKE
WITH 6 SPLINES
SEE FIGURE 11A

3.484 in	OE	88.50 mm
1.250 in	BD	31.75 mm
1.375 in	SB	34.93 mm
4.563 in	EC	115.89 mm

AEC 46035
ALLOY 0751
BORG-WARNR 3-4-5251X
BORG-WARNR 62101
G & G MFG 182-3506
HAYES 3-4-5251X
HUB CITY 0335-02102
MCQUAY-NOR U03-72
NEAPCO 21101X
NEAPCO 211D7X
NEAPCO 22-1101
NEAPCO 22-1319
NEW IDEA K2964A
PRECISION 1852
REX CHAIN 760086
ROCKWELL L16SYS22-205A
SPICER 3-4-5181
SPICER 3-4-5181X
SPICER 3-4-5251X
TRW 21262
WESCO 35N0220-20
WESCO 35N220-20
WESCO 471223

IDLI GROUP 84532

QUICK DISCONNECT YOKE
WITH 6 SPLINES
SEE FIGURE 11A

3.484 in	OE	88.50 mm
1.250 in	BD	31.75 mm
1.375 in	SB	34.93 mm
5.063 in	EC	128.59 mm

ROCKWELL L16SYS22-30A

IDLI GROUP 84533

QUICK DISCONNECT YOKE
WITH 6 SPLINES
SEE FIGURE 11A

3.484 in	OE	88.50 mm
1.250 in	BD	31.75 mm
1.375 in	SB	34.93 mm
5.313 in	EC	134.94 mm

PRECISION 1852
ROCKWELL L16SYS22-3A

IDLI GROUP 84534

QUICK DISCONNECT YOKE
WITH 6 SPLINES
SEE FIGURE 11A

3.484 in	OE	88.50 mm
1.250 in	BD	31.75 mm
1.750 in	SB	44.45 mm
5.313 in	EC	134.94 mm

ROCKWELL L16SYS28-2A

IDLI GROUP 84535

QUICK DISCONNECT YOKE
WITH 6 SPLINES
SEE FIGURE 11A

4.188 in	OE	106.36 mm
1.000 in	BD	25.40 mm
1.125 in	SB	28.58 mm
5.375 in	EC	136.53 mm

ROCKWELL 3DRYS18-13A

IDLI GROUP 84536

QUICK DISCONNECT YOKE
WITH 6 SPLINES
SEE FIGURE 11A

4.188 in	OE	106.36 mm
1.000 in	BD	25.40 mm
1.375 in	SB	34.93 mm
4.688 in	EC	119.06 mm

ROCKWELL 3DRYS22-22A

IDLI GROUP 84537

QUICK DISCONNECT YOKE
WITH 6 SPLINES
SEE FIGURE 11A

4.188 in	OE	106.36 mm
1.000 in	BD	25.40 mm
1.375 in	SB	34.93 mm
5.375 in	EC	136.53 mm

PRECISION 1951
ROCKWELL 3DRYS22-2A

IDLI GROUP 84538

QUICK DISCONNECT YOKE
WITH 21 INVOLUTED SPLINES
SEE FIGURE 11A

4.625 in	OE	117.48 mm
1.531 in	BD	38.90 mm
1.375 in	SB	34.93 mm
4.938 in	EC	125.41 mm

G & G MFG 182-5521
ROCKWELL 55NYS22-56A
WESCO 56N220TQD

BD = BEARING DIAMETER (outside)   BW = BEARING WIDTH   CB = CROSS LENGTH WITH BEARINGS   CC = CENTER TO CENTER   CD = CROSS DIAMETER
CL = CROSS LENGTH WITHOUT BEARINGS   EC = END TO CENTER or FACE TO FACE   EE = END TO END   EL = EFFECTIVE LENGTH
HD = HUB DIAMETER (or insert)   IE = INSIDE OF EARS (or recess)   OD = OUTSIDE DIAMETER   OE = OUTSIDE OF EARS   SB = SPLINE OR BORE SIZE

IDLI GROUP 84539

QUICK DISCONNECT YOKE
WITH 21 INVOLUTED SPLINES
SEE FIGURE 11A

4.625 in	OE	117.48 mm
1.531 in	BD	38.90 mm
1.375 in	SB	34.93 mm
5.313 in	EC	134.94 mm

ROCKWELL 55NYS22-19A

IDLI GROUP 84540

QUICK DISCONNECT YOKE
WITH 21 INVOLUTED SPLINES
slight variation from above group
SEE FIGURE 11A

4.625 in	OE	117.48 mm
1.531 in	BD	38.90 mm
1.375 in	SB	34.93 mm
5.313 in	EC	134.94 mm

ROCKWELL 55NYS22-25A

IDLI GROUP 84541

QUICK DISCONNECT YOKE
WITH 6 SPLINES
SEE FIGURE 11A

4.625 in	OE	117.48 mm
1.531 in	BD	38.90 mm
1.375 in	SB	34.93 mm
5.813 in	EC	147.64 mm

G & G MFG 182-5506
ROCKWELL 55NYS22-7A
WESCO 56N-220-20

IDLI GROUP 84542

QUICK DISCONNECT YOKE
WITH 6 SPLINES
SEE FIGURE 11A

4.625 in	OE	117.48 mm
1.531 in	BD	38.90 mm
1.375 in	SB	34.93 mm
5.813 in	EC	147.64 mm

ROCKWELL 55NYS22-9A

IDLI GROUP 84543

QUICK DISCONNECT YOKE
WITH 6 SPLINES
SEE FIGURE 11A

4.625 in	OE	117.48 mm
1.531 in	BD	38.90 mm
1.750 in	SB	44.45 mm
5.625 in	EC	142.88 mm

ROCKWELL 55NYS28-1A

IDLI GROUP 84544

QUICK DISCONNECT YOKE
WITH 20 INVOLUTED SPLINES
SEE FIGURE 11A

4.625 in	OE	117.48 mm
1.531 in	BD	38.90 mm
1.750 in	SB	44.45 mm
6.375 in	EC	161.93 mm

ROCKWELL 55NYS28-50A

IDLI GROUP 84545

QUICK DISCONNECT YOKE
WITH 27 INVOLUTED SPLINES
SEE FIGURE 11A

4.625 in	OE	117.48 mm
1.531 in	BD	38.90 mm
1.750 in	SB	44.45 mm
6.375 in	EC	161.93 mm

ROCKWELL 55NYS28-48A

FOR YOUR CONVENIENCE
NOTES ON OTHER INFORMATION

We urge you to send us any data that
you feel should be included with the
future editions of the I.D.L.I. Guide.

IDLI GROUP 84601

FLANGE YOKE
SEE FIGURE 16

1.438 in	IE	36.51 mm
.969 in	BD	24.61 mm
2.000 in	EC	50.80 mm
4.000 in	CC	101.60 mm
5.250 in	OD	133.35 mm

ROCKWELL L6NF123

IDLI GROUP 84602

FLANGE YOKE
SEE FIGURE 16

1.438 in	IE	36.51 mm
.969 in	BD	24.61 mm
2.250 in	EC	57.15 mm
3.344 in	CC	84.93 mm
3.875 in	OD	98.43 mm

ROCKWELL L6NF134

IDLI GROUP 84603

FLANGE YOKE
SEE FIGURE 16

1.438 in	IE	36.51 mm
.969 in	BD	24.61 mm
2.281 in	EC	57.95 mm
2.750 in	CC	69.85 mm
3.625 in	OD	92.08 mm

ROCKWELL L6NF115

IDLI GROUP 84604

FLANGE YOKE
SEE FIGURE 15

1.438 in	IE	36.51 mm
.969 in	BD	24.61 mm
2.281 in	EC	57.95 mm
3.922 in	CC	99.62 mm
4.813 in	OD	122.24 mm

ROCKWELL L6NF43

IDLI GROUP 84605

FLANGE YOKE
SEE FIGURE 16

1.438 in	IE	36.51 mm
.969 in	BD	24.61 mm
2.313 in	EC	58.74 mm
3.125 in	CC	79.38 mm
3.938 in	OD	100.01 mm

ROCKWELL L6NF53

IDLI GROUP 84606

FLANGE YOKE
SEE FIGURE 15

1.438 in	IE	36.51 mm
.969 in	BD	24.61 mm
2.406 in	EC	61.12 mm
3.000 in	CC	76.20 mm
3.875 in	OD	98.43 mm

ROCKWELL L6NF20

IDLI GROUP 84607

FLANGE YOKE
SEE FIGURE 16

1.813 in	IE	46.04 mm
1.063 in	BD	26.99 mm
2.500 in	EC	63.50 mm
3.813 in	CC	96.84 mm
4.750 in	OD	120.65 mm

ROCKWELL L12NF100

IDLI GROUP 84608

FLANGE YOKE
SEE FIGURE 16

1.813 in	IE	46.04 mm
1.063 in	BD	26.99 mm
2.750 in	EC	69.85 mm
3.797 in	CC	96.44 mm
4.688 in	OD	119.06 mm

ROCKWELL L12NF81

IDLI GROUP 84609

FLANGE YOKE
SEE FIGURE 16

1.813 in	IE	46.04 mm
1.063 in	BD	26.99 mm
2.875 in	EC	73.03 mm
4.000 in	CC	101.60 mm
5.250 in	OD	133.35 mm

ROCKWELL L12NF138

IDLI GROUP 84610

FLANGE YOKE
SEE FIGURE 16

1.813 in	IE	46.04 mm
1.063 in	BD	26.99 mm
2.938 in	EC	74.61 mm
2.406 in	CC	61.12 mm
3.250 in	OD	82.55 mm

ROCKWELL L12NF123A

IDLI GROUP 84611

FLANGE YOKE
SEE FIGURE 16

1.813 in	IE	46.04 mm
1.063 in	BD	26.99 mm
3.375 in	EC	85.73 mm
8.375 in	OD	212.73 mm

ROCKWELL L12NF130

IDLI GROUP 84612

FLANGE YOKE
SEE FIGURE 16

2.797 in	IE	71.04 mm
1.313 in	BD	33.34 mm
1.063 in	SB	26.99 mm
5.063 in	EC	128.59 mm
8.000 in	OD	203.20 mm

ROCKWELL 44RF22

IDLI GROUP 84613

FLANGE YOKE
SEE FIGURE 16

2.797 in	IE	71.04 mm
1.313 in	BD	33.34 mm
1.063 in	SB	26.99 mm
5.313 in	EC	134.94 mm

ROCKWELL 44RF12

IDLI GROUP 84614

FLANGE YOKE
SEE FIGURE 28

3.125 in	IE	79.38 mm
1.188 in	BD	30.16 mm
1.688 in	EC	42.86 mm
4.750 in	OD	120.65 mm

CLEVE.STL. O55-2
ROCKWELL 141NF1
SPICER 3-2-159

IDLI GROUP 84615

FLANGE YOKE
SEE FIGURE 28

3.125 in	IE	79.38 mm
1.188 in	BD	30.16 mm
1.813 in	EC	46.04 mm
5.813 in	OD	147.64 mm

CLEVE.STL. O55-2-2
ROCKWELL 141NF2
SPICER 3-2-549

IDLI GROUP 84616

FLANGE YOKE
SEE FIGURE 28

4.188 in	IE	106.36 mm
1.375 in	BD	34.93 mm
2.000 in	EC	50.80 mm
5.813 in	OD	147.64 mm

CLEVE.STL. D76-2
ROCKWELL 148NF2
SPICER 3-2-479

IDLI GROUP 84617

FLANGE YOKE
SEE FIGURE 28

4.188 in	IE	106.36 mm
1.375 in	BD	34.93 mm
2.375 in	EC	60.33 mm
5.813 in	OD	147.64 mm

ROCKWELL 148NF4

IDLI GROUP 84618

FLANGE YOKE
SEE FIGURE 28

4.969 in	IE	126.21 mm
1.375 in	BD	34.93 mm
2.000 in	EC	50.80 mm
5.875 in	OD	149.23 mm

CLEVE.STL. U76-2
ROCKWELL 155NF2
SPICER 4-2-669

IDLI GROUP 84619

FLANGE YOKE
SEE FIGURE 27

5.313 in	IE	134.94 mm
5.625 in	OE	142.88 mm
1.109 in	EC	28.18 mm
6.875 in	OD	174.63 mm

ROCKWELL 58WBF42

IDLI GROUP 84620

FLANGE YOKE
SEE FIGURE 27

5.313 in	IE	134.94 mm
5.625 in	OE	142.88 mm
1.188 in	EC	30.16 mm
6.875 in	OD	174.63 mm

ROCKWELL 58WBF34

IDLI GROUP 84621

FLANGE YOKE
slight variation from above group
SEE FIGURE 27

5.313 in	IE	134.94 mm
5.625 in	OE	142.88 mm
1.188 in	EC	30.16 mm
6.875 in	OD	174.63 mm

ROCKWELL 58WBF6

IDLI GROUP 84622

FLANGE YOKE
SEE FIGURE 27

5.313 in	IE	134.94 mm
5.625 in	OE	142.88 mm
1.250 in	EC	31.75 mm
8.000 in	OD	203.20 mm

ROCKWELL 58WBF29

IDLI GROUP 84623

FLANGE YOKE
SEE FIGURE 27

5.313 in	IE	134.94 mm
5.625 in	OE	142.88 mm
1.266 in	EC	32.15 mm
6.875 in	OD	174.63 mm

ROCKWELL 58WBF40

IDLI GROUP 84624

FLANGE YOKE
SEE FIGURE 27

5.313 in	IE	134.94 mm
5.625 in	OE	142.88 mm
1.266 in	EC	32.15 mm
7.125 in	OD	180.98 mm

ROCKWELL 58WBF35

IDLI GROUP 84625

FLANGE YOKE
SEE FIGURE 27

5.313 in	IE	134.94 mm
5.625 in	OE	142.88 mm
1.438 in	EC	36.51 mm
8.000 in	OD	203.20 mm

ROCKWELL 58WBF28

BD=BEARING DIAMETER (outside)   **BW**=BEARING WIDTH   **CB**=CROSS LENGTH WITH BEARINGS   **CC**=CENTER TO CENTER   **CD**=CROSS DIAMETER
CL=CROSS LENGTH WITHOUT BEARINGS   **EC**=END TO CENTER or FACE TO FACE   **EE**=END TO END   **EL**=EFFECTIVE LENGTH
HD=HUB DIAMETER (or insert)   **IE**=INSIDE OF EARS (or recess)   **OD**=OUTSIDE DIAMETER   **OE**=OUTSIDE OF EARS   **SB**=SPLINE OR BORE SIZE

CAUTION: BE SURE TO REFER TO ENGINEERING CATALOGS FOR SPECIAL APPLICATIONS THAT REQUIRE SPECIFIC MATERIAL CONTENT, TOLERANCES, ETC. SEE FOOTNOTE.

IDLI GROUP 84626

FLANGE YOKE
SEE FIGURE 27

5.313 in	IE	134.94 mm
5.625 in	OE	142.88 mm
1.438 in	EC	36.51 mm
8.125 in	OD	206.38 mm

ROCKWELL 58WBF36

IDLI GROUP 84627

FLANGE YOKE
SEE FIGURE 27

5.313 in	IE	134.94 mm
5.625 in	OE	142.88 mm
1.500 in	EC	38.10 mm
5.750 in	OD	146.05 mm

ROCKWELL 58WBF12

IDLI GROUP 84628

FLANGE YOKE
SEE FIGURE 27

5.313 in	IE	134.94 mm
5.625 in	OE	142.88 mm
2.188 in	EC	55.56 mm
8.000 in	OD	203.20 mm

ROCKWELL 58WBF1

IDLI GROUP 84629

FLANGE YOKE
slight variation from above group
SEE FIGURE 27

5.313 in	IE	134.94 mm
5.625 in	OE	142.88 mm
2.188 in	EC	55.56 mm
8.000 in	OD	203.20 mm

ROCKWELL 58WBF2

IDLI GROUP 84630

FLANGE YOKE
SEE FIGURE 27

5.313 in	IE	134.94 mm
5.625 in	OE	142.88 mm
2.188 in	EC	55.56 mm
10.000 in	OD	254.00 mm

ROCKWELL 58WBF3

IDLI GROUP 84631

FLANGE YOKE
SEE FIGURE 27

5.313 in	IE	134.94 mm
5.625 in	OE	142.88 mm
2.250 in	EC	57.15 mm
7.625 in	OD	193.68 mm

ROCKWELL 58WBF20

IDLI GROUP 84632

FLANGE YOKE
SEE FIGURE 27

5.313 in	IE	134.94 mm
5.625 in	OE	142.88 mm
2.313 in	EC	58.74 mm
5.750 in	OD	146.05 mm

ROCKWELL 58WBF38

IDLI GROUP 84633

FLANGE YOKE
SEE FIGURE 27

5.313 in	IE	134.94 mm
5.625 in	OE	142.88 mm
2.313 in	EC	58.74 mm
6.875 in	OD	174.63 mm

ROCKWELL 58WBF4

IDLI GROUP 84634

FLANGE YOKE
SEE FIGURE 27

5.313 in	IE	134.94 mm
5.625 in	OE	142.88 mm
2.375 in	EC	60.33 mm
5.969 in	OD	151.61 mm

ROCKWELL 58WBF8

IDLI GROUP 84635

FLANGE YOKE
SEE FIGURE 27

5.313 in	IE	134.94 mm
5.625 in	OE	142.88 mm
2.375 in	EC	60.33 mm
6.000 in	OD	152.40 mm

ROCKWELL 58WBF19

IDLI GROUP 84636

FLANGE YOKE
SEE FIGURE 27

5.313 in	IE	134.94 mm
5.625 in	OE	142.88 mm
2.375 in	EC	60.33 mm
7.750 in	OD	196.85 mm

ROCKWELL 58WBF32

IDLI GROUP 84637

FLANGE YOKE
SEE FIGURE 27

5.313 in	IE	134.94 mm
5.625 in	OE	142.88 mm
2.375 in	EC	60.33 mm
8.750 in	OD	222.25 mm

ROCKWELL 58WBF10

IDLI GROUP 84638

FLANGE YOKE
SEE FIGURE 27

5.313 in	IE	134.94 mm
5.625 in	OE	142.88 mm
2.375 in	EC	60.33 mm
10.000 in	OD	254.00 mm

ROCKWELL 58WBF11

IDLI GROUP 84639

FLANGE YOKE
SEE FIGURE 27

5.313 in	IE	134.94 mm
5.625 in	OE	142.88 mm
2.438 in	EC	61.91 mm
7.000 in	OD	177.80 mm

ROCKWELL 58WBF14

IDLI GROUP 84640

FLANGE YOKE
SEE FIGURE 27

5.313 in	IE	134.94 mm
5.625 in	OE	142.88 mm
2.688 in	EC	68.26 mm
5.375 in	OD	136.53 mm

ROCKWELL 58WBF25

IDLI GROUP 84641

FLANGE YOKE
slight variation from above group
SEE FIGURE 27

5.313 in	IE	134.94 mm
5.625 in	OE	142.88 mm
2.688 in	EC	68.26 mm
5.375 in	OD	136.53 mm

ROCKWELL 58WBF41

IDLI GROUP 84642

FLANGE YOKE
SEE FIGURE 27

5.313 in	IE	134.94 mm
5.625 in	OE	142.88 mm
2.688 in	EC	68.26 mm
7.500 in	OD	190.50 mm

ROCKWELL 58WBF21

IDLI GROUP 84643

FLANGE YOKE
SEE FIGURE 27

5.313 in	IE	134.94 mm
5.625 in	OE	142.88 mm
2.688 in	EC	68.26 mm
7.938 in	OD	201.61 mm

ROCKWELL 58WBF16

IDLI GROUP 84644

FLANGE YOKE
SEE FIGURE 27

5.313 in	IE	134.94 mm
5.625 in	OE	142.88 mm
2.750 in	EC	69.85 mm
8.500 in	OD	215.90 mm

ROCKWELL 58WBF31

IDLI GROUP 84645

FLANGE YOKE
SEE FIGURE 27

5.313 in	IE	134.94 mm
5.625 in	OE	142.88 mm
2.813 in	EC	71.44 mm
8.000 in	OD	203.20 mm

ROCKWELL 58WBF5

IDLI GROUP 84646

FLANGE YOKE
SEE FIGURE 27

5.313 in	IE	134.94 mm
5.625 in	OE	142.88 mm
3.375 in	EC	85.73 mm
7.625 in	OD	193.68 mm

ROCKWELL 58WBF7

IDLI GROUP 84647

FLANGE YOKE
slight variation from above group
SEE FIGURE 27

5.313 in	IE	134.94 mm
5.625 in	OE	142.88 mm
3.375 in	EC	85.73 mm
7.625 in	OD	193.68 mm

ROCKWELL 58WBF43

IDLI GROUP 84648

FLANGE YOKE
SEE FIGURE 27

5.313 in	IE	134.94 mm
5.625 in	OE	142.88 mm
5.688 in	EC	144.46 mm
8.500 in	OD	215.90 mm

ROCKWELL 58WBF26

IDLI GROUP 84649

FLANGE YOKE
SEE FIGURE 27

5.844 in	IE	148.43 mm
6.219 in	OE	157.96 mm
1.125 in	EC	28.58 mm
6.219 in	OD	157.96 mm

BORG-WARNR 15866J
ROCKWELL 72NF1

IDLI GROUP 84650

FLANGE YOKE
SEE FIGURE 27

5.844 in	IE	148.43 mm
6.219 in	OE	157.96 mm
1.313 in	EC	33.34 mm
6.375 in	OD	161.93 mm

ROCKWELL 72NF56

IDLI GROUP 84651

FLANGE YOKE
SEE FIGURE 27

5.844 in	IE	148.43 mm
6.219 in	OE	157.96 mm
1.438 in	EC	36.51 mm
6.219 in	OD	157.96 mm

BORG-WARNR 20431J
ROCKWELL 72NF13

ENGINEERING CATALOGS MUST BE CONSULTED FOR DETAILS NOT INCLUDED
IN THIS GUIDE. SPECIFIC DESIGNS, MATERIAL CONTENT, TOLERANCES
LUBE FITTINGS AND OTHER DIMENSIONS ARE INTENTIONALLY OMITTED HERE.

IDLI GROUP 84652

FLANGE YOKE
SEE FIGURE 27

5.844 in	IE	148.43 mm
6.219 in	OE	157.96 mm
1.625 in	EC	41.28 mm
6.875 in	OD	174.63 mm

BORG-WARNR 6397J
ROCKWELL 72NF6

IDLI GROUP 84653

FLANGE YOKE
SEE FIGURE 27

5.844 in	IE	148.43 mm
6.219 in	OE	157.96 mm
2.125 in	EC	53.98 mm
6.875 in	OD	174.63 mm

ROCKWELL 72NF24

IDLI GROUP 84654

FLANGE YOKE
SEE FIGURE 27

5.844 in	IE	148.43 mm
6.219 in	OE	157.96 mm
2.188 in	EC	55.56 mm
6.875 in	OD	174.63 mm

ROCKWELL 72NF3
ROCKWELL 72NF50

IDLI GROUP 84655

FLANGE YOKE
SEE FIGURE 27

5.844 in	IE	148.43 mm
6.219 in	OE	157.96 mm
2.500 in	EC	63.50 mm
6.875 in	OD	174.63 mm

BORG-WARNR 23397J
ROCKWELL 72NF26

IDLI GROUP 84656

FLANGE YOKE
slight variation from above group
SEE FIGURE 27

5.844 in	IE	148.43 mm
6.219 in	OE	157.96 mm
2.500 in	EC	63.50 mm
6.875 in	OD	174.63 mm

BORG-WARNR 4687J
ROCKWELL 72NF8

IDLI GROUP 84657

FLANGE YOKE
slight variation from above group
SEE FIGURE 27

5.844 in	IE	148.43 mm
6.219 in	OE	157.96 mm
2.500 in	EC	63.50 mm
6.875 in	OD	174.63 mm

BORG-WARNR 23393J
ROCKWELL 72NF30

IDLI GROUP 84658

FLANGE YOKE
slight variation from above group
SEE FIGURE 27

5.844 in	IE	148.43 mm
6.219 in	OE	157.96 mm
2.500 in	EC	63.50 mm
6.875 in	OD	174.63 mm

ROCKWELL 72NF52

IDLI GROUP 84659

FLANGE YOKE
SEE FIGURE 27

5.844 in	IE	148.43 mm
6.219 in	OE	157.96 mm
2.563 in	EC	65.09 mm
6.219 in	OD	157.96 mm

ROCKWELL 72NF51

IDLI GROUP 84660

FLANGE YOKE
SEE FIGURE 27

5.844 in	IE	148.43 mm
6.219 in	OE	157.96 mm
2.563 in	EC	65.09 mm
6.875 in	OD	174.63 mm

ROCKWELL 72NF53

IDLI GROUP 84661

FLANGE YOKE
SEE FIGURE 27

5.844 in	IE	148.43 mm
6.219 in	OE	157.96 mm
2.563 in	EC	65.09 mm
8.000 in	OD	203.20 mm

ROCKWELL 72NF28

IDLI GROUP 84662

FLANGE YOKE
slight variation from above group
SEE FIGURE 27

5.844 in	IE	148.43 mm
6.219 in	OE	157.96 mm
2.563 in	EC	65.09 mm
8.000 in	OD	203.20 mm

BORG-WARNR 4222J
ROCKWELL 72NF18

IDLI GROUP 84663

FLANGE YOKE
SEE FIGURE 27

5.844 in	IE	148.43 mm
6.219 in	OE	157.96 mm
2.625 in	EC	66.68 mm
6.875 in	OD	174.63 mm

ROCKWELL 72NF11

IDLI GROUP 84664

FLANGE YOKE
slight variation from above group
SEE FIGURE 27

5.844 in	IE	148.43 mm
6.219 in	OE	157.96 mm
2.625 in	EC	66.68 mm
6.875 in	OD	174.63 mm

BORG-WARNR 23395J
ROCKWELL 72NF27

IDLI GROUP 84665

FLANGE YOKE
slight variation from above group
SEE FIGURE 27

5.844 in	IE	148.43 mm
6.219 in	OE	157.96 mm
2.625 in	EC	66.68 mm
6.875 in	OD	174.63 mm

ROCKWELL 72NF34

IDLI GROUP 84666

FLANGE YOKE
slight variation from above group
SEE FIGURE 27

5.844 in	IE	148.43 mm
6.219 in	OE	157.96 mm
2.625 in	EC	66.68 mm
8.000 in	OD	203.20 mm

ROCKWELL 72NF4

IDLI GROUP 84667

FLANGE YOKE
SEE FIGURE 27

5.844 in	IE	148.43 mm
6.219 in	OE	157.96 mm
2.625 in	EC	66.68 mm
8.000 in	OD	203.20 mm

ROCKWELL 72NF9

IDLI GROUP 84668

FLANGE YOKE
slight variation from above group
SEE FIGURE 27

5.844 in	IE	148.43 mm
6.219 in	OE	157.96 mm
2.625 in	EC	66.68 mm
8.000 in	OD	203.20 mm

BORG-WARNR 23396J
ROCKWELL 72NF25

IDLI GROUP 84669

FLANGE YOKE
slight variation from above group
SEE FIGURE 27

5.844 in	IE	148.43 mm
6.219 in	OE	157.96 mm
2.625 in	EC	66.68 mm
8.000 in	OD	203.20 mm

ROCKWELL 72NF32

IDLI GROUP 84670

FLANGE YOKE
SEE FIGURE 27

5.844 in	IE	148.43 mm
6.219 in	OE	157.96 mm
2.688 in	EC	68.26 mm
5.375 in	OD	136.53 mm

ROCKWELL 72NF31

IDLI GROUP 84671

FLANGE YOKE
SEE FIGURE 27

5.844 in	IE	148.43 mm
6.219 in	OE	157.96 mm
2.688 in	EC	68.26 mm
7.125 in	OD	180.98 mm

ROCKWELL 72NF14

IDLI GROUP 84672

FLANGE YOKE
SEE FIGURE 27

5.844 in	IE	148.43 mm
6.219 in	OE	157.96 mm
2.688 in	EC	68.26 mm
8.750 in	OD	222.25 mm

ROCKWELL 72NF2

IDLI GROUP 84673

FLANGE YOKE
SEE FIGURE 27

5.844 in	IE	148.43 mm
6.219 in	OE	157.96 mm
3.375 in	EC	85.73 mm
5.750 in	OD	146.05 mm

ROCKWELL 72NF41

IDLI GROUP 84674

FLANGE YOKE
SEE FIGURE 27

6.500 in	IE	165.10 mm
6.875 in	OE	174.63 mm
2.188 in	EC	55.56 mm
8.000 in	OD	203.20 mm

ROCKWELL 85WBF2

IDLI GROUP 84675

FLANGE YOKE
slight variation from above group
SEE FIGURE 27

6.500 in	IE	165.10 mm
6.875 in	OE	174.63 mm
2.188 in	EC	55.56 mm
8.000 in	OD	203.20 mm

ROCKWELL 85WBF3

IDLI GROUP 84676

FLANGE YOKE
SEE FIGURE 27

6.500 in	IE	165.10 mm
6.875 in	OE	174.63 mm
2.500 in	EC	63.50 mm
7.125 in	OD	180.98 mm

ROCKWELL 85WBF1

BD=BEARING DIAMETER (outside)   **BW**=BEARING WIDTH   **CB**=CROSS LENGTH WITH BEARINGS   **CC**=CENTER TO CENTER   **CD**=CROSS DIAMETER
CL=CROSS LENGTH WITHOUT BEARINGS   **EC**=END TO CENTER or FACE TO FACE   **EE**=END TO END   **EL**=EFFECTIVE LENGTH
HD=HUB DIAMETER (or insert)   **IE**=INSIDE OF EARS (or recess)   **OD**=OUTSIDE DIAMETER   **OE**=OUTSIDE OF EARS   **SB**=SPLINE OR BORE SIZE

IDLI GUIDE   COPYRIGHT ® INTERCHANGE, INC.   ST. LOUIS PARK, MN. 55416 USA

CAUTION: BE SURE TO REFER TO ENGINEERING CATALOGS FOR SPECIAL APPLICATIONS THAT REQUIRE SPECIFIC MATERIAL CONTENT, TOLERANCES, ETC. SEE FOOTNOTE.

IDLI GROUP 84677
FLANGE YOKE
SEE FIGURE 27

6.500 in	IE	165.10 mm
6.875 in	OE	174.63 mm
2.875 in	EC	73.03 mm
6.875 in	OD	174.63 mm
ROCKWELL		85WBF4

IDLI GROUP 84678
FLANGE YOKE
SEE FIGURE 27

6.500 in	IE	165.10 mm
6.875 in	OE	174.63 mm
3.000 in	EC	76.20 mm
8.000 in	OD	203.20 mm
ROCKWELL		85WBF5

IDLI GROUP 84679
FLANGE YOKE
SEE FIGURE 27

6.500 in	IE	165.10 mm
6.875 in	OE	174.63 mm
4.750 in	EC	120.65 mm
6.875 in	OD	174.63 mm
ROCKWELL		85WBF6

IDLI GROUP 84680
FLANGE YOKE
SEE FIGURE 27

8.125 in	IE	206.38 mm
8.500 in	OE	215.90 mm
2.375 in	EC	60.33 mm
8.000 in	OD	203.20 mm
ROCKWELL		82NF22

IDLI GROUP 84681
FLANGE YOKE
slight variation from above group
SEE FIGURE 27

8.125 in	IE	206.38 mm
8.500 in	OE	215.90 mm
2.375 in	EC	60.33 mm
8.000 in	OD	203.20 mm
ROCKWELL		82NF24

IDLI GROUP 84682
FLANGE YOKE
SEE FIGURE 27

8.125 in	IE	206.38 mm
8.500 in	OE	215.90 mm
2.938 in	EC	74.61 mm
6.875 in	OD	174.63 mm
ROCKWELL		82NF13

IDLI GROUP 84683
FLANGE YOKE
SEE FIGURE 27

8.125 in	IE	206.38 mm
8.500 in	OE	215.90 mm
2.938 in	EC	74.61 mm
8.000 in	OD	203.20 mm
ROCKWELL		82NF6

IDLI GROUP 84684
FLANGE YOKE
slight variation from above group
SEE FIGURE 27

8.125 in	IE	206.38 mm
8.500 in	OE	215.90 mm
2.938 in	EC	74.61 mm
8.000 in	OD	203.20 mm
BORG-WARNR		23394J
ROCKWELL		82NF14

IDLI GROUP 84685
FLANGE YOKE
slight variation from above group
SEE FIGURE 27

8.125 in	IE	206.38 mm
8.500 in	OE	215.90 mm
2.938 in	EC	74.61 mm
8.000 in	OD	203.20 mm
ROCKWELL		82NF16

IDLI GROUP 84686
FLANGE YOKE
SEE FIGURE 27

8.125 in	IE	206.38 mm
8.500 in	OE	215.90 mm
3.000 in	EC	76.20 mm
10.000 in	OD	254.00 mm
ROCKWELL		82NF15

IDLI GROUP 84687
FLANGE YOKE
SEE FIGURE 27

8.125 in	IE	206.38 mm
8.500 in	OE	215.90 mm
3.063 in	EC	77.79 mm
8.000 in	OD	203.20 mm
BORG-WARNR		16954J
ROCKWELL		82NF2

IDLI GROUP 84688
FLANGE YOKE
slight variation from above group
SEE FIGURE 27

8.125 in	IE	206.38 mm
8.500 in	OE	215.90 mm
3.063 in	EC	77.79 mm
8.000 in	OD	203.20 mm
BORG-WARNR		23635J
ROCKWELL		82NF19

IDLI GROUP 84689
FLANGE YOKE
SEE FIGURE 27

8.125 in	IE	206.38 mm
8.500 in	OE	215.90 mm
3.125 in	EC	79.38 mm
7.750 in	OD	196.85 mm
ROCKWELL		82NF4

IDLI GROUP 84690
FLANGE YOKE
SEE FIGURE 27

8.125 in	IE	206.38 mm
8.500 in	OE	215.90 mm
3.125 in	EC	79.38 mm
10.875 in	OD	276.23 mm
ROCKWELL		82NF25

IDLI GROUP 84691
FLANGE YOKE
SEE FIGURE 27

8.125 in	IE	206.38 mm
8.500 in	OE	215.90 mm
3.250 in	EC	82.55 mm
9.250 in	OD	234.95 mm
ROCKWELL		82NF3

IDLI GROUP 84692
FLANGE YOKE
SEE FIGURE 27

8.125 in	IE	206.38 mm
8.500 in	OE	215.90 mm
4.063 in	EC	103.19 mm
8.750 in	OD	222.25 mm
ROCKWELL		82NF18

IDLI GROUP 84693
FLANGE YOKE
SEE FIGURE 27

8.125 in	IE	206.38 mm
8.500 in	OE	215.90 mm
4.313 in	EC	109.54 mm
14.000 in	OD	355.60 mm
ROCKWELL		82NF7

IDLI GROUP 84694
FLANGE YOKE
SEE FIGURE 27

8.250 in	IE	209.55 mm
8.625 in	OE	219.08 mm
2.250 in	EC	57.15 mm
8.000 in	OD	203.20 mm
ROCKWELL		92NF12

IDLI GROUP 84695
FLANGE YOKE
SEE FIGURE 27

8.250 in	IE	209.55 mm
8.625 in	OE	219.08 mm
2.750 in	EC	69.85 mm
8.000 in	OD	203.20 mm
ROCKWELL		92NF14

IDLI GROUP 84696
FLANGE YOKE
slight variation from above group
SEE FIGURE 27

8.250 in	IE	209.55 mm
8.625 in	OE	219.08 mm
2.750 in	EC	69.85 mm
8.000 in	OD	203.20 mm
ROCKWELL		92NF7

IDLI GROUP 84697
FLANGE YOKE
SEE FIGURE 27

8.250 in	IE	209.55 mm
8.625 in	OE	219.08 mm
2.813 in	EC	71.44 mm
6.875 in	OD	174.63 mm
ROCKWELL		92NF8

IDLI GROUP 84698
FLANGE YOKE
SEE FIGURE 27

8.250 in	IE	209.55 mm
8.625 in	OE	219.08 mm
2.813 in	EC	71.44 mm
8.000 in	OD	203.20 mm
ROCKWELL		92NF2

IDLI GROUP 84699
FLANGE YOKE
SEE FIGURE 27

8.250 in	IE	209.55 mm
8.625 in	OE	219.08 mm
2.813 in	EC	71.44 mm
10.000 in	OD	254.00 mm
ROCKWELL		92NF16

IDLI GROUP 84700
FLANGE YOKE
SEE FIGURE 27

8.250 in	IE	209.55 mm
8.625 in	OE	219.08 mm
2.938 in	EC	74.61 mm
8.000 in	OD	203.20 mm
ROCKWELL		92NF17

IDLI GROUP 84701
FLANGE YOKE
slight variation from above group
SEE FIGURE 27

8.250 in	IE	209.55 mm
8.625 in	OE	219.08 mm
2.938 in	EC	74.61 mm
8.000 in	OD	203.20 mm
BORG-WARNR		20655J
ROCKWELL		92NF4

ENGINEERING CATALOGS MUST BE CONSULTED FOR DETAILS NOT INCLUDED
IN THIS GUIDE. SPECIFIC DESIGNS, MATERIAL CONTENT, TOLERANCES
LUBE FITTINGS AND OTHER DIMENSIONS ARE INTENTIONALLY OMITTED HERE.

IDLI GROUP 84702

FLANGE YOKE
slight variation from above group
SEE FIGURE 27

8.250 in	IE	209.55 mm
8.625 in	OE	219.08 mm
2.938 in	EC	74.61 mm
8.000 in	OD	203.20 mm

BORG-WARNR 15589J
ROCKWELL 92NF5

IDLI GROUP 84703

FLANGE YOKE
slight variation from above group
SEE FIGURE 27

8.250 in	IE	209.55 mm
8.625 in	OE	219.08 mm
2.938 in	EC	74.61 mm
8.000 in	OD	203.20 mm

ROCKWELL 92NF6

IDLI GROUP 84704

FLANGE YOKE
slight variation from above group
SEE FIGURE 27

8.250 in	IE	209.55 mm
8.625 in	OE	219.08 mm
2.938 in	EC	74.61 mm
8.000 in	OD	203.20 mm

ROCKWELL 92NF26

IDLI GROUP 84705

FLANGE YOKE
SEE FIGURE 27

8.250 in	IE	209.55 mm
8.625 in	OE	219.08 mm
2.938 in	EC	74.61 mm
8.750 in	OD	222.25 mm

ROCKWELL 92NF18

IDLI GROUP 84706

FLANGE YOKE
SEE FIGURE 27

8.250 in	IE	209.55 mm
8.625 in	OE	219.08 mm
2.938 in	EC	74.61 mm
10.500 in	OD	266.70 mm

ROCKWELL 92NF22

IDLI GROUP 84707

FLANGE YOKE
SEE FIGURE 27

8.250 in	IE	209.55 mm
8.625 in	OE	219.08 mm
2.938 in	EC	74.61 mm
11.500 in	OD	292.10 mm

ROCKWELL 92NF3

IDLI GROUP 84708

FLANGE YOKE
SEE FIGURE 27

8.250 in	IE	209.55 mm
8.625 in	OE	219.08 mm
2.938 in	EC	74.61 mm
12.500 in	OD	317.50 mm

ROCKWELL 92NF11

IDLI GROUP 84709

FLANGE YOKE
SEE FIGURE 27

8.250 in	IE	209.55 mm
8.625 in	OE	219.08 mm
3.000 in	EC	76.20 mm
8.000 in	OD	203.20 mm

ROCKWELL 92NF23

IDLI GROUP 84710

FLANGE YOKE
SEE FIGURE 27

8.250 in	IE	209.55 mm
8.625 in	OE	219.08 mm
3.000 in	EC	76.20 mm
12.500 in	OD	317.50 mm

ROCKWELL 92NF10

IDLI GROUP 84711

FLANGE YOKE
SEE FIGURE 27

8.250 in	IE	209.55 mm
8.625 in	OE	219.08 mm
3.688 in	EC	93.66 mm
8.750 in	OD	222.25 mm

ROCKWELL 92NF20

IDLI GROUP 84712

FLANGE YOKE
SEE FIGURE 27

8.250 in	IE	209.55 mm
8.625 in	OE	219.08 mm
4.125 in	EC	104.78 mm
6.875 in	OD	174.63 mm

ROCKWELL 92NF15

IDLI GROUP 84713

FLANGE YOKE
SEE FIGURE 27

8.250 in	IE	209.55 mm
8.625 in	OE	219.08 mm
5.250 in	EC	133.35 mm
11.000 in	OD	279.40 mm

ROCKWELL 92NF1

IDLI GROUP 84714

FLANGE YOKE
SEE FIGURE 27

8.250 in	IE	209.55 mm
8.625 in	OE	219.08 mm
7.250 in	EC	184.15 mm
8.000 in	OD	203.20 mm

ROCKWELL 92NF13

BD=BEARING DIAMETER (outside)   **BW**=BEARING WIDTH   **CB**=CROSS LENGTH WITH BEARINGS   **CC**=CENTER TO CENTER   **CD**=CROSS DIAMETER
CL=CROSS LENGTH WITHOUT BEARINGS   **EC**=END TO CENTER or FACE TO FACE   **EE**=END TO END   **EL**=EFFECTIVE LENGTH
HD=HUB DIAMETER (or insert)   **IE**=INSIDE OF EARS (or recess)   **OD**=OUTSIDE DIAMETER   **OE**=OUTSIDE OF EARS   **SB**=SPLINE OR BORE SIZE

Other Valuable Interchange Publications

The International Bearing Interchange Guide

New Ninth Edition

The I.B.I. Guide has over 365,000 listings of 25,326 ball and roller bearings available worldwide and dating back to 1918. It includes manufacturers, original equipment users and government part numbers, with THOUSANDS of new listings not included in earlier editions!

I.S.B.N. #916966-17-8

The International Seal Interchange Guide

New Sixth Edition

The I.S.I. Guide has over 150,000 listings of 12,632 different seals and "O" rings. It includes government numbers as well as producers and users throughout the world—over 700 pages of information to save you time and money!

I.S.B.N. #916966-16-X

The International Drive Belt Interchange Guide

New Fourth Edition

The I.D.B.I. Guide has more than 115,000 listings for nearly 5,863 types and sizes of drive and V-belts. Listings include manufacturers, original equipment and government users from all over the world. 760 pages.

I.S.B.N. #916966-15-1

IDLI GROUP 84901

FLANGE YOKE
SEE FIGURE 18

2.906 in	OE	73.82 mm
1.125 in	BD	28.58 mm
2.375 in	EC	60.33 mm
3.625 in	CC	92.08 mm
4.375 in	OD	111.13 mm

ROCKWELL L14SF184

IDLI GROUP 84902

FLANGE YOKE
SEE FIGURE 17

2.906 in	OE	73.82 mm
1.125 in	BD	28.58 mm
1.063 in	SB	26.99 mm
4.531 in	EC	115.10 mm
7.188 in	CC	182.56 mm
8.000 in	OD	203.20 mm

ROCKWELL L14SF89

IDLI GROUP 84903

FLANGE YOKE
SEE FIGURE 19

2.906 in	OE	73.82 mm
1.125 in	BD	28.58 mm
1.125 in	SB	28.58 mm
4.875 in	EC	123.83 mm
3.906 in	CC	99.22 mm
4.875 in	OD	123.83 mm

ROCKWELL L14SF113

IDLI GROUP 84904

FLANGE YOKE
SEE FIGURE 18

2.906 in	OE	73.82 mm
1.125 in	BD	28.58 mm
1.250 in	SB	31.75 mm
2.375 in	EC	60.33 mm
5.750 in	CC	146.05 mm
7.125 in	OD	180.98 mm

ROCKWELL L14SF159

IDLI GROUP 84905

FLANGE YOKE
SEE FIGURE 20

2.906 in	OE	73.82 mm
1.125 in	BD	28.58 mm
1.375 in	SB	34.93 mm
4.500 in	EC	114.30 mm
3.125 in	CC	79.38 mm
4.000 in	OD	101.60 mm

ROCKWELL L14SF168A

IDLI GROUP 84906

FLANGE YOKE
SEE FIGURE 25

3.484 in	OE	88.50 mm
1.250 in	BD	31.75 mm
1.375 in	EC	34.93 mm
4.750 in	OD	120.65 mm

ROCKWELL 3NF20

IDLI GROUP 84907

FLANGE YOKE
SEE FIGURE 23

3.484 in	OE	88.50 mm
1.250 in	BD	31.75 mm
1.375 in	EC	34.93 mm
6.000 in	OD	152.40 mm

ROCKWELL 3NF33

IDLI GROUP 84908

FLANGE YOKE
SEE FIGURE 23

3.484 in	OE	88.50 mm
1.250 in	BD	31.75 mm
1.625 in	EC	41.28 mm
6.000 in	OD	152.40 mm

ROCKWELL 3NF5

IDLI GROUP 84909

FLANGE YOKE
SEE FIGURE 25

3.484 in	OE	88.50 mm
1.250 in	BD	31.75 mm
1.938 in	EC	49.21 mm
5.000 in	OD	127.00 mm

ROCKWELL L16SF20

IDLI GROUP 84910

FLANGE YOKE
slight variation from above group
SEE FIGURE 25

3.484 in	OE	88.50 mm
1.250 in	BD	31.75 mm
1.938 in	EC	49.21 mm
5.000 in	OD	127.00 mm

ROCKWELL L16SF4

IDLI GROUP 84911

FLANGE YOKE
SEE FIGURE 17

3.484 in	OE	88.50 mm
1.250 in	BD	31.75 mm
1.063 in	SB	26.99 mm
5.000 in	OD	127.00 mm

ROCKWELL L16SF61A

IDLI GROUP 84912

FLANGE YOKE
SEE FIGURE 25

3.484 in	OE	88.50 mm
1.250 in	BD	31.75 mm
1.969 in	EC	50.01 mm
5.000 in	OD	127.00 mm

ROCKWELL 3NF6

IDLI GROUP 84913

FLANGE YOKE
SEE FIGURE 19

3.484 in	OE	88.50 mm
1.250 in	BD	31.75 mm
5.250 in	EC	133.35 mm
5.000 in	OD	127.00 mm

ROCKWELL L16SF238A

IDLI GROUP 84914

FLANGE YOKE
SEE FIGURE 21

3.484 in	OE	88.50 mm
1.250 in	BD	31.75 mm
1.063 in	SB	26.99 mm
4.188 in	EC	106.36 mm
6.125 in	OD	155.58 mm

ROCKWELL L16SF102A

IDLI GROUP 84915

FLANGE YOKE
SEE FIGURE 17

3.484 in	OE	88.50 mm
1.250 in	BD	31.75 mm
1.063 in	SB	26.99 mm
5.000 in	EC	127.00 mm
8.000 in	OD	203.20 mm

ROCKWELL L16SF47

IDLI GROUP 84916

FLANGE YOKE
slight variation from above group
SEE FIGURE 17

3.484 in	OE	88.50 mm
1.250 in	BD	31.75 mm
1.063 in	SB	26.99 mm
5.000 in	EC	127.00 mm
8.000 in	OD	203.20 mm

ROCKWELL L16SF98

IDLI GROUP 84917

FLANGE YOKE
SEE FIGURE 17

3.484 in	OE	88.50 mm
1.250 in	BD	31.75 mm
1.063 in	SB	26.99 mm
5.313 in	EC	134.94 mm
9.500 in	OD	241.30 mm

ROCKWELL L16SF59A

IDLI GROUP 84918

FLANGE YOKE
SEE FIGURE 19

3.484 in	OE	88.50 mm
1.250 in	BD	31.75 mm
1.125 in	SB	28.58 mm
4.750 in	EC	120.65 mm
5.000 in	OD	127.00 mm

ROCKWELL L16SF195A

IDLI GROUP 84919

FLANGE YOKE
SEE FIGURE 18

3.484 in	OE	88.50 mm
1.250 in	BD	31.75 mm
1.188 in	SB	30.16 mm
3.250 in	EC	82.55 mm
7.125 in	OD	180.98 mm

ROCKWELL L16SF201

IDLI GROUP 84920

FLANGE YOKE
SEE FIGURE 24

3.484 in	OE	88.50 mm
1.250 in	BD	31.75 mm
1.250 in	SB	31.75 mm
3.500 in	EC	88.90 mm
4.750 in	OD	120.65 mm

ROCKWELL L16SF220

IDLI GROUP 84921

FLANGE YOKE
SEE FIGURE 20

3.484 in	OE	88.50 mm
1.250 in	BD	31.75 mm
1.375 in	SB	34.93 mm
4.500 in	EC	114.30 mm
4.000 in	OD	101.60 mm

ROCKWELL L16SF181A

IDLI GROUP 84922

FLANGE YOKE
SEE FIGURE 17

3.484 in	OE	88.50 mm
1.250 in	BD	31.75 mm
1.500 in	SB	38.10 mm
5.000 in	EC	127.00 mm
8.000 in	OD	203.20 mm

ROCKWELL L16SF227

IDLI GROUP 84923

FLANGE YOKE
SEE FIGURE 18

3.484 in	OE	88.50 mm
1.250 in	BD	31.75 mm
1.938 in	SB	49.21 mm
4.094 in	EC	103.98 mm
8.500 in	OD	215.90 mm

BORG-WARNR 64005
NEAPCO 211G5X19-10
NEAPCO 46-1910
ROCKWELL L16SF88

IDLI GROUP 84924

FLANGE YOKE
SEE FIGURE 22

3.484 in	OE	88.50 mm
1.250 in	BD	31.75 mm
2.000 in	SB	50.80 mm
3.875 in	EC	98.43 mm
5.125 in	OD	130.18 mm

ROCKWELL L16SF206

IDLI GROUP 84925

FLANGE YOKE
SEE FIGURE 26

4.625 in	OE	117.48 mm
1.531 in	BD	38.90 mm
4.625 in	EC	117.48 mm

ROCKWELL 55NF29

ENGINEERING CATALOGS MUST BE CONSULTED FOR DETAILS NOT INCLUDED
IN THIS GUIDE. SPECIFIC DESIGNS, MATERIAL CONTENT, TOLERANCES
LUBE FITTINGS AND OTHER DIMENSIONS ARE INTENTIONALLY OMITTED HERE.

CAUTION: BE SURE TO REFER TO ENGINEERING CATALOGS FOR SPECIAL APPLICATIONS THAT REQUIRE SPECIFIC MATERIAL CONTENT, TOLERANCES, ETC. SEE FOOTNOTE.

IDLI GROUP 84926

FLANGE YOKE
SEE FIGURE 18

4.625 in	OE	117.48 mm
1.531 in	BD	38.90 mm
.750 in	SB	19.05 mm
3.875 in	EC	98.43 mm
5.875 in	OD	149.23 mm

ROCKWELL 55NF36

IDLI GROUP 84927

FLANGE YOKE
SEE FIGURE 26

4.625 in	OE	117.48 mm
1.531 in	BD	38.90 mm
1.000 in	SB	25.40 mm
4.125 in	EC	104.78 mm
5.000 in	OD	127.00 mm

ROCKWELL 55NF44

RD = BEARING DIAMETER (outside)　　BW = BEARING WIDTH　　CB = CROSS LENGTH WITH BEARINGS　　CC = CENTER TO CENTER　　CD = CROSS DIAMETER
CL = CROSS LENGTH WITHOUT BEARINGS　　EC = END TO CENTER or FACE TO FACE　　EE = END TO END　　EL = EFFECTIVE LENGTH
HD = HUB DIAMETER (or insert)　　IE = INSIDE OF EARS (or recess)　　OD = OUTSIDE DIAMETER　　OE = OUTSIDE OF EARS　　SB = SPLINE OR BORE SIZE

IDLI GUIDE COPYRIGHT ® INTERCHANGE, INC. ST. LOUIS PARK, MN. 55416 USA

IDLI GROUP 85001

SLIP YOKE WITH ROUND BORE
SEE FIGURE 38B

1.438 in	IE	36.51 mm
.969 in	BD	24.61 mm
.750 in	SB	19.05 mm
5.750 in	EC	146.05 mm
1.500 in	HD	38.10 mm

ROCKWELL L6-109

IDLI GROUP 85002

SLIP YOKE WITH SQUARE BORE
SEE FIGURE 38D

1.438 in	IE	36.51 mm
.969 in	BD	24.61 mm
.750 in	SB	19.05 mm
5.750 in	EC	146.05 mm
1.500 in	HD	38.10 mm

ROCKWELL L6-117

IDLI GROUP 85003

SLIP YOKE WITH ROUND BORE
SEE FIGURE 35B

1.438 in	IE	36.51 mm
.969 in	BD	24.61 mm
.813 in	SB	20.64 mm
5.750 in	EC	146.05 mm

ROCKWELL L6NYR13-17A

IDLI GROUP 85004

SLIP YOKE WITH ROUND BORE
SEE FIGURE 38B

1.438 in	IE	36.51 mm
.969 in	BD	24.61 mm
.813 in	SB	20.64 mm
5.750 in	EC	146.05 mm
1.500 in	HD	38.10 mm

ROCKWELL L6-110

IDLI GROUP 85005

SEE FIGURE 38

1.438 in	IE	36.51 mm
.969 in	BD	24.61 mm
.875 in	SB	22.23 mm
4.750 in	EC	120.65 mm
1.625 in	HD	41.28 mm

ROCKWELL L6-124

IDLI GROUP 85006

SLIP YOKE WITH ROUND BORE
SEE FIGURE 35B

1.438 in	IE	36.51 mm
.969 in	BD	24.61 mm
.875 in	SB	22.23 mm
5.750 in	EC	146.05 mm

ROCKWELL L6NYR14-22A

IDLI GROUP 85007

SLIP YOKE WITH RECTANGULAR BORE
maximum SB dimension shown
SEE FIGURE 37C

1.438 in	IE	36.51 mm
.969 in	BD	24.61 mm
.875 in	SB	22.23 mm
5.750 in	EC	146.05 mm

ROCKWELL L6NYP12-28A

IDLI GROUP 85008

SLIP YOKE WITH SQUARE BORE
SEE FIGURE 38D

1.438 in	IE	36.51 mm
.969 in	BD	24.61 mm
.875 in	SB	22.23 mm
5.750 in	EC	146.05 mm

ROCKWELL L6NYQ14-2A

IDLI GROUP 85009

SLIP YOKE WITH ROUND BORE
SEE FIGURE 38B

1.438 in	IE	36.51 mm
.969 in	BD	24.61 mm
.875 in	SB	22.23 mm
5.750 in	EC	146.05 mm
1.500 in	HD	38.10 mm

ROCKWELL L6-111

IDLI GROUP 85010

SLIP YOKE WITH SQUARE BORE
SEE FIGURE 38D

1.438 in	IE	36.51 mm
.969 in	BD	24.61 mm
.875 in	SB	22.23 mm
5.750 in	EC	146.05 mm
1.500 in	HD	38.10 mm

ROCKWELL L6-58

IDLI GROUP 85011

SLIP YOKE WITH ROUND BORE
SEE FIGURE 38B

1.438 in	IE	36.51 mm
.969 in	BD	24.61 mm
.938 in	SB	23.81 mm
5.750 in	EC	146.05 mm
1.500 in	HD	38.10 mm

ROCKWELL L6-112

IDLI GROUP 85012

SLIP YOKE WITH SQUARE BORE
SEE FIGURE 38D

1.438 in	IE	36.51 mm
.969 in	BD	24.61 mm
.938 in	SB	23.81 mm
5.750 in	EC	146.05 mm
2.000 in	HD	50.80 mm

ROCKWELL L6-118

IDLI GROUP 85013

SLIP YOKE WITH ROUND BORE
SEE FIGURE 35B

1.438 in	IE	36.51 mm
.969 in	BD	24.61 mm
1.000 in	SB	25.40 mm
4.250 in	EC	107.95 mm

ROCKWELL L6NYR16-38A

IDLI GROUP 85014

SEE FIGURE 38A

1.438 in	IE	36.51 mm
.969 in	BD	24.61 mm
1.000 in	SB	25.40 mm
4.250 in	EC	107.95 mm
1.625 in	HD	41.28 mm

ROCKWELL L6-56

IDLI GROUP 85015

SLIP YOKE WITH ROUND BORE
SEE FIGURE 35B

1.438 in	IE	36.51 mm
.969 in	BD	24.61 mm
1.000 in	SB	25.40 mm
4.500 in	EC	114.30 mm

ROCKWELL L6NYR16-201

IDLI GROUP 85016

SLIP YOKE WITH SQUARE BORE
SEE FIGURE 38D

1.438 in	IE	36.51 mm
.969 in	BD	24.61 mm
1.000 in	SB	25.40 mm
5.500 in	EC	139.70 mm

ROCKWELL L6NYQ16-1A

IDLI GROUP 85017

SLIP YOKE WITH SQUARE BORE
SEE FIGURE 38D

1.438 in	IE	36.51 mm
.969 in	BD	24.61 mm
1.000 in	SB	25.40 mm
5.500 in	EC	139.70 mm
2.000 in	HD	50.80 mm

ROCKWELL L6-119

IDLI GROUP 85018

SLIP YOKE WITH ROUND BORE
SEE FIGURE 35B

1.438 in	IE	36.51 mm
.969 in	BD	24.61 mm
1.000 in	SB	25.40 mm
5.750 in	EC	146.05 mm

ROCKWELL L6NYR16-44A

IDLI GROUP 85019

SLIP YOKE WITH ROUND BORE
SEE FIGURE 38B

1.438 in	IE	36.51 mm
.969 in	BD	24.61 mm
1.000 in	SB	25.40 mm
5.750 in	EC	146.05 mm
1.625 in	HD	41.28 mm

ROCKWELL L6-113

IDLI GROUP 85020

SLIP YOKE WITH 10 SPLINES
SEE FIGURE 36A

1.438 in	IE	36.51 mm
.969 in	BD	24.61 mm
1.125 in	SB	28.58 mm
4.125 in	EC	104.78 mm

ROCKWELL L6NLS18-3

IDLI GROUP 85021

SLIP YOKE WITH 10 SPLINES
SEE FIGURE 38A

1.438 in	IE	36.51 mm
.969 in	BD	24.61 mm
1.125 in	SB	28.58 mm
5.125 in	EC	130.18 mm
1.500 in	HD	38.10 mm

ROCKWELL L6-120

IDLI GROUP 85022

SLIP YOKE WITH 10 SPLINES
SEE FIGURE 36A

1.438 in	IE	36.51 mm
.969 in	BD	24.61 mm
1.125 in	SB	28.58 mm
5.188 in	EC	131.76 mm

ROCKWELL L6NLS18DC

IDLI GROUP 85023

SLIP YOKE WITH ROUND BORE
SEE FIGURE 38B

1.438 in	IE	36.51 mm
.969 in	BD	24.61 mm
1.125 in	SB	28.58 mm
5.500 in	EC	139.70 mm
1.875 in	HD	47.63 mm

ROCKWELL L6-114

IDLI GROUP 85024

SLIP YOKE WITH ROUND BORE
SEE FIGURE 38B

1.438 in	IE	36.51 mm
.969 in	BD	24.61 mm
1.125 in	SB	28.58 mm
5.500 in	EC	139.70 mm
1.875 in	HD	47.63 mm

ROCKWELL L6-115

IDLI GROUP 85025

SLIP YOKE WITH 6 SPLINES
SEE FIGURE 36A

1.438 in	IE	36.51 mm
.969 in	BD	24.61 mm
1.125 in	SB	28.58 mm
5.750 in	EC	146.05 mm

ROCKWELL L6NYS18-6A

ENGINEERING CATALOGS MUST BE CONSULTED FOR DETAILS NOT INCLUDED
IN THIS GUIDE. SPECIFIC DESIGNS, MATERIAL CONTENT, TOLERANCES
LUBE FITTINGS AND OTHER DIMENSIONS ARE INTENTIONALLY OMITTED HERE.

CAUTION: BE SURE TO REFER TO ENGINEERING CATALOGS FOR SPECIAL APPLICATIONS THAT REQUIRE SPECIFIC MATERIAL CONTENT, TOLERANCES, ETC. SEE FOOTNOTE.

IDLI GROUP 85026

SLIP YOKE WITH 6 SPLINES
SEE FIGURE 38A

1.438 in	IE	36.51 mm
.969 in	BD	24.61 mm
1.125 in	SB	28.58 mm
5.750 in	EC	146.05 mm
1.563 in	HD	39.69 mm

ROCKWELL L6-61

IDLI GROUP 85027

SEE FIGURE 38

1.438 in	IE	36.51 mm
.969 in	BD	24.61 mm
1.125 in	SB	28.58 mm
5.750 in	EC	146.05 mm
1.875 in	HD	47.63 mm

ROCKWELL L6-125

IDLI GROUP 85028

SLIP YOKE WITH 10 SPLINES
SEE FIGURE 36A

1.438 in	IE	36.51 mm
.969 in	BD	24.61 mm
1.125 in	SB	28.58 mm
6.250 in	EC	158.75 mm

ROCKWELL L6NLS18-1DC

IDLI GROUP 85029

SLIP YOKE WITH 32 INVOLUTED SPLINES
SEE FIGURE 36A

1.438 in	IE	36.51 mm
.969 in	BD	24.61 mm
1.250 in	SB	31.75 mm
4.000 in	EC	101.60 mm

ROCKWELL L6NLS20-31DC

IDLI GROUP 85030

SLIP YOKE WITH 32 INVOLUTED SPLINES
SEE FIGURE 36A

1.438 in	IE	36.51 mm
.969 in	BD	24.61 mm
1.250 in	SB	31.75 mm
5.375 in	EC	136.53 mm

ROCKWELL L6NLS20-22DC

IDLI GROUP 85031

SLIP YOKE WITH ROUND BORE
SEE FIGURE 35B

1.438 in	IE	36.51 mm
.969 in	BD	24.61 mm
1.250 in	SB	31.75 mm
5.500 in	EC	139.70 mm

ROCKWELL L6NYR20-9A

IDLI GROUP 85032

SLIP YOKE WITH ROUND BORE
SEE FIGURE 38B

1.438 in	IE	36.51 mm
.969 in	BD	24.61 mm
1.250 in	SB	31.75 mm
5.500 in	EC	139.70 mm
1.875 in	HD	47.63 mm

ROCKWELL L6-116

IDLI GROUP 85033

SLIP YOKE WITH 32 INVOLUTED SPLINES
SEE FIGURE 36A

1.438 in	IE	36.51 mm
.969 in	BD	24.61 mm
1.250 in	SB	31.75 mm
5.750 in	EC	146.05 mm

ROCKWELL L6NLS20-71DC

IDLI GROUP 85034

SLIP YOKE WITH 6 SPLINES
SEE FIGURE 38A

1.438 in	IE	36.51 mm
.969 in	BD	24.61 mm
1.250 in	SB	31.75 mm
5.750 in	EC	146.05 mm
1.625 in	HD	41.28 mm

ROCKWELL L6-122

IDLI GROUP 85035

SLIP YOKE WITH 10 SPLINES
SEE FIGURE 38A

1.438 in	IE	36.51 mm
.969 in	BD	24.61 mm
1.250 in	SB	31.75 mm
5.750 in	EC	146.05 mm
1.875 in	HD	47.63 mm

ROCKWELL L6-121

IDLI GROUP 85036

SLIP YOKE WITH 32 INVOLUTED SPLINES
SEE FIGURE 36A

1.438 in	IE	36.51 mm
.969 in	BD	24.61 mm
1.250 in	SB	31.75 mm
6.750 in	EC	171.45 mm

ROCKWELL L6NLS20-1DC2

IDLI GROUP 85037

SLIP YOKE WITH 32 INVOLUTED SPLINES
SEE FIGURE 36A

1.438 in	IE	36.51 mm
.969 in	BD	24.61 mm
1.250 in	SB	31.75 mm
7.250 in	EC	184.15 mm

ROCKWELL L6NLS20-65DC1

IDLI GROUP 85038

SLIP YOKE WITH 32 INVOLUTED SPLINES
slight variation from above group
SEE FIGURE 36A

1.438 in	IE	36.51 mm
.969 in	BD	24.61 mm
1.250 in	SB	31.75 mm
7.250 in	EC	184.15 mm

ROCKWELL L6NLS20-65DC

IDLI GROUP 85039

SLIP YOKE WITH 32 INVOLUTED SPLINES
SEE FIGURE 36A

1.438 in	IE	36.51 mm
.969 in	BD	24.61 mm
1.250 in	SB	31.75 mm
9.125 in	EC	231.78 mm

ROCKWELL L6NLS20-55DC

IDLI GROUP 85040

SLIP YOKE WITH 32 INVOLUTED SPLINES
SEE FIGURE 36A

1.438 in	IE	36.51 mm
.969 in	BD	24.61 mm
1.250 in	SB	31.75 mm
9.750 in	EC	247.65 mm

ROCKWELL L6NLS20-77DC

IDLI GROUP 85041

SLIP YOKE WITH 32 INVOLUTED SPLINES
SEE FIGURE 36A

1.438 in	IE	36.51 mm
.969 in	BD	24.61 mm
1.250 in	SB	31.75 mm
10.313 in	EC	261.94 mm

ROCKWELL L6NLS20-35DC

IDLI GROUP 85042

SLIP YOKE WITH 32 INVOLUTED SPLINES
slight variation from above group
SEE FIGURE 36A

1.438 in	IE	36.51 mm
.969 in	BD	24.61 mm
1.250 in	SB	31.75 mm
10.313 in	EC	261.94 mm

ROCKWELL L6NLS20-63DC1

IDLI GROUP 85043

SLIP YOKE WITH 32 INVOLUTED SPLINES
SEE FIGURE 36A

1.438 in	IE	36.51 mm
.969 in	BD	24.61 mm
1.250 in	SB	31.75 mm
10.750 in	EC	273.05 mm

ROCKWELL L6NLS20-6DC

IDLI GROUP 85044

SLIP YOKE WITH 32 INVOLUTED SPLINES
SEE FIGURE 36A

1.438 in	IE	36.51 mm
.969 in	BD	24.61 mm
1.250 in	SB	31.75 mm
11.734 in	EC	298.05 mm

ROCKWELL L6NLS20-94DC

IDLI GROUP 85045

SLIP YOKE WITH 32 INVOLUTED SPLINES
SEE FIGURE 36A

1.438 in	IE	36.51 mm
.969 in	BD	24.61 mm
1.250 in	SB	31.75 mm
12.813 in	EC	325.44 mm

ROCKWELL L6NLS20-25DC

IDLI GROUP 85046

SLIP YOKE WITH 32 INVOLUTED SPLINES
SEE FIGURE 36A

1.438 in	IE	36.51 mm
.969 in	BD	24.61 mm
1.250 in	SB	31.75 mm
15.750 in	EC	400.05 mm

ROCKWELL L6NLS20-50DC

IDLI GROUP 85047

SLIP YOKE WITH 6 SPLINES
SEE FIGURE 38A

1.438 in	IE	36.51 mm
.969 in	BD	24.61 mm
1.375 in	SB	34.93 mm
5.125 in	EC	130.18 mm
1.875 in	HD	47.63 mm

ROCKWELL L6-123

IDLI GROUP 85048

SLIP YOKE WITH ROUND BORE
SEE FIGURE 38B

1.813 in	IE	46.04 mm
1.063 in	BD	26.99 mm
.750 in	SB	19.05 mm
4.750 in	EC	120.65 mm
1.750 in	HD	44.45 mm

PRECISION 1260
ROCKWELL L12-39

IDLI GROUP 85049

SLIP YOKE WITH SQUARE BORE
SEE FIGURE 38D

1.813 in	IE	46.04 mm
1.063 in	BD	26.99 mm
.750 in	SB	19.05 mm
5.250 in	EC	133.35 mm
1.875 in	HD	47.63 mm

BPC H12SY-3/4
PRECISION 1270
ROCKWELL L12-45

IDLI GROUP 85050

SLIP YOKE WITH ROUND BORE
SEE FIGURE 38B

1.813 in	IE	46.04 mm
1.063 in	BD	26.99 mm
.875 in	SB	22.23 mm
4.750 in	EC	120.65 mm
1.750 in	HD	44.45 mm

BPC H12SY-7/8R-1/4KW
PRECISION 1261
ROCKWELL L12-40

BD = BEARING DIAMETER (outside)   BW = BEARING WIDTH   CB = CROSS LENGTH WITH BEARINGS   CC = CENTER TO CENTER   CD = CROSS DIAMETER
CL = CROSS LENGTH WITHOUT BEARINGS   EC = END TO CENTER or FACE TO FACE   EE = END TO END   EL = EFFECTIVE LENGTH
HD = HUB DIAMETER (or insert)   IE = INSIDE OF EARS (or recess)   OD = OUTSIDE DIAMETER   OE = OUTSIDE OF EARS   SB = SPLINE OR BORE SIZE

IDLI GROUP 85051

SLIP YOKE WITH SQUARE BORE
SEE FIGURE 38D

1.813 in	IE	46.04 mm
1.063 in	BD	26.99 mm
.875 in	SB	22.23 mm
4.750 in	EC	120.65 mm
1.750 in	HD	44.45 mm

PRECISION 1271
ROCKWELL L12-46

IDLI GROUP 85052

SLIP YOKE WITH ROUND BORE
SEE FIGURE 38B

1.813 in	IE	46.04 mm
1.063 in	BD	26.99 mm
.938 in	SB	23.81 mm
5.250 in	EC	133.35 mm
2.000 in	HD	50.80 mm

PRECISION 1262
ROCKWELL L12-41

IDLI GROUP 85053

SEE FIGURE 38A

1.813 in	IE	46.04 mm
1.063 in	BD	26.99 mm
1.000 in	SB	25.40 mm
4.500 in	EC	114.30 mm
1.625 in	HD	41.28 mm

PRECISION 1280
ROCKWELL L12-48

IDLI GROUP 85054

SLIP YOKE WITH ROUND BORE
SEE FIGURE 38B

1.813 in	IE	46.04 mm
1.063 in	BD	26.99 mm
1.000 in	SB	25.40 mm
4.750 in	EC	120.65 mm
1.750 in	HD	44.45 mm

BPC H12SY-1R-1/4KW
MUNCIE MSY54
NEAPCO 111E5
NEAPCO 12-1316
PRECISION 1263
ROCKWELL L12-42
WESCO 12N160SL
WESCO 12N164L
WESCO 750049

IDLI GROUP 85055

SLIP YOKE WITH SQUARE BORE
SEE FIGURE 38D

1.813 in	IE	46.04 mm
1.063 in	BD	26.99 mm
1.000 in	SB	25.40 mm
4.750 in	EC	120.65 mm
1.750 in	HD	44.45 mm

BORG-WARNR 12-1314
BORG-WARNR 61252
BPC H12SY-1
MUNCIE MSY69
NEAPCO 111E2
PRECISION 1272
ROCKWELL L12-47
WESCO 12N160LQ

WESCO 750048

IDLI GROUP 85056

SLIP YOKE WITH 10 SPLINES
SEE FIGURE 36A

1.813 in	IE	46.04 mm
1.063 in	BD	26.99 mm
1.125 in	SB	28.58 mm
5.250 in	EC	133.35 mm

ROCKWELL L12NLS18DC

IDLI GROUP 85057

SLIP YOKE WITH 6 SPLINES
SEE FIGURE 38A

1.813 in	IE	46.04 mm
1.063 in	BD	26.99 mm
1.125 in	SB	28.58 mm
5.250 in	EC	133.35 mm
1.875 in	HD	47.63 mm

BPC H12SY-1-1/8
PRECISION 1281
ROCKWELL L12-49

IDLI GROUP 85058

SEE FIGURE 38

1.813 in	IE	46.04 mm
1.063 in	BD	26.99 mm
1.125 in	SB	28.58 mm
5.250 in	EC	133.35 mm
1.875 in	HD	47.63 mm

BORG-WARNR 61173
BPC H12SY-1-1/8
MUNCIE MSY60
NEAPCO 11173
NEAPCO 12-1173
PRECISION 1275
ROCKWELL L12-53
WESCO 12N180HSL
WESCO 12N180LH
WESCO 750050

IDLI GROUP 85059

SLIP YOKE WITH ROUND BORE
SEE FIGURE 38B

1.813 in	IE	46.04 mm
1.063 in	BD	26.99 mm
1.125 in	SB	28.58 mm
5.250 in	EC	133.35 mm
2.000 in	HD	50.80 mm

PRECISION 1264
ROCKWELL L12-43

IDLI GROUP 85060

SEE FIGURE 38

1.813 in	IE	46.04 mm
1.063 in	BD	26.99 mm
1.125 in	SB	28.58 mm
5.250 in	EC	133.35 mm
2.000 in	HD	50.80 mm

PRECISION 1276
ROCKWELL L12-54

IDLI GROUP 85061

SLIP YOKE WITH 6 SPLINES
SEE FIGURE 38A

1.813 in	IE	46.04 mm
1.063 in	BD	26.99 mm
1.250 in	SB	31.75 mm
5.250 in	EC	133.35 mm
1.875 in	HD	47.63 mm

PRECISION 1282
ROCKWELL L12-50

IDLI GROUP 85062

SLIP YOKE WITH 10 SPLINES
SEE FIGURE 38A

1.813 in	IE	46.04 mm
1.063 in	BD	26.99 mm
1.250 in	SB	31.75 mm
5.250 in	EC	133.35 mm
1.875 in	HD	47.63 mm

BPC H12SY-1-1/4
PRECISION 1283
ROCKWELL L12-51

IDLI GROUP 85063

SLIP YOKE WITH ROUND BORE
SEE FIGURE 38B

1.813 in	IE	46.04 mm
1.063 in	BD	26.99 mm
1.250 in	SB	31.75 mm
5.250 in	EC	133.35 mm
2.000 in	HD	50.80 mm

BPC H12SY-1-1/4R-5/16KW
MUNCIE MSY57
NEAPCO 11174
NEAPCO 111E4
NEAPCO 12-1174
PRECISION 1265
ROCKWELL L12-44
WESCO 12N200SL
WESCO 12N205L
WESCO 750052

IDLI GROUP 85064

SLIP YOKE WITH 6 SPLINES
SEE FIGURE 38A

1.813 in	IE	46.04 mm
1.063 in	BD	26.99 mm
1.375 in	SB	34.93 mm
4.500 in	EC	114.30 mm
2.000 in	HD	50.80 mm

BPC H12SY-1-3/8
NEAPCO 111AF
PRECISION 1284
ROCKWELL L12-52

IDLI GROUP 85065

SLIP YOKE WITH 16 SPLINES
SEE FIGURE 43

3.125 in	IE	79.38 mm
1.188 in	BD	30.16 mm
1.500 in	SB	38.10 mm
5.250 in	EC	133.35 mm

CLEVE.STL. O55-54-5
ROCKWELL 141NLS24-3DC
SPICER 3-3-2041KX
SPICER 3-3-2041X

IDLI GROUP 85066

SLIP YOKE WITH 32 INVOLUTED SPLINES
SEE FIGURE 44

3.125 in	IE	79.38 mm
1.188 in	BD	30.16 mm
1.500 in	SB	38.10 mm
5.688 in	EC	144.46 mm

CLEVE.STL. O55-54-1
ROCKWELL 141NLS24DC

IDLI GROUP 85067

SLIP YOKE WITH 32 INVOLUTED SPLINES
SEE FIGURE 44

3.125 in	IE	79.38 mm
1.188 in	BD	30.16 mm
1.500 in	SB	38.10 mm
8.500 in	EC	215.90 mm

CLEVE.STL. O55-54-240
ROCKWELL 141NLS24-6DC

IDLI GROUP 85068

SLIP YOKE WITH 16 SPLINES
SEE FIGURE 43

4.188 in	IE	106.36 mm
1.375 in	BD	34.93 mm
1.563 in	SB	39.69 mm
6.813 in	EC	173.04 mm

CLEVE.STL. D76-54
ROCKWELL 148NLS25DC
SPICER 3-3-1601KX
SPICER 3-3-1601X

IDLI GROUP 85069

SLIP YOKE WITH 16 SPLINES
slight variation from above group
SEE FIGURE 43

4.188 in	IE	106.36 mm
1.375 in	BD	34.93 mm
1.563 in	SB	39.69 mm
6.813 in	EC	173.04 mm

ROCKWELL 148NLS25DC1

IDLI GROUP 85070

SLIP YOKE WITH 32 INVOLUTED SPLINES
SEE FIGURE 44

4.188 in	IE	106.36 mm
1.375 in	BD	34.93 mm
1.625 in	SB	41.28 mm
6.375 in	EC	161.93 mm

CLEVE.STL. D56-54
ROCKWELL 148NLS26DC

IDLI GROUP 85071

SLIP YOKE WITH 32 INVOLUTED SPLINES
SEE FIGURE 44

4.188 in	IE	106.36 mm
1.375 in	BD	34.93 mm
1.625 in	SB	41.28 mm
8.938 in	EC	227.01 mm

CLEVE.STL. D56-54-140
ROCKWELL 148NLS26-3DC

ENGINEERING CATALOGS MUST BE CONSULTED FOR DETAILS NOT INCLUDED
IN THIS GUIDE. SPECIFIC DESIGNS, MATERIAL CONTENT, TOLERANCES
LUBE FITTINGS AND OTHER DIMENSIONS ARE INTENTIONALLY OMITTED HERE.

IDLI GUIDE   COPYRIGHT ® INTERCHANGE, INC.   ST. LOUIS PARK, MN. 55416 USA

CAUTION: BE SURE TO REFER TO ENGINEERING CATALOGS FOR SPECIAL APPLICATIONS THAT REQUIRE SPECIFIC MATERIAL CONTENT, TOLERANCES, ETC. SEE FOOTNOTE.

IDLI GROUP 85072

SLIP YOKE WITH 16 SPLINES

4.188 in	IE	106.36 mm
1.375 in	BD	34.93 mm
2.000 in	SB	50.80 mm
16.188 in	EC	411.16 mm

ROCKWELL 148NLS32DC

IDLI GROUP 85073

SLIP YOKE WITH 16 SPLINES
SEE FIGURE 43

4.969 in	IE	126.21 mm
1.375 in	BD	34.93 mm
1.750 in	SB	44.45 mm
6.875 in	EC	174.63 mm

CLEVE.STL. U76-54
ROCKWELL 155NLS28DC
SPICER 4-3-1241KX
SPICER 4-3-1241X

IDLI GROUP 85074

SLIP YOKE WITH 16 SPLINES
slight variation from above group
SEE FIGURE 43

4.969 in	IE	126.21 mm
1.375 in	BD	34.93 mm
1.750 in	SB	44.45 mm
6.875 in	EC	174.63 mm

ROCKWELL 155NLS28DC1

IDLI GROUP 85075

SLIP YOKE WITH 32 INVOLUTED SPLINES
SEE FIGURE 44

4.969 in	IE	126.21 mm
1.375 in	BD	34.93 mm
1.750 in	SB	44.45 mm
7.625 in	EC	193.68 mm

CLEVE.STL. U56-54
ROCKWELL 155NLS28-6DC

IDLI GROUP 85076

SLIP YOKE WITH 32 INVOLUTED SPLINES
SEE FIGURE 43

4.969 in	IE	126.21 mm
1.375 in	BD	34.93 mm
1.750 in	SB	44.45 mm
10.063 in	EC	255.59 mm

CLEVE.STL. U76-54L
ROCKWELL 155NLS28-3DC
SPICER 4-3-1431KX

IDLI GROUP 85077

SLIP YOKE WITH 16 SPLINES
SEE FIGURE 41

5.313 in	IE	134.94 mm
5.625 in	OE	142.88 mm
1.750 in	SB	44.45 mm
5.250 in	EC	133.35 mm

ROCKWELL 58WBLS28-8

IDLI GROUP 85078

SLIP YOKE WITH 16 SPLINES
SEE FIGURE 41

5.313 in	IE	134.94 mm
5.625 in	OE	142.88 mm
1.750 in	SB	44.45 mm
6.750 in	EC	171.45 mm

ROCKWELL 58WBLS28-1

IDLI GROUP 85079

SLIP YOKE WITH 16 SPLINES
slight variation from above group
SEE FIGURE 41

5.313 in	IE	134.94 mm
5.625 in	OE	142.88 mm
1.750 in	SB	44.45 mm
6.750 in	EC	171.45 mm

ROCKWELL 58WBLS28-12

IDLI GROUP 85080

SLIP YOKE WITH 16 SPLINES
SEE FIGURE 41

5.313 in	IE	134.94 mm
5.625 in	OE	142.88 mm
1.750 in	SB	44.45 mm
7.000 in	EC	177.80 mm

ROCKWELL 58WBLS28-15

IDLI GROUP 85081

SLIP YOKE WITH 16 SPLINES
SEE FIGURE 41

5.313 in	IE	134.94 mm
5.625 in	OE	142.88 mm
1.750 in	SB	44.45 mm
8.750 in	EC	222.25 mm

ROCKWELL 58WBLS28-2

IDLI GROUP 85082

SLIP YOKE WITH 16 SPLINES
SEE FIGURE 42

5.313 in	IE	134.94 mm
5.625 in	OE	142.88 mm
1.750 in	SB	44.45 mm
8.750 in	EC	222.25 mm

ROCKWELL 58WBLS28-21

IDLI GROUP 85083

SLIP YOKE WITH 16 SPLINES
slight variation from above group
SEE FIGURE 41

5.313 in	IE	134.94 mm
5.625 in	OE	142.88 mm
1.750 in	SB	44.45 mm
8.750 in	EC	222.25 mm

ROCKWELL 58WBLS28-10

IDLI GROUP 85084

SLIP YOKE WITH 16 SPLINES
slight variation from above group
SEE FIGURE 42

5.313 in	IE	134.94 mm
5.625 in	OE	142.88 mm
1.750 in	SB	44.45 mm
8.750 in	EC	222.25 mm

ROCKWELL 58WBLS28-11

IDLI GROUP 85085

SLIP YOKE WITH 16 SPLINES
SEE FIGURE 41

5.313 in	IE	134.94 mm
5.625 in	OE	142.88 mm
2.500 in	SB	63.50 mm
3.688 in	EC	93.66 mm

ROCKWELL 58WBLS40-1

IDLI GROUP 85086

SLIP YOKE WITH 16 SPLINES
slight variation from above group
SEE FIGURE 41

5.531 in	IE	140.50 mm
5.841 in	OE	148.37 mm
2.000 in	SB	50.80 mm
7.000 in	EC	177.80 mm

ROCKWELL 62NLS32-7

IDLI GROUP 85087

SLIP YOKE WITH 10 SPLINES
SEE FIGURE 41

5.531 in	IE	140.50 mm
5.844 in	OE	148.43 mm
1.750 in	SB	44.45 mm
5.625 in	EC	142.88 mm

BORG-WARNR 3593J
BORG-WARNR 66000
BORG-WARNR 6A6000
FORD TGAA4841A
IHC 38777HX
NEAPCO 05175X5
ROCKWELL 62NLS28

IDLI GROUP 85088

SLIP YOKE WITH 10 SPLINES
SEE FIGURE 42

5.531 in	IE	140.50 mm
5.844 in	OE	148.43 mm
1.750 in	SB	44.45 mm
5.625 in	EC	142.88 mm

ROCKWELL 62NLS28-43

IDLI GROUP 85089

SLIP YOKE WITH 10 SPLINES
slight variation from above group
SEE FIGURE 41

5.531 in	IE	140.50 mm
5.844 in	OE	148.43 mm
1.750 in	SB	44.45 mm
5.625 in	EC	142.88 mm

ROCKWELL 62NLS28-1

IDLI GROUP 85090

SLIP YOKE WITH 10 SPLINES
SEE FIGURE 41

5.531 in	IE	140.50 mm
5.844 in	OE	148.43 mm
1.750 in	SB	44.45 mm
5.750 in	EC	146.05 mm

ROCKWELL 62NLS28-41

IDLI GROUP 85091

SLIP YOKE WITH 10 SPLINES
SEE FIGURE 41

5.531 in	IE	140.50 mm
5.844 in	OE	148.43 mm
1.750 in	SB	44.45 mm
6.500 in	EC	165.10 mm

ROCKWELL 62NLS28-35

IDLI GROUP 85092

SLIP YOKE WITH 10 SPLINES
SEE FIGURE 41

5.531 in	IE	140.50 mm
5.844 in	OE	148.43 mm
1.750 in	SB	44.45 mm
7.375 in	EC	187.33 mm

ROCKWELL 62NLS28-28

IDLI GROUP 85093

SLIP YOKE WITH 10 SPLINES
SEE FIGURE 42

5.531 in	IE	140.50 mm
5.844 in	OE	148.43 mm
1.750 in	SB	44.45 mm
7.375 in	EC	187.33 mm

BORG-WARNR 16991J
BORG-WARNR 5096J
BORG-WARNR 6-6007
BORG-WARNR 6-6067
BORG-WARNR 6A6007
BORG-WARNR 6A6067
FORD B4QH4841C
FORD TWAA4841B
ROCKWELL 62NLS28-3

IDLI GROUP 85094

SLIP YOKE WITH 16 SPLINES
SEE FIGURE 41

5.531 in	IE	140.50 mm
5.844 in	OE	148.43 mm
1.750 in	SB	44.45 mm
9.500 in	EC	241.30 mm

BORG-WARNR 17070J
BORG-WARNR 6-6047
BORG-WARNR 6-6068
BORG-WARNR 6A6047
BORG-WARNR 6A6068
FORD C3TZ4841G
ROCKWELL 62NLS28-2

IDLI GROUP 85095

SLIP YOKE WITH 10 SPLINES
SEE FIGURE 42

5.531 in	IE	140.50 mm
5.844 in	OE	148.43 mm
1.750 in	SB	44.45 mm
12.438 in	EC	315.91 mm

BORG-WARNR 6-6025
BORG-WARNR 6-6057
BORG-WARNR 6-6069
BORG-WARNR 6A6025
BORG-WARNR 6A6057
BORG-WARNR 6A6069
BORG-WARNR 9519J
ROCKWELL 62NLS28-4

BD = BEARING DIAMETER (outside)   **BW** = BEARING WIDTH   **CB** = CROSS LENGTH WITH BEARINGS   **CC** = CENTER TO CENTER   **CD** = CROSS DIAMETER
CL = CROSS LENGTH WITHOUT BEARINGS   **EC** = END TO CENTER or FACE TO FACE   **EE** = END TO END   **EL** = EFFECTIVE LENGTH
HD = HUB DIAMETER (or insert)   **IE** = INSIDE OF EARS (or recess)   **OD** = OUTSIDE DIAMETER   **OE** = OUTSIDE OF EARS   **SB** = SPLINE OR BORE SIZE

IDLI GROUP 85096

SLIP YOKE WITH 10 SPLINES
SEE FIGURE 41

5.531 in	IE	140.50 mm
5.844 in	OE	148.43 mm
2.000 in	SB	50.80 mm
5.750 in	EC	146.05 mm

ROCKWELL 62NLS32-3

IDLI GROUP 85097

SLIP YOKE WITH 16 SPLINES
SEE FIGURE 41

5.531 in	IE	140.50 mm
5.844 in	OE	148.43 mm
2.000 in	SB	50.80 mm
7.000 in	EC	177.80 mm

ROCKWELL 62NLS32-5

IDLI GROUP 85098

SLIP YOKE WITH 16 SPLINES
SEE FIGURE 41

5.531 in	IE	140.50 mm
5.844 in	OE	148.43 mm
2.000 in	SB	50.80 mm
8.500 in	EC	215.90 mm

ROCKWELL 62NLS32-2

IDLI GROUP 85099

SLIP YOKE WITH 10 SPLINES
SEE FIGURE 41

5.844 in	IE	148.43 mm
6.219 in	OE	157.96 mm
1.969 in	SB	50.01 mm
4.781 in	EC	121.45 mm

ROCKWELL 72NLS31-1

IDLI GROUP 85100

SLIP YOKE WITH 10 SPLINES
SEE FIGURE 42

5.844 in	IE	148.43 mm
6.219 in	OE	157.96 mm
2.000 in	SB	50.80 mm
3.000 in	EC	76.20 mm

ROCKWELL 72NLS32-89

IDLI GROUP 85101

SLIP YOKE WITH 10 SPLINES
SEE FIGURE 41

5.844 in	IE	148.43 mm
6.219 in	OE	157.96 mm
2.000 in	SB	50.80 mm
4.000 in	EC	101.60 mm

BORG-WARNR 12972J
BORG-WARNR 67032
BORG-WARNR 6-7077
BORG-WARNR 6A7032
BORG-WARNR 6A7077
NEAPCO 26-229
NEAPCO 26-239
NEAPCO 26-251
NEAPCO 68-2639
ROCKWELL 72NLS32-29
SPICER 228606-3000X
TRW 21483

IDLI GROUP 85102

SEE FIGURE 41

5.844 in	IE	148.43 mm
6.219 in	OE	157.96 mm
2.000 in	SB	50.80 mm
4.938 in	EC	125.41 mm

BORG-WARNR 5506J
BORG-WARNR 67005
BORG-WARNR 67018
BORG-WARNR 67040
BORG-WARNR 67049
BORG-WARNR 6A7005
BORG-WARNR 6A7018
BORG-WARNR 6A7040
BORG-WARNR 6A7049
FORD C1TZ4841C
NEAPCO 06-251
NEAPCO 12-139
NEAPCO 18-328
NEAPCO 26-329
NEAPCO 6-251
NEAPCO 64-1828
NEAPCO 64-2629
NEAPCO 66-1239
NEAPCO 68-0651
NEAPCO UT26-002900
ROCKWELL 72NLS32-13
SPICER 228626-3010X
TRW 21403
WESCO 06NSSM48
WESCO 920-302

IDLI GROUP 85103

SLIP YOKE WITH 10 SPLINES
SEE FIGURE 42

5.844 in	IE	148.43 mm
6.219 in	OE	157.96 mm
2.000 in	SB	50.80 mm
5.188 in	EC	131.76 mm

ROCKWELL 72NLS32-113

IDLI GROUP 85104

SLIP YOKE WITH 10 SPLINES
SEE FIGURE 41

5.844 in	IE	148.43 mm
6.219 in	OE	157.96 mm
2.000 in	SB	50.80 mm
5.313 in	EC	134.94 mm

ROCKWELL 72NLS32-4

IDLI GROUP 85105

SLIP YOKE WITH 10 SPLINES
slight variation from above group
SEE FIGURE 41

5.844 in	IE	148.43 mm
6.219 in	OE	157.96 mm
2.000 in	SB	50.80 mm
5.313 in	EC	134.94 mm

ROCKWELL 72NLS32-8

IDLI GROUP 85106

SLIP YOKE WITH 10 SPLINES
slight variation from above group
SEE FIGURE 41

5.844 in	IE	148.43 mm
6.219 in	OE	157.96 mm
2.000 in	SB	50.80 mm
5.313 in	EC	134.94 mm

ROCKWELL 72NLS32-78

IDLI GROUP 85107

SLIP YOKE WITH 10 SPLINES
SEE FIGURE 41

5.844 in	IE	148.43 mm
6.219 in	OE	157.96 mm
2.000 in	SB	50.80 mm
5.875 in	EC	149.23 mm

BORG-WARNR 5147J
BORG-WARNR 67004
BORG-WARNR 6-7073
BORG-WARNR 6A7004
BORG-WARNR 6A7073
NEAPCO 12-127
NEAPCO 66-1227
ROCKWELL 72NLS32-24

IDLI GROUP 85108

SLIP YOKE WITH 10 SPLINES
slight variation from above group
SEE FIGURE 41

5.844 in	IE	148.43 mm
6.219 in	OE	157.96 mm
2.000 in	SB	50.80 mm
5.875 in	EC	149.23 mm

BORG-WARNR 6-7086
BORG-WARNR 6A7086
ROCKWELL 72NLS32-61

IDLI GROUP 85109

SLIP YOKE WITH 10 SPLINES
SEE FIGURE 41

5.844 in	IE	148.43 mm
6.219 in	OE	157.96 mm
2.000 in	SB	50.80 mm
6.063 in	EC	153.99 mm

ROCKWELL 72NLS32-7

IDLI GROUP 85110

SLIP YOKE WITH 10 SPLINES
SEE FIGURE 41

5.844 in	IE	148.43 mm
6.219 in	OE	157.96 mm
2.000 in	SB	50.80 mm
6.563 in	EC	166.69 mm

ROCKWELL 72NLS32-52

IDLI GROUP 85111

SLIP YOKE WITH 10 SPLINES
slight variation from above group
SEE FIGURE 41

5.844 in	IE	148.43 mm
6.219 in	OE	157.96 mm
2.000 in	SB	50.80 mm
6.563 in	EC	166.69 mm

ROCKWELL 72NLS32-33

IDLI GROUP 85112

SLIP YOKE WITH 10 SPLINES
slight variation from above group
SEE FIGURE 41

5.844 in	IE	148.43 mm
6.219 in	OE	157.96 mm
2.000 in	SB	50.80 mm
7.063 in	EC	179.39 mm

BORG-WARNR 67059
BORG-WARNR 6A7059
NEAPCO 18-440
NEAPCO 65-1840
ROCKWELL 72NLS32-1

IDLI GROUP 85113

SLIP YOKE WITH 10 SPLINES
SEE FIGURE 41

5.844 in	IE	148.43 mm
6.219 in	OE	157.96 mm
2.000 in	SB	50.80 mm
7.063 in	EC	179.39 mm

ROCKWELL 72NLS32-77

IDLI GROUP 85114

SLIP YOKE WITH 10 SPLINES
slight variation from above group
SEE FIGURE 41

5.844 in	IE	148.43 mm
6.219 in	OE	157.96 mm
2.000 in	SB	50.80 mm
7.063 in	EC	179.39 mm

BORG-WARNR 3555J
BORG-WARNR 6-7000
BORG-WARNR 6-7022
BORG-WARNR 67022
BORG-WARNR 67023
BORG-WARNR 6-7023
BORG-WARNR 6-7036
BORG-WARNR 67036
BORG-WARNR 67047
BORG-WARNR 6-7047
BORG-WARNR 67048
BORG-WARNR 6-7048
BORG-WARNR 6-7079
BORG-WARNR 6A7000
BORG-WARNR 6A7022
BORG-WARNR 6A7023
BORG-WARNR 6A7036
BORG-WARNR 6A7047
BORG-WARNR 6A7048
BORG-WARNR 6A7079
BORG-WARNR 7990J
FORD B4QH4841B
FORD C3TZ4841E
FORD CITT4841E
FORD TWAA4841A
MUNCIE A302
NEAPCO 06-351
NEAPCO 18-228
NEAPCO 18-240
NEAPCO 22-339
NEAPCO 22-351
NEAPCO 64-0651
NEAPCO 64-2239
NEAPCO 64-2251
NEAPCO 68-1828

CAUTION: BE SURE TO REFER TO ENGINEERING CATALOGS FOR SPECIAL APPLICATIONS THAT REQUIRE SPECIFIC MATERIAL CONTENT, TOLERANCES, ETC. SEE FOOTNOTE.

NEAPCO 68-1840
ROCKWELL 72NLS32
ROCKWELL 72NLS32-18

IDLI GROUP 85115

SLIP YOKE WITH 10 SPLINES
slight variation from above group
SEE FIGURE 41

5.844 in	IE	148.43 mm
6.219 in	OE	157.96 mm
2.000 in	SB	50.80 mm
7.063 in	EC	179.39 mm

ROCKWELL 72NLS32-55

IDLI GROUP 85116

SLIP YOKE WITH 10 SPLINES
slight variation from above group
SEE FIGURE 41

5.844 in	IE	148.43 mm
6.219 in	OE	157.96 mm
2.000 in	SB	50.80 mm
7.063 in	EC	179.39 mm

ROCKWELL 72NLS32-80

IDLI GROUP 85117

SLIP YOKE WITH 16 SPLINES
SEE FIGURE 41

5.844 in	IE	148.43 mm
6.219 in	OE	157.96 mm
2.000 in	SB	50.80 mm
7.063 in	EC	179.39 mm

ROCKWELL 72NLS32-90

IDLI GROUP 85118

SLIP YOKE WITH 16 SPLINES
slight variation from above group
SEE FIGURE 41

5.844 in	IE	148.43 mm
6.219 in	OE	157.96 mm
2.000 in	SB	50.80 mm
7.063 in	EC	179.39 mm

ROCKWELL 72NLS32-86

IDLI GROUP 85119

SLIP YOKE WITH 16 SPLINES
SEE FIGURE 41

5.844 in	IE	148.43 mm
6.219 in	OE	157.96 mm
2.000 in	SB	50.80 mm
8.250 in	EC	209.55 mm

ROCKWELL 72NLS32-74

IDLI GROUP 85120

SLIP YOKE WITH 16 SPLINES
slight variation from above group
SEE FIGURE 41

5.844 in	IE	148.43 mm
6.219 in	OE	157.96 mm
2.000 in	SB	50.80 mm
8.250 in	EC	209.55 mm

ROCKWELL 72NLS32-72

IDLI GROUP 85121

SLIP YOKE WITH 10 SPLINES
SEE FIGURE 42

5.844 in	IE	148.43 mm
6.219 in	OE	157.96 mm
2.000 in	SB	50.80 mm
8.563 in	EC	217.49 mm

ROCKWELL 72NLS32-27

IDLI GROUP 85122

SLIP YOKE WITH 16 SPLINES
SEE FIGURE 41

5.844 in	IE	148.43 mm
6.219 in	OE	157.96 mm
2.000 in	SB	50.80 mm
8.813 in	EC	223.84 mm

ROCKWELL 72NLS32-124

IDLI GROUP 85123

SLIP YOKE WITH 10 SPLINES
SEE FIGURE 41

5.844 in	IE	148.43 mm
6.219 in	OE	157.96 mm
2.000 in	SB	50.80 mm
9.125 in	EC	231.78 mm

BORG-WARNR 16974J
BORG-WARNR 6-7075
BORG-WARNR 6A7075
ROCKWELL 72NLS32-2

IDLI GROUP 85124

SLIP YOKE WITH 10 SPLINES
slight variation from above group
SEE FIGURE 41

5.844 in	IE	148.43 mm
6.219 in	OE	157.96 mm
2.000 in	SB	50.80 mm
9.125 in	EC	231.78 mm

BORG-WARNR 6-7080
BORG-WARNR 6-7096
BORG-WARNR 6A7080
BORG-WARNR 6A7096
ROCKWELL 72NLS32-34

IDLI GROUP 85125

SLIP YOKE WITH 10 SPLINES
SEE FIGURE 41

5.844 in	IE	148.43 mm
6.219 in	OE	157.96 mm
2.000 in	SB	50.80 mm
9.563 in	EC	242.89 mm

ROCKWELL 72NLS32-71

IDLI GROUP 85126

SLIP YOKE WITH 10 SPLINES
SEE FIGURE 42

5.844 in	IE	148.43 mm
6.219 in	OE	157.96 mm
2.000 in	SB	50.80 mm
9.625 in	EC	244.48 mm

BORG-WARNR 19403J
BORG-WARNR 67039
BORG-WARNR 67061
BORG-WARNR 6A7039
BORG-WARNR 6A7061
FORD B8QH4841D
NEAPCO 12-351
NEAPCO 20-428
NEAPCO 64-1251
NEAPCO 65-2028
ROCKWELL 72NLS32-15
SPICER 2-82-82-2706X

IDLI GROUP 85127

SLIP YOKE
WITH 15 SPLINES IN 16 SPLINE SPACES
SEE FIGURE 41

5.844 in	IE	148.43 mm
6.219 in	OE	157.96 mm
2.000 in	SB	50.80 mm
9.750 in	EC	247.65 mm

ROCKWELL 72NLS32-21

IDLI GROUP 85128

SLIP YOKE WITH 16 SPLINES
SEE FIGURE 41

5.844 in	IE	148.43 mm
6.219 in	OE	157.96 mm
2.000 in	SB	50.80 mm
9.750 in	EC	247.65 mm

ROCKWELL 72NLS32-51

IDLI GROUP 85129

SLIP YOKE WITH 16 SPLINES
SEE FIGURE 41

5.844 in	IE	148.43 mm
6.219 in	OE	157.96 mm
2.000 in	SB	50.80 mm
9.813 in	EC	249.24 mm

ROCKWELL 72NLS32-19

IDLI GROUP 85130

SLIP YOKE WITH 16 SPLINES
SEE FIGURE 41

5.844 in	IE	148.43 mm
6.219 in	OE	157.96 mm
2.000 in	SB	50.80 mm
10.063 in	EC	255.59 mm

ROCKWELL 72NLS32-6

IDLI GROUP 85131

SLIP YOKE WITH 16 SPLINES
SEE FIGURE 41

5.844 in	IE	148.43 mm
6.219 in	OE	157.96 mm
2.000 in	SB	50.80 mm
10.875 in	EC	276.23 mm

BORG-WARNR 6-7093
BORG-WARNR 6A7093
ROCKWELL 72NLS32-20

IDLI GROUP 85132

SLIP YOKE WITH 16 SPLINES
slight variation from above group
SEE FIGURE 41

5.844 in	IE	148.43 mm
6.219 in	OE	157.96 mm
2.000 in	SB	50.80 mm
10.875 in	EC	276.23 mm

BORG-WARNR 6-7099
BORG-WARNR 6A7099
ROCKWELL 72NLS32-35

IDLI GROUP 85133

SLIP YOKE
WITH 15 SPLINES IN 16 SPLINE SPACES
SEE FIGURE 41

5.844 in	IE	148.43 mm
6.219 in	OE	157.96 mm
2.000 in	SB	50.80 mm
12.000 in	EC	304.80 mm

ROCKWELL 72NLS32-22

IDLI GROUP 85134

SLIP YOKE WITH 16 SPLINES
SEE FIGURE 41

5.844 in	IE	148.43 mm
6.219 in	OE	157.96 mm
2.000 in	SB	50.80 mm
12.188 in	EC	309.56 mm

ROCKWELL 72NLS32-129

IDLI GROUP 85135

SLIP YOKE WITH 16 SPLINES
SEE FIGURE 41

5.844 in	IE	148.43 mm
6.219 in	OE	157.96 mm
2.500 in	SB	63.50 mm
2.875 in	EC	73.03 mm

ROCKWELL 72NLS40-3

IDLI GROUP 85136

SLIP YOKE WITH 16 SPLINES
slight variation from above group
SEE FIGURE 41

5.844 in	IE	148.43 mm
6.219 in	OE	157.96 mm
2.500 in	SB	63.50 mm
2.875 in	EC	73.03 mm

ROCKWELL 72NLS40-8

BD = BEARING DIAMETER (outside)　　**BW** = BEARING WIDTH　　**CB** = CROSS LENGTH WITH BEARINGS　　**CC** = CENTER TO CENTER　　**CD** = CROSS DIAMETER
CL = CROSS LENGTH WITHOUT BEARINGS　　**EC** = END TO CENTER or FACE TO FACE　　**EE** = END TO END　　**EL** = EFFECTIVE LENGTH
HD = HUB DIAMETER (or insert)　　**IE** = INSIDE OF EARS (or recess)　　**OD** = OUTSIDE DIAMETER　　**OE** = OUTSIDE OF EARS　　**SB** = SPLINE OR BORE SIZE

IDLI GROUP 85137

SLIP YOKE WITH 16 SPLINES
SEE FIGURE 41

5.844 in	IE	148.43 mm
6.219 in	OE	157.96 mm
2.500 in	SB	63.50 mm
3.125 in	EC	79.38 mm

ROCKWELL 72NLS40-6

IDLI GROUP 85138

SLIP YOKE WITH 16 SPLINES
SEE FIGURE 41

5.844 in	IE	148.43 mm
6.219 in	OE	157.96 mm
2.500 in	SB	63.50 mm
3.500 in	EC	88.90 mm

ROCKWELL 72NLS40-12

IDLI GROUP 85139

SLIP YOKE WITH 16 SPLINES
SEE FIGURE 42

5.844 in	IE	148.43 mm
6.219 in	OE	157.96 mm
2.500 in	SB	63.50 mm
3.625 in	EC	92.08 mm

ROCKWELL 72NLS40-14

IDLI GROUP 85140

SLIP YOKE WITH 16 SPLINES
SEE FIGURE 41

5.844 in	IE	148.43 mm
6.219 in	OE	157.96 mm
2.500 in	SB	63.50 mm
3.875 in	EC	98.43 mm

ROCKWELL 72NLS40-5

IDLI GROUP 85141

SLIP YOKE WITH 16 SPLINES
slight variation from above group
SEE FIGURE 41

5.844 in	IE	148.43 mm
6.219 in	OE	157.96 mm
2.500 in	SB	63.50 mm
3.875 in	EC	98.43 mm

BORG-WARNR 6-7095
BORG-WARNR 6A7095
ROCKWELL 72NLS40-7

IDLI GROUP 85142

SLIP YOKE WITH 16 SPLINES
SEE FIGURE 41

5.844 in	IE	148.43 mm
6.219 in	OE	157.96 mm
2.500 in	SB	63.50 mm
4.250 in	EC	107.95 mm

ROCKWELL 72NLS40-16

IDLI GROUP 85143

SLIP YOKE WITH 16 SPLINES
SEE FIGURE 42

5.844 in	IE	148.43 mm
6.219 in	OE	157.96 mm
2.500 in	SB	63.50 mm
7.063 in	EC	179.39 mm

ROCKWELL 72NLS40-10

IDLI GROUP 85144

SLIP YOKE WITH 10 SPLINES
SEE FIGURE 41

5.844 in	IE	148.43 mm
6.219 in	OE	157.96 mm
2.500 in	SB	63.50 mm
8.063 in	EC	204.79 mm

ROCKWELL 72NLS40-1

IDLI GROUP 85145

SLIP YOKE WITH 22 INVOLUTED SPLINES
IN 23 INVOLUTED SPLINE SPACES
SEE FIGURE 41

5.844 in	IE	148.43 mm
6.219 in	OE	157.96 mm
3.000 in	SB	76.20 mm
11.500 in	EC	292.10 mm

ROCKWELL 72NLS48-4

IDLI GROUP 85146

SLIP YOKE WITH 16 SPLINES
SEE FIGURE 41

6.500 in	IE	165.10 mm
6.875 in	OE	174.63 mm
2.500 in	SB	63.50 mm
3.250 in	EC	82.55 mm

ROCKWELL 85WBLS40-2

IDLI GROUP 85147

SLIP YOKE WITH 16 SPLINES
SEE FIGURE 42

6.500 in	IE	165.10 mm
6.875 in	OE	174.63 mm
2.500 in	SB	63.50 mm
8.500 in	EC	215.90 mm

ROCKWELL 85WBLS40-3

IDLI GROUP 85148

SLIP YOKE WITH 10 SPLINES
SEE FIGURE 41

6.500 in	IE	165.10 mm
6.875 in	OE	174.63 mm
2.500 in	SB	63.50 mm
9.250 in	EC	234.95 mm

ROCKWELL 85WBLS40-1

IDLI GROUP 85149

SLIP YOKE WITH 10 SPLINES
SEE FIGURE 41

6.500 in	IE	165.10 mm
6.875 in	OE	174.63 mm
3.000 in	SB	76.20 mm
9.750 in	EC	247.65 mm

ROCKWELL 85WBLS48-1

IDLI GROUP 85150

SLIP YOKE WITH 10 SPLINES
SEE FIGURE 42

6.500 in	IE	165.10 mm
6.875 in	OE	174.63 mm
3.000 in	SB	76.20 mm
10.750 in	EC	273.05 mm

ROCKWELL 85WBLS48

IDLI GROUP 85151

SLIP YOKE
WITH 22 SPLINES IN 23 SPLINE SPACES
SEE FIGURE 41

6.500 in	IE	165.10 mm
6.875 in	OE	174.63 mm
3.000 in	SB	76.20 mm
11.625 in	EC	295.28 mm

ROCKWELL 85WBLS48-4

IDLI GROUP 85152

SLIP YOKE WITH 16 SPLINES
SEE FIGURE 41

8.125 in	IE	206.38 mm
8.500 in	OE	215.90 mm
2.500 in	SB	63.50 mm
3.188 in	EC	80.96 mm

ROCKWELL 82NLS40-12

IDLI GROUP 85153

SLIP YOKE WITH 16 SPLINES
SEE FIGURE 42

8.125 in	IE	206.38 mm
8.500 in	OE	215.90 mm
2.500 in	SB	63.50 mm
3.938 in	EC	100.01 mm

ROCKWELL 82NLS40-30

IDLI GROUP 85154

SLIP YOKE WITH 16 SPLINES
SEE FIGURE 42

8.125 in	IE	206.38 mm
8.500 in	OE	215.90 mm
2.500 in	SB	63.50 mm
4.500 in	EC	114.30 mm

ROCKWELL 82NLS40-17

IDLI GROUP 85155

SLIP YOKE WITH 16 SPLINES
slight variation from above group
SEE FIGURE 42

8.125 in	IE	206.38 mm
8.500 in	OE	215.90 mm
2.500 in	SB	63.50 mm
4.500 in	EC	114.30 mm

ROCKWELL 82NLS40-31

IDLI GROUP 85156

SLIP YOKE WITH 16 SPLINES
SEE FIGURE 42

8.125 in	IE	206.38 mm
8.500 in	OE	215.90 mm
2.500 in	SB	63.50 mm
5.688 in	EC	144.46 mm

ROCKWELL 82NLS40-54

IDLI GROUP 85157

SLIP YOKE WITH 10 SPLINES
SEE FIGURE 42

8.125 in	IE	206.38 mm
8.500 in	OE	215.90 mm
2.500 in	SB	63.50 mm
6.000 in	EC	152.40 mm

BORG-WARNR 6-8003
BORG-WARNR 6A8003
ROCKWELL 82NLS40-32

IDLI GROUP 85158

SLIP YOKE WITH 10 SPLINES
SEE FIGURE 41

8.125 in	IE	206.38 mm
8.500 in	OE	215.90 mm
2.500 in	SB	63.50 mm
6.500 in	EC	165.10 mm

BORG-WARNR 4658J
BORG-WARNR 6-8003
BORG-WARNR 6A8003
ROCKWELL 82NLS40-3

IDLI GROUP 85159

SLIP YOKE WITH 10 SPLINES
slight variation from above group
SEE FIGURE 41

8.125 in	IE	206.38 mm
8.500 in	OE	215.90 mm
2.500 in	SB	63.50 mm
6.500 in	EC	165.10 mm

BORG-WARNR 6-8020
BORG-WARNR 6-8026
BORG-WARNR 6A8020
BORG-WARNR 6A8021
BORG-WARNR 6A8026
ROCKWELL 82NLS40

IDLI GROUP 85160

SLIP YOKE WITH 10 SPLINES
SEE FIGURE 41

8.125 in	IE	206.38 mm
8.500 in	OE	215.90 mm
2.500 in	SB	63.50 mm
6.688 in	EC	169.86 mm

ROCKWELL 82NLS40-36

IDLI GROUP 85161

SLIP YOKE WITH 10 SPLINES
SEE FIGURE 41

8.125 in	IE	206.38 mm
8.500 in	OE	215.90 mm
2.500 in	SB	63.50 mm
6.813 in	EC	173.04 mm

ROCKWELL 82NLS40-52

IDLI GROUP 85162

SLIP YOKE WITH 10 SPLINES
SEE FIGURE 41

8.125 in	IE	206.38 mm
8.500 in	OE	215.90 mm
2.500 in	SB	63.50 mm
7.000 in	EC	177.80 mm

ROCKWELL 82NLS40-25

ENGINEERING CATALOGS MUST BE CONSULTED FOR DETAILS NOT INCLUDED
IN THIS GUIDE. SPECIFIC DESIGNS, MATERIAL CONTENT, TOLERANCES
LUBE FITTINGS AND OTHER DIMENSIONS ARE INTENTIONALLY OMITTED HERE.

CAUTION: BE SURE TO REFER TO ENGINEERING CATALOGS FOR SPECIAL APPLICATIONS THAT REQUIRE SPECIFIC MATERIAL CONTENT, TOLERANCES, ETC. SEE FOOTNOTE.

IDLI GROUP 85163

SLIP YOKE WITH 10 SPLINES
SEE FIGURE 41

8.125 in	IE	206.38 mm
8.500 in	OE	215.90 mm
2.500 in	SB	63.50 mm
8.000 in	EC	203.20 mm

ROCKWELL 82NLS40-21

IDLI GROUP 85164

SLIP YOKE WITH 10 SPLINES
slight variation from above group
SEE FIGURE 41

8.125 in	IE	206.38 mm
8.500 in	OE	215.90 mm
2.500 in	SB	63.50 mm
8.000 in	EC	203.20 mm

BORG-WARNR 6-8021
BORG-WARNR 6-8023
BORG-WARNR 6A8021
BORG-WARNR 6A8023
ROCKWELL 82NLS40-1

IDLI GROUP 85165

SLIP YOKE WITH 10 SPLINES
slight variation from above group
SEE FIGURE 41

8.125 in	IE	206.38 mm
8.500 in	OE	215.90 mm
2.500 in	SB	63.50 mm
8.000 in	EC	203.20 mm

BORG-WARNR 3578J
BORG-WARNR 6201J
BORG-WARNR 6-8002
BORG-WARNR 6-8015
BORG-WARNR 6-8036
BORG-WARNR 6A8002
BORG-WARNR 6A8015
BORG-WARNR 6A8036
FORD B8QH4841C
IHC 38025HX
ROCKWELL 82NLS40-2

IDLI GROUP 85166

SLIP YOKE WITH 10 SPLINES
SEE FIGURE 41

8.125 in	IE	206.38 mm
8.500 in	OE	215.90 mm
2.500 in	SB	63.50 mm
8.500 in	EC	215.90 mm

ROCKWELL 82NLS40-44

IDLI GROUP 85167

SLIP YOKE WITH 10 SPLINES
SEE FIGURE 41

8.125 in	IE	206.38 mm
8.500 in	OE	215.90 mm
2.500 in	SB	63.50 mm
9.438 in	EC	239.71 mm

BORG-WARNR 6-8041
BORG-WARNR 6A8041

ROCKWELL 82NLS40-16

IDLI GROUP 85168

SLIP YOKE WITH 10 SPLINES
SEE FIGURE 42

8.125 in	IE	206.38 mm
8.500 in	OE	215.90 mm
2.500 in	SB	63.50 mm
9.750 in	EC	247.65 mm

BORG-WARNR 6-8012
BORG-WARNR 6A8012
ROCKWELL 82NLS40-6

IDLI GROUP 85169

SLIP YOKE WITH 10 SPLINES
slight variation from above group
SEE FIGURE 41

8.125 in	IE	206.38 mm
8.500 in	OE	215.90 mm
2.500 in	SB	63.50 mm
9.750 in	EC	247.65 mm

BORG-WARNR 6-8045
BORG-WARNR 6-8054
BORG-WARNR 6A8045
BORG-WARNR 6A8054
ROCKWELL 82NLS40-26

IDLI GROUP 85170

SLIP YOKE WITH 10 SPLINES
slight variation from above group
SEE FIGURE 41

8.125 in	IE	206.38 mm
8.500 in	OE	215.90 mm
2.500 in	SB	63.50 mm
9.750 in	EC	247.65 mm

ROCKWELL 82NLS40-35

IDLI GROUP 85171

SLIP YOKE WITH 10 SPLINES
SEE FIGURE 41

8.250 in	IE	209.55 mm
8.625 in	OE	219.08 mm
3.000 in	SB	76.20 mm
6.250 in	EC	158.75 mm

ROCKWELL 92NLS48-2

IDLI GROUP 85172

SLIP YOKE WITH 10 SPLINES
slight variation from above group
SEE FIGURE 41

8.250 in	IE	209.55 mm
8.625 in	OE	219.08 mm
3.000 in	SB	76.20 mm
6.250 in	EC	158.75 mm

ROCKWELL 92NLS48-2DC1

IDLI GROUP 85173

SLIP YOKE WITH 10 SPLINES
SEE FIGURE 41

8.250 in	IE	209.55 mm
8.625 in	OE	219.08 mm
3.000 in	SB	76.20 mm
6.750 in	EC	171.45 mm

BORG-WARNR 6-9008
BORG-WARNR 6A9008
ROCKWELL 92NLS48-3

IDLI GROUP 85174

SLIP YOKE WITH 10 SPLINES
SEE FIGURE 41

8.250 in	IE	209.55 mm
8.625 in	OE	219.08 mm
3.000 in	SB	76.20 mm
7.000 in	EC	177.80 mm

ROCKWELL 92NLS48-7

IDLI GROUP 85175

SLIP YOKE WITH 10 SPLINES
SEE FIGURE 41

8.250 in	IE	209.55 mm
8.625 in	OE	219.08 mm
3.000 in	SB	76.20 mm
8.125 in	EC	206.38 mm

ROCKWELL 92NLS48-10

IDLI GROUP 85176

SLIP YOKE WITH 10 SPLINES
SEE FIGURE 41

8.250 in	IE	209.55 mm
8.625 in	OE	219.08 mm
3.000 in	SB	76.20 mm
8.750 in	EC	222.25 mm

ROCKWELL 92NLS48-9

IDLI GROUP 85177

SLIP YOKE WITH 10 SPLINES
SEE FIGURE 42

8.250 in	IE	209.55 mm
8.625 in	OE	219.08 mm
3.000 in	SB	76.20 mm
9.125 in	EC	231.78 mm

ROCKWELL 92NLS48-17

IDLI GROUP 85178

SLIP YOKE WITH 10 SPLINES
slight variation from above group
SEE FIGURE 42

8.250 in	IE	209.55 mm
8.625 in	OE	219.08 mm
3.000 in	SB	76.20 mm
9.125 in	EC	231.78 mm

ROCKWELL 92NLS48-17DC

IDLI GROUP 85179

SLIP YOKE WITH 10 SPLINES
SEE FIGURE 41

8.250 in	IE	209.55 mm
8.625 in	OE	219.08 mm
3.000 in	SB	76.20 mm
9.625 in	EC	244.48 mm

BORG-WARNR 6-9005
BORG-WARNR 6A9005
ROCKWELL 92NLS48-1

IDLI GROUP 85180

SLIP YOKE WITH 10 SPLINES
slight variation from above group
SEE FIGURE 41

8.250 in	IE	209.55 mm
8.625 in	OE	219.08 mm
3.000 in	SB	76.20 mm
9.625 in	EC	244.48 mm

BORG-WARNR 6-9001
BORG-WARNR 6-9013
BORG-WARNR 6A9001
BORG-WARNR 6A9013
ROCKWELL 92NLS48

IDLI GROUP 85181

SLIP YOKE WITH 10 SPLINES
slight variation from above group
SEE FIGURE 41

8.250 in	IE	209.55 mm
8.625 in	OE	219.08 mm
3.000 in	SB	76.20 mm
9.625 in	EC	244.48 mm

BORG-WARNR 6-9005
BORG-WARNR 6A9005
ROCKWELL 92NLS48-1DC1

IDLI GROUP 85182

SLIP YOKE WITH 10 SPLINES
slight variation from above group
SEE FIGURE 41

8.250 in	IE	209.55 mm
8.625 in	OE	219.08 mm
3.000 in	SB	76.20 mm
9.625 in	EC	244.48 mm

BORG-WARNR 6-9001
BORG-WARNR 6-9013
BORG-WARNR 6A9001
BORG-WARNR 6A9013
ROCKWELL 92NLS48DC

IDLI GROUP 85183

SLIP YOKE WITH 10 SPLINES
slight variation from above group
SEE FIGURE 42

8.250 in	IE	209.55 mm
8.625 in	OE	219.08 mm
3.000 in	SB	76.20 mm
9.625 in	EC	244.48 mm

ROCKWELL 92NLS48DC1

DD = BEARING DIAMETER (outside)　　**BW** = BEARING WIDTH　　**CB** = CROSS LENGTH WITH BEARINGS　　**CC** = CENTER TO CENTER　　**CD** = CROSS DIAMETER
CL = CROSS LENGTH WITHOUT BEARINGS　　**EC** = END TO CENTER or FACE TO FACE　　**EE** = END TO END　　**EL** = EFFECTIVE LENGTH
HD = HUB DIAMETER (or insert)　　**IE** = INSIDE OF EARS (or recess)　　**OD** = OUTSIDE DIAMETER　　**OE** = OUTSIDE OF EARS　　**SB** = SPLINE OR BORE SIZE

IDLI GUIDE COPYRIGHT ® INTERCHANGE, INC. ST. LOUIS PARK, MN. 55416 USA

IDLI GROUP 85184

SLIP YOKE WITH 16 SPLINES
SEE FIGURE 41

8.250 in	IE	209.55 mm
8.625 in	OE	219.08 mm
3.000 in	SB	76.20 mm
9.625 in	EC	244.48 mm

BORG-WARNR 6-9033
BORG-WARNR 6A9033
ROCKWELL 92NLS48-47DC

IDLI GROUP 85185

SLIP YOKE WITH 16 SPLINES
slight variation from above group
SEE FIGURE 41

8.250 in	IE	209.55 mm
8.625 in	OE	219.08 mm
3.000 in	SB	76.20 mm
9.625 in	EC	244.48 mm

BORG-WARNR 6-9034
BORG-WARNR 6A9034
ROCKWELL 92NLS48-49DC

IDLI GROUP 85186

SLIP YOKE
WITH 22 SPLINES IN 23 SPLINE SPACES
SEE FIGURE 41

8.250 in	IE	209.55 mm
8.625 in	OE	219.08 mm
3.000 in	SB	76.20 mm
9.625 in	EC	244.48 mm

ROCKWELL 92NLS48-52DC

IDLI GROUP 85187

SLIP YOKE
WITH 22 SPLINES IN 23 SPLINE SPACES

8.250 in	IE	209.55 mm
8.625 in	OE	219.08 mm
3.000 in	SB	76.20 mm
15.500 in	EC	393.70 mm

ROCKWELL 92NLS48-12

IDLI GROUP 85188

SLIP YOKE
WITH 22 SPLINES IN 23 SPLINE SPACES

8.250 in	IE	209.55 mm
8.625 in	OE	219.08 mm
3.000 in	SB	76.20 mm
15.500 in	EC	393.70 mm

ROCKWELL 92NLS48-12DC

IDLI GROUP 85189

SLIP YOKE
WITH 22 SPLINES IN 23 SPLINE SPACES

8.250 in	IE	209.55 mm
8.625 in	OE	219.08 mm
3.000 in	SB	76.20 mm
16.188 in	EC	411.16 mm

ROCKWELL 92NLS48-37DC

ENGINEERING CATALOGS MUST BE CONSULTED FOR DETAILS NOT INCLUDED
IN THIS GUIDE. SPECIFIC DESIGNS, MATERIAL CONTENT, TOLERANCES
LUBE FITTINGS AND OTHER DIMENSIONS ARE INTENTIONALLY OMITTED HERE.

FOR YOUR CONVENIENCE

NOTES ON OTHER INFORMATION

We urge you to send us any data that
you feel should be included with the
future editions of the I.D.L.I. Guide.

IDLI GROUP 85301

SLIP YOKE WITH 10 SPLINES
SEE FIGURE 39A

2.906 in	OE	73.82 mm
1.125 in	BD	28.58 mm
1.250 in	SB	31.75 mm
4.688 in	EC	119.06 mm

ROCKWELL L14SLS20-4DC

IDLI GROUP 85302

SLIP YOKE WITH 14 INVOLUTED SPLINES
SEE FIGURE 39A

3.484 in	OE	88.50 mm
1.250 in	BD	31.75 mm
1.250 in	SB	31.75 mm
3.250 in	EC	82.55 mm

ROCKWELL 3NYS20-12A

IDLI GROUP 85303

SLIP YOKE WITH 21 INVOLUTED SPLINES
SEE FIGURE 39A

3.484 in	OE	88.50 mm
1.250 in	BD	31.75 mm
1.375 in	SB	34.93 mm
3.938 in	EC	100.01 mm

ROCKWELL L16SYS22-194

IDLI GROUP 85304

SLIP YOKE WITH 14 INVOLUTED SPLINES
SEE FIGURE 39A

3.484 in	OE	88.50 mm
1.250 in	BD	31.75 mm
1.500 in	SB	38.10 mm
3.500 in	EC	88.90 mm

ROCKWELL 3NLS24-8DC

IDLI GROUP 85305

SLIP YOKE WITH 10 SPLINES
SEE FIGURE 39A

3.484 in	OE	88.50 mm
1.250 in	BD	31.75 mm
1.500 in	SB	38.10 mm
4.188 in	EC	106.36 mm

ROCKWELL 3NLS24-6DC

IDLI GROUP 85306

SLIP YOKE WITH 10 SPLINES
SEE FIGURE 39A

3.484 in	OE	88.50 mm
1.250 in	BD	31.75 mm
1.500 in	SB	38.10 mm
6.375 in	EC	161.93 mm

ROCKWELL 3NLS24-7A

IDLI GROUP 85307

SLIP YOKE WITH 10 SPLINES
SEE FIGURE 39A

3.484 in	OE	88.50 mm
1.250 in	BD	31.75 mm
1.500 in	SB	38.10 mm
6.500 in	EC	165.10 mm

ROCKWELL 3NLS24DC

IDLI GROUP 85308

SLIP YOKE WITH 10 SPLINES
SEE FIGURE 39A

3.484 in	OE	88.50 mm
1.250 in	BD	31.75 mm
1.500 in	SB	38.10 mm
7.750 in	EC	196.85 mm

ROCKWELL 3NLS24-10DC

IDLI GROUP 85309

SLIP YOKE WITH 10 SPLINES
slight variation from above group
SEE FIGURE 39A

3.484 in	OE	88.50 mm
1.250 in	BD	31.75 mm
1.500 in	SB	38.10 mm
7.750 in	EC	196.85 mm

ROCKWELL L16SLS24-6DC

IDLI GROUP 85310

SLIP YOKE WITH 13 INVOLUTED SPLINES
SEE FIGURE 39A

3.484 in	OE	88.50 mm
1.250 in	BD	31.75 mm
1.750 in	SB	44.45 mm
3.250 in	EC	82.55 mm

ROCKWELL L16SYS28-25

IDLI GROUP 85311

SLIP YOKE WITH 10 SPLINES
SEE FIGURE 39A

3.484 in	OE	88.50 mm
1.250 in	BD	31.75 mm
1.750 in	SB	44.45 mm
6.375 in	EC	161.93 mm

ROCKWELL L16SLS28A

IDLI GROUP 85312

SLIP YOKE WITH 16 SPLINES
SEE FIGURE 40

5.313 in	OE	134.94 mm
1.875 in	BD	47.63 mm
2.000 in	SB	50.80 mm
7.813 in	EC	198.44 mm

ROCKWELL 16N3-108X
SPICER 5-3-108KX
SPICER 5-3-108X

IDLI GROUP 85313

SLIP YOKE WITH 16 SPLINES
SEE FIGURE 40

5.313 in	OE	134.94 mm
1.875 in	BD	47.63 mm
2.000 in	SB	50.80 mm
9.313 in	EC	236.54 mm

ROCKWELL 16N3-288X
SPICER 5-3-288KX
SPICER 5-3-288X

IDLI GROUP 85314

SLIP YOKE WITH 16 SPLINES
SEE FIGURE 40

5.313 in	OE	134.94 mm
1.875 in	BD	47.63 mm
2.000 in	SB	50.80 mm
9.813 in	EC	249.24 mm

ROCKWELL 16N3-128X
SPICER 5-3-128K-X
SPICER 5-3-128X

IDLI GROUP 85315

SLIP YOKE WITH 16 SPLINES
SEE FIGURE 40

5.313 in	OE	134.94 mm
1.875 in	BD	47.63 mm
2.000 in	SB	50.80 mm
10.313 in	EC	261.94 mm

ROCKWELL 16N3-1411X
SPICER 5-3-1411KX
SPICER 5-3-1411X

IDLI GROUP 85316

SLIP YOKE WITH 16 SPLINES
SEE FIGURE 40

5.313 in	OE	134.94 mm
1.875 in	BD	47.63 mm
2.000 in	SB	50.80 mm
10.813 in	EC	274.64 mm

ROCKWELL 16N3-2261X
SPICER 5-3-2261KX

IDLI GROUP 85317

SLIP YOKE WITH 10 SPLINES
SEE FIGURE 40

6.094 in	OE	154.78 mm
1.938 in	BD	49.21 mm
2.500 in	SB	63.50 mm
1.688 in	EC	42.86 mm
9.000 in	HD	228.60 mm

ROCKWELL 17N3-1521X
SPICER 6-3-1521KX
SPICER 6-3-1521X
SPICER 6-3-2171
SPICER 6-3-2171KX
SPICER 6-3-398X
SPICER 6-3-408X

IDLI GROUP 85318

SLIP YOKE WITH 16 SPLINES
SEE FIGURE 40

6.094 in	OE	154.78 mm
1.938 in	BD	49.21 mm
2.500 in	SB	63.50 mm
6.719 in	EC	170.66 mm

ROCKWELL 17N3-2761X
SPICER 6-3-2761KX

IDLI GROUP 85319

SLIP YOKE WITH 16 SPLINES
SEE FIGURE 40

6.094 in	OE	154.78 mm
1.938 in	BD	49.21 mm
2.500 in	SB	63.50 mm
7.969 in	EC	202.41 mm

ROCKWELL 17NLS40-50

IDLI GROUP 85320

SLIP YOKE WITH 16 SPLINES
SEE FIGURE 40

6.094 in	OE	154.78 mm
1.938 in	BD	49.21 mm
2.500 in	SB	63.50 mm
8.469 in	EC	215.11 mm

ROCKWELL 17NLS40-85

IDLI GROUP 85321

SLIP YOKE WITH 10 SPLINES
SEE FIGURE 40

6.094 in	OE	154.78 mm
1.938 in	BD	49.21 mm
2.500 in	SB	63.50 mm
8.844 in	EC	224.63 mm

ROCKWELL 17N3-1461X
SPICER 6-3-1461KX
SPICER 6-3-1461X
SPICER 6-3-158X

IDLI GROUP 85322

SLIP YOKE WITH 16 SPLINES
SEE FIGURE 40

6.094 in	OE	154.78 mm
1.938 in	BD	49.21 mm
2.500 in	SB	63.50 mm
8.844 in	EC	224.63 mm

ROCKWELL 17N3-2241
ROCKWELL 17N3-2241X
SPICER 6-3-2241
SPICER 6-3-2241KX
SPICER 6-3-2241X
SPICER 6-3-2631KX

IDLI GROUP 85323

SLIP YOKE WITH 16 SPLINES
SEE FIGURE 40

6.094 in	OE	154.78 mm
1.938 in	BD	49.21 mm
2.500 in	SB	63.50 mm
8.938 in	EC	227.01 mm

ROCKWELL 17NLS40-40

IDLI GROUP 85324

SLIP YOKE WITH 16 SPLINES
SEE FIGURE 40

6.094 in	OE	154.78 mm
1.938 in	BD	49.21 mm
2.500 in	SB	63.50 mm
9.219 in	EC	234.16 mm

ROCKWELL 17NLS40-47

BD = BEARING DIAMETER (outside)   **BW** = BEARING WIDTH   **CB** = CROSS LENGTH WITH BEARINGS   **CC** = CENTER TO CENTER   **CD** = CROSS DIAMETER
CL = CROSS LENGTH WITHOUT BEARINGS   **EC** = END TO CENTER or FACE TO FACE   **EE** = END TO END   **EL** = EFFECTIVE LENGTH
HD = HUB DIAMETER (or insert)   **IE** = INSIDE OF EARS (or recess)   **OD** = OUTSIDE DIAMETER   **OE** = OUTSIDE OF EARS   **SB** = SPLINE OR BORE SIZE

IDLI GUIDE　　　COPYRIGHT ® INTERCHANGE, INC.　　　ST. LOUIS PARK, MN. 55416 USA

CAUTION: BE SURE TO REFER TO ENGINEERING CATALOGS FOR SPECIAL APPLICATIONS THAT REQUIRE SPECIFIC MATERIAL CONTENT, TOLERANCES, ETC. SEE FOOTNOTE.

IDLI GROUP 85325

SLIP YOKE WITH 16 SPLINES
SEE FIGURE 40

6.094 in	OE	154.78 mm
1.938 in	BD	49.21 mm
2.500 in	SB	63.50 mm
9.250 in	EC	234.95 mm

ROCKWELL 17NLS40-49

IDLI GROUP 85326

SLIP YOKE WITH 10 SPLINES
SEE FIGURE 40

6.094 in	OE	154.78 mm
1.938 in	BD	49.21 mm
2.500 in	SB	63.50 mm
9.281 in	EC	235.75 mm

ROCKWELL 17N3-1541X
SPICER 6-3-1531
SPICER 6-3-1531KX
SPICER 6-3-1541KX
SPICER 6-3-1541X
SPICER 6-3-268X

IDLI GROUP 85327

SLIP YOKE WITH 16 SPLINES
SEE FIGURE 40

6.094 in	OE	154.78 mm
1.938 in	BD	49.21 mm
2.500 in	SB	63.50 mm
9.594 in	EC	243.68 mm

ROCKWELL 17N3-2341X
SPICER 6-3-2331KX
SPICER 6-3-2331X
SPICER 6-3-2341KX
SPICER 6-3-2341X

IDLI GROUP 85328

SLIP YOKE WITH 16 SPLINES
slight variation from above group
SEE FIGURE 40

6.094 in	OE	154.78 mm
1.938 in	BD	49.21 mm
2.500 in	SB	63.50 mm
9.594 in	EC	243.68 mm

ROCKWELL 17N3-2671X
SPICER 6-3-2671KX
SPICER 6-3-2671X

IDLI GROUP 85329

SLIP YOKE WITH 16 SPLINES
SEE FIGURE 40

6.094 in	OE	154.78 mm
1.938 in	BD	49.21 mm
2.500 in	SB	63.50 mm
9.813 in	EC	249.24 mm

ROCKWELL 17NLS40-90

IDLI GROUP 85330

SLIP YOKE WITH 16 SPLINES
SEE FIGURE 40

6.094 in	OE	154.78 mm
1.938 in	BD	49.21 mm
2.500 in	SB	63.50 mm
10.406 in	EC	264.32 mm

ROCKWELL 17N3-3021X
SPICER 6-3-3021KX

IDLI GROUP 85331

SLIP YOKE WITH 16 SPLINES
SEE FIGURE 40

6.094 in	OE	154.78 mm
1.938 in	BD	49.21 mm
2.500 in	SB	63.50 mm
11.000 in	EC	279.40 mm

ROCKWELL 17NLS40-91

IDLI GROUP 85332

SLIP YOKE WITH 16 SPLINES
SEE FIGURE 40

6.094 in	OE	154.78 mm
1.938 in	BD	49.21 mm
2.500 in	SB	63.50 mm
11.406 in	EC	289.72 mm

ROCKWELL 17NLS40-82

IDLI GROUP 85333

SLIP YOKE WITH 10 SPLINES
slight variation from above group
SEE FIGURE 40

6.094 in	OE	154.78 mm
1.938 in	BD	49.21 mm
2.500 in	SB	63.50 mm
11.688 in	EC	296.86 mm

ROCKWELL 17N3-1501X
SPICER 6-3-1491
SPICER 6-3-1491KX
SPICER 6-3-1501KX
SPICER 6-3-1501X
SPICER 6-3-368X

IDLI GROUP 85334

SLIP YOKE WITH 16 SPLINES
SEE FIGURE 40

6.094 in	OE	154.78 mm
1.938 in	BD	49.21 mm
2.500 in	SB	63.50 mm
11.688 in	EC	296.86 mm

ROCKWELL 17N3-2261X
SPICER 6-3-2251KX
SPICER 6-3-2251X
SPICER 6-3-2261KX
SPICER 6-3-2261X

IDLI GROUP 85335

SLIP YOKE WITH 16 SPLINES
SEE FIGURE 40

6.094 in	OE	154.78 mm
1.938 in	BD	49.21 mm
2.500 in	SB	63.50 mm
11.688 in	EC	296.86 mm

ROCKWELL 17N3-2651X
SPICER 6-3-2651KX

IDLI GROUP 85336

SLIP YOKE WITH 16 SPLINES
SEE FIGURE 40

6.094 in	OE	154.78 mm
1.938 in	BD	49.21 mm
2.500 in	SB	63.50 mm
11.719 in	EC	297.66 mm

ROCKWELL 17NLS40-86
SPICER 6-3-2571KX

IDLI GROUP 85337

SLIP YOKE WITH 16 SPLINES
SEE FIGURE 40

6.094 in	OE	154.78 mm
1.938 in	BD	49.21 mm
2.500 in	SB	63.50 mm
11.969 in	EC	304.01 mm

ROCKWELL 17NLS40-56
SPICER 6-3-2591KX

IDLI GROUP 85338

SLIP YOKE WITH 16 SPLINES
SEE FIGURE 40

6.094 in	OE	154.78 mm
1.938 in	BD	49.21 mm
2.500 in	SB	63.50 mm
12.719 in	EC	323.06 mm

ROCKWELL 17NLS40-72

IDLI GROUP 85339

SLIP YOKE WITH 16 SPLINES
slight variation from above group
SEE FIGURE 40

6.094 in	OE	154.78 mm
1.938 in	BD	49.21 mm
2.500 in	SB	63.50 mm
12.719 in	EC	323.06 mm

ROCKWELL 17N3-2401X
SPICER 6-3-2401KX
SPICER 6-3-2401X
SPICER 6-3-2551KX

IDLI GROUP 85340

END YOKE WITH 16 SPLINES
SEE FIGURE 40

7.000 in	OE	177.80 mm
1.938 in	BD	49.21 mm
2.500 in	SB	63.50 mm
11.156 in	EC	283.37 mm

ROCKWELL 176N3-21X
SPICER 6.3-3-21KX

IDLI GROUP 85341

END YOKE WITH 16 SPLINES
SEE FIGURE 40

7.547 in	OE	191.69 mm
1.938 in	BD	49.21 mm
3.000 in	SB	76.20 mm
7.063 in	EC	179.39 mm

ROCKWELL 18NYS48-4

IDLI GROUP 85342

END YOKE WITH 16 SPLINES
SEE FIGURE 40

7.547 in	OE	191.69 mm
1.938 in	BD	49.21 mm
3.000 in	SB	76.20 mm
8.938 in	EC	227.01 mm

ROCKWELL 18N3-1431X
SPICER 6.5-3-1431KX
SPICER 6.5-3-1431X

IDLI GROUP 85343

END YOKE WITH 16 SPLINES
SEE FIGURE 40

7.547 in	OE	191.69 mm
1.938 in	BD	49.21 mm
3.000 in	SB	76.20 mm
10.250 in	EC	260.35 mm

ROCKWELL 18N3-1351X
SPICER 6.5-3-1351KX
SPICER 6.5-3-1351X

IDLI GROUP 85344

END YOKE WITH 16 SPLINES
SEE FIGURE 40

7.547 in	OE	191.69 mm
1.938 in	BD	49.21 mm
3.000 in	SB	76.20 mm
11.219 in	EC	284.96 mm

ROCKWELL 18NLS48-10

IDLI GROUP 85345

END YOKE WITH 16 SPLINES
SEE FIGURE 40

7.547 in	OE	191.69 mm
1.938 in	BD	49.21 mm
3.000 in	SB	76.20 mm
11.875 in	EC	301.63 mm

ROCKWELL 18N3-1371X
SPICER 6.5-3-1371KX
SPICER 6.5-3-1371X

SPICER 6-3-2651X

ENGINEERING CATALOGS MUST BE CONSULTED FOR DETAILS NOT INCLUDED
IN THIS GUIDE. SPECIFIC DESIGNS, MATERIAL CONTENT, TOLERANCES
LUBE FITTINGS AND OTHER DIMENSIONS ARE INTENTIONALLY OMITTED HERE.

IDLI GUIDE   COPYRIGHT © INTERCHANGE, INC.   ST. LOUIS PARK, MN. 55416 USA

IDLI GROUP 85501

SOLID WELD YOKE
SEE FIGURE 45

1.438 in	IE	36.51 mm
.969 in	BD	24.61 mm
1.250 in	EC	31.75 mm
1.316 in	HD	33.43 mm

ROCKWELL L6NY21-1

IDLI GROUP 85502

SOLID WELD YOKE
SEE FIGURE 45

1.438 in	IE	36.51 mm
.969 in	BD	24.61 mm
1.875 in	EC	47.63 mm
1.016 in	HD	25.81 mm

ROCKWELL L6NY16-1

IDLI GROUP 85503

SOLID WELD YOKE
SEE FIGURE 45

1.438 in	IE	36.51 mm
.969 in	BD	24.61 mm
1.875 in	EC	47.63 mm
1.266 in	HD	32.16 mm

ROCKWELL L6NY20

IDLI GROUP 85504

HOLLOW WELD YOKE
SEE FIGURE 45

1.438 in	IE	36.51 mm
.969 in	BD	24.61 mm
1.875 in	EC	47.63 mm
1.316 in	HD	33.43 mm

ROCKWELL L6NY21

IDLI GROUP 85505

SOLID WELD YOKE
SEE FIGURE 45

1.438 in	IE	36.51 mm
.969 in	BD	24.61 mm
1.875 in	EC	47.63 mm
1.570 in	HD	39.88 mm

ROCKWELL L6NY25

IDLI GROUP 85506

SOLID WELD YOKE
SEE FIGURE 45

1.438 in	IE	36.51 mm
.969 in	BD	24.61 mm
1.875 in	EC	47.63 mm
1.628 in	HD	41.35 mm

ROCKWELL L6NY26

IDLI GROUP 85507

HOLLOW WELD YOKE
SEE FIGURE 45

1.438 in	IE	36.51 mm
.969 in	BD	24.61 mm
2.813 in	EC	71.44 mm
1.465 in	HD	37.21 mm

ROCKWELL L6NY24-6

IDLI GROUP 85508

HOLLOW WELD YOKE
SEE FIGURE 45

1.438 in	IE	36.51 mm
.969 in	BD	24.61 mm
2.875 in	EC	73.03 mm
1.316 in	HD	33.43 mm

ROCKWELL L6NY21-8

IDLI GROUP 85509

SOLID WELD YOKE
SEE FIGURE 45

1.813 in	IE	46.04 mm
1.063 in	BD	26.99 mm
2.000 in	EC	50.80 mm
1.270 in	HD	32.26 mm

ROCKWELL L12NY20

IDLI GROUP 85510

SOLID WELD YOKE
SEE FIGURE 45

1.813 in	IE	46.04 mm
1.063 in	BD	26.99 mm
2.000 in	EC	50.80 mm
1.880 in	HD	47.75 mm

ROCKWELL L12NY30-3

IDLI GROUP 85511

HOLLOW WELD YOKE
SEE FIGURE 45

1.813 in	IE	46.04 mm
1.063 in	BD	26.99 mm
2.000 in	EC	50.80 mm
1.880 in	HD	47.75 mm

ROCKWELL L12NY30-8

IDLI GROUP 85512

SOLID WELD YOKE
SEE FIGURE 45

1.813 in	IE	46.04 mm
1.063 in	BD	26.99 mm
2.375 in	EC	60.33 mm
1.844 in	HD	46.84 mm

ROCKWELL L12NY30-26

IDLI GROUP 85513

SOLID WELD YOKE
SEE FIGURE 45

1.813 in	IE	46.04 mm
1.063 in	BD	26.99 mm
2.375 in	EC	60.33 mm
1.880 in	HD	47.75 mm

ROCKWELL L12NY30-5

IDLI GROUP 85514

HOLLOW WELD YOKE
SEE FIGURE 45

1.813 in	IE	46.04 mm
1.063 in	BD	26.99 mm
3.125 in	EC	79.38 mm
1.844 in	HD	46.84 mm

NEAPCO 111B8
NEAPCO 12-1308
ROCKWELL L12NY30-23
SPICER 1-28-277

TRW 21108
WESCO 12NWY1834
WESCO 850-044

IDLI GROUP 85515

HOLLOW WELD YOKE
SEE FIGURE 45

1.813 in	IE	46.04 mm
1.063 in	BD	26.99 mm
3.125 in	EC	79.38 mm
1.880 in	HD	47.75 mm

ROCKWELL L12NY30-14

IDLI GROUP 85516

SOLID WELD YOKE
SEE FIGURE 45

2.797 in	IE	71.04 mm
1.313 in	BD	33.34 mm
4.313 in	EC	109.54 mm
2.240 in	HD	56.90 mm

G & G MFG 181-4400
NEAPCO 44-2200
NEAPCO N44NWY2200
PRECISION 1455
ROCKWELL 44RY36-5
SPICER 3.4-26-11
SPICER 3-4-26-17
WESCO 44N2100
WESCO 44NW4-1200
WESCO 481218

IDLI GROUP 85517

HOLLOW WELD YOKE
SEE FIGURE 48

3.125 in	IE	79.38 mm
1.188 in	BD	30.16 mm
2.000 in	EC	50.80 mm
2.604 in	HD	66.14 mm

ROCKWELL 141NY42

IDLI GROUP 85518

HOLLOW WELD YOKE
SEE FIGURE 48

3.125 in	IE	79.38 mm
1.188 in	BD	30.16 mm
2.000 in	EC	50.80 mm
2.844 in	HD	72.24 mm

CLEVE.STL. O55-3-3-14
ROCKWELL 141NY45-1

IDLI GROUP 85519

HOLLOW WELD YOKE
SEE FIGURE 48

3.125 in	IE	79.38 mm
1.188 in	BD	30.16 mm
2.000 in	EC	50.80 mm
3.344 in	HD	84.94 mm

ROCKWELL 141NY54

IDLI GROUP 85520

HOLLOW WELD YOKE
SEE FIGURE 48

3.125 in	IE	79.38 mm
1.188 in	BD	30.16 mm
2.000 in	EC	50.80 mm
3.392 in	HD	86.16 mm

ROCKWELL 141NY54-2

IDLI GROUP 85521

HOLLOW WELD YOKE
SEE FIGURE 48

3.125 in	IE	79.38 mm
1.188 in	BD	30.16 mm
2.125 in	EC	53.98 mm
2.820 in	HD	71.63 mm

ROCKWELL 141NY45

IDLI GROUP 85522

HOLLOW WELD YOKE
SEE FIGURE 48

3.125 in	IE	79.38 mm
1.188 in	BD	30.16 mm
2.125 in	EC	53.98 mm
2.844 in	HD	72.24 mm

CLEVE.STL. O55-3-3-14S
ROCKWELL 141NY45-3
SPICER 3-28-97

IDLI GROUP 85523

HOLLOW WELD YOKE
SEE FIGURE 48

4.188 in	IE	106.36 mm
1.375 in	BD	34.93 mm
2.031 in	EC	51.60 mm
3.344 in	HD	84.94 mm

ROCKWELL 148NY54-2

IDLI GROUP 85524

HOLLOW WELD YOKE
SEE FIGURE 48

4.188 in	IE	106.36 mm
1.375 in	BD	34.93 mm
2.031 in	EC	51.60 mm
3.846 in	HD	97.69 mm

CLEVE.STL. D76-3-4-14
ROCKWELL 148NY62-1
SPICER 3-28-507

IDLI GROUP 85525

HOLLOW WELD YOKE
SEE FIGURE 48

4.188 in	IE	106.36 mm
1.375 in	BD	34.93 mm
2.313 in	EC	58.74 mm
3.344 in	HD	84.94 mm

ROCKWELL 148NY54

BD=BEARING DIAMETER (outside)   **BW**=BEARING WIDTH   **CB**=CROSS LENGTH WITH BEARINGS   **CC**=CENTER TO CENTER   **CD**=CROSS DIAMETER
CL=CROSS LENGTH WITHOUT BEARINGS   **EC**=END TO CENTER or FACE TO FACE   **EE**=END TO END   **EL**=EFFECTIVE LENGTH
HD=HUB DIAMETER (or insert)   **IE**=INSIDE OF EARS (or recess)   **OD**=OUTSIDE DIAMETER   **OE**=OUTSIDE OF EARS   **SB**=SPLINE OR BORE SIZE

IDLI GUIDE   COPYRIGHT © INTERCHANGE, INC.   ST. LOUIS PARK, MN. 55416 USA

CAUTION: BE SURE TO REFER TO ENGINEERING CATALOGS FOR SPECIAL APPLICATIONS THAT REQUIRE SPECIFIC MATERIAL CONTENT, TOLERANCES, ETC. SEE FOOTNOTE.

IDLI GROUP 85526

HOLLOW WELD YOKE
SEE FIGURE 48

4.969 in	IE	126.21 mm
1.375 in	BD	34.93 mm
2.188 in	EC	55.56 mm
3.320 in	HD	84.33 mm

ROCKWELL 155NY53-4

IDLI GROUP 85527

HOLLOW WELD YOKE
SEE FIGURE 48

4.969 in	IE	126.21 mm
1.375 in	BD	34.93 mm
2.188 in	EC	55.56 mm
3.846 in	HD	97.69 mm

CLEVE.STL. U76-3-4-14
ROCKWELL 155NY62
SPICER 4-28-277

IDLI GROUP 85528

HOLLOW WELD YOKE
SEE FIGURE 48

4.969 in	IE	126.21 mm
1.375 in	BD	34.93 mm
2.688 in	EC	68.26 mm
3.320 in	HD	84.33 mm

ROCKWELL 155NY53-5

IDLI GROUP 85529

SOLID WELD YOKE
SEE FIGURE 47

5.313 in	IE	134.94 mm
5.625 in	OE	142.88 mm
1.063 in	EC	26.99 mm
5.844 in	HD	148.44 mm

ROCKWELL 58WBY94

IDLI GROUP 85530

HOLLOW WELD YOKE
SEE FIGURE 47

5.313 in	IE	134.94 mm
5.625 in	OE	142.88 mm
1.813 in	EC	46.04 mm
3.320 in	HD	84.33 mm

ROCKWELL 58WBY53-8

IDLI GROUP 85531

HOLLOW WELD YOKE
slight variation from above group
SEE FIGURE 47

5.313 in	IE	134.94 mm
5.625 in	OE	142.88 mm
1.813 in	EC	46.04 mm
3.320 in	HD	84.33 mm

ROCKWELL 58WBY53-1

IDLI GROUP 85532

HOLLOW WELD YOKE
slight variation from above group
SEE FIGURE 47

5.313 in	IE	134.94 mm
5.625 in	OE	142.88 mm
1.813 in	EC	46.04 mm
3.320 in	HD	84.33 mm

ROCKWELL 58WBY53-4

IDLI GROUP 85533

HOLLOW WELD YOKE
SEE FIGURE 47

5.313 in	IE	134.94 mm
5.625 in	OE	142.88 mm
1.813 in	EC	46.04 mm
3.694 in	HD	93.83 mm

ROCKWELL 58WBY59

IDLI GROUP 85534

HOLLOW WELD YOKE
SEE FIGURE 47

5.313 in	IE	134.94 mm
5.625 in	OE	142.88 mm
1.813 in	EC	46.04 mm
3.844 in	HD	97.64 mm

ROCKWELL 58WBY62-3

IDLI GROUP 85535

HOLLOW WELD YOKE
slight variation from above group
SEE FIGURE 47

5.313 in	IE	134.94 mm
5.625 in	OE	142.88 mm
1.813 in	EC	46.04 mm
3.844 in	HD	97.64 mm

ROCKWELL 58WBY62

IDLI GROUP 85536

HOLLOW WELD YOKE
SEE FIGURE 47

5.313 in	IE	134.94 mm
5.625 in	OE	142.88 mm
2.063 in	EC	52.39 mm
3.320 in	HD	84.33 mm

ROCKWELL 58WBY53-9

IDLI GROUP 85537

HOLLOW WELD YOKE
SEE FIGURE 47

5.313 in	IE	134.94 mm
5.625 in	OE	142.88 mm
2.125 in	EC	53.98 mm
3.320 in	HD	84.33 mm

ROCKWELL 58WBY53-6

IDLI GROUP 85538

SOLID WELD YOKE
SEE FIGURE 47

5.313 in	IE	134.94 mm
5.625 in	OE	142.88 mm
2.438 in	EC	61.91 mm
3.320 in	HD	84.33 mm

ROCKWELL 58WBY53-5

IDLI GROUP 85539

HOLLOW WELD YOKE
SEE FIGURE 47

5.313 in	IE	134.94 mm
5.625 in	OE	142.88 mm
2.750 in	EC	69.85 mm
4.320 in	HD	109.73 mm

ROCKWELL 58WBY69-1

IDLI GROUP 85540

HOLLOW WELD YOKE
SEE FIGURE 47

5.531 in	IE	140.50 mm
5.844 in	OE	148.43 mm
1.594 in	EC	40.48 mm
2.820 in	HD	71.63 mm

ROCKWELL 62NY45-3

IDLI GROUP 85541

HOLLOW WELD YOKE
SEE FIGURE 47

5.531 in	IE	140.50 mm
5.844 in	OE	148.43 mm
1.750 in	EC	44.45 mm
3.270 in	HD	83.06 mm

ROCKWELL 62NY52-7

IDLI GROUP 85542

HOLLOW WELD YOKE
SEE FIGURE 47

5.531 in	IE	140.50 mm
5.844 in	OE	148.43 mm
1.875 in	EC	47.63 mm
2.635 in	HD	66.93 mm

BORG-WARNR 16542J
ROCKWELL 62NY42

IDLI GROUP 85543

HOLLOW WELD YOKE
SEE FIGURE 47

5.531 in	IE	140.50 mm
5.844 in	OE	148.43 mm
1.875 in	EC	47.63 mm
2.770 in	HD	70.36 mm

BORG-WARNR 18160J
BORG-WARNR 4227J
BORG-WARNR 5140J
BORG-WARNR 6805J
BORG-WARNR 9668J
ROCKWELL 62NY44

IDLI GROUP 85544

HOLLOW WELD YOKE
slight variation from above group
SEE FIGURE 47

5.531 in	IE	140.50 mm
5.844 in	OE	148.43 mm
1.875 in	EC	47.63 mm
2.770 in	HD	70.36 mm

ROCKWELL 62NY44-1

IDLI GROUP 85545

HOLLOW WELD YOKE
SEE FIGURE 47

5.531 in	IE	140.50 mm
5.844 in	OE	148.43 mm
1.875 in	EC	47.63 mm
2.820 in	HD	71.63 mm

BORG-WARNR 3669J
BORG-WARNR 4441J
IHC 38780H
ROCKWELL 62NY45-2

IDLI GROUP 85546

HOLLOW WELD YOKE
slight variation from above group
SEE FIGURE 47

5.531 in	IE	140.50 mm
5.844 in	OE	148.43 mm
1.875 in	EC	47.63 mm
2.820 in	HD	71.63 mm

BORG-WARNR 13301J
BORG-WARNR 15238J
BORG-WARNR 9668J
ROCKWELL 62NY45

IDLI GROUP 85547

HOLLOW WELD YOKE
SEE FIGURE 47

5.531 in	IE	140.50 mm
5.844 in	OE	148.43 mm
1.875 in	EC	47.63 mm
3.270 in	HD	83.06 mm

ROCKWELL 62NY52-3

IDLI GROUP 85548

HOLLOW WELD YOKE
slight variation from above group
SEE FIGURE 47

5.531 in	IE	140.50 mm
5.844 in	OE	148.43 mm
1.875 in	EC	47.63 mm
3.270 in	HD	83.06 mm

BORG-WARNR 13780J
ROCKWELL 62NY52

IDLI GROUP 85549

HOLLOW WELD YOKE
SEE FIGURE 47

5.531 in	IE	140.50 mm
5.844 in	OE	148.43 mm
1.875 in	EC	47.63 mm
3.320 in	HD	84.33 mm

BORG-WARNR 15365J
BORG-WARNR 9138J
ROCKWELL 62NY53-1

IDLI GROUP 85550

HOLLOW WELD YOKE
SEE FIGURE 47

5.531 in	IE	140.50 mm
5.844 in	OE	148.43 mm
1.875 in	EC	47.63 mm
3.792 in	HD	96.32 mm

BORG-WARNR 16206J
ROCKWELL 62NY61-1

ENGINEERING CATALOGS MUST BE CONSULTED FOR DETAILS NOT INCLUDED IN THIS GUIDE. SPECIFIC DESIGNS, MATERIAL CONTENT, TOLERANCES LUBE FITTINGS AND OTHER DIMENSIONS ARE INTENTIONALLY OMITTED HERE.

IDLI GROUP 85551

HOLLOW WELD YOKE
SEE FIGURE 47

5.531 in	IE	140.50 mm
5.844 in	OE	148.43 mm
1.875 in	EC	47.63 mm
3.844 in	HD	97.64 mm

BORG-WARNR 9135J
ROCKWELL 62NY61

IDLI GROUP 85552

HOLLOW WELD YOKE
SEE FIGURE 47

5.531 in	IE	140.50 mm
5.844 in	OE	148.43 mm
1.906 in	EC	48.42 mm
2.635 in	HD	66.93 mm

BORG-WARNR 20319J
ROCKWELL 62NY42-1

IDLI GROUP 85553

HOLLOW WELD YOKE
SEE FIGURE 47

5.531 in	IE	140.50 mm
5.844 in	OE	148.43 mm
1.969 in	EC	50.01 mm
2.820 in	HD	71.63 mm

BORG-WARNR 17441J
ROCKWELL 62NY45-4

IDLI GROUP 85554

HOLLOW WELD YOKE
slight variation from above group
SEE FIGURE 47

5.531 in	IE	140.50 mm
5.844 in	OE	148.43 mm
2.000 in	EC	50.80 mm
3.270 in	HD	83.06 mm

ROCKWELL 62NY52-5

IDLI GROUP 85555

HOLLOW WELD YOKE
SEE FIGURE 47

5.531 in	IE	140.50 mm
5.844 in	OE	148.43 mm
2.250 in	EC	57.15 mm
3.020 in	HD	76.71 mm

ROCKWELL 62NY48

IDLI GROUP 85556

HOLLOW WELD YOKE
SEE FIGURE 47

5.531 in	IE	140.50 mm
5.844 in	OE	148.43 mm
2.250 in	EC	57.15 mm
3.320 in	HD	84.33 mm

BORG-WARNR 16008J
ROCKWELL 62NY53

IDLI GROUP 85557

SOLID WELD YOKE
SEE FIGURE 47

5.844 in	IE	148.43 mm
6.219 in	OE	157.96 mm
1.125 in	EC	28.58 mm
3.320 in	HD	84.33 mm

BORG-WARNR 23850J
ROCKWELL 72NY53-7

IDLI GROUP 85558

SOLID WELD YOKE
SEE FIGURE 47

5.844 in	IE	148.43 mm
6.219 in	OE	157.96 mm
2.125 in	EC	53.98 mm
2.510 in	HD	63.75 mm

ROCKWELL 72NY40

IDLI GROUP 85559

SOLID WELD YOKE
SEE FIGURE 47

5.844 in	IE	148.43 mm
6.219 in	OE	157.96 mm
2.125 in	EC	53.98 mm
2.697 in	HD	68.50 mm

BORG-WARNR 3747J
BORG-WARNR 4165J
BORG-WARNR 7428J
ROCKWELL 72NY43

IDLI GROUP 85560

HOLLOW WELD YOKE
SEE FIGURE 47

5.844 in	IE	148.43 mm
6.219 in	OE	157.96 mm
2.125 in	EC	53.98 mm
3.135 in	HD	79.63 mm

BORG-WARNR 22168J
ROCKWELL 72NY50-3

IDLI GROUP 85561

SOLID WELD YOKE
SEE FIGURE 47

5.844 in	IE	148.43 mm
6.219 in	OE	157.96 mm
2.125 in	EC	53.98 mm
3.270 in	HD	83.06 mm

BORG-WARNR 10012J
ROCKWELL 72NY52-2

IDLI GROUP 85562

HOLLOW WELD YOKE
slight variation from above group
SEE FIGURE 47

5.844 in	IE	148.43 mm
6.219 in	OE	157.96 mm
2.125 in	EC	53.98 mm
3.270 in	HD	83.06 mm

BORG-WARNR 13922J
ROCKWELL 72NY52-1

IDLI GROUP 85563

HOLLOW WELD YOKE
slight variation from above group
SEE FIGURE 47

5.844 in	IE	148.43 mm
6.219 in	OE	157.96 mm
2.125 in	EC	53.98 mm
3.270 in	HD	83.06 mm

ROCKWELL 72NY52-4

IDLI GROUP 85564

HOLLOW WELD YOKE
SEE FIGURE 47

5.844 in	IE	148.43 mm
6.219 in	OE	157.96 mm
2.125 in	EC	53.98 mm
3.320 in	HD	84.33 mm

BORG-WARNR 8655J
ROCKWELL 72NY53

IDLI GROUP 85565

HOLLOW WELD YOKE
slight variation from above group
SEE FIGURE 47

5.844 in	IE	148.43 mm
6.219 in	OE	157.96 mm
2.125 in	EC	53.98 mm
3.320 in	HD	84.33 mm

ROCKWELL 72NY53-1

IDLI GROUP 85566

HOLLOW WELD YOKE
SEE FIGURE 47

5.844 in	IE	148.43 mm
6.219 in	OE	157.96 mm
2.125 in	EC	53.98 mm
3.742 in	HD	95.05 mm

ROCKWELL 72NY60-3

IDLI GROUP 85567

HOLLOW WELD YOKE
SEE FIGURE 47

5.844 in	IE	148.43 mm
6.219 in	OE	157.96 mm
2.125 in	EC	53.98 mm
3.792 in	HD	96.32 mm

BORG-WARNR 13434J
BORG-WARNR 17898J
ROCKWELL 72NY61

IDLI GROUP 85568

HOLLOW WELD YOKE
slight variation from above group
SEE FIGURE 47

5.844 in	IE	148.43 mm
6.219 in	OE	157.96 mm
2.125 in	EC	53.98 mm
3.792 in	HD	96.32 mm

ROCKWELL 72NY61-4

IDLI GROUP 85569

HOLLOW WELD YOKE
SEE FIGURE 47

5.844 in	IE	148.43 mm
6.219 in	OE	157.96 mm
2.125 in	EC	53.98 mm
3.844 in	HD	97.64 mm

BORG-WARNR 10804J
ROCKWELL 72NY61-1

IDLI GROUP 85570

HOLLOW WELD YOKE
slight variation from above group
SEE FIGURE 47

5.844 in	IE	148.43 mm
6.219 in	OE	157.96 mm
2.125 in	EC	53.98 mm
3.844 in	HD	97.64 mm

ROCKWELL 72NY61-10

IDLI GROUP 85571

HOLLOW WELD YOKE
SEE FIGURE 47

5.844 in	IE	148.43 mm
6.219 in	OE	157.96 mm
2.250 in	EC	57.15 mm
3.270 in	HD	83.06 mm

ROCKWELL 72NY52-23

IDLI GROUP 85572

SOLID WELD YOKE
SEE FIGURE 47

5.844 in	IE	148.43 mm
6.219 in	OE	157.96 mm
2.375 in	EC	60.33 mm
2.697 in	HD	68.50 mm

ROCKWELL 72NY43-1

IDLI GROUP 85573

SOLID WELD YOKE
SEE FIGURE 47

5.844 in	IE	148.43 mm
6.219 in	OE	157.96 mm
2.375 in	EC	60.33 mm
2.697 in	HD	68.50 mm

ROCKWELL 72NY43-3

IDLI GROUP 85574

HOLLOW WELD YOKE
SEE FIGURE 47

5.844 in	IE	148.43 mm
6.219 in	OE	157.96 mm
2.375 in	EC	60.33 mm
2.697 in	HD	68.50 mm

ROCKWELL 72NY43-5

IDLI GROUP 85575

HOLLOW WELD YOKE
SEE FIGURE 47

5.844 in	IE	148.43 mm
6.219 in	OE	157.96 mm
2.563 in	EC	65.09 mm
3.320 in	HD	84.33 mm

ROCKWELL 72NY53-3

BD=BEARING DIAMETER (outside)   **BW**=BEARING WIDTH   **CB**=CROSS LENGTH WITH BEARINGS   **CC**=CENTER TO CENTER   **CD**=CROSS DIAMETER
CL=CROSS LENGTH WITHOUT BEARINGS   **EC**=END TO CENTER or FACE TO FACE   **EE**=END TO END   **EL**=EFFECTIVE LENGTH
HD=HUB DIAMETER (or insert)   **IE**=INSIDE OF EARS (or recess)   **OD**=OUTSIDE DIAMETER   **OE**=OUTSIDE OF EARS   **SB**=SPLINE OR BORE SIZE

CAUTION: BE SURE TO REFER TO ENGINEERING CATALOGS FOR SPECIAL APPLICATIONS THAT REQUIRE SPECIFIC MATERIAL CONTENT, TOLERANCES, ETC. SEE FOOTNOTE.

IDLI GROUP 85576

HOLLOW WELD YOKE
SEE FIGURE 47

5.844 in	IE	148.43 mm
6.219 in	OE	157.96 mm
2.688 in	EC	68.26 mm
2.770 in	HD	70.36 mm

ROCKWELL 72NY44-2

IDLI GROUP 85577

HOLLOW WELD YOKE
slight variation from above group
SEE FIGURE 47

5.844 in	IE	148.43 mm
6.219 in	OE	157.96 mm
2.688 in	EC	68.26 mm
2.770 in	HD	70.36 mm

ROCKWELL 72NY44

IDLI GROUP 85578

SOLID WELD YOKE
SEE FIGURE 47

5.844 in	IE	148.43 mm
6.219 in	OE	157.96 mm
2.875 in	EC	73.03 mm
2.697 in	HD	68.50 mm

ROCKWELL 72NY43-2

IDLI GROUP 85579

HOLLOW WELD YOKE
SEE FIGURE 47

5.844 in	IE	148.43 mm
6.219 in	OE	157.96 mm
3.000 in	EC	76.20 mm
3.198 in	HD	81.23 mm

ROCKWELL 72NY51-1

IDLI GROUP 85580

HOLLOW WELD YOKE
SEE FIGURE 47

5.844 in	IE	148.43 mm
6.219 in	OE	157.96 mm
3.000 in	EC	76.20 mm
3.242 in	HD	82.35 mm

ROCKWELL 72NY52-9

IDLI GROUP 85581

SOLID WELD YOKE
SEE FIGURE 47

5.844 in	IE	148.43 mm
6.219 in	OE	157.96 mm
3.000 in	EC	76.20 mm
3.270 in	HD	83.06 mm

BORG-WARNR 14089J
ROCKWELL 72NY52

IDLI GROUP 85582

HOLLOW WELD YOKE
slight variation from above group
SEE FIGURE 47

5.844 in	IE	148.43 mm
6.219 in	OE	157.96 mm
3.000 in	EC	76.20 mm
3.270 in	HD	83.06 mm

BORG-WARNR 22902J
ROCKWELL 72NY52-5

IDLI GROUP 85583

HOLLOW WELD YOKE
SEE FIGURE 47

5.844 in	IE	148.43 mm
6.219 in	OE	157.96 mm
3.000 in	EC	76.20 mm
3.320 in	HD	84.33 mm

BORG-WARNR 16787J
ROCKWELL 72NY53-2

IDLI GROUP 85584

HOLLOW WELD YOKE
SEE FIGURE 47

5.844 in	IE	148.43 mm
6.219 in	OE	157.96 mm
3.000 in	EC	76.20 mm
3.792 in	HD	96.32 mm

ROCKWELL 72NY61-3

IDLI GROUP 85585

HOLLOW WELD YOKE
SEE FIGURE 47

6.500 in	IE	165.10 mm
6.875 in	OE	174.63 mm
1.688 in	EC	42.86 mm
3.270 in	HD	83.06 mm

ROCKWELL 85WBY52-2

IDLI GROUP 85586

SOLID WELD YOKE
SEE FIGURE 47

6.500 in	IE	165.10 mm
6.875 in	OE	174.63 mm
1.875 in	EC	47.63 mm
3.320 in	HD	84.33 mm

ROCKWELL 85WBY53

IDLI GROUP 85587

SOLID WELD YOKE
slight variation from above group
SEE FIGURE 47

6.500 in	IE	165.10 mm
6.875 in	OE	174.63 mm
1.875 in	EC	47.63 mm
3.320 in	HD	84.33 mm

ROCKWELL 85WBY53-1

IDLI GROUP 85588

SOLID WELD YOKE
SEE FIGURE 47

6.500 in	IE	165.10 mm
6.875 in	OE	174.63 mm
1.875 in	EC	47.63 mm
3.792 in	HD	96.32 mm

ROCKWELL 85WBY60-6

IDLI GROUP 85589

SOLID WELD YOKE
SEE FIGURE 47

6.500 in	IE	165.10 mm
6.875 in	OE	174.63 mm
2.500 in	EC	63.50 mm
3.792 in	HD	96.32 mm

ROCKWELL 85WBY60

IDLI GROUP 85590

SOLID WELD YOKE
SEE FIGURE 47

6.500 in	IE	165.10 mm
6.875 in	OE	174.63 mm
2.500 in	EC	63.50 mm
4.135 in	HD	105.03 mm

ROCKWELL 85WBY66

IDLI GROUP 85591

SOLID WELD YOKE
SEE FIGURE 47

6.500 in	IE	165.10 mm
6.875 in	OE	174.63 mm
2.625 in	EC	66.68 mm
3.742 in	HD	95.05 mm

ROCKWELL 85WBY60-3

IDLI GROUP 85592

HOLLOW WELD YOKE
SEE FIGURE 47

8.125 in	IE	206.38 mm
8.500 in	OE	215.90 mm
2.313 in	EC	58.74 mm
3.792 in	HD	96.32 mm

ROCKWELL 82NY60-3

IDLI GROUP 85593

HOLLOW WELD YOKE
SEE FIGURE 47

8.125 in	IE	206.38 mm
8.500 in	OE	215.90 mm
2.375 in	EC	60.33 mm
3.792 in	HD	96.32 mm

ROCKWELL 82NY60-6

IDLI GROUP 85594

HOLLOW WELD YOKE
SEE FIGURE 47

8.125 in	IE	206.38 mm
8.500 in	OE	215.90 mm
2.438 in	EC	61.91 mm
2.635 in	HD	66.93 mm

BORG-WARNR 7623J
ROCKWELL 82NY42

IDLI GROUP 85595

SOLID WELD YOKE
SEE FIGURE 47

8.125 in	IE	206.38 mm
8.500 in	OE	215.90 mm
2.438 in	EC	61.91 mm
3.135 in	HD	79.63 mm

ROCKWELL 82NY50

IDLI GROUP 85596

HOLLOW WELD YOKE
SEE FIGURE 47

8.125 in	IE	206.38 mm
8.500 in	OE	215.90 mm
2.438 in	EC	61.91 mm
3.635 in	HD	92.33 mm

BORG-WARNR 12710J
BORG-WARNR 16752J
BORG-WARNR 18114J
ROCKWELL 82NY58

IDLI GROUP 85597

HOLLOW WELD YOKE
slight variation from above group
SEE FIGURE 47

8.125 in	IE	206.38 mm
8.500 in	OE	215.90 mm
2.438 in	EC	61.91 mm
3.635 in	HD	92.33 mm

ROCKWELL 82NY58-2

IDLI GROUP 85598

HOLLOW WELD YOKE
SEE FIGURE 47

8.125 in	IE	206.38 mm
8.500 in	OE	215.90 mm
2.438 in	EC	61.91 mm
3.792 in	HD	96.32 mm

BORG-WARNR 10473J
BORG-WARNR 10908J
BORG-WARNR 8022J
BORG-WARNR 8402J
ROCKWELL 82NY60-1

IDLI GROUP 85599

HOLLOW WELD YOKE
slight variation from above group
SEE FIGURE 47

8.125 in	IE	206.38 mm
8.500 in	OE	215.90 mm
2.438 in	EC	61.91 mm
3.792 in	HD	96.32 mm

BORG-WARNR 12812J
BORG-WARNR 9728J
ROCKWELL 82NY60

IDLI GROUP 85600

HOLLOW WELD YOKE
SEE FIGURE 47

8.125 in	IE	206.38 mm
8.500 in	OE	215.90 mm
2.438 in	EC	61.91 mm
4.135 in	HD	105.03 mm

ROCKWELL 82NY66-1

ENGINEERING CATALOGS MUST BE CONSULTED FOR DETAILS NOT INCLUDED IN THIS GUIDE. SPECIFIC DESIGNS, MATERIAL CONTENT, TOLERANCES LUBE FITTINGS AND OTHER DIMENSIONS ARE INTENTIONALLY OMITTED HERE.

IDLI GROUP 85601

HOLLOW WELD YOKE
SEE FIGURE 47

8.125 in	IE	206.38 mm
8.500 in	OE	215.90 mm
2.563 in	EC	65.09 mm
3.742 in	HD	95.05 mm

ROCKWELL 82NY60-5

IDLI GROUP 85602

HOLLOW WELD YOKE
SEE FIGURE 47

8.125 in	IE	206.38 mm
8.500 in	OE	215.90 mm
2.875 in	EC	73.03 mm
3.742 in	HD	95.05 mm

ROCKWELL 82NY60-20

IDLI GROUP 85603

HOLLOW WELD YOKE
slight variation from above group
SEE FIGURE 47

8.125 in	IE	206.38 mm
8.500 in	OE	215.90 mm
2.875 in	EC	73.03 mm
3.742 in	HD	95.05 mm

BORG-WARNR 23327J
ROCKWELL 82NY60-16

IDLI GROUP 85604

HOLLOW WELD YOKE
slight variation from above group
SEE FIGURE 47

8.125 in	IE	206.38 mm
8.500 in	OE	215.90 mm
2.875 in	EC	73.03 mm
3.742 in	HD	95.05 mm

ROCKWELL 82NY60-19

IDLI GROUP 85605

SOLID WELD YOKE
SEE FIGURE 47

8.250 in	IE	209.55 mm
8.625 in	OE	219.08 mm
2.500 in	EC	63.50 mm
4.135 in	HD	105.03 mm

BORG-WARNR 11756J
BORG-WARNR 11853J
BORG-WARNR 12714J
BORG-WARNR 1730J
BORG-WARNR 17320J
ROCKWELL 92NY66-5

IDLI GROUP 85606

SOLID WELD YOKE
slight variation from above group
SEE FIGURE 47

8.250 in	IE	209.55 mm
8.625 in	OE	219.08 mm
2.250 in	EC	57.15 mm
4.135 in	HD	105.03 mm

ROCKWELL 92NY66-1

IDLI GROUP 85607

SOLID WELD YOKE
slight variation from above group
SEE FIGURE 47

8.250 in	IE	209.55 mm
8.625 in	OE	219.08 mm
2.250 in	EC	57.15 mm
4.135 in	HD	105.03 mm

ROCKWELL 92NY66-6

IDLI GROUP 85608

SOLID WELD YOKE
slight variation from above group
SEE FIGURE 47

8.250 in	IE	209.55 mm
8.625 in	OE	219.08 mm
2.250 in	EC	57.15 mm
4.135 in	HD	105.03 mm

BORG-WARNR 8012J
ROCKWELL 92NY66

IDLI GROUP 85609

SOLID WELD YOKE
slight variation from above group
SEE FIGURE 47

8.250 in	IE	209.55 mm
8.625 in	OE	219.08 mm
2.250 in	EC	57.15 mm
4.135 in	HD	105.03 mm

ROCKWELL 92NY66-12

IDLI GROUP 85610

SOLID WELD YOKE
SEE FIGURE 47

8.250 in	IE	209.55 mm
8.625 in	OE	219.08 mm
2.250 in	EC	57.15 mm
4.242 in	HD	107.75 mm

ROCKWELL 92NY68-3

IDLI GROUP 85611

SOLID WELD YOKE
SEE FIGURE 47

8.250 in	IE	209.55 mm
8.625 in	OE	219.08 mm
2.438 in	EC	61.91 mm
4.135 in	HD	105.03 mm

ROCKWELL 92NY66-8

IDLI GROUP 85612

SOLID WELD YOKE
slight variation from above group
SEE FIGURE 47

8.250 in	IE	209.55 mm
8.625 in	OE	219.08 mm
2.438 in	EC	61.91 mm
4.135 in	HD	105.03 mm

ROCKWELL 92NY66-10

IDLI GROUP 85613

SOLID WELD YOKE
SEE FIGURE 47

8.250 in	IE	209.55 mm
8.625 in	OE	219.08 mm
2.750 in	EC	69.85 mm
3.152 in	HD	80.06 mm

ROCKWELL 92NY51

IDLI GROUP 85614

SOLID WELD YOKE
slight variation from above group
SEE FIGURE 47

8.250 in	IE	209.55 mm
8.625 in	OE	219.08 mm
2.750 in	EC	69.85 mm
3.152 in	HD	80.06 mm

ROCKWELL 92NY51-2

IDLI GROUP 85615

SOLID WELD YOKE
SEE FIGURE 47

8.250 in	IE	209.55 mm
8.625 in	OE	219.08 mm
2.750 in	EC	69.85 mm
4.002 in	HD	101.65 mm

ROCKWELL 92NY64

IDLI GROUP 85616

SOLID WELD YOKE
SEE FIGURE 47

8.250 in	IE	209.55 mm
8.625 in	OE	219.08 mm
2.750 in	EC	69.85 mm
4.135 in	HD	105.03 mm

ROCKWELL 92NY66-11

IDLI GROUP 85617

SOLID WELD YOKE
SEE FIGURE 47

8.250 in	IE	209.55 mm
8.625 in	OE	219.08 mm
2.750 in	EC	69.85 mm
4.242 in	HD	107.75 mm

ROCKWELL 92NY68-1

IDLI GROUP 85618

SOLID WELD YOKE
SEE FIGURE 47

8.250 in	IE	209.55 mm
8.625 in	OE	219.08 mm
2.813 in	EC	71.44 mm
3.755 in	HD	95.38 mm

ROCKWELL 92NY60-6

IDLI GROUP 85619

SOLID WELD YOKE
SEE FIGURE 47

8.250 in	IE	209.55 mm
8.625 in	OE	219.08 mm
2.813 in	EC	71.44 mm
4.135 in	HD	105.03 mm

ROCKWELL 92NY66-14

IDLI GROUP 85620

SOLID WELD YOKE
slight variation from above group
SEE FIGURE 47

8.250 in	IE	209.55 mm
8.625 in	OE	219.08 mm
2.813 in	EC	71.44 mm
4.135 in	HD	105.03 mm

ROCKWELL 92NY66-13

BD = BEARING DIAMETER (outside)   **BW** = BEARING WIDTH   **CB** = CROSS LENGTH WITH BEARINGS   **CC** = CENTER TO CENTER   **CD** = CROSS DIAMETER
CL = CROSS LENGTH WITHOUT BEARINGS   **EC** = END TO CENTER or FACE TO FACE   **EE** = END TO END   **EL** = EFFECTIVE LENGTH
HD = HUB DIAMETER (or insert)   **IE** = INSIDE OF EARS (or recess)   **OD** = OUTSIDE DIAMETER   **OE** = OUTSIDE OF EARS   **SB** = SPLINE OR BORE SIZE

Other Valuable Interchange Publications

The International Bearing Interchange Guide

New Ninth Edition

The I.B.I. Guide has over 365,000 listings of 25,326 ball and roller bearings available worldwide and dating back to 1918. It includes manufacturers, original equipment users and government part numbers, with THOUSANDS of new listings not included in earlier editions!

I.S.B.N. #916966-17-8

The International Seal Interchange Guide

New Sixth Edition

The I.S.I. Guide has over 150,000 listings of 12,632 different seals and "O" rings. It includes government numbers as well as producers and users throughout the world—over 700 páges of information to save you time and money!

I.S.B.N. #916966-16-X

The International Drive Belt Interchange Guide

New Fourth Edition

The I.D.B.I. Guide has more than 115,000 listings for nearly 5,863 types and sizes of drive and V-belts. Listings include manufacturers, original equipment and government users from all over the world. 760 pages.

I.S.B.N. #916966-15-1

IDLI GROUP 85701

HOLLOW WELD YOKE
SEE FIGURE 45A

2.906 in	OE	73.82 mm
1.125 in	BD	28.58 mm
2.000 in	EC	50.80 mm
1.844 in	HD	46.84 mm

ROCKWELL L14SY29-10

IDLI GROUP 85702

SOLID WELD YOKE
SEE FIGURE 45A

2.906 in	OE	73.82 mm
1.125 in	BD	28.58 mm
2.625 in	EC	66.68 mm
1.844 in	HD	46.84 mm

ROCKWELL L14SY29-1

IDLI GROUP 85703

HOLLOW WELD YOKE
SEE FIGURE 45A

2.906 in	OE	73.82 mm
1.125 in	BD	28.58 mm
2.625 in	EC	66.68 mm
1.969 in	HD	50.01 mm

ROCKWELL L14SY32-6

IDLI GROUP 85704

HOLLOW WELD YOKE
SEE FIGURE 45A

2.906 in	OE	73.82 mm
1.125 in	BD	28.58 mm
3.563 in	EC	90.49 mm
1.844 in	HD	46.84 mm

ROCKWELL L14SY29-18
SPICER 2-28-1127
WESCO 14NWY1834

IDLI GROUP 85705

HOLLOW WELD YOKE
with blind hole
SEE FIGURE 45A

2.906 in	OE	73.82 mm
1.125 in	BD	28.58 mm
3.563 in	EC	90.49 mm
1.844 in	HD	46.84 mm

NEAPCO 19131
NEAPCO 20-9131
PRECISION 1375
ROCKWELL L14SY30-43
WESCO 860062

IDLI GROUP 85706

HOLLOW WELD YOKE
SEE FIGURE 45A

2.906 in	OE	73.82 mm
1.125 in	BD	28.58 mm
3.563 in	EC	90.49 mm
1.888 in	HD	47.96 mm

ROCKWELL L14SY30-39

IDLI GROUP 85707

HOLLOW WELD YOKE
SEE FIGURE 45A

2.906 in	OE	73.82 mm
1.125 in	BD	28.58 mm
3.563 in	EC	90.49 mm
1.969 in	HD	50.01 mm

ROCKWELL L14SY32-17

IDLI GROUP 85708

SOLID WELD YOKE
SEE FIGURE 45A

3.484 in	OE	88.50 mm
1.250 in	BD	31.75 mm
2.625 in	EC	66.68 mm
1.969 in	HD	50.01 mm

ROCKWELL 3NY32-2

IDLI GROUP 85709

SOLID WELD YOKE
SEE FIGURE 45A

3.484 in	OE	88.50 mm
1.250 in	BD	31.75 mm
2.625 in	EC	66.68 mm
2.021 in	HD	51.33 mm

ROCKWELL 3NY32

IDLI GROUP 85710

HOLLOW WELD YOKE
SEE FIGURE 45A

3.484 in	OE	88.50 mm
1.250 in	BD	31.75 mm
2.625 in	EC	66.68 mm
2.021 in	HD	51.33 mm

ROCKWELL 3NY32-9

IDLI GROUP 85711

SOLID WELD YOKE
SEE FIGURE 45A

3.484 in	OE	88.50 mm
1.250 in	BD	31.75 mm
2.625 in	EC	66.68 mm
2.242 in	HD	56.95 mm

ROCKWELL L16SY36-3

IDLI GROUP 85712

SOLID WELD YOKE
SEE FIGURE 45A

3.484 in	OE	88.50 mm
1.250 in	BD	31.75 mm
3.250 in	EC	82.55 mm
1.844 in	HD	46.84 mm

NEAPCO 211F2
NEAPCO 211Z3
NEAPCO 22-1325
NEAPCO 22-1343
PRECISION 1855
ROCKWELL L16SY30-3
SPICER 3-2-667
WESCO 471885

IDLI GROUP 85713

SOLID WELD YOKE
SEE FIGURE 45A

3.484 in	OE	88.50 mm
1.250 in	BD	31.75 mm
3.250 in	EC	82.55 mm
1.969 in	HD	50.01 mm

ROCKWELL L16SY32

IDLI GROUP 85714

SOLID WELD YOKE
SEE FIGURE 45A

3.484 in	OE	88.50 mm
1.250 in	BD	31.75 mm
3.250 in	EC	82.55 mm
2.021 in	HD	51.33 mm

ROCKWELL L16SY32-10

IDLI GROUP 85715

HOLLOW WELD YOKE
SEE FIGURE 45A

3.484 in	OE	88.50 mm
1.250 in	BD	31.75 mm
3.750 in	EC	95.25 mm
1.969 in	HD	50.01 mm

ROCKWELL L16SY32-52

IDLI GROUP 85716

HOLLOW WELD YOKE
SEE FIGURE 45A

3.484 in	OE	88.50 mm
1.250 in	BD	31.75 mm
3.875 in	EC	98.43 mm
1.844 in	HD	46.84 mm

ROCKWELL L16SY30-19

IDLI GROUP 85717

HOLLOW WELD YOKE
SEE FIGURE 45A

3.484 in	OE	88.50 mm
1.250 in	BD	31.75 mm
3.875 in	EC	98.43 mm
1.888 in	HD	47.96 mm

ROCKWELL L16SY30-31

IDLI GROUP 85718

HOLLOW WELD YOKE
SEE FIGURE 45A

3.484 in	OE	88.50 mm
1.250 in	BD	31.75 mm
3.875 in	EC	98.43 mm
1.969 in	HD	50.01 mm

ROCKWELL L16SY32-35
SPICER 3-28-667

IDLI GROUP 85719

HOLLOW WELD YOKE
SEE FIGURE 45A

3.484 in	OE	88.50 mm
1.250 in	BD	31.75 mm
3.875 in	EC	98.43 mm
2.242 in	HD	56.95 mm

ROCKWELL L16SY36-4

IDLI GROUP 85720

HOLLOW WELD YOKE
SEE FIGURE 45A

4.625 in	OE	117.48 mm
1.531 in	BD	38.90 mm
4.750 in	EC	120.65 mm
2.240 in	HD	56.90 mm

G & G MFG 181-5500
ROCKWELL 55NY36-4

IDLI GROUP 85721

HOLLOW WELD YOKE
SEE FIGURE 46

5.313 in	OE	134.94 mm
1.875 in	BD	47.63 mm
3.000 in	EC	76.20 mm
3.198 in	HD	81.23 mm

ROCKWELL 16N28-207
SPICER 5-28-207
SPICER 5-28-407

IDLI GROUP 85722

HOLLOW WELD YOKE
SEE FIGURE 46

5.313 in	OE	134.94 mm
1.875 in	BD	47.63 mm
3.000 in	EC	76.20 mm
3.242 in	HD	82.35 mm

ROCKWELL 16N28-627
SPICER 5-28-627

IDLI GROUP 85723

HOLLOW WELD YOKE
SEE FIGURE 46

5.313 in	OE	134.94 mm
1.875 in	BD	47.63 mm
3.000 in	EC	76.20 mm
3.320 in	HD	84.33 mm

ROCKWELL 16N28-167
SPICER 5-28-167
SPICER 5-28-217

IDLI GROUP 85724

HOLLOW WELD YOKE
SEE FIGURE 46

6.094 in	OE	154.78 mm
1.938 in	BD	49.21 mm
3.031 in	EC	77.00 mm
3.198 in	HD	81.23 mm

ROCKWELL 17N28-137
SPICER 6-28-137
SPICER 6-28-17
SPICER 6-28-197
SPICER 6-28-87

IDLI GROUP 85725

HOLLOW WELD YOKE
SEE FIGURE 46

6.094 in	OE	154.78 mm
1.938 in	BD	49.21 mm
3.031 in	EC	77.00 mm
3.270 in	HD	83.06 mm

ROCKWELL 17NY52

CAUTION: BE SURE TO REFER TO ENGINEERING CATALOGS FOR SPECIAL APPLICATIONS THAT REQUIRE SPECIFIC MATERIAL CONTENT, TOLERANCES, ETC. SEE FOOTNOTE.

IDLI GROUP 85726

HOLLOW WELD YOKE
SEE FIGURE 46

6.094 in	OE	154.78 mm
1.938 in	BD	49.21 mm
3.031 in	EC	77.00 mm
3.635 in	HD	92.33 mm

ROCKWELL 17NY58

IDLI GROUP 85727

HOLLOW WELD YOKE
SEE FIGURE 46

6.094 in	OE	154.78 mm
1.938 in	BD	49.21 mm
3.031 in	EC	77.00 mm
3.742 in	HD	95.05 mm

ROCKWELL 17N28-347
SPICER 6-28-347

IDLI GROUP 85728

HOLLOW WELD YOKE
slight variation from above group
SEE FIGURE 46

6.094 in	OE	154.78 mm
1.938 in	BD	49.21 mm
3.031 in	EC	77.00 mm
3.742 in	HD	95.05 mm

ROCKWELL 17NY60-7

IDLI GROUP 85729

HOLLOW WELD YOKE
SEE FIGURE 46

6.094 in	OE	154.78 mm
1.938 in	BD	49.21 mm
3.031 in	EC	77.00 mm
3.770 in	HD	95.76 mm

ROCKWELL 17N28-277
SPICER 6-28-277

IDLI GROUP 85730

HOLLOW WELD YOKE
SEE FIGURE 46

6.094 in	OE	154.78 mm
1.938 in	BD	49.21 mm
3.031 in	EC	77.00 mm
4.242 in	HD	107.75 mm

ROCKWELL 17NY68

IDLI GROUP 85731

HOLLOW WELD YOKE
SEE FIGURE 46

6.094 in	OE	154.78 mm
1.938 in	BD	49.21 mm
3.063 in	EC	77.79 mm
3.742 in	HD	95.05 mm

ROCKWELL 17NY60-3

IDLI GROUP 85732

SOLID WELD YOKE
SEE FIGURE 46

6.094 in	OE	154.78 mm
1.938 in	BD	49.21 mm
3.750 in	EC	95.25 mm
3.742 in	HD	95.05 mm

ROCKWELL 17NY60-8

IDLI GROUP 85733

SOLID WELD YOKE
SEE FIGURE 46

6.094 in	OE	154.78 mm
1.938 in	BD	49.21 mm
4.688 in	EC	119.06 mm
3.742 in	HD	95.05 mm

ROCKWELL 17N28-367
SPICER 6-28-367

IDLI GROUP 85734

HOLLOW WELD YOKE
SEE FIGURE 46

7.000 in	OE	177.80 mm
1.938 in	BD	49.21 mm
3.031 in	EC	77.00 mm
3.742 in	HD	95.05 mm

ROCKWELL 176N28-17
SPICER 6.3-28-17

IDLI GROUP 85735

HOLLOW WELD YOKE
SEE FIGURE 46

7.547 in	OE	191.69 mm
1.938 in	BD	49.21 mm
3.375 in	EC	85.73 mm
4.002 in	HD	101.65 mm

ROCKWELL 18NY64-1

IDLI GROUP 85736

HOLLOW WELD YOKE
SEE FIGURE 46

7.547 in	OE	191.69 mm
1.938 in	BD	49.21 mm
3.375 in	EC	85.73 mm
4.242 in	HD	107.75 mm

ROCKWELL 18N28-117
SPICER 6.5-28-117

BD=BEARING DIAMETER (outside)   **BW**=BEARING WIDTH   **CB**=CROSS LENGTH WITH BEARINGS   **CC**=CENTER TO CENTER   **CD**=CROSS DIAMETER
CL=CROSS LENGTH WITHOUT BEARINGS   **EC**=END TO CENTER or FACE TO FACE   **EE**=END TO END   **EL**=EFFECTIVE LENGTH
HD=HUB DIAMETER (or insert)   **IE**=INSIDE OF EARS (or recess)   **OD**=OUTSIDE DIAMETER   **OE**=OUTSIDE OF EARS   **SB**=SPLINE OR BORE SIZE

IDLI GROUP 85801

BRAKE DRUM YOKE
WITH 32 INVOLUTED SPLINES
SEE FIGURE 31

3.125 in	IE	79.38 mm
1.188 in	BD	30.16 mm
1.391 in	SB	35.32 mm
8.719 in	EC	221.46 mm
2.125 in	HD	53.98 mm

ROCKWELL 14NYSB22-8

IDLI GROUP 85802

BRAKE DRUM YOKE
WITH 35 INVOLUTED SPLINES
SEE FIGURE 30

3.125 in	IE	79.38 mm
1.188 in	BD	30.16 mm
1.531 in	SB	38.90 mm
5.063 in	EC	128.59 mm
2.500 in	HD	63.50 mm

ROCKWELL 141NYSB24-2

IDLI GROUP 85803

BRAKE DRUM YOKE
WITH 10 SPLINES
SEE FIGURE 30

3.125 in	IE	79.38 mm
1.188 in	BD	30.16 mm
1.750 in	SB	44.45 mm
5.063 in	EC	128.59 mm
2.500 in	HD	63.50 mm

ROCKWELL 141NYSB28-2

IDLI GROUP 85804

BRAKE DRUM YOKE
WITH 34 INVOLUTED SPLINES
SEE FIGURE 30

3.125 in	IE	79.38 mm
1.188 in	BD	30.16 mm
1.781 in	SB	45.25 mm
6.188 in	EC	157.16 mm
2.563 in	HD	65.09 mm

ROCKWELL 141NYSB29

IDLI GROUP 85805

BRAKE DRUM YOKE
WITH 10 SPLINES
SEE FIGURE 30

4.188 in	IE	106.36 mm
1.375 in	BD	34.93 mm
1.500 in	SB	38.10 mm
3.656 in	EC	92.87 mm
2.125 in	HD	53.98 mm

ROCKWELL 148NYSB24-10

IDLI GROUP 85806

BRAKE DRUM YOKE
WITH 35 INVOLUTED SPLINES
SEE FIGURE 30

4.188 in	IE	106.36 mm
1.375 in	BD	34.93 mm
1.531 in	SB	38.90 mm
5.063 in	EC	128.59 mm
2.500 in	HD	63.50 mm

ROCKWELL 148NYSB24-7

IDLI GROUP 85807

BRAKE DRUM YOKE
WITH 10 SPLINES
SEE FIGURE 30

4.188 in	IE	106.36 mm
1.375 in	BD	34.93 mm
1.750 in	SB	44.45 mm
3.719 in	EC	94.46 mm
2.625 in	HD	66.68 mm

ROCKWELL 148NYSB28-26

IDLI GROUP 85808

BRAKE DRUM YOKE
WITH 10 SPLINES
SEE FIGURE 30

4.188 in	IE	106.36 mm
1.375 in	BD	34.93 mm
1.750 in	SB	44.45 mm
5.063 in	EC	128.59 mm
2.500 in	HD	63.50 mm

ROCKWELL 148NYSB28-17

IDLI GROUP 85809

BRAKE DRUM YOKE
WITH 34 INVOLUTED SPLINES
SEE FIGURE 30

4.188 in	IE	106.36 mm
1.375 in	BD	34.93 mm
1.781 in	SB	45.25 mm
6.188 in	EC	157.16 mm
2.563 in	HD	65.09 mm

ROCKWELL 148NYSB29

IDLI GROUP 85810

BRAKE FLANGE YOKE
WITH 10 SPLINES
SEE FIGURE 29A

5.313 in	IE	134.94 mm
5.625 in	OE	142.88 mm
1.375 in	SB	34.93 mm
3.750 in	EC	95.25 mm
5.813 in	OD	147.64 mm

ROCKWELL 58WBYS22-2

IDLI GROUP 85811

BRAKE FLANGE YOKE
WITH 10 SPLINES
slight variation from above group
SEE FIGURE 29A

5.313 in	IE	134.94 mm
5.625 in	OE	142.88 mm
1.375 in	SB	34.93 mm
3.750 in	EC	95.25 mm
5.813 in	OD	147.64 mm

ROCKWELL 58WBYS22-1

IDLI GROUP 85812

BRAKE FLANGE YOKE
WITH 10 SPLINES
SEE FIGURE 29A

5.313 in	IE	134.94 mm
5.625 in	OE	142.88 mm
1.500 in	SB	38.10 mm
3.750 in	EC	95.25 mm

ROCKWELL 58WBYS24

IDLI GROUP 85813

BRAKE FLANGE YOKE
WITH 10 SPLINES
SEE FIGURE 29A

5.313 in	IE	134.94 mm
5.625 in	OE	142.88 mm
1.750 in	SB	44.45 mm
3.188 in	EC	80.96 mm
5.813 in	OD	147.64 mm

ROCKWELL 58WBYS28-8

IDLI GROUP 85814

BRAKE FLANGE YOKE
WITH 10 SPLINES
SEE FIGURE 29A

5.313 in	IE	134.94 mm
5.625 in	OE	142.88 mm
1.750 in	SB	44.45 mm
3.938 in	EC	100.01 mm
5.813 in	OD	147.64 mm

ROCKWELL 58WBYS28-81

IDLI GROUP 85815

BRAKE FLANGE YOKE
WITH 10 SPLINES
SEE FIGURE 29A

5.313 in	IE	134.94 mm
5.625 in	OE	142.88 mm
1.750 in	SB	44.45 mm
4.250 in	EC	107.95 mm
5.813 in	OD	147.64 mm

ROCKWELL 58WBYS28-54

IDLI GROUP 85816

BRAKE FLANGE YOKE
WITH 10 SPLINES
SEE FIGURE 29A

5.313 in	IE	134.94 mm
5.625 in	OE	142.88 mm
2.000 in	SB	50.80 mm
5.125 in	EC	130.18 mm
6.875 in	OD	174.63 mm

ROCKWELL 58WBYS32-9

IDLI GROUP 85817

BRAKE FLANGE YOKE
WITH 10 SPLINES
SEE FIGURE 29A

5.313 in	IE	134.94 mm
5.625 in	OE	142.88 mm
2.500 in	SB	63.50 mm
4.719 in	EC	119.86 mm

ROCKWELL 58WBYS40-1

IDLI GROUP 85818

BRAKE FLANGE YOKE
WITH 10 SPLINES
SEE FIGURE 29A

5.531 in	IE	140.50 mm
5.844 in	OE	148.43 mm
1.500 in	SB	38.10 mm
2.875 in	EC	73.03 mm
5.750 in	OD	146.05 mm

ROCKWELL 62NYS24-21

IDLI GROUP 85819

BRAKE FLANGE YOKE
WITH 10 SPLINES
SEE FIGURE 29B

5.844 in	IE	148.43 mm
6.219 in	OE	157.96 mm
1.750 in	SB	44.45 mm
4.609 in	EC	117.08 mm
8.125 in	OD	206.38 mm

ROCKWELL 72NYS24-24

IDLI GROUP 85820

BRAKE FLANGE YOKE
WITH 34 SLANTED SPLINES
SEE FIGURE 29B

5.844 in	IE	148.43 mm
6.219 in	OE	157.96 mm
1.766 in	SB	44.85 mm
3.844 in	EC	97.63 mm
8.125 in	OD	206.38 mm

ROCKWELL 72NYS28-163

IDLI GROUP 85821

BRAKE FLANGE YOKE
WITH 27 INVOLUTED SPLINES
SEE FIGURE 29B

5.844 in	IE	148.43 mm
6.219 in	OE	157.96 mm
1.781 in	SB	45.25 mm
6.625 in	EC	168.28 mm
6.875 in	OD	174.63 mm

ROCKWELL 72NYS28-105

IDLI GROUP 85822

BRAKE FLANGE YOKE
WITH 10 SPLINES
SEE FIGURE 29B

5.844 in	IE	148.43 mm
6.219 in	OE	157.96 mm
1.875 in	SB	47.63 mm
4.063 in	EC	103.19 mm
8.250 in	OD	209.55 mm

ROCKWELL 72NYS30-18

IDLI GROUP 85823

BRAKE FLANGE YOKE
WITH 10 SPLINES
slight variation from above group
SEE FIGURE 29B

5.844 in	IE	148.43 mm
6.219 in	OE	157.96 mm
1.875 in	SB	47.63 mm
4.063 in	EC	103.19 mm
8.250 in	OD	209.55 mm

BORG-WARNR 7169J
ROCKWELL 72NYS30-4

IDLI GROUP 85824

BRAKE FLANGE YOKE
WITH 30 INVOLUTED SPLINES
SEE FIGURE 29B

5.844 in	IE	148.43 mm
6.219 in	OE	157.96 mm
1.938 in	SB	49.21 mm
3.875 in	EC	98.43 mm
8.000 in	OD	203.20 mm

ENGINEERING CATALOGS MUST BE CONSULTED FOR DETAILS NOT INCLUDED
IN THIS GUIDE. SPECIFIC DESIGNS, MATERIAL CONTENT, TOLERANCES
LUBE FITTINGS AND OTHER DIMENSIONS ARE INTENTIONALLY OMITTED HERE.

CAUTION: BE SURE TO REFER TO ENGINEERING CATALOGS FOR SPECIAL APPLICATIONS THAT REQUIRE SPECIFIC MATERIAL CONTENT, TOLERANCES, ETC. SEE FOOTNOTE.

ROCKWELL 72NYS31-89

IDLI GROUP 85825

BRAKE FLANGE YOKE
WITH 10 SPLINES
SEE FIGURE 29B

5.844 in	IE	148.43 mm
6.219 in	OE	157.96 mm
1.969 in	SB	50.01 mm
3.875 in	EC	98.43 mm
8.000 in	OD	203.20 mm

BORG-WARNR 21660J
ROCKWELL 72NYS31-33

IDLI GROUP 85826

BRAKE FLANGE YOKE
WITH 10 SPLINES
slight variation from above group
SEE FIGURE 29B

5.844 in	IE	148.43 mm
6.219 in	OE	157.96 mm
1.969 in	SB	50.01 mm
3.875 in	EC	98.43 mm
8.000 in	OD	203.20 mm

ROCKWELL 72NYS31-69

IDLI GROUP 85827

BRAKE FLANGE YOKE
WITH 10 SPLINES
SEE FIGURE 29B

5.844 in	IE	148.43 mm
6.219 in	OE	157.96 mm
1.969 in	SB	50.01 mm
3.875 in	EC	98.43 mm
8.250 in	OD	209.55 mm

ROCKWELL 72NYS31-79

IDLI GROUP 85828

BRAKE FLANGE YOKE
WITH 6 SPLINES
SEE FIGURE 29B

5.844 in	IE	148.43 mm
6.219 in	OE	157.96 mm
1.969 in	SB	50.01 mm
4.625 in	EC	117.48 mm

ROCKWELL 72NYS31-64

IDLI GROUP 85829

BRAKE FLANGE YOKE
WITH 10 SPLINES
SEE FIGURE 29B

5.844 in	IE	148.43 mm
6.219 in	OE	157.96 mm
2.000 in	SB	50.80 mm
3.500 in	EC	88.90 mm
8.000 in	OD	203.20 mm

ROCKWELL 72NYS32-167

IDLI GROUP 85830

BRAKE FLANGE YOKE
WITH 15 INVOLUTED SPLINES
SEE FIGURE 29B

5.844 in	IE	148.43 mm
6.219 in	OE	157.96 mm
2.000 in	SB	50.80 mm
4.875 in	EC	123.83 mm
8.000 in	OD	203.20 mm

BORG-WARNR 19767J
ROCKWELL 72NYS32-35

IDLI GROUP 85831

BRAKE FLANGE YOKE
WITH 10 SPLINES
SEE FIGURE 29B

5.844 in	IE	148.43 mm
6.219 in	OE	157.96 mm
2.000 in	SB	50.80 mm
5.750 in	EC	146.05 mm
6.438 in	OD	163.51 mm

ROCKWELL 72NYS32-174

IDLI GROUP 85832

BRAKE FLANGE YOKE
WITH 10 SPLINES
SEE FIGURE 29B

5.844 in	IE	148.43 mm
6.219 in	OE	157.96 mm
2.000 in	SB	50.80 mm
5.750 in	EC	146.05 mm
8.000 in	OD	203.20 mm

ROCKWELL 72NYS31-171A

IDLI GROUP 85833

BRAKE FLANGE YOKE
WITH 6 SPLINES
SEE FIGURE 29B

5.844 in	IE	148.43 mm
6.219 in	OE	157.96 mm
2.219 in	SB	56.36 mm
4.438 in	EC	112.71 mm
8.000 in	OD	203.20 mm

ROCKWELL 72NYS36-63

IDLI GROUP 85834

BRAKE FLANGE YOKE
WITH 10 SPLINES
SEE FIGURE 29B

5.844 in	IE	148.43 mm
6.219 in	OE	157.96 mm
2.250 in	SB	57.15 mm
4.313 in	EC	109.54 mm
8.000 in	OD	203.20 mm

ROCKWELL 72NYS36-78

IDLI GROUP 85835

BRAKE FLANGE YOKE
WITH 17 INVOLUTED SPLINES
SEE FIGURE 29B

5.844 in	IE	148.43 mm
6.219 in	OE	157.96 mm
2.281 in	SB	57.95 mm
3.563 in	EC	90.49 mm
8.125 in	OD	206.38 mm

ROCKWELL 72NYS36-7

IDLI GROUP 85836

BRAKE FLANGE YOKE
WITH 17 INVOLUTED SPLINES
SEE FIGURE 29B

5.844 in	IE	148.43 mm
6.219 in	OE	157.96 mm
2.281 in	SB	57.95 mm
4.188 in	EC	106.36 mm
8.125 in	OD	206.38 mm

ROCKWELL 72NYS36-4

IDLI GROUP 85837

BRAKE FLANGE YOKE
WITH 46 INVOLUTED SPLINES
SEE FIGURE 29B

5.844 in	IE	148.43 mm
6.219 in	OE	157.96 mm
2.375 in	SB	60.33 mm
5.219 in	EC	132.56 mm
8.000 in	OD	203.20 mm

ROCKWELL 72NYS38-10

IDLI GROUP 85838

BRAKE FLANGE YOKE
WITH 28 INVOLUTED SPLINES
SEE FIGURE 29B

5.844 in	IE	148.43 mm
6.219 in	OE	157.96 mm
2.438 in	SB	61.91 mm
4.750 in	EC	120.65 mm
8.000 in	OD	203.20 mm

BORG-WARNR 19884J
ROCKWELL 72NYS39-1

IDLI GROUP 85839

BRAKE FLANGE YOKE
WITH 10 SPLINES
SEE FIGURE 29B

5.844 in	IE	148.43 mm
6.219 in	OE	157.96 mm
2.500 in	SB	63.50 mm
4.719 in	EC	119.86 mm
8.000 in	OD	203.20 mm

ROCKWELL 72NYS40-21

IDLI GROUP 85840

BRAKE FLANGE YOKE
WITH 10 SPLINES
SEE FIGURE 29B

6.500 in	IE	165.10 mm
6.875 in	OE	174.63 mm
2.250 in	SB	57.15 mm
5.563 in	EC	141.29 mm
8.000 in	OD	203.20 mm

ROCKWELL 85WBYS36-4

IDLI GROUP 85841

BRAKE FLANGE YOKE
WITH 17 INVOLUTED SPLINES
SEE FIGURE 29B

6.500 in	IE	165.10 mm
6.875 in	OE	174.63 mm
2.281 in	SB	57.95 mm
4.688 in	EC	119.06 mm
8.125 in	OD	206.38 mm

ROCKWELL 85WBYS36

IDLI GROUP 85842

BRAKE FLANGE YOKE
WITH 28 INVOLUTED SPLINES
SEE FIGURE 29B

6.500 in	IE	165.10 mm
6.875 in	OE	174.63 mm
2.438 in	SB	61.91 mm
4.750 in	EC	120.65 mm
8.000 in	OD	203.20 mm

BORG-WARNR 19951J
ROCKWELL 85WBYS39-1

IDLI GROUP 85843

BRAKE FLANGE YOKE
WITH 16 INVOLUTED SPLINES
SEE FIGURE 29B

8.125 in	IE	206.38 mm
8.500 in	OE	215.90 mm
1.750 in	SB	44.45 mm
4.813 in	EC	122.24 mm
8.000 in	OD	203.20 mm

BORG-WARNR 13446J
ROCKWELL 82NYS28-3

IDLI GROUP 85844

BRAKE FLANGE YOKE
WITH 10 SPLINES
SEE FIGURE 29B

8.125 in	IE	206.38 mm
8.500 in	OE	215.90 mm
1.875 in	SB	47.63 mm
4.875 in	EC	123.83 mm
8.000 in	OD	203.20 mm

ROCKWELL 82NYS30

IDLI GROUP 85845

BRAKE FLANGE YOKE
WITH 10 SPLINES
SEE FIGURE 29B

8.125 in	IE	206.38 mm
8.500 in	OE	215.90 mm
1.969 in	SB	50.01 mm
3.719 in	EC	94.46 mm
9.500 in	OD	241.30 mm

ROCKWELL 82NYS31-42

IDLI GROUP 85846

BRAKE FLANGE YOKE
WITH 10 SPLINES
SEE FIGURE 29B

8.125 in	IE	206.38 mm
8.500 in	OE	215.90 mm
1.969 in	SB	50.01 mm
5.281 in	EC	134.15 mm
8.000 in	OD	203.20 mm

BORG-WARNR 8657J
ROCKWELL 82NYS31-1

BD = BEARING DIAMETER (outside)　　**BW** = BEARING WIDTH　　**CB** = CROSS LENGTH WITH BEARINGS　　**CC** = CENTER TO CENTER　　**CD** = CROSS DIAMETER
CL = CROSS LENGTH WITHOUT BEARINGS　　**EC** = END TO CENTER or FACE TO FACE　　**EE** = END TO END　　**EL** = EFFECTIVE LENGTH
HD = HUB DIAMETER (or insert)　　**IE** = INSIDE OF EARS (or recess)　　**OD** = OUTSIDE DIAMETER　　**OE** = OUTSIDE OF EARS　　**SB** = SPLINE OR BORE SIZE

IDLI GROUP 85847

BRAKE FLANGE YOKE
WITH 6 SPLINES
SEE FIGURE 29B

8.125 in	IE	206.38 mm
8.500 in	OE	215.90 mm
1.969 in	SB	50.01 mm
6.500 in	EC	165.10 mm
10.000 in	OD	254.00 mm

BORG-WARNR 14217J
ROCKWELL 82NYS31-2

IDLI GROUP 85848

BRAKE FLANGE YOKE
WITH 10 SPLINES
SEE FIGURE 29B

8.125 in	IE	206.38 mm
8.500 in	OE	215.90 mm
2.000 in	SB	50.80 mm
4.281 in	EC	108.75 mm
8.000 in	OD	203.20 mm

ROCKWELL 82NYS32-45

IDLI GROUP 85849

BRAKE FLANGE YOKE
WITH 10 SPLINES
SEE FIGURE 29B

8.125 in	IE	206.38 mm
8.500 in	OE	215.90 mm
2.000 in	SB	50.80 mm
4.813 in	EC	122.24 mm
10.000 in	OD	254.00 mm

BORG-WARNR 22100J
ROCKWELL 82NYS32-15

IDLI GROUP 85850

BRAKE FLANGE YOKE
WITH 15 INVOLUTED SPLINES
SEE FIGURE 29B

8.125 in	IE	206.38 mm
8.500 in	OE	215.90 mm
2.000 in	SB	50.80 mm
4.875 in	EC	123.83 mm
8.000 in	OD	203.20 mm

BORG-WARNR 12074J
ROCKWELL 82NYS32-2

IDLI GROUP 85851

BRAKE FLANGE YOKE
WITH 10 SPLINES
SEE FIGURE 29B

8.125 in	IE	206.38 mm
8.500 in	OE	215.90 mm
2.000 in	SB	50.80 mm
5.281 in	EC	134.15 mm
8.000 in	OD	203.20 mm

BORG-WARNR 13458J
ROCKWELL 82NYS32-5

IDLI GROUP 85852

BRAKE FLANGE YOKE
WITH 10 SPLINES
SEE FIGURE 29B

8.125 in	IE	206.38 mm
8.500 in	OE	215.90 mm
2.000 in	SB	50.80 mm
5.375 in	EC	136.53 mm
8.000 in	OD	203.20 mm

ROCKWELL 82NYS32-21

IDLI GROUP 85853

BRAKE FLANGE YOKE
WITH 10 SPLINES
SEE FIGURE 29B

8.125 in	IE	206.38 mm
8.500 in	OE	215.90 mm
2.000 in	SB	50.80 mm
5.375 in	EC	136.53 mm
8.000 in	OD	203.20 mm

ROCKWELL 82NYS32-25

IDLI GROUP 85854

BRAKE FLANGE YOKE
WITH 10 SPLINES
SEE FIGURE 29B

8.125 in	IE	206.38 mm
8.500 in	OE	215.90 mm
2.000 in	SB	50.80 mm
5.688 in	EC	144.46 mm
8.000 in	OD	203.20 mm

BORG-WARNR 14104J
ROCKWELL 82NYS32-12

IDLI GROUP 85855

BRAKE FLANGE YOKE
WITH 6 SPLINES
SEE FIGURE 29B

8.125 in	IE	206.38 mm
8.500 in	OE	215.90 mm
2.219 in	SB	56.36 mm
6.875 in	EC	174.63 mm
8.000 in	OD	203.20 mm

ROCKWELL 82NYS36-51

IDLI GROUP 85856

BRAKE FLANGE YOKE
WITH 6 SPLINES
SEE FIGURE 29B

8.125 in	IE	206.38 mm
8.500 in	OE	215.90 mm
2.219 in	SB	56.36 mm
7.000 in	EC	177.80 mm
8.000 in	OD	203.20 mm

BORG-WARNR 11598J
BORG-WARNR 7774J
BORG-WARNR 8408J
ROCKWELL 82NYS36-2

IDLI GROUP 85857

BRAKE FLANGE YOKE
WITH 6 SPLINES
slight variation from above group
SEE FIGURE 29B

8.125 in	IE	206.38 mm
8.500 in	OE	215.90 mm
2.219 in	SB	56.36 mm
7.000 in	EC	177.80 mm
8.000 in	OD	203.20 mm

ROCKWELL 82NYS36-77

IDLI GROUP 85858

BRAKE FLANGE YOKE
WITH 10 SPLINES
SEE FIGURE 29B

8.125 in	IE	206.38 mm
8.500 in	OE	215.90 mm
2.250 in	SB	57.15 mm
4.750 in	EC	120.65 mm
9.500 in	OD	241.30 mm

ROCKWELL 82NYS36-97

IDLI GROUP 85859

BRAKE FLANGE YOKE
WITH 10 SPLINES
SEE FIGURE 29B

8.125 in	IE	206.38 mm
8.500 in	OE	215.90 mm
2.250 in	SB	57.15 mm
4.781 in	EC	121.45 mm
8.000 in	OD	203.20 mm

ROCKWELL 82NYS36-31

IDLI GROUP 85860

BRAKE FLANGE YOKE
WITH 10 SPLINES
slight variation from above group
SEE FIGURE 29B

8.125 in	IE	206.38 mm
8.500 in	OE	215.90 mm
2.250 in	SB	57.15 mm
4.781 in	EC	121.45 mm
8.000 in	OD	203.20 mm

ROCKWELL 82NYS36-11

IDLI GROUP 85861

BRAKE FLANGE YOKE
WITH 10 SPLINES
slight variation from above group
SEE FIGURE 29B

8.125 in	IE	206.38 mm
8.500 in	OE	215.90 mm
2.250 in	SB	57.15 mm
4.781 in	EC	121.45 mm
8.000 in	OD	203.20 mm

BORG-WARNR 11938J
ROCKWELL 82NYS36-78

IDLI GROUP 85862

BRAKE FLANGE YOKE
WITH 10 SPLINES
SEE FIGURE 29B

8.125 in	IE	206.38 mm
8.500 in	OE	215.90 mm
2.250 in	SB	57.15 mm
5.125 in	EC	130.18 mm
8.000 in	OD	203.20 mm

ROCKWELL 82NYS36-16

IDLI GROUP 85863

BRAKE FLANGE YOKE
WITH 10 SPLINES
SEE FIGURE 29B

8.125 in	IE	206.38 mm
8.500 in	OE	215.90 mm
2.250 in	SB	57.15 mm
5.188 in	EC	131.76 mm
8.000 in	OD	203.20 mm

ROCKWELL 82NYS36-6

IDLI GROUP 85864

BRAKE FLANGE YOKE
WITH 10 SPLINES
SEE FIGURE 29B

8.125 in	IE	206.38 mm
8.500 in	OE	215.90 mm
2.250 in	SB	57.15 mm
5.281 in	EC	134.15 mm
8.000 in	OD	203.20 mm

BORG-WARNR 12750J
BORG-WARNR 16257J
ROCKWELL 82NYS36-22

IDLI GROUP 85865

BRAKE FLANGE YOKE
WITH 10 SPLINES
slight variation from above group
SEE FIGURE 29B

8.125 in	IE	206.38 mm
8.500 in	OE	215.90 mm
2.250 in	SB	57.15 mm
5.281 in	EC	134.15 mm
8.000 in	OD	203.20 mm

BORG-WARNR 8433J
ROCKWELL 82NYS36-1

IDLI GROUP 85866

BRAKE FLANGE YOKE
WITH 10 SPLINES
slight variation from above group
SEE FIGURE 29B

8.125 in	IE	206.38 mm
8.500 in	OE	215.90 mm
2.250 in	SB	57.15 mm
5.281 in	EC	134.15 mm
8.000 in	OD	203.20 mm

BORG-WARNR 11408J
ROCKWELL 82NYS36-8

ENGINEERING CATALOGS MUST BE CONSULTED FOR DETAILS NOT INCLUDED
IN THIS GUIDE. SPECIFIC DESIGNS, MATERIAL CONTENT, TOLERANCES
LUBE FITTINGS AND OTHER DIMENSIONS ARE INTENTIONALLY OMITTED HERE.

CAUTION: BE SURE TO REFER TO ENGINEERING CATALOGS FOR SPECIAL APPLICATIONS THAT REQUIRE SPECIFIC MATERIAL CONTENT, TOLERANCES, ETC. SEE FOOTNOTE.

IDLI GROUP 85867

BRAKE FLANGE YOKE
WITH 10 SPLINES
SEE FIGURE 29B

8.125 in	IE	206.38 mm
8.500 in	OE	215.90 mm
2.250 in	SB	57.15 mm
5.688 in	EC	144.46 mm
8.000 in	OD	203.20 mm

BORG-WARNR 21283J
BORG-WARNR 5126J
ROCKWELL 82NYS36-14

IDLI GROUP 85868

BRAKE FLANGE YOKE
WITH 10 SPLINES
slight variation from above group
SEE FIGURE 29B

8.125 in	IE	206.38 mm
8.500 in	OE	215.90 mm
2.250 in	SB	57.15 mm
5.688 in	EC	144.46 mm
8.000 in	OD	203.20 mm

ROCKWELL 82NYS36-29

IDLI GROUP 85869

BRAKE FLANGE YOKE
WITH 10 SPLINES
slight variation from above group
SEE FIGURE 29B

8.125 in	IE	206.38 mm
8.500 in	OE	215.90 mm
2.250 in	SB	57.15 mm
5.688 in	EC	144.46 mm
8.000 in	OD	203.20 mm

ROCKWELL 82NYS36-36

IDLI GROUP 85870

BRAKE FLANGE YOKE
WITH 10 SPLINES
SEE FIGURE 29B

8.125 in	IE	206.38 mm
8.500 in	OE	215.90 mm
2.250 in	SB	57.15 mm
7.344 in	EC	186.53 mm
10.250 in	OD	260.35 mm

BORG-WARNR 20627J
ROCKWELL 82NYS36-27

IDLI GROUP 85871

BRAKE FLANGE YOKE
WITH 10 SPLINES
SEE FIGURE 29B

8.125 in	IE	206.38 mm
8.500 in	OE	215.90 mm
2.250 in	SB	57.15 mm
7.375 in	EC	187.33 mm
8.000 in	OD	203.20 mm

ROCKWELL 82NYS36-82

IDLI GROUP 85872

BRAKE FLANGE YOKE
WITH 6 SPLINES
SEE FIGURE 29B

8.125 in	IE	206.38 mm
8.500 in	OE	215.90 mm
2.266 in	SB	57.55 mm
7.375 in	EC	187.33 mm
8.000 in	OD	203.20 mm

ROCKWELL 82NYS36-83

IDLI GROUP 85873

BRAKE FLANGE YOKE
WITH 17 INVOLUTED SPLINES
SEE FIGURE 29B

8.125 in	IE	206.38 mm
8.500 in	OE	215.90 mm
2.281 in	SB	57.95 mm
4.250 in	EC	107.95 mm
8.000 in	OD	203.20 mm

BORG-WARNR 15049J
ROCKWELL 82NYS36-12

IDLI GROUP 85874

BRAKE FLANGE YOKE
WITH 28 INVOLUTED SPLINES
SEE FIGURE 29B

8.125 in	IE	206.38 mm
8.500 in	OE	215.90 mm
2.438 in	SB	61.91 mm
4.750 in	EC	120.65 mm
8.000 in	OD	203.20 mm

BORG-WARNR 17280J
ROCKWELL 82NYS39-1

IDLI GROUP 85875

BRAKE FLANGE YOKE
WITH 10 SPLINES
SEE FIGURE 29B

8.125 in	IE	206.38 mm
8.500 in	OE	215.90 mm
2.500 in	SB	63.50 mm
4.969 in	EC	126.21 mm
8.000 in	OD	203.20 mm

BORG-WARNR 20523J
ROCKWELL 82NYS40-34

IDLI GROUP 85876

BRAKE FLANGE YOKE
WITH 25 INVOLUTED SPLINES
SEE FIGURE 29B

8.125 in	IE	206.38 mm
8.500 in	OE	215.90 mm
2.625 in	SB	66.68 mm
6.563 in	EC	166.69 mm
10.188 in	OD	258.76 mm

BORG-WARNR 13323J
ROCKWELL 82NYS42-1

IDLI GROUP 85877

BRAKE FLANGE YOKE
WITH 21 INVOLUTED SPLINES
SEE FIGURE 29B

8.125 in	IE	206.38 mm
8.500 in	OE	215.90 mm
2.859 in	SB	72.63 mm
6.563 in	EC	166.69 mm
10.188 in	OD	258.76 mm

BORG-WARNR 17123J
ROCKWELL 82NYS44

IDLI GROUP 85878

BRAKE FLANGE YOKE
WITH 6 SPLINES
SEE FIGURE 29B

8.250 in	IE	209.55 mm
8.625 in	OE	219.08 mm
2.219 in	SB	56.36 mm
6.500 in	EC	165.10 mm
8.000 in	OD	203.20 mm

ROCKWELL 92NYS36-10

IDLI GROUP 85879

BRAKE FLANGE YOKE
WITH 10 SPLINES
SEE FIGURE 29B

8.250 in	IE	209.55 mm
8.625 in	OE	219.08 mm
2.250 in	SB	57.15 mm
5.563 in	EC	141.29 mm
8.000 in	OD	203.20 mm

ROCKWELL 92NYS36-12

IDLI GROUP 85880

BRAKE FLANGE YOKE
WITH 10 SPLINES
SEE FIGURE 29B

8.250 in	IE	209.55 mm
8.625 in	OE	219.08 mm
2.250 in	SB	57.15 mm
6.375 in	EC	161.93 mm
8.000 in	OD	203.20 mm

ROCKWELL 92NYS36-7

IDLI GROUP 85881

BRAKE FLANGE YOKE
WITH 10 SPLINES
SEE FIGURE 29B

8.250 in	IE	209.55 mm
8.625 in	OE	219.08 mm
2.250 in	SB	57.15 mm
6.938 in	EC	176.21 mm
8.000 in	OD	203.20 mm

ROCKWELL 92NYS36-39

IDLI GROUP 85882

BRAKE FLANGE YOKE
WITH 10 SPLINES
SEE FIGURE 29B

8.250 in	IE	209.55 mm
8.625 in	OE	219.08 mm
2.250 in	SB	57.15 mm
7.344 in	EC	186.53 mm
10.250 in	OD	260.35 mm

BORG-WARNR 20630J

ROCKWELL 92NYS36-4

IDLI GROUP 85883

BRAKE FLANGE YOKE
WITH 21 INVOLUTED SPLINES
SEE FIGURE 29B

8.250 in	IE	209.55 mm
8.625 in	OE	219.08 mm
2.859 in	SB	72.63 mm
4.938 in	EC	125.41 mm
8.000 in	OD	203.20 mm

ROCKWELL 92NYS44-7

IDLI GROUP 85884

BRAKE FLANGE YOKE
WITH 28 INVOLUTED SPLINES
SEE FIGURE 29B

8.250 in	IE	209.55 mm
8.625 in	OE	219.08 mm
2.938 in	SB	74.61 mm
4.625 in	EC	117.48 mm
10.250 in	OD	260.35 mm

ROCKWELL 92NYS47-1

IDLI GROUP 85885

BRAKE FLANGE YOKE
WITH 29 INVOLUTED SPLINES
SEE FIGURE 29B

8.250 in	IE	209.55 mm
8.625 in	OE	219.08 mm
3.000 in	SB	76.20 mm
5.000 in	EC	127.00 mm
10.188 in	OD	258.76 mm

ROCKWELL 92NF24

BD=BEARING DIAMETER (outside)　　**BW**=BEARING WIDTH　　**CB**=CROSS LENGTH WITH BEARINGS　　**CC**=CENTER TO CENTER　　**CD**=CROSS DIAMETER
CL=CROSS LENGTH WITHOUT BEARINGS　　**EC**=END TO CENTER or FACE TO FACE　　**EE**=END TO END　　**EL**=EFFECTIVE LENGTH
HD=HUB DIAMETER (or insert)　　**IE**=INSIDE OF EARS (or recess)　　**OD**=OUTSIDE DIAMETER　　**OE**=OUTSIDE OF EARS　　**SB**=SPLINE OR BORE SIZE

IDLI GROUP 86001

CONSTANT VELOCITY SERVICE KIT

AMER.MOTOR	22004R
AUDI	321407271D
MIDAS	CV1000
MOOG	CV2200
NEAPCO	89-8005
PRECISION	2200
TRW	22004R
VOLKSWAGEN	321407271AG
VOLKSWAGEN	321407271D

IDLI GROUP 86002

CONSTANT VELOCITY SERVICE KIT

AMER.MOTOR	22006R
AUDI	321407271C
MIDAS	CV1001
MOOG	CV2201
NEAPCO	89-8007
PRECISION	2201
TRW	22006R
VOLKSWAGEN	321407271AE
VOLKSWAGEN	321407271C

IDLI GROUP 86003

CONSTANT VELOCITY SERVICE KIT

AMER.MOTOR	22005R
AUDI	321407272B
MIDAS	CV1002
MOOG	CV2202
NEAPCO	89-8006
PRECISION	2202
TRW	22005R
VOLKSWAGEN	321407272B
VOLKSWAGEN	321407272P

IDLI GROUP 86004

CONSTANT VELOCITY SERVICE KIT

MIDAS	CV1003
MOOG	CV2203
NEAPCO	89-8048
PRECISION	2203
TRW	22047R
VOLKSWAGEN	171407271
VOLKSWAGEN	171407271R

IDLI GROUP 86005

CONSTANT VELOCITY SERVICE KIT

MIDAS	CV1004
MOOG	CV2204
NEAPCO	89-8047
PRECISION	2204
TRW	22046R
TRW	22048R
VOLKSWAGEN	171407272
VOLKSWAGEN	171407272M

IDLI GROUP 86006

CONSTANT VELOCITY SERVICE KIT

MIDAS	CV1005
MOOG	CV2205
PRECISION	2205
TRW	22049R
VOLKSWAGEN	171407271G

IDLI GROUP 86007

CONSTANT VELOCITY SERVICE KIT

MIDAS	CV1006
MOOG	CV2206
PRECISION	2206
TRW	22046R
VOLKSWAGEN	171407272G

IDLI GROUP 86008

CONSTANT VELOCITY SERVICE KIT

FORD	E2GZ3B436A
MIDAS	CV1007
MOOG	CV2209
NEAPCO	89-8077
PRECISION	2209
TRW	22015R

IDLI GROUP 86009

CONSTANT VELOCITY SERVICE KIT

FORD	E2GZ3B436B
MIDAS	CV1008
MOOG	CV2210
NEAPCO	89-8078
PRECISION	22014R
PRECISION	2210
TRW	22014R

IDLI GROUP 86010

CONSTANT VELOCITY SERVICE KIT

FORD	78TT3A425AA
FORD	D8RZ3A426B
MIDAS	CV1009
MOOG	CV2211
NEAPCO	89-8044
PRECISION	2211
TRW	22043R

IDLI GROUP 86011

CONSTANT VELOCITY SERVICE KIT

CHRYSLER	5212604
CHRYSLER	5212654
MIDAS	CV1010
MOOG	CV2212
NEAPCO	89-8040
NEAPCO	89-8070
PRECISION	2212
TRW	22039R
TRW	22072R

IDLI GROUP 86012

CONSTANT VELOCITY SERVICE KIT

CHRYSLER	5212605
CHRYSLER	5212655
MIDAS	CV1011
MOOG	CV2212
MOOG	CV2213
NEAPCO	89-8041
NEAPCO	89-8071
PRECISION	2213
TRW	22040R
TRW	22073R

IDLI GROUP 86013

CONSTANT VELOCITY SERVICE KIT

CHRYSLER	5212602
CHRYSLER	5212652
MIDAS	CV1012
MOOG	CV2214
NEAPCO	89-8068
PRECISION	2214
TRW	22037R
TRW	22070R
TRW	89-8038

IDLI GROUP 86014

CONSTANT VELOCITY SERVICE KIT

CHRYSLER	5212603
CHRYSLER	5212653
MIDAS	CV1013
MOOG	CV2215
NEAPCO	89-8039
NEAPCO	89-8069
PRECISION	2215
TRW	22038R
TRW	22071R

IDLI GROUP 86015

CONSTANT VELOCITY SERVICE KIT

CHRYSLER	5212592
MIDAS	CV1014
MOOG	CV2216
NEAPCO	89-8030
PRECISION	2216
TRW	22029R

IDLI GROUP 86016

CONSTANT VELOCITY SERVICE KIT

CHRYSLER	5212593
MIDAS	CV1015
MOOG	CV2217
NEAPCO	89-8031
PRECISION	2217
TRW	22030R

IDLI GROUP 86017

CONSTANT VELOCITY SERVICE KIT

CHRYSLER	5212594
MIDAS	CV1016
MOOG	CV2218
NEAPCO	89-8032
PRECISION	2218
TRW	22031R

IDLI GROUP 86018

CONSTANT VELOCITY SERVICE KIT

CHRYSLER	5212595
MIDAS	CV1017
MOOG	CV2219
NEAPCO	89-8033
PRECISION	2219
TRW	22032R

IDLI GROUP 86019

CONSTANT VELOCITY SERVICE KIT

CHRYSLER	5212552
MIDAS	CV1018
MOOG	CV2220
NEAPCO	89-8026
PRECISION	2220
TRW	22004R
TRW	22025R

IDLI GROUP 86020

CONSTANT VELOCITY SERVICE KIT

CHRYSLER	5212553
MIDAS	CV1019
MOOG	CV2221
NEAPCO	89-8027
PRECISION	2221
TRW	22026R

IDLI GROUP 86021

CONSTANT VELOCITY SERVICE KIT

CHRYSLER	5212554
MIDAS	CV1020
MOOG	CV2222
NEAPCO	89-8028
PRECISION	2222
TRW	22027R

IDLI GROUP 86022

CONSTANT VELOCITY SERVICE KIT

CHRYSLER	5212555
MIDAS	CV1021
MOOG	CV2223
NEAPCO	89-8029
PRECISION	2223
TRW	22028R

ENGINEERING CATALOGS MUST BE CONSULTED FOR DETAILS NOT INCLUDED
IN THIS GUIDE. SPECIFIC DESIGNS, MATERIAL CONTENT, TOLERANCES
LUBE FITTINGS AND OTHER DIMENSIONS ARE INTENTIONALLY OMITTED HERE.

CAUTION: BE SURE TO REFER TO ENGINEERING CATALOGS FOR SPECIAL APPLICATIONS THAT REQUIRE SPECIFIC MATERIAL CONTENT, TOLERANCES, ETC. SEE FOOTNOTE.

IDLI GROUP 86023

CONSTANT VELOCITY SERVICE KIT

CHRYSLER	5212622
CHRYSLER	5212640
MIDAS	CV1022
MOOG	CV2224
NEAPCO	89-8034
NEAPCO	89-8066
PRECISION	2224
TRW	22033R
TRW	22068R

IDLI GROUP 86024

CONSTANT VELOCITY SERVICE KIT

CHRYSLER	5212623
CHRYSLER	5212641
MIDAS	CV1023
MOOG	CV2225
NEAPCO	89-8035
NEAPCO	89-8067
PRECISION	2225
TRW	22034R
TRW	22069R

IDLI GROUP 86025

CONSTANT VELOCITY SERVICE KIT

CHRYSLER	5212624
MIDAS	CV1024
MOOG	CV2226
NEAPCO	89-8036
PRECISION	2226
TRW	22035R

IDLI GROUP 86026

CONSTANT VELOCITY SERVICE KIT

CHRYSLER	5212625
MIDAS	CV1025
MOOG	CV2227
NEAPCO	89-8037
PRECISION	2227
TRW	22036R

IDLI GROUP 86027

CONSTANT VELOCITY SERVICE KIT

AUDI	321407271AJ
MIDAS	CV1026
MOOG	CV2228
NEAPCO	89-8055
PRECISION	2228
TRW	22054R
VOLKSWAGEN	321407271AJ

IDLI GROUP 86028

CONSTANT VELOCITY SERVICE KIT

AUDI	803407271A
MIDAS	CV1027
MOOG	CV2229
NEAPCO	89-8004
PRECISION	2229
TRW	22003R

IDLI GROUP 86029

CONSTANT VELOCITY SERVICE KIT

AUDI	803407271
MIDAS	CV1028
MOOG	CV2230
NEAPCO	89-8003
PRECISION	2230
TRW	22002R

IDLI GROUP 86030

CONSTANT VELOCITY SERVICE KIT

AUDI	431407271C
MIDAS	CV1029
MOOG	CV2231
NEAPCO	89-8018
PRECISION	2231
TRW	22017R

IDLI GROUP 86031

CONSTANT VELOCITY SERVICE KIT

AUDI	431407272A
MIDAS	CV1030
MOOG	CV2232
NEAPCO	89-8017
NEAPCO	89-8076
PRECISION	2232
TRW	22016R

IDLI GROUP 86032

CONSTANT VELOCITY SERVICE KIT

AUDI	431407271B
MIDAS	CV1031
MOOG	CV2233
NEAPCO	89-8056
PRECISION	2233
TRW	22055R

IDLI GROUP 86033

CONSTANT VELOCITY SERVICE KIT

MIDAS	CV1032
MOOG	CV2234
NEAPCO	89-8042
PRECISION	2234
TRW	22041R

IDLI GROUP 86034

CONSTANT VELOCITY SERVICE KIT

MIDAS	CV1033
MOOG	CV2235
NEAPCO	89-8043
PRECISION	2235
TRW	22042R

IDLI GROUP 86035

CONSTANT VELOCITY SERVICE KIT

AMER.MOTOR	321407272D
MIDAS	CV1034
MOOG	CV2236
NEAPCO	89-8009
PRECISION	2236
TRW	22008R

IDLI GROUP 86036

CONSTANT VELOCITY SERVICE KIT

AUDI	321407271K
MIDAS	CV1035
MOOG	CV2237
NEAPCO	89-8010
PRECISION	2237
TRW	22009R

IDLI GROUP 86037

CONSTANT VELOCITY SERVICE KIT

AUDI	321407271P
MIDAS	CV1036
MOOG	CV2238
NEAPCO	89-8008
NEAPCO	89-8009
PRECISION	2238
TRW	22007R

IDLI GROUP 86038

CONSTANT VELOCITY SERVICE KIT

AUDI	437407271E
MIDAS	CV1037
MOOG	CV2239
NEAPCO	89-8024
PRECISION	2239
TRW	22023R

IDLI GROUP 86039

CONSTANT VELOCITY SERVICE KIT

AUDI	437407272A
MIDAS	CV1038
MOOG	CV2240
NEAPCO	89-8023
PRECISION	2240
TRW	22022R

IDLI GROUP 86040

CONSTANT VELOCITY SERVICE KIT

MIDAS	CV1039
NEAPCO	89-8025
PRECISION	2241

IDLI GROUP 86041

CONSTANT VELOCITY SERVICE KIT

AUDI	431407276H
MIDAS	CV1040
MOOG	CV2242
NEAPCO	89-8022
PRECISION	2242
TRW	22021R

IDLI GROUP 86042

CONSTANT VELOCITY SERVICE KIT

AUDI	431407272L
MIDAS	CV1041
MOOG	CV2243
NEAPCO	89-8020
PRECISION	2243
TRW	22019R

IDLI GROUP 86043

CONSTANT VELOCITY SERVICE KIT

AUDI	431407276G
MIDAS	CV1042
MOOG	CV2244
NEAPCO	89-8021
PRECISION	2244
TRW	22020R

IDLI GROUP 86044

CONSTANT VELOCITY SERVICE KIT

AUDI	431407276F
MIDAS	CV1043
NEAPCO	89-8019
PRECISION	2245
TRW	22018R

IDLI GROUP 86045

CONSTANT VELOCITY SERVICE KIT

AUDI	321407272R
MIDAS	CV1044
MOOG	CV2246
NEAPCO	89-8072
NEAPCO	89-8083
PRECISION	2246
TRW	22061R
VOLKSWAGEN	321407272R

IDLI GROUP 86046

CONSTANT VELOCITY SERVICE KIT

AUDI	321407271AK
MIDAS	CV1045
MOOG	CV2247
NEAPCO	89-8073
PRECISION	2247
TRW	22062R
VOLKSWAGEN	321407271AK

BD=BEARING DIAMETER (outside) **BW**=BEARING WIDTH **CB**=CROSS LENGTH WITH BEARINGS **CC**=CENTER TO CENTER **CD**=CROSS DIAMETER
CL=CROSS LENGTH WITHOUT BEARINGS **EC**=END TO CENTER or FACE TO FACE **EE**=END TO END **EL**=EFFECTIVE LENGTH
HD=HUB DIAMETER (or insert) **IE**=INSIDE OF EARS (or recess) **OD**=OUTSIDE DIAMETER **OE**=OUTSIDE OF EARS **SB**=SPLINE OR BORE SIZE

IDLI GUIDE   COPYRIGHT © INTERCHANGE, INC.   ST. LOUIS PARK, MN. 55416 USA

IDLI GROUP 86047

CONSTANT VELOCITY SERVICE KIT

AUDI	803407271D
MIDAS	CV1046
MOOG	CV2248
NEAPCO	89-8079
PRECISION	2248
TRW	22063R

IDLI GROUP 86048

CONSTANT VELOCITY SERVICE KIT

MIDAS	CV1047
MOOG	CV2249
NEAPCO	89-8046
PRECISION	2249
RENAULT	7701348968
TRW	22045R

IDLI GROUP 86049

CONSTANT VELOCITY SERVICE KIT

MIDAS	CV1048
MOOG	CV2250
NEAPCO	89-8045
PRECISION	2250
RENAULT	7701348159
RENAULT	7701348759
TRW	22044R

IDLI GROUP 86050

CONSTANT VELOCITY SERVICE KIT

MIDAS	CV1049
PRECISION	2251

IDLI GROUP 86051

CONSTANT VELOCITY SERVICE KIT

AEC	3237089
AEC	3238675
AMER.MOTOR	3237089
AMER.MOTOR	3238675
AMER.MOTOR	3239621
MIDAS	CV1050
MOOG	CV2253
MOOG	CV2254
NEAPCO	89-8001
PRECISION	2253
TRW	22000R
TRW	22001R

IDLI GROUP 86052

CONSTANT VELOCITY SERVICE KIT

AMER.MOTOR	3238675
MIDAS	CV1051
NEAPCO	89-8002
PRECISION	2254
TRW	22001R

IDLI GROUP 86053

CONSTANT VELOCITY SERVICE KIT

AMER.MOTOR	8933001363
MOOG	CV2255
NEAPCO	89-8104
PRECISION	2255
TRW	22057R

IDLI GROUP 86054

CONSTANT VELOCITY SERVICE KIT

MOOG	CV2257
NEAPCO	89-8082
PRECISION	2257
TRW	22065R
TRW	803407271E

IDLI GROUP 86055

CONSTANT VELOCITY SERVICE KIT

AUDI	803407272A
MOOG	CV2258
NEAPCO	89-8081
PRECISION	2258
TRW	22064R

IDLI GROUP 86056

CONSTANT VELOCITY SERVICE KIT

AUDI	321407272AA
MIDAS	CV1052
MOOG	CV2260
NEAPCO	89-8013
PRECISION	2260
TRW	22012R

IDLI GROUP 86057

CONSTANT VELOCITY SERVICE KIT

AUDI	321407271AN
MIDAS	CV1053
MOOG	CV2261
NEAPCO	89-8014
PRECISION	2261
TRW	22013R

IDLI GROUP 86058

CONSTANT VELOCITY SERVICE KIT

AUDI	321407272S
MIDAS	CV1054
MOOG	CV2262
NEAPCO	89-8011
PRECISION	2262
TRW	22010R

IDLI GROUP 86059

CONSTANT VELOCITY SERVICE KIT

AUDI	321407271AP
MIDAS	CV1055
MOOG	CV2263
NEAPCO	89-8012
PRECISION	2263
TRW	22011R

IDLI GROUP 86060

CONSTANT VELOCITY SERVICE KIT

AUDI	857407272A
MIDAS	CV1056
MOOG	CV2264
NEAPCO	89-8085
PRECISION	2264
TRW	22066R

IDLI GROUP 86061

CONSTANT VELOCITY SERVICE KIT

AUDI	857407271A
MIDAS	CV1057
MOOG	CV2265
NEAPCO	89-8086
PRECISION	2265
TRW	22067R

IDLI GROUP 86062

CONSTANT VELOCITY SERVICE KIT

CHRYSLER	5212670
MOOG	CV2270
NEAPCO	89-8087
PRECISION	2270
TRW	22074R

IDLI GROUP 86063

CONSTANT VELOCITY SERVICE KIT

CHRYSLER	5212671
MOOG	CV2271
NEAPCO	89-8088
PRECISION	2271
TRW	22075R

IDLI GROUP 86064

CONSTANT VELOCITY SERVICE KIT

CHRYSLER	5212853
MOOG	CV2273
MOOG	CV2275
NEAPCO	89-8113
PRECISION	2273
TRW	22081R

IDLI GROUP 86065

CONSTANT VELOCITY SERVICE KIT

CHRYSLER	5212852
MOOG	CV2274
NEAPCO	89-8115
PRECISION	2274
TRW	22082R

IDLI GROUP 86066

CONSTANT VELOCITY SERVICE KIT

CHRYSLER	5212877
NEAPCO	89-8114
PRECISION	2275
TRW	22081R

IDLI GROUP 86067

CONSTANT VELOCITY SERVICE KIT

CHRYSLER	5212854
MOOG	CV2276
NEAPCO	89-8105
PRECISION	2276
TRW	22076R

IDLI GROUP 86068

CONSTANT VELOCITY SERVICE KIT

CHRYSLER	5212855
CHRYSLER	CV2277
NEAPCO	89-8106
PRECISION	2277
TRW	22077R

IDLI GROUP 86069

CONSTANT VELOCITY SERVICE KIT

GMB	GV11312
LUDWIG	21-1010
MIDAS	CV1058
MOOG	6-200X
MOOG	CV2300
NEAPCO	86-0908
PRECISION	2300
SPICER	6-2001X
SPICER	6-200X
SPICER	6-200X(8)
TRW	20150
TRW	22311
TRW	22512
VOLKSWAGEN	113501331
WORLDPARTS	W62-751
WORLDPARTS	W62-766

IDLI GROUP 86070

CONSTANT VELOCITY SERVICE KIT

GMB	GV12310
LUDWIG	21-1004
MCQUAY-NOR	U01-68
MIDAS	CV1059
MOOG	CV2301
NEAPCO	86-0909
PRECISION	2301
SPICER	6-201X
SPICER	6-201X(8)
SPICER	970168X
TRW	20152
VOLKSWAGEN	211-501-331B
VOLKSWAGEN	211-598-101
VOLKSWAGEN	221501149
VOLKSWAGEN	221501331B
WORLDPARTS	W62-752
WORLDPARTS	W62-764

ENGINEERING CATALOGS MUST BE CONSULTED FOR DETAILS NOT INCLUDED
IN THIS GUIDE. SPECIFIC DESIGNS, MATERIAL CONTENT, TOLERANCES
LUBE FITTINGS AND OTHER DIMENSIONS ARE INTENTIONALLY OMITTED HERE.

CAUTION: BE SURE TO REFER TO ENGINEERING CATALOGS FOR SPECIAL APPLICATIONS THAT REQUIRE SPECIFIC MATERIAL CONTENT, TOLERANCES, ETC. SEE FOOTNOTE.

IDLI GROUP 86071

CONSTANT VELOCITY SERVICE KIT

GMB	GV11312
LUDWIG	21-1011
MIDAS	CV1060
MOOG	CV2302
NEAPCO	86-0939
PORSCHE	113501331D
PORSCHE	411501149A
PRECISION	2302
SPICER	6-202X
SPICER	6-202X(8)
SPICER	970169X
TRW	20151
VOLKSWAGEN	113501331D
VOLKSWAGEN	411501149A
WORLDPARTS	W62-753

IDLI GROUP 86072

CONSTANT VELOCITY SERVICE KIT

AUDI	113501149
AUDI	321407331
AUDI	321498201
MIDAS	CV1061
MOOG	CV2303
NEAPCO	86-0902
PRECISION	2303
SPICER	6-203X
SPICER	6-203X(8)
TRW	20153
TRW	20154
TRW	22512
TRW	22515
VOLKSWAGEN	171407331F
VOLKSWAGEN	171-498-103
VOLKSWAGEN	321407331
VOLKSWAGEN	321-498-103
VOLKSWAGEN	CV2303
WORLDPARTS	W62-769

IDLI GROUP 86073

CONSTANT VELOCITY SERVICE KIT

MIDAS	CV1062
PRECISION	2304
VOLKSWAGEN	171407282B
VOLKSWAGEN	171407331C

IDLI GROUP 86074

CONSTANT VELOCITY SERVICE KIT

GMC	7837827
GMC	7840072
GMC	7845016
GMC	7845171
PRECISION	2307
PRECISION	5307
TRW	22572

IDLI GROUP 86075

CONSTANT VELOCITY SERVICE KIT

AEC	26001
ALLOY	10019
GMC	7545016
GMC	7833098
GMC	7837827
GMC	7837828
GMC	7840070
GMC	7840072
GMC	7845016
GMC	7845019
MOOG	CV2307
MOOG	CV2308
MOOG	CV2407
NEAPCO	86-0991
PRECISION	2308
PRECISION	5307
SPICER	6-221X
SPICER	6-285X
TRW	22573

IDLI GROUP 86076

CONSTANT VELOCITY SERVICE KIT

FORD	E1F2-3A331B
FORD	E2GZ3B414A
MIDAS	CV1063
MOOG	CV2310
NEAPCO	22549
NEAPCO	86-0935
PRECISION	2310
SPICER	6-2271X
SPICER	6-227X(8)
TRW	22549

IDLI GROUP 86077

CONSTANT VELOCITY SERVICE KIT

FORD	6058776
FORD	D8RZ3249B
FORD	D8RZ3A331B
MIDAS	CV1064
MOOG	CV2311
NEAPCO	86-0931
PRECISION	2311
SPICER	6-208X(8)
TRW	22545
WORLDPARTS	W62-781

IDLI GROUP 86078

CONSTANT VELOCITY SERVICE KIT

CHRYSLER	5212608
CHRYSLER	5212612
MIDAS	CV1065
NEAPCO	86-0973
PRECISION	2312
TRW	22526

IDLI GROUP 86079

CONSTANT VELOCITY SERVICE KIT

CHRYSLER	5212608
CHRYSLER	5212613
MIDAS	CV1066
MOOG	CV2312
MOOG	CV2313
NEAPCO	86-0974
PRECISION	2313
SPICER	6-245X(8)
TRW	22587

IDLI GROUP 86080

CONSTANT VELOCITY SERVICE KIT

AEC	26111
CHRYSLER	4205733
CHRYSLER	4205852
CHRYSLER	5212859
GMC	7832255
GMC	7845171
MOOG	CV2315
MOOG	CV2417
NEAPCO	86-0956
PRECISION	2315
SPICER	6-268X
TRW	22588

IDLI GROUP 86081

CONSTANT VELOCITY SERVICE KIT

CHRYSLER	4057032
CHRYSLER	4205852
CHRYSLER	5212567
CHRYSLER	5212574
CHRYSLER	5212627
MIDAS	CV1067
MOOG	CV2316
NEAPCO	86-0961
PRECISION	2316
SPICER	6-206X(8)
TRW	22314
TRW	22517

IDLI GROUP 86082

CONSTANT VELOCITY SERVICE KIT

CHRYSLER	22584
CHRYSLER	4057033
CHRYSLER	4205733
CHRYSLER	5212567
CHRYSLER	5212575
MIDAS	CV1068
MOOG	CV2317
NEAPCO	86-0962
PRECISION	2317
SPICER	6-207X(8)
TRW	22316

IDLI GROUP 86083

CONSTANT VELOCITY SERVICE KIT

CHRYSLER	5212567
CHRYSLER	5212568
MIDAS	CV1069
MOOG	CV2318
NEAPCO	86-0963
PRECISION	2318
SPICER	6-205X(8)
TRW	22323
TRW	22585

IDLI GROUP 86084

CONSTANT VELOCITY SERVICE KIT

CHRYSLER	5212567
CHRYSLER	5212569
MIDAS	CV1070
MOOG	CV2319
NEAPCO	86-0964
PRECISION	2319
SPICER	6-205X(8)
TRW	22315
TRW	22586
WORLDPARTS	W62-805

IDLI GROUP 86085

CONSTANT VELOCITY SERVICE KIT

AEC	26111
CHRYSLER	4205852
CHRYSLER	5212859
GMC	7832255
GMC	7845171
MOOG	CV2417
NEAPCO	86-0955
PRECISION	2320
SPICER	6-268X
TRW	22532

IDLI GROUP 86086

CONSTANT VELOCITY SERVICE KIT

CHRYSLER	5212614
MIDAS	CV1071
MOOG	CV2321
NEAPCO	86-0975
PRECISION	2321
SPICER	6-246X
TRW	22527

IDLI GROUP 86087

CONSTANT VELOCITY SERVICE KIT

CHRYSLER	5212615
MIDAS	CV1072
MOOG	CV2322
NEAPCO	86-0976
PRECISION	2322
SPICER	6-247X
TRW	22528

BD = BEARING DIAMETER (outside)   BW = BEARING WIDTH   CB = CROSS LENGTH WITH BEARINGS   CC = CENTER TO CENTER   CD = CROSS DIAMETER
CL = CROSS LENGTH WITHOUT BEARINGS   EC = END TO CENTER or FACE TO FACE   EE = END TO END   EL = EFFECTIVE LENGTH
HD = HUB DIAMETER (or insert)   IE = INSIDE OF EARS (or recess)   OD = OUTSIDE DIAMETER   OE = OUTSIDE OF EARS   SB = SPLINE OR BORE SIZE

IDLI GROUP 86088

CONSTANT VELOCITY SERVICE KIT

CHRYSLER	5212572
MIDAS	CV1073
MOOG	CV2323
NEAPCO	86-0967
PRECISION	2323
SPICER	6-205-1X
SPICER	6-205X1
TRW	22520
WORLDPARTS	W62-803

IDLI GROUP 86089

CONSTANT VELOCITY SERVICE KIT

CHRYSLER	4205852
CHRYSLER	5212576
MIDAS	CV1074
MOOG	CV2324
NEAPCO	86-0965
PRECISION	2324
SPICER	6-206-1X
SPICER	6-206X1
TRW	22518

IDLI GROUP 86090

CONSTANT VELOCITY SERVICE KIT

CHRYSLER	5212577
MIDAS	CV1075
MOOG	CV2325
NEAPCO	86-0943
NEAPCO	86-0966
PRECISION	2325
SPICER	6-206X1
SPICER	6-207-1X
SPICER	6-207X1
TRW	22519
WORLDPARTS	W62-801
WORLDPARTS	W62-802

IDLI GROUP 86091

CONSTANT VELOCITY SERVICE KIT

CHRYSLER	5212644
MIDAS	CV1076
NEAPCO	86-0970
PRECISION	2326
SPICER	6-254
SPICER	6-254X
TRW	22523

IDLI GROUP 86092

CONSTANT VELOCITY SERVICE KIT

CHRYSLER	5212645
CHRYSLER	5212897
MIDAS	CV1077
MOOG	CV2327
NEAPCO	86-0971
PRECISION	2327
SPICER	6-255X
TRW	22524

IDLI GROUP 86093

CONSTANT VELOCITY SERVICE KIT

CHRYSLER	5212656
CHRYSLER	5212898
MIDAS	CV1078
MOOG	CV2328
NEAPCO	86-0977
PRECISION	2328
SPICER	6-256X
TRW	22529

IDLI GROUP 86094

CONSTANT VELOCITY SERVICE KIT

CHRYSLER	5212657
CHRYSLER	5212899
MIDAS	CV1079
MOOG	CV2329
NEAPCO	86-0978
PRECISION	2329
SPICER	6-257X
TRW	22530

IDLI GROUP 86095

CONSTANT VELOCITY SERVICE KIT

AUDI	431-407-331C
AUDI	431407331G
AUDI	431498201A
AUDI	803407283C
MIDAS	CV1080
MOOG	CV2331
NEAPCO	86-0912
PRECISION	2331
SPICER	6-212X(8)
TRW	22507

IDLI GROUP 86096

CONSTANT VELOCITY SERVICE KIT

AUDI	431-407-331G
MIDAS	CV1081
MOOG	CV2332
MOOG	CV2333
NEAPCO	86-0915
PRECISION	2332
SPICER	6-241X(8)
TRW	22508
WORLDPARTS	W62-775

IDLI GROUP 86097

CONSTANT VELOCITY SERVICE KIT

MIDAS	CV1082
PRECISION	2333
TRW	22508

IDLI GROUP 86098

CONSTANT VELOCITY SERVICE KIT

FIAT	4204117
FIAT	4247393
FIAT	4311356
FIAT	4314783
FIAT	4333088
FIAT-ALLIS	4247393
FIAT-ALLIS	4313087
FIAT-ALLIS	4314783
MIDAS	CV1083
MOOG	CV2334
NEAPCO	86-0921
PRECISION	2334
SPICER	6-215X(8)
TRW	22538
TRW	22539
WORLDPARTS	W62-782

IDLI GROUP 86099

CONSTANT VELOCITY SERVICE KIT

AUDI	321-407-311B
MIDAS	CV1084
MOOG	CV2337
NEAPCO	86-0916
PRECISION	2337
SPICER	6-242X(8)
TRW	22515
VOLKSWAGEN	171-498-103C

IDLI GROUP 86100

CONSTANT VELOCITY SERVICE KIT

BMW	33211205745
MIDAS	CV1085
MOOG	CV2338
NEAPCO	86-1010
PRECISION	2338

IDLI GROUP 86101

CONSTANT VELOCITY SERVICE KIT

BMW	33211205746
MIDAS	CV1086
MOOG	CV2339
NEAPCO	86-1011
PRECISION	2339
SPICER	6-213X(8)
WORLDPARTS	W62-785

IDLI GROUP 86102

CONSTANT VELOCITY SERVICE KIT

BMW	33211205748
BMW	33211205840
BMW	33211207037
MIDAS	CV1087
MOOG	CV2340
NEAPCO	86-1012
PRECISION	2340
SPICER	6-214X(8)
WORLDPARTS	W62-786

IDLI GROUP 86103

CONSTANT VELOCITY SERVICE KIT

BMW	33-211-207-722
MIDAS	CV1088
MOOG	CV2341
NEAPCO	86-1013
PRECISION	2341
SPICER	6-238X(8)

IDLI GROUP 86104

CONSTANT VELOCITY SERVICE KIT

BMW	33-211-208-439
GMC	7842379
MIDAS	CV1089
MOOG	CV2342
MOOG	CV5387
NEAPCO	86-1014
PRECISION	2342
SPICER	6-239X(8)
SPICER	6-279X

IDLI GROUP 86105

CONSTANT VELOCITY SERVICE KIT

BMW	33-211-209-551
BMW	33-211-209-552
MIDAS	CV1090
MOOG	CV2343
NEAPCO	86-1015
PRECISION	2343
SPICER	6-2401X
SPICER	6-240X

IDLI GROUP 86106

CONSTANT VELOCITY SERVICE KIT

MIDAS	CV1091
NEAPCO	86-1020
PORSCHE	90133203110
PRECISION	2344

IDLI GROUP 86107

CONSTANT VELOCITY SERVICE KIT

MIDAS	CV1092
MOOG	CV2345
NEAPCO	86-1021
PORSCHE	90133203200
PORSCHE	90133229312
PRECISION	2345
SPICER	6-216X(8)
WORLDPARTS	W62-788

IDLI GROUP 86108

CONSTANT VELOCITY SERVICE KIT

MIDAS	CV1093
MOOG	CV2346
NEAPCO	86-1022
PORSCHE	90133229312
PORSCHE	91133203202
PRECISION	2346
SPICER	6-217X(8)
WORLDPARTS	W62-779

IDLI GROUP 86109

CONSTANT VELOCITY SERVICE KIT

MIDAS	CV1094
MOOG	CV2347
PORSCHE	92333203200
PORSCHE	92333203700
PRECISION	2347
SPICER	6-218X(8)
WORLDPARTS	W62-778

ENGINEERING CATALOGS MUST BE CONSULTED FOR DETAILS NOT INCLUDED
IN THIS GUIDE. SPECIFIC DESIGNS, MATERIAL CONTENT, TOLERANCES
LUBE FITTINGS AND OTHER DIMENSIONS ARE INTENTIONALLY OMITTED HERE.

CAUTION: BE SURE TO REFER TO ENGINEERING CATALOGS FOR SPECIAL APPLICATIONS THAT REQUIRE SPECIFIC MATERIAL CONTENT, TOLERANCES, ETC. SEE FOOTNOTE.

IDLI GROUP 86110

CONSTANT VELOCITY SERVICE KIT

MIDAS CV1095
MOOG CV2348
NEAPCO 86-1023
PORSCHE 90133229312
PORSCHE 93033203400
PRECISION 2348
SPICER . . . . 6-219X(8)
WORLDPARTS W62-777

IDLI GROUP 86111

CONSTANT VELOCITY SERVICE KIT

MIDAS CV1096
MOOG CV2349
NEAPCO 86-1024
PORSCHE 91433202903
PRECISION 2349
SPICER . . . . 6-220X(8)
WORLDPARTS W62-790

IDLI GROUP 86112

CONSTANT VELOCITY SERVICE KIT

MIDAS CV1097
NEAPCO 86-1025
PORSCHE 92833202901
PRECISION 2350
SPICER 6-258X

IDLI GROUP 86113

CONSTANT VELOCITY SERVICE KIT

MOOG CV2351
PRECISION 2351
RENAULT 7701201045
RENAULT 7701460779
SPICER . . . . 6-108X(8)
SPICER 86-0940
TRW 22578

IDLI GROUP 86114

CONSTANT VELOCITY SERVICE KIT

AMER.MOTOR 8132434
AMER.MOTOR 8132435
AMER.MOTOR 813-3581
AMER.MOTOR 813-3582
MIDAS CV1098
MOOG CV2353
NEAPCO 86-0951
PRECISION 2353
SPICER . . . . 2-225X(8)
SPICER 6-225X
SPICER 6-225X8
TRW 22302
TRW 22500
TRW 22502
TRW 22504

IDLI GROUP 86115

CONSTANT VELOCITY SERVICE KIT

AMER.MOTOR 813-2437
AMER.MOTOR 8133581
AMER.MOTOR 813-3583
MIDAS CV1099
MOOG CV2354
NEAPCO 86-0952
PRECISION 2354
SPICER 6-226X
TRW 22503
TRW 22505

IDLI GROUP 86116

CONSTANT VELOCITY SERVICE KIT

AMER.MOTOR 813-2434
AMER.MOTOR 813-2435
MIDAS CV1100
MOOG CV2355
NEAPCO 86-0953
PRECISION 2355
SPICER . . . . 6-223X(8)
TRW 22504

IDLI GROUP 86117

CONSTANT VELOCITY SERVICE KIT

MIDAS CV1101
MOOG CV2356
NEAPCO 86-0954
PRECISION 2356
SPICER 6-224X
TRW 22505

IDLI GROUP 86118

CONSTANT VELOCITY SERVICE KIT

CHRYSLER 5212644
CHRYSLER 5212896
MOOG CV2326
MOOG CV2358
MOOG CV2359
NEAPCO 86-0922
PRECISION 2358
SPICER 6-254X
TRW 22518

IDLI GROUP 86119

CONSTANT VELOCITY SERVICE KIT

CHRYSLER 5212897
NEAPCO 22519
NEAPCO 86-0924
PRECISION 2359
SPICER 6-254X

IDLI GROUP 86120

CONSTANT VELOCITY SERVICE KIT

GMC 7832258
GMC 7832259
GMC 7844997
GMC 7845009
MOOG CV5347
MOOG CV5348
MOOG CV5349

IDLI GROUP 86121

CONSTANT VELOCITY SERVICE KIT

GMC 7832259
GMC 7836405
GMC 7845009
GMC 7845015
MOOG CV2360
MOOG CV2361
NEAPCO 86-1050
PRECISION 2361
SPICER . . . . 6-109X(8)
TRW 22583

Top of this column (above IDLI GROUP 86121):

NEAPCO 86-0990
PRECISION 2360
SPICER . . . . 6-109X(8)
TRW 22583
TRW 22595

IDLI GROUP 86122

CONSTANT VELOCITY SERVICE KIT

AUDI 171407311D
AUDI 321407285B
AUDI 321498203
MIDAS CV1102
MOOG CV2304
MOOG CV2370
NEAPCO 86-0905
PRECISION 2370
SPICER 6-1001X
SPICER 6-100X
SPICER . . . . 6-100X(8)
SPICER 970164X
TRW 20195
TRW 22511
VOLKSWAGEN 171407331D
WORLDPARTS W62-771
WORLDPARTS W62-772

IDLI GROUP 86123

CONSTANT VELOCITY SERVICE KIT

MIDAS CV1103
MOOG CV2371
NEAPCO 86-0901
PRECISION 2371
TRW 22580
VOLKSWAGEN 171407311

IDLI GROUP 86124

CONSTANT VELOCITY SERVICE KIT

AUDI 321-407-311L
MIDAS CV1104
MOOG CV2372
NEAPCO 86-0918
PRECISION 2372
SPICER . . . . 6-123X(8)
TRW 22513
VOLKSWAGEN 321-407-311L

IDLI GROUP 86125

CONSTANT VELOCITY SERVICE KIT

MIDAS CV1105
MOOG CV2373
NEAPCO 86-0903
PRECISION 2373
SPICER . . . . 6-126X(8)
TRW 22581
VOLKSWAGEN 171-498-99B

IDLI GROUP 86126

CONSTANT VELOCITY SERVICE KIT

FORD 22328
FORD 77TT3K187AA
FORD D8RZ3A331A
FORD D8RZ3K187A
MIDAS CV1106
MOOG CV2376
NEAPCO 86-0930
PRECISION 2376
SPICER . . . . 6-105X(8)
TRW 22544
WORLDPARTS W62-780

IDLI GROUP 86127

CONSTANT VELOCITY SERVICE KIT

FORD E1FZ3A331A
FORD E2GZ3B413A
FORD E2GZ3B413B
MIDAS CV1107
MOOG CV2377
NEAPCO 86-0933
PRECISION 2377
SPICER . . . . 6-111X(8)
TRW 22547

IDLI GROUP 86128

CONSTANT VELOCITY SERVICE KIT

CHRYSLER 3780299
CHRYSLER 4205853
CHRYSLER 5212558
CHRYSLER 5212607
MIDAS CV1108
MOOG CV2379
NEAPCO 86-0960
PRECISION 2379
SPICER . . . . 6-103X(8)
TRW 22516
WORLDPARTS W62-804

IDLI GROUP 86129

CONSTANT VELOCITY SERVICE KIT

CHRYSLER 4205853
CHRYSLER 5212591
CHRYSLER 5212607
MIDAS CV1109
MOOG CV2380
NEAPCO 86-0968
PRECISION 2380
SPICER . . . . 6-104X(8)
TRW 22521
WORLDPARTS W62-804

BD = BEARING DIAMETER (outside)   **BW** = BEARING WIDTH   **CB** = CROSS LENGTH WITH BEARINGS   **CC** = CENTER TO CENTER   **CD** = CROSS DIAMETER
CL = CROSS LENGTH WITHOUT BEARINGS   **EC** = END TO CENTER or FACE TO FACE   **EE** = END TO END   **EL** = EFFECTIVE LENGTH
HD = HUB DIAMETER (or insert)   **IE** = INSIDE OF EARS (or recess)   **OD** = OUTSIDE DIAMETER   **OE** = OUTSIDE OF EARS   **SB** = SPLINE OR BORE SIZE

IDLI GROUP 86130

CONSTANT VELOCITY SERVICE KIT

CHRYSLER	4186890
CHRYSLER	4205853
CHRYSLER	5212607
CHRYSLER	5212621
MIDAS	CV1110
MOOG	CV2381
NEAPCO	86-0969
PRECISION	2381
SPICER	6-127X(8)
TRW	22522
WORLDPARTS	W62-804

IDLI GROUP 86131

CONSTANT VELOCITY SERVICE KIT

CHRYSLER	4186891
CHRYSLER	4205853
CHRYSLER	5212606
CHRYSLER	5212607
MIDAS	CV1111
MOOG	CV2382
NEAPCO	86-0972
PRECISION	2382
SPICER	6-128X(8)
TRW	22525
WORLDPARTS	W62-804

IDLI GROUP 86132

CONSTANT VELOCITY SERVICE KIT

CHRYSLER	5212856
CHRYSLER	5212858
NEAPCO	86-0947
PRECISION	2383
SPICER	6-144X
TRW	22531

IDLI GROUP 86133

CONSTANT VELOCITY SERVICE KIT

MIDAS	CV1112
NEAPCO	86-0910
PRECISION	2384
TRW	22510

IDLI GROUP 86134

CONSTANT VELOCITY SERVICE KIT

AUDI	803-407-275
AUDI	803407275A
AUDI	803498203A
MIDAS	CV1113
MOOG	CV2384
MOOG	CV2385
NEAPCO	86-0911
PRECISION	2385
SPICER	6-100X
SPICER	6-102X(8)
SPICER	970163X
TRW	22506
VOLKSWAGEN	171-407-311D
VOLKSWAGEN	803-407-275
VOLKSWAGEN	803-407-275A

IDLI GROUP 86135

CONSTANT VELOCITY SERVICE KIT

AUDI	431407275
AUDI	431498203C
MIDAS	CV1114
MOOG	CV2386
PRECISION	2386
SPICER	6-106X(8)
TRW	22506
WORLDPARTS	W62-774
WORLDPARTS	W62-776

IDLI GROUP 86136

CONSTANT VELOCITY SERVICE KIT

AUDI	431-407-311C
MIDAS	CV1115
MOOG	CV2387
NEAPCO	86-0917
PRECISION	2387
SPICER	6-122X(8)
TRW	22509

IDLI GROUP 86137

CONSTANT VELOCITY SERVICE KIT

AUDI	321-407-311M
MIDAS	CV1116
MOOG	CV2388
NEAPCO	86-0919
PRECISION	2388
SPICER	6-124X(8)
TRW	22514

IDLI GROUP 86138

CONSTANT VELOCITY SERVICE KIT

FIAT	4215015
FIAT	4247399
FIAT	4247400
FIAT	4287138
FIAT	4308412
FIAT	4308650
FIAT	4313087
FIAT	4328468
FIAT	4383758
FIAT-ALLIS	4247399
FIAT-ALLIS	4247400
FIAT-ALLIS	4287138
FIAT-ALLIS	4308650
FIAT-ALLIS	4328468
MIDAS	CV1117
MOOG	CV2389
NEAPCO	86-0920
PRECISION	2389
SPICER	6-107X(8)
TRW	22538
WORLDPARTS	W62-783

IDLI GROUP 86139

CONSTANT VELOCITY SERVICE KIT

AMER.MOTOR	8132433
AMER.MOTOR	8132441
AMER.MOTOR	8983500976
MIDAS	CV1118
MOOG	CV2393
NEAPCO	86-0950
PRECISION	2393
SPICER	6-110X(8)
TRW	22300
TRW	22500

IDLI GROUP 86140

CONSTANT VELOCITY SERVICE KIT

AMER.MOTOR	8983500495
AMER.MOTOR	8983500497
MOOG	CV2394
MOOG	CV2494
NEAPCO	86-1038
PRECISION	2394
SPICER	6-1451X
SPICER	6-145X
TRW	22501

IDLI GROUP 86141

CONSTANT VELOCITY SERVICE KIT

MOOG	CV2398
NEAPCO	86-0945
PRECISION	2398
TRW	22577

IDLI GROUP 86142

CONSTANT VELOCITY SERVICE KIT

MOOG	CV2399
NEAPCO	86-0946
PRECISION	2399
TRW	22576

IDLI GROUP 86143

CONSTANT VELOCITY SERVICE KIT

AEC	26005
ALLOY	10015
AUDI	113501149
AUDI	321498201
MIDAS	CV1119
MOOG	CV2300-2303
MOOG	CV2337
MOOG	CV2400
NEAPCO	85-0905
PRECISION	2300-2303
PRECISION	2400
SPICER	6-2001X
SPICER	970167X
TRW	10015
TRW	15010
TRW	22311
VOLKSWAGEN	113501149
VOLKSWAGEN	171-498-103
VOLKSWAGEN	171-498-103C
VOLKSWAGEN	211-598-101
VOLKSWAGEN	321-498-103
VOLKSWAGEN	321498201
VOLKSWAGEN	321-498-201A
VOLKSWAGEN	321498203
WESCO	10015
WORLDPARTS	62-751
WORLDPARTS	W62-717
WORLDPARTS	W62-751
WORLDPARTS	W62-754

IDLI GROUP 86144

CONSTANT VELOCITY SERVICE KIT

MIDAS	CV1120
MOOG	CV2401
NEAPCO	85-0913
PRECISION	2401
SPICER	6-2011X
SPICER	970168X
VOLKSWAGEN	211-598-201
VOLKSWAGEN	221501149
WORLDPARTS	62-752
WORLDPARTS	W62-718
WORLDPARTS	W62-752
WORLDPARTS	W62-756
WORLDPARTS	W62-898

IDLI GROUP 86145

CONSTANT VELOCITY SERVICE KIT

MIDAS	CV1121
MOOG	CV2402
NEAPCO	85-0939
PORSCHE	411501149A
PORSCHE	94433190300
PRECISION	2402
PRECISION	2498
SPICER	6-1121X
SPICER	6-1141X
SPICER	6-1161X
SPICER	6-1171X
SPICER	6-2021X
SPICER	6-2301X
SPICER	6-2311X
SPICER	6-2321X
SPICER	6-2341X
SPICER	6-2351X
SPICER	970169X
VOLKSWAGEN	411501149A
VOLKSWAGEN	411-598-201
WORLDPARTS	62-753
WORLDPARTS	W62-716
WORLDPARTS	W62-736
WORLDPARTS	W62-740
WORLDPARTS	W62-741
WORLDPARTS	W62-746
WORLDPARTS	W62-753
WORLDPARTS	W62-756
WORLDPARTS	W62-898

ENGINEERING CATALOGS MUST BE CONSULTED FOR DETAILS NOT INCLUDED
IN THIS GUIDE. SPECIFIC DESIGNS, MATERIAL CONTENT, TOLERANCES
LUBE FITTINGS AND OTHER DIMENSIONS ARE INTENTIONALLY OMITTED HERE.

CAUTION: BE SURE TO REFER TO ENGINEERING CATALOGS FOR SPECIAL APPLICATIONS THAT REQUIRE SPECIFIC MATERIAL CONTENT, TOLERANCES, ETC. SEE FOOTNOTE.

IDLI GROUP 86146
CONSTANT VELOCITY SERVICE KIT
MIDAS	CV1122
MOOG	CV2404
PRECISION	2404
TRW	22311
VOLKSWAGEN	171407282B

IDLI GROUP 86147
CONSTANT VELOCITY SERVICE KIT
AEC	26007
ALLOY	10021
GMC	7836928
GMC	7840071
GMC	7840072
MIDAS	CV1123
MOOG	CV2308
MOOG	CV2406
MOOG	CV5308
NEAPCO	85-0907
PRECISION	2406
SPICER	6-2221X
SPICER	6-222X
SPICER	6-285X
TRW	22345
WESCO	10021

IDLI GROUP 86148
CONSTANT VELOCITY SERVICE KIT
AEC	26001
ALLOY	10019
GMC	7545016
GMC	7832257
GMC	7833098
GMC	7837827
GMC	7837828
GMC	7840070
GMC	7845019
MIDAS	CV1124
MOOG	CV2307
MOOG	CV2407
NEAPCO	85-0901
PRECISION	2407
SPICER	6-2211X
SPICER	6-221X
TRW	22344
WESCO	10019

IDLI GROUP 86149
CONSTANT VELOCITY SERVICE KIT
AEC	26003
ALLOY	10018
FORD	E1FZ3A331B
FORD	E2GZ3B414B
MIDAS	CV1125
MOOG	CV2410
MOOG	CV8306
NEAPCO	85-0903
PRECISION	2410

SPICER	6-2271X
SPICER	6-262X
TRW	22331
WESCO	10018

IDLI GROUP 86150
CONSTANT VELOCITY SERVICE KIT
AEC	26116
FORD	D8RZ3A331B
MIDAS	CV1126
MOOG	CV2411
NEAPCO	85-0912
PRECISION	2411
SPICER	6-2081X
TRW	22329
TRW	22349
WORLDPARTS	W62-710
WORLDPARTS	W62-781

IDLI GROUP 86151
CONSTANT VELOCITY SERVICE KIT
AEC	26104
CHRYSLER	5212612
MIDAS	CV1127
MOOG	CV2312
MOOG	CV2412
NEAPCO	85-0925
PRECISION	2412
SPICER	6-2461X
TRW	22317
WESCO	10061

IDLI GROUP 86152
CONSTANT VELOCITY SERVICE KIT
AEC	26105
CHRYSLER	5212613
MIDAS	CV1128
MOOG	CV2313
MOOG	CV2413
NEAPCO	85-0926
PRECISION	2413
SPICER	6-2471X
TRW	22318
WESCO	10062

IDLI GROUP 86153
CONSTANT VELOCITY SERVICE KIT
AEC	26110
MIDAS	CV1129
MOOG	CV2320
MOOG	CV2416
NEAPCO	85-0921
PRECISION	2416
SPICER	6-2061X
TRW	22314
WESCO	10058

IDLI GROUP 86154
CONSTANT VELOCITY SERVICE KIT
AEC	26111
CHRYSLER	4205852
CHRYSLER	5212859
GMC	7832255
GMC	7845171
MIDAS	CV1130
MOOG	CV2417
NEAPCO	85-0922
PRECISION	2417
SPICER	6-2071X
SPICER	6-268X
TRW	22316
WESCO	10057

IDLI GROUP 86155
CONSTANT VELOCITY SERVICE KIT
AEC	26108
MIDAS	CV1131
MOOG	CV2418
NEAPCO	85-0923
PRECISION	2418
SPICER	6-2052X
TRW	22323
WESCO	10055

IDLI GROUP 86156
CONSTANT VELOCITY SERVICE KIT
AEC	26109
MIDAS	CV1132
MOOG	CV2419
NEAPCO	85-0924
PRECISION	2419
SPICER	6-2051X
TRW	22315
WESCO	10054

IDLI GROUP 86157
CONSTANT VELOCITY SERVICE KIT
AEC	26008
ALLOY	10022
GMC	7834238
GMC	7834241
GMC	7845014
MOOG	CV2423
MOOG	CV5301
MOOG	CV5302
MOOG	CV5401
NEAPCO	85-0908
PRECISION	2423
SPICER	6-2361X
SPICER	6-274X
SPICER	6-286X
TRW	22350
WESCO	10022

IDLI GROUP 86158
CONSTANT VELOCITY SERVICE KIT
CHRYSLER	MB109012
CHRYSLER	MB109184
DATSUN	39741-11M25
DATSUN	40089M0820
DATSUN	40089M0826
DATSUN	40089M0827
HONDA	44315-634-003
HONDA	44315SA2-300
MIDAS	CV1133
MITSUBISHI	MB109012
MITSUBISHI	MB109184
MOOG	CV2428
NEAPCO	85-0910
PRECISION	2428
PRECISION	2498
SPICER	6-2341X
TRW	22355
WESCO	10013

IDLI GROUP 86159
CONSTANT VELOCITY SERVICE KIT
AUDI	803-498-203B
MIDAS	CV1134
MOOG	CV2429
NEAPCO	85-0932
PRECISION	2429
SPICER	6-1251X
TRW	22308

IDLI GROUP 86160
CONSTANT VELOCITY SERVICE KIT
AEC	26123
AUDI	803407283B
AUDI	803407331
AUDI	803498201A
MIDAS	CV1135
MOOG	6-2401X
MOOG	CV2330
MOOG	CV2430
NEAPCO	85-0933
PRECISION	2330
PRECISION	2430
SPICER	6-2041X
SPICER	6-204X
SPICER	6-204X(8)
SPICER	970162X
TRW	22309
WORLDPARTS	W62-702
WORLDPARTS	W62-773

IDLI GROUP 86161
CONSTANT VELOCITY SERVICE KIT
AEC	26127
AUDI	431498201A
AUDI	803407283C
MIDAS	CV1136
MOOG	CV2431
NEAPCO	85-0931
NEAPCO	85-0938
PRECISION	2431
SPICER	6-2121X
TRW	22305

BD=BEARING DIAMETER (outside)　　**BW**=BEARING WIDTH　　**CB**=CROSS LENGTH WITH BEARINGS　　**CC**=CENTER TO CENTER　　**CD**=CROSS DIAMETER
CL=CROSS LENGTH WITHOUT BEARINGS　　**EC**=END TO CENTER or FACE TO FACE　　**EE**=END TO END　　**EL**=EFFECTIVE LENGTH
HD=HUB DIAMETER (or insert)　　**IE**=INSIDE OF EARS (or recess)　　**OD**=OUTSIDE DIAMETER　　**OE**=OUTSIDE OF EARS　　**SB**=SPLINE OR BORE SIZE

WORLDPARTS W62-704

IDLI GROUP 86162

CONSTANT VELOCITY SERVICE KIT

AEC	26103
AUDI	431-498-201D
MIDAS	CV1137
MOOG	CV2332
MOOG	CV2333
MOOG	CV2432
NEAPCO	85-0934
NEAPCO	85-0936
PRECISION	2432
SPICER	6-2411X
TRW	22306
WORLDPARTS	W62-775

IDLI GROUP 86163

CONSTANT VELOCITY SERVICE KIT

AEC	26115
ALLOY	10067
FIAT	4204117
FIAT	4311356
FIAT	4330 88
FIAT	4333088
FIAT-ALLIS	4333088
MIDAS	CV1138
MOOG	CV2434
NEAPCO	85-0957
PRECISION	2434
SPICER	6-2151X
TRW	22325
WESCO	10067
WORLDPARTS	W62-711
WORLDPARTS	W62-782

IDLI GROUP 86164

CONSTANT VELOCITY SERVICE KIT

BMW	33-211-105-446
BMW	33211205816
MIDAS	CV1139
MOOG	CV2338-2339
MOOG	CV2438
NEAPCO	85-1010
PRECISION	2338-2339
PRECISION	2438
SPICER	6-2131X
WORLDPARTS	W62-713
WORLDPARTS	W62-785

IDLI GROUP 86165

CONSTANT VELOCITY SERVICE KIT

AEC	26124
BMW	33211205748
BMW	33211207037
MIDAS	CV1140
MOOG	CV2440
NEAPCO	85-1012
PRECISION	2440
SPICER	6-2141X
WORLDPARTS	W62-714
WORLDPARTS	W62-788

IDLI GROUP 86166

CONSTANT VELOCITY SERVICE KIT

AEC	26121
AEC	26125
AEC	26129
AEC	26130
GMC	94027768
GMC	94027769
MIDAS	CV1141
MOOG	CV2341
MOOG	CV2441
NEAPCO	85-1013
PRECISION	2441
SPICER	6-1121X
SPICER	6-1141X
SPICER	6-1171X
SPICER	6-2301X
SPICER	6-2311X
SPICER	6-2321X
SPICER	6-2351X
SPICER	6-2381X
SUBARU	26125
SUBARU	6232-07050
SUBARU	6232-07051
SUBARU	6232-07052
SUBARU	6232-07091
SUBARU	6232-07190
SUBARU	7232-22020
WORLDPARTS	W620744
WORLDPARTS	W62-735
WORLDPARTS	W62-737
WORLDPARTS	W62-738
WORLDPARTS	W62-739
WORLDPARTS	W62-743
WORLDPARTS	W62-851
WORLDPARTS	W62-898

IDLI GROUP 86167

CONSTANT VELOCITY SERVICE KIT

BMW	33-211-208-443
MIDAS	CV1142
MOOG	CV2342
MOOG	CV2442
NEAPCO	85-1014
PRECISION	2442
SPICER	6-2391X

IDLI GROUP 86168

CONSTANT VELOCITY SERVICE KIT

MIDAS	CV1143
NEAPCO	85-1015
PRECISION	2330
PRECISION	2443
SPICER	6-2401X

IDLI GROUP 86169

CONSTANT VELOCITY SERVICE KIT

AEC	26126
MIDAS	CV1144
MOOG	CV2349
MOOG	CV2444
NEAPCO	85-1020
PORSCHE	90133229310
PRECISION	2444

IDLI GROUP 86170

CONSTANT VELOCITY SERVICE KIT

MIDAS	CV1145
MOOG	CV2344-2349
MOOG	CV2445
NEAPCO	85-1021
PORSCHE	90133229310
PRECISION	2344-2349
PRECISION	2445
SPICER	6-2161X
WORLDPARTS	W62-779
WORLDPARTS	W62-788

IDLI GROUP 86171

CONSTANT VELOCITY SERVICE KIT

MIDAS	CV1146
MOOG	CV2447
PORSCHE	92333203700
PRECISION	2447
SPICER	6-2181X
WORLDPARTS	W62-778

IDLI GROUP 86172

CONSTANT VELOCITY SERVICE KIT

MIDAS	CV1147
MOOG	CV2350
MOOG	CV2448
NEAPCO	85-1023
PORSCHE	90133229312
PRECISION	2448
SPICER	6-2191X
WORLDPARTS	W62-707
WORLDPARTS	W62-777

IDLI GROUP 86173

CONSTANT VELOCITY SERVICE KIT

MIDAS	CV1148
MOOG	CV2450
NEAPCO	85-0955
PRECISION	2450
RENAULT	7701450617
TRW	22358

IDLI GROUP 86174

CONSTANT VELOCITY SERVICE KIT

AEC	26117
MIDAS	CV1149
MOOG	CV2351
MOOG	CV2451
NEAPCO	85-0951
NEAPCO	85-0954
NEAPCO	86-0940
PRECISION	2451
RENAULT	7701200922
RENAULT	7701457538
SPICER	6-1081X
TRW	22356

IDLI GROUP 86175

CONSTANT VELOCITY SERVICE KIT

ALLOY	10111
AMER.MOTOR	8132435
AMER.MOTOR	8133582
MIDAS	CV1150
MOOG	CV2453
NEAPCO	85-0916
PRECISION	2453
SPICER	6-2251X
TRW	22301
WESCO	10121

IDLI GROUP 86176

CONSTANT VELOCITY SERVICE KIT

MIDAS	CV1151
MOOG	CV2355
MOOG	CV2455
NEAPCO	85-0917
PRECISION	2455
SPICER	6-2231X
TRW	22302
WESCO	10111

IDLI GROUP 86177

CONSTANT VELOCITY SERVICE KIT

AEC	26112
GMC	7800666
MOOG	CV2456
NEAPCO	85-0970
PRECISION	2456
SPICER	6-1005X
TRW	22353
WESCO	10104

IDLI GROUP 86178

CONSTANT VELOCITY SERVICE KIT

AEC	26009
ALLOY	10023
GMC	7831589
GMC	7834239
GMC	7835070
GMC	7845005
MOOG	CV2457
MOOG	CV5341
MOOG	CV5342
MOOG	CV5441
NEAPCO	85-0909
PRECISION	2457
SPICER	6-1201X
SPICER	6-120X
SPICER	6-148X
TRW	22352
WESCO	10023

ENGINEERING CATALOGS MUST BE CONSULTED FOR DETAILS NOT INCLUDED
IN THIS GUIDE. SPECIFIC DESIGNS, MATERIAL CONTENT, TOLERANCES
LUBE FITTINGS AND OTHER DIMENSIONS ARE INTENTIONALLY OMITTED HERE.

IDLI GUIDE — COPYRIGHT © INTERCHANGE, INC. — ST. LOUIS PARK, MN. 55416 USA

CAUTION: BE SURE TO REFER TO ENGINEERING CATALOGS FOR SPECIAL APPLICATIONS THAT REQUIRE SPECIFIC MATERIAL CONTENT, TOLERANCES, ETC. SEE FOOTNOTE.

IDLI GROUP 86179

CONSTANT VELOCITY SERVICE KIT

AEC	26002
ALLOY	10020
GMC	7832258
GMC	7836405
GMC	7838306
GMC	7840950
GMC	7842034
GMC	7842037
GMC	7843919
GMC	7844747
GMC	7844785
GMC	7844786
GMC	7844997
GMC	7845015
GMC	7845018
GMC	7845020
GMC	7845021
GMC	7845024
GMC	7845026
GMC	7845028
GMC	7845029
GMC	7845030
GMC	7845031
GMC	7845915
MIDAS	CV1152
MOOG	CV2360
MOOG	CV2361
MOOG	CV2460
MOOG	CV5347
MOOG	CV5348
MOOG	CV5349
MOOG	CV5352
MOOG	CV5353
MOOG	CV5354
MOOG	CV5358
MOOG	CV5362
MOOG	CV5366
MOOG	CV5367
MOOG	CV5447
MOOG	CV5448
MOOG	CV5449
MOOG	CV5462
MOOG	CV5467
NEAPCO	85-0902
NEAPCO	85-0975
PRECISION	2460
PRECISION	5447
SPICER	6-1091X
SPICER	6-149X
SPICER	6-150X
SPICER	6-152X
SPICER	6-1531X
TRW	22343
TRW	22346
WESCO	10020

IDLI GROUP 86180

CONSTANT VELOCITY SERVICE KIT

AEC	26006
ALLOY	10016
AUDI	321407285B
AUDI	321498203
MIDAS	CV1153
MOOG	CV2370-2371
MOOG	CV2470
NEAPCO	85-0906
PRECISION	2370-2371
PRECISION	2470
SPICER	6-1001X
SPICER	970164X
TRW	10016
TRW	22310
TRW	22511
VOLKSWAGEN	22511
VOLKSWAGEN	22580
VOLKSWAGEN	321407285B
VOLKSWAGEN	321498203
WESCO	10016
WORLDPARTS	W62-701
WORLDPARTS	W62-771

IDLI GROUP 86181

CONSTANT VELOCITY SERVICE KIT

AUDI	171-498-203
MIDAS	CV1154
MOOG	CV2372
MOOG	CV2373
MOOG	CV2471
NEAPCO	85-0937
PRECISION	2471
SPICER	6-1231X
TRW	22312
VOLKSWAGEN	171-498-099B
VOLKSWAGEN	171-498-203
WORLDPARTS	W62-772

IDLI GROUP 86182

CONSTANT VELOCITY SERVICE KIT

ALLOY	10017
FORD	D8RZ3A331A
FORD	E1FZ3A331A
FORD	E3FZ3B413B
MIDAS	CV1155
MOOG	CV2377
MOOG	CV2476
MOOG	CV2477
MOOG	CV8346
NEAPCO	85-0914
PRECISION	2476
SPICER	6-1051X
SPICER	6-1111X
SPICER	6-140X
TRW	22328
WESCO	10017

IDLI GROUP 86183

CONSTANT VELOCITY SERVICE KIT

AEC	26004
CHRYSLER	5212809
CHRYSLER	5212830
FORD	E1FZ3A331A
MIDAS	CV1156
NEAPCO	85-0904
NEAPCO	85-0944
NEAPCO	86-0949
PRECISION	2477
SPICER	6-1051X
SPICER	6-143X
TRW	22321
TRW	22330
TRW	22536
WORLDPARTS	W62-709
WORLDPARTS	W62-780

IDLI GROUP 86184

CONSTANT VELOCITY SERVICE KIT

AEC	26011
ALLOY	10056
CHRYSLER	4205853
CHRYSLER	5212858
CHRYSLER	5212869
MIDAS	CV1157
MOOG	CV2381
MOOG	CV2382
MOOG	CV2383
MOOG	CV2479
MOOG	CV2482
MOOG	CV2483
NEAPCO	85-0920
NEAPCO	86-0948
PRECISION	2479
SPICER	6-1031X
SPICER	6-1271X
SPICER	6-142X
SPICER	6-144X
TRW	22313
TRW	22533
WESCO	10056
WORLDPARTS	W62-804

IDLI GROUP 86185

CONSTANT VELOCITY SERVICE KIT

MIDAS	CV1158
PRECISION	2482
TRW	22313

IDLI GROUP 86186

CONSTANT VELOCITY SERVICE KIT

NEAPCO	85-0943
PRECISION	2483
TRW	22319

IDLI GROUP 86187

CONSTANT VELOCITY SERVICE KIT

AUDI	803498203A
MIDAS	CV1159
MOOG	CV2384
MOOG	CV2386
MOOG	CV2485
NEAPCO	85-0930
PRECISION	2485
SPICER	6-1021X
SPICER	970163X
TRW	22304
VOLKSWAGEN	431-498-203C
WORLDPARTS	W62-703
WORLDPARTS	W62-705
WORLDPARTS	W62-774
WORLDPARTS	W62-776

IDLI GROUP 86188

CONSTANT VELOCITY SERVICE KIT

AUDI	431498203C
MIDAS	CV1160
MOOG	CV2486
PRECISION	2486
TRW	22304
TRW	22506

IDLI GROUP 86189

CONSTANT VELOCITY SERVICE KIT

AEC	26102
AUDI	431-498-203E
MIDAS	CV1161
MOOG	CV2487
NEAPCO	85-0935
PRECISION	2487
SPICER	6-1221X
TRW	22307

IDLI GROUP 86190

CONSTANT VELOCITY SERVICE KIT

AEC	26114
ALLOY	10066
FIAT	4215015
FIAT	4383758
MIDAS	CV1162
MOOG	CV2389
MOOG	CV2489
NEAPCO	85-0956
PRECISION	2489
SPICER	6-1071X
TRW	22324
WESCO	10066
WORLDPARTS	W62-712
WORLDPARTS	W62-783

BD = BEARING DIAMETER (outside) BW = BEARING WIDTH CB = CROSS LENGTH WITH BEARINGS CC = CENTER TO CENTER CD = CROSS DIAMETER CL = CROSS LENGTH WITHOUT BEARINGS EC = END TO CENTER or FACE TO FACE EE = END TO END EL = EFFECTIVE LENGTH HD = HUB DIAMETER (or insert) IE = INSIDE OF EARS (or recess) OD = OUTSIDE DIAMETER OE = OUTSIDE OF EARS SB = SPLINE OR BORE SIZE

IDLI GROUP 86191

CONSTANT VELOCITY SERVICE KIT

AEC	26101
AEC	26128
ALLOY	10110
AMER.MOTOR	8132433
AMER.MOTOR	8983500976
MIDAS	CV1163
MOOG	CV2493
NEAPCO	85-0915
PRECISION	2493
SPICER	6-1101X
TRW	22300
WESCO	10110

IDLI GROUP 86192

CONSTANT VELOCITY SERVICE KIT

MIDAS	CV1164
NEAPCO	85-0979
PRECISION	2494
SPICER	6-1451X
TRW	22303

IDLI GROUP 86193

CONSTANT VELOCITY SERVICE KIT

AEC	26010
ALLOY	10013
ALLOY	10014
CHRYSLER	MB109183
CHRYSLER	MB176162
DATSUN	39241-11M25
DATSUN	40088M0820
DATSUN	40088M0825
DATSUN	40088M0826
DATSUN	40089M0826
HONDA	44315-634-013
HONDA	44315SA2-310
HONDA	44333-634-013
HONDA	44333SA2-310
MIDAS	CV1165
MITSUBISHI	MB109183
MITSUBISHI	MB176162
MOOG	CV2499
NEAPCO	85-0911
PRECISION	2498
PRECISION	2499
SPICER	6-1161X
SPICER	6-1191X
SPICER	6-2341X
SPICER	6-2351X
TRW	10014
TRW	2235
TRW	22354
WESCO	10014
WORLDPARTS	W62-733
WORLDPARTS	W62-734
WORLDPARTS	W62-742
WORLDPARTS	W62-747
WORLDPARTS	W62-850
WORLDPARTS	W62-898

IDLI GROUP 86194

CONSTANT VELOCITY SERVICE KIT

GMC	7834245
GMC	7834303
MOOG	CV5203
PRECISION	5203
TRW	22131R

IDLI GROUP 86195

CONSTANT VELOCITY SERVICE KIT

GMC	7834244
GMC	7834304
MOOG	CV5204
PRECISION	5204
TRW	22130R

IDLI GROUP 86196

CONSTANT VELOCITY SERVICE KIT

GMC	7832254
GMC	7840074
GMC	7840152
MOOG	CV5207
NEAPCO	89-8103
PRECISION	5207
TRW	22096R

IDLI GROUP 86197

CONSTANT VELOCITY SERVICE KIT

GMC	7832253
GMC	7840073
GMC	7840151
MOOG	CV5208
NEAPCO	89-8102
PRECISION	5208
TRW	22095R

IDLI GROUP 86198

CONSTANT VELOCITY SERVICE KIT

GMC	7832252
GMC	7840075
GMC	7840150
MOOG	CV5209
NEAPCO	89-8101
PRECISION	5209
TRW	22097R

IDLI GROUP 86199

CONSTANT VELOCITY SERVICE KIT

GMC	7832251
GMC	7840076
GMC	7840149
MOOG	CV5210
NEAPCO	89-8100
PRECISION	5210
TRW	22098R

IDLI GROUP 86200

CONSTANT VELOCITY SERVICE KIT

GMC	7842336
GMC	7842347
MOOG	CV5211
NEAPCO	89-8205
PRECISION	5211
TRW	22099R

IDLI GROUP 86201

CONSTANT VELOCITY SERVICE KIT

GMC	7842335
GMC	7842346
MOOG	CV5212
NEAPCO	89-8206
PRECISION	5212
TRW	22100R

IDLI GROUP 86202

CONSTANT VELOCITY SERVICE KIT

GMC	7844748
GMC	7844749
MOOG	CV5213
NEAPCO	89-8220
PRECISION	5213
TRW	22114R

IDLI GROUP 86203

CONSTANT VELOCITY SERVICE KIT

GMC	7844745
GMC	7844746
MOOG	CV5214
NEAPCO	89-8221
PRECISION	5214
TRW	22115R

IDLI GROUP 86204

CONSTANT VELOCITY SERVICE KIT

GMC	7844756
GMC	7844757
MOOG	CV5215
NEAPCO	89-8222
PRECISION	5215
TRW	22116R

IDLI GROUP 86205

CONSTANT VELOCITY SERVICE KIT

GMC	7844754
GMC	7844755
MOOG	CV5216
NEAPCO	89-8224
PRECISION	5216
TRW	22117R

IDLI GROUP 86206

CONSTANT VELOCITY SERVICE KIT

GMC	7844752
NEAPCO	89-8060
PRECISION	5217
TRW	22103R

IDLI GROUP 86207

CONSTANT VELOCITY SERVICE KIT

GMC	7844750
GMC	7844751
MOOG	CV5218
NEAPCO	89-8080
PRECISION	5218
TRW	22104R

IDLI GROUP 86208

CONSTANT VELOCITY SERVICE KIT

GMC	7836413
GMC	7836414
MOOG	CV5225
NEAPCO	89-8015
PRECISION	5225
TRW	22111R

IDLI GROUP 86209

CONSTANT VELOCITY SERVICE KIT

GMC	7836440
GMC	7836441
MOOG	CV5226
NEAPCO	89-8016
PRECISION	5226
TRW	22112R

IDLI GROUP 86210

CONSTANT VELOCITY SERVICE KIT

GMC	7836462
GMC	7836463
MOOG	CV5227
NEAPCO	89-8053
PRECISION	5227
TRW	22113R

IDLI GROUP 86211

CONSTANT VELOCITY SERVICE KIT

GMC	7836409
GMC	7836411
MOOG	CV5228
NEAPCO	89-8054
PRECISION	5228
TRW	22118R

IDLI GROUP 86212

CONSTANT VELOCITY SERVICE KIT

GMC	7841256
GMC	7842332
GMC	7842345
MOOG	CV5229
NEAPCO	89-8051
NEAPCO	89-8064
PRECISION	5229

ENGINEERING CATALOGS MUST BE CONSULTED FOR DETAILS NOT INCLUDED
IN THIS GUIDE. SPECIFIC DESIGNS, MATERIAL CONTENT, TOLERANCES
LUBE FITTINGS AND OTHER DIMENSIONS ARE INTENTIONALLY OMITTED HERE.

CAUTION: BE SURE TO REFER TO ENGINEERING CATALOGS FOR SPECIAL APPLICATIONS THAT REQUIRE SPECIFIC MATERIAL CONTENT, TOLERANCES, ETC. SEE FOOTNOTE.

TRW 22119R

IDLI GROUP 86213

CONSTANT VELOCITY SERVICE KIT

GMC	7841257
GMC	7842331
GMC	7842344
MOOG	CV5230
NEAPCO	89-8052
NEAPCO	89-8065
PRECISION	5230
TRW	22120R

IDLI GROUP 86214

CONSTANT VELOCITY SERVICE KIT

GMC	7842258
GMC	7842334
MOOG	CV5231
NEAPCO	89-8049
PRECISION	5231
TRW	22121R

IDLI GROUP 86215

CONSTANT VELOCITY SERVICE KIT

GMC	7842256
GMC	7842333
MOOG	CV5232
NEAPCO	89-8050
PRECISION	5232
TRW	22122R

IDLI GROUP 86216

CONSTANT VELOCITY SERVICE KIT

GMC	7844539
PRECISION	5233

IDLI GROUP 86217

CONSTANT VELOCITY SERVICE KIT

GMC	7844536
PRECISION	5234

IDLI GROUP 86218

CONSTANT VELOCITY SERVICE KIT

GMC	7844766
GMC	7844767
MOOG	CV5235
NEAPCO	89-8062
PRECISION	5235
TRW	22123R

IDLI GROUP 86219

CONSTANT VELOCITY SERVICE KIT

GMC	7844764
GMC	7844765
MOOG	CV5236
NEAPCO	89-8063
PRECISION	5236
TRW	22124R

IDLI GROUP 86220

CONSTANT VELOCITY SERVICE KIT

GMC	7844752
GMC	7844753
MOOG	CV5217
MOOG	CV5237
NEAPCO	89-8060
PRECISION	5237
TRW	22103R
TRW	89-8209

IDLI GROUP 86221

CONSTANT VELOCITY SERVICE KIT

GMC	7844762
GMC	7844763
MOOG	CV5238
NEAPCO	89-8061
PRECISION	5238
TRW	22125R

IDLI GROUP 86222

CONSTANT VELOCITY SERVICE KIT

GMC	7843916
GMC	CV5251
MOOG	CV5251
NEAPCO	89-8211
PRECISION	5251
TRW	22101R
TRW	22105R

IDLI GROUP 86223

CONSTANT VELOCITY SERVICE KIT

GMC	7843892
GMC	7843915
GMC	7844759
MOOG	CV5252
MOOG	CV5258
NEAPCO	89-8212
PRECISION	5252
TRW	22102R
TRW	22106R

IDLI GROUP 86224

CONSTANT VELOCITY SERVICE KIT

GMC	7843887
GMC	7843912
MOOG	CV5253
NEAPCO	89-8207
PRECISION	5253
TRW	22101R
TRW	22109R

IDLI GROUP 86225

CONSTANT VELOCITY SERVICE KIT

GMC	7843886
GMC	7843911
MOOG	CV5254
NEAPCO	89-8208
PRECISION	5254
TRW	22102R
TRW	22110R

IDLI GROUP 86226

CONSTANT VELOCITY SERVICE KIT

GMC	7843914
NEAPCO	82-8209
PRECISION	5255
TRW	22107R

IDLI GROUP 86227

CONSTANT VELOCITY SERVICE KIT

GMC	7843888
GMC	7843913
MOOG	CV5256
NEAPCO	89-8210
PRECISION	5256
TRW	22104R
TRW	22108R

IDLI GROUP 86228

CONSTANT VELOCITY SERVICE KIT

GMC	7843891
GMC	7844760
GMC	7844761
MOOG	CV5255
MOOG	CV5257
NEAPCO	89-8213
PRECISION	5257
TRW	22105R
TRW	22107R

IDLI GROUP 86229

CONSTANT VELOCITY SERVICE KIT

GMC	7844758
NEAPCO	89-8214
PRECISION	5258
TRW	22106R
TRW	22108R

IDLI GROUP 86230

CONSTANT VELOCITY SERVICE KIT

GMC	7844542
GMC	7844671
MOOG	CV5271
PRECISION	5271
TRW	22126R

IDLI GROUP 86231

CONSTANT VELOCITY SERVICE KIT

GMC	7844541
GMC	7844672
MOOG	CV5272
PRECISION	5272
TRW	22127R

IDLI GROUP 86232

CONSTANT VELOCITY SERVICE KIT

GMC	7844804
GMC	7844805
GMC	CV5273
PRECISION	5273
TRW	22129R

IDLI GROUP 86233

CONSTANT VELOCITY SERVICE KIT

GMC	7842340
GMC	7842351
MOOG	CV5274
PRECISION	5274
TRW	22128R

IDLI GROUP 86234

CONSTANT VELOCITY SERVICE KIT

CHRYSLER	5212810
CHRYSLER	5212811
GMC	7831590
GMC	7831591
GMC	7845006
GMC	7845007
NEAPCO	86-0957
NEAPCO	86-0958
PRECISION	5301
SPICER	6-2671X
SPICER	6-269X
SPICER	6-286X
TRW	22320
TRW	22322
TRW	22534
TRW	22568
TRW	22596

IDLI GROUP 86235

CONSTANT VELOCITY SERVICE KIT

GMC	7834238
GMC	7834241
GMC	7845012
GMC	7845014
PRECISION	5302
SPICER	6-274X
TRW	22569

BD = BEARING DIAMETER (outside)   **BW** = BEARING WIDTH   **CB** = CROSS LENGTH WITH BEARINGS   **CC** = CENTER TO CENTER   **CD** = CROSS DIAMETER
CL = CROSS LENGTH WITHOUT BEARINGS   **EC** = END TO CENTER or FACE TO FACE   **EE** = END TO END   **EL** = EFFECTIVE LENGTH
HD = HUB DIAMETER (or insert)   **IE** = INSIDE OF EARS (or recess)   **OD** = OUTSIDE DIAMETER   **OE** = OUTSIDE OF EARS   **SB** = SPLINE OR BORE SIZE

IDLI GROUP 86236

CONSTANT VELOCITY SERVICE KIT

AEC	26007
ALLOY	10021
GMC	7836928
GMC	7837828
GMC	7840071
GMC	7840072
MOOG	CV2308
MOOG	CV2406
MOOG	CV5308
NEAPCO	86-0992
NEAPCO	86-0993
PRECISION	5307
PRECISION	5308
SPICER	6-222X
SPICER	6-285X
TRW	22574

IDLI GROUP 86237

CONSTANT VELOCITY SERVICE KIT

GMC	7838298
GMC	7838299
GMC	7838304
GMC	7842152
MOOG	CV5311
MOOG	CV5312
MOOG	CV5411
MOOG	CV5412
NEAPCO	86-0995
PRECISION	5311
SPICER	6-2601X
SPICER	6-2611X
SPICER	6-271X
TRW	22563

IDLI GROUP 86238

CONSTANT VELOCITY SERVICE KIT

GMC	7838288
GMC	7838298
GMC	7838299
GMC	7838304
GMC	7842152
MOOG	CV5311
MOOG	CV5312
MOOG	CV5411
MOOG	CV5412
NEAPCO	86-0998
PRECISION	5312
SPICER	6-2601X
SPICER	6-2611X
SPICER	6-271X
TRW	22589

IDLI GROUP 86239

CONSTANT VELOCITY SERVICE KIT

GMC	7842382
GMC	7843576
GMC	7843577
GMC	7845022
MOOG	CV5315
MOOG	CV5316
MOOG	CV5415
MOOG	CV5416
NEAPCO	86-1001
PRECISION	5315
SPICER	6-272X
SPICER	6-2781X
SPICER	6-2791X
TRW	22564

IDLI GROUP 86240

CONSTANT VELOCITY SERVICE KIT

GMC	7842382
GMC	7843576
GMC	7843577
GMC	7845022
MOOG	CV5315
MOOG	CV5316
MOOG	CV5415
MOOG	CV5416
NEAPCO	86-1003
PRECISION	5316
SPICER	6-272X
SPICER	6-2781X
SPICER	6-2791X
TRW	22590

IDLI GROUP 86241

CONSTANT VELOCITY SERVICE KIT

GMC	7842036
GMC	7842039
GMC	7842040
GMC	7842129
MOOG	CV5319
MOOG	CV5320
MOOG	CV5419
MOOG	CV5420
NEAPCO	86-1006
PRECISION	5319
SPICER	6-270X
SPICER	6-2771X
TRW	22565

IDLI GROUP 86242

CONSTANT VELOCITY SERVICE KIT

GMC	7842036
GMC	7842039
GMC	7842040
GMC	7842129
MOOG	CV5319
MOOG	CV5320
MOOG	CV5419
MOOG	CV5420
NEAPCO	86-1031
PRECISION	5320
SPICER	6-270X
SPICER	6-2761X
TRW	22591

IDLI GROUP 86243

CONSTANT VELOCITY SERVICE KIT

GMC	7842380
GMC	7843917
GMC	7843921
GMC	7843923
GMC	7843924
GMC	7845025
MOOG	CV5323
MOOG	CV5324
MOOG	CV5325
MOOG	CV5391
MOOG	CV5423
NEAPCO	86-1007
PRECISION	5323
SPICER	6-273X
SPICER	6-2831X
SPICER	6-284X
TRW	22566

IDLI GROUP 86244

CONSTANT VELOCITY SERVICE KIT

GMC	7843917
GMC	7843921
GMC	7843924
GMC	7845025
MOOG	CV5323
MOOG	CV5324
MOOG	CV5325
MOOG	CV5391
MOOG	CV5423
MOOG	CV5424
NEAPCO	86-1030
PRECISION	5324
SPICER	6-273X
SPICER	6-2821X
SPICER	6-2831X
TRW	22592

IDLI GROUP 86245

CONSTANT VELOCITY SERVICE KIT

GMC	7843917
GMC	7843921
GMC	7843924
GMC	7844538
GMC	7845025
MOOG	CV5323
MOOG	CV5324
MOOG	CV5325
MOOG	CV5391
MOOG	CV5423
PRECISION	5325
SPICER	6-273X
SPICER	6-2811X
SPICER	6-2821X
SPICER	6-2831X
TRW	22594

IDLI GROUP 86246

CONSTANT VELOCITY SERVICE KIT

AEC	26009
ALLOY	10023
GMC	7831589
GMC	7831592
GMC	7834239
GMC	7835070
GMC	7845005
GMC	7845008
MOOG	CV2457
MOOG	CV5341
MOOG	CV5342
MOOG	CV5441
PRECISION	5341
SPICER	6-120X
SPICER	6-148X
TRW	22554

IDLI GROUP 86247

CONSTANT VELOCITY SERVICE KIT

AEC	26009
ALLOY	10023
GMC	7831589
GMC	7834239
GMC	7835070
GMC	7845005
GMC	7845013
MOOG	CV2457
MOOG	CV5341
MOOG	CV5342
MOOG	CV5441
PRECISION	5342
SPICER	6-120X
SPICER	6-148X
TRW	22555

IDLI GROUP 86248

CONSTANT VELOCITY SERVICE KIT

GMC	7832259
GMC	7838306
GMC	7845009
GMC	7845018
NEAPCO	86-0994
PRECISION	5348
TRW	22552

IDLI GROUP 86249

CONSTANT VELOCITY SERVICE KIT

GMC	7832259
GMC	7840950
GMC	7845009
GMC	7845020
NEAPCO	86-1034
PRECISION	5349
TRW	22582

CAUTION: BE SURE TO REFER TO ENGINEERING CATALOGS FOR SPECIAL APPLICATIONS THAT REQUIRE SPECIFIC MATERIAL CONTENT, TOLERANCES, ETC. SEE FOOTNOTE.

IDLI GROUP 86250

CONSTANT VELOCITY SERVICE KIT

GMC	7844747
GMC	7844785
GMC	7844786
GMC	7845028
GMC	7845029
GMC	7845030
NEAPCO	86-0999
PRECISION	5352
SPICER	6-150X
TRW	22583

IDLI GROUP 86251

CONSTANT VELOCITY SERVICE KIT

AEC	26002
ALLOY	10020
GMC	7832258
GMC	7836405
GMC	7838306
GMC	7840950
GMC	7842034
GMC	7842037
GMC	7843919
GMC	7844747
GMC	7844785
GMC	7844786
GMC	7844997
GMC	7845015
GMC	7845018
GMC	7845020
GMC	7845021
GMC	7845024
GMC	7845026
GMC	7845027
GMC	7845028
GMC	7845029
GMC	7845030
GMC	7845031
GMC	7845915
GMC	CV5362
GMC	CV5447
GMC	CV5448
GMC	CV5449
GMC	CV5462
GMC	CV5467
MOOG	CV2360
MOOG	CV2361
MOOG	CV2460
MOOG	CV5347
MOOG	CV5348
MOOG	CV5349
MOOG	CV5352
MOOG	CV5353
MOOG	CV5354
MOOG	CV5358
MOOG	CV5359
MOOG	CV5362
MOOG	CV5366
MOOG	CV5367
NEAPCO	86-1035
PRECISION	5353
SPICER	6-149X
SPICER	6-1501X
SPICER	6-150X
SPICER	6-152X
SPICER	6-1531X
TRW	22582
TRW	22583

IDLI GROUP 86252

CONSTANT VELOCITY SERVICE KIT

AEC	26002
ALLOY	10020
GMC	7832258
GMC	7836405
GMC	7838306
GMC	7840950
GMC	7842034
GMC	7842037
GMC	7843919
GMC	7844747
GMC	7844785
GMC	7844786
GMC	7844997
GMC	7845015
GMC	7845018
GMC	7845020
GMC	7845021
GMC	7845024
GMC	7845026
GMC	7845027
GMC	7845028
GMC	7845029
GMC	7845030
GMC	7845031
GMC	7845915
MOOG	CV2360
MOOG	CV2361
MOOG	CV2460
MOOG	CV5347
MOOG	CV5348
MOOG	CV5349
MOOG	CV5352
MOOG	CV5353
MOOG	CV5354
MOOG	CV5358
MOOG	CV5359
MOOG	CV5362
MOOG	CV5366
MOOG	CV5367
MOOG	CV5447
MOOG	CV5448
MOOG	CV5449
MOOG	CV5462
MOOG	CV5467
PRECISION	5354
SPICER	6-149X
SPICER	6-1501X
SPICER	6-150X
SPICER	6-152X
SPICER	6-1531X
TRW	22593

IDLI GROUP 86253

CONSTANT VELOCITY SERVICE KIT

NEAPCO	86-1033
PRECISION	5358
TRW	22564
TRW	22582

IDLI GROUP 86254

CONSTANT VELOCITY SERVICE KIT

AEC	26002
ALLOY	10020
GMC	7832258
GMC	7836405
GMC	7838306
GMC	7840950
GMC	7842034
GMC	7842037
GMC	7843919
GMC	7844747
GMC	7844785
GMC	7844786
GMC	7844997
GMC	7845015
GMC	7845018
GMC	7845020
GMC	7845021
GMC	7845024
GMC	7845026
GMC	7845027
GMC	7845028
GMC	7845029
GMC	7845030
GMC	7845031
GMC	7845915
MOOG	CV2360
MOOG	CV2361
MOOG	CV2460
MOOG	CV5347
MOOG	CV5348
MOOG	CV5349
MOOG	CV5352
MOOG	CV5353
MOOG	CV5354
MOOG	CV5358
MOOG	CV5359
MOOG	CV5362
MOOG	CV5366
MOOG	CV5367
MOOG	CV5447
MOOG	CV5448
MOOG	CV5449
MOOG	CV5462
MOOG	CV5467
NEAPCO	86-1032
PRECISION	5359
SPICER	6-149X
SPICER	6-1501X
SPICER	6-150X
SPICER	6-152X
SPICER	6-1531X
TRW	22564
TRW	22582

IDLI GROUP 86255

CONSTANT VELOCITY SERVICE KIT

AEC	26002
ALLOY	10020
GMC	7832258
GMC	7836405
GMC	7838306
GMC	7840950
GMC	7842034
GMC	7842037
GMC	7843919
GMC	7844747
GMC	7844785
GMC	7844786
GMC	7844997
GMC	7845015
GMC	7845018
GMC	7845020
GMC	7845021
GMC	7845024
GMC	7845026
GMC	7845027
GMC	7845028
GMC	7845029
GMC	7845030
GMC	7845031
GMC	7845915
MOOG	CV2360
MOOG	CV2361
MOOG	CV2460
MOOG	CV5347
MOOG	CV5348
MOOG	CV5349
MOOG	CV5352
MOOG	CV5353
MOOG	CV5354
MOOG	CV5358
MOOG	CV5359
MOOG	CV5362
MOOG	CV5366
MOOG	CV5367
MOOG	CV5447
MOOG	CV5448
MOOG	CV5449
NEAPCO	86-1004
PRECISION	5362
SPICER	6-149X
SPICER	6-1501X
SPICER	6-150X
SPICER	6-152X
SPICER	6-1531X
TRW	22551

IDLI GROUP 86256

CONSTANT VELOCITY SERVICE KIT

AEC	26002
ALLOY	10020
GMC	7832258
GMC	7836405
GMC	7838306
GMC	7840950
GMC	7842034
GMC	7842037
GMC	7843919
GMC	7844747

BD = BEARING DIAMETER (outside)   **BW** = BEARING WIDTH   **CB** = CROSS LENGTH WITH BEARINGS   **CC** = CENTER TO CENTER   **CD** = CROSS DIAMETER
CL = CROSS LENGTH WITHOUT BEARINGS   **EC** = END TO CENTER or FACE TO FACE   **EE** = END TO END   **EL** = EFFECTIVE LENGTH
HD = HUB DIAMETER (or insert)   **IE** = INSIDE OF EARS (or recess)   **OD** = OUTSIDE DIAMETER   **OE** = OUTSIDE OF EARS   **SB** = SPLINE OR BORE SIZE

Column 1

GMC	7844785
GMC	7844786
GMC	7844997
GMC	7845015
GMC	7845018
GMC	7845020
GMC	7845021
GMC	7845024
GMC	7845026
GMC	7845027
GMC	7845028
GMC	7845029
GMC	7845030
GMC	7845031
GMC	7845915
MOOG	CV2360
MOOG	CV2361
MOOG	CV2460
MOOG	CV5347
MOOG	CV5348
MOOG	CV5349
MOOG	CV5352
MOOG	CV5353
MOOG	CV5354
MOOG	CV5358
MOOG	CV5359
MOOG	CV5362
MOOG	CV5366
MOOG	CV5367
MOOG	CV5447
MOOG	CV5448
MOOG	CV5449
MOOG	CV5462
MOOG	CV5467
PRECISION	5366
SPICER	6-149X
SPICER	6-1501X
SPICER	6-150X
SPICER	6-152X
SPICER	6-1531X

IDLI GROUP 86257
CONSTANT VELOCITY SERVICE KIT

AEC	26002
ALLOY	10020
GMC	7832258
GMC	7836405
GMC	7838306
GMC	7840950
GMC	7842034
GMC	7842037
GMC	7843919
GMC	7844747
GMC	7844785
GMC	7844786
GMC	7844997
GMC	7845015
GMC	7845018
GMC	7845020
GMC	7845021
GMC	7845024
GMC	7845026
GMC	7845027
GMC	7845028
GMC	7845029

Column 2

GMC	7845030
GMC	7845031
GMC	7845915
MOOG	CV2360
MOOG	CV2361
MOOG	CV2460
MOOG	CV5347
MOOG	CV5348
MOOG	CV5349
MOOG	CV5352
MOOG	CV5353
MOOG	CV5354
MOOG	CV5358
MOOG	CV5359
MOOG	CV5362
MOOG	CV5366
MOOG	CV5367
MOOG	CV5447
MOOG	CV5448
MOOG	CV5449
MOOG	CV5462
MOOG	CV5467
PRECISION	5367
SPICER	6-149X
SPICER	6-1501X
SPICER	6-150X
SPICER	6-152X
SPICER	6-1531X
TRW	22553

IDLI GROUP 86258
CONSTANT VELOCITY SERVICE KIT

GMC	7834082
MOOG	CV5381
PRECISION	5381
TRW	22571

IDLI GROUP 86259
CONSTANT VELOCITY SERVICE KIT

GMC	7830112
GMC	7831588
GMC	7837071
MOOG	CV5382
PRECISION	5382
SPICER	6-280X
TRW	22570

IDLI GROUP 86260
CONSTANT VELOCITY SERVICE KIT

GMC	7837135
GMC	7840364
MOOG	CV5383
NEAPCO	86-0996
PRECISION	5383
SPICER	6-277X
TRW	22556

Column 3

IDLI GROUP 86261
CONSTANT VELOCITY SERVICE KIT

GMC	7837071
MOOG	CV5384
NEAPCO	86-0997
PRECISION	5384
SPICER	6-276X
TRW	22557

IDLI GROUP 86262
CONSTANT VELOCITY SERVICE KIT

GMC	7842380
MOOG	CV5385
NEAPCO	86-1000
PRECISION	5385
SPICER	6-281X
TRW	22558

IDLI GROUP 86263
CONSTANT VELOCITY SERVICE KIT

GMC	7842378
MOOG	CV5386
NEAPCO	86-1002
PRECISION	5386
SPICER	6-278X
TRW	22559
TRW	22589

IDLI GROUP 86264
CONSTANT VELOCITY SERVICE KIT

BMW	33-211-208-439
GMC	7842379
MOOG	CV2342
MOOG	CV5387
NEAPCO	86-1008
PRECISION	5387
SPICER	6-279X
TRW	22560

IDLI GROUP 86265
CONSTANT VELOCITY SERVICE KIT

GMC	7842380
GMC	7843923
MOOG	CV5389
NEAPCO	86-1005
PRECISION	5389
SPICER	6-284X
TRW	22567

IDLI GROUP 86266
CONSTANT VELOCITY SERVICE KIT

GMC	7843920
MOOG	CV5390
NEAPCO	86-1009
PRECISION	5390
SPICER	6-282X
TRW	22561

Column 4

IDLI GROUP 86267
CONSTANT VELOCITY SERVICE KIT

GMC	7842380
GMC	7843922
GMC	7843923
GMC	7843924
MOOG	CV5323
MOOG	CV5389
MOOG	CV5391
MOOG	CV5423
NEAPCO	86-1037
NEAPCO	86-1039
PRECISION	5391
SPICER	6-2831X
SPICER	6-283X
SPICER	6-284X
TRW	22562

IDLI GROUP 86268
CONSTANT VELOCITY SERVICE KIT

AEC	26113
ALLOY	10105
GMC	7802092
MOOG	CV2422
NEAPCO	85-0971
PRECISION	2422
PRECISION	5401
SPICER	6-2481X
TRW	22351
WESCO	10105

IDLI GROUP 86269
CONSTANT VELOCITY SERVICE KIT

GMC	7838298
GMC	7838299
GMC	7838304
GMC	7842152
MOOG	CV5311
MOOG	CV5312
MOOG	CV5411
MOOG	CV5412
NEAPCO	85-0974
PRECISION	5411
SPICER	6-2601X
SPICER	6-2611X
SPICER	6-271X
TRW	22333

IDLI GROUP 86270
CONSTANT VELOCITY SERVICE KIT

GMC	7838298
GMC	7838299
GMC	7838304
GMC	7842152
MOOG	CV5311
MOOG	CV5312
MOOG	CV5411
MOOG	CV5412
NEAPCO	85-0973
PRECISION	5412
SPICER	6-2601X
SPICER	6-2611X
SPICER	6-271X
TRW	22334

CAUTION: BE SURE TO REFER TO ENGINEERING CATALOGS FOR SPECIAL APPLICATIONS THAT REQUIRE SPECIFIC MATERIAL CONTENT, TOLERANCES, ETC. SEE FOOTNOTE.

IDLI GROUP 86271

CONSTANT VELOCITY SERVICE KIT

GMC	7842382
GMC	7843576
GMC	7843577
GMC	7845022
MOOG	CV5315
MOOG	CV5316
MOOG	CV5415
MOOG	CV5416
NEAPCO	85-0976
PRECISION	5415
SPICER	6-272X
SPICER	6-2791X
TRW	22335

IDLI GROUP 86272

CONSTANT VELOCITY SERVICE KIT

GMC	7842382
GMC	7843576
GMC	7843577
GMC	7845022
MOOG	CV5315
MOOG	CV5316
MOOG	CV5415
MOOG	CV5416
NEAPCO	85-0977
PRECISION	5416
SPICER	6-272X
SPICER	6-2781X
SPICER	6-2791X
TRW	22336

IDLI GROUP 86273

CONSTANT VELOCITY SERVICE KIT

GMC	7842036
GMC	7842039
GMC	7842040
GMC	7842129
MOOG	CV5319
MOOG	CV5419
MOOG	CV5420
NEAPCO	85-0919
PRECISION	5419
SPICER	6-270X
SPICER	6-2771X
TRW	22338

IDLI GROUP 86274

CONSTANT VELOCITY SERVICE KIT

GMC	7842036
GMC	7842039
GMC	7842040
GMC	7842129
MOOG	CV5319
MOOG	CV5320
MOOG	CV5419
MOOG	CV5420
NEAPCO	85-0928

(continued)

PRECISION	5420
SPICER	6-270X
SPICER	6-2761X
SPICER	6-2771X
TRW	22339

IDLI GROUP 86275

CONSTANT VELOCITY SERVICE KIT

GMC	7843917
GMC	7843921
GMC	7843923
GMC	7843924
GMC	7845025
MOOG	CV5323
MOOG	CV5324
MOOG	CV5325
MOOG	CV5391
MOOG	CV5423
NEAPCO	85-0927
PRECISION	5423
SPICER	6-273X
SPICER	6-2831X
SPICER	6-284X
TRW	22340

IDLI GROUP 86276

CONSTANT VELOCITY SERVICE KIT

GMC	7843917
GMC	7843921
GMC	7843924
GMC	7845025
MOOG	CV5323
MOOG	CV5324
MOOG	CV5325
MOOG	CV5391
MOOG	CV5423
MOOG	CV5424
NEAPCO	85-0940
PRECISION	5424
SPICER	6-273X
SPICER	6-2821X
SPICER	6-2831X
TRW	22341

IDLI GROUP 86277

CONSTANT VELOCITY SERVICE KIT

GMC	7844538
MOOG	CV5325
MOOG	CV5425
PRECISION	5425
SPICER	6-2811X
TRW	22349

IDLI GROUP 86278

CONSTANT VELOCITY SERVICE KIT

PRECISION	5441
TRW	22353

IDLI GROUP 86279

CONSTANT VELOCITY SERVICE KIT

NEAPCO	85-0972
PRECISION	5448
TRW	22332

IDLI GROUP 86280

CONSTANT VELOCITY SERVICE KIT

NEAPCO	85-0929
PRECISION	5449
TRW	22342

IDLI GROUP 86281

CONSTANT VELOCITY SERVICE KIT

MOOG	CV5454
PRECISION	5454
TRW	22348

IDLI GROUP 86282

CONSTANT VELOCITY SERVICE KIT

AEC	26002
ALLOY	10020
GMC	7832258
GMC	7836405
GMC	7838306
GMC	7840950
GMC	7842034
GMC	7842037
GMC	7843919
GMC	7844747
GMC	7844785
GMC	7844786
GMC	7844997
GMC	7845015
GMC	7845018
GMC	7845020
GMC	7845021
GMC	7845024
GMC	7845026
GMC	7845027
GMC	7845028
GMC	7845029
GMC	7845030
GMC	7845031
GMC	7845915
MOOG	CV2360
MOOG	CV2361
MOOG	CV2460
MOOG	CV5347
MOOG	CV5348
MOOG	CV5349
MOOG	CV5352
MOOG	CV5353
MOOG	CV5354
MOOG	CV5358
MOOG	CV5359
MOOG	CV5362
MOOG	CV5366
MOOG	CV5367
MOOG	CV5447
MOOG	CV5448
MOOG	CV5449
MOOG	CV5462
MOOG	CV5467

(continued)

NEAPCO	85-0918
PRECISION	5462
SPICER	6-149X
SPICER	6-1501X
SPICER	6-150X
SPICER	6-152X
SPICER	6-1531X
TRW	22337

IDLI GROUP 86283

CONSTANT VELOCITY SERVICE KIT

PRECISION	5467
TRW	22347

IDLI GROUP 86284

CONSTANT VELOCITY SERVICE KIT

FORD	E3FZ3B436D
MOOG	CV8207
NEAPCO	89-8091
PRECISION	8207
TRW	22084R

IDLI GROUP 86285

CONSTANT VELOCITY SERVICE KIT

FORD	E3FZ3B436G
MOOG	CV8209
NEAPCO	89-8093
PRECISION	8209
TRW	22086R

IDLI GROUP 86286

CONSTANT VELOCITY SERVICE KIT

FORD	E3FZ3B436F
MOOG	CV8210
NEAPCO	89-8090
PRECISION	8210
TRW	22083R

IDLI GROUP 86287

CONSTANT VELOCITY SERVICE KIT

FORD	E3FZ3B436H
MOOG	CV8211
NEAPCO	89-8092
PRECISION	8211
TRW	22085R

IDLI GROUP 86288

CONSTANT VELOCITY SERVICE KIT

FORD	E3FZ3B436C
FORD	E3FZ3B436E
MOOG	CV8212
NEAPCO	89-8094
PRECISION	8212
TRW	22087R

BD = BEARING DIAMETER (outside)   **BW** = BEARING WIDTH   **CB** = CROSS LENGTH WITH BEARINGS   **CC** = CENTER TO CENTER   **CD** = CROSS DIAMETER
CL = CROSS LENGTH WITHOUT BEARINGS   **EC** = END TO CENTER or FACE TO FACE   **EE** = END TO END   **EL** = EFFECTIVE LENGTH
HD = HUB DIAMETER (or insert)   **IE** = INSIDE OF EARS (or recess)   **OD** = OUTSIDE DIAMETER   **OE** = OUTSIDE OF EARS   **SB** = SPLINE OR BORE SIZE

IDLI GROUP 86289

CONSTANT VELOCITY SERVICE KIT

FORD	E43Z3B436B
MOOG	CV8213
NEAPCO	89-8074
PRECISION	8213
TRW	22093R

IDLI GROUP 86290

CONSTANT VELOCITY SERVICE KIT

FORD	E43Z3B436A
MOOG	CV8214
NEAPCO	89-8075
PRECISION	8214
TRW	22094R

IDLI GROUP 86291

CONSTANT VELOCITY SERVICE KIT

FORD	E43Z3B436D
MOOG	CV8215
NEAPCO	89-8095
PRECISION	8215
TRW	22089R

IDLI GROUP 86292

CONSTANT VELOCITY SERVICE KIT

FORD	E43Z3B436C
MOOG	CV8216
NEAPCO	89-8096
PRECISION	8216
TRW	22090R

IDLI GROUP 86293

CONSTANT VELOCITY SERVICE KIT

FORD	E4GZ3B436D
MOOG	CV8217
NEAPCO	89-8098
PRECISION	8217
TRW	22092R

IDLI GROUP 86294

CONSTANT VELOCITY SERVICE KIT

FORD	E4GZ3B436C
MOOG	CV8218
NEAPCO	89-8097
PRECISION	8218
TRW	22091R

IDLI GROUP 86295

CONSTANT VELOCITY SERVICE KIT

AEC	26003
ALLOY	10018
FORD	E1FZ3A331B
FORD	E2GZ3B414B
MOOG	CV2410
MOOG	CV8306
NEAPCO	86-0936
PRECISION	8306
SPICER	6-2271X
SPICER	6-262X
TRW	22550

IDLI GROUP 86296

CONSTANT VELOCITY SERVICE KIT

FORD	E1FZ3A331B
FORD	E3FZ3B414C
MOOG	CV8307
NEAPCO	86-0937
PRECISION	8307
SPICER	6-2271X
SPICER	6-263X
TRW	22540

IDLI GROUP 86297

CONSTANT VELOCITY SERVICE KIT

FORD	E43Z3A331B
FORD	E43Z3B414B
MOOG	CV8308
MOOG	CV8309
MOOG	CV8406
NEAPCO	22543
NEAPCO	86-0927
PRECISION	8308
SPICER	6-2641X
SPICER	6-264X

IDLI GROUP 86298

CONSTANT VELOCITY SERVICE KIT

FORD	E43Z3A331B
FORD	E43Z3B414C
MOOG	CV8308
MOOG	CV8309
MOOG	CV8406
NEAPCO	86-0929
PRECISION	8309
SPICER	6-2641X
SPICER	6-264X
SPICER	6-265X
TRW	22542

IDLI GROUP 86299

CONSTANT VELOCITY SERVICE KIT

ALLOY	10017
FORD	E1FZ3A331A
FORD	E3FZ3B413B
MOOG	CV2377
MOOG	CV2477
MOOG	CV8346
NEAPCO	86-0934
PRECISION	8346
SPICER	6-140X
TRW	22548

IDLI GROUP 86300

CONSTANT VELOCITY SERVICE KIT

FORD	E43Z3A331A
FORD	E43Z3B413B
MOOG	CV8347
MOOG	CV8446
NEAPCO	86-0928
PRECISION	8347
SPICER	6-1411X
SPICER	6-141X
TRW	22541

IDLI GROUP 86301

CONSTANT VELOCITY SERVICE KIT

FORD	E43Z3A331B
FORD	E43Z3B414B
MOOG	CV8308
MOOG	CV8309
MOOG	CV8406
NEAPCO	85-0942
PRECISION	8406
SPICER	6-2641X
SPICER	6-264X
TRW	22327
TRW	22543

IDLI GROUP 86302

CONSTANT VELOCITY SERVICE KIT

FORD	E43Z3A331A
FORD	E43Z3B413B
MOOG	CV8347
MOOG	CV8446
NEAPCO	85-0941
PRECISION	8446
SPICER	6-1411X
SPICER	6-141X
TRW	22326

ENGINEERING CATALOGS MUST BE CONSULTED FOR DETAILS NOT INCLUDED
IN THIS GUIDE. SPECIFIC DESIGNS, MATERIAL CONTENT, TOLERANCES
LUBE FITTINGS AND OTHER DIMENSIONS ARE INTENTIONALLY OMITTED HERE.

FOR YOUR CONVENIENCE
NOTES ON OTHER INFORMATION

We urge you to send us any data that
you feel should be included with the
future editions of the I.D.L.I. Guide.

IDLI GUIDE   COPYRIGHT © INTERCHANGE, INC.   ST. LOUIS PARK, MN. 55416 USA

IDLI GROUP 86601

DOUBLE CENTER
SEE FIGURE 58

1.438 in	IE	36.51 mm
.969 in	BD	24.61 mm
1.875 in	CC	47.63 mm

PRECISION 1925
ROCKWELL L6N170-1

IDLI GROUP 86602

DOUBLE CENTER
SEE FIGURE 58

1.438 in	IE	36.51 mm
.969 in	BD	24.61 mm
2.500 in	CC	63.50 mm

PRECISION 1926
ROCKWELL L6N171-1

IDLI GROUP 86603

DOUBLE CENTER
SEE FIGURE 59

1.438 in	IE	36.51 mm
.969 in	BD	24.61 mm
3.500 in	CC	88.90 mm

ROCKWELL L6NDCA22

IDLI GROUP 86604

DOUBLE CENTER
SEE FIGURE 59

1.438 in	IE	36.51 mm
.969 in	BD	24.61 mm
5.000 in	CC	127.00 mm

ROCKWELL L6NDCA3

IDLI GROUP 86605

CLAMP YOKE WITH 26 SERRATIONS
SEE FIGURE 12C

1.438 in	IE	36.51 mm
.969 in	BD	24.61 mm
.750 in	SB	19.05 mm
2.250 in	EC	57.15 mm

ROCKWELL L6NYS12-39

IDLI GROUP 86606

CLAMP YOKE WITH 26 SERRATIONS
slight variation from above group
SEE FIGURE 12C

1.438 in	IE	36.51 mm
.969 in	BD	24.61 mm
.750 in	SB	19.05 mm
2.250 in	EC	57.15 mm

ROCKWELL L6NYS12-27

IDLI GROUP 86607

CLAMP YOKE WITH 26 SERRATIONS
slight variation from above group
SEE FIGURE 12C

1.438 in	IE	36.51 mm
.969 in	BD	24.61 mm
.750 in	SB	19.05 mm
2.250 in	EC	57.15 mm

ROCKWELL L6NYS12-42

IDLI GROUP 86608

CLAMP YOKE WITH 26 SERRATIONS
slight variation from above group
SEE FIGURE 12C

1.438 in	IE	36.51 mm
.969 in	BD	24.61 mm
.750 in	SB	19.05 mm
2.250 in	EC	57.15 mm

ROCKWELL L6NYS12-48

IDLI GROUP 86609

CLAMP YOKE
WITH ROUND BORE
SEE FIGURE 12B

1.438 in	IE	36.51 mm
.969 in	BD	24.61 mm
.750 in	SB	19.05 mm
2.500 in	EC	63.50 mm

ROCKWELL L6NYR12-146

IDLI GROUP 86610

CLAMP YOKE WITH ROUND BORE
slight variation from above group
SEE FIGURE 12B

1.438 in	IE	36.51 mm
.969 in	BD	24.61 mm
.750 in	SB	19.05 mm
2.500 in	EC	63.50 mm

ROCKWELL L6NYR12-140

IDLI GROUP 86611

CLAMP YOKE
WITH ROUND BORE
SEE FIGURE 12B

1.438 in	IE	36.51 mm
.969 in	BD	24.61 mm
.750 in	SB	19.05 mm
2.625 in	EC	66.68 mm

ROCKWELL L6NYR12-30

IDLI GROUP 86612

CLAMP YOKE WITH 18 SERRATIONS
SEE FIGURE 12C

1.438 in	IE	36.51 mm
.969 in	BD	24.61 mm
.813 in	SB	20.64 mm
2.141 in	EC	54.37 mm

ROCKWELL L6NYS13-25

IDLI GROUP 86613

CLAMP YOKE WITH 36 SERRATIONS
SEE FIGURE 12C

1.438 in	IE	36.51 mm
.969 in	BD	24.61 mm
.813 in	SB	20.64 mm
2.141 in	EC	54.37 mm

ROCKWELL L6NYS14-27

IDLI GROUP 86614

CLAMP YOKE WITH 18 SERRATIONS
SEE FIGURE 12C

1.438 in	IE	36.51 mm
.969 in	BD	24.61 mm
.813 in	SB	20.64 mm
2.156 in	EC	54.77 mm

ROCKWELL L6NYS13-31

IDLI GROUP 86615

CLAMP YOKE WITH 29 SERRATIONS
SEE FIGURE 12C

1.438 in	IE	36.51 mm
.969 in	BD	24.61 mm
.813 in	SB	20.64 mm
2.500 in	EC	63.50 mm

ROCKWELL L6NYS14-39

IDLI GROUP 86616

CLAMP YOKE WITH 18 SERRATIONS
SEE FIGURE 12C

1.438 in	IE	36.51 mm
.969 in	BD	24.61 mm
.813 in	SB	20.64 mm
2.703 in	EC	68.66 mm

ROCKWELL L6NYS13-6

IDLI GROUP 86617

CLAMP YOKE WITH 18 SERRATIONS
SEE FIGURE 12C

1.438 in	IE	36.51 mm
.969 in	BD	24.61 mm
.813 in	SB	20.64 mm
3.297 in	EC	83.74 mm

ROCKWELL L6NYS13-21

IDLI GROUP 86618

CLAMP YOKE WITH 36 SERRATIONS
SEE FIGURE 12C

1.438 in	IE	36.51 mm
.969 in	BD	24.61 mm
.813 in	SB	20.64 mm
3.297 in	EC	83.74 mm

ROCKWELL L6NYS14-16

IDLI GROUP 86619

CLAMP YOKE WITH 18 SERRATIONS
SEE FIGURE 12C

1.438 in	IE	36.51 mm
.969 in	BD	24.61 mm
.813 in	SB	20.64 mm
4.063 in	EC	103.19 mm

ROCKWELL L6NYS13-20

IDLI GROUP 86620

CLAMP YOKE WITH 20 SERRATIONS
SEE FIGURE 12C

1.438 in	IE	36.51 mm
.969 in	BD	24.61 mm
.875 in	SB	22.23 mm
2.156 in	EC	54.77 mm

ROCKWELL L6NYS14-123

IDLI GROUP 86621

CLAMP YOKE WITH 30 SERRATIONS
SEE FIGURE 12C

1.438 in	IE	36.51 mm
.969 in	BD	24.61 mm
.875 in	SB	22.23 mm
2.250 in	EC	57.15 mm

ROCKWELL L6NYS14-100

IDLI GROUP 86622

CLAMP YOKE WITH 36 SERRATIONS
SEE FIGURE 12C

1.438 in	IE	36.51 mm
.969 in	BD	24.61 mm
.875 in	SB	22.23 mm
2.250 in	EC	57.15 mm

ROCKWELL L6NYS14-17

IDLI GROUP 86623

CLAMP YOKE WITH 36 SERRATIONS
slight variation from above group
SEE FIGURE 12C

1.438 in	IE	36.51 mm
.969 in	BD	24.61 mm
.875 in	SB	22.23 mm
2.250 in	EC	57.15 mm

ROCKWELL L6NYS14-122

IDLI GROUP 86624

CLAMP YOKE WITH 30 SERRATIONS
SEE FIGURE 12C

1.438 in	IE	36.51 mm
.969 in	BD	24.61 mm
.875 in	SB	22.23 mm
2.500 in	EC	63.50 mm

ROCKWELL L6NYS14-65

IDLI GROUP 86625

CLAMP YOKE
WITH ROUND BORE
SEE FIGURE 12B

1.438 in	IE	36.51 mm
.969 in	BD	24.61 mm
.875 in	SB	22.23 mm
2.500 in	EC	63.50 mm

ROCKWELL L6NYR14-96

IDLI GROUP 86626

CLAMP YOKE
WITH ROUND BORE
SEE FIGURE 12B

1.438 in	IE	36.51 mm
.969 in	BD	24.61 mm
.875 in	SB	22.23 mm
2.625 in	EC	66.68 mm

ROCKWELL L6NYR14-7

BD = BEARING DIAMETER (outside)   **BW** = BEARING WIDTH   **CB** = CROSS LENGTH WITH BEARINGS   **CC** = CENTER TO CENTER   **CD** = CROSS DIAMETER
CL = CROSS LENGTH WITHOUT BEARINGS   **EC** = END TO CENTER or FACE TO FACE   **EE** = END TO END   **EL** = EFFECTIVE LENGTH
HD = HUB DIAMETER (or insert)   **IE** = INSIDE OF EARS (or recess)   **OD** = OUTSIDE DIAMETER   **OE** = OUTSIDE OF EARS   **SB** = SPLINE OR BORE SIZE

IDLI GUIDE　　COPYRIGHT ® INTERCHANGE, INC.　　ST. LOUIS PARK, MN. 55416 USA

CAUTION: BE SURE TO REFER TO ENGINEERING CATALOGS FOR SPECIAL APPLICATIONS THAT REQUIRE SPECIFIC MATERIAL CONTENT, TOLERANCES, ETC. SEE FOOTNOTE.

IDLI GROUP 86627

CLAMP YOKE
WITH 13 INVOLUTED SPLINES
SEE FIGURE 12A

1.438 in	IE	36.51 mm
.969 in	BD	24.61 mm
.875 in	SB	22.23 mm
2.750 in	EC	69.85 mm

ROCKWELL L6NYS14-109

IDLI GROUP 86628

CLAMP YOKE
WITH ROUND BORE
SEE FIGURE 12A

1.438 in	IE	36.51 mm
.969 in	BD	24.61 mm
.875 in	SB	22.23 mm
2.750 in	EC	69.85 mm

ROCKWELL L6NYR14-149

IDLI GROUP 86629

CLAMP YOKE
WITH ROUND BORE
SEE FIGURE 12B

1.438 in	IE	36.51 mm
.969 in	BD	24.61 mm
.875 in	SB	22.23 mm
2.750 in	EC	69.85 mm

ROCKWELL L6NYR14-16

IDLI GROUP 86630

CLAMP YOKE
WITH 6 SPLINES
SEE FIGURE 12A

1.438 in	IE	36.51 mm
.969 in	BD	24.61 mm
.984 in	SB	25.00 mm
3.000 in	EC	76.20 mm

ROCKWELL L6NYS16-59

IDLI GROUP 86631

CLAMP YOKE
WITH ROUND BORE
SEE FIGURE 13

1.438 in	IE	36.51 mm
.969 in	BD	24.61 mm
1.000 in	SB	25.40 mm
2.250 in	CC	57.15 mm

ROCKWELL L6NYR16-114

IDLI GROUP 86632

CLAMP YOKE WITH 36 SERRATIONS
SEE FIGURE 12C

1.438 in	IE	36.51 mm
.969 in	BD	24.61 mm
1.000 in	SB	25.40 mm
2.250 in	EC	57.15 mm

ROCKWELL L6NYS16-77

IDLI GROUP 86633

CLAMP YOKE WITH 36 SERRATIONS
slight variation from above group
SEE FIGURE 12C

1.438 in	IE	36.51 mm
.969 in	BD	24.61 mm
1.000 in	SB	25.40 mm
2.250 in	EC	57.15 mm

ROCKWELL L6NYS16-18

IDLI GROUP 86634

CLAMP YOKE
WITH ROUND BORE
SEE FIGURE 12B

1.438 in	IE	36.51 mm
.969 in	BD	24.61 mm
1.000 in	SB	25.40 mm
2.250 in	EC	57.15 mm

ROCKWELL L6NYR16-54

IDLI GROUP 86635

CLAMP YOKE WITH 30 SERRATIONS
SEE FIGURE 12C

1.438 in	IE	36.51 mm
.969 in	BD	24.61 mm
1.000 in	SB	25.40 mm
2.500 in	EC	63.50 mm

ROCKWELL L6NYS16-34

IDLI GROUP 86636

CLAMP YOKE WITH 30 SERRATIONS
slight variation from above group
SEE FIGURE 12C

1.438 in	IE	36.51 mm
.969 in	BD	24.61 mm
1.000 in	SB	25.40 mm
2.500 in	EC	63.50 mm

ROCKWELL L6NYS16-46

IDLI GROUP 86637

CLAMP YOKE WITH 36 SERRATIONS
SEE FIGURE 12C

1.438 in	IE	36.51 mm
.969 in	BD	24.61 mm
1.000 in	SB	25.40 mm
2.500 in	EC	63.50 mm

ROCKWELL L6NYS16-64

IDLI GROUP 86638

CLAMP YOKE
WITH ROUND BORE
SEE FIGURE 12B

1.438 in	IE	36.51 mm
.969 in	BD	24.61 mm
1.000 in	SB	25.40 mm
2.500 in	EC	63.50 mm

ROCKWELL L6NYR16-287

IDLI GROUP 86639

CLAMP YOKE WITH 36 SERRATIONS
SEE FIGURE 12C

1.438 in	IE	36.51 mm
.969 in	BD	24.61 mm
1.000 in	SB	25.40 mm
2.703 in	EC	68.66 mm

ROCKWELL L6NYS16-14

IDLI GROUP 86640

CLAMP YOKE
WITH 15 INVOLUTED SPLINES
SEE FIGURE 12A

1.438 in	IE	36.51 mm
.969 in	BD	24.61 mm
1.000 in	SB	25.40 mm
2.750 in	EC	69.85 mm

ROCKWELL L6NYS16-66

IDLI GROUP 86641

CLAMP YOKE
WITH ROUND BORE
SEE FIGURE 12B

1.438 in	IE	36.51 mm
.969 in	BD	24.61 mm
1.000 in	SB	25.40 mm
2.750 in	EC	69.85 mm

ROCKWELL L6NYR16-78

IDLI GROUP 86642

CLAMP YOKE WITH ROUND BORE
slight variation from above group
SEE FIGURE 12B

1.438 in	IE	36.51 mm
.969 in	BD	24.61 mm
1.000 in	SB	25.40 mm
2.750 in	EC	69.85 mm

ROCKWELL L6NYR16-86

IDLI GROUP 86643

CLAMP YOKE
WITH THREADED BORE
SEE FIGURE 12D

1.438 in	IE	36.51 mm
.969 in	BD	24.61 mm
1.000 in	SB	25.40 mm
3.188 in	EC	80.96 mm

ROCKWELL L6NYD16-5

IDLI GROUP 86644

CLAMP YOKE
WITH ROUND BORE
SEE FIGURE 12B

1.438 in	IE	36.51 mm
.969 in	BD	24.61 mm
1.125 in	SB	28.58 mm
3.063 in	EC	77.79 mm

ROCKWELL L6NYR18-62

IDLI GROUP 86645

CLAMP YOKE
WITH ROUND BORE
SEE FIGURE 12B

1.438 in	IE	36.51 mm
.969 in	BD	24.61 mm
1.125 in	SB	28.58 mm
3.063 in	EC	77.79 mm

ROCKWELL L6NYR18-104

IDLI GROUP 86646

CLAMP YOKE
WITH ROUND BORE
SEE FIGURE 12B

1.438 in	IE	36.51 mm
.969 in	BD	24.61 mm
1.188 in	SB	30.16 mm
2.625 in	EC	66.68 mm

ROCKWELL L6NYR19-3

IDLI GROUP 86647

YOKE SHAFT
WITH 32 INVOLUTED SPLINES
SEE FIGURE 69

1.438 in	IE	36.51 mm
.969 in	BD	24.61 mm
1.250 in	SB	31.75 mm
10.875 in	EC	276.23 mm

ROCKWELL L6NYSM20-18

IDLI GROUP 86648

YOKE SHAFT
WITH 32 INVOLUTED SPLINES
SEE FIGURE 68

1.438 in	IE	36.51 mm
.969 in	BD	24.61 mm
1.250 in	SB	31.75 mm
17.125 in	EC	434.98 mm

ROCKWELL L6NYSM20-61

IDLI GROUP 86649

YOKE SHAFT
WITH 21 INVOLUTED SPLINES
SEE FIGURE 68

1.438 in	IE	36.51 mm
.969 in	BD	24.61 mm
1.375 in	SB	34.93 mm
7.719 in	EC	196.06 mm

ROCKWELL L6NYSM22-5

IDLI GROUP 86650

DOUBLE CENTER
SEE FIGURE 58

1.813 in	IE	46.04 mm
1.063 in	BD	26.99 mm
4.438 in	CC	112.71 mm

PRECISION 1295
ROCKWELL L12N170-1

ENGINEERING CATALOGS MUST BE CONSULTED FOR DETAILS NOT INCLUDED
IN THIS GUIDE. SPECIFIC DESIGNS, MATERIAL CONTENT, TOLERANCES
LUBE FITTINGS AND OTHER DIMENSIONS ARE INTENTIONALLY OMITTED HERE.

IDLI GROUP 86651

CLAMP YOKE
WITH SQUARE BORE

1.813 in	IE	46.04 mm
1.063 in	BD	26.99 mm
.875 in	SB	22.23 mm
2.750 in	EC	69.85 mm

ROCKWELL L12NYQ14

IDLI GROUP 86652

CLAMP YOKE
WITH 9 INVOLUTED SPLINES
SEE FIGURE 12A

1.813 in	IE	46.04 mm
1.063 in	BD	26.99 mm
1.000 in	SB	25.40 mm
2.750 in	EC	69.85 mm

ROCKWELL L12NYS16

IDLI GROUP 86653

CLAMP YOKE
WITH ROUND BORE
SEE FIGURE 12B

1.813 in	IE	46.04 mm
1.063 in	BD	26.99 mm
1.000 in	SB	25.40 mm
2.750 in	EC	69.85 mm

ROCKWELL L12NYR16-55

IDLI GROUP 86654

CLAMP YOKE WITH ROUND BORE
slight variation from above group
SEE FIGURE 12B

1.813 in	IE	46.04 mm
1.063 in	BD	26.99 mm
1.000 in	SB	25.40 mm
2.750 in	EC	69.85 mm

ROCKWELL L12NYR16

IDLI GROUP 86655

CLAMP YOKE
WITH ROUND BORE
SEE FIGURE 12B

1.813 in	IE	46.04 mm
1.063 in	BD	26.99 mm
1.125 in	SB	28.58 mm
2.750 in	EC	69.85 mm

ROCKWELL L12NYR18-60

IDLI GROUP 86656

CLAMP YOKE
WITH ROUND BORE
SEE FIGURE 12B

1.813 in	IE	46.04 mm
1.063 in	BD	26.99 mm
1.125 in	SB	28.58 mm
2.750 in	EC	69.85 mm

ROCKWELL L12NYR18-73

IDLI GROUP 86657

CLAMP YOKE
WITH 19 INVOLUTED SPLINES
SEE FIGURE 12A

1.813 in	IE	46.04 mm
1.063 in	BD	26.99 mm
1.250 in	SB	31.75 mm
3.438 in	EC	87.31 mm

ROCKWELL L12NYS20-34

IDLI GROUP 86658

CLAMP YOKE
WITH ROUND BORE
SEE FIGURE 12B

1.813 in	IE	46.04 mm
1.063 in	BD	26.99 mm
1.250 in	SB	31.75 mm
3.438 in	EC	87.31 mm

ROCKWELL L12NYR20-61

IDLI GROUP 86659

DOUBLE CENTER
SEE FIGURE 59

2.797 in	IE	71.04 mm
1.313 in	BD	33.34 mm
7.125 in	CC	180.98 mm

ROCKWELL 44RDCA1

IDLI GROUP 86660

CLAMP YOKE
WITH HEX BORE
SEE FIGURE 12E

2.797 in	IE	71.04 mm
1.313 in	BD	33.34 mm
1.250 in	SB	31.75 mm
4.750 in	EC	120.65 mm

ROCKWELL 44RYH20

IDLI GROUP 86661

CLAMP YOKE
WITH 21 INVOLUTED SPLINES
SEE FIGURE 12A

2.797 in	IE	71.04 mm
1.313 in	BD	33.34 mm
1.375 in	SB	34.93 mm
4.563 in	EC	115.89 mm

ROCKWELL 44RYS22-14

IDLI GROUP 86662

CLAMP YOKE
WITH 21 INVOLUTED SPLINES
SEE FIGURE 12A

2.797 in	IE	71.04 mm
1.313 in	BD	33.34 mm
1.375 in	SB	34.93 mm
4.750 in	EC	120.65 mm

ROCKWELL 44RYS22-36

IDLI GROUP 86663

CLAMP YOKE
WITH ROUND BORE
SEE FIGURE 12B

2.979 in	IE	75.67 mm
1.313 in	BD	33.34 mm
1.500 in	SB	38.10 mm
5.063 in	EC	128.59 mm

ROCKWELL 44RYR24-5

IDLI GROUP 86664

CLAMP YOKE
WITH 20 INVOLUTED SPLINES
SEE FIGURE 12A

2.797 in	IE	71.04 mm
1.313 in	BD	33.34 mm
1.750 in	SB	44.45 mm
4.750 in	EC	120.65 mm

ROCKWELL 44RYS28-21

IDLI GROUP 86665

CLAMP YOKE
WITH 20 INVOLUTED SPLINES
SEE FIGURE 12A

2.797 in	IE	71.04 mm
1.313 in	BD	33.34 mm
1.750 in	SB	44.45 mm
5.063 in	EC	128.59 mm

ROCKWELL 44RYS28-9
SPICER 3.4-4-61X

IDLI GROUP 86666

CLAMP YOKE
WITH 20 INVOLUTED SPLINES
SEE FIGURE 12A

2.797 in	IE	71.04 mm
1.313 in	BD	33.34 mm
1.750 in	SB	44.45 mm
5.063 in	EC	128.59 mm

ROCKWELL 44RYS28-18

IDLI GROUP 86667

CLAMP YOKE
WITH 6 SPLINES
SEE FIGURE 12A

2.797 in	IE	71.04 mm
1.313 in	BD	33.34 mm
1.750 in	SB	44.45 mm
5.063 in	EC	128.59 mm

ROCKWELL 44RYS28-3

IDLI GROUP 86668

TUBE AND WELD YOKE ASSEMBLY
SEE FIGURE 51

3.125 in	IE	79.38 mm
1.188 in	BD	30.16 mm
64.000 in	EC	1625.60 mm
3.000 in	OD	76.20 mm

CLEVE.STL. O55-43-175
ROCKWELL 991-10-2
SPICER 3-27-3-6414X

IDLI GROUP 86669

TUBE AND WELD YOKE ASSEMBLY
SEE FIGURE 51

3.125 in	IE	79.38 mm
1.188 in	BD	30.16 mm
74.000 in	EC	1879.60 mm
3.500 in	OD	88.90 mm

CLEVE.STL. O55-43-169
ROCKWELL 991-11-2
SPICER 3-27-8-7317X

IDLI GROUP 86670

TUBE AND WELD YOKE ASSEMBLY
SEE FIGURE 52

3.125 in	IE	79.38 mm
1.188 in	BD	30.16 mm
74.000 in	EC	1879.60 mm
4.000 in	OD	101.60 mm

CLEVE.STL. O55-43-176
ROCKWELL 991-12-2
SPICER 3-27-16-7120X

IDLI GROUP 86671

TUBE AND WELD YOKE ASSEMBLY
SEE FIGURE 51

4.188 in	IE	106.36 mm
1.375 in	BD	34.93 mm
74.000 in	EC	1879.60 mm
3.500 in	OD	88.90 mm

CLEVE.STL. D56-43-102
ROCKWELL 992-10-2

IDLI GROUP 86672

TUBE AND WELD YOKE ASSEMBLY
slight variation from above group
SEE FIGURE 51

4.188 in	IE	106.36 mm
1.375 in	BD	34.93 mm
74.000 in	EC	1879.60 mm
3.500 in	OD	88.90 mm

CLEVE.STL. D76-43-158
ROCKWELL 992-12-2
SPICER 3-27-22-6221X

IDLI GROUP 86673

TUBE AND WELD YOKE ASSEMBLY
SEE FIGURE 51

4.188 in	IE	106.36 mm
1.375 in	BD	34.93 mm
74.000 in	EC	1879.60 mm
4.000 in	OD	101.60 mm

CLEVE.STL. D76-43-159
ROCKWELL 992-13-2
SPICER 3-27-20-7311X

IDLI GROUP 86674

TUBE AND WELD YOKE ASSEMBLY
SEE FIGURE 52

4.188 in	IE	106.36 mm
1.375 in	BD	34.93 mm
74.000 in	EC	1879.60 mm
4.000 in	OD	101.60 mm

CLEVE.STL. D56-43-926
ROCKWELL 992-14-2

BD = BEARING DIAMETER (outside)   **BW** = BEARING WIDTH   **CB** = CROSS LENGTH WITH BEARINGS   **CC** = CENTER TO CENTER   **CD** = CROSS DIAMETER
CL = CROSS LENGTH WITHOUT BEARINGS   **EC** = END TO CENTER or FACE TO FACE   **EE** = END TO END   **EL** = EFFECTIVE LENGTH
HD = HUB DIAMETER (or insert)   **IE** = INSIDE OF EARS (or recess)   **OD** = OUTSIDE DIAMETER   **OE** = OUTSIDE OF EARS   **SB** = SPLINE OR BORE SIZE

IDLI GUIDE   COPYRIGHT ® INTERCHANGE, INC.   ST. LOUIS PARK, MN. 55416 USA

CAUTION: BE SURE TO REFER TO ENGINEERING CATALOGS FOR SPECIAL APPLICATIONS THAT REQUIRE SPECIFIC MATERIAL CONTENT, TOLERANCES, ETC. SEE FOOTNOTE.

IDLI GROUP 86675

TUBE AND WELD YOKE ASSEMBLY
SEE FIGURE 51

4.969 in	IE	126.21 mm
1.375 in	BD	34.93 mm
74.000 in	EC	1879.60 mm
3.500 in	OD	88.90 mm

CLEVE.STL. U56-43-154
ROCKWELL 990-11-2

IDLI GROUP 86676

TUBE AND WELD YOKE ASSEMBLY
slight variation from above group
SEE FIGURE 51

4.969 in	IE	126.21 mm
1.375 in	BD	34.93 mm
74.000 in	EC	1879.60 mm
3.500 in	OD	88.90 mm

CLEVE.STL. U76-43-199
ROCKWELL 990-12-2
SPICER 4-27-16-6200X

IDLI GROUP 86677

TUBE AND WELD YOKE ASSEMBLY
SEE FIGURE 51

4.969 in	IE	126.21 mm
1.375 in	BD	34.93 mm
74.000 in	EC	1879.60 mm
3.500 in	OD	88.90 mm

CLEVE.STL. U76-43-198
ROCKWELL 990-14-2
SPICER 4-27-18-6116X

IDLI GROUP 86678

TUBE AND WELD YOKE ASSEMBLY
SEE FIGURE 51

4.969 in	IE	126.21 mm
1.375 in	BD	34.93 mm
74.000 in	EC	1879.60 mm
4.000 in	OD	101.60 mm

CLEVE.STL. U76-43-200
ROCKWELL 990-13-2

IDLI GROUP 86679

DOUBLE CENTER
SEE FIGURE 62

5.313 in	IE	134.94 mm
5.625 in	OE	142.88 mm
1.375 in	EC	34.93 mm

ROCKWELL 58WB170

IDLI GROUP 86680

DOUBLE CENTER
slight variation from above group
SEE FIGURE 62

5.313 in	IE	134.94 mm
5.625 in	OE	142.88 mm
1.375 in	EC	34.93 mm

ROCKWELL 58WB170-1

IDLI GROUP 86681

DOUBLE CENTER
SEE FIGURE 61

5.313 in	IE	134.94 mm
5.625 in	OE	142.88 mm
5.250 in	EC	133.35 mm

ROCKWELL 58WBDCA4

IDLI GROUP 86682

YOKE SHAFT WITH 16 SPLINES
SEE FIGURE 71

5.313 in	IE	134.94 mm
5.625 in	OE	142.88 mm
1.750 in	SB	44.45 mm
6.250 in	EC	158.75 mm

ROCKWELL 58WBYSM28-22

IDLI GROUP 86683

YOKE SHAFT WITH 16 SPLINES
SEE FIGURE 71

5.313 in	IE	134.94 mm
5.625 in	OE	142.88 mm
1.750 in	SB	44.45 mm
6.500 in	EC	165.10 mm

ROCKWELL 58WBYSM28-11

IDLI GROUP 86684

YOKE SHAFT WITH 16 SPLINES
SEE FIGURE 71

5.313 in	IE	134.94 mm
5.625 in	OE	142.88 mm
1.750 in	SB	44.45 mm
6.750 in	EC	171.45 mm

ROCKWELL 58WBYSM28-4

IDLI GROUP 86685

YOKE SHAFT WITH 16 SPLINES
SEE FIGURE 71

5.313 in	IE	134.94 mm
5.625 in	OE	142.88 mm
1.750 in	SB	44.45 mm
7.625 in	EC	193.68 mm

ROCKWELL 58WBYSM28-38

IDLI GROUP 86686

YOKE SHAFT WITH 16 SPLINES
SEE FIGURE 71

5.313 in	IE	134.94 mm
5.625 in	OE	142.88 mm
2.500 in	SB	63.50 mm
5.938 in	EC	150.81 mm

ROCKWELL 58WBYSM40-3

IDLI GROUP 86687

YOKE SHAFT WITH 16 SPLINES
SEE FIGURE 71

5.313 in	IE	134.94 mm
5.625 in	OE	142.88 mm
2.500 in	SB	63.50 mm
6.688 in	EC	169.86 mm

ROCKWELL 58WBYSM40-21

IDLI GROUP 86688

YOKE SHAFT WITH 16 SPLINES
SEE FIGURE 71

5.313 in	IE	134.94 mm
5.625 in	OE	142.88 mm
2.500 in	SB	63.50 mm
7.188 in	EC	182.56 mm

ROCKWELL 58WBYSM40-5

IDLI GROUP 86689

YOKE SHAFT WITH 16 SPLINES
SEE FIGURE 71

5.313 in	IE	134.94 mm
5.625 in	OE	142.88 mm
2.500 in	SB	63.50 mm
7.438 in	EC	188.91 mm

ROCKWELL 58WBYSM40-9

IDLI GROUP 86690

YOKE SHAFT WITH 16 SPLINES
SEE FIGURE 71

5.313 in	IE	134.94 mm
5.625 in	OE	142.88 mm
2.500 in	SB	63.50 mm
7.688 in	EC	195.26 mm

ROCKWELL 58WBYSM40-12

IDLI GROUP 86691

YOKE SHAFT WITH 16 SPLINES
SEE FIGURE 71

5.313 in	IE	134.94 mm
5.625 in	OE	142.88 mm
2.500 in	SB	63.50 mm
8.125 in	EC	206.38 mm

ROCKWELL 58WBYSM40

IDLI GROUP 86692

YOKE SHAFT WITH 16 SPLINES
SEE FIGURE 71

5.313 in	IE	134.94 mm
5.625 in	OE	142.88 mm
2.500 in	SB	63.50 mm
8.188 in	EC	207.96 mm

ROCKWELL 58WBYSM40-10

IDLI GROUP 86693

YOKE SHAFT WITH 16 SPLINES
SEE FIGURE 72

5.313 in	IE	134.94 mm
5.625 in	OE	142.88 mm
2.500 in	SB	63.50 mm
9.125 in	EC	231.78 mm
4.375 in	HD	111.13 mm

ROCKWELL 58WBYSM40-6

IDLI GROUP 86694

YOKE SHAFT WITH 16 SPLINES
SEE FIGURE 71

5.313 in	IE	134.94 mm
5.625 in	OE	142.88 mm
2.500 in	SB	63.50 mm
9.188 in	EC	233.36 mm

ROCKWELL 58WBYSM40-8

IDLI GROUP 86695

YOKE SHAFT WITH 16 SPLINES
SEE FIGURE 71

5.313 in	IE	134.94 mm
5.625 in	OE	142.88 mm
2.500 in	SB	63.50 mm
9.750 in	EC	247.65 mm

ROCKWELL 58WBYSM40-7

IDLI GROUP 86696

TUBE AND WELD YOKE ASSEMBLY
SEE FIGURE 50

5.313 in	IE	134.94 mm
5.625 in	OE	142.88 mm
3.320 in	SB	84.33 mm
49.813 in	EC	1265.24 mm
3.500 in	OD	88.90 mm

ROCKWELL 959-10-2

IDLI GROUP 86697

TUBE AND WELD YOKE ASSEMBLY
SEE FIGURE 50

5.313 in	IE	134.94 mm
5.625 in	OE	142.88 mm
3.320 in	SB	84.33 mm
61.813 in	EC	1570.04 mm
3.500 in	OD	88.90 mm

ROCKWELL 959-11-2

IDLI GROUP 86698

TUBE AND WELD YOKE ASSEMBLY
SEE FIGURE 50

5.313 in	IE	134.94 mm
5.625 in	OE	142.88 mm
3.840 in	SB	97.54 mm
61.813 in	EC	1570.04 mm
4.000 in	OD	101.60 mm

ROCKWELL 959-10-2

IDLI GROUP 86699

YOKE SHAFT WITH 10 SPLINES
slight variation from above group
SEE FIGURE 71

5.531 in	IE	140.50 mm
5.841 in	OE	148.37 mm
1.750 in	SB	44.45 mm
5.563 in	EC	141.29 mm

BORG-WARNR 13928J
ROCKWELL 62NYSM28-63

IDLI GROUP 86700

DOUBLE CENTER
SEE FIGURE 62

5.531 in	IE	140.50 mm
5.844 in	OE	148.43 mm
1.375 in	EC	34.93 mm

BORG-WARNR 5567J
ROCKWELL 62N170-11

ENGINEERING CATALOGS MUST BE CONSULTED FOR DETAILS NOT INCLUDED
IN THIS GUIDE. SPECIFIC DESIGNS, MATERIAL CONTENT, TOLERANCES
LUBE FITTINGS AND OTHER DIMENSIONS ARE INTENTIONALLY OMITTED HERE.

IDLI GROUP 86701

DOUBLE CENTER
SEE FIGURE 61

5.531 in	IE	140.50 mm
5.844 in	OE	148.43 mm
2.563 in	EC	65.09 mm

ROCKWELL 62NDCA2

IDLI GROUP 86702

DOUBLE CENTER
SEE FIGURE 61

5.531 in	IE	140.50 mm
5.844 in	OE	148.43 mm
3.938 in	EC	100.01 mm

BORG-WARNR 119-16985-1
ROCKWELL 62NDCA1

IDLI GROUP 86703

DOUBLE CENTER
SEE FIGURE 61

5.531 in	IE	140.50 mm
5.844 in	OE	148.43 mm
4.313 in	EC	109.54 mm

ROCKWELL 62NDCA11

IDLI GROUP 86704

DOUBLE CENTER
SEE FIGURE 61

5.531 in	IE	140.50 mm
5.844 in	OE	148.43 mm
4.500 in	EC	114.30 mm

BORG-WARNR 119-15562-1
BORG-WARNR 119-16985-10
ROCKWELL 62NDCA10

IDLI GROUP 86705

DOUBLE CENTER
SEE FIGURE 61

5.531 in	IE	140.50 mm
5.844 in	OE	148.43 mm
4.688 in	EC	119.06 mm

BORG-WARNR 119-19369-1
ROCKWELL 62NDCA3

IDLI GROUP 86706

YOKE SHAFT WITH 10 SPLINES
SEE FIGURE 71

5.531 in	IE	140.50 mm
5.844 in	OE	148.43 mm
1.750 in	SB	44.45 mm
4.750 in	EC	120.65 mm

ROCKWELL 62NYSM28-16

IDLI GROUP 86707

YOKE SHAFT WITH 10 SPLINES
SEE FIGURE 71

5.531 in	IE	140.50 mm
5.844 in	OE	148.43 mm
1.750 in	SB	44.45 mm
5.563 in	EC	141.29 mm

BORG-WARNR 5571J
ROCKWELL 62NYSM28-13

IDLI GROUP 86708

YOKE SHAFT WITH 10 SPLINES
SEE FIGURE 71

5.531 in	IE	140.50 mm
5.844 in	OE	148.43 mm
1.750 in	SB	44.45 mm
5.625 in	EC	142.88 mm

BORG-WARNR 10617J
ROCKWELL 62NYSM28-37

IDLI GROUP 86709

YOKE SHAFT WITH 10 SPLINES
SEE FIGURE 71

5.531 in	IE	140.50 mm
5.844 in	OE	148.43 mm
1.750 in	SB	44.45 mm
6.000 in	EC	152.40 mm

ROCKWELL 62NYSM28-72

IDLI GROUP 86710

YOKE SHAFT WITH 10 SPLINES
SEE FIGURE 72

5.531 in	IE	140.50 mm
5.844 in	OE	148.43 mm
1.750 in	SB	44.45 mm
6.156 in	EC	156.37 mm

ROCKWELL 62NYSM28-89

IDLI GROUP 86711

YOKE SHAFT WITH 10 SPLINES
SEE FIGURE 71

5.531 in	IE	140.50 mm
5.844 in	OE	148.43 mm
1.750 in	SB	44.45 mm
6.375 in	EC	161.93 mm

ROCKWELL 62NYSM28-9

IDLI GROUP 86712

YOKE SHAFT WITH 10 SPLINES
SEE FIGURE 71

5.531 in	IE	140.50 mm
5.844 in	OE	148.43 mm
1.750 in	SB	44.45 mm
7.375 in	EC	187.33 mm

ROCKWELL 62NYSM28-109

IDLI GROUP 86713

YOKE SHAFT WITH 10 SPLINES
SEE FIGURE 71

5.531 in	IE	140.50 mm
5.844 in	OE	148.43 mm
1.750 in	SB	44.45 mm
7.500 in	EC	190.50 mm

BORG-WARNR 105-2752
BORG-WARNR 13062J
FORD B8Q4816C
FORD B8TZ4385B
FORD TVBV44627A
ROCKWELL 62NYSM28-7

IDLI GROUP 86714

YOKE SHAFT WITH 10 SPLINES
SEE FIGURE 71

5.531 in	IE	140.50 mm
5.844 in	OE	148.43 mm
1.750 in	SB	44.45 mm
8.625 in	EC	219.08 mm

ROCKWELL 62NYSM28-23

IDLI GROUP 86715

YOKE SHAFT WITH 10 SPLINES
SEE FIGURE 71

5.531 in	IE	140.50 mm
5.844 in	OE	148.43 mm
1.750 in	SB	44.45 mm
9.000 in	EC	228.60 mm

ROCKWELL 62NYSM28-64

IDLI GROUP 86716

YOKE SHAFT WITH 10 SPLINES
slight variation from above group
SEE FIGURE 71

5.531 in	IE	140.50 mm
5.844 in	OE	148.43 mm
1.750 in	SB	44.45 mm
9.000 in	EC	228.60 mm

ROCKWELL 62NYSM28-51

IDLI GROUP 86717

YOKE SHAFT WITH 10 SPLINES
SEE FIGURE 72

5.531 in	IE	140.50 mm
5.844 in	OE	148.43 mm
1.750 in	SB	44.45 mm
18.188 in	EC	461.96 mm

BORG-WARNR 105-12768-2
ROCKWELL 62NYSM28-18

IDLI GROUP 86718

YOKE SHAFT WITH 10 SPLINES
SEE FIGURE 72

5.531 in	IE	140.50 mm
5.844 in	OE	148.43 mm
1.750 in	SB	44.45 mm
33.438 in	EC	849.31 mm

BORG-WARNR 105-10752-1
IHC 37499R11
ROCKWELL 62NYSM28-2

IDLI GROUP 86719

YOKE SHAFT WITH 10 SPLINES
SEE FIGURE 72

5.531 in	IE	140.50 mm
5.844 in	OE	148.43 mm
1.750 in	SB	44.45 mm
36.438 in	EC	925.51 mm

BORG-WARNR 105-16197-2
ROCKWELL 62NYSM28-30

IDLI GROUP 86720

YOKE SHAFT WITH 14 SPLINES
SEE FIGURE 72

5.531 in	IE	140.50 mm
5.844 in	OE	148.43 mm
1.875 in	SB	47.63 mm
3.875 in	EC	98.43 mm

ROCKWELL 62NYSM30-8

IDLI GROUP 86721

YOKE SHAFT WITH 10 SPLINES
SEE FIGURE 71

5.531 in	IE	140.50 mm
5.844 in	OE	148.43 mm
2.000 in	SB	50.80 mm
7.375 in	EC	187.33 mm

ROCKWELL 62NYSM32-13

IDLI GROUP 86722

YOKE SHAFT WITH 16 SPLINES
SEE FIGURE 71

5.531 in	IE	140.50 mm
5.844 in	OE	148.43 mm
2.000 in	SB	50.80 mm
9.313 in	EC	236.54 mm

ROCKWELL 62NYSM32-17

IDLI GROUP 86723

YOKE SHAFT WITH 4 SPLINES
FOR ROLL JOINT ASSEMBLY
SEE FIGURE 72

5.531 in	IE	140.50 mm
5.844 in	OE	148.43 mm
2.125 in	SB	53.98 mm
13.375 in	EC	339.73 mm

ROCKWELL 62NYSM34

IDLI GROUP 86724

TUBE AND WELD YOKE ASSEMBLY
SEE FIGURE 50

5.531 in	IE	140.50 mm
5.844 in	OE	148.43 mm
2.810 in	SB	71.37 mm
57.000 in	EC	1447.80 mm
3.000 in	OD	76.20 mm

BORG-WARNR 42-6001
ROCKWELL 965-10-2

IDLI GROUP 86725

TUBE AND WELD YOKE ASSEMBLY
SEE FIGURE 50

5.531 in	IE	140.50 mm
5.844 in	OE	148.43 mm
3.260 in	SB	82.80 mm
61.875 in	EC	1571.63 mm
3.500 in	OD	88.90 mm

BORG-WARNR 42-6006
ROCKWELL 965-12-2

BD = BEARING DIAMETER (outside)   **BW** = BEARING WIDTH   **CB** = CROSS LENGTH WITH BEARINGS   **CC** = CENTER TO CENTER   **CD** = CROSS DIAMETER
CL = CROSS LENGTH WITHOUT BEARINGS   **EC** = END TO CENTER or FACE TO FACE   **EE** = END TO END   **EL** = EFFECTIVE LENGTH
HD = HUB DIAMETER (or insert)   **IE** = INSIDE OF EARS (or recess)   **OD** = OUTSIDE DIAMETER   **OE** = OUTSIDE OF EARS   **SB** = SPLINE OR BORE SIZE

IDLI GUIDE   COPYRIGHT ® INTERCHANGE, INC.   ST. LOUIS PARK, MN. 55416 USA

CAUTION: BE SURE TO REFER TO ENGINEERING CATALOGS FOR SPECIAL APPLICATIONS THAT REQUIRE SPECIFIC MATERIAL CONTENT, TOLERANCES, ETC. SEE FOOTNOTE.

IDLI GROUP 86726

TUBE AND WELD YOKE ASSEMBLY
SEE FIGURE 50

5.531 in	IE	140.50 mm
5.844 in	OE	148.43 mm
3.310 in	SB	84.07 mm
73.875 in	EC	1876.43 mm
3.500 in	OD	88.90 mm

BORG-WARNR 42-6003
ROCKWELL 965-11-2

IDLI GROUP 86727

TUBE AND WELD YOKE ASSEMBLY
SEE FIGURE 50

5.531 in	IE	140.50 mm
5.844 in	OE	148.43 mm
3.782 in	SB	96.06 mm
67.875 in	EC	1724.03 mm
4.000 in	OD	101.60 mm

BORG-WARNR 42-6005
ROCKWELL 965-14-2

IDLI GROUP 86728

TUBE AND WELD YOKE ASSEMBLY
SEE FIGURE 50

5.531 in	IE	140.50 mm
5.844 in	OE	148.43 mm
3.834 in	SB	97.38 mm
67.875 in	EC	1724.03 mm
4.000 in	OD	101.60 mm

BORG-WARNR 42-6007
ROCKWELL 965-13-2

IDLI GROUP 86729

DOUBLE CENTER
SEE FIGURE 62

5.844 in	IE	148.43 mm
6.219 in	OE	157.96 mm
1.250 in	EC	31.75 mm

ROCKWELL 72N170-9

IDLI GROUP 86730

DOUBLE CENTER
slight variation from above group
SEE FIGURE 62

5.844 in	IE	148.43 mm
6.219 in	OE	157.96 mm
1.250 in	EC	31.75 mm

BORG-WARNR 20643J
ROCKWELL 72N175-1

IDLI GROUP 86731

DOUBLE CENTER
SEE FIGURE 62

5.844 in	IE	148.43 mm
6.219 in	OE	157.96 mm
1.500 in	EC	38.10 mm

BORG-WARNR 11655J
ROCKWELL 72N174-1

IDLI GROUP 86732

DOUBLE CENTER
slight variation from above group
SEE FIGURE 62

5.844 in	IE	148.43 mm
6.219 in	OE	157.96 mm
1.500 in	EC	38.10 mm

BORG-WARNR 7398J
BORG-WARNR 7981J
ROCKWELL 72N170

IDLI GROUP 86733

DOUBLE CENTER
SEE FIGURE 62

5.844 in	IE	148.43 mm
6.219 in	OE	157.96 mm
1.625 in	EC	41.28 mm

BORG-WARNR 20956J
ROCKWELL 72N176-1

IDLI GROUP 86734

DOUBLE CENTER
SEE FIGURE 62

5.844 in	IE	148.43 mm
6.219 in	OE	157.96 mm
1.750 in	EC	44.45 mm

BORG-WARNR 5586J
ROCKWELL 72N171-1

IDLI GROUP 86735

DOUBLE CENTER
slight variation from above group
SEE FIGURE 62

5.844 in	IE	148.43 mm
6.219 in	OE	157.96 mm
1.750 in	EC	44.45 mm

BORG-WARNR 9403J
ROCKWELL 72N172-1

IDLI GROUP 86736

DOUBLE CENTER
SEE FIGURE 61

5.844 in	IE	148.43 mm
6.219 in	OE	157.96 mm
3.938 in	EC	100.01 mm

ROCKWELL 72N173-1

IDLI GROUP 86737

DOUBLE CENTER
SEE FIGURE 71

5.844 in	IE	148.43 mm
6.219 in	OE	157.96 mm
4.188 in	EC	106.36 mm

BORG-WARNR 119-13961-1
ROCKWELL 72NDCA1

IDLI GROUP 86738

DOUBLE CENTER
SEE FIGURE 61

5.844 in	IE	148.43 mm
6.219 in	OE	157.96 mm
4.906 in	EC	124.62 mm

BORG-WARNR 119-17900-1
ROCKWELL 72NDCA2

IDLI GROUP 86739

DOUBLE CENTER
slight variation from above group
SEE FIGURE 61

5.844 in	IE	148.43 mm
6.219 in	OE	157.96 mm
4.906 in	EC	124.62 mm

BORG-WARNR 25228J
ROCKWELL 72N177-1

IDLI GROUP 86740

DOUBLE CENTER
SEE FIGURE 61

5.844 in	IE	148.43 mm
6.219 in	OE	157.96 mm
6.828 in	EC	173.43 mm

ROCKWELL 72N170-14

IDLI GROUP 86741

YOKE SHAFT WITH 16 INVOLUTED SPLINES
SEE FIGURE 72

5.844 in	IE	148.43 mm
6.219 in	OE	157.96 mm
1.875 in	SB	47.63 mm
9.938 in	EC	252.41 mm

ROCKWELL 72NYSM30-8

IDLI GROUP 86742

YOKE SHAFT WITH 10 SPLINES
SEE FIGURE 71

5.844 in	IE	148.43 mm
6.219 in	OE	157.96 mm
2.000 in	SB	50.80 mm
4.875 in	EC	123.83 mm

BORG-WARNR 10946J
ROCKWELL 72NYSM32-65

IDLI GROUP 86743

YOKE SHAFT WITH 10 SPLINES
SEE FIGURE 71

5.844 in	IE	148.43 mm
6.219 in	OE	157.96 mm
2.000 in	SB	50.80 mm
7.000 in	EC	177.80 mm

BORG-WARNR 14742J
ROCKWELL 72NYSM32-3

IDLI GROUP 86744

YOKE SHAFT WITH 10 SPLINES
slight variation from above group
SEE FIGURE 71

5.844 in	IE	148.43 mm
6.219 in	OE	157.96 mm
2.000 in	SB	50.80 mm
7.000 in	EC	177.80 mm

BORG-WARNR 5505J
FORD CITT4385A
ROCKWELL 72NYSM32-11

IDLI GROUP 86745

YOKE SHAFT WITH 10 SPLINES
slight variation from above group
SEE FIGURE 71

5.844 in	IE	148.43 mm
6.219 in	OE	157.96 mm
2.000 in	SB	50.80 mm
7.000 in	EC	177.80 mm

ROCKWELL 72NYSM32-161

IDLI GROUP 86746

YOKE SHAFT WITH 10 SPLINES
SEE FIGURE 71

5.844 in	IE	148.43 mm
6.219 in	OE	157.96 mm
2.000 in	SB	50.80 mm
7.063 in	EC	179.39 mm

BORG-WARNR 105-3730
ROCKWELL 72NYSM32-1

IDLI GROUP 86747

YOKE SHAFT WITH 10 SPLINES
SEE FIGURE 71

5.844 in	IE	148.43 mm
6.219 in	OE	157.96 mm
2.000 in	SB	50.80 mm
7.688 in	EC	195.26 mm

BORG-WARNR 105-4499
ROCKWELL 72NYSM32-59

IDLI GROUP 86748

YOKE SHAFT WITH 10 SPLINES
SEE FIGURE 71

5.844 in	IE	148.43 mm
6.219 in	OE	157.96 mm
2.000 in	SB	50.80 mm
8.000 in	EC	203.20 mm

ROCKWELL 72NYSM32-164

IDLI GROUP 86749

YOKE SHAFT WITH 10 SPLINES
SEE FIGURE 71

5.844 in	IE	148.43 mm
6.219 in	OE	157.96 mm
2.000 in	SB	50.80 mm
8.125 in	EC	206.38 mm

BORG-WARNR 105-12399-1
ROCKWELL 72NYSM32-74

IDLI GROUP 86750

YOKE SHAFT WITH 10 SPLINES
SEE FIGURE 71

5.844 in	IE	148.43 mm
6.219 in	OE	157.96 mm
2.000 in	SB	50.80 mm
8.813 in	EC	223.84 mm

ROCKWELL 72NYSM32-122

IDLI GROUP 86751

YOKE SHAFT WITH 10 SPLINES
SEE FIGURE 71

5.844 in	IE	148.43 mm
6.219 in	OE	157.96 mm
2.000 in	SB	50.80 mm
9.125 in	EC	231.78 mm

BORG-WARNR 105-2510
ROCKWELL 72NYSM32-58

IDLI GROUP 86752

YOKE SHAFT WITH 10 SPLINES
slight variation from above group
SEE FIGURE 71

5.844 in	IE	148.43 mm
6.219 in	OE	157.96 mm
2.000 in	SB	50.80 mm
9.125 in	EC	231.78 mm

ROCKWELL 72NYSM32-67

IDLI GROUP 86753

YOKE SHAFT WITH 10 SPLINES
slight variation from above group
SEE FIGURE 71

5.844 in	IE	148.43 mm
6.219 in	OE	157.96 mm
2.000 in	SB	50.80 mm
9.125 in	EC	231.78 mm

ROCKWELL 72NYSM32-118

IDLI GROUP 86754

YOKE SHAFT WITH 10 SPLINES
SEE FIGURE 71

5.844 in	IE	148.43 mm
6.219 in	OE	157.96 mm
2.000 in	SB	50.80 mm
9.688 in	EC	246.06 mm

ROCKWELL 72NYSM32-73

IDLI GROUP 86755

YOKE SHAFT WITH 10 SPLINES
SEE FIGURE 71

5.844 in	IE	148.43 mm
6.219 in	OE	157.96 mm
2.000 in	SB	50.80 mm
10.313 in	EC	261.94 mm

BORG-WARNR 29056J
ROCKWELL 72NYSM32-144

IDLI GROUP 86756

YOKE SHAFT WITH 10 SPLINES
SEE FIGURE 71

5.844 in	IE	148.43 mm
6.219 in	OE	157.96 mm
2.000 in	SB	50.80 mm
10.500 in	EC	266.70 mm

ROCKWELL 72NYSM32-161

IDLI GROUP 86757

YOKE SHAFT WITH 10 SPLINES
SEE FIGURE 71

5.844 in	IE	148.43 mm
6.219 in	OE	157.96 mm
2.000 in	SB	50.80 mm
10.938 in	EC	277.81 mm

BORG-WARNR 29047J
ROCKWELL 72NYSM32-42

IDLI GROUP 86758

YOKE SHAFT WITH 10 SPLINES
SEE FIGURE 71

5.844 in	IE	148.43 mm
6.219 in	OE	157.96 mm
2.000 in	SB	50.80 mm
11.125 in	EC	282.58 mm

ROCKWELL 72NYSM32-54

IDLI GROUP 86759

YOKE SHAFT WITH 10 SPLINES
SEE FIGURE 72

5.844 in	IE	148.43 mm
6.219 in	OE	157.96 mm
2.000 in	SB	50.80 mm
11.313 in	EC	287.34 mm

BORG-WARNR 105-11905-1
ROCKWELL 72NYSM32-57

IDLI GROUP 86760

YOKE SHAFT WITH 16 SPLINES
SEE FIGURE 71

5.844 in	IE	148.43 mm
6.219 in	OE	157.96 mm
2.000 in	SB	50.80 mm
12.375 in	EC	314.33 mm

ROCKWELL 72NYSM32-216

IDLI GROUP 86761

YOKE SHAFT WITH 16 SPLINES
SEE FIGURE 71

5.844 in	IE	148.43 mm
6.219 in	OE	157.96 mm
2.000 in	SB	50.80 mm
12.750 in	EC	323.85 mm

ROCKWELL 72NYSM32-240

IDLI GROUP 86762

YOKE SHAFT WITH 10 SPLINES
SEE FIGURE 71

5.844 in	IE	148.43 mm
6.219 in	OE	157.96 mm
2.000 in	SB	50.80 mm
13.375 in	EC	339.73 mm

ROCKWELL 72NYSM32-131

IDLI GROUP 86763

YOKE SHAFT WITH 16 SPLINES
SEE FIGURE 71

5.844 in	IE	148.43 mm
6.219 in	OE	157.96 mm
2.000 in	SB	50.80 mm
13.375 in	EC	339.73 mm

ROCKWELL 72NYSM32-128

IDLI GROUP 86764

YOKE SHAFT WITH 10 SPLINES
SEE FIGURE 71

5.844 in	IE	148.43 mm
6.219 in	OE	157.96 mm
2.000 in	SB	50.80 mm
15.063 in	EC	382.59 mm

ROCKWELL 72NYSM32-137

IDLI GROUP 86765

YOKE SHAFT
WITH 15 SPLINES IN 16 SPLINE SPACES
SEE FIGURE 71

5.844 in	IE	148.43 mm
6.219 in	OE	157.96 mm
2.000 in	SB	50.80 mm
18.375 in	EC	466.73 mm

ROCKWELL 72NYSM32-30

IDLI GROUP 86766

YOKE SHAFT WITH 10 SPLINES
SEE FIGURE 71

5.844 in	IE	148.43 mm
6.219 in	OE	157.96 mm
2.000 in	SB	50.80 mm
19.313 in	EC	490.54 mm

ROCKWELL 72NYSM32-121

IDLI GROUP 86767

YOKE SHAFT
WITH 15 SPLINES IN 16 SPLINE SPACES
SEE FIGURE 71

5.844 in	IE	148.43 mm
6.219 in	OE	157.96 mm
2.000 in	SB	50.80 mm
21.250 in	EC	539.75 mm

ROCKWELL 72NYSM32-32

IDLI GROUP 86768

YOKE SHAFT WITH 10 SPLINES
SEE FIGURE 71

5.844 in	IE	148.43 mm
6.219 in	OE	157.96 mm
2.000 in	SB	50.80 mm
28.063 in	EC	712.79 mm

ROCKWELL 72NYSM32-88

IDLI GROUP 86769

YOKE SHAFT WITH 10 SPLINES
SEE FIGURE 72

5.844 in	IE	148.43 mm
6.219 in	OE	157.96 mm
2.000 in	SB	50.80 mm
42.563 in	EC	1081.09 mm

ROCKWELL 72NYSM32-120

IDLI GROUP 86770

YOKE SHAFT WITH 4 SPLINES
FOR ROLL JOINT ASSEMBLY
SEE FIGURE 72

5.844 in	IE	148.43 mm
6.219 in	OE	157.96 mm
2.109 in	SB	53.58 mm
12.875 in	EC	327.03 mm

ROCKWELL 72NYSM34-20

IDLI GROUP 86771

YOKE SHAFT WITH 4 SPLINES
FOR ROLL JOINT ASSEMBLY
SEE FIGURE 72

5.844 in	IE	148.43 mm
6.219 in	OE	157.96 mm
2.109 in	SB	53.58 mm
13.000 in	EC	330.20 mm

ROCKWELL 72NYSM34-15

IDLI GROUP 86772

YOKE SHAFT WITH 4 SPLINES
FOR ROLL JOINT ASSEMBLY
slight variation from above group
SEE FIGURE 72

5.844 in	IE	148.43 mm
6.219 in	OE	157.96 mm
2.109 in	SB	53.58 mm
13.000 in	EC	330.20 mm

ROCKWELL 72NYSM34-19

IDLI GROUP 86773

YOKE SHAFT WITH 16 SPLINES
SEE FIGURE 71

5.844 in	IE	148.43 mm
6.219 in	OE	157.96 mm
2.500 in	SB	63.50 mm
3.250 in	EC	82.55 mm

ROCKWELL 72NYSM40-49

IDLI GROUP 86774

YOKE SHAFT WITH 16 SPLINES
SEE FIGURE 71

5.844 in	IE	148.43 mm
6.219 in	OE	157.96 mm
2.500 in	SB	63.50 mm
4.063 in	EC	103.19 mm

ROCKWELL 72NYSM40-12

IDLI GROUP 86775

YOKE SHAFT WITH 16 SPLINES
SEE FIGURE 71

5.844 in	IE	148.43 mm
6.219 in	OE	157.96 mm
2.500 in	SB	63.50 mm
4.125 in	EC	104.78 mm

BORG-WARNR 22093J
ROCKWELL 72NYSM40-40

BD=BEARING DIAMETER (outside)　　**BW**=BEARING WIDTH　　**CB**=CROSS LENGTH WITH BEARINGS　　**CC**=CENTER TO CENTER　　**CD**=CROSS DIAMETER
CL=CROSS LENGTH WITHOUT BEARINGS　　**EC**=END TO CENTER or FACE TO FACE　　**EE**=END TO END　　**EL**=EFFECTIVE LENGTH
HD=HUB DIAMETER (or insert)　　**IE**=INSIDE OF EARS (or recess)　　**OD**=OUTSIDE DIAMETER　　**OE**=OUTSIDE OF EARS　　**SB**=SPLINE OR BORE SIZE

IDLI GUIDE COPYRIGHT ® INTERCHANGE, INC. ST. LOUIS PARK, MN. 55416 USA

CAUTION: BE SURE TO REFER TO ENGINEERING CATALOGS FOR SPECIAL APPLICATIONS THAT REQUIRE SPECIFIC MATERIAL CONTENT, TOLERANCES, ETC. SEE FOOTNOTE.

IDLI GROUP 86776

YOKE SHAFT WITH 16 SPLINES
SEE FIGURE 72

5.844 in	IE	148.43 mm
6.219 in	OE	157.96 mm
2.500 in	SB	63.50 mm
4.375 in	EC	111.13 mm

ROCKWELL 72NYSM40-53

IDLI GROUP 86777

YOKE SHAFT WITH 16 SPLINES
SEE FIGURE 71

5.844 in	IE	148.43 mm
6.219 in	OE	157.96 mm
2.500 in	SB	63.50 mm
4.750 in	EC	120.65 mm

ROCKWELL 72NYSM40-14

IDLI GROUP 86778

YOKE SHAFT WITH 16 SPLINES
slight variation from above group
SEE FIGURE 71

5.844 in	IE	148.43 mm
6.219 in	OE	157.96 mm
2.500 in	SB	63.50 mm
4.750 in	EC	120.65 mm

ROCKWELL 72NYSM40-35

IDLI GROUP 86779

YOKE SHAFT WITH 16 SPLINES
SEE FIGURE 72

5.844 in	IE	148.43 mm
6.219 in	OE	157.96 mm
2.500 in	SB	63.50 mm
4.875 in	EC	123.83 mm

ROCKWELL 72NYSM40-54

IDLI GROUP 86780

YOKE SHAFT WITH 16 SPLINES
SEE FIGURE 71

5.844 in	IE	148.43 mm
6.219 in	OE	157.96 mm
2.500 in	SB	63.50 mm
5.375 in	EC	136.53 mm

ROCKWELL 72NYSM40-66

IDLI GROUP 86781

YOKE SHAFT WITH 16 SPLINES
slight variation from above group
SEE FIGURE 72

5.844 in	IE	148.43 mm
6.219 in	OE	157.96 mm
2.500 in	SB	63.50 mm
5.375 in	EC	136.53 mm

ROCKWELL 72NYSM40-55

IDLI GROUP 86782

YOKE SHAFT WITH 16 SPLINES
SEE FIGURE 71

5.844 in	IE	148.43 mm
6.219 in	OE	157.96 mm
2.500 in	SB	63.50 mm
5.875 in	EC	149.23 mm

ROCKWELL 72NYSM40-76

IDLI GROUP 86783

YOKE SHAFT WITH 16 SPLINES
SEE FIGURE 72

5.844 in	IE	148.43 mm
6.219 in	OE	157.96 mm
2.500 in	SB	63.50 mm
5.875 in	EC	149.23 mm

ROCKWELL 72NYSM40-56

IDLI GROUP 86784

YOKE SHAFT WITH 16 SPLINES
SEE FIGURE 71

5.844 in	IE	148.43 mm
6.219 in	OE	157.96 mm
2.500 in	SB	63.50 mm
6.000 in	EC	152.40 mm

ROCKWELL 72NYSM40-93

IDLI GROUP 86785

YOKE SHAFT WITH 16 SPLINES
SEE FIGURE 71

5.844 in	IE	148.43 mm
6.219 in	OE	157.96 mm
2.500 in	SB	63.50 mm
6.188 in	EC	157.16 mm

ROCKWELL 72NYSM40-88

IDLI GROUP 86786

YOKE SHAFT WITH 16 SPLINES
slight variation from above group
SEE FIGURE 71

5.844 in	IE	148.43 mm
6.219 in	OE	157.96 mm
2.500 in	SB	63.50 mm
6.188 in	EC	157.16 mm

ROCKWELL 72NYSM40-17

IDLI GROUP 86787

YOKE SHAFT WITH 16 SPLINES
slight variation from above group
SEE FIGURE 71

5.844 in	IE	148.43 mm
6.219 in	OE	157.96 mm
2.500 in	SB	63.50 mm
6.188 in	EC	157.16 mm

ROCKWELL 72NYSM40-36

IDLI GROUP 86788

YOKE SHAFT WITH 16 SPLINES
SEE FIGURE 72

5.844 in	IE	148.43 mm
6.219 in	OE	157.96 mm
2.500 in	SB	63.50 mm
6.375 in	EC	161.93 mm

ROCKWELL 72NYSM40-57

IDLI GROUP 86789

YOKE SHAFT WITH 16 SPLINES
SEE FIGURE 71

5.844 in	IE	148.43 mm
6.219 in	OE	157.96 mm
2.500 in	SB	63.50 mm
6.688 in	EC	169.86 mm

ROCKWELL 72NYSM40-16

IDLI GROUP 86790

YOKE SHAFT WITH 16 SPLINES
slight variation from above group
SEE FIGURE 71

5.844 in	IE	148.43 mm
6.219 in	OE	157.96 mm
2.500 in	SB	63.50 mm
6.688 in	EC	169.86 mm

ROCKWELL 72NYSM40-37

IDLI GROUP 86791

YOKE SHAFT WITH 16 SPLINES
slight variation from above group
SEE FIGURE 71

5.844 in	IE	148.43 mm
6.219 in	OE	157.96 mm
2.500 in	SB	63.50 mm
6.688 in	EC	169.86 mm

ROCKWELL 72NYSM40-87

IDLI GROUP 86792

YOKE SHAFT WITH 16 SPLINES
SEE FIGURE 71

5.844 in	IE	148.43 mm
6.219 in	OE	157.96 mm
2.500 in	SB	63.50 mm
7.000 in	EC	177.80 mm

ROCKWELL 72NYSM40-13

IDLI GROUP 86793

YOKE SHAFT WITH 16 SPLINES
slight variation from above group
SEE FIGURE 71

5.844 in	IE	148.43 mm
6.219 in	OE	157.96 mm
2.500 in	SB	63.50 mm
7.000 in	EC	177.80 mm

ROCKWELL 72NYSM40-34

IDLI GROUP 86794

YOKE SHAFT WITH 16 SPLINES
slight variation from above group
SEE FIGURE 71

5.844 in	IE	148.43 mm
6.219 in	OE	157.96 mm
2.500 in	SB	63.50 mm
7.000 in	EC	177.80 mm

ROCKWELL 72NYSM40-84

IDLI GROUP 86795

YOKE SHAFT WITH 16 SPLINES
SEE FIGURE 71

5.844 in	IE	148.43 mm
6.219 in	OE	157.96 mm
2.500 in	SB	63.50 mm
7.688 in	EC	195.26 mm

ROCKWELL 72NYSM40-71

IDLI GROUP 86796

YOKE SHAFT WITH 16 SPLINES
SEE FIGURE 71

5.844 in	IE	148.43 mm
6.219 in	OE	157.96 mm
2.500 in	SB	63.50 mm
8.125 in	EC	206.38 mm

ROCKWELL 72NYSM40-15

IDLI GROUP 86797

YOKE SHAFT WITH 16 SPLINES
slight variation from above group
SEE FIGURE 71

5.844 in	IE	148.43 mm
6.219 in	OE	157.96 mm
2.500 in	SB	63.50 mm
8.125 in	EC	206.38 mm

ROCKWELL 72NYSM40-39

IDLI GROUP 86798

YOKE SHAFT WITH 16 SPLINES
SEE FIGURE 71

5.844 in	IE	148.43 mm
6.219 in	OE	157.96 mm
2.500 in	SB	63.50 mm
9.125 in	EC	231.78 mm

ROCKWELL 72NYSM40-69

IDLI GROUP 86799

YOKE SHAFT WITH 16 SPLINES
SEE FIGURE 71

5.844 in	IE	148.43 mm
6.219 in	OE	157.96 mm
2.500 in	SB	63.50 mm
9.375 in	EC	238.13 mm

ROCKWELL 72NYSM40-18

IDLI GROUP 86800

TUBE AND WELD YOKE ASSEMBLY
SEE FIGURE 50

5.844 in	IE	148.43 mm
6.219 in	OE	157.96 mm
2.760 in	SB	70.10 mm
54.813 in	EC	1392.24 mm
3.000 in	OD	76.20 mm

BORG-WARNR 42-7000
ROCKWELL 954-13-2

ENGINEERING CATALOGS MUST BE CONSULTED FOR DETAILS NOT INCLUDED
IN THIS GUIDE. SPECIFIC DESIGNS, MATERIAL CONTENT, TOLERANCES
LUBE FITTINGS AND OTHER DIMENSIONS ARE INTENTIONALLY OMITTED HERE.

IDLI GUIDE   COPYRIGHT © INTERCHANGE, INC.   ST. LOUIS PARK, MN. 55416 USA

IDLI GROUP 86801

YOKE SHAFT WITH 22 INVOLUTED SPLINES IN 23 SPLINE SPACES
SEE FIGURE 71

5.844 in	IE	148.43 mm
6.219 in	OE	157.96 mm
3.000 in	SB	76.20 mm
11.688 in	EC	296.86 mm

BORG-WARNR 105-21424J
BORG-WARNR 105-23569J
ROCKWELL 72NYSM48-4

IDLI GROUP 86802

TUBE AND WELD YOKE ASSEMBLY
SEE FIGURE 50

5.844 in	IE	148.43 mm
6.219 in	OE	157.96 mm
3.260 in	SB	82.80 mm
62.125 in	EC	1577.98 mm
3.500 in	OD	88.90 mm

BORG-WARNR 42-7004
ROCKWELL 954-12-2

IDLI GROUP 86803

TUBE AND WELD YOKE ASSEMBLY
SEE FIGURE 50

5.844 in	IE	148.43 mm
6.219 in	OE	157.96 mm
3.310 in	SB	84.07 mm
64.125 in	EC	1628.78 mm
3.500 in	OD	88.90 mm

BORG-WARNR 42-7001
ROCKWELL 954-10-2

IDLI GROUP 86804

TUBE AND WELD YOKE ASSEMBLY
SEE FIGURE 50

5.844 in	IE	148.43 mm
6.219 in	OE	157.96 mm
3.782 in	SB	96.06 mm
62.125 in	EC	1577.98 mm
4.000 in	OD	101.60 mm

BORG-WARNR 42-7003
ROCKWELL 954-11-2

IDLI GROUP 86805

DOUBLE CENTER
SEE FIGURE 62

6.500 in	IE	165.10 mm
6.875 in	OE	174.63 mm
2.000 in	EC	50.80 mm

BORG-WARNR 15706J
BORG-WARNR 18639J
ROCKWELL 85WB170

IDLI GROUP 86806

YOKE SHAFT WITH 10 SPLINES
SEE FIGURE 71

6.500 in	IE	165.10 mm
6.875 in	OE	174.63 mm
2.500 in	SB	63.50 mm
10.000 in	EC	254.00 mm

ROCKWELL 85WBYSM40-5

IDLI GROUP 86807

YOKE SHAFT WITH 23 INVOLUTED SPLINES
SEE FIGURE 71

6.500 in	IE	165.10 mm
6.875 in	OE	174.63 mm
3.000 in	SB	76.20 mm
12.313 in	EC	312.74 mm

ROCKWELL 85WBYSM48-7

IDLI GROUP 86808

DOUBLE CENTER
SEE FIGURE 62

8.125 in	IE	206.38 mm
8.500 in	OE	215.90 mm
1.750 in	EC	44.45 mm

BORG-WARNR 10925J
BORG-WARNR 5461J
BORG-WARNR 5481J
ROCKWELL 82N170-1

IDLI GROUP 86809

DOUBLE CENTER
slight variation from above group
SEE FIGURE 62

8.125 in	IE	206.38 mm
8.500 in	OE	215.90 mm
1.750 in	EC	44.45 mm

BORG-WARNR 7936J
BORG-WARNR 9699J
ROCKWELL 82N170

IDLI GROUP 86810

DOUBLE CENTER
SEE FIGURE 62

8.125 in	IE	206.38 mm
8.500 in	OE	215.90 mm
2.563 in	EC	65.09 mm

BORG-WARNR 10033J
ROCKWELL 82N174-1

IDLI GROUP 86811

DOUBLE CENTER
SEE FIGURE 61

8.125 in	IE	206.38 mm
8.500 in	OE	215.90 mm
3.375 in	EC	85.73 mm

BORG-WARNR 11221J
ROCKWELL 82NDCA1

IDLI GROUP 86812

YOKE SHAFT WITH 4 SPLINES FOR ROLL JOINT ASSEMBLY
SEE FIGURE 72

8.125 in	IE	206.38 mm
8.500 in	OE	215.90 mm
2.109 in	SB	53.58 mm
15.813 in	EC	401.64 mm

ROCKWELL 82NYSM34-9

IDLI GROUP 86813

YOKE SHAFT WITH 16 SPLINES
SEE FIGURE 71

8.125 in	IE	206.38 mm
8.500 in	OE	215.90 mm
2.500 in	SB	63.50 mm
5.813 in	EC	147.64 mm

ROCKWELL 82NYSM40-19

IDLI GROUP 86814

YOKE SHAFT WITH 16 SPLINES
SEE FIGURE 71

8.125 in	IE	206.38 mm
8.500 in	OE	215.90 mm
2.500 in	SB	63.50 mm
5.938 in	EC	150.81 mm

ROCKWELL 82NYSM40-48

IDLI GROUP 86815

YOKE SHAFT WITH 10 SPLINES
SEE FIGURE 71

8.125 in	IE	206.38 mm
8.500 in	OE	215.90 mm
2.500 in	SB	63.50 mm
6.188 in	EC	157.16 mm

ROCKWELL 82NYSM40-14

IDLI GROUP 86816

YOKE SHAFT WITH 16 SPLINES
SEE FIGURE 72

8.125 in	IE	206.38 mm
8.500 in	OE	215.90 mm
2.500 in	SB	63.50 mm
7.625 in	EC	193.68 mm

ROCKWELL 82NYSM40-54

IDLI GROUP 86817

YOKE SHAFT WITH 10 SPLINES
SEE FIGURE 71

8.125 in	IE	206.38 mm
8.500 in	OE	215.90 mm
2.500 in	SB	63.50 mm
8.125 in	EC	206.38 mm

ROCKWELL 82NYSM40-4

IDLI GROUP 86818

YOKE SHAFT WITH 10 SPLINES
slight variation from above group
SEE FIGURE 71

8.125 in	IE	206.38 mm
8.500 in	OE	215.90 mm
2.500 in	SB	63.50 mm
8.125 in	EC	206.38 mm

ROCKWELL 82NYSM40

IDLI GROUP 86819

YOKE SHAFT WITH 10 SPLINES
SEE FIGURE 71

8.125 in	IE	206.38 mm
8.500 in	OE	215.90 mm
2.500 in	SB	63.50 mm
9.625 in	EC	244.48 mm

ROCKWELL 82NYSM40-8

IDLI GROUP 86820

YOKE SHAFT WITH 10 SPLINES
SEE FIGURE 71

8.125 in	IE	206.38 mm
8.500 in	OE	215.90 mm
2.500 in	SB	63.50 mm
9.813 in	EC	249.24 mm

ROCKWELL 82NYSM40-49

IDLI GROUP 86821

YOKE SHAFT WITH 10 SPLINES
SEE FIGURE 71

8.125 in	IE	206.38 mm
8.500 in	OE	215.90 mm
2.500 in	SB	63.50 mm
10.250 in	EC	260.35 mm

ROCKWELL 82NYSM40-6

IDLI GROUP 86822

YOKE SHAFT WITH 10 SPLINES
SEE FIGURE 71

8.125 in	IE	206.38 mm
8.500 in	OE	215.90 mm
2.500 in	SB	63.50 mm
10.875 in	EC	276.23 mm

ROCKWELL 82NYSM40-45

IDLI GROUP 86823

YOKE SHAFT WITH 10 SPLINES
SEE FIGURE 71

8.125 in	IE	206.38 mm
8.500 in	OE	215.90 mm
2.500 in	SB	63.50 mm
11.625 in	EC	295.28 mm

ROCKWELL 82NYSM40-9

IDLI GROUP 86824

YOKE SHAFT WITH 10 SPLINES
SEE FIGURE 71

8.125 in	IE	206.38 mm
8.500 in	OE	215.90 mm
2.500 in	SB	63.50 mm
11.875 in	EC	301.63 mm

ROCKWELL 82NYSM40-21

IDLI GROUP 86825

YOKE SHAFT WITH 10 SPLINES
SEE FIGURE 71

8.125 in	IE	206.38 mm
8.500 in	OE	215.90 mm
2.500 in	SB	63.50 mm
12.250 in	EC	311.15 mm

ROCKWELL 82NYSM40-53

IDLI GROUP 86826

YOKE SHAFT WITH 10 SPLINES
SEE FIGURE 71

8.125 in	IE	206.38 mm
8.500 in	OE	215.90 mm
2.500 in	SB	63.50 mm
12.375 in	EC	314.33 mm

ROCKWELL 82NYSM40-7

BD = BEARING DIAMETER (outside)   **BW** = BEARING WIDTH   **CB** = CROSS LENGTH WITH BEARINGS   **CC** = CENTER TO CENTER   **CD** = CROSS DIAMETER
CL = CROSS LENGTH WITHOUT BEARINGS   **EC** = END TO CENTER or FACE TO FACE   **EE** = END TO END   **EL** = EFFECTIVE LENGTH
HD = HUB DIAMETER (or insert)   **IE** = INSIDE OF EARS (or recess)   **OD** = OUTSIDE DIAMETER   **OE** = OUTSIDE OF EARS   **SB** = SPLINE OR BORE SIZE

IDLI GUIDE   COPYRIGHT ® INTERCHANGE, INC.   ST. LOUIS PARK, MN. 55416 USA

CAUTION: BE SURE TO REFER TO ENGINEERING CATALOGS FOR SPECIAL APPLICATIONS THAT REQUIRE SPECIFIC MATERIAL CONTENT, TOLERANCES, ETC. SEE FOOTNOTE.

IDLI GROUP 86827

YOKE SHAFT WITH 10 SPLINES
SEE FIGURE 72

8.125 in	IE	206.38 mm
8.500 in	OE	215.90 mm
2.500 in	SB	63.50 mm
12.938 in	EC	328.61 mm

ROCKWELL 82NYSM40-52

IDLI GROUP 86828

YOKE SHAFT WITH 10 SPLINES
SEE FIGURE 72

8.125 in	IE	206.38 mm
8.500 in	OE	215.90 mm
2.500 in	SB	63.50 mm
14.188 in	EC	360.36 mm

ROCKWELL 82NYSM40-75

IDLI GROUP 86829

TUBE AND WELD YOKE ASSEMBLY
SEE FIGURE 50

8.125 in	IE	206.38 mm
8.500 in	OE	215.90 mm
3.626 in	SB	92.10 mm
68.438 in	EC	1738.31 mm
4.000 in	OD	101.60 mm

BORG-WARNR 42-8003
ROCKWELL 955-11-2

IDLI GROUP 86830

TUBE AND WELD YOKE ASSEMBLY
SEE FIGURE 50

8.125 in	IE	206.38 mm
8.500 in	OE	215.90 mm
3.782 in	SB	96.06 mm
64.438 in	EC	1636.71 mm
4.000 in	OD	101.60 mm

BORG-WARNR 42-8002
ROCKWELL 955-10-2

IDLI GROUP 86831

DOUBLE CENTER
SEE FIGURE 62

8.250 in	IE	209.55 mm
8.625 in	OE	219.08 mm
2.000 in	EC	50.80 mm

BORG-WARNR 7307J
BORG-WARNR 7985J
ROCKWELL 92N170

IDLI GROUP 86832

DOUBLE CENTER
SEE FIGURE 61

8.250 in	IE	209.55 mm
8.625 in	OE	219.08 mm
3.125 in	EC	79.38 mm

ROCKWELL 92NDCA4

IDLI GROUP 86833

DOUBLE CENTER
SEE FIGURE 61

8.250 in	IE	209.55 mm
8.625 in	OE	219.08 mm
5.000 in	EC	127.00 mm

ROCKWELL 92NDCA2

IDLI GROUP 86834

DOUBLE CENTER
SEE FIGURE 61

8.250 in	IE	209.55 mm
8.625 in	OE	219.08 mm
5.188 in	EC	131.76 mm

ROCKWELL 92NDCA8

IDLI GROUP 86835

DOUBLE CENTER
SEE FIGURE 61

8.250 in	IE	209.55 mm
8.625 in	OE	219.08 mm
5.250 in	EC	133.35 mm

ROCKWELL 92NDCA6

IDLI GROUP 86836

DOUBLE CENTER
SEE FIGURE 61

8.250 in	IE	209.55 mm
8.625 in	OE	219.08 mm
6.250 in	EC	158.75 mm

ROCKWELL 92NDCA5

IDLI GROUP 86837

DOUBLE CENTER
SEE FIGURE 61

8.250 in	IE	209.55 mm
8.625 in	OE	219.08 mm
8.500 in	EC	215.90 mm

ROCKWELL 92NDCA1

IDLI GROUP 86838

YOKE SHAFT
WITH 23 INVOLUTED SPLINES

8.250 in	IE	209.55 mm
8.625 in	OE	219.08 mm
2.250 in	SB	57.15 mm
31.500 in	EC	800.10 mm

ROCKWELL 92NYSM34-3

IDLI GROUP 86839

YOKE SHAFT
WITH 24 INVOLUTED SPLINES

8.250 in	IE	209.55 mm
8.625 in	OE	219.08 mm
2.375 in	SB	60.33 mm
18.125 in	EC	460.38 mm

ROCKWELL 92NYSM38-9

IDLI GROUP 86840

YOKE SHAFT
WITH 24 INVOLUTED SPLINES

8.250 in	IE	209.55 mm
8.625 in	OE	219.08 mm
2.375 in	SB	60.33 mm
19.188 in	EC	487.36 mm

ROCKWELL 92NYSM38-23

IDLI GROUP 86841

YOKE SHAFT
WITH 24 INVOLUTED SPLINES

8.250 in	IE	209.55 mm
8.625 in	OE	219.08 mm
2.375 in	SB	60.33 mm
34.938 in	EC	887.41 mm

ROCKWELL 92NYSM38-13

IDLI GROUP 86842

YOKE SHAFT
WITH 24 INVOLUTED SPLINES

8.250 in	IE	209.55 mm
8.625 in	OE	219.08 mm
2.375 in	SB	60.33 mm
39.438 in	EC	1001.71 mm

ROCKWELL 92NYSM38-8

IDLI GROUP 86843

YOKE SHAFT
WITH 24 INVOLUTED SPLINES
SEE FIGURE 72

8.250 in	IE	209.55 mm
8.625 in	OE	219.08 mm
2.469 in	SB	62.71 mm
7.688 in	EC	195.26 mm

BORG-WARNR 24513J
ROCKWELL 92NYSM40-19

IDLI GROUP 86844

YOKE SHAFT
WITH 26 INVOLUTED SPLINES

8.250 in	IE	209.55 mm
8.625 in	OE	219.08 mm
2.625 in	SB	66.68 mm
12.750 in	EC	323.85 mm

ROCKWELL 92NYSM42-8

IDLI GROUP 86845

YOKE SHAFT
WITH 21 INVOLUTED SPLINES
SEE FIGURE 72

8.250 in	IE	209.55 mm
8.625 in	OE	219.08 mm
2.750 in	SB	69.85 mm
10.750 in	EC	273.05 mm

ROCKWELL 92NYSM44-6

IDLI GROUP 86846

YOKE SHAFT WITH 10 SPLINES
SEE FIGURE 72

8.250 in	IE	209.55 mm
8.625 in	OE	219.08 mm
3.000 in	SB	76.20 mm
7.500 in	EC	190.50 mm

ROCKWELL 92NYSM48-30

IDLI GROUP 86847

YOKE SHAFT WITH 10 SPLINES
SEE FIGURE 71

8.250 in	IE	209.55 mm
8.625 in	OE	219.08 mm
3.000 in	SB	76.20 mm
8.813 in	EC	223.84 mm

ROCKWELL 92NYSM48-23

IDLI GROUP 86848

YOKE SHAFT WITH 10 SPLINES
SEE FIGURE 71

8.250 in	IE	209.55 mm
8.625 in	OE	219.08 mm
3.000 in	SB	76.20 mm
9.000 in	EC	228.60 mm

ROCKWELL 92NYSM48-4

IDLI GROUP 86849

YOKE SHAFT WITH 10 SPLINES
SEE FIGURE 71

8.250 in	IE	209.55 mm
8.625 in	OE	219.08 mm
3.000 in	SB	76.20 mm
9.688 in	EC	246.06 mm

ROCKWELL 92NYSM48-27

IDLI GROUP 86850

YOKE SHAFT WITH 10 SPLINES
SEE FIGURE 71

8.250 in	IE	209.55 mm
8.625 in	OE	219.08 mm
3.000 in	SB	76.20 mm
10.500 in	EC	266.70 mm

ROCKWELL 92NYSM48-15

IDLI GROUP 86851

YOKE SHAFT WITH 10 SPLINES
SEE FIGURE 71

8.250 in	IE	209.55 mm
8.625 in	OE	219.08 mm
3.000 in	SB	76.20 mm
11.438 in	EC	290.51 mm

ROCKWELL 92NYSM48-11

IDLI GROUP 86852

YOKE SHAFT WITH 16 SPLINES
SEE FIGURE 71

8.250 in	IE	209.55 mm
8.625 in	OE	219.08 mm
3.000 in	SB	76.20 mm
11.438 in	EC	290.51 mm

ROCKWELL 92NYSM48-60

ENGINEERING CATALOGS MUST BE CONSULTED FOR DETAILS NOT INCLUDED
IN THIS GUIDE. SPECIFIC DESIGNS, MATERIAL CONTENT, TOLERANCES
LUBE FITTINGS AND OTHER DIMENSIONS ARE INTENTIONALLY OMITTED HERE.

IDLI GROUP 86853

**YOKE SHAFT WITH 22 INVOLUTED
SPLINES IN 23 SPLINE SPACES
SEE FIGURE 71**

8.250 in	IE	209.55 mm
8.625 in	OE	219.08 mm
3.000 in	SB	76.20 mm
13.313 in	EC	338.14 mm

ROCKWELL 92NYSM48-21

IDLI GROUP 86854

**YOKE SHAFT WITH 10 SPLINES
SEE FIGURE 71**

8.250 in	IE	209.55 mm
8.625 in	OE	219.08 mm
3.000 in	SB	76.20 mm
14.125 in	EC	358.78 mm

BORG-WARNR 105-4494
ROCKWELL 92NYSM48-19

IDLI GROUP 86855

**YOKE SHAFT WITH 22 INVOLUTED
SPLINES IN 23 SPLINE SPACES
SEE FIGURE 71**

8.250 in	IE	209.55 mm
8.625 in	OE	219.08 mm
3.000 in	SB	76.20 mm
15.375 in	EC	390.53 mm

ROCKWELL 92NYSM48-55

IDLI GROUP 86856

**TUBE AND WELD YOKE ASSEMBLY
SEE FIGURE 50**

8.250 in	IE	209.55 mm
8.625 in	OE	219.08 mm
3.982 in	SB	101.14 mm
4.500 in	OD	114.30 mm

ROCKWELL 960-11-2

IDLI GROUP 86857

**TUBE AND WELD YOKE ASSEMBLY
SEE FIGURE 50**

8.250 in	IE	209.55 mm
8.625 in	OE	219.08 mm
4.125 in	SB	104.78 mm
4.500 in	OD	114.30 mm

ROCKWELL 960-10-2

BD=BEARING DIAMETER (outside)   **BW**=BEARING WIDTH   **CB**=CROSS LENGTH WITH BEARINGS   **CC**=CENTER TO CENTER   **CD**=CROSS DIAMETER
CL=CROSS LENGTH WITHOUT BEARINGS   **EC**=END TO CENTER or FACE TO FACE   **EE**=END TO END   **EL**=EFFECTIVE LENGTH
HD=HUB DIAMETER (or insert)   **IE**=INSIDE OF EARS (or recess)   **OD**=OUTSIDE DIAMETER   **OE**=OUTSIDE OF EARS   **SB**=SPLINE OR BORE SIZE

Other Valuable Interchange Publications

The International Bearing Interchange Guide

New Ninth Edition

The I.B.I. Guide has over 365,000 listings of 25,326 ball and roller bearings available worldwide and dating back to 1918. It includes manufacturers, original equipment users and government part numbers, with THOUSANDS of new listings not included in earlier editions!

I.S.B.N. #916966-17-8

The International Seal Interchange Guide

New Sixth Edition

The I.S.I. Guide has over 150,000 listings of 12,632 different seals and "O" rings. It includes government numbers as well as producers and users throughout the world—over 700 pages of information to save you time and money!

I.S.B.N. #916966-16-X

The International Drive Belt Interchange Guide

New Fourth Edition

The I.D.B.I. Guide has more than 115,000 listings for nearly 5,863 types and sizes of drive and V-belts. Listings include manufacturers, original equipment and government users from all over the world. 760 pages.

I.S.B.N. #916966-15-1

IDLI GROUP 87001

DOUBLE CENTER
SEE FIGURE 60

2.906 in	OE	73.82 mm
1.125 in	BD	28.58 mm
1.875 in	CC	47.63 mm

PRECISION 1380
ROCKWELL L14S170-1

IDLI GROUP 87002

DOUBLE CENTER
SEE FIGURE 59

2.906 in	OE	73.82 mm
1.125 in	BD	28.58 mm
4.500 in	CC	114.30 mm

ROCKWELL L14SDCA8

IDLI GROUP 87003

CLAMP YOKE
WITH ROUND BORE
SEE FIGURE 14B

2.906 in	OE	73.82 mm
1.125 in	BD	28.58 mm
1.125 in	SB	28.58 mm
3.219 in	EC	81.76 mm

ROCKWELL L14SYR18-72

IDLI GROUP 87004

CLAMP YOKE
WITH HEX BORE
SEE FIGURE 14E

2.906 in	OE	73.82 mm
1.125 in	BD	28.58 mm
1.125 in	SB	28.58 mm
3.375 in	EC	85.73 mm

ROCKWELL L14SYH18-5

IDLI GROUP 87005

CLAMP YOKE
WITH 17 INVOLUTED SPLINES
SEE FIGURE 14A

2.906 in	OE	73.82 mm
1.125 in	BD	28.58 mm
1.125 in	SB	28.58 mm
3.438 in	EC	87.31 mm

ROCKWELL L14SYS18-23

IDLI GROUP 87006

CLAMP YOKE
WITH 6 SPLINES
SEE FIGURE 14A

2.906 in	OE	73.82 mm
1.125 in	BD	28.58 mm
1.125 in	SB	28.58 mm
3.688 in	EC	93.66 mm

ROCKWELL L14SYS18-20

IDLI GROUP 87007

CLAMP YOKE
WITH 6 SPLINES
SEE FIGURE 14A

2.906 in	OE	73.82 mm
1.125 in	BD	28.58 mm
1.125 in	SB	28.58 mm
3.875 in	EC	98.43 mm

ROCKWELL L14SYS18-11

IDLI GROUP 87008

CLAMP YOKE
WITH ROUND BORE
SEE FIGURE 14B

2.906 in	OE	73.82 mm
1.125 in	BD	28.58 mm
1.250 in	SB	31.75 mm
3.438 in	EC	87.31 mm

ROCKWELL L14SYR20-114

IDLI GROUP 87009

CLAMP YOKE
WITH HEX BORE
SEE FIGURE 14E

2.906 in	OE	73.82 mm
1.125 in	BD	28.58 mm
1.250 in	SB	31.75 mm
3.563 in	EC	90.49 mm

ROCKWELL L14SYH20

IDLI GROUP 87010

CLAMP YOKE
WITH ROUND BORE
SEE FIGURE 14B

2.906 in	OE	73.82 mm
1.125 in	BD	28.58 mm
1.250 in	SB	31.75 mm
4.000 in	EC	101.60 mm

ROCKWELL L14SYR20-29

IDLI GROUP 87011

CLAMP YOKE
WITH ROUND BORE
SEE FIGURE 14B

2.906 in	OE	73.82 mm
1.125 in	BD	28.58 mm
1.375 in	SB	34.93 mm
3.438 in	EC	87.31 mm

ROCKWELL L14SYR22-65

IDLI GROUP 87012

CLAMP YOKE WITH 43 SERRATIONS
SEE FIGURE 14C

2.906 in	OE	73.82 mm
1.125 in	BD	28.58 mm
1.375 in	SB	34.93 mm
3.688 in	EC	93.66 mm

ROCKWELL L14SYS22-26
SPICER 2-4-2731

IDLI GROUP 87013

CLAMP YOKE
WITH 21 INVOLUTED SPLINES
SEE FIGURE 14A

2.906 in	OE	73.82 mm
1.125 in	BD	28.58 mm
1.375 in	SB	34.93 mm
3.688 in	EC	93.66 mm

PRECISION 1371
ROCKWELL L14SYS22-43
SPICER 2-4-3061
SPICER 2-4-3061X

IDLI GROUP 87014

CLAMP YOKE
WITH ROUND BORE
SEE FIGURE 14B

2.906 in	OE	73.82 mm
1.125 in	BD	28.58 mm
1.375 in	SB	34.93 mm
4.000 in	EC	101.60 mm

ROCKWELL L14SYR22-44

IDLI GROUP 87015

CLAMP YOKE WITH ROUND BORE
slight variation from above group
SEE FIGURE 14B

2.906 in	OE	73.82 mm
1.125 in	BD	28.58 mm
1.375 in	SB	34.93 mm
4.000 in	EC	101.60 mm

ROCKWELL L14SYS22-28

IDLI GROUP 87016

CLAMP YOKE
WITH 6 SPLINES
SEE FIGURE 14A

2.906 in	OE	73.82 mm
1.125 in	BD	28.58 mm
1.375 in	SB	34.93 mm
4.375 in	EC	111.13 mm

PRECISION 1370
ROCKWELL L14SYS22-2

IDLI GROUP 87017

CLAMP YOKE
WITH ROUND BORE
SEE FIGURE 14B

3.125 in	OE	79.38 mm
.875 in	BD	22.23 mm
1.000 in	SB	25.40 mm
3.688 in	EC	93.66 mm

ROCKWELL 1FRYR16-57

IDLI GROUP 87018

CLAMP YOKE
WITH 6 SPLINES
SEE FIGURE 14B

3.125 in	OE	79.38 mm
.875 in	BD	22.23 mm
1.125 in	SB	28.58 mm
3.688 in	EC	93.66 mm

ROCKWELL 1FRYS18-1

IDLI GROUP 87019

CLAMP YOKE
WITH 6 SPLINES
slight variation from above group
SEE FIGURE 14B

3.125 in	OE	79.38 mm
.875 in	BD	22.23 mm
1.125 in	SB	28.58 mm
3.688 in	EC	93.66 mm

PRECISION 1940
ROCKWELL 1FRYS22-2

IDLI GROUP 87020

CLAMP YOKE
WITH ROUND BORE
SEE FIGURE 14B

3.125 in	OE	79.38 mm
.875 in	BD	22.23 mm
1.125 in	SB	28.58 mm
3.688 in	EC	93.66 mm

ROCKWELL 1FRYR18-69

IDLI GROUP 87021

CLAMP YOKE
WITH ROUND BORE
SEE FIGURE 14B

3.125 in	OE	79.38 mm
.875 in	BD	22.23 mm
1.250 in	SB	31.75 mm
3.750 in	EC	95.25 mm

ROCKWELL 1FRYR20-64

IDLI GROUP 87022

CLAMP YOKE WITH ROUND BORE
slight variation from above group
SEE FIGURE 14B

3.125 in	OE	79.38 mm
.875 in	BD	22.23 mm
1.250 in	SB	31.75 mm
3.750 in	EC	95.25 mm

ROCKWELL 1FRYR20-82

IDLI GROUP 87023

CLAMP YOKE
WITH 21 INVOLUTED SPLINES
SEE FIGURE 14B

3.125 in	OE	79.38 mm
.875 in	BD	22.23 mm
1.375 in	SB	34.93 mm
3.688 in	EC	93.66 mm

ROCKWELL 1FRYS22-47

IDLI GROUP 87024

CLAMP YOKE
WITH 6 SPLINES
SEE FIGURE 14B

3.125 in	OE	79.38 mm
.875 in	BD	22.23 mm
1.375 in	SB	34.93 mm
4.938 in	EC	125.41 mm

ROCKWELL 1FRYS22

ENGINEERING CATALOGS MUST BE CONSULTED FOR DETAILS NOT INCLUDED
IN THIS GUIDE. SPECIFIC DESIGNS, MATERIAL CONTENT, TOLERANCES
LUBE FITTINGS AND OTHER DIMENSIONS ARE INTENTIONALLY OMITTED HERE.

IDLI GUIDE   COPYRIGHT ® INTERCHANGE, INC.   ST. LOUIS PARK, MN. 55416 USA

CAUTION: BE SURE TO REFER TO ENGINEERING CATALOGS FOR SPECIAL APPLICATIONS THAT REQUIRE SPECIFIC MATERIAL CONTENT, TOLERANCES, ETC. SEE FOOTNOTE.

IDLI GROUP 87025

CLAMP YOKE
WITH 6 SPLINES
SEE FIGURE 14B

3.125 in	OE	79.38 mm
.875 in	BD	22.23 mm
1.750 in	SB	44.45 mm
3.750 in	EC	95.25 mm

ROCKWELL 1FRYS28-1

IDLI GROUP 87026

DOUBLE CENTER
SEE FIGURE 60

3.484 in	OE	88.50 mm
1.250 in	BD	31.75 mm
2.250 in	CC	57.15 mm

ROCKWELL 3N171-1

IDLI GROUP 87027

DOUBLE CENTER
SEE FIGURE 60

3.484 in	OE	88.50 mm
1.250 in	BD	31.75 mm
3.000 in	CC	76.20 mm

ROCKWELL 3N170-1

IDLI GROUP 87028

CLAMP YOKE
WITH 6 SPLINES
SEE FIGURE 14A

3.484 in	OE	88.50 mm
1.250 in	BD	31.75 mm
1.125 in	SB	28.58 mm
5.250 in	EC	133.35 mm

ROCKWELL L16SYS18-3

IDLI GROUP 87029

CLAMP YOKE
WITH HEX BORE
SEE FIGURE 14C

3.484 in	OE	88.50 mm
1.250 in	BD	31.75 mm
1.250 in	SB	31.75 mm
3.875 in	EC	98.43 mm

ROCKWELL L16SYH20

IDLI GROUP 87030

CLAMP YOKE
WITH ROUND BORE
SEE FIGURE 14B

3.484 in	OE	88.50 mm
1.250 in	BD	31.75 mm
1.250 in	SB	31.75 mm
4.438 in	EC	112.71 mm

ROCKWELL L16SYR20-52

IDLI GROUP 87031

CLAMP YOKE
WITH 21 INVOLUTED SPLINES
SEE FIGURE 14A

3.484 in	OE	88.50 mm
1.250 in	BD	31.75 mm
1.375 in	SB	34.93 mm
3.750 in	EC	95.25 mm

PRECISION 1851
ROCKWELL L16SYS22-187

IDLI GROUP 87032

CLAMP YOKE
WITH 21 INVOLUTED SPLINES
SEE FIGURE 14A

3.484 in	OE	88.50 mm
1.250 in	BD	31.75 mm
1.375 in	SB	34.93 mm
4.000 in	EC	101.60 mm

ALLOY 0801
ALLOY 801
BORG-WARNR 62105
NEAPCO 21105
NEAPCO 22-1105
NEAPCO 25105
NEAPCO 26-5105
PRECISION 1851
ROCKWELL L16SYS22-46
SPICER 3-4-5131X

IDLI GROUP 87033

CLAMP YOKE
WITH 6 SPLINES
SEE FIGURE 14A

3.484 in	OE	88.50 mm
1.250 in	BD	31.75 mm
1.375 in	SB	34.93 mm
4.750 in	EC	120.65 mm

ROCKWELL 3NYS22-10

IDLI GROUP 87034

CLAMP YOKE
WITH ROUND BORE
SEE FIGURE 14B

3.484 in	OE	88.50 mm
1.250 in	BD	31.75 mm
1.375 in	SB	34.93 mm
4.750 in	EC	120.65 mm

ROCKWELL 3NYR22-25

IDLI GROUP 87035

CLAMP YOKE
WITH HEX BORE
SEE FIGURE 14C

3.484 in	OE	88.50 mm
1.250 in	BD	31.75 mm
1.375 in	SB	34.93 mm
5.000 in	EC	127.00 mm

ROCKWELL L16SYH22

IDLI GROUP 87036

CLAMP YOKE WITH 43 SERRATIONS
SEE FIGURE 14A

3.484 in	OE	88.50 mm
1.250 in	BD	31.75 mm
1.375 in	SB	34.93 mm
5.125 in	EC	130.18 mm

ROCKWELL L16SYS22-48

IDLI GROUP 87037

CLAMP YOKE
WITH 6 SPLINES
SEE FIGURE 14A

3.484 in	OE	88.50 mm
1.250 in	BD	31.75 mm
1.375 in	SB	34.93 mm
5.250 in	EC	133.35 mm

ALLOY 0767
BORG-WARNR 62140
NEAPCO 21140
NEAPCO 22-1140
PRECISION 1850
ROCKWELL 3NYS22-8

IDLI GROUP 87038

CLAMP YOKE
WITH ROUND BORE
SEE FIGURE 14B

3.484 in	OE	88.50 mm
1.250 in	BD	31.75 mm
1.438 in	SB	36.51 mm
4.750 in	EC	120.65 mm

ROCKWELL L16SYR23-18

IDLI GROUP 87039

YOKE SHAFT WITH 10 SPLINES
SEE FIGURE 70

3.484 in	OE	88.50 mm
1.250 in	BD	31.75 mm
1.500 in	SB	38.10 mm
10.125 in	EC	257.18 mm

ROCKWELL 3NYSM24-35

IDLI GROUP 87040

YOKE SHAFT WITH 10 SPLINES
SEE FIGURE 70

3.484 in	OE	88.50 mm
1.250 in	BD	31.75 mm
1.500 in	SB	38.10 mm
11.125 in	EC	282.58 mm

ROCKWELL 3NYSM24-37

IDLI GROUP 87041

YOKE SHAFT WITH 10 SPLINES
SEE FIGURE 70

3.484 in	OE	88.50 mm
1.250 in	BD	31.75 mm
1.500 in	SB	38.10 mm
11.500 in	EC	292.10 mm

ROCKWELL 3NYSM24-39

IDLI GROUP 87042

YOKE SHAFT WITH 10 SPLINES
SEE FIGURE 70

3.484 in	OE	88.50 mm
1.250 in	BD	31.75 mm
1.500 in	SB	38.10 mm
15.500 in	EC	393.70 mm

ROCKWELL 3NYSM24-42

IDLI GROUP 87043

YOKE SHAFT WITH 10 SPLINES
SEE FIGURE 70

3.484 in	OE	88.50 mm
1.250 in	BD	31.75 mm
1.500 in	SB	38.10 mm
16.875 in	EC	428.63 mm

ROCKWELL 3NYSM24-43

IDLI GROUP 87044

CLAMP YOKE
WITH 27 INVOLUTED SPLINES
SEE FIGURE 14A

3.484 in	OE	88.50 mm
1.250 in	BD	31.75 mm
1.750 in	SB	44.45 mm
5.000 in	EC	127.00 mm

ROCKWELL L16SYS28-31

IDLI GROUP 87045

CLAMP YOKE
WITH 6 SPLINES
SEE FIGURE 14A

3.484 in	OE	88.50 mm
1.250 in	BD	31.75 mm
1.750 in	SB	44.45 mm
5.313 in	EC	134.94 mm

ROCKWELL L16SYS28

IDLI GROUP 87046

CLAMP YOKE
WITH 6 SPLINES
SEE FIGURE 14A

4.188 in	OE	106.36 mm
1.000 in	BD	25.40 mm
1.125 in	SB	28.58 mm
4.938 in	EC	125.41 mm

ROCKWELL 3DRYS18-9

IDLI GROUP 87047

CLAMP YOKE
WITH 6 SPLINES
SEE FIGURE 14A

4.188 in	OE	106.36 mm
1.000 in	BD	25.40 mm
1.125 in	SB	28.58 mm
5.250 in	EC	133.35 mm

ROCKWELL 3DRYS18-6

BD — BEARING DIAMETER (outside)   BW — BEARING WIDTH   CB — CROSS LENGTH WITH BEARINGS   CC — CENTER TO CENTER   CD — CROSS DIAMETER
CL = CROSS LENGTH WITHOUT BEARINGS   EC = END TO CENTER or FACE TO FACE   EE = END TO END   EL = EFFECTIVE LENGTH
HD = HUB DIAMETER (or insert)   IE = INSIDE OF EARS (or recess)   OD = OUTSIDE DIAMETER   OE = OUTSIDE OF EARS   SB = SPLINE OR BORE SIZE

IDLI GUIDE   COPYRIGHT ® INTERCHANGE, INC.   ST. LOUIS PARK, MN. 55416 USA

IDLI GROUP 87048

CLAMP YOKE
WITH 6 SPLINES
slight variation from above group
SEE FIGURE 14A

4.188 in	OE	106.36 mm
1.000 in	BD	25.40 mm
1.250 in	SB	31.75 mm
5.250 in	EC	133.35 mm

ROCKWELL 3DRYS20-2

IDLI GROUP 87049

CLAMP YOKE
WITH ROUND BORE
SEE FIGURE 14B

4.188 in	OE	106.36 mm
1.000 in	BD	25.40 mm
1.250 in	SB	31.75 mm
5.250 in	EC	133.35 mm

ROCKWELL 3DRYR20-45

IDLI GROUP 87050

CLAMP YOKE
WITH 6 SPLINES
slight variation from above group
SEE FIGURE 14A

4.188 in	OE	106.36 mm
1.000 in	BD	25.40 mm
1.375 in	SB	34.93 mm
5.250 in	EC	133.35 mm

PRECISION 1950
ROCKWELL 3DRYS22-5

IDLI GROUP 87051

CLAMP YOKE
WITH ROUND BORE
SEE FIGURE 14B

4.188 in	OE	106.36 mm
1.000 in	BD	25.40 mm
1.438 in	SB	36.51 mm
5.250 in	EC	133.35 mm

ROCKWELL 3DRYS23-12

IDLI GROUP 87052

CLAMP YOKE
WITH ROUND BORE
SEE FIGURE 14B

4.188 in	OE	106.36 mm
1.000 in	BD	25.40 mm
1.500 in	SB	38.10 mm
5.250 in	EC	133.35 mm

ROCKWELL 3DRYS24-30

IDLI GROUP 87053

CLAMP YOKE
WITH 6 SPLINES
slight variation from above group
SEE FIGURE 14A

4.188 in	OE	106.36 mm
1.000 in	BD	25.40 mm
1.750 in	SB	44.45 mm
5.250 in	EC	133.35 mm

ROCKWELL 3DRYS28

IDLI GROUP 87054

DOUBLE CENTER
SEE FIGURE 60

4.625 in	OE	117.48 mm
1.531 in	BD	38.90 mm
3.125 in	CC	79.38 mm

ROCKWELL 55N170-1

IDLI GROUP 87055

CLAMP YOKE
WITH 21 INVOLUTED SPLINES
SEE FIGURE 14A

4.625 in	OE	117.48 mm
1.531 in	BD	38.90 mm
1.375 in	SB	34.93 mm
5.125 in	EC	130.18 mm

ROCKWELL 55NYS22-37

IDLI GROUP 87056

CLAMP YOKE
WITH 21 INVOLUTED SPLINES
SEE FIGURE 14A

4.625 in	OE	117.48 mm
1.531 in	BD	38.90 mm
1.375 in	SB	34.93 mm
5.313 in	EC	134.94 mm

ROCKWELL 55NYS22-33

IDLI GROUP 87057

CLAMP YOKE
WITH 6 SPLINES
SEE FIGURE 14A

4.625 in	OE	117.48 mm
1.531 in	BD	38.90 mm
1.375 in	SB	34.93 mm
5.813 in	EC	147.64 mm

ROCKWELL 55NYS22-6

IDLI GROUP 87058

CLAMP YOKE
WITH ROUND BORE
SEE FIGURE 14B

4.625 in	OE	117.48 mm
1.531 in	BD	38.90 mm
1.500 in	SB	38.10 mm
5.563 in	EC	141.29 mm

ROCKWELL 55NYR24-15

IDLI GROUP 87059

CLAMP YOKE
WITH 20 INVOLUTED SPLINES
SEE FIGURE 14A

4.625 in	OE	117.48 mm
1.531 in	BD	38.90 mm
1.750 in	SB	44.45 mm
5.125 in	EC	130.18 mm

ROCKWELL 55NYS28-36

IDLI GROUP 87060

CLAMP YOKE
WITH 6 SPLINES
SEE FIGURE 14A

4.625 in	OE	117.48 mm
1.531 in	BD	38.90 mm
1.750 in	SB	44.45 mm
5.438 in	EC	138.11 mm

ROCKWELL 55NYS28-3

IDLI GROUP 87061

TUBE AND WELD YOKE ASSEMBLY
SEE FIGURE 49

5.313 in	OE	134.94 mm
1.875 in	BD	47.63 mm
61.938 in	EC	1573.21 mm
3.500 in	OD	88.90 mm

ROCKWELL 983-13-2
SPICER 5-27-348X
SPICER 5-27-9-6100X

IDLI GROUP 87062

TUBE AND WELD YOKE ASSEMBLY
SEE FIGURE 49

5.313 in	OE	134.94 mm
1.875 in	BD	47.63 mm
68.625 in	EC	1743.08 mm
3.500 in	OD	88.90 mm

ROCKWELL 983-10-2
SPICER 5-27-48X
SPICER 5-27-7-2726X
SPICER 5-27-7-6722X

IDLI GROUP 87063

TUBE AND WELD YOKE ASSEMBLY
SEE FIGURE 49

6.094 in	OE	154.78 mm
1.938 in	BD	49.21 mm
27.031 in	EC	686.60 mm
3.500 in	OD	88.90 mm

ROCKWELL 984-21-2
SPICER 6-27-278X
SPICER 6-27-7-2602X
SPICER 6-27-948X

IDLI GROUP 87064

TUBE AND WELD YOKE ASSEMBLY
SEE FIGURE 49

6.094 in	OE	154.78 mm
1.938 in	BD	49.21 mm
27.031 in	EC	686.60 mm
4.000 in	OD	101.60 mm

ROCKWELL 984-22-2
SPICER 6-27-17-2602X

IDLI GROUP 87065

TUBE AND WELD YOKE ASSEMBLY
SEE FIGURE 49

6.094 in	OE	154.78 mm
1.938 in	BD	49.21 mm
27.031 in	EC	686.60 mm
4.094 in	OD	103.98 mm

ROCKWELL 984-26-2

IDLI GROUP 87066

TUBE AND WELD YOKE ASSEMBLY
SEE FIGURE 49

6.094 in	OE	154.78 mm
1.938 in	BD	49.21 mm
37.094 in	EC	942.18 mm
3.500 in	OD	88.90 mm

ROCKWELL 984-10-2
SPICER 6-27-48X
SPICER 6-27-708X
SPICER 6-27-7-3604X

IDLI GROUP 87067

TUBE AND WELD YOKE ASSEMBLY
SEE FIGURE 49

6.094 in	OE	154.78 mm
1.938 in	BD	49.21 mm
51.031 in	EC	1296.20 mm
3.500 in	OD	88.90 mm

ROCKWELL 984-19-2

IDLI GROUP 87068

TUBE AND WELD YOKE ASSEMBLY
SEE FIGURE 49

6.094 in	OE	154.78 mm
1.938 in	BD	49.21 mm
51.031 in	EC	1296.20 mm
4.000 in	OD	101.60 mm

ROCKWELL 984-23-2
SPICER 6-27-17-5002X

IDLI GROUP 87069

TUBE AND WELD YOKE ASSEMBLY
SEE FIGURE 49

6.094 in	OE	154.78 mm
1.938 in	BD	49.21 mm
51.031 in	EC	1296.20 mm
4.094 in	OD	103.98 mm

ROCKWELL 984-27-2

IDLI GROUP 87070

TUBE AND WELD YOKE ASSEMBLY
SEE FIGURE 49

6.094 in	OE	154.78 mm
1.938 in	BD	49.21 mm
56.313 in	EC	1430.34 mm
3.500 in	OD	88.90 mm

ROCKWELL 984-11-2
SPICER 6-27-608X
SPICER 6-27-7-5511X

IDLI GROUP 87071

TUBE AND WELD YOKE ASSEMBLY
SEE FIGURE 49

6.094 in	OE	154.78 mm
1.938 in	BD	49.21 mm
63.031 in	EC	1601.00 mm
4.000 in	OD	101.60 mm

ROCKWELL 984-24-2
SPICER 6-27-17-6202X

ENGINEERING CATALOGS MUST BE CONSULTED FOR DETAILS NOT INCLUDED
IN THIS GUIDE. SPECIFIC DESIGNS, MATERIAL CONTENT, TOLERANCES
LUBE FITTINGS AND OTHER DIMENSIONS ARE INTENTIONALLY OMITTED HERE.

IDLI GUIDE   COPYRIGHT © INTERCHANGE, INC.   ST. LOUIS PARK, MN. 55416 USA

CAUTION: BE SURE TO REFER TO ENGINEERING CATALOGS FOR SPECIAL APPLICATIONS THAT REQUIRE SPECIFIC MATERIAL CONTENT, TOLERANCES, ETC. SEE FOOTNOTE.

IDLI GROUP 87072

TUBE AND WELD YOKE ASSEMBLY
SEE FIGURE 49

6.094 in	OE	154.78 mm
1.938 in	BD	49.21 mm
64.938 in	EC	1649.41 mm
3.500 in	OD	88.90 mm

ROCKWELL 984-13-2
SPICER 6-27-18X
SPICER 6-27-38X
SPICER 6-27-618X
SPICER 6-27-7-6331X

IDLI GROUP 87073

TUBE AND WELD YOKE ASSEMBLY
SEE FIGURE 49

6.094 in	OE	154.78 mm
1.938 in	BD	49.21 mm
68.031 in	EC	1728.00 mm
4.000 in	OD	101.60 mm

ROCKWELL 984-16-2

IDLI GROUP 87074

TUBE AND WELD YOKE ASSEMBLY
SEE FIGURE 49

6.094 in	OE	154.78 mm
1.938 in	BD	49.21 mm
68.031 in	EC	1728.00 mm
4.094 in	OD	103.98 mm

ROCKWELL 984-28-2

IDLI GROUP 87075

TUBE AND WELD YOKE ASSEMBLY
SEE FIGURE 49

6.094 in	OE	154.78 mm
1.938 in	BD	49.21 mm
75.000 in	EC	1905.00 mm
4.000 in	OD	101.60 mm

ROCKWELL 984-25-2
SPICER 6-27-17-7401X

IDLI GROUP 87076

YOKE SHAFT WITH 16 SPLINES
SEE FIGURE 73

6.094 in	OE	154.78 mm
1.938 in	BD	49.21 mm
2.500 in	SB	63.50 mm
8.156 in	EC	207.17 mm

ROCKWELL 17NYSM40-89

IDLI GROUP 87077

YOKE SHAFT WITH 16 SPLINES
SEE FIGURE 73

6.094 in	OE	154.78 mm
1.938 in	BD	49.21 mm
2.500 in	SB	63.50 mm
9.625 in	EC	244.48 mm

ROCKWELL 17NYSM40-40

IDLI GROUP 87078

YOKE SHAFT WITH 16 SPLINES
SEE FIGURE 73

6.094 in	OE	154.78 mm
1.938 in	BD	49.21 mm
2.500 in	SB	63.50 mm
10.031 in	EC	254.80 mm

ROCKWELL 17NYSM40-64

IDLI GROUP 87079

YOKE SHAFT WITH 16 SPLINES
slight variation from above group
SEE FIGURE 73

6.094 in	OE	154.78 mm
1.938 in	BD	49.21 mm
2.500 in	SB	63.50 mm
10.031 in	EC	254.80 mm

ROCKWELL 17NYSM40-50

IDLI GROUP 87080

YOKE SHAFT WITH 16 SPLINES
SEE FIGURE 73

6.094 in	OE	154.78 mm
1.938 in	BD	49.21 mm
2.500 in	SB	63.50 mm
10.656 in	EC	270.67 mm

ROCKWELL 17N82-1171-12
SPICER 6-82-1171-12

IDLI GROUP 87081

YOKE SHAFT WITH 16 SPLINES
SEE FIGURE 73

6.094 in	OE	154.78 mm
1.938 in	BD	49.21 mm
2.500 in	SB	63.50 mm
10.906 in	EC	277.02 mm

ROCKWELL 17NYSM40-55

IDLI GROUP 87082

YOKE SHAFT WITH 16 SPLINES
SEE FIGURE 73

6.094 in	OE	154.78 mm
1.938 in	BD	49.21 mm
2.500 in	SB	63.50 mm
11.000 in	EC	279.40 mm

ROCKWELL 17NYSM40-85

IDLI GROUP 87083

YOKE SHAFT WITH 16 SPLINES
SEE FIGURE 73

6.094 in	OE	154.78 mm
1.938 in	BD	49.21 mm
2.500 in	SB	63.50 mm
11.156 in	EC	283.37 mm

ROCKWELL 17NYSM40-81

IDLI GROUP 87084

YOKE SHAFT WITH 16 SPLINES
SEE FIGURE 73

6.094 in	OE	154.78 mm
1.938 in	BD	49.21 mm
2.500 in	SB	63.50 mm
11.250 in	EC	285.75 mm

ROCKWELL 17NYSM40-50

IDLI GROUP 87085

YOKE SHAFT WITH 16 SPLINES
SEE FIGURE 73

6.094 in	OE	154.78 mm
1.938 in	BD	49.21 mm
2.500 in	SB	63.50 mm
11.781 in	EC	299.25 mm

ROCKWELL 17NYSM40-41

IDLI GROUP 87086

YOKE SHAFT WITH 16 SPLINES
slight variation from above group
SEE FIGURE 73

6.094 in	OE	154.78 mm
1.938 in	BD	49.21 mm
2.500 in	SB	63.50 mm
11.781 in	EC	299.25 mm

ROCKWELL 17NYSM40-58

IDLI GROUP 87087

YOKE SHAFT WITH 16 SPLINES
SEE FIGURE 73

6.094 in	OE	154.78 mm
1.938 in	BD	49.21 mm
2.500 in	SB	63.50 mm
12.156 in	EC	308.77 mm

ROCKWELL 17N82-1091-1
SPICER 6-82-1091-1

IDLI GROUP 87088

YOKE SHAFT WITH 16 SPLINES
SEE FIGURE 73

6.094 in	OE	154.78 mm
1.938 in	BD	49.21 mm
2.500 in	SB	63.50 mm
12.656 in	EC	321.47 mm

ROCKWELL 17NYSM40-59

IDLI GROUP 87089

YOKE SHAFT WITH 16 SPLINES
slight variation from above group
SEE FIGURE 73

6.094 in	OE	154.78 mm
1.938 in	BD	49.21 mm
2.500 in	SB	63.50 mm
12.656 in	EC	321.47 mm

ROCKWELL 17NYSM40-46

IDLI GROUP 87090

CROSS AND LOW OR HIGH WING TYPE
DRILLED OR THREADED BEARINGS
SEE FIGURE 73

6.094 in	OE	154.78 mm
1.938 in	BD	49.21 mm
2.500 in	SB	63.50 mm
13.000 in	EC	330.20 mm

ROCKWELL 17NYSM40-71

IDLI GROUP 87091

YOKE SHAFT WITH 16 SPLINES
slight variation from above group
SEE FIGURE 73

6.094 in	OE	154.78 mm
1.938 in	BD	49.21 mm
2.500 in	SB	63.50 mm
13.000 in	EC	330.20 mm

ROCKWELL 17NYSM40-88

IDLI GROUP 87092

YOKE SHAFT WITH 16 SPLINES
SEE FIGURE 73

6.094 in	OE	154.78 mm
1.938 in	BD	49.21 mm
2.500 in	SB	63.50 mm
13.094 in	EC	332.58 mm

ROCKWELL 17N82-1281
SPICER 6-28-1281

IDLI GROUP 87093

YOKE SHAFT WITH 16 SPLINES
SEE FIGURE 73

6.094 in	OE	154.78 mm
1.938 in	BD	49.21 mm
2.500 in	SB	63.50 mm
13.156 in	EC	334.17 mm

ROCKWELL 17NYSM40-60

IDLI GROUP 87094

YOKE SHAFT WITH 16 SPLINES
SEE FIGURE 73

6.094 in	OE	154.78 mm
1.938 in	BD	49.21 mm
2.500 in	SB	63.50 mm
13.406 in	EC	340.52 mm

ROCKWELL 17NYSM40-86

IDLI GROUP 87095

YOKE SHAFT WITH 16 SPLINES
SEE FIGURE 73

6.094 in	OE	154.78 mm
1.938 in	BD	49.21 mm
2.500 in	SB	63.50 mm
13.438 in	EC	341.31 mm

ROCKWELL 17N82-1091-4
SPICER 6-82-1091-4

IDLI GROUP 87096

YOKE SHAFT WITH 16 SPLINES
SEE FIGURE 73

6.094 in	OE	154.78 mm
1.938 in	BD	49.21 mm
2.500 in	SB	63.50 mm
13.563 in	EC	344.49 mm

ROCKWELL 17NYSM40-42

BD = BEARING DIAMETER (outside)   **BW** = BEARING WIDTH   **CB** = CROSS LENGTH WITH BEARINGS   **CC** = CENTER TO CENTER   **CD** = CROSS DIAMETER
CL = CROSS LENGTH WITHOUT BEARINGS   **EC** = END TO CENTER or FACE TO FACE   **EE** = END TO END   **EL** = EFFECTIVE LENGTH
HD = HUB DIAMETER (or insert)   **IE** = INSIDE OF EARS (or recess)   **OD** = OUTSIDE DIAMETER   **OE** = OUTSIDE OF EARS   **SB** = SPLINE OR BORE SIZE

IDLI GROUP 87097

YOKE SHAFT WITH 16 SPLINES
SEE FIGURE 73

6.094 in	OE	154.78 mm
1.938 in	BD	49.21 mm
2.500 in	SB	63.50 mm
13.625 in	EC	346.08 mm

ROCKWELL 17NYSM40-61

IDLI GROUP 87098

YOKE SHAFT WITH 16 SPLINES
SEE FIGURE 73

6.094 in	OE	154.78 mm
1.938 in	BD	49.21 mm
2.500 in	SB	63.50 mm
13.688 in	EC	347.66 mm

ROCKWELL 17NYSM40-104

IDLI GROUP 87099

YOKE SHAFT WITH 16 SPLINES
SEE FIGURE 73

6.094 in	OE	154.78 mm
1.938 in	BD	49.21 mm
2.500 in	SB	63.50 mm
13.938 in	EC	354.01 mm

ROCKWELL 17NYSM40-102

IDLI GROUP 87100

YOKE SHAFT WITH 16 SPLINES
SEE FIGURE 73

6.094 in	OE	154.78 mm
1.938 in	BD	49.21 mm
2.500 in	SB	63.50 mm
14.000 in	EC	355.60 mm

ROCKWELL 17NYSM40-54

IDLI GROUP 87101

YOKE SHAFT WITH 16 SPLINES
SEE FIGURE 73

6.094 in	OE	154.78 mm
1.938 in	BD	49.21 mm
2.500 in	SB	63.50 mm
14.063 in	EC	357.19 mm

ROCKWELL 17NYSM40-47

IDLI GROUP 87102

YOKE SHAFT WITH 16 SPLINES
SEE FIGURE 73

6.094 in	OE	154.78 mm
1.938 in	BD	49.21 mm
2.500 in	SB	63.50 mm
14.156 in	EC	359.57 mm

ROCKWELL 17NYSM40-62

IDLI GROUP 87103

YOKE SHAFT WITH 16 SPLINES
SEE FIGURE 73

6.094 in	OE	154.78 mm
1.938 in	BD	49.21 mm
2.500 in	SB	63.50 mm
14.281 in	EC	362.75 mm

ROCKWELL 17NYSM40-96

IDLI GROUP 87104

YOKE SHAFT WITH 16 SPLINES
SEE FIGURE 73

6.094 in	OE	154.78 mm
1.938 in	BD	49.21 mm
2.500 in	SB	63.50 mm
14.563 in	EC	369.89 mm

ROCKWELL 17NYSM40-80

IDLI GROUP 87105

YOKE SHAFT WITH 16 SPLINES
SEE FIGURE 73

6.094 in	OE	154.78 mm
1.938 in	BD	49.21 mm
2.500 in	SB	63.50 mm
14.750 in	EC	374.65 mm

ROCKWELL 17NYSM40-95

IDLI GROUP 87106

YOKE SHAFT WITH 16 SPLINES
SEE FIGURE 73

6.094 in	OE	154.78 mm
1.938 in	BD	49.21 mm
2.500 in	SB	63.50 mm
14.813 in	EC	376.24 mm

ROCKWELL 17NYSM40-63

IDLI GROUP 87107

YOKE SHAFT WITH 16 SPLINES
SEE FIGURE 73

6.094 in	OE	154.78 mm
1.938 in	BD	49.21 mm
2.500 in	SB	63.50 mm
14.906 in	EC	378.62 mm

ROCKWELL 17NYSM40-68

IDLI GROUP 87108

YOKE SHAFT WITH 16 SPLINES
SEE FIGURE 73

6.094 in	OE	154.78 mm
1.938 in	BD	49.21 mm
2.500 in	SB	63.50 mm
15.031 in	EC	381.80 mm

ROCKWELL 17N82-971-1X
SPICER 6-82-971-1X

IDLI GROUP 87109

YOKE SHAFT WITH 16 SPLINES
slight variation from above group
SEE FIGURE 73

6.094 in	OE	154.78 mm
1.938 in	BD	49.21 mm
2.500 in	SB	63.50 mm
15.031 in	EC	381.80 mm

ROCKWELL 17NYSM40-56

IDLI GROUP 87110

YOKE SHAFT WITH 16 SPLINES
SEE FIGURE 73

6.094 in	OE	154.78 mm
1.938 in	BD	49.21 mm
2.500 in	SB	63.50 mm
16.406 in	EC	416.72 mm

ROCKWELL 17NYSM40-33

IDLI GROUP 87111

YOKE SHAFT WITH 16 SPLINES
SEE FIGURE 73

6.094 in	OE	154.78 mm
1.938 in	BD	49.21 mm
2.500 in	SB	63.50 mm
16.688 in	EC	423.86 mm

ROCKWELL 17NYSM40-92

IDLI GROUP 87112

YOKE SHAFT WITH 16 SPLINES
SEE FIGURE 73

6.094 in	OE	154.78 mm
1.938 in	BD	49.21 mm
2.500 in	SB	63.50 mm
16.875 in	EC	428.63 mm

ROCKWELL 17NYSM40-91

IDLI GROUP 87113

YOKE SHAFT WITH 16 SPLINES
SEE FIGURE 73

6.094 in	OE	154.78 mm
1.938 in	BD	49.21 mm
2.500 in	SB	63.50 mm
17.031 in	EC	432.60 mm

ROCKWELL 17NYSM40-82

IDLI GROUP 87114

YOKE SHAFT WITH 16 SPLINES
SEE FIGURE 73

6.094 in	OE	154.78 mm
1.938 in	BD	49.21 mm
2.500 in	SB	63.50 mm
17.500 in	EC	444.50 mm

ROCKWELL 17NYSM40-94

IDLI GROUP 87115

YOKE SHAFT WITH 4 SPLINES
FOR ROLL JOINT ASSEMBLY
SEE FIGURE 74

6.094 in	OE	154.78 mm
1.938 in	BD	49.21 mm
2.500 in	SB	63.50 mm
17.750 in	EC	450.85 mm

ROCKWELL 17NYSM36-28

IDLI GROUP 87116

TUBE AND WELD YOKE ASSEMBLY
SEE FIGURE 49

7.547 in	OE	191.69 mm
1.938 in	BD	49.21 mm
39.125 in	EC	993.78 mm
4.500 in	OD	114.30 mm

ROCKWELL 18N27-9-3813X
SPICER 6.5-27-9-3813X

IDLI GROUP 87117

TUBE AND WELD YOKE ASSEMBLY
SEE FIGURE 49

7.547 in	OE	191.69 mm
1.938 in	BD	49.21 mm
52.375 in	EC	1330.33 mm
4.500 in	OD	114.30 mm

ROCKWELL 18N27-9-5013X
SPICER 6.5-27-9-5013X

IDLI GROUP 87118

TUBE AND WELD YOKE ASSEMBLY
SEE FIGURE 49

7.547 in	OE	191.69 mm
1.938 in	BD	49.21 mm
64.375 in	EC	1635.13 mm
4.500 in	OD	114.30 mm

ROCKWELL 18N27-9-6213X
SPICER 6.5-27-9-6213X

ENGINEERING CATALOGS MUST BE CONSULTED FOR DETAILS NOT INCLUDED
IN THIS GUIDE. SPECIFIC DESIGNS, MATERIAL CONTENT, TOLERANCES
LUBE FITTINGS AND OTHER DIMENSIONS ARE INTENTIONALLY OMITTED HERE.

FOR YOUR CONVENIENCE

NOTES ON OTHER INFORMATION

We urge you to send us any data that
you feel should be included with the
future editions of the I.D.L.I. Guide.

IDLI GROUP 87201

ROUND BEARING STUB
SEE FIGURE 67A

.750 in	SB	19.05 mm
2.656 in	EC	67.47 mm
2.162 in	HD	54.91 mm

ROCKWELL BSA-R12-1

IDLI GROUP 87202

ROUND BEARING STUB
SEE FIGURE 67A

.750 in	SB	19.05 mm
3.406 in	EC	86.52 mm
2.162 in	HD	54.91 mm

ROCKWELL BSA-R12-2

IDLI GROUP 87203

SQUARE TUBING
SEE FIGURE 76

.781 in	SB	19.85 mm
72.000 in	EE	1828.80 mm
1.031 in	OD	26.20 mm

ROCKWELL TU-Q17-1-72

IDLI GROUP 87204

SQUARE TUBING
SEE FIGURE 76

.781 in	SB	19.85 mm
108.000 in	EE	2743.20 mm
1.031 in	OD	26.20 mm

ROCKWELL TU-Q17-1-108

IDLI GROUP 87205

SLEEVE WITH RECTANGULAR BORE
maximum SB dimension shown
SEE FIGURE 84

.875 in	SB	22.23 mm
1.880 in	HD	47.75 mm
4.000 in	EL	101.60 mm
2.125 in	OD	53.98 mm

ROCKWELL SL-P12-32

IDLI GROUP 87206

ROUND BEARING STUB
SEE FIGURE 67A

.875 in	SB	22.23 mm
4.938 in	EC	125.41 mm
1.844 in	HD	46.84 mm

ROCKWELL BSA-R14-1

IDLI GROUP 87207

RECTANGULAR TUBING
maximum SB and OD dimensions shown
SEE FIGURE 76

.891 in	SB	22.62 mm
72.000 in	EE	1828.80 mm
1.125 in	OD	28.58 mm

ROCKWELL TU-P16-5-72

IDLI GROUP 87208

RECTANGULAR TUBING
maximum SB and OD dimensions shown
SEE FIGURE 76

.891 in	SB	22.62 mm
108.000 in	EE	2743.20 mm
1.125 in	OD	28.58 mm

ROCKWELL TU-P16-5-108

IDLI GROUP 87209

SLEEVE WITH SQUARE BORE
SEE FIGURE 84

1.000 in	SB	25.40 mm
1.880 in	HD	47.75 mm
3.500 in	EL	88.90 mm
2.000 in	OD	50.80 mm

ROCKWELL SL-Q16-64

IDLI GROUP 87210

ROUND BEARING STUB
SEE FIGURE 67A

1.000 in	SB	25.40 mm
3.375 in	EC	85.73 mm
1.880 in	HD	47.75 mm

ROCKWELL BS-R16-2

IDLI GROUP 87211

ROUND BEARING STUB
SEE FIGURE 67A

1.000 in	SB	25.40 mm
5.750 in	EC	146.05 mm
2.820 in	HD	71.63 mm

ROCKWELL BSA-R16-3

IDLI GROUP 87212

ROUND BEARING STUB
SEE FIGURE 67A

1.000 in	SB	25.40 mm
6.250 in	EC	158.75 mm
1.844 in	HD	46.84 mm

ROCKWELL BSA-R16-1

IDLI GROUP 87213

ROUND TUBING
SEE FIGURE 75

1.010 in	SB	25.65 mm
72.000 in	EE	1828.80 mm
1.250 in	OD	31.75 mm

ROCKWELL RT125-11-72

IDLI GROUP 87214

ROUND TUBING
SEE FIGURE 75

1.010 in	SB	25.65 mm
108.000 in	EE	2743.20 mm
1.250 in	OD	31.75 mm

ROCKWELL RT125-11-108

IDLI GROUP 87215

SQUARE TUBING
SEE FIGURE 76

1.063 in	SB	26.99 mm
72.000 in	EE	1828.80 mm
1.313 in	OD	33.34 mm

ROCKWELL SQT13125-72

IDLI GROUP 87216

SQUARE TUBING
SEE FIGURE 76

1.063 in	SB	26.99 mm
108.000 in	EE	2743.20 mm
1.313 in	OD	33.34 mm

ROCKWELL SQT13125-108

IDLI GROUP 87217

SPLINED PLUG WITH 10 SPLINES
SEE FIGURE 63

1.125 in	SB	28.58 mm
1.018 in	HD	25.86 mm
4.375 in	EL	111.13 mm

ROCKWELL L6NPS18-3

IDLI GROUP 87218

SPLINED PLUG WITH 10 SPLINES
SEE FIGURE 63

1.125 in	SB	28.58 mm
1.018 in	HD	25.86 mm
4.875 in	EL	123.83 mm

ROCKWELL L6NPS18

IDLI GROUP 87219

SPLINED PLUG WITH 10 SPLINES
SEE FIGURE 63

1.125 in	SB	28.58 mm
1.018 in	HD	25.86 mm
5.750 in	EL	146.05 mm

ROCKWELL L6NPS18-1

IDLI GROUP 87220

SLEEVE WITH RECTANGULAR BORE
maximum SB dimension shown
SEE FIGURE 84

1.125 in	SB	28.58 mm
1.464 in	HD	37.19 mm
3.938 in	EL	100.01 mm
1.500 in	OD	38.10 mm

ROCKWELL SL-P12-31

IDLI GROUP 87221

SLEEVE WITH RECTANGULAR BORE
maximum SB dimension shown
SEE FIGURE 84

1.125 in	SB	28.58 mm
1.464 in	HD	37.19 mm
4.313 in	EL	109.54 mm
1.500 in	OD	38.10 mm

ROCKWELL SL-P12-38

IDLI GROUP 87222

SLEEVE WITH RECTANGULAR BORE
maximum SB dimension shown
SEE FIGURE 84

1.125 in	SB	28.58 mm
1.628 in	HD	41.35 mm
4.000 in	EL	101.60 mm
1.750 in	OD	44.45 mm

ROCKWELL SL-P12-1

IDLI GROUP 87223

SPLINED PLUG WITH 10 SPLINES
SEE FIGURE 63

1.125 in	SB	28.58 mm
1.880 in	HD	47.75 mm
4.750 in	EL	120.65 mm

ROCKWELL L12NPS18

IDLI GROUP 87224

SPLINED PLUG WITH 10 SPLINES
SEE FIGURE 63

1.125 in	SB	28.58 mm
1.630 in	HD	41.40 mm
4.875 in	EL	123.83 mm

ROCKWELL L6NPS18-8

IDLI GROUP 87225

SLEEVE WITH RECTANGULAR BORE
maximum SB dimension shown
SEE FIGURE 84

1.125 in	SB	28.58 mm
1.880 in	HD	47.75 mm
3.500 in	EL	88.90 mm
2.000 in	OD	50.80 mm

ROCKWELL SL-P16-40

IDLI GROUP 87226

SLEEVE WITH RECTANGULAR BORE
maximum SB dimension shown
SEE FIGURE 84

1.125 in	SB	28.58 mm
1.880 in	HD	47.75 mm
3.500 in	EL	88.90 mm
2.125 in	OD	53.98 mm

ROCKWELL SL-P16-47

IDLI GROUP 87227

SLEEVE WITH RECTANGULAR BORE
maximum SB dimension shown
SEE FIGURE 84

1.125 in	SB	28.58 mm
1.880 in	HD	47.75 mm
7.000 in	EL	177.80 mm
2.125 in	OD	53.98 mm

ROCKWELL SLA-P16-14
SPICER 2-55-102
WESCO 14NSA
WESCO 760-053

BD = BEARING DIAMETER (outside)   BW = BEARING WIDTH   CB = CROSS LENGTH WITH BEARINGS   CC = CENTER TO CENTER   CD = CROSS DIAMETER
CL = CROSS LENGTH WITHOUT BEARINGS   EC = END TO CENTER or FACE TO FACE   EE = END TO END   EL = EFFECTIVE LENGTH
HD = HUB DIAMETER (or insert)   IE = INSIDE OF EARS (or recess)   OD = OUTSIDE DIAMETER   OE = OUTSIDE OF EARS   SB = SPLINE OR BORE SIZE

CAUTION: BE SURE TO REFER TO ENGINEERING CATALOGS FOR SPECIAL APPLICATIONS THAT REQUIRE SPECIFIC MATERIAL CONTENT, TOLERANCES, ETC. SEE FOOTNOTE.

IDLI GROUP 87228
ROUND BEARING STUB
SEE FIGURE 67A

1.125 in	SB	28.58 mm
2.875 in	EC	73.03 mm
1.880 in	HD	47.75 mm

ROCKWELL BS-R18-1

IDLI GROUP 87229
ROUND BEARING STUB
SEE FIGURE 67A

1.125 in	SB	28.58 mm
3.563 in	EC	90.49 mm
1.880 in	HD	47.75 mm

ROCKWELL BS-R18-5

IDLI GROUP 87230
SPLINED BEARING STUB
WITH 6 SPLINES
SEE FIGURE 66

1.125 in	SB	28.58 mm
3.875 in	EC	98.43 mm
2.630 in	HD	66.80 mm

ROCKWELL BSA-S18-1

IDLI GROUP 87231
SPLINED BEARING STUB
WITH 6 SPLINES
SEE FIGURE 66

1.125 in	SB	28.58 mm
3.938 in	EC	100.01 mm
1.880 in	HD	47.75 mm

ROCKWELL BSA-S18-2

IDLI GROUP 87232
ROUND BEARING STUB
SEE FIGURE 67A

1.125 in	SB	28.58 mm
4.000 in	EC	101.60 mm
1.880 in	HD	47.75 mm

ROCKWELL BSA-R18-6

IDLI GROUP 87233
ROUND BEARING STUB
SEE FIGURE 67A

1.125 in	SB	28.58 mm
4.375 in	EC	111.13 mm
1.880 in	HD	47.75 mm

ROCKWELL BS-R18-25

IDLI GROUP 87234
ROUND BEARING STUB
SEE FIGURE 67B

1.125 in	SB	28.58 mm
4.438 in	EC	112.71 mm
1.844 in	HD	46.84 mm

ROCKWELL BS-R18-23

IDLI GROUP 87235
ROUND BEARING STUB
SEE FIGURE 67B

1.125 in	SB	28.58 mm
4.625 in	EC	117.48 mm
1.844 in	HD	46.84 mm

ROCKWELL BS-R18-20

IDLI GROUP 87236
ROUND BEARING STUB
SEE FIGURE 67A

1.125 in	SB	28.58 mm
4.625 in	EC	117.48 mm
1.880 in	HD	47.75 mm

ROCKWELL BSA-R18-7

IDLI GROUP 87237
ROUND BEARING STUB
SEE FIGURE 67A

1.125 in	SB	28.58 mm
5.625 in	EC	142.88 mm
1.880 in	HD	47.75 mm

ROCKWELL BS-R18-32

IDLI GROUP 87238
CENTER BEARING
SEE FIGURE 57

1.181 in	SB	30.00 mm
3.563 in	EC	90.49 mm
1.500 in	CC	38.10 mm
.625 in	BW	15.88 mm

AMER.PARTS	 2-6035
ANCHOR	 35-6028
ANCHOR	 35-6035
ANCHORDOAN	 35-6035
BEAR.INC.	. . . . B1-101K
BEAR.INC.	. . . . BI101
BEAR.INC.	. . . . BI101K
BORG-WARNR	 35-6035
BUTLER	. . . . 510527X
CLEVE.MOT.	. . . . DS101K
CLEVE.STL.	. . . . SCB76
CMP	. . . . DS101
CMP	. . . . DS101K
CMP	. . . . DS101K2
DETROIT	. . . . MB600
DORMAN	 662-013
DORMAN	. . . . DS10
DORMAN	. . . . DS13B
EATON	. . . . MM6035
FED.BEAR.	. . . . HB1206LL
FEDERAL	. . . . HB1206LL
FED.MOGUL	. . . . HB20600
FED.MOGUL	. . . . HB206DD
FED.MOGUL	. . . . HB206FF
GMC	 907794
L & S BRG.	. . . . PBS6
MCQUAY-NOR	. . . . MM6035
MOPAR	 600
MOPAR	. . . . D600
MOTOR MAST	. . . . DS13B
NEAPCO	 81-1020
NEAPCO	 81-1022
NEAPCO	. . . . DS101K
NEAPCO	. . . . DS101K2

REPUBLIC	 35-6035
REPUBLIC	 85-6035
ROCKWELL	. . . . CB210527X
SKF	. . . . PBS6
SPICER	. . . . 210527-1X
SPICER	. . . . 210527X
TRW	 20801
WESCO	 960-012
WESCO	 960-013
WESCO	. . . . DS13B

IDLI GROUP 87239
SLEEVE WITH SQUARE BORE
SEE FIGURE 84

1.188 in	SB	30.16 mm
1.969 in	HD	50.01 mm
4.500 in	EL	114.30 mm
2.000 in	OD	50.80 mm

ROCKWELL SL-Q19-83

IDLI GROUP 87240
SLEEVE WITH SQUARE BORE
SEE FIGURE 84

1.188 in	SB	30.16 mm
1.969 in	HD	50.01 mm
5.000 in	EL	127.00 mm
2.125 in	OD	53.98 mm

ROCKWELL SL-Q19-53

IDLI GROUP 87241
SLEEVE WITH SQUARE BORE
SEE FIGURE 84

1.188 in	SB	30.16 mm
2.020 in	HD	51.31 mm
4.500 in	EL	114.30 mm
2.125 in	OD	53.98 mm

ROCKWELL SL-Q19-102

IDLI GROUP 87242
SLEEVE WITH RECTANGULAR BORE
maximum SB dimension shown
SEE FIGURE 84

1.188 in	SB	30.16 mm
2.020 in	HD	51.31 mm
5.000 in	EL	127.00 mm
2.125 in	OD	53.98 mm

ROCKWELL SL-Q19-82

IDLI GROUP 87243
ROUND BEARING STUB
SEE FIGURE 67A

1.188 in	SB	30.16 mm
2.375 in	EC	60.33 mm
2.320 in	HD	58.93 mm

ROCKWELL BSA-R19-4

IDLI GROUP 87244
ROUND BEARING STUB
SEE FIGURE 67A

1.188 in	SB	30.16 mm
5.063 in	EC	128.59 mm
1.969 in	HD	50.01 mm

ROCKWELL BSA-R20-1

IDLI GROUP 87245
RECTANGULAR TUBING
maximum SB and OD dimensions shown
SEE FIGURE 76

1.203 in	SB	30.56 mm
72.000 in	EE	1828.80 mm
1.406 in	OD	35.72 mm

ROCKWELL TU-P20-11-72

IDLI GROUP 87246
RECTANGULAR TUBING
maximum SB and OD dimensions shown
SEE FIGURE 76

1.203 in	SB	30.56 mm
108.000 in	EE	2743.20 mm
1.406 in	OD	35.72 mm

ROCKWELL TU-P20-11-108

IDLI GROUP 87247
SHAFTING WITH 6 SPLINES
SEE FIGURE 86

| 1.250 in | SB | 31.75 mm |
| 12.000 in | EE | 304.80 mm |

ROCKWELL SH-S20-88-12

IDLI GROUP 87248
SHAFTING WITH 10 SPLINES
SEE FIGURE 86

| 1.250 in | SB | 31.75 mm |
| 12.000 in | EE | 304.80 mm |

ROCKWELL SH-S20-90-12

IDLI GROUP 87249
SHAFTING WITH 10 SPLINES
SEE FIGURE 86

| 1.250 in | SB | 31.75 mm |
| 24.000 in | EE | 609.60 mm |

ROCKWELL SH-S20-90-24

IDLI GROUP 87250
KEYED SHAFTING WITH 6 SPLINES
SEE FIGURE 87

| 1.250 in | SB | 31.75 mm |
| 44.000 in | EE | 1117.60 mm |

ROCKWELL SH-S20-26

IDLI GROUP 87251
KEYED SHAFTING WITH 6 SPLINES
SEE FIGURE 87

| 1.250 in | SB | 31.75 mm |
| 65.000 in | EE | 1651.00 mm |

ROCKWELL SH-S20-27

IDLI GUIDE   COPYRIGHT © INTERCHANGE, INC.   ST. LOUIS PARK, MN. 55416 USA

IDLI GROUP 87252

SLEEVE WITH 10 SPLINES
SEE FIGURE 82

1.250 in	SB	31.75 mm
1.844 in	HD	46.84 mm
3.500 in	EL	88.90 mm

ROCKWELL SL-S20-15

IDLI GROUP 87253

SPLINED PLUG WITH 10 SPLINES
SEE FIGURE 63

1.250 in	SB	31.75 mm
1.844 in	HD	46.84 mm
4.250 in	EL	107.95 mm

ROCKWELL L14SPS20-2

IDLI GROUP 87254

SPLINED PLUG WITH 10 SPLINES
SEE FIGURE 63

1.250 in	SB	31.75 mm
1.844 in	HD	46.84 mm
5.125 in	EL	130.18 mm

ROCKWELL PS20-10-1

IDLI GROUP 87255

SPLINED PLUG WITH 10 SPLINES
SEE FIGURE 63

1.250 in	SB	31.75 mm
1.880 in	HD	47.75 mm
4.250 in	EL	107.95 mm

ROCKWELL 1FSP20-1

IDLI GROUP 87256

SPLINED BEARING STUB
WITH 6 SPLINES
SEE FIGURE 66

1.250 in	SB	31.75 mm
3.313 in	EC	84.14 mm
2.320 in	HD	58.93 mm

ROCKWELL BS-S20-9

IDLI GROUP 87257

SPLINED BEARING STUB
WITH 6 SPLINES
SEE FIGURE 66

1.250 in	SB	31.75 mm
3.563 in	EC	90.49 mm
1.844 in	HD	46.84 mm

ROCKWELL BS-S20-3

IDLI GROUP 87258

SPLINED BEARING STUB
WITH 6 SPLINES
SEE FIGURE 66

1.250 in	SB	31.75 mm
3.563 in	EC	90.49 mm
1.880 in	HD	47.75 mm

ROCKWELL BS-S20-1

IDLI GROUP 87259

SPLINED BEARING STUB
WITH 6 SPLINES
SEE FIGURE 66

1.250 in	SB	31.75 mm
3.688 in	EC	93.66 mm
1.844 in	HD	46.84 mm

ROCKWELL BS-S20-11

IDLI GROUP 87260

SPLINED BEARING STUB
WITH 6 SPLINES
SEE FIGURE 66

1.250 in	SB	31.75 mm
4.250 in	EC	107.95 mm
1.317 in	HD	33.45 mm

ROCKWELL BS-S20-13

IDLI GROUP 87261

SPLINED BEARING STUB
WITH 6 SPLINES
SEE FIGURE 66

1.250 in	SB	31.75 mm
4.625 in	EC	117.48 mm
1.844 in	HD	46.84 mm

ROCKWELL BS-S20-23

IDLI GROUP 87262

SPLINED BEARING STUB WITH 6
SPLINES
slight variation from above group
SEE FIGURE 66

1.250 in	SB	31.75 mm
4.625 in	EC	117.48 mm
1.844 in	HD	46.84 mm

ROCKWELL BS-S20-15

IDLI GROUP 87263

SPLINED BEARING STUB WITH 6
SPLINES
slight variation from above group
SEE FIGURE 66

1.250 in	SB	31.75 mm
4.625 in	EC	117.48 mm
1.844 in	HD	46.84 mm

ROCKWELL BS-S20-17

IDLI GROUP 87264

SPLINED BEARING STUB
WITH 6 SPLINES
SEE FIGURE 66

1.250 in	SB	31.75 mm
4.625 in	EC	117.48 mm
1.880 in	HD	47.75 mm

ROCKWELL BS-S20-14

IDLI GROUP 87265

SPLINED BEARING STUB
WITH 6 SPLINES
SEE FIGURE 66

1.250 in	SB	31.75 mm
4.875 in	EC	123.83 mm
1.880 in	HD	47.75 mm

ROCKWELL BS-S20-24

IDLI GROUP 87266

SPLINED BEARING STUB WITH 6
SPLINES
slight variation from above group
SEE FIGURE 66

1.250 in	SB	31.75 mm
4.875 in	EC	123.83 mm
1.880 in	HD	47.75 mm

ROCKWELL BS-S20-16

IDLI GROUP 87267

ROUND BEARING STUB
SEE FIGURE 67A

1.250 in	SB	31.75 mm
5.063 in	EC	128.59 mm
1.880 in	HD	47.75 mm

ROCKWELL BSA-R20-2

IDLI GROUP 87268

ROUND BEARING STUB
SEE FIGURE 67A

1.250 in	SB	31.75 mm
5.375 in	EC	136.53 mm
1.880 in	HD	47.75 mm

ROCKWELL BSA-R20-6

IDLI GROUP 87269

ROUND BEARING STUB
SEE FIGURE 67B

1.250 in	SB	31.75 mm
6.125 in	EC	155.58 mm
1.880 in	HD	47.75 mm

ROCKWELL BSA-R20-10

IDLI GROUP 87270

ROUND BEARING STUB
SEE FIGURE 67A

1.250 in	SB	31.75 mm
6.188 in	EC	157.16 mm
1.880 in	HD	47.75 mm

ROCKWELL BSA-R20-19

IDLI GROUP 87271

ROUND BEARING STUB
SEE FIGURE 67A

1.250 in	SB	31.75 mm
7.000 in	EC	177.80 mm
1.969 in	HD	50.01 mm

ROCKWELL BSA-R20-3

IDLI GROUP 87272

ROUND TUBING
SEE FIGURE 75

1.260 in	SB	32.00 mm
72.000 in	EE	1828.80 mm
1.500 in	OD	38.10 mm

ROCKWELL RT15-11-72

IDLI GROUP 87273

ROUND TUBING
SEE FIGURE 75

1.260 in	SB	32.00 mm
108.000 in	EE	2743.20 mm
1.500 in	OD	38.10 mm

ROCKWELL RT15-11-108

IDLI GROUP 87274

ROUND TUBING
SEE FIGURE 75

1.310 in	SB	33.27 mm
72.000 in	EE	1828.80 mm
1.500 in	OD	38.10 mm

ROCKWELL RT15-13-72

IDLI GROUP 87275

ROUND TUBING
SEE FIGURE 75

1.310 in	SB	33.27 mm
108.000 in	EE	2743.20 mm
1.500 in	OD	38.10 mm

ROCKWELL RT15-13-108

IDLI GROUP 87276

SLEEVE WITH SQUARE BORE
SEE FIGURE 83

1.313 in	SB	33.34 mm
2.240 in	HD	56.90 mm
5.000 in	EL	127.00 mm

G & G MFG 194-4400
NEAPCO 44NS06
PRECISION 1490
ROCKWELL SL-Q21-8
SPICER 3-4-55-12
WESCO 44N-SO6
WESCO 780-059

IDLI GROUP 87277

SLEEVE WITH SQUARE BORE
SEE FIGURE 83

1.313 in	SB	33.34 mm
2.240 in	HD	56.90 mm
13.000 in	EL	330.20 mm

ROCKWELL SLA-Q21-7

IDLI GROUP 87278

ROUND BEARING STUB
SEE FIGURE 67A

1.313 in	SB	33.34 mm
2.875 in	EC	73.03 mm
1.880 in	HD	47.75 mm

ROCKWELL BS-R21-3

IDLI GROUP 87279

ROUND BEARING STUB
SEE FIGURE 67A

1.313 in	SB	33.34 mm
2.875 in	EC	73.03 mm
1.969 in	HD	50.01 mm

ROCKWELL BS-R21-1

BD = BEARING DIAMETER (outside)   **BW** = BEARING WIDTH   **CB** = CROSS LENGTH WITH BEARINGS   **CC** = CENTER TO CENTER   **CD** = CROSS DIAMETER
CL = CROSS LENGTH WITHOUT BEARINGS   **EC** = END TO CENTER or FACE TO FACE   **EE** = END TO END   **EL** = EFFECTIVE LENGTH
HD = HUB DIAMETER (or insert)   **IE** = INSIDE OF EARS (or recess)   **OD** = OUTSIDE DIAMETER   **OE** = OUTSIDE OF EARS   **SB** = SPLINE OR BORE SIZE

CAUTION: BE SURE TO REFER TO ENGINEERING CATALOGS FOR SPECIAL APPLICATIONS THAT REQUIRE SPECIFIC MATERIAL CONTENT, TOLERANCES, ETC. SEE FOOTNOTE.

IDLI GROUP 87280

ROUND BEARING STUB
SEE FIGURE 67A

1.313 in	SB	33.34 mm
3.250 in	EC	82.55 mm
1.969 in	HD	50.01 mm

ROCKWELL BS-R21-2

IDLI GROUP 87281

ROUND BEARING STUB
SEE FIGURE 67A

1.313 in	SB	33.34 mm
3.250 in	EC	82.55 mm
2.240 in	HD	56.90 mm

ROCKWELL BS-R21-14

IDLI GROUP 87282

ROUND BEARING STUB
SEE FIGURE 67B

1.313 in	SB	33.34 mm
4.250 in	EC	107.95 mm
1.844 in	HD	46.84 mm

ROCKWELL BS-R21-19

IDLI GROUP 87283

ROUND BEARING STUB
SEE FIGURE 67B

1.313 in	SB	33.34 mm
4.500 in	EC	114.30 mm
1.844 in	HD	46.84 mm

ROCKWELL BS-R21-22

IDLI GROUP 87284

ROUND BEARING STUB
SEE FIGURE 67B

1.313 in	SB	33.34 mm
4.500 in	EC	114.30 mm
1.969 in	HD	50.01 mm

ROCKWELL BS-R21-23

IDLI GROUP 87285

CENTER BEARING
SEE FIGURE 55

1.375 in	SB	34.93 mm
3.500 in	OD	88.90 mm
.750 in	BW	19.05 mm

AEC 22
AEC AE22
AMER.PARTS 2-6508
ANCHORDOAN 35-6050
BEAR.INC. B1-203KC
BORG-WARNR 35-6050
BORG-WARNR 36-6050
CLEVE.MOT. DS203KC
CLEVE.STL. RCB100L
DORMAN 662-022
DORMAN DS22
EATON MM6053
FED.BEAR. HB101
FED.MOGUL HB1

FORD D0TZ4840B
GREEN BALL RSB13
GREEN BALL TSB12
IHC 223572R93
IHC 233572R93
KENBAR 203KC
L & S BRG. PBS1
LEMPCO 36-2030
MCQUAY-NOR MM6053
MCQUAY-NOR MM6054
MOPAR D604
MOTOR MAST DS22
NEAPCO 81-2030
NEAPCO DS203KC
PILOT H-1005
REPUBLIC 35-6053
ROCKWELL CB310008-1X
SPICER 211302X
TRW 20805

IDLI GROUP 87286

CENTER BEARING
slight variation from above group
SEE FIGURE 55

1.375 in	SB	34.93 mm
3.500 in	OD	88.90 mm
.750 in	BW	19.05 mm

CLEVE.STL. RCB100LS
ROCKWELL CB310008-2X

IDLI GROUP 87287

SHAFTING WITH 6 SPLINES
SEE FIGURE 86

| 1.375 in | SB | 34.93 mm |
| 12.000 in | EE | 304.80 mm |

ROCKWELL SH-S22-88-12

IDLI GROUP 87288

SQUARE TUBING
SEE FIGURE 76

1.375 in	SB	34.93 mm
72.000 in	EE	1828.80 mm
1.625 in	OD	41.28 mm

ROCKWELL SQT1625-72

IDLI GROUP 87289

SQUARE TUBING
SEE FIGURE 76

1.375 in	SB	34.93 mm
108.000 in	EE	2743.20 mm
1.625 in	OD	41.28 mm

ROCKWELL SQT1625-108

IDLI GROUP 87290

SPLINED BEARING STUB
WITH 16 INVOLUTED SPLINES
SEE FIGURE 67

1.375 in	SB	34.93 mm
2.844 in	HD	72.24 mm
8.094 in	EL	205.58 mm

ROCKWELL 141N53-161
SPICER 3-53-161

IDLI GROUP 87291

SPLINED BEARING STUB
WITH 21 INVOLUTED SPLINES
SEE FIGURE 66

1.375 in	SB	34.93 mm
3.188 in	EC	80.96 mm
2.320 in	HD	58.93 mm

ROCKWELL BS-S22-44

IDLI GROUP 87292

SPLINED BEARING STUB
WITH 21 INVOLUTED SPLINES
SEE FIGURE 66

1.375 in	SB	34.93 mm
3.313 in	EC	84.14 mm
2.320 in	HD	58.93 mm

ROCKWELL BS-S22-16

IDLI GROUP 87293

SPLINED BEARING STUB
WITH 21 INVOLUTED SPLINES
SEE FIGURE 66

1.375 in	SB	34.93 mm
3.313 in	EC	84.14 mm
2.542 in	HD	64.57 mm

ROCKWELL BS-S22-40

IDLI GROUP 87294

SPLINED BEARING STUB
WITH 21 INVOLUTED SPLINES
SEE FIGURE 66

1.375 in	SB	34.93 mm
3.375 in	EC	85.73 mm
1.844 in	HD	46.84 mm

ROCKWELL BS-S22-39

IDLI GROUP 87295

SPLINED BEARING STUB
WITH 6 SPLINES
SEE FIGURE 66

1.375 in	SB	34.93 mm
3.375 in	EC	85.73 mm
1.969 in	HD	50.01 mm

ROCKWELL BS-S22-1

IDLI GROUP 87296

SPLINED BEARING STUB
WITH 21 INVOLUTED SPLINES
SEE FIGURE 66

1.375 in	SB	34.93 mm
3.375 in	EC	85.73 mm
1.969 in	HD	50.01 mm

ROCKWELL BS-S22-38

IDLI GROUP 87297

SPLINED BEARING STUB
WITH 21 INVOLUTED SPLINES
SEE FIGURE 66

1.375 in	SB	34.93 mm
3.375 in	EC	85.73 mm
2.240 in	HD	56.90 mm

ROCKWELL BS-S22-60

IDLI GROUP 87298

SPLINED BEARING STUB
WITH 21 INVOLUTED SPLINES
slight variation from above group
SEE FIGURE 66

1.375 in	SB	34.93 mm
3.375 in	EC	85.73 mm
2.240 in	HD	56.90 mm

ROCKWELL BS-S22-65

IDLI GROUP 87299

SPLINED BEARING STUB
WITH 21 INVOLUTED SPLINES
SEE FIGURE 66

1.375 in	SB	34.93 mm
3.375 in	EC	85.73 mm
2.270 in	HD	57.66 mm

ROCKWELL BS-S22-49

IDLI GROUP 87300

SPLINED BEARING STUB
WITH 6 SPLINES
SEE FIGURE 66

1.375 in	SB	34.93 mm
3.375 in	EC	85.73 mm
2.320 in	HD	58.93 mm

ROCKWELL BS-S22-4

IDLI GROUP 87301

SPLINED BEARING STUB
WITH 21 INVOLUTED SPLINES
SEE FIGURE 66

1.375 in	SB	34.93 mm
3.438 in	EC	87.31 mm
1.844 in	HD	46.84 mm

ROCKWELL BS-S22-34

IDLI GROUP 87302

ROUND BEARING STUB
SEE FIGURE 67A

1.375 in	SB	34.93 mm
3.500 in	EC	88.90 mm
1.844 in	HD	46.84 mm

ROCKWELL BS-R22-15

IDLI GROUP 87303

SPLINED BEARING STUB
WITH 6 SPLINES
SEE FIGURE 66

1.375 in	SB	34.93 mm
3.500 in	EC	88.90 mm
1.880 in	HD	47.75 mm

ROCKWELL BS-S22-13

IDLI GROUP 87304

ROUND BEARING STUB
SEE FIGURE 67A

1.375 in	SB	34.93 mm
3.500 in	EC	88.90 mm
1.880 in	HD	47.75 mm

ROCKWELL BS-R22-14

ENGINEERING CATALOGS MUST BE CONSULTED FOR DETAILS NOT INCLUDED
IN THIS GUIDE. SPECIFIC DESIGNS, MATERIAL CONTENT, TOLERANCES
LUBE FITTINGS AND OTHER DIMENSIONS ARE INTENTIONALLY OMITTED HERE.

IDLI GUIDE   COPYRIGHT ® INTERCHANGE, INC.   ST. LOUIS PARK, MN. 55416 USA

IDLI GROUP 87305

SPLINED BEARING STUB
WITH 6 SPLINES
SEE FIGURE 66

1.375 in	SB	34.93 mm
3.500 in	EC	88.90 mm
1.969 in	HD	50.01 mm

ROCKWELL BS-S22-41

IDLI GROUP 87306

SPLINED BEARING STUB
WITH 6 SPLINES
SEE FIGURE 66

1.375 in	SB	34.93 mm
3.563 in	EC	90.49 mm
1.844 in	HD	46.84 mm

ROCKWELL BS-S22-25

IDLI GROUP 87307

SPLINED BEARING STUB
WITH 21 INVOLUTED SPLINES
SEE FIGURE 66

1.375 in	SB	34.93 mm
3.563 in	EC	90.49 mm
1.880 in	HD	47.75 mm

ROCKWELL BS-S22-29

IDLI GROUP 87308

SPLINED BEARING STUB
WITH 6 SPLINES
SEE FIGURE 66

1.375 in	SB	34.93 mm
3.625 in	EC	92.08 mm
3.270 in	HD	83.06 mm

ROCKWELL BS-S22-17

IDLI GROUP 87309

SPLINED BEARING STUB
WITH 6 SPLINES
SEE FIGURE 66

1.375 in	SB	34.93 mm
3.625 in	EC	92.08 mm
3.320 in	HD	84.33 mm

ROCKWELL BS-S22-11

IDLI GROUP 87310

SPLINED BEARING STUB
WITH 21 INVOLUTED SPLINES
SEE FIGURE 66

1.375 in	SB	34.93 mm
4.063 in	EC	103.19 mm
3.844 in	HD	97.64 mm

ROCKWELL BS-S22-12

IDLI GROUP 87311

SPLINED BEARING STUB
WITH 6 SPLINES
SEE FIGURE 66

1.375 in	SB	34.93 mm
4.188 in	EC	106.36 mm
2.320 in	HD	58.93 mm

ROCKWELL BS-S22-12A

IDLI GROUP 87312

SPLINED BEARING STUB
WITH 6 SPLINES
SEE FIGURE 66

1.375 in	SB	34.93 mm
4.500 in	EC	114.30 mm
1.969 in	HD	50.01 mm

ROCKWELL BS-S22-50

IDLI GROUP 87313

SPLINED BEARING STUB WITH 6
SPLINES
slight variation from above group
SEE FIGURE 66

.1.375 in	SB	34.93 mm
4.500 in	EC	114.30 mm
1.969 in	HD	50.01 mm

ROCKWELL BS-S22-57

IDLI GROUP 87314

SPLINED BEARING STUB
WITH 10 SPLINES
SEE FIGURE 66

1.375 in	SB	34.93 mm
4.531 in	EC	115.10 mm
3.792 in	HD	96.32 mm

ROCKWELL BS-S28-12

IDLI GROUP 87315

SPLINED BEARING STUB
WITH 21 INVOLUTED SPLINES
SEE FIGURE 66

1.375 in	SB	34.93 mm
4.563 in	EC	115.89 mm
1.844 in	HD	46.84 mm

ROCKWELL BS-S22-43

IDLI GROUP 87316

SPLINED BEARING STUB
WITH 6 SPLINES
SEE FIGURE 66

1.375 in	SB	34.93 mm
4.625 in	EC	117.48 mm
1.844 in	HD	46.84 mm

ROCKWELL BS-S22-10

IDLI GROUP 87317

SPLINED BEARING STUB
WITH 6 SPLINES
SEE FIGURE 66

1.375 in	SB	34.93 mm
4.625 in	EC	117.48 mm
1.969 in	HD	50.01 mm

ROCKWELL BS-S22-36

IDLI GROUP 87318

SPLINED BEARING STUB WITH 6
SPLINES
slight variation from above group
SEE FIGURE 66

1.375 in	SB	34.93 mm
4.625 in	EC	117.48 mm
1.969 in	HD	50.01 mm

ROCKWELL BS-S22-53

IDLI GROUP 87319

SPLINED BEARING STUB WITH 6
SPLINES
slight variation from above group
SEE FIGURE 66

1.375 in	SB	34.93 mm
4.625 in	EC	117.48 mm
1.969 in	HD	50.01 mm

ROCKWELL BS-S22-82

IDLI GROUP 87320

SPLINED BEARING STUB
WITH 21 INVOLUTED SPLINES
SEE FIGURE 66

1.375 in	SB	34.93 mm
4.625 in	EC	117.48 mm
1.969 in	HD	50.01 mm

ROCKWELL BS-S22-51

IDLI GROUP 87321

ROUND BEARING STUB
SEE FIGURE 67A

1.375 in	SB	34.93 mm
4.688 in	EC	119.06 mm
1.969 in	HD	50.01 mm

ROCKWELL BS-R22-1

IDLI GROUP 87322

SPLINED BEARING STUB
WITH 5 SPLINES IN 6 SPACES
SEE FIGURE 66

1.375 in	SB	34.93 mm
5.250 in	EC	133.35 mm
1.844 in	HD	46.84 mm

ROCKWELL BS-S22-72

IDLI GROUP 87323

SPLINED BEARING STUB
WITH 6 SPLINES
SEE FIGURE 66

1.375 in	SB	34.93 mm
5.375 in	EC	136.53 mm
1.844 in	HD	46.84 mm

ROCKWELL BS-S22-9

IDLI GROUP 87324

SPLINED BEARING STUB
WITH 6 SPLINES
SEE FIGURE 66

1.375 in	SB	34.93 mm
7.875 in	EC	200.03 mm
1.969 in	HD	50.01 mm

ROCKWELL BS-S22-64

IDLI GROUP 87325

CENTER BEARING
SEE FIGURE 55

1.378 in	SB	35.00 mm
3.500 in	OD	88.90 mm
.984 in	BW	25.00 mm

AEC	9
AEC	....................AE9
ALLOY	4010
AMER.PARTS	2-6510
ANCHORDOAN	35-6054

BORG-WARNR 35-6054
BUTLER 51085X
CLEVE.MOT. DS205KG
CLEVE.STL. SCB85
CMP DS205KG
DORMAN 662-026
DORMAN DS26
FED.BEAR. HGB88107
FEDERAL HBG88107
FED.MOGUL HB5
GMC 2309277
KENBAR 9277-2
MOPAR D614
MOTOR MAST DS9
NEAPCO 81-2050
NEAPCO DS205KG
PILOT H1007
ROCKWELL CB310011-1X
SKF 1700947
WESCO 960-010

IDLI GROUP 87326

CENTER BEARING
SEE FIGURE 56

1.378 in	SB	35.00 mm
2.250 in	EC	57.15 mm
6.625 in	CC	168.28 mm
.984 in	BW	25.00 mm

AEC 12
AEC AE12
ALLOY 4002
AMER.PARTS 2-6036
ANCHOR 35-6036
ANCHOR 3-6036
ANCHOR 6036
ANCHORDOAN 35-6036
ANCHORDOAN 35-6056
BORG-WARNR 35-6036
BORG-WARNR 35-6036A
BUTLER 510367-1X
BUTLER 510367-1XWF
BUTLER 510370-1X
CLEVE.MOT. DS201KG
CLEVE.STL. SCB103
CMP DS201KG
CMP DS201KG1
CMP DS201KG3
DETROIT MB602
DORMAN 662-012
DORMAN 662-027
DORMAN DS12
EATON MM6036
FED.BEAR. HB88107
FEDERAL HB88107
FED.MOGUL HB88107A
GMC 2343325
GMC 2765513
GMC 3765513
GMC 3990291
GMC 960-002
GREEN BALL RSB3
KENBAR 4513-2
KENBAR 5513-2
LEMPCO 35-6036
MCQUAY-NOR MM6036
MCQUAY-NOR MM6042

BD=BEARING DIAMETER (outside)   **BW**=BEARING WIDTH   **CB**=CROSS LENGTH WITH BEARINGS   **CC**=CENTER TO CENTER   **CD**=CROSS DIAMETER
CL=CROSS LENGTH WITHOUT BEARINGS   **EC**=END TO CENTER or FACE TO FACE   **EE**=END TO END   **EL**=EFFECTIVE LENGTH
HD=HUB DIAMETER (or insert)   **IE**=INSIDE OF EARS (or recess)   **OD**=OUTSIDE DIAMETER   **OE**=OUTSIDE OF EARS   **SB**=SPLINE OR BORE SIZE

CAUTION: BE SURE TO REFER TO ENGINEERING CATALOGS FOR SPECIAL APPLICATIONS THAT REQUIRE SPECIFIC MATERIAL CONTENT, TOLERANCES, ETC. SEE FOOTNOTE.

MOPAR	D602
MOTOR MAST	DS12
NEAPCO	81-2070
NEAPCO	81-2120
NEAPCO	81-2123
NEAPCO	DS201KG
NEAPCO	DS207KF
PILOT	H-1003
PILOT	H1003
PILOT	H-1003
PILOT	H1003
REPUBLIC	35-6036
ROCKWELL	CB310005-1X
SKF	1700547
SKF	1700547A
SKF	PBS7
SPICER	210087-1X
SPICER	210089-1X
SPICER	210091-1X
SPICER	210367-1X
SPICER	210370-1X
SPICER	210851-1X
TRW	20803
TRW	20817
TRW	20818
WESCO	960-002

IDLI GROUP 87327

CENTER BEARING
slight variation from above group
SEE FIGURE 56

1.378 in	SB	35.00 mm
2.250 in	EC	57.15 mm
6.625 in	CC	168.28 mm
.984 in	BW	25.00 mm

CHRYSLER	3004542
CHRYSLER	3621457
CHRYSLER	3632989
CHRYSLER	3632990
CLEVE.STL.	SCB102
FORD	D2TZ4840B
MOPAR	MB202
NEAPCO	81-2010
NEAPCO	81-4012
NEAPCO	DS201KDS
NEAPCO	DS201KF
NEAPCO	DS401KF2
ROCKWELL	CB210088-1X
SPICER	210088-1X
WESCO	960-001
WHITE	694297

IDLI GROUP 87328

CENTER BEARING
slight variation from above group
SEE FIGURE 56

1.378 in	SB	35.00 mm
2.250 in	EC	57.15 mm
6.625 in	CC	168.28 mm
.984 in	BW	25.00 mm

BUTLER	510088-1X
CLEVE.STL.	SCB107
IHC	383182C91
IHC	427349C91
NEAPCO	81-2012
NEAPCO	DS201KF1
ROCKWELL	CB210090-1X
SPICER	210090-1X
SPICER	210939CX
SPICER	210939X
WESCO	960-001

IDLI GROUP 87329

CENTER BEARING
slight variation from above group
SEE FIGURE 56

1.378 in	SB	35.00 mm
2.250 in	EC	57.15 mm
6.625 in	CC	168.28 mm
.984 in	BW	25.00 mm

CLEVE.STL.	SCB104
NEAPCO	81-2121
NEAPCO	DS201KG1
ROCKWELL	CB210367-1X
SPICER	210367-1X

IDLI GROUP 87330

CENTER BEARING
slight variation from above group
SEE FIGURE 56

1.378 in	SB	35.00 mm
2.250 in	EC	57.15 mm
6.625 in	CC	168.28 mm
.984 in	BW	25.00 mm

AEC	12
BUTLER	510367-1XWF
BUTLER	510370-1X
CLEVE.STL.	SCB106
CMP	DS201KG
CMP	DS201KG2
CMP	DS201KG3
GMC	2473119
GMC	3765513
IHC	432377C91
MOPAR	602
NEAPCO	81-2122
NEAPCO	81-2123
NEAPCO	DS201KG
NEAPCO	DS201KG2
ROCKWELL	CB210370-1X
SPICER	210370-1X
SPICER	210851-1X
WESCO	960-002

IDLI GROUP 87331

CENTER BEARING
slight variation from above group
SEE FIGURE 56

1.378 in	SB	35.00 mm
2.250 in	EC	57.15 mm
6.625 in	CC	168.28 mm
.984 in	BW	25.00 mm

AEC	14
AEC	14S
AEC	15
AEC	AE14S
AEC	AE15
ALLOY	4001
AMER.PARTS	2-6023
AMER.PARTS	2-6037
ANCHOR	35-6037
ANCHOR	6037
ANCHORDOAN	35-6037
ANCHORDOAN	35-6057
BEAR.INC.	B1-202KD
BORG-WARNR	35-6037
BORG-WARNR	35-6047
BUTLER	510088-1X
BUTLER	510088-1X2
BUTLER	510088-1XWF
BUTLER	510090-1X
CHECKER	03568
CHECKER	64575
CHRYSLER	2514402
CHRYSLER	2908812
CHRYSLER	4117147
CLEVE.MOT.	DS201KF
CLEVE.MOT.	DS202KD
CLEVE.STL.	SCB101
CMP	DS201KF
CMP	DS201KF1
CMP	DS201KF2
CMP	DS201KF4
DORMAN	662-014
DORMAN	662-015
DORMAN	662-027
DORMAN	DS15
EATON	MM6037
FED.BEAR.	HBD88107
FED.BEAR.	HBF88107
FED.MOGUL	HB88107
FED.MOGUL	HBD88107
FORD	D2TZ4840B
FORD	D4TZ4800A
FORD	D4TZ4800B
FORD	D4TZ4800D
FORD	D7TZ4800A
FORD	D8TZ4800A
FORD	D9TZ4800A
FORD	DSVZ4800A
GMC	3990291
GMC	7830355
GREEN BALL	RSB4
GREEN BALL	RSB5
IHC	386418C91
KENBAR	5407-1
KENBAR	8812
L & S BRG.	PBS7
L & S BRG.	PBS9
LEMPCO	35-6037
MCQUAY-NOR	MM6037
MOPAR	601
MOPAR	D-601
MOPAR	D603
MOPAR	MB202
MOTOR MAST	DS14
MOTOR MAST	DS145
MOTOR MAST	DS14S
MOTOR MAST	DS15
NEAPCO	35-6024
NEAPCO	81-2010
NEAPCO	81-2011
NEAPCO	81-2020
NEAPCO	81-2021
NEAPCO	81-2060
NEAPCO	81-2061
NEAPCO	81-2062
NEAPCO	81-2071
NEAPCO	DS201BF
NEAPCO	DS201KDS
NEAPCO	DS201KF
NEAPCO	DS202KD
NEAPCO	DS202KD1
NEAPCO	DS206KF
NEAPCO	DS206KF1
NEAPCO	DS206KF2
NEAPCO	DS207KF1
PILOT	H-1002
PILOT	H-1004
PILOT	H1004
REPUBLIC	35-6037
REPUBLIC	35-6052
ROCKWELL	CB310006-1X
SKF	1700537
SKF	1700537A
SKF	PBS8
SPICER	210088-1X
SPICER	210090-1X
SPICER	210137-1X
SPICER	210939X
SPICER	211036-2X
SPICER	211139X
SPICER	211187X
TRW	20802
TRW	20804
TRW	20817
TRW	20818
WESCO	960-001

IDLI GROUP 87332

CENTER BEARING
slight variation from above group
SEE FIGURE 56

1.378 in	SB	35.00 mm
2.250 in	EC	57.15 mm
6.625 in	CC	168.28 mm
.984 in	BW	25.00 mm

BEAR.INC.	BI202KD
CLEVE.STL.	SCB110
CMP	DS202KD
DETROIT	MB202
DETROIT	MB601
DETROIT	MB603
DORMAN	DS14
FEDERAL	HBD88107
FEDERAL	HBF88107
FORD	D4TJ4800BA
NEAPCO	81-2014
NEAPCO	DS201KF4
ROCKWELL	CB310006-2X
SPICER	210088-1X
SPICER	210090-1X
SPICER	210797-2X
WESCO	960-001

ENGINEERING CATALOGS MUST BE CONSULTED FOR DETAILS NOT INCLUDED IN THIS GUIDE. SPECIFIC DESIGNS, MATERIAL CONTENT, TOLERANCES LUBE FITTINGS AND OTHER DIMENSIONS ARE INTENTIONALLY OMITTED HERE.

IDLI GROUP 87333

ROUND BEARING STUB
SEE FIGURE 67A

1.438 in	SB	36.51 mm
7.625 in	EC	193.68 mm
3.844 in	HD	97.64 mm

ROCKWELL BSA-R23-5

IDLI GROUP 87334

ROUND BEARING STUB
SEE FIGURE 67A

1.438 in	SB	36.51 mm
8.750 in	EC	222.25 mm
1.969 in	HD	50.01 mm

ROCKWELL BSA-R23-6

IDLI GROUP 87335

ROUND TUBING
SEE FIGURE 75

1.459 in	SB	37.06 mm
72.000 in	EE	1828.80 mm
1.625 in	OD	41.28 mm

ROCKWELL RT1625-14-72

IDLI GROUP 87336

ROUND TUBING
SEE FIGURE 75

1.459 in	SB	37.06 mm
108.000 in	EE	2743.20 mm
1.625 in	OD	41.28 mm

ROCKWELL RT1625-14-108

IDLI GROUP 87337

CENTER BEARING
SEE FIGURE 55

1.500 in	SB	38.10 mm
3.500 in	OD	88.90 mm
.750 in	BW	19.05 mm

AEC	 23
AEC	.... AE23
AMER.PARTS	 2-6511
ANCHORDOAN	 35-6051
BEAR.INC.	.... B1-303KC
BORG-WARNR	 35-6051
CLEVE.MOT.	.... DS303KC
CLEVE.STL.	.... OCB100L
DORMAN	 662-023
DORMAN	.... DS23
EATON	.... MM6054
FED.BEAR.	.... HB102
FED.MOGUL	.... HB2
FORD	.... D0TZ4840A
GREEN BALL	.... RSB12
IHC	.................223573R93
L & S BRG.	.... PBS2
MCQUAY-NOR	.... MM6054
MOPAR	.... D607
MOTOR MAST	.... DS23
NEAPCO	 81-3030
NEAPCO	.............DS303KC
REPUBLIC	 35-6054
ROCKWELL	.... CB310010-1X
SPICER	...............211303X
TRW	 20808

IDLI GROUP 87338

CENTER BEARING
slight variation from above group
SEE FIGURE 55

1.500 in	SB	38.10 mm
3.500 in	OD	88.90 mm
.750 in	BW	19.05 mm

CLEVE.STL. OCB100LS
ROCKWELL CB310010-2X

IDLI GROUP 87339

SHAFTING WITH 10 SPLINES
SEE FIGURE 86

1.500 in	SB	38.10 mm
12.000 in	EE	304.80 mm

ROCKWELL SH-S24-53-12

IDLI GROUP 87340

SHAFTING WITH 10 SPLINES
SEE FIGURE 86

1.500 in	SB	38.10 mm
24.000 in	EE	609.60 mm

ROCKWELL SH-S24-53-24

IDLI GROUP 87341

SPLINED PLUG WITH 10 SPLINES
SEE FIGURE 63

1.500 in	SB	38.10 mm
1.969 in	HD	50.01 mm
5.750 in	EL	146.05 mm

ROCKWELL 3NPS24

IDLI GROUP 87342

SLEEVE WITH 10 SPLINES
SEE FIGURE 82

1.500 in	SB	38.10 mm
2.020 in	HD	51.31 mm
3.000 in	EL	76.20 mm
2.250 in	OD	57.15 mm

ROCKWELL SL-S24-3

IDLI GROUP 87343

SPLINED PLUG WITH 10 SPLINES
SEE FIGURE 63

1.500 in	SB	38.10 mm
2.021 in	HD	51.33 mm
5.250 in	EL	133.35 mm

ROCKWELL 4NPS24-29

IDLI GROUP 87344

SPLINED PLUG WITH 10 SPLINES
SEE FIGURE 64

1.500 in	SB	38.10 mm
2.021 in	HD	51.33 mm
6.500 in	EL	165.10 mm

ROCKWELL 3NPS24-3

IDLI GROUP 87345

SPLINED PLUG
WITH 23 INVOLUTED SPLINES
SEE FIGURE 63

1.500 in	SB	38.10 mm
2.021 in	HD	51.33 mm
6.500 in	EL	165.10 mm

ROCKWELL 4NPS24-26

IDLI GROUP 87346

SPLINED PLUG WITH 10 SPLINES
SEE FIGURE 63

1.500 in	SB	38.10 mm
2.021 in	HD	51.33 mm
7.500 in	EL	190.50 mm

ROCKWELL 4NPS24

IDLI GROUP 87347

SLEEVE WITH 10 SPLINES
SEE FIGURE 82

1.500 in	SB	38.10 mm
2.240 in	HD	56.90 mm
5.000 in	EL	127.00 mm
2.375 in	OD	60.33 mm

ROCKWELL SL-S24-30

IDLI GROUP 87348

SLEEVE WITH 10 SPLINES
SEE FIGURE 82

1.500 in	SB	38.10 mm
2.240 in	HD	56.90 mm
7.000 in	EL	177.80 mm
2.500 in	OD	63.50 mm

ROCKWELL SL-S24-32

IDLI GROUP 87349

SPLINED PLUG WITH 10 SPLINES
SEE FIGURE 64

1.500 in	SB	38.10 mm
2.542 in	HD	64.57 mm
7.875 in	EL	200.03 mm

ROCKWELL 4NPS24-23

IDLI GROUP 87350

SPLINED BEARING STUB
WITH 10 SPLINES
SEE FIGURE 66

1.500 in	SB	38.10 mm
2.823 in	HD	71.70 mm
3.875 in	EL	98.43 mm

ROCKWELL BS-S24-36

IDLI GROUP 87351

SPLINED BEARING STUB
WITH 16 INVOLUTED SPLINES
SEE FIGURE 67

1.500 in	SB	38.10 mm
2.823 in	HD	71.70 mm
7.063 in	EL	179.39 mm

ROCKWELL BS-S24-37

IDLI GROUP 87352

SPLINED BEARING STUB
WITH 10 SPLINES
SEE FIGURE 66

1.500 in	SB	38.10 mm
2.844 in	HD	72.24 mm
3.875 in	EL	98.43 mm

ROCKWELL 141N53-1071
SPICER 3-53-1071

IDLI GROUP 87353

SPLINED BEARING STUB
WITH 16 INVOLUTED SPLINES
SEE FIGURE 67

1.500 in	SB	38.10 mm
2.844 in	HD	72.24 mm
7.063 in	EL	179.39 mm

ROCKWELL 141N53-1031
SPICER 3-53-1031

IDLI GROUP 87354

SPLINED PLUG
WITH 32 INVOLUTED SPLINES
SEE FIGURE 67

1.500 in	SB	38.10 mm
2.844 in	HD	72.24 mm
9.375 in	EL	238.13 mm

CLEVE.STL. O55-129
ROCKWELL BS-S24-39

IDLI GROUP 87355

SPLINED BEARING STUB
WITH 10 SPLINES
SEE FIGURE 66

1.500 in	SB	38.10 mm
3.320 in	HD	84.33 mm
4.031 in	EL	102.40 mm

ROCKWELL 155N53-71
SPICER 4-53-71

IDLI GROUP 87356

SPLINED BEARING STUB
WITH 10 SPLINES
SEE FIGURE 66

1.500 in	SB	38.10 mm
3.320 in	HD	84.33 mm
4.063 in	EL	103.19 mm

ROCKWELL 155N53-261
SPICER 4-53-261

IDLI GROUP 87357

SPLINED BEARING STUB
WITH 10 SPLINES
SEE FIGURE 66

1.500 in	SB	38.10 mm
3.344 in	HD	84.94 mm
4.031 in	EL	102.40 mm

NEAPCO N3-53-431
ROCKWELL 148N53-431
SPICER 3-53-431

BD=BEARING DIAMETER (outside)   **BW**=BEARING WIDTH   **CB**=CROSS LENGTH WITH BEARINGS   **CC**=CENTER TO CENTER   **CD**=CROSS DIAMETER
CL=CROSS LENGTH WITHOUT BEARINGS   **EC**=END TO CENTER or FACE TO FACE   **EE**=END TO END   **EL**=EFFECTIVE LENGTH
HD=HUB DIAMETER (or insert)   **IE**=INSIDE OF EARS (or recess)   **OD**=OUTSIDE DIAMETER   **OE**=OUTSIDE OF EARS   **SB**=SPLINE OR BORE SIZE

CAUTION: BE SURE TO REFER TO ENGINEERING CATALOGS FOR SPECIAL APPLICATIONS THAT REQUIRE SPECIFIC MATERIAL CONTENT, TOLERANCES, ETC. SEE FOOTNOTE.

IDLI GROUP 87358

**SPLINED PLUG
WITH 32 INVOLUTED SPLINES
SEE FIGURE 67**

1.500 in	SB	38.10 mm
3.344 in	HD	84.94 mm
9.375 in	EL	238.13 mm

CLEVE.STL. O55-132
ROCKWELL BS-S24-40

IDLI GROUP 87359

**SPLINED PLUG
WITH 32 INVOLUTED SPLINES
SEE FIGURE 67**

1.500 in	SB	38.10 mm
3.392 in	HD	86.16 mm
9.375 in	EL	238.13 mm

CLEVE.STL. O55-245
ROCKWELL BS-S24-38

IDLI GROUP 87360

**SPLINED BEARING STUB
WITH 10 SPLINES
SEE FIGURE 66**

1.500 in	SB	38.10 mm
3.846 in	HD	97.69 mm
4.031 in	EL	102.40 mm

ROCKWELL 148N53-241
SPICER 3-53-241

IDLI GROUP 87361

**SPLINED BEARING STUB
WITH 10 SPLINES
SEE FIGURE 66**

1.500 in	SB	38.10 mm
3.846 in	HD	97.69 mm
4.063 in	EL	103.19 mm

ROCKWELL 155N53-271
SPICER 4-53-271

IDLI GROUP 87362

**SPLINED PLUG
WITH 32 INVOLUTED SPLINES
SEE FIGURE 67**

1.500 in	SB	38.10 mm
4.320 in	HD	109.73 mm
9.375 in	EL	238.13 mm

ROCKWELL BS-S24-40A

IDLI GROUP 87363

**SPLINED BEARING STUB
WITH 10 SPLINES
SEE FIGURE 66**

1.500 in	SB	38.10 mm
4.250 in	EC	107.95 mm
3.270 in	HD	83.06 mm

BORG-WARNR 13782J
ROCKWELL BS-S24-11

IDLI GROUP 87364

**SPLINED BEARING STUB
WITH 10 SPLINES
SEE FIGURE 66**

1.500 in	SB	38.10 mm
4.250 in	EC	107.95 mm
3.320 in	HD	84.33 mm

BORG-WARNR 10503J
ROCKWELL BS-S24-8

IDLI GROUP 87365

**ROUND BEARING STUB
SEE FIGURE 67A**

1.500 in	SB	38.10 mm
4.875 in	EC	123.83 mm
1.844 in	HD	46.84 mm

ROCKWELL BS-R24-5

IDLI GROUP 87366

**ROUND TUBING
SEE FIGURE 75**

1.510 in	SB	38.35 mm
72.000 in	EE	1828.80 mm
1.750 in	OD	44.45 mm

ROCKWELL RT175-11-72

IDLI GROUP 87367

**ROUND TUBING
SEE FIGURE 75**

1.510 in	SB	38.35 mm
108.000 in	EE	2743.20 mm
1.750 in	OD	44.45 mm

ROCKWELL RT175-11-108

IDLI GROUP 87368

**ROUND TUBING
SEE FIGURE 75**

1.560 in	SB	39.62 mm
72.000 in	EE	1828.80 mm
1.750 in	OD	44.45 mm

ROCKWELL RT175-13-72

IDLI GROUP 87369

**ROUND TUBING
SEE FIGURE 75**

1.560 in	SB	39.62 mm
108.000 in	EE	2743.20 mm
1.750 in	OD	44.45 mm

ROCKWELL RT175-13-108

IDLI GROUP 87370

**SPLINED PLUG WITH 16 SPLINES
SEE FIGURE 63**

1.563 in	SB	39.69 mm
3.344 in	HD	84.94 mm
6.750 in	EL	171.45 mm

NEAPCO N3-40-971
ROCKWELL 148N40-971
SPICER 3-40-1571
SPICER 3-40-971

IDLI GROUP 87371

**SPLINED BEARING STUB
WITH 16 INVOLUTED SPLINES
SEE FIGURE 67**

1.563 in	SB	39.69 mm
3.344 in	HD	84.94 mm
8.438 in	EL	214.31 mm

ROCKWELL 148N53-451
SPICER 3-53-151
SPICER 3-53-341
SPICER 3-53-451

IDLI GROUP 87372

**SPLINED BEARING STUB
WITH 16 INVOLUTED SPLINES
SEE FIGURE 67**

1.563 in	SB	39.69 mm
3.846 in	HD	97.69 mm
8.438 in	EL	214.31 mm

ROCKWELL 148N53-291
SPICER 3-53-291

IDLI GROUP 87373

**CENTER BEARING
SEE FIGURE 56**

1.575 in	SB	40.00 mm
2.250 in	EC	57.15 mm
6.625 in	CC	168.28 mm
.859 in	BW	21.83 mm

AEC 11
AEC AE11
ALLOY 4011
AMER.PARTS 2-6509
ANCHORDOAN 35-6053
BORG-WARNR 35-6053
BUTLER 510866-1X
BUTLER 511016X
BUTLER 511016X3
CLEVE.MOT. DS204KF
CLEVE.STL. OCB103
CMP DS204KF
CMP DS204KF1
CMP DS204KF3
CMP DS240KF
DORMAN 662-025
DORMAN DS25
FED.BEAR. HBF88507A
FEDERAL HBF88507A
FED.MOGUL HB88508A
FORD DSTZ4800A
GMC 3730827
GMC 694987
GMC 719413
GMC 720161
KENBAR 9413-1
MOPAR 613
MOPAR D613
MOTOR MAST DS11
NEAPCO 81-2040
NEAPCO DS204KF
ROCKWELL CB211016X
SKF 1700946
SPICER 210866-1X
SPICER 210873-1X
SPICER 211016
SPICER 211016X

TRW 20816
WESCO 960-011

IDLI GROUP 87374

**CENTER BEARING
slight variation from above group
SEE FIGURE 56**

1.575 in	SB	40.00 mm
2.250 in	EC	57.15 mm
6.625 in	CC	168.28 mm
.859 in	BW	21.83 mm

BUTLER 510866-1X
CLEVE.STL. OCB108
CLEVE.STL. OCB109
FORD D4TA4800BA
IHC 468656C91
NEAPCO 81-2041
NEAPCO DS204KF1
ROCKWELL CB210866-1X
SPICER 210866-1X
WESCO 960-011

IDLI GROUP 87375

**CENTER BEARING
slight variation from above group
SEE FIGURE 56**

1.575 in	SB	40.00 mm
2.250 in	EC	57.15 mm
6.625 in	CC	168.28 mm
.859 in	BW	21.83 mm

BUTLER 510873-1X
BUTLER 510873X
BUTLER 511016X
BUTLER 511016X3
CLEVE.STL. OCB107
CMP DS204KF
CMP DS204KF2
CMP DS204KF3
FORD D4TA4800AA
IHC 468666C91
NEAPCO 81-2042
NEAPCO DS204KF2
ROCKWELL CB210873-1X
SPICER 210873-1X
SPICER 21162X
WESCO 960-011

IDLI GROUP 87376

**CENTER BEARING
slight variation from above group
SEE FIGURE 56**

1.575 in	SB	40.00 mm
2.250 in	EC	57.15 mm
6.625 in	CC	168.28 mm
.859 in	BW	21.83 mm

AEC 11
BUTLER 511016X3
CLEVE.STL. OCB106
FORD D5TZ4800A
NEAPCO 81-2043
NEAPCO DS204KF3
ROCKWELL CB310003-1X
SPICER 210866-1X
SPICER 210873-1X

ENGINEERING CATALOGS MUST BE CONSULTED FOR DETAILS NOT INCLUDED
IN THIS GUIDE. SPECIFIC DESIGNS, MATERIAL CONTENT, TOLERANCES
LUBE FITTINGS AND OTHER DIMENSIONS ARE INTENTIONALLY OMITTED HERE.

IDLI GROUP 87377

CENTER BEARING
SEE FIGURE 56

1.575 in	SB	40.00 mm
2.438 in	EC	61.91 mm
6.625 in	CC	168.28 mm
1.063 in	BW	26.99 mm

AEC	16
AEC	AE16
ALLOY	4004
AMER.PARTS	2-6038
ANCHOR	35-6038
ANCHOR	6038
ANCHORDOAN	35-6038
BORG-WARNR	35-6038
BUTLER	510391-1X3
BUTLER	510391-1XWF
BUTLER	510433-1X
CLEVE.MOT.	DS301KF
CLEVE.STL.	DCB101
CMP	DS301KF
CMP	DS301KF2
CMP	DS301KF3
DETROIT	MB605
DORMAN	662-016
DORMAN	DS16
EATON	MM6038
FED.BEAR.	HBF88508
FEDERAL	HBF88508
FED.MOGUL	HB8805
FED.MOGUL	HB88508
FORD	315442C91
FORD	D4TJ4800AA
GREEN BALL	RSB7
L & S BRG.	PBS10
MCQUAY-NOR	MM6038
MOPAR	D605
MOTOR MAST	DS16
NEAPCO	81-3010
NEAPCO	81-310
NEAPCO	DS301KF
REPUBLIC	35-6047
ROCKWELL	CB310001-1X
SKF	PBS10
SPICER	210083-1X
SPICER	210391-1X
SPICER	210433-1X
SPICER	211098-1X
TRW	20806
WESCO	960-004

IDLI GROUP 87378

CENTER BEARING
slight variation from above group
SEE FIGURE 56

1.575 in	SB	40.00 mm
2.438 in	EC	61.91 mm
6.625 in	CC	168.28 mm
1.063 in	BW	26.99 mm

BUTLER	510140-1X
CLEVE.STL.	DCB106
NEAPCO	81-3021
NEAPCO	DS301KG1
ROCKWELL	CB210140-1X
SPICER	210140-1X

IDLI GROUP 87379

CENTER BEARING
slight variation from above group
SEE FIGURE 56

1.575 in	SB	40.00 mm
2.438 in	EC	61.91 mm
6.625 in	CC	168.28 mm
1.063 in	BW	26.99 mm

CLEVE.STL.	DCB107
NEAPCO	81-3022
NEAPCO	DS301KG2
ROCKWELL	CB210144-1X
SPICER	210144-1X

IDLI GROUP 87380

CENTER BEARING
slight variation from above group
SEE FIGURE 56

1.575 in	SB	40.00 mm
2.438 in	EC	61.91 mm
6.625 in	CC	168.28 mm
1.063 in	BW	26.99 mm

BUTLER	510391-1X
CLEVE.STL.	DCB102
IHC	315442C91
NEAPCO	81-3011
NEAPCO	DS301KF1
ROCKWELL	CB210391-1X
SPICER	210391-1X
WESCO	960-004

IDLI GROUP 87381

CENTER BEARING
slight variation from above group
SEE FIGURE 56

1.575 in	SB	40.00 mm
2.438 in	EC	61.91 mm
6.625 in	CC	168.28 mm
1.063 in	BW	26.99 mm

CLEVE.STL.	DCB103
IHC	463236C91
NEAPCO	81-3012
NEAPCO	DS301KF2
ROCKWELL	CB210433-1X
SPICER	210433-1X

IDLI GROUP 87382

CENTER BEARING
slight variation from above group
SEE FIGURE 56

1.575 in	SB	40.00 mm
2.438 in	EC	61.91 mm
6.625 in	CC	168.28 mm
1.063 in	BW	26.99 mm

AEC	16
BUTLER	510391-1X
BUTLER	510391-1X3
BUTLER	510391-1XWF
CHRYSLER	1663975
CHRYSLER	1663979
CHRYSLER	1663981
CHRYSLER	1667747
CLEVE.STL.	DCB108
CMP	DS301KF
CMP	DS301KF1
CMP	DS301KF3

FORD	D4TZ4800C
FORD	D4TZ4800D
MOPAR	605
NEAPCO	81-3013
NEAPCO	DS301KF
NEAPCO	DS301KF3
ROCKWELL	CB310001-2X
SPICER	210391-1X
SPICER	210433-1X
SPICER	211037-1X
WESCO	960-004
WHITE	02-7011949
WHITE	689434
WHITE	692313

IDLI GROUP 87383

CENTER BEARING
slight variation from above group
SEE FIGURE 56

1.575 in	SB	40.00 mm
2.438 in	EC	61.91 mm
6.625 in	CC	168.28 mm
1.063 in	BW	26.99 mm

AEC	17
AEC	AE17
ALLOY	4003
AMER.PARTS	2-6039
ANCHOR	6039
ANCHORDOAN	35-6039
BORG-WARNR	35-6039
CLEVE.MOT.	DS301KG
CLEVE.STL.	DCB105
DETROIT	MB606
DORMAN	662-017
DORMAN	DS17
EATON	MM6039
FED.BEAR.	HB88508
FEDERAL	HB88508
GMC	2351948
GMC	3766233
GREEN BALL	RSB6
MCQUAY-NOR	MM6039
MOPAR	D606
MOTOR MAST	DS17
NEAPCO	81-3020
NEAPCO	DS301KG
REPUBLIC	35-6045
ROCKWELL	CB310002-1X
SPICER	210128-1X
SPICER	210140-1X
SPICER	210144-1X
SPICER	210410-1X
TRW	20807
WESCO	960-003

IDLI GROUP 87384

CENTER BEARING
slight variation from above group
SEE FIGURE 56

1.575 in	SB	40.00 mm
2.438 in	EC	61.91 mm
6.625 in	CC	168.28 mm
1.063 in	BW	26.99 mm

AEC	17
BUTLER	510140-1X
BUTLER	510140-1X3

BUTLER	510140-1XWF
BUTLER	510144-1X
CLEVE.STL.	DCB109
CMP	DS301KG
CMP	DS301KG1
CMP	DS301KG2
CMP	DS301KG3
GMC	2021805
GMC	2351948
MOPAR	606
NEAPCO	81-3023
NEAPCO	DS301KG
ROCKWELL	CB310002-2X
SPICER	210140-1X
SPICER	210144-1X
WESCO	960-003

IDLI GROUP 87385

ROUND TUBING
SEE FIGURE 75

1.583 in	SB	40.21 mm
72.000 in	EE	1828.80 mm
1.750 in	OD	44.45 mm

ROCKWELL	RT175-14-72

IDLI GROUP 87386

ROUND TUBING
SEE FIGURE 75

1.583 in	SB	40.21 mm
108.000 in	EE	2743.20 mm
1.750 in	OD	44.45 mm

ROCKWELL	RT175-14-108

IDLI GROUP 87387

ROUND TUBING
SEE FIGURE 75

1.620 in	SB	41.15 mm
72.000 in	EE	1828.80 mm
1.750 in	OD	44.45 mm

ROCKWELL	RT175-16-72

IDLI GROUP 87388

ROUND TUBING
SEE FIGURE 75

1.620 in	SB	41.15 mm
108.000 in	EE	2743.20 mm
1.750 in	OD	44.45 mm

ROCKWELL	RT175-16-108

IDLI GROUP 87389

SPLINED PLUG
WITH 32 INVOLUTED SPLINES
SEE FIGURE 64

1.625 in	SB	41.28 mm
3.344 in	HD	84.94 mm
7.438 in	EL	188.91 mm

ROCKWELL	PS26-32-1

BD = BEARING DIAMETER (outside)   BW = BEARING WIDTH   CB = CROSS LENGTH WITH BEARINGS   CC = CENTER TO CENTER   CD = CROSS DIAMETER
CL = CROSS LENGTH WITHOUT BEARINGS   EC = END TO CENTER or FACE TO FACE   EE = END TO END   EL = EFFECTIVE LENGTH
HD = HUB DIAMETER (or insert)   IE = INSIDE OF EARS (or recess)   OD = OUTSIDE DIAMETER   OE = OUTSIDE OF EARS   SB = SPLINE OR BORE SIZE

IDLI GUIDE   COPYRIGHT ® INTERCHANGE, INC.   ST. LOUIS PARK, MN. 55416 USA

CAUTION: BE SURE TO REFER TO ENGINEERING CATALOGS FOR SPECIAL APPLICATIONS THAT REQUIRE SPECIFIC MATERIAL CONTENT, TOLERANCES, ETC. SEE FOOTNOTE.

IDLI GROUP 87390

SPLINED PLUG WITH 32 INVOLUTED SPLINES
SEE FIGURE 64

1.625 in	SB	41.28 mm
3.344 in	HD	84.94 mm
8.813 in	EL	223.84 mm

CLEVE.STL. D56-7-138
ROCKWELL PS26-32-2

IDLI GROUP 87391

SPLINED PLUG WITH 32 INVOLUTED SPLINES
SEE FIGURE 67

1.625 in	SB	41.28 mm
3.344 in	HD	84.94 mm
10.375 in	EL	263.53 mm

CLEVE.STL. D56-120
ROCKWELL BS-S26-16

IDLI GROUP 87392

SPLINED PLUG WITH 32 INVOLUTED SPLINES
SEE FIGURE 64

1.625 in	SB	41.28 mm
3.990 in	HD	101.35 mm
8.000 in	EL	203.20 mm

ROCKWELL PS26-32-1A

IDLI GROUP 87393

SPLINED PLUG WITH 32 INVOLUTED SPLINES
SEE FIGURE 64

1.625 in	SB	41.28 mm
3.990 in	HD	101.35 mm
9.375 in	EL	238.13 mm

ROCKWELL PS26-32-2A

IDLI GROUP 87394

SPLINED PLUG WITH 32 INVOLUTED SPLINES
SEE FIGURE 67

1.625 in	SB	41.28 mm
3.990 in	HD	101.35 mm
11.000 in	EL	279.40 mm

ROCKWELL BS-S26-16A

IDLI GROUP 87395

SPLINED BEARING STUB WITH 10 SPLINES
SEE FIGURE 65

1.625 in	SB	41.28 mm
2.250 in	EC	57.15 mm
2.021 in	HD	51.33 mm

ROCKWELL BS-S26-3

IDLI GROUP 87396

COMPANION FLANGE WITH 10 SPLINES
SEE FIGURE 33

1.625 in	SB	41.28 mm
3.000 in	EC	76.20 mm
2.375 in	HD	60.33 mm
6.000 in	OD	152.40 mm

ROCKWELL 2WCS26-13

IDLI GROUP 87397

SPLINED BEARING STUB WITH 10 SPLINES
SEE FIGURE 65

1.625 in	SB	41.28 mm
5.875 in	EC	149.23 mm
2.240 in	HD	56.90 mm

ROCKWELL BS-S26-6

IDLI GROUP 87398

SPLINED BEARING STUB WITH 10 SPLINES
SEE FIGURE 65

1.625 in	SB	41.28 mm
5.875 in	EC	149.23 mm
2.395 in	HD	60.83 mm

ROCKWELL BS-S26-12

IDLI GROUP 87399

CENTER BEARING
SEE FIGURE 55

1.750 in	SB	44.45 mm
3.875 in	OD	98.43 mm
1.000 in	BW	25.40 mm

AEC	. . . .	24
AEC	. . . .	AE24
AMER.PARTS	. . . .	2-6512
AMER.PARTS	. . . .	26512
AMER.PARTS	. . . .	2-6512
AMER.PARTS	. . . .	26512
ANCHORDOAN	. . . .	35-6052
BEAR.INC.	. . . .	B1403KC
BORG-WARNR	. . . .	35-6052
CLEVE.MOT.	. . . .	DS403KC
CLEVE.STL.	. . . .	UCB100L
CLEVE.STL.	. . . .	VCB100L
DORMAN	. . . .	662-024
DORMAN	. . . .	DS24
EATON	. . . .	MM6055
FED.BEAR.	. . . .	HB103
FED.MOGUL	. . . .	HB3
GREEN BALL	. . . .	RSB14
IHC	. . . .	879829R92
IHC	. . . .	879849R91
L & S BRG.	. . . .	PBS3
MCQUAY-NOR	. . . .	MM6055
MOPAR	. . . .	D610
MOTOR MAST	. . . .	DS24
NEAPCO	. . . .	81-4030
NEAPCO	. . . .	DS403KC
REPUBLIC	. . . .	35-6055
ROCKWELL	. . . .	CB310009-1X
SPICER	. . . .	211304X
TRW	. . . .	20811

IDLI GROUP 87400

CENTER BEARING
slight variation from above group
SEE FIGURE 55

1.750 in	SB	44.45 mm
3.875 in	OD	98.43 mm
1.000 in	BW	25.40 mm

CLEVE.STL. UCB100LS
ROCKWELL CB310009-2X

IDLI GROUP 87401

SHAFTING WITH 10 SPLINES
SEE FIGURE 86

| 1.750 in | SB | 44.45 mm |
| 12.000 in | EE | 304.80 mm |

ROCKWELL SH-S28-48-12

IDLI GROUP 87402

SHAFTING WITH 10 SPLINES
SEE FIGURE 86

| 1.750 in | SB | 44.45 mm |
| 24.000 in | EE | 609.60 mm |

ROCKWELL SH-S28-48-24

IDLI GROUP 87403

SPLINED PLUG WITH 16 SPLINES
SEE FIGURE 63

1.750 in	SB	44.45 mm
2.542 in	HD	64.57 mm
6.875 in	EL	174.63 mm

ROCKWELL 50NPS28-27

IDLI GROUP 87404

SPLINED PLUG WITH 10 SPLINES
SEE FIGURE 63

1.750 in	SB	44.45 mm
2.542 in	HD	64.57 mm
7.500 in	EL	190.50 mm

ROCKWELL 5NPS28

IDLI GROUP 87405

SPLINED PLUG WITH 16 SPLINES
SEE FIGURE 63

1.750 in	SB	44.45 mm
2.542 in	HD	64.57 mm
7.500 in	EL	190.50 mm

ROCKWELL 50NPS28-25

IDLI GROUP 87406

SPLINED PLUG WITH 16 SPLINES
SEE FIGURE 63

1.750 in	SB	44.45 mm
2.542 in	HD	64.57 mm
9.750 in	EL	247.65 mm

ROCKWELL 5NPS28-36

IDLI GROUP 87407

SPLINED PLUG WITH 10 SPLINES
SEE FIGURE 63

1.750 in	SB	44.45 mm
2.542 in	HD	64.57 mm
11.500 in	EL	292.10 mm

ROCKWELL 5NPS28-1

IDLI GROUP 87408

SPLINED PLUG WITH 10 SPLINES
SEE FIGURE 63

1.750 in	SB	44.45 mm
2.635 in	HD	66.93 mm
7.000 in	EL	177.80 mm

BORG-WARNR 16545J
ROCKWELL 62NPS28-32

IDLI GROUP 87409

SPLINED PLUG WITH 16 SPLINES
SEE FIGURE 63

1.750 in	SB	44.45 mm
2.770 in	HD	70.36 mm
6.875 in	EL	174.63 mm

ROCKWELL 50NPS28-31

IDLI GROUP 87410

SPLINED PLUG WITH 10 SPLINES
SEE FIGURE 63

1.750 in	SB	44.45 mm
2.770 in	HD	70.36 mm
7.000 in	EL	177.80 mm

BORG-WARNR 18872J
BORG-WARNR 3308J
BORG-WARNR 5004J
ROCKWELL 62NPS28-13

IDLI GROUP 87411

SPLINED PLUG WITH 10 SPLINES
SEE FIGURE 64

1.750 in	SB	44.45 mm
2.770 in	HD	70.36 mm
13.000 in	EL	330.20 mm

ROCKWELL 5NPS28-24

IDLI GROUP 87412

SPLINED PLUG WITH 16 SPLINES
SEE FIGURE 63

1.750 in	SB	44.45 mm
2.820 in	HD	71.63 mm
6.875 in	EL	174.63 mm

ROCKWELL 50NPS28-26

IDLI GROUP 87413

SPLINED PLUG WITH 10 SPLINES
SEE FIGURE 63

1.750 in	SB	44.45 mm
2.820 in	HD	71.63 mm
7.000 in	EL	177.80 mm

BORG-WARNR 3670J
BORG-WARNR 4151J
BORG-WARNR 4642J
IHC 38781H
ROCKWELL 62NPS28-14

ENGINEERING CATALOGS MUST BE CONSULTED FOR DETAILS NOT INCLUDED IN THIS GUIDE. SPECIFIC DESIGNS, MATERIAL CONTENT, TOLERANCES LUBE FITTINGS AND OTHER DIMENSIONS ARE INTENTIONALLY OMITTED HERE.

IDLI GROUP 87414

SPLINED PLUG WITH 16 SPLINES
SEE FIGURE 63

1.750 in	SB	44.45 mm
3.020 in	HD	76.71 mm
7.375 in	EL	187.33 mm

ROCKWELL 50NPS28-28

IDLI GROUP 87415

SPLINED PLUG WITH 10 SPLINES
SEE FIGURE 63

1.750 in	SB	44.45 mm
3.020 in	HD	76.71 mm
8.000 in	EL	203.20 mm

ROCKWELL 5NPS28-13

IDLI GROUP 87416

SPLINED BEARING STUB
WITH 10 SPLINES
SEE FIGURE 65

1.750 in	SB	44.45 mm
3.242 in	HD	82.35 mm
4.344 in	EL	110.33 mm

ROCKWELL 16N53-141
SPICER 5-53-141

IDLI GROUP 87417

SPLINED PLUG WITH 10 SPLINES
SEE FIGURE 63

1.750 in	SB	44.45 mm
3.270 in	HD	83.06 mm
7.125 in	EL	180.98 mm

BORG-WARNR 13781J
ROCKWELL 62NPS28-15

IDLI GROUP 87418

SPLINED PLUG WITH 16 SPLINES
SEE FIGURE 63

1.750 in	SB	44.45 mm
3.320 in	HD	84.33 mm
6.750 in	EL	171.45 mm

ROCKWELL 50NPS28-44
SPICER 4-40-141
SPICER 4-40-461

IDLI GROUP 87419

SPLINED PLUG WITH 16 SPLINES
slight variation from above group
SEE FIGURE 63

1.750 in	SB	44.45 mm
3.320 in	HD	84.33 mm
6.750 in	EL	171.45 mm

ROCKWELL 50NPS28-59

IDLI GROUP 87420

SPLINED PLUG WITH 16 SPLINES
SEE FIGURE 63

1.750 in	SB	44.45 mm
3.320 in	HD	84.33 mm
6.781 in	EL	172.25 mm

ROCKWELL 155N40-461
SPICER 4-40-461
SPICER 4-40-761

IDLI GROUP 87421

SPLINED PLUG WITH 10 SPLINES
SEE FIGURE 63

1.750 in	SB	44.45 mm
3.320 in	HD	84.33 mm
7.125 in	EL	180.98 mm

BORG-WARNR 9146J
ROCKWELL 62NPS28-8

IDLI GROUP 87422

SPLINED PLUG WITH 16 SPLINES
SEE FIGURE 63

1.750 in	SB	44.45 mm
3.320 in	HD	84.33 mm
7.500 in	EL	190.50 mm

ROCKWELL 50NPS28-38

IDLI GROUP 87423

SPLINED PLUG WITH 10 SPLINES
SEE FIGURE 63

1.750 in	SB	44.45 mm
3.320 in	HD	84.33 mm
7.938 in	EL	201.61 mm

BORG-WARNR 17233J
ROCKWELL 62NPS28-5

IDLI GROUP 87424

SPLINED BEARING STUB
WITH 16 INVOLUTED SPLINES
SEE FIGURE 67

1.750 in	SB	44.45 mm
3.320 in	HD	84.33 mm
8.563 in	EL	217.49 mm

ROCKWELL 155N53-61
SPICER 4-53-61

IDLI GROUP 87425

SPLINED PLUG WITH 16 SPLINES
SEE FIGURE 63

1.750 in	SB	44.45 mm
3.320 in	HD	84.33 mm
8.750 in	EL	222.25 mm

ROCKWELL 50NPS28-45
SPICER 4-40-481

IDLI GROUP 87426

SPLINED PLUG WITH 16 SPLINES
slight variation from above group
SEE FIGURE 63

1.750 in	SB	44.45 mm
3.320 in	HD	84.33 mm
8.750 in	EL	222.25 mm

ROCKWELL 5NPS28-40

IDLI GROUP 87427

SPLINED PLUG WITH 16 SPLINES
slight variation from above group
SEE FIGURE 63

1.750 in	SB	44.45 mm
3.320 in	HD	84.33 mm
8.750 in	EL	222.25 mm

ROCKWELL 50NPS28-60

IDLI GROUP 87428

SPLINED PLUG WITH 16 SPLINES
SEE FIGURE 63

1.750 in	SB	44.45 mm
3.320 in	HD	84.33 mm
9.625 in	EL	244.48 mm

BORG-WARNR 17072J
ROCKWELL 62NPS28-4

IDLI GROUP 87429

SPLINED PLUG
WITH 32 INVOLUTED SPLINES
SEE FIGURE 67

1.750 in	SB	44.45 mm
3.303 in	HD	83.90 mm
11.438 in	EL	290.51 mm

CLEVE.STL. U56-102
ROCKWELL BS-S28-67

IDLI GROUP 87430

SPLINED PLUG WITH 16 SPLINES
SEE FIGURE 64

1.750 in	SB	44.45 mm
3.320 in	HD	84.33 mm
9.813 in	EL	249.24 mm

ROCKWELL 50NPS28-43
SPICER 4-53-111

IDLI GROUP 87431

SPLINED PLUG WITH 16 SPLINES
SEE FIGURE 63

1.750 in	SB	44.45 mm
3.844 in	HD	97.64 mm
6.938 in	EL	176.21 mm

ROCKWELL 50NPS28-47
SPICER 4-40-431

IDLI GROUP 87432

SPLINED PLUG WITH 10 SPLINES
SEE FIGURE 63

1.750 in	SB	44.45 mm
3.844 in	HD	97.64 mm
7.250 in	EL	184.15 mm

BORG-WARNR 9147J
ROCKWELL 62NPS28-1

IDLI GROUP 87433

SPLINED PLUG WITH 16 SPLINES
SEE FIGURE 64

1.750 in	SB	44.45 mm
3.844 in	HD	97.64 mm
8.063 in	EL	204.79 mm

ROCKWELL 50NPS28-51

IDLI GROUP 87434

SPLINED PLUG WITH 10 SPLINES
SEE FIGURE 64

1.750 in	SB	44.45 mm
3.844 in	HD	97.64 mm
8.313 in	EL	211.14 mm

BORG-WARNR 11454J
ROCKWELL 62NPS28-12

IDLI GROUP 87435

SPLINED PLUG WITH 16 SPLINES
SEE FIGURE 63

1.750 in	SB	44.45 mm
3.844 in	HD	97.64 mm
8.875 in	EL	225.43 mm

ROCKWELL 50NPS28-48

IDLI GROUP 87436

SPLINED BEARING STUB
WITH 16 INVOLUTED SPLINES
SEE FIGURE 67

1.750 in	SB	44.45 mm
3.846 in	HD	97.69 mm
8.563 in	EL	217.49 mm

CLEVE.STL. U76-103
ROCKWELL BS-S28-62

IDLI GROUP 87437

SPLINED PLUG WITH 16 SPLINES
SEE FIGURE 64

1.750 in	SB	44.45 mm
4.320 in	HD	109.73 mm
8.500 in	EL	215.90 mm

ROCKWELL 50NPS28-57

IDLI GROUP 87438

SPLINED PLUG
WITH 32 INVOLUTED SPLINES
SEE FIGURE 67

1.750 in	SB	44.45 mm
4.320 in	HD	109.73 mm
12.063 in	EL	306.39 mm

ROCKWELL BS-S28-67A

IDLI GROUP 87439

COMPANION FLANGE
WITH 10 SPLINES
SEE FIGURE 33

1.750 in	SB	44.45 mm
2.750 in	EC	69.85 mm
2.250 in	HD	57.15 mm
6.000 in	OD	152.40 mm

ROCKWELL 2WCS28-78

IDLI GROUP 87440

COMPANION FLANGE
WITH TAPERED BORE
SEE FIGURE 34

1.750 in	SB	44.45 mm
3.500 in	EC	88.90 mm
2.500 in	HD	63.50 mm
4.563 in	OD	115.89 mm

ROCKWELL 2WCT28-15

IDLI GROUP 87441

SPLINED BEARING STUB
WITH 20 INVOLUTED SPLINES
SEE FIGURE 66

1.750 in	SB	44.45 mm
4.438 in	EC	112.71 mm
2.380 in	HD	60.45 mm

ROCKWELL BSA-S28-5

BD=BEARING DIAMETER (outside)   **BW**=BEARING WIDTH   **CB**=CROSS LENGTH WITH BEARINGS   **CC**=CENTER TO CENTER   **CD**=CROSS DIAMETER
CL=CROSS LENGTH WITHOUT BEARINGS   **EC**=END TO CENTER or FACE TO FACE   **EE**=END TO END   **EL**=EFFECTIVE LENGTH
HD=HUB DIAMETER (or insert)   **IE**=INSIDE OF EARS (or recess)   **OD**=OUTSIDE DIAMETER   **OE**=OUTSIDE OF EARS   **SB**=SPLINE OR BORE SIZE

IDLI GUIDE · COPYRIGHT © INTERCHANGE, INC. · ST. LOUIS PARK, MN. 55416 USA

CAUTION: BE SURE TO REFER TO ENGINEERING CATALOGS FOR SPECIAL APPLICATIONS THAT REQUIRE SPECIFIC MATERIAL CONTENT, TOLERANCES, ETC. SEE FOOTNOTE.

IDLI GROUP 87442

SPLINED BEARING STUB WITH 10 SPLINES
SEE FIGURE 66

1.750 in	SB	44.45 mm
4.500 in	EC	114.30 mm
2.820 in	HD	71.63 mm

ROCKWELL BS-S28-36

IDLI GROUP 87443

SPLINED BEARING STUB WITH 10 SPLINES
SEE FIGURE 66

1.750 in	SB	44.45 mm
4.531 in	EC	115.10 mm
3.270 in	HD	83.06 mm

BORG-WARNR 13818J
ROCKWELL BS-S28-12

IDLI GROUP 87444

SPLINED BEARING STUB WITH 10 SPLINES
SEE FIGURE 66

1.750 in	SB	44.45 mm
4.531 in	EC	115.10 mm
3.320 in	HD	84.33 mm

BORG-WARNR 10504J
ROCKWELL BS-S28-9

IDLI GROUP 87445

SPLINED BEARING STUB WITH 10 SPLINES
SEE FIGURE 65

1.750 in	SB	44.45 mm
4.813 in	EC	122.24 mm
2.770 in	HD	70.36 mm

ROCKWELL BS-S28-33

IDLI GROUP 87446

SPLINED BEARING STUB WITH 6 SPLINES
SEE FIGURE 66

1.750 in	SB	44.45 mm
5.375 in	EC	136.53 mm
2.380 in	HD	60.45 mm

ROCKWELL BSA-S28-3

IDLI GROUP 87447

SPLINED BEARING STUB WITH 10 SPLINES
SEE FIGURE 65

1.750 in	SB	44.45 mm
6.563 in	EC	166.69 mm
3.320 in	HD	84.33 mm

ROCKWELL BS-S28-17

IDLI GROUP 87448

SPLINED BEARING STUB WITH 10 SPLINES
SEE FIGURE 65

1.750 in	SB	44.45 mm
7.875 in	EC	200.03 mm
3.844 in	HD	97.64 mm

BORG-WARNR 14059J
ROCKWELL BS-S28-28

IDLI GROUP 87449

SPLINED BEARING STUB WITH 10 SPLINES
SEE FIGURE 65

1.750 in	SB	44.45 mm
7.906 in	EC	200.82 mm
2.820 in	HD	71.63 mm

BORG-WARNR 5814J
ROCKWELL BS-S28-10
SPICER 4-54-381

IDLI GROUP 87450

SPLINED BEARING STUB WITH 10 SPLINES
SEE FIGURE 65

1.750 in	SB	44.45 mm
8.063 in	EC	204.79 mm
3.320 in	HD	84.33 mm

BORG-WARNR 9283J
ROCKWELL BS-S28-8

IDLI GROUP 87451

SPLINED BEARING STUB WITH 16 SPLINES
SEE FIGURE 65

1.750 in	SB	44.45 mm
8.063 in	EC	204.79 mm
3.320 in	HD	84.33 mm

BORG-WARNR 12089J
ROCKWELL BS-S28-4

IDLI GROUP 87452

SPLINED BEARING STUB WITH 16 SPLINES
SEE FIGURE 65

1.750 in	SB	44.45 mm
8.313 in	EC	211.14 mm
3.792 in	HD	96.32 mm

BORG-WARNR 13457J
ROCKWELL BS-S28-6

IDLI GROUP 87453

SPLINED BEARING STUB WITH 10 SPLINES
SEE FIGURE 65

1.750 in	SB	44.45 mm
8.313 in	EC	211.14 mm
3.844 in	HD	97.64 mm

BORG-WARNR 10834J
ROCKWELL BS-S28-11

IDLI GROUP 87454

SPLINED BEARING STUB WITH 20 INVOLUTED SPLINES
SEE FIGURE 66

1.750 in	SB	44.45 mm
8.938 in	EC	227.01 mm
2.380 in	HD	60.45 mm

ROCKWELL BS-S28-42

IDLI GROUP 87455

SPLINED BEARING STUB WITH 16 SPLINES
SEE FIGURE 66

1.750 in	SB	44.45 mm
9.813 in	EC	249.24 mm
3.320 in	HD	84.33 mm

ROCKWELL 50NPS28-43

IDLI GROUP 87456

SPLINED BEARING STUB WITH 6 SPLINES
SEE FIGURE 66

1.750 in	SB	44.45 mm
9.875 in	EC	250.83 mm
2.380 in	HD	60.45 mm

ROCKWELL BSA-S28-2

IDLI GROUP 87457

CENTER BEARING
SEE FIGURE 56

1.772 in	SB	45.00 mm
2.688 in	EC	68.26 mm
7.625 in	CC	193.68 mm
1.063 in	BW	26.99 mm

AEC	. . . .	18
AEC	. . . .	AE18
ALLOY	. . . .	4006
AMER.PARTS	. . . .	2-6040
ANCHOR	. . . .	35-6040
ANCHOR	. . . .	6040
ANCHORDOAN	. . . .	35-6040
AUTOCAR	. . . .	3AD0970
BORG-WARNR	. . . .	35-6040
CLEVE.MOT.	. . . .	DS401KF
CLEVE.STL.	. . . .	UCB103
DETROIT	. . . .	MB608
DORMAN	. . . .	662-018
DORMAN	. . . .	DS18
EATON	. . . .	MM6040
FED.BEAR.	. . . .	HBF88509
FEDERAL	. . . .	HBF88509
FED.MOGUL	. . . .	HB88509
FORD	. . . .	B9TT4800A
FORD	. . . .	D4TZ4800F
GREEN BALL	. . . .	RSB9
IHC	. . . .	283067C91
L & S BRG.	. . . .	PBS11
MCQUAY-NOR	. . . .	MM6040
MOPAR	. . . .	D608
MOTOR MAST	. . . .	DS18
REPUBLIC	. . . .	35-6049
ROCKWELL	. . . .	CB310007-1X
SKF	. . . .	PBS11
SKF	. . . .	PNS11
SPICER	. . . .	210084-2X
SPICER	. . . .	210207-1X

SPICER	. . . .	210969X
TRW	. . . .	20809
WESCO	. . . .	960-006

IDLI GROUP 87458

CENTER BEARING
slight variation from above group
SEE FIGURE 56

1.772 in	SB	45.00 mm
2.688 in	EC	68.26 mm
7.625 in	CC	193.68 mm
1.063 in	BW	26.99 mm

BUTLER	. . . .	510084-2X
CHRYSLER	. . . .	1667743
CLEVE.STL.	. . . .	UCB104
FORD	. . . .	D4TZ4800E
FORD	. . . .	D6TJ4800CA
NEAPCO	. . . .	81-4011
NEAPCO	. . . .	DS401KF
NEAPCO	. . . .	DS401KF1
ROCKWELL	. . . .	CB210084-2X
SPICER	. . . .	210084-2X
WESCO	. . . .	960-006
WHITE	. . . .	692296
WHITE	. . . .	692297

IDLI GROUP 87459

CENTER BEARING
slight variation from above group
SEE FIGURE 56

1.772 in	SB	45.00 mm
2.688 in	EC	68.26 mm
7.625 in	CC	193.68 mm
1.063 in	BW	26.99 mm

AEC	. . . .	19
AEC	. . . .	AE19
ALLOY	. . . .	4005
AMER.PARTS	. . . .	2-6041
ANCHOR	. . . .	35-6039
ANCHOR	. . . .	35-6041
ANCHOR	. . . .	6041
ANCHORDOAN	. . . .	35-6041
BORG-WARNR	. . . .	35-6041
BUTLER	. . . .	510230-1X
CLEVE.MOT.	. . . .	DS401KG
CLEVE.STL.	. . . .	UCB107
CMP	. . . .	DS401KG
DETROIT	. . . .	MB609
DORMAN	. . . .	662-019
DORMAN	. . . .	DS19
EATON	. . . .	MM6041
FED.BEAR.	. . . .	HB88509
FEDERAL	. . . .	HB88509
GMC	. . . .	2351949
GREEN BALL	. . . .	RSB8
KENBAR	. . . .	5841-5
LEMPCO	. . . .	35-6041
MCQUAY-NOR	. . . .	MM6041
MOPAR	. . . .	D609
MOTOR MAST	. . . .	DS19
NEAPCO	. . . .	81-4020
NEAPCO	. . . .	81-4021
NEAPCO	. . . .	DS401KG
NEAPCO	. . . .	DS401KG1
PILOT	. . . .	H-1012
REPUBLIC	. . . .	35-6046

ENGINEERING CATALOGS MUST BE CONSULTED FOR DETAILS NOT INCLUDED IN THIS GUIDE. SPECIFIC DESIGNS, MATERIAL CONTENT, TOLERANCES LUBE FITTINGS AND OTHER DIMENSIONS ARE INTENTIONALLY OMITTED HERE.

IDLI GUIDE   COPYRIGHT ® INTERCHANGE, INC.   ST. LOUIS PARK, MN. 55416 USA

Column 1

ROCKWELL CB210130-1X
SKF 1700549
SPICER 210130-1X
SPICER 210209-1X
TRW 20810
WESCO 960-005

IDLI GROUP 87460

CENTER BEARING
slight variation from above group
SEE FIGURE 56

1.772 in	SB	45.00 mm
2.688 in	EC	68.26 mm
7.625 in	CC	193.68 mm
1.063 in	BW	26.99 mm

BUTLER 510084-2X3
BUTLER 510084-2XWF
BUTLER 510207-1X
CLEVE.STL. UCB105
CMP DS401KF
CMP DS401KF2
CMP DS401KF3
IHC 466505C91
NEAPCO 81-4012
NEAPCO DS401KF2
ROCKWELL CB210207-1X
SPICER 210207-1X
WESCO 960-006

IDLI GROUP 87461

CENTER BEARING
slight variation from above group
SEE FIGURE 56

1.772 in	SB	45.00 mm
2.688 in	EC	68.26 mm
7.625 in	CC	193.68 mm
1.063 in	BW	26.99 mm

AEC 18
BUTLER 510084-2X
BUTLER 510084-2X3
BUTLER 510084-2XWF
CHRYSLER 1667748
CHRYSLER 1667749
CLEVE.STL. UCB109
CMP DS401KF
CMP DS401KF1
CMP DS401KF3
CRANE CAR. 53-17
FORD D9HZ4800A
IHC 283067
MOPAR 608
NEAPCO 81-4013
NEAPCO DS401KF3
ROCKWELL CB310007-2X
SPICER 210084-2X
SPICER 210207-1X
SPICER 211172-1X
WESCO 960-006
WHITE 622296

Column 2

IDLI GROUP 87462

ROUND TUBING
SEE FIGURE 75

1.808 in	SB	45.91 mm
72.000 in	EE	1828.80 mm
1.938 in	OD	49.21 mm

ROCKWELL RT1937-16-72

IDLI GROUP 87463

ROUND TUBING
SEE FIGURE 75

1.808 in	SB	45.91 mm
108.000 in	EE	2743.20 mm
1.938 in	OD	49.21 mm

ROCKWELL RT1937-16-108

IDLI GROUP 87464

ROUND TUBING
SEE FIGURE 75

1.834 in	SB	46.58 mm
72.000 in	EL	1828.80 mm
2.000 in	OD	50.80 mm

BORG-WARNR 4365
BORG-WARNR 4365J-7200
ROCKWELL RT2-14-72

IDLI GROUP 87465

ROUND TUBING
SEE FIGURE 75

1.834 in	SB	46.58 mm
108.000 in	EL	2743.20 mm
2.000 in	OD	50.80 mm

BORG-WARNR 4365J10800
ROCKWELL RT2-14-108

IDLI GROUP 87466

ROUND TUBING
SEE FIGURE 75

1.870 in	SB	47.50 mm
72.000 in	EL	1828.80 mm
2.000 in	OD	50.80 mm

BORG-WARNR 4367J7200
CLEVE.STL. 82-16-74
ROCKWELL RT2-16-72

IDLI GROUP 87467

ROUND TUBING
SEE FIGURE 75

1.870 in	SB	47.50 mm
108.000 in	EL	2743.20 mm
2.000 in	OD	50.80 mm

BORG-WARNR 4367J10800
CLEVE.STL. 82-16-108
ROCKWELL RT2-16-108

IDLI GROUP 87468

ROUND TUBING
SEE FIGURE 75

1.885 in	SB	47.88 mm
72.000 in	EE	1828.80 mm
2.125 in	OD	53.98 mm

ROCKWELL RT2125-11-72

Column 3

IDLI GROUP 87469

ROUND TUBING
SEE FIGURE 75

1.885 in	SB	47.88 mm
108.000 in	EE	2743.20 mm
2.125 in	OD	53.98 mm

ROCKWELL RT2125-11-108

IDLI GROUP 87470

SPLINED BEARING STUB
WITH 10 SPLINES
SEE FIGURE 65

1.953 in	SB	49.61 mm
4.438 in	EC	112.71 mm
3.198 in	HD	81.23 mm

ROCKWELL BS-S31-19

IDLI GROUP 87471

SPLINED BEARING STUB
WITH 10 SPLINES
SEE FIGURE 65

1.953 in	SB	49.61 mm
4.438 in	EC	112.71 mm
3.242 in	HD	82.35 mm

ROCKWELL BS-S31-12

IDLI GROUP 87472

SPLINED BEARING STUB
WITH 10 SPLINES
SEE FIGURE 65

1.953 in	SB	49.61 mm
4.438 in	EC	112.71 mm
3.320 in	HD	84.33 mm

ROCKWELL BS-S31-14
SPICER 4-54-481

IDLI GROUP 87473

SPLINED BEARING STUB
WITH 10 SPLINES
SEE FIGURE 65

1.953 in	SB	49.61 mm
4.438 in	EC	112.71 mm
3.742 in	HD	95.05 mm

ROCKWELL BS-S31-13

IDLI GROUP 87474

SPLINED BEARING STUB
WITH 10 SPLINES
SEE FIGURE 65

1.953 in	SB	49.61 mm
5.156 in	EC	130.97 mm
3.792 in	HD	96.32 mm

ROCKWELL BS-S31-9

IDLI GROUP 87475

SPLINED BEARING STUB
WITH 10 SPLINES
SEE FIGURE 66

1.953 in	SB	49.61 mm
5.313 in	EC	134.94 mm
3.635 in	HD	92.33 mm

BORG-WARNR 16757J
ROCKWELL BS-S31-4

Column 4

IDLI GROUP 87476

SPLINED BEARING STUB
WITH 10 SPLINES
SEE FIGURE 66

1.953 in	SB	49.61 mm
5.313 in	EC	134.94 mm
3.742 in	HD	95.05 mm

ROCKWELL BS-S31-21

IDLI GROUP 87477

SPLINED BEARING STUB
WITH 10 SPLINES
SEE FIGURE 66

1.953 in	SB	49.61 mm
5.313 in	EC	134.94 mm
3.792 in	HD	96.32 mm

BORG-WARNR 16418J
ROCKWELL BS-S31-3

IDLI GROUP 87478

SPLINED BEARING STUB
WITH 10 SPLINES
SEE FIGURE 65

1.953 in	SB	49.61 mm
10.750 in	EC	273.05 mm
3.792 in	HD	96.32 mm

BORG-WARNR 8424J
ROCKWELL BS-S31-1

IDLI GROUP 87479

ROUND TUBING
SEE FIGURE 75

1.959 in	SB	49.76 mm
72.000 in	EE	1828.80 mm
2.125 in	OD	53.98 mm

ROCKWELL RT2125-14-72

IDLI GROUP 87480

ROUND TUBING
SEE FIGURE 75

1.959 in	SB	49.76 mm
108.000 in	EE	2743.20 mm
2.125 in	OD	53.98 mm

ROCKWELL RT2125-14-108

IDLI GROUP 87481

CENTER BEARING
SEE FIGURE 56

1.969 in	SB	50.00 mm
2.781 in	EC	70.65 mm
7.625 in	CC	193.68 mm
1.188 in	BW	30.16 mm

AEC 20
AEC 326
AEC AE20
ALLOY 4007
AMER.PARTS............2-6513
ANCHORDOAN 35-6048
AUTOCAR 3AA
AUTOCAR 3AA0970
BEAR.INC. B1-501KF
BORG-WARNR 114-644
BORG-WARNR 35-6048
BUTLER 510121-1X
CHRYSLER 1932710

BD = BEARING DIAMETER (outside)   BW = BEARING WIDTH   CB = CROSS LENGTH WITH BEARINGS   CC = CENTER TO CENTER   CD = CROSS DIAMETER
CL = CROSS LENGTH WITHOUT BEARINGS   EC = END TO CENTER or FACE TO FACE   EE = END TO END   EL = EFFECTIVE LENGTH
HD = HUB DIAMETER (or insert)   IE = INSIDE OF EARS (or recess)   OD = OUTSIDE DIAMETER   OE = OUTSIDE OF EARS   SB = SPLINE OR BORE SIZE

IDLI GUIDE   COPYRIGHT ® INTERCHANGE, INC.   ST. LOUIS PARK, MN. 55416 USA

CAUTION: BE SURE TO REFER TO ENGINEERING CATALOGS FOR SPECIAL APPLICATIONS THAT REQUIRE SPECIFIC MATERIAL CONTENT, TOLERANCES, ETC. SEE FOOTNOTE.

CLEVE.MOT.	DS501KF
CLEVE.STL.	CCB101
CMP	DS501KF
CRANE CAR.	53-18
DORMAN	662-020
DORMAN	DS20
EATON	MM6050
FED.BEAR.	BF88510
FED.BEAR.	HBF88510
FED.MOGUL	HB88510
FORD	C1TT4840A
FORD	C8HA4800B
FORD	D4TZ4800G
FORD	D4TZ4800H
GMC	2406989
GREEN BALL	RSB10
IHC	283069C91
IHC	283069C92
KENBAR	6060-10
L & S BRG.	PBS12
LEMPCO	35-6025
MACK	35MU31
MACK	35MU32P1
MACK	35MUA31
MCQUAY-NOR	MM6050
MOPAR	611
MOPAR	D611
MOTOR MAST	7244DC
MOTOR MAST	DS20
NEAPCO	81-5010
NEAPCO	82-5010
NEAPCO	DS501KF
PILOT	H-1014
REPCO	5-326X
REPUBLIC	35-6050
REPUBLIC	CB1880C
ROCKWELL	CB210121-1X
SKF	1700810
SPICER	210121-1X
SPICER	5-326X
TRW	20182
TRW	20812
WHITE	515693
WHITE	515697

IDLI GROUP 87482

SPLINED BEARING STUB WITH 10 SPLINES
SEE FIGURE 65

1.969 in	SB	50.01 mm
3.198 in	HD	81.23 mm
5.156 in	EL	130.97 mm

ROCKWELL 17N53-151
SPICER 6-53-151

IDLI GROUP 87483

SPLINED BEARING STUB WITH 10 SPLINES
SEE FIGURE 65

1.969 in	SB	50.01 mm
3.742 in	HD	95.05 mm
5.156 in	EL	130.97 mm

ROCKWELL 17N53-201
SPICER 6-53-201

IDLI GROUP 87484

ROUND TUBING
SEE FIGURE 75

1.995 in	SB	50.67 mm
72.000 in	EE	1828.80 mm
2.125 in	OD	53.98 mm

ROCKWELL RT2125-16-72

IDLI GROUP 87485

ROUND TUBING
SEE FIGURE 75

1.995 in	SB	50.67 mm
108.000 in	EE	2743.20 mm
2.125 in	OD	53.98 mm

ROCKWELL RT2125-16-108

IDLI GROUP 87486

SPLINED PLUG WITH 10 SPLINES
SEE FIGURE 63

2.000 in	SB	50.80 mm
2.542 in	HD	64.57 mm
8.000 in	EL	203.20 mm

ROCKWELL 5NPS32-12

IDLI GROUP 87487

SPLINED PLUG WITH 16 SPLINES
SEE FIGURE 63

2.000 in	SB	50.80 mm
2.742 in	HD	69.65 mm
8.000 in	EL	203.20 mm

ROCKWELL 60NPS32-26

IDLI GROUP 87488

SPLINED PLUG WITH 10 SPLINES
SEE FIGURE 63

2.000 in	SB	50.80 mm
2.770 in	HD	70.36 mm
8.500 in	EL	215.90 mm

BORG-WARNR 2176J
ROCKWELL 72NPS32-53

IDLI GROUP 87489

SPLINED PLUG WITH 16 SPLINES
SEE FIGURE 63

2.000 in	SB	50.80 mm
2.814 in	HD	71.48 mm
8.000 in	EL	203.20 mm

ROCKWELL 60NPS32-25

IDLI GROUP 87491

SPLINED PLUG WITH 16 SPLINES
SEE FIGURE 63

2.000 in	SB	50.80 mm
3.020 in	HD	76.71 mm
8.000 in	EL	203.20 mm

ROCKWELL 6NPS32-15

IDLI GROUP 87492

SPLINED PLUG WITH 10 SPLINES
SEE FIGURE 63

2.000 in	SB	50.80 mm
3.020 in	HD	76.71 mm
9.188 in	EL	233.36 mm

ROCKWELL 6NPS32-31

IDLI GROUP 87493

SPLINED PLUG WITH 16 SPLINES
SEE FIGURE 63

2.000 in	SB	50.80 mm
3.020 in	HD	76.71 mm
10.750 in	EL	273.05 mm

ROCKWELL 6NPS32-29

IDLI GROUP 87494

SPLINED PLUG WITH 16 SPLINES
SEE FIGURE 63

2.000 in	SB	50.80 mm
3.020 in	HD	76.71 mm
10.875 in	EL	276.23 mm

ROCKWELL 6NPS32-32

IDLI GROUP 87495

SPLINED PLUG WITH 10 SPLINES
SEE FIGURE 63

2.000 in	SB	50.80 mm
3.020 in	HD	76.71 mm
11.500 in	EL	292.10 mm

BORG-WARNR 18613J
ROCKWELL 6NPS32-10

IDLI GROUP 87496

SPLINED PLUG WITH 10 SPLINES
SEE FIGURE 63

2.000 in	SB	50.80 mm
3.135 in	HD	79.63 mm
8.500 in	EL	215.90 mm

BORG-WARNR 22167J
ROCKWELL 72NPS32-100

IDLI GROUP 87497

SPLINED PLUG WITH 16 SPLINES
SEE FIGURE 63

2.000 in	SB	50.80 mm
3.198 in	HD	81.23 mm
8.563 in	EL	217.49 mm

ROCKWELL 16N40-481
SPICER 5-40-1121
SPICER 5-40-481
SPICER 5-40-701

IDLI GROUP 87498

SPLINED PLUG WITH 16 SPLINES
SEE FIGURE 63

2.000 in	SB	50.80 mm
3.198 in	HD	81.23 mm
9.438 in	EL	239.71 mm

ROCKWELL 72NPS32-83

IDLI GROUP 87499

SPLINED PLUG WITH 16 SPLINES
SEE FIGURE 63

2.000 in	SB	50.80 mm
3.020 in	HD	76.71 mm
8.000 in	EL	203.20 mm

ROCKWELL 16N40-511
ROCKWELL 6NPS32
SPICER 5-40-511

IDLI GROUP 87500

SPLINED PLUG WITH 16 SPLINES
SEE FIGURE 63

2.000 in	SB	50.80 mm
3.198 in	HD	81.23 mm
10.875 in	EL	276.23 mm

ROCKWELL 72NPS32-84

IDLI GROUP 87501

SPLINED PLUG WITH 16 SPLINES
SEE FIGURE 63

2.000 in	SB	50.80 mm
3.200 in	HD	81.28 mm
9.063 in	EL	230.19 mm

ROCKWELL 60NPS32-30

IDLI GROUP 87502

SPLINED PLUG WITH 16 SPLINES
SEE FIGURE 63

| 2.000 in | SB | 50.80 mm |
| 3.242 in | HD | 82.35 mm |

ROCKWELL 72NPS32-87

IDLI GROUP 87503

SPLINED PLUG WITH 16 SPLINES
SEE FIGURE 63

2.000 in	SB	50.80 mm
3.242 in	HD	82.35 mm
8.563 in	EL	217.49 mm

ROCKWELL 16N40-671
SPICER 5-40-1151
SPICER 5-40-671

IDLI GROUP 87504

SPLINED PLUG WITH 16 SPLINES
SEE FIGURE 63

2.000 in	SB	50.80 mm
3.242 in	HD	82.35 mm
9.688 in	EL	246.06 mm

ROCKWELL 16N40-741
SPICER 5-40-1011
SPICER 5-40-741

ENGINEERING CATALOGS MUST BE CONSULTED FOR DETAILS NOT INCLUDED IN THIS GUIDE. SPECIFIC DESIGNS, MATERIAL CONTENT, TOLERANCES LUBE FITTINGS AND OTHER DIMENSIONS ARE INTENTIONALLY OMITTED HERE.

IDLI GUIDE   COPYRIGHT ® INTERCHANGE, INC.   ST. LOUIS PARK, MN. 55416 USA

IDLI GROUP 87505

SPLINED PLUG WITH 16 SPLINES
SEE FIGURE 63

2.000 in	SB	50.80 mm
3.242 in	HD	82.35 mm
9.688 in	EL	246.06 mm

ROCKWELL 16N40-1011

IDLI GROUP 87506

SPLINED PLUG WITH 16 SPLINES
SEE FIGURE 63

2.000 in	SB	50.80 mm
3.242 in	HD	82.35 mm
10.688 in	EL	271.46 mm

ROCKWELL 72NPS32-96

IDLI GROUP 87507

SPLINED PLUG WITH 16 SPLINES
SEE FIGURE 63

2.000 in	SB	50.80 mm
3.242 in	HD	82.35 mm
14.500 in	EL	368.30 mm

ROCKWELL 72NPS32-87

IDLI GROUP 87508

SPLINED PLUG WITH 16 SPLINES
SEE FIGURE 63

2.000 in	SB	50.80 mm
3.270 in	HD	83.06 mm
7.938 in	EL	201.61 mm

ROCKWELL PS32-16-31

IDLI GROUP 87509

SPLINED PLUG WITH 10 SPLINES
SEE FIGURE 63

2.000 in	SB	50.80 mm
3.270 in	HD	83.06 mm
8.000 in	EL	203.20 mm

ROCKWELL 72NPS32-14

IDLI GROUP 87510

SPLINED PLUG WITH 16 SPLINES
SEE FIGURE 63

2.000 in	SB	50.80 mm
3.270 in	HD	83.06 mm
8.000 in	EL	203.20 mm

ROCKWELL 60NPS32-24

IDLI GROUP 87511

SPLINED PLUG WITH 10 SPLINES
SEE FIGURE 63

2.000 in	SB	50.80 mm
3.270 in	HD	83.06 mm
8.500 in	EL	215.90 mm

BORG-WARNR 10011J
ROCKWELL 72NPS32-13

IDLI GROUP 87512

SPLINED PLUG WITH 10 SPLINES
SEE FIGURE 64

2.000 in	SB	50.80 mm
3.270 in	HD	83.06 mm
9.375 in	EL	238.13 mm

BORG-WARNR 13819J
ROCKWELL 72NPS32-17

IDLI GROUP 87513

SPLINED PLUG
WITH 15 SPLINES IN 16 SPLINE SPACES
SEE FIGURE 63

2.000 in	SB	50.80 mm
3.270 in	HD	83.06 mm
9.375 in	EL	238.13 mm

BORG-WARNR 14785J
ROCKWELL 72NPS32-35

IDLI GROUP 87514

SPLINED PLUG WITH 10 SPLINES
SEE FIGURE 63

2.000 in	SB	50.80 mm
3.270 in	HD	83.06 mm
9.438 in	EL	239.71 mm

BORG-WARNR 21535J
BORG-WARNR 22898J
ROCKWELL 72NPS32-73

IDLI GROUP 87515

SPLINED PLUG WITH 10 SPLINES
SEE FIGURE 64

2.000 in	SB	50.80 mm
3.270 in	HD	83.06 mm
10.750 in	EL	273.05 mm

BORG-WARNR 14-114J
BORG-WARNR 14144J
ROCKWELL 72NPS32-64

IDLI GROUP 87516

SPLINED PLUG WITH 16 SPLINES
SEE FIGURE 63

2.000 in	SB	50.80 mm
3.270 in	HD	83.06 mm
10.875 in	EL	276.23 mm

ROCKWELL 6NPS32-38

IDLI GROUP 87517

SPLINED PLUG WITH 16 SPLINES
slight variation from above group
SEE FIGURE 63

2.000 in	SB	50.80 mm
3.270 in	HD	83.06 mm
10.875 in	EL	276.23 mm

BORG-WARNR 23332J
ROCKWELL 72NPS32-74

IDLI GROUP 87518

SPLINED PLUG WITH 10 SPLINES
SEE FIGURE 64

2.000 in	SB	50.80 mm
3.270 in	HD	83.06 mm
11.500 in	EL	292.10 mm

BORG-WARNR 19402J
BORG-WARNR 21665J
ROCKWELL 72NPS32-29

IDLI GROUP 87519

SPLINED PLUG WITH 16 SPLINES
SEE FIGURE 63

2.000 in	SB	50.80 mm
3.270 in	HD	83.06 mm
11.625 in	EL	295.28 mm

ROCKWELL PS32-16-16

IDLI GROUP 87520

SPLINED PLUG WITH 16 SPLINES
slight variation from above group
SEE FIGURE 63

2.000 in	SB	50.80 mm
3.270 in	HD	83.06 mm
11.625 in	EL	295.28 mm

ROCKWELL 72NPS32-25

IDLI GROUP 87521

SPLINED PLUG WITH 16 SPLINES
SEE FIGURE 63

2.000 in	SB	50.80 mm
3.320 in	HD	84.33 mm
8.000 in	EL	203.20 mm

ROCKWELL 60NPS32-27

IDLI GROUP 87522

SPLINED PLUG WITH 10 SPLINES
SEE FIGURE 63

2.000 in	SB	50.80 mm
3.320 in	HD	84.33 mm
8.500 in	EL	215.90 mm

BORG-WARNR 9281J
ROCKWELL 72NPS32

IDLI GROUP 87523

SPLINED PLUG WITH 16 SPLINES
SEE FIGURE 63

2.000 in	SB	50.80 mm
3.320 in	HD	84.33 mm
8.563 in	EL	217.49 mm

ROCKWELL 16N40-471
SPICER 5-40-471

IDLI GROUP 87524

SPLINED PLUG WITH 10 SPLINES
SEE FIGURE 64

2.000 in	SB	50.80 mm
3.320 in	HD	84.33 mm
9.375 in	EL	238.13 mm

BORG-WARNR 9276J
ROCKWELL 72NPS32-9

IDLI GROUP 87525

SPLINED PLUG WITH 10 SPLINES
SEE FIGURE 63

2.000 in	SB	50.80 mm
3.320 in	HD	84.33 mm
9.438 in	EL	239.71 mm

BORG-WARNR 17015J
ROCKWELL 72NPS32-5

IDLI GROUP 87526

SPLINED PLUG WITH 16 SPLINES
SEE FIGURE 63

2.000 in	SB	50.80 mm
3.320 in	HD	84.33 mm
9.563 in	EL	242.89 mm

ROCKWELL 16N40-501
SPICER 5-40-1031
SPICER 5-40-501

IDLI GROUP 87527

SPLINED PLUG WITH 16 SPLINES
SEE FIGURE 63

2.000 in	SB	50.80 mm
3.320 in	HD	84.33 mm
10.875 in	EL	276.23 mm

BORG-WARNR 18613J
ROCKWELL 72NPS32-10

IDLI GROUP 87528

SPLINED PLUG WITH 16 SPLINES
SEE FIGURE 63

2.000 in	SB	50.80 mm
3.344 in	HD	84.94 mm
15.375 in	EL	390.53 mm

ROCKWELL PS32-16-43

IDLI GROUP 87529

SPLINED PLUG WITH 10 SPLINES
SEE FIGURE 63

2.000 in	SB	50.80 mm
3.792 in	HD	96.32 mm
8.625 in	EL	219.08 mm

BORG-WARNR 13451J
ROCKWELL 72NPS32-22

IDLI GROUP 87530

SPLINED PLUG WITH 10 SPLINES
SEE FIGURE 63

2.000 in	SB	50.80 mm
3.792 in	HD	96.32 mm
9.438 in	EL	239.71 mm

BORG-WARNR 17016J
ROCKWELL 72NPS32-6

IDLI GROUP 87531

SPLINED PLUG WITH 10 SPLINES
SEE FIGURE 64

2.000 in	SB	50.80 mm
3.792 in	HD	96.32 mm
9.563 in	EL	242.89 mm

BORG-WARNR 13435J
ROCKWELL 72NPS32-19

BD=BEARING DIAMETER (outside)   **BW**=BEARING WIDTH   **CB**=CROSS LENGTH WITH BEARINGS   **CC**=CENTER TO CENTER   **CD**=CROSS DIAMETER
CL=CROSS LENGTH WITHOUT BEARINGS   **EC**=END TO CENTER or FACE TO FACE   **EE**=END TO END   **EL**=EFFECTIVE LENGTH
HD=HUB DIAMETER (or insert)   **IE**=INSIDE OF EARS (or recess)   **OD**=OUTSIDE DIAMETER   **OE**=OUTSIDE OF EARS   **SB**=SPLINE OR BORE SIZE

IDLI GUIDE   COPYRIGHT ® INTERCHANGE, INC.   ST. LOUIS PARK, MN. 55416 USA

CAUTION: BE SURE TO REFER TO ENGINEERING CATALOGS FOR SPECIAL APPLICATIONS THAT REQUIRE SPECIFIC MATERIAL CONTENT, TOLERANCES, ETC. SEE FOOTNOTE.

IDLI GROUP 87532

SPLINED PLUG WITH 10 SPLINES
SEE FIGURE 64

2.000 in	SB	50.80 mm
3.792 in	HD	96.32 mm
11.000 in	EL	279.40 mm

BORG-WARNR 21665J
ROCKWELL 72NPS32-65

IDLI GROUP 87533

SEE FIGURE 64

2.000 in	SB	50.80 mm
3.792 in	HD	96.32 mm
11.625 in	EL	295.28 mm

ROCKWELL 72NPS32-28

IDLI GROUP 87534

SPLINED PLUG WITH 16 SPLINES
SEE FIGURE 63

2.000 in	SB	50.80 mm
3.820 in	HD	97.03 mm
8.000 in	EL	203.20 mm

ROCKWELL 60NPS32-35

IDLI GROUP 87535

SPLINED PLUG WITH 10 SPLINES
SEE FIGURE 63

2.000 in	SB	50.80 mm
3.844 in	HD	97.64 mm
8.625 in	EL	219.08 mm

BORG-WARNR 10803J
ROCKWELL 72NPS32-21

IDLI GROUP 87536

SPLINED PLUG WITH 10 SPLINES
SEE FIGURE 64

2.000 in	SB	50.80 mm
3.844 in	HD	97.64 mm
9.563 in	EL	242.89 mm

BORG-WARNR 11451J
ROCKWELL 72NPS32-26

IDLI GROUP 87537

COMPANION FLANGE WITH 10 SPLINES
SEE FIGURE 32

2.000 in	SB	50.80 mm
3.250 in	EC	82.55 mm
2.938 in	HD	74.61 mm
8.000 in	OD	203.20 mm

ROCKWELL 5WCS32-37
SPICER 6-1-2141

IDLI GROUP 87538

SPLINED BEARING STUB WITH 10 SPLINES
SEE FIGURE 65

2.000 in	SB	50.80 mm
6.125 in	EC	155.58 mm
3.635 in	HD	92.33 mm

ROCKWELL BS-S32-9

IDLI GROUP 87539

SPLINED BEARING STUB WITH 10 SPLINES
SEE FIGURE 65

2.000 in	SB	50.80 mm
7.000 in	EC	177.80 mm
2.885 in	HD	73.28 mm

ROCKWELL BS-S32-8

IDLI GROUP 87540

SPLINED BEARING STUB WITH 10 SPLINES
SEE FIGURE 65

2.000 in	SB	50.80 mm
7.000 in	EC	177.80 mm
3.135 in	HD	79.63 mm

ROCKWELL BS-S32-19

IDLI GROUP 87541

SPLINED BEARING STUB WITH 10 SPLINES
SEE FIGURE 65

2.000 in	SB	50.80 mm
7.719 in	EC	196.06 mm
3.152 in	HD	80.06 mm

ROCKWELL BS-S32-12

IDLI GROUP 87542

SPLINED BEARING STUB WITH 10 SPLINES
SEE FIGURE 65

2.000 in	SB	50.80 mm
7.719 in	EC	196.06 mm
4.002 in	HD	101.65 mm

ROCKWELL BS-S32-16

IDLI GROUP 87543

SPLINED BEARING STUB WITH 10 SPLINES
SEE FIGURE 65

2.000 in	SB	50.80 mm
7.719 in	EC	196.06 mm
4.135 in	HD	105.03 mm

ROCKWELL BS-S32-15

IDLI GROUP 87544

ROUND TUBING
SEE FIGURE 75

2.010 in	SB	51.05 mm
72.000 in	EE	1828.80 mm
2.250 in	OD	57.15 mm

ROCKWELL RT225-11-72

IDLI GROUP 87545

ROUND TUBING
SEE FIGURE 75

2.010 in	SB	51.05 mm
108.000 in	EE	2743.20 mm
2.250 in	OD	57.15 mm

ROCKWELL RT225-11-108

IDLI GROUP 87546

ROUND TUBING
SEE FIGURE 75

2.084 in	SB	52.93 mm
72.000 in	EE	1828.80 mm
2.250 in	OD	57.15 mm

BORG-WARNR 4871J7200
ROCKWELL RT225-14-72

IDLI GROUP 87547

ROUND TUBING
SEE FIGURE 75

2.084 in	SB	52.93 mm
108.000 in	EE	2743.20 mm
2.250 in	OD	57.15 mm

BORG-WARNR 4871J10800
ROCKWELL RT225-14-108

IDLI GROUP 87548

SPLINED BEARING STUB WITH 16 INVOLUTED SPLINES
SEE FIGURE 66

2.094 in	SB	53.18 mm
7.781 in	EC	197.65 mm
2.770 in	HD	70.36 mm

ROCKWELL BS-S33-1

IDLI GROUP 87549

ROUND TUBING
SEE FIGURE 75

2.172 in	SB	55.17 mm
72.000 in	EE	1828.80 mm
2.250 in	OD	57.15 mm

ROCKWELL RT225-18-72

IDLI GROUP 87550

ROUND TUBING
SEE FIGURE 75

2.172 in	SB	55.17 mm
108.000 in	EE	2743.20 mm
2.250 in	OD	57.15 mm

ROCKWELL RT225-18-108

IDLI GROUP 87551

ROUND TUBING
SEE FIGURE 75

2.209 in	SB	56.11 mm
72.000 in	EE	1828.80 mm
2.375 in	OD	60.33 mm

ROCKWELL RT2375-14-72

IDLI GROUP 87552

ROUND TUBING
SEE FIGURE 75

2.209 in	SB	56.11 mm
108.000 in	EE	2743.20 mm
2.375 in	OD	60.33 mm

ROCKWELL RT2375-14-108

IDLI GROUP 87553

ROUND TUBING
SEE FIGURE 75

2.232 in	SB	56.69 mm
72.000 in	EL	1828.80 mm
2.500 in	OD	63.50 mm

ROCKWELL RT25-10-72
SPICER 27-93-27-5400

IDLI GROUP 87554

ROUND TUBING
SEE FIGURE 75

2.232 in	SB	56.69 mm
108.000 in	EL	2743.20 mm
2.500 in	OD	63.50 mm

ROCKWELL RT25-10-108
SPICER 27-93-27-10800

IDLI GROUP 87555

SPLINED PLUG WITH 10 SPLINES
SEE FIGURE 63

2.250 in	SB	57.15 mm
3.270 in	HD	83.06 mm
8.750 in	EL	222.25 mm

ROCKWELL 7NPS36

IDLI GROUP 87556

SPLINED PLUG WITH 16 SPLINES
SEE FIGURE 63

2.250 in	SB	57.15 mm
3.270 in	HD	83.06 mm
8.750 in	EL	222.25 mm

ROCKWELL 70NPS36-18

IDLI GROUP 87557

SPLINED BEARING STUB WITH 10 SPLINES
SEE FIGURE 65

2.250 in	SB	57.15 mm
5.281 in	EC	134.15 mm
3.152 in	HD	80.06 mm

ROCKWELL BS-S36-12
SPICER 6-1/2-53-11

IDLI GROUP 87558

SPLINED BEARING STUB WITH 10 SPLINES
SEE FIGURE 65

2.250 in	SB	57.15 mm
5.281 in	EC	134.15 mm
4.002 in	HD	101.65 mm

ROCKWELL BS-S36-10
SPICER 8-53-41

ENGINEERING CATALOGS MUST BE CONSULTED FOR DETAILS NOT INCLUDED IN THIS GUIDE. SPECIFIC DESIGNS, MATERIAL CONTENT, TOLERANCES LUBE FITTINGS AND OTHER DIMENSIONS ARE INTENTIONALLY OMITTED HERE.

IDLI GROUP 87559

SPLINED BEARING STUB WITH 10 SPLINES
SEE FIGURE 65

2.250 in	SB	57.15 mm
5.938 in	EC	150.81 mm
3.792 in	HD	96.32 mm

ROCKWELL BS-S36-3

IDLI GROUP 87560

ROUND TUBING
SEE FIGURE 75

2.282 in	SB	57.96 mm
72.000 in	EE	1828.80 mm
2.500 in	OD	63.50 mm

ROCKWELL RT25-12-72

IDLI GROUP 87561

ROUND TUBING
SEE FIGURE 75

2.282 in	SB	57.96 mm
108.000 in	EE	2743.20 mm
2.500 in	OD	63.50 mm

ROCKWELL RT25-12-108

IDLI GROUP 87562

ROUND TUBING
SEE FIGURE 75

2.310 in	SB	58.67 mm
72.000 in	EE	1828.80 mm
2.500 in	OD	63.50 mm

BORG-WARNR 4369J7200
ROCKWELL RT25-13-72

IDLI GROUP 87563

ROUND TUBING
SEE FIGURE 75

2.310 in	SB	58.67 mm
108.000 in	EE	2743.20 mm
2.500 in	OD	63.50 mm

BORG-WARNR 4369J10800
ROCKWELL RT25-13-108

IDLI GROUP 87564

SPLINED BEARING STUB WITH 16 SPLINES
SEE FIGURE 65

2.344 in	SB	59.53 mm
3.742 in	HD	95.05 mm
5.313 in	EL	134.94 mm

ROCKWELL 17N53-311
SPICER 6-53-311

IDLI GROUP 87565

SPLINED BEARING STUB WITH 18 INVOLUTED SPLINES
SEE FIGURE 66

2.344 in	SB	59.53 mm
7.500 in	EC	190.50 mm
3.270 in	HD	83.06 mm

ROCKWELL BS-S38-1

IDLI GROUP 87566

SPLINED BEARING STUB WITH 18 INVOLUTED SPLINES
SEE FIGURE 66

2.344 in	SB	59.53 mm
7.563 in	EC	192.09 mm
3.135 in	HD	79.63 mm

ROCKWELL BS-S38-4

IDLI GROUP 87567

SPLINED BEARING STUB WITH 16 SPLINES
SEE FIGURE 65

2.349 in	SB	59.66 mm
4.242 in	HD	107.75 mm
5.313 in	EL	134.94 mm

ROCKWELL 18N53-91
SPICER 6.5-53-91

IDLI GROUP 87568

SPLINED BEARING STUB WITH 16 SPLINES
SEE FIGURE 65

2.349 in	SB	59.66 mm
5.313 in	EC	134.94 mm
4.002 in	HD	101.65 mm

ROCKWELL BS-S38-6

IDLI GROUP 87569

SPLINED BEARING STUB WITH 16 SPLINES
SEE FIGURE 65

2.349 in	SB	59.66 mm
5.313 in	EC	134.94 mm
4.242 in	HD	107.75 mm

ROCKWELL BS-S38-5

IDLI GROUP 87570

CENTER BEARING
SEE FIGURE 56

2.362 in	SB	60.00 mm
3.375 in	EC	85.73 mm
8.625 in	CC	219.08 mm
1.406 in	BW	35.72 mm

AEC 21
AEC AE21
ALLOY 4008
AMER.PARTS 2-6514
ANCHORDOAN 35-6049
AUTOCAR 3AB0970
BORG-WARNR 35-6049
BUTLER 510661-1X
CLEVE.MOT. DS601KF
CLEVE.STL. HCB101
CMP DS601KF
DORMAN 662-021
EATON MM6051
FED.BEAR. HBF88512A
FED.MOGUL HB88512A
FORD D4TZ4800K
L & S BRG. PBS13
MACK 203SJ13
MACK 35MU32P3
MCQUAY-NOR MM6051
MOPAR D612
MOTOR MAST DS21
NEAPCO 81-6010
NEAPCO DS601KF
REPUBLIC 35-6051
ROCKWELL CB210661-1X
SPICER 210661-1X
SPICER 210874-1X
SPICER 210939X
TRW 20813
WHITE 02-7010593
WHITE 991790

IDLI GROUP 87571

ROUND TUBING
SEE FIGURE 75

2.385 in	SB	60.58 mm
72.000 in	EE	1828.80 mm
2.625 in	OD	66.68 mm

ROCKWELL RT2625-11-72

IDLI GROUP 87572

ROUND TUBING
SEE FIGURE 75

2.385 in	SB	60.58 mm
108.000 in	EE	2743.20 mm
2.625 in	OD	66.68 mm

ROCKWELL RT2625-11-108

IDLI GROUP 87573

ROUND TUBING
SEE FIGURE 75

2.402 in	SB	61.01 mm
72.000 in	EE	1828.80 mm
2.500 in	OD	63.50 mm

ROCKWELL RT25-18-72

IDLI GROUP 87574

ROUND TUBING
SEE FIGURE 75

2.402 in	SB	61.01 mm
108.000 in	EE	2743.20 mm
2.500 in	OD	63.50 mm

ROCKWELL RT25-18-108

IDLI GROUP 87575

SPLINED PLUG WITH 10 SPLINES
SEE FIGURE 63

| 2.500 in | SB | 63.50 mm |
| 10.344 in | EL | 262.73 mm |

ROCKWELL 17N42-1081
SPICER 6-42-1081
SPICER 6-42-1101

IDLI GROUP 87576

SPLINED PLUG WITH 10 SPLINES
SEE FIGURE 63

2.500 in	SB	63.50 mm
2.885 in	HD	73.28 mm
9.750 in	EL	247.65 mm

ROCKWELL 82NPS40-27

IDLI GROUP 87577

SPLINED PLUG WITH 16 SPLINES
SEE FIGURE 63

2.500 in	SB	63.50 mm
3.020 in	HD	76.71 mm
8.000 in	EL	203.20 mm

ROCKWELL 70NPS40-10

IDLI GROUP 87578

SPLINED PLUG WITH 10 SPLINES
SEE FIGURE 63

2.500 in	SB	63.50 mm
3.135 in	HD	79.63 mm
10.344 in	EL	262.73 mm

ROCKWELL 17N42-1081

IDLI GROUP 87579

SPLINED PLUG WITH 10 SPLINES
SEE FIGURE 63

2.500 in	SB	63.50 mm
3.198 in	HD	81.23 mm
8.344 in	EL	211.93 mm

ROCKWELL 17N40-51
SPICER 6-40-141
SPICER 6-40-51
SPICER 6-40-821

IDLI GROUP 87580

SPLINED PLUG WITH 10 SPLINES
SEE FIGURE 63

2.500 in	SB	63.50 mm
3.198 in	HD	81.23 mm
9.156 in	EL	232.57 mm

ROCKWELL 17N40-161
SPICER 6-40-161

IDLI GROUP 87581

SPLINED PLUG WITH 10 SPLINES
SEE FIGURE 63

2.500 in	SB	63.50 mm
3.198 in	HD	81.23 mm
10.349 in	EL	262.86 mm

ROCKWELL 17N40-61
SPICER 6-40-101
SPICER 6-40-61

IDLI GROUP 87582

SPLINED PLUG WITH 16 SPLINES
SEE FIGURE 63

2.500 in	SB	63.50 mm
3.270 in	HD	83.06 mm
4.938 in	EL	125.41 mm

ROCKWELL 72NPS40-5

IDLI GROUP 87583

SPLINED PLUG WITH 16 SPLINES
SEE FIGURE 63

2.500 in	SB	63.50 mm
3.320 in	HD	84.33 mm
4.188 in	EL	106.36 mm

BORG-WARNR 23526J
ROCKWELL 72NPS40-2

BD = BEARING DIAMETER (outside)   BW = BEARING WIDTH   CB = CROSS LENGTH WITH BEARINGS   CC = CENTER TO CENTER   CD = CROSS DIAMETER
CL = CROSS LENGTH WITHOUT BEARINGS   EC = END TO CENTER or FACE TO FACE   EE = END TO END   EL = EFFECTIVE LENGTH
HD = HUB DIAMETER (or insert)   IE = INSIDE OF EARS (or recess)   OD = OUTSIDE DIAMETER   OE = OUTSIDE OF EARS   SB = SPLINE OR BORE SIZE

CAUTION: BE SURE TO REFER TO ENGINEERING CATALOGS FOR SPECIAL APPLICATIONS THAT REQUIRE SPECIFIC MATERIAL CONTENT, TOLERANCES, ETC. SEE FOOTNOTE.

IDLI GROUP 87584

SPLINED PLUG WITH 10 SPLINES
SEE FIGURE 63

2.500 in	SB	63.50 mm
3.635 in	HD	92.33 mm
10.000 in	EL	254.00 mm

BORG-WARNR 16821J
BORG-WARNR 18870J
ROCKWELL 82NPS40-16

IDLI GROUP 87585

SPLINED PLUG WITH 10 SPLINES
SEE FIGURE 64

2.500 in	SB	63.50 mm
3.635 in	HD	92.33 mm
10.563 in	EL	268.29 mm

BORG-WARNR 12711J
BORG-WARNR 16822J
ROCKWELL 82NPS40-5

IDLI GROUP 87586

SPLINED PLUG WITH 10 SPLINES
SEE FIGURE 64

2.500 in	SB	63.50 mm
3.635 in	HD	92.33 mm
12.750 in	EL	323.85 mm

BORG-WARNR 21667J
ROCKWELL 82NPS40-23

IDLI GROUP 87587

SPLINED PLUG WITH 16 SPLINES
SEE FIGURE 63

2.500 in	SB	63.50 mm
3.742 in	HD	95.05 mm
8.000 in	EL	203.20 mm

ROCKWELL 7NPS40-9

IDLI GROUP 87588

SPLINED PLUG WITH 16 SPLINES
slight variation from above group
SEE FIGURE 63

2.500 in	SB	63.50 mm
3.742 in	HD	95.05 mm
8.000 in	EL	203.20 mm

ROCKWELL 72NPS40-9

IDLI GROUP 87589

SPLINED PLUG WITH 16 SPLINES
SEE FIGURE 63

2.500 in	SB	63.50 mm
3.742 in	HD	95.05 mm
8.938 in	EL	227.01 mm

ROCKWELL 17NPS40

IDLI GROUP 87590

SPLINED PLUG WITH 16 SPLINES
slight variation from above group
SEE FIGURE 63

2.500 in	SB	63.50 mm
3.742 in	HD	95.05 mm
8.938 in	EL	227.01 mm

ROCKWELL PS40-16-18

IDLI GROUP 87591

SPLINED PLUG WITH 16 SPLINES
SEE FIGURE 63

2.500 in	SB	63.50 mm
3.742 in	HD	95.05 mm
9.250 in	EL	234.95 mm

ROCKWELL 17N40-401
SPICER 6-40-401
SPICER 6-40-461
SPICER 6-40-541

IDLI GROUP 87592

SPLINED PLUG WITH 16 SPLINES
slight variation from above group
SEE FIGURE 63

2.500 in	SB	63.50 mm
3.742 in	HD	95.05 mm
9.250 in	EL	234.95 mm

ROCKWELL 17N40-541
SPICER 6-40-541

IDLI GROUP 87593

SPLINED PLUG WITH 10 SPLINES
SEE FIGURE 63

2.500 in	SB	63.50 mm
3.742 in	HD	95.05 mm
10.000 in	EL	254.00 mm

ROCKWELL 7NPS40-17

IDLI GROUP 87594

SPLINED PLUG WITH 10 SPLINES
slight variation from above group
SEE FIGURE 63

2.500 in	SB	63.50 mm
3.742 in	HD	95.05 mm
10.000 in	EL	254.00 mm

BORG-WARNR 23328J
ROCKWELL 82NPS40-17

IDLI GROUP 87595

SPLINED PLUG WITH 10 SPLINES
slight variation from above group
SEE FIGURE 63

2.500 in	SB	63.50 mm
3.742 in	HD	95.05 mm
10.000 in	EL	254.00 mm

ROCKWELL 82NPS40-59

IDLI GROUP 87596

SPLINED PLUG WITH 16 SPLINES
SEE FIGURE 63

2.500 in	SB	63.50 mm
3.742 in	HD	95.05 mm
10.563 in	EL	268.29 mm

CLEVE.STL. C67-7-4-10
ROCKWELL 17N40-451
SPICER 6-40-451
SPICER 6-40-521

IDLI GROUP 87597

SPLINED PLUG WITH 10 SPLINES
SEE FIGURE 63

2.500 in	SB	63.50 mm
3.742 in	HD	95.05 mm
10.563 in	EL	268.29 mm

ROCKWELL 82NPS40-70

IDLI GROUP 87598

SPLINED PLUG WITH 16 SPLINES
slight variation from above group
SEE FIGURE 63

2.500 in	SB	63.50 mm
3.742 in	HD	95.05 mm
10.563 in	EL	268.29 mm

ROCKWELL 17N40-521
SPICER 6-40-521

IDLI GROUP 87599

SPLINED PLUG WITH 10 SPLINES
SEE FIGURE 63

2.500 in	SB	63.50 mm
3.742 in	HD	95.05 mm
10.594 in	EL	269.08 mm

ROCKWELL 82NPS40-60

IDLI GROUP 87600

SPLINED PLUG WITH 10 SPLINES
SEE FIGURE 64

2.500 in	SB	63.50 mm
3.742 in	HD	95.05 mm
12.750 in	EL	323.85 mm

BORG-WARNR 23330J
ROCKWELL 82NPS40-48

IDLI GROUP 87601

SPLINED PLUG WITH 10 SPLINES
SEE FIGURE 63

2.500 in	SB	63.50 mm
3.770 in	HD	95.76 mm
9.469 in	EL	240.51 mm

ROCKWELL 17N40-311
SPICER 6-40-311

IDLI GROUP 87602

SPLINED PLUG WITH 10 SPLINES
SEE FIGURE 63

2.500 in	SB	63.50 mm
4.135 in	HD	105.03 mm
9.438 in	EL	239.71 mm

ROCKWELL 82NPS40-22

IDLI GROUP 87603

SPLINED BEARING STUB WITH 10 SPLINES
SEE FIGURE 65

2.500 in	SB	63.50 mm
6.688 in	EC	169.86 mm
3.792 in	HD	96.32 mm

ROCKWELL BS-S40-4

IDLI GROUP 87604

ROUND TUBING
SEE FIGURE 75

2.532 in	SB	64.31 mm
72.000 in	EE	1828.80 mm
2.750 in	OD	69.85 mm

ROCKWELL RT275-12-72

IDLI GROUP 87605

ROUND TUBING
SEE FIGURE 75

2.532 in	SB	64.31 mm
108.000 in	EE	2743.20 mm
2.750 in	OD	69.85 mm

ROCKWELL RT275-12-108

IDLI GROUP 87606

ROUND TUBING
SEE FIGURE 75

2.560 in	SB	65.02 mm
72.000 in	EE	1828.80 mm
2.750 in	OD	69.85 mm

ROCKWELL RT275-13-72

IDLI GROUP 87607

ROUND TUBING
SEE FIGURE 75

2.560 in	SB	65.02 mm
108.000 in	EE	2743.20 mm
2.750 in	OD	69.85 mm

ROCKWELL RT275-13-108

IDLI GROUP 87608

ROUND TUBING
SEE FIGURE 75

2.625 in	SB	66.68 mm
72.000 in	EE	1828.80 mm
3.000 in	OD	76.20 mm

BORG-WARNR 4379J7200
ROCKWELL RT3-187-72

IDLI GROUP 87609

ROUND TUBING
SEE FIGURE 75

2.625 in	SB	66.68 mm
108.000 in	EE	2743.20 mm
3.000 in	OD	76.20 mm

BORG-WARNR 4379J10800
ROCKWELL RT3-187-108

ENGINEERING CATALOGS MUST BE CONSULTED FOR DETAILS NOT INCLUDED
IN THIS GUIDE. SPECIFIC DESIGNS, MATERIAL CONTENT, TOLERANCES
LUBE FITTINGS AND OTHER DIMENSIONS ARE INTENTIONALLY OMITTED HERE.

IDLI GROUP 87610

ROUND TUBING
SEE FIGURE 75

2.688 in	SB	68.26 mm
72.000 in	EL	1828.80 mm
3.000 in	OD	76.20 mm

BORG-WARNR 4378J7200
ROCKWELL RT3-156-72

IDLI GROUP 87611

ROUND TUBING
SEE FIGURE 75

2.688 in	SB	68.26 mm
108.000 in	EL	2743.20 mm
3.000 in	OD	76.20 mm

BORG-WARNR 4378J10800
ROCKWELL RT3-156-108

IDLI GROUP 87612

ROUND TUBING
SEE FIGURE 75

2.745 in	SB	69.72 mm
72.000 in	EE	1828.80 mm
2.875 in	OD	73.03 mm

ROCKWELL RT2875-16-72

IDLI GROUP 87613

ROUND TUBING
SEE FIGURE 75

2.745 in	SB	69.72 mm
108.000 in	EE	2743.20 mm
2.875 in	OD	73.03 mm

ROCKWELL RT2875-16-108

IDLI GROUP 87614

ROUND TUBING
SEE FIGURE 75

2.760 in	SB	70.10 mm
72.000 in	EL	1828.80 mm
3.000 in	OD	76.20 mm

BORG-WARNR 4375J7200
ROCKWELL RT3-11-72

IDLI GROUP 87615

ROUND TUBING
SEE FIGURE 75

2.760 in	SB	70.10 mm
108.000 in	EL	2743.20 mm
3.000 in	OD	76.20 mm

BORG-WARNR 4375J10800
ROCKWELL RT3-11-108

IDLI GROUP 87616

ROUND TUBING
SEE FIGURE 75

2.810 in	SB	71.37 mm
72.000 in	EE	1828.80 mm
3.000 in	OD	76.20 mm

BORG-WARNR 4376J7200
ROCKWELL RT3-13-72

IDLI GROUP 87617

ROUND TUBING
SEE FIGURE 75

2.810 in	SB	71.37 mm
108.000 in	EE	2743.20 mm
3.000 in	OD	76.20 mm

BORG-WARNR 4376J10800
ROCKWELL RT3-13-108

IDLI GROUP 87618

ROUND TUBING
SEE FIGURE 75

2.834 in	SB	71.98 mm
72.000 in	EE	1828.80 mm
3.000 in	OD	76.20 mm

CLEVE.STL. 83-14-74
ROCKWELL RT3-14-72

IDLI GROUP 87619

ROUND TUBING
SEE FIGURE 75

2.834 in	SB	71.98 mm
108.000 in	EE	2743.20 mm
3.000 in	OD	76.20 mm

CLEVE.STL. 83-14-108
ROCKWELL RT3-14-108

IDLI GROUP 87620

SPLINED BEARING STUB
WITH 22 INVOLUTED SPLINES
SEE FIGURE 66

2.844 in	SB	72.23 mm
8.313 in	EC	211.14 mm
4.135 in	HD	105.03 mm

ROCKWELL BS-S46-1

IDLI GROUP 87621

ROUND TUBING
SEE FIGURE 75

2.870 in	SB	72.90 mm
72.000 in	EE	1828.80 mm
3.000 in	OD	76.20 mm

BORG-WARNR 5792J7200
CLEVE.STL. 83-16-74
ROCKWELL RT3-16-72

IDLI GROUP 87622

ROUND TUBING
SEE FIGURE 75

2.870 in	SB	72.90 mm
108.000 in	EE	2743.20 mm
3.000 in	OD	76.20 mm

BORG-WARNR 5792J10800
CLEVE.STL. 83-16-108
ROCKWELL RT3-16-108

IDLI GROUP 87623

ROUND TUBING
SEE FIGURE 75

2.875 in	SB	73.03 mm
72.000 in	EL	1828.80 mm
3.250 in	OD	82.55 mm

ROCKWELL RT325-187-72

IDLI GROUP 87624

ROUND TUBING
SEE FIGURE 75

2.875 in	SB	73.03 mm
108.000 in	EL	2743.20 mm
3.250 in	OD	82.55 mm

ROCKWELL RT325-187-108

IDLI GROUP 87625

ROUND TUBING
SEE FIGURE 75

2.935 in	SB	74.55 mm
72.000 in	EE	1828.80 mm
3.125 in	OD	79.38 mm

ROCKWELL RT3125-13-72

IDLI GROUP 87626

ROUND TUBING
SEE FIGURE 75

2.935 in	SB	74.55 mm
108.000 in	EE	2743.20 mm
3.125 in	OD	79.38 mm

ROCKWELL RT3125-13-108

IDLI GROUP 87627

SHAFTING WITH 10 SPLINES
SEE FIGURE 86

3.000 in	SB	76.20 mm
12.000 in	EE	304.80 mm

ROCKWELL SH-S48-21-12

IDLI GROUP 87628

SHAFTING WITH 10 SPLINES
SEE FIGURE 86

3.000 in	SB	76.20 mm
24.000 in	EE	609.60 mm

ROCKWELL SH-S48-21-24

IDLI GROUP 87629

SPLINED PLUG WITH 10 SPLINES
SEE FIGURE 64

3.000 in	SB	76.20 mm
3.152 in	HD	80.06 mm
10.750 in	EL	273.05 mm

ROCKWELL 92NPS48-21

IDLI GROUP 87630

SPLINED PLUG WITH 16 SPLINES
SEE FIGURE 63

3.000 in	SB	76.20 mm
3.152 in	HD	80.06 mm
11.250 in	EL	285.75 mm

ROCKWELL PS48-16-9

IDLI GROUP 87631

SPLINED PLUG WITH 22 INVOLUTED
SPLINES IN 23 SPLINE SPACES
SEE FIGURE 63

3.000 in	SB	76.20 mm
3.755 in	HD	95.38 mm
11.188 in	EL	284.16 mm

ROCKWELL PS48-23-4

IDLI GROUP 87632

SPLINED PLUG WITH 10 SPLINES
SEE FIGURE 63

3.000 in	SB	76.20 mm
3.792 in	HD	96.32 mm
11.188 in	EL	284.16 mm

BORG-WARNR 14045J
ROCKWELL 92NPS48-12

IDLI GROUP 87633

SPLINED PLUG WITH 16 SPLINES
SEE FIGURE 63

3.000 in	SB	76.20 mm
4.002 in	HD	101.65 mm
9.469 in	EL	240.51 mm

ROCKWELL PS48-16-19

IDLI GROUP 87634

SPLINED PLUG WITH 10 SPLINES
SEE FIGURE 64

3.000 in	SB	76.20 mm
4.002 in	HD	101.65 mm
11.188 in	EL	284.16 mm

ROCKWELL 92NPS48-25

IDLI GROUP 87635

SPLINED PLUG WITH 22 INVOLUTED
SPLINES IN 23 SPLINE SPACES
SEE FIGURE 63

3.000 in	SB	76.20 mm
4.002 in	HD	101.65 mm
11.188 in	EL	284.16 mm

ROCKWELL PS48-23-3

IDLI GROUP 87636

SPLINED PLUG WITH 16 SPLINES
SEE FIGURE 63

3.000 in	SB	76.20 mm
4.002 in	HD	101.65 mm
11.250 in	EL	285.75 mm

ROCKWELL PS48-16-1

IDLI GROUP 87637

SPLINED PLUG WITH 16 SPLINES
SEE FIGURE 63

3.000 in	SB	76.20 mm
4.002 in	HD	101.65 mm
11.250 in	EL	285.75 mm

BORG-WARNR 25806J
ROCKWELL PS48-16-10

BD=BEARING DIAMETER (outside)   **BW**=BEARING WIDTH   **CB**=CROSS LENGTH WITH BEARINGS   **CC**=CENTER TO CENTER   **CD**=CROSS DIAMETER
CL=CROSS LENGTH WITHOUT BEARINGS   **EC**=END TO CENTER or FACE TO FACE   **EE**=END TO END   **EL**=EFFECTIVE LENGTH
HD=HUB DIAMETER (or insert)   **IE**=INSIDE OF EARS (or recess)   **OD**=OUTSIDE DIAMETER   **OE**=OUTSIDE OF EARS   **SB**=SPLINE OR BORE SIZE

IDLI GUIDE   COPYRIGHT © INTERCHANGE, INC.   ST. LOUIS PARK, MN. 55416 USA

CAUTION: BE SURE TO REFER TO ENGINEERING CATALOGS FOR SPECIAL APPLICATIONS THAT REQUIRE SPECIFIC MATERIAL CONTENT, TOLERANCES, ETC. SEE FOOTNOTE.

IDLI GROUP 87638

SPLINED PLUG WITH 16 SPLINES
slight variation from above group
SEE FIGURE 63

3.000 in	SB	76.20 mm
4.002 in	HD	101.65 mm
11.250 in	EL	285.75 mm

ROCKWELL PS48-16-1

IDLI GROUP 87639

SPLINED PLUG WITH 10 SPLINES
SEE FIGURE 63

3.000 in	SB	76.20 mm
4.135 in	HD	105.03 mm
9.125 in	EL	231.78 mm

ROCKWELL 92NPS48-14

IDLI GROUP 87640

SPLINED PLUG WITH 10 SPLINES
SEE FIGURE 63

3.000 in	SB	76.20 mm
4.135 in	HD	105.03 mm
11.188 in	EL	284.16 mm

ROCKWELL 92NPS48-10

IDLI GROUP 87641

SPLINED PLUG WITH 10 SPLINES
slight variation from above group
SEE FIGURE 63

3.000 in	SB	76.20 mm
4.135 in	HD	105.03 mm
11.188 in	EL	284.16 mm

BORG-WARNR 11757J
BORG-WARNR 11854J
BORG-WARNR 15262J
ROCKWELL 92NPS48-1

IDLI GROUP 87642

SPLINED PLUG WITH 10 SPLINES
slight variation from above group
SEE FIGURE 64

3.000 in	SB	76.20 mm
4.135 in	HD	105.03 mm
11.188 in	EL	284.16 mm

ROCKWELL 92NPS48-2

IDLI GROUP 87643

SPLINED PLUG WITH 16 SPLINES
SEE FIGURE 63

3.000 in	SB	76.20 mm
4.135 in	HD	105.03 mm
11.250 in	EL	285.75 mm

ROCKWELL PS48-16-17

IDLI GROUP 87644

SPLINED PLUG WITH 16 SPLINES
slight variation from above group
SEE FIGURE 63

3.000 in	SB	76.20 mm
4.242 in	HD	107.75 mm
9.469 in	EL	240.51 mm

ROCKWELL 18N40-151
SPICER 6.5-40-151

IDLI GROUP 87645

SPLINED PLUG WITH 16 SPLINES
slight variation from above group
SEE FIGURE 63

3.000 in	SB	76.20 mm
4.242 in	HD	107.75 mm
9.469 in	EL	240.51 mm

ROCKWELL 18N40-201
SPICER 6.5-40-201

IDLI GROUP 87646

SPLINED PLUG WITH 10 SPLINES
SEE FIGURE 64

3.000 in	SB	76.20 mm
4.242 in	HD	107.75 mm
10.750 in	EL	273.05 mm

ROCKWELL PS48-10-13

IDLI GROUP 87647

SPLINED PLUG WITH 10 SPLINES
SEE FIGURE 63

3.000 in	SB	76.20 mm
4.242 in	HD	107.75 mm
11.188 in	EL	284.16 mm

BORG-WARNR 8019J
ROCKWELL 92NPS48-8

IDLI GROUP 87648

SPLINED PLUG WITH 16 SPLINES
slight variation from above group
SEE FIGURE 63

3.000 in	SB	76.20 mm
4.242 in	HD	107.75 mm
11.250 in	EL	285.75 mm

ROCKWELL 18N40-161
SPICER 6.5-40-161

IDLI GROUP 87649

SPLINED PLUG WITH 16 SPLINES
slight variation from above group
SEE FIGURE 63

3.000 in	SB	76.20 mm
4.242 in	HD	107.75 mm
11.250 in	EL	285.75 mm

ROCKWELL 18N40-191
SPICER 6.5-40-191

IDLI GROUP 87650

ROUND TUBING
SEE FIGURE 75

3.010 in	SB	76.45 mm
72.000 in	EL	1828.80 mm
3.250 in	OD	82.55 mm

ROCKWELL RT325-11-72

IDLI GROUP 87651

ROUND TUBING
SEE FIGURE 75

3.010 in	SB	76.45 mm
108.000 in	EL	2743.20 mm
3.250 in	OD	82.55 mm

ROCKWELL RT325-11-108

IDLI GROUP 87652

ROUND TUBING
SEE FIGURE 75

3.060 in	SB	77.72 mm
72.000 in	EL	1828.80 mm
3.250 in	OD	82.55 mm

ROCKWELL RT325-13-72

IDLI GROUP 87653

ROUND TUBING
SEE FIGURE 75

3.060 in	SB	77.72 mm
108.000 in	EL	2743.20 mm
3.250 in	OD	82.55 mm

ROCKWELL RT325-13-108

IDLI GROUP 87654

ROUND TUBING
SEE FIGURE 75

3.188 in	SB	80.98 mm
72.000 in	EE	1828.80 mm
3.500 in	OD	88.90 mm

ROCKWELL RT35-156-72

IDLI GROUP 87655

ROUND TUBING
SEE FIGURE 75

3.188 in	SB	80.98 mm
108.000 in	EE	2743.20 mm
3.500 in	OD	88.90 mm

ROCKWELL RT35-156-108

IDLI GROUP 87656

SPLINED PLUG WITH 18 INVOLUTED SPLINES
SEE FIGURE 63

3.250 in	SB	82.55 mm
3.198 in	HD	81.23 mm
4.250 in	EL	107.95 mm

ROCKWELL 17N40-181
SPICER 6-40-181
SPICER 6-40-38X

IDLI GROUP 87657

ROUND TUBING
SEE FIGURE 75

3.260 in	SB	82.80 mm
72.000 in	EE	1828.80 mm
3.500 in	OD	88.90 mm

BORG-WARNR 10000J7200
ROCKWELL RT35-11-72

IDLI GROUP 87658

ROUND TUBING
SEE FIGURE 75

3.260 in	SB	82.80 mm
108.000 in	EE	2743.20 mm
3.500 in	OD	88.90 mm

BORG-WARNR 10000J10800
ROCKWELL RT35-11-108

IDLI GROUP 87659

ROUND TUBING
SEE FIGURE 75

3.310 in	SB	84.07 mm
72.000 in	EE	1828.80 mm
3.500 in	OD	88.90 mm

ROCKWELL RT35-13-72

IDLI GROUP 87660

ROUND TUBING
SEE FIGURE 75

3.310 in	SB	84.07 mm
108.000 in	EE	2743.20 mm
3.500 in	OD	88.90 mm

ROCKWELL RT35-13-108

IDLI GROUP 87661

ROUND TUBING
SEE FIGURE 75

3.320 in	SB	84.33 mm
72.000 in	EE	1828.80 mm
3.500 in	OD	88.90 mm

ROCKWELL RT35-10-72

IDLI GROUP 87662

ROUND TUBING
SEE FIGURE 75

3.320 in	SB	84.33 mm
108.000 in	EE	2743.20 mm
3.500 in	OD	88.90 mm

ROCKWELL RT35-10-108

IDLI GROUP 87663

ROUND TUBING
SEE FIGURE 75

3.334 in	SB	84.68 mm
72.000 in	EE	1828.80 mm
3.500 in	OD	88.90 mm

ROCKWELL RT35-14-72

ENGINEERING CATALOGS MUST BE CONSULTED FOR DETAILS NOT INCLUDED IN THIS GUIDE. SPECIFIC DESIGNS, MATERIAL CONTENT, TOLERANCES LUBE FITTINGS AND OTHER DIMENSIONS ARE INTENTIONALLY OMITTED HERE.

IDLI GROUP 87664

ROUND TUBING
SEE FIGURE 75

3.334 in	SB	84.68 mm
108.000 in	EE	2743.20 mm
3.500 in	OD	88.90 mm

ROCKWELL RT35-14-108

IDLI GROUP 87665

ROUND TUBING
SEE FIGURE 75

3.626 in	SB	92.10 mm
72.000 in	EE	1828.80 mm
4.000 in	OD	101.60 mm

BORG-WARNR 16745J7200
ROCKWELL RT4-187-72

IDLI GROUP 87666

ROUND TUBING
SEE FIGURE 75

3.626 in	SB	92.10 mm
108.000 in	EE	2743.20 mm
4.000 in	OD	101.60 mm

BORG-WARNR 16745J10800
ROCKWELL RT4-187-108

IDLI GROUP 87667

ROUND TUBING
SEE FIGURE 75

3.732 in	SB	94.79 mm
72.000 in	EE	1828.80 mm
4.000 in	OD	101.60 mm

CLEVE.STL. 84-10-74
ROCKWELL RT4-10-72
SPICER 32-50-52

IDLI GROUP 87668

3.732 in	SB	94.79 mm
108.000 in	EE	2743.20 mm
4.000 in	OD	101.60 mm

CLEVE.STL. 84-10-108
ROCKWELL RT4-10-108
SPICER 32-30-52

IDLI GROUP 87669

ROUND TUBING
SEE FIGURE 75

3.734 in	SB	94.84 mm
72.000 in	EE	1828.80 mm
4.094 in	OD	103.98 mm

ROCKWELL RT4095-7-72

IDLI GROUP 87670

ROUND TUBING
SEE FIGURE 75

3.734 in	SB	94.84 mm
108.000 in	EE	2743.20 mm
4.094 in	OD	103.98 mm

ROCKWELL RT4095-7-108

IDLI GROUP 87671

ROUND TUBING
SEE FIGURE 75

3.750 in	SB	95.25 mm
72.000 in	EL	1828.80 mm
4.500 in	OD	114.30 mm

ROCKWELL RT45-375-72

IDLI GROUP 87672

ROUND TUBING
SEE FIGURE 75

3.750 in	SB	95.25 mm
108.000 in	EL	2743.20 mm
4.500 in	OD	114.30 mm

ROCKWELL RT45-375-108

IDLI GROUP 87673

ROUND TUBING
SEE FIGURE 75

3.782 in	SB	96.06 mm
72.000 in	EE	1828.80 mm
4.000 in	OD	101.60 mm

BORG-WARNR 8128J7200
ROCKWELL RT4-12-72

IDLI GROUP 87674

ROUND TUBING
SEE FIGURE 75

3.782 in	SB	96.06 mm
108.000 in	EE	2743.20 mm
4.000 in	OD	101.60 mm

BORG-WARNR 8128J10800
ROCKWELL RT4-12-108

IDLI GROUP 87675

ROUND TUBING
SEE FIGURE 75

3.834 in	SB	97.38 mm
72.000 in	EE	1828.80 mm
4.000 in	OD	101.60 mm

BORG-WARNR 9113J7200
CLEVE.STL. 84-14-74
ROCKWELL RT4-14-72

IDLI GROUP 87676

ROUND TUBING
SEE FIGURE 75

3.834 in	SB	97.38 mm
108.000 in	EE	2743.20 mm
4.000 in	OD	101.60 mm

BORG-WARNR 9113J10800
CLEVE.STL. 84-14-108
ROCKWELL RT4-14-108

IDLI GROUP 87677

ROUND TUBING
SEE FIGURE 75

3.982 in	SB	101.14 mm
72.000 in	EE	1828.80 mm
4.500 in	OD	114.30 mm

ROCKWELL RT45-3-72

IDLI GROUP 87678

ROUND TUBING
SEE FIGURE 75

3.982 in	SB	101.14 mm
108.000 in	EE	2743.20 mm
4.500 in	OD	114.30 mm

ROCKWELL RT45-3-108

IDLI GROUP 87679

ROUND TUBING
SEE FIGURE 75

4.125 in	SB	104.78 mm
72.000 in	EL	1828.80 mm
4.500 in	OD	114.30 mm

BORG-WARNR 11855J7200
ROCKWELL RT45-187-72

IDLI GROUP 87680

ROUND TUBING
SEE FIGURE 75

4.125 in	SB	104.78 mm
108.000 in	EL	2743.20 mm
4.500 in	OD	114.30 mm

BORG-WARNR 11855J10800
ROCKWELL RT45-187-108

IDLI GROUP 87681

ROUND TUBING
SEE FIGURE 75

4.232 in	SB	107.49 mm
72.000 in	EE	1828.80 mm
4.500 in	OD	114.30 mm

ROCKWELL RT45-10-72

IDLI GROUP 87682

ROUND TUBING
SEE FIGURE 75

4.232 in	SB	107.49 mm
108.000 in	EE	2743.20 mm
4.500 in	OD	114.30 mm

ROCKWELL RT45-10-108

IDLI GROUP 87683

ROUND TUBING
SEE FIGURE 75

4.310 in	SB	109.47 mm
72.000 in	EE	1828.80 mm
4.500 in	OD	114.30 mm

ROCKWELL RT45-13-72

IDLI GROUP 87684

ROUND TUBING
SEE FIGURE 75

4.310 in	SB	109.47 mm
108.000 in	EE	2743.20 mm
4.500 in	OD	114.30 mm

ROCKWELL RT45-13-108

IDLI GROUP 87685

ROUND TUBING
SEE FIGURE 75

5.834 in	SB	148.18 mm
72.000 in	EE	1828.80 mm
6.000 in	OD	152.40 mm

ROCKWELL RT6-14-72

IDLI GROUP 87686

ROUND TUBING
SEE FIGURE 75

5.834 in	SB	148.18 mm
108.000 in	EE	2743.20 mm
6.000 in	OD	152.40 mm

ROCKWELL RT6-14-108

BD = BEARING DIAMETER (outside)   BW = BEARING WIDTH   CB = CROSS LENGTH WITH BEARINGS   CC = CENTER TO CENTER   CD = CROSS DIAMETER
CL = CROSS LENGTH WITHOUT BEARINGS   EC = END TO CENTER or FACE TO FACE   EE = END TO END   EL = EFFECTIVE LENGTH
HD = HUB DIAMETER (or insert)   IE = INSIDE OF EARS (or recess)   OD = OUTSIDE DIAMETER   OE = OUTSIDE OF EARS   SB = SPLINE OR BORE SIZE

Other Valuable Interchange Publications

The International Bearing Interchange Guide

New Ninth Edition

The I.B.I. Guide has over 365,000 listings of 25,326 ball and roller bearings available worldwide and dating back to 1918. It includes manufacturers, original equipment users and government part numbers, with THOUSANDS of new listings not included in earlier editions!

I.S.B.N. #916966-17-8

The International Seal Interchange Guide

New Sixth Edition

The I.S.I. Guide has over 150,000 listings of 12,632 different seals and "O" rings. It includes government numbers as well as producers and users throughout the world—over 700 páges of information to save you time·and money!

I.S.B.N. #916966-16-X

The International Drive Belt Interchange Guide

New Fourth Edition

The I.D.B.I. Guide has more than 115,000 listings for nearly 5,863 types and sizes of drive and V-belts. Listings include manufacturers, original equipment and government users from all over the world. 760 pages.

I.S.B.N. #916966-15-1

IDLI GROUP 87801

UNWELDED CENTER ASSEMBLY
extended EE dimension shown
SEE FIGURE 85
63.625 in   EE   1616.08 mm
3.500 in   OD   88.90 mm
ROCKWELL 959-10-1

IDLI GROUP 87802

UNWELDED CENTER ASSEMBLY
extended EE dimension shown
SEE FIGURE 85
67.375 in   EE   1711.33 mm
3.000 in   OD   76.20 mm
ROCKWELL 965-10-1

IDLI GROUP 87803

UNWELDED CENTER ASSEMBLY
extended EE dimension shown
SEE FIGURE 85
69.688 in   EE   1770.06 mm
3.000 in   OD   76.20 mm
ROCKWELL 954-13-1

IDLI GROUP 87804

SQUARE SHAFTING
SEE FIGURE 88
72.000 in   EE   1828.80 mm
.750 in   OD   19.05 mm
ROCKWELL SQS75-72

IDLI GROUP 87805

RECTANGULAR SHAFTING
maximum OD dimension shown
SEE FIGURE 88
72.000 in   EE   1828.80 mm
.875 in   OD   22.23 mm
ROCKWELL SH-P12-11-72

IDLI GROUP 87806

SQUARE SHAFTING
SEE FIGURE 88
72.000 in   EE   1828.80 mm
1.000 in   OD   25.40 mm
ROCKWELL SQS100-72

IDLI GROUP 87807

RECTANGULAR SHAFTING
maximum OD dimension shown
SEE FIGURE 88
72.000 in   EE   1828.80 mm
1.125 in   OD   28.58 mm
ROCKWELL SH-P16-7-72

IDLI GROUP 87808

SQUARE SHAFTING
SEE FIGURE 88
72.000 in   EE   1828.80 mm
1.188 in   OD   30.16 mm
ROCKWELL SQS1187-72

IDLI GROUP 87809

SQUARE SHAFTING
SEE FIGURE 88
72.000 in   EE   1828.80 mm
1.313 in   OD   33.34 mm
ROCKWELL SQS13125-72

IDLI GROUP 87810

UNWELDED CENTER ASSEMBLY
extended EE dimension shown
SEE FIGURE 85
72.375 in   EE   1838.33 mm
3.500 in   OD   88.90 mm
ROCKWELL 965-12-1

IDLI GROUP 87811

UNWELDED CENTER ASSEMBLY
extended EE dimension shown
SEE FIGURE 85
75.625 in   EE   1920.88 mm
3.500 in   OD   88.90 mm
ROCKWELL 959-11-1

IDLI GROUP 87812

UNWELDED CENTER ASSEMBLY
extended EE dimension shown
SEE FIGURE 85
75.750 in   EE   1924.05 mm
4.000 in   OD   101.60 mm
ROCKWELL 959-12-1

IDLI GROUP 87813

UNWELDED CENTER ASSEMBLY
extended EE dimension shown
SEE FIGURE 85
77.000 in   EE   1955.80 mm
3.500 in   OD   88.90 mm
ROCKWELL 954-12-1

IDLI GROUP 87814

UNWELDED CENTER ASSEMBLY
extended EE dimension shown
SEE FIGURE 85
77.125 in   EE   1958.98 mm
4.000 in   OD   101.60 mm
ROCKWELL 954-11-1

IDLI GROUP 87815

UNWELDED CENTER ASSEMBLY
extended EE dimension shown
SEE FIGURE 85
78.500 in   EE   1993.90 mm
4.000 in   OD   101.60 mm
ROCKWELL 965-13-1

IDLI GROUP 87816

UNWELDED CENTER ASSEMBLY
extended EE dimension shown
SEE FIGURE 85
79.000 in   EE   2006.60 mm
3.500 in   OD   88.90 mm
ROCKWELL 954-10-1

IDLI GROUP 87817

UNWELDED CENTER ASSEMBLY
extended EE dimension shown
SEE FIGURE 85
81.188 in   EE   2062.16 mm
4.000 in   OD   101.60 mm
ROCKWELL 955-10-1

IDLI GROUP 87818

UNWELDED CENTER ASSEMBLY
extended EE dimension shown
SEE FIGURE 85
81.750 in   EE   2076.45 mm
4.500 in   OD   114.30 mm
ROCKWELL 960-10-1

IDLI GROUP 87819

UNWELDED CENTER ASSEMBLY
extended EE dimension shown
SEE FIGURE 85
82.188 in   EE   2087.56 mm
4.000 in   OD   101.60 mm
ROCKWELL 955-11-1

IDLI GROUP 87820

UNWELDED CENTER ASSEMBLY
extended EE dimension shown
SEE FIGURE 85
83.188 in   EE   2112.96 mm
4.500 in   OD   114.30 mm
ROCKWELL 960-11-1

IDLI GROUP 87821

UNWELDED CENTER ASSEMBLY
extended EE dimension shown
SEE FIGURE 85
84.375 in   EE   2143.13 mm
3.500 in   OD   88.90 mm
ROCKWELL 965-11-1

IDLI GROUP 87822

SQUARE SHAFTING
SEE FIGURE 88
108.000 in   EE   2743.20 mm
.750 in   OD   19.05 mm
ROCKWELL SQS75-108

IDLI GROUP 87823

RECTANGULAR SHAFTING
maximum OD dimension shown
SEE FIGURE 88
108.000 in   EE   2743.20 mm
.875 in   OD   22.23 mm
ROCKWELL SH-P12-11-108

IDLI GROUP 87824

SQUARE SHAFTING
SEE FIGURE 88
108.000 in   EE   2743.20 mm
1.000 in   OD   25.40 mm
ROCKWELL SQS100-108

IDLI GROUP 87825

RECTANGULAR SHAFTING
maximum OD dimension shown
SEE FIGURE 88
108.000 in   EE   2743.20 mm
1.125 in   OD   28.58 mm
ROCKWELL SH-P16-7-108

IDLI GROUP 87826

SQUARE SHAFTING
SEE FIGURE 88
108.000 in   EE   2743.20 mm
1.188 in   OD   30.16 mm
ROCKWELL SQS1187-108

IDLI GROUP 87827

SQUARE SHAFTING
SEE FIGURE 88
108.000 in   EE   2743.20 mm
1.313 in   OD   33.34 mm
ROCKWELL SQS13125-108

ENGINEERING CATALOGS MUST BE CONSULTED FOR DETAILS NOT INCLUDED
IN THIS GUIDE. SPECIFIC DESIGNS, MATERIAL CONTENT, TOLERANCES
LUBE FITTINGS AND OTHER DIMENSIONS ARE INTENTIONALLY OMITTED HERE.

FOR YOUR CONVENIENCE
NOTES ON OTHER INFORMATION

We urge you to send us any data that
you feel should be included with the
future editions of the I.D.L.I. Guide.

IDLI GROUP 87901

SLIP ASSEMBLY
WITH RECTANGULAR SHAFT
maximum dimension is shown
SEE FIGURE 80

40.000 in   CC   1016.00 mm
ROCKWELL 907-13-1

IDLI GROUP 87902

SLIP ASSEMBLY WITH SHIELDED
RECTANGULAR TUBE AND SHAFT
extended dimension is shown
SEE FIGURE 77

40.125 in   CC   1019.18 mm
ROCKWELL 982-50-1

IDLI GROUP 87903

SLIP ASSEMBLY WITH SHIELDED
RECTANGULAR TUBE AND SHAFT
extended dimension is shown
SEE FIGURE 77B

40.125 in   CC   1019.18 mm
ROCKWELL 982-50-5

IDLI GROUP 87904

SLIP ASSEMBLY WITH SHIELDED
RECTANGULAR TUBE AND SHAFT
extended dimension is shown
SEE FIGURE 77

40.875 in   CC   1038.23 mm
ROCKWELL 927-50-1

IDLI GROUP 87905

SLIP ASSEMBLY WITH SHIELDED
RECTANGULAR TUBE AND SHAFT
extended dimension is shown
SEE FIGURE 77B

40.875 in   CC   1038.23 mm
ROCKWELL 927-50-5

IDLI GROUP 87906

SLIP ASSEMBLY
WITH RECTANGULAR SHAFT
maximum dimension is shown
SEE FIGURE 80

41.500 in   CC   1054.10 mm
ROCKWELL 902-15-1

IDLI GROUP 87907

SLIP ASSEMBLY
WITH RECTANGULAR SHAFT
maximum dimension is shown
SEE FIGURE 80

41.500 in   CC   1054.10 mm
ROCKWELL 903-15-1

IDLI GROUP 87908

SLIP ASSEMBLY
WITH RECTANGULAR SHAFT
maximum dimension is shown
SEE FIGURE 79

41.875 in   CC   1063.63 mm
ROCKWELL 907-15-1

IDLI GROUP 87909

SLIP ASSEMBLY
WITH RECTANGULAR SHAFT
maximum dimension is shown
SEE FIGURE 79

43.313 in   CC   1100.14 mm
ROCKWELL 902-17-1

IDLI GROUP 87910

SLIP ASSEMBLY
WITH RECTANGULAR SHAFT
maximum dimension is shown
SEE FIGURE 79

43.313 in   CC   1100.14 mm
ROCKWELL 903-17-1

IDLI GROUP 87911

SLIP ASSEMBLY WITH SHIELDED
RECTANGULAR TUBE AND SHAFT
extended dimension is shown
SEE FIGURE 77

44.063 in   CC   1119.19 mm
ROCKWELL 922-50-1
SPICER 208655-1

IDLI GROUP 87912

SLIP ASSEMBLY WITH SHIELDED
RECTANGULAR TUBE AND SHAFT
extended dimension is shown
SEE FIGURE 77A

44.063 in   CC   1119.19 mm
NEAPCO ST22-002900P
ROCKWELL 922-50-4

IDLI GROUP 87913

SLIP ASSEMBLY WITH SHIELDED
RECTANGULAR TUBE AND SHAFT
extended dimension is shown
SEE FIGURE 77B

44.063 in   CC   1119.19 mm
BORG-WARNR 64500
NEAPCO 60-2229
NEAPCO ST22-002900
PRECISION 1848
ROCKWELL 922-50-5

IDLI GROUP 87914

SLIP ASSEMBLY WITH SHIELDED
RECTANGULAR TUBE AND SHAFT
extended dimension is shown
SEE FIGURE 77

47.563 in   CC   1208.09 mm
ROCKWELL 909-50-1
SPICER 208955-1
SPICER 208955-2

IDLI GROUP 87915

SLIP ASSEMBLY WITH SHIELDED
RECTANGULAR TUBE AND SHAFT
extended dimension is shown
SEE FIGURE 77A

47.563 in   CC   1208.09 mm
NEAPCO ST20-002800P
ROCKWELL 909-50-4

IDLI GROUP 87916

SLIP ASSEMBLY WITH SHIELDED
RECTANGULAR TUBE AND SHAFT
extended dimension is shown
SEE FIGURE 77B

47.563 in   CC   1208.09 mm
BORG-WARNR 66128
NEAPCO 60-2028
NEAPCO ST20-002800
ROCKWELL 909-50-5

IDLI GROUP 87917

SLIP ASSEMBLY WITH SHIELDED
RECTANGULAR TUBE AND SHAFT
extended dimension is shown
SEE FIGURE 77

48.000 in   CC   1219.20 mm
ROCKWELL 907-50-1
SPICER 208555-1
SPICER 208555-2

IDLI GROUP 87918

SLIP ASSEMBLY WITH SHIELDED
RECTANGULAR TUBE AND SHAFT
extended dimension is shown
SEE FIGURE 77A

48.000 in   CC   1219.20 mm
NEAPCO STR12-002700P
ROCKWELL 907-50-4
SPICER 208565-4

IDLI GROUP 87919

SLIP ASSEMBLY WITH SHIELDED
RECTANGULAR TUBE AND SHAFT
extended dimension is shown
SEE FIGURE 77B

48.000 in   CC   1219.20 mm
BORG-WARNR 60370
NEAPCO 60-1227
NEAPCO ST12-002700
NEAPCO STR12-002700
ROCKWELL 907-50-5

IDLI GROUP 87920

SLIP ASSEMBLY WITH SHIELDED
RECTANGULAR TUBE AND SHAFT
extended dimension is shown
SEE FIGURE 77

49.125 in   CC   1247.78 mm
BORG-WARNR 61768
NEAPCO 60-0627
NEAPCO STR06-002700
ROCKWELL 903-50-1
SPICER 208805-1

IDLI GROUP 87921

SLIP ASSEMBLY WITH SHIELDED
RECTANGULAR TUBE AND SHAFT
extended dimension is shown
SEE FIGURE 77A

49.125 in   CC   1247.78 mm
ROCKWELL 903-50-4

IDLI GROUP 87922

SLIP ASSEMBLY WITH SHIELDED
RECTANGULAR TUBE AND SHAFT
extended dimension is shown
SEE FIGURE 77B

49.125 in   CC   1247.78 mm
ROCKWELL 903-50-5

IDLI GROUP 87923

SLIP ASSEMBLY
WITH SPLINED SHAFT
maximum dimension is shown
SEE FIGURE 78

54.188 in   CC   1376.36 mm
BORG-WARNR 60448
NEAPCO 63-1254
NEAPCO UTP12-005400
ROCKWELL 907-10-1

IDLI GROUP 87924

SLIP ASSEMBLY
WITH SPLINED SHAFT
maximum dimension is shown
SEE FIGURE 78

55.875 in   CC   1419.23 mm
ROCKWELL 902-10-1

IDLI GROUP 87925

SLIP ASSEMBLY
WITH SPLINED SHAFT
maximum dimension is shown
SEE FIGURE 78

55.875 in   CC   1419.23 mm
ROCKWELL 902-12-1

IDLI GROUP 87926

SLIP ASSEMBLY
WITH SPLINED SHAFT
maximum dimension is shown
SEE FIGURE 78

55.875 in   CC   1419.23 mm
ROCKWELL 903-10-1

IDLI GROUP 87927

SLIP ASSEMBLY
WITH SPLINED SHAFT
maximum dimension is shown
SEE FIGURE 78

55.875 in   CC   1419.23 mm
ROCKWELL 903-12-1

BD = BEARING DIAMETER (outside)   **BW** = BEARING WIDTH   **CB** = CROSS LENGTH WITH BEARINGS   **CC** = CENTER TO CENTER   **CD** = CROSS DIAMETER
CL = CROSS LENGTH WITHOUT BEARINGS   **EC** = END TO CENTER or FACE TO FACE   **EE** = END TO END   **EL** = EFFECTIVE LENGTH
HD = HUB DIAMETER (or insert)   **IE** = INSIDE OF EARS (or recess)   **OD** = OUTSIDE DIAMETER   **OE** = OUTSIDE OF EARS   **SB** = SPLINE OR BORE SIZE

CAUTION: BE SURE TO REFER TO ENGINEERING CATALOGS FOR SPECIAL APPLICATIONS THAT REQUIRE SPECIFIC MATERIAL CONTENT, TOLERANCES, ETC. SEE FOOTNOTE.

IDLI GROUP 87928

SLIP ASSEMBLY WITH SHIELDED RECTANGULAR TUBE AND SHAFT
extended dimension is shown
SEE FIGURE 77
60.125 in   CC   1527.18 mm
ROCKWELL 982-51-1

IDLI GROUP 87929

SLIP ASSEMBLY WITH SHIELDED RECTANGULAR TUBE AND SHAFT
extended dimension is shown
SEE FIGURE 77B
60.125 in   CC   1527.18 mm
ROCKWELL 982-51-5
SPICER 949238-1

IDLI GROUP 87930

SLIP ASSEMBLY WITH SHIELDED RECTANGULAR TUBE AND SHAFT
extended dimension is shown
SEE FIGURE 77
60.875 in   CC   1546.23 mm
ROCKWELL 927-51-1

IDLI GROUP 87931

SLIP ASSEMBLY WITH SHIELDED RECTANGULAR TUBE AND SHAFT
extended dimension is shown
SEE FIGURE 77B
60.875 in   CC   1546.23 mm
ROCKWELL 927-51-5

IDLI GROUP 87932

SLIP ASSEMBLY WITH SHIELDED RECTANGULAR TUBE AND SHAFT
extended dimension is shown
SEE FIGURE 77
64.063 in   CC   1627.19 mm
PRECISION 1848
ROCKWELL 922-51-1
SPICER 208655-2

IDLI GROUP 87933

SLIP ASSEMBLY WITH SHIELDED RECTANGULAR TUBE AND SHAFT
extended dimension is shown
SEE FIGURE 77A
64.063 in   CC   1627.19 mm
NEAPCO 62-2239
NEAPCO 62-2251
NEAPCO ST22-003900P
NEAPCO UT22-005100
NEAPCO UT22-3900
NEAPCO UT22-5100
PRECISION 1848
ROCKWELL 922-51-4
SPICER 208663-4
TRW 21465
TRW 21470

WESCO 35N48
WESCO 920140

IDLI GROUP 87934

SLIP ASSEMBLY WITH SHIELDED RECTANGULAR TUBE AND SHAFT
extended dimension is shown
SEE FIGURE 77B
64.063 in   CC   1627.19 mm
BORG-WARNR 64600
NEAPCO 60-2239
NEAPCO ST22-003900
ROCKWELL 922-51-5

IDLI GROUP 87935

SLIP ASSEMBLY WITH SPLINED SHAFT
maximum dimension is shown
SEE FIGURE 78
66.188 in   CC   1681.16 mm
BORG-WARNR 64602
NEAPCO 63-1266
NEAPCO UTP12-006600
ROCKWELL 907-11-1

IDLI GROUP 87936

SLIP ASSEMBLY WITH SPLINED SHAFT
maximum dimension is shown
SEE FIGURE 78
66.875 in   CC   1698.63 mm
ROCKWELL 903-11-1

IDLI GROUP 87937

SLIP ASSEMBLY WITH SHIELDED RECTANGULAR TUBE AND SHAFT
extended dimension is shown
SEE FIGURE 77
67.563 in   CC   1716.09 mm
ROCKWELL 909-51-1
SPICER 208955-2

IDLI GROUP 87938

SLIP ASSEMBLY WITH SHIELDED RECTANGULAR TUBE AND SHAFT
extended dimension is shown
SEE FIGURE 77A
67.563 in   CC   1716.09 mm
NEAPCO 62-2040
NEAPCO 62-2052
NEAPCO ST20-004000P
NEAPCO UT20-005200
NEAPCO UT20-4000
PRECISION 1348
ROCKWELL 909-51-4
SPICER 208963-4
TRW 21451
TRW 21456
WESCO 14N48TB
WESCO 62-2040
WESCO 920-118
WESCO 920119

IDLI GROUP 87939

SLIP ASSEMBLY WITH SHIELDED RECTANGULAR TUBE AND SHAFT
extended dimension is shown
SEE FIGURE 77B
67.563 in   CC   1716.09 mm
BORG-WARNR 66140
NEAPCO 60-2040
NEAPCO ST20-004000
ROCKWELL 909-51-5

IDLI GROUP 87940

SLIP ASSEMBLY WITH SPLINED SHAFT
maximum dimension is shown
SEE FIGURE 78
67.875 in   CC   1724.03 mm
ROCKWELL 902-11-1

IDLI GROUP 87941

SLIP ASSEMBLY WITH SPLINED SHAFT
maximum dimension is shown
SEE FIGURE 78
67.875 in   CC   1724.03 mm
ROCKWELL 902-13-1

IDLI GROUP 87942

SLIP ASSEMBLY WITH SPLINED SHAFT
maximum dimension is shown
SEE FIGURE 78
67.875 in   CC   1724.03 mm
ROCKWELL 903-13-1

IDLI GROUP 87943

SLIP ASSEMBLY WITH SHIELDED RECTANGULAR TUBE AND SHAFT
extended dimension is shown
SEE FIGURE 77
68.000 in   CC   1727.20 mm
ROCKWELL 907-51-1
SPICER 208555-2
WESCO 920-113

IDLI GROUP 87944

SLIP ASSEMBLY WITH SHIELDED RECTANGULAR TUBE AND SHAFT
extended dimension is shown
SEE FIGURE 77A
68.000 in   CC   1727.20 mm
BORG-WARNR 208565-4
NEAPCO STR12-003900P
ROCKWELL 907-51-4
SPICER 208565-4

IDLI GROUP 87945

SLIP ASSEMBLY WITH SHIELDED RECTANGULAR TUBE AND SHAFT
extended dimension is shown
SEE FIGURE 77B
68.000 in   CC   1727.20 mm
BORG-WARNR 60371
NEAPCO 60-1239
NEAPCO ST12-003900
NEAPCO STR12-003900
ROCKWELL 907-51-5
SPICER 208555-4

IDLI GROUP 87946

SLIP ASSEMBLY WITH SHIELDED RECTANGULAR TUBE AND SHAFT
extended dimension is shown
SEE FIGURE 77
69.125 in   CC   1755.78 mm
BORG-WARNR 61774
NEAPCO 60-0639
NEAPCO STR06-003900
ROCKWELL 903-51-1
SPICER 208805-2

IDLI GROUP 87947

SLIP ASSEMBLY WITH SHIELDED RECTANGULAR TUBE AND SHAFT
extended dimension is shown
SEE FIGURE 77A
69.125 in   CC   1755.78 mm
ROCKWELL 903-51-4

IDLI GROUP 87948

SLIP ASSEMBLY WITH SHIELDED RECTANGULAR TUBE AND SHAFT
extended dimension is shown
SEE FIGURE 77B
69.125 in   CC   1755.78 mm
ROCKWELL 903-51-5

IDLI GROUP 87949

SLIP ASSEMBLY WITH RECTANGULAR SHAFT
maximum dimension is shown
SEE FIGURE 80
70.000 in   CC   1778.00 mm
ROCKWELL 907-12-1

IDLI GROUP 87950

SLIP ASSEMBLY WITH RECTANGULAR SHAFT
maximum dimension is shown
SEE FIGURE 80
71.500 in   CC   1816.10 mm
ROCKWELL 902-14-1

IDLI GUIDE   COPYRIGHT © INTERCHANGE, INC.   ST. LOUIS PARK, MN. 55416 USA

IDLI GROUP 87951

**SLIP ASSEMBLY
WITH RECTANGULAR SHAFT**
maximum dimension is shown
SEE FIGURE 80

71.500 in   CC   1816.10 mm
ROCKWELL 903-14-1

IDLI GROUP 87952

**SLIP ASSEMBLY
WITH RECTANGULAR SHAFT**
maximum dimension is shown
SEE FIGURE 79

71.875 in   CC   1825.63 mm
ROCKWELL 907-14-1

IDLI GROUP 87953

**SLIP ASSEMBLY
WITH RECTANGULAR SHAFT**
maximum dimension is shown
SEE FIGURE 79

73.313 in   CC   1862.14 mm
ROCKWELL 902-16-1

IDLI GROUP 87954

**SLIP ASSEMBLY
WITH RECTANGULAR SHAFT**
maximum dimension is shown
SEE FIGURE 79

73.313 in   CC   1862.14 mm
ROCKWELL 903-16-1

IDLI GROUP 87955

**SLIP ASSEMBLY WITH SHIELDED
RECTANGULAR TUBE AND SHAFT**
extended dimension is shown
SEE FIGURE 77

100.188 in   CC   2544.76 mm
ROCKWELL 922-52-1

IDLI GROUP 87956

**SLIP ASSEMBLY WITH SHIELDED
RECTANGULAR TUBE AND SHAFT**
extended dimension is shown
SEE FIGURE 77A

100.188 in   CC   2544.76 mm
ROCKWELL 922-52-4

IDLI GROUP 87957

**SLIP ASSEMBLY WITH SHIELDED
RECTANGULAR TUBE AND SHAFT**
extended dimension is shown
SEE FIGURE 77B

100.188 in   CC   2544.76 mm
ROCKWELL 922-52-5

IDLI GROUP 87958

**SLIP ASSEMBLY WITH SHIELDED
RECTANGULAR TUBE AND SHAFT**
extended dimension is shown
SEE FIGURE 77

102.750 in   CC   2609.85 mm
ROCKWELL 982-52-1

IDLI GROUP 87959

**SLIP ASSEMBLY WITH SHIELDED
RECTANGULAR TUBE AND SHAFT**
extended dimension is shown
SEE FIGURE 77B

102.750 in   CC   2609.85 mm
ROCKWELL 982-52-5

IDLI GROUP 87960

**SLIP ASSEMBLY WITH SHIELDED
RECTANGULAR TUBE AND SHAFT**
extended dimension is shown
SEE FIGURE 77

103.500 in   CC   2628.90 mm
ROCKWELL 927-52-1

IDLI GROUP 87961

**SLIP ASSEMBLY WITH SHIELDED
RECTANGULAR TUBE AND SHAFT**
extended dimension is shown
SEE FIGURE 77B

103.500 in   CC   2628.90 mm
ROCKWELL 927-52-5

IDLI GROUP 87962

**SLIP ASSEMBLY WITH SHIELDED
RECTANGULAR TUBE AND SHAFT**
extended dimension is shown
SEE FIGURE 77

111.000 in   CC   2819.40 mm
ROCKWELL 907-52-1

IDLI GROUP 87963

**SLIP ASSEMBLY WITH SHIELDED
RECTANGULAR TUBE AND SHAFT**
extended dimension is shown
SEE FIGURE 77A

111.000 in   CC   2819.40 mm
ROCKWELL 907-52-4

IDLI GROUP 87964

**SLIP ASSEMBLY WITH SHIELDED
RECTANGULAR TUBE AND SHAFT**
extended dimension is shown
SEE FIGURE 77B

111.000 in   CC   2819.40 mm
ROCKWELL 907-52-5

IDLI GROUP 87965

**SLIP ASSEMBLY WITH SHIELDED
RECTANGULAR TUBE AND SHAFT**
extended dimension is shown
SEE FIGURE 77

112.750 in   CC   2863.85 mm
ROCKWELL 909-52-1

IDLI GROUP 87966

**SLIP ASSEMBLY WITH SHIELDED
RECTANGULAR TUBE AND SHAFT**
extended dimension is shown
SEE FIGURE 77A

112.750 in   CC   2863.85 mm
ROCKWELL 909-52-4

IDLI GROUP 87967

**SLIP ASSEMBLY WITH SHIELDED
RECTANGULAR TUBE AND SHAFT**
extended dimension is shown
SEE FIGURE 77B

112.750 in   CC   2863.85 mm
ROCKWELL 909-52-5

IDLI GROUP 87968

**SLIP ASSEMBLY WITH SHIELDED
RECTANGULAR TUBE AND SHAFT**
extended dimension is shown
SEE FIGURE 77

114.750 in   CC   2914.65 mm
ROCKWELL 903-54-1

IDLI GROUP 87969

**SLIP ASSEMBLY WITH SHIELDED
RECTANGULAR TUBE AND SHAFT**
extended dimension is shown
SEE FIGURE 77A

114.750 in   CC   2914.65 mm
ROCKWELL 903-54-4

IDLI GROUP 87970

**SLIP ASSEMBLY WITH SHIELDED
RECTANGULAR TUBE AND SHAFT**
extended dimension is shown
SEE FIGURE 77B

114.750 in   CC   2914.65 mm
ROCKWELL 903-54-5

BD = BEARING DIAMETER (outside)   **BW** = BEARING WIDTH   **CB** = CROSS LENGTH WITH BEARINGS   **CC** = CENTER TO CENTER   **CD** = CROSS DIAMETER
CL = CROSS LENGTH WITHOUT BEARINGS   **EC** = END TO CENTER or FACE TO FACE   **EE** = END TO END   **EL** = EFFECTIVE LENGTH
HD = HUB DIAMETER (or insert)   **IE** = INSIDE OF EARS (or recess)   **OD** = OUTSIDE DIAMETER   **OE** = OUTSIDE OF EARS   **SB** = SPLINE OR BORE SIZE

IDLI GROUP 88000

BORG-WARNR	61112
BORG-WARNR	61212
NEAPCO	11112
NEAPCO	12-1112
NEAPCO	211A2
NEAPCO	22-1301
PRECISION	1202
SPICER	1-4-2923
SPICER	1-4-2953
WESCO	12N134
WESCO	850011

IDLI GROUP 88000.01

ALLOY	0601
ALLOY	601
BORG-WARNR	61113
NEAPCO	11113
NEAPCO	12-1113
PRECISION	1205
SPICER	1-4-2973
TRW	21100
TRW	21100D
WESCO	12N0144
WESCO	12N144
WESCO	850-016
WESCO	855016

IDLI GROUP 88000.02

BORG-WARNR	12-1114
BORG-WARNR	61114
NEAPCO	11114
PRECISION	1208
SPICER	1-4-2983
WESCO	12N154
WESCO	850-018
WESCO	855018

IDLI GROUP 88000.03

BORG-WARNR	61116
NEAPCO	11116
NEAPCO	12-1116
PRECISION	1215
SPICER	1-4-3023

IDLI GROUP 88000.04

ALLOY	0603
ALLOY	603
BORG-WARNR	61117
G & G MFG	184-1218
HAYES	1-4-3043
HAYES	1-4-3063
NEAPCO	11117
NEAPCO	111A4
NEAPCO	111C9
NEAPCO	12-1117
PRECISION	1219
REX CHAIN	760035
SPICER	1-4-3063

IDLI GROUP 88000.05

ALLOY	0604
ALLOY	604
BORG-WARNR	61118
NEAPCO	11118
NEAPCO	12-1118
PRECISION	1222
SPICER	1-4-3083
TRW	21107

IDLI GROUP 88000.06

G & G MFG	184-1220
HAYES	1-4-3093
HAYES	1-4-3113
NEAPCO	111B9
NEAPCO	111E9
NEAPCO	111F2
NEAPCO	111F4
NEAPCO	111P7
NEAPCO	12-1319
PRECISION	1228

IDLI GROUP 88000.07

G & G MFG	192-1222
HAYES	1-4-981
NEAPCO	111Z7
PRECISION	1259

IDLI GROUP 88000.08

BORG-WARNR	61166
MUNCIE	MJY27
NEAPCO	11166
NEAPCO	12-1166
PRECISION	1290
WESCO	12N180-22
WESCO	850027

IDLI GROUP 88000.09

BORG-WARNR	60101
BORG-WARNR	60102
G & G MFG	128-1200
HAYES	228162X
HAYES	228397X
NEAPCO	1859X
NEAPCO	56-1859
NEAPCO	56-2299
NEAPCO	56-2300
NEAPCO	N1859X
NEAPCO	N2004X
NEAPCO	N2299-1X
NEAPCO	N2299X
NEAPCO	N-2299X-1
PRECISION	1299
SPICER	228305X
SPICER	228397X
TRW	21501
WESCO	12NNB2
WESCO	251015

IDLI GROUP 88000.10

NEAPCO	2034A
NEAPCO	53-2034
PRECISION	1390
SPICER	2-55-92
WESCO	14N-S13
WESCO	560-101
WESCO	560-102
WESCO	760054

IDLI GROUP 88000.11

PRECISION	1448
SPICER	949238-1
WESCO	44N48
WESCO	920151

IDLI GROUP 88000.12

G & G MFG	186-0600
PRECISION	1449
SPICER	229116-3227X
WESCO	44N48F
WESCO	920152

IDLI GROUP 88000.13

PRECISION	1499
SPICER	229152X
WESCO	281045
WESCO	44N-NB4

IDLI GROUP 88000.14

AEC	571
ALCO	1782
ALLOY	1782
ALLOY	498
BORG-WARNR	114-146
BPC	PB7A
CHRYSLER	D300
DIAMOND R	100336K1
GMB	G5-103X
MCQUAY-NOR	U03-9
MOPAR	D300
MOTOR MAST	2207-P
MOTOR MAST	2207PE
MUNCIE	MK1X
NEAPCO	0505X
NEAPCO	1-0505
NEAPCO	15052X
NEAPCO	1-605
NEAPCO	1-675
NEAPCO	605X
NEAPCO	675X
PILOT	1135R
PRECISION	1500
REPUBLIC	CB140P
TRW	20018
WESCO	201-571
ZELLER	2207P

IDLI GROUP 88000.15

AEC	510
AEC	570
AEC	570-6N8
AEC	570P
AEC	576
AEC	AE575
ALCO	1783
ALCO	499
ALLIS-CHLM	61709
ALLIS-CHLM	6179097
ALLIS-CHLM	645053
ALLIS-CHLM	645110
ALLOY	1070
ALLOY	1783
ALLOY	1786
ALLOY	576
ALLOY	J5-170
BORG-WARNR	114-145
BORG-WARNR	114-148
BORG-WARNR	114-570
BORG-WARNR	114-575
BPC	NB7A
BWE	114-570
CATERPILLR	709715
CATERPILLR	801508
CLEVE.MOT.	CM5-103X
CLEVE.MOT.	CM5555X
DIAMOND R	859107T2
G & G MFG	302-0675
GMB	G499
GMB	G5-103X
GMB	G5-170X
GMC	2296407
GMC	612072
IHC	975663R92
LEMPCO	15-70X
MASTERS	MS570X
MOTOR MAST	2107A
MOTOR MAST	2107-A
MOTOR MAST	2107A
MOTOR MAST	2107-A
MOTOR MAST	2207B
MOTOR MAST	2207C
MOTOR MAST	2207E
MUNCIE	MK2X
MUNCIE	MK3X
NEAPCO	0505X1
NEAPCO	0605X
NEAPCO	0675X
NEAPCO	0675X1A
NEAPCO	06L6
NEAPCO	1-0170
NEAPCO	1-0605
NEAPCO	1-0616
NEAPCO	1-0675
NEAPCO	1-0677
NEAPCO	1-170
NEAPCO	1-605
NEAPCO	1605X
NEAPCO	1-675
NEAPCO	1675X
NEAPCO	1675X1A
NEAPCO	605X
NEAPCO	675X
NEW HOLL.	139842

CAUTION: BE SURE TO REFER TO ENGINEERING CATALOGS FOR SPECIAL APPLICATIONS THAT REQUIRE SPECIFIC MATERIAL CONTENT, TOLERANCES, ETC. SEE FOOTNOTE.

PERFECT-CI	430
PERFECTION	RG570K
PILOT	1135R
PILOT	11575
PRECISION	1501
REPUBLIC	CB140R
REPUBLIC	CB14DR
SPICER	5-170X
SPICER	5-430X
TRU-CROSS	110
TRU-CROSS	170
TRU-CROSS	430
TRW	20017
TRW	20018
WESCO	201-570
WESCO	N1570
WHITE	1717518
ZELLER	2107A
ZELLER	2207A
ZELLER	2207B

IDLI GROUP 88000.16

AEC	510
AEC	AE572
BORG-WARNR	114-148
BORG-WARNR	114-503
GMB	G5-110X
GMB	GU7010
HUSCO	1605XBA
HUSCO	5-110XBA
HUSCO	NB7
HUSCO	NB751
HUSCO	NB751A
MOPAR	2207C
MOTOR MAST	2207C
MOTOR MAST	2207E
MUNCIE	MK3X
NEAPCO	0605X1
NEAPCO	0675X1
NEAPCO	1-0606
NEAPCO	1-0676
NEAPCO	1-1556
NEAPCO	1605X1
NEAPCO	1675X1
NEAPCO	1-676
NEAPCO	281556
NEAPCO	675X1
PRECISION	1502
TRW	20019
TRW	20198
WESCO	201-128

IDLI GROUP 88000.17

ALLOY	1192
BORG-WARNR	66251
BORG-WARNR	68516
NEAPCO	1225X
NEAPCO	1856X
NEAPCO	228311X
NEAPCO	56-1225
NEAPCO	56-1856
PRECISION	1503
SPICER	228317X
TRW	21351
TRW	21352
WESCO	201102
WESCO	2QDK

IDLI GROUP 88000.18

ALLOY	1190
BORG-WARNR	66251
MCQUAY-NOR	U03-110
NEAPCO	0625X
NEAPCO	56-0625
NEAPCO	625X
PRECISION	1504
SPICER	228163X
SPICER	228317X
TRW	21350
TRW	21351
WESCO	1QDK
WESCO	201205

IDLI GROUP 88000.19

ALCO	0506
ALLOY	506
BORG-WARNR	65344
BPC	H7EY-5/8R-3/16KW
HUB CITY	0335-00045
MOPAR	D307
MUNCIE	EY100
NEAPCO	051D4
NEAPCO	151D4
NEAPCO	151P1
NEAPCO	15-5336
NEAPCO	51D4
PRECISION	1510
SPICER	10-4-0373
SPICER	10-4-373
TRW	21000
WESCO	1A103
WESCO	1A103-1
WESCO	510001

IDLI GROUP 88000.20

AEC	0-13
AEC	40013
AEC	AE013
AEC	AEO13
ALCO	0516
ALLOY	516
BORG-WARNR	100-13
BORG-WARNR	65111
BPC	H7EY-3/4R-3/16KW
CHRYSLER	D318
CLEVE.MOT.	CM13ST
FLEETRITE	ZAD50Y13
G & G MFG	184-0612
LEMPCO	1-13
MCQUAY-NOR	U03-47
MOPAR	D318
MOTOR MAST	MY0-13
MUNCIE	EY101
NEAPCO	05103
NEAPCO	05111
NEAPCO	10-4133
NEAPCO	15111
NEAPCO	151Z2
NEAPCO	15-5103
NEAPCO	15-5111
NEAPCO	5111
NEAPCO	N104133
PILOT	50Y13
PRECISION	1511
REPUBLIC	HY362-1
REPUBLIC	HY362-2
REPUBLIC	HY363-8
SPICER	10-4-133
SPICER	10-4-303
SPICER	10-4-483
TRW	20602
WESCO	1A123
WESCO	1A123-13
WESCO	1A123-2
WESCO	1A123-4
WESCO	510005
WESCO	51-1230
WESCO	YA123

IDLI GROUP 88000.21

AEC	AE40013
ALCO	0525
ALCO	516
ALLOY	525
BORG-WARNR	2B0001
BORG-WARNR	2B1
BORG-WARNR	65123
CHELSEA	Y123Y123
CMP	CM13SH
CMP	CM13ST
FLEETRITE	ZAD5P2
G & G MFG	184-0512
G & G MFG	184-0612
HUB CITY	03-35-00008
LEMPCO	1-13
LEMPCO	12-13
LEMPCO	2L1
LEMPCO	2L4
MCQUAY-NOR	U03-47
MOPAR	D326
MOTOR MAST	2B1
MOTOR MAST	2B2
MOTOR MAST	MY0-13
MOTOR MAST	MYO13
MUNCIE	101-1-101
MUNCIE	101-1-120
MUNCIE	101-2-101
MUNCIE	101-2-120
MUNCIE	10-4-13
MUNCIE	EY102
NEAPCO	05111
NEAPCO	05123
NEAPCO	051BH
NEAPCO	051CM
NEAPCO	051G1
NEAPCO	051R4
NEAPCO	061111
NEAPCO	061113
NEAPCO	062323
NEAPCO	06B8B8
NEAPCO	10-0413
NEAPCO	11-2005
NEAPCO	11-2017
NEAPCO	15123
NEAPCO	151Z1
NEAPCO	15-5123
NEAPCO	161111
NEAPCO	5123
NEAPCO	65111
NEAPCO	N10242-17
NEAPCO	N10242-5
NEAPCO	N10413
PRECISION	1512
REPUBLIC	2790-100M
REPUBLIC	HY362-1
REPUBLIC	HY362-2
SPICER	10242-17SF
SPICER	10242-18SF
SPICER	10242-5SF
SPICER	10-4-0013
SPICER	10-4-0133
SPICER	10-4-13
SPICER	10-4-133
TRW	20501
TRW	20502
TRW	20509
TRW	20602
WESCO	1A123-1
WESCO	510006
WESCO	51-1230
WESCO	A123-1
WESCO	A123A123
WESCO	A144-1

IDLI GROUP 88000.22

AEC	0-33
AEC	0-43
AEC	40043
AEC	AE043
AEC	AE243
AEC	AEO43
ALCO	0502
ALCO	502
ALLOY	502
BORG-WARNR	100-43
BORG-WARNR	2B13
BORG-WARNR	65112
BORG-WARNR	65132
BORG-WARNR	65162
BORG-WARNR	65167
BORG-WARNR	65219
BORG-WARNR	65261
BORG-WARNR	65262
BPC	H7EY-13/16R-1/4KW
CHELSEA	Y134Y134

BD = BEARING DIAMETER (outside) BW = BEARING WIDTH CB = CROSS LENGTH WITH BEARINGS CC = CENTER TO CENTER CD = CROSS DIAMETER
CL = CROSS LENGTH WITHOUT BEARINGS EC = END TO CENTER or FACE TO FACE EE = END TO END EL = EFFECTIVE LENGTH
HD = HUB DIAMETER (or insert) IE = INSIDE OF EARS (or recess) OD = OUTSIDE DIAMETER OE = OUTSIDE OF EARS SB = SPLINE OR BORE SIZE

CHELSEA	Y134Y144
CHELSEA	Y134Y154
CHELSEA	Y134Y164
CHRYSLER	D303
CLEVE.MOT.	CM-33ST
CLEVE.MOT.	CM43ST
CMP	CM43SH
CMP	CM43ST
FLEETRITE	ZAD50Y43
FLEETRITE	ZAD5P10
FLEETRITE	ZAD5P12
G & G MFG	184-0513
LEMPCO	12-43
LEMPCO	1-33
LEMPCO	1-43
LEMPCO	2L10
LEMPCO	2L11
LEMPCO	2L12
LEMPCO	2L13
MCQUAY-NOR	U03-49
MOPAR	D303
MOTOR MAST	2B10
MOTOR MAST	MYO-33
MOTOR MAST	MYO-43
MUNCIE	10-4-43
MUNCIE	110-1-110
MUNCIE	110-1-120
MUNCIE	110-1-130
MUNCIE	110-1-140
MUNCIE	110-2-110
MUNCIE	110-2-120
MUNCIE	110-2-130
MUNCIE	110-2-140
MUNCIE	EY102
MUNCIE	EY110
MUNCIE	EY112
MUNCIE	EY163
NEAPCO	05112
NEAPCO	05132
NEAPCO	05142
NEAPCO	05162
NEAPCO	05167
NEAPCO	051A2
NEAPCO	051A9
NEAPCO	051B1
NEAPCO	051B2
NEAPCO	051CL
NEAPCO	051G2
NEAPCO	051KZ
NEAPCO	061212
NEAPCO	061213
NEAPCO	061214
NEAPCO	061215
NEAPCO	10-0443
NEAPCO	10-4173
NEAPCO	10-4453
NEAPCO	11-2003
NEAPCO	11-2030
NEAPCO	15112
NEAPCO	151B2
NEAPCO	15-5112
NEAPCO	15-5132
NEAPCO	15-5162
NEAPCO	15-5167
NEAPCO	15-5306
NEAPCO	15-5313
NEAPCO	15-5314
NEAPCO	161213
NEAPCO	161214
NEAPCO	161215
NEAPCO	5112
NEAPCO	51B1
NEAPCO	51B2
NEAPCO	.5P-12
NEAPCO	N10242-3
NEAPCO	N10242-30
NEAPCO	N104173
NEAPCO	N10443
NEAPCO	N104453
PILOT	50Y43
PILOT	.5P-10
PRECISION	1513
REPUBLIC	2790-127M
REPUBLIC	HY362-3A
REPUBLIC	HY362-4
SPICER	10242-21SF
SPICER	10242-30SF
SPICER	10242-3SF
SPICER	10242-40SF
SPICER	10242-45SF
SPICER	10-4-0033
SPICER	10-4-0043
SPICER	10-4-0173
SPICER	10-4-173
SPICER	10-4-43
SPICER	10-4-453
TRW	20503
TRW	20510
TRW	20511
TRW	20512
TRW	20605
TRW	20606
TRW	21001
WESCO	1A134
WESCO	1A134-1
WESCO	1A134-16
WESCO	1A134-17
WESCO	1A134-2
WESCO	1A134-20
WESCO	1A134-21
WESCO	1A134-4
WESCO	510010
WESCO	510011
WESCO	510-106
WESCO	51-1340
WESCO	A134-1
WESCO	A134A164
WESCO	A134A205
WESCO	A144-1
WESCO	A154-1
WESCO	A164-1
WESCO	YA134

IDLI GROUP 88000.23

ALCO	0528
ALLOY	528
BORG-WARNR	65146
BORG-WARNR	65150
MOPAR	D329
MUNCIE	EY111
NEAPCO	05146
NEAPCO	05150
NEAPCO	15146
NEAPCO	15150
NEAPCO	15-5146
NEAPCO	15-5150
PRECISION	1514
WESCO	1A130-17
WESCO	1A130-7
WESCO	1A134-15
WESCO	1A134-18
WESCO	510008
WESCO	510012

IDLI GROUP 88000.24

ALCO	0502
PRECISION	1515

IDLI GROUP 88000.25

AEC	0-63
AEC	40063
AEC	43063
AEC	5363
AEC	AE2B17
AEC	AE2B49
AEC	AEO63
ALCO	0503
ALLOY	503
BORG-WARNR	2B0017
BORG-WARNR	2B17
BORG-WARNR	65113
BORG-WARNR	66141
BPC	H7EY-7/8R-1/4KW
CHELSEA	Y134Y144
CHELSEA	Y144Y144
CHRYSLER	D304
CLEVE.MOT.	CM63ST
CMP	CM63ST
FLEETRITE	ZAD50Y63
FLEETRITE	ZAD5P4
G & G MFG	184-0514
HUB CITY	03-35-00046
LEMPCO	1-63
LEMPCO	2L11
LEMPCO	2L17
LEMPCO	2L4
MCQUAY-NOR	U03-52
MOPAR	D304
MOTOR MAST	2B17
MOTOR MAST	MYO-5363
MOTOR MAST	MYO-63
MUNCIE	101-1-120
MUNCIE	101-2-120
MUNCIE	110-1-120
MUNCIE	110-2-120
MUNCIE	120-1-120
MUNCIE	120-2-120
MUNCIE	EY12
MUNCIE	EY120
NEAPCO	05113
NEAPCO	0513A
NEAPCO	051B3
NEAPCO	051B5
NEAPCO	051B6
NEAPCO	051EZ
NEAPCO	051GV
NEAPCO	051HG
NEAPCO	061113
NEAPCO	061213
NEAPCO	061313
NEAPCO	06DYDY
NEAPCO	11-2016
NEAPCO	15113
NEAPCO	15190
NEAPCO	151B3
NEAPCO	15-5113
NEAPCO	15-5213
NEAPCO	161213
NEAPCO	161313
NEAPCO	5113
NEAPCO	513A
NEAPCO	N10242-16
PILOT	50Y63
PILOT	5P-4
PRECISION	1516
REPUBLIC	2790-150M
REPUBLIC	HY362-6
REPUBLIC	HY362-6A
SPICER	10242-16SF
SPICER	10242-17SF
SPICER	10242-45SF
SPICER	10242-57SF
SPICER	10-4-253
SPICER	10-4-453
TRW	20504
TRW	20509
TRW	20510
TRW	20608
TRW	21002
WESCO	1A144
WESCO	1A144-2
WESCO	1A144-24
WESCO	1A144-4
WESCO	510016
WESCO	510018
WESCO	820-003
WESCO	A123-1
WESCO	A134-1
WESCO	A144-1
WESCO	A144A144

IDLI GROUP 88000.26

AEC	40033
AEC	AE033
AEC	AEO-33
BORG-WARNR	100-33
BORG-WARNR	65167
CMP	CM33ST
LEMPCO	1-33
MOTOR MAST	MYO-33
MUNCIE	10-4-33
NEAPCO	05167

ENGINEERING CATALOGS MUST BE CONSULTED FOR DETAILS NOT INCLUDED
IN THIS GUIDE. SPECIFIC DESIGNS, MATERIAL CONTENT, TOLERANCES
LUBE FITTINGS AND OTHER DIMENSIONS ARE INTENTIONALLY OMITTED HERE.

CAUTION: BE SURE TO REFER TO ENGINEERING CATALOGS FOR SPECIAL APPLICATIONS THAT REQUIRE SPECIFIC MATERIAL CONTENT, TOLERANCES, ETC. SEE FOOTNOTE.

NEAPCO	10-0433
NEAPCO	15162
NEAPCO	15-5167
NEAPCO	N10433
PILOT	50Y33
PRECISION	1517
REPUBLIC	HY362-3
REPUBLIC	HY362-7
SPICER	10-4-0033
SPICER	10-4-33
TRW	20605
WESCO	1A133
WESCO	1A133-1
WESCO	510009
WESCO	YA133

IDLI GROUP 88000.27

AEC	AE063
AEC	AEO63
ALCO	503
ALCO	517
ALLOY	503
BORG-WARNR	100-63
BORG-WARNR	65263
CHELSEA	Y134Y144
CHELSEA	Y144Y144
CMP	CM63ST
FLEETRITE	ZAD5P11
FLEETRITE	ZAD5P17
G & G MFG	184-0514
HUB CITY	03-35-00046
LEMPCO	1-63
LEMPCO	2L11
LEMPCO	2L17
LEMPCO	2L4
MCQUAY-NOR	U03-52
MOTOR MAST	MYO-63
MUNCIE	101-1-120
MUNCIE	101-2-120
MUNCIE	10-463
MUNCIE	110-1-120
MUNCIE	110-2-120
MUNCIE	120-1-120
MUNCIE	120-2-120
MUNCIE	EY12
MUNCIE	EY121
MUNCIE	EY122
NEAPCO	05113
NEAPCO	051B3
NEAPCO	051CJ
NEAPCO	061113
NEAPCO	061213
NEAPCO	061313
NEAPCO	10-0463
NEAPCO	11-2045
NEAPCO	15113
NEAPCO	151B3
NEAPCO	151B5
NEAPCO	151B6
NEAPCO	15-5315
NEAPCO	15-5326
NEAPCO	161213
NEAPCO	51B3
NEAPCO	N10242-45
NEAPCO	N10463
PILOT	50Y63
PILOT	5P-11
PILOT	5P-17
PRECISION	1518
REPUBLIC	2790-129M
REPUBLIC	HY362-6
SPICER	10242-16SF
SPICER	10242-17SF
SPICER	10242-45SF
SPICER	10-4-0063
SPICER	10-4-0073
SPICER	10-4-63
TRW	20504
TRW	20509
TRW	20510
TRW	20608
TRW	20609
WESCO	1A144-1
WESCO	1A144-12
WESCO	1A144-13
WESCO	510017
WESCO	51-1440
WESCO	A123-1
WESCO	A134-1
WESCO	A144-1
WESCO	YA144

IDLI GROUP 88000.28

AEC	0-73
AEC	40073
AEC	AE073
AEC	AE40073
AEC	AEO73
ALCO	0517
ALLOY	517
BORG-WARNR	100-73
BORG-WARNR	65114
BORG-WARNR	65337
BPC	H7EY-15/16R-1/4KW
CHELSEA	Y134Y154
CHELSEA	Y154Y154
CHRYSLER	D319
CLEVE.MOT.	CM73ST
CMP	CM73ST
FLEETRITE	ZAD50Y73
FLEETRITE	ZAD5P22
G & G MFG	184-0615
HUB CITY	03-35-00013
LEMPCO	1-73
LEMPCO	2L12
LEMPCO	2L22
MCQUAY-NOR	U03-53
MOPAR	D319
MOTOR MAST	MYO-73
MUNCIE	10-4-73
MUNCIE	110-1-130
MUNCIE	110-2-130
MUNCIE	130-1-130
MUNCIE	130-2-130
MUNCIE	EY130
NEAPCO	05114
NEAPCO	061214
NEAPCO	061414
NEAPCO	10-0473
NEAPCO	10-473
NEAPCO	11-2006
NEAPCO	15114
NEAPCO	151C7
NEAPCO	151M9
NEAPCO	15-5114
NEAPCO	161214
NEAPCO	5114
NEAPCO	N10242-6
NEAPCO	N10473
PILOT	50Y73
PILOT	5P-22
PRECISION	1519
REPUBLIC	2790-131M
REPUBLIC	2790-169M
REPUBLIC	HY362-7
SPICER	10242-40SF
SPICER	10242-6SF
SPICER	10-4-73
TRW	20511
TRW	20609
WESCO	1A154
WESCO	1A15401
WESCO	1A154-1
WESCO	1A154-4
WESCO	510020
WESCO	51-1540
WESCO	A134-1
WESCO	A154-1
WESCO	YA154

IDLI GROUP 88000.29

AEC	40076
AEC	AE076
AEC	AEO76
ALCO	0505
ALLOY	505
BORG-WARNR	100-76
BORG-WARNR	65338
BPC	H7EY15/16R
CMP	CM76ST
LEMPCO	1-76
MOPAR	D306
MOTOR MAST	MYO-76
MUNCIE	EY131
NEAPCO	051C8
NEAPCO	15153
NEAPCO	151C8
NEAPCO	15-5324
PILOT	50Y76
PRECISION	1520
SPICER	10-4-643
WESCO	1A150-26
WESCO	1A150-9
WESCO	1A15926
WESCO	510019

IDLI GROUP 88000.30

AEC	40093
AEC	43093
AEC	8393
AEC	AE093
ALCO	0518
ALLOY	518
BORG-WARNR	65115
BORG-WARNR	66187
BPC	H7EY-1R-1/4KW
CHRYSLER	D320
CLEVE.MOT.	CM93ST
FLEETRITE	ZAD50Y93
G & G MFG	184-0516
G & G MFG	184-0616
LEMPCO	1-93
MCQUAY-NOR	U03-56
MOPAR	D320
MOTOR MAST	MYO-8393
MUNCIE	EY140
NEAPCO	05115
NEAPCO	0515A
NEAPCO	10-4573
NEAPCO	15115
NEAPCO	15164
NEAPCO	151A6
NEAPCO	151K5
NEAPCO	15-5115
NEAPCO	15-5215
NEAPCO	5115
NEAPCO	515A
NEAPCO	N104573
PRECISION	1521
REPUBLIC	HY362-10
SPICER	10-4-493
SPICER	10-4-573
SPICER	10-4-853
SPICER	10-4-93
TRW	20612
TRW	21007
WESCO	1A164
WESCO	1A164-2
WESCO	1A164-4
WESCO	510024
WESCO	51-1640

IDLI GROUP 88000.31

AEC	0-93
AEC	40093
AEC	AE40093
BORG-WARNR	100-93
BORG-WARNR	2B0027
BORG-WARNR	2B27
BORG-WARNR	65124
CHELSEA	SS12Y164
CHELSEA	Y134Y164
CHELSEA	Y164Y164
CMP	CM93ST
FLEETRITE	ZAD5P13
FLEETRITE	ZAD5P27
FLEETRITE	ZAD5P76
G & G MFG	184-0516
G & G MFG	184-0616
HUB CITY	03-35-00006
LEMPCO	1-93

LEMPCO	2L13
LEMPCO	2L27
LEMPCO	2L76
MCQUAY-NOR	U03-56
MOTOR MAST	2B27
MOTOR MAST	MYO-93
MUNCIE	10-4-93
MUNCIE	110-1-140
MUNCIE	110-2-140
MUNCIE	140-1-140
MUNCIE	140-1-550
MUNCIE	140-2-140
MUNCIE	140-2-550
MUNCIE	EY140
MUNCIE	EY141
NEAPCO	05115
NEAPCO	05124
NEAPCO	061215
NEAPCO	061515
NEAPCO	061576
NEAPCO	062424
NEAPCO	065454
NEAPCO	10-0493
NEAPCO	11-2007
NEAPCO	11-2021
NEAPCO	15124
NEAPCO	15-5124
NEAPCO	161515
NEAPCO	161576
NEAPCO	5124
NEAPCO	N10242-21
NEAPCO	N10242-7
NEAPCO	N10493
PILOT	50Y93
PILOT	5P-13
PILOT	5P-27
PILOT	5P-76
PRECISION	1522
REPUBLIC	2790-133M
REPUBLIC	2790-190M
REPUBLIC	HY362-10
SPICER	10242-21SF
SPICER	10242-7SF
SPICER	10243-8SF
SPICER	10-4-0083
SPICER	10-4-0853
SPICER	10-4-493
SPICER	10-4-523
SPICER	10-4-93
TRW	20505
TRW	20512
TRW	20517
TRW	20611
TRW	20612
WESCO	1A164-1
WESCO	510025
WESCO	51-1640
WESCO	6N164-5
WESCO	6N164-7
WESCO	820-004
WESCO	A134-1
WESCO	A164-1
WESCO	A164A164
WESCO	B120-26
WESCO	YA164

IDLI GROUP 88000.32

ALCO	0531
ALLOY	531
BORG-WARNR	65154
BPC	H7EY-1R-3/8PH
G & G MFG	184-0516
MOPAR	D332
MUNCIE	EY142
NEAPCO	05154
NEAPCO	10-4443
NEAPCO	15154
NEAPCO	15160
NEAPCO	151L6
NEAPCO	15-5154
NEAPCO	5154
NEAPCO	N104443
PRECISION	1523
SPICER	10-4-443

IDLI GROUP 88000.33

ALLOY	0558
ALLOY	558
BORG-WARNR	65164
NEAPCO	05164
NEAPCO	051GJ
NEAPCO	051RP
NEAPCO	15164
NEAPCO	151A6
NEAPCO	151Z5
NEAPCO	15-5164
NEAPCO	5164
PRECISION	1524
SPICER	10-4-1203
SPICER	10-4-853
WESCO	1A164-11
WESCO	1A164-28

IDLI GROUP 88000.34

AEC	0-95
AEC	40095
AEC	AE095
AEC	AEO95
ALCO	0550
ALLOY	550
BORG-WARNR	100-95
BORG-WARNR	65191
BPC	H7EY-1-1/16R-1/4KW
CMP	CM95ST
FLEETRITE	ZAD50Y95
G & G MFG	184-0518
LEMPCO	1-95
MCQUAY-NOR	U03-57
MCQUAY-NOR	U03-59
MOPAR	D351
MOTOR MAST	MYO-95
MUNCIE	EY150
NEAPCO	05191
NEAPCO	15191
NEAPCO	15-5191
NEAPCO	5191
PILOT	50Y95
PRECISION	1525
REPUBLIC	HY363-1
REPUBLIC	HY363-1A

SPICER	10-4-163
TRW	21011
WESCO	1A174
WESCO	1A174-1
WESCO	510027

IDLI GROUP 88000.35

AEC	0-113
AEC	103113
AEC	40113
AEC	43113
AEC	AE0113
AEC	AEO113
ALCO	0519
ALCO	519
ALLOY	519
BORG-WARNR	100-113
BORG-WARNR	65116
BORG-WARNR	65197
BORG-WARNR	66210
BPC	H7EY-1-1/8R-5/16KW
CHELSEA	Y185Y185
CHELSEA	Y185Y205
CHRYSLER	D321
CLEVE.MOT.	CM113ST
CMP	CM113ST
FLEETRITE	ZAD50Y113
FLEETRITE	ZAD5P30
G & G MFG	184-0518
HUB CITY	03-35-00026
LEMPCO	1-113
LEMPCO	2L30
LEMPCO	2L31
MCQUAY-NOR	U03-59
MOPAR	D321
MOTOR MAST	2B30
MOTOR MAST	MYO-103113
MOTOR MAST	MYO-113
MUNCIE	EY160
MUNCIE	EY162
NEAPCO	05116
NEAPCO	0516A
NEAPCO	05191
NEAPCO	05197
NEAPCO	061616
NEAPCO	061618
NEAPCO	10-4113
NEAPCO	10-4473
NEAPCO	11-2009
NEAPCO	11-2023
NEAPCO	15116
NEAPCO	15197
NEAPCO	15-5116
NEAPCO	15-5197
NEAPCO	15-5216
NEAPCO	5116
NEAPCO	516A
NEAPCO	5197
NEAPCO	N10242-23
NEAPCO	N10242-9
NEAPCO	N104113
NEAPCO	N1041473
PILOT	50Y113
PILOT	5P-30
PRECISION	1526

REPUBLIC	2790-195M
REPUBLIC	HY363-1
SPICER	10242-23SF
SPICER	10242-9SF
SPICER	10-4-0113
SPICER	10-4-113
SPICER	10-4-1473
TRW	20506
TRW	20514
TRW	20614
TRW	21012
WESCO	1A185
WESCO	1A185-1
WESCO	510031
WESCO	510032
WESCO	A185-1
WESCO	A205-1
WESCO	YA185

IDLI GROUP 88000.36

AEC	0-103
AEC	40103
AEC	AE0103
AEC	AE40103
AEC	AEO103
ALCO	0552
ALCO	552
ALLOY	552
BORG-WARNR	100-103
BORG-WARNR	65196
BORG-WARNR	65348
BPC	H7EY-1-1/8R-1/4KW
CHRYSLER	D353
CLEVE.MOT.	CM103ST
CMP	CM103ST
FLEETRITE	ZAD50Y103
G & G MFG	184-0518
G & G MFG	184-0618
HUB CITY	03-35-00025
LEMPCO	1-103
MCQUAY-NOR	U03-58
MOPAR	D353
MOTOR MAST	MYO-103
MUNCIE	EY161
NEAPCO	05196
NEAPCO	051D8
NEAPCO	10-4103
NEAPCO	10-4143
NEAPCO	15196
NEAPCO	15198
NEAPCO	151A9
NEAPCO	151D8
NEAPCO	151D9
NEAPCO	15-5196
NEAPCO	15-5339
NEAPCO	5196
NEAPCO	51D8
NEAPCO	N104103
NEAPCO	N104143
PILOT	50Y103
PRECISION	1527
REPUBLIC	HY363-2
REPUBLIC	HY363-3A
SPICER	10-4-0103
SPICER	10-4-0143

ENGINEERING CATALOGS MUST BE CONSULTED FOR DETAILS NOT INCLUDED
IN THIS GUIDE. SPECIFIC DESIGNS, MATERIAL CONTENT, TOLERANCES
LUBE FITTINGS AND OTHER DIMENSIONS ARE INTENTIONALLY OMITTED HERE.

CAUTION: BE SURE TO REFER TO ENGINEERING CATALOGS FOR SPECIAL APPLICATIONS THAT REQUIRE SPECIFIC MATERIAL CONTENT, TOLERANCES, ETC. SEE FOOTNOTE.

SPICER	10-4-103
SPICER	10-4-143
TRW	20613
WESCO	1A184
WESCO	1A184-1
WESCO	510030
WESCO	6N185-5
WESCO	YA184

IDLI GROUP 88000.37

AEC	0-124
AEC	124125
AEC	40124
AEC	43125
AEC	AE0124
AEC	AEO124
ALCO	0520
ALCO	520
ALLOY	520
BORG-WARNR	100-124
BORG-WARNR	65117
BORG-WARNR	66233
BPC	H7EY-1-3/16R-5/16KW
CHRYSLER	D322
CLEVE.MOT.	CM115ST
CMP	CM115ST
FLEETRITE	ZAD50Y124
HUB CITY	0335-00014
LEMPCO	1-124
LEMPCO	1-24
MCQUAY-NOR	U03-60
MOPAR	D322
MOTOR MAST	MYO-124
MOTOR MAST	MYO-124125
MUNCIE	EY170
NEAPCO	05117
NEAPCO	0517A
NEAPCO	15117
NEAPCO	15-5117
NEAPCO	15-5217
NEAPCO	5117
NEAPCO	517A
PILOT	50Y124
PRECISION	1528
REPUBLIC	HY363-5
SPICER	10-3-0122X
SPICER	10-4-0153
SPICER	10-4-153
TRW	20617
TRW	21016
WESCO	1A195
WESCO	1A195-1
WESCO	510033

IDLI GROUP 88000.38

AEC	0-123
AEC	115123
AEC	40123
AEC	43123
AEC	AE0123
AEC	AEO123
ALCO	0521
ALLOY	521
BORG-WARNR	100-123
BORG-WARNR	2B0032
BORG-WARNR	2B32
BORG-WARNR	65118
BORG-WARNR	66256
BPC	H7EY-1-1/4R-5/16KW
CHELSEA	Y185Y205
CHELSEA	Y205Y205
CHRYSLER	D323
CLEVE.MOT.	CM124ST
CMP	CM124ST
FLEETRITE	ZAD50Y123
FLEETRITE	ZAD5P31
FLEETRITE	ZAD5P32
HUB CITY	03-35-00061
LEMPCO	1-123
LEMPCO	2L31
LEMPCO	2L32
MCQUAY-NOR	U03-62
MOPAR	D323
MOTOR MAST	2B32
MOTOR MAST	MYO-115
MOTOR MAST	MYO-115123
MOTOR MAST	MYO-123
MUNCIE	EY180
NEAPCO	05118
NEAPCO	0518A
NEAPCO	061618
NEAPCO	061818
NEAPCO	062525
NEAPCO	10-4193
NEAPCO	11-2008
NEAPCO	11-2032
NEAPCO	14-0601
NEAPCO	15118
NEAPCO	15-5118
NEAPCO	15-5218
NEAPCO	161818
NEAPCO	5118
NEAPCO	518A
NEAPCO	N10242-32
NEAPCO	N10242-8
NEAPCO	N104193
PILOT	50Y123
PILOT	5P-31
PILOT	5P-32
PRECISION	1529
REPUBLIC	2790-197M
REPUBLIC	2790-202M
REPUBLIC	HY363-6A
SPICER	10242-23SF
SPICER	10242-32SF
SPICER	10242-8SF
SPICER	10-4-0183
SPICER	10-4-193
TRW	20507
TRW	20514

TRW	20615
TRW	21017
WESCO	1A205
WESCO	510036
WESCO	A185-1
WESCO	A205-1
WESCO	A205A205
WESCO	YA205

IDLI GROUP 88000.39

AEC	AE2B72
AEC	AE40123
ALCO	521
BORG-WARNR	65125
CHELSEA	S144Y123
CHELSEA	Y123Y123
CMP	CM124ST
FLEETRITE	ZAD5P1
G & G MFG	184-0520
HUB CITY	03-35-00061
LEMPCO	1-123
LEMPCO	2L1
LEMPCO	2L4
LEMPCO	2L52
MCQUAY-NOR	U03-62
MOTOR MAST	MYO-123
MOTOR MAST	MYO-13
MUNCIE	101-1-101
MUNCIE	101-1-120
MUNCIE	101-1-510
MUNCIE	101-2-101
MUNCIE	101-2-120
MUNCIE	101-2-510
MUNCIE	EY181
NEAPCO	05118
NEAPCO	05125
NEAPCO	061111
NEAPCO	061113
NEAPCO	061184
NEAPCO	10-4123
NEAPCO	11-2018
NEAPCO	11-3003
NEAPCO	15125
NEAPCO	15-5125
NEAPCO	5125
NEAPCO	N10242-18
NEAPCO	N10243-3
NEAPCO	N104123
PILOT	5P-1
PRECISION	1530
REPUBLIC	2790-100M
REPUBLIC	2790-105M
REPUBLIC	HY363-3
REPUBLIC	HY363-6A
SPICER	10242-17SF
SPICER	10242-18SF
SPICER	10242-5SF
SPICER	10243-3SF
SPICER	10-4-0123
SPICER	10-4-0193
SPICER	10-4-123
TRW	20501
TRW	20509
TRW	20516
TRW	20616

WESCO	1A205-1
WESCO	510037
WESCO	51-2050
WESCO	A123-1
WESCO	A144-1
WESCO	B144-1

IDLI GROUP 88000.40

ALCO	0536
ALLOY	536
BORG-WARNR	65165
MOPAR	D337
MUNCIE	EY184
NEAPCO	05165
NEAPCO	10-4363
NEAPCO	15165
NEAPCO	15-5165
NEAPCO	N1041363
PRECISION	1531
SPICER	10-4-1363
WESCO	1A200-35
WESCO	1A200-7
WESCO	510034

IDLI GROUP 88000.41

ALCO	0504
ALLOY	504
BORG-WARNR	65357
MOPAR	D305
MUNCIE	EY190
NEAPCO	051E7
NEAPCO	151E7
NEAPCO	15-5344
PRECISION	1532

IDLI GROUP 88000.42

PRECISION	1533

IDLI GROUP 88000.43

AEC	0-22
AEC	40022
AEC	AE022
AEC	AEO22
ALCO	0522
ALCO	522
ALLOY	0557
ALLOY	522
BORG-WARNR	100-22
BORG-WARNR	65119
BORG-WARNR	65171
BORG-WARNR	65355
BPC	H7EY-3/4
CHRYSLER	D324
CLEVE.MOT.	CM12ST
CMP	CM12ST
FLEETRITE	ZAD50Y22
HUB CITY	03-35-00017
LEMPCO	1-22
MCQUAY-NOR	U03-46
MOPAR	D324
MOTOR MAST	MYO-22
MUNCIE	10-4-22
MUNCIE	EY202

BD = BEARING DIAMETER (outside)   BW = BEARING WIDTH   CB = CROSS LENGTH WITH BEARINGS   CC = CENTER TO CENTER   CD = CROSS DIAMETER
CL = CROSS LENGTH WITHOUT BEARINGS   EC = END TO CENTER or FACE TO FACE   EE = END TO END   EL = EFFECTIVE LENGTH
HD = HUB DIAMETER (or insert)   IE = INSIDE OF EARS (or recess)   OD = OUTSIDE DIAMETER   OE = OUTSIDE OF EARS   SB = SPLINE OR BORE SIZE

MUNCIE	EY-340
MUNCIE	EY340
MUNCIE	EY-340
MUNCIE	EY340
NEAPCO	05149
NEAPCO	05171
NEAPCO	051E5
NEAPCO	051H4
NEAPCO	10-0422
NEAPCO	15119
NEAPCO	15129
NEAPCO	15171
NEAPCO	151Z9
NEAPCO	15-5149
NEAPCO	15-5171
NEAPCO	15-5342
NEAPCO	5171
NEAPCO	N10422
PILOT	50Y22
PRECISION	1534
PREC.TOR.	1533
REPUBLIC	HY364-1
SPICER	10-4-0022
SPICER	10-4-182
SPICER	10-4-22
TRW	20603
WESCO	1A120-10
WESCO	1A120-26
WESCO	1AS120
WESCO	510002
WESCO	510003
WESCO	510004
WESCO	YAS120

IDLI GROUP 88000.44

AEC	40023
AEC	AE023
AEC	AE40023
AEC	AE40043
ALLOY	542
BORG-WARNR	65182
BORG-WARNR	65268
CLEVE.MOT.	CM23ST
CMP	CM23ST
FLEETRITE	ZAD50Y23
HUB CITY	03-35-00019
LEMPCO	1-23
MOPAR	D343
MOPAR	D-343
MOPAR	D343
MOPAR	D-343
MUNCIE	10-4-23
MUNCIE	EY103
MUNCIE	EY-210
MUNCIE	EY210
MUNCIE	EY-210
MUNCIE	EY210
NEAPCO	05182
NEAPCO	051B8
NEAPCO	10-0423
NEAPCO	15-5182
NEAPCO	15-5319
NEAPCO	51B8
NEAPCO	N10423
PILOT	50Y23

PRECISION	1535
REPUBLIC	HY362-1A
SPICER	10-4-23
TRW	20604
WESCO	1A124
WESCO	510007

IDLI GROUP 88000.45

AEC	0-12
AEC	40012
AEC	AE012
AEC	AEO12
ALCO	523
ALCO	549
ALLOY	0549
BORG-WARNR	100-12
BORG-WARNR	65120
BORG-WARNR	65189
BORG-WARNR	65432
BPC	H7EY7/8
CHRYSLER	D350
CLEVE.MOT.	CM22ST
CMP	CM22ST
FLEETRITE	ZAD50Y12
HUB CITY	03-35-00005
LEMPCO	1-12
MCQUAY-NOR	U03-50
MOPAR	D350
MOTOR MAST	MYO-12
MOTOR MAST	MYO12
MUNCIE	10-4-12
MUNCIE	EY220
MUNCIE	EY221
NEAPCO	05189
NEAPCO	051H5
NEAPCO	051L2
NEAPCO	10-0412
NEAPCO	15120
NEAPCO	15189
NEAPCO	15-5189
NEAPCO	15-5363
NEAPCO	5189
NEAPCO	N10412
PILOT	50Y12
PRECISION	1537
PREC.TOR.	1536
REPUBLIC	HY364-3
SPICER	10-4-0012
SPICER	10-4-0062
SPICER	10-4-12
SPICER	10-4-192
SPICER	10-4-62
TRW	20601
TRW	21003
WESCO	1A140-10
WESCO	1A140-25
WESCO	1A140-26
WESCO	1A160-30
WESCO	1A160-33
WESCO	1AS140
WESCO	510013
WESCO	51-140-26
WESCO	YAS140

IDLI GROUP 88000.46

AEC	0-15D
AEC	40150
AEC	AEO150
ALCO	551
ALLOY	0551
BORG-WARNR	65193
BPC	H7EY-1
CHRYSLER	D352
FLEETRITE	ZAD50Y52
MCQUAY-NOR	U03-54
MCQUAY-NOR	U03-55
MOPAR	D352
MOTOR MAST	MYO-150
MUNCIE	10-4-52
MUNCIE	EY240
NEAPCO	05139
NEAPCO	05193
NEAPCO	05195
NEAPCO	10-0452
NEAPCO	15122
NEAPCO	15193
NEAPCO	15195
NEAPCO	15-5193
NEAPCO	5193
NEAPCO	N10452
PRECISION	1539
REPUBLIC	HY364-4
SPICER	10-4-0052
SPICER	10-4-52
SPICER	10-4-72
TRW	20618
TRW	21009
WESCO	1A160
WESCO	1A160-10
WESCO	1A160-26
WESCO	1AS160
WESCO	510021
WESCO	51-160-26

IDLI GROUP 88000.47

ALCO	0538
ALLOY	538
BORG-WARNR	65172
BPC	H7EY-1-1/8
MOPAR	D339
MUNCIE	10-4-32
MUNCIE	EY325
NEAPCO	05172
NEAPCO	10-0432
NEAPCO	15172
NEAPCO	15-5172
NEAPCO	5172
NEAPCO	N10432
PRECISION	1540
SPICER	10-4-0032
SPICER	10-4-32
TRW	21014
WESCO	1A180-11
WESCO	1A180-32
WESCO	1AH180
WESCO	510029
WESCO	51-1850

IDLI GROUP 88000.48

ALCO	0533
ALLOY	533
BORG-WARNR	65158
MOPAR	D334
MOPAR	D336
MUNCIE	EY370
NEAPCO	05158
NEAPCO	05163
NEAPCO	15158
NEAPCO	15-5158
NEAPCO	5158
PRECISION	1541
SPICER	10-4-101
WESCO	1A200-21

IDLI GROUP 88000.49

AEC	0-81
AEC	40081
AEC	AE081
AEC	AEO81
ALCO	0526
ALLOY	526
BORG-WARNR	100-81
BORG-WARNR	65134
BPC	H7EY-1-1/8
CMP	CM81ST
HUB CITY	03-35-00015
LEMPCO	1-81
MOPAR	D327
MOTOR MAST	MYO-81
MUNCIE	EY380
NEAPCO	05134
NEAPCO	15134
NEAPCO	15188
NEAPCO	15-5134
NEAPCO	5134
PRECISION	1542
SPICER	10-4-0011
SPICER	10-4-11
SPICER	10-4-171
SPICER	10-4-211
SPICER	1A180-21
TRW	21015
WESCO	151028
WESCO	1A180-21
WESCO	510-028
WESCO	A1809-6B

IDLI GROUP 88000.50

ALCO	527
BORG-WARNR	65141
BPC	H7EY-1-1/4
MUNCIE	EY390
NEAPCO	15141
NEAPCO	15-4141
PRECISION	1544
SPICER	10-4-21

CAUTION: BE SURE TO REFER TO ENGINEERING CATALOGS FOR SPECIAL APPLICATIONS THAT REQUIRE SPECIFIC MATERIAL CONTENT, TOLERANCES, ETC. SEE FOOTNOTE.

IDLI GROUP 88000.51

ALCO	0547
ALLOY	547
BORG-WARNR	65145
BORG-WARNR	65187
BPC	H7EY-1-3/8
HUB CITY	0335-0004
MOPAR	D348
MUNCIE	EY395
MUNCIE	EY-722
NEAPCO	05145
NEAPCO	05187
NEAPCO	15145
NEAPCO	15187
NEAPCO	15-5145
NEAPCO	15-5187
NEAPCO	5187
PRECISION	1545
SPICER	10-4-0031
SPICER	10-4-31
TRW	21018
WESCO	1A220-21
WESCO	1A2209-6B
WESCO	1B220-21
WESCO	510038
WESCO	720-022

IDLI GROUP 88000.52

AEC	0-53
AEC	40053
AEC	AE053
AEC	AEO53
BORG-WARNR	100-53
BORG-WARNR	65136
BORG-WARNR	65264
BPC	H7EY-7/8R-3/16KW
CLEVE.MOT.	CM53ST
CLEVE.MOT.	CM-63SL
CMP	CM53ST
FLEETRITE	ZAD50Y53
G & G MFG	184-0514
G & G MFG	184-0614
HUB CITY	03-35-00007
LEMPCO	1-53
MCQUAY-NOR	U03-51
MOTOR MAST	MY0-53
MUNCIE	10-4-53
MUNCIE	EY123
NEAPCO	05105
NEAPCO	05136
NEAPCO	051A1
NEAPCO	051B4
NEAPCO	10-0453
NEAPCO	11-2002
NEAPCO	15136
NEAPCO	151A1
NEAPCO	151B4
NEAPCO	151Z6
NEAPCO	15-5105
NEAPCO	15-5136

NEAPCO	15-5316
NEAPCO	5136
NEAPCO	N10242-2
NEAPCO	N10453
PILOT	50Y53
PRECISION	1546
REPUBLIC	HY362-5
REPUBLIC	HY363-9
SPICER	10242-2SF
SPICER	10-4-0053
SPICER	10-4-203
SPICER	10-4-53
TRW	20607
WESCO	1A141-1A
WESCO	1A143
WESCO	1A143-1
WESCO	1A143-1B
WESCO	1A143-22
WESCO	1A143-23
WESCO	1A143-24
WESCO	510015
WESCO	YA143

IDLI GROUP 88000.53

AEC	0-83
AEC	40083
AEC	AE083
AEC	AEO83
ALCO	518
BORG-WARNR	100-83
BORG-WARNR	65139
BPC	H7EY-1R-3/16KW
CHELSEA	Y163Y163
CLEVE.MOT.	CM83ST
CMP	CM83ST
FLEETRITE	ZAD50Y83
FLEETRITE	ZAD5P26
G & G MFG	184-0516
HUB CITY	0335-00063
HUB CITY	03-35-0063
LEMPCO	1-83
LEMPCO	2L26
MCQUAY-NOR	U03-55
MOTOR MAST	MY0-83
MUNCIE	10-4-83
MUNCIE	EY143
NEAPCO	05139
NEAPCO	05143
NEAPCO	05160
NEAPCO	051T3
NEAPCO	063939
NEAPCO	10-0483
NEAPCO	11-2001
NEAPCO	15139
NEAPCO	15170
NEAPCO	15-5139
NEAPCO	5139
NEAPCO	N10242-1
NEAPCO	N10483
PILOT	50Y83
PILOT	5P-26
PRECISION	1548
REPUBLIC	2790-184M
REPUBLIC	HY362-9
SPICER	10242-1SF

SPICER	10-4-0093
SPICER	10-4-0853
SPICER	10-4-83
TRW	20513
TRW	20611
TRW	20612
WESCO	1A163
WESCO	1A163-1
WESCO	510023
WESCO	A163-1
WESCO	YA163

IDLI GROUP 88000.54

BORG-WARNR	65448
NEAPCO	051E3
NEAPCO	151M8
NEAPCO	15-5340
PRECISION	1550
SPICER	10-4-12253
WESCO	1A164-14
WESCO	1A164-18
WESCO	1A164-29
WESCO	1A164-30
WESCO	1A164-31
WESCO	1A164-7
WESCO	510026

IDLI GROUP 88000.55

AEC	0-115
AEC	40115
AEC	AE0115
AEC	AEO115
BORG-WARNR	100-115
BORG-WARNR	65156
BORG-WARNR	65342
BPC	H7EY-1-1/4R-1/4KW
CLEVE.MOT.	CM123ST
CMP	CM123ST
FLEETRITE	ZAD50Y115
G & G MFG	184-0521
G & G MFG	184-0620
HUB CITY	03-35-00016
LEMPCO	1-115
MCQUAY-NOR	U03-61
MOTOR MAST	MY0-115
MUNCIE	EY183
NEAPCO	05156
NEAPCO	051D2
NEAPCO	10-4183
NEAPCO	15156
NEAPCO	15174
NEAPCO	151D2
NEAPCO	15-5156
NEAPCO	15-5335
NEAPCO	5156
NEAPCO	51D2
NEAPCO	N104183
PILOT	50Y115
PRECISION	1552
REPUBLIC	HY363-6
SPICER	10-4-0183
SPICER	10-4-183
TRW	20615
WESCO	1A204

WESCO	1A204-1
WESCO	1A204-11
WESCO	1A204-34
WESCO	1A204-37
WESCO	1A204-7
WESCO	510035
WESCO	6N200-7
WESCO	6N205-5
WESCO	YA204

IDLI GROUP 88000.56

ALLOY	0562
ALLOY	562
BORG-WARNR	65499
G & G MFG	184-0520
NEAPCO	051R9
NEAPCO	151R2
NEAPCO	151R9
NEAPCO	15-5374
NEAPCO	51R9
PRECISION	1553
REPUBLIC	HY363-7
SPICER	10-4-293

IDLI GROUP 88000.57

ALLOY	557
BORG-WARNR	65103
BORG-WARNR	65149
BORG-WARNR	65150
MUNCIE	EY203
NEAPCO	05149
NEAPCO	15149
NEAPCO	15-5149
PRECISION	1554
SPICER	10-4-303
WESCO	1A120-14
WESCO	1A120-25
WESCO	510002
WESCO	51-120-26

IDLI GROUP 88000.58

BORG-WARNR	65121
BORG-WARNR	65478
HUB CITY	0335-00012
MOPAR	D-325
MOPAR	D325
MOPAR	D-325
MOPAR	D325
MUNCIE	EY230
NEAPCO	05121
NEAPCO	151P8
NEAPCO	15-5121
PRECISION	1555
SPICER	10-4-152
SPICER	10-4-172

BD = BEARING DIAMETER (outside)　　BW = BEARING WIDTH　　CB = CROSS LENGTH WITH BEARINGS　　CC = CENTER TO CENTER　　CD = CROSS DIAMETER
CL = CROSS LENGTH WITHOUT BEARINGS　　EC = END TO CENTER or FACE TO FACE　　EE = END TO END　　EL = EFFECTIVE LENGTH
HD = HUB DIAMETER (or insert)　　IE = INSIDE OF EARS (or recess)　　OD = OUTSIDE DIAMETER　　OE = OUTSIDE OF EARS　　SB = SPLINE OR BORE SIZE

IDLI GROUP 88000.59

ALLOY	0559
ALLOY	559
BORG-WARNR	65195
NEAPCO	05195
NEAPCO	15-5195
PRECISION	1556
SPICER	10-4-1193
WESCO	1A160-27
WESCO	1A160-32
WESCO	510022

IDLI GROUP 88000.60

AEC	0-23
AEC	40023
AEC	AEO23
ALCO	0538
ALLOY	0556
ALLOY	556
BORG-WARNR	65138
HUB CITY	0335-00042
MCQUAY-NOR	U03-48
MOTOR MAST	MY0-23
MUNCIE	EY250
NEAPCO	05138
NEAPCO	15138
NEAPCO	15-5138
NEAPCO	5138
PRECISION	1557
WESCO	1A180-26

IDLI GROUP 88000.61

ALLOY	0560
ALLOY	560
BORG-WARNR	65356
MUNCIE	EY345
NEAPCO	051E6
NEAPCO	151E6
NEAPCO	151S6
NEAPCO	15-5343
PRECISION	1559

IDLI GROUP 88000.62

ALLOY	0561
ALLOY	561
BORG-WARNR	65919
BPC	H7EY-7/8
MUNCIE	EY330
NEAPCO	051R8
NEAPCO	10-4282
NEAPCO	151R8
NEAPCO	15-5373
NEAPCO	51R8
NEAPCO	N104282
PRECISION	1560
SPICER	10-4-282
TRW	21004
WESCO	1A140-11
WESCO	1A140-32
WESCO	1AH140
WESCO	510014

IDLI GROUP 88000.63

BORG-WARNR	65177
NEAPCO	05177
NEAPCO	15177
NEAPCO	15-5177
PRECISION	1562
WESCO	1B154-19

IDLI GROUP 88000.64

BORG-WARNR	65166
HUB CITY	0335-00003
MUNCIE	EY725
NEAPCO	05166
NEAPCO	15166
NEAPCO	15-5166
PRECISION	1563
WESCO	1B180-22
WESCO	720014

IDLI GROUP 88000.65

NEAPCO	151U7
PRECISION	1564

IDLI GROUP 88000.66

BORG-WARNR	65180
MUNCIE	EY721
NEAPCO	05180
NEAPCO	15180
NEAPCO	151P9
NEAPCO	15-5180
PRECISION	1565
WESCO	1B220-22
WESCO	720023

IDLI GROUP 88000.67

AEC	46001
ALLOY	0500
ALLOY	500
BORG-WARNR	65101
BPC	H7EY-1-3/8
G & G MFG	182-0606
HUB CITY	0335-00002
MUNCIE	EY720
NEAPCO	05101X
NEAPCO	10-4141
NEAPCO	15101X
NEAPCO	15102X
NEAPCO	15-5101
NEAPCO	5101X
NEAPCO	N104141X
PRECISION	1566
SPICER	10-4-141X
TRW	21019
WESCO	1B220-20
WESCO	820001

IDLI GROUP 88000.68

AEC	2-13
AEC	42013
AEC	AE213
AEC	AEZ13
ALCO	0507
ALCO	0512
ALLOY	507
ALLOY	512
BORG-WARNR	102-13
BORG-WARNR	65371
BORG-WARNR	65376
BPC	H751EY-3/4R-3/16KW
FLEETRITE	ZAD52Y13
LEMPCO	12-13
MOPAR	D308
MOTOR MAST	MY2-13
MUNCIE	EY400
NEAPCO	051G1
NEAPCO	051G6
NEAPCO	151G1
NEAPCO	15-5345
NEAPCO	15-5350
NEAPCO	51G1
PILOT	52Y13
PRECISION	1700
REPUBLIC	HY284-1M
SPICER	10-4-693
SPICER	1-4-2543
TRW	20645
WESCO	1C123
WESCO	1C1232
WESCO	530040
WESCO	YC1232

IDLI GROUP 88000.69

ALCO	0512
BORG-WARNR	65376
CHELSEA	CY123CY123
LEMPCO	06G1G1
LEMPCO	12-13
MOPAR	D-313
MOPAR	D313
MOPAR	D-313
MOPAR	D313
MOTOR MAST	MY2-13
MUNCIE	400-3-400
NEAPCO	051G6
NEAPCO	151G6
NEAPCO	151J2
NEAPCO	151J5
NEAPCO	151J6
NEAPCO	15-5350
PRECISION	1701
REPUBLIC	2920-100M
SPICER	10280-3SF
SPICER	10-4-0693
SPICER	207056-3
TRW	20645
WESCO	1B123-1
WESCO	1C123-1
WESCO	530041
WESCO	720005
WESCO	C123-1

IDLI GROUP 88000.70

AEC	2-43
AEC	42043
AEC	AEZ43
ALCO	0508
ALLOY	508
BORG-WARNR	102-43
BORG-WARNR	65372
BPC	H751EY-13/16R-1/4KW
CHELSEA	CY134CY134
CLEVE.MOT.	CM43SH
FLEETRITE	ZAD52Y43
FLEETRITE	ZAD5P45
LEMPCO	12-43
LEMPCO	2L45
MOPAR	D309
MOTOR MAST	MY2-43
MUNCIE	405-3-405
MUNCIE	EY405
NEAPCO	051G2
NEAPCO	06G2G2
NEAPCO	151G2
NEAPCO	151G7
NEAPCO	151H2
NEAPCO	15-5346
NEAPCO	51G2
PILOT	52Y43
PILOT	5P-45
PRECISION	1702
REPUBLIC	2920-139M
REPUBLIC	HY284-4M
SPICER	10280-2SF
SPICER	10-4-713
SPICER	1-4-2553
SPICER	207056-2
TRW	20646
WESCO	1C134
WESCO	1C134-13
WESCO	1C1342
WESCO	1C134-43
WESCO	530042
WESCO	C134
WESCO	YC1342

IDLI GROUP 88000.71

AEC	2-63
AEC	42063
AEC	AE263
AEC	AEZ63
ALCO	0509
ALLOY	509
BORG-WARNR	102-63
BORG-WARNR	65373
BORG-WARNR	65418
BPC	H751EY-7/8R-1/4KW
CHELSEA	CY144CY144
CHRYSLER	D310
CLEVE.MOT.	CM63SH
CMP	CM63SH
FLEETRITE	ZAD52Y63
FLEETRITE	ZAD5P49
LEMPCO	12-63
LEMPCO	2L49
MOPAR	D310
MOTOR MAST	MY2-63

CAUTION: BE SURE TO REFER TO ENGINEERING CATALOGS FOR SPECIAL APPLICATIONS THAT REQUIRE SPECIFIC MATERIAL CONTENT, TOLERANCES, ETC. SEE FOOTNOTE.

MUNCIE	410-3-410
MUNCIE	EY410
MUNCIE	EY470
NEAPCO	051G3
NEAPCO	051J8
NEAPCO	06G3G3
NEAPCO	10-4703
NEAPCO	151G3
NEAPCO	151G8
NEAPCO	151J3
NEAPCO	151J4
NEAPCO	151J9
NEAPCO	15-5347
NEAPCO	15-5359
NEAPCO	51G3
NEAPCO	N104703
PILOT	52Y63
PILOT	5P-49
PRECISION	1703
REPUBLIC	2920-160M
REPUBLIC	HY284-6M
SPICER	10280-1SF
SPICER	10-4-0703
SPICER	10-4-703
SPICER	1-4-2493
SPICER	207056-1
TRW	20508
TRW	20647
WESCO	1C144
WESCO	1C1442
WESCO	530-044
WESCO	530-045
WESCO	C144

IDLI GROUP 88000.72

AEC	2-73
AEC	42073
AEC	AE273
AEC	AEZ73
AEC	AEZ76
ALCO	0510
ALLOY	510
BORG-WARNR	102-73
BORG-WARNR	102-76
BORG-WARNR	65374
BORG-WARNR	65388
BORG-WARNR	65389
BPC	H751EY-15/16R-1/4KW
CHELSEA	CY154CY154
CMP	CM73SH
FLEETRITE	ZAD5P52
LEMPCO	12-73
LEMPCO	12-76
LEMPCO	2L52
MOPAR	D311
MOTOR MAST	MY2-73
MOTOR MAST	MY2-76
MUNCIE	415-3-415
MUNCIE	EY415
MUNCIE	EY416
NEAPCO	051G4

NEAPCO	06G4G4
NEAPCO	151G4
NEAPCO	151H8
NEAPCO	151H9
NEAPCO	15-5348
NEAPCO	51G4
PILOT	52Y73
PILOT	5P-52
PRECISION	1704
REPUBLIC	2920-169M
REPUBLIC	HY284-8M
SPICER	10280-7SF
SPICER	10-4-0683
SPICER	10-4-683
SPICER	1-4-2783
SPICER	207056-7
TRW	20515
TRW	20648
TRW	20649
TRW	21006
WESCO	1C150-10
WESCO	1C15045
WESCO	1C154
WESCO	1C1542
WESCO	530046
WESCO	530047
WESCO	C154
WESCO	YC1542

IDLI GROUP 88000.73

AEC	2-93
AEC	42093
AEC	AE293
AEC	AEZ93
ALCO	0511
ALLOY	511
BORG-WARNR	102-93
BORG-WARNR	104-93
BORG-WARNR	65375
BORG-WARNR	65387
BPC	H751EY-1R-1/4KW
CHRYSLER	D312
CLEVE.MOT.	CM93SH
CMP	CM93SH
LEMPCO	12-93
MCQUAY-NOR	U03-70
MOPAR	D312
MOTOR MAST	MY2-93
MUNCIE	EY420
MUNCIE	EY421
NEAPCO	051G5
NEAPCO	151G5
NEAPCO	151G9
NEAPCO	15-5349
NEAPCO	51G5
PILOT	52Y93
PRECISION	1705
REPUBLIC	HY284-10M
SPICER	10-4-0723
SPICER	10-4-723
TRW	20648
WESCO	1C160-46
WESCO	1C160-7
WESCO	1C164
WESCO	1C1642

WESCO	530-048

IDLI GROUP 88000.74

ALCO	0513
ALLOY	513
BORG-WARNR	65384
BORG-WARNR	65386
BPC	H751EY-3/4
MOPAR	D314
MUNCIE	EY450
MUNCIE	EY451
NEAPCO	051H4
NEAPCO	051H6
NEAPCO	151H4
NEAPCO	151H6
NEAPCO	15-5354
NEAPCO	15-5356
NEAPCO	51H4
PRECISION	1706
WESCO	1C120
WESCO	1C120-27
WESCO	1C120-29
WESCO	1C120-40
WESCO	1C120-41
WESCO	530-039

IDLI GROUP 88000.75

ALCO	0514
ALLOY	514
BORG-WARNR	65385
BPC	H751EY-7/8
MOPAR	D315
MUNCIE	EY460
NEAPCO	051H5
NEAPCO	151E5
NEAPCO	151H5
NEAPCO	15-5355
NEAPCO	51H5
PRECISION	1707
WESCO	1C140-29
WESCO	1C140-44
WESCO	1CS140
WESCO	530-043

IDLI GROUP 88000.76

BORG-WARNR	65421
NEAPCO	051K1
NEAPCO	151K1
NEAPCO	15-5361
PRECISION	1708

IDLI GROUP 88000.77

BORG-WARNR	65422
MUNCIE	EY430
NEAPCO	051K2
NEAPCO	151K2
NEAPCO	15-5362
PRECISION	1711

IDLI GROUP 88000.78

NEAPCO	151H4
PRECISION	1740

IDLI GROUP 88000.79

AEC	1-13
AEC	41113
AEC	AE113
AEC	AEI13
ALCO	0543
ALCO	543
ALLOY	543
BORG-WARNR	101-13
BORG-WARNR	65183
BPC	H7SY-3/4R-3/16KW
CMP	CM13SL
FLEETRITE	ZAD51Y13
LEMPCO	11-13
MCQUAY-NOR	U03-63
MOPAR	D344
MOTOR MAST	MY1-13
MUNCIE	SY500
NEAPCO	05183
NEAPCO	15183
NEAPCO	151D7
NEAPCO	151G6
NEAPCO	15-5183
NEAPCO	5183
PILOT	51Y13
PRECISION	1750
REPUBLIC	SY360-1
SPICER	10-3-0013X
SPICER	10-3-13X
TRW	20631
WESCO	1B123-1
WESCO	720005
WESCO	72-1231
WESCO	YB123

IDLI GROUP 88000.80

AEC	1-23
AEC	41123
AEC	AE123
AEC	AEI23
ALCO	0532
ALCO	532
ALLOY	532
BORG-WARNR	101-23
BORG-WARNR	65157
BPC	H7SY-13/16R-1/4KW
CHRYSLER	D333
CLEVE.MOT.	CM43SL
CMP	CM43SL
FLEETRITE	ZAD51Y23
LEMPCO	11-23
LEMPCO	1-23
MCQUAY-NOR	U03-64
MOPAR	D333
MOTOR MAST	MY1-23
MUNCIE	10-3-23X
MUNCIE	SY505
NEAPCO	05157
NEAPCO	10-0323
NEAPCO	15157

BD = BEARING DIAMETER (outside) BW = BEARING WIDTH CB = CROSS LENGTH WITH BEARINGS CC = CENTER TO CENTER CD = CROSS DIAMETER
CL = CROSS LENGTH WITHOUT BEARINGS EC = END TO CENTER or FACE TO FACE EE = END TO END EL = EFFECTIVE LENGTH
HD = HUB DIAMETER (or insert) IE = INSIDE OF EARS (or recess) OD = OUTSIDE DIAMETER OE = OUTSIDE OF EARS SB = SPLINE OR BORE SIZE

NEAPCO	151R5
NEAPCO	15-5157
NEAPCO	5157
NEAPCO	N10323X
PILOT	50Y23
PILOT	51Y23
PRECISION	1751
REPUBLIC	SY360-2
SPICER	10-3-0023X
SPICER	10-3-23X
TRW	20633
WESCO	1B134-1
WESCO	720006
WESCO	72-1441
WESCO	YB134

IDLI GROUP 88000.81

AEC	1-33
AEC	41133
AEC	AE133
AEC	AEI33
ALCO	0544
ALCO	544
ALLOY	544
BORG-WARNR	101-33
BORG-WARNR	65184
BORG-WARNR	65264
BPC	H7SY-7/8R-1/4KW
CHELSEA	S144Y123
CHRYSLER	D345
CMP	CM63SL
FLEETRITE	ZAD51Y33
FLEETRITE	ZAD5P72
LEMPCO	11-33
LEMPCO	2L72
MCQUAY-NOR	U03-66
MOPAR	D345
MOTOR MAST	MY1-33
MUNCIE	101-1-510
MUNCIE	101-2-510
MUNCIE	10-3-33X
MUNCIE	SY510
NEAPCO	05184
NEAPCO	061184
NEAPCO	10-0333
NEAPCO	15184
NEAPCO	151R3
NEAPCO	15-5184
NEAPCO	5184
NEAPCO	N10333X
PILOT	51Y33
PILOT	5P-72
PRECISION	1752
REPUBLIC	SY360-3
SPICER	10243-3SF
SPICER	10-3-0033X
SPICER	10-3-33X
SPICER	10-4-1043
SPICER	10-4-44
TRW	20516
TRW	20635
WESCO	1B144-1
WESCO	720009
WESCO	72-1441
WESCO	A123-1
WESCO	B144-1
WESCO	YB144

IDLI GROUP 88000.82

AEC	1-83
AEC	41183
AEC	AE183
AEC	AEI83
ALCO	0545
BORG-WARNR	65185
BPC	H7SY-15/16R-1/4KW
FLEETRITE	ZAD51Y83
LEMPCO	11-83
MOPAR	D346
MOTOR MAST	MY1-83
MUNCIE	SY515
NEAPCO	05185
NEAPCO	15185
NEAPCO	15-5185
NEAPCO	5185
PRECISION	1753
SPICER	10-3-103X
SPICER	10-3-203X
WESCO	1B154-1
WESCO	1B154-19
WESCO	1B154-51
WESCO	720010
WESCO	72-1541

IDLI GROUP 88000.83

AEC	1-53
AEC	41153
AEC	AE153
AEC	AEI53
ALCO	0546
ALCO	546
ALLOY	546
BORG-WARNR	101-53
BORG-WARNR	65186
BPC	H7SY-1R-1/4KW
CHRYSLER	D347
CLEVE.MOT.	CM93SL
CMP	CM93SL
FLEETRITE	ZAD51Y53
HUB CITY	0335-00116
LEMPCO	11-53
MCQUAY-NOR	U03-67
MOPAR	D347
MOTOR MAST	MY1-53
MUNCIE	10-3-83X
MUNCIE	SY520
NEAPCO	05186
NEAPCO	10-0383
NEAPCO	15186
NEAPCO	15-5186
NEAPCO	5186
NEAPCO	N10383X
PILOT	51Y53
PRECISION	1754
REPUBLIC	SY360-5
SPICER	10-3-0083X
SPICER	10-3-83X
TRW	20636
WESCO	1B164-1
WESCO	720012
WESCO	72-1641
WESCO	YB164

IDLI GROUP 88000.84

AEC	1-93
AEC	41193
AEC	AEI93
ALCO	0553
ALLOY	553
BORG-WARNR	101-93
BORG-WARNR	65199
BPC	H7SY-1-1/8R-5/16KW
CLEVE.MOT.	CM113SL
FLEETRITE	ZAD51Y93
LEMPCO	11-93
MOPAR	D354
MOTOR MAST	MY1-93
MUNCIE	10-3-183X
MUNCIE	SY530
NEAPCO	05199
NEAPCO	10-3183
NEAPCO	15199
NEAPCO	15-5199
NEAPCO	5199
NEAPCO	N103183X
PILOT	51Y93
PRECISION	1755
SPICER	10-3-0183X
SPICER	10-3-163X
TRW	21013
WESCO	1B185-1
WESCO	720019
WESCO	YC1642

IDLI GROUP 88000.85

AEC	1-103
AEC	41103
AEC	AE1193
AEC	AE41103
AEC	AEI103
ALCO	0515
ALCO	537
ALLOY	515
BORG-WARNR	65107
BORG-WARNR	65347
BPC	H7SY-1-1/4R-5/16KW
CHRYSLER	D317
MCQUAY-NOR	U03-69
MOPAR	D317
MOTOR MAST	MY1-103
MUNCIE	10-3-163X
MUNCIE	SY540
NEAPCO	051D7
NEAPCO	10-3163
NEAPCO	151D7
NEAPCO	15-5338
NEAPCO	51D7
NEAPCO	N103163X
PRECISION	1756
REPUBLIC	SY360-23
SPICER	10-3-0051X
SPICER	10-3-183X
TRW	20638

WESCO	1B205-1
WESCO	720021

IDLI GROUP 88000.86

AEC	1-12
AEC	41112
AEC	AE112
ALCO	0529
ALLOY	529
BORG-WARNR	65147
BORG-WARNR	65176
BPC	H7SY-3/4
CLEVE.MOT.	CM12SL
FLEETRITE	ZAD51Y12
LEMPCO	11-12
MOTOR MAST	MY1-12
MUNCIE	MYI12
MUNCIE	SY550
NEAPCO	05147
NEAPCO	05176
NEAPCO	15147
NEAPCO	15-5147
NEAPCO	15-5176
NEAPCO	5176
PILOT	51Y12
PRECISION	1757
REPUBLIC	SY360-11
SPICER	10-3-12X
TRW	20630
WESCO	1B120-10
WESCO	1B120-25
WESCO	1B120-26
WESCO	1B120-47
WESCO	1BS120
WESCO	720002
WESCO	720-003
WESCO	72-120-126

IDLI GROUP 88000.87

ALCO	0530
ALLOY	530
BORG-WARNR	65148
MOPAR	D331
MUNCIE	EY550
MUNCIE	SY550
NEAPCO	05148
NEAPCO	15148
NEAPCO	15-5148
NEAPCO	5148
PRECISION	1758
TRW	21008
WESCO	1B120-24
WESCO	1B120-48
WESCO	720001

IDLI GROUP 88000.88

AEC	41112
AEC	AE112
AEC	AEI12
ALCO	0529
ALCO	530
ALLOY	0529
BORG-WARNR	101-12

ENGINEERING CATALOGS MUST BE CONSULTED FOR DETAILS NOT INCLUDED
IN THIS GUIDE. SPECIFIC DESIGNS, MATERIAL CONTENT, TOLERANCES
LUBE FITTINGS AND OTHER DIMENSIONS ARE INTENTIONALLY OMITTED HERE.

CAUTION: BE SURE TO REFER TO ENGINEERING CATALOGS FOR SPECIAL APPLICATIONS THAT REQUIRE SPECIFIC MATERIAL CONTENT, TOLERANCES, ETC. SEE FOOTNOTE.

BORG-WARNR	65147
CHELSEA	SS12Y164
CMP	CM12SL
LEMPCO	11-12
LEMPCO	2L76
MOTOR MAST	MY1-12
MUNCIE	10-3-12X
MUNCIE	140-1-550
MUNCIE	140-2-550
NEAPCO	05147
NEAPCO	05148
NEAPCO	05176
NEAPCO	05E4
NEAPCO	061576
NEAPCO	10-0312
NEAPCO	11-3008
NEAPCO	15176
NEAPCO	15-5147
NEAPCO	15-5148
NEAPCO	161576
NEAPCO	N10243-8
NEAPCO	N10312X
PILOT	51Y12
PRECISION	1759
REPUBLIC	SY360-11
SPICER	10243-8SF
SPICER	10-3-0012X
SPICER	10-3-12X
TRW	20517
TRW	20630
WESCO	720-002
WESCO	72-120-26
WESCO	A164-1
WESCO	B120-26
WESCO	YBX120

IDLI GROUP 88000.89

AEC	1-22
AEC	41122
AEC	AE122
AEC	AEI122
AEC	AEI22
ALCO	0541
ALCO	541
ALLOY	541
BORG-WARNR	101-22
BORG-WARNR	65178
BPC	H7SY-7/8
CHRYSLER	D342
CLEVE.MOT.	CM22SL
CMP	CM22SL
FLEETRITE	ZAD51Y22
LEMPCO	11-22
MCQUAY-NOR	U03-65
MOPAR	D342
MOTOR MAST	MY1-22
MUNCIE	10-3-22X
MUNCIE	SY560
NEAPCO	05178
NEAPCO	10-0322
NEAPCO	15158

NEAPCO	15178
NEAPCO	15-5178
NEAPCO	5178
NEAPCO	N10322X
PILOT	51Y22
PRECISION	1760
REPUBLIC	SY360-13
SPICER	10-3-0022X
SPICER	10-3-22X
TRW	20632
WESCO	1B140-26
WESCO	1B140-28
WESCO	720007
WESCO	72-140-28
WESCO	YBS140

IDLI GROUP 88000.90

AEC	1-150
AEC	41150
ALCO	0501
ALLOY	501
BORG-WARNR	65217
BPC	H7SY-1
CHRYSLER	D302
FLEETRITE	ZAD51Y52
MOPAR	D302
MOTOR MAST	MY1-150
MUNCIE	10-3-122X
MUNCIE	SY570
NEAPCO	051A7
NEAPCO	10-3122
NEAPCO	151A7
NEAPCO	15-5305
NEAPCO	51A7
NEAPCO	N103122X
PRECISION	1761
REPUBLIC	SY360-15
SPICER	10-3-122X
TRW	20639
TRW	21010
WESCO	1A160-24
WESCO	1B160-24
WESCO	1B160-52
WESCO	720011
WESCO	72-160-24

IDLI GROUP 88000.91

ALCO	0535
ALLOY	535
BORG-WARNR	65163
BPC	H7SY-1R
MOPAR	D336
MOPAR	D-336
MOPAR	D336
MOPAR	D-336
MUNCIE	SY670
NEAPCO	05163
NEAPCO	15163
NEAPCO	15-5163
PRECISION	1762
WESCO	1B160-63

IDLI GROUP 88000.92

AEC	41181
AEC	AE181
ALCO	0534
ALCO	534
ALLOY	534
BORG-WARNR	101-83
BORG-WARNR	65159
BPC	H7SY-1-1/8
CMP	CM81SL
LEMPCO	11-81
MOPAR	D335
MOTOR MAST	MY1-81
MUNCIE	10-3-81X
NEAPCO	05159
NEAPCO	05159X
NEAPCO	05175
NEAPCO	05175X
NEAPCO	05175XS
NEAPCO	10-0381
NEAPCO	15159
NEAPCO	15-5159
NEAPCO	5168
NEAPCO	N10381X
PILOT	51Y81
PRECISION	1763
REPUBLIC	SY370-22
SPICER	10-3-0081X
SPICER	10-3-81X
SPICER	D2A
TRW	20637
WESCO	1B18054
WESCO	72-180-23
WESCO	YB154

IDLI GROUP 88000.93

ALCO	0540
ALLOY	540
BORG-WARNR	101-81
BORG-WARNR	65175
CLEVE.MOT.	CM81SL
LEMPCO	11-81
MOPAR	D341
MUNCIE	SY675
MUNCIE	SY676
NEAPCO	05159X
NEAPCO	05175
NEAPCO	05175X
NEAPCO	05175XS
NEAPCO	15175
NEAPCO	15-5159
NEAPCO	15-5175
NEAPCO	5175
PILOT	50Y81
PILOT	51Y81
PRECISION	1764
SPICER	10-3-81X
WESCO	1B180-23
WESCO	1B180-23A
WESCO	1B18054-10C
WESCO	1B18056-10C
WESCO	720015
WESCO	720016

IDLI GROUP 88000.94

AEC	AE01103
ALCO	0537
ALLOY	0527
ALLOY	537
BORG-WARNR	101-103
BORG-WARNR	65169
BPC	H7SY-1-1/4
CLEVE.MOT.	CM124SL
CMP	CM124SL
FLEETRITE	ZAD51Y103
LEMPCO	11-103
LEMPCO	12-103
MOPAR	D338
MUNCIE	SY680
NEAPCO	05141
NEAPCO	05169
NEAPCO	15169
NEAPCO	15-5169
NEAPCO	5169
PILOT	51Y103
PRECISION	1765
REPUBLIC	SY360-23
SPICER	10-3-51X
WESCO	1B200-21
WESCO	1B20058-6B
WESCO	1BP200
WESCO	720020

IDLI GROUP 88000.95

AEC	1-32
AEC	41132
AEC	AEI32
ALCO	0539
ALCO	539
ALLOY	539
BORG-WARNR	101-32
BORG-WARNR	65173
BPC	H7SY-1-1/8
CHRYSLER	D340
CLEVE.MOT.	CM32SL
CMP	CM32SL
FLEETRITE	ZAD51Y32
LEMPCO	11-32
MCQUAY-NOR	U03-68
MOPAR	D340
MOTOR MAST	MY1-32
MUNCIE	10-3-32X
MUNCIE	EY132
MUNCIE	SY630
NEAPCO	05173
NEAPCO	10-0332
NEAPCO	15173
NEAPCO	15-5173
NEAPCO	5173
NEAPCO	N10332X
PILOT	51Y32
PRECISION	1766
REPUBLIC	SY360-31
SPICER	10-3-0032
SPICER	10-3-32X
TRW	20634
WESCO	1B140-50
WESCO	1B180-32
WESCO	1B180-57

WESCO 1BS140
WESCO 720017
WESCO YBH180

IDLI GROUP 88000.96

AEC 1-63
AEC 41163
AEC AEI63
BORG-WARNR 101-63
BORG-WARNR 65472
BPC H7SY-1-1/8R-1/4KW
CLEVE.MOT. CM103SL
LEMPCO 11-63
MOTOR MAST MY1-63
NEAPCO 051P2
NEAPCO 151M5
NEAPCO 151P2
NEAPCO 15-5369
PILOT 51Y63
PRECISION 1767
WESCO 1B184-1
WESCO 720018
WESCO YC1442

IDLI GROUP 88000.97

PRECISION 1768

IDLI GROUP 88000.98

BORG-WARNR 65354
NEAPCO 051E4
NEAPCO 151E4
NEAPCO 15-5341
NEAPCO 51E4
PRECISION 1769
SPICER 10-4-182
WESCO 1B120-34
WESCO 1B120-49
WESCO 720004

IDLI GROUP 88000.99

ALLOY 564
BORG-WARNR 65129
BORG-WARNR 65355
MUNCIE EY340
NEAPCO 05129
NEAPCO 051E5
NEAPCO 151Z9
NEAPCO 15-5129
NEAPCO 15-5342
PRECISION 1770
WESCO 1A120-33
WESCO 510004
WESCO 510-105

IDLI GROUP 88001

AEC 0-76
AEC 1-81
AEC 40076
AEC 41181
AEC AEI81
BORG-WARNR 65168
BPC H7SY-1-1/8
MOTOR MAST MY1-81
MUNCIE EY131
MUNCIE SY677
NEAPCO 05168
NEAPCO 15168
NEAPCO 15-5168
PRECISION 1771
REPUBLIC SY360-21
SPICER 10-3-41X
TRW 20637
WESCO 1B180-21
WESCO 1B18055-6B
WESCO 1B1890-6B
WESCO 1BP180
WESCO 720013
WESCO 720-014

IDLI GROUP 88001.01

ALLOY 0563
ALLOY 563
BORG-WARNR 65528
BPC H7SY-7/8
MUNCIE SY635
NEAPCO 051S8
NEAPCO 10-3162
NEAPCO 151S8
NEAPCO 15-5377
NEAPCO 51S8
NEAPCO N103162X
PRECISION 1772
SPICER 10-3-162X
TRW 21005
WESCO 1B140-11
WESCO 1B140-32
WESCO 1BH140
WESCO 720008

IDLI GROUP 88001.02

NEAPCO 05175XS
NEAPCO 15-5275
PRECISION 1773
REPUBLIC SY360-22

IDLI GROUP 88001.03

PRECISION 1774

IDLI GROUP 88001.04

PRECISION 1775

IDLI GROUP 88001.05

PRECISION 1776
WESCO 1BH180

IDLI GROUP 88001.06

BORG-WARNR 67046
G & G MFG 185-3500
HAYES 228217-3316X
NEAPCO 22-329
NEAPCO 22-339
NEAPCO 22-529
NEAPCO 22-539
NEAPCO 22-551
NEAPCO 64-2229
NEAPCO 64-2529
NEAPCO 64-2539
NEAPCO 64-2551
NEAPCO UT22-002900
NEAPCO UT22-003900
PRECISION 1849
SPICER 228217-3316X
TRW 21461
TRW 21466
TRW 21471
WESCO 35N048F
WESCO 35N48F
WESCO 35N60F
WESCO 35N72
WESCO 920-138
WESCO 920141
WESCO 920-145
WESCO 920-146

IDLI GROUP 88001.07

BORG-WARNR 64353
G & G MFG 194-3500
HAYES 3-55-42
NEAPCO 2434-13A
NEAPCO 53-2434
PRECISION 1890
SPICER 3-53-831
SPICER 3-55-42
WESCO 35N-SE138
WESCO 770058

IDLI GROUP 88001.08

BORG-WARNR 60103
NEAPCO 2459X
NEAPCO 56-2300
NEAPCO 56-2459
NEAPCO N-2299X-1
NEAPCO N2459X
PRECISION 1899
SPICER 228305X
TRW 21502
WESCO 12N-NB2
WESCO 252-015
WESCO 271038
WESCO 35NNB3
WESCO 35N-NB3
WESCO 35NNB3
WESCO 35N-NB3

IDLI GROUP 88001.09

NEAPCO 16-6217
NEAPCO L617A
PRECISION 1914
TRW 21061
WESCO 06N1945-37
WESCO 540089

IDLI GROUP 88001.10

MCQUAY-NOR U03-109
NEAPCO 16-6302
NEAPCO L61L2
PRECISION 1916
TRW 21057
WESCO 06N0140-30
WESCO 540061

IDLI GROUP 88001.11

NEAPCO 16-6131
NEAPCO L6131
PRECISION 1935
WESCO 540085

IDLI GROUP 88001.12

BORG-WARNR 67042
BORG-WARNR 67054
G & G MFG 188-0600
MUNCIE A402
MUNCIE A-402
NEAPCO 06-451
NEAPCO 18-352
NEAPCO 64-1852
NEAPCO 6-451
NEAPCO 65-0651
NEAPCO 65-1651
NEAPCO L6-451
PRECISION 1947
SPICER 10-60-19-3625X
TRW 21402
TRW 21431
WESCO 06N048M
WESCO 6N48M
WESCO 920103

IDLI GROUP 88001.13

BORG-WARNR 66642
MCQUAY-NOR U03-75
MUNCIE 78-2
MUNCIE TA78-2
NEAPCO 62-0651
NEAPCO 62-1651
NEAPCO UTP06-005100
NEAPCO UTR06-002700
NEAPCO UTR06-003900
NEAPCO UTR06-005100
NEAPCO UTR06-5100
NEAPCO UTRL6-005100
PRECISION 1948
SPICER 204192-1
TRW 21400
WESCO 06N048
WESCO 6N48

ENGINEERING CATALOGS MUST BE CONSULTED FOR DETAILS NOT INCLUDED
IN THIS GUIDE. SPECIFIC DESIGNS, MATERIAL CONTENT, TOLERANCES
LUBE FITTINGS AND OTHER DIMENSIONS ARE INTENTIONALLY OMITTED HERE.

CAUTION: BE SURE TO REFER TO ENGINEERING CATALOGS FOR SPECIAL APPLICATIONS THAT REQUIRE SPECIFIC MATERIAL CONTENT, TOLERANCES, ETC. SEE FOOTNOTE.

WESCO	920100

IDLI GROUP 88001.14

MUNCIE	A-302
NEAPCO	06-351
NEAPCO	6-351
NEAPCO	64-0651
NEAPCO	64-1651
NEAPCO	L6-351
PRECISION	1949
TRW	21401
WESCO	06N048F
WESCO	6N48F
WESCO	920102

IDLI GROUP 88001.15

BORG-WARNR	60100
G & G MFG	128-0600
NEAPCO	1666X
NEAPCO	56-0666
NEAPCO	N-0666X
NEAPCO	N1666X
NEAPCO	N666X
PRECISION	1999
SPICER	228161X
TRW	21500
WESCO	06NNB1
WESCO	241008
WESCO	6N-NB1

IDLI GROUP 88001.16

CATERPILLR	1H8499
CATERPILLR	3H2972
CATERPILLR	3H301
PRECISION	2100

IDLI GROUP 88001.17

CATERPILLR	3H8141(2)
CATERPILLR	5H1816
CATERPILLR	5H7701(2)
PRECISION	2102

IDLI GROUP 88001.18

CATERPILLR	4H8461
CATERPILLR	5H2312(2)
PRECISION	2103

IDLI GROUP 88001.19

CATERPILLR	5H3754(2)
CATERPILLR	5H625
PRECISION	2104

IDLI GROUP 88001.20

CATERPILLR	5H6609(2)
CATERPILLR	5H6610
PRECISION	2105

IDLI GROUP 88001.21

CATERPILLR	9H5107(2)
CATERPILLR	9H5108
PRECISION	2106

IDLI GROUP 88001.22

CATERPILLR	1M4410(2)
CATERPILLR	1M8110
PRECISION	2107

IDLI GROUP 88001.23

CATERPILLR	2M2378(2)
CATERPILLR	2M5001
PRECISION	2108

IDLI GROUP 88001.24

CATERPILLR	1M4411(2)
CATERPILLR	2M215
PRECISION	2109

IDLI GROUP 88001.25

CATERPILLR	3H1945
CATERPILLR	3H9099
PRECISION	2110

IDLI GROUP 88001.26

CATERPILLR	2H798(2)
CATERPILLR	9F3761
PRECISION	2111

IDLI GROUP 88001.27

CATERPILLR	4M5961(2)
CATERPILLR	4M5962(2)
PRECISION	2112

IDLI GROUP 88001.28

AEC	2L
AEC	AE2L102B
ALLOY	1747
DARLINGTON	813-744
DIGIO-LIFT	3-0M22
FIAT-ALLIS	4707800
GALION	D63084
LA PORTA	LA7957
LIFT TRUCK	3-0M22
MOTOR MAST	2L102B
NEAPCO	F00200
PRECISION	2113
TOWMOTOR	B1729
WESCO	401-113
ZELLER	2L102B

IDLI GROUP 88001.29

CATERPILLR	5F1083
CATERPILLR	6F4339
PRECISION	2114

IDLI GROUP 88001.30

CATERPILLR	3F1518
CATERPILLR	5F2551
CATERPILLR	6F5321
PRECISION	2115

IDLI GROUP 88001.31

CATERPILLR	5F6332
PRECISION	2116

IDLI GROUP 88001.32

CATERPILLR	1J8584
PRECISION	2117

IDLI GROUP 88001.33

CATERPILLR	4F7024
CATERPILLR	5F2552
CATERPILLR	6F5322
PRECISION	2118

IDLI GROUP 88001.34

CATERPILLR	5M5072(2)
CATERPILLR	5M5073
PRECISION	2119

IDLI GROUP 88001.35

CATERPILLR	5M7882
CATERPILLR	9M2036(2)
PRECISION	2120

IDLI GROUP 88001.36

CATERPILLR	1S8753(2)
CATERPILLR	1S8754
CATERPILLR	5M7962
CATERPILLR	5S5684
PRECISION	2121

IDLI GROUP 88001.37

CATERPILLR	5K2005
CATERPILLR	5K3171(2)
PRECISION	2122

IDLI GROUP 88001.38

CATERPILLR	3K8807
CATERPILLR	8H6473
PRECISION	2123

IDLI GROUP 88001.39

CATERPILLR	3H2852
PRECISION	2124

IDLI GROUP 88001.40

CATERPILLR	2S471
CATERPILLR	2S476(2)
PRECISION	2125

IDLI GROUP 88001.41

PRECISION	2126

IDLI GROUP 88001.42

PRECISION	2127

IDLI GROUP 88001.43

PRECISION	2128

IDLI GROUP 88001.44

PRECISION	2129

IDLI GROUP 88001.45

PRECISION	2130

IDLI GROUP 88001.46

PRECISION	2131

IDLI GROUP 88001.47

PRECISION	2132

IDLI GROUP 88001.48

PRECISION	2133

IDLI GROUP 88001.49

PRECISION	2134

IDLI GROUP 88001.50

CATERPILLR	6H9294
PRECISION	2135

IDLI GROUP 88001.51

PRECISION	2136

IDLI GROUP 88001.52

CATERPILLR	1P1569
PRECISION	2137

BD=BEARING DIAMETER (outside)   **BW**=BEARING WIDTH   **CB**=CROSS LENGTH WITH BEARINGS   **CC**=CENTER TO CENTER   **CD**=CROSS DIAMETER
CL=CROSS LENGTH WITHOUT BEARINGS   **EC**=END TO CENTER or FACE TO FACE   **EE**=END TO END   **EL**=EFFECTIVE LENGTH
HD=HUB DIAMETER (or insert)   **IE**=INSIDE OF EARS (or recess)   **OD**=OUTSIDE DIAMETER   **OE**=OUTSIDE OF EARS   **SB**=SPLINE OR BORE SIZE

IDLI GROUP 88001.53

CATERPILLR	5S723
PRECISION	2138

IDLI GROUP 88001.54

CATERPILLR	9S7184
PRECISION	2139

IDLI GROUP 88001.55

CHRYSLER	1653469
PRECISION	3111

IDLI GROUP 88001.56

CATERPILLR	2K8780
PRECISION	3136

IDLI GROUP 88001.57

BORG-WARNR	114-7006
IHC	133507HA
LOBRO	U672
PRECISION	9317

IDLI GROUP 88001.58

ALCO	72525-2
BORG-WARNR	114-252
BORG-WARNR	114-254
GMB	G5-1311X
PRECISION	294
WESCO	201256
WESCO	D256

IDLI GROUP 88001.59

ALCO	72625-3
BORG-WARNR	114-255
PERFECTION	RG255
PERFECTION	RG266
PRECISION	295
SPICER	7225 5
SPICER	72625-3
SPICER	73525-3
WESCO	201255
WESCO	D255

IDLI GROUP 88001.60

ALCO	J51525-2
BORG-WARNR	114-154
BORG-WARNR	114-156
BORG-WARNR	114-354
CHRYSLER	1403764
CHRYSLER	1532930
CHRYSLER	572489
CHRYSLER	575787
CHRYSLER	588572
CHRYSLER	599250
CHRYSLER	663349
CHRYSLER	915141
CHRYSLER	947548
DETROIT	515251-2
DETROIT	51525-2
DETROIT	515253-2
DETROIT	535251-2
GMB	G51525-2
LEMPCO	U417
NEAPCO	281100
NEAPCO	284300
PRECISION	297
PRECISION	735251-2
REPUBLIC	CB51525
REPUBLIC	CB515251
REPUBLIC	CB515252
REPUBLIC	CB515253
STUDEBAKER	382852
WESCO	201258
WESCO	D154
WESCO	D156
WESCO	D258

IDLI GROUP 88001.61

ALCO	735251-2
BORG-WARNR	114-324
GMB	G5-1310X
PRECISION	298
WESCO	201257
WESCO	D257

IDLI GROUP 88001.62

ALCO	J53525-2
BORG-WARNR	114-325
BORG-WARNR	114-352
BORG-WARNR	114-354
CHRYSLER	663366
CHRYSLER	947549
DETROIT	535251-2
DETROIT	53525-2
GMB	G5-1313X
GMB	G53525-2
LEMPCO	U418
MASTERS	MD4252
NEAPCO	281800
PRECISION	299
REPUBLIC	CB53525
REPUBLIC	CB535251
WESCO	301354
WESCO	D352
WESCO	D354

IDLI GROUP 88001.63

ALCO	72625-2
BORG-WARNR	114-225
BORG-WARNR	114-265
CLEVE.MOT.	CM72625-2
DETROIT	72625-2
DETROIT	J7225-2
DETROIT	J72525-3
HABERLE	H7753
MASTERS	MD2625
MOPAR	D132
PERFECTION	RG225
PERFECTION	RG261
PERFECTION	RG265
PRECISION	300
REPUBLIC	CB7225

IDLI GROUP 88001.64

ALCO	73252-2
BORG-WARNR	114-325
GMB	G5-1313X
PERFECTION	RG262
PERFECTION	RG325
PRECISION	301
REPUBLIC	CB7325
REPUBLIC	CB73252-2
WESCO	201735
WESCO	D7325

IDLI GROUP 88001.65

ALCO	53625-2
BORG-WARNR	114-362
MOPAR	D127
PERFECTION	RG362
PERFECTION	RG373
PRECISION	303
SPICER	53625-2

IDLI GROUP 88001.66

ALCO	7354A
ALCO	J52A
ALCO	J54A
ALCO	JR54A
BORG-WARNR	114-54A
DETROIT	54A
DETROIT	R54
DETROIT	R54A
LEMPCO	U421
NEAPCO	D4141X
PERFECTION	RG54A
PILOT	11362
PRECISION	221R
PRECISION	306
REPUBLIC	AD5400A
REPUBLIC	RG52A
SPICER	5-1402X
SPICER	54A
WESCO	206443
WESCO	301441
WESCO	D4410
WILLYS	115828
WILLYS	117164

IDLI GROUP 88001.67

BORG-WARNR	114-53A
CHRYSLER	986956
DETROIT	53A
LEMPCO	U423
LEMPCO	U423R
NEAPCO	D4441X
PRECISION	307
REPUBLIC	AD5300A
WESCO	D4310

IDLI GROUP 88001.68

ALCO	72625-2
BORG-WARNR	114-261
PRECISION	310
WESCO	201725
WESCO	D7225

IDLI GROUP 88001.69

ALCO	73252-2
BORG-WARNR	114-262
PRECISION	312
WESCO	201735
WESCO	D7325

IDLI GROUP 88001.70

AEC	106
AEC	6
AEC	AE266BCK2
ALLOY	5821
AMER.MOTOR	940034
BORG-WARNR	114-448SK
CHRYSLER	72620
D&D MACH.	UB253K
GMB	G3-266S
JAMCO	UJ2387
MCQUAY-NOR	U04-27
MOPAR	D134A
MOTOR MAST	2387BCK
MOTOR MAST	2387BCK2
MOTOR MAST	2387UBK2T
NEAPCO	1-0023
NEAPCO	35023X
PRECISION	318-10
SPICER	2-70-38X
TRU-CROSS	9728
TRU-CROSS	9738
TRW	20704
WESCO	255-90
WESCO	306255
WESCO	306265
ZELLER	2387BCK2

IDLI GROUP 88001.71

BORG-WARNR	114-573
FORD	XB4635DA
FORD	XW4635B
PILOT	11573
PRECISION	322
REPCO	RUJ2029
REPUBLIC	5-260X
WYLIE	56
ZELLER	3200D

IDLI GROUP 88001.72

ALCO	7352A
BORG-WARNR	114-52A
MCQUAY-NOR	U04-19
PRECISION	323
REPUBLIC	AD5200A
SPICER	52A

CAUTION: BE SURE TO REFER TO ENGINEERING CATALOGS FOR SPECIAL APPLICATIONS THAT REQUIRE SPECIFIC MATERIAL CONTENT, TOLERANCES, ETC. SEE FOOTNOTE.

IDLI GROUP 88001.73

ALCO	7352
ALCO	JR60
ALCO	U52
BORG-WARNR	114-52
GMB	GR60
PERFECTION	G52
PRECISION	324
REPUBLIC	AD5200
SPICER	52
WESCO	D5200HD

IDLI GROUP 88001.74

AEC	AE54R
ALCO	7354
ALCO	JR54
ALCO	R7354
ALCO	U54
ALCO	UR54
AMER.MOTOR	117164
BORG-WARNR	114-54
BORG-WARNR	114-54R
DETROIT	54
GMB	GR54
GMB	GUR54
LEMPCO	U420
LEMPCO	U426R
MOTOR MAST	MR54
NEAPCO	D4190
NEAPCO	D41904
NEAPCO	D4190R
PERFECTION	G54
PILOT	11364
PILOT	11364R
PRECISION	221R
PRECISION	325
REPUBLIC	AD5400
REPUBLIC	AD5400R
SPICER	5-1401X
SPICER	54
TEDDY TORQ	401C
TEDDY TORQ	GUR54
TEDDY TORQ	R401C
WESCO	401440
WESCO	D4400

IDLI GROUP 88001.75

BORG-WARNR	114-53
DETROIT	53
LEMPCO	U422
NEAPCO	D4490
PRECISION	326
REPCO	114-54
REPUBLIC	AD5300
WESCO	D4300

IDLI GROUP 88001.76

ALCO	2032
ALMETAL	CB2017
BALKAMP	M2017
BALKAMP	M2032
BORG-WARNR	114-2017
CLEVE.MOT.	CM2017
GMC	602796
HABERLE	H2017
HABERLE	H2032
MASTERS	MM2032
MOPAR	D3
NEAPCO	28052X
PERFECTION	RG2017
PERFECTION	RG2023
PRECISION	327
REPCO	114-2032
REPUBLIC	CB2017
ROCKWELL	CP2017
WESCO	201217
WESCO	S2017

IDLI GROUP 88001.77

FORD	C1TZ3249A
PRECISION	331(6)

IDLI GROUP 88001.78

GMC	2301544
PRECISION	333-2

IDLI GROUP 88001.79

GMC	2301520
GMC	2366104
PRECISION	334-2
WESCO	306517

IDLI GROUP 88001.80

CURTISS-WR	2-41485
CURTISS-WR	2-41485(5)
CURTISS-WR	7-21453
CURTISS-WR	7-21453(5)
PRECISION	334-5

IDLI GROUP 88001.81

GMC	2301775
PRECISION	335-2

IDLI GROUP 88001.82

AEC	586
AEC	AE586
CLEVE.MOT.	CM5-186X
CMP	CM5-186X
GMB	G5-121B
GMB	G5-186X
HARDY SPCR	K5GB206
LEMPCO	U108
LEMPCO	U110
MACK	201SJ46
MACK	201ST46
MOTOR MAST	7244B
NEAPCO	28186X

IDLI GROUP 88001.83 (continued)

NEAPCO	6-0186
PILOT	11522
PRECISION	336
REPCO	5-186X
REPUBLIC	CB1855B
SPICER	5-186X
SPICER	5-234X
SPICER	5-276X

IDLI GROUP 88001.83

AEC	587
AEC	AE587
CLEVE.MOT.	CM5-187X
CMP	CM5-187X
GMB	G5-187X
LEMPCO	U109
MOTOR MAST	7244C
NEAPCO	28187X
NEAPCO	6-0187
PRECISION	337
REPCO	5-187X
SPICER	5-187X

IDLI GROUP 88001.84

GMB	G7225-2
PRECISION	341G

IDLI GROUP 88001.85

ALCO	154
ALCO	J5-4
AMER.MOTOR	28075A
BORG-WARNR	114-502
HERCULES	405
PERFECTION	RG5-2X
PERFECTION	RG5-4X
PRECISION	343
WESCO	201504
WESCO	N1504

IDLI GROUP 88001.86

GLAENZER	165-004-0
GMB	GU730
MIPER	C3116
PRECISION	367

IDLI GROUP 88001.87

ALCO	JS55-55
ALCO	S55-55
BORG-WARNR	S55-55-1
PERFECTION	RS55-55-1
PRECISION	424

IDLI GROUP 88001.88

DEERE,JOHN	AT3427
DEERE,JOHN	AT64485
PRECISION	471

IDLI GROUP 88001.89

CATERPILLR	7V4077
PRECISION	475

IDLI GROUP 88001.90

CATERPILLR	7V3842
PRECISION	476

IDLI GROUP 88001.91

DEERE,JOHN	AR95974
PRECISION	482

IDLI GROUP 88001.92

CATERPILLR	8P8132
CATERPILLR	8V1540
PRECISION	485

IDLI GROUP 88001.93

AEC	6417
CATERPILLR	6S706
CLEVE.MOT.	CM206417
MOTOR MAST	7087T
PRECISION	490
TRW	20178
TWIN DISC	X206417

IDLI GROUP 88001.94

AEC	108
AEC	AE3105BCK2
BORG-WARNR	114-458SK
CLEVE.MOT.	CM5-013
GMB	G3-3105S
GMB	G5-3105S
MCQUAY-NOR	U04-29
MCQUAY-NOR	UO4-29
MOTOR MAST	3031BCK2
PRECISION	492-10
SPICER	2-70-28X
TRU-CROSS	9728

IDLI GROUP 88001.95

ALCO	1750
BORG-WARNR	114-1762
PERFECTION	RG2053
PRECISION	500

IDLI GROUP 88001.96

AEC	AE2126
ALCO	1523
ALLIS-CHLM	178461
ALLIS-CHLM	178544
ALLIS-CHLM	178554
ALLIS-CHLM	3006546
ALLIS-CHLM	4984504
ALLOY	J2100
BALKAMP	M2010
BALKAMP	M2031
BALKAMP	M2036
BALKAMP	M2040
BALKAMP	M2045

BALKAMP	M2048
BORG-WARNR	114-2100
CLEVE.MOT.	CM2002
CLEVE.MOT.	CM2062
CLEVE.MOT.	CM2100
HABERLE	H2000
HABERLE	H2010
HABERLE	H2046
HABERLE	H2062
LEMPCO	U239
MOPAR	D12
MOPAR	D4
NEAPCO	283500
PERFECTION	RG2045
PILOT	11463
PRECISION	501
REPUBLIC	CB2063
SPICER	5-2031X
TRW	20006
WESCO	N20456
ZELLER	2060C

IDLI GROUP 88001.97

ALCO	1501
ALCO	2006
ALCO	J1501
BORG-WARNR	114-1501
BORG-WARNR	114-2006
CLEVE.MOT.	CM1501
LEMPCO	U244
MOPAR	D1
PERFECTION	RG2006
PILOT	11406
PRECISION	502
TRW	20002D
WABCO	2055743

IDLI GROUP 88001.98

AEC	AE2127
AEC	AE2727
ALCO	1071
ALCO	2046
ALLIS-CHLM	20268
ALLIS-CHLM	631383
ALLOY	J2101
AMER.MOTOR	8126606
BALKAMP	M2011
BALKAMP	M2039
BALKAMP	M2041
BALKAMP	M2046
BALKAMP	M2063
BALKAMP	M2064
BORG-WARNR	114-2101
CLEVE.MOT.	CM2063
CLEVE.MOT.	CM2101
CLEVE.MOT.	CM5-108X
CMP	CM2063
EXCEL	U218
EXCEL	U239
FORD	1M7042A
HABERLE	H2011
HABERLE	H2016
HABERLE	H2036
HABERLE	H2039

HABERLE	H2045
HABERLE	H2064
LEMPCO	U218
MASTERS	MM2063
MASTERS	MM2101
MOPAR	D13
MOPAR	D5
MOTOR MAST	2060LA
NEAPCO	2-350
NEAPCO	283600
PERFECT-CI	2011
PERFECTION	RG2046
PILOT	11446
PRECISION	503
PRECISION	525
REPUBLIC	CB2046
SPICER	5-2011X
TRW	20007
WESCO	201-259
WESCO	N2046G
ZELLER	2060D

IDLI GROUP 88001.99

ALCO	1750
PERFECTION	RG2026
PRECISION	506
WESCO	201253
WESCO	N2053

IDLI GROUP 88002

PRECISION	509
WESCO	201249
WESCO	N2049

IDLI GROUP 88002.01

PILOT	11410
PRECISION	510
WESCO	201245
WESCO	N2045G

IDLI GROUP 88002.02

PRECISION	511
WESCO	201245
WESCO	N2045G

IDLI GROUP 88002.03

ALCO	2114
PERFECTION	RG2049
PRECISION	512
WESCO	201249
WESCO	N2049

IDLI GROUP 88002.04

ALCO	1077
PRECISION	513G

IDLI GROUP 88002.05

AMER.MOTOR	3113002
AMER.MOTOR	3115912
AMER.MOTOR	3119649
AMER.MOTOR	3172840
AMER.MOTOR	3200332
AMER.MOTOR	3200538
AMER.MOTOR	320085
AMER.MOTOR	3203992
AMER.MOTOR	3205371
AMER.MOTOR	3207078
PRECISION	515(3)

IDLI GROUP 88002.06

ALCO	2045
LEMPCO	U206
PERFECTION	RG2045
PRECISION	520
WESCO	201245
WESCO	G2045G
WESCO	N2045G

IDLI GROUP 88002.07

ALCO	2032
PRECISION	521
WESCO	201217
WESCO	S2017

IDLI GROUP 88002.08

ALCO	2046
PRECISION	524
PRECISION	525
WESCO	201246
WESCO	N2046G

IDLI GROUP 88002.09

ALCO	2045
BORG-WARNR	114-2000
BORG-WARNR	114-2010
PRECISION	52G
REPUBLIC	CB2000
REPUBLIC	CB2010
REPUBLIC	CB2014
WESCO	201245
WESCO	N2045G

IDLI GROUP 88002.10

ALCO	2002
BORG-WARNR	114-2002
BORG-WARNR	114-2008
BORG-WARNR	114-2103
GMB	G2103
GMC	1309468
GMC	509295
ITALCARDAN	50-812-000
LEMPCO	U203
MCQUAY-NOR	UO2-32
MOPAR	D6
NEAPCO	280200
PERFECTION	RG2002
PILOT	11402
PILOT	11403

PRECISION	527
REPUBLIC	CB2002
WESCO	N2002-1

IDLI GROUP 88002.11

ALCO	2045
PRECISION	528

IDLI GROUP 88002.12

ALCO	2045
PRECISION	529

IDLI GROUP 88002.13

AEC	105
AEC	AE5-1204UBK2
BORG-WARNR	114-438UB
GMB	G3-1204UB
MCQUAY-NOR	U04-26
MOTOR MAST	3437UBK2
MOTOR MAST	3437UBK2T
PRECISION	534-10
REPUBLIC	CB-P55-5UB
SPICER	2-94-58X
SPICER	3-94-58X
TRU-CROSS	9458
TRW	20703

IDLI GROUP 88002.14

ALCO	3012
BORG-WARNR	114-3113
PRECISION	537
WESCO	301312
WESCO	N3012G

IDLI GROUP 88002.15

ALCO	4102
PRECISION	547
WESCO	301414
WESCO	N4014

IDLI GROUP 88002.16

BORG-WARNR	114-4021
PRECISION	548
WESCO	301414
WESCO	N4014

IDLI GROUP 88002.17

PRECISION	555
WESCO	601500
WESCO	N5000

IDLI GROUP 88002.18

BORG-WARNR	114-5012
PRECISION	556
WESCO	301507
WESCO	N5007

ENGINEERING CATALOGS MUST BE CONSULTED FOR DETAILS NOT INCLUDED
IN THIS GUIDE. SPECIFIC DESIGNS, MATERIAL CONTENT, TOLERANCES
LUBE FITTINGS AND OTHER DIMENSIONS ARE INTENTIONALLY OMITTED HERE.

CAUTION: BE SURE TO REFER TO ENGINEERING CATALOGS FOR SPECIAL APPLICATIONS THAT REQUIRE SPECIFIC MATERIAL CONTENT, TOLERANCES, ETC. SEE FOOTNOTE.

IDLI GROUP 88002.19

PRECISION	560-1
WESCO	301601

IDLI GROUP 88002.20

PRECISION	560-3
WESCO	201603

IDLI GROUP 88002.21

PRECISION	563-3B
WESCO	201604
WESCO	201606

IDLI GROUP 88002.22

PRECISION	568-3
WESCO	201607

IDLI GROUP 88002.23

PRECISION	570-1
WESCO	301701

IDLI GROUP 88002.24

PRECISION	570-3
WESCO	201703
WESCO	201705

IDLI GROUP 88002.25

PRECISION	572-3A
WESCO	201709

IDLI GROUP 88002.26

PRECISION	574-3B
WESCO	201704

IDLI GROUP 88002.27

PRECISION	581

IDLI GROUP 88002.28

PRECISION	581-1
WESCO	601801

IDLI GROUP 88002.29

PRECISION	581-3
WESCO	201803
WESCO	201805

IDLI GROUP 88002.30

PRECISION	582-3A
WESCO	201809

IDLI GROUP 88002.31

PRECISION	582-3B

IDLI GROUP 88002.32

NEAPCO	6-9015
PRECISION	591

IDLI GROUP 88002.33

AEC	AECVK3
AEC	AE-CVK4
AEC	AECVK4
AEC	AECVK5
AEC	AE-CVK6
AEC	CVK003
AEC	CVK005
AEC	CVK3
AEC	CVK4
AEC	CVK 4
AEC	CVK5
AEC	CVK6
ALCO	1176
ALCO	1177
ALCO	1180
ALCO	1186
ALCO	J4-3
ALCO	J4-5
ALLOY	1176
ALLOY	1177
ALLOY	1179
ALLOY	1180
ALLOY	J4-3
ALLOY	J4-4
ALLOY	J4-5
ALLOY	J4-6
BORG-WARNR	117-113
BORG-WARNR	117-114
BORG-WARNR	117-115
BORG-WARNR	117-115A
BORG-WARNR	117-116
CLEVE.MOT.	CMCVK20
CLEVE.MOT.	CMCVK20A
CLEVE.MOT.	CM-CVK30
CLEVE.MOT.	CM-CVK30A
FLEETRITE	ZAD314
FLEETRITE	ZAD316
GMB	GCVK3
GMB	GCVK4
GMB	GCVK-4
GMB	GCVK4
GMB	GCVK5
GMB	GCVK6
GMB	GCVK7
GMC	1389864
GMC	1396838
GMC	7808979
GOODYEAR	219-2523
LEMPCO	U603K
LEMPCO	U-603K
LEMPCO	U603K
LEMPCO	U604K
LEMPCO	U605K
LEMPCO	U606K
MCQUAY-NOR	U01-48
MCQUAY-NOR	U01-49
MCQUAY-NOR	U01-53
MCQUAY-NOR	U01-54
MCQUAY-NOR	UO1-49
MOPAR	D209
MOPAR	D210
MOTOR MAST	CVK3
MOTOR MAST	CVK4
MOTOR MAST	CVK5
MOTOR MAST	CVK5A
MOTOR MAST	CVK6
NEAPCO	2-9203
NEAPCO	2-9205
NEAPCO	2-9303
NEAPCO	2-9305
NEAPCO	9203
NEAPCO	9203B
NEAPCO	9205
NEAPCO	9205B
PILOT	314
PILOT	315
PILOT	316
PILOT	317
PRECISION	603
REPCO	117-113
REPCO	117-114
REPCO	117-115A
REPCO	117-116
REPUBLIC	CVK3
REPUBLIC	CVK-4
REPUBLIC	CVK4
REPUBLIC	CVK5
REPUBLIC	CVK5A
REPUBLIC	CVK6A
SPICER	211010X
SPICER	211011X
SPICER	211012
SPICER	211012X
SPICER	211013
SPICER	211013X
TRW	20726
TRW	20727
TRW	20730
TRW	20731
WESCO	117-114
WESCO	208838
WESCO	208839
WESCO	208-864
WESCO	208-865
WESCO	408-838
WESCO	920-838
ZELLER	CVK-3
ZELLER	CVK5
ZELLER	CVK5-115A
ZELLER	CVK6

IDLI GROUP 88002.34

AEC	AE-CVK1
AEC	AE-CVK2
AEC	CVK001
AEC	CVK1
AEC	CVK2
ALCO	1175
ALCO	1181
ALCO	J4-1
ALLOY	1175
ALLOY	1178
ALLOY	J4-1
ALLOY	J4-2
BORG-WARNR	117-110
BORG-WARNR	117-111
CLEVE.MOT.	CM-CPK10
CLEVE.MOT.	CM-CVK10
CLEVE.MOT.	CM-CVK10A
D&D MACH.	MCVK1
D&D MACH.	MCVK2
FLEETRITE	ZAD311
FORD	C8SZ4A031
FORD	C8SZ4A031A
GMB	GCVK1
GMB	GCVK2
GMC	7809156
GMC	7816006
GOODYEAR	219-2521
LEMPCO	U601K
LEMPCO	U602K
MCQUAY-NOR	U01-46
MCQUAY-NOR	U01-65
MCQUAY-NOR	UO1-46
MOPAR	D208
MOTOR MAST	3031BCK2T
MOTOR MAST	CVK1
MOTOR MAST	CVK2
NEAPCO	2-9201
NEAPCO	2-9301
NEAPCO	9201
NEAPCO	9201B
PILOT	311
PILOT	312
PRECISION	604
REPCO	117-110
REPCO	117-111
REPUBLIC	CVK1
REPUBLIC	CVK2
SPICER	211007
SPICER	211007X
SPICER	211008
SPICER	211008X
TRW	20705
TRW	20725
TRW	20729
WESCO	208-156
WESCO	208-157
WESCO	208308
WESCO	208309
WESCO	408-308
WESCO	920-308
ZELLER	CVK1
ZELLER	CVK1-111
ZELLER	CVK2

BD = BEARING DIAMETER (outside)   **BW** = BEARING WIDTH   **CB** = CROSS LENGTH WITH BEARINGS   **CC** = CENTER TO CENTER   **CD** = CROSS DIAMETER
CL = CROSS LENGTH WITHOUT BEARINGS   **EC** = END TO CENTER or FACE TO FACE   **EE** = END TO END   **EL** = EFFECTIVE LENGTH
HD = HUB DIAMETER (or insert)   **IE** = INSIDE OF EARS (or recess)   **OD** = OUTSIDE DIAMETER   **OE** = OUTSIDE OF EARS   **SB** = SPLINE OR BORE SIZE

IDLI GROUP 88002.35

AEC	AE-CVK1
BORG-WARNR	117-111
GMC	7814308
MCQUAY-NOR	U01-46
MOTOR MAST	CVK1
NEAPCO	9201
PRECISION	605
REPCO	117-111
WESCO	208308
ZELLER	CVK1-111

IDLI GROUP 88002.36

AEC	AE-CVK2A
AEC	CVK002A
AEC	CVK2A
ALCO	1178
ALLOY	1181
ALLOY	J4-7
BORG-WARNR	117-112
CHRYSLER	4030849
CHRYSLER	4049697
CLEVE.MOT.	CM-CVK10B
D&D MACH.	MCVK2A
FLEETRITE	ZAD313
GMB	GCVK2
GMB	GCVK2A
GMC	7815848
GMC	7828953
GMC	7829456
LEMPCO	U602A
LEMPCO	U602AK
MCQUAY-NOR	U01-47
MCQUAY-NOR	UO1-47
MOTOR MAST	CVK2A
NEAPCO	2-9302
NEAPCO	9202B
PILOT	313
PRECISION	606
REPCO	117-112
REPUBLIC	CVK2A
SPICER	211009
SPICER	211009X
TRU-CROSS	9119
TRW	20728
WESCO	208-157
WESCO	208-308
WESCO	408-309
WESCO	920-309
ZELLER	CVK2A
ZELLER	CVK2A-112

IDLI GROUP 88002.37

AEC	SCV082
AEC	SCV82
ALLOY	40782
AMER.MOTOR	8130303
AMER.MOTOR	994823
BORG-WARNR	117-1782J
CHRYSLER	3827050
FORD	C9TZ4782A
FORD	D2TZ4782A
FORD	D2TZ4782Z
GMB	G210782X

IDLI GROUP 88002.38

GMC	6258549
LEMPCO	U607T
MCQUAY-NOR	U02-35
MOTOR MAST	SCV82
NEAPCO	28CV82
NEAPCO	7-0082
PRECISION	607
REPUBLIC	CVK8
SPICER	210782X
SPICER	211007
SPICER	211177X
SPICER	211355X
TRU-CROSS	9182
TRW	20186
TRW	20186D
WESCO	520782
ZELLER	SCV82

IDLI GROUP 88002.38

ALLOY	40179
FORD	D9TZ4782A
FORD	D9TZ4782Z
MOTOR MAST	SCV79
PRECISION	608
SPICER	211179X
TRW	20215
WESCO	530179
ZELLER	SCV79

IDLI GROUP 88002.39

FORD	E3TZ4782A
PRECISION	609
SPICER	211342X

IDLI GROUP 88002.40

PRECISION	610
WESCO	950003

IDLI GROUP 88002.41

PRECISION	611
SPICER	2-86-348
WESCO	201800

IDLI GROUP 88002.42

AEC	2T2911
AEC	AE2T2911
ALCO	1920
ALCO	6642
ALCO	U2T2911
ALLOY	1920
ALLOY	U2T2911
BORG-WARNR	2T2911
BORG-WARNR	60611
CLEVE.MOT.	CM591642
CMP	CM591642
FLEETRITE	ZAD11042
GMB	2T2911
GMB	GU2T2911
GMC	591642
GMC	606779
HABERLE	H6642
HERCULES	729

IDLI GROUP 88002.42

LEMPCO	U10
LEMPCO	U11
MASTERS	MT1642
MCQUAY-NOR	U01-61
MCQUAY-NOR	UO1-61
MOPAR	D181
MOTOR MAST	26-101
NEAPCO	7-9061
NEAPCO	9061
PERFECTION	G2911
PILOT	11042
PILOT	11079
PRECISION	660
REPUBLIC	C1109
REPUBLIC	CB1109
SPICER	212-3X
TRW	20710
UJS	55-212-3
WESCO	401642
WESCO	S1642
ZELLER	26-101

IDLI GROUP 88002.43

AEC	AE2T2910
ALCO	6779
ALCO	U2T2910
BORG-WARNR	2T2910
BORG-WARNR	2T4295
CMP	CM606779
GMB	GU2T2910
GMC	1288621
GMC	606779
HERCULES	727
HERCULES	728
LEMPCO	U10
LEMPCO	U8
MOPAR	D180
MOTOR MAST	26-100
NEAPCO	9056
NEAPCO	9060
PERFECTION	G2910
PERFECTION	G4295
PILOT	11079
PRECISION	661
REPCO	5-212X
REPUBLIC	C1009
REPUBLIC	C809
SPICER	212-1X
UJS	55-212-1
WESCO	AS6779
WESCO	SN6779
WESCO	SR6779

IDLI GROUP 88002.44

AEC	2T2912
AEC	AE2T2912
ALCO	1921
ALCO	6232
ALCO	U2T2912
ALCO	UT2912
ALLOY	1921
ALLOY	U2T2912
BORG-WARNR	2T2912
BORG-WARNR	2T2914

IDLI GROUP 88002.44 (cont.)

GMB	2T2912
GMB	GU2T2912
GMC	3693692
GMC	3704232
LEMPCO	U12
LEMPCO	U21
MASTERS	MT3692
MOPAR	D182
MOTOR MAST	26-102
NEAPCO	9063
PERFECTION	G2912
PILOT	11032
PILOT	11092
PRECISION	662
REPUBLIC	C1209
REPUBLIC	C1309
SPICER	212-4X
SPICER	212-5X
TORQUE	U2T2914
WESCO	401424
WESCO	S3692
WESCO	S4232
ZELLER	26-102

IDLI GROUP 88002.45

ALCO	U3T2222
BORG-WARNR	3T2222
GMB	GU3T2222
LEMPCO	U9B
NEAPCO	9058
PRECISION	663
REPUBLIC	C901
REPUBLIC	C910
SPICER	302-8X

IDLI GROUP 88002.46

ALCO	U3T2894
BORG-WARNR	3T2894
GMB	GU3T2894
GMC	604373
LEMPCO	U5A
NEAPCO	9044
PRECISION	664
REPUBLIC	C1410
SPICER	302-10X

IDLI GROUP 88002.47

ALCO	U3T1595
BORG-WARNR	3T2595
GMB	GU3T1595
GMC	602795
LEMPCO	U9C
NEAPCO	9059
PRECISION	665
REPUBLIC	C1110
SPICER	302-9X
WESCO	401953
WESCO	S9531

CAUTION: BE SURE TO REFER TO ENGINEERING CATALOGS FOR SPECIAL APPLICATIONS THAT REQUIRE SPECIFIC MATERIAL CONTENT, TOLERANCES, ETC. SEE FOOTNOTE.

IDLI GROUP 88002.48

AEC	3T2895
AEC	AE3T2895
ALCO	6531
ALCO	U3T2895
BORG-WARNR	3T2893
BORG-WARNR	3T2895
CLEVE.MOT.	CM609531
GMB	GU3T2895
GMC	3653379
GMC	602795
GMC	609531
HABERLE	H6531
LEMPCO	U5B
MASTERS	MT9531
MOPAR	D185
MOTOR MAST	26-105
NEAPCO	9046
NEAPCO	9059
PERFECTION	G2895
PILOT	11031
PRECISION	666
REPUBLIC	C1610
SPICER	302-9X
SPICER	332-3X
TORQUE	U3T2893
TRW	20711
WESCO	401953
WESCO	S9531

IDLI GROUP 88002.49

ALCO	6232
GMB	2T2911
PRECISION	667
REPUBLIC	CB1209
WESCO	401424
WESCO	S4232

IDLI GROUP 88002.50

AEC	3T2896
AEC	AE3T2896
ALCO	9266
ALCO	U3T2896
BORG-WARNR	3T2896
CLEVE.MOT.	CM3689266
FORD	3705572
GMB	GU3T2896
GMC	3689266
GMC	3705572
HABERLE	H9266
LEMPCO	U22
MOPAR	D186
MOTOR MAST	26-106
NEAPCO	9065
PERFECTION	G2896
PILOT	11066
PRECISION	668
REPUBLIC	C1810
REPUBLIC	C1910
SPICER	266-1X

SPICER	288-1X
TRW	20712
WESCO	411926
WESCO	S9266

IDLI GROUP 88002.51

AEC	3T2897
AEC	AE3T2897
ALCO	6646
ALCO	U3T2897
BORG-WARNR	3T2897
GMB	GU3T2897
GMC	591646
LEMPCO	U24
MOPAR	D187
NEAPCO	9067
PERFECTION	G2897
PILOT	11046
PRECISION	669
REPUBLIC	C1709
REPUBLIC	C1710
SPICER	302-13X
WESCO	401646
WESCO	S1646

IDLI GROUP 88002.52

AEC	3T2898
AEC	AE3T2898
ALCO	9288
ALCO	U3T2898
BORG-WARNR	3T2898
FORD	3689288
GMB	GU3T2898
GMC	3609288
GMC	3689288
LEMPCO	U23
MOPAR	D188
NEAPCO	9066
PERFECTION	G2898
PILOT	11088
PRECISION	670
REPUBLIC	C1910
SPICER	288-1X
WESCO	411928
WESCO	S9288

IDLI GROUP 88002.53

ALCO	U3T2899
BORG-WARNR	3T2899
FORD	81C7090
GMB	GU3T2899
PRECISION	673
REPUBLIC	F115
SPICER	202-8X

IDLI GROUP 88002.54

ALCO	U3T2266
ALMETAL	F111
BORG-WARNR	3T2266
FORD	74-7090
GMB	GU3T2266
PRECISION	674
REPUBLIC	F111
SPICER	202-7X
WESCO	S7409

IDLI GROUP 88002.55

AEC	3T1328
AEC	AE3T1328
ALCO	B7090
ALCO	U3T1328
BORG-WARNR	3T1328
CLEVE.MOT.	CM-B7090
FORD	B7090
GMB	GU3T1328
HABERLE	H7090
HERCULES	380
HERCULES	757
LEMPCO	U17
LEMPCO	U2
MASTERS	3T-1328
MASTERS	MT7090
MOPAR	D183
MOTOR MAST	26-128
MOTOR MAST	27-128
NEAPCO	9005
NEAPCO	9005X
NEAPCO	9025
PERFECTION	G1328
PILOT	11090
PRECISION	675
REPUBLIC	F100
SPICER	202-6X
TEDDY TORQ	F1
TRW	20713
WESCO	401709
WESCO	S7090

IDLI GROUP 88002.56

ALCO	6174
ALCO	U3T1602
BORG-WARNR	3T1602
GMB	GU3T1602
GMC	600174
HERCULES	731
LEMPCO	U3
NEAPCO	9031
PERFECTION	G1602
PILOT	11074
PRECISION	676
REPUBLIC	C409
SPICER	332-1X

IDLI GROUP 88002.57

ALCO	6889
ALCO	U3T2007
BORG-WARNR	3T2007
GMB	GU3T2007
GMC	601889
HERCULES	735
LEMPCO	U4
NEAPCO	9035
PERFECTION	G2007
PILOT	11089
PRECISION	677
REPUBLIC	C609
SPICER	232-1X
WESCO	S1889

IDLI GROUP 88002.58

ALCO	6462
ALCO	6646
ALCO	U3T2221
BORG-WARNR	3T2221
GMB	GU3T2221
GMC	602462
HERCULES	736
LEMPCO	U5
NEAPCO	9036
PERFECTION	G2221
PILOT	11062
PRECISION	678
REPUBLIC	C709
SPICER	332-1X
SPICER	332-2X
WESCO	S2462

IDLI GROUP 88002.59

ALCO	6779
BORG-WARNR	2T2910
NEAPCO	9060
PRECISION	679
REPUBLIC	C809
WESCO	401642
WESCO	S1642

IDLI GROUP 88002.60

ALCO	6118
ALCO	U3T2531
ALCO	U3T2538
BORG-WARNR	3T2531
GMB	GU3T2531
GMB	GU3T2538
GMC	3653380
GMC	605118
HERCULES	737
LEMPCO	U9
LEMPCO	U9A
MOPAR	D184
NEAPCO	9054
NEAPCO	9057
PERFECTION	G2531
PILOT	11018
PRECISION	680
REPUBLIC	C1210
REPUBLIC	CR1210

BD = BEARING DIAMETER (outside)　　**BW** = BEARING WIDTH　　**CB** = CROSS LENGTH WITH BEARINGS　　**CC** = CENTER TO CENTER　　**CD** = CROSS DIAMETER
CL = CROSS LENGTH WITHOUT BEARINGS　　**EC** = END TO CENTER or FACE TO FACE　　**EE** = END TO END　　**EL** = EFFECTIVE LENGTH
HD = HUB DIAMETER (or insert)　　**IE** = INSIDE OF EARS (or recess)　　**OD** = OUTSIDE DIAMETER　　**OE** = OUTSIDE OF EARS　　**SB** = SPLINE OR BORE SIZE

SPICER 302-11X
WESCO S1018

IDLI GROUP 88002.61

ALCO U3T1328
BORG-WARNR 114-1328
BORG-WARNR 2T2910
FORD B7084
GMB GU3T1328
LEMPCO U1
PRECISION 681
SPICER 5-44X

IDLI GROUP 88002.62

ALCO 6779
BORG-WARNR 2T2910
PERFECTION G4295
PRECISION 682
REPUBLIC C1009
WESCO 401642
WESCO S1642

IDLI GROUP 88002.63

AEC AE3T3501
ALCO 6717
ALCO U3T2501
ALCO U3T3501
BORG-WARNR 3T3501
GMB GU3T2501
GMB GU3T3501
GMC 1393717
HERCULES 758
LEMPCO U6
PERFECTION G3501
PERFECTION G3502
PILOT 11917
PRECISION 683
REPUBLIC BU200
SPICER 202-5X
SPICER 5-47X-JA
TEDDY TORQ B374
TORQUE U3T3502
WESCO S3717

IDLI GROUP 88002.64

ALCO 6164
ALCO U3T3501
ALCO U3T3502
BORG-WARNR 3T3502
BORG-WARNR 3T3504
GMB GU3T3502
GMC 1393718
HERCULES 759
LEMPCO U7
NEAPCO 9052
PILOT 11918
PILOT 18208
PRECISION 684
REPUBLIC BU201
SPICER 202-13X
SPICER 5-47X-JA
WESCO S1164

IDLI GROUP 88002.65

ALCO U3T3505
BORG-WARNR 3T3505
GMB GU3T3505
GMC 1393716
LEMPCO U7B
NEAPCO 9055
PRECISION 685
REPUBLIC BU203
SPICER 302-3X

IDLI GROUP 88002.66

ALCO 5939
BORG-WARNR 3T3503
BORG-WARNR 3T3506
PRECISION 686
WESCO S1385

IDLI GROUP 88002.67

AEC AE3T3501
ALCO 5939
ALCO U3T3501
ALCO U3T3506
BORG-WARNR 3T3503
BORG-WARNR 3T3506
GMB GU3T3506
GMC 1391262
GMC 1393718
GMC 1393719
GMC 5661385
GMC 5665939
LEMPCO U15
LEMPCO U18
LEMPCO U20
LEMPCO U7A
NEAPCO 9051
NEAPCO 9053
NEAPCO 9062
PERFECTION G3506
PILOT 11985
PRECISION 687
REPUBLIC BU202
REPUBLIC BU204
REPUBLIC BU207
SPICER 302-14X
TEDDY TORQ B262
TEDDY TORQ B385
TORQUE U3T3502
TORQUE U3T3508
WESCO S1385

IDLI GROUP 88002.68

AMER.MOTOR 3131428
AMER.MOTOR 3138316
BORG-WARNR 2B3053
BORG-WARNR B7200-20
DETROIT 7200-20
GMC 600865
LEMPCO U14
PRECISION 688
REPUBLIC N7200-20

IDLI GROUP 88002.69

AMER.MOTOR 3127013
AMER.MOTOR 3131449
AMER.MOTOR 3131517
AMER.MOTOR 3144950
AMER.MOTOR 3145689
BORG-WARNR B7100-1A
BORG-WARNR B7140-2A
DETROIT 7100-1A
DETROIT 7140-2A
LEMPCO U13
PRECISION 689
REPUBLIC N7100-1
REPUBLIC N7140-2A

IDLI GROUP 88002.70

ALCO 6605
ALCO U3T3509
BORG-WARNR 3T3509
GMB GU3T3509
GMC 5665605
LEMPCO U20
NEAPCO 9069
PERFECTION G3509
PILOT 11905
PRECISION 692
REPUBLIC BU206
WESCO S5605

IDLI GROUP 88002.71

AEC AE3T3507
ALCO 6164
ALCO U3T3407
ALCO U3T3507
BORG-WARNR 3T3507
GMB GU3T3407
GMB GU3T3507
GMC 1391164
GMC 1393719
GMC 5665608
LEMPCO U19
LEMPCO U8A
NEAPCO 9050
PERFECTION G3507
PILOT 11918
PILOT 11964
PRECISION 694
REPUBLIC BU205
SPICER 202-13X
SPICER 202-15X
TEDDY TORQ B663

IDLI GROUP 88002.72

ALCO 5939
ALCO U3T3508
BORG-WARNR 3T3508
GMB GU3T3508
GMC 1391262
GMC 5665939
LEMPCO U18
PERFECTION 3506
PERFECTION G3508
PRECISION 695

REPUBLIC BU207

REPUBLIC BU207
TORQUE U3T3503
TORQUE U3T3508

IDLI GROUP 88002.73

ALCO U3T3510
BORG-WARNR 3T3510
GMB GU3T3510
GMC 5670292
LEMPCO U25
NEAPCO 9072
PRECISION 696
REPUBLIC BU208
WESCO SO174
WESCO SO292

IDLI GROUP 88002.74

ALCO U3T3511
BORG-WARNR 3T3511
GMB GU3T3511
GMC 5670233
GMC 5670293
LEMPCO U26
NEAPCO 9071
PRECISION 697
REPUBLIC BU209
WESCO SO293

IDLI GROUP 88002.75

GMC 5670615
GMC 5672334
LEMPCO U28
LEMPCO U29
PRECISION 698
REPUBLIC BU210

IDLI GROUP 88002.76

GMC 5671108
GMC 5672333
LEMPCO U27
LEMPCO U30
PRECISION 699
REPUBLIC BU211

IDLI GROUP 88002.77

ALLOY 1061
ARMSTRONG 5410-7010
ARMSTRONG 5410-7150
ARMSTRONG 5410-7280
AUST.MOT 04371-22010C
AUST.MOT T106-01-39
AUST.MOT T61-01-85
BMC HYL3090
BORG-WARNR 114-1750
BORG-WARNR 114-1762
CHRYSLER 2630272
CHRYSLER 4156466
DATSUN 37125H1025A
FLEETRITE ZAD715
FLEETRITE ZAD715G
FORD ARC4DZ4635A
FORD CF4635A

ENGINEERING CATALOGS MUST BE CONSULTED FOR DETAILS NOT INCLUDED
IN THIS GUIDE. SPECIFIC DESIGNS, MATERIAL CONTENT, TOLERANCES
LUBE FITTINGS AND OTHER DIMENSIONS ARE INTENTIONALLY OMITTED HERE.

CAUTION: BE SURE TO REFER TO ENGINEERING CATALOGS FOR SPECIAL APPLICATIONS THAT REQUIRE SPECIFIC MATERIAL CONTENT, TOLERANCES, ETC. SEE FOOTNOTE.

FORD	CH4635A
FORD	CIUZ4635A
FORD	CODZ7039A
GELENKWEL	7.175.30.06.13.060
GELENKWEL	7.185.10.06.17.060
GLAENZER	065-257-8
GLAENZER	065-935-8
GLAENZER	065-959-8
GLAENZER	065-960-8
GLAENZER	065-980-8
GMB	G1750G
GMB	G5-1767
GMB	GU6120
GMB	GU7150
GMC	7405065
GMC	7421879
GMC	7423286
GMC	7423287
GMC	9044923
GMC	VS10604
GMC	VS10605
HOLDEN	7405065
HOLDEN	7421879
HOLDEN	7423286
HOLDEN	7423287
LEMPCO	U262
LEMPCO	U262G
LEYLAND	HYL3826
LOBRO	U592
LOBRO	U595
LOBRO	U597
LOBRO	U701
LOBRO	U715
LOBRO	U728
LOBRO	U728/3
MATSUBA	UJ030
MCQUAY-NOR	U01-1
MCQUAY-NOR	U01-12
MERCEDES	100 410 01 31
MERCEDES	111 410 00 31
MERCEDES	180 410 03 31
MIPER	C3318-6
MITSUBISHI	4156466
MOPAR	D202
MOTOR MAST	2025A
NEAPCO	1-105
NEAPCO	1-1555
NEAPCO	281555
OPEL	1604073
OPEL	458011
OPEL	45802
OPEL	45807
OPEL	TB94402
PILOT	715
PILOT	715G
PITT.VIOL.	40.746.000
PITT.VIOL.	40.752.000
PITT.VIOL.	40.814.000
PRECISION	715
QUINT.HAZ.	QL11007
QUINT.HAZ.	QL1761

QUINT.HAZ.	QL1762
QUINT.HAZ.	QL1765
QUINT.HAZ.	QL1786
QUINT.HAZ.	QL203
QUINT.HAZ.	QL204
QUINT.HAZ.	QL212
QUINT.HAZ.	QL300
QUINT.HAZ.	QL312
RENAULT	SB539-8695
REPCO	114-1750
REPCO	114-1762
REPCO	114-1767
REPCO	7.175.30.06-1R
REPCO	7.185.10.06-1R
REPCO	7.185.10.06-2R
REPCO	RUJ1760
REPCO	RUJ1761
REPCO	RUJ1762
REPCO	RUJ1765
REPCO	RUJ1773
REPCO	RUJ1786
RUEDARSA	1291
RUEDARSA	2104-1291-0-0
SPICER	5-2053
TOYOTA	04371-22010C
TRW	20003
TRW	20003D
WYLIE	50
WYLIE	9

IDLI GROUP 88002.78

PRECISION	716
RENAULT	9812055
REPCO	SKA11654

IDLI GROUP 88002.79

PRECISION	717
RENAULT	9816522
REPCO	SKA11655

IDLI GROUP 88002.80

PRECISION	718
RENAULT	9816566
REPCO	SKA11653

IDLI GROUP 88002.81

ALCO	J1755G
GMB	G1755G
GMC	7406001
GMC	7421881
GMC	7425515
HOLDEN	7406001
HOLDEN	7421881
LOBRO	U591
MATSUBA	UJ031
PRECISION	719
QUINT.HAZ.	QL1758
REPCO	RUJ1758
REPUBLIC	CB1755G
WESCO	201251
WYLIE	31

IDLI GROUP 88002.82

ALCO	J7200
GMB	G7200
PRECISION	740
TEDDY TORQ	1470

IDLI GROUP 88002.83

KOMATSU	141-10-00012(4)
PRECISION	794

IDLI GROUP 88002.84

KOMATSU	150-11-12360
PRECISION	795

IDLI GROUP 88002.85

KOMATSU	175-10-00191(4)
PRECISION	796

IDLI GROUP 88002.86

AEC	AE6A2046
AEC	AE-Y1001
ALCO	1662
ALCO	4851
BORG-WARNR	6-2046
BORG-WARNR	6-2074
BORG-WARNR	6A2046
CHRYSLER	1134584
FORD	1M4841
FORD	1M4841A
FORD	1N4841
FORD	8M4841
FORD	AF4841A
FORD	B5A4841A
GMB	G1662
LEMPCO	UY1
NEAPCO	NF4841A
PERFECTION	R424
PERFECTION	R425
PILOT	14841
PRECISION	841
REPUBLIC	HY311
REPUBLIC	HY365
TEDDY TORQ	841
TEDDY TORQ	851
WESCO	401484
WESCO	Y4841

IDLI GROUP 88002.87

AEC	AE6A2046
AEC	AE-Y1001
ALCO	1661
ALCO	4841
BORG-WARNR	6A2046
BORG-WARNR	Y1001
FORD	0A4841
FORD	AB4851
FORD	AB4851A
FORD	0A4851
GMB	G1661
LEMPCO	UF1
NEAPCO	NF4851A
PERFECTION	R424

PERFECTION	R425
PILOT	14851
PRECISION	851
REPUBLIC	HY311
REPUBLIC	HY365
WESCO	920018
WESCO	G1661
WESCO	Y4851

IDLI GROUP 88002.88

CLARK EQU.	873726
CLARK EQU.	885386
FEDERAL	16B6050
GAR WOOD	3000-100
GAR WOOD	304626
GMB	G5-264X
PRECISION	854
REPCO	CP1FRN4
REPCO	SCP1FR
REPUBLIC	CB1FRN4

IDLI GROUP 88002.89

FLEETRITE	ZAD83DR
GMB	G5-265X
PRECISION	857
WESCO	201300
WESCO	3DR

IDLI GROUP 88002.90

CATERPILLR	3H8257(5)
CATERPILLR	3H8262(5)
CATERPILLR	3H8263(5)
PRECISION	861(5)

IDLI GROUP 88002.91

PRECISION	862
REPUBLIC	CB-L6N5

IDLI GROUP 88002.92

CATERPILLR	3K9532(5)
CATERPILLR	4F3056(5)
CATERPILLR	4F3057(5)
CATERPILLR	4F3058(5)
CATERPILLR	4K1917(5)
CATERPILLR	9H3772(5)
PRECISION	870(5)

IDLI GROUP 88002.93

AEC	AE-CPL10N2
CATERPILLR	9H3766(2)
GMB	GGPL10N
PRECISION	871

IDLI GROUP 88002.94

OLIVER	AC17-77
PRECISION	881
WEST.LAND	AS60190P

IDLI GROUP 88002.95

COCKSHUTT	FH1131
FOX RIVER	F4CX5
PRECISION	885
REPUBLIC	CB-L16S
WHITE	876347

IDLI GROUP 88002.96

BORG-WARNR	CP5N
PRECISION	891
TRW	20088D
WESCO	601515
WESCO	N1515

IDLI GROUP 88002.97

AEC	AE7105
ALCO	J7105
ALCO	J7112
ALCO	J7113
ALCO	J7117
ALCO	J7122
BORG-WARNR	114-7105
BORG-WARNR	114-7113
BORG-WARNR	114-7119
BORG-WARNR	114-7122
PRECISION	901

IDLI GROUP 88002.98

BORG-WARNR	114-3108
MOPAR	D18
PRECISION	902

IDLI GROUP 88002.99

GMC	9025904
PRECISION	905-3B

IDLI GROUP 88003

GMC	9024661
GMC	9024787
GMC	9026499
PRECISION	906-3

IDLI GROUP 88003.01

EATON	2017732(5)
EATON	2017744
EATON	2017758
EATON	T698-1
EATON	T698-7(5)
EATON	T699-1
PRECISION	906(5)

IDLI GROUP 88003.02

GMC	9024789
GMC	9025557
GMC	9026534
GMC	9046312
PRECISION	907-3A

IDLI GROUP 88003.03

EATON	2017732(5)
EATON	2017744
EATON	T698-1
EATON	T698-7(5)
PRECISION	908-5

IDLI GROUP 88003.04

GMC	9021250
GMC	9024792
GMC	9025559
GMC	9046303
PRECISION	909-3

IDLI GROUP 88003.05

GMC	9021691
GMC	9022353
GMC	9024793
GMC	9025560
GMC	9026500
PRECISION	909-3B

IDLI GROUP 88003.06

GMC	9026541
PRECISION	911-3B

IDLI GROUP 88003.07

GMC	9024796
PRECISION	913-3A

IDLI GROUP 88003.08

GMB	G8113
PRECISION	914

IDLI GROUP 88003.09

AEC	AE9002
AEC	AE9026
ALCO	J9002
ALLOY	J9017
BORG-WARNR	114-9002
BORG-WARNR	114-9026
GMB	G9002
NEAPCO	289026
PILOT	19026
PRECISION	919
REPUBLIC	CB719
SPICER	5-9002
SPICER	5-9002X
TRW	20114
WESCO	N9002
ZELLER	7075HA

IDLI GROUP 88003.10

CATERPILLR	6F2426(5)
CATERPILLR	6F4341(5)
CATERPILLR	6F4342(5)
CATERPILLR	6F4343(5)
IHC	104397
IHC	104398
IHC	104401
IHC	104401(5)
IHC	104402
IHC	111048
IHC	111050
IHC	111051(5)
PRECISION	934-5

IDLI GROUP 88003.11

BORG-WARNR	114-6100
GMB	G6143
PRECISION	935

IDLI GROUP 88003.12

BORG-WARNR	114-6101
GMB	G6152
PRECISION	936

IDLI GROUP 88003.13

BALKAMP	M4014
BORG-WARNR	114-4102
GMB	G4152
PRECISION	938

IDLI GROUP 88003.14

BORG-WARNR	114-4006
GMB	G4123
PRECISION	939

IDLI GROUP 88003.15

BORG-WARNR	114-4100
GMB	G4138
PRECISION	940

IDLI GROUP 88003.16

GMB	G4113
IHC	127290
IHC	127290H1
MCQUAY-NOR	U03-17
NEAPCO	284103
PRECISION	941
REPCO	114-4103
REPUBLIC	CB741

IDLI GROUP 88003.17

AEC	AE4113
BORG-WARNR	114-4143
NEAPCO	284143
PILOT	14143
PRECISION	942
REPUBLIC	CB742
TRW	20073
WESCO	N4113

ZELLER ... 4000E

IDLI GROUP 88003.18

BORG-WARNR	114-6106
BORG-WARNR	114-6110
PRECISION	943

IDLI GROUP 88003.19

GMB	GCP35N3
PRECISION	945
REPCO	CP35N

IDLI GROUP 88003.20

AEC	AE8105
ALCO	J8105
ALCO	J8107
ALCO	J8113
BORG-WARNR	114-8105
BORG-WARNR	114-8111
GMB	G8113
PRECISION	946

IDLI GROUP 88003.21

BORG-WARNR	114-7119
GMB	G7126
PRECISION	947

IDLI GROUP 88003.22

PRECISION	953
REPUBLIC	CB753

IDLI GROUP 88003.23

GMB	G6152
INBRA	CT11-045
PRECISION	956

IDLI GROUP 88003.24

GMB	G5132
PRECISION	959

IDLI GROUP 88003.25

GMB	G6140
PRECISION	960

IDLI GROUP 88003.26

ATHEY PROD	P20261
ATHEY PROD	P20331
CATERPILLR	2D3765
CATERPILLR	4D8212
CATERPILLR	4H7808
CATERPILLR	5R56
CATERPILLR	D2X2202
FIAT-ALLIS	0178544-3
FIAT-ALLIS	082073
FIAT-ALLIS	178544
FIAT-ALLIS	178544-3
FIAT-ALLIS	70082073
FIAT-ALLIS	70178544
FIAT-ALLIS	70178544-3

CAUTION: BE SURE TO REFER TO ENGINEERING CATALOGS FOR SPECIAL APPLICATIONS THAT REQUIRE SPECIFIC MATERIAL CONTENT, TOLERANCES, ETC. SEE FOOTNOTE.

GMB	G5-2001X
GMC	9020268
GMC	9025561
INBRA	CT13-015
MASS.FERG.	840-302M92
MCQUAY-NOR	U01-3
PRECISION	961
REPUBLIC	CB761
WABCO	472196
WABCO	472197
WABCO	472200
WABCO	474041

IDLI GROUP 88003.27

CATERPILLR	1H8022(5)
CATERPILLR	1H8024(5)
CATERPILLR	1H8025(5)
PRECISION	961(5)

IDLI GROUP 88003.28

AEC	AE8500
ALCO	J8500
BORG-WARNR	114-8500
BORG-WARNR	114-8504
GMB	G5-8500X
GMB	G8500
MICHIGAN	944686
PRECISION	962

IDLI GROUP 88003.29

PRECISION	965
REPUBLIC	CB765

IDLI GROUP 88003.30

PRECISION	970
REPCO	5-280X
REPUBLIC	CB770

IDLI GROUP 88003.31

PRECISION	971
REPCO	CP35N
REPUBLIC	CB771

IDLI GROUP 88003.32

GLAENZER	065-164-8
PRECISION	973
REPUBLIC	CB773

IDLI GROUP 88003.33

GLAENZER	065-149-8
PRECISION	974
REPUBLIC	CB774

IDLI GROUP 88003.34

PRECISION	975
REPUBLIC	CB775

IDLI GROUP 88003.35

PRECISION	976
REPUBLIC	CB776

IDLI GROUP 88003.36

PRECISION	977
REPUBLIC	CB777

IDLI GROUP 88003.37

PRECISION	978
REPUBLIC	CB778

IDLI GROUP 88003.38

BORG-WARNR	114-1002
BORG-WARNR	114-1003
BORG-WARNR	114-1007
BORG-WARNR	114-1009
BPC	B600N
DEERE,JOHN	AJ7331H
GMB	G1003
MACK	793XX1
PRECISION	981

IDLI GROUP 88003.39

GLAENZER	065-154-8
PRECISION	984

IDLI GROUP 88003.40

BORG-WARNR	114-10000
CLARK EQU.	943424
CLEVE.MOT.	CM10002
GLAENZER	065-172-8
MOTOR MAST	7087C
PRECISION	986
REPUBLIC	CB786
TRW	20177

IDLI GROUP 88003.41

GMB	G3111
PRECISION	989

IDLI GROUP 88003.42

ALCO	7050
ALCO	7622
ALCO	J7622
BORG-WARNR	42021
BORG-WARNR	B42021
CHRYSLER	5794345
CHRYSLER	579445
DIVCO	16260
GMB	G7622
HERCULES	42021
KAISER	201767
NEAPCO	D2022
PRECISION	N12
REPUBLIC	D42021
STUDEBAKER	364669
WESCO	206422
WESCO	D4222

IDLI GROUP 88003.43

ALCO	7627
BORG-WARNR	4203-1
BORG-WARNR	B42031
BORG-WARNR	B4303-1
CHRYSLER	1550466
CHRYSLER	1739927
CHRYSLER	629770
GMB	G7627
HERCULES	4203-1
NEAPCO	D20311
PRECISION	N13
REPUBLIC	D42031
REPUBLIC	D-4303-1
WESCO	206421
WESCO	206453
WESCO	D4221

IDLI GROUP 88003.44

ALCO	7206
BORG-WARNR	4257-1
BORG-WARNR	B4257-1
CHRYSLER	1532907
CHRYSLER	629768
DIVCO	16262
DIVCO	16271
HERCULES	4257
NEAPCO	D4257-1
NEAPCO	D4257-2
PRECISION	N14
REPUBLIC	D4257-1
WESCO	206429
WESCO	D4229
WESCO	D4229-30

IDLI GROUP 88003.45

ALCO	7060
ALCO	7604
ALCO	J7604
BORG-WARNR	4204
BORG-WARNR	B4204
BORG-WARNR	B42042
CHRYSLER	1616185
CHRYSLER	602819
DIVCO	16253
GMB	G7604
HERCULES	4204
KAISER	201773
NEAPCO	D204
PRECISION	N15
REPUBLIC	D4204
REPUBLIC	D42042

IDLI GROUP 88003.46

ALCO	7075
ALCO	7608
ALCO	76-8
ALCO	J7608
BORG-WARNR	4261
BORG-WARNR	B4261
CHRYSLER	605004
DIVCO	16259
GMB	G7608
HERCULES	4261
KAISER	201768
NEAPCO	D4261
PRECISION	N16
REPUBLIC	D4261
STUDEBAKER	364674
WESCO	D4230

IDLI GROUP 88003.47

ALMETAL	D42083
AMER.MOTOR	116035
BORG-WARNR	41083
BORG-WARNR	44086
BORG-WARNR	B41083
BORG-WARNR	B42083
BORG-WARNR	B44086
CHRYSLER	391437
CHRYSLER	858017
CHRYSLER	916767
DETROIT	42083
HERCULES	42083
NEAPCO	B42083
NEAPCO	D42083
PRECISION	N17
REPUBLIC	D41083
REPUBLIC	D42083
REPUBLIC	D42086
REPUBLIC	D44086
WESCO	206423
WESCO	D4223
WILLYS	116035

IDLI GROUP 88003.48

BORG-WARNR	42051
BORG-WARNR	B42051
CHRYSLER	639937
CHRYSLER	639973
GMB	G7605
HERCULES	42051
NEAPCO	D42051
PRECISION	N18
REPUBLIC	D42051
REPUBLIC	D440811-1
STUDEBAKER	436426
WESCO	206428
WESCO	D4228

Top of column 4:

WESCO	206427
WESCO	D4227

BD = BEARING DIAMETER (outside) **BW** = BEARING WIDTH **CB** = CROSS LENGTH WITH BEARINGS **CC** = CENTER TO CENTER **CD** = CROSS DIAMETER
CL = CROSS LENGTH WITHOUT BEARINGS **EC** = END TO CENTER or FACE TO FACE **EE** = END TO END **EL** = EFFECTIVE LENGTH
HD = HUB DIAMETER (or insert) **IE** = INSIDE OF EARS (or recess) **OD** = OUTSIDE DIAMETER **OE** = OUTSIDE OF EARS **SB** = SPLINE OR BORE SIZE

IDLI GROUP 88003.49

ALCO	7080
ALCO	7607
ALCO	J7607
BORG-WARNR	42591
BORG-WARNR	B42591
CHRYSLER	602820
DIVCO	16252
GMB	G7607
HERCULES	42591
KAISER	201774
NEAPCO	D4259-1
PRECISION	N19
REPUBLIC	D42591
WESCO	206426
WESCO	D4226

IDLI GROUP 88003.50

ALCO	7512
ALCO	7612
ALCO	J7612
ALMETAL	D420819
AMER.MOTOR	490801
BORG-WARNR	420822-1
BORG-WARNR	B420821
BORG-WARNR	B420822-1
BORG-WARNR	B420825-1
BORG-WARNR	B42083
CHRYSLER	1238571
CHRYSLER	1288571
CHRYSLER	1603397
CHRYSLER	391437
CHRYSLER	697320
CHRYSLER	858017
CHRYSLER	870311
CHRYSLER	870538
DETROIT	41085-1
DETROIT	420111
DETROIT	420112
DETROIT	420819
DETROIT	420822-1
DETROIT	420825-1
DETROIT	42083
DIVCO	16258
GMB	G7609
GMB	G7612
GMC	1288574
KAISER	201765
LEMPCO	U401
LEMPCO	U411
NEAPCO	D42083
NEAPCO	D428
PILOT	C420825-1
PILOT	C42083
PRECISION	N20
REPUBLIC	D-420819
REPUBLIC	D420822-1
REPUBLIC	D420825-1
REPUBLIC	D42086
WESCO	D4223
WESCO	D4225

IDLI GROUP 88003.51

ALMETAL	D44023
BORG-WARNR	B44023
BORG-WARNR	B440811-1
CHRYSLER	1450448
CHRYSLER	2275514
CHRYSLER	915757
DETROIT	440811-1
LEMPCO	U401B
NEAPCO	D240811X
PRECISION	N24
REPUBLIC	D-44023
REPUBLIC	D440811-1
STUDEBAKER	436426
WESCO	406428

IDLI GROUP 88003.52

PRECISION	N42
REPUBLIC	D42017
WESCO	401423
WESCO	D4220HD
WILLYS	201776

IDLI GROUP 88003.53

PRECISION	N47
REPUBLIC	B42017

IDLI GROUP 88003.54

CATERPILLR	6F9192
GMC	9021527
PRECISION	N5
WABCO	VJ4057
WABCO	VJ4180
WABCO	VJ6789

IDLI GROUP 88003.55

DETROIT	41
PRECISION	P41

IDLI GROUP 88003.56

ALCO	7012
ALCO	7601
ALCO	7605
ALCO	J7605
ALMETAL	D42013
BORG-WARNR	B42012
BORG-WARNR	B42013
BORG-WARNR	B42019
CHRYSLER	1534402
CHRYSLER	1534428
CHRYSLER	1550456
CHRYSLER	602813
CHRYSLER	636362
CHRYSLER	668504
CHRYSLER	668594
CHRYSLER	952545
DETROIT	42012
DETROIT	42017
DETROIT	42019
DIVCO	16254
DIVCO	16264
GMB	G7601
GMB	G7605
GMC	639973
GMC	668504
HERCULES	42012
HERCULES	42012C
KAISER	201766
KAISER	201772
NEAPCO	D2012X
PRECISION	P42
REPUBLIC	D4015
REPUBLIC	D42012
REPUBLIC	D-42013
REPUBLIC	D42017
REPUBLIC	D42019
WESCO	401423
WESCO	406451
WESCO	D4220HD

IDLI GROUP 88003.57

BORG-WARNR	B4301
CHRYSLER	1739920
DETROIT	4301
PRECISION	P43
REPUBLIC	D4301
WESCO	406461

IDLI GROUP 88003.58

BORG-WARNR	B420111
CHRYSLER	1734159
DETROIT	420111
DETROIT	420112
PRECISION	P57
REPUBLIC	D420111

IDLI GROUP 88003.59

MASS.FERG.	3406270M1
VAPORMATIC	VTE2000
WALTERSCHD	75-06-60

IDLI GROUP 88003.60

MASS.FERG.	3406280M1
VAPORMATIC	VTE2001
WALTERSCHD	75-12-00

IDLI GROUP 88003.61

HARDY SPCR	00-32-30
MASS.FERG.	3406271M1
VAPORMATIC	VTE2002
WALTERSCHD	75-11-00

IDLI GROUP 88003.62

HARDY SPCR	00-32-40
MASS.FERG.	3406281M1
VAPORMATIC	VTE2003
WALTERSCHD	75-15-00

IDLI GROUP 88003.63

HARDY SPCR	00-32-50
MASS.FERG.	340627M1
VAPORMATIC	VTE2004
WALTERSCHD	75-25-00

IDLI GROUP 88003.64

HARDY SPCR	00-32-60
MASS.FERG.	3406282M1
VAPORMATIC	VTE2005
WALTERSCHD	75-35-00

IDLI GROUP 88003.65

BYPY	2-03
MASS.FERG.	3406212M1
VAPORMATIC	VTE2010

IDLI GROUP 88003.66

BYPY	2-04
MASS.FERG.	3406211M1
VAPORMATIC	VTE2011

IDLI GROUP 88003.67

BYPY	3-03
MASS.FERG.	3406221M1
VAPORMATIC	VTE2012

IDLI GROUP 88003.68

BYPY	4-04
MASS.FERG.	3406213M1
VAPORMATIC	VTE2013

IDLI GROUP 88003.69

BYPY	5-03
MASS.FERG.	3406222M1
VAPORMATIC	VTE2014

IDLI GROUP 88003.70

BYPY	5-04
MASS.FERG.	3406214M1
VAPORMATIC	VTE2015

IDLI GROUP 88003.71

BYPY	6-03
MASS.FERG.	3406223M1
VAPORMATIC	VTE2016

IDLI GROUP 88003.72

BYPY	7-04
MASS.FERG.	3406215M1
VAPORMATIC	VTE2017

ENGINEERING CATALOGS MUST BE CONSULTED FOR DETAILS NOT INCLUDED
IN THIS GUIDE. SPECIFIC DESIGNS, MATERIAL CONTENT, TOLERANCES
LUBE FITTINGS AND OTHER DIMENSIONS ARE INTENTIONALLY OMITTED HERE.

CAUTION: BE SURE TO REFER TO ENGINEERING CATALOGS FOR SPECIAL APPLICATIONS THAT REQUIRE SPECIFIC MATERIAL CONTENT, TOLERANCES, ETC. SEE FOOTNOTE.

IDLI GROUP 88003.73

**QUICK DISCONNECT YOKE
WITH 6 SPLINES
SEE FIGURE 10**

BYPY 1-05-07
MASS.FERG. 3406011M91
VAPORMATIC VTE3000
WALTERSCHD 10-10-00

IDLI GROUP 88003.74

**QUICK DISCONNECT YOKE
WITH 6 SPLINES
SEE FIGURE 10**

BYPY 2-05-07
HARDY SPCR 14-04-813
MASS.FERG. 3406021M91
VAPORMATIC VTE3001
WALTERSCHD 20-10-00

IDLI GROUP 88003.75

**QUICK DISCONNECT YOKE
WITH 6 SPLINES
SEE FIGURE 10**

BYPY 3-05-07
MASS.FERG. 3406031M91
VAPORMATIC VTE3002
WALTERSCHD 11-10-00

IDLI GROUP 88003.76

**QUICK DISCONNECT YOKE
WITH 6 SPLINES
SEE FIGURE 10**

BYPY 4-05-07
HARDY SPCR 30-04-813
MASS.FERG. 3406041M91
VAPORMATIC VTE3003
WALTERSCHD 21-10-00

IDLI GROUP 88003.77

**QUICK DISCONNECT YOKE
WITH 6 SPLINES
SEE FIGURE 10**

BYPY 5-05-07
MASS.FERG. 3406061M91
VAPORMATIC VTE3004

IDLI GROUP 88003.78

**QUICK DISCONNECT YOKE
WITH 6 SPLINES
SEE FIGURE 10**

MASS.FERG. 3406051M91
VAPORMATIC VTE3005
WALTERSCHD 35-10-00

IDLI GROUP 88003.79

**QUICK DISCONNECT YOKE
WITH 6 SPLINES
SEE FIGURE 10**

BYPY 6-05-07
HARDY SPCR 35-04-813
MASS.FERG. 3406081M91
VAPORMATIC VTE3006
WALTERSCHD 22-10-00

IDLI GROUP 88003.80

**QUICK DISCONNECT YOKE
WITH 6 SPLINES
SEE FIGURE 10**

MASS.FERG. 3406071M91
VAPORMATIC VTE3007
WALTERSCHD 12-10-00

IDLI GROUP 88003.81

**QUICK DISCONNECT YOKE
WITH 6 SPLINES
SEE FIGURE 10**

BYPY 7-05-07
HARDY SPCR KS4GB3458
MASS.FERG. 3406091M91
VAPORMATIC VTE3008
WALTERSCHD 23-51-00

IDLI GROUP 88003.82

**QUICK DISCONNECT YOKE
WITH 6 SPLINES
SEE FIGURE 10**

BYPY 8-05-07
MASS.FERG. 3406101M91
VAPORMATIC VTE3009
WALTERSCHD 24-10-00

IDLI GROUP 88003.83

**QUICK DISCONNECT YOKE
WITH 6 SPLINES
SEE FIGURE 10**

MASS.FERG. 3406111M91
VAPORMATIC VTE3010
WALTERSCHD 37-10-00

IDLI GROUP 88003.84

**QUICK DISCONNECT YOKE
WITH 21 INVOLUTED SPLINES
SEE FIGURE 10**

BYPY 5-05-08
MASS.FERG. 3406062M91
VAPORMATIC VTE3020

IDLI GROUP 88003.85

**QUICK DISCONNECT YOKE
WITH 21 INVOLUTED SPLINES
SEE FIGURE 10**

MASS.FERG. 3406052M91
VAPORMATIC VTE3021
WALTERSCHD 35-10-02

IDLI GROUP 88003.86

**QUICK DISCONNECT YOKE
WITH 21 INVOLUTED SPLINES
SEE FIGURE 10**

BYPY 6-05-08
MASS.FERG. 3406082M91
VAPORMATIC VTE3022
WALTERSCHD 22-19-00

IDLI GROUP 88003.87

**QUICK DISCONNECT YOKE
WITH 21 INVOLUTED SPLINES
SEE FIGURE 10**

MASS.FERG. 3406072M91
VAPORMATIC VTE3023
WALTERSCHD 12-50-00

IDLI GROUP 88003.88

**QUICK DISCONNECT YOKE
WITH 21 INVOLUTED SPLINES
SEE FIGURE 10**

BYPY 8-05-08
MASS.FERG. 3046102M91
VAPORMATIC VTE3025
WALTERSCHD 24-18-00

IDLI GROUP 88003.89

**QUICK DISCONNECT YOKE
WITH 21 INVOLUTED SPLINES
SEE FIGURE 10**

MASS.FERG. 3406112M91
VAPORMATIC VTE3026
WALTERSCHD 36-10-02

IDLI GROUP 88003.90

MASS.FERG. 3406015M1
VAPORMATIC VTE3100
WALTERSCHD 10-48-00

IDLI GROUP 88003.91

VAPORMATIC VTE3101
WALTERSCHD 20-32-00

IDLI GROUP 88003.92

HARDY SPCR 14-26-000
MASS.FERG. 3406025M1
VAPORMATIC VTE3102
WALTERSCHD 20-11-00

IDLI GROUP 88003.93

MASS.FERG. 3406035M1
VAPORMATIC VTE3103
WALTERSCHD 11-38-00

IDLI GROUP 88003.94

HARDY SPCR 30-26-000
MASS.FERG. 3406045M1
VAPORMATIC VTE3104
WALTERSCHD 21-11-00

IDLI GROUP 88003.95

HARDY SPCR 30-26-009
VAPORMATIC VTE3105
WALTERSCHD 21-22-100

IDLI GROUP 88003.96

MASS.FERG. 3406055M1
VAPORMATIC VTE3106
WALTERSCHD 35-13-00

IDLI GROUP 88003.97

HARDY SPCR 35-26-000
MASS.FERG. 3406085M1
VAPORMATIC VTE3107
WALTERSCHD 22-11-00

IDLI GROUP 88003.98

MASS.FERG. 3406075M1
VAPORMATIC VTE3108
WALTERSCHD 12-38-00

IDLI GROUP 88003.99

MASS.FERG. 3406115M1
VAPORMATIC VTE3109
WALTERSCHD 36-13-00

IDLI GROUP 88004

MASS.FERG. 3406018M1
VAPORMATIC VTE3110

IDLI GROUP 88004.01

VAPORMATIC VTE3111
WALTERSCHD 20-33-00

IDLI GROUP 88004.02

MASS.FERG. 3406028M1
VAPORMATIC VTE3112
WALTERSCHD 20-12-00

IDLI GROUP 88004.03

MASS.FERG. 3406038M1
VAPORMATIC VTE3113
WALTERSCHD 11-39-00

BD=BEARING DIAMETER (outside)   **BW**=BEARING WIDTH   **CB**=CROSS LENGTH WITH BEARINGS   **CC**=CENTER TO CENTER   **CD**=CROSS DIAMETER
CL=CROSS LENGTH WITHOUT BEARINGS   **EC**=END TO CENTER or FACE TO FACE   **EE**=END TO END   **EL**=EFFECTIVE LENGTH
HD=HUB DIAMETER (or insert)   **IE**=INSIDE OF EARS (or recess)   **OD**=OUTSIDE DIAMETER   **OE**=OUTSIDE OF EARS   **SB**=SPLINE OR BORE SIZE

IDLI GROUP 88004.04

HARDY SPCR	30-26-001
MASS.FERG.	3406048M1
VAPORMATIC	VTE3114
WALTERSCHD	21-12-00

IDLI GROUP 88004.05

HARDY SPCR	30-26-010
VAPORMATIC	VTE3115
WALTERSCHD	21-22-101

IDLI GROUP 88004.06

MASS.FERG.	3406058M1
VAPORMATIC	VTE3116
WALTERSCHD	35-14-00

IDLI GROUP 88004.07

HARDY SPCR	35-26-001
MASS.FERG.	3406088M1
VAPORMATIC	VTE3117

IDLI GROUP 88004.08

MASS.FERG.	3406078M1
VAPORMATIC	VTE3118
WALTERSCHD	12-39-00

IDLI GROUP 88004.09

MASS.FERG.	3406118M1
VAPORMATIC	VTE3119
WALTERSCHD	36-14-00

IDLI GROUP 88004.10

BYPY	2-07
MASS.FERG.	3406023M1
VAPORMATIC	VTE3200

IDLI GROUP 88004.11

BYPY	3-07
MASS.FERG.	3406033M1
VAPORMATIC	VTE3201

IDLI GROUP 88004.12

BYPY	4-07
MASS.FERG.	3406043M1
VAPORMATIC	VTE3202

IDLI GROUP 88004.13

BYPY	5-07
MASS.FERG.	3406063M1
VAPORMATIC	VTE3203

IDLI GROUP 88004.14

BYPY	6-07
MASS.FERG.	3406083M1
VAPORMATIC	VTE3204

IDLI GROUP 88004.15

BYPY	7-07
MASS.FERG.	3406093M1
VAPORMATIC	VTE3205

IDLI GROUP 88004.16

BYPY	8-07
MASS.FERG.	3406103M1
VAPORMATIC	VTE3206

IDLI GROUP 88004.17

BYPY	2-06
MASS.FERG.	3406026M1
VAPORMATIC	VTE3207

IDLI GROUP 88004.18

BYPY	3-06
MASS.FERG.	3406036M1
VAPORMATIC	VTE3208

IDLI GROUP 88004.19

BYPY	4-06
MASS.FERG.	3406046M1
VAPORMATIC	VTE3209

IDLI GROUP 88004.20

BYPY	5 06
MASS.FERG.	3406066M1
VAPORMATIC	VTE3210

IDLI GROUP 88004.21

BYPY	6-06
MASS.FERG.	3406086M1
VAPORMATIC	VTE3211

IDLI GROUP 88004.22

BYPY	7-06
MASS.FERG.	3406096M1
VAPORMATIC	VTE3212

IDLI GROUP 88004.23

BYPY	8-06
MASS.FERG.	3406106M1
VAPORMATIC	VTE3213

IDLI GROUP 88004.24

HARDY SPCR	SK3281E
VAPORMATIC	VTE3900

IDLI GROUP 88004.25

MASS.FERG.	3406453M91
VAPORMATIC	VTE3901
WALTERSCHD	00-80-00

IDLI GROUP 88004.26

BYPY	1-08-25
MASS.FERG.	3406451M91
VAPORMATIC	VTE3902

IDLI GROUP 88004.27

MASS.FERG.	3406454M91
VAPORMATIC	VTE3903
WALTERSCHD	00-08-00

IDLI GROUP 88004.28

BYPY	1-08-25-01
VAPORMATIC	VTE3904

IDLI GROUP 88004.29

CROSS + 4 ROUND BEARINGS + OUTSIDE SNAP RINGS SEE FIGURE 1A

.866 in	BD	22.00 mm	
2.166 in	CB	55.00 mm	

MASS.FERG.	3406010M91
VAPORMATIC	VTE5000
WALTERSCHD	10-01-00

IDLI GROUP 88004.30

CROSS + 4 ROUND BEARINGS + OUTSIDE SNAP RINGS SEE FIGURE 1A

1.063 in	BD	27.00 mm	
2.756 in	CB	70.00 mm	

BYPY	3-41
MASS.FERG.	3406030M91
VAPORMATIC	VTE5001
WALTERSCHD	11-03-00

IDLI GROUP 88004.31

CROSS + 4 ROUND BEARINGS + OUTSIDE SNAP RINGS SEE FIGURE 1A

1.339 in	BD	34.00 mm	
3.544 in	CB	90.00 mm	

MASS.FERG.	3406070M91
VAPORMATIC	VTE5002
WALTERSCHD	12-00-00

IDLI GROUP 88004.32

CROSS + 4 ROUND BEARINGS + OUTSIDE SNAP RINGS SEE FIGURE 1A

1.189 in	BD	30.20 mm	
4.193 in	CB	106.50 mm	

BYPY	7-41
HARDY SPCR	KS-GB179
MASS.FERG.	3406090M91
VAPORMATIC	VTE5006
WALTERSCHD	23-00-00

IDLI GROUP 88004.33

CROSS + 4 ROUND BEARINGS + OUTSIDE SNAP RINGS SEE FIGURE 1A

1.189 in	BD	30.20 mm	
3.150 in	CB	80.00 mm	

BYPY	5-41
MASS.FERG.	3406060M91
VAPORMATIC	VTE5007

IDLI GROUP 88004.34

CROSS + 4 ROUND BEARINGS + OUTSIDE SNAP RINGS SEE FIGURE 1A

1.260 in	BD	32.00 mm	
2.992 in	CB	76.00 mm	

MASS.FERG.	3406050M91
VAPORMATIC	VTE5008
WALTERSCHD	35-00-00

IDLI GROUP 88004.35

CROSS + 4 ROUND BEARINGS + OUTSIDE SNAP RINGS SEE FIGURE 1A

1.378 in	BD	35.00 mm	
4.193 in	CB	106.50 mm	

BYPY	8-41
MASS.FERG.	3406100M91
VAPORMATIC	VTE5009
WALTERSCHD	24-01-00

IDLI GROUP 88004.36

CROSS + 4 ROUND BEARINGS + OUTSIDE SNAP RINGS SEE FIGURE 1A

1.417 in	BD	36.00 mm	
3.504 in	CB	89.00 mm	

MASS.FERG.	3406110M91
VAPORMATIC	VTE5010
WALTERSCHD	36-00-00

IDLI GROUP 88004.37

CLEVE.STL.	T55-54-170
ROCKWELL	126NLS18DC

ENGINEERING CATALOGS MUST BE CONSULTED FOR DETAILS NOT INCLUDED
IN THIS GUIDE. SPECIFIC DESIGNS, MATERIAL CONTENT, TOLERANCES
LUBE FITTINGS AND OTHER DIMENSIONS ARE INTENTIONALLY OMITTED HERE.

CAUTION: BE SURE TO REFER TO ENGINEERING CATALOGS FOR SPECIAL APPLICATIONS THAT REQUIRE SPECIFIC MATERIAL CONTENT, TOLERANCES, ETC. SEE FOOTNOTE.

IDLI GROUP 88004.38
CLEVE.STL. S55-2
ROCKWELL 131NF3

IDLI GROUP 88004.39
CLEVE.STL. S55-2-1
ROCKWELL 131NF6

IDLI GROUP 88004.40
CLEVE.STL. S55-54-1L
ROCKWELL 131NLS20-3DC

IDLI GROUP 88004.41
CLEVE.STL. S55-54-960
ROCKWELL 131NLS20-6DC

IDLI GROUP 88004.42
CLEVE.STL. S55-54-1
ROCKWELL 131NLS20DC

IDLI GROUP 88004.43
CLEVE.STL. S55-54-1167
ROCKWELL 131NLS24DC

IDLI GROUP 88004.44
CLEVE.STL. S55-3-2-16
ROCKWELL 131NY30

IDLI GROUP 88004.45
CLEVE.STL. S55-3-2-16D
ROCKWELL 131NY30-1

IDLI GROUP 88004.46
CLEVE.STL. S55-3-3-14
ROCKWELL 131NY46

IDLI GROUP 88004.47
CLEVE.STL. S55-3-3-16
ROCKWELL 131NY46-1

IDLI GROUP 88004.48
CLEVE.STL. S55-3-3-16D
ROCKWELL 131NY46-2

IDLI GROUP 88004.49
CLEVE.STL. S55-1-1551
ROCKWELL 131NYR12

IDLI GROUP 88004.50
CLEVE.STL. S55-1-1431
ROCKWELL 131NYR12-1

IDLI GROUP 88004.51
CLEVE.STL. S55-1-129
ROCKWELL 131NYR20

IDLI GROUP 88004.52
CLEVE.STL. S55-1-132
ROCKWELL 131NYS18-2

IDLI GROUP 88004.53
CLEVE.STL. S55-1-133
ROCKWELL 131NYS19

IDLI GROUP 88004.54
CLEVE.STL. S55-1-153
ROCKWELL 131NYS19-1

IDLI GROUP 88004.55
CLEVE.STL. S55-1-218
ROCKWELL 131NYS19-2

IDLI GROUP 88004.56
CLEVE.STL. S55-1-130
ROCKWELL 131NYS20

IDLI GROUP 88004.57
CLEVE.STL. S55-1-131
ROCKWELL 131NYS20-1

IDLI GROUP 88004.58
CLEVE.STL. S55-1-229
ROCKWELL 131NYS20-2

IDLI GROUP 88004.59
CLEVE.STL. S55-1-225
ROCKWELL 131NYS22-4

IDLI GROUP 88004.60
CLEVE.STL. S55-1-1170
ROCKWELL 131NYS22-5

IDLI GROUP 88004.61
CLEVE.STL. R55-19
ROCKWELL 135N170
SPICER 3-26-317

IDLI GROUP 88004.62
CLEVE.STL. R55-2
ROCKWELL 135NF2
SPICER 3-2-119

IDLI GROUP 88004.63
CLEVE.STL. R55-2-1
ROCKWELL 135NF3

IDLI GROUP 88004.64
CLEVE.STL. R55-2-4
ROCKWELL 135NF4

IDLI GROUP 88004.65
CLEVE.STL. R55-54-16
ROCKWELL 135NLS22-3DC

IDLI GROUP 88004.66
CLEVE.STL. R55-54-1
ROCKWELL 135NLS22DC

IDLI GROUP 88004.67
CLEVE.STL. R55-3-3-14
ROCKWELL 135NY46
SPICER 3-28-57

IDLI GROUP 88004.68
CLEVE.STL. R55-3-3-14D
ROCKWELL 135NY46-1

IDLI GROUP 88004.69
CLEVE.STL. R55-1-1434
ROCKWELL 135NYR14

IDLI GROUP 88004.70
CLEVE.STL. R55-1-544
ROCKWELL 135NYR26

IDLI GROUP 88004.71
CLEVE.STL. R55-1-669
ROCKWELL 135NYR28

IDLI GROUP 88004.72
CLEVE.STL. R55-1-324
ROCKWELL 135NYS22

IDLI GROUP 88004.73
CLEVE.STL. R55-1-446
ROCKWELL 135NYS22-1

IDLI GROUP 88004.74
CLEVE.STL. R55-1-1226
ROCKWELL 135NYS22-4

IDLI GROUP 88004.75
CLEVE.STL. R55-1-179
ROCKWELL 135NYS22-5

IDLI GROUP 88004.76
CLEVE.STL. R55-1-101
ROCKWELL 135NYS24

IDLI GROUP 88004.77
CLEVE.STL. R55-1-104
ROCKWELL 135NYS24-1

IDLI GROUP 88004.78
CLEVE.STL. R55-1-112
ROCKWELL 135NYS24-2

IDLI GROUP 88004.79
CLEVE.STL. R55-1-172
ROCKWELL 135NYS24-3

IDLI GROUP 88004.80
CLEVE.STL. R55-1-107
ROCKWELL 135NYS24-4

IDLI GROUP 88004.81
CLEVE.STL. R55-1-240
ROCKWELL 135NYS26

IDLI GROUP 88004.82
CLEVE.STL. R55-1-188
ROCKWELL 135NYS28

IDLI GROUP 88004.83
CLEVE.STL. O55-3-3-14D
ROCKWELL 141NY45-2

IDLI GROUP 88004.84
CLEVE.STL. O55-1-961
ROCKWELL 141NYS22-1

IDLI GROUP 88004.85
CLEVE.STL. O55-1-961A
ROCKWELL 141NYS22-1A

BD=BEARING DIAMETER (outside)   **BW**=BEARING WIDTH   **CB**=CROSS LENGTH WITH BEARINGS   **CC**=CENTER TO CENTER   **CD**=CROSS DIAMETER
CL=CROSS LENGTH WITHOUT BEARINGS   **EC**=END TO CENTER or FACE TO FACE   **EE**=END TO END   **EL**=EFFECTIVE LENGTH
HD=HUB DIAMETER (or insert)   **IE**=INSIDE OF EARS (or recess)   **OD**=OUTSIDE DIAMETER   **OE**=OUTSIDE OF EARS   **SB**=SPLINE OR BORE SIZE

IDLI GROUP 88004.86
CLEVE.STL. O55-1-1305
ROCKWELL 141NYS22-3

IDLI GROUP 88004.87
CLEVE.STL. O55-1-1306
ROCKWELL 141NYS22-3A

IDLI GROUP 88004.88
CLEVE.STL. O55-1-1186
ROCKWELL 141NYS22-5

IDLI GROUP 88004.89
CLEVE.STL. O55-1-1255
ROCKWELL 141NYS22-6

IDLI GROUP 88004.90
CLEVE.STL. O55-1-1548
ROCKWELL 141NYS22-8

IDLI GROUP 88004.91
CLEVE.STL. O55-1-1409
ROCKWELL 141NYS24-1
SPICER 3-4-6561

IDLI GROUP 88004.92
CLEVE.STL. O55-1-1705
ROCKWELL 141NYS24-2

IDLI GROUP 88004.93
CLEVE.STL. O55-1-877
ROCKWELL 141NYS24-4

IDLI GROUP 88004.94
CLEVE.STL. O55-1-877A
ROCKWELL 141NYS24-4A

IDLI GROUP 88004.95
CLEVE.STL. O55-1-812
ROCKWELL 141NYS28-2

IDLI GROUP 88004.96
CLEVE.STL. O55-1-812A
ROCKWELL 141NYS28-2A

IDLI GROUP 88004.97
CLEVE.STL. O55-1-1714
ROCKWELL 141NYS28-4

IDLI GROUP 88004.98
CLEVE.STL. O55-1-1188
ROCKWELL 141NYS29

IDLI GROUP 88004.99
CLEVE.STL. O55-1-1188A
ROCKWELL 141NYS29A

IDLI GROUP 88005
CLEVE.STL. O55-1259
ROCKWELL 141NYS29A1

IDLI GROUP 88005.01
CLEVE.STL. O55-1259A
ROCKWELL 141NYS29A2

IDLI GROUP 88005.02
CLEVE.STL. O55-1-1694
ROCKWELL 141NYT22

IDLI GROUP 88005.03
ROCKWELL 148-53-291
SPICER 3-53-291

IDLI GROUP 88005.04
CLEVE.STL. D56-2-3
ROCKWELL 148NF3

IDLI GROUP 88005.05
CLEVE.STL. D76-3-4-14D
ROCKWELL 148NY62-2

IDLI GROUP 88005.06
CLEVE.STL. D76-1-100
ROCKWELL 148NYS24-4
SPICER 3-4-3091

IDLI GROUP 88005.07
CLEVE.STL. D56-1-713
ROCKWELL 148NYS24-5

IDLI GROUP 88005.08
CLEVE.STL. D56-1-1215
ROCKWELL 148NYS24-7

IDLI GROUP 88005.09
CLEVE.STL. D56-1-1075
ROCKWELL 148NYS25

IDLI GROUP 88005.10
CLEVE.STL. D56-1-135-1
ROCKWELL 148NYS26-3

IDLI GROUP 88005.11
CLEVE.STL. D56-1-135-1A
ROCKWELL 148NYS26-3A

IDLI GROUP 88005.12
CLEVE.STL. D56-1-141-1
ROCKWELL 148NYS28-12

IDLI GROUP 88005.13
CLEVE.STL. D56-1-141-1A
ROCKWELL 148NYS28-12A

IDLI GROUP 88005.14
CLEVE.STL. D56-1-743
ROCKWELL 148NYS28-14

IDLI GROUP 88005.15
CLEVE.STL. D56-1-1642
CLEVE.STL. D56-1-642
ROCKWELL 148NYS28-15

IDLI GROUP 88005.16
CLEVE.STL. D56-1-1220
ROCKWELL 148NYS28-17

IDLI GROUP 88005.17
CLEVE.STL. D56-1-785
ROCKWELL 148NYS29

IDLI GROUP 88005.18
CLEVE.STL. D56-1-887
ROCKWELL 148NYS29-1

IDLI GROUP 88005.19
CLEVE.STL. D56-888
ROCKWELL 148NYS29-1A

IDLI GROUP 88005.20
CLEVE.STL. D56-888A
ROCKWELL 148NYS29-1A1

IDLI GROUP 88005.21
CLEVE.STL. D56-884
ROCKWELL 148NYS29A

IDLI GROUP 88005.22
CLEVE.STL. D56-1-139-1
ROCKWELL 148NYS31-1

IDLI GROUP 88005.23
CLEVE.STL. D56-1-139-1A
ROCKWELL 148NYS31-1A

IDLI GROUP 88005.24
CLEVE.STL. D56-1-142-1
ROCKWELL 148NYS31-3

IDLI GROUP 88005.25
CLEVE.STL. D56-1-142-1A
ROCKWELL 148NYS31-3A

IDLI GROUP 88005.26
CLEVE.STL. D76-1-129
ROCKWELL 148NYS31-5
SPICER 3-4-3561

IDLI GROUP 88005.27
CLEVE.STL. U56-2-1
ROCKWELL 155NF3

IDLI GROUP 88005.28
CLEVE.STL. U56-54-109
ROCKWELL 155NLS28-10DC

IDLI GROUP 88005.29
CLEVE.STL. U76-3-4-14D
ROCKWELL 155NY62-1

IDLI GROUP 88005.30
CLEVE.STL. U76-1-100
ROCKWELL 155NYS24-1
SPICER 4-4-2051

IDLI GROUP 88005.31
CLEVE.STL. U56-1-108-1
ROCKWELL 155NYS26

IDLI GROUP 88005.32
CLEVE.STL. U56-1-108-1A
ROCKWELL 155NYS26A

IDLI GROUP 88005.33
CLEVE.STL. U76-1-132
ROCKWELL 155NYS28-2
SPICER 4-4-2041
SPICER 4-4-2091

IDLI GUIDE   COPYRIGHT © INTERCHANGE, INC.   ST. LOUIS PARK, MN. 55416 USA

CAUTION: BE SURE TO REFER TO ENGINEERING CATALOGS FOR SPECIAL APPLICATIONS THAT REQUIRE SPECIFIC MATERIAL CONTENT, TOLERANCES, ETC. SEE FOOTNOTE.

IDLI GROUP 88005.34

CLEVE.STL. U76-1-133
ROCKWELL 155NYS28-3

IDLI GROUP 88005.35

ROCKWELL 16N14-19
SPICER 4-14-19

IDLI GROUP 88005.36

ROCKWELL 16N15-23
SPICER 5-15-23

IDLI GROUP 88005.37

ROCKWELL 16N16-23
SPICER 5-16-23

IDLI GROUP 88005.38

ROCKWELL 16N4-2401

IDLI GROUP 88005.39

ROCKWELL 16N4-5061
SPICER 5-4-5061

IDLI GROUP 88005.40

ROCKWELL 16N4-6351
SPICER 5-4-6351

IDLI GROUP 88005.41

ROCKWELL 176N4-231X
SPICER 6.3-4-231X

IDLI GROUP 88005.42

ROCKWELL 17N1-101X
SPICER 1701-101X

IDLI GROUP 88005.43

ROCKWELL 17N1-104X
SPICER 1701-104X

IDLI GROUP 88005.44

ROCKWELL 17N1-108X
SPICER 1701-108X

IDLI GROUP 88005.45

ROCKWELL 17N1-113X
SPICER 1701-113X

IDLI GROUP 88005.46

ROCKWELL 17N1-116X
SPICER 1701-116X

IDLI GROUP 88005.47

ROCKWELL 17N1-118X
SPICER 1701-118X

IDLI GROUP 88005.48

ROCKWELL 17N14-29
SPICER 6-14-29

IDLI GROUP 88005.49

ROCKWELL 17N15-33
SPICER 6-15-33

IDLI GROUP 88005.50

ROCKWELL 17N15-73
SPICER 6-15-73

IDLI GROUP 88005.51

ROCKWELL 17N15-93
SPICER 6-15-93

IDLI GROUP 88005.52

ROCKWELL 17N16-123
SPICER 6-16-123

IDLI GROUP 88005.53

ROCKWELL 17N16-53
SPICER 6-16-53

IDLI GROUP 88005.54

ROCKWELL 17N16-93
SPICER 6-16-93

IDLI GROUP 88005.55

ROCKWELL 17N1-9101X
SPICER 1701-9101X

IDLI GROUP 88005.56

ROCKWELL 17N1971
SPICER 6-4-1971

IDLI GROUP 88005.57

ROCKWELL 17N3-1501
SPICER 6-3-1501

IDLI GROUP 88005.58

ROCKWELL 17N3-1521
SPICER 6-3-1521

IDLI GROUP 88005.59

ROCKWELL 17N3-2261
SPICER 6-3-2261

IDLI GROUP 88005.60

ROCKWELL 17N3-2341
SPICER 6-3-2341

IDLI GROUP 88005.61

CLEVE.STL. C67-7-4-105
ROCKWELL 17N40-461
SPICER 6-40-461

IDLI GROUP 88005.62

ROCKWELL 17N42-108
SPICER 6-42-1081

IDLI GROUP 88005.63

ROCKWELL 17N4-3231
SPICER 6-4-1211
SPICER 6-4-2011
SPICER 6-4-3231

IDLI GROUP 88005.64

ROCKWELL 17N4-3321
SPICER 6-4-3311X

IDLI GROUP 88005.65

ROCKWELL 17N4-3621
SPICER 6-4-3621

IDLI GROUP 88005.66

ROCKWELL 17N4-3901
SPICER 6-4-4761X

IDLI GROUP 88005.67

ROCKWELL 17N4-4411
SPICER 6-4-4411

IDLI GROUP 88005.68

ROCKWELL 17N4-4881
SPICER 6-4-4881

IDLI GROUP 88005.69

ROCKWELL 17N4-4971
SPICER 6-4-4971

IDLI GROUP 88005.70

ROCKWELL 17N4-5041X
SPICER 6-4-5041X

IDLI GROUP 88005.71

ROCKWELL 17N4-5461X
SPICER 6-4-5461X

IDLI GROUP 88005.72

ROCKWELL 17N4-5661
SPICER 6-4-5661

IDLI GROUP 88005.73

ROCKWELL 17N4-6451
SPICER 6-4-6451

IDLI GROUP 88005.74

ROCKWELL 17N4-6481
SPICER 6-4-6481

IDLI GROUP 88005.75

ROCKWELL 17N82-1061-3
SPICER 6-82-1061-3

IDLI GROUP 88005.76

ROCKWELL 17N82-1171-5
SPICER 6-82-1171-5

IDLI GROUP 88005.77

ROCKWELL 17NYS28-5
SPICER 6-4-2261

IDLI GROUP 88005.78

ROCKWELL 18N1-211
SPICER 6.5-1-211

IDLI GROUP 88005.79

ROCKWELL 18N4-2011

IDLI GROUP 88005.80

ROCKWELL 18N82-461-1
SPICER 6.5-82-461-1

IDLI GROUP 88005.81

ROCKWELL 18N82-461-2
SPICER 6.5-82-461-2

IDLI GROUP 88005.82

ROCKWELL 18N82-461-4
SPICER 6.5-82-461-4

BD = BEARING DIAMETER (outside)   **BW** = BEARING WIDTH   **CB** = CROSS LENGTH WITH BEARINGS   **CC** = CENTER TO CENTER   **CD** = CROSS DIAMETER
CL = CROSS LENGTH WITHOUT BEARINGS   **EC** = END TO CENTER or FACE TO FACE   **EE** = END TO END   **EL** = EFFECTIVE LENGTH
HD = HUB DIAMETER (or insert)   **IE** = INSIDE OF EARS (or recess)   **OD** = OUTSIDE DIAMETER   **OE** = OUTSIDE OF EARS   **SB** = SPLINE OR BORE SIZE

IDLI GROUP 88005.83
ROCKWELL 18N82-461-8
SPICER 6.5-82-461-8

IDLI GROUP 88005.84
ROCKWELL 2CP92N5

IDLI GROUP 88005.85
CLEVE.STL. R96-1-814
ROCKWELL 2WCS22-19

IDLI GROUP 88005.86
CLEVE.STL. R96-1-792
ROCKWELL 2WCS22-23

IDLI GROUP 88005.87
CLEVE.STL. R96-1-812
ROCKWELL 2WCS22-75

IDLI GROUP 88005.88
CLEVE.STL. R96-1-811
ROCKWELL 2WCS22-76

IDLI GROUP 88005.89
ROCKWELL 2WCS24-41
SPICER 3-1-3521

IDLI GROUP 88005.90
CLEVE.STL. R96-1-793
ROCKWELL 2WCS24-48

IDLI GROUP 88005.91
ROCKWELL 2WCS24-52
SPICER 4-1-2761

IDLI GROUP 88005.92
CLEVE.STL. D56-1-645
ROCKWELL 2WCS24-76

IDLI GROUP 88005.93
CLEVE.STL. D56-1-631
ROCKWELL 2WCS26-19

IDLI GROUP 88005.94
CLEVE.STL. R96-1-809
ROCKWELL 2WCS26-20

IDLI GROUP 88005.95
ROCKWELL 2WCS28-73
SPICER 4-1-4791

IDLI GROUP 88005.96
ROCKWELL 3DRYS22-5

IDLI GROUP 88005.97
PRECISION 1860
ROCKWELL 3N173-1

IDLI GROUP 88005.98
PRECISION 1861
ROCKWELL 3NDCA2

IDLI GROUP 88005.99
PRECISION 1806
ROCKWELL 3NYR20-10

IDLI GROUP 88006
ROCKWELL 3WCS28-93
SPICER 5-1-6121

IDLI GROUP 88006.01
ROCKWELL 3WCS30-4
SPICER 5-1-2841

IDLI GROUP 88006.02
ROCKWELL 3WCS31-15
SPICER 5-1-2121

IDLI GROUP 88006.03
ROCKWELL 3WCS32-58
SPICER 5-1-3161

IDLI GROUP 88006.04
ROCKWELL 3WCS36-21
SPICER 5-1-4511

IDLI GROUP 88006.05
PRECISION 1408
ROCKWELL 44RY24-2

IDLI GROUP 88006.06
PRECISION 1408
ROCKWELL 44RYR24-2

IDLI GROUP 88006.07
ROCKWELL 44RYR28-2

IDLI GROUP 88006.08
REXNORD 762349
ROCKWELL 44RYS22-28
SPICER 3.4-4-221

IDLI GROUP 88006.09
ROCKWELL 55NYR24-13

IDLI GROUP 88006.10
ROCKWELL 5WCS30-2
SPICER 6-1-1641

IDLI GROUP 88006.11
ROCKWELL 5WCS31-7
SPICER 6-1-1461

IDLI GROUP 88006.12
ROCKWELL 5WCS36-76
SPICER 6-1-1421

IDLI GROUP 88006.13
ROCKWELL 5WCS38-4
SPICER 6-1-5081

IDLI GROUP 88006.14
ROCKWELL 62NLS28

IDLI GROUP 88006.15
BORG-WARNR 6-6026
BORG-WARNR 6-6032
BORG-WARNR 6-6036
BORG-WARNR 6A6026
BORG-WARNR 6A6032
BORG-WARNR 6A6036
ROCKWELL 62NLS28-1

IDLI GROUP 88006.16
BORG-WARNR 6-6038
BORG-WARNR 6A6038
ROCKWELL 62NLS28-12

IDLI GROUP 88006.17
ROCKWELL 62NLS28-22

IDLI GROUP 88006.18
BORG-WARNR 6-6084
BORG-WARNR 6A6084
ROCKWELL 62NLS28-25

IDLI GROUP 88006.19
ROCKWELL 62NLS28-27

IDLI GROUP 88006.20
BORG-WARNR 6-6010
BORG-WARNR 6-6019
BORG-WARNR 6-6023
BORG-WARNR 6-6035
BORG-WARNR 6-6072
BORG-WARNR 6-6082
BORG-WARNR 6A6010
BORG-WARNR 6A6019
BORG-WARNR 6A6023
BORG-WARNR 6A6035
BORG-WARNR 6A6072
BORG-WARNR 6A6082
ROCKWELL 62NLS28-8

IDLI GROUP 88006.21
ROCKWELL 62NLS28-9

IDLI GROUP 88006.22
BORG-WARNR 17072J
ROCKWELL 62NPA28-4

IDLI GROUP 88006.23
BORG-WARNR 14082J
ROCKWELL 62NPS28-11

IDLI GROUP 88006.24
ROCKWELL 62NPS28-13

IDLI GROUP 88006.25
BORG-WARNR 13084J
ROCKWELL 62NPS28-19

IDLI GROUP 88006.26
BORG-WARNR 15016J
ROCKWELL 62NY44-2

IDLI GROUP 88006.27
BORG-WARNR 21324J
ROCKWELL 62NY45-6

IDLI GROUP 88006.28
BORG-WARNR 5520J
ROCKWELL 62NYS24-14

IDLI GROUP 88006.29
BORG-WARNR 11935
BORG-WARNR 11935J
ROCKWELL 62NYS24-16

IDLI GROUP 88006.30
BORG-WARNR 11998J
ROCKWELL 62NYS24-5

IDLI GROUP 88006.31
BORG-WARNR 28315J
ROCKWELL 62NYS28-190

CAUTION: BE SURE TO REFER TO ENGINEERING CATALOGS FOR SPECIAL APPLICATIONS THAT REQUIRE SPECIFIC MATERIAL CONTENT, TOLERANCES, ETC. SEE FOOTNOTE.

IDLI GROUP 88006.32

BORG-WARNR 26181J
ROCKWELL 62NYS28-193

IDLI GROUP 88006.33

BORG-WARNR 7901J
ROCKWELL 62NYS31-20

IDLI GROUP 88006.34

BORG-WARNR 11344J
BORG-WARNR 119-11339-1
ROCKWELL 72NDCA3

IDLI GROUP 88006.35

BORG-WARNR 4685J
ROCKWELL 72NF15

IDLI GROUP 88006.36

BORG-WARNR 60078
BORG-WARNR 67002
BORG-WARNR 6A7002
NEAPCO 06-127
NEAPCO 06-139
NEAPCO 66-0639
ROCKWELL 72NLS32-25
SPICER 210400-3900X

IDLI GROUP 88006.37

BORG-WARNR 6344J
BORG-WARNR 6-7014
BORG-WARNR 6A7014
NEAPCO 26-129
NEAPCO 26-139
NEAPCO 66-2629
NEAPCO N2604X
NEAPCO N2691X
ROCKWELL 72NLS32-50
SPICER 228596X
TRW 21479

IDLI GROUP 88006.38

BORG-WARNR 6-7177
ROCKWELL 72NLS32-86DC

IDLI GROUP 88006.39

BORG-WARNR 17218J
ROCKWELL 72NPS32-44

IDLI GROUP 88006.40

BORG-WARNR 17150J
ROCKWELL 72NPS32-45

IDLI GROUP 88006.41

BORG-WARNR 16515J
ROCKWELL 72NY51

IDLI GROUP 88006.42

BORG-WARNR 19605J
ROCKWELL 72NYS24-16

IDLI GROUP 88006.43

BORG-WARNR 20658J
ROCKWELL 72NYS26-5

IDLI GROUP 88006.44

BORG-WARNR 16720J
ROCKWELL 72NYS26-6

IDLI GROUP 88006.45

BORG-WARNR 7141J
ROCKWELL 72NYS28-12

IDLI GROUP 88006.46

BORG-WARNR 29983J
ROCKWELL 72NYS28-180

IDLI GROUP 88006.47

BORG-WARNR 12309J
BORG-WARNR 16946J
ROCKWELL 72NYS28-20

IDLI GROUP 88006.48

BORG-WARNR 20664J
ROCKWELL 72NYS28-31

IDLI GROUP 88006.49

BORG-WARNR 22708J
ROCKWELL 72NYS28-55

IDLI GROUP 88006.50

BORG-WARNR 23398J
ROCKWELL 72NYS28-62

IDLI GROUP 88006.51

ROCKWELL 72NYS28-79

IDLI GROUP 88006.52

BORG-WARNR 24064J
ROCKWELL 72NYS28-81

IDLI GROUP 88006.53

BORG-WARNR 15528J
ROCKWELL 72NYS30

IDLI GROUP 88006.54

ROCKWELL 72NYS31-33

IDLI GROUP 88006.55

BORG-WARNR 7133J
ROCKWELL 72NYS31-47

IDLI GROUP 88006.56

BORG-WARNR 17272J
ROCKWELL 72NYS32-186

IDLI GROUP 88006.57

BORG-WARNR 12308J
ROCKWELL 72NYS32-38

IDLI GROUP 88006.58

BORG-WARNR 20657J
ROCKWELL 72NYS32-66

IDLI GROUP 88006.59

BORG-WARNR 11965J
ROCKWELL 72NYS32-80

IDLI GROUP 88006.60

BORG-WARNR 29984J
ROCKWELL 72NYS33-3

IDLI GROUP 88006.61

BORG-WARNR 8572J
ROCKWELL 72NYS34-2

IDLI GROUP 88006.62

ROCKWELL 72NYS36-21

IDLI GROUP 88006.63

BORG-WARNR 15718J
ROCKWELL 72NYS38-3

IDLI GROUP 88006.64

BORG-WARNR 6-7107
BORG-WARNR 6A7107
ROCKWELL 72NYS40-59A

IDLI GROUP 88006.65

BORG-WARNR 105-16717-1
ROCKWELL 72NYSM32-21

IDLI GROUP 88006.66

BORG-WARNR 105-16513-1
FORD C1TZ4557C
ROCKWELL 72NYSM32-23

IDLI GROUP 88006.67

BORG-WARNR 105-16512-1
ROCKWELL 72NYSM32-24

IDLI GROUP 88006.68

BORG-WARNR 105-16514-1
ROCKWELL 72NYSM32-27

IDLI GROUP 88006.69

BORG-WARNR 105-16717-1
ROCKWELL 72NYSM32-31

IDLI GROUP 88006.70

BORG-WARNR 105-16511-1
BORG-WARNR 106-16511-1
FORD C1TZ4557F
ROCKWELL 72NYSM32-33

IDLI GROUP 88006.71

BORG-WARNR 105-2304
ROCKWELL 72NYSM32-61

IDLI GROUP 88006.72

ROCKWELL 72NYSM48-4

IDLI GROUP 88006.73

BORG-WARNR 18750J
ROCKWELL 72NYT26-1

IDLI GROUP 88006.74

ROCKWELL 82N170

IDLI GROUP 88006.75

BORG-WARNR 13455J
ROCKWELL 82NF1

IDLI GROUP 88006.76

BORG-WARNR 15263J
BORG-WARNR 8021J
ROCKWELL 82NPS40

BD=BEARING DIAMETER (outside)   BW=BEARING WIDTH   CB=CROSS LENGTH WITH BEARINGS   CC=CENTER TO CENTER   CD=CROSS DIAMETER
CL=CROSS LENGTH WITHOUT BEARINGS   EC=END TO CENTER or FACE TO FACE   EE=END TO END   EL=EFFECTIVE LENGTH
HD=HUB DIAMETER (or insert)   IE=INSIDE OF EARS (or recess)   OD=OUTSIDE DIAMETER   OE=OUTSIDE OF EARS   SB=SPLINE OR BORE SIZE

IDLI GROUP 88006.77

BORG-WARNR 8425J
ROCKWELL 82NPS40-1

IDLI GROUP 88006.78

BORG-WARNR 19550J
ROCKWELL 82NPS40-19

IDLI GROUP 88006.79

BORG-WARNR 5288J
ROCKWELL 82NPS40-21

IDLI GROUP 88006.80

BORG-WARNR 13368J
ROCKWELL 82NPS40-3

IDLI GROUP 88006.81

ROCKWELL 82NPS40-5

IDLI GROUP 88006.82

BORG-WARNR 5290J
ROCKWELL 82NPS40-9

IDLI GROUP 88006.83

ROCKWELL 82NY58

IDLI GROUP 88006.84

BORG-WARNR 18114J
ROCKWELL 82NY58-2

IDLI GROUP 88006.85

BORG-WARNR 22353J
ROCKWELL 82NY58-3

IDLI GROUP 88006.86

BORG-WARNR 22353J
ROCKWELL 82NY58-4

IDLI GROUP 88006.87

ROCKWELL 82NYS32-47

IDLI GROUP 88006.88

BORG-WARNR 32046J
ROCKWELL 82NYS33-5

IDLI GROUP 88006.89

BORG-WARNR 15753J
ROCKWELL 82NYS34-1

IDLI GROUP 88006.90

BORG-WARNR 12595J
ROCKWELL 82NYS34-6

IDLI GROUP 88006.91

BORG-WARNR 29982J
ROCKWELL 82NYS36-122

IDLI GROUP 88006.92

BORG-WARNR 15534J
BORG-WARNR 7140J
ROCKWELL 82NYS36-18

IDLI GROUP 88006.93

ROCKWELL 82NYS36-50

IDLI GROUP 88006.94

NEAPCO 12-827
ROCKWELL 907-50-3

IDLI GROUP 88006.95

NEAPCO 12-727
ROCKWELL 907-50-7

IDLI GROUP 88006.96

ROCKWELL 907-51-10
WESCO 920-825

IDLI GROUP 88006.97

BORG-WARNR 67019
BORG-WARNR 67020
BORG-WARNR 67026
BORG-WARNR 67027
NEAPCO 12-227
NEAPCO 12-239
NEAPCO 12-627
NEAPCO 12-651
NEAPCO 20-240
NEAPCO 20-252
NEAPCO 20-628
NEAPCO 20-640
NEAPCO 66-2651
ROCKWELL 907-51-11
SPICER 210405-3800X
SPICER 228417X
TRW 21419
TRW 21449
TRW 21454
TRW 21499
TRW 67026
TRW 67027
WESCO 12N-SSF48
WESCO 12NSSF48
WESCO 12N-SSF72
WESCO 920-306
WESCO 920-307

IDLI GROUP 88006.98

BORG-WARNR 1-82-22-3715X
NEAPCO 12-839
PRECISION 1241
ROCKWELL 907-51-3
SPICER 1-82-82-3715X

IDLI GROUP 88006.99

BORG-WARNR 1-60-34-3519X
BORG-WARNR 67037
BORG-WARNR 67038
G & G MFG 186-1200
HAYES 1-60-34-3519X
MCQUAY-NOR U03-94
MOTOR MAST C-302
MOTOR MAST C302
NEAPCO 12-327
NEAPCO 12-339
NEAPCO 12-739
NEAPCO 12-751
NEAPCO 64-2739
NEAPCO 64-2751
NEAPCO 65-2751
NEAPCO UTR12-002700
NEAPCO UTR12-003900
PRECISION 1249
ROCKWELL 907-51-7
SPICER 1-60-34-3515X
SPICER 1-60-34-3519X
TRW 21416
WESCO 12N048F
WESCO 12N48F
WESCO 12N72
WESCO 920111

IDLI GROUP 88007

ROCKWELL 907-52-10
WESCO 920-310

IDLI GROUP 88007.01

ROCKWELL 907-52-11
WESCO 920-311

IDLI GROUP 88007.02

ROCKWELL 907-52-3
SPICER 1-82-22-3515X
WESCO 920-115

IDLI GROUP 88007.03

ROCKWELL 907-52-7
SPICER 1-60-34-3513X
WESCO 920-114

IDLI GROUP 88007.04

NEAPCO 20-828
NEAPCO 65-2828
ROCKWELL 909-50-3
SPICER 2-82-22-2706X
TRW 21448

IDLI GROUP 88007.05

BORG-WARNR 67043
NEAPCO 20-328
NEAPCO 20-728
NEAPCO 64-2728
NEAPCO UT20-002800
ROCKWELL 909-50-6
SPICER 228226-2502X
TRW 21446
TRW 21447

IDLI GROUP 88007.06

NEAPCO 12-551
NEAPCO 68-2551
ROCKWELL 909-51-10
SPICER 210406-3700X
TRW 21418
WESCO 12NSSM48
WESCO 920-306
WESCO 920-825

IDLI GROUP 88007.07

ROCKWELL 909-51-11
WESCO 920-307

IDLI GROUP 88007.08

BORG-WARNR 67062
G & G MFG 188-1400
G & G MFG 188-1472
HAYES 2-82-82-3714X
NEAPCO 20-440
NEAPCO 20-840
NEAPCO 65-2040
NEAPCO UT20-004000
PRECISION 1347
ROCKWELL 909-51-3
SPICER 2-82-82-3714X
TRW 21453
WESCO 14N48M
WESCO 14N72M
WESCO 65-2840
WESCO 920123
WESCO 920-127

IDLI GROUP 88007.09

BORG-WARNR 67044
G & G MFG 185-1400
G & G MFG 185-4400
HAYES 228226-3502X
NEAPCO 20-340
NEAPCO 20-740
NEAPCO 20-752
NEAPCO 64-2040
NEAPCO 64-2740
NEAPCO 64-2752

CAUTION: BE SURE TO REFER TO ENGINEERING CATALOGS FOR SPECIAL APPLICATIONS THAT REQUIRE SPECIFIC MATERIAL CONTENT, TOLERANCES, ETC. SEE FOOTNOTE.

NEAPCO UT20-004000
PRECISION 1349
ROCKWELL 909-51-6
SPICER 228226-35
SPICER 228226-3502X
TRW 21452
TRW 21457
WESCO 14N048F
WESCO 14N48F
WESCO 14N48FS
WESCO 14N72F
WESCO 14N72SF
WESCO 920-122
WESCO 920126
WESCO 920-130

IDLI GROUP 88007.10

ROCKWELL 909-52-10
SPICER 210406-3700X
WESCO 12N-SSM48
WESCO 12N-SSM72
WESCO 920-310

IDLI GROUP 88007.11

ROCKWELL 909-52-11
SPICER 210405-3800X
WESCO 920-311

IDLI GROUP 88007.12

ROCKWELL 909-52-3
WESCO 920-131

IDLI GROUP 88007.13

ROCKWELL 909-52-6
WESCO 920-134

IDLI GROUP 88007.14

BORG-WARNR 67065
NEAPCO 22-629
NEAPCO 26-429
NEAPCO 65-2629
ROCKWELL 922-50-3
SPICER 3-82-448-3230X
TRW 21462

IDLI GROUP 88007.15

BORG-WARNR 67063
BORG-WARNR 67064
FORD C3TZ4841H
G & G MFG 188-3572
NEAPCO 22-429
NEAPCO 22-439
NEAPCO 22-639
NEAPCO 62-2229
NEAPCO 65-2229
NEAPCO 65-2239
NEAPCO UT22-002900

NEAPCO UT22-003900
NEAPCO UT22-2900
ROCKWELL 922-51-3
SPICER 3-82-338-3512X
TRW 21460
WESCO 35N72M
WESCO 920-147

IDLI GROUP 88007.16

BORG-WARNR 67066
G & G MFG 188-3500
HAYES 3-82-338-3512X
NEAPCO 26-439
NEAPCO 65-2639
PRECISION 1847
ROCKWELL 922-51-6
SPICER 228217-3316X
TRW 21467
WESCO 35N48M
WESCO 920142

IDLI GROUP 88007.17

BORG-WARNR 6-9033
ROCKWELL 92NLS48-55DC

IDLI GROUP 88007.18

BORG-WARNR 7176J
ROCKWELL 92NYS31-3

IDLI GROUP 88007.19

ROCKWELL 92NYS36-2

IDLI GROUP 88007.20

BORG-WARNR 16128J
ROCKWELL 92NYSM44-1

IDLI GROUP 88007.21

BORG-WARNR 42-8001
ROCKWELL 955-12-2

IDLI GROUP 88007.22

PREC.TOR. 1439
ROCKWELL 982-51-10
SPICER 229092-3020X

IDLI GROUP 88007.23

PREC.TOR. 1437
ROCKWELL 982-51-11
SPICER 229091-3012X

IDLI GROUP 88007.24

PRECISION 1447
ROCKWELL 982-51-3
SPICER 3-4-82-78-3227X
WESCO 44N48M
WESCO 920153

IDLI GROUP 88007.25

ROCKWELL 982-51-6
SPICER 229116-3227

IDLI GROUP 88007.26

ROCKWELL 982-52-3
SPICER 3.4-82-78-5923X

IDLI GROUP 88007.27

ROCKWELL 982-52-6
SPICER 229116-5923X

IDLI GROUP 88007.28

ROCKWELL 984-10-1
SPICER 204574-1
SPICER 9506SF

IDLI GROUP 88007.29

ROCKWELL 984-11-1
SPICER 204572-1
SPICER 7710-SF

IDLI GROUP 88007.30

ROCKWELL 984-12-1
SPICER 204989-1
SPICER 9296SF

IDLI GROUP 88007.31

ROCKWELL 984-13-1
SPICER 204572-2
SPICER 7710-1SF

IDLI GROUP 88007.32

ROCKWELL 984-22-2

IDLI GROUP 88007.33

CLEVE.STL. U56-43-131
ROCKWELL 990-10-2

IDLI GROUP 88007.34

CLEVE.STL. O55-43-171
ROCKWELL 991-13-2

IDLI GROUP 88007.35

CLEVE.STL. D56-43-321
ROCKWELL 992-11-2

IDLI GROUP 88007.36

CLEVE.STL. TS55-43-181
ROCKWELL 993-10-2

IDLI GROUP 88007.37

CLEVE.STL. TS55-43-182
ROCKWELL 993-11-2

IDLI GROUP 88007.38

CLEVE.STL. S55-43-201
ROCKWELL 993-12-2

IDLI GROUP 88007.39

CLEVE.STL. S55-43-202
ROCKWELL 993-13-2

IDLI GROUP 88007.40

CLEVE.STL. S55-43-203
ROCKWELL 993-14-2

IDLI GROUP 88007.41

CLEVE.STL. S55-43-204
ROCKWELL 993-15-2

IDLI GROUP 88007.42

CLEVE.STL. S55-43-469
ROCKWELL 993-16-2

IDLI GROUP 88007.43

CLEVE.STL. S55-43-209
ROCKWELL 993-17-2

IDLI GROUP 88007.44

CLEVE.STL. S55-43-210
ROCKWELL 993-18-2

IDLI GROUP 88007.45

CLEVE.STL. R55-43-311
ROCKWELL 994-10-2

IDLI GROUP 88007.46

CLEVE.STL. R55-43-313
ROCKWELL 994-11-2

IDLI GROUP 88007.47

CLEVE.STL. R55-43-314
ROCKWELL 994-12-2

IDLI GROUP 88007.48

CLEVE.STL. R55-43-593
ROCKWELL 994-13-2

BD=BEARING DIAMETER (outside)   **BW**=BEARING WIDTH   **CB**=CROSS LENGTH WITH BEARINGS   **CC**=CENTER TO CENTER   **CD**=CROSS DIAMETER
CL=CROSS LENGTH WITHOUT BEARINGS   **EC**=END TO CENTER or FACE TO FACE   **EE**=END TO END   **EL**=EFFECTIVE LENGTH
HD=HUB DIAMETER (or insert)   **IE**=INSIDE OF EARS (or recess)   **OD**=OUTSIDE DIAMETER   **OE**=OUTSIDE OF EARS   **SB**=SPLINE OR BORE SIZE

IDLI GROUP 88007.49

CLEVE.STL.	R55-43-315
ROCKWELL	994-14-2

IDLI GROUP 88007.50

CLEVE.STL.	R55-43-316
ROCKWELL	994-15-2

IDLI GROUP 88007.51

CLEVE.STL.	HCB1-4
ROCKWELL	BHR110-1
SPICER	230750

IDLI GROUP 88007.52

CLEVE.STL.	RCB1-2
ROCKWELL	BHR56-1

IDLI GROUP 88007.53

CLEVE.STL.	RCB1-10
FORD	D0TZ4808A
IHC	455004C1
ROCKWELL	BHR56-2

IDLI GROUP 88007.54

CLEVE.STL.	UCB1-5
IHC	455005C1
ROCKWELL	BHR62-3

IDLI GROUP 88007.55

CLEVE.STL.	SCB1-4
FORD	D0TZ4808B
GMC	3990292
IHC	392952C1
IHC	914236R1
ROCKWELL	BHR72-1
SPICER	230122

IDLI GROUP 88007.56

CLEVE.STL.	UCB1-4
IHC	291872C1
ROCKWELL	BHR86-1
SPICER	230148

IDLI GROUP 88007.57

CLEVE.STL.	CCB1-4
GMC	2402705
IHC	291873C1
ROCKWELL	BHR90-1
SPICER	230164

IDLI GROUP 88007.58

AEC	13B
ALLOY	4013
AMER.PARTS	2-6035
ANCHORDOAN	35-6035
BUTLER	510527X
DORMAN	662-013
FED.MOGUL	HB206FF
GREEN BALL	RSB2
IHC	328384C91
KENBAR	6030B
LEMPCO	35-6035
MCQUAY-NOR	MM6035
NEAPCO	81-1020
NEAPCO	DS101K
PILOT	H-1001
PILOT	H1001
PILOT	H-1001
PILOT	H-1001
ROCKWELL	BRAR19-2
SKF	1700536

IDLI GROUP 88007.59

AEC	9
ALLOY	4010
AMER.PARTS	2-6510
ANCHORDOAN	35-6054
BUTLER	51085X
DORMAN	662-026
GMC	2309277
KENBAR	9277-2
MOPAR	D-614
NEAPCO	81-2050
NEAPCO	DS205KG
PILOT	H-1007
ROCKWELL	BRAR22-2
SKF	1700947

IDLI GROUP 88007.60

NEAPCO	81-2012
NEAPCO	DS201KF1
ROCKWELL	BRAR22-4

IDLI GROUP 88007.61

BUTLER	510367-1X
NEAPCO	81-2121
NEAPCO	DS201KG1
ROCKWELL	BRAR22-6

IDLI GROUP 88007.62

NEAPCO	81-2122
NEAPCO	DS201KG2
ROCKWELL	BRAR22-7

IDLI GROUP 88007.63

BUTLER	510090-1X
CHRYSLER	3637710
NEAPCO	81-2013
NEAPCO	DS201KF2
ROCKWELL	BRAR22-8

IDLI GROUP 88007.64

BUTLER	510088-1X2
NEAPCO	81-2014
NEAPCO	DS201KF4
ROCKWELL	BRAR22-9
SPICER	210797-2X
SPICER	211036-2X

IDLI GROUP 88007.65

AEC	11
BUTLER	511016X3
NEAPCO	81-2043
NEAPCO	DS204KF3
ROCKWELL	BRAR25-12

IDLI GROUP 88007.66

BUTLER	510873X
NEAPCO	81-2042
NEAPCO	DS204KF2
ROCKWELL	BRAR25-13
SPICER	21162X

IDLI GROUP 88007.67

BUTLER	510866-1X
NEAPCO	81-2041
NEAPCO	DS204KF1
ROCKWELL	BRAR25-14

IDLI GROUP 88007.68

NEAPCO	81-3021
NEAPCO	DS301KG1
ROCKWELL	BRAR25-3

IDLI GROUP 88007.69

NEAPCO	81-3011
NEAPCO	DS301KF1
ROCKWELL	BRAR25-4

IDLI GROUP 88007.70

BUTLER	510433-1X
NEAPCO	81-3012
NEAPCO	DS301KF2
ROCKWELL	BRAR25-5

IDLI GROUP 88007.71

BUTLER	510144-1X
NEAPCO	81-3022
NEAPCO	DS301KG2
ROCKWELL	BRAR25-6

IDLI GROUP 88007.72

NEAPCO	81-3013
NEAPCO	DS301KF3
ROCKWELL	BRAR25-7

IDLI GROUP 88007.73

NEAPCO	81-3023
ROCKWELL	BRAR25-8

IDLI GROUP 88007.74

BUTLER	510084-2X
NEAPCO	81-4011
NEAPCO	DS401KF1
ROCKWELL	BRAR28-1

IDLI GROUP 88007.75

BUTLER	510207-1X
NEAPCO	81-4012
NEAPCO	DS401KF2
ROCKWELL	BRAR28-2

IDLI GROUP 88007.76

AEC	18
BUTLER	510084-2X3
FORD	D9HZ4800A
NEAPCO	81-4013
NEAPCO	DS401KF3
ROCKWELL	BRAR28-5
SPICER	211172-1X

IDLI GROUP 88007.77

AEC	20
ALLOY	4007
AMER.PARTS	2-6513
ANCHORDOAN	35-6048
AUTOCAR	3AA0970
BEAR.INC.	BI501KF
BORG-WARNR	35-6048
BUTLER	510121-1X
CMP	DS501KF
CRANE CAR.	53-18
DETROIT	MB611
DORMAN	662-020
DORMAN	DS20
FEDERAL	HBF88510
FORD	C8HA4800B
FORD	D4TZ4800G
FORD	D4TZ4800H
GREEN BALL	RSB10
IHC	283069C92
KENBAR	6060-10
LEMPCO	35-6025
MCQUAY-NOR	MM6050
NEAPCO	81-5010
NEAPCO	DS501KF
PILOT	H1014
ROCKWELL	BRAR32-1
SKF	1700810
WESCO	960-007

CAUTION: BE SURE TO REFER TO ENGINEERING CATALOGS FOR SPECIAL APPLICATIONS THAT REQUIRE SPECIFIC MATERIAL CONTENT, TOLERANCES, ETC. SEE FOOTNOTE.

IDLI GROUP 88007.78

AEC	21S
KENDAR	6161-12S
MACK	35MU32P4
MOTOR MAST	BS21S
MOTOR MAST	DS21S
NEAPCO	81-6011
NEAPCO	DS601KFS
ROCKWELL	BRAR38-1
SKF	1700945

IDLI GROUP 88007.79

AEC	21
ALLOY	4008
AMER.PARTS	2-6514
ANCHORDOAN	35-6049
AUTOCAR	3AB0970
BEAR.INC.	B1-601KF
BEAR.INC.	BI601KF
BORG-WARNR	35-6049
BUTLER	510661-1X
CMP	DS601KF
CMP	DS701KF
CRANE CAR.	53-17
DETROIT	MB612
DORMAN	662-021
DORMAN	DS21
FEDERAL	HBF88512A
FORD	C8HA4800A
FORD	D4TZ4800J
FORD	D4TZ4800K
GMC	662044
GREEN BALL	RSB11
GREEN BALL	TSB11
IHC	283071C91
IHC	283071C92
KENDAR	6161-12
LEMPCO	35-6026
MACK	35MU32PC
MCQUAY-NOR	MM6051
NEAPCO	81-6010
NEAPCO	DS601KF
PILOT	H1016
PILOT	H-1016
PILOT	H1016
PILOT	H-1016
ROCKWELL	BRAR38-2
SKF	1700912
SPICER	210128-1X
SPICER	210786-1X
SPICER	210874-1X
WESCO	960-008
WHITE	991790A

IDLI GROUP 88007.80

CLEVE.STL.	SCB61-1
ROCKWELL	BRBAR19-1

IDLI GROUP 88007.81

CLEVE.STL.	SCB61
ROCKWELL	BRBAR22-6

IDLI GROUP 88007.82

CLEVE.STL.	DCB61
ROCKWELL	BRBAR25-4

IDLI GROUP 88007.83

CLEVE.STL.	OCB61-1
ROCKWELL	BRBAR25-5

IDLI GROUP 88007.84

CLEVE.STL.	UCB61-1
ROCKWELL	BRBAR28-3

IDLI GROUP 88007.85

CLEVE.STL.	CC61
CLEVE.STL.	CCB61
ROCKWELL	BRBAR32-6

IDLI GROUP 88007.86

CLEVE.STL.	HCB61
ROCKWELL	BRBAR38-2

IDLI GROUP 88007.87

CLEVE.STL.	OCB61
ROCKWELL	BRNR24-2

IDLI GROUP 88007.88

CLEVE.STL.	UCB61
ROCKWELL	BRNR28-2

IDLI GROUP 88007.89

BORG-WARNR	13782J
ROCKWELL	BS524-11

IDLI GROUP 88007.90

BORG-WARNR	10503J
ROCKWELL	BS524-8

IDLI GROUP 88007.91

BORG-WARNR	5814J
ROCKWELL	BS528-10
SPICER	4-54-381

IDLI GROUP 88007.92

BORG-WARNR	10834J
ROCKWELL	BS528-11

IDLI GROUP 88007.93

BORG-WARNR	13818J
ROCKWELL	BS528-12

IDLI GROUP 88007.94

BORG-WARNR	16728J
ROCKWELL	BS528-16

IDLI GROUP 88007.95

BORG-WARNR	14059J
ROCKWELL	BS528-28

IDLI GROUP 88007.96

BORG-WARNR	12089J
ROCKWELL	BS528-4

IDLI GROUP 88007.97

BORG-WARNR	13457J
ROCKWELL	BS528-6

IDLI GROUP 88007.98

BORG-WARNR	12802J
ROCKWELL	BS528-7

IDLI GROUP 88007.99

BORG-WARNR	10504J
ROCKWELL	BS528-9

IDLI GROUP 88008

ROCKWELL	BS531-14
SPICER	4-54-481

IDLI GROUP 88008.01

BORG-WARNR	16418J
ROCKWELL	BS531-3

IDLI GROUP 88008.02

BORG-WARNR	16757J
ROCKWELL	BS531-4

IDLI GROUP 88008.03

CLEVE.STL.	O55-130
CLEVE.STL.	O55-130
ROCKWELL	BS-S22-114

IDLI GROUP 88008.04

CLEVE.STL.	O55-131
ROCKWELL	BS-S22-115

IDLI GROUP 88008.05

CLEVE.STL.	O55-135
ROCKWELL	BS-S22-116

IDLI GROUP 88008.06

CLEVE.STL.	R55-449
ROCKWELL	BS-S22-117

IDLI GROUP 88008.07

CLEVE.STL.	R55-450
ROCKWELL	BS-S22-118

IDLI GROUP 88008.08

CLEVE.STL.	R55-452
ROCKWELL	BS-S22-119

IDLI GROUP 88008.09

CLEVE.STL.	O55-1295
ROCKWELL	BS-S22-120
SPICER	3-53-161

IDLI GROUP 88008.10

CLEVE.STL.	S55-1168
ROCKWELL	BS-S22-121

IDLI GROUP 88008.11

CLEVE.STL.	S55-1169
ROCKWELL	BS-S22-122

IDLI GROUP 88008.12

CLEVE.STL.	R55-1233
ROCKWELL	BS-S22-123

IDLI GROUP 88008.13

CLEVE.STL.	O55-1303
ROCKWELL	BS-S22-124
SPICER	3-54-611

IDLI GROUP 88008.14

CLEVE.STL.	D76-101
ROCKWELL	BS-S24-30
SPICER	3-53-431

IDLI GROUP 88008.15

CLEVE.STL.	D76-102
ROCKWELL	BS-S24-31
SPICER	3-53-241

BD = BEARING DIAMETER (outside)　　**BW** = BEARING WIDTH　　**CB** = CROSS LENGTH WITH BEARINGS　　**CC** = CENTER TO CENTER　　**CD** = CROSS DIAMETER
CL = CROSS LENGTH WITHOUT BEARINGS　　**EC** = END TO CENTER or FACE TO FACE　　**EE** = END TO END　　**EL** = EFFECTIVE LENGTH
HD = HUB DIAMETER (or insert)　　**IE** = INSIDE OF EARS (or recess)　　**OD** = OUTSIDE DIAMETER　　**OE** = OUTSIDE OF EARS　　**SB** = SPLINE OR BORE SIZE

IDLI GROUP 88008.16

CLEVE.STL.	U76-101
NEAPCO	N4-53-71
ROCKWELL	BS-S24-32
SPICER	4-53-71

IDLI GROUP 88008.17

CLEVE.STL.	O55-1410
ROCKWELL	BS-S24-41
SPICER	3-53-1071

IDLI GROUP 88008.18

CLEVE.STL.	O55-1408
ROCKWELL	BS-S24-42
SPICER	3-53-1031

IDLI GROUP 88008.19

CLEVE.STL.	D76-125
ROCKWELL	BS-S25-3
SPICER	3-53-451

IDLI GROUP 88008.20

CLEVE.STL.	D76-126
ROCKWELL	BS-S25-4
SPICER	3-53-291

IDLI GROUP 88008.21

CLEVE.STL.	D56-104
ROCKWELL	BS-S26-14

IDLI GROUP 88008.22

CLEVE.STL.	D56-125
ROCKWELL	BS-S26-15

IDLI GROUP 88008.23

BORG-WARNR	16728J
ROCKWELL	BS-S28-16

IDLI GROUP 88008.24

BORG-WARNR	5916J
ROCKWELL	BS-S28-49

IDLI GROUP 88008.25

CLEVE.STL.	U76-102
ROCKWELL	BS-S28-61
SPICER	4-53-61

IDLI GROUP 88008.26

CLEVE.STL.	U56-106
ROCKWELL	BS-S28-68

IDLI GROUP 88008.27

BORG-WARNR	12802J
ROCKWELL	BS-S28-7

IDLI GROUP 88008.28

CLEVE.STL.	C67-110
ROCKWELL	BS-S31-18

IDLI GROUP 88008.29

BORG-WARNR	22263J
ROCKWELL	BS-S32-7A

IDLI GROUP 88008.30

CLEVE.STL.	OCB51-2
ROCKWELL	CASR30-1

IDLI GROUP 88008.31

AEC	18
ALLOY	4006
AMER.PARTS	2-6040
ANCHORDOAN	35-6040
AUTOCAR	3AD0970
BUTLER	510084-2XWF
DORMAN	662-018
GREEN BALL	RSB9
IHC	283067C92
KENBAR	5954-6
LEMPCO	35-6040
MCQUAY-NOR	MM6040
NEAPCO	81-4010
NEAPCO	DS401KF
PILOT	H-1011
ROCKWELL	CB130007-1X
SKF	1700539
SPICER	210969X

IDLI GROUP 88008.32

ROCKWELL	CB210084-1X
WHITE	02-7011870

IDLI GROUP 88008.33

IHC	291869C91
ROCKWELL	CB210094-2X

IDLI GROUP 88008.34

AEC	AE13B
ROCKWELL	CB21057X

IDLI GROUP 88008.35

AEC	10
AEC	AE10
ALLOY	4012
AMER.PARTS	2-6028
ANCHORDOAN	35-6028
BORG-WARNR	35-6028
BUTLER	510580X
CLEVE.MOT.	DS101
CLEVE.STL.	SCB75
DORMAN	662-010
GMC	3860801
GMC	465286
GREEN BALL	RS2
KENBAR	27424
KENBAR	7424
LEMPCO	35-6028
MCQUAY-NOR	MM6028
MOTOR MAST	DS10
NEAPCO	81-1010
NEAPCO	81-1011
NEAPCO	81-1012
NEAPCO	DS101
NEAPCO	DS101-1
NEAPCO	DS101-2
PILOT	H1000
REPUBLIC	35-6028
ROCKWELL	CB210580X
SPICER	210471-X
SPICER	210580X
TRW	20815
WESCO	960-012

IDLI GROUP 88008.36

CMP	DS601KFS
ROCKWELL	CB210875-1X
SPICER	210875-1X

IDLI GROUP 88008.37

ROCKWELL	CB58WB17

IDLI GROUP 88008.38

IHC	392953
ROCKWELL	CB5R40-1

IDLI GROUP 88008.39

AEC	15
ALLOY	4001
AMER.PARTS	2-6037
ANCHORDOAN	35-6037
BUTLER	510088-1XWF
CHECKER	03568
CHECKER	64575
DORMAN	662-015
GREEN BALL	RSB4
KENBAR	5407-1
LEMPCO	35-6037
MCQUAY-NOR	MM6037
MOPAR	D601
NEAPCO	DS201KF
PILOT	H1002
ROCKWELL	CBAR22-1
SKF	1700537
SPICER	210137-1X
SPICER	210939-X
SPICER	211036-2X
TRW	20817

IDLI GROUP 88008.40

NEAPCO	81-2120
NEAPCO	DS201KG
ROCKWELL	CBAR22-2

IDLI GROUP 88008.41

AEC	16
ALLOY	4004
AMER.PARTS	2-6038
ANCHORDOAN	35-6038
BUTLER	510391-1XWF
DORMAN	662-016
GMC	904570
GREEN BALL	RSB7
IHC	484002C91
KENBAR	5732-3
LEMPCO	35-6038
MCQUAY-NOR	MM6038
NEAPCO	81-3010
NEAPCO	DS301KF
PILOT	H-1008
PILOT	H1008
PILOT	H-1008
PILOT	H1008
ROCKWELL	CBAR25-1
SKF	1700538
SPICER	210083-1X
SPICER	210139-1X
SPICER	211098-1X

IDLI GROUP 88008.42

AEC	17
ALLOY	4003
AMER.PARTS	2-6039
ANCHORDOAN	35-6039
BUTLER	510140-1XWF
DORMAN	662-017
GMC	2351948
GREEN BALL	RSB6
KENBAR	5620-4
LEMPCO	35-6039
MCQUAY-NOR	MM6039
NEAPCO	81-3020
NEAPCO	DS301KG
PILOT	H-1009
PILOT	H1009
PILOT	H-1009
PILOT	H1009
ROCKWELL	CBAR25-2
SKF	1700548
SPICER	210128-1X

IDLI GROUP 88008.43

AEC	11
ALLOY	4011
AMER.PARTS	2-6509
ANCHORDOAN	35-6053
BUTLER	511016X
DORMAN	662-025
GMC	719413
KENBAR	9413-1
MOPAR	D-613
NEAPCO	81-2040

CAUTION: BE SURE TO REFER TO ENGINEERING CATALOGS FOR SPECIAL APPLICATIONS THAT REQUIRE SPECIFIC MATERIAL CONTENT, TOLERANCES, ETC. SEE FOOTNOTE.

NEAPCO DS204KF
REPUBLIC 35-6058
ROCKWELL CBAR25-3
SKF 1700946
TRW 20816

IDLI GROUP 88008.44

AEC 18
ALLOY 4006
AMER.PARTS 2-6040
ANCHORDOAN 35-6040
AUTOCAR 3AD0970
BUTLER 510084-2XWF
DORMAN 662-018
GREEN BALL RSB9
IHC 283067C92
KENBAR 5954-6
LEMPCO 35-6040
MCQUAY-NOR MM6040
NEAPCO 81-4010
NEAPCO DS401KF
PILOT H1011
ROCKWELL CBAR28-1
SKF 1700539
SPICER 210969X

IDLI GROUP 88008.45

AEC 19
ALLOY 4005
AMER.PARTS 2-6041
ANCHORDOAN 35-6041
BUTLER 510230-1X
DORMAN 662-019
GREEN BALL TSB8
KENBAR 5841-5
LEMPCO 35-6041
MCQUAY-NOR MM6041
NEAPCO 81-4020
NEAPCO DS401KG
PILOT H1012
ROCKWELL CBAR28-2
SKF 1700549
SPICER 210209-1X

IDLI GROUP 88008.46

ROCKWELL CBAR32-1
SPICER 210120-1X

IDLI GROUP 88008.47

BUTLER 510086-1X
CLEVE.STL. SCB109
FORD B9TZ4840A
FORD D2TZ4840A
LEMPCO 35-6037B
NEAPCO 82-2010
NEAPCO DS201BF
ROCKWELL CBRR22-1
SPICER 210086-1X

SPICER 210086X
SPICER 210921X

IDLI GROUP 88008.48

AEC 22
AMER.PARTS 2-6508
ANCHORDOAN 35-6050
DORMAN 662-022
GREEN BALL RSB12
IHC 223572R93
KENBAR 203KC
LEMPCO 36-2030
MCQUAY-NOR MM6053
MCQUAY-NOR MM6054
NEAPCO 81-2030
NEAPCO DS203KC
PILOT H1005
ROCKWELL CBRR22-2

IDLI GROUP 88008.49

NEAPCO 81-2030
NEAPCO DS203KC
ROCKWELL CBRR22-2A

IDLI GROUP 88008.50

BUTLER 510228X
CLEVE.STL. SCB108
GMC 2470889
LEMPCO 35-6036B
NEAPCO 82-2011
NEAPCO DS201BG
ROCKWELL CBRR22-5
SPICER 210087-1X
SPICER 210228-1X
SPICER 210228X

IDLI GROUP 88008.51

AEC 23
AMER.PARTS 2-6511
ANCHORDOAN 35-6051
DORMAN 662-023
GREEN BALL RSB13
KENBAR 303KC
LEMPCO 36-3030
NEAPCO 81-3030
NEAPCO DS303KC
PILOT H-1010
ROCKWELL CBRR24-1

IDLI GROUP 88008.52

AEC 23
AMER.PARTS 2-6511
ANCHORDOAN 35-6051
DORMAN 662-023
GREEN BALL RSB13
KENBAR 303KC
LEMPCO 36-3030
NEAPCO 81-3030
NEAPCO DS303KC
PILOT H1010
ROCKWELL CBRR24-1A

IDLI GROUP 88008.53

BUTLER 510085-1X
CLEVE.STL. DCB100
FORD B5TZ4840B
IHC 918837R91
LEMPCO 35-6038B
NEAPCO 82-3010
NEAPCO DS301BF
ROCKWELL CBRR25-1
SPICER 210085-1X

IDLI GROUP 88008.54

BUTLER 510119-1X
CLEVE.STL. DCB104
GMC 2333914
GMC 9422980
LEMPCO 35-6039B
NEAPCO 82-3011
NEAPCO DS301BG
ROCKWELL CBRR25-2
SPICER 210119-1X

IDLI GROUP 88008.55

CLEVE.STL. OCB105
IHC 470682C91
NEAPCO 82-2040
NEAPCO DS204BF
ROCKWELL CBRR25-3
SPICER 210865-1X

IDLI GROUP 88008.56

BUTLER 510092-1X
CLEVE.STL. UCB106
FORD B5TZ4840A
GMC 2338929
IHC 281869C91
LEMPCO 35-6040B
NEAPCO 82-4010
NEAPCO DS401BF
ROCKWELL CBRR28-1
SPICER 210092-1X

IDLI GROUP 88008.57

BUTLER 510129-1X
CLEVE.STL. UCB108
LEMPCO 35-6041B
NEAPCO 510129-1X
NEAPCO 82-4011
NEAPCO DS401BG
ROCKWELL CBRR28-2
SPICER 210129-1X

IDLI GROUP 88008.58

NEAPCO 81-4030
NEAPCO DS403KC
ROCKWELL CBRR28-4

IDLI GROUP 88008.59

AEC 24
AMER.PARTS 2-6512
ANCHORDOAN 35-6052
DORMAN 662-024
GREEN BALL RSB14
IHC 879849R92
KENBAR 403KC
LEMPCO 36-4030

MCQUAY-NOR MM6055
NEAPCO 81-4030
NEAPCO DS403KC
PILOT H1013
PILOT H-1013
PILOT H1013
PILOT H-1013
ROCKWELL CBRR28-4A

IDLI GROUP 88008.60

BUTLER 510120-1X
CLEVE.STL. CCB100
GMC 2402704
IHC 291870C91
LEMPCO 35-6025B
NEAPCO 82-5010
NEAPCO DS501BF
ROCKWELL CBRR32-1
SPICER 210120-1X

IDLI GROUP 88008.61

CLEVE.STL. OCB51-2
ROCKWELL CBSR30-1

IDLI GROUP 88008.62

CLEVE.STL. UCB51-2
ROCKWELL CBSR38-1R

IDLI GROUP 88008.63

ANCHORDOAN 35-6027
BORG-WARNR 35-6027
BUTLER 530119-1
CHRYSLER 3632993
CLEVE.STL. SCB50-1
GMC 3730827
IHC 914238R1
LEMPCO 35-6037R
MCQUAY-NOR MM6027
NEAPCO 82-2110
NEAPCO DS201R
REPUBLIC 35-6027
ROCKWELL CBSR40-1
SPICER 230119-1
SPICER 230119-1X
TRW 20814

BD=BEARING DIAMETER (outside)   **BW**=BEARING WIDTH   **CB**=CROSS LENGTH WITH BEARINGS   **CC**=CENTER TO CENTER   **CD**=CROSS DIAMETER
CL=CROSS LENGTH WITHOUT BEARINGS   **EC**=END TO CENTER or FACE TO FACE   **EE**=END TO END   **EL**=EFFECTIVE LENGTH
HD=HUB DIAMETER (or insert)   **IE**=INSIDE OF EARS (or recess)   **OD**=OUTSIDE DIAMETER   **OE**=OUTSIDE OF EARS   **SB**=SPLINE OR BORE SIZE

IDLI GROUP 88008.64

AMER.PARTS	2-6025
BUTLER	530126-1
CHRYSLER	1663979
CLEVE.STL.	DCB50-1
FORD	B5QH4826B
GMC	2333919
GMC	3730828
IHC	918836R91
LEMPCO	35-6038R
NEAPCO	82-3110
NEAPCO	DS301R
ROCKWELL	CBSR45-1
SPICER	230126-1

IDLI GROUP 88008.65

AMER.PARTS	2-6026
BUTLER	530147-3
CHRYSLER	1667747
CLEVE.STL.	UCB50-1
FORD	B9TT4826A
GMC	2336424
IHC	291866C1
LEMPCO	35-6040R
NEAPCO	82-4110
NEAPCO	DS401R
ROCKWELL	CBSR48-1
SPICER	230147-3

IDLI GROUP 88008.66

BUTLER	530167-1
CLEVE.STL.	CCB50-1
FORD	C1TT4826A
GMC	2402707
IHC	291867C1
LEMPCO	35-6025R
NEAPCO	82-5110
NEAPCO	DS501R
ROCKWELL	CBSR51-1
SPICER	230167-1

IDLI GROUP 88008.67

BUTLER	530549
CLEVE.STL.	HCB50-1
FORD	C8TZ4826A
GMC	2236223
LEMPCO	35-6026R
NEAPCO	82-6110
NEAPCO	DS601R
ROCKWELL	CBSR64-3
SPICER	230749-1

IDLI GROUP 88008.68

ALCO	1581
ALCO	RG581X
ALLOY	15-181
MOPAR	D8Z
NEAPCO	28181X
PRECISION	349
ROCKWELL	CP101X

IDLI GROUP 88008.69

ROCKWELL	CP105X

IDLI GROUP 88008.70

ROCKWELL	CP111X

IDLI GROUP 88008.71

ROCKWELL	CP115X

IDLI GROUP 88008.72

ROCKWELL	CP116X

IDLI GROUP 88008.73

ROCKWELL	CP1200X

IDLI GROUP 88008.74

ROCKWELL	CP1201X

IDLI GROUP 88008.75

ROCKWELL	CP1203X

IDLI GROUP 88008.76

ROCKWELL	CP1204X

IDLI GROUP 88008.77

ROCKWELL	CP1205X

IDLI GROUP 88008.78

ROCKWELL	CP1206X

IDLI GROUP 88008.79

ROCKWELL	CP1207X

IDLI GROUP 88008.80

ROCKWELL	CP1208X

IDLI GROUP 88008.81

ROCKWELL	CP1209X

IDLI GROUP 88008.82

ROCKWELL	CP1210X

IDLI GROUP 88008.83

ROCKWELL	CP1211X

IDLI GROUP 88008.84

ROCKWELL	CP1214X

IDLI GROUP 88008.85

ROCKWELL	CP121X

IDLI GROUP 88008.86

ROCKWELL	CP124X

IDLI GROUP 88008.87

ROCKWELL	CP131N

IDLI GROUP 88008.88

ROCKWELL	CP131N1

IDLI GROUP 88008.89

ROCKWELL	CP133N

IDLI GROUP 88008.90

CLEVE.STL.	P55-55-675
ROCKWELL	CP133N1

IDLI GROUP 88008.91

ROCKWELL	CP134X

IDLI GROUP 88008.92

ROCKWELL	CP135N

IDLI GROUP 88008.93

ROCKWELL	CP141N

IDLI GROUP 88008.94

ROCKWELL	CP148N

IDLI GROUP 88008.95

ROCKWELL	CP153X

IDLI GROUP 88008.96

ROCKWELL	CP155N

IDLI GROUP 88008.97

MICHIGAN	835846
PRECISION	328
REO MOTORS	10057P2
ROCKWELL	CP160X
WHITE	3W01172

IDLI GROUP 88008.98

ROCKWELL	CP165X

IDLI GROUP 88008.99

ROCKWELL	CP16N

IDLI GROUP 88009

ROCKWELL	CP16N1

IDLI GROUP 88009.01

ROCKWELL	CP16N2

IDLI GROUP 88009.02

ROCKWELL	CP16N3

IDLI GROUP 88009.03

ROCKWELL	CP170X

IDLI GROUP 88009.04

ROCKWELL	CP176N
SPICER	5-407X

IDLI GROUP 88009.05

ROCKWELL	CP177X

IDLI GROUP 88009.06

PRECISION	345
ROCKWELL	CP178X

IDLI GROUP 88009.07

ROCKWELL	CP17N

IDLI GROUP 88009.08

ROCKWELL	CP17N12

IDLI GROUP 88009.09

ROCKWELL	CP17N20

IDLI GROUP 88009.10

ROCKWELL	CP17N22

IDLI GROUP 88009.11

REPCO	5-280X
ROCKWELL	CP17N3

IDLI GROUP 88009.12

ROCKWELL	CP17N7

IDLI GROUP 88009.13

ROCKWELL	CP188X

IDLI GUIDE　　　COPYRIGHT ® INTERCHANGE, INC.　　　ST. LOUIS PARK, MN. 55416 USA

CAUTION: BE SURE TO REFER TO ENGINEERING CATALOGS FOR SPECIAL APPLICATIONS THAT REQUIRE SPECIFIC MATERIAL CONTENT, TOLERANCES, ETC. SEE FOOTNOTE.

IDLI GROUP 88009.14
ROCKWELL CP18N1

IDLI GROUP 88009.15
ROCKWELL CP18N3

IDLI GROUP 88009.16
ROCKWELL CP18N8

IDLI GROUP 88009.17
GMB G5-324X
ROCKWELL CP18N95
SPICER 5-324X

IDLI GROUP 88009.18
ROCKWELL CP1F6

IDLI GROUP 88009.19
ROCKWELL CP1FN12

IDLI GROUP 88009.20
ROCKWELL CP1FN7

IDLI GROUP 88009.21
ROCKWELL CP1FN8

IDLI GROUP 88009.22
ROCKWELL CP1FN9

IDLI GROUP 88009.23
ROCKWELL CP1FR

IDLI GROUP 88009.24
ROCKWELL CP1FR1

IDLI GROUP 88009.25
ROCKWELL CP1FR2

IDLI GROUP 88009.26
ROCKWELL CP1FR4

IDLI GROUP 88009.27
ROCKWELL CP1FR5

IDLI GROUP 88009.28
ROCKWELL CP1FR7

IDLI GROUP 88009.29
ROCKWELL CP1FRN

IDLI GROUP 88009.30
ROCKWELL CP1FRN2

IDLI GROUP 88009.31
ROCKWELL CP1FRN4

IDLI GROUP 88009.32
ROCKWELL CP1FRN5

IDLI GROUP 88009.33
ROCKWELL CP1FRN5

IDLI GROUP 88009.34
ROCKWELL CP1FS1

IDLI GROUP 88009.35
ROCKWELL CP1FS2

IDLI GROUP 88009.36
ROCKWELL CP1FS5

IDLI GROUP 88009.37
ROCKWELL CP1FS8

IDLI GROUP 88009.38
ROCKWELL CP1FS9

IDLI GROUP 88009.39
ROCKWELL CP2002

IDLI GROUP 88009.40
ROCKWELL CP200X

IDLI GROUP 88009.41
ROCKWELL CP2011

IDLI GROUP 88009.42
ROCKWELL CP2031

IDLI GROUP 88009.43
ROCKWELL CP2033

IDLI GROUP 88009.44
ROCKWELL CP2116

IDLI GROUP 88009.45
ROCKWELL CP213X

IDLI GROUP 88009.46
ROCKWELL CP2140

IDLI GROUP 88009.47
ROCKWELL CP248X

IDLI GROUP 88009.48
ROCKWELL CP3000

IDLI GROUP 88009.49
PRECISION 534
ROCKWELL CP3008

IDLI GROUP 88009.50
ROCKWELL CP3012

IDLI GROUP 88009.51
IHC 69-550R91
ROCKWELL CP3021

IDLI GROUP 88009.52
ROCKWELL CP3120

IDLI GROUP 88009.53
ROCKWELL CP3121

IDLI GROUP 88009.54
MICHIGAN 854639
REPCO CP35N
ROCKWELL CP35N

IDLI GROUP 88009.55
ROCKWELL CP35N10

IDLI GROUP 88009.56
ROCKWELL CP35N2

IDLI GROUP 88009.57
ROCKWELL CP35N20

IDLI GROUP 88009.58
ROCKWELL CP35N3

IDLI GROUP 88009.59
ROCKWELL CP35N4

IDLI GROUP 88009.60
ROCKWELL CP35N5

IDLI GROUP 88009.61
ROCKWELL CP35N8

IDLI GROUP 88009.62
ROCKWELL CP35N9

IDLI GROUP 88009.63
ROCKWELL CP35R

IDLI GROUP 88009.64
ROCKWELL CP35R17

IDLI GROUP 88009.65
ROCKWELL CP35R3

IDLI GROUP 88009.66
ROCKWELL CP3D

IDLI GROUP 88009.67
ROCKWELL CP3D1

IDLI GROUP 88009.68
ROCKWELL CP3DN3

IDLI GROUP 88009.69
ROCKWELL CP3DN4

IDLI GROUP 88009.70
ROCKWELL CP3DR

IDLI GROUP 88009.71
ROCKWELL CP3DRN

IDLI GROUP 88009.72
ROCKWELL CP3DRN3

IDLI GROUP 88009.73
ROCKWELL CP3DS1

IDLI GROUP 88009.74
ROCKWELL CP3DS4

BD=BEARING DIAMETER (outside)　　BW=BEARING WIDTH　　CB=CROSS LENGTH WITH BEARINGS　　CC=CENTER TO CENTER　　CD=CROSS DIAMETER
CL=CROSS LENGTH WITHOUT BEARINGS　　EC=END TO CENTER or FACE TO FACE　　EE=END TO END　　EL=EFFECTIVE LENGTH
HD=HUB DIAMETER (or insert)　　IE=INSIDE OF EARS (or recess)　　OD=OUTSIDE DIAMETER　　OE=OUTSIDE OF EARS　　SB=SPLINE OR BORE SIZE

IDLI GROUP 88009.75
ROCKWELL CP4002

IDLI GROUP 88009.76
ROCKWELL CP4007

IDLI GROUP 88009.77
ROCKWELL CP4008

IDLI GROUP 88009.78
ROCKWELL CP4012

IDLI GROUP 88009.79
ROCKWELL CP4015

IDLI GROUP 88009.80
ROCKWELL CP4016

IDLI GROUP 88009.81
ROCKWELL CP4101

IDLI GROUP 88009.82
ROCKWELL CP4102

IDLI GROUP 88009.83
ROCKWELL CP4113

IDLI GROUP 88009.84
ROCKWELL CP4130

IDLI GROUP 88009.85
ROCKWELL CP4156

IDLI GROUP 88009.86
ROCKWELL CP44R

IDLI GROUP 88009.87
ROCKWELL CP45N

IDLI GROUP 88009.88
ROCKWELL CP45N1

IDLI GROUP 88009.89
ROCKWELL CP4N

IDLI GROUP 88009.90
ROCKWELL CP4N1

IDLI GROUP 88009.91
ROCKWELL CP4N5

IDLI GROUP 88009.92
ROCKWELL CP4N7

IDLI GROUP 88009.93
ROCKWELL CP5000

IDLI GROUP 88009.94
ROCKWELL CP5007

IDLI GROUP 88009.95
ROCKWELL CP5008

IDLI GROUP 88009.96
ROCKWELL CP5015

IDLI GROUP 88009.97
ROCKWELL CP5101

IDLI GROUP 88009.98
ROCKWELL CP5109

IDLI GROUP 88009.99
ROCKWELL CP5122

IDLI GROUP 88010
ROCKWELL CP5154

IDLI GROUP 88010.01
ROCKWELL CP51625-2

IDLI GROUP 88010.02
ROCKWELL CP5177

IDLI GROUP 88010.03
ROCKWELL CP52625-2

IDLI GROUP 88010.04
ROCKWELL CP53625-2

IDLI GROUP 88010.05
ROCKWELL CP55N

IDLI GROUP 88010.06
ROCKWELL CP55N1

IDLI GROUP 88010.07
ROCKWELL CP55N2

IDLI GROUP 88010.08
ROCKWELL CP55N3

IDLI GROUP 88010.09
ROCKWELL CP55N4

IDLI GROUP 88010.10
ROCKWELL CP58N

IDLI GROUP 88010.11
ROCKWELL CP58N3

IDLI GROUP 88010.12
ROCKWELL CP58N4

IDLI GROUP 88010.13
ALCO CP58NBW-4
ROCKWELL CP58NBW4

IDLI GROUP 88010.14
ROCKWELL CP58WB

IDLI GROUP 88010.15
ROCKWELL CP58WB1

IDLI GROUP 88010.16
ROCKWELL CP58WB17
TRW 20142D

IDLI GROUP 88010.17
ROCKWELL CP58WB18

IDLI GROUP 88010.18
ROCKWELL CP58WB2

IDLI GROUP 88010.19
ROCKWELL CP58WB20

IDLI GROUP 88010.20
ROCKWELL CP58WB21

IDLI GROUP 88010.21
ROCKWELL CP58WB29

IDLI GROUP 88010.22
ROCKWELL CP58WB3

IDLI GROUP 88010.23
ROCKWELL CP58WB37

IDLI GROUP 88010.24
ROCKWELL CP58WB38

IDLI GROUP 88010.25
ROCKWELL CP58WB4

IDLI GROUP 88010.26
ROCKWELL CP58WB5

IDLI GROUP 88010.27
ROCKWELL CP58WB6

IDLI GROUP 88010.28
ROCKWELL CP58WB7

IDLI GROUP 88010.29
ROCKWELL CP58WB9

IDLI GROUP 88010.30
ROCKWELL CP5N

IDLI GROUP 88010.31
ROCKWELL CP5N1

IDLI GROUP 88010.32
ROCKWELL CP5N14

IDLI GROUP 88010.33
ROCKWELL CP5N4

IDLI GROUP 88010.34
ROCKWELL CP5N9

IDLI GROUP 88010.35
IOWA MFG. 45375-500-03
ROCKWELL CP6

IDLI GROUP 88010.36
ROCKWELL CP6107

ENGINEERING CATALOGS MUST BE CONSULTED FOR DETAILS NOT INCLUDED
IN THIS GUIDE. SPECIFIC DESIGNS, MATERIAL CONTENT, TOLERANCES
LUBE FITTINGS AND OTHER DIMENSIONS ARE INTENTIONALLY OMITTED HERE.

CAUTION: BE SURE TO REFER TO ENGINEERING CATALOGS FOR SPECIAL APPLICATIONS THAT REQUIRE SPECIFIC MATERIAL CONTENT, TOLERANCES, ETC. SEE FOOTNOTE.

IDLI GROUP 88010.37
ROCKWELL CP6135

IDLI GROUP 88010.38
BORG-WARNR 114-6149
PRECISION 925
ROCKWELL CP62N
WESCO N6119

IDLI GROUP 88010.39
ROCKWELL CP62N10

IDLI GROUP 88010.40
ROCKWELL CP62N11

IDLI GROUP 88010.41
ROCKWELL CP62N12

IDLI GROUP 88010.42
BORG-WARNR 114-6012
BORG-WARNR 114-6028
ROCKWELL CP62N12

IDLI GROUP 88010.43
ROCKWELL CP62N13

IDLI GROUP 88010.44
PRECISION 925
ROCKWELL CP62N14

IDLI GROUP 88010.45
ROCKWELL CP62N15

IDLI GROUP 88010.46
ROCKWELL CP62N16

IDLI GROUP 88010.47
ROCKWELL CP62N17

IDLI GROUP 88010.48
ROCKWELL CP62N18

IDLI GROUP 88010.49
ROCKWELL CP62N23

IDLI GROUP 88010.50
ROCKWELL CP62N25

IDLI GROUP 88010.51
ROCKWELL CP62N26

IDLI GROUP 88010.52
ROCKWELL CP62N3

IDLI GROUP 88010.53
ROCKWELL CP62N4

IDLI GROUP 88010.54
CLARK EQU. 1265587
ROCKWELL CP62N46

IDLI GROUP 88010.55
ROCKWELL CP62N47

IDLI GROUP 88010.56
ROCKWELL CP62N48

IDLI GROUP 88010.57
ROCKWELL CP62N49

IDLI GROUP 88010.58
ROCKWELL CP62N50

IDLI GROUP 88010.59
ROCKWELL CP62N51

IDLI GROUP 88010.60
EIMCO 912A15034
ROCKWELL CP62N56

IDLI GROUP 88010.61
ROCKWELL CP62N6

IDLI GROUP 88010.62
ROCKWELL CP62N8

IDLI GROUP 88010.63
ROCKWELL CP62N9

IDLI GROUP 88010.64
A.E.LEE 3881-1
NATL.MINE 3881-1
NATL.MINE 7601171
ROCKWELL CP65N

IDLI GROUP 88010.65
ROCKWELL CP68WB

IDLI GROUP 88010.66
ROCKWELL CP68WB1

IDLI GROUP 88010.67
ROCKWELL CP68WB10

IDLI GROUP 88010.68
ROCKWELL CP68WB2

IDLI GROUP 88010.69
ROCKWELL CP68WB6

IDLI GROUP 88010.70
ROCKWELL CP68WB64

IDLI GROUP 88010.71
ROCKWELL CP68WB7

IDLI GROUP 88010.72
ROCKWELL CP68WB8

IDLI GROUP 88010.73
ROCKWELL CP68WB9

IDLI GROUP 88010.74
ROCKWELL CP6N

IDLI GROUP 88010.75
ROCKWELL CP6N1

IDLI GROUP 88010.76
ROCKWELL CP6N10

IDLI GROUP 88010.77
ROCKWELL CP6N5

IDLI GROUP 88010.78
ROCKWELL CP6N8

IDLI GROUP 88010.79
ROCKWELL CP6N9

IDLI GROUP 88010.80
ROCKWELL CP726251-2

IDLI GROUP 88010.81
ROCKWELL CP72625-2

IDLI GROUP 88010.82
ROCKWELL CP72625-3

IDLI GROUP 88010.83
ROCKWELL CP72625-4

IDLI GROUP 88010.84
ROCKWELL CP72N1

IDLI GROUP 88010.85
ROCKWELL CP72N10

IDLI GROUP 88010.86
ROCKWELL CP72N103

IDLI GROUP 88010.87
ROCKWELL CP72N104

IDLI GROUP 88010.88
ROCKWELL CP72N11

IDLI GROUP 88010.89
ROCKWELL CP72N112

IDLI GROUP 88010.90
ROCKWELL CP72N113

IDLI GROUP 88010.91
ROCKWELL CP72N12

IDLI GROUP 88010.92
ROCKWELL CP72N13

IDLI GROUP 88010.93
ROCKWELL CP72N14

IDLI GROUP 88010.94
ROCKWELL CP72N14A

IDLI GROUP 88010.95
ROCKWELL CP72N15

IDLI GROUP 88010.96
ROCKWELL CP72N16

BD=BEARING DIAMETER (outside)   **BW**=BEARING WIDTH   **CB**=CROSS LENGTH WITH BEARINGS   **CC**=CENTER TO CENTER   **CD**=CROSS DIAMETER
CL=CROSS LENGTH WITHOUT BEARINGS   **EC**=END TO CENTER or FACE TO FACE   **EE**=END TO END   **EL**=EFFECTIVE LENGTH
HD=HUB DIAMETER (or insert)   **IE**=INSIDE OF EARS (or recess)   **OD**=OUTSIDE DIAMETER   **OE**=OUTSIDE OF EARS   **SB**=SPLINE OR BORE SIZE

IDLI GROUP 88010.97
ROCKWELL CP72N17

IDLI GROUP 88010.98
ROCKWELL CP72N18

IDLI GROUP 88010.99
ROCKWELL CP72N19

IDLI GROUP 88011
ROCKWELL CP72N2

IDLI GROUP 88011.01
ROCKWELL CP72N20

IDLI GROUP 88011.02
ROCKWELL CP72N21

IDLI GROUP 88011.03
ROCKWELL CP72N22

IDLI GROUP 88011.04
ROCKWELL CP72N23

IDLI GROUP 88011.05
ROCKWELL CP72N24

IDLI GROUP 88011.06
ROCKWELL CP72N25

IDLI GROUP 88011.07
ROCKWELL CP72N26

IDLI GROUP 88011.08
ROCKWELL CP72N27

IDLI GROUP 88011.09
ROCKWELL CP72N28

IDLI GROUP 88011.10
ROCKWELL CP72N3

IDLI GROUP 88011.11
BORG-WARNR 114-7013
ROCKWELL CP72N31

IDLI GROUP 88011.12
BORG-WARNR 114-7011
BORG-WARNR 114-7012
BORG-WARNR 114-7015
ROCKWELL CP72N32

IDLI GROUP 88011.13
ROCKWELL CP72N33

IDLI GROUP 88011.14
BORG-WARNR 114-7001
ROCKWELL CP72N34

IDLI GROUP 88011.15
ROCKWELL CP72N35

IDLI GROUP 88011.16
ROCKWELL CP72N36

IDLI GROUP 88011.17
ROCKWELL CP72N37

IDLI GROUP 88011.18
ROCKWELL CP72N38

IDLI GROUP 88011.19
ROCKWELL CP72N39

IDLI GROUP 88011.20
ROCKWELL CP72N4

IDLI GROUP 88011.21
ROCKWELL CP72N40

IDLI GROUP 88011.22
ROCKWELL CP72N42

IDLI GROUP 88011.23
ROCKWELL CP72N43

IDLI GROUP 88011.24
ROCKWELL CP72N44

IDLI GROUP 88011.25
ROCKWELL CP72N45

IDLI GROUP 88011.26
ROCKWELL CP72N47

IDLI GROUP 88011.27
ROCKWELL CP72N48

IDLI GROUP 88011.28
ROCKWELL CP72N49

IDLI GROUP 88011.29
ROCKWELL CP72N5

IDLI GROUP 88011.30
ROCKWELL CP72N55

IDLI GROUP 88011.31
ROCKWELL CP72N57

IDLI GROUP 88011.32
ROCKWELL CP72N6

IDLI GROUP 88011.33
ROCKWELL CP72N68

IDLI GROUP 88011.34
ROCKWELL CP72N7

IDLI GROUP 88011.35
EIMCO 920M940043
ROCKWELL CP72N72

IDLI GROUP 88011.36
ROCKWELL CP72N73

IDLI GROUP 88011.37
BALKAMP 3-910
ROCKWELL CP72N76

IDLI GROUP 88011.38
ROCKWELL CP72N8

IDLI GROUP 88011.39
ROCKWELL CP72N9

IDLI GROUP 88011.40
ROCKWELL CP74X

IDLI GROUP 88011.41
ROCKWELL CP750N

IDLI GROUP 88011.42
ROCKWELL CP750N1

IDLI GROUP 88011.43
ROCKWELL CP75N

IDLI GROUP 88011.44
ROCKWELL CP75N1

IDLI GROUP 88011.45
ROCKWELL CP75N2

IDLI GROUP 88011.46
ROCKWELL CP75N4

IDLI GROUP 88011.47
ROCKWELL CP75WB13

IDLI GROUP 88011.48
ROCKWELL CP78WB12

IDLI GROUP 88011.49
ROCKWELL CP78WB13

IDLI GROUP 88011.50
ROCKWELL CP78WB14

IDLI GROUP 88011.51
ROCKWELL CP78WB15

IDLI GROUP 88011.52
ROCKWELL CP78WB16

IDLI GROUP 88011.53
ROCKWELL CP78WB2

IDLI GROUP 88011.54
ROCKWELL CP78WB5

IDLI GROUP 88011.55
ROCKWELL CP78WB6

IDLI GROUP 88011.56
ROCKWELL CP78X

IDLI GROUP 88011.57
ROCKWELL CP7N

ENGINEERING CATALOGS MUST BE CONSULTED FOR DETAILS NOT INCLUDED
IN THIS GUIDE. SPECIFIC DESIGNS, MATERIAL CONTENT, TOLERANCES
LUBE FITTINGS AND OTHER DIMENSIONS ARE INTENTIONALLY OMITTED HERE.

CAUTION: BE SURE TO REFER TO ENGINEERING CATALOGS FOR SPECIAL APPLICATIONS THAT REQUIRE SPECIFIC MATERIAL CONTENT, TOLERANCES, ETC. SEE FOOTNOTE.

IDLI GROUP 88011.58
ROCKWELL CP7N1

IDLI GROUP 88011.59
ROCKWELL CP7N10

IDLI GROUP 88011.60
ROCKWELL CP7N2

IDLI GROUP 88011.61
ROCKWELL CP7N5

IDLI GROUP 88011.62
ROCKWELL CP7N6

IDLI GROUP 88011.63
ROCKWELL CP7N7

IDLI GROUP 88011.64
ROCKWELL CP7N8

IDLI GROUP 88011.65
ROCKWELL CP7N9

IDLI GROUP 88011.66
ROCKWELL CP82N1

IDLI GROUP 88011.67
ROCKWELL CP82N10

IDLI GROUP 88011.68
ROCKWELL CP82N11

IDLI GROUP 88011.69
ROCKWELL CP82N12

IDLI GROUP 88011.70
ROCKWELL CP82N13

IDLI GROUP 88011.71
ROCKWELL CP82N14

IDLI GROUP 88011.72
ROCKWELL CP82N14A

IDLI GROUP 88011.73
ROCKWELL CP82N15

IDLI GROUP 88011.74
ROCKWELL CP82N16

IDLI GROUP 88011.75
ROCKWELL CP82N17

IDLI GROUP 88011.76
ROCKWELL CP82N18

IDLI GROUP 88011.77
ROCKWELL CP82N19

IDLI GROUP 88011.78
ROCKWELL CP82N2

IDLI GROUP 88011.79
ROCKWELL CP82N20

IDLI GROUP 88011.80
ROCKWELL CP82N22

IDLI GROUP 88011.81
ROCKWELL CP82N24

IDLI GROUP 88011.82
ROCKWELL CP82N25

IDLI GROUP 88011.83
ROCKWELL CP82N27

IDLI GROUP 88011.84
ROCKWELL CP82N28

IDLI GROUP 88011.85
ROCKWELL CP82N29

IDLI GROUP 88011.86
ROCKWELL CP82N3

IDLI GROUP 88011.87
ROCKWELL CP82N30

IDLI GROUP 88011.88
ROCKWELL CP82N35

IDLI GROUP 88011.89
ROCKWELL CP82N36

IDLI GROUP 88011.90
ROCKWELL CP82N37

IDLI GROUP 88011.91
ROCKWELL CP82N4

IDLI GROUP 88011.92
ROCKWELL CP82N47

IDLI GROUP 88011.93
ROCKWELL CP82N5

IDLI GROUP 88011.94
ROCKWELL CP82N6

IDLI GROUP 88011.95
ROCKWELL CP82N72

IDLI GROUP 88011.96
ROCKWELL CP82N75

IDLI GROUP 88011.97
ROCKWELL CP82N76

IDLI GROUP 88011.98
ROCKWELL CP82N9

IDLI GROUP 88011.99
ROCKWELL CP83N75

IDLI GROUP 88012
ROCKWELL CP850N

IDLI GROUP 88012.01
ROCKWELL CP850N4

IDLI GROUP 88012.02
ROCKWELL CP85WB1

IDLI GROUP 88012.03
ROCKWELL CP85WB10

IDLI GROUP 88012.04
ROCKWELL CP85WB12

IDLI GROUP 88012.05
ROCKWELL CP85WB13

IDLI GROUP 88012.06
ROCKWELL CP85WB14

IDLI GROUP 88012.07
ROCKWELL CP85WB19

IDLI GROUP 88012.08
ROCKWELL CP85WB2

IDLI GROUP 88012.09
ROCKWELL CP85WB20

IDLI GROUP 88012.10
ROCKWELL CP85WB22

IDLI GROUP 88012.11
ROCKWELL CP85WB23

IDLI GROUP 88012.12
ROCKWELL CP85WB3

IDLI GROUP 88012.13
ROCKWELL CP85WB46

IDLI GROUP 88012.14
ROCKWELL CP85WB47

IDLI GROUP 88012.15
ROCKWELL CP85WB48

IDLI GROUP 88012.16
ROCKWELL CP85WB5

IDLI GROUP 88012.17
ROCKWELL CP85WB6

IDLI GROUP 88012.18
ROCKWELL CP85WB8

IDLI GROUP 88012.19
ROCKWELL CP88WB

BD=BEARING DIAMETER (outside)   BW=BEARING WIDTH   CB=CROSS LENGTH WITH BEARINGS   CC=CENTER TO CENTER   CD=CROSS DIAMETER
CL=CROSS LENGTH WITHOUT BEARINGS   EC=END TO CENTER or FACE TO FACE   EE=END TO END   EL=EFFECTIVE LENGTH
HD=HUB DIAMETER (or insert)   IE=INSIDE OF EARS (or recess)   OD=OUTSIDE DIAMETER   OE=OUTSIDE OF EARS   SB=SPLINE OR BORE SIZE

IDLI GUIDE   COPYRIGHT ® INTERCHANGE, INC.   ST. LOUIS PARK, MN. 55416 USA

IDLI GROUP 88012.20
ROCKWELL CP88WB10

IDLI GROUP 88012.21
ROCKWELL CP88WB6

IDLI GROUP 88012.22
ROCKWELL CP88WB9

IDLI GROUP 88012.23
IOWA MFG. 4537501706
NEAPCO 28CP8N
NEAPCO 5-0080
REPUBLIC CB-8N
ROCKWELL CP8N
WESCO 811008

IDLI GROUP 88012.24
ROCKWELL CP92N

IDLI GROUP 88012.25
ROCKWELL CP92N10

IDLI GROUP 88012.26
ROCKWELL CP92N103

IDLI GROUP 88012.27
ROCKWELL CP92N11

IDLI GROUP 88012.28
ROCKWELL CP92N12

IDLI GROUP 88012.29
ROCKWELL CP92N13

IDLI GROUP 88012.30
ROCKWELL CP92N14

IDLI GROUP 88012.31
PRECISION 904
ROCKWELL CP92N15

IDLI GROUP 88012.32
ROCKWELL CP92N16

IDLI GROUP 88012.33
ROCKWELL CP92N17

IDLI GROUP 88012.34
ALCO 9000
MOPAR MM9000
PRECISION 590
PRECISION 903
ROCKWELL CP92N18
SCOOPMOBLE 708686

IDLI GROUP 88012.35
ROCKWELL CP92N19

IDLI GROUP 88012.36
ALCO J9016
GMB G9016
ROCKWELL CP92N20

IDLI GROUP 88012.37
ROCKWELL CP92N21

IDLI GROUP 88012.38
ROCKWELL CP92N22

IDLI GROUP 88012.39
ROCKWELL CP92N24

IDLI GROUP 88012.40
ROCKWELL CP92N25

IDLI GROUP 88012.41
ROCKWELL CP92N26

IDLI GROUP 88012.42
ROCKWELL CP92N3

IDLI GROUP 88012.43
ROCKWELL CP92N4

IDLI GROUP 88012.44
ROCKWELL CP92N49

IDLI GROUP 88012.45
ROCKWELL CP92N5

IDLI GROUP 88012.46
ROCKWELL CP92N50

IDLI GROUP 88012.47
ROCKWELL CP92N6

IDLI GROUP 88012.48
ROCKWELL CP92N7

IDLI GROUP 88012.49
ROCKWELL CP92N71

IDLI GROUP 88012.50
ROCKWELL CP92N72

IDLI GROUP 88012.51
ROCKWELL CP92N73

IDLI GROUP 88012.52
ROCKWELL CP92N8

IDLI GROUP 88012.53
ROCKWELL CP92N9

IDLI GROUP 88012.54
ROCKWELL CP92N93

IDLI GROUP 88012.55
ROCKWELL CPL105

IDLI GROUP 88012.56
ROCKWELL CPL10N1

IDLI GROUP 88012.57
ROCKWELL CPL10N2

IDLI GROUP 88012.58
ROCKWELL CPL10N3

IDLI GROUP 88012.59
ROCKWELL CPL10N4

IDLI GROUP 88012.60
ROCKWELL CPL10S

IDLI GROUP 88012.61
ROCKWELL CPL10S1

IDLI GROUP 88012.62
ROCKWELL CPL10S10

IDLI GROUP 88012.63
ROCKWELL CPL10S3

IDLI GROUP 88012.64
ROCKWELL CPL10S5

IDLI GROUP 88012.65
ROCKWELL CPL10S7

IDLI GROUP 88012.66
ROCKWELL CPL10S8

IDLI GROUP 88012.67
ROCKWELL CPL10S9

IDLI GROUP 88012.68
PRECISION 972
ROCKWELL CPL12N

IDLI GROUP 88012.69
PRECISION 972
ROCKWELL CPL12N1

IDLI GROUP 88012.70
ROCKWELL CPL12N2

IDLI GROUP 88012.71
ROCKWELL CPL12N3

IDLI GROUP 88012.72
ROCKWELL CPL12N4

IDLI GROUP 88012.73
ROCKWELL CPL12N5

IDLI GROUP 88012.74
ROCKWELL CPL12N7

IDLI GROUP 88012.75
ROCKWELL CPL12N8

IDLI GROUP 88012.76
ROCKWELL CPL12R

IDLI GROUP 88012.77
ROCKWELL CPL12R12

IDLI GROUP 88012.78
ROCKWELL CPL14N

ENGINEERING CATALOGS MUST BE CONSULTED FOR DETAILS NOT INCLUDED
IN THIS GUIDE. SPECIFIC DESIGNS, MATERIAL CONTENT, TOLERANCES
LUBE FITTINGS AND OTHER DIMENSIONS ARE INTENTIONALLY OMITTED HERE.

IDLI GUIDE   COPYRIGHT ® INTERCHANGE, INC.   ST. LOUIS PARK, MN. 55416 USA

CAUTION: BE SURE TO REFER TO ENGINEERING CATALOGS FOR SPECIAL APPLICATIONS THAT REQUIRE SPECIFIC MATERIAL CONTENT, TOLERANCES, ETC. SEE FOOTNOTE.

IDLI GROUP 88012.79
ROCKWELL CPL14N1

IDLI GROUP 88012.80
ROCKWELL CPL14N2

IDLI GROUP 88012.81
ROCKWELL CPL14N4

IDLI GROUP 88012.82
ROCKWELL CPL14R

IDLI GROUP 88012.83
ROCKWELL CPL14R1

IDLI GROUP 88012.84
ROCKWELL CPL14S

IDLI GROUP 88012.85
ROCKWELL CPL14S1

IDLI GROUP 88012.86
ROCKWELL CPL14S2

IDLI GROUP 88012.87
ROCKWELL CPL16S

IDLI GROUP 88012.88
ROCKWELL CPL16S1

IDLI GROUP 88012.89
PRECISION 982
ROCKWELL CPL6N

IDLI GROUP 88012.90
ROCKWELL CPL6N1

IDLI GROUP 88012.91
PRECISION 982
ROCKWELL CPL6N10

IDLI GROUP 88012.92
PRECISION 982
ROCKWELL CPL6N11

IDLI GROUP 88012.93
ROCKWELL CPL6N12

IDLI GROUP 88012.94
ROCKWELL CPL6N13

IDLI GROUP 88012.95
PRECISION 982
ROCKWELL CPL6N14

IDLI GROUP 88012.96
ROCKWELL CPL6N17

IDLI GROUP 88012.97
ROCKWELL CPL6N19

IDLI GROUP 88012.98
PRECISION 982
ROCKWELL CPL6N2

IDLI GROUP 88012.99
ROCKWELL CPL6N20

IDLI GROUP 88013
ROCKWELL CPL6N4

IDLI GROUP 88013.01
PRECISION 982
ROCKWELL CPL6N5

IDLI GROUP 88013.02
ROCKWELL CPL6N6

IDLI GROUP 88013.03
ROCKWELL CPL6N7

IDLI GROUP 88013.04
PRECISION 982
ROCKWELL CPL6N8

IDLI GROUP 88013.05
ROCKWELL CPL6N9

IDLI GROUP 88013.06
ROCKWELL CPL6R

IDLI GROUP 88013.07
ROCKWELL CPL6R30

IDLI GROUP 88013.08
ROCKWELL CPL6R34

IDLI GROUP 88013.09
ROCKWELL CPL6R39

IDLI GROUP 88013.10
ROCKWELL CPL6R40

IDLI GROUP 88013.11
PRECISION 982
ROCKWELL CPL6S10

IDLI GROUP 88013.12
PRECISION 982
ROCKWELL CPL6S11

IDLI GROUP 88013.13
ROCKWELL CPL6S12

IDLI GROUP 88013.14
ROCKWELL CPL6S13

IDLI GROUP 88013.15
ROCKWELL CPL6S14

IDLI GROUP 88013.16
ROCKWELL CPL6S15

IDLI GROUP 88013.17
ROCKWELL CPL6S16

IDLI GROUP 88013.18
ROCKWELL CPL6S19

IDLI GROUP 88013.19
ROCKWELL CPL6S20

IDLI GROUP 88013.20
PRECISION 982
ROCKWELL CPL6S22

IDLI GROUP 88013.21
ROCKWELL CPL6S26

IDLI GROUP 88013.22
ROCKWELL CPL6S3

IDLI GROUP 88013.23
ROCKWELL CPL6S36

IDLI GROUP 88013.24
ROCKWELL CPL6S4

IDLI GROUP 88013.25
ROCKWELL CPL6S48

IDLI GROUP 88013.26
PRECISION 982
ROCKWELL CPL6S5

IDLI GROUP 88013.27
PRECISION 982
ROCKWELL CPL6S7

IDLI GROUP 88013.28
ALLOY J1004
PRECISION 982
ROCKWELL CPL6S8

IDLI GROUP 88013.29
PRECISION 982
ROCKWELL CPL6S9

IDLI GROUP 88013.30
ROCKWELL CPL7S48

IDLI GROUP 88013.31
ROCKWELL CPLSN10

IDLI GROUP 88013.32
CLEVE.STL. X1405
ROCKWELL CS-H12-16-6
SPICER 500415-9

IDLI GROUP 88013.33
CLEVE.STL. D96-13
ROCKWELL CS-H4-20-13

IDLI GROUP 88013.34
CLEVE.STL. D56-13
ROCKWELL CS-H5-18-27

IDLI GROUP 88013.35
CLEVE.STL. U96-13
ROCKWELL CS-H5-18-28

BD = BEARING DIAMETER (outside)   BW = BEARING WIDTH   CB = CROSS LENGTH WITH BEARINGS   CC = CENTER TO CENTER   CD = CROSS DIAMETER
CL = CROSS LENGTH WITHOUT BEARINGS   EC = END TO CENTER or FACE TO FACE   EE = END TO END   EL = EFFECTIVE LENGTH
HD = HUB DIAMETER (or insert)   IE = INSIDE OF EARS (or recess)   OD = OUTSIDE DIAMETER   OE = OUTSIDE OF EARS   SB = SPLINE OR BORE SIZE

IDLI GROUP 88013.36
```
CLEVE.STL. .............. C67-13
ROCKWELL ......... CS-H6-24-49
SPICER ............... 6-73-109
```

IDLI GROUP 88013.37
```
CLEVE.STL. .............. T55-17
ROCKWELL ........... DC126N
```

IDLI GROUP 88013.38
```
CLEVE.STL. .............. S55-17
ROCKWELL ........... DC131N
```

IDLI GROUP 88013.39
```
CLEVE.STL. ............ S55-17-1
ROCKWELL ........... DC131N1
```

IDLI GROUP 88013.40
```
CLEVE.STL. ............. R55-17
ROCKWELL ............ DC135N
```

IDLI GROUP 88013.41
```
CLEVE.STL. ............. O55-17
ROCKWELL ............ DC141N
```

IDLI GROUP 88013.42
```
CLEVE.STL. ............ O55-17-1
ROCKWELL ........... DC141N1
SPICER .................. D3K
```

IDLI GROUP 88013.43
```
CLEVE.STL. ............. D56-17
ROCKWELL ............ DC148N
```

IDLI GROUP 88013.44
```
CLEVE.STL. ............. D76-17
ROCKWELL ............ DC148N1
SPICER .................. D3H
```

IDLI GROUP 88013.45
```
CLEVE.STL. ............ D76-17-1
ROCKWELL ........... DC148N2
SPICER .................. D3J
```

IDLI GROUP 88013.46
```
CLEVE.STL. ............. U56-17
ROCKWELL ............ DC155N
```

IDLI GROUP 88013.47
```
CLEVE.STL. ............. U76-17
ROCKWELL ........... DC155N1
SPICER .................. D4J
```

IDLI GROUP 88013.48
```
CLEVE.STL. ............. U76-17-1
ROCKWELL ........... DC155N2
```

IDLI GROUP 88013.49
```
ROCKWELL .............. DC16N
SPICER ................... D5A
```

IDLI GROUP 88013.50
```
ROCKWELL .............. DC17N
SPICER ................... D6A
SPICER ................... D6B
```

IDLI GROUP 88013.51
```
ROCKWELL ............ DC17N2
SPICER ................... D6F
```

IDLI GROUP 88013.52
```
ROCKWELL ............ DC17N3
SPICER .................. D10A
```

IDLI GROUP 88013.53
```
CLEVE.STL. ............. O55-9-3
ROCKWELL ............ DCD24-1
SPICER .............. 3.5-14-39
```

IDLI GROUP 88013.54
```
CLEVE.STL. ............. D76-9
ROCKWELL ............ DCD25-1
SPICER ................ 3-14-59
```

IDLI GROUP 88013.55
```
CLEVE.STL. ............. U76-9
ROCKWELL ............ DCD28-1
SPICER ................ 4-14-69
```

IDLI GROUP 88013.56
```
CLEVE.STL. ............. S55-9-1
ROCKWELL ............ DCR20-2
```

IDLI GROUP 88013.57
```
CLEVE.STL. ............. O96-9
ROCKWELL ............ DCS22-1
```

IDLI GROUP 88013.58
```
CLEVE.STL. ............. R55-9
ROCKWELL ............ DCS22-2
```

IDLI GROUP 88013.59
```
CLEVE.STL. ............. D96-9
ROCKWELL ............ DCS24-1
```

IDLI GROUP 88013.60
```
CLEVE.STL. ............. O55-9
ROCKWELL ............ DCS24-2
```

IDLI GROUP 88013.61
```
CLEVE.STL. ............. D56-9
ROCKWELL ............ DCS26-1
```

IDLI GROUP 88013.62
```
CLEVE.STL. ............. U56-9
ROCKWELL ............ DCS28-1
```

IDLI GROUP 88013.63
```
CLEVE.STL. ............. U96-9
ROCKWELL ............ DCS30-1
```

IDLI GROUP 88013.64
```
CLEVE.STL. ............. SCB2-5
GMC ................. 3705198
IHC .................. 474339C1
ROCKWELL ........... DEFR22-4
SPICER ............... 230104
```

IDLI GROUP 88013.65
```
CLEVE.STL. ............. SCB2-6
FORD ............... B5T4835A
ROCKWELL ........... DEFR22-5
SPICER ............... 230121
```

IDLI GROUP 88013.66
```
CLEVE.STL. ............. DCB2-1
FORD ............... C1TT4835A
IHC .................. 918842R1
ROCKWELL ........... DEFR25-1
SPICER ............... 230209
```

IDLI GROUP 88013.67
```
CLEVE.MOT. ............ DCB2-3
GMC ................. 3723615
ROCKWELL ........... DEFR25-2
SPICER ............... 230208
```

IDLI GROUP 88013.68
```
CLEVE.STL. ............. OCB2-1
GMC ................. 694981
IHC .................. 470683C1
ROCKWELL ........... DEFR25-4
SPICER ............... 231000
```

IDLI GROUP 88013.69
```
CLEVE.STL. ............. UCB2-3
IHC .................. 922078R1
ROCKWELL ........... DEFR28-3
SPICER ............... 230363
```

IDLI GROUP 88013.70
```
CLEVE.STL. ............. SCB2-3
GMC ................. 3714168
IHC .................. 474340C1
ROCKWELL ........... DEFR28-4
SPICER ............... 230141
```

IDLI GROUP 88013.71
```
CLEVE.STL. ............. RCB2-1
FORD .............. C5TZ4805A
ROCKWELL ........... DEFR30-1
```

IDLI GROUP 88013.72
```
CLEVE.STL. ............. UCB2-1
IHC .................. 236092R1
ROCKWELL ........... DEFR32-3
```

IDLI GROUP 88013.73
```
CLEVE.STL. ............. SCB2-2
FORD ............... B5T4805A
IHC .................. 386417C1
IHC .................. 914243R1
ROCKWELL ........... DEFR33-1
SPICER ............... 230118
```

IDLI GROUP 88013.74
```
CLEVE.STL. ............... OCB2
GMC ................. 694973
IHC .................. 470684C1
ROCKWELL ........... DEFR34-1
SPICER ............... 231004
```

IDLI GROUP 88013.75
```
CLEVE.STL. ............... CCB2
GMC ................. 2402706
IHC .................. 283068C1
ROCKWELL ........... DEFR35-1
SPICER ............... 230166
```

IDLI GROUP 88013.76
```
CLEVE.STL. ............... DCB2
FORD ............... B5Q4805B
IHC .................. 918834R1
ROCKWELL ........... DEFR37-1
SPICER ............... 230124
```

IDLI GROUP 88013.77
```
CLEVE.MOT. ............ DCB2-2
GMC ................. 3723609
ROCKWELL ........... DEFR40-5
SPICER ............... 230207
```

ENGINEERING CATALOGS MUST BE CONSULTED FOR DETAILS NOT INCLUDED
IN THIS GUIDE. SPECIFIC DESIGNS, MATERIAL CONTENT, TOLERANCES
LUBE FITTINGS AND OTHER DIMENSIONS ARE INTENTIONALLY OMITTED HERE.

CAUTION: BE SURE TO REFER TO ENGINEERING CATALOGS FOR SPECIAL APPLICATIONS THAT REQUIRE SPECIFIC MATERIAL CONTENT, TOLERANCES, ETC. SEE FOOTNOTE.

IDLI GROUP 88013.78

CLEVE.STL.	UCB2-2
IHC	283066C1
ROCKWELL	DEFR41-1
SPICER	230151

IDLI GROUP 88013.79

ROCKWELL	DEFR42-1
SPICER	230909

IDLI GROUP 88013.80

CLEVE.STL.	HCB2
ROCKWELL	DEFR55-1
SPICER	230746

IDLI GROUP 88013.81

ROCKWELL	DEFR56-2
SPICER	231047

IDLI GROUP 88013.82

BORG-WARNR	62112
HUB CITY	0335-01254
HUB CITY	0355-02154
NEAPCO	N2299X1
ROCKWELL	KT105
TRW	21502

IDLI GROUP 88013.83

ROCKWELL	KT109
SPICER	3.4-16-33

IDLI GROUP 88013.84

BORG-WARNR	62251
G & G MFG	02010006
ROCKWELL	KT110
SPICER	228311X

IDLI GROUP 88013.85

ALLOY	1191
BORG-WARNR	68561
G & G MFG	02010021
G & G MFG	02010221
NEAPCO	1856X
ROCKWELL	KT111
SPICER	228317X
SPICER	229153X

IDLI GROUP 88013.86

AEC	102
AEC	103
AEC	2
AEC	AE2-94-28X2
ALCO	4996
ALCO	C4996
ALLOY	4996
BORG-WARNR	114-418UB
BORG-WARNR	114-428
BORG-WARNR	114-428UB
CLEVE.MOT.	CM5-007
CMP	CM5-7
D&D MACH.	UB200K
GMB	G3-121XU
MASTERS	MS1210
MCQUAY-NOR	UO4-23
MOTOR MAST	2290UBK2T
MOTOR MAST	3200UBK2
MOTOR MAST	3290UBK2
NEAPCO	1-0089
NEAPCO	28089
NEAPCO	28099
PRECISION	329-10
REPUBLIC	CB1270UB
ROCKWELL	KT116
SPICER	2-94-28X
TRU-CROSS	9428
TRW	20701
WESCO	1521-90
WESCO	206-521
WESCO	306521
ZELLER	3287UBK2

IDLI GROUP 88013.87

AEC	102
AEC	103
AEC	3
AEC	AE3-94-18X2
ALCO	4999
ALCO	C4999
ALLOY	4999
BORG-WARNR	114-418UB
BORG-WARNR	114-428UB
BWE	114-418UB
CLEVE.MOT.	CM5-008
CMP	CM5-8
D&D MACH.	UB560K
GMB	G3-178UB
GMB	G3-178XU
MCQUAY-NOR	UO4-24
MCQUAY-NOR	UO4-24
MOTOR MAST	3200UBK2
MOTOR MAST	3290UBK2
MOTOR MAST	3290UBK2T
NEAPCO	1-0099
NEAPCO	28089
NEAPCO	28099
PRECISION	330-10
REPUBLIC	1350UB
REPUBLIC	CB1350UB
ROCKWELL	KT117
SPICER	2-94-18X
SPICER	3-94-18X
TRU-CROSS	9418

TRW	20702
WESCO	1578-90
WESCO	306-578
ZELLER	3290UBK2

IDLI GROUP 88013.88

ROCKWELL	KT118
SPICER	3-94-28X

IDLI GROUP 88013.89

PRECISION	1226
ROCKWELL	L12N2424

IDLI GROUP 88013.90

ALLOY	0605
ALLOY	605
BORG-WARNR	61119
BORG-WARNR	61176
EATON	6421591
HUB CITY	0335-01226
MCQUAY-NOR	U03-85
MUNCIE	MJY15A
NEAPCO	11119
NEAPCO	11176
NEAPCO	1118A
NEAPCO	12-1119
NEAPCO	12-1176
NEAPCO	12-1218
PRECISION	1226
REX CHAIN	760038
REX CHAIN	760040
ROCKWELL	L12N2438
WESCO	12N2045-37
WESCO	850042

IDLI GROUP 88013.91

PRECISION	1226
ROCKWELL	L12N2453

IDLI GROUP 88013.92

ROCKWELL	L12NY30-23

IDLI GROUP 88013.93

PRECISION	1276
ROCKWELL	L12NYH20

IDLI GROUP 88013.94

PRECISION	1270
ROCKWELL	L12NYQ12-15

IDLI GROUP 88013.95

PRECISION	1241
ROCKWELL	L12NYQ13-2

IDLI GROUP 88013.96

BPC	H12SY-7/8
PRECISION	1271
ROCKWELL	L12NYQ14-4

IDLI GROUP 88013.97

PRECISION	1272
ROCKWELL	L12NYQ16-16

IDLI GROUP 88013.98

ROCKWELL	L12NYR12-3

IDLI GROUP 88013.99

BPC	H12SY-1R
PRECISION	1280
ROCKWELL	L12NYR16-15

IDLI GROUP 88014

BORG-WARNR	61185
NEAPCO	11185
NEAPCO	12-1185
PRECISION	1212
ROCKWELL	L12NYR16-36

IDLI GROUP 88014.01

BORG-WARNR	61312
G & G MFG	184-1216
NEAPCO	111J2
NEAPCO	12-1331
PRECISION	1213
ROCKWELL	L12NYR16-5
SPICER	1-4-3253
WESCO	12N164
WESCO	850022

IDLI GROUP 88014.02

PRECISION	1209
ROCKWELL	L12NYR16-9

IDLI GROUP 88014.03

PRECISION	1226
ROCKWELL	L12NYR20-37

IDLI GROUP 88014.04

PRECISION	1225
ROCKWELL	L12NYR20-40

IDLI GROUP 88014.05

BORG-WARNR	61176
BORG-WARNR	61213
HUB CITY	0335-01216
NEAPCO	111A3
NEAPCO	12-1304
PRECISION	1227
REX CHAIN	760032
ROCKWELL	L12NYR20-42
SPICER	1-4-3373

IDLI GROUP 88014.06

AEC	46002
BORG-WARNR	61154
BORG-WARNR	61221
G & G MFG	182-0621
NEAPCO	05102X
NEAPCO	111B1
PRECISION	1291
PRECISION	1292
REX CHAIN	760004
ROCKWELL	L12NYR22-16
SPICER	10-4-331X
TRW	21020
WESCO	12N220TQB
WESCO	12N220TQD
WESCO	12N222-0
WESCO	451220
WESCO	6N220TQD

IDLI GROUP 88014.07

PRECISION	1291
ROCKWELL	L12NYR22A

IDLI GROUP 88014.08

ROCKWELL	L12NYR24-7

IDLI GROUP 88014.09

BPC	H12SY-3/4R-3/16KW
PRECISION	1260
ROCKWELL	L12NYS12-2

IDLI GROUP 88014.10

PRECISION	1261
ROCKWELL	L12NYS14-11

IDLI GROUP 88014.11

BPC	H12SY-15/16R-1/4KW
PRECISION	1262
ROCKWELL	L12NYS15-4

IDLI GROUP 88014.12

PRECISION	1263
ROCKWELL	L12NYS16-13

IDLI GROUP 88014.13

PRECISION	1281
ROCKWELL	L12NYS18-10

IDLI GROUP 88014.14

PRECISION	1251
ROCKWELL	L12NYS18-15

IDLI GROUP 88014.15

ROCKWELL	L12NYS18-17
SPICER	1-4-931

IDLI GROUP 88014.16

PRECISION	1264
ROCKWELL	L12NYS18-25

IDLI GROUP 88014.17

ROCKWELL	L12NYS18-8
SPICER	1-4-921

IDLI GROUP 88014.18

PRECISION	1283
ROCKWELL	L12NYS20-15

IDLI GROUP 88014.19

BORG-WARNR	61171
BPC	H12SY-1-1/4
MUNCIE	MSY62
NEAPCO	11171
NEAPCO	12-1171
PRECISION	1282
ROCKWELL	L12NYS20-7
WESCO	12N200SP

IDLI GROUP 88014.20

PRECISION	1255
ROCKWELL	L12NYS20-9

IDLI GROUP 88014.21

NEAPCO	19132
NEAPCO	19187
NEAPCO	20-9187
PRECISION	1350
ROCKWELL	L14SY18-5
SPICER	2-4-2971
SPICER	2-4-3681
WESCO	14N180-21
WESCO	860048

IDLI GROUP 88014.22

PRECISION	1323
ROCKWELL	L14SYQ16-11

IDLI GROUP 88014.23

PRECISION	1323
ROCKWELL	L14SYQ16-18

IDLI GROUP 88014.24

PRECISION	1321
ROCKWELL	L14SYQ16-4

IDLI GROUP 88014.25

PRECISION	1329
ROCKWELL	L14SYQ19-15
WESCO	860070

IDLI GROUP 88014.26

BORG-WARNR	69130
PRECISION	1324
ROCKWELL	L14SYQ20-1

IDLI GROUP 88014.27

BORG-WARNR	69127
NEAPCO	19127
NEAPCO	19129
NEAPCO	20-9127
PRECISION	1322
ROCKWELL	L14SYQ20-5
TRW	21205
WESCO	14N0200-26
WESCO	14N200-26
WESCO	860053

IDLI GROUP 88014.28

BORG-WARNR	69117
HUB CITY	0335-01481
PRECISION	1306
ROCKWELL	L14SYR18-41

IDLI GROUP 88014.29

NEAPCO	1916A
PRECISION	1304
ROCKWELL	L14SYR18-82

IDLI GROUP 88014.30

PRECISION	1311
ROCKWELL	L14SYR20-43

IDLI GROUP 88014.31

PRECISION	1311
ROCKWELL	L14SYR20-51

IDLI GROUP 88014.32

AEC	46014
ALLOY	0750
CASE,J.I.	B13272
G & G MFG	182-1406
MCQUAY-NOR	U03-86
MCQUAY-NOR	U03-87
NEAPCO	19101
NEAPCO	19101X
NEAPCO	19132
NEAPCO	19137
NEAPCO	19138
NEAPCO	20-9101
NEAPCO	20-9132
NEAPCO	20-9137
PRECISION	1372
ROCKWELL	L14SYR22-1A
SPICER	2-4-3041X
SPICER	2-4-3481X
SPICER	3-4-3481X
TRW	21207
TRW	21209
WESCO	14N0220-20
WESCO	14N220-20
WESCO	461221
WESCO	860-056

IDLI GROUP 88014.33

PRECISION	1352
ROCKWELL	L14SYS16

IDLI GROUP 88014.34

ALLOY	750
BORG-WARNR	69101
G & G MFG	184-1406
HUB CITY	0335-01402
NEAPCO	19137X
PRECISION	1372
ROCKWELL	L14SYS22-127A

IDLI GROUP 88014.35

PRECISION	1370
REX CHAIN	760049
REX CHAIN	760058
ROCKWELL	L14SYS22-4

IDLI GROUP 88014.36

PRECISION	1835
ROCKWELL	L16SYQ16-3

IDLI GROUP 88014.37

ROCKWELL	L16SYQ20-3

IDLI GROUP 88014.38

ALLOY	0756
ALLOY	756
BORG-WARNR	62122
NEAPCO	21122
NEAPCO	22-1122
PRECISION	1836
ROCKWELL	L16SYQ20-7

IDLI GROUP 88014.39

ROCKWELL	L16SYQ26

IDLI GROUP 88014.40

G & G MFG	184-3524
PRECISION	1808
ROCKWELL	L16SYR20-25
SPICER	3-4-803
SPICER	3-4-823
SPICER	3-4-873
SPICER	3-4-883
SPICER	3-4-943

CAUTION: BE SURE TO REFER TO ENGINEERING CATALOGS FOR SPECIAL APPLICATIONS THAT REQUIRE SPECIFIC MATERIAL CONTENT, TOLERANCES, ETC. SEE FOOTNOTE.

IDLI GROUP 88014.41

BORG-WARNR 62158
BORG-WARNR 62175
FORD C5ZZ4841A
PRECISION 1810
ROCKWELL L16SYR22-115

IDLI GROUP 88014.42

ROCKWELL L16SYR28-8

IDLI GROUP 88014.43

ROCKWELL L16SYS223A

IDLI GROUP 88014.44

BORG-WARNR 62134
PRECISION 1842
ROCKWELL L16SYS22-40

IDLI GROUP 88014.45

PRECISION 1851
ROCKWELL L16SYS22-46A

IDLI GROUP 88014.46

ROCKWELL L6-107

IDLI GROUP 88014.47

PRECISION 1910
ROCKWELL L6-23

IDLI GROUP 88014.48

PRECISION 1903
ROCKWELL L6-28

IDLI GROUP 88014.49

ROCKWELL L6N15-15

IDLI GROUP 88014.50

PRECISION 1912
ROCKWELL L6NYH14-3

IDLI GROUP 88014.51

PRECISION 1901
ROCKWELL L6NYR12-10

IDLI GROUP 88014.52

BORG-WARNR 7238J
BORG-WARNR 7522J
BORG-WARNR 8338J
BORG-WARNR 9487J
NEAPCO 16-6111
NEAPCO L6111
PRECISION 1900

ROCKWELL L6NYR12-9
TRW 21051
WESCO 06N0123
WESCO 540053
WESCO 540054
WESCO 6N123
WESCO 6N123-5

IDLI GROUP 88014.53

PRECISION 1903
ROCKWELL L6NYR14-10

IDLI GROUP 88014.54

PRECISION 1905
ROCKWELL L6NYR16-8

IDLI GROUP 88014.55

CLEVE.STL. C67-12
ROCKWELL LP-HC39-3

IDLI GROUP 88014.56

CLEVE.STL. X23
ROCKWELL NUSL16-20-1

IDLI GROUP 88014.57

CLEVE.STL. X1394
ROCKWELL NUSL16-20-2

IDLI GROUP 88014.58

ROCKWELL NU-SL20-18-1
SPICER 332J

IDLI GROUP 88014.59

ROCKWELL NU-SL20-18-3
SPICER 20-74-11

IDLI GROUP 88014.60

ROCKWELL PL6R40

IDLI GROUP 88014.61

CLEVE.STL. X1391
ROCKWELL PNCO2-32-6

IDLI GROUP 88014.62

ROCKWELL PN-CO2-40-4
SPICER 500024-16

IDLI GROUP 88014.63

CLEVE.STL. T55-7-2-16
ROCKWELL PS18-24-2

IDLI GROUP 88014.64

CLEVE.STL. S55-7-2-16
ROCKWELL PS20-32-1

IDLI GROUP 88014.65

CLEVE.STL. S55-7-3-16LN
ROCKWELL PS20-32-13

IDLI GROUP 88014.66

CLEVE.STL. S55-7-2-16L
ROCKWELL PS20-32-2

IDLI GROUP 88014.67

CLEVE.STL. S55-7-3-16
ROCKWELL PS20-32-5

IDLI GROUP 88014.68

CLEVE.STL. S55-7-3-16L
ROCKWELL PS20-32-6

IDLI GROUP 88014.69

CLEVE.STL. R55-7-3-14
ROCKWELL PS22-32-3

IDLI GROUP 88014.70

CLEVE.STL. R55-7-3-14L
ROCKWELL PS22-32-4

IDLI GROUP 88014.71

CLEVE.STL. O55-7-3-14
ROCKWELL PS24-32-1

IDLI GROUP 88014.72

CLEVE.STL. O55-7-241
ROCKWELL PS24-32-5

IDLI GROUP 88014.73

CLEVE.STL. O55-7-242
ROCKWELL PS24-32-6

IDLI GROUP 88014.74

CLEVE.STL. S55-7-1166
ROCKWELL PS24-32-7

IDLI GROUP 88014.75

CLEVE.STL. D76-7-4-14N
ROCKWELL PS25-16-2
SPICER 3-40-1381
SPICER 3-40-881

IDLI GROUP 88014.76

CLEVE.STL. U76-7-4-14N
ROCKWELL PS28-16-6
SPICER 4-40-431
SPICER 4-40-781

IDLI GROUP 88014.77

CLEVE.STL. U56-7-110
ROCKWELL PS28-32-4

IDLI GROUP 88014.78

CLEVE.STL. U56-7-111
ROCKWELL PS28-32-5

IDLI GROUP 88014.79

CLEVE.STL. SCB2-4
GMC 3685958
NEAPCO 98-1473
ROCKWELL RBR28-1
SPICER 98-1473

IDLI GROUP 88014.80

CLEVE.STL. O55-6
ROCKWELL REBC141

IDLI GROUP 88014.81

CLEVE.STL. D56-6
ROCKWELL REBC148

IDLI GROUP 88014.82

CLEVE.STL. S55-6
ROCKWELL RECB131

IDLI GROUP 88014.83

CLEVE.STL. D96-23
ROCKWELL RECP148

IDLI GROUP 88014.84

CLEVE.STL. U56-23
ROCKWELL RECP155

IDLI GROUP 88014.85

CLEVE.STL. SCB1-4A
FORD B5T4818A
GMC 3714195
IHC 392951C1
IHC 914237R1
ROCKWELL REP26-2
SPICER 3-95-59

DD = BEARING DIAMETER (outside)   BW = BEARING WIDTH   CB = CROSS LENGTH WITH BEARINGS   CC = CENTER TO CENTER   CD = CROSS DIAMETER
CL = CROSS LENGTH WITHOUT BEARINGS   EC = END TO CENTER or FACE TO FACE   EE = END TO END   EL = EFFECTIVE LENGTH
HD = HUB DIAMETER (or insert)   IE = INSIDE OF EARS (or recess)   OD = OUTSIDE DIAMETER   OE = OUTSIDE OF EARS   SB = SPLINE OR BORE SIZE

IDLI GROUP 88014.86

CLEVE.STL. DCB1-4A
FORD B5QH4818B
GMC 2333918
IHC 918835R1
ROCKWELL REP29-1
SPICER 3-95-69
SPICER 3-95-96

IDLI GROUP 88014.87

CLEVE.STL. CCB1-4A
GMC 2402703
IHC 291876C1
ROCKWELL REP31-1
SPICER 6-95-29

IDLI GROUP 88014.88

CLEVE.STL. UCB1-4A
IHC 291875C1
ROCKWELL REP31-2
SPICER 5-95-29

IDLI GROUP 88014.89

CLEVE.STL. HCB-1-4A
ROCKWELL REP35-4
SPICER 8-95-49

IDLI GROUP 88014.90

CLEVE.STL. S55-12S
ROCKWELL RR1R19-6
SPICER 2-7-29

IDLI GROUP 88014.91

CLEVE.STL. O55-12S
ROCKWELL RR1R22-6
SPICER 3-7-39

IDLI GROUP 88014.92

CLEVE.STL. D56-12S
ROCKWELL RR1R24-2
SPICER 3-7-89

IDLI GROUP 88014.93

CLEVE.STL. SS55-12
ROCKWELL RRER16-5

IDLI GROUP 88014.94

CLEVE.STL. O55-12
ROCKWELL RRER17-3

IDLI GROUP 88014.95

ROCKWELL RT125-11
SPICER 10-32-92

IDLI GROUP 88014.96

ROCKWELL RT175-14
SPICER 14-30-52

IDLI GROUP 88014.97

ROCKWELL RT175-16
SPICER 14-30-12

IDLI GROUP 88014.98

ROCKWELL RT2125-14
SPICER 17-30-12

IDLI GROUP 88014.99

BORG-WARNR 4365J
ROCKWELL RT2-14
SPICER 16-30-62

IDLI GROUP 88015

BORG-WARNR 4367J
ROCKWELL RT2-16
SPICER 16-30-32

IDLI GROUP 88015.01

ROCKWELL RT225-11
SPICER 18-32-22

IDLI GROUP 88015.02

BORG-WARNR 4871J
ROCKWELL RT225-14

IDLI GROUP 88015.03

BORG-WARNR 4368J
ROCKWELL RT25-12

IDLI GROUP 88015.04

BORG-WARNR 4369J
ROCKWELL RT25-13
SPICER 20-30-52

IDLI GROUP 88015.05

ROCKWELL RT25-14
SPICER 20-30-22

IDLI GROUP 88015.06

ROCKWELL RT25-16
SPICER 20-30-12

IDLI GROUP 88015.07

ROCKWELL RT2625-11
SPICER 21-32-22

IDLI GROUP 88015.08

ROCKWELL RT275-14
SPICER 22-30-22

IDLI GROUP 88015.09

BORG-WARNR 4375J
ROCKWELL RT3-11
SPICER 24-32-142

IDLI GROUP 88015.10

BORG-WARNR 4376J
ROCKWELL RT3-13
SPICER 24-30-12
SPICER 24-32-62

IDLI GROUP 88015.11

ROCKWELL RT3-14
SPICER 24-30-42

IDLI GROUP 88015.12

BORG-WARNR 4378J
ROCKWELL RT3-156

IDLI GROUP 88015.13

BORG-WARNR 5792J
ROCKWELL RT3-16
SPICER 24-30-32

IDLI GROUP 88015.14

BORG-WARNR 4379J
ROCKWELL RT3-187

IDLI GROUP 88015.15

BORG-WARNR 24710J
ROCKWELL RT35-10
SPICER 28-30-92

IDLI GROUP 88015.16

BORG-WARNR 10000J
ROCKWELL RT35-11
SPICER 28-32-42

IDLI GROUP 88015.17

BORG-WARNR 4380J
ROCKWELL RT35-13
SPICER 28-30-22
SPICER 28-32-22

IDLI GROUP 88015.18

ROCKWELL RT35-14
SPICER 28-30-62

IDLI GROUP 88015.19

BORG-WARNR 6892J
ROCKWELL RT35-156
SPICER 28-30-52

IDLI GROUP 88015.20

ROCKWELL RT35-16
SPICER 28-30-42

IDLI GROUP 88015.21

ROCKWELL RT4095-7
SPICER 32-30-72

IDLI GROUP 88015.22

BORG-WARNR 8128J
ROCKWELL RT4-10
SPICER 32-30-52

IDLI GROUP 88015.23

BORG-WARNR 8128J
ROCKWELL RT4-12

IDLI GROUP 88015.24

BORG-WARNR 9113J
ROCKWELL RT4-14
SPICER 32-30-22

IDLI GROUP 88015.25

BORG-WARNR 16745J
ROCKWELL RT4-187

IDLI GROUP 88015.26

ROCKWELL RT45-10
SPICER 36-30-62

IDLI GROUP 88015.27

ROCKWELL RT45-13
SPICER 36-30-12

IDLI GROUP 88015.28

BORG-WARNR 11855J
ROCKWELL RT45-187
SPICER 36-32-12

IDLI GROUP 88015.29

ROCKWELL RT45-3
SPICER 36-30-22

IDLI GROUP 88015.30

CLEVE.STL. D76-10
ROCKWELL SECR20-7
SPICER 3-16-103

ENGINEERING CATALOGS MUST BE CONSULTED FOR DETAILS NOT INCLUDED
IN THIS GUIDE. SPECIFIC DESIGNS, MATERIAL CONTENT, TOLERANCES
LUBE FITTINGS AND OTHER DIMENSIONS ARE INTENTIONALLY OMITTED HERE.

CAUTION: BE SURE TO REFER TO ENGINEERING CATALOGS FOR SPECIAL APPLICATIONS THAT REQUIRE SPECIFIC MATERIAL CONTENT, TOLERANCES, ETC. SEE FOOTNOTE.

IDLI GROUP 88015.31

CLEVE.STL. U76-10
ROCKWELL SECR22-8
SPICER 4-16-143

IDLI GROUP 88015.32

CLEVE.STL. O55-10-4
ROCKWELL SECS24-7
SPICER 3-16-51

IDLI GROUP 88015.33

CLEVE.STL. D76-10-1
ROCKWELL SECS25-3
SPICER 3-16-41

IDLI GROUP 88015.34

CLEVE.STL. U76-10-1
ROCKWELL SECS28-2

IDLI GROUP 88015.35

ROCKWELL SH-P16-7
SPICER 230565

IDLI GROUP 88015.36

CLEVE.STL. U56-7-140
ROCKWELL SHS28-91

IDLI GROUP 88015.37

CLEVE.STL. U56-7-137
ROCKWELL SHS28-92

IDLI GROUP 88015.38

ROCKWELL SLA-P16-14

IDLI GROUP 88015.39

BORG-WARNR 68343
G & G MFG 194-1400
HAYES 2-55-62
NEAPCO 1834-3A
NEAPCO 2034A
NEAPCO 53-1834
ROCKWELL SL-P16-47
SPICER 2-55-202
SPICER 2-55-62
WESCO 560-101
WESCO 560-102

IDLI GROUP 88015.40

ROCKWELL SQS100
SPICER 230566

IDLI GROUP 88015.41

G & G MFG 136-1900
G & G MFG 137-1900
ROCKWELL SQS1187
SPICER 278269

IDLI GROUP 88015.42

G & G MFG 136-2100
G & G MFG 137-2100
ROCKWELL SQS13125
SPICER 278635

IDLI GROUP 88015.43

BORG-WARNR 68810
NEAPCO 72-0750
NEAPCO N230590
ROCKWELL SQS75
SPICER 230590
WESCO 922-013
WESCO 922-014
WESCO 922-015
WESCO 922-016
WESCO 922-017
WESCO 922-018

IDLI GROUP 88015.44

CLEVE.STL. X1345
ROCKWELL SSFL5-18-3

IDLI GROUP 88015.45

ROCKWELL TU-P20-11
SPICER 278814

IDLI GROUP 88015.46

CLEVE.STL. S55-10-1
ROCKWELL WAFR17-7

IDLI GROUP 88015.47

CLEVE.STL. T55-10
ROCKWELL WAFS18-2

IDLI GROUP 88015.48

CLEVE.STL. S55-10
ROCKWELL WAFS20-6

IDLI GROUP 88015.49

CLEVE.STL. R55-10
ROCKWELL WAFS22-1

IDLI GROUP 88015.50

CLEVE.STL. D96-10
ROCKWELL WAFS24-3

IDLI GROUP 88015.51

CLEVE.STL. O55-10
ROCKWELL WAFS24-4

IDLI GROUP 88015.52

CLEVE.STL. D56-10
ROCKWELL WAFS26-4

IDLI GROUP 88015.53

CLEVE.STL. U56-10
ROCKWELL WAFS28-5

IDLI GROUP 88015.54

CLEVE.STL. U96-10
ROCKWELL WAFS30-1

IDLI GROUP 88015.55

CLEVE.STL. S55-11
ROCKWELL WAR18-15

IDLI GROUP 88015.56

CLEVE.STL. D76-11
ROCKWELL WAR21-4
SPICER 3-15-63

IDLI GROUP 88015.57

CLEVE.STL. U76-11
ROCKWELL WAR23-4
SPICER 4-15-13

IDLI GROUP 88015.58

CLEVE.STL. O55-11-1
ROCKWELL WARS24-5
SPICER 3-15-51

IDLI GROUP 88015.59

CLEVE.STL. D76-11-1
ROCKWELL WASR25-4
SPICER 3-15-41

BD=BEARING DIAMETER (outside)   BW=BEARING WIDTH   CB=CROSS LENGTH WITH BEARINGS   CC=CENTER TO CENTER   CD=CROSS DIAMETER
CL=CROSS LENGTH WITHOUT BEARINGS   EC=END TO CENTER or FACE TO FACE   EE=END TO END   EL=EFFECTIVE LENGTH
HD=HUB DIAMETER (or insert)   IE=INSIDE OF EARS (or recess)   OD=OUTSIDE DIAMETER   OE=OUTSIDE OF EARS   SB=SPLINE OR BORE SIZE

FOR YOUR CONVENIENCE
NOTES ON OTHER INFORMATION

We urge you to send us any data that
you feel should be included with the
future editions of the I.D.L.I. Guide.

SERIES CROSS REFERENCE CHART ON UNIVERSAL JOINTS AND DRIVE LINE COMPONENTS

YOKE DIMENSIONS IN INCHES			AGITRANS	APC	CHAINBELT	G & G
INSIDE OF EARS	OUTSIDE OF EARS	BEARING OUTSIDE DIAMETER				
1.4375	2.1875	.9688	6	6N		6N
1.8125		1.0625	12	12N	R12N, R31N/120	12N
2.3750	2.6875	1.0625				
2.5625		1.1875				
2.7969		1.3125	44	44N	440	44R
	2.9063	1.1250	14	14N	R14N/140	14N
3.1250	4.4375	1.1875				
	3.4844	1.2500	35	35N	R35N/350	35N
	4.0625	1.2500				
4.1875	4.4375	1.3750				
	4.5625	1.5625				
	4.6250	1.5313	55			55N
	4.9688	1.3750				
	5.1250	1.9375				
	5.3125	1.8750				
	5.9375	2.0000				
	6.0938	1.9375				
	7.0000	1.9375				
	7.5469	1.9375				
5.3125	5.6250					
5.5313	5.8438					
5.8438	6.2188					
6.5000	6.8750					
8.1250	8.5000					
8.2500	8.6250					
2.3750		1.0625				
2.5625		1.1875				
3.1250	4.1875	1.1875				
	4.4375	1.3750				
	4.9688	1.3750				

Left margin labels: ROUND BEARING TYPE (rows 1–19), WING BEARING TYPE (rows 20–25), U-BOLT TYPE (rows 26–30).

MECHANICS (BORG-WARNER)	NEAPCO	PRECISION	REXNORD	CLEVE. STEEL PROD. ROCKWELL	HAYES-DANA SPICER	WEASLER	WESCO
100●	1500/1600● L600	1500/1700 1900		L6N	1000●	6R L6W/6RW	1A, 1B, 1C,● 6N
SERIES 2 1½RA●	1200	1200	120	L12N	1100● 1140	12R, 12RW	12N
				131N✳			
				135N✳			
	44R		440	44R	1440	44R	44N, 45N
SERIES 3.5	2000		140	L14N, L14S	1240	14R, GN/14NW	14N
				141N✳	1410		
SERIES 4 3LRA●	2200	1800	350	35N, 16S, 3N, L16S	1270/1600 1340/1610	35R,35RW,DNW, DN/35NW	35N
				4N			
SERIES 5	2600			148N✳	1480		
				5N			
	55N			55N		L55/55NW	55N
				155N✳			
				6N			
				16N			
				7N			
				17N	1700/1710		
				176N	1760		
				18N	1810		
				58WB			
6C				62N			
7C				72N			
8½C				85WB			
8C				82N			
9C				92N			
				131N✳			
				135N✳			
				141N✳			
				148N✳			
				155N✳			

● COMPARABLE CAPACITY   ✳ IN TWO TYPES

EQUIVALENT CHART

ADD COMPONENTS TO DETERMINE THE DECIMAL EQUIVALENT OVER 1" OR 25 MM.

FRACTIONS	MILLIMETERS	DECIMALS
1/64		.0156
1/32		.0313
	1	.0394
3/64		.0469
1/16		.0625
5/64		.0781
	2	.0787
3/32		.0938
7/64		.1094
	3	.1181
1/8		.1250
9/64		.1406
5/32		.1563
	4	.1575
11/64		.1719
3/16		.1875
	5	.1969
13/64		.2031
7/32		.2188
15/64		.2344
	6	.2362
1/4		.2500
17/64		.2656
	7	.2756
9/32		.2813
19/64		.2969
5/16		.3125
	8	.3150
21/64		.3281
11/32		.3438

FRACTIONS	MILLIMETERS	DECIMALS
	9	.3543
23/64		.3594
3/8		.3750
25/64		.3906
	10	.3937
13/32		.4063
27/64		.4219
	11	.4331
7/16		.4375
29/64		.4531
15/32		.4688
	12	.4724
31/64		.4844
1/2		.5000
	13	.5118
33/64		.5156
17/32		.5313
35/64		.5469
	14	.5512
9/16		.5625
37/64		.5781
	15	.5906
19/32		.5938
39/64		.6094
5/8		.6250
	16	.6299
41/64		.6406
21/32		.6563
	17	.6693
43/64		.6719

FRACTIONS	MILLIMETERS	DECIMALS
11/16		.6875
45/64		.7031
	18	.7087
23/32		.7188
47/64		.7344
	19	.7480
3/4		.7500
49/64		.7656
25/32		.7813
	20	.7874
51/64		.7969
13/16		.8125
	21	.8268
53/64		.8281
27/32		.8438
55/64		.8594
	22	.8661
7/8		.8750
57/64		.8906
	23	.9055
29/32		.9063
59/64		.9219
15/16		.9375
	24	.9449
61/64		.9531
31/32		.9688
	25	.9843
63/64		.9844
1		1.0000

We need your help.

If there are others in your organization who should be added to our mailing list, or if you have branches, warehouses or distributors that might be included, please send us their names and addresses and we will oblige. On the other hand, if you prefer, we can send our circulars to you to be included with your periodic mailings with a cover letter if desired. Advise how many copies you require and we will send whatever material we have available. Your assistance in this regard will be greatly appreciated.

ORDER FORM

PLEASE SEND:

☐ **NINTH EDITION (1986-87) INTERNATIONAL BEARING INTERCHANGE**
I.B.I. GUIDE(S) @ $120.00 Each Plus Shipping Charges

☐ **SIXTH EDITION (1986-87) INTERNATIONAL SEAL INTERCHANGE**
I.S.I. GUIDE(S) @ $90.00 Each Plus Shipping Charges

☐ **FOURTH EDITION (1986-87) INTERNATIONAL DRIVE BELT INTERCHANGE**
I.D.B.I. GUIDE(S) @ $90.00 Each Plus Shipping Charges

☐ **FIRST EDITION (1986-87) INTERNATIONAL DRIVE LINE INTERCHANGE**
I.D.L.I. GUIDE(S) @ $90.00 Each Plus Shipping Charges

SHIPPING AND HANDLING CHARGES PER BOOK

Approximate Delivery Shown Is Dependent Upon Distance, Customs, etc.
Quantity Discounts Available (on 150 or more books)–Please Inquire

Destination	I.B.I.-I.S.I.–I.D.B.I. or I.D.L.I. By Surface Shipping & Handling Each		I.B.I. Guide By Air Shipping & Handling Each		I.S.I. Guide By Air Shipping & Handling Each		I.D.B.I. Guide By Air Shipping & Handling Each		I.D.L.I. Guide By Air Shipping & Handling Each	
United States Canada	$4.00	10-20 Days	By Zone	1 Week	By Zone	1 Week	By Zone	1 Week	By Zone	1 Week
Mexico	$4.00	1-2 Months	$13.00	1 Week	$ 9.00	1 Week	$ 9.00	1 Week	$ 9.00	1 Week
Central America, Columbia Venezuela & Caribbean Islands	$4.00	3-4 Months	$22.00	2-4 Weeks	$14.00	2-4 Weeks	$14.00	2-4 Weeks	$14.00	2-4 Weeks
Europe & South America Except Columbia & Venezuela	$4.00	4-5 Months	$36.00	3-5 Weeks	$22.00	3-5 Weeks	$22.00	3-5 Weeks	$22.00	3-5 Weeks
All Others	$4.00	5-6 Months	$50.00	4-6 Weeks	$30.00	4-6 Weeks	$30.00	4-6 Weeks	$30.00	4-6 Weeks

Subject To Change—Large Quantities Shipped Via Freight or Steamer Are Reduced Accordingly. U.S. Dollars Shown

Date ____________________Order No. ____________________Phone No. (_______)________________

Ship To __________________________________Attention ____________________________

Address ____________________________________City ____________________________

Postal (Zip) Code _________________State or Province _________________Country ________________

SPECIAL SHIPPING OR BILLING INSTRUCTIONS:

Ship Via: ☐ **Surface**   ☐ **Air**   ☐ **As Noted Below:**

Minnesota Residents Add 6% Sales Tax

Make checks payable to INTERCHANGE, INC.– all payments due in U.S. dollars drawn on U.S. banks.

To expedite shipments overseas, it is suggested that you send payment with your order. Pro-forma invoices will be sent on request.

IS THE RIGHT PERSON IN YOUR ORGANIZATION ON OUR MAILING LIST? IF NOT, PLEASE ADVISE.

INTERCHANGE INCORPORATED

P.O. Box 16012-D, St. Louis Park, MN 55416, U.S.A.
Telephone 612-929-6669   Telex 29-1081

We need your help.

If there are others in your organization who should be added to our mailing list, or if you have branches, warehouses or distributors that might be included, please send us their names and addresses and we will oblige. On the other hand, if you prefer, we can send our circulars to you to be included with your periodic mailings with a cover letter if desired. Advise how many copies you require and we will send whatever material we have available. Your assistance in this regard will be greatly appreciated.

ORDER FORM

PLEASE SEND:

☐ **NINTH EDITION (1986-87) INTERNATIONAL BEARING INTERCHANGE**
I.B.I. GUIDE(S) @ $120.00 Each Plus Shipping Charges

☐ **SIXTH EDITION (1986-87) INTERNATIONAL SEAL INTERCHANGE**
I.S.I. GUIDE(S) @ $90.00 Each Plus Shipping Charges

☐ **FOURTH EDITION (1986-87) INTERNATIONAL DRIVE BELT INTERCHANGE**
I.D.B.I. GUIDE(S) @ $90.00 Each Plus Shipping Charges

☐ **FIRST EDITION (1986-87) INTERNATIONAL DRIVE LINE INTERCHANGE**
I.D.L.I. GUIDE(S) @ $90.00 Each Plus Shipping Charges

SHIPPING AND HANDLING CHARGES PER BOOK

Approximate Delivery Shown Is Dependent Upon Distance, Customs, etc.
Quantity Discounts Available (on 150 or more books)–Please Inquire

Destination	I.B.I.-I.S.I.–I.D.B.I. or I.D.L.I. By Surface Shipping & Handling Each		I.B.I. Guide By Air Shipping & Handling Each		I.S.I. Guide By Air Shipping & Handling Each		I.D.B.I. Guide By Air Shipping & Handling Each		I.D.L.I. Guide By Air Shipping & Handling Each	
United States Canada	$4.00	10-20 Days	By Zone	1 Week	By Zone	1 Week	By Zone	1 Week	By Zone	1 Week
Mexico	$4.00	1-2 Months	$13.00	1 Week	$ 9.00	1 Week	$ 9.00	1 Week	$ 9.00	1 Week
Central America, Columbia Venezuela & Caribbean Islands	$4.00	3-4 Months	$22.00	2-4 Weeks	$14.00	2-4 Weeks	$14.00	2-4 Weeks	$ 14.00	2-4 Weeks
Europe & South America Except Columbia & Venezuela	$4.00	4-5 Months	$36.00	3-5 Weeks	$22.00	3-5 Weeks	$22.00	3-5 Weeks	$22.00	3-5 Weeks
All Others	$4.00	5-6 Months	$50.00	4-6 Weeks	$30.00	4-6 Weeks	$30.00	4-6 Weeks	$30.00	4-6 Weeks

Subject To Change—Large Quantities Shipped Via Freight or Steamer Are Reduced Accordingly. U.S. Dollars Shown

Date _______________ Order No. _______________ Phone No. (______)______________

Ship To _______________________________ Attention _______________________

Address _________________________________ City ______________________

Postal (Zip) Code ______________ State or Province ______________ Country ______________

SPECIAL SHIPPING OR BILLING INSTRUCTIONS:

Ship Via: ☐ **Surface**   ☐ **Air**   ☐ **As Noted Below:**

Minnesota Residents Add 6% Sales Tax

Make checks payable to INTERCHANGE, INC.– all payments due in U.S. dollars drawn on U.S. banks.

To expedite shipments overseas, it is suggested that you send payment with your order. Pro-forma invoices will be sent on request.

IS THE RIGHT PERSON IN YOUR ORGANIZATION ON OUR MAILING LIST? IF NOT, PLEASE ADVISE.

P.O. Box 16012-D, St. Louis Park, MN 55416, U.S.A.
Telephone 612-929-6669   Telex 29-1081